GENETIC VARIANTS AND STRAINS
OF THE LABORATORY MOUSE

GENETIC VARIANTS AND STRAINS
OF THE
LABORATORY MOUSE

Second Edition

Edited by
MARY F. LYON
MRC Radiobiology Unit

and

ANTONY G. SEARLE
MRC Radiobiology Unit

for the

International Committee on
Standardized Genetic Nomenclature for
Mice

1989

OXFORD UNIVERSITY PRESS · OXFORD · NEW YORK · TOKYO
GUSTAV FISCHER VERLAG · STUTTGART

Oxford University Press, Walton Street, Oxford OX2 6DP
Oxford New York Toronto
Delhi Bombay Calcutta Madras Karachi
Petaling Jaya Singapore Hong Kong Tokyo
Nairobi Dar es Salaam Cape Town
Melbourne Auckland
and associated companies in
Berlin Ibadan

Oxford is a trade mark of Oxford University Press

Published in the United States
by Oxford University Press, New York

Distribution for
Continental Europe and
European Socialist Countries
by:
Gustav Fischer Verlag
Wollgrasweg 49,
D-7000 Stuttgart 70,
Federal Republic of Germany

British Library Cataloguing in Publication Data
Genetic variants and strains of the laboratory
mouse—2nd ed.
1. Laboratory animals: Mice. Genetic aspects
I. Lyon, Mary F. II. Searle, Antony G.
636'.93233
ISBN 0-19-854204-6

Library of Congress Cataloging-in-Publication Data
Genetic variants and strains of the laboratory mouse/edited by M. F.
Lyon and Antony G. Searle.—2nd ed.
Includes index.
1. Mice—Genetics. 2. Mice—Variation. 3. Mice as laboratory
animals. 4. Mammals—Genetics. 5. Mammals—Variation. I. Lyon,
M. F. (Mary F.) II. Searle, A. G. (Antony Gilbert)
QL737.R638G45 1989 599.32'32—dc 19 88–39796
ISBN 0-19-854204-6

Deutsche Bibliothek Cataloguing-in-Publication Data
Genetic variants and strains of the laboratory mouse/ed. by
Mary F. Lyon and Antony G. Searle for the Internat.
Committee on Standardized Genet. Nomenclature for Mice.—
2. ed.—Oxford; New York; Tokyo: Oxford Univ. Press;
Stuttgart: Fischer, 1989
ISBN 3-437-11268-6
NE: Lyon, Mary F. [Hrsg.]

Set by
Promenade Graphics, Cheltenham
Printed in Great Britain by
Butler & Tanner Ltd
Frome, Somerset

Dedication

This book is dedicated to

LESLIE C. DUNN, HANS GRÜNEBERG, and GEORGE D. SNELL,

whose pioneering efforts produced the first rules
for mouse genetic nomenclature and set an example for the
habits of consultation and co-operation among
mouse geneticists that still endure.

Extract from foreword to the first edition

This book constitutes an attempt to list and briefly describe the known genetic variants and inbred strains of the mouse, for the benefit not only of those working in mouse genetics *per se*, but also of those in diverse fields such as immunology, oncology, developmental biology, and endocrinology who now find the mouse and its variants such a valuable source of research material.

The consistent and orderly system of nomenclature used in the book for genes, chromosomal variants, and strains is the result of the work of the International Committee on Standardized Genetic Nomenclature for Mice. Mouse genetics owes a great debt to those scientists, including those to whom this book is dedicated, who realized at an early stage the need for a committee to co-ordinate genetic nomenclature for the species. The first committee, called the *Committee on Mouse Genetics Nomenclature*, was formed in 1939 when there were only 31 known gene loci and seven linkage groups, and consisted of Drs L. C. Dunn, H. Grüneberg, and G. D. Snell. They formulated the basic rules for gene nomenclature on which the present rules are founded, exhibiting and encouraging the general co-operative spirit which has been so important for the success of the nomenclature system. This was followed in 1952 by a *Committee on Standardized Nomenclature for Inbred Strains of Mice* in the formation of which G. D. Snell was again a leading figure. This committee laid down not only the rules for nomenclature of inbred strains but also the genetic criteria for the establishment of inbred strains and congenic strains. Later still, in 1958, the first two committees were merged into the present *International Committee on Standardized Genetic Nomenclature for Mice* which deals with mutant genes, inbred strains, and all types of genetic variation. G. D. Snell was its first chairman.

From the outset, the various committees realized the importance of the dissemination of information concerning nomenclature. The first step was the founding of *Mouse News Letter*, first issued in 1949 and still a major organ for the dissemination of information concerning committee rulings and nomenclature of genes and chromosomal variants. Information concerning inbred strains and their nomenclature is regularly promulgated in *Inbred Strains of Mice*, and by listings in *Cancer Research*, carried out by Joan Staats.*

Naturally, as advances in knowledge have occurred, it has been necessary to modify and extend the rules of nomenclature. This has been done by consultation between the committee members, the formation of ad hoc subcommittees, and by seeking the views of scientists concerned in the field. The new rules have been published in *Mouse News Letter* and elsewhere. Included among the items tackled in this way have been the standard karyotype of the mouse, guidelines for nomenclature of biochemical variants, rules for designation of chromosome anomalies, and nomenclature for inbred strains preserved by freezing.

M. F. Lyon
20 May, 1981

* At the time of the second edition, this information is now included in the November issue of *Mouse News Letter*.

Preface to the second edition

THE years since the first edition of this book have seen not only a great increase in knowledge of mouse genetics, but also the opening up of new fields, through the application of recombinant DNA technology. In addition, there is a new interest in mouse genetics, resulting from the recent increasingly detailed knowledge of homologies between mouse and human genes and chromosomes.

Thus, the second edition includes both updated and expanded versions of the chapters in the first edition, and also six entirely new chapters. As in the first edition, the book is intended as a work of reference and provides catalogues and lists of the different classes of genetic information. The most important catalogue, that of known genes of the mouse, by Dr M. C. Green, now includes about 1500 loci. About 1000 of these are genetically mapped, and other chapters give the overall map, the data on which this is based, and some specialized maps of particular regions or types of genes. Knowledge of the correlation between the genetic map and the physical chromosomal map is becoming more detailed, both through increasing numbers of known chromosome aberrations, which provide this information, and through the technique of *in situ* hybridization. The chapter on chromosomal variants has been expanded, and there are comprehensive maps showing breakpoint positions. The normal karyotypes for G-bands, Q-bands, pachytene, and early replicating bands are also given, together with a new G-band idiogram, at a considerably higher resolution.

A new chapter shows in detail the homologies of mouse and human genetic maps, by depicting those segments of mouse chromosomes with homology to particular human chromosome segments.

As in the earlier edition, the characteristics of inbred, recombinant inbred and congenic strains are listed, in considerably expanded chapters. Restriction fragment length polymorphisms (RFLPs) are dealt with in a separate chapter giving details of the probes and the known variants. Up-to-date information on oncogenes and tumour viruses, and on the various classes of repeated DNA sequences in the mouse genome is given in other new chapters. Also new is a chapter on wild mice, which have become so important to mouse genetics, in view of the variants they carry, particularly the high frequency of RFLPs found in crosses between *Mus* species.

Finally, but importantly, the book provides three chapters on nomenclature. Use of the correct nomenclature becomes ever more essential as mouse genetics becomes more complex. Workers are urged to abide closely by the rules for nomenclature.

It is hoped that, like the first edition, this book will become a standard work of reference. Inevitably, however, new information will accrue very rapidly, and additional sources of information will be needed. Fortunately, a well organized system of information for mouse genetics exists, the main sources being as follows:

1. *Mouse News Letter*, sponsored by the International Committee on Standardized Genetic Nomenclature for Mice, is published by Oxford University Press. February and July issues are currently edited by Dr J. Peters, and the November issue by Dr M. F. W. Festing (for affiliations see list of Committee members). It publishes an annual gene list, a range of annual genetic maps, lists of chromosomal variants, lists of new gene symbols,

new linkage and gene mapping data, and lists of available DNA probes and libraries. The November issue contains details of inbred strains, congenic strains, recombinant inbred strains and their polymorphisms, and the list of substrain holders' symbols. All issues carry contributed items giving research results on mutants, linkages and other mouse genetic data.

2. *Linkage Map of the Mouse* is prepared by Dr M. T. Davisson and Dr T. H. Roderick by use of information culled from the literature and from personal communications, and is available from them.

3. The lists of DNA probes and libraries and their holders and availability, which appear in *Mouse News Letter*, are prepared by Dr J. Eppig, Jackson Laboratory, Bar Harbor, Maine 04609, USA.

4. *List of Mutations and Mutant Stocks of the Mouse* prepared by Mrs P. W. Lane, The Jackson Laboratory, contains a list of the mutant stocks, chromosome aberrations and other variants kept at the Jackson Laboratory.

The editors would like to express their gratitude to all who have helped in the preparation of this edition, particularly Mrs D. Badger, who patiently dealt with a great deal of secretarial work, and all the authors, especially Dr Margaret Green, for undertaking the heavy task of preparing the main catalogue. The editorial staff at Oxford University Press have always been ready to help us overcome the many editorial problems. Since June 1987 A. G. Searle has been helped by the award of an Emeritus Fellowship from the Leverhulme Trust.

Harwell
August 1988

M. F. LYON
A. G. SEARLE

Contents

Contents

Contributors

MR COLIN V. BEECHEY, MRC Radiobiology Unit, Chilton, Didcot, Oxon OX11 0RD, England

DR FRANCOIS BONHOMME, Laboratoire de Génétique, Institut des Sciences de l'Evolution, USTL-Pl. E. Bataillon, 34060 Montpellier CEDEX, France

DR MURIEL T. DAVISSON, The Jackson Laboratory, 600 Main Street, Bar Harbor, Maine 04609, USA

DR DONALD P. DOOLITTLE, The Jackson Laboratory, 600 Main Street, Bar Harbor, Maine 04609, USA

DR ROSEMARY W. ELLIOTT, Department of Molecular and Cellular Biology, Roswell Park Memorial Institute, 666 Elm Street, Buffalo, New York 14263, USA

DR EDWARD P. EVANS, Sir William Dunn School of Pathology, University of Oxford, South Parks Road, Oxford OX1 3RE, England

DR MICHAEL F. W. FESTING, MRC Toxicology Unit, Woodmansterne Road, Carshalton, Surrey SM5 4EF, England

DR ANNA-MARIA FRISCHAUF, Imperial Cancer Research Fund, Lincoln's Inn Fields, London WC2A 3PX, England

DR MARGARET C. GREEN, 8 Seely Road, Bar Harbor, Maine 04609, USA

DR DOROTHY D. GREENHOUSE, Institute of Laboratory Animal Resources, National Research Council, 2101 Constitution Avenue, N.W., Washington, DC 20418, USA

DR JEAN-LOUIS GUÉNET, Unité de Génétique des Mammifères, Institut Pasteur, 28 rue du Dr Roux, 75724 Paris, CEDEX 15, France

MR JOHN H. GUIDI, The Jackson Laboratory, 600 Main Street, Bar Harbor, Maine 04609, USA

DR HORST HAMEISTER, Universität Ulm, Abt. Klinische Genetik, Oberer Eselsberg, D-7900 Ulm, Germany (FRG)

DR NICHOLAS D. HASTIE, MRC Human Genetics Unit, Western General Hospital, Crewe Road, Edinburgh EH4 2XU, UK

DR NORMAN L. HAWES, The Jackson Laboratory, 600 Main Street, Bar Harbor, Maine 04609, USA

DR ALAN L. HILLYARD, The Jackson Laboratory, 600 Main Street, Bar Harbor, Maine 04609, USA

PROF JAN KLEIN, Department of Immunogenetics, Max Planck Institute for Biology, Corrensstrasse 42, D-7400 Tübingen 1, Germany (FRG)

DR CHRISTINE A. KOZAK, Laboratory of Molecular Microbiology, National Institute of Allergy and Infectious Diseases, 9000 Rockville Pike, Bethesda, Md 20892, USA

MRS PRISCILLA W. LANE, The Jackson Laboratory, 600 Main Street, Bar Harbor, Maine 04609, USA

DR MARY F. LYON, MRC Radiobiology Unit, Chilton, Didcot, Oxon OX11 0RD, England

DR JOSEPH H. NADEAU, The Jackson Laboratory, 600 Main Street, Bar Harbor, Maine 04609, USA

Contributors

Mr Andrew H. Reiner, The Jackson Laboratory, 600 Main Street, Bar Harbor, Maine 04609, USA

Dr Thomas H. Roderick, The Jackson Laboratory, 600 Main Street, Bar Harbor, Maine 04609, USA

Dr Antony G. Searle, MRC Radiobiology Unit, Chilton, Didcot, Oxon OX11 0RD, England

Dr Imre E. Somssich, Universität Ulm, Abt. Klinische Genetik, Oberer Eselsberg, D-7900 Ulm, Germany (FRG)

Dr Benjamin A. Taylor, The Jackson Laboratory, 600 Main Street, Bar Harbor, Maine 04609, USA

Members of
International Committee on Standardized Genetic Nomenclature for Mice

1 RULES AND GUIDELINES FOR GENE NOMENCLATURE

COMMITTEE ON STANDARDIZED GENETIC NOMENCLATURE FOR MICE

Chairman: MARY F. LYON

These rules were adopted by the Committee in 1984. They are the most recent version of rules first published by Dunn, Grüneberg, and Snell in 1940 (3), revised by the Committee in 1963 (1), amplified for biochemical variants by the Committee in 1973 (2), and revised again in 1979 (5).

1.1 Rules for gene nomenclature

1.1.1 Names of gene loci

Names of gene loci should be brief and should be chosen so as to convey as accurately as possible the character by which the gene is usually recognized. Such a character may range from a coat colour or morphological effect to change in an enzyme or other protein, disease susceptibility or resistance, resemblance to a human syndrome, or a DNA sequence identified by a DNA probe for the gene.

1.1.2 Symbols for gene loci

Symbols for gene loci should typically be two-, three-, or four-letter abbreviations of the name. For convenience in alphabetical listings, the initial letters of names and symbols should where possible be the same. Arabic numbers may be included for proteins in which a number is part of the recognized name or abbreviation, but the symbol should always begin with a letter, e.g. *G6pd*, glucose-6-phosphate dehydrogenase; *B2m*, β2-microglobulin; *C3* and *C4*, third and fourth components of complement, respectively. Roman numbers and Greek letters should not be used. Names of persons or places are in general not suitable for gene names or symbols.

Except in the case of loci first discovered because of a recessive mutation (see Section 1.1.5), the initial letter of the locus symbol should be a capital, and all others lower case.

In published articles gene symbols should be set in italics, e.g. *dw*, dwarf; *Hbb*, haemoglobin β-chain.

Identification of new loci should not be assumed from the discovery of variation, whether morphological, biochemical, or antigenic. Appropriate genetic tests should be made to show Mendelian segregation, and identity or not with known loci should be established as far as possible by mapping or tests for allelism. Loci may also be identified by somatic cell genetics, or studies of DNA.

A proposed new symbol must not duplicate one already used for another locus, even if the gene effect is very different. Listing of a gene symbol in *Mouse News Letter* establishes priority.

1.1.3 Loci which are members of a series

Loci which are members of a series specifying similar proteins or other characters (e.g. isoenzymes, lymphocyte antigens, histocompatibility antigens) should be designated by the same letter symbol with the addition of a hyphen and a distinguishing number: e.g. *H-1*, *H-2*, etc., histocompatibility loci; *Es-1*, *Es-2*, etc., for esterase loci.

For morphological or 'visible' loci with similar effects (e.g. 'waltzing' genes, hair-waving genes) distinctive names should be given since the gene actions and gene products may ultimately prove to be very different, e.g. *v*, waltzer, and *kr* kreisler; *Sl*, steel, and *W*, dominant spotting.

1.1.4 Homology with other organisms

It is highly desirable that terminology for homologous genes should be standardized among species. Therefore, when choosing a gene symbol an attempt should first be made to discover and use any symbols already adopted for this gene in other species. However, care should be taken that such symbols *do not duplicate any already in use in the mouse* for other loci. If duplication would occur, then the symbol should be modified to one resembling that used in the other species but not duplicating that used for a different gene in that (or

1

another) species, e.g. carbonic anhydrase: man *CA*, mouse, *Car-1* and *Car-2*; catalase: man *CAT*, mouse, *Cas-1*.

Where possible the numbering of homologous loci in a series should be made concordant in various species, with locus *1* in the mouse corresponding to the locus *A* in the other species, locus *2* with locus *B*, and so on.

1.1.5 Alleles

Alleles should be designated by the locus symbol with an added superscript (in italics when printed). In computerized symbols the superscript may be denoted by prefixing an asterisk, e.g. Hbb^d or $Hbb*d$.

1. Exceptions occur for the first discovered allele in those cases in which there is clearly a wild type. No superscript is then used, e.g. *nu*, nude; *Ca*, caracul. When further alleles are discovered the first mutant allele may still be without a superscript.

2. Recessive alleles should be indicated by the use of a lower case initial letter for a mutant gene, e.g. *a*, nonagouti; *nu*, nude.

All other alleles, whether dominant, codominant, or having dominance relationships which vary with method of assessment, should be indicated by the use of a capital initial letter followed by lower case letters, as in the locus symbol, e.g. *Re*, rex; *Ta*, tabby.

3. Allele superscripts should typically be one or two lower case letters and, if possible, should convey additional information about the allele, e.g. c^{ch}, chinchilla allele of *c* or albino; Mi^{wh}, white allele at the microphthalmia locus; Hbb^d, haemoglobin β-chain allele giving a *diffuse* band after electrophoresis. If information is too complex to be conveyed conveniently in the symbol (e.g. biochemical properties, antigenic specificities), the alleles are given superscripts and the information concerning the allelic properties is shown in catalogues or tables, e.g. $Pgm-1^a$, $Pgm-1^b$; $H-2^a$, $H-2^b$, etc.

4. Wild type alleles should be designated by a + sign, with the locus symbol as a superscript e.g. $+^d$, $+^c$. Reversions from a mutant allele to wild type should be distinguished from the original wild type allele by designating them by the locus symbol, with a + sign as superscript e.g. d^+, pe^+. A + sign only may be used when the context leaves no doubt as to the locus represented e.g. in genetic formulae.

5. Indistinguishable alleles of independent origin (e.g. reoccurrences, reversions to wild type) should be designated by the existing gene symbol with a series symbol (see below) appended as a superscript in italics. If the gene symbol already has a superscript, this should be separated from the series symbol by a hyphen. The series symbol should consist of an Arabic numeral corresponding to the serial number of the variant in any given laboratory, plus an abbreviation indicating the discoverer or laboratory of origin. Where an abbreviation for the designation of inbred substrains or sublines has already been assigned, this should be used. Otherwise, the abbreviations used should follow the rules for substrain designation and not duplicate an existing symbol in the standard list of abbreviations. To avoid the confusion of the numeral *1* and the letter *l*, a first-discovered variant may be left unnumbered, and the second variant numbered 2. Examples: c^{4Rl}, the fourth reoccurrence of *c* found by Russell; a^{t-7J}, the seventh reoccurrence of a^t found at the Jackson Laboratory; d^{+J} and d^{+2J}, the first and second reversions from *d* to d^+ found at the Jackson Laboratory.

6. Mutations or other variations occurring in known alleles may be denoted by a superscript *m* followed by an appropriate series symbol (as above) and separate from the original allele symbol by a hyphen; e.g. $Mod-1^{a-m1Lws}$, the first mutant allele of $Mod-1^a$ found by Lewis.

For known deletions of all or part of an allele the superscript *m* may be replaced with *dl*.

Information on the allele of origin of mutations may be valuable in elucidating changes in DNA sequence.

1.1.6 Phenotype symbols

Phenotype symbols, where these are necessary (e.g. antigen loci, enzyme loci), should be the same as genotype symbols except that symbols for phenotypes should be in capitals, not italicized, and with superscripts lowered to the line. The phenotypes of heterozygotes should be written as in the following example: GPI-1A, GPI-1B, and GPI-1AB are phenotypes associated with the *Gpi-1* locus.

In those cases in which information concerning the subunit structure of a protein is available, phenotype symbols should reflect the subunit composition. Where multiple loci exist for a single enzyme (genetic isozymes), locus *1* should be considered to synthesize the A subunit, locus *2* the B subunit, etc. Phenotypes are then designated according to the rules of the International Union of Biochemistry (6), e.g.

Locus	Isozyme phenotype
Pgm-1	PGM-A
Pgm-2	PGM-B
Adh-1	ADH-A$_2$

$Adh\text{-}2$ $ADH\text{-}B_2$
$Ldh\text{-}1$ $LDH\text{-}A_4$

Phenotype symbols for allelic variants reflect the subunit structure; i.e. allele 1^a synthesizes the A^a subunit; allele 1^b the A^b subunit, etc.; e.g.

Genotype	Phenotype
$Ldh\text{-}1^a/Ldh\text{-}1^a$	$LDH\text{-}A_4^a$
$Ldh\text{-}1^b/Ldh\text{-}1^b$	$LDH\text{-}A_4^a$
$Ldh\text{-}1^a/Ldh\text{-}1^b$	$LDH\text{-}A_4^aA, A_3^aA^b, A_2^aA_2^b, A^aA_3^b, A_4^b$

1.1.7 Gene complexes

Gene complexes are considered to exist when a number of apparently functionally related loci are genetically closely linked. Alternative states of complexes are referred to as *haplotypes* rather than alleles.

Known complexes are of two main types: (a) less extensive complexes involving duplicate loci or in which operators or *cis*-acting regulators of structural genes for protein show little or no recombination with the loci on which they act: and (b) very extensive complexes, possibly involving hundreds of related loci, for which special rules may be necessary. The *H-2* and the immunoglobin complexes are in category (b).

Less extensive complexes involving operators, cis-acting regulators, or duplicate loci

The existence of a gene complex, rather than multiple types of variation in a structural gene, should not be postulated unless there is good evidence. It should be remembered that different mutations in a structural gene may affect not only electrophoretic mobility but also activity and stability. In addition, changes in 5′ or 3′ regulatory sequences may cause apparent changes in tissue specificity or inducibility. Thus, such changes in effect should be attributed to mutations in the structural gene unless there is good evidence otherwise.

To distinguish different loci of a complex, the basic symbol should have appended a single lower case letter designating the presumed function or means of identification of the locus, such as *s* (structural), *e* (electrophoretic), *r* (regulatory), or *t* (temporal). This letter should be set off by a hyphen, as in *Bgl-e* (beta-galactosidase electrophoretic), except in the case of numbered unlinked loci in a series, in which case the letter should follow the number without a second hyphen, as in *Adh-3t*, (alcohol dehydrogenase-3 temporal). When it is discovered that a previously described locus is part of a complex, a letter indicative of its function or means of identification should be added to the basic symbol to represent the already known locus, as well as a different letter for the newly discovered locus. An example is: the *Adh-3* locus, after discovery of a temporal regu-

lator, becomes *Adh-3e* (*Adh-3* electrophoretic) and the regulator is called *Adh-3t* (*Adh-3* temporal). The basic symbol then represents the entire complex; if necessary for clarity, the complex may be additionally indicated by enclosing the basic symbol in parentheses or in brackets. Haplotypes are designated by the symbol for the complex with a superscript small letter. The components of the haplotype can be briefly indicated as in the following example: $Adh\text{-}3^a$ or $(Adh\text{-}3)^a = Adh\text{-}3e^a$ $Adh\text{-}3t^a = Adh\text{-}3e^at^a$;. In the case where two or more closely linked and functionally related structural loci have been given serial numbers, the complex, loci, and haplotypes should be indicated as in the following example: complex, *Amy* or *(Amy)*; loci, *Amy-1*, *Amy-2*; haplotypes, Amy^a or $(Amy)^a = Amy\text{-}1^a \; Amy\text{-}2^a = Amy\text{-}1^a2^a$.

Distantly acting regulators should be given locus symbols different from but related to the locus they regulate and preferably with the same initial letter, for example, *Ldr* (a regulator of the lactate dehydrogenase locus *Ldh*).

Suffixes so far used, either for loci in gene complexes, or for subdividing series of loci, include:

-C Constant region (immunoglobulins),
-e Electrophoretic,
-m Mitochondrial,
-r Regulatory,
-s Structural,
-t Temporal,
-V Variable region (immunoglobulins).

Very extensive complexes with special rules

The list of extensive complexes with special rules continually increases. The need for special rules for each arises because the various complexes differ widely in their structure and no suitable single nomenclature system has yet been found that is adequate for all these complexes.

Some complexes with special rules include:

(a) *H-2* complex;

(b) immunoglobulin complexes;

(c) globin gene complexes;

(d) homeobox-containing gene complexes;

(e) *t*-complex.

(a) *The histocompatibility-2 or H-2 region complex.* Symbols for haplotypes should be superscripts, consisting of small letters and Arabic numerals, assigned according to the standing rules for the *H-2* complex (7), such as $H\text{-}2^b$, $H\text{-}2^{bm1}$, $H\text{-}2^{ap4}$. Parts of the complex that are shown to be separable from each other by recombination are referred to as *regions*, and may consist of many loci. The regions are denoted by symbols on the line in capital letters, for example, *K*, *I*, *S*, and *D*. In referring to alternative states or 'alleles' of particular regions, the symbol of the region is appended by a superscript denoting the haplotype from which the region originates, for example K^k, S^d, D^b. The individual loci have symbols reflecting their characteristics, which are chosen according to the valid 'Rules for gene nomenclature in mice', the rules for nomenclature of the *H-2* gene complex (7), and rules for the nomenclature of the *I* region (13). Examples: *H-2K*, *Ia-1*, *C4*, *H-2L*. The allelic forms of these loci are designated by superscript letters denoting the haplotype of origin, for example, $H\text{-}2K^b$, $C4^d$. Rules for nomenclature of mutant haplotypes were summarized by Kohn *et al.* (8) and by Shreffler (14).

(b) *The immunoglobin complexes.* The heavy-, kappa-, and lambda-chain regions are designated *Igh*, *Igk*, and *Igl*, respectively. Haplotypes are designated by the symbol for the region with a superscript lower case letter. The constant subregions are designated *Igh-C*, *Igk-C*, and *Igl-C*, and individual loci in these subregions are designated by Arabic numbers following the hyphen and assigned chronologically, for example, *Igh-1*, *Igh-2*, etc., *Igl-1*. Allelic symbols are superscript lower case letters denoting the haplotype, or one of the haplotypes, in which the allele occurs. The variable subregions are designated *Igh-V*, *Igk-V*, and *Igl-V*, and individual loci in these subregions are designated by a two- or three-letter symbol or by two letters and a number following the hyphen, for example, *Igh-Dex*, *Igh-Pc*, *Igk-Ef1*. The symbol following the hyphen for the variable-region loci should be related to the antigen for which the immunoglobulin is specific or to the method used for recognizing the variant. Allelic symbols are superscript lower case letters such as: *a* and *b* in the case where allelic markers are well established, for example, $Igk\text{-}Ef1^a$ and $Igk\text{-}Ef1^b$ or *a* and *o* in the case where the allelic nature of markers is in doubt and the alleles are postulated to determine a marker and its absence, for example, $Igh\text{-}Dex^a$ and $Igh\text{-}Dex^o$. These rules are given in greater detail by Green (4).

The nomenclature systems for (c) globin gene complexes and (d) homeobox-containing gene complexes are summarized at the end of these rules.

1.1.8 Pseudogenes

Pseudogenes located away from the main gene complex should be denoted by the suffix *ps*, separated from the locus symbol by a hyphen, and followed by an appropriate serial number; e.g. *Hba-ps3*, *Hba-ps4*, pseudogenes of α-globin located away from the *Hba* complex.

1.1.9 Lethals

Appropriate locus symbols for recessive lethals with no known heterozygous effect and unidentified function consist of a lower case letter *l* followed by the chromosome number of location in parentheses, and by a hyphen and series symbol indicating the serial number of the lethal in the laboratory of origin, e.g. *l(5)-1Rk*, the first lethal on chr. 5 found by Roderick; *l(17)-2Pas*, the second lethal on chr. 17 found at the Pasteur Institute.

Such symbols should be considered as provisional. The lethal should be renamed if found to be allelic with a known gene, or if the underlying defect becomes understood.

1.1.10 Viruses

Nomenclature for genes related to the expression of viral antigens, or to sensitivity or resistance to viruses, should follow the standard rules for gene nomenclature, i.e. symbols should be italicized, with the initial letter a capital and all others lower case. Where possible and appropriate, the letters of the symbol should be those by which the virus is usually known; e.g. *Mtv-1*, a locus concerned in induction of mammary tumour virus, MTV. Successive loci concerned with the same virus should be distinguished by appending a number separated by a hyphen; e.g. *Fv-1*, *Fv-2*, loci concerned with resistance to Friend virus.

Locus symbols ending in *v* should be reserved for virological loci.

1.1.11 Oncogenes

Nomenclature for mouse cellular oncogene sequences should follow the standard nomenclature for oncogenes. However in lists of symbols and maps the prefix *c-* denoting cellular sequence should be omitted and the initial letter of the symbol should be capitalized; e.g. *c-myc* becomes *Myc*, myelocytoma oncogene; *c-Hras-1* becomes *Hras-1*, Harvey rat sarcoma-1 oncogene; *c-erba* becomes *Erba*, avian erythroblastosis oncogene.

The names and symbols of oncogenes should be regarded as provisional until the true functions of the genes become known, when they should be renamed, e.g. *Erbb* becomes epidermal growth factor receptor, *Egfr*; *Sis* becomes platelet derived growth factor, beta polypeptide, *Pdgfb*.

1.1.12 Mitochondrial genome

Loci in the mitochondrial genome should be denoted by the prefix *mt-* set off from the main symbol by a hyphen.

1.1.13 Restriction fragment length polymorphisms

There are a number of different situations in which restriction fragment length polymorphism may occur.

(a) variation in DNA sequence within exons of a known gene;

(b) variation in DNA sequence within introns and/or flanking sequences of a known gene;

(c) variation in DNA sequence outside situations in (a) or (b) but detected by a probe for the known gene, e.g. the *Hpa* site variation 5 kb from the 3′ end of the human β-globin structural gene;

(d) variation in DNA sequence detected using an arbitrary DNA sequence as a probe.

The type of variation detected in (a) and (b) should be described using current rules for nomenclature of gene loci and alleles. Thus these variants could be listed both in a compilation of restriction fragment length variants and in lists of gene loci.

For (c) symbols for the restriction fragments should begin with D (for DNA), followed by the gene symbol, followed by a number, e.g. the *Hpa* site variant cited above would be symbolized *DHbb1*, the 1 indicating that this was the first probe found. The variation in possession of the *Hpa* site can be described in terms of 'alleles'. Thus the presence of the site would be designated *DHbb1^a* and the absence *DHbb1^b*. If the allele in which the variation occurs is known this should be indicated in the symbol; e.g. *DHbb^{d1a}*.

For (d) it is not possible to ascertain if the variation fits into categories (a), (b) or (c). The nomenclature should follow that in human gene mapping for provisional nomenclature (15).

An arbitrary probe is given a name composed of four parts:

(i) D for DNA.

(ii) 0, 1 . . . . . 19, XY for the chromosomal assignment, with 0 for unassigned segments.

(iii) a symbol indicating the laboratory or scientist describing the probe. This should be the abbreviation

used to designate inbred substrains, e.g. Pas for Pasteur Institute.

(iv) a number to give uniqueness to the probe; e.g. *D1Pas5*, the fifth chromosome 1 probe, developed at the Pasteur Institute; *D17Leh48*, a chromosome 17 probe designated no. 48 by Lehrach.

Alleles of arbitrary DNA sequences could be designated as follows: *D1Pas5^a* could indicate possession of a restriction site for a particular enzyme and *D1Pas5^b* its absence.

If the arbitrary sequence is later identified to be at a known locus, the nomenclature should be altered to take this into account.

Anonymous DNA segments from the human genome which hybridize with mouse DNA and are mapped to a mouse chromosome should retain their human symbol, and this should be followed by a lower case h to denote the human origin e.g. *D21S56h*, a DNA segment from human chromosome 21 (which hybridizes to mouse chromosome 17).

1.1.14 Antigenic variants

Symbols adopted for loci concerned in cell-membrane alloantigens should be based on the method of demonstrating such loci:

1. Loci primarily demonstrable by transplantation techniques should be designated by an initial H, e.g. *H-1*, *H-2*, etc.

2. Loci demonstrable by red-cell agglutination should be designated by the letters *Ea*, e.g. *Ea-1*, *Ea-2*, etc.

3. Loci of genes coding for a cell surface molecule on lymphocytes, or shared by lymphocytes and other cell types, and detected by serological or biochemical methods, and for which there is a demonstrable polymorphism, should be designated by the letter *Ly*, e.g. *Ly-4*, *Ly-5*, etc. More details of nomenclature for *Ly* antigen genes are given at the end of these rules.

4. Similarly, other loci involving other cell types should be denoted by symbols indicating the cell type, e.g. *Pca*, plasma cell antigen; *Tla*, thymus leukaemia antigen.

Appropriate genetic nomenclature for the H-2 complex has been put forward by Klein *et al.* (7), and for the *I* region of this complex by Shreffler *et al.* (13); this nomenclature should be followed.

A registry of symbols for *H-2* haplotypes and antigenic specificities is maintained and additions to the register are listed in Mouse Membrane Alloantigen News (MMAN) incorporated in *Mouse News Letter*. New variants should be registered with the editor of MMAN, at present Dr P. Demant.

1.2 Nomenclature for special classes of genes and gene complexes

1.2.1 Guidelines for nomenclature of biochemical variants

Biochemical nomenclature should be in accord with the rules of the International Union of Biochemistry, Commission on Biochemical Nomenclature. The nomenclature recommended by the Commission is published periodically in major international biochemical journals, such as the *Journal of Biological Chemistry* and the *Biochemical Journal*. Enzymes and other biochemicals have both trivial and formal names. The correct formal name should be given the first time a substance is mentioned in a publication; trivial or abbreviated names can be used subsequently. Example: G6PD, GPD, or Gd, abbreviations of glucose-6-phosphate dehydrogenase (E.C. 1.1.1.49); D-glucose 6-phosphate:NADP oxidoreductase). This nomenclature is used in periodicals, reference works, and textbooks of biochemistry.

Symbols of structural loci should typically be two-, three-, or four-letter abbreviations (italic type) of the official Commission name of the enzyme, protein, or other substance affected. The initial letter of the symbol should be capitalized. Example: *Gpi-1*, the first identified structural locus of glucosephosphate isomerase (E.C. 5.3.1.9; D-glucose 6-phosphate ketol-isomerase). In the case of biochemical variants, the use of the locus symbol with a lower case initial letter to indicate recessive mutant genes and with a capital initial letter to indicate dominant mutant genes should generally be avoided. Such nomenclature is not suited to polymorphic systems of alleles, and the dominance–recessive relationship usually varies and depends on the method used to assess it.

Greek letters preceding the name of an enzyme or other protein, should be changed to an appropriate English letter and placed at the end of the locus symbol; e.g. *Fuca*, α-fucosidase. This permits a rational alphabetic ordering of locus symbols.

Similarly, adjectives describing the tissue specificity or other property of an enzyme or protein should normally be placed after the noun, again in order to allow appropriate alphabetic ordering of symbols, e.g. *Actc*, actin, cardiac; *Acts*, actin, skeletal.

A series of loci specifying the structure of isoenzymes that catalyse the same or similar reactions but are structurally different, or the different polypeptide chains of a protein, can be designated by the same letter symbol for the structural locus with the addition of a hyphen and a distinguishing number. Example: *Pgm-1* and *Pgm-2*, loci of structurally different isoenzymes of phosphoglucomutase (E.C. 2.7.5.1; phosphotransferase).

Homology with other organisms. It is highly desirable that terminology for homologous genes should be standardized among species. Therefore, as in the standard rules, when choosing a gene symbol an attempt should first be made to discover and use any symbols already adopted for this locus in other species. However, care should be taken that such symbols *do not duplicate any already in use in the mouse* for other loci. If duplication would occur, then the symbol should be modified to one resembling that used in the other species but not duplicating that used for a different gene in that (or another) species, e.g. carbonic anhydrase: man CA; mouse, *Car-1* and *Car-2*.

Where possible the numbering of homologous loci in a series should be made concordant in various species, with locus *1* in the mouse corresponding to the locus *A* in other species, locus *2* with locus *B*, and so on.

It is not appropriate to insert the letter m or M (for mouse) as the first letter of the symbol for a locus with homologues in other species since this would lead to all mouse locus symbols beginning with the same letter.

Alleles should be designated by the locus symbol with an added superscript as in the standard rules. In describing alleles, whether found in inbred strains or in the wild, it is desirable that the phenotype of a number of widely used inbred strains be reported. One strain should arbitrarily be designated the prototype strain for each allele, since variation that has not been detected by the methods used may be present within each allelic class. If an apparently identical allele in other strains is found by new methods to be different from that in the prototype strain, it should be assigned a new alphabetical symbol as a superscript and a prototype strain designated. This system permits the orderly assignment of symbols to newly identified alleles and allows ready comparisons of new variants with previously reported variants.

Locus and allele symbols are necessarily brief and cannot contain more than a small fraction of the known information. Additional information may be contained in gene descriptions which in some cases can be collected in catalogues or tables. The haemoglobin α-chain locus *Hba*, for example, specifies at least four different polypeptides, with the additional complication that in some strains two different polypeptides are both produced. The alleles can be assigned letter designations, and information about the amino acid composition of the chains produced by the alleles can be shown in tables. This is somewhat similar to the method already in use to record the specificities determined by alleles at loci for antigenic variants.

Proteins detected as spots on 2D-gels but not identified. Locus symbols for proteins detected as spots on 2D-gels should only be given if genetic variation or gene location is established, the gene behaves in a Mendelian fashion, and, as far as possible, the protein is known to be distinct from those already named.

Such symbols should consist of four parts: (i) the capital letter P for protein; (ii) a number indicating the chromosome on which the coding gene is located (using the number 0 to indicate unknown location); (iii) an abbreviation for the laboratory or scientist discovering the protein; (iv) a number distinguishing the protein from others found in the same laboratory, e.g. *P0Pas1* the first 2D protein in a Pasteur Institute series.

Wherever possible, the abbreviation used should be the same as that assigned for designating inbred substrains. The number of digits in the distinguishing number should be kept as low as possible for convenience in listing.

When the protein is identified the locus should be given a new and appropriate symbol.

Phenotype symbols, where these are necessary, should be the same as genotype symbols, except that symbols for phenotypes should be in capitals, not italicized, and with superscripts lowered to the line. Example: GPI-1A, GPI-1B, phenotypes associated with the *Gpi-1* locus.

In those cases in which information concerning subunit structure is available, phenotype symbols should reflect the subunit composition, according to the rules of the International Union of Biochemistry (4) by use of capital letters. Details are given in Rules for gene nomenclature (Section 1.1.6).

Identification of loci should not be assumed from the discovery of phenotypic structural variation; crosses should be made to show Mendelian segregation of the alleles. Official gene symbols should not be assigned to variants found in wild mice unless appropriate genetic tests for allelism with known similar variants are carried out. In the absence of genetic tests, phenotypic symbols (as in the standard rules) should be used, together with a description of the criteria used to establish identity with phenotypes of inbred strains.

Genetic variants affecting enzyme activity may do so for reasons other than a direct change in the catalytic activity per molecule of the enzyme under study. Presumptive mutations in this group include those producing activity differences with no discernible alteration in physical or chemical properties of the enzyme and those producing tissue-specific differences in activity. Mutations producing this type of quantitative variation may or may not prove to be allelic or forming a gene complex with the structural locus of the enzyme in question. When allelic with the structural locus, they should be designated following the standard rules. Even when not allelic, or when the structural locus has not been identified, the new locus should be named on the basis of its discernible phenotype, following the above rules. Examples: *Lv*, levulinate dehydrase, a locus affecting the amount of enzyme present; *Ah*, aromatic hydrocarbon responsiveness, a locus affecting the level of hydroxylase induced by aromatic hydrocarbons.

1.2.2 Nomenclature for loci controlling mouse lymphocyte antigens

Recently, research into mouse lymphocyte antigens has been very active, with many new loci and alleles being named. This high activity has unfortunately led to considerable confusion in the nomenclature, with duplication of Ly numbers, and allocation of different names for the same antigen.

A small group of workers in the field met recently and drafted a revised nomenclature as a means of overcoming the various problems (11). A summary of the suggestions is presented here.

Definition of an Ly locus

The designation '*Ly*' (italicized) should be given to a gene coding for a cell surface molecule on lymphocytes, or shared by lymphocytes and other cell types. The antigenic molecule is designated by the same symbol unitalicized. The criteria for recognizing such genes have been broadened considerably since the first *Ly* loci were described, in order to accommodate new information obtained from major technical advances, such as molecular genetic techniques, which have led to the identification of new loci for lymphocyte surface antigens. In particular, (a) somatic cell hybridization techniques have led to the mapping of several loci not exhibiting genetic polymorphism, and must be considered with conventional segregation analysis to identify and map genes; (b) biochemical techniques not involving the use of antibody can be used to identify cell-surface molecules; (c) molecular cloning can be used to identify *Ly* genes not showing genetic polymorphism; and (d) genes can be identified by use of cDNA probes from other species. Genes for which no cell membrane molecule has been found in the mouse can be isolated by this technique using cDNA probes for genes encoding membrane molecules identified in other species. These genes can be given the *Ly* designation followed by the gene name (in brackets) in the original species from which the gene was isolated. For example, human T3 can be detected by antibody but has no known mouse counterpart. However, the gene coding for a human T3 chain was used to isolate a corresponding mouse *T3* gene. This gene should be known as *Ly[T3]*. Placing the name in brackets indicates that the symbol is reserved pending evidence that the gene encodes a lymphocyte cell-surface molecule in the mouse.

It should be noted that these criteria do not distinguish between loci expressed only on lymphocytes, e.g. *Ly-1*, *Ly-2*, and *Ly-7*, and those shared with other cell types, e.g. *Ly-5* and *Ly-6*. It is not practical to make this distinction, as experience has taught us that, with time and more extensive tissue distribution analysis, many molecules have a wider tissue distribution than first reported.

Specificities and alleles

Alleles are designated with lower case letter superscripts and specificities with Arabic numerals separated from the antigen symbol by a period. The allelic superscript *a* is used for specificity .1, *b* is used for specificity .2, etc. The uniform practice of designating the C57BL/6 allele with the superscript *b* and specificity with the number .2 is recommended, e.g., for *Ly-1*, C57BL/6 mice have the allele *Ly-1^b* and the specificity *Ly-1.2*. This convention requires a change of two alleles and specificities, for the *Ly-5* and *Ly-7* loci.

Monoclonal antibodies

Since the advent of monoclonal antibodies, it is possible to produce many different antibodies recognizing the same molecule, including antibodies detecting different epitopes present on one molecule. In the absence of any co-ordinated effort to determine if such epitope differences have any significance in terms of function or differentiation, we do not consider it necessary to distinguish between specificities defined by independently derived monoclonal antibodies. However, where possible, a monoclonal antibody should be used to define an antigen. Some *Ly* antigens defined by monoclonal antibodies have been designated *Ly-m*,

e.g. *Ly-m10*. Since the vast majority of antigens are now defined by monoclonal antibodies, the 'm' should no longer be used. Thus *Ly-m10* is now *Ly-10*.

Xenogeneic antibodies

Lymphocyte antigens can often be detected by xenogeneic antibodies. In some cases, these antibodies can distinguish between specificities determined by different alleles at the particular locus; in others, they react equally well with all specificities. In the latter case, e.g. with an antibody recognizing the *Ly-1* antigen, the antigenic determinant recognized by the antibody is not assigned a specificity number, but can be referred to as a 'framework' determinant or as a molecule carrying the *Ly-1* specificity.

Requirements for designating a new Ly locus

A cell-surface molecule must be identified and shown to be present on lymphocytes. The molecule will usually be identified by serological methods, preferably by a monoclonal antibody, often combined with biochemical characterization.

Evidence for genetic control of the molecule must be shown. If genetic variation is found, crosses should be made to show that the variation is controlled by a single gene and, if possible, to find the chromosomal location of the gene. A unique strain distribution of the alleles or a unique chromosomal location can be taken as evidence for a new *Ly* locus. If variation in the antigen is not found, it may be possible to identify the gene controlling the antigen by somatic cell hybridization, by molecular cloning, or by discovery of other variation in the gene such as a restriction fragment length polymorphism.

If the presumed new locus shows no recombination with a previously known locus, it should not be given a new locus symbol until more definitive evidence is available that the two genes are indeed distinct. Such evidence may include different tissue distributions or different molecular weights. Neither are entirely reliable, however, since apparently different tissue distributions may result from use of different antibodies or different methods of quantifying presence of the antigen, and different molecular weights may result from posttranslational modifications of the antigen.

Ly nomenclature registry

The registry has been established at the National Institutes of Health and will be administered by H.C. Morse III. New *Ly* gene numbers will be allocated in order of receipt of requests, which should be accompanied by a copy of the evidence demonstrating a new *Ly* locus. All applications should be addressed to Dr H.C. Morse III, Laboratory of Immunopathology, National Institute of Allergy and Infectious Diseases, Building 7, Room 304, National Institutes of Health, Bethesda, MD 20892, USA.

Rationalization of existing nomenclature

In order to remove duplications and various other ambiguities some existing nomenclature has been changed. For a comparison of old and new nomenclature please see Morse *et al.* (11). The following points should be noted:

1. In the course of eliminating duplications etc., some loci have been renumbered.

2. The *Lyt* designation has been abandoned, but *Lyb-2* to *Lyb-8* remain unchanged.

3. Various loci already well known by non-conforming names have been left unchanged to avoid confusion. These include *Thy-1*, *Thy-2*, and *Lna-1*. However, it is recommended that, if similar loci are described in the future, they should receive an *Ly* number.

1.2.3 Mouse globin gene nomenclature

A meeting to discuss mouse globin gene nomenclature, organized by Dr Jane Barker, was held at Bar Harbor on 21–24 May 1984.

The need for revision of the nomenclature had arisen largely as a result of increasing knowledge of the α- and β-globin gene complexes, and also to obtain a consistent nomenclature of spontaneous or induced mutations or other variants of already known alleles or haplotypes.

The aim was to devise a system which would as far as possible be consistent with other mouse genetic nomenclature, and nomenclature used for globin genes in other species, and which would be practicable for use with the types of genetic variation found in the mouse.

The following is an extract from the proposals published more fully elsewhere (9, 12).

The proposed system of nomenclature is as follows.

Gene complexes

The α- and β-globin genes should be considered as constituting gene complexes and should be given the names and symbols—haemoglobin-alpha, *Hba* and haemoglobin-beta, *Hbb*.

Haplotypes

The different forms of the complexes should be considered as haplotypes, and designated by superscript lower case letters, e.g. Hba^a, Hba^b, Hbb^d, Hbb^s.

The letters *m* and *o* should be omitted as they might be confused with 'mutant' or 'null', and the letter *w* should be reserved for wild-derived haplotypes.

If the alphabet becomes exhausted then a series of two-letter symbols should be used, beginning *aa*, *bb*, etc.

Loci within complexes

The individual gene loci within the *Hba* and *Hbb* complexes should be denoted by lower case letters, in some cases followed by numbers and set off from the main symbol by a hyphen, e.g. *Hba-x* (previously *Hbx*) and *Hbb-y* (previously *Hbby*). The numbers should run from the 5' end (in contrast to comparable genes in man, but in line with previous numbering for the mouse).

Alleles

The alleles of genes within the complexes should be denoted by superscript lower case letters, indicating the haplotype of origin, e.g. $Hbb\text{-}y^s$, $Hbb\text{-}y^d$, the alleles of *Hbb-y* occurring in haplotypes Hbb^s and Hbb^d.

Variant haplotypes and alleles

When new haplotypes or alleles arise by mutation or other changes in already known haplotypes the new haplotype or allele should be denoted by appending a serial number to the haplotype superscript. If the change is known to be due to a mutation or deletion within a particular allele(s) this should be indicated by adding an appropriate letter symbol and serial number to the allele superscript, set off by a hyphen, e.g. Hbb^{d2}, Hbb^{d3}, the first two variants of the haplotype Hbb^d; $Hbb\text{-}b1^{d\text{-}m1}$, the first mutant allele of the gene $Hbb\text{-}b1^d$; $Hbb\text{-}b1^{d\text{-}d11}$, the first deletion found in the allele $Hbb\text{-}b1^d$.

Pseudogenes

Pseudogenes located at a distance from the main complexes should be given a locus symbol consisting of the main haplotype symbol followed by a hyphen, a serial number, and the lower case letters *ps*, e.g. *Hba-ps3*, *Hba-ps4*, α-globin pseudogenes located away from the main *Hba* complex.

Globin gene registry

In order to maintain an orderly sequence of designation of letters and serial numbers to newly discovered and variant haplotypes and alleles it is proposed that there should be a Mouse Globin Gene Registry. Dr Ray Popp has agreed to be the keeper of the Registry.

Those wishing to name new haplotypes or alleles should write to

> Dr R.A. Popp,
> Biology Division,
> Oak Ridge National Laboratory,
> Oak Ridge, TN 37830,
> USA

with details of the new variant. Appropriate letters and/or numbers can then be assigned without duplication or confusion.

1.2.4 Nomenclature for homoeobox-containing genes

The following is an extract from recommendations on nomenclature for homoeobox-containing genes, drawn up at a meeting on such genes and published by Martin (10).

In accord with the rules of the International Committee for Standardized Nomenclature for Mice the following modified suggestions are proposed: First, any homoeobox-containing gene or genomic fragment may be given the designation *Hox* (Homoeobox) providing that a significant fraction of the amino acids it encodes are identical to those of the homoeobox in the *Drosophila Antennapedia* gene. Second, the criterion for designating a new *Hox* locus is that it occupies a different map position (that is, it is physically distinct) from all other known *Hox* loci. Until this criterion is met, a new homoeobox-containing gene or genomic sequence should be designated by a 'laboratory' name. Third, the designation of any new *Hox* locus or group of loci (complex), will be determined as follows: (a) If it is not apparently closely linked to any previously described *Hox* locus, it should be numbered by accession, using the next available number in the series (*Hox-1*, *Hox-2*, *Hox-3*, and *Hox-4* have already been used to designate *Hox* loci or complexes). If two or more homoeobox-containing loci are present these will be designated by decimal numbers (for example, *Hox-2.1*, *Hox-2.2*). These decimal subdesignations should, where possible, reflect the linear order of the *Hox* loci along the chromosome; (b) If the new gene is known to be closely linked to a previously designated *Hox* locus, it will be given the numerical designation of that locus or complex, and the next available decimal subdesignation in the series. Whenever possible, this should reflect the linear order of the *Hox* loci along the chromosome; (c) Before any new *Hox* designation is used in a publication, the authors are asked to contact Dr Thomas Roderick at the Jackson Laboratory, Bar Harbor, Maine 04609, USA, to ensure that this designation has

not already been assigned and to inform Dr J. Peters (editor of *Mouse News Letter*) of the MRC Radiobiology Unit, Chilton, Didcot, Oxon OX11 0RD, UK, so that the symbols can be added to the annual gene list.

1.2.5 Nomenclature for transgenic mice

A group met at the Jackson Laboratory, Bar Harbor, Maine on 21–24 May 1984 to discuss mouse α- and β-globin gene nomenclature. During the meeting the question of nomenclature for inserted gene sequences was also discussed. A proposed general form for the nomenclature was suggested and the views of a number of experts in the field of transgenic mice were then sought. The majority of those replying favoured the proposals. Various comments and suggestions made have been incorporated in the revised proposals which are now put forward.

The proposed general form is as follows, and is of the type used for chromosome anomalies, or DNA polymorphisms. The symbol would have several parts.

1. A genetic symbol for the type of anomaly, namely Tg—indicating transgenic. (NB The symbol Is is already used for insertions of sizeable chromosome segments into another chromosome, and should not be used for inserted genes.)

2. Enclosed in brackets, the number of the chromosome into which the gene is inserted, e.g. Tg(12). If the chromosome is not known, then the number zero should be used, e.g. Tg(0).

3. Also enclosed in the same brackets, details of the inserted gene. The most important of these details would be the gene symbol of the structural sequence of most interest, e.g. *Hba* for an inserted α-globin sequence. If desired the species of origin of the gene may be denoted by the three-letter code worked out by human geneticists for origin of chromosomes, e.g. HSA, man; MMU, mouse, e.g. Tg(12OCU*Hba*) a rabbit α-globin gene inserted into chromosome 12.

   More detailed information may exist, such as nature and origin of the promoter sequence, insertion of a gene constructed from two or more parts of other genes, and so on. However, inclusion of such details in the symbol would probably make it too unwieldy to use. Hence, such information should be given separately, in the text or in the form of tables.

4. After the brackets, a serial number for the laboratory or investigator performing the insertion, and a symbol used as that investigator's symbol, e.g. Tg(12OCU*Hba*)N33, the 33rd insertion performed at NIH.

The serial number obviously refers to all insertions performed and not to all those in Chr 12 or involving globin genes. The laboratory or investigator's symbol should be that used to designate inbred substrains.

5. The symbol may be so long that abbreviation is required. It is more or less standard practice with chromosome anomalies to give the full symbol only in the 'Materials and Methods' section of a paper and thereafter to abbreviate, e.g. Tg33 or TgN33 might be suitable.

References

1. Committee on Standardized Genetic Nomenclature for Mice. 1963. A revision of the standardized genetic nomenclature for mice. J. Hered. 54:159–162.
2. Committee on Standardized Genetic Nomenclature for Mice. 1973. Guidelines for nomenclature of genetically determined biochemical variants in the house mouse, *Mus musculus*. Biochem. Genet. 9:369–374.
3. Dunn, L.C., H. Grüneberg, and G.D. Snell. 1940. Report of the committee on mouse genetics nomenclature. J. Hered. 31:505–506.
4. Green, M.C. 1979. Genetic nomenclature for the immunoglobulin loci of the mouse. Immunogenetics 8:89–97.
5. Green, M.C. 1981. *Genetic Variants and Strains of the Laboratory Mouse*, 1st edn. Gustav Fischer, Stuttgart.
6. IUPAC-IUB Commission on Biochemical Nomenclature (CBN). 1978. Nomenclature of multiple forms of enzymes. Recommendations (1976). Arch. Biochem. Biophys. 185:1–3.
7. Klein, J., F.H. Bach, F. Festenstein, H.O. McDevitt, D.C. Shreffler, G.D. Snell, and J.H. Stimpfling. 1974. Genetic nomenclature for the H-2 complex of the mouse. Immunogenetics 1:184–188.
8. Kohn, H.,I., J. Klein, R.W. Melvold, S.G. Nathenson, D. Pious, and D.C. Shreffler. 1978. The first H-2 mutant workshop. Immunogenetics 1:279–284.
9. Lyon, M.F., J.E. Barker, and R.A. Popp. 1988. Mouse globin gene nomenclature. J. Hered. 79:93–95.
10. Martin, G.R. 1987. Nomenclature for homoeobox containing genes. Nature 325:21–22.
11. Morse, H.C. III, F.-W. Shen, and U. Hämmerling. 1987. Genetic nomenclature for loci controlling mouse lymphocyte antigens. Immunogenetics 25:71–78.
12. Mouse News Letter. 1985. Mouse News Lett. 72:23–26.
13. Shreffler, D.C., C. David, D. Gotze, J. Klein, H.O. McDevitt, and D. Sachs. 1974. Genetic nomenclature for new lymphocyte antigens controlled by the *I* region of the *H-2* complex. Immunogenetics 1:189–190.
14. Shreffler, D.C. 1979. Report on recommendations for new mutant *H-2* haplotype nomenclature. Mouse News Lett. 60:34–36.
15. Skolnick, M.H., and U. Francke. 1981. Report of the committee on human gene mapping by recombinant DNA techniques. Cytogenet Cell Genet 32:194–204.

2 CATALOG OF MUTANT GENES AND POLYMORPHIC LOCI

MARGARET C. GREEN

More than 1300 genetic loci have been described in the mouse. Most of them have been recognized by the occurrence of mutations causing visible effects such as changed color, morphology, behavior, etc., or by discovery of differences between strains in protein structure, enzyme activity, antigenic determinants, immune responses, DNA sequences, etc. A growing number of loci for which no variants are known in the mouse are being identified in somatic cell hybrids with Chinese hamster, rat, or human cell lines. This catalog is an attempt to list and describe all of these loci of the mouse. It does not include loci identified only as proteins or DNA sequences, and for which there are no genetic variants or no information on chromosomal location. Since the information about individual loci and their alleles varies considerably in reliability, the choice of genes to be included in the list has been somewhat arbitrary. With some exceptions, I have included those listed in the main Mouse Gene List in *Mouse News Letter* and excluded those in the list of Provisional Gene Symbols. Except in the case of highly polymorphic loci such as the *H-2* and *T* complexes, I have tried to list all known alleles.

A number of traits in the mouse that appear to be under single-locus control have been described, but the loci concerned have not been named. A list with descriptions of the unnamed loci known to me is given at the end of the list of named loci.

The catalog is arranged alphabetically by locus *symbol*. Locus and gene *names* are not listed separately in the catalog but may all be found in the index, which gives the symbol under which each is described.

The chromosomal assignment of each locus is given, when known, immediately after the name of the locus. When not given, this indicates that the information is not known. References for the chromosome assignments are given in many but not all cases. References for all chromosome assignments can be found in Chapter 6. 'Chromosome' is abbreviated 'Chr' throughout.

For a number of cases in which authors have given names and symbols for loci but not for alleles, I have adopted the convention of designating the allele for the presence of the marker trait by a superscript a, and the allele for absence of the trait by a superscript b. In the case of loci governing resistance and susceptibility to infectious organisms or other noxious agents, when authors have not assigned allelic symbols, I have used superscripts r for resistance and s for susceptibility.

For the convenience of readers, the references cited for each locus, gene complex, or group of related loci are listed immediately after the catalog entry. The lists of references are not intended to be complete, but are selected so as to give access to the previous literature.

One purpose of the catalog is to allow searches for genes of particular types or with particular phenotypic effects. To facilitate such searches, the genes have been roughly classified according to their effects as shown in Table 2.1. The classification is modified from that used in *Mouse News Letter*. To identify genes with more specific effects than those used in this classification, the index should be consulted.

The assembly of this large amount of information would have been nearly impossible without the facilities of the Jackson Laboratory Library. The Library's collections provide virtually complete coverage of the literature on subjects related to mouse genetics, and have the added advantage of easy and convenient accessibility. It is a pleasure to express my gratitude to Joan Staats, the librarian responsible for assembling and organizing the collection until her retirement in 1984, and to the present staff, Alison Baker, Moyha Lennon-Pierce, and Ann Jordan, for tolerating my use of office space in the library over a period of several years and for ever ready assistance in tracking down elusive references.

I am also deeply indebted for review of all or parts of the catalog to Drs George A. Carlson, Muriel T. Davisson, Eva M. Eicher, Joseph H. Nadeau, Leonard D. Shultz, Christine A. Kozak, and Mary F. Lyon. I owe a special debt of gratitude to Muriel Davisson for alerting me to many references I would otherwise have missed and for resolving dozens of nomenclatural problems. Finally, I am grateful to Dr Earl L. Green for manning the dictionary during my struggles with English and scientific spelling and for much other help in preparing the 'diskscript'.

Table 2.1 Classification of loci by their phenotypic effects. For loci with multiple alleles, one or more of the alleles have the effect in that category.

Color and white spotting

a	coa	f	le	mu	ru-2	U
ash	Crm	fc	Li	Och	Rw	uw
b	d	fe	ln	p	s	Va
bf	da	Fk	ls	pa	sea	vl
bg	Dfp	Ga	m	pe	si	W
Bst	dp	gc	md	Ph	Sl	Ym
bt	dr	gl	mg	Rn	slt	
bt-2	dsu	gm	mh	rp	Sp	
c	e	gr	mi	rs	Sta	
cht	ep	hs	Mo	ru	tp	

Skin and hair texture

ab	cr	fs	hr	Ng	sa	tf
Al	crh	ft	ic	nu	sch	thd
ao	cw	fz	lm	olt	sf	Tsk
ap	dep	fzt	lt	pf	Sha	Tsk-2
at	dl	go	ma	pk	Sk	Ve
ba	Eh	Gs	mc	Ra	soc	wa-1
bal	Er	Hct	md	rc	spc	wa-2
Bda	exf	hid	me	Re	spf	wal
Bpa	fd	hl	mk	rhg	Str	Wc
Bsk	fr	Hpt	Mo	ro	Ta	we
Ca	Frl	Hq	N	rst	Td	Xpa

Skeleton

Aft	ct	fl	lx	Po	srn	Xpl
am	cy	fro	mdac	pr	st	Xt
ank	Dac	Fu	mdg	Ps	stb	
Bda	Dcp	gl	Mdmg-1	Pt	stm	
Bhd	de	Gy	mea	pu	sy	
bm	Dh	Hd	mi	px	t	
Bn	di	Hk	Mo	py	Tal	
bp	dm	hl	Mv	Q	tc	
Br	Dmm	hop	myd	Rf	tk	
can	dpy	Hx	oc	rh	tl	
cby	dr	Hyp	ol	rv	Ts	
ccd	Dre	It	op	sb	Tsk	
Cd	Ds	jg	Os	sc	Tsk-2	
ch	Dsh	Kw	oto	Sd	tu	
cho	Ed	ky	p	se	Ul	
cl	f	ld	pc	sho	un	
cmd	far	Lp	Pdn	sm	ur	
cn	fhd	lst	Ph	smc	us	
crn	fi	lu	pn	sno	vt	

Tail and other appendages

Aft	Dsh	heb	lu	pu	sy	vt
bp	eb	Hk	lx	px	t	Xpl
bl	Ed	Hm	mdac	py	Tal	Xt
Bn	Eh	hop	mea	Q	tc	
Bst	f	Ht	Mp	Rf	Tht	
Cd	fh	Hx	my	rh	tint	
cl	fi	Ie	ol	sb	tk	
ct	fl	jg	Os	sc	Ts	
cy	ft	jt	Pdn	Sd	tu	
Dac	Fu	Kw	Po	se	Ul	
de	fzt	ld	pr	sm	un	
Dfp	gr	Lp	Ps	srn	ur	
Dre	Hd	lst	Pt	st	us	

Table 2.1—cont.

Eye

Acc	Cat	Eo	j	Mp	or	Vlm
ak	ch	ey-1,-2	Len-1,-2	my	pcd	Xcat
Alm	Cm	fh	lg	nct	Phil	
Anc	cri	fhd	lm	Nop	Pscr	
Apoc	Crya-1	fi	Lop	nr	Pscs	
Apyc	Cts	gp	lop-2	nuc	rd	
bl	eb	heb	Lop-4	Nuca	rds	
Bld	ec	Iac	lr	Nzc	Sey	
Bru	Elo	Idc	Lse	oe	Sig	
cac	Em	Ie	mi	oel	vl	

Inner ear and circling behavior

av	dv	het	Kw	pi	sh-2	th
bv	ecl	jc	mk	pr	sr	Tw
co	ecs	je	mu	Pv	st	v
Dc	fi	js	nv	Q	su	Va
dn	Fu	jv	oc	rg	sv	wl
dr	Gy	kr	pa	sh-1	sy	Wt

Neurological and neuromuscular

ac	ct	hpc	mdg	oto	st	trm
adr	d	hpy	mdm	par	stu	twi
ag	dfw	hy-1 to	mdx	pcd	sw	twt
anx	Do	hy-3	mea	pma	Swl	ty
asp-1	dt	hyh	med	qk	sy	vb
ax	du	ji	Mnd	qv	tb	vc
bc	dy	jp	Mo	rl	tg	vl
bh	El	Lc	mto	sg	ti	wd
cb	epf	lcsd	myd	shi	tip	wh
ch	Exnm	lh	nr	shm	tlt	wl
cla	fa	lm	oc	Slf	tm	wr
Cm	gt	Lp	oh	Sp	tn	wv
cod	Hld	lz	opt	spa	tor	xn
cri	ho	mdf	ot	spm	Tr	

Other behavioral

Aal	Cpz	Phr	Ri-1,-2	Sco
Aap	Eam	Pp	Rua	Sip
anx	Exa	Qui	Sac	Soa

Hematological

Aia-1,-2	Ema	hbd	ja	nb	Spna-1
an	f	hea	lst	Sl	Spnb-1
cri	Hba	hpx	mk	sla	Ts
dm	Hbb	hub	Nan	sph	W

Endocrine defects, hormones, growth, obesity

a	cho	dw	hg	Ins-1 to	ob	stm
Ad	cmd	Ex	Hnl	Ins-3	pg	Tgn
ald	cn	Ezg	Hom-1	Lhb	Pth	Tsha
bm	cog	fat	hpg	lit	Smst	Tshb
Calc	df	Fshb	hyt	mn	Spl	tub
can	dm	Gcg	Idd-1,-2	Mo	stb	

Reproductive organs, sterility

an	hpg	jsd	p	sph	Tda-1
at	hop	lst	px	Sxr	Tdy
azh	Hst-1,-2	lu	qk	sy	Tfm
bs	hyt	olt	qv	t	W
H-Y	jg	om	Sl	Tas	wr

Table 2.1—*cont.*

Internal defects of viscera, etc.

bd	*cpk*	*far*	*ld*	*nu*	*Sd*	*vi*
bl	*Dh*	*fm*	*lec*	*ol*	*se*	*Xpl*
Br	*Dom*	*hf*	*ls*	*Os*	*srn*	
cab	*dr*	*iv*	*lx*	*pnc*	*Tsk*	
ch	*Ds*	*j*	*Mo*	*px*	*Tsz-1*	
Chy	*eb*	*jpk*	*my*	*Ra*	*ur*	
cph	*epi*	*kd*	*Nil*	*s*	*us*	

Enzymes

Aco-1	*Atpa*	*Es-27*	*Gsta*	*Lzp*	*Pfk-1*	*Sdr-1,-2*
Acp-1	*Atpb*	*Esr*	*Gus*	*Lzm*	*Pgam*	*Sod-1,-2*
Acy-1	*Bcd*	*Fbp-1,-2*	*Hao-1,-2*	*Map-1*	*Pgd*	*spf*
Ada	*bg*	*For-1* to	*Hat-1*	*Mep-1*	*pge*	*Spi-1,-2*
Adh-1,-3	*Bgl*	*For-5*	*Hdc*	*Mod-1,-2*	*Pgm-1* to	*Sto-1*
Adk	*Blvr*	*Fpgs*	*Hex-2*	*Mor-1*	*Pgm-3*	*Sts*
Adr-1 to	*c*	*Fuca*	*Hiomt*	*Mpi-1*	*Phk*	*Suc-1*
Adr-4	*Caa*	*G6pd*	*his*	*Nat-1,-2*	*Pk-1,-3*	*Tag*
Ags	*Car-1,-2*	*Galt*	*Hk-1*	*Neu-1*	*Plat*	*Tam-1*
Ah	*Cas-1*	*Gapd*	*Hma*	*Np-1,-2*	*Plau*	*Tat*
Ahd-1 to	*Ce-1,-2*	*Gda*	*hph-1*	*Oat*	*Por*	*Tdt*
Ahd-6	*cld*	*Gdc-1,-2*	*Hprt*	*Odc-1* to	*Pre-1,-2*	*Thd*
Ahr-1 to	*Coh*	*Gdcr-1,-2*	*Hsd*	*Odc-11*	*Prgs*	*Tk-1*
Ahr-4	*Cpa*	*Gdr-1,-2*	*Hti*	*Oh21-1,-2*	*Pro-1*	*Tpi-1*
Ak-1,-2	*Cs*	*Ggc*	*Idh-1,-2*	*Otc*	*Prt-1* to	*Trp-1*
Akp-1 to	*Ctrb*	*Ggm-1,-2*	*Ipo-1*	*Ox-1,-2*	*Prt-5*	*Tse-1,-2*
Akp-3	*Dao*	*Ggtb-1*	*Itp*	*P450-1,-3*	*Psp*	*Udpgt-3*
Aldh	*Dhfr*	*Gk*	*Kal*	*P450-2D*	*Psph*	*Umph-2*
Amy-1,-2	*Dia-1*	*Glk*	*Lap-1,-2*	*pa*	*Ren-1,-2*	*Upg*
Aox-1,-2	*Dlb*	*Glo-1*	*Lck*	*Pck-1*	*Rib-1*	*Ups*
Apk	*Ela-1*	*Got-1,-2*	*Ldh-1,-2*	*Pcn*	*Rpo2-1*	*Xld-1*
Aprt	*Eno-1*	*Gpd-1,-2*	*Ldr-1,-2*	*Pd*	*Rrm2*	*Xpa*
As-1,-2	*ep*	*Gpi-1*	*le*	*pe*	*ru*	
Asl	*Eph-1*	*Gpt-1*	*Lip-1*	*Pep-1* to	*ru-2*	
Ass	*Es-1* to	*Gr-1*	*Lv*	*Pep-7*	*Sdh-1*	

Receptors

Abp	*Acrg*	*Gbp-1*	*Micrl*	*Rmc-1*	*Tcrb*
Acra	*C4bp*	*Grl-1*	*Mtvr*	*Ssp*	*Tcrd*
Acrb	*Crl-1*	*Ifcr*	*Ram-1*	*T3d*	*Tcrg*
Acrd	*Fcr*	*Ifgr*	*Rev-1*	*Tcra*	

Other proteins

Acf-1	*cdm*	*Gsa*	*Kth-1,-2*	*Mylc*	*Prm-1,-2*	*t* complex
Afp	*Cola-1,-2*	*Hdl-1*	*Lamb-1,-2*	*Mylf*	*Prn*	*Tcn-2*
Afr-1,-2	*Cp*	*Hist1*	*Len-1,-2*	*Mylpf*	*Prp*	*Timp*
Alb-1	*Crbp*	*Hist2*	*Lfo-1,-2*	*Ncam*	*Q10*	*Tnfa*
Ap2	*Crip*	*Hba*	*Lth-1,-2*	*Ngfa*	*Rpl*	*Tnfb*
Aph-1	*Crya-1*	*Hbb*	*Ltw-1* to	*Ngfb*	*Rps*	*Trf*
Apoa-1,-2	*Csn*	*Hp*	*Ltw-6*	*Ngfg*	*Saa*	*Trp53*
Apob	*Eg*	*Hrt-1*	*Lvp-1*	*Orm-1,-2*	*Sap*	*Tse-3*
Apoe	*Egf*	*Ifa*	*Mt-1,-2*	*Pan-1*	*Sas-1,-2*	*Ucp*
App	*Elo*	*Ifb*	*Mtp-1,-3*	*Pgy-1*	*Sinc*	*Wap*
Ath-1	*Erp-1*	*Ifbip-1*	*Mup-1*	*Plp*	*Sparc*	
Badm	*Fabph-1* to	*Ifg*	*Myhc*	*Pnd*	*Svp-1* to	
Brp-1 to	*Fabph-4*	*Kfo-1*	*Myhs*	*Pomc-1,-2*	*Svp-3*	
Brp-14	*Gia*	*Krt-1*	*Myla*	*Pre-1,-2*	*Syn-1*	

Table 2.1—*cont.*

Cell surface antigens

B2m	Grl-1	Hya	Lna-1	Lyx-1 to	Pca-1	Thb
Cxv-1-2	Gt-1,-2	Hye	Lsd	Lyx-3	Qa-1 to	Thy-1,-2
Dtc-1	Gv-1,-2	Ia	Ly-1 to	Meta	Qa-11	Tind
Ea-1 to	H-1 to	Ii	Ly-34	Mls	Qb-1	Tla
Ea-8	H-45	Jt-1	Ly-37	Mph-1	Sgp-1,-2	Tpre
Epa-1	Hmt	Lm-1	Lyb-1 to	Nel-1	Skn-1,-2	Tsu
Gm-3	Hxa	Lmp	Lyb-8	Nk-1,-2	Spl-1	Tthy

Other immunological

Air1	Crl-1	If-1 to	Ifg	Il-1	Rig-1
Bmfr-1,-2	Csfgm	If-4	Ifrc	Ir-1 to	Sas-1
C2 to C7	Csfmu	If-X	Igh	Ir-7	Sr-1
C4bp	Ctl-1	Ifa	Igj	Lus	Tol-1
Cfh	Ctla-1	Ifb	Igk	Lyx-1 to	xid
Cfhe		Ifbip-1	Igl	Lyx-3	Xlr

Viral, disease, and tumor resistance

Ack	Flv	Lsr-1	Rcs-1	Rmcf	Rv-2,-3	Tgfb
Av-1,-2	Fv-1 to	Mal-1	Rfv-1 to	Rmp-1	Rvil-1	
Bcg	Fv-6	Mx	Rfv-3	Rmv-1 to	Scl-1,-2	
Bcga	Hv-1,-2	Pas-1 to	Rgv-1,-2	Rmv-3	Srlv-1	
Cms	Ihe	Pas-3	Rhv-1,-2	Rrs	Srv	
crz	Ity	Pgy-1	Ric	Rrv-1	Sxv	
Fhe	Lsh	Ptv-1	Ril-1	Rsm-1	ter	

Immune defects

A^y	Agp-1,-2	Dh	kd	lpr	med	W
an	Aia-1,-2	gld	lh	Lps	nu	wst
Adr-1,-2	Arp	hr	Lpn-1 to	Lus	scid	xid
Aem-1	bg	Idd-1,-2	Lpn-3	me	Sl	Yaa

Oncogenes and viral integration sites

Abl	Erbb	Hras-1	Mlvi-1	Pim-1	Sis
Araf	Ets-1,-2	Int-1 to	Mos	Pvt-1	Src
Bcl-2	Fes	Int-3	Myb	Raf	Srch
Dsi-1	Fis-1	Kras-2	Myc	Rel	
Erba	Fos	Mis-1	Nras	Rras	

Endogenous viruses

Bbv	Fgv-1,-2	Mlv-1	Mov-24	Nfxv-1	Xmmv-1 to
Bdv-1	Emv-1 to	Mmv-1 to	Mov-34	Nzv-1,-2	Xmmv-76
Bxv-1	Emv-18	Mmv-13	Mtv-1 to	Prn	Xmv-1 to
C58v-1 to	Inb-1	Mov-1 to	Mtv-25	Ptv-1	Xmv-5
C58v-4	Inc-1	Mov-15	Mxv-1	Sinc	

DNA sequences, transgenes

D7Ag2	D12Nyu1 to	D4Rp1 to	Hgh-3, -4	Mgpt-1	Ms6hm	t complex
D9Ag3	D12Nyu5	D12Rp54	Hox-1 to	Mov-14	Ms15-1 to	Y DNA
D12Mcg1	DXPas1 to	DXS32	Hox-3	Ms6-1 to	Ms15-8	
	DxPas5	En-1	Mdm-1, -2	Ms6-5	Myk-103	

RNA

Rn7s-6	Rnr12	Rnu1a	Rnu1b

Table 2.1—*cont.*

Homozygous lethality or sublethality

a	cp	Fk	Lp	ot	Sl	ur
ag	cpk	Frl	ls	pc	Sp	us
am	crn	gl	mdg	pf	sph	vb
bd	d	Hd	me	Ph	srn	Ve
Bhd	Dc	Ht	med	Ps	Sta	W
Bld	Dfp	hy-1 to	mi	Ra	Str	Wc
Bpa	Dh	hy-3	mn	Rf	sy	wh
Br	Dmm	j	Mo	Rw	t	wl
c	Do	ji	N	s	Td	Wt
cab	dt	jp	Nil	Sd	tn	xn
can	dy	jpk	oc	Sey	Ts	Xt
cb	Eh	l(17)-1 to	oed	sf	Tsk	Ym
ch	Er	l(17)-4	oel	sg	Tsk-2	
cho	fa	Lc	opt	sho	Tw	
cmd	fh	lec	Os	Sig	ty	

Miscellaneous

Ath-1	cab	fh	lm	Oua-1	tx
bl	Caf	heb	Lse	Oubr	War
Bru	cdm	ic	my	Segr	Xce
Bsp	Ds	lcsd	oed	Sxa	

A

a locus, Chr 2

The agouti locus controls the relative amount and distribution of yellow pigment (phaeomelanin) and black pigment (eumelanin) in hairs of the coat. Experimental evidence indicates that the genes at this locus act only in the hair follicles. Pigment produced at sites other than the hair follicles is always eumelanin regardless of the genotype at the a locus (25, 39). Separation and recombination of embryonic dermis and epidermis have been used to determine the site of action of agouti locus alleles in the skin. These experiments have shown that all alleles studied act in the dermis but that the A^y, A^{vy}, a^m, and A^s alleles act also in the epidermis (26, 32–5). They may do so by interfering with a message from the dermis to the follicular melanocytes (33, 34). In genotypes having both black and yellow pigment in the hair, the melanocytes can switch back and forth between black and yellow, but the mechanism controlling the switch is not known (12, 13, 20, 28). Administration of melanocyte-stimulating hormone *in vivo* or in cultured skin to a- locus genotypes making yellow pigment (A^y/-, A/-, etc.) causes a shift from phaeomelanin to eumelanin formation (17). The shift is probably mediated by cAMP and is dependent on gene transcription and translation (42). The melanin granules forming phaeomelanin are different in shape and internal structure from those forming eumelanin (38). Mutations from a to a^t and A^w are very common, but other mutations within the locus appear to be rare. The dominance relations among the alleles are complex and are given, when known, with the descriptions of the various alleles.

+ or A, agouti. Usually regarded as the wild-type allele. Hairs are black with a subapical yellow band, the typical agouti pattern. The general appearance is yellowish gray, slightly lighter on the belly than on the back.

A^i, intermediate agouti. Arose spontaneously in the C57BL/6J strain. A^i/A^i, A^i/A, and A^i/a^{td} resemble A^w/-. A^i/a^t and A^i/a mice have a dark back and light belly with agouti hairs along the side (4).

a

A^{iy}, intermediate yellow, dominant. Arose spontaneously in the C3H/HeJ strain. Heterozygotes (A^{iy}/a) are sootier than A^y/a and are often indistinguishable from A/a. They do not have the mottling often observed in A^{vy}/a. Homozygotes are viable, as are A^{iy}/A^y and A^{iy}/A^{vy} (6).

A^s, agouti suppressor, semidominant. Found among the offspring of an irradiated (C3H × 101)F1 male. A^s shows 0.6 per cent crossing over with A^w and a^t, but is regarded as part of the agouti locus because it shows a *cis–trans* position effect with other alleles at the locus. When homozygous, A^s reduces the amount of yellow pigment produced by whatever other alleles are present, particularly in the hairs of the pinna and belly. In heterozygotes, it affects only the allele situated on the same chromosome (30). The effect in heterozygotes with other agouti-locus alleles ($A^s A^w/+ x$, where $x A^w$, A, a^t, or a) is more pronounced in the pinna, tail, and hindfoot than in the body pelage (15). A^s is associated with a long inversion, In(2)2H, which suppresses crossing over between *pa* and A^s and increases it between *a* and *bp* (11).

A^{sy}, sienna yellow, semidominant. Arose spontaneously in the C57BL/6J strain. Homozygotes are viable and are a clear yellow. A^{sy}/a mice are a dark sooty yellow. A^{sy}/A^y mice are viable and fertile (7).

A^{vy}, viable yellow, dominant. Arose spontaneously in the C3H/HeJ strain. A^{vy} resembles A^y except that homozygotes are viable and resemble heterozygotes with lower members of the series. Both homozygotes (A^{vy}/A^{vy}) and heterozygotes (A^{vy}/A and A^{vy}/a) show considerable variation in appearance, ranging from clear yellow, through mottling with dark patches, to a complete agouti-like coat (14). The variation is strongly influenced by the agouti-locus genotype and strain genome of the dam (46). A larger proportion of homozygotes than heterozygotes are clear yellow (5). In combination with the dominant spotting allele A^{2J}, A^{vy} heterozygotes that have the agouti phenotype have much larger white spots than those of yellow or mottled yellow phenotype (21). Homozygotes and heterozygotes tend to become obese, and the degree of obesity is correlated with the amount of yellow in the coat (43), except that E^{so} prevents synthesis of yellow pigment in $A^{vy}/-$ mice but does not prevent obesity (46). A^{vy} resembles A^y in causing greater tumor susceptibility and lower graft vs. host reactivity (16) and higher hepatic malic enzyme activity (18). The greater tumor susceptibility as well as several altered immune responses occur in A^{vy}/a mice of mottled phenotype but not in those of agouti phenotype (36).

A^w, white-bellied agouti. Like A except that the belly is white or cream. Dominant to A and all alleles with lower case symbols. Since mutations to this allele are common, the A^w alleles found in many laboratory stocks may not be of common origin.

A^y, yellow, semidominant. An old mutant of the mouse fancy. In heterozygotes, all the hair pigment is yellow, but eyes are black. In combination with spotting genes, A^y usually causes reduction in size of white spots (9, 22). Heterozygotes usually become obese and sterile after the first few months. Increased adipose tissue mass is due to fat-cell hypertrophy. The amount of total body fat is higher than normal even when body weight is maintained at normal levels by diet restriction (18). Heterozygotes are more susceptible to several kinds of tumors than normal mice, and their spleen cells cause a significantly lower graft vs. host reaction (16). The level of malic enzyme in the liver is elevated (45). Homozygotes die before implantation or shortly thereafter, the time of death and type of abnormality being in part dependent on the genetic background (27). In embryos affected early, there is exclusion of some blastomeres from the embryo after the eight-cell stage (1); in embryos affected later, abnormalities begin at implantation with failure of trophoblast giant cell development (10). No single ultrastructural alteration characteristic of A^y/A^y pre-implantation embryos has been found (2). Probable recombination has been reported between A^y and the *a* and a^x alleles. This suggests that A^y is a pseudoallele of the *a* locus. The yellow color is not separable from lethality and may result from an interaction of the lethal gene with A to produce the yellow phenotype. A^y is probably on the proximal side of the *a* locus (41). An ecotropic provirus (*Emv-15*) is closely associated with the A^y mutation (3), but has been separated from it in the YBR-A^y/a strain (40). See *Emv-15* locus.

a, nonagouti. An old mutant of the mouse fancy. *a/a* mice are plain black on back and belly, with some yellow hairs on the ears and perineum. Recessive to A, A^w, a^t, a^{td}, and all the yellow alleles.

a^{16H}, nonagouti-16H. ENU-induced in spermatogonial stem cells of a (C3H/HeH × 101/H)F1 male. a^{16H}/a mice resemble A/A on the belly and flanks and *a/a* on the back. Homozygotes die prenatally. a^{16H}/A^y and a^{16H}/a^t mice are viable, indicating that the lethal site of a^{16H} differs from that of the other two alleles. a^{16H}/a^l and a^{16H}/a^e mice are alike and have a black back and pinna hairs and agouti belly and flanks (24).

a^{da}, nonagouti with dark agouti belly. Recovered from an irradiation experiment. Recessive to A. Homozygotes and a^{da}/a mice have a black or very dark back, a dark agouti belly, and yellow pinna hairs (31).

a^e, extreme nonagouti. Found among descendants of an irradiated mouse (19). Homozygotes are very dark all over with no yellow hairs in the ears or around the nipples and perineum. Recessive to all other alleles except a^l.

a^l, nonagouti lethal. Radiation-induced in a (C3H/HeH × 101/H)F1 female. Recessive to A, dominant to a. a^l/a^e mice resemble a^e/a^e mice. Homozygotes die prenatally. a^l/A^y mice also die prenatally, indicating that a^l and A^y may be deletions that overlap at a site causing prenatal lethality. The a^l deletion does not extend as far as the closely linked bp locus (24).

a^m, mottled agouti. Derived from an irradiation experiment (37). Homozygotes and a^m/a mice vary from all black, indistinguishable from a^e/a^e, through various degrees of yellow and black mottling to a full agouti pattern. The variation is strongly influenced by the agouti-locus genotype and strain genome of the dam (44).

a^t, black and tan. Found by Dunn (8) in a strain obtained from an English fancier. Black back and cream or yellow belly. Recessive to A on back but dominant on belly. A/a^t mice resemble $A^w/-$. Dermal–epidermal recombination experiments have shown that the regional differences in color are determined by gene action in the dermis (35).

a^{td}, tanoid. Arose spontaneously in the C57BL/6J strain. Homozygotes, a^{td}/a^t, and a^{td}/a all have a light belly and a very dark agouti back. a^{td}/a resembles $A^w/-$ (23).

a^u, agouti umbrous. Recovered from an irradiation experiment. Recessive to A. Homozygotes have a dark agouti back, belly, and pinna hairs; a^u/a mice have an umbrous back, dark agouti belly, and yellow pinna hairs; and a^u/a^t mice have an umbrous back, tan belly, and yellow pinna hairs. a^u resembles A^s, but does not cross over with A (31).

a^x, lethal nonagouti. Induced by irradiation of spermatogonia. Homozygotes are abnormal at the blastocyst stage but implant and develop to varying degrees. By 7.5 to 8.5 days, only small disorganized clumps of cells remain. a^x is recessive to A^y, A^w, A, and a^t, and dominant to a. Heterozygotes with a or a^e have a lighter belly than a/a or a^e/a^e animals. a^x shows about 0.5 per cent recombination with A^y (29).

References

1. Calarco, P.G., and R.A. Pedersen. 1976. Ultrastructural observations of lethal yellow (A^y/A^y) mouse embryos. J. Embryol. Exp. Morphol. 35:73–80.
2. Cizaldo, G.R., and N.H. Granholm. 1978. Ultrastructural analysis of preimplantation lethal yellow (A^y/A^y) mouse embryos. J. Embryol. Exp. Morphol. 45:13–24.
3. Copeland, N.G., N.A. Jenkins, and B.K. Lee. 1982. Association of the lethal yellow (A^y) coat color mutation with an ecotropic murine leukemia virus genome. Proc. Natl. Acad. Sci. USA 80:247–249.
4. Dickie, M.M. 1962. Mouse News Lett. 27:37.
5. Dickie, M.M. 1962. A new viable yellow mutation in the house mouse. J. Hered. 53:84–86.
6. Dickie, M.M. 1966. Mouse News Lett. 34:30.
7. Dickie, M.M. 1969. Mouse News Lett. 41:31.
8. Dunn, L.C. 1928. A fifth allelomorph in the agouti series of the house mouse. Proc. Natl. Acad. Sci. USA 14:816–819.
9. Dunn, L.C., E.C. MacDowell, and G.A. Lebedeff. 1937. Studies on spotting patterns. III. Interactions between genes affecting white spotting and those affecting color in the house mouse. Genetics 22:307–318.
10. Eaton, G.J., and M.M. Green. 1963. Giant cell differentiation and lethality of homozygous *yellow* mouse embryos. Genetics 34:155–161.
11. Evans, E.P., and R.J.S. Phillips. 1978. Mouse News Lett. 58:44–35.
12. Galbraith, D.B., and R.J. Arceci. 1974. Melanocyte populations of yellow and black hair bulbs in the mouse. J. Hered. 65:381–382.
13. Galbraith, D.B., and A.M. Patrignani. 1976. Sulfhydryl compounds in melanocytes of yellow (A^y/a), nonagouti (a/a), and agouti (A/A) mice. Genetics 84:587–591.
14. Galbraith, D.B., and G.L. Wolff. 1974. Aberrant regulation of the agouti pigment pattern in the viable yellow mouse. J. Hered. 65:137–140.
15. Galbraith, D.B., G.L. Wolff, and N.L. Brewer. 1980. Hair pigment patterns in different integumental environments of the mouse. Influence of the agouti suppressor (A^s) mutation on expression of agouti locus alleles. J. Hered. 71:229–234.
16. Gasser, D.L., and T. Fischgrund. 1973. Genetic control of the immune response in mice. IV. Relationship between graft vs host reactivity and possession of the high tumor genotypes A^{y}a and A^{vy}a. J. Immunol. 110:305–308.
17. Geschwind, I.I., R.A. Huseby, and R. Nishioka. 1972. The effect of melanocyte–stimulating hormone on coat color in the mouse. Rec. Prog. Horm. Res. 28:91–130.
18. Herberg, L., and D.L. Coleman. 1977. Laboratory animals exhibiting obesity and diabetes syndromes. Metabolism 26:59–99.
19. Hollander, W.F., and J.W. Gowen. 1956. An extreme non-agouti mutant in the mouse. J. Hered. 47:221–224.
20. Knisely, A.S., D.L. Gasser, and W.K. Silvers. 1975. Expression in organ culture of agouti locus genes of the mouse. Genetics 79:471–475.

21. Lamoreux, M.L. 1983. Developmental interaction in the pigmentary system of mice. II. Interaction in genetically identical mice. J. Hered. 74:440–442.

22. Lamoreux, L., and E.S. Russell. 1971. Effects of agouti-locus alleles (A^y/a vs a/a) on amount of white spotting when combined with two different spotting genotypes (s/s^l and W^{j2}/+). Genetics 68:s36 (Abstr.).

23. Loosli, R. 1963. Tanoid—a new agouti mutant in the mouse. J. Hered. 54:26–29.

24. Lyon, M.F., G. Fisher, and P.H. Glenister. 1985. A recessive allele of the mouse agouti locus showing lethality with yellow, A^y. Genet. Res. 46:95–99.

25. Markert, C.L., and W.K. Silvers. 1956. The effects of genotype and cell environment on melanoblast differentiation in the house mouse. Genetics 41:429–450.

26. Mayer, T.C., and J.L. Fishbane. 1972. Mesoderm–ectoderm interaction in the production of the agouti pigmentation pattern in mice. Genetics 71:297–303.

27. McLaren, A. 1976. Genetics of the early mouse embryo. Ann. Rev. Genet. 10:361–388.

28. Moyer, F.H. 1966. Genetic variation in the fine structure and ontogeny of mouse melanin granules. Am. Zool. 6:43–66.

29. Papaioannou, V.E., and H. Mardon. 1984. Lethal nonagouti (a^x): description of a second embryonic lethal at the agouti locus. Dev. Genet. 4:21–29.

30. Phillips, R.J.S. 1966. A cis–trans position effect at the A locus of the house mouse. Genetics 54:485–495.

31. Phillips, R.J.S. 1976. Mouse News Lett. 55:14.

32. Poole, T.W. 1975. Dermal–epidermal interactions and the action of alleles at the agouti locus in the mouse. Dev. Biol. 42:203–210.

33. Poole, T.W. 1980. Dermal–epidermal interactions and the action of alleles at the agouti locus in the mouse. II. The viable yellow (A^{vy}) and mottled agouti (a^m) alleles. Dev. Biol. 80:495–500.

34. Poole, T.W. 1982. The agouti suppressor (A^s) coat color mutation in mice: developmental effects on the expression of agouti locus alleles. J. Exp. Zool. 220:57–64.

35. Poole. T.W., and W.K. Silvers. 1976. The development of regional pigmentation patterns in black and tan (a^t) mice. J. Exp. Zool. 197:115–120.

36. Roberts, D.W., G.L. Wolff, and W.L. Campbell. 1984. Differential effects of mottled yellow and pseudoagouti phenotypes on immunocompetence in A^{vy}/a mice. Proc. Natl. Acad. Sci. USA 81:2152–2156.

37. Russell, L.B. 1964. Genetic and functional mosaicism in the mouse. In M. Locke, ed., Role of Chromosomes in Development, 153–181. Academic Press, New York.

38. Sakurai, T., H. Ochiai, and T. Takeuchi. 1975. Ultrastructural change of melanosomes associated with agouti pattern formation in mouse hair. Dev. Biol. 47:466–471.

39. Silvers, W.K. 1958. An experimental approach to action of genes at the agouti locus in the mouse. III. Transplants of newborn A^w-, A-, and a^t-skin to A^y-, A^w-, and aa hosts. J. Exp. Zool. 137:189–196.

40. Siracusa, L.D., L.B. Russell, N.A. Jenkins, and N.G. Copeland. 1987. Allelic variation within the Emv-15 locus defines genomic sequences closely linked to the agouti locus on mouse chromosome 2. Genetics 117:85–92.

41. Siracusa, L.D., L.B. Russell, E.M. Eicher, D.J. Corrow, N.G. Copeland, and N.A. Jenkins. 1987. Genetic organization of the agouti region of the mouse. Genetics 117:93–100.

42. Tamate, H.B., and T. Takeuchi. 1981. Induction of the shift in melanin synthesis in lethal yellow (A^y/a) mice in vitro. Dev. Genet. 2:349–356.

43. Wolff, G.L. 1965. Body composition and coat color correlation in different phenotypes of "viable yellow" mice. Science 147:1145–1147.

44. Wolff, G.L. 1978. Influence of maternal phenotype on metabolic differentiation of agouti locus mutants in the mouse. Genetics 88:529–539.

45. Wolff, G.L., and H.C. Pitot. 1973. Influence of background genome on enzymatic characteristics of yellow (A^y/-. A^{vy}/-) mice. Genetics 73:109–123.

46. Wolff, G.L., D.B. Galbraith, O.E. Domon, and J.M. Row. 1978. Phaeomelanin synthesis and obesity in mice: interaction of the viable yellow (A^{vy}) and sombre (E^{so}) mutations. J. Hered. 69:295–298.

Aal locus, active avoidance learning, Chr 1

Active avoidance learning, measured in a shuttle box apparatus, is high in mice of the BALB/cBy strain carrying the recessive Aal^h allele, and low in mice of the C57BL/6By strain carrying the dominant Aal^l allele. Evidence from recombinant inbred strains showed that the difference was due to a single locus, linked to *H-25* and *H-35* (1), known to be located on Chr 1 (D.W. Bailey 1986, personal communication).

References

1. Oliverio, A., B.E. Eleftheriou, and D.W. Bailey. 1973. A gene influencing active avoidance performance in mice. Physiol. Behav. 11:497–501.

Aap locus, active avoidance performance

Performance in an active avoidance jump-up task is poor (Aap^d) in mice of the DBA/2J strain and good (Aap^b) in mice of the C57BL/6J strain. Crosses between the two strains supported a single-locus mode of inheritance, with Aap^d dominant to Aap^b (1).

References

1. Stavnes, K.L., and R.L. Sprott. 1975. Genetic analysis of active avoidance performance in mice. Psychol. Rep. 36:515–521.

Aat locus, α-1 antitrypsin, Chr 12

Probably the same as *Pre-1*. See *Pre-1*, *Pre-2* loci.

ab, asebia, recessive, Chr 19

Arose as a spontaneous mutation in the BALB/cCrglGa strain. Homozygotes may be recognizable at 7 days of age by retarded growth of the coat. Alopecia increases, and at adulthood the hair is very sparse. Homozygotes are viable and fertile, but fertility of females is somewhat reduced (1). Hyperplasia of the cellular layers of the skin is apparent at birth and increases markedly with age. Fibroblasts in the dermis are morphologically abnormal, and collagen and elastin show alterations (2). Hair follicles are initiated normally, but soon become abnormal, with cells of the inner root sheath plugging the hair canal and adhering to the emerging hair (3). Sebaceous glands are present but differentiate abnormally, with abnormal smooth endoplasmic reticulum and mitochondria, and few lipid droplets. Meibomian glands are also defective (4). Results of reciprocal transplantation of skin between normal and mutant mice suggest that some diffusible substance synthesized by normal skin can stimulate hair growth and alleviate the hyperkeratosis characteristic of the mutant skin (1). The skin is deficient in esterified sterols and waxes, but rich in free sterols (5).

References

1. Gates, A.H., and M. Karasek. 1965. Hereditary absence of sebaceous glands in the mouse. Science 148:1471–1473.
2. Josefowicz, W.J., and M.H. Hardy. 1978. The expression of the gene asebia in the laboratory mouse. 1. Epidermis and dermis. Genet. Res. 31:53–65.
3. Josefowicz, W.J., and M.H. Hardy. 1978. The expression of the gene asebia in the laboratory mouse. 2. Hair follicles. Genet. Res. 31:145–155.
4. Josefowicz, W.J., and M.H. Hardy. 1978. The expression of the gene asebia in the laboratory mouse. 3. Sebaceous glands. Genet. Res. 31:157–166.
5. Wilkinson, D.I., and M.A. Karasek. 1966. Skin lipids of a normal and a mutant (asebic) mouse strain. J. Invest. Dermatol. 47:449–455.

Abl locus (*c-abl*), Abelson leukemia oncogene, Chr 2

This is the cellular homolog of the transforming gene (v-*abl*) of the Abelson leukemia virus. No genetic variants have been described. The location of *Abl* on Chr 2 was found using mouse–Chinese hamster somatic cell hybrids screened with a v-*abl* DNA probe (1). The viral v-*abl* gene was shown to have protein kinase activity specific for tyrosine (4, 6). The tyrosine kinase coding sequences in the *Abl* gene reside in the 5' end, distributed in 10 clusters along 25 kb of the total of 40 kb (7). The *Abl* gene is actively transcribed in embryos at all stages and in all postnatal tissues. Two size classes of mRNA transcripts are found in embryos and in postnatal thymus, liver, and spleen and a third class in adult testis (2, 5). The testicular class is not present in immature testis; its appearance is coincident with the entry of germ cells into the haploid state (3).

References

1. Goff, S.P., P. D'Eustachio, F.H. Ruddle, and D. Baltimore. 1982. Chromosomal assignment of the endogenous proto- oncogene C-*abl*. Science 218:1317–1319.
2. Muüller, R., D.J. Shamon, J.M. Tremblay, M.J. Cline, and I.M. Verma. 1982. Differential expression of cellular oncogenes during pre- and postnatal development of the mouse. Nature 299:640–644.
3. Ponzetto, C., and D.J. Wolfgemuth. 1985. Haploid expression of a unique *c-abl* transcript in the mouse male germ line. Mol. Cell. Biol. 5:1791–1794.
4. Van de Vin, W.J.M., F.H. Reynold, Jr., and J.R. Stephenson. 1980. The nonstructural components of polyproteins encoded by replication-defective mammalian transforming retroviruses are phosphorylated and have associated protein kinase activity. Virology 101:185–197.
5. Wang, J.Y.J., and D. Baltimore. 1983. Cellular RNA homologous to the Abelson murine leukemia virus transforming gene: expression and relationship to the viral sequence. Mol. Cell. Biol. 3:773–779.
6. Wang, J.Y.J., C. Queen, and D. Baltimore. 1982. Expression of an Abelson murine leukemia virus-encoded protein in Escherichia coli causes extensive phosphorylation of tyrosine residues. J. Biol. Chem. 257:13,181–13,184.
7. Wang, J.Y.J., F. Ledley, S. Goff, R. Lee, Y. Groner, and D. Baltimore. 1984. The mouse c-*abl* locus: molecular cloning and characterization. Cell 36:349–356.

Abp complex (formerly also *Sal-1*), androgen-binding protein, Chr 7

This complex contains at least two and probably three genes for androgen-binding protein (ABP) of the saliva. After electrophoresis and isoelectric focusing, the protein is seen as three bands. Two haplotypes are known: *Abp*^a determines a more basic form of the three bands and occurs in strains A/J, AKR, BALB/c, C3H/St, C57BL/6, and many others; *Abp*^b determines a more acidic form and occurs in strains DBA/1, DBA/2, DW/J, MOLD/Rk, NZB, RIIIS, and SM. ABP is produced in the submandibular gland. It binds androgen and has a dimeric structure. The protein is also influenced by the *Ssp* locus which controls susceptibility to the effect of androgen on electrophoretic mobility of the protein (1). *Abp* and *Ssp* are not linked (2). See *Ssp* locus. There are thought to be three genes that code for ABP subunits, *Abpa* coding for the ABP-α subunit,

Abp

Abpg coding for the ABP-γ subunit, and *Abpb* coding for the ABP-β subunit. The dimeric ABP protein is formed by random combinations of the A subunit with the G or B subunits. *Abpa* and *Abpg* both have two alleles, *Abpa^a* and *Abpg^a* in the *Abp^a* haplotype, and *Abpa^b* and *Abpg^b* in the *Abp^b* haplotype. The six bands seen in heterozygotes are the result of combinations of the B subunit with the allelic forms of the A and G subunits. *Abpa* and *Abpg* are very closely linked and are part of the *Abp* complex. *Abpb* is not polymorphic and its location is not known, but it is probably also in the *Abp* complex. The *Abp* complex was shown to be on Chr 7, 1.6 cM proximal to *Gpi-1* by use of recombinant inbred strains and by a backcross (3).

References

1. Dlouhy, S.R., and R.C. Karn. 1983. The tissue source and cellular control of the apparent size of androgen binding protein (Abp), a mouse salivary protein whose electrophoretic mobility is under the control of *sex-limited saliva pattern (Ssp)*. Biochem. Genet. 21:1057–1070.
2. Dlouhy, S.R., and R.C. Karn. 1984. Multiple gene action determining a mouse salivary protein phenotype: identification of the structural gene for androgen binding protein (Abp). Biochem. Genet. 22:657–667.
3. Dlouhy, S.R., B.A. Taylor, and R.C. Karn. 1987. The genes for mouse salivary androgen binding protein (ABP) subunits alpha and gamma are located on chromosome 7. Genetics 115:535–543.

ac, absent corpus callosum, recessive

Discovered accidentally in the course of studies of the brain anatomy of mice with rodless retina. The corpus callosum is partially or completely absent, but a tract of fibers, the longitudinal callosal bundle, which does not occur in normal mice, is present (1). A condition similar to that in *ac/ac* mice occurs in some, but not all mice of the BALB/cJ and 129/J strains (2,3).

References

1. King, L.S. 1936. Hereditary defects of the corpus callosum in the mouse, *Mus musculus*. J. Comp. Neurol. 64:337–363.
2. Wahlsten, D. 1974. Heritable aspects of anomalous myelinated fibre tracts in the forebrain of the laboratory mouse. Brain Res. 68:1–18.
3. Wimer, R.E. 1965. Personal communication.

Acc, anterior capsular cataract, probably semidominant

Arose in a (101 × C3H)F1 hybrid after irradiation with ^{137}Cs gamma rays. Homozygotes have not been recovered. The lenses of heterozygotes show a dustlike capsular opacity of the anterior pole and an opaque area in the region of the anterior embryonic suture. Manifestation is uni- or bilateral and penetrance is incomplete (1).

References

1. Kratochvilova, J. 1981. Dominant cataract mutations detected in offspring of gamma-irradiated male mice. J. Hered. 72:302–307.

Acf-1 locus, albumin conformation factor-1, Chr 1

This locus controls the presence of a factor in liver capable of converting a large proportion of mouse, bovine, and horse albumin from an electrophoretically slow to a faster migrating form. The factor is not present in plasma (1). The conversion can be reversed by markedly lowering the pH, and the change is thought to be conformational (2). Absence of the factor is determined by the recessive allele *Acf-1^a*, found in C3H/HeJ and 10 other strains, and presence by the dominant allele *Acf-1^b*, found in SWR/J and 16 other strains (3).

References

1. Wilcox, F.H. 1973. Genetic differences in alteration of albumin in the house mouse, *Mus musculus*. Biochem. Genet. 10:69–78.
2. Wilcox, F.H. 1973. Genetic differences in conformational changes of albumin in the house mouse. Nature 246:35.
3. Wilcox, F.H. 1975. Genetic variation in plasma prealbumin of the house mouse. J. Hered. 66:19–22.

Ack locus, anti-*C. kutscheri*

This locus controls the microbicidal efficiency of mononuclear phagocytes against *Corynebacterium kutscheri*. The allele *Ack^r* (formerly *Ack*) determines high microbicidal efficiency and resistance to the bacterium and is found in the C57BL/6 and C57BL/10 strains; the allele *Ack^s* (formerly *ack*) determines low microbicidal efficiency and sensitivity to the bacterium, and is found in the Swiss Lynch (SWR?), CBA, A/J, C3H/He, and BALB/c strains. *Ack^r* is dominant (3). The difference between C57BL/6 and Swiss Lynch in mononuclear phagocytic microbicidal efficiency is probably due to ability of the phagocytes of the liver to destroy the bacteria, and not to their ability to ingest the bacteria or to a bactericidal serum factor. Resistance could be transferred with spleen or bone marrow cells (2). A single locus determining susceptibility and resistance to *Listeria monocytogenes* (1,4) has the same strain distribution of alleles as *Ack* and may be identical. Linkage tests with both these loci have shown no linkage with genes on Chrs 2, 4, 7, 9, 17, or with *H-7*, *H-8*, or *Igh*.

References

1. Cheers, C., and I.F.C. McKenzie. 1978. Resistance and susceptibility of mice to bacterial infection: genetics of listeriosis. Infect. Immun. 19:755–762.
2. Hirst, R.G., and R. Campbell. 1977. Mechanisms of resistance to *Corynebacterium kutscheri* in mice. Infect. Immun. 17:319–324.
3. Hirst, R.G., and M.E. Wallace. 1976. Inherited resistance to *Corynebacterium kutscheri* in mice. Infect. Immun. 14:475–482.
4. Skamene, E., and P.A.L. Kongshavn. 1978. Resistance to Listeria monocytes in mice is genetically controlled by non-H-2 linked genes. Fed. Proc. 37:1783 (Abstr.).

Aco-1 locus, aconitase-1, Chr 4

This locus controls electrophoretic variation in soluble aconitase (E.C. 4.2.1.3), an enzyme found in liver, kidney, brain, heart, lung, muscle, spleen, testis, and thymus, but not in red blood cells or plasma. The allele *Aco-1^a* determines a variant of intermediate mobility and occurs in strains A, AKR, CBA/J, C57BL/6, SM, and most other strains; the allele *Aco-1^b* determines the most cathodally migrating variant and occurs in a wild-derived strain, CALIF/Ei, and a strain derived from wild *M. castaneus*, CAST/Ei; the allele *Aco-1^c* determines the most anodally migrating variant and occurs in the ST/b and PERU-ATTECK strains. Heterozygotes have the two parental bands, indicating a monomeric structure of the enzyme (1).

References

1. Nadeau, J.H., and E.M. Eicher. 1982. Conserved linkage of soluble aconitase and galactose-1-phosphate uridyl transferase in mouse and man: assignment of the genes to mouse chromosome 4. Cytogenet. Cell Genet. 34:271–281.

Acp-1 locus (*Acp-1* was formerly used for *Apl* which is now *Neu-1*), acid phosphatase-1, Chr 12

This is probably the structural locus for an acid phosphatase isozyme (ACP-1; E.C. 3.1.3.2) that occurs primarily in red cells but also in other tissues. No genetic variants are known. The locus was shown to be on Chr 12 by use of mouse–Chinese hamster somatic cell hybrids segregating mouse chromosomes. The enzyme forms two bands of activity after electrophoresis on starch gels and behaves as a monomer in hybrid clones (1).

References

1. Francke, U., P.A. Lalley, W. Moss, J. Ivy, and D. Minna. 1977. Gene mapping in *Mus musculus* by interspecific cell hybridization: assignment of the genes for tripeptidase-1 to chromosome 10, dipeptidase-2 to chromosome 18, acid phosphatase-1 to chromosome 12, and adenylate kinase-1 to chromosome 2. Cytogenet. Cell Genet. 19:57–84.

Acp-2 locus (*Acp-2* was formerly used for *Apk*), acid phosphatase-2, Chr 2

This is probably the structural locus for the tissue form of acid phosphatase (ACP-2; E.C. 3.1.3.2). No genetic variants are known. The gene was located on Chr 2 using mouse–Chinese hamster somatic cell hybrids segregating mouse chromosomes (1).

References

1. Lalley, P.A. 1981. Mouse News Lett. 64:79–81.

Acra, *Acrb*, *Acrd*, *Acrg* loci

These loci encode the four subunits of muscle nicotinic acetylcholine receptor. The subunits are assembled into the $\alpha_2\beta\gamma\delta$ pentamer of the complete molecule. DNA restriction fragment length variants for the four genes were found between the DBA/2 strain and a strain (SPE) derived from wild *Mus spretus*. The variants were detected with cDNA clones from mouse muscle cell lines used to probe genomic DNA digested with several restriction enzymes. A backcross of DBA/2 × (DBA/2 × SPE)F1 was used to look for linkage of the four genes with other markers (1).

Acra locus, acetylcholine receptor-α, Chr 17. DNA digested with *Hinc*II shows a 6-kb band in DBA/2 and a 3-kb band in SPE. *Acra* is loosely linked to *Actc* (19 per cent recombination) on Chr 17 and both loci are probably near the distal end of the chromosome.

Acrb locus, acetylcholine receptor-β, Chr 11. DNA digested with *Hinc*II shows a 2.5-kb band in DBA/2 and a 1.5-kb band in SPE. *Acrb* shows no recombination with *Myha* on Chr 11.

Acrd locus, acetylcholine receptor-δ, Chr 1. DNA digested with *Pvu*II shows a 5-kb band in DBA/2 and a 4.2-kb band in SPE.

Acrg locus, acetylcholine receptor-γ, Chr 1. DNA digested with *Eco*RI shows a 9-kb band in DBA/2 and a 4-kb band in SPE. *Acrd* and *Acrg* show no recombination with each other and both are closely linked (5 per cent recombination) to *Mylf* on Chr 1 (1).

Acra

References

1. Heidmann, D., A. Buonanno, B. Geoffroy, B. Robert, J.-L. Guénet, J.P. Merlie, and J.-P Changeux. 1986. Chromosomal localization of muscle nicotinic acetylcholine receptor genes in the mouse. Science 234:866–868.

Acry-1 locus

See *Crya-1* locus.

Actb, Actc, Acts loci

These loci code for cytoplasmic and muscle actins. No genetic variants at these loci have been described.

Actb locus, cytoplasmic β-actin, Chr 5. This locus codes for β-actin, one of the two actins (the other is γ-actin) found in the cytoplasm of many non-muscle cell types. The gene was found to be on Chr 5 using mouse–Chinese hamster somatic cell hybrids screened with a DNA fragment of a cloned rat β-actin gene (2).

Actc locus, cardiac muscle actin, Chr 17. This locus codes for cardiac muscle actin. This is one of two vertebrate striated muscle actins, the other being skeletal muscle actin (see *Acts* locus). *Actc* was found to be on Chr 17 by use of mouse–Chinese hamster somatic cell hybrids screened with a cDNA probe made from a rat heart mRNA template (2). Heidmann *et al.* (3) described linkage of *Actc* with *Acra* (acetylcholine receptor-α) in a cross between DBA/2 and a strain derived from wild *Mus spretus*. This implies existence of a variant of *Actc* in *M. spretus* but the variant was not described. *Actc* and *Acra* show no linkage with *Glo-1* which is near the centromeric end of Chr 17, and are therefore probably near the distal end of the chromosome.

Acts locus, skeletal muscle actin, Chr 3. This locus codes for skeletal muscle α-actin. In experiments with mouse–Chinese hamster somatic cell hybrids designed to determine the location of the skeletal muscle actin genes, 12 mouse-specific *Eco*RI-restricted DNA fragments were identified using a rat DNA probe coding for skeletal muscle actin. The probe hybridizes only with the skeletal muscle actin gene under strigent conditions and with other muscle and non-muscle actin genes under less stringent conditions. Under less stringent conditions, five to seven of the 12 fragments were assigned to Chr 3, one to Chr 2, and three to Chr 17. Under stringent conditions, only one fragment was detected and it was assigned to Chr 3 (1). Thus most actin genes detected by this rat probe, including *Acts*, are on Chr 3, with some others on Chrs 2 and 17 (see *Actc* locus). It is not known whether the Chr 3 actin genes are located in a single cluster.

References

1. Czosnek, H., U. Nudel, M. Shani, P.E. Barker, D.D. Pravtcheva, F.H. Ruddle, and D. Yaffe. 1982. The genes coding for the muscle contractile proteins, myosin heavy chain, myosin light chain 2, and skeletal muscle actin, are located on three different mouse chromosomes. EMBO J. 1:1299–1305.
2. Czosnek, H., U. Nadel, Y. Mayer, P.E. Barker, D.E. Pravtcheva, F.H. Ruddle, and D. Yaffe. 1983. The genes coding for the cardiac muscle actin, the skeletal muscle actin, and the cytoplasmic β-actin are located on three different chromosomes. EMBO J. 2:1977–1979.
3. Heidmann, D., A. Buonanno, B. Geoffroy, B. Robert, J.-L. Guénet, J.P. Merlie, and J.-P Changeux. 1986. Chromosomal localization of muscle nicotinic acetylcholine receptor genes in the mouse. Science 234:866–868.

Acy-1 locus, aminoacylase-1, Chr 9

This is the structural locus for aminoacylase-1 (ACY-1; E.C. 3.5.1.14), an enzyme catalyzing degradation of acylated amino acids. It is found in all tissues, the highest activity occurring in kidney and small intestine. Three alleles recognized by differences in electrophoretic mobility of the enzyme are known: $Acy-1^a$ determines a slow anodally moving band and occurs in strain ST/b; $Acy-1^b$ determines a band of intermediate mobility and occurs in strains C57BL/6, CBA/J, and AU/Ss; and $Acy-1^c$ determines a fast band and occurs in strains AKR, BALB/c, C3H/He, LP, NZB, and many others (3). *Acy-1* was first assigned to Chr 9 using mouse–Chinese hamster somatic cell hybrids segregating mouse chromosomes (1). It was later found by use of recombinant inbred strains (4) and backcrosses (2) to be located between *Mod-1* and *Bgl-s* on Chr 9.

References

1. Lalley, P.A, S.L. Naylor, R.W. Elliott, and T.B. Shows. 1982. Assignment of *Acy-1* to mouse chromosome 9: evidence for homologous linkage groups in man and mouse. Cytogenet. Cell Genet. 32:293 (Abstr.).
2. Nadeau, J.H. 1987. A chromosomal segment conserved since divergence of lineages leading to man and mouse: the gene order of aminoacylase-1, transferrin, and beta-galactosidase on mouse chromosome 9. Genet. Res. 48:175–178.
3. Naylor, S.L., T.B. Shows, and R.J. Klebe. 1979. Bioautographic visualization of aminoacylase-1: assignment of the structural gene *ACY-1* to chromosome 3 in man. Somat. Cell Genet. 5:11–21.
4. Naylor, S.L., R.W. Elliott, J.A. Brown, and T.B. Shows. 1982. Mapping of aminoacylase-1 and beta-galactosidase-A to homologous regions of human chromosome 3 and mouse chromosome 9 suggests location of additional genes. Am. J. Hum. Genet. 34:235–244.

Ad, adult obesity and diabetes, semidominant, Chr 7

Found in a cross between a random-bred stock descended from wild mice trapped in the Plant Breeding Institute, Cambridge, and a chromosome 7 marker stock. It was later found in two other wild populations nearby. *Ad/+* and *Ad/Ad* mice of both sexes are normally fertile and appear to have normal activity. The obesity can be recognized at 4 to 6 months and is greater in homozygotes than in heterozygotes. Penetrance is probably complete in homozygotes. In heterozygotes classified at 4 months, it was found to be nearly complete in males but severely reduced in females (2). In a study of 6-month-old offspring of matings between obese and lean mice of an *Ad* stock, it was found that the obese mice were 50 per cent heavier than the lean mice, and that they were hyperinsulinemic but had normal blood glucose levels. The increase in weight was due entirely to increased amount of adipose tissue. The increase in fat occurred in all fat depots and resulted from increase in cell size in males and increase in both cell size and cell number in females (1). *Ad* is located at the distal end of Chr 7 (2).

References

1. Trayhurn, P., W.P.T. James, and M.I. Gurr. 1979. Studies on the body composition, fat distribution and fat cell size and number of *Ad*, a new obese mutant mouse. Br. J. Nutr. 41:211–221.
2. Wallace, M.E., and F.M. MacSwiney. 1979. An inherited mild middle-aged adiposity in wild mice. J. Hyg. 82:309–317.

ad

See *db*.

Ada locus, adenosine deaminase, Chr 2

This locus codes for adenosine deaminase (E.C. 3.5.4.4). The allele *Ada^b* determines a slow electrophoretic variant and was found in wild mice on the Orkney island of Eday; the allele *Ada^a* determines a faster band and occurs in other wild mice and in all inbred strains tested (2). By use of mouse–Chinese hamster somatic cell hybrids, *Ada* was found to be on Chr 2 close to *Itp* in the C1—ter region (1, 3).

References

1. Lalley, P.A., and J.A. Diaz. 1984. Comparative gene mapping in the mouse involving genes assigned to human chromosomes 7 and 20. Cytogenet. Cell Genet. 37:514–515 (Abstr.).
2. Nash, H.R. 1984. Mouse News Lett. 71:33.
3. Siciliano, M.J., R.E.K. Fournier, and R.L. Stallings. 1984. Regional assignment of *ADA* and *ITPA* to mouse chromosome 2 (C1—ter). A demonstration of the conserved linkage of enzyme and proto-oncogene loci. J. Hered. 75:175–180.

Adh-1 complex, Chr 3

This complex consists of a structural and a closely linked regulatory locus for an isozyme of alcohol dehydrogenase (ADH-A$_2$ or ADH-1; E.C. 1.1.1.1). The *Adh-1* complex is closely linked to the *Adh-3* complex on Chr 3 (1–3).

Adh-1e locus (formerly *Adh-1*), alcohol dehydrogenase-1- electrophoretic. This locus controls electrophoretic mobility of ADH-1 present principally in liver but also in kidney, intestine, adrenals, and other tissues (4). The allele *Adh-1e^a* determines a fast cathodally migrating band and occurs in all inbred strains examined; the allele *Adh-1e^b* determines a slower band and occurs in wild populations of *M. m. musculus* and *M. m. domesticus* in Denmark (3, 6). A similar slow variant occurs in wild *M. spretus* (1). Heterozygotes have three bands, consistent with a dimeric structure of the enzyme (4).

Adh-1t locus, alcohol dehydrogenase-1-temporal. This locus controls level of activity of ADH-1 in kidney. The allele *Adh-1t^a* determines high activity and occurs in a strain descended from wild Danish mice; the allele *Adh-1t^b* determines low activity and occurs in the CBA/H strain. *Adh-1t* is closely linked to *Adh-1e* and appears to be *cis*-acting. Its effect is specific to kidney (5).

References

1. Bonhomme, F, F. Benmehdi, J. Britton-Davidian, and S. Martin. 1979. Analyse génétique de croisements interspécifiques *Mus musculus* L. × *Mus spretus* Lataste: liason de *Adh-1* avec *Amy-1* sur le chromosome 3 et de *Es-14* avec *Mod-1* sur le chromosome 9. C. R. Acad. Sci. Paris 289:545–548.
2. Holmes, R.S., and J.A. Duley. 1982. Mouse News Lett. 66:60.
3. Holmes, R.S., R. Albanese, F.D. Whitehead, and J.A. Duley. 1981. Mouse alcohol dehydrogenase isozymes: products of closely localized duplicated genes exhibiting divergent kinetic properties. J. Exp. Zool. 217:151–157.
4. Holmes, R.S., J.A. Duley, and J. Hilgers. 1982. Sorbitol dehydrogenase genetics in the mouse: a "null" mutant in a "European" C57BL strain. Anim. Bld. Grps. Biochem. Genet. 13:263–272.
5. Holmes, R.S., J.A. Duley, and S. Imai. 1983. Mouse News Lett. 68:67.
6. Selander, R.K., W.G. Hunt, and S.Y. Yang. 1969. Pro-

tein polymorphism and gene heterozygosity in two European subspecies of the house mouse. Evolution 23:379–390.

Adh-1-t

This is a non-standard symbol for a locus that controls level of activity of alcohol dehydrogenase-1 (ADH-A_2 or ADH-1; E.C. 1.1.1.1) in liver. The locus has no effect in kidney and is therefore clearly different from *Adh-1t*. Activity of the enzyme is low in all strains at birth and rises beginning at 5 days to a maximum at about 40 days. The allele designated *Adh-1-t^a* determines a low maximum activity and occurs in strains C3H/He, A/J, DDS, DBA/2, and BALB/c; the allele designated *Adh-1-t^b* determines a maximum about twice that of the low strains and occurs in strain C57BL/6. Heterozygotes have intermediate activity. The enzymes in the high and low strains appear to be identical, and evidence indicates that the difference is due to a difference in rate of synthesis (1, 2). The chromosomal location of *Adh-1-t* is not known (1).

References

1. Balak, K.J., R.H. Keith, and M.R. Felder. 1982. Genetic and developmental regulation of mouse liver alcohol dehydrogenase. J. Biol. Chem. 257:15000–15007.
2. Rex, D.K., W.F. Bosson, and T.-K. Li. 1984. Purification and characterization of mouse alcohol dehydrogenase from two inbred strains that differ in total liver enzyme activity. Biochem. Genet. 22:115–124.

Adh-3 complex, Chr 3

This complex includes a structural and a regulatory locus for an isozyme of alcohol dehydrogenase (ADH-C_2 or ADH-3; E.C. 1.1.1.1). The two loci are closely linked but show some recombination (1/156 backcross progeny 0.39 ± 0.05 per cent recombination) (4).

Adh-3e locus (formerly *Adh-3*), alcohol dehydrogenase-3-electrophoretic. This locus controls electrophoretic mobility of ADH-3. The enzyme occurs in stomach, adrenals, lung, epididymis, and other tissues and uses a wide range of alcohols as substrates. The allele *Adh-3e^a* determines a band of intermediate mobility and occurs in strains NZC, CBA/Ca, C3H/He, DBA/2, and others; the allele *Adh-3e^b* determines a slow band and occurs in strain C57BL/6J and in *M. m. castaneus*; the allele *Adh-3e^c* determines a fast band and was found in wild mice in Queensland, Australia. Heterozygotes have the two parental bands and an intermediate band (1, 3, 5).

Adh-3t locus (formerly *Adt-1*), alcohol dehydrogenase-3-temporal. This *cis*-acting regulator locus, which is closely linked to *Adh-3e*, controls expression of ADH-3 in seminal vesicles, epididymis, and ovaries, but has little effect in other tissues (5). The allele *Adh-3t^a* determines presence of the enzyme in the affected tissues; the allele *Adh-3t^b* determines its absence. Heterozygotes express only the enzyme determined by the *Adh-3e* allele on the same chromosome as *Adh-3t^a*. Most inbred strains carry the *Adh-3e^{a}t^a* haplotype; C57BL/Go and C57BL/10 carry *Adh-3e^{b}t^b*; and some Peru-Coppock mice carry *Adh-3e^{a}t^b* (2).

References

1. Holmes, R.S. 1977. The genetics of α-hydroxyacid oxidase and alcohol dehydrogenase in the mouse: evidence for multiple gene loci and linkage between *Hao-2* and *Adh-3*. Genetics 87:709–716.
2. Holmes, R.S. 1979. Genetics and ontogeny of alcohol dehydrogenase in the mouse: evidence for a *cis*-acting regulator gene (*Adt-1*) controlling C_2 isozyme expression in reproductive tissues and close linkage of *Adh-3* and *Adt-1* on chromosome 3. Biochem Genet. 17:461–472.
3. Holmes, R.S., J.T. Jones, and J. Peters. 1978. Genetic variation, cellular distribution and ontogeny of sorbitol dehydrogenase and alcohol dehydrogenase isozymes in male reproductive tissues of the mouse. J. Exp. Zool. 206:279–288.
4. Holmes, R.S., S.J. Andrews, and C.V. Beechey. 1981. Genetic regulation of alcohol dehydrogenase C2 in the mouse. Developmental consequences of the temporal locus (Adh-3t) and positioning of Adh-3 on chromosome 3. Dev. Genet. 2:89–98.
5. Mather, P.B., J.A. Duley, and R.S. Holmes. 1983. Aldehyde oxidase and alcohol dehydrogenase genetics in the mouse. New alleles at the *Aox-2* and *Adh-3* loci. Anim. Bld. Grps. Biochem. Genet. 14:279–287.

ADH inducibility regulator

One or two unnamed loci control *in vivo* inducibility by ingested ethanol of activity of alcohol dehydrogenase (ADH; E.C. 1.1.1.1) in liver. In strains BALB/c and 'S.W.', the enzyme level shows induction, i.e. the level rises after 2 weeks of exposure; in strains C57BL/6 and 129/Re, the level shows repression, i.e. the levels falls after the same treatment. The F1 offspring of crosses between inducible and repressible strains are inducible. A set of recombinant inbred strains between C57BL/6 and BALB/c showed segregation into eight inducible and four repressible strains, indicating a possible single gene control of the difference. However, a cross between the two repressible strains, C57BL/6 and 129/Re, gave inducible F1 offspring, suggesting that two complementing loci control the variation in inducibility. The liver ADH isozymes affected by alcohol treatment are ADA-1 (ADH-A_2) and ADH-2 (ADH-B_2).

No genetic variants of ADA-2 are known in the mouse (1).

References

1. Wang, C.H., and S.M. Singh. 1984. Genetic considerations in the effects of ethanol in mice. II. A *trans*-acting inducibility regulator(s) affecting alcohol dehydrogenase (ADH) activity. Biochem. Genet. 22:597–609.

Adk locus, adenosine kinase, Chr 14

This is probably the structural locus for adenosine kinase (ADK; E.C. 2.7.1.20). No genetic variants are known. The locus was shown to be on Chr 14 by use of mouse–Chinese hamster somatic cell hybrids (1). In similar hybrids carrying radiation-induced chromosome rearrangements, *Adk* was shown to be in the 14A2–B region (2).

References

1. Leinwand, L., R.E.K. Fournier, E.A. Nichols, and F.H. Ruddle. 1978. Assignment of the gene for adenosine kinase to chromosome 14 in *Mus musculus* by somatic cell hybridization. Cytogenet. Cell Genet. 21:77–85.
2. Samuelson, L.C., and R.A. Farber. 1985. Cytological localization of adenosine kinase, nucleoside phosphorylase-1, and esterase-10 genes on mouse chromosome 14. Somat. Cell Mol. Genet. 11:157–165.

adr locus

adr, arrested development of righting response, recessive. Arose spontaneously in the A2G strain. Homozygotes are recognizable at 9 days by their retarded righting response. From about 4 weeks, they show stunted growth and stiffened rear limbs. Females may produce small litters but males have not bred. Life span is 7 to 12 weeks. The defect appears to be confined to type II muscle fibers, which fail to undergo a normal increase in activity of several enzymes, including succinate dehydrogenase, AMP deaminase, and 1,6-biphosphate aldolase (5). The concentration of parvalbumin in type II fibers of the gastrocnemius is drastically reduced (4). There is a 200 per cent increase in alanine and aspartate aminotransferases, which may indicate enhanced protein degradation (3). Functional studies of *adr*/*adr* fast twitch muscle (tibialis anterior) showed only slightly prolonged contraction times, but long-lasting after-contractions following repeated direct stimulation (4).

*adr^mto^*, myotonia, recessive. Arose spontaneously in the SWR/J strain (1). It was shown to be allelic with *adr* by Jockusch and Bertram (2). Homozygotes are recog-

nizable at 2 weeks of age or earlier by prolonged stiff extension postures of the limbs when the cage is shaken or the mouse is dropped from about 10 cm. This behavior persists throughout life. When undisturbed, affected animals walk almost normally, but somewhat stiffly. They grow more slowly and weigh about 40 per cent less than controls in adulthood. On an outbred background they may live a year or more and may be fertile. Electromyographic studies revealed changes characteristic of myotonia, i.e. repetitive firing at varying amplitude and frequency in all skeletal muscles tested. These discharges do not originate in peripheral nerves, and there is no evidence of muscle fiber necrosis. The *mto* mutant exhibits classical myotonia similar to that described in man (1).

References

1. Heller, A.H., E.M. Eicher, M. Hallett, and R.L. Sidman. 1982. Myotonia, a new inherited muscle disease in mice. J. Neurosci. 2:924–933.
2. Jockusch, H., and K. Bertram. 1986. "Arrested development of righting response" (adr) and "myotonia" (mto) are allelic. Mouse News Lett. 75:19.
3. Soothill, P.W., F. Kousebati, R.L. Watts, and D.C. Watts. 1981. Glycolytic, pentose-phosphate shunt and transaminase enzymes in gastrocnemius muscle, liver, heart, and brain of two mouse mutants, 129J-dy and A2G-adr, with abnormal muscle function. J. Neurochem. 32:506–510.
4. Stuhlfauth, I., J. Reininghaus, H. Jockusch, and C.W. Heizmann. 1984. Calcium-binding protein, parvalbumin, is reduced in mutant mammalian muscle with abnormal contractile properties. Proc. Natl. Acad. Sci. USA 81:4814–4818.
5. Watkins, J., R.L. Watts, and D.C. Watts. 1979. Deficiency of fructose 1,6-biphosphate aldolase in type II muscle fibers of the A2G-*adr* mouse mutant with abnormal muscle function. Biochem. Soc. Trans. 7:907–908 (Abstr.).

Ads-1 to Ads-4 loci

Ads-1, *-3* loci, anti-dsDNA antibodies-1, -3, Chr 17

Ads-2, *-4* loci, anti-dsDNA antibodies-2, -4

These four loci are postulated to affect the level of double-stranded DNA (dsDNA) in NZB and NZW mice and their crosses. The dominant alleles of the independently segregating *Ads-1* and *Ads-2* loci cause high level of IgM with anti-dsDNA specificity and occur in the NZB strain; the recessive alleles for low level occur in the NZW strain which produces very little anti-dsDNA antibody. The dominant alleles of the independently segregating *Ads-3* and *Ads-4* loci occur in the

Ads-1

NZW strain and cause conversion of the anti-dsDNA antibodies of (NZB × NZW)F1 mice from IgM to IgG; the recessive alleles allowing production of IgM antibodies occur in the NZB strain. (NZB × NZW)F1 mice thus have high levels of anti-dsDNA antibodies, mostly composed of IgG. *Ads-1* is probably linked to *H-2* but not within the *H-2* complex; *Ads-3* is probably within the *H-2* complex. The locations of *Ads-2* and *Ads-4* are not known (1, 2). Several other manifestations of autoimmunity in NZB and NZW mice are also controlled by genes on Chr 17. Two loci, *Agp-1* and *Agp-3*, are postulated to control level of retroviral gp70 immune complexes in NZB and NZW mice and crosses between them (1, 3). The *Agp* loci parallel the strain distribution and effects of *Ads-1* and *Ads-3*, respectively (1). It is therefore possible that *Ads-1* and *Agp-1* are identical or parts of a complex and that *Ads-3* and *Agp-3* are similarly related. The incidence of lupus nephritis in NZB and NZW and their crosses is also controlled by two genes on Chr 17, *Lpn-1* and *Lpn-2*, that may be parts of the *Ads/Agp* complexes (1). See *Agp-1*, *-3* loci, *Lpn-1*, *-2* loci.

References

1. Hirose, S., R. Nagasawa, I. Sekikawa, M. Hamaoki, Y. Ishida, H. Sato, and T. Shirai. 1983. Enhancing effect of H-2-linked NZW gene(s) on the autoimmune traits of (NZB × NZW)F1 mice. J. Exp. Med. 158:228–233.
2. Kohno, A., H. Yoshida, K. Sekita, N. Maruyama, S. Ozaki, S. Hirose, and T. Shirai. 1983. Genetic regulation of the class conversion of dsDNA-specific antibodies in (NZB × NZW)F₁ hybrid. Immunogenetics 18:513–524.
3. Maruyama, N., F. Furukawa, Y. Nakai, Y. Sasake, K. Ohta, S. Ozaki, S. Hirose, and T. Shirai. 1983. Genetic studies of autoimmunity in New Zealand mice. IV. Contribution of NZB and NZW genes to the spontaneous occurrence of retroviral gp70 immune complexes in (NZB × NZW)F₁ hybrid and correlation to renal disease. J. Immunol. 130:740–746.

Aem-1 locus, anti-erythrocyte autoantibody modifier-1, Chr 4?

This locus modifies the incidence of anti-erythrocyte autoantibodies (AEA) in crosses between the NZB and C57BL/6 strains. The recessive allele present in NZB allows 100 per cent incidence of AEA and the dominant allele present in C57BL/6 prevents production of AEA. *Aem-1* does not affect spontaneous production of dsDNA-specific antibodies, retroviral gp70–anti-gp70 immune complexes, natural thymocytotoxic autoantibodies, or serum levels of retroviral gp70. The locus may affect serum IgM level; there was a significantly higher IgM level (like NZB) in AEA-positive than in AEA-negative mice in the backcross of (C57BL/6 × NZB)F1 × NZB. *Aem-1* showed loose linkage with *Mup-1* and may be on Chr 4 (1).

References

1. Ozaki, S., H. Honda, N. Maruyama, S. Hirose, M. Hamaoki, H. Sato, and T. Shirai. 1983. Genetic regulation of erythrocyte autoantibody production in New Zealand black mice. Immunogenetics 18:241–254.

Afp locus, α-fetoprotein, Chr 5

This is the structural locus for α-fetoprotein (AFP) the major serum protein of the developing fetus. It is synthesized by the embryonic liver and yolk sac. No genetic variants are known. The locus was shown to be on Chr 5 by use of mouse–Chinese hamster somatic cell hybrids segregating mouse chromosomes (2). The rate of synthesis of AFP declines at birth at a time when the rate of synthesis of serum albumin, the major adult serum protein, is increasing (5). The histone H1°, which is preferentially associated with nuclease-resistant chromatin, is found associated with the *Afp* gene in adult liver, indicating that control of transcription of *Afp* may be mediated by H1° (4). The serum albumin locus (*Alb-1*) is known to be on Chr 5. The two proteins of the mouse have similar amino acid sequences, and cloned DNA segments of the two genes reveal identically organized coding and intervening sequences. *Afp* and *Alb-1* thus probably arose in evolution by duplication of a common ancestral gene. This conclusion is supported by the isolation and characterization of a recombinant DNA clone whose sequences span the distance between the two genes, thus demonstrating that they are closely linked in tandem (3). The DNA of the C57BL/6 strain carries an *EcoRI* restriction enzyme variant, located between the *Alb-1* and *Afp* genes, that can be used in crosses as a genetic marker for the invariant *Afp* locus (1). The level of AFP in plasma is regulated by two loci, *Afr-1* and *Afr-2*, unlinked to *Afp* or to each other. See *Afr-1*, *Afr-2*.

References

1. Belayew, A., and S.M. Tilghman. 1982. Genetic analysis of α-fetoprotein synthesis in mice. Mol. Cell. Biol. 2:1427–1435.
2. D'Eustachio, P., R.S. Ingram, S.M. Tilghman, and F.H. Ruddle. 1981. Murine alpha-fetoprotein and albumin: two evolutionarily linked proteins encoded in the same mouse chromosome. Somat. Cell Genet. 7:289–294.
3. Ingram, R.S., R.W. Scott, and S.M. Tilghman. 1981. Alpha fetoprotein and albumin genes are in tandem in the mouse genome. Proc. Natl. Acad. Sci. USA 78:4694–4698.

4. Roche, J., C. Gorka, P. Goeltz, and J.J. Lawrence. 1985. Association of histone H1° with a gene repressed during liver development. Nature 314:197–198.
5. Sellem, C.H., M. Frain, T. Erdos, and J.M. Sala-Trepat. 1984. Different expression of albumin and α-fetoprotein genes in fetal tissues of mouse and rat. Dev. Biol. 102:51–60.

Afr-1, *Afr-2* loci

Afr-1 locus (formerly *Raf*), α-fetoprotein regulation-1. The level of α-fetoprotein (AFP) in plasma declines precipitously after birth to a level 10^3-fold lower than at birth in most strains. The *Afr-1* locus controls a difference in the rate of decrease. The allele *Afr-1^b*, found only in the BALB/cJ substrain, determines a slow rate of decline and consequently a high adult level (about 10-fold higher than normal) of AFP; the allele *Afr-1^a* determines a fast rate of decline of AFP and low adult level and occurs in the C3H/He, DBA/2, and all other strains tested including two other BALB/c substrains, as well as in *M. m. castaneus*. Heterozygotes have low adult levels. The increased level of AFP in BALB/cJ may be due to a high rate of synthesis (2). *Afr-1* also regulates the rate of production of an mRNA termed H19, the gene for which is on Chr 7 unlinked to *Afp*, *Afr-1*, or *Afr-2* (3).

Afr-2 locus (formerly *Rif*), α-fetoprotein regulation-2. The synthesis of AFP in liver, which is very low in adults of most strains, can resume during liver regeneration induced by partial hepatectomy, poisoning, or other damaging conditions. Rate of induction of AFP during liver regeneration is controlled by the *Afr-2* locus. The allele *Afr-2^a* determines a high rate of induction and occurs in the BALB/c and C3H/He strains; the allele *Afr-2^b* determines a 10-fold lower rate of induction and occurs in the C57BL/6 strain. Heterozygotes have a low rate of induction. Interaction between the *Afr-1* and *Afr-2* loci results in a high rate of AFP induction in BALB/c, an intermediate level in C3H/He, and a low level in C57BL/6, and in a 1:2:1 segregation of high, medium, and low induction in the backcross of (BALB/c × C57BL/6)F1 to BALB/c. *Afr-2* is not linked to *Afp* or to *Afr-1* (1). *Afr-2*, like *Afr-1*, also regulates rate of production of the H19 mRNA (3).

References

1. Belayew, A., and S.M. Tilghman. 1982. Genetic analysis of α-fetoprotein synthesis in mice. Mol. Cell. Biol. 2:1427–1435.
2. Olsson, M., G. Lindahl, and E. Ruoslahti. 1977. Genetic control of alpha-fetoprotein synthesis in the mouse. J. Exp. Med. 145:819–827.

3. Pachnis, V., A. Belayew, and S.M. Tilghman. 1984. Locus unlinked to α-fetoprotein under the control of the murine *raf* and *Rif* genes. Proc. Natl. Acad. Sci. USA 81:5523–5527.

Aft, abnormal feet and tail, probably semidominant, Chr 9

Arose spontaneously in a subline of strain 129/Sv. Heterozygotes have syndactyly of digits 3 and 4 of the hindfeet and a kinky tail. Penetrance in heterozygotes is low, about 50 per cent. Homozygotes have not been described. *Aft* is on Chr 9 about 10 cM from *se*, probably on the centromeric side (1).

References

1. Lane, P.W. 1987. Abnormal feet and tail (Aft). Mouse News Lett. 78:56–57.

Afuc locus

See *Fuca* locus.

ag, agitans, recessive, Chr 14

Arose spontaneously in a moderately inbred stock. Homozygotes can be recognized from about 10 days onward. They show retarded growth, generalized tremor, and ataxia. They usually die at weaning, but may live as long as 3 months. Atrophy of the Purkinje cells has been found in some regions of the cerebellum (1,2).

References

1. Hoecker, G., A. Martinez, S. Markovic, and O. Pizzaro. 1954. Agitans, a new mutation in the house mouse with neurological effects. J. Hered. 45:10–14.
2. Martinez, A., and J.L. Sirlin. 1955. Neurohistology of the agitans mouse. J. Comp. Neurol. 103:131–137.

Agp-1, *-3* loci, anti-gp70 immune complex-1, -3, Chr 17

These two loci control the formation of retroviral gp70-anti-gp70 immune complexes (gp70 IC) in the NZB and NZW strains and their crosses. The level of gp70 IC is low in the two strains but high in the F1 generation. A dominant allele at the *Agp-1* locus in the NZB strain determines the formation of the immune complexes; the recessive allele preventing their formation occurs in the NZW strain. A dominant allele at the *Agp-3* locus in the NZW strain enhances the level of the immune complexes; the recessive allele causing low level occurs in the NZB strain. The level of gp70 is high in all these mice, indicating that the effect of the Agp loci is on the production of anti-gp70 antibodies (2). *Agp-1* is loosely linked to *H-2* and *Agp-3* is probably within the *H-2* complex (1). The *Agp* loci may be related to the *Ads* and *Lpn* loci. See *Ads-1, -3* loci, *Lpn-1, -2* loci.

References

1. Hirose, S., R. Nagasawa, I. Sekikawa, M. Hamaoki, Y. Ishida, H. Sato, and T. Shirai. 1983. Enhancing effect of H-2-linked NZW gene(s) on the autoimmune traits of (NZB × NZW)F1 mice. J. Exp. Med. 158:228–233.
2. Maruyama, N., F. Furukawa, Y. Nakai, Y. Sasake, K. Ohta, S. Ozaki, S. Hirose, and T. Shirai. 1983. Genetic studies of autoimmunity in New Zealand mice. IV. Contribution of NZB and NZW genes to the spontaneous occurrence of retroviral gp70 immune complexes in (NZB × NZW)F$_1$ hybrid and correlation to renal disease. J. Immunol. 130:740–746.

Agp-1, -2 loci

See *Orm-1, -2* loci.

Ags locus, α-galactosidase, Chr X

This is probably the structural locus for α-galactosidase (AGS; E.C. 3.2.1.22), which is found in all tissues, including pre-implantation embryos (1). The allele *Ags^m*, found in *M. m. molossinus*, determines a thermolabile form of the enzyme, and the allele *Ags^h*, found in C3H/HeJ and about 100 other strains examined, determines the more stable form. Female heterozygotes have intermediate stability (4). A third allele *Ags^s*, found in *M. spretus*, determines an enzyme with the same thermostability as that determined by *Ags^m* but with about twice the activity determined by both *Ags^h* and *Ags^m*. *Ags^s* is fully dominant to the other two alleles in males and nearly so in females (2). *Ags* was shown to be distal to *Pgk-1* on the X chromosome (4) and was also shown to be in the same region of the chromosome by somatic cell hybridization using an X-autosome translocation (3).

References

1. Adler, D.A., J.D. West, and V.M. Chapman. 1977. Expression of α-galactosidase in preimplantation mouse embryos. Nature 267:838–839.
2. Chapman, V.M., P.G. Kratzer, and B.A. Quarantillo. 1983. Electrophoretic variation for X chromosome-linked hypoxanthine phosphoribosyl transferase (HPRT) in wild-derived mice. Genetics 103:785–795.
3. Francke, U., and R. Taggart. 1980. Comparative gene mapping: order of loci on the X chromosome is different in mice and humans. Proc. Natl. Acad. Sci. USA 77:3595–3599.
4. Lusis, A.J., and J.D. West. 1976. X-linked inheritance of a structural gene for α-galactosidase in *Mus musculus*. Biochem. Genet. 14:849–855.

Ah locus (formerly *In, Ahh*), aromatic hydrocarbon responsiveness, Chr 12

This locus controls inducibility, by 3-methylcholanthrene and numerous other polycyclic aromatic compounds, of at least 14 monooxygenase activities and associated cytochromes P$_1$-450, including aryl hydrocarbon hydroxylase (AHH; E.C. 1.14.14.2). Regulation of responsiveness may involve more then one locus, but differences between C57BL/6 (responsive, *Ah^b*) and DBA/2 (non-responsive, *Ah^d*) can be almost completely explained by the difference at the *Ah* locus. Heterozygotes (*Ah^b/Ah^d*) are responsive (7). Some other responsive strains are C3H/He, BALB/c, CBA/H, and PL/J; some other non-responsive strains are 129/J, RF/J, AKR/J, and AU/SsJ. Taylor (11) has summarized the available information on identification of *Ah* alleles and their strain distribution. Responsiveness in *Ah^b* mice occurs not only in liver but in numerous non-hepatic tissues such as lung, kidney, bowel, skin, lymph nodes, retinal pigmented epithelium, bone marrow, ovary, testis, cultured fetal cells (reviewed by Thorgeirson and Nebert, 12; Nebert *et al.*, 9), and brain (5). The monooxygenase activities are associated with an electron transport chain whose terminal oxidases are numerous forms of cytochrome P-450. Increases in enzymatic activities after polycyclic aromatic inducer treatment are accompanied by *de novo* synthesis of cytochrome P$_1$-450 (also termed P-448) (3). Compared with *Ah^d/Ah^d* mice, *Ah^b/Ah^b* individuals have: a high inflammatory response to cutaneous application of 7,12-dimethylbenz[a]anthracene; a high susceptibility to 3-methylcholanthrene- and benzo[a]pyrene-induced subcutaneous sarcomas and 3-methylcholanthrene-induced lung tumors; an increased resistance to zoxazolamine-induced paralysis, lindane toxicity, and benzo[a]pyrene-induced aplastic anemia and leukemia; a high susceptibility to acetaminophen-induced hepatic necrosis and cataract formation; and an increased susceptibility to polycyclic hydrocarbon-induced birth defects, stillbirths, resorptions, decreased body weight, ovarian primordial oocyte depletion, and spermatozoal aberrations (reviewed by Thorgeirson and Nebert, 12; Nebert *et al.*, 9). The *Ah^b* allele is associated with increases in numerous metabolites of chemical carcinogens binding to DNA nucleotides (1). The effectiveness of several mutagens for *Salmonella in vitro* is enhanced by presence of a liver fraction from *Ah^b/Ah^b* mice treated with polycyclic hydrocarbons, but not from similarly treated Ah^d/Ah^d mice (2). In contrast, oral doses of benzo[a]pyrene cause a high rate of leukemia in Ah^d/Ah^d but not in Ah^b/Ah^d mice, probably because the carcinogenic metabolites produced in the responsive Ah^b/Ah^d mice are rapidly degraded in the intestine and excreted in the feces (8).

The product of the *Ah* locus is a cytosolic receptor which combines with the polycyclic aromatic inducers. Translocation into the nucleus of the receptor with its bound inducer is necessary for the formation of membrane-bound P-450 cytochromes. The receptor is not detectable in Ah^d strains (10). A regulatory gene similar to the *Ah* locus was found to be on Chr 17 by use of mouse–Chinese hamster somatic cell hybrids screened for inducibility of AHH (4). However, examination of the strain distribution pattern of *Ah* alleles in several sets of recombinant inbred strains clearly shows that *Ah* is very closely linked to *Ly-18* on Chr 12 and lies between the two DNA polymorphisms *D12Nyu1* and *D12Nyu2* near the centromeric end (6). See also *Ox-1*, *Ox-2 loci*, and *P450-1*, *P450-3* loci.

References

1. Boobis, A.R., and D.W. Nebert. 1977. Genetic differences in the metabolism of carcinogens and in the binding of benzo[a]pyrene metabolites to DNA. Adv. Enz. Regul. 15:339–362.
2. Felton, J.S., and D.W. Nebert. 1975. Mutagenesis of certain activated carcinogens *in vitro* associated with genetically mediated increases in monooxygenase activity and cytochrome P_1-450. J. Biol. Chem. 250:6769–6778.
3. Haugen, D.A., M.J. Coon, and D.W. Nebert. 1976. Induction of multiple forms of mouse liver P-450. Evidence for genetically controlled *de novo* protein synthesis in response to treatment with β-naphthoflavone or phenobarbital. J. Biol. Chem. 251:1817–1827.
4. Legraverend, C., S.O. Kärenlampi, S.W. Bigelow, P.A. Lalley, C.A. Kozak, J.E. Womack, and D.W. Nebert. 1984. Aryl hydrocarbon hydroxylase induction by benzo[a]anthracene: regulatory gene localized to the distal portion of mouse chromosome 17. Genetics 107:447–461.
5. Levitt, R.C., J.M. Fysh, N.M. Jensen, and D.W. Nebert. 1979. The *Ah* locus: biochemical basis for genetic differences in brain tumor formation in mice. Genetics 92:1205–1210.
6. Lusis, A.J., B.A. Taylor, D. Quon, S. Zollman, and R.C. LeBoeuf. 1987. Genetic factors controlling structure and expression of apolipoproteins B and E in mice. J. Biol. Chem. 262:7594–7504.
7. Nebert, D.W., and J.E. Gielen. 1972. Genetic regulation of aryl hydrocarbon hydroxylase induction in the mouse. Fed. Proc. 31:1315–1325.
8. Nebert, D.W., and N.M. Jensen. 1979. Benzo[a]pyrene-initiated leukemia in mice. Association with allelic differences at the *Ah* locus. Biochem. Pharmacol. 28:149–151.
9. Nebert, D.W., S.A. Atlas, T.M. Guenthner, and R.E. Kouri. 1978. The *Ah* locus: genetic regulation of the enzymes which metabolize polycyclic hydrocarbons and the risk of cancer. *In* P.O.P. T'so and H.V. Gelboin, eds., Polycyclic Hydrocarbons and Cancer: Chemistry, Molecular Biology and Environment, 345. Academic Press, New York.
10. Okey, A.B., G.P. Bondy, M.E. Mason, G.F. Kahl, H.J. Eisen, T.M. Guenthner, and D.W. Nebert. 1979. Regulatory gene product of the *Ah* locus. Characterization of the cytosolic inducer–receptor complex and evidence for its nuclear translocation. J. Biol. Chem. 254:11636–11648.
11. Taylor, B.A. 1984. The aryl hydrocarbon hydroxylase inducibility locus (Ah) of the mouse, a genetic perspective. *In* A. Poland and R.D. Kimbrough, eds., Biological Mechanisms of Dioxin Action, Banbury Report 18:97–108. Cold Spring Harbor Laboratory, New York.
12. Thorgeirson, S.S., and D.W. Nebert. 1977. The *Ah* locus and the metabolism of chemical carcinogens and other foreign compounds. Adv. Cancer Res. 25:149–193.

Ahd-1 to *Ahd-6* loci

These loci control electrophoretic or activity variation in isozymes of aldehyde dehydrogenase (AHD; E.C. 1.2.1.3), as determined on cellulose acetate gel using proprionaldehyde as substrate and NAD as cofactor (2).

Ahd-1 locus, aldehyde dehydrogenase-1, Chr 4. This locus controls electrophoretic variation in the mitochondrial AHD isozyme, AHD-1, present in liver, kidney, and lung. The allele $Ahd\text{-}1^a$ determines a slow migrating band and occurs in strains A/J, C57BL/6, NZB, and others; the allele $Ahd\text{-}1^b$ determines a faster band and occurs in strains AKR, BALB/c, CBA/Ca, and others. Heterozygotes have a three-banded pattern indicating a dimeric structure of the enzyme (2, 4). A third allele $Ahd\text{-}1^c$ determines a band slower than the other two bands and occurs in *M. m. castaneus* (3). *Ahd-1* is located on the distal part of Chr 4 (2).

Ahd-2 locus, aldehyde dehydrogenase-2, Chr 19. This locus controls electrophoretic mobility of a cytoplasmic AHD isozyme, AHD-2, present in liver, pancreas, and to a lesser extent in testis, stomach, and lung. The isozyme migrates more slowly that AHD-1. The allele $Ahd\text{-}2^a$ determines a slow band and occurs in most inbred strains; the allele $Ahd\text{-}2^b$ determines a faster band with greater activity, and occurs in the NZB, NZC, NZW, and other NZ strains. Homozygotes have a band intermediate in position and activity (4). A third allele $Ahd\text{-}2^c$ determines a band slower than the other two bands and occurs in *M. m. castaneus* (3). *Ahd-2* is located near the centromere on Chr 19 (4).

Ahd-3r locus, aldehyde dehydrogenase-3 regulator. AHD-3 is a microsomal isozyme of AHD that migrates faster than AHD-1. It is present in liver. No electrophoretic variants have been found (4). However, variation in inducibility of AHD-3 by phenobarbital in liver occurs and is controlled by a single locus designated *Ahd-3r*. The allele in strains A/J and BALB/c determines lack of inducibility; the allele in strains 101/H, C3H/He, C57BL6/J, and others determines inducibility (a threefold increase in activity under the conditions used). The alleles are codominant (5).

Ahd-4 locus, aldehyde dehydrogenase-4. This locus controls electrophoretic mobility of an AHD isozyme, AHD-4, present in stomach. The allele *Ahd-4ᵃ* determines a slow band and occurs in most inbred strains; the allele *Ahd-4ᵇ* determines a faster band and occurs in strains AL/NiA, O20/A, LIII, and a strain descended from wild Danish mice. *Ahd-4* is linked to *Ahd-6* with about 7 per cent recombination but the chromosomal location is not known (3).

AHD-5. This isozyme of AHD is found in testis. No variants are known (1).

Ahd-6 locus, aldehyde dehydrogenase-6. This locus controls activity level of an AHD isozyme, AHD-6, present in testis. The allele *Ahd-6ᵃ* determines low activity and occurs in most inbred strains; the allele *Ahd-6ᵇ* determines high activity and occurs in strains C57BL/LiA, ddN, DD/HeAf, SF/Cam, and GRS/A. *Ahd-6* is linked to *Ahd-4* with about 7 per cent recombination but the chromosomal location is not known (3).

References

1. Algar, E.M., and R.S. Holmes. 1985. Mouse mitochondrial aldehyde dehydrogenase isozymes: purification and molecular properties. Int. J. Biochem. 17:51–60.
2. Holmes, R.S. 1978. Genetics and ontogeny of aldehyde dehydrogenase isozymes in the mouse: localization of *Ahd-1* encoding the mitochondrial isozyme on chromosome 4. Biochem. Genet. 16:1207–1218.
3. Mather, P.B., and R.S. Holmes. 1984. Biochemical genetics of aldehyde dehydrogenase isozymes in the mouse: evidence for stomach- and testis-specific isozymes. Biochem. Genet. 22:981–995.
4. Timms, G.P., and R.S. Holmes. 1981. Genetics of aldehyde dehydrogenase isozymes in the mouse: evidence for multiple loci and localization of *Ahd-2* on chromosome 19. Genetics 97:327–336.
5. Timms, G.P., and R.S. Holmes. 1982. Mouse News Lett. 66:60–61.

Ahh locus

See *Ah* locus.

Ahr-1 to *Ahr-4* loci

These loci control electrophoretic pattern and activity of several isozymes of aldehyde reductase (AHR; E.C. 1.1.1.2).

Ahr-1 locus, aldehyde reductase-1, Chr 3. This locus controls electrophoretic pattern and activity of an isozyme of aldehyde reductase (AHD-1) that shows specificity toward *p*-nitrobenzaldehyde as a substrate. Three alleles are known: *Ahr-1ᵃ* determines a single anodally migrating band and occurs in many strains; *Ahr-1ᵇ* determines absence of an enzyme band and occurs in all C57 strains as well as in AQR, O20/A, SF/Cam, and several others; *Ahr-1ᶜ* determines a slower moving band and occurs in *M. m. castaneus*. *Ahr-1ᵃ/Ahr-1ᵇ* mice show three bands, consistent with a dimeric structure of the enzyme. The enzyme occurs in the cytoplasmic fraction, predominantly in liver with some activity also in small intestine. *Ahr-1* shows no recombination with *Adh-3* and pronounced linkage disequilibrium with it; it may be the product of a gene duplication event during evolution with considerable subsequent divergence in catalytic properties (1).

AHR-2. This member of the aldehyde reductase group of enzymes is a hexonate dehydrogenase. No genetic variants are known (1, 2).

Ahr-3 locus, aldehyde reductase-3, Chr 7. This locus controls activity of an isozyme of aldehyde reductase (AHR-3) that migrates faster anodally than AHR-2 or AHR-4 on cellulose acetate. It is found in liver and kidney but not in brain. The allele *Ahr-3ᵇ* determines low activity and occurs in the SJL/J strain; the allele *Ahr-3ᵃ* determines high activity and occurs in all other strains examined. Heterozygotes have high activity. *Ahr-3* is on Chr 7 about 27 cM from albino (*c*) (2).

Ahr-4 locus, aldehyde reductase-4. This locus controls activity of an isozyme of aldehyde reductase (AHD-4) that migrates slower anodally than AHR-2 or AHR-3 on cellulose acetate. It is found in liver and kidney but not in brain. The allele *Ahr-4ᵃ* determines high activity and occurs in most inbred strains; the allele *Ahr-4ᵇ* determines low activity and occurs in strains C57BL/6J, DBA/2, DD/Tbr, LTS/A, LIII, and one or two others. Heterozygotes have high activity (2).

References

1. Duley, J.A., and R.S. Holmes. 1982. Biochemical genetics of aldehyde reductase in the mouse: *Ahr-1*—a new locus linked to the alcohol dehydrogenase gene complex on chromosome 3. Biochem. Genet. 20:1067–1083.
2. Mather, P.B., and R.S. Holmes. 1985. Aldehyde reductase isozymes in the mouse: evidence for two new loci and

localization of *Ahr-3* on chromosome 7. Biochem. Genet. 23:483–496.

Aia-1, *Aia-2* loci

Aia-1 locus, autoimmune hemolytic anemia-1, Chr 4.

Aia-2 locus, autoimmune hemolytic anemia-2

These two loci have been postulated to explain the occurrence of autoimmune hemolytic anemia in the NZB strain. The dominant allele of *Aia-1* determining presence of the anemia occurs in the NZB strain; the recessive allele determining absence of the anemia occurs in the NZC strain and probably most other strains (1). Slightly different dominant alleles at the *Aia-2* locus are said to occur in the NZB and NZC strains to account for a slight difference in the time of onset of the anemia in backcrosses to the two strains, but the evidence for existence of this locus is not strong.

References

1. Knight, J.G., and D.D. Adams. 1981. Genes determining autoimmune disease in New Zealand mice. J. Clin. Lab. Immunol. 5:165–170.

Air1 locus, activator of *Ir-1*, Chr 16

This locus controls an activator for the expression of class II or Ia antigens of the major histocompatibility complex. The locus was identified in mouse–human somatic cell hybrids. Class II antigens were not expressed in the human cell line from which the hybrids were derived but were expressed in some of the hybrid lines. Expression of human as well as of mouse class II antigens in the hybrid lines was shown to depend on the presence of mouse Chr 16 (1).

References

1. Accolla, R.S., M. Jotterand-Bellomo, L. Scarpellino, A. Maffei, G. Carra, and J. Guardiola. 1986. *aIr-1*, a newly found locus on mouse chromosome 16 encoding a *trans*-acting activator factor for MHC class II gene expression. J. Exp. Med. 164:367–374.

ak, aphakia, recessive, Chr 19

Arose spontaneously in inbred strain 129/Sv-*Sl^J*. Homozygotes have small eyes lacking a lens, and the eyelids are closed. Newborn homozygotes can be recognized by their small eyes. Affected mice are often slightly smaller than their normal littermates, and their breeding ability may be reduced, especially in females. Heterozygotes appear normal (1). In homozygotes the abnormality is first detectable at 10 days of gestation during the early lens vesicle stage when the epithelium of the lens rudiment becomes disorganized and the lumen of the vesicle fills with rounded cells. In the lens placode more mitoses are oriented with their long axis perpendicular or oblique to the surface then in normal mice, in which most mitoses are parallel to the surface (4). The extracellular matrix material that forms a firm attachment between the lens cup and the optic cup during invagination of the lens cup is abnormal in affected embryos, and there are abnormal deposits of this material between the cells of the lens epithelium (5). The lens does not develop beyond 11 days, and the rest of the eye secondarily becomes abnormal (3). The degenerating lens remains as a clump of cells or as a cyst that persists into adulthood. Lens crystallins and sulfated materials similar to those in normal lens are found in these lens-derived units, indicating that *ak* does not primarily affect the synthesis of these substances (2).

References

1. Varnum, D.S., and L.C. Stevens. 1968. Aphakia, a new mutation in the mouse. J. Hered. 59:147–150.
2. Webster, E.H. Jr., J. Zwaan, and P. Cooper. 1986. Abnormal accumulation of sulphated materials in lens tissue of mice with the aphakia mutation. J. Embryol. Exp. Morphol. 92:85–101.
3. Zwaan, J. 1975. Immunofluorescent studies on aphakia, a mutation of a gene involved in the control of lens differentiation in the mouse embryo. Dev. Biol. 44:306–312.
4. Zwaan, J., and B.M. Kirkland. 1975. Malorientation of mitotic figures in the early lens rudiment of aphakia mouse embryos. Anat. Rec. 182:345–354.
5. Zwaan, J., and E.H. Webster Jr. 1984. Histochemical analysis of extracellular matrix material during embryonic mouse lens morphogenesis in an aphakia strain of mice. Dev. Biol. 104:380–389.

Ak-1 locus, adenylate kinase-1, Chr 2

This locus is probably the structural locus for an isozyme of adenylate kinase (AK-1; E.C. 2.7.4.3) that occurs in red blood cells and in tissues. The enzyme migrates rapidly toward the anode as a single band on starch gels. An electrophoretic variant has been briefly reported (1). The locus was shown to be located on Chr 2 by use of mouse–Chinese hamster somatic cell hybrid clones segregating mouse chromosomes (2).

References

1. Burow, K., H.J. Hedrick, and K.G. Rapp. 1977. Mouse News Lett. 56:65.
2. Francke, U., P.A. Lalley, W. Moss, J. Ivy, and J.D.

Ak-1

Minna. 1977. Gene mapping in *Mus musculus* by interspecific cell hybridization: assignment of the genes for tripeptidase-1 to chromosome 10, dipeptidase-2 to chromosome 18, acid phosphatase-1 to chromosome 12, and adenylate kinase-1 to chromosome 2. Cytogenet. Cell Genet. 19:57–84.

Ak-2 locus, adenylate kinase-2, Chr 4

This is probably the structural locus for an isozyme of adenylate kinase (AK-2; E.C. 2.7.4.3) that migrates as a double-banded pattern on starch gel and migrates less anodally than the AK-1 band (1). No genetic variants are known. The locus was shown to be located on Chr 4 by use of mouse–Chinese hamster somatic cell hybrid clones segregating mouse chromosomes. The synteny of *Ak-2* with *Eno-1*, *Pgd*, and *Pmg-2* on Chr 4 in the mouse is paralleled by the synteny of the homologous loci on the short arm of human Chr 1 (2).

References

1. Francke, U., P.A. Lalley, W. Moss, J. Ivy, and J.D. Minna. 1977. Gene mapping in *Mus musculus* by interspecific cell hybridization: assignment of the genes for tripeptidase-1 to chromosome 10, dipeptidase-2 to chromosome 18, acid phosphatase-1 to chromosome 12, and adenylate kinase-1 to chromosome 2. Cytogenet. Cell Genet. 19:57–84.
2. Lalley, P.A., U. Francke, and J.D. Minna. 1978. Homologous genes for enolase, phosphogluconate dehydrogenase, phosphoglucomutase, and adenylate kinase are syntenic on mouse chromosome 4 and human chromosome 1p. Proc. Natl. Acad. Sci. USA 75:2382–2386.

Akp-1 to Akp-3 loci

These loci control variation in isozymes of alkaline phosphatase (AKP; E.C. 3.1.3.1).

Akp-1 locus, alkaline phosphatase-1, Chr 1. This locus controls electrophoretic variation of a manganese-requiring isozyme of alkaline phosphatase, AKP-1, in liver and kidney. In liver, the enzyme is found in three fast anodally migrating bands numbered 1 to 3 from fastest to slowest. The allele *Akp-1^a* determines presence of bands 1 and 2 and occurs in strains C57BL/6, C57L, DBA/2, and others; the allele *Akp-1^b* determines presence of bands 2 and 3 and occurs in strains C3H/He, SJL, AKR/J, and many others, and in several wild subspecies. Heterozygotes have all three bands. In kidney, *Akp-1* determines the presence of different single fast anodally migrating bands in the two homozygotes, with both bands in heterozygotes (3).

Akp-2 locus, alkaline phosphatase-2, Chr 4. This locus controls electrophoretic variation in an isozyme of alkaline phosphatase, AKP-2, that does not require magnesium and that migrates more slowly than the isozyme controlled by *Akp-1*. It is present in liver, kidney, bone, placenta, and blood plasma. The allele *Akp-2^a* determines a relatively fast anodally migrating band and occurs in many inbred strains; the allele *Akp-2^b* determines a slightly slower faint band and occurs in the C57–C58 family of strains and a few others, including several wild-derived strains. Heterozygotes are intermediate but not always easy to distinguish from homozygotes. *Akp-2* is on Chr 4 very close to *Ahd-1* (2, 3).

Akp-3 locus, alkaline phosphatase-3, Chr 1. This locus controls electrophoretic mobility of an isozyme of alkaline phosphatase, AKP-3, in the small intestine. The allele *Akp-3^a* determines a fast migrating band and occurs in the DBA/2, LP, and probably also in the SWR strains; the allele *Akp-3^b* determines a slower band and occurs in the PL/J strain. Other strains like DBA/2 include A/J, AKR, BALB/c, CBA/J, C3H/He, NZB, and others. Other strains like PL/J are BDP, BUB, MA/My, P, SEA, and SM. The enzyme in SWR has a faster migration rate than that in the other *Akp-3^a* strains, and the genetic control of the difference is not clear. *Akp-3* is on Chr 1 near the center (1).

References

1. Wilcox, F.H., 1983. Genetics of alkaline phosphatase of the small intestine of the house mouse *(Mus musculus)*. Biochem. Genet. 21:641–6522.
2. Wilcox, F.H. and B.A. Taylor. 1981. Genetics of the *Akp-2* locus for alkaline phosphatase of liver, kidney, bone, and placenta in the mouse. J. Hered. 72:387–390.
3. Wilcox, F.H., L. Hirschhorn, B.A. Taylor, J.E. Womack, and T.H. Roderick. 1979. Genetic variation in alkaline phosphatase of the house mouse *(Mus musculus)* with emphasis on a manganese-requiring isozyme. Biochem. Genet. 17:1093–1107.

Akv-1 locus

See *Emv-11* locus.

Akv-2 locus

See *Emv-12* locus.

Akv-2J locus

See *Emv-14* locus.

Akv-3 locus

See *Emv-13* locus.

Akv-4 locus

See *Emv-14* locus.

Akvp locus

See *Emv-13* locus.

Akvr locus

See *Fv-4* locus.

Al, alopecia, dominant, Chr 11

Appeared as a spontaneous mutation in the first generation of an outcross. Hair loss is first noticeable at 25 to 30 days. *Al/Al* mice lose all hair over the body with the exception of some guard hairs. *Al/+* have more hair, but the distinction between *Al/Al* and *Al/+* is not always clear. After the first moult, hair grows in, but is lost again at the next moult. Mice appear more patchy as they grow older (1).

References

1. Dickie, M.M. 1955. Alopecia, a dominant mutation in the house mouse. J. Hered. 46:31–34.

Ala-1 locus

See *Ly-6E* locus.

Alb-1 locus, serum albumin variant, Chr 5

The alleles at this locus control electrophoretic mobility of serum albumin. All laboratory mice and nearly all wild mice examined have the *Alb-1ᵃ* allele. A variant with a slower migrating band (*Alb-1ᵇ*) was found in wild mice in Denmark (4) and another slow variant (*Alb-1ᶜ*), not known to be identical with *Alb-1ᵇ*, was found in wild mice near Windsor, Ontario. *Alb-1ᵃ/Alb-1ᶜ* mice have both bands (1). The albumin product of *Alb-1ᶜ* differs from that of *Alb-1ᵃ* by a lysine substituted for a glutamic acid (2). *Alb-1* is very closely linked to the *Afp* (α-fetoprotein) locus on Chr 5. The two genes show considerable homology. Both are transcribed in fetuses but after birth the rate of transcription of *Afp* falls rapidly to a very low level and the rate of transcription of *Alb-1* increases and remains high (5). The histone H1°, which is preferentially associated with nuclease-resistant chromatin and which is found associated with the *Afp* gene in adult mouse liver, is not found associated with the *Alb-1* gene (3). See *Afp* locus.

References

1. Petras, M.L. 1972. An inherited albumin variant in the mouse, *Mus musculus*. Biochem. Genet. 7:273–277.
2. Petras, M.L., and I.A. MacLaren. 1976. Structural differences between albumin A and albumin C of the house mouse *Mus musculus*. Biochem. Genet. 14:67–73.
3. Roche, J., C. Gorka, P. Goeltz, and J.J. Lawrence. 1985. Association of histone H1° with a gene repressed during liver development. Nature 314:197–198.
4. Selander, R.K. 1970. Biochemical polymorphism in populations of the house mouse and the old-field mouse. Symp. Zool. Soc. Lond. 26:73.
5. Sellam, C.H., M. Frain, T. Erdos, and J.M. Sala-Trepat. 1984. Differential expression of albumin and α-fetoprotein genes in fetal tissues of mouse and rat. Dev. Biol. 102:51–60.

ald, adrenocortical lipid depletion, recessive, Chr 1

Found in the AKR/O inbred strain by Arnesen (1). In homozygotes, spontaneous lipid depletion in the adrenal cortex occurs at the time of puberty. In adult males the lipid depletion depends on the presence of gonads in both sexes and on the presence of the hypophysis (5). There is a higher than normal turnover of RNA per cell in all three zones of the cortex in *ald/ald* mice (4). The depleted lipid consists of cholesterol esters, but not of free cholesterol or triglycerides (3). The site of action of *ald* is not in the adrenal gland as shown by normal lipid content of AKR adrenals after transplantation into normal (AKR × C57L)F1 mice (6). The adrenal cortex of the DBA/2 strain closely resembles that of AKR and of two related substrains, AC and CS (2), but apparently is caused by a variant at a different locus (3).

References

1. Arnesen, K. 1963. The cytology of the adrenal cortex in mice with spontaneous adrenocortical lipid depletion. Acta Pathol. Microbiol. Scand. 58:212–218.
2. Arnesen, K. 1974. Adrenocortical lipid depletion and leukemia in mice. Acta Pathol. Microbiol. Scand., Sect. A, Suppl. 248:15–19.
3. Doering, C.H., J.G.M. Shire, S. Kessler, and R.B. Clayton. 1973. Genetic and biochemical studies of the adrenal lipid depletion phenotype in mice. Biochem. Genet. 8:101–111.
4. Molne, K. 1969. Autoradiographic studies on the incorporation of ³H-uridine into the adrenal glands of mice with spontaneous lipid depletion. Acta Pathol. Microbiol. Scand. 77:369–378.
5. Molne, K., and G. Braband. 1968. Spontaneous adrenocortical lipid depletion in mice. Relation to general growth, degeneration of adrenal × zone, and maturation of seminiferous epithelium. Acta Pathol. Microbiol. Scand. 72:478–490.
6. Taylor, B.A., H. Meier, and W.K. Whitten. 1974. Chro-

mosomal location and site of action of the adrenal lipid depletion gene of the mouse. Genetics 77:s65 (Abstr.).

Aldh locus, aldehyde dehydrogenase

This locus described by Lush (1) does not seem to be the same as any of the *Ahd* loci described by Holmes and his co-workers (see *Ahd-1* to *Ahd-6* loci). It controls electrophoretic mobility on starch gel of a supernatant isozyme of aldehyde dehydrogenase (ALDH; E.C. 1.2.1.3) that is detected using salicylaldehyde as substrate and NAD as cofactor. The allele *Aldh^a* determines a single cathodally migrating band and occurs in the C3H/He and many other inbred strains; the allele *Aldh^b* determines three additional slower bands and occurs in the NZB, IS/Cam, and a few other strains. The slow bands stain with greater intensity in females than in males. Homozygotes are indistinguishable from *Aldh^b/Aldh^b* (1).

References

1. Lush, I.E. 1978. Genetic variation of some aldehyde-oxidizing enzymes in the mouse. Anim. Blood Grps. Biochem. Genet. 9:85–96.

Alm, anterior lenticonus with microphthalmia, probably semidominant

Arose in a (101 × C3H)F1 hybrid after irradiation with ^{137}Cs gamma rays. Homozygotes have not been recovered. In heterozygotes, *Alm* is probably semilethal, and survivors are often sterile. The lens is opaque at the anterior pole, abnormal in shape, and adhered to the cornea. Heterozygotes are characterized by microphthalmia, a white belly spot, and reduced body size (1).

References

1. Kratochvilova, J. 1981. Dominant cataract mutations detected in offspring of gamma-irradiated male mice. J. Hered. 72:302–307.

Alp-1, *Alp-2* loci

See *Apoa-1*, *Apoa-2* loci.

am, amputated, recessive, Chr 8

Found among the fetuses of dissected pregnant females in an irradiation experiment. Homozygotes die at birth. Fetuses lack a tail, and at late stages of gestation the limbs appear to have been severed close to the trunk. The head is short and the body edematous dorsally.

Vertebrae in the lumbosacral region are absent, and there is disorganization of the remainder of the spinal column and some rib fusion. In the forelimbs most of the bones are present but abnormal (6). The body axis is shortened from the eighth day of gestation, and the newly formed somites are only one-half normal size (5). By 9 1/2 days the cells of the sclerotomes are aggregated in small clumps with greater areas of cell contact than normal and with filopodia that form an abnormal tangled web instead of being stretched out between cells. Similar abnormalities occur in the mesoderm of the facial region and are already present in the presomatic mesoderm (2, 3). Clumping of mesoderm cells in the palatal shelves, rather than separation and spreading out as in normal mesoderm, causes failure of the shelves to grow and results in cleft palate (1). The mesodermal abnormalities occur in cultured cells as well as *in vivo*, suggesting that the abnormal cell behavior is intrinsic to the cells (4).

References

1. Flint, O.P. 1980. Cell behavior and cleft palate in the mutant mouse, *amputated*. J. Embryol. Exp. Morphol. 58:131–142.
2. Flint, O.P., and D.A. Ede. 1978. Cell interactions in the developing somite; *in vivo* comparisons between *amputated* (*am/am*) and normal embryos. J. Cell Sci. 31:275–291.
3. Flint, O.P., and D.A. Ede. 1978. Facial development in the mouse; a comparison between normal and mutant (*amputated*) mouse embryos. J. Embryol. Exp. Morphol. 48:249–267.
4. Flint, O.P., and D.A. Ede. 1981. Cell interactions in the developing somite: *in vitro* comparison between *amputated* (*am/am*) and normal mouse embryos. J. Embryol. Exp. Morphol. 67:113–125.
5. Flint, O.P., D.A. Ede, O.K. Wilby, and J. Proctor. 1978. Control of somite number in normal and *amputated* mutant mouse embryos: an experimental and theoretical analysis. J. Embryol. Exp. Morphol. 45:189–202.
6. Meredith, R. 1964. Mouse News Lett. 31:25.

Amy complex, amylase, Chr 3

This complex includes structural loci for salivary and pancreatic α-amylase (AMY; E.C. 3.2.1.1) and regulatory elements affecting rate of expression of the structural genes. There is one gene for salivary amylase and at least four for pancreatic amylase (5). They have not shown recombination in crosses (7).

Amy-1 locus, salivary amylase. This locus controls electrophoretic mobility of α-amylase found in saliva and parotid gland. The amount of enzyme present is also controlled by this locus or by a very closely linked element, which acts by controlling rate of synthesis of the enzyme and is *cis*-acting. At least four codominant alleles are known (3). *Amy-1ᵃ* determines a slow migrating component (SAL A) and an intermediate enzyme level and occurs in strain C3H/As and most other inbred strains. *Amy-1ᵇ* determines a fast migrating component (SAL B) and an intermediate enzyme level and occurs in *M. m. castaneus*. *Amy-1ᶜ* determines the SAL B component and high enzyme level (2.7 times that of *Amy-1ᵃ*) and occurs in strain YBR. *Amy-1ᵈ* determines the SAL B component and low enzyme level (0.72 times that of *Amy-1ᵃ*) and occurs in strain CE (3, 7). Heterozygotes have both parental electrophoretic components and intermediate levels of enzyme. Other alleles determining other enzyme levels may be present in wild populations (3). An apparent mutation from *Amy-1ᵃ* to *Amy-1ᵇ* was found in the BXD-16 recombinant inbred strain. The enzyme determined by the mutation has a slightly higher anodal electrophoretic mobility than that determined by *Amy-1ᵇ* (4). In the *Amy-1ᵃ* strain, A/J, it was shown that salivary amylase is also produced in the liver, but at a much lower level than in the parotid. In both organs, the amylase mRNA is transcribed from the same gene but with two promoters of different strength, only the weaker one in the liver and both in the parotid. The strong promoter is activated at about 2 weeks of age and is expressed in all secreting cells of the parotid in adulthood (8).

Amy-2 loci, pancreatic amylase. This group of loci controls electrophoretic mobility and rate of synthesis of pancreatic α-amylase. At least four electrophoretic forms of the enzyme occur. In the terminology of Bloor *et al.* (1), the four isozymes are A₁, A₂, B₁, and B₂. Different haplotypes occur among inbred strains and wild mice, which may have one or more of the forms. The haplotype of common inbred strains, including C3H/He and A/J, has been designated *Amy-2ᵃ*; these strains have A₂. The haplotype of YBR is designated *Amy-2ᵇ* and this strain has A₁B₁. The CE strain has A₁A₂B₁B₂, a strain derived from *M. m. castaneus* has B₁, and various wild-derived strains have other combinations, e.g. A₁A₂, A₁B₁B₂, and A₂B₁ (1, 5, 9). The quantitative differences always co-segregate with the isozyme phenotypes. DNA restriction enzyme analysis has shown that haplotypes with identical isozymes but different total or relative amounts of the different isozymes as well as haplotypes with different isozymes have different restriction enzyme patterns, suggesting that the quantitative differences are coded by regulatory sequences closely linked to the structural genes (11). In the CE strain (A₁A₂B₁B₂), Tosi *et al.* (12) found five

distinct mRNA sequences specifying *Amy-2* amylases, two of which specify the same protein. In the A/J strain (A₂), four *Amy-2*-related sequences in the genomic DNA have been found (2). In the YBR strain (A₁B₁), two *Amy-2* genes determining enzymes that differ at two amino acids are expressed, but the total enzyme activity is less than half that of most other strains (10). In this strain there are two complete genes probably coding for the two proteins and two truncated genes (6).

References

1. Bloor, J.H., M.H. Meisler, and J.T. Nielsen. 1981. Genetic determination of amylase synthesis in the mouse. J. Biol. Chem. 256:373–377.
2. Hagenbüchle, O, P.K. Wellauer, D.L. Cribbs, and U. Schibler. 1984. Termination of transcription in the mouse α-amylase gene *Amy-2ᵃ* occurs at multiple sites downstream of the polyadenylation site. Cell 38:737–744.
3. Hjorth, J.P. 1979. Genetic variation in mouse salivary amylase rate of synthesis. Biochem. Genet. 17:665–682.
4. Hjorth, J.P. 1982. Altered salivary amylase gene in the mouse strain BXD-16. Heredity 48:127–135.
5. Hjorth, J.P., A.J. Lusis, and J.T. Nielsen. 1980. Multiple structural genes for mouse amylase. Biochem. Genet. 18:281–302.
6. Mikkelsen, B.M., M.E. Clark, G. Christiansen, O.M. Klintebaek, J.T. Nielsen, K.K. Thomsen, and J.P. Hjorth. 1985. The structure of two distinct pancreatic amylase genes in mouse strain YBR. Biochem. Genet. 23:511–524.
7. Nielsen, J.T., and K. Sick. 1975. Genetic polymorphism of amylase isozymes in feral populations of the house mouse. Hereditas 79:279–286.
8. Shaw, P., B. Sordat, and U. Schibler. 1985. The two promoters of the mouse α-amylase gene *Amy-1ᵃ* are differentially activated during parotid gland differentiation. Cell 40:907–912.
9. Sick, K., and J.T. Nielsen. 1964. Genetics of amylase isozymes in the mouse. Hereditas 51:291–296.
10. Strahler, J.R., and M. Meisler. 1982. Two distinct pancreatic amylase genes are active in YBR mice. Genetics 101:91–102.
11. Thomsen, K.K., J.P. Hjorth, and J.T. Nielsen. 1984. Genetic variation in restriction patterns among mouse amylase gene complexes. Biochem. Genet. 22:257–273.
12. Tosi, M., R. Bovey, S. Astolfi, S. Bodary, M. Meisler, and P.K. Wallauer. 1984. Multiple non-allelic genes encoding pancreatic alpha-amylase of mouse are expressed in a strain-specific fashion. EMBO J. 12:2809–2816.

an, Hertwig's anemia, recessive, Chr 4

Probably X-ray-induced. Homozygotes are anemic at birth, are small, and usually die within a few days (6).

Survival has been greatly prolonged in a selected strain, BAN/Re (7). Survivors are almost completely sterile (8), with low germ cell number beginning at 12 to 13 days of gestation (9). The anemia is the normochromic macrocytic type. It can be seen as early as 12 days of gestation and persists throughout life. The white cell count is also much reduced (1). Phytohemagglutinin-stimulated proliferation of spleen cells and immune response to sheep red blood cells are about half normal in *an/an* mice (5). Transplantation studies have shown that the anemia results from an intrinsic stem cell defect (3). Examination of the number of cells in various precursor hemopoietic stem cell populations shows that there is a progressive loss of *an/an* precursor cells during maturation (2, 10). In adult bone marrow, kidney cell cultures, and fetal liver, there is a 5 to 15 per cent incidence of abnormal mitotic figures consisting mostly of hyperploid cells. This could explain the loss of progenitor cells during erythroid maturation and the depletion of germ cells during ontogeny (4, 10).

References

1. Bannerman, R.M., J.A. Edwards, and P.H. Pinkerton. 1973. Hereditary disorders of the red cells in animals. Prog. Hematol. 8:131–179.
2. Barker, J.E., and S.E. Bernstein. 1983. Hertwig's anemia: characterization of the stem cell defect. Blood 61:765–769.
3. Bernstein, S.E. 1975. Bone marrow implantation therapy in Hertwig's anemic mice. Jackson Lab. Ann. Rep. 46:87.
4. Eppig, J.T., and J.E. Barker. 1984. Chromosome abnormalities in mice with Hertwig's anemia. Blood 64:727–732.
5. Harrison, D.E. 1976. Stem cell multiplication rate in W-anemic mice, and immune response of mice with Hertwig's anemia. Jackson Lab. Ann. Rep. 47:59.
6. Hertwig, P. 1942. Neue Mutationen und Koppelungsgruppen bei der Hausmaus. Z. Indukt. Abstammungs-Vererbungsl. 80:220–246.
7. Russell, E.S. 1970. Abnormalities of erythropoiesis associated with mutant genes in mice. *In* A.L. Gordon, ed., Regulation of Hematopoiesis, Vol. I, 649–675. Appleton-Century-Crofts, New York.
8. Russell, E.S., and H. Meier. 1966. Constitutional diseases. *In* E.L. Green, ed., Biology of the Laboratory Mouse, 2nd ed., 571–587. McGraw-Hill, New York.
9. Russell, E.S., W.K. Whitten, W.G. Beamer, E.C. McFarland, and H. Peters. 1976. Germ cell defect in Hertwig's anemic mice. Jackson Lab. Ann. Rep. 47:89.
10. Russell, E.S., E.C. McFarland, and H. Peters. 1985. Genetic and pleiotropic defects in mouse fetuses with Hertwig's anemia. Dev. Biol. 110:331–337.

Anc, anisocoria, probably semidominant

Arose in a (101 × C3H)F1 hybrid after irradiation with ^{137}Cs gamma rays. Homozygotes have not been recovered. In heterozygotes, the pupils are ovoid and dilated abnormally toward the nasal side after atropine treatment. The pupillary anomaly is frequently accompanied by remnants of the pupillary membrane (70 per cent) or anterior polar cataract (20 per cent). Penetrance is incomplete (1).

References

1. Kratochvilova, J. 1981. Dominant cataract mutations detected in offspring of gamma-irradiated male mice. J. Hered. 72:302–307.

Anf locus

See *Pnd* locus.

ank, progressive ankylosis, recessive, Chr 15

Arose spontaneously in a balanced stock carrying buff (*bf*) and jagged tail (*jg*). Homozygotes have a severe joint disease that affects nearly all the movable joints. Affected mice are first recognizable at about 5 weeks of age by stiffness of the toes of the forefeet. Loss of mobility soon spreads to other joints, so that the mice become quite rigid and seldom survive beyond 5 months. Both sexes can breed but do not produce more than two litters. There is increased deposition of calcium in the calcified cartilages (2). Development of the joint disorder studied in the joints of the forefeet is characterized by an initial distention of joints with a milky fluid containing large macrophage-like mononuclear cells and suspended calcium hydroxyapatite, accompanied by proliferation of the synovial cells with subsequent roughness of the cartilaginous articular surfaces and eventual fusion of the articular cartilages, and by formation of periarticular cartilage bridges with subsequent ossification leading to complete joint fusion. The *ank* disease somewhat resembles the calcium hydroxyapatite-associated arthropathies in man but is considerably more severe (1). *ank* is closely linked to underwhite (*uw*) near the centromere of Chr 15 (2).

References

1. Hakim, F.T., R. Cranley, K.S. Brown, E.D. Eanes, L. Haine, and J.T. Oppenheim. 1984. Hereditary joint disorder in progressive ankylosis (ank/ank) mice. I. Association of calcium hydroxyapatite deposition with inflammatory arthropathy. Arth. Rheum. 27:1411–1420.
2. Sweet, H.O., and M.C. Green. 1981. Progressive ankylosis, a new skeletal mutation in the mouse. J. Hered. 72:87–93.

anx, anorexia, recessive, Chr 2

Arose in an F2 of a cross between the DW/J strain and an inbred strain derived from a cross of *M. m. poschiavinus* to an inbred Swiss stock. Homozygotes are characterized by marked reduction in body weight, emaciated appearance, and abnormal behavior including head weaving and body tremors, uncoordinated gait, hyperactivity, and poor appetite. Stomach contents are reduced at about 5 days of age and continue so until death at about 22 days. Injecting 15-day-old normal mice with serotonin produced the body tremors and head weaving typical of 18-day anorexic mice, and injecting 20-day-old anorexic mice with a serotonin antagonist diminished the severity of the neurological symptoms. The data suggest that the mice are anorexic, failing to ingest the amount of food necessary to sustain life. *anx* is located close to pallid (*pa*) on Chr 2 (1).

References

1. Maltais, L.J., P.W. Lane, and W.G. Beamer. 1984. Anorexia, a recessive mutation causing starvation in preweanling mice. J. Hered. 75:468–472.

ao, apampischo, recessive

Arose spontaneously in a stock of sib-mated mice at Greensboro College. In homozygotes, hair loss begins at 10 to 15 days on the head and proceeds toward the ventrum and tail. A narrow strip of hair is retained along the dorsal midline. Regrowth begins at 3 to 4 weeks. The new coat is rougher, thinner, and darker colored than normal, but in males and non-pregnant females there is no further hair loss until about a year of age. In breeding females, however, hair loss occurs at every pregnancy, beginning at about 8 days and becoming maximum at birth, with regrowth starting at the end of lactation unless the female is already pregnant again (1).

References

1. VanPelt, A.F., R. Knorr, and K. Cain. 1969. A new hairless mutant in the house mouse. J. Hered. 60:79–96.

Aox-1, Aox-2 loci, Chr 1

Aox-1 locus, aldehyde oxidase-1. Alleles at this locus probably determine variation in the molecular structure of aldehyde oxidase (AOX-1; E.C. 1.2.3.1; originally referred to as N^1-methylnicotinamide oxidase). The C57BL/6 strain (*Aox-1^a*) and the DBA/2 strain (*Aox-1^b*) differ by a 10-fold higher liver enzyme activity in C57BL/6, and a twofold higher K_m of the DBA/2 liver enzyme (4). Most other strains, including DBA/1, C3H/He, and CBA/J, have low activity (3). In segregating generations of crosses between C57BL/6 and DBA/2, the activity difference segregated as though controlled by a single locus (1). Heterozygotes have intermediate activity. An electrophoretic variant found in the CE strain and at first thought to be a product of a third *Aox-1* allele (5) is now known to be the product of *Aox-2* (3). A survey of 30 inbred strains showed two classes of strains with respect to K_m values, but an intermediate class in addition to the high and low classes with respect to activity level (7). Studies to clarify the genetic basis for the intermediate activity have not been reported. *Aox-1* and *Aox-2* are closely linked on Chr 1 (2).

Aox-2 locus, aldehyde oxidase-2. This locus controls activity and electrophoretic mobility of a soluble liver isozyme of aldehyde oxidase (AOX-2) that migrates more slowly electrophoretically than the AOX-1 isozyme. Most strains carry the allele determining high activity, *Aox-2^a*; SWR carries the allele determining low activity, *Aox-2^b*; the allele, *Aox-2^c*, determining high activity and a variant electrophoretic form occurs in the CE strain and in wild Danish mice (3); a fourth allele, *Aox-2^d*, determines a slower electrophoretic form than *Aox-2^c* and was found in wild mice from Darling Downs, Queensland. *Aox-2^c*/*Aox-2^d* mice have a three-banded pattern indicating a dimeric structure of the enzyme (6). *Aox-2* has shown no recombination with *Aox-1* (3).

References

1. Gluecksohn-Waelsch, S., P. Greengard, G.P. Quinn, and L.S. Teicher. 1967. Genetic variations of an oxidase in mammals. J. Biol. Chem. 242:1271–1273.
2. Holmes, R.S. 1979. Genetics, ontogeny, and testosterone inducibility of aldehyde oxidase isozymes in the mouse: evidence for two genetic loci (*Aox-1* and *Aox-2*) closely linked on chromosome 1. Biochem. Genet. 17:517–527.
3. Holmes, R.S., L.R. Leijten, and J.A. Duley. 1981. Liver aldehyde oxidase and xanthine oxidase genetics in the mouse. Anim. Blood Grps. Biochem. Genet. 12:193–199.
4. Huff, S.D., and S. Chaykin. 1967. Genetic and androgenic control of N^1-methylnicotinamide oxidase activity in mice. J. Biol. Chem. 242:1265–1270.
5. Lush, I.E. 1978. Genetic variation of some aldehyde-oxidizing enzymes in the mouse. Anim. Blood Grps. Biochem. Genet. 9:85–96.
6. Mather, P.B., J.A. Duley, and R.S. Holmes. 1983. Aldehyde oxidase and alcohol dehydrogenase genetics in the mouse. New alleles at the *Aox-2* and *Adh-3* loci. Anim. Blood Grps. Biochem. Genet. 14:279–287.
7. Watson, J.G., T.J. Higgins, P.B. Collins, and S. Chaykin.

1972. The mouse liver aldehyde oxidase locus *(Aox)*. Biochem. Genet. 6:195–204.

ap, alopecia periodica, recessive

Arose spontaneously in an inbred stock at the National Institute of Genetics, Japan. Homozygotes are recognizable at 3 to 4 days of age by their short whiskers and short, sparse first coat. The first coat is shed almost completely between 13 to 15 days and 20 to 24 days. Thereafter the hair is lost and regenerated in cycles which begin at the anterior end and spread toward the tail. During growth of the first coat the skin may have thick scales, and part of the tail may be lost. All four kinds of hair are present, but they are somewhat abnormal (1). This mutant is not allelic with furless *(fs)* nor linked to it (2).

References

1. Tutikawa, K. 1953. Studies on an apparently new mutant, üalopecia periodica,' found in the mouse. Ann. Rep. Natl. Inst. Genet. Jpn. (1952) 3:9–10.
2. Tutikawa, K. 1955. Test for allelism of üalopecia periodica' and üfurless' in the house mouse. Ann. Rep. Natl. Inst. Genet. Jpn. (1954) 5:16.

Ap2 locus, adipocyte-specific protein aP2, Chr 3

This locus codes for the adipocyte-specific protein aP2, a member of a family of proteins that includes fatty acid binding proteins and cellular retinol binding proteins among others. Restriction fragment length differences in *TaqI* digested DNA occur between strains AKR and C57L and between strains C57BL/6 and DBA/2. By use of these differences in the AKXL and BXD recombinant inbred strains as well as by use of mouse–Chinese hamster somatic cell hybrids, *Ap2* was shown to be very closely linked to *Car-2* on Chr 3 (1).

References

1. Heuckeroth, R.D., E.H. Birkenmeier, M.S. Levin, and J.I. Gordon. 1987. Analysis of the tissue-specific expression, developmental regulation, and linkage relationships of a rodent gene encoding heart fatty acid binding protein. J. Biol. Chem. 262:9709–9717.

Aph-1 locus, alpha protein-1, Chr 2

This locus controls a serum protein that migrates electrophoretically in the α globulin region. It was found by reciprocal immunization between strain BALB/c and Mol-A, an inbred strain derived from wild *M. musculus molossinus* in Japan. The allele *Aph-1^a* determines an electrophoretic mobility that is slighter faster than that determined by *Aph-1^b*. *Aph-1^b* occurs in all tested strains derived from *M. m. molossinus*; *Aph-1^a* occurs in strain BALB/c and in all other strains tested. The alleles are codominant. *Aph-1* is linked to *a* on Chr 2 with 38.7 ± 4.2 per cent recombination (1).

References

1. Harada, Y., J. Hayakawa, and T. Tomita. 1986. Antigenic polymorphism of a serum protein migrating electrophoretically in the α region detected by using a strain derived from the Japanese wild mouse *(Mus musculus molossinus)*. Immunogenetics 24:47–50.

Apk locus (formerly *Acp-2*), acid phosphatase-kidney, Chr 10

This locus controls electrophoretic mobility of an acid phosphatase (E.C. 3.1.3.2) found in kidney, but not in red cells, liver, spleen, heart, lung, brain, skeletal muscle, stomach, or testis. The enzyme band is detectable with naphthol AS-MX phosphoric acid as substrate. The allele *Apk^a* determines a cathodally migrating band and occurs in strain C57BL/6 and all other inbred strains examined; the allele *Apk^m* determines a faster cathodally migrating band and occurs in *M. m. molossinus*. Heterozygotes have both parental bands (1).

References

1. Womack, J.E., and S.B. Auerbach. 1978. An acid phosphatase locus expressed in mouse kidney *(Apk)* and its genetic location on chromosome 10. Biochem. Genet. 16:239–245.

Apl locus

See *Neu-1* locus.

Apoa-1, Apoa-2 loci (formerly *Alp-1, Alp-2*)

These loci control variation in isoelectric point (pI) revealed by isoelectric focusing of apolipoproteins that are major constituents of mouse high density lipoprotein (HDL). The *Apoa* nomenclature was suggested by Lusis *et al.* (4) to conform with that for apoliproteins B and E.

Apoa-1 locus (formerly also *Sep-1* and *Ltw-1*), apolipoprotein A-1, Chr 9. This locus controls variation in pI of apolipoprotein A-I (apoA-I). After isoelectric focusing, apoA-I consists of two major bands and one or more minor bands at pI 5.6. Two alleles are known, *Apoa-1^a* in strains AKR, CBA/Ca, the C57 family, NZB, and others, and *Apoa-1^b* in strains A, BALB/c,

C3H/He, DBA/2, and others. The two alleles determine the same pattern of bands but the pI of the bands determined by *Apoa-1^a* is more acidic than that of those determined by *Apoa-1^b*. Heterozygotes show all the parental bands (3). The same locus was identified by electrophoretic mobility *(Sep-1)* (1) and was also identified by two-dimensional electrophoresis *(Ltw-1)* (2).

Apoa-2 locus, apolipoprotein A-2, Chr 1. This locus controls variation in pI of apolipoprotein A-II (apoA-II). After isoelectric focusing, apoA-II consists of a single major band at pI approximately 4.9. The allele *Apoa-2^a* determines a band that is slightly more acidic than that determined by *Apoa-2^b*. Heterozygotes have both parental bands. *Apoa-2^a* occurs in strains AKR, the C57 family, DBA/2, and others; *Apoa-2^b* occurs in strains A, BALB/c, CBA/Ca, C3H/He, NZB, and others. *Apoa-2* is very closely linked to the high-density lipoprotein-1 locus, *Hdl-1*, but the distribution of alleles of the two loci shows some lack of concordance among inbred strains. The two loci are probably not identical but may be part of a cluster of HDL genes on Chr 1 (3). See also *Hdl-1* locus.

References

1. Eicher, E.M., B.A. Taylor, S.C. Leighton, and J.E. Womack 1980. A serum protein polymorphism determinant on chromosome 9 of *Mus musculus*. Mol. Gen. Genet. 177:571–576.
2. Elliott, R.W. 1979. Use of two-dimensional electrophoresis to identify and map new mouse genes. Genetics 91:295–308.
3. Lusis, A.J., B.A. Taylor, R.W. Wangenstein, and R.C. LeBoeuf. 1983. Genetic control of lipid transport in mice. II. Genes controlling structure of high density lipoproteins. J. Biol. Chem. 258:5071–5078.
4. Lusis, A.J., B.A. Taylor, D. Quon, S. Zollman, and R.C. LeBoeuf. 1987. Genetic factors controlling structure and expression of apoliproteins B and E in mice. J. Biol. Chem. 262:7594–7604.

Apob, *Apoe* loci

These loci code for apolipoproteins B and E (apoB and apoE), the major proteins of low-density lipoproteins (LDL) and very-low-density lipoproteins (VLDL). Although most strains of mice have large increases in plasma LDL and VLDL in response to a high fat diet, the levels of apoB and apoE mRNA are not increased, suggesting that the increase in apolipoproteins is due principally to a decrease in rate of catabolism.

Apob locus, apolipoprotein B, Chr 12. The apolipoprotein (apoB) coded for by this locus occurs in two forms, apoB100 with a molecular weight of 550 kDa and apoB48 with a molecular weight of 210 kDa. Both forms are secreted by liver. Both species are probably derived from the *Apob* gene, but the mechanism by which this is accomplished is not clear. A DNA restriction fragment length polymorphism detected with a rat apoB cDNA probe and a combination of *Eco*RV and *Hind*III endonucleases occurs among inbred strains: the allele *Apob^a* occurs in strains C57BL/6, C57L, DBA/2, LT/Sv, SWR, and 129/J; the allele *Apob^b* occurs in strains AKR and C3H/HeJ; the allele *Apob^c* occurs in strains BALB/c and NZB. The distribution of these alleles in five sets of recombinant inbred strains showed that *Apob* is located on Chr 12 close to *Rnr12* and the centromere. The *Apob^b* strains, AKR and C3H/He, differ from the other strains in failing to show an increase in either apoB48 or in LDL/VLDL lipid when on a high fat diet. Whether this is an effect of the *Apob^b* allele is not known (1).

Apoe locus, apolipoprotein E, Chr 7. A restriction fragment length polymorphism detected with a mouse apoE cDNA probe hybridized to DNA cut with *Kpn*I occurs among inbred strains: the allele *Apoe^b* occurs in strains AKR, BALB/c, C57BL/6, C57L, C58, DBA/2, and IS/Cam; the allele *Apoe^h* occurs in strains C3H/HeJ, LT/Sv, NZB, 129/J, and SJL. Distribution of these alleles in two sets of recombinant inbred strains showed that *Apoe* is located on Chr 7 near *Es-8* and *Svp-2* (1).

References

1. Lusis, A.J., B.A. Taylor, D. Quon, S. Zollman, and R.C. LeBoeuf. 1987. Genetic factors controlling structure and expression of apolipoproteins B and E in mice. J. Biol. Chem. 262:7594–7604.

Apoc, anterior polar cataract, probably semidominant

Arose in a (101 × C3H)F1 hybrid after irradiation with ^{137}Cs gamma rays. Viable homozygotes have not been recovered. In heterozygotes, the lens shows irregular concentric opaque layers at the anterior pole. Remnants of the pupillary membrane and adhesions of the iris to the cornea occur. Manifestation is uni- or bilateral. The gene is probably semilethal in heterozygotes (1).

References

1. Kratochvilova, J. 1981. Dominant cataract mutations detected in offspring of gamma-irradiated male mice. J. Hered. 72:302–307.

App

App locus (formerly Cvap), amyloid β (A4) protein, Chr 16

This locus is the mouse homologue of the human gene coding for amyloid β (A4) protein. This protein is a major constituent of the amyloid filaments found in brains of patients with Alzheimer's disease. The locus in the mouse was found to be on Chr 16 by use of mouse–Chinese hamster somatic cell hybrids screened with a clone isolated from a human brain cDNA library. Somatic cell hybrids made with mouse cells carrying the T(2;6)28H reciprocal translocation showed that the locus is on the distal part of Chr 16 (1). It was found to be closely linked to Sod-1 and Ets-2 on this chromosome by use of a restriction fragment length polymorphism found in M. spretus (2). This region contains several loci homologous to loci on human Chr 21, the chromosome that is trisomic in patients with Down syndrome and that carries the human amyloid β gene (1, 2).

References

1. Lovett, M., O Goldgaber, P. Ashley, D.R. Cox, D.C. Gajdusek, and C.J. Epstein. 1987. The mouse homolog of the human amyloid β protein gene is located on the distal end of mouse chromosome 16: further extension of the homology between human chromosome 21 and mouse chromosome 16. Biochem. Biophys. Res. Comm. 144:1069–1075.
2. Reeves, R.H., N.K. Robakis, M.L. Oster-Granite, H.M. Wisniewski, J.T. Coyle, and J.D. Gearhart. 1987. Genetic linkage in the mouse of genes involved in Down syndrome and Alzheimer's disease in man. Mol. Brain Res. 2:215–221.

Aprt loci

Aprt locus, adenine phosphoribosyltransferase, Chr 8. This is probably the structural locus for adenine phosphoribosyltransferase (APRT; E.C. 2.4.2.7). In mouse–Chinese hamster somatic cell hybrid cultures, the mouse electrophoretic form of the enzyme was found to segregate with Gr-1 on Chr 8 (2). A slow electrophoretic variant determined by the allele designated $Aprt^a$ was found in wild mice from South Turkemia; the allele $Aprt^b$ determines the fast form found in many inbred strains. Heterozygotes have the three-banded pattern characteristic of a dimeric protein. The use of this variant in crosses showed that Aprt was located about 25 cM distal to Es-1 (3).

Aprt-ps locus, adenine phosphoribosyltransferase pseudogene, Chr 8. Of several Aprt pseudogenes in the mouse, one has been cloned and sequenced. A restriction fragment length polymorphism associated with this pseudogene segregated concordantly with Chr 8 in APRT⁻ mutants of a near-diploid cell line that had lost one copy of Chr 8 (1).

References

1. Farber, R.A., and D. Zielinski. 1986. Assignment of a processed mouse Aprt pseudogene to the same chromosome as the functional gene. Cytogenet. Cell Genet. 42:198–201.
2. Kozak, C., E. Nichols, and F.H. Ruddle. 1975. Gene linkage analysis in the mouse by somatic cell hybridization: assignment of adenine phosphoribosyltransferase to chromosome 8 and α-galactosidase to the X chromosome. Somat. Cell Genet. 1:371–382.
3. Nesterova, T.B., P.M. Borodin, S.M. Zakian, and O.L. Serov. 1987. Assignment of the gene for anenine phosphoribosyltransferase on the genetic map of mouse chromosome 8. Biochem Genet. 25:563–568.

Apyc, anterior pyramidal cataract, semidominant

Arose in a (101 × C3H)F1 hybrid after irradiation with ^{137}Cs gamma rays. In heterozygotes, the lens shows a pyramidal opacity at the anterior pole which protrudes into the anterior chamber and attaches to the cornea. Manifestation is uni- or bilateral and penetrance is incomplete. All homozygotes show anophthalmia and are viable and fertile (1).

References

1. Kratochvilova, J. 1981. Dominant cataract mutations detected in offspring of gamma-irradiated male mice. J. Hered. 72:302–307.

Araf locus, Raf-related oncogene, Chr X

This gene shows 85 per cent homology with the mouse Raf oncogene, but has a more restricted tissue distribution, being expressed in epidydimis, liver, and intestine (3) and also in spleen, fibroblasts, and cells of myeloid and T-cell lineage (2). Its function is unknown but it shows strong homology with the human X-linked A-raf-1 gene which has serine–threonine-specific protein kinase activity. Araf was first shown to be on Chr X by use of mouse–Chinese hamster somatic cell hybrids screened with a spleen cDNA probe (2). A restriction fragment length difference in XbaI-restricted DNA between laboratory strains and an inbred strain derived from Mus spretus was later found and used in a backcross to locate *Araf* to a region 10 to 17 cM proximal to *Hprt* (1).

References

1. Avner, P., M. Bućan, D. Arnaud, H. Lehrach, and U. Rapp. 1987. *A-raf* oncogene localizes on mouse × chromosome to region some 10–17 centimorgans proximal to hypoxanthine phosphoribosyltransferase gene. Somat. Cell Mol. Genet. 13:267–272.
2. Huebner, K., A. ar-Rushdi, C.A. Griffin, M. Isobe, C. Kozak, B.S. Emanuel, L. Nagarajan, J.L. Cleveland, T.I. Bonner, M.D. Goldsborough, C.M. Croce, and U. Rapp. 1986. Actively transcribed genes in the *raf* oncogene group, located on the × chromosome in mouse and human. Proc. Natl. Acad. Sci. USA 83:3934–3938.
3. Huleihel, M., M. Goldsborough, J. Cleveland, M. Gunnell, T. Bonner, and U.R. Rapp. 1986. Characterization of murine A-*raf*, a new oncogene related to v-*raf* oncogene. Mol. Cell. Biol. 6:2655–2662.

Arp locus, Arp lymphoid/erythroid hyperplasia, Chr 5

This locus controls presence or absence of gross hyperplasia of the thymus, spleen, Peyer's patches, and lymph nodes. There are also lymphocytic and metamyelocytic accumulations in the liver and hyperplasia of the bone marrow. Aspects of the pathological picture are consistent with the disorder being due to a primary hemolytic process or autoimmune disorder with secondary hyperplasia (1). The recessive allele responsible for the disorder, *Arp*[b], occurred as a mutation in a substrain of the BALB/c strain now designated BALB/c-*Arp*[b]; the dominant allele preventing the disorder, *Arp*[a], occurs in normal BALB/c substrains and in all other strains tested including the B10.D2 strain (2). Occurrence of the disorder also requires presence of the dominant allele *Lus*[a] present in the B10.D2 strain. *Lus* and *Arp* assort independently. See also *Lus* locus (1). The *Arp* locus appears to code for a surface antigen in lymphocytes, since mice of the two homozygous genotypes on the BALB/c background show a range of alloimmune reactions including mixed lymphocyte reaction, host-vs.-graft and graft-vs.-host reactions, and development of weak cytotoxic but strong cytostatic effector lymphocytes which are allo- as well as autoreactive. *Arp* shows no recombination with *Pep-7* on Chr 5 (2).

References

1. DeGiorgi, L., A. Matossian-Rogers, and H. Festenstein. 1981. Two genes interact to control development of a lymphoid/erythroid hyperplastic disorder of mice. Nature 292:545–547.
2. Matossian-Rogers, A., L. DeGiorgi and S. Povey. 1983. Alloimmune interactions of a lymphoproliferative disease-inducer gene *Arp* and linkage to *Pep-7*. Immunogenetics 18:639–648.

As-1 complex, Chr 13

This gene complex comprises structural and regulatory loci for arylsulfatase B (AS-B; E.C. 3.1.6.1), an enzyme found in most tissues, that catalyzes hydrolysis of sulfate ester bonds in glycosaminoglycans. The complex is on Chr 13 near *pe* (6). Different combinations of the structural and regulatory loci occur in different strains, but no recombination has been observed between structural and regulatory loci in crosses. The structural locus was originally called *As-1* and the regulatory loci were originally called *Asr-1* and *Ast-1*, but since all three are closely linked, they are part of a complex and more properly called *As-1s*, *As-1r*, and *As-1t*, respectively, according to the rules for gene nomenclature.

As-1s locus (formerly *As-1*), arylsulfatase B structural. This locus controls heat stability and is probably the structural locus for the enzyme. The allele *As-1s*[a] determines the heat-stable form and occurs in the SWR, MA/My, RF, C3H/He, and other strains; the allele *As-1s*[b] determines the heat-labile form and occurs in the A/He, LG, C57BL/6 and other C57 strains, and in *M. m. molossinus* (2). Heterozygotes produce both enzyme forms (1). The enzyme in SWR has a slightly higher electrophoretic mobility at pH 4.0 than that in C57BL/6 and A/J (5).

As-1r locus (formerly *Asr-1*), arylsulfatase B regulation. This locus controls activity level of the enzyme in all tissues. The allele *As-1r*[b] determines high activity and occurs in the C57BL/6, SWR, MA, and many other strains; the allele *As-1r*[a] determines low activity and occurs in the A, LG, and IS/Cam strains. *As-1r* is *cis*-acting, affecting activity of the *As-1s* allele on the same chromosome (3).

As-1t locus (formerly *Ast-1*), arylsulfatase B temporal. This locus controls the developmental pattern of activity of arylsulfatase B in liver. The allele *As-1t*[b] in the C57BL/6 strain determines a twofold higher activity of the enzyme in liver than the allele *As-1t*[a] which occurs in strains SWR/J and A/J. *As-1t*[b] homozygotes have a more prominent spike of activity at 12 days postnatal and maintain a higher activity in adults than *As-1t*[a] homozygotes. Heterozygotes are intermediate. The locus is *trans*-acting, affecting activity of the *As-1s* allele on the same chromosome and on its homolog (4).

References

1. Daniel, W.L. 1976. Genetic control of heat sensitivity and activity level of murine arylsulfatase B. Biochem. Genet. 14:1003–1019.
2. Daniel, W.L., and M.S. Caplan. 1980. Comparative biochemistry of murine arylsulfatase B. Biochem. Genet. 18:625–642.
3. Daniel. W.L., K. Abedin, and R.E. Langelan. 1980. Murine arylsulfatase B: evidence favoring control of liver and kidney activity by two regulatory elements. J. Hered. 71:161–167.
4. Daniel, W.L., B.W. Harrison, and K. Nelson. 1982. Genetic regulation of murine hepatic arylsulfatase B activity during development. J. Hered. 73:24–28.
5. Daniel, W.L., B.M. Ruoff, D.B. Thompson, and J.B. Gore. 1985. Purification and properties of arylsulfatase B from high-and low-activity mouse strains. Biochem. Genet. 23:771–786.
6. Elliott, R.W., W.L. Daniel, B.A. Taylor, and E.K. Novak. 1985. Linkage of loci affecting murine liver protein and arylsulfatase B to chromosome 13. J. Hered. 78:243–246.

As-2 locus, arylsulfatase A, Chr 15

This is the structural locus for arylsulfatase A (AS-A; E.C. 3.1.5.1). No genetic variants are known. The locus was assigned to Chr 15 by use of mouse–Chinese hamster somatic cell hybrids segregating mouse chromosomes. The mouse enzyme was recognized in the hybrid cells by its electrophoretic pattern (1).

References

1. Francke, U., P. Tetri, R.T. Taggart, and N. Oliver. 1981. Conserved autosomal syntenic groups on mouse (MMU) chromosome 15 and human (HSA) chromosome 22: assignment of a gene for arylsulfatase A to MMU 15 and regional mapping of DIA1, ARSA, and ACO2 on HSA 22. Cytogenet. Cell Genet. 31:58–69.

asd

See *Gus* complex, *gus^mps*.

ash, ashen, recessive, Chr 9

Arose in strain C3H/HeSn. Homozygotes have a coat color similar to that of *d/d* and *ln/ln* in combination with *a/a*, but greater dilution of yellow pigment than either *d/d* or *ln/ln* in combination with *A/A*. The pigment granules of *ash/ash* melanocytes are clumped around the nucleus like those of *d/d* and *ln/ln*, rather than being widely distributed in the dendrites as in wild-type melanocytes. Ashen has no effect on level of lysosomal enzymes in liver (1). A second mutation at this locus similar to *ash* occurred in the B10.F strain (2).

References

1. Lane, P.W., and J.E. Womack. 1979. Ashen, a new color mutation on chromosome 9 of the mouse. J. Hered. 70:133–135.
2. Langdon, W.Y., T.S. Theodore, C.E. Buckler, J.H. Stimpfling, M.A. Martin, and H.C. Morse III. 1984. Relationship between a retroviral germ line reintegration and a new mutation at the ashen locus in B10.F mice. Virology 133:183–190.

Asl locus, argininosuccinate lyase, Chr 5

This is the structural locus for argininosuccinate lyase (ASL; E.C. 4.3.2.1). No genetic variants are known. The locus was assigned to Chr 5 using mouse–Chinese hamster somatic cell hybrids segregating mouse chromosomes. Comparison with the known location of homologous human genes suggests that *Asl* may be in the distal half of Chr 5 (1).

References

1. Lalley, P.A., S.L. Naylor, and T.B. Shows. 1979. Gene assignment of arginosuccinate lyase to mouse chromosome 5. Cytogenet. Cell Genet. 25:178 (Abstr.).

asp-1 locus (formerly *Ias*), audiogenic seizure prone-1, recessive, Chr 12

Occurs in the DBA/2 inbred strain. Homozygotes are susceptible to sound-induced convulsions upon initial exposure to a loud electric bell at 3 weeks of age. Mice of the C57BL/6 strain have the wild-type allele for resistance. Heterozygotes are generally resistant. The gene was first described by Collins and Fuller (2). Susceptibility is also influenced by two or more other genes and by non-genetic factors. Collins (1) found loose linkage of *asp-1* to brown (*b*) on Chr 4, but Seyfried *et al.* (5) were unable to confirm this finding. Seyfried and Glaser (4) found that the gene controlling the major difference in seizure susceptibility between the DBA/2 and C57BL/6 strains (formerly *Ias*) (3) was linked to the Ah locus which is known to be on Chr 12. See *Ah* locus.

References

1. Collins, R.L. 1970. A new genetic locus mapped from behavioral variation in mice: audiogenic seizure prone (*asp*). Behav. Genet. 1:99–109.
2. Collins, R.L., and J.L. Fuller. 1968. Audiogenic seizure prone (asp): a gene affecting behavior in linkage group VIII of the mouse. Science 162:1137–1139.
3. Roderick, T.H., and M.T. Davisson. 1987. Nomenclature change: Ias to asp-1. Mouse News Lett. 77:106.
4. Seyfried, T.N., and G.H. Glaser. 1981. Genetic linkage

between the *Ah* locus and a major gene that inhibits susceptibility to audiogenic seizures in mice. Genetics 99:117–126.

5. Seyfried, T.N., R.K. Yu, and G.H. Glaser. 1980. Genetic analysis of audiogenic seizure susceptibility in C57BL/6J × DBA/2J recombinant inbred strains of mice. Genetics 94:701–718.

Asr-1 locus

See *As-1* complex.

Ass gene family

Ass locus, argininosuccinate synthetase, Chr 2. This locus codes for argininosuccinate synthetase (ASS; E.C. 6.3.4.5). It was found to be on Chr 2 by use of mouse–Chinese hamster somatic cell hybrids screened with a cDNA clone of the human ASS gene. In the same study, 10 to 15 ASS-related pseudogenes were found and assigned to two other chromosomes (1).

Ass-1ps, argininosuccinate synthetase pseudogene-1, Chr 7.

Ass-2ps, argininosuccinate synthetase pseudogene-2, Chr 10(1).

References

1. Nakamura, D., R. Popp, E. Eicher, and P.A. Lalley. 1985. Comparison of the arginosuccinate synthetase gene family in mouse and man. Cytogenet. Cell Genet. 40:710 (Abstr.).

Ast-1 locus

See *As-1* complex.

at, atrichosis, recessive. Chr 10

Arose spontaneously in strain DBA/1₀Hu. Homozygotes have sparsely distributed hair, especially on the trunk. The amount of hair is variable, but affected mice can be easily recognized at 8 to 10 days. Vibrissae are not affected. Both sexes are sterile, having very small gonads containing few germ cells (2). In the testis, there is a great deficiency of primary spermatogonia (1). In the ovary, a quantitative study of maturation of oocytes showed that the proportion of resident oocytes recruited into the first wave of growth was much higher than in normal ± controls (J.J. Eppig, personal communication).

References

1. Handel, M.A., and J.J. Eppig. 1979. Sertoli cell differentiation in the testes of mice genetically deficient in germ cells. Biol. Reprod. 20:1031–1038.
2. Hummel, K.P. 1966. Mouse News Lett. 34:31.

Ath-1 locus, atherosclerosis susceptibility-1, Chr 1

Inbred strains differ in response to a high-fat diet. In some strains, the level of high-density lipoprotein (HDL) decreases by about 50 per cent and lipid-staining lesions form in the aorta after 3 months. In other strains, HDL level decreases only slightly and few or no aortic lesions develop. The difference is controlled by the *Ath-1* locus. The allele *Ath-1ˢ* determines a 50 per cent decrease in HDL and susceptibility to aortic lesions and occurs in the C57BL/6 strain; the allele *Ath-1ʳ* determines a small decrease in HDL and resistance to aortic lesions and occurs in the C3H/He and BALB/c strains. HDL levels in (C57BL/6 × BALB/c)F1 mice resemble those of BALB/c (*Ath-1ʳ*) and HDL levels in (C57BL/6 × C3H/He)F1 are intermediate; number of aortic lesions is intermediate in both crosses. By use of CXB and BXH recombinant inbred strains and backcrosses of F1 mice to the susceptible parent (C57BL/6), *Ath-1* was shown to be on Chr 1, near but distinct from *Alp-2* and *Hdl*, genes determining apolipoprotein A-II and high-density lipoprotein, respectively (1).

References

1. Paigen, B., D. Mitchell, K. Rene, A. Morrow, A.J. Lusis, and R.C. LeBoeuf. 1987. *Ath-1*, a gene determining atherosclerosis susceptibility and high density lipoprotein levels in mice. Proc. Natl. Acad. Sci. USA 84:3763–3767.

Atpa, Atpb loci

These loci code for the α and β subunits of Na⁺,K⁺-activated ATP phosphohydrolase (E.C. 3.6.1.3). The enzyme consists of two subunits, α and β. The α subunit is a polypeptide of 100 kDa and exists in three isoforms which are encoded by three unlinked genes. The β subunit is a glycosylated polypeptide of 55 kDa encoded by a single gene. Their chromosome location was determined by use of mouse–Chinese hamster somatic cell hybrids screened with rat cDNA probes and their position on the chromosomes determined by use of restriction fragment length differences between *Mus musculus* and *M. spretus* in a backcross of the F1 hybrid between these species to *M. musculus*.

Atpa-1 locus, Na,K-ATPaseα-1, Chr 3. This locus is expressed in all tissues examined. It is about 28 cM from *Egf* (9/32 recombinants).

Atpa-2 locus, Na,K-ATPaseα-2, Chr 7. This locus is expressed predominantly in brain and fetal heart. It is very close to *Coh* (0/32 recombinants).

Atpa-3 locus, Na,K-ATPaseα-3, Chr 1. This locus encodes two mRNA species, the smaller expressed in a variety of tissues, the larger only in brain, heart, and skeletal muscle. It is very close to *Spna-1* (0/31 recombinants).

Atpb locus, Na,K-ATPaseβ, Chr 1. This locus has a complex pattern of tissue expression. It is on the same chromosome as *Atpa-3* (2/31 recombinants) but not tightly linked to it (1).

References

1. Kent, R.B., D.A. Fallows, E. Geissler, T. Glaser, J.R. Emanuel, P.A. Lalley, R. Levenson, and D.E. Housman. 1987. Genes encoding α and β subunits of N^+,K^+-ATPase are located on three different chromosomes in the mouse. Proc. Natl. Acad. Sci. USA 84:5369–5373.

av, Ames waltzer, recessive, Chr 10

Arose as a mutation in the K strain (2). Viability and fertility are about normal, but females are poor mothers. Homozygotes show the typical circling, head-tossing, deafness, and hyperactivity of the circling mutants. In the membranous labyrinth the fluid spaces in the Organ of Corti fail to develop. Later, hair cells and spiral ganglion cells degenerate. The permeability of the stria vascularis appears to be normal (1).

References

1. Osako, S., and D.A. Hilding. 1971. Electron microscopic studies of capillary permeability in normal and Ames waltzer deaf mice. Acta Otolaryngol. 71:365–376.
2. Schaible, R.H. 1956. Mouse News Lett. 15:29.

Av-1, *Av-2* loci, Abelson virus susceptibility-1, -2

These loci control susceptibility to induction of non-thymic lymphomas by Abelson leukemia virus (A-MuLV). The alleles determining susceptibility, *Av-1^s* and *Av-2^s*, occur in the BALB/c strain; the alleles determining resistance, *Av-1^r* and *Av-2^r*, occur in the C57BL/6 strain. Other susceptible strains are SEA and DBA/2; most other strains are resistant. The alleles for susceptibility are dominant. Analysis of C57BL/6 × BALB/c recombinant inbred strains and hybrids between them has shown that presence of the allele for susceptibility at one locus results in intermediate susceptibility and that *Av-1^s* determines greater susceptibility (about 75 per cent) than *Av-2^s* (about 33 per cent). A-MuLV is a complex of defective virus and non-defective Moloney leukemia virus, and resistance to the defective virus is responsible for the resistance of C57BL/6 mice (2). An antigen present on A-MuLV is also present on uninfected adult bone marrow and spleen cells of BALB/c and is amplified on cells of regenerating bone marrow and spleen of BALB/c but not of C57BL/6 mice (3). It is also present on bone marrow cells of the recombinant inbred strains carrying *As-2^s*. The function of the *As-2^s* gene product in hematopoietic differentiation may be related to its role in sensitivity to A-MuLV disease (1, 2).

References

1. Risser, R. 1979. Abelson antigen is expressed on hematopoietic spleen colony-forming cells from mice carrying the *Av-2^s* virus sensitivity gene. Proc. Natl. Acad. Sci USA 76:5350–5354.
2. Risser, R., M. Potter, and W.P. Rowe. 1978. Abelson virus-induced lymphomagenesis in mice. J. Exp. Med. 148:714–726.
3. Risser, R., E. Stockert, and L.J. Old. 1978. Abelson antigen: a viral tumor antigen that is also a differentiation antigen of BALB/c mice. Proc. Natl. Acad. Sci. USA 75:3918–3922.

ax, ataxia, recessive, Chr 18

Arose as a spontaneous mutation in the CBA/H strain. Viability is low; both sexes are sterile. A mutation at the same locus (*ax^J*, at first called 'paralytic') occurred independently at the Jackson Laboratory. Homozygotes are hyperactive during the second week of life. From the third week on they show a steadily increasing paralysis up to 3 months of age. The paralysis is characterized by tremor, and loss of co-ordination and weakness of limbs (3). In the central nervous system there is underdevelopment of the corpus callosum, hippocampus, dentate gyrus, some other forebrain structures associated with the limbic system and some nuclei and tracts of the brainstem, and degeneration in Purkinje cell axons and dendrites (1, 2). The dendritic trees of hippocampal cells show marked reduction in height and lateral spread between 19 and 41 days and the dendritic trees of granule cells of the dentate gyrus are shorter than normal (1). No significant differences from normal were found in the major brain lipid classes, fatty acids, nucleic acids, or 'neurokeratin' amino acid content, suggesting that *ax* causes a generalized retardation of brain development (4).

References

1. Burt, A.M. 1980. Morphologic abnormalities in the postnatal differentiation of CA1 pyramidal cells and granule cells in the hippocampal formation of the *ataxic* mouse. Anat. Rec. 196:61–69.
2. d'Amato, C.J., and S.P. Hicks. 1965. Neuropathologic alterations in the ataxia (paralytic) mouse. Arch. Pathol. 80:604–612.

3. Lyon, M.F. 1955. Ataxia—a new recessive mutant of the house mouse. J. Hered. 46:77–80.
4. Weber, M.B. 1968. A study of brain lipids, nucleic acids and proteins in the ataxic (ax^J) mouse. Neurology 18:243–249.

azh, abnormal spermatozoon head shape, recessive

Probably arose spontaneously during a radiation experiment. In homozygous males all sperm are affected, having a characteristic ladle shape. Up to 40 per cent of sperm lack a flagella. Most males breed and litter sizes are either normal or about half that of litter mates. The karyotype appears to be normal. Sperm-head packaging of the chromatin appears to be affected, and acrosomal and flagellar development may also be affected (1).

References

1. Hugenholtz, A.P. 1984. Mouse News Lett. 71:35.

B

b locus, Chr 4

b, brown, recessive. An old mutant of the mouse fancy. The eumelanin of the hair and eyes is brown rather than black. The pigment granules also appear brown rather than black and are spheroid rather than ovoid in shape (5). The fine structure of the developing pigment granules is fibrillar, like that of +/+ mice, but the appearance of the mature granule may be more coarsely granular (2,7,10). The granules incorporate twice as much C^{14}-tyrosine as normal (1). Mutations from + to b have been recorded frequently.

b^c, cordovan. Arose as a spontaneous mutation in an F1 hybrid between C57BL and DBA/1. Recessive to +, dominant to b. The hair color is a rich deep brown (6).

B^{lt}, light, semidominant. Arose as a spontaneous mutation in the C58 inbred strain (4). B^{lt}/B^{lt} mice have almost white hair except for the tips which are brown. The pigment granules are the same shape as in *b/b*, but somewhat darker in color (5). The granules are clumped (8,9). $B^{lt}/+$ mice have darker hair tips, and the pigment extends much further down the hair. With advancing age the hair of both B^{lt}/B^{lt} and $B^{lt}/+$ mice becomes progressively lighter. Hair follicle melanocytes of both genotypes have centrally located melanin granules and few dendrites, and tend to be incorporated whole into hair cells, so that few melanocytes are left at the time the lower part of the hair is being formed (11).

B^w, white-based brown, semidominant. Causes pigment reduction at base of the hair, more extreme in homozygotes than in heterozygotes. In $B^w/+$, the hair is almost black except for the extreme base which is light; in B^w/b, the outer two-thirds to three-quarters of the hair is brown, and the base is white. The eye pigment resembles that of *b/b*. In B^w/B^w, the hair resembles that of B^w/b except that the light base is about twice as wide (3).

References

1. Coleman, D.L. 1962. Effect of genic substitution on the incorporation of tyrosine into the melanin of mouse skin. Arch. Biochem. Biophys. 96:562–568.
2. Hearing, V.J., P. Phillips, and M.A. Lutzner. 1973. The fine structure of melanogenesis in coat color mutants of the mouse. J. Ultrastruct. Res. 43:88–106.
3. Hunsicker, P.R. 1969. Mouse News Lett. 40:41.
4. MacDowell, E.C. 1950. "Light"—a new mouse color. J. Hered. 41:35–36.
5. Markert, C.L., and W.K. Silvers. 1956. The effects of genotype and cell environment on melanoblast differentiation in the house mouse. Genetics 41:429–450.
6. Miller, D.S., and M.Z. Potas. 1955. Cordovan, a new allele of black and brown color in the mouse. J. Hered. 46:293–296.
7. Moyer, F.H. 1966. Genetic variations in the fine structure and ontogeny of mouse melanin granules. Am. Zool. 6:43–66.
8. Pierro, L.J. 1963. Effects of the *light* mutation of mouse coat color on eye pigmentation. J. Exp. Zool. 153:81–87.
9. Quevedo, W.C., Jr., and H.B. Chase. 1958. An analysis of the light mutation of coat color in mice. J. Morphol. 102:329–345.
10. Rittenhouse, E. 1968. Genetic effects on fine structure and development of pigment granules in mouse hair bulb melanocytes. I. The *b* and *d* loci. Dev. Biol. 17:351–365.
11. Sweet, S.E., and W.C. Quevedo, Jr. 1968. Role of melanocyte morphology in pigmentation of mouse hair. Anat. Rec. 162:243–254.

B2m

B2m locus (formerly also *Ly-m11*, *4*), β2-microglobulin, Chr 2

This is the structural locus for β2-microglobulin (β2M), a 12-kDa protein that is associated with the 40 to 50 kDa Class I product of the *H-2* major histocompatibility complex. Some of the β2M-associated gene products are the antigens H-2K, H-2D, QA-1, QA-2, and TL (6). The β2M molecules show considerable homology with a single immunoglobulin constant-region domain (2). In addition to its association with cell surface molecules, β2M is also found in serum (4). Three alleles are known, $B2m^a$ found in the A, BALB/c, and most other strains, $B2m^b$ found in the C57 family of strains and a few others (6), and $B2m^c$ found in the B10-pa a^t strain (5). Several alleles that differ from these three have been found in wild mice (7). The products of $B2m^a$ and $B2m^b$ are distinguished electrophoretically, $B2m^b$ determining a faster variant and $B2m^a$ a slower one (6). $B2m^c$ determines a protein that is indistinguishable electrophoretically from $B2m^b$ but differs in reaction to monoclonal antibodies (5). The three alleles differ from each other by a single amino acid at position 85 (1, 6). *B2m* is closely linked to *H-3* and *H-42* on Chr 2 (3, 5).

References

1. Gasser, D.L., K.A. Klein, E. Choi, and J.G. Seidman. 1985. A new beta-2 microglobulin allele in mice defined by DNA sequencing. Immunogenetics 22:413–416.
2. Gates, F.T., J.E. Coligan, and T.J. Kindt. 1981. Complete amino acid sequence of murine 2-microglobulin: structural evidence for strain-related polymorphism. Proc. Natl. Acad. Sci. USA 78:554–558.
3. Graff, R.J., D. Martin-Morgan, and M.E. Kurtz. 1985. Allograft rejection-defined antigens of the *B2m*, *H-3* region. J. Immunol. 135:2842–2846.
4. Kimura, S., N. Tada, Y. Liu, and U. Hämmerling. 1981. The presence of Ly-m11 alloantigen in mouse body fluids. Immunogenetics 14:445–447.
5. Kurtz, M.E., R.J. Graff, A. Adelman, D. Martin-Morgan, and R.E. Clark. 1985. CTL and serologically defined antigens of *B2m*, *H-3* region. J. Immunol. 135:2847–2852.
6. Michaelson, J. 1983. Genetics of beta-2 microglobulin in the mouse. Immunogenetics 17:219–260.
7. Robinson, P.J., M. Steinmetz, K. Moriwaki, and K. Fischer Lindahl. 1984. Beta-2 microglobulin types in mice of wild origin. Immunogenetics 20:655–665.

ba, bare, recessive

Arose in a strain of Swiss albino mice in 1953 at the Poona Virus Research Center (1). Homozygotes have no vibrissae at birth. The first coat is much delayed in appearance (13 to 14 days), and the hairs are very thin and short. The coat is shed at about 30 days. A new hair cycle starts at 45 days, but the hairs are thin and are soon shed. Affected mice are entirely naked at 6 months. The number of hair follicles is not reduced, but the follicles contain keratinized globular masses instead of straight hair. The hairs formed are very thin with no regular internal structure. May be extinct.

References

1. Randelia, H.P., and L. D. Sanghvi. 1961. "Bare," a new hairless mutant in the mouse—genetics and histology. Genet. Res. 2:283–289.

Badm locus, β-adrenergic receptor magnesium effect

This locus controls the effect of magnesium on binding ability of β-adrenergic receptors. Binding was measured with the potent β-adrenergic antagonist dihydroalprenolol. The allele in the A/J strain, $Badm^h$, determines a marked increase in level of binding with increase in magnesium concentration; the allele in the C57BL/6J strain, $Badm^l$, determines little increase in level of binding with increase in magnesium concentration. Heterozygotes have not been described. The evidence for a single-gene effect comes from eight recombinant inbred (RI) strains derived from the A/J and C57BL/6J progenitor strains. Four of the RI strains were high and four were low with no intermediates (1).

References

1. Markovac, J., and R.P. Erickson. 1983. Genetic variation in β-adrenergic receptors in mice: a magnesium effect determined by a single gene. Genet. Res. 42:159–168.

bal, balding, recessive

Arose spontaneously in the C57BL/6J strain. Homozygotes are smaller than their normal littermates at birth. Prior to weaning hair loss is evident. Later, bare patches appear on top of the head and in other areas (1).

References

1. Fox, S., and E.M. Eicher. 1977. Mouse News Lett. 56:40.

Bbv locus, B10.BR B-tropic virus, Chr 11

This locus controls inducibility of a B-tropic virus in mice of the B10.BR/SnLi strain and is probably the site of the integrated virus genome. B10.BR mice show a high level of expression of the virus (2). The B10.BR strain was not examined by Jenkins *et al.* (1), and this locus has not been assigned an *Emv* number. See *Emv* loci.

References

1. Jenkins, N.A., N.G. Copeland, B.A. Taylor, and B.K. Lee. 1982. Organization, distribution, and stability of endogenous ecotropic murine leukemia virus DNA sequences in chromosomes of *Mus musculus*. J. Virol. 43:26–36.
2. Kozak, C.A., and W.P. Rowe. 1982. Genetic mapping of ecotropic murine leukemia virus-inducing loci in six inbred strains. J. Exp. Med. 155:524–534.

bc, bouncy, recessive, Chr 18

Found in a linkage cross involving *tc* and *wa-1*. Remutations at this locus have occurred in the C57BL/6 (*bc²ʲ*) and C3H/HeSn (*bc³ʲ*) strains. Homozygotes can be recognized at 10 days by their small size and by a severe tremor and bouncy gait when moving. Homozygous females usually breed and raise their litters. Affected *bc* and *bc³ʲ* males do not breed but *bc²ʲ* males do. Some homozygotes die before weaning but survivors live as long as normal sibs. *bc* is on Chr 18 between *Tw* and *sy* (1).

References

1. Lane, P.W., A.G. Searle, C.V. Beechey, and E.M. Eicher. 1981. Chromosome 18 of the mouse. J. Hered. 72:409–412.

Bcd-1 locus, butyryl CoA dehydrogenase-1, Chr 5

This locus controls electrophoretic variation in butyryl CoA dehydrogenase (BCD-1; E.C. 1.3.99.2), one of a group of mammalian fatty acyl CoA dehydrogenases. The allele *Bcd-1ᵃ* determines a fast anodally migrating band and occurs in strains A/J, C57BL/6, SJL, SM, and 101/H; the allele *Bcd-1ᵇ* determines a slower band and occurs in strains BALB/c, CBA/H, NZB, NZC, and 129/J (3); the allele *Bcd-1ᶜ* determines absence of a band and occurs in the BALB/cByJ subline where it apparently arose as a mutation at some time after 1975. *Bcd-1ᵃ/Bcd-1ᵇ* mice have a band of intermediate mobility; *Bcd-1ᶜ* is recessive to both the other alleles (2). The enzyme is found in the mitochondrial fraction of extracts of liver, kidney, heart, and intestine. *Bcd-1* showed no recombination with *rd* on Chr 5 in 14 STS × BALB/c recombinant inbred strains (1, 3).

References

1. Hilgers, J., and R. van Nie. 1984. Mouse News Lett. 70:65.
2. Prochazka, M., and E.H. Leiter. 1985. A null activity variant found at the butyryl CoA dehydrogenase (Bcd-1) locus in BALB/cByJ subline. Mouse News Lett. 75:31.
3. Seeley, T.-L., and R.S. Holmes. 1981. Genetics and ontogeny of butyryl CoA dehydrogenase in the mouse and linkage of Bcd-1 with Dao-1. Biochem. Genet. 19:333–345.

Bcg locus, BCG resistance, Chr 1

This locus controls natural resistance or susceptibility to multiplication of *Mycobacterium bovis* (BCG) in the spleen, determined 4 weeks after injection of a low dose of bacteria. The allele *Bcgˢ* determines susceptibility and occurs in the C57BL/6 and related strains and in BALB/c and DBA/1; the allele *Bcgʳ* determines resistance and occurs in strains A, C3H/He, DBA/2, CBA/J, C57BR, and AKR (3). *Bcg* is expressed in a radiation-resistant mature cell of the mononuclear macrophage lineage that is susceptible to prolonged exposure to silica. Susceptibility can be transferred with bone marrow-derived precursors (4). *Bcgʳ* macrophages, both *in vitro* and *in vivo*, significantly reduce the metabolic activity of BCG as measured by rate of RNA synthesis when compared to *Bcgˢ* macrophages from congenic mice (8). Resistance and susceptibility to *M. lepraemurium* (1, 5, 7) and *M. intracellulare* (2) are probably also controlled by the *Bcg* locus. The strain distribution pattern of *Bcg* alleles in inbred and recombinant inbred strains matches exactly that of *Lsh* and *Ity*. All three of these loci may therefore be identical (6). See *Ity* locus, *Lsh* locus.

References

1. Brown, I.N., A.A. Glynn, and J. Plant. 1982. Inbred mouse strain resistance to *Mycobacterium lepraemurium* follows the *Ity/Lsh* pattern. Immunology 47:149–156.
2. Goto, Y., R.M. Nakamura, H. Takahashi, and T. Tokunaga. 1984. Genetic control of resistance to *Mycobacterium intracellulare* infection in mice. Infect. Immun. 46:135–140.
3. Gros, P., E. Skamene, and A. Forget. 1981. Genetic control of natural resistance to *Mycobacterium bovis* (BCG) in mice. J. Immunol. 127:2417–2421.
4. Gros, P., E. Skamene, and A. Forget. 1983. Cellular mechanisms of genetically controlled host resistance to *Mycobacterium bovis* (BCG). J. Immunol. 131:1966–1972.
5. Hoffenbach, A., P.H. Lagrange, and M.A. Bach. 1985. Strain variation of lymphokine production and specific antibody secretion in mice infected with *Mycobacterium lepraemurium*. Cell. Immunol. 91:1–11.
6. Skamene, E., P. Gros, A. Forget, P.A.L. Kongshaavn, C. St. Charles, and B.A. Taylor. 1982. Genetic regulation of resistance to intracellular pathogens. Nature 297:506–509.

Bcg

7. Skamene, E., P. Gros, A. Forget, P.J. Patel, and M.N. Nesbitt. 1984. Regulation of resistance to leprosy by chromosome 1 locus in the mouse. Immunogenetics 19:117–124.
8. Stach, J.-L., P. Gros, A. Forget, and E. Skamene. 1984. Phenotypic expression of genetically-controlled natural resistance to Mycobacterium bovis (BCG). J. Immunol. 132:888–892.

Bcga locus, BCG-induced anergy, Chr 12

This locus determines presence or absence of immunological unresponsiveness induced by killed *Mycobacterium bovis* (BCG), and evaluated by delayed type hypersensitivity to sheep red blood cells. Anergy, or unresponsiveness, is recessive and occurs in strains C57BL/6, C57BL/10, and B10.BR; the allele in these strains can be designated $Bcga^b$. Strains that do not develop BCG-induced anergy are C57BR, BALB/c, C3H, CBA, and C3H.SW; the allele in these strains can be designated $Bcga^a$. Bcga is linked to the immunoglobulin heavy chain complex, Igh, probably in the genetic region between Lyb-7 and Igh-Src (1, 2). The degree of anergy is also influenced by haplotypes of the H-2 complex (1).

References

1. Callis, A.H., D.J. Schrier, C.S. David, and V.L. Moore. 1983. Immunogenetics of BCG-induced anergy in mice. Control by Igh- and H-2-linked genes. Immunology 49:601–616.
2. Schrier, D.J., J.L. Sternick, E.M. Allen, and V.L. Moore. 1982. Immunogenetics of BCG-induced anergy in mice: control by genes linked to the Igh complex. J. Immunol. 128:1466–1469.

Bcl-2 locus, B-cell leukemia/lymphoma-2, Chr 1

This locus is the mouse homolog of the human bcl-2 gene that is close to the breakpoint of a translocation occurring in more than 85 per cent of human follicular lymphomas. The mouse gene has two exons and produces two species of mRNA, one coding for a polypeptide of 236 amino acids, Bcl-2α, and the other for a polypeptide of 199 amino acids, Bcl-2β. In adults, expression of Bcl-2 transcripts is tissue specific, the highest levels occurring in spleen and thymus. No genetic variants of Bcl-2 are known. It was shown to be located on Chr 1 by use of mouse–Chinese hamster somatic cell hybrids screened with a mouse genomic DNA clone (1).

References

1. Negrini, M., E. Silini, C. Kozak, Y. Tsujimoto, and C.M. Croce. 1987. Molecular analysis of mbcl-2: structure and expression of the murine gene homologous to the human gene involved in follicular lymphoma. Cell 49:455–463.

bd, bradypneic, recessive, Chr 12

Arose spontaneously in the AEJ inbred strain. Homozygotes have variable weight reduction beginning at 3 to 5 days. They may die early or survive up to weaning or to adulthood and may be fertile. Breathing rate is about half normal, and the lungs show moderate emphysema. Large amounts of gas may be found in the gastrointestinal tracts of preweaning mice, and some distal tubules of the kidney are dilated. The cause of the breathing rate defect is not known (1). bd is probably on Chr 12, 16 to 27 cM from Pre-1 (2).

References

1. Green, M.C., and C.L. Jahn. 1977. Bradypneic, a new mutation in mice causing slow breathing, runting, and early death. J, Hered. 68:150–156.
2. Sweet, H.O., S. Langley, and M.C. Green. 1982. Mouse News Lett. 66:66.

Bda, bald arthritic, semidominant, Chr 11

Arose spontaneously in a laboratory strain. Heterozygotes have sparse hair and homozygotes are hairless (1). There is considerable overlap between the two genotypes at weaning but not at 2 months. The legs of homozygotes become stiff beginning at about 2 months. Bda is closely linked to Re on Chr 11 (2).

References

1. Ferguson, J.M., and M.E. Wallace. 1977. Mouse News Lett. 57:11.
2. Wallace, M.E., and J.M. Ferguson. 1984. Mouse News Lett. 71:19.

Bdv-1, BALB/c defective provirus-1

This locus controls inducibility in lymphocytes of a reverse transcriptase-positive defective retrovirus. The BALB/cTif strain carries the allele for inducibility; the 129/RrJ strain carries the allele for non-inducibility. The virus is induced in lymphocytes by treatment with lipopolysaccharide (LPS), and the induction is slightly amplified by the addition of bromodeoxyuridine (BrdU). Bdv-1 segregates independently of Bxv-1, a locus that controls inducibility of a complete xenotropic virus. The two loci are also regulated differently; Bxv-1 is induced in lymphocytes mainly by BrdU, only marginally by LPS (1).

References

1. Stoye, J.P., and C. Moroni. 1983. Endogenous retrovirus expression in stimulated murine lymphocytes. Identification of a new locus controlling mitogen induction of a defective virus. J. Exp. Med. 157:1660–1674.

bf, buff, recessive, Chr 5

Arose spontaneously in the C57BL/6J strain. Non-agouti buff mice (*a/a bf/bf*) are khaki-colored. Eye color does not appear to be affected (1). Activity levels of β-galactosidase, β-glucuronidase, and N-acetyl-β-hexoseaminidase are usually higher in kidneys of *bf/bf* than in *bf/+* controls (2).

References

1. Dickie, M.M. 1964. Mouse News Lett. 30:30.
2. Hakansson, E.M., and L.G. Lundin. 1977. Effect of a coat color locus on kidney lysosomal glycosidases in the house mouse. Biochem. Genet. 15:75–85.

Bf locus

See *H-2S*.

Bfo locus, bell-flash ovulation, Chr 4

Virgin female mice of two inbred strains differ in the number of ova released following exposure to an intermittently ringing bell or a flashing high-intensity light preceded by an injection of PMSG. Mice of the BALB/cBy strain have a high response (*Bfo^h*), and mice of the C57BL/6By strain have a low response (*Bfo^l*). The difference appears to be due to a single locus on Chr 4 (1).

References

1. Eleftheriou, B.E., and M.B. Kristal. 1974. A gene controlling bell- and photically-induced ovulation in mice. J. Reprod. Fertil. 38:41–47.

bg, beige (formerly *slt*, slate), recessive, Chr 13

Several very similar alleles of independent origin are in wide use, including *bg*, which was probably radiation-induced at Oak Ridge National Laboratory, and *bg^J*, which occurred spontaneously in the C57BL/6J strain at the Jackson Laboratory (10). In homozygotes, eye color is light at birth and varies from ruby to almost black in adults. Ear and tail pigmentation are reduced, and coat color is lighter. The condition in homozygotes closely resembles Chediak–Higashi disease in man and similar conditions in mink and cattle (11).

Pigment granules of both optic cup and neural crest derivatives are reduced in number and much enlarged. They have normal fine structure (7). Abnormal giant lysosomal granules occur in all tissues with granule-containing cells, including granulocytes, lymphocytes, liver parenchyma, kidney proximal tubules, brain and spinal cord, exocrine and endocrine pancreas, the ducts of most glands, thyroid follicle cells, and others (16, 17); type II pneumocytes (18); mast cells (3); and retinal pigment epithelium (19). The giant lysosomes are formed by fusion of normal-sized ones (27). Granulocytes of peritoneal exudate from *bg/bg* mice show defective chemotaxis and reduced bactericidal activity (6). Beige mice are more susceptible than controls to pneumonitis (10) and to various viral (23), bacterial (5), and parasitic (9) infections. They have a severe deficiency of natural killer (NK) cells (20) and a defective cytotoxic T-cell and cytotoxic antibody response to allogeneic tumor cells (2, 22). Syngeneic tumor cells grow better in *bg/bg* mice than in littermate controls, an effect thought to be due to the deficiency of NK cells (8, 26). The lysosomal serine proteases, elastase and cathepsin G, are profoundly decreased in peritoneal neutrophils of beige mice (25), and the amount of plasminogen activator secreted by beige macrophages in culture is increased (21). In the SB/Le-*bg* strain, development of an autoimmune lupus-like syndrome in males is greatly retarded in *bg/bg* compared to rapid onset of autoimmunity in *+/bg* controls (4). Whether this is a result of the immune defects cited above is not known.

Beige mice have platelet storage pool deficiency, i.e. a prolonged bleeding time associated with deficiency in number of platelet dense granules and the granule components serotonin, adenosine triphosphate, and adenosine diphosphate (13). The prolonged bleeding time and low dense granule content can be cured by transplantation of bone marrow from normal mice, and transferred to normal mice by transplantation of *bg/bg* bone marrow (14). Beige mice that survive to 17 months of age show a progressive neurological disorder accompanied by a complete or nearly complete loss of cerebellar Purkinje cells (12).

The abnormal lysosomes of kidney proximal tubules of *bg/bg* mice show a striking accumulation of three lysosomal enzymes after androgen treatment, and the level of these enzymes in urine is lower than normal. Release of lysosomal contents into the tubule lumen appears to be defective (1, 24). It has been shown by use of concanavalin A cap formation that *bg/bg* granulocytes have defective membrane-related microtubular function, and that this as well as the granule abnormality can be corrected by promoters of cyclic GMP generation (15).

References

1. Brandt, E.J., R.W. Elliott, and R.T. Swank. 1975. Defective lysosomal enzyme secretion in kidneys of Chediak–Higashi (beige) mouse. J. Cell. Biol. 67:774–788.
2. Carlson, G.A., S.T. Marshall, and A.T. Truesdale. 1984. Adaptive immune defects and delayed rejection of allogeneic tumor cells in beige mice. Cell. Immunol. 87:348–356.
3. Chi, E.Y., E. Ignácio, and D. Lagunoff. 1978. Mast cell granule formation in the beige mouse. J. Histochem. Cytochem. 26:131–137.
4. Clark, E.A., J.B. Roths, E.D. Murphy, J.A. Ledbetter, and J.A. Clagett. 1982. The beige (*bg*) gene influences the development of autoimmune disease in SB/Le male mice. *In* R.B. Herberman, ed., NK Cells and Other Natural Effector Cells, 301–306. Academic Press, New York.
5. Elin, R.J., J.B. Edelin, and S.M. Wolff. 1974. Infection and immunoglobulin concentrations in Chediak–Higashi mice. Infect. Immun. 10:88–91.
6. Gallin, J.I., J.S. Bujak, E. Patten, and S.M. Wolff. 1974. Granulocyte function in the Chediak–Higashi syndrome of mice. Blood 43:201–206.
7. Hearing, V.J., P. Phillips, and M.A. Lutzner. 1973. The fine structure of melanogenesis in coat color mutants of the mouse. J. Ultrastruct. Res. 43:88–106.
8. Kärre, K., G.O, Klein, R. Kiessling, G. Klein, and J.C. Roder. 1980. Low natural *in vivo* resistance to syngeneic leukaemias in natural killer-deficient mice. Nature 284:624–626.
9. Kirkpatrick, C.E., and J.P. Farrell. 1982. Leishmaniasis in beige mice. Infect. Immun. 38:1208–1216.
10. Lane, P.W., and E.D. Murphy. 1972. Susceptibility to spontaneous pneumonitis in an inbred strain of beige and satin mice. Genetics 72:451–460.
11. Lutzner, M.A., C.T. Lowrie, and H.W. Jordan. 1967. Giant granules in leukocytes of the beige mouse. J. Hered. 58:299–300.
12. Murphy, E.D., and J.B. Roths. 1978. Purkinje cell degeneration, a late effect of beige mutations in mice. Jackson Lab. Ann. Rep. 49:108–109.
13. Novak, E.K., S.-W. Hui, and R.T. Swank. 1984. Platelet storage pool deficiency in mouse pigment mutations associated with several distinct genetic loci. Blood 63:536–544.
14. Novak, E.K, M.P. McGarry, and R.T. Swank. 1985. Correction of symptoms of platelet storage pool deficiency in animal models for Chediak–Higashi syndrome and Hermansky–Pudlak syndrome. Blood 66:1196–1201.
15. Oliver, J.M. 1976. Impaired microtubule function correc-
16. Oliver, C., and E. Essner. 1973. Distribution of anomalous lysosomes in the beige mouse. A homologue of Chediak–Higashi syndrome. J. Histochem. Cytochem. 21:218–228.
17. Oliver, C., and E. Essner. 1975. Formation of anomalous lysosomes in monocytes, neutrophils, and eosinophils from bone marrow of mice with Chediak–Higashi syndrome. Lab. Invest. 32:17–27.
18. Prueitt, J.L., E.Y. Chi, and D. Lagunoff. 1978. Pulmonary surface active materials in the Chediak–Higashi syndrome. J. Lipid Res. 19:410–415.
19. Robison, W.G. Jr., and T. Kuwabara. 1976. Light-induced alterations of retinal pigment epithelium in black, albino, and beige mice. Exp. Eye Res. 22:549–557.
20. Roder, J.C. 1979. The beige mutation in the mouse. I. A stem cell predetermined impairment in natural killer cell function. J. Immunol. 123:2168–2173.
21. de Saint Basile, G., A. Fischer, M.D. Dautzenberg, A. Duransky, F. Le Deist, E. Angles-Cano, and C. Griscelli. 1985. Enhanced plasminogen-activator production by leukocytes in the human and murine Chediak–Higashi syndrome. Blood 65:1275–1281.
22. Saxena, R.K., Q.B. Saxena, and W.H. Adler. 1982. Defective T-cell response in beige mutant mice. Nature 295:240–241.
23. Skellam, G.R., J.E. Allan, J.M. Papadimitriou, and G.J. Bancroft. 1981. Increased susceptibility to cytomegalovirus infection in beige mutant mice. Proc. Natl. Acad. Sci. USA 78:5104–5108.
24. Swank, R.T., and E.J. Brandt. 1978. Turnover of kidney glucuronidase in normal and Chediak–Higashi (beige) mice. Am. J. Pathol. 92:755–772.
25. Takeuchi, K., M. Wood, and R.T. Swank. 1986. Lysosomal elastase and cathepsin G in beige mice. Neutrophils of beige (Chediak–Higashi) mice selectively lack lysosomal elastase and cathepsin G. J. Exp. Med. 163:665–677.
26. Talmadge, J.E., K.M. Meyers, D.J. Prieur, and J.R. Starkey. 1980. Role of NK cells in tumour growth and metastasis in *beige* mice. Nature 284:622–624.
27. Willingham, M.C., S.S. Spicer, and R.A. Vincent Jr. 1981. The origin and fate of large dense bodies in beige mouse fibroblasts. Exp. Cell Res. 136:157–168.

table by cyclic GMP and cholinergic agonists in the Chediak–Higashi syndrome. Am. J. Pathol. 85:395–418.

Bge locus

See *Bgl* complex.

Bgl complex, β-galactosidase, Chr 9

This complex, which can be symbolized [*Bgl*], includes three postulated loci, described below, which have not shown recombination in crosses. There are three known haplotypes: $[Bgl]^b$ (Bgl-$e^b s^b t^b$) in the C57BL/6J and other strains; $[Bgl]^d$ (Bgl-$e^b s^d t^d$) in the DBA/2J, CBA/J, and many other strains; and $[Bgl]^c$ (Bgl-$e^a s^h t^b$) in the BALB/c, C3H/HeJ, and other strains (4, 5).

Bgl-e locus (formerly *Bge*), β-galactosidase electrophoretic. This locus determines an electrophoretic variation in β-galactosidase (B-GAL; E.C. 3.2.1.23) among common inbred strains. The allele Bgl-e^a determines a fast variant and occurs in strains A/J, BALB/c, C3H/He, and others; the allele Bgl-e^b determines a slow variant and occurs in strains AU/Ss, DBA/2, C57BL/6, and many others (5). The fast variant enzyme also has low thermal stability, and the slow enzyme has high thermal stability (1, 9). Heterozygotes are intermediate for both characteristics. The strain differences in both characteristics are found in all tissues examined. No recombinants between *Bgl-s* and *Bgl-e* were observed in the F2 and backcross generations of a cross between *AU/Ss* and *A/J* (5).

Bgl-s locus (formerly *Bgs*), β-galactosidase systemic regulator. This locus determines activity levels of β-galactosidase in brain, kidney, heart, and liver from birth to adulthood (8) and in the embryo beginning at pre-implantation stages (7). The allele Bgl-s^h, which occurs in the C3H/He strain, determines tissue levels about twice those determined by Bgl-s^d, which occurs in the DBA/2 strain. Heterozygotes are intermediate (8, 10). *Bgl-s* is *cis*-acting, controlling expression of the structural gene on the same chromosome (2). Activity level of B-GAL in the salivary gland and pancreas varies independently of *Bgl-s* (6). The *Bgl-s*-associated activity difference between strains is due to differences in specific rate of synthesis of the enzyme (3).

Bgl-t locus (formerly *Bgt*), β-galactosidase temporal. This locus determines a difference in adult level of activity of β-galactosidase in liver, but not in other tissues. The allele Bgl-t^b, which occurs in the C57BL/6 and related strains, determines adult liver activity levels about twice as high as those determined by the allele Bgl-t^d, which occurs in strain DBA/2, CBA/J, and many others. The major part of the difference arises after 30 days of age through an increase in the specific rate of synthesis in C57BL/6 livers (3). Heterozygotes have intermediate activity in liver. *Bgl-t* has shown no recombination with *Bgl-s* in F2 and backcross generations from a cross between C57BL/6 and DBA/2 (5, 11). Activity of B-GAL in adult liver is affected also by regulator genes unlinked to *Bgl*, the alleles for high activity at these loci being necessary, in addition to Bgl-t^b, for high activity in adult liver (4).

References

1. Berger, F.G., and A.J. Lusis. 1978. Relationship between genetic variation in thermal stability and electrophoretic mobility of mouse β-galactosidase. Biochem. Genet. 16:1007–1014.
2. Berger, F.G., and K. Paigen. 1979. *Cis*-active control of mouse β-galactosidase biosynthesis by a systemic regulatory locus. Nature 282:314–316.
3. Berger, F.G., K. Paigen, and M.M. Meisler. 1978. Regulation of the rate of β-galactosidase synthesis by the *Bgs* and *Bgt* loci in the mouse. J. Biol. Chem. 253:5280–5282.
4. Berger, F.G., G.A.M. Breen, and K. Paigen. 1979. Genetic determination of the developmental program for mouse liver β-galactosidase: involvement of sites proximate to and distant from the structural gene. Genetics 92:1187–1203.
5. Breen, G.A.M., A.J. Lusis, and K. Paigen. 1977. Linkage of genetic determinants for mouse β-galactosidase electrophoresis and activity. Genetics 85:73–84.
6. Dewey, M.J., and B. Mintz. 1978. Genetic control of cell-type-specific levels of mouse β-galactosidase. Dev. Biol. 66:560–563.
7. Esworthy, S., and V.M. Chapman. 1981. The expression of β-galactosidase during preimplantation mouse embryogenesis. Dev. Genet. 2:1–12.
8. Felton, J., M. Meisler, and K. Paigen. 1974. A locus determining β-galactosidase activity in the mouse. J. Biol. Chem. 249:3267–3272.
9. Li, I.C., and W.L. Daniel. 1976. Correlation between structural variation and activity of murine kidney β-galactosidase: implications for genetic control. Biochem. Genet. 14:933–952.
10. Lundin, L.-G. and R. Seyedyazdani. 1973. Mendelian inheritance of variations in β-galactosidase activities in the house mouse. Biochem. Genet. 10:351–361.
11. Paigen, K., M. Meisler, J. Felton, and V. Chapman. 1976. Genetic determination of the β-galactosidase developmental program in mouse liver. Cell 9:533–539.

Bgs locus

See *Bgl* complex.

Bgt locus

See *Bgl* complex.

bh, brain-hernia, recessive, probably Chr 7

Appeared as a spontaneous mutation in the second generation following an outcross. Viability of homozygotes is reduced, but some survive and may breed well. Penetrance is probably variable and dependent on the genetic background. At birth homozygotes may have cerebral hernia, hydrocephaly, anophthalmia, or microphthalmia. Later in life they all develop polycystic kidneys (1). The polycystic disease is preceded by the appearance of PAS-positive plugs in the tubules of the kidney. Analysis of the urine shows that *bh/bh* mice have higher than normal levels of free amino acids and protein prior to the development of the disease (2).

References

1. Bennett, D. 1959. Brain hernia, a new recessive mutation in the mouse. J. Hered. 50:265–268.
2. Bennett, D., 1961. A chromatographic study of abnormal urinary amino acid excretion in mutant mice. Ann. Hum. Genet. 25:1–6.

Bhd, broad-headed, semidominant, Chr X

Found among offspring of an irradiated female. Hemizygous males have broad heads and noses and usually die before weaning. Heterozygous females also have broad heads and noses, but are viable and fertile (1).

References

1. Phillips, R.J.S., and G. Fisher. 1978. Mouse News Lett. 58:43–44.

bl, blebbed, recessive, Chr 5

Found among the F3 offspring of a male irradiated with neutrons. Closely resembles *my*. Homozygotes usually have reduced eyes, many have clubbed feet, and one or both kidneys may be absent. Blebs under the skin are often still visible at birth (1).

References

1. Phillips, R.J.S. 1970. Mouse News Lett. 42:26.

Bld, blind, semidominant, Chr 15

Arose spontaneously in the BALB/Ci-*Fu* strain. Heterozygotes are born with their eyelids open, and the cornea becomes damaged (3). Homozygotes die at the end of the seventh day of gestation. The trophectoderm fails to fuse with the uterine decidua. Mesoderm is missing or fails to proliferate, and cells of the proximal extra-embryonic ectoderm become enlarged and polyploid (2). Penetrance in the heterozygote is incomplete (3), but may be increased on certain genetic backgrounds (1). The mutation recurred at Harwell, England (3).

References

1. Teicher, L.S., and E.W. Caspari. 1978. The genetics of blind—a lethal factor in mice. J. Hered. 69:86–90.
2. Vankin, G.L., and E.W. Caspari. 1979. Developmental studies of the lethal gene *Bld* in the mouse. I. Post-implantation development of the lethal homozygote. J. Embryol. Exp. Morphol. 49:1–12.
3. Watson, M.L. 1968. Blind—a dominant mutation in mice. J. Hered. 59:60–64.

Blvr locus, biliverdin reductase, Chr 2

This locus controls electrophoretic mobility of biliverdin reductase. The allele *Blvr*^{a} determines a slowly migrating band and occurs in the C3H/He, C57BL/H, and 101/H strains and in several Harwell linkage testing stocks; the allele *Blvr*^{b} determines a fast band and occurs in the BALB/c, CBA/Ca, TFH/H, and SM/J strains. Heterozygotes have both parental bands. *Blvr* is on Chr 2 about 14 cM from *a* (1).

References

1. von Deimling, D., J. Peters, and S. Povey. 1984. Mouse News Lett. 70:81.

bm, brachymorphic, recessive, Chr 19

Found in an inbred mahogany (*mg*) stock in 1964. Homozygotes have disproportionate dwarfing; overall length and length of long bones are reduced to about three-quarters of normal, and bone width is very little affected. Homozygotes can be recognized at about 5 days by their short domed skulls and short tails. Malocclusion occurs occasionally. Viability is good, and both sexes are fertile. Mutant epiphyseal growth plates are smaller than normal (1). The appearance and orientation of the cells are said to be abnormal by Miller and Flynn-Miller (2), but normal by Orkin *et al.* (3). The matrix contains normal collagen, but it is deficient in sulfated glycosaminoglycans. Fine structure studies show that the proteoglycan granules are small and few (3, 4). The deficiency in sulfated glycosaminoglycans is due to defective conversion of ATP sulfurylase (APS) to the sulfate donor 3'-phosphoadenosine 5'-phosphosulfate (PAPS), resulting from a marked deficiency of APS kinase (E.C. 2.7.1.25) (7). The deficiency in synthesis of PAPS occurs in other tissues besides cartilage, but is manifested only in cartilage because of the large amount of the sulfated glycosaminoglycan, chondroitin sulfate, synthesized there (5). Homozygous *bm/bm* mice have a 95 per cent incidence of cleft palate induced by hydrocortisone administered on days 11 to 14 of gestation in contrast to congenic C57BL/6 mice which have a 20 per cent incidence (6).

References

1. Lane, P.W., and M.M. Dickie. 1968. Three recessive mutations producing disproportionate dwarfing in mice:

achondroplasia, brachymorphic, and stubby. J. Hered. 59:300–308.

2. Miller, W.A., and K.L. Flynn-Miller. 1976. Achondroplastic, brachymorphic and stubby chondrodystrophies in mice. J. Comp. Pathol. 86:349–363.

3. Orkin, R.W., R.M. Pratt, and G.R. Martin. 1976. Undersulfated chondroitin sulfate in the cartilage matrix of brachymorphic mice. Dev. Biol. 50:82–94.

4. Orkin, R.W., B.R. Williams, R.E. Cranley, D.C. Poppke, and K.S. Brown. 1977. Defects in the cartilaginous growth plates of brachymorphic mice. J. Cell Biol. 73:287–299.

5. Pennypacker, J.P., K. Kimata, and K.S. Brown. 1981. Brachymorphic mice (bm/bm): a generalized biochemical defect expressed primarily in cartilage. Dev. Biol. 81:280–287.

6. Pratt, R.M., D.S. Salomon, V.M. Diewert, R.P. Erickson, R. Burns, and K.S. Brown. 1980. Cortisone-induced cleft palate in the brachymorphic mouse. Teratogen. Carcinogen. Mutagen. 1:15–23.

7. Sugahara, K., and N.B. Schwartz. 1979. Defect in 3'-phosphoadenosine 5'-phosphosulfate formation in brachymorphic mice. Proc. Natl. Acad. Sci. USA 76:6615–6618.

Bmfr-1, *Bmfr-2* loci

Bmfr-1 locus, B-cell maturation factor responsiveness-1, Chr 4. This locus controls ability of splenic B-cells to differentiate into antibody-secreting cells in response to several B-cell maturation factors (BMF) when tested at one month of age or younger. The allele *Bmfr-1^{nr}* determines non-responsiveness and occurs in the DBA/2Ha strain; the allele *Bmfr-1^r* determines responsiveness and occurs in strains C57BL/6, BALB/cBy, DBA/2J, C3H/HeJ, and CBA/N. *Bmfr-1^r* is dominant. *Bmfr-1* is on Chr 4 about 15 cM distal to brown (*b*) (1).

Bmfr-2 locus, B-cell maturation factor responsiveness-2, Chr 9. This locus controls ability of splenic B-cells to differentiate into antibody-secreting cells in response to BMF when tested at 3 to 6 months of age. The alleles *Bmfr-2^{nr}* determining non-responsiveness and *Bmfr-2^r* determining responsiveness have the same strain distribution, respectively, as *Bmfr-1^{nr}* and *Bmfr-1^r*. *Bmfr-2^r* is dominant. *Bmfr-1^{nr}*/*Bmfr-1^{nr}* *Bmfr-2^r*/- mice are nonresponsive during the first month after birth but begin to convert to responsiveness after that time. Conversion to responsiveness is significantly reduced by ovariectomy in females but is unaffected by castration in males. *Bmfr-2* is closely linked to dilute (*d*) on Chr 9 (1).

References

1. Sidman, C.L., J.D. Marshall, W.G. Beamer, J.H. Nadeau, and E.R. Unanue. 1986. Two loci affecting B cell response to B cell maturation factors. J. Exp. Med. 163:116–128.

Bn, bent-tail, semidominant, Chr X

Appeared as a spontaneous mutation in a crossbred male. Tails of heterozygous females are more or less shortened, with one or several bends. Penetrance is incomplete. Tails of homozygous females and hemizygous males are shorter and more kinked and not always clearly distinguishable from the tails of heterozygotes. Heterozygous females have normal viability and fertility, but homozygotes and hemizygotes may be less viable and fertile than normal (1,2). Homozygotes and heterozygotes may have open neural tubes in the sacral region and cranioschisis *in utero* (J. Butler and M.F. Lyon quoted as personal communication in Johnson, 3). In the skulls of homozygotes and heterozygotes an interfrontal bone is often present, and the skull is slightly wider than normal (3).

References

1. Garber, E.D. 1952. "Bent tail," a dominant sex-linked mutation in the mouse. Proc. Natl. Acad. Sci. USA 38:876–879.

2. Grüneberg, H. 1955. Genetical studies on the skeleton of the mouse. XVII. Bent-tail. J. Genet. 53:551–562.

3. Johnson, D.R. 1976. The interfrontal bone and mutant genes in the mouse. J. Anat. 121:507–513.

bp, brachypodism, recessive, Chr 2

Arose in a stock of Swiss mice at Harvard University. Fully viable and fertile. The long bones of the limbs are significantly shortened, and the metacarpals, metatarsals, and basal and middle phalanges are severely shortened. Anomalies of the digital blastemata are detectable at 12 days of gestation, before the onset of chondrification (4). A similar allele, *bp^H*, arose independently at Harwell, England, in the PCT multiple recessive stock (5) and has been extensively studied. Dissociated mesoderm cells from 12-day hindlimb buds of *bp^H*/*bp^H* embryos make fewer and smaller condensations than those of control embryos and chondrogenesis is delayed (2). Cells of the limb mesenchyme are more adhesive than in normal embryos (3). From experiments in which cell aggregation and cartilaginous nodule formation were determined in cultures of various combinations of cells from different stages and parts of limb buds of 10- to 13-day embryos, Owens and Solursh (6, 7) showed that in *bp^H*/*bp^H* limb buds, there is reduced ability of a specific mesenchyme cell population to provide inductive stimulus for chondrogenesis, and that this cell–cell interaction results in a lower proportion of chondrogenic cells and a higher proportion of more mature non-chondrogenic cells. Konyukhov and his colleagues (8) have described a growth-inhibiting factor produced by the femur and fibula or by the postaxial part of the hindlimb bud of 13-day bp^H/bp^H embryos. The factor inhibits the growth *in vitro* of tibias of 13-day +/+ embryos. They have also shown that, while 13-day bp^H/bp^H tibias grow normally in standard cultures, addition of a factor from the postaxial part of 11- to 12-day +/+ limb buds causes reduced growth of bp^H/bp^H but not of +/+ 13-day tibias (1).

References

1. Bugrilova, R.S., and B.V. Konyukhov. 1978. Regulator of the expression of the gene brachypodism-H in the mouse. Soviet Genet. 14:1272–1278.
2. Duke, J., and W.A. Elmer. 1977. Effect of the brachypod mutation on cell adhesion and chondrogenesis in aggregates of mouse limb bud mesenchyme. J. Embryol. Exp. Morphol. 42:209–217.
3. Duke, J., and W.A. Elmer. 1978. Cell adhesion and chondrogenesis in brachypod mouse limb bud mesenchyme: fragment fusion studies. J. Embryol. Exp. Morphol. 48:161–168.
4. Grüneberg, H., and A.J. Lee. 1973. The anatomy and development of brachypodism in the mouse. J. Embryol. Exp. Morphol. 30:119–141.
5. Lyon, M.F. 1958. Mouse News Lett. 19:22.
6. Owens, E.M., and M. Solursh. 1982. Cell–cell interaction by mouse limb bud cells during *in vitro* chondrogenesis: analysis of the brachypod mutation. Dev. Biol. 91:376–388.
7. Owens, E.M., and M. Solursh. 1983. Accelerated maturation of limb mesenchyme by the brachypod^H mouse mutation. Differentiation 24:145–148.
8. Pleskova, M.N., V.M. Rodionov, R.S. Bugrilova, and B.V. Konyukhov. 1974. The partial purification of growth-inhibiting factor of the brachypodism-H mouse embryos. Dev. Biol. 37:417–421.

Bpa, bare patches, Chr X

Found among offspring of an irradiated (C3H/HeH × 101/H)F1 male. Hemizygous males die before birth, probably at the small mole stage. Heterozygous females can be recognized by narrow transverse dark stripes due to bare stripes of mutant skin which allow the dark base of the immediately posterior hair to show. Heterozygotes are first identifiable at 5 days by patches of bare skin visible among the emerging hairs. As adults, they may look somewhat scruffy as some of the larger bare patches become visible (4). They are smaller than their normal sibs, have short bent tails, and shortened limbs with flexion contractures. Closer

examination shows various abnormalities with patchy distribution. These include punctate calcification in the epiphyses of the vertebrae and long bones, opacities in the lens, and thickened hyperkeratotic skin, These characteristics closely resemble those of human females with the X-linked trait chondrodysplasia punctata, suggesting that these X-linked genes in mouse and man are homologous (2). The mutation to *Bpa* occurred coincidentally with a long X-chromosome inversion, In(X)1H, but has been separated from it (1, 3).

References

1. Evans, E.P., and R.J.S. Phillips. 1975. Inversion heterozygosity and the origin of XO daughters of *Bpa*/+ female mice. Nature 256:40–41.
2. Happle, R., R.J.S. Phillips, A. Roessner, and G. Jünemann. 1983. Homologous genes for X-linked chondrodysplasia punctata in man and mouse. Human Genet. 63:24–27.
3. Phillips, R.J.S., and M.H. Kaufman. 1974. Bare-patches, a new sex-linked gene in the mouse, associated with a high production of *XO* females. II. Investigation into the nature and mechanism of the *XO* production. Genet. Res. 24:27–41.
4. Phillips, R.J.S., S.G. Hawker, and H.J. Moseley. 1973. Bare-patches, a new sex-linked gene in the mouse, associated with a high production of *XO* females. I. A preliminary report of breeding experiments. Genet. Res. 22:91–99.

Br, brachyrrhine, semidominant

Arose in a neutron irradiation experiment. Heterozygotes have a much shortened snout classifiable at birth, and a deeper than usual median cleft in the upper lip. The kidneys are small at birth and, in a number of *Br*/+ mice dying in early maturity, have been extremely pale with very few glomeruli. Heterozygotes are smaller than normal, and there is a shortage of them at birth in outcrosses. Homozygotes have not been identified (1).

References

1. Searle, A.G. 1966. Mouse News Lett. 35:27.

Brp-1 to *Brp-14* loci, brain protein-1 to -14

These loci code for polypeptides isolated from brain and visualized by two-dimensional electrophoresis. Among approximately 200 brain polypeptides examined, variation for these 14 was found in inbred strains. Examination of C57BL/6 × DBA/2 recombinant inbred strains has revealed the chromosomal locations of four of the loci. Two of the four are probably identical to previously described loci. The distribution of

Table 2.2

Locus	C57BL	C3H	AKR	IS	BALB/c	A/J	DBA/2	SS	LS	Identity	Chr	MW (kDa)	IEP acidic
Brp-1	b	a	b	b	b	b	a	b	a	–	6	81	5.6
Brp-2	b	b	b	a	b	b	b	b	a	–	–	48	5.6
Brp-3	a	a	a	b	b	b	a	a,b	b	–	–	30	5.6
Brp-4	b	a	b	–	b	a	a	a,b	a	–	–	112	5.7
Brp-5	b	b	b	–	b	b	a	b	b	–	–	122	5.8
Brp-6	a	a	a	–	a	a	a	a,b	a	–	–	72	5.8
Brp-7	a	b	a	–	a	a	a	a	a	–	–	71	5.4
Brp-8	b	a	a	–	a	a	a	a	a	–	–	31	5.2
Brp-9	a	b	b	–	b	b	b	a	a	–	–	33	5.8
Brp-10	a	a	b	–	a	a	a	a	a	–	–	23	5.4
Brp-11	a	a	b	–	a	b	a	b	a	–	–	65	5.2
Brp-12	b	a	b	–	a	b	a	a	a	Ltw-4	1	28	5.6
Brp-13	a	–	–	–	–	–	b	–	–	–	2	62	5.9
Brp-14	a	b	a	–	b	b	b	–	–	Apoa-1	9	26	5.3

alleles in seven inbred strains and two strains selected for short (SS) and long (LS) sleeping time after i.p. injection of ethanol (*a* acidic isoelectric point, *b* basic IEP) are given in Table 2.2, as well as chromosome locations, molecular weight in kilodaltons (kDa) and IEP of the acidic variant. All loci tested show codominance except *Brp-1* in which the *Brp-1*^b allele is fully dominant (1, 2). No association of strain distribution of alleles with long and short sleeping time after ethanol administration was found in seven inbred and two selected strains (2). *Brp-12* (*Ltw-4*) is also expressed in liver and kidney. It is associated with variation in level of ethanol intake. Among 15 BXD recombinant inbred strains and 19 other inbred strains, those strains carrying the basic (*b*) allele showed higher intake than those carrying the acidic (*a*) allele (3).

References

1. Goldman, D., and H.J. Pikus. 1986. Fourteen genetically variant proteins of mouse brain: discovery of two new variants and chromosomal mapping of four loci. Biochem. Genet. 24:183–194.
2. Goldman, D., R. Nelson, R.A. Deitrich, R.C. Baker, K. Spuhler, H. Markley, M. Ebert, and C.A. Morril. 1985. Genetic brain polypeptide variants in inbred mice and in mouse strains with high and low sensitivity to alcohol. Brain Res. 341:130–138.
3. Goldman, D., R.G. Lister, and J.C. Crabbe. 1987. Mapping of a putative genetic locus determining ethanol intake in the mouse. Brain Res. 420:220–226.

Brs-1 to *-11* loci

See *Odc-1* to *-11* loci.

Bru, bruised, semidominant

Found among offspring of a male treated with ethylnitrosourea. Heterozygotes are recognizable at birth by apparent bruising about the head or around the hindquarters. Some have a pale anemic appearance. Viability before weaning is reduced. Survivors have enlarged eyes by 1 month. Slit lamp examination shows that cataracts are present and that strands of tissue connect the iris, lens, and cornea. The animals remain small and some are sterile. Homozygotes die prenatally as large or small moles (1).

References

1. Lyon, M.F., P.H. Glenister, and J.D. West. 1984. Mouse News Lett. 71:26.

bs, blind-sterile, recessive, Chr 2

Arose spontaneously in the AKR/J strain. Homozygotes have bilateral nuclear cataracts, eyes that are slightly smaller than normal, and glossy coats. Females are normally fertile but males are sterile. The cataracts are present in 16 day fetuses in the center of the lens. They can be seen through the eyelids at birth in albino mice and are easily recognized in adult albinos. In older pigmented animals they are difficult to see unless the pupils are dilated (2). Adult males copulate normally but have smaller than normal testes. There is a reduced number of sperm, and those that are formed completely lack acrosomes (1).

References

1. Sotomayer, R.E., and M.A. Handel. 1986. Failure of acrosome assembly in a male sterile mouse mutant. Biol. Reprod. 34:171–182.

2. Varnum, D.S. 1983. Blind-sterile: a new mutation on chromosome 2 of the house mouse. J. Hered. 74:206–207.

Bsk, bareskin, Chr 11

Found among offspring of a male treated with ENU. Heterozygotes are detectable at about 2 weeks of age by somewhat thickened skin thrown into folds. At 3 weeks the coat appears sparse and becomes progressively thinner with age, so that adults are nearly bare except for some hair on the snout. Corneal opacities develop in some animals beginning at 2 months. Homozygotes resemble heterozygotes and are viable and fertile (1). *Bsk* shows no recombination with *Re* (rex) on Chr 11 (2). *Bsk* closely resembles the rex allele denuded, *Re^{den}*. *Re*, *Re^{den}*, and *Bsk* may be mutations at the same locus or may be parts of a complex. See *Re* locus.

References

1. Lyon, M.F., and P.H. Glenister. 1984. Bareskin (Bsk). Mouse News Lett. 71:26.
2. Lyon, M.F., and J.F. Zenthon. 1986. Close linkage of bareskin (Bsk) and rex (Re). Mouse News Lett. 74:96.

Bsp, black spleen, dominant

All sublines of C57BL examined apparently are homozygous for this gene, which determines the presence of black pigment (lipofuscin) in the anterior end of the spleen in 4 to 50 per cent of the mice. The percentage of lipofuscinosis differs between sublines but not between sexes. Other strains examined, including BALB/c, CBA/Ca, C3H/He, DBA/2, and 129/J, show no blackening of the spleen and apparently carry the *bsp* allele. (C57BL/6 × BALB/c)F1 mice have black spleens in about the same percentage as the C57BL/6 parent subline or somewhat greater. Observations on backcrosses and recombinant inbred strains and their intercrosses showed that lipofuscinosis did not occur in homozygotes for albino (*c*) or pink-eyed dilution (*p*), two linked genes known to cause deficiency of melanin pigment, but did occur in the expected frequencies in the pigmented mice. Lipofuscin has been thought to be unrelated to melanin. These results show either that lipofuscin is affected by the same genes as those that affect melanin or that there is a gene in the *p*–*c* region of Chr 7, in addition to the independent *Bsp*$_+$ gene, that causes a deficiency of lipofuscin (1). Lipofuscin occurs in structures resembling lysosomes. Two pigmentation genes, beige (*bg*) and reduced pigmentation (*rp*), which cause abnormal lysosomes, prevent or greatly reduce lipofuscinosis in C57BL mice (2).

References

1. Crichton, D.N., and J.G.M. Shire. 1982. Genetic basis of susceptibility to splenic lipofuscinosis in mice. Genet. Res. 39:275–285.
2. Ahmed, F., and J.G.M. Shire. 1985. Lysosomal mutations inhibit lipofuscinosis of the spleen in C57BL mice. J. Hered. 76:311–312.

Bst, belly spot and tail, semidominant, Chr 16

Arose spontaneously in the C57BL/Ks strain. Homozygotes die *in utero*. Heterozygotes have a short kinked tail, ventral spotting, and occasionally reduced body size, malocclusion, white feet, anomalies of the spine and eyes, and polydactyly. Both sexes are viable and fertile. Penetrance in heterozygotes is incomplete after crosses to the C3H/HeJ strain. Bst is on Chr 16 about 16 cM distal to *Igl-1* (1).

References

1. Epstein, R., M. Davisson, K. Lehmann, E.C. Akeson, and M. Cohn. 1986. Position of *Igl-1*, *md*, and *Bst* loci on chromosome 16 of the mouse. Immunogenetics 23:78–83.

bt, belted, recessive, Chr 15

Arose as a spontaneous mutation in the DBA inbred strain (2). Homozygotes have a white belt across the back in the midtrunk region and a white belly patch. The two may coalesce to form a white belt. The white spotting appears to be due to a defect in the hair follicles which prevents pigment cells from entering the hair follicles or from developing there (1).

References

1. Mayer, T.C., and E. Maltby. 1964. An experimental investigation of pattern development in lethal spotting and belted mouse embryos. Dev. Biol. 9:269–286.
2. Murray, J.M., and G.D. Snell. 1945. Belted, a new sixth chromosome mutation in the mouse. J. Hered. 36:266–268.

bt-2, belted-2, recessive

Found in progeny of a male treated with MNNG. Not allelic with *bt* or *s*. Homozygotes have a narrow white belt (1).

References

1. Kelly, E.M. 1975. Mouse News Lett. 52:46.

bv, Bronx waltzer, recessive

Found in a stock at Albert Einstein College of Medicine. Homozygotes show typical waltzing behavior. Most are completely deaf, but some hear for a brief period shortly after 13 days of age. The inner hair cells of the inner ear are missing or abnormal at birth and thereafter, and the cells that innervate them, the spiral ganglion cells, degenerate almost completely, beginning by 6 days of age. The outer hair cells are normal (2). Electron microscopy of the vestibular receptors and ganglia revealed a considerable lack of hair bundles in the utricular macula by 3 days of age and early degeneration of many sensory cells. Many of the remaining hair bundles and sensory cells survive but show abnormalities and delayed maturation. Immature features are also seen in the vestibular ganglion (1).

References

1. Demémes, D., and A. Sans. 1985. Pathological changes during the development of the vestibular sensory and ganglion cells of the Bronx waltzer mouse: scanning and transmission electron microscopy. Dev. Brain Res. 18:285–295 (Brain Res. 350).
2. Deol, M.S., and S. Gluecksohn-Waelsch. 1979. The role of the inner ear cells in hearing. Nature 278:250–252.

Bv or *Bv-1* locus

See *Emv-2* locus.

Bxv-1 locus, B10 xenotropic virus-1, Chr 1

This locus controls inducibility by 5-iododeoxyuridine of xenotropic leukemia virus in cultures of tail-biopsy tissues. The allele *Bxv-1^a* determines inducibility and occurs in strains C57BL/10, BALB/c, C57L, AKR, LPT/Le (2), and MA/My (3); the allele *Bxv-1^b* determines non-inducibility and occurs in strains A/J, NFS, SEA, and SWR. Heterozygotes are inducible (2). Virus is produced shortly after induction but the inducibility is transient, in contrast to the persistent induction of virus by loci for inducibility of ecotropic viruses (1). *Bxv-1* is on Chr 1 about 20 cM distal to *Pep-3* (2). A xenotropic virus-inducibility gene in strain F/St maps on Chr 1 but significantly closer to *Pep-3* than *Bxv-1* (9.3 ± 3 per cent vs. 19.5 ± 2.2 per cent). It may or may not be identical to *Bxv-1* (4).

References

1. Kozak, C.A. 1984. Differential expression of murine leukemia virus loci in chemically induced hybrid cells. J. Virol. 51:876–879.
2. Kozak, C.A., and W.P. Rowe. 1980. Genetic mapping of xenotropic murine leukemia virus-inducing loci in five mouse strains. J. Exp. Med. 152:219–228.
3. Kozak, C.A., J.W. Hartley, and H.C. Morse III. 1984. Laboratory and wild-derived mice with multiple loci for production of xenotropic murine leukemia virus. J. Virol. 51:77–80.
4. Morse, H.C. III, C.A. Kozak, R.A. Yetter, and J.W. Hartley. 1982. Unique features of retrovirus expression in F/St mice. J. Virol. 43:1–7.

C

c locus, Chr 7

The albino locus affects the amount of tyrosinase in pigment cells, but does not interfere with the production of pigment cells themselves (2, 34). It may be the structural locus for tyrosinase (40), but there is evidence that the enzyme structure is not affected by several mutations at the locus (37), and that the locus controls an inhibitor of tyrosinase (18). All the mutant alleles are recessive to wild type in appearance, but heterozygotes with wild type produce intermediate amounts of tyrosinase (2). Albino-locus mutants with lightly pigmented eyes have a reduced number of fibers of the optic nerve going to the ipsilateral lateral geniculate nucleus of the brain. This is probably a secondary effect of reduced tyrosinase activity or amount of pigment in the pigment epithelium, since genes at other loci that reduce eye pigmentation also cause the same anomaly (17, 22). Spontaneous mutations at the *c* locus are relatively infrequent.

c, albino. A very old mutant, already known in Greek and Roman times. Hair and eyes are completely devoid of pigment. Melanocytes with melanosomes showing normal fine structure occur in the retina and hair follicles. The granules are smaller and fewer than normal and completely lack melanin (18, 28). Tyrosinase is almost absent (2). Homozygotes do not perform as well as normal in a number of behavioral tests. It is likely that this effect is mediated, at least in part, by defective vision resulting from lack of retinal pigment (19, 36, 39).

c^{ch}, chinchilla. First described by Feldman (9). Agouti chinchilla mice are gray rather than brownish gray. The yellow pigment in the hair is greatly reduced and the black pigment slightly so. Eyes are black. The number and size of pigment granules in retinal melanocytes is normal (28). Tyrosinase activity in the skin is about one-third that of normal (2). Heterozygotes with c^h, c^e, and c are intermediate both in color and in tyrosinase activity.

c^e, extreme dilution. Found in the wild by Detlefsen (4). Hair is very light gray, and eyes are black. Melanin granules of both retinal and neural-crest derived melanocytes are pigmented but are smaller and fewer than normal (25, 28). Tyrosinase activity in the skin is very low (2). c^e/c mice are almost white with black eyes.

c^h, himalayan. Found in the offspring of a cross between the DBA/2 and AKR/J strains. In homozygotes, the first coat is a uniform light tan. At the first molt, the body hair becomes lighter and the ears, nose, tail, and scrotum become dark as in Siamese cats. Body hair may be very light or quite sooty, depending on the genetic background. Eyes are slightly pigmented and appear red. c^h/c and c^h/c^e mice are intermediate between the homozygotes (16). Melanin granules of the retina are indistinguishable from those of c^{ch}/c^{ch} or of $+/+$ (28). Pigment synthesis in himalayan mice is temperature-sensitive; pigment develops in growing hairs of mice housed at 15°C but not in those at 30°C (20). In explanation of this effect, it has been shown that tyrosinase in these mice is heat-labile (2), being maximally active at well below body temperature. Tyrosinase of c^h/c^h mice is strongly bound by an inhibitor at 37°C but not at lower temperatures (21).

c^m, chinchilla-mottled. Found in the progeny of a neutron-irradiated male. Homozygotes and c^m/c^{ch} have patches of normal chinchilla-type fur and patches of lighter fur similar to that of c/c^{ch}. The two genotypes are distinguishable by the whiter belly of c^m/c^m mice (30).

c^p, platinum. Occurred in the DBA/2 strain and was found in an F1 hybrid with AKR/J. Homozygotes are lighter than c^e/c^e and have pink eyes. c^p/c mice are intermediate between c^p/c^p and c/c (5). Platinum homozygotes have almost as much tyrosinase in their skin as c^{ch}/c^{ch} mice but a much larger proportion of it is in solu-ble form where it is not available for production of melanin in the melanosomes (37).

c^r, ruby-eyed dilute. Found in a nonagouti chinchilla (a/a c^{ch}/c^{ch}) stock. Homozygotes have reduced black pigment and no yellow pigment. They are lighter that c^{ch}/c^{ch} but darker than c^e/c^e mice. Eyes are ruby in adults. Heterozygotes with other albino locus mutations are intermediate (10).

Radiation-induced lethal c-alleles. Six of these mutations have been intensively studied. They are probably small deletions. In their effects on proteins other than those on tyrosinase known to be caused by mutation at the c locus itself, they are completely recessive to wild type, indicating that they are probably deletions of regulatory loci rather than of loci affecting protein structure (12). The structural genes for some of these proteins are known to be located elsewhere (3).

c^{3H}. Found at Harwell. This allele resembles c^{65K} and c^{112K} in most respects studied, including neonatal death, effect on liver enzymes and serum proteins (13), and defective rough endoplasmic reticulum of liver (38). Liver cells of homozygotes have reduced numbers of receptors for epidermal growth factor and glucagon (33), and three specific liver polypeptides recognizable in two-dimensional gels are lacking (1). Kidneys are severely reduced in size, but there are very few thymus abnormalities (7). The deletion includes the *Mod-2* locus (6). c^{3H}/c^{6H} mice are viable, but runted and sterile (24). They have been used to study the effects of absence of mitochondrial malic enzyme caused by the absence of the *Mod-2* locus (27).

c^{6H}. Found at Harwell. Homozygotes die soon after implantation (23). Complementation studies with other lethal c-alleles show that c^{6H} probably does not affect liver enzyme production or structure of rough endoplasmic reticulum (13). Homozygotes lack a specific liver polypeptide recognizable in two-dimensional gels (1). The deletion includes the *Mod-2* locus (8).

c^{25H}. Found at Harwell. Homozygotes die during early embryogenesis at the three- to six-cell stage. Before death, the nuclei become extremely aberrant in shape, but other organelles appear normal (29). Complementation studies with other lethal c-alleles show that c^{25H} probably also affects liver enzyme production and structure of rough endoplasmic reticulum (13). The deletion includes the *Mod-2* locus (8) and is probably the longest of the six deletions. It is detectable in banded mitotic chromosomes by the absence of a band in the middle region of chromosome 7 (26).

c^{14CoS}. Found at Oak Ridge National Laboratory. Homozygotes die neonatally (7), and have deficiencies in activity of at least five liver enzymes, glucose-6-phosphatase, tyrosine aminotransferase, serine dehydratase, glutamine synthetase, and UDP-glucuronyltransferase, as well as low levels of liver microsomal cytochrome P-450 and the plasma proteins, albumin, α-fetoprotein, and transferrin (11, 14, 35). Control of the level of two enzymes occurs at the transcriptional level; control of the level of the three plasma proteins occurs at the posttranscriptional level (31, 32). There is a marked decrease in number of receptors for insulin and glucocorticoids (15) and for epidermal growth factor and glucagon (33). Two specific liver polypeptides recognizable in two-dimensional gels are lacking (1). The rough endoplasmic reticulum of the liver, and to a lesser extent of the kidneys, is dilated and vesiculated (38). Complementation studies with other alleles have shown that this is probably the shortest of the six deletions (13). It is known not to include the *Mod-2* locus (8).

References

1. Baier, L.J., S.M. Hanash, and R.P. Erickson. 1984. Mice homozygous for chromosomal deletions at the albino locus region lack specific polypeptides in two-dimensional gels. Proc. Natl. Acad. Sci. USA 81:2132–2136.
2. Coleman, D.L. 1962. Effect of genic substitution on the incorporation of tyrosine into the melanin of mouse skin. Arch. Biochem. Biophys. 96:562–568.
3. Cori, C.F., S. Gluecksohn-Waelsch, P.A. Shaw, and C. Robinson. 1983. Correction of a genetically caused enzyme defect by somatic cell hybridization. Proc. Natl. Acad. Sci. USA 80:6611–6614.
4. Detlefsen, J.A. 1921. A new mutation in the house mouse. Am. Nat. 55:469–473.
5. Dickie, M.M. 1966. Mouse News Lett. 34:30.
6. Eicher, E.M., S.E. Lewis, H.A. Turchin, and S. Gluecksohn-Waelsch. 1978. Absence of mitochondrial malic enzyme in mice carrying two complementing lethal albino alleles. Genet. Res. 32:1–7.
7. Erickson, R.P., S. Gluecksohn-Waelsch, and C.F. Cori. 1968. Glucose-6-phosphate deficiency caused by radiation-induced alleles at the albino locus in the mouse. Proc. Natl. Acad. Sci. USA 59:437–444.
8. Erickson, R.P., E.M. Eicher, and S. Gluecksohn-Waelsch. 1974. Demonstration in mouse of X-ray induced deletions for a known enzyme structural locus. Nature 248:416–418.
9. Feldman, H.W. 1922. A fourth allelomorph in the albino series in mice. Am. Nat. 56:573–574.
10. Fisher, G. 1984. Mouse News Lett. 70:81.
11. Garland, R.C., J. Satrustegui, S Gluecksohn-Waelsch, and C.F. Cori. 1976. Deficiency in plasma protein synthesis caused by X-ray induced lethal albino alleles in mouse. Proc. Natl. Acad. Sci. USA 73:3376–3380.
12. Gluecksohn-Waelsch, S. 1979. Genetic control of mor-

13. Gluecksohn-Waelsch, S., M.B. Schiffman, J. Thorndike, and C.F. Cori. 1974. Complementation studies of lethal alleles in the mouse causing deficiencies of glucose-6-phosphatase, tyrosine aminotransferase, and serine dehydratase. Proc. Natl. Acad. Sci. USA 71:825–829.
14. Gluecksohn-Waelsch, S., M.B. Schiffman, and M.H. Moscona. 1975. Glutamine synthetase in newborn mice homozygous for lethal albino alleles. Dev. Biol. 45:369–371.
15. Goldfeld, A.E., G.L. Firestone, P.A. Shaw, and S. Gluecksohn-Waelsch. 1983. Recessive lethal deletion on mouse chromosome 7 affects glucocorticoid receptor binding activities. Proc. Natl. Acad. Sci. USA 80:1431–1434.
16. Green, M.C. 1961. Himalayan, a new allele of albino in the mouse. J. Hered. 52:73–75.
17. Guillery, R.W. 1974. Visual pathways in albinos. Sci. Amer. 230:44–54.
18. Hearing, V.J., P. Phillips, and M.A. Lutzner. 1973. The fine structure of melanogenesis in coat color mutants of the mouse. J. Ultrastruct. Res. 43:88–106.
19. Hegmann, J.P., R.A. Kieso, and H.B. Hartman. 1974. Gene differences influencing visual system function and behavior. Behav. Genet. 4:165–170.
20. Kidson, S., and B. Fabian. 1979. Pigment synthesis in the himalayan mouse. J. Exp. Zool. 210:145–152.
21. Kidson, S.H., and B.C. Fabian. 1981. The effect of temperature on tyrosinase activity in himalayan mouse skin. J. Exp. Zool. 215:91–97.
22. LaVail, J.H., R.A. Nixon, and R.L. Sidman. 1978. Genetic control of retinal ganglion cell projections. J. Comp. Neurol. 182:399–422.
23. Lewis, S.E., H.A. Turchin, and S. Gluecksohn-Waelsch. 1976. The developmental analysis of an embryonic lethal (c^{6H}) in the mouse. J. Embryol. Exp. Morphol. 36:363–371.
24. Lewis, S.E., H.A. Turchin, and T.E. Wojtowicz. 1978. Fertility studies of complementing genotypes at the albino locus of the mouse. J. Reprod. Fert. 53:197–202.
25. Markert, C.L., and W.K. Silvers. 1956. The effects of genotype and cell environment on melanoblast differentiation in the house mouse. Genetics 41:429–430.
26. Miller, D.A., V.G. Dev, R. Tantravahi, O.J. Miller, M.B. Schiffman, R.A. Yates, and S. Gluecksohn-Waelsch. 1974. Cytological detection of the c^{25H} deletion involving the albino (c) locus on chromosome 7 in the mouse. Genetics 78:905–910.
27. Mohrenweiser, H.W., and R.P. Erickson. 1979. Enzyme changes associated with mitochondrial malic enzyme deficiency in mice. Biochim. Biophys. Acta 587:313–323.
28. Moyer, F.H. 1966. Genetic variation in the fine structure and ontogeny of mouse pigment granules. Am. Zool. 6:43–66.
29. Nadijcka, M.D., N. Hillman, and S. Gluecksohn-

phogenetic and biochemical differentiation: lethal albino deletions in the mouse. Cell 16:225–237.

c

Waelsch. 1979. Ultrastructural studies of lethal c²⁵ᴴ/c²⁵ᴴ mouse embryos. J. Embryol. Exp. Morphol. 52:1–11.

30. Phillips, R.J.S. 1970. Mouse News Lett. 42:26.
31. Sala-Trepat, J.M., M. Poiret, C.H. Sellem, R. Bessada, T. Erdos, and S. Gluecksohn-Waelsch. 1985. A lethal deletion on mouse chromosome 7 affects regulation of liver-cell-specific functions: post transcriptional control of serum protein and transcriptional control of aldolase B synthesis. Proc. Natl. Acad. Sci. USA 82:2442–2446.
32. Schmid, W., G. Müller, G. Schütz, and S. Gluecksohn-Waelsch. 1985. Deletions near the albino locus on chromosome 7 of the mouse affect the level of tyrosine aminotransferase mRNA. Proc. Natl. Acad. Sci. USA 82:2866–2869.
33. Shaw, P.A., and S. Gluecksohn-Waelsch. 1983. Epidermal growth factor and glucagon receptors in mice homozygous for a lethal chromosomal deletion. Proc. Natl. Acad. Sci. USA 80:5379–5382. 34.Silvers, W.K., 1958. Origin and identity of clear cells found in the hair bulbs of albino mice. Anat. Rec. 130:135–144.
35. Thaler, M.M., R.P. Erickson, and A. Pelger. 1976. Genetically determined abnormalities of microsomal enzymes in liver of mutant newborn mice. Biochem. Biophys. Res. Comm. 72:1244–1250.
36. Thiessen, D.D., G. Lindsey, and K. Owen. 1970. Behavior and allelic variations in enzyme activity and coat color at the *c* locus of the mouse. Behav. Genet. 1:257–267.
37. Townsend, D., C.T. Witkop Jr., and J. Mattson. 1981. Tyrosinase subcellular distribution and kinetic parameters in wild type and *c*-locus mutant C57BL/6J mice. J. Exp. Zool. 216:113–119.
38. Trigg, M.J., and S. Gluecksohn-Waelsch. 1973. Ultrastructural basis of biochemical effects in a series of lethal alleles in the mouse. Neonatal and developmental studies. J. Cell. Biol. 58:549–563.
39. Tyler, P.A. 1970. Coat color differences and runway learning in mice. Behav. Genet. 1:149–155.
40. Wolfe, H.G., and D.L. Coleman. 1966. Pigmentation. *In* E.L. Green, ed., Biology of the Laboratory Mouse, 2nd ed., 405–425. McGraw-Hill, New York.

C2

See *H-2S*.

C3 locus (formerly *C3-1*, C3, and *Plp*), complement component-3, Chr 17

This locus controls variation in complement component 3 in serum, detected by cellulose acetate electrophoresis (*Plp*) (1, 5), by high-voltage electrophoresis in agarose gels (C3) (4), or by analytical isoelectric focusing and antigenicity (*C3-1*) (2, 3). Three alleles are known. *C3ᵃ* (*C3-1ᵃ*, *Plpᶜ*) determines a form with an isoelectric point (pI) of 6.0 and an intermediate migration rate on cellulose acetate and occurs in *M. m.*

molossinus and several Japanese strains derived from that subspecies. *C3ᵇ* (*C3-1ᵇ*, *Plpᵃ*, C3-S) determines a form with a pI of 6.1 and a slow migration rate, and occurs in strains C57BL/6, BALB/c, CBA/J, and many others. *C3ᶜ* (*C3-1ᶜ*) determines a pI slightly lower than 6.0 and the protein differs antigenically from that of *C3ᵃ*. It occurs in strain SWR/J and is probably the same allele as *Plpᵃ* and C3-F which determine fast electrophoretic forms and occur in strains WB/Re, SF/Cam, SWR, SJL, and others, and in GRS/A, STS/A, and others, respectively.

References

1. Hoppe, P.C., and K. Illmensee. 1977. Microsurgically produced homozygous-diploid uniparental mice. Proc. Natl. Acad. Sci. USA 74:5657–5661.
2. Natsuume-Sakai, S., J.I. Hayakawa, and M. Takahashi. 1978. Genetic polymorphism of murine C3 controlled by a single co-dominant locus on chromosome 17. J. Immunol. 121:491–498.
3. Natsuume-Sakai, S., S. Amano, J.I. Hayakawa, and M. Takahashi. 1978. Preparation of an alloantiserum to murine C3 and demonstration of multiple alleles. J. Immunol. 121:2025–2029.
4. da Silva, F.P., G.F. Hoecker, N.K. Day, K. Vienne, and P. Rubenstein. 1978. Murine complement component 3: genetic variation and linkage to *H-2*. Proc. Natl. Acad. Sci. USA 75:963–965.
5. Whitney, J.B., B.A. Taylor, and M. Cherry. 1978. Mouse News Lett. 58:49.

C3-1

See *C3*.

C4

See *H-2S*.

C4bp locus, C4 binding protein, Chr 17?

This locus was reported to control variation in isoelectric point (IEP) of C4-BP, a protein that binds to complement component 4 and is an essential regulatory protein in the classical complement pathway. Two genetic variants were reported, *C4bpᵃ* determining a broadly diffuse band with a IEP of 6.5 to 7.0 and occurring in the A/J, CBA, C3H/He, DBA/2, BALB/c, C57BR, and HTI strains, and *C4bpᵇ* determining a similar band with an IEP of 6.3 to 6.6 and occurring in the C57BL/6, C57BL/10, AKR, C58, HTG, and HTH strains. By use of backcrosses and *H-2* congenic and recombinant strains, the locus was mapped to the *H-2D–Qa-2* interval of Chr 17 about 1.7 cM distal to

H-2S (2). However, de Cordoba *et al.* (1) reported that they were unable to detect this polymorphism for purified C4-BP and conclude that the location of *C4bp* is as yet undetermined.

References

1. de Cordoba, S.R., A. Ferreira, and P. Rubenstein. 1985. Does the mouse C4-binding protein gene (*C4BP*) map in the *H-2* region? Immunogenetics 21:257–265.
2. Kaidoh, T., S. Natsuume-Sakai, and M. Takahashi. 1981. Murine binding protein of the fourth component of complement: structural polymorphism and its linkage to the major histocompatibility complex. Proc. Natl. Acad. Sci. USA 78:3794–3798.

C5

See *Hc* locus.

C6, C7 loci

C6 locus, complement component-6, Chr 15. This locus controls variation in isoelectric focusing (IEP) of the sixth component of complement C6. Immunofixation after IEF shows that C6 focuses as two major bands and one or more minor bands (1). Orren *et al.* (3, 4) found that strains CBA/Ca and AKR had two molecular forms of different molecular weight (90 kDa and 100 kDa) but identical IEF patterns; all other strains examined, including BALB/c and DBA/2, had only the 90 kDa form. The allelic symbols here used are modified from those used by the authors and are in accordance with the rules for mouse genetic nomenclature. The allele (or haplotype) *C6A^a* determines a 90 kDa molecule with an isoelectric point (pI) range of <6.3 and occurs in strains A/J, BALB/c, DBA/2, C57BL/6, NZB, and others; the allele *C6A^m* determines a 90 kDa molecule with a pI range of <6.2 and occurs in wild *M. m. molossinus* in Japan (Hayakawa *et al.*, 2); the allele *C6A^{a}B^a* determines presence of both 90 and 100 kDa molecules with pIs like that of *C6A^a* and occurs in strain CBA/Ca; the allele *C6A^{b}B^b* determines presence of both low and high molecular weight molecules with more acidic pIs than those determined by *C6A^{a}B^a* and occurs in the AKR strain (2, 4). Heterozygotes have all parental bands. In most strains, activity of C6 is strong in males but very weak in females. In the IS/Cam (*C6A^a*) strain, however, activity in females is strong. In female offspring of a cross of IS/Cam with AKR, the *C6A^a* allele was much more strongly expressed than the *C6A^{b}B^b* allele, indicating that IS/Cam carries an allele of a putative regulatory locus causing high expression in females that acts in a *cis* manner. C6 is on Chr 15 close to the centromere (2) and is also closely linked to complement component 7 (*C7*) (4).

C7 locus, complement component-7, Chr 15. This locus controls variation in isoelectric focusing of the seventh component of complement C7. The AKR strain carries an allele here designated *C7^b* that determines a pattern of about five bands; 26 other strains examined carry an allele here designated *C7^a* that determines presence of an extra cathodal band. Heterozygotes are not distinguishable from *C7^a* homozygotes. *C7* showed no recombination with *C6* in 29 mice tested (4).

References

1. Hayakawa, J., H. Nikaido, and T. Koizumi. 1984. Genetic polymorphism of the sixth component of complement (C6) in mice. Immunogenetics 20:633–638.
2. Hayakawa, J., H. Nikaido, and T. Koizumi. 1985. Assignment of the gene locus for the sixth component (C6) of complement in mice to chromosome 15. Immunogenetics 22:637–642.
3. Orren, A., C.J. Preese, and E.B. Dowdle. 1985. Genetically determined molecular weight differences in murine complement component C6. Eur. J. Immunol. 15:100–103.
4. Orren, A., M.J. Hobart, H.R. Nash, and P.J. Lachmann. 1985. Close linkage between mouse genes determining the two forms of complement component 6 and component 7 and *cis* action of a *C6* regulator gene. Immunogenetics 21:591–599.

C58v-1 to *C58v-4* loci

The C58 strain is a high-leukemia high-virus strain that may carry at least six integrated ecotropic virus sites (2).

C58v-1 locus, C58 ecotropic virus-1, Chr 8.

C58v-2 locus, C58 ecotropic virus-2. These two viruses have been isolated in NFS congenic strains. In crosses in which presence of the viral genome was recognized by its inducibility in cultured cells, C58v-1 was shown to be on Chr 8 very close to but probably not identical with *Emv-2* (*Bv*). The location of *C58v-2* is not known; it is not linked to markers on Chrs 1, 5, 7, 8, and 11 (3).

C58v-3 locus (also *Emv-x1*), C58 ecotropic virus-3, Chr 4.

C58v-4 locus (also *Emv-x2*), C58 ecotropic virus-4, Chr 7. These two loci were recognized as integrated proviruses. *C58v-3* was found to be on the distal part of Chr 4, and *C58v-4* on Chr 7 (1). The *C58v-4* provirus was also found by Silver (4) in the NFS.C58-*c$^+$* congenic strain. It is not expressed spontaneously and is not inducible.

Silver found close linkage to *c* on Chr 7 (0/12 recombinants).

References

1. Jenkins, N.A., and N.G. Copeland. 1987. Retroviral DNA content of the mouse genome. *In* V.A. McKusick, T.H. Roderick, J. Mori, and N.W. Paul, eds, Medical and Experimental Mammalian Genetics: a Perspective. Birth Defects: Original Articles Series 23:109–122.
2. Jenkins, N.A., N.G. Copeland, B.A. Taylor, and B.K. Lee. 1982. Organization, distribution, and stability of endogenous ecotropic murine leukemia virus DNA sequences in chromosomes of *Mus musculus*. J. Virol. 43:26–36.
3. Kozak, C.A., and W.P. Rowe. 1982. Genetic mapping of ecotropic murine leukemia virus-inducing loci in six inbred strains. J. Exp. Med. 155:524–534.
4. Silver, J. 1985. A defective ecotropic provirus closely linked to the albino locus. J. Virol. 55:494–496.

Ca locus, Chr 15

Ca, caracul, dominant. Arose in a stock of Swiss mice (2). Homozygotes are indistinguishable from heterozygotes. Both are fully viable and fertile. Vibrissae are curved, and hair is wavy from time of first appearance until about 4 weeks of age. After that the waves disappear, but the hair has a plush-like look. Several recurrences of mutations similar to *Ca* have been reported at this locus.

Ca^d, directional caracul, semidominant. Arose spontaneously in a cross of (101 × C3H)F1 by the T multiple recessive stock. *Ca^d*/+ resembles *Ca*/+ except for a clear ridge of hair down the back which disappears at weaning age. *Ca^d*/*Ca^d* mice have extremely curly hair and develop baldish patches over the upper back that persist throughout life. Homozygotes and heterozygotes are both fully viable and fertile (1).

References

1. Bangham, J.W. 1969. Mouse News Lett. 40:41.
2. Dunn, L.C. 1937. Caracul, a dominant mutation. J. Hered. 28:334.

Caa locus, Ca^{2+}-ATPase activity, Chr 12

This locus controls part of the difference in activity of Ca^{2+}-ATPase between the C57BL/6J and DBA/2J strains. The allele *Caa^b* in C57BL/6 partly determines high activity of the enzyme; the *Caa^d* in DBA/2 partly determines low activity. Activity appears to be controlled by more than one locus as shown by the continuous rather than bimodal distribution of activity levels in 21 B6 × D2 recombinant inbred strains (BXD RI). In these RI strains there is a high correlation between Ca^{2+}-ATPase activity and resistance to audiogenic seizures. High activity is also associated with presence of the *Ah^b* (aromatic hydrocarbon responsiveness) allele from C57BL/6 and low activity with the *Ah^d* allele from DBA/2. *Caa* appears to be different from *asp-1* (audiogenic seizure prone-1) which is also closely linked to the *Ah* locus. Thus, susceptibility to audiogenic seizures is influenced by two major genes, both linked to the *Ah* locus, which is known to be located on Chr 12.

References

1. Palayoor, S.T., and T.N. Seyfried. 1984. Genetic association between Ca^{2+}-ATPase activity and audiogenic seizures in mice. J. Neurochem. 42:1771–1774.

cab, cardiac abnormality, recessive

Arose spontaneously in a closed colony carrying *mdg*. Homozygotes die at birth or half a day earlier. Late stage fetuses show greatly reduced spontaneous or induced movements. In the heart muscle there are numerous vacuolated cells and prominent foci of mesenchyme-like cells. In contrast to the condition in normal littermates, there is no histologically detectable glycogen in the liver, lung, heart, or skeletal muscle (1). In liver cells from late fetal stages, there is severe hypoglycemia and low ATP levels, low specific activity of glycogen synthase and phosphorylase, and low output of CO_2 from glucose metabolism. The glucose deficit is not due to defective placental transport (2).

References

1. Essien, F.B. 1979. A lethal mutation (*cab*) affecting heart function in the mouse. Genet. Res. 33:57–59.
2. Tyson, F.L., and F.B. Essien. 1985. Prenatal expression of a lethal defect in carbohydrate metabolism in mice. Proc. Natl. Acad. Sci. USA 82:2101–2105.

cac, recessive cataract

Arose in an inbred strain carrying *Fu^ki*. Penetrance is incomplete. Homozygotes develop cataracts at about one month of age, with degenerative changes in the posterior part of the lens capsule and subcapsular material (2; *cat* used as gene symbol). Konyukhov and Wachtel (1) found differences from normal in the relative amounts of α-, β-, and γ-crystallins, but Moser and Gluecksohn-Waelsch (2) were unable to repeat these results.

References

1. Konyukhov, B.V., and A.W. Wachtel. 1963. Electrophoretic studies of proteins in normal lenses and cataracts of inbred and mutant mice. Exp. Eye Res. 2:325–330.

2. Moser, G.C., and S. Gluecksohn-Waelsch. 1967. Electrophoretic patterns of lens protein from genetically caused cataracts in the mouse. Exp. Eye Res. 6:297–298.

Caf locus, caffeine susceptibility

This locus determines relative susceptibility to the acute toxic effects of intraperitoneally administered caffeine in SWR and CBA mice. The allele Caf^s occurs in the CBA/J strain and determines occurrence of clonic seizures and death within 30 seconds after an intraperitoneal dose of 187 mg/kg; the allele Caf^r occurs in the SWR strain and determines absence of seizures and survival after a similar dose. Heterozygotes are sensitive. The difference in response does not appear to be due to a difference in caffeine metabolism, since plasma levels of caffeine are comparable in the two strains (1).

References
1. Searle, T.W., P. Johnson, T.H. Roderick, J.M. Carney, and O.M. Rennert. 1985. A single gene difference determines relative susceptibility to caffeine-induced lethality in SWR and CBA inbred mice. Pharmacol. Biochem. Behav. 23:275–278.

Calc locus, calcitonin, Chr 7

This locus codes for the hormone calcitonin secreted by the parafollicular cells of the thyroid gland. No genetic variants are known. The locus was found to be on Chr 7 using mouse–Chinese hamster somatic cell hybrids and a rat cDNA clone coding for calcitonin (1).

References
1. Naylor, S.L., A.Y. Sakaguchi, W.W. Chin, J. Jacobs, L.P. Shen, W.J. Rutter, and P.A. Lalley. 1984. Chromosomal mapping of mouse hormone genes: somatostatin, calcitonin, thyrotropin alpha subunit, and luteinizing hormone beta subunit. Cytogenet. Cell Genet. 37:551 (Abstr.).

can, cartilage anomaly, recessive

Probably extinct. Arose as a spontaneous mutation in 1965 in a stock carrying *dn*. Homozygotes are small at birth and have shortened domed skulls and short limbs. Growth is retarded, and death usually occurs at about 10 days. Homozygotes can be recognized as early as 17 days of gestation by their size and shape (3). Collagen fibrils in cartilage are well formed, but appear more crowded than normal. At 17 days of gestation there is reduced protein synthesis in cartilage (2). Explanted cartilage from 3-day mutants shows higher than normal protein and mucopolysaccharide synthesis (1).

References
1. Johnson, D.R. 1974. The *in vivo* behaviour of achondroplastic cartilage from the cartilage anomaly (*can/can*) mouse. J. Embryol. Exp. Morphol. 31:313–318.
2. Johnson, D.R., and D.M. Hunt. 1974. Biochemical observations on the cartilage of achondroplastic (*can*) mice. J. Embryol. Exp. Morphol. 31:319–328.
3. Johnson, D.R., and J.M. Wise. 1971. Cartilage anomaly (*can*); a new mutant gene in the mouse J. Embryol. Exp. Morphol. 25:21–31.

Car-1, *Car-2* loci, Chr 3

These loci control electrophoretic variation in two isozymes of carbonic anhydrase (E.C. 4.2.1.1) in red blood cells. The loci are very closely linked; all four combinations of the *a* and *b* alleles at the two loci have been found in inbred strains but recombination has not been observed in offspring of crosses (1).

Car-1 locus, carbonic anhydrase-1. Electrophoretic variation in the enzyme, CA-I, controlled by this locus is demonstrated in red blood cells at alkaline pH. The allele $Car-1^a$ determines a slow form present in C57BL/6, C3H/He, and many other strains; the allele $Car-2^b$ determines a fast form present in IS/Cam, WB/Re, SM, and a few related strains, and in *M. m. castaneus* and *M. m. molossinus*. Heterozygotes have both parental bands (1).

Car-2 locus (formerly *Pro-1*, *Car-1*), carbonic anhydrase-2. Electrophoretic variation in the enzyme, CA-II, controlled by this locus is demonstrated in red blood cells at slightly acid pH. The allele $Car-2^a$ determines a slow form with a high isoelectric point (pI) and is present in AKR/J, C57BL/6, and a number of other strains; the allele $Car-2^b$ determines a fast form with a lower pI and is present in BALB/c, C3H/He, SM, and a number of other strains; a third allele $Car-2^c$ determines a form with a still lower pI and was found in wild mice, *Mus spretus*, of Spanish origin. $Car-2^a/Car-2^b$ heterozygotes have both parental bands (1, 3). The soluble liver polypeptide called LTW-5 by Elliott (2) is probably determined by the *Car-2* locus.

References
1. Eicher, E.M., R.H. Stern, J.E. Womack, M.T. Davisson, T.H. Roderick, and S.C. Reynolds. 1976. Evolution of mammalian carbonic anhydrase loci by tandem duplication: close linkage of *Car-1* and *Car-2* to the centromere region of chromosome 3 of the mouse. Biochem. Genet. 14:651–660.
2. Elliott, R.W. 1979. Use of two-dimensional electrophoresis to identify and map new mouse genes. Genetics 91:295–308.

3. Whitney, J.B. III. 1984. Novel carbonic anhydrase (*Car-2*) allele in Spanish mice. Biochem. Genet. 22:913–922.

Cas-1 locus (formerly *Cs*, *Cs-1*, see Lyon, 9), catalase-1, Chr 2

This locus controls molecular structure of catalase found in erythrocytes, liver, and other solid tissues. Variants at the locus were first found among offspring of a cross of X-irradiated (101 × C3H)F1 males to females of a multiple recessive stock, by use of a rapid blood catalase assay (2). The mutant alleles are *Cas-1*^{b} which determines acatalasemia and *Cas-1*^{c} to *Cas-1*^{f} which determine several grades of severe hypocatalasemia. There is considerable variation among inbred strains in erythrocyte catalase activity, and Hoffman and Grieshaber (7) have shown that the difference between the high level of BALB/c (*Cas-1*^{a} allele) and the low level of C57BL/An (*Cas-1*^{g} allele) is also determined by the *Cas-1* locus. It is not certain whether the wild-type allele of Feinstein (originally called *Cs*^{a}) is identical with *Cas-1*^{a} in the BALB/c strain. An eighth allele, *Cas-1*^{v}, determining a very high level of erythrocyte catalase with a slight electrophoretic variation was found in all tested *M. m. castaneus* (10). Heterozygotes have catalase levels intermediate between those of the parental homozygotes. The X-ray-induced mutants all produce normal amounts of catalase protein, but in modified form (3). The catalase molecule appears to have at least two antigenic sites, one of which is at or near the mutated site of *Cas-1*^{b} and hence probably near the enzyme active site. It may be near the histidyl residue 74 (4). For a brief review of catalase genetics see Holmes and Masters (8).

Cas-1^b/*Cas-1*^{b} mice have a high peroxidase activity in liver and kidneys, thought to result from degradation of unstable catalase. Peroxidase is known to lower serum triglycerides and cholesterol, and these are both found to be lower in the mutant mice (1, 6). On the C3H/He background under certain conditions, *Cas-1*^{b} causes greater susceptibility to liver tumors than does *Cas-1*^{a}, possibly because of the diminished rate of degradation of H_2O_2 in *Cas-1*^{b}/*Cas-1*^{b} mice (5).

References

1. Cuadrado, R.R., and L.A. Bricker, 1973. An abnormality of hepatic lipogenesis in a mutant strain of acatalasemic mice. Biochim. Biophys. Acta 306:168–172.
2. Feinstein, R.N., J.B. Howard, J.T. Braun, and J.E. Seaholm. 1966. Acatalasemic and hypocatalasemic mouse mutants. Genetics 53:923–933.
3. Feinstein, R.N., H. Suter, and B.N. Jaroslow. 1968. Blood catalase polymorphism: some immunological aspects. Science 159:638–640.
4. Feinstein, R.N., B.N. Jaroslow, and J.B. Howard. 1972. Molecular location of the genetic lesion in the acatalasemic mouse. Biochem. Genet. 6:263–273.
5. Feinstein, R.N., R.J.M. Fry, and E.F. Staffeldt. 1978. Carcinogenic and antitumor effects of aminotriazole on acatalasemic and normal catalase mice. J. Natl. Cancer Inst. 60:1113–1116.
6. Goldfischer, S., P.S. Roheim, D. Edelstein, and E. Essner. 1971. Hypolipidemia in a mutant strain of "acatalasemic" mice. Science 173:65–66.
7. Hoffman, H.A., and C.K. Grieshaber. 1974. Genetic studies of murine catalase: liver and erythrocyte catalase controlled by independent loci. J. Hered. 65:277–279.
8. Holmes, R.S., and C.J. Masters. 1978. Genetic control and ontogeny of microbody enzymes: a review. Biochem. Genet. 16:171–190.
9. Lyon, M.F. 1983. Mouse News Lett. 68:50.
10. Whitney, J.B., and D.L. Schlais. 1978. Mouse News Lett. 58:48.

Cat, dominant cataract, semidominant, Chr 10

Arose in an albino strain of unknown origin (5). A probably identical allele, *Cat*^{Fr} (originally called shrivelled, *Svl*), was found in the A/J strain (6). Homozygotes and heterozygotes are fully viable and fertile. Homozygotes have lens opacities which are visible when the eyes are first open at 14 days. The lens is smaller than normal, the main lens fiber mass degenerates, and the epithelial cells form multiple layers of atypical cells (8). By light microscopy, the earliest effect occurs at 14 days of gestation when the centrally located nuclei of the lens fibers are seen to be abnormally small and dark with indistinct nucleoli. Nuclear degeneration is followed by swelling and vacuolization of the lens fibers (7). By electron microscopy, changes can be seen in the nuclear envelope of the primary lens fibers in 12-day embryos (3). A study of the lens crystallins showed a decrease in the proportion of γ-crystallins and an increase in the proportion of α-crystallins (1). At birth, γ-crystallin mRNAs are present in nearly normal amounts and functional capacity but are selectively lost by 40 days of age. The effect is more severe in homozygotes than in heterozygotes. The γ-crystallin cellular DNA does not differ from normal (2). *Cat*^{Fr} was shown to be allelic with *Lop*, known to be located on Chr 10 (4).

References

1. Garber, A.T., L. Stirk, and R.J.M. Gold. 1983. Abnormalities of crystallins in the lens of the Cat^{Fraser} mouse. Exp. Eye Res. 36:165–169.

2. Garber, A.T., C. Winkler, T. Shinohara, C.R. King, G. Inana, J. Piatigorsky, and R.J.M. Gold. 1984. Selective loss of a family of gene transcripts in a hereditary murine cataract. Science 227:74–77.
3. Konyukhov, B.V., and N.A. Kolesova. 1976. Ultra-structural analysis of effects of dominant cataract-Fr gene in mouse embryos. Ontogenez. 7:271–277. (Russian) Abstract in Genet. Abst. 8 (9):160, 1976.
4. Muggleton-Harris, A.L., M.F.W. Festing, and M. Hall. 1987. A gene location for the inheritance of the cataract Fraser (*Cat^Fr*) mouse congenital cataract. Genet. Res. 49:235–238.
5. Paget, O.E. 1953. *Cataracta hereditaria subcapsularis*: ein neues, dominantes Allel bei der Hausmaus. Z. Indukt. Abstammungs-Vererbungsl. 85:238–244.
6. Verrusio, A.C., and F.C. Fraser. 1966. Identity of mutant genes "shrivelled" and cataracta congenita subcapsularis in the mouse. Genet. Res. 8:377–378.
7. Zwaan, J., and R.M. Williams. 1968. Morphogenesis of the eye lens in a mouse strain with hereditary cataracts. J. Exp. Zool. 169:407–422.
8. Zwaan, J., and R.M. Williams. 1969. Cataracts and abnormal proliferation of the lens epithelium in mice carrying the *Cat^Fr* gene. Exp. Eye Res. 8:161–167.

Cat^Na

See *nct*.

cat

See *cac*, *nct*.

cb, cerebral degeneration, recessive

Found by Deol and Truslove (1). Homozygotes usually die before reaching maturity. If they live on, they are sterile. The degeneration produces an *ex vacuo* hydro-cephalus visible in the living animal and is usually recognizable at birth. The most susceptible parts of the brain are the cerebral hemispheres and the olfactory lobes. Later the epithelium of the nose and trachea also degenerates.

References

1. Deol, M.S., and G.M. Truslove. 1963. A new gene causing cerebral degeneration in the mouse. *In* S.J. Geerts, ed., Proc. XI Int. Congr. Genet. Vol. 1, 183–184. Pergamon Press, New York (Abstr.).

cby, chubby, recessive

Arose in the Peru-Coppock stock. Homozygotes can be recognized at 15 days of age by their short heads and shortened bodies, limbs, and tails. Both sexes are fer-tile and the gene appears to be fully penetrant. The growth cartilage of the proximal tibial epiphyseal plate shows increased height and pronounced morphological abnormalities in the proliferative zone. There is exten-sive cell death and accumulations of matrix vesicles in the extracellular matrix. In the hypertrophic zone, mineralization is irregular. The chondroitin sulfate chains of *cby/cby* cartilage are considerably longer than normal (1).

References

1. Wikström, B., M.E. Wallace, A. Hjerpe, and B. Eng-feldt. 1987. Chubby; a new autosomal recessive mutation producing dwarfism in the mouse. J. Hered. 78:8–14.

Ccd, cleidocranial dysplasia, semidominant

Induced by gamma irradiation in a 101 strain male. Heterozygotes show striking homology to cleidocranial dysplasia, a skeletal disorder in man. The mutation is fully penetrant in heterozygotes. It causes partial or complete failure of ossification of clavicles, symphysis pubis, and cranial fontanelles. *Ccd* is lethal in homozy-gotes (1).

References

1. Selby, P. 1985. Mouse News Lett. 72:123.

Cd, crooked, semidominant, Chr 6

Arose spontaneously in the A inbred strain (5). Heterozygotes have crooked tails with abnormal caudal vertebrae and often have abnormal sacral and lumbar vertebrae as well. Some have no abnormalities of the skeleton at all. Homozygotes are more severely affec-ted and show no normal overlapping. They may die early in development, or later with exencephaly, or, if they survive, are smaller than normal from about 15 days of age and may have crooked tails, microphthal-mia, small or non-erupted lower incisors, absence of third molars, irregular tail rings with few hairs, and ner-vous movement of the head (2, 3, 5). Heterozygous embryos are more susceptible to insulin-induced exen-cephaly than normal controls (1). The vertebral anoma-lies can be traced back to the somite stage when the somites are seen to be irregular in size and often partly fused with their neighbors (4). Small size of the molars can be traced back to a transient proliferation of the dental lamina which may competitively inhibit normal molar growth (6).

Cd

References

1. Cole, W.A., and D.G. Trasler. 1980. Gene teratogen interaction in insulin-induced mouse exencephaly. Teratology 22:125–139.
2. Grewal, M.S. 1962. The development of an inherited tooth defect in the mouse. J. Embryol. Exp. Morphol. 10:202–211.
3. Grüneberg, H. 1965. Genes and genotypes affecting the teeth of the mouse. J. Embryol. Exp. Morphol. 14:137–159.
4. Matter, H. 1957. Die normale Genese einer vererbten Wirbelsäulenmissbildung am Beispiel der Mutante Crooked-tail der Maus. Rev. Suisse Zool. 64:1–38.
5. Morgan, W.C. 1954. A new crooked tail mutation involving distinctive pleiotropism. J. Genet. 52:354–373.
6. Sofaer, J.A. 1977. Tooth development in the "crooked" mouse. J. Embryol. Exp. Morphol. 41:279–287.

CD-2 locus

See *Ly-37* locus.

cdm, cadmium resistance, recessive, Chr 3

Inbred strains differ markedly in sensitivity to cadmium-induced necrosis of the testis. Resistance is determined by homozygosity for a single recessive gene (*cdm/cdm*). Homozygotes show undamaged testes after a single subcutaneous dose of $CdCl_2$, whereas *cdm/+* and +/+ mice show testes with hemorrhagic and necrotic areas. *cdm* is found in strains A/J, BALB/c, C3H/He, and C57BL/6; most other strains are wild type (5). The level of cadmium in testes of DBA/2 (+/+) mice 2 hours after treatment is twice as high as in C3H and BALB/c (*cdm/cdm*) mice. Paradoxically, the *cdm/cdm* resistant strains are more susceptible to the lethal effects of cadmium treatment than the +/+ susceptible strains (1) and also more susceptible to the teratogenic effects of cadmium (3). Induction by cadmium of metallothionein (MT) in liver of the +/+ DBA/2 strain is higher than in the *cdm/cdm* C3H and C57BL/6 strains. The low level in C3H appears to be due to increased rate of degradation of MT (4). The difference between DBA/2 and C57BL/6 in cadmium-induced MT level is controlled by a single locus with codominant alleles that shows close linkage with *cdm* and may be the same locus (2).

References

1. Hata, A., H. Tsunoo, H. Nakajima, K. Shintaku, and M. Kimura. 1980. Acute cadmium intoxication in inbred mice: a study on strain differences. Chem.-Biol. Interactions 32:29–39.
2. Hunt, D.M., and T. Mhlanga. 1983. Genetic studies on metallothionein synthesis in the mouse: the induction of metallothionein by cadmium in inbred strains. Biochem. Genet. 21:609–625.
3. Layton, W.M. Jr., and M.W. Layton. 1979. Cadmium induced limb defects in mice: strain associated differences in sensitivity. Teratology 19:229–236.
4. Piletz, J.E., R.D. Andersen, W. Berry, and H.R. Herschman. 1983. Synthesis and degradation of hepatic metallothionein in mice differing in susceptibility to cadmium mortality. Biochem. Genet. 21:561–578.
5. Taylor, B.A., H.J. Heiniger, and H. Meier. 1973. Genetic analysis of resistance to cadmium-induced testicular damage in mice. Proc. Soc. Exp. Biol. Med. 143:629–633.

Ce-1 locus (formerly Ce), liver catalase

This locus controls the rate of degradation of catalase (E.C. 1.11.1.6) in liver. The allele $Ce-1^b$ determines a slow rate of degradation and therefore high activity level and is present in the C57BL/An, C57BL/He, and C57BL/Ha strains. The allele $Ce-1^c$ determines a fast rate of degradation and therefore low activity level and is present in all other strains of the C57 and C58 group and, possibly, in DBA/2 and other strains. $Ce-1^c$ is completely dominant to $Ce-1^b$. *Ce-1* has no effect on kidney catalase activity (1,3,4). *Ce-1* is not linked to the structural locus for catalase, *Cs-1* (2).

References

1. Ganschow, R.E., and R.T. Schimke. 1970. Murine catalase phenotypes. Biochem. Genet. 4:157–167.
2. Hoffman, H.A., and C.K. Grieshaber. 1974. Genetic studies of murine catalase: liver and erythrocyte catalase controlled by independent loci. J. Hered. 65:277–279.
3. Paigen, K. 1971. The genetics of enzyme realization. In Enzyme Synthesis and Degradation in Mammalian Systems, 1–46. Karger, Basel, New York.
4. Rechcigl, M. Jr., and W.E. Heston. 1967. Genetic regulation of enzyme activity in mammalian systems by the alteration of the rates of enzyme degradation. Biochem. Biophys. Res. Comm. 27:119–124.

Ce-2 locus, kidney catalase, Chr 17

In kidney and various tissues, mouse catalase can exist in five major electrophoretic forms, Ct-1 to Ct-5. The variation is associated with progressive removal of sialic acid from the Ct-1 form (3). The *Ce-2* locus controls the multiplicity of catalase in kidney. The allele $Ce-2^a$, present in C57BL/6 and other strains, determines a single band pattern and the allele $Ce-2^b$, present in C3H/He and other strains, determines a multiple band pattern. $Ce-2^a$ is fully dominant. Mixture of kidney homogenates of various genotypes indicates that the kidneys of $Ce-2^a/Ce-2^a$ and $Ce-2^a/Ce-2^b$ mice contain a factor capable of converting the multiple banded pattern to a single band. Liver catalase has the multiple banded pattern in all strains, and the kidney factor also converts liver catalase to a single banded pattern (1,2).

References

1. Hoffman, H.A., and C.K. Grieshaber. 1976. Genetic studies of murine catalase: regulation of multiple molecular forms of kidney catalase. Biochem. Genet. 14:59–66.
2. Hoffman, H.A., and C.K. Grieshaber. 1977. Mouse News Lett. 56:51.
3. Holmes, R.S., and C.J. Masters. 1978. Genetic control and ontogeny of microbody enzymes: a review. Biochem. Genet. 16:171–190.

Cfh locus, complement factor H, Chr 1

This locus codes for the complement component, factor H. It was identified in *Eco*RI-digested genomic DNA by hybridization with a factor H cDNA clone isolated from a mouse liver cDNA library. Southern blots show 16 bands, with two patterns found among inbred strains. A pattern containing a 4-kb fragment occurs in strains AKR, C57L, C57BL/6, and BALB/c; this fragment is not found in strains DBA/2, C3H/He, and SJL. By use of mouse–Chinese hamster and mouse–rat somatic cell hybrids and two panels of recombinant inbred (RI) strains (AKXL and BXD), *Cfh* was shown to be on Chr 1 very close to *Sas-1*, a locus controlling a serum antigenic substance. The two loci showed no recombination among 36 RI strains, suggesting that *Sas-1* may be the factor H structural locus (1).

References

1. D'Eustachio, P., T. Kristensen, R.A. Wetsel, R. Riblet, B.A. Taylor, and B.F. Tack. 1986. Chromosomal location of the genes encoding complement components C5 and factor H in the mouse. J. Immunol. 137:3990–3995.

Cfhe locus, complement factor H-electrophoretic, Chr 2

Factor H protein is a constituent of plasma that serves as a cofactor in the cleavage of complememt component C3b. Polymorphism of factor H was detected by three different methods. Factor H.1, determined by the allele *Cfhe^a*, moves faster anodally during prolonged agarose electrophoresis than factor H.2, determined by the allele *Cfhe^b*; the two factors differ at five spots by two-dimensional peptide mapping; and alloantiserum against each factor was produced by repeated injection of purified factor H of one type into mice of the other type. Heterozygotes have both parental phenotypes by all three of these methods. The allele *Cfhe^a* occurs in strains BALB/c, A/J, C57BL/10, and most other inbred strains; the allele *Cfhe^b* occurs in strain STR and in *M. m. molossinus*. *Cfhe* is probably not the structural locus for factor H since a locus identified by a cDNA clone for factor H has been found to be located on Chr 1 (1) and *Cfhe* is on Chr 2. It lies about 17 cM from hemolytic complement, *Hc*, probably on the distal side (2).

References

1. D'Eustachio, P., T. Kristensen, R.A. Wetsel, R. Riblet, B.A. Taylor, and B.F. Tack. 1986. Chromosomal location of the genes encoding complement components C5 and factor H in the mouse. J. Immunol. 137:3990–3995.
2. Natsuume-Sakai, S., K. Sudoh, T. Kaidoh, J.-I Hayakawa, and M. Takahashi. 1985. Structural polymorphism of murine complement factor H controlled by a locus located between Hc and the *β*2M locus on the second chromosome of the mouse. J. Immunol. 134:2600–2606.

Cg locus

See *Xce* locus.

ch, congenital hydrocephalus, recessive, Chr 13

Appeared as a spontaneous mutation in descendants of an outcross of the CBA inbred strain. Homozygotes die at birth, probably from inability to breathe. They have bulging hemorrhagic cerebral hemispheres and open eyelids, urogenital abnormalities, and severely abnormal skull, cervical vertebrae, sternum, laryngeal cartilages, and hyoid bone (2,3). Many other parts of the skeleton are mildly affected, and there are numerous defects of glands and their ducts in the head region (4). Some heterozygotes show the same type of defect as the homozygotes, but much less severe (2). The urogenital defects trace to the formation in 10-day embryos of many excess mesonephric tubules posterior to the normal mesonephros. This is accompanied by a deficiency of celiac ganglion cells in the same region. The hydrocephalus may trace to lack of formation of subarachnoid space, which is first detectable in 11-day embryos (2). Most of the defects in the head region are probably secondary to the severe displacements created by the hydrocephalus. The basic defect responsible for the syndrome is not known, but may be a very early defect of the mesenchyme which prevents proper aggregation of mesenchyme cells to form bone and cartilage rudiments, prevents proper movement of neural crest cells and epithelial ducts through the mesenchyme, and prevents proper dispersal of cells to form subarachnoid space (2,4). Breen *et al.* (1) describe a deficiency of glycosaminoglycans in developing cartilage of *ch/ch* mice.

References

1. Breen, M., R. Richardson, W. Bondareff, and H.G. Weinstein. 1973. Acidic glycosaminoglycans in developing sterno-costal cartilage of the hydrocephalic (*ch +/ch +*) mouse. Biochim. Biophys. Acta 304:828–836.
2. Green, M.C. 1970. The developmental of congenital hydrocephalus (*ch*) in the mouse. Dev. Biol. 23:585–608.
3. Grüneberg, H. 1953. Genetical studies on the skeleton of the mouse. VII. Congenital hydrocephalus. J. Genet. 51:327–358.
4. Grüneberg, H., and G.A.DeS. Wickramaratne. 1974. A re-examination of two skeletal mutants of the mouse, vestigial-tail (*vt*) and congenital hydrocephalus (*ch*). J. Embryol. Exp. Morphol. 31:207–222.

cho, chondrodysplasia, recessive

Arose spontaneously in the C57BL/6 strain. Homozygotes have disproportionately shortened limbs, shortened heads, protruding tongues, cleft palates, and short mandibles, and die at birth from asphyxia due to defective tracheal rings. In mutant cartilage the collagen fibrils are large and banded rather than fine and unbanded as in normal cartilage (1). There is reduction of acid mucopolysaccharides in the matrix (2). These substances are synthesized at a normal rate, but are more easily extracted with water than in normal controls, probably because of defective interaction with some other matrix component, possible collagen (4). Examination of development of the secondary palate and lower jaw in homozygotes suggests that growth of Meckel's cartilage is necessary for normal palate formation (3).

References

1. Seegmiller, R., F.C. Fraser, and H. Sheldon. 1971. A new chondrodystrophic mutant in mice. Electron microscopy of normal and abnormal chondrogenesis. J. Cell. Biol. 48:580–593.
2. Seegmiller, R., C.C. Ferguson, and H. Sheldon. 1972. Studies on cartilage. VI. A genetically determined defect in tracheal cartilage. J. Ultrastruct. Res. 38:288–301.
3. Seegmiller, R.E., and F.C. Fraser. 1977. Mandibular growth retardation as a cause of cleft palate in mice homozygous for the chondrodysplasia gene. J. Embryol. Exp. Morphol. 38:227–238.
4. Stephens, T.D., and R.E. Seegmiller. 1976. Normal production of cartilage glycosaminoglycan in mice homozygous for the chondrodysplasia gene. Teratology 13:317–326.

cht, chocolate, recessive, Chr 7

Arose spontaneously in the C57BL/6 strain. Nonagouti chocolate (*a/a cht/cht*) mice are a rich dark brown (2). *cht* is on Chr 7 probably near *p* (1).

References

1. Davisson, M.T., and H.P. Bunker. 1986. Linkage of chocolate (cht) on Chr 7. Mouse News Lett. 75:31.
2. MacPike, A.D., and L.E. Mobraaten. 1984. Mouse News Lett. 70:86.

Chy, chylous ascites, probably semidominant, Chr 11

Found among offspring of a male treated with ethylnitrosourea. Heterozygotes can be recognized after they begin to suckle by the appearance of a milky fluid in the abdomen. At weaning, the hindfeet are noticeably swollen and edematous. Both sexes are fertile. Homozygotes have not been seen. The chylous ascites resembles that seen in ragged (*Ra*) mice. *Chy* is on Chr 11 proximal to *vt* (1, 2).

References

1. Lyon, M.F., and P.H. Glenister. 1984. Mouse News Lett. 71:26.
2. Lyon, M.F., and P.H. Glenister. 1986. Mouse News Lett. 74:96.

Cis locus

See *Cs* locus.

ck

See *cpk*.

cl, clubfoot, recessive

Appeared in a stock of unknown ancestry. Homozygotes are much less viable and fertile than normal. The clubbing is a simple calcaneum (dorsiflexed) type, always affecting both hindfeet and sometimes one or both forefeet. Muscles of the lower limb are smaller than normal, and there is some carpal and tarsal fusion (1).

References

1. Robins, M.W. 1959. A mutation causing congenital clubfoot in the house mouse. J. Hered. 50:188–192.

cla, clasper, recessive, Chr 4

Arose spontaneously in the C57BL/6ByJ strain. Homozygotes can be identified at 14 days of age by smaller size and a fine whole-body tremor, with clasping of both forefeet and hindfeet when held up by the tail. Adults hold their hindfeet close together and frequently clasped, resulting in a hobbling gait. This condition persists throughout life. Both sexes are fertile. *cla* is on Chr 4 about 10 cM distal to *Pt* (1).

References

1. Sweet, H.O. 1985. Mouse News Lett. 73:18.

cld, combined lipase deficiency, recessive, Chr 17

This mutation was found as a 'parasitic lethal' in the *T*-complex haplotype carrying t^{w73}, a recessive embryonic lethal. It was recovered from the haplotype by recombination. Homozygotes are normal at birth but show poor weight gain, paleness, and progressive cyanosis leading to death within 36 to 48 hours. Plasma triglycerides and cholesterol are normal at birth but rise dramatically as soon as nursing starts. Chylomicrons accumulate and the serum becomes opaque. This probably results in coalescence of particles and obstruction of capillary fields in vital organs. Homozygotes are deficient in activity of two lipolytic enzymes, lipoprotein lipase (LPL; E.C. 3.1.3.34) present in the capillary endothelium of most extrahepatic tissues, and hepatic triglyceride lipase (HTGL; E.C. 3.1.3.3) present in the capillary endothelium of liver. The deficiency is present *in utero*, but apparently the enzymes are not needed until intake of dietary fat occurs. Heterozygotes have normal enzyme activity (2). Homozygotes synthesize a lipoprotein lipase-like protein in amounts two to six times greater than normal, and with normal electrophoretic and immunological properties. It is possible that *cld* affects the two lipase enzymes by determining a factor required for maturation of the enzymes to their active form (1).

References

1. Olivecrona, T., S.S. Chernick, G. Bengtsson-Olivecrona, J.R. Paterniti Jr., W.V. Brown, and R.O. Skow. 1985. Combined lipase deficiency (*cld/cld*) in mice. Demonstration that an inactive form of lipoprotein lipase is synthesized. J. Biol. Chem. 260:2552–2557.
2. Paterniti, J.R. Jr., W.V. Brown, H.N. Ginsberg, and K. Artzt. 1983. Combined lipase deficiency (*cld*): a lethal mutation on chromosome 17 of the mouse. Science 221:167–169.

Cm, coloboma, semidominant, Chr 2

Found among offspring of an irradiated (C3H/HeH × 101/H)F1 male. In heterozygotes the eyes look smaller at birth, and a ventral segment of the choroid comprising about 25 per cent of the whole is absent. In adults the eyeballs are rotated ventrally so that the pupil may be partly hidden by the lower lid. There is also abnormal posture and some head-shaking and circling. The homozygote has not been described (1). See also *Sey*, which may be an allele.

References

1. Searle, A.G. 1966. Mouse News Lett. 35:27.

cmd, cartilage matrix deficiency, recessive, probably Chr 7

Arose in a stock carrying *T*, t^{low}, and *tf*. Homozygotes are dwarfed at birth and have an abnormally short trunk, limbs, tail, and snout, and a protruding tongue, cleft palate, and distended abdomen. They die at birth from inability to breathe. Cartilage matrix is defective, with abnormally large and closely packed collagen fibers and an altered form and distribution of mucopolysaccharides. In sections, the chondrocytes appear tightly packed with very little matrix showing between them. The syndrome closely resembles the rare lethal condition, achondrogenesis, in man (5). Biochemical and immunofluorescent examination of *cmd/cmd* cartilage has shown that the core protein of the large major cartilage proteoglycan is absent, but that smaller proteoglycans are present in normal amounts (2, 4). Homozygotes can be recognized at 15 days of gestation by reduced limb length and difference in shape and contour of forelimb flexures. In organ culture of limb buds from mutant and normal embryos, synthesis of chondroitin sulfate, a sulfated glycosaminoglycan (s-GAG), was markedly lower in *cmd/cmd* cultures than in normal controls. Heterozygotes were intermediate between the two homozygotes in amount of s-GAG synthesized. Among the normal controls, heterozygotes for *cmd* could be distinguished from homozygous wild type by segregation of a closely linked gene causing pink eyes at birth, presumably albino (*c*), i.e. *cmd* +/+ *c* (dark eyes) vs. + *c*/+ *c* (unpigmented eyes), indicating that *cmd* is probably on Chr 7 near *c* (3). A second mutation at the *cmd* locus, cmd^{Bc}, occurred in the BALB/cGaBc strain. It closely resembles *cmd*, and *cmd/cmd^{Bc}* mice are indistinguishable from either homozygote (1).

References

1. Bell, L., D.M. Juriloff, and M.J. Harris. 1986. A new mutation at the *cmd* locus in the mouse. J. Hered. 77:205–206.
2. Brennan, M.J., A. Oldberg, E. Ruoslahti, K. Brown, and

N. Schwartz. 1983. Immunological evidence for two distinct chondroitin sulfate proteoglycan core proteins: differential expression in cartilage matrix deficiency mice. Dev. Biol. 98:139–147.

3. Kochhar, D.M. 1985. Cellular expression of a mutant gene (cmd/cmd) causing limb and other defects in mouse embryos. *In* M. Marois, ed., Prevention of Physical and Mental Congenital Defects, 131–144. Alan Liss, New York.

4. Kimata, K., H.-J. Barrach, K.S. Brown, and J.P. Pennypacker. 1981. Absence of proteoglycan core protein in cartilage from cmd/cmd (cartilage matrix deficiency) mice. J. Biol. Chem. 256:6961–6968.

5. Rittenhouse, E., L.C. Dunn, J. Cookingham, C. Calo, M. Spiegelman, G.B. Dooher, and D. Bennett. 1978. Cartilage matrix deficiency (*cmd*): a new autosomal recessive mutation in the mouse. J. Embryol. Exp. Morphol. 43:71–84.

Cms locus, *Coccidioides immitis* resistance

This locus controls resistance and susceptibility to intraperitoneal infection with *C. immitis*. The eight inbred strains tested show variation in LD_{50}: BALB/c, AKR/N, A/J, and C57BL/6 are all susceptible ($LD_{50} < 2.8$); DBA/1, C3H/HeN, and CBA/J are intermediate ($LD_{50} = 3.32$–3.85); and DBA/2 is resistant ($LD_{50} = 5.25$). The difference between DBA/2 and both BALB/c and C57BL/6 behaves as though controlled by a single locus designated *Cms*, with the allele Cms^r in DBA/2 and Cms^s in BALB/c and C57BL/6. Heterozygotes are resistant. It is not known whether intermediate susceptibility is due to a third allele of *Cms* or to modifying genes. Males are more susceptible than females, but *Cms* is not X-linked. Resistance is radiosensitive, indicating that it depends on a dividing cell. *Cms* appears to be active in spleen cells, since resistance can be transferred to irradiated susceptible hosts by injection of spleen cells from resistant donors (1).

References

1. Kirkland, T.N., and J. Fierer. 1985. Genetic control of resistance to *Coccidioides immitis*: a single gene that is expressed in spleen cells determines resistance. J. Immunol. 135:548–552.

cn, achondroplasia, recessive, Chr 4

Arose in strain AKR/J in 1960. Homozygotes are recognizable at birth by a short domed skull and short thick tail. As adults they are square rumped with short head, body, limbs, and tail, bulging abdomen, and often with malocclusion. Many die before weaning from malocclusion, and others die by 3 months from cyanosis (6). Pituitary growth hormone is normal in amount. Growth rate is lower than normal during the first 3 weeks, but equal to that of normal littermates from the fourth week on (4). The epiphyseal plates are reduced in thickness and have fewer chondrocytes per capsule (5). In the skull, the endochondral bones of the cranial base are markedly reduced in length, but not in width, and the cartilage of the condylar process of the mandible is reduced to a lesser extent (2). The cells and matrix of cartilage of the mutant long bones show no striking histological abnormalities (7) except larger than normal accumulation of glycogen in chondrocytes and failure of many hypertrophic chondrocytes to degenerate (1), and changes which have been interpreted as increased vulnerability of chondrocytes to regressive changes (8). Thurston *et al*. (9) found evidence in their stock for two non-overlapping classes of *cn/cn* mice distinguishable at 22 days, one with more severe disruption of the proximal tibial growth plate than the other. The cause of the difference is not known. The amount and composition of proteoglycans and collagen are unaffected by *cn* (3).

References

1. Bonucci, E., G. Gherardi, A. Del Marco, B. Nicoletti, and P. Petrinelli. 1977. An electron microscope investigation of cartilage and bone in achondroplastic *cn/cn* mice. J. Submicr. Cytol. 9:299–306.

2. Brewer, A.K., D.R. Johnson, and W.J. Moore. 1977. Further studies on skull growth in achondroplastic (*cn*) mice. J. Embryol. Exp. Morphol. 39:59–70.

3. Kleinman, H.K., J.P. Pennypacker, and K.S. Brown. 1977. Proteoglycan and collagen of "achondroplastic" (*cn/cn*) neonatal mouse cartilage. Growth 41:171–177.

4. Konyukhov, B.V., and Y.V. Paschin. 1967. Experimental study of the achondroplasia gene effects in the mouse. Acta Biol. Acad. Sci. Hungar. 18:285–294.

5. Konyukhov, B.V., and Y.V. Paschin. 1970. Abnormal growth of the body, internal organs, and skeleton in the achondroplastic mice. Acta Biol. Acad. Sci. Hungar. 21:347–354.

6. Lane, P.W., and M.M. Dickie. 1968. Three recessive mutations producing disproportionate dwarfing in mice: achondroplasia, brachymorphic, and stubby. J. Hered. 59:300–308.

7. Miller, W.A., and K.L. Flynn-Miller. 1976. Achondroplastic, brachymorphic and stubby chondrodystrophies in mice. J. Comp. Pathol. 86:349–363.

8. Silberberg, R., M. Hasler, and P. Lesker. 1976. Ultrastructure of articular cartilage of achondroplastic mice. Acta Anat. 96:162–175.

9. Thurston, M.N., D.R. Johnson, and N.F. Kember. 1985. Cell kinetics of growth cartilage of achondroplastic (*cn*) mice. J. Anat. 140:425–434.

co, cocked, recessive, Chr 11

Arose spontaneously in a line being inbred for *oe*. Homozygotes have a gross deficiency of the utricular otolith, consisting either of one or more distinctively large crystals or of total absence of utricular otolith (1,2).

References

1. Peterson, A.C. 1970. The genetics of cocked, a new behavioural mutant in the house mouse. Can. J. Genet. Cytol. 12:391–392 (Abstr.).
2. Peterson, A., and F. Biddle. 1970. Mouse News Lett. 43:19.

coa, cocoa, recessive, Chr 3

Arose spontaneously in the C57BL/6J strain. Cocoa resembles ruby (*ru*), and *coa*/*coa* mice cannot be distinguished from either *ru*/*ru* or *coa*/*coa ru*/*ru* mice. *coa* is on Chr 3 close to *Car-2* near the centromere (1).

References

1. Sweet, H.O., and M. Prochaska. 1985. Mouse News Lett. 73:18.

cod, cerebellar outflow degeneration, recessive

Arose in a stock at Iowa State University. Manifestation in homozygotes is variable on some genetic backgrounds but is more uniform after crossing to the BALB/c inbred strain. Affected mice show loss of balance beginning at 12 days of age. By 45 days, lethargy, ataxia, nodding of the head and trunk, malpositioning of hindlegs, and poor swimming are the principal features, and these remain unchanged throughout life. Neither sex is fertile. At about 11 days, there is chromatolysis in vestibular and red nuclei accompanied by degenerative changes in their axons, and followed closely by neuron degeneration in deep cerebellar nuclei and brainstem reticular formation. Many additional nuclear groups and axonal systems become involved later (1, 2).

References

1. Sidman, R.L. 1967. Mouse News Lett. 36:33.
2. Sidman, R.L. 1980. Mouse News Lett. 63:13.

cog, congenital goiter, recessive, Chr 15

Arose spontaneously in the AKR/J strain. Homozygotes are recognizable by their small size as early as 15 days of age. Later, their growth rate increases so that it becomes difficult to distinguish them from their normal sibs. They have enlarged thyroid glands. Thyroid hormone therapy reduces the size of the thyroid and repairs the growth deficiency. In untreated animals at 7 to 8 weeks, the thyroids show complete absence of colloid, obliteration of follicular lumens, hypertrophy and hyperplasia of follicular cells, and some degenerating follicles. Serum thyroxine is significantly reduced, serum thyroid stimulating hormone is markedly elevated, and thyroglobulin (TG) is markedly reduced in the thyroid gland and slightly elevated in serum. These findings suggest that *cog* may cause defective synthesis or processing of TG. *cog* was found to be on Chr 15 by close linkage with *Eh*, a mutation that suppresses recombination in the distal half of Chr 15 (1), and later found to be located close to *Gpt-1* near the middle of that chromosome. It also shows no recombination with *Tgn*, the TG structural locus. This suggests that *cog* is probably a mutation in the *Tgn* gene. No difference between *cog*/*cog* and +/+ DNA in Southern blots or in size and abundance of TG mRNA was found. The thyroids of mutants may synthesize an abnormal TG polypeptide that is released directly into the circulation (2). See *Tgn* locus.

References

1. Beamer, W.G., L.J. Malthais, M.H. DeBaets, and E.M. Eicher. 1987. Inherited congenital goiter in mice. Endocrinology 120: 838–840.
2. Taylor, B.A., and L. Rowe. 1987. The congenital goiter mutation is linked to the thyroglobulin gene in the mouse. Proc. Natl. Acad. Sci. USA 84:1986–1990.

Coh locus, coumarin hydroxylase, Chr 7

This locus controls basal and phenobarbital-induced levels of coumarin hydroxylase in microsomes of liver. The allele *Coh^h* determines high activity and occurs in strains DBA/2J, C57BR/cd, and others; the allele *Coh^l* determines low activity and occurs in strains AKR/J, C57BL/6J, C3H/HeJ, and others. Heterozygotes are intermediate (6). The same strain difference exists for activity of the enzyme in lung and kidney, although these tissues have less than 1 per cent of the enzyme activity in liver (5). The enzyme from high activity strains is preferentially inhibited by aniline; the enzyme from low activity strains is preferentially inhibited by metyrapone (6). The difference between high- and low-activity strains is also detected as a difference in rate of metabolism of coumarin to umbelliferone (2). DBA/2J mice are less sensitive to the lethal anticoagulant effect of coumarin derivatives, possibly because of the rapid

Coh

hydroxylation in this strain (1,5). Coumarin hydroxylase is believed to be a cytochrome P-450. A cDNA probe for phenobarbital-induced P-450 revealed several polymorphisms detected with three different restriction enzymes between DBA/2 and C57BL/6, and these were all shown in BXD recombinant inbred strains to cosegregate completely with *Coh*. It is likely that *Coh* is the site of a cluster of P-450 genes (3, 4).

References

1. Endell, W., and G. Seidel. 1978. Coumarin toxicity in different strains of mice. Agents and Actions 8:299–302.
2. Lush, I.E., and K.M. Andrews. 1978. Genetic variation between mice in their metabolism of coumarin and its derivatives. Genet. Res. 31:177–186.
3. Simmons, D.L., and C.B. Kasper. 1983. Genetic polymorphisms for a phenobarbital-inducible cytochrome P-450 map to the *Coh* locus in mice. J. Biol. Chem. 258:9585–9588.
4. Simmons, D.L., P.A. Lalley, and C.B. Kasper. 1985. Chromosomal assignments of genes coding for components of the mixed-function oxidase system in mice. Genetic localization of the cytochrome P-450PCN and P-450PB gene families and the NADPH-cytochrome P-450 oxidoreductase and epoxide hydratase genes. J. Biol. Chem. 260:515–521.
5. Wood, A.W., and A.H. Conney. 1974. Genetic variation in coumarin hydroxylase activity in the mouse (Mus musculus). Science 185:612–614.
6. Wood, A.W., and B.A. Taylor. 1979. Genetic regulation of coumarin hydroxylase activity in mice: evidence for single locus control on chromosome 7. J. Biol. Chem. 254:5647–5651.

Cola-1, *Cola-2* loci

Cola-1 locus, procollagen type 1 α-1, Chr 11
Cola-2 locus, procollagen type 1 α-2, Chr 16
These loci code for the $\alpha 1$ and $\alpha 2$ chains of type I procollagen, the precursor molecules for the type I collagen $\alpha 1$ and $\alpha 2$ chains. The *Mov-13* locus is a Moloney leukemia virus integration into the *Cola-1* gene which produced a lethal mutation in this gene. By use of a unique mouse DNA sequence flanking the *Mov-13* virus, *Cola-1* was mapped to Chr 11 using mouse–Chinese hamster and mouse–rat somatic cell hybrids and by *in situ* hybridization (1). The type I procollagen genes were originally assigned to Chr 16 by use of mouse–Chinese hamster somatic cell hybrids in which the mouse collagens were detected by ELISA and 'Western blotting' immunochemical analyses (2). In view of the results of Münke *et al.* (1), it is likely that only *Cola-2* is on Chr 16.

References

1. Münke, M., K. Harbers, R. Jaenisch, D.R. Cox, and U. Francke. 1985. Assignment of the alpha 1(I) collagen gene to mouse chromosome 11. Cytogenet. Cell Genet. 40:706 (Abstr.).
2. Shupp Byrne, D.E., and R.L. Church. 1983. Assignment of the genes for mouse type I procollagen to chromosome 16 using mouse fibroblast–Chinese hamster somatic cell hybrids. Somat. Cell Genet. 9:313–331.

Cp locus, ceruloplasmin, Chr 9

This locus codes for ceruloplasmin, a glycoprotein of plasma that plays a leading role in processes of transport, intracellular metabolism, and excretion of copper. By *in situ* hybridization of rat ceruloplasmin DNA probes with bone marrow cells carrying various Robertsonian translocations, *Cp* was shown to be on Chr 9, band D, near the center of the chromosome (1).

References

1. Baranov, D.S., M. Shvartsman, V.N. Gorbunova, V.S. Gaitskhoki, and S.A. Neifakh. 1985. Mapping of the ceruloplasmin gene on the chromosomes of humans and laboratory mice by the method of direct hybridization in situ. Soviet Genet. 21:315–324.

Cpa locus, carboxypeptidase A, Chr 6

This locus codes for the exopeptidase, carboxypeptidase A (E.C. 3.4.17.1). No genetic variants are known among standard laboratory inbred strains. The locus was first found to be on Chr 6 using mouse–Chinese hamster somatic cell hybrids and a cDNA probe isolated from a rat cDNA library (2). A restriction fragment length difference between laboratory strains and *Mus spretus* detected in *Eco*RI digested DNA was later found and used to map *Cpa* to about 5 cM proximal to *Tcrb* on Chr 6 (1).

References

1. Bućan, M., T. Yang-Feng, A.M. Colberg-Poley, D.J. Wolgemuth, J.-L. Guénet, U. Francke, and H. Lehrach. 1986. Genetic and cytogenetic localisation of the homeo box containing genes on mouse chromosome 6 and human chromosome 7. EMBO J. 5:2899–2905.
2. Honey, N.K., A.Y. Sakaguchi, P.A. Lalley, C. Quinto, R.J. MacDonald, W.J. Rutter, and S.L. Naylor. 1984. Assignment in mouse of the genes for chymotrypsinogen B, elastase, trypsin, and carboxypeptidase A. Cytogenet. Cell Genet. 37:492–493 (Abstr.).

cph (formerly *jpk*), congenital progressive hydronephrosis, recessive, Chr 15

Arose spontaneously in the C57BL/6J strain in 1973. Homozygotes can be recognized by 2 weeks of age by abdominal swelling accompanied by small size, alopecia, and lethargic behavior. By 3 weeks, both kidneys are massively hydropic; ureters and bladder appear normal. Most homozygotes die by 1 month; a few survive to adulthood and breed. At 1 day of age, the kidneys appear grossly normal but the papilla is slightly smaller than that of normal littermates. The calyx becomes progressively distended with fluid and the papilla, medulla, and cortex erode. By 20 days, little kidney tissue remains within the distended kidney capsule. The disease appears to be due to obstruction at the ureteropelvic junction. *cph* is on Chr 15 about 5 cM from *Ca* (1).

References

1. Horton, C.E. Jr., M.T. Davisson, J.B. Jacobs, G.T. Bernstein, A.B. Retik, and J. Mandell. 1988. Congenital progressive hydronephrosis in mice: a new recessive mutation. J. Urol. 140:1310–1315.

cpk (formerly *ck*), congenital polycystic kidneys, recessive

Arose in the C57BL/6J-*pa* strain. Homozygotes have bilateral polycystic kidneys which are detectable in living animals at 10 to 13 days by the enlarged abdomen. Death occurs at about 3 weeks. Histological examination shows that the cysts occur first in the proximal convoluted tubules and at later stages, probably secondarily, in the collecting tubules. No casts or obstructing lesions have been seen. Mild cystic changes were observed as early as the 17th day of gestation (1, 2). A similar mutation occurred in the PM/Se strain at the Division of Cancer Research, Perugia. Polycystic kidneys are present at birth, and death occurs by 5 weeks (3). The two mutations have not been tested for allelism with each other.

References

1. Mandell, J., W.K. Koch, R. Nidess, G.M. Preminger, and E. McFarland. 1983. Congenital polycystic kidney disease: genetically transmitted infantile polycystic kidney disease in C57BL/6J mice. Am. J. Pathol. 113:112–114.
2. Preminger, G.M., W.E. Koch, F.A. Fried. E. McFarland, E.D. Murphy, and J. Mandell. 1982. Murine congenital polycystic kidney disease: a model for studying development of cystic disease. J. Urol. 127:556–560.
3. Ribacchi, R. 1975. Rene policistico congenito in topi PM/Se. Lav. Anat. Pat. Perugia 35:81–88.

Cpz locus, chlorpromazine avoidance

This is one of two postulated loci thought to control a difference in impairment by chlorpromazine of performance of mice pre-trained in an active avoidance task. Strain BALB/cBy mice have the allele for high effect of chlorpromazine and strain C57BL/6By the allele for low effect (1). *Cpz* is not on Chr 9 as originally reported, since *H(w106)*, to which it was found to be linked, is now known not to be on that chromosome.

References

1. Castellano, C., B.E. Eleftheriou, D.W. Bailey, and A. Oliverio. 1974. Chlorpromazine and avoidance: a genetic analysis. Psychopharmacologia 34:309–316.

cr, crinkled, recessive, Chr 13

Appeared in the progeny of a male treated with nitrogen mustard. It is an almost exact mimic of *dl* and of *Ta* in the homozygous condition. Homozygotes have coats that are thinner than normal and contain only one type of hair resembling an abnormal awl, rather than the four types of hair of the normal coat. There is a bald patch behind the ears, bald tail, kinks at the tail tip, absence of Meibomian glands, a respiratory disorder, modification of the agouti pattern (1), abnormal molars and incisors (2, 3, 8), and probably abnormalities of many exocrine glands (4). Viability and breeding performance are somewhat reduced (1). Myelin abnormalities have been described in the brains of old crinkled mice (9). Hair follicle initiation is restricted to the period between 17 days of gestation and birth, rather than extending from 14 days of gestation to several days after birth as in normal mice (1). Heterozygotes have normal coats, but many have slight abnormalities of the upper molars (3). The site of gene action in skin of *cr/cr* mice has been shown, by means of dermal–epidermal recombination grafts of embryonic skin, to be in the epidermis (7). The level of copper in liver is low in young *cr/cr* mice and the level of activity in liver of the copper-containing enzyme superoxide dismutase (SOD) is also low in both young and adult mice (6). Copper supplementation of the diet of mothers during pregnancy and lactation increases survival of homozygotes, improves coat development, and raises SOD activity and liver copper concentration to normal (5, 6).

References

1. Falconer, D.S., A.S. Fraser, and J.W.B. King. 1951. The genetics and development of "crinkled", a new mutant in the mouse. J. Genet. 50:324–344.
2. Grüneberg, H. 1965. Genes and genotypes affecting the

teeth of the mouse. J. Embryol. Exp. Morphol. 14:137–159.

3. Grüneberg, H. 1966. The molars of the tabby mouse, and a test of the single-active X-chromosome hypothesis. J. Embryol. Exp. Morphol. 15:223–244.
4. Grüneberg, H. 1971. The glandular aspects of the tabby syndrome in the mouse. J. Embryol. Exp. Morphol. 25:1–19.
5. Hurley, L.S., and I.T. Bell. 1975. Amelioration by copper supplementation of mutant gene effects in the crinkled mouse. Proc. Soc. Exp. Biol. Med. 149:830–834.
6. Keen, C.L., and L.S. Hurley. 1979. Superoxide dismutase activity in the crinkled mutant mouse: ameliorative effects of dietary copper supplementation. Proc. Soc. Exp. Biol. Med. 162:152–156.
7. Mayer, T.C., C.K. Miller, and M.C. Green. 1977. Site of action of the crinkled (*cr*) locus in the mouse. Dev. Biol. 55:397–401.
8. Sofaer, J.A. 1969. Aspects of the tabby-crinkled-downless syndrome. II. Observations on the reaction to changes in genetic background. J. Embryol. Exp. Morphol. 22:207–227.
9. Theriault, L.L., D.D. Dugan, S. Simons, C.L. Keen, and L.S. Hurley. 1977. Lipid and myelin abnormalities of brain in the crinkled mouse. Proc. Soc. Exp. Biol. Med. 155:549–553.

Crbp complex, Chr 9

This complex contains genes that code for two highly homologous cellular retinol-binding proteins, CRBP and CRBP II. The proteins have been described in rats but not in mice. CRBP II is found primarily in the enterocytes of the rat small intestine and may play an important role in the uptake of the vitamin A precursor retinol from the intestinal lumen. CRBP is present in a wide variety of adult rat tissues but is most abundant in liver and kidney. The two proteins belong to a family of low molecular weight proteins with at least eight known members that include fatty acid and retinoic acid binding proteins and have diverse tissue distributions (1).

Crbp, cellular retinol-binding protein (CRBP). Two alleles distinguished by a *Kpn*I restriction fragment length difference detected with a rat CRBP cDNA probe are known at this locus, *Crbp*^{b} in strains C57BL/6 and C57L, and *Crbp*^{d} in strains DBA/2 and AKR. In 17 BXD and 13 AKXL recombinant inbred strains, no recombinants between *Crbp* and *Crbp-2* were found (1).

Crbp-2, cellular retinol-binding protein-2 (CRBP II). Three alleles distinguished by *Taq*I restriction fragment length differences detected with a rat CRBP II cDNA probe are known, *Crbp-2*^{b} in strains C57BL/6 and C57L, *Crbp-2*^{d} in strain DBA/2, and *Crbp-2*^{a} in

strain AKR. Segregation of these alleles in 25 BXD and 18 AKXL recombinant inbred strains showed that *Crbp-2* is located on Chr 9 near *Pgm-3* (1).

References

1. Demmer, L.A., E.H. Birkenmeier, D.A. Sweetser, M.S. Levin, S. Zollman, R.S. Sparkes, T. Mohandas, A.J. Lusis, and J.I. Gordon. 1987. The cellular retinol binding protein II gene: sequence analysis of the rat gene, chromosomal location in mice and humans, and documentation of its close linkage to the cellular retinol binding protein gene. J. Biol. Chem. 262:2458–2467.

crh, cryptothrix, recessive

Arose as a spontaneous mutation in the non-inbred CD-1 line at Charles River Breeding Laboratories. Homozygotes do not grow a normal first coat. There are numerous short fuzzy hairs which persist throughout life but are less frequent in older mice. The coat apparently has a full complement of hair follicles, but most of the hairs disintegrate at the level of the sebaceous glands and fail to erupt (1,2).

References

1. Mann, S.J. 1971. Varieties of hairless-like mutant mice. J. Invest. Dermatol. 56:170–173.
2. Mann, S.J. 1972. Mouse News Lett. 47:64.

cri, cribriform degeneration, recessive, Chr 4

Arose spontaneously in the DBA/2J inbred strain. Homozygotes have abnormal behavior, anemia, and an electrolyte imbalance. Many die before weaning and most are dead before 3 months. None have bred. They are small, weak, and show ataxic behavior beginning at 2.5 to 3 weeks of age. Light and electron microscopy reveal severe vacuolar degeneration in white and gray matter of the spinal cord and brainstem and in the inner nuclear layer of the retina. The vacuoles are almost exclusively intracellular and occur in cell bodies, axons, and myelin sheaths. The anemia is often visible as pallor at birth. It is due to a deficiency of normocytic red blood cells, which becomes less severe with age, but does not disappear (1). The electrolyte imbalance of homozygotes is manifested by their preference for physiological saline over water, decreased concentrations of potassium in plasma and of sodium and potassium in urine, increased concentration of sodium in feces and hair and of chloride in sweat (3), and marked elevation of sodium concentration in the secretion of the parotid gland (2). DBA/2-*cri/cri* mice have more neutrophils recoverable from lung washings than normal controls, and a greater increase in neutrophils after exposure to an aerosol of *Staphlococcus aureus* (4). Homozygotes on the DBA/2 genetic background develop clonic seizures like their normal sibs after exposure to loud noise, but, in contrast to their normal sibs, do not develop tonic seizures or die. Presumably they are protected by their severe neurological defect (5).

References

1. Green, M.C., R.L. Sidman, and O.H. Pivetta. 1972. Cribriform degeneration (cri): a new recessive neurological mutation in the mouse. Science 176:800–803.
2. Kaiser, D., O. Pivetta, and O.M. Rennert. 1974. Autosomal recessively inherited electrolytic defect in the parotid of the 'cribriform degeneration' mouse mutant—possible analogy to cystic fibrosis. Life Sci. 15:803–810.
3. Pivetta, O.H., and M.C. Green. 1977. Apetencia salina y anormalidades electroliticas en ratones con degeneracion cribriforme. Medicina (Buenos Aires) 37:379–384.
4. Pivetta, O.H., R.J.J. Cassino, and D.O. Sordelli. 1980. Pulmonary cell response to bacterial challenge in mutant mice with some hereditary alterations resembling cystic fibrosis. Life Sci. 26:1349–1357.
5. Seyfried, T.N., G.H. Glaser, and R.K. Yu. 1979. Influence of the cribriform degeneration (*cri*) mutation on audiogenic seizures in DBA/2J mice. Exp. Neurol. 63:643–646.

Crip locus, cysteine-rich intestinal protein, Chr 12

This locus codes for a cysteine-rich protein (CRIP) in the intestine. The cDNA coding for CRIP was identified among cDNAs transcribed from mRNAs from mouse and rat intestine whose concentration changed during transition from suckling to weaning age. In the rat, CRIP mRNA is produced in the small and large intestines and at lower levels in lung, spleen, adrenal, and testis. In the rat intestine, it is present at low levels at birth, rises dramatically at the mid-suckling stage to a peak at weaning, and decreases to adult levels thereafter. CRIP is highly conserved, being found in most vertebrates, as well as in the sea squirt. CRIP is homologous to certain ferrodoxins. In genomic DNA screened with a rat cDNA probe for *Crip*, strain C57BL/6J was found to differ from other strains in *Kpn*I and *Taq*I restriction fragment lengths, and AKR to differ from others at a *Taq*I fragment. Among 62 recombinant inbred strains involving these strains as progenitors (25 BXD, 12 BXH, 7 CXB, and 18 AKXL), there were no recombinants between *Crip* and the immunoglobulin heavy chain locus, *Igh-C*, on Chr 12 (1).

References

1. Birkenmeier, E.H., and J.L. Gordon. 1986. Developmental regulation of a gene that encodes a cysteine-rich intestinal protein and maps near the murine immunoglobulin heavy chain locus. Proc. Natl. Acad. Sci. USA 83:2516–2520.

Crl-1 locus, complement receptor lymphocytes-1, Chr 17

This locus influences time of appearance of receptors for the C3 component of complement on B lymphocytes. The allele in strain AKR/J determines a high frequency at 2 weeks of age of complement receptor lymphocytes; the allele in DBA/2J determines a low frequency. Analysis of *H-2* recombinant haplotypes indicates that *Crl-1* is between the *Ss-Slp* and *Tla* loci. At least one other non-*H-2* linked gene influences level of C3 receptors (1).

References

1. Gelfand, M.C., D.H. Sachs, R. Lieberman, and W.E. Paul. 1974. Ontogeny of B lymphocytes. III. *H-2* linkage of a gene controlling the rate of appearance of complement receptor lymphocytes. J. Exp. Med. 139:1142–1153.

Crm, cream, semidominant, Chr X

Found in an outbred albino TO colony (3). Hemizygous males and homozygous females are pale yellow. This color is expressed even in the unpigmented areas of mice with white spotting and in albino mice. The color thus does not seem to be connected with pigment cell activity. *Crm* causes fluorescence of the coat in long-wave UV light, which produces a clear mosaic pattern of fluorescent and non-fluorescent areas in the coat of *Crm/+* females (4), thus confirming the observation of pale yellow patches in these mice (3). *Crm* is near *Gy* close to the distal end of Chr X (1) and, in female heterozygotes, shows about 10 per cent recombination with *Sts*, a locus in the pairing region of the X and Y chromosomes (2).

References

1. Beechey, C.V., and A.G. Searle. 1979. Mouse News Lett. 60:47.
2. Cattanach, B.M., and A.J.M. Crocker. 1986. X chromosomal location of Sts. Mouse News Lett. 74:94–95.
3. Hetherington, C.M. 1977. Mouse News Lett. 56:35.

4. Peters, J., and A.G. Searle. 1979. Mouse News Lett. 61:37.

crn, cranioschisis, recessive

Arose in a strain carrying *oel* (open eyelids with cleft palate). Homozygotes die shortly after birth. They have anencephalus varying from a small skin and frontal cranium defect to absence of the calvarium and associated overgrowth of neural tissue. In litters segregating for both *crn* and *oel*, segregation of the two genes is compatible with the hypothesis that they are alleles at the same locus (1). The defect caused by *crn* appears to reside in the supporting mesenchyme around the neural tube. The mesenchyme loses its normal stellate cellular interconnections, causing the cells to round up and perhaps involving detachment from the matrix. This condition is not associated with cell death and persists only 24 hours (2).

References

1. Brown, K.S., and L.C. Horne. 1973. Recessive anencephalus in the oel strain of mice. Genetics 74:s31–s32 (Abstr.).
2. Brown, K.S., S.C. Hetzel, and L.C. Horne. 1984. Craniofacial consequences of connective tissue disorders in mice. Birth Defects: Orig. Art. Ser. 20:113–136.

Crya-1 locus (formerly *Acry-1*), α-crystallin-1, Chr 17

This locus codes for αA-crystallin, one of the two polypeptide chains of α-crystallin. Variation was recognized by restriction fragment length polymorphism using a number of different restriction endonucleases. Six alleles were recognized: *Crya-1^a* in 11 strains including BALB/c, DBA/2, MA/My, and NZB; *Crya-1^b* in 11 strains including A/J, AKR, CBA/J, C3H/He, and C57BL/6; *Crya-1^c* in PL/J; *Crya-1^d* in MOLD/Rk; *Crya-1^e* in CASO/Rk, and *Crya-1^f* in SJL (1). A seventh allele, *Crya-1^t*, was found in nine of 17 *t* haplotypes examined (2). *Crya-1* is on Chr 17 between *Glo-1* and *H-2K*, within the distal inverted region of *t* haplotypes (1, 2).

References

1. Skow, L.C., and M.A. Donner. 1985. The locus encoding αA-crystallin is closely linked to *H-2K* on mouse chromosome 17. Genetics 110:723–732.
2. Skow, L.C., J.N. Nadeau, J.C. Ahn, H.-S. Shin, K. Artzt, and D. Bennett. 1987. Polymorphism and linkage of the αA-crystallin gene in *t*-haplotypes of the mouse. Genetics 116:107–111.

crz, susceptibility to *Trypanosoma cruzi*, recessive

Arose spontaneously in the BXH-2 (C57BL/6 × C3H/HeJ) recombinant inbred strain. Homozygotes are unable to control early proliferation of the Brazil strain of *T. cruzi*, as determined by parasite number in the blood 16 days after intraperitoneal infection. Both progenitor strains have very little proliferation within that period and F1 offspring of BXH-2 with both progenitor strains also have low proliferation. *crz* is distinct from *lpr*, a gene that also causes susceptibility to *T. cruzi* (1).

References

1. Trischmann, T.M. 1984. Single locus in BXH-2 mice responsible for inability to control early proliferation of *Trypanosoma cruzi*. Infect. Immun. 46:658–662.

Cs locus (formerly *Cis*; the symbol *Cs* was formerly used for the *Cts* locus), citrate synthase, Chr 10

This is the structural locus for mitochondrial citrate synthase (E.C. 4.1.3.7). No genetic variants are known. The locus was found to be on Chr 10 using mouse–Chinese hamster somatic cell hybrids. The synteny of *Cs* and *Pep-2* on Chr 10 is similar to the synteny of the homologous loci on human chromosome 12 (1).

References

1. Cox, D.R., D. Goldblatt, and C.J. Epstein. 1981. Comparative gene mapping in man and mouse: assignment of the gene for citrate synthase (Cis) to mouse chromosome 10. Cytogenet. Cell Genet. 32:259 (Abstr.).

Cs locus

See *Cas-1* locus, *Cts*.

Cs-1 locus

See *Cas-1* locus.

Csfgm locus, colony-stimulating factor, granulocyte-macrophage-specific, Chr 11

This locus codes for a colony-stimulating factor, GM-CSF, that acts on the granulocyte–macrophage lineage. It is produced by multiple cell types in many different organs. It has a molecular weight of 23 000, is active in low concentration, is required continuously for the differentiation and proliferation of granulocytes and macrophages, and can regulate the functional activity of mature end cells. *Csfgm* is a single gene and has no structural similarity to the gene for the functionally related interleukin-3 (colony-stimulating factor-multi, *Csfmu*). Like *Csfmu*, *Csfgm* is on Chr 11 (1, 2).

References

1. Gough, N.M., J. Gough, D. Metcalf, A. Kelso, D. Grail, N.A. Nicola, A.W. Burgess, and A.R. Dunn. 1984. Molecular cloning of cDNA encoding a murine haematopoietic growth regulator, granulocyte–macrophage colony stimulating factor. Nature 309:763–767.
2. Metcalf, D. 1985. The granulocyte–macrophage colony stimulating factors. Cell 43:5–6.

Csfmu locus, colony-stimulating factor-multi, Chr 11

This locus codes for a lymphokine, also called interleukin-3 (IL-3), that is produced by the monocytic cell line WEHI-3 and by lectin-activated T-cells. It acts as a colony-stimulating factor for diverse kinds of cells of hematopoietic lineage (1, 3). *Csfmu* is a single gene composed of five exons and four introns (4). A restriction fragment length polymorphism was found by hybridization of genomic DNA with a cDNA clone for IL-3. Strains BALB/c and C57BL/6 have an *Eco*RI 8.5-kb fragment and strains NFS and A/J have an *Eco*RI 10.8-kb fragment. By use of mouse–Chinese hamster somatic cell hybrids and genetic crosses between NFS and C57BL/6 mice, *Csfmu* was found to be located on Chr 11 (2).

References

1. Ihle, J.N. 1985. Biochemical and biological properties of interleukin-3, a lymphokine mediating the differentiation of a lineage of cells that includes prothymocytes and mast-like cells. Contemp. Top. Mol. Immunol. 10:93–119.
2. Ihle, J.N., J. Silver, and C.A. Kozak. 1987. Genetic mapping of the mouse interleukin 3 gene to chromosome 11. J. Immunol. 138:3051–3054.
3. Metcalf, D. 1985. The granulocyte–macrophage colony stimulating factors. Cell 43:5–6.
4. Miyatake, S., T. Yokota, F. Lee, and K. Arai. 1985. Structure of the chromosomal gene for interleukin 3. Proc. Natl. Acad. Sci. USA 82:316–320.

Csn complex, casein gene family, Chr 5

This complex contains the structural genes for the hormone-inducible α-, β-, and γ-caseins of mouse milk. No genetic variants are known. The genes were found to be on Chr 5 using mouse–Chinese hamster somatic cell hybrids that were classified for presence or absence of the casein genes with cDNA clones encoding the three caseins (1).

References

1. Gupta, P., J.M. Rosen, P. D'Eustachio, and F.H. Ruddle. 1982. Localization of the casein gene family to a single mouse chromosome. J. Cell Biol. 93:199–204.

ct, curly-tail, recessive

Arose spontaneously in the GFF inbred strain. Genetic data indicate that *ct* is probably a recessive with incomplete penetrance, but do not absolutely exclude the possibility that it is a dominant with reduced penetrance. Presumed homozygotes, if abnormal, usually have some degree of spina bifida and, occasionally, exencephaly. If the mice survive, the spina bifida usually results in a curly or kinky tail (4). Among the exencephalic embryos, there is a great preponderance of females (3). The posterior defects may be due to delay in closure of the posterior neuropore, which occurs in about 50 per cent of homozygotes *in vivo* and can also be demonstrated in embryos cultivated *in vitro* from the headfold stage (2). The level of α-fetoprotein is raised in the amniotic fluid of *ct/ct* fetuses with exencephaly or open spina bifida (1). Administration of vitamin A to mothers on day 8 of pregnancy increases the frequency of neural tube defects in *ct/ct* offspring, but vitamin A administration on day 9 causes a marked decrease (5). Administration of vitamin A on day 8 also causes neural tube defects in a proportion of F1 hybrids between *ct/ct* and strain A mice. More embryos are affected when the mother is *ct/ct* than when she is strain A (6).

References

1. Adinolfi, M., S. Beck, S. Embury, P. Polani, and M.J. Seller. 1976. Levels of α-fetoprotein in amniotic fluids of mice (curly-tail) with neural tube defects. J. Med. Genet. 13:511–513.
2. Copp, A.J., M.J. Seller, and P.E. Polani. 1982. Neural tube development in mutant (curly tail) and normal mouse embryos: the timing of posterior neuropore closure *in vivo* and *in vitro*. J. Embryol. Exp. Morphol. 69:151–167.
3. Embury, S., M.J. Seller, M. Adinolfi, and P.E. Polani. 1979. Neural tube defects in curly-tail mice. I. Incidence, expression and similarity to the human condition. Proc. R. Soc. Lond. (Biol.) 206:85–94.
4. Grüneberg, H. 1954. Genetical studies on the skeleton of the mouse. VIII. Curly-tail. J. Genet. 52:52–67.
5. Seller, M.J., S. Embury, P.E. Polani, and M. Adinolfi. 1979. Neural tube defects in curly-tail mice. II. Effect of maternal administration of vitamin A. Proc. R. Soc. Lond. (Biol.) 206:95–107.
6. Seller, M.J., K.J. Perkins, and M. Adinolfi. 1983. Differential response of heterozygous curly-tail mouse embryos

to vitamin A teratogenesis depending on maternal geno-type. Teratology 28:123–129.

Ctl-1 locus, cytotoxic T lymphocyte response-1, Chr 1

This is one of two loci controlling primary *in vitro* cyto-toxic T lymphocyte (CTL) response to *H-2*-compatible cells. The allele determining responsiveness occurs in the NZB ($H-2^d$) strain. It is detected by the ability of NZB spleen cells to generate a strong CTL response to BALB/c ($H-2^d$) cells. The allele determining non-responsiveness occurs in the B10.D2 ($H-2^d$) strain. These mice do not produce a CTL response to BALB/c cells. (NZB × B10.D2)F1 mice have an intermediate response. The response is not dependent on differences at the *Qa-1* locus (1). *Ctl-1* is located on Chr 1 near *Ly-9* (2, 3).

References

1. Davidson, W.F., T.M. Chused, and H.C. Morse III. 1981. Genetic and functional analysis of the primary in vitro CTL:response of NZB lymphocytes to *H-2*-compat-ible cells. Immunogenetics 12:445–463.
2. Davidson, W.F., B.J. Mathieson, C.A. Kozak, T.M. Chused, and H.C. Morse III. 1982. Chromosome 1 locus required for induction of CTL to *H-2*-compatible cells in NZB mice. Immunogenetics 15:321–325.
3. Kozak, C.A., W.F. Davidson, and H.C. Morse III. 1984. Genetic and functional relationships of the retroviral and lymphocyte alloantigen loci on mouse chromosome 1. Immunogenetics 19:163–168.

Ctla-1 locus, cytotoxic T-lymphocyte-associated protein-1, Chr 14

This locus was recognized as a cDNA clone of a gene expressed in cytotoxic T lymphocytes but not (or less so) in brain cells or in a range of noncytotoxic lymphoid cells. The amino acid sequence deduced from the cDNA sequence shows significant homology, but not identity, with known serine esterases. *Ctla-1* was found to be located in the D segment of Chr 14 by *in situ* hybridization (1).

References

1. Brunet, J.-F., M. Dosseto, F. Denizot, M.-G. Mattei, W.R. Clark, T.M. Haqqi, P. Ferrier, M. Nabholz, A.-M. Schmitt-Verholst, M.-F. Luciani, and P. Golstein. 1986. The inducible cytotoxic T-lymphocyte-associated gene transcript CTLA-1 sequence and gene localization to mouse chromosome 14. Nature 322:268–271.

Ctrb locus, chymotrypsin B, Chr 8

This locus codes for the endopeptidase chymotrypsin B (E.C. 3.4.4.6). It was found to be on Chr 8 using mouse–Chinese hamster somatic cell hybrids and a cDNA probe isolated from a rat cDNA library (1). The locus *Prt-2*, pancreatic proteinase-2, which controls electrophoretic mobility of chymotrypsin is also on Chr 8 (2), and it seem likely that the two loci are identical. See *Prt-2*.

References

1. Honey, N.K., A.Y. Sakaguchi, P.A. Lalley, C. Quinto, R.J. MacDonald, C. Craik, G.I. Bell, W.J. Rutter, and S.L. Naylor. 1984. Chromosomal assignment of genes for trypsin, chymotrypsin B, and elastase in mouse. Somat. Cell Molec. Genet. 10:377–383.
2. Watanabe, T., N. Ogasawara, and H. Goto. 1976. Gene-tic study of pancreatic proteinase in mice (*Mus musculus*): linkage of the *Prt-2* locus on chromosome 8. Biochem. Genet. 14:999–1002.

Cts, cataract and small eye (formerly *Cs*), semidominant

Arose spontaneously in the JCL:ICR non-inbred stock. Heterozygotes have cataracts of varying degree of severity, depending in part on the genetic background. Penetrance in heterozygotes appears to be complete. Homozygotes have more severe cataracts and microph-thalmia. In homozygotes, swelling of lens fibers can first be detected at 15 days of gestation and is followed by degeneration of lens tissue, particularly in the anter-ior pole (1,2). *Cts* resembles *Cat*, but has not been tested for allelism with it.

References

1. Harata, M., R. Shoji, and R. Semba. 1978. Genetic back-ground and expressivity of congenital cataract in mice. Jpn. J. Genet. 53:147–152.
2. Ohotori, H., T. Yoshida, and T. Inuta. 1968. 'Small eye and cataracts,' a new dominant mutation in the mouse. Exp. Anim. 17:91–96 (Japanese, English summary).

Cv or *Cv-1* locus

See *Emv-1* locus.

Cvap locus

See *App* locus.

cw locus, Chr 9

cw, curly-whiskers, recessive. Arose in the CBA strain maintained at the Chester Beatty Research Institute, London. Homozygotes have strongly curled whiskers easily recognizable soon after birth. Their coats are not waved, but slightly abnormal in appearance, and there is a slight darkening on the CBA wild type background. Heterozygotes have slightly bent whiskers, but cannot be reliably identified on this basis (1).

cw^thd^, tail hair depletion. Arose spontaneously in the SJL/J strain. Homozygotes have abnormal or depleted whiskers and tail hair beginning at 1 to 2 weeks of age. Hair is also depleted on the feet and genitalia and around the nipples. Homozygotes are reported to have an uncharacteristic immune response to certain polysaccharide antigens. Mutant mice accept tail skin grafts from normal mice, but normal mice reject tail skin grafts from mutant mice over a period of 12 weeks. Response to sheep red blood cells is normal (2,3).

References

1. Falconer, D.S., and J.H. Isaacson. 1966. Curly-whiskers and its linkage with tail-kinks in linkage group II of the mouse. Genet. Res. 8:111–113.
2. Les, E.P., and J.B. Roths. 1975. Mouse News Lett. 53:34.
3. Roths, J.B. 1978. Mouse News Lett. 58:50.

Cxv-1 locus, control of xenotropic virus-1, Chr 17

This locus controls level of production by lymphocytes of infectious xenotropic leukemia virus and expression of the cell surface antigen XenCSA on lymphocytes. The recessive allele *Cxv-1*^s^ determines high level of virus and XenCSA and occurs in the F/St strain; the dominant allele *Cxv-1*^r^ determines low level of virus and XenCSA and occurs in the AKR and C57BL/10 strains. The expression of the high-virus phenotype is also dependent on presence of the *Bxv-1*^a^ allele on Chr 1 determining presence of the integrated xenotropic virus genome. *Cxv-1* is either within or tightly linked to the *H-2* complex. The effect of *Cxv-1* on expression of virus and XenCSA is greater in thymocytes than in spleen cells (1).

References

1. Yetter, R.A., J.W. Hartley, and H.C. Morse, III. 1983. *H-2*-linked regulation of xenotropic murine leukemia virus expression. Proc. Natl. Acad. Sci. USA 80:505–509.

Cxv-2 locus, control of xenotropic virus-2, Chr 4

This locus controls level of expression of an antigen, XenCSA, related to the major glycoprotein gp70 of the xenotropic murine leukemia virus (MuLV) envelope. Strains NZB, NZW, and DBA/2 carrying the allele *Cxv-2*^a^ express high levels of XenCSA on thymus and spleen cells; strains C57BL/6, NFS, MA, and many others carrying the allele *Cxv-2*^b^ express low levels on both types of cells. The antigen is recognized by a rabbit antiserum made against xenotropic MuLV-infected cells. The difference between C57BL/6 and DBA/2 depends on more than one gene but is primarily controlled by a locus linked to *Fv-1* on Chr 4. Expression in the F1 hybrids differs between reciprocal crosses. When the female parent is C57BL/6, the F1 mice are intermediate between the two parents; when the female parent is DBA/2, the F1 mice resemble the C57BL/6 parent. This locus may also affect spontaneous xenotropic virus production by spleen cells. The locus shows some recombination with *Fv-1* but it is not certain that these are true recombinants and that the locus differs from *Fv-1* (1, 2).

References

1. Morse, H.C., Jr., T.M. Chused, J.W. Hartley, B.M. Mathieson, S.O. Sharrow, and B.A. Taylor. 1979. Expression of xenotropic murine leukemia viruses as cell surface gp70 in genetic crosses between strains DBA/2 and C57BL/6. J. Exp. Med. 149:1183–1196.
2. Morse, H.C. III, B.A. Taylor, C.A. Kozak, T.M. Chused, S.O. Sharrow, J.W. Hartley, and E. Stockert. 1982. Expression on normal lymphocytes of two cell surface antigens, XenCSA and G_{IX}, related to the major glycoproteins (gp70) of murine leukemia viruses. J. Immunol. 128:2111–2115.

cy, crinkly-tail, recessive, Chr 4

Arose in the randombred 'Parke's albino' stock in the Department of Investigative Medicine, Cambridge, England. Penetrance varies from near 0 per cent to 100 per cent, depending on the genetic background. The only abnormality found in homozygotes is abnormal shape of caudal vertebrae 9 to 14. In 11-day embryos, the abnormality can be recognized as a duplicated tail gut in this region (1).

References

1. Johnson, D.R., and M.E. Wallace. 1979. Crinkly-tail, a mild skeletal mutant in the mouse. J. Embryol. Exp. Morphol. 53:327–333.

D

d locus (see also *Emv-3* locus), Chr 9

Mutations at this locus alter coat color; some alleles are lethal and cause a severe neuromuscular disorder. There is conflicting evidence on whether they affect incidence of audiogenic seizures and phenylalanine hydroxylase activity (15). Spontaneous and radiation-induced mutations from + to *d^l* and spontaneous mutations from *d* to + have been reported repeatedly.

d, dilute, recessive. An old mutation of the mouse fancy. The blue-gray color of the hair produced by this mutation in nonagouti (*a/a*) mice is caused by clumping of the melanin pigment into a few large masses (11). The clumping is due to the shape of the melanocytes which have fewer and thinner dendritic processes than wild-type melanocytes, with the result that melanin granules are largely clumped around the nucleus (8). Incorporation of tyrosine into melanin proceeds at a normal rate (1), and the fine structure of the melanin granules is normal (4). The original *d* allele which first identified the locus was found to have been caused by insertion of an ecotropic murine leukemia virus (called *Emv-3*) (6, 7). All other mutations at the locus examined lack the virus. Reversions of *d* to + are caused by excision of the virus leaving exactly one long terminal repeat of the virus in place (2). The virus is integrated into a non-coding region of the DNA (5). The *d* locus is very closely linked to the *se* locus on Chr 9, and many mutagen-induced mutations at one or the other or both loci have been produced. By means of complementation mapping of these mutations, Russell (12) has identified eight or nine functional units in the region. Molecular mapping with the unique DNA probes produced with the aid of the viral probe have shown co-linearity of the physical map with the complementation map (10).

d^l, dilute-lethal, recessive to both + and *d*. Arose in the C57BL/Go strain. The color of *d^l/d^l* is identical to that of *d/d* but the mice develop a severe neuromuscular disorder characterized by convulsions and opisthotonos (arching upward of the head and tail), and they usually die at about 3 weeks (13). No evidence of myelin breakdown products in the central nervous system has been found (14).

d^s, slight dilution, recessive to +. Discovered in a C57BL/6J × DBA/2 hybrid. Homozygotes are darker

than *d/d* mice. *d^s/d* mice are intermediate between *d^s/d^s* and *d/d* (3).

d^{15}, dilute-15, recessive to +. Homozygotes have slightly diluted coats, darker than *d/d*. They develop behavioral abnormalities similar to those of *d^l/d^l* at about 20 days, but some may live and one male survived to breed. *d^{15}/d* mice are similar in color to *d^{15}/d^{15}* but behave normally (9).

References

1. Coleman, D.L. 1962. Effect of genic substitution on the incorporation of tyrosine into the melanin of mouse skin. Arch. Biochem. Biophys. 96:562–568.
2. Copeland, N.G., K.W. Hutchison, and N.A. Jenkins. 1983. Excision of the DBA ecotropic provirus in dilute coat-color revertants of mice occurs by homologous recombination involving the viral LTRs. Cell 33:379–387.
3. Dickie, M.M. 1965. Mouse News Lett. 32:43.
4. Hearing, V.J., P. Phillips, and M.A. Lutzner. 1973. The fine structure of melanogenesis in coat color mutants of the mouse. J. Ultrastruct. Res. 43:88–106.
5. Hutchison, K.W., N.G. Copeland, and N.A. Jenkins. 1984. Dilute-coat-color locus of mice: nucleotide sequence analysis of the *d^{+2J}* and *d^{+Ha}* revertant alleles. Mol. Cell. Biol. 4:2899–2904.
6. Jenkins, N.A., N.G. Copeland, B.A. Taylor, and B.K. Lee. 1981. Dilute (*d*) coat color mutation of DBA/2J mice is associated with the site of integration of an ecotropic MuLV genome. Nature 293:370–374.
7. Jenkins, N.A., N.G. Copeland, B.A. Taylor, and B.K. Lee. 1982. Organization, distribution, and stability of endogenous ecotropic murine leukemia virus DNA sequences in chromosomes of *Mus musculus*. J. Virol. 43:26–36.
8. Markert, C.L., and W.K. Silvers. 1956. The effects of genotype and cell environment on melanoblast differentiation in the house mouse. Genetics 41:429–450.
9. Phillips, R.J.S. 1962. Mouse News Lett. 27:34.
10. Rinchik, E.M., L.B. Russell, N.G. Copeland, and N.A. Jenkins. 1986. Molecular genetic analysis of the *dilute-short ear* (*d-se*) region of the mouse. Genetics 112:321–342.
11. Russell, E.S. 1949. A quantitative histological study of the pigment found in the coat-color mutants of the house mouse. IV. The nature of the effects of genic substitution in five major allelic series. Genetics 34:146–166.
12. Russell, L.B. 1971. Definition of functional units in a small chromosomal segment of the mouse and its use in

interpreting the nature of radiation-induced mutations. Mutat. Res. 11:107–123.

13. Searle, A.G. 1952. A lethal allele of dilute in the house mouse. Heredity 6:395–401.
14. Winterbourn, C.C., F. Woolf, and L.I. Woolf. 1971. Brain lipids of mice homozygous for the gene "dilute lethal" (d^l). J. Neurochem. 18:1077–1086.
15. Woolf, L.I., A. Jakubovic, F. Woolf, and P. Bory. 1970. Metabolism of phenylalanine in mice homozygous for the gene "dilute lethal". Biochem. J. 119:895–903.

D7Ag-2, *D7Ag-3* DNA segments

D7Ag2, DNA segment Agaron-2, Chr 7
D9Ag3, DNA segment Agaron-3, Chr 9

These DNA segments, identified at the Agaron Institute, La Jolla, are recognized by rat cDNA probes prepared from brain-specific mRNAs. They detect a restriction site difference in genomic DNA between C57BL/6 and DBA/2 and between A/J and C57BL/6. By use of BXD and AXB recombinant inbred strains, D7Ag2 was found to be near the neurological mutation quivering (*qv*) on Chr 7 and D9Ag3 to be near another neurological mutation Snell's waltzer (*sv*) on Chr 9. It is possible that the DNA differences identify the two neurological genetic loci (1).

References

1. Blatt, C., L. Weiner, J.G. Sutcliffe, M.N. Nesbitt, and M.I. Simon. Chromosome mapping of murine brain specific genes. Cytogenet. Cell Genet. 40:583 (Abstr.).

D17Leh loci

See *t* complex.

D12Mcg1, DNA segment MCG-1, Chr 12

This DNA segment, identified at the Medical College of Georgia, was recognized by a unique sequence DNA probe. In *Hind*III digested DNA, the probe hybridizes with a 7.6-kb fragment in the C57BL/6 and related strains and with an 11.2-kb fragment in all other strains tested. *Sac*I and *Eco*RV also detect a polymorphism in this region. By use of BXD recombinant inbred strains, *D12Mcg1* was shown to be very closely linked to *D12Nyu2* on Chr 12 (1).

References

1. Cobb, R.R., T.A. Stoming, and B.J. Whitney III. 1987. The arylhydrocarbon hydroxylase (*Ah*) locus and a novel restriction-fragment length polymorphism (RFLP) are located on mouse chromosome 12. Biochem. Genet. 25:401–413.

D12Nyu1 to *D12Nyu5*, DNA segments NYU-1 to -5, Chr 12

These DNA segments, identified at New York University, were recognized by probes prepared from mouse genomic DNA isolated from a mouse–Chinese hamster somatic cell hybrid line containing mouse chromosomes 12 and X only. The probes detect restriction site polymorphisms in DNA digested with several different restriction endonucleases. The five loci are all on Chr 12 in the order from the centromere of 2, 5, (1, 3, *Lyb-7*), (4, *Fos*), *Igh-C*.

D12Nyu1. Three alleles are distinguishable in DNA digested in *Eco*RI, *Taq*I, or *Msp*I: *D12Nyu1^a* in BALB/c, C3H/He, DBA/2, and AKR; *D12Nyu1^b* in C57BL/6, C57L, and others; and *D12Nyu1^c* in SJL and SWR.

D12Nyu2. Two alleles are distinguishable in DNA digested in *Taq*I: *D12Nyu2^a* in BALB/c, C3H/He, DBA/2, AKR, C57L, and others; and *D12Nyu2^b* in C57BL/6.

D12Nyu3. Two alleles are distinguishable in DNA digested in *Eco*RI or *Taq*I: *D12Nyu3^a* in SWR, SJL, and SM; and *D12Nyu3^b* in C57BL/6, C3H/He, DBA/2, BALB/c, C57L, and others.

D12Nyu4. Two alleles are distinguishable in DNA digested in *Taq*I: *D12Nyu4^a* in SWR; and *D12Nyu4^b* in C57BL/6, SJL, C3H/He, DBA/2, AKR, BALB/c, C57L, and others.

D12Nyu5. Two alleles are distinguishable in DNA digested with *Msp*I: *D12Nyu5^a* in BALB/c and AKR; and *D12Nyu5^b* in C57BL/6, SWR, and C57L (1).

References

1. D'Eustachio, P. 1984. A genetic map of murine chromosome 12 composed of polymorphic DNA segments. J. Exp. Med. 160:827–838.

DXPas1 to *DXPas5*, DNA segments Pasteur-1 to -5, Chr X

These X-linked DNA segments found at the Pasteur Institute are polymorphic between the SPE/Pas inbred strain derived from wild *Mus spretus* and the laboratory strains C57BL/6 and BALB/c. They were detected in *Taq*I-restricted DNA using genomic DNA probes prepared from X-chromosome-enriched material from an established cell line. Fragments *DXPas1* to *DXPas5* were recognized by probes numbered 45, 52, 66, 87, and 100, respectively. The fragments are located in the order from the centromere *spf*, 3, 4, 5, *Hprt*, *Ta*, 2, *jp*, 1 (1).

References

1. Amar, L.C., D. Armaud, J. Cambrou, J.-L. Guénet, and P.R. Avner. 1985. Mapping of the mouse X chromosome using random genomic probes and an interspecific mouse cross. EMBO J. 4:3695–3700.

D4Rp1 to D12Rp54, DNA segments Roswell Park-1 to -54

These DNA polymorphisms were found at Roswell Park Memorial Institute. They were detected in restriction endonuclease fragments of DNA from various mouse strains hybridized to recombinant plasmid cDNA made from mRNA from kidney or other tissue. The known chromosome locations were found by use of recombinant inbred strains.

D4Rp1 locus (formerly *RP1*), Chr 4 (1).

D7Rp2 locus (formerly *RP2*), Chr 7. Detected by a cDNA probe made from an androgen-inducible kidney mRNA. After *Hind*III digestion of genomic DNA, the allele $D7Rp2^b$ produces a 1.9-kb fragment; it occurs in the C57BL/6, A/J, AKR, BALB/c, C3H/He, and other strains. The allele $D7R2^d$ produces a 2.1-kb fragment and occurs in the DBA/2, C57L, C58, RF, SJL, SWR, 129/J, and other strains (2). Three mRNAs of different sizes are transcribed in the kidney by *D7Rp2*. They differ in the length of their 3′ untranslated region. The two smaller ones are the same in all strains examined but the larger one is 2150 nucleotides long in DBA/2 and 1950 long in C57BL/6 and BALB/c. The size difference in the large mRNA is due to insertion of a member of the B1 family of repeats into the 3′ untranslated region of the gene in DBA/2 mice (4). The mRNA polymorphism has a strain distribution pattern in both inbred and recombinant inbred strains identical to that of *D7Rp2*. *D7Rp2* is closely linked to *Gpi-1* on Chr 7 (2).

D0Rp3 locus (formerly *RP3*) (1).

D14Rp4 locus (formerly *RP4*), Chr 14 (1).

D0Rp5 locus (formerly *RP5*) (1).

D0Rp6 locus (formerly *RP6*) (1).

D17Rp10 locus (formerly *RP10*), Chr 17. Detected by a human phosphoglycerate kinase probe, pHPGK-7e. It is probably a PGK pseudogene. The locus lies 8 cM distal to *D17Rp11*. Allelic variants segregate in the BXD and AKXL recombinant inbred strains (5).

D17Rp11 (formerly *RP11*), Chr 17. Detected by a cDNA probe made from kidney mRNA. The locus lies between *D17Rp17* and *D17Rp10*. Allelic variants segregate in the AKXL and SWXL recombinant inbred strains (5).

D17Rp17 (formerly *RP17*), Chr 17. Detected by a cDNA probe made from testosterone-induced kidney mRNA. The locus was mapped to the *t*-complex region using BXD and BXH recombinant inbred strains (5).

D12Rp54 locus (formerly *RP54*), Chr 12. This locus, discovered as an mRNA polymorphism, controls a liver-specific protein that is regulated both developmentally and by glucocorticoids. The two restriction enzyme patterns of the DNA are related to the two forms of the protein differing in molecular weight and to the two electrophoretic patterns of the mRNA. Analysis of protein and DNA variation in BXD recombinant inbred strains showed that the locus was on Chr 12 near *Pre-1* (3).

References

1. Elliott, R. 1981. Mouse News Lett. 64:87.
2. Elliott, R.W., and F.G. Berger. 1983. DNA sequence homology in an androgen-regulated gene is associated with alteration in the encoded RNAs. Proc. Natl. Acad. Sci. USA 80:501–504.
3. Hill. R.E., R.K. Barth, and N.D. Hastie. 1982. Mouse News Lett. 67:36.
4. King, D., L.D. Snider, and J.B. Lingrel. 1986. Polymorphism in an androgen-regulated mouse gene is the result of the insertion of a B1 repetitive element into the transcriptional unit. Mol. Cell. Biol. 6:209–217.
5. Mann, E., R.W. Elliott, and C. Rohman. 1984. Mouse News Lett. 71:48.

DXS32 locus, DNA segment S32, Chr X

This locus was identified by hybridization with a probe for the human X-linked DXS32 hypervariable locus that is closely linked to the hemophilia A gene. The segment has a 500 base pair region of very high homology between the two species. In the mouse, *DXS32* is located about 10 cM from *spf* (*Otc*) (1).

References

1. Mandel, J.L., B. Arveiler, G. Camerino, A. Hanauer, R. Heilig, M. Koenig, and L. Oberlé. 1986. Genetic mapping of the human X chromosome: linkage analysis of the q26–q28 region that includes the fragile X locus and isolation of expressed sequences. Cold Spring Harbor Symp. Quant. Biol. 51:195–203.

da, dark, recessive, Chr 7

Arose in the CBA/Fa inbred strain. Probably extinct. Homozygotes are smaller than normal. In combination with *A* or A^y the yellow pigment on the back is replaced by black so that both look like *a/a* except on the flanks. The darkening of the back of *A/-* lessens as the animals get older (1).

References

1. Falconer, D.S. 1956. Mouse News Lett. 15:23.

Dac, dactylaplasia, semidominant

Arose in the SM7B/SC strain, a congenic subline of SM. Heterozygotes show absence of digits, particularly the three middle digits, on each foot. Often the metatarsal and metacarpal bones are partially or completely missing or fused. The forefeet are more severely affected than the hindfeet. The long bones are normal. The defects are clearly recognizable in 15-day embryos. Homozygotes die before birth. The abnormal phenotype is dependent also on the homozygous presence of a recessive gene, *mdac*, that is present in the SM strain, and also in BALB/c, LG, and 129/J. There is no visible effect in *mdac/mdac* mice except in combination with *Dac/+* (1). See *mdac*.

References

1. Chai, C.K. 1981. Dactylaplasia in mice, a two-locus model for developmental anomalies. J. Hered. 72:234–237.

Dao-1 locus, D-amino acid oxidase, Chr 5

This locus controls electrophoretic variation in the peroxisomal enzyme D-amino acid oxidase (DAOX; E.C. 1.4.3.3). The allele *Dao-1^a* determines a single band and occurs in the BALB/c, CBA/H, C57L, and C3H strains; the allele *Dao-1^b* determines a single more anodal band and occurs in the NZC strain. Heterozygotes (*Dao-1^a/Dao-1^b*) have two fainter bands (2). *Dao-1* is located about 21 cM from *Bcd-1* on Chr 5 (1, 3).

References

1. Hilgers, J., and R. van Nie. 1984. Mouse News Lett. 70:65.
2. Holmes, R.S. 1976. Genetics of peroxisomal enzymes in the mouse: nonlinkage of D-amino acid oxidase locus (*Dao*) to catalase (*Cs*) and L-α-hydroxyacid oxidase (*Hao-1*) loci on chromosome 2. Biochem. Genet. 14:981–987.
3. Seeley, T.-L., and R.S. Holmes. 1981. Genetics and ontogeny of butyryl CoA dehydrogenase in the mouse and linkage of *Bcd-1* with *Dao-1*. Biochem. Genet. 19:333–345.

db, diabetes, recessive, Chr 4

Arose spontaneously in the C57BL/Ks strain. Several alleles of independent origin, *db^{2J}*, *db^{ad}*, *db^{3J}*, and *db^{Pas}*, have been studied and are probably very similar, any differences most likely being due to differences in genetic background (7, 8, 12). Homozygotes are sterile. The closely linked mutant gene misty (*m*) has been incorporated into stocks for maintenance of *db*, in repulsion (*db +/+ m*) to facilitate identification of heterozygotes for breeding, and in coupling (*db m/+ +*) to allow identification of homozygotes before onset of clinical symptoms. There is a large and growing literature concerning this locus (see Staats, 19, for early references). The literature has been critically reviewed by Herberg and Coleman (8), Coleman (1), and Coleman and Brodoff (4).

Briefly, homozygous *db/db* mice are first visibly recognizable at about 3 to 4 weeks by their obesity in comparison to normal littermates. Elevation of plasma insulin begins at 10 to 14 days and elevation of blood sugar at 4 to 8 weeks. Affected mice show polyphagia, polydipsia, and polyuria. The course of the disease is markedly influenced by genetic background, the C57BL/Ks background allowing uncontrolled rise in blood sugar, severe depletion of the insulin-producing B-cells of the pancreatic islets, and death before 10 months, and the C57BL/6 background causing compensatory hyperplasia of the islet B-cells, and continued hyperinsulinemia throughout an 18- to 20-month life span. In the severely diabetic homozygotes on the C57BL/Ks background, exogenous insulin fails to control blood glucose levels. There is increased activity of gluconeogenic enzymes. In cultures of pancreatic cells of 4-day C57BL/*Ks-db/db* mice, there is increased synthesis and secretion of glucagon preceding any hyperfunction of the insulin-secreting B-cells (11). Cultured fibroblasts of C57BL/Ks-*db/db* mice have about 40 per cent as many insulin receptors as controls and this difference is maintained over many generations in culture (16).

Many complications similar to those of human diabetes are minor or absent in *db/db* mice, but peripheral neuropathy (15, 17) and myocardial disease (6) have been described in C57BL/Ks mice. There is accelerated thickening of the glomerular basement membrane associated with proteinuria (13), and large quantities of immunoglobulin and complement in the mesangium (14). An autoimmune process may be involved in the severe diabetes of the C57BL/Ks-*db/db* mice. Spleen cells and serum of these mice show a specific anti-pancreatic immune activity at 10 days of age (5). The B-cells of the pancreatic islets contain intracisternal type A particles and show constitutive expression of p73, the core protein antigen of the particles, which may act as a 'neoantigen' and generate an autoimmune response (10).

Homozygotes show increased metabolic efficiency which may be due in part to a defect in thermoregulation leading to a decreased expenditure of energy on thermogenesis (18). Increased metabolic efficiency also occurs in heterozygotes (*db*/+) which, although normal in body weight, blood glucose, and plasma insulin, survive a prolonged fast longer than congenic +/+ controls (2). This might be explained by the observation that fasting induces conversion of acetone to lactate to a significantly higher degree in *db*/+ than in +/+ mice (3).

Parabiosis of homozygotes with normal mice causes the normal partner to stop eating and die of starvation within 2 weeks. This suggests that *db*/*db* mice are unable to respond to a circulating satiety factor which they therefore produce in excess and which causes the normal partner to stop eating. This and other experiments involving destruction of the ventromedial nucleus of the hypothalamus support the suggestion that *db* may cause a defective hypothalamus (8). A defective hypothalamus may also be involved in the sterility of *db*/*db* females which have been shown to have defective release of gonadotropin-releasing hormone (9).

References

1. Coleman, D.L. 1978. Obese and diabetes: two mutant genes causing diabetes–obesity syndromes in mice. Diabetologia 14:141–148.
2. Coleman, D.L. 1979. Obesity genes: beneficial effects in heterozygous mice. Science 203:663–665.
3. Coleman, D.L. 1980. Acetone metabolism in mice: increased activity in mice heterozygous for obesity genes. Proc. Natl. Acad. Sci. USA 77:290–293.
4. Coleman, D.L., and B.N. Brodoff. 1982. Spontaneous diabetes and obesity in rodents. In B.N. Brodoff and S. Bleicher, eds., Diabetes Mellitus and Obesity, 283–293. Williams and Wilkins, Baltimore.
5. Debray-Sachs, M., P. Saï, C. Boitard, R. Assan, and J. Hamburger. 1983. Anti-pancreatic immunity in genetically diabetic mice. Clin. Exp. Immunol. 51:1–7.
6. Giacomelli, F., and J. Wiener. 1979. Primary myocardial disease in the diabetic mouse. An ultrastructural study. Lab. Invest. 40:460–473.
7. Guenet, J.L., R. Aubert, and D. Lemonnier. 1984. Mouse News Lett. 70:95.
8. Herberg, L., and D.L. Coleman. 1977. Laboratory animals exhibiting obesity and diabetes syndromes. Metabolism 26:59–99.
9. Johnson, L.M., and R.L. Sidman. 1979. A reproductive endocrine profile in the diabetes (*db*) mutant mouse. Biol. Reprod. 20:552–559.
10. Leiter, E.H., and E.L. Kuff. 1984. Intracisternal type A particles in murine pancreatic B cells. Immunocytochemical demonstration of increased antigen (p73) in genetically diabetic mice. Am. J. Pathol. 114:46–55.
11. Leiter, E.H., D.L. Coleman, and J.J. Eppig. 1979. Endocrine pancreatic cells of postnatal 'diabetes' (*db*) mice in cell culture. In Vitro 15:507–521.
12. Leiter, E.H., D.L. Coleman, A.B. Eisenstein, and I. Strack. 1980. A new mutation (*db²ᴶ*) at the diabetes locus in strain 129/J mice. I. Physiological and histological characterization. Diabetologia 19:58–65.
13. Like, A.A., R.L. Lavine, P.L. Poffenbarger, and W.L. Chick. 1972. Studies in the diabetic mutant mouse. VI. Evolution of glomerular lesions and associated proteinuria. Am. J. Pathol. 66:193–324.
14. Mauer, S.M., M.W. Steffes, A.F. Michael, and D.W. Brown. 1976. Studies of diabetic nephropathy in animals and man. Diabetes 25 (Suppl. 2):850–857.
15. Norido, F., R. Canella, R. Zanoni, and A. Gorio. 1984. Development of diabetic neuropathy in the C57BL/Ks (db/db) mouse and its treatment with gangliosides. Exp. Neurol. 83:221–232.
16. Raizada, M.K., G. Tan, and R.E. Fellows. 1980. Fibroblastic cultures from diabetic db/db mouse. Demonstration of decreased insulin receptors and impaired responses to insulin. J. Biol. Chem. 255:9149–9155.
17. Robertson, D.M., and A.A.F. Sima. 1980. Diabetic neuropathy in the mutant mouse [C57BL/Ks(db/db)]: a morphometric study. Diabetes 29:60–67.
18. Trayhurn, P. 1979. Thermoregulation in the diabetic–obese (*db/db*) mouse. The role of non-shivering thermogenesis in energy balance. Pflügers Arch. 380:227–232.
19. Staats, J. 1975. Diabetes in the mouse due to two mutant genes—a bibliography. Diabetologia 11:325–327.

Dc, dancer, semidominant, Chr 19

Arose as a spontaneous mutation in a heterogeneous stock. Not allelic with *Tw*, which it somewhat resembles. Viability and fertility of heterozygotes are normal. Penetrance is incomplete in heterozygotes. Homozygotes die at birth with cleft lip and cleft palate (1). On some genetic backgrounds (after crosses to C57BL/6 and A/J), some heterozygotes also have cleft lip (2). On other backgrounds (after crosses to C3H/He), cleft lip does not occur spontaneously but can be induced in heterozygotes by treatment of pregnant mothers with 6-aminonicotinamide (3). Cleft lip probably results from reduction in the median nasal processes (2). Heterozygotes exhibit circling and head-tossing behavior, but are not deaf. They usually have a small patch of white hairs on the forehead. There is complete absence of the macula of the utriculus, accompanied by gross defects of the bony and membranous labyrinths in the vestibular region. These defects probably result from a reduction in the vestibular ganglion and some of its nerves which occurs about 2 days earlier (1).

References

1. Deol, M.S., and P.W. Lane. 1966. A new gene affecting the morphogenesis of the vestibular part of the inner ear in the mouse. J. Embryol. Exp. Morphol. 16:543–558.

2. Trasler, D.G., and S. Leong. 1982. Mitotic index in mouse embryos with 6-aminonicotinamide-induced and inherited cleft lip. Teratology 25:259–265.

3. Trasler, D.G., D. Kemp, and T.A. Trasler. 1984. Increased susceptibility to 6-aminonicotinamide-induced cleft lip of heterozygous dancer mice. Teratology 29:101–104.

Dcp loci, dexamethasone-induced cleft palate, Chr 17

These *H-2*-linked loci influence susceptibility to dexamethasone-induced cleft palate. There are probably two or three such loci. Analysis of *H-2* congenic strains shows that the genes are located in the *H-2* region on Chr 17 and gives some information about their location within that region. On a two-locus model, the genes map to locations somewhat different from those on a three-locus model. Susceptibility was determined by the incidence of cleft palate in 18-day embryos gestated in mothers that received an intraperitoneal injection of dexamethasone (a synthetic glucocorticoid) on day 12 of pregnancy (1).

References

1. Bonner, J.J., and M.L. Tyan. 1984. Glucocorticoid-induced cleft palate in chromosome 17 genetic linkage and mapping analyses. Immunogenetics 20:169–183.

de, droopy ear, recessive, Chr 3

Arose spontaneously in the J stock at the Institute of Animal Genetics, Edinburgh. In homozygotes the ears are set low on the sides of the head and the pinnae project laterally. On an *a^t* background the light belly hair comes up farther than normal around the sides of the body and face. Homozygotes tend to be smaller than their normal sibs, but they breed well, except that some females are poor mothers. The skeleton of the occiput and shoulder girdle shows immaturity of form with disproportionate shortening of the limb bones. The skeletal abnormalities can be traced back to disturbed mesenchymal condensations. It is possible that growth of cartilage itself is abnormal (1). *de* is on Chr 3 about 6 cM distal to matted (*ma*) (2, 3).

References

1. Curry, G.A. 1959. Genetical and developmental studies on droopy-eared mice. J. Embryol. Exp. Morphol. 7:39–65.

2. Lane, P.W. 1980. Mouse News Lett. 62:56.

3. Lane, P.W., and E.M. Eicher. 1979. Gene order in linkage group XVI of the house mouse. J. Hered. 70:239–244.

Den

See *Re* locus.

dep, depilated, recessive, Chr 4

Arose spontaneously in a stock carrying *T* and *tf*. Homozygotes are viable and fertile. They are identifiable at about 1 week when the first coat appears. The hair is abnormally thin and short. By 3 weeks some mice are nearly hairless; others possess a rather substantial coat. The hair has a matted greasy appearance. Many hair follicles are misshapen, and others are completely disoriented. Clumps of pigment that appear to be the remains of degenerating follicles are numerous. In addition, about 20 per cent of the hairs of the coat show an irregular arrangement of septa in the shaft or a complete interruption of septal pattern. By dermal–epidermal recombination grafts of embryonic skin, *dep* has been shown to act in the epidermis (1).

References

1. Mayer, T.C., N.J. Kleiman, and M.C. Green. 1976. Depilated (*dep*), a mutant gene that affects the coat of the mouse and acts in the epidermis. Genetics 84:59–65.

Dey

See *Sey*.

df, Ames dwarf, recessive, Chr 11

Arose spontaneously in descendants of a cross of Goodale giant mice to a pink-eyed stock. Homozygotes resemble *dw/dw* mice. Growth retardation occurs after the first week, and weight at 2 months is only about one-half normal. Females and all but an occasional male are sterile (2). The anterior hypophysis lacks identifiable growth hormone (GH)-producing (4) and prolactin (PRL)-producing cells (1), and contains no detectable GH or PRL at any stage from birth to 6 weeks (7). Dopamine is severely deficient in the median eminence (6). Treatment of affected mice with GH or PRL or both greatly improves growth rate and fertility (3). Ames dwarf mice have an immune deficiency presumably resulting from the lack of GH (see review by Duquesnoy and Pederson, 5). The defects reported for *df/df* mice also occur in *dw/dw* mice, indicating that they are probably secondary to the as yet unknown primary effects of the genes.

References

1. Barkley, M.S., A. Bartke, D.S. Gross, and Y.N. Sinha. 1982. Prolactin status of hereditary dwarf mice. Endocrinology 110:2088–2096.

2. Bartke, A. 1964. Histology of the anterior hypophysis, thyroid and gonads of two types of dwarf mice. Anat Rec. 149:225–235.
3. Bartke, A., B.D. Goldman, F. Bex, and S. Dalterio. 1977. Effects of prolactin (PRL) on pituitary and testicular function in mice with hereditary PRL deficiency. Endocrinology 101:1760–1766.
4. Cheng, T.C., W.G. Beamer, S. Phillips III, A. Bartke, R.L. Mallonee, and C. Dowling. 1983. Etiology of growth hormone deficiency in little, Ames, and Snell dwarf mice. Endocrinology 113:1669–1678.
5. Duquesnoy, R.J., and G.M. Pedersen. 1981. Immunologic and hematologic deficiencies of the hypopituitary dwarf mouse. *In* M.E. Gershwin and B. Merchant, eds., Immunologic defects in Laboratory Animals, Vol. 1, 309–324. Plenum Press, New York.
6. Morgan, W.W., A. Bartke, and K. Pfeil. 1981. Deficiency of dopamine in the median eminence of Snell dwarf mice. Endocrinology 109:2069–2075.
7. Slabaugh, M.B., M.E. Lieberman, J.J. Rutledge, and J. Gorski. 1981. Growth hormone and prolactin synthesis in normal and homozygous Snell and Ames dwarf mice. Endocrinology 109:1040–1046.

Dfp, dark foot pads, semidominant

First observed in a later generation of descendants of an irradiated female. Homozygotes die before implantation. In heterozygotes, the pads of the plantar surface of the feet are black or markedly darker than the rest of the foot, but there is some variation, with one foot occasionally unaffected or some pads on all feet unaffected (1, 2).

References

1. Kelly, E.M. 1968. Mouse News Lett. 38:31.
2. Kelly, E.M. 1970. Mouse News Lett. 43:59.

dfw, deaf waddler, recessive, Chr 6

Arose spontaneously in strain C3H/HeJ. Homozygotes are deaf, they walk with a hesitant wobbly gain, and they bob their heads. Both sexes are fertile. *dfw* is on Chr 6 about 7 cM distal to *Mi^{wh}* (1).

References

1. Lane, P.W. 1987. Deaf waddler (dfw). Mouse News Lett. 77:129.

Dh, dominant hemimelia, semidominant, Chr 1

Arose spontaneously in a crossbred stock carrying a translocation at the Institute of Animal Genetics, Edinburgh (2). Heterozygotes show preaxial polydactyly or oligodactyly of the hindlimbs, tibial hemimelia, and sometimes reduction of the femur and pubic element of the pelvic girdle. Number of ribs, sternebrae, and presacral vertebrae tends to be reduced. The spleen is entirely lacking; the stomach is somewhat smaller and the alimentary tract shorter than normal. Homozygotes usually die within a few days after birth. Abnormalities are similar to those of heterozygotes but much more severe and include also abnormalities of the urogenital system (9). The defects of both heterozygotes and homozygotes are traceable to early $9\frac{1}{2}$ day embryos in which the normal columnar or cuboidal architecture of the epithelium of the splanchnic mesoderm is defective or lacking, leading to defective morphogenesis of the gut and associated structures (6). Studies on the effect of the spleenless condition on the immune system have been reviewed by Welles and Battisto (10). These studies have shown that in *Dh*/+ mice: the lymph nodes are enlarged; most *Dh*/+ mice have elevated levels of lymphocytes, granulocytes, and thrombocytes (8); there are decreased serum levels of IgM and IgG_2 but normal IgG_1 (1); and there is normal cellular immunity (7). *Dh*/+ mice have defective B- and T-cell co-operation in the production of IgM, probably due to defective T helper cell function (10). Fletcher *et al.* (3) concluded that absence of the spleen in *Dh*/+ mice significantly reduces the rate of T-cell maturation. The naturally occurring autoimmunity of strain NZB mice is significantly retarded in congenic NZB-*Dh*/+ mice (5). *Dh* has been combined with *nu* (nude) to produce *Dh*/+ *nu*/*nu* mice lacking both spleen and thymus for use in immunological research (4).

References

1. Battisto, J.R., L.C. Cantor, F. Borek, A.L. Goldstein, and E. Cabrerra. 1969. Immunoglobulin synthesis in hereditarily spleenless mice. Nature 222:1196–1198.
2. Carter, T.C. 1954. Mouse News Lett. 11:16.
3. Fletcher, M.P., R.M. Ikeda, and M.E. Gershwin. 1977. Splenic influence of T cell function: the immunobiology of the inbred hereditarily asplenic mouse. J. Immunol. 119:110–117.
4. Gershwin, M.E., A. Ahmed, R.M. Ikeda, M. Shifrine, and F. Wilson. 1978. Immunobiology of congenitally athymic–asplenic mice. Immunology 34:631–642.
5. Gershwin, M.E., J.J. Castles, R.M. Ikeda, K. Erickson, and J. Montero. 1979. Studies on congenitally immunologic mutant New Zealand mice. I. Autoimmune features of hereditarily asplenic (Dh/+) NZB mice, reduction of naturally occurring thymocytotoxic antibody and normal suppressor function. J. Immunol. 122:710–717.
6. Green, M.C. 1967. A defect of the splanchnic mesoderm caused by the mutant gene dominant hemimelia in the mouse. Dev. Biol. 15:62–89.

7. Lozzio, B.B., and L.B. Wargon. 1974. Immune competence of hereditarily spleenless mice. Immunology 27:167–178.
8. Machado, E.A., and B.B. Lozzio. 1976. Animal model: hereditary asplenia in mice. Am. J. Pathol. 85:515–518.
9. Searle, A.G. 1964. The genetics and morphology of two 'luxoid' mutants in the house mouse. Genet. Res. 5:171–197.
10. Welles, W.L., and J.R. Battisto. 1981. The significance of hereditary asplenia for immunologic competence. *In* M.E. Gershwin and B. Merchant, eds., Immunologic Defects in Laboratory Animals, Vol. 1, 191–212. Plenum Press, New York.

Dhfr locus, dihydrofolate reductase, Chr 13

This locus codes for dihydrofolate reductase (DHFR; E.C. 1.5.1.3), an enzyme essential in DNA synthesis. It was found to be on Chr 13 by fusion of microcells prepared from mouse fibroblasts with Chinese hamster cells lacking the DHFR gene. Hybrids that selectively retained Chr 13 expressed wild-type levels of DHFR enzyme activity. The assignment to Chr 13 was confirmed by use of Southern blots of DNA prepared from these hybrids and from independently derived rat–mouse microcell hybrids (1).

References

1. Killary, A.N., R.J. Leach, R.G. Moran, and R.E.K. Fournier. 1986. Assignment of the genes encoding dihydrofolate reductase and hexosaminidase B to *Mus musculus* chromosome 13. Somat. Cell Mol. Genet. 12:641–648.

di, duplicate incisors, recessive

Found in the F2 generation from mice treated *in utero* with nitrogen mustard. Penetrance is high in the original stock, but low and variable in descendants of outcrosses. Affected mice are characterized by the appearance on one or both sides of supernumerary lower incisors located directly behind, or slightly medial to, the corresponding normal teeth. The extra teeth usually erupt a few days later than the normal incisors. Usually they become deflected and are broken off, after which it may be difficult to say whether an extra tooth has ever been present (1).

References

1. Danforth, C.H. 1958. The occurrence and genetic behavior of duplicate lower incisors in the mouse. Genetics 43:139–148.

Dia-1 locus, diaphorase-1, Chr 15

This is the structural locus for NADH-dependent diaphorase (DIA-1; E.C. 1.6.2.2) which catalyzes the oxidation of NADH. No genetic variants are known. The locus was assigned to Chr 15 using mouse–Chinese hamster somatic cell hybrids segregating mouse chromosomes. The mouse and Chinese hamster enzymes were distinguishable by isoelectric focusing in polyacrylamide gels (1).

References

1. Taggart, R.T., P. Tetri, and U. Francke. 1980. Assignment of the gene for NADH diaphorase *Dia-1* to mouse chromosome 15. Somat. Cell Genet. 6:769–776.

Dip-1 locus

See *Pep-3* locus.

Dip-2 locus

See *Pep-1* locus.

dl locus, Chr 10

dl, downless, recessive. Arose spontaneously in strain A/H. Homozygotes closely resemble crinkled (*cr/cr*) mice and tabby homozygotes (*Ta/Ta*) and hemizygotes (*Ta/Y*). The coat is thin and shiny with bald patches behind the ears and a bald tail, and the incisors may be abnormal or absent and the molars abnormal (3). Instead of the normal four types of hair, only one type, resembling an abnormal awl, is found (2). By means of dermal–epidermal recombination grafts of tail skin, the site of action of *dl* has been shown to be in the epidermis (4, 5).

Dl^{slk}, sleek, dominant. Arose spontaneously in a stock carrying the Robertsonian translocation *Rb(4.6)2Bnr*. Homozygotes are indistinguishable from *dl/dl* mice. Heterozygotes are like homozygotes except for the presence of some hair on the tail and a smaller reduction in number of vibrissae. Dl^{slk}/dl mice resemble *dl/dl* or Dl^{slk}/Dl^{slk} mice (1). $Dl^{slk}/+$ mice have the same syndrome of dental abnormalities as *dl/dl*, including small or absent incisors and third molars, and the first and second molars are usually small (6).

References

1. Crocker, M., and B.M. Cattanach. 1979. The genetics of *sleek*: a possible regulatory mutation of the *tabby-crinkled-downless* syndrome. Genet. Res. 34:231–238.
2. Green, M.C., D. Durham, T.C. Mayer, and P.C. Hoppe. 1977. Evidence from chimaeras for the pattern of proliferation of epidermis in the mouse. Genet. Res. 29:279–284.

3. Sofaer, J.A. 1969. Aspects of the tabby-crinkled-downless syndrome. II. Observations on the reaction to changes in the genetic background. J. Embryol. Exp. Morphol. 22:207–227.
4. Sofaer, J.A. 1973. Hair follicle initiation in reciprocal recombinations of downless homozygote and heterozygote mouse tail epidermis and dermis. Dev. Biol. 34:289–296.
5. Sofaer, J.A. 1974. Differences between *tabby* and *downless* epidermis and dermis in culture. Genet. Res. 23:219–225.
6. Sofaer, J.A. 1977. The teeth of the "sleek" mouse. Arch. Oral Biol. 22:299–300.

Dlb-1 locus, *Dolichos* lectin-binding-1, Chr 11

This locus controls the binding pattern of the N-acetyl-galactosamine-binding lectins from *Dolichos biflorus* (DBA) to vascular endothelium or intestinal epithelium. In some strains, DBA binds to vascular endothelium but not to intestinal epithelium; in other strains it binds to intestinal epithelium but not to vascular endothelium. The allele *Dlb-1^a* determines binding of DBA to vascular endothelium and not to intestinal epithelium and occurs in the RIII/Ro, DDK/Ro, MAS, SM, LIS/Af, SWR, GRS, and STS strains; the allele *Dlb-1^b* determines the opposite pattern and occurs in the C57BL/6, CBA/Ca, BALB/c, AKR, C3H/He, DBA/2, 101, HTG, SEA, SJL, O20, and other strains, Heterozygotes show binding in both tissues. *Dlb-1* also controls expression of binding sites for lectins from *Helix pomatia* and *Wisteria floribunda* but not for eight other lectins tested. In RIII/Ro and C57BL/6 embryos, the polymorphism is only partially expressed at 9.5 days but has acquired the adult type by 11 days. It is likely that the polymorphism is due to tissue specific pattern of expression of N-acetylgalactosaminosyl transferase (2). By use of BALB/c × STS recombinant inbred strains and backcrosses, *Dlb-1* was shown to be on Chr 11 about 3 cM proximal to *Re* (1).

References

1. Uiterdijk, H.G., B.A.J. Ponder, M.F.W. Festing, J. Hilgers, L. Skow, and R. van Nie. 1986. The gene controlling the binding sites of *Dolichos biflorus* agglutinin, *Dlb-1*, is on chromosome 11 of the mouse. Genet. Res. 47:125–129.
2. Ponder, B.A.J., M.F.W. Festing, and M.M. Wilkinson, 1985. An allelic difference determines reciprocal patterns of expression of binding sites for *Dolichos biflorus* lectin in inbred strains of mice. J. Embryol. Exp. Morphol. 87:229–239.

dm, diminutive, recessive, Chr 2

Appeared in the descendants of a cross of strains 129 and C57L/J. Viability of homozygotes is low, but many surviving to maturity are fertile. Homozygotes are consistently smaller than normal and have macrocytic anemia which becomes slightly less severe with age. Preliminary studies indicate that the bone marrow contains only half the normal number of colony-forming units (1). Homozygotes have short kinked tails and tend to have an additional rib at both the cervical and lumbar ends of the thorax, and an additional presacral vertebra. The vertebrae are malformed, and ribs may be fused (2).

References

1. Bannerman, R.M., J.A. Edwards, and P.H. Pinkerton. 1973. Hereditary disorders of the red cell in animals. Prog. Hematol. 8:131–179.
2. Stevens, L.C., and J.A. Mackensen. 1958. The inheritance and expression of a mutation in the mouse affecting blood formation, the axial skeleton, and body size. J. Hered. 49:153–160.

Dmm, disproportionate micromelia, semidominant

Found among the offspring of an irradiated male of strain 101. Homozygotes are disproportionately short, have cleft palates, and die at birth. The long bones are shorter and wider than normal. In 18-day fetuses the epiphyseal growth plates of homozygotes are shorter and wider than normal and the inner region of the cartilaginous epiphysis shows loss of structural integrity of the matrix. The amounts of both type II collagen and proteoglycans in the matrix are greatly reduced. The type II collagen that remains is located near the cell surface of the chondrocytes rather than being distributed throughout the matrix as in normal mice. The proteoglycans of the homozygote are normal in size and glycosaminoglycan composition but are more easily extractable from the matrix than normal. Heterozygotes appear normal at birth but by 1 week are recognizable by short nose, tail, and limbs. The dwarfism increases during growth. The epiphyseal plates of the long bones are somewhat shorter than normal but show normal organization. The amount of proteoglycan in the matrix is intermediate between that of homozygotes and normals. Penetrance in the heterozygote may not be fully complete (1).

References

1. Brown, K.S., R.E. Cranley, R. Greene, H.K. Kleinman, and J.P. Pennypacker. 1981. Disproportionate microme-

lia (*Dmm*): an incomplete dominant mouse dwarfism with abnormal cartilage matrix. J. Embryol. Exp. Morphol. 62:165–182.

dn, deafness, recessive

Discovered in a stock at University College, London, in the course of a systematic search for uncomplicated deafness. Fertility of homozygotes is normal. Homozygotes are deaf their entire life, and a few of them show slight head-tossing (2). The hair cells of the organ of Corti appear to be intact between 13 and 17 days of age but the fluid spaces around the outer hair cells and in the tunnel of Corti are abnormal. Most of the hair cells have degenerated by 40 days (3). The macula of the sacculus may degenerate in both the head-tossing and normally behaving mice, but remains histologically normal in many of them (2). Stimulus-induced action potential in the auditory nerve is absent at all ages tested from 12 days on (3). However, central auditory function is preserved as shown by evoked action potentials in the inferior colliculus in response to direct stimulation of the cochlear nerve (1).

References

1. Bock, G.R., M.P. Frank, and K.P. Steel. 1982. Preservation of central auditory function in the *deafness* mouse. Brain Res. 239:608–612.
2. Deol, M.S., and W. Kocher. 1958. A new gene for deafness in the mouse. Heredity 12:463–466.
3. Steel, K.P., and G.R. Bock. 1980. The nature of inherited deafness in *deafness* mice. Nature 288:159–161.

Do, disoriented, semidominant, extinct

Arose spontaneously in the BALB/c strain. Heterozygotes exhibit a wide range of unusual behavior from mere hesitation, through heaving motions, to a tendency to attempt handstands. Behavior in any one individual is consistent throughout life. Most heterozygotes breed well. Homozygotes are completely disoriented in motion and position sense and die prior to weaning, probably from malnutrition (1).

References

1. Dickie, M.M. 1968. Mouse News Lett. 38:27.

Dom, dominant megacolon, semidominant, Chr 15

Arose in a hybrid stock carrying bouncy, B6C3-*a*/*a-bc*. Heterozygotes have a white belly spot of variable size and white feet. On the B6C3 hybrid background, about 30 per cent develop severe megacolon before or shortly after weaning and die; on the inbred C57BL/6 background, few survive to weaning. Homozygotes die *in utero* before 13 days. The megacolon is associated with a deficiency of enteric ganglion cells in the colon, the deficiency being minimal at the anterior end of the colon and complete or nearly so at the posterior end. *Dom* is located in the central region of Chr 15 (1).

References

1. Lane, P.W., and H.M. Liu. 1984. Association of megacolon with a new dominant spotting gene (*Dom*) in the mouse. J. Hered. 75:435–439.

dp, dilution-Peru, recessive, Chr 15

Arose in descendants of a cross between CBA/Fa and a Peru-Harland female. Homozygotes have diluted coat color, slightly darker than that of misty (*m*/*m*). They are viable and fertile (1).

References

1. Wallace, M.E. 1971. Mouse News Lett. 44:18.

dpy, dumpy, recessive, Chr 13

Found among the F2 mice of a cross between BALB/c-*go* and a stock carrying *bp*. Homozygotes are small with shortened snout and hind feet. The inner metatarsals and the hallux are markedly reduced. Homozygotes of both sexes are fertile (1).

References

1. Hollander, W.F. 1981. Linkage relations of dumpy, a recessive mutant on chromosome 13 of the mouse. J. Hered. 72:358–359.

dr locus, Chr 1

dr, dreher, recessive. Arose in the wild, presumably as a spontaneous mutation, and was obtained in a litter of four, all homozygous, captured in a factory in Detmold, Germany (3). Homozygotes show the circling, head-tossing, deafness, and hyperactivity characteristic of the circling mutants. They may also have a partial or complete white belt (5). The bony and membranous labyrinths develop abnormally, and there may be hydrocephalus of the hindbrain (1, 4). The defects of the ear and hindbrain trace back to an abnormality of the neural tube at 9 days of gestation. The roof of the posterior part of the rhombencephalon, normally very thin, in *dr*/*dr* embryos is thick like that of the neural tube in the posterior region, as though joining of the lateral plates had extended more anteriorly than normal (2).

dr^sst, shaker short-tail, recessive. Arose in the BALB/cCF strain. Homozygotes are recognizable at birth by a short or blunt tail with a small filament at the tip and often by a small blood bleb on the top of the head. About one quarter of pigmented homozygotes develop a white belly spot. Beginning at 10 days of age, the righting reflex is defective, some homozygotes never acquiring the ability to right themselves. Affected animals sit with widely splayed hindlegs and usually fall over when trying to groom or eat. Locomotion in aroused animals is incessant and chaotic. Hearing appears to be impaired. Most homozygotes die by 30 days of age. Examination of the brain shows that the cerebellum is smaller than normal, sometimes markedly so, with abnormal patterns of foliation. The cellular organization of the lobes appears normal, however. The hippocampus and the corpus callosum also may be abnormal. The array of abnormalities suggests a defect in development of the neural tube (7). A very similar mutation at the same locus, *dr^J* (formerly *sst^J*) occurred at the Jackson Laboratory in a C3B6-*A^{w-J}*/*A* hybrid stock (6).

dr^{2J}, dreher-2J. Arose in the AKR/J strain. Homozygotes show the behavioral abnormalities and white belly spot characteristic of *dr*/*dr* mice. In addition, homozygous females have grossly underdeveloped Fallopian tubes, a condition not described for the other *dr* alleles (8).

References

1. Bierwolf, D. 1958. Die Embryogenese des Hydrocephalus und der Kleinhirnmissbildungen beim Dreherstamm der Hausmaus. Morphol. Jahr. 99:542–612.
2. Deol, M.S. 1964. The origin of the abnormalities of the inner ear in dreher mice. J. Embryol. Exp. Morphol. 12:727–733.
3. Falconer, D.S., and U. Sierts-Roth. 1951. Dreher, ein neues Gen der Tanzmausgruppe bei der Hausmaus. Z. Indukt. Abstammungs-Vererbungsl. 84:71–73.
4. Fischer, H. 1958. Die Embryogenese der Innenohrmissbildungen bei der spontanmutierten Dreherstamm der Hausmaus. Z. Mikroskop.-Anat. Forsch. 64:476–497.
5. Lyon, M.F. 1961. Linkage relations and some pleiotropic effects of the dreher mutant of the house mouse. Genet. Res. 2:92–95.
6. Sweet, H.O., and D. Wahlsten. 1983. Mouse News Lett. 69:26.
7. Wahlsten, D., J.P. Lyons, and W. Zagaja. 1983. Shaker short-tail, a spontaneous neurological mutant in the mouse. J. Hered. 74:421–425.

8. Washburn, L., and E.M. Eicher. 1986. A mutation at the dreher locus (*dr^{2J}*). Mouse News Lett. 75:28–29.

Dre, dominant reduced ear, semidominant, Chr 4

Found in offspring of an X-irradiated female. Heterozygotes are small, the pinna is slightly reduced, and the head may be blunt and rounded. The homozygotes have not been identified (1).

References

1. Kelly, E.M. 1968. Mouse News Lett. 38:31.

Ds, disorganization, semidominant, Chr 14

Arose spontaneously in the inbred DA/Hu strain. *Ds* disrupts the orderly processes of development. Most homozygotes probably die *in utero* after implantation and before term (2), but some may possibly be normal and viable (1). Heterozygotes are normal, deviate slightly, or are monstrous individuals with multiple defects. The defects show great variety, range of severity, and random distribution among derivatives of all germ layers. Penetrance varied between 9 and 89 per cent on different genetic backgrounds observed by Hummel (2,3).

References

1. Guénet, J.L. 1976. Mouse News Lett. 54:51.
2. Hummel, K.P. 1958. The inheritance and expression of disorganization, an unusual mutation in the mouse. J. Exp. Zool. 137:389–423.
3. Hummel, K.P. 1959. Developmental anomalies in mice resulting from the action of the gene, disorganization, a semi-dominant lethal. Pediatrics 23:212–221.

Dsh, short digits, semidominant

Found among offspring of an X-irradiated (101 × C3H)F1 male. Heterozygotes have shortening of all digits on both fore- and hindfeet. One phalanx is absent in digits 2 through 5. Other effects include a large hole between the frontals, dyssymphyses between many cervical vertebrae, and bony plates over many ribs. Penetrance is complete but expression is variable. Homozygotes are short limbed dwarfs lacking forefeet and hindfeet. They usually lack eyes, ears, external genitalia, and a mouth. Many die before embryonic day 12 but some have survived to day 18 (1).

References

1. Selby, P.B., Rayner, and McKinley Jr. 1986. Short digits, a new mutation. Mouse News Lett. 75:43.

Dsi-1 locus, David Steffen integration-1, Chr 4

This locus is the mouse homolog of a 13.5-kb region of DNA that contains proviral insertions in 10 to 20 per cent of Moloney virus-induced thymomas in rats. In rat thymomas, the sequence is associated with two clusters of DNase I hypersensitive sites. It appears not to be transcribed. Homologous regions occur also in man, hamsters, and chickens. In the mouse, the locus was found to be on Chr 4 by use of mouse–Chinese hamster somatic cell hybrids screened with a genomic DNA clone isolated from a virus-induced rat lymphoma (1).

References

1. Vijaya, S., D.L. Steffen, C. Kozak, and H.L. Robinson. 1987. *Dsi-1*, a region with frequent proviral insertions in Moloney murine leukemia virus-induced rat thymomas. J. Virol. 61:1164–1170.

dsu, dilute suppressor, recessive, Chr 1

Arose spontaneously in a partially inbred stock homozygous for nonagouti (*a*) and dilute (*d*) and carrying curly whiskers (*cw*). In mice homozygous for *d* and *dsu*, expression of *d* is suppressed, and the coat color resembles that of non-dilute mice. The effect of *dsu* on coat color is due to its effects on melanocyte morphology and clumping of pigment. In *d/d* mice, melanocytes have fewer and thinner dendritic processes which results in clumped pigment granules and the dilute color. In *d/d dsu/dsu* mice, the morphology of the melanocytes and pigment granules is restored to normal and the number of melanocytes is increased, although not to the fully normal number. *dsu* has no effect on the neurological abnormalities of the lethal alleles at the *d* locus. *dsu* is not linked to *d* and is located near *Idh-1* on Chr 1 (1).

References

1. Sweet, H.O. 1983. Dilute suppressor, a new suppressor gene in the house mouse. J. Hered. 74:305–306.

dt locus, Chr 1

Several mutations at this locus have occurred at the Jackson Laboratory and elsewhere. Two have been extensively studied.

dt, dystonia musculorum, recessive. Arose spontaneously in mice bred at the Institute of Animal genetics, Edinburgh. Homozygotes can first be recognized at 7 to 10 days of age by clasping of the hindlimbs when mice are lifted by the tail. There is increasing incoordi-nation with alternating hyperextension and hyperflexion of the limbs. There is no obvious paralysis. Many affected animals die before weaning, but some survive several months. The clinical condition becomes relatively stationary after the first phase of deterioration. Histologically the nervous system shows degenerative changes and progressive loss of nerve fibers in the central and peripheral branches of the sensory ganglion cells of the spinal and cranial nerves, in the central sensory pathways, and in peripheral sensory structures such as skin, Pacinian corpuscles, and muscle spindles. In longer surviving animals there is also evidence of lower motor neuron involvement (1, 2). In the peripheral nervous system, there is some segmental demyelination and other abnormalities of the myelin sheaths. Transplantation experiments with sciatic nerve showed that the defect resides in the Schwann cells, not in the axons (7). The histochemical and ultrastructural changes in the degenerating sensory nerves have been described by Janota (4) and Hanker and Peach (3).

dt^{alb}, dystonia musculorum-Albany. This allele at the *dt* locus arose in a BALB/cByJ strain at the New York State Department of Health laboratories in Albany, New York. In homozygotes, the pathology of the peripheral and central sensory pathways is similar to that described for *dt/dt* mice. In addition, there are pathological changes in the brain. The coarse neurons of the red nucleus show distorted shapes, eccentric nuclei, and varying degrees of central chromatolysis, and there are patches of small, darkly staining, and abnormally shaped cells in the striatum (6). There is loss of almost 50 per cent of γ-amino butyric acid (GABA) synthetic capacity in whole tissue samples of striatum and substantia nigra (5). Tyrosine hydroxylase activity and the level of a norepinephrin metabolite, MHPG, are elevated in the cerebellum, although no abnormalities of granule or Purkinje cells are seen. The increase may be a response to decreased afferent information which provokes a compensatory increase in noradrenergic tonus (8).

References

1. Duchen, L.W. 1976. *Dystonia musculorum*—an inherited disease of the nervous system in the mouse. Adv. Neurol. 14:353–365.
2. Duchen, L.W., and S.J. Strich (with appendix by D.S. Falconer). 1964. Clinical and pathological studies of an hereditary neuropathy in mice (*dystonia musculorum*). Brain 87:367–378.
3. Hanker, J.S., and R. Peach. 1976. Histochemical and ultrastructural studies on primary sensory neurons in mice with dystonia musculorum. I. Acetylcholinesterase and

lysosomal hydrolases. Neuropathol. Appl. Neurobiol. 2:79–97.

4. Janota. I. 1972. Ultrastructural studies of an hereditary sensory neuropathy in mice (dystonia musculorum). Brain 95:529–536.
5. Messer, A., and D. Gordon. 1979. Changes in whole tissue biosynthesis of γ-amino butyric acid (GABA) in basal ganglia of the dystonia (*dt^alb*) mouse. Life Sci. 25:2217–2221.
6. Messer, A., and N.L. Strominger. 1980. An allele of the mouse mutant dystonia musculorum exhibits lesions in red nucleus and striatum. Neuroscience 5:543–549.
7. Moss, T.H. 1981. Schwann cell involvement in the neurological lesion of the dystonic mutant mouse. A nerve grafting study. J. Neurol. Sci. 49:207–232.
8. Ricker, D.K., A. Messer, and R.H. Roth. 1981. Increased noradrenergic metabolism in the cerebellum of the mouse mutant *dystonia musculorum*. J. Neurochem. 37:649–654.

Dtc-1 locus (formerly *tra-1*), detected by T-cells antigen-1

This locus controls an antigen present on Con A- and LPS-stimulated lymphoblasts, peritoneal macrophages, and SV40-transformed embryo fibroblasts. The antigen is detected by cell-mediated lympholysis (CTL) techniques. Lysis by CTL of cells bearing the antigen is H-$2D^k$-restricted. The allele *Dtc-1^b* determines presence of the antigen and occurs in the C57BL/6 and all inbred strains tested except *B6-H-2^k* which carries the *Dtc-1^m* allele determining absence of the antigen. *Dtc-1^m* apparently arose as a mutation in *B6-H-2^k*. The locus is not linked to *T* or *H-2* on Chr 17 (1).

References

1. Rinchik, E.M., and D.B. Amos. 1986. A mutation in a non-MHC murine cell surface antigen detected by cytotoxic T lymphocytes. J. Immunol. 136:1992–1998.

du locus, Chr 9

du, ducky, recessive. Arose as a spontaneous mutation in a non-inbred stock (5). Viability is somewhat less than normal. Males living to maturity may be fertile, but are poor breeders. Females rarely breed. Homozygotes show a waddling or reeling gait and a tendency to fall to one side. They are slightly smaller than normal and may occasionally have seizures. Histologically they show severe dysgenesis of hindbrain and spinal cord, myelin deficiency which is more marked the more caudad the CNS region, and demyelination and axonal dystrophy in selective fiber systems including the spinocerebellar and vestibulospinal tracts. There is a deficit of cerebrosides in the hindbrain and spinal cord, but other lipid classes are present in normal amounts relative to the size of the CNS (4). Non-specific esterases of liver and kidney show developmental changes that are different from normal (6).

du^td, torpid, recessive. Arose in the DBA/2J strain. Homozygotes move very slowly. They may lie on their backs for minutes at a time, sometimes with the mouth and feet quivering. When standing, one foot is often raised and quivers. Homozygotes are viable. Both sexes are fertile, but males are more productive (1,2). It is not certain whether the behavioral differences between *du/du* and *du^td/du^td* are due to real differences between the alleles or to differences in genetic background. The pituitary of female *du^td/du^td* mice is deficient in prolactin cells (3).

References

1. Dickie, M.M. 1965. Mouse News Lett. 32:45.
2. Dickie, M.M. 1969. Mouse News Lett. 41:31.
3. Dung, H.C. 1975. Evidence of prolactin cell deficiency in connection with low reproductive efficiency of female 'torpid' mice. J. Reprod. Fertil. 45:91–99.
4. Meier, H., and A.D. MacPike. 1970. Ducky, a neurological mutation in mice characterized by deficiency of cerebrosides. Exp. Med. Surg. 28:256–269.
5. Snell, G.D. 1955. Ducky, a new second chromosome mutation in the mouse. J. Hered. 46:27–29.
6. Tsuji, S., and H. Meier. 1970. Esterase alterations in liver and kidneys of ducky, a neurological mutation in mice. Biochem. Genet. 4:539–547.

dv, dervish, recessive

Possibly radiation-induced. Homozygotes show circling behavior and are viable (1). No tests for allelism with other waltzing mutants have been reported.

References

1. Russell, L.B. 1961. Mouse News Lett. 25:64.

dw, dwarf, recessive, Chr 16

Arose in a stock of silver mice obtained from an English fancier (14). Another mutation at the same locus, *dw^J*, occurred in the C3H/HeJ strain (8). Homozygous *dw/dw* mice are about one-fourth to one-third normal size and are sterile. The small size is due to a defective anterior pituitary in which there is a great deficiency of growth hormone (GH)-producing, prolactin (PRL)-producing, and thyrotropic hormone (TSH)-producing cells (1, 5, 9,). The anterior pituitary is already abnormal at birth with no identifiable GH or PRL cells (15). GH and PRL synthesis is not detectable at any stages from birth to 6 weeks (13), and there is probably also a deficiency of TSH and corticotropin (10). The defects in growth and fertility may be corrected by pituitary implants (4) or by administration of pituitary hormones (2, 3). Dwarf mice have been reported to have a defective immune response that primarily affects the T-cell system (7), but Schneider (12) and Dumont *et al.* (6) have been unable to confirm these findings and attribute the previous results to the secondary effects of dwarfing on overall vigor and nutritional status. Dwarf homozygotes have a severe deficiency of dopamine in the median eminence (11). There is a very large literature on the use of the dwarf mouse as a naturally GH-deficient animal.

References

1. Barkley, M.S., A. Bartke, D.S. Gross, and Y.N. Sinha. 1982. Prolactin status of hereditary dwarf mice. Endocrinology 110:2088–2096.
2. Bartke, A. 1967. Prolactin deficiency in genetically sterile dwarf mice. *In* S.G. Spickett and J.G.M. Shire, eds., Endocrine Genetics, 193–197. Cambridge Univ. Press, Cambridge.
3. Bartke, A. 1968. The response of dwarf mice to murine thyroid-stimulating hormone. Gen. Comp. Endocrinol. 11:246–247.
4. Carsner, R.L., and E.G. Rennels. 1960. Primary site of gene action in anterior pituitary dwarf mice. Science 131:829.
5. Cheng, T.C., W.G. Beamer, A. Phillips III, A. Bartke, R.L. Mallonee, and C. Dowling. 1983. Etiology of growth hormone deficiency in little, Ames, and Snell dwarf mice. Endocrinology 113:1669–1678.
6. Dumont, F., F. Robert, and P. Bischoff. 1979. T and B lymphocytes in pituitary dwarf Snell–Bagg mice. Immunology 38:23–31.
7. Duquesnoy, R.J., and G.M. Pedersen. 1981. Immunologic and hematologic deficiencies of the hypopituitary dwarf mouse. *In* M.E. Gershwin and B. Merchant, eds., Immunologic defects in Laboratory Animals, Vol. 1, 309–324. Plenum Press, New York.
8. Eicher, E.M., and W.G. Beamer. 1980. New mouse *dw* allele: genetic location and effects on lifespan and growth hormone levels. J. Hered. 71:187–190.
9. Elftman, H., and O. Wegelius. 1959. Anterior pituitary cytology of the dwarf mouse. Anat. Rec. 135:43–49.
10. Lewis, U.J. 1967. Growth hormone of normal and dwarf mice. *In* S.G. Spickett and J.G.M. Shire, eds., Endocrine Genetics, 179–188. Cambridge Univ. Press, Cambridge.
11. Morgan, W.W., A. Bartke, and K. Pfeil. 1981. Deficiency of dopamine in the median eminence of Snell dwarf mice. Endocrinology 109:2069–2075.
12. Schneider, G.B. 1976. Immunological competence in Snell–Bagg pituitary dwarf mice: response to the contact-sensitizing agent oxazalone. Am. J. Anat. 145:371–394.
13. Slabaugh, M.B., M.E. Lieberman, J.J. Rutledge, and J. Gorski. 1981. Growth hormone and prolactin synthesis in normal and homozygous Snell and Ames dwarf mice. Endocrinology 109:1040–1046.
14. Snell, G.D. 1929. 'Dwarf', a new mendelian recessive character in the house mouse. Proc. Natl. Acad. Sci. USA 15:733–734.
15. Wilson, D.B., and E. Christensen. 1981. Fine structure of somatotrophs and mammotrophs during development of the dwarf (*dw*) mutant mouse. J. Anat. 133:407–417.

dy locus, Chr 10

dy, dystrophia muscularis, recessive. Arose as a spontaneous mutation in the 129/Re inbred strain. Homozygotes are characterized by progressive weakness and paralysis beginning at about $3\frac{1}{2}$ weeks of age. The hindlimbs are affected first, later the axial and forelimb musculature. Death usually occurs before 6 months of age, and the mice are usually sterile. Skeletal muscle shows degenerative changes with proliferation of sarcolemmal nuclei, increase in amount of interstitial tissue, and size variation among individual muscle fibers (14). In the dorsal and ventral roots of the peripheral nerves, both spinal and cranial, Schwann cells fail to separate and ensheathe axons so that groups of closely apposed naked axons, normally seen only in early stages, persist into adulthood (1). At birth there is a deficiency of Schwann cells in the spinal roots, but presence of many undifferentiated cells capable of cell division, which are probably uncommitted Schwann cells that can differentiate when transplanted to a normal environment (18). In the rest of the peripheral nervous system, the basement membrane of Schwann cells is interrupted by gaps (11), the internodal gap in the nodes of Ranvier is lengthened (2, 8), and there is delayed onset of myelination with fewer myelinated axons and shorter internode length (9). There is reduced conduction velocity in spinal roots and peripheral nerves (20) and evidence of defective axonal transport in the sciatic nerve (3, 12). It is probable that the myelination defects of the peripheral nerves are due to an intrinsic defect in the dystrophic Schwann cells but the evidence from transplantation experiments, chimeras, and cell cultures is somewhat conflicting (5, 15, 17, 19,). There is some evidence for defective myelination in the central nervous system also (21). In culture, myogenic cells from *dy/dy* mice develop muscle properties just as well as cells from control mice (23) and there is no peripheral block of neuromuscular transmission (6, 7). Muscle of *dy/dy* mice transplanted into normal hosts is indistinguishable from transplanted control muscle by 100 days (4). There is a very large literature on morphological and biochemical defects in *dy/dy* muscle.

dy

dy^{2J}, dystrophia muscularis-2J, recessive. This allele arose in the WK/Re strain in 1969 (13). Its effects on that genetic background appeared to be milder than those previously described for dy, but it is similar clinically and histologically to dy when on the same genetic background (10). It is not certain whether all the observed differences between dy^{2J}/dy^{2J} and dy/dy are due to differences in genetic background rather than to real differences between the alleles. Abnormalities of the nerve roots and peripheral nerves are like those of dy (16, 22).

References

1. Bradley, W.G., and M. Jenkison, 1973. Abnormalities of peripheral nerves in murine muscular dystrophy. J. Neurol. Sci. 18:227–247.
2. Bradley, W.G., E. Jaros, and M. Jenkison. 1977. The nodes of Ranvier in the nerves of mice with muscular dystrophy. J. Neuropathol. Exp. Neurol. 36:797–806.
3. Brimijoin, S., and C. Jablecki, 1976. Reduced axonal transport of dopamine-β-hydroxylase activity in dystrophic mice: evidence for abnormality of adrenergic nerve cells. Exp. Neurol. 53:454–464.
4. Bourke, D.L., and M. Ontell. 1986. Modification of the phenotypic expression of murine dystrophy: a morphological study. Anat. Rec. 214:17–24.
5. Bunge, R.P., M.B. Bunge, E. Okada, and C.J. Cornbrooks. 1980. Abnormalities expressed in cultures prepared from peripheral nerve tissues of trembler and dystrophic mice. In N. Baumann, ed., Neurological Mutations Affecting Myelination, 433–446. Elsevier/North Holland Biomedical Press, Amsterdam.
6. Carbonetto, S. 1977. Neuromuscular transmission in dystrophic mice. J. Neurophysiol. 40:836–843.
7. Harris, J.B., and R.R. Ribchester. 1978. Neuromuscular transmission is adequate in identified abnormal dystrophic muscle fibers. Nature 271:362–364.
8. Jaros, E., and W.G. Bradley. 1979. Atypical axon–Schwann cell relationships in the common peroneal nerve of the dystrophic mouse: an ultrastructural study. Neuropathol. Appl. Neurobiol. 5:133–147.
9. Jaros, E., and M. Jenkison. 1983. Defective differentiation of peripheral nerves in the dystrophic mouse. Dev. Brain Res. 6:231–242.
10. MacPike, A.D., and H. Meier. 1976. Comparison of dy and dy^{2J}, two alleles expressing forms of muscular dystrophy in the mouse. Proc. Soc. Exp. Biol. Med. 151:670–672.
11. Madrid, R.E., E. Jaros, M.J. Cullen and W.G. Bradley. 1975. Genetically determined defect of Schwann cell basement membrane in dystrophic mice. Nature 257:319–321.
12. McLane, J.A., and W.O. McClure. 1977. Rapid axoplasmic transport in dystrophic mice. J. Neurochem. 29:865–872.
13. Meier, H., and J.L. Southard. 1970. Muscular dystrophy in the mouse caused by an allele at the dy locus. Life Sci. 9:137–144.
14. Michelson, A.M., E.S. Russell, and P.J. Harman. 1955. Dystrophia muscularis: a hereditary primary myopathy in the house mouse. Proc. Natl. Acad. Sci. USA 41:1079–1084.
15. Montgomery, A. 1975. Parabiotic reinnervation in normal and dystrophic mice. Part I. Muscle weight and physiological studies. J. Neurol. Sci. 26:401–423.
16. Montgomery, A., and L. Swenarchuk. 1977. Dystrophic mice show age related muscle fiber and myelinated axon losses. Nature 267:167–168.
17. Neerunjun, J.S., and V. Dubowitz. 1977. Regeneration of muscles transplanted between normal and dystrophic mice: a quantitative study of early transplants. J. Anat. 124:459–467.
18. Perkins, C.S., G.M. Bray, and A.J. Aguayo. 1981. Ongoing block of Schwann cell differentiation and deployment in dystrophic mouse spinal roots. Dev. Brain Res. 1:213–220.
19. Peterson, A.C., and G.M. Bray. 1984. Normal basal laminas are realized on dystrophic Schwann cells in dystrophic $\longleftrightarrow$ shiverer chimera nerves. J. Cell Biol. 99:1831–1837.
20. Rasminsky, M., R.E. Kearney, A.J. Aguayo, and G.M. Bray. 1978. Conduction of nerve impulses in spinal roots and peripheral nerves of dystrophic mice. Brain Res. 143:71–85.
21. Tsuji, S., and H. Matsushita. 1985. Evidence on hypomyelination of central nervous system in murine muscular dystrophy. J. Neurol. Sci. 68:175 184.
22. Weinberg, H.J., P.S. Spencer, and C.S. Raine. 1975. Aberrant PNS development in dystrophic mice. Brain Res. 88:532–537.
23. Yaffe, D., and O. Saxel. 1977. Serial passaging and differentiation of myogenic cells isolated from dystrophic mouse muscle. Nature 270:725–727.

dyl, dysgenetic lens, recessive, Chr 4

Arose in the BALB/cLiA strain. Homozygotes are recognizable by smaller than normal eyes. The pupil is markedly smaller than normal and irregular in outline and the cornea is opaque in varying degrees. In sections of the eye, the lens is much reduced in size and irregular in shape, and there is a persistent connection between the lens and the corneal epithelium. The interior of the lens is disorganized, with large irregular vacuoles. Affected mice of both sexes are fully viable and fertile. The defect is first recognizable at 10 days of gestation when the lens vesicle of affected embryos fails to separate from the ectoderm. By day 13, the lens is markedly smaller than normal. Vacuolar structures soon appear in the lens and by day 16 the lens is catar-actous in all mutants. Part of the lens protrudes to the exterior through the persistent connection and materials from the lens are expelled, resulting in the greatly reduced and distorted lens seen in young ani-mals (2). Lens crystallins are produced in normal amounts in the early stages before degeneration becomes severe (1). *dyl* is on Chr 4 about 12 cM distal to the *b* locus (3).

References

1. Brahma, S.K., and S. Sanyal. 1984. Immunohistochemi-cal studies of lens crystallins in the dysgenetic lens (*dyl*) mutant mice. Exp. Eye Res. 38:305–311.
2. Sanyal, S., and R.K. Hawkins. 1979. *Dysgenetic lens (dyl)*—a new gene in the mouse. Invest. Ophthalmol. Vis. Sci. 18:642–645.
3. Sanyal, S., R. van Nie, J. de Moes, and R.K. Hawkins. 1986. Map position of dysgenetic lens (*dyl*) locus on chro-mosome 4 in the mouse. Genet. Res. 48:199–200.

E

e locus, Chr 8

This locus is probably homologous with the extension locus which occurs in several other mammals and gov-erns the extension of eumelanin with the opposite effects on phaeomelanin. In this series of alleles, in contrast to those of the agouti locus, the dominant alleles produce partial or full extension of black, the wild-type allele produces normal extension of yellow and black as in the agouti pattern, and the recessive alleles produce partial or full extension of yellow (6).

E^{so}, sombre, semidominant. Arose spontaneously in the C3H inbred strain (1). Heterozygotes resemble nonagouti (*a/a*) except that they have darker ears and nipples and have a stronger tendency to develop yellow hairs on their sides when mature. Homozygotes are black all over. Except for a few yellowed hairs on the perineum, they exactly resemble extreme nonagoutis (a^e/a^e). Viability and fertility are normal in both homo-zygotes and heterozygotes.

E^{tob}, tobacco darkening. Mice of the species *Mus pos-chiavinus*, the tobacco mouse, are homozygous for this allele. They are black in color until the eighth week, after which the flanks become agouti. In descendants of crosses with *Mus musculus*, heterozygotes were recovered and shown to have an agouti pattern with darkened back. E^{tob}/e *a/a* mice are black (8, 9).

e, recessive yellow. Arose spontaneously in the C57BL/6Ha strain. Homozygotes are a clear yellow with black eyes. Slight sootiness on the back may occur before weaning but disappears with successive molts. Viability, fertility, body weight, and tumor incidence are normal. In combination with either piebald (*s/s*) or belted (*bt/bt*), *e/e* decreases the amount of white spot-ting. Leaden (*ln/ln*) prevents the expression of *e/e*, the double mutant being indistinguishable from *ln/ln* (3). By use of embryonic neural tube-skin recombination grafts, it was shown that the site of action of *e* is in the melanocyte. However, *e/e* melanocytes produce yellow pigment only in the hair follicles, the pigment produced in the skin, eye, and elsewhere being black (4, 5). In contrast to $A^y/$- mice, *e/e* mice do not darken after injection of α-melanocyte-stimulating hormone (α-MSH) (2), nor do their skin cells in culture produce eumelanin after treatment with α-MSH. However, the skin cells do produce eumelanin after treatment with dibutyryl cyclic AMP, suggesting that the *e* locus may control the function of an α-MSH receptor (7).

References

1. Bateman, N. 1961. Sombre, a viable dominant mutant in the house mouse. J. Hered. 52:186–189.
2. Geschwind, I.I., R.A. Huseby, and R. Nishioka. 1972. The effect of melanocyte-stimulating hormone on coat color in the mouse. Rec. Prog. Horm. Res. 28:91–130.
3. Hauschka, T.S., B.B. Jacobs, and B.A. Holdridge. 1968. Recessive yellow and its interaction with belted in the mouse. J. Hered. 59:339–341.
4. Lamoreux, M.L., and T.C. Mayer. 1975. Site of action in the development of hair pigment in recessive yellow (*e/e*) mice. Dev. Biol. 46:160–166.
5. Poole, T.W., and W.K. Silvers. 1976. An experimental analysis of the recessive yellow coat color mutant in the mouse. Dev. Biol. 48:377–381.
6. Searle, A.G. 1968. An extension series in the mouse. J. Hered. 59:341–342.
7. Tamate, H.B., and T. Takeuchi. 1984. Action of the *e* locus of mice in the response of phaeomelanic hair folli-

cles to α-melanocyte-stimulating hormone in vitro. Science 224:1241–1242.

8. von Lehmann, E. 1973. Mouse News Lett. 48:23.
9. von Lehmann, E. 1974. Mouse News Lett. 50:26–27.

Ea-1 locus, erythrocyte antigen-1, Chr 8

Wild populations of house mice have been found to contain four phenotypes with respect to erythrocyte antigens, designated A, B, AB, and O. The antigens are under the control of a system of three alleles: *Ea-1ᵃ* determines the A antigen, *Ea-1ᵇ* determines the B antigen, and *Ea-1ᵒ* is a null allele determining absence of an antigen. *Ea-1ᵃ* and *Ea-1ᵇ* are co-dominant, and both are dominant to *Ea-1ᵒ*. Thirteen inbred strains surveyed were all homozygous for *Ea-1ᵒ* (1,2).

References

1. Foster, M., M.L. Petras, and D.L. Gassner. 1968. The *Ea-1* blood group locus of the house mouse: inheritance, linkage, polymorphism and control of antibody synthesis. Proc. 12th Int. Cong. Genet. 1:245 (Abstr.).
2. Singer, M.F., M. Foster, M.L. Petras, P. Tomlin, and R.W. Sloane. 1964. A new case of blood group inheritance in the house mouse, *Mus musculus*. Genetics 50:285–286 (Abstr.).

Ea-2 locus (formerly rho, R-Z, H-14), erythrocyte antigen-2

This locus regulates a cellular antigen system that is serologically detectable on erythrocytes, spleen, thymus, liver, lung, brain, and testis (3). The allele *Ea-2ᵃ* determines the Ea-2.1 antigen and is found in the F/St, RF/J, FM/Un, and RIII strains; the allele *Ea-2ᵇ* determines the Ea-2.2 antigen and is found in most other strains (4). Linkage of *Ea-2* to *H-2* in Chr 17 has been reported (2), but others have been unable to confirm this linkage (5) (see *Ea-?*). Hemagglutinating antibody response against Ea-2.1 is much more vigorous in mice bearing *H-2ʳ* than in mice bearing *H-2ᵇ* (6). Incompatibility at the *Ea-2* locus does not prevent survival of wild-type marrow cells injected intravenously into *W/Wᵛ* anemic hosts (1).

References

1. Harrison, D.E., and M. Cherry. 1975. Survival of marrow allografts in *W/Wᵛ* anemic mice: effect of disparity at the *Ea-2* locus. Immunogenetics 2:219–229.
2. Pizarro, O., and U. Vergera. 1973. Relationship between locus R (Ea-2) and other loci of the ninth linkage group of the house mouse. Folia Biol (Praha) 19:89–94.
3. Popp, D.M. 1969. *Histocompatibility-14*: correlation of the isoantigen rho and the R-Z locus. Transplantation 7:233–241.

4. Snell, G.D., and M. Cherry. 1972. Loci determining cell surface alloantigens. *In* P. Emmelot and P. Bentvelzen, eds., RNA Viruses and Host Genome in Oncogenesis, 221–228. North-Holland, Amsterdam.
5. Stimpfling, J.H. 1974. The Ea-2 (H-14) cellular antigen system in the mouse. Transplantation 18:350–356.
6. Stimpfling, J.H., A.E. Reichert, and S. Blanchard. 1976. Genetic control of the immune response to the Ea-2 alloantigen system of the mouse. I. The humoral antibody response. J. Immunol. 116:1096–1098.

Ea-3 locus (formerly lambda), erythrocyte antigen-3

This locus controls an antigen present on erythrocytes. The allele *Ea-3ᵃ* determines presence of the antigen and occurs in strain C57L/J; the allele *Ea-3ᵇ* determines absence of the antigen and occurs in strains A/Mv, BALB/cDe, C3H/He, and C57BL/10 (1,2).

References

1. Egorov, I.K. 1965. A new isoantigen of mouse erythrocytes. Genetika. Mosk. 1(6):80–85.
2. Snell, G.D., and M. Cherry. 1972. Loci determining cell surface alloantigens. *In* P. Emmelot and P. Bentvelzen, eds., RNA Viruses and Host Genome in Oncogenesis, 221–228. North-Holland, Amsterdam.

Ea-4 locus (formerly D), erythrocyte antigen-4, Chr 19 (P.M. Hogarth, personal communication)

This locus controls an antigen present in high concentration on erythrocytes, in moderate or lesser amounts on lymph node cells, lung, and kidney, and absent or nearly so on thymus, gut, and other tissues and in serum (3). The allele *Ea-4ᵃ* determines the antigen Ea-4.1 and is present in most inbred strains; the allele *Ea-4ᵇ* determines the antigen Ea-4.2 and is present in strains of the C57BL group (1, 4). The presence of the Ea-4.2 antigen on erythrocytes (or presence of the product of a dominant gene closely linked to *Ea-4*) influences the interaction of anti-H-2K.31 antibodies with H-2K.31 sites on the same erythrocytes in such a way that the erythrocytes are agglutinated only at high antibody titers (2).

References

1. Klein, J. and J. Martinkova. 1968. A new non-H-2 antigen of C57BL mice. Folia Biol. (Praha) 14:237–238.
2. Lilly, F. 1974. Genetic regulation of the expression of antigen specificity H-2K.31 on mouse erythrocytes. Immunogenetics 1:22–32.
3. Shreffler, D.C. 1966. A new erythrocyte antigen in the house mouse. Genetics 54:362 (Abstr.).

4. Snell, G.D., and M. Cherry. 1972. Loci determining cell surface alloantigens. *In* P. Emmelot and P. Bentvelzen, eds., RNA Viruses and Host Genome in Oncogenesis, 221–228. North-Holland, Amsterdam.

Ea-5 locus (formerly *H-5*), erythrocyte antigen-5

This locus controls a serologically detectable antigen present on red cells, kidney, testis, and lung, and in smaller amounts on spleen, brain, and muscle (1). The allele *Ea-5ᵃ* determines the antigen Ea-5.1 and is present in the C3H/St, 129, and A strains; the allele *Ea-5ᵇ* determines absence of the Ea-5.1 antigen and is present in the C3H/He, DBA, C57BL, and several other strains (2,3).

References

1. Amos, D.B., M. Zumpft, and P. Armstrong. 1963. H-5.A and H-6.A, two mouse isoantigens on red cells and tissues detected serologically. Transplantation 1:270–283.
2. Snell, G.D., and M. Cherry. 1972. Loci determining cell surface alloantigens. *In* P. Emmelot and P. Bentvelzen, eds., RNA Viruses and Host Genome in Oncogenesis, 221–228. North-Holland, Amsterdam.
3. Snell, G.D., and J.H. Stimpfling. 1966. Genetics of tissue transplantation. *In* E.L. Green, ed., Biology of the Laboratory Mouse, 2nd ed., 457–491. McGraw-Hill, New York.

Ea-6 locus (formerly *H-6*), Chr 2, erythrocyte antigen-6

This locus controls a serologically detectable antigen present on red cells, testis, brain, gut, and spleen, and in smaller amounts on lung, liver, kidney, and muscle (1). The allele *Ea-6ᵃ* determines the antigen Ea-6.1 and is present in the C3H/He, C57BL, A, and 129 strains (3,4); the allele *Ea-6ᵇ* determines the antigen Ea-6.2 and is present in the BALB/c, DBA, and other strains (2).

References

1. Amos, D.B., M. Zumpft, and P. Armstrong. 1963. H-5.A and H-6.A, two mouse isoantigens on red cells and tissues detected serologically. Transplantation 1:270–283.
2. Lilly, F. 1974. Genetic regulation of the expression of antigen specificity H-2K.31 on mouse erythrocytes. Immunogenetics 1:22–32.
3. Snell, G.D., and M. Cherry. 1972. Loci determining cell surface alloantigens. *In* P. Emmelot and P. Bentvelzen, eds., RNA Viruses and Host Genome in Oncogenesis, 221–228. North-Holland, Amsterdam.
4. Snell, G.D., and J.H. Stimpfling. 1966. Genetics of tissue transplantation. *In* E.L. Green, ed., Biology of the

Laboratory Mouse, 2nd ed., 457–491. McGraw-Hill, New York.

Ea-7 locus (formerly *T*), erythrocyte antigen-7

This locus controls a serologically detectable antigen on erythrocytes. The allele *Ea-7ᵃ* determines the antigen Ea-7.1 and is present in strains 129, LP, and C3H/He; the allele *Ea-7ᵇ* determines the antigen Ea-7.2 and is present in strains C57BL, BALB/c, DBA, A, and others (1,2).

References

1. Snell, G.D., and M. Cherry. 1972. Loci determining cell surface alloantigens. *In* P. Emmelot and P. Bentvelzen, eds., RNA Viruses and Host Genome in Oncogenesis, 221–228. North-Holland, Amsterdam.
2. Stimpfling, J.H., and G.D. Snell. 1968. Detection of a non-H-2 blood group system with the aid of B10.129(5M) mice. Transplantation 6:468–475.

Ea-8 locus, erythrocyte antigen-8

This locus controls a serologically detectable antigen present on erythrocytes. The *Ea-8ᵃ* allele determines the antigen Ea-8.1 and is present in strains A, C57BL/10, C3H/He, NZB, some wild mice, and in *Mus poschiavinus*. The *Ea-8ᵇ* allele determines absence of the antigen and is present in strains BALB/c, DBA/2, 129, and LP (1).

References

1. Klein, J. 1972. Mouse News Lett. 46:19.

Ea-9 locus, erythrocyte antigen-9, Chr 4

This locus, recently designated *Ea-9* (2), controls a serologically detected antigen on red cells. The allele *Ea-9ᵇ* determines absence of the antigen and occurs in the C57BL/10Sn and related congenic strains; the allele *Ea-9ᵃ* determines presence of the antigen and occurs in the C57BL/6J and all other inbred strains tested. *Ea-9ᵇ* appears to be closely associated with a histocompatibility allele from C57BL/10 that was introduced into the C57BL/6J background by 10 generations of backcrosses (1). Walker and Phillips-Quagliata (3) have described a locus for an antigen present on red cells of C57BL/10Sn and absent from C57BL/6J and all other inbred strains tested. Noting the antithetical strain distribution of their antigen with that described by Cherry *et al.* (1), these authors suggested that the genes determining the two antigens might be alleles. *Ea-9* is on Chr 4, about 24 cM from *Gpd-1* (2).

References

1. Cherry, M.. D.W. Bailey, and G.D. Snell. 1978. Differences between C57BL strains at an erythrocyte antigen locus. *In* H.C. Morse III, ed., Origins of Inbred Mice, 481–482. Academic Press, New York.
2. Fox, R.R., and P.M. Black. 1987. Erythrocyte antigen-9 (Ea-9). Mouse News Lett. 78:57.
3. Walker, M.C., and J.M. Phillips-Quagliata. 1985. A new erythrocyte antigen of C57BL/10 (B10) mice. Proc. Soc. Exp. Biol. Med. 178:402–406.

Ea-? locus (formerly *Ea-2*, R-Z), erythrocyte antigen-?, Chr 17

This locus was found to be linked to *H-2*, *T*, and *tf* on Chr 17 by Pizarro and Vergara (1). It controls a serologically defined erythrocyte antigen and was discovered by use of apparently the same strain combinations as those used for the *Ea-2* locus. Others have been unable to find linkage between *Ea-2* and *H-2* (see *Ea-2*).

References

1. Pizarro, O., and U. Vergara. 1973. Relationship between locus R (Ea-2) and other loci of the ninth linkage group of the house mouse. Folia Biol. (Praha) 19:89–94.

Eam locus, ethanol activity modifier, Chr 4

This locus exerts some control over the degree of decline in basal activity level induced by alcohol. The allele *Eam^h* determines a high decrement in activity and is found in the C57BL/6By strain; the allele *Eam^l* determines a lower decrement in activity and is found in the BALB/cBy strain. Heterozygotes are similar to *Eam^l* homozygotes. From analysis of CXB recombinant inbred strains, *Eam* appears to be on Chr 4 near *Exa* (1).

References

1. Oliverio, A., and B.E. Eleftheriou. 1976. Motor activity and alcohol; genetic analysis in the mouse. Physiol. Behav. 16:577–581.

eb, eye-blebs, recessive, Chr 10

Arose in a non-inbred stock of hairless mice. It resembles *my* but is not allelic to it. Penetrance is incomplete. Homozygotes may be normal or have defects of the eyes, kidneys, or feet including: anophthalmia; small, missing, or cystic kidneys; and clubbed feet, webbed toes, or extra digits (2). The eye and foot abnormalities trace to clear blisters of fluid in the hypodermis of the eye and apex of the limbs at the 12th and 13th days of gestation. The blisters enlarge and become hemorrhagic and persist until birth (1).

References

1. Beasley, A.B., and F.L. Crutchfield. 1969. Development of a mutant with abnormalities of the eye and extremities. Anat. Rec. 163:293 (Abstr.).
2. Hummel, K.P., and D.B. Chapman. 1963. Mouse News Lett. 28:32.

ec, ectopic, recessive

Found among the descendants of mice exposed to 1000 R of X-rays. Homozygotes are first distinguishable from normal at 28 to 30 days, when there is a slight increase in the size of the eyeball. The increase disappears after 6 weeks. At 5 weeks an opacity of the lens develops. The nucleus of the lens is gradually displaced through the posterior pole of the ruptured capsule. Histologically the lens epithelium is abnormal at about 4 weeks. In normal mice, it is a single layer with no mitotic figures, but in *ec/ec* mice the epithelium is irregularly stratified with many mitotic figures. The posterior part of the lens capsule is much thinner than normal (1). *ec* is similar to *lr*, but allelism tests have not been made.

References

1. Beasley, A.B. 1963. Inheritance and development of a lens abnormality in the mouse. J. Morphol. 112:1–11.

ecl, *ecs* loci

ecl, epistatic circling of C57L/J, probably Chr 4.
ecs, epistatic circling of SWR/J.
Simultaneous homozygosity for these two recessive genes causes circling behavior. C57L/J is homozygous for *ecl*, which is probably located on Chr 4 near *Gpd-1*, and SWR/J is homozygous for *ecs*. Homozygosity for either gene alone does not cause circling (1).

References

1. Taylor, B.A. 1976. Mouse News Lett. 55:17.

Ed, extra digit

This mutation arose at Oak Ridge National Laboratory. The homozygote appears to be viable, but the effects of the gene have not been described (1,2).

References

1. Russell, L.B. 1961. Mouse News Lett. 25:64.
2. Russell, L.B. 1972. Mouse News Lett. 47:58.

Ee-1 locus

See *Es-1* locus.

Ee-2 locus

See *Es-3* locus.

Eg locus, endoplasmic β-glucuronidase, Chr 8

This locus controls the ability to integrate β-glucuronidase into the endoplasmic reticulum of liver (3). It does not affect localization of the enzyme in lysosomes and is specific for β-glucuronidase (1). Mice homozygous for the recessive allele *Eg°*, which occurs in the YBR strain, lack glucuronidase in the microsomal fraction (endoplasmic reticulum) of liver cells. Nearly all other strains have the *Egᵃ* or wild-type allele and have about 10 to 50 per cent of their glucuronidase in microsomes. Heterozygotes are like *Egᵃ/Egᵃ* in this respect. *Egᵃ* has been shown to determine the presence of a protein, egasyn, with a molecular weight of 64 000, which is necessary to stabilize the binding of β-glucuronidase to the membrane of the endoplasmic reticulum. *Eg°/Eg°* mice do not make egasyn, and *Egᵃ/Eg°* mice have half normal levels (4, 5). Egasyn binding of glucuronidase to microsomes occurs in liver, kidney, and lung, but not in spleen, heart, and brain (2). Only about 10 per cent of the egasyn in liver cells is bound to glucuronidase, suggesting that egasyn may also be complexed to other proteins (4). Egasyn is also an esterase with multiple electrophoretic forms and is probably identical with *Es-22* (6).

References

1. Ganschow, R., and K. Paigen. 1967. Separate genes determining the structure and intracellular location of hepatic glucuronidase. Proc. Natl. Acad. Sci. USA 58:938–945.
2. Lusis, A.J., and K. Paigen. 1977. Relationships between levels of membrane-bound glucuronidase and the associated protein egasyn in mouse tissues. J. Cell. Biol. 73:728–735.
3. Lusis, A.J., and K. Paigen. 1977. Mechanisms involved in the intracellular localization of mouse glucuronidase. Curr. Top. Biol. Med. Res. 2:63–106.
4. Lusis, A.J., S. Tomino, and K. Paigen. 1976. Isolation, characterization, and radioimmunoassay of murine egasyn, a protein stabilizing glucuronidase membrane binding. J. Biol. Chem. 251:7753–7760.
5. Lusis, A.J., S. Tomino, and K. Paigen. 1977. Inheritance in mice of the membrane anchor protein egasyn: the *Eg* locus determines egasyn levels. Biochem. Genet. 15:115–122.
6. Medda, S., O. von Deimling, and R.T. Swank. 1986. Identity of esterase-22 and egasyn, the protein which complexes with microsomal β-glucuronidase. Biochem. Genet. 24:229–243.

Egf locus, epidermal growth factor, Chr 3

This locus codes for epidermal growth factor (EGF), a 53 amino acid polypeptide with diverse biological activities that probably functions in the regulation of growth and function of various tissues throughout life. EGF is abundant in the male submandibular gland and to a lesser extent in the female submandibular gland and is sparse in other tissues. It is synthesized as a much larger precursor, pre-pro-EGF, which may produce other biologically active peptides in addition to EGF (2). The level of pre-pro-EGF mRNA produced in kidney is only twofold lower than that in submandibular gland, but apparently is not processed into EGF in this tissue (1). No genetic variants of *Egf* are known. The locus was found to be on Chr 3 by use of mouse–Chinese hamster somatic cell hybrids screened with a probe consisting of a fragment from male submandibular gland pre-pro-EGF cDNA (2).

References

1. Rall, L.B., J. Scott, and G.I. Bell. 1985. Mouse prepro-epidermal growth factor synthesis by the kidney and other tissues. Nature 313:228–231.
2. Zabel, B.U., R.L. Eddy, P.A. Lalley, J. Scott, G.I. Bell, and T.B. Shows. 1985. Chromosomal locations of the human and mouse genes for precursors of epidermal growth factor and the β subunit of nerve growth factor. Proc. Natl. Acad. Sci. USA 82:469–473.

Eh, hairy ears, semidominant, Chr 15

Found among offspring of a neutron-irradiated male. Heterozygotes have a small pinna with a tuft of hair on the inner surface. The edge of the pinna is not crinkled as in *se/se* mice (1). Homozygotes die *in utero* (2), but *Eh/Eh* is probably not cell-lethal, since viable and fertile *Eh/Eh* ⟷ +/+ aggregation chimeras can be formed (3). *Eh* suppresses recombination in the region of Chr 15 between *Dom* and *Ca* and is probably associated with a chromosomal rearrangement (4).

References

1. Bangham, J.W. 1965. Mouse News Lett. 33:68.
2. Bangham, J.W. 1968. Mouse News Lett. 38:32.
3. Guenet, J.L. 1978. Mouse News Lett. 58:67.
4. Lane, P.W., and H.M. Liu. 1984. Association of megacolon with a new dominant spotting gene (*Dom*) in the mouse. J. Hered. 75:435–439.

El, epilepsy (formerly *ep*), dominant

Found in 1954 in a strain being studied for hydrocephalus in Tokyo. Mice of a strain (EL) that breeds true for the defect and presumably is homozygous (*El/El*) have a brief violent convulsion after being thrown up gently about 10 to 15 cm several times. Susceptibility begins at about 7 weeks. Most F1 offspring of outcrosses convulse; most F2 offspring also convulse (2, 3). Conclusive evidence for single-locus segregation has not been published. Numerous differences between the EL strain and non-convulsing strains in brain chemistry have been described (4–6). The strain has been used extensively to study factors precipitating the convulsions (8), effects of the convulsions on various aspects of brain metabolism (7, 9), and the effect of anticonvulsants on incidence of seizures (1).

References

1. Honda, T. 1984. Amino acid metabolism in the brain with convulsive disorders. Part 2: the effects of anticonvulsants on convulsions and free amino acid patterns in the brain of El mouse. Brain Dev. 1:22–26.
2. Imaizumi, K. 1964. Mouse News Lett. 31:57.
3. Imaizumi, K., S. Ito, G. Kutukake, T. Takizawa, K. Fujiwara, and K. Tutikawa. 1959. Epilepsy-like anomaly in mice. Bull. Exp. Anim. 8:6–10 (English summary).
4. Kurokawa, M., M. Kato, and Y. Machiyama. 1961. Choline acetylase activity in a convulsive strain of mouse. Biochim. Biophys. Acta 50:385–386.
5. Kurokawa, M., H. Naruse, and M. Kato. 1966. Metabolic studies on *ep* mouse, a special strain with convulsive predisposition. Prog. Brain Res. 21A:112–130.
6. Naruse, H., M. Kato, M. Kurokawa, R. Haba, and T. Yabe. 1960. Metabolic defects in a convulsive strain of mouse. J. Neurochem. 5:359–369.
7. Onishi, H., S. Yamagami, K. Mori, and Y. Kawakita. 1984. Effect of convulsions on the synthesis of heterogeneous nuclear RNA associated with polyadenylate and oligoadenylate sequences from El mouse brain as a convulsive strain. Exp. Neurol. 83:98–107.
8. Suzuki, J., and Y. Nakamoto. 1982. Abnormal plastic phenomena of sensory-precipitated epilepsy in the mutant El mouse. Exp. Neurol. 75:440–452.
9. Suzuki, J., Y. Nakamoto, and Y. Shinkawa. 1983. Local cerebral glucose utilization in epileptic seizures of the mutant El mouse. Brain Res. 266:359–363.

Ela-1 locus, elastase-1, Chr 4, 15

Two chromosomal locations have been reported for this locus which codes for the endopeptidase, elastase-1 (E.C. 3.4.21.11). One was found to be on Chr 15 using mouse–Chinese hamster somatic cell hybrids and a cDNA probe isolated from a rat pancreatic cDNA library (2). The other was found to be on Chr 4 by use of a DNA restriction fragment length polymorphism segregating in recombinant inbred strains (1). Whether one of these findings is in error or one of the loci is a pseudogene has not been reported.

References

1. Harper, M.E., C. Blatt, S. Marks, M.N. Nesbitt, and M.I. Simon. 1984. Gene mapping in the mouse by analysis of RFLP segregation in recombinant inbred strains. Cytogenet. Cell Genet. 37:488 (Abstr.).
2. Honey, N.K., A.Y. Sakaguchi, P.A. Lalley, C. Quinto, R.J. MacDonald, C. Craik, G.I. Bell, W.J. Rutter, and S.L. Naylor. 1984. Chromosomal assignments of genes for trypsin, chymotrypsin B, and elastase in mouse. Somat. Cell Mol. Genet. 10:377–383.

Elo, eye lens obsolescence, dominant or semidominant, Chr 1

Found in descendants of a cross between C57BL/6 and a pink-eyed-dilution (*p/p*) mouse. In both heterozygotes and homozygotes the lens is underdeveloped at birth and has completely degenerated by 30 days. Other eye structures are normal but the eyes are smaller than normal and somewhat collapsed. Penetrance appears to be complete. Both homozygotes and heterozygotes are viable and fertile (1). In homozygotes, impairment in elongation of the central lens fibers can be detected at 12 days of gestation. These cells become necrotic, so that by late fetal stages the lens is small and deformed (2). Whether heterozygotes are as severely affected as homozygotes has not been reported. Lenses from 10- and 11-day homozygous embryos develop their typical abnormalities in culture, indicating that the defect is intrinsic to the lens. Mutant lenses from 11 days to birth, even though severely necrotic, accumulate immunohistochemically detectable γ-crystallins, indicating that *Elo* does not act by preventing the synthesis of γ-crystallins (5). However, *Elo* is on Chr 1 close to *Len-1*, the structural locus for a family of γ-crystallins (1, 3), and Skow and Donner (4) have found that the DNA of C3H-*Elo* mice has a different pattern of restriction fragments that hybridize to a γ-crystallin probe from that of the congenic C3H/He strain. The pattern in C3H-*Elo* is unique to that strain. These results indicate that the *Elo* mutation is an altered gene that probably produces an abnormal γ-crystallin immunologically similar to the normal protein but with a defective function.

References

1. Oda, S., T. Watanabe, and K. Kondo. 1980. A new mutation, eye lens obsolescence, Elo, on chromosome 1 in the mouse. Jpn. J. Genet. 55:71–75.
2. Oda, S., K. Watanabe, H. Fujisawa, and Y. Kameyama.

1980. Impaired development of lens fibers in genetic microphthalmia, eye lens obsolescence, *Elo*, of the mouse. Exp. Eye Res. 31:673–681.

3. Skow, L.C. 1982. Location of a gene controlling electrophoretic variation in mouse γ-crystallins. Exp. Eye Res. 39:509–516.
4. Skow, L.C., and M. Donner. 1986. Mouse News Lett. 74:119.
5. Watanabe, K., H. Fujisawa, S. Oda, and Y. Kameyama. 1980. Organ culture and immunohistochemistry of the genetically malformed lens, in eye lens obsolescence, *Elo*, of the mouse. Exp. Eye Res. 31:683–689.

Em, Emory cataract, dominant or semidominant

Bilateral cataracts were found in an 11-month-old male of the CFW stock. Selective breeding of descendants of this and related mice gave rise to a strain of mice with a high probability of cataract formation. The mode of inheritance is said to be autosomal dominant, but it is not clear to what extent modifiers for high cataract incidence have been selected. Lens opacities appear as early as 6 to 8 months and progress to a dense opaque central zone, with a shrunken appearance and loss of wet and dry weight of the lens (3). The most striking and consistent changes accompanying development of the cataracts are conversion of soluble to insoluble proteins, with lower total protein and lower glutathione (2). Changes in levels of soluble crystallins with age show only minor differences from normal controls, chiefly a greater than normal decrease in γ-crystallins (1).

References

1. Barron, B.C., J.F.R. Kuck, and K.P. Kuck. 1984. The Emory mouse cataract: changes in the β and γ-crystallins during aging and cataractogenesis as revealed by isoelectric focusing of the native soluble proteins. Curr. Eye Res. 3:1365–1372.
2. Kuck, J.F.R., and K.D. Kuck. 1983. The Emory mouse cataract: loss of soluble protein, glutathione, protein sulfhydryl and other changes. Exp. Eye Res. 36:351–362.
3. Kuck, J.F.R., T. Kuwabara, and K.D. Kuck. 1981/1982. The Emory mouse cataract: an animal model for human senile cataract. Curr. Eye Res. 1:643–649.

Ema locus, electrophoretic mobility and agglutinability

This locus controls both the degree of agglutinability and the electrophoretic mobility of red blood cells. The allele *Ema^h* determining fast mobility and high agglutinability is found in strains C57BL, BALB/c, SWR, and others; the allele *Ema^l* determining slow mobility and low agglutinability is found in strains A, DBA, AKR, C3H/He, and others. Heterozygotes are intermediate for both characteristics (1,2).

References

1. Rubenstein, P. 1975. Mouse News Lett. 52:45.
2. Rubenstein, P., N. Liu, E.W. Streun, and F. Decary. 1974. Electrophoretic mobility and agglutinability of red blood cells: a "new" polymorphism in mice. J. Exp. Med. 139:313–322.

Emv-1 to *Emv-18* loci

These loci are the sites of integration of the proviral genome of endogenous murine leukemia viruses (MuLV). Jenkins *et al.* (10) have used a hybridization probe specific for ecotropic MuLV DNA sequences to determine the presence of proviral sequences in restriction enzyme fragments of DNA in many inbred strains. The number of loci was found to vary among inbred strains from none to as many as five separate sites. Some of these loci have previously been recognized as genes causing virus inducibility or spontaneous expression and assigned names according to the strain in which they were identified.

Emv-1 locus (formerly *Cv* or *Cv-1*), ecotropic murine leukemia virus-1, Chr 5. The proviral genome is present in strains A, BALB/c, CBA/J, C3H/He, SEA/Gn, HRS, SM/J, and others (10). It was first identified as a locus controlling inducibility of the N-tropic leukemia virus (BALB:virus-1) in BALB/c mice. The virus is expressed only sporadically and late in life in BALB/c mice, but is inducible by IUdR in embryo cell cultures (22). It is expressed earlier in C3H/He mice (6), and is also expressed in HRS mice. The *Emv-1* virus differs slightly in structure from the *Emv-3* virus also present in HRS mice. In this strain, recombinants between the two viruses are found in thymocytes (24). In the RF/J strain, a gene for low efficiency virus induction by IUdR in cell cultures, *Rjv-2*, is probably identical to the *Emv-1* proviral locus (14). A gene in the BALB/c strain that is either very closely linked to *Emv-1* or identical with it, *Inc-1*, has been shown to interact with a similar gene closely linked to *Emv-2* in the C57BL/6 strain, *Inb-1*, to produce a dramatic increase in the rate of induction of virus in cultured cells (13). See *Inb-1* locus, *Inc-1* locus.

Emv-2 locus (formerly *Bv* or *Bv-1*), ecotropic murine leukemia virus-2, Chr 8. Provirus is present at this site in strains C57BL/6, C57BL/10, and C57BR/cd (10). The virus is N-tropic. It was previously detected in strains C57BL/6 and C57BL/10 by the presence of viral antibodies that develop at low titer at 4 to 5 months of age (5). The virus is not expressed in these *Fv-1^b* strains and is only poorly inducible. In genetic crosses it was recognized by inducibility by IdU in cultured cells, but with incomplete penetrance (12). A gene in the C57BL/6 strain that is either very closely linked to *Emv-2* or identical with it, *Inb-1*, has been shown to interact with a similar gene closely linked to *Emv-1* in the BALB/c strain, *Inc-1*, to produce a dramatic increase in the rate of induction of virus in cultured cells (13). See *Inb-1* locus, *Inc-1* locus.

Emv-3 locus (formerly also *Sev-1* (12)), ecotropic murine leukemia virus-3, Chr 9. Provirus is present at this site in the DBA and SEA strains and all other strains carrying the *d* (dilute) mutation (10). In the HRS strain the virus is expressed in thymocytes where it often recombines with the *Emv-1* virus also expressed in these cells (24). *Emv-3* shows no recombination with *d*. Spontaneous *d$^+$* revertant mutations in the DBA strain lack the *Emv-3* proviral DNA, suggesting that the *d* mutation resulted from integration of an ecotropic provirus into the mouse genome (9). The revertant DNA has been shown to differ from wild type by having a long terminal repeat derived from the provirus at the *Emv-3* site (4).

Emv-4 locus, ecotropic murine leukemia virus-4. Provirus is present at this site in the LG strain (10).

Emv-5 locus, ecotropic murine leukemia virus-5. Provirus is present at this site in the LP strain (10).

Emv-6 locus. ecotropic murine leukemia virus-6. Provirus is present at this site in the ST/b strain (10).

Emv-7 locus, ecotropic murine leukemia virus-7. Provirus is present at this site in the DA/Hu strain (10).

Emv-8 locus, ecotropic murine leukemia virus-8. Provirus is present at this site in the MA/My strain (10).

Emv-9 locus, ecotropic murine leukemia virus-9. Provirus is present at this locus in the MA/J and SJL strains (10). In the SJL strain, *Emv-9* virus is not produced spontaneously and is non-inducible. It contains a 3.1-kb deletion, possibly in the polymerase region (25).

Emv-10 locus, ecotropic murine leukemia virus-10. Provirus is present at this locus in the SJL strain. (10). The SJL strain produces low levels of infectious virus by 10 weeks of age. This was shown to be due to expression of the *Emv-10* proviral locus, with no contribution from *Emv-9* (25).

Emv-11 locus (originally *Akv-1*), ecotropic murine leukemia virus-11, Chr 7. Provirus is present at this site in strains AKR/N and AKR/J (10). This is one of two loci, *Akv-1* and *Akv-2*, first identified as controlling

expression of N-tropic AKR leukemia viruses in crosses between the AKR high leukemia strain and several low leukemia strains (17). *Emv-11* is on Chr 7 close to the centromere (18).

Emv-12 locus (originally *Akv-2*), ecotropic murine leukemia virus-12, Chr 16. See *Emv-11*. The locus was found to be on Chr 16 by Kozak and Rowe (11). Provirus is present at this site in AKR/N but not in AKR/J (10, 20).

Emv-13 locus (formerly *Akv-3*, *Akvp*), ecotropic murine leukemia virus-13, Chr 2. Provirus is present at this locus in strains AKR/N and AKR/J (10). The *Emv-13* virus, unlike the other AKR viruses, is not expressed as infectious virus (1). This locus is probably the same as the *Akvp* locus, which is expressed only as group-specific antigen p30 (an internal viral protein) and as the G_{IX} viral antigen gp69/71 (a viral envelope glycoprotein) (7). *Emv-13* is near the agouti locus on Chr 2 (23).

Emv-14 locus (formerly *Akv-4*, *Akv-2J*), ecotropic murine leukemia virus-14, Chr 11. Provirus is present at this locus in the AKR/J strain (10, 21). The virus at this locus is expressed spontaneously as infectious virus as determined by the XC-plaque assay (1). *Emv-14* is near *Tr* on Chr 11 (23).

Emv-15 locus, ecotropic murine leukemia virus-15, Chr 2. This provirus is closely associated with the *A^y* (lethal yellow) mutation at the agouti locus (3). By use as a probe of a unique mouse sequence flanking the *Emv-15* provirus, three alleles of the *Emv-15* locus were identified: *Emv-15^a* occurs in strains 129/Sv-*A^w* and C3H/He and in all *a*-locus mutations derived from these strains; *Emv-15^b* occurs in strain C57BL/6 and in all *a*-locus mutations derived from this strain; *Emv-15^v* is always associated with the *A^y* mutation except in the *YBR-A^y* strain in which a recombination between *Emv-15* and *A^y* has presumably occurred. *Emv-15^v* is always associated with presence of the provirus (19). See *A^y*.

Emv-16 locus, ecotropic murine leukemia virus-16, Chr 1.

Emv-17 locus, ecotropic murine leukemia virus-17, Chr 1. These two proviruses occur in the RF/J strain. No recombinants between them have been found in 895 backcross offspring. The RF strain does not express leukemia virus early in life because of a combination of the *Fv-1^{nr}* semirestrictive allele and a maternal resistance factor transmitted in the milk. However, when RF males are crossed to SWR/J strain females (*Fv-1^n*) and the offspring backcrossed to SWR, females carrying *Emv-16 Emv-17* in the N2 and later generations were found to produce offspring with a very high incidence of newly integrated germline proviruses. The viruses are probably acquired early in development by virus infection. A combination of *Emv-16* and *Emv-17* with the SWR genetic background appears to be necessary for the high incidence of integration (8). *Emv-16* and *Emv-17* are located on Chr 1 close to *ln* (2). These proviral loci are probably identical with the *Rjv-1* gene determining high efficiency virus induction by IUdR in cultured cells of the RF/J strain (14).

Emv-18 locus, ecotropic murine leukemia virus-18, Chr 9. This is the locus of a newly integrated provirus found in the C57BL/6.SIM-*Fv-2ˢ* congenic strain. Since SIM, the *Fv-2ˢ* donor strain, does not contain proviruses, the new provirus presumably came from the low-virus C57BL/6 strain and was integrated into the germline during the construction of the congenic strain. The virus is expressed in spleen, bone marrow, and kidney. *Emv-18* is linked to *Fv-2* on Chr 9 with about 11 per cent recombination (15, 16).

References

1. Bedigian, H.G., N.G. Copeland, N.A. Jenkins, K. Salvatore, and S. Rodick. 1983. *Emv-13* (*Akv-3*): a noninducible endogenous ecotropic provirus of AKR/J mice. J. Virol. 46:490–497.
2. Buchberg, A.M., B.A. Taylor, N.A. Jenkins, and N.G. Copeland, 1986. Chromosomal localization of *Emv-16* and *Emv-17*, two closely linked ecotropic proviruses of RF/J mice. J. Virol. 60:1175–1178.
3. Copeland, N.G., N.A. Jenkins, and B.K. Lee. 1982. Association of the lethal yellow (*Aʸ*) coat color mutation with an ecotropic murine leukemia virus genome. Proc. Natl. Acad. Sci. USA 80:247–249.
4. Copeland, N.G., K.W. Hutchison, and N.A. Jenkins. 1983. Excision of the DBA ecotropic provirus in dilute coat-color revertants of mice occurs by homologous recombination involving the viral LTRs. Cell 33:379–387.
5. Ihle, J.N., and D.R. Joseph. 1978. Genetic analysis of the endogenous C3H murine leukemia virus genome: evidence for one locus unlinked to the endogenous murine leukemia virus genome of C57BL/6 mice. Virology 87:298–306.
6. Ihle, J.N., D.R. Joseph, and J.J. Domotor Jr. 1979. Genetic linkage of C3H/HeJ and BALB/c endogenous ecotropic C-type viruses to phosphoglucomutase-1 on chromosome 5. Science 204:71–73.
7. Ikeda, H., W.P. Rowe, E.A. Boyse, E. Stockert, H. Sato, and S. Jacobs. 1976. Relationship of infectious murine leukemia virus and virus-related antigens in genetic crosses between AKR and *Fv-1* compatible strain C57L. J. Exp. Med. 143:32–46.
8. Jenkins, N.A., and N.G. Copeland. 1985. High frequency germline acquisition of ecotropic MuLV proviruses in SWR/J-RF/J hybrid mice. Cell 43:811–819.
9. Jenkins, N.A., N.G. Copeland, B.A. Taylor, and B.K. Lee. 1981. Dilute (*d*) coat color mutation of DBA/2J mice is associated with the site of integration of an ecotropic MuLV genome. Nature 293:370–374.
10. Jenkins, N.A., N.G. Copeland, B.A. Taylor, and B.K. Lee. 1982. Organization, distribution, and stability of endogenous ecotropic murine leukemia virus DNA sequences in chromosomes of *Mus musculus*. J. Virol. 43:26–36.
11. Kozak, C.A., and W.P. Rowe. 1980. Genetic mapping of the ecotropic virus-inducing locus *Akv-2* of the AKR mouse. J. Exp. Med. 152:1419–1423.
12. Kozak, C.A., and W.P. Rowe. 1982. Genetic mapping of ecotropic murine leukemia virus-inducing loci in six inbred strains. J. Exp. Med. 155:524–534.
13. McCubrey, J., and R. Risser. 1982. Allelism and linkage studies of murine leukemia virus activation genes in low leukemic strains of mice. J. Exp. Med. 155:1233–1238.
14. McCubrey, J., J.M. Horowitz, and R. Risser. 1982. Structure and expression of endogenous ecotropic murine leukemia viruses in RF/J mice. J. Exp. Med. 156:1461–1474.
15. Mowat, M., and A. Bernstein. 1983. Linkage of the *Fv-2* gene to a newly inserted ecotropic retrovirus in *Fv-2* congenic mice. J. Hered. 47:471–477.
16. Robson, I.B., M. Mowat, and A. Bernstein. 1985. Tissue-specific expression of the newly acquired ecotropic *Emv-18* provirus in *Fv-2* congenic mice. J. Virol. 55:54–59.
17. Rowe, W.P., and J.W. Hartley. 1972. Studies of genetic transmission of murine leukemia virus by AKR mice. II. Crosses with *Fv-1ᵇ* strains of mice. J. Exp. Med. 136:1286–1301.
18. Rowe, W.P., J.W. Hartley, and T. Bremmer. 1972. Genetic mapping of a murine leukemia virus-inducing locus of AKR mice. Science 168:860–862.
19. Siracusa, L.D., L.B. Russell, N.A. Jenkins, and N.G. Copeland. 1987. Allelic variation within the *Emv-15* locus defines genomic sequences closely linked to the *agouti* locus on mouse chromosome 2. Genetics 117:85–92.
20. Steffen, D., S. Bird, W.P. Rowe, and R.A. Weinberg. 1979. Identification of DNA fragments carrying ecotropic proviruses of AKR mice. Proc. Natl. Acad. Sci. USA 76:4554–4558.
21. Steffen, D.L., B.A. Taylor, and R.A. Weinberg. 1982. Continued germline integration of AKR proviruses during the breeding of AKR mice and derivative recombinant inbred strains. J. Virol. 42:165–175.
22. Stephenson, J.R., and S.A. Aaronson. 1973. Segregation of loci for C-type virus induction in strains of mice with high and low incidence of leukemia. Science 180:865–866.
23. Taylor, B.A., L. Rowe, N.A. Jenkins, and N.G. Copeland. 1985. Chromosomal assignment of two endogenous ecotropic murine leukemia virus proviruses of the AKR/J mouse strain. J. Virol. 56:172–175.
24. Thomas, C.Y., R. Khiroya, R.S. Schwartz, and J.M. Coffin. 1984. Role of recombinant ecotropic and polytropic viruses in the development of spontaneous thymic lymphomas in HRS/J mice. J. Virol. 50:397–407.

25. Yetter, R.A., W.Y. Langdon, and H.C. Morse III. 1985. Characterization of ecotropic murine leukemia viruses in SJL/J mice. Virology 141:319–321.

Emv-x1, *-x2* loci

See *C58v-3*, *-4* loci.

En-1, *En-2* loci

En-1 locus, engrailed-1, Chr 1. This locus was identified by Southern blot analysis using as a probe a cDNA clone of the Drosophila homeo box-containing *engrailed* gene. Homology of *En-1* with the *engrailed* gene is localized in a 180-bp *engrailed*-like homeo box and 63 nucleotides immediately 3′ to it. *En-1* is transcribed at all stages of embryogenesis beginning at 4.5 to 6.5 days of development, with a peak at 10.5 to 12.5 days (3). Strains C57BL/6 and DBA/2 differ in length of *Taq*I DNA restriction fragments that hybridize with a mouse genomic *En-1* sequence 3′ to the homeo box: C57BL/6 has fragments of 3.5 and 3.3 kb; DBA/2 has fragments of 4.4 and 3.3 kb (2). *En-1* was first shown to be on Chr 1, probably distal to band 1C2, by use of mouse–Chinese hamster somatic cell hybrids (2). By use of the DNA polymorphism in BXD recombinant inbred strains, it was later found to be close to *Ren-1* in about the same position as dominant hemimelia (*Dh*), a gene that causes defective development beginning at about 9 days of gestation. The *Dh* mutation is thus a candidate for a mutation at the *En-1* locus (1, 2).

En-2 locus, engrailed-2, Chr 5. This locus is highly homologous with *En-1* and with the *engrailed* and *invected* genes in Drosophila, and, like them, contains a centrally located homeo box sequence. *En-2* is expressed in embryos from 9.5 through 17.5 days of development, predominantly in the posterior brain and to a much smaller extent in other regions and tissues. RNA transcripts of several different sizes are produced and their relative abundance varies during this period. The gene is also transcribed in cultured teratocarcinoma cells. Strain C57BL/6 differs from strains C3H/He and BALB/c in length of an *Msp*I DNA restriction fragment that hybridizes with a mouse genomic DNA probe; C57BL/6 has fragments of 1.5 and 1.3 kb, and the other two strains have a fragment of 2.8 kb. By use of 12 BXH and 7 CXB recombinant inbred strains segregating for this fragment length difference, *En-2* was shown to be on the proximal portion of Chr 5 close to *Hm* (hammer toe) and *Hx* (hemimelic extra toes), two loci that affect development of the limbs (2).

References

1. Hill, R.E., A.E. Hall, C.M. Sime, and N.D. Hastie. 1987. A mouse homeo box-containing gene maps near a developmental mutation. Cytogenet. Cell Genet. 44:171–174.
2. Joyner, A.L., and G.R. Martin. 1987. *En-1* and *En-2*, two mouse genes with sequence homology to the *Drosophila engrailed* gene: expression during embryogenesis. Genes Dev. 1:29–38.
3. Joyner, A.L., T. Kornberg, K.G. Coleman, D.R. Cox, and G.R. Martin. 1985. Expression during embryogenesis of a mouse gene with sequence homology to the Drosophila *engrailed* gene. Cell 43:29–37.

Eno-1 locus, enolase-1, Chr 4

This locus is probably the structural locus for enolase (E.C. 4.2.1.11). No variants are known in the mouse. The locus was shown to be located on Chr 4 by somatic cell hybridization using mouse peritoneal macrophages and human fibrosarcoma cells or Chinese hamster cells. The synteny of *Eno-1*, *Pgd*, *Pgm-2*, and *Ak-2* on Chr 4 in the mouse is paralleled by the synteny of the homologous loci on the short arm of human Chr 1 (1,2).

References

1. D'Ancona, G.G., and C.M. Croce. 1977. Assignment of the gene for enolase to mouse chromosome 4 using somatic cell hybrids. Cytogenet. Cell. Genet. 19:1–6.
2. Lalley, P.A., U. Francke, and J.D. Minna. 1978. Homologous genes for enolase, phosphogluconate dehydrogenase, phosphoglucomutase, and adenylate kinase are syntenic on mouse chromosome 4 and human chromosome 1p. Proc. Natl. Acad. Sci. USA 75:2382–2386.

Env-1 to *-40* loci

See *Xmmv-1* to *-40* loci.

Eo, eye-opacity, dominant

Radiation-induced. Eye opacity and small eyes and occasionally open lids at birth (1). Probably extinct (2).

References

1. Gower, J.S. 1953. Mouse News Lett. 8, Suppl:14.
2. Juriloff, D.M., M.J. Harris, and J.R. Miller. 1983. The lidgap defect in mice: update and hypotheses. Can. J. Genet. Cytol. 25:246-254.

eo

See *wa-1*.

ep, pale ear, recessive, Chr 19

Arose as a spontaneous mutation in the C3HeB/FeJ inbred strain. The ears and tail are pale in color. The juvenile coat is slightly diluted in color but the adult coat is almost normal. Eyes are pale at birth but darken with age. *ep* is almost an exact mimic of *le* (2). Homozygotes, in common with some other mutations that reduce pigmentation, have a reduced number of projections of retinal ganglion cells to the ipsilateral lateral geniculate nucleus (3). In homozygotes, *ep* is similar to beige (*bg*) and several other pigment mutations in causing a more than twofold increase in the kidney lysosomal enzymes, β-glucuronidase, β-galactosidase, and α-mannosidase, accompanied by lowered lysosomal enzyme excretion in the urine, and elevated serum levels of glucuronidase and galactosidase (5). Lysosomal enzymes are synthesized in the endoplasmic reticulum as large precursors which are passed through the Golgi apparatus to the lysosomes where they are processed into smaller mature enzymes before secretion. Brown *et al.* (1) have presented evidence that the *ep* locus functions at the stage of secretion of the enzymes from the lysosomes, the mutant *ep* allele inhibiting secretion. In addition to their defect in lysosomal secretion, pale ear mice resemble *bg* and several other pigment mutations in two other characteristics: (1) They have a platelet storage pool deficiency characterized by prolonged bleeding time and a deficiency of platelet dense granules and the dense-granule components, serotonin, ATP, and ADP (6, 7). The deficiency can be corrected by transplantation of bone marrow from normal donors (4); and (2) They have reduced natural (NK) cell activity (8).

References

1. Brown, J.A., E.K. Novak, and R.T. Swank. 1985. Effects of ammonia on processing and secretion of precursor and mature lysosomal enzyme from macrophages of normal and pale ear mice: evidence for two distinct pathways. J. Cell Biol. 100:1894–1904.
2. Lane, P.W., and E.L. Green. 1967. Pale ear and light ear in the house mouse. Mimic mutations in linkage groups XII and XVII. J. Hered. 58:17–20.
3. LaVail, J.H., R.A. Nixon, and R.L. Sidman. 1978. Genetic control of retinal ganglion cell projections. J. Comp. Neurol. 182:399–422.
4. McGarry, M.P., E.K. Novak, and R.T. Swank. 1986. Progenitor cell defect correctable by bone marrow transplantation in five independent mouse models of platelet storage pool deficiency. Exp. Hematol. 14:261–265.
5. Novak, E.K., and R.T. Swank. 1979. Lysosomal dysfunctions associated with mutations at mouse pigment genes. Genetics 92:189–204.
6. Novak, E.K., S.-W. Hui, and R.T. Swank. 1981. The mouse pale ear pigment mutant as a possible animal model for human platelet storage pool deficiency. Blood 57:38–43.
7. Novak, E.K., S.-W. Hui, and R.T. Swank. 1984. Platelet storage pool deficiency in mouse pigment mutations associated with seven distinct genetic loci. Blood 63:536–544.
8. Örn, A., E.M. Hakansson, M. Gidlund, U. Ramstedt, I. Axberg, H. Wigzell, and L.-G. Lundin. 1982. Pigment mutations in the mouse which also affect lysosomal functions lead to suppressed natural killer cell activity. Scand. J. Immunol. 15:305–310.

Epa-1 locus, epidermal antigen-1

This locus controls presence or absence of an antigen on epidermal cells. The antigen induces potent cytotoxic T lymphocytes (CTL) in *H-2* compatible hosts lacking the antigen. The CTLs react strongly against epidermal cells, skin fibroblasts, and activated macrophages, but not against fresh peritoneal macrophages or lymphocytes (1–3). The *Epa-1^a* allele determining presence of the antigen occurs in strains AKR, BALB/c, CBA/J, C3H/Bi, C57BL, DBA, and others; the *Epa-1^b* allele determining absence of the antigen occurs in strains C3H/He, C3H/An, A/J, A/He, LG, and ST/b (5). *Epa-1* also acts like a minor histocompatibility locus, as demonstrated by delayed rejection of skin grafts between *Epa-1* identical pairs of mice as compared to *Epa-1* non-identical pairs in mice identical at the *H-2* locus but segregating for other minor histocompatibility loci (4).

References

1. Burlingham, W.J., M.E. Snyder, J.D. Tyler, and D. Steinmuller. 1984. Lysis of mouse macrophages, fibroblasts, and epidermal cells by epidermal alloantigen-specific cytotoxic T lymphocytes: effect of culture and inflammatory agents on Epa-1 expression. Cell. Immunol. 87:553–565.
2. Steinmuller, D., J.D. Tyler, and S.S. David. 1981. Cell-mediated cytotoxicity to non-MHC alloantigens on mouse epidermal cells. I. H-2 restricted reactions among strains sharing the *H-2^k* haplotype. J. Immunol. 126:1747–1753.
3. Steinmuller, D., J.D. Tyler, and S.S. David. 1981. Cell-mediated cytotoxicity to non-MHC alloantigens on mouse epidermal cells. II. Genetic basis of the response of C3H mice. J. Immunol. 126:1754–1758.
4. Steinmuller, D., J.D. Tyler, K.G. Waddick, and W.J. Burlingham. 1982. Epidermal alloantigen and the survival of mouse skin allografts. Transplantation 33:308–313.
5. Steinmuller, D., J.D. Tyler, and A.R. Zinsmeister. 1985. Strain distribution of the new tissue-restricted alloantigen Epa-1. Transplant. Proc. 17:749–753.

epf, epileptiform, recessive

Arose in a strain of black and white (*a/a s/s*) mice. Homozygotes are recognizable by the occurrence of seizures but otherwise appear quite normal. They have a normal life span and normal fertility, except that females have slightly smaller litters than their normal sibs and about 5 per cent die within the first 3 to 6 months, apparently as the result of a seizure. Seizures may begin as early as 20 days of age or as late as 14 months. It is possible that as many as 8 per cent of homozygotes escape detection. Seizures may be induced by a range of mild stimuli, including physical, visual, and auditory ones. The seizure pattern involves a tonic phase, a clonic phase, and a post-ictal recovery phase (1).

References

1. Hare, J.E., and A.S. Hare. 1979. Epileptiform mice, a new neurological mutant. J. Hered. 70:417–420.

Eph-1 locus, epoxide hydrolase, Chr 1

This locus controls pH optimum and thermolability of hepatic microsomal epoxide hydrolase (E.C. 3.3.2.3) activity. The allele *Eph-1^b* determines a pH optimum of 9.5 and heat stability and is found in the C57BL/6, AKR, BALB/c, C3H/He, and numerous other strains; the allele *Eph-1^d* determines a pH optimum of 8.7 and marked heat lability and occurs in the DBA/2, A/J, CBA/J, and a few related strains. *Eph-1* was found to be on Chr 1 between *Ltw-4* and *Mls* (1).

References

1. Lyman, S.D., A. Poland, and B.A. Taylor. 1980. Genetic polymorphism of microsomal epoxide hydrolase activity in the mouse. J. Biol. Chem. 255:8650–8654.

epi, exocrine pancreatic insufficiency, recessive

This condition was found in the CBA/J strain in 1972. Affected mice are small at 2 to 3 weeks and have yellow fatty feces. The exocrine pancreas tissue is almost completely replaced by fat, causing functional starvation. The islets of Langerhans appear normal. Life span is greatly reduced and affected mice do not breed. Segregation data from crosses within the CBA/J strain led Pivetta and Green (4) to postulate that the condition was caused by a recessive gene *epi*. However, a very similar condition can be caused by various viruses, and subsequent genetic tests have failed to confirm the hypothesis of a single recessive mutation (1, 2). The disease in CBA/J mice appears to be initiated by premature activation of proteolytic enzymes within the exocrine cells, which causes autodigestion of the cells (3).

References

1. Eppig, J.J., and E.H. Leiter. 1977. Exocrine pancreatic insufficienty syndrome in CBA/J mice. I. Ultrastructural study. Am. J. Pathol. 86:17–30.
2. Leiter, E.H., and T. Cunliffe-Beamer. 1977. Exocrine pancreatic insufficiency syndrome in CBA/J mice. III. Pathological and genetic analysis. Gastroenterology 73:260–266.
3. Leiter, E.H., E.C. Dempsey, and J.J. Eppig. 1977. Exocrine pancreatic insufficiency syndrome in CBA/J mice. II. Biochemical studies. Am. J. Pathol. 86:31–46.
4. Pivetta, O.H., and E.L. Green. 1973. Exocrine pancreatic insufficiency: a new recessive mutation in mice. J. Hered. 64:301–302.

Er, repeated epilation, semidominant, Chr 4

Probably radiation-induced. There is complete penetrance in heterozygotes. They grow a normal appearing but probably somewhat dry first coat until the age of about 13 days; then hair loss begins and continues until the fur becomes sparse. Repeated growth and re-epilation follow without a definite pattern. Heterozygotes are slightly reduced in size and some may die before weaning, but adults are fully viable and fertile. Homozygotes die at birth from inability to breathe because of a closed oral cavity. In 17-day embryos the skin is extremely thin and smooth with few vibrissae and hair follicles. The snout is truncated and the mouth closed. The limbs and tail are greatly shortened and held close to the trunk and the anal and urogenital orifices are closed (2). There are marked skeletal abnormalities and cleft palate. Homozygotes can be recognized at 13 days of gestation by their blunt limbs and stumpy tail. Between 13 and 15 days the nares and oral opening close, resulting in marked compression and abnormality of the structures in the area (5). Heterozygotes are recognizable at 18 days of gestation by edema of the feet and hemorrhagic tail tip. Histologically, at 19 days of gestation, *Er/+* skin does not differ from +/+ skin. *Er/Er* skin is characterized by wide intercellular spaces and highly variable numbers and distribution of cells of the spinous, granular, and superficial layers. A 26.5 kDa protein, filaggrin, is absent, but filaggrin precursors appear to be present. The mutation is probably responsible for a defect in some regulatory step important in many processes of differentiation and development (3). In *Er/+* mice, there is a high incidence of multiple cutaneous papillomas and squamous cell carcinomas (4). *Er* is on Chr 4 close to pupoid fetus (*pf*), a mutant which it closely resembles. The two mutations may be alleles (1). See *pf*.

References

1. Fisher, C., B.A. Dale, and E.J. Kollar. 1984. Abnormal keratinization in the pupoid fetus (*pf/pf*) mutant mouse epidermis. Dev. Biol. 102:290–299.
2. Guénet, J.L., B. Salzburger, and M.T. Tassin. 1979. Repeated epilation: a genetic epidermal syndrome in mice. J. Hered. 70:90–94.
3. Holbrook, K.A., B.A. Dale, and K.S. Brown. 1982. Abnormal epidermal keratinization in the repeated epilation mutant mouse. J. Cell Biol. 92:387–397.
4. Lutzner, M.A., J.L. Guénet, and F. Breitburd. 1985. Multiple cutaneous papillomas and carcinomas that develop spontaneously in a mouse mutant, the repeated epilation heterozygote *Er/+*. J. Natl. Cancer Inst. 75:161–166.
5. Tassin, M.T., B. Salzgeber, and J.-L. Guénet. 1983. Studies of "repeated epilation" mouse mutant embryos. I. Development of facial malformations. J. Craniofac. Genet. Dev. Biol. 3:289–307.

Erba, *Erbb* loci, Chr 11

These loci are the cellular homologs in the mouse of the v-*erbA* and v-*erbB* transforming genes of the avian erythroblastosis virus (AEV). The two loci were found to be on Chr 11 by use of mouse–Chinese hamster somatic cell hybrids screened with a viral oncogene probe. It is not known how close together they are; they were separated in two of six hybrid cell lines with broken Chrs 11 (5).

Erba locus (c-*erbA*), avian erythroblastosis oncogene-A. No genetic variants are known. The v-*erbA* protein shows no homology with other known oncogene products but is related to carbonic anhydrase (1).

Erbb locus (c-*erbB*), avian erythroblastosis oncogene-B. A DNA polymorphism at this locus occurs among inbred strains. The allele in AKR, BALB/c, and SEC/Re (suggested symbol *Erbb^a*) determines a 5.3-kb *Pst1* fragment; the allele in C57BL/6, C57L, CE, 129/J, SM, I/Ln, BDP, P, and WB/Re (suggested symbol *Erbb^b*) determines a 5.5-kb *Pst1* fragment. *Erbb* was found to be closely linked to *Hba* on Chr 11 by use of the DNA polymorphism segregating in the AKXL and CXB recombinant inbred strains (4). The v-*erbB* gene product is a membrane glycoprotein that shows close homology with the cytoplasmic domain of human nerve growth factor receptor (2, 3). *Erbb* and the gene for nerve growth factor receptor may be the same.

References

1. Debuire, B., C. Henry, M. Benaissa, G. Besirte, J.M. Claverie, S. Saule, P. Martin, and D. Stehelin. 1984. Sequencing the *erbA* gene of avian erythroblastosis virus reveals a new type of oncogene. Science 224: 1456–1459.
2. Downward, J., Y. Yarden, E. Mayes, G. Scrace, N. Totty, P. Stockwell, A. Ullrich, J. Schlessinger, and M.D. Waterfield. 1984. Close similarity of epidermal growth factor receptor and v-*erb-B* oncogene protein sequences. Nature 307:521–527.
3. Hunter, T. 1984. The epidermal growth factor receptor gene and its product. Nature 311:414–416.
4. Silver, J., J.B. Whitney III, C.A. Kozak, G. Hollis, and I. Kirsch. 1985. *Erbb* is linked to the alpha-globin locus on mouse chromosome 11. Mol. Cell. Biol. 5:1784–1786.
5. Zabel, B.U., R.E.K. Fournier, P.A. Lalley, S.L. Naylor, and A.Y. Sakaguchi. 1984. Cellular homologs of the avian erythroblastosis virus *erb-A* and *erb-B* genes are syntenic in mouse but asyntenic in man. Proc. Natl. Acad. Sci. USA 81:4874–4878.

Erp-1 locus, erythrocyte protein-1, Chr 8

This locus controls the electrophoretic mobility of a protein band anodal to hemoglobin in alkaline starch gels. The protein is not hemoglobin or carbonic anhydrase. The allele *Erp-1^a* determines a fast band and is found in most wild mice and in all inbred strains examined; the allele *Erp-1^b* determines a slower band and has a frequency of about 1 per cent in wild mice in South Australia. Heterozygotes show three bands (1).

References

1. Kirby, G.C. 1974. The genetics of an electrophoretic variant of an erythrocytic protein in the house mouse (*Mus musculus*). Anim. Blood Grps. Biochem. Genet. 5:153–157.

Es-1 locus (formerly *Es-4*, *Ee-1*), esterase-1, Chr 8

This locus controls electrophoretic mobility of an esterase present in serum, lung, testis, heart, and other organs and absent from erythrocytes. The esterase is detectable using either naphthyl or 4-methylumbelliferyl esters as substrates (5). The allele *Es-1^a* determines a fast anodally migrating band and is present in the C57BL family of strains. The allele *Es-1^b* determines a slower migrating band and is present in the C58, DBA/2, BALB/c, and many other strains. Heterozygotes have both bands (6, 8). The ES-1A isozyme is more heat-stable that the ES-1B isozyme (1). Two more alleles occur in wild mice, *Es-1^c*, a null allele found in southern California and northern Denmark, and *Es-1^d* determining a fast band found in southern California (9). The allele *Es-1^e* arose in strain DBA/2 as a spontaneous mutation. The ES-1E esterase migrates to a position between those of ES-1A and ES-1B and has much less activity (10). The allele *Es-1^f* determines a strong slowly migrating band and is found in MOL3/JA and other strains derived from *M. m. molossinus* (3). Further alleles have been observed in wild mice (2). The locus designated *Ee-1* by Pelzer (4) is probably identical with *Es-1* (7).

References

1. Aotsoka, T., and K. Moriwaki. 1977. A comparison of heat sensitivity of serum esterase (Es-1) in different inbred mice, *Mus musculus*. Ann. Rep. Natl. Inst. Genet. Jpn. 27:32–33.
2. von Deimling, O. 1984. Esterase-23 (ES-23): characterization of a new carboxylesterase isozyme (E.C. 3.1.1.1) of the house mouse, genetically linked to ES-2 on chromosome 8. Biochem. Genet. 22:769–782.
3. Otto, J., A. Ronai, and O. von Deimling. 1981. Purification and characterization of esterase 1F, the albuminesterase of the house mouse (*Mus musculus*). Eur. J. Biochem. 116:285–291.
4. Pelzer, C.F. 1965. Genetic control of erythrocyte esterase forms in *Mus musculus*. Genetics 52:819–828.
5. Peters, J. and H.R. Nash. 1977. Polymorphism of esterase 11 in *Mus musculus*, a further esterase locus on chromosome 8. Biochem. Genet. 15:217–226.
6. Popp, R.A., and D.M. Popp. 1962. Inheritance of serum esterases having different electrophoretic patterns among inbred strains of mice. J. Hered. 53:111–114.
7. Ruddle, F.H., and T.H. Roderick. 1966. The genetic control of two types of esterases in inbred strains of the mouse. Genetics 54:191–202.
8. Ruddle, F.H., and T.H. Roderick. 1968. Allelically determined isozyme polymorphisms in laboratory populations of mice. Ann. NY Acad. Sci. 151:531–539.
9. Selander, R.K., S.Y. Yang, and W.G. Hunt. 1969. Polymorphism in esterases and hemoglobin in wild populations of the house mouse (*Mus musculus*). Stud. Genet. V, Univ. Texas Publ. 6918:271–338.
10. Soares, E.R. 1979. Identification of a new allele of *Es-1* segregating in an inbred strain of mice. Biochem. Genet. 17:577–583.

Es-2 locus, esterase-2, Chr 8

This locus controls electrophoretic variation of a fast anodally migrating carboxylesterase (E.C. 3.1.1.1) of kidney and serum. The esterase is detectable using either naphthyl or 4-methylumbelliferyl esters as substrates (2). The allele *Es-2^a* determines absence of an esterase band and occurs in the RF and RFM strains and in some populations of wild mice; the allele *Es-2^b* determines a fast migrating band and occurs in most inbred strains and some wild mice; the allele *Es-2^c* determines a slower band and occurs in the PL/J strain and in the wild (3, 4). Further phenotypes have been described by von Deimling (1). Heterozygotes with *Es-2^a* have lighter staining bands than homozygotes for the *Es-2^b* and *Es-2^c* alleles, and *Es-2^b*/*Es-2^c* mice have two bands. In the electrophoretic patterns from kidney, the enzyme bands show three subbands, whereas those from serum are single bands (3, 4).

References

1. von Deimling, O. 1984. Esterase-23 (ES-23): characterization of a new carboxylesterase isozyme (E.C. 3.1.1.1) of the house mouse, genetically linked to ES-2 on chromosome 8. Biochem. Genet. 22:769–782.
2. Peters, J. and H.R. Nash. 1977. Polymorphism of esterase 11 in *Mus musculus*, a further esterase locus on chromosome 8. Biochem. Genet. 15:217–226.
3. Petras, M.L. 1963. Genetic control of a serum esterase component in *Mus musculus*. Proc. Natl. Acad. Sci. USA 50:112–116.
4. Ruddle, F.H., T.B. Shows, and T.H. Roderick. 1969. Esterase genetics in *Mus musculus*: expression, linkage, and polymorphism of locus *Es-2*. Genetics 62:393–399.

Es-3 locus (formerly also *Ee-2*), esterase-3, Chr 11

This locus controls electrophoretic variation of an eserine-sensitive esterase present in kidney, liver, and erythrocytes. Four alleles are known: *Es-3^a* determines a faintly staining band of intermediate mobility and occurs in strains C57BL, BALB/c, SEC, and others (5); *Es-3^b* determines a fast anodally migrating band and occurs in strains RF and RFM; *Es-3^c* determines a slow band and occurs in strains CBA, C3H, SJL, AKR, and others (3, 4); *Es-3^d* determines a band that is faster than that of *Es-3^b* and occurs at low frequency in wild mice near Windsor, Ontario (1). Heterozygotes show the two bands of the parental types. The locus designated *Ee-2* by Pelzer (2) is probably identical with *Es-3* (5).

References

1. Martin, J.E., and M.L. Petras. 1971. Es-3 esterases in erythrocytes of *Mus musculus*. Can J. Genet. Cytol. 13:777–781.
2. Pelzer, C.F. 1965. Genetic control of erythrocyte esterase forms in *Mus musculus*. Genetics 52:819–828.
3. Popp, R.A. 1966. Inheritance of an erythrocyte and kidney esterase in the mouse. J. Hered. 57:197–201.
4. Ruddle, F.H., and T.H. Roderick. 1966. The genetic control of two types of esterases in inbred strains of the mouse. Genetics 54:191–202.

5. Ruddle, F.H., and T.H. Roderick. 1968. Allelically determined isozyme polymorphisms in laboratory populations of mice. Ann. NY Acad. Sci. 151:531–539.

Es-4 locus

See *Es-1* locus.

Es-5 locus, esterase-5, Chr 8

This locus controls electrophoretic variation in a postalbumin serum esterase. The esterase is detectable using either naphthyl or 4-methylumbelliferyl esters as substrates (2). The allele *Es-5^a* determines absence of the enzyme band and is found in relatively high frequency in wild mice in southwestern Ontario (3) and in strains derived from other wild populations (von Deimling, personal communication); the allele *Es-5^b* determines presence of the enzyme band and is found in all laboratory strains examined and in some wild mice. Heterozygotes resemble *Es-5^b/Es-5^b* (3). A third allele *Es-5^c* that determines a band slower than that of *Es-5^b* occurs in *M. m. molossinus* in Japan (1).

References

1. Minezawa, M., K. Moriwaki, and K. Kondo. 1976. Studies on protein polymorphism of Japanese wild mouse, *Mus musculus molossinus*. Ann. Rep. Natl. Inst. Genet. Jpn. 26:23–25.
2. Peters, J. and H.R. Nash. 1977. Polymorphism of esterase 11 in *Mus musculus*, a further esterase locus on chromosome 8. Biochem. Genet. 15:217–226.
3. Petras, M.L., and F.G. Biddle, 1967. Serum esterases in the house mouse, *Mus musculus*. Can. J. Genet. Cytol. 9:704–710.

Es-6 locus, esterase-6, Chr 8

This locus controls electrophoretic variation of a carboxylesterase (E.C. 3.1.1.1) present in kidney, liver, lung, heart, and most other organs but absent from serum and erythrocytes. The esterase is detectable using naphthyl and naphthol AS esters as substrates but not 4-methylumbelliferyl esters and it stains only poorly if 5-bromoindoxyl acetate is used (4). The ES-6 content is highest in organs with active fat metabolism. The banding pattern is complex and secondary bands occur in varying degrees in different organs (2). Parts of the pattern originally described by Peters and Nash (5) and Womack (7) have been ascribed by Nash and von Deimling (4) to variation in ES-20. The alleles as defined by Nash and von Deimling (4) are as follows: *Es-6^a* determines a multiple banding pattern and occurs in the C57BL, C3H, and most other laboratory strains

and in many wild-derived strains; *Es-6^b* determines a strong single band and occurs in the Peru/1Fr and other wild-derived strains; *Es-6^c* determines a single band, more intense than that of *Es-6^b*, and occurs in wild-derived Peru-Coppock strains (the variant from Skokholm Island ascribed to the *Es-6^c* allele by Peters and Nash (5) is due entirely to a variant of ES-20); *Es-6^d* determines a multiple banded pattern more anodal than that of *Es-6^a* and occurs in the A/WySnA strain (von Deimling, personal communication); a fifth allele *Es-6^e* occurs in wild *M. m. molossinus* and *M. m. castaneus* (1). A variant with a slow single band which may be the same as *Es-6^c* has been described by Minezawa *et al.* (3) in *M. m. molossinus*. In the cases described, heterozygotes have all parental bands. The ES-20 esterase is probably a hybrid of the *Es-6* and *Es-9* gene products (6). See ES-20.

References

1. Eisenhardt, E., and O. von Deimling. 1982. Interstrain variation of esterase-22, a new isozyme of the house mouse. Comp. Biochem. Physiol. 73B:719–724.
2. de Looze, S.M., A. Ronai, and O. v. Deimling. 1982. Organ specific expression of esterase-6 in the house mouse, *Mus musculus*. Histochemistry 74:553–561.
3. Minezawa, M., K. Moriwaki, and K. Kondo. 1976. Studies of protein polymorphism of Japanese wild mouse, *Mus musculus molossinus*. Ann. Rep. Natl. Inst. Genet. Jpn. 26:23–25.
4. Nash, H.R., and O. von Deimling. 1982. Kidney esterases in *Mus musculus*: further polymorphism of esterase-6, esterase-9, and a new esterase, esterase-20. Biochem. Genet. 20:537–554.
5. Peters, J., and H.R. Nash. 1978. Esterases of *Mus musculus*: substrate and inhibition characteristics, new isozymes, and homologies with man. Biochem. Genet. 16:553–569.
6. Wassmer, B., S.M. de Looze, and O.H. von Deimling. 1985. Biochemistry and genetics of esterase 20 (ES-20), a second trimeric carboxylesterase of the house mouse (*Mus musculus*). II. A unique recombination reveals ES-20 is a hybrid enzyme. Biochem. Genet. 23:759–770.
7. Womack, J.E. 1975. Esterase-6 (*Es-6*) in laboratory mice: hormone-influenced expression and linkage relationship to oligosyndactylism (*Os*) and esterase-2 (*Es-2*) in chromosome 8. Biochem. Genet. 13:311–322.

Es-7 locus, esterase-7, Chr 8

This locus controls electrophoretic variation of a carboxylesterase (E.C. 3.1.1.1) present in erythrocyte lysates and membranes and in lung, tongue, testis, and most other tissues, but not in plasma or brain (2). The allele *Es-7^b* determines a pattern of three or four bands in gels stained with 4-methylumbelliferyl heptanoate and occurs in most inbred strains (3); the allele *Es-7^a*

determines a similar pattern of lesser anodal mobility and occurs in *M. m. molossinus*; the allele *Es-7ᶜ* determines a similar pattern of greater anodal mobility than that of *Es-7ᵇ* and occurs in *M. m. molossinus* and *M. m. castaneus* (2). An allele described by Peters and Nash (3) determines three bands of similar mobility to those of *Es-7ᵇ* but with greatly diminished staining ability and occurs in the IS/Cam strain and in the Peru-maus wild stock. *Es-7* shows no recombination with *Es-2* on Chr 8 but has a different tissue distribution and different substrate specificity (1, 3).

References

1. Chapman, V.M., and F.H. Ruddle. 1973. Mouse News Lett. 48:45.
2. Lipps, A., A. Ronai, and O. von Deimling. 1979. Esterase-7, a common constituent of numerous mouse tissues. Comp. Biochem. Physiol. 62B:201–206.
3. Peters, J., and H.R. Nash. 1977. Polymorphism of esterase 11 in *Mus musculus*, a further esterase locus on chromosome 8. Biochem. Genet. 15:217–226.

Es-8 locus, esterase-8, Chr 7

This locus controls electrophoretic variation of an erythrocyte esterase specific for the substrates α- and β-naphthyl acetate. The enzyme is probably not carbonic anhydrase. The allele *Es-8ᵃ* determines a more anodal band and is present in C57BL, SWR, and all other strains examined; the allele *Es-8ᵇ* determines a more cathodal band and is found in the subspecies *M. m. castaneus*. Heterozygotes show both bands (1).

References

1. Chapman, V.M., E.A. Nichols, and F.H. Ruddle. 1974. Esterase-8 (*Es-8*): characterization, polymorphism, and linkage of an erythrocyte esterase locus on chromosome 7 of *Mus musculus*. Biochem. Genet. 11:347–358.

Es-9 locus, esterase-9, Chr 8

This locus controls electrophoretic variation of a carboxylesterase (E.C. 3.1.1.1) in kidney (4). The allele *Es-9ᵃ* determines a pattern of three prominent cathodally migrating bands and occurs in strains NMRI, C3H, C57BL/Gr, and some Peru-Coppock mice; the allele *Es-9ᵇ* determines three fainter bands with slightly less mobility than those of *Es-9ᵃ* and occurs in the SK/Cam strain, in some Peru-Coppock mice, and in all wild-caught mice from Skokholm Island (3, 4); the allele *Es-9ᶜ* determines three faint bands migrating between those of *Es-9ᵃ* and *Es-9ᵇ* and occurs in *M. m. molossinus*; the allele *Es-9ᵈ* determines three very

strongly staining bands and occurs in some Peru-Coppock mice; the allele *Es-9ᵉ* determines three faint bands more anodal than those of *Es-9ᶜ* and occurs in an inbred Peru-Coppock strain, Peru/1Fr (2). Heterozygotes have intermediate patterns. ES-9 is found in kidney, liver, diaphragm, submandibular gland, brain, and testis, but not in red cells or serum. Activity is low in kidneys of females, but is induced to male levels by testosterone. A study of cellular localization of ES-9 in the kidneys of NMRI (*Es-9ᵃ*) and SK/Cam (*Es-9ᵇ*) strains showed that the esterase was fully repressed in the cortical proximal tubules in females or in the absence of testosterone, and was constitutive in the medullary part of the proximal tubules. The esterase is probably located in the mitochondria (1). The ES-20 esterase is probably a hybrid of the *Es-6* and *Es-9* gene products (5). See ES-20.

References

1. Bocking, A., J. Hansert, and O. von Deimling. 1976. Esterase. XXII. Cellular and subcellular localization of the Es-9 esterase in mouse kidney. Histochemistry 46:177–188.
2. Nash, H.R., and O. von Deimling. 1982. Kidney esterases of *Mus musculus*: further polymorphism of esterase-6, esterase-9, and a new esterase, esterase-20. Biochem. Genet. 20:537–554.
3. Peters, J., and H.R. Nash. 1978. Esterases of *Mus musculus*: substrate and inhibition characteristics, new isozymes, and homologies with man. Biochem. Genet. 16:553–569.
4. Schollen, J., K. Bender, and O. von Deimling. 1975. Esterase. XXI. *Es-9*, a possibly new polymorphic esterase in *Mus musculus* genetically related to *Es-2*. Biochem. Genet. 13:369–377.
5. Wassmer, B., S.M. de Looze, and O.H. von Deimling. 1985. Biochemistry and genetics of esterase 20 (ES-20), a second trimeric carboxylesterase of the house mouse (*Mus musculus*). II. A unique recombination reveals ES-20 is a hybrid enzyme. Biochem. Genet. 23:759–770.

Es-10 locus, esterase-10, Chr 14

This locus controls electrophoretic variation of a slowly migrating esterase present in erythrocytes, kidney, liver, and many other tissues. The esterase is detectable with 4-methylumbelliferyl ester as a substrate, but not with naphthyl esters. Three alleles are known: *Es-10ᵇ* determines the most anodal band and occurs in strains AKR, C3H, DBA, and others; *Es-10ᵃ* determines a slightly more cathodal band and occurs in strains A/He, BALB/c, C57BL, and others, and in *M. m. castaneus* and some *M. m. molossinus*; *Es-10ᶜ* determines the most cathodal band and occurs in strain BUB/BnJ and in some *M. m. molossinus*. Heterozygotes have the three bands indicative of a dimeric enzyme (1, 2).

References

1. Peters, J., and H.R. Nash. 1976. Polymorphism of esterase-10 in *Mus musculus*. Biochem. Genet. 14:119–125.
2. Womack, J.E., M.T. Davisson, E.M. Eicher, and D.A. Kendall. 1977. Mapping of nucleotide phosphorylase (*Np-1*) and esterase-10 (*Es-10*) on mouse chromosome 14. Biochem. Genet. 15:347–355.

Es-11 locus, esterase-11, Chr 8

This locus controls electrophoretic variation of an eserine-sensitive carboxylesterase (E.C. 3.1.1.1) found in liver, kidney, and small intestine, and detectable with naphthyl or 4-methylumbelliferyl esters. The allele *Es-11ᵃ* determines the presence of two anodally located bands and occurs in C57BL, BALB/c, and many other inbred strains; the allele *Es-11ᵇ* determines two less anodal bands and occurs in the IS/Cam strain and in many wild mice (2); the allele *Es-11ᶜ* determines two weak bands migrating between those of *Es-11ᵃ* and *Es-11ᵇ* and occurs in *M. m. molossinus* and *M. spretus*; the allele *Es-11ᵈ* determines two strong bands and occurs in *M. m. molossinus* (1). In heterozygotes the phenotype resembles a mixture of the parental types (2). *Es-11* does not recombine with *Es-2*, *Es-7*, or *Es-23* on Chr 8 (1, 2).

References

1. von Deimling, O. 1984. Esterase 23 (ES-23): characterization of a new carboxylesterase isozyme (E.C. 3.1.1.1) of the house mouse, genetically linked to ES-2 on chromosome 8. Biochem. Genet. 22:769–782.
2. Peters, J., and H.R. Nash. 1977. Polymorphism of esterase-11 in *Mus musculus*, a further esterase locus on chromosome 8. Biochem. Genet. 15:217–226.

Es-12 locus, esterase-12

This locus controls electrophoretic variation, detected on polyacrylamide gels, of a serum esterase. The allele *Es-12ᵃ* determines a single band and is present in strains SWR/J, C3H/HeJ, C57BL/6J, and DBA/2J. The allele *Es-12ᵇ* determines two bands, the more anodal of which comigrates with the single band of *Es-12ᵃ*; it is present in strains C57BR/cdJ and C57L/J. Heterozygotes show both bands. *Es-12* is not closely linked to either *Es-1* on Chr 8 or *Es-3* on Chr 11 (1, 2).

References

1. Taylor, B.A. 1977. Mouse News Lett. 56:41.
2. Taylor, B.A. 1977. Mouse News Lett. 57:19.

Es-13 locus, esterase-13, Chr 9

This locus controls the isoelectric focusing pattern in kidney of the ES-18 esterase coded for by the *Es-18* locus (1). The allele *Es-13ᵇ* determines a fast migrating band and is found in strains AEJ/GnRk, LG/J, SJL/J, and SWR/J; the allele *Es-13ᵃ* determines a slower band and is found in all other strains examined. *Es-13ᵇ* is completely recessive, suggesting that the locus acts by posttranslational modification of the enzyme. *Es-13* is near *Mod-1* on Chr 9 (4) unlinked to *Es-18* on Chr 19. See *Es-18* locus. *Es-21* described by Nash (3) has the same strain distribution of alleles as *Es-13* in inbred strains and in BALB/c × STS/A recombinant inbred strains and is probably identical to *Es-13* (2).

References

1. von Deimling, O., and J. Hilgers. 1983. Mouse News Lett. 68:65.
2. von Deimling, O., and M. Müller. 1985. Mouse News Lett. 72:101.
3. Nash, H.R. 1982. Mouse News Lett. 66:80.
4. Womack, J.E., B.A. Taylor, and J.E. Barton. 1978. Esterase-13, a new mouse esterase locus with recessive expression and its location on chromosome 9. Biochem. Genet. 16:1107–1112.

Es-14 locus, esterase-14, Chr 9

This locus controls variation in electrophoretic migration and isoelectric focusing pattern of a butyrylesterase present in erythrocytes and many tissues but absent from plasma. ES-14 is active on several substrates (2, 3). All laboratory and wild strains of *Mus musculus* carry the *Es-14ʳ* allele which manifests itself as a single rapidly migrating band and focuses at pH 4.29; all *Mus spretus* mice carry the *Es-14ˢ* allele which manifests as a single slower band and focuses at pH 4.42 (2, 3). Heterozygotes have a five-banded pattern indicative of a tetrameric structure (3). *Es-14* shows loose linkage with *Mod-1* on Chr 9 (1).

References

1. Bonhomme, F., F. Benmehdi, J. Britton-Davidian, and S. Martin. 1979. Analyse génétique de croisements interspécifiques *Mus musculus* L. × *Mus spretus* Lataste: liaison de *Adh-1* avec *Amy-1* sur le chromosome 3 et de *Es-14* avec *Mod-1* sur le chromosome 9. C. R. Acad. Sci. Paris 289:545–548.
2. Britton-Davidian, J., and F. Bonhomme. 1979. Deux nouveaux locus d'estérases *Es-14* et *Es-15* chez les souris (genre *Mus* L.): caractérisation par différents substrats et inhibiteurs. C. R. Acad. Sci. Ser. D 288:1419–1422.
3. von Deimling, O., and S. de Looze. 1983. Human red cell

butyrylesterase, and its homologies in thirteen other mammalian species. Hum. Genet. 63:241–346.

Es-15 locus, esterase-15

This locus controls electrophoretic variation in an esterase present in muscle, stomach, heart, and testis. ES-15 is active on naphthyl substrates. The allele *Es-15^r* determines a rapidly migrating enzyme band and is present in *Mus specilegus*; *Es-15^m* determines a slower migrating band and is present in inbred strains; *Es-15^l* determines a slow band and is present in wild *Mus musculus* in Greece. The enzyme is probably a dimer, since presumed heterozygotes have three bands (1).

References

1. Britton-Davidian, J., and F. Bonhomme. 1979. Deux nouveaux locus d'estérases *Es-14* et *Es-15* chez les souris (genre *Mus* L.): caractérisation par différents substrats et inhibiteurs. C. R. Acad. Sci. Ser. D 288:1419–1422.

Es-16 locus, esterase-16, Chr 3

This locus controls a polymorphism for an arylesterase (E.C. 3.1.1.2) detected in kidney, spleen, and heart by means of isoelectric focusing and disc electrophoresis. The allele *Es-16^a* determines an enzyme with an isoelectric point (IEP) of 5.9 and occurs in the C57BL/10 and many other inbred strains; the allele *Es-16^b* determines an IEP of 6.1 and occurs in *M. m. molossinus*; the allele *Es-16^c* determines an enzyme with very weak activity and an IEP of 5.9 and occurs in the Peru-Coppock stock. *Es-16^a/Es-16^b* mice have both parental bands. ES-16 is completely insoluble in water. It is active on indoxyl acetate substrate but reacts only weakly with β-naphthyl acetate. *Es-16* is located near *Car-1* on Chr 3 (1).

References

1. von Deimling, O.H., P. Schupp, and L. Otto. 1981. Esterase-16 (*Es-16*): characterization, polymorphism, and linkage to chromosome 3 of a kidney esterase locus of the house mouse. Biochem. Genet. 19:1091–1099.

Es-17 locus, esterase-17, Chr 9

This locus determines variation in isoelectric focusing of an enzyme tentatively classified as an acetyl esterase (E.C. 3.1.1.6) and present in all tissues examined and in fetal organs but not in hemolysate or serum. The allele *Es-17^a* determines presence of three bands with isoelectric points (IEP) between 5.55 and 5.90 and occurs in the C57BL/10 and many other strains; the

allele *Es-17^b* determines three bands with IEPs between 5.05 and 5.55 and occurs in strain LP/J, in *M. m. molossinus* and *M. m. castaneus* (2), and in some wild populations of *Mus musculus* in central Europe (1). Heterozygotes show all the parental bands. *Es-17* is located on Chr 9 between the centromere and *Mpi-1* (2).

References

1. Antonucci, T.K., O.H. von Deimling, B.B. Rosenblum, L.C. Skow, and M.H. Meisler. 1984. Conserved linkage within a 4 cM region of mouse chromosome 9 and human chromosome 11. Genetics 107:463–475.
2. Otto, J., and O.H. von Deimling. 1983. Esterase 17 (ES-17): characterization and genetic location on chromosome 9 of a bis-p-nitrophenyl phosphate-resistant esterase of the house mouse *Mus musculus*. Biochem. Genet. 21:37–48.

Es-18 locus, esterase-18, Chr 19

This locus controls a polymorphism detectable by isoelectric focusing of an arylesterase (E.C. 3.1.1.2) active with β-naphthyl butyrate as substrate and present in kidney, spleen, and lung. The esterase shows an isoelectric point (IEP) in the range of 7.2 to 7.8. The allele *Es-18^a* determines an esterase with the higher IEP and is present in C57BL/10 and many other laboratory strains; the allele *Es-18^b* determines an esterase with the lower IEP and occurs in *M. m. molossinus* (1). The esterase is regulated in kidney by the unlinked *Es-13* locus; the recessive *Es-13^b* allele in AEJ, SWR, and some other strains not closely related to the more widely used standard inbred strains determines what is probably a posttranslational modification of the enzyme (3). *Es-18* is on Chr 19 close to *Got-1* (2). See *Es-13* locus.

References

1. von Deimling, O. 1981. Mouse News Lett. 65:13.
2. von Deimling, O. 1985. Mouse News Lett. 73:15.
3. von Deimling, O., and J. Hilgers. 1983. Mouse News Lett. 68:65.

Es-19 locus, esterase-19

This symbol is reserved for an arylesterase (E.C. 3.1.1.2) that differs significantly from other known mouse esterases but for which no polymorphism detectable by disc electrophoresis is known (1). The esterase occurs in many organs including testis and adrenals and in brain of newborn animals but not in red cells or plasma. It is detectable with α-naphthyl acetate after inhibition of the comigrating ES-7 (2).

References

1. von Deimling, O. 1981. Mouse News Lett. 65:13.
2. Münz, M., and O. von Deimling. 1985. Electrophoretic characterization of esterase-19 (ES-19), a new arylesterase of the house mouse (*Mus musculus*). Electrophoresis 6:175–178.

ES-20, esterase-20, Chr 8

Nash and von Deimling (2) have described five phenotypes of this carboxylesterase (E.C. 3.1.1.1) detectable in liver and kidney by polyacrylamide and starch gel electrophoresis. The phenotypes are different from those determined by five variants at the Es-9 locus, but there is complete association of variants of the two esterases. The five phenotypes of ES-20 are as follows: ES-20A shows the slowest anodal migration and weakest activity and occurs in C57BL/10, C3H/He, and the P/3 stock derived from Peru-Coppock mice; ES-20B is slightly faster, shows tightly grouped and very active bands, and occurs in SK/Cam and *M. m. castaneus*; ES-20C shows a very active major band and a more anodal minor band and occurs in *M. m. molossinus*; ES-20D shows no activity and occurs in the P/2 strain derived from the Peru-Coppock stock; and ES-20E is marginally faster and stains more intensely than ES-20B and occurs in the Peru/1Fr strain derived from the Peru-Coppock stock (1). The enzyme is a trimer. A rare recombinant between the haplotypes *Es-6ᵃ Es-9ᵃ* (ES-20A) and *Es-6ˀ Es-9ᵈ* (ES-20D), *Es-6ᵃ Es-9ᵈ*, was neither ES-20A nor ES-20D but a new type. In addition, *in vitro* incubation of purified ES-6A and ES-9A generated a product identical to ES-20A. These results are evidence that ES-20 is a hybrid of products of the *Es-6* and *Es-9* loci and not the product of a separate locus (3).

References

1. Berning, W., S.M. deLooze, and O. von Deimling. 1985. Identification and development of a genetically closely-linked carboxylesterase family of the mouse liver. Comp. Biochem. Physiol. 80B:859–865.
2. Nash, H.R., and O. von Deimling. 1982. Kidney esterases of *Mus musculus*: further polymorphism of esterase-6, esterase-9, and a new esterase, esterase-20. Biochem. Genet. 20:537–554.
3. Wassmer, B., S.M. de Looze, and O. von Deimling. 1985. Biochemistry and genetics of esterase-20 (ES-20), a second trimeric carboxylesterase of the house mouse (*Mus musculus*) II. A unique recombination reveals ES-20 as a hybrid enzyme. Biochem. Genet. 23:759–770.

Es-21 locus

See *Es-13* locus.

Es-22 locus, esterase-22, Chr 8

This locus controls electrophoretic mobility of a carboxylesterase (E.C. 3.1.1.1) present in submandibular gland and also in liver, lung, and tongue. The enzyme shows three bands on disc electrophoresis and about five bands on isoelectric focusing between pH 5.2 and 5.7. Eight alleles have been distinguished. In order of decreasing anodal mobility, they are: *Es-22ᶠ* in *M. spretus*; *Es-22ᵍ* in *M. hortulanus*; *Es-22ᵃ* in *M. m. molossinus*; *Es-22ᵇ* in *M. m. castaneus*; *Es-22ᶜ* in IS/Cam; and *Es-22ᵈ* in C57BL/10 and many other inbred strains; *Es-22ᵉ* in Peru-Coppock determines very weak ES-22 bands; *Es-22ʰ* in STS and YBR determines absence of ES-22 bands (1, 2). *Es-22* is identical with *Eg*, egasyn, a locus controlling a protein, egasyn, that stabilizes the binding of glucuronidase to microsomes and that also has esterase activity. *Eg* and *Es-22* both map near *Es-9* on Chr 8. The *Egᵒ* allele determining absence of egasyn is the same as *Es-22ʰ* (2). See *Eg* locus.

References

1. Eisenhardt, E., and O. von Deimling. 1982. Interstrain variation of esterase-22, a new isozyme of the house mouse. Comp. Biochem. Physiol. 73B:719–724.
2. Medda, S., O. von Deimling, and R.T. Swank. 1986. Identity of esterase-22 and egasyn, the protein which complexes with microsomal β-glucuronidase. Biochem. Genet. 24:229–243.

Es-23 complex, Chr 8

This complex includes a structural and a temporal locus for a carboxylesterase (E.C. 3.1.1.1), ES-23, detected by disc electrophoresis and isoelectric focusing and present in kidney, liver, and intestine. It is active with substrates 5-bromoindoxyl acetate and α-naphthyl butyrate.

Es-23e locus, esterase-23 electrophoretic. Five alleles are known: *Es-23eᵃ* in *M. m. molossinus* determines two or three fast anodally migrating bands; *Es-23eᵇ* in *M. m. castaneus* determines slightly faster bands; *Es-23eᶜ* in C57BL/10 and many other laboratory strains determines bands slower than those of *Es-23eᵃ*; *Es-23eᵈ* in SK/Cam is a null allele; *Es-23eᵉ* in *Mus spretus* determines faint bands migrating like those of *Es-23eᵇ*. The *Es-23* complex is near *Es-2*, *Es-7*, and *Es-11* on Chr 8, but the esterase is clearly different from ES-2, ES-7, and ES-11 (1).

Es-23t locus, esterase-23 temporal. This locus affects lung-specific expression of ES-23 and was recognized as a variant in *M. m. castaneus* and some *M. m. molossinus*. In lung, ES-23 is present only in mice carrying

Es-23e^b. This allele, designated *Es-23t^b*, determines expression of ES-23 in lung. In crosses, it remains associated with *Es-23e^b* and in heterozygotes acts only on expression of the *Es-23e^b* allele. *Es-23t* is thus *cis*-acting. The allele determining lack of expression in lung is designated *Es-23t^a* and occurs in association with all the other *Es-23e* alleles expressing ES-23 (1).

References

1. von Deimling, O. 1984. Esterase 23 (ES-23): characterization of a new carboxylesterase isozyme (E.C. 3.1.1.1) of the house mouse, genetically linked to ES-2 on chromosome 8. Biochem. Genet. 22:769–782.

Es-24 locus, esterase-24, Chr 8

This locus controls electrophoretic mobility of a non-specific esterase present in lung, liver, and possibly kidney and absent from all other organs investigated (2). Six phenotypes were distinguished: ES-24A in C57BL/10 and many inbred strains; ES-24B in *M. m. castaneus*; ES-24C in *M. m. molossinus*; ES-24D in Peru-Coppock; ES-24E in Peru-Coppock; and ES-24F in strains AKR, NMRI, NFS, and 129. ES-24 has a relatively high molecular weight (180 000). It may be a hybrid aggregate of the products of two other undetermined Chr 8 esterase loci (1). The locus or loci determining ES-24 are located on Chr 8 near *Es-6* and *Es-9* (3).

References

1. Berning, W., S.M. de Looze, and O. von Deimling. 1985. Identification and development of a genetically closely-linked carboxylesterase family of the mouse liver. Comp. Biochem. Physiol. 80B:859–865.
2. von Deimling, O., and J. Hilgers. 1982. Mouse News Lett. 67:16.
3. von Deimling, O., M. Müller, and E. Eisenhardt. 1983. The non-specific esterases of mouse lung. Histochemistry 78:271–284.

Es-25 locus, esterase-25, Chr 12

This locus controls electrophoretic mobility of an arylesterase (E.C. 3.1.1.2) present in kidney and absent from other organs. The allele *Es-25^a* determines a fast migrating double band and occurs in the BALB/c, AKR, and A/J strains; the allele *Es-25^b* determines a slower double band and occurs in DBA/2, C3H, the various C57 strains, and many other laboratory strains. The esterase appears to be a monomeric protein (1). *Es-25* is on Chr 12, probably proximal to *Ah* (B.A. Taylor, personal communication).

References

1. Wassmer, B., and O. von Deimling. 1983. Mouse News Lett. 69:20.

Es-26 locus, esterase-26, Chr 3

This locus controls variation in isoelectric focusing of an esterase present in liver, stomach, and small intestine but not in kidney, lung, testis, plasma, or red blood cells. The esterase is active on a variety of substrates including choline esters. The allele *Es-26^a* determines a band focusing at pH 8.2 and occurs in the C57BL/10, BALB/c, CBA/J, C3H, and other inbred and wild-derived strains; the allele *Es-26^b* determines a more diffuse band focusing at pH 7.8–7.9 and occurs in the A/J, DD/He, and NZW strains, and in *M. spretus*; the allele *Es-26^c* determines two bands focusing at 8.1 and 8.3 and occurs in the AKR, SJL, and FVB/N strains. The esterase is extremely sensitive to eserine, is particle-bound, and has the properties of a cholinesterase (E.C. 3.1.1.8). *Es-26* is on Chr 3 near the center (1).

References

1. von Deimling, O.H., B. Wassmer, and M. Müller. 1984. Esterase 26 (ES-26): characterization and genetic location on chromosome 3 of an eserine-sensitive esterase of the house mouse (*Mus musculus*). Biochem. Genet. 22:1119–1126.

Es-27 locus, esterase-27, Chr 3

This locus controls heat stability of a serum cholinesterase (E.C. 3.1.1.8). The allele *Es-27^a* determines a heat-stable form and occurs in the C57BL/10, BALB/c, DBA/2, and many other laboratory strains and in *M. m. molossinus* and *M. m. castaneus*; the allele *Es-27^b* determines a heat-labile form and occurs in the AKR and SJL strains and in some *M. m. castaneus*. Heterozygotes have intermediate heat stability (1, 2). *Es-27* is on Chr 3, about 9 cM distal to *Es-26*, the other known mouse cholinesterase locus (3).

References

1. Bonhomme, F., and R.K. Selander. 1978. Estimating total genic diversity in the house mouse. Biochem. Genet. 16:287–297.
2. von Deimling, O. 1985. Mouse News Lett. 73:15.
3. von Deimling, O. 1986. Mouse News Lett. 74:89.

Esr locus, esterase-5 regulator, Chr 6

This locus controls presence or absence of the ES-5B esterase in plasma. The allele *Esr^o* occurs in *M. m.*

molossinus and, when homozygous, prevents the expression of ES-5B; the allele *Esr*^{a} occurs in strain C57BL/10, SK/Cam, and NMRI/Fr and allows expression of ES-5B. *Esr*^{a} is dominant to *Esr*^{o}. *Esr*^{o} has no effect on the ES-5A esterase or on enzymes determined by the *Es-2*, *Es-7*, and *Es-11* loci. *Esr* is located on Chr 6 between *Ly-3* and *Ldr-1* (1).

References

1. von Deimling, O.H., J. Otto, and A. Reske-Kunz. 1981. Esr, a second locus in the house mouse controlling ES-5. Biochem. Genet. 20:351–358.

Ets-1, *Ets-2* loci

These loci are cellular homologs of the *ets* sequence (v-*ets*) of the oncogenic avian E26 replication-defective retrovirus. The coding sequence of the gene in the virus is disrupted in the mouse and other mammals, with the 5′ and 3′ ends on different chromosomes (2).

Ets-1 locus, E26 oncogene-1, Chr 9. This locus is homologous with the 5′ end of the v-*ets* gene. No genetic variants are known. The locus was found to be on Chr 9 by use of mouse–Chinese hamster somatic cell hybrids screened with a human genomic probe homologous with the 5′ end of v-*ets* (2).

Ets-2 locus, E26 oncogene-2, Chr 16. This locus is homologous with the 3′ end of the v-*ets* gene. It was first found to be on Chr 16 by use of mouse–Chinese hamster somatic cell hybrids screened with a human genomic probe homologous with the 3′ end of v-*ets* (2). A DNA fragment length variant was found in the Czech II strain of wild mice from Czechoslovakia. Czech II mice have a *Bam*HI fragment of 17 kb; BALB/c mice have a *Bam*HI fragment of 9.4 kb. In a cross between these strains, *Ets-2* was mapped to the distal end of Chr 16. This region of the chromosome is homologous with the Down syndrome region of human Chr 21 (1).

References

1. Reeves, R.H., D. Gallahan, B.F. O'Hara, R. Callahan, and J.D. Gearhart. 1987. Genetic mapping of *Prm-1*, *Igl-1*, *Smst*, *Mtv-6*, *Sod-1* and *Ets-2*, and localization of the Down syndrome-region of mouse chromosome 16. Cytogenet. Cell Genet. 44:76–81.
2. Watson, D.K., M.J. McWilliams-Smith, C. Kozak, R. Reeves, J. Gearhart, M.F. Nunn, W. Nash, J.R. Fowle III, P. Duesberg, T.S. Papas, and S.J. O'Brien. 1986. Conserved chromosomal positions of dual domains of the *ets* protooncogene in cats, mice, and humans. Proc. Natl. Acad. Sci. USA 83:1792–1796.

Ex locus, earlier X-zone degeneration, Chr 7

Discovered as a difference between the A/Cam and CBA/FaCam strains in time of fatty degeneration of the X zone of the adrenal gland of females. A/Cam is homozygous for the dominant allele which causes well advanced degeneration by 8 weeks. In CBA females and females of some other strains homozygous for the recessive allele, degeneration has hardly begun at this age. Ovariectomy does not prevent fatty degeneration in A strain females or in F1 hybrids with CBA females (1). The authors suggest the symbols *Ex* and *ex* for the dominant and recessive alleles, respectively, but since there is not a clearly defined wild type allele, this nomenclature is inappropriate. I suggest *Ex*^{e} for the dominant 'early' allele in A/Cam, and *Ex*^{d} for the recessive 'delayed' allele in CBA.

References

1. Shire, J.G.M., and S.G. Spickett. 1968. A strain difference in the time and mode of regression of the adrenal X zone in female mice. Gen. Comp. Endocrinol. 11:355–365.

ex

See *exf*.

Exa locus, exploratory activity, Chr 4

This locus was discovered as a difference in short-term exploratory activity between the C57BL/6By and BALB/cBy inbred strains. The allele *Exa*^{h}, present in C57BL/6, determines high short-term exploratory activity; the allele *Exa*^{l}, present in BALB/c, determines low activity. Heterozygotes have high activity like the *Exa*^{h}/*Exa*^{h} parent. Analysis of CXB recombinant inbred strains shows that *Exa* is probably on Chr 4 near *Eam* (1).

References

1. Oliverio, A., B.E. Eleftheriou, and D.W. Bailey. 1973. Exploratory activity: genetic analysis of its modification by scopolamine and amphetamine. Physiol. Behav. 10:893–899.

exf (formerly *ex*), exfoliative, recessive

Arose spontaneously in the 'A^y viable-yellow inbred' strain at the National Institutes of Health, USA. Homozygotes have a transient purulent conjunctivitis beginning soon after the eyes open at 13 days and lasting 5 or 6 days. During the fourth week, hair loss and exfoliation of the outer layer of skin begins and spreads over the entire body. Regrowth of a dense, short, rough coat occurs, but adult mice are subject to recurring patches of hair loss and transient eyelid swelling. Gram-negative bacilli are present in blood and cerebrospinal fluid during the conjunctivitis stage and in skin during the exfoliative period. Mortality is high at 18 or 19 days, but survivors grow to normal size and are fertile (1).

References

1. Kent, R.L., M.A. Lutzner, and K.P. Smith. 1976. Skin exfoliation and purulent conjunctivitis in a new mutant—the exfoliative mouse. J. Invest. Dermatol. 66:248–251.

Exnm locus, neuromuscular excitability, probably Chr 1

This locus controls a difference in threshold of neuromuscular excitability between the C57BL/6By and BALB/cBy strains. The threshold, measured in volts, was determined in calf muscle electrically stimulated for 2.00 or 0.01 ms. The allele *Exnm^l* in BALB/c determines low threshold (about 0.3 V at stimulation for 2 ms); the allele *Exnm^h* in C57BL/6 determines high threshold (about 0.6 V at stimulation for 2 ms). Examination of the strain distribution pattern in the CXB recombinant inbred strains and the B6.C-H-35 (H(w56)) congenic strain shows that *Exnm* is linked to H-25 and H-35 (1). Contrary to the statement by Oliverio *et al.* (4) cited by the author that these loci are on Chr 9, they are on Chr 1 (2, 3, D.W. Bailey, personal communication).

References

1. Dmitriev, Y.S. 1981. Mapping the genes responsible for strain differences in thresholds of neuromuscular excitability in mice. Doklady Biol. Sci. USSR 261:204–206 (Russian).
2. Durda, P.J., S.C. Boos, and P.D. Gottlieb. 1979. T100: a new murine cell surface glycoprotein detected by anti-Lyt-2.1 serum. J. Immunol. 122:1407–1412.
3. Mathieson, B.J., S.O. Sharrow, K. Bottomly, and B.J. Fowlkes. 1980. Ly-9, an alloantigenic marker of lymphocyte differentiation. J. Immunol. 125:2127–2136.
4. Oliverio, A., B.E. Eleftheriou, and D.W. Bailey. 1973. A gene influencing active avoidance performance in mice. Physiol. Behav. 11:497–501.

ey-1, *ey-2* loci, eyeless-1, -2

Chase (5) postulated that these two recessive mutant genes were responsible for the microphthalmia observed in nearly 90 per cent of the mice in certain selected strains. It is not certain that two discrete loci

with large effects exist (2, 3). If they exist, they are not distinguishable from each other either by their effects or by known linkage map position. Mice heterozygous at one or both of these loci have a higher incidence of abnormal eyes than homozygous normal mice after injection of trypan blue into their mothers during gestation (1, 4). In mice of the ZRDCT/Ch strain (70 per cent eyeless, 30 per cent microphthalmic), lens invagination at 10 days of gestation is abnormal, the lens being smaller than normal and often improperly centered in the optic cup. When improperly centered, it may fail to be enclosed in the cup, which then eventually collapses (6). There is a deficiency of sulfated glycosaminoglycans in the basal lamina of the optic vesicle at its interface with the lens cup during lens cup invagination (8). In eyeless animals of this strain, the nucleus suprachiasmaticus (SCN) of the mediolateral hypothalamus is abnormal, a condition thought to be due to absence of the eye and subjacent optic axons, which are probably necessary at least as early as day 13 of gestation for normal development of the SCN (7).

References

1. Barber, A.N. 1957. The effects of maternal hypoxia on inheritance of recessive blindness in mice. Am. J. Ophthalmol. 44:94–101.
2. Beck, S.L. 1963. The anophthalmic mutant of the mouse. I. Genetic contribution to the anophthalmic phenotype. J. Hered. 54:39–44.
3. Beck. S.L. 1963. The anophthalmic mutant of the mouse. II. An association of anophthalmia and polydactyly. J. Hered. 54:79–83.
4. Beck, S.L. 1963. Frequencies of teratologies among homozygous normal mice compared with those heterozygous for anophthalmia. Nature 200:810–811.
5. Chase, H.B. 1944. Studies on an anophthalmic strain of mice. IV. A second major gene for anophthalmia. Genetics 29:264–269.
6. Harch, C., H.B. Chase, and N.I. Gonsalves. 1978. Studies on an anophthalmic strain of mice. VI. Lens and cup interaction. Dev. Biol. 63:352–357.
7. Silver, J. 1977. Abnormal development of the suprachiasmatic nuclei of the hypothalamus in a strain of genetically anophthalmic mice. J. Comp. Neurol. 176:589–606.
8. Webster, E.H. Jr., A.F. Silver, and N.I. Gonsalves. 1984. The extracellular matrix between the optic vesicle and presumptive lens during lens morphogenesis in an anophthalmic strain of mice. Dev. Biol. 103:142–150.

Ezg locus, extent of zona glomerulosa

This locus was discovered as a difference between the CBA/Cam and Peru strains in width of the zona glomerulosa of the adrenal gland. The dominant allele determines a wide well defined zona, and is present in the CBA strain; the recessive allele determines a narrow poorly defined zona, and is present in the Peru strain. The A/Cam strain resembles CBA and the SF/Cam strain resembles Peru with respect to width of the zona. Sodium deficiency results in glomerular hypertrophy in CBA but not in Peru mice (1). The two strains do not differ in rate of aldosterone secretion (2). Shire (1) suggested the symbols *Ezg* and *ezg* for the dominant and recessive alleles, respectively, but since there is not a clearly defined wild-type allele, this nomenclature is inappropriate. I suggest *Ezg^w* for the dominant 'wide' allele of CBA, and *Ezg^n* for the recessive 'narrow' allele in Peru.

References

1. Shire, J.G.M. 1969. A strain difference in the adrenal zona glomerulosa determined by one gene-locus. Endocrinology 85:415–422.
2. Stewart, J., R. Fraser, V. Papaioannou, and A. Tait. 1972. Aldosterone production and the zona glomerulosa: a genetic study. Endocrinology 90:968–972.

F

f, flexed-tail, recessive, Chr 13

Appeared in a stock maintained by Hunt (8). Homozygotes are small at birth and have a transitory hypochromic, microcytic anemia characterized by a large number of siderocytes containing non-haem iron granules. Most homozygotes also have flexed tail and a belly spot, but these are not constant manifestations of the mutant. Because of the anemia there is probably greater postnatal mortality among *f/f* than among normal mice. The anemia begins on the 12th day of embryonic life when the liver first starts to produce blood cells (7). It is most intense at 15 days of gestation and still severe at birth. By 2 weeks of age the anemia has disappeared. Although adults have normal blood values, their response to hemopoietic stress is defective (1,10). The results of numerous studies have led to the conclusion that the prenatal deficiency in number of erythrocytes and the defective response of adult erythropoietic cells are due to a delay in maturation of already committed erythroid stem cells, and that earlier uncommitted precursors are unaffected by *f* (1,5,6). An additional effect of *f* in homozygotes is defective heme synthesis, which occurs in fetal reticulocytes but not in adult reticulocytes nor in erythroblasts at earlier stages of maturation. In fetal reticulocytes there is normal uptake of iron but poor incorporation into hemoglobin (1), probably as a result of reduced activity of δ-aminolevulinate synthetase and dehydratase (4). Fetal erythrocytes of *f/f* mice have more α than β globin chains. In both *f/f* and +/+ fetal erythrocytes there is more α- than β-chain mRNA; probably some regulatory mechanism bringing about equal α- and β-chain synthesis exists in +/+ mice but is defective in *f/f*. (2,3). The tail abnormalities are first noticeable on the 14th day of gestation as abnormal differentiation of the intervertebral discs (9). The possibility that abnormal heme synthesis could cause the tail and pigment defects in *f/f* mice is discussed by Cole *et al*. (4).

References

1. Bannerman, R.M., J.A. Edwards, and P.H. Pinkerton. 1973. Hereditary disorders of the red cell in animals. Prog. Hematol. 8:131–179.
2. Chui, D.H.K., M. Patterson, and S.T. Bayley. 1977. Unequal α and β globin mRNA in reticulocytes of normal and mutant *f/f* fetal mice. Blood 50 Suppl. 1:104 (Abstr.).
3. Chui, D.H.K., G.D. Sweeney, M. Patterson, and E.S. Russell. 1977. Hemoglobin synthesis in siderocytes of flexed-tailed mutant (*f/f*) fetal mice. Blood 50:165–177.
4. Cole, R.J., J. Garlick, and E.M. Cheek. 1975. Activities of haem synthetic enzymes in blood cells of pre-natal flexed-tailed (*f/f*) anaemic mice. J. Embryol. Exp. Morphol. 34:373–386.
5. Cole, R.J., and T. Regan. 1976. Haemopoietic progenitor cells in prenatal congenitally anaemic "flexed-tailed" (*f/f*) mice. Brit. J. Haematol. 33:387–394.
6. Gregory, C.J., E.A. McCulloch, and J.E. Till. 1975. The cellular basis for the defect in haemopoiesis in flexed-tailed mice. III. Restriction of the defect to erythropoietin progenitors capable of transient colony formation *in vivo*. Brit. J. Haematol. 30:401–410.
7. Grüneberg, H. 1942. The anaemia of flexed-tailed mice. II. Siderocytes. J. Genet. 44:246–271.

f

8. Hunt, H.R., R. Mixter, and D. Permar. 1933. Flexed tail in the mouse, *Mus musculus*. Genetics 18:335–366.
9. Kamenoff, R.J. 1935. Effects of the flexed-tailed gene on the development of the house mouse. J. Morphol. 58:117–155.
10. Russell, E.S. 1970. Abnormalities of erythropoiesis associated with mutant genes in mice. *In* A.L. Gordon, ed., Regulation of Hematopoiesis, Vol. I, 649–675. Appleton- Century-Crofts, New York.

fa, falter, recessive

Arose as a spontaneous mutation in a DBA subline. Homozygotes have an abnormal side-to-side swaying walk beginning at about 10 days. The abnormality becomes more pronounced with age and all mice die at about 20 days. *fa* has not been tested for allelism with similar mutants (1).

References

1. Yosida, T.H. 1960. Study of the new mutant "falter" found in the house mouse. Ann. Rep. Natl. Inst. Genet. Jpn. (1959) 10:27–28.

Fabph-1 to *Fabph-4* loci

These loci are mouse homologs of a rat gene for heart fatty acid binding protein (H-FBAP). In the rat, the gene is expressed most abundantly in heart, and in diminishing amounts in slow-twitch skeletal muscle, fast-twitch skeletal muscle, brain, kidney, and adrenal. Four mouse genes were identified by hybridization of *Hind*III digested mouse DNA with a rat H-FBAP probe. The chromosomal locations of the genes were found by use of mouse–Chinese hamster somatic cell hybrids screened with the rat probe.

Fabph-1 locus, fatty acid binding protein, heart-1, Chr 4.

Fabph-2 locus, fatty acid binding protein, heart-2, Chr 8.

Fabph-3 locus, fatty acid binding protein, heart-3, Chrs 10 or 15.

Fabph-4 locus, fatty acid binding protein, heart-4, Chr 4.

References

1. Heuckeroth, R.D., E.H. Birkenmeier, M.S. Levin, and J.I. Gordon. 1987. Analysis of the tissue-specific expression, developmental regulation, and linkage relationships of a rodent gene encoding heart fatty acid binding protein. J. Biol. Chem. 262:9709–9717.

far, first arch malformation, recessive

Arose spontaneously in the BALB/cGa strain. Affected mice die within 24 hours of birth. Homozygous newborns have a cleft secondary palate, a pointed snout, and extensive abnormalities of bones derived from the first branchial arch. These abnormalities involve the zygomatic, squamosal, sphenoid, and palatine bones, the stylohyoid cartilage, the premaxilla, maxilla, malleus, and mandible (2), abnormally arranged or defective vibrissae, and open eyelids. All the defects can be accounted for by abnormalities present on day 12 of gestation, which include deficient and irregular palatal shelves and orbital ridges, and an abnormal maxillary branch of the trigeminal nerve associated with an abnormal pattern of mesenchymal cells between the vibrissae and palate (1).

References

1. Juriloff, D.M., and M.J. Harris. 1983. Abnormal facial development in the mouse mutant first arch. J. Craniofac. Genet. Dev. Biol. 3:317–337.
2. McLeod, M.J., M.J. Harris, G.F. Chernoff, and J.R. Miller. 1983. First arch malformation: a new craniofacial mutant in the mouse. J. Hered. 71:331–335.

fat, fat, recessive

Arose spontaneously in the HRS/J strain. Homozygotes are recognizable at 4 to 5 weeks by the deposition of abnormal amounts of fat. They become fatter with age, often reaching 60 g by 6 months. They are hyperinsulinemic throughout life. Polyuria and glycosuria have not been observed, although some fluctuations in blood sugar occur. Hyperglycemia as high as 300 mg/100 ml is infrequent, but hypoglycemia is common in affected mice older than 6 months. These observations are not reliable indicators of differences or similarities of *fat* and *db* or *ob* because the genetic backgrounds are different (1).

References

1. Hummel, K.P., and D.L. Coleman. 1974. Mouse News Lett. 50:43.

Fbp-1, *Fbp-2* loci

Fbp-1 locus, fructose biphosphatase-1. This locus controls electrophoretic variation of the isozymes of fructose biphosphatase (FBP-1; E.C. 3.1.3.11) found in muscle. The isozymes of kidney, liver, and testis are not affected. The allele *Fbp-1^a* determines a slow migrating variant and is found in the P, SEA, and SWR strains and in the Peru-Coppock and other wild-derived stocks; the allele *Fbp-1^b* determines a faster variant and is found in the SM, C3H/He, C57BL/Go, CE, DBA/2, and many other strains. Heterozygotes have an enzyme of intermediate mobility (1).

Fbp-2 locus, fructose biphosphatase-2. This enzyme (FBP-2) is found in kidney, liver, and testis. Laboratory strains and all wild *M. musculus* examined are monomorphic for electrophoretic pattern of FBP-2. A low activity variant with slightly slower mobility occurs in *M. spretus*. However, crosses of *M. spretus* with several inbred strains failed to produce offspring and the mode of inheritance could therefore not be determined (1).

References

1. Nash, H.R., and P.R. Challinor. 1985. Fructose biphosphatase isozymes of the mouse. I. Inheritance. Biochem. Genet. 23:499–510.

fc, flecking, recessive, Chr 2

Found in an outbred laboratory stock. The phenotype found in this stock, consisting of a head spot and occasional belly spot, was shown by outcrossing to be due to two mimic genes, either of which alone when homozygous could cause the spotting. One of the genes, *fc*, was shown to be linked to Chr 2 markers and is located between *Ra* and *a*. The other gene is not on Chr 2 and has not been named (1).

References

1. Wallace, M.E. 1979. Mapping with mimic genes in the mouse. J. Hered. 70:216.

Fcr locus, Fc receptor, Chr 12

This locus controls degree of affinity of Fc receptors on B-cells for several subclasses of rat IgG. The allele here designated *Fcr^a* determines low affinity and occurs in strains A/J, A/He, NZB, and AL/N; the allele here designated *Fcr^b* determines high affinity and occurs in strains C57BL/6, BALB/c, DBA/2, AKR, and others. Heterozygotes have high-affinity receptors. The same receptors bind mouse IgG1, IgG2a, and IgG2b subclasses with no differences in affinity between subclasses or strains. *Fcr* is on Chr 12 about 20 cM proximal to *Lm-1* (1).

References

1. Baum, C.M., J.P. McKearn, R. Riblet, and J.M. Davie. 1985. Polymorphism of Fc receptor on murine B cells is Igh-linked. J. Exp. Med. 162:282–296.

fd, fur deficient, recessive, Chr 9

Arose spontaneously in the DBA/2J strain. Homozygotes are usually recognizable before weaning by sparse underfur which gives the remaining fur the appearance of being longer than normal. Adults have a sparse coat but all types of hair are present. Viability and fertility are normal (1).

References

1. Lane, P.W., and H.O. Sweet. 1973. Mouse News Lett. 48:34–35.

fe, faded, recessive, Chr 6

Arose spontaneously in the KSB inbred strain at Nagoya University. Homozygotes can be distinguished at 15 days of age by a slight dilution of coat color but are more easily classified at 30 days. The coat color appears faded and the underfur is white. Loss of pigment occurs progressively with age, so that homozygotes are almost white by 1 year of age. Skin lesions occurred in about a quarter of the *fe/fe* mice, in contrast to an incidence of only 2 per cent in normal controls, and were often followed by loss of weight and death (2). *fe* shows loose linkage with *Ggc* on Chr 6 (1).

References

1. Oh, Y.-S., and T. Tomita. 1987. Linkage of faded (*fe*) to chromosome 6 of the mouse. Exp. Anim. 36:73–77.
2. Oh, Y.-S, T. Tomita, and K. Kondo. 1986. Faded, a mutation in the KSB strain of mouse which shows age-related pigment change. Exp. Anim. 35:131–138.

Fes locus (c-*fes*), feline sarcoma oncogene, Chr 7

This is the cellular homolog in the mouse of the transforming gene (v-*fes*) of feline sarcoma virus found in the independent Snyder–Theiler and Gardner–Arnstein isolates. The c-*fes* gene is closely related to the v-*fps* oncogene of several avian transforming viruses (3). The product of the *fes* gene in viruses and humans and presumably also of the mouse cellular gene is a protein kinase with specificity for tyrosinase (1, 5). In the mouse it is found almost exclusively in myeloid cells including granulocyte–macrophage progenitors and the mononuclear fraction of peripheral blood (4). *Fes* was first shown to be on Chr 7 by use of mouse–Chinese hamster somatic cell hybrids screened with a viral DNA probe (3). Later, a DNA restriction site difference at this locus was found between the C57BL/6 and A/J strains. The allele in C57BL/6, *Fes^b*, generates an *Eco*RI restriction fragment of 12 kb; the allele in

Fes

A/J, Fes^a, generates an EcoRI fragment of 13 kb. By use of 30 AXB and BXA recombinant inbred strains, Fes was found to be near Gpi-1 on Chr 7 (2).

References

1. Barbacid, M., K. Beemon, and S.G. Devare. 1980. Origin and functional properties of the major gene product of the Snyder–Theiler strain of feline sarcoma virus. Proc. Natl. Acad. Sci. USA 77:5158–5162.
2. Blatt, C., M.E. Harper, G. Franchini, M.N. Nesbitt, and M.I. Simon. 1984. Chromosome mapping of murine c-fes and c-src genes. Mol. Cell Biol. 4:978–981.
3. Kozak, C.A., J.F. Sears, and M.D. Hoggan. 1983. Genetic mapping of the mouse oncogenes c-Ha-ras-1 and c-fes to chromosome 7. J. Virol. 47:217–220.
4. MacDonald, I., J. Jevy, and T. Pawson. 1985. Expression of the mammalian c-fes protein in hematopoietic cells and identification of a distinct fes-related protein. Mol. Cell. Biol. 5:2543–2551.
5. Van de Vin, W.J.M., F.H. Reynold, Jr., and J.R. Stephenson. 1980. The nonstructural components of polyproteins encoded by replication-defective mammalian transforming retroviruses are phosphorylated and have associated protein kinase activity. Virology 101:185–187.

fg

See myd.

Fgv-1, Fgv-2 loci

The C3H/FgLw strain is a high leukemia and high virus strain. It has been shown to carry three independent N-tropic high-virus loci (2). Two of these have been isolated in NFS congenic lines.

Fgv-1 locus, C3H/Fg virus-1, Chr 7, closely linked to Hbb (2) .

Fgv-2 locus, C3H/Fg virus-2. Not allelic with Emv-1 (Cv), which is the locus of a C3H/He virus on Chr 5 (1).

References

1. Kozak, C.A., and W.P. Rowe. 1982. Genetic mapping of ecotropic murine leukemia virus-inducing loci in six inbred strains. J. Exp. Med. 155:524–534.
2. Rowe, W.P. 1973. Genetic factors in the natural history of murine leukemia virus infection: G.H.A. Clowes Memorial Lecture. Cancer Res. 33:3061–3068.

fh, fetal hematoma, recessive

Found in a stock descended from a cross of C57BL/6 by Stanford-J mice. Penetrance is incomplete in homozygotes, being about 60 per cent in a stock observed by Center. Affected mice may have reduced and fused digits on one or more feet, and one or both eyes may be small with an opaque cornea. Abnormalities may be seen as early as 12 or 13 days of gestation, and consist of clear or hemorrhagic blebs on the head and feet. The blebs interfere with normal development of the feet and eyes. Adults may be fully fertile. fh closely resembles my but is not allelic to it. Allelism has not been tested with eb and bl (1).

References

1. Center, E.M. 1977. Genetical and embryological comparison of two mutations which cause foetal blebs in mice. Genet. Res. 29:147–157.

fhd, fathead, Chr 2

Found in the descendants of (C57BL/6 × DBA/2)F1 mice carrying an ENU-induced null mutation at the Pep-7 locus. Homozygotes are recognizable at 3 weeks of age by their deformed skulls. The nasal and maxillary bones are shortened and twisted to the left, the orbits are abnormally shaped, and the junction between the frontal and parietal bones is misaligned. The fhd locus is on Chr 2 linked to agouti (1).

References

1. Whitmore, S.P. 1986. Mouse News Lett. 74:107.

Fhe locus, Friend helper virus erythroblastosis

This locus controls susceptibility and resistance to erythroblastosis induced by the helper virus component of Friend NB-tropic virus. The allele Fhe^r determines resistance and occurs in the C57BL/6 and B10.D2 strains; the allele Fhe^s determines susceptibility and occurs in the BALB/c strain. F1 mice are resistant. The virus replicates equally well in the two strains inoculated as neonates, but causes erythroblastosis only in BALB/c. The erythroblastosis develops with a mean latency of 85 days. In C57BL/6 the virus causes non-erythroid leukemias but with a greater latency, 119 to 168 days. Fhe is not linked to b or to the viral-susceptibility genes H-2, Fv-1, Fv-2, Rfv-3, Rmcf, Emv-1, and Emv-2. It is phenotypically distinct from Fv-4 (1, 2).

References

1. Silver, J.E., and T.N. Fredrickson. 1983. A new gene that controls the type of leukemia induced by Friend murine leukemia virus. J. Exp. Med. 158:493–505.
2. Silver, J.E., and T.N. Fredrickson. 1983. Susceptibility to Friend helper virus leukemias in CXB recombinant inbred mice. J. Exp. Med. 158:1693–1702.

fi, fidget, recessive, Chr 2

Discovered by Grüneberg (1) in a heterogeneous stock. Viability of homozygotes is usually less than normal and may be as low as 50 per cent on some genetic backgrounds. Fertility is poor, but some males are reliable breeders. Homozygotes toss their heads from side to side and tend to run in circles but can hear throughout life. The bony labyrinth is quite defective. There are other abnormalities including small eyes, absence of lachrymal glands, dislocation of the hip, increased incidence of polydactylism, and displaced parafloccular lobes of the cerebellum (3). The defects of the eye are due to prolongation of the presynthetic period of the cell cycle in the retinal anlage beginning at 11 days of gestation and leading to retardation in transition of retinal cells to the differentiated state (2). All of the normal lens crystallins are synthesized, but synthesis is delayed (4).

References

1. Gruneberg, H. 1943. Two new mutant genes in the house mouse. J. Genet. 45:22–28.
2. Konyukhov, B.V., and M.V. Sazhina. 1976. The cell cycle and retinal histogenesis in *fidget* mutant mice. Dev. Biol. 54:13–22.
3. Truslove, G.M. 1956. The anatomy of development of the fidget mouse. J. Genet. 54:64–86.
4. Yakovlev, M.I., E.S. Platonov, and B.V. Konyukhov. 1977. A study of the effect of mutant genes on crystallin synthesis in the developing mouse lens. II. Genes fidget and ocular retardation. Ontogenez 8:115–120 (English summary).

Fis-1 locus, Friend virus integration site-1, Chr 7

This locus is a common proviral integration site of Friend murine leukemia virus in lymphoid and myeloid tumors induced by the virus. Provirus was found at this site in 4/35 tumors examined. All proviruses were inserted in the same 1.5-kb region and all were in the same orientation. *Fis-1* was first shown to be on Chr 7 by use of mouse–Chinese hamster somatic cell hybrids (1). Later, a DNA restriction fragment length polymorphism was found among inbred strains. In *Hinc*II digested DNA, the allele *Fis-1ᵃ* determines a fragment of 2.1 kb and occurs in strain NFS; the allele *Fis-1ᵇ* determines a fragment of 2.5 kb and occurs in strains C57BL/6, C57BL/10, and C57L; the allele *Fis-1ᶜ* determines a fragment of 2.3 kb and occurs in strains AKR, BALB/c, and DBA/2. In 51 recombinant inbred strains and 31 progeny of a backcross from these progenitor strains, no recombinants were found between *Fis-1* and *Int-2* on Chr 7. However, the two genes are distinct and are separated by at least 25 to 30 kb (2).

References

1. Silver, J., and C. Kozak. 1986. Common proviral integration region on mouse chromosome 7 in lymphomas and myelologenous leukemias induced by Friend murine leukemia virus. J. Virol. 57:526–533.
2. Silver, J., and C.E. Buckler. 1986. A preferred region for integration of Friend murine leukemia virus in hematopoietic neoplasms is closely linked to the *Int-2* oncogene. J. Virol. 61:1156–1158.

Fk, fleck, semidominant

Arose spontaneously in the CBA strain. Heterozygotes have white spotting on the belly, tail tip, and often on the back paws. Tests of fertility showed both male and female heterozygotes were less productive breeders than their normal sibs. Postnatal survival to weaning was normal in heterozygotes. Homozygotes die *in utero* at about the time of implantation. *Fk* is not allelic with *Sp*, *Wᵛ*, or *Sltᵈ* (1).

References

1. Sheridan, W. 1968. The dominant effects of a recessive lethal in the mouse. Mutat. Res. 5:323–328.

Fkl

See *Rnᶠᵏˡ*.

fl, flipper-arm, recessive

Possibly radiation-induced. Viability of homozygotes is poor. Males are sterile, females only occasionally fertile. Homozygotes are smaller than their normal sibs. The ulna and radius bend outward sharply at the wrist. The hindlimbs are normal (1).

References

1. Kelly, E.M. 1957. Mouse News Lett. 16:36.

Fla locus (called *Flp* (7) and *Laf* (3)), F liver antigen, Chr 5

F antigen is a soluble antigen that occurs in the livers of all mice and exists in two antigenic forms. The allele *Flaᵃ* determines the F.1 form and occurs in strains CBA, C3H, DBA/2, SM, and AKR; the allele *Flaᵇ* determines the F.2 form and occurs in strains A/J, A2G, BALB/c, C57BL/10, and many others. F antigen from the F.2 strains can induce precipitating antibodies in the F.1 strains but not in the F.2 strains, and vice versa. The anti-F antibodies so produced, however, can react with F antigen from all strains, including the strain that produced the antibody (2, 4). The two forms of the antigen differ in isoelectric point; that of the F.1 form is 7.3, that of the F.2 form is 6.8 (1). By this method a large number of strains were typed and all shown to have either the F.1 or F.2 forms. *Fla* was found to be on Chr 5 about 3 cM proximal to *Gus* by Winchester *et al.* (7) using C57BL/6 × DBA/2 recombinant inbred strains and an immunological method of classification, and in approximately the same position by Hayakawa and Nikaido (3) using both C57BL/6 × C3H/He recombinant inbred strains and a backcross between the F1 of these two strains and C57BL/6 and classification by isoelectric point. The F antigen is a protein of 43 kDa with the electrophoretic mobility of a β-globulin. It was found to be a strong inhibitor of DNA synthesis by Con A-stimulated lymphocytes (4). Immune responsiveness to F antigen requires presence of the *H-2K^k* or *H-2I-A^k* alleles, and the magnitude of the response is controlled by a non-*H-2*-linked locus (5). The immune response also occurs *in vitro* (6).

References

1. Anders, R.F., P.C. Cooper, M.F. Coffey, and I.R. MacKay. 1978. Autoimmune response to F antigen. *In* N.R. Rose, P.E. Bigazzi, and N.L. Warner, eds., Genetic Control of Autoimmune Disease, 393–398. Elsevier North Holland, Amsterdam.
2. Fravi, G., and J. Lindenmann. 1968. Induction by allogeneic extracts of liver-specific precipitating autoantibodies in the mouse. Nature 218:141–143.
3. Hayakawa, J., and H. Nikaido. 1987. Two types of liver-specific F antigen are encoded by a locus located on chromosome 5 in mice. Immunogenetics 26:366–369.
4. Lane, D.P., and D.M. Silver. 1977. Inhibition of lymphocyte proliferation by murine liver-specific F antigen. Transplant. Proc. 9:1063–1065.
5. Silver, D.M., and D.P. Lane. 1981. Polygenic control of the immune response to F antigen. Immunogenetics 12:237–251.
6. Sunshine, G.H., M. Cyrus, and G. Winchester. 1982. *In vitro* responses to the liver antigen F. Immunology 45:357–363.
7. Winchester, G., N.A. Mitchison, and B.A. Taylor. 1987. The structural gene for F liver protein (*Flp*) maps to chromosome 5 of the mouse. Immunogenetics 26:356–358.

Flp locus

See *Fla* locus.

Flv locus, flavivirus resistance, Chr 5?

The symbol has recently been assigned to this locus (5). The locus controls susceptibility and resistance to the lethal effects of the flaviviruses (group B arboviruses or togaviruses). The allele *Flv^r* determines resistance and occurs in the BRVR and PRI strains, the C3H.PRI-*Flv^r* (C3H/RV) congenic strain, and in many wild mice; the allele *Flv^s* determines susceptibility and occurs in the C3H/He strain. All other laboratory strains tested are also susceptible (3, 6). Heterozygotes are resistant (9). Resistant mice produce lower amounts of virus after infection, and this difference is also observed in cultured macrophages, brain tissue, spleen, and embryo fibroblasts (4, 7). Resistant and susceptible mice show no difference in production of neutralizing antibodies to West Nile virus (4), but immunodepressants cause severe reduction in resistance of C3H/RV mice to another flavivirus (1). Both in cell culture and *in vivo*, resistant mice (C3H/RV) show a greater inhibitory effect of interferon on production of flavivirus than do susceptible mice (C3H/He) (7), but antibody against NDV-induced interferon does not lower resistance of C3H/RV mice to the lethal effects of these viruses (2). After intraperitoneal inoculation with Banzai virus, the brains of C3H/RV but not of C3H/He mice produce a component that interferes with virus replication (10). The C3H/He and C3H/RV strains carry different alleles at the *Ric* (resistance to *Rickettsia tsutsugamushi*) locus, *Ric^s* and *Ric^r*, respectively. This suggests that *Flv* may be closely linked to *Ric*, which is on Chr 5. See *Ric* locus. *Flv* segregates independently of the genes for salmonella resistance (*Ity*) on Chr 1 (11) and for hepatitis virus susceptibility-1 (*Hv-1*) (8).

References

1. Bhatt, P.N., and R.O. Jacoby. 1976. Genetic resistance to lethal flavivirus encephalitis. II. Effect of immunosuppression. J. Infect. Dis. 134:166–173.
2. Brinton, M.A., H. Arnheiter, and O. Haller. 1982. Interferon independence of genetically controlled resistance to flaviviruses. Infect. Immun. 36:284–288.
3. Darnell, M.B., H. Koprowski, and K. Lagerspetz. 1974. Genetically determined resistance to infection with group B arboviruses. I. Distribution of the resistance gene among various mouse populations and characteristics of gene expression *in vivo*. J. Infect. Dis. 129:240–247.
4. Goodman, G.T., and H. Koprowski. 1962. Study of the mechanism of innate resistance to virus infection. J. Cell. Comp. Physiol. 59:333–373.
5. Green, M.C. 1979. Mouse News Lett. 60:29.
6. Groschel, D., and H. Koprowski. 1965. Development of a virus-resistant inbred mouse strain for the study of innate resistance to arbo B viruses. Arch. Gesamte Virusforsch. 17:379–391.
7. Hanson, B., H. Koprowski, S. Baron, and C.E. Buckler. 1969. Interferon-mediated natural resistance of mice to arbo B virus infection. Microbios 1B:51–68.

8. Kantoch, M., A. Warwick, and F.B. Bang. 1963. The cellular nature of genetic susceptibility to a virus. J. Exp. Med. 117:781–798.

9. Sabin, A.B. 1952. Nature of inherited resistance to viruses affecting the nervous system. Proc. Natl. Acad. Sci USA 38:540–546.

10. Smith, A.L., R.O. Jacoby, and P.N. Bhatt. 1980. Genetic resistance to lethal flavivirus infection: Detection of interfering virus produced *in vivo*. *In* E. Skamene, P.A.L. Kongshavn, and M. Tandy, eds., Genetic Control of Natural Resistance to Infection and Malignancy, 305–312. Academic Press, New York.

11. Webster, L.T. 1937. Inheritance of resistance of mice to enteric bacterial and neurotropic virus infection. J. Exp. Med. 65:261–286.

fm, foam-cell reticulosis, recessive

Arose spontaneously in the CBA/H strain. Homozygotes are first recognizable at 10 to 14 weeks when they become thin and less active. They continue to lose weight and usually die before 6 months. Histological examination shows that lipid-containing foam cells replace the lymphoid tissue of the thymus, Peyer's patches, lymph nodes, and spleen, and occur in the liver as well (2,4). Electron microscopy demonstrated occasional electron-lucent inclusions in the neurons of the central nervous system (1). There is an increase in sphingomyelin and cholesterol in the liver, spleen, and thymus, but not in brain, and an increase in lysolecithin in the thymus. Sphingomyelinase is not increased (1,3). The disease resembles a group of lipid storage diseases in man and is most similar to Niemann–Pick disease, type C (2).

References

1. Adachi, M., C.Y. Tsai, L.M. Hoffman, L. Schneck, and B.W. Volk. 1974. The central nervous system, liver, and spleen of FM mice. Arch. Pathol. 97:232–238.

2. Adachi, M., B.W. Volk, and L. Schneck. 1976. Niemann–Pick disease type C. Animal model: mouse Niemann–Pick disease. Am. J. Pathol. 85:229–231.

3. Frederickson, D.S., H.R. Sloan, and C.T. Hansen. 1969. Lipid abnormalities in foam-cell reticulosis of mice, an analogue of human sphingomyelin lipidosis. J. Lipid Res. 10:288–293.

4. Lyon, M.F., E.V. Hulse, and C.E. Rowe. 1965. Foam-cell reticulosis of mice: an inherited condition resembling Gaucher's and Niemann–Pick diseases. J. Med. Genet. 2:99–106.

For-1, *-3*, *-4*, *-5* loci

These loci control molecular structure or activity level or both of isozymes of formamidase (FOR; E.C. 3.5.1.9) in various tissues.

For-1 locus, formamidase-1. This locus controls level of activity and molecular structure of kynurenine formamidase (FOR-1) in liver. The affected enzyme is found in the second or major peak of enzyme activity eluting from DEAE cellulose columns. The allele *For-1^a* determines high activity and high heat stability and is found in strain C3Hf/Rl and probably also in 129, DBA, 101, and others; the allele *For-1^b* determines low activity and low heat stability and is found in strain C57BL/10ScSn and probably also in BALB/c, SEC, C57BL/6, and others. Heterozygotes have intermediate levels of activity. Formamidase activity begins after birth and increases slowly, reaching adult levels at 12 days in low strains (C57BL/10) and at 50 days in high strains (C3Hf) (1).

For-3, *For-4* loci, formamidase-3, -4. These loci control two formamidase isozymes (FOR-3, FOR-4) in kidney that chromatograph together and can be separated from each other by ammonium sulfate fractionation. In inbred strains there are allelic forms of both loci that can be distinguished from each other by heat stability and substrate affinity (1).

For-5 locus, formamidase-5, Chr 14. This locus controls heat sensitivity of an isozyme of formamidase (FOR-5) that is found in brain and is different from formamidases of other tissues. The allele *For-5^b* determines heat stability and occurs in strain C57BL/6J; the allele *For-5^d* determines heat sensitivity and occurs in strain DBA/2J. *For-5* has shown no recombination with *Es-10*, and there is an exact correspondence of *For-5^b* with *Es-10^a* and *Es-10^c* and of *For-5^d* with *Es-10^b* among 46 inbred strains tested (2).

References

1. Cumming, R.B., M.F. Walton, and F.H. Gaertner. 1978. Mouse News Lett. 59:47.

2. Cumming, R.B., M.F. Walton, J.C. Fuscoe, B.A. Taylor, J.E. Womack, and F.H. Gaertner. 1979. Genetics of formamidase-5 (brain formamidase) in the mouse: localization of the structural gene on chromosome 14. Biochem. Genet. 17:415–431.

Fos locus (c-*fos*), FBJ osteosarcoma oncogene, Chr 12

This is the cellular homolog of the transforming gene (v-*fos*) of the FBJ murine osteosarcoma virus. Three DNA variants distinguishable with a combination of *Eco*RI and *Msp*I nucleases are known: *Fos^a* in SWR, *Fos^b* in C57BL/6, NZB, C3H/He, DBA/2, and other strains, and *Fos^c* in SJL. By use of these markers in various recombinant inbred strains, D'Eustachio (3) mapped *Fos* to a position between *D12Nyu3* and *Igh-C*

Fos

on Chr 12. The function of the gene in the mouse is not known, but its expression in various tissues and in response to various treatments has been examined. *Fos* is actively transcribed in early embryos, in the placenta after the ninth day of gestation, and in bone and skin and to a lesser extent in other tissues of early postnatal mice (8, 9). A 55 to 60 kDa protein product of *Fos* is found in the nuclei of the same tissues (1). In differentiating monocyte–macrophage cells *in vitro*, *Fos* is transcribed principally at late stages of differentiation (4, 10). Stimulation of quiescent fibroblast cultures with growth factors results in induction within minutes of *Fos* mRNA and protein (5, 6). Transfection of the *Fos* gene into F9 teratocarcinoma stem cells results in a marked change in morphology to greatly enlarged flat cells growing in epithelial fashion (7). These results suggest that *Fos* plays some important role in differentiation. The *Fos* gene is 3397 nucleotides in length and contains three introns. The amino acid sequence of the protein predicted from the nucleotide sequence has been compared with those of other known protein sequences and no homology has been found (11). The protein was found to be localized in the nucleus (2, 6).

References

1. Adamson, E.D., J. Meek, and S.A. Edwards. 1985. Product of the cellular oncogene, c-*fos*, observed in mouse and human tissues using an antibody to a synthetic peptide. EMBO J. 4:941–947.
2. Curran, T., A.D. Miller, L. Zokas, and I.M. Verma. 1984. Viral and cellular *fos* proteins: a comparative analysis. Cell 36:259–268.
3. D'Eustachio, P. 1984. A genetic map of murine chromosome 12 composed of polymorphic DNA fragments. J. Exp. Med. 160:827–838.
4. Gonda, T.J., and D. Metcalf. 1984. Expression of *myb*, *myc* and *fos* proto-oncogenes during the differentiation of a murine myeloid leukaemia. Nature 310:249–251.
5. Greenberg, M.E., and E.B. Ziff. 1984. Stimulation of 3T3 cells induces transcription of the c-*fos* proto-oncogene. Nature 311:433–438.
6. Kruijer, W., J.A. Cooper, T. Hunter, and I.M. Verma. 1984. Platelet-derived growth factor induces rapid but transient expression of the c-*fos* gene and protein. Nature 312:711–716.
7. Müller, R., and E.F. Wagner. 1984. Differentiation of F9 teratocarcinoma stem cells after transfer of c-*fos* proto-oncogenes. Nature 311:438–442.
8. Müller, R., D.J. Shamon, J.M. Tremblay, M.J. Cline, and I.M. Verma. 1982. Differential expression of cellular oncogenes during pre- and postnatal development of the mouse. Nature 299:640–644.
9. Müller, R., I.M. Verma, and E.D. Adamson. 1983. Expression of c-*onc* genes: c-*fos* transcripts accumulate to high levels during development of mouse placenta, yolk sac, and amnion. EMBO J. 2:679–684.
10. Müller, R., D. Müller, and L. Guilbert. 1984. Differential expression of c-*fos* in hematopoietic cells: correlation with differentiation of monomyelocytic cells *in vitro*. EMBO J. 3:1887–1890.
11. Van Beveren, C., F. van Straaten, T. Curran, R. Müller, and I.M. Verma. 1983. Analysis of FBJ-MuSV provirus and c-*fos* (mouse) gene reveals that viral and cellular *fos* gene products have different carboxy termini. Cell 32:1241–1255.

Fpgs locus, folylpolyglutamyl synthetase, Chr 2

This is probably the structural locus for the enzyme folylpolyglutamyl synthetase, which catalyzes the sequential addition of glutamate residues to the folate nucleus. No genetic variants are known in the mouse. A line of Chinese hamster cells carrying a mutation resulting in a deficiency of the enzyme was used for complementation mapping of *Fpgs*. Microcells from mouse embryo fibroblasts were fused with the Chinese hamster cells. Cultures of the hybrid cells selected for the active enzyme were all found to possess Chr 2 or a fragment consisting of its centromeric portion. *Fpgs* is therefore located on Chr 2 near the centromeric end (1).

References

1. Fournier, R.E.K., and R.G. Moran. 1983. Complementation mapping in microcell hybrids: localization of *Fpgs* and *Ak-1* on *Mus musculus* chromosome 2. Somat. Cell Genet. 9:69–84.

fr, frizzy, recessive, Chr 7

Arose in a stock of mixed origin at the Jackson Laboratory. Homozygotes have wavy or curly vibrissae at 1 or 2 days after birth. The coat is short and rough at first and then may become normal at 6 to 7 weeks. It is usually short and thin in older animals (1). In some outcrosses the proportion of *fr/fr* mice identifiable in the F2 was low, an indication that penetrance may be low on some backgrounds.

References

1. Falconer, D.S., and G.D. Snell. 1952. Two new hair mutants, rough and frizzy, in the house mouse. J. Hered. 43:53–57.

Frl locus

Three dominant alleles of independent origin have been reported.

Frl^a, furloss-a. Arose spontaneously in a stock carrying *d* and *se* (1). Heterozygotes can be recognized at 7 to 10 days by abnormally wrinkled skin. Loss of fur begins at 13 to 15 days, resulting in bare patches on the head and then on the back and abdomen. Several cycles of regrowth and hair loss occur, but by 12 to 14 weeks the animals are completely bare except for the muzzle. Heterozygotes are smaller than normal beginning at about 16 days but are vigorous and fertile. Homozygotes have not been recovered from matings between heterozygotes (2).

Frl^b, furloss-b. Found among offspring of an X-irradiated (101 × C3H)F1 male mated to a T-stock female. Heterozygotes closely resemble *Frl^a*/+ mice. Homozygotes are identical in appearance to heterozygotes but occur in much lower than expected frequencies in offspring of matings between heterozygotes, probably as a result of prenatal death (2).

Frl^c, furloss-c. Arose spontaneously in a stock many generations removed from X-irradiated ancestors. Closely resembles *Frl^b* except that homozygotes appear to be normally viable (2).

References

1. Rayner, D. 1971. Mouse News Lett. 45:39.
2. Stelzner, K.F. 1983. Four dominant autosomal mutations affecting skin and hair development in the mouse. J. Hered. 74:193–196.

fro, fragilitas ossium, recessive

Arose in a random bred stock after treatment of spermatids with the chemical mutagen tris(1-aziridinyl) phosphinesulphide. Homozygotes are moderately runted at birth and have deformities of all four limbs. The long bones of the limbs are short and curved and show fractures. No deformities are apparent in the tail, spine, or ribs. Up to 90 per cent of homozygotes die before day 3. Those that survive remain small and deformed throughout life but have normal breeding behavior and life span (2). Cartilage formation and mineralization appears to be normal but osteoid, the mineralized bone matrix of normal developing bones, is severely undermineralized in *fro/fro* mice. Blood calcium levels are normal but parathyroid hormone is markedly elevated, probably in response to the reduced calcium stores in bone (1). The condition shows roentgenological and pathological features similar to those of severe human osteogenesis imperfecta.

References

1. Brown, K.S., S.C. Hetzel, and L.C. Harne. 1984. Craniofacial consequences of connective tissue disorders in mice. Birth Defects: Orig. Art. Series 20:113–136.
2. Guénet, J.L., R. Stanescu, P. Maroteaux, and V. Stanescu. 1981. Fragilitas ossium: a new autosomal recessive mutation in the mouse. J. Hered. 72:440–441.

fs, furless, recessive, Chr 13

Appeared in an unpedigreed stock maintained in the Department of Zoology, Ohio State University. Homozygotes have short or missing vibrissae 2 days after birth. The first coat grows normally but begins to thin out at 19 days, starting on the head. A new coat grows but persists only a short time. Mature mice are partly devoid of hair at all times. The hairs present are shorter than normal (1). The viability up to weaning of furless is considerably reduced, but survivors are fertile. Heterozygotes have normal reproductive fitness and radiation sensitivity (2). Reciprocal skin grafts between normal and furless mice demonstrate some curative influence of the normal host on grafted furless skin and of grafted normal skin on furless skin posterior to the graft (3).

References

1. Green, E.L. 1954. The genetics of a new hair deficiency, furless, in the house mouse. J. Hered. 45:115–118.
2. Green, E.L. 1971. Fitness of heterozygotes of deleterious recessive mutations in the mouse. Mutat. Res. 12:281–289.
3. Tsuji, S., and T.H. Yosida. 1965. Reciprocal skin transplantation between normal and hereditary hairless mice. Jpn. J. Genet. 40:55–62.

Fshb locus, follicle-stimulating hormone-β, Chr 2

This is the structural locus for the β subunit of follicle-stimulating hormone (FSH-β). It is said to be on Chr 2 but no supporting evidence has been published (1).

References

1. Glaser, T., D. Gerhard, C. Jones, L. Albritton, P. Lalley, and D. Housman. 1985. A fine structure deletion map of chromosome 11p. Cytogenet. Cell Genet. 40:643 (Abstr.).

ft, flaky tail, recessive, Chr 3

Discovered in a heterogeneous stock. Homozygotes are recognizable at 2 to 4 days by the stretched appearance of the skin on the feet and dorsum. Between 5 and 14 days the tail becomes constricted and flaky and the pinna becomes thickened and shortened. At this stage, homozygotes are often smaller than normal. After 14 days the flakiness disappears and the mice appear normal except for slightly smaller ears and occasionally an amputated tail. There is some inviability prior to weaning, but survivors are normally viable and fertile. During the first postnatal week the stratum granulosum of the skin is thinner than normal, presumably resulting in slower than normal proliferation of the stratum corneum and keratin (1).

References

1. Lane, P.W. 1972. Two new mutations in Linkage Group XVI of the house mouse. Flaky tail and varitint-waddler-J. J. Hered. 63:135–140.

Fu locus, Chr 17

Fu, fused, semidominant. Arose in stocks at the Bussy Institution prior to 1931. Expression of this mutant shows great variability. Both homozygotes and heterozygotes may have shortened and kinked tails or they may both be normal. Homozygotes tend to be more severely affected than heterozygotes. Expression is not dependent on the residual genotype but offspring of *Fu/+* or *Fu/Fu* mothers are less likely to express the character than offspring of +/+ mothers (8). Homozygotes and heterozygotes occasionally show an abnormal behavior similar to that of the circling mutants and are deaf (3). Embryos of homozygotes show some overgrowth and duplication of the posterior part of the neural tube (9).

Fu^{ki}, kinky, semidominant. Arose in stocks of a Florida mouse fancier (1). Heterozygotes are very similar to fused in abnormalities of the skeleton and may show the same behavioral abnormalities. Dunn and Caspari (3) found five probable crossovers between *Fu* and *Fu^{ki}* among 505 gametes tested, but Dunn and Glucksohn-Waelsch (4) found none in 971 and concluded that the previous study was in error and that kinky is an allele of *Fu*. Unlike *Fu*, kinky homozygotes are inviable. They show tissue hyperplasia and twinning at 7 days and die between 8 and 10 days of embryonic life (5). Many heterozygotes show morphogenetic abnormalities of the inner ear, the severity of which is correlated with the severity of the behavioral defects, although a particular behavioral defect is not invariably associated with a particular abnormality (2).

Fu^{kb}, knobbly, semidominant. Found in a mutagenesis experiment at Harwell, England. The *T*-locus allele *t^{h20}* is a deletion including the *Fu^{kb}* locus (7). Heterozy-

gotes have short bent tails. They occur in less than expected proportion in offspring of male but not of female heterozygotes, suggesting either reduced penetrance or low transmission ratio. Homozygotes closely resemble *Fu^{ki}* homozygotes and die by 9 days of gestation (6).

References

1. Caspari, E., and P.R. David. 1940. The inheritance of a tail abnormality in the house mouse. J. Hered. 31:427–431.
2. Deol, M.S. 1966. The probable mode of gene action in the circling mutants of the mouse. Genet. Res. 7:363–371.
3. Dunn, L.C., and E. Caspari. 1945. A case of neighboring loci with similar effects. Genetics 30:543–568.
4. Dunn, L.C., and S. Glucksohn-Waelsch. 1954. A genetical study of the mutation "fused" in the house mouse, with evidence concerning its allelism with a similar mutation "kink". J. Genet. 53:383–391.
5. Glucksohn-Schoenheimer, S. 1949. The effect of a lethal mutation responsible for duplication and twinning in mouse embryos. J. Exp. Zool. 110:47–76.
6. Jacobs-Cohen, R.J., M. Spiegelman, J.C. Cookingham, and D. Bennett. 1984. Knobbly, a new dominant mutation in the mouse that affects embryonic ectoderm organization. Genet. Res. 43:43–50.
7. Lyon, M.F., and K.B. Bechtol. 1977. Derivation of mutant t-haplotypes of the mouse by presumed duplication or deletion. Genet. Res. 30:63–76.
8. Reed, S.C. 1937. The inheritance and expression of fused, a new mutation in the house mouse. Genetics 22:1–13.
9. Theiler, K., and S. Glucksohn-Waelsch. 1956. The morphological effects and the development of the fused mutation in the mouse. Anat. Rec. 125:83–104.

Fuca locus (formerly *Afuc*), α-L-fucosidase, Chr 4

This is the structural locus for an α-L-fucosidase (FUCA: E.C. 3.2.1.51) found in somatic cells. The locus was found to be on Chr 4 using mouse–Chinese hamster somatic cell hybrids segregating mouse chromosomes (1). Johnson and Hong (2) found a strain difference in several properties of FUCA among 28 inbred strains. Strains A/J, BDP, LP, P, SEA/Gn, and 129/J have high activity of FUCA, and 22 other strains, including C57BL/6, C3H/He, DBA/2, and BALB/c, have low activity. The enzyme from high activity strains has high heat stability and a pH activity curve with high relative activity at pH 2.8; the enzyme from the low activity strains has low heat stability and a pH activity curve with low relative activity at pH 2.8. The heat-stability differences occur in brain, liver, kidney, spleen, heart, skeletal muscle lung, and testis. The strain difference is thought to be due to a structural variant in the FUCA gene but no genetic tests have been made. An activity variant of FUCA in sperm was described by Self *et al.* (3) and ascribed to a single locus *Afuc-2*. Strain B10.M had high activity and BALB/c had low activity. How these variants are related to each other and to the gene described by Lalley (1) is not known.

References

1. Lalley, P.A. 1981. Mouse News Lett. 64:80.
2. Johnson, W.G., and J.L. Hong. 1986. Variation in alpha-L-fucosidase properties among 28 inbred mouse strains: six strains have high enzyme activity and heat stabile enzyme with a variant *p*H-activity curve; twenty-two strains have low activity and heat-labile enzyme. Biochem. Genet. 24:469–483.
3. Self, S.J., B.G. Winchester, and J.R. Archer. 1978. Strain variation in spermatozoal glycosidases in inbred mice. Genet. Res. 32:183–193.

Fv-1 locus, Friend virus susceptibility-1, Chr 4

Studies on this locus have been extensively reviewed by Jolicoeur (3). Briefly, a strain difference in susceptibility to splenomegaly induced by inoculation with Friend leukemia virus was first discovered by Odaka (9). In 1970, Lilly showed that susceptibility was controlled by two independent loci, *Fv-1* and *Fv-2*. *Fv-1* controls susceptibility to the lymphatic leukemia virus component of the Friend virus complex, as assayed either by splenomegaly in the presence of the recessive *Fv-2* allele for susceptibility ($Fv-2^s$) (6), or probably also as assayed by amount of Friend virus in the spleen, even in the presence of the dominant *Fv-2* allele for resistance ($Fv-2^r$) (11). The allele $Fv-1^n$ causes susceptibility to the F-S strain of Friend virus (now known to be N-tropic, see below), and the allele $Fv-1^b$ causes resistance to this strain. Resistance is dominant. *Fv-1* also controls susceptibility to naturally occurring ecotropic leukemia viruses. These viruses can be divided into two classes which will grow preferentially in NIH Swiss embryo cells (N-tropic) or BALB/c embryo cells (B-tropic). Cells of $Fv-1^n$ strains are susceptible to N-tropic and resistant to B-tropic viruses; cells of $Fv-1^b$ strains are susceptible to B-tropic and resistant to N-tropic viruses. Resistance is dominant, so that heterozygotes ($Fv-1^n/Fv-1^b$) are resistant to both kinds of viruses. $Fv-1^n$ occurs in the NIH Swiss strain and in AKR, C57L, and other strains; $Fv-1^b$ occurs in BALB/c, A, C57BL/6, and other strains (7). A third allele, $Fv-1^{nr}$, causes at least fourfold greater resistance to N-tropic viruses than $Fv-1^n$ and occurs in strains RF, 129, NZB, and NZW. It is dominant to $Fv-1^n$ (8, 14). A fourth allele, $Fv-1^o$, allows replication of both N- and B-tropic viruses and is found in wild *M. spretus* and *M. m. praetextus* and probably in other wild species of *Mus* (5). *Fv-1* also affects the expression in living animals of endogenous viruses; endogenously transmitted N- and B-tropic viruses are expressed only or more readily if the mouse is of the appropriate $Fv-1^n$ or $Fv-1^b$ genotype (10, 13). *Fv-1* is expressed in erythroblasts, lymphocytes, fibroblasts, and myocytes (3).

Studies on the mechanism of action of the *Fv-1* locus have shown that the restriction does not affect entry of the virus into the cell or prevent formation of linear double-stranded DNA (form III) from the viral RNA, but does interfere with the change from form III to closed circular double-stranded DNA (form I) and reduces integration of form I into the cellular DNA (1, 4). It does not affect transfection with integrated viral DNA from chronically infected SC-1 cells (2). The restrictive activity can be transferred by RNA from embryo liver cells applied to the test cultures shortly after infection (16). The N-tropic viruses appear to carry a specific determinant that renders them sensitive to restriction by an $Fv-1^b$ product, B-tropic viruses carry a determinant sensitive to the $Fv-1^n$ product, and NB-tropic viruses lack both determinants (12). A locus *Rv-1*, resistance to Rauscher virus-1, described by Tóth *et al.* (15) is very similar to *Fv-1* and may be identical.

References

1. Chinsky, J., R. Soeiro, and J. Kopchick. 1984. *Fv-1* host cell restriction of Friend leukemia virus: microinjection of unintegrated viral DNA. J. Virol. 50:271–274.
2. Hsu, I.C., W.K. Yang, R.W. Tennant, and A. Brown. 1978. Transfection of Fv-1 permissive and restrictive mouse cells with integrated DNA of murine leukemia viruses. Proc. Natl. Acad. Sci. USA 75:1451–1455.
3. Jolicoeur, P. 1979. The Fv-1 gene of the mouse and its control of murine leukemia virus replication. Curr. Top. Microbiol. Immunol. 86:67–122.
4. Jolicoeur, P., and E. Rassart. 1980. Effect of the *Fv-1* gene product on synthesis of linear and supercoiled viral DNA in cells infected with murine leukemia virus. J. Virol. 33:183–196.
5. Kozak, C.A. 1985. Analysis of wild-derived mice for *Fv-1* and *Fv-2* murine leukemia virus restriction loci: a novel wild mouse *Fv-1* allele responsible for lack of host range restriction. J. Virol. 55:281–285.
6. Lilly, F. 1970. Fv-2: identification and location of a second gene governing spleen focus response to Friend leukemia virus in mice. J. Natl. Cancer Inst. 45:163–169.
7. Lilly, F., and T. Pincus. 1973. Genetic control of murine viral leukemogenesis. Adv. Cancer Res. 17:231–277.
8. Mayer, A., M.L. Duran-Reynals, and F. Lilly. 1978. *Fv-1*

regulation of lymphoma development and of thymic, ecotropic and xenotropic MuLV expression in mice of the AKR/J × RF/J cross. Cell 15:429–435.

9. Odaka, T. 1969. Inheritance of susceptibility to Friend mouse leukemia virus. V. Introduction of a gene responsible for susceptibility in the genetic complement of resistant mice. J. Virol. 3:543–548.

10. Odaka, T. 1975. Genetic transmission of endogenous N- and B-tropic murine leukemia viruses in low-leukemic strain C57BL/6. J. Virol. 15:332–337.

11. Odaka, T., and S. Hino. 1973. Inheritance of susceptibility to Friend mouse leukemia virus. IX. Genetic control of virus multiplication. *In* W.S. Ceglowski and H. Friedman, eds., Virus Tumorigenesis and Immunogenesis, 223–238. Academic Press, New York.

12. Rein, A., S.V.S. Kashmiri, R.H. Bassin, B.I. Gerwin, and G. Duran-Troise. 1976. Phenotypic mixing between N- and B-tropic murine leukemia viruses: infectious particles with dual sensitivity to Fv-1 restriction. Cell 7:373–379.

13. Rowe, W.P. 1973. Genetic factors in the natural history of murine leukemia virus infection. Cancer Res. 33:3061–3068.

14. Steeves, R., amd F. Lilly. 1977. Interactions between host and viral genomes in mouse leukemia. Ann. Rev. Genet. 11:277–296.

15. Tóth, F.D., L. Vácsi, and M. Balogh. 1973. Inheritance of susceptibility and resistance to Rauscher leukemia virus. Acta Microbiol, Acad. Sci. Hung. 20:183–189.

16. Yang, W.K., R.W. Tennant, R.J. Rascati, J.A. Otten, B. Schluter, J.O. Kiggans, F.E. Myer, and A. Brown. 1978. Transfer of Fv-1 locus-specific resistance to murine N-tropic and B-tropic retroviruses by cytoplasmic RNA. J. Virol. 27:288–299.

Fv-2 locus (formerly *Fv*), Friend virus susceptibility-2, Chr 9

This locus controls susceptibility to splenomegaly or spleen focus formation induced by inoculation with Friend leukemia virus (6). Two alleles are known, $Fv-2^s$ for susceptibility to the virus, and $Fv-2^r$ for resistance. Heterozygotes are susceptible. $Fv-2^s$ occurs in strains NIH Swiss, DDD, BALB/c, DBA, AKR, and others; $Fv-2^r$ occurs only in strain C57BL/6 and other members of the C57 and C58 family and in wild *M. spretus* (2). Lilly (3) showed that *Fv-2* segregates independently of *Fv-1* and that the presence of $Fv-2^r$ in single or double dose prevents the detection of *Fv-1* differences by the splenomegaly assay. In contrast to *Fv-1*, *Fv-2* has no effect on susceptibility of cultured cells. It appears to act by controlling susceptibility to the spleen focus-forming component (SFFV) of the Friend virus complex (4). *Fv-2* is expressed in hemopoietic cells as can be shown by transplantation of bone marrow

between $Fv-2^s$ and $Fv-2^r$ strains (7). Bone marrow and spleen cells of infected mice of $Fv-2^s$ strains bear a cell-surface Friend virus antigen which is not present in $Fv-2^r$ strains (8). It was shown that virus-specific RNA sequences related to the SFFV component of Friend virus are expressed in normal hemopoietic cells of $Fv-2^s$ but not of $Fv-2^r$ strains (5). This suggests that these sequences may be involved in normal hemopoiesis, a conclusion that is supported by evidence that $Fv-2^s$ mice have a much higher rate of DNA synthesis in BFU-E or CFU-E hemopoietic cells than do $Fv-2^r$ mice (1, 9). *Fv-2* probably influences susceptibility to Rauscher leukemia virus in a manner similar to its effect on Friend virus (11), and may be the same as *Rv-2*, resistance to Rauscher virus-2, described by Tóth *et al.* (10).

References

1. Behringer, R.R., and M.J. Dewey. 1985. Cellular site and mode of *Fv-2* gene action. Cell 40:441–447.

2. Kozak, C.A. 1985. Analysis of wild-derived mice for *Fv-1* and *Fv-2* murine leukemia virus restriction loci: a novel wild mouse *Fv-1* allele responsible for lack of host range restriction. J. Virol. 55:281–285.

3. Lilly, F. 1970. *Fv-2*: identification and location of a second gene governing spleen focus response to Friend leukemia virus in mice. J. Natl. Cancer Inst. 45:163–169.

4. Lilly, F., and T. Pincus. 1973. Genetic control of murine viral leukemogenesis. Adv. Cancer Res. 17:231–277.

5. Mak, T.W., A.A. Axelrad, and A. Bernstein. 1979. *Fv-2* locus controls expression of Friend spleen focus-forming virus-specific sequences in normal and infected mice. Proc. Natl. Acad. Sci. USA 76:5809–5812.

6. Odaka, T. 1969. Inheritance of susceptibility to Friend mouse leukemia virus. V. Introduction of a gene responsible for susceptibility in the genetic complement of resistant mice. J. Virol. 3:543–548.

7. Odaka, T., and M. Matsukura. 1969. Inheritance of susceptibility to Friend mouse leukemia virus. VI. Reciprocal alterations of innate resistance or susceptibility by bone marrow transplantation between congenic strains. J. Virol. 4:837–843.

8. Risser, R. 1979. Friend erythroleukemia antigen. A viral antigen specified by spleen focus-forming virus and differentiation antigen controlled by the *Fv-2* locus. J. Exp. Med. 149:1152–1167.

9. Suzuki, S. and A.A. Axelrad. 1980. *Fv-2* locus controls the propagation of erythropoietic progenitor cells (BFU-E) synthesizing DNA in normal mice. Cell 19:225–236.

10. Tóth, F.D., L. Vácsi, and M. Balogh. 1973. Inheritance of susceptibility and resistance to Rauscher leukemia virus. Acta Microbiol. Acad. Sci. Hung. 20:183–189.

11. Varet, B., A. Cannat, and S. Gisselbrecht. 1977. Genetic control of antinuclear antibodies in mice infected with Rauscher leukemia virus. Cancer Res. 37:1115–1118.

Fv-3 locus, Friend virus susceptibility-3

Infection with Friend leukemia virus suppresses humoral antibody synthesis *in vivo* and suppresses the proliferative response of normal lymphocytes to mitogens *in vitro*. The *Fv-3* locus controls susceptibility to both these effects of NB-tropic virus. The allele for susceptibility *Fv-3ˢ* occurs in strains 129, DBA/2, and others; the allele for resistance *Fv-3ʳ* occurs in strains C57BL/6, C57BL/10, and C58. *Fv-3* segregates independently of *Fv-2* and *H-2*. *Fv-3* may mediate its effects by regulating the activity of T-suppressor cells (1, 2).

References

1. Kumar, V., L. Goldschmidt, J.W. Eastcott, and M. Bennett. 1978. Mechanism of genetic resistance to Friend virus leukemia in mice. IV. Identification of a gene (*Fv-3*) regulating immunosuppression *in vitro*, and its distinction from *Fv-2* and genes regulating marrow allograft reactivity. J. Exp. Med. 147:422–433.
2. Kumar, V., P. Resnick, J.W. Eastcott, and M. Bennett. 1978. Mechanism of genetic resistance to Friend leukemia virus in mice. V. Relevance of *Fv-3* gene in the regulation of *in vitro* suppression. J. Natl. Cancer Inst. 61:1117–1123.

Fv-4 locus (formerly also *Akvr*), Friend virus susceptibility-4, Chr 12

This locus was identified by a variant found in a new strain, G, selected for resistance to N-tropic Friend virus in Japan. The allele *Fv-4ʳ* found in the G strain determines resistance to splenomegaly induced by Friend virus; the allele *Fv-4ˢ* determines susceptibility and occurs in all other strains tested including DBA/2 and BALB/c. Heterozygotes are resistant (5, 9). A locus designated *Akvr-1* controlling presence or absence of viremia and virus-mediated lymphoma induced by AKR N-tropic virus was described by Gardner *et al.* (1). The dominant allele determining resistance to the virus was found in wild mice in California. *Akvr-1* was subsequently shown to be identical to *Fv-4*. The locus is polymorphic in modern wild mouse populations and the dominant allele *Fv-4ʳ* is effective against AKR virus, NB-tropic Friend leukemia virus, and NB-tropic Moloney leukemia virus, but ineffective against amphotropic leukemia virus (8). Cells of uninfected *Fv-4ʳ* mice have on their cell membranes a leukemia virus *env*-gene product related to gp70 that is unique to *Fv-4ʳ* mice. This gp70 protein occurs on cells from thymus, spleen, lymph nodes, bone marrow, and on embryo fibroblasts (2). The DNA of *Fv-4ʳ* mice contains a unique sequence not found in *Fv-4ˢ* mice that hybridizes with an ecotropic *env*-gene probe. It has been partially sequenced by Ikeda *et al.* (4). It is likely that this sequence codes for the gp70 protein. The gp70 protein may be part of a defective virus that causes resistance by binding to exogenous viruses (6, 7). *Fv-4* is located on Chr 12 near the center (3).

References

1. Gardner, M.B., S. Rasheed, B.K. Pal, J.D. Estes, and S.J. O'Brien. 1980. *Akvr-1*, a dominant murine leukemia virus restriction gene, is polymorphic in leukemia-prone wild mice. Proc. Natl. Acad. Sci. USA 77:531–535.
2. Ikeda, H., and T. Odaka. 1984. A cell membrane "gp70" associated with *Fv-4* gene: immunological characterization, and tissue and strain distribution. Virology 133:65–76.
3. Ikeda, H., H. Sato, and T. Odaka. 1981. Mapping of the *Fv-4* mouse gene controlling resistance to murine leukemia viruses. Int. J. Cancer 28:237–240.
4. Ikeda, H., F. F. Laigret, M.A. Martin, and R. Repaske. 1985. Characterization of a molecularly cloned retroviral sequence associated with *Fv-4* resistance. J. Virol. 55:768–777.
5. Kai, K., H. Ikeda, Y. Yuasa, S. Suzuki, and T. Odaka. 1976. Mouse strain resistant to N-, B-, and NB-tropic murine leukemia viruses. J. Virol. 20:436–440.
6. Kai, K., H. Sato, and T. Odaka. 1986. Relationship between the cellular resistance to Friend murine leukemia virus infection and the expression of murine leukemia virus-gp70-related glycoprotein on cell surface of BALB/c-*Fv-4wʳ* mice. Virology 150:509–512.
7. Kozak, C.A., N.J. Gromet, H. Ikeda, and C.E. Buckler. 1984. A unique sequence related to the ecotropic murine leukemia virus is associated with the *Fv-4* resistance gene. Proc. Natl. Acad. Sci. USA 81:834–837.
8. O'Brien, S.J., E.J. Berman, J.D. Estes, and M.B. Gardner. 1983. Murine retroviral restriction genes *Fv-4* and *Akvr-1* are alleles at a single locus. J. Virol. 47:649–651.
9. Suzuki, S. 1975. *Fv-4*: a new gene affecting the splenomegaly induction by Friend leukemia virus. Jpn. J. Exp. Med. 45:473–478.

Fv-5 locus, Friend virus susceptibility-5

This locus controls induction of early anemia or polycythemia by the FV-P strain of Friend leukemia virus. The allele *Fv-5ᵖ* determines an early polycythemia (high hematocrit) response to infection with FV-P and occurs in strains DBA/2, BALB/c, and AKR; the allele *Fv-5ᵃ* determines an early anemic response (low hematocrit) to FV-P and occurs in strains CBA/J and C3H. Heterozygotes (CBA × DBA/2 F1) have a slight polycythemia with hematocrits intermediate between those of the parent strains (1). *Fv-5* appears not to affect the early stages of erythropoiesis, but to control the rate of proliferation of a late erythroid cell. The Fv-P strain of

Friend virus is different from the original strain isolated by Friend (FV-A) which induces an early anemic response apparently in all strains tested. The difference between the two viral strains resides in the spleen focus-forming viral component SFFV (2).

References

1. Shibuya, T., and T.W. Mak. 1982. A host gene controlling early anaemia or polycythaemia induced by Friend erythroleukaemia virus. Nature 296:577–579.
2. Shibuya, T., Y. Niho, and T.W. Mak. 1982. Erythroleukemia induction by Friend leukemia virus. A host gene locus controlling early anemia or polycythemia and the rate of proliferation of late erythroid cells. J. Exp. Med. 156:398–414.

Fv-6 locus, Friend virus susceptibility-6

This locus controls susceptibility and resistance to early erythroleukemia induction by helper virus-independent NB-tropic Friend murine leukemia virus. The allele *Fv-6^s* determines susceptibility and occurs in strain BALB/c (2) and probably also in NFS and C3H/He (1); the allele *Fv-6^r* determines resistance and occurs in strain DBA/2 (2) and probably also in DBA/1, C57BL/6, C57BL/10, C57L, CBA/N, and AKR/N (1). Heterozygotes are resistant. The disease induced by neonatal injection of virus is characterized by rapid splenomegaly at 5 to 10 weeks with severe anemia and death by 15 weeks. Resistant strains show no such symptoms by 20 weeks. *Fv-6* is distinct from *Fv-1*, *Fv-2*, *Fv-4*, *Fv-5*, and from *H-2*-linked genes (1, 2).

References

1. Ruscetti, S., L. Davis, J. Field, and A. Oliff. 1981. Friend murine leukemia virus-induced leukemia is associated with the formation of mink cell focus-inducing viruses and is blocked in mice expressing endogenous mink cell focus-forming xenotropic viral envelope genes. J. Exp. Med. 154:907–920.
2. Shibuya, T., and T.W. Mak. 1982. Host control of susceptibility to erythroleukemia and to the types of leukemia induced by Friend murine leukemia virus: initial and late stages. Cell 31:483–493.

fz locus, Chr 1

fz, fuzzy, recessive. Discovered in 1945 in the CFW stock at Carworth Farms. Homozygotes are viable and fertile. They are classifiable at 2 days of age by their thin, wavy vibrissae and later by their uneven coat. In the adult the coat is thin and wavy or curly (1). All four hair types are present but they are thin and curly and less easily distinguishable from each other than normal (4, 7). In the hair follicle of *fz/fz* mice, the dermal papilla becomes short and abnormal during the stage of hair shaft proliferation, and the abnormal papilla may be responsible for the failure of normal hair growth (7). An abnormality of the epidermal hair bulb may in turn be responsible for the abnormal dermal papilla, since the site of action of the *fz* locus has been shown to be in the epidermis and not in the dermis (5).

fz^fy, frowzy, recessive. Arose spontaneously in either the C3H or 101 inbred strain. Homozygotes are viable and usually fertile. The skin is wrinkled, and the hair is sparse and frizzy. Vibrissae are curly. Ear tufts are absent, and feet, nose, and tail are white (2, 3, 6).

References

1. Dickie, M.M., and G.W. Woolley. 1950. Fuzzy mice. J. Hered. 41:193–196.
2. Hollander, W.F. 1966. Mouse News Lett. 35:30.
3. Major, M.H., and M.S. Hawkins. 1958. Mouse News Lett. 19:37.
4. Mann, S.J. 1964. The hair of fuzzy mice. J. Hered. 55:121–123.
5. Mayer, T.C., J.A. Mittelberger, and M.C. Green. 1974. The site of action of the fuzzy locus (*fz*) in the mouse, as determined by dermal–epidermal recombinations. J. Embryol. Exp. Morphol. 32:707–713.
6. Russell, L.B. 1960. Mouse News Lett. 22:49.
7. Trigg, M.J. 1972. Hair growth in mouse mutants affecting coat texture. J. Zool. Lond. 168:165–198.

fzt, fuzzy tail, recessive

Found in strain DBA/2J. Homozygotes have curved hairs on the tail beginning with the first coat and continuing into adulthood. The body hair shows slight waving before weaning but is normal in adults (1).

References

1. Varnum, D.S. 1976. Mouse News Lett. 55:17.

G

G6pd locus (formerly *Gpdx*), glucose-6-phosphate dehydrogenase, Chr X

This locus, controlling the major isozyme of glucose-6-phosphate dehydrogenase (G6PD; E.C. 1.1.1.49), was first found to be on the X chromosome by Epstein (2), who showed that the enzyme was synthesized in oocytes and that oocytes of X/O females contained half as much enzyme as those of X/X females. No naturally occurring variants have been found in laboratory mice or in wild *Mus musculus*. A mutant allele with low G6PD activity was induced in a (101/El × C3H/El)F1 male mouse by ENU. In hemizygous males, heterozygous females, and homozygous females, the activity level in blood was 20 per cent, 60 per cent, and 15 per cent of normal, respectively (1). Using this mutation, Peters and Ball (5) found *G6pd* to be close to *Hq* in the proximal region of the × chromosome. A variant allele occurs in *Mus spretus*. In an interspecific backcross with a laboratory strain of *M. mus domesticus*, the locus was found to lie distal to *Hprt* (3). However, Martin-DeLeon *et al*. (4) found it to be near the centromere in band A by use of *in situ* hybridization. The homologous locus in man is near the distal end of the × chromosome.

References

1. Charles, D.J., and W. Pretsch. 1984. Mouse News Lett. 71:37–38.
2. Epstein, C.J. 1969. Mammalian oocytes: × chromosome activity. Science 163:1078–1079.
3. Heilig, R., C. Lemaire, J.-L Mandel, L. Dandolo, L. Amar, and P. Avner. 1987. Localization of the region homologous to the Duchenne muscular dystrophy locus on the mouse × chromosome. Nature 328:168–170.
4. Martin-DeLeon, P.A., S.F. Wolf, G. Persico, D. Tonido, G. Martini, and B.R. Migeon. 1985. Localization of glucose-6-phosphate dehydrogenase in mouse and man by in situ hybridization: evidence for a single locus and transposition of homologous X-linked genes. Cytogenet. Cell Genet. 39:87–92.
5. Peters, J., and S.T. Ball. 1985. Mouse News Lett. 73:17–18.

Ga, graying with age

Mice that become gray with age were found among wild mice in South Australia in 1970. The condition was ascribed to a dominant mutation with variation depending in part on genetic background and in part on maternal influence. All offspring of presumed heterozygous females mated to C57BL males were gray by 10 months of age, whereas only 6/65 offspring of the reciprocal cross were gray by that age (1). Morse *et al*. (2) have described a graying with age traceable to a maternally transmitted leukemia virus that causes melanocyte dysfunction subsequent to infection at pre- or early postnatal stages. The virus is not endogenous and is transmitted principally in milk. It is likely that viral transmission is responsible for the condition ascribed to *Ga* and the symbol should be withdrawn.

References

1. Kirby, G.C. 1974. Greying with age: a coat-color variant in wild Australian populations of mice. J. Hered. 65:126–128.
2. Morse, H.C. III, R.A. Yetter, J.H. Stimpfling, O.M. Pitts, T.N. Fredrickson, and J.W. Hartley. 1985. Greying with age in mice: relation to expression of murine leukemia virus. Cell 41:439–448.

Galt locus, galactose-1-phosphate uridyl transferase, Chr 4

This locus controls electrophoretic variation of an enzyme (GALT; E.C. 2.7.7.12) present in liver, spleen, heart, kidney, lung, thymus, brain, muscle, red blood cells, and plasma, but not in testis. Three alleles are known: *Galt^a* determines two anodal bands of equal staining intensity and occurs in the AKR, BDP, CE, NZB, RF, SEA, and ST/b strains and a few wild-derived stocks; *Galt^b* determines a darkly staining band at the position of the cathodal band of *Galt^a* and occurs in most inbred strains; and *Galt^c* determines two darkly staining bands at the same positions as those of *Galt^a* and occurs in the BUB and SF/Cam strains. The enzyme in heterozygotes appears to consist of homopolymers only (1).

References

1. Nadeau, J.H., and E.M. Eicher. 1982. Conserved linkage of soluble aconitase and galactose-1-phosphate uridyl transferase in mouse and man: assignment of the genes to mouse chromosome 4. Cytogenet. Cell Genet. 34:271–281.

Gapd

Gapd locus, glyceraldehyde-3-phosphate dehydrogenase (E.C. 1.2.1.12), Chr 6

No genetic variants at this locus are known. The locus was shown to be on Chr 6 by use of mouse–Chinese hamster somatic cell hybrids in which the mouse enzyme was identified by electrophoresis (1).

References

1. Bruns, G., P.S. Gerald, P. Lalley, U. Francke, and J. Minna. 1979. Gene mapping of the mouse by somatic cell hybridization. Cytogenet. Cell Genet. 25:139 (Abstr.).

Gbp-1 locus, guanine nucleotide-binding protein-1, Chr 3

This locus controls inducibility by interferons (both types I and II) of a protein with high binding affinity for guanylates. The protein (GBP-1) has a molecular weight of 65 000. The allele $Gbp-1^a$ determines inducibility of GBP-1 and occurs in strains A/J, BALB/c, C3H/He, and others and probably also in wild mice; the allele $Gbp-1^b$ determines non-inducibility and occurs in strains DBA/2, AKR, C57BL/6, CBA/J, STS, and many others and probably also in wild mice. Heterozygotes have an intermediate rate of synthesis of GBP-1 (1, 2). Gbp-1 is on Chr 3 near Adh-3 (1).

References

1. Prochazka, M., P. Staeheli, R.S. Holmes, and O. Haller. 1985. Interferon-induced guanylate-binding protein: mapping of the murine Gbp-1 locus to chromosome 3. Virology 145:273–279.
2. Staeheli, P., M. Prochazka, P.A. Steigmeier, and O. Haller. 1984. Genetic control of interferon action: mouse strain distribution and inheritance of an induced protein with guanylate-binding property. Virology 137:135–142.

gc, grey coat, recessive, Chr 5

Appeared within the first three generations of sib mating of mice trapped in Peru. The coat color resembles that of Va. The gene is not allelic with si. It is on Chr 5 about 20 cM from Hm (1).

References

1. Wallace, M.E., and J. Ferguson. 1984. Mouse News Lett. 71:18.

Gcg locus, glucagon, Chr 2

This is the structural locus for the polypeptide hormone glucagon. It was found to be on Chr 2 by use of mouse–Chinese hamster somatic cell hybrids screened with a full length cDNA clone of the Syrian hamster glucagon gene (1).

References

1. Lalley, P.A., A.Y. Sakaguchi, R.L. Eddy, N.H. Honey, G.I. Bell, L.-P. Shen, W.J. Rutter, J.W. Jacobs, G. Heinrich, W.W. Chin, and S.L. Naylor. 1987. Mapping polypeptide hormone genes in the mouse: somatostatin, glucagon, calcitonin, and parathyroid hormone. Cytogenet. Cell Genet. 44:92–97.

Gda locus, guanine deaminase

This locus controls electrophoretic mobility of guanine deaminase (E.C. 3.5.4.3). The allele Gda^b determines a fast migrating enzyme band and was found in wild mice from the Orkney island of Eday; the allele Gda^a determines a slower band and occurs in inbred strains and in other wild mouse populations. Presumptive heterozygotes have a three-banded pattern typical of a dimeric protein (1).

References

1. Nash, H.R. 1984. Mouse News Lett. 71:33.

Gdc-1, Gdc-2 loci

These two independent loci code for the adult and embryonic forms, respectively, of sn-glycerol-3-phosphate dehydrogenase (GPDH; E.C. 1.1.1.8). The embryonic form predominates during fetal and neonatal stages and is superseded after 10 days by the adult form. Mouse tumors may possess both embryonic and adult forms. Kozak and Murphy (6) have shown that in solid tumors the adult form predominates and in ascites tumors the embryonic form predominates, and that the enzyme form changes with conversion from solid to ascites form and back. Kozak and Fisher (5) have critically summarized the molecular and genetic aspects of GPDH development in the cerebellum.

Gdc-1 locus, glycerol-3-phosphate dehydrogenase-1, Chr 15. Variation at this locus is detected by heat stability and electrophoretic mobility. Five alleles are known: $Gdc-1^b$ determines high heat stability and slow cathodal electrophoretic mobility and occurs in C57BL/6 and most inbred strains; $Gdc-1^c$ determines heat sensitivity and slow electrophoretic mobility and occurs in BALB/c and CBA; $Gdc-1^d$ determines high heat stability and fast electrophoretic mobility and occurs in M. m. castaneus (4). $Gdc-1^e$ determines very high heat stability and very slow electrophoretic mobility and was induced by ethylnitrosourea in a (101 × C3H)F1 male (8); $Gdc-1^f$ determines high heat stability and occurs in the DDS strain (1). Heterozygotes have intermediate heat stability and the three-banded electrophoretic pattern characteristic of a dimeric protein. In addition, a *Hind*III DNA polymorphism has been found; *Gdc-1^d* DNA has a 5.9-kb fragment that hybridizes with a GPDH cDNA probe, and *Gdc-1^b* and *Gdc-1^c* DNA have a 7.9-kb fragment (3). *Gdc-1* acts in cerebellum, cerebral cortex, kidney, liver, muscle, and heart (4). In the cerebellum, GPDH is localized in Bergmann glia and oligodendrocytes where it begins to accumulate at 10 days. Very little activity is found in neuronal cells, but a study of the distribution of the enzyme in mice bearing mutant genes that cause deficiency of Purkinje cells indicates that expression of GPDH in glial cells is dependent on presence of Purkinje cells (5). *Gdc-1* is on Chr 15, probably distal to *Gpt-1* (2).

Gdc-2 locus, glycerol-3-phosphate dehydrogenase-2, Chr 9. Variation at this locus is detected by electrophoretic mobility of GPDH that is expressed chiefly in undifferentiated tissue. The allele *Gdc-2^b* determines a fast anodally migrating enzyme band and occurs in all inbred strains examined and several wild populations; the allele *Gdc-1^d* determines a more slowly migrating band and occurs in *M. m. castaneus*. Heterozygotes have an intermediate band plus the two parental bands. The level of expression of *Gdc-2* is about 1000-fold less than that of *Gdc-1*. *Gdc-2* is on Chr 9 distal to *Mod-1* (7).

References

1. Chen, A.K., and E.A. Thompson. 1981. Appearance of a second form of hepatic glycerol-3-phosphate dehydrogenase during neonatal development in the mouse. Arch. Biochem. Biophys. 207:96–102.
2. Hogarth, P.M., I.F.C. McKenzie, V.R. Sutton, K.M. Curnow, B.K. Lee, and E.M. Eicher. 1987. Mapping of the murine *Ly-6*, *Xp-14*, and *Gdc-1* loci to chromosome 15. Immunogenetics 25:21–27.
3. Kozak, L.P., and E.H. Berkenmeier. 1983. Mouse *sn*-glycerol-3-phosphate dehydrogenase: molecular cloning and genetic mapping of a cDNA sequence. Proc. Natl. Acad. Sci. USA 80:3020–3024.
4. Kozak, L.P., and K.J. Erdelsky. 1975. The genetic and developmental regulation of L-glycerol-3-phosphate dehydrogenase. J. Cell. Physiol. 85:437–448.
5. Kozak, L.P., and M. Fisher. 1985. The molecular genetic analysis of *sn*-glycerol-3-phosphate dehydrogenase development in mouse cerebellum. *In* C. Zomzely-Neurath and W.A. Walker, eds., Gene Expression in Brain, 173–204. John Wiley and Sons, New York.
6. Kozak, L.P., and E.D. Murphy. 1976. The reversible expression of an adult isozyme locus, *Gdc-1*, in tumors of the mouse. Cancer Res. 36:3711–3717.
7. Kozak, L.P., D.L. Burkart, and J.P. Hjorth. 1982. Unlinked structural genes for the developmentally regulated isozymes of *sn*-glycerol-3-phosphate dehydrogenase in mice. Dev. Genet. 3:1–6.
8. Pretsch, W., and D.J. Charles. 1984. An inherited variant of mouse *sn*-glycerol-3-phosphate dehydrogenase detected by isoelectric focusing: genetical and biochemical analysis. Biochem. Genet. 22:419–428.

Gdcr-1, *Gdcr-2* loci, glycerol phosphate dehydrogenase regulator-1 and -2

The level of expression of *sn*-glycerol-3-phosphate dehydrogenase (GPDH: E.C. 1.1.1.8) in cerebellum of BALB/c mice is 2.5 times that of C57BL/6 mice. To explain the genetic control of this difference, Kozak (2) has postulated the existence of two loci, *Gdcr-1* and *Gdcr-2*, unlinked to each other or to *Gdc-1*. The *Gdcr-1^c* and *Gdcr-2^c* alleles cause enhancement of expression of GPDH and occur in BALB/c. *Gdcr-1^c* acts only on *Gdc-1^c* expression, but *Gdcr-2^c* acts on whatever alleles of *Gdc-1* are present. The alleles in C57BL/6 and also in *M. m. castaneus* are *Gdcr-1^b* and *Gdcr-2^b*. Differences in cerebellar activity of GPDH between BALB/c and C57BL/6 are determined by the amount of translatable GPDH mRNA (3). The activity difference can also be detected in reaggregating cultures of 3-day cerebellar cells (1).

References

1. Kozak, L.P. 1978. Increased synthesis of L-glycerol-3-phosphate dehydrogenase during *in vitro* differentiation of reaggregating cerebellar cells. Dev. Biol. 66:593–600.
2. Kozak, L.P. 1985. Interacting genes control glycerol-3-phosphate dehydrogenase expression in developing cerebellum of the mouse. Genetics 110:123–143.
3. Kozak, L.P., and P.L. Ratner. 1980. Genetic regulation of translatable mRNA levels for mouse *sn*-glycerol-3-phosphate dehydrogenase during development of the cerebellum. J. Biol. Chem. 255:7589–7594.

Gdr-1, *Gdr-2* loci, G6PD regulator-1, -2

These two loci are postulated to govern the level of activity of glucose-6-phosphate dehydrogenase in erythrocytes. Inbred strains fall into three classes: those with high activity, A and A/He; those with low activity, C57L and C57BR/cd; and those with intermediate activity, AKR, BALB/c, C57BL/6, and numerous others. Activity levels in segregating generations from crosses between high and intermediate strains and between intermediate and low strains were compatible with single-locus control in each cross. Activity levels in a segregating generation from a cross between a high and a low strain were compatible with two-locus control. The symbol *Gdr-1* has tentatively been assigned to the locus by which AKR (intermediate) and C57L (low) differ, and the symbol *Gdr-2* to the locus by which A (high) and C57BL/6 (intermediate) differ. No structural difference in the enzyme was detected between the strains. Activity of the enzyme in tissues other than red cells is not significantly affected by these loci (2). A single-locus difference was found between C57BL/6 (intermediate) and C57L (low) as judged by the distribution of G6PD level in four recombinant inbred strains derived from these two strains. The strain distribution pattern indicated a possible linkage to brown (*b*) on Chr 4 (1). This locus may be the same as *Gdr-1*.

References

1. Erickson, R.P., and K. Harper. 1980. A major autosomal gene effect on activity of glucose-6-phosphate dehydrogenase segregating between recombinant inbred lines of mice. Genet. Res. 36:91–97.
2. Hutton, J.J. 1971. Genetic regulation of glucose-6-phosphate dehydrogenase activity in the inbred mouse. Biochem. Genet. 5:315–331.

Ggc locus, γ-glutamyl cyclotransferase, Chr 6

This locus controls electrophoretic variation of γ-glutamyl cyclotransferase (E.C. 2.3.2.4), an enzyme found in red cells, liver, and kidney of the mouse and probably in many other tissues as well. The allele *Ggc^a* determines a pattern of two bands, the minor one faster anodally migrating than the major one; the allele *Ggc^b* determines a pattern of banding similar to that of *Ggc^a* but more slowly migrating. Heterozygotes have all four parental bands. *Ggc^a* occurs in the A/J strain and in most other inbred strains; *Ggc^b* occurs in the C57BL, C58, and some related strains and in the SK/Cam strain. *Ggc* is near the *Ly-2* and *Igk* loci on Chr 6(1).

References

1. Tulchin, N. and B.A. Taylor. 1981. γ-glutamyl transferase: a new genetic polymorphism in the mouse (*Mus musculus*) linked to *Lyt-2*. Genetics 99:109–116.

Ggm-1 locus, ganglioside expression-1, Chr 17

This locus controls expression of gangliosides GM1 and GD1a in liver and erythrocytes. The allele *Ggm-1^a* determines presence of GM1 and GD1a and occurs in the SJL, SWR, PL, RFM, SWM, SL, and WHT strains

and in *M. m. castaneus*; the allele *Ggm-1^b* determines absence or very low level of these gangliosides and occurs in BALB/c, DBA/2, C57BL/6, and most other inbred strains and in wild mice in China and Japan (1, 2). Heterozygotes express the gangliosides. Examination of ganglioside expression in *H-2*-congenic strains showed that *Ggm-1* is located on the proximal side of *H-2K* on Chr 17 (1).

References

1. Hashimoto, Y., A. Suzuki, T. Yamakawa, N. Miyashita, and K. Moriwaki. 1983. Expression of GM1 and GD1a in mouse liver is linked to the H-2 complex on chromosome 17. J. Biochem. 94:2043–2048.
2. Hashimoto, Y., A. Suzuki, T. Yamakawa, C.-H. Wang, F. Bonhomme, N. Miyashita, and K. Moriwaki. 1984. Expression of GM1 and Gd1a in liver of wild mice. J. Biochem. 95:7–12.

Ggm-2 locus, ganglioside expression-2

This locus controls expression of ganglioside GM2 in liver. The allele *Ggm-2^a* determines presence of GM2 and occurs in all inbred strains examined except one; the allele *Ggm-2^b* determines absence of GM2 and occurs in the WHT/Ht strain. Heterozygotes express GM2 (1). GM2 is synthesized from GM3 by the activity of UDP-*N*-acetylgalactosamine: GM3 *N*-acetylgalactosaminyltransferase (E.C. 2.4.1.92). This enzyme is present in liver of BALB/c and absent from liver of WHT mice. Crosses between these two strains showed that GM2 expression always segregated with activity of the enzyme. F1 hybrids have intermediate levels of the enzyme. It seems likely that *Ggm-2* codes for the enzyme (2).

References

1. Hashimoto, Y., H. Otsuka, K. Suzuki, K. Sudo, A. Suzuki, and T. Yamakawa. 1983. Genetic regulation of GM2 expression in liver of mouse. J. Biochem. 93:895–901.
2. Hashimoto, Y., M. Abe, Y. Kiuchi, A. Suzuki, and T. Yamakawa. 1984. Genetically regulated expression of UDP-*N*-acetylgalactosamine: GM3(NeuGc) *N*-acetylgalactosaminyltransferase (E.C. 2.4.1.92) activity in mouse liver. J. Biochem. 95:1543–1549.

Ggtb-1 locus, galactosyltransferase-1, Chr 4

This is the structural locus for galactosyltransferase (E.C. 2.4.1.38), an enzyme that catalyzes the transfer of galactose from UDP-galactose to N-acetylglucosamine. The enzyme is located in the plasma membrane of a variety of cells including that of the acrosome of mature sperm. It has been hypothesized to play a role in the segregation distortion caused by many *t*-complex haplotypes, since *t*-bearing sperm, which have a high transmission ratio, have higher galactosyltransferase activity than sperm bearing the wild-type haplotype. However, *Ggtb-1* is on Chr 4 and thus cannot be a part of the *t* complex on Chr 17. See *t* complex. No genetic variants of *Ggtb-1* are known. Its chromosome location was found by use of mouse–Chinese hamster somatic cell hybrids screened with a cDNA clone for bovine galactosyltransferase (1).

References

1. Shaper, N.L., J.H. Shaper, G.F. Hollis, H. Chang, J.R. Kirsch, and C.A. Kozak. 1987. The gene for galactosyltransferase maps to mouse chromosome 4. Cytogenet. Cell Genet. 44:18–21.

Gia, *Gsa* loci

These loci code for two members of the G protein family of transmembrane signalling molecules. Their chromosomal positions were determined by use of cloned cDNA probes prepared from mouse mRNAs, in conjunction with mouse–Chinese hamster somatic cell hybrids.

Gia locus, G_i protein α, Chr 9. The G_i protein is the inhibitory regulator of adenylate cyclase.

Gsa locus, G_s protein α, Chr 2. The G_s protein is the stimulatory regulator of adenylate cyclase (1).

References

1. Ashley, P.L., J. Ellison, W.A. Sullivan, H.R. Bourne, and D.R. Cox. 1987. Chromosomal assignment of the murine G_i and G_s genes. Am. J. Hum. Genet. 46:A155 (Abstr.).

Gk locus, glucokinase activity

This locus controls activity level of the enzyme glucokinase (E.C. 2.7.1.2) which is found exclusively in liver. Inbred strains fall into three groups with high, intermediate, and low activity. The difference between the high strain C3H/HeJ and the low strains RF/J and C58 is controlled by the *Gk* locus, with the allele Gk^a (high) in C3H and the allele Gk^b (low) in RF and C58. Heterozygotes have intermediate activity. The genetic basis for the strains with intermediate activity has not been reported. Glucokinase is considered the rate-limiting enzyme in the glucose metabolism pathway, and its activity is regulated by insulin. Gk^a/- and Gk^b/Gk^b mice are equally healthy, but RF and C58 (Gk^b/Gk^b) mice cannot withstand a 48-hour fast (1, 2). Plasma insulin is

significantly higher in C3H/He than in C58 mice under various conditions of feeding and fasting but plasma glucose concentrations are not significantly different (2). The glucokinase molecules of the two strains appear to be identical, indicating that *Gk* is probably a regulatory locus (3).

References

1. Coleman D.L. 1977. Genetic control of glucokinase activity in mice. Biochem. Genet. 15:297–305.
2. Lavender, F.L., P.A. James, and D.G. Walker. 1983. Hepatic glucokinase activity and circulating insulin concentrations in two inbred mouse strains. Diabetologia 25:114–119.
3. James, P.A., F.L. Lavender, G.M. Lawrence, and D.G. Walker. 1985. Comparison of glucokinase in C3H/He and C58 mice that differ in their hepatic activity. Biochem. Genet. 23:525–538.

gl, grey-lethal, recessive, Chr 10

Arose spontaneously in a stock segregating for c^e (2). Homozygotes die between 20 and 30 days of age. The yellow parts of the coat appear white as a result of clumping of the phaeomelanin granules (3). The color of *a*/*a* mice is not noticeably influenced by *gl*. Homozygotes lack the power of secondary bone resorption. Consequently the bones cannot grow normally, they no not form normal marrow cavities, and the teeth do not erupt. Homozygotes can be recognized at 18 days of gestation by their short lower incisors which do not extend caudally beneath the first lower molar tooth germs (4). An atlas of the skeletal anatomy of grey-lethal and normal sibs at 3 weeks of age has been prepared by Bateman (1). Grey-lethals have lower than normal serum calcium concentration, higher rate of bone deposition, and an eightfold increase in number of calcitonin- secreting cells of the thyroid. Administration of parathyroid hormone does not cure the disease and raises the serum calcium level only slightly (5). The plasma of grey-lethals contains a hypocalcemic factor, probably calcitonin (6). Transplantation of bone marrow or spleen cells between grey-lethals and normal sibs, using lethally irradiated hosts, transmits the osteopetrosis to normal mice or cures it in grey-lethals (7,8). These results suggest that the basic defect caused by *gl* is either in osteoclasts or some other cells which can develop from bone marrow or spleen and which control the activity of osteoclasts. The low serum calcium resulting from decreased bone resorption could secondarily cause the elevated production of calcitonin. The relation of the pigment defect to the bone defects is not known.

References

1. Bateman, N. 1954. Bone growth: a study of the grey-lethal and microphthalmic mutants of the mouse. J. Anat. 88:212–262.
2. Grüneberg, H. 1935. A new sub-lethal colour mutation in the house mouse. Proc. R. Soc. Lond. (Biol.) 118:321–342.
3. Grüneberg, H. 1936. Grey-lethal, a new mutation in the house mouse. J. Hered. 27:105–109.
4. Hollinshead, M.B., L.C. Schneider, and M.E. Smith. 1975. Prenatal development of the grey-lethal mouse. I. Teeth and jaws. Anat. Rec. 182:305–320.
5. Marks, S.C., and D.G. Walker. 1969. The role of the parafollicular cell of the thyroid gland in pathogenesis of congenital osteopetrosis in mice. Am. J. Anat. 126:299–314.
6. Murphy, H.M., 1972. Calcitonin-like activity in the circulation of osteopetrotic grey-lethal mice. J. Endocrinol. 53:139–150.
7. Walker, D.G. 1975. Bone resorption restored in osteopetrotic mice by transplants of normal bone marrow and spleen cells. Science 190:784–785.
8. Walker, D.G. 1975. Spleen cells transmit osteopetrosis in mice. Science 190:785–787.

gld, generalized lymphoproliferative disease, recessive, Chr 1

Arose spontaneously in the C3H/HeJ strain. Homozygotes have enlargement of all lymph nodes to 50 times the control weight by 20 weeks of age and fourfold enlargement of the spleen. Thymus weight is unaffected. The number of T, B, and null lymphocytes is increased 6-, 15-, and 33-fold, respectively. There is a broad-based hypergammaglobulinemia with increases in levels of IgA, IgG, and IgM, and high titers of antinuclear autoantibodies. Homozygotes live only one-half as long as controls. Virtually all homozygotes had interstitial pneumonitis at autopsy. At 22 weeks or older, all had deposits of immune complexes in the glomeruli (4). Spleens and lymph nodes of *gld/gld* mice contain a population of aberrant T-cells that bear the Ly-5(B220) cell surface antigen normally found only on B-cells (2). Activation of B-cells does not occur until significant numbers of Ly-5(B220)$^+$ cells are present in spleen and lymph nodes, suggesting that B-cell activation may be secondary to the T-cell abnormality. Spleen and lymph node cells of *gld/gld* mice have a decreased ability to produce IL-2 *in vitro* (1). The abnormal T-cells of lymph nodes from C3H mice homozygous for *gld* and *lpr* were compared for cell surface antigens, lectin binding sites, and for functional and other characteristics. No differences were found, suggesting that these mutations affect a common pathway important in T-cell differentiation and function

(3). The *gld* locus is on Chr 1 between *Pep-3* and *Lp* (4).

References

1. Davidson, W.F., J.B. Roths, and H.C. Morse III. 1984. Single gene mutations that cause SLE-like autoimmune disease in mice. Clin. Immunol. Newsletter 5:17–20.
2. Davidson, W.F., K.L. Holmes, J.B. Roths, and H.C. Morse III. 1985. Immunologic abnormalities of mice bearing the *gld* mutation suggest a common pathway for murine nonmalignant lymphoproliferative disorders with autoimmunity. Proc. Natl. Acad. Sci. USA 82:1219–1223.
3. Davidson, W.F., F.J. Dumont, H.G. Bedigian, B.F. Fowlkes, and H.C. Morse III. 1986. Phenotypic, functional, and molecular genetic comparisons of the abnormal lymphoid cells of C3H-*lpr/lpr* and C3H-*gld/gld* mice. J. Immunol. 136:4075–4084.
4. Roths, J.B., E.D. Murphy, and E.M. Eicher. 1984. A new mutation, *gld*, that produces lymphoproliferation and autoimmunity in C3H/HeJ mice. J. Exp. Med. 159:1–20.

Glk locus, galactokinase, Chr 11

This locus controls the level of activity of galactokinase (E.C. 2.7.1.6) in whole blood. The allele *Glk*^{a} determines low activity and is found in 21 strains including C57BL/6, BALB/c, and C57L; the allele *Glk*^{b} determines high activity, and is found in strains C3H/He, A, AKR, and others. Heterozygotes have intermediate activity (2). Galactokinase is present in cultured peritoneal macrophages and embryo fibroblasts. Its unvarying association with Chr 11 in hybrid cultures of these cells with Chinese hamster cells is evidence that *Glk* is the structural locus for galactokinase (1).

References

1. Kozak, C.A., and F.H. Ruddle. 1977. Assignment of the genes for thymidine kinase and galactokinase to *Mus musculus* Chromosome 11, and the preferential segregation of this chromosome in Chinese hamster/mouse somatic cell hybrids. Somat. Cell Genet. 3:121–133.
2. Mishkin, J.D., B.A. Taylor, and W.J. Mellman. 1976. *Glk*: a locus controlling galactokinase activity in the mouse. Biochem. Genet. 14:635–640.

Glo-1 complex, Chr 17

This complex comprises a structural locus and a regulatory element for glyoxalase-1 (GLO-1; E.C. 4.4.1.5). The close linkage of *Glo-1* to *H-2* (3 per cent recombination) resembles the close linkage in man of the glyoxalase 1 locus to the HLA major histocompatibility locus (1).

Glo-1s locus (formerly *Glo-1*), glyoxalase-1-structural. This locus controls electrophoretic mobility of glyoxalase-1 in erythrocytes. The allele *Glo-1s^a* determines a fast anodally migrating band and occurs in strains AKR, CBA/H, C57BL/6, and many others; the allele *Glo-1s^b* determines a slower band and occurs in strain MA/MyJ. Heterozygotes have the two parental bands and an intermediate band, suggesting a dimeric structure of the enzyme (1). An allele designated *Glo-1s^c* determines activity level of GLO-1 three- to eightfold less than that in *Glo-1s^a* strains and was found in all *t*-complex haplotypes examined (2).

Glo-1r locus (originally designated *QGlo-1*), glyoxalase-1 regulatory. This locus controls quantitative variation in GLO-1 content of red cells. It shows no recombination with *Glo-1s*. At least six different alleles, designated *Glo-1r^a* through *Glo-1r^f* are found in different inbred strains, varying from a GLO-1 activity of 3.7 in the NZB/Bn strain (*Glo-1r^a*) to an activity of 18 to 19 in the BALB/c strains of E.A. Boyse (*Glo-1r^f*) (3). It is possible that the *Glo-1s^c* allele of Nadeau (2) may be a regulatory variant and assignable to the *Glo-1r* locus.

References

1. Meo, T., T. Douglas, and A.M. Rijnbeek. 1977. Glyoxalase I polymorphism in the mouse: a new genetic marker linked to *H-2*. Science 198:311–313.
2. Nadeau, J.H. 1986. A glyoxalase-1 variant associated with the *t*-complex in house mice. Genetics 113:91–99.
3. Rubenstein, P., and K. Vienne. 1982. Genetic control of the quantitative variation of erythrocyte glyoxalase-1 (GLO-1) in mice. Biochem. Genet. 20:153–163.

gm, gunmetal, recessive, Chr 14

Arose spontaneously in the C57BL/6J inbred strain. Homozygotes have a diluted coat color somewhat similar to that of dilute (*d/d*). Eye color does not appear to be affected. Homozygotes have high mortality and do not breed well (1).

References

1. Dickie, M.M. 1964. Mouse News Lett. 30:30.

Gm-2 locus

See *Ly-6B* locus.

Gm-3 locus, granulocyte-macrophage antigen-3, Chr 2

This locus controls a cell-surface antigen found on mature neutrophils in peritoneal exudates and in bone marrow, and on activated macrophages, but absent from lymphoid and red cells and from kidney, liver, and heart. The allele *Gm-3^b* determines presence of the antigen and occurs in strains A/WySn, the C57 strain family, CBA, and RF; the allele *Gm-3^a* determines absence of the antigen and occurs in strains AKR, BALB/c, C3H/He, and many others. Gm-3 was shown to be on Chr 2 near *Ly-4*, *B2m*, and *H-3* by use of BALB/c × C57BL/6 recombinant inbred strains and an *H-3* congenic strain (1).

References

1. Hibbs, M.L., P.M. Hogarth, R.A. Harris, and I.F.C. McKenzie. 1985. Gm-3.2, a new granulocyte/macrophage alloantigen. Immunogenetics 21:61–70.

go, angora, recessive, Chr 5

Arose spontaneously in the BALB/cJ inbred strain. Homozygotes are recognizable at about 18 days by the extra length of the guard hairs. At weaning the guard hairs are more than twice normal length. Vibrissae are also long (1).

References

1. Dickie, M.M. 1963. Mouse News Lett. 29:39.

Got-1 locus, glutamate oxaloacetate transaminase-1, Chr 19

This locus controls electrophoretic variation in the supernatant form of glutamate oxaloacetate transaminase (GOT-1; E.C. 2.6.1.1). The enzyme is found in most tissues with the exception of erythrocytes, and in cultured fibroblasts (2). The allele *Got-1^a* determines a fast anodally migrating band and occurs in strains C57BL/6, SJL, SWR, and all other inbred strains examined and in wild mice from North America and Europe; the allele *Got-1^b* determines a slower anodally migrating band and occurs in *M. m. castaneus* from Thailand and wild house mice from Calcutta. Heterozygotes have three major bands, evidence for dimeric structure of the enzyme (1). *Got-1* is on Chr 19 near *ru* (3). Its assignment to Chr 19 has been confirmed by use of mouse–Chinese hamster somatic cell hybrids classified for mouse GOT-1 by isoelectric focusing (4).

References

1. Chapman, V.M., and F.H. Ruddle. 1972. Glutamate oxaloacetate transaminase (GOT) genetics in the mouse: polymorphism of GOT-1. Genetics 70:299–305.
2. DeLorenzo, R.J., and F.H. Ruddle. 1970. Glutamate oxalate transaminase (GOT) genetics in *Mus musculus*: linkage, polymorphism, and phenotypes of the *Got-2* and *Got-1* loci. Biochem. Genet. 4:259–273.

3. Eicher, E.M., S. Reynolds, and J.L. Southard. 1977. Mouse News Lett. 56:42.
4. Schäfer, R., J. Doehmer, I. Rademacher, and K. Willecke. 1980. Assignment of the gene for cytoplasmic glutamate oxaloacetate transaminase to mouse chromosome 19 using Chinese hamster × mouse somatic cell hybrids. Somat. Cell Genet. 6:709–717.

Got-2 locus, glutamate oxaloacetate transaminase-2, Chr 8

This locus controls electrophoretic variation in the mitochondrial form of glutamate oxaloacetate transaminase (GOT-2; E.C. 2.6.1.1) found in most tissues with the exception of erythrocytes and in cultured fibroblasts. The allele *Got-2ᵃ* determines a slow cathodally migrating band and occurs in strain SWR and in strains derived from wild populations in Peru, Israel, and Skokholm Island; the allele *Got-2ᵇ* determines a faster cathodally migrating band and occurs in strains C57BL/6, A/He, DBA/2, and many others. Heterozygotes have the parental bands plus an intermediate band (1).

References

1. DeLorenzo, R.J., and F.H. Ruddle. 1970. Glutamate oxalate transaminase (GOT) genetics in *Mus musculus*: linkage, polymorphism, and phenotypes of the *Got-2* and *Got-1* loci. Biochem. Genet. 4:259–273.

gp, gaping lids, recessive

Not allelic with *oe* or *lg* (2) but has not been tested for allelism with *lo*. Arose in a stock carrying *ax*. In homozygotes the eyes are open at birth and become opaque with age. Homozygotes are viable and fertile. The gene appears to have full penetrance. From 15 days of gestation onward the lens is about twice normal size (1).

References

1. Kelton, D.E., and V. Smith. 1964. Gaping, a new open-eyelid mutation in the house mouse. Genetics 50:261–262 (Abstr.).
2. Miller, J.R. 1964. Mouse News Lett. 30:16.

Gpd-1 locus, glucose phosphate dehydrogenase-1, Chr 4

This locus controls electrophoretic variation of glucose-6-phosphate dehydrogenase (GPD-1; E.C. 1.1.1.49) found in kidney and many other tissues and in cultured cells, but not in erythrocyte lysates. The enzyme controlled by *Gpd-1* is a minor isozyme with this activity and is probably a general hexose-6-phosphate dehydrogenase rather than a specific glucose-6-phosphate dehydrogenase (2). The allele *Gpd-1ᵃ* determines a slow migrating form and is found in strains C57BL/6, 129/J, RF, and others; the allele *Gpd-1ᵇ* determines a fast migrating form and is found in strains SJL, DBA/2, AKR, and many others. Heterozygotes show three bands, evidence that the enzyme is a dimer (3). A third allele, *Gpd-1ᶜ*, which determines a band migrating more slowly than that of *Gpd-1ᵃ*, occurs in wild mice in Denmark (4) and in *M. m. castaneus* (1).

References

1. Chapman, V.M., and F.H. Ruddle. 1972. Glutamate oxaloacetate transaminase (GOT) genetics in the mouse; polymorphism of GOT-1. Genetics 70:299–305.
2. Hutton, J.J., and T.H. Roderick. 1970. Linkage analyses using biochemical variants in mice. III. Linkage relationships of eleven biochemical markers. Biochem. Genet. 4:339–350.
3. Ruddle, F.H., T.B. Shows, and T.H. Roderick. 1968. Autosomal control of an electrophoretic variant of glucose-6-phosphate dehydrogenase in the mouse (*Mus musculus*). Genetics 58:599–606.
4. Selander, R.K., W.G. Hunt, and S.Y. Yang. 1969. Protein polymorphism and genic heterozygosity in two European subspecies of the house mouse. Evolution 23:379–390.

Gpd-2 locus, glucose phosphate dehydrogenase-2, Chr 5

This locus controls electrophoretic pattern of glucose-6-phosphate dehydrogenase (GPD-2; E.C. 1.1.1.49) in testis extracts. The enzyme migrates cathodal to the X-linked enzyme (G6PD) and has substrate and cofactor preferences quite different from GPD-1. The allele *Gpd-2ᵃ* determines a fast single band and occurs in strains CE/J, C57BL/Go, IS/Cam, and P/J; the allele *Gpd-2ᵇ* determines a slower band and occurs in strains DBA/2 and SM/J. Heterozygotes have a three-banded pattern typical of a dimeric protein. *Gpd-2* is on Chr 5 very close to *Pgm-1* (1).

References

1. Nash, H.R. 1984. Mouse News Lett. 71:33.

Gpdx locus

See *G6pd* locus.

Gpi-1 complex, Chr 7

This complex includes a structural and a putative regulatory locus that have not shown recombination in crosses. Five distinct haplotypes have been proposed by West and Fisher (11), but more recent evidence suggests a larger number of distinct types (8).

Gpi-1s locus (formerly *Gpi-1*), glucosephosphate isomerase-1 structural. This locus controls electrophoretic variation in glucosephosphate isomerase (GPI-1; E.C. 5.3.1.9), an enzyme widely distributed in mouse tissues and found also in cultured fibroblasts and in several tumors. The allele *Gpi-1s^a* determines a slow cathodally migrating band and occurs in strains SJL, PL, BALB/c, DBA/2, and many others (4); it has an isoelectric point (IEP) of 8.4 and is thermostable (3). The allele *Gpi-1s^b* determines a faster migrating band and occurs in the wild and in strains C57BL/6, CBA, C3H, and others (4); it has an IEP of 8.7 and is thermolabile (3). The allele *Gpi-1s^c* determines an enzyme that migrates faster than GPI-1B and has reduced activity and heat stability and altered pH profile, and occurs in wild mice in Somerset, UK. The allele *Gpi-1s^d* determines an enzyme that migrates more slowly than GPI-1A and occurs with low frequency in wild mice in Somerset (6). Heterozygotes have the two parental bands plus an intermediate band, evidence for dimeric structure of the enzyme. A TEM-induced null allele of *Gpi-1s* has been described by Soares (9). GPI-1 of maternal type is found in 1- to 4-day embryos; at 5 days GPI-1 of paternal type appears (2, 12). Because GPI-1 occurs so early in development and because the intermediate hybrid band is formed only in heterozygous cells (*Gpi-1s^a*/*Gpi-1s^b*) or in heterokaryons and not in mixtures of the two kinds of homozygous cells, this locus has been extremely useful in studying the origin of multinucleated and polyploid cells in embryos (1). The ratio of GPI-1A to GPI-1B has been used to determine the proportion of cells of each genotype in chimeric mice, but this method may be subject to inaccuracy since it has been shown that red cells and peripheral lymph nodes of A/J (*Gpi-1s^a*) and C57BL/6 (*Gpi-1s^b*) differ significantly in GPI-1 activity (10).

Gpi-1t locus (formerly *Org*, *Gpi-1r*), glucosephosphate isomerase-1 temporal. This locus controls level of activity of GPI-1 in oocytes. Three alleles were originally described, *Gpi-1t^a* determining low activity in strain DBA/2 and NZW, *Gpi-1t^b* determining intermediate activity in strains A/He, C57BL, BALB/c, LP, and most other strains, and *Gpi-1t^c* determining high activity in strain 101/H and probably in AKR (11). However, Peterson *et al.* (8) have found multiple strain differences in GPI-1 activity in oocytes among 45 inbred strains and substrains tested, with a fairly continuous distribution from low to high. Closely related strains tend to have similar levels of activity. The authors believe that there may be more than three distinct alleles. There is no association of level of expression in oocytes with the *Gpi-1s^a* or *Gpi-1s^b* allele. Heterozygotes for *Gpi-1t* have activity that is intermediate between that of the parents and shows expression of the parental *Gpi-1s* alleles in the proportions expected if *Gpi-1t* is *cis*-acting, i.e. controls expression of the *Gpi-1s* allele on the same chromosome (7, 8). Expression of *Gpi-1t* is first detected in the oocytes of females at 6 to 7 days of age about a week after the onset of meiosis when the oocytes are just beginning to grow (5).

References

1. Ansell, J.D., P.W. Barlow, and A. McLaren. 1974. Binucleate and polyploid cells in the decidua of the mouse. J. Embryol. Exp. Morphol. 31:223–227.
2. Chapman, V.M., W.K. Whitten, and F.H. Ruddle. 1971. Expression of paternal glucose phosphate isomerase-1 (Gpi-1) in preimplantation stages of mouse embryos. Dev. Biol. 26:153–158.
3. Charles, D.J., and C.-Y. Lee. 1980. Biochemical and immunological characterization of genetic variants of phosphoglucose isomerase from mouse. Biochem. Genet. 18:153–169.
4. DeLorenzo, R.J., and F.H. Ruddle. 1969. Genetic control of two electrophoretic variants of glucosephosphate isomerase in the mouse (*Mus musculus*). Biochem. Genet. 3:151–162.
5. McLaren, A., and M. Buehr. 1981. GPI expression in female germ cells of the mouse. Genet. Res. 37:303–309.
6. Padua, R.A., G. Bulfield, and J. Peters. 1978. Biochemical genetics of a new glucosephosphate isomerase allele (*Gpi-1^c*) from wild mice. Biochem. Genet. 16:127–143.
7. Peterson, A.C., and G.G. Wang. 1978. Genetic regulation of glucosephosphate isomerase in mouse oocytes. Nature 276:267–269.
8. Peterson, A., F. Choy, G. Wong, S. Clapoff, and P. Frair. 1985. Glucosephosphate isomerase (GPI-1) expression in mouse ova: cis regulation of monomer realization. Biochem. Genet. 23:827–846.
9. Soares, E.R. 1979. TEM-induced gene mutations at enzyme loci in the mouse. Environ. Mutagen. 1:19–25.
10. Warner, C.M., T.E. Meyer, D. Balinsky, and C.J. Briggs. 1985. Variation in the amount of glucose phosphate isomerase in lymphocytes and erythrocytes from A/J and C57BL/6J mice. Biochem. Genet. 23:815–825.
11. West, J.D., and G. Fisher. 1984. A new allele of the *Gpi-1t* temporal gene that regulates the expression of glucosephosphate isomerase in mouse oocytes. Genet. Res. 44:169–181.

12. West, J.D., and J.F. Green. 1983. The transition from oocyte-coded to embryo-coded glucosephosphate isomerase in the early mouse embryo. J. Embryol. Exp. Morphol. 78:127–140.

Gpt-1 locus, glutamic-pyruvic transaminase-1, Chr 15

This locus controls electrophoretic variation in glutamic-pyruvic transaminase (GPT-1; E.C. 2.6.1.2) found in liver and, in lesser amounts, in heart, stomach, brain, skin, and several other tissues. Three alleles are known: *Gpt-1ᵃ* determines a pattern of four distinct bands migrating with intermediate speed, and occurs in many strains including C57BL/6, BALB/c, and DBA/2; *Gpt-1ᵇ* determines a pattern of four slower migrating bands and occurs in strains MA/J and NZB/Bl and in some *M. m. molossinus* and *M. m. castaneus*; *Gpt-1ᶜ* determines a pattern of four fast migrating bands and occurs in some *M. m. molossinus* and some *M. m. castaneus*. Heterozygotes show a pattern of five or six distinct bands (1,2). In liver the enzyme activity is low at birth and rises dramatically between 12 and 19 days to the adult level (1). The enzyme controlled by the locus is the soluble, not the mitochondrial, form (2).

References

1. Chen, S.H., R.P. Donahue, and C.R. Scott. 1973. The genetics of glutamic-pyruvic transaminase in mice: inheritance, electrophoretic phenotypes, and postnatal changes. Biochem. Genet. 10:23–28.
2. Eicher, E.M., and J.E. Womack. 1977. Chromosomal location of soluble glutamic-pyruvic transaminase-1 (*Gpt-1*) in the mouse. Biochem. Genet. 15:1–8.

gr, grizzled, recessive, Chr 10

Discovered in 1948 in descendants of a cross between the A strain and a stock carrying *fz*. Homozygotes resemble *cᶜʰ/cᶜʰ* mice, but the yellow pigment in the hair is more diluted and the black pigment is unaffected. In combination with nonagouti the hair on the ears and genitalia is white instead of yellow. Homozygotes are about 20 to 30 per cent smaller than their normal littermates at birth and about 10 per cent smaller as adults. Some have kinky tails, the degree of kinkiness being very variable. Viability up to the time of classification is about 50 to 60 per cent of normal, and death occurs at all stages from about 10 days of gestation on. The cause of the mortality is not known (1).

References

1. Bloom, J.L., and D.S. Falconer. 1966. "Grizzled," a mutant in linkage group X of the mouse. Genet. Res. 7:159–167.

Gr-1 locus, glutathione reductase-1, Chr 8

This locus controls electrophoretic variation in glutathione reductase (E.C. 1.6.4.2) as assayed in kidney samples. It is present also in erythrocytes, liver, and brain. The allele *Gr-1ᵃ* determines a fast anodally migrating prominent band with one or two subbands on either side and low enzyme activity; it occurs in strains C57BL/6, CE, and many others and in some wild mice in Ontario. The allele *Gr-1ᵇ* determines a slower migrating prominent band and high enzyme activity; it occurs in strains SJL, SWR, and others. Heterozygotes have both parental prominent bands plus a band of equal prominence between them, evidence that the enzyme is a dimer (1, 2).

References

1. Firth, S., J. Peters, and G. Bulfield. 1979. Activity and electrophoretic mobility of glutathione reductase allozymes in different tissues of the mouse. Biochem. Genet. 17:229–232.
2. Nichols, E.A., and F.H. Ruddle. 1975. Polymorphism and linkage of glutathione reductase in *Mus musculus*. Biochem. Genet. 13:323–329.

Grl-1 locus, glucocorticoid receptor-1, Chr 18

This locus is probably the structural gene for the glucocorticoid receptor present in the S49.1 myeloma cell line. No genetic variants are known in intact mice. However, the lymphoma cell line EL4 does not possess this receptor. The S49.1 receptor is recognized by sensitivity to dexamethazone. Hybrids between the two cell lines were sensitive. Segregants resistant to dexamethazone were isolated from the hybrids and all found to be lacking a specific Chr 18 from S49.1 that was recognizable by a structural difference from the Chr 18 of EL4, thus demonstrating that the *Grl-1* locus is on Chr 18 (1).

References

1. Francke, U., and U. Gehring. 1980. Chromosome assignment of a mouse glucocorticoid receptor gene (*Grl-1*) using intraspecific somatic cell hybrids. Cell 22:657–664.

Gs, greasy, semidominant, Chr X

Discovered among later descendants of an irradiated male but is probably of spontaneous origin. *Gs/+* resembles *Ta/+*. Homozygous females and hemizygous males have shiny fur like the corresponding *Ta* genotypes but lack the bare patches behind the ears, the dark middorsal stripe in agouti mice, and the characteristic sticky feel of the tail (1). *Gs* is very close to *Ta* on the X chromosome, but a few crossovers have occurred, showing that the two genes are not alleles (2).

References

1. Larsen, M.M. 1964. Mouse News Lett. 30:47.
2. Russell, L.B., and M.M. Larsen. 1965. Mouse News Lett. 33:69.

Gsa locus

See *Gia* locus.

Gsta complex, glutathione S-transferase Ya subunit gene family, Chr 9

This gene family consists of at least four genes coding for the Ya subunit of glutathione S-transferase. No genetic variants are known. The genes were found to be on Chr 9 by use of mouse–Chinese hamster somatic cell hybrids and a DNA probe of Ya, the major glutathione S-transferase subunit of rat liver. At least some of the genes are clustered. The glutathione S-transferases of mouse liver have not been extensively characterized, but in rat liver, there are at least six, and all are dimers of four subunits, Ya, Yb^1, Yb^2, and Yc. The glutathione S-transferases constitute a tissue protection system by catalyzing conjugation of glutathione with a variety of compounds, most of which derive from oxidative metabolism of xenobiotics (1).

References

1. Czosnek, H., S. Sarid, P.E. Barker, F.H. Ruddle, and V. Daniel. 1984. Glutathione S-transferase Ya subunit is coded by a multigene family located on a single mouse chromosome. Nucl. Acids Res. 12:4825–4833.

gt, gray tremor, recessive, Chr 15

Arose in strain HYIII/Le in 1977. Homozygotes have a reduced amount of yellow pigment in the coat, a white blaze on the head, and white feet, tail, and belly. They have a marked tremor beginning at 8 days of age and are subject to convulsions. Most homozygotes die by 3 months of age, but on a heterogeneous genetic background, many survive and reproduce. Histological examination of the nervous system reveals intense vacuolation in most areas of the CNS gray matter, vacuolation and hypomyelination in many CNS white matter tracts, very deficient myelin formation of nerves at dorsal and ventral root entry zones, and slow rate of myelination in the peripheral nervous system generally (2). Vacuolation is present in the first prenatal week and spreads by the second month to involve gray and white matter throughout the neuraxis. The basic defect may be a membrane disorder (1). Heterozygotes resemble homozygous wild-type mice except that they develop mild vacuolation from about 7 weeks onward in the same gray matter areas as those affected in *gt/gt* mice. The disorder closely resembles the spongiform encephalopathies caused by slow viruses. Possible transmissibility of the disease was indicated by the occurrence of spongiform encephalopathy in all seven mice of three inbred strains that survived to 22 to 24 months of age after neonatal intracerebral inoculation with *gt/gt* brain homogenates. The *gt* locus is on Chr 15 about 21 cM from *Ca* (2).

References

1. Kinney, H.C., and R.L. Sidman. 1986. Pathology of the spongiform encephalopathy in the *gray tremor* mutant mouse. J. Neuropathol. Exp. Neurol. 45:108–126.
2. Sidman, R.L., H.C. Kinney, and H.O. Sweet. 1985. Transmissible spongiform encephalopathy in the gray tremor mutant mouse. Proc. Natl. Acad. Sci. USA 82:253–257.

Gt-1, *-2* loci, teratocarcinoma graft antigen-1, -2, Chr 17

These two loci have been postulated to control strong transplantation antigens on embryonal teratocarcinoma (EC) cells that are demonstrated in grafts of the tumors into allogeneic mice. The antigens are present also on adult tissues (1). Several alleles have been described for *Gt-1*: *Gt-1^{sv}* occurs in the 129/Sv and C57BL/6 strains (2); *Gt-1^a* occurs in the A strain (4); the alleles in BALB/c and C3H/He differ from *Gt-1^{sv}* and may or may not be identical with each other (5); the allele in A.BY/SnJ differs from *Gt-1^a* and *Gt-1^{sv}* (1). The loci are near *H-2* and were at first thought to be outside the *H-2* complex, *Gt-1* on the proximal side and *Gt-2* on the distal side (2, 4). However, Johnson *et al.* (1) concluded that there was no convincing evidence for recombination with *H-2* and that the two loci probably lie within the *H-2* complex, one at the *K* end and one at the *D* end. More recently, Moser *et al.* (3) have shown that mice bearing mutations of *H-2K^b* reject grafts of EC cells of the *H-2K^b* genotype and have concluded that the *Gt* specificity probably resides in the H-2K molecule. A weaker specificity may be associated with H-2D.

References

1. Johnson, L.L., L.J. Clipson, W.F. Dove, J. Feilbach. L.J. Maher, and A. Shedlovsky. 1983. Teratocarcinoma trans-

plantation antigens are encoded in the *H-2* region. Immunogenetics 18:137–145.

2. Levine, A.J., and A.K. Teresky. 1981. Teratocarcinoma transplantation rejection loci: genetic localization of the *Gt-1* locus on chromosome 17 and the expression of alternate alleles. Immunogenetics 13:405–412.

3. Moser, A.R., A. Shedlovsky, and L.L. Johnson. 1986. H-2 class I and Gt(H-2) antigens are identical: evidence from H-2 mutant mice. Immunogenetics 23:271–273.

4. Shedlovsky, A., L.J. Clipson, J.L. VandeBerg, and W.F. Dove. 1981. Strong teratocarcinoma transplantation loci, *Gt-1* and *Gt-2*, flank *H-2*. Immunogenetics 13:413–419.

5. Siegler, E.L., N. Tick, A.K. Teresky, M. Rosenstraus, and A.J. Levine. 1979. Teratocarcinoma transplantation rejection loci: an *H-2*-linked tumor rejection locus. Immunogenetics 9:207–220.

Gur locus

See *Gus* complex.

Gus complex, Chr 5

This complex contains structural and regulatory genes for β-glucuronidase (GUS; E.C. 3.2.1.31). It includes four loci which have not shown confirmed recombination in crosses. Three haplotypes are known in inbred strains: $[Gus]^a$ ($Gus\text{-}s^a r^a t^a u^b$) in type strain A/J; $[Gus]^b$ ($Gus\text{-}s^b r^b t^a u^b$) in type strain C57BL/6; and $[Gus]^h$ ($Gus\text{-}s^h r^h t^h u^h$) in type strain C3H/He. Several other haplotypes have been found in wild mice (7, 10). $[Gus]^a$ differs from the other two haplotypes by a *Pst*I restriction fragment pattern (14). A recessive mutation at the *Gus* locus, described below, causes a condition similar to human mucopolysaccharidosis type VII.

Gus-s locus (formerly *g*, *Gus*), β-glucuronidase-structural. This locus was first recognized by a difference in activity level of β-glucuronidase and later found to affect heat stability (4) and electrophoretic mobility (2) of the enzyme. Three alleles are known. $Gus\text{-}s^a$ determines heat stability and fast anodal migration; it occurs in strains A/J, BALB/c, LG/J and others. $Gus\text{-}s^b$ determines heat stability and slow anodally migrating electrophoretic pattern; it occurs in C57BL/6 and other members of the C57 family of strains and in numerous other strains. $Gus\text{-}s^h$ determines heat lability and slow anodal migration; it occurs in strains C3H/He, CBA/J, and AKR/J (3, 9, 11). Other alleles have been found in wild mice (10). Heterozygotes have intermediate electrophoretic pattern and heat stability. The structural differences between alleles do not alter the catalytic properties of GUS either in substrate affinity or catalytic efficiency per molecule (11). The differences among the enzymes determined by the three alleles are restricted to a few amino acids of the polypeptide chain (4). Glucuronidase is found in many different tissues and at all ages including early embryos, and the structural type of the enzyme present is determined by the *Gus-s* genotype (9). In heterozygotes, the paternal gene becomes effective beginning at 57 hours post-coitum (15). Within cells, glucuronidase is found both in lysosomes and in membranes of the endoplasmic reticulum (microsomes). The enzymes at the two sites are affected identically by different *Gus-s* alleles. However, the relative amount of enzyme at the two sites is affected by the the $Gus\text{-}s^h$ allele. The localization of glucuronidase in the microsomal fraction is also affected by the *Eg* locus (9).

Gus-r locus (formerly *Gur*), β-glucuronidase-regulator. This locus controls the rate of induction by androgens of β-glucuronidase in kidney proximal tubule cells; the effect is confined to these cells and to the lysosomal form of the enzyme (2). The allele $Gus\text{-}r^a$ determines high inducibility and occurs in strains A/J, BALB/c, SM/J, and others; the allele $Gus\text{-}r^b$ determines intermediate inducibility and occurs in strains C57BL/6, DBA/2, SWR, and others; low inducibility occurs in strain C3H/He (13, 14), but it is not certain that this is an effect of the *Gus-r* locus. Heterozygotes have intermediate inducibility. In backcrosses and F2 from a cross of C57BL/6 to A, two apparent recombinants between *Gus-r* and *Gus-s* were found (0.7 per cent), but it was not possible to confirm that they were true crossovers. In heterozygotes the *Gus-r* allele affects the induced rate of synthesis only of the glucuronidase coded for by the *Gus-s* allele on the same chromosome (12). The *Gus-r* locus acts by regulating the rate of production of mRNA after induction (14). In the C3H/He strain ($[Gus]^k$), GUS mRNA is induced at the same rate as in C57BL/6 ($[Gus]^b$) and the low inducibility of the enzyme is presumably due to reduced translation capacity of the mRNA (14).

Gus-t locus (formerly *Gut*), β-glucuronidase-temporal. The developmental pattern of glucuronidase activity in liver is different in mice of $Gus\text{-}s^h$ strains from that in mice of other strains. In liver of C3H/He mice ($Gus\text{-}s^h$), enzyme activity level rises to a peak at about 10 to 12 days and falls off rapidly thereafter to a lower level at 30 days; in liver of strains with other *Gus-s* alleles, enzyme activity rises to a much higher peak at 10 days and remains high. This difference in developmental pattern is attributed to variation at a temporal locus, *Gus-t*, with alleles $Gus\text{-}t^h$ determining the pattern associated with $Gus\text{-}s^h$, and $Gus\text{-}t^a$ determining the pattern associated with the other two *Gus-s* alleles.

Heterozygotes have an intermediate pattern (5). The low level of glucuronidase determined by *Gus-t^h* is due to decreased synthesis of the enzyme (9). The action of *Gus-t* is confined to liver cells (8). Meredith and Ganschow (6) have shown that in *Gus-s^{b}t^a/Gus-s^{h}t^h* heterozygotes the products of the two haplotypes, as determined by heat stability, are synthesized at about the same rate, indicating that the temporal regulation is not *cis*-acting, but affects the two structural alleles equally.

Gus-u locus, *β*-glucuronidase-systemic regulator. This locus influences the systemic activity level of *β*-glucuronidase. An allele *Gus-u^c* found in *M. m. castaneus* reduces the activity level in all organs studied to approximately half that in the strains carrying the *Gus-u^b* allele, C57BL/6, DBA/2, A, and BALB/c. The androgen-inducible rate of synthesis in the kidney is also lowered by one-half (7, 10). The allele *Gus-u^h* in strain C3H/He reduces activity level to about 30 per cent that in DBA/2 and C57BL/6. Heterozygotes have intermediate activity (8).

gus^{mps}, mucopolysaccharidosis type VII, recessive. Arose spontaneously in the C57BL/6By strain and was originally called adipose storage deficiency (*asd*). Homozygotes are essentially devoid of visible adipose tissue except for brown fat. They are smaller than normal and their bones are shorter and thicker than normal. Males are sterile but have apparently normal reproductive tracts. About half the females produce one or more litters, but do not lactate adequately. Affected mice die between 150 and 200 days (1). The mutation is a null allele at the *Gus-s* locus. Homozygotes lack *β*-glucuronidase activity and produce no glucuronidase mRNA. In Southern blots, the gene appears normal, indicating that the mutation is not a large deletion. *gus^{mps}* appears to be an exact model for human mucopolysaccharidosis type VII (E.H. Birkenmeier and M.T. Davisson, personal communication).

References

1. Beamer, W.G., and D.L. Coleman. 1982. Mouse News Lett. 67:21.
2. Dofuku, R., U. Tettenborn, and S. Ohno. 1971. Further characterization of *o^s* mutation of mouse *β*-glucuronidase locus. Nature New Biol. 234:259–261.
3. Lalley, P.A., and T.B. Shows. 1974. Lysosomal and microsomal glucuronidase: genetic variant alters electrophoretic mobility of both hydrolases. Science 195:442–444.
4. Lusis, A.J., and K. Paigen. 1979. The large scale isolation of mouse *β*-glucuronidase and comparison of allozymes. J. Biol. Chem. 253:7336–7345.
5. Lusis, A.J., V.M. Chapman, R.W. Wangenstein, and K. Paigen. 1983. *Trans*-acting temporal locus within the *β*-glucuronidase gene complex. Proc. Natl. Acad. Sci. USA 90:4398–4402.
6. Meredith, S.A., and R.E. Ganschow, 1978. Apparent *trans* control of murine *β*-glucuronidase synthesis by a temporal element. Genetics 90:715–734.
7. Paigen, K. 1979. Acid hydrolases as models of genetic control. Ann. Rev. Genet. 13:417–466.
8. Paigen, K., and A. Jakubowski. 1985. Cell specificity in the developmental regulation of acid hydrolases by temporal genes. Dev. Genet. 5:83–91.
9. Paigen, K., R.T. Swank, S. Tomino, and R.E. Ganschow. 1975. The molecular genetics of mammalian glucuronidase. J. Cell. Physiol. 85:379–392.
10. Pfister, K., V. Chapman, and K. Paigen. 1981. A systemic regulatory locus associated with two new structural alleles of murine *β*-glucuronidase in *M. musculus castaneus*. Hereditas 94:18 (Abstr.).
11. Pfister, K., K. Paigen, G. Watson, and V. Chapman. 1982. Expression of *β*-glucuronidase haplotypes in prototype and congenic mouse strains. Biochem. Genet. 20:519–536.
12. Swank, R.T., K. Paigen, and R.E. Ganschow. 1973. Genetic control of glucuronidase induction in mice. J. Mol. Biol. 81:225–243.
13. Watson, G., R.A. Davey, C. Labarca, and K. Paigen. 1981. Genetic determination of kinetic parameters in *β*-glucuronidase induction by androgen. J. Biol. Chem. 256:3005–3011.
14. Watson, G., M. Felder, L. Rabinow, K. Moore, C. Labarca, C. Tietze, G. Vander Molen, L. Bracey, M. Brabant, J. Cai, and K. Paigen. 1985. Properties of rat and mouse *β*-glucuronidase mRNA and cDNA, including evidence for sequence polymorphism and genetic regulation of mRNA levels. Gene 36:15–25.
15. Wudl, L., and V. Chapman. 1976. The expression of *β*-glucuronidase during preimplantation development of mouse embryos. Dev. Biol. 48:104–109.

Gut locus

See *Gus* complex.

Gv-1 locus, Gross virus antigen-1

This locus and *Gv-2* control expression of the G$_{IX}$ antigen, a component of gp70, the major glycoprotein of the murine leukemia virus envelope. G$_{IX}$ is found on thymocytes, the positive alleles at both loci being required for expression of the antigen. At the *Gv-1* locus, two alleles are known, *Gv-1^a* determining presence of G$_{IX}$ and *Gv-1^b* determining its absence. Heterozygotes have half as much antigen as homozygotes for *Gv-1^a*. Many strains have been typed for G$_{IX}$ antigen. Of these it is known that *Gv-1^a* occurs in 129, CE, and AKR. Other strains that are G$_{IX}^+$ and therefore presumably *Gv-1^a* are SJL, DBA/2, C3H/An, 101, C58, A, and I. These strains show smaller amounts of the antigen on their thymocytes than 129 or CE. G_{IX}^- strains that are known to be *Gv-1^b* are C57BL/6, C57BR/cd, BALB/c, and CBA/J. Other G_{IX}^- strains are C57BL/10, RF/J, DBA/2, MA/J, and SWR (4, 5). In linkage crosses in which strain AKR is the source of the *Gv-1^a* allele, *Gv-1* behaves as if linked to *Gpd-1* on Chr 4 (1). In crosses in which strain 129 is the source of the *Gv-1^a* allele, *Gv-1* behaves as if linked to both *Gpd-1* and *H-2* on Chr 17 (5). It is now thought that both these linkages are spurious and due to the effects of the *Fv-1* (Chr 4) and *H-2* (Chr 17) loci in these crosses (5). The true chromosomal location of *Gv-1* is as yet unknown.

The G_{IX} antigen is located mostly on thymocytes in the low leukemia 129 strain except in old age. However, productive infection with murine leukemia virus causes expression on other lymphatic cells and leukemic cells in both G_{IX}^+ and G_{IX}^- strains. In strain 129 mice, G_{IX}-gp70 occurs in serum and in various epithelia associated with the digestive tract and the male reproductive tract (3). By use of strain 129 (*Gv-1^a*) and a 129 congenic strain carrying the *Gv-1^b* allele, it was shown that the *Gv-1* locus regulates the expression of multiple endogenous xenotropic retroviral sequences present in strain 129. The viral transcripts are derived from distinct proviruses, many exhibiting extensive deletions, and they are regulated in a tissue-specific manner. The multiple viral antigens produced in mice carrying *Gv-1^a* (and *Gv-2^a*) give rise to the G_{IX}^+ phenotype (2).

References

1. Ikeda, H., E. Stockert, W.P. Rowe, E.A. Boyse, F. Lilly, H. Sato, and L.J. Old. 1973. Relation of chromosome 4 (Linkage Group VIII) to murine leukemia virus-associated antigens in AKR mice. J. Exp. Med. 137:1103–1107.
2. Levy, D.E., R.A. Lerner, and M.C. Wilson. 1985. The Gv-1 locus coordinately regulates the expression of multiple endogenous murine retroviruses. Cell 41:289–299.
3. Obata, Y., E. Stockert, M. Yamaguchi, and E.A. Boyse. 1978. Source and hormone-dependence of G_{IX}-gp70 in mouse serum. J. Exp. Med. 148:793–798.
4. Stockert, E., L.J. Old, and E.A. Boyse. 1971. The G_{IX} system. A cell surface allo-antigen associated with murine leukemia virus: implications regarding chromosomal integration of the viral genome. J. Exp. Med. 133:1334–1353.
5. Stockert, E., E.A. Boyse, H. Sato, and K. Itakura. 1976. Heredity of the G_{IX} thymocyte antigen associated with murine leukemia virus: segregation data simulating genetic linkage. Proc. Natl. Acad. Sci. USA 73:2077–2081.

Gv-2 locus, Gross virus antigen-2, Chr 7

This locus and *Gv-1* control expression of the G_{IX} antigen on thymocytes, the positive alleles at both loci being required for expression of the antigen. At the *Gv-2* locus the positive allele (*Gv-2^a*) is fully dominant over the negative allele (*Gv-2^b*). G_{IX} positive strains known to carry *Gv-2^a* are 129, AKR, and CE. G_{IX} negative strains known to carry *Gv-2^b* are C57BL/6 and C57BR/cd (2, 4, 5). Because of full dominance at the *Gv-2* locus and lack of a strain known to carry *Gv-1^a* and *Gv-2^b*, the assignment of *Gv-2* to Chr 7 is based on results of only one cross in which *Gv-1* was also segregating, and is thus not too firmly established (3). However, *Gv-2* closely resembles *Sgp-2* which is known to be linked to *Hbb* on Chr 7 and the two loci may be identical (1). This supports the conclusion that *Gv-2* is on Chr 7. See *Sgp-2* locus.

References

1. Maruyama, N., C.O. Lindstrom, H. Sato, and F.J. Dixon. 1983. Serum gp70 production regulated by a gene on murine chromosome 7. Immunogenetics 18:365–371.
2. Stockert, E., L.J. Old, and E.A. Boyse. 1971. The G_{IX} system. A cell surface allo-antigen associated with murine leukemia virus; implications regarding chromosomal integration of the viral genome. J. Exp. Med. 133:1334–1355.
3. Stockert, E., H. Sato, K. Itakura, E.A. Boyse, L.J. Old, and J.J. Hutton. 1972. Location of the second gene required for expression of the leukemia-associated mouse antigen G_{IX}. Science 178:862–863.
4. Stockert, E., E.A. Boyse, Y. Obata, H. Ikeda, N. Sarkar, and H.A. Hoffman. 1975. New mutant and congenic mouse stocks expressing the murine leukemia virus-associated thymocyte surface antigen G_{IX}. J. Exp. Med. 142:512–517.
5. Stockert, E., E.A. Boyse, H. Sato, and K. Itakura. 1976. Heredity of the G_{IX} thymocyte antigen associated with murine leukemia virus: segregation data simulating genetic linkage. Proc. Natl. Acad. Sci. USA 73:2077–2081.

Gy, gyro, semidominant, Chr X

Found among offspring of an irradiated female. Hemizygous males have hypophosphatemia, rickets/osteomalacia, circling behavior, inner ear abnormalities, and sterility; heterozygous females have milder symptoms. *Gy* is located close (0.4 to 0.8 per cent recombination) to the X-linked gene *Hyp* that also causes hypophosphatemia. The two genes have very similar effects on serum phosphorus levels and on phosphate transport, but *Hyp* does not affect the inner ear (1). See *Hyp*.

References

1. Lyon, M.F., C.R. Scriver, L.R.I. Baker, H.S. Tenenhouse, J. Kronick, and S. Mandla. 1986. The Gy mutation: another cause of X-linked hypophosphatemia in mouse. Proc. Natl. Acad. Sci. USA 83:4899–4903.

H

H-1 locus

See after *H-2* complex.

H-2 complex, histocompatibility-2, Chr 17

The *H-2* genes are part of the major histocompatibility complex (MHC) of the mouse. The complexities of the system are too great to be adequately described here, but they are well covered in reviews by Bell *et al.* (2), Campbell *et al.* (3), Flavell *et al.* (7), Hansen *et al.* (9), Hood *et al.* (11), Klein (12), Nathenson *et al.* (18), Schwartz (22), and Snell *et al.* (21). Lists of standard *H-2* haplotypes with their strain distribution, *H-2* congenic strains including strains congenic for recombinant and mutant haplotypes, alleles of the loci within each haplotype, class I and class II antigenic specificities, etc., are given by Klein *et al.* (14) and Murphy (17). Similar lists of the complement loci of the *H-2S* region, as well as other *H-2*-associated loci are given in Klein *et al.* (13). Because of the great complexity of the *H-2* complex, special rules have been devised for its nomenclature. These are summarized elsewhere in this volume. The number of mutant and recombinant haplotypes are now so numerous that new haplotype superscript symbols are assigned through a registry (currently maintained by Dr Peter Démant, The Netherlands Cancer Institute, Plesmanlaan 121, 1066 CX Amsterdam, The Netherlands).

The MHC of the mouse is a multigene cluster containing three major classes of genes, class I (located in the *H-2K*, *H-2D*, *Qa*, and *Tla* regions), class II (in the *H-2I* region), and class III (in the *H-2S* region). The *Qa* and *Tla* genes are described elsewhere in this catalog. A number of genes controlling susceptibility to virus-induced leukemia also map to the MHC; these loci have been assigned other names and are described elsewhere in this catalog; see *Rfv-1*, *Rfv-2*, *Rgv-1*, *Rrs*, and *Rv-1*. The MHC regions are located in the order from the centromere, *H-2K*, *H-2I*, *H-2S*, *H-2D*, *Qa*, *Tla*; they span approximately 2 cM of genetic length. Most class I gene products are tightly bound to the cell surface membrane and are heterodimers consisting of a heavy chain of 45 kDa associated non-covalently with a light chain of 11-kDa (β2-microglobulin, a product of the *B2m* locus on Chr 2). The class II gene products (Ia molecules) are also tightly bound to the cell surface and are heterodimers consisting of an α-chain of 35 kDa and a β-chain of 29 kDa. Antigens determined by the class I genes of the H-2K and H-2D regions are found on practically all cells except in very early embryos. They function in cytolytic immune responses. Allogeneic differences at these loci induce vigorous graft rejection and strong primary *in vitro* cytotoxic responses. In responses against non-*H-2* or viral antigens, the receptor of cytotoxic T-cells recognizes antigen only in association with a class I molecule. This phenomenon is termed *H-2* restriction. Products of class II genes of the *H-2I* region, unlike the class I gene products, are restricted primarily to B lymphocytes, macrophages, and dendritic cells. The antigen-specific receptors of helper T-cells that are required for the generation of cytotoxic T-cells and for antibody production by B-cells recognize the foreign antigen only when it is associated with class II (Ia) molecules. Class III genes determine several complement components and are diverse in structure.

The class I and class II genes of the *H-2* region are highly polymorphic, the class III genes much less so. Mutations in the *H-2K* and *H-2D* regions are thought to have been generated by a gene conversion mechanism involving exchange of sequences, in most cases with class I genes of the *Qa* region (18, 23). Such mechanisms may have played a role in generating the high degree of polymorphism seen in the *H-2K* and *H-2D* regions.

Following is a brief summary of information on the regions of the *H-2* complex. In most cases the information is derived from the sources cited above, and they are usually not cited again. A few original sources not cited in the reviews above are given. The regions are listed in the order of their arrangement on Chr 17 beginning at the centromeric end, even though this does not correspond with their functional grouping. The antigenic loci within the complex were mapped with the aid of a series of recombinant haplotypes and the order has been more recently confirmed by molecular DNA mapping.

K. This region contains the *H-2K* locus which determines antigens detectable serologically and by graft rejection. For reviews, see Hansen *et al.* (9) and Hood *et al.* (11). In most cases alleles are codominant. The H-2K antigen is the principal product of this region, but a second product of the *H-2K^d* region, designated K^l, has been described (25). *H-2K* differences can cause graft-vs-host reactions and they can also cause mixed lymphocyte reactions and cell-mediated lysis when lymphocytes of different *H-2K* genotype are cultured together. The DNA of the *K* region of the C57BL/10 and BALB/c strains is known to contain two genes, one *H-2K* and the other designated *K1* (8). Many mutations at the *H-2K* locus have been found; their structure is described by Nathenson *et al.* (18).

I. The literature on this region is well reviewed by Bell *et al.* (2), Flavell *et al.* (8), Hansen *et al.* (9), Hood *et al.* (11), Murphy (17), and Schwartz (22). The I region was originally defined as containing immune response (*Ir-1*) loci that determined the level of response to many antigens, including synthetic polypeptides, ovalbumin, thyroglobulin, immunoglobulins, and alloantigens, and probably susceptibility to viral oncogenesis. The *I* region has been subdivided into several subregions, *I-A*, *I-B*, *I-J*, *I-E*, and *I-C*, by genetic mapping of immune responses and by serological analysis of *I* region-encoded alloantigens (Ia antigens) in congenic strains carrying recombinant *H-2* haplotypes. The *I-A* and *I-E* subregions were serologically defined and code for Ia antigens. These two class II molecules each consist of an α-and a β-chain. Analysis of Ia antigens in *H-2* recombinant strains has demonstrated that the A_α, A_β, and E_β chains are encoded in the *I-A* subregion and the E_α chain is encoded in the *I-E* subregion. Each of the β-chains is encoded by two distinct but adjacent genes, each of the α-chains by only one gene. In addition, the region contains a β pseudogene for both the I-A and I-E molecules. Polymorphisms of the *I-A* and *I-E* genes at the molecular level are reviewed by Bell *et al.* (2). The *I-B* and *I-C* regions were defined as immune response loci; the *I-J* locus was defined by antibodies against determinants unique to suppressor T-cells. No structural genes for I-B, I-J, and I-C have been identified. The *I-B* subregion can be explained by interaction between *I-E*-restricted suppressor cells and *I-A*-restricted helper cells. The *I-C* subregion can be explained by complementation between two immune response genes, one mapping in the *I-A* subregion. Although specific anti-I-J antibodies identify determinants unique to suppressor T-cells, it has not been possible to characterize the I-J molecule. Restriction site polymorphisms and DNA sequence analysis of *H-2* recombinant mice localized the I-J subregion to a 1-kb region within the E_β gene, but the two *H-2* recombinant strains used to define the *I-J* subregion were found

to have identical DNA sequences in this region (15). Hayes *et al.* (10) have presented evidence that I-J is encoded by a gene on Chr 4 (*Jt-1*). The Ia molecules may function as immune response genes by serving as restricting elements for helper T-cells, although the exact mechanism is not clear. The E_β gene contains one of several hotspots of recombination that have been localized in the *H-2* complex (15, 26).

S. This region contains the structural loci for the complement components C2, C4 (formerly Ss), Slp, and Bf. The organization and function of genes in the region are well reviewed by Campbell *et al.* (3), Chaplin (5), Nonaka *et al.* (19), Ogata (20), and Tosi *et al.* (24). C4 and Slp are 200 kDa proteins composed of three subunits: an α-chain of 100 kDa, a β-chain of 70 kDa, and a γ-chain of 30 kDa. C4 is a functional complement component but the function of Slp is unknown. C2 and Bf are 100 kDa and 92 kDa single-chain glycoproteins, respectively, and are analogous in structure and function. C2 functions in the classical pathway of complement activation and Bf functions in the alternative pathway. These four genes are expressed in liver as soluble proteins that are secreted into the plasma, and they are also expressed in macrophages. They have been mapped by molecular methods in the order *C2, Bf, Slp, C4*. Two or three additional *Slp* genes have been reported in wild-derived strains. The region also contains two steroid 21-hydroxylase genes which are located one each in association with *Slp* and *C4*. See *Oh21-1, -2* loci.

C2 variants are recognized by micropeptide mapping and as DNA polymorphisms; *Bf* variants are recognized by electrophoretic mobility, hemolytic activity, and as DNA polymorphisms; and *Slp* and *C4* variants were recognized by specific antibodies and peptide mapping. *Slp* and *C4* show very limited DNA sequence variation. *Slp* expression is in most cases androgen-dependent. Alleles may determine either absence of Slp protein (*Slp^o*), sex-limited expression (*Slp^a*), or constitutive expression in both males and females (*Slp^c*). Alleles of *C4* determine amount of C4 protein formed, with a 20-fold difference between highest and lowest values. The genetic variants of these complement molecules have been described and are summarized by Atkinson *et al.* (1).

D. The major characteristics of this region are identical to those of the *H-2K* region. The two regions share some antigenic specificities but differ in many others. The *D* region contains the *H-2D* locus controlling the expression of most of the specificities of the *D* region, and, at least in some strains, the *H-2L* locus. The number of genes in the *H-2D* region appears to differ in different strains. There is one, *H-2D^b* in C57BL/6 and *H-2D^k* in AKR, and probably five in BALB/c, the most proximal determining the H-2D^d antigen and the most distal the H-2L^d antigen (23).

The *H-2* complex contains several hemopoietic histo-compatibility (*Hh*) loci (formerly called hybrid histo-compatibility). Loci controlling this trait have been reported at other locations on Chr 17, but their assignment is less certain. *Hh* incompatibility causes natural resistance of F1 hybrid mice to bone marrow or leukemia cells from the parental strains. The alleles do not act codominantly as in other histocompatibility systems, but appear to produce their antigenic products only when homozygous (6). F1 hosts may resist *Hh*-incompatible grafts even after irradiation, but this effect is controlled by non-*H-2* genes. *Hh-1* maps within the *H-2D* region and *Hh-3* maps within the H-2K region. It is not known whether the class I *H-2K* and *H-2D* gene products themselves serve as Hh antigens, but results with *H-2* mutants suggest that *Hh-1* may be identical with *H-2L* (16). Natural resistance against *Hh* incompatible cells is not due to conventional immune responses. Members of the natural killer (NK) cell family are involved in the rejection, but NK cells do not exhibit the immunogenetic specificity of the *Hh* system and cannot alone account for the rejection phenomenon (4).

Numerous other traits reported to be associated with the *H-2* complex are summarized by Klein *et al.* (13). In addition, a group of low molecular weight antigens are controlled by a gene that appears, by typing of *H-2* congenic and recombinant strains, to lie between the *H-2K* and *I-A* subregions. See *Lmp* locus.

References

1. Atkinson, J.P., D.R. Karp, E.P. Seeskin, C.C. Killion, P.A. Rosa, S.L. Newell, and D.C. Shreffler. 1982. H-2S region determined polymorphic variants of C4, Slp, C2, and B comple ment proteins: a compilation. Immunogenetics 16:617–623.

2. Bell, J.I., D.W. Denny Jr., and H.O. McDevitt. 1985. Structure and polymorphism of murine and human class II major histocompatibility antigens. Immunol. Rev. 84:51–71.

3. Campbell, R.D., M.C. Carroll, and R.R. Porter. 1986. The molecular genetics of components of complement. Adv. Immunol. 38:203–244.

4. Carlson, G.A., B.A. Taylor, S.T. Marshall, and A.H. Greenberg. 1984. A genetic analysis of natural resistance to nonsyngeneic cells: the role of *H-2*. Immunogenetics 20:287–300.

5. Chaplin, D.D. 1985. Molecular organization and *in vitro*

6. Cudkowicz, G., and I. Nakamura. 1983. Genetics of the murine hemopoietic-histocompatibility system: an overview. Transplant. Proc. 15:2059–2063.

7. Flavell, R.A., H. Allen, B. Huber, C. Wake, and G. Widera. 1985. Organization and expression of the MHC of the C57Black/10 mouse. Immunol. Rev. 84:29–50.

8. Flavell, R.A., H. Allen, L.C. L.C. Burkly, D.H. Sherman, G.L. Waneck, and G. Widera. 1986. Molecular biology of the H-2 histocompatibility complex. Science 233:437–443.

9. Hansen, T.H., D.G. Spinella, D.R. Lee, and D.C. Shreffler. 1984. The immunogenetics of the mouse major histocompatibility gene complex. Ann. Rev. Genet. 18:99–129.

10. Hayes, C.E., K. Klyczek, D.P. Krum, R.M. Whitcomb, D.A. Hullett, and H. Cantor. 1984. Chromosome 4 *Jt* gene controls murine cell surface I-J expression. Science 223:559–563.

11. Hood, L., M. Steinmetz, and B. Malissen. 1983. Genes of the major histocompatibility complex of the mouse. Ann. Rev. Immunol. 1:529–568.

12. Klein, J. 1975. Biology of the mouse histocompatibility-2 complex. Springer-Verlag, New York. 620 p.

13. Klein, J., F. Figueroa, and D. Klein. 1982. *H-2* haplotypes, genes, and antigens: second listing. I. Non-*H-2* loci on chromosome 17. Immunogenetics 16:285–317.

14. Klein, J., F. Figueroa, and C.S. David. 1983. *H-2* haplotypes, genes, and antigens: second listing. II. The *H-2* complex. Immunogenetics 17:553–596.

15. Kobori, J.A., E. Strauss, K. Minard, and L. Hood. 1986. Molecular analysis of the hotspot of recombination in the murine major histocompatibility complex. Science 234:173–179.

16. Morgan, G.M., and I.F.C. McKenzie. 1981. Implication of the *H-2L* locus in hybrid histocompatibility (*Hh-1*). Transplantation 31:417–422.

17. Murphy, D.B. 1986. Overview: the murine MHC. *In* D.M. Weir, ed., Handbook of Experimental Immunology, vol 3, Genetic and Molecular Immunology, 100.1–100.19. Blackwell Scientific Publ., Oxford.

18. Nathenson, S.G., J. Geliebter, G.M. Pfaffenbach, and R.A. Zeff. 1986. Murine major histocompatibility complex class-I mutants: molecular analysis and structure-function implications. Ann. Rev. Immunol. 4:471–502.

19. Nonaka, M., K. Nakayama, Y.D. Yeul, A. Shimizu, and M. Takahashi. 1985. Molecular cloning and characterization of complementary and genomic DNA clones for mouse C4 and SLP. Immunol. Rev. 87:81–99.

20. Ogata, R.T. 1985. Structure and expression of murine fourth complement component (C4) and sex-limited protein (Slp). Immunol. Rev. 87:101–122.

21. Snell, G.D., J. Dausset, and S. Nathenson. 1976. Histocompatibility. Academic Press, New York. 401 p.

expression of murine class III genes. Immunol. Rev. 87:61–80.

22. Schwartz, R.H. 1986. Immune response (*Ir*) genes of the murine major histocompatibility complex. Adv. Immunol. 38:31–201.

23. Stephan, D., H. Sun, K. Fischer Lindahl, E. Meyer, G. Hämmerling, L. Hood, and M. Steinmetz. 1986. Organization and evolution of the D region class I genes in the mouse major histocompatibility complex. J. Exp. Med. 163:1227–1244.

24. Tosi, M., M. Lévi-Strauss, S. Georgatsou, M. Amor, and T. Meo. 1985. Duplications of complement and noncomplement genes in the *H-2S* region: evolutionary aspects of the C4 isotypes and molecular analysis of their variants. Immunol. Rev. 87:151–183.

25. Tryphonas, M., D.P. King, and P.P. Jones. 1983. Identification of a second class I antigen controlled by the *K* end of the *H-2* complex and its selective cellular expression. Proc. Natl. Acad. Sci. USA 80:1445–1448.

26. Uematsu, Y., H. Kiefer, R. Schulze, K. Fischer-Lindahl, and M. Steinmetz. 1986. Molecular characterization of a meiotic recombinant hotspot enhancing homologous equal crossing-over. EMBO J. 5:2123–2129.

H-1, *H-3* to *H-13* loci

These loci were isolated and identified by Snell and his co-workers by development of congenic resistant strains in which histocompatibility gene differences were identified by resistance to transplantable tumors.

H-1 locus, histocompatibility-1, Chr 7. *H-1ᵃ* and *H-1ᵇ* were identified by the congenic strains C3H and C3H.K, respectively (13). At least six alleles are known (5,6). The antigen is known to be present on transplantable tumors, skin, thymocytes (7), macrophages (1), and many other tissues (10). Zink and Heyner (21) have prepared alloantibodies that recognize *H-1ᵃ* antigens on lymphocytes.

H-3 locus, histocompatibility-3, Chr 2. *H-3ᵃ* and *H-3ᵇ* were identified by the congenic strains C57BL/10 and B10.LP, respectively (13). At least six alleles are known (3, 4). The antigen is present on transplantable tumors, skin, thymocytes (7), bone marrow, liver, newborn heart (see Johnson *et al.*, 10 for references), cultured kidney cells (20), and possibly on macrophages (1), ova (12), and sperm (19). *H-3* allelic differences can also be recognized by cytotoxic T lymphocyte responses, and these are H-2D-restricted (11). *H-3* is closely linked on Chr 2 to two other loci controlling cell-surface antigens, *B2m* and *H-42* (8, 14).

H-4 locus, histocompatibility-4, Chr 7. *H-4ᵃ* and *H-4ᵇ* were identified by congenic strains C57BL/10 and B10.129(21M), respectively (16). At least three alleles are known. The antigen is present on transplantable tumors, thymocytes (7), bone marrow liver, and newborn heart (see Johnson *et al.*, 10 for references).

H-5 locus. See *Ea-5* locus.

H-6 locus. See *Ea-6* locus.

H-7 locus, histocompatibility-7, Chr 9. *H-7ᵃ* and *H-7ᵇ* were identified by congenic strains C57BL/10 and B10.C(47N), respectively (15). At least three alleles are known. The antigen is present on a transplantable leukemia, skin, thymocytes (7), liver, and heart (10).

H-8 locus, histocompatibility-8, Chr 14. *H-8ᵃ* and *H-8ᵇ* were identified by congenic strains C57BL/10 and B10.D2(57N), respectively (15). At least three alleles are known. The antigen is present on a transplantable leukemia, skin, thymocytes (7), heart, and testis (10). Zink and Heyner (20) were able to produce antibodies to H-8ᵃ antigen, which they used to detect antigens on cultured kidney cells and lymphocytes. Linkage of *H-8* to *Np-2* on Chr 14 was found by Dembić *et al.* (2).

H-9 locus, histocompatibility-9. *H-9ᵃ* and *H-9ᵇ* were identified by congenic strains C57BL/10 and B10.C(45N), respectively (15). At least three alleles are known. The antigen is present on a transplantable leukemia, skin, and thymocytes (7).

H-10 locus, histocompatibility-10. *H-10ᵃ* and *H-10ᵇ* were identified by congenic strains C57BL/10 and B10.129(9M), respectively (15). Two alleles are known. The antigen is present on a transplantable leukemia, skin, and thymocytes (7).

H-11 locus, histocompatibility-11. *H-11ᵃ* and *H-11ᵇ* were identified by congenic strains C57BL/10 and B10.129(10M), respectively. A third allele may be present in strain DBA/2 (15). The antigen is present on a transplantable leukemia, skin, and thymocytes (7).

H-12 locus, histocompatibility-12. *H-12ᵃ* and *H-12ᵇ* were identified by congenic strains C57BL/10 and B10.129(12M), respectively (18). At least three alleles are known. The antigen is present on a transplantable leukemia, skin, thymocytes, and bone marrow (9).

H-13 locus, histocompatibility-13, Chr 2. *H-13ᵃ* and *H-13ᵇ* were identified by congenic strains C57BL/10 and B10.129(14M), respectively (17). At least three alleles are known. The antigen is present on macrophages (1), ova (12), and sperm (19). Zink and Heyner (21) were able to produce antibodies that detected H-13ᵃ antigen on lymphocytes.

References

1. DeAngelis, W.J., and G. Haughton. 1971. Detection of weak histocompatibility loci on mouse peritoneal macrophages. Transplant. Proc. 3:202–206.

2. Dembić, W. Bannwarth, B.A. Taylor, and M. Steinmetz. 1985. The gene coding for the T-cell receptor α-chain maps close to the *Np-2* locus on mouse chromosome 14. Nature 314:271–273.

3. Gasser, D.L. 1976. Genetic studies of the H-3 region of mice and implications for polymorphism of histocompatibility loci. Immunogenetics 3:271–276.

4. Graff, R.J., and D.W. Bailey. 1973. The non-H-2 histocompatibility loci and their antigens. Transplant. Rev. 15:26–49.

5. Graff, R.J., and G.D. Snell. 1968. Histocompatibility genes of mice. VIII. The alleles of the *H-1* locus. Transplantation 6:598–617.

6. Graff, R.J., D.H. Brown, and G.D. Snell. 1979. Thirteen new chromosome-7 congenic lines. Immunogenetics 8:245–256.

7. Graff, R.J., W.H. Hildemann, and G.D. Snell. 1966. Histocompatibility genes of mice. VI. Allografts in mice congenic at various non-*H-2* histocompatibility loci. Transplantation 4:425–437.

8. Graff, R.J., D. Martin-Morgan, and M.E. Kurtz. 1985. Allograft rejection-defined antigens of the *B2m*, *H-3* region. J. Immunol. 135:2842–2846.

9. Harrison, D.E., and J.W. Doubleday. 1976. Marrow allograft success in *W/Wᵛ* anemic mice: relation to skin graft survival times. Immunogenetics 3:289–297.

10. Johnson, L.L., D.W. Bailey, and L.E. Mobraaten. 1981. Genetics of histocompatibility in mice. IV. Detection of certain minor (non-H-2) H antigens in selected organs by the popliteal node test. Immunogenetics 14:63–71.

11. Kurtz, M.E., R.J. Graff, A. Adelman, D. Martin-Morgan, and R.E. Click. 1985. CTL and serologically defined antigens of *B2m H-3* region. J. Immunol. 135:2847–2852.

12. Palm, J., S. Heyner, and R.L. Brinster. 1971. Differential immunofluorescence of fertilized mouse eggs with H-2 and non-H-2 antibody. J. Exp. Med. 133:1282–1293.

13. Snell, G.D. 1958. Histocompatibility genes of the mouse. II. Production and analysis of isogenic resistant lines. J. Natl. Cancer Inst. 21:843–877.

14. Snell, G.D., and H.P. Bunker. 1964. Histocompatibility genes of mice. IV. The position of *H-3* in the fifth linkage group. Transplantation 2:743–751.

15. Snell, G.D., and H.P. Bunker. 1965. Histocompatibility genes of mice. V. Five new histocompatibility loci identified by congenic resistant lines on a C57BL/10 background. Transplantation 3:235–252.

16. Snell, G.D., and L.C. Stevens. 1961. Histocompatibility genes of mice. III. H-1 and H-4, two histocompatibility loci in the first linkage group. Immunology 4:366–379.

17. Snell, G.D., G. Cudkowicz, and H.P. Bunker. 1967. Histocompatibility genes of mice. VII. H-13, a new histocompatibility locus in the fifth linkage group. Transplantation 5:492–503.

18. Snell, G.D., R.J. Graff, and M. Cherry. 1971. Histocompatibility genes of mice. XI. Evidence establishing a new histocompatibility locus, *H-12* and a new *H-2* allele, *H-2ᵇᶜ*. Transplantation 11:525–530.

19. Vojtisková, M., M. Poláčková, and Z. Pokorná. 1968. Histocompatibility antigens on mouse spermatozoa. Folia Biol. (Praha) 15:322–332.

20. Zink, G.L., and S. Heyner. 1977. The production of non-*H-2* histocompatibility antibodies in the mouse. Immunogenetics 4:257–266.

21. Zink, G.L., and S. Heyner. 1978. Further studies on non-*H-2* histocompatibility alloantibodies in the mouse. Immunogenetics 6:269–276.

H-14 locus

See *Ea-2* locus.

H-15 to *H-30* and *H-34* to *H-38* loci

These loci were all identified in congenic strains made by crossing *H*-locus alleles from BALB/cBy into C57BL/6By. Each has two known alleles, *b* in C57BL/6 and *c* in the particular B6.C congenic strain and in BALB/c. The antigens are known to be present on tail skin in all cases (1) and on various other tissues in *H-25*, *-29*, *-34*, *-36*, and *-38* (4). Except where noted, the chromosome locations were all determined by D.W. Bailey (personal communication).

H-15 locus, histocompatibility-15, Chr 4.

H-16 locus, histocompatibility-16, Chr 4.

H-17 locus, histocompatibility-17.

H-18 locus, histocompatibility-18, Chr 4.

H-19 locus, histocompatibility-19, Chr 8.

H-20 locus, histocompatibility-20, Chr 4.

H-21 locus, histocompatibility-21, Chr 4.

H-22 locus, histocompatibility-22, Chr 7.

H-23 locus, histocompatibility-23, Chr 3 (7).

H-24 locus, histocompatibility-24, Chr 7.

H-25 locus, histocompatibility-25, Chr 1 (2, 5).

H-26 locus, histocompatibility-26.

H-27 locus, histocompatibility-27, Chr 5.

H-28 locus, histocompatibility-28, Chr 3 (7).

H-29 locus, histocompatibility-29, Chr 8.

H-30 locus, histocompatibility-30, Chr 15 (3, 6).

H-31 to *H-33* loci. See after *H-38* locus.

H-34 locus, histocompatibility-34.

H-35 locus, histocompatibility-35, Chr 1. Linked to *H-25* with 12 per cent recombination.

H-36 locus, histocompatibility-36.

H-37 locus, histocompatibility-37.

H-38 locus, histocompatibility-38.

References

1. Bailey, D.W. 1975. Genetics of histocompatibility in mice. I. New loci and congenic lines. Immunogenetics 2:249–256.
2. Durda, P.J., S.C. Boos, and P.D. Gottlieb. 1979. T100: a new murine cell surface glycoprotein detected by anti-Lyt-2.1 serum. J. Immunol. 122:1407–1412.
3. Hogarth, P.M., I.F.C. McKenzie, V.R. Sutton, K.M. Curnow, B.K. Lee, and E.M. Eicher. 1987. Mapping of the murine *Ly-6*, *Xp-14*, and *Gdc-1* loci to chromosome 15. Immunogenetics 25:21–27.
4. Johnson, L.L., D.W. Bailey, and L.E. Mobraaten. 1981. Genetics of histocompatibility in mice. IV. Detection of certain minor (non-H-2) H antigens in selected organs by the popliteal node test. Immunogenetics 14:63–71.
5. Mathieson, B.J., S.O. Sharrow, K. Bottomly, and B.J. Fowlkes. 1980. Ly9, an alloantigenic marker of lymphocyte differentiation. J. Immunol. 125:2127–2136.
6. Meruelo, D., M. Offer, and A. Rossomando. 1982. Evidence for a major cluster of lymphocyte differentiation antigens in murine chromosome 2. Proc. Natl. Acad. Sci. USA 79:7460–7464.
7. Mobraaten, L.E., H.P. Bunker, J. DeMaeyer-Guignard, E. DeMaeyer, and D.W. Bailey. 1984. Location of histocompatibility and interferon loci on chromosome 3 of the mouse. J. Hered. 75:233–234.

H-31 to *H-33* loci

These three loci are all on Chr 17 near *H-2*.

H-31 locus, histocompatibility-31. *H-31^b* and *H-31^a* were identified in congenic strains C57BL/6 and B6.A-*Tla^a*, respectively. A third allele, *H-31^c*, occurs in B6.AKR-*H-2^k*. The antigen is present on skin and tumors (2, 3).

H-32 locus, histocompatibility-32. *H-32^a* and *H-32^b* were identified in congenic strains A and A.B6-*Tla^b*, respectively. The antigen is present on skin and tumors (2, 3).

H-33 locus, histocompatibility-33. *H-33^b* and *H-33^a* were identified in congenic strains BALB/c and BALB.TTF, respectively. The antigen is present on skin (1).

References

1. Flaherty, L. 1975. *H-33*—a histocompatibility locus to the left of the *H-2* complex. Immunogenetics 2:325–329.
2. Flaherty, L., and S.S. Wachtel. 1975. *H(Tla)* system: identification of two new loci, *H-31* and *H-32*, and alleles. Immunogenetics 2:81–85.
3. Flaherty, L., and S.S. Wachtel. 1975. H(Tla) system: allelism and linkage studies. Transplant. Proc. 7: Suppl. 1:143–145.

H-34 to *H-38* loci

See *H-15*.

H-39 locus, histocompatibility-39, Chr 17

The two known alleles at this locus are present in the congenic strains BTBRTF/Nev (abbreviated BT/Nev) and BT/Nev-*t^{w18}*/+, and were detected by rejection of tail skin grafts between these strains. *H-39* showed no recombination with *t^{w18}* in 71 gametes classified, although *t^{w18}* allows normal recombination in this chromosome region (1).

References

1. Artzt, K., L. Hamburger, and L. Flaherty. 1977. *H-39*, a histocompatibility locus closely linked to the *T/t* complex. Immunogenetics 5:477–480.

H-40 locus, histocompatibility-40, Chr 12

This locus controls a minor histocompatibility antigen present on surface immunoglobulin-positive B-cells and on tumors. It may also be present on normal tissues. Its expression on B-cells correlates with the expression of surface IgM. The antigen is recognized by cytotoxic T lymphocytes and by tumor rejection. Three alleles are known: *H-40^a* in strains BALB/c, C57L, DBA/2, C.AL-20, and AKR; *H-40^b* in strains C.B-20 and B10.D2; and *H-40^c*, possibly a null allele, in NZB. *H-40* is on Chr 12 proximal to *Tsu* in the region of *Pre-1* (1–3).

References

1. Forman, J., R. Riblet, K. Brooks, E.S. Vitetta, and L.A. Henderson. 1984. H-40, an antigen controlled by an *Igh* linked gene and recognized by toxic T lymphocytes. I. Genetic analysis of *H-40* and distribution of its products on B cell tumors. J. Exp. Med. 159:1724–1740.
2. Forman, J., R. Ciavarra, and L.A. Henderson. 1985. Recognition of an *Igh*-linked histocompatibility antigen, *H-40*, on B-cell tumors by cytotoxic T lymphocytes. Surv. Immunol. Res. 4:41–47.
3. Henderson, L.A., R. Ciavarra, R. Riblet, and J. Forman. 1984. H-40, an antigen controlled by an *Igh*-linked gene and recognized by cytotoxic T lymphocytes. II. Recognition of H-40 as a tumor antigen in leukemic animals. J. Immunol. 133: 2778–2785.

H-41 locus, histocompatibility-41

This locus controls a minor histocompatibility antigen recognized by skin graft rejection. The antigen can also generate a cytotoxic lymphocyte response after one *in vivo* immunization. Presence of the *H-2D^p* allele is necessary for the response. The allele *H-41ᵃ* occurs in the wild and in the B10.STA12 congenic strain where it was introduced along with the *H-2ʷ¹³* haplotype which carries the *H-2Dᵖ* specificity; *H-41ᵃ* is also present in B10.P and C3H.NB. The allele *H-41ᵇ* occurs in the similarly derived B10.STA10 and B10.LIB55 *H-2* congenic strains as well as in C57BL/10, C57BL/6, C57L, BALB/c, A, AKR, WB, DBA/1, and DBA/2 (1).

References

1. Juretić, A., I. Vucak, B. Malenica, Z.A. Nagy, and J. Klein. 1984. *H-41*, a new minor histocompatibility locus. I. Histogenetic analysis. J. Immunol. 133:2950–2954.

H-42 locus, histocompatibility-42, Chr 2

This locus controls a minor histocompatibility antigen recognized by rejection of skin grafts between the C57BL/10 and B10.FS-a strains, which are congenic at the *B2m*, *H-3*, and *H-13* loci on Chr 2. The *H-42ᵃ* allele occurs in C57BL/10; the *H-42ᵇ* allele occurs in B10.FS-a and presumably also in the FS/SnEi strain. *H-42* is on Chr 2, probably distal to *B2m* and *H-3*, and near *we* (1, 2).

References

1. Graff, R.J., D. Martin-Morgan, and M.E. Kurtz. 1985. Allograft rejection-defined antigens of the *B2m*, *H-3* region. J. Immunol. 135:2842–2846.
2. Kurtz, M.E., R.J. Graff, A. Adelman, D. Martin-Morgan, and R.E. Click. 1985. CTL and serologically defined antigens of *B2m H-3* region. J. Immunol. 135:2847–2852.

H-43 locus (formerly *H-42*), histocompatibility-43, Chr 17

This locus controls a minor histocompatibility antigen recognized by cytotoxic T lymphocyte (CTL) responsiveness. The allele *H-43ᵃ* determines presence of the antigen and occurs in the C3H strain and the C3H-*H-2ᵇ* congenic strain CSW; the allele *H-43ᵇ* determines absence of the antigen and occurs in the C57BL/10 and B10.BR strains and in the C3H-*H-2ᵇ Ighᵇ* congenic strain CWB (1). The CTL response to *H-43ᵃ* is generated in *H-43ᵇ* mice primed with *H-2I-A* semicompatible spleen cells or with *H-2I-A* compatible skin grafts. However, priming with *H-2I-A* compatible spleen cells induces tolerance, i.e. no CTL response (2). *H-43* is linked to *H-2* on Chr 17 with about 8 per cent recombination (1).

References

1. Ishikawa, H., E. Kubota, H. Suzuki, and K. Saito. 1985. The target minor H antigen for F₁ cytotoxic T lympho-

cytes induced by Igh-congenic parental spleen cells is coded for by gene linked to H-2. J. Immunol. 134:2953–2959.
2. Ishikawa, H., H. Suzuki, T. Hino, E. Kubota, and K. Saito. 1985. *In vivo* priming of mouse CTL precursors directed to product of a newly defined minor H-42 locus is under a novel control of class II MHC gene. J. Immunol. 135:3681–3685.

H-44, 45 loci, histocompatibility-44, 45, Chr 2

These two histocompatibility loci on Chr 2, in addition to the formerly described *H-3*, *H-13*, and *H-42*, were detected by Graff *et al.* (1) in skin grafting experiments using strain C57BL/10 and a number of B10 strains congenic for Chr 2 loci in combinations and F1 tests not previously used. The *H-44ᵃ* allele occurs in C57BL/10, the *H-44ᵇ* allele in B10.LP-*a H-44*. The *H-45ᵃ* allele occurs in C57BL/10, the *H-45ᵇ* allele in B10.KR-*a* and B10.KR-*Aʷ*.

References

1. Graff, R.J., D. Martin-Morgan, and M.E. Kurtz. 1987. Multiplicity of chromosome 2 histocompatibility genes: new loci, *H-44* and *H-45*. Immunogenetics 26:111–114.

H-X locus

See *Hxa* locus.

H-Y locus

See *Hya* locus.

Histocompatibility loci linked to visible markers

Several histocompatibility loci have been found by looking for graft rejection between the C57BL/6 strain and congenic strains made by introducing visible marker genes on to the C57BL/6 background. Five *H*-loci with known chromosomal location have been identified by this means. Because it is not certain that they are different from all the previously numbered *H*-loci, they are given provisional symbols incorporating the symbol of the marker gene (1).

H(Eh) locus, histocompatibility (Eh), Chr 15.

H(ep) locus, histocompatibility (ep), Chr 19.

H(go) locus, histocompatibility (go), Chr 5. Graft rejection associated with another locus on Chr 5, *pi*, may also be an effect of this locus.

H(js)

H(js) locus, histocompatibility (js), Chr 11. Graft rejection associated with two other loci on Chr 11, lt and tn, may also be an effect of this locus.

H(ln) locus, histocompatibility (ln), Chr 1.

References

1. Bailey, D.W., and H.P. Bunker. 1972. Mouse News Lett. 47:18.

H(Igh) locus, histocompatibility (Igh), Chr 12

One or possibly two minor histocompatibility genes have been shown to be linked to the Igh complex on Chr 12, one on the distal and the other on the proximal side of Igh. They are both detectable by cytotoxic lymphocytes in mixed lymphocyte cultures but the cytotoxicity is thought to be directed against minor histocompatibility antigens rather than against serologically defined allotypic determinants. The loci were recognized as a difference between BALB/c and several BALB/c congenics in which the Igh complex was derived from C57BL/6 (2). The proximal locus may be the same as that described by Riblet and Congleton (1).

References

1. Riblet, R., and C. Congleton. 1977. A possible allotype-linked histocompatibility gene. Immunogenetics 5:511 (Abstr.).
2. Rolink, T., K. Eichmann, and M.M. Simon. 1978. Detection of two allotype-(Ig-1)-linked minor histocompatibility loci by the use of H-2-restricted cytotoxic lymphocytes in congenic mice. Immunogenetics 7:321–336.

H19 locus

See Afr-1 locus.

ha

See sph locus.

Hao-1 locus (formerly GOX), hydroxyacid oxidase-1, Chr 2

This locus controls electrophoretic variation of an isozyme of α-hydroxyacid oxidase (HAOX-A; E.C. 1.1.3.1) that preferentially oxidizes short-chain hydroxyacids and is present in liver. The allele Hao-1^a determines a fast anodally migrating band and is found in strains BALB/c, CBA/H, C3H/H, C57BL/6, and many others; the allele Hao-1^b determines a slower band and is found in the NZC strain; the allele Hao-1^c deter-

mines a band faster than that of Hao-1^a and occurs in strains DBA/Li, NFS/N, STS, 101/H, 129, and a few others. Heterozygotes have three intermediate bands in addition to the parental bands, consistent with a tetrameric structure of the enzyme (1, 2).

References

1. Duley, J., and R.S. Holmes. 1974. α-hydroxyacid oxidase genetics in the mouse: evidence for two genetic loci and a tetrameric subunit structure for the liver isozyme. Genetics 76:93–97.
2. Holmes, R.S., J.A. Duley, and J. Hilgers. 1982. Sorbitol dehydrogenase genetics in the mouse: a 'null' mutant in a 'European' C57BL strain. Anim. Bld. Grps. Biochem. Genet. 13:263–272.

Hao-2 locus, hydroxyacid oxidase-2, Chr 3

This locus controls electrophoretic variation of an L-α-hydroxyacid oxidase (HAOX-B; E.C. 1.1.3.1) present in kidney. The allele Hao-2^a determines a slow anodally migrating band and is present in all inbred strains tested; the allele Hao-2^b determines a fast band and is present in M. m. castaneus. Heterozygotes have a single broad intermediate band (1).

References

1. Holmes, R.S. 1977. The genetics of α-hydroxyacid oxidase and alcohol dehydrogenase in the mouse: evidence for multiple loci and linkage between Hao-2 and Adh-3. Genetics 87:709–716.

Hat-1 locus, histidine aminotransferase-1

This locus controls variation in activity and heat stability of liver cytosolic histidine aminotransferase (HAT-1; E.C. 2.6.1.38). The allele Hat-1^a determines high activity and high heat stability and occurs in strain SM/J; the allele Hat-1^b determines low activity and low heat stability and occurs in strain C57BL/6J. Heterozygotes have intermediate activity and heat stability. Hat-1 does not affect the mitochondrial form of the enzyme (1).

References

1. Bulfield, G. 1978. Genetic variation in the activity of the histidine catabolic enzymes between inbred strains of mice: a structural locus for a cytosol histidine aminotransferase isozyme (Hat-1). Biochem. Genet. 16:1233–1241.

Hb locus

See Hbb locus.

Hba gene family

This family consists of three closely linked genes coding for the α-chain of adult and embryonic hemoglobin and two pseudogenes which are located on different chromosomes from each other and from the active genes (3).

Hba complex (formerly *Sol*), hemoglobin α-chain, Chr 11. Variants of this complex were first recognized as differences in solubility of carbon monoxyhemoglobin (5). It was later shown that variants could be differentiated on the basis of peptide differences revealed by electrophoresis or chromatography, isoelectric focusing, or by amino acid sequencing (6, 9, 16). There are at least five different α-chains differing at one or more amino acid positions, and an allele (or haplotype) may determine one chain or two different chains. The alleles known in inbred strains are: *Hba*^{a} (chain 1) in prototype strain C57BL/6; *Hba*^{b} (chains 2, 3) in BALB/c; *Hba*^{c} (chains 1, 4) in C3H/Cum; *Hba*^{d} chains (1, 2) in C3H/B; *Hba*^{e} (chain 4) in NB, now extinct (11, 12); *Hba*^{f} (chain 5) in AKR/J; *Hba*^{g} (chains 1, 5) in DBA/2; and *Hba*^{h} (chains 4, 5) in P/J (15, 16). A sixth chain and other combinations have been found in wild mice (17). A mutant allele, *Hba*^{g2} (originally *Hba*^{y9}) was induced in the DBA/2 strain with ethylnitrosourea (10). Three mutations that are probably deletions of the Hba complex have been induced, two by X-rays in (101 × C3H)F1 mice, *Hba*$^{b2(th)}$ and *Hba*$^{b3(th)}$ (formerly 352HB and 27HB, respectively), and one by triethylenemelamine, *Hba*$^{th-J}$. Mice homozygous for these mutations die *in utero*. In heterozygotes. the absence of one α-chain gene results in α-thalassemia (7, 13). The *Hba* complex of strain BALB/c has been shown by recombinant DNA techniques to consist of two genes coding for adult α-chain and one coding for embryonic X chain, arranged with the X chain sequence 5′ to the two adult chain sequences (3). The two α-chains of BALB/c are presumably coded for by the two adult genes. The arrangement of the *Hba* DNA in strains with single α chains has not been reported.

The *Hba-x* locus is invariant. It codes for an α-like chain that is expressed only in embryonic nucleated red cells and is homologous with the human embryonic α-like zeta chain (4). It was first shown to be part of the *Hba* complex by determining that it was not active in the *Hba*$^{th-J}$ deletion (14).

Hba-3ps locus, hemoglobin α-chain pseudogene 3, Chr 15. This is the locus for a non-functional α-globin gene. It lacks the intervening sequences present in the functional genes of the Hba complex. No genetic variants are known. The gene was shown to be on Chr 15 by use of mouse–Chinese hamster somatic cell hybrids and a molecular probe to detect presence of the gene in mouse DNA (3, 8).

Hba-4ps locus, hemoglobin α-chain pseudogene 4, Chr 17. This is the locus for a non-functional α-globin-like gene. It differs from *Hba-ps3* in that it retains the intervening sequences of the functional *Hba* genes. The gene was first shown to be on Chr 17 by use of mouse–Chinese hamster somatic cell hybrids classified with a molecular probe (3, 8). There are four known variants of *Hba-4ps* DNA revealed by digestion with *Taq*I endonuclease. The allele *Hba-4ps*^{a} produces two fragments of 3.5 and 1.3 kb and occurs in BALB/c, AKR, and a few other strains; the allele *Hba-4ps*^{b} produces a 3.1-kb fragment and occurs in C57BL/6, DBA/2, and several other strains; the allele *Hba-4ps*^{c} produces a 4.7-kb fragment and occurs in the SM/J strain; the allele *Hba-4ps*^{d} produces a different pattern and occurs in the MOL/Ei strain (1). Another allele *Hba-4ps*^{t} (5.2 kb) has been found in all *t*-haplotypes examined (2). By use of recombinant inbred strains derived from progenitor strains carrying different alleles, *Hba-4ps* was found to be located on the proximal side of *H-2*, probably close to *tf* (1).

References

1. D'Eustachio, P., B. Fein, J. Michaelson, and B.A. Taylor. 1984. The α-globin pseudogene on mouse chromosome 17 is closely linked to H-2. J. Exp. Med. 159:958–963.
2. Fox, H.S., L.M. Silver, and G.R. Martin. 1984. An alpha globin pseudogene is located within the mouse *t* complex. Immunogenetics 19:125–130.
3. Leder, A., D. Swan, F. Ruddle, P. D'Eustachio, and P. Leder. 1981. Dispersion of α-like globin genes of the mouse to three different chromosomes. Nature 293:196–200.
4. Leder, A., L. Weir, and P. Leder. 1985. Characterization, expression, and evolution of the mouse embryonic zeta-globin gene. Mol. Cell. Biol. 5:1025–1033.
5. Popp, R.A. 1962. Studies on the mouse hemoglobin loci. II, III, IV. J. Hered. 53:73–80.
6. Popp, R.A. 1965. Hemoglobin variants in mice. Fed. Proc. 24:1252–1257.
7. Popp, R.A., B.S. Bradshaw, and L.C. Skow. 1980. Effects of alpha thalassemia on mouse development. Differentiation 17:205–210.
8. Popp, R.A., P.A. Lalley, J.B. Whitney III, and W.F. Anderson. 1981. Mouse α-globin genes and α-globin-like pseudogenes are not syntenic. Proc. Natl. Acad. Sci. USA 78: 6362–6366.
9. Popp, R.A., E.G. Bailiff, L.C. Skow, and J.B. Whitney III. 1982. The primary structure of genetic variants of mouse hemoglobin. Biochem. Genet. 20:199–208.

10. Popp, R.A., E.G. Bailiff, L.C. Skow, F.M. Johnson, and S.E. Lewis. 1983. Analysis of a mouse α-globin gene mutation induced by ethylnitrosourea. Genetics 105:157–167.
11. Russell, E.S., and E.C. McFarland. 1974. Genetics of mouse hemoglobins. Ann. NY Acad. Sci. 241:25–38.
12. Russell, E.S., S.L. Blake, and E.C. McFarland. 1972. Characterization and strain distribution of four alleles at the hemoglobin α-chain locus in the mouse. Biochem. Genet. 7:313–330.
13. Whitney, J.B. III, and R.A. Popp. 1984. Alpha-thalassemia in laboratory mice. Am. J. Pathol. 116:523–525.
14. Whitney, J.B. III, and E.S. Russell. 1980. Linkage of genes for adult α-globin and embryonic α-like globin chains. Proc. Natl. Acad. Sci. USA 77:1087–1090.
15. Whitney, J.B. III, G.T. Copeland, L.C. Skow, and E.S. Russell. 1978. Mouse News Lett. 59:26.
16. Whitney, J.B. III, G.T. Copeland, L.C. Skow, and E.S. Russell. 1979. Resolution of products of the duplicated hemoglobin α-chain loci by isoelectric focusing. Proc. Natl. Acad. Sci. USA 76:867–871.
17. Whitney, J.B. III, R.R. Cobb, R.A. Popp, and T.W. O'Rourke. 1985. Detection of neutral amino acid substitutions in proteins. Proc. Natl. Acad. Sci. USA 82:7646–7650.

Hbb complex (formerly *Hb*), hemoglobin β-chain, Chr 7

The nomenclature here used is that recommended by Lyon *et al.* (4). This complex comprises seven β-chain genes as determined by DNA sequencing. The components are arranged from 5' to 3' in the order *y*, *bh0*, *bh1*, *bh2*, *bh3*, *b1*, *b2* (1). *Hbb-y* codes for embryonic Y globin, *Hbb-bh1* codes for embryonic Z globin, *Hbb-bh0* probably codes for a minor embryonic β-globin, *Hbb-b1* and *Hbb-b2* code for the adult β-major and β-minor globins, respectively, and *Hbb-bh2* and *Hbb-bh3* are probably pseudogenes (1, 2). No variants are known at the *Hbb-bh0*, *-bh1*, *-bh2*, and *-bh3* loci. The allelic forms of the embryonic Y and adult hemoglobins are distinguished by electrophoretic patterns. Three codominant haplotypes are found among inbred strains. *Hbb^s (Hbb-y^s b1^s b2^s)* produces a fast Y globin and a single band of adult globin; it occurs in strains C57BL/6, SEC/1, SWR, and others. *Hbb^d (Hbb-y^d b1^d b2^d)* produces a slow Y globin and a diffuse band of adult globin comprising a major and a minor component; it occurs in strains A, AKR, BALB/c, C3H/He, DBA/2, RFM, and others (8). *Hbb^p (Hbb-y^d b1^d b2^p)* produces a slow Y globin and a diffuse adult globin with a different minor component from that of *Hbb^d*; it occurs in the AU/Ss strain and in wild Asiatic mice (12, 13). No recombinant haplotypes have been found. Four mutant haplotypes have been described, all induced by ethylnitrosourea. *Hbb^{s2}* was induced in a C57BL/6 female and is probably a mutation in the *b1^s* gene. It is detectable as an electrophoretically fast band in addition to the normal single band and for the first time allows distinction of the two otherwise identical *Hbb-b1^s* and *-b2^s* genes (3). Wawrzyniak and Popp (14) have used this distinction to study the developmental regulation of expression of *Hbb-b1^s* and *-b2^s*. *Hbb^{d2}* was induced in a C3H/He male and is probably a mutation in the *b2^d* gene. It causes slower mobility of the minor band of the *Hbb^d* haplotype (5). *Hbb$^{d3(th)}$* was induced in a DBA/2 male and is a 3-kb deletion of the *b1^d* gene. It causes a mild reticulocytosis in heterozygotes and hypochromic microcytic anemia in homozygotes, most of which nevertheless survive and reproduce (11). *Hbb^{d4}* was induced in a laboratory stock and is a mutation in the *b1^d* gene. It causes increased anodal electrophoretic mobility of the diffuse major hemoglobin band. In both homozygotes and heterozygotes, oxygen affinity of hemoglobin is increased. Oxygen is given up less readily and this leads to a marked increase in red cell number (polycythemia). Heterozygotes with *Hbb^d* and *Hbb^s* have intermediate oxygen affinities (6). The two adult genes in the *Hbb^s* haplotype, *b1^s* and *b2^s*, produce indistinguishable β-globins which are different from the major and minor β-globins produced by the *Hbb-b1^d* and *-b2^d* genes (7). The *Hbb^d* and *Hbb^s* haplotypes also differ in DNA structure between the *b1* and *b2* genes by a rearrangement and an insertion (10). In *Hbb^p/Hbb^p* and probably also in *Hbb^d/Hbb^d* mice a much greater proportion of the minor β-chain is synthesized in fetuses than in adults, with the switch to adult proportions occurring between birth and 3 weeks of age (15). An X-ray-induced tandem duplication of the *Hbb^s* allele has been found. It involves about 20 per cent of Chr 7 from the SEC strain and includes the *c* and presumably the *Mod-2* loci (9).

References

1. Jahn, C.L., C.A. Hutchison III, S.J. Phillips, S. Weaver, N.L. Haigwood, C.F. Voliva, and M.H. Edgell. 1980. DNA sequence organization of the β-globin complex in the BALB/c mouse. Cell 21:159–168.
2. Hill, A., S.C. Hardies, S.J. Phillips, M.G. David, C.A. Hutchison III, and M.H. Edgell. 1984. Two mouse early embryonic β-globin gene sequences: evolution of the nonadult β-globins. J. Biol. Chem. 259:3739–3747.
3. Lewis, S.E., F.M. Johnson, L.C. Skow, D. Popp, L.B. Barnett, and R.A. Popp. 1985. A mutation in the β-globin gene detected among the progeny of a female mouse treated with ethylnitrosourea. Proc. Natl. Acad. Sci. USA 82:5829–5831.

4. Lyon, M.F., J.E. Barker, and R.A. Popp. 1988. Mouse globin gene nomenclature. J. Hered. 79:93–95.

5. Murota, T., T. Shibuya, and K. Tutikawa. 1982. Genetic analysis of an *N*-ethyl-*N*-nitrosourea-induced mutation at the hemoglobin β-chain locus in mice. Mutat. Res. 104:317–321.

6. Peters, J., J.F. Loutit, and S.J. Andrews. 1982. Mouse News Lett. 67:19.

7. Popp, R.A. 1965. Hemoglobin variants in mice. Fed. Proc. 24:1252–1257.

8. Russell, E.S., and E.C. McFarland. 1974. Genetics of mouse hemoglobins. Ann. NY Acad. Sci. 241:25–38.

9. Russell, L.B., W.L. Russell, R.A. Popp, C. Vaughan, and K.B. Jacobson. 1976. Radiation-induced mutations at mouse hemoglobin loci. Proc. Natl. Acad. Sci. USA 73:2843–2846.

10. Shyman, S., and S. Weaver. 1985. Chromosomal rearrangements associated with LINE elements in the mouse genome. Nucl. Acids Res. 13:5085–5093.

11. Skow, L.C., B.A. Burkhart, F.M. Johnson, R.A. Popp, D.M. Popp, S.Z. Goldberg, W.F. Anderson, L.B. Barnett, and S.E. Lewis. 1983. A mouse model for β-thalassemia. Cell 34:1043–1052.

12. Vulpis, G., E. Ballardin, and S. Quaggia. 1970. Characteristics of mouse hemoglobin: a new type of hemoglobin in the Shanghai strain (*Mus musculus bactrianus*). Biochim. Biophys. Acta 221:672–674.

13. Watanabe, T., K. Shimizu, H. Goto, and N. Ogasawara. 1976. Genetic study of hemoglobin variants in mice (*Mus musculus*). Jpn. J. Genet. 51:377–383.

14. Wawrzyniak, C.J., and R.A. Popp. 1986. Independent expression of the two mouse adult β-globin genes. Biochem. Genet. 24:259–272.

15. Whitney, J.B. III. 1977. Differential control of the synthesis of two hemoglobin β chains in normal mice. Cell 12:863–871.

hbd, hemoglobin deficit, recessive

Arose spontaneously in an inbred strain carrying A^y. Homozygotes have moderate anemia when young, with some improvement, but not complete remission, by adult life (1, 3). The deficit is characterized by anemia, red cell hyperchromia and microcytosis, and reticulocytosis. It is not due to excess hemolysis, iron deficiency, or ferrochelatase deficiency, and is not a thalassemia. Free red cell protoporphyrin is markedly elevated, suggesting that the anemia is due to a deficit in the erythroid cell iron procurement mechanism leading to diminished heme and hemoglobin synthesis (2).

References

1. Bannerman, R.M., J.A. Edwards, and P.H. Pinkerton. 1973. Hereditary disorders of the red cells in animals. Prog. Hematol. 8:131–179.

2. Bannerman, R.M., L.M. Garrick, P. Rusniak-Smalley, J.E. Hoke, and J.A. Edwards. 1986. Hemoglobin deficit: an inherited hypochromic anemia in the mouse. Proc. Soc. Exp. Biol. Med. 182:52–57.

3. Scheufler, H. 1969. Eine weitere Mutante der Hausmaus mit Anämie (hbd). Z. Versuchstierk. 11:348–353.

Hbx locus

See *Hba* gene family.

Hby locus

See *Hbb* complex.

Hc locus, hemolytic complement, Chr 2

The locus was independently discovered by immunological methods, and the antigen produced by Hc^1 was called MuB1 (3). Two alleles are known: Hc^1 determines presence of hemolytic complement in serum and is found in strains DBA/1, BALB/c, C57BL/6, and others; Hc^0 determines absence of detectable hemolytic complement in serum and is found in strain DBA/2 and others (5). The complement component missing in Hc^0/Hc^0 mice is C5 (6). Hc^1/Hc^0 mice have half the amount of complement of Hc^1/Hc^1 mice. C5 is produced by macrophages of Hc^1/Hc^1 mice but not by other spleen cells, by kidney, brain, or liver cells, or by skin fibroblasts. C5 is synthesized in macrophages, probably as pro-C5 of molecular weight 210 kDa, which is secreted and converted to the two-chain (α-chain, 125 kDa, and β-chain, 83 kDa) C5 protein. Pro-C5 is made in macrophages of Hc^0 strains but is not secreted (7). C5 deficiency has little apparent effect on viability but some differences from normal have been shown in several immunological functions (1). Hc^1 strains are more resistant to certain infections than Hc^0 strains (2, 9). The effect of the *Hc* locus on resistance to *Listeria monocytogenes*, however, does not correlate with presence or absence of C5 in serum, but does correlate with presence of macrophages of the Hc^1/Hc^1 or Hc^0/Hc^0 genotype (8). A DNA polymorphism for C5 was found to segregate completely concordantly with *Hc* in 48 recombinant inbred strains. *Eco*RI-digested genomic DNA shows 11 bands that hybridize with a C5 cDNA clone isolated from a mouse liver cDNA library. A 0.9-kb fragment occurs in the Hc^1 strains C57L, C57BL/6. C3H/He, SJL, and BALB/c; a 6.0-kb fragment occurs in the Hc^0 strains AKR, SWR, and DBA/2 (4).

References

1. Alper, C.A., and F.S. Rosen. 1971. Genetic aspects of the complement system. Adv. Immunol. 14:251–290.

2. Cerquetti, M.C., D.O. Sordelli, R.A. Ortegon, and J.A. Bellanti. 1983. Impaired lung defenses against *Staphylococcus aureus* in mice with hereditary deficiency of the fifth component of complement. Infect. Immun. 41:1071–1076.
3. Cinader, B., S. Dubiski, and A.C. Wardlaw. 1964. Distribution, inheritance, and properties of an antigen, MuB1, and its relation to hemolytic complement. J. Exp. Med. 120:897–924.
4. D'Eustachio, P., T. Kristensen, R.A. Wetsel, R. Riblet, B.A. Taylor, and B.F. Tack. 1986. Chromosomal location of the genes encoding complement components C5 and factor H in the mouse. J. Immunol. 137:3990–3995.
5. Herzenberg, L.A., D.K. Tatchibana, L.A. Herzenberg, and L.T. Rosenberg. 1963. A gene locus concerned with hemolytic complement in *Mus musculus*. Genetics 48:711–715.
6. Nilsson, U.R., and H.J. Müller-Eberhard. 1967. Deficiency of the fifth component of complement in mice with an inherited complement defect. J. Exp. Med. 125:1–16.
7. Ooi, Y.M., and H.R. Colten. 1979. Genetic defect in secretion of complement C5 in mice. Nature 282:207–208.
8. Petit, J.-C. 1980. Resistance to listeriosis in mice that are deficient in the fifth component of complement. Infect. Immun. 27:61–67.
9. Rhodes, J.C., L.S. Wicker, and W.J. Urba. 1980. Genetic control of susceptibility to *Cryptococcus neoformans* in mice. Infect. Immun. 29:494–499.

Hct locus, hair constriction

This locus controls a difference in frequency of hairs with single constrictions on the caudal mid-dorsum of mice of strains C57BL/10 (9.41 per cent) and CBA/St (0.13 per cent). F1 mice have an intermediate frequency (4.41 per cent). No allelic designations have been assigned (1).

References

1. Mann, S.J., and W.E. Straile. 1972. Inheritance of hair constrictions in mice. Genetics 70:631–637.

Hd, hypodactyly, semidominant, Chr 6

Arose spontaneously in the MYA strain. Heterozygotes have a uniform loss of the terminal phalanx of the first digit of the hindfoot. Sometimes the first phalanx is either shortened or missing. The forefeet are normal. Both sexes are viable and fertile. Homozygotes have a single digit on each foot and greatly reduced carpals, metacarpals, tarsals, and metatarsals. The surviving digit is probably the fifth. Most homozygotes die before birth, but a few have survived to maturity. The cause of death is not obvious (1).

References

1. Hummel, K.P. 1970. Hypodactyly, a semidominant lethal mutation in mice. J. Hered. 61:219–220.

Hdc complex, Chr 2

This complex comprises four loci affecting structure and regulation of kidney histidine decarboxylase (HDC; E.C. 4.1.1.22) that have not shown recombination in crosses. Four haplotypes have been described: $[Hdc]^{B10}$ $(Hdc\text{-}a^b c^b e^b s^b)$ in strains C57BL/10, BALB/c, and SWR; $[Hdc]^D$ $(Hdc\text{-}a^b c^d e^d s^d)$ in strains DBA/2 and C3H/He; $[Hdc]^{B6}$ $(Hdc\text{-}a^b c^b e^d s^b)$ in strains C57BL/6, A, and CBA/Ca (1–3); and $[Hdc]^{DAN}$ $Hdc\text{-}a^w c^d e^b s^d$ in the DAN strain derived from wild Danish mice (5). The complex is on Chr 2 about 2 cM distal to *pa* (4).

Hdc-a locus, histidine decarboxylase-testosterone sensitivity. This locus controls sensitivity to the suppressive effects of testosterone on level of HDC in the kidney. The allele $Hdc\text{-}a^b$ determines high sensitivity and occurs in the C57BL/10, DBA/2, and probably other laboratory strains; the allele $Hdc\text{-}a^w$ determines low sensitivity and occurs in the DAN strain derived from wild Danish mice. Heterozygotes have intermediate sensitivity (5).

Hdc-c locus (formerly *Hdc*), histidine decarboxylase-concentration. This locus controls concentration of HDC in kidney. The allele $Hdc\text{-}c^b$ determines low concentration and occurs in C57BL/10, BALB/c, and most other strains; the allele $Hdc\text{-}c^d$ determines high concentration and occurs in strains DBA/2, C3H/He, and I. The difference becomes apparent between 2 and 3 weeks of age. Heterozygotes have intermediate levels (4).

Hdc-e locus, histidine decarboxylase-estrogen. This locus controls estrogen inducibility or repression of HDC in kidney. The allele $Hdc\text{-}e^b$ determines inducibility by estrogen and occurs in strains C57BL/10, A2G, BALB/c, CE, NZB, RIII, and SWR; the allele $Hdc\text{-}e^d$ determines repression by estrogen and occurs in strains C57BL/6, DBA/2, DBA/1, A, C3H/He, CBA/Ca, and F/St. Heterozygotes are intermediate (2, 3).

Hdc-s locus, histidine decarboxylase-structural. This locus affects heat stability of kidney HDC. The allele $Hdc\text{-}s^b$ determines heat stability and occurs in strains C57BL/6, C57BL/10, A, BALB/c, and SWR; the allele $Hdc\text{-}s^d$ determines heat lability and occurs in strains DBA/2 and C3H/He. Heterozygotes have intermediate heat stability (1).

References

1. Martin, S.A.M., and G. Bulfield. 1984. A structural gene *Hdc-s* for mouse kidney histidine decarboxylase. Biochem. Genet. 22:645–656.
2. Martin, S.A.M., and G. Bulfield. 1984. A regulatory locus, *Hdc-e*, determines the response of mouse kidney histidine decarboxylase to estrogen. Biochem. Genet. 22:1037–1046.
3. Martin, S.A.M., and G. Bulfield. 1986. Genetic analysis of a new haplotype of the histidine decarboxylase gene complex in C57BL/6 mice. Genet. Res. 47:131–134.
4. Martin, S.A.M., B.A. Taylor, T. Watanabe, and G. Bulfield. 1984. Histidine decarboxylase phenotypes of inbred mouse strains: a regulatory locus (*Hdc*) determines kidney enzyme concentration. Biochem. Genet. 22:305–322.
5. Middleton, R.J., S.A.M. Martin, and G. Bulfield. 1987. A new regulatory gene in the histidine decarboxylase gene complex determines the responsiveness of the mouse kidney to testosterone. Genet. Res. 49:61–67.

Hdl-1 locus, high-density lipoprotein-1, Chr 1

This locus determines variation in electrophoretic pattern of intact high-density lipoprotein. The electrophoretic patterns are complex but can be grouped into three types: A, in strain PERU, is a single rapidly migrating band; B, present in many strains including C57BL/6J, consists of two major bands migrating more slowly than A; and C, present in several strains including BALB/c and C3H/He, consists of two major bands migrating more slowly than B. Genetic tests between strains with B and C patterns reveal the existence of a single locus with the two alleles $Hdl-1^b$ and $Hdl-1^c$. The A pattern may be determined by a third allele $Hdl-1^a$ but this has not been tested. *Hdl-1* shows very close linkage with apolipoprotein-2, *Apoa-2* (1). See also *Apoa-2* locus.

References

1. Lusis, A.J., B.A. Taylor, R.W. Wangenstein, and R.C. LeBoeuf. 1983. Genetic control of lipid transport in mice. II. Genes controlling structure of high density lipoproteins. J. Biol. Chem. 258:5071–5078.

hea, hereditary erythroblastic anemia, recessive

Arose in the inbred CFO strain derived from outbred CF-1 mice. Homozygotes are distinguishable at birth by their yellow body color. They die between 15 and 20 days of age. Red cells of homozygotes show polychromasia, anisocytosis, and poikilocytosis. There are large numbers of nucleated erythroblasts and myeloid cells in the circulation, and many naked nuclei are found in smears in circulating blood. By 17 days, hematocrit and hemoglobin content are about 50 per cent of normal (1).

References

1. Shimizu, K., H. Keino, N. Ogasawara, and K. Esaki. 1983. Hereditary erythroblastic anaemia in the laboratory mouse. Lab. Anim. 17:198–202.

heb, head bleb, recessive, Chr 4

Arose spontaneously in strain AKR/J. Homozygotes usually have open eyelids at birth, resulting in abnormal eyes with closed eyelids in adults. A small proportion have extra digits. At 12 to 14 days of gestation, blebs are formed on the head, usually on or near the nose or eye, and occasionally on a foot. The blebs become hemorrhagic. Ninety-seven per cent of affected 17- and 19-day fetuses had open eyelids and 12 per cent had extra toes. Penetrance is probably complete, but there is some fetal death of affected animals leading to low apparent segregation ratios of the mutant in some crosses. The *heb* locus is on Chr 4 near Pt (1).

References

1. Varnum, D.S., and S.C. Fox. 1981. Head blebs: a new mutation on chromosome 4 of the mouse. J. Hered. 72:293.

het, head tilt, recessive, Chr 17

Arose in the GL/Le strain in 1976. Homozygotes have a tilted head, circling behavior, hyperactivity, and inability to swim, and are not deaf. The locus is near *T* on Chr 17 (1).

References

1. Sweet, H.O. 1980. Mouse News Lett. 63:19.

Hexb locus, hexosaminidase B, Chr 13

This locus codes for the B form of hexosaminidase (E.C. 3.2.1.30), a lysosomal hydrolase involved in the catabolism of sphingolipids (1). It was found to be on Chr 13 by use of mouse–Chinese hamster microcell hybrids screened with a human hexosaminidase B cDNA clone (2).

References

1. Gilbert, F., R. Kucherlapati, R.P. Creagan, M.J. Murnane, G.J. Darlington, and F.H. Ruddle. 1975. Tay-Sachs' and Sandhoff's diseases: the assignment of genes

for hexosaminidase A and B to individual human chromosomes. Proc. Natl. Acad. Sci. USA 72:263–267.

2. Killary, A.M., R.J. Leach, R.G. Moran, and R.E.K. Fournier. 1986. Assignment of the genes encoding dihydrofolate reductase and hexosaminidase B to *Mus musculus* chromosome 13. Somat. Cell Mol. Genet. 12:641–648.

hf, hepatic fusion, recessive, Chr 7

Discovered in the MA/MyJ inbred strain in which the mutant is homozygous. Homozygotes have varying degrees of fusion between the left portion of the central lobe and the adjacent left lateral lobe. Penetrance decreases on outcrossing, suggesting that the residual genotype of the MA strain is favorable for the complete expression of the gene. Similar fusion has not been discovered in other strains routinely autopsied at the Jackson Laboratory (1).

References

1. Bunker, L.E., Jr. 1959. Hepatic fusion, a new gene in linkage group I of the mouse. J. Hered. 50:40–44.

hg, high growth, recessive

Discovered in mice three or more standard deviations above the mean in generation 25 in a line selected for high 21- to 42-day weight gain. Weight gain in this interval is 30 to 50 per cent higher in *hg/hg* mice than in their normal sibs and mature weight is much higher, but weaning weight is slightly lower. Fertility in both sexes may be normal or as much as 40 per cent lower than that of +/− controls. +/+ and +/*hg* mice are indistinguishable but there is some overlap in weight gain between +/− and *hghg* mice (1). Energetic efficiency of growth is greater in *hg/hg* mice than in controls (2).

References

1. Bradford, G.E., and T.R. Famula. 1984. Evidence for a major gene for rapid postweaning growth in mice. Genet Res. 44:293–308.
2. Calvert, C.C., T.R. Famula, J.F. Bernier, N. Khalat, and G.E. Bradford. 1986. Efficiency of growth in mice with a major gene for rapid postweaning growth. J. Anim. Sci. 162:77–85.

Hgh-3, *Hgh-4* loci, human growth hormone-3, -4

These two loci are the sites of integration of the gene for human growth hormone into the DNA of mice obtained from eggs injected with the hormone gene in a plasmid vector. Heterozygotes for both insertions are normal, but homozygotes die prenatally. It is thought that disruption of native gene sequences by the insertions has inactivated the affected genes, producing recessive lethal mutations. The two mutations are not at the same locus (1).

References

1. Wagner, E.F., L. Covarrubias, T.A. Stewart, and B. Mintz. 1983. Prenatal lethalities in mice homozygous for human growth hormone gene sequences integrated in the germ line. Cell 35:647–655.

Hh loci

See *H-2* complex.

hid, hair interior defect, recessive

Discovered in the AKR strain in which it is homozygous. Homozygotes appear completely normal externally and can be recognized only by microscopic examination of the hair. All hair types are affected and are more abnormal distally than proximally. The medullary cells are large and elongated, and irregularly packed into the medullary space. Normal medullary cells become closely packed and rectangular in shape directly above the dermal papilla shortly after they form. Cells of the *hid/hid* medulla remain rounded and never attain the normal regular arrangement, compactness, or rectangular shape (1).

References

1. Trigg, M.J. 1972. Hair growth in mouse mutants affecting coat texture. J. Zool. Lond. 168:165–198.

Hiomt locus, hydroxyindole-*O*-methyltransferase

This locus controls presence or absence of hydroxyindole-*O*-methyltransferase (HIOMT) in the pineal gland. This enzyme and *N*-acetyltransferase (NAT) are necessary for the synthesis of pineal melatonin, a substance that plays an important role in regulation of circadian rhythm. The pineal glands of C57BL/6, BALB/c, AKR/J, NZB/Bl, IS/Cam, and CAST/Ei completely lack melatonin; the pineal glands of SK/Cam, SF/Cam, PERU-Coppock/Cam, and a wild-derived stock (FDS) from Alberta contain normal amounts of melatonin. Analysis of the pineals of C57BL/6 and FDS mice shows that the two enzymes necessary for melatonin synthesis, NAT and HIOMT, occur in normal amounts in FDS but are absent from C57BL/6 mice. The two enzymes are controlled by two independently segregating autosomal genes. *Hiomt*ᵃ in FDS determines presence of HIOMT; *Hiomt^b* in C57BL/6 and probably also in NZB determines absence of the enzyme. Heterozygotes have intermediate levels (1). See *Nat-2* locus.

References

1. Ebihara, S., T. Marks, D.J. Hudson, and M. Menaker. 1986. Genetic control of melatonin synthesis in the pineal gland of the mouse. Science 231:491–493.

his, histidinemia, recessive, Chr 10

Discovered in a stock derived from wild mice from Peru. The stock had been selected at Cambridge University for increased incidence of a balance defect (4). Homozygotes are histidinemic with high levels of histidine in liver, brain, plasma, urine, and skin, and high levels of derived imidazoles, particularly in urine (2). The histidinemic effect is almost fully recessive, but the liver concentration of histidine in heterozygotes is slightly but significantly lower than in +/+ mice (1). Homozygous females may produce offspring with a balance defect due to malformation of the inner ear caused by the high level of histidine or its imidazole derivatives. The proportion of offspring affected varies with genetic background and with genotype of the offspring, *his/his* offspring being more susceptible than *his/+* offspring (3). Homozygotes show greatly reduced activity of skin and liver histidine ammonia-lyase (HAL; E.C. 4.3.1.3). Liver HAL in homozygotes is heat- and salt-labile and is inhibited by high substrate concentrations. In normal mouse liver there are two HAL components, a minor component with properties similar to those of the HAL of *his/his* mice and a major component that is heat- and salt-stable and insensitive to substrate inhibition. The livers of mutant mice contain none of the major component. The *his* mutation, therefore, is probably a structural or regulatory locus for the major HAL component. It is linked to *Sl* on Chr 10 with about the same recombination frequency as the *Hsd* locus which controls a difference between inbred strains in the rate of HAL synthesis. It is thus likely that the two loci are either identical or part of a gene complex controlling HAL synthesis and regulation (5). See also *Hsd* locus.

References

1. Bulfield, G., and H. Kacser. 1974. Histidinaemia in mouse and man. Arch. Dis. Child. 49:545–552.
2. Kacser, H., G. Bulfield, and M.E. Wallace. 1973. Histidinaemic mutant in the mouse. Nature 244:77–79.
3. Kacser, H., K.M. Mya, M. Dunker, A.F. Wright, G. Bulfield, A. McLaren, and M.F. Lyon. 1977. Maternal histi-

dine metabolism and its effect on foetal development in the mouse. Nature 265:262–266.
4. Wallace, M.E. 1970. Mouse News Lett. 42:20.
5. Wright, A.F., G. Bulfield, S.M. Arfin, and H. Kacser. 1982. Comparison of the properties of histidine ammonia-lyase in normal and histidinemic mutant mice. Biochem. Genet. 20:245–263.

Hist1, *Hist2* complexes

These complexes code for the class of histone genes known as replication variants that are predominant in rapidly growing cells and whose synthesis is tightly linked to DNA replication (2). *Hist1* also contains genes for spermatocyte-specific histones (3). There are probably several genes coding for each histone protein; they differ in the untranslated regions and give rise to different mRNAs (2).

Hist1 complex, histone gene complex-1, Chr 13. This complex, identified by DNA clone MM221, contains genes for the replication-dependent histones H3.1, H3.2, H2b.1, and probably also H2a and H4. The genes are transcribed at low levels in cultured cells and fetal mice. The complex was shown to be on Chr 13 using mouse–Chinese hamster and mouse–rat somatic cell hybrids and a cloned DNA fragment from MM221 as a probe (1). A difference in electrophoretic mobility of the spermatocyte-specific histone H1.S in this complex occurs among inbred strains. The *Hist1^b* haplotype contains an allele for slow mobility and occurs in strain C57BL/6; the *Hist1^a* haplotype contains an allele for fast mobility and occurs in strains DBA/2, C3H/He, and all other strains tested. Another spermatocyte-specific histone, H2B.S, with a variant in wild *M. m. domesticus*, is also controlled by a gene in the *Hist1* complex. The alleles of both these loci are codominant (4). By use of the H1.S marker in BXD and BXH recombinant inbred strains, *Hist1* was shown to be very closely linked to *Tcrg* near the centromere of Chr 13 (3).

Hist2 complex, histone gene complex-2, Chr 3. This complex, identified by DNA clone 614, codes for histones H3 and H2a. These genes are expressed at high levels in cultured cells and fetal mice. The 614 clone is said by Cox (personal communication) to be derived from Chr 3 (2).

A genetic variant of the replication-defective somatic histone H2A.1 occurs in the Icr subline of the BALB/c strain. The gene controlling this variant is not linked to the *Hist1* complex. H2A.1 appears to be controlled by 10 or 11 genes, different arrays of which are active in different tissues. Only one of these genes is mutated and responsible for the BALB/cIcr variant (4).

Hist1, Hist2

References

1. Cox, D.R., V.E. Groppi, D. Buber, D.B. Sittman, P. Coffino, and W.F. Marzluff. 1984. Assignment of murine histone genes to mouse chromosome 13. Cytogenet. Cell Genet. 37:443 (Abstr.).
2. Graves, R.A., S.E. Wellman, I.-M. Chiu, and W.F. Marzluff. 1985. Differential expression of two clusters of mouse histone genes. J. Mol. Biol. 183:179–194.
3. Owen, F.L., B.A. Taylor, A. Zweidler, and J.G. Seidman. 1986. The murine γ-chain of the T cell receptor is closely linked to a spermatocyte-specific histone gene and the beige coat color locus on chromosome 13. J. Immunol. 137:1044–1046.
4. Zweidler, A. 1984. Core histone variants of the mouse: primary structure and differential expression. *In* G.S. Stein, J.L. Stein, and W.F. Marzluff, eds., Histone Genes: Structure, Organization, and Regulation, 339–371. John Wiley and Sons, New York.

Hk, hook, semidominant, Chr 8

Arose spontaneously in a piebald shaker stock. Homozygotes are viable and fertile, and at least on some genetic backgrounds are more severely affected than heterozygotes. Heterozygotes may not be detectably abnormal, or they may have shortened, crooked, or hook-shaped tails, and an abnormal anus. The anus tends to be slit-shaped rather than round, and to be bare rather than surrounded by hair. It may be very much elongated and displaced toward the tail, sometimes running well out on to the tail (1).

References

1. Holman, S.P. 1951. The hook-tailed mouse. J. Hered. 42:305–306.

Hk-1 locus, hexokinase-1, Chr 10

This is probably the structural locus for an isozyme of hexokinase (HK-1; E.C. 2.7.1.1). No genetic variants are known. The locus was found to be located on Chr 10 by use of mouse–Chinese hamster somatic cell hybrid clones segregating mouse chromosomes (1). Discovery of an electrophoretic variant (not described) allowed Peters *et al.* (2) to localize *Hk-1* to a position 1 cM distal to *dl* (2).

References

1. Lalley, P.A., J.D. Minna, and U. Francke. 1978. Conservation of autosomal gene synteny groups in mouse and man. Nature 274:160–162.
2. Peters, J., S.J. Andrews, and C.V. Beechey. 1980. Position of Hk-1 on chromosome 10. Mouse News Lett. 63:17.

hl, hair-loss, recessive, Chr 15

Arose spontaneously. Homozygotes show progressive degeneration of hair follicles over the entire body, loss of hair becoming obvious after about 5 weeks. Some specimens never lose all hair; others become very wrinkled and resemble rhino (2). An interesting peculiarity of this mutant is that the *hl*/+ offspring of *hl*/*hl* mothers have extremely fragile bones that break easily and lead to excessive mortality until the age of 2 weeks. The antagonism is probably prenatal. It is not immunological. The suggestion has been made that *hl*/*hl* mice may maintain and require a smaller pool of calcium than normal mice. Their *hl*/*hl* offspring are able to obtain enough to satisfy their smaller requirement, but their *hl*/+ offspring obtain too little (1).

References

1. Hollander, W.F. 1960. Genetics in relation to reproductive physiology in mammals. J. Cell. Comp. Physiol. 56, Suppl.1:61–72.
2. Hollander, W.F., and J.W. Gowen. 1959. A single-gene antagonism between mother and fetus in the mouse. Proc. Soc. Exp. Biol. Med. 101:425–428.

Hld, hippocampal lamination defect, dominant or semidominant

This gene causes abnormal laminar organization of the pyramidal layer of the cerebellum, particularly in the proximal segment of the layer. It occurs in the BALB/cJ strain. In normal strains, the latest formed or youngest neurons migrate past the earlier formed or older neurons to a position in the pyramidal layer that is superficial to that of the older cells. In the BALB/c strain, the positions are reversed, with the older cells lying deep to the younger ones (4). Since mossy fibers form synapses primarily with the older cells, this aberrant pattern of cell migration in BALB/c leads to a different pattern of mossy-fiber synapses, easily visualized with Timm's stain (1). The dendritic excrescences induced by contact with mossy fibers on late-generated pyramidal cells in +/+ mice occur at sites on both the apical and basal dendrites; in *Hld*/*Hld* mice, they occur in two sites on the apical dendrites only (3). Strains having the normal pattern include C57BL/6, SM, LP, CE, DBA/2, and 129/J. *Hld* was identified as a single gene in the F2 generation from a cross of BALB/c with C57BL/6 and in the CXB recombinant inbred strains (2).

References

1. Barber, R.P., J.E. Vaughn, R.E. Wimer, and C.C. Wimer. 1974. Genetically associated variations in the dis-

tribution of dentate granule cell synapses upon the pyramidal cell dendrites in mouse hippocampus. J. Comp. Neurol. 156:417–434.

2. Nowakowski, R.S. 1984. Mouse News Lett. 71:35.
3. Nowakowski, R.S., and T.L. Davis. 1985. Dendritic arbors and dendritic excrescences of abnormally positioned neurons in area CA3c of mice carrying the mutation "hippocampal lamination defect." J. Comp. Neurol. 239:267–275.
4. Vaughn, J.E., D.A. Matthews, R.P. Barber, C.C. Wimer, and R.E. Wimer. 1977. Genetically-associated variations in the development of hippocampal pyramidal neurons may produce differences in mossy fiber connectivity. J. Comp. Neurol. 173:41–52.

Hm, hammer-toe, semidominant, Chr 5

Found in a linkage cross. Homozygotes are viable and fertile. Both homozygotes and heterozygotes are characterized by feet in which the second phalanx of digits 2, 3, 4, and 5 is strongly flexed. All four feet are affected. Homozygotes have webbing between digits 2, 3, 4, and 5, extending to the nails in the hindfeet and to the base of the distal phalanx in the forefeet. Heterozygotes also show webbing but it does not extend so far on the toes (1; Green and Kaufer, unpublished). The condition is detectable at $14\frac{1}{2}$ days of gestation in homozygotes and at 15 days in heterozygotes by failure of the webbing between the toes to undergo normal regression. Persistence of the webbing prevents the toes from elongating independently as they do in normal mice and results in the marked flexion seen at birth (Green and Winikoff, unpublished).

References

1. Green, M.C. 1964. Mouse News Lett. 31:27.

Hma locus, HPRT mobility alteration, Chr 7

This locus controls variation in electrophoretic mobility of the enzyme hypoxanthine-guanine phosphoribosyltransferase (E.C.2.4.2.8) in erythrocyte lysates but not in other tissues. The allele Hma^a determines a three-banded pattern in erythrocytes only and occurs in strains A/J, DBA/2J, and CBA/J; the allele Hma^b determines a two-banded pattern (not identical with any of the three bands of Hma^a), and occurs in strains C57BL/6J, SJL/J, and LT/Sv. Other tissues of all strains have the two-banded pattern. Hma^a is fully dominant. The strain distribution of Hma^a and Hma^b alleles is completely concordant with that of Hbb^d and Hbb^s, respectively, suggesting that the HPRT mobility alteration may be an effect of the hemoglobin β-chain difference (1).

References

1. Nesbitt, M.N., B. Bakay, M.B. Gardner, and C. Day. 1979. Isoenzyme pattern of HPRT in murine erythrocytes: control by an autosomal locus. Biochem. Genet. 17:957–964.

Hmt locus, histocompatibility maternally transmitted, Chr 17

A maternally transmitted antigen, Mta, is a weak histocompatibility antigen found on the lymphoid cells and fibroblasts of most strains of mice. It is also recognized by cytotoxic T lymphocytes and has many similarities to MHC class I genes (see Rodgers *et al.*, 3, for an extensive review of the literature on Mta). Presence of the antigen is dependent on a factor Mtf^α that is transmitted only through the oocyte and is probably carried in the mitochondria. Absence of the antigen depends on the maternally transmitted Mtf^β factor, known to occur in the NZO and NMRI strains. Presence of the Mta antigen is also dependent on presence of at least one Hmt^a allele, known to occur in the C57BL/6, C3H-*Pgk-1^a*, NZO, and NMRI strains. Homozygosity for the Hmt^b allele, found in *M. m. castaneus*, results in absence of the Mta antigen (1). A third allele Hmt^c determining a specificity different from that of Hmt^a occurs in *M. spretus*, which also carries Mtf^α (2). *Hmt* is closely linked to *H-2* in the *Tla* region (1).

References

1. Fischer Lindahl, K., B. Hausmann, and V.M. Chapman. 1983. A new *H-2*-linked class I gene whose expression depends on a maternally inherited factor. Nature 306:383–385.
2. Fischer Lindahl, K., B. Hausmann, P.J. Robinson, J.-L. Guénet, D.C. Wharton, and H. Winking. 1986. Mta, the maternally transmitted antigen, is determined jointly by the chromosomal *Hmt* and the extrachromosomal *Mtf* genes. J. Exp. Med. 163:334–346.
3. Rodgers, J.P., R. Smith III, M.M. Huston, and R.R. Rich. 1986. Maternally transmitted antigen. Adv. Immunol. 38:313–359.

Hnl locus, hypothalamic norepinephrine level, Chr 3

This locus controls level of norepinephrine in the hypothalamus. The allele Hnl^h determines a high level and occurs in the BALB/cBy strain; the allele Hnl^l determines a level about one-third that of the high level and occurs in the C57BL/6By strain. Heterozygotes have intermediate levels. Analysis of CXB recombinant inbred strains shows that *Hnl* is probably linked to H-23 on Chr 3 (1).

References

1. Eleftheriou, B.E. 1974. A gene influencing hypothalamic norepinephrine levels in mice. Brain Res. 70:538–540.

ho, hotfoot, recessive, Chr 6

This mutation first occurred in strain C57BL/Ks in 1964. It was subsequently discontinued, but other similar mutations at the same locus have been found at the Jackson Laboratory in the AKR, C3H/He, DBA/2, and C57BL/6 strains. Homozygotes show a quick patting motion of their feet. Females are fertile, but males have not bred (1). An allele, *ho^{cpr}* (creeper), was found in the B10.D2 strain at Duke University. In behavior and fertility, creepers are similar to hot foot mice. On the B10.D2 background, they can be identified at 14 to 25 days; on an outbred background, onset of symptoms is later. Survival to weaning is poor, but life span is normal in those that survive weaning (4). *ho^{cpr}/ho^{4J}* males may breed (3). *ho* is on Chr 6 near *Hd* (2).

References

1. Dickie, M.M. 1966. Mouse News Lett. 34:30.
2. Southard, J.L. 1981. Mouse News Lett. 64:60.
3. Whitmore, S.P. 1986. Mouse News Lett. 74:106–107.
4. Whitmore, S.V.P., and J. Bresnahan. 1983. Mouse News Lett. 69:18.

Hom-1 locus, hormone metabolism-1, Chr 17

This locus, which may be identical with some part of the *H-2* system, controls differences in weights of testis, seminal vesicle, thymus, and lymph nodes between the C57BL/10 and A/Ph strains. The allele *Hom-1^a*, present in strain A, determines lower seminal vesicle weight, higher testis, thymus, and lymph node weights, and higher plasma testosterone level than does *Hom-1^b*, present in C57BL/10. *Hom-1* may be involved in androgen hormone metabolism since it affects the weight of these primarily or partially sex-hormone dependent organs (1–3).

References

1. Iványi, P., S. Gregorová, and M. Micková. 1972. Genetic differences in thymus, lymph node, testes and vesicular gland weights among inbred mouse strains. Association with the major histocompatibility (H-2) system. Folia Biol. 18:81–97.
2. Ivaányi, P., R. Hampl, L. Stárka, and M. Micková. 1972. Genetic association between H-2 gene and testosterone metabolism in mice. Nature New Biol. 238:280–281.
3. Iványi, P., S. Gregorová, M. Marková, R. Hampl, and L. Stárka. 1973. Genetic association between a histocompatibility gene (H-2) and androgen metabolism in mice. Transplant. Proc. 5:189–191.

hop locus, Chr 6

hop, hop-sterile, recessive. Arose spontaneously in 1967 at the MRC Radiobiology Unit at Harwell. Homozygotes walk with a hopping gait using the hindlegs simultaneously. There is preaxial polydactyly of both fore- and hindfeet. Females are fully fertile, but males are completely sterile. There is no gross abnormality of the male reproductive system, but tailed sperm are completely absent from the lumen of the testicular tubules. The second meiotic division is frequently abnormal or incomplete, often with four centrioles per cell. These centrioles usually fail to form flagella, and tail development is arrested (5).

hop^{hpy}, hydrocephalic–polydactyl, recessive. Found among descendants of X-irradiated mice. Homozygotes resemble *hop/hop* mice in behavior and sperm abnormalities and, in addition, may have hydrocephalus. Adults, particularly females, often develop scoliosis. Most females are fertile, at least for a time (4). The sterility of males is associated with sperm abnormalities which develop during postmeiotic stages of spermiogenesis and are most severe in the sperm tails (1). The hydrocephalus is the non-obstructive type. The entire ventricular system is dilated, and the ependymal cells are destroyed or damaged over wide areas (2). *hop/hop^{hpy}* males are sterile and females are fertile (3).

References

1. Bryan, J.H.D. 1977. Spermatogenesis revisited. IV. Abnormal spermiogenesis in mice homozygous for another male-sterility-inducing mutation, *hyp* (hydrocephalic–polydactyl). Cell Tiss. Res. 180:187–201.
2. Bryan, J.H.D., R.L. Hughes, and T.J. Bates. 1977. Brain development in hydrocephalic-polydactyl, a recessive pleiotropic mutant in the mouse. Virchows Arch. A Pathol. Anat. Histol. 374:205–214.
3. Handel, M.A. 1985. Mouse News Lett. 72:124.
4. Hollander, W.F. 1976. Hydrocephalic polydactyl, a recessive pleiotropic mutant in the mouse, and its location in chromosome 6. Iowa State J. Res. 51:13–23.
5. Johnson, D.R., and D.M. Hunt. 1971. Hop-sterile, a mutant gene affecting sperm tail development in the mouse. J. Embryol. Exp. Morphol. 25:223–236.

Hox-1 to *Hox-3* loci

These three loci are DNA segments homologous to sequences of 180 base pairs that occur in several complexes of homeotic genes of Drosophila. These 'homeo box' sequences are widely conserved throughout the animal kingdom and are thought, from their function in Drosophila, to be concerned with regulation of segmental differentiation. In the mouse, there are numerous separate restriction fragments that hybridize with homeo box probes (9, 10). Three chromosomal locations containing some of these are known. The nomenclature used here is that described by Martin *et al.* (8).

Hox-1 locus, homeo box-1, Chr 6. This locus was recognized as a 180-bp sequence detected with a DNA clone, MO-10 (*Hox-1.5*), isolated from the mouse genome. The sequence codes for a protein domain homologous to the conserved domain of Drosophila homeotic genes (10). *Hox-1* marks a cluster of 6 to 10 homeo box sequences. Six of these have been identified and designated *Hox-1.1, Hox-1.2 . . . Hox-1.6* (8). Restriction fragment length polymorphisms detected with at least three different *Hox-1* probes and two different restriction enzymes have been found among inbred strains (3, 11). The allele *Hox-1.5^a* (*Eco*RV 12.5 kb) occurs in many strains including A/J, BALB/c, CBA/N, C3H/He, DBA/2, O20, and NZB; the allele *Hox-1.5^b* (*Eco*RV 16.4 kb) occurs in strains AKR, C57BL/6, C57L, I/Ln, STS, and several others (11); a third allele occurs in *Mus spretus* (3). *Hox-1* was first shown to be on Chr 6 proximal to *Igk* by use of mouse–Chinese hamster somatic cell hybrids (10). Later, Mock *et al.* (11) showed that *Hox-1.5* is located very close to the developmental mutation hypodactyly (*Hd*) but is not allelic with it (1/118 recombinants). Colberg-Poley *et al.* (4) have shown that *Hox-1.1* (m6) and *Hox-1.2* (m5) are expressed in embryonal carcinoma cells and in 12-day embryos; *Hox-1.1* is also expressed in almost all other stages of prenatal development. Rubin *et al.* (14) have found that a gene in this region identified as *Hox-1.4* (MH-3) is expressed in male germ cells at late meiotic stages and in 11.5-day embryos before differentiation of the male and female gonads.

Hox-2 locus, homeo box-2, Chr 11. This locus contains a cluster of four homeo box sequences (5). They were found to be on Chr 11 by use of mouse–Chinese hamster somatic cell hybrids (7, 13). By *in situ* hybridization, the locus was found to be in band 11D, a region containing *T*s (tail-short), a mutation causing skeletal abnormalities in heterozygotes and early death in homozygotes (12). *Hox-2* is transcribed during embryogenesis from as early as 7.5 days post-coitum onward, with a peak level at 11.5 to 12.5 days. The level is high in embryonic spinal cord and brain and low elsewhere, and is also high in adult kidney but low in other tissues (6).

Hox-3 locus, homeo box-3, Chr 15. This 180-bp sequence was detected with a DNA clone (pMo-EA) isolated from mouse genomic DNA. It was found to be on Chr 15 by use of mouse–Chinese hamster somatic cell hybrids. *Hox-3* shows spatially restricted expression in the newborn central nervous system. It is expressed at high levels in posterior cervical and anterior thoracic regions of the spinal cord, but is not expressed anterior to this region and only weakly posterior to it. Expression occurs at all stages of development from 11 days of development to adulthood (1). A gene in this cluster that may be the same as that detected by pMo-EA is expressed in 12-day embryos and in F9 teratocarcinoma cells differentiating into parietal endoderm (2).

References

1. Awgulewitsch, A., M.F. Utset, C.P. Hart, W. McGinnis, and F.H. Ruddle. 1986. Spatial restriction in expression of a mouse homeo box locus within the central nervous system. Nature 320:328–335.
2. Breier, G., M. Bućan, U. Francke, A. Colberg-Poley, and P. Gruss. 1986. Sequential expression of murine homeo box genes during F9 EC cell differentiation. EMBO J. 5:2209–2215.
3. Bućan, M., T. Yang-Feng, A.M. Colberg-Poley, D.J. Wolgemuth, J.-L. Guénet, U. Francke, and H. Lehrach. 1986. Genetic and cytogenetic localisation of the homeo box containing genes on mouse chromosome 6 and human chromosome 7. EMBO J. 5:2899–2905.
4. Colberg-Poley, A.M., S.D. Voss, K. Chowdhury, C.L. Stewart, E.F. Wagner, and P. Gruss. 1985. Clustered homeo boxes are differentially expressed during murine development. Cell 43:39–45.
5. Hart, C.P., A. Awgulewitsch, A. Feinsod, W. McGinnis, and F.H. Ruddle. 1985. Homeo box gene complex on mouse chromosome 11: molecular cloning, expression in embryogenesis, and homology to a human homeo box gene. Cell 43:9–18.
6. Jackson, I.J., P. Schofield, and B. Hogan. 1985. A mouse homoeo box gene is expressed during embryogenesis and in adult kidney. Nature 317:745–748.
7. Joyner, A.L., R.V. Lebo, Y.W. Kan, R. Tjian, D.R. Cox, and G.R. Martin. 1985. Comparative chromosome mapping of a conserved homoeo box region in mouse and human. Nature 314:173–175.
8. Martin, G.R., E. Boncinelli, D. Duboule, P. Gruss, I. Jackson, R. Krumlauf, P. Lonai, W. McGinnis, F. Ruddle, and D. Wolgemuth. 1987. Nomenclature for homeo-box-containing genes. Nature 325:21–22.
9. McGinnis, W., R.L. Garber, J. Wirz, A. Kuroiwa, and W.J. Gehring. 1984. A homologous protein-coding sequence in Drosophila homeotic genes and its conservation in other metazoans. Cell 37:403–408.
10. McGinnis, W., C.P. Hart, W.J. Gehring, and F.H. Rud-

dle. 1984. Molecular cloning and chromosome mapping of a mouse DNA sequence homologous to homeotic genes of Drosophila. Cell 38:675–680.

11. Mock, B.A., L.A. D'Hoostelaere, R. Matthai, and K. Huppi. 1987. A mouse homeo box gene, *Hox-1.5*, and the morphological locus, *Hd*, map to within 1 cM on chromosome 6. Genetics 116:607–612.

12. Münke, M., D.R. Cox, I.J. Jackson, B.L.M. Hogan, and U. Francke. 1986. The murine *Hox-2* cluster of homeo box containing genes maps distal on chromosome 11 near the tail-short (*Ts*) locus. Cytogenet. Cell Genet. 42:236–240.

13. Rabin, M., C.P. Hart, A. Ferguson-Smith, W. McGinnis, M. Levine, and F.H. Ruddle. 1985. Two homoeo box loci mapped in evolutionarily related mouse and human chromosomes. Nature 314:175–178.

14. Rubin, M.R., L.E. Toth, M.D. Patel, P. D'Eustachio, and M.C. Nguyen-Huu. 1986. A mouse homeo box gene is expressed in spermatocytes and embryos. Science 233:663–667.

Hp locus, haptoglobin, Chr 8

This locus is the structural gene for haptoglobin, an acute phase protein synthesized in liver in response to a variety of systemic injuries. A variant determined by the allele *Hp^b* and identified by two-dimensional acrylamide electrophoresis of cell-free translation product of liver mRNA occurs in wild *M. spretus*. It has a slightly lower molecular weight and is more basic than the protein determined by the allele *Hp^a*, found in all inbred strains examined, including C57BL/6, C3H/He, AKR, BALB/c, and others (2). *Hp* is 7 cM from *Es-1* on Chr 8 (1).

References

1. Baumann, H., and F.G. Berger. 1985. Genetics and evolution of the acute phase proteins in mice. Mol. Gen. Genet. 201:505–512.

2. Baumann, H., W.A. Held, and F.G. Berger. 1984. The acute phase response of mouse liver: genetic analysis of the major acute phase reactants. J. Biol. Chem. 259:566–573.

hpc, hyperspiny Purkinje cell, recessive

Found in an experiment to detect mutations, after treatment with the mutagen Thiotepa, of postmeiotic cells of males of a moderately inbred stock carrying *ru* and *f*. Homozygotes can first be recognized at 6 to 8 days by incoordination and lack of a righting response, symptoms typical of cerebellar defects. Affected mice grow slowly and some die at weaning. With extra care,

most survive to a rather advanced age. Both sexes are sterile. The cerebellum is slightly reduced in size with selective defects in the Purkinje cells. These cells have atrophic dendritic trees, and most of their axon terminals and cell bodies are studded with spines. Most of the axon terminals undergo progressive degeneration but there is little death of Purkinje cells. The synapses between climbing fibers and Purkinje cells are functionally normal. Some heterozygotes are affected by a mild neurological syndrome in old age (1).

References

1. Guénet, J.-L, C. Sotelo, and J. Mariani. 1983. Hyperspiny Purkinje cell, a new neurological mutation in the mouse. J. Hered. 74:105–108.

hpg, hypogonadal, recessive

Found in a stock segregating for *Rb(1.3)1Bnr* on a (C3H/HeH × 101/H) genetic background. Homozygotes of both sexes have underdeveloped reproductive tracts. Males have a small penis and scrotum and short anogenital distance. All male reproductive organs are present but immature. The testes are small and located in the abdomen. Spermatogenesis is arrested usually by the diplotene stage. In females the vagina does not open fully, the uterus is thread-like, and the ovaries are very small. The primary defect is a deficiency of hypothalamic gonadotrophin-releasing hormone (GnRH) with a consequent reduction in pituitary content and circulating levels of luteinizing hormone and follicle-stimulating hormone (3). The gonadotrophic cells of the pituitary are less abundant and smaller than normal (10), but respond normally when stimulated by exogenous GnRH (5, 11). Both ovaries and testes can develop normally when transplanted into normal hosts (2). Implantation of normal fetal preoptic area grafts into the third ventricle of both male and female *hpg/hpg* mice increases GnRH production to nearly normal but does not completely restore normal reproductive functions (6, 7). The *hpg* mutant gene has a 33.5-kb deletion of the distal half of the gene for the common precursor of GnRH and GnRH-associated peptide (GAP), a hormone that is also absent from *hpg/hpg* brains (8). Mason *et al.* (9) were able to produce normal mice transgenic for an additional *hpg^+* gene and, by a mating scheme starting with crosses to *hpg/+* mice, to produce *hpg/hpg* mice carrying the *hpg^+* transgene. These mice, both males and females, were fully fertile and had normal regulation of GnRH and GAP. GnRH-deficient *hpg/hpg* mice can be used to study the function of GnRH and its interrelationship with pituitary and gonadal hormones (1, 4).

References

1. Amador, A., T. Parkening, W. Beamer, A. Bartke, and T.J. Collins. 1984. Autoregulation of testicular luteinizing hormone receptors in hypogonadal (*hpg/hpg*) mice. Biochem. Genet. 22:395–401.
2. Bamber, S., C.A. Iddon, H.M. Charlton, and B.J. Ward. 1980. Transplantation of the gonads of hypogonadal (*hpg*) mice. J. Reprod. Fertil. 58:249–252.
3. Cattanach, B.M., C.A. Iddon, H.M. Charlton, S.A. Chiappa, and G. Fink. 1977. Gonadotrophin-releasing hormone deficiency in a mutant mouse with hypogonadism. Nature 269:338–340.
4. Charlton, H.M., A. Speight, D.M.G. Halpin, A. Bramwell, W.J. Sheward, and G. Fink. 1983. Prolactin measurements in normal and hypogonadal (*hpg*) mice: developmental and experimental studies. Endocrinology 113:545–548.
5. Fink, G., W.J. Sheward, and H.M. Charlton. 1982. Priming effect of luteinizing hormone releasing hormone in the hypogonadal mouse. J. Endocrinol. 94:283–287.
6. Gibson, M.J., D.T. Krieger, H.M. Charlton, E.A. Zimmerman, A.-J. Silverman, and M.J. Perlow. 1984. Mating and pregnancy in genetically hypogonadal mice with preoptic area brain grafts. Science 225:949–951.
7. Krieger, D.T., M.J. Perlow, M.J. Gibson, T.F. Davies, E.A. Zimmerman, M. Ferin, and H.M. Charlton. 1982. Brain grafts reverse hypogonadism of gonadotropin releasing hormone deficiency. Nature 298:468–471.
8. Mason, A.J., J.S. Hayflick, R.T. Zoeller, W.S. Young III, H.S. Phillips, K. Nikolics, and P.H. Seeburg. 1986. A deletion truncating the gonadotropin-releasing hormone gene is responsible for hypogonadism in the *hpg* mouse. Science 234:1366–1371.
9. Mason, A.J., S.L. Pitts, K. Nikolics, E. Szonyi, J.N. Wilcox, P.H. Seeburg, and T.A. Stewart. 1986. The hypogonadal mouse: reproductive functions restored by gene therapy. Science 234:1372–1378.
10. McDowell, I.F.W., J.F. Morris, and H.M. Charlton. 1982. Characterization of the pituitary gonadotroph cells of hypogonadal (*hpg*) male mice: comparison with normal mice. J. Endocrinol. 95:321–330.
11. McDowell, I.F.W., J.F. Morris, H.M. Charlton, and G. Fink. 1982. Effects of luteinizing hormone releasing hormone on the gonadotrophs of hypogonadal (*hpg*) mice. J. Endocrinol. 95:331–340.

hph-1, hyperphenylalaninemia, recessive, Chr 14

Found in an ENU mutagenesis experiment by screening with the Guthrie test for serum phenylalanine. In an interspecific backcross with *M. spretus*, *hph-1* was shown to be closely linked to *Np-1* on Chr 14 (0/29 recombinants) (1).

References

1. Bode, V., F. Bonhomme, and J.-L. Guénet. 1986. Mouse News Lett. 74:97.

Hprt locus, hypoxanthine phosphoribosyltransferase, Chr X

This locus controls electrophoretic mobility of HPRT (E.C. 2.4.2.8) and is probably the structural locus for the enzyme. The allele *Hprt^b* determines a slow anodal band and occurs in strain C3H/He and all other inbred strains tested; the allele *Hprt^a* determines a fast band and was found in wild *Mus spretus* and *M. m. castaneus*. After isoelectric focusing, the enzyme is distributed in five bands, those of HPRT-A being about 0.4 pH units more acidic than those of HPRT-B (2). The wild-derived allele produces a DNA restriction fragment that is different from that of *Hprt^b* (1). *Hprt* was first found to be on the X chromosome by Epstein (3) who showed that the enzyme was synthesized in oocytes and 1-day embryos and that oocytes and 1-day embryos of X/O females had half as much enzyme activity as those of X/X females. The gene was later found, by conventional crosses, to be located about 20 cM proximal to *Pgk-1* on the X chromosome (2). *Hprt* is expressed in all tissues. The embryonic genes probably begin to be expressed in 3-day embryos (3).

References

1. Berger, S. and V. Chapman. 1984. Mouse News Lett. 71:50.
2. Chapman, V.M., P.G. Kratzer, and B.A. Quarantillo. 1983. Electrophoretic variation for X chromosome-linked hypoxanthine phosphoribosyl transferase (HPRT) in wild-derived mice. Genetics 103:785–795.
3. Epstein, C.J. 1972. Expression of the mammalian X chromosome before and after fertilization. Science 175:1467–1468.

Hpt, hair patches, dominant, Chr 4

Arose in a C57BL/6 × C3H hybrid stock in 1979. Both homozygotes and heterozygotes show great variation in expression; some have a permanent coat of very little hair and extensive bare patches, others have an almost normal coat. Hair follicles are abnormal in the bare patches and the skin is littered with heavily pigmented debris. Both sexes are fertile. *Hpt* is located near *b* on Chr 4 (1).

References

1. Lane, P.W. 1981. Mouse News Lett. 65:27.

hpx, hypotransferrinemia with hemochromatosis, recessive, Chr 9

Arose spontaneously in the BALB/cJ strain. Homozygotes usually die within 2 weeks after birth with hypochromic anemia and very low serum transferrin. Injection of transferrin prolongs life for a year or more. The mutant condition is evident in 13-day embryos, which have severe transferrin deficiency and hepatic iron loading. Heterozygotes have normal blood values but half normal concentrations of transferrin. Transferrin from affected mice appears normal in molecular weight, isoelectric point, and antigenic specificity. The condition closely resembles human atransferrinemia. *hpx* is on Chr 9 very close to the transferrin locus *Trf*, but has shown 0.5 per cent (1/196) recombination with *Trf* (1).

References

1. Bernstein, S.E. 1987. Hereditary hypotransferrinemia with hemosiderosis, a murine disorder resembling human atransferrinemia. J. Lab. Clin. Med. 110:690–705.

hpy

See *hop*.

Hq, harlequin, semidominant, Chr X

Arose spontaneously in the outbred CF1 stock. Hemizygous males and homozygous females are almost completely bald. They are fully viable and fertile, but are smaller than normal. In heterozygous females there are bald patches with varying distribution. The bald patches are very much less than 50 per cent of the total area and are not always easy to see (1,2).

References

1. Barber, B.R. 1971. Mouse News Lett. 45:34.
2. Falconer, D.S., and J.H. Isaacson. 1972. Mouse News Lett. 47:28.

hr locus, Chr 14

hr, hairless, recessive. Found in a mouse caught in an aviary in London (1). Hairless mice are fertile, but most females do not nurse their young well. Homozygotes develop a normal coat up to the age of about 10 days, at which time they begin to lose their hair. The complete hair is lost from the follicle. After a short time a few thin fuzzy hairs grow in but are soon lost. Waves of similar growth occur at intervals of about one month thereafter, but the animals eventually appear hairless (2). The vibrissae are repeatedly shed and become

more abnormal with age. The toenails are excessively long and curved. Histologically there is hyperkeratosis of the stratified epithelium and upper part of the hair canals beginning at about 14 days. Hair club formation is abnormal, with the internal root sheath coalescing around the terminal part of the hair shaft so that the lower part of the external root sheath fails to follow the ascending hair club and becomes stranded in the dermis. Cysts develop from two sources, the hyperkeratotic upper part of the hair canals, and the sheaths of the abnormal follicles stranded in the dermis (3, 10). Some cysts arise from isolated sebaceous glands. Regardless of their origin, all cysts undergo a sebaceous transformation and later a keratinization (12). Homozygous hairless mice of the inbred HRS/J strain have a higher incidence of leukemia with earlier onset than their normal sibs, a 13-fold higher titer of ecotropic virus in tail extracts at 6 months of age (5), a 100-fold higher titer of xenotropic virus in the thymus at 8 months of age (6), and a lower cellular immune response to tumor viruses that may be responsible for the increased tumor susceptibility of the mice (8). The incidence of leukemia in *hr/hr* mice can be significantly reduced by passive immunization with virus-specific antiserum (14). The lower cellular immune response of *hr/hr* mice is characteristic of spleen cells but not of lymph node cells and is due to a deficiency of T helper (Ly-1$^+$) cells (13, 15).

hr^{ba}, bald, recessive, probably extinct. Arose spontaneously in a stock of albino mice of unknown origin. This allele is intermediate between *hr* and *hr^{rh}* in its effects (4).

hr^{rh}, rhino, recessive. Arose spontaneously in the descendants of a cross between two inbred lines. Homozygotes are similar to *hr/hr* except that no hair regeneration occurs and the skin becomes thickened and enormously wrinkled (7). The hyperkeratosis in the hair follicles is very extensive, producing hair canal cysts containing balls of keratin in such large numbers that the skin area is greatly increased. In addition, the subepidermal cysts are much larger than in *hr/hr* (3, 10). Rhino females do not nurse their young well. *hr^{rh}/hr* mice resemble *hr/hr* but histologically are somewhat intermediate. Massive doses of vitamin A prevent enlargement of the subepidermal and sebaceous cysts but do not prevent depilation (11). Rhino skin transplanted to normal hosts grows some normal hair around the edges, and normal skin transplanted to rhino hosts grows normal hair except for a narrow hairless strip around the edges (3), possibly indicating an inductive effect of host dermal tissue on follicle formation in the graft epidermis. Rhino mice have spleen cells with a defective response to T-dependent antigens (17) and they develop an autoimmune disease characterized by hypergammaglobulinemia, immunoglobulin deposits in basement membrane of skin, spleen, liver, and kidney, and presence of antinuclear antibodies which occur in young mice and increase with age (9).

Hr^n, near naked, semidominant. Arose spontaneously in a stock many generations removed from an irradiated ancestor. Heterozygotes are recognizable at 5 days of age by their thin slick skin and short curly vibrissae. A thin fuzzy coat develops at about 10 days. Adults undergo repeated hair loss and regrowth but are never entirely bare. Homozygotes are completely naked throughout life. The skin is very thin and becomes deeply wrinkled in adults. Both sexes are fertile but females are poor mothers. *Hr^n/hr* mice resemble *Hr^n/+* mice (16).

References

1. Brooke, H.C. 1926. Hairless mice. J. Hered. 17:173–174.
2. Crew, F.A.E., and L. Mirskaia. 1931. The character 'hairless" in the mouse. J. Genet. 25:17–24.
3. Fraser, F.C. 1946. The expression and interaction of hereditary factors producing hypotrichosis in the mouse: histology and experimental results. Can. J. Res. D 24:10–25.
4. Garber, E.D. 1952. Bald, a second allele of hairless in the mouse. J. Hered. 43:45–46.
5. Heiniger, H.J., R.J. Huebner, and H. Meier. 1976. Effect of allelic substitutions at the hairless locus on endogenous ecotropic murine leukemia virus titers and leukemogenesis. J. Natl. Cancer Inst. 56:1073–1074.
6. Hiai, H., P. Morrissey, R. Khiroya, and R.S. Schwartz. 1977. Selective expression of xenotropic virus in congenic HRS/J (hairless) mice. Nature 270:247–249.
7. Howard, A. 1940. "Rhino," an allele of hairless in the mouse. J. Hered. 31:467–470.
8. Johnson, D.A., and H. Meier. 1981. Immune responsiveness of HRS/J mice to syngeneic lymphoma cells. J. Immunol. 127:461–464.
9. Kawaji, H., R. Tsukuda, and T. Nakaguchi. 1980. Immunopathology of rhino mouse, an autosomal recessive mutant with murine lupus-like disease. Acta Pathol. Jpn. 30:515–530.
10. Mann, S.J. 1971. Hair loss and cyst formation in hairless and rhino mutant mice. Anat. Rec. 170:485–500.
11. Mauer, I. 1961. The effect of vitamin A in hyperkeratotic mouse mutants. J. Exp. Zool. 146:181–207.
12. Montagna, W., H.B. Chase, and H.P. Melaragno. 1952. The skin of hairless mice. I. The formation of cysts and the distribution of lipids. J. Invest. Dermatol. 19:83–94.
13. Morrissey, P.J., D.R. Parkinson, R.S. Schwartz, and S.D. Waksal. 1980. Immunological abnormalities in HRS/J mice. I. Specific deficit in T lymphocyte helper function in a mutant mouse. J. Immunol. 125:1558–1562.
14. Pottathil, R., H. Meier, and R.J. Huebner. 1978. Sup-

pression of leukemogenesis in hairless mice by anti-type C viral gamma globulins. Naturwissenschaften 65:443–444.
15. Reske-Kunz, A.B., M.P. Schéid, and E.A. Boyse. 1979. Disproportion in T-cell subpopulations in immunodeficient mutant *hr/hr* mice. J. Exp. Med. 149:228–233.
16. Stelzner, K.F. 1983. Four dominant autosomal mutations affecting skin and hair development in the mouse. J. Hered. 74:193–196.
17. Takaoki, M., and K. Kawaji. 1980. Impaired antibody response against T-dependent antigens in rhino mice. Immunology 40:27–32.

Hras-1 locus (c-*Hras-1*, c-Ha-*ras-1*), Harvey sarcoma oncogene-1, Chr 7

This is a cellular homolog in the mouse of the transforming gene of the Harvey murine sarcoma virus (v-Ha-*ras*). In this recombinant virus, the *ras* transforming gene was derived from a rat cellular DNA sequence (3). A very similar and perhaps identical gene occurs in a murine sarcoma virus from the BALB/c mouse strain (6). The protein product of *Hras-1* has a molecular weight of 21 000 (3). The virally encoded protein has guanine nucleotide-binding activity and GTP-dependent autophosphorylating activity at a threonine residue (7). *Hras-1* is expressed at all stages of prenatal development in both embryonic and extraembryonic tissues, particularly in bone, brain, kidney, and skin (4). No genetic variants are known. The locus was found to be on Chr 7 by use of mouse–Chinese hamster somatic cell hybrids screened with a viral or a human oncogene probe (2, 5). *Hras-1* and *Kras-2* contain a common DNA sequence of 0.35 kb, indicating that they probably derive from a common ancestral gene (1).

References

1. Ellis, R.W. D. DeFeo, T.Y. Shih, M.A. Gonda, H.A. Young, N. Tsuchida, D.R. Lowy, and E.M. Scolnick. 1981. The p21 *src* genes of Harvey and Kirsten sarcoma viruses originate from divergent members of a family of normal vertebrate genes. Nature 292:506–511.
2. Kozak, C.A., J.F. Sears, and M.D. Hoggan. 1983. Genetic mapping of the mouse oncogenes c-Ha-*ras-1* and c-*fes* to chromosome 7. J. Virol. 47:217–220.
3. Langbeheim, H., T.Y. Shih, and E.M. Scolnick. 1980. Identification of a normal vertebrate cell protein related to the p21 *src* of Harvey murine sarcoma virus. Virology 106:292–300.
4. Müller, R., D.J. Shamon, J.M. Tremblay, M.J. Cline, and I.M. Verma. 1982. Differential expression of cellular oncogenes during pre- and postnatal development of the mouse. 299:640–644.
5. Sakaguchi, A.Y., P.A. Lalley, B.U. Zabel, R.W. Ellis, E.M. Scolnick, and S.L. Naylor. 1984. Chromosome

assignments of four mouse cellular homologs of sarcoma and leukemia virus oncogenes. Proc. Natl. Acad. Sci. USA 81:525–529.
6. Santos, E., S.R. Tronick, S.A. Aronson, S. Pulciani, and M. Barbacid. 1982. T24 human bladder carcinoma oncogene is an activated form of the normal human homologue of BALB-and Harvey-MSV transforming genes. Nature 298:343–347.
7. Shih, T.Y., A.G. Papageorge, P.E. Stokes, M.D. Weeks, and E.M. Scolnick. 1980. Guanine nucleotide-binding and autophosphorylatiny activities associated with the p21^{src} protein of Harvey murine sarcoma virus. Nature 287:686–691.

Hrt-1 locus, heart protein-1

This locus controls variation in pattern of isoelectric focusing in polyacrylamide gels of a basic protein (pI = 8.0) in heart supernatant. The allele *Hrt-1^a* occurs in the DBA/2J strain; the allele *Hrt-1^b* occurs in Peru mice (1, 2).

References

1. Skow, L.C. 1979. Mouse News Lett. 61:55.
2. Skow, L.C. 1983. Mouse News Lett. 69:47.

hs, head spot, recessive

This gene occurs in the GI stock which was derived from an outcross of Goodale's head spot strain and has been selected for large size of white head spots. Homozygotes have a conspicuous white spot on the forehead. The size of the head spot is under the control of multiple modifying genes, but the *hs* locus appears to have a major effect on presence or absence of the head spot. A sufficient number of plus or minus modifying genes can override the effect of *hs*, producing no head spot in *hs/hs* mice or a head spot in +/− mice. No linkage has been found with loci on Chrs 1, 4, 6, or 18 (1).

References

1. Wildman, J., and D.P. Doolittle. 1986. Single gene inheritance of occurrence of head spots in mice. J. Hered. 77:136–138.

Hsd locus, histidase synthetic rate, Chr 10

This locus controls differences between inbred strains in rate of histidine synthesis (2). The allele *Hsd^b* determines high rate of synthesis and occurs in the C57 family of strains; the allele *Hsd^h* determines low rate of synthesis and occurs in strains C3H/He, BALB/c, AKR, and many others. Heterozygotes are intermediate. *Hsd* is on Chr 10 close to Sl (1). This locus is

probably either identical to histidinemia (*his*) or part of a gene complex controlling histidine synthesis (3). See also *his*.

References

1. Arfin, S.M., W.C. Hanford and B.A. Taylor. 1979. Assignment of histidase regulating locus to chromosome 10 of the mouse. Biochem. Genet. 17:529–535.
2. Hanford, W.C., and S.M. Arfin. 1977. Genetic differences in the rate of histidase synthesis in inbred mice. J. Biol. Chem. 252:6695–6699.
3. Wright, A.F., G. Bulfield, S.M. Arfin, and H. Kacser. 1982. Comparison of the properties of histidine ammonia-lyase in normal and histidinemic mutant mice. Biochem. Genet. 20:245–263.

Hst-1 locus, hybrid sterility-1, Chr 17

A particular combination of alleles at this locus determines sterility of heterozygous males. The fertility of females is unaffected. In laboratory mice, the allele *Hst-1^s* occurs in the C57BL/10 strain, and the allele *Hst-1^f* occurs in the C3H/Di strain. A locus in wild mice, which interacts with *Hst-1*, which is located at the same position on Chr 17, and which is probably allelic with *Hst-1*, is tentatively called *Hst^w*. Two alleles have been found, *Hst^{ws}* in wild mice from central Bohemia and Jutland, and *Hst^{wf}* from the same populations and from Great Gull Island, New York. Males of the combination *Hst-1^s/Hst^{ws}* are sterile; all other heterozygotes and all homozygotes are fertile. Testis weight is sufficiently reduced in sterile males so that it can be used as a means of classifying for sterility. In the sterile males, spermatogenesis usually breaks down at the spermatogonial stage. No chromosomal abnormalities that could account for the hybrid sterility have been found. Although *Hst-1* and *Hst^w* are closely linked to the sterility factors of the *t*-complex, they are apparently not identical with these factors. *Hst^{ws}/Hst^{ws}* males yield sterile male offspring when crossed to females not only of the C57BL/10 strain but also of the A/Ph, BALB/c, DBA/1, and AKR/J strains, but the genetic factor responsible for the effect is known to be independent of *Hst-1* in the case of the A/Ph strain and may be different for the other strains also. Strains CBA/J, PL/J, and F/St produced fertile hybrids with *Hst^{ws}/Hst^{ws}* mice (1).

References

1. Forejt, J., and P. Iványi. 1975. Genetic studies on male sterility of hybrids between laboratory and wild mice (*Mus musculus* L.). Genet. Res. 24:189–206.

Hst-2 locus, hybrid sterility-2, Chr 9

F1 hybrid males of a cross between *Mus musculus* and *M. spretus* are sterile. The sterility depends in large part upon heterozygosity at a locus designated *Hst-2*. The allele in a partially inbred strain, SPE derived from *M. spretus*, is designated *Hst-2ˢ*; the allele in the BALB/c strain of *M. musculus* is designated *Hst-2ᵐ*. Female *Hst-2ˢ*/*Hst-2ᵐ* mice are fertile. In a backcross of F1 hybrids between the two strains to BALB/c, *Hst-2* showed close linkage to *Mod-1* on Chr 9 (1).

References
1. Bonhomme, F., J.-L. Guénet, and J. Catalan. 1982. Présence d'un facteur stèrilité mâle, *Hst-2*, segrégeant dans les croisements interspècifiques, *M. musculus* L. × *M. spretus* Lataste et lié à *Mod-1* et *Mpi-1* sur le chromosome 9. C. R. Acad. Sci. Paris 294:691–693.

Ht, hightail, semidominant, Chr 15

Radiation-induced. Homozygotes die before birth. In heterozygotes the tail emerges high, is short and thick at the base, but is never kinked. *Ht*/+ mice are usually smaller than their normal littermates (1).

References
1. Gower, J.S. 1957. Mouse News Lett. 16:37.

Hti locus, HDC thyroxine-inducibility

This locus controls inducibility by thyroxine of histidine decarboxylase (HDC; E.C. 4.1.1.22) activity in kidney. The allele *Htiˢ* determines low inducibility (about 50 per cent of that in C57BL/6) and occurs in the SWR strain; the allele *Htiᵇ* determines high inducibility and occurs in the C57BL/6 and other inbred strains. *Htiˢ* is fully recessive. The induced change in activity is due to a change in HDC concentration (1).

References
1. Martin, S.A., R.J. Middleton, and G. Bulfield. 1984. Mouse News Lett. 70:75.

hub, hyperunconjugated bilirubinemia, recessive

Arose in the randomly bred ICR albino strain at North Carolina State University. Homozygotes appear jaundiced at about 24 hours post partum. They have a major increase in unconjugated bilirubin in serum. The condition improves with age; adults are not jaundiced and have normal bilirubin levels. Body weights are normal but liver and spleen weights and hematocrit values are greater than normal. Adult males have an enlarged scrotum due to inguinal hernia. They are sterile with reduced testis size and no viable sperm, possibly as a result of increased temperature in the scrotum caused by contact with the herniated intestine. Females are fertile but have somewhat reduced maternal performance (1).

References
1. Saxton, A.M., E.J. Eisen, B.H. Johnson, and J.G. Burkhart. 1985. New mutation causing jaundice in mice. J. Hered. 76:441–446.

Hv-1 locus (formerly *Hv*), hepatitis virus susceptibility-1

This locus controls susceptibility to lethal infection by mouse hepatitis virus-2 (MHV-2). The allele *Hv-1ˢ* determines susceptibility and occurs in the PRI strain and the C3H.PRI-*Hv-1ˢ* (C3HSS) congenic strain (5); the allele *Hv-1ʳ* determines resistance and occurs in the C3H/An strain. Heterozygotes are susceptible (1). Resistance or susceptibility is a property of the macrophages. The virus is taken up equally well by resistant and susceptible macrophages but fails to multiply in resistant ones (3). Resistance and susceptibility can be modified by various treatments which have parallel effects in intact mice and in cultured macrophages (4). The *Hv-1* locus is independent of the locus for resistance to flavivirus, *Flv* (2), and it has a different strain distribution from *Hv-2* and controls susceptibility to a different hepatitis virus. See *Hv-2* locus.

References
1. Kantoch, M., and F.B. Bang. 1962. Conversion of genetic resistance of mammalian cells to susceptibility to a virus infection. Proc. Natl. Acad. Sci. USA 48:1553–1559.
2. Kantoch, M., A. Warwick, and F.B. Bang. 1963. The cellular nature of genetic susceptibility to a virus. J. Exp. Med. 117:781–798.
3. Shif, I., and F.B. Bang. 1970. *in vitro* interaction of mouse hepatitis virus and macrophages from genetically resistant mice. I. Adsorption of virus and growth curves. J. Exp. Med. 131:843–850.
4. Weiser, W.Y., and F.B. Bang. 1977. Blocking of *in vitro* and *in vivo* susceptibility to mouse hepatitis virus. J. Exp. Med. 146:1467–1472.
5. Weiser, W., I. Vellisto, and F.B. Bang. 1976. Congenic strains of mice susceptible and resistant to mouse hepatitis virus. Proc. Soc. Exp. Biol. Med. 152:499–502.

Hv-2 locus (also called *Mhv-1*), hepatitis virus susceptibility-2, Chr 7

This locus controls resistance of neurons and macrophages to infection by the JHM and A59 strains of mouse hepatitis virus MHV-4. The allele $Hv\text{-}2^r$ determines resistance and has been found only in the SJL/J strain; the allele $Hv\text{-}2^s$ determines susceptibility and occurs in strains A/J, C57BL/10, BALB/c, NZB, DBA/2, AKR/J, CBA/J, C3H/He, and SWR. Heterozygotes are fully susceptible. *Hv-2* is located on Chr 7 closely linked to *Svp-2* (1–3). Stohlman and Frelinger (4) have concluded that two loci (*Rhv-1, Rhv-2*) are necessary to explain the difference between SJL and other strains in susceptibility to the lethal effects of the JHM strain of mouse hepatitis virus. The cause of this discrepancy is not apparent. See *Rhv-1, Rhv-2* loci.

References

1. Knobler, R.L., M.V. Harpel, and M.B.A. Oldstone. 1981. Mouse hepatitis virus type 4 (JHM strain)-induced fatal central nervous system disease. I. Genetic control and the murine neuron as the susceptible site of disease. J. Exp. Med. 153:832–843.
2. Knobler, R.L., B.A. Taylor, M.K. Woodell, W.G. Beamer, and M.B.A. Oldstone. 1984. Host genetic control of mouse hepatitis virus type-4 (JHM strain) replication. II. The gene locus for susceptibility is linked to the *Svp-1* locus on mouse chromosome 7. Exp. Clin. Immunol. 1:217–222.
3. Smith, M.S., R.E. Click, and B.P.G.W. Plagemann. 1984. Control of mouse hepatitis virus replication in macrophages by a recessive gene on chromosome 7. J. Immunol. 133:428–432.
4. Stohlman, S.A., and J.A. Frelinger. 1978. Resistance to fatal central nervous system disease by mouse hepatitis virus, strain JHM. I. Genetic analysis. Immunogenetics 6:277–281.

Hx, hemimelic extra toes, semidominant, Chr 5

Arose in strain B10.D2oSn. Heterozygotes have preaxial polydactyly on all four feet. They are fertile but some males do not breed, possibly because their crippled limbs prevent them from mating successfully (1). Homozygotes probably die early in embryogenesis. In heterozygotes, hindlimbs are always more severely affected than forelimbs. Typical expression includes shortening of the radius, tibia, and talus with supernumerary metacarpals, metatarsals, and digits. The supernumerary digits are always preaxial. The fibula and ulna are normal in size but often bowed. The humerus, femur, and limb girdles are normal (2). Heterozygotes can be identified by day-11 or -12 of gestation. Knudsen and Kochhar (2) have described embryonic development of the hindlimbs in detail. *Hx* is located very close to hammer-toe, *Hm*, but the two genes are not allelic. They have very different effects on the feet, and one

recombinant has been found in 1664 offspring of double backcrosses (3).

References

1. Dickie, M.M. 1968. Mouse News Lett. 38:24.
2. Knudsen, T.B., and D.M. Kochhar. 1981. The role of morphogenetic cell death during abnormal limb bud outgrowth in mice heterozygous for the dominant mutation *Hemimelic-extra toe* (*Hm^x*). J. Embryol. Exp. Morphol. 65 (Suppl.):289–307.
3. Sweet, H.O. 1982. Mouse News Lett. 66:66.

Hxa locus (formerly *H-X*), H-X antigen, Chr X

Identified by rejection of skin grafts between male offspring of reciprocal crosses between strains C57BL/6 and BALB/c (1, 2). At least four alleles are known. Different alleles occur in BALB/c, DBA/2, A/J, and C57BL/6 and C57BL/10 (3). Strong (4) described an antigen, present in a tumor, that was determined by an X-linked gene which was very probably *Hxa*.

References

1. Bailey, D.W. 1963. Histocompatibility associated with the X chromosome in mice. Transplantation 1:70–74.
2. Bailey, D.W. 1964. Genetically modified survival times of grafts from mice bearing X-linked histocompatibility. Transplantation 2:203–206.
3. Berryman, P.L., and W.K. Silvers. 1979. Studies on the *H-X* locus of mice. I. Analysis of polymorphism. Immunogenetics 9:363–367.
4. Strong, L.C. 1929. Transplantation studies on tumors arising spontaneously in heterozygous individuals. I. Experimental evidence for the theory that the tumor cell has deviated from a definitive somatic cell by a process analagous to genetic mutation. J. Cancer Res. 13:103–115.

hy-1, hydrocephalus-1, recessive, probably extinct

Discovered by Clark (2). Penetrance is incomplete. Homozygotes with recognizable hydrocephalus usually die, but a few reach maturity and breed. The abnormality is recognized externally as a dome-shaped head that may sometimes be seen at birth and in other cases develops during the first 2 weeks (3). Internally, the whole ventricular system is dilated, but the cause has not been determined (1).

References

1. Bonnevie, K., and A. Brodal. 1946. Hereditary hydrocephalus in the house mouse. IV. The development of the cerebellar anomalies during foetal life with notes on the

normal development of the mouse cerebellum. Skr. Norske Vidensk.-Akad. Oslo. I. Mat.-Natur. Kl. 1946. (4) 60p.
2. Clark, F.H. 1932. Hydrocephalus, a hereditary character in the house mouse. Proc. Natl. Acad. Sci. USA 18:654–656.
3. Clark, F.H. 1934. Anatomical basis of a hereditary hydrocephalus in the house mouse. Anat. Rec. 58:225–233.

hy-2, hydrocephalus-2, recessive, probably extinct

Discovered by Zimmermann (2) in the offspring of mice caught in the wild. Penetrance was reported to be complete, but homozygotes never bred. Not allelic with *hy-1* (1). The anomaly resembles that of *hy-1* but is more severe.

References
1. Clark, F.H. 1935. Two hereditary types of hydrocephalus in the house mouse (*Mus musculus*). Proc. Natl. Acad. Sci. USA 21:150–152.
2. Zimmermann, K. 1933. Eine neue Mutation der Hausmaus: "hydrocephalus." Z. Indukt. Abstammungs-Vererbungsl. 64:176–180.

hy-3, hydrocephalus-3, recessive, Chr 8

Found by inbreeding a heterogeneous stock of laboratory mice (2). It has not been tested for allelism with *hy-1* or *hy-2*. Penetrance is incomplete. Homozygotes are usually detectable at 3 to 5 days or sometimes later. Those with frank hydrocephalus die by 4 or 5 weeks of age. The lateral ventricles and the third ventricle are enlarged, the aqueduct of Sylvius and the fourth ventricle are only slightly affected, and there is some dilatation of the ventral subarachnoid cistern. The hydrocephalus seems to be due to a defect in the subarachnoid space under the calvarium caused by an abnormal postnatal differentiation of the arachnoid mater and pia mater which prevents their separation (1). As the hydrocephalus progresses, there is an increase in size of the extracellular space in the telencephalic choroid plexus and ependyma and in the cerebral cortex (3,4). This apparently is a secondary result of the increased fluid pressure in the ventricles.

References
1. Berry, R.J. 1961. The inheritance and pathogenesis of hydrocephalus-3. J. Pathol. Bacteriol. 81:157–167.
2. Grüneberg, H. 1943. Two new mutant genes in the house mouse. J. Genet. 45:22–28.
3. McLone, D.G., W. Bondareff, and A.J. Raimondi. 1973. Hydrocephalus-3, a murine mutant: II. Changes in the brain extracellular space. Surg. Neurol. 1:233–242.
4. Raimondi, A.J., O.T. Bailey, D.G. McLone, R.F. Lawson, and A. Echeverry. 1973. The pathophysiology and morphology of murine hydrocephalus in hy-3 and ch mutants. Surg. Neurol. 1:50–55.

Hya locus (formerly *H-Y*), H-Y antigen, Chr Y

The existence of this locus was first proposed to explain the rejection by females of skin grafts from males of the same strain. The extensive literature on the H-Y antigen has been reviewed by Wachtel (8). Briefly, *Hya* is probably not polymorphic; it appears to be the same in all strains tested (1). The immune response of females to the H-Y antigen is affected by the alleles they carry at the *H-2* complex (3), and the speed of rejection of grafts is affected by *H-2* alleles of the donor male (9). The H-Y antigen has been shown by tissue transplantation techniques to be present on nearly all tissues including eight-cell embryos (8). H-Y antigen on lymphocytes can also be detected by cell-mediated cytotoxicity tests (4) and on sperm by reaction with anti-H-Y antibodies (2). However, it is not certain that the antigen recognized by skin grafts is identical to the one recognized by antiserum (7). Since the Y chromosome is necessary for normal male development, the simplest and most likely explanation for genetic control of the H-Y antigen is that the structural gene is located on the Y chromosome. This conclusion is supported by evidence from the study of the *Sxr* (sex reversed) mutation which has demonstrated that *Hya* is probably near the centromere on the Y chromosome. It has been hypothesized that *Hya* is the primary testis determining gene and that embryos lacking *Hya* develop into females (6). However, a variant *Sxr* mutation was found that caused male development but absence of H-Y antigen. From this observation it was concluded that the *Hya* gene is different from the testis-determining gene (*Tdy*) but probably closely linked to it (5). Two autosomal loci (*Tas*, *Tda*) also affect sexual differentiation, mutant alleles at these loci promoting female development in the presence of H-Y antigen. See *Sxr*, *Tdy*, *Tas*, and *Tda* loci.

References
1. Billingham, R.E., and W.K. Silvers. 1960. Studies on tolerance of the Y chromosome antigen in mice. J. Immunol. 85:14–26.
2. Goldberg, E.H., E.A. Boyse, D. Bennett, M. Scheid, and E.A. Carswell. 1971. Serological demonstration of H-Y (male) antigen on mouse sperm. Nature 232:478–480.
3. Gordon, R.D., and E. Simpson. 1977. Immune response gene control of cytotoxic T-cell responses to H-Y. Transplant. Proc. 9:885–888.

4. Gordon, R.D., E. Simpson, and L.E. Samuelson. 1975. *in vitro* cell-mediated immune responses to the male specific (H-Y) antigen in mice. J. Exp. Med. 142:1108–1120.
5. McLaren, A., E. Simpson, K. Tomonari, P. Chandler, and H. Hogg. 1984. Male sexual differentiation in mice lacking H-Y antigen. Nature 312:552–555.
6. Ohno, S. 1976. Major regulatory genes for mammalian sexual development. Cell 7:315–321.
7. Simpson, E., P. Chandler, L.L. Washburn, H.P. Bunker, and E.M. Eicher. 1983. H-Y typing of karyotypically abnormal mice. Differentiation 23(Suppl.):S116–S120.
8. Wachtel, S.S. 1983. H-Y antigen and the biology of sex determination. Grune and Stratton, New York. 302p.
9. Wachtel, S.S., D.L. Gasser, and W.K. Silvers. 1973. Male-specific antigen: modification of potency by the H-2 locus in mice. Science 181:862–863.

Hye locus, H-Y antigen expression, Chr 17

This locus controls the level of expression of the H-Y antigen on thymocytes as measured by immunogenicity of parental strain thymocytes in F1 hybrids in a host-versus-graft popliteal lymph node enlargement assay. The allele *Hye^h* determines high expression and occurs in the C57BL/10Sn strain and the B10.A(5R) and B10.T congenic strains; the allele *Hye^l* determines low expression and occurs in the C3H/Di strain and the B10.BR and B10.D2 congenic strains. *Hye* is on Chr 17 between *T* and *H-2*, probably somewhat closer to *T* (1).

References

1. Králová, J., and A Lengerová. H-Y antigen: genetic control of the expression as detected by host-versus-graft popliteal lymph node enlargement assay maps between T and H-2 complexes. J. Immunogenet. 6:429–438.

hyh, hydrocephaly with hop gait, recessive, Chr 7

Arose spontaneously in the C57BL/10J strain. Homozygotes have hydrocephalus and a hop gait. They usually die by 2 months of age but on a B6C3 hybrid background some have survived and bred. *hyh* is on Chr 7 near *Gpi-1* (1).

References

1. Lane, P.W. 1985. Mouse News Lett. 73:18.

Hyp, hypophosphatemia, semidominant, Chr X

Found in a linkage cross in 1966. Hemizygous males and heterozygous females can be recognized at 20 to 30 days of age by their shortened hindlimbs and tail. They have reduced body size which persists throughout life

and skeletal changes resembling rickets, both characteristics more marked in *Hyp*/Y males than in *Hyp*/+ females. *Hyp*/*Hyp* females have not been described, but presumably are as severely affected as *Hyp*/Y males. Viability is normal in both sexes, but heterozygous females show better fertility than hemizygous males (4). A bone-inducing substance that induces cartilage followed by ossification in normal mice induced only cartilage followed by unmineralized osteoid in *Hyp*/Y mice (15). Plasma phosphate in hypophosphatemic mice is markedly reduced and plasma calcium is slightly reduced. Phosphate supplementation of drinking water after weaning reduces the severity of the skeletal abnormalities. Fractional phosphate excretion is increased, probably because of low net tubular reabsorption (3, 4, 8). The defect in phosphate transport in the kidney is due to a defect in the brush border membrane of the cortical tubules (13). Activity of protein kinase C is enhanced in the supernatant fraction of *Hyp*/Y kidneys (12). Intestinal transport of phosphate (1, 11) and calcium (2, 7) is significantly reduced. Plasma levels of the vitamin D metabolites 25 hydroxyvitamin D and 1,25-dihydroxyvitamin D_3 (1,25-$(OH)_2D_3$) are normal but the level of 1,25-$(OH)_2D_3$ does not rise in response to low plasma phosphate as it does in normal mice, although it shows a normal elevation in response to a low calcium diet (9, 10). Treatment of 3- to 7-week old *Hyp*/Y males by continuous infusion of 1,25-$(OH)_2D_3$ raised serum phosphate level to normal, presumably through increase in intestinal transport, and corrected the skeletal defects, but did not correct the renal phosphate leak (6). A search for effects of *Hyp* on systems other than skeletal, renal, endocrine, or gastrointestinal unexpectedly showed that *Hyp* causes increased food and O_2 consumption (14). *Hyp* is very closely linked to the X-linked gene *Gy* (gyro) which has very similar effects on serum phosphorus levels and on phosphate transport, but also causes circling behavior (5). See *Gy*.

References

1. Beamer, W.G., M.C. Wilson, and H.F. DeLuca. 1980. Successful treatment of genetically hypophosphatemic mice by 1α-hydroxyvitamin D_5 but not by 1,25-dihydroxyvitamin D_3. Endocrinology 106:1949–1955.
2. Bruns, M.E., R.A. Meyer Jr., and M.H. Meyer. 1984. Low levels of intestinal vitamin D-dependent calcium-binding protein in juvenile X-linked hypophosphatemic mice. Endocrinology 115:1459–1463.
3. Cowgill, L.D., S. Goldfarb, K. Lau, E. Slatopolsky, and Z.A. Agus. 1979. Evidence for an intrinsic renal tubular defect in mice with genetic hypophosphatemic rickets. J. Clin. Invest. 63:1203–1210.

4. Eicher, E.M., J.L. Southard, C.R. Scriver, and F.H. Glorieux. 1976. Hypophosphatemia: mouse model for human familial hypophosphatemic (vitamin D-resistant) rickets. Proc. Natl. Acad. Sci. USA 73:4667–4671.

5. Lyon, M.F., C.R. Scriver, L.R.I. Baker, H.S. Tenenhouse, J. Kronick, and D. Mandla. 1986. The *Gy* mutation: another cause of X-linked hypophosphatemia in mouse. Proc. Natl. Acad. Sci. USA 83:4899–4903.

6. Marie, P.J., R. Travers, and F.H. Glorieux. 1982. Healing of bone lesions with 1,25-dihydroxyvitamin D_3 in the young X-linked hypophosphatemic male mouse. Endocrinology 111:904–911.

7. Meyer, M.H., R.A. Meyer Jr., and R.J. Iorio. 1984. A role for the intestine in the bone disease of juvenile X-linked hypophosphatemic mice: malabsorption of calcium and reduced skeletal mineralization. Endocrinology 115:1464–1470.

8. Meyer, R.A. Jr. 1985. X-linked hypophosphatemia (familial or sex-linked vitamin-D-resistant rickets): X-linked hypophosphatemic (*Hyp*) mice. Am. J. Pathol. 118:340–342.

9. Meyer, R.A. Jr., R.W. Gray, and M.H. Meyer. 1980. Abnormal vitamin D metabolism in the X-linked hypophosphatemic mouse. Endocrinology 107:1577–1581.

10. Meyer, R.A. Jr., R.W. Gray, B.A. Roos, and G.M. Kiebzak. 1982. Increased plasma 1.25-dihydroxyvitamin D after low calcium challenge in X-linked hypophosphatemic mice. Endocrinology 111:174–177.

11. O'Doherty, P.J.A., H.F. DeLuca, and E.M. Eicher. 1977. Lack of effect of vitamin D and its metabolites on intestinal transport in familial hypophosphatemia of mice. Endocrinology 101:1325–1330.

12. Tenenhouse, H.S., and H.L. Henry. 1985. Protein kinase activity and protein kinase inhibitor in mouse kidney: effect of the X-linked *Hyp* mutation and vitamin D status. Endocrinology 117:1719–1726.

13. Tenenhouse, H.S., and C.R. Scriver. 1978. The defect in transcellular transport is located in brush-border membranes in X-linked hypophosphatemia (*Hyp* mouse model). Can. J. Biochem. 56:640–646.

14. Vaughn, L.K., R.A. Meyer Jr., and M.H. Meyer. 1986. Increased metabolic rate in X-linked hypophosphatemic mice. Endocrinology 118:441–445.

15. Yoshikawa, H., K. Masuhara, K. Takaoka, K. Ono, H. Tanaka, and Y. Seino. 1985. Abnormal bone formation induced by implantation of osteosarcoma-derived bone-inducing substance in the X-linked hypophosphatemic mouse. Bone 6:235–239.

hyt (formerly *pet*), hypothyroidism, recessive, Chr 12

Arose spontaneously in the RF/J strain. Homozygotes show retarded growth, very low serum thyroid hormones, elevated serum thyroid-stimulating hormone (TSH), and hypoplastic thyroid glands. Mice of both sexes are infertile but fertility as well as growth can be restored by feeding desiccated thyroid powder. The mice do not respond to exogenous TSH. The defect is thought to be due to unresponsiveness of the mutant thyroid glands to TSH. The ability of the thyroid gland of mutants to concentrate iodine was found to be only 5 to 10 per cent that of normal. Deficient iodine-concentrating ability of the thyroid could be demonstrated as early as 18 days of gestation (1, 2). *hyt/hyt* mice have been used as thyroid-deficient animals in several investigations of the effects of thyroxin on development of various systems. The *hyt* locus is on Chr 12 about 30 cM from the centromere (2).

References

1. Beamer, W.G., and L.A. Cresswell. 1982. Defective thyroid ontogenesis in fetal hypothyroid (*hyt/hyt*) mice. Anat. Rec. 202:387–393.

2. Beamer, W.G., E.M. Eicher, L.J. Malthais, and J.L. Southard. 1981. Inherited primary hypothyroidism in mice. Science 212:61–62.

hz

See *ru-2*.

I

Ia loci

See *H-2* complex.

Iac, iris anomaly with cataract, semidominant

Arose in a (101 × C3H)F1 hybrid after irradiation with ^{137}Cs gamma rays. Homozygotes have not been recovered. In heterozygotes, the shape of the pupil varies from oval to slit-like and pupillary reaction is minimal or absent. The lens shows cortical or total opacity. Manifestation is bilateral. Heterozygotes occasionally have long teeth, leading to a twisted skull, reduced body size, and high mortality (1).

References

1. Kratochvilova, J. 1981. Dominant cataract mutations detected in offspring of gamma-irradiated male mice. J. Hered. 72:302–307.

Ias locus

See *asp-1*.

ic, ichthyosis, recessive, Chr 1

Arose spontaneously in a sib-mated stock (1). This mutation has a generalized effect on nuclear morphology and DNA content, although it was recognized and long known by its effects on skin and hair. In homozygotes the chromatin in leukocytes and many other cell types is abnormally clumped (2, 4). Nuclei of *ic/ic* mice contain about 10 per cent more DNA than those of +/+ mice, with +/*ic* mice being intermediate (6). Homozygotes are recognizable at 2 days of age by their shorter vibrissae. The first coat is delayed and is short, thin, and somewhat curly. During the third and fourth weeks the skin sometimes develops large hard scales. Often the tail has constrictions resulting in loss of the tip. Adults may be bare or have a thin fuzzy coat. Viability and fertility are considerably reduced (1). The skin and hair abnormalities have been described in more detail by Spearman (7) and Jensen and Esterly

(5). The effect of *ic* on skin and hair is exerted through action in the epidermis (3).

References

1. Carter, T.C., and R.S. Phillips. 1950. Ichthyosis, a new recessive mutant in the house mouse. J. Hered. 41:297–300.
2. Goldowitz, D., and R.J. Mullen. 1982. Nuclear morphology of ichthyosis mutant mice as a cell marker in chimeric brain. Dev. Biol. 89:261–267.
3. Green, M.C., B.N. Alpert, and T.C. Mayer. 1974. The site of action of the ichthyosis locus (*ic*) in the mouse, as determined by dermal–epidermal recombinations. J. Embryol. Exp. Morphol. 32:715–721.
4. Green, M.C., L.D. Shultz, and L.A. Nedzi. 1975. Abnormal nuclear morphology of leukocytes in the mouse mutant ichthyosis: a possible transplantation marker. Transplantation 20:172–175.
5. Jensen, J.E., and N.B. Esterly. 1977. The ichthyosis mouse: histological, histochemical, ultrastructural, and autoradiagraphic studies of interfollicular epidermis. J. Invest. Dermatol. 68:23–31.
6. Meyers, R.S., A.S. Klein, J.J. Eppig, and R.A. Eckhardt. 1976. Altered DNA content and chromatin distribution associated with the mouse ichthyosis gene. Genetics 83:s50 (Abstr.).
7. Spearman, R.I. 1960. The skin abnormality of "ichthyosis", a mutant of the house mouse. J. Embryol. Exp. Morphol. 8:387–395.

Id-1 locus

See *Idh-1* locus.

Idc, iris dysplasia with cataract, semidominant

Arose in a (101 × C3H)F1 hybrid irradiated with ^{137}Cs gamma rays. Homozygotes have not been recovered. In heterozygotes, the pupils are irregularly shaped and have no pupillary reaction. The cataractous lenses are attached to the opaque cornea. The eyes are microphthalmic with inflammation of the lids. Manifestation is bilateral. There may be a white belly spot, reduced body size, and high mortality. Penetrance is complete (1).

References

1. Kratochvilova, J. 1981. Dominant cataract mutations detected in offspring of gamma-irradiated male mice. J. Hered. 72:302–307.

Idd-1, *Idd-2* loci

Idd-1 locus, insulin-dependent diabetes-1, Chr 17
Idd-2 locus, insulin-dependent diabetes-2, Chr 9

These two loci control susceptibility to an insulin-dependent diabetes (IDD) which is probably of autoimmune etiology. The NOD strain, derived from the ICR stock by selection and inbreeding, is characterized by spontaneous development of IDD. Genetic crosses to an ICR-derived non-diabetic strain NON showed that NOD is homozygous for two recessive alleles for susceptibility $Idd-1^s$ and $Idd-2^s$. The two dominant alleles for non-susceptibility to IDD, $Idd-1^r$ and $Idd-2^r$, occur in the NON strain. It is likely that at least one other as yet unidentified locus affecting susceptibility to IDD differentiates the NOD and NON strains. Homozygosity for the recessive alleles at all three (or more) loci is necessary for the development of IDD (1).

References

1. Prochazka, M., E.H. Leiter, D.V. Serreze, and D.L. Coleman. 1987. Three recessive loci required for insulin-dependent diabetes in non-obese diabetic mice. Science 237:286–289.

Idh-1 locus (formerly Id-1), isocitrate dehydrogenase-1, Chr 1

This locus controls electrophoretic mobility of the isozyme of NADP-specific isocitrate dehydrogenase (IDH-1; E.C. 1.1.1.41) found mainly in the supernatant fraction of tissue homogenates. The enzyme occurs in liver, kidney, spleen, and muscle (4), and in embryos as young as 9 days (2). The allele $Idh-1^a$ determines a slow migrating band and is found in strains C57BL/6, C3H/He, and many others; the allele $Idh-1^b$ determines a faster band and is found in strain DBA/2 and others (4); the allele $Idh-1^c$ determines a band slower than that of $Idh-1^a$ and is found in *M. m. castaneus* (1) and in *M. m. molossinus* in Japan (6); a fourth allele $Idh-1^d$ determines a band faster than that of $Idh-1^b$ and is found in very low frequency in *M. m. molossinus* in Japan (5, 6). Heterozygotes have the two parental bands and an intermediate band, indicating a dimeric structure of the enzyme. Presence of the hybrid band in skeletal muscle of chimeric mice was used by Mintz and Baker (7) to demonstrate that skeletal muscle forms by fusion of cells. Epstein *et al.* (3) showed that, in mice trisomic for Chr 1 with three *Idh-1* genes, the enzyme activity was 1.5 times normal.

References

1. Chapman, V.M., and F.H. Ruddle. 1972. Glutamate oxaloacetate transaminase (GOT) genetics in the mouse: polymorphism of GOT-1. Genetics 70:299–305.
2. Epstein, C.J., J.A. Weston, W.K. Whitten, and E.S. Russell. 1972. The expression of the isocitrate dehydrogenase locus (*Id-1*) during mouse embryogenesis. Dev. Biol. 27:430–433.
3. Epstein, C.J., G. Tucker, and B. Travis. 1977. Gene dosage for isocitrate dehydrogenase in mouse embryos trisomic for chromosome 1. Nature 267:615–616.
4. Henderson, N.S. 1965. Isozymes of isocitrate dehydrogenase: subunit structure and intracellular location. J. Exp. Zool. 158:263–273.
5. Minezawa, M., K. Moriwaki, and K. Kondo. 1976. Studies on protein polymorphism of Japanese wild mouse, *Mus musculus molossinus*. Ann. Rep. Natl. Inst. Genet. Jpn. 26:23–25.
6. Minezawa, M., K. Moriwaki, and K. Kondo. 1978. Geographical distribution of isocitrate dehydrogenase-1^c in the Japanese wild mouse, *Mus musculus molossinus*. Ann. Rep. Natl. Inst. Genet. Jpn. 28:34–35.
7. Mintz, B., and W.W. Baker. 1967. Normal mammalian muscle differentiation and gene control of isocitrate dehydrogenase synthesis. Proc. Natl. Acad. Sci. USA 58:592–598.

Idh-2 locus, isocitrate dehydrogenase-2, Chr 7

This is probably the structural locus for mitochondrial isocitrate dehydrogenase (IDH-2; E.C. 1.1.1.42). By use of mouse–Chinese hamster somatic cell hybrid clones segregating mouse chromosomes, the locus was found to be on Chr 7 distal to band 7E1 (2). By use of a variant segregating in an interspecific cross, *Idh-2* was found to lie 5 to 10 cM proximal to *Hbb* (1).

References

1. Britton-Davidian, J. 1983. Mouse News Lett. 68:79.
2. Lalley, P.A., J.A. Diaz, and D.S. Woodward. 1979. Comparative mapping in man and mouse involving genes assigned to human chromosomes 11, 15, 19, and X. Cytogenet. Cell Genet. 25:177 (Abstr.).

Ie, eye–ear reduction, semidominant, Chr X

Arose following irradiation of spermatogonia. Homozygous females and hemizygous males are anophthalmic and the external ear is very small with thickened crinkled edges. Heterozygous females may be normal, or they may have varying degrees of abnormality including the typical hemi- or homozygous phenotype (1).

References

1. Hunsicker, P. 1974. Mouse News Lett. 50:51–52.

If-1 locus, NDV-induced circulating interferon, Chr 3

This locus controls circulating level of interferon induced by intravenous injection of Newcastle disease virus. This virus is not pathogenic in mice. The allele *If-1ʰ* determines high level of production of interferon and occurs in strain C57BL/6 and many others; the allele *If-1ˡ* determines a 10-fold lower level and occurs in strain BALB/c and many others. Heterozygotes have levels slightly higher then those of BALB/c (1–3). The interferon phenotype can be transferred with bone marrow cells, indicating that the interferon is probably produced by a cell type present in bone marrow (3). There is no difference in molecular weight of interferon from the two genotypes (4). *If-1* is on Chr 3 very close to the distal end (5).

References

1. DeMaeyer, E., and J. DeMaeyer-Guignard. 1969. Gene with quantitative effect on circulating interferon induced by Newcastle disease virus. J. Virol. 3:506–512.
2. DeMaeyer, E., J. DeMaeyer-Guignard, and D.W. Bailey. 1975. Effect of mouse genotype on interferon production. I. Lines congenic at the *If-1* locus. Immunogenetics 1:438–443.
3. DeMaeyer, E., P. Jullien, J. DeMaeyer-Guignard, and P. Demant. 1975. Effect of mouse genotype on interferon production. II. Distribution of *If-1* alleles among inbred strains and transfer of phenotype by grafting bone marrow cells. Immunogenetics 2:151–160.
4. Martelly, I., D.W. Bailey, and J. DeMaeyer-Guignard. 1972. Poids molécular de l'interféron sérique induit par le virus de la maladie de Newcastle dans des lignées de sourie hautes et basses productrices. Ann. Inst. Pasteur 123:835–848.
5. Mobraaten, L.E., H.P. Bunker, J. DeMaeyer-Guignard, E. DeMaeyer, and D.W. Bailey. 1984. Location of histocompatibility and interferon loci on chromosome 3 of the mouse. J. Hered. 75:233–234.

If-2 locus, MTV-induced circulating interferon

This locus controls level of circulating interferon induced by intravenous inoculation of mouse mammary tumor virus. The allele *If-2ʰ* determines high level of interferon production and occurs in the C57BL/6 strain; the allele *If-2ˡ* determines a lower level and occurs in the BALB/c strain. Heterozygotes do not differ from the low strain. The C57BL/6 strain is resistant to MTV-induced tumor formation, and BALB/c is susceptible, but it is not known whether the difference in interferon level is responsible for the difference in susceptibility (1).

References

1. DeMaeyer, E., J. DeMaeyer-Guignard, W.T. Hall, and D.W. Bailey. 1974. A locus affecting circulating interferon levels induced by mouse mammary tumor virus. J. Gen. Virol. 23:209–211.

If-3, *If-4* loci, Sendai virus-induced interferon-3, -4

These two loci have been postulated to control the level of Sendai virus-induced circulating interferon. The level in C57BL/6By mice is 10 times that in BALB/cBy mice. The difference is due to at least two loci as shown by the intermediate levels in the CXB recombinant inbred strains derived from a cross between these two strains (1).

References

1. DeMaeyer, E., and J. DeMaeyer-Guignard. 1979. Considerations on mouse genes influencing interferon production and action. *In* I. Gresser, ed., Interferon, Vol. I, 75–100. Academic Press, New York.

If-X locus, early interferon induction, Chr X

Level of early interferon (IFN) induction (2 to 3 hours after treatment) by Newcastle disease virus (NDV) and herpes simplex virus (HSV) is controlled predominantly by X-linked loci, probably identical for the two viruses (1, 5). Strain C57BL/6 carries the allele for high early IFN induction and strain BALB/c the allele for low early induction. F1 females have levels nearly as high as C57BL/6 females. Early IFN induction is very radio-resistant, suggesting that macrophages may be involved. This is in contrast to late IFN induction by NDV, controlled by the *If-1* locus on Chr 3, which is very radiation-sensitive (1). The *If-X* locus may also be responsible for a strain difference in susceptibility to lethal hepatitis induced by HSV. C57BL/6 and GR are resistant and BALB/c is susceptible and the difference is due in part to an X-linked gene (2, 3). The resistance of C57BL/6 is greater at high doses than at low or intermediate doses and this effect seems to be due to early interferon production induced by high doses of the virus. The similar pattern of inheritance of resistance to

HSV and induction of IFN by HSV and NDV suggests that a single X-linked locus, *If-X*, is involved in all three phenomena (3, 4).

References

1. DeMaeyer-Guignard, J., R. Zawatsky, F. Dandoy, and E. DeMaeyer. 1983. An X-linked locus influences early serum interferon levels in the mouse. J. Interferon Res. 3:241–252.
2. Mogensen, S.C. 1980. Genetics of macrophage-controlled natural resistance to hepatitis induced by herpes simplex virus type 2 in mice. *In* E. Skamene, P.A.L. Kongshavn, and M. Landy, eds., Genetic Control of Natural Resistance to Infection and Malignancy, 291–296. Academic Press, New York.
3. Pedersen, E.B., S. Haahr, and S.C. Mogensen. 1983. X-linked resistance of mice to high doses of herpes simplex virus type 2 correlates with early interferon production. Infect. Immun. 42:740–746.
4. Zawatsky, R., I. Gresser, E. DeMaeyer, and H. Kirschner. 1982. The role of interferon in the resistance of C57BL/6 mice to various doses of herpes simplex virus type 1. J. Infect. Dis. 146:405–410.
5. Zawatsky, R., H. Kirschner, J. DeMaeyer-Guignard, and E. DeMaeyer. 1982. An X-linked locus influences the amount of circulating interferon induced in the mouse by herpes simplex virus type 1. J. Gen. Virol. 63:325–332.

Ifa, *Ifb* loci, Chr 4

Ifa complex, interferon-α gene family. This complex codes for the family of mouse α-interferon (IFA) genes that may contain 12 or more loci (1, 5, 6). Several of these genes have been isolated and characterized. They do not contain introns (2, 6, 7). By use of mouse–Chinese hamster somatic cell hybrids, the complex was shown to be on Chr 4 in several laboratories (3–5). Several restriction fragment length polymorphisms (RFLP) have been found and have permitted mapping of *Ifa* to the central portion of Chr 4 near *H-15*. Haplotypes for these RFLPs showing minor differences occur among inbred strains. Different haplotypes occur in BALB/c and C57BL/6 (1); four different haplotypes distinguish DBA/2 / SWR / ICR, AKR / and C3H, C57BL/6 (5). The haplotype of *M. spretus* shows major differences from those of inbred strains (1). The *Ifa* complex is very closely linked to *Ifb*.

Ifb locus, interferon-β. This is the locus of the single interferon-β gene present in the mouse genome. Like *Ifa*, *Ifb* contains no introns. *Ifb* was shown to be on Chr 4 by use of mouse–Chinese hamster somatic cell hybrids (4). No genetic variants have been found among inbred strains but a restriction fragment length variant occurs in *M. spretus*. This was used in an inter-specific cross which showed that no recombination occurred between *Ifa* and *Ifb* (1).

References

1. Dandoy, F., E. deMaeyer, F. Bonhomme, J.-L. Guénet, and J. DeMaeyer-Guignard. 1985. Segregation of restriction fragment length polymorphism in an interspecific cross of laboratory and wild mice indicates tight linkage of the murine IFN-β gene to the murine IFN-α genes. J. Virol. 56:216–220.
2. Kelley, K.A., and P.M. Pitha. 1985. Characterization of a mouse interferon gene locus. I. Isolation of a cluster of four α-interferon genes. Nucl. Acids Res. 13:805–823.
3. Kelley, K.A., C.A. Kozak, F. Dandoy, F. Sor, J.D. Windass, J. DeMaeyer-Guignard, P.M. Pitha, and E. DeMaeyer. 1983. Mapping of mouse interferon-α genes to chromosome 4. Gene 26:181–188.
4. van der Korput, J.A.G.M., J. Hilgers, V. Kroezen, E.C. Swarthoff, and J. Trapman. 1985. Mouse interferon alpha and beta genes are linked at the centromere proximal region of chromosome 4. J. Gen. Virol. 66:493–502.
5. Lovett, M., D.R. Cox, D. Yee, W. Boll, C. Weissmann, C.J. Epstein, and L.B. Epstein. 1984. The chromosomal location of mouse interferon α genes. EMBO J. 3:1643–1646.
6. Shaw, G.D., W. Boll, M. Taira, N. Mantei, P. Lengyel, and C. Weissmann. 1983. Structure and expression of cloned murine IFN-α genes. Nucl. Acids Res. 11:555–573.
7. Zwarthoff, E.C., A.T.A. Mooren, and J. Trapman. 1985. Organization, structure and expression of murine interferon alpha genes. Nucl. Acids Res. 13:791–804.

Ifbip-1 locus (gene 202), interferon-beta-induced protein-1, Chr 1

This locus codes for a 56.5-kDa protein that can be induced in Ehrlich ascites tumor cells by beta interferon. The function of the protein is unknown. The locus was found to be on Chr 1 by use of mouse–Chinese hamster somatic cell hybrids screened with a cDNA probe complementary to an mRNA translatable *in vitro* into the 56.5-kDa protein (1).

References

1. Samanta, H., D.D. Pravtcheva, F.H. Ruddle, and P. Lengyel. 1984. Chromosomal location of mouse gene 202 which is induced by interferons and specifies a 56.5 kD protein. J. Interferon Res. 4:295–300.

Ifg locus, interferon-γ, Chr 10

This locus codes for immune or γ-interferon. No genetic variants are known. The locus was shown to be on Chr 10 using mouse–Chinese hamster somatic cell hybrids and a cDNA clone specific for mouse gamma interferon (1).

References

1. Naylor, S.L., P.W. Gray, and P.A. Lalley. 1984. Mouse immune interferon (IFN-γ) gene is on chromosome 10. Somat. Cell Genet. 10:531–534.

Ifgr locus, interferon γ receptor, Chr 10

This is the structural locus for the receptor for immune or γ interferon. When cells are incubated with labeled recombinant immune interferon and subsequently crosslinked with disuccinimidyl suberate, a major complex of interferon and receptor is formed and can be recognized by SDS polyacrylamide gel electrophoresis. By use of this technique to classify mouse–Chinese hamster somatic cell hybrids, *Ifgr* was found to be on Chr 10 (1).

References

1. Mariano, T.M., C.A. Kozak, J.A. Langer, and S. Pestka. 1987. The mouse immune interferon receptor gene is located on chromosome 10. J. Biol. Chem. 262:5812–5814.

Ifrc locus, interferon receptor, Chr 16

This locus controls presence of the receptor for α and β interferon on cells as measured by ability of the cells to be protected by mouse interferon against challenge by vesicular stomatitis virus. No genetic variants are known. *Ifrc* was shown to be on Chr 16 by use of mouse–Chinese hamster (1) and mouse–human (3) somatic cell hybrids. *Ifrc* may be homologous with a similar locus IFRC in man known to be located on human chromosome 21 (2).

References

1. Cox, D.R., L.B. Epstein, and C.J. Epstein. 1980. Genes coding for sensitivity to interferon (IfRec) and soluble superoxide dismutase (SOD-1) are linked in mouse and man and map to mouse chromosome 16. Proc. Natl. Acad. Sci. USA 77:2168–2172.
2. Epstein, C.J., and L.B. Epstein. 1982. Genetic control of the response to interferon. Texas Rep. Biol. Med. 41:324–331.
3. Lin, P.-F., D.L. Slate, F.C. Lawyer, and F.H. Ruddle. 1980. Assignment of the murine interferon sensitivity and cytoplasmic superoxide dismutase genes to chromosome 16. Science 209:285–287.

Igh complex, immunoglobulin heavy chain, Chr 12

The nomenclature used here is that of Green (13). This chromosome region contains loci controlling both the constant and variable regions of the immunoglobulin heavy chains. Molecular genetic studies have shown that in the germline the number of genes and their order from 5′ to 3′ is approximately 100 or more *V* (variable) genes, 12 or more *D* (diversity) genes, about four *J* (joining) genes, and eight *C* (constant) genes (8, 15, 34). During differentiation of B lymphocytes, in which the immunoglobulin genes are expressed, the DNA is rearranged, bringing together one of each of the four classes, so that an immunoglobulin heavy chain consisting of a particular combination of the four parts is synthesized in one piece. The *Igh* complex was found to be on Chr 12 almost simultaneously by standard genetic crosses (25) and by use of somatic cell hybrids (7).

Igh-C, immunoglobulin heavy-chain constant region (formerly *Ig-1* to *Ig-6*). The eight constant region loci are located in a tightly linked cluster distal to the variable region loci (38). The loci are very closely linked, no recombinants having been observed in over 5000 offspring of crosses designed to look for them. The order of the *Igh-C* genes as determined by molecular cloning is *Igh-2* (α), *Igh-7* (ε), *Igh-1* (γ2a), *Igh-3* (γ2b), *Igh-4* (γ1), *Igh-8* (γ3), *Igh-5* (δ), *Igh-6* (μ), *Igh-V* (33). Variants are known for all eight loci. The greatest number of alleles occurs at the *Igh-1* locus, and most strains having the same *Igh-1* allele have the same allele at the other loci. It has therefore been convenient to designate most of the haplotypes by superscripts corresponding to those of the *Igh-1* alleles. Table 2.3 gives the distribution of the alleles at the eight loci in the 15 haplotypes (13, 14, 16, 20, 26, 29).

Igh-1 locus (γ2a), immunoglobulin heavy chain-1. Variants were detected in serum IgG2a immunoglobulin by use of precipitating antisera prepared in mice of one inbred strain against immunoglobulin of another strain (alloantisera) (14). Thirteen alleles are known (Table 2.3) (16, 20). Alleles *Igh-1ᵃ* of BALB/c and *Igh-1ᵇ* of C57BL/6 have multiple amino acid sequence differences, probably arising from gene conversion events (32).

Igh-2 locus (α), immunoglobulin heavy chain-2. Variants were detected in serum IgA immunoglobulin by use of alloantisera. Five alleles are known (Table 2.3) (14, 20).

Igh-3 locus (γ2b), immunoglobulin heavy chain-3. Variants were detected in serum IgG2b immunoglobulin by use of alloantisera. Six alleles are known (Table 2.3) (14, 20).

Igh-4 locus (γ1), immunoglobulin heavy chain-4. Variants were originally detected in serum IgG1 immunoglobulin by a difference in electrophoretic mobility and were later defined by alloantisera as well. Two alleles are known (Table 2.3) (14, 20).

Table 2.3 Distribution of alleles of the *Igh* loci in the *Igh* haplotypes

Haplotype	Prototype strain	Locus and chain									
		Igh-1 γ2a	*-2* α	*-3* γ2b	*-4* γ1	*-5* δ	*-6* μ	*-7* ε	*-8* γ3	*-V*	*-D*
a	BALB/c	a	a	a	a	a	a	a	a	a	a
b	C57BL/6	b	b	b	b	b	b	b	a	b	b
c	DBA/2	c	c	a	a	a	–	–	a	c	c
d	AKR	d	d	d	a	a	–	a	a	d,j	a
e	A	e	d	e	a	e	e	a	a	e	e
f	CE	f	f	f	a	a	–	–	a	f	f
g	RIII	g	c	g	a	a	–	–	a	a	a
h	SEA	h	a	a	a	a	–	–	a	–	–
j	CBA/H, C3H/He	j	a	a	a	a	a	a	a	j	g
k	KH-1 (wild)	k	c	a	a	–	–	–	–	–	–
l	KH-2 (wild)	l	c	a	a	–	–	–	–	–	–
m	Ky (wild)	m	b	b	b	–	–	–	–	–	–
n	NZB	e	d	e	a	a	e	–	a	d,j	a
o	AL/N	e	–	d	–	e	–	–	a	e	e
p	SWR/J	p	–	–	–	–	–	–	a	f,g	g

Igh-5 locus (δ), immunoglobulin heavy chain-5. This locus controls the heavy chain of an IgD cell surface immunoglobulin. Variants were originally detected by alloantisera prepared in C57BL/6 against CBA/H spleen cells (12). Three alleles are known (Table 2.3) (20, 29). IgD occurs in two molecular forms, the four-chain form typical of other immunoglobulins with two heavy and two light chains (IgD$_I$) and a two-chain form with one heavy and one light chain (IgD$_{II}$) The ratio of I/II varies among inbred strains; a low ratio occurs in IgD-5^e strains and a high ratio in IgD-5^a or IgD-5^b strains (29). *Igh-5*-determined immunoglobulin is found on lymphocytes from spleen, mesenteric lymph nodes, and Peyer's patches, but not from bone marrow or thymus (12).

Igh-6 locus (μ), immunoglobulin heavy chain-6. This locus controls the heavy chain of IgM immunoglobulin. Variants were detected by alloantisera raised against serum immunoglobulins or against spleen cells. Three alleles are known (Table 2.3). Alleles *Igh-6^a* and *Igh-6^b* in BALB/c and C57BL/6, respectively, differ at only one nucleotide, leading to an arginine in the product of *Igh-6^a* and a lysine in that of *Igh-6^b* (32). *Igh-6* determinants occur on 19S IgM of serum and on cell surface 8S IgM of spleen cells (12, 20, 36).

Igh-7 locus (ε), immunoglobulin heavy chain-7. Variation was detected by monoclonal antisera produced by reciprocal immunization of C57BL/6 and BALB/c with the IgE produced in large quantities by hybridomas derived from the two strains. Two alleles are known (Table 2.3) (4).

Igh-8 locus (γ3), immunoglobulin heavy chain-8. No variants of IgG3 are known among laboratory strains or in most wild populations, but the wild *Mus spretus* sub-species contains a variant recognized by alloantiserum prepared in *M. spretus* against BALB/c myeloma IgG3 or by rat anti-BALB/c monoclonal antibodies (1, 16).

Igh-V, immunoglobulin heavy-chain variable region. The variable region of the heavy chain is synthesized from a DNA segment consisting of a *V*, a *D*, and a *J* gene. Molecular genetic studies have shown that the 100 or more *V* genes of inbred strains are grouped into seven families within which there is more than 80 per cent sequence homology and between which there is less than 70 per cent homology (6). Polymorphism of the *V* and *D* sequences allows the assignment of 'allelic' symbols to these regions (Table 2.3) (6, 35). In classical genetic studies, the variable region loci have been identified by variants found among inbred strains in capacity to produce antibodies of particular specificities. The variants are detected either by variation in fine specificity of the induced antibodies or by antibodies raised against myeloma proteins (homogeneous immunoglobulins secreted by plasma cell tumors) that detect antigenic determinants (idiotypes) in the heavy-chain variable region of the antibody molecules. Molecular analysis of some of these variants has shown that they depend on a single or on very few *Igh-V* genes (summary and references in Rajewsky and Takemori, 30, p. 577 ff). Table 2.4 lists the known variants and the loci postulated to determine them. Some of the loci have not shown recombination and may in

Table 2.4 Immunoglobulin heavy-chain variable-region loci, *Igh-V*.

Ab = Antibody, MP = Myeloma protein

Locus symbol	Antigen	Idiotype	Reference immunoglobulin	Reference strain	Reference
Igh-Aa1	ABA-HOP	ABA-HOP-b	Ab	C57BL	23
Igh-Aa2	ABA-HOP	ABA-HOP-e	Ab	A/J	23
Igh-Aa3	ABA-HOP	ABA-HOP-j	Ab	CBA/H	23
Igh-Ars	Arsanyl	ARS	Ab	A/J	27
Igh-Dex	1-3-dextran	J558	MP	BALB/c	3
Igh-Inu	Inulin	Inuldx	MP	BALB/c	22
Igh-Lev	Levan	U10-173	MP	BALB/c	5
Igh-Nbp	NBrP	NBrP-a	Ab	BALB/c	18
Igh-Np	NP	NP	Ab	C57BL	17
Igh-Ns1	Nuclease (1–99)	Nase-A	Ab	A/J	28
Igh-Ns2	Nuclease (1–99)	Nase-B	Ab	SJL	28
Igh-Ns3	Nuclease (1–99)	Nase-C	Ab	C57BL/10	28
Igh-Ns4	Nuclease (99–149)	Nase-D	Ab	A/J	28
Igh-Ns5	Nuclease (99–149)	Nase-E	Ab	SJL	28
Igh-Pc	PC	T15	MP	BALB/c	21,11
Igh-Sa1	Strep-A-CHO	A5A	Ab	A/J	9
Igh-Sa2	Strep-A-CHO	A5Acr	Ab	C57L	2
Igh-Sa3	Strep-A-CHO	S117	MP	BALB/c	10
Igh-Sa4	Strep-A-CHO	S117cr	MP	DBA/2	2
Igh-Sa5	Strep-A-CHO	ACH[d]	Ab	AKR	19
Igh-Src	SRBC	ESE	Ab	C57BL/6	24

fact be alleles. The loci determining these variants are closely linked to the *Igh-C* loci but show some recombination with them and with each other. Recombination frequencies of the order of 0.5 to 5 per cent have been observed (2, 9, 37). There is a high frequency of double recombination in the *Igh-V*, *Igh-C*, *Pre-1* interval (9, 31).

References

1. Amor, M., F. Bonhomme, J.-L. Guénet, F. Potter, and P.-A. Cazanave. 1984. Polymorphism of heavy chain immunoglobulin isotypes in *Mus* subgenus. I. Limited $\gamma 3$ polymorphism revealed by antibodies raised in SPE wild-derived inbred strain. Immunogenetics 20:577–581.

2. Berek, C., B.A. Taylor, and K. Eichmann. 1976. Genetics of the idiotype of BALB/c myeloma S117: multiple chromosomal loci for V_H genes encoding specificity for group A streptococcal carbohydrate. J. Exp. Med. 144:1164–1174.

3. Blomberg, B., W.R. Geckeler, and M. Weigert. 1972. Genetics of the response to dextran in mice. Science 177:178–180.

4. Borges, M.S., Y. Kumagai, K. Okumura, N. Hirayama, Z. Ovary, and T. Tada. 1981. Allelic polymorphism of murine IgE controlled by the seventh immunoglobulin heavy chain allotype locus. Immunogenetics 13:499–507.

5. Bosma, M.J., C. De Witt, S.J. Hausman, R. Marks, M. Potter, and B. Taylor. 1977. A new heavy chain marker in mice. Immunogenetics 4:418–419 (Abstr.).

6. Brodeur, P.H., and R. Riblet. 1984. The immunoglobulin heavy chain variable region (Igh-V) locus in the mouse. I. One hundred Igh-V genes comprise seven families of homologous genes. Eur. J. Immunol. 14:922–930.

7. D'Eustachio, P., D. Pravtcheva, K. Marcu, and F.H. Ruddle. 1980. Chromosomal location of the structural gene cluster encoding murine immunoglobulin heavy chains. J. Exp. Med. 151:1545–1550.

8. Early, P., H. Huang, M. Davis, K. Calame, and L. Hood. 1980. An immunoglobulin heavy chain variable region gene is generated from three segments of DNA: V_H, D, and J_H. Cell 19:981–992.

9. Eichmann, K. 1973. Idiotype expression and inheritance of mouse antibody clones. J. Exp. Med. 137:603–621.

10. Eichmann, K. 1975. Genetic control of antibody specificity in the mouse. Immunogenetics 2:491–506.

11. Gearhart, P.J., and J.J. Cebra. 1978. Idiotype sharing by murine strains differing in immunoglobulin allotype. Nature 272:264–265.

12. Goding, J.W., D.W. Scott, and J.E. Layton. 1977. Genetics, cellular expression, and function of IgD and IgM receptors. Immunol. Rev. 37:152–186.

13. Green, M.C. 1979. Genetic nomenclature for the immunoglobulin loci of the mouse. Immunogenetics 8:89–97.

14. Herzenberg, L.A., H.O. McDevitt, and L.A. Herzenberg. 1968. Genetics of antibodies. Ann. Rev. Genet. 2:209–244.

15. Honjo, T. 1983. Immunoglobulin genes. Ann. Rev. Immunol. 1:499–528.

16. Huang, C.-M., H.-J.S. Huang, and S.-C. Lee. 1984. Detection of immunoglobulin heavy chain IgG$_3$ polymorphism in wild mice with xenogeneic monoclonal antibodies. Immunogenetics 20:565–575.

17. Imanishi, T., and O. Mäkelä. 1974. Inheritance of antibody specificity. I. Anti-(4-hydroxy-3-nitrophenyl) acetyl of the mouse primary response. J. Exp. Med. 140:1498–1510.

18. Imanishi, T., and O. Mäkelä. 1975. Inheritance of antibody specificity. II. Anti-(4-hydroxy-3-nitrophenyl) acetyl in the mouse. J. Exp. Med. 141:840–854.

19. Krammer, P., B.A. Taylor, and K. Eichmann. 1978. Genetics of the idiotype of strain AKR antibodies to group A streptococcal carbohydrate: further evidence for a low degree of homology in the V$_H$ chromosomal region. Z. Immun.-Forsch. 154:284–295.

20. Lieberman, R. 1978. Genetics of the IgCH (allotype) locus in the mouse. Springer Seminars Immunopathol. 1:7–30.

21. Lieberman, R., M. Potter, E.C. Mushinski, W. Humphrey Jr., and S. Rudikoff. 1974. Genetics of a new IgV$_H$ (T15 idiotype) marker in the mouse regulating natural antibody to phosphorylcholine. J. Exp. Med. 139:983–1001.

22. Lieberman, R., M. Potter, and W. Humphrey Jr. 1977. Genetics of $\beta2 \rightarrow 1$ fructosan (inulin) antibodies in mice. Immunogenetics 4:416 (Abstr.).

23. Mäkelä, O., M. Julin, and M. Becker. 1976. Inheritance of antibody specificity. III. A new V$_H$ gene controls fine specificity of anti-*p*-azobenzenearsonate coupled to the carbon 5 of hydroxyphenylacetic acid in the mouse. J. Exp. Med. 143:316–328.

24. McCarthy, M.M., and R.W. Dutton. 1975. The humoral response of mouse spleen cells to two types of sheep erythrocytes. III. A new V$_H$ region marker. J. Immunol. 115:1327–1329.

25. Meo, T., J. Johnson, C.V. Beechey, S.J. Andrews, J. Peters, and A.G. Searle. 1980. Linkage analysis of murine immunoglobulin heavy chain and serum prealbumin genes establish their location on chromosome 12 proximal to the *T(5;12)31H* breakpoint in band 12F1. Proc. Natl. Acad. Sci. USA 77:550–553.

26. Parsone, M., L.A. Herzenberg, A.M. Stall, and L.A. Herzenberg. 1986. Mouse immunoglobulin allotypes. *In* D.M. Weir, ed., Handbook of Experimental Immunology vol. 3, Genetic and Molecular Immunology, 97.1–97.17. Blackwell Scientific Publ., Oxford.

27. Pawlak, L.L., E.B. Mushinski, A. Nisonaff, and M. Potter. 1973. Evidence for linkage of the IgC$_H$ locus to a gene controlling the idiotype specificity of anti-*p*-azophenylarsonate antibodies in strain A mice. J. Exp. Med. 137:22–31.

28. Pisetsky, D.S., and D.H. Sachs. 1978. Genetic control of the immune response to staphylococcal nuclease. VIII. Mapping of genes for antibodies to different antigenic regions of nuclease. J. Exp. Med. 147:1517–1525.

29. Pollock, R.R., M.E. Dorf, and M.F. Mescher. 1980.

30. Rajewsky, K., and T. Takemori. 1983. Genetics, expression, and function of idiotypes. Ann. Rev. Immunol. 1:569–607.

31. Riblet, R., and D. Boyer. 1977. Unusual genetic features of the heavy-chain complex in the mouse. Immunogenetics 5:512 (Abstr.).

32. Schreier, P.H., S. Quester, and A. Bothwell. 1986. Allotype differences in murine μ genes. Nucl. Acids Res. 14:2381–2389.

33. Shimizu, A., T. Kataoka, and T. Honjo. 1981. Ordering of mouse immunoglobulin heavy chain genes by molecular cloning. Nature 289:149–153.

34. Tonegawa, S. 1983. Somatic generation of antibody diversity. Nature 302:575–581.

35. Trepicchio, W. Jr., and K.J. Barrett. 1985. The IGH-V locus of MRL mice: restriction fragment length polymorphism in eleven strains of mice as determined with V$_H$ and D gene probes. J. Immunol. 134:2734–2739.

36. Warner, N.L., J.W. Goding, G.A. Gutman, G.W. Warr, L.A. Herzenberg, B.A. Osborne, S. van der Loo, S.J. Black, and M.R. Loken. 1977. Allotypes of mouse IgM immunoglobulin. Nature 265:447–449.

37. Weigert, M., and M. Potter. 1977. Antibody variable-region genetics: summary and abstracts of the Homogeneous Immunoglobulin Workshop VII. Immunogenetics 4:401–435.

38. Weigert, M., and R. Riblet. 1978. The genetic control of antibody variable regions of the mouse. Springer Seminars Immunopathol. 1:133–169.

Igj locus, immunoglobulin joining chain, Chr 5

This locus controls the structure of the J chain necessary for the polymeric assembly of IgM and IgA. No genetic variants are known. The locus was found to be on Chr 5 by use of mouse–Chinese hamster somatic cell hybrids screened with a ^{32}P-labeled cDNA probe (1).

References

1. Yagi, M., P. D'Eustachio, F.H. Ruddle, and M.E. Koshland. 1982. J chain is encoded by a single gene unlinked to other immunoglobulin structural genes. J. Exp. Med. 155:647–654.

Igk complex, immunoglobulin kappa chain, Chr 6

This complex in the germline consists of a single *C* gene, a cluster of five *J* segments, and a cluster of 90 to 300 different *V* segments. In lymphocytes the *C* gene joins with one *J* and one *V* gene to form a complete *Igk* gene (12, 17). By use of *Igk-V* region variants, the com-

Genetic control of murine IgD structural heterogeneity. Proc. Natl. Acad. Sci. USA 77:4256–4259.

plex was shown to be linked with 0.30 per cent recombination to the *Ly-2,3* locus on Chr 6 (8). Three recombinants in the *Igk-V* region have been found in *Igk* congenic strains. Analysis of the DNA of these strains showed that the order of loci on Chr 6 is (*Ly-2*, *Igk-V21*); *Igk-Ef1*; (*Igk-V4*, *-V8*, *-V10*); (*Igk-V11*, *-V24*, *-Ef2*) (4).

Igk-C, immunoglobulin κ-chain constant region. No genetic variants in kappa-chain constant-region polypeptides have been identified, but a DNA polymorphism detectable with *Bam*HI, *Igk-C^a*, was found in strains SJL, STA, RB156, MA/My, and Peru-Atteck. The allele *Igk-C^b* occurs in all other strains examined (13).

Igk-V, immunoglobulin κ-chain variable region. The variants controlled by this region have been recognized by isoelectric focusing or molecular structure of immunoglobulin κ-chains, by variation in level of expression of immunoglobulin idiotypes that segregate with the *Ly-3* locus to which *Igk* is very closely linked, or as restriction fragment length polymorphisms. *Igk-Ef1*, *-Pc*, and *-Trp* have the same strain distribution of alleles as *Ly-3*, but one recombination between *Igk-Ef1* and *Ly-3* has been found (8).

Igk-Ars locus, immunoglobulin κ-chain-Ars. This locus was recognized as presence or absence of the idiotype of anti-p-azophenylarsonate antibodies produced in strain A/J mice. The allele *Igk-Ars^b* determines presence of determinants of the idiotype on the $V_κ$ chains of these antibodies and occurs in the A/J, BALB/c, C57BL/10, CBA/J, C3H/St, SJL, and 129 strains; the allele *Igk-Ars^a* determines absence of these determinants and occurs in the PL, AKR, and C58 strains. Expression of this idiotype is also affected by *Igh-V* genes (1, 14).

Igk-Dex locus, immunoglobulin κ-chain-Dex. This locus controls presence or absence of a $V_κ$ response to α(1,3) dextran after hyperimmunization in strains that do not make an *Igl-1* response to this antigen. The allele *Igk-Dex^b* determines responsiveness and occurs in the C57BL/6 strain; the allele *Igk-Dex^a* determines unresponsiveness and occurs in the BALB/c strain. F1 hybrids are responsive (3).

Igk-Ef1 locus, immunoglobulin κ-chain-Ef1. This locus was recognized by variation in isoelectric focusing of the light chain of normal immunoglobulin. The allele *Igk-Ef1^a* determines presence of bands 58, 61, and 66 and occurs in strains AKR, RF, PL, and C58; the allele *Igk-Ef1^b* determines presence of bands 25 and 60 and occurs in all other inbred strains tested (6). The light chains determined by *Igk-Ef1^a* comprise the $V_κ$-Ser

subgroup. This subgroup is also associated with the *Igk-Trp^a* allele (11).

Igk-Ef2 locus, immunoglobulin κ-chain-Ef2. This locus determines variation in isoelectric focusing of the variable region of a kappa light chain of normal immunoglobulin. The allele *Igk-Ef2^a* determines presence of two major and several minor bands in the isoelectric focusing profile and occurs in strains BALB/c, AKR, C57BL/6, and most other inbred strains; the allele *Igk-Ef2^b* determines absence of these bands and occurs in strains NZB, BDP, CE, I/Ln, P/J, and C58 (7). Amino acid sequencing of the variable region of light chains of BALB/c and NZB myelomas showed that the four chains from BALB/c were all members of the same subgroup (designated $V_κ$-1A) and that the two closely related chains from NZB were members of a distinct subgroup ($V_κ$-1B) differing by four substitutions from the $V_κ$-1A subgroup (15). By restriction fragment analysis of DNA with a $V_κ$-1A probe, it was shown that all *Igk-Ef2^a* strains had identical *Bam*HI fragments and that two of the *Igk-Ef2^b* strains (NZB and C58) had a different pattern of fragments. However, four other *Igk-Ef2^b* strains (BDP, CE, I/Ln, and P) did not differ from the *Igk-Ef2^a* pattern. The basis for this difference among the *Igk-Ef2^b* strains is not known (16). *Igk-Ef2* has shown 0.45 per cent recombination with *Igk-Ef1* (9).

Igk-Id460 locus, immunoglobulin κ-chain-Id460. This locus controls level of expression of an anti-DNP-ovalbumin antibody with an idiotype, Id460, like that of BALB/c myeloma protein MOPC 460. The allele *Igk-Id460^a* determines a high level of expression and occurs in strains AKR, BALB/c, and CBA/J; the allele *Igk-Id460^b* determines a low level of expression and occurs in the C58 strain. The level of expression is also affected by *Igh-V* genes. The $V_κ$ chain of the MOPC 460 protein belongs to the $V_κ$-1 group, like those of *Igk-Ef2*. The strain distribution and $V_κ$-chain structure of *Igk-Id460* is the same of that of *Igk-Ef2* (5).

Igk-Pc locus, immunoglobulin κ-chain-Pc. This locus was recognized by variation in isoelectric focusing of the light chain of anti-phosphorylcholine immunoglobulin. The allele *Igk-Pc^a* determines lower isoelectric point and occurs in strains AKR, RF, PL, and C58; the allele *Igk-Pc^b* determines higher isoelectric point and occurs in all other strains tested (2).

Igk-Trp locus (formerly VK-1), immunoglobulin κ-chain-Trp. This locus was recognized as a difference in molecular structure of light chains detected by presence or absence of the I_B tryptic peptide. The allele *Igk-Trp^a* determines presence of the peptide and occurs in strains AKR, RF, PL, and C58; the allele *Igk-Trp^b* determines its absence and occurs in all other strains tested (10). The V_x chain determined by *Igk-Trp^a* is associated with the same V_x subgroup, V_x-Ser, as the chain determined by *Igk-Ef1^a* (11).

Igk-V19, immunoglobulin *x*-chain V19 group. A *Bam*HI DNA polymorphism distinguishes two alleles, *Igk-V19^a* in strains AKR, RF, PL, and C58, and *Igk-V19^b* in all other strains examined (13).

Igk-V21, immunoglobulin *x*-chain V21 group. A *Bam*HI DNA polymorphism distinguishes three alleles; *Igk-V21^a* in strains AKR, RF, PL, and C58, *Igk-V21^b* in most inbred strains, and *Igk-V21^c* in strains SJL, STA, RB156, MA/My, and Peru-Atteck (13).

References

1. Brown, A.R., P.D. Gottlieb, and A. Nisonoff. 1981. Role and strain distribution of genes controlling light chains needed for the expression of an intrastrain cross-reactive idiotype. Immunogenetics 14:85–99.
2. Claflin, J.L. 1976. Genetic marker in the variable region of kappa chains of mouse anti-phosphorylcholine antibodies. Eur. J. Immunol. 6:666–668.
3. DePreval. C., B. Blomberg, and M. Cohn. 1979. Determination of the kappa anti-α(1,3) dextran immune response difference by a gene(s) in the V_x locus of mice. J. Exp. Med. 149:1265–1270.
4. D'Hoostelaere, L.A., and D.M. Gibson. 1986. The organization of immunoglobulin variable kappa chain genes on mouse chromosome 6. Immunogenetics 23:260–265.
5. Dzierzak, E.A., C.A. Janeway Jr., R.W. Rosenstein, and P.D. Gottlieb. 1980. Expression of an idiotype (Id-460) during *in vivo* anti-dinitrophenyl antibody response. J. Exp. Med. 152:720–729.
6. Gibson, D. 1976. Genetic polymorphism of mouse immunoglobulin light chains revealed by isoelectric focusing. J. Exp. Med. 144:298–303.
7. Gibson, D.M., and S.J. MacLean. 1979. Ef-2: a new *Ly-3* linked light-chain marker expressed in normal mouse serum immunoglobulin. J. Exp. Med. 149:1477–1486.
8. Gibson, D.M., S.J. MacLean, and M. Cherry. 1983. Recombination between kappa chain genetic markers and the *Lyt-3* locus. Immunogenetics 18:111–116.
9. Gibson, D.M., S.J. MacLean, D. Anctil, and B.J. Mathieson. 1984. Recombination between kappa chain genetic markers in the mouse. Immunogenetics 20:493–501.
10. Gottlieb, P.D., and P.J. Durda. 1977. The I_B peptide marker and the Ly-3 surface alloantigen: structural studies of a V_x-region polymorphism and a T-cell marker determined by linked genes. Cold Spring Harbor Symp. Quant. Biol. 41:805–815.
11. Goldrick, M.M., R.T. Boyd, P.D. Ponetti, S.-Y. Lou, and P.D. Gottlieb. Molecular genetic analysis of the V_xSer group associated with two mouse light chain genetic markers. J. Exp. Med. 162:713–728.
12. Honjo, T. 1983. Immunoglobulin genes. Ann. Rev. Immunol. 1:499–528.
13. Huppi, K., E. Jouvin-Marche, C. Scott, M. Potter, and M. Weigert. 1985. Genetic polymorphism at the *x* chain locus in mice: comparison of restriction enzyme hybridization fragments of variable and constant region genes. Immunogenetics 21:445–457.
14. Laskin, J.A., A. Gray, A. Nisonoff, N.R. Klinman, and P.D. Gottlieb. 1977. Segregation at a locus determining an immunoglobulin genetic marker for the light chain variable region affects inheritance of expression of an idiotype. Proc. Natl. Acad. Sci. USA 74:4600–4604.
15. Lazure, C., W.-T. Hum, and D.M. Gibson. 1981. Sequence diversity within a subgroup of mouse immunoglobulin kappa chains controlled by the *Igk-Ef2* locus. J. Exp. Med. 154:146–155.
16. Moynet, D., S.J. MacLean, K.H. Ng, D. Anctil, and D.M. Gibson. 1985. Polymorphism of *x* variable region (V_x-1) genes in inbred mice: relationship to the Igk-Ef2 serum light chain marker. J. Immunol. 135:727–732.
17. Tonegawa, S. 1983. Somatic generation of antibody diversity. Nature 302:575–581.

Igl complex, immunoglobulin lambda chain, Chr 16

This complex has been shown by recombinant DNA techniques to contain four constant (*C*) region genes, four joining (*J*) genes, and two variable (V) region genes. They appear to be arranged in two clusters, $V_1 J_3 C_3 J_1 C_1$ and $V_2 J_2 C_2 J_4 C_4$, which may have arisen in evolution by duplication. C_4 is probably a pseudogene (3). By use of mouse–Chinese hamster somatic cell hybrids screened with cloned lambda DNA probes the complex was found to be on Chr 16 in the region proximal to band B5 (4, 8). By use of genetic crosses it was later found to be 10 cM distal to *md* (7).

Igl-C, immunoglobulin *λ*-chain constant region. Three immunoglobulin *λ* chains occur in the mouse but genetic variants are known only for λ_1.

Igl-1 locus, immunoglobulin *λ*-chain-1. This locus was recognized by alloantisera prepared in strain SJL against BALB/c myeloma proteins containing λ_1 light chains. The locus controls variation in the constant region of the λ_1 chain. Three alleles are known; *Igl-1^a* which determines presence of the BALB/c *Igl-1* specificity and occurs in BALB/c and most other inbred strains, *Igl-1^b* which determines absence of the specificity and occurs in strains SJL, YBR, BSVS, and DE (11), and *Igl-1^c* which determines a specificity present in all inbred strains tested including SJL and absent in many populations of wild *Mus musculus* and *M. spretus*

Igl

(1). Analysis of the structural gene for λ_1 in the SJL and BALB/c strains has shown one base pair difference that leads to an amino acid substitution of valine for glycine. The difference also causes elimination in SJL of a *Kpn*I restriction enzyme site that is present in BALB/c (2).

Igl-1r locus (formerly *Igl-Lo*), immunoglobulin λ-chain-1 regulator. This locus controls expression of the λ_1 immunoglobulin light chain and is *cis*-acting. The variant was detected as a difference in the amount of λ_1 chain in normal immunoglobulin (9). Mice of strain SJL have about one-fiftieth as much λ_1 in their serum and one-fiftieth as many λ_1-bearing spleen cells as mice of strain BALB/c. F1 heterozygotes are intermediate for both variables. The low allele *Igl-1r*^{lo} occurs in inbred strains SJL and BSVS and in random-bred CFW and CD-1 mice; the normal allele *Igl-1r*^{hi} occurs in BALB/c, C57BL/6, AKR/N, and most inbred strains (5, 6). A low activity variant also occurs in various wild populations but may be different from the SJL allele (10). No recombination has been found between *Igl-1* and *Igl-1r* and they have the same strain distribution. It is therefore possible that the apparent regulatory defect may be due to the *Igl-1* structural variant (6).

Igl-V, immunoglobulin λ-chain variable region.

Igl-V1 locus, immunoglobulin λ-chain-V1. This locus was detected by an antiserum against a BALB/c myeloma protein J558 raised in wild-derived *Mus spretus*. It detects an *Igl-V1* specificity present in all inbred and wild mice tested except *Mus spretus* (1).

References

1. Amor, M., J.-L. Guénet, F. Bonhomme, and P.-A. Cazenave. 1983. Genetic polymorphism of $\lambda 1$ and $\lambda 3$ immunoglobulin light chains in the *Mus* subgenus. Eur. J. Immunol 13:312–317.
2. Arp, B., M.D. McMullen, and U. Storb. 1982. Sequences of immunoglobulin λ_1 genes in a λ_1 defective mouse strain. Nature 298:184–187.
3. Blomberg, B., A. Traunecker, H. Eisen, and S. Tonegawa. 1981. Organization of four mouse λ light chain immunoglobulin genes. Proc. Natl. Acad. Sci. USA 78:3765–3769.
4. D'Eustachio, P., A.L.M. Bothwell, T.K. Takaro, D. Baltimore, and F.H. Ruddle. 1981. Chromosomal location of structural genes encoding murine immunoglobulin λ light chains. Genetics of murine λ light chains. J. Exp. Med. 153:793–800.
5. Epstein, R., K. Lehmann, and M. Cohn. 1983. Induction of λ_1-immunoglobulin is determined by a regulatory gene ($r_{\lambda 1}$) linked (or identical) to the structural ($c_{\lambda 1}$) gene. J. Exp. Med. 157:1681–1686.
6. Epstein, R., K. Lehmann, M. Cohn, C. Buckler, W. Rowe, and M. Davisson. 1984. Linkage of the *Igl-1* struc-

tural and regulatory genes to *Akv-2* on chromosome 16. Immunogenetics 19:527–537.
7. Epstein, R., M. Davisson, K. Lehmann, E.C. Akeson, and M. Cohn. 1986. Position of *Igl*, *md*, and *Bst* loci on chromosome 16 of the mouse. Immunogenetics 23:78–83.
8. Francke, U., B. de Martinville, P. D'Eustachio, and F.H. Ruddle. 1982. Comparative gene mapping: murine lambda light chain genes are located in region cen $\rightarrow$ B5 of mouse chromosome 16 not homologous to human chromosome 21. Cytogenet. Cell Genet. 33:267–271.
9. Geckeler, W., J. Faversham, and M. Cohn. 1978. On a regulatory locus controlling the expression of the murine λ_1 light chain. J. Exp. Med. 148:1122–1136.
10. Ju, S.-T., E. Selsing, M.-C. Huang, K. Kelly, and M.E. Dorf. 1986. Genetic analysis of low $\lambda 1$ chain production in mice. J. Immunol. 136:2684–2688.
11. Lieberman, R., W. Humphrey Jr., and C.C. Chien. 1977. Identification and genetics of λ_1 light chain allotype in the mouse. Immunogenetics 5:515 (Abstr.).

Ihe-1 locus, intestinal helminth expulsion-1

This locus controls the immune response to a challenge dose of the parasitic helminth *Trichinella spiralis* that results in rapid expulsion of the parasite. Immunity is generated by a standard dose of the parasite 17 days before challenge, followed by treatment to eliminate all infestation. The allele determining immune responsiveness, here called *Ihe-1*^{r}, occurs in strains NFR/N, NFS/N, and an English strain designated NIH, all originating from outbred Swiss mice; the allele determining non-response, here called *Ihe-1*^{s}, occurs in strains C3H/He, B10.BR, SJL, SWR, BALB/c, CBA, A, C57BL/6, DBA/1, and DBA/2. No linkage to *H-2* on Chr 17 or to *c* on Chr 7 was found (1).

References

1. Bell R.G., L.S. Adams, and R.W. Ogden. 1984. A single gene determines rapid expulsion of *Trichinella spiralis* in mice. Infect. Immun. 45:273–275.

Ii locus, Ia-associated invariant chain, Chr 18

This locus was first identified by electrophoretic variation in the invariant chain of mouse Ia antigens. This chain is associated with the intracellular forms of the $A_\alpha{:}A_\beta$ and $E_\alpha{:}E_\beta$ Ia complexes. The allele *Ii*^{a} occurs in all inbred strains tested; a variant allele *Ii*^{b} was found in *M. spretus* mice trapped in Cadiz, Spain (1). In addition, restriction fragment length polymorphisms have been found among inbred strains. After *Pst*I digestion, AKR gave a 10-kb fragment and C57L, DBA/2, and BALB/c gave two fragments of 9.0 and 1.1 kb. After *Eco*RI digestion, C57BL/6 and BALB/c gave fragments of 10.0 and 12.5 kb, respectively (2). *Ii*, unlike the genes for the Ia chains, is not linked to the *H-2* complex on Chr 17; it was found to be on Chr 18 by use of mouse–Chinese hamster somatic cell hybrids screened with a mouse cDNA probe (2, 4). The Ii polypeptide has a molecular weight of 31K. Four other polypeptides are also complexed with Ia molecules. Of these, a 41K chain is serologically and structurally closely related to Ii and both chains are encoded by the same gene, *Ii* (5). Ia molecules are formed intracellularly and transported to the cell membrane at which site they are found to be associated with a chondroitin sulfate proteoglycan (CSPG). The core protein of the CSPG is the Ii molecule, which may be converted to CSPG before the complex reaches the membrane (3).

References

1. Day, C.E., and P.P. Jones. 1983. The gene controlling the Ia antigen-associated invariant chain (I_i) is not linked to the *H-2* complex. Nature 302:157–159.
2. Richards, J.E., D.D. Pravtcheva, C. Day, F.H. Ruddle, and P.P. Jones. 1985. Murine invariant chain gene: chromosomal assignment and segregation in recombinant inbred strains. Immunogenetics 22:193–199.
3. Sant, A.J., S.E. Cullen, K.S. Giacoletto, and B.D. Schwartz. 1985. Invariant chain is the core protein of the Ia-associated chondroitin sulfate proteoglycan. J. Exp. Med. 162:1916–1934.
4. Yamamoto, K., G. Floyd-Smith, U. Francke, N. Koch, W. Lauer, B. Dobberstein, R. Shäfer, and G.J. Hämmerling. 1985. The gene encoding the Ia-associated invariant chain is located on chromosome 18 in the mouse. Immunogenetics 21:83–90.
5. Yamamoto, K., N. Koch, M. Steinmetz, and G.J. Hämmerling. 1985. One gene encodes two distinct Ia-associated invariant chains. J. Immunol. 134:3461–3467.

Il-1 complex, interleukin-1, Chr 2

This complex includes two genes, one for each of the two interleukin-1 (IL-1) chains, α and β. IL-1 is a lymphokine released from macrophages and monocytes and probably from many other cells, and it exhibits many diverse biological actions. The complex is on Chr 2 as shown by use of mouse–Chinese hamster somatic cell hybrids screened with mouse genomic clones. The two genes are tightly linked and are located about 5 cM distal to *B2m* as shown by use of five sets of recombinant inbred strains.

Il-1a locus, interleukin-1α. DNA restriction fragment length polymorphisms among inbred strains were found by use of a combination of *Msp*I and *Taq*I restriction enzymes. *Il-1a*^{a} occurs in BALB/c, *Il-1a*^{b} in C57BL/6 and C57L, *Il-1a*^{c} in AKR and SWR, *Il-1a*^{d} in DBA/2, *Il-1a*^{e} in C3H/He, and *Il-1a*^{f} in SJL.

Il-1b locus, interleukin-1β. A DNA restriction fragment length variant was detected with *Taq*I in the BALB/c strain. The allele in this strain was designated *Il-1b*^{a}; *Il-1b*^{b} occurs in strains C57BL/6, C57L, AKR, SWR, DBA/2, C3H/He, and SJL (1).

References

1. D'Eustachio, P., S. Jadidi, R.C. Fuhlbrigge, P.W. Gray, and D.D. Chaplin. 1987. Interleukin-1 α and β genes: linkage on chromosome 2 in the mouse. Immunogenetics 26:339–343.

IL-3
See *Csfmu* locus.

In locus
See *Ah* locus.

Inb-1, Inc-1 loci

Inb-1 locus, inducibility of MuLV in C57BL, Chr 8

Inc-1 locus, inducibility of MuLV in BALB/c, Chr 5

The *Inb-1* allele from the low leukemic strain C57BL (*Inb-1*^{b}) interacts in F1 hybrids with the *Inc-1* allele from low leukemic strain BALB/c (*Inc-1*^{c}) to enhance inducibility of virus in cultured embryo cells by a factor of 10 to 50 (2). Various crosses have shown that the *Inb-1*^{b} allele is present also in strain C57BR and that the *Inc-1*^{c} allele is present also in strains A/J, C3H/He, and SEC. *Inb-1* is located close to *Emv-2* (*Bv-1*) on Chr 8, and *Inc-1* is located close to *Emv-1* (*Cv-1*) on Chr 5. *Inb-1* and *Inc-1* may either be part of the virus-inducing genes *Bv-1* and *Cv-1*, respectively, or they may be tightly linked regulatory genes (1). See *Emv-1* and *Emv-2* loci.

References

1. McCubrey, J., and R. Risser. 1982. Allelism and linkage studies of murine leukemia virus activation genes in low leukemic strains of mice. J. Exp. Med. 155:1233–1238.
2. McCubrey, J., and R. Risser. 1982. Genetic interactions in the spontaneous production of endogenous murine leukemia virus in low leukemic mouse strains. J. Exp. Med. 156:337–349.

Ins-1, -2, -3 loci

The chromosomal location of these loci was determined in mouse–Chinese hamster somatic cell hybrids by restriction enzyme analysis of DNA.

Ins-1, -2, -3

Ins-1, *Ins-3* loci. One of these loci is the active gene for insulin I and the other is probably a pseudogene. One is on Chr 6 and the other on Chr 15, but it is not known which one is the active gene. A DNA restriction fragment length difference for a locus designated *Ins-3* ('a novel insulin-like gene') between the A/J and C57BL/6 strains was found by Meruelo *et al.* (2). By use of AXB and BXA recombinant inbred strains, this gene was found to be located near *Ly-6* and *Sis* on Chr 15.

Ins-2 locus, insulin II, Chr 7. This locus codes for the insulin II protein and is probably homologous with the human and rat insulin II genes (1).

References

1. Lalley, P.A., and J.M. Chirgwin. 1984. Mapping of mouse insulin genes. Cytogenet. Cell Genet. 37:515 (Abstr.).
2. Meruelo, D., A. Rossomando, S. Scandalis, P. D'Eustachio, R.E.K. Fournier, D.R. Roop, D. Saxe, C. Blatt, and M.N. Nesbitt. 1987. Assignment of the *Ly-6–Ril-1–Sis–H-30–Pol-5/Xmmv-72–Ins-3–Krt-1–Int-1–Gdc-1* region to mouse chromosome 15. Immunogenetics 25:361–372.

Int-1, -2, -3 loci

These loci are the sites of mammary tumor proviral insertion. They lack homology with known proviral oncogenes. They are transcribed in mammary carcinomas but not in normal tissues. It is thought that the long latency of induction of tumors by mammary tumor viruses (MTVs) occurs because the viruses do not themselves contain oncogenes but, when integrated at specific sites, activate adjacent cellular oncogenes. Activation of each of the *Int* loci appears to be associated primarily with a particular strain of virus. In tumor DNA, the C3H MTV more frequently occupies *Int-1*, the RIII MTV more frequently occupies *Int-2*, and the Czech II MTV more frequently occupies *Int-3* (2). In *Int-1* and *Int-2*, viral integration occurs near or outside the 5′ and 3′ ends of the genes, usually oriented away from the gene (1, 7). The location of the three genes was first found by use of mouse–Chinese hamster somatic cell hybrids and cloned mouse DNA probes (2, 4, 5).

Int-1 locus, mammary tumor integration site-1, Chr 15. A DNA restriction fragment length difference occurs between the A/J and C57BL/6 strains and was used in the AXB and BXA recombinant inbred strains to map *Int-1* close to *Gdc-1* on Chr 15 (3).

Int-2 locus, mammary tumor integration site-2, Chr 7. A DNA restriction fragment length polymorphism occurs among inbred strains. In *Pst*I digested DNA, the allele *Int-2ᵃ* determines a fragment of 4.4 kb and occurs

in strains C57BL/6, C57BL/10, and C57L; the allele *Int-2ᵇ* determines a fragment of 2.1 kb and occurs in strains NFS, AKR. BALB/c, and DBA/2. In 51 recombinant inbred strains and 31 progeny of a backcross from these progenitors, no recombination was found between *Int-2* and *Fis-1* on Chr 7. However, the two loci are distinct and are separated by at least 25 to 30 kb (6).

Int-3 locus, mammary tumor integration site-3, Chr 17 (2).

References

1. Dickson, C., R. Smith, S. Brookes, and G. Peters. 1984. Tumorigenesis by mouse mammary tumor virus: proviral activation of a cellular gene in the common integration region *int-2*. Cell 37:529–536.
2. Gallahan, D., C. Kozak, and R. Callahan. 1987. A new common integration region (*int-3*) for mouse mammary tumor virus on mouse chromosome 17. J. Virol. 61:218–220.
3. Meruelo, D., A. Rossomando, S. Scandalis, P. D'Eustachio, R.E.K. Fournier, D.R. Roop, D. Saxe, C. Blatt, and M.N. Nesbitt. 1987. Assignment of the *Ly-6–Ril-1–Sis–H-30–Pol-5/Xmmv-72–Ins-3–Krt-1–Int-1–Gdc-1* region to mouse chromosome 15. Immunogenetics 25:361–372.
4. Nusse, R., A. van Ooyen, D. Cox, Y.K.T. Fung, and A. Varmus. 1984. Mode of proviral activation of a putative mammary oncogene (*int-1*) on mouse chromosome 15. Nature 307:131–136.
5. Peters, G., C. Kozak, and C. Dickson. 1984. Mouse mammary tumor virus integration regions *int-1* and *int-2* map on different mouse chromosomes. Mol. Cell. Biol. 4:375–378.
6. Silver, J, and C.E. Buckler. 1986. A preferred region for integration of Friend murine leukemia virus in hematopoietic neoplasms is closely linked to the *Int-2* oncogene. J. Immunol. 60:1156–1158.
7. van Ooyen, A., and R. Nusse. 1984. Structure and nucleotide sequences of the putative mammary oncogene *int-1*: proviral insertions leave the protein-encoding domain intact. Cell 39:233–240.

Ipo-1 locus

See *Sod-1* locus.

Ir-1 locus

See *H-2* complex.

Ir-2 locus, immune response-2, Chr 2

This locus controls ability to produce agglutinating antibodies to the Ea-1 antigens in crosses between the YBR strain bearing the *Ir-2ᵇ* allele causing non-responsiveness and either BALB or CBA bearing the *Ir-2ᵃ* allele causing responsiveness. *Ir-2* is unusual in that responsiveness is recessive. If matings are made between several other strains, multiple genetic and non-genetic factors control the response (2). There is evidence that non-responsiveness is due to shared or cross-reactive antigens (3). Genes in the *Ir-2* region, which may in fact be *Ir-2*, were shown to control acquired resistance to *Leishmania donovani* (1) and to reverse the unresponsiveness to hen egg-white lysozyme caused by the *H-2ᵇ* immune response genes (4).

References

1. DeTolla, L.J. Jr., L.H. Semprevivo, N.C. Palczuk, and H.C. Passmore. 1980. Genetic control of acquired resistance to visceral leishmaniasis in mice. Immunogenetics 10:353–361.
2. Gasser, D.L. 1976. Genetic studies of the *H-3* region of mice and implications for polymorphism of histocompatibility loci. Immunogenetics 3:271–276.
3. Gasser, D.L. 1976. H-3-linked unresponsiveness to *Ea-1* and H-13 antigens. *In* D.H. Katz and B. Benacerraf, eds., The Role of Products of the Histocompatibility Gene Complex in Immune Responses, 289–295. Academic Press, New York.
4. Sadegh-Nasseri, S., D.E. Kipp, B.A. Taylor, A. Miller, and E. Sercarz. 1984. Selective reversal of *H-2* linked genetic nonresponsiveness to lysozyme. I. Non-*H-2* gene(s) closely linked to the *Ir-2* locus on chromosome 2 permit(s) an antilysozyme response in *H-2ᵇ* mice. Immunogenetics 20:535–546.

Ir-3 locus, immune response-3

This locus determines ability to make antibody to the Pro–L determinant of the synthetic polypeptides (Phe,G)-Pro–L [poly-**L**-(Phe,Glu)-poly-**L**–Pro–poly-**L**-Lys] and (T,G)-Pro–L [poly(Tyr,Glu)-poly-**L**-Pro–poly-**L**-Lys]. Strain SJL has the allele for responsiveness *Ir-3ᵃ*, and strain DBA/1 has the allele for non-responsiveness *Ir-3ᵇ*. Response to these polypeptides requires both a T-cell co-operative signal and a B-cell reaction to the signal. The response to the Pro–L determinant controlled by the *Ir-3* locus is mediated by B-cells. Response to the (Phe,G) and (T,G) determinants is controlled by the *H-2*-linked *Ir-1* locus and is mediated by T-cells. *Ir-3* is not linked to *H-2*, and its location is unknown. Strain SWR has the responsive *Ir-3ᵃ* allele but is non-responsive because it has the non-responsive *Ir-1* allele causing defective T-cell function. (SWR × DBA)F1 hybrids are responsive because they receive the high response *Ir-1* allele permitting normal

T-cell function from the DBA/1 parent and the high response *Ir-3ᵃ* allele permitting normal B-cell function from the SWR parent (1–3).

References

1. Gasser, D.L., and W.K. Silvers. 1973. The genetic determinants of immunologic responsiveness. Adv. Immunol. 18:1–66.
2. Isac, R., E. Mozes, and M. Taussig. 1976. Antigen-specific T-cell factors in the immune response to poly(Tyr,Glu)-poly(Pro)-poly(Lys). Immunogenetics 3:409–416.
3. Mozes, E. 1976. The nature of antigen specific T cell factors involved in the genetic regulation of immune response. *In* D.H. Katz and B. Benacerraf, eds., The Role of the Products of the Histocompatibility Gene Complex in Immune Response, 485–505. Academic Press, New York.

Ir-4 locus, immune response-4

This locus determines ability to produce cytotoxic or agglutinating antibodies to the H-2.2 antigenic determinant of the *H-2Dᵇ* allele. Strains HTI/Go, B10.A, and C57BL have the *Ir-4ᵇ* allele and are non-responsive; strains BALB/c and A/J have the *Ir-4ᵃ* allele and are responsive. Heterozygotes are intermediate. *Ir-4* is not linked to *H-2* or *a* (1). Alleles at the *Ir-1* locus also affect ability to respond to the H-2.2 antigen (2).

References

1. Lilly, F., J.S. Jacoby, and R.C. Coley. 1971. Immunologic unresponsiveness to the H-2.2 antigen. *In* A. Lengerová and M. Vojtisková, eds., Immunogenetics of the H-2 System, 197–199. Karger, Basel.
2. Lilly, F., H. Graham, and R. Coley. 1973. Genetic control of the antibody response to the H-2.2 alloantigen in mice. Transplant. Proc. 5:193–196.

Ir-5 locus, immune response-5, Chr 17

This locus controls, in part, the immune response to the Thy-1.1 antigen of AKR thymocytes, as measured by the number of spleen cells producing antibodies lytic for AKR thymocytes. The response is also controlled by one or more loci in the *Ir-1* region of the *H-2* system. Strains DBA/2, CBA, C3H, and C57BR have the *Ir-5ᵃ* allele for high responsiveness, and strains C57BL/6 and C57BL/10 have the *Ir-5ᵇ* allele for low responsiveness. High responsiveness is dominant. *Ir-5* is on the same chromosome as *Ir-1*, but is located about 19 map units distal to it (1–3).

Ir-5

References

1. Zaleski, M.B. 1975. Immune response of mice to the Thy-1.1 antigen: effect of the *Ir-5* alleles studied in 129/J and B10.129(6M) mice. Immunogenetics 2:241–248.
2. Zaleski, M., and J. Klein. 1974. Immune response of mice to the Thy-1.1 antigen: genetic control by alleles at the *Ir-5* locus loosely linked to the *H-2* complex. J. Immunol. 113:1170–1177.
3. Zaleski, M., and J. Klein. 1978. Genetic control of the immune response to Thy-1 antigens. Immunol. Rev. 38:120–162.

Ir-6 locus

See *Ir-7* locus.

Ir-7 locus (formerly *Ir-A-CHO*, *Ir-6*) (2), immune response-7

This locus controls the serum antibody response to cell-wall polysaccharide of group A streptococci (A-CHO). Three alleles have been described. *Ir-7^a* determines high response and occurs in strain BALB/cJ. *Ir-7^b* determines low response and is fully recessive to *Ir-7^a*; it occurs in strains AKR/J, C57BL/10, C57BL/6, SJL, and CBA/J. *Ir-7^c* determines low response and produces intermediate heterozygotes with *Ir-7^a*; it occurs in strains A/J, DBA/2, and C3H/He. *Ir-7* is not linked to *a*, *c*, *H-2*, or *Igh-1* (1).

References

1. Cramer, M., and D.G. Braun. 1975. Genetics of restricted antibodies to streptococcal group polysaccharides in mice. II. The Ir-A-CHO gene determines antibody levels, and regulatory genes influence the restriction of the response. Eur. J. Immunol. 5:823–830.
2. Green, M.C. 1979. Mouse News Lett. 61:16.

Ir-A-CHO locus

See *Ir-7* locus.

It, irregular teeth, semidominant, Chr X

Arose in a radiation experiment but was probably spontaneous. Heterozygous females have one or both lower incisors markedly reduced or absent. Upper incisors may be absent also. Expression is variable, and normal overlaps occur. Viability and fertility are low even when females are kept on powdered food. Hemizygous males die before birth (1).

References

1. Phipps, E.L. 1969. Mouse News Lett. 40:41–42.

190

Itp locus, inosine triphosphatase, Chr 2

This locus codes for inosine triphosphatase (E.C. 3.6.1.19), an enzyme found in all tissues. It was first found to be on Chr 2 using mouse–human or mouse–Chinese hamster somatic cell hybrids (1, 2). Later, a variant allele, *Itp^a*, determining slower electrophoretic mobility was found in strains BALB/cJ, BALB/cBy, DBA/1, PL/J, RIIIS/J, CAST/Ei, and five strains descended from outcrosses of BALB/c; the allele *Itp^b* determining faster mobility occurs in C57BL/6, DBA/2 and all other strains tested. Heterozygotes have an intermediate diffuse band, consistent with the known dimeric structure of the enzyme. *Itp* is located about 6 cM distal to *pa* (pallid) on Chr 2 (3).

References

1. Lalley, P.A., and J.A Diaz. 1984. Comparative gene mapping in the mouse involving genes assigned to human chromosomes 7 and 20. Cytogenet. Cell Genet. 37:514–515 (Abstr.).
2. Siciliano, M.J., R.E.K. Fournier, and R.L. Stallings. 1984. Regional assignment of *ADA* and *ITPA* to mouse chromosome 2 (C1-ter): a demonstration of the conserved linkage of enzyme and proto-oncogene loci. J. Hered. 75:175–180.
3. Taylor, B.A., D.M. Walls, and R.J. Wimsatt. 1987. Localization of the inosine triphosphatase locus (*Itp*) on chromosome 2 of the mouse. Biochem. Genet. 25:267–274.

Ity locus, immunity to *S. typhimurium*, Chr 1

This locus controls resistance to the lethal effects of *Salmonella typhimurium* given subcutaneously. The allele *Ity^r* determines resistance (LD$_{50}$ > 10) and occurs in strains CBA, A/J, C3H/He, and DBA/2; the allele *Ity^s* determines susceptibility (LD$_{50}$ < 10) and occurs in strains BALB/c, C57BL, B10.D2, and DBA/1 (4). Heterozygotes are resistant. *Ity* was the first of three loci (the others are *Bcg* and *Lsh*) assigned to Chr 1 that control resistance to infection by various pathogens and are probably identical. All three have the same strain distribution of alleles and no recombination has been observed between them (5, 6). The locus is on Chr 1 between *Idh-1* and *ln* (6, 8). *Ity* is probably the same locus as that controlling resistance to *Salmonella* in the BRVR and A/J strains and susceptibility in the BSVS strain (1, 10). *Ity* appears to act through macrophages at a very early stage after infection. There is no difference between resistant and susceptible mice in the rate of phagocytosis of the bacteria by macrophages, but the rate of intracellular killing of bacteria is 1.7 times greater in resistant DBA/2 than in susceptible C57BL/10 mice (2, 7, 9). The macrophages of Ity^r mice are also better able than those of Ity^s mice to kill the extracellular bacteria *Escherichia coli*, *Staphylococcus aureus*, and *Corynebacterium diphtheriae* (3). See *Bcg* locus, *Lsh* locus.

References

1. Briles, D.E., R.H. Benjamin Jr., C.A. Williams, and J.M. Davie. 1981. A genetic locus responsible for salmonella susceptibility in BSVS mice is not responsible for the limited T-dependent immune responsiveness of BSVS mice. J. Immunol. 127:906–911.
2. Lissner, C.R., R.N. Swanson, and A.D. O'Brien. 1983. Genetic control of the innate resistance of mice to *Salmonella typhimurium*: expression of the *Ity* gene in peritoneal and splenic macrophages isolated *in vitro*. J. Immunol. 131:3006–3013.
3. Lissner, C.F., D.L. Weinstein, and A.D. O'Brien. 1985. Mouse chromosome 1 *Ity* locus regulates microbicidal activity of isolated peritoneal macrophages against a diverse group of intracellular and extracellular bacteria. J. Immunol. 135:544–547.
4. Plant, J., and A.A. Glynn. 1976. Genetics of resistance to infection with *Salmonella typhimurium* in mice. J. Infect. Dis. 133:72–78.
5. Plant, J., J.M. Blackwell, A.D. O'Brien, D.J. Bradley, and A.A. Glynn. 1982. Are the *Lsh* and *Ity* disease resistance genes at one locus on mouse chromosome 1? Nature 297:510–511.
6. Skamene, E., P. Gros, A. Forget, P.A.L. Kongshavn, C. St. Charles, and B.A. Taylor. 1982. Genetic regulation of resistance to intracellular pathogens. Nature 297:506–509.
7. Swanson, R.N., and A.D. O'Brien. 1983. Genetic control of the innate resistance of mice to *Salmonella typhimurium*: *Ity* gene is expressed *in vivo* by 24 hours after infection. J. Immunol. 131:3014–3020.
8. Taylor, B.A., and A.D. O'Brien. 1982. Position on mouse chromosome 1 of a gene that controls resistance to *Salmonella typhimurium*. Infect Immun. 36:1257–1260.
9. van Dissel, J.T., P.C.J. Leijh, and R. van Furth. 1985. Differences in initial rate of intracellular killing of *Salmonella typhimurium* by resident peritoneal macrophages from various mouse strains. J. Immunol. 134:3404–3410.
10. Webster, L.T. 1937. Inheritance of resistance of mice to enteric bacterial and neurotropic virus infection. J. Exp. Med. 65:261–286.

iv, situs inversus viscerum, recessive

Arose spontaneously in a non-inbred stock (1). About 50 per cent of homozygotes showed left–right transposition of stomach and abdominal viscera. About half of all the homozygotes showed discordance in asymmetry between the stomach and the major abdominal and thoracic veins, the discordance occurring with similar frequency in mice with normal and reversed viscera. In the stock examined by Hummel and Chapman (1) penetrance was found to be 71 per cent when mice were classified by both these criteria. Layton (2) has postulated that the wild-type allele of the *iv* locus is necessary for development of the normal asymmetrical configuration of the viscera. In *iv/iv* mice, this control is absent, allowing random direction of the asymmetry and thus accounting for the fact that only 50 per cent of the homozygotes show reversed asymmetry.

References

1. Hummel, K.P., and D.B. Chapman. 1959. Visceral inversion and associated anomalies in the mouse. J. Hered. 50:9–13.
2. Layton, W.M., Jr. 1976. Random determination of a developmental process: reversal of normal visceral asymmetry in the mouse. J. Hered. 67:336–338.

J

j, jaw-lethal, recessive, probably extinct

Found in the descendants of both irradiated and control mice by Little and Bagg (2). Homozygotes are still-born, death resulting from abnormalities of the head including shortening of the head, reduced or absent lower jaw, microphthalmia, anophthalmia or cyclopia, abnormal or missing tongue, lack of openings between oral and nasal passages and the trachea and pharynx, and various abnormalities of the brain (1).

References

1. Johnson, P.L. 1926. An anatomical study of abnormal jaws in the progeny of X-rayed mice. Am. J. Anat. 38:281–317.
2. Little, C.C., and H.J. Bagg. 1924. The occurrence of four inheritable morphological variations in mice and their possible relation to treatment with X-rays. J. Exp. Zool. 41:45–91.

ja, jaundiced, recessive

Arose in the 129 inbred strain. Homozygotes are anemic at birth and become jaundiced within a few hours. The anemia is present in 15-day fetuses. Examination of the blood of newborn jaundiced mice shows that they have a severe microcytic anemia with about half the normal number of red cells. There is considerable erythrocyte destruction and a compensating release of immature blood cells into the blood (4). Most homozygotes die within a day or two of birth, but a few survive to adulthood. The red cell membrane in homozygotes lacks the membrane skeletal protein spectrin. The cells synthesize α spectrin but do not synthesize β spectrin, the component that appears to be necessary for incorporation of spectrin into the red cell membrane (1). It is possible that *ja* is a mutation at the β spectrin locus, *Spnb*, on Chr 12. See *Spnb* locus. The survival time of red cells in heterozygotes is slightly reduced, indicating that heterozygotes have a mild, well compensated anemia. They have a higher concentration of protoporphyrin in red cells than +/+ mice and are identifiable with virtual certainty by this method (3). Transplantation experiments show that the gene defect is expressed in the red cells or their progenitors (2).

References

1. Bodine, D.M. IV, C.S. Berkenmeier, and J.E. Barker. 1984. Spectrin deficient inherited hemolytic anemias in the mouse: characterization by spectrin synthesis and mRNA activity in reticulocytes. Cell 37:721–729.
2. Russell, E.S. 1970. Abnormalities of erythropoiesis associated with mutant genes in mice. *In* A.L. Gordon, ed., Regulation of Hematopoiesis, Vol. 1, 649–675. Appleton-Century-Crofts, New York.
3. Sassa, S., and S.E. Bernstein. 1978. Studies of erythrocyte protoporphyrin in anemic mice: use of modified hemato-fluorimeter for the detection of heterozygotes for hemolytic disease. Exp. Hematol. 6:479–487.
4. Stevens, L.C., J.A. Mackensen, and S.E. Bernstein. 1959. A mutation causing neonatal jaundice in the house mouse. J. Hered. 50:35–39.

jc, Jackson circler, recessive, Chr 10

Arose in strain C57BL/6J in 1963. Homozygotes show erratic circling behavior and hyperactivity which becomes more pronounced with age. *jc* is not allelic with *av* or *v*, other circling mutants on Chr 10 (1).

References

1. Southard, J.L., and M.M. Dickie. 1970. Mouse News Lett. 42:30.

je, jerker, recessive, Chr 4

Originated as a dancing mouse found by a fancier and given to Grüneberg (2). Homozygotes show the typical behavior of the circling mutants, head-tossing, circling, hyperactivity, and deafness. The abnormal behavior is associated with postnatal degeneration of the sensory cells of the cochlea and the sacculus and utriculus (1). Affected animals apparently are deaf from birth and have no detectable stimulus-related cochlear potential at any stage (3).

References

1. Deol, M.S. 1954. The anomalies of the labyrinth of the mutants varitint-waddler, shaker-2, and jerker in the mouse. J. Genet. 52:562–588.
2. Grüneberg, H., J.B. Burnett, and G.D. Snell. 1941. The origin of jerker, a new gene mutation of the house mouse, and linkage studies made with it. Proc. Natl. Acad. Sci. USA 27:562–565.

3. Steel, K.P., and G.R. Bock. 1983. Cochlear dysfunction in the jerker mouse. Behav. Neurosci. 97:381–391.

jg, jagged-tail, recessive, Chr 5

Arose spontaneously in the C3H/HeJ inbred strain. Homozygotes have tails with a normal tapering tip but usually much shortened and kinked. In the most severely affected individuals other parts of the vertebral column may be abnormal; in the less severely affected it appears nearly normal. Testes are much reduced in size, and males are probably sterile. Females have small ovaries. They may breed, but litters are few and small. Viability of both sexes is considerably reduced (1).

References

1. Green, M.C. 1964. Mouse News Lett. 30:32.

ji, jittery, recessive, Chr 10

Discovered by DeOme (1) as a spontaneous mutation in the Bagg albino strain. Homozygotes usually die by 4 weeks of age. They show muscular incoordination beginning at 10 to 16 days. As seizures and incoordination become more severe, the mice lose ability to obtain food and apparently die from starvation and thirst. Polycystic alterations in white matter of the brain have been reported (2).

References

1. DeOme, K.B. 1945. A new recessive lethal mutation in mice. Univ. Calif. Publ. Zool. 53:41–66.
2. Harman, P.J. 1950. Polycystic alterations in the white matter of the brain of the "jittery" mouse. Anat. Rec. 106:304 (Abstr.).

jp locus, Chr X

jp, jimpy, recessive. Arose spontaneously in an inbred line. Hemizygous males die between 20 and 40 days of age. Affected males appear normal when sitting quietly, but beginning at about 3 weeks they show a marked tremor of the hindquarters when attempting movement. From 3 weeks on convulsions may occur. Heterozygous females have normal behavior (15). In affected males, the central nervous system is very deficient in myelin, but the peripheral nervous system is normally myelinated (6, 16). Some poorly myelinated axons occur in the spinal cord and optic nerves, but the distribution is patchy and myelin usually does not extend more than one internodal length (9). Oligodendroglia are reduced in number in the early period of myelination and become abnormal as myelination

advances, containing lipid inclusions and multimembranous tubes (10). A large number of studies seeking to discover the basic defect causing the jimpy disease have found deficiencies in various myelin components including galactose cerebrosides, sulfatides, and myelin basic protein (6, 8, 13). These deficiencies are probably all secondary, however. The jimpy mutant gene appears to be the result of a mutation at the X-linked locus *Plp* coding for myelin proteolipid protein (PLP). *Plp* is in the same position on the X chromosome as *jp*. In *jp*/Y males, there is no alteration in *Plp* genomic DNA detectable with several restriction enzymes, but PLP mRNA is severely reduced in abundance (3) and contains a 74-base deletion. Probes specific for the deleted sequence hybridize with equal efficiency to wild-type and *jp* genomic DNA, suggesting that an incorrectly spliced RNA transcript is responsible for the jimpy defect (14). See *Plp* locus. The jimpy defect is expressed in cultured cerebellar tissue from newborn mice, very little myelin being formed by mutant cells in comparison to abundant myelin in control cultures (17). Myelination is greatly improved in jimpy cultures by the addition of normal astroglial and oligodendroglial cells (2). Young heterozygous females (*jp*/+) show mosaic expression of *jp* in optic nerve myelination (5). Eicher and Hoppe (4) were able to produce a fertile male chimera by combining a *jp*/Y and a +/− embryo, and used it to sire *jp*/*jp* females, which were found to resemble *jp*/Y males. The jimpy disease may be homologous to the X-linked Pelizaeus–Merzbacher disease in man (11).

jp^{msd}, myelin synthesis deficiency, recessive. Arose spontaneously in the B10.C(47N) strain. The mutation was shown to be an allele of *jp* by mating a fertile male *jp*/Y ⟷ +/− chimera to a *jp^{msd}*/+ female and producing *jp*/*jp^{msd}* females which had the typical jimpy phenotype (4). Hemizygous males resemble *jp*/Y males in behavior, early death, and other aspects (12). They have a deficiency of myelin in the central nervous system, but the myelination defect is less severe than in jimpy males (18). Myelin of the peripheral nervous system is normal (1, 7).

References

1. Billings-Gagliardi, S. and L.H. Adcock. 1981. Hypomyelinated mutant mice. IV. Peripheral myelin in *jp^{msd}*. Brain Res. 225:309–317.
2. Billings-Gagliardi, S., L.H. Adcock, E.D. Lamperti, G. Schwing-Stanhope, and M.K. Wolf. 1983. Myelination of *jp*, *jp^{msd}*, and *qk* axons by normal glia in vitro: ultrastructural and autoradiographic evidence. Brain Res. 268:255–266.

3. Dautigny, A., M.-G. Mattei, D. Morello, P.M. Alliel, D. Pham-Dinh, L. Amar, D. Arnaud, D. Simon, J.-F. Mattei, J.-L. Guénet, P. Jollès, and P. Avner. 1986. The structural gene coding for myelin-associated proteolipid protein is mutated in *jimpy* mice. Nature 321:867–869.
4. Eicher, E.M., and P.C. Hoppe. 1973. Use of chimeras to transmit lethal genes in the mouse and to demonstrate allelism of the two X-linked lethal genes *jp* and *msd*. J. Exp. Zool. 183:181–184.
5. Hatfield, J.S., and R.P. Skoff. 1982. GFAP immunoreactivity reveals astrogliosis in females heterozygous for jimpy. Brain Res. 250:123–131.
6. Jacque, C., A. Delassalle. M. Raoul, and N. Baumann. 1983. Myelin basic protein deposition in the optic and sciatic nerves of demyelinating mutants quaking, jimpy, trembler, MLD, and shiverer during development. J. Neurochem. 41:1335–1340.
7. Kirschner, D.A., and R.L. Sidman. 1976. X-ray diffraction study of myelin structure in immature and mutant mice. Biochim. Biophys. Acta 448:73–87.
8. Matthieu, J.M., S. Widmer, and N. Herschkowitz. 1973. Jimpy, an anomaly of myelin maturation. Biochemical study of myelination phases. Brain Res. 55:403–412.
9. Meier, C., and A. Bischoff. 1974. Dysmyelination in "jimpy" mouse. Electron microscopic study. J. Neuropathol. Exp. Neurol. 33:343–353.
10. Meier, C., and A. Bischoff. 1975. Oligodendroglial cell development in jimpy mice and controls: an electron-microscopic study in the optic nerve. J. Neurol. Sci. 26:517–528.
11. Meier, C., N. Herschkowitz, and A. Bischoff. 1974. Morphological and biochemical observations in the jimpy spinal cord. Acta Neuropathol. 17:349–362.
12. Meier, H., and A.D. MacPike. 1970. A neurological mutation (msd) of the mouse causing a deficiency of myelin synthesis. Exp. Brain Res. 10:512–525.
13. Mikoshiba, K., S. Kohsaka, T. Hayakawa, T. Takamatsu, and Y. Tsukada. 1985. Immunohistochemical localization of 2′,3′-cyclic nucleotide 3′-phosphodiesterase and myelin basic protein in the central nervous system of the *jimpy* and the normal mouse. J. Neurochem. 44:686–691.
14. Nave, K.-A., C. Lai, F.E. Bloom, and R.J. Milner. 1986. Jimpy mutant mice: a 74-base deletion in the mRNA for myelin proteolipid protein and evidence for a primary defect in RNA splicing. Proc. Natl. Acad. Sci. USA 83:9264–9268.
15. Phillips, R.J.S. 1954. *Jimpy*, a new totally sex-linked gene in the mouse. Z. Indukt. Abstammungs-Vererbungsl. 86:322–326.
16. Sidman, R.L., M.M. Dickie, and S.H. Appel. 1964. Mutant mice (*quaking* and *jimpy*) with deficient myelination in the central nervous system. Science 144:309–311.
17. Wolf, M.K., and A.B. Holden. 1969. Tissue culture analysis of the inherited defect of central nervous myelination in jimpy mice. J. Neuropathol. Exp. Neurol. 28:195–213.
18. Wolf, M.K., G.B. Kardon, L.H. Adcock, and S. Billings-Gagliardi. 1983. Hypomyelinated mutant mice. V. Relationship between *jp* and *jp^{msd}* re-examined on identical genetic backgrounds. Brain Res. 271:121–129.

jpk

See *cph*.

js, Jackson shaker, recessive, Chr 11

Arose in strain A/J in 1963. Homozygotes show circling, head-shaking, and hyperactivity, and are indistinguishable behaviorally from other shaker–waltzer mutants. Pathologically, they have few if any distinguishing features (1).

References

1. Dickie, M.M., and M.S. Deol. 1967. Mouse News Lett. 36:39.

jsd, juvenile spermatogonial depletion, recessive, Chr 1

Arose in the C57BL/6J strain. Homozygous females are normal and fertile. Homozygous males have small testes, normal serum testosterone, elevated serum follicle-stimulating hormone, azoospermia, and normal body and seminal vesicle weights. At 3 to 4 weeks, there is a reduction in number of spermatogenic cells that proceeds at different rates in different tubules, but by 8 to 10 weeks most tubules are devoid of germ cells, leaving only Sertoli cells. *jsd* is on Chr 1 about 6 cM proximal to *Idh-1* (1).

References

1. Beamer, W.G., T.L. Cunliffe-Beamer, K.L. Shultz, S.H. Langley, and T.H. Roderick. 1988. Juvenile spermatogonial depletion (*jsd*): a genetic defect of germ cell proliferation of male mice. Biol. Reprod. 38:899-908.

jt, joined-toes, recessive

Arose spontaneously in the 129 inbred strain. Homozygotes show fusion of the soft tissue of the digits, more commonly on the hindfeet than on the forefeet. Fusions are more common between the second and third and between the third and fourth digits (2, as 'syndactylism'). A similar mutation at the same locus occurred at Stanford University in descendants of Swiss-albino mice that had been treated with nitrogen mustard. In the developing footplate there is loss of both preaxial and postaxial tissue prior to and during the condensation of mesenchyme into the precartilage of the digital elements (1).

References

1. Center, E.M. 1966. Genetical and embryological studies of the *jt* form of syndactylism in the mouse. Genet. Res. 8:33–40.
2. Stevens, L.C. 1955. Mouse News Lett. 13:41.

Jt-1 locus (originally *Jt*), I-J antigen expression, Chr 4

The I-J antigen was originally thought to the product of a gene in the *H-2I* region of Chr 17, but molecular studies of this region failed to find such a gene. Expression of the I-J antigen on the surface of T-cells was found to be controlled in part by the *Jt-1* gene on Chr 4. It acts in complementation with a gene in the *H-2I* region, probably at or near *I-E*. The dominant allele *Jt-1^b* that allows expression of the *I-J^k* antigen when in combination with *H-2^k* occurs in the C57BL/10, CBA/J, C3H/HeJ, C58/J, MA/My, MRL/Mp, RF/J, and B10.A(5R) strains. The recessive allele *Jt-1^a* that does not allow expression of the I-J^k antigen in the presence of *H-2^k* occurs in the AKR/J, CE/J, and B10.A(3R) strains (2). However, Waltenbaugh *et al.* (5) found that *I-J^k* can be expressed in AKR spleen cells, and it is also possible that a Chr 12 gene influences I-J expression (1, 4). Klyczek *et al.* (3) suggest that the non-*H-2* gene (presumably *Jt-1*) may be the structural gene for the I-J^k antigen whose transcription or translation is controlled by a gene in the *H-2I-E* region.

References

1. Flood, P.M. 1985. Investigation into the nature of Igh-V region-restricted T cell interactions by using antibodies to antigen on methylcholanthrene-induced sarcomas. I. Analysis of Igh-V-restricted suppressor-inducer factor. J. Immunol. 134:1665–1672.
2. Hayes, C.E., K.K. Klyczek, D.P. Krum, R.M. Whitcomb, D.A. Hullett, and H. Cantor. 1984. Chromosome 4 *Jt* gene controls murine T cell surface I-J expression. Science 223:559–563.
3. Klyczek, K.K., H. Cantor, and C.E. Hayes. 1984. T cell surface I-J glycoprotein: concerted action of chromosome-4 and -17 genes forms an epitope dependent on α-D-mannosyl residues. J. Exp. Med. 159:1604–1617.
4. Murphy. D.B. 1985. Commentary on the genetic basis for control of I-J determinants. J. Immunol. 135:1543–1547.
5. Waltenbaugh, C., L. Sun, and H.-Y. Lei. 1985. I-J expression is not associated with murine chromosome 4. Eur. J. Immunol. 15:922–926.

jv, Jackson waltzer, recessive

Arose in strain C57BL/6J in 1961. Homozygotes show circling and head-shaking but are not deaf. The abnormalities of the inner ear are unlike those of any other shaker–waltzer mutant. The lateral semicircular canal and its crista are missing, and the morphogenesis of the sacculus and utriculus may be abnormal (1).

References

1. Dickie, M.M., and M.S. Deol. 1966. Mouse News Lett. 35:31.

K

Kal complex, kallikrein gene family, Chr 7

The mouse kallikrein gene family consists of 25 to 30 highly homologous genes encoding specific proteases that process biologically active peptides such as nerve growth factor and epidermal growth factor. By use of mouse–Chinese hamster somatic cell hybrids and a mouse submaxillary gland cDNA clone that codes for a member of the kallikrein family and cross-reacts with many other members, it was found that all the kallikrein genes hybridizing with the clone were located on Chr 7 (2). A polymorphism for five DNA fragments recognized by a different cDNA probe was found between the DBA/2 and C57BL/6 strains and shown to be closely linked to the tamase-1 (*Tam-1*) locus on Chr 7 (1). Two members of the *Kal* complex, nerve growth factor-α (NGF-α) and nerve growth factor-γ (NGF-γ) have been assigned individual gene symbols. See *Ngfa*, *Ngfg* loci.

References

1. Howles, P.N., D.P. Dickinson, L.L. DiCaprio, M. Woodworth-Gutai, and K.W. Gross. 1984. Use of a cDNA recombinant for the γ-subunit of mouse nerve growth factor to localize members of this multigene family near the

TAM-1 locus on chromosome 7. Nucl. Acids Res. 12:2791–2805.

2. Mason, A.J., B.A. Evans, D.R. Cox, J. Shine, and R.I. Richards. 1983. Studies of mouse kallikrein gene family suggests a role in specific processing of biological active peptides. Nature 303:300–307.

kd, kidney disease, recessive, Chr 10

Arose spontaneously in the CBA/H strain. Homozygotes develop nephrosis recognizable at about 10 weeks of age by increased proteinuria. This is followed by excessive drinking, loss of weight, anemia, and death usually at 5 to 7 months. Pathologically, the kidneys have areas of tubular atrophy alternating with areas of dilated tubules often containing hyaline casts. Glomeruli are sclerotic in the atrophied areas, and cells of some of the atrophied tubules contain hemosiderin. The disease strongly resembles human nephronophthisis (3). The kidney disease of *kd/kd* mice can be strikingly inhibited and life span more than doubled by restriction of food intake to 8 calories per day (1). The disease can be transferred by transfer of lymph node cells from nephritic *kd/kd* mice to irradiated thymectomized congenic normal mice. The disease appears to be due to an autoimmune process mediated by an antigen-specific, H-2^k-restricted effector cell (2, 4).

References

1. Fernandes, G., E.Y. Yunis, M. Miranda, J. Smith, and R.A. Good. 1978. Nutritional inhibition of genetically determined renal disease and autoimmunity with prolongation of life in *kd/kd* mice. Proc. Natl. Acad. Sci. USA 75:2888–2892.
2. Kelly, C.J., R. Korngold, R. Mann, M. Clayman, T. Haverty, and E.G. Neilson. 1986. Spontaneous interstitial nephritis in kdkd mice. II. Characterization of a tubular antigen-specific, H-2K-restricted Lyt-2+ effector T cell that mediates destructive tubulointerstitial injury. J. Immunol. 136:526–531.
3. Lyon, M.F., and E.V. Hulse. 1971. An inherited kidney disease of mice resembling human nephronophthisis. J. Med. Genet. 8:41–48.
4. Neilson, E.G., E. McCafferty, A. Feldman, M.D. Clayman, B. Zakheim, and R. Korngold. 1984. Spontaneous interstitial nephritis in kdkd mice. I. An experimental model of autoimmune renal disease. J. Immunol. 133:2560–2565.

Kfo-1, *Kth-1*, *-2* loci

These loci control variation in soluble kidney proteins revealed by two-dimensional electrophoresis. The symbols *th* and *fo* indicate the range of molecular weight of the proteins in thousands (thirties and forties, respectively). Chromosomal locations were determined by use of recombinant inbred strains.

Kfo-1 locus, kidney 40–50 thousand MW protein-1, Chr 9. The allele *Kfo-1ᵃ* determines a protein with an isoelectric point (pI) of 6.0 and occurs in the C57BL/6 and related strains; the allele *Kfo-1ᵇ* determines a pI of 6.1 and occurs in the DBA/2, C3H/He, and many other strains. No recombinants with *Ltw-3* have been found and all strains carrying *Ltw-3ᵃ* also carry *Kfo-1ᵇ* (1).

Kth-1 locus, kidney 30–40 thousand MW protein-1, Chr 7 (?). The allele *Kth-1ᵃ* determines a protein with a pI of 5.5 and occurs in most inbred strains; the allele *Kth-1ᵒ* determines absence of the protein and occurs in AKR/J and C3H/He (1).

Kth-2 locus, kidney 30–40 thousand MW protein-2. The allele *Kth-2ᵃ* determines a protein with a pI of 6.8 or higher; the allele *Kth-2ᵒ* determines absence of the protein. The two alleles are randomly distributed among inbred strains (1).

References

1. Elliott, R.W. 1979. Mouse News Lett. 61:59.

kr, kreisler, recessive, Chr 2

Found by Hertwig (2) in the descendants of an irradiated male. Viability and fertility are less than normal. Males are more likely to breed than females. In homozygotes both the bone and membranous labyrinths of the inner ear are very abnormal. The abnormality traces back to faulty segmentation of the neural tube in the region of the rhombencephalon in $8\frac{3}{4}$-day embryos, followed by degeneration of cells in the fourth rhombomere (1). The abnormalities of the neural tube are thought to lead to an abnormal position of the ear vesicles, which in 9-day kreisler embryos are not directly in contact with the neural tube as they are in normal embryos. An imperfect capsule forms around the vesicle, and extracapsular cysts result from growth of the vesicle through gaps in the capsule. The cysts are often quite large and located under the brain (1,3). Shortening of the cell cycle in otocysts of 11-day embryos has been described (4). The abnormal behavior resulting from these defects is very similar to that of the degenerative group of circling mutants.

References

1. Deol, M.S. 1964. The abnormalities of the inner ear in *kreisler* mice. J. Embryol. Exp. Morphol. 12:475–490.
2. Hertwig, P. 1942. Neue Mutationem und Koppelungsgruppen bei der Hausmaus. Z. Indukt. Abstammungs-Vererbungsl. 80:220–246.

3. Hertwig, P. 1944. Die Genese der Hirn- und Gehörorganmissbildungen bei röntgenmutierten Kreisler-Mäusen. Z. Mensch. Vererb. Konst. 18:327–354.
4. Ruben, R.J. 1973. Development and cell kinetics of the kreisler (*kr/kr*) mouse. Laryngoscope 83:1440–1468.

7. Scolnick, E.M., A.G. Papageorge, and T.Y. Shih. 1979. Guanine nucleotide-binding as an assay for *src* protein of rat-derived murine sarcoma virus. Proc. Natl. Acad. Sci. USA 76:5355–5359.

Kras-2 locus (c-*Kras*-2, c-Ki-*ras*-2), Kirsten sarcoma oncogene-2, Chr 6

This is the cellular homolog of the transforming gene of the Kirsten sarcoma virus, v-Ki-*ras*. It is found in normal mouse cells and in amplified condition in Y1 adrenal tumor cells (6). It is homologous to the human KRAS2 gene (3). The protein product of the viral oncogene has the property of binding to guanine nucleotides (7). *Kras-2* was found to be on Chr 6 by use of mouse–Chinese hamster somatic cell hybrids and DNA probes of viral and cellular origin (3, 5). *In situ* hybridization localized the gene to the distal portion of the chromosome (1). The *Kras-2* DNA is polymorphic; an *Eco*RI fragment of 0.55 kb (*Kras-2^a*) occurs in strains A/J, A/He, SWR, BALB/c, CBA/J, RF, and 129/Rr; a fragment of 0.7 kb (*Kras-2^b*) occurs in strains AKR, C58, C57L, C3H/He, SJL, DBA/2, and C57BL/6 (4). *Kras-2* and *Hras-1* contain a common DNA sequence of 0.35 kb, indicating that they probably derive from a common ancestral gene (2).

References

1. Cahilly, L.A., and D.L. George. 1985. Regional mapping of the cKi-*ras* proto-oncogene on mouse chromosome 6 by in situ hybridization. Cytogenet. Cell Genet. 39:140–144.
2. Ellis, R.W., D. DeFeo, T.Y. Shih, M.A. Gonda, H.A. Young, N. Tsuchida, D.R. Lowy, and E.M. Scolnick. 1981. The p21 *src* genes of Harvey and Kirsten sarcoma viruses originate from divergent members of a family of normal vertebrate genes. Nature 292:505–511.
3. George, D.L., A.F. Scott., B. de Martinville, and U. Francke. 1984. Amplified DNA in Y1 mouse adrenal tumor cells: isolation of cDNAs complementary to an amplified c-Ki-*ras* gene and localization of homologous sequences to mouse chromosome 6. Nucl. Acids Res. 12:2731–2743.
4. Ryan, J., P.E. Barker, and F.H. Ruddle. 1986. An *Eco*RI restriction fragment length polymorphism at the KRAS2 locus on mouse chromosome 6. Nucl. Acids Res. 14:9222.
5. Sakaguchi, A.Y., P.A. Lalley, B.U. Zabel, R.W. Ellis, E.M. Scolnick, and S.L. Naylor. 1984. Chromosome assignments of four mouse cellular homologs of sarcoma and leukemia virus oncogenes. Proc. Natl. Acad. Sci. USA 81:525–529.
6. Schwab, M., K. Alitalo, H.E. Varmus, and J.M. Bishop. 1983. A cellular oncogene (c-Ki-*ras*) is amplified, overexpressed, and located within karyotypic abnormalities in mouse adrenocortical cells. Nature 303:497–501.

Krt-1 locus, keratin-1, Chr 15

This locus contains a cluster of at least three keratin genes coding for the keratins Krtb1, Krta1, and Krtb2 of molecular mass 67 000, 59 000, and 60 000, respectively. A DNA restriction fragment length difference between strains A/J and C57BL/6 was found by Meruelo *et al.* (1) and used in AXB and BXA recombinant inbred strains to locate *Krt-1* close to *Gdc-1* on Chr 15.

References

1. Meruelo, D., A. Rossomando, S. Scandalis, P. D'Eustachio, R.E.K. Fournier, D.R. Roop, D. Saxe, C. Blatt, and M.N. Nesbitt. 1987. Assignment of the *Ly-6–Ril-1–Sis–H-30–Pol-5/Xmmv-72–Ins-3–Krt-1–Int-1–Gdc-1* region to mouse chromosome 15. Immunogenetics 25:361–372.

Kth-1, *-2* loci

See *Kfo-1*, etc.

Kw, kinky-waltzer, dominant or semidominant

Probably radiation-induced. Heterozygotes are viable and fertile. Penetrance is incomplete in heterozygotes. No description of homozygotes has been published. In heterozygotes the amount of kinking of the tail is variable and may or may not be accompanied by a balance defect. The balance defect has not been seen without the kinky tail (1, as 'kinky tail balance defect'). This mutant resembles *Q*, *Fu*, and *Fu^{ki}*, but tests for allelism have not been reported.

References

1. Gower, J.S., and M.B. Cupp. 1958. Mouse News Lett. 19:37.

ky, kyphoscoliosis, recessive

Arose in a noninbred stock. Most homozygotes are identifiable by postural changes soon after weaning. When held by the tail and placed on a surface, they show a defective placing reflex, tending to land on their noses instead of on their forefeet. There is progressive

ky

erosion of the articulating thoracic vertebral surfaces, giving a complete S-shaped kyphosis at the extreme. Breathing is difficult, and body weight is reduced. Both sexes are fertile, but breeding performance is somewhat reduced, especially in females (1). *ky* somewhat resembles *myd* (myodystrophy) but has not been tested

for allelism with it, and has not been reported to have a defect in muscle.

References

1. Dickinson, A.G., and V.M.H. Meikle. 1973. Genetic kyphoscoliosis in mice. Lancet 1973 i:1186.

L

L3T4 locus

See *Ly-4* locus.

l(17)-1 to *-4* loci, Chr 17 lethal-1 to -4, recessive

These mutations were induced by ENU in the inbred BTBR strain and isolated by a classical breeding scheme for detecting recessive lethals. They all lie in the *T–H-2* region of Chr 17 and are known to be mutations at different loci. *l(17)-1* and *l(17)-2* are close to *T*, and the other two loci are further away. *l(17)-2* probably is proximal to *T* since it is not within the T^{Orl} or T^{hp} deletions (1).

References

1. Shedlovsky, E., J.-L. Guénet, L.L. Johnson, and W.F. Dove. 1986. Induction of recessive lethal mutations of the T/t–H-2 region of the mouse genome by a point mutagen. Genet. Res. 47:135–142.

la

See *tg^{la}*.

Laf locus

See *Fla* locus.

Lamb-1, *Lamb-2* loci

Lamb-1 locus, laminin B1 subunit, Chr 12, 19?
Lamb-2 locus, laminin B2 subunit, Chr 1
These loci code for the B1 and B2 light polypeptide chains of the extracellular matrix protein laminin synthesized by the mouse embryo parietal endoderm cells. Restriction fragment length polymorphisms for both

loci were found in genomic DNA by use of cDNA probes and a number of different restriction enzymes. For *Lamb-1*, a variant detected with *Pst*I (allele designated *Lamb-1^l*) was found in strains C57L, C57BR/cd, and C58; all other strains carry the allele *Lamb-1^a*. For *Lamb-2*, three alleles were detected by a combination of the enzymes *Hinf*I, *Eco*RI, *Bal*I, and *Pvu*II; *Lamb-2^a* in strains 129/J, AKR, AU/Ss, NZB, SM, and MOLD; *Lamb-2^b* in most other strains; and *Lamb-2^d* in DBA/1, DBA/2, O20/A, and ST/b. In 18 AKR × C57L recombinant inbred strains, there were no recombinants between the two loci and both appeared to be on Chr 1 (2). This location was later confirmed for *Lamb-2* but not for *Lamb-1*. Two fragments hybridizing with *Lamb-1* cDNA were found, one on Chr 12 and the other probably on Chr 19. Segregation of the Chr 12 locus in recombinant inbred strains suggests that it is near *Ly-18* (1).

References

1. Elliott, R.W. 1987. Map locations for laminin subunits B1 and B2. Mouse News Lett. 78:74.
2. Elliott, R.W., D. Barlow, and B.L.M. Hogan. 1985. Linkage of genes for laminin B1 and B2 subunits on chromosome 1 in mouse. In vitro Cell. Dev. Biol. 21:477–484.

Lap-1 locus, leucine arylaminopeptidase-1, Chr 9

This locus controls electrophoretic variation of leucine arylaminopeptidase, an enzyme found in the the intestine that catalyzes the hydrolysis of L-leucyl-β-naphthylamine. The allele *Lap-1^a* determines presence of an electrophoretic band and is found in the C57 family of strains: the allele *Lap-1^b* determines absence of the band and is found in strains DBA/2, BALB/c, C3H/He, and many others. Heterozygotes are indistinguishable from *Lap-1ᵃ* homozygotes (2). Electrophoretic differences in the intestine due to *Lap-1* can also be demonstrated with a staining method for alkaline phosphatase. *Lap-1ᵃ* determines presence of a series of subbands, and *Lap-1ᵇ* determines presence of a single wide band (1).

References

1. Wilcox. F.H., L. Hirschhorn, B.A. Taylor, J.E. Womack, and T.H. Roderick. 1979. Genetic variation in alkaline phosphatase of the house mouse (*Mus musculus*) with emphasis on a manganese-requiring enzyme. Biochem. Genet. 17:1093–1107.
2. Womack, J.E., M.A. Lynes, and B.A. Taylor. 1975. Genetic variation of an intestinal leucine arylaminopeptidase (*Lap-1*) in the mouse and its location on chromosome 9. Biochem. Genet. 13:511–518.

Lap-2 locus, leucine aminopeptidase-2

This locus controls electrophoretic mobility and activity level of leucine aminopeptidase (E.C. 3.4.11.1) in serum. The allele *Lap-2ᵇ* determines presence of two bands and high activity and occurs in the DD/S strain; the allele *Lap-2ᵃ* determines a single band and an activity level about half that in DD/S and occurs in strains C57BL/6, BALB/c, and DBA/2. F1 hybrids between DD/S and C57BL/6 are identical to C57BL/6 in both electrophoretic pattern and activity level. This suggests that *Lap-2* is a regulatory rather than a structural locus. *Lap-2* segregates independently of *Lap-1* on Chr 9, but is linked to an intestinal alkaline phosphatase (AKP) variant identified by the authors (1). This AKP variant may be the same as that determined by the *Akp-3* locus on Chr 1 (2).

References

1. Finlay, M.F., and L.L. Huang. 1985. A new variant of serum leucine aminopeptidase in the mouse: its development and possible regulation. Biochem. Genet. 23:169–180.
2. Wilcox, F.H. 1983. Genetics of alkaline phosphatase of the small intestine of the house mouse (*Mus musculus*). Biochem. Genet. 21:641–652.

Lc, lurcher, semidominant, Chr 6

Arose as a spontaneous mutation in a male homozygous for white (*Miʷʰ*). Heterozygotes show a characteristic swaying of the hindquarters and a jerky up and down movement. They are identifiable with sureness by their behavior at 12 to 14 days of age. They are smaller than normal at maturity but fertile and have a normal life span. Homozygotes die shortly after birth but have no visible abnormalities (2). The cerebellum of heterozygotes is much smaller than normal and shows severe postnatal loss of Purkinje cells and granule cells. Virtually no Purkinje cells are found in adults and granule cells are reduced to about 10 per cent of normal. The number of neurons in the inferior olivary nucleus falls to about 25 per cent of normal. Other cell populations are normal (1, 3). Wetts and Herrup (4, 5) have used chimeras of *Lc*/+ and wild-type mice in which the *Lc*/+ and +/+ cells were distinguishable by cytoplasmic or nuclear markers to show that the action of *Lc* is intrinsic to Purkinje cells and that the loss of granule cells and olive cells is probably secondary to the loss of Purkinje cells. The same authors (6, 7) have concluded, from the numbers of +/+ Purkinje cells found in different *Lc*/+ ⟷ +/+ chimeras, that the number of Purkinje cell progenitors is very small, perhaps 8 to 11 in each brain half, and that each progenitor gives rise to about 10 000 Purkinje cells. *Lc* has been used as a model reproducibly deficient in defined cell populations to study cerebellar function and the distribution of various brain components on cerebellar cells.

References

1. Caddy, K.W.T., and T.J. Biscoe. 1979. Structural and quantitative studies on the normal C3H and lurcher mutant mouse. Phil. Trans. R. Soc. Lond. (Biol.) 287:167–201.
2. Phillips, R.J.S. 1960. "Lurcher", a new gene in linkage group XI of the house mouse. J. Genet. 57:35–42.
3. Swisher, D.A., and D.B. Wilson. 1977. Cerebellar histogenesis in the lurcher (*Lc*) mutant mouse. J. Comp. Neurol. 173:205–218.
4. Wetts, R., and K. Herrup. 1982. Interaction of granule, Purkinje and inferior olivary neurons in lurcher chimaeric mice. I. Qualitative studies. J. Embryol. Exp. Morphol. 68:87–98.
5. Wetts, R., and K. Herrup. 1982. Interaction of granule, Purkinje and inferior olivary neurons in lurcher chimaeric mice. II. Granule cell death. Brain Res. 250:358–362.
6. Wetts, R., and K. Herrup. 1982. Cerebellar Purkinje cells are descended from a small number of progenitors committed during early development: quantitative analysis of lurcher chimeric mice. J. Neurosci. 2:1494–1498.
7. Wetts, R., and K. Herrup. 1983. Direct correlation between Purkinje and granule cell number in the cerebella of lurcher chimeras and wild-type mice. Dev. Brain Res. 10:41–47.

Lck locus, lymphocyte-specific protein tyrosine kinase, Chr 4

This locus codes for a lymphocyte-specific protein tyrosine kinase that is a member of a family of protein tyrosine kinases that are membrane-associated but lack a

Lck

transmembrane domain. The human homolog is located at a site that is frequently rearranged in human lymphomas, and the mouse gene is rearranged and overexpressed in a murine lymphoma. *Lck* was found to be on the distal end of Chr 4 by use of mouse–Chinese hamster somatic cell hybrids screened with an *Lck* clone isolated from a thymus cDNA library (1).

References

1. Marth, J.D., C. Disteche, D. Pravtcheva, F. Ruddle, E.G. Krebs, and R.M. Perlmutter. 1986. Localization of a lymphocyte-specific protein tyrosine kinase gene (*lck*) at a site of frequent chromosomal abnormalities in human lymphomas. Proc. Natl. Acad. Sci. USA 83:7400–7404.

lcsd, lysosomal cholesterol storage disorder, recessive

Arose in a BALB/c substrain maintained at the National Center for Toxicological Research, Jefferson, Arkansas. Homozygotes develop normally until about 4 weeks, after which they show tremor followed at 7 to 8 weeks by jerky movements and staggering gait. Body weight is nearly normal until about 40 days and then declines. Death occurs between 80 and 120 days. Large numbers of foam cells are found in most visceral tissues with concomitant elevation of cholesterol and sphingomyelin (2). The foam cells are transformed reticuloendothelial cells with lysosomes greatly enlarged by accumulated cholesterol (5) and sphingomyelin (1). In cultured fibroblasts, the mutation specifically disrupts processes controlling esterification of exogenous cholesterol subsequent to its delivery to the microsomes. In *lcsd*/+ fibroblasts, the level of esterification is intermediate between that of mutant and normal cells. The cholesterol-esterification disorder of fibroblasts closely resembles that found in fibroblasts of Niemann–Pick disease in man (4). There is a severe myelin deficiency in the central nervous system which is probably due to a defect in myelinogenesis rather than to myelin breakdown (6). In cultured neuroglial cells of homozygotes, as in fibroblasts, cholesterol esterification is defective, with an intermediate level in heterozygotes (3).

References

1. Bhuvaneswaran, C., M.D. Morris, H. Shio, and S. Fowler. 1982. Lysosome lipid storage disorder in NCTR-BALB/c mice. III. Isolation and analysis of storage inclusions from liver. Am. J. Pathol. 108:160–170.
2. Morris, M.D., C. Bhuvaneswaran, H. Shio, and S. Fowler. 1982. Lysosome lipid storage disorder in NCTR-BALB/c mice. I. Description of the disease and genetics. Am. J. Pathol. 108:140–149.
3. Patel, S.C., S. Suresh, H. Weintroub, R.O. Brady, and P.G. Pentchev. 1987. Impaired cholesterol esterification in primary brain cultures of the lysosomal cholesterol storage disorder (LCSD) mouse mutant. Biochem. Biophys. Res. Comm. 143:233–240.
4. Pentchev, P.G., M.C. Comly, H.S. Kruth, S. Patel, M. Proestel, and H. Weintroub. 1986. The cholesterol storage disorder of the mutant BALB/c mouse. A primary genetic lesion closely linked to defective esterification of exogenously derived cholesterol and its relationship to human type C Niemann–Pick disease. J. Biol. Chem. 261:2772–2777.
5. Shio, H., S. Fowler, C. Bhuvaneswaran, and M.D. Morris. 1982. Lysosome lipid storage disorder in NCTR-BALB/c mice. II. Morphological and cytochemical studies. Am. J. Pathol. 108:150–159.
6. Weintraub, H., A. Abramovici, U. Sandbank, P.G. Pentchev, R.O. Brady, M. Sekine, A. Suzuki, and B. Sela. 1985. Neurological mutation characterized by dysmyelination in NCTR-BALB/c mouse with lysosomal storage disease. J. Neurochem. 45:665–672.

ld, limb deformity, recessive, Chr 2

Arose at Oak Ridge National Laboratory, possibly radiation-induced. Homozygotes have reduced viability, but often live and breed. Skeletal deformities appear to be confined to the portions of the limbs below the elbow and knee. The radius and ulna appear fused into a broad flat bone, usually triangular. The fore- and hindfeet are radically reduced and disorganized (1). A similar mutation (*ld^J*) occurred in the CBA/Ca strain at the Jackson Laboratory. The tibia and fibula and the ulna and radius are replaced by single long bones in homozygotes of this allele, and the viability appears to be lower. Whether the difference is an effect of the residual genotype or due to a real difference in the alleles is not known (2, 3). A third allele (*ld^{TgHd}*) was induced by injection of the c-*myc* oncogene into the pronucleus of eggs from a cross of outbred CD-1 × inbred C57BL/6 mice. Its effects on limb morphology appear to resemble those of *ld^J* (5). These authors showed that the three alleles differ from wild type and from each other in DNA restriction fragment lengths revealed by digestion with *Bam*HI and *Bgl*II restriction enzymes. Homozygotes for *ld^J* very often have small or absent kidneys (4). Examination of 11- to 15-day embryos showed that short or absent ureters were always associated with the small or absent kidneys (M.C. Green and C. Masiak, unpublished).

References

1. Cupp. M.B. 1960. Mouse News Lett. 22:50.
2. Cupp, M.B. 1962. Mouse News Lett. 26:51.
3. Green, M.C. 1962. Mouse News Lett. 26:34.

4. Kleinbrecht, J. 1974. Mouse News Lett. 51:19.
5. Woychik, R.P., T.A. Stewart, L.G. Davis, P. D'Eustachio, and P. Leder. 1985. An inherited limb deformity created by insertional mutagenesis in a transgenic mouse. Nature 318:36–40.

Ldh-1 locus, lactate dehydrogenase-1, Chr 7

This locus controls electrophoretic mobility of the A subunit of lactate dehydrogenase (LDH-A; E.C. 1.1.1.27). It was found to be on Chr 7 by use of mouse–human and mouse–Chinese hamster somatic cell hybrids (2, 3) and later found by classical genetic crosses to be located between *Gpi-1* and *Hbb* (5). The allele *Ldh-1^a* determines a pattern of multiple bands after isoelectric focusing on polyacrylamide gels and occurs in all strains and stocks tested. The codominant allele *Ldh-1^b* occurred spontaneously in a cross between mice of the DBA/2J and C57BL/6J strains (4). The zymogram pattern of homozygotes for this allele shows considerably less staining than that of wild-type mice. Homozygotes are viable and fertile (5). The allele *Ldh-1^c* was found in an offspring of a (101 × C3H)F1 male mouse treated with the mutagen procarbazine hydrochloride. In heterozygotes with *Ldh-1^a*, there is a pronounced shift of LDH bands from the acidic to the basic pole. In homozygotes, there is very low staining intensity of all bands. Homozygotes are viable and fertile (1). *Ldh-1^b* and *Ldh-1^c* appear to have very similar effects but direct comparisons have not been reported.

References

1. Charles, D.J., and W. Pretsch. 1981. A mutation affecting the lactate dehydrogenase locus *Ldh-1* in the mouse. I. Genetical and electrophoretical characterization. Biochem. Genet. 19:301–309.
2. Lalley, P.A., J.D. Minna, and U. Francke. 1978. Conservation of autosomal gene synteny groups in mouse and man. Nature 274:160–162.
3. O'Brien, D., A. Linnenbach, and C.M. Croce. 1978. Assignment of the gene for lactic dehydrogenase A to mouse chromosome 7 using mouse–human hybrids. Cytogenet. Cell Genet. 21:72–76.
4. Soares, E.R. 1977. Mouse News Lett. 57:33.
5. Soares, E.R. 1978. Mouse News Lett. 59:12.

Ldh-2 locus, lactate dehydrogenase-2, Chr 6

This locus controls electrophoretic variation in the B subunit of lactate dehydrogenase (LDH-B; E.C. 1.1.1.27). The allele *Ldh-2^a* determines a B subunit with fast anodal mobility and occurs in all inbred strains tested; the allele *Ldh-2^b* determines a subunit with slower mobility and was found in the Peru–Coppock stock (2); three variant alleles occur in wild *Mus spretus* in southern France and in Spain, *Ldh-2^l* determining a slow form, *Ldh-2^m* determining a form with intermediate mobility, and *Ldh-2^p* determining a variant found in Pratx in southern France (1, 3). The alleles are codominant. *Ldh-2* is on Chr 6 near the distal end (2).

References

1. Britton-Davidian, J., A.R. Bustos, L. Thaler, and M. Topal. 1978. Lactate dehydrogenase polymorphism in *Mus musculus* L. and *Mus spretus* Lataste. Experientia 34:1144–1145.
2. Peters, J., and S.J. Andrews. 1985. Linkage of lactate dehydrogenase-2, *Ldh-2*, in the mouse. Biochem. Genet. 23:217–225.
3. Thaler, L., F. Bonhomme, and J. Britton-Davidian. 1979. Mouse News Lett. 60:62.

Ldr-1 locus, lactate dehydrogenase regulator-1, Chr 6

This locus controls presence or absence of the B subunits in lactate dehydrogenase of erythrocytes. The allele *Ldr-1^a* determines absence of the B subunit and is found in strains C57BL/6, C3HeB/Fe, CBA/J, and many others; the allele *Ldr-1^b* determines presence of B and is found in the European substrains of CBA and in a few other strains (2–4). The alleles are recognized by the electrophoretic pattern of hemolysates. *Ldr-1^a* produces a single band (LDH-5) composed of the tetrameric isozyme with four A subunits, and *Ldr-1^b* produces three additional bands (LDR-4, LDR-3, and LDR-2) containing respectively one, two, and three B subunits. Heterozygotes are like *Ldr-1^b* homozygotes. The B phenotype is not present at birth, but appears at about 2 weeks and increases in intensity with age (3). The variant enzyme is present only in erythrocytes, not in other tissues. There is evidence that *Ldr-1^a* determines the presence in erythrocytes of a factor that associates with the A subunits and prevents them from combining with B subunits (1).

References

1. Glass, R.D., and D. Doyle. 1972. Genetic control of lactate dehydrogenase expression in mammalian tissues. Science 176:180–181.
2. Martin, J.E., and M.L. Petras. 1971. Two erythrocyte lactate dehydrogenase variants in the house mouse, Mus musculus. Anim. Blood Grps. Biochem. Genet. 2:229–237.
3. Riles, L. 1965. A lactate dehydrogenase variant in the mouse. Nature 208:814–815.
4. Shows, T.B., and F.H. Ruddle. 1968. Function of the lac-

tate dehydrogenase B gene in mouse erythrocytes: evidence for control by a regulatory gene. Proc. Natl. Acad. Sci. USA 61:574–581.

Ldr-2 locus, lactate dehydrogenase regulator-2, Chr 6

This locus regulates expression of the B subunit of lactate dehydrogenase (LDH-B) in liver. The allele *Ldr-2ᵇ* determines high activity of LDH-B and occurs in the CBA/Lac strain; the allele *Ldr-2ᵃ* determines low activity of LDH-B and occurs in the DBA/1, DBA/2, C57BL/6, and C3H/He strains. Heterozygotes resemble the *Ldr-2ᵇ* parent as determined by starch-gel electrophoretic pattern, but are intermediate when the amount of LDH-B is determined by radial immunodiffusion. *Ldr-2* appears also to have a slight effect on the amount of liver LDH-A as measured by radial immunodiffusion. The effect of *Ldr-2* on LDH-B is confined to the liver; it becomes manifest beginning at 6 to 8 days after birth. *Ldr-2* is on Chr 6 about 20 cM from *Ldr-1* (1).

References

1. Khlebodarova, T.M., and O.L. Serov. 1980. A new locus regulating the expression of the *Ldh-2* gene in mouse liver. Biochem. Genet. 18:1027–1039.

le, light ear, recessive, Chr 5

Arose spontaneously in the C3H/HeJ strain. It closely resembles *ep*. Homozygotes have light ears, tail, and feet, and a slightly diluted juvenile coat which becomes darker in the adult. Eyes are pale at birth and darken with age (1). The light ears are conspicuous by 3 to 4 days. In melanocytes of the choroid of the eye, the melanin granules are fewer than normal, three to four times larger, and more heterogeneous in size (2). Like *bg* and *ep*, *le* in homozygotes causes an increase in concentration of the lysosomal enzyme β-galactosidase in the kidney cortex and a decrease in urinary excretion of the enzyme (5). Lysosomes of the kidney proximal tubules are enlarged and of reduced density (4). Homozygotes have a platelet storage pool deficiency characterized by prolonged bleeding time accompanied by normal platelet counts, decreased platelet dense granule content and function, and reduced number of platelet dense granules (6). The platelet abnormality can be corrected by transplantation of normal bone marrow, or transferred by transplantation of mutant marrow to normal mice (3).

References

1. Lane, P.W., and E.L. Green. 1967. Pale ear and light ear in the house mouse: mimic mutations in linkage groups XII and XVII. J. Hered. 58:17–20.
2. LaVail, M.M., and R.L. Sidman. 1974. C57BL/6J mice with inherited retinal degeneration. Arch. Ophthalmol. 91:394–400.
3. McGarry, M.P., E.K. Novak, and R.T. Swank. 1986. Progenitor cell defect correctable by bone marrow transplantation in five independent mouse models of platelet storage pool deficiency. Exp. Hematol. 14:261–265.
4. Meisler, M., J. Levy, F. Sansone, and M. Gordon. 1980. Morphologic and biochemical abnormalities of kidney lysosomes in mice with an inherited albinism. Am. J. Pathol. 101:581–594.
5. Meisler, M.H., L. Wanner, and J. Strahler. 1984. Pigmentation and lysosomal phenotypes in mice doubly homozygous for both light-ear and pale-ear mutant alleles. J. Hered. 75:103–106.
6. Novak, E.K., S.-W. Hui, and R.T. Swank. 1984. Platelet storage pool deficiency in mouse pigment mutations associated with seven distinct genetic loci. Blood 63:536–544.

lec, laryngotracheo-esophageal cleft, recessive

Found in a stock carrying the *cab* (cardiac abnormality) mutation. Homozygotes can be recognized shortly after birth by bloating caused by distention of the stomach or abdominal cavity with air. Some affected newborns remain quite active and may live up to 10 to 20 hours, but all experience frequent cyanotic bouts and finally die. These symptoms are caused by failure of division of the foregut into totally separate digestive and respiratory ducts. This allows excessive air to pass into the stomach, which at more advanced stages may rupture, allowing air to pass into the abdominal cavity. The cleft between the esophagus and trachea extends from the pharynx about 1 to 2 mm through the larynx into the trachea. The *lec* mutation is not linked to *cab* (1).

References

1. Essien, F.B., and A. Maderious. 1981. A genetic factor controlling morphogenesis of the laryngotracheo-esophageal complex in the mouse. Teratology 24:235–239.

Len-1, *Len-2* loci

Len-1 locus, lens protein-1, Chr 1. This locus encodes a major lens protein of the γ-crystallin class. It has a molecular weight of about 22 000, occurs primarily in the lens nucleus, and is synthesized from about day 14 of gestation to about 4 weeks of age (2, 5). The alleles are distinguishable by isoelectric focusing in polyacrylamide gels. The LEN-1 proteins are distributed in six major bands in the gel with isoelectric points (pI) of 7.1 to 8.5. The allele *Len-1ᵃ* determines five bands with pIs of 7.1, 7.5, 7.6, 8.0, and 8.1; it occurs in DBA/2 and most other inbred strains. The allele *Len-1ᵇ* determines four bands with pIs of 7.1, 7.5, 7.6, and 8.1; it occurs in the BALB/c, C57BL/6, C57BL/10, and SEC/1 strains. The allele *Len-1ᶜ* determines five bands with pIs of 7.1, 7.5, 7.6, 8.1, and 8.5; it occurs in *M. m. molossinus* and *M. m. castaneus*. The alleles are codominantly expressed in heterozygotes. No hybrid bands are formed, indicating that the proteins are monomeric. *Len-1* is located on Chr 1 near *Idh-1* (3, 5). Lok *et al.* (1) have identified the genomic DNA of a family of at least four γ-crystalline genes, and Skow (4), by use of restriction fragment length polymorphisms, has found that the four genes map near *Len-1*. Thus, the *Len-1* variation probably marks a complex of γ-crystallin genes.

Len-2 locus, lens protein-2. This locus encodes a major lens protein that has many similarities to β-crystallin. It has a molecular weight of about 23 000, occurs primarily in the lens nucleus, and is synthesized from about day 14 of gestation to about 14 days of age (2). The alleles are distinguished by isoelectric focusing in polyacrylamide gels. The allele *Len-2ᵃ* determines a band that focuses at pH 7.1; it occurs in the O20/A inbred strain, in wild mice from California and Peru, and in *M. m. molossinus*. The allele *Len-2ᵇ* determines a band focusing at a little higher pH; it occurs in all other strains examined (5).

References

1. Lok, S., L.-C. Tsui, T. Shinohara, J. Piatigorsky, R. Gold, and M. Breitman. 1984. Analysis of the mouse γ-crystallin gene family: assignment of multiple cDNAs to discrete genome sequences and characterization of a representative gene. Nucl. Acids Res. 12:4517–4529.
2. Skow, L. 1981. Mouse News Lett. 64:77.
3. Skow, L. 1982. Location of a gene controlling electrophoretic variation in mouse γ-crystallins. Exp. Eye Res. 34:509–516.
4. Skow, L.C. 1985. Mouse News Lett. 73:32.
5. Skow, L.C., M.E. Donner, R.A. Popp, and E.G. Bailiff. 1985. A second polymorphic lens crystallin (LEN-2) in the mouse: genetic and biochemical analysis of LEN-1 and LEN-2. Biochem. Genet. 23:181–189.

Lfo-1, *-2*, *Lth-1*, *-2*, *Ltn-1*, *-2*, *Ltw-1* to *-6* loci

These loci control variation in liver cytosol polypeptides detected by high resolution two-dimensional elec-

trophoresis (2). For most of the loci, differences were first found between strains C57BL/6By and BALB/cBy. Linkage was determined in recombinant inbred strains derived from these and other progenitors. Some of the loci were found to be identical to previously described loci but the functions of the polypeptide products of the other loci are not known. All polypeptides except LTH-1 and LTH-2 occur also in kidney (4). In the locus symbols, *tn*, *tw*, *th*, and *fo* indicate the range of molecular weights of the polypeptides in thousands (teens, twenties, thirties, and forties, respectively).

Lfo-1 locus, liver 40–50 thousand MW protein-1, Chr 7. The polypeptide controlled by this locus represents 1 to 2 per cent of the total liver supernatant protein. The allele *Lfo-1ᵃ* determines a more acidic form and occurs in strain C57BL/6; the allele *Lfo-1ᵇ* determines a more basic form and occurs in strain BALB/c. Heterozygotes have both forms in decreased intensity (1, 3).

Lfo-2 locus, liver 40–50 thousand MW protein-2, Chr 7. Different alleles occur in the C57BL/6 and DBA/2 strains (1).

Lth-1 locus, liver 30–40 thousand MW protein-1, Chr 13. The polypeptide controlled by this locus has a molecular weight of about 35 000. The allele *Lth-1ᵃ* determines presence of the protein and occurs in the BALB/c strain; the allele *Lth-1ᵒ* determines absence of the protein and occurs in strain C57BL/6. Heterozygotes show a very faint spot on electrophoretograms (3). *Lth-1* is on Chr 13 near *pe* (6).

Lth-2 locus, liver 30–40 thousand MW protein-2. Different alleles occur in the AKR and C57L strains (1).

Ltn-1 locus. Same as *Mup-1* locus.

Ltn-2 locus, liver 10–20 thousand MW protein-2. Different alleles occur in the AKR and C57L strains (1).

Ltw-1 locus. Same as *Apoa-1* locus.

Ltw-2 locus, liver 20–30 thousand MW protein-2, Chr 12. The allele *Ltw-2ᵃ* determines presence of the polypeptide and occurs in the C57BL/6 strain; the allele *Ltw-2ᵒ* determines its absence and occurs in BALB/c. Heterozygotes show a faint spot on electrophoretograms (3). *Ltw-2* was shown to be on the proximal part of Chr 12 by use of AKXL and BXH recombinant inbred strains (B.A. Taylor, personal communication).

Ltw-3 locus, liver 20–30 thousand MW protein-3, Chr 9. This locus controls electrophoretic variation of a polypeptide of about 23 000 molecular weight. The allele *Ltw-3ᵇ* determines a more basic polypeptide and occurs in strains C57BL/6, C3H/St, AU/SsJ, and a few others; the allele *Ltw-3ᵃ* determines a more acidic polypeptide and occurs in strains BALB/c, C3H/He, DBA/

2, A/J, AKR/J, and many others. Heterozygotes have both forms in decreased intensity. *Ltw-3* is located near *Trf* (transferrin) on Chr 9, but determines a different protein (3, 5).

Ltw-4 locus, liver 20–30 thousand MW protein-4, Chr 1. The allele *Ltw-4^b* determines a polypeptide seen as a prominent basic spot and occurs in strain BALB/c; the allele *Ltw-4^a* determines a more acidic prominent spot widely separated from that of *Ltw-4^b* and occurs in strain C57BL/6. Both spots are present in reduced intensity in heterozygotes (1, 3). This locus is probably identical with brain protein-12 (*Brp-12*) and is associated with variation in level of ethanol intake. See *Brp-12* locus.

Ltw-5 locus. See *Car-2* locus.

Ltw-6 locus, liver 20–30 thousand MW protein-6. Most strains carry the *Ltw-6^a* allele. The SWR/J, BUB, PL, and MA/My strains carry the *Ltw-6^b* allele (4).

References

1. Elliott, R.W. 1978. Mouse News Lett. 59:59.
2. Elliott, R.W. 1979. Mouse News Lett. 60:76.
3. Elliott, R.W. 1979. Use of two-dimensional electrophoresis to identify and map new mouse genes. Genetics 91:295–308.
4. Elliott, R.W. 1979. Mouse News Lett. 61:59.
5. Elliott, R.W., C. Hohman, C. Romejko, P. Louis, and F. Lilly. 1978. Use of high resolution two dimensional electrophoresis of liver cytosol proteins in the discovery and mapping of a new mouse variant. *In* N. Catsimpoolas, ed., Electrophoresis '78, 261–274. Elsevier North Holland, Amsterdam.
6. Elliott, R.W., W.L. Daniel, B.A. Taylor, and E.K. Novak. 1985. Linkage of loci affecting murine liver protein and arylsulfatase B to chromosome 13. J. Hered. 78:243–246.

lg locus, lid gap

Mutations at this locus were found by Strong (*lgSt*), Russell (*lg*), Miller (*lg^{Ml}*), and Stein (*lg^{Stn}*) (4, 6). A fifth allele, *lg^{Ga}*, occurred at Stanford University (3). Homozygotes may have one or both eyes open at birth. They are viable and fertile. The alleles show incomplete penetrance, varying from about 70 per cent to 100 per cent depending on the genetic background (4). In overall severity of the defect, the alleles rank in the order *lg* > *lg^{Stn}* = *lgSt* > *lg^{Ga}* > *lg^{Ml}* (3). Homozygotes for *lg^{Ml}* and *lg^{Stn}* are cured wholly or in part by administration of cortisone on day 14 of pregnancy (1, 4, 5) or by thyroxine on day 10 to 11 (2). In *lg^{Stn}*/*lg^{Stn}* embryos there is extensive vacuolization of the lenses, which is

first seen as fine vacuolization at 14 days of gestation. Later, defects of the cornea and retina occur (6).

References

1. Harris, M.J., D.M. Juriloff, and F.G. Biddle. 1984. Cortisone cure of the lidgap defect in fetal mice: a dose-response and time-response study. Teratology 29:287–295.
2. Juriloff, D.M. 1985. Prevention of the eye closure defect in *lg^{Ml}*/*lg^{Ml}* fetal mice by thyroxine. Teratology 32:73–86.
3. Juriloff, D.M., M.J. Harris, and J.R. Miller. 1983. The lidgap defect in mice: update and hypothesis. Can. J. Genet. Cytol. 25:246–254.
4. Ricardo, N.S., and J.R. Miller. 1967. Further observations on *lg^{Ml}* (lid-gap Miller) and other open-eyelid mutants in the house mouse. Can. J. Genet. Cytol. 9:596–605.
5. Stein, K.F., and N.C. Kettyle. 1973. Further data on the slit-lid mutant in mice and the effect of cortisone and doca on its development. Teratology 8:51–54.
6. Stein, K.F., B.E. Norris, and J. Mason. 1967. Development of an open eyelid mutant in *Mus musculus*. Dev. Biol. 16:315–330.

lh, lethargic, recessive, Chr 2

Arose spontaneously in the BALB/cGn strain. Homozygotes are first recognizable at 15 days by their lethargic behavior with instability of gait and occasional seizures. They are smaller and weaker than their normal littermates and often die before 2 months. Survivors of both sexes may breed, but their reproductivity is low. No pathological changes have been found in the central nervous system or the skeletal muscles (3), but peripheral motor nerves show reduced conduction velocity and prolonged distal latency (5). There is early thymic involution at 3 to 4 weeks, accompanied by decreased lymphocyte count, decreased cell-mediated immunity (1), and increased levels of serum IgG1 (4). The defects in the immune system tend to disappear by 2 months of age (1, 2).

References

1. Dung, H.C. 1977. Deficiency in the thymus-dependent immunity in "lethargic" mutant mice. Transplantation 23:39–43.
2. Dung, H.C. 1981. Lethargic mice. *In* M.E. Gershwin and B. Merchant, eds., Immunologic Defects in Laboratory Animals, Vol. 2, 17–37. Plenum Press, New York .
3. Dung, H.C., and R.H. Swigart. 1972. Histo-pathologic observations of the nervous and lymphoid tissues of "lethargic" mutant mice. Tex. Rep. Biol. Med. 30:23–39.
4. Dung, H.C., R.L. Lawson, and M. Stevens. 1977. A study of the increased serum level of IgG1 in lethargic mice combined with a depressed thymus-dependent lymphoid system. J. Immunogenet. 4:287–293.

5. Herring, J.M., H.C. Dung, J.H. Yoo, and J. Yu. 1981. Chronological studies of peripheral motor nerve condition in lethargic mice. Electromygr. Clin. Neurophysiol. 21:121–134.

Lhb locus, luteinizing hormone β, Chr 7

This locus codes for the luteinizing hormone β subunit (LH-β) secreted by the anterior pituitary. No genetic variants are known. The locus was found to be on Chr 7 using mouse–Chinese hamster somatic cell hybrids and a rat cDNA clone coding for LH-β (1, 2).

References

1. Naylor, S.L., W.W. Chin, H.M. Goodman, P.A. Lalley, K.-H Grzeschik, and A.Y. Sakaguchi. 1983. Chromosome assignments of genes encoding α and β subunits of glycoprotein hormones in man and mouse. Somat. Cell Genet. 9:757–770.
2. Naylor, S.L., A.Y. Sakaguchi, W.W. Chin, J. Jacobs, L.P. Shen, W.J. Rutter, and P.A. Lalley. 1984. Chromosomal mapping of mouse hormone genes: somatostatin, calcitonin, thyrotropin alpha subunit, and luteinizing hormone beta subunit. Cytogenet. Cell Genet. 37:551 (Abstr.).

Li, lined, semidominant, Chr X

Found among descendants of an X-irradiated male. Hemizygous males die before birth. *Li*/+ females are probably fully viable. They show a very fine striping, often seen only on the dorsal body area, and it is possible that some escape detection. The predominance of wildtype coat is due to a non-random X chromosome expression, since heterozygotes with other X-linked genes including *Blo*, *Ta*, *Hq*, and *Pgk-1* show a predominant expression of the alleles on the chromosome not carrying *Li*. The X chromosome bearing *Li* appears to be inactive in at least 90 per cent of cells (1). *Li* is located in the distal portion of Chr X near *Hyp* (2).

References

1. Cattanach, B.M., A.J.M. Crocker, and J. Peters. 1984. Mouse News Lett. 70:80–81.
2. Cattanach, B.M. 1985. Mouse News Lett. 73:17.

Lip-1 locus, lysosomal acid lipase, Chr 19

This is the structural locus for lysosomal acid lipase A (LIPA), an enzyme that functions in the hydrolysis of triglycerides and cholesterol esters. No genetic variants are known. The locus was found to be on Chr 19 by use of mouse–Chinese hamster somatic cell hybrids segregating mouse chromosomes. The synteny of *Lip-1* and *Got-1* on Chr 19 is similar to the synteny of genes for the homologous enzymes on human chromosome 10 (1).

References

1. Koch, G., P.A. Lalley, M. McAvoy, and T.B. Shows. 1981. Assignment of LIPA, associated with human lipase deficiency, to human chromosome 10 and comparative assignment to mouse chromosome 19. Somat. Cell Genet. 7:345–358.

lit, little, recessive, Chr 6

Arose spontaneously in the C57BL/6 strain. Homozygotes are smaller than normal beginning at about 2 weeks and continuing thereafter. Adult weight is about two-thirds that of controls. Females are fully fertile, but usually fail to nurse their first litters; males have reduced fertility. Females grow during pregnancy and after growth hormone (GH) treatment (4). The pituitary is deficient in GH and prolactin (PRL) (1), but plasma is deficient only in GH (5). The GH gene is grossly intact (7), but the GH-producing cells (somatotrophs) of the anterior pituitary have few or no secretory granules (3), and there is a deficiency of GH precursor RNA and mRNA (2). The pituitaries of little mice do not release GH in response to hypothalamic GH-releasing factor (GRF). This is not due to a defect in cAMP-mediation of the response, but is probably due to a defect in GRF-receptor binding or in function of the hormone–receptor complex (6).

References

1. Beamer, W.G., and E.M. Eicher. 1976. Stimulation of growth in the little mouse. J. Endocrinol. 71:37–45.
2. Cheng, T.C., W.G. Beamer, A. Phillips III, A. Bartke, R.L. Mallonee, and C. Dowling. 1983. Etiology of growth hormone deficiency in little, Ames dwarf and Snell dwarf mice. Endocrinology 113:1669–1678.
3. Christensen, E., and D.B. Wilson. 1981. Fine structure of somatotrophs and mammotrophs in the pituitary pars distalis of the little (*lit*) mutant mouse. Virchows Arch. (Cell. Pathol.) 37:89–96.
4. Eicher, E.M., and W.G. Beamer. 1976. Inherited ateliotic dwarfism in mice: characteristics of the mutation, little, on chromosome 6. J. Hered. 67:87–91.
5. Herington, H.C., D. Harrison, and J. Graystone. 1983. Hepatic binding of human and bovine growth hormones and ovine prolactin in the dwarf "little" mouse. Endocrinology 112:2032–2038.
6. Jansson, J.-O, J.R. Downs, W.G. Beamer, and L.A. Frohman. 1986. Receptor-associated resistance to growth hormone releasing factor in dwarf "little" mice. Science 232:511–512.
7. Phillips, J.A. III, W.G. Beamer, and A. Bartke. 1982.

Analysis of growth hormone genes in mice with genetic defects of growth hormone expression. J. Endocrinol. 92:405–407.

lm, lethal milk, recessive, Chr 2

Occurred spontaneously in the C57BL/6J strain. Homozygous females produce milk that is lethal to mice nursed during the first week of life and detrimental to mice nursed thereafter. The lethal and detrimental effects are due to a deficiency of zinc in the milk (2). If foster nursed on normal dams, *lm/lm* mice survive and become reproductively mature. Zinc supplementation of the drinking water of dams enables them to nurse their young successfully. All homozygotes lack utricular otoliths and this defect is not prevented by zinc supplementation. They have varying loss of saccular otoliths and show mild behavioral abnormalities related to the otolith defect. Homozygotes over 8 months of age show progressive hair loss, dermatitis, and skin lesions, symptoms of zinc deficiency (1). Administration of zinc at low doses to *lm/lm* mice induces in liver an increase in the zinc-binding protein metallothionein (MT) that is about two to three times higher than in controls. Old *lm/lm* mice with skin lesions have about three times as much zinc bound to MT in liver as *lm/lm* mice of the same age without lesions or as controls. While there is no difference in younger mice between untreated *lm/lm* mice and controls in zinc concentration in liver and other tissues, it is thought that *lm* may cause an abnormality of MT metabolism that results in sequestration of zinc during pregnancy and in old age and a consequent zinc deficiency in milk and elsewhere in old age (3). How an MT defect could be related to the otolith abnormality is not known.

References

1. Erway, L.C., and A. Grider Jr. 1984. Zinc metabolism in lethal-milk mice. Otolith, lactation, and aging effects. J. Hered. 75:480–484.
2. Piletz, J.E., and R.E. Ganschow. 1978. Zinc deficiency in murine milk underlies expression of the lethal-milk (*lm*) mutation. Science 199:181–183.
3. Piletz, J.E., and H.R. Herschman. 1982. Induction of metallothionein by zinc in *lethal milk* mutant mice. Biochem. Genet. 20:1221–1233.

Lm-1 locus, lymphomyeloid antigen-1, Chr 12

This locus controls an antigen, present on thymocytes, T lymphoblasts, some plasmacytomas, and some myeloid cells, that is recognized with a cytotoxic T-cell assay. The allele *Lm-1ᵃ* determines the Lm-1.1 specificity and occurs in the BALB/c strain; the allele *Lm-1ᵇ* determines the Lm-1.2 specificity and occurs in the C57BL/6 and NZB strains and in the *Igh*-congenic strains C.B-17, C.B-23, and C.B-26. O'Toole *et al.* (2) found differential expression of at least two distinct Lm-1 antigenic determinants, one on T- and B-cell blasts and the other on pre-B-cells and B-cell blasts. Fibroblasts may express a third determinant. *Lm-1* may also be a histocompatibility locus. It is located on Chr 12 proximal to *Igh* and *Pre-1*, in the same region as *Ly-7*, *H-40*, and *H(Igh)* (1). It is probably different from *Ly-7*, since it has a different strain distribution. However, it has many similarities to *H-40* and *H(Igh)*. *Lm-1* may be a lymphocyte–macrophage family of genes that includes those designated *Ly-40* and *H(Igh)* (2).

References

1. O'Toole, M.M., R. Riblet, and M.J. Bosma. 1984. Identification and mapping of *Lm-1*, an *Igh*-linked locus for a murine alloantigen. Immunogenetics 20:265–275.
2. O'Toole, M.M., G.C. Bosma, and M.J. Bosma. 1985. Expression of murine *Lm-1* locus. Lm-1 determinants on lymphocytes and macrophages, and effects of Lm-1 incompatibility on bone marrow grafts. J. Exp. Med. 162:607–624.

Lmp locus, low molecular weight polypeptides, Chr 17

This locus codes for an intracellular antigen or antigens present at high levels in macrophages and at lower levels in lymphocytes and fibroblasts. Strains positive for the antigen (*Lmpᵈ*, *Lmpᵃ*) are BALB/c, A/J, and strains carrying the *H-2* haplotypes *d, a, f, ja, k, s, u, v*, and *z*; strains negative for the antigen (*Lmpᵇ*) are C57BL/6, DBA/1, and strains carrying the *H-2* haplotypes *b, p, q*, and *r*. The LMP antigens comprise a large number of non-covalently linked low molecular weight polypeptide subunits in a large intracellular complex of unknown function. By two-dimensional gel electrophoresis, polymorphism for two of the subunits (2 and 7) was detected. The subunit 2 difference corresponds exactly to the antigenic difference. However, the *H-2ᵈ* haplotype differs from that of the other LMP₊ strains by a subunit 7 difference, thereby defining a third allele *Lmpᵈ*. LMP typing of *H-2* congenic and recombinant strains provides evidence that *Lmp* lies within the *H-2* complex between the *H-2K* and *I-A* subregions (1, 2).

References

1. Monaco, J.J., and H.O. McDevitt. 1982. Identification of a fourth class of proteins linked to the murine major histo-

compatibility complex. Proc. Natl. Acad. Sci. USA 79:3001–3005.

2. Monaco, J.J., and H.O. McDevitt. 1986. The LMP antigens: a stable MHC-controlled multisubunit protein complex. Hum. Immunol. 15:416–426.

ln, leaden, recessive, Chr 1

Arose spontaneously in the C57BR inbred strain (2). In its effect on coat color it is indistinguishable from *d*. Like *d* it causes clumping of melanin granules into larger masses, but no change in color of the pigment. The clumping is due to the shape of the melanocytes, which have fewer and thinner dendritic processes than wild-type melanocytes (1). These melanocytes are more easily dislodged from fixed sites in the hair bulb and incorporated into the developing hair, resulting in large clumps of pigment in the hair shaft (4). The site of action of *ln* was shown by use of chimeras and dermal–epidermal recombination grafts to be in the melanocytes (3).

References

1. Markert, C.L., and W.K. Silvers. 1956. The effects of genotype and cell environment on melanoblast differentiation in the house mouse. Genetics 41:429–450.
2. Murray, J.M. 1933. "Leaden", a recent color mutation in the house mouse. Am. Nat. 67:278–283.
3. Stephenson, D.A., P.H. Glenister, and J.E. Hornby. 1985. Site of beige (*bg*) and leaden (*ln*) pigment gene expression determined by recombinant embryonic grafts and aggregation mouse chimaeras employing sash (*W*_{sh}) homozygotes. Genet. Res. 46:193–205.
4. Sweet, S.E., and W.C. Quevedo Jr. 1968. Role of melanocyte morphology in pigmentation of mouse hair. Anat. Rec. 162:243–354.

Lna-1 locus, lymph node antigen-1

This locus controls an antigen present on the cells of lymph nodes but not of other lymphoid organs. It occurs on about 30 to 40 per cent of both T- and B-cells of the lymph nodes. The allele *Lna-1^b* determines presence of the Lna-1.2 antigen and occurs in the C57, C58 family, and in the GR/A, HRS, and I/St strains; the allele *Lna-1^a* determines absence of the antigen and occurs in the A, AKR, BALB/c, CBA/J, C3H/An, and many other strains. The antigen can be induced *in vitro* with Con A in spleen and thymus cells of *Lna-1^b* strains but not in bone marrow. It can be induced *in vivo* in *Lna-1^b* mice by inoculation 5 days previously with sheep red blood cells (1, 2).

References

1. Shen, E.-W., G. Viamontes, and E.A. Boyse. 1982. A system of alloantigens that selectively identifies lymph node lymphocytes. Immunogenetics 15:17–21.
2. Shen, F.-W., H. Yakura, and J.-S. Tung. 1983. Some compartments of B cell differentiation. Immunol. Rev. 69:69–80.

Lop, lens opacity, semidominant, Chr 10

Arose spontaneously in a stock of mice homozygous for the Robertsonian translocation Rb(6.15)1Ald. Most heterozygotes have total lens opacity but in a few the opacity is less than total. The opacity is present at the time the eyes first open. It is more easily seen with the naked eye in albino than in pigmented mice. Size of the eye is slightly reduced in heterozygotes. Homozygotes have very small eyes with opaque lenses. The lenses are also much smaller than normal and there is no regular arrangement of layers of lens fibers. *Lop* was shown to be on Chr 10 near the distal end (1). It is probably an allele of *Cat* and is therefore more properly designated *Cat* ^{lop} (2).

References

1. Lyon, M.F., S.E. Jarvis, I. Sayers, and R.S. Holmes. 1981. Lens opacity: a new gene for congenital cataract on chromosome 10 of the mouse. Genet. Res. 38:337–341.
2. Muggleton-Harris, A.L., M.F.W. Festing, and M. Hall. 1987. A gene location for the inheritance of the cataract Fraser (*Cat*_{Fr}) mouse congenital cataract. Genet. Res. 49:235–238.

lop-2, lens opacity-2, recessive

Mice of the 101/H inbred strain are all homozygous for this gene which causes a colored cataract revealed by examination with a slit lamp. The cataract is present by weaning age. The 129/Sv strain which is related to 101/H also probably carries *lop-2*. Strains not carrying *lop-2* include C3H/He, C57BL/Ola, C57BL/10, DBA/2, JU/Fa, and 101/El. Several sublines of CBA/Ca have a bright white cataract which appears to be controlled by a single semidominant gene tentatively designated *Lop-3*. F1 mice of crosses to C3H/He have very mild cataracts. However, crosses between CBA/Ca and 101/H to determine whether *lop-2* and *Lop-3* are alleles gave ambiguous results. Neither *lop-2* nor *Lop-3* is allelic with *Lop*. No tests for allelism with any of the numerous other genes causing cataracts in mice have been made (1).

References

1. West, J.D., and G. Fisher. 1985. Inherited cataracts in inbred mice. Genet. Res. 46:45–56.

Lop-4

Lop-4, lens opacity-4, semidominant, Chr 2

Found among the offspring of a C3H/HeH male whose spermatogonia had been treated with ENU. Heterozygotes have cataracts of variable severity that typically involve clusters or arcs of reflectant specks in the nuclear or perinuclear region, or, less frequently, nuclear or perinuclear opacities extending into the cortex. Penetrance is probably incomplete. Homozygotes have not been positively identified, but mice with bright white nuclear cataracts were found only in progeny of crosses that could have produced Lop-4/Lop-4 offspring. Lop-4 is on Chr 2, probably about 7 cM distal to the a locus (1).

References

1. West, J.D., and G. Fisher. 1986. Further experience of the mouse dominant cataract mutation test from an experiment with ethylnitrosourea. Mut. Res. 164:127–136.

Lp, loop-tail, semidominant, Chr 1

Arose as a spontaneous mutation in the A strain. Lethal when homozygous, probably not fully penetrant in heterozygotes. Heterozygotes have looped or crooked tails and often show a wobbling of the head. In females the vagina may be imperforate (3). Many heterozygotes show enlarged lateral ventricles of the brain (5) and numerous slight behavioral abnormalities (4). The head wobbling was shown to be associated with enlargement of the ventricles (9). Homozygotes have open neural folds in the region of the brain and may or may not have open spinal cords as well. They are usually alive at birth but never survive thereafter (3). The axial skeleton and ribs of homozygotes are abnormal, apparently as a secondary result of the abnormalities of the nervous system (2). In Lp/Lp embryos of 9 to 9.5 days, the shortening or regression of the primitive streak is retarded, the neural plate and notochord are shorter than normal (1), there are gaps in the neural basal lamina (6), and the cells and basal lamina of the otocyst are abnormal (7). At 10 and 11 days, there is prolongation of the cell cycle, particularly in the M and G_1 periods (8), and a marked flattening of the surface cells of the brain (10).

References

1. Smith, L.J., and K.F. Stein. 1962. Axial elongation in the mouse and its retardation in homozygous looptail mice. J. Embryol. Exp. Morphol. 10:73–87.
2. Stein, K.F., and J.A. Mackensen. 1957. Abnormal development of the thoracic skeleton in mice homozygous for looped-tail. Am. J. Anat. 100:205–223.
3. Strong, L.C., and W.F. Hollander. 1949. Hereditary loop-tail in the house mouse accompanied by inperforate vagina and craniorachischisis when homozygous. J. Hered. 40:329–334.
4. Van Abeelen, J.H.F. 1968. Behavioural ontogeny of looptail mice. Anim. Behav. 16:1–4.
5. Van Abeelen, J.H.F., and S.M.J. Raven. 1968. Enlarged ventricles in the cerebrum of loop-tail mice. Experientia 24:191–192.
6. Wilson, D.B. 1985. Ultrastructural analysis of basal neuroepithelial cells in dysraphic mice. Virchows Arch. (Cell Pathol.) 48:9–17.
7. Wilson, D.B. 1985. An ultrastructural analysis of abnormal otic development in exencephalic mutant mice. Acta Otolaryngol. 241:203–208.
8. Wilson, D.B., and E.M. Center. 1974. The neural cell cycle in the looptail (Lp) mutant mouse. J. Embryol. Exp. Morphol. 32:697–705.
9. Wilson, D.B., and E.M. Center. 1977. Differences in cerebral morphology in 2 stocks of mutant mice heterozygous for the loop-tail (Lp) gene. Experientia 33:1502–1503.
10. Wilson, D.B., and S.D. Michael. 1975. Surface defects in ventricular cells of brains of mouse embryos homozygous for the loop-tail gene: scanning electron microscopic study. Teratology 11:87–98.

Lpn-1 to Lpn-3 loci

Lpn-1, -2 loci, NZ lupus nephritis-1, -2, Chr 17
Lpn-3 locus, NZ lupus nephritis-3

Lupus nephritis occurs in the F1 of crosses between the NZB and NZW strains and in the offspring of backcrosses to the parent strains, but not in the parent strains themselves (2). To explain these results, the authors postulate three loci with dominant alleles determining susceptibility to nephritis, the dominant alleles of all three loci being necessary for expression of the disease. The dominant allele of Lpn-1 occurs in NZB, and those of Lpn-2 and Lpn-3 occur in NZW. Lpn-1 shows about 33 per cent recombination with H-2, and Lpn-2 shows no recombination with H-2 (2). Lpn-1 and Lpn-2 have the same strain distribution and chromosomal location as Ads-1, Agp-1 and Ads-3, Agp-3, respectively, and may be parts of the two gene complexes or may be identical with one or the other component (1). See Ads-1 to -4 loci, Agp-1, -3 loci.

References

1. Hirose, S., R. Nagasawa, I. Sekikawa, M. Hamaki, Y. Ishida, H. Sato, and T. Shirai. 1983. Enhancing effect of H-2-linked NZW gene(s) on the autoimmune traits of (NZB × NZW)F1 mice. J. Exp. Med. 158:228–233.
2. Knight, J.G., and D.D. Adams. 1978. Three genes for

lupus nephritis in NZB × NZW mice. J. Exp. Med. 147:1653–1660.

lpr, lymphoproliferation, recessive

Found during inbreeding in a stock derived from strains LG, AKR, C3H, and C57BL/6. In the derived inbred strain, MRL/Mp-*lpr/lpr*, compared with its congenic normal strain, MRL/Mp-+/+, homozygotes develop an autoimmune syndrome resembling systemic lupus erythematosus with massive lymphoproliferation and immune complex glomerulonephrosis beginning by 8 weeks of age. Females die at an average age of 17 weeks and males at 22 weeks. Hypergammaglobuline-mia develops with twofold increases in IgA, IgM, and IgG2b, and 10-fold increases in IgG1 and IgG2a (12). The excess lymphocyte population consists largely of abnormal T-cells with the T-cell surface markers Thy-1^+, Lyt-1^+2^-, and with the B-cell marker Ly-5^+ in addition (3, 10). There is enhanced T-helper cell activity (15) and early-onset generalized polyclonal B-cell activation. The polyclonal B-cell activation occurs only after appearance of the abnormal Ly-5^+ T-cells in the spleen and lymph nodes (1). The abnormal T-cells are descended from abnormal bone-marrow stem cells that require processing in the thymus to acquire their abnormal surface markers (7, 16, 17). There is a progressive defect in ability of lymphocytes to produce and respond to interleukin 2 (19) and an increase in production of interleukin 3 (13). Transfer of spleen or bone marrow cells from +/+ to *lpr/lpr* mice delays the onset of autoimmunity and increases longevity to almost normal; transfer of *lpr/lpr* cells to +/+ mice causes severe graft-vs.-host disease, wasting, and early death. A possible explanation of these results is that *lpr/lpr* cells intrinsically lack an antigen which is present on +/+ cells and is recognized as foreign by the transplanted *lpr/lpr* lymphocytes (18). The disease symptoms are ameliorated in *lpr/lpr* mice bearing the X-linked immune deficiency gene *xid*, which causes absence of a class of mature B-cells (6, 14). Disease symptoms are also ameliorated by a restricted-calorie diet (5) or by a fish oil diet that reduces cyclooxygenase metabolites (9). The *lpr* gene has been introduced into a number of different inbred strains. On different backgrounds it causes varying degrees of lymphoproliferation, autoimmune antibody production, earliness of death, and incidence of severe lupus-like disease (1, 8, 11). On strain backgrounds positive for the plasma cell antigen PC.1 (*Pca-1^a*), there is a very high level of PC.1 on the abnormal T-cells of *lpr/lpr* mice in contrast to a low level on T-cells of +/+ mice (4). The mutation *gld* (generalized lymphoproliferative disease) produces an abnormal T-cell population indistinguishable from that of *lpr* (2).

References

1. Davidson, W.F., J.B. Roths, K.L. Holmes, E. Rudikoff, and H.C. Morse III. 1984. Dissociation of severe lupus-like disease from polyclonal B cell activation and IL 2 deficiency in C3H-*lpr/lpr* mice. J. Immunol. 133:1048–1056.
2. Davidson, W.F., F.J. Dumont, H.G. Bedigian, B.J. Fowlkes, and H.C. Morse III. 1986. Phenotypic, functional, and molecular genetic comparisons of the abnormal lymphoid cells of C3H-*lpr/lpr* and C3H-*gld/gld* mice. J. Immunol. 126:4075–4084.
3. Dumont, F.J., R.C. Habbessett, E.A. Nichols, J.A. Treffinger, and A.S. Tung. 1983. A monoclonal antibody (100C5) to the Lyt-2^- T cell population expanding in MRL/Mp-*lpr/lpr* mice detects a surface antigen normally expressed on Lyt-2^+ cells and B cells. Eur. J. Immunol. 13:455–459.
4. Dumont, F.J., R.C. Habberset, L.Z. Coker, E.A. Nichols, and J.A. Treffinger. 1985. High level expression of the plasma cell antigen PC.1 on the T-cell subset expanding in MRL/MpJ-*lpr/lpr* mice: detection with a xenogeneic monoclonal antibody and alloantisera. Cell. Immunol. 96:327–337.
5. Fernandes, G., and R.A. Good. 1984. Inhibition by restricted-calorie diet of lymphoproliferation disease and renal damage in MRL/*lpr* mice. Proc. Natl. Acad. Sci. USA 81:6144–6148.
6. Fieser, T.M., M.E. Gershwin, A.D. Steinberg, F.J. Dixon, and A.N. Theofilopoulos. 1984. Abrogation of murine lupus by the *xid* gene is associated with reduced responsiveness of B cells to T-cell-helper signals. Cell. Immunol. 87:708–713.
7. Fujiwara, M., and A. Kariyone. 1984. One-way occurrence of graft-versus-host disease in bone marrow chimaeras between congenic MRL mice. Immunology 53:251–256.
8. Izui, S., V.E. Kelley, K. Masuda, H. Yoshida, J.B. Roths, and E.D. Murphy. 1984. Induction of various autoantibodies by mutant gene *lpr* in several strains of mice. J. Immunol. 133:227–233.
9. Kelley, V.E., A. Ferretti, S. Izui, and T.B. Strom. 1985. A fish oil diet rich in eicosapentaenoic acid reduces cyclooxygenase metabolites and suppresses lupus in MRL-*lpr* mice. J. Immunol. 134:1914–1919.
10. Morse, H.C. III, W.F. Davidson, R.A. Yetter, E.D. Murphy, J.B. Roths, and R.L. Coffman. 1982. Abnormalities induced by the mutant gene *lpr*: expansion of a unique lymphocyte subset. J. Immunol. 129:2612–2615.
11. Morse, H.C. III, J.B. Roths, W.F. Davidson, W.Y. Langdon, T.N. Fredrickson, and J.W. Hartley. 1985. Abnormalities induced by the mutant gene *lpr*: patterns of disease and expression of murine leukemia virus in SJL/J mice homozygous and heterozygous for *lpr*. J. Exp. Med. 161:602–616.

12. Murphy, E.D., and J.B. Roths. 1978. Autoimmunity and lymphoproliferation: induction by mutant gene *lpr*, and alteration by a male-associated factor in strain BXSB mice. *In* N.R. Rose, P.E. Bigazzi, and N.L. Warner, eds., Genetic Control of Immune Disease, 207–221. Elsevier North Holland, Amsterdam.

13. Palacios, R. 1984. Spontaneous production of interleukin 3 by T lymphocytes from autoimmune MRL/Mp-*lpr/lpr* mice. Eur. J. Immunol. 14:599–605.

14. Steinberg, E.B., T.J. Santoro, T.M. Chused, P.A. Smathers, and A.D. Steinberg. 1983. Studies of congenic MRL-*lpr/lpr.xid* mice. J. Immunol. 131:2789–2795.

15. Theofilopoulos, A.N., D.L. Shawler, R.A. Eisenberg, and F.J. Dixon. 1980. Splenic immunoglobulin-secreting cells and their regulation in autoimmune mice. J. Exp. Med. 151:446–466.

16. Theofilopoulos, A.N., R. Balderas, D.L. Shawler, S. Izui, B.L. Kotzin, A. Strober, and F.J. Dixon. 1980. Inhibition of T cell proliferation and SLE-like syndrome of MRL/1 mice by whole body or total lymphoid irradiation. J. Immunol. 125:2137–2142.

17. Theofilopoulos, A.N., R.S. Balderas, D.L. Shawler, S. Lee, and F.J. Dixon. 1981. Influence of thymus genotype on the systemic lupus erythematosus-like disease and T cell proliferation of MRL/Mp-*lpr/lpr* mice. J. Exp. Med. 153:1405–1414.

18. Theofilopoulos, A.N., R.S. Balderas, Y. Gozes, M.T. Aguado, L. Hang, P.R. Morrow, and F.J. Dixon. 1985. Association of *lpr* gene with graft-vs-host disease-like syndrome. J. Exp. Med. 162:1–18.

19. Wofsy, D., E.D. Murphy, J.B. Roths, M.J. Dauphinée, S.B. Kipper, and N. Talal. 1981. Deficient interleukin 2 activity in MRL/Mp and C57BL/6J mice bearing the *lpr* gene. J. Exp. Med. 154:1671–1680.

Lps locus, lipopolysaccharide response, Chr 4

This locus controls response to treatment with the lipid A moiety of endotoxin, the lipopolysaccharide (LPS) from the cell wall of Gram-negative bacteria. The allele for defective response, Lps^d, occurs in the C3H/HeJ strain where it probably originated by mutation between 1960 and 1968 (8); the allele for normal response, Lps^n, occurs in most other C3H substrains and other inbred strains. The effects of LPS which are absent or modified in C3H/HeJ, but occur in closely related substrains and are therefore probably dependent on the *Lps* locus include: toxicity (C3H/HeJ is 10 to 40 times more resistant to lethal effects of LPS then Lps^n strains) (14, 20); mitogenic stimulation of B lymphocytes (5); polyclonal activation of B lymphocytes (27); maturation of B-cells with acquisition of Ia antigens (26), surface immunoglobulin, and complement receptor (12); induction of Thy-1 antigen on a subpopulation of T lymphocytes (12); hypothermia in a cool environment; elevation of serum precursor of secondary amyloid protein; elevation of colony-stimulating activity in serum (28); adjuventicity (19); induction of interferon (1); toxicity to macrophages (9), and inhibition of phagocytosis by macrophages *in vitro* (25); tumoricidal activity of macrophages (16); and increase of glucose utilization in fibroblasts (17). In addition, LPS causes a more rapid and greater influx of neutrophils and macrophages into the peritoneal cavity in C3H/HeJ mice than in mice of Lps^n strains, a response that may be related to the greater LPS resistance of C3H/HeJ mice (14). Lps^d exerts some effects in the absence of LPS. These include: defective tumoricidal activity of macrophages and rate of glucose utilization of fibroblasts (16, 17); defective response of macrophages to macrophage inhibition factor (21); lower electrophoretic mobility of B-cells (7); resistance of spleen cells to replication of herpes simplex virus (11); 50 per cent drop in number of peripheral leukocytes (22); failure of macrophages in culture to maintain their phagocytic ability (23) and increased susceptibility to *Salmonella typhimurium* and other Gram-negative bacteria (10, 15). C3H/HeN (Lps^n) mice have a higher level of transcription of mammary tumor provirus DNA than C3H/HeJ (Lps^d) mice and respond to injection of LPS with a 3- to 4-fold increase in transcription, in contrast to C3H/HeJ mice which show no increase after injection (3). Various combinations of *Lps*-associated traits have been followed in crosses between C3H/HeJ and other C3H substrains, and the traits have in all cases segregated together (2, 5, 19, 28). Some of the traits show dominance of the Lps^n allele; others show codominance. The C57BL/10ScCr and C57BL/10ScN sublines of C57BL/10 are like C3H/HeJ in failure to respond to the B-cell mitogenic effect of LPS (6, 24) and to the LPS stimulation of acute phase serum amyloid (13), and the defect is probably caused by a mutation at the *Lps* locus similar or identical to Lps^d (4, 24). Since the *Lps* locus affects the response to LPS of B-cells, macrophages, fibroblasts, and probably also T-cells, it probably acts by changing a membrane component that has a similar function in the different cell types (18), and this function may be important in at least some of the cell types even in the absence of LPS. *Lps* is on Chr 4 between *Mup-1* and *Ps* (29).

References

1. Apte, R.N., O. Ascher, and D.H. Pluznik. 1977. Genetic analysis of generation of serum interferon by bacterial lipopolysaccharide. J. Immunol. 119:1898–1902.

2. Apte, R.N., O. Ascher, and D.H. Pluznik. 1977. Rela-

tionship between lipopolysaccharide-induced serum colony stimulating factor and interferon in regulation of murine granulopoiesis. Exp. Hematol. 5 Suppl. 2:12 (Abstr.).

3. Clark, J.K., V.L. Train-Dorge, and J.C. Cohen. 1985. Mouse mammary tumor virus gene expression regulated *in trans* by *Lps* locus. Virology 147:210–213.

4. Coutinho, A., and T. Meo. 1978. Genetic basis for unresponsiveness to lipopolysaccharide in C57BL/10Cr mice. Immunogenetics 7:17–24.

5. Coutinho, A., G. Müller, and E. Gronowicz. 1975. Genetical control of B-cell responses. IV. Inheritance of the unresponsiveness to lipopolysaccharides. J. Exp. Med. 142:253–258.

6. Coutinho, A., L. Forni, F. Melchers, and T. Watanabe. 1977. Genetic defect in responsiveness to the B cell mitogen lipopolysaccharide. Eur. J. Immunol. 7:325–328.

7. Dumont, F., and R. Barrois. 1976. Electrokinetic properties of splenic lymphocytes from the low-polysaccharide responder C3H/HeJ mice. Folia Biol. Praha 22:145–150.

8. Glode, L.M., and D.L. Rosenstreich. 1976. Genetic control of B cell activation by bacterial lipopolysaccharide is mediated by multiple distinct genes or alleles. J. Immunol. 117:2061–2066.

9. Glode, L.M., A. Jacques, S.E. Mergenhagen, and D.L. Rosenstreich. 1977. Resistance of macrophages from C3H/HeJ mice to the *in vitro* cytotoxic effects of endotoxin. J. Immunol. 119:162–166.

10. Hagberg, L., D.E. Briles, and C.S. Eden. 1985. Evidence for separate defects in C3H/HeJ and C3HeB/FeJ mice, that affect susceptibility to gram-negative infection. J. Immunol. 134:4118–4122.

11. Kirchner, H., H.M. Hirt, D.L. Rosenstreich, and S.E. Mergenhagen. 1978. Resistance of C3H/HeJ mice to lethal challenge with herpes simplex virus. Proc. Soc. Exp. Biol. Med. 157:29–32.

12. Koenig, S., M.K. Hoffmann, and L. Thomas. 1977. Induction of phenotypic lymphocyte differentiation in LPS unresponsive mice by an LPS-induced serum factor and by lipid-A-associated protein. J. Immunol. 118:1910–1911.

13. McAdam, K.P.W.J., and J.L. Ryan. 1978. C57BL/10/CR mice: nonresponders to activation by the lipid A moiety of bacterial lipopolysaccharide. J. Immunol. 120:249–253.

14. Moeller, G.R., L. Terry, and R. Snyderman. 1978. The inflammatory response and resistance to endotoxin in mice. J. Immunol. 120:116–123.

15. O'Brien, A.D., D.A. Weinstein, M.Y. Soliman, and D.L. Rosenstreich. 1985. Additional evidence that the *Lps* gene locus regulates native resistance to *S. typhimurium* in mice. J. Immunol. 134:2820–2823.

16. Ruco, L.P., M.S. Meltzer, and D.L. Rosenstreich. 1978. Macrophage activation for tumor cytotoxicity: control of macrophage tumoricidal capacity by the Lps gene. J. Immunol. 121:543–548.

17. Ryan, J.L., and K.P.W.J. McAdam. 1977. Genetic non-responsiveness of murine fibroblasts to bacterial endotoxin. Nature 269:153–155.

18. Scibienski, R.J. 1981. Defects in murine responsiveness to bacterial lipopolysaccharide. The C3H/HeJ and C57BL/10ScCr strains. *In* M.E. Gershwin and B. Merchant, eds., Immunologic Defects in Laboratory Animals, Vol. 2, 241–258. Plenum Press, New York.

19. Skidmore, B.J., J.M. Chiller, W.O. Weigle, R. Riblet, and J. Watson. 1976. Immunologic properties of bacterial lipopolysaccharide (LPS). III. Genetic linkage between the *in vitro* mitogenic and *in vivo* adjuvant properties of LPS. J. Exp. Med. 143:143–150.

20. Sultzer, B.M. 1968. Genetic control of leukocyte responses to endotoxin. Nature 219:1253–1254.

21. Tagliabue, A., J.L. McCoy, and R.B. Herberman. 1978. Refractoriness to migration inhibitory factor of macrophages of LPS nonresponder mouse strains. J. Immunol. 121:1223–1226.

22. Verghese, M.W., M. Prince, and R. Snyderman. 1980. Genetic control of peripheral leukocyte response to endotoxin in mice. J. Immunol. 124:2468–2473.

23. Vogel, S.N., and D.L. Rosenstreich. 1979. Defective Fc receptor-mediated phagocytosis in C3H/HeJ macrophages. I. Correction by lymphokine-induced stimulation. J. Immunol. 123:2842–2850.

24. Vogel, S.N., C.T. Hansen, and D.L. Rosenstreich. 1979. Characterization of a congenitally LPS-resistant, athymic mouse strain. J. Immunol. 122:619–622.

25. Vogel, S.N., S.T. Marshall, and D.L. Rosenstreich. 1979. Analysis of the effects of lipopolysaccharide on macrophages; differential phagocytic responses of C3H/HeN and C3H/HeJ macrophages *in vitro*. Infect. Immun. 25:328–336.

26. Watson, J. 1977. Differentiation of B lymphocytes in C3H/HeJ mice: the induction of Ia antigens by lipopolysaccharide. J. Immunol. 118:1103–1108.

27. Watson, J., and R. Riblet. 1975. Genetic control of responses to bacterial lipopolysaccharides in mice. II. A gene that influences a membrane component involved in the activation of bone marrow-derived lymphocytes by lipopolysaccharide. J. Immunol. 114:1462–1468.

28. Watson, J., M. Largen, and K.P.W.J. McAdam. 1978. Genetic control of endotoxic response in mice. J. Exp. Med. 147:39–49.

29. Watson, J., K. Kelly, M. Largen, and B.A. Taylor. 1978. The genetic mapping of a defective LPS response gene in C3H/HeJ mice. J. Immunol. 120:422–424.

lr, lens rupture, recessive

Arose spontaneously in an inbred albino strain. Penetrance is complete on some genetic backgrounds, but may be reduced on others. In homozygotes, beginning at about 3 weeks of age, there is cataractous degeneration of the lens with rupture of the capsule and expulsion of the lens nucleus into the vitreous chamber. The lens shrinks, tearing the suspensory ligaments, and may pass through the pupil into the anterior chamber (1,2). This mutant resembles *ec*, but has not been tested for allelism with it.

References

1. Fraser, F.C., and M.L. Herer. 1948. Lens rupture, a new recessive gene in the house mouse. J. Hered. 39:149.
2. Fraser, F.C., and M.L. Herer. 1950. The inheritance and expression of the "lens rupture" gene in the house mouse. J. Hered. 41:3–7.

Lr or *Lr-1* locus

See *Lsr-1* locus.

ls, lethal spotting, recessive, Chr 2

Arose in a subline of an inbred strain, C57BL-*a^t*. Homozygotes resemble piebald mice in having considerable white spotting. They die usually in the third week of life with megacolon associated with deficiency of intrinsic ganglion cells in the lower colon. Some survive and are fertile (1, 2). The narrowed distal segment of the colon shows absence of the serotonergic inhibitory system which functions normally to cause the relaxation phase of peristalsis and relaxation of the internal anal sphincter (5, 6). The *ls* mutation seems to exert its effect on pigmentation by reducing the number of melanoblasts, possibly through an effect on the neural crest (3). Since melanoblasts and intrinsic ganglion cells are both neural crest derivatives, it is possible that *ls* acts by causing defective migration or function of these cells or by reducing their number. However, Rothman and Gershon (6) have obtained indirect evidence that, in the lower bowel, the gene acts by causing an intrinsic abnormality of the terminal 2 mm of the gut. The *ls* locus is on Chr 2 close to *Ra* near the distal end (4).

References

1. Bolande, R.P., and W.F. Towler. 1972. Ultrastructure and histochemical studies of murine megacolon. Am. J. Pathol. 69:139–162.
2. Lane, P.W. 1966. Association of megacolon with two recessive spotting genes in the mouse. J. Hered. 57:29–31.
3. Mayer, T.C., and E. Maltby. 1964. An experimental investigation of pattern development in lethal spotting and belted mouse embryos. Dev. Biol. 9:269–286.
4. Phillips, R.J.S. 1966. Mouse News Lett. 34:27.
5. Richardson, J. 1975. Pharmacological studies of Hirschsprung's disease on a murine model. J. Pediat. Surg. 10:875–884.
6. Rothman, T.P., and M.D. Gershon. 1984. Regionally defective colonization of the terminal bowel by the pre-

cursors of enteric neurons in lethal spotted mutant mice. Neuroscience 12:1293–1311.

Lsd locus, lymphocyte-stimulating determinant, Chr 1

This locus controls a lymphocyte-stimulating determinant on spleen cells. In mixed lymphocyte cultures, the determinant stimulates proliferation and effector function of T cells not having the determinant. The allele here designated *Lsd^a* determines presence of the determinant and occurs in strains DBA/2, AKR, and C3H/Tif; the allele here designated *Lsd^b* determines absence of the determinant and occurs in strains C57BL/6, C57BL/10, BALB/c, DBA/1, and C3H/DiSn. *Lsd* is on Chr 1 very close to *Mls* which it closely resembles, but the two loci have shown about 1 per cent recombination (1).

References

1. Opalka, B., and E. Kölsch. 1983. Evidence for a new lymphocyte-stimulating determinant (Lsd) detected by alloreactive T cell lines. Eur. J. Immunol. 13:24–30.

Lse, low-set ears, probably semidominant

Found among offspring of a cross of a mutant stock carrying *dr^J* to C3HeB/JLe-a/a. Heterozygotes can be recognized at birth; the external auditory meatus which is sealed over by the pinna in normal mice remains open. Adults have abnormally shaped external ears which appear to be set lower on the head than normal. The pinna is smaller than normal and the tragus and antitragus are set wider apart, widely exposing the entrance of the auditory meatus. The cornea is opaque and the mutants are smaller than normal and suffer high mortality before weaning. No inner-ear defects or cleft palate have been observed. Homozygotes have not been identified. Heterozygotes can be recognized at 13 days of gestation by the position of the primitive pinna which is placed more posteriorly than in controls (1).

References

1. Theiler, K., and H.O. Sweet. 1986. Low set ears (*Lse*), a new mutation of the house mouse. Anat. Embryol. 175:241–246.

Lsh locus, leishmaniasis resistance, Chr 1

This locus controls resistance to infection by the intracellular parasite *Leishmania donovani*, as measured by the mean liver parasite load on day 15 after infection. Two alleles are known: *Lsh^s* determines susceptibility and occurs in strains BALB/c and B10.D2n; *Lsh^r* determines resistance and occurs in strains DBA/2, C3H/He, A/J, and CBA/Ca, and in all wild mice examined (1). Heterozygotes are resistant. Resistance and susceptibility can be transferred with bone marrow into irradiated hosts (2) and the trait is expressed by macrophages in vitro (3). Infection with L. donovani causes profound suppression of immune response to parasite antigen in *Lsh^s* mice and the suppression appears to be due to action of the macrophages of these mice (4). The strain distribution of *Lsh* alleles corresponds exactly to that for immunity to *typhimurium* (*Ity*) and for *Mycobacterium bovis* resistance (*Bcg*) and the genes for the three traits all map to the same position on Chr 1, indicating that the three loci may be identical (5, 6). See *Bcg* locus, *Ity* locus.

References

1. Bradley, D.J. 1977. Regulation of *Leishmania* populations within the host. II. Genetic control of acute susceptibility of mice to *Leishmania donovani* infection. Clin. Exp. Immunol. 30:130–140.
2. Crocker, P.R., J.M. Blackwell, and D.J. Bradley. 1984. Transfer of innate resistance and susceptibility to *Leishmania donovani* infection in mouse radiation bone marrow chimaeras. Immunology 52:417–422.
3. Crocker, P.R., J.M. Blackwell, and D.J. Bradley. 1984. Expression of the natural resistance gene *Lsh* in resident macrophages. Infect. Immunol. 43:1033–1040.
4. Nichol, A.D., and P.F. Bonventre. 1985. Visceral leishmaniasis in congenic mice of susceptible and resistant phenotypes: immunosuppression by adherent spleen cells. Infect. Immun. 50:160–168.
5. Plant, J., J.M. Blackwell, A.D. O'Brien, D.J. Bradley, and A.A. Glynn. 1982. Are the *Lsh* and *Ity* disease resistance genes at one locus on mouse chromosome 1? Nature 297:510–511.
6. Skamene, E., P. Gros, A. Forget, P.A.L. Kongshavn, C. St. Charles, and B.A. Taylor. 1982. Genetic regulation of resistance to intracellular pathogens. Nature 297:506–509.

Lsr-1 locus (originally *Lr* or *Lr-1*), listeria resistance, Chr 2

This locus controls natural resistance or susceptibility to infection by *Listeria monocytogenes* as measured by lethal dose, or by bacterial proliferation in liver or spleen. The allele *Lsr-1^r* determining resistance occurs in the C57BL/6, C57BL/10, NZB, and SJL strains; the allele *Lsr-1^s* determining susceptibility occurs in the BALB/c, CBA, DBA/1, DBA/2, A/J, LP, RIII, WB, 129, and C3H strains (1). Heterozygotes are almost fully resistant. The cellular basis of resistance is an augmented pool size of mononuclear phagocytes and a prompt mobilization of these cells to foci of infection

(6). The mechanism of resistance is highly radio-resistant (5). It does not reside in the macrophages themselves, but in the level of environmental factors in the resistant mice that regulate the proliferation or maturation of the monocyte–macrophage cell line (4). Ability to recruit inflammatory macrophages and neutrophils to the peritoneal cavity after intraperitoneal injection of *Listeria* is enhanced in susceptible A/J mice after transfer of plasma from resistant C57BL/6 mice (2). In the C57BL/6 × A/J (AXB) and reciprocal (BXA) recombinant inbred strains, the strain distribution pattern of *Lsr-1* is fully concordant with that of *Hc* which controls level of the C5 component of complement (3). This is the result of genetic linkage between *Hc* and *Lsr-1* and does not indicate a causal relationship between C5 and *Listeria* resistance, since *Hc* and *Lsr-1* have quite different distributions of alleles among inbred strains. Another bacterial resistance locus, *Ack* (anti-*Corynebacterium kutscheri*), has the same strain distribution as *Lsr* and may be identical with it. See *Ack* locus.

References

1. Cheers, C., and I.F.C. McKenzie. 1980. A single gene (Lr) controlling natural resistance to murine listeriosis. In E. Skamene, P.A.L. Kongshavn, and M. Landy, eds., Genetic Control of Natural Resistance to Infection and Malignancy, 141–147. Academic Press, NY.
2. Czuprynski, C.T., B.P. Canono, P.M. Henson, and P.A. Campbell. 1985. Genetically determined resistance to listeriosis is associated with increased accumulation of inflammatory neutrophils and macrophages which have enhanced bactericidal activity. Immunology 55:511–518.
3. Gervais, F., M. Stevenson, and E. Skamene. 1984. Genetic control of resistance to *Listeria monocytogenes*: regulation of leukocyte inflammatory responses by the Hc locus. J. Immunol. 132:2078–2083.
4. Kongshavn, P.A.L., C. Sandarangani, and E. Skamene. 1980. Genetically determined differences in antibacterial activity of macrophages are expressed in the environment in which the macrophage precursors mature. Cell. Immunol. 53:341–349.
5. Sandarangani, C., E. Skamene, and P.A.L. Kongshavn. 1980. Cellular basis for genetically determined enhanced resistance of certain strains to listeriosis. Infect. Immun. 28:381–386.
6. Stevenson, M.M., P.A.L. Kongshavn, and E. Skamene. 1981. Genetic linkage of resistance to *Listeria monocytogenes* with macrophage inflammatory responses. J. Immunol. 127:402–407.

lst, Strong's luxoid, semidominant, Chr 2

Arose in a line treated with methylcholanthrene for 22 generations. Heterozygotes show preaxial abnormalities of the hindfeet and also, very rarely, of the forefeet. Penetrance is incomplete and dependent on the genetic background. Homozygotes show preaxial polydactyly of all four feet, reductions and duplications of the radius, reductions and rarely duplications of the tibia, reduction of the pubis, modifications of the skull, temporary dorsal alopecia, and a posterior shift of the umbilicus. Males are sterile, but females sometimes breed (1,3). Embryonic development of parts of the limbs, head, integument, and belly is retarded (2). Certain teratogens increase the effect of *lst* on the skeleton, and modifying genes that increase the effects of *lst* also increase the effects of the teratogens (4).

References

1. Forsthoefel, P.F. 1962. Genetics and manifold effects of Strong's luxoid gene in the mouse, including its interactions with Green's luxoid and Carter's luxate genes. J. Morphol. 110:391–420.
2. Forsthoefel, P.F. 1963. The embryological development of the effects of Strong's luxoid gene in the mouse. J. Morphol. 113:427–452.
3. Forsthoefel, P.F. 1968. Responses to selection for plus and minus modifiers of some effects of Strong's luxoid gene on the mouse skeleton. Teratology 1:339–352.
4. Forsthoefel, P.F., and M.L. Williams. 1975. The effects of 5- fluorouracil and 5-fluorodeoxyuridine used alone and in combination with normal nucleic acid precursors on development of mice selected for low and high expression of Strong's luxoid gene. Teratology 11:1–20.

lt, lustrous, recessive, Chr 11

Arose spontaneously in the DBA/2 strain. Homozygotes are recognizable at birth by their short curled whiskers, which gradually lose their curliness, but remain slightly irregular throughout life. The coat has a glossy appearance similar to that of satin mice. All hair types are present, but they may show irregularities in the diameter of the shaft and irregularly spaced medullary cells. The air spaces of the medulla are filled with fatty fluid. This changes the optical properties of the shaft and accounts for the lustrous appearance of the coat (1).

References

1. Trigg, M.J. 1972. Hair growth in mouse mutants affecting coat texture. J. Zool. Lond. 168:165–198.

Lth-1, *-2* loci

See *Lfo-1*, etc.

Ltn-1, *-2* loci

See *Lfo-1*, etc.

Ltw-1 to *-6* loci

See *Lfo-1*, etc.

lu, luxoid, semidominant, Chr 9

Arose spontaneously in the C3H/He inbred strain (6). Heterozygotes show preaxial polydactyly or hyperphalangy of the hindfeet. Penetrance is close to zero in the C3H/He strain and about 90 per cent in the C57BL/10 strain. Homozygotes show preaxial polydactyly or oligodactyly of the hindfeet, preaxial polydactyly of the forefeet, tibial hemimelia, and occasionally radial hemimelia and tail kinks. The gene tends to increase the number of presacral vertebrae, ribs, sternebrae, and total number of vertebrae, the effect being greater in homozygotes than in heterozygotes (4). Homozygous females may occasionally breed, but homozygous males are sterile, with a complete lack of spermatogonia after 6 weeks of age (3). The developmental effects of *lu* in homozygotes can be traced back to the $10\frac{1}{2}$-day stage when the posterior end of the coelom is found to be more caudal than in normal controls. The somites of the tail may be abnormal, and there are more somites than in controls (5). Cartilage from the hindlimbs of 13-day *lu/lu* embryos retains its defective morphology when grown in culture (1). 5-fluorouracil administered to pregnant females on day 10 increases the expression of the effects of *lu* on hindlimbs of heterozygotes (2). A very similar mutant has been described by Kobozieff and Pomriaskinsky-Kobozieff (7 and earlier). It has not been tested for allelism with *lu*.

References

1. Burda, D.J., and E.M. Center. 1969. Development of luxoid (*lu*) skeletal defects *in vitro*. J. Embryol. Exp. Morphol. 21:347–360.
2. Dagg, C.P. 1967. Combined action of fluorouracil and two mutant genes on limb development in the mouse. J. Exp. Zool. 164:479–490.
3. Elkins, D., and P.F. Forsthoefel, cited as personal communication in Johnson, D.R., and D.M. Hunt. 1971. Hop-sterile, a mutant gene affecting sperm tail development in the mouse. J. Embryol. Exp. Morphol. 25:223–236.
4. Forsthoefel, P.F. 1958. The skeletal effects of the luxoid gene in the mouse, including its interactions with the luxate gene. J. Morphol. 102:247–287.
5. Forsthoefel, P.F. 1959. The embryological development of the skeletal effects of the luxoid gene in the mouse, including its interactions with the luxate gene. J. Morphol. 104:89–141.

6. Green, M.C. 1955. Luxoid, a new hereditary leg and foot abnormality in the house mouse. J. Hered. 46:91–99.
7. Kobozieff, N., and N.A. Pomriaskinsky-Kobozieff. 1962. Hémimélie chez la souri. Rec. Méd. Vét. 138:671–686.

Lus locus, lus cytostasis suppressor

This locus controls the ability of lymphoid cells injected into allogeneic mice to induce sensitized T- and B-cells that can cause cytostasis of target tumor cells. The dominant allele *Lus^a*, present in strain B10.D2, causes suppression of ability to produce cytostatic T and B effector cells; the recessive allele *Lus^b*, present in strain BALB/c, enhances the ability to produce cytostatic effector cells. *Lus^a* interacts with the recessive allele at the *Arp* locus, *Arp^a*, to produce a lymphoid/erythroid hyperplastic disorder that may be of autoimmune origin. See also *Arp* locus (1).

References

1. DeGiorgi, L., A. Matossian-Rogers, and H. Festenstein. 1981. Two genes interact to control development of lymphoid/erythroid hyperplastic disorder of mice. Nature 292:545–547.

Lv locus, delta-aminolevulinate dehydratase, Chr 4

An interstrain difference in hepatic activity of δ-aminolevulinate dehydratase is controlled by this locus. There are at least two alleles, *Lv^a* determining high activity and occurring in strains AKR/J, DBA/2J, C57BR/cdJ, and others, and *Lv^b* determining low activity and occurring in strains C57BL/6J, SM/J, and others (4). A number of other strains have an intermediate level of activity and are said to carry the allele *Lv^c*, but no crosses have been reported between these strains and strains carrying either *Lv^a* or *Lv^b*. *Lv^b*/*Lv^b* mice have lower levels of enzyme than *Lv^a*/*Lv^a* mice at all stages from 8 days before birth to adulthood. Heterozygotes are intermediate (5). The *Lv* locus appears also to affect the level of the enzyme in kidney and spleen. The liver enzyme from the *Lv^a* strains, AKR and DBA/2, was identical in all respects tested to the enzyme from the *Lv^b* strain, C57BL/6 (1), and the difference in activity was shown to be due to difference in the rate of enzyme synthesis (3). However, Coleman (2) found differences in heat stability of the enzyme between C57BR/cd (*Lv^a*, low heat stability), C57BL/6 (*Lv^b*, intermediate heat stability), and SM (*Lv^b*, high heat stability). In a cross between C57BR/cd and SM, heat stability showed no recombination with the *Lv* alleles.

References

1. Coleman, D.L. 1966. Purification and properties of δ-aminolevulinate dehydratase from tissues of two strains of mice. J. Biol. Chem. 241:5511–5517.
2. Coleman, D.L. 1971. Linkage of genes controlling the rate of synthesis and structure of aminolevulinate dehydratase. Science 173:1245–1246.
3. Doyle, D., and R.T. Schimke. 1969. The genetic and developmental regulation of hepatic δ-aminolevulinate dehydratase in mice. J. Biol. Chem. 244:5449–5459.
4. Hutton, J.J., and D.L. Coleman. 1969. Linkage analyses using biochemical variants in mice. II. Levulinate dehydratase and autosomal glucose 6-phosphate dehydrogenase. Biochem. Genet. 3:517–523.
5. Russell, R.L., and D.L. Coleman. 1963. Genetic control of hepatic δ-aminolevulinate dehydratase in mice. Genetics 48:1033–1039.

Lvp-1 locus, major liver protein-1, Chr 6

This locus controls electrophoretic mobility of a low molecular weight (about 18 000) protein of the soluble liver fraction. The protein migrates as the fastest of the major soluble proteins of liver and is probably not an enzyme, a hemoglobin polypeptide, or a histone. The allele *Lvp-1^a* determines a fast cathodally migrating band and occurs in strains SWR/J, BALB/c, C57BL/6, and others; the allele *Lvp-1^b* determines a slower band and occurs in strains C3H/HeJ, DBA/2, and others (1,2).

References

1. Wilcox, F.H. 1972. Genetic variation of a major liver protein in the mouse. J. Hered. 63:60–62.
2. Wilcox, F.H., T.H. Roderick, B.A. Taylor, and J.E. Womack. 1978. Mouse News Lett. 59:23.

lx, luxate, semidominant, Chr 5

Found in descendants of a silver mouse obtained from a fancier. A similar mutant had been reported by Rabaud but is extinct and cannot now be tested for allelism with *lx*. Heterozygous *lx* mice show preaxial polydactyly (including hyperphalangy of the first digit) of the hindfeet. Penetrance is incomplete and dependent on the genetic background. Homozygotes show preaxial polydactyly or oligodactyly of the hindfeet, reduction of the tibia, loss of part of the femur and pubis, decrease in number of presacral vertebrae, and anomalies of the urogenital system including horseshoe kidney, hydronephrosis, and hydroureter (1,2). The abnormalities of homozygotes can be traced back to the 10-day stage when the posterior end of the coelom is nearly a full segment anterior to its position in normal sibs. A day later the right umbilical artery is highly abnormal and, with the left umbilical artery, forms a ring which is often too small to allow free passage of the developing kidneys as they migrate forward. The posterior limb buds of homozygotes are narrower than normal from a very early stage (3). 5-fluorouracil administered to pregnant females on day 10 increases the expression of the effects of *lx* on the hindlimbs of heterozygotes (4,5).

References

1. Carter, T.C. 1951. The genetics of luxate mice. I. Morphological abnormalities of heterozygotes and homozygotes. J. Genet. 50:277–299.
2. Carter, T.C. 1953. The genetics of luxate mice. III. Horseshoe kidney, hydronephrosis and lumbar reduction. J. Genet. 51:441–457.
3. Carter, T.C. 1954. The genetics of luxate mice. IV. Embryology. J. Genet. 52:1–35.
4. Dagg, C.P. 1967. Combined action of fluorouracil and two mutant genes on limb development in the mouse. J. Exp. Zool. 164:479–490.
5. Long, S.Y., and E.M. Johnson. 1968. Enzyme ontogeny in normal and hemimelic limbs of mice. J. Embryol. Exp. Morphol. 20:415–430.

3. Cherry, M., and P.W. Lane. 1973. Mouse News Lett. 49:33.
4. Hayakawa, K., R.R. Hardy, D.R. Parks, and L.A. Herzenberg. 1983. The "Ly-1 B" cell subpopulation in normal, immunodefective, and autoimmune mice. J. Exp. Med. 157:202–218.
5. Huber, B., O. Devinsky, R.K. Gershon, and H. Cantor. 1976. Cell-mediated immunity: delayed-type hypersensitivity and cytotoxic responses are mediated by different T-cell subclasses. J. Exp. Med. 143:1534–1539.
6. Ledbetter, J.A., R.V. Rouse, H.S. Micklem, and L.A. Herzenberg. 1980. T cell subsets defined by expression of Lyt-1,2,3, and Thy-1 antigens: two parameter immunofluorescence and cytotoxicity analysis with monoclonal antibodies modifies current views. J. Exp. Med. 152:280–295.
7. Morse, H.C. III, F.-W. Shen, and U. Hämmerling. 1987. Genetic nomenclature for loci controlling mouse lymphocyte antigens. Immunogenetics 25:71–78.
8. Shiku, H., P. Kisielow, E.A. Boyse, and H.F. Oettgen. 1976. Immunogenetic identification of functional T-cell subsets. Transplant. Proc. 8:381–385.
9. Snell, G.D., and M. Cherry. 1972. Loci determining cell surface antigens. *In* P. Emmelot and P. Bentvelsen, eds., RNA Viruses and Host Genome in Oncogenesis, 221–228. North-Holland, Amsterdam.

Ly-1 locus (formerly *Ly-A*, *Lyt-1*), lymphocyte antigen-1, Chr 19

This locus controls an antigen present on lymphocytes of the T-cell class. Two alleles are known, *Ly-1ᵃ* which determines presence of the Ly-1.1 specificity and occurs in strains C3H/An, DBA/2, and others, and *Ly-1ᵇ* which determines presence of the Ly-1.2 specificity and occurs in strains AKR, C58, C57BL/6, and others (2, 7, 9). Ly-1 antigen is present on all cortical thymocytes but markedly declines on some of these cells when they pass into the peripheral circulation and acquire differentiated functions. The antigen is retained on helper cells and cells involved in delayed-type hypersensitivity and declines on suppressor-cytotoxic cells (1, 5, 6, 8). Ly-1 is absent from most B-cells but is found on a small population of splenic B-cells that comprise about 2 per cent of B-cells in most normal strains and about 5 to 10 per cent in NZB mice (4). *Ly-1* is on Chr 19 very closely linked to *Dc* near the centromeric end (3).

References

1. Beverly, P.C.L., J. Woody, M. Dunkley, and M. Feldmann. 1976. Separation of suppressor and killer T cells by surface phenotype. Nature 262:495–497.
2. Boyse, E.A., M. Miyazawa, T. Aoki, and L.J. Old. 1968. Ly-A and Ly-B: two systems of lymphocyte isoantigens in the mouse. Proc. R. Soc. Lond. (Biol.) 170:175–193.

Ly-2/3 complex, Chr 6

These loci control a heterodimeric antigenic molecule present on T lymphocytes. The molecule consists of two polypeptides, an α-chain of 38 kDa (α) or 35 kDa (α') and a β-chain of 30 kDa, bound by disulfide bonds. The Ly-2 epitope resides on the α-chain, the Ly-3 epitope resides on the β-chain (7). The Ly-2/3 antigen in present on all cortical thymocytes and on peripheral cytotoxic killer cells and suppressor cells (1, 6). The *Ly-2* and *Ly-3* loci have shown no recombination in crosses, but alleles at the two loci have different strain distributions. The complex is on Chr 6 very close to *Igk* (4).

Ly-2 locus (formerly *Ly-B*, *Lyt-2*), lymphocyte antigen-2. Two alleles are known, *Ly-2ᵃ* which determines specificity Ly-2.1 and occurs in strains C3H/An, DBA/2, C58, and others, and *Ly-2ᵇ* which determines specificity Ly-2.2 and occurs in strains A, C57BL/6, SJL/J, and others (2). Thymocytes synthesize both the α- and α'-chains bearing the Ly-2 epitope (7). The two chains arise from a single gene by alternate modes of mRNA splicing (8).

Ly-3 locus (formerly *Ly-C*, *Lyt-3*), lymphocyte antigen-3. Two alleles are known: *Ly-3ᵃ* determines specificity Ly-3.1 and occurs in strains C58, AKR, PL, and RF; *Ly-3ᵇ* determines specificity Ly-3.2 and occurs in C3H/He, DBA/2, C57BL/6, and others (3). The *Ly-3* alleles have the same strain distribution as the alleles of three of the *Igk-V* loci (immunoglobulin kappa chain variable region) but 0.30 per cent recombination was found between them and *Ly-3* (5).

References

1. Beverly, P.C.L., J. Woody, M. Dunkley, and M. Feldman. 1976. Separation of suppressor and killer T cells by surface phenotype. Nature 262:495–497.
2. Boyse, E.A., M. Miyazawa, T. Aoki, and L.J. Old. 1968. Ly-A and Ly-B: two systems of lymphocyte isoantigens in the mouse. Proc. R. Soc. Lond. (Biol.) 170:175–193.
3. Boyse, E.A., K. Itakura, E. Stockert, C.A. Iritani, and M. Miura. 1971. Ly-C: a third locus specifying alloantigens expressed only on thymocytes and lymphocytes. Transplantation 11:351–353.
4. Gibson, D.M., B.A. Taylor, and M. Cherry. 1978. Evidence for close linkage of a mouse light chain marker with the *Ly-2,3* locus. J. Immunol 121:1585–1590.
5. Gibson, D.M., S.J. MacLean, and M. Cherry. 1983. Recombination between kappa chain genetic markers and the *Lyt-3* locus. Immunogenetics 18:111–116.
6. Shiku, H., P. Kisielow, E.A. Boyse, and H.F. Oettgen. 1976. Immunogenetic identification of functional T-cell subsets. Transplant. Proc. 8:381–385.
7. Walker, I.D., P.M. Hogarth, B.J. Murray, K.E. Lovering, B.J. Classon, G.W. Chambers, and I.F.C. McKenzie. 1984. Ly antigens associated with T-cell recognition and effector function. Immunol. Rev. 82:47–77.
8. Zamoyska, R., A.C. Vollmer, K.C. Sizer, C.W. Liaw, and J.R. Parnes. 1985. Two Lyt-2 polypeptides arise from a single gene by alternative splicing of mRNA. Cell 43:153–163.

Ly-4 locus, lymphocyte antigen-4, Chr 6

This locus codes for the L3T4 T-cell differentiation antigen (2). The locus formerly called *Ly-4* (and before that *Lyb-1*) and located on Chr 2 (3) has essentially the same strain distribution of alleles and tissue distribution as *B2m* (β-2 microglobulin) and is probably the same locus (2). T-cells bearing L3T4 generally mediate T-helper or -inducer function and are characterized by their interaction with cells bearing class II MHC molecules. *Ly-4* was found to be on Chr 6 by use of mouse–Chinese hamster and mouse–rat somatic cell hybrids screened with a mouse cDNA clone (1).

References

1. Field, E.H., B. Tourvieille, P. D'Eustachio, and J.R. Parnis. 1987. The gene encoding the mouse T-cell differentiation antigen L3T4 is located on chromosome 6. J. Immunol. 138:1968–1970.
2. Morse, H.C. III, F.-W. Shen, and U. Hämmerling. 1987. Genetic nomenclature for loci controlling mouse lymphocyte antigens. Immunogenetics 25:71–78.

3. Snell, G.D., M. Cherry, I.F.C. McKenzie, and D.W. Bailey. 1973. *Ly-4*, a new locus determining a lymphocyte cell-surface alloantigen in mice. Proc. Natl. Acad. Sci. USA 70:1108–1111.

Ly-5 locus (formerly *Lyt-4*), lymphocyte antigen-5, Chr 1

This locus controls an antigen present on most cells of hematopoietic lineage but not on other cell types. Two alleles are known, *Ly-5^b* which determines specificity Ly-5.2 and is found in strains C57BL/6, C3H/An, DBA/2, AKR, and many others, and *Ly-5^a* which determines specificity Ly-5.1 and is found only in strains SJL/J, STS/A, and DA (1, 3–5). Ly-5 antigens are present on all T-cells and on prothymocytes, B-cells, myeloid cells, monocytes, macrophages, and NK cells (6). The antigen is found in 8-day yolk sac cells and on 13-day fetal liver cells (8). The size of the antigenic molecule is different on different cell types: on macrophages it is 205 kDa, on B-cells 220 kDa, on Ly-1$^+$ T-cells 200 kDa, and on Ly-2$^+$3$^+$ T-cells either 210 or 215 kDa (9). The 200 and 220 kDa molecules appear to be different proteins (10). The *Ly-5* gene product may play a role in several important T-cell functions, since antiserum to Ly-5 blocks these functions (2). *Ly-5* is on Chr 1 linked to *Pep-3* (7).

References

1. Ewald, S.J., and I.F.C. McKenzie. 1979. Immunochemical characterization of the Ly-5 alloantigens on thymocytes. Immunogenetics 8:583–587.
2. Harp, J.A., B.S. Davis, and S.J. Ewald. 1984. Inhibition of T cell responses to alloantigens and polyclonal activation by Ly-5 antisera. J. Immunol. 133:10–15.
3. Kasai, M., J.C. Leclerc, F.-W. Shen, and H. Cantor. 1979. Identification of Ly-5 on the surface of "natural killer" cells in normal and athymic inbred mouse strains. Immunogenetics 8:153–159.
4. Komura, K., K. Itakura, E.A. Boyse, and M. John. 1975. Ly-5: a new lymphocyte antigen system. Immunogenetics 1:452–456.
5. Morse, H.C. III, F.-W. Shen, and U. Hämmerling. 1987. Genetic nomenclature for loci controlling mouse lymphocyte antigens. Immunogenetics 25:71–78.
6. Scheid, M.P., and D. Triglia. 1979. Further description of the *Ly-5* system. Immunogenetics 9:423–433.
7. Scheid, M.P., K.S. Landreth, J.-S. Tung, and P.W. Kincade. 1983. Preferential but nonexclusive expression of macromolecular antigens on B-lineage cells. Immunol. Rev. 69:141–159.
8. Triglia, D. 1980. Expression of Ly-5 on yolk sac and fetal liver cells of the mouse. Immunogenetics 11:303–307.
9. Tung, J.-S., M.P. Scheid, and M.A. Palladino. 1983. Dif-

ferent forms of Ly-5 within the T-cell lineage. Immunogenetics 17:649–654.

10. Tung, L.-S., M.C. Deere, and E.A. Boyse. 1984. Evidence that Ly-5 product of T and B cells differ in protein structure. Immunogenetics 19:149–154.

Ly-6 complex, Chr 15

This complex controls a number of antigens determined by loci that are either very closely linked or identical. The alleles at the loci have identical or very similar strain distributions but the antigens have different tissue distributions. The antigens were defined at first by conventional antibodies and later by monoclonal antibodies. All strains examined except NZB have either one of two haplotypes, *Ly-6*^{b} which occurs in strain C57BL/6 and related strains and in AKR, DBA/2, and many others, and *Ly-6*^{a} which occurs in strains A, BALB/c, C3H/He, and many others. The NZB strain has the *Ly-6B*^{b} allele but the *a* alleles at the other loci. A cDNA clone encoding an Ly-6 antigen hybridizes with multiple DNA fragments which display a fragment length polymorphism between *Ly-6*^{a} and *Ly-6*^{b} (8). The complex is located on Chr 15, probably near *Sis* (5, 9). The literature has been summarized by Kimura *et al.* (7) who have devised the following classification based on monoclonal antibodies.

Ly-6A locus, lymphocyte antigen-6A. This is the classical *Ly-6* locus described by McKenzie *et al.* (9) and Kimura *et al.* (6). It is probably identical with *Ly-8* (2). The antigen is present on peripheral T- and B-cells of spleen and lymph nodes and on activated lymphocytes, but is virtually absent from thymus and bone marrow. The allele *Ly-6A*^{b} which is part of the *Ly-6*^{b} haplotype determines presence of specificity Ly-6A.2. No antigenic specificity has been demonstrated for the *Ly-6A*^{a} allele. However, Frelinger and Murphy (2) were able to demonstrate two specificities of the Ly-8 antigen, Ly-8.2 and Ly-8.1. Antigens with a tissue distribution similar to that of Ly-6A.2, and precipitated by a monoclonal antibody, were found to comprise two molecules of different molecular weight, 34 000 and 56 000, translatable from separable mRNA molecules. Both were present on spleen cells, but only the 34 000 molecular weight molecule was found on thymus cells (13).

Ly-6B locus, lymphocyte antigen-6B. The antigen controlled by this locus is present on bone marrow cells, on a very small proportion of lymph node and spleen cells, and on neutrophils. The locus is probably identical with *Gm-2* (3, 10). The allele *Ly-6B*^{b}, part of the *Ly-6*^{b} haplotype, determines presence of specificity Ly-6B.2. No antigenic specificity has been demonstrated for the *Ly-6B*^{a} allele.

Ly-6C locus, lymphocyte antigen-6C. The antigen is present on lymph node, spleen, and bone marrow cells and on activated lymphocytes. It is probably identical with the H9/25 antigen described by Takei *et al.* (14) and with Ly-28 (4, 10). The allele *Ly-6C*^{b}, part of the *Ly-6*^{b} haplotype, determines presence of specificity Ly-6C.2. No antigenic specificity has been demonstrated for the *Ly-6C*^{a} allele.

Ly-6D locus, lymphocyte antigen-6D. The antigen is present on thymus, lymph node, and spleen cells, on activated lymphocytes, and in lower proportion on bone marrow cells. It is probably identical with Ly-27 (4, 10). The allele *Ly-6D*^{b}, part of the *Ly-6*^{b} haplotype, determines presence of specificity Ly-6D.2. No antigenic specificity has been demonstrated for the *Ly-6D*^{a} allele.

Ly-6E locus, lymphocyte antigen-6E. The antigen is present almost exclusively on activated lymphocytes. It is thus probably identical with the Ala-1 antigen described by Feeney and Hämmerling (1). The allele *Ly-6E*^{a}, part of the *Ly-6*^{a} haplotype, determines presence of specificity Ly-6E.1. No antigenic specificity has been demonstrated for the *Ly-6E*^{b} allele. However, Feeney and Hämmerling (1) described two specificities for the Ala-1 antigen, Ala-1.1 and Ala-1.2. Palfrey *et al.* (11) have presented evidence that *Ly-6E*^{a} and *Ly-6A*^{b} may be alleles rather than the antigen-positive alleles at separate loci.

A T-cell-activating protein TAP is controlled by a locus with alleles that have a strain distribution identical to that of the *Ly-6* haplotypes. The locus does not recombine with *Ly-6* but the TAP molecule has a different structure and molecular properties from other Ly-6 molecules (12).

References

1. Feeney, A.J., and U. Hämmerling. 1977. Ala-1: a murine alloantigen of activated lymphocytes. II. T and B effector cells express Ala-1. J. Immunol. 118:1488–1494.

2. Frelinger, J.A., and D.B. Murphy. 1976. A new alloantigen, Ly-8, recognized by C3H anti-AKR serum. Immunogenetics 3:481–487.

3. Hibbs, M.L., P.M. Hogarth, B.M. Scott, R.A. Harris, and I.F. C. McKenzie. 1984. Monoclonal antibody to murine neutrophils: identification of the Gm-2.2 specificity. J. Immunol. 133:2619–2623.

4. Hogarth, P.M., B.A. Houlden, S.E. Latham, V.R. Sutton, and I.F.C. McKenzie. 1984. Definition of new alloantigens encoded by genes in the *Ly-6* complex. Immunogenetics 20:57–69.

5. Hogarth, P.M., I.F.C. McKenzie, V.R. Sutton, K.M. Curnow, B.K. Lee, and E.M. Eicher. 1987. Mapping of

the murine *Ly-6*, *Xp-14*, and *Gdc-1* loci to chromosome 15. Immunogenetics 25:21–27.

6. Kimura, S., N. Tada, E. Nakayama, and U. Hämmerling. 1980. Studies of the mouse *Ly-6* alloantigen system. I. Serological characterization of the mouse *Ly-6* alloantigen by monoclonal antibodies. Immunogenetics 11:373–381.

7. Kimura, S., N. Tada, Y. Liu-Lam, and U. Hämmerling. 1984. Studies of the mouse *Ly-6* alloantigen system. II. Complexities of the *Ly-6* region. Immunogenetics 20:47–56.

8. LeClair, K.P., M. Rabin, M.N. Nesbitt, D. Pravtcheva, F.H. Ruddle, R.C.E. Palfree, and A. Bothwell. 1987. Murine *Ly-6* multigene family is located on chromosome 15. Proc. Natl. Acad. Sci. USA 84:1638–1642.

9. McKenzie, I.F.C., M. Cherry, and G.D. Snell. 1977. Ly-6.2: a new lymphocyte specificity of peripheral T cells. Immunogenetics 5:25–32.

10. Morse, H.C. III, F.-W. Shen, and U. Hämmerling. 1987. Genetic nomenclature for loci controlling mouse lymphocyte antigens. Immunogenetics 25:71–78.

11. Palfree, R.G.E., F.J. Dumont, and U. Hämmerling. 1986. Ly-6A.2 and Ly-6E.1 molecules are antithetical and identical to MALA-1. Immunogenetics 23:197–207.

12. Reiser, H., E.T.H. Yeh, D.F. Gramm, B. Benacerraf, and K.L. Rock. 1986. Gene encoding T-cell activating protein TAP maps to the *Ly-6* locus. Proc. Natl. Acad. Sci. USA 83:2954–2958.

13. Sutton, V.R., C.A. Mickelson, G.H. Tobias, P.M. Hogarth, and I.F.C. McKenzie. 1985. Identification of two forms of the Ly-6 antigen showing differential expression. Immunogenetics 21:539–547.

14. Takei, F., G. Galfrè, T. Alderson, E.S. Lennox, and C. Milstein. 1980. H9/25 monoclonal antibody recognizes a new allospecificity of mouse lymphocyte subpopulations: strain and tissue distribution. Eur. J. Immunol. 10:241–246.

Ly-7 locus, lymphocyte antigen-7

This locus controls an antigen present on lymphocytes. It is expressed in very low amounts on thymocytes but is readily detectable on peripheral lymphocytes, with greatest expression on B-cells (4). Three alleles but only two specificities are known. The allele *Ly-7ᵃ* determines presence of Ly-7.1 and absence of Ly-7.3 and occurs in strains NZB, C3H/He, CBA/J, CBA/N, and SJL (2, 3); the allele *Ly-7ᵇ* determines absence of both specificities and occurs in the C57BL/6, C57BL/10, and C58 strains; the allele *Ly-7ᶜ* determines presence of both specificities Ly-7.1 (formerly Ly-7.2) and Ly-7.3 and occurs in strains BALB/c, DBA/1, DBA/2, MRL, CE, AKR/J, and RIIIS/J (1). A linkage backcross failed to confirm the location of *Ly-7* on Chr 12 (R. Riblet, personal communication).

References

1. Hogarth, P.M., I.F.C. Mckenzie, L. Lanier, D.W. Bailey, and B.A. Taylor. 1984. Location of *Ly-7* on mouse chromosome 12. Immunogenetics 19:539–543.

2. Lanier, L.L., and N.L. Warner, 1983. Lym-7.3 a new murine specificity defining third allele of the *Ly-7* locus. Hybridoma 2:177–155.

3. Morse, H.C. III, F.-W. Shen, U. Hämmerling. 1987. Genetic nomenclature for loci controlling mouse lymphocyte antigens. Immunogenetics 25:71–78.

4. Potter, T.A., G.M. Morgan, and I.F.C. Mckenzie. 1980. Murine lymphocyte alloantigens. II. The *Ly-7* locus. J. Immunol. 125:546–550.

Ly-8, locus (formerly *Ly-11* (1)), lymphocyte antigen-8

This locus controls an alloantigen present on thymocytes and kidney cells and detectable by cytotoxicity. The allele *Ly-8ᵃ* determines presence of the Ly-8.1 antigen and occurs in strains 129, 101, and LP.RIII; the allele *Ly-8ᵇ* determines presence of the antigen Ly-8.2 and occurs in the C57 family of strains and in many others. The Ly-8.2 antigen was detected by a 129 anti-C57L antiserum, the Ly-8.1 antigen by a B10.RIII(71NS) anti-LP.RIII antiserum (2).

References

1. Morse, H.C. III, F.-W. Shen, and U. Hämmerling. 1987. Genetic nomenclature for loci controlling mouse lymphocyte antigens. Immunogenetics 25:71–78.

2. Potter, T.A., and I.F.C. McKenzie. 1981. Identification of new murine lymphocyte alloantigens: antisera prepared between C57L, 129, and related strains define new loci. Immunogenetics 12:351–369.

Ly-8 locus

See *Ly-6A* locus.

Ly-9 locus (formerly T100, Lgp100), lymphocyte antigen-9, Chr 1

This locus controls an antigen detected by flow microfluorometry on thymocytes, peripheral lymphocytes, and bone marrow cells, including both T- and B-cells. It is not present on red cells or other tissues. The allele *Ly-9ᵃ* determines presence of the Ly-9.1 antigen and occurs in the A/J, 129/Re, BALB/c, C3H/He, and many other strains; the allele *Ly-9ᵇ* determines presence of the Ly-9.2 antigen and occurs in the C57 family of strains and in HTI/Go, MA/My, F/St, and C58/Lw (3). *Ly-9* is probably the same locus as that controlling the glycoproteins called T100 (1) and Lgp100 (2). It is closely linked to *Mls* on Chr 1 (2, 3).

Ly-9

References

1. Durda, P.J., S.C. Boos, and P.D. Gottlieb. 1979. T100: a new murine cell surface glycoprotein detected by anti-Ly-2.1 serum. J. Immunol. 122:1407–1412.
2. Ledbetter, J.A., J.W. Goding, T.T. Tsu, and L.A. Herzenberg. 1979. A new mouse lymphoid alloantigen (Lgp100) recognized by a monoclonal rat antibody. Immunogenetics 8:347–360.
3. Mathieson, B.J., S.O. Sharrow, K. Bottomly, and B.J. Fowlkes. 1980. Ly-9, an alloantigenic marker of lymphocyte differentiation. J. Immunol. 125:2127–2136.

Ly-10 locus (formerly Ly-m10), Chr 19

This locus controls an antigen present on thymocytes, splenic T- and B-cells, bone marrow, brain, kidney, and liver. The allele $Ly-10^a$ determines presence of the Ly-10.1 antigen and occurs in C57BR/cd, C57L, C3Hf/Bi, DBA/1, DBA/2, CBA/J, and CBA/H; the allele $Ly-10^b$ determines absence of the antigen and occurs in C3H/He, MA/J, C57BL/6, A, AKR, BALB/c, and others. Ly-10.1 is detected by a monoclonal antibody made with spleen cells from (BALB/c × C57BL/6)F1 mice immunized against CBA/J spleen cells (1, 2). Ly-10 is closely linked to Ly-1 on Chr 19 (1).

References

1. Kimura, S., N. Tada, and U. Hämmerling. 1980. A new lymphocyte antigen (Ly-10) controlled by a gene linked to the Lyt-1 locus. Immunogenetics 10:363–372.
2. Kimura, S., N. Tada, Y. Liu-Lam, and U. Hämmerling. 1983. Further genetic characterization of mouse Ly-m10 alloantigen. Immunogenetics 17:211–214.

Ly-11 locus, lymphocyte antigen-11, Chr 2

This locus controls an antigen on lymphocytes of the T-cell lineage but not on B-cells or non-lymphoid cells. The allele $Ly-11^a$ determines absence of the Ly-11.2 antigen and occurs in strains A/J, BALB/c, C3H/DiSn, CE, NZB, CBA/J, and SWR; the allele $Ly-11^b$ determines presence of the Ly-11.2 antigen and occurs in the C57 strains and in AKR, C58, DBA/2, SJL, 129/J, PL, and MA/My (1). Functional studies have shown that Ly-11.2 antigen is present on prothymocytes and natural killer cells but not on certain other subpopulations of T-cells. In the early stages of leukemogenesis, there is a marked increase in the expression of Ly-11.2 in thymus and bone marrow, and in the overt and late stages a dramatic drop (2). Ly-11 is located on Chr 2 proximal to pa (3, 4).

References

1. Meruelo, D., A. Paolino, N. Flieger, and M. Offer. 1980. Definition of a new T lymphocyte cell surface antigen, Ly 11.2. J. Immunol. 125:2713–2718.
2. Meruelo, D., A. Paolino, N. Flieger, J. Dworkin, M. Offer, N. Hirayama, and Z. Ovary. 1980. Functional properties of Ly 11.2 lymphocytes: a role for these cells in leukemia. J. Immunol. 125:2719–2726.
3. Meruelo, D., M. Offer, and A. Rossomando. 1982. Evidence for a major cluster of lymphocyte differentiation antigens on murine chromosome 2. Proc. Natl. Acad. Sci. USA 79:7460–7464.
4. Meruelo, D., A. Rossomando, S. Scandalis, P. D'Eustachio, R.E.K. Fournier, D.R. Roop, D. Saxe, C. Blatt, and M.N. Nesbitt. 1987. Assignment of the Ly-6–Ril-1–Sis–H-30–Pol-5/Xmmv-72–Ins-3–Krt-1–Int-1–Gdc-1 region to mouse chromosome 15. Immunogenetics 25:361–372.

Ly-11 locus

See B2m, Ly-8 loci.

Ly-12 locus, lymphocyte antigen-12, Chr 19

This locus controls an antigen present on thymus, lymph nodes, and spleen cells and detectable by cytotoxicity and by SAM rosetting. The allele $Ly-12^a$ determines presence of the antigen Ly-12.1 and occurs in the C57L, C57BR, and SWR strains; the allele $Ly-12^b$ determines absence of Ly-12.1 and occurs in the C57BL/10, C58, 129, CBA, and many other strains. The Ly-12.1 antigen was detected by a 129 anti-C57L antiserum (1). Ly-12 is on Chr 19 linked to Ly-10 (B.A. Taylor, personal communication).

References

1. Potter, T.A., and I.F.C. McKenzie. 1981. Identification of new murine lymphocyte alloantigens: antisera prepared between C57L, 129, and related strains define new loci. Immunogenetics 12:351–369.

Ly-13 locus, lymphocyte antigen-13

This locus controls an antigen present on thymus, lymph node, and spleen cells, and on brain, liver, kidney, and red blood cells and detectable by cytotoxicity and by SAM rosetting. The allele $Ly-13^a$ determines presence of the antigen Ly-13.1 and occurs in strains C57L and C57BR; the allele $Ly-13^b$ determines absence of the antigen and occurs in the C57BL/10, C58, CBA, 129, and many other strains. The antigen was detected by a 129 anti-C57L antiserum (1).

References

1. Potter, T.A., and I.F.C. McKenzie. 1981. Identification of new murine lymphocyte alloantigens: antisera prepared between C57L, 129, and related strains define new loci. Immunogenetics 12:351–369.

Ly-14 locus, lymphocyte antigen-14, Chr 7

This locus controls an antigen present on thymus, lymph node, and spleen cells, and on brain, liver, and kidney and detectable by SAM rosetting. The allele *Ly-14^b* determines presence of the antigen Ly-14.2 and occurs in the C57L, C57BR, C57BL/10, A, CE, C58, DBA/1, DBA/2, CBA, and C3H strains; the allele *Ly-14^a* determines absence of the antigen and occurs in the 129, SWR, BALB/c, SJL, and AKR strains. The Ly-14.2 antigen was detected by a 129 anti-C57L antiserum. *Ly-14* is linked to *c* on Chr 7 with 17 per cent recombination (1).

References

1. Potter, T.A., and I.F.C. McKenzie. 1981. Identification of new murine lymphocyte alloantigens: antisera prepared between C57L, 129, and related strains define new loci. Immunogenetics 12:351–369.

Ly-15 locus, lymphocyte antigen-15, Chr 7

This locus controls an antigen present on all thymocytes, lymphocytes, neutrophils, and on liver. The Ly-15.1 antigen was detected by antisera produced in several donor–host combinations of BALB/c × C57BL/6 (BXD) recombinant inbred (RI) strains; it is determined by the *Ly-15^a* allele and occurs in the BALB/c, SWR, SJL, AKR, C58, NZB, and 129/Re strains. The Ly-15.2 antigen was detected by a monoclonal antibody produced from 129/Re anti-B6-*Ly-1^a* spleen cells; it is determined by the *Ly-15^b* allele and occurs in the C57BL/6, C57BL/10, C57L, C57BR, A, DBA/1, DBA/2, CE, RF, 129/J, LP.RIII, C3H/He, and CBA strains (3). Ly-15.2 is identical with the non-polymorphic LFA-1 (lymphocyte function-associated antigen-1) molecule. Both are dimers of an α-chain (180 kDa) and a β-chain (94 kDa) and they have identical tissue distributions (1). The Ly-15 specificity resides on the β-chain (4). *Ly-15* is on Chr 7 near *Mod-2* and *Hbb* (2).

References

1. Hogarth, P.M., I.D. Walker, I.F.C. McKenzie, and J.A. Springer. 1985. The Ly-15 alloantigenic system: a genetically determined polymorphism of the murine function-associated antigen-1 molecule. Proc. Natl. Acad. Sci. USA 82:526–530.

2. Hogarth, P.M., E.M. Eicher, and I.F.C. McKenzie. 1986. Mapping of the murine *Ly-15* (LFA-1) locus to chromosome 7. Immunogenetics 23:348–349.
3. Potter, T.A., P.M. Hogarth, and I.F.C. McKenzie. 1981. *Ly-15*: a new murine lymphocyte alloantigenic locus. Transplantation 31:339–342.
4. Walker, I.D., P.M. Hogarth, B.J. Murray, K.E. Lovering, B.J. Classon, G.W. Chambers, and I.F.C. McKenzie. 1984. Ly antigens associated with T cell recognition and effector function. Immunol. Rev. 82:47–77.

Ly-16 locus (formerly *Ly-18* (2)), lymphocyte antigen-16, Chr 12

This locus controls an antigen expressed on cytotoxic T-cells. It is present on a subpopulation of activated T-cells and is probably a differentiation antigen. The allele *Ly-16^a* determines presence of the antigen Ly-16.1 and occurs in the BALB/c, C57L, C58, and several *Igh-1* congenic strains; the allele *Ly-16^b* determines absence of the antigen and occurs in the CBA/Tu, C57BL/6, SJL, LP and some congenic strains. The association of *Ly-16^a* with the *Igh-1^a* allele in BALB/c, C57L, and C58 and of *Ly-16^b* with the *Igh-1^b* allele in C57BL/6, SJL, and LP, suggests that *Ly-16* is closely linked to *Igh-1* on Chr 12. The *Ly-16* type of the *Igh-1* congenics suggests that *Ly-16* is probably on the centromeric side of *Igh-1* (1).

References

1. Finnegan, A., and F.L. Owen. 1981. Ly-18, a new alloantigen on cytotoxic T cells and controlled by a gene(s) linked to the *Igh-V* locus. J. Immunol. 127:1947–1953.
2. Morse, H.C. III, F.-W. Shen, and U. Hämmerling. 1987. Genetic nomenclature for loci controlling mouse lymphocyte antigens. Immunogenetics 25:71–78.

Ly-17 locus (formerly also *Ly-m20*, *LyM-1*), lymphocyte antigen-17, Chr 1

This locus controls an antigen preferentially on B-cells but also on liver and kidney cells. It was first described by Shen and Boyse (8). It was subsequently shown to be the same as *LyM-1* (2, 9) and *Ly-m20* (1, 6). The allele *Ly-17^a* determines presence of the Ly-17.1 antigen (8) and occurs in strains A, CBA/H, C3H/He, RF, SJL, 129, and others; the allele *Ly-17^b* determines presence of the Ly-17.2 antigen (6) and occurs in strains AKR, BALB/c, CBA/J, the C57 strains, DBA/2, and others. *Ly-17* determines a polymorphism of the Fc receptor (4, 5). The locus is closely linked to *Mls* on Chr 1 and shows about 6 per cent recombination with it (3, 7).

Ly-17

References

1. Davidson, W.F., H.C. Morse III, B.J. Mathieson, C.A. Kozak, and F.-W. Shen. 1983. The B cell alloantigen Ly-17.1 is controlled by a gene closely linked to Ly-20 and Ly-9 on chromosome 1. Immunogenetics 17:325–329.
2. Dickler, H.B., A. Ahmed, and D.H. Sachs. 1977. B-lymphocyte F:c receptor-associated non-H-2 antigens are determined by a single polymorphic locus which is linked to the Mls locus. J. Exp. Med. 146:1678–1692.
3. Dickler, H.B., D.L. Rosenstreich, A. Ahmed, and D.H. Sachs. 1980. Genetic linkage between Lym-1, Sas-1, and Mls loci. Immunogenetics 10:93–96.
4. Hibbs, M.L., P.M. Hogarth, and I.F.C. McKenzie. 1985. The mouse Ly-17 locus identifies a polymorphism of the Fc receptor. Immunogenetics 22:335–348.
5. Holmes, K.L., R.G.E. Palfree, U. Hämmerling, and H.C. Morse III. 1985. Alleles of the Ly-17 alloantigen define polymorphisms of the murine IgG Fc receptor. Proc. Natl. Acad. Sci. USA 82:7706–7710.
6. Kimura, S., N. Tada, E. Nakayama, Y. Liu, and U. Hämmerling. 1981. A new mouse cell-surface antigen (Ly-m20) controlled by a gene linked to Mls locus and defined by monoclonal antibodies. Immunogenetics 14:3–14.
7. Sato, H., S. Natsuume-Sakai, M. Kalagiri, and K. Itakura. 1981. Location of LyM-1 on Mus musculus chromosome 1. J. Immunogenet. 8:185–189.
8. Shen, F.-W., and E.A. Boyse. 1980. An alloantigen selective for B cells: Ly-17.1. Immunogenetics 11:315–317.
9. Tonkonogy, S.L., and H.J. Winn. 1977. Further genetic and serological analysis of the LyM-1 alloantigenic system. Immunogenetics 5:57–64.

Ly-18 locus (formerly Ly-m18), lymphocyte antigen-18, Chr 12

This locus controls an antigen present on 90 per cent of thymus cells, 55 per cent of spleen cells, and 45 per cent of lymph node or bone marrow cells, on 50 per cent of either peripheral T- or B-cells, some tumor cell lines, and also on brain, kidney, and liver. The allele $Ly-18^b$ determines presence of the antigen Ly-18.2 and occurs in the C57 strains, C58, CE, DBA/2, GR/A, SJL, SWR, and others; the allele $Ly-18^a$ determines absence of the antigen and occurs in strains A, AKR, BALB/c, C3H/He, CBA/H, and CBA/J (1). Ly-18 is closely linked to Ah on Chr 12 (2).

References

1. Kimura, S., N. Tada, Y. Liu, and U. Hämmerling. 1981. A new mouse cell surface antigen (Ly-m18) defined by a monoclonal antibody. Immunogenetics 13:547–554.
2. Lusis, A.J., B.A. Taylor, D. Quon, S. Zollman, and R.C. LeBoeuf. 1987. Genetic factors controlling structure and expression of apolipoproteins B and C. J. Biol. Chem. 262:7594–7604.

Ly-18 locus

See Ly-16 locus.

Ly-19 locus (formerly Ly-m19), lymphocyte antigen-19, Chr 4

This locus controls an antigen present on both T- and B-cells. By cytotoxicity assays, it was found that the antigen occurs on 75 per cent of spleen and lymph node cells, 35 per cent of bone marrow cells, and 15 per cent of thymus cells. The allele $Ly-19^a$ determines absence of the Ly-19.2 antigen and occurs in the AKR, CE, RF, GR/A, SJL, P, BDP, and LG strains; the allele $Ly-19^b$ determines presence of the Ly-19.2 antigen and occurs in all other strains tested. Ly-19 has shown no recombination with Lyb-2 on Chr 4 and the two loci have the same strain distribution, but they are thought not to be identical because Lyb-2 is not expressed on T-cells (1).

References

1. Tada, N., S. Kimura, Y. Liu, B.A. Taylor, and U. Hämmerling. 1981. Ly-m19: the Lyb-2 region of mouse chromosome 4 controls a new cell surface antigen. Immunogenetics 13:539–546.

Ly-20 locus (formerly Ly-22 (2)), lymphocyte antigen-20, Chr 4

This locus controls an antigen present in varying amounts in all lymphoid organs but primarily in T-cells. It is not present on brain, kidney, lung, liver, or red cells. The antigen Ly-20.2 is found in all strains examined including C57BL/6 but not in C57BL/10 and its congenics. The amount of antigen in the $Ly-20.2^+$ strains shows considerable variation. The F1 and F2 of a cross between A/J and C57BL/10 gave evidence that expression of the antigen was controlled by two dominant genes, at least one dominant allele at both loci being necessary for Ly-20.2 expression on the cell surface. One of these loci, Ly-20 (Ly-22) is on Chr 4 near Fv-1. The allele $Ly-20^a$ determines presence of the Ly-20.2 antigen in C57BL/10; the allele $Ly-20^b$ determines absence the antigen in all other strains (1).

References

1. Meruelo, D., M. Offer, and N. Flieger. 1983. A new lymphocyte cell surface antigen, Ly-22.2, controlled by a locus on chromosome 4 and a second unlinked locus. J. Immunol. 130:946–950.
2. Morse, H.C. III, F.-W. Shen, and U. Hämmerling. 1987. Genetic nomenclature for loci controlling mouse lymphocyte antigens. Immunogenetics 25:71–78.

Ly-21 locus, lymphocyte antigen-21, Chr 7

This locus controls an antigen, Ly-21.2, present in varying amounts in all lymphoid tissue, in trace amounts in liver, but not detectable in brain, kidney, liver, lung, and red cells. The allele *Ly-21ᵃ* determines absence of Ly-21.2 and occurs in the A/J strain; the allele *Ly-21ᵇ* determines presence of the antigen and occurs in all other strains examined. Ly-21.2 expression increases dramatically in mice with overt leukemia regardless of the etiological agent (1). *Ly-21* is on the distal portion of Chr 7 (2).

References

1. Kennard, J., and D. Meruelo. 1982. A new murine lymphocyte alloantigen, Ly-21.2, mapping to the seventh chromosome. Immunogenetics 15:239–250
2. Kennard, J., and D. Meruelo. 1982. Precise mapping of the gene for Ly-21.2 on the 7th chromosome. J. Immunogenet. 9:389–395.

Ly-22 locus (formerly *Ly-m22*), lymphocyte antigen-22, Chr 1

This locus controls an antigen expressed predominantly on T-cells and detectable on lymph node, spleen, thymus, and bone marrow cells. The allele *Ly-22ᵃ* determines absence of the Ly-22.2 antigen and occurs in the DBA/2, RF/J, SJL, and 14 other strains; the allele *Ly-22ᵇ* determines presence of the antigen and occurs in the A, BALB/c, CBA/J, C3H/He, and many other strains. *Ly-22* is located on the distal part of Chr 1 in a cluster that includes the other antigenic loci *Ly-9*, *Ly-17*, and *Mls* (1).

References

1. Tada, N., S. Kimura, Y. Liu-Lam, and U. Hämmerling. 1983. Mouse alloantigen system Ly-m22 predominantly expressed on T lymphocytes and controlled by a gene linked to Mls region on chromosome 1. Hybridoma 2:29–38.

Ly-22 locus

See *Ly-20* locus.

Ly-23 locus, lymphocyte antigen-23, Chr 2

This locus controls an antigen present on lymphocytes, liver, and kidney. The antigen was detected by a BALB/c anti-B10.D2 monoclonal antibody. The allele *Ly-23ᵃ* determines absence of the antigen and occurs in the A/J, C3H/OuJ, NZB, BALB/c, and other strains; the allele *Ly-23ᵇ* determines presence of the antigen

and occurs in the C57 strains and in DBA/2, SJL, C58, 129/Sv, and others. *Ly-23* is closely linked to *B2m* on Chr 2 and has shown at least one recombination with it (1).

References

1. Gasser, D.L., J. Ziebur, and K. Matsumoto. 1983. A lymphocyte antigen encoded by a locus closely linked to *H-3* on the second chromosome of the mouse. Proc. Natl. Acad. Sci. USA 80:6620–6623.

Ly-24 locus (formerly *Pgp-1* (3)), lymphocyte antigen-24, Chr 2

This locus controls an antigen present on fibroblasts, macrophages, bone marrow cells of the granulocyte series, thymocytes, lung, kidney, brain, and liver, and on hematopoietic tumor cell lines and some lymphomas. The antigen is a cell-surface glycoprotein of molecular weight 95 000 (1, 4). The allele *Ly-24ᵃ* determines the antigen Ly-24.1 and occurs in the BALB/c, DBA/1, DBA/2, and CBA/J strains; the allele *Ly-24ᵇ* determines the antigen Ly-24.2 and occurs in the C57 strains and in C58, A/J, AKR/J, C3H/He, MA/My, NZB, and SJL (2). *Ly-24* is on Chr 2 near *Ea-6* (1).

References

1. Colombatti, A., E.N. Hughes, B.A. Taylor, and J.T. August. 1982. Gene for a major cell surface glycoprotein of mouse macrophages and other phagocytic cells on chromosome 2. Proc. Natl. Acad. Sci. USA 79:1926–1929.
2. Lesley, J., and I.S. Trowbridge. 1982. Genetic characterization of a polymorphic murine cell-surface glycoprotein. Immunogenetics 15:313–320.
3. Morse, H.C. III, F.-W. Shen, and U. Hämmerling. 1987. Genetic nomenclature for loci controlling mouse lymphocyte antigens. Immunogenetics 25:71–78.
4. Trowbridge, I.S., J. Lesley, R. Schulte, R. Hyman, and J. Trotter. 1982. Biochemical characterization and cellular distribution of a polymorphic murine cell-surface glycoprotein expressed on lymphoid tissues. Immunogenetics 15:299–312.

Ly-25 locus, lymphocyte antigen-25, Chr 2

This locus controls an antigen present on lymphocytes and non-lymphoid cells of hematopoietic origin. The antigen was recognized by a monoclonal antibody made by immunizing (C3H × DBA/2)F1 mice with CE/J thymocytes. The allele *Ly-25ᵃ* determines presence of the Ly25.1 antigen and occurs in the CE/J and STS/A strains; the allele *Ly-25ᵇ* determines absence of the Ly-25.1 antigen and occurs in the C3H, DBA/2, and BALB/c strains. In 14 BALB/c × STS recombinant inbred strains, *Ly-25* showed only one discordance with *Ly-24* (formerly *Pgp-1*) on Chr 2 (1).

References

1. Hogarth, P.A., A. Rigby, V.R. Sutton, I.F.C. McKenzie, and J. Hilgers. 1984. The Ly-25.1 specificity: definition with a monoclonal antibody. Immunogenetics 19:83–86.

Ly-26 locus, lymphocyte antigen-26

This locus controls an antigen present on B lymphocytes and other non-lymphoid cells of bone marrow origin, possibly including monocytes, macrophages, and granulocytes. It was detected by three different monoclonal antibodies. The allele *Ly-26ᵃ* determines presence of the Ly-26.1 antigenic specificity and occurs in strains LP.RIII, CE, and NZB; the allele *Ly-26ᵇ* determines absence of Ly-26.1 and occurs in all other strains tested including C57BL/6, BALB/c, DBA/2, and many others. Surprisingly, both LP and RIII, the progenitors of LP.RIII, are both *Ly-26ᵇ*. Crosses to test for segregation at the *Ly-26* locus have not been described (1).

References

1. Hogarth, P.M., S.E. Latham, and I.F.C. McKenzie. 1984. A new murine alloantigen: Ly-26.1. Immunogenetics 19:355–358.

Ly-27 locus, lymphocyte antigen-27 (*Ly-27* formerly designated a locus now called *Ly-6D*)

This locus determines an antigen detected by a monoclonal antibody in a subpopulation of T-cells and in low quantities in a small fraction of B-cells. It is not found in bone marrow cells. The allele *Ly-27ᵇ* determines presence of the Ly-27.2 antigen and occurs in C57BL/6 and most other strains; the allele *Ly-27ᵃ* determines absence of the antigen and occurs in the BALB/c, A/J, C3H/He, CBA/Ca, CE, and SM strains (1).

References

1. Matsushima, G.K., R.C. Harmon, and J.A. Frelinger. 1986. A new murine lymphocyte alloantigen Ly-27.2. Immunogenetics 23:406–408.

Ly-28 locus, lymphocyte antigen-28 (*Ly-28* formerly designated a locus now called *Ly-6C*)

This locus controls an antigen present on thymus, spleen, and lymph node cells but not on bone marrow cells, and on both T- and B-cells. It is detected by monoclonal antibodies. The allele *Ly-28ᵇ* determines presence of the Ly-28.2 antigen and occurs in most inbred strains, including BALB/c, C57BL/6, C3H/He, and DBA/2; the allele *Ly-28ᵃ* determines absence of the antigen and occurs in strains AKR/N, RF, and RIIIS (1).

References

1. Tada, N., N. Tamaoki, S. Ikegami, S. Nakamura, K. Dairiki, and T. Fujimori. 1986. A new mouse lymphocyte alloantigen (Ly-28) defined by monoclonal antibodies. Immunogenetics 24:275–277.

Ly-29 locus, lymphocyte antigen-29, Chr 4

This locus controls an antigen recognized by a monoclonal antibody on T- and B lymphocytes and not on other cells and tissues. The allele *Ly-29ᵇ* determines the antigen Ly-29.2 and occurs in strains BALB/c, CBA/J, C57BL/6, C57BL/10, NZB, and 129; the allele *Ly-29ᵃ* determines lack of the antigen and occurs in strains A, AKR, CE, C57L, C57BR/cd, C58, and others. *Ly-29* showed no recombinants with *Mtv-17* (formerly *Mtv-20*) on Chr 4 among 14 BALB/c × STS recombinant inbred strains (1).

References

1. Gomez, L.J., M.S. Sandrin, M.M. Henning, J. Hilgers, and I.F.C. McKenzie. 1985. Ly-29: a locus closely linked to Mtv-20 on chromosome 4 codes for a new mouse lymphocyte surface alloantigen. Immunogenetics 22:305–308.

Ly-30 locus, lymphocyte antigen-30

This locus controls an antigen recognized by a monoclonal antibody on some T- and B-cells, liver, and kidney, but not on thymocytes. The allele *Ly-30ᵇ* determines the Ly-30.2 antigen and occurs in strains C57BL/10, BALB/c, AKR, DBA/2, and many others; the allele *Ly-30ᵃ* determines absence of the antigen and occurs in strains CE, C58, and SJL (1).

References

1. Gomez, L.J., M.S. Sandrin, M.M. Henning, and I.F.C. McKenzie. 1985. Ly-30: a new mouse lymphocyte surface alloantigen defined by a monoclonal antibody. Immunogenetics 22:301–303.

Ly-31 locus, lymphocyte antigen-31, Chr 4

This locus controls a cell surface antigen present on lymphocytes of the thymus, spleen, lymph nodes, bone marrow, and peripheral blood. It is present on both T- and B-cells and on many myelomas and leukemias. The allele *Ly-31ᵃ* determines the Ly-31.1 antigen and occurs in the A/He, AKR, BALB/c, C3H/He, DBA/2, SWR, and many other strains; the allele *Ly-31ᵇ* determines absence of the Ly-31.1 antigen and occurs in the C57BL family of strains and in strains DBA/1, SM, 129/J, and others. The strain distributions of *Ly-31* and *Akp-2* are completely concordant in recombinant inbred strains and in 30 standard inbred strains, indicating very close linkage of the two loci on Chr 4 (1).

References

1. Tada, N., S. Kimura, Y. Liu-Lam, and U. Hämmerling. 1984. A new lymphocyte surface antigen (Ly-m31) controlled by a gene closely linked to the *Akp-2* locus on mouse chromosome 4. Immunogenetics 20:589–592.

Ly-32 locus, lymphocyte antigen-32, Chr 4

This locus controls an antigen present on spleen, lymph node, and bone marrow cells and to a lesser extent on thymus cells. It is found on most B-cells and on a proportion of T-cells. The antigen was detected by a monoclonal antibody produced by a cell line derived from spleen cells of an AKR mouse immunized with a BALB/c tumor and fused with myeloma cells. The allele *Ly-32ᵇ* determines presence of the antigen and occurs in strains A/J, BALB/c, CBA/Ca, C3H/He, C57BL/6, NZB, and others; the allele *Ly-32ᵃ* determines absence of the antigen and occurs in strains AKR, CBA/J, C57BR/cd, DBA/1, DBA/2, RF, SJL, and others. *Ly-32* alleles have the same strain distribution as alleles of *Lyb-2*, a locus on Chr 4, and close linkage is confirmed by presence of the *Ly-32ᵃ* allele in the B6.1-*Lyb-2ᵃ* congenic strain. *Ly-32ᵇ* is probably different from *Ly-29ᵇ* which is also in this chromosomal region, since the A/WySn strain is *Ly-32ᵇ* and *Ly-29ᵃ* (1).

References

1. Tada, N., N. Tamaoki, S. Ikegami, S. Nakamura, K. Dairiki, and T. Fujimori. 1987. The *Lyb-2–Ly-19* region of mouse chromosome 4 controls a new cell-surface alloantigen: Ly-32. Immunogenetics 25:269–270.

Ly-33 locus, lymphocyte antigen-33, Chr 1

This locus controls an antigen present on lymphocytes of thymus, lymph nodes, spleen, and bone marrow, and also on brain, liver, and kidney cells. The allele determining presence of the antigen, *Ly-33ᵃ*, occurs in strains GR/A, RF/J, STS/A, LIS/A, and NZW/LacJ; the allele determining absence of the antigen, *Ly-33ᵇ*, occurs in strains A/J, C57BL/6, BALB/c, and all other inbred strains tested. *Ly-33* is on Chr 1 about 20 cM distal to *Ly-17* (1).

References

1. Kimura, S., N. Tada, J. Furze, and H. Festenstein. 1987. Ly-33: a new lymphocyte alloantigen controlled by a gene linked to *Ly-17* locus on mouse chromosome 1. Immunogenetics 26:315–316.

Ly-34 locus, lymphocyte antigen-34

This locus controls an antigen present on 100 per cent, 50 per cent, 30 per cent, and 20 per cent of cells of thymus, lymph nodes, spleen, and bone marrow, predominantly on the T-cell population of these tissues. It is also present in brain, liver, and kidney. The antigen was detected by a monoclonal antibody prepared from spleen cells from HRS/J mice hyperimmunized with C57BL/6-*H-2ᵏ* thymocytes. The allele *Ly-34ᵇ* determines presence of the antigen and occurs in all the C57 strains tested except C57BL/Ks, and in strains C58, C3H/He, DBA/2, NZB, P, CE, and others; the allele *Ly-34ᵃ* determines absence of the antigen and occurs in strains C57BL/Ks, A/He, AKR, CBA/J, DBA/1, BALB/c, HRS, STS, and others. No linkage was found with loci on Chrs 2, 4, 6, 9, and 12 (1).

References

1. Kimura, S., J. Furze, and H. Festenstein. 1987. Ly-34: a new mouse lymphocyte alloantigen. Immunogenetics 26:381–382.

Ly-37 locus, lymphocyte antigen-37, Chr 3

This locus is homologous with the locus in man coding for the CD2 antigen. It was identified among clones isolated from a mouse T helper cDNA library by hybridization with a human CD2 cDNA probe. The human antigen is a glycoprotein present in all T-cells except the most immature thymocytes and also in natural killer (NK) cells. The murine antigen has not been isolated, but CD2 mRNA is expressed in mouse thymus and spleen and in T-cell line EL4, and a rabbit antiserum against human CD2 precipitated from thymocytes a glycoprotein similar in size to CD2. *Ly-37* was assigned to Chr 3 by use of mouse–Chinese hamster somatic cell hybrids screened with a cDNA probe (1).

Ly-37

References

1. Sewell, W.A., M.H. Brown, M.J. Owen, P.J. Fink, C.A. Kozak, and M.J. Crumpton. 1987. The murine homologue of the T lymphocyte CD2 antigen: molecular cloning, chromosome assignment and cell surface expression. Eur. J. Immunol. 17:1015–1020.

Ly-A locus

See *Ly-1* locus.

Ly-B locus

See *Ly-2* locus.

Ly-C locus

See *Ly-3* locus.

Lyb-1 locus

See *Ly-4* locus.

Lyb-2 locus, B-lymphocyte antigen-2, Chr 4

This locus controls an antigen present on B lymphocytes. Four alleles are known: *Lyb-2ᵃ* determines the specificities Lyb-2.1, 4, and 5 and occurs in strains I/St, C57BR/cd, CBA/J, DBA/2, and others; *Lyb-2ᵇ* determines the specificities Lyb-2.2 and 4 and occurs in strains A, C57BL/6, C3H/An, CBA/H, and others; *Lyb-2ᶜ* determines the specificity Lyb-2.3 and occurs in strains CE, AKR, SJL, and others; *Lyb-2ᵈ* determines specificity Lyb-2.1 and occurs in strain STS/A (3, 5). The Lyb-2 antigen is present on spleen and bone marrow cells but absent from thymocytes (2). It is expressed on relatively mature B-cells (1). The cell surface component specified by *Lyb-2* is similar to the components of molecular weight 40 and 45 kDa specified by the *H-2* and *Tla* loci but differs from them in lacking the β2-microglobulin unit (4).

References

1. Brown, L.D., F.W. Shen, J.W. Uhr, and E.S. Vitetta. 1978. The expression of Lyb-2.1 on murine B lymphocytes. Fed. Proc. 37:1584 (Abstr.).
2. Sato, H., and E.A. Boyse. 1976. A new alloantigen expressed on B cells: the *Lyb-2* system. Immunogenetics 3:565–572.
3. Shen, F.W., M. Spanondis, and E.A. Boyse. 1977. Multiple alleles at the *Lyb-2* locus. Immunogenetics 5:481–484.

4. Tung, J.-S., J. Michaelson, H. Sato, E.S. Vitetta, and E.A. Boyse. 1977. Properties of the Lyb-2 molecule. Immunogenetics 5:485–488.
5. Tung, J.-S, F.-W. Shen, V. LaRegina, and E.A. Boyse. 1986. Antigenic complexity and protein-structural polymorphism in the *Lyb-2* system. Immunogenetics 23:208–210.

Lyb-3 locus, B-lymphocyte antigen-3

This antigen occurs on a late maturing portion of splenic B-cells of all strains tested except CBA/N. It is not present on bone marrow cells, thymocytes, or peripheral T-cells. Anti-Lyb-3 serum is not cytotoxic, but it binds to B-cells and enhances the antibody response to low doses of thymus-dependent antigens. The enhancement is antigen-specific. In an Lyb-3-secreting myeloma, MOPC-315, anti-Lyb-3 enhances secretory differentiation but not clonal proliferation of myeloma cells (3), indicating that *Lyb-3* plays some role in terminal differentiation of B-cells. The Lyb-3 antigen bound by the antiserum is a single molecular species of 68 kDa (1). It is thought that lack of the antigen in CBA/N mice is due to absence of the class of late maturing B-cells that bear the antigen. Absence of this class of cells results from action of the X-linked immune deficiency gene (*xid*) in that strain. The evidence does not preclude the possibility that the Lyb-3 antigen is a product of the *xid* locus (2).

References

1. Cone, R.E., B. Huber, H. Cantor, and R.K. Gershon. 1978. Molecular identification of a surface structure on B cells (Lyb-3) and its relationship to B cell triggering. J. Immunol. 120:1733–1740.
2. Huber, B., R.K. Gershon, and H. Cantor. 1977. Identification of a B-cell surface structure involved in antigen-dependent triggering: absence of this structure on B cells from CBA/N mutant mice. J. Exp. Med. 145:10–20.
3. Kemp, J.D., J.W. Rohrer, and B.T. Huber. 1983. Lyb-3: a B cell surface antigen associated with triggering secretory differentiation. Immunol. Rev. 69:127–140.

Lyb-4 locus, B-lymphocyte antigen-4, Chr 4

This locus controls an antigen detectable by a cytotoxic antiserum and present on B-cells of spleen, lymph nodes, and Peyer's patches but not on peripheral T-cells or on most bone marrow cells. The allele *Lyb-4ᵃ* determines presence of the Lyb-4.1 antigen and occurs in strains DBA/2, C3H/He, and others; the allele *Lyb-4ᵇ* determines absence of the Lyb-4.1 antigen and occurs in strains C57BL/Ks, C57BL/6, BALB/c, and others. Heterozygotes express the antigen (1). Anti-Lyb-4.1 blocks the mixed lymphocyte culture (MLC) response across *H-2* and *Mls* disparities. In the response of parental cells to F1 lymphocytes from crosses between strains with and without the Lyb-4 antigen, MLC stimulation was blocked only when the responding parental cell recognized on the F1 cell *H-2* and *Mls* disparities derived from the parent which possessed the Lyb-4.1 antigen (2). *Lyb-4ᵃ*, when derived from the DBA/2 strain, shows close linkage with *Mup-1* on Chr 4. When derived from the C3H/HeJ strain, it shows no linkage with *Mup-1*. The Lyb-4.1 antigenic determinants in the two strains are indistinguishable (3). *Lyb-4* is also closely linked to *Lyb-2* and *Lyb-6* on Chr 4 (4).

References

1. Freund, J.G., A. Ahmed, R.E. Budd, M.E. Dorf, K.W. Sell, W.E. Vannier, and R.E. Humphreys. 1976. The L1210 leukemia cell bears a B lymphocyte specific, non-H-2 linked alloantigen. J. Immunol. 117:1903–1905.
2. Freund, J.G., A. Ahmed, M.E. Dorf, K.W. Sell, and R.E. Humphreys. 1977. Inhibition of the mixed lymphocyte culture response with anti-Lyb-4.1 serum. J. Immunol. 118:1143–1149.
3. Howe, R.C., A. Ahmed, T.J. Faldetta, J.E. Byrnes, K.M. Rogan, M.E. Dorf, B.A. Taylor, and R.E. Humphreys. 1979. Mapping of the *Lyb-4* gene to different chromosomes in DBA/2 and C3H/HeJ mice. Immunogenetics 9:221–232.
4. Kessler, S., A. Ahmed, and I. Scher. 1979. Murine B cell alloantigens Lyb-2, Lyb-4, and Lyb-6 are products of distinct genes clustered on chromosome 4. Fed. Proc. 38:1460 (Abstr.).

Lyb-5 locus, B-lymphocyte antigen-5

This locus controls an antigen present on the late maturing class of B-cells which is deficient in the CBA/N strain bearing the *xid* gene. The allele *Lyb-5ᵃ* determines presence of the Lyb-5.1 antigen and occurs in the DBA/2 strain; the allele *Lyb-5ᵇ* determines absence of the Lyb-5.1 antigen and occurs in the C57BL/6 strain (1–3).

References

1. Ahmed, A., I. Scher, S.O. Sharrow, A.H. Smith, W.E. Paul., D.H. Sachs, and K.W. Sell. 1977. B-lymphocyte heterogeneity: development and characterization of an alloantiserum which distinguishes B-lymphocyte differentiation alloantigens. J. Exp. Med. 145:101–110.
2. Ahmed, A., A.H. Smith, S.W. Kessler, and I. Scher. 1978. Ontogeny of murine B-cell surface antigens and function. *In* M.E. Gershwin and R.L. Cooper, eds., Animal Models of Comparative and Developmental Aspects of Immunity and Disease, 175–200. Pergamon Press, New York.
3. Smith, H.R., L.J. Yaffe, D.L. Kastner, and A.D. Steinberg. 1986. Evidence that Lyb-5 is a differentiation antigen in normal and *xid* mice. J. Immunol. 136:1194–1200.

Lyb-6 locus, B-lymphocyte antigen-6, Chr 4

This locus controls an antigen present on B-cells of the spleen, Peyer's patches, and lymph nodes. The allele *Lyb-6ᵃ* determines presence of the antigen and occurs in strain CBA/J and others; the allele *Lyb-6ᵇ* determines absence of the antigen and occurs in strain C3H/HeJ and others (1, 2). The antigen is a glycoprotein of 45 000 molecular weight that has been identified by its mobility in SDS-polyacrylamide gel electrophoresis of immune precipitates from radioiodinated cell lysates (2). *Lyb-6* is closely linked to *Lyb-2* and *Lyb-4* on Chr 4 (3).

References

1. Ahmed, A., A.H. Smith, S.W. Kessler, and I. Scher. 1978. Ontogeny of murine B-cell surface antigens and function. *In* M.E. Gershwin and R.L. Cooper, eds., Animal Models of Comparative and Developmental Aspects of Immunity and Disease, 175–200. Pergamon Press, New York.
2. Kessler, S., A. Ahmed, and I. Scher. 1978. A new murine B cell alloantigen: biochemical characteristics and distribution. Fed. Proc. 37:1584 (Abstr.).
3. Kessler, S., A. Ahmed, and I. Scher. 1979. Murine B cell alloantigens Lyb-2, Lyb-4, and Lyb-6 are products of distinct genes clustered on chromosome 4. Fed. Proc. 38:1460 (Abstr.).

Lyb-7 locus, B-lymphocyte antigen-7, Chr 12

This locus controls an antigen present on mature B lymphocytes of the class that is absent in *xid*-bearing mice of the CBA/N strain. The antigen is recognized by an antibody, anti-Lyb-7.1, that specifically inhibits the *in vitro* primary antibody response to TNP-Ficoll and some other TI-2 antigens (1). The allele *Lyb-7ᵃ* determines susceptibility to the blocking activity of anti-Lyb-7.1 and occurs in strains DBA/2 and C3H/He; the allele *Lyb-7ᵇ* determines lack of susceptibility to the blocking activity and occurs in strains C57BL/6 and AL/N. Heterozygotes are as susceptible as the susceptible parent. *Lyb-7* is loosely linked to *Igh* on Chr-12 (2).

References

1. Subbarao, B., and D.E. Mosier. 1983. Lyb antigens and their role in B lymphocyte activation. Immunol. Rev. 69:81–97.

2. Subbarao, B., A. Ahmed, W.E. Paul, I. Scher, R. Lieberman, and D.E. Mosier. 1979. *Lyb-7*, a new B cell alloantigen controlled by genes linked to the *IgC_H* locus. J. Immunol. 122:2279–2285.

Lyb-8 locus, B-lymphocyte antigen-8, Chr 7

This locus controls an antigen present on pre-B-cells and on all mature B lymphocytes. The antigen is a glycoprotein with a single polypeptide chain of 95 000 to 105 000 molecular weight. It is found on spleen, bone marrow, and lymph node cells, and on B-cell lymphomas. The allele *Lyb-8^b* determines presence of the Lyb-8.2 antigen and occurs in the C57 and C58 strains and in strains CBA/J, CBA/Ca, SWR, BALB/c, CE, C3H/He, SJL, A/J, and AL/N; the allele *Lyb-8^a* determines absence of the antigen and occurs in strains DBA/1, DBA/2, NZB, AKR, and PL. *Lyb-8* is on Chr 7 between *Gpi-1* and *D7Rp2* (1).

References

1. Symington, F.W., B. Subbarao, D.E. Mosier, and J. Sprent. 1982. Lyb-8.2: a new B cell antigen defined and characterized with a monoclonal antibody. Immunogenetics 16:381–391.

Lyb-X

See *Lyx-2* locus.

Lyt-1 locus

See *Ly-1* locus.

Lyt-2 locus

See *Ly-2* locus.

Lyt-3 locus

See *Ly-3* locus.

Lyt-4 locus

See *Ly-5* locus.

Lyt-X

See *Lyx-2* locus.

Lyx-1 to *Lyx-3* loci, Chr X

These three loci control presence of antigens on T lymphocytes and to some extent also on B lymphocytes. They are all X-linked but have different strain distributions. The three T-cell antigens are closely associated with level of immune response to three different thymus-independent antigens (3, 4). Response to a T-dependent antigen, lactate dehydrogenase isozyme-C_4 (LDH-C_4), is also controlled by an X-linked gene. SJL/J mice have a strong immediate and secondary response, C3H/HeJ mice have a low immediate and a delayed secondary response (1). It is not known whether the genes for all these effects are identical or even whether they are closely linked.

Lyx-1 locus, lymphocyte antigen X-1. The antigen controlled by this locus is present on T-cells of the BALB/c and C3H/He strains and absent from T-cells of the DBA/2, SJL, A/J, C57BL/6, and AKR strains. Immune response to type III pneumococcal polysaccharide (sIII) is high in strains positive for the antigen (BALB/c, C3H/He) and low in strains negative for it (DBA/2, SJL) (3).

Lyx-2 locus, lymphocyte antigen X-2. The antigen controlled by this locus is present on T-cells of the DBA/2, BALB/c, and A/J strains and absent from T-cells of the SJL, C57BL/6, A/J, and AKR strains. Immune response to synthetic double-stranded RNA [poly(I)-poly(C)] is high in strains positive for the antigen and low in strains negative for it (3). Zeicher *et al.* (4) later found an antigen present on B-cells that was detected by the same antiserum that recognized Lyx-2. It was shown to be a different molecule from the T-cell antigen but had the same strain distribution and the same X-linked inheritance. The authors tentatively called the two antigens Lyt-X and Lyb-X.

Lyx-3 locus, lymphocyte antigen X-3. The antigen controlled by this locus is present on T-cells of the SJL, C57BL/6, and AKR strains and absent from T-cells of the DBA/2, BALB/c, C3H/He, and A/J strains. Immune response to denatured DNA is high in strains positive for the antigen and low in strains negative for it (3). The immune response shows X-linked inheritance, with the allele for high responsiveness in SJL dominant to the allele for low responsiveness in DBA/2 (2).

References

1. Marsh, J.A., T.E. Wheat, and E. Goldberg. 1977. Temporal regulation of the immune response to LDH-C_4 by an X-linked gene in C3H/HeJ and SJL/J mice. J. Immunol. 118:2293–2298.
2. Mozes, E., and S. Fuchs. 1974. Linkage between immune response potential to DNA and X chromosome. Nature 249:167–168.

3. Zeicher, M., E. Mozes, and P. Lonai. 1977. Lymphocyte alloantigens associated with X-chromosome-linked immune response genes. Proc. Natl. Acad. Sci. USA 74:721–724.
4. Zeicher, M., E. Mozes, Y. Reisner, and P. Lonai. 1979. Selective expression of murine lymphocyte alloantigens controlled by the X chromosome. Immunogenetics 9:119–124.

lz, lizard, recessive, Chr 15

Probably extinct. Found in a mutagenesis experiment among the progeny of crosses involving four inbred strains and probably X-ray induced (1). Homozygotes show partial paralysis of the hindlimbs and marked tremor of the whole musculature and are recognizable at about 2 weeks. Females breed only rarely and do not rear their young; males never breed although normal spermatozoa appear to be formed. Preliminary studies (by L.W. Duchen and S. Strich) suggest that the syndrome is muscular rather than nervous in origin (2).

References

1. Lyon, M.F., R.J.S. Phillips, and A.G. Searle. 1964. The overall rates of dominant and recessive lethal and visible mutation induced by spermatogonial X-irradiation in mice. Genet. Res. 5:448–467.
2. Morris, T. 1966. Mouse News Lett. 34:27.

Lzp, *Lzm* complex

This complex consists of closely linked structural and regulatory loci affecting two highly homologous forms of intestinal lysozyme c, the P form which comprises 97 per cent and the M form which comprises 3 per cent of the total in the C58/J and C3H/He strains. The two forms differ in structure and pattern of expression. No confirmed recombinants between any of the loci have been found.

Lzp-r locus, lysozyme P-regulatory. This locus controls level of activity of the P form of intestinal lysozyme. Activity is high but with a wide range of variation in most laboratory inbred strains and low in strains derived from *Mus castaneus*. The difference in crosses between C3H/He and the CASA strain derived from *M. castaneus* was shown to be due to a single locus withthe allele $Lzp-r^d$ in C3H/He and $Lzp-r^c$ in CASA. Heterozygotes have intermediate activity.

Lzp-s locus, lysozyme P-structural. This locus controls heat stability of lysozyme P. The allele $Lzp-s^d$ determines low heat stability and occurs in the C3H/He strain, and the allele $Lzp-s^c$ determines high heat stability and probably occurs in the CASA strain. The level of P lysozyme is so low in CASA that heat stability cannot be determined, but presence of the $Lzp-s^c$ allele is inferred from the high heat stability of the F1 hybrid. In addition, the P lysozymes of strain BALB/c and *Mus spretus* differ in electrophoretic mobility. BALB/c has a relative mobility of 84 and *M. spretus* one of 82. The trait segregates as if controlled by a single gene closely linked to *Lzm-s* and is presumably due to a difference at the *Lzp-s* locus.

Lzm-s locus, lysozyme M-structural. This locus controls electrophoretic mobility of the M form of intestinal lysozyme. The allele $Lzm-s^d$ in strain C3H/He determines an intermediate relative mobility of 68, the allele $Lzm-s^c$ in the CASA strain determines a fast relative mobility of 73, and the allele $Lzm-s^s$ in *M. spretus* determines a slow relative mobility of 62. There is evidence from wild *M. castaneus* populations that *Lzm-s* may consist of two closely linked loci, *Lzm-s1* and *Lzm-s2*, each of which has a c (73) and/or a d (68) allele (1).

References

1. Hammer, M.F., and A.C. Wilson. 1987. Regulatory and structural genes for lysozymes of mice. Genetics 115:521–533.

M

M locus

See *Mls* locus.

m, misty, recessive, Chr 4

Arose spontaneously in the DBA/J strain. Homozygotes are lighter in color than normal and usually have a white belly spot and tail tip. The color of *m/m* mice is not as diluted as that of *d/d* or *ln/ln*, and the hairs of the coat have more cortical pigment than in *d/d* or *ln/ln* (1,2).

References

1. Woolley, G.W. 1941. "Misty," a new coat color dilution in the mouse, *Mus musculus*. Am. Nat. 75:507–508.
2. Woolley, G.W. 1945. Misty dilution in the mouse. J. Hered. 36:269–270.

ma, matted, recessive, Chr 3

Arose spontaneously in the CBA/Gr inbred strain. Homozygotes can first be identified between 2 and 4 weeks of age. The hairs are more erect than normal and are matted in clumps, there is a tendency to baldness depending on areas of friction and resulting hair breakage, and a color change of black hair to russet happens toward the end of each hair cycle. Hairs are normal in morphology but are brittle and tend to split longitudinally. Viability and fertility are normal (2). The defect in matted hairs is thought to be in the cuticle (1).

References

1. Jarrett, A., and R.I. Spearman. 1957. The keratin defect and hair-cycle of a new mutant (matted) in the house-mouse. J. Embryol. Exp. Morphol. 5:103–110.
2. Searle, A.G., and R.I. Spearman. 1957. "Matted," a new hair-mutant in the house-mouse: genetics and morphology. J. Embryol. Exp. Morphol. 5:93–102.

Mal-1 locus, malaria resistance, Chr 1

This locus controls resistance to *Plasmodium chabaudi* as measured by survival after infection. Differences in resistance between C57BL/6 and DBA/2 may be due to several loci, one of which, *Mal-1*, is located on the distal part of Chr 1 (1).

References

1. Lundin, L.-G., P. Borwell, and G. Holmquist. 1983. Mouse News Lett. 68:89.

Map-1 locus, mannosidase processing-1, Chr 5

This locus controls electrophoretic mobility of an acid lysosomal α-mannosidase (E.C. 3.2.1.24) in liver. The allele *Map-1^b* determines a fast anodally migrating band and occurs in strains C57BL/6J and many others; the allele *Map-1^h* determines a slower band and occurs in strains CBA/J, C3H/HeJ, and AKR/J. Heterozygotes are indistinguishable from *Map-1^b* homozygotes. The gene is expressed only in liver and appears to control post-translational sialylation of the enzyme. The phenotype of *Map-1^b/Map-1^b* mice at 17 days of gestation resembles a mixture of enzymes of both homozygotes. This changes gradually to the adult expression by 20 days after birth. *Map-1* is very closely linked to *Gus* (1).

References

1. Dizik, M., and R.W. Elliott. 1977. A gene apparently determining the extent of sialylation of lysosomal α-mannosidase in mouse liver. Biochem. Genet. 15:31–46.

Map-2 locus

See *Neu-1* locus.

Mbp locus

See *shi* locus.

mc, marcel, recessive, Chr 5

Arose in the C57BL/10 strain. Homozygotes have pronounced waves in the coat at 14 to 21 days. Female homozygotes are sterile (1).

References

1. Foerster, E. 1956. Mouse News Lett. 14:23.

md locus, Chr 16

md, mahoganoid, recessive. Arose in the C3H/HeJ strain. This gene is similar to mahogany (*mg*) in its effects on coat color, but causes less darkening of the back, ears, and tail of agouti mice (1,4).

*md^nc^*, nonagouti curly, recessive. Found in the F2 generation of a mutagenesis experiment with caffeine. Homozygotes are black with black pinna hairs and curly whiskers and have reduced viability (2,3). *md^nc^/md* mice have straight whiskers and are darker than *md/md* but not as dark as *md^nc^/md^nc^* mice (4).

References

1. Lane, P.W. 1960. Mouse News Lett. 22:35.
2. Phillips, R.J.S. 1963. Mouse News Lett. 29:38.
3. Phillips, R.J.S. 1971. Mouse News Lett. 45:25.
4. Phillips, R.J.S., and G. Fisher. 1979. Mouse News Lett. 60:46–47.

mdac, modifier of *Dac*, recessive

Homozygosity for this gene is necessary for expression of dactylaplasia in *Dac/+* mice. It is homozygous in strains SM, BALB/c, LG, and 129/J. The wild-type allele occurs in strains C57BL/6, DBA/2, CBA/J, AKR, and SWR. There is no visible effect on the feet of *mdac/mdac* in the absence of *Dac/+* (1). See *Dac*.

References

1. Chai, C.K. 1981. Dactylaplasia in mice, a two-locus model for developmental anomalies. J. Hered. 72:234–237.

mdf, muscle deficient, recessive, Chr 19

Arose in a C57BL/6J-*ln fz* stock. Homozygotes can first be detected at 5 to 6 weeks of age. They are slightly smaller than normal, have a waddling gait, and often have a nervous tremor. By 12 weeks, they can progress only by pulling themselves forward with their forelimbs. Fertility is low in both males and females. Homozygotes have a marked reduction of muscle mass in the sartorius, vastus lateralis, and rectus femoris, and a lesser reduction in forelimb muscles. Histologically, there is atrophy of both type I and type II muscle fibers. Serum creatine kinase activity is not affected (2). The mutation is on Chr 19 close to the centromere (1).

References

1. Sweet, H.O. 1983. Mouse News Lett. 68:72.
2. Womack, J.E., A. MacPike, and H. Meier. 1980. Muscle deficient, a new mutation in the mouse. J. Hered. 71:68.

mdg, muscular dysgenesis, recessive

Arose spontaneously in a tailless line. Homozygotes never develop normal skeletal muscle. The first observable effects occur at 13 days of gestation when there is a marked edema followed by failure of myoblasts to differentiate into striated myotubes. Subsequently all muscle cells degenerate. There are numerous skeletal abnormalities in homozygotes, including a short lower jaw and usually a cleft palate, which may be secondary to the abnormalities of muscle. Homozygotes do not survive birth (3). The earliest abnormality found in muscle was dilated sarcoplasmic reticulum at 14 days of gestation (5). No differences were detected in proteins of sarcoplasmic reticulum except for an 80 per cent reduction in the binding of phosphorylase (2). Skeletal muscle of homozygotes never develops contractility either in culture or *in vivo*. In culture the myoblasts may fuse and differentiate into cross-striated myotubes (12) with normal electrical properties and action potentials (6), but no contractility. Innervation of muscle is abnormal beginning as early as $13\frac{1}{2}$ to 14 days of gestation. There is extensive overgrowth and generalized sprouting of nerve endings and focal accumulations of acetylcholine esterase and multiple clusters of acetylcholine receptors over the entire surface of the myotubes (8–10). Mutant myotubes in culture become contractile when cocultured with normal nerve cord, an effect due to formation of normal muscle–nerve synapses (7). Further evidence that *mdg* may have an extramuscular expression comes from *mdg/mdg* ⟷ +/− chimeras in which the sites of abnormal innervation of diaphragm muscle were not correlated with the *mdg/mdg* genotype of the abnormally innervated muscle (11). However, Peterson and Pina (4) showed that, in mosaic myotubes formed by coculturing mutant and normal muscle tissue, normal contractility could occur when only a very small proportion of nuclei in a myotube were of +/− genotype. A possible effect of *mdg* in heterozygotes (*mdg/+*) was discovered by Atchley *et al*. (1) who found that the dimensions and shape of the mandible showed greater right–left asymmetry in *mdg/+* than in *+/+* mice and that the parts of the mandible affected were those with most extensive muscle attachment.

References

1. Atchley, W.R., S.W. Herring, B. Riska, and A.A. Plummer. 1984. Effects of the muscular dysgenesis gene on developmental stability in the mouse mandible. J. Craniofac. Genet. Dev. Biol. 4:179–189.
2. Essien, F.B., A.O. Jorgensen, V.I. Kalmins, E. Zubrzycka, and D.H. MacLennan. 1977. The biosynthesis and localization of sarcoplasmic reticulum proteins in dysgenic (*mdg/mdg*) mouse cells. Lab. Invest. 37:562–568.
3. Pai, A.C. 1965. Developmental genetics of a lethal mutation, muscular dysgenesis (*mdg*), in the mouse. II. Developmental analysis. Dev. Biol. 11:93–109.

4. Peterson, A., and S. Pina. 1984. Relationship of genotype and *in vitro* contractility in *mdg/mdg* ⟷ +/+ "mosaic" myotubes. Muscle News 7:194–203.
5. Platzner, A.C., and S. Gluecksohn-Waelsch. 1972. Fine structure of mutant (muscular dysgenesis) embryonic mouse muscle. Dev. Biol. 28:242–252.
6. Powell, J.A., and D.M. Fambrough. 1973. Electrical properties of normal and dysgenic mouse skeletal muscle in culture. J. Cell. Physiol. 82:21–38.
7. Powell, J.A., A. Peterson, and C.V. Paul. 1984. Neurons induce contractions in myotubes containing only muscular dysgenic nuclei. Muscle Nerve. 7:204–210.
8. Powell, J.A., F. Rieger, B. Blondet, P. Dreyfus, and M. Pinçon-Raymond. 1984. Distribution and quantification of ACh receptors and innervation in diaphragm muscle of normal and *mdg* mouse embryos. Dev. Biol. 101:168–180.
9. Rieger, F., and M. Pinçon-Raymond. 1981. Muscle and nerve in muscular dysgenesis in the mouse at birth: sprouting and multiple innervation. Dev. Biol. 87:85–101.
10. Rieger, F., J.A. Powell, and M. Pinçon-Raymond. 1984. Extensive nerve overgrowth and paucity of the tailed asymmetric form (16S) of acetylcholinesterase in the developing skeletal neuromuscular system of the dysgenic (*mdg/mdg*) mouse. Dev. Biol. 101:181–191.
11. Rieger, F., D. Cross, A. Peterson, M. Pinçon-Raymond, and I. Tretjakoff. 1984. Disease expression in +/+ ⟷ *mdg/mdg* mouse chimeras: evidence for an extra-muscular component in the pathogenesis of both dysgenic abnormal diaphragm innervation and skeletal muscle 16S acetylcholinesterase deficiency. Dev. Biol. 106:296–306.
12. Yao, M.I., and F.B. Essien. 1975. DNA synthesis and the cell cycle in cultures of normal and mutant (*mdg/mdg*) embryonic mouse muscle cells. Dev. Biol. 45:166–175.

Mdh-1 locus

See *Mod-1* locus.

mdm, muscular dystrophy with myositis, recessive, Chr 2

Arose spontaneously in the C57BL/6J strain. Homozygotes can be recognized at 12 days by an unusual stiff and humpbacked posture. They become very stiff and immobile and most are dead by 2 months. Histological sections show severe muscular degeneration with acute chronic myositis. *mdm* is 26 cM from *Sd* on Chr 2 (1).

References
1. Lane, P.W. 1985. Mouse News Lett. 73:18.

Mdm-1, *-2* loci, mouse double minute-1, -2, Chr 10

These loci were identified by use of a cDNA library from a transformed line of mouse 3T3 cells. The trans-

formed cells are characterized by presence of a large number of double minute chromatin bodies thought to result from gene amplification. The sequences isolated from the library represent sequences that are amplified and overexpressed in the transformed cells. Among these sequences, two distinct genes were identified, *Mdm-1* and *Mdm-2*. Both genes were found to be on Chr 10 in the C1–C3 region by use of mouse–Chinese hamster somatic cell hybrids and by *in situ* hybridization (1).

References
1. Cahilly-Snyder, L., T. Yang-Feng, U. Francke, and D.L. George. 1987. Molecular analysis and chromosomal mapping of amplified genes isolated from a transformed mouse 3T3 cell line. Somat. Cell Mol. Genet. 13:235–244.

Mdmg-1 locus, mandibular-morphogenesis-1, Chr 17

This locus controls a difference in elongation of the dorsal edge of the condylar process of the mandible between strains C57BL/6 and the congenic B6.C-*H-2^d*/*a* strain. Comparison with two other B6.C-*H-2^d* congenic strains shows that *Mdmg-1* lies outside *H-2* on the centromeric side. C57BL/cBy carries the *Mdmg-1^b* allele which determines a shorter dorsal edge; BALB/cBy carries the *Mdmg 1^c* allele which determines a longer dorsal edge. Heterozygotes are intermediate (1).

References
1. Bailey, D.W. 1986. Mouse News Lett. 74:99.

Mdr locus

See *Mod-2r* locus.

mdx, X-linked muscular dystrophy, recessive, Chr X

Arose spontaneously in the C57BL/10 strain. Hemizygous males and homozygous females have elevated plasma levels of muscle creatine kinase and pyruvate kinase. Heterozygous females are indistinguishable from normal. No clinical symptoms appear in affected mice until about 12 months of age when the mice may develop tremors and mild incoordination (2). Affected mice show dystrophic changes in muscle histopathology beginning at 3 weeks. Massive muscle fiber destruction followed by complete regeneration begins at this time and recurs at intervals. Even after repeated episodes of muscle fiber necrosis and regeneration, fibrosis is minimal and there is no apparent fiber loss (4). The *mdx* locus may be homologous with either the Duchenne (DMD) or Emery Dreifuss (EMD) muscular dystrophy loci in man. The two loci are on different arms of the human X chromosome. In the mouse, *mdx* maps close to the *G6pd* locus, an association similar to that of EMD and G6PD in man. However, a DNA sequence on the mouse X chromosome homologous to the human DMD probe also maps close to the *mdx* locus. It is possible that there are two loci in the mouse homologous to EMD and DMD but lying close together rather than far apart as in man (1, 3).

References

1. Brockdorff, N., G.S. Cross, J.S. Cavanna, E.M.C. Fisher, M.F. Lyon, K.E. Davies, and S.D.M. Brown. 1987. The mapping of a cDNA from the human X-linked Duchenne muscular dystrophy gene to the mouse X chromosome. Nature 238:166–168.
2. Bulfield, G., W.G. Siller, P.A.L. Wight, and K.J. Moore. 1984. X chromosome-linked muscular dystrophy (*mdx*) in the mouse. Proc. Natl. Acad. Sci. USA 81:1189–1192.
3. Heilig, R., C. Lemaire, J.-L. Mandel, L. Dandolo, L. Amar, and P. Avner. 1987. Localization of the region homologous to the Duchenne muscular dystrophy locus on the mouse X chromosome. Nature 328:168–170.
4. Tanabe, Y., K. Esaki, and T. Nomura. 1986. Skeletal muscle pathology in X chromosome-linked muscular dystrophy (*mdx*) mouse. Acta Neuropathol. 69:91–95.

me locus, Chr 6

This locus has two known alleles. It is about 16 cM from *mi* near the distal end of Chr 6 (3).

me, motheaten, recessive. Arose spontaneously in the C57BL/6J strain. Homozygotes are characterized by early onset autoimmunity and severe immunodeficiency. They can often be detected at 1 day of age by neutrophilic lesions in the epidermis. A few days later, subepidermal lesions occur; they disrupt hair follicles and pigmentation and give the mice their characteristic motheaten appearance (3). Death occurs before 6 to 8 weeks, even under SPF conditions (4), and is apparently due to an unusual pneumonia which begins as early as 3 days and progresses to a massive accumulation of crystal-containing macrophages in the alveoli (14). Spleen macrophages of *me/me* mice have an accelerated rate of maturation which may contribute to the pulmonary disease (5). Serum immunoglobulins are elevated, particularly IgM, which may reach 25 times the control level at 6 weeks (7). The number of surface-Ig-bearing lymphocytes (B-cells) is greatly reduced,

and the B-cells present are of the immature rather that adult type (11). There is polyclonal activation of B-cells with an increased frequency of cells with cytoplasmic Ig (plasma cells) as early as 8 days (2, 6). The number of T-cells is normal, but several T-cell functions are defective. Natural killer cell activity is virtually absent (1). There are abnormal immunoglobulin deposits in the kidney, thymus, skin, and lung. The serum contains factors cytotoxic to thymus and skin. Bone marrow contains normal hematopoietic stem cells with a normal number of colony-forming units and capable of curing the anemia of genetically anemic *W/W^v* mice, but also contains abnormal stem cells capable of transmitting the motheaten disease to sublethally irradiated mice and to *W/W^v* mice (8, 11). Normal bone marrow cells injected into sublethally irradiated mutants increase life span to 10 months (9).

me^v, viable motheaten, recessive. Arose spontaneously in the C57BL/6J strain. The disease of homozygotes is very similar to that of *me/me* mice but progresses more slowly. Mean life span is 61 days versus 22 days for *me/me* mice. The cause of death appears to be pneumonitis as in *me/me*, but the pneumonitis develops more slowly. The life span of *me^v/me* mice is as long as that of *me^v/me^v* mice (10). The B-cells of *me^v/me^v* mice produce a B-cell maturation factor (BMF, a factor that drives B-cells to a state of active Ig secretion) different from previously described BMFs and unique to mice of this genotype (12). In addition, there is a factor in *me^v/me^v* serum that increases the activity of *me^v/me^v*-derived BMF by up to three orders of magnitude. The action of these two factors is probably responsible for the severe immunodeficiency of motheaten mice by causing the observed great increase in number of plasma cells, leaving few or no resting B-cells able to respond to specific antigens (13).

References

1. Clark, E.A., L.D. Shultz, and S.B. Pollack. 1981. Mutations in mice that influence natural killer (NK) cell activity. Immunogenetics 12:601–613.
2. Davidson, W.F., H.C. Morse III, S.O. Sharrow, and T.M. Chused. 1979. Phenotypic and functional effects of the motheaten gene on murine T and B lymphocytes. J. Immunol. 122:884–891.
3. Green, M.C., and L.D. Shultz. 1975. Motheaten, an immunodeficient mutant of the mouse. I. Genetics and pathology. J. Hered. 66:250–258.
4. Lutzner, M.A., and C.T. Hansen. 1976. Motheaten: an immunodeficient mouse with markedly less ability to survive than the nude mouse in a germfree environment. J. Immunol. 116:1496–1497.

5. McCoy, K.L., K. Nielson, and J. Clagett. 1984. Spontaneous production of colony-stimulating activity by splenic Mac-1 antigen-positive cells from autoimmune motheaten mice. J. Immunol. 132:272–276.

6. McCoy, K.L., J. Clagett, and C. Rosse. 1985. Effects of the motheaten gene on murine B-cell production. Exp. Hematol. 13:554–559.

7. Shultz, L.D., and M.C. Green. 1976. Motheaten, an immunodeficient mutant of the mouse. II. Depressed immune response and elevated serum immunoglobulins. J. Immunol. 116:936–943.

8. Shultz, L.D., and R.B. Zurier. 1978. "Motheaten", a single gene model for stem cell dysfunction and early onset autoimmunity. *In* N.R. Rose, P.E. Bigazzi, and N.L. Warner, eds., Genetic Control of Autoimmune Disease, 229–240. Elsevier North Holland, Amsterdam.

9. Shultz, L.D., C.L. Bailey, and D.R. Coman. 1983. Hematopoietic stem cell function in motheaten mice. Exp. Hematol. 11:667–680.

10. Shultz, L.D., D.R. Coman, C.L. Bailey, W.G. Beamer, and C.L. Sidman. 1984. "Viable motheaten", a new allele at the motheaten locus. I. Pathology. Am. J. Pathol. 116:179–192.

11. Sidman, C.L., L.D. Shultz, and E.R. Unanue. 1978. The mouse mutant "motheaten". II. Functional studies of the immune system. J. Immunol. 121:2399–2404.

12. Sidman C.L., J.D. Marshall, N.C. Masiello, J.B. Roths, and L.D. Shultz. 1984. Novel B-cell maturation factor from spontaneously autoimmune viable motheaten mice. Proc. Natl. Acad. Sci. USA 81:7199–7202.

13. Sidman, C.L., L.D. Shultz, and R. Evans. 1985. A serum-derived molecule from antoimmune viable motheaten mice potentiates the action of a B cell maturation factor. J. Immunol. 135:870–872.

14. Ward, J.M. 1978. Pulmonary pathology of the motheaten mouse. Vet. Pathol. 15:170–178.

mea, meander tail, recessive, Chr 4

Arose spontaneously in a moderately inbred multiple marker stock. Homozygotes have variably shortened and kinked tails. At about 2 weeks an unsteadiness of gait, often amounting to a tremor, appears. In the F2 from an outcross, some homozygotes showed partial paralysis of one or both hindlegs. Penetrance is probably incomplete, especially if mice are classified before 2 weeks of age. Homozygotes are smaller and less vigorous than normal sibs but are fertile. The anterior lobe of the cerebellum is abnormal (1).

References

1. Hollander, W.F., and K.S. Waggie. 1977. Meander tail: a recessive mutant located in chromosome 4 of the mouse. J. Hered. 68:403–406.

med locus, Chr 15

This locus has three known alleles. It is located very close to caracul (*Ca*) on Chr 15 (10).

med, motor end-plate disease, recessive. Arose as a spontaneous mutation in a stock homozygous for six recessive mutations. Homozygotes show progressive weakness of skeletal muscle beginning at 8 to 10 days and usually die within 2 weeks of onset. There is progressive atrophy of skeletal muscle, which is particularly severe in proximal limb muscles. Motor nerves show marked terminal sprouting, the sprouts growing beyond the normal confines on the motor end-plates (5). With longer survival, an increasing proportion of muscle fibers fail to show end-plate potentials or action potentials in response to nerve stimulation (6). Motor nerves show slower conduction velocity and prolonged refractory period. The non-myelinated gaps at the nodes of Ranvier are wider in *med*/*med* than in normal mice, even before the onset of clinical symptoms, and there is an increased number of Na^+ channels at the nodes (1, 9). Minced muscle from *med*/*med* mice implanted into normal hosts regenerates normally, suggesting that the genetic defect is expressed in factors extrinsic to the muscle, presumably in the motor nerves or their sheaths (12). Because functional and morphological abnormalities of Purkinje cells were found in the allele *med^{jo}*, these were looked for in *med*/*med* mice. No morphological abnormalities could be detected, but in *med*/*med* cerebella, there is a marked reduction in the high-frequency spike discharges characteristic of normal Purkinje cells (3).

med^J, motor end-plate disease-J, recessive. Occurred at the Jackson Laboratory on the same chromosome as the very closely linked gene *Ca* in a stock carrying that gene (11). Homozygotes resemble *med*/*med* mice. Heterozygotes may show mild manifestations of the disease during the first 2 weeks but they recover. Both homozygotes and heterozygotes exhibit immunological aberrations: reduced PFC response to sheep red blood cells in 14- to 16-day old mice, with reduced suppressor cell function and precocious maturation of the cytotoxic response to allogeneic cells at 21 to 23 days. The reduced PFC response disappears in older heterozygotes but remains in the few homozygotes that survive beyond 6 weeks (8).

med^{jo}, jolting, recessive. Arose in the DBA/2WyDi strain. Homozygotes show unsteady gait and tremor of the head and trunk beginning at about 3 weeks of age. They often survive to adulthood and some have bred (4). The peripheral nervous system is indistinguishable from normal and neuromuscular transmission is unimpaired (7). Anatomical and electrophysiological examination of the cerebellum of homozygotes revealed a progressive loss of Purkinje cells from the age of about 6 months and conspicuously less spontaneous extracellular electrical activity at 1 to 9 months (2).

References

1. Angaut-Petit, D., J.J. McArdle, A. Mallart, R. Bournard, M. Pinçon-Raymond, and F. Rieger. 1982. Electrophysiological and morphological studies of a motor nerve in "motor endplate disease" of the mouse. Proc. R. Soc. Lond. (Biol.) 215:117–125.
2. Boakes, R.J., J.M. Candy, D.J. Dick, J.B. Harris, S. Pollard, and J. Thompson. 1984. Abnormal electrophysiological and anatomical characteristics of cerebellar Purkinje cells in the mutant mouse *jolting*. J. Physiol. 346:27P (Abstr.).
3. Dick, D.J., R.J. Boakes, and J.B. Harris. 1985. A cerebellar abnormality in the mouse with motor end-plate disease. Neuropathol. Appl. Neurol. 11:141–147.
4. Dickie, M.M. 1965. Mouse News Lett. 32:44.
5. Duchen, L.W. 1970. Hereditary motor end-plate disease in the mouse: light and electron microscope studies. J. Neurol. Neurosurg. Psychiat. 33:238–250.
6. Duchen, L.W., and E. Stefani. 1971. Electrophysiological studies of neuromuscular transmission in hereditary "motor end-plate disease" of the mouse. J. Physiol. 212:535–548.
7. Harris, J.B., and S. Pollard. 1982. Motor end plate disease and the mutant jolting (*med^{jo}*) mouse. J. Physiol. 326:50–51P.
8. Papiernik, M., F. Rieger, S. Ezine, and M. Pinçon-Raymond. 1982. Impairment of T lymphocyte functions in mice with motor-end-plate disease. Clin. Exp. Immunol. 48:429–436.
9. Rieger, F., M. Pinçon-Raymond, A. Lambet, G. Ponzio, M. Lazdunski, and R.L. Sidman. 1984. Paranodal dysmyelination and increase in tetrodotoxin binding sites in the sciatic nerve of the motor end-plate disease (*med/med*) mouse during postnatal development. Dev. Biol. 101:401–409.
10. Searle, A.G., and C.V. Beechey. 1975. Mouse News Lett. 52:35.
11. Sidman, R.L., J.S. Cowen, and E.M. Eicher. 1979. Inherited muscle and nerve diseases in mice: a tabulation and commentary. Ann. NY Acad. Sci. 317:497–505.
12. Zachs, S.L., and M.F. Sheff. 1977. Regeneration and differentiation of minced anterior tibial muscle explants from mice with MED myopathy. Lab. Invest. 36:303–309.

Mep-1 locus, meprin activity-1, Chr 17

This locus controls activity level of meprin, a glycoprotein with potent metalloendopeptidase activity found in the brush border membrane of kidney proximal tubules. The allele *Mep-1^a* determines high activity and occurs in strains C57BL/10, C57L, C57BR/cd, C58, DBA/1, DBA/2, A, AU, 129/J, and LP; the allele *Mep-1^b* determines low activity and occurs in strains CBA/J, C3H/He, AKR/J, CE, MA/My, RF, and CHI. Meprin has little specificity and will hydrolyze small peptides as well as large proteins (1). *Mep-1* is on Chr 17, 2.1 cM distal to *H-2D* (2).

References

1. Bond, J.S., R.J. Beynon, J.F. Reckelhoff, and C.S. David. 1984. Mep-1 gene controlling a kidney metalloendopeptidase is linked to the major histocompatibility complex in mice. Proc. Natl. Acad. Sci. USA 81:5542–5545.
2. Rickelhoff, J.F., J.S. Bond, R.J. Beynon, S. Sevarirayan, and C.S. David. 1985. Proximity of the *Mep-1* gene to *H-2D* on chromosome 17 in mice. Immunogenetics 22:617–623.

Meta locus, MethA tumor antigen, Chr 12

This locus codes for a tumor-specific antigen present on the methylcholanthrene (MC)-induced BALB/c sarcoma, MethA. Tumor-specific antigens are detected as transplantation antigens causing immunization of syngeneic hosts against subsequent challenge with the same tumor. It has not usually been possible to produce serological reagents to detect these antigens, but a syngeneic antiserum that recognizes MethA antigen on tumor cells has been produced. By use of this antiserum to classify mouse–Chinese hamster somatic cell hybrids, it was shown that the *Meta* gene is located on Chr 12 (3). A protein very similar to MethA in molecular weight and amino acid composition was isolated from another MC-induced BALB/c sarcoma, CI-4. The CI-4 protein produces immunity only against CI-4 challenge but is recognized by a rabbit antibody against the MethA protein. The location of the CI-4 gene is not known, but it is possible that such tumor-specific antigens are the products of a single multigene family (1). *Meta* is located in the same chromosomal region as *Lm-1* and *H-40* (2). The antigens produced by all three loci are recognized by T-cells (1).

References

1. DuBois, G.C., L.W. Law, and E. Appella. 1982. Purification and biochemical properties of tumor-associated transplantation antigens from methylcholanthrene-induced murine sarcomas. Proc. Natl. Acad. Sci. USA 79:7669–7673.
2. O'Toole, M.M., R. Riblet, and M.J. Bosma. 1984. Identification and mapping of *Lm-1*, an *Igh*-linked locus for a murine alloantigen. Immunogenetics 20:265–275.
3. Pravtcheva, D.D., A.B. DeLeo, F.H. Ruddle, and L.J. Old. 1981. Chromosome assignment of the tumor-specific

antigen of a 3-methylcholanthrene-induced mouse sarcoma. J. Exp. Med. 154:964–977.

mg, mahogany, recessive, Chr 2

Appeared in an agouti stock of unknown origin. In homozygous condition *mg* darkens the back, ears, and tail of agouti mice and darkens the ears and tail of nonagouti mice. *a/a mg/mg* mice resemble a^e/a^e mice (1).

References

1. Lane, P.W., and M.C. Green. 1960. Mahogany, a recessive color mutation in linkage group V of the mouse. J. Hered. 51:228–230.

Mgpt-1 locus, *gpt*-insertion-1

This locus is the site of insertion of a retrovirus carrying the *gpt* gene of *Escherichia coli*. It was produced by cocultivation of cells producing the virus with strain 129 pre-implantation embryos. Both heterozygotes and homozygotes are normal. The virus is highly methylated and is not expressed (1).

References

1. Jähner, D., K. Haase, R. Mulligan, and R. Jaenisch. 1985. Insertion of the bacterial *gpt* gene into the germ line of mice by retroviral infection. Proc. Natl. Acad. Sci. USA 82:6927–6931.

mh, mocha, recessive, Chr 10

Arose spontaneously in the C57BL/6J-*pi* stock. Homozygotes are recognizable at birth by absence of visible pigment in the eyes, which darken to deep red in adults. Both yellow and black pigment in the coat are diluted. The hairs have considerably smaller and fewer melanin granules than normal. Behavior is abnormal. Homozygotes are hyperactive, many hold their heads tilted to one side, and some are unable to swim on the surface. They can hear. All homozygotes show degenerative changes in the organ of Corti, stria vascularis, and spiral ganglion, and most show abnormalities of the otoliths in the saccule and utricle. Homozygotes may be fertile but are poor breeders (1). The otolith defects can be prevented or reduced by supplementing the diet of pregnant dams with manganese or zinc. Perinatal mortality is not reduced by this treatment (2). The *mh* gene shows no recombination with grizzled (*gr*) on Chr 10 (1).

References

1. Lane, P.W., and M.S. Deol. 1974. Mocha, a new coat color and behavior mutation on chromosome 10 of the mouse. J. Hered. 65:362–364.
2. Rolfsen, R.M., and L.C. Erway. 1984. Trace metals and otolith defects in mocha mice. J. Hered. 75:158–162.

Mhv-1 locus

See *Hv-2* locus.

mi locus, Chr 6

Mutations at this locus affect one or more of three different traits: size of eye, pigmentation, and capacity for secondary bone resorption. Some combinations of alleles (Mi^{wh}/mi, Mi^{wh}/Mi^b, Mi^{wh}/mi^{ws}) show interallelic complementation, in that the heterozygote of the two alleles is more normal than either homozygote. The *mi* alleles, therefore, are probably part of a complex locus (9, 10). The locus is on Chr 6 near *wa-1* (1).

Mi^b, microphthalmia-brownish, semidominant. Appeared spontaneously in offspring of a (101 × C3H)F1 × T (multiple recessive) cross. Heterozygotes have fur of a diluted brownish color with paler ears and tail. Homozygotes are white with reduced eye pigment, are probably not microphthalmic, and have normal teeth. Both sexes are fertile. Mi^b/Mi^{wh} A/A mice are light cream with white spots and ruby eyes. Mi^b/Mi^{or} mice are indistinguishable from Mi^b/Mi^b (12).

Mi^{or}, microphthalmia-Oak Ridge, semidominant. Appeared among offspring of an irradiated male. Heterozygotes have slightly diluted fur color with pale pinkish ears and tail and often a white belly streak or head spot. Eye pigmentation is slightly reduced at birth but appears normal in adults. Homozygotes are white. Their eyes are small or absent, and incisors may fail to erupt. Homozygous males are fertile, but homozygous females have not bred (25). Mi^{or}/mi mice resemble Mi^{or}/Mi^{or}. Mi^{or}/Mi^b mice resemble Mi^b/Mi^b (26).

Mi^{wh}, white, semidominant. Found among offspring of a cross between the DBA and C57BL strains (4). $Mi^{wh}/+$ mice are somewhat lighter than *d/d* mice and have light ears. They also have a white belly spot and rarely a dorsal spot. The number of epidermal melanocytes and the rate of melanosome production within the melanocytes are reduced (21). The dermis, in contrast to that of +/+ mice, contains no melanocytes (20). Homozygotes are white with very slightly pigmented eyes. They have no pigment cells except a few in the retina (14).

Dermal–epidermal recombination grafts have shown that the defect in both $Mi^{wh}/+$ and Mi^{wh}/Mi^{wh} is expressed in the melanocytes, but in Mi^{wh}/Mi^{wh} may also be expressed in the dermis (20). Homozygotes are slightly microphthalmic but have normal skeletons. The microphthalmia is due to increased rate of proliferation of the outer (pigment) layer of the retina, which leads to failure of the choroid fissure to close normally (19). Measurements of some neural crest derivatives other than pigment cells (spinal ganglia, adrenal medulla, dermis) have shown them to be smaller in Mi^{wh}/Mi^{wh} than in normal controls (10). $Mi^{wh}/+$ mice have abnormalities of the inner ear in both the cochlear and vestibular portions (2). Mi^{wh}/mi mice closely resemble Mi^{wh}/Mi^{wh} mice, but their eyes are slightly more pigmented and more nearly normal in size (3, 6). Konyukhov and Osipov (10) have shown that these mice are also more normal with respect to width of the dermis and size of the spinal ganglia and adrenal medulla than either homozygote.

mi, microphthalmia, semidominant. Found among descendants of an irradiated male (7). Heterozygotes have eyes with less iris pigment than normal and often have white spotting on the belly, head, and tail. Homozygotes are devoid of pigment in the eyes and coat, and the eyes are small (5). There is a deficiency of mast cells in the gut and spleen (27). Most homozygotes die at about weaning age, but they may live several months longer. There is a deficiency of secondary bone resorption (osteopetrosis) with failure of the incisors to erupt (5). The eye abnormality traces to hyperplasia of the pigment layer of the retina (19). Some neural crest derivatives (spinal ganglia, adrenal medulla, dermis) are smaller than in normal controls (10). In Mi^{wh}/mi mice some complementation occurs with respect to these effects and to size and pigmentation of the eyes (3, 10).

Numerous investigations of the cause of the osteopetrosis are summarized in Marks and Walker (16) and Marks (15). The defect can be transmitted with transplanted spleen or bone marrow cells, probably through progenitors of osteoclasts (17, 29). These cells from *mi/mi* mice show a generalized defect in function and hormone response (22) and a fusion disability. Normal osteoclasts are multinucleate and have ruffled borders. They are formed by fusion of embryonic mononuclear precursors. In *mi/mi* mice, embryonic precursors do not fuse and do not develop ruffled borders. They are also not capable of forming multinucleate foreign-body giant cells (28). It is likely that the osteopetrosis is due to a basic defect in osteoclasts, but how this effect is related to the effects on pigmentation and the eye is not known.

mi^{bw}, black-eyed white, recessive to wild type. Arose spontaneously in the C3H inbred strain. Homozygotes have white coats and black eyes. Occasionally a few pigmented hairs may be found on the back (11). Melanoblasts cannot be demonstrated in the skin of embryos of these mice by transplantation to the eyes of albino hosts (13). Mi^{wh}/mi^{bw} mice have pigmented eyes and are white with some pale yellow spots which become white in the adult (23).

mi^{di}, microphthalmia-defective iris. Found among progeny of a male treated with ENU. Homozygotes have white coats, mildly defective bone resorption, and small eyes with very little pigmentation. The iris and retina are abnormal. Heterozygotes appear normal except for slightly reduced choroidal pigment. mi^{di}/mi mice are white with small eyes and small teeth; mi^{di}/Mi^{wh} mice are white with only slightly affected irises (30).

mi^{rw}, red-eyed white, recessive. Arose spontaneously in the CBA/J strain. Homozygotes have some pigmentation around the neck and have small red eyes. mi^{rw}/mi mice are all white with small or absent eyes; mi^{rw}/mi^{sp} mice are fully pigmented except for a white belly spot; and mi^{rw}/Mi^{wh} are mostly white with tan spots and red eyes (24).

mi^{sp}, mi-spotted. Found in a C57BL/6J-Mi^{wh} stock. Homozygotes and $mi^{sp}/+$ mice are not detectably different from $+/+$ mice in color but have slightly less tyrosinase activity in the skin. Mi^{wh}/mi^{sp} mice are light yellow with dorsal and ventral white spots and pigmented eyes. Medullary pigment granules in the hair show much clumping and are yellowish brown. mi/mi^{sp} mice are white with some pigmentation in the eyes and some flecks of pigmented hair on the back. All these combinations are viable and fertile (31).

mi^{ws}, white spot, semidominant. Arose in the C57BL/6 strain. $mi^{ws}/+$ mice have a white diamond on the belly. mi^{ws}/mi^{ws} mice are white with unpigmented eyes, they are viable and fertile, and some have microphthalmia. mi^{ws}/mi^{bw} mice are black-eyed white (8, 18). Mi^{wh}/mi^{ws} mice have more pigmentation and more normal eyes than homozygotes of either allele, having dark eyes and a patchy distribution of yellowish gray pigment in the coat (9).

References

1. Bunker, H, and G.D. Snell. 1948. Linkage of white and waved-1. J. Hered. 39:28.
2. Deol, M.S. 1967. The neural crest and acoustic ganglion. J. Embryol. Exp. Morphol. 17:533–541.
3. Deol, M.S. 1973. The role of the tissue environment in the

expression of spotting genes in the mouse. J. Embryol. Exp. Morphol. 30:483–489.

4. Grobman, A.B., and D.R. Charles. 1947. Mutant white mice. A new dominant autosomal mutant affecting coat color in *Mus musculus*. J. Hered. 38:381–384.

5. Grüneberg, H. 1948. Some observations on the microphthalmia gene in the mouse. J. Genet. 49:1–13.

6. Grüneberg, H. 1953. The relations of microphthalmia and white in the mouse. J. Genet. 51:359–362.

7. Hertwig, P. 1942. Neue Mutationen und Koppelungsgruppen bei der Hausmaus. Z. Indukt. Abstammungs-Vererbungsl. 80:220–246.

8. Hollander, W.F. 1964. Mouse News Lett. 30:29.

9. Hollander, W.F. 1968. Complementary alleles at the mi–locus in the mouse. Genetics 60:189 (Abstr.).

10. Konyukhov, B.V., and V.V. Osipov. 1968. Interallelic complementation of microphthalmia and white genes in mice. Soviet Genet. 4:1457–1465.

11. Kreitner, P.C. 1957. Linkage studies in a new black–eyed–white mutation in the mouse (not *W*). J. Hered. 48:300–304.

12. Larsen, M.M. 1966. Mouse News Lett. 34:41.

13. Markert, C.L. 1960. Biochemical embryology and genetics. *In* N. Kaliss, ed., Symposium on Normal and Abnormal Differentiation and Development. Natl. Cancer Inst. Monogr. 2:3–17.

14. Markert, C.L., and W.K. Silvers. 1956. The effect of genotype and cell environment on melanoblast differentiation in the house mouse. Genetics 41:429–450.

15. Marks, S.C. Jr. 1977. Pathogenesis of osteopetrosis in the microphthalmic mouse: reduced bone resorption. Am. J. Anat. 149:269–276.

16. Marks, S.C. Jr., and D.G. Walker. 1976. Mammalian osteopetrosis—a model for studying cellular and humoral factors in bone resorption. *In* G.H. Bourne, ed., The Biochemistry and Physiology of Bone, Vol. IV, 227–301. Academic Press, New York.

17. Marks, S.C. Jr., and D.G. Walker. 1981. The hematogenous origin of osteoclasts: experimental evidence from osteopetrotic (microphthalmic) mice treated with spleen cells from beige mouse donors. Am. J. Anat. 161:1–10.

18. Miller, D.S. 1963. Coat color and behavior mutations in inbred mice under chronic low level γ-radiation. Radiat. Res. 19:184–185.

19. Packer, S.O. 1967. The eye and skeletal effects of two mutant alleles at the microphthalmia locus of *Mus musculus*. J. Exp. Zool. 165:21–45.

20. Pratt, B.M. 1982. Site of gene action of the white allele (*Mi^{wh}*) of the microphthalmia locus: a dermal–epidermal recombination study. J. Exp. Zool. 220:93–101.

21. Pratt, B.M. 1983. Effects of microphthalmic white (*Mi^{wh}*) on the number and function of cutaneous melanocytes. J. Exp. Zool. 225:329–335.

22. Raisz, L.G., H.A. Simmons, S.C. Gworek, and G. Eilon. 1977. Studies on congenital osteopetrosis in microphthalmic mice using organ cultures: impairment of bone resorption in response to physiological stimulators. J. Exp. Med. 145:857–865.

23. Schaible, R.H. 1963. Developmental genetics of spotting patterns in the mouse. PhD Dissertation, Iowa State Univ., Ames. 225 p.

24. Southard, J.L. 1974. Mouse News Lett. 51:23.

25. Stelzner, K.F. 1964. Mouse News Lett. 31:40.

26. Stelzner, K.F. 1966. Mouse News Lett. 34:41.

27. Stevens, J., and J.F. Loutit. 1982. Mast cells in spotted mutant mice (*W*, *Ph*, *mi*). Proc. R. Soc. Lond. (Biol.) 215:405–409.

28. Thesingh, C.W., and J.P. Scherft. 1985. Fusion disability of embryonic osteoclast precursor cells and macrophages in the microphthalmic osteopetrotic mouse. Bone 6:43–52.

29. Walker, D.G. 1975. Spleen cells transmit osteopetrosis in mice. Science 190:785–787.

30. West, J.D., G. Fisher, J.F. Loutit, M.J. Marshall, N.W. Nisbet, and V.H. Perry. 1986. A new allele of microphthalmia induced in the mouse: microphthalmia-defective iris (*mi^{di}*). Genet. Res. 46:309–324.

31. Wolfe, H.G., and D.L. Coleman. 1964. Mi-spotted: a mutation in the mouse. Genet. Res. 5:432–440.

Micrl locus, microwave-induced increase in complement-receptor B-cells, Chr 5

This locus controls susceptibility to induction by a single exposure to 2450-MHz microwaves of a transitory increase in percentage of B lymphocytes bearing the receptor for the third component of complement. Mice of strains AKR, CBA/J, CBA/N, and C3H/He are responders (*Micrl^r*); mice of strains A/J, BALB/c, C57BL/6, C58, C57BL/10, and C3HFeB/Fe are nonresponders (*Micrl^n*). Heterozygotes are responsive. Examination of 10 BXH recombinant inbred strains and two C57BL/6.C3H congenic strains showed that *Micrl* is located on Chr 5 between *Pgm-1* and *rd* (1).

References

1. Schagel, C.J., and A. Ahmed. 1982. Evidence for genetic control of microwave-induced augmentation of complement receptor-bearing B lymphocytes. J. Immunol. 129:1530–1533.

Mis-1 locus

See *Pvt-1* locus.

mk, microcytic anemia, recessive, Chr 15

Found among descendants of a cross between C57BL/6J and DBA/2J. Homozygotes are recognizable by their pallor at birth. The red cells are smaller than normal and hypochromic from 15 days of gestation onward, but they are present in normal or greater than normal numbers (5). Viability and fertility may be reduced and, on some genetic backgrounds, skin lesions occur early in life (4). The anemia appears to be due to a generalized impairment of cellular iron uptake involving transfer of iron from the intestinal lumen to the mucosa as well as from plasma to erythroblasts (1–3).

References

1. Bannerman, R.M., J.A. Edwards, M. Kreimer-Birnbaum, E. McFarland, and E.S. Russell. 1972. Hereditary microcytic anemia of the mouse; studies in iron distribution and metabolism. Br. J. Haematol. 23:235–245.
2. Edwards, J.A., and J.E. Hoke. 1975. Red cell iron uptake in hereditary microcytic anemia. Blood 46:381–388.
3. Harrison, D.E. 1972. Marrow transplantation and iron therapy in mouse hereditary microcytic anemia. Blood 40:893–901.
4. Russell, E.S., E.C. McFarland, and E.L. Kent. 1970. Low viability, skin lesions, and reduced fertility associated with microcytic anemia in the mouse. Transplant. Proc. 2:144–151.
5. Russell, E.S., D.J. Nash, S.E. Bernstein, E.L. Kent, E.C. McFarland, S.M. Matthews, and M.S. Norwood. 1970. Characterization and genetic studies of microcytic anemia in house mouse. Blood 35:838–850.

mld

See *shi* locus.

Mls locus (formerly *M*), minor MLC-stimulating, Chr 1

This locus controls presence on B lymphocytes and on macrophages of lymphocyte-activating determinants which stimulate strong proliferative reactions of T lymphocytes in mixed lymphocyte cultures (MLC) of cells from *H-2*-compatible strains. Four alleles are known: Mls^a in strains DBA/2, AKR, and others; Mls^b in strains CBA/H, C57BL/6, and others; Mls^c in strains C3H/He, A/J, and SJL; and Mls^d in CBA/J. Mls^a and Mls^d are strongly stimulatory, Mls^c is weakly stimulatory, and Mls^b is non-stimulatory (3). The mixed lymphocyte reaction in *Mls*-disparate cultures is characteristically one way: determinants of all three other alleles stimulate Mls^b, and Mls^d determinants stimulate Mls^a and Mls^c (4). There is, however, conflicting evidence that Mls^b determinants can stimulate Mls^a and Mls^d cells and that Mls^c can stimulate Mls^a and Mls^d (2). *Mls*-locus determinants probably also act *in vivo* in causing proliferation of T lymphocytes injected into irradiated *Mls*-disparate hosts (8). Tissue grafting is affected only weakly at best by *Mls* incompatibility, and

Mls determinants are not serologically detectable. The cells bearing the Mls determinants as determined in MLCs are present at 4 to 5 weeks of age and cannot be detected from birth to 2 weeks. They are adult-type B lymphocytes with a high density of surface immunoglobulin and complement (C3) receptor (1). The Ia determinants of the *H-2* complex must be centrally involved in the unidirectional T-cell response to Mls^a and Mls^d cells, since anti-Ia antibodies readily inhibit the response (6). B-cells bearing the strongly stimulatory Mls^a and Mls^d alleles also cause increased proliferative response of selected foreign protein-specific Ia-restricted T-cell lines derived from Mls^b strains. This and other evidence led Katz and Janeway (7) to propose that the *Mls* locus regulates the expression of a cell surface molecule that promotes interaction of the receptor on T-cells for antigen-self-Ia with the relatively non-polymorphic region of Ia glycoprotein on the antigen-presenting cell. The level of expression of the molecule varies between *Mls*-disparate strains, being greater in the more strongly stimulating Mls^a and Mls^d strains and weaker in the weakly stimulating strains. *Mls* is on Chr 1 near the distal end (5).

References

1. Ahmed, A., I. Scher, and K.W. Sell. 1977. Studies on non-H-2-linked lymphocyte-activating determinants. IV. Ontogeny of the Mls product on murine B cells. Cell. Immunol. 30:122–134.
2. Click, R.E., A.M. Adelmann, and M.M. Azar. 1985. Immune response *in vitro*. XIV. MLR detectability of Mls^a-, Mls^b-, Mls^c-, and Mls^d-encoded products. J. Immunol. 134:2948–2952.
3. Festenstein, H. 1974. Pertinent features of *M* locus determinants including revised nomenclature and strain distribution. Transplantation 18:555–557.
4. Festenstein, H. 1976. The Mls system. Transplant. Proc. 8:339–342.
5. Festenstein, H., C. Bishop, and B.A. Taylor. 1977. Location of *Mls* locus on mouse chromosome 1. Immunogenetics 5:357–361.
6. Janeway, C.A. Jr. and M.E. Katz. 1985. The immunobiology of the T cell response to Mls-locus-disparate stimulator cells. I. Unidirectionality, new strain combinations, and the role of Ia antigens. J. Immunol. 134:2057–2063.
7. Katz, M.E., and C.A. Janeway Jr. 1985. The immunobiology of T cell responses to Mls-locus-disparate stimulator cells. II. Effects of Mls-locus-disparate stimulator cells on cloned, protein antigen-specific, Ia-restricted T cell clones. J. Immunol. 134:2064–2070.
8. Sprent, J., and H. von Boehmer. 1976. Activation of T lymphocytes to M locus determinants *in vivo*. I. Quantitation of T cell proliferation and migration into thoracic duct lymph. Eur. J. Immunol. 6:352–358.

Mlv-1

Mlv-1 locus, murine leukemia virus-1

This locus controls expression of group-specific antigen (gs-AG) of type C RNA tumor virus. The allele $Mlv-1^a$ determines presence of the antigen on spleen cells and occurs in strains DBA/2J and C57BL/10.D2(58N); the allele $Mlv-1^b$ determines absence of the antigen and occurs in strains C57BL/10Sn and C57BL/10.129(21M). $Mlv-1^b$, determining absence of the antigen, is dominant to $Mlv-1^a$, suggesting that Mlv-1 is a regulatory locus, with $Mlv-1^b$ acting to inhibit expression of the viral antigen. Presence of gs-AG in $Mlv-1^a/Mlv-1^a$ mice may be unaccompanied by detectable complete replicating leukemia virus (1).

References

1. Taylor, B.A., H. Meier, and R.J. Huebner. 1973. Genetic control of the group-specific antigen of murine leukemia virus. Nature New Biol. 241:184–186.

Mlvi-1 to Mlvi-3 loci

Mlvi-1, -2 loci, Moloney leukemia virus integration site-1, -2, Chr 15
Mlvi-3 locus, Moloney leukemia virus integration site-3

These loci are chromosomal sites at which Moloney virus integrates in cells that develop into thymic lymphomas. In the rat and mouse there are three such sites. No genetic variants are known in the mouse. The cellular sequences of two of the sites, Mlvi-1 and Mlvi-2, have been mapped to Chr 15 by use of mouse–Chinese hamster somatic cell hybrids. The two loci are on separate parts of the chromosome with Mlvi-2 probably proximal to Mlvi-1 (1, 2).

References

1. Kozak, C.A., P.G. Strauss, and P.N. Tsichlis. 1985. Genetic mapping of a cellular DNA region involved in induction of thymic lymphomas (Mlvi-1) to mouse chromosome 15. Mol. Cell. Biol. 5:894–897.
2. Tsichlis, P.N., P.G. Strauss, and C.A. Kozak. 1984. Cellular DNA region involved in induction of thymic lymphomas (Mlvi-2) maps to chromosome 15. Mol. Cell. Biol. 4:997–1000.

Mmv-1 to Mmv-13 loci, MCF murine leukemia virus-1 to -13

These loci (Table 2.5) are sites of integration of the proviral genomes of mink-cell focus-forming (MCF) leukemia viruses in the BALB/c strain. They were recognized in HindIII digested genomic DNA hybridized to a probe specific for the env gene of MCF vir-

Table 2.5

Locus	Fragment size (kb)	Chromosome
Mmv-1	27.5	5
Mmv-2	22.9	3
Mmv-3	20.8	7
Mmv-4	15.1	7
Mmv-5	13.7	1
Mmv-6	13.7	1
Mmv-7	12.3	1
Mmv-8	10.8	11
Mmv-9	10.4	1
Mmv-10	8.9	1
Mmv-11	6.6	11
Mmv-12	6.3	3
Mmv-13	4.2	11

uses. The chromosomal location of the loci was found by use of mouse–Chinese hamster somatic cell hybrids made with BALB/c mouse cells (1).

References

1. Hoggan, M.D., R.R. O'Neill, and C.A. Kozak. 1986. Nonecotropic murine leukemia viruses in BALB/c and NFS/N mice: characterization of the BALB/c Bxv-1 provirus and the single NFS endogenous xenotrope. J. Virol. 60:980–986.

mn, miniature, recessive, Chr 15

Appeared in the descendants of an outcross of AKR to a stock at Columbia University. Homozygotes are smaller than their normal sibs at birth and have a dorsoventral flattening of the skull. They grow more slowly and are only about one-fourth to one-third normal size at 75 days. They do not seem to differ much from normal in body proportions or in other anatomical features or behavior, but mortality is high at all times up to 2 months when 96 per cent were dead. A few that lived to the age of 4 months all proved infertile (1).

References

1. Bennett, D. 1961. Miniature, a new gene for small size in the mouse. J. Hered. 52:95–98.

Mnd, motoneuron degeneration, dominant

Found in the C57BL/6.KB2/Rn congenic inbred strain, in which Mnd was homozygous. Onset of the disease occurs between 5 and 11 months. It is characterized by hindlimb weakness and ataxia which progresses to severe spastic paralysis of all limbs, with death usually by 9 to 14 months. Males and females are equally affected and both are fertile, although breeding efficiency is reduced. Heterozygotes resemble homozygotes in age of onset of mild symptoms and incidence of moderate symptoms. Histological examination of the nervous system of affected animals shows degeneration of the upper and lower motoneurons of the spinal cord and cranial nerves and of some areas of the brain (1).

References

1. Messer, A., and L. Flaherty. 1986. Autosomal dominance in a late-onset motor neuron disease in the mouse. J. Neurogenet. 3:345–355.

Mo locus, Chr X

Mutations at the *Mo* locus are very common. Numerous mutations similar in phenotype have been reported, and although allelism is difficult to prove, it is likely that most of them are mutations at this locus (23, 36, 40). Mutants at the *Mo* locus cause varying degrees of defective pigmentation, wavy vibrissae and coat, slight skeletal abnormalities, tremor and incoordination, and defective collagen and elastin. Some alleles cause death of hemizygous males *in utero*. Heterozygous females have mottled coats. The variegated phenotype is due to the inactivation of one or the other of the X chromosomes, the mutant gene being active in the light patches and the wild-type gene in the dark patches (24). The basic cause of the abnormalities is thought to be a defect in copper transport or copper binding which affects activity of many copper-dependent or copper-containing enzymes (16, 31). The *Mo* locus may be the homologue of the X-linked locus causing Menkes' kinky hair syndrome in man.

Mo, mottled, semidominant. Arose spontaneously in a cross segregating for several color factors (11). Heterozygous females have irregular patches of full colored and very lightly colored fur over the whole coat. Patches with well defined edges very rarely cross the middorsal or midventral line. Vibrissae are curly, but the coat is not waved. Viability is reduced, some heterozygotes dying prenatally and some postnatally. Survivors are usually fertile. Hemizygous males die *in utero* at about 11 days of gestation with no visible abnormality (8).

Mo^{blo}, blotchy, semidominant. Arose spontaneously. Mo^{blo} shows no recombination with Mo^{br} and is therefore considered to be an allele at the *Mo* locus (13). Heterozygous females have irregular patches of light-colored fur. Expression is occasionally poor at weaning age but is complete in adulthood. Viability and fertility are normal. Hemizygous males and homozygous females are light all over with no blotching, are usually

small, and occasionally have deformed hindlegs. Vibrissae are kinked at birth but straight at weaning. Hemizygotes and homozygotes have reduced viability and many are infertile (34). Most hemizygotes and homozygotes have defective elastin in the aorta and usually die with rupture of the aorta (1, 12). Hemizygous males have enlarged air spaces in the lung (emphysema), probably because of defective elastin and collagen (10). Skin collagen and aortic elastin have defective crosslinking at the step at which lysine residues are converted to aldehydes (32). This defect is probably due to a deficiency in activity of the copper-dependent enzyme, lysyl oxidase, known to occur in the skin (33), in the lung, and in cultured fibroblasts of blotchy mice (37). Hemizygous males have a deficiency of noradrenalin in the brain (15), probably caused by defective activity of the enzyme dopamine-beta-hydroxylase, which may be due to a deficiency of copper shown to exist in the brain (16). Copper absorption from the gut and hepatic copper concentration are reduced to 64 per cent and 56 per cent of normal, respectively (38). In heterozygotes with the X-linked gene tabby (*Ta*), the effect of Mo^{blo} on coat color was shown to be expressed in the same cell population as *Ta*, i.e. the skin, demonstrating that the mutant gene acts through an effect of hair follicles on pigmentation of the hairs (25). Mo^{blo}/Mo^{br} females are viable; Mo^{blo}/Mo^{dp} and Mo^{blo}/Mo^{to} females die by 15 days of age (12).

Mo^{br}, brindled, semidominant. Arose spontaneously in the C57BL inbred strain (11). Heterozygous females are very similar to *Mo*/+ females in appearance but have normal viability. They have curly vibrissae, but the coat is not noticeably waved. Hemizygous males are almost devoid of pigment except in the eyes and ears. The vibrissae are strongly curled, and the coat is wavy. The abnormalities of hair structure have been described by Grüneberg (14). Males usually die when 2 weeks old, but a few have lived and been fertile. They have a behavioral abnormality consisting of a slight tremor, uncoordinated gait, and clasping of the hindfeet when held up by the tail (8, 9). Homozygous females and also Mo^{br}/Mo and Mo^{br}/Mo^{to} females are identical in phenotype to hemizygous males and die at the same age (9, 12). Mo^{br}/Mo^{blo} females are viable. Histological examination of the brain of brindled males shows widespread neuronal degeneration in the cerebral cortex and thalamic nuclei and scattered degeneration in the cerebellum (42). In contrast to mice bearing other *Mo* alleles, brindled mice have no aortic lesions (12) and no defect in crosslinking of collagen and elastin. Brindled males do, however, have a 30 to 40 per cent reduction in lysyl oxidase activity in skin (33). They have reduced synthesis of noradrenalin by dopamine-beta-hydroxy-lase in the brain and peripheral nervous system (19), and decreased activity of cytochrome C oxidase and superoxide dismutase (17). These enzymes are all copper-dependent or copper-containing, and their reduced activity may be due to defective copper transport in cells of all affected tissues (7, 15, 16, 35). Brindled males have defective placental transport and defective intestinal absorption of copper. Concentration of copper is high in gut mucosa, kidney, and testis and low in liver, brain, plasma, and most other organs (2, 27). Concentration of the copper-binding protein metallothionien in various tissues is correlated with the concentration of copper (18). Parenteral injection of copper at 7 to 10 days of age has a striking therapeutic effect; it prevents tremor and early death, allows normal pigmentation, improves growth, produces normal concentrations of copper in previously deficient organs except liver, produces normal activity of some copper-dependent enzymes (26, 41), and prevents neuronal degeneration (29). In heterozygotes with tabby (*Ta*), *Mo*^{br} resembles *Mo*^{blo}, and thus appears to act on coat color through an effect in hair follicles (25).

Mo^{dp}, dappled, semidominant. Arose in a low-dosage γ-irradiation experiment. Heterozygous females are similar in color and curliness of vibrissae to *Mo*/+ females. Some have clubbing of the forefeet at birth or, at weaning, a tendency to walk on the dorsal surfaces of the hindfeet. With age, calcified lumps may appear in the region of the periosteum of the thoracic and lumbar vertebrae. Hemizygous males die at about 17 days of gestation. They show bending and thickening of the ribs and distortion of the pectoral and pelvic girdles and limb bones (30). About 10 per cent of *Mo*^{dp}/+ females show lesions of the aorta. No homozygous females have been produced, but *Mo*^{dp}/*Mo*^{blo} females live to about 15 days of age and *Mo*^{dp}/*Mo*^{to} females die *in utero* (12). The effect of *Mo*^{dp} on pigmentation, in contrast to that of *Mo*^{blo} and *Mo*^{br}, has been shown to be expressed in a different cell population from *Ta* and in the same cell population as pink-eyed dilution (*p*), i.e. melanocytes (25). Evidence from chimeras shows that *Mo*^{dp} acts independently in melanocytes and hair follicles (5).

Mo^{ms}, mosaic, semidominant. Arose spontaneously in an outbred stock. This mutant has not been tested for allelism with other *Mo* alleles, but it is similar to them genetically and phenotypically and should probably be regarded as an *Mo* allele. In hemizygous males the coat color is considerably diluted and the vibrissae are curly. Affected males die when 2 to 3 weeks old. Heterozy-

gous females have normal-colored and mutant hair arranged in an irregular pattern of transverse bars. They have curly vibrissae which become straight in adults (21). Hemizygous males have abnormal hair structure. They have a neurological disorder consisting of progressive paresis of the hindlimbs beginning at 8 days. As in other mutants at the *Mo* locus, they have defective copper transport as shown by impaired absorption of copper from the intestine, low copper levels in brain and liver, low serum ceruloplasmin activity, and low activity in the eyeball of o-diphenol oxidase, a copper-dependent enzyme involved in the synthesis of melanin (39).

Mo^{to}, tortoiseshell, semidominant. Arose spontaneously in an obese stock. Because of its interaction with other alleles at the locus (see below) and because it shows about the same linkage relation with *Bn* as *Mo* (22), it is considered to be an allele at the *Mo* locus. Heterozygous females resemble *Mo*/+ females in color. The vibrissae are slightly wavy and the coat has a slightly silky texture. The distribution of patches of wild-type and white color is different from the distribution of patches of wild-type and wavy hair, suggesting that *Mo*^{to} probably acts independently in melanoblasts and hair follicles (28). Most hemizygous males die before birth (6), but a few are stillborn (12). Using as parent a male presumed to be chimeric for *Mo*^{to}/Y and +/Y, Grahn *et al.* (12) showed that *Mo*^{to}/*Mo*^{to} and *Mo*^{to}/*Mo*^{dp} females die *in utero*, and that *Mo*^{to}/*Mo*^{blo} and *Mo*^{to}/*Mo*^{br} females reach term and die by 15 days of age. The elastic lamina of the aorta is defective in stillborn *Mo*^{to} males and in 40 per cent of *Mo*^{to}/+ females.

Mo^{vbr}, viable brindled, semidominant. Found among mice in a selection experiment with Cattanach's translocation *T(X;7)1Ct*, now known as *Is(In7;X)1Ct* (3). Heterozygous females have a mottled coat of colored and whitish hair in a pattern of somewhat transversely arranged stripes. The vibrissae are curled, and the coat is slightly rippled. Hemizygous males are white, have somewhat reduced viability, and are sterile (4). They have aortic aneurysms, reduced breaking strength of skin, and defective crosslinking of skin collagen and aortic elastin (32). Lysyl oxidase activity in skin is markedly lower than in controls (33). There is decreased level of norepinephrine in the brain as in brindled and blotchy males (20).

References

1. Andrews, E.J., W.J. White, and L.P. Bullock. 1975. Spontaneous aortic aneurysms in blotchy mice. Am. J. Pathol. 78:190–210.

2. Camarkis, J., J.R. Mann, and D.M. Danks. 1979. Copper metabolism in mottled mouse mutants. Copper concentrations in tissues during development. Biochem. J. 180:597–604.

3. Cattanach, B.M., and J.H. Isaacson. 1968. Mouse News Lett. 38:17.

4. Cattanach, B.M., C.E. Pollard, and J.N. Perez. 1969. Controlling elements in the mouse X-chromosome. I. Interaction with the X-linked genes. Genet. Res. 14:223–235.

5. Cattanach, B.M., H.G. Wolfe, and M.F. Lyon. 1972. A comparative study of the coats of chimaeric mice and those heterozygous for X-linked genes. Genet Res. 19:213–228.

6. Dickie, M.M. 1954. The tortoise shell house mouse. J. Hered. 45:158, 190.

7. Evans, G.W., and B.L. Reis. 1978. Impaired copper homeostasis in neonatal male and adult female brindled (*Mo^{br}*) mice. J. Nutr. 108:554–560.

8. Falconer, D.S. 1953. Total sex linkage in the house mouse. Z. Indukt. Abstammungs-Vererbungsl. 85:210–219.

9. Falconer, D.S. 1956. Mouse News Lett. 15:24.

10. Fisk, D.E., and C. Kuhn. 1976. Emphysema-like changes in the lungs of the blotchy mouse. Am. Rev. Resp. Dis. 113:787–797.

11. Fraser, A.S., S. Sobey, and C.C. Spicer. 1953. Mottled, a sex-modified lethal in the house mouse. J. Genet. 51:217–221.

12. Grahn, D., R.J.M. Fry, and K.F. Hamilton. 1969. Genetic and pathological analysis of the sex-linked allelic series, mottled, in the mouse. Genetics 61:s22 (Abstr.).

13. Grahn, D., R.J.M. Fry, and K. Allen. 1972. Mouse News Lett. 47:20.

14. Grüneberg, H. 1969. Threshold phenomena versus cell heredity in the manifestation of sex-linked genes in mammals. J. Embryol. Exp. Morphol. 22:145–179.

15. Hunt, D.M. 1974. Primary defect in copper transport underlies mottled mutants in the mouse. Nature 249:852–854.

16. Hunt, D.M. 1976. A study of copper treatment and tissue copper levels in the murine copper deficiency, mottled. Life Sci. 19:1913–1920.

17. Hunt, D.M. 1977. Catecholamine biosynthesis and the activity of a number of copper-dependent enzymes in the copper-deficient mottled mouse mutants. Comp. Biochem. Physiol. 57:79–83.

18. Hunt, D.M., and R. Clarke. 1983. Metallothionein and the development of the mottled disorder in the mouse. Biochem. Genet. 21:1175–1194.

19. Hunt, D.M., and D.R. Johnson. 1972. An inherited deficiency in noradrenalin biosynthesis in the brindled mouse. J. Neurochem. 19:2811–2819.

20. Hunt, D.M., and D.R. Johnson. 1972. Aromatic acid metabolism in brindled (*Mo^{br}*) and viable brindled (*Mo^{vbr}*), two alleles at the mottled locus in the mouse. Biochem. Genet. 6:31–40.

21. Krzanowska, H. 1966. Mouse News Lett. 35:44.

22. Lane, P.W. 1960. Mouse News Lett. 23:36.

23. Lyon, M.F. 1960. A further mutation of the mottled type in the house mouse. J. Hered. 51:116–121.

24. Lyon, M.F. 1961. Gene action in the X-chromosome of the mouse (*Mus musculus* L.). Nature 190:372–373.

25. Lyon, M.F. 1970. Genetic activity of sex chromosomes in somatic cells of mammals. Phil. Trans. R. Soc. Lond. (Biol.) 259:41–52.

26. Mann, J.R., J. Camarkis, D.M. Danks, and E.G. Walleczek. 1979. Copper metabolism in mottled mouse mutants. Copper therapy of brindled (*Mo^{br}*) mice. Biochem. J. 180:605–612.

27. Mann, J.R., J. Camarkis, and D.M. Danks. 1979. Copper metabolism in mottled mouse mutants. Distribution of ^{64}Cu in brindled (*Mo^{br}*) mice. Biochem. J. 180:613–619.

28. Mintz, B. 1970. Gene expression in allophenic mice. Symp. Int. Soc. Cell Biol. 9:14–42.

29. Nagara, H., K. Yajima, and K. Suzuki. 1981. The effect of copper supplementation on the brindled mouse, a clinico-pathological study. J. Neuropath. Exp. Neurol. 40:428–446.

30. Phillips, R.J.S. 1961. "Dappled", a new allele at the mottled locus in the house mouse. Genet. Res. 2:290–295.

31. Prins, H.W., and C.J.A. Van den Hamer. 1979. Primary biochemical defect in copper metabolism in mice with a recessive X-linked mutation analogous to Menkes' disease in man. J. Inorg. Biochem. 10:19–27.

32. Rowe, D.W., E.B. McGoodwin, G.R. Martin, M.D. Sussman, D. Grahn, B. Faris, and C. Fransblau. 1974. A sex-linked defect in the cross-linking of collagen and elastin associated with the mottled locus in mice. J. Exp. Med. 139:180–192.

33. Rowe, D.W., E.B. McGoodwin, G.R. Martin, and D. Grahn. 1977. Decreased lysyl oxidase activity in the aneurysm-prone, mottled mouse. J. Biol. Chem. 252:939–942.

34. Russell, L.B. 1960. Mouse News Lett. 23:58.

35. Sayed, A.K., J.A. Edwards, and R.M. Bannerman. 1978. ^{64}Cu uptake by cultured fibroblasts from *Mo^{br}*/Y mice. Fed. Proc. 37:746 (Abstr.).

36. Schröder, J.H. 1975. *Mottled Neuherberg* (Mo^{br}), a new male-lethal coat colour mutation of the house mouse (*Mus musculus*). Theor. Appl. Genet. 46:135–142.

37. Starcher, B.C., J.A. Madaras, and A.S. Tepper. 1977. Lysyl oxidase activity in lung and fibroblasts from mice with hereditary emphysema. Biochem. Biophys. Res. Comm. 78:706–712.

38. Starcher, B., J.A. Madaras, D. Fisk, E.F. Perry, and C.H. Hill. 1978. Abnormal cellular copper metabolism in the blotchy mouse. J. Nutr. 108:1229–1233.

39. Styrna, J. 1977. Analysis of causes of lethality in mice with *Ms (mosaic)* gene. Genet. Polon. 18:61–79.

40. Welshons, W.J., and L.B. Russell. 1959. The Y-chromosome as the bearer of male determining factors in the mouse. Proc. Natl. Acad. Sci. USA 45:560–566.

41. Wenk, G., and K. Suzuki. 1983. Congenital copper

deficiency: copper therapy and dopamine-β-hydroxylase activity in the mottled (brindled) mouse. J. Neurochem. 41:1648–1652.

42. Yajima, K., and K. Suzuki. 1979. Neuronal degeneration in the brain of the brindled mouse—a light microscopic study. J. Neuropathol. Exp. Neurol. 38:35–46.

Mod-1 locus (formerly *Mdh-1*), malic enzyme-supernatant, Chr 9

This locus controls electrophoretic mobility of soluble malic enzyme (E.C. 1.1.1.40) present in many tissues from embryonic stages on. The allele *Mod-1^a* determines a slow cathodally migrating band and occurs in strains SJL/J, C3H/HeJ, DBA/2J, and many others; the allele *Mod-1^b* determines a faster band and occurs in strains C57BL/6J, AKR/J, CBA/J, and many others. Both alleles occur in wild populations. Heterozygotes have five bands, indicating a tetrameric structure of the enzyme (2–4). A mutation to a null allele, *Mod-1^n*, occurred spontaneously in the C57BL/6J strain. There appears to be some reduction in viability of homozygous as well as heterozygous males (1).

References

1. Johnson, F.M., F. Charalow, S.E. Lewis, L. Barnett, and C.-Y. Lee. 1981. A null allele at the *Mod-1* locus of the mouse. J. Hered. 72:134–136.
2. Shows, T.B., and F.H. Ruddle. 1968. Malate dehydrogenase: evidence for tetrameric structure in Mus musculus. Science 160:1356–1357.
3. Shows, T.B., V.M. Chapman, and F.H. Ruddle. 1970. Mitochondrial malate dehydrogenase and malic enzyme: mendelian inherited electrophoretic variants in the mouse. Biochem. Genet. 4:707–718.
4. Wolf, U., and W. Engel. 1972. Gene activation during early development in mammals. Humangenetik 15:99–118.

Mod-2 complex, Chr 7

This complex comprises a structural and a closely linked *cis*-acting regulatory locus.

Mod-2s locus (formerly *Mod-2*), malic enzyme-mitochondrial-structural. This locus controls electrophoretic mobility of the mitochondrial form of malic enzyme (E.C. 1.1.1.40) present in many tissues. The allele *Mod-2s^a* determines a slow anodally migrating band and occurs in strain SM/J and in wild mice in Ohio; the allele *Mod-2s^b* determines a faster band and occurs in all other inbred strains examined. Heterozygotes have a single band in an intermediate position (7). In studies in which enzyme activity was determined in mice with deletions or duplications for the *Mod-2s* locus, it was

shown that enzyme activity was proportional to gene dose (4–6), and that mice completely lacking the *Mod-2s* gene were viable (4).

Mod-2r locus (formerly *Mdr-1*), malic enzyme-mitochondrial-regulatory. This locus controls level of mitochondrial malic enzyme in brain. The allele for high activity, *Mod-2r^a*, occurs in strains C3H/Rl, A/Rl, SM/J, 129/Rl, DBA/Rl, and 101/Rl; the allele for low activity occurs in strains C57BL/10 and SEC/Rl (1). The difference in activity is due to a difference in rate of synthesis of the enzyme. *Mod-2r* shows no recombination with *Mod-2s* (3). It is *cis*-acting, i.e. it controls the rate of synthesis of the enzyme determined by the *Mod-2s* gene on the same chromosome (2).

References

1. Bernstine, E.G. 1979. Genetic control of mitochondrial malic enzyme in mouse brain. J. Biol. Chem. 254:83–87.
2. Bernstine, E.G., and C. Koh. 1980. A *cis*-acting regulatory gene in the mouse: direct demonstration of *cis*-active control of the rate of enzyme subunit synthesis. Proc. Natl. Acad. Sci. USA 77:4193–4195.
3. Bernstine, E.G., C. Koh, and C.C. Lovelace. 1979. Regulation of mitochondrial malic enzyme synthesis in mouse brain. Proc. Natl. Acad. Sci. USA 76:6539–6541.
4. Bernstine, E.G., L.B. Russell, and C.S. Cain. 1978. Effect of gene dosage on expression of mitochondrial malic enzyme activity in the mouse. Nature 271:748–750.
5. Diamond, R.P., and R.P. Erickson. 1974. Gene dosage in a deletion for a nuclear-coded mitochondrial enzyme. Nature 248:418–419.
6. Eicher, E.M., and D.L. Coleman. 1977. Influence of gene duplication and X-inactivation on mouse malic enzyme activity and electrophoretic patterns. Genetics 85:647–658.
7. Shows, T.B., V.M. Chapman, and F.H. Ruddle. 1970. Mitochondrial malate dehydrogenase and malic enzyme: mendelian inherited electrophoretic variants in the mouse. Biochem. Genet. 4:707–718.

Mor-1 locus, mitochondrial malate dehydrogenase, Chr 5

This locus controls electrophoretic mobility of the enzyme malate dehydrogenase (E.C. 1.1.1.37) in mitochondria of most tissues. The allele *Mor-1^a* determines the presence of four cathodally migrating bands and occurs in most wild mice and in 40 inbred strains; the allele *Mor-1^b* determines presence of four slower bands and occurs in some wild mice in Ohio. The major band in both homozygotes is the most cathodal band; the other three bands may be conformational variants. Heterozygotes show both parental major bands and an intermediate band in addition to three other more diffuse bands. This is in agreement with a dimeric structure of the enzyme (1).

References

1. Shows, T.B., V.M. Chapman, and F.H. Ruddle. 1970. Mitochondrial malate dehydrogenase and malic enzyme: Mendelian inherited electrophoretic variants in the mouse. Biochem. Genet. 4:707–718.

Mos locus (c-*mos*), Moloney sarcoma oncogene, Chr 4

This locus is the cellular homolog of the transforming gene (v-*mos*) of Moloney sarcoma virus. By use of a sensitive S_1 nuclease assay, the gene was shown to be transcribed at very low levels in embryonic cells, testes, and ovaries. The transcription products are of different sizes in these different tissues but are all transcribed from the same single-copy gene (3). The amino acid sequence inferred from part of the DNA sequence of *Mos* shows homology with bovine cAMP-dependent protein kinase, a characteristic shared by several other murine oncogenes (1). When cloned *Mos* sequences were covalently linked to cloned long terminal repeats (LTR) from the Moloney virus, they transformed as efficiently in DNA transfection assays as cloned viral sequences containing the viral *mos* sequences and LTR (2). No genetic variants of *Mos* are known. It was shown to be on Chr 4 by use of mouse–Chinese hamster somatic cell hybrids screened with a cellular DNA probe (4).

References

1. Barker, W.C., and M.O. Dayhoff. 1982. Viral *src* gene products are related to the catalytic chain of mammalian cAMP-dependent protein kinase. Proc. Natl. Acad. Sci. USA 79:2836–2839.
2. Blair, D.G., M. Oskarsson, T.G. Wood, W.L. Clements, P.J. Fischinger, and G.G. VandeWoude. 1981. Activation of the transforming potential of a normal cell sequence: a molecular model for oncogenesis. Science 212:941–943.
3. Propst, F., and G.F. VandeWoude. 1985. Expression of *cmos* proto-oncogene transcripts in mouse tissues. Nature 315:516–518.
4. Swan, D., M. Oskarsson, D. Keithly, F. H. Ruddle, P. D'Eustachio, and G.F. VandeWoude. 1982. Chromosomal localization of the Moloney sarcoma virus mouse cellular (c-*mos*) sequence. J. Virol. 44:752–754.

Mov-1 to *Mov-15* loci

These loci are the sites of integration of Moloney leukemia virus produced by infecting embryos at various stages with virus and testing the resulting mice for transmission of viremia or of viral DNA. In the first generation, transmission was usually less than 50 per cent but in succeeding generations it approximated 50 per cent, indicating that the integrated virus behaved like a dominant gene. The majority of the integrated viruses do not cause viremia but some do. Those with viremia vary in the time of virus activation. The chromosomal region at which the viral genome is integrated influences its expression (4).

Mov-1 locus, Moloney leukemia virus-1, Chr 6. Integration of the virus at this locus was produced by infecting four- to eight-cell (BALB/c × 129)F1 embryos with virus and gestating them in foster mothers. Homozygotes are fully viable and fertile and are no more leukemic than heterozygotes (3). Virus activation occurs in the lymphatic target tissues, spleen and thymus, soon after birth (4). *Mov-1* was shown to be on Chr 6 by use of somatic cell hybrids and found to show 31 per cent recombination with *wa-1* by a standard genetic cross (1).

Mov-2 locus, Moloney leukemia virus-2. Integration of the virus at this locus was produced by cocultivating four- to eight-cell strain ICR embryos overnight with Cl-1A cells producing Moloney leukemia virus. A single male developing from one of these embryos proved to be carrying integrated virus at three different sites distinguished by restriction enzyme Southern blotting analysis. They were named *Mov-2*, *Mov-3*, and *Mov-4*. Mice carrying *Mov-2* do not develop viremia early in life, but about 20 per cent of them become viremic at 2 to 4 months of age (6). The proviral genome carries at least two alterations, probably in the *gag-pol* region, which may be the cause of the failure to develop viremia (9).

Mov-3 locus, Moloney leukemia virus-3. See *Mov-2* for origin, Mice carrying *Mov-3* develop viremia at 3 weeks of age or younger, followed by rapid development of fatal leukemia at 2 to 4 months (6).

Mov-4 locus, Moloney leukemia virus-4. See *Mov-2* for origin. The integrated virus at this locus probably has a deletion in the 3' half of the genome (4). Mice carrying *Mov-4* do not develop viremia (6).

Mov-5 locus, Moloney leukemia virus-5. Integration of virus at this locus was produced by cocultivating four- to 16-cell strain 129 embryos overnight with Cl-1A cells producing Moloney virus. Seven resulting lines, *Mov-5, -6, -7, -8, -10, -11,* and *-12,* which were not viremic and could not be detected by this criterion, were identified by the characteristic restriction enzyme pattern of Moloney virus-specific sequences in DNA obtained from liver biopsies (4).

Mov-6 locus, Moloney leukemia virus-6. See *Mov-5* for origin. The integrated virus at this locus has a deletion in the 3' half of the genome similar to that of *Mov-4* (4).

Mov-7 locus, Moloney leukemia virus-7, Chr 1. See *Mov-5* for origin. The proviral genome at this locus carries at least two alterations, probably in the *gag-pol* region. These may be responsible for the failure of the virus to cause viremia (9). *Mov-7* was found to be on Chr 1 by use of mouse–hamster somatic cell hybrids screened with a single-copy mouse DNA sequence flanking the proviral DNA and by *in situ* hybridization (7).

Mov-8 locus, Moloney leukemia virus-8. See *Mov-5*.

Mov-9 locus, Moloney leukemia virus-9, Chr 11. Integration of virus at this locus was produced by cocultivating four- to 16-cell strain 129 embryos with Cl-1A cells producing Moloney leukemia virus. Mice carrying *Mov-9* invariably activate virus and subsequently develop leukemia (4). *Mov-9* was found to be on Chr 11 by use of mouse–rat and mouse–hamster somatic cell hybrids screened with a single-copy mouse DNA sequence flanking the proviral DNA and by *in situ* hybridization (7).

Mov-10 locus, Moloney leukemia virus-10, Chr 3. See *Mov-5* for origin. The proviral genome at this locus has an alteration in the 5' end of the *gag* region, which may account for the failure of mice bearing *Mov-10* to develop viremia (9). *Mov-10* was found to be on Chr 3 by use of mouse–rat and mouse–hamster somatic cell hybrids screened with a single-copy mouse DNA sequence flanking the proviral DNA and by *in situ* hybridization. The DNA sequence is polymorphic; in C57BL/6 it produces a 7.0-kb EcoRI fragment and, in BALB/c and AKR, a 9.0-kb *Eco*RI fragment (7).

Mov-11 locus, Moloney leukemia virus-11. See *Mov-5*.

Mov-12 locus, Moloney leukemia virus-12. See *Mov-5*.

Mov-13 locus, Moloney leukemia virus-13, Chr 11 or 16. Integration of the virus at this locus was produced by microinjection of Moloney virus into 8-day embryos of the C57BL/6 strain (4). In heterozygotes the virus begins to be expressed at day 12 of gestation and is subsequently found in most if not all organs. Homozygotes are arrested in development at 11 to 12 days of gestation and die a day or two later (5). It has been found that the site of integration of the virus is in the $\alpha1(I)$ collagen gene near its 5' end. This apparently prevents transcription of the gene, resulting in a complete deficiency of the protein, a necessary constituent of most organs. The time of arrest of development of the homozygous embryos coincides with onset of high expression of the $\alpha1(I)$ collagen gene (2, 8). *Mov-13* thus serves as a marker for the gene, which may be the same as *Cola-1*. A unique mouse DNA sequence flanking the *Mov-13* provirus was mapped to Chr 11 by use of mouse–rat and mouse–hamster somatic cell hybrids and *in situ* hybridization (7). See *Cola-1* locus.

Mov-14 locus, Moloney leukemia virus-14, Chr X. Integration of the virus at this locus was produced by microinjection of a genomic clone derived from the *Mov-3* strain and consisting of the entire Moloney virus genome and some moderately repetitive adjacent sequences of mouse DNA. The clone was injected into one pronucleus of fertilized C57BL/6 eggs. An animal was derived that was viremic and carried 10 to 20 copies of proviral DNA in a tandem array. The moderately repetitive sequences of mouse DNA were not integrated. Subsequent breeding showed that the site of integration was the X chromosome. The viral gene was expressed in all mice carrying *Mov-14* (10).

Mov-15 locus, Moloney leukemia virus-15, Chrs X and Y pairing region (7).

References

1. Breindl, M., J. Doehmer, K. Willecke, J. Dausman, and R. Jaenisch. 1979. Germ line integration of Moloney leukemia virus: identification of the chromosomal integration site. Proc. Natl. Acad. Sci. USA 76:1938–1942.
2. Hartnung, S., R. Jaenisch, and M. Breindl. 1986. Retrovirus insertion inactivates mouse $\alpha1(I)$ collagen gene by blocking initiation of transcription. Nature 320:365–367.
3. Jaenisch, R. 1977. Germ line integration of Moloney leukemia virus: effects of homozygosity at the M-Mulv locus. Cell 12:691–696.
4. Jaenisch, R., D. Jähner, P. Nobis, I. Simon, J. Löhler, K. Harbers, and D. Grotkopp. 1981. Chromosomal position and activation of retroviral genomes inserted into the germline of mice. Cell 24:519–529.
5. Jaenisch, R., K. Harbers, A. Schnieke, J. Löhler, I. Chumakov, D. Jähner, D. Grotkopp, and E. Hoffmann. 1983. Germline integration of Moloney leukemia virus at the *Mov13* locus leads to recessive lethal mutation and early embryonic death. Cell 32:209–216.
6. Jähner, D., and R. Jaenisch. 1980. Integration of Moloney leukemia virus into the germ line of mice: correlation between site of integration and virus activation. Nature 287:456–458.
7. Münke, M., K. Harbers, R. Jaenisch, and U. Francke. 1986. Chromosomal mapping of four different integration sites of Moloney murine leukemia virus including the locus for $\alpha1(I)$ collagen in mouse. Cytogenet. Cell Genet. 43:140–149.
8. Schnieke, A., K. Harbers, and R. Jaenisch. 1983.

Embryonic lethal mutation in mice induced by retrovirus insertion into the 1(I) collagen gene. Nature 304:315–320.

9. Schnieke, A., H. Stuhlmann, K. Harbers, I. Chumakov, and R. Jaenisch. 1983. Endogenous Moloney leukemia virus in nonviremic Mov substrains of mice carries defects in the proviral genome. J. Virol. 45:505–513.
10. Stewart, C., K. Harbers, D. Jähner, and R. Jaenisch. 1983. × chromosome-linked transmission and expression of retroviral genomes microinjected into mouse zygotes. Science 221:760–762.

Mov-24, -34 loci

These loci are sites of integration of Moloney leukemia virus induced by infection of pre-implantation embryos with the virus. They are two of the 34 such integrations induced in addition to the *Mov-1* to *Mov-14* loci previously described. All were autosomal and homozygous viable except these two loci.

Mov-24 locus, Moloney leukemia virus-24, Chr Y.

Mov-34 locus, Moloney leukemia virus-34. Homozygotes die at an early embryonic stage. The provirus is integrated on the 5′ side of an abundantly and ubiquitously transcribed gene, which is probably inactivated by the integration (1).

References

1. Soriano, P., T. Gridley, and R. Jaenisch. 1987. Retroviruses and insertional mutagenesis in mice: proviral integration at the *Mov-34* locus leads to early embryonic death. Genes Dev. 1:366–375.

Mp, micropinna-microphthalmia, semidominant

Found in the offspring of an irradiated male. Heterozygotes invariably have microphthalmia and often have reduced external ears. Homozygotes appear to be eyeless and have oligodactyly of the hindfeet. They are very small and die at about 12 days. The pinna is present and very probably of reduced size, but, because homozygotes die before the ears open and the pinna expands, no accurate comparison with heterozygotes can be made (1,2).

References

1. Phipps, E.L. 1964. Mouse News Lett. 31:41.
2. Phipps, E.L. 1965. Mouse News Lett. 33:68.

Mph-1 locus, macrophage antigen-1, Chr 7

This locus controls an antigen on a subpopulation of unstimulated peritoneal exudate cells. The allele *Mph-*

1^a determines the antigen Mph-1.1 on cells (probably macrophages) of strains F/St and I/StDa; the allele *Mph-1^b* determines an antigen Mph-1.2 on cells of strains A/J, C57BL/10, C57BL/6, 129, DBA/2, and others. Heterozygotes have both antigens. The antigen is not present on lymph node cells (1).

References

1. Archer, J.R. 1975. Inheritance of the macrophage alloantigen marker (Mph-1) in inbred mice. Genet. Res. 26:213–219.

Mpi-1 locus, mannosephosphate isomerase-1, Chr 9

This locus controls electrophoretic mobility of the enzyme mannosephosphate isomerase (MPI-1; E.C. 5.3.1.8) present in many tissues and in fibroblast cultures. The allele *Mpi-1^a* determines a fast anodally migrating band and occurs in strain MA/J and in *M. m. castaneus*; the allele *Mpi-1^b* determines a slower band and occurs in C57BL/6J and many other strains. Heterozygotes have both parental bands (1).

References

1. Nichols, E.A., V.M. Chapman, and F.H. Ruddle. 1973. Polymorphism and linkage for mannosephosphate isomerase in *Mus musculus*. Biochem. Genet. 8:47–53.

mr

See *ru-2*.

Ms

See *Mo^{ms}*.

Ms6, Ms15 loci

Ms6-1 to *-5* loci, minisatellite 6-1 to -5
Ms6hm locus, minisatellite 6 hypermutable
Ms15-1 to *-8* loci, minisatellite 15-1 to -8
These loci are variable regions of DNA consisting of tandem repeats of short sequences (minisatellites), the variability or polymorphism resulting from allelic differences in the number of repeats. In man, some of the repeat units share a common core sequence which may promote the unequal exchanges that create allelic variability. Two human minisatellite probes, 33.6 and 33.15, that contain tandem repeats of different versions of the core sequence were used to detect similar sequences in *Hinf*I-digested mouse DNA. Examination of the pattern of DNA fragments in Southern blots of

Table 2.6

Locus	Chromosome	Number of fragments present in	
		C57BL/6	DBA/2
Ms6-1	–	0	1
Ms6-2	4	1	1
Ms6-3	–	1	1
Ms6-4	6	0	1
Ms6-5	14	0	1
Ms6hm	–	1	0[a]
Ms15-1	4	10	2
Ms15-2	–	0	2
Ms15-3	17	0	1
Ms15-4	–	1	0
Ms15-5	5	0	1
Ms15-6	5	1	0
Ms15-7	14	0	1
Ms15-8	–	1	0

[a] See text.

the C57BL/6 and DBA/2 strains and in 25 BXD recombinant inbred strains allowed identification of 14 autosomal minisatellite loci and the chromosomal location of eight of them (Table 2.6). Most of the loci were identified by a single fragment in one or the other progenitor strain, but three determined allelic fragments in the two strains and one of these determined multiple fragments in both strains. The sequences varied in length between 2 and 20 kb. The total length of the sequences determined by *Ms15-1* is about 90 kb in C57BL/6 and 19 kb in DBA/2.

The *Ms6hm* locus appears to mutate frequently, to the extent that none of the BXD strains show the 7-kb fragment found in the progenitor C57BL/6 strain but instead contain novel fragments at least some of which represent different length alleles of the C57BL/6 fragment. The locus is so unstable that new mutant alleles have arisen in most or all of the 25 BXD strains during the 60 generations since their origin (1).

References

1. Jeffreys, A.J., V. Wilson, R. Kelly, B.A. Taylor, and G. Bulfield. 1987. Mouse DNA "fingerprints": analysis of chromosome localization and germ-line stability of hypervariable loci in recombinant inbred strains. Nucl. Acids Res. 15:2823–2836.

Msp-1 locus

See *Ssp* locus.

Mt-1, *Mt-2* loci, metallothionein-1, -2, Chr 8

These loci code for the metal-binding proteins MT-1 and MT-2. No genetic variants for either gene are known. *Mt-1* was found to be on Chr 8 by use of mouse–Chinese hamster somatic cell hybrids screened with a cloned DNA probe for *Mt-1* (2), and *Mt-2* was subsequently found to be located 6 kb upstream from *Mt-1* (4). *Mt-1* is a 'housekeeping' gene. It is expressed in most cells and regulated by a variety of stimuli, including heavy metals and glucocorticoid hormones. The promoter DNA sequences of *Mt-1* can be fused to the structural genes for other substances and microinjected into mouse fertilized eggs. If the fused gene is successfully incorporated into the host genome, it can be expressed in all cells in which *Mt-1* is normally expressed. This process has been used with the gene for thymidine kinase from herpes simplex virus (1) and with the genes for rat and human growth hormone (3). In the first case, thymidine kinase was produced in liver and kidney of the transfected mouse; in the second two cases, rat and human growth hormone were found to be produced in all tissues examined (3). *Mt-1* and *Mt-2* are probably the only genes for metallothionein in the mouse. The two proteins are very similar and are coordinately regulated (4).

References

1. Brinster, R.L., H.Y. Chen, M. Trumbauer, A.W. Senear, R. Warren, and R.D. Palmiter. 1981. Somatic expression of herpes thymidine kinase in mice following injection of a fusion gene into eggs. Cell 27:223–231.
2. Cox, D.R., and R.D. Palmiter. 1983. The metallothioncin gene maps to mouse chromosome 8: implications for human Menkes' disease. Hum. Genet. 64:61–64.
3. Palmiter, R.D., G. Norstedt, R.E. Gelinas, R.E. Hammer, and R.L. Brinster. 1983. Metallothionein–human GH fusion genes stimulate growth of mice. Science 222:809–814.
4. Searle, R.F., B.L. Davison, G.W. Stewart, T.M. Wallace, G. Norstadt, and R.D. Palmiter. 1984. Regulation, linkage, and sequence of mouse metallothionein I and II genes. Mol. Cell. Biol. 4:1221–1230.

mto

See *adr* locus.

Mtp-1 locus, mouse tear protein-1, Chr 7

This locus controls electrophoretic variation of a protein in tears that migrates in the region designated region II. The allele *Mtp-1^f* determines a fast migrating band and occurs in strains C3H/He, CBA/N, IC/Le, KK, SWM/MS, WB/Re, and WC/Re; the allele *Mtp-1^s* determines a slow band and occurs in strains DBA/1, DBA/2, C57L, NZB, NZW, and SM; the null allele *Mtp-1°* determines absence of a band and occurs in strains BALB/c, A/He, AKR, CL/Fr, DDD, and NC. *Mtp-1ᶠ/Mtp-1ˢ* mice have both parental bands and *Mtp-1ᵒ* is recessive to both the other alleles (1). *Mtp-1* is on Chr 7 very close to *Gpi-1* (2).

References

1. Matsushima, Y., and S. Ikemoto. 1983. Genetic analysis of mouse tear protein system-1 (*Mtp-1*). Jpn. J. Zootech. Sci. 54:269–274.
2. Matsushima, Y., and S. Ikemoto. 1984. Genetic analysis of mouse tear protein: linkage of the *Mtp-1* locus on chromosome 7. Jpn. J. Zootech. Sci. 55:197–198.

Mtp-3 locus, mouse tear protein-3, Chr X

This locus controls presence or absence of a protein of tears that migrates to a position cathodal to the protein determined by the *Mtp-1* locus. The allele *Mtp-3ᵃ* determines presence of the protein band and occurs in the NZB strain; the allele *Mtp-3ᵇ* determines absence of the band and occurs in 19 other strains including A/He, BALB/c, C57BL/6, and NZW. Most heterozygous females lacked the MTP-3 protein but a few showed mosaic expression when tear samples were collected separately from the right and left eyes, probably as a result of X-inactivation. Segregation was abnormal in both males and females in the F2 generations with a marked deficiency of *Mtp-3ᵃ/Y* males and *Mtp-3ᵃ/Mtp-3ᵃ* females (1). This suggests that other loci may also affect expression of the MTP-3 protein.

References

1. Matsushima, Y., T. Imai, and S. Ikemoto. 1984. Evidence for X linkage of the mouse tear protein system-3 (*Mtp-3*) with mosaic expression in heterozygous females. Biochem. Genet. 22:577–585.

Mtv-1 to *Mtv-26* loci

These loci are the sites of integration of mouse mammary tumor viruses (MTV) in various strains of mice. A few of them were first described as loci controlling expression of mammary tumors or of MTV protein in mammary glands and milk. The others have been described as presence of fragments of *Eco*RI-digested DNA detected by hybridization with cloned probes of the MTV genome. Since *Eco*RI cleaves the MTV genome once internally, two cell–virus junction fragments are formed for each complete provirus. The occurrence of only one fragment probably indicates an incomplete proviral genome. There has been considerable confusion in the literature resulting in part from similar sizes of fragments at different integration sites and from lack of a system for assigning new *Mtv*

numbers. The nomenclature and numbering system for the loci used here is that proposed by Kozak *et al.* (8) who have attempted to resolve the inconsistencies present in the literature. There is some variation in size of *Eco*RI fragments given by various authors for individual loci. The sizes given here follow Kozak *et al.* (8).

Mtv-1 locus, mammary tumor virus-1, Chr 7. This locus controls expression of antigens of the nodule-inducing mammary tumor virus MTV-L (low) in the milk, as detected by a sensitive radioimmunoassay. C3H/Hef mice have the allele *Mtv-1ᵃ* determining presence of the viral antigens, and BALB/c mice have the allele *Mtv-1ᵇ* determining absence of these antigens. *Mtv-1ᵃ* is dominant and is transmitted through both males and females (20). The *Mtv-1* locus is the site of integration of an MTV provirus recognized as 6.5- and 4.5-kb *Eco*RI restriction fragments. The provirus also occurs in the DBA/2 strain (17).

Mtv-2 locus, mammary tumor virus-2, Chr 18. This locus controls occurrence of mammary tumors and expression of a mammary tumor virus, MTV-P, that occurs in the GRS/A (GR) strain. The virus is transmitted not only through the milk but also by male and female germ cells and cannot be eliminated by foster nursing. A dominant allele (*Mtv-2ᵃ*) is present in the GR strain and determines the presence of MTV antigen and susceptibility to hormone-induced mammary tumors; a recessive allele (*Mtv-2ᵇ*) occurs in strains DBAf, C57BL, BALB/c, and C3Hf, and determines absence of MTV antigens and of mammary tumors (19, 21). *Mtv-2* is the site of integration of an MTV provirus recognized as 11.7 and 6.9-kb *Eco*RI restriction fragments and found to be located on Chr 18 (9).

Mtv-3 locus, mammary tumor virus-3, Chr 11. This locus codes for partial MTV expression in the GR strain. It was discovered in the congenic GR-*Mtv-2ᵇ* strain not expressing MTV-2. By radio immunoassay for individual viral proteins in extracts of mammary glands, substantial amounts of *gag* proteins but essentially no *env* proteins were detected. *Mtv-3ᵃ* behaved as a dominant gene in a cross to BALB/c and was shown to be on Chr 11 near *Es-3* (11). The locus was called *Mtv-18* by Gray *et al.* (3), who identified it by means of *Eco*RI proviral fragments of 17.4, 6.9, and 0.9 kb, which are very similar to those identified for *Mtv-3* by Michalides *et al.* (9).

Mtv-4 locus, mammary tumor virus-4. This locus was found in the SHN strain by Imai *et al.* (7). It resembles *Mtv-2* in the GR strain in causing a high incidence (90–95 per cent) of mammary tumors early in life. It cannot be identical with *Mtv-2*, however, because the

DNA of the SHN strain does not show the characteristic *Eco*RI fragments of the *Mtv-2* locus (7). The proviral fragments of this strain have not been described, and the relationship of *Mtv-4* to other *Mtv* loci is not known.

Mtv-5 locus, mammary tumor virus-5. This locus was found in the SL/NiA strain by Imai and Hilgers (6), but not named by them. Mice of this strain produce low levels of viral p27 *gag* protein but no gp52 *env* protein and they show a low incidence of mammary tumors. F1 hybrids with the BALB/c strain were all positive for p27 *gag* protein (6).

Mtv-6 locus, mammary tumor virus-6, Chr 16. This is one of at least three sites of integration of MTV in the BALB/c strain. It was also called *Mtv-19* (3). It is recognized as a single 16.7-kb fragment in *Eco*RI digests of BALB/c DNA and is probably an incomplete viral genome. It was found to be on Chr 16 by use of BALB/c–Chinese hamster somatic cell hybrids. Further, in a genetic cross of BALB/c with the wild-derived CL stock, *Mtv-6* was shown to be located about 16 cM from the *Igl* complex which was also recognized by a DNA polymorphism (1). MTV is integrated at the same site in the C3H/HeJ strain (17).

Mtv-7 locus, mammary tumor virus-7, Chr 1. There were at first thought to be several *Mtv* proviral loci on Chr 1, including *Mtv-10* (16), but it is now known that there is only one such locus, called *Mtv-7* (8). It is a site of integration of MTV in the DBA/2, NFS, GRS/A, and STS/A strains. It is recognized as two *Eco*RI restriction fragments of 11.7 and 16.7 kb (5, 9). *Mtv-7* was found to be on Chr 1 close to *Mls* by its strain distribution in C57BL/6 × DBA/2 and BALB/c × STS recombinant inbred strains (12, 16).

Mtv-8 locus (formerly also *Mtv-16*), mammary tumor virus-8, Chr 6. This is one of at least three sites of integration of MTV in the BALB/c strain. It is recognized as two fragments of 8.5 and 6.7 kb in *Eco*RI digests of BALB/c DNA (1) and was found to be on Chr 6 by use of mouse–Chinese hamster somatic cell hybrids (15). MTV is integrated at the same site in strains C57BL/6, C3H/He, DBA/2, GR, and others (17), and was called *Mtv-16* by Stoye and Coffin (16).

Mtv-9 locus, mammary tumor virus-9, Chr 12. This is one of at least three sites of integration of MTV virus in the BALB/c strain. It is recognized as two fragments of 10.0 and 7.8 kb in *Eco*RI digests of BALB/c DNA. It was found to be on Chr 12 by use of mouse–Chinese hamster somatic cell hybrids (1). *Mtv-9* is also present in AKR, C57BL/6, CBA, and NZB (12, 13).

Mtv-10 locus. See *Mtv-7*, *Mtv-17*.

Mtv-11 locus, mammary tumor virus-11, Chr 14. This is a site of integration of MTV in the DBA/2 and C3H/He strains. It is recognized as fragments of 5.8 and 15.0 kb in *Eco*RI digests of DNA of these strains (4, 16). It is probably the same as the locus provisionally called *Mtv-7a* by Prakash *et al.* (14) and found to be on Chr 14, and the same as the locus called *Mtv-12* by Traina *et al.* (17).

Mtv-12 locus. See *Mtv-11*.

Mtv-13 locus, mammary tumor virus-13, Chr 4. This is a site of integration of MTV in the A/ST, DBA/2, and GRS strains. It is recognized as fragments of 5.8 and 9.0 kb (3, 10, 16) and is located on Chr 4 (10).

Mtv-14 locus, mammary tumor virus-14. This is the site of integration of MTV in the C3H/He and DBA/2 strains. It is recognized as a single fragment of 1.7 kb. It was at first thought to be on Chr 6 from its strain distribution pattern in C57BL/6 × C3H/He and C57BL/6 × DBA/2 recombinant inbred strains (17), but later found to show no linkage with three loci on this chromosome (15).

Mtv-15 locus. See *Mtv-17*.

Mtv-16 locus. See *Mtv-8*.

Mtv-17 locus, mammary tumor virus-17, Chr 4. This locus is the site of integration of an MTV provirus in the C57BL/6 strain (13) and also in the GRS, DBA/2, NFS, AKR, NZB, and BR6 strains (3, 12). It is recognized by two *Eco*RI fragments of 10.0 and 8.3 kb (12) and was formerly designated *Mtv-10* (3), *Mtv-15* (17), and *Mtv-20* (9). It is located on Chr 4 (12).

Mtv-18 locus, mammary tumor virus-18 (this designation was formerly used for a provirus identical to *Mtv-3* (3)). This is a site of integration recognized as a 9.2-kb *Eco*RI fragment that occurs exclusively in the BALB/c × C57BL/6 recombinant inbred strain CXBJ. It is not present in either progenitor strain and therefore must be regarded as a recent germline integration (5).

Mtv-19 locus. Same as *Mtv-10*.

Mtv-20 locus, mammary tumor virus-20. This is the site of integration of MTV in some C57BL strains but not in C57BL/6 or C57BL/10. The provirus is probably transcribed in lactating mammary glands but is not translated into viral protein. The provirus is detected as 13.0- and 5.3-kb fragments in *Eco*RI digests of DNA (18).

Mtv-21 locus, mammary tumor virus-21, Chr 8. This is a site of integration of MTV only in the BR6 strain among 10 strains examined. It is recognized as 8.0- and 8.1-kb fragments in *Eco*RI digests of DNA. It was found to be on Chr 8 by use of mouse–Chinese hamster somatic cell hybrids (12).

Mtv-22 to *Mtv-25* loci, mammary tumor virus-22 to -25. These four loci were defined by comparison of the DNA fragment patterns in a panel of inbred strains. Their chromosomal location is not known. Their *Eco*RI fragment size and strain distribution are given in Table

Table 2.7

Locus	Fragment	size (kb)	Strains
Mtv-22	17	11.5	AKR, NZB
Mtv-23	11.8	4.6	AKR, A/St
Mtv-24	20	6.6	DDSio, NZB
Mtv-25	5.9	6.1	DDSio

2.7 (12).

Mtv-26 locus (formerly *Mtv-22*), mammary tumor virus-26. This locus is the integration site of a mammary tumor virus in B6.C-KH-84, a strain carrying a 'gain' mutation at a minor histocompatibility locus. 21 other such mutations were examined for integrated MTVs and none found. Neither parental strain, BALB/c nor C57BL/6, carries this viral integration. The site in B6.C-KH-84 is recognized as a 5.5-kb *Eco*RI fragment with an *env* probe and as two *Eco*RI fragments of 9.2 and 5.5 kb with an LTR probe. The histocompatibility effect may be due to the presence of the MTV (2).

References

1. Callahan, R., D. Gallahan, and C. Kozak. 1984. Two genetically transmitted BALB/c mouse mammary tumor virus genomes located on chromosomes 12 and 16. J. Virol. 49:1005–1008.
2. Colombo, M.P., R.W. Melvold, and P.J. Wettstein. 1987. Inheritance of a mutant histocompatibility gene and a new mammary tumor virus genome in the B6.KH-84 mouse strain. Immunogenetics 26:99–104.
3. Gray, D.A., E.C.M.L. Chan, J.I. McInnes, and R.L. Morris. 1986. Restriction endonuclease map of endogenous mouse mammary tumor virus loci in GR, DBA, and NFS mice. Virology 148:237–242.
4. Hilgers, J. 1986. Mtv-11. Mouse News Lett. 74:80–81.
5. Hilgers, J., and R. Michalides. 1984. Mouse News Lett. 70:63–64.
6. Imai, S., and J. Hilgers. 1979. Levels of mammary tumor virus proteins (MTV p27 and MTV gp52) in the milk of low and high mammary cancer strains of Japanese origin compared with European and American strains. Int. J. Cancer 24:359–364.
7. Imai, S., Y. Tsubura, J. Hilgers, and R. Michalides. 1983. A new locus (Mtv-4) for endogenous mammary tumor virus expres sion and early mammary tumor development in the SHN mouse strain. J. Natl. Cancer Inst. 71:517–521.
8. Kozak, C., G. Peters, R. Pauley, V. Morris, R. Michalides, and 16 others. 1987. A standardized nomenclature for endogenous mouse mammary tumor viruses. J. Virol. 61:1651–1654.
9. Michalides, R., R. Verstraeten, F.W. Shen, and J. Hilgers. 1985. Characterization and chromosomal distribution of endogenous mouse mammary tumor viruses of European mouse strains STS/A and GR/A. Virology 142:278–290.
10. Morris, V.L., C. Kozak, J.C. Cohen, P.R. Shank, P. Jolicoeur, F. Ruddle, and H.E. Varmus. 1979. Endogenous mouse mammary tumor virus DNA is distributed among multiple mouse chromo somes. Virology 92:46–55.
11. Nusse, R., J. deMoes, J. Hilkens, and R. van Nie. 1980. Localization of a gene for expression of mouse mammary tumor virus antigens in the GR/Mtv-2⁻ mouse strain. J. Exp. Med. 152:712–719.
12. Peters, G., M. Placzek, S. Brookes, C. Kozak, R. Smith, and C. Dickson. 1986. Characterization, chromosomal assignment and segregation analysis of endogenous proviral units of mouse mammary tumor viruses. J. Virol. 59:535–544.
13. Peterson, D.O., K.G. Kriz, J.E. Marich, and M.G. Toohey. 1985. Sequence organization and molecular cloning of mouse mammary tumor virus DNA endogenous to C57BL/6 mice. J. Virol. 54:525–531.
14. Prakash, O., C. Kozak, and N.H. Sarkar. 1985. Molecular cloning, characterization, and genetic mapping of an endogenous mammary tumor virus proviral unit I of C3H/He mice. J. Virol. 54:285–294.
15. Robbins, J.M., D. Gallahan, E. Hogg, C. Kozak, and R. Callahan. 1986. An endogenous mouse mammary tumor virus genome common in inbred mouse strains is located on chromosome 6. J. Virol. 57:709–713.
16. Stoye, J., and J. Coffin. 1985. Endogenous viruses, In R. Weiss, N. Teich, H. Varmus, and J. Coffin, eds., RNA Tumor Viruses, 2nd ed, 2/Suppl. Append., 357–404. Cold Spring Harbor Lab., Cold Spring Harbor, New York.
17. Traina, V.L., B.A. Taylor, and J.C. Cohen. 1981. Genetic mapping of endogenous mammary tumor viruses: locus character ization, segregation, and chromosomal distribution. J. Virol. 40:735–744.
18. Vaidya, A.B., N.E. Taraschi, S.L. Tancin, and C.A. Long. 1983. Regulation of endogenous murine mammary tumor virus expression in C57BL mouse lactating mammary glands: transcription of functional mRNA with a block at the translational level. J. Virol. 46:818–828.
19. van Nie, R., and J. Hilgers. 1976. Genetic analysis of mammary tumor induction and expression of mammary tumor virus antigen in hormone-treated and ovariectomized GR mice. J. Natl. Cancer Inst. 56:23–32.
20. van Nie, R., and A.A. Verstraeten. 1975. Studies of genetic transmission of mammary tumor virus by C3Hf mice. Int. J. Cancer 16:922–931.
21. van Nie, R., A.A. Verstraeten, and J. de Moes. 1977.

Genetic transmission of mammary tumour virus in GR mice. Int. J. Cancer 19:383–390.

Mtvr-1 locus, mammary tumor virus receptor-1, Chr 16

This locus controls a mammary tumor virus receptor present on mouse cells of strains GRS/A and C3H. It was detected by ability of cells to bind a pseudotype of vesicular stomatitis virus containing envelope proteins provided by C3H mammary tumor virus. In somatic cell hybrids of Chinese hamster cells and spontaneous leukemia or mammary tumor cells from mouse strain GRS/A, binding of the pseudotype virus always segregated with Chr 16 (1).

References

1. Hilkens, J., B. vander Zeijst, F. Buijs, V. Kroezen, N. Bleumink, and J. Hilgers. 1983. Identification of a cellular receptor for mouse mammary tumor virus and mapping of its gene to chromosome 16. J. Virol. 45:140–147.

mu, muted, recessive, Chr 13

Arose in a stock carrying t-alleles. Homozygotes have light eyes at birth and fur of a muted brown shade, often with white underfur. Some have a balance defect, similar to that of pa, which is due to absence of otoliths from the sacculus and utriculus of one or both ears. The bony labyrinth is normal. The mice are not deaf and do not circle (1).

References

1. Lyon, M.F., and R. Meredith. 1969. Muted, a new mutant affecting coat colour and otoliths of the mouse, and its position in linkage group XIV. Genet. Res. 14:163–166.

MuB1

See Hc locus.

Mud-1 locus

See Sas-1 locus.

Mup-1 complex, major urinary protein, Chr 4

This complex was originally recognized as a locus controlling the major proteins of the urine (MUP) as revealed by electrophoresis. The proteins secreted in the urine are produced in the liver, and MUPs are also produced in the lachrymal, mammary, and submaxill-

ary glands (14). Two codominant alleles are known in laboratory mice, Mup-1^a in strains BALB/c, DBA/2, C57BR/cd, and others, and Mup-1^b in strains C57BL/6, C57L, C58, and others (5). Wild populations probably contain other alleles (8, 12). In the urine, Mup-1^a determines presence of a slow migrating band and Mup-1^b determines presence of a faster band. A third band is present in all mice (5). The electrophoretic pattern of females differs from that of males in some strains, but is converted to the male pattern by androgen administration (11). Molecular studies have shown that the MUPs are coded for by a cluster of about 35 genes, all tightly linked on Chr 4 (1, 9), the previously known location of Mup-1 (6). Distinct sets of MUP genes are transcribed in each tissue that produces these proteins (13). Transcription is regulated by testosterone, thyroxine, and growth hormone, and different hormones are effective in different tissues (4, 10, 14). Two patterns of MUP DNA restriction fragment lengths are found among inbred strains and they correspond exactly to the presence of Mup-1^a and Mup-1^b (1). The 35 MUP genes are largely composed of two groups, group 1 and group 2, of about 15 genes each, and a few others belonging to neither group. The group 1 genes have seven exons and the group 2 genes have six (3). They are organized on the chromosome in pairs of one group 1 and one group 2 gene, oriented head to head (2). The group 2 genes appear to be pseudogenes (7).

References

1. Bennett, K., P. Lalley, R. Barth, and N. Hastie. 1982. Mapping the structural genes coding for the major urinary proteins in the mouse: combined use of recombinant inbred strains and somatic cell hybrids. Proc. Natl. Acad. Sci. USA 79:1220–1224.

2. Bishop, J.O., G.G. Selman, J. Hickman, L. Black, R.D.P. Saunders, and A.J. Clark. 1985. The 45-kb unit of major urinary protein gene organization is a gigantic imperfect palindrome. Molec. Cell. Biol. 5:1591–1600.

3. Clark, A.J., P.M. Clissold, R.A. Shawi, P. Beattie, and J. Bishop. 1984. Structure of mouse major urinary protein genes: different splicing configurations in the 3'-non-coding region. EMBO J. 3:1045–1052.

4. Clissold, P.M., S. Hainey, and J.O. Bishop. 1984. Messenger RNAs coding for mouse major urinary proteins are differentially induced by testosterone. Biochem. Genet. 22:379–387.

5. Finlayson, J.S., M. Potter, and C.C. Runner. 1963. Electrophoretic variation and sex dimorphism of the major urinary protein complex of inbred mice: a new genetic marker. J. Natl. Cancer Inst. 31:91–107.

6. Finlayson, J.S., D.M. Hudson, and B.L. Armstrong.

1969. Location of the *Mup-a* locus on mouse linkage group VIII. Genet. Res. 14:329–331.

7. Ghazal, P., A.J. Clark, and J. Bishop. 1985. Mouse News Lett. 72:97.

8. Hayakawa, J., H. Nikaido, and T. Koizumi. 1983. Components of major urinary proteins (MUP's) in the mouse: Sex, strain, and subspecies differences. J. Hered. 74:453–456.

9. Krauter, K., L. Leinwand, P. D'Eustachio, F. Ruddle, and J.E. Darnell Jr. 1982. Structural genes of the mouse major urinary protein are on chromosome 4. J. Cell Biol. 94:414–424.

10. Kuhn, N.J., M. Woodworth-Gutai, K.W. Gross, and W.A. Held. 1984. Subfamilies of the mouse major urinary protein (MUP) multigene family: sequence analysis of cDNA clones and differential regulation in the liver. Nucl. Acids Res. 12:6073–6090.

11. Ozawa, S., and S. Tomino. 1977. Regulation by androgen of mRNA level for the major urinary protein complex of mouse liver. Biochem. Biophys. Res. Comm. 77:628–633.

12. Sampsell, B.M., and W.A. Held. 1985. Variation in the major urinary protein multigene family in wild-derived mice. 1985. Genetics 109:549–568.

13. Shahan, K., and E. Derman. 1984. Tissue-specific expression of major urinary protein (MUP) genes in mice: characterization of MUP mRNAs by restriction mapping of cDNA and by *in vitro* translation. Mol. Cell. Biol. 4:2259–2265.

14. Shaw, P.H., W.A. Held, and N.D. Hastie. 1983. The gene family for major urinary proteins: expression in several secretory tissues of the mouse. Cell 32:755–761.

Mv, malformed vertebrae, semidominant

Arose spontaneously in the AKR/J strain. Heterozygotes have kinking of the tail tip and an occasional fused rib. Penetrance is incomplete. Homozygotes have very short tails and extreme disorganization of the axial skeleton, and they die at birth. The appendicular skeleton is normal. The skeletal malformations can be traced to disturbed somite formation (1).

References

1. Theiler, K., D.S. Varnum, J.L. Southard, and L.C. Stevens. 1975. Malformed vertebrae: a new mutant with the "Wirbel-Rippen-Syndrom" in the mouse. Anat. Embryol. 147:161–166.

Mx, myxovirus resistance, dominant, Chr 16

This gene occurs in the A2G strain and determines resistance to the lethal effects of various myxoviruses including neurotropic avian influenza A virus injected intracerebrally, pneumotropic strains injected intranasally, and a hepatotropic strain injected intraperitoneally (5, 10). Strains A/J, C3H/HeJ, BALB/c, and many others are susceptible. Heterozygotes (*Mx*/+) are resistant. Resistance is not dependent on presence of the thymus and is not abolished by immunosuppression or by inhibitors of macrophage function (1, 4, 5). Resistance is specific for the orthomyxoviruses (7). It is dependent on the presence of interferon α and β but not γ (12). Resistance is expressed by macrophages and other cells *in vitro* (3, 11), but resistance could not be transferred to susceptible animals by transfer of *Mx*-bearing macrophages (6). The *Mx* gene controls production of a unique 75 000 molecular weight protein, Mx, which is synthesized by cells of *Mx*/- but not of +/+ mice in response to interferon α and β (8, 13). Interferon is induced by viral infection and it then induces the Mx protein. The protein is located in the nucleus (2) and produces its antiviral effect by preventing synthesis of viral mRNA in the nucleus (9). By use of mouse–Chinese hamster somatic cell hybrids, *Mx* was found to be on Chr 16, probably near the distal end (14).

References

1. Arnheiter, H., O. Haller, and J. Lindenmann. 1976. Pathology of influenza hepatitis in susceptible and genetically resistant mice. Exp. Cell Biol. 44:95–107.

2. Dreiding, P., P. Staeheli, and O. Haller. 1985. Interferon-induced protein Mx accumulates in nuclei of mouse cells expressing resistance to influenza viruses. Virology 140:192–196.

3. Haller, O. 1981. Interferon resistance to myxoviruses. Curr. Top. Microbiol. 92:25–52.

4. Haller, O., and J. Lindenmann. 1974. Athymic (nude) mice express gene for myxovirus resistance. Nature 250:679–680.

5. Haller, O., H. Arnheiter, and J. Lindenmann. 1976. Genetically determined resistance to infection by hepatotropic influenza A virus in mice: effect of immunosuppression. Infect. Immun. 13:844–854.

6. Haller, O., H. Arnheiter, and J. Lindenmann. 1979. Natural, genetically determined resistance toward influenza virus in hemopoietic mouse chimeras: role of mononuclear phagocytes. J. Exp. Med. 150:117–126.

7. Haller, O., H. Arnheiter, J. Lindenmann, and I. Gresser. 1980. Host gene influences sensitivity to interferon action selectively for influenza virus. Nature 283:660–662.

8. Horisberger, M.A., P. Staeheli, and O. Haller. 1983. Interferon induces a unique protein in mouse cells bearing a gene for resistance to influenza virus. Proc. Natl. Acad. Sci. USA 80:1910–1914.

9. Krug, R.M., M. Shaw, B. Broni, G. Shapiro, and O. Haller. 1985. Inhibition of influenza viral mRNA synthesis in cells expressing the interferon-induced *Mx* gene product. J. Virol. 56:201–206.

10. Lindenmann, J. 1964. Inheritance of resistance to

influenza virus in mice. Proc. Soc. Exp. Biol. Med. 116:506–509.

11. Lindenmann, J., E. Deuel, S. Fanconi, and O. Haller. 1978. Inborn resistance of mice to myxoviruses: macrophages express phenotype *in vitro*. J. Exp. Med. 147:531–540.
12. Staeheli, P., M.A. Horisberger, and O. Haller. 1984. *Mx*-dependent resistance to influenza viruses is induced by mouse interferons α and β but not γ. Virology 132:456–461.
13. Staeheli, P., P. Dreiding, O. Haller, and J. Lindenmann. 1985. Polyclonal and monoclonal antibodies to the interferon-inducible protein Mx of influenza virus-resistant mice. J. Biol. Chem. 260:1821–1825.
14. Staeheli, P., D. Pravtcheva, L.-G. Lundin, M. Acklin, F. Ruddle, J. Lindenmann, and O. Haller. 1986. Interferon-regulated influenza virus resistance gene *Mx* is localized on mouse chromosome 16. J. Virol. 58:967–969.

Mxv-1 locus, murine xenotropic virus-1

This locus controls inducibility of xenotropic virus by 5-iododeoxyuridine in tail fibroblasts. The MA/MyJ strain carries the allele for inducibility; the SEA/Gn and NFS/N strains carry the allele for noninducibility. Heterozygotes are inducible. In the MA strain, xenotropic virus is not produced spontaneously in mice younger than 6 months. *Mxv-1* is not closely linked to markers on Chrs 4, 5, 7, 9, 11, and 17. It is not known whether this inducibility locus is identical with any of the known sites of xenotropic virus integration in the mouse genome (1).

References

1. Kozak, C.A., J.W. Hartley, and H.C. Morse III. 1984. Laboratory and wild-derived mice with multiple loci for production of xenotropic murine leukemia virus. J. Virol. 51:77–80.

my, blebs, recessive, Chr 3

Found among the descendants of irradiated mice (4). Effects are variable, and penetrance is incomplete. This mutant has been the subject of a large number of investigations reviewed by Grüneberg (3). The genetics and development were reinvestigated by Carter (1,2) who concluded that the effects of *my* include: embryonic subepidermal blebs leading to abnormalities of the eyes, skin, and hair, and clubbing of the feet; split sternum, pseudencephaly, acrania, renal agenesis, preaxial polydactyly, and syndactyly of the middle digits; and probably also hydronephrosis, ectopia viscerum, and midcerebral lesions.

References

1. Carter, T.C. 1956. Genetics of the Little and Bagg X-rayed mouse stock. J. Genet. 54:311–326.
2. Carter, T.C. 1959. Embryology of the Little and Bagg X-rayed mouse stock. J. Genet. 56:401–435.
3. Grüneberg, H. 1952. The Genetics of the Mouse, 2nd ed. Martinus Nijhoff, The Hague. 650 p.
4. Little, C.C., and H.J. Bagg. 1923. The occurrence of two heritable types of abnormality among descendants of X-rayed mice. Am. J. Roentgenol. 10:975–989.

Myb locus (c-*myb*), myeloblastosis oncogene, Chr 10

This is the cellular homolog in the mouse of the transforming gene (v-*myb*) of the avian myeloblastosis virus. Homologous DNA sequences are found in all the higher vertebrates (1). The gene is expressed in small amounts in most normal cells and at much higher levels in thymic lymphocytes and cells of the erythroid lineage. The level in T-cells decreases as they mature (3, 4). *Myb* occurs in rearranged form associated with viral insertions in virus-induced plasmacytoid lymphosarcomas and gives rise to abnormal mRNA transcripts (5). The abnormal T-cells of autoimmune MRL-*lpr/lpr* (lymphoproliferation) mice have increased expression of *Myb* (2). No genetic variants of *Myb* are known. It was found to be on Chr 10 by use of mouse–Chinese hamster somatic cell hybrids screened with a viral gene probe (6).

References

1. Bergmann, D.G., L.M. Souza, and M.A. Baluda. 1981. Vertebrate DNAs contain nucleotide sequences related to the transforming gene of avian myeloblastosis virus. J. Virol. 40:450–455.
2. Mountz, J.D., A.D. Steinberg, D.M. Klinman, H.R. Smith, and J.F. Mushinski. 1984. Autoimmunity and increased c-*myb* transcription. Science 226:1087–1089.
3. Mushinski, J.F., M. Potter, S.R. Bauer, and E.P. Reddy. 1983. DNA rearrangement and altered RNA expression of the c-*myb* oncogene in mouse plasmacytoid lymphosarcomas. Science 220:795–798.
4. Sheiness, D., and M. Gardinier. 1984. Expression of a proto-oncogene (proto-*myb*) in hemopoietic tissues of mice. Mol. Cell. Biol. 4:1206–1212.
5. Shen-Ong, G.L.C., M. Potter, J.F. Mushinski, S. Lavu, and E. Prekumar Reddy. 1984. Activation of the c-*myb* locus by viral insertional mutagenesis in plasmacytoid lymphosarcomas. Science 226:1077–1080.
6. Sakaguchi, A.Y., P.A. Lalley, B.U. Zabel, R.W. Ellis, E.M. Scolnick, and S.L. Naylor. 1984. Chromosome assignments of four mouse cellular homologs of sarcoma and leukemia virus oncogenes. Proc. Natl. Acad. Sci. USA 81:525–529.

Myc locus (c-*myc*), myelocytomatosis oncogene, Chr 15

This is the cellular homolog in the mouse of the transforming gene (v-*myc*) of avian myelocytomatosis virus. No genetic variants are known. Translocations between Chr 15 and Chr 12 are a regular feature of plasma cell tumors (plasmacytomas, PC) of the mouse (8). Molecular studies of PC DNA have shown that in the T(12^15) translocations, the immunoglobulin heavy chain (*Igh*) region on Chr 12 is brought into association with the *Myc* gene on Chr 15. The exchange points differ in different translocations (2, 3, 5, 7). Three plasmacytomas lacking translocations were found to have deletions in the D region of Chr 15 in which *Myc* is located (10). In about 10 to 20 per cent of T-cell lymphomas, a proviral integration was found 1 to 2 kb 5′ to *Myc* (1, 6). *Myc* is transcribed in most somatic tissues but not in germ cells (9). It is strongly inducible in lymphocytes by lipopolysaccharide and concanavalin A and in fibroblasts by platelet-derived growth factor (4). See *Pvt-1* locus for the role of *Myc* in the variant plasmacytomas bearing (6^15) translocations involving the *Igk* locus.

References

1. Corcoran, L.M., J.M. Adams, A.R. Dunn, and S. Cory. 1984. Murine T lymphomas in which the cellular *myc* oncogene has been activated by retroviral insertion. Cell 37:113–122.
2. Cory, S., S. Gerondakis, and J.M. Adams. 1983. Interchromosomal recombination of the cellular oncogene c-*myc* with the immunoglobulin heavy chain in murine plasmacytomas is a reciprocal exchange. EMBO J. 2:697–703.
3. Crews, S., R. Barth, L. Hood, J. Prehn, and K. Calame. 1982. Mouse c-*myc* oncogene is located on chromosome 15 and translocated to chromosome 12 in plasmacytomas. Science 218:1319–1321.
4. Kelly, K., B.H. Cochran, C.D. Stiles, and P. Leder. 1983. Cell-specific regulation of the c-*myc* gene by lymphocyte mitogens and platelet-derived growth factor. Cell 35:603–610.
5. Klein, G. 1983. Specific chromosomal translocations and the genesis of B-cell derived tumors in mice and men. Cell 32:311–315.
6. Li, Y., C.A. Holland, J.W. Hartley, and N. Hopkins. 1984. Viral integrations near c-*myc* in 10–20% of MCF 247-induced AKR lymphomas. Proc. Natl. Acad. Sci. USA 81:6808–6811.
7. Neuberger, M.S., and F. Calabi. 1983. Reciprocal chromosome translocation between c-*myc* and immunoglobulin 2b genes. Nature 305:240–243.
8. Ohno, S., M. Babonits, F. Wiener, J. Spira, G. Klein, and M. Potter. 1979. Nonrandom chromosome changes involving the Ig gene-carrying chromosomes 12 and 6 in pristane-induced mouse plasmacytomas. Cell 18:1001–1007.
9. Stewart, T.A., A.R. Bellvé, and P. Leder. 1984. Transcription and promoter usage of the *myc* gene in normal somatic and spermatogenic cells. Science 226:707–710.
10. Wiener, F., S. Ohno, M. Babonits, J. Sümegi, Z. Wirschubsky, G. Klein, J.M. Mushinski, and M. Potter. 1984. Hemizygous interstitial deletion of chromosome 15 (band D) in three translocation-negative murine plasmacytomas. Proc. Natl. Acad. Sci. USA 81:1159–1163.

myd (formerly *fg*), myodystrophy, recessive, Chr 8

Arose spontaneously in mice carrying lethal spotting (*ls*). Homozygotes have a diffuse progressive myopathy resulting in abnormal posture, particularly a frog-like position of the hindlimbs, and severe thoracic kyphosis. They can usually be recognized at 12 to 15 days. Both sexes may be fertile, but breeding performance is very poor. The average age at death is about 4 months. Histological examination shows widely distributed focal lesions in all skeletal muscle (1). The calcium content of the diaphragm averages about 100 times that of normal mice and the excess calcium occurs in focal deposits in the muscle. There is moderate elevation of calcium content in other muscles (4). Basal ATPase activity in sarcoplasmic reticulum of *myd/myd* mice is elevated above normal (3) and skinned muscle fibers accumulate calcium ions in the sarcoplasmic reticulum at a slower rate than normal (2). This mutant resembles *dy* in that the dorsal and ventral spinal roots, particularly in the lumbar region, contain groups of axons completely devoid of ensheathing Schwann cells or myelin (5).

References

1. Lane, P.W., T.C. Beamer, and D.D. Myers. 1976. Myodystrophy, a new myopathy on chromosome 8 of the mouse. J. Hered. 67:135–138.
2. Mobley, B.A. 1985. Ca^{2+} capacity and uptake rate in abnormal skinned fibers of myodystrophic muscle. Exp. Neurol. 87:137–146.
3. Neymark, M.A., S.J. Kopacz, and C.-P. Lee. 1980. Characterization of ATPase in sarcoplasmic reticulum from two strains of dystrophic mice. Muscle Nerve 3:316–325.
4. Nutting, D.F., A.D. MacPike, and H. Meier. 1980. The calcium content of various tissues from myodystrophic and dystrophic mice. J. Hered. 71:15–18.
5. Rayburn, H.B., and A.C. Peterson. 1978. Naked axons in myodystrophic mice. Brain Res. 146:380–384.

Myhc, *Myhs* complexes

Myhc complex, myosin heavy chain, cardiac muscle, Chr 14. This complex includes the adult and fetal genes for cardiac myosin heavy chain (M. Buckingham and J. Peters, personal communication). No genetic variants for the two genes among inbred strains are known.

Myhc-a locus, myosin heavy chain, cardiac muscle, adult. In an interspecific backcross, *Myhc-a* showed very close linkage with *Np-1* on Chr 14 (1).

Myhc-b locus, myosin heavy chain, cardiac muscle, fetal.

Myhs complex, myosin heavy chain, skeletal muscle, Chr 11. This complex includes most, if not all, of the mouse genes for sarcomeric myosin heavy chains (MHC). These genes are actively transcribed during terminal differentiation of muscle cells (1). No genetic variants are known among inbred strains. All detectable mouse MHC genes were assigned to Chr 11 using mouse- Chinese hamster or mouse-rat somatic cell hybrids and cDNA clones of rat MHC genes (1–3). By use of an interspecific backcross, the three members of the complex listed below were found to be located near *nu* on Chr 11 (4).

Myhs-e locus, myosin heavy chain, skeletal muscle, embryonic.

Myhs-p locus, myosin heavy chain, skeletal muscle, perinatal.

Myhs-f locus, myosin heavy chain, skeletal muscle, adult fast.

References

1. Czosnek, H., U. Nudel, M. Shani, P.E. Barker, D.D. Pravtcheva, F.H. Ruddle, and D. Yaffe. 1982. The genes coding for the muscle contractile proteins, myosin heavy chain, myosin light chain 2, and skeletal muscle actin, are located on three different mouse chromosomes. EMBO J. 1:1299–1305.
2. Leinwand, L.A., R.E.K. Fournier, B. Nadal-Ginard, and T.B. Shows. 1983. Multigene family for sarcomeric myosin heavy chain in mouse and human DNA: localization on a single chromosome. Science 221:766–768.
3. Leinwand, L.A., R.E.K. Fournier, B. Nadal-Ginard, and T.B. Shows. 1984. Assignment of the sarcomeric myosin heavy chain multigene family to chromosome 17 in humans and chromosome 11 in the mouse. Cytogenet. Cell Genet. 37:521–522 (Abstr.).
4. Weydert, A., P. Daubas, I. Lagaritis, P. Barton, I. Garner, D.P. Leader, F. Bonhomme, J. Catalan, D. Simon, J.L. Guénet, F. Gros, and M.E. Buckingham. 1985. Genes for skeletal muscle myosin heavy chains are clustered and are not located on the same mouse chromosome as a cardiac myosin heavy chain gene. Proc. Natl. Acad. Sci. USA 82:7183–7187.

Myk-103 locus, transgenic sperm-lethality

This is the integration site of a plasmid containing a mouse metallothionein (*Mt-1*) promoter fused to the structural gene of thymidine kinase from herpes simplex virus (HSV). The insert consists of two copies of the plasmid oriented as inverted repeats. It is stably transmitted in a mendelian manner through females but expression of HSV thymidine kinase is extremely variable among mice carrying the insert. Males carrying the insert produce normal proportions of male and female offspring but do not transmit the insert. It is postulated that the insertion has disrupted a gene necessary for normal spermiogenesis in the haploid stages. The site of integration must be autosomal but its location is not known (1).

References

1. Palmiter, R.D., T.M. Wilkie, H.Y. Chen, and R.L. Brinster. 1984. Transmission distortion and mosaicism in an unusual transgenic mouse pedigree. Cell 36:869–877.

Myla, *Mylc*, *Mylf* loci

Myla locus, myosin light chain, alkali, cardiac atria, Chr 11. This locus codes for a myosin alkali light chain expressed in cardiac atrial muscle ($MLC1_A$) and in fetal skeletal and ventricular muscle ($MLC1_{emb}$). *Myla* was shown to cosegregate with *Es-3* (3/37 recombinants) on Chr 11 in an interspecific backcross (4).

Mylc locus, myosin light chain, alkali, cardiac ventricles, Chr 9. This locus codes for a myosin alkali light chain that is present in both heart ventricular ($MLC1_V$) and slow skeletal muscle ($MLC1_S$). The chains in the two tissues are indistinguishable and are coded for by the same gene (1). *Mylc* was shown to cosegregate with *Es-14* (0/37 recombinants) on Chr 9 in an interspecific backcross (4).

Mylf locus, myosin light chain, alkali, fast skeletal muscle, Chr 1. This locus codes for two different myosin alkali light chains ($MLC1_F$, $MLC3_F$) present in fast skeletal muscle. The common part of the two chains is encoded by four exons, the different parts by four other exons upstream from the common ones, two for chain 1 and two for chain 3 (3). *Mylf* was shown to be on Chr 1 near *Idh-1* by use of an interspecific backcross (4).

Mylf-ps locus, myosin light chain, alkali, fast skeletal muscle, pseudogene, Chr 12. This processed pseudogene was found to be on Chr 12 by use of mouse–Chinese hamster somatic cell hybrids (2).

References

1. Barton, P.J.R., A. Cohen, B. Robert, M.Y. Fiszman, F. Bonhomme, J.-L. Guénet, D.P. Leader, and M.E. Buckingham. 1985. The myosin alkali light chains of mouse ventricular and slow skeletal muscle are indistinguishable and are encoded by the same gene. J. Biol. Chem. 260:8578–8584.
2. Czosnek, H., P.E. Barker, F.H. Ruddle, and B. Robert. 1985. Chromosomal distribution of genes coding for fast twitch skeletal myosin light chains. Somat. Cell Mol. Genet. 11:533–540.
3. Robert, B., P. Daubas, M.-A. Akimenko, A. Cohen, I. Garner, J.-L. Guénet, and M. Buckingham. 1984. A single locus in the mouse encodes both myosin light chains 1 and 3, a second locus corresponds to a related pseudogene. Cell 39:129–140.
4. Robert, B., P. Barton, A. Minty, P. Daubas, A. Weydert, F. Bonhomme, J. Catalan, D. Chazottes, J.-L. Guénet, and M. Buckingham. 1985. Investigation of genetic

linkage between myosin and actin genes using an interspecific mouse backcross. Nature 314:181–183.

Mylpf locus, myosin light chain, phosphorylatable, fast skeletal muscle, Chr 7

This locus codes for the phosphorylatable myosin light chain (MLC 2_f) polypeptide of fast skeletal muscle. It is expressed throughout fetal and adult life. *Mylpf* was shown to be on Chr 7 using mouse–Chinese hamster somatic cell hybrids screened by Southern blot analysis with a rat cDNA probe (1).

References

1. Czosnek, H., P.E. Barker, F.H. Ruddle, and B. Robert. 1985. Chromosomal distribution of genes coding for fast twitch skeletal myosin light chains. Somat. Cell Mol. Genet. 11:533–540.

N

N, naked, semidominant, Chr 15

Arose as a spontaneous mutation in a stock at the Latvian University of Riga (3). Heterozygotes grow a nearly normal first coat. At 10 to 14 days the hairs begin to break off because of weakness due to incomplete keratinization. Breaking off and regeneration occur in cycles and produce animals with irregular bare and haired patches which change as the cycle progresses. Heterozygotes are viable and fertile. Homozygotes lack vibrissae at birth. They often die before 10 days, but some live and breed. The effect on the coat is much more severe than in heterozygotes. Homozygotes never grow a complete coat. They have cycles of hair regeneration and loss, but the new hairs break off almost as soon as they are formed (1, 2). The hair of heterozygotes is deficient in a protein fraction with high glycine and tyrosine content (6). This and other quantitative differences in proteins and amino acids of the hair of normal and naked mice are probably only indirect effects of the *N* gene (5). By use of dermal–epidermal recombination grafts, *N* has been shown to act in the epidermis (4).

References

1. David, L.T. 1932. The external expression and comparative dermal histology of hereditary hairlessness in mammals. Z. Zellforsch. 14:616–719.
2. Fraser, F.C. 1946. The expression of hereditary factors producing hypotrichosis in the mouse: histology and experimental results. Can. J. Res. D 24:10–25.
3. Lebedinsky, N.G., and A. Dauwart. 1927. Atrichosis und ihre Vererbung bei der albinotischen Hausmaus. Biol. Zentralbl. 47:746–752.
4. Raphael, K.A., and P.R. Pennycuik. 1980. The site of action of the naked locus (*N*) in the mouse as determined by dermal–epidermal recombinations. J. Embryol. Exp. Morphol. 57:143–153.
5. Raphael, K.A., R.C. Marshall, and P.R. Pennycuik. 1984. Protein and amino acid composition of hair from mice carrying the naked (*N*) gene. Genet. Res. 44:29–38.
6. Tenenhouse, H.S., R.J.M. Gold, Z. Kachra, and F.C. Fraser. 1974. Biochemical marker in dominantly inherited ectodermal malformation. Nature 251:431–432.

Nan, neonatal anemia, semidominant, Chr 8

Found in an offspring of a male mouse treated with the mutagen ethylnitrosourea. Homozygotes die at 10 to 11 days of gestation with severe deficiency of hemopoiesis. Heterozygotes have reasonably normal viability and fertility. They have a marked hemolytic anemia at birth which persists throughout life. The RBC count is moderately reduced with anisocytosis and many nucleated cells. The spleen is enlarged and there are iron deposits in spleen, liver, and kidney. The anemia can be transmitted with spleen cells and is therefore of the 'intrinsic' type (2). *Nan* is on Chr 8 near *Os* (1).

References

1. Lyon, M.F., and P.H. Glenister. 1986. Position of neonatal anemia (Nan) on chromosome 8. Mouse News Lett. 74:95.
2. Lyon, M.F., P.H. Glenister, J.F. Loutit, and J. Peters. 1983. Mouse News Lett. 68:68.

Nat-1 locus, *N*-acetyltransferase-1

This locus controls activity of the enzyme *N*-acetyltransferase (NAT) found in liver and also in blood. The enzyme catalyzes the *N*-acetylation of benzidine or the alternative substrate aminofluorene. The allele *Nat-1ʳ* determines rapid acetylation and occurs in the C57BL/6J strain; the allele *Nat-1ˢ* determines a 10-fold slower acetylation and occurs in the A/J and A/HeJ strains. Heterozygotes have intermediate activity (1).

References

1. Glowinski, J., and W. Weber. 1981. Mouse News Lett. 64:71.

Nat-2 locus, *N*-acetyltransferase-2

This locus controls presence or absence of *N*-acetyltransferase (NAT), an enzyme necessary for the synthesis of melatonin in the pineal gland. The allele *Nat-2ᵃ* determines presence of the enzyme in the pineal and occurs in the FDS stock derived from wild mice from Alberta and probably also in SK/Cam, SF/Cam, PERU-Atteck, and NZB; the allele *Nat-2ᵇ* determines absence of the enzyme and is known to occur in C57BL/6. Other strains that lack melatonin are BALB/c, AKR/J, IS/Cam, and CAST/Ei, but it has not been reported whether this results from lack of NAT or from lack of HIOMT, another enzyme necessary for melatonin synthesis, or possibly from other causes (1). See *Hiomt* locus.

References

1. Ebihara, S., T. Marks, D.J. Hudson, and M. Menaker. 1986. Genetic control of melatonin synthesis in the pineal gland of the mouse. Science 231:491–493.

nb, normoblastic anemia, recessive

Arose spontaneously in a heterogeneous stock. Homozygotes are recognizable shortly after birth by their bright orange color. Most of them survive to adulthood. As adults they are not obviously jaundiced although the urine, serum, and intestines are highly colored. Affected mice are smaller than normal and appear to be functionally sterile. The anemia is normocytic and hypochromic, with low red cell number and total hemoglobin, and a high proportion of immature red cells and nucleated red cell precursors. The affected mice show marked splenomegaly and also enlarged liver, heart, and kidneys (2). After 6 months of age, 57 per cent of *nb/nb* mice develop pigment gallstones. The frequency is twice as high in females as in males (7). Heme catabolism is greatly increased and is much greater during the period of erythroid maturation than after the cells have been released into the circulation (5). Fecal excretion of urobilinogen is elevated about 20-fold above normal (1, 4). Protoporphyrin concentration is higher in red cells of *nb/+* than of *+/+* mice, and 100 per cent of *nb/+* mice can be identified by this method (6). The red cells of *nb/nb* mice contain no ankyrin, a red cell membrane skeletal protein, suggesting that *nb* may be the structural locus for this protein (3).

References

1. Bannerman, R.M., J.A. Edwards, and P.H. Pinkerton. 1973. Hereditary disorders of the red cell in animals. Prog. Hematol. 8:131–179.
2. Bernstein, S.E. 1969. Hereditary disorders of the rodent erythron. *In* J.R. Lindsay, ed., Genetics in Laboratory Animal Medicine. Natl. Acad. Sci. Publ. 1697:9–33. Washington, DC.
3. Bodine, D.M. IV, C.S. Berkenmeier, and J.E. Barker. 1984. Spectrin deficient inherited anemias in the mouse: characterization by spectrin synthesis and mRNA activity in reticulocytes. Cell 37:721–729.
4. Kreimer-Birnbaum, M., R.M. Bannerman, E.S. Russell, and S.E. Bernstein. 1972. Pyrrole pigments in normal and congenitally anaemic mice. Comp. Biochem. Physiol. 43A:21–30.
5. Landaw, S.A. 1970. Kinetic aspects of endogenous carbon monoxide production in experimental animals. Ann. NY Acad. Sci.174:32–48.
6. Sassa, S. and S.E. Bernstein. 1978. Studies of erythrocyte protoporphyrin in anemic mutant mice: use of modified hematofluorometer for the detection of heterozygotes for hemolytic disease. Exp. Hematol. 6:479–487.
7. Trotman, B.W., S.E. Bernstein, K.E. Bove, and G.D. Wirt. 1980. Studies on the pathogenesis of pigment gall-

stones in hemolytic anemia. Description and characterization of a mouse model. J. Clin. Invest. 65:1301–1308.

Ncam locus, neural cell adhesion molecule, Chr 9

This locus encodes the neural cell adhesion molecule, N-CAM, a protein produced early in development and important in morphogenesis of the nervous system. There are embryonic and adult forms of N-CAM, which are probably the products of two mRNAs derived from the same gene by differential splicing of a single transcript. A DNA restriction fragment length polymorphism is revealed by digestion with *Msp*I. The allele *Ncam*^{a} determines three fragments of 0.7, 0.8, and 1.0 kb and occurs in strains DBA/2, C3H/He, and BALB/c; the allele *Ncam*^{b} determines two fragments of 1.7 and 2.7 kb and occurs in strains AKR, C57L, C57BL/6, SWR, and SJL. *Ncam* was shown to be on Chr 9 by use of somatic cell hybrids and to be close to *Thy-1* and *sg* by use of recombinant inbred strains. Both of these loci are associated with neural development, and *sg*, in particular, causes defective switching from the embryonic to adult form of N-CAM. *Ncam* recombines with *Thy-1* but has not been tested for recombination with *sg* (1).

References

1. D'Eustachio, P., G.C. Owens, G.M. Edelman, and B.A. Cunningham. 1985. Chromosomal location of the gene encoding the neural cell adhesion molecule (N-CAM) in the mouse. Proc. Natl. Acad. Sci. USA 82:7631–7635.

nct (previously *cat*, *Cat*^{Na}), Nakano cataract, recessive

Arose in an inbred strain at the National Institute of Health, Tokyo. Homozygotes are viable and fertile and are first identifiable on gross examination at 16 to 22 days of age by an opaque spot in the lens. By 4 to 5 weeks the whole lens is opaque. Histologically, the earliest abnormalities are seen at about 7 days and consist of retardation in loss of nuclei by the lens cells, predominantly in the bow area. Many lens fibers swell and become hydropic (1, 2). The histological changes are accompanied by an abnormally low sodium–potassium ATPase activity and an abnormal increase in sodium and water uptake by the lens (3). There is a decrease in rate of crystallin synthesis and leakage of crystallins from the lens (4). However, the lens fiber cells possess a full complement of crystallin mRNAs, indicating that poor utilization of the mRNAs must be responsible for the reduction in crystallin synthesis (5).

References

1. Brown, E.R., T. Nakano, and G.L. Vankin. 1970. Early development of an inherited cataract in mice. Exp. Anim. 19:95–99.
2. Hamai, Y., H.N. Fukui, and T. Kuwabara. 1974. Morphology of hereditary mouse cataract. Exp. Eye Res. 18:537–546.
3. Iwata, S., and J.H. Kinoshita. 1971. Mechanism of development of hereditary cataract in mice. Invest. Ophthalmol. 10:504–512.
4. Piatigorsky, J., H.N. Fukui, and J.H. Kinoshita. 1978. Differential metabolism and leakage of protein in an inherited cataract and a normal lens cultured with ouabain. Nature 274:558–562.
5. Shinohara, T., and J. Piatigorsky. 1980. Persistence of crystallin messenger RNA's with reduced translation in hereditary cataracts in mice. Science 210:914–916.

Nel-1 locus, Nel lymphocyte antigen-1, Chr 17

This locus controls an H-2-restricted T-cell antigen that elicits strong immunity in *H-2*-congenic recipients and can act as a target antigen in CML experiments. It has no effect on skin graft rejection and is therefore probably not the product of a minor histocompatibility gene. The allele *Nel-1*^{a} determines presence of the antigen Nel-1.1 and occurs in the DBA/2, A, CBA, C3H/He, NZO, and 129 strains; the allele *Nel-1*^{b} determines absence of the antigen and occurs in the AKR, BALB/c, C58, HTI, NZW, and SWR strains. The locus was shown to be on Chr 17 proximal to *H-2K* by presence or absence of the antigen in various *H-2*-congenic strains (1).

References

1. Hetherington, C.M., E.P. Blankenhorn, and E. Simpson. 1979. An H-2-restricted CML target antigen controlled by a gene linked to the *H-2* complex. Immunogenetics 9:255–260.

Neu-1 locus, neuraminidase-1, Chr 17

This locus controls activity level of neuraminidase (E.C. 3.2.1.18) in the liver. Low activity determined by the *Neu-1*^{a} allele occurs in the SM/J inbred strain and in wild mice in the area of Ann Arbor, Michigan; all other inbred strains have high activity determined by the *Neu-1*^{b} allele. Heterozygotes have intermediate activity. The neuraminidase deficiency in the SM/J strain causes defective posttranslational processing of at least four liver lysosomal enzymes, acid phosphatase, α-mannosidase, arylsulfatase B, and acid α-glucosidase. Defective processing of two of these, acid phosphatase and α-mannosidase, was previously ascribed to effects of the loci *Apl* (7) and *Map-2* (2), respectively, both located on Chr 17 near *H-2*. The processing of the other two enzymes is also controlled by a gene in the same region (1, 5). Since *Neu-1* maps either close to the *H-2* complex or within it (3, 6) and no recombination has been found between *Neu-1* and the loci controlling the other enzymes, it is very likely that the variation in all the enzymes is controlled by a single locus, *Neu-1*. Neuraminidase removes extra sialic acid residues from these enzymes. The defective neuraminidase of *Neu-1ᵃ* mice fails to remove the extra sialic acid and so changes their electrophoretic mobility (8). Level of activity of neuraminidase in activated T lymphocytes is also depressed in $Neu-1^a/Neu-1^a$ mice (4).

References

1. Daniel, W.L., J.E. Womack, and P.S. Henthorn. 1981. Murine arylsulfatase B processing by region on chromosome 17. Biochem. Genet. 19:211–225.
2. Dizik, M., and R.W. Elliott. 1978. A second gene affecting the sialylation of lysosomal α-mannosidase in mouse liver. Biochem. Genet. 16:247–260.
3. Figueroa, F., D. Klein, S. Tewarson, and J. Klein. 1982. Evidence for placing the *Neu-1* locus within the mouse *H-2* complex. J. Immunol. 129:2089–2093.
4. Landolfi, N.F., J. Leone, J.E. Womack, and R.G. Cook. 1985. Activation of T lymphocytes results in an increase in *H-2*-encoded neuraminidase. Immunogenetics 22:159–167.
5. Peters, J., D.M. Swallow, S.J. Andrews, and L. Evans. 1981. A gene (*Neu-1*) on chromosome 17 of the mouse affects acid α-mannosidase and codes for neuraminidase. Genet. Res. 38:47–55.
6. Womack, J.E., and C.S. David. 1982. Mouse gene for neuraminidase activity (*Neu-1*) maps to the *D* end of *H-2* Immunogenetics 16:177–180.
7. Womack, J.E., and E.M. Eicher. 1977. Liver-specific lysosomal acid phosphatase deficiency (*Apl*) on mouse chromosome 17. Mol. Gen. Genet. 155:315–317.
8. Womack, J.E., D.L. Shun Yan, and M. Potier. 1981. Gene for neuraminidase activity on mouse chromosome 17 near H-2: pleiotropic effects on multiple hydrolases. Science 212:63–65.

Nfxv-1 locus, NFS xenotropic virus-1

This is the site of integration of a xenotropic *env* viral sequence that is probably not associated with an intact xenotropic virus. It is present in the low-virus NFS strain in which it is not possible to induce virus. The structure of *Nfxv-1* is substantially different from that of all known infectious xenotropic viruses (1)

References

1. Hoggan, M.D., R.R. O'Neill, and C.A. Kozak. 1986. Nonecotropic murine leukemia viruses in BALB/c and NFS/N mice: characterization of the BALB/c *Bxv-1* provirus and the single NFS endogenous xenotrope. J. Virol.60:980–986.

Ng, nackig, semidominant

Arose spontaneously in the NMRI strain. Heterozygotes have a sparse first coat, and homozygotes are completely naked. The interfollicular epidermis of *Ng/Ng* mice is hyperplastic with marked hyperkeratosis, and the lower part of the hair follicles is greatly reduced (1).

References

1. Franz, E., K. Bosse, and E. Kreutzer. 1978. Uüber den Hautcyclus einer Spontanmutante (Nackig) mit mangelhafter Haarkeratinisierung bei der Maus. Arch. Dermatol. Res. 262:63–71.

Ngfa, *Ngfb*, *Ngfg* loci

These loci code for the α, β, and γ subunits of nerve growth factor (NGF), a high molecular weight complex (7s NGF) composed of six polypeptides ($\alpha_2\beta_2\gamma_2$). NGF is produced at high levels in the submandibular gland of male mice. It appears to be essential for growth and development of the peripheral nervous system. The β subunit (NGF-β) is responsible for the growth-promoting activity. The γ subunit (NGF-γ) is believed to cleave pro-β-NGF to generate the active growth factor. The α subunit (NGF-α) has no enzymatic activity but may be necessary for formation of a stable 7S complex (1).

Ngfa locus, nerve growth factor-α, Chr 7.

Ngfg locus, nerve growth factor-γ, Chr 7. NGF-α and NGF-γ are members of a family of closely related serine proteases that includes the kallikrein (*Kal*) family, tamases (*Tam*), and the esteroprotease loci, *Prt-4* and *Prt-5*, all located on Chr 7. *Ngfg* was found to be on Chr 7 by use of mouse–Chinese hamster somatic cell hybrids screened with a cDNA clone isolated from a male submandibular gland cDNA library (2). *Ngfa* was found to be contiguous with *Ngfg*, transcribed from the same DNA strand, and separated from *Ngfg* by 5.3 kb of intergenic DNA (1). *Ngfg* of strain DBA/2 shows a difference in a single base pair from the gene in strains BALB/c and Swiss Webster, producing aspartate at position 175 in DBA/2 and glutamate at that position in BALB/c and Swiss Webster (1, 2). A polymorphism for five DNA fragments recognized by an NGF-γ probe occurs between the DBA/2 and C57BL/6 strains. By use of BXD recombinant inbred strains, these differences were all found to cosegregate with *Tam-1* on Chr 7 (2). See *Kal* complex.

Ngfb locus, nerve growth factor-β, Chr 3. This locus codes for the precursor of NGF-β, pre-pro-NGF-β. NGF-β is generated by processing of the precursor, probably by action of NGF-γ (1). No genetic variants of *Ngfb* are known. The locus was found to be on Chr 3 using mouse–Chinese hamster somatic cell hybrids and a cDNA probe encoding mouse pre-pro-NGF-β (3).

References

1. Evans, B.A., and R.I. Richards. 1985. Genes for the alpha and gamma subunits of mouse nerve growth factor are contiguous. EMBO J. 4:133–138.
2. Howles, P.N., D.P. Dickinson, L.L. DiCaprio, M. Woodworth-Gutai, and R.W. Gross. 1984. Use of a cDNA recombinant for the γ-subunit of mouse nerve growth factor to localize members of this multigene family near the TAM-1 locus on chromosome 7. Nucl. Acids Res. 12:2791–2805.
3. Zabel, B.U., R.L. Eddy, P.A. Lalley, J. Scott, G.I. Bell, and T.B. Shows. 1985. Chromosomal locations of the human and mouse genes for precursors of epidermal growth factor and the β subunit of nerve growth factor. Proc. Natl. Acad. Sci. USA 82:469–473.

Nil, neonatal intestinal lipidosis, semidominant, Chr 7, probably extinct

Found in 1962 in the A/Cam strain. Heterozygotes are detectable shortly after suckling begins by whiteness of parts of the small intestine. The abnormal portions of the intestinal wall are thicker than normal and contain abundant lipid, particularly in the submucosa. The condition often persists until after weaning. Penetrance in heterozygotes is incomplete and dependent on the genetic background. Heterozygotes have slightly reduced viability and fertility and tend to be small as adults. Homozygotes die at or before 11 days of gestation (1).

References

1. Wallace, M.E., and B.M. Herbertson. 1969. Neonatal intestinal lipidosis in mice. An inherited disorder of the intestinal lymphatic vessels. J. Med. Genet. 6:361–375.

Nk-1, Nk-2 loci

Nk-1 locus, natural killer cell antigen-1. This locus controls an antigen on the surface of natural killer cells. It was detected by use of an antibody produced in (BALB/c × C3H)F1 mice against CE lymphoid cells

(2). The allele *Nk-1ᵃ* determines the antigen, designated Nk-1.1, thus detected; it occurs in strains CE, C57BL/6, C57BR/cd, C57L, C58, DBA/1, MA/My, NZB, SJL, SM, and B10.D2 (3). The allelic antigen was detected by an antibody produced in CE mice against CBA cells (1). The allele *Nk-1ᵇ* determines this antigen, designated Nk-1.2; it occurs in strains CBA/J, BALB/c, C3H/He, A/J, DBA/2, LP, and 129 and possibly some others (1, 3). Pollack and Emmons (3) have shown that *Nk-1* segregates independently of *Nk-2*. It is possible that some of the antisera used in studying these two loci are not specific for only one of them.

Nk-2 locus, natural killer cell antigen-2, Chr 2. This locus controls an antigen on natural killer cells. It was detected by use of an antibody produced in strain NZB/J against BALB/c spleen cells. The antibody causes complement-dependent depletion of natural killer (NK) cell activity in antigen-positive mice as measured in an assay with Yac-1 tumor cells. The allele *Nk-2ᵃ* determines presence of the Nk-2.1 antigen and occurs in strains BALB/c, C3H/He, C57BL/6, DBA/1, DBA/2, and LP; the allele *Nk-2ᵇ* determines absence of the Nk-2.1 antigen and occurs in strains CE, C57BR/cd, C57L, MA/My, NZB, SJL, SM, 129, and B10.A. (BALB/c × NZB)F1 hybrids were slightly more sensitive to the antiserum than NZB mice, but in the F2 generation heterozygotes were apparently classified as not sensitive (negative for the Nk-2.1 antigen), so that the negative allele (*Nk-2ᵇ*) acts as a dominant. *Nk-2* segregates independently of *Nk-1* and is loosely linked to *a* on Chr 2 (1).

References

1. Burton, R.C., and H.J. Winn. 1981. Studies on natural killer (NK) cells. I. NK cell specific antibodies in CE anti-CBA serum. J. Immunol. 126:1985–1989.
2. Glimcher, L., F.-W. Shen, and H. Cantor. 1977. Identification of a cell-surface antigen selectively expressed on natural killer cells. J. Exp. Med. 145:1–9.
3. Pollack, S.B., and S.L. Emmons. 1982. NK-2.1; an NK-associated antigen detected with NZB anti-BALB/c serum. J. Immunol. 129:2277–2281.

Nop, nuclear opacity, dominant

Found among untreated (101 × C3H)F1 hybrids. Both homozygotes and heterozygotes are fully viable and fertile, and penetrance is complete in both. The nuclear opacity of the lens is seen as a white spot in the eye. The cataractous lens shows a decrease in activity of four glycolytic enzymes, a decreased amount of glutathione disulfide, and loss of a single protein, probably a γ-crystallin, as revealed by isoelectric focusing. No tests for allelism with other cataract mutations have been made, but the absence of γ-crystallin suggests possible allelism with *Len-1* (1).

References

1. Graw, J., J. Kratochvilova, and K.-H. Summer. 1984. Genetical and biochemical studies of a dominant cataract mutant in mice. Exp. Eye Res. 39:37–45.

Np-1, *Np-2* loci, Chr 14

These loci control electrophoretic and quantitative variation of the enzyme purine nucleoside phosphorylase (NP; E.C. 2.4.2.1). They have shown no recombination in crosses. They are located about 12 cM proximal to *Es-10* on Chr 14 (2).

Np-1 locus, nucleoside phosphorylase-1. This locus controls electrophoretic mobility of NP occurring in a variety of tissues. The allele *Np-1ª* determines a fast anodally migrating band and occurs in all inbred strains tested; the allele *Np-1ᵇ* determines a slower band and occurs in *M. m. castaneus* and *M. m. molossinus*. Heterozygotes have a blurred intermediate band (4).

Np-2 locus, nucleoside phosphorylase-2. This locus controls presence or absence of an isozyme of NP that is present mostly on erythrocytes and is expressed as an electrophoretic band just cathodal to that of *Np-1ª*. Presence of the band is always associated with about a 2.5-fold increase in enzyme activity (3). The allele *Np-2ᵇ* determines presence of the band and high activity and occurs in the C57, C58 family of strains; the allele *Np-2ª* determines absence of the band and low activity and occurs in BALB/c, NZB/Bl, 129/J and most other inbred strains. Heterozygotes resemble *Np-2ᵇ*/*Np-2ᵇ* (1). No difference in biochemical characteristics of NP was found between the *Np-2ᵇ* strain C57BL/6J and the *Np-2ª* strain DBA/2J (3).

References

1. Bremmer, T.A., E. Premkumar-Reddy, K. Nayar, and R.E. Kouri. 1978. Nucleoside phosphorylase 2 (*Np-2*) in mice. Biochem. Genet. 16:1143–1151.
2. Lukey, T., K. Neote, J.E. Loman, A.E. Unger, F.G. Biddle, and F.F. Snyder. 1985. Purine nucleoside phosphorylase (Np) in the mouse: linkage relationship of *Np-2* to esterase-10 (*Es-10*) and *Np-1* on chromosome 14. Biochem. Genet. 23:347–356.
3. Snyder, F.F., F.G. Biddle, T. Lukey, and M.J. Sparling. 1983. Genetic variability of purine nucleoside phosphorylase activity in the mouse: relationship to NP-1 and NP-2. Biochem. Genet. 21:323–332.
4. Womack, J.E., M.T. Davisson, E.M. Eicher, and D.A. Kendall. 1977. Mapping of nucleoside phosphorylase

(*Np-1*) and esterase-10 (*Es-10*) on mouse chromosome 14. Biochem. Genet. 15:347–355.

nr, nervous, recessive, Chr 8

Arose spontaneously in a BALB/cGr subline carrying *tk* (tail-kinks). Homozygotes may be recognized at 2 to 3 weeks by their smaller size and hyperactive ataxic behavior. Their gait is hesitant and wobbly and remains so throughout life. Viability is nearly normal, and both sexes are fertile but are very poor breeders (9). There is a 90 per cent loss of Purkinje cells between 23 and 50 days. The mitochondria of all Purkinje cells become large and rounded beginning at 9 days. Subsequently most Purkinje cells degenerate. A few reacquire normal-appearing mitochondria and survive longer (2), but some degeneration of Purkinje cells continues throughout life (10). Several mitochondrial dehydrogenases were found to be histochemically normal up until the time of mitochondrial rounding (4). Purkinje cell abnormalities are preceded by partial degeneration of cells of the external granular layer (EGL) at 6 to 8 days. Since normal maturation of Purkinje cells is dependent on proliferation and inward migration of EGL cells, the Purkinje cell degeneration may be secondary to the EGL defect (7). Some neurons in other regions of the brain also transiently contain spherical mitochondria, but these cells probably do not subsequently degenerate (3). In the retina there is degeneration of the photoreceptors, which is already present at 13 days. Whorls of outer segment membranes form and the outer segments eventually disappear completely (6). The nearly total lack of Purkinje cells in nervous mice has been used as a tool in several investigations of the chemistry, physiology, and development of these cells (1, 5, 8, 10).

References

1. Atlas, D., V.I. Teichberg, and J.P. Changeux. 1977. Direct evidence for beta-adrenoreceptors on the Purkinje cells of mouse cerebellum. Brain Res. 128:532–536.
2. Landis, S.C. 1973. Ultrastructural changes in the mitochondria of cerebellar Purkinje cells of nervous mutant mice. J. Cell Biol. 57:782–797.
3. Landis, S.C. 1973. Changes in neuronal mitochondrial shape in brains of nervous mutant mice. J. Hered. 64:193–196.
4. Landis, S.C. 1975. Histochemical demonstration of mitochondrial dehydrogenases in developing normal and nervous mutant mouse Purkinje cells. J. Histochem. Cytochem. 23:136–143.
5. Mallet, J., M. Huchet, R. Pougeois, and J.P. Changeux. 1976. Anatomical, physiological and biochemical studies on the cerebellum from mutant mice. III. Protein differ-

ences associated with the weaver, staggerer and nervous mutations. Brain Res. 103:291–312.

6. Mullen, R.J., and M.M. LaVail. 1975. Two new types of retinal degeneration in cerebellar mutant mice. Nature 258:528–530.

7. Mishra, P.R., V.K. Jain, V. Bijlani, and M.S. Grewal. 1983. DNA loss in the developing cerebellum of nervous mice: a flow cytometric study. Dev. Brain Res. 6:193–196.

8. Schmidt, M.J., and N.S. Nadi. 1977. Cyclic nucleotide accumulation *in vitro* in the cerebellum of "nervous" neurologically mutant mice. J. Neurochem. 29:87–90.

9. Sidman, R.L., and M.C. Green. 1970. "Nervous", a new mutant mouse with cerebellar disease. *In* M. Sabourdy, ed., Les mutants pathologiques chez l'animal, 69–79. Centre National de la Recherche Scientifique, Paris.

10. Sotelo, C., and A. Triller. 1979. Fate of presynaptic afferents to Purkinje cells in the adult nervous mutant mouse: a model to study presynaptic stabilization. Brain Res. 175:11–36.

Nras locus, neuroblastoma oncogene, Chr 3

This is the homolog of the N-*ras* oncogene originally found in a human neuroblastoma (3). Unlike the *Hras* and *Kras* oncogenes, it has not been found in transforming retroviruses. In the mouse, the activated *Nras* gene was found in thymic lymphomas induced by N-methylnitrosourea. No genetic variants of *Nras* are known. By use of mouse–Chinese hamster somatic cell hybrids, it was found to be on Chr 3 unlinked to *Hras* or *Kras* (1, 2). Sequences homologous to 5′ and 3′ flanking regions of *Nras*, which may represent a complete *Nras*-related gene, were found to be located on Chr 14 (2).

References

1. Guerrero, I., A. Villasante, P. D'Eustachio, and A. Pellicer. 1984. Isolation, characterization, and chromosome assignment of a mouse N-*ras* gene from carcinogen-induced thymic lymphoma. Science 225:1041–1043.

2. Ryan, J., C.P. Hart, and F.H. Ruddle. 1984. Molecular cloning and chromosome assignment of murine N-*ras*. Nucl. Acids Res. 12:6063–6072.

3. Shimizu, K., M. Goldfarb, Y. Suard, M. Perucho, Y. Li, T. Kamata, J. Feramisco, E. Stavnezer, J. Fogh, and M.H. Wigler. 1983. Three human transforming genes are related to the viral *ras* oncogenes. Proc. Natl. Acad. Sci. USA 80:2112–2116.

nu, nude, recessive, Chr 11

Found in mice in the Virus Laboratory, Ruchill Hospital, Glasgow. Homozygotes are hairless from birth throughout life and completely lack a thymus. Viability and fertility are severely reduced (3, 8). Hair follicles are normal at birth, but keratinization, which normally occurs in the middle third of the follicle, is faulty and the hairs do not erupt (3). Absence of the thymus is due to failure of development of the thymic anlage which arises from the ectoderm of the third pharyngeal pouch. The rudiment remains small and cystic. Another derivative of the third pharyngeal pouch, the parathyroid, is unaffected (2). Differentiation of the stroma of the normal thymus into cortical and medullary parts, as recognized by monoclonal antibodies, occurs by embryonic day 13. This differentiation does not occur in *nu/nu* mice (16). Transplantation studies have shown that the thymic rudiment of *nu/nu* mice fails to attract lymphoid cells, but that lymphoid cells of *nu/nu* mice are quite normal in their ability to populate implanted normal thymuses (9). Heterozygotes (*nu/+*) have a thymus about 50 to 80 per cent as large as that of *+/+* congenic controls, the size of the difference depending on strain background. The two genotypes can be distinguished by this criterion (7). Lack of the thymus in homozygotes leads to many defects of the immune system, including depletion of lymphocytes from thymus-dependent areas of lymph nodes and spleen, a much reduced lymphocyte population composed almost entirely of B-cells, relatively normal IgM response to thymus-independent antigens, very poor response to thymus-dependent antigens including failure to reject allogeneic and xenogeneic skin and tumor grafts, and greatly increased susceptibility to infection (5, 9, 11, 13). This work has been reviewed by Kindred (6). Viability and fertility of nude mice can be greatly increased by protecting them from infectious organisms, as under specific pathogen-free or germ-free conditions (12). Nude mice are widely used as naturally thymectomized mice in investigations of the role of the thymus in immune reactions. Rygaard and Povlsen (14) published an extensive bibliography of the literature on nude, and descriptions of more recent work can be found in Fogh and Giovanella (4) and Reed (10).

A remutation at the *nu* locus (*nu^{str}*, streaker) occurred in the AKR/J strain. Homozygotes resemble *nu/nu* mice, the difference between the two alleles being no greater than reported variation among different experiments with *nu* (15). AKR/J mice have a very high incidence of lymphatic leukemia which is abolished by absence of the thymus in AKR/J-*nu^{str}/nu^{str}* mice (1).

References

1. Bedigian, H.G., L.D. Shultz, and H. Meier. 1979. Expression of endogenous murine leukemia viruses in AKR/J streaker mice. Nature 279:434–436.

2. Cordier, A.C., and S.M. Haumont. 1980. Development of thymus, parathyroids, and ultimo-branchial bodies in NMRI and nude mice. Am. J. Anat. 157:227–263.

3. Flanagan, S.P. 1966. "Nude", a new hairless gene with pleiotropic effects in the mouse. Genet. Res. 8:295–309.

4. Fogh, J., and B.C. Giovanella, eds. 1982. The Nude Mouse in Experimental and Clinical Research, Vol. 2. Academic Press, New York, 587p.

5. Gershwin, M.E., B. Merchant, M.C. Gelfand, J. Vickers, A.D. Steinberg, and C.T. Hansen. 1975. The natural history and immunopathology of outbred athymic (nude) mice. Clin. Immunol. Immunopathol. 4:324–340.

6. Kindred, B. 1981. Deficient and sufficient immune systems in the nude mouse. *In* M.E. Gershwin and B. Merchant, eds., Immunological Defects in Laboratory Animals, Vol. 1, 215–265. Plenum Press, New York.

7. Kojima, A., M. Saito, K. Hioki, K. Shimanura, and S. Habu. 1984. NFS/N-nu/+ mice can macroscopically be distinguished from NFS/N-+/+ littermates by their thymic size and shape. Exp. Cell Biol. 52:107–110.

8. Pantelouris, E.M. 1968. Absence of thymus in a mouse mutant. Nature 217:370–371.

9. Pantelouris, E.M. 1973. Athymic development in the mouse. Differentiation 1:437–450.

10. Reed, N.D., ed. 1982. Proceedings of the Third International Workshop on Nude Mice, 2 vols. Gustav Fischer, New York.

11. Rygaard, J. 1973. Thymus and self. Immunobiology of the mouse mutant *nude*. F.A.D.L. Copenhagen.

12. Rygaard, J., and C.W. Friis. 1974. The husbandry of mice with congenital absence of the thymus (*nude* mice). Z. Versuchstierk. 16:1–10.

13. Rygaard, J., and C.O. Povlsen. 1974. Effects of homozygosity of the *nude* (*nu*) gene in three inbred strains of mice. A detailed study of mice of three genetic backgrounds (BALB/c, C3H, C57BL/6) with congenital absence of the thymus (*nude* mice) at a stage in the gene transfer. Acta Pathol. Microbiol. Scand. A 82:48–70.

14. Rygaard, J. and C.O. Povlsen. 1977. Bibliography of the nude mouse. 1966–1976. Gustav Fischer, Stuttgart, New York.

15. Shultz, L.D., H.J. Heiniger, and E.M. Eicher. 1978. Immunopathology of streaker mice: a remutation to nude in the AKR/J strain. *In* M.E. Gershwin and E.L. Cooper, eds., Animal Models of Comparative and Developmental Aspects of Immunity and Disease, 211–222. Pergamon Press, New York.

16. Van Vliet, E., E.J. Jenkinson, R. Kingston, J.J.T. Owen, and W.V. Van Ewijk. 1985. Stromal cell types in the developing thymus of the normal and nude mouse embryo. Eur. J. Immunol. 15:675–681.

nuc, nuclear cataract, recessive

Arose in the 129/Sv-ter strain. Homozygotes can be identified at 3 to 4 weeks of age by a white area in the center of the lens. The cortex of the lens is not affected, and the eyes are of normal size. Manifestation is bilateral. The mutation resembles vacuolated lens, *vl*, but is not allelic with it or with blind sterile, *bs* (1).

References

1. Varnum, D. 1981. Mouse News Lett. 64:59.

Nuca (originally *Nuc*), dominant nuclear cataract, probably semidominant

Arose in a (101 × C3H)F1 hybrid after irradiation with ^{137}Cs gamma rays. Homozygotes have not been obtained. In heterozygotes, a granular opacity of the embryonic nucleus protrudes into the anterior part of the lens. There is also an opacity of the posterior embryonic suture. Manifestation is bilateral and penetrance is complete. Most heterozygotes are sterile or have greatly reduced litter size with considerable prenatal mortality. Matings between heterozygotes have not produced offspring (1).

References

1. Kratochvilova, J. 1981. Dominant cataract mutations detected in offspring of gamma-irradiated male mice. J. Hered. 72:302–307.

nv, Nijmegen waltzer, recessive, Chr 7

Arose spontaneously in a non-inbred Swiss albino stock. Homozygotes show circling behavior and head shaking but are not deaf. There is considerable variability in degree of abnormal behavior. Mutants are best recognized at 4 weeks. Both sexes are viable and fertile (1). In a detailed study of behavioral development of *nv*/*nv* and control mice, van Abeelen and Kalkhoven (2) found that incoordination and hyperkinesis developed at about 1 week and that, otherwise, aberrations of reflexes in the mutants were not very severe.

References

1. van Abeelen, J.H.F., and P.H.W. van der Kroon. 1967. *Nijmegen waltzer*—a new neurological mutant in the mouse. Genet. Res. 10:117–118.

2. van Abeelen, J.H.F., and J.Th.R. Kalkhoven. 1970. Behavioural ontogeny of the Nijmegen waltzer, a neurological mutant in the mouse. Anim. Behav. 18:711–718.

Nzc, nuclear and zonular cataract, semidominant

Arose in a (101 × C3H)F1 hybrid after irradiation with ^{137}Cs gamma rays. In heterozygotes, there is a fine granular opacity of the nucleus surrounded by a less opaque concentric area. In homozygotes, a dense opacity involves the whole lens except the peripheral layer. Manifestation is bilateral and penetrance is complete in both heterozygotes and homozygotes. Homozygotes are viable and fertile (1).

References

1. Kratochvilova, J. 1981. Dominant cataract mutations detected in offspring of gamma-irradiated male mice. J. Hered. 72:302–307.

Nzv-1, *Nzv-2* loci, NZB virus-1, -2

These loci control expression of two infectious xenotropic viruses found in the NZB strain (2). *Nzv-1* controls expression of a virus spontaneously produced in high titer and detectable as infectious centers in a direct focus assay on mink S^+L^-cells; *Nzv-2* controls expression of a virus produced in low titer and detectable only in the absence of *Nzv-1* virus by induction with IdU in fibroblasts or by immunofluorescence assay in uninfected mink cells (1, 5, 6). Strains carrying neither *Nzv-1* nor *Nzv-2* are SWR (2), 129/J and SEA/Gn (6), and NFS (1). The chromosomal location of these loci is not known, but no linkage was detected with loci on Chrs 4, 5, 7, and 9 (6). *Nzv-1* and *Nzv-2* are inherited independently of the autoimmune disease of NZV mice and independently of the level of envelope glycoprotein gp70 (3, 4).

References

1. Chused, T.M., and H.C. Morse III. 1978. Expression of XenCSA, a cell surface antigen related to the major glycoprotein (gp70) of xenotropic murine leukemia virus, by lymphocytes of inbred mouse strains. *In* H.C. Morse III, ed., Origins of Inbred Mice, 297–319. Academic Press, New York.
2. Datta, S.K., and R.S. Schwartz. 1977. Mendelian segregation of loci controlling xenotropic virus production in NZB crosses. Virology 83:449–452.
3. Datta, S.K., N. Manny, C. Andrzejewski, J. Andreé-Schwartz, and R.S. Schwartz. 1978. Genetic studies of autoimmunity and retrovirus expression in crosses of New Zealand black mice. I. Xenotropic virus. J. Exp. Med. 147:854–871.
4. Datta, S.K., P.J. McConahey, N. Manny, A.N. Theofilopoulos, F.J. Dixon, and R.S. Schwartz. 1978. Genetic studies of autoimmunity and retrovirus expression in crosses of New Zealand black mice. II. The viral envelope glycoprotein gp70. J. Exp. Med. 147:872–881.
5. Elder, J.H., J.W. Gautsch, F.C. Jensen, R.A. Lerner, T.H. Chused, H.C. Morse, J.W. Hartley, and W.P. Rowe. 1980. Differential expression of two distinct xenotropic viruses in NZB mice. Clin. Immunol. Immunopathol. 15:493–501.
6. Kozak, C.A., J.W. Hartley, and H.C. Morse III. 1984. Laboratory and wild-derived mice with multiple loci for production of xenotropic murine leukemia virus. J. Virol. 51:77–80.

O

Oat locus, ornithine aminotransferase, Chr 7

This is probably the structural locus for the enzyme ornithine aminotransferase (OAT; E.C. 2.6.1.13). *Oat* was found to be on Chr 7 by use of mouse–Chinese hamster somatic cell hybrids classified for presence of mouse OAT by starch gel electrophoresis (1).

References

1. O'Donnell, J.J., K.M. Vannas-Sulonen, T.B. Shows, and D.E. Cox. 1985. Ornithine aminotransferase (OAT) maps to human chromosome 10 and mouse chromosome 7. Cytogenet. Cell Genet. 40:716 (Abstr.).

ob, obese, recessive, Chr 6

Arose as a spontaneous mutation in a multiple recessive stock (7). Homozygotes are first recognizable at about 4 weeks. They increase in weight rapidly and may reach three times the normal weight. Females are always sterile, but an occasional male may breed, particularly if maintained on a restricted diet. As in the case of *db* (diabetes), progress of the disease is strikingly dependent on the genetic background: on the C57BL/6 background, homozygotes develop a well compensated diabetes with only temporary and moderate hyperglycemia, marked hyperinsulinemia, and enlarged islets of Langerhans; on the C57BL/Ks background, they become severely diabetic with regression of the islets and early death (5). The obesity is due in part to increase in number of adipocytes; this is in contrast to other genetic obesities in the mouse in which increase in fat depots is due to cell enlargement (8,9). Hyperinsulinemia does not develop until after the increase in food intake and is probably the result of it (6). Heterozygotes gain excess weight and deposit excess fat even when restricted to a diet sufficient for normal lean mice, thus demonstrating an increased metabolic efficiency (12). They have an impaired capacity for non-shivering thermogenesis which can be demonstrated at low temperatures as early as 10 days (14). At 4°C they rapidly become hypothermic and die within a few hours (13). It has been suggested that a reduction in the energy requirement normally used for thermoregulatory heat production could be responsible for the increased metabolic efficiency. That is unlikely to be the major cause, however, since brief exposure of obese mice to 10°C produces cold adaptation and allows indefinite survival at 4°C with maintenance of nearly normal body temperature (4). Coleman (4) has suggested that the hyperinsulinemia of obese mice, which increases synthetic processes and decreases degradation, might spare the energy normally spent on tissue turnover and account at least in part for the increased efficiency. Some contribution to the effect may also come from a low basal N^+,K^+ATPase activity, shown to occur in muscle of 14-day old mutants (10). In parabiosis with normal mice, *ob/ob* mice eat less and lose weight. In parabiosis with *db/db* mice, they cease eating completely and die of starvation. This is taken to mean that they have a normal response to a 'satiety' factor but are unable to produce enough of it to maintain normal food intake (2). Efforts to identify the satiety factor have produced several promising suspects but all have failed of confirmation (1, 11). Heterozygotes (*ob/+*) have normal body weight, blood glucose, and plasma insulin, but survive a prolonged fast longer than congenic (*+/+*) controls, suggesting that increased metabolic efficiency is expressed to some extent in heterozygotes (3). There is a vast literature on the biochemical and physiological alterations in *ob/ob* mice which has been reviewed by Herberg and Coleman (6) and more recently by Charlton (1). A large number of these studies were performed in adult mice and may therefore involve the secondary effects of obesity rather than the primary causes of it.

References

1. Charlton, H.M. 1984. Mouse mutants as models in endocrine research. Quart. J. Exp. Physiol. 69:655–676.
2. Coleman, D.L. 1973. Effects of parabiosis of obese with diabetes and normal mice. Diabetologia 9:294–298.
3. Coleman, D.L. 1979. Obesity genes: beneficial effects in heterozygous mice. Science 203:663–665.
4. Coleman, D.L. 1982. Thermogenesis in diabetes–obesity syndromes in mutant mice. Diabetologia 22:205–211.
5. Coleman, D.L., and K.P. Hummel. 1973. The influence of genetic background on the expression of the obese (*ob*) gene in the mouse. Diabetologia 9:287–293.
6. Herberg, L., and D.L. Coleman. 1977. Laboratory animals exhibiting obesity and diabetes syndromes. Metabolism 26:59–99.
7. Ingalls, A.M., M.M. Dickie, and G.D. Snell. 1950. Obese, a new mutation in the house mouse. J. Hered. 41:317–318.
8. Johnson, P.R., and J. Hirsch. 1972. Cellularity of adipose tissue depots in six strains of genetically obese mice. J. Lipid. Res. 13:2–11.
9. Kaplan, M.L., J.R. Trout, and G.A. Leveille. 1976. Adipocyte size distribution in *ob/ob* mice during preobese and obese phases of development. Proc. Soc. Exp. Biol. Med. 153:476–482.
10. Lin, M.H., D.R. Romsos, T. Akera, and G.A. Leveille. 1979. Na^+,K^+ATPase enzyme units in skeletal muscle and liver of 14-day-old lean and obese (*ob/ob*) mice. Proc. Soc. Exp. Biol. Med. 161:235–238.
11. Taylor, I.L., and R. Garcia. 1985. Effects of pancreatic polypeptide, cerulein, and bombesin on satiety in obese mice. Am. J. Physiol. 248:G277–G280.
12. Thurlby, P.L., and P. Trayhurn. 1979. The role of thermoregulatory thermogenesis in the development of obesity in genetically-obese (*ob/ob*) mice pair-fed with lean siblings. Br. J. Nutr. 42:377–385.
13. Trayhurn, P., and W.P.T. James. 1978. Thermoregulation and non-shivering thermogenesis in the genetically obese (*ob/ob*) mouse. Pflügers Arch. 373:189–193.
14. Trayhurn, P., P.L. Thurlby, and W.P.T. James. 1977. Thermogenic defect in pre-obese *ob/ob* mice. Nature 226:60–61.

oc, osteosclerotic, recessive, Chr 19

Arose in the C57BL/6J-*bf* strain. Homozygotes can be recognized at about 10 days by failure of eruption of the incisors. Affected mice may have a kinked tail and clubbed hindfeet. They grow normally for about the first 10 days, then lose weight or fail to gain, and usually die on or before 20 days. They show mild circling behavior beginning at about 2 weeks. Long bones show the appearance typical of osteopetrotic bones with failure of secondary bone resorption and lack of normal marrow cavities. The hypertrophic region of the epiphyseal plate is markedly thickened and the metaphyses have excessive unmineralized bone matrix (osteoid), a rickets-like condition not seen in other mouse osteopetrotic mutants (*gl*, *mi*, *op*). Osteoclasts are numerous but tend to be smaller than normal, they lack cytoplasmic vacuolization next to the surface of the trabeculae, and they either lack a ruffled border or have one that is less elaborate than normal. The rachitic condition, reduced bone resorption, and undermineralized matrix in this mutant are a unique combination of characteristics (1, 2).

References

1. Marks, S.C. Jr., M.F. Seifert, and P.W. Lane. 1985. Osteosclerosis, a recessive skeletal mutation on chromosome 19 in the mouse. J. Hered. 76:171–176.
2. Seifert, M.F., and S.C. Marks Jr. 1985. Morphological evidence of reduced bone resorption in the osteosclerotic (*oc*) mouse. Am. J. Anat. 172:141–153.

Och, ochre, semidominant, Chr 4

Found among progeny of an irradiated male. Heterozygotes look very light at weaning age (like $Mi^{wh}/+$ except for darker ears) but become darker with age (like *d/d*). Homozygotes are yellower than heterozygotes and smaller; many die before or soon after weaning, but some have lived to maturity (2). From about 8 days of age they exhibit an abnormal landing reaction, tending to tip their heads ventrally. The gross structure of the ears and otoliths is normal (1).

References

1. Phillips, R.J.S., and M.F. Lyon. 1975. Mouse News Lett. 53:29.
2. Phillips, R.J.S., S.G. Hawkes, and H.J. Moseley. 1973. Mouse News Lett. 48:30.

Oct locus

See *Otc* locus.

Odc-1 to *Odc-11* loci (formerly BRS-1 to -11)

These loci are probably a family of genes and pseudogenes for the enzyme ornithine decarboxylase (E.C. 4.1.1.17), a testosterone-inducible enzyme of mouse kidney (1). Numerous DNA restriction fragment length polymorphisms occur among inbred strains. The distribution of these polymorphisms in recombinant inbred strains has been used to determine the chromosomal location of nine of the loci (2–4).

Odc-1 locus, ornithine decarboxylase-1, Chr 1 (2). Different alleles occur in AKR and C57L.

Odc-2 locus, ornithine decarboxylase-2, Chr 2 (4). Different alleles occur in C57BL/6 and DBA/2.

Odc-3 locus, ornithine decarboxylase-3, Chr 3 (3). Different alleles occur in C57BL/6 and C3H/He.

Odc-4 locus, ornithine decarboxylase-4, Chr 4 (4). Different alleles occur in C57BL/6 and C3H/He.

Odc-5 locus, ornithine decarboxylase-5, Chr 6 (4). Different alleles occur in AKR and C57L.

Odc-6 locus, ornithine decarboxylase-6, Chr 7 (4). Different alleles occur in C57BL/6 and DBA/2 and in SWR and C57L.

Odc-7 locus, ornithine decarboxylase-7, Chr 7 (4). Different alleles occur in C57BL/6 and DBA/2, in AKR and C57L, in C57BL/6 and BALB/c, and in SWR and C57L.

Odc-8 locus, ornithine decarboxylase-8, Chr 12 (4). Different alleles occur in C57BL/6 and DBA/2, in AKR and C57L, in C57BL/6 and C3H/He, and in C57BL/6 and BALB/c.

Odc-9 locus, ornithine decarboxylase-9, Chr 14 (4). Different alleles occur in C57BL/6 and DBA/2 and in C57BL/6 and C3H/He.

Odc-10 locus, ornithine decarboxylase-10. Different alleles occur in C57BL/6 and DBA/2.

Odc-11 locus, ornithine decarboxylase-11. Different alleles occur in C57BL/6 and DBA/2 (4).

References

1. Berger, F.G., P. Szymanski, E. Read, and G. Watson. 1984. Androgen-regulated ornithine decarboxylase mRNAs of mouse kidney. J. Biol. Chem. 259:7941–7946.
2. Elliott, R.W., D. Barlow, and B.L. Hogan. 1985. Linkage of genes for laminin B1 and B2 subunits on chromosome 1 in mouse. *in vitro* Cell. Dev. Biol. 21:477–484.
3. Paul, P.R., and R.W. Elliott. 1987. Analysis of the mouse *Amy* locus in recombinant inbred mouse strains. Biochem. Genet. 25:569–579.
4. Richards-Smith, B., and R.W. Elliott. 1984. Mouse News Lett. 71:46–47.

oe, open eyelids, recessive, Chr 11

Arose spontaneously in the 129/ReSv strain. Not allelic with *gp* or *lg* (2). Homozygotes have open eyelids at birth. The cornea usually becomes opaque in adults, and the eyes are usually smaller than normal. A corneal staphyloma may be present. Homozygous embryos can be distinguished at 17 days when the eyelids normally close. Some homozygotes at this stage have a protruding lens, folded retina, and constricted cornea (1).

References

1. Mackensen, J.S. 1960. "Open eyelids" in newborn mice. J. Hered. 51:188–190.
2. Miller, J.R. 1964. Mouse News Lett. 30:16.

oed, edematous, recessive

Arose spontaneously in a strain carrying *pc* (phocomelia). Homozygotes die shortly after birth. They have a bloated appearance with shiny brittle skin and distal hematomata of the extremities. There is a threefold increase in leukocyte count, primarily due to increase in number of granulocytes. The plasma is deficient in high-, low-, and very-low-density lipoproteins, and in total and free cholesterol, triglycerides, and phospholipids. In liver, triglyceride concentration is reduced, and rate of synthesis of triglycerides, total fatty acids, and cholesterol esters is also reduced. The primary defect as well as the relationship between the abnormalities of lipid metabolism and those of skin and blood are unknown (1).

References

1. Schiffman, M.B., M.L. Santorineou, S.E. Lewis, H.A. Turchin, and S. Gluecksohn-Waelsch. 1975. Lipid deficiencies, leukocytosis, brittle skin—a lethal syndrome caused by a recessive mutation, edematous (*oed*), in the mouse. Genetics 81:525–536.

oel, open eyelids with cleft palate, recessive

Appeared in a phocomelic strain. Homozygotes have a cleft palate and die soon after birth (1). A mutation *crn*, cranioschisis, arose in this strain and segregates as if allelic or closely linked to *oel*. See *crn*.

References

1. Gluecksohn-Waelsch, S. 1961. Mouse News Lett. 25:12.

oh, obstructive hydrocephalus, recessive

Arose spontaneously in a non-inbred stock carrying *tb* (tumbler). Homozygotes develop a generalized ventricular dilatation a few days after birth. After 2 weeks there is a superimposed aqueductal stenosis that appears to be secondary to compression of the midbrain by the expanding cerebral hemispheres. Affected animals usually die by 1 month of age. No evidence for obstruction to flow of cerebrospinal fluid was found in the early stage of the disease, and the cause of the early low-grade generalized hydrocephalus is not known (1).

References

1. Borit, A., and R.L. Sidman. 1972. New mutant mouse with communicating hydrocephalus and secondary aqueductal stenosis. Acta Neuropathol. 21:316–331.

Oh21-1, *Oh21-2* loci, steroid 21-hydroxylase-1 and -2, Chr 17

These loci are structural genes for steroid 21-hydroxylase (21-OH; E.C. 1.14.99.10), a cytochrome P-450 enzyme active in adrenal glands and necessary for normal cortisol synthesis. No genetic variants are known. In man, the 21-OH genes are known to be associated with the HLA-linked complement genes. In the mouse, the loci were shown to be on Chr 17 within the *H-2S* region. Two complement-related and the two 21-OH genes were found to be located in the order *Slp*, *Oh21-1*, . . . *C4*, *Oh21-2*. *Oh21-2* appears to be inactive (1–3).

References

1. Amor, M., M. Tosi, C. Duponchel, M. Steinmetz, and T. Meo. 1985. Liver mRNA probes disclose two cytochrome P-450 genes duplicated in tandem with the complement *C-4* loci of the mouse *H-2S* region. Proc. Natl. Acad. Sci. USA 82:4453–4457.
2. Parker, K.L., D.D. Chaplin, M. Wong, J.G. Seidman, and J.A. Smith. 1985. Expression of murine 21-hydroxylase in mouse adrenal glands and in transfected Y1 adrenocortical tumor cells. Proc. Natl. Acad. Sci. USA 82:7860–7864.
3. White, P.C., M.I. New, and B. Dupont. 1985. Adrenal 21-hydroxylase cytochrome P-450 genes within the MHC class III region. Immunol. Rev. 87:123–150.

ol, oligodactyly, recessive, Chr 7

Appeared among the descendants of an X-rayed male (1,2). Homozygotes show reduction of the postaxial digits of all four limbs in varying degrees of severity. There may also be reduction or absence of the ulna or fibula. The tail may be shortened and kinked, the last rib is reduced in size or absent, and the ribs and sternebrae may show fusions. The spleen may be reduced in size and deformed, and there may be horseshoe kidney or the kidneys may be cystic or absent (3). Few homozygotes live beyond 1 month. Survivors are small and do not breed.

References

1. Hertwig, P. 1939. Zwei subletale recessive Mutationen in der Nachkommenschaft von röntgenbestrahlten Mäusen. Erbarzt 6:41–43.
2. Hertwig, P. 1942. Neue Mutationen und Koppelungsgruppen bei der Hausmaus. Z. Indukt. Abstammungs-Vererbungsl. 80:220–246.
3. Freye, H. 1954. Anatomische und entwicklungsgeschichtliche Untersuchungen am Skelett normaler und oligodactyler Mäuse. Wiss. Z. Martin-Luther-Univ. 3:801–824.

olt, oligotriche, recessive

Arose in the C3H/HeOrl strain. Homozygotes show retarded growth of the ventral coat, and adults have sparse coats, females being more severely affected than males. Females are fertile; males are sterile. Spermatogenesis is normal until the spermatid stage, then becomes abortive and mature spermatozoa are not found in either the seminiferous tubules or the epididymis (1).

References

1. Moutier, R. 1976. New mutations causing sterility restricted to the male in rats and mice. *In* The Laboratory Animal in the Study of Reproduction, 115–117. Gustav Fischer, Stuttgart.

om, ovum mutant, semidominant

Wakasugi (2) postulated the existence of this mutant gene in the DDK strain to explain the much reduced litter size of DDK females mated to males of strains KK, NC, and C57BL. Reciprocal crosses and matings within the DDK strain were fully fertile. Reduced litter size was due to a defect in the eggs of the DDK females, not in the uterine environment. To explain the results of breeding tests, Wakasugi assumed that the DDK strain is homozygous for *om*, a gene that produces a substance o in ova. A closely linked or identical gene produces the substance s in sperm. Homozygotes for the wild-type allele produce the substances O and S in ova and sperm, respectively. Substance o in ova reacts with S in fertilizing sperm to produce a factor lethal to the embryo. All other combinations are normal. Heterozygous females (*om/+*) are thought to produce o and O in equal amounts, and one or the other at random interacts irreversibly with the spermatozoon in any given fertilized egg (1, 2).

References

1. McLaren, A. 1976. Genetics of the early mouse embryo. Ann. Rev. Genet. 10:361–388.
2. Wakasugi, N. 1974. A genetically determined incompatibility system between spermatozoa and eggs leading to embryonic death in mice. J. Reprod. Fert. 41:85–96.

op, osteopetrosis, recessive, Chr 3

Arose spontaneously in the C57BL/6J-*dw* stock. Homozygotes can be recognized at 10 days of age by absence of incisors and by a domed skull. They survive weaning if provided with soft food, but viability is somewhat reduced and breeding performance is very poor. Young homozygotes have excessive accumulations of bone with lack of marrow cavities, increases in bone matrix formation and in parafollicular (calcitonin-secreting) cells of the thyroid, and have normal

serum calcium but low serum phosphate. The skeletal signs of the disease slowly disappear beginning at 6 weeks, when bone matrix formation begins to change from above normal (145 per cent of normal) to below normal (20 per cent of normal). Osteoclasts are small and few in number and have abnormal distribution of acid phosphatase. These deficiencies persist into old age. Unusual lipoid masses occur in the soft tissue spaces in the skeleton. The main skeletal defect appears to be a severely restricted capacity for bone remodelling that is capable, however, when combined with the decline in rate of bone formation to one-fifth of normal after 6 weeks, of removing excess skeletal mass to produce nearly normal bones (1). The linkage group containing *op*, now known to be on Chr 3, was at first mistakenly assigned to Chr 12.

References

1. Marks, S.C., and P.W. Lane. 1976. Osteopetrosis, a new recessive skeletal mutation on chromosome 12 of the mouse. J. Hered. 67:11–18.

opt, opisthotonos, recessive, Chr 6

Arose in the C57BL/Ks-*db²ᴶ* (N3) strain. Homozygotes can be recognized at about 10 days by their loss of balance when standing or moving. Typical behavior between 15 and 20 days consists of falling over, struggling to get up, becoming agitated, and developing severe opisthotonos (arching upward of the head and tail). Death occurs by weaning age or before (1).

References

1. Lane, P.W. 1972. Mouse News Lett.47:36–37.

or, ocular retardation, recessive

Appeared in a stock segregating for patch (*Ph*) (7). Homozygotes have small eyes at birth and can be easily classified. The eyes develop normally up to the age of 11 days of gestation, when there is retardation in growth of the retinal anlage. The central artery and vein fail to establish a pathway along the choroid fissure, which closes completely. In the adult there is no optic nerve or optic chiasma. The space between the small eyeball and the orbit is filled with hypertrophied Harderian and lachrymal glands (1,2,7). Eyes of *or/or* embryos cultured in the anterior chamber of eyes of adult mice develop more normal lenses than when left *in situ* (3). A repeat mutation at the *or* locus, *orᴶ*, occurred in the 129/Sv-*Sl c⁺ p⁺* strain. Adult homozygotes resemble adult *or/or* mice. Beginning at 10.5 days of gestation there is much cell death in the normal retina but none in the *or^J*/*or^J* retina (4,6). Intercellular channels form in the normal retina and continue along the optic stalk, and the optic nerve grows out of the eye through these channels. Presumably as a result of lack of cell death the intercellular channels are much reduced in the *or^J*/*or^J* eyes. This may be responsible for absence of the optic nerve (5).

References

1. Konyukhov, B.V., and D.M. Glukharev. 1962. The genetic and morphological characteristics of microphthalmic mice (blind mutants). Bull. Exp. Biol. Med. 52:1437–1440 (English translation).
2. Konyukhov, B.V., and M.V. Sazhina. 1975. Inhibition of DNA synthesis in embryonic mouse retina as a result of gene interaction. Dev. Biol. 45:1–6.
3. Konyukhov, B.V., O.G. Stroeva, M.V. Sazhina, and T.A. Lipgart. 1963. Retinal injury as a cause of microphthalmia in mice of the ocular retardation mutant line. Arkh. Anat. Gistol. Embryol. 44:36–43 (English summary).
4. Robb, R.M., J. Silver, and R.T. Sullivan. 1978. Ocular retardation (or) in the mouse. Invest. Ophthalmol. Vis. Sci. 17:468–473.
5. Silver, J., and R.M. Robb. 1979. Studies on the development of the eye cup and optic nerve in normal mice and in mutants with congenital optic nerve aplasia. Dev. Biol. 68:175–190.
6. Theiler, K., D.S. Varnum, J.H. Nadeau, L.C. Stevens, and B. Cagianut. 1976. A new allele of ocular retardation: early development and morphogenetic cell death. Anat. Embryol. 150:85–97.
7. Truslove, G.M. 1962. A gene causing ocular retardation in the mouse. J. Embryol. Exp. Morphol. 10:652–660.

Org locus

See *Gpi-1* complex.

Orm-1, *Orm-2* loci (formerly *Agp-1*, *Agp-2*), Chr 4

These are the structural loci for two forms of the acute phase reactant α_1-acid glycoprotein or orosomucoid (ORM), which differ in size and charge. The proteins are inducible in liver by LPS or turpentine. The two loci are closely linked on Chr 4 near *b* (1, 3).

Orm-1 locus, orosomucoid-1. This locus controls a protein, ORM-1, that is more acidic in charge than ORM-2 and contains five N-glycans. The allele *Orm-1^a* occurs in strains AKR, C57L, SWR, and DE/Cv; the allele *Orm-1^b* determines a slightly more basic protein and occurs in strains C57BL/6, DBA/2, C3H/He, BALB/c, and A/J, and in wild *M. spretus* (2).

Orm-2 locus, orosomucoid-2. This locus controls a protein, ORM-2, that is more basic than Orm-1 and contains six N-glycans. The allele *Orm-2^b*, found in *M. spretus*, determines a protein that is more basic and of higher molecular weight than that determined by the allele *Orm-2^a*, found in all inbred strains examined (2). *Orm-1* and *Orm-2* have shown no recombination in 58 backcross mice tested (1).

References

1. Baumann, H., and F.G. Berger. 1984. Mouse News Lett. 71:49.
2. Baumann, H., W.A. Held, and F.G. Berger. 1984. The acute phase response of mouse liver: genetic analysis of the major acute phase reactants. J. Biol. Chem. 259:566–573.
3. Nadeau, J.H., F.G. Berger, K.A. Kelley, P.M. Pitha, C.L. Sidman, and N. Worrall. 1986. Rearrangement of genes located on homologous chromosomal segments in mouse and man: the location of alpha- and beta-interferon, alpha-1 acid glycoprotein-1 and -2, and aminolevulinate dehydratase on mouse chromosome 4. Genetics 114:1239–1255.

Os, oligosyndactylism, semidominant, Chr 8

Arose in an irradiation experiment and was probably X-ray induced. Homozygotes die by the fifth day of embryonic life, shortly after the 64-cell stage, as a result of abnormalities occurring during the seventh and eighth divisions (9). There is a very high mitotic index, more than a third of the cells containing mitotic figures. *Os* in homozygotes may exert its primary effect on the mitotic apparatus (5). Heterozygotes are affected on all four feet. Fusion usually occurs between the second and third digits and occasionally involves the fourth (2). The muscles of the forearms and lower legs as well as of the feet show anomalous arrangements not necessarily correlated with the skeletal changes (4). At 11 days of gestation the preaxial border of the limbs can be seen to be reduced (3), and a histological examination at this time shows that there is a small amount of cellular degeneration in the preaxial part of the footplate mesoderm, leading to coalescence of the second and third digital rudiments (6). *Os*/+ mice have a mild diabetes insipidus present at 5 weeks and increasing with age. In combination with one or more recessive modifying genes in the selected DI stock, *Os*/+ mice have a severe diabetes insipidus (1). The cause of the diabetes is a 45 per cent reduction in size of the kidneys with an 80 per cent reduction in number of glomeruli. Compensatory hypertrophy of the nephrons is not sufficient to restore normal urine-concentrating ability (7,8). It is not known how the kidney and foot defects are related, or how either is related to the early death of the homozygote.

References

1. Falconer, D.S., M. Latyszewski, and J.H. Isaacson. 1964. Diabetes insipidus associated with oligosyndactylism in the mouse. Genet. Res. 5:473–488.
2. Grüneberg, H. 1956. Genetical studies on the skeleton of the mouse. XVIII. Three genes for syndactylism. J. Genet. 54:113–145.
3. Grüneberg, H. 1961. Genetical studies on the skeleton of the mouse. XXVII. The development of oligosyndactylism. Genet. Res. 2:33–42.
4. Kadam, K.M. 1962. Genetical studies on the skeleton of the mouse. XXXI. The muscular anatomy of syndactylism and oligosyndactylism. Genet. Res. 3:139–156.
5. McLaren, A. 1976. Genetics of the early mouse embryo. Ann. Rev. Genet. 10:361–388.
6. Milaire, J. 1967. Histochemical observations on the developing foot of normal, oligosyndactylous (*Os/+*) and syndactylous (*sm/sm*) mouse embryos. Arch. Biol. 78:223–288.
7. Naik, D.V., and H. Valtin. 1969. Hereditary vasopressin-resistant urinary concentrating defects in mice. Am. J. Physiol. 217:1183–1190.
8. Stewart, A.D., and J. Stewart. 1969. Studies on syndrome of diabetes insipidus associated with oligosyndactylism in mice. Am. J. Physiol. 217:1191–1198.
9. Van Valen, P. 1966. Oligosyndactylism, an early embryonic lethal in the mouse. J. Embryol. Exp. Morphol. 15:119–124.

ot, oscillator, recessive

Arose in the C57BL/6J-*rc* strain. Homozygotes can be recognized at 14 days by violent rapid trembling which increases in severity daily. At 19 to 21 days homozygotes may have prolonged periods of very rapid tremor, producing great rigor and stiffness, in which they appear to be dead. They usually die at 22 to 23 days. Therapy with aminooxyacetic acid is helpful, but only temporarily (1).

References

1. Dickie, M.M. 1967. Mouse News Lett. 36:39.

Otc locus (also called *Oct*), ornithine transcarbamylase, Chr X

This is the structural locus for ornithine transcarbamylase (ornithine carbamoyltransferase, OTC; E.C. 2.1.3.3). It is identical to the sparse-fur (*spf*) locus; a defective mutant allele is responsible for the sparse-fur phenotype. See *spf*. Five DNA polymorphisms for *Otc* have been found among several wild *Mus* subspecies (1).

References

1. Mullins, L.J., and V. Chapman. 1986. Polymorphism of the ornithine transcarbamylase locus and control of gene expression. Mouse News Lett. 74:114.

oto, otocephaly, recessive, Chr 1

Induced by X irradiation of a male homozygous for the Chr 1 inversion In(1)1Rk. Segregation in descendants of this male showed that *oto* is either within or closely linked to the inversion. Homozygotes examined after 12 days of gestation may show microphthalmia, reduced or absent jaws, cyclopia, fused maxillary prominences, or complete absence of the face. Some mutant embryos can be identified as early as the 0 to 4 somite stage when they show extreme truncation of the anterior neural area. Penetrance in homozygotes is variable, being nearly complete in offspring of crosses to C57BL/6 and nearly zero in offspring of crosses to SWR (1).

References

1. Juriloff, D.M., K.K. Smith, T.H. Roderick, and B.K. Hogan. 1985. Genetic and developmental studies of a new mouse mutation that produces otocephaly. J. Craniofac. Genet. Dev. Biol. 5:121–145.

Oua-1 locus, ouabain resistance-1, Chr 3

This locus controls resistance of cells to the growth-inhibiting effect of the cardiac glycoside ouabain. No genetic variants are known, but mouse cells are more resistant than human cells. Using cultures of human HeLa cells fused with microcells prepared from mouse embryo fibroblasts, Kozak *et al.* (1) showed that resistance to ouabain was always associated with presence of mouse Chr 3. There is evidence that this locus may code for a component of the Na^+ K^+ Mg^{2+} activated membrane ATPase (1).

References

1. Kozak, C.A., R.E.K. Fournier, L.A. Leinwand, and F.H. Ruddle. 1979. Assignment of the gene governing cellular ouabain resistance to *Mus musculus* chromosome 3 using human/mouse microcell hybrids. Biochem. Genet. 17:23–34.

Oubr locus, ouabain resistance, Chr X

The gene at this locus in the mouse determines resistance of cultured cells to killing by ouabain. No genetic variants are known. In somatic cell hybrids with human cells, which are susceptible to killing by ouabain, resistance was concordant with X-linked mouse *Hprt* in 6/6 cases of untreated cells and in 10/12 cases in which the mouse cells had been irradiated. This indicates that *Oubr* is probably on the X chromosome (1).

References

1. Law, M.L., X. Mo, X. Zhang, and F.T. Kao. 1984. Genes coding for ouabain resistance (OUBR) and HPRT are syntenic in the mouse genome. Cytogenet. Cell Genet. 37:518 (Abstr.).

Ox-1, Ox-2 loci

Ox-1 locus, menadione oxidoreductase-1, Chr 12
Ox-2 locus, menadione oxidoreductase-2
Variation at these two loci is postulated to explain the difference between strains C57BL/6 and DBA/2 in inducibility in liver cytosol of reduced-NAD(P):menadione oxidoreductase (E.C. 1.6.99.2) by 3-methylcholanthrene and TCDD. The alleles for high inducibility occur in the C57BL/6 strain and those for low inducibility in DBA/2. Inducibility is nearly as high in the F1 as in C57BL/6. Enzyme activity is not found in liver microsomes. In F1, F2, and backcrosses of C57BL/6 and DBA/2 there is association between induced oxidoreductase activity and induced aryl hydrocarbon hydroxylase (controlled by the Ah locus) activity, but not a strict correlation. To explain the results, the authors postulate two loci, *Ox-1* and *Ox-2*, with *Ox-1* linked to *Ah* on Chr 12 with 2 to 23 per cent recombination, and *Ox-2* independent of *Ah* (1).

References

1. Kumaki, K., N.M. Jensen, J.G.M. Shire, and D.W. Nesbit. 1977. Genetic differences in induction of cytosol reduced-NAD(P):menadione oxidoreductase and microsomal aryl hydrocarbon hydroxylase in the mouse. J. Biol. Chem. 252:157–165.

P

p locus, Chr 7

p, pink-eyed dilution, recessive. This is a very old mutant carried in many varieties of fancy mice. Homozygotes have pink eyes with pigmentation very much reduced but not completely absent in both the retina and choroid. The black pigment of the hair is very much diluted, but the yellow pigment is only slightly affected. Pigment granules are irregular and shred-like in shape. The small amount of pigment they contain is of wild-type color (11, 20). The fine structure of the pigment granules was said to be disrupted by Moyer (14), but Hearing *et al.* (5) found the structure to be normal, with premature termination of the melanization process. In tissue culture of the eye, the amount of pigment formed can be increased by increasing the concentration of tyrosine. This suggests that *p* may block the melanin-synthesizing pathway by interference with tyrosine supply (22). The site of gene action is in the melanocytes and not in either the dermis or the epidermis (24).

p'. See *p^{un}*.

p^{bs} (formerly *p^{24H}*), p-black-eyed sterile, recessive. Found in the progeny of an irradiated male. Homozygotes have dark eyes at birth. The coat color in adults is slightly lighter than that of *p^d/p^d* and darker than *p^{dn}/p^{dn}*. At weaning age homozygotes show a slightly jerky behavior (15, 18). Males are sterile and females have reduced fertility. Males have a high proportion of abnormal sperm (26) and a low proportion of pituitary gonadotropin cells, and females have a high proportion of polyovular follicles and no corpora lutea (13). The sperm of *p^{bs}* heterozygotes, as well as those of *p^{6H}* and *p^{25H}*, show a loss of negative charge along the whole length of the tail, suggesting to the authors that the defect in spermiogenesis might involve the Golgi apparatus (4).

p^{cp} (formerly *p^{11}*), p-cleft palate, recessive. Found in the progeny of a neutron-irradiated male. Most homozygotes die soon after birth with cleft palate, but a few survive to maturity, presumably with unaffected or slightly affected palates, and are fertile. *p^{cp}/p^{cp}* and *p^{cp}/p* mice have eyes and coats resembling those of *p/p* (16–18).

p^d, dark pink eye, recessive. Probably X-ray-induced. Eyes of homozygotes are slightly pigmented at birth and darken in the next few days; the coat is only slightly diluted. Eyes of p^d/p are colorless at birth but darken during the next 2 weeks; the coat is diluted but darker than that of p/p (2). Eyes of p^d/p^{bs} are dark at birth, and the coat is slightly lighter and the ears slightly darker than those of p^d/p^d (18).

p^{dn}, p-darkening, recessive. EMS-induced. Homozygotes have pink eyes at birth which become darker by weaning. The coat is slightly darker than that of p/p, and the ear color is light (18).

p^m. See p^{un}.

p^{m1}, pink-eyed mottled-1.

p^{m2}. pink-eyed mottled-2. These two mutants arose in different radiation experiments. They have identical effects. In p^m/p mice, the coat consists of wild-type and typically p/p color intermingled in about equal proportions. Segregation is normal in p^m/p mice. Probably p^m acts by, at random, producing wild-type pigment in some cells and p/p pigment in others (21).

p^r, Japanese ruby, recessive. Discovered in a stock of Japanese waltzers (23). The eyes of homozygotes are very variable in color and may be different on the two sides of the same animal. Coat color is intermediate between wild type and p/p. p^r/p mice resemble p/p. This mutant may be extinct.

p^s, p-sterile, recessive. Found in the progeny of an X-irradiated male of the S strain. This mutant is like p in its effect on pigmentation, but homozygotes are small and males are almost completely sterile. They have a high proportion of abnormal sperm and produce very few vaginal plugs when caged with normal females. Females may be fertile but are poor mothers. Both sexes show slightly uncoordinated behavior and have incisors that wear abnormally (6, 7). The sperm abnormalities develop after meiosis is completed and involve the head, the tail being normal and motile (1).

p^{un} (formerly p', p^m), pink-eyed unstable, recessive. Arose spontaneously in the C57BL/6J strain. Homozygotes resemble p/p mice phenotypically, but p^{un} reverts somatically to wild type with high frequency. The reversion rate varies with age, being highest at 2 and at 10 days of gestation, and with genotype of parents, being higher in p^{un}/p^{un} progeny of heterozygous parents, $p^{un}/+$ or p^{un}/p, mated *inter se*, than of homozygous parents, p^{un}/p^{un}, mated *inter se* (12). Whitney and Lamoreux (25) have suggested that p^{un} may have been produced by insertion of a transposable element into the p^+ gene, frequent excision of the element being responsible for the high reversion rate. In the pigmented retinal epithelium, darkly pigmented revertant cells occur with greater frequency in the more distal or anterior part of the epithelial layer than in the proximal part near the optic nerve, indicating that the rate of reversion is conditioned by the tissue environment (3). Homozygous p^{un} mice, in common with some other mutations that reduce pigmentation, have a reduced number of projections of retinal ganglion cells to the ipsilateral lateral geniculate nucleus (10).

p^x, p-extra dark, recessive. Arose spontaneously in the C3H strain. Homozygotes have dark eyes at birth, and the coat color and ears of adults are only slightly lighter than wild type. p^x/p mice have dark eyes at birth, the coat color is slightly lighter than that of p^d/p^d, and the ear color is slightly darker (18).

p^{6H}, p-6H, recessive. Radiation-induced. Homozygotes resemble p/p mice in eye and coat color, but are small and show a nervous jerky behavior (19). No dental abnormalities were found. Males are sterile, and females have greatly reduced fertility. Males have a high proportion of abnormal sperm (8, 26) and a reduced proportion of gonadotropic cells in the pituitary. Females have an increased proportion of polyovular follicles and no corpora lutea (13). See also p^{bs}.

p^{25H}, p-25H, recessive. Radiation-induced. Homozygotes are very similar to p^{6H}/p^{6H}, except that sperm morphology is different, p^{25H}/p^{25H} males having a lower proportion of normal sperm (8, 13, 18, 26). In addition, in p^{25H}/p^{25H} mice of both sexes, the pars nervosa of the pituitary contains degenerating nerve axons, and the hypothalamus has reduced binding capacity for estradiol-17β (9). See also p^{bs}.

References

1. Bryan, J.H.D. 1977. Spermatogenesis revisited. III. The course of spermatogenesis in a male-sterile pink-eyed mutant type in the mouse. Cell Tiss. Res. 180:173–180.
2. Carter, T.C. 1959. Mouse News Lett. 21:40.
3. Deol, M.S., and G.M. Truslove. 1983. The effect of the pink-eyed unstable gene on the retinal pigmented epithelium of the mouse. J. Embryol. Exp. Morphol. 78:291–298.
4. Hash, D.C., and H.G. Wolfe. 1979. Pink-eyed dilution alleles affect negative surface charges on mouse spermatozoa. Dev. Genet. 1:61–68.
5. Hearing, V.J., P. Phillips, and M.A. Lutzner. 1973. The fine structure of melanosomes in coat color mutants of the mouse. J. Ultrastruct. Res. 43:88–106.
6. Hollander, W.F., J.H.D. Bryan, and J.W. Gowen. 1960. Pleiotropic effects of a mutant at the p locus from X-irradiated mice. Genetics 45:413–418.
7. Hollander, W.F., J.H.D. Bryan, and J.W. Gowen. 1960. A male-sterile pink-eyed mutant type in the mouse. Fertil. Steril. 11:316–324.
8. Hunt, D.M., and D.R. Johnson. 1971. Abnormal sper-

miogenesis in two pink-eyed sterile mutants in the mouse. J. Embryol. Exp. Morphol. 26:111–121.

9. Johnson, D.R., and D.M. Hunt. 1975. Endocrinological findings in pink-eyed mice. J. Reprod. Fertil. 42:51–58.

10. LaVail, J.H., R.A. Nixon, and R.L. Sidman. 1978. Genetic control of retinal ganglion cell projections. J. Comp. Neurol. 182:399–422.

11. Markert, C.L., and W.K. Silvers. 1956. The effects of genotype and cell environment on melanoblast differentiation in the house mouse. Genetics 41:429–450.

12. Melvold, R.W. 1971. Spontaneous somatic reversion in mice. Effects of parental genotype on stability at the *p*-locus. Mutat. Res. 12:171–174.

13. Melvold, R.W. 1974. The effects of mutant *p*-alleles on the reproductive system in mice. Genet. Res. 23:319–325.

14. Moyer, F.H. 1966. Genetic variations in the fine structure and ontogeny of mouse melanin granules. Am. Zool. 6:43–66.

15. Phillips, R.J.S. 1965. Mouse News Lett. 32:39.

16. Phillips, R.J.S. 1973. Mouse News Lett. 48:30.

17. Phillips, R.J.S. 1975. Mouse News Lett. 53:29.

18. Phillips, R.J.S. 1977. Mouse News Lett. 56:38.

19. Phillips, R.J.S. 1977. Mouse News Lett. 57:18.

20. Russell. E.S. 1949. A quantitative histological study of the pigment found in the coat-color mutants of the house mouse. IV. The nature of the effects of gene substitution in five major allelic series. Genetics 34:146–166.

21. Russell. L.B. 1964. Genetic and functional mosaicism in the mouse. *In* M. Locke, ed., The Role of Chromosomes in Development, 153–181. Academic Press, New York.

22. Sidman, R.L., and R. Pearlstein. 1965. Pink-eyed dilution (*p*) in rodents: increased pigmentation in tissue culture. Dev. Biol. 12:93–116.

23. Sô, M., and Y. Imai. 1926. On the inheritance of ruby eye in mice. Jpn. J. Genet. 4:1–9.

24. Stephenson, D.A., and J.E. Hornby. 1985. Gene expression at the pink-eyed dilution (*p*) locus is confirmed to be pigment-cell autonomous using recombinant skin grafts. J. Embryol. Exp. Morphol. 87:65–73.

25. Whitney, J.B. III, and M.L. Lamoreux. 1982. Transposable elements controlling genetic instabilities in mammals. J. Hered. 73:12–18.

26. Wolfe, H.G., R.P. Erickson, and L.C. Schmidt. 1977. Effects on sperm morphology by alleles at the pink-eyed dilution locus in mice. Genetics 85:303–308.

P450-1, P450-3 loci

P450-1, cytochrome P_1-450, Chr 9
P450-3, cytochrome P_3-450, Chr 9

These loci are the structural genes for two members of the cytochrome P-450 family of enzymes. P-450s have the capacity to oxidize a wide variety of structurally unrelated compounds. P_1-450 and P_3-450 are defined as those forms of 3-methylcholanthrene-induced P-450 having the highest turnover number for induced aryl

hydrocarbon hydroxylase activity and acetanilide 4-hydroxylase activity, respectively (2). P_3-450 is also correlated with estradiol 2-hydroxylase activity (3). The two P-450 enzymes are coordinately regulated by a product of the *Ah* locus, which controls a cytosolic receptor for both of them. Although the two genes are homologous and are regulated by the same locus, there are striking differences between them in tissue-specific constitutive and induced levels of transcription and mRNA stabilization and in developmental patterns (1, 3). By use of mouse–Chinese hamster somatic cell hybrids and cDNA probes for the P_1-450 and P_3-450 genes, it was shown that both structural genes are located on Chr 9 (2).

References

1. Kimura, S., F.J. Gonzalez, and D.W. Nebert. 1986. Tissue-specific expression of the mouse dioxin-inducible $P_1$450 and $P_3$450 genes: differential transcriptional activation and mRNA stability in liver and extrahepatic tissues. Mol. Cell. Biol. 6:1471–1477.

2. Tukey, R.H., P.A. Lalley, and D.W. Nebert. 1984. Localization of the P_1-450 and P_3-450 genes to mouse chromosome 9. Proc. Natl. Acad. Sci. USA 81:3163–3166.

3. Tuteja, N., F.J. Gonzalez, and D.W. Nebert. 1985. Developmental and tissue-specific differential regulation of the mouse dioxin-inducible P_1-450 and P_3-450 genes. Dev. Biol. 112:177–184.

P450-2D complex, cytochrome P450IID, Chr 15

This complex is homologous with loci in the rat that are members of the P450IID subfamily. In the rat, the P450s in this subfamily are P450db1 (debrisoquine 4-hydroxylase), and another protein, P450db2, without 4-hydroxylating properties. Both are produced in rat liver but are differentially regulated during development. The mouse complex, *P450-2D*, was shown to be located on Chr 15 by use of mouse–Chinese hamster somatic cell hybrids screened with a rat cDNA P450db1 probe (1).

References

1. Gonzalez, F.J., T. Matsunaga, K. Nagata, U.A. Meyer, D.W. Nebert, J. Pastewka, C.A. Kozak, J. Gillette, H.V. Gelboin, and J.P. Hardwick. 1987. Debrisoquine 4-hydroxylase: characterization of a new P450 gene subfamily, regulation, chromosomal mapping, and molecular analysis of the DNA rat polymorphism. DNA 6:149–161.

pa, pallid, recessive, Chr 2

Found in a mouse caught in the wild (9). Eyes of homozygotes are pink, and the coat is a little lighter than that of *p/p*. Both black and yellow pigments are diluted. Homozygotes have slightly reduced viability, and some have slightly abnormal behavior. The head is tilted to one side or the other, and the mice may be unable to orient themselves if submerged in water. The abnormal behavior is due to the absence of otoliths in the sacculus and utriculus which occurs in many but not in all *pa/pa* mice (4, 5, 11). The effect of *pa* on behavior and otolith morphology can be duplicated by deficiency of manganese in the mother's diet during pregnancy, and supplementation of the mother's diet with about a hundredfold increase in the normal amount of manganese cures both the behavioral and otolith defects. In homozygotes there is defective mucopolysaccharide synthesis in the otolith matrix and slower rate of transport of manganese, L-dopa, and L-tryptophane in the brain (1, 3, 10). Homozygotes have elevated basal and testosterone-induced levels of the kidney lysosomal enzymes β-glucuronidase, β-galactosidase, and α-mannosidase, accompanied by lowered enzyme excretion in the urine (6). They have prolonged bleeding time due to platelet storage pool deficiency (SPD) characterized by a normal platelet number but a deficiency in the number of platelet dense granules and in the serotonin, ATP, and ADP content of the granules (7). The SPD could be corrected by bone marrow transplants from normal mice (8). Graff *et al.* (2) reported a cross in which there was an unusual interaction between *pa* and the closely linked wellhaarig (*we*) locus that produced, in the F2, pallid wellhaarig offspring in greater than expected frequency, all with a severe skeletal abnormality. See *we*.

References

1. Cotzias, G.C., L.C. Tang, S.T. Miller, D. Sladic-Simic, and L.S. Hurley. 1972. A mutation influencing the transportation of manganese, L-dopa, and L-tryptophane. Science 176:410–412.
2. Graff, R.J., D. Simmons, J. Meyer, D. Martin-Morgan, and M. Kurtz. 1986. Abnormal bone production associated with mutant mouse genes *pa* and *we*. J. Hered. 77:109–113.
3. Hurley, L.S. 1976. Interaction of genes and metals in development. Fed. Proc. 35:2271–2275.
4. Lyon, M.F. 1953a. Absence of otoliths in the mouse: an effect of the pallid mutant. J. Genet. 51:638–650.
5. Lyon, M.F. 1953b. The developmental origin of hereditary absence of otoliths in mice. J. Embryol. Exp. Morphol. 3:230–241.
6. Novak, E.K., and R.T. Swank. 1979. Lysosomal dysfunctions associated with mutation at mouse pigment genes. Genetics 92:189–204.
7. Novak, E.K., S.-W. Hui, and R.T. Swank. 1984. Platelet storage pool deficiency in mouse pigment mutations associated with seven distinct genetic loci. Blood 63:536–544.
8. Novak, E.K., M.P. McGarry, and R.T. Swank. 1985. Correction of symptoms of platelet storage pool deficiency in animal models for Chediak–Higashi syndrome and Hermansky–Pudlak syndrome. Blood 66:1196–1201.
9. Roberts, E. 1931. A new mutation in the house mouse (*Mus musculus*). Science 74:569.
10. Shrader, R.E., L.C. Erway, and L.S. Hurley. 1973. Mucopolysaccharide synthesis in the developing inner ear of manganese-deficient and pallid mutant mice. Teratology 8:257–266.
11. Trune, D.R., and D.J. Lim. 1983. The behavior and vestibular nuclear morphology of otoconia-deficient pallid mutant mice. J. Neurogenetics 1:53–69.

Pan-1 locus, pancreas protein-1, Chr 4

This locus encodes a basic (pI = 9.5) monomeric protein of unknown function in the pancreas. Different alleles occur in the B10.129 and DBA/2 strains. *Pan-1* is located on Chr 4 near the distal end (1).

References

1. Skow, L.C. 1983. Mouse News Lett. 69:46.

par, paralysé, recessive

Homozygotes can be recognized by 8 days of age after which they develop progressive muscular weakness and are clearly smaller than their normal littermates. They rarely survive beyond 16 days when they are almost totally paralyzed. The neuromuscular junctions, after developing normally, show progressive atrophy and eventual loss of motor nerve terminals. Occasional degenerating intramuscular nerve fibers are found but nerve roots and anterior horn cells appear normal (1).

References

1. Duchen, L.W., and J.-L. Guénet. 1983. Mouse News Lett. 69:34.

Pas-1, -2, -3 loci, pulmonary adenoma susceptibility-1, -2, -3

Strain differences in resistance to urethan-induced lung tumors were first noted by Cowen (2) who showed that the major part of the difference between the susceptible A and resistant C57BL strains was due to a single locus with susceptibility dominant. The locus was not linked to *a*, *b*, or *c*. In a more extensive analysis, Bloom and Falconer (1) found the same strain distribution of susceptibility and confirmed the single-locus difference between strains A and C57BL. The recessive allele for resistance in C57BL was designated *ptr*. Malkinson *et al.* (4) used a series of 46 AXB and BXA recombinant inbred strains to clarify the genetic basis of the difference in susceptibility to urethan between the two strains. They found that the difference was consistent with a three- or possibly only two-locus difference. *Pas-1* had a considerably stronger effect than *Pas-2* or *Pas-3* (4) and is possibly identical with the *ptr* locus. In another study, Malkinson and Beer (3) found that SWR mice are also very susceptible, that BALB/cBy and several other strains have an intermediate degree of resistance, and that DBA/2, C3H/He, and some other strains are very resistant. The resistance of BALB/c is dominant in crosses with A/J and SWR.

References

1. Bloom, J.L., and D.S. Falconer. 1964. A gene with major effect on susceptibility to induced lung tumors in mice. J. Natl. Cancer Inst.33:607–618.
2. Cowen, P.N. 1950. Strain differences in mice to the carcinogenic action of urethane and its noncarcinogenicity in chicks and guinea-pigs. Br. J. Cancer 4:245–253.
3. Malkinson, A.M., and D.S. Beer. 1983. Major effect on susceptibility to urethan-induced pulmonary adenoma by a single gene in BALB/cBy mice. J. Natl. Cancer Inst. 70:931–936.
4. Malkinson, A.M., M.N. Nesbitt, and E. Skamene. 1985. Susceptibility to urethan-induced pulmonary adenomas between A/J and C57BL/6 mice: use of AXB and BXA recombinant inbred lines indicating a three-locus genetic model. J. Natl. Cancer Inst.75:971–974.

pc, phocomelic, recessive

Found in a tailless stock. Homozygotes show disproportionate dwarfism and die shortly after birth as a result of a large median cleft palate. The skull is narrow and pointed, the upper incisors are small or absent, and the limbs are disproportionately shortened. Both polydactyly and syndactyly may occur (2). Embryos show retardation of precartilage formation in the limbs and head at 12 days of gestation, and chondrification and ossification of the extremities are retarded throughout development (1,3). Extra pieces of cartilage are often found in the head and limbs. Two abnormal bars of cartilage in the area where palatine closure normally occurs are thought to interfere mechanically with closure.

References

1. Fitch, N. 1957. An embryological analysis of two mutants in the house mouse, both producing cleft palate. J. Exp. Zool. 136:329–357.
2. Gluecksohn-Waelsch, S., D. Hagedorn, and B.F. Sisken. 1956. Genetics and morphology of a recessive mutation in the house mouse, affecting head and limb skeleton. J. Morphol. 99:465–479.
3. Sisken, B.F., and S. Gluecksohn-Waelsch. 1959. A developmental study of the mutation "phocomelia" in the mouse. J. Exp. Zool. 142:623–642.

Pca-1 locus, plasma cell antigen-1

This locus controls presence or absence of an antigen (PC-1) on plasma cells and myeloma cells, and to a lesser extent on liver, kidney, brain, spleen, and lymph node cells. It is not found on thymocytes, peritoneal or blood lymphocytes, bone marrow cells, or erythrocytes. The allele *Pca-1^a* determines presence of the antigen and occurs in strains BALB/c, C3H/He, AKR, NZB, and others; the allele *Pca-1^b* determines absence of the antigen and occurs in strains C3Hf/Bi, C57BL/6, DBA/2, and others. The antigen is probably not an immunoglobulin (3). Natural antibodies to an antigen indistinguishable from PC-1 and from a viral envelope antigen are found in the sera of all strains tested (2). PC-1 consists of two similar or identical disulfidebonded polypeptide chains, each of molecular weight 115 000. It is similar to transferrin receptor in molecular weight, charge, subunit composition, disulfide bonding, and developmental regulation, suggesting that the two proteins may have a common evolutionary origin (1). A PC-1 cDNA probe hybridized to genomic DNA digested with *Bam*HI and *Pst*I revealed that BALB/c showed a restriction fragment length difference from C57BL/6 and DBA/2. There is probably only a single *Pca-1* gene (4).

References

1. Goding, J.W., and F.-W. Shen, 1982. Structure of the murine plasma cell alloantigen PC-1: comparison with the receptor for transferrin. J. Immunol. 129:2636–2640.
2. Herberman, R.B., and T. Aoki. 1972. Immune and natural antibodies to syngeneic murine plasma cell tumors. J. Exp. Med. 136:94–111.
3. Takahashi, T., L.J. Old, and E.A. Boyse. 1970. Surface alloantigens of plasma cells. J. Exp. Med. 131:1325–1341.
4. van Driel, I.R., A.P. Welles, G.A. Pietersz, and J.W. Goding. 1985. Murine plasma cell membrane antigen PC-1: molecular cloning of cDNA and analysis of expression. Proc. Natl. Acad. Sci. USA 82:8619–8623.

pcd, Purkinje cell degeneration, recessive, Chr 13

Arose spontaneously in the C57BR/cdJ strain. Homozygotes show a moderate ataxia beginning at 3 to 4 weeks. They are somewhat smaller than normal but may live a fairly normal life span. Males have abnormal sperm and are sterile. Females are fertile but are poor breeders. There is rapid degeneration of nearly all Purkinje cells beginning at 15 to 18 days, and a slower degeneration of the photoreceptor cells of the retina and mitral cells of the olfactory bulb (8). Degeneration of Purkinje cells is followed by partial loss of granule cells (2). Several discrete populations of thalamic neurons also degenerate, some severely, between 50 and 60 days of age; these populations are a diverse assortment of target cells with no discernible common denominator (9, 10). Purkinje cell death occurs during the third and fourth weeks and is preceded by abnormal retention of the basal accumulation of perikaryonal polysomes and abnormal configuration of the endoplasmic reticulum (5). By use of an independent cell marker, β-glucuronidase (GUS), in fusion chimeras between *pcd/pcd* and +/+ embryos, Mullen (7) was able to show that *pcd* acts directly in the Purkinje cells. In the retina of *pcd/pcd* mice, pycnotic nuclei begin to appear in the photoreceptor cells between 18 and 25 days, and the outer rod segments become disorganized. Degeneration of the photoreceptor cells proceeds slowly to completeness over the course of a year (1, 6). In *pcd/pcd* males, sperm count is reduced 100-fold, and sperm are non-motile and have structural abnormalities of heads and/or tails which develop during the later stages of spermiogenesis (4). This mutation has been used extensively to study the functions of Purkinje cells and mitral cells and their interrelationships with other neurons (3, 11, 12).

References

1. Blanks, J.C., R.J. Mullen, and M.M. LaVail. 1982. Retinal degeneration in the *pcd* cerebellar mutant mouse. II. Electron microscopic analysis. J. Comp. Neurol. 212:231–246.
2. Ghetti, B., C.J. Alyea, and J. Muller. 1978. Studies on the Purkinje cell degeneration (pcd) mutant: primary pathology and transneuronal changes. J. Neuropath. Exp. Neurol. 37:667 (Abstr.).
3. Greer, C.A., and G.M. Shepherd. 1982. Mitral cell degeneration and sensory function in the neurological mutant mouse Purkinje cell degeneration (PCD). Brain Res. 235:156–161.
4. Handel, M.A., and M. Dawson. 1981. Effects on spermiogenesis in the mouse of a male sterile neurological mutation, Purkinje cell degeneration. Gam. Res. 4:185–192.
5. Landis, S.C., and R.J. Mullen. 1978. The development and degeneration of Purkinje cells in *pcd* mutant mice. J. Comp. Neurol. 177:125–143.
6. LaVail, M.M., J.C. Blanks, and R.J. Mullen. 1982. Retinal degeneration in the *pcd* cerebellar mutant mouse. I. Light microscopic and autoradiographic analysis. J. Comp. Neurol. 212:217–230.
7. Mullen, R.J. 1977. Site of *pcd* gene action and Purkinje cell mosaicism in cerebella of chimaeric mice. Nature 270:245–247.
8. Mullen, R.J., E.M. Eicher, and R.L. Sidman. 1976. Purkinje cell degeneration, a new neurological mutation in the mouse. Proc. Natl. Acad. Sci. USA 73:208–212.
9. O'Gorman, S. 1985. Degeneration of thalamic nuclei in "Purkinje cell degeneration" mutant mice. II. Cytology of neuron loss. J. Comp. Neurol. 234:298–316.
10. O'Gorman, S., and R.L. Sidman. 1985. Degeneration of thalamic nuclei in "Purkinje cell degeneration" mutant mice. I. Distribution of neuron loss. J. Comp. Neurol. 234:277–297.
11. Roffler-Tarlov, S., S.C. Landis, and M.J. Zigmond. 1984. Effects of Purkinje cell degeneration on the noradrenergic projection to mouse cerebellar cortex. Brain Res. 298:303–311.
12. Vaccarino, F.M., B. Ghetti, and J.I. Nurnberger. 1985. Residual benzodiazepine (BZ) binding in the cortex of *pcd* mutant cerebella and qualitative BZ binding in the deep cerebellar nuclei of control and mutant mice: an autoradiographic study. Brain Res. 343:70–78.

Pck-1 locus, phosphoenolpyruvate carboxykinase-1, Chr 2

This locus codes for cytosolic phosphoenolpyruvate carboxykinase (PEPCK; E.C. 4.1.1.32), a key enzyme in mammalian gluconeogenesis. The enzyme is synthesized primarily in liver and kidney. By use of mouse–Chinese hamster and mouse–rat somatic cell hybrids screened with a rat kidney cDNA probe, *Pck-1* was assigned to Chr 2 (1).

References

1. Lem, J., and R.E.K. Fournier. 1985. Assignment of the gene encoding cytosolic phosphoenolpyruvate carboxykinase (GPT) to *Mus musculus* chromosome 2. Somat. Cell Mol. Genet. 11:633–638.

Pcn complex, PCN-inducible cytochrome P-450, Chr 6

This complex codes for a family of cytochromes P-450 inducible by pregnenolone-16α-carbonitrile (PCN). There are probably several genes in this family but it is not known how many are functional or whether they are all inducible by PCN. *Pcn* was shown to be located on Chr 6 by use of mouse–Chinese hamster somatic cell hybrids. Strains DBA/2 and C57BL/6 show a *Hind*III DNA restriction fragment length difference, with C57BL/6 having a 1.4-kb fragment not present in DBA/

2. In the BXD recombinant inbred strains, the fragment shows no linkage with markers in the central region of Chr 6, indicating that *Pcn* must be near either the centromeric or telomeric end of the chromosome (1).

References

1. Simmons, D.L., P.A. Lalley, and C.B. Kasper. 1985. Chromosomal assignment of genes coding for components of the mixed-function oxidase system in mice: genetic localization of the cytochrome P-450PCN and P-450PB gene families and the NADPH-cytochrome P-450 oxidoreductase and epoxide hydratase genes. J. Biol. Chem. 260:515–521.

Pd locus, pyrimidine degrading

Two alleles are known at this locus: Pd^a determines a low level of pyrimidine degrading enzyme activity and occurs in strains C57BL/6J and BDP/J; Pd^b determines a high level and occurs in strains SJL/J and RF/J. Three enzyme activities appear to be affected simultaneously; they catalyze the stepwise reactions, uracil → dihydrouracil → β-ureido-proprionic acid → CO_2 + NH_3 + β-alanine (1). However, Sanno *et al*. (2) have shown that only the last enzyme in the pathway, β-ureidopropionase (E.C. 3.5.1.6), is defective in C57BL/6, with 16 per cent of the activity of RF. *Pd* probably affects structure of the enzyme, since the enzyme from C57BL/6 is more thermolabile and has lower affinity for substrate than that of RF. Heterozygotes have intermediate enzyme as measured in liver slices (1).

References

1. Dagg, C.P., D.L. Coleman, and G.M. Fraser. 1964. A gene affecting the rate of pyrimidine degradation in mice. Genetics 49:979–989.
2. Sanno, Y., M. Holzer, and R.T. Shimke. 1970. Studies of a mutation affecting pyrimidine degradation in inbred mice. J. Biol. Chem. 245:5668–5676.

Pdn, polydactyly Nagoya, semidominant

Arose in the JCL:ICR stock. Heterozygotes are fully viable and fertile. At birth they have an enlarged first digit on the forelimb, often with a tab on the postaxial side of the forefoot, and an extra digit on the preaxial side of the hindfoot. The tab on the forefoot disappears and the extra digit on the hindfoot becomes rudimentary in adults. Homozygotes die within 2 days after birth. They have two or three extra preaxial digits and a tab on the postaxial side of the forefoot and one to three extra preaxial digits on the hindfoot. Exencephaly, cleft palate, and open eyelids were found in 21

per cent, 12 per cent, and 12 per cent respectively. *Pdn* resembles *Xt* (extra toes) in some respects but has not been tested for allelism with it (1).

References

1. Hayasaka, I., T. Nakatsuka, T. Fujii, I. Naruse, and S.-I. Oda. 1981. Polydactyly Nagoya, *Pdn*: a new mutant gene in the mouse. Exp. Anim. 29:391–395.

pe, pearl, recessive, Chr 13

Arose spontaneously in the C3H/He strain. Homozygotes have a smaller amount of pigment in the eyes at birth than normal, but the eye color cannot be distinguished from normal in adults. The yellow and black pigments of the coat are diluted. Viability until maturity is good, but females have a tendency to die during pregnancy and lactation, and the survivors are poor mothers (7). Somatic reverse mutations to wild type are frequent, but the frequency is different on different genetic backgrounds and probably also at different stages of development (6). Whitney and Lamoreux (8) have suggested that the *pe* mutation may have resulted from integration of a transposable element into the pe^+ gene, the high rate of reversion being due to frequent excision of the element. Homozygotes have higher levels than controls of the lysosomal enzymes β-glucuronidase, β-galactosidase, and α-mannosidase in the kidney, and lower rates of excretion of glucuronidase and galactosidase. They also have a 50 per cent elevation in rate of glucuronidase synthesis in the kidney (3). Homozygotes have a bleeding abnormality due to platelet storage pool deficiency (SPD) characterized by normal platelet number but reduced number of platelet dense granules and reduction in amount of the dense-granule contents, serotonin, ATP, and ADP (4). The platelet defect is corrected by transplantation of bone marrow from normal mice (5). Pearl mice have reduced ipsilateral projection from the retina to the lateral geniculate nucleus, the superior colliculus, and the visual cortex, and reduced sensitivity of the dark-adapted retina (1, 2).

References

1. Balkema, G.W., L.H. Pinto, U.C. Dräger, and J.W. Vanable Jr. 1981. Characterization of abnormality in the visual system of the mutant mouse pearl. J. Neurochem. 1:1320–1329.
2. Balkema, G.W., N.J. Mangini, and L.H. Pinto. 1983. Discrete visual defects in pearl mutant mice. Science 219:1085–1087.
3. Novak, E.K., and R.T. Swank. 1979. Lysosomal dysfunctions associated with mutations at mouse pigment genes. Genetics 92:189–204.

4. Novak, E.K., S.-W. Hui, and R.T. Swank. 1984. Platelet storage pool deficiency in mouse pigment mutations associated with several distinct genetic loci. Blood 63:536–544.
5. Novak. E., R. Swank, and M. McGarry. 1984. Mouse News Lett. 71:46.
6. Russell. L.B. 1964. Genetic and functional mosaicism in the mouse. *In* M. Locke, ed., The Role of the Chromosomes in Development, 153–181. Academic Press, New York.
7. Sarvella, P.A. 1954. Pearl, a new spontaneous coat and eye color mutation in the house mouse. J. Hered. 45:19–20.
8. Whitney, J.B. III, and M.L. Lamoreux. 1982. Transposable elements controlling genetic instabilities in mammals. J. Hered. 73:12–18.

Ped locus

See *Qa-2* locus.

Pep-1 locus (formerly *Dip-2*), peptidase-1, Chr 18

This is probably the structural locus for a peptidase (PEP-A; E.C. 3.4.11.-) that uses valyl-l-leucine as a substrate. The enzyme migrates as two bands of activity following electrophoresis in a Tris-EDTA-borate buffer system. No genetic variants are known, but the locus was found to be on Chr 18 by use of mouse–Chinese hamster somatic cell hybrids segregating mouse chromosomes (1, 2).

References

1. Francke, U., P.A. Lalley, W. Moss, J. Ivy, and J.D. Minna. 1977. Gene mapping in *Mus musculus* by interspecific cell hybridization: assignment of the genes for tripeptidase-1 to chromosome 10, dipeptidase-2 to chromosome 19, acid phosphatase-1 to chromosome 12, and adenylate kinase-1 to chromosome 2. Cytogenet. Cell Genet. 19:57–84.
2. Leinwand, L.A., and F.H. Ruddle. 1978. Assignment of the gene for dipeptidase 2 to *Mus musculus* chromosome 18 by somatic cell hybridization. Biochem. Genet. 16:477–484.

Pep-2 locus (formerly *Trip-1*), peptidase-2, Chr 10

This is probably the structural locus for a tripeptidase (PEP-B; E.C. 3.4.11.-) that cleaves l-leucyl-l-glycylglycine. The enzyme migrates as a single band of activity following electrophoresis in a Tris-EDTA-borate buffer system. The locus was first shown to be on Chr 10 by use of mouse–Chinese hamster somatic cell hybrids segregating mouse chromosomes (1, 2). Two ENU-induced mutations at this locus were found by Lewis and Johnson (3). Using one of these, designated *Pep-2^b*, which determines a less anodal enzyme than the common allele, *Pep-2^a*, Womack *et al.* (4) found that *Pep-2* is located about 4 cM proximal to *Sl* on Chr 10.

References

1. Francke, U., P.A. Lalley, W. Moss, J. Ivy, and J.D. Minna. 1977. Gene mapping in *Mus musculus* by interspecific cell hybridization: assignment of the genes for tripeptidase-1 to chromosome 10, dipeptidase-2 to chromosome 19, acid phosphatase-1 to chromosome 12, and adenylate kinase-1 to chromosome 2. Cytogenet. Cell Genet. 19:57–84.
2. Leinwand, L.A., C.A. Kozak, and F.H. Ruddle. 1978. Assignment of genes for triose phosphate isomerase to chromosome 6 and tripeptidase-1 to chromosome 10 in *Mus musculus* by somatic cell hybridization. Somat. Cell Genet. 4:233–240.
3. Lewis, S.E., and F.M. Johnson. 1983. Dominant and recessive effects of electrophoretically detected specific locus mutations. *In* F.J. de Serres and W. Sheridan, eds., Utilization of Mammalian Specific Locus Studies in Hazard Evaluation and Estimation of Genetic Risk, 267–278. Plenum Publ. Corp.
4. Womack, J.E., S. Ashley, L.B. Barnett, and S.E. Lewis. 1986. Linkage of *Pep-2* and *Apk* on mouse chromosome 10. Biochem. Genet. 24:721–727.

Pep-3 locus (formerly *Pep-C*, *Dip-1*), peptidase-3, Chr 1

This locus controls electrophoretic variation of peptidase C (PEP-C; E.C. 3.4.11.-) of red blood cells. The allele *Pep-3^a* determines a slow anodally migrating band and is found in the C57BL/6 strain; the allele *Pep-3^b* determines a faster migrating band and is found in the CBA/J and A/He strains (1, 3); the allele *Pep-3^c* determines an even faster band and is found in strains NZB, ST/b, and RIIIS/J (4). Heterozygotes have both parental bands. Several mutations to null alleles of *Pep-3* induced by ENU and procarbazine have been reported by Lewis and Johnson (2). All were homozygous viable. *Pep-3* is also expressed on embryonic cells and in the placenta (3).

References

1. Chapman, V.M., F.H. Ruddle, and T.H. Roderick. 1971. Linkage of isozyme loci in the mouse: phosphoglucomutase-2 (*Pgm-2*), mitochondrial NADP malate dehydrogenase (*Mod-2*), and dipeptidase-1 (*Dip-1*). Biochem. Genet. 5:101–110.
2. Lewis, S.E., and F.M. Johnson. 1983. Dominant and

recessive effects of electrophoretically detected specific locus mutations. *In* F.J. de Serres and W. Sheridan, eds., Utilization of Mammalian Specific Locus Studies in Hazard Evaluation and Estimation of Genetic Risk, 267–278. Plenum Publ. Corp.

3. Lewis, W.H.P., and G.M. Truslove. 1969. Electrophoretic heterogeneity of mouse erythrocytes peptidase. Biochem. Genet. 3:493–498.
4. Taylor, B.A. 1979. Mouse News Lett. 61:42.

Pep-4 locus, peptidase-4, Chr 7

This locus codes for prolidase (PEP-4; E.C. 3.4.13.9), a dimeric enzyme with high specificity for peptides with carboxy-terminal proline residues. Allelic variants were recognized by cellulose acetate electrophoresis. The allele *Pep-4^a* determines a fast anodally migrating enzyme band and occurs in the IS/Cam and CASOP-6 (*M. m. castaneus*) strains; the allele *Pep-4^b* determines a slightly slower band and occurs in the C57 strains and in many other inbred strains. Heterozygotes have three bands, consistent with the dimeric structure of the enzyme. By use of mouse–Chinese hamster somatic cell hybrids, *Pep-4* was first found to be on Chr 7 proximal to band 7E1 (1, 2) and later mapped near *Gpi-1* on that chromosome (3).

References

1. Lalley, P.A., J.D. Minna, and U. Francke. 1978. Conservation of autosomal gene synteny groups in mouse and man. Nature 274:160–162.
2. Lalley, P.A., J.A. Diaz, and D.S. Woodward. 1979. Comparative mapping in man and mouse involving genes assigned to human chromosomes 11, 15, 19, and X. Cytogenet. Cell Genet. 25:177 (Abstr.).
3. Skow, L.C. 1981. Genetic variation for prolidase (PEP-4) in the mouse maps near the gene for glucosephosphate isomerase (GPI-1) on chromosome 7. Biochem. Genet. 19:695–700.

Pep-7 locus, peptidase-7, Chr 5

This locus codes for peptidase S (PEP-S; E.C. 3.4.11.-), an enzyme found in gut, liver, kidney, and spleen, but not in red cells. It hydrolyzes a wide range of substrates. An electrophoretic variant was discovered in the BALB/c strain maintained in the Immunology Department of the London Hospital Medical College, The allele *Pep-7^a* determines a slow migrating enzyme band and occurs in the BALB/c and all other strains tested; the allele *Pep-7^b* determines the faster variant found in BALB/c. Heterozygotes have a band of intermediate mobility. *Pep-7* was first found to be on Chr 5 by use of mouse–Chinese hamster somatic cell

hybrids (1) and later mapped near *Rw* (rump white) on that chromosome (2).

References

1. Lalley, P.A., J.D. Minna, and U. Francke. 1978. Conservation of autosomal gene synteny groups in mouse and man. Nature 274:160–162.
2. Peters, J., S. Povey, S. Jeremiah, and L. DeGiorgi. 1983. Linkage relationships of peptidase-7, *Pep-7*, in the mouse. Biochem. Genet. 21:801–807.

Pep-C locus

See *Pep-3* locus.

pet

See *hyt*.

pf, pupoid fetus, recessive, Chr 4

Found among fetuses of dissected pregnant females in an irradiation experiment. Homozygotes die at birth. The skin of late fetuses is stretched in an anteroposterior direction and lacks hair follicles. Fore- and hindlimbs and tail are seen as smooth stubs covered and held bound to the body by an abnormally overdeveloped epidermis. The cartilaginous skeleton appears grossly normal (4). At 13 to 14 days of gestation when normal epidermis is developing a multilayered structure, the epidermis of *pf/pf* embryos develops abnormally with thickness varying from 1 to 10 layers. The dermis is more cellular than normal and soon begins invading the epidermis (3). By birth, the epidermis is greatly thickened and is permeated by fibroblasts, blood vessels, nerve fibers, and Schwann cells. A basal lamina always occurs between the epidermal cells and the invading cells (2). The epidermis never forms a stratum corneum and the only keratins found are those of lower molecular weight characteristic of the basal cell layer of the epidermis (1). Skin of *pf/pf* embryos transplanted to normal hosts becomes completely normal, leading to the conclusion that *pf* acts either systemically or very transiently in the dermis or epidermis (1, 3). The effects of *pf* are very similar to those of *Er* (repeated epilation). Since *pf* maps to the same region of Chr 4 (5) as *Er*, the two genes may be alleles.

References

1. Fisher, C. 1987. Abnormal development in the skin of the pupoid fetus (*pf/pf*) mutant mouse: abnormal keratinization, recovery of a normal phenotype, and relationship to the repeated epilation (*Er*) mutant mouse. Curr. Top. Dev. Biol. 22:209–234.

2. Fisher, C., and E.J. Kollar. 1985. Abnormal skin development in pupoid fetus (pf/pf) mutant mice. J. Embryol. Exp. Morphol. 87:47–64.
3. Fisher, C., B.A. Dale, and E.J. Kollar. 1984. Abnormal keratinization in the pupoid fetus (pf/pf) mutant mouse epidermis. Dev. Biol. 102:290–299.
4. Meredith, R. 1964. Mouse News Lett. 31:25.
5. Meredith, R. 1965. Mouse News Lett. 32:39.

Pfk-4 locus, phosphofructokinase-4, Chr 15

This locus codes for a member of the phosphofructokinase (PFK; E.C. 2.7.1.-) gene family. It was identified by a mouse PFK cDNA clone and mapped to Chr 15 by use of a panel of mouse–Chinese hamster somatic cell hybrids. High levels of this PFK mRNA occur in brain and testis, and lower amounts in kidney, spleen, and liver. The mRNA of testis is slightly larger than that of other tissues. The mouse PFK probe hybridizes to a human DNA fragment that maps to human Chr 12 and is therefore different from the previously described enzymes of muscle (PFKM, Chr 1), platelets (PFKP, Chr 10) and liver (PFKL, Chr 21) (1).

References

1. Ashley, P.L., R.R. Flandermeyer, and D.R. Cox. 1986. Identification of novel phosphofructokinase loci in mouse and man. Am. J. Hum. Genet. 39:Suppl. A186 (Abstr.).

pg, pygmy, recessive, Chr 10

Appeared as undersized segregants in a strain selected for small size (3). Homozygotes are recognizably smaller than normal at birth. Their growth rate is low, and they weigh about one-third as much as normal when adult (1). They are healthy and active, but some are sterile. Fertility may be almost normal on genetic backgrounds favoring large size. The dwarfing is not due to a pituitary deficiency, and the thyroid and adrenals appear normal (2). Somatomedin activity is normal (4). Sinha *et al.* (6) found that concentrations of growth hormone and prolactin were normal in serum but below normal in the pituitary. Homozygotes are similar to human pygmies in being unresponsive to administration of growth hormone (5). The gene increased in frequency during five generations of selection for small size, indicating that *pg* probably reduces the size of heterozygotes ((7).

References

1. King, J.W.B. 1950. Pygmy, a dwarfing gene in the house mouse. J. Hered. 41:249–252.
2. King, J.W.B. 1955. Observations on the mutant "pygmy" in the house mouse. J. Genet. 53:487–497.

3. MacArthur, J.W. 1944. Genetics of body size and related characters. Am. Nat. 78:142–157, 224–237.
4. Nissly, S.P., R.A. Knazek, and G.L. Wolff. 1980. Somatomedin activity in sera of genetically small mice. Horm. Metabol. Res. 12:158–164.
5. Rimoin, D.L., and L. Richmond. 1972. The pigmy (pg) mutant of the mouse—a model of the human pygmy. J. Clin. Endocrinol. Metab. 35:467–468.
6. Sinha, Y.N., G.L. Wolff, S.R. Baxter, and O.E. Domon. 1979. Serum and pituitary concentrations of growth hormone and prolactin in pygmy mice. Proc. Soc. Exp. Biol. Med. 162:221–223.
7. Warwick, E.J., and W.L. Lewis. 1954. Increase in frequency of a deleterious recessive gene in mice. J. Hered. 45:143–145.

Pgam-1 locus, phosphoglyceromutase-1, Chr 19

This locus codes for the A subunit, one of the two different subunits leading to organ-specific isozyme patterns, of the dimeric enzyme phosphoglyceromutase (E.C. 2.7.5.3). No genetic variants are known. The locus was assigned to Chr 19 by showing that, in mice trisomic for Chr 19, the activity of PGAM-A was about 50 per cent higher than in controls. The gene dosage effects were fully expressed in liver, brain, and erythrocytes (AA-type isozyme) but not in skeletal muscle (BB-type). After electrophoretic separation of the AA, AB, and BB enzyme bands in developing heart muscle, the activity difference could be demonstrated in AA and AB bands but not in BB bands (1).

References

1. Fundele, R., T. Bücher, A. Gropp, and H. Winking. 1981. Enzyme patterns in trisomy 19 of the mouse. Dev. Genet. 2:291–303.

Pgd locus, 6-phosphogluconate dehydrogenase, Chr 4

This locus controls electrophoretic variation of the enzyme 6-phosphogluconate dehydrogenase (PGD; E.C. 1.1.1.43) in kidney and erythrocytes. The allele *Pgd^a* determines a fast anodally migrating band and occurs in wild mice in California and in the *M. m. musculus* subspecies in Denmark; the allele *Pgd^b* determines a slower band and occurs in all inbred strains tested. Heterozygotes have the two parental bands and a third intermediate band indicative of dimeric structure of the enzyme (1). Several ENU-induced mutations at the *Pgd* locus causing altered electrophoretic mobility of the enzyme have been reported (2). One of these may be the same as *Pgd^b* (3). *Pgd* is closely linked to *Gpd-1* (recombination = 1.7 per cent), and the enzymes controlled by these loci catalyse sequential steps in metabolism (1).

References

1. Chapman, V.M. 1975. 6-Phosphogluconate dehydrogenase (PGD) genetics in the mouse: linkage with metabolically related enzyme loci. Biochem. Genet. 13:849–856.
2. Lewis, S.E., and F.M. Johnson. 1983. Dominant and recessive effects of electrophoretically detected specific locus mutations. *In* F.J. de Serres and W. Sheridan, eds., Utilization of Mammalian Specific Locus Studies in Hazard Evaluation and Estimation of Genetic Risk, 267–278. Plenum Press Corp.
3. Lewis, S.E., and F.M. Johnson. 1986. The nature of spontaneous and induced electrophoretically detected mutations in the mouse. Prog. Clin. Biol. Res. 209B:359–365.

pge, phosphoglyceromutase expression, recessive

Found among offspring of a C3H/HeH mouse treated with ENU. Homozygotes are fully viable and fertile. They have 25- to 50-fold higher red cell phosphoglyceromutase (PGAM; E.C. 2.7.5.3) activity than normal and the enzyme shows decreased heat stability. Electrophoretic mobility is normal. This regulatory locus is not linked to *ru* on Chr 19, the chromosome to which *Pgam-1*, the structural locus for the enzyme, has been assigned by a gene dosage method (1).

References

1. Peters, J., and S.T. Bell. 1986. Phosphoglyceromutase expression (pge). Mouse News Lett. 74:94.

Pgk-1 locus, phosphoglycerate kinase-1, Chr X

This locus codes for the isozyme of phosphoglycerate kinase (PGK-; E.C. 2.7.2.3) found in somatic tissues. It was found to be on the X chromosome by Kozak *et al.* (4) who showed that the activity level in X/O one-celled embryos was half that in X/X embryos. This location was confirmed by Chapman and Shows (2) who found an electrophoretically different form of the enzyme in *M. caroli*. Further confirmation was provided by crosses with an electrophoretic variant found in wild mice in Denmark. The allele in these mice is designated *Pgk-1^a*, and it determines a fast anodally migrating band; the allele in inbred mice and other wild mice, *Pgk-1^b*, determines a slower band. Heterozygotes have both bands (7). PGK-1B is more heat stable and has higher specific activity than PGK-1A (6). In growing

oocytes but not in other tissues, the expression of *Pgk-1^b* is greater by 50 per cent or more than that of *Pgk-1^a*, both in heterozygotes and in homozygotes (5). *Pgk-1^a* and *Pgk-1^b* show a restriction fragment length difference (1). Because PGK-1 is expressed in early embryos, it has been used to study dosage compensation through X-inactivation (3), and the allelic differences have been used to study preferential inactivation of maternal or paternal × chromosomes (8).

References

1. Berger, S., and V. Chapman. 1984. Mouse News Lett. 71:50.
2. Chapman, V.M., and T.B. Shows. 1976. Somatic cell genetic evidence for X-chromosome linkage of three enzymes in the mouse. Nature 259:665–667.
3. Kozak, L.P., and P.J. Quinn. 1975. Evidence for dosage compensation of an X-linked gene in the 6-day embryo of the mouse. Dev. Biol. 45:65–73.
4. Kozak. L.P., G.K. McLean, and E.M. Eicher. 1974. X linkage of phosphoglycerate kinase in the mouse. Biochem. Genet. 11:41–47.
5. McMahon, A. 1983. Oocyte specific regulation of PGK-1 isozyme activity in female germ cells of the mouse. Genet. Res. 42:77–89.
6. Muhlbacher, C., G.W.K. Kuntz, G.A. Haedenkamp, and G.W.K. Krietsch. 1983. Comparison of the purified allozymes (1B and 1A) of X-linked phosphoglycerate kinase in the mouse. Biochem. Genet. 21:487–496.
7. Nielsen, J.T., and V.M. Chapman. 1977. Electrophoretic variation for X-chromosome-linked phosphoglycerate kinase (PGK-1) in the mouse. Genetics 87:319–325.
8. West, J.D., W.I. Frels, V.M. Chapman, and V.E. Papaioannou. 1977. Preferential expression of the maternally derived X chromosome in the mouse yolk sac. Cell 12:873–882.

Pgk-2 locus (also called *Tcp-2*), phosphoglycerate kinase-2, Chr 17

This locus controls electrophoretic variation, heat stability, and activity level of the isozyme of phosphoglycerate kinase (PGK-2; E.C. 2.7.2.3) present in sperm. This isozyme migrates faster anodally than PGK-1. The allele *Pgk-2^b* determines a slow migrating band and high heat stability and occurs in strains AKR, C3H/He, CBA/J, and others; the allele *Pgk-2^a* determines a faster band and intermediate heat stability and occurs in strains BALB/c, C57BL/6, DBA/2, and many others; the allele *Pgk-2^c* determines a faint band faster than that of *Pgk-2^a* and low heat stability and occurs in strains C57L, LP, and 129/Sv (2, 5). This allelic form shows lower activity than the other two. Heterozygotes show both parental bands (1, 4). PGK-2 apparently is present only on male germ cells and is first produced in the early spermatid stage (6). The *Tcp-2* (*t*-complex protein-2) locus has been shown to be identical to *Pgk-2* (3).

References

1. Eicher, E.M., M. Cherry, and L. Flaherty. 1978. Autosomal phosphoglycerate kinase linked to mouse major histocompatibility complex. Mol. Gen. Genet. 158:225–228.
2. Lee, C.-Y., and B. Pegoraro. 1979. Biochemical and immunological studies of three genetic variants of 3-phosphoglycerate kinase 2 from the mouse. Biochem. Genet. 17:631–644.
3. Silver, L.M., J. Uman, J. Danska, and J.I. Garrels. 1983. A diversified set of testicular cell proteins specified by genes within the mouse *t* complex. Cell 35:35–45.
4. VandeBerg, J.L., and S.V. Blohm. 1977. An allelic isozyme of Pgk-2 with low activity. J. Exp. Zool. 201:479–483.
5. VandeBerg, J.L., D.W. Cooper, and P.J. Close. 1976. Testis specific phosphoglycerate kinase B in mouse. J. Exp. Zool. 198:231–240.
6. VandeBerg, J.L., C.-Y. Lee, and E. Goldberg. 1981. Immunochemical localization of phosphoglycerate kinase isozymes in mouse testes. J. Exp. Zool. 217:435–441.

Pgm-1 locus, phosphoglucomutase-1, Chr 5

This locus controls electrophoretic variation of a fast anodally migrating isozyme of the enzyme phosphoglucomutase (PGM-1; E.C. 2.7.5.1). The enzyme is present in most tissues, erythrocytes, and in cultured fibroblasts. The allele *Pgm-1^a* determines a pattern of five fast migrating bands and occurs in strains C57BL/6, BALB/c, AKR, and others; the allele *Pgm-1^b* determines five slower bands and occurs in strains DBA/2, SJL, and others (3); the allele *Pgm-1^c* determines a pattern similar to that of *Pgm-1^b* but with smearing in the faster region and occurs in strain C3H/He; the allele *Pgm-1^d* determines a pattern similar to that of *Pgm-1^a* but also with smearing in the faster region and occurs in strain 129/Re. Heterozygotes show a mixture of parental bands (2). Several ENU-induced null mutations at this locus have been reported by Lewis and Johnson (1). Those tested were normally viable in homozygotes.

References

1. Lewis, S.E., and F.M. Johnson. 1983. Dominant and recessive effects of electrophoretically detected specific locus mutations. *In* F.J. de Serres and W. Sheridan, eds., Utilization of Mammalian Specific Locus Studies in Hazard Evaluation and Estimation of Genetic Risk, 267–278. Plenum Press Corp.
2. Miner, G.D., and H.G. Wolfe. 1972. Phosphoglucomutase isozymes of red cells in *Mus musculus*: additional

polymorphism at PGM-1 compared by two electrophoretic systems. Biochem Genet. 7:247–252.
3. Shows, T.B., F.H. Ruddle, and T.H. Roderick. 1969. Phosphoglucomutase electrophoretic variants in the mouse. Biochem. Genet. 3:25–35.

Pgm-2 locus, phosphoglucomutase-2, Chr 4

This locus controls electrophoretic variation in a slow anodally migrating isozyme of the enzyme phosphoglucomutase (PGM-2; E.C. 2.7.5.1). The enzyme is present in most tissues and to a lesser extent in erythrocytes, and is also present in cultured fibroblasts. The allele *Pgm-2^a* determines a pattern of at least two fast migrating bands and occurs in most inbred strains tested; the allele *Pgm-2^b* determines two slower migrating bands and occurs in strain SM/J and probably also in some wild mice from Ohio. Heterozygotes have a mixture of the parental bands (1, 3). An ENU-induced mutation causing an electrophoretic pattern indistinguishable from that of PGM-2B was found by Lewis and Johnson (2).

References

1. Chapman, V.M., F.H. Ruddle, and T.H. Roderick. 1971. Linkage of isozyme loci in the mouse. I. Phosphoglucomutase-2 (*Pgm-2*), mitochondrial NADP malate dehydrogenase (*Mod-2*), and dipeptidase-1 (*Dip-1*). Biochem. Genet. 5:101–110.
2. Lewis, S.E., and F.M. Johnson. 1983. Dominant and recessive effects of electrophoretically detected specific locus mutations. *In* F.J. de Serres and W. Sheridan, eds., Utilization of Mammalian Specific Locus Studies in Hazard Evaluation and Estimation of Genetic Risk, 267–278. Plenum Press Corp.
3. Shows, T.B., F.H. Ruddle, and T.H. Roderick. 1969. Phosphoglucomutase electrophoretic variants in the mouse. Biochem. Genet. 3:25–35.

Pgm-3 locus, phosphoglucomutase-3, Chr 9

This locus controls electrophoretic pattern of an isozyme of phosphoglucomutase (PGM-3; E.C. 2.7.5.1) that migrates between PGM-1 and PGM-2. The isozyme is present in muscle, brain, kidney, liver, lung, stomach, testis, red cells, and serum. The variants were most easily classified in hemolysates. The allele *Pgm-3^a* determines a fast anodally migrating band and occurs in the C57BL/6, A/J, BALB/c, C3H/He, and many other strains; the allele *Pgm-3^b* determines a slower band and occurs in the LP/Pas, CE, DW/J, FS/Ei, LG, SF/Cam, and 129/J strains; the allele *Pgm-3^c* determines absence of both bands and occurs in the DBA/2, C57L, C58, and several other strains. *Pgm-3^a*/*Pgm-3^b* mice have

Pgm-3

both parental bands. Both these alleles appear fully dominant over *Pgm-3ᶜ*. All three alleles are common in wild populations. From its strain distribution pattern in 74 BXD, AKXL, AKXD, and NX129 recombinant inbred strains, *Pgm-3* was found to be on Chr 9 between *d* and *Mod-1* (2). The *Pgm-3* locus was also described by Johnson *et al.* (1), who found it to be closely linked to *Mod-1* in the F2 and backcross generations of a cross between strains C57BL/6 and DBA/2. The authors used different methods and described effects of the alleles in different tissues from those used by Nadeau *et al.* (2). The effects of these differences have been discussed by Peters and Andrews (3).

References

1. Johnson, F.M., R.W. Hendren, F. Charalow, L.B. Barnett, and S.E. Lewis. 1981. A null mutation at the mouse *phosphoglucomutase-1* locus and a new locus, *Pgm-3*. Biochem. Genet. 19:599–615.
2. Nadeau, J.H., J. Kömpf, G. Siebert, and B.A. Taylor. 1981. Linkage of *Pgm-3* in the house mouse and homologies of three phosphoglucomutase loci in mouse and man. Biochem. Genet. 19:465–474.
3. Peters, J., and S.J. Andrews. 1984. The *Pk-3* gene determines both the heart, M₁, and the kidney, M₂, pyruvate kinase isozyme in the mouse; and a simple electrophoretic method for separating phosphoglucomutase-3. Biochem. Genet. 22:1047–1063.

Pgp-1 locus

See *Ly-24* locus.

Pgy-1 locus, P-glycoprotein-1, Chr 5

This is the structural locus for the 170-kDa P-glycoprotein associated with multidrug resistance. Multidrug resistance is thought to be a major cause of the frequent failure of cancer chemotherapy. *Pgy-1* was found to be on Chr 5 in the A2-A3 region by *in situ* hybridization with a P-glycoprotein cDNA probe. In cell lines selected for multidrug resistance, the gene is greatly amplified and is associated with homogeneously staining regions of the chromosomes. The cDNA probe hybridized with these regions in different chromosomal locations in each of the cell lines. None of these locations were on Chr 5 (1).

References

1. Martinsson, T., and G. Levan. 1987. Localization of the multidrug resistance-associated 170 kDa P-glycoprotein gene to mouse chromosome 5 and to homogeneously staining regions in multidrug-resistant mouse cells by *in situ* hybridization. Cytogenet. Cell Genet. 45:99–101.

Ph locus, Chr 5

Ph, patch, semidominant. Arose spontaneously in the C57BL strain. Heterozygotes have sharply defined white spotting in the belt region. The spot may vary from a large belly spot to a wide complete belt which includes the fore- and hindlegs. The skull is a little wider and shorter than normal and has a large interfrontal bone. Homozygotes die *in utero* from about 10 days of gestation onward. At 9 days they have excessive amounts of fluid in the circulation, the pericardium, the tissues, and under the epidermis. Those that survive the 10-day period have a large bleb in the middle of the face which interferes with the development of the nose and palate (1). Double heterozygotes of *Ph* and the allele *Wᶜᵗ* at the closely linked *W* locus (*Ph* +/+ *Wᶜᵗ*) are more anemic than *Wᶜᵗ* heterozygotes alone (+ *Wᶜᵗ* + +), have significantly raised leucocyte counts, and are unduly sensitive to whole body X-irradiation, indicating that the *Ph* locus exerts some control over hematopoiesis (2).

Phᵉ, patch-extended, semidominant. Arose in a stock carrying *Ph*. *Phᵉ*/+ mice are very uniform in appearance in contrast to *Ph*/+, and have a white trunk, legs, and tail, and a pigmented head and shoulders. Most *Phᵉ*/*Phᵉ* mice probably die *in utero*, but one that survived to birth had a split face. Most *Phᵉ*/*Ph* embryos die before 14 days, but one that survived to that age had a split face. Double heterozygotes of *Phᵉ* with the closely linked genes *Wᵛ* and *Rw* were different in color from those of *Ph*. In particular, *Phᵉ* +/+ *Wᵛ* mice resembled + *Wᵛ*/+ *Wᵛ* and were sterile (3).

References

1. Grüneberg, H., and G.M. Truslove. 1960. Two closely linked genes in the mouse. Genet. Res. 1:69–90.
2. Loutit, J.F., and B.M. Cattanach. 1983. Hematopoietic role for patch (*Ph*) revealed by a new *W* mutant (*Wᶜᵗ*) in mice. Genet. Res. 42:29–39.
3. Truslove, G.M. 1977. A new allele at the patch locus in the mouse. Genet. Res. 29:183–186.

Phil (tentative symbol), Philly cataract

No gene symbol has been assigned by the authors. The condition was found in a mouse of the Swiss-Webster non-inbred stock from Taconic Farms. The mouse's cataractous offspring were mated and the stock was 'inbred until all offspring developed hereditary cataracts.' Whether the trait is dominant, recessive, or polygenic has not been reported. All mice of the strain develop cataracts visible to the naked eye at 5 to 6 weeks after birth. The mice are otherwise normal. Slit lamp observation shows that a faint anterior opacity is visible at 15 days. Subsequently, opacity develops in the nucleus. At 45 days, there is total dense nuclear opacity along with striking anterior and posterior subcapsular opacity. Dry weight is markedly reduced. Normal lenses maintain a high potassium–low sodium content. In Philly lenses this ratio is reversed, beginning at about 15 days. The change appears to be due to a change in membrane permeability rather than to a defect in the cation pump mechanism (2). The content of α-crystallins is about normal but β- and γ-crystallins are strikingly reduced (3). In particular, there is a total absence of a 27kDa β-crystallin and its messenger RNA, although other β-crystallins occur throughout the lens. This protein normally becomes abundant by 16 days and is a major component of the adult lens. Its absence, preceding cataract formation, may be the basic defect in the Philly cataract (1).

References

1. Carper, D., S.J. Smith-Gill, and J.H. Kinoshita. 1986. Immunochemical localization of the 27K β-crystallin polypeptide in the mouse lens during development using a specific monoclonal antibody: implications for cataract formation in the Philly mouse. Dev. Biol. 113:104–109.
2. Kador, P.F., H.N. Fukui, S. Fukushi, H.M. Jernigan, Jr., and J.H. Kinoshita. 1980. Philly mouse: a new model of hereditary cataract. Exp. Eye Res. 30:59–68.
3. Piatigorsky, J., P.F. Kador, and J.H. Kinoshita. 1980. Differential synthesis and degradation of protein in the hereditary Philly mouse cataract. Exp. Eye Res. 30:69–78.

Phk locus, phosphorylase kinase, Chr X

This locus controls activity in skeletal muscle of the enzyme phosphorylase kinase (PHK; E.C. 2.7.1.38), which catalyzes the conversion of phosphorylase b to phosphorylase a. The allele Phk^b, found only in the I strain, determines virtual absence of PHK in skeletal muscle and a reduced amount in brain, heart, and kidney; the allele Phk^a, found in all other strains, determines normal levels of PHK in muscle and other organs. Heterozygous females have about 85 per cent as much activity as in normal mice (4, 5). PHK is not completely absent from muscle in the I strain; Lyon (5) found a small amount at birth which disappears by 3 weeks; Huijing (3) found about 5 per cent of normal activity when whole homogenates rather than supernatants were assayed, and Varsànyi et al. (6) found nearly normal activity in sarcoplasmic reticulum isolated from I-strain muscle. Gross et al. (2) have found evidence that the Phk^b allele produces normal amounts of an enzyme with abnormal structure and defective function, but this has been disputed by Cohen et al. (1) who were unable by three methods to demonstrate in I-strain muscle any of the four polypeptide chains that make up PHK. They concluded that the *Phk* locus on the X chromosome may control the expression of structural genes for the four chains which are probably located elsewhere. An X-linked gene causing partial deficiency of PHK and presumed to be an allele at this locus (Phk^c) was found by Varsànyi et al. (7) and propagated in a strain designated V. Phk^c is fully dominant to both Phk^a and Phk^b. Mice carrying one or two Phk^c genes have about 25 per cent of normal PHK activity.

References

1. Cohen, P.T.W., A. Burchell, and P. Cohen. 1976. The molecular basis of skeletal muscle phosphorylase kinase deficiency. Eur. J. Biochem. 66:347–356.
2. Gross, S.R., M.A. Longshore, and S. Pangburn. 1975. The phosphorylase kinase deficiency (*Phk*) locus in the mouse: evidence that the mutant allele codes for an enzyme with an abnormal structure. Biochem. Genet. 13:567–584.
3. Huijing, F. 1970. Phosphorylase kinase deficiency in mice. FEBS Lett. 10:328–332.
4. Huijing, F., E.M. Eicher, and D.L. Coleman. 1973. Location of phosphorylase kinase (*Phk*) in the mouse X-chromosome. Biochem. Genet. 9:193–196.
5. Lyon, J.B. 1970. The X-chromosome and the enzymes controlling muscle glycogen: phosphorylase kinase. Biochem. Genet. 4:169–185.
6. Varsànyi, M., U. Gröschel-Stewart, and L.M.G. Heilmeyer Jr. 1978. Characterization of a Ca^{2+}-dependent protein kinase in skeletal muscle membranes of I-strain and wild-type mice. Eur. J. Biochem. 87:331–340.
7. Varsànyi, M., A. Vrbica, and L.M.G. Heilmeyer Jr. 1980. X-linked dominant inheritance of partial phosphorylase kinase deficiency in mice. Biochem. Genet. 18:247–261.

Phr locus, pheromonal response

This locus controls production in males of a pheromone that induces ovulation in females previously treated with pregnant mare serum gonadotropin. The allele Phr^h determines production of the pheromone and occurs in strain SWR/J; the allele Phr^l determines lack of the pheromone and occurs in strain C57BL/6J. Phr^h is dominant to Phr^l. The effect of these alleles in females in not known (1).

References

1. Eleftheriou, B.E., D.W. Bailey, and M.X. Zarrow. 1972. A gene controlling male pheromonal facilitation of PMSG-induced ovulation in mice. J. Reprod. Fertil. 31:155–158.

pi, pirouette, recessive, Chr 5

Found by Woolley and Dickie (3) as a spontaneous mutation in the C3H strain. Somewhat less viable and fertile than normal. Males are more reliable breeders than females. Homozygotes show the typical behavior of the circling mutants: circling, head-tossing, hyperactivity, and deafness. Deafness may be present from the beginning or may be delayed until 1 to 4 months of age depending on the genetic background (2). Behavioral abnormalities are due to degenerative changes in the inner ear including degeneration of Corti's organ, the spiral ganglion, stria vascularis, saccular macula, and (after 9 months) the cristae ampullaris (1,2).

References

1. Deol, M.S. 1956. The anatomy and development of the mutants pirouette, shaker-1 and waltzer in the mouse. Proc. R. Soc. Lond. (Biol.) 145:206–213.
2. Kocher, W. 1960. Untersuchungen zur Genetik und Pathologie der Entwicklung von 8 Labyrinth-mutanten (*deaf–waltzer–shaker–Mutanten*) der Maus (*Mus musculus*). Z. Vererb. 91:114–140.
3. Woolley, G.W., and M.M. Dickie. 1945. Pirouetting mice. J. Hered. 36:281–284.

Pim-1 locus, proviral integration, MCF, Chr 17

This locus is a preferred site of integration of murine leukemia viruses in cells that develop into lymphomas after long exposure to the virus. Integrations at the *Pim-1* site were found in 24 of 93 lymphomas examined. The lymphomas were from three strains, AKR, BALB/c, and C57BL, and integrations at the *Pim-1* site occurred in all three strains. The proportion of *Pim-1* integrations was higher in the lymphomas with shorter latencies (14 to 30 weeks) than in those with longer latencies (1). Viral insertion at the *Pim-1* locus causes activation of the gene with an increase in *Pim-1* mRNA, normally expressed at low levels in lymphoid tissues. The mRNAs are variable in size, some lacking up to 1300 bases of *Pim-1*-specific sequences (4). *Pim-1* shows no homology with the *Mlvi* and *Int* loci (1). It was first shown to be on Chr 17 by use of mouse–Chinese hamster somatic cell hybrids; two other regions containing *Pim-1* homologous sequences were localized to Chrs 6 and 16 (2). Restriction fragment length differences among inbred strains were later found. Six patterns were revealed in DNA cleaved with the three enzymes, *Bam*HI, *Hinc*II, and *Taq*I. The allele *Pim-1ᵃ* occurs in strains A/J, AKR, CBA/Ca, C3H, and others; *Pim-1ᵇ* occurs in strains BALB/c, C57BL/Ks, C57L, C58, DBA/2, LP, and others, and in seven *t* haplo-

types; *Pim-1ᶜ* occurs in strains NZB and RIIIS; *Pim-1ᵈ* occurs in strains BDP, C57BL/6, IS/Cam, MA/My, SJL, SWR, and others; *Pim-1ᵉ* occurs in strains CAST/EI, MOLF/EI, and RF; *Pim-1ᶠ* occurs in *Mus spretus*; and *Pim-1ᵗ* occurs in three *t* haplotypes. Segregation of *Pim-1* in three sets of recombinant inbred strains showed it to be located between *Hba-4ps* and *Crya-1* within the region covered by the t complex (3).

References

1. Cuypers, H.J., G. Selten, W. Quint, *et al.* 1984. Murine leukemia virus-induced T-cell lymphomagenesis: integration of provirus in a distinct chromosomal region. Cell 37:141–150.
2. Hilkins, J., H.T. Cuypers, G. Selten, V. Kroezen, J. Hilgers, and A. Berns. 1986. Genetic mapping of *Pim-1* putative oncogene to mouse chromosome 17. Somat. Cell Mol. Genet. 12:81–88.
3. Nadeau, J.H., and S.J. Phillips. 1987. The putative oncogene *Pim-1* in the mouse: its linkage and variation among *t* haplotypes. Genetics 117:533–541.
4. Selten, G., H.T. Cuypers, and A. Berns. 1985. Proviral activation of the putative oncogene *Pim-1* in MuLV induced T cell lymphomas. EMBO J. 4:1793–1798.

pk, plucked, recessive, Chr 18

Arose in the DBA/2J strain. The development and morphology of homozygotes closely resembles that of bare (*ba*), but the two genes have not been tested for allelism. Plucked mice have short or absent whiskers at birth. The skin becomes thickened, folded, and darkly pigmented, and the first coat is delayed until about 11 days, when a thick coat of short irregular hairs appears. The same sequence is repeated in later hair cycles. The distal ends of all hair types are missing, and the proximal parts are variably irregular in structure. In development of normal hairs the external root sheath breaks down just below the level of the sebaceous gland, leaving the hair shaft free. This does not happen in *pk/pk* mice; the sebaceous glands enlarge abnormally, and the internal root sheath regresses prematurely. As a result of these defects, passage of the hair shaft up the follicle is disrupted, with consequent abnormalities of the shaft and delay in eruption through the epidermis (2). The *pk* locus is on Chr 18, 3 cM proximal to *Tw* (1).

References

1. Lane, P.W., and E.M. Eicher. 1985. The location of plucked, (*pk*) on chromosome 18 of the mouse. J. Hered. 76:476–477.
2. Trigg, M.J. 1972. Hair growth in mouse mutants affecting coat texture. J. Zool. Lond. 168:165–198.

Pk-1 locus, pyruvate kinase-1

This locus controls activity level and heat stability of an isozyme of pyruvate kinase (PK-1; E.C. 2.7.1.40) found in erythrocytes and liver. A variant allele, $Pk-1^b$, was discovered in a wild-caught male in Edinburgh. Homozygotes have less than half the enzyme activity of C57BL mice ($Pk-1^a/Pk-1^a$) in both erythrocytes and liver, and the enzyme has lower heat stability. Isoelectric point is unaffected. Heterozygotes have intermediate activity (2). A mutation induced by ENU in a (101 × C3H/H)F1 male was found by Charles and Pretsch (1). It causes high activity of PK in erythrocytes and liver and is probably a mutation at the *Pk-1* locus. Heterozygotes with $Pk-1^a$ have 160 per cent and homozygotes have 240 per cent of normal activity. Heat stability and electrophoretic mobility are unaffected.

References

1. Charles, D.J., and W. Pretsch. 1984. A new pyruvate kinase mutation and hyperactivity in the mouse. Biochem. Genet. 22:1103–1117.
2. Moore, K.J., and G. Bulfield. 1981. An allele ($Pk-1^b$) from wild-caught mice that affects the activity and kinetics of erythrocyte and liver pyruvate kinase. Biochem. Genet. 19:771–781.

Pk-2 locus

See *Pk-3* locus.

Pk-3 locus, pyruvate kinase-3, Chr 9

This locus controls electrophoretic variation in two isozymes of pyruvate kinase (PK-3; E.C. 2.7.1.40), M_1 present in muscle, and M_2 present in kidney (4, 5). M_2 was previously thought to be controlled by a separate locus, *Pk-2*. M_2 migrates a little more slowly than M_1. The allele $Pk-3^a$ determines slow anodally migrating forms of the two isozymes and occurs in the C57BL and most other inbred strains; the allele $Pk-3^b$ determines faster migrating forms and occurs in the IS/Cam strain and in some wild mice from Peru. $Pk-3^a/Pk-3^b$ mice show five bands of both isozymes, in accord with the known tetrameric structure of the enzyme (5). A third allele, $Pk-3^r$, arose spontaneously in the C57BL strain. It is prenatally lethal in homozygotes. In heterozygotes it causes about half normal activity of PK-3 in kidney and heart (1, 3). *Pk-3* was first found to be on Chr 9 by use of mouse–Chinese hamster somatic cell hybrids (2) and later found to be located between *Alp-1* and *Mod-1* (5).

References

1. Johnson, F.M., F. Chasalow, G. Anderson, P. MacDougal, R.W. Hendren, and S.E. Lewis. 1981. A variation in mouse kidney pyruvate kinase activity determined by a mutant gene on chromosome 9. Genet Res. 37:123–131.
2. Lalley, P.A., J.D. Minna, and U. Francke. 1978. Conservation of autosomal gene synteny groups in mouse and man. Nature 274:160–162.
3. Lewis, S.E., and F.M. Johnson. 1983. Dominant and recessive effects of electrophoretically detected specific locus mutations. *In* F.J. de Serres and W. Sheridan, eds., Utilization of Mammalian Specific Locus Studies in Hazard Evaluation and Estimation of Genetic Risk, 267–278. Plenum Press Corp.
4. Peters, J., and S.J. Andrews. 1984. The *Pk-3* gene determines both the heart, M_1, and the kidney, M_2, pyruvate kinase isozymes in the mouse; and a simple electrophoretic method for separating phosphoglucomutase-3. Biochem. Genet. 22:1047–1063.
5. Peters, J., H.R. Nash, E.M. Eicher, and G. Bulfield. 1981. Polymorphism of kidney pyruvate kinase in the mouse is determined by a gene, *Pk-3*, on chromosome 9. Biochem. Genet. 19:757–769.

Plat, *Plau* loci

These loci code for two distinct plasminogen activators which convert plasminogen, an inactive proenzyme present in plasma and many other extracellular fluids, to plasmin, a serine protease of broad specificity. The chromosomal locations of the two loci were determined by use of mouse–Chinese hamster somatic cell hybrids screened with specific mouse cDNA probes.

Plat locus, tissue plasminogen activator (t-PA), Chr 8.

Plau locus, urokinase (u-PA), Chr 14 (1).

References

1. Rajput, B., A. Marshall, A.M. Killary, P.A. Lalley, S.L. Naylor, D. Belin, R.J. Rickler, and S. Strickland. 1987. Chromosomal assignments of genes for tissue plasminogen activator and urokinase in mouse. Somat. Cell Mol. Genet. 13:581–586.

Plp locus, proteolipid protein, Chr X

This locus codes for the proteolipid protein found in the central nervous system. By use of somatic cell hybrids and a cDNA probe, the locus was found to be on the X chromosome in both mouse and man (2). In mice, an *Eco*RI restriction fragment length polymorphism (RFLP) occurs in *M. spretus*. By use of this RFLP, *Plp* was found to be located 30 cM distal to *Ta* (tabby) in the same position as *jp* (jimpy). Further evidence showed that *jp* is a mutation in the *Plp* gene (1). See *jp*.

References

1. Dautigny, A., M.-G. Mattei, D. Morello, P.M. Alliel, D. Pham-Dinh, L. Amar, D. Arnaud, D. Simon, J.-F. Mattei, J.-L. Guenet, P. Jollès, and P Avner. 1986. The structural gene coding for myelin-associated proteolipid protein is mutated in *jimpy* mice. Nature 321:867–869.
2. Willard, H.F., and J.R. Riordan. 1985. Assignment of the gene for myelin proteolipid protein to the X chromosome: implications for X-linked myelin disorders. Science 230:940–942.

Plp locus

See *C3* locus.

pma, peroneal muscular atrophy, recessive

Found in the CF1 stock during the process of inbreeding. Affected mice show clubbed hindfeet at birth and no progression of the defect. Growth is slightly delayed but is almost normal in both sexes. Expression is sometimes unilateral. In affected limbs there is nearly complete atrophy of the peroneal muscles which show the histochemical and electronmicroscopic characteristics of fetal muscle. The common peroneal nerve of affected limbs is absent, but there is no abnormality of the sciatic nerve or contributing spinal roots (1, 2). Penetrance and expressivity are complete in the original and related strains but probably incomplete on other genetic backgrounds (1).

References

1. Esaki, K., Y. Yasuda, M. Nakamura, H. Hayashi, and K. Ono. 1981. A new mutant in the mouse: peroneal muscular atrophy. Exp. Anim. 30:151–155 (English abstract).
2. Nonaka, I., A. Kikuchi, T. Suzuki, and K. Esaki. 1986. Hereditary peroneal muscular atrophy in the mouse: an experimental model for congenital contractures (arthrogryposis). Exp. Neurol. 91:571–579.

pn, pugnose, recessive, Chr 14

Arose spontaneously in the S strain at the Rockefeller Institute for Medical Research. Homozygotes are distinguishable at 4 weeks by the relatively shorter wider head. The forehead protrudes slightly and the skin over the nose is wrinkled. Adults are 2 to 4 g lighter than their normal littermates. The parietal, frontal, and nasal bones and the mandible are shorter and wider than normal. The xiphisternum is bent ventrally. There is reduced survival to weaning. Males are normally fertile, but females are poor breeders, often failing to deliver their young or care for them properly (1).

References

1. Kidwell, J.F., J.W. Gowen, and J. Stadler. 1961. Pugnose—a recessive mutation in linkage group 3 of mice. J. Hered. 52:145–148.

pnc, pancreas defect, recessive

Arose in the Peru-Coppock stock. In homozygotes at 15 to 21 days, the abdomen is distended with fluid. Affected mice usually die before weaning, some probably before classification. One male alive at 46 days had a plasma insulin level elevated more than 20-fold (1).

References

1. Wallace, M.E., and J. Ferguson. 1981. Mouse News Lett. 65:11.

Pnd locus (also called *Anf*), pronatriodilatin, Chr 4

This locus codes for a precursor, pronatriodilatin, from which are synthesized a number of atrial natriuretic factors having natriuretic, diuretic, and smooth muscle relaxing activities. There is good evidence that *Pnd* is a single gene. It was found to be on Chr 4 by use of mouse–Chinese hamster somatic cell hybrids screened with a rat cDNA probe (1).

References

1. Yang-Feng, T.L., G. Floyd-Smith, M. Nemer, J. Drosein, and U. Francke. 1985. The pronatriodilatin gene is located on the distal short arm of human chromosome 1 and on mouse chromosome 4. Am. J. Hum. Genet. 37:1117–1128.

Po, postaxial polydactyly, dominant

Found in mice obtained from a fancier in Japan. This abnormality is characterized by an extra toe on the ulnar side of one or both forefeet. The character depends on a major dominant gene (*Po*) and possibly some minor genes which may modify its expression (1).

References

1. Nakamura, A., H.Sakamoto, and K. Moriwaki. 1963. Genetical studies of post-axial polydactyly in the house mouse. Ann. Rep. Nat. Inst. Genet. Jap. (1962) 13:31.

Pomc-1, *Pomc-2* loci

One or both of these loci code for pro-opiomelanocortin (POMC), the 30 000 molecular weight protein precursor of both adrenocorticotropin (ACTH) and β-lipotropic hormone (β-LPH) formed in the anterior pituitary and the α-melanocyte stimulating hormone (α-MSH) and β-endorphin formed in the intermediate pituitary. POMC is also expressed in the hypothalamus and other areas of the central nervous system, in the placenta, and in the gut. The mouse has two *Pomc* genes but all the evidence indicates that only one of them, *Pomc-1*, codes for the POMC produced in the anterior and intermediate pituitary lobes. The other is probably a pseudogene (see *Pomc-2*). In the anterior pituitary, POMC is proteolytically cleaved into three fragments, ACTH, β-LPH, and an NH_2-terminal glyco-peptide; in the intermediate lobe, the polypeptides are again cleaved at sites where pairs of basic amino acids occur, ACTH to yield α-MSH and β-LPH to yield β-endorphin. Processing of the POMC produced in other tissues has not been studied, and it is even possible that they are coded for by *Pomc-2*.

Pomc-1 locus, pro-opiomelanocortin-1, Chr 12. No genetic variants of this gene are known. It was found to be on Chr 12 by use of mouse–Chinese hamster somatic cell hybrids screened with a POMC cDNA probe (1).

Pomc-2 locus, pro-opiomelanocortin-2, Chr 19?. The DNA of this gene is similar to that of *Pomc-1* but differs from it in several respects, including the presence of a transcription stop signal in place of the codon for the first amino acid of β-endorphin. It has features in common with the pseudogene of the β-globin family. It is possible that *Pomc-2* codes for the POMC produced in the central nervous system, placenta, or gut, but more likely that it is a pseudogene. No genetic variants are known. *Pomc-2* is probably on Chr 19 as shown by use of mouse–Chinese hamster somatic cell hybrids screened with a POMC cDNA probe (1).

References

1. Uhler, M., E. Herbert, P. D'Eustachio, and F.H. Ruddle. 1983. The mouse genome contains two nonallelic pro-opiomelanocortin genes. J. Biol. Chem. 258:9444–9453.

Por locus, NADPH-cytochrome P-450 oxidoreductase, Chr 6

This locus codes for NADPH-cytochrome P-450 oxidoreductase. It was shown to be on Chr 6 by use of mouse–Chinese hamster somatic cell hybrids screened with a cDNA probe (1).

References

1. Simmons, D.L., P.A. Lalley, and C.B. Kasper. 1985. Chromosomal assignment of genes coding for components of the mixed-function oxidase system in mice: genetic localization of the cytochrome P-450PCN and P-450PB gene families and the NADPH-cytochrome P-450 oxido-reductase and epoxide hydratase genes. J. Biol. Chem. 260:515–521.

Pp locus, passive performance

This locus controls part of the variation in passive avoidance learning under conditions in which mice were trained to avoid a foot shock by remaining on a shelf. The dominant allele Pp^b determines poor passive avoidance learning and occurs in the C57BL/6J strain; the recessive allele Pp^d determines good passive avoidance learning and occurs in DBA/2J (1).

References

1. Sprott, R.L. 1974. Passive-avoidance performance in mice: evidence for single-locus inheritance. Behav. Biol. 11:231–237.

pr, porcine tail, recessive

Arose spontaneously in a stock carrying *qv* (quivering) (1). Penetrance in homozygotes is variable and dependent on the genetic background. Abnormalities of the tail vary from a slight bend to a short tightly coiled stump. The defect is apparently due to unequal growth in the primitive streak beginning at about the level of the seventh caudal vertebra (3). Homozygotes with quite short tails may also exhibit waltzing behavior which is associated with varying degrees of abnormality of the membranous and bony labyrinths of the inner ear (2).

References

1. McNutt, W. 1967. Porcine tail, a new mutation in the house mouse. Anat. Rec. 157:286 (Abstr.).
2. McNutt, W. 1968. Abnormalities of the inner ear in choreic porcine tail (*pr*) mice. Anat. Rec. 160:392–393 (Abstr.).
3. McNutt, W. 1969. Developmental anomalies associated with the porcine (*pr*) gene in the mouse. Anat. Rec. 163:340 (Abstr.).

Pre-1, *Pre-2* loci, Chr 12

These two closely linked loci control electrophoretic pattern of two minor serum proteins migrating as prealbumins. The bands for the two loci are separated by a major prealbumin band (PRE-3) for which no genetic variants have been found. The alleles at the two loci have different strain distributions, but they have shown no recombination in 282 gametes tested (5, 6). They also show very close linkage with *Spi-1* and *Spi-2*, two multigene families coding for the two plasma protease inhibitors, α_1-protease inhibitor (also called α_1-antitrypsin) and contrapsin, respectively (2). It is possible that the prealbumin proteins are identical with the protease inhibitors. If so, *Pre-1* may correspond to *Spi-1* and *Pre-2* to *Spi-2*, since there has been one recombinant between *Spi-2* and (*Spi-1 Pre-1*) and none between *Spi-1* and *Pre-1* (2). The loci are on Chr 12 about 10 cM proximal to *Igh-C* (4).

Pre-1 locus (formerly *Pre*, *Pre-2*), prealbumin-1. This locus determines electrophoretic variation in a serum prealbumin. The allele *Pre-1^a* determines presence of a fast migrating band and occurs in strains AKR/J, BALB/c, DBA/2J, CBA/J, and others; the allele *Pre-1^b* determines a slightly slower band and occurs in strains SWR/J, C57BR/cdJ, and LP/J; the allele *Pre-1^c* determines a slow band and occurs in strains C3H/HeJ, A/J, and C57BL/6J. Heterozygotes have both parental bands (3, 5). Strains classified as having the *Pre-1^b* and *Pre-1^c* alleles by Wilcox (5), as well as a number of other strains, were originally thought by Shreffler (3) to lack prealbumin-1, and classified as *Pre-1^o* (6). Classification is most easily made as presence or absence of the *Pre-1^a* band. Density of the band is hormone-dependent and is most easily classified in sexually mature males (4). The locus for a prealbumin variant later found to be controlled by the *Mup-1* locus was at first called *Pre-1* (1).

Pre-2 locus (formerly *Pre-4*), prealbumin-2. This locus controls presence and absence of a serum prealbumin migrating just ahead of albumin. The allele *Pre-2^a* determines absence of the band and occurs in strains SWR/J, AKR/J, BALB/c, and others; the allele *Pre-2^b* determines presence of the band and occurs in strains C3H/HeJ, A/J, C57BL/6J, and CE/J. *Pre-2^b* is dominant to *Pre-2^a*. The *Pre-2^b* band is easily seen in males but is absent in most females (5).

References

1. Claxton, L., H.V. Malling, and H.E. Walburg. 1974. Serum prealbumin in inbred strains of *Mus musculus*. Biochem. Genet. 11:97–102.
2. Hill, R.E., P.H. Shaw, R.K. Barth, and N.D. Hastie. 1985. A genetic locus closely linked to a protease inhibitor gene complex controls the level of multiple RNA transcripts. Mol. Cell. Biol. 5:2114–2122.
3. Shreffler, D.C. 1964. Inheritance of a serum pre-albumin variant in the mouse. Genetics 49:629–634.
4. Taylor, B.A., D.W. Bailey, M. Cherry, R. Riblet, and M. Weigert. 1975. Genes for immunoglobulin heavy chain and serum prealbumin protein are linked in mouse. Nature 256:644–646.
5. Wilcox, F.H. 1975. Genetic variation in plasma prealbumin of the house mouse. J. Hered. 66:19–22.
6. Wilcox, F.H., and D.C. Shreffler. 1976. Terminology for prealbumin loci in the house mouse. J. Hered. 67:113.

Pre-4 locus

See *Pre-2* locus.

Prgs locus, phosphoribosylglycineamide synthetase, Chr 16

This locus codes for the enzyme phosphoribosylglycineamide synthetase (PRGS; E.C. 6.3.4.13), the third enzyme in the pathway of *de novo* purine biosynthesis. No genetic variants are known. *Prgs* was shown to be on Chr 16 by use of mouse–Chinese hamster somatic cell hybrids and electrophoretic identification of mouse PRGS (1).

References

1. Cox, D.R., D. Goldblatt, and C.J. Epstein. 1981. Chromosomal assignment of mouse *PRGS*: further evidence for homology between mouse chromosome 16 and human chromosome 21. Am. J. Hum. Genet. 33:145A (Abstr.).

Prm-1, *Prm-2* loci, protamine-1, -2, Chr 16

These loci code for two protamines, nuclear proteins that replace histones late in the haploid phase of spermatogenesis and are believed to be essential for sperm head condensation and DNA stabilization. Both protamines are determined by single-copy genes and both were shown to be on Chr 16 by use of mouse–Chinese hamster somatic cell hybrids screened with cDNA probes of the two mouse protamines. The nucleotide sequences of the two genes predict markedly different amino acid sequences for the two proteins.

Prm-1 locus, protamine-1. This locus codes for a tyrosine-containing protamine. A DNA fragment length variant was found in the Czech II strain derived from wild *M. m. musculus* in Czechoslovakia. These mice have a *Bst*EII fragment of 6.8 kb; BALB/c mice have a *Bst*EII fragment of 16.8 kb. This difference was used in a cross of Czech II with BALB/c to locate *Prm-1* about 7 cM proximal to *Igl-1* on Chr 16 (2).

Prm-2 locus, protamine-2. This locus codes for a histamine-containing protamine (1).

References

1. Hecht, N.B., K.C. Kleene, P.C. Yelick, P.A. Johnson, D.D. Pravtcheva, and F.H. Ruddle. 1986. Mapping of haploid expressed genes: genes for both mouse prota-

mines are located on Chr 16. Somat. Cell. Mol. Genet. 12:203–208.

2. Reeves, R.H., D. Gallahan, B.F. O'Hara, R. Callahan, and J.D. Gearhart. 1987. Genetic mapping of *Prm-1*, *Igl-1*, *Smst*, *Mtv-6*, *Sod-1* and *Ets-2*, and localization of the Down syndrome-region of mouse chromosome 16. Cytogenet. Cell Genet. 44:76–81.

Prn complex, prion gene complex, Chr 2

This is the locus of the gene or genes that code for the scrapie prion protein and regulate scrapie incubation time. It was assigned to Chr 2 by use of somatic cell hybrids (6) and later found, by use of restriction fragment length polymorphisms, to lie about 6 cM distal to *B2m* (1). Scrapie is a degenerative neurological disease that is transmitted by an infectious agent and occurs after long incubation periods. It occurs naturally in sheep and can be transmitted to mice by intracerebral or intraperitoneal inoculation. The scrapie agent is unusual among infective agents in that it contains protein but no detectable nucleic acid. The gene coding for the protein has been shown to reside in the genome of the host (3, 5). The exact nature of the scrapie agent and the mechanism of its infectivity are unresolved questions (4).

Prn-p locus, prion protein. This gene codes for the protein of the scrapie prion. Six alleles have been distinguished by the pattern of restriction fragment length polymorphisms detected with five different restriction endonucleases: $Prn-p^a$ in the BALB/c, CBA/J, C57BL/6, DBA/2, and many other strains; $Prn-p^b$ in I/Ln, BDP, P, and IM; $Prn-p^c$ in RIIIS; $Prn-p^d$ in MA/My; $Prn-p^e$ in CAST/Ei; and $Prn-p^f$ in MOLF/Ei (1, 2).

Prn-i locus, scrapie incubation period. This locus determines length of scrapie incubation period after intracerebral inoculation with the Chandler strain of scrapie. The allele $Prn-i^i$ determines a long incubation period (359 ± 19 days) and occurs only in the I/Ln and other $Prn-p^b$ strains; the allele $Prn-i^n$ determines a short incubation period (113 ± 2 days) and occurs in the NZW/Lac strain and probably in most other strains. Heterozygotes are intermediate (223 ± 3 days). *Prn-i* shows no recombination with *Prn-p*; length of incubation time may possibly be a pleiotropic effect of *Prn-p*. *Prn-i* closely resembles *Sinc* and may be identical with it (1, 2). See *Sinc* locus.

References

1. Carlson, G.A., D. Westaway, P.A. Goodman, M. Peterson, S.T. Marshall, and S.B. Prusiner. 1988. Genetic control of prion incubation period in mice. *In* Ciba Foundation Symposium 135, Novel Infectious Agents and the Central Nervous system, 84–99. J. Wiley and Sons, London.

2. Carlson, G.A., D.T. Kingsbury, P.A. Goodman, S. Coleman, S.T. Marshall, S. DeArmond, D. Westaway, and S.B. Prusiner. 1986. Linkage of prion protein and scrapie incubation time genes. Cell 46:503–511.

3. Chesbro, B., R. Race, K. Wehrly, J. Nishio, M. Bloom, D. Lechner, S. Bergstrom, K. Robbins, L. Mayer, J.M. Keith, C. Garon, and A. Haase. 1985. Identification of scrapie prion protein-specific mRNA in scrapie-infected and uninfected brain. Nature 315:331–333.

4. Diener, T.O. 1987. PrP and the nature of the scrapie agent. Cell 49:719–721.

5. Oesch, B., D. Westaway, M. Wälchli, M.P. McKinley, S.B.H. Kent, R. Aebersold, R.R. Barry, P. Tempst, D.B. Taplow, L.E. Hood, S.B. Prusiner, and C. Weissmann. 1985. A cellular gene encodes scrapie PrP 27-30 protein. Cell 40:735–746.

6. Sparkes, R.S., M. Simon, V.H. Cohn, R.E.K. Fournier, J. Lem, I. Klisak, C. Heinzmann, C. Blatt, M. Lucero, T. Mohandas, S.J. DeArmond, D. Westaway, S.B. Prusiner, and L.P. Weiner. 1986. Assignment of the human and mouse prion protein genes to homologous chromosomes. Proc. Natl. Acad. Sci. USA 83:7358–7362.

Pro-1 locus, proline oxidase-1 (the symbol *Pro-1* was formerly used for *Car-2*)

This locus controls activity level of proline oxidase in liver, kidney, and brain. The allele $Pro-1^b$ determines low activity of the enzyme (less than 20 per cent of normal) and occurs in the PRO/Re strain where it apparently occurred as a mutation; the allele $Pro-1^a$ determines normal activity of the enzyme and occurs in all other strains tested (1). This enzyme is said by Holmes (2) to be the same as D-aminoacid oxidase (E.C. 1.4.3.3), an enzyme with a wide specificity for D-amino acids. Holmes has observed an electrophoretic variant (suggested symbol $Pro-1^c$) in the NZC/Bl strain. Low activity of proline oxidase in mice of the PRO/Re strain results in a sevenfold elevation of proline in blood and a 50-fold elevation in urine. In heterozygotes ($Pro-1^a/Pro-1^b$) enzyme activity is intermediate, but blood and urine levels of proline are normal. The exaggerated renal clearance of proline in $Pro-1^b/Pro-1^b$ mice is due to impaired proline oxidation in nephron cells, which causes a back flux in the urine (3). The enzyme deficiency in $Pro-1^b/Pro-1^b$ mice is due to the absence of component 1, one of two electrophoretically separable components with proline dehydrogenase activity found in normal liver mitochondria (1).

References

1. Blake, R.L., J.G. Hall, and E.S. Russell. 1976. Mitochondrial proline dehydrogenase deficiency in hyperproli-

nemic PRO/Re mice: genetic and enzymatic analyses. Biochem. Genet. 14:739–757.

2. Holmes, R.S. Mouse News Lett. 54:33.

3. Scriver, C.R., R.R. McInnes, and F. Mohyuddin. 1975. Role of epithelial architecture and intracellular metabolism in proline uptake and transtubular reclamation in PRO/Re mouse kidney. Proc. Natl. Acad. Sci. USA 72:1431–1435.

Prp complex, proline-rich protein gene family, Chr 8 (?)

This complex comprises all loci of the mouse for the proline-rich proteins (PRP) of the salivary glands that are detectable with two rat PRP cDNA probes. The BALB/cBy and C57BL/6By strains have different *Prp* restriction enzyme patterns. In mouse–Chinese hamster somatic cell hybrids, the whole *Prp* complex appeared to segregate concordantly with Chr 8. However, two C57BL/6.BALB/c congenic strains carrying the BALB/c alleles of *Es-1*, *H-29*, and *H-19* known to be in the distal two-thirds of Chr 8 did not also carry the BALB/c allele of *Prp*, casting some doubt on the assignment of *Prp* to Chr 8 (1).

References

1. Azen, E.A., D.M. Carlson, S. Clements, P.A. Lalley, and E. Vanin. 1984. Salivary proline-rich protein genes on chromosome 8 of the mouse. Science 226:967–969.

Prt-1, *Prt-3* loci, Chr 6 (?)

These two closely linked loci control electrophoretic patterns and activity level of trypsin (E.C. 3.4.21.4) and trypsinogen. Their chromosomal location has not been determined but *Prt-1* is probably identical to *Try-1* which codes for trypsin and was found to be on Chr 6 by use of mouse–Chinese hamster somatic cell hybrids (1).

Prt-1 locus, pancreatic proteinase-1. Two codominant alleles at this locus control electrophoretic patterns and activity level of trypsin I, one of three isozymes of trypsin described by Watanabe (3, 4). *Prt-1^a* determines presence of two bands and occurs in strains CBA/J, DBA/2, C57BL/6, and others; *Prt-1^b* determines presence of only the most anodal of the two bands and occurs in strains C57L, DDK, NC, MOL-A, and others (2, 3). *Prt-1* has a parallel effect on the electrophoretic properties of trypsinogen and is probably the structural locus for trypsinogen (3).

Prt-3 locus, pancreatic proteinase-3. This locus controls activity level of trypsin I and trypsinogen I. The allele *Prt-3^b* determines high level of activity and occurs

in the MOL-A strain; the allele *Prt-3^a* determines low level of activity and occurs in all other strains tested. The difference is due to a difference in specific activity of the enzymes determined by the two alleles. Heterozygotes have intermediate activity (2, 3).

References

1. Honey, N.K., A.Y. Sakaguchi, P.A. Lalley, C. Quinto, R.J. MacDonald, C. Craik, G.I. Bell, W.J. Rutter, and S.L. Naylor. 1984. Chromosomal assignments of genes for trypsin, chymotrypsinogen B, and elastase in mouse. Somat. Cell Mol. Genet. 10:377–383.

2. Watanabe, T., N. Ogasawara, and H. Goto. 1976. Genetic study of pancreatic proteinase in mice (*Mus musculus*): genetic variants of trypsin and chymotrypsin. Biochem. Genet. 14:697–707.

3. Watanabe, T. 1981. Genetic study of pancreatic proteinase in mice (*Mus musculus*): genetic variation at the *Prt-1* locus as trypsinogen. Biochem. Genet. 19:191–197.

4. Watanabe, T. 1983. Purification and properties of genetic variants of mouse trypsinogen. Biochem. Genet. 21:761–772.

Prt-2 locus, pancreatic proteinase-2, Chr 8

This locus controls electrophoretic mobility of chymotrypsin (E.C. 3.4.21.1) in the pancreas. The allele *Prt-2^a* determines a fast anodally migrating band and occurs in all laboratory strains tested; the allele *Prt-2^b* determines a slower band and occurs in MOL-A, a strain derived from *M. m. molossinus* in Japan. Heterozygotes have both bands (2). *Prt-2* is on Chr 8, 4.5 cM distal to *Es-2* (3). It is probably identical with *Ctrb* which codes for chymotrypsinogen B and is also on Chr 8, found to be there by use of somatic cell hybrids (1).

References

1. Honey, N.K., A.Y. Sakaguchi, P.A. Lalley, C. Quinto, R.J. MacDonald, W.J. Rutter, and S.L. Naylor. 1984. Assignment in mouse of the genes for chymotrypsinogen B, elastase, trypsin, and carboxypeptidase A. Cytogenet. Cell Genet. 37:492–493 (Abstr.).

2. Watanabe, T., N. Ogasawara, and H. Goto. 1976. Genetic study of pancreatic proteinase in mice (*Mus musculus*): genetic variants of trypsin and chymotrypsin. Biochem. Genet. 14:697–707.

3. Watanabe, T., N. Ogasawara, and H. Goto. 1976. Genetic study of pancreatic proteinase in mice (*Mus musculus*): linkage of the *Prt-2* locus on chromosome 8. Biochem. Genet. 14:999–1002.

Prt-4, *Prt-5* loci (formerly *Smg-1*, *Smg-2*), Chr 7

These loci control electrophoretic mobility of two esteroproteases inducible by testosterone in submandibular glands, probably in the secretory tubules. The two loci are on Chr 7 and show no recombination with each other or with *Tam-1*, another esteroprotease expressed in the submandibular gland. It is unlikely that the three loci are identical, however, as they have different numbers of alleles, *Tam-1* has a different strain distribution, and the gene products have different chemical properties (1).

Prt-4 locus, protease-4. The enzyme controlled by this locus is designated protease A. The allele *Prt-4^a* determines a pattern of bands after isoelectric focusing (IEF) consisting of a main band (IEP = 5.2) and two satellite bands; it occurs in strain DBA/2 and in *M. m. molossinus* (strain kl/oci) and *M. m. castaneus*. The allele *Prt-4^b* determines a pattern of bands after IEF consisting of a main band (IEP = 6.0) and two other satellite bands; it occurs in strains C57BL/10, A/J, NMRI/F, and SK/Cam. Heterozygotes have all parental bands (1).

Prt-5 locus, protease-5. The enzyme controlled by this locus is designated protease E. The allele *Prt-5^a* determines a single basic band (IEP = 9.5) and occurs in *M. m. molossinus* (strain kl/oci); the allele *Prt-5^b* determines a single more acidic band and occurs in strains C57BL/10, A/J, and NMRI/F; the allele *Prt-5^c* determines two even more acidic bands and occurs in strain DBA/2. Heterozygotes have all parental bands. A null allele, *Prt-5^d*, occurs in *M. m. castaneus* and SK/Cam, but no crosses with mice bearing other alleles have been described (1).

References

1. Otto, J., and O. von Deimling. 1981. *Prt-4* and *Prt-5*: new constituents of a gene cluster on chromosome 7 coding for esteroproteases in the submandibular gland of the house mouse (*Mus musculus*). Biochem. Genet. 19:431–444.

Ps, polysyndactyly, semidominant, Chr 4

Found among offspring of a (C3H/HeH × 101/H)F1 male irradiated with low-intensity neutrons (1). In heterozygotes, the hindfeet commonly show soft-tissue syndactylism, shortening of the nails and terminal phalanges, shortening and thickening of the hallux, and sometimes an extra digit between digits III and IV. Forefeet show occasional soft-tissue syndactyly, the nails and terminal phalanges are usually shortened, and there is often duplication of digit V. Heterozygotes are slightly smaller than their normal sibs. Homozygotes die at or before birth. The feet end in a broad pad and are edematous as far as the knee or elbow. Homozygotes can be recognized at 13 days of gestation by lack

of division of the footplates into digits and by a hypertrophied and wavy apical ectodermal ridge. At 14 days the footplates show no interdigital cell death, and there is subcutaneous edema of the limbs and trunk (2).

References

1. Batchelor, A.L., R.J.S. Phillips, and A.G. Searle. 1966. A comparison of the mutagenic effectiveness of chronic neutron- and γ-irradiation of mouse spermatogonia. Mutat. Res. 3:218–229.
2. Johnson, D.R. 1969. Polysyndactyly, a new mutant gene in the mouse. J. Embryol. Exp. Morphol. 21:285–294.

Pscr-1, Pscr-2, Pscs loci

Pscr-1, -2 loci, corneal resistance to *P. aeruginosa*-1, -2. These loci are postulated to explain inheritance of resistance and susceptibility to infection of the cornea by *Pseudomonas aeruginosa*. In resistant mice, intracorneal infection remains localized and spontaneously heals within 3 to 4 weeks; in susceptible mice, the infection leads to extensive necrosis, eye shrinkage, and blindness within 12 to 14 days. Mice of strains BALB/c and C57BL/6 are both susceptible but their F1 hybrid is resistant. This suggests that the two strains each carry a different dominant gene, both of which are necessary for resistance. The locus for which C57BL/6 carries the dominant allele is *Pscr-1*, that for BALB/c is *Pscr-2* (2).

Pscs locus, corneal susceptibility to *P. aeruginosa*. This additional locus, is postulated to be responsible for the difference between the resistant DBA/2 and susceptible C3H/He strains. In this case, resistance is recessive (1).

References

1. Beisel, K.W., L.D. Hazlett, and R.S. Berk. 1983. Dominant susceptibility effect on the murine corneal response to *Pseudomonas aeruginosa*. Proc. Soc. Exp. Biol. Med. 172:488–491.
2. Berk, R.S., K. Beisel, and L.D. Hazlett. 1981. Genetic studies on the murine corneal response to *Pseudomonas aeruginosa*. Infect. Immun. 34:1–5.

Psp locus, parotid secretory protein, Chr 2

This locus controls electrophoretic mobility and sensitivity to cyanogen bromide (CNBr) degradation of a protein found only in the parotid gland and secreted in the saliva. The allele *Psp^a* determines fast electrophoretic mobility and sensitivity to CNBr and occurs in the YBR/WiHa strain; the allele *Psp^b* determines slow electrophoretic mobility and insensitivity to CNBr and occurs in the C3H/A strain. The alleles are codominant (1).

References

1. Owerbach, D., and J.P. Hjorth. 1978. Genetic determination of a parotid secretory protein in mouse. Hereditas 89:146 (Abstr.).

Psph locus, phosphoserine phosphatase, Chr 5

This locus codes for the enzyme phosphoserine phosphatase (E.C. 3.1.3.3). No genetic variants are known. The locus was found to be on Chr 5 by use of mouse–Chinese hamster somatic cell hybrids screened for expression of the enzyme (1).

References

1. Lalley, P.A., and J.A. Diaz. 1984. Comparative gene mapping in the mouse involving genes assigned to human chromosomes 7 and 20. Cytogenet. Cell Genet. 37:514–515 (Abstr.).

Pt, pintail, semidominant, Chr 4

Arose in a strain protractedly treated with methylcholanthrene. Heterozygotes have tails of variable length, usually characterized by kinks near the end and a thin threadlike tip. Homozygotes have tails similar to those of heterozygotes but usually much shorter. Homozygotes are smaller than normal and have a high preweaning mortality rate, but survivors are healthy and fertile (3). Examination of the skeletons of *Pt*/+ and *Pt*/*Pt* mice shows that there is a progressive reduction in the nucleus pulposus of the intervertebral discs in a cephalocaudad direction. The reduction is more severe in *Pt*/*Pt* than in *Pt*/+. The defect can be traced back to the 10th day of embryonic development where it appears as a reduced rate of cell division of the notochord. This leads to a smaller than normal notochord, and this in turn to smaller intervertebral discs (1, 2).

References

1. Berry, R.J. 1960. Genetical studies on the skeleton of the mouse. XXVI. Pintail. Genet. Res. 1:439–451.
2. Berry, R.J. 1961. Genetically controlled degeneration of the nucleus pulposus in the mouse. J. Bone Joint Surg. 43B:387–393.
3. Hollander. W.F., and L.C. Strong. 1951. Pintail, a dominant mutation linked to brown in the house mouse. J. Hered. 42:179–182.

Pth locus, parathyroid hormone, Chr 7

This is the structural locus for parathyroid hormone (PTH). It was found to be on Chr 7 by use of mouse–Chinese hamster somatic cell hybrids screened with a genomic clone of the rat PTH gene (1).

References

1. Lalley, P.A., A.Y. Sakaguchi, R.L. Eddy, N.H. Honey, G.I. Bell, L.-P. Shen, W.J. Rutter, J.W. Jacobs, G. Heinrich, W.W. Chin, and S.L. Naylor. 1987. Mapping polypeptide hormone genes in the mouse: somatostatin, glucagon, calcitonin, and parathyroid hormone. Cytogenet. Cell Genet. 44:92–97.

ptr

See *Pas-1*, *-2*, *-3* loci.

Ptv-1 locus, polytropic virus-1

This locus controls susceptibility to infection by a polytropic leukemia virus PTV-1 isolated from the HRS/J strain, as determined by an assay for infected thymus cells 4 to 6 weeks after injection of the virus. The dominant allele *Ptv-1ˢ* determines susceptibility and occurs in the CBA/J strain; the recessive allele *Ptv-1ʳ* determines resistance and occurs in the NFS strain. Polytropic viruses are formed by recombination between the viral genomes of endogenous ecotropic and xenotropic viruses in the *env* region of the viral genomes. This region codes for the envelope glycoprotein, gp70. Polytropic viruses are leukemogenic whereas their precursors are not, or only weakly so. Susceptibility to leukemogenesis by these viruses, however, is restricted to certain strains. The HRS/J isolate is leukemogenic in strains HRS/J and CBA/J but not in NFS and SWR/J. It seems likely that *Ptv-1ˢ* also occurs in the HRS/J strain and is responsible for the occurrence of leukemia in both HRS/J and CBA/J, and that *Ptv-1ʳ* also occurs in SWR/J and is responsible for absence of leukemia in both NFS and SWR/J. Resistance to infection by PTV-1 does not seem to be immunological but may be due to lack of a receptor for recombinant gp70, the receptor being present in susceptible strains (1).

References

1. Schwartz, R.S., and R.H. Khiroya. 1981. A single dominant gene determines susceptibility to a leukemogenic recombinant retrovirus. Nature 292:245–246.

pu, pudgy, recessive, Chr 7

Appeared in descendants of an X-rayed male. Homozygotes are identifiable at birth by their extremely short tails and shortened trunk region. The whole vertebral column is greatly shortened and highly irregular. Ribs and sternum are also abnormal. The rest of the skeleton is normal. Viability is somewhat reduced. Females are very poor breeders, but many males, surprisingly, breed well. The axial abnormalities arise from defective segmentation. Somite tissue with an epithelially arranged outer layer is formed, but segmentation is abortive or absent (1). *pu* is possibly a recurrence of stub (*sb*).

References

1. Grüneberg, H. 1961. Genetical studies on the skeleton of the mouse. XXIX. Pudgy. Genet. Res. 2:384–393.

Pv, pivoter, probably semidominant

Arose spontaneously in the amputated (*am*) stock. Heterozygotes have variable behavior ranging from hyperactivity with typical waltzing motion to normal behavior. They are deaf and have structural abnormalities of the inner ear, including constricted or absent semicircular canals. Both sexes are fertile (1).

References

1. Meredith, R. 1969. Mouse News Lett. 40:36.

Pvt-1 locus, plasmacytoma variant translocation-1, Chr 15

About 10 to 25 per cent of B-cell tumors (plasmacytomas) involve translocations between Chrs 6 and 15 that bring an oncogene on Chr 15 into association with the immunoglobulin kappa chain gene on Chr 6 (4). These tumors are referred to as 'variant plasmacytomas.' The breakpoint on Chr 15 occurs at the *Pvt-1* locus, which is located at least 72 kb away from the oncogene *Myc* (2), in a position distal to *Myc* (1). Both loci are in chromosome band 15D. In these tumors, the *Myc* gene is activated, presumably through some long-range mechanism, either direct or through a *Pvt-1* gene product (2). Some murine T-cell lymphomas have proviral inserts at the *Pvt-1* locus. *Pvt-1* altered by viral insertion may act on *Myc* expression in a manner similar to that of the translocation-altered *Pvt-1* gene (3). The role of *Pvt-1* in (6^15) variant plasmacytomas is in some ways comparable to that of *Myc* in (12;15) plasmacytomas. See *Myc* locus. A locus originally called *Mis-1*, Moloney virus integration site, on Chr 15 is now known to be identical to *Pvt-1* (5).

References

1. Banerjee, M., F. Wiener, J. Spira, M. Babonits, M.-G. Nilsson, J. Sumegi, and G. Klein. 1985. Mapping of the c-myc, pvt-1, and immunoglobulin kappa genes in relation to the mouse plasmacytoma-associated variant (6, 15) translocation breakpoint. EMBO J. 4:3183–3188.
2. Cory, S., M. Graham, E. Webb, L. Corcoran, and J.M. Adams. 1985. Variant (6:15) translocations in murine plasmacytomas involve a chromosome 15 locus at least 72 kb from the c-myc oncogene. EMBO J. 4:675–681.
3. Graham, M., J.M. Adams, and S. Cory. 1985. Murine T lymphomas with retroviral inserts in the chromosomal 15 locus for plasmacytoma variant translocations. Nature 314:740–743.
4. Ohno, S., M. Babonits, F. Wiener, J. Spira, G. Klein, and M. Potter. 1979. Nonrandom chromosome changes involving the Ig gene-carrying chromosomes 12 and 6 in pristane-induced mouse plasmacytomas. Cell 18:1001–1007.
5. Villeneuve, L., E. Bassart, P. Jolicoeur, M. Graham, and J.M. Adams. 1986. Proviral integration site Mis-1 in rat thymomas corresponds to the pvt-1 translocation breakpoint in murine plasmacytomas. Mol. Cell Biol. 6:1834–1837.

px locus, Chr 6

px, postaxial hemimelia, recessive. Arose spontaneously in a stock carrying *Ra*. The forelimbs are regularly affected. There may be absence of digits 5, 4, and sometimes 3, and reduction or absence of the ulna. There is always a large oval foramen in the scapula. The hindlimbs are usually normal, but digit 5 may be absent, and occasionally the fibula is reduced. On the dorsal surface of both forefeet and hindfeet, there are epidermal papillae, situated a little behind the base of the digits, which sometimes develop a rudimentary claw. Mice with more severely affected limbs tend to have an extra pair of ribs and a slight reduction in number of presacral vertebrae. Both sexes are sterile and show anomalies of the Müllerian ducts including a partly or wholly double vagina and uncoiled oviducts in the female, and persistent Müllerian ducts in the male (3).

px^r, postaxial hemimelia-Russell, recessive. Arose in the HR stock at Oak Ridge National Laboratory (1). It is similar to *px* except that homozygous males are fertile and the scapula does not have the large foramen present in *px/px* mice (2).

References

1. Russell, L.B. 1957. Mouse News Lett. 17:84.
2. Russell, L.B. 1972. Mouse News Lett. 47:61.
3. Searle, A.G. 1964. The genetics and morphology of two "luxoid" mutants in the house mouse. Genet. Res. 5:171–197.

py, polydactyly, recessive, Chr 1

Polydactyly unassociated with other major defects has been described several times in the mouse. In one such case, described by Holt (2), the polydactyly is on the preaxial side and affects the hindfeet almost exclusively. Penetrance is variable and may be increased by selection. In appropriate stocks it has been possible to follow the segregation of this gene sufficiently to determine its linkage (1). In stocks selected for regular manifestation of *py/py*, heterozygotes (*py/+*) were occasionally polydactylous on one foot (4). After outcrossing *py/py* to CBA or C57BL, Kocher (3) was able to produce by selection stocks of mice with polydactyly on all four feet. When polydactyly was single-sided, it occurred predominantly on the right side on the hindfeet and predominantly on the left side on the forefeet.

References

1. Fisher, R.A. 1953. The linkage of *polydactyly* with *leaden* in the house mouse. Heredity 7:91–95.
2. Holt, S.B. 1945. A polydactyl gene in mice capable of nearly regular manifestation. Ann. Eugen. 12:220–249.
3. Kocher, W. 1966. Genetisch und chemisch verursachte praeaxial Polydactylie bei der Maus. Zool. Anzeiger Suppl. 30:412–422.
4. Parsons, P.A. 1958. A balanced four-point linkage experiment for linkage group XIII of the house mouse. Heredity 12:77–95.

Pyp locus, pyrophosphatase, Chr 10

This is probably the structural locus for inorganic pyrophosphatase (PYP; E.C. 3.6.1.1). No genetic variants are known. By use of mouse–Chinese hamster somatic cell hybrid clones segregating mouse chromosomes, the locus was found to be located on Chr 10. A heteropolymeric band is formed in hybrid cells expressing mouse enzyme, indicating that the enzyme is a dimer (1).

References

1. Lalley, P.A., J.D. Minna, and U. Francke. 1978. Conservation of autosomal gene synteny groups in mouse and man. Nature 274:160–162.

Q

Q, quinky, semidominant, Chr 8

Arose as a spontaneous mutation in the Q strain. Heterozygotes are viable and fertile, but penetrance is incomplete. Heterozygotes may have short kinked tails and shaking or circling behavior. Most homozygotes die soon after implantation (1).

References

1. Schaible, R.H. 1961. Mouse News Lett. 24:38.

Q1–Q9

See *Qa-2* locus.

Q10, Q region gene 10, Chr 17

This is one of 10 genes in the *Qa-2/3* region of Chr 17. Unlike other class I genes, it is expressed only in liver cells and its product is secreted into the serum (1). The *Q10* gene from strain C57BL/10 was isolated and found to have an unusual structure in its exon 5 which encodes the transmembrane domain. The gene may encode a functional Class I polypeptide, but one that fails to be anchored in the cell membrane. The *Q10* of C57BL/6 is almost identical (99.5 per cent homology) with a cDNA clone isolated from an SWR/J liver cDNA library. It is thus possible that the C57BL/10 and SWR genes are alleles (3). The amount of Q10 protein in serum varies among inbred strains with different *H-2* haplotypes and is completely absent in the *H-2^f* strain B10.M. The genetic control of this variation maps distal to *H-2S* (2). The function of Q10 is unknown, but it is conserved in distantly related members of the genus *Mus* (4).

References

1. Kress, M., D. Cosman, G. Khoury, and G. Jay. 1983. Secretion of a transplantation-related antigen. Cell 34:189–196.
2. Lew, A.M., W.L. Maloy, and J.E. Coligan. 1986. Characteristics of the expression of the murine soluble class I molecule (Q10). J. Immunol. 136:254–263.
3. Mellor, A.L., E.H. Weiss, M. Kress, G. Jay, and R.A. Flavell. 1984. A nonpolymorphic class I gene in the murine major histocompatibility complex. Cell 36:139–144.
4. Siwarski, D.F., Y. Barra, G. Jay, and M.J. Rogers. 1985. Occurrence of a unique MHC class I gene in distantly

related members of the genus *Mus*. Immunogenetics 21:267–276.

Qa-1 locus (formerly also *Qed-1*), Qa lymphocyte antigen-1, Chr 17

This locus controls antigenic determinants on lymphocytes recognizable by antisera and by cytotoxic T lymphocytes (3, 5, 7, 8). Four alleles are known expressing different combinations of at least four antigenic determinants. *Qa-1ᵃ* expresses determinants Qa-1.1 and -1.3 and occurs in strains A, C57BR/cd, SWR, and NZB; *Qa-1ᵇ* expresses Qa-1.1 and -1.4 and occurs in strains C57BL/6, C57BL/10, C57L, BALB/c, CBA/J, DBA/2, AKR, and C3H/He; *Qa-1ᶜ* expresses Qa-1.2 and occurs in strain RIII; and *Qa-1ᵈ* expresses Qa-1.1 and -1.2 and occurs in strains A.CA and B10.M (4, 7). In a later study, Aldrich *et al.* (1) generated cytotoxic T lymphocytes against the determinants of *Qa-1ᵃ*, *Qa-1ᵇ*, and *Qa-1ᵈ* and were able to show 12 different reactivity patterns among target cells carrying the four alleles. The locus encodes a class I molecule of about 46 000 molecular weight that is associated with β_2-microglobulin (2, 9). It is expressed on thymus cells and on peripheral T- and B-cells, including cells that respond to LPS, PHA, and ConA. The degree of expression on different cell types is markedly affected by the method used for detection (4). The products of the *a*, *b*, and *d* alleles all have different isoelectric points (6). *Qa-1* is part of the *H-2 Tla* complex and is located distal to *Tla* (10).

References

1. Aldrich, C.J., R.N. Jenkins, and R.R. Rich. 1986. Clonal analysis of the anti-Qa-1 cytotoxic T lymphocyte repertoire: definition of the Qa-1ᵈ and Qa-1ᶜ alloantigens and cross reactivity with H-2. J. Immunol. 136:383–388.
2. Cook, R.G., R.N. Jenkins, L. Flaherty, and R.R. Rich. 1983. The Qa-1 alloantigens. II. Evidence for the expression of two Qa-1 molecules by the *Qa-1ᵈ* genotype and for cross-reactivity between Qa-1 and H-2Kᶠ. J. Immunol. 130:1293–1299.
3. Fischer Lindahl, K., B. Hausmann, and L. Flaherty. 1982. Polymorphism of a Qa-1-associated antigen defined by cytotoxic T cells. I. Qed-1ᵃ and Qed-1ᵈ. Eur. J. Immunol. 12:159–166.
4. Harris, R.A., P.M. Hogarth, D.G. Pennington, and I.F.C. McKenzie. 1984. Qa antigens and their differential distribution on lymphoid, myeloid, and stem cells. J. Immunogenet. 11:265–281.
5. Jenkins, R.N., C.J. Aldrich, N.F. Landolfi, and R.R. Rich. 1985. Correlation of Qa-1 determinants defined by antisera and by cytotoxic T lymphocytes. Immunogenetics 21:215–225.
6. Landolfi, N.F., R.R. Rich, and R.G. Cook. 1985. The Qa-1 alloantigens. III. Biochemical analysis of the structure and extent of polymorphism of the Qa-1 allelic products. J. Immunol. 135:1264–1270.
7. Nell, L.J., R.G. Cook, and R.R. Rich. 1982. Qa-1-associated antigens. IV. Evidence for additional Qa-1 polymorphism defined biochemically and by cytotoxic T lymphocyte recognition. Transplantation 34:54–59.
8. Stanton, T.H., and E.A. Boyse. 1976. A new serologically defined locus, *Qa-1*, in the *Tla*-region of the mouse. Immunogenetics 3:525–531.
9. Stanton, T.H., and L. Hood. 1980. Biochemical identification of the Qa-1 alloantigen. Immunogenetics 11:309–314.
10. Stanton, T.H., S. Carbon, and M. Maynard. 1981. Recognition of alternate alleles and mapping of the *Qa-1* locus. J. Immunol. 127:1640–1643.

Qa-2 locus, Qa lymphocyte antigen-2, Chr 17

This locus controls a lymphocyte antigen detected serologically or by cytotoxic T lymphocytes (4). The allele *Qa-2ᵃ* determines presence of the antigen and occurs in strains A, C57BL/6, BALB/cJ, DBA/1, DBA/2, SWR, and others; the allele *Qa-2ᵇ* determines absence of the antigen and occurs in strains AKR, C3H/An, BALB/cAn, BALB/cBy, GR/A, and others (1). The antigen is widely distributed on cells of hemopoietic lineage, including most normal T lymphocytes, a portion of thymocytes, T and B lymphoblasts, multipotential stem cells, progenitors of granulocytes and macrophages, and some natural killer cells (4). Most T-cell functions require the presence of Qa-2⁺ cells (5). The antigen also occurs on oocytes and on two-cell, eight-cell, and blastocyst embryos. In strains expressing the antigen (*Qa-2ᵃ*), the rate of cleavage division is fast and, in strains not expressing the antigen (*Qa-2ᵇ*), the rate is slow. This effect was ascribed to a gene *Ped*, pre-implantation-embryo-development, but may be an effect of the *Qa-2* locus (8). *Qa-2* is a class I antigen consisting of molecules of about 40 kDa associated with β2-microglobulin, a composition similar to that of the H-2K, H-2D, and TL antigens controlled by genes closely linked to *Qa-2* (4). Different clones of cytotoxic T lymphocytes from the same strain have been found to bear Qa-2 molecules that vary in molecular weight, isoelectric point, and in number of different Qa-2 molecules present (one or two). The cause of this variability is not known (7). Molecular studies have shown that there are 10 class I genes in the *Qa-2* region numbered Q1 to Q10 in order from 5′ to 3′ (9). Q6, Q7, Q8, and Q9 code for polypeptides that are precipitated with anti-Qa-2/3 antiserum. All are present in C57BL/10, a strain in which Qa-2 expression is high. In BALB/cJ in which Qa-2 expression is low, Q6 and Q7 are present but there is a deletion and fusion between Q8 and Q9. In BALB/cBy which is *Qa-2*^{b} (*Qa-2*$^-$), there is also a deletion and fusion between Q6 and Q7 (6). The DNA of *Qa-2*^{a} strains has a 3.7-kb *Xba*I fragment that is lacking in *Qa-2*^{b} strains (3). The *Qa-2* region of Chr 17 lies between *H-2D* and *Tla* (2). See also *Qa-3*.

References

1. Flaherty, L. 1976. The *Tla* region of the mouse: identification of a new serologically defined locus, *Qa-2*. Immunogenetics 3:533–549.
2. Flaherty L. 1978. Genes of the *Tla* system. The new *Qa* system of antigens. *In* H.C. Morse III, ed., Origins of Inbred Strains of Mice, 409–422. Academic Press, New York.
3. Flaherty, L., K. DiBiase, M.A. Lynes, J.G. Seidman, O. Weinberger, and E.M. Rinchik. 1985. Characterization of a *Q* subregion gene in the murine major histocompatibility complex. Proc. Natl. Acad. Sci. USA 82:1503–1507.
4. Forman, J., J. Trial, S. Tonkonogy, and L. Flaherty. 1982. The *Qa-2* subregion controls the expression of two antigens recognized by *H-2*-unrestricted cytotoxic T cells. J. Exp. Med. 155:749–767.
5. Hogarth, P.M., A. Basten, H. Prichard-Briscoe, M.H. Henning, V.R. Sutton, and I.F.C., McKenzie. 1985. The distribution of the Qa-2 alloantigen on functional T lymphocytes. J. Immunol. 135:1632–1636.
6. Mellor, A.L., J. Antoniou, and P.J. Robison. 1985. Structure and expression of genes encoding murine Qa-2 class I antigens. Proc. Natl. Acad. Sci. USA 82:5920–5924.
7. Sherman, D.H., D.M. Kranz, and H.N. Eisen. 1984. Expression of structurally diverse Qa-2-encoded molecules on the surface of cytotoxic T lymphocytes. J. Exp. Med. 160:1421–1430.
8. Warner, C.M., S.O. Gollnick, L. Flaherty, and S.B. Goldbard. 1987. Analysis of Qa-2 antigen expression by preimplantation mouse embryos: possible relationship to the preimplantation-embryo-development (*Ped*) gene product. Biol. Reprod. 36:611–616.
9. Weiss, E.H., L. Golden, K. Fahrner, A.L. Mellor, J.J. Devlin, H. Bulman, H. Tiddens, H. Bud, and R.A. Flavell. 1984. Organization and evolution of the class I gene family in the major histocompatibility complex of the C57BL/10 mouse. Nature 310:650–655.

Qa-3 locus, Qa lymphocyte antigen-3, Chr 17

This locus controls a serologically detected antigen of lymphocytes. The allele *Qa-3*^{a} determines presence of the antigen and occurs in most *Qa-2*^{a} strains including BALB/cJ, C56BL/6, and A; *Qa-3*^{b} determines absence of the antigen and occurs in the *Qa-2*^{b} strains as well as in DBA/1 and SWR. The antigen is limited to lymph node and spleen cells and predominantly to Thy-1$^+$ cells (1). Qa-2 and Qa-3 antigens occur on lymphocytes that respond to Con A and PHA (T-cells), but not on lymphocytes that respond to LPS (B-cells). Among Thy-1$^+$ cells, there are three classes with respect to these antigens, Qa-2$^-$ Qa-3$^-$, Qa-2$^+$ Qa-3$^-$, and Qa-2$^+$ Qa-3$^+$ (2). See also *Qa-2*.

References

1. Flaherty, L., D. Zimmerman, and T.H. Hansen. 1978. Further serological analysis of the Qa antigens: analysis of an anti-H-2.28 serum. Immunogenetics 6:245–251.
2. Flaherty, L., D. Zimmerman, and K.A. Sullivan. 1978. Qa-2 and Qa-3 antigens on lymphocyte subpopulations. I. Mitogen responsiveness. J. Immunol. 121:1640–1643.

Qa-4 locus (formerly *Qat-4*), Qa lymphocyte antigen-4, Chr 17

This locus controls an antigen present on peripheral T-cells. It is expressed on 70 per cent of spleen and lymph node cells but not on thymocytes. It was identified with a monoclonal AKR anti-C57BL/6 antibody (2). It can also be recognized by cytotoxic T lymphocytes (1). The allele *Qa-4*^{a} determines presence of the antigen and occurs in strains C57BL/6, C57BL/10, BALB/cJ, 129, C58, C57L, A/J, SJL, and DBA/2; the allele *Qa-4*^{b} determines absence of the antigen and occurs in strains AKR, CBA/J, CBA/H, C3H/An, C57BR/cd, BALB/cBy, SEC/1, P, BDP, BUB, and SWR. *Qa-4* is located between *H-2D* and *Tla* on Chr 17 (2). *Qa-4* has the same strain distribution as *Qa-2*. The tissue distributions are similar except that Qa-2 is present on thymocytes and activated T lymphocytes and Qa-4 is not present on either. Anti-Qa-2 monoclonal antibody completely blocks binding of anti-Qa-4 to resting spleen cells but anti-Qa-4 only partially blocks binding of Qa-2 to these cells. It therefore seems likely that Qa-4 is a weak determinant on the Qa-2 molecule (1, 3).

References

1. Forman, J., J. Trial, S. Tonkonogy, and L. Flaherty. 1982. The *Qa-2* subregion controls the expression of two antigens recognized by *H-2*-unrestricted cytotoxic T cells. J. Exp. Med. 155:749–767.
2. Hämmerling, G.J., U. Hämmerling, and L. Flaherty. 1979. Qat-4 and Qat-5, new murine T-cell antigens governed by the *Tla* region and identified by monoclonal antibodies. J. Exp. Med. 150:108–116.
3. Landolfi, N.F., R.R. Rich, M.J. Bevan, and R.C. Cook. 1985. Analysis of the Qa-4 determinant. Immunogenetics 21:511–516.

Qa-5 locus (formerly *Qat-5*), Qa lymphocyte antigen-5, Chr 17

This locus controls an antigen present on peripheral T-cells. It is expressed on 30 per cent of lymph node and spleen cells. It was identified with a monoclonal AKR anti-C57BL/6 antibody. The allele *Qa-5ᵃ* determines presence of the antigen and occurs in strains C57BL/6, C57BL/10, BALB/c, 129, C58, and C57L; the allele *Qa-5ᵇ* determines absence of the antigen and occurs in strains AKR, A/J, SJL, DBA/2, CBA/J, CBA/H, C3H/An, C57BR/cd, SEC/1, P, BDP, BUB, and SWR. *Qa-5* is located between *H-2D* and *Tla* on Chr 17 (2). *Qa-5* can also be recognized by cytotoxic T lymphocytes (CTL). However, it is possible that the CTL target antigen may not be identical with that recognized by the monoclonal antibody (1).

References

1. Forman, J., J. Trial, S. Tonkonogy, and L. Flaherty. 1982. The *Qa-2* subregion controls the expression of two antigens recognized by *H-2*-unrestricted cytotoxic T cells. J. Exp. Med. 155:749–767.
2. Haämmerling, G.J., U. Hämmerling, and L. Flaherty. 1979. Qat-4 and Qat-5, new murine T-cell antigens governed by the *Tla* region and identified by monoclonal antibodies. J. Exp. Med. 150:108–116.

Qa-6 locus, Qa lymphocyte antigen-6, Chr 17

This locus controls an antigen present on approximately 40 per cent of lymph node and splenic lymphocytes. The allele *Qa-6ᵃ* determines presence of the antigen and occurs in strains C57BL/6, C57BL/10, B10.D2, C57BL/Ks, C57L, A/J, BALB/cJ, LG, NZB, SEA, 129/J, DBA/1, and DBA/2; the allele *Qa-6ᵇ* determines absence of the antigen and occurs in strains AKR, BALB/cBy, BUB, CBA/J, C3H/He, C57BR/cd, C58, I/Ln, MA/My, P, RIIIS, SJL, and SWR. The antigen was detected by an antiserum produced by injecting BALB/cBy mice with the BALB/cBy tumor ORA1-a, which anomalously was Qa-6⁺. The distribution of *Qa-6* alleles in a number of congenic strains provides evidence that the locus is within the *Tla* region of Chr 17. *Qa-6* is the latest of four loci on the distal side of *H-2* (the others are *Qa-2*, *Qa-3*, and *Tla*) controlling antigens expressed anomalously on tumors from genetically antigen-negative strains (1). Although *Qa-6* has a strain distribution different from that of *Qa-2* and is known to be located distal to *Qa-2* on Chr 17, the molecules controlled by the two loci are both associated with β2-microglobulin and are indistinguishable by molecular weight (about 41 000), two-dimensional gel electrophoresis, peptide map analysis, and by immunodepletion experiments (2).

References

1. Flaherty, L., M. Karl, and C.L. Reinisch. 1982. Syngeneic tumor immunizations produce Qa antibodies. Discovery of a new Qa antigen, Qa-6. Immunogenetics 16:329–337.
2. Widacki, S.M., L. Flaherty, and R.C. Cook. 1985. Biochemical characterization of the molecules reactive with Qa-6 antiserum and the monoclonal antibody 20-8-4: evidence for structural similarity with the Qa-2 molecule. J. Immunol. 135:3333–3339.

Qa-7 locus, Qa lymphocyte antigen-7, Chr 17

This locus controls a lymphocyte antigen recognized by a monoclonal antibody. It is present on all lymph node and spleen cells, 67 per cent of bone marrow cells, and absent from the thymus. Strains that are positive for the antigen are C57BL/6, C57BL/10, C57L, NZB, DBA/1, DBA/2, SWR, SJL, and 129; negative strains are A, AKR, C3H, CBA, C58, CE, BALB/c, and RIII. *Qa-7* maps in the *Qa-2* region between *H-2D* and *Tla*. The Qa-7 determinant appears to be located on a 40 000 molecular weight class I molecule associated with β2-microglobulin (1).

References

1. Sandrin, M.S., P.M. Hogarth, and I.F.C. McKenzie. 1983. Two "Qa" specificities: Qa-m7 and Qa-m8 defined by monoclonal antibodies. J. Immunol. 131:546–547.

Qa-8 locus, Qa lymphocyte antigen-8, Chr 17

This locus controls a lymphocyte antigen recognized by a monoclonal antibody. It is present on 37 per cent of lymph node and spleen cells, on 24 per cent of bone marrow cells, and absent from the thymus. It is found only on T-cells, and is absent from B-cells. Strains positive for the antigen are C57BL/6, C57BL/10, C57L, 129, NZB, and SJL; all other strains tested were negative. Like *Qa-7*, *Qa-8* maps in the *Qa-2* region between *H-2D* and *Tla* and appears to determine specificity of a 40 000 molecular weight class I molecule (1).

References

1. Sandrin, M.S., P.M. Hogarth, and I.F.C. McKenzie. 1983. Two "Qa" specificities: Qa-m7 and Qa-m8 defined by monoclonal antibodies. J. Immunol. 131:546–547.

Qa-9 locus, Qa lymphocyte antigen-9, Chr 17

The lymphocyte antigenic determinant Qa-9 is determined by an allele at the *Qa-9* locus in the *Qa-2* region (between *H-2D* and *Tla*) of Chr 17 in combination with a gene linked to $H\text{-}3^a$ (probably $B2m^b$) on Chr 2. Strains with the positive allele, $Qa\text{-}9^a$, are C57BL/6, C57BL/10, C57L, and 129/J; all other strains tested carried the negative allele, $Qa\text{-}9^b$. Strains that are $Qa\text{-}9^a$ $B2m^b$ (C57BL/6, C57BL/10, C57L) have high levels of Qa-9 antigen; strains that are $Qa\text{-}9^a$ $B2m^a$ (129/J and several *H-3* and *H-2* congenic strains) have low levels of Qa-9 antigen (20 to 50 per cent expression); strains that are $Qa\text{-}9^b$ have no Qa-9 antigen regardless of the *B2m* genotype. In addition to *Qa-9* and *B2m*, another gene closely linked to *H-2D* controls level of expression of Qa-9. $H\text{-}2D^b$ strains have > 95 per cent expression; $H\text{-}2D^d$ and $H\text{-}2D^q$ strains have 75 per cent expression. Qa-9 is found on T-and B-cells of spleen and lymph nodes and on a small proportion of bone marrow cells, and is absent from the thymus (1).

References

1. Sutton, V.R., P.M. Hogarth, and I.F.C. McKenzie. 1983. Description of a new Qa antigenic specificity, "Qa-m9," whose expression is under complex genetic control. J. Immunol. 131:1363–1367.

Qa-11 locus, Qa lymphocyte antigen-11, Chr 17

The Qa-11 antigen is similar to the Qa-9 antigen in that its presence requires simultaneous presence of an *H-2* haplotype bearing the positive allele for the antigen ($Qa\text{-}11^a$) and the *b* allele at the *B2m* locus ($B2m^b$). Since the C57BL strains are $B2m^b$ and most other strains are $B2m^a$, the genotype of other strains with respect to *Qa-11* was determined from the Qa-11 phenotype of C57BL/6 and C57BL/10 congenic strains with *H-2* haplotypes derived from the other strains. The $Qa\text{-}11^a$ allele ($Qa\text{-}11^+$) occurs in the C57BL strains and in DBA/1, DBA/2, A, and O20; the $Qa\text{-}11^b$ allele ($Qa\text{-}11^-$) occurs in C57BR/cd and in several congenic strains with *H-2* haplotypes of non-inbred or recombinant origin. Examination of the recombinant congenic strains shows that *Qa-11* is in the *Tla* region of Chr 17. *Qa-11* has a strain distribution different from that of *Qa-9* and is therefore probably a different locus. The Qa-11 molecule is a 40 kDa molecule that is indistinguishable from Qa-2 by molecular weight, isoelectric focusing, or peptide mapping, but it does not react with anti-Qa-2 antiserum (1).

References

1. van de Meugheuvel, W., G. van Seventer, and P. Demant. 1985. A new Tla region antigen Qa-11, similar to Qa-2 and associated with B-type β2-microglobulin. J. Immunol. 134:2507–2512.

Qat-4 locus

See *Qa-4* locus.

Qat-5 locus

See *Qa-5* locus.

Qb-1 locus, Qb antigen-1, Chr 17

This locus determines an antigen present on spleen, lymph node, and bone marrow cells, but not on thymus cells. The allele $Qb\text{-}1^a$ determines a more acidic protein and occurs in the A/J, C3H/HeJ, and BALB/c strains and in the B6.K1 and several other C57BL congenic strains; the allele $Qb\text{-}1^b$ determines a more basic protein and occurs in the C57BL/6, C57BL/10, and 129/J strains and in the B6.K2 and other C57BL congenic strains; the allele $Qb\text{-}1^c$ is apparently a null allele and occurs in B10.RIII and two other B10 congenics. *Qb-1* is within the *Qa-2* region centromeric to *Tla*. The Qb-1 molecule is different from other biochemically characterized *Qa* region gene products. It may be encoded by one of the *Q1* to *Q5* genes (1).

References

1. Robinson, P.J. 1985. Qb-1, a new class I polypeptide encoded by the *Qa* region of the mouse *H-2* complex. Immunogenetics 22:285–289.

Qed-1 locus

See *Qa-1* locus.

QGlo-1 locus

See *Glo-1* locus.

qk, quaking, recessive, Chr 17

Arose spontaneously in the DBA/2J strain. Homozygotes have marked rapid tremor which disappears when they are at rest and in contact with bedding, but increases during locomotion. It begins at about 10 days and is fully developed by 3 weeks. Mature mice may have seizures in which motionless posture is maintained many seconds. Females are viable and fertile; males are sterile. The entire central nervous system is very deficient in myelin at all ages (15), and there is a less severe myelin deficiency in the peripheral nervous system (14). In the central nervous system myelin sheaths are present, but they are thinner than normal, some consisting of only one to four myelin lamellae. The sheaths are usually loosely wound, with patches of oligodendroglial cell cytoplasm between the lamellae, and there are abnormal inclusions and vacuoles in the processes and perikarya of oligodendrocytes. Development of the myelin sheaths appears to be arrested in a stage characteristic of very young animals (5, 17, 18). There is variable hyperplasia of oligodendrocytes, greatest in the tracts with the greatest degree of myelination (8). Axons have normal morphology but there is abnormally high proteolysis in the axons of the optic nerve (11). There is evidence that the myelination defect in the central nervous system is due to defective oligodendrocytes (10). In the peripheral nervous system, thinly myelinated fibers and unmyelinated fibers have been described in the sciatic nerve and in the intracranial portion of the trigeminal nerve (5, 17). The sheaths may be structurally abnormal with regions of uncompacted myelin lamellae similar to those of the central nervous system (16). Orthotopic transplantation of pieces of sciatic nerve between quaking and normal mice has shown that the genetic defect is expressed in Schwann cells (1). It is possible that qk causes defective myelinogenesis through a mechanism common to oligodendrocytes and Schwann cells (6). There is an extensive literature on biochemical defects related to the deficiency of myelin in quaking mice (reviewed by Baumann *et al.*, 3). A consistent finding is a severe deficiency of the myelin lipids, sphingomyelin, cerebrosides, and sulfatides, particularly those containing long-chain fatty acids. The normal increase in these fatty acids which occurs between 15 and 20 days does not occur in *qk/qk* mice, so that adult mutants tend to resemble very young controls (2). Brain proteolipids in adult quaking mice retain the relative proportions found in 10-day controls (12). The myelin-associated glycoproteins of different molecular weight in the brains of quaking mice 15 days of age and older are expressed in abnormal proportions (7). Synthesis of myelin basic protein and proteolipids is normal in quaking brains but their incorporation into myelin is defective (9). Quaking mice may have abnormal levels of copper and zinc in the brain but the evidence on this point is conflicting (13). The sterility of *qk/qk* mice is due to defective spermatid differentiation, the details of which have been described by Bennett *et al.* (4).

References

1. Aguayo, A.J., K. Mizuno, and G.M. Bray. 1977. Schwann cell transplantation: evidence for a primary sheath cell disorder causing hypomyelination in quaking mice. J. Neuropathol. Exp. Neurol. 36:595 (Abstr.).

2. Baumann N.A., M.L. Harper, and J.M. Bourre. 1970. Long chain fatty acid formation: key step in myelination studied in mutant mice. Nature 227:960–961.

3. Baumann, N.A., J.M. Bourre, C. Jacque, and S. Pollet. 1972. Genetic disorders of myelination. *In* Ciba Found. Symp., Lipids, Malnutrition and the Developing Brain, 91–100. ASP (Elsevier, Excerpta Medica, North-Holland), Amsterdam.

4. Bennett, W.I., A.M. Gall, J.L. Southard, and R.L. Sidman. 1971. Abnormal spermiogenesis in quaking, a myelin deficient mutant mouse. Biol. Reprod. 5:30–58.

5. Berger, B. 1971. Quelques aspects ultrastructuraux de la substance blanche chez la souris quaking. Brain Res. 25:35–53.

6. Fagg, G.E. 1979. The quaking mouse: regional variation in the content and protein composition of myelin isolated from the central nervous system. Neuroscience 4:973–978.

7. Frail, D.E., and P.E. Braun. 1985. Abnormal expression of the myelin-associated glycoprotein in the central nervous system of dysmyelinating mutant mice. J. Neurochem. 45:1071–1075.

8. Friedrich, V.L. 1975. Hyperplasia of oligodendrocytes in quaking mice. Anat. Embryol. 147:259–271.

9. Greenfield, S., N.I. Williams, M. White, S.W. Brostoff, and E.L. Hogan. 1979. Proteolipid protein: synthesis and assembly into quaking mouse myelin. J. Neurochem. 32:1647–1651.

10. Mikoshiba, K., K. Nagaike, E. Aoki, and Y. Tsukada. 1979. Biochemical and immunohistochemical studies on dysmyelination of quaking mutant mice *in vivo* and *in vitro*. Brain Res. 177:287–299.

11. Nixon, R.A. 1982. Increased axonal proteolysis in myelin-deficient mutant mice. Science 215:999–1001.

12. Nussbaum, J.L., and P. Mandel. 1973. Brain proteolipids in neurological mutant mice. Brain Res. 61:295–310.

13. Rosenfeld, J., A.W. Zimmerman, and V.L. Friedrich Jr. 1983. Altered brain copper and zinc in quaking mice. Exp. Neurol. 82:55–63.

14. Samorajski, T., R.I. Friede, and P.R. Reiner. 1970. Hypomyelination in the quaking mouse. A model for the analysis of disturbed myelin formation. J. Neuropathol. Exp. Neurol. 29:507–523.

15. Sidman, R.L., M.M. Dickie, and S.H. Appel. 1964. Mutant mice (*quaking* and *jimpy*) with deficient myelination in the central nervous system. Science 144:309–311.

16. Suzuki, K., and J.C. Zagoren. 1977. Quaking mouse: an ultrastructural study of the peripheral nerves. J. Neurocytol. 6:71–84.

17. Watanabe, I., and G.J. Bingle. 1972. Dysmyelination in "quaking" mouse. Electron microscopic study. J. Neuropathol. Exp. Neurol. 31:352–369.

18. Wisniewski, H., and P. Morell. 1971. Quaking mouse: ultrastructural evidence for arrest of myelinogenesis. Brain Res. 29:63–73.

Qui

Qui locus, quinine sensitivity, Chr 8 (?)

This locus controls sensitivity to the taste of quinine sulphate as measured by aversion to consuming it in drinking water when presented with distilled water as an alternative. There are large strain differences in aversion to quinine (interpreted as ability to taste it), with A2G, DBA/2, and BALB/c showing little aversion and 129/Sv and C57BL/6 showing strong aversion. The differences are probably due to a single locus with three alleles, Qui^a, and Qui^b determining strong aversion in 129/Sv and C57BL/6, respectively, and Qui^c determining little aversion in A2G, DBA/2, and BALB/c. Different alleles in 129/Sv and C57BL/6 are postulated because C57BL/6 mice show an unusual pattern of aversion, with aversion high and low on alternate days. In backcrosses with segregation at both the Qui and Soa (sucrose octo-acetate aversion) loci, no linkage was evident, but all Soa non-tasters were also Qui non-tasters, a result which was not satisfactorily explained (1). Qui is tightly linked to the gene for raffinose undecaacetate sensitivity, Rua (0/23 recombinants in CXB and BXD recombinant inbred strains) (2), and to Prp, proline-rich protein, (0/20 recombinants in BXD RI strains) (3). The assignment of Prp to Chr 8 is in some doubt. See Prp locus.

References

1. Lush, I.E. 1984. The genetics of tasting in mice. III. Quinine. Genet. Res. 44:151–160.
2. Lush, I.E. 1986. The genetics of tasting in mice. IV. The acetates of raffinose, galactose, and β-lactose. Genet. Res. 47:117–123.
3. Lush, I.E. 1986. Mouse News Lett. 74:121.

qv, quivering, recessive, Chr 7

Arose as a spontaneous mutation in a non-inbred stock (7). Viability of homozygotes at weaning is normal, but the life span is short, the majority dying before 5 months of age. Males are sterile, but females may be fertile and nurse their litters. Homozygotes are characterized by locomotor instability, pronounced quivering, deafness, varying degrees of paralysis of the hindlegs, clasping of the hindlegs when held up by the tail, and priapism in most males (4, 7). Serial sections of the brain, cord, and nerve roots reveal no abnormalities, and urinary amino acids are normal (3). The inner ear is histologically normal and has normal thresholds for compound action potentials at the round window. However, the thresholds for potentials in the inferior colliculus are twice as high as normal, showing that the deafness, unlike that of any other deaf mutants in the mouse, is of central origin (1). The raised thresholds are characteristic of all single units recorded. It is not certain whether the defect in the inferior colliculus is primary or the result of dysfunction of the superior olivary complex and lateral limniscus (2). Studies of the effect of qv on serum proteins and serum cholinesterase have been made, but it is not clear how the results are related to the effects of the gene on behavior (5, 6).

References

1. Deol, M.S., M.P. Frank, K.P. Steel, and G.R. Bock. 1983. Genetic deafness of central origin. Brain Res. 258:177–179.
2. Horner, K.C., and G.R. Bock. 1985. Combined electrophysiological and autoradiographic delimitation of retrocochlear dysfunction in a mouse mutant. Brain Res. 331:217–223.
3. McNutt, W. 1962. Urinary amino acid excretion in quivering mice (qvqv). Anat. Rec. 142:257 (Abstr.).
4. Yoon, C.H. 1960. Genetic and non-genetic factors affecting the quivering condition in mice. Am. Nat. 94:435–440.
5. Yoon, C.H. 1961. Electrophoretic analysis of the serum proteins of neurological mutants in mice. Science 134:1009–1010.
6. Yoon, C.H., and S.R. Harris. 1962. Cholinesterase studies of neurological mutants in mice. I. Alterations in serum cholinesterase levels. Neurology 12:423–426.
7. Yoon, C.H., and E.P. Les. 1957. Quivering, a new first chromosome mutation in mice. J. Hered. 48:176–180.

R

r

See *rd*.

Ra locus, Chr 2

Ra, ragged, semidominant. Arose spontaneously in a crossbred stock. In heterozygotes the first coat develops a little more slowly than normal. The coat contains guard hairs and awls but no auchenes and very few zigzags. This gives the coat a thin ragged appearance. The agouti pattern is modified, the entire coat being unusually dark. Heterozygotes are normally viable and fertile. Homozygotes are almost completely naked. Many are edematous at birth, and almost all die before weaning. A few survive and may breed (1). Developmental studies have shown that in *Ra/+* mice, growth of the late differentiating hair follicles which produce auchenes and zigzags is very retarded or arrested (5). A low percentage of *Ra/+* mice in some stocks were found by Herbertson and Wallace (3) to have a white chylous fluid in the abdomen from shortly after birth until a week or so of age. The incidence of chylous ascites in these mice is affected by one or more genes unlinked to *Ra* and also by two mutant genes linked to *Ra* (*fi*, *we*) and one on a different chromosome (*py*, Chr 1) (6).

*Ra^op^*, opossum, semidominant. Appeared spontaneously in a hybrid between the C57BL/6J and DBA/2J strains (2). Heterozygotes have very sparse fur consisting mostly of guard hairs and awls. The anterior vibrissae are absent. Viability is low, many heterozygotes dying shortly after birth in an edematous condition. Some develop a white fluid in the peritoneal cavity as in *Ra/+* and usually die before weaning. The surviving females are very poor breeders, but males may be normally fertile. Homozygotes die *in utero*, probably before 11 days. *Ra/Ra^op^* mice also die *in utero*, probably before 11 days (4).

References

1. Carter, T.C., and R.J.S. Phillips. 1954. Ragged, a semidominant coat texture mutant in the house mouse. J. Hered. 45:151–154.
2. Green, E.L., and S.J. Mann. 1961. Opossum, a semidominant lethal mutation affecting hair and other characteristics of mice. J. Hered. 52:223–227.
3. Herbertson, B.M., and M.E. Wallace. 1964. Chylous ascites in newborn mice. J. Med. Genet. 1:10–23.
4. Mann, S.J. 1963. The phenogenetics of hair mutants in the house mouse: opossum and ragged. Genet. Res. 4:1–11.
5. Slee, J. 1962. Developmental morphology of the skin and hair follicles in normal and in ragged mice. J. Embryol. Exp. Morphol. 10:507–529.
6. Wallace, M.E. 1979. Analysis of genetic control of chylous ascites in ragged mice. Heredity 43:9–18.

Raf locus

See *Afr-1* locus.

Raf-1 locus (*c-raf-1*), ras-related fibrosarcoma oncogene, Chr 6

This is the cellular homolog of the transforming gene (*v-raf*) of murine sarcoma virus 3611. The function of the cellular gene has not been reported, and no protein kinase activity has been found in two polyproteins produced by the viral v-*raf* gene (5). *Raf-1* is expressed at high levels in 14-day embryos, heart, lung, testis, epididymis, brain, and intestine and at lower levels in 18-day embryos, spleen, liver, and placenta (2). No genetic variants are known among standard laboratory inbred strains. *Raf-1* was first found to be on Chr 6 by use of mouse–Chinese hamster somatic cell hybrids screened with a v-*raf* probe (4). A restriction fragment length difference between laboratory strains and *Mus spretus* detected in *Taq*I digested DNA was later found and used to map *Raf-1* about 13 cM distal to *Igk* on Chr 6 (1). *Raf-1* in the mouse is 95 per cent homologous with a cellular gene in chickens that gave rise to the oncogene in avian carcinoma virus MH2 (3).

References

1. Bućan, M., T. Yang-Feng, A.M. Colberg-Poley, D.J. Wolgemuth, J.-L. Guénet, U. Francke, and H. Lehrach. 1986. Genetic and cytogenetic localisation of the homeo box containing genes on mouse chromosome 6 and human chromosome 7. EMBO J. 5:2899–2905.
2. Huleihel, M., M. Goldsborough, J. Cleveland, M. Gunnell, T. Bonner, and U.R. Rapp. 1986. Characterization of murine A-*raf*, a new oncogene related to v-*raf* oncogene. Mol. Cell. Biol. 6:2655–2662.
3. Kan, N.C., C.S. Flordellis, G.E. Mark, P.H. Duesberg,

Raf-1

and T.S. Papas. 1984. A common *onc* gene sequence transduced by avian carcinoma virus MH2 and by murine sarcoma virus 3611. Science 223:813–816.

4. Kozak, C., M.A. Gunnell, and U.R. Rapp. 1984. A new oncogene, c-*raf*, is located on mouse chromosome 6. J. Virol. 49:297–299.
5. Rapp, U.R., M.D. Goldsborough, G.E. Mark, T.I. Bonner, J. Groffen, F.H. Reynolds, Jr., and J.R. Stephenson. 1983. Structure and biological activity of v-*raf*, a unique oncogene transduced by a retrovirus. Proc. Natl. Acad. Sci. USA 80:4218–4222.

Ram-1 locus, replication of amphotropic virus-1, Chr 8

No genetic variant is known for this locus. It was discovered in Chinese hamster–mouse somatic cell hybrid clones segregating mouse chromosomes. Ability of the clones to support replication of two strains of exogenous amphotropic murine leukemia virus was shown to segregate with Chr 8 and with the enzyme locus *Aprt* known to be on Chr 8. Because Chrs 8 and 19 tended to segregate together, there is a possibility that *Ram-1* or a similar locus may be on Chr 19 (1).

References

1. Gazdar, A.F., H. Oie, P. Lalley, W.W. Moss, J.D. Minna, and U. Francke. 1977. Identification of mouse chromosomes required for murine leukemia virus reproduction. Cell 11:949–956.

rc, rough coat, recessive, Chr 9

Arose in the C57BL/6J strain. Homozygotes have unkempt coats and look identical to matted (*ma/ma*) mice. Both sexes are fertile (1). *rc* is on Chr 9 close to *Mpi-1* (2).

References

1. Dickie, M.M. 1966. Mouse News Lett. 34:30.
2. Eicher, E.M., S. Fox, and S. Reynolds. 1977. Mouse News Lett. 56:42.

Rcs-1 locus, reticulum cell sarcoma-1

This locus controls a difference between strains A.SW/Sn and SJL/J in spontaneous development of reticulum cell sarcoma (RCS). The allele *Rcs-1^s* determines an RCS incidence of over 95 per cent by 2 years of age and occurs in the SJL strain; the allele *Rcs-1^r* determines an incidence of less than 1 per cent by that time and occurs in the A.SW strain. F1 mice resemble A.SW. *Rcs-1^r* probably suppresses RCS expression at a point relatively early in tumor initiation rather than later at a point affecting tumor growth. *Rcs-1* is not an *H-2* gene and is not linked to markers on Chrs 1, 5, 7, 8, 12, or X (1).

References

1. Bubbers, J.E. 1984. Identification and linkage analysis of a gene, *Rcs-1*, suppressing spontaneous SJL/J lymphoma expression. J. Natl. Cancer Inst. 72:441–446.

rd, retinal degeneration, recessive, Chr 5

Described by Bruckner (3) and Tansley (17) in various stocks, and later found to be present in many inbred strains including C3H/He, C3H/St, C3HeB/Fe, P/J, BDP/J, PL/J, ST/J, and CBA/J (16). It is probably identical with rodless retina, *r* (11). Homozygotes are fully viable and fertile. Eyes develop normally up to the seven- to 10-day stage. At this stage the outer segment of the rod cell has begun to form, and in normal mice it elongates rapidly during the 10th to 15th days. In *rd/rd* nice the nascent outer segments and the rod cells degenerate rapidly so that by 15 days there is only a thin layer of rod cells left, and they have disappeared completely by 35 days (4, 12). The inner nuclear layer and the retinal ganglion cells appear normal but may show slight quantitative reduction (1, 10). Although the eyes of *rd/rd* mice are devoid of normal rods, the mice have some visual capacity (2, 6). Carter-Dawson *et al.* (5) were able to distinguish about 3 per cent of cones among the visual cells and showed that they degenerated at a much slower rate in *rd/rd* mice than rods, so that a few were still present at 18 months. By use of fusion chimeras between *rd/rd* and +/+ embryos, LaVail and Mullen (12) have shown that the *rd* locus does not act in the pigment epithelium of the retina and probably does act in the photoreceptor cells. At a stage before degeneration can be seen, the photoreceptor cells are deficient in a phosphodiesterase that hydrolyzes cyclic GMP. During the period of degeneration, the level of cyclic GMP in the photoreceptor layer rises to several times normal (7, 14, 15). At approximately the same time, the levels of adenylate cyclase and cyclic AMP increase above normal and remain so in adulthood. This is partly due to loss of the photoreceptor cell layer which is low in these factors, but is probably also due to increased levels of the factors in the inner nuclear layer (8, 13). Unexpectedly, the rate of retinal degeneration in mutants doubly homozygous with retinal degeneration-slow (*rd/rd rds/rds*) is intermediate between those of the two homozygotes (9).

References

1. Blanks, J.C., and D. Bok. 1977. An autoradiographic analysis of postnatal cell proliferation in the normal and

degenerative mouse retina. J. Comp. Neurol. 174:317–328.

2. Bonaventure, N., and P. Karli. 1961. Sensibilité visuelle spectrale chez des souris à rétine entièrement dépourvue de cellules visuelles photoréceptrices. C.R. Soc. Biol. 155:2015–2018.

3. Brückner, R. 1951. Spaltlampenmikroskopie und Ophthalmoskopie am Auge von Ratte und Maus. Doc. Ophthalmol. 5–6:452–554.

4. Caley, D.W., C. Johnson, and R.A. Liebelt. 1972. The postnatal development of the retina in the normal and rodless CBA mouse: a light and electron microscopic study. Am. J. Anat. 133:179–212.

5. Carter-Dawson, L.D., M.M. LaVail, and R.L. Sidman. 1978. Differential effect of the *rd* mutation on rods and cones in the mouse retina. Invest. Ophthalmol. Vis. Sci. 17:489–498.

6. Dräger, U.C., and D.H. Hubel. 1978. Studies of visual function and its decay in mice with hereditary retinal degeneration. J. Comp. Neurol. 180:85–114.

7. Farber, D.B., and R.N. Lolley. 1974. Cyclic guanosine monophosphate: elevation in degenerating photoreceptor cells in the C3H mouse. Science 186:449–451.

8. Ferrendelli, J.A., K.M. Campan, and G.W. De Vries. 1982. Adenylate cyclases in vertebrate retina: enzymatic characteristics in normal and dystrophic mouse retina. J. Neurochem. 38:753–758.

9. Fletcher, R.T., S. Sanyal, G. Krishna, G. Aguirre, and G.J. Chader. 1986. Genetic expression of cyclic GMP phosphodiesterase activity defines abnormal photoreceptor differentiation in neurological mutants of inherited retinal degeneration. J. Neurochem. 46:1240–1245.

10. Grafstein, B., M. Murray, and N.A. Ingoglia. 1972. Protein synthesis and axonal transport in retinal ganglion cells of mice lacking visual receptors. Brain Res. 44:37–48.

11. Keeler, C. 1966. Retinal degeneration in the mouse is rodless retina. J. Hered. 57:47–50.

12. LaVail, M.M., and R.J. Mullen. 1976. Role of the pigment epithelium in inherited retinal degeneration analyzed with experimental mouse chimeras. Exp. Eye Res. 23:227–245.

13. Lolley, R.N., S.Y. Schmidt, and D.B. Farber. 1974. Alterations in cyclic AMP metabolism associated with photoreceptor cell degeneration in the C3H mouse. J. Neurochem. 22:701–707.

14. Robb, R.M. 1979. Cyclic nucleotide phosphodiesterase activity in normal mice and mice with retinal degeneration. Invest. Ophthalmol. Vis. Sci. 18:1097–1100.

15. Schmidt, S.Y., and R.N. Lolley. 1973. Cyclic-nucleotide phosphodiesterase: an early defect in inherited retinal degeneration of C3H mice. J. Cell. Biol. 57:117–123.

16. Sidman, R.L., and M.C. Green. 1965. Retinal degeneration in the mouse: location of the *rd* locus in linkage group XVII. J. Hered. 56:23–29.

17. Tansley, K. 1954. An inherited retinal degeneration in the mouse. J. Hered. 45:123–127.

rds, retinal degeneration-slow, recessive, Chr 17

Found to be carried by the O20/A inbred strain (6). Homozygotes have retinal degeneration of early onset and slow progression. They lack outer segments of the receptor layer of the retina which normally begin to develop at 7 days of age. At 2 weeks the outer nuclear layer begins to decrease and continues to degenerate slowly. It is gone by 9 months in the peripheral part of the retina and by 12 months in the central part (1, 4). The electroretinogram decreases over the same period and is absent by 1 year (3). The level of cyclic AMP in the retina is higher than normal and that of cyclic GMP is lower than normal after 10 days. Activity of cAMP phosphodiesterase is somewhat higher than normal and that of cGMP phosphodiesterase is considerably lower than normal beginning at the same age. This contrasts with the case of *rd* in which cGMP level is much elevated, thought to be a consequence of low activity of cGMP phosphodiesterase (5). In heterozygotes (*rds*/ +), the outer segments formed are abnormal and there is a very slow rate of photoreceptor cell loss in old animals. Activity of cGMP phosphodiesterase in the retina is intermediate between that of +/+ and *rds*/*rds* retinas (2).

References

1. Jansen, H.G., and S. Sanyal. 1984. Development and degeneration of retina in *rds* mutant mice: electron microscopy. J. Comp. Neurol. 224:71–84.

2. Fletcher, R.T., S. Sanyal, G. Krishna, G. Aguirre, and G.J. Chader. 1986. Genetic expression of cyclic GMP phosphodiesterase activity defines abnormal photoreceptor differentiation in neurological mutants of inherited retinal degeneration. J. Neurochem. 46:1240–1245.

3. Reuter, J.H., and S. Sanyal. 1984. Development and degeneration of retina in *rds* mutant mice: the electroretinogram. Neurosci. Lett. 48:231–237.

4. Sanyal, S., A. De Reuter, and R.K. Hawkins. 1980. Development and degeneration of retina in *rds* mutant mice: light microscopy. J. Comp. Neurol. 194:193–207.

5. Sanyal, S., R. Fletcher, J.P. Liu, A. Aguirre, and G. Chador. 1984. Cyclic nucleotide content and phosphodiesterase activity in the *rds* mouse (O20/A) retina. Exp. Eye Res. 38:247–256.

6. van Nie, R., D. Iványi, and P. Démant. 1978. A new H-2-linked mutation, *rds*, causing retinal degeneration in the mouse. Tissue Antigens 12:106–108.

Re locus, Chr 11

Re, rex, dominant. First described by Crew and Auerbach (2) who obtained it from a commercial breeder. Both homozygotes and heterozygotes are fully viable and fertile. Homozygotes have a slightly more extreme expression than heterozygotes when young (1). Both have curly whiskers and wavy coats. The waviness of the coat disappears in adults, but the vibrissae and guard hairs remain curly. In both homozygotes and heterozygotes, all hair types are present, but the shaft diameter is irregular with variation in width of the cortex, and with some bends and twists. The hair defects appear to be due to a defect in the internal root sheath which is irregular in shape and unable to support the hair shaft and control its diameter (5).

Re^{wc}, wavy coat, dominant. Arose in a chronic neutron irradiation experiment. *Re^{wc}* resembles *Re* very closely except that its effects are more severe. Waviness of the juvenile coat is visible a little earlier, and the adult coat appears rougher. *Re^{wc}/Re^{wc}* are the most severely affected, *Re^{wc}/+* and *Re/Re* a little less so, and *Re/+* least affected. Morphology of the individual hairs and hair development are also affected in the same order of severity (5).

Re^{den}, denuded, semidominant. Arose spontaneously in the congenic strain B10.129(11M). Homozygotes and heterozygotes are viable and healthy. Heterozygotes can be distinguished by folds in the body skin at about one week. The coat grows normally until about 4 weeks when it becomes sparse on the head and neck. After 6 weeks loss is progressive, and older animals are almost naked and usually blind by 6 months (4). Homozygotes are similar to heterozygotes except that they usually become completely naked. The whiskers of both genotypes are normal. *Re^{den}/Re* mice are naked and have curly whiskers (3).

References

1. Carter, T.C. 1951. Wavy-coated mice: phenotypic interactions and linkage tests between rex and (a) waved-1, (b) waved-2. J. Genet. 50:268–276.
2. Crew, F.A.E., and C. Auerbach. 1939. Rex: a dominant autosomal monogenic coat texture character in the mouse. J. Genet. 38:341–344.
3. Eicher, E.M., and D. Varnum. 1986. Allelism of Den and Re. Mouse News Lett. 75:29–30.
4. Snell, G.D., and H.P. Bunker. 1968. Mouse News Lett. 39:28.
5. Trigg, M.J. 1972. Hair growth in mouse mutants affecting coat texture. J. Zool. Lond. 168:165–198.

Rec-1 locus (formerly also *Rev-1*), receptor for ecotropic virus-1, Chr 5

This locus controls a cell-surface receptor for ecotropic murine leukemia virus. No genetic variants are known. In a number of experiments using mouse–Chinese ham-

ster somatic cell hybrids, the gene for the receptor was found to be on Chr 5. The mouse strains from which the hybrid cells were derived include several different laboratory strains and two wild subspecies. The viruses used to infect the cells include Rauscher, Moloney, and Friend leukemia viruses, a vesicular stomatitis virus pseudotype, an N-tropic virus, and a B-tropic virus (1–5). Presence of the receptor was assayed by ability to bind gp70 envelope protein (2–4) or by ability to support replication of virus (1, 5). It is likely that both assays detect presence of the receptor.

References

1. Gazdar, A.F., H. Oie, P. Lalley, W.W. Moss, J.D. Minna, and U. Francke. 1977. Identification of mouse chromosomes for murine leukemia virus reproduction. Cell 11:949–956.
2. Hilkens, J., A. Colombatti, M. Strand, E. Nichols, F.H. Ruddle, and J. Hilgers. 1979. Identification of a mouse gene required for binding of Rauscher MuLV envelope gp70. Somat. Cell Genet. 5:39–49.
3. Marshall, T.H., and U.R. Rapp. 1979. Genes controlling receptors for ecotropic and xenotropic type C virus in *Mus cervicolor* and *Mus musculus*. J. Virol. 29:501–506.
4. Oie, H.K., A.F. Gazdar, P.A. Lalley, E.K. Russell, J.D. Minna, J. DeLarco, G.J. Todaro, and U. Francke. 1978. Mouse chromosome 5 codes for ecotropic murine leukemia virus cell-surface receptor. Nature 274:60–62.
5. Ruddle, N.H., B.S. Conta, L. Leinwand, C. Kozak, F. Ruddle, P. Besmer, and D. Baltimore. 1978. Assignment of the receptor for ecotropic murine leukemia virus to mouse chromosome 5. 1978. J. Exp. Med. 148:451–465.

Rel locus, reticuloendotheliosis oncogene, Chr 11

This is the cellular homolog of the transforming gene (v-*rel*) of the avian reticuloendotheliosis virus strain T, a gene that probably was derived from a turkey cellular gene c-*rel*. Homologous sequences occur in man and cats. In the mouse, *Rel* was shown to be on Chr 11 by use of mouse–Chinese hamster somatic cell hybrids (1).

References

1. Brownell, E., C.A. Kozak, J.R Fowle III, and W.S. Modi, N.R. Rice, and S.J. O'Brien. 1986. Comparative genetic mapping of cellular *rel* sequences in man, mouse, and the domestic cat. Am. J. Hum. Genet. 39:194–202.

Ren-1, *Ren-2* loci (formerly *Rnr*), Chr 1

These two loci, which are very closely linked, were thought to be a single locus controlling basal and induced activity of renin (E.C. 3.4.99.19) in the submandibular gland (SMG) when the polymorphism was first discovered (7). Low basal and induced activity was found in strains C57BL/6, C57L, A/J, BALB/c, C3H/He, and many others and ascribed to the allele Rnr^b; high basal and induced activity was found in strains SWR, DBA/2, AKR, Swiss, and many others and ascribed to the allele Rnr^s. Rnr was found to be on Chr 1 by Wilson *et al.* (8). Wilson and Taylor (9) found that renin in Rnr^s strains was thermolabile while renin in Rnr^b strains was thermostable and proposed the hypothesis, soon confirmed by others (3–5), that there were two renin genes, one controlling the thermostable renin and expressed primarily in the kidney, and the other controlling the thermolabile renin and expressed primarily in the SMG. Strains with high levels of SMG renin have both genes, strains with low levels have only one. The two genes show substantial homology (3), but *Ren-2* contains a 3-kb insert of a sequence coding for an intracisternal A particle that is not present in *Ren-1* (1). The nomenclature used below was proposed by Wilson (6).

Ren-1 locus, renin-1. This locus codes for renin produced primarily in the kidney but also at a very low level in the SMG. No genetic variants have been discovered in many inbred strains examined. The allele in these strains has been assigned the symbol $Ren-1^a$. The *Ren-1* renin is thermostable and is not inducible by androgen. *Ren-1* was shown to cosegregate with *Ren-2* on Chr 1 in the BXD recombinant inbred strains (5).

Ren-2 locus, renin-2. This locus codes for the renin produced primarily in the SMG. The allele in strains producing high levels of SMG renin is designated $Ren-2^s$. This renin is thermolabile and is highly inducible by androgen. The allele in strains producing low levels of SMG renin is designated $Ren-2^n$ for a null allele or absence of the gene. Field and Gross (2) found that, in the kidneys of $Ren-2^s$ strains, *Ren-2* mRNA accumulates at approximately the same rate as *Ren-1* mRNA, but how the difference in kidney *Ren-1* and *Ren-2* proteins arises is not known. *Ren-1* and *Ren-2* are probably the result of a duplication event occurring during evolution (5).

References

1. Burt, D.W., A.D. Reith, and W.J. Brammar. 1984. A retroviral provirus closely associated with the *Ren-2* gene of DBA/2 mice. Nucl. Acids Res. 12:8579–8593.
2. Field, L.J., and K.W. Gross. 1985. *Ren-1* and *Ren-2* loci are expressed in mouse kidney. Proc. Natl. Acad. Sci. USA 82:6196–6200.
3. Mullins, J.J., D.W. Burt, J.D. Windass, P. McTush, H. George, and W.J. Brammar. 1982. Molecular cloning of two distinct renin genes from the DBA/2 mouse. EMBO J. 1:1461–1466.
4. Panthier, J.-J., and F. Rougeon. 1983. Kidney and submaxillary gland renins are encoded by two non-allelic genes in Swiss mice. EMBO J. 2:675–678.
5. Piccini, N., J.L. Knopf, and K.W. Gross. 1982. A DNA polymorphism, consistent with gene duplication, correlates with high renin levels in mouse submaxillary gland. Cell 30:205–213.
6. Wilson, C.M. 1983. Mouse News Lett. 69:53–54.
7. Wilson, C.M., E.G. Erdös, J.F. Dunn, and J.D. Wilson. 1977. Genetic control of renin activity in the submaxillary gland of the mouse. Proc. Natl. Acad. Sci. USA 74:1185–1189.
8. Wilson, C.M., E.G. Erdös, J.D. Wilson, and B.A. Taylor. 1978. Location on chromosome 1 of *Rnr*, a gene that regulates renin in the submaxillary gland of the mouse. Proc. Natl. Acad. Sci. USA 75:5623–5626.
9. Wilson, C.M., and B.A. Taylor. 1982. Genetic regulation of thermostability of mouse submaxillary gland renin. J. Biol. Chem. 257:217–223.

Rev-1 locus

See *Rec-1* locus.

Rf, rib fusions, semidominant

Arose spontaneously in strain 129/RrSv. Heterozygotes have various skeletal anomalies, the commonest of which is fusion of ribs, usually somewhat distal to the vertebrae. The vertebrae may also be abnormal, and the tail may be kinked. Homozygotes may have very abnormal shortened vertebral columns and a short wide thoracic basket with extensive neural arch and rib fusions. They die at or before birth (2). Homozygous embryos fail to form somites. The rib fusions of heterozygotes are probably the result of abnormalities of the ventrolateral extensions of normal somites (3). *Rf*/+ embryos are significantly more susceptible than their normal sibs to exencephaly induced by insulin administered to their mothers during gestation (1).

References

1. Cole, W.A., and D.G. Trasler. 1980. Gene teratogen interaction in insulin-induced mouse exencephaly. Teratology 22:125–139.
2. Mackensen, J.A., and L.C. Stevens. 1960. Rib fusions, a new mutation in the mouse. J. Hered. 51:264–268.
3. Theiler, K., and L.C. Stevens. 1960. The development of rib fusions, a mutation in the house mouse. Am J. Anat. 106:171–183.

Rfv-1, Rfv-2 loci, Chr 17

Rfv-1 locus, recovery from Friend virus-1. This *H-2*-linked locus controls relative susceptibility to splenomegaly induced by Friend leukemia virus. The allele for susceptibility to splenomegaly and low rate of recovery, $Rfv-1^s$, occurs in association with the $H-2^d$ haplotype in strains BALB/c and DBA/2; the allele for resistance to splenomegaly and high rate of recovery, $Rfv-1^r$, occurs in association with the $H-2^b$ haplotype in strain C57BL/6. Heterozygotes have intermediate susceptibility to splenomegaly and low rate of recovery. The effect of *Rfv-1* is demonstrable only in *Fv-2*-susceptible mice ($Fv-2^s/-$) (4). *Rfv-1* is located in the *H-2G* or *H-2D* region (2, 3). Recovery appears to be mediated, at least in part, by the Friend virus-induced immune response of splenic T lymphocytes (1).

Rfv-2 locus, recovery from Friend virus-2. This second *H-2*-linked locus influences rate of recovery from splenomegaly induced by Friend virus. By use of F1 hybrids between C57BL/10 and A strain congenic lines bearing *H-2* recombinant haplotypes, *Rfv-2* was shown to be located either in the *H-2K* or *H-2I* regions or in the *Tla* region. Low recovery ($Rfv-2^s$) is associated with *d* alleles of the *H-2K–H-2I* region, high recovery ($Rfv-2^r$) with *b* alleles. The *Rfv-2* effect was seen in mice heterozygous at the *Rfv-1* locus but not in homozygotes for the high recovery $Rfv-1^r$ allele (2). Recovery appears to be mediated at least in part by the Friend virus-induced immune response of splenic T lymphocytes (1).

References

1. Britt, W.J., and B. Chesebro. 1983. H-2D control of recovery from Friend virus leukemia: H-2D region influences the kinetics of the T lymphocyte response to Friend virus. J. Exp. Med. 157:1736–1745.
2. Chesebro, B., and K. Wehrly. 1978. *Rfv-1* and *Rfv-2*, two *H-2*-associated genes that influence recovery from Friend leukemia virus-induced splenomegaly. J. Immunol. 120:1081–1085.
3. Chesebro, B., K. Wehrly, and J. Stimpfling. 1974. Host genetic control of recovery from Friend leukemia virus-induced splenomegaly. Mapping of a gene within the major histocompatibility complex. J. Exp. Med. 140:1457–1467.
4. Lilly, F. 1968. The effect of histocompatibility-2 type on response to the Friend leukemia virus in mice. J. Exp. Med. 127:465–473.

Rfv-3 locus, recovery from Friend virus-3

This locus controls recovery from viremia and leukemia induced by Friend virus (FV). The dominant allele required for recovery, $Rfv-3^r$, occurs in the C57BL/10 strain and allows recovery from both viremia and leukemia in $H-2^b/H-2^b$ mice. In $H-2^a/H-2^a$ mice, $Rfv-3^r$ allows recovery from viremia only; these mice lack the

appropriate *H-2* genotype for recovery from leukemia. The recessive allele $Rfv-3^s$ determines lack of recovery from viremia and leukemia and occurs in the A and A.BY strains (1). Leukemic $Rfv-3^r/Rfv-3^s$ mice 30 days after virus inoculation have lower levels of virus in spleen cells and much lower release of virus from cells than $Rfv-3^s/Rfv-3^s$ mice (2). The higher rate of recovery in heterozygotes is associated with loss of FV-induced antigen on the surface of leukemic spleen cells and high levels of anti-FV cytotoxic antibodies possibly responsible for the loss of FV-induced antigen (3). *Rfv-3* is not linked to *a*, *b*, *c*, *Igh-1*, *H-2*, or *Fv-2* (2).

References

1. Chesebro, B., and K. Wehrly. 1979. Identification of a non-*H-2* gene (*Rfv-3*) influencing recovery from viremia and leukemia induced by Friend virus complex. Proc. Natl. Acad. Sci. USA 76:425–429.
2. Chesebro, B., K. Wehrly, D. Doig, and J. Nishio. 1979. Antibody-induced modulation of Friend virus cell surface antigens decreases virus production by persistent erythroleukemia cells: influence of the *Rfv-3* gene. Proc. Natl. Acad. Sci. USA 76:5784–5788.
3. Doig, D., and B. Chesebro. 1979. Anti-Friend virus antibody is associated with recovery from viremia and loss of viral leukemia cell-surface antigens in leukemic mice. Identification of Rfv-3 as a gene locus influencing antibody production. J. Exp. Med. 150:10–19.

rg, rotating, recessive

Arose spontaneously in the AKR/J strain. Homozygotes show hyperactivity, circling, and some head-shaking from about 10 days on, but are not deaf. Expression is variable, and some homozygotes are difficult to distinguish from normal. A very small belly spot was found in about half of the homozygotes observed and a slight kinkiness of the tail in about one-fifth of them. Most homozygotes are smaller than normal throughout life. The abnormalities of the inner ear are of both degenerative and morphogenetic types: the saccular macula degenerates with age, and the semicircular ducts and canals have prominent constrictions. The organ of Corti is unaffected (1).

References

1. Deol, M.S., and M.M. Dickie. 1967. Rotating, a new gene affecting behavior and the inner ear in the mouse. J. Hered. 58:69–72.

Rgv-1 locus, resistance to Gross virus-1, Chr 17

This is an *H-2* linked locus that controls resistance to leukemogenesis induced by Gross murine leukemia virus. The allele *Rgv-1ˢ* determining susceptibility is associated with the *H-2* haplotypes, *k*, *a*, and *d* which occur in strains C3H, A, and DBA/2 respectively; the allele *Rgv-1ʳ* determining resistance is associated with *H-2ᵇ* which occurs in strain C57BL. Heterozygotes are resistant. By use of recombinant *H-2* haplotypes *Rgv-1* was shown to reside in the *H-2K* region at the left end of the *H-2* complex. The mechanism of action of the *Rgv-1* locus has been investigated extensively, but the results are inconclusive (1,2).

References

1. Lilly, F., and T. Pincus. 1973. Genetic control of murine viral leukemogenesis. Adv. Cancer Res. 17:231–277.
2. Steeves, R., and F. Lilly. 1977. Interactions between host and viral genomes in mouse leukemia. Ann. Rev. Genet. 11:277–296.

Rgv-2 locus, resistance to Gross virus-2

This non-*H-2*-linked locus, in addition to *Rgv-1*, was postulated to account for segregation of resistance to leukemogenesis induced by Gross leukemia virus in a cross between strains C3Hf/Bi and C57BL/6. The allele *Rgv-2ˢ* determines sensitivity and occurs in strain C3Hf/Bi; the allele *Rgv-2ʳ* determines resistance and occurs in strains C57BL/6, 129, and I. Heterozygotes are resistant. Susceptibility results from homozygosity for the *s* alleles at either the *Rgv-1* or *Rgv-2* loci. It is possible that *Rgv-2* is identical with *Fv-1* (1,2).

References

1. Lilly, F. 1966. The inheritance of susceptibility to Gross leukemia virus in mice. Genetics 53:529–539.
2. Lilly, F., and T. Pincus. 1973. Genetic control of murine viral leukemogenesis. Adv. Cancer Res. 17:231–277.

rh, rachiterata, recessive, Chr 2

Arose spontaneously in a subline of strain A/He. Homozygotes are small at birth and have short tails with distal kinks. Affected mice have a missing or abnormal axis, fused ribs, and fused thoracic and lumbar vertebrae (2). They have six cervical vertebrae instead of the normal seven and occasionally have cervical ribs on the sixth cervical vertebra. The malformations of the thoraco-lumbar region are due to a disturbed arrangement of somites first detectable in 11-day embryos (1).

References

1. Theiler, K., D. Varnum, and L.C. Stevens. 1974. Development of rachiterata, a mutation in the house mouse
with 6 cervical vertebrae. Z. Anat. Entwickl-Gesch. 145:75–80.
2. Varnum, D.S., and L.C. Stevens. 1974. Rachiterata: a new skeletal mutation on chromosome 2 of the mouse. J. Hered. 65:91–93.

rhg, retarded hair growth, recessive

Arose in strain AKR/J. Homozygotes are smaller than their normal littermates at birth. Hair development is noticeably retarded at 3 weeks but appears normal by 10 weeks (1).

References

1. Fox, S., and E.M. Eicher. 1978. Mouse News Lett. 58:47.

Rhv-1, Rhv-2 loci, resistance to hepatitis virus-1, -2

These two loci have been postulated to account for the pattern of inheritance of resistance to the JHM neurotropic strain of mouse hepatitis virus. In most strains of mice including C57BL/10 and BALB/c, intracerebral inoculation of JHM causes a lethal acute encephalomyelitis. Strain SJL/J is resistant. It is postulated that at the *Rhv-1* locus, the allele for resistance, *Rhv-1ʳ*, is dominant over the allele for susceptibility, *Rhv-1ˢ*, and at the *Rhv-2* (formerly *rhv-2*) locus, the allele for resistance, *Rhv-2ʳ*, is recessive to the allele for susceptibility, *Rhv-2ˢ*. Resistance thus requires the genotype *Rhv-2ʳ/Rhv-2ʳ Rhv-1ʳ/-*. A more complicated multigene model is not ruled out (1). *Rhv-2* may be identical to *Hv-2*. See *Hv-2* locus.

References

1. Stohlman, S.A., and J.A. Frelinger. 1978. Resistance to fatal nervous system disease by mouse hepatitis virus, strain JHM. I. Genetic analysis. Immunogenetics 6:277–281.

Ri-1, Ri-2 loci, recognition of identity-1, -2, Chr 17

These loci are postulated to explain the inheritance of variation in mating preference associated with differences in the region of the *H-2* complex. *Ri-1* is expressed in females and appears to be situated distal to *H-2D* in the *Qa-Tla* region (1). *Ri-2* is expressed in males and appears to be situated in the *H-2K* region (2). The variation was first demonstrated by caging a male with two congenic females differing at the *H-2* complex and noting a difference in the frequency of matings with the two females (3). It was later shown that both males and females could be trained in a Y maze to recognize the odor of congenic mice that differed only at the *H-2* complex (4). The odor of urine produces the same effect as the odor of whole animals. The critical odor of urine is derived from the hematopoietic system as shown by the finding that urine from bone-marrow chimeras acquires the odor of the donor marrow (5).

References

1. Andrews, P.W., and E.A. Boyse. 1978. Mapping of an *Il-2*-linked gene that influences mating preference in mice. Immunogenetics 6:265–268.
2. Yamaguchi, M., K. Yamazaki, and E.A. Boyse. 1978. Mating preference tests with the recombinant congenic strain BALB.HTG. Immunogenetics 6:261–264.
3. Yamazaki, K., M. Yamaguchi, P.W. Andrews, B. Peake, and E.A. Boyse. 1978. Mating preferences of F₂ segregants of crosses between MHC-congenic mouse strains. Immunogenetics 6:253–259.
4. Yamazaki, K., M. Yamaguchi, L. Baranoski, J. Bard, E.A. Boyse, and L. Thomas. 1979. Recognition among mice. Evidence from the use of a Y-maze differentially scented by congenic mice of different major histocompatibility types. J. Exp. Med. 150:755–760.
5. Yamazaki, K., G.K. Beauchamp, L. Thomas, and E.A. Boyse. 1985. The hematopoietic system is a source of odorants that distinguish major histocompatibility types. J. Exp. Med. 162:1377–1380.

Rib-1 locus, ribonuclease-1, Chr 14

This locus codes for pancreatic ribonuclease (E.C. 3.1.27.5). Genomic DNA of 43 inbred strains digested with *Bam*HI and probed with a clone isolated from mouse pancreatic cDNA showed a fragment of 3.7 kb in 34 strains including DBA/2, A/J, AKR, C3H/He, and BALB/c, and a fragment of 5.8 kb in nine strains including C57BL/6 and C57L. Segregation of the fragment size in BXD, BXH, and CXB recombinant inbred strains showed complete concordance of *Rib-1* with *Tcra* and *Np-2* on Chr 14 and looser linkage to other Chr 14 loci (1).

References

1. Elliott, R.W., L.C. Samuelson, M.S. Lambert, and M.H. Meisler. 1986. Assignment of pancreatic ribonuclease gene to mouse chromosome 14. Cytogenet. Cell Genet. 42:110–112.

Ric locus, resistance to *Rickettsia tsutsugamushi*, Chr 5

This locus controls natural resistance to the lethal effects of intraperitoneal injection of the Gilliam strain of *R. tsutsugamushi*. The allele *Ric^r* determines resistance and occurs in strains AKR, BALB/c, C57BL/6, SWR, and C3H/RV (a strain congenic with C3H/He, originally produced by crossing the gene for flavivirus resistance on to the C3H/He background and more exactly designated C3H.PRI-*Flv^r*); the allele *Ric^s* determines susceptibility and occurs in strains A/He, C3H/He, CBA/J, DBA/1, DBA/2, and SJL. Heterozygotes are resistant (1, 4). The mechanism of resistance has been studied in the congenic C3H/He and C3H/RV strains. After injection, susceptible C3H/He mice have an early influx of polymorphonuclear cells into the peritoneal cavity followed by a mononuclear influx which persists until death. Resistant C3H/RV mice have a small response consisting predominantly of mononuclear cells (4). The resistance can be abolished by irradiation, consistent with the explanation that resistance is due to the action of peripheral blood monocytes derived from bone marrow (5). Macrophages produced in response to intraperitoneal infection in C3H/RV mice contain a high proportion of I region associated (Ia) antigen-positive cells, whereas macrophages produced in C3H/He mice are very deficient in these cells. These results indicate that a factor regulating the presence of Ia antigen on macrophages may be responsible for the *Ric* effect (3). *Ric* was shown to be on Chr 5 near *rd* (retinal degeneration) by its strain distribution in BXD recombinant inbred strains (2).

References

1. Groves, M.G., and J.V. Osterman. 1978. Host defenses in experimental scrub typhus: genetics of natural resistance to infection. Infect. Immun. 19:583–588.
2. Groves, M.G., D.L. Rosenstreich, B.A. Taylor, and J.V. Osterman. 1980. Host defenses in experimental scrub typhus: mapping the gene that controls natural resistance in mice. J. Immunol. 125:1395–1399.
3. Jerrells, T.R. 1983. Association of an inflammatory I region-associated antigen-positive macrophage influx and genetic resistance of inbred mice to *Rickettsia tsutsugamushi*. Infect. Immun. 42:549–557.
4. Jerrells, T.R., and J.V. Osterman. 1981. Host defenses in experimental scrub typhus: inflammatory response of congenic C3H mice differing at the *Ric* gene. Infect. Immun. 31:1014–1022.
5. Jerrells, T.R., and J.V. Osterman. 1982. Role of macrophages in innate and acquired host resistance to experimental scrub typhus infection of inbred mice. Infect. Immun. 37:1066–1073.

rif locus

See *Afr-2* locus.

Rig-1 locus, regulation of Igh-1b-1, Chr 17

This locus regulates the level of Igh-1b immunoglobulin in serum. This immunoglobulin is of the IgG_{2a} class and is controlled by the *Igh-1* locus. The *Igh-1^b* allele occurs in the C57BL/6 strain and has been crossed into the BALB/c strain to produce the BAB/14 congenic strain. The BAB/14 mice have aberrantly low levels of Igh-1b immunoglobulin. This has been shown to be due to presence in the BALB/c genome of a recessive allele designated *Rig-1^d* which suppresses Igh-1b production. The allele *Rig-1^b* occurs in the C57BL/10 strain and allows normal Igh-1b production. Heterozygotes have normal Igh-1b production. There is only a slight effect of *Rig-1^d* on Igh-1a, the allotypic immunoglobulin of the BALB/c strain. Igh-4 immunoglobulins are not affected. *Rig-1* is either closely linked to or within the *H-2* complex, probably centromeric to the I-C subregion. There is probably at least one other gene in the BALB/c genome causing suppression of Igh-1b. It is dominant and not linked to H-2 (1).

References

1. Dowsett, A.P., L.A. Herzenberg, and L.A. Herzenberg. 1981. Regulation of the production of murine IgG_{2a} by an *H-2*-linked gene and other unlinked genes. Immunogenetics 13:237–245.

Ril-1 locus, radiation-induced leukemia-1, Chr 15

This locus controls susceptibility to leukemia induced by fractionated doses of gamma radiation (FX) given at weekly intervals. A/J mice carry the allele *Ril-1^r* and have a very low incidence of leukemia after FX; C57BL/10 mice carry the allele *Ril-1^s* and develop a high incidence of leukemia after FX. F1 hybrids are intermediate (1). *Ril-1^r* probably also occurs in strains DBA/2 and BALB/cBy (2). How *Ril-1* functions in the control of FX-induced leukemia is not known, but it may interact with loci on Chrs 1 and 4 that code for xenotropic viruses. It may control incidence of radiation-induced recombination between the xenotropic viruses and an ecotropic virus, such recombination being needed to produce a leukemogenic virus (2). *Ril-1* is on Chr 15 near *Ly-6* (3).

References

1. Meruelo, D., M. Offer, and N. Flieger. 1981. Genetics of susceptibility to radiation-induced leukemia. Mapping of genes involved to chromosomes 1, 2, and 4, and implications for a viral etiology of the disease. J. Exp. Med. 154:1201–1211.

2. Meruelo, D., M. Offer, and A. Rossomando. 1983. Induction of leukemia by both fractionated X-irradiation and radiation leukemia virus involves loci in the chromosome 2 segment H-30–A. Proc. Natl. Acad. Sci. USA 80:462–466.

3. Meruelo D., A. Rossomando, S. Scandalis, P. D'Eustachio, R.E.K. Fournier, D.R.Roop, D. Saxe, C. Blatt, and M.N. Nesbitt. Assignment of the *Ly-6–Ril-1–Sis–H-30–Pol-5/Xmmv-72–Ins-3–Krt-1–Int-1–Gdc-1* region to mouse chromosome 15. Immunogenetics 25:361–372.

Rjv-1 locus

See *Emv-16*, *Emv-17* loci.

Rjv-2 locus

See *Emv-1* locus.

rl, reeler, recessive, Chr 5

Found by Falconer (2) as a spontaneous mutation in a mildly inbred stock. Homozygotes are unable to keep their hindquarters upright and frequently fall over on their sides when walking or running. Viability and fertility are much reduced, particularly when the gene is on an inbred genetic background, but viability is greatly improved on a hybrid background, and an occasional female or rarely a male may breed (1). Healthy reeler mice have fairly normal behavior except for difficulties in locomotion (5). The neuropathology of *rl/rl* mice has been studied very extensively. These studies have been summarized and critically reviewed by Goffinet (3). Briefly, the cerebellum is greatly reduced in size, and the typical organization and lamination of the cerebellar cortex, the cerebral cortex, and the hippocampus are altered. Abnormal arrangement of neurons is also seen in the inferior olivary nucleus, the facial nerve nucleus, the tectum and lateral geniculate nucleus, and to some extent in the olfactory bulb. The neurons of the normal cortex originate in the ependymal layer and migrate out, with the later formed cells migrating past the earlier formed cells. Autoradiographic studies of development of the cerebral cortex in reelers have shown that different classes of neurons take their origin from the ependymal layer at the normal time but migrate abnormally, so that they come to rest in abnormal relations to each other. Neurons migrate out from the ependymal layer along radial glial fibers and separate from them when they reach their normal destination. In *rl/rl* mice, there appears to be close adhesion between postmigratory cells and the glial fibers, which may interfere with migration of the later formed neurons (6). In all other affected parts of the brain studied, the abnormal arrangement of neurons is the result of a similar abnormal pattern of migration. In spite of abnormal location of the neurons and also their greatly reduced number in the cerebellum, relatively normal cell connections are established. Some evidence on the cell type in which *rl* is expressed in the cerebellum is available from examination of chimeras produced by fusion between reeler and normal embryos, which showed that factors extrinsic to the abnormally positioned Purkinje cells were defective in reeler (4).

References

1. Caviness, V.S. Jr., D.K. So, and R.L. Sidman. 1972. The hybrid reeler mouse. J. Hered. 63:241–246.
2. Falconer, D.S. 1951. Two new mutants, "trembler" and "reeler," with neurological actions in the house mouse. J. Genet. 50:192–201.
3. Goffinet, A.M. 1984. Events governing organization of postmigratory neurons: studies on brain development in normal and reeler mice. Brain Res. Rev. 7:261–296.
4. Mullen, R.J. 1977. Genetic dissection of the CNS with mutant-normal mouse and rat chimeras. Soc. Neurosci. Symp. 2:47–65.
5. Myers, W.A. 1970. Some observations on "reeler," a neuromuscular mutation in mice. Behav. Genet. 1:225–234.
6. Pinto-Lord, M.C., P. Everard, and V.S. Caviness Jr. 1982. Obstructed neuronal migration along radial glial fibers in the neocortex of the reeler mouse; a Golgi-EM analysis. Dev. Brain Res. 4:379–393.

Rmc-1 locus, receptor for MCF virus-1, Chr 1

This locus controls capacity of mouse cells to support replication of several lines of exogenous mink cell focus-forming virus, probably because the locus codes for a cell-surface virus receptor. The cells tested and found to support replication were from the BALB/c, A, and NFS strains. No genetic variants were found. The locus was shown to be on Chr 1, probably at its distal end, by use of mouse–Chinese hamster somatic cell hybrids (1). *Rmc-1* is in the same region of Chr 1 as *Sxv*, a locus that controls susceptibility to xenotropic viruses and it is possible that the two genes may control allelic receptors to MCF and xenotropic viruses (2). See *Sxv* locus.

References

1. Kozak, C.A. 1983. Genetic mapping of a mouse chromosomal locus required for mink cell focus-forming virus replication. J. Virol. 48:300–303.
2. Kozak, C.A. 1985. Susceptibility of wild mouse cells to exogenous infection with xenotropic leukemia virus: con-

trol by a single dominant locus on chromosome 1. J. Virol. 55:690–695.

Rmcf locus, resistance to MCF virus, Chr 5

This locus controls resistance and susceptibility of cells in tissue culture to infection by mink cell focus-forming murine leukemia viruses. The allele *Rmcf^r* determines resistance and occurs in strains DBA/1, DBA/2, and CBA/Ca; the allele *Rmcf^s* determines susceptibility and occurs in strains AKR/J, C57BL/6, BALB/c, CBA/ J, NFS, NZB, 129/J, and many others. Heterozygotes are as resistant as the resistant parent or nearly so. *Rmcf* is different from and independent of *Fv-1*, a locus that controls susceptibility to infection by ecotropic viruses. *Rmcf* is located on Chr 5 close to *Hm* near the centromeric end (2). *Rmcf^r* protects (AKR × CBA/ Ca)F1 and (AKR × DBA/2)F1 hybrids from development of spontaneous thymic lymphomas and reduces the incidence of MCF-induced thymic lymphomas (4), and it reduces susceptibility of cells of *Sxv^s/Sxv^r* mice to exogenous xenotropic viruses (3). In addition, in strains susceptible to Friend virus-induced erythroleukemia, a condition thought to be due to the replication of MCF virus, *Rmcf^r* increases resistance to the virus-induced erythroleukemia. It may cause resistance by coding for or regulating the production of an MCF-related envelope glycoprotein that blocks the receptor for MCF viruses (5). This conclusion is reinforced by the findings of Buller *et al.* (1), who showed that the *Rmcf^r* allele contains an endogenous MCF gp70 *env* gene and that the *Rmcf^s* allele, at least in some strains (C57BL/6, CBA/ J, and A/WySn), contains a xenotropic gp70 *env* gene. Presumably the MCF gp70 inhibits exogenous MCF infection *in vitro* by a mechanism of viral interference.

References

1. Buller, R.S., A. Ahmed, and J.L. Portis. 1987. Identification of two forms of an endogenous murine retroviral *env* gene linked to the *Rmcf* locus. J. Virol. 61:29–34.
2. Hartley, J.W., R.A. Yetter, and H.C. Morse III. 1983. A mouse gene on chromosome 5 that restricts infectivity of mink cell focus-forming recombinant murine leukemia viruses. J. Exp. Med. 158:16–24.
3. Kozak, C.A. 1985. Susceptibility of wild mouse cells to exogenous infection with xenotropic leukemia viruses: control by a single dominant locus on chromosome 1. J. Virol. 55:690–695.
4. Rowe, W.P., and J.W. Hartley. 1983. Genes affecting mink cell focus-inducing (MCF) murine leukemia virus infection and spontaneous lymphoma in AKR F1 hybrids. J. Exp. Med. 158:353–364.
5. Ruscetti, S., R. Matthai, and M. Potter. 1985. Susceptibility of BALB/c mice carrying various DBA/2 genes to

development of Friend murine leukemia virus-induced erythroleukemia. J. Exp. Med. 162:1579–1587.

Rmp-1 locus, resistance to mousepox-1

This is one of probably several loci controlling resistance to mortality from infection with ectromelia (mousepox) virus. C57BL/6J mice bear the allele for resistance *Rmp-1^r* and A/J mice bear the allele for susceptibility *Rmp-1^s*. Heterozygotes are resistant and resistance segregates normally in both sexes of backcrosses to A/J. Mice of strain AKR/J are also resistant, and mice of strains BALB/cBy and DBA/2J are susceptible, but crosses involving these strains give evidence of multiple-locus inheritance of resistance and susceptibility. Resistance to mortality is apparently not related to resistance to infection, since the 50 per cent infectious doses of virus for all strains are similar (1).

References

1. Wallace, G.D., R.M.L. Buller, and H.C. Morse III. 1985. Genetic determinants of resistance to ectromelia (mousepox) virus-induced mortality. J. Virol. 55:890–891.

Rmv-1, *-2*, *-3* loci, resistance to Moloney virus-1, -2, -3, Chr 17

There is evidence that resistance to viremia induced by Moloney leukemia virus, an exogenous NB tropic virus, is controlled by three genes in the *H-2* complex. The loci were mapped by use of *H-2* congenic and recombinant strains. *Rmv-1* and *Rmv-2* appear to be located near the left and right ends of the *I* region, respectively, and *Rmv-3* at the *D* end of the complex. Resistance appears to be dominant for *Rmv-2* and semidominant for *Rmv-1* and *Rmv-3*. There is a good but not perfect correlation between viremia and the development of leukemia (1).

References

1. Debre, P., S. Gisselbrecht, F. Pozo, and J.P. Levy. 1979. Genetic control of sensitivity to Moloney leukemia virus in mice. II. Mapping of three resistance genes within the H-2 complex. J. Immunol. 123:1806–1812.

Rn locus, Chr 14

Rn, roan, semidominant. Arose spontaneously in a stock carrying *ld* and *mg*. Heterozygotes have white or partly pigmented hairs distributed throughout the coat. Expression is very variable and dependent on the genetic background, some heterozygotes being indistinguishable from normal. Homozygotes are lighter than heterozygotes and are viable and fertile (3).

Rn^fkl, freckled, semidominant. Arose spontaneously in a random bred stock carrying *t*-alleles at Harwell. Heterozygotes are variegated with small white or light colored patches in the coat and whitish pinna hairs. The amount of white increases with age. Homozygotes are viable and fertile and considerably lighter colored than heterozygotes (1,2). As with *Rn*, expression is variable and dependent on the genetic background. However, *Rn* and *Rn^fkl* have different expressions on the 101/H background, *Rn^fkl*/+ mice being much lighter than *Rn*/+. The alleles are therefore, not identical (4). They show the same recombination with piebald (5).

References

1. Butler, J., and M.F. Lyon. 1969. Mouse News Lett. 40:25.
2. Butler, J., and M.F. Lyon. 1969. Mouse News Lett. 41:28.
3. Green, M.C. 1966. Mouse News Lett. 34:31.
4. Lyon, M.F., and P.H. Glenister. 1973. Mouse News Lett. 48:31.
5. Lyon, M.F., and P.H. Glenister. 1978. Mouse News Lett. 59:18.

Rn7s loci

These loci code for 7S RNA, a small abundant cytoplasmic species that is a constituent of a ribonucleoprotein, the signal recognition particle that is responsible for the transport of secretory proteins from the ribosomes to the endoplasmic reticulum. Multiple 7S RNA genomic sequences are transcribed, even in single tissues. Numerous DNA restriction fragment variants among inbred strains were identified by a mouse 7S RNA cDNA plasmid. Eleven of these that have been assigned chromosomal locations by use of recombinant inbred strains and congenic strains are listed below.

Chr	Linked gene	Ref.	Chr	Linked gene	Ref.
1	*Ly-1*	1	6	*Igk*	4
2	*Hc*	2	8	*Emv-2*	2
3	*Amy-1*, *Adh-1*	2	9	*Pgm-3*	3
4	*Mup-1*, *b*	2	12	*Igh*	2
	(2 variants)		13	*As-1*	2
5	*Pgm-1*, *Ric*	3	X	–	1

The X-linked gene was identified by a twofold greater intensity of hybridization to a single band in females than in males. Only the Chr 6 locus has been assigned a symbol and allelic designations.

Rn7s-6 locus, 7S RNA-6, Chr 6. This locus determines presence or absence of a fragment detected with at least five different restriction enzymes. The allele designated *Rn7s-6^b* determines presence of the fragment and occurs in the C57BL/6 and most other inbred strains; the allele *Rn7s-6^a* determines absence of the fragment and occurs in strains AKR, C58, PL, and RF, and in wild mice from many locations. These four inbred strains are also the only laboratory strains carrying the closely linked *Ly-3^a* and *Igk-Ef1^a* alleles. Evidence from recombinant inbred and congenic strains shows that *Rn7s-6* is closely linked to *Igk* on Chr 6, probably in the order (*Ly-2, Ly-3*)–(*Rn7s-6, Igk-Ef1*)–*Igk-Ef2* (4).

References

1. Taylor, B.A., and L. Rowe. 1984. Mapping 7S RNA sequences. Mouse News Lett. 71:31.
2. Taylor, B.A., and L. Rowe. 1985. Mapping additional 7S RNA sequences. Mouse News Lett. 72:110.
3. Taylor, B.A., L. Rowe, and K. Gibbons. 1984. 7S RNA sequences mapped. Mouse News Lett. 70:84.
4. Taylor, B.A., L. Rowe, D.M. Gibson, R. Riblet, R. Yetter, and P.D. Gottlieb. 1985. Linkage of a 7S RNA sequence and kappa light chain genes in the mouse. Immunogenetics 22:471–481.

Rnr locus

See *Ren-1, Ren-2* loci.

Rnr12 gene cluster, ribosomal RNA-12, Chr 12

These are the ribosomal RNA genes (rDNA) that are located near the centromere of Chr 12. Such clusters have previously been called nucleolus organizer (NO) regions. *Rnr12* was identified as two size classes of rDNA hybridizing with a cloned variable length rDNA segment. The two classes are present in strain C57BL/6J (*Rnr12^b*) and absent in strain C3H/HeJ (*Rnr12^h*). In 13 BXH recombinant inbred strains, the two classes show no recombination. *Rnr12* was shown to be linked to the Robertsonian translocation Rb(8.12)5Bnr, and is known to be on Chr 12 since this chromosome has an NO region whereas Chr 8 does not (1, 2).

References

1. Arnheim, N., D. Treco, B. Taylor, and E.M. Eicher. 1982. Distribution of ribosomal gene length variants among mouse chromosomes. Proc. Natl. Acad. Sci. USA 79:4677–4680.
2. Taylor, B.A., E.M. Eicher, and N. Arnheim. 1987. Locus symbol for rRNA genes. Mouse News Lett. 77:130.

Rnu1 loci, U1 small nuclear RNAs

In the mouse, there are two molecular species of U1 small nuclear RNA (snRNA), U1a and U1b, and probably several genes and/or pseudogenes for both. Three *Rnu1* genes, not identified as coding for either U1a or U1b, were mapped by Harper *et al.* (1) using C57BL/6 × DBA/2 and A/J × C57BL/6 recombinant inbred strains classified with DNA probes for restriction fragment length polymorphisms.

Rnu1-1, U1 small nuclear RNA-1, Chr 1 near *Pep-3*.

Rnu1-2, U1 small nuclear RNA-2, Chr 9 near *Apoa-1*.

Rnu1-3, U1 small nuclear RNA-3, chromosome not identified.

Rnu1a-1, U1a small nuclear RNA-1, Chr 11. This locus codes for a U1a snRNA. A single-copy fragment from the 3′ flanking region of the gene was used as a probe in linkage tests. Strains BALB/c and STS were found to differ in a *Hind*III fragment length, 5.1 kb in BALB/c and 4.5 kb in STS. In BALB/c × STS recombinant inbred strains, in mouse–Chinese hamster somatic cell hybrids, and in a backcross with a strain carrying *Re* (rex), *Rnu1a-1* was found to be on Chr 11, probably very close to *Re* and *Dlb* (4). A DNA region coding for a U1b gene was found to consist of two identical genes 5 kb apart and located on opposite DNA strands as inverted repeats (3). Several U1a and U1b variant snRNAs that probably reflect polymorphisms of the *Rnu1* genes were found among strains BALB/c, C57BL/6, C3H/He, A, 129, and LT by Lund *et al.* (2), who also showed that U1a and U1b occur at comparable levels in fetal tissue, but that only U1a occurs in differentiated tissues.

References

1. Harper, M.E., C. Blatt, S. Marks, M.N. Nesbitt, and M.I. Simon. 1984. Gene mapping in the mouse by analysis of RFLP segregation in recombinant inbred strains. Cytogenet. Cell Genet. 37:488 (Abstr.).
2. Lund, E., B. Kahan, and J.E. Dahlberg. 1985. Differential control of U1 small nuclear RNA expression during mouse development. Science 229:1271–1274.
3. Marzluff, W.F., D.T. Brown, S. Lobo, and S.-S. Wang. 1983. Isolation and characterization of two linked mouse U1b small nuclear RNA genes. Nucl. Acids Res. 11:6255–6270.
4. Michael, S.K., J. Hilgers, C. Kozak, J.B. Whitney III, and E.F. Howard. 1986. Characterization and mapping of DNA sequence homologous to mouse U1a1 snRNA: localization on chromosome 11 near *Dlb* and *Re* loci. Somat. Cell Mol. Genet. 12:215–223.

ro, rough, recessive, Chr 2

Arose spontaneously in the RIII inbred strain. Homozygotes have wavy vibrissae, apparent a few days after birth and still visible, though reduced, in the adult. The hair of the coat is not wavy, but the hairs tend to stick together in bundles as if wet or greasy. The septa between the air spaces of the hair are thicker and the air spaces smaller than normal. Many of the air spaces are filled with fluid which does not dry out because the hairs are greasier than normal (1).

References

1. Falconer, D.S., and G.D. Snell. 1952. Two new hair mutants, rough and frizzy, in the house mouse. J. Hered. 43:53–57.

rp, reduced pigmentation, recessive, Chr 7

Arose spontaneously in the C57BL/Tb strain. Homozygotes have light ears and tails, diluted fur color, and dark eyes. They are fully viable and fertile. Pigment granules in the hairs are smaller than normal and not clumped. Homozygotes have significantly higher activity in kidney of the lysosomal enzymes, β-galactosidase, β-glucuronidase, and N-acetyl-β-hexoseaminidase than their heterozygous littermates (1). They also have reduced activity of splenic NK cells (2). In these characteristics they resemble a number of other mutants that dilute coat color in the mouse.

References

1. Gibb, S., E.M. Hakensson, L.-G. Lundin, and J.G.M. Shire. 1981. Reduced pigmentation (*rp*), a new colour gene with effects on kidney lysosomal glycosidases in the mouse. Genet. Res. 37:95–103.
2. Orn, A., E.M. Hakansson, M. Gidlund, U. Ramstedt, I. Axberg, H. Wigzell, and L.-G. Lundin. 1982. Pigment mutations in the mouse which also affect lysosomal functions lead to suppressed natural killer cell activity. Scand. J. Immunol. 15:305–310.

Rpl, *Rps* loci

Rpl, ribosomal protein-large
Rps, ribosomal protein-small

The mouse has about 70 different ribosomal proteins encoded by multigene families with an average of about 10 members each. Members of a particular family are often located on more than one chromosome and there is no evidence for clustering of many families on a few chromosomes. Screening of mouse–Chinese hamster somatic cell hybrids with ribosomal protein cDNA probes for four *Rpl* and one *Rps* gene families showed that three *Rpl18* genes are probably on Chr 12 and two *Rpl18* genes are probably on Chr 7. One gene of the *Rps16* family is probably on Chr 5 and one or more genes of the other families could be assigned to each of Chrs 3, 5, and 6. Other genes could tentatively be assigned to Chrs 2, 11, 16, and X (1). A DNA polymorphism occurs between C3H and BALB/c, which differ by one A $\rightarrow$ G substitution in an *Rpl32* gene (2). There is evidence that only one gene in each family is expressed and that the others are pseudogenes (2–4).

References

1. D'Eustachio, P., O. Meynhas, F. Ruddle, and R.P. Perry. 1981. Chromosomal distribution of ribosomal protein genes in the mouse. Cell 24:307–313.
2. Oudov, K.P., and R.P. Perry. 1984. The gene family encoding the mouse ribosomal protein L32 contains a uniquely expressed intron-containing gene and an unmutated processed gene. Cell 37:457–468.
3. Wagner, M., and R.P. Perry. 1985. Characterization of the multigene family encoding the mouse S16 ribosomal protein: strategy for distinguishing an expressed gene from its processed pseudogene counterparts by an analysis of total genomic DNA. Mol. Cell. Biol. 5:3560–3576.
4. Wiedemann, L.M., and R.P. Perry. 1984. Characterization of the expressed gene and several processed pseudogenes for the mouse ribosomal protein L30 gene family. Mol. Cell. Biol. 4:2518–2528.

Rpo2-1 locus, RNA polymerase II-1, Chr 11

This locus encodes the largest subunit of RNA polymerase II (RpII). RpII is the DNA-dependent RNA polymerase responsible for transcription of messenger RNA precursors and is composed of at least 10 different subunits, whose individual functions have not been determined. No genetic variants of *Rpo2-1* are known. It was found to be on Chr 11 by use of mouse–Chinese hamster somatic cell hybrids screened with a genomic DNA clone of the mouse RNA polymerase II large subunit locus, and by *in situ* hybridization (1).

References

1. Pravtcheva, D., M. Rabin, M. Bartolomei, J. Corden, and F.H. Ruddle. 1986. Chromosomal assignment of gene encoding the largest subunit of RNA polymerase II in the mouse. Somat. Cell Mol. Genet. 12:523–528.

Rps loci

See *Rpl* loci.

Rras locus, ras-related oncogene, Chr 7

This gene was detected in mouse DNA with a probe derived from a human genomic library by low stringency screening with viral *Ha-ras*. The human R-*ras* gene shares several highly conserved domains with the other *ras* genes (Ha-*ras*, K-*ras*, N-*ras*) but is clearly different in size and in location of introns. The mouse *Rras* gene has 88 per cent nucleotide similarity with the human R-*ras* gene. *Rras* was found to be on Chr 7 by use of mouse–Chinese hamster somatic cell hybrids (1).

References

1. Sakaguchi, A.Y., D.F. Lowe, G. Goeddel, P.A. Lalley, and S.L. Naylor. 1986. R-*ras*: a member of the *ras* family on human chromosome 19. Am J. Hum. Genet. 39 (Suppl.):A167.

Rrm2 loci, ribonucleotide reductase-M2

At least one of these loci codes for protein M2, one of two non-identical subunits of ribonucleotide reductase, an enzyme that catalyzes the formation of deoxyribonucleotides from corresponding ribonucleotides. Southern blot analysis of mouse–human somatic cell hybrids identified M2 sequences on Chrs 4, 7, 12, and 13. It is not known which of these sites is the active one (1).

References

1. Tang-Feng, T.L., L. Thelander, W.H. Lewis, F.R. Srinivasan, and U. Francke. 1986. Gene localization of the ribonucleotide reductase M2 subunit, and of related and co-amplified sequences on human and mouse chromosomes. Am. J. Hum. Genet. 39:Suppl. A174 (Abstr.).

Rrs locus, resistance to Rous sarcoma, Chr 17

This locus controls resistance and susceptibility to primary sarcomas induced by neonatal injection of Rous sarcoma virus-induced chicken tumor material. *Rrs2* is located in the *H-2* complex, probably near the *K* end of the *I* region. The allele *Rrs^r* determines resistance and occurs in the *H-2^s* haplotype in strain A.SW and probably also in the *H-2^d* haplotype in strain B10.D2/oSn; the allele *Rrs^s* determines susceptibility and occurs in the *H-2^k*, *H-2^a*, and *H-2^b* haplotypes in strains CBA/J, A/WySn, and C57BL/10, respectively. Heterozygotes are resistant or intermediate in susceptibility. Mice from resistant strains develop stronger cellular anti-Rous tumor immunity after immunization with allogeneic Rous sarcoma than do mice from susceptible strains (1).

References

1. Whitmore, A.C., G.F. Babcock, and G. Haughton. 1978. Genetic control of susceptibility of mice to Rous sarcoma virus tumorigenesis. II. Segregation analysis of strain A.SW-associated resistance to primary tumor induction. J. Immunol. 121:213–220.

Rrv-1 locus, resistance to RadLV-1, Chr 17

This is an *H-2*-linked locus controlling resistance to leukemogenesis induced by a B-tropic radiation leukemia virus A-RadLV in *Fv-1^b* strains. The allele for resistance *Rrv-1^r* occurs in strains with the *H-2^s* haplotype; the allele for susceptibility *Rrv-1^s* occurs in strains carrying *H-2* haplotypes r, b, k, and d. No genetic crosses have been made, but analysis of *H-2* recombinants indicates that *Rrv-1* is in the left part of the *I* region. The existence of resistant recombinants derived from two sensitive haplotypes suggests that at least one other similar locus is present and that the alleles for resistance are dominant (1). Using similar methods, Meruelo *et al*. (2) found that resistance to leukemia induced by a RadLV virus was determined by a dominant allele of a locus in the D end of the *H-2* complex, resistance being associated with the *H-2D^d* allele and susceptibility with the *H-2D^q* and *H-2D^s* alleles.

References

1. Lonai, P., and N. Haran-Ghera. 1977. Resistance genes to murine leukemia in the *I* immune response region of the *H-2* complex. J. Exp. Med. 146:1164–1168.
2. Meruelo, D., M. Lieberman, N. Ginzton, B. Deak, and H.O. McDevitt. 1977. Genetic control of radiation leukemia virus-induced tumorigenesis. I. Role of the major histocompatibility complex, *H-2*. J. Exp. Med. 146:1079–1087.

rs, recessive spotting, recessive, Chr 5

Arose spontaneously in the C3H/HeJ strain. Homozygotes have a large head blaze and belly spot and diluted color of the dark fur on the belly. Some heterozygotes may have a long thin white spot on the ventrum. Homozygotes are viable and fertile (1). *W^v* +/+ *rs* mice are black-eyed white and are fertile and not anemic. However, *rs* shows no recombination with *W^v* and may be an allele at the *W* locus (2).

References

1. Dickie, M.M. 1966. Mouse News Lett. 35:31.
2. Southard, J.L., and M.C. Green. 1971. Mouse News Lett. 45:29.

Rsm-1 locus, resistance to *Schistosoma mansoni*-1

This locus controls resistance to vaccine-induced immunity to the trematode parasite, *Schistosoma mansoni*. The fully recessive allele *Rsm-1^s* determines lack of protective immunity after exposure to the vaccine and occurs in the P (P/N or P/J) strain; the dominant allele *Rsm^r* determines presence of protective immunity after exposure to the vaccine and occurs in the C57BL/6 and probably most other inbred strains. The lack of immunity determined by *Rsm-1^s* appears to result from defective antigen-specific cell-mediated immunity. *Rsm-1* is not linked to *Igh* or *H-2* (1).

References

1. Correa-Oliveira, R., S.L. James, D. McCall, and A. Sher. 1986. Identification of a genetic locus, Rsm-1, controlling protective immunity against *Schistosoma mansoni*. J. Immunol. 137:2014–2019.

rst, rosette, recessive

Found in a strain from an English fancier. Penetrance in homozygotes appears to be complete, but expression is variable. The commonest expression is two whorls, each centered on a hindleg and extending mid-dorsally and almost mid-ventrally (1).

References

1. Wallace, M.E. 1971. Mouse News Lett. 44:18.

ru, ruby-eye, recessive, Chr 19

Found by Dunn (1) in a silver piebald stock of Danforth. Homozygotes at birth have unpigmented eyes which later darken to a ruby color. The black pigment of the coat is diluted to a dark slate color, and the yellow pigment is diluted slightly. Ruby-eye in homozygous condition greatly reduces the number of melanocytes in the retina, ear skin, Harderian gland, nictitans (4), and retinal pigment epithelium (3). It has the same effect on shape and color of pigment granules as *b* (brown), i.e. it makes the granules spheroidal rather than ovoid as in wild type, and it changes the color of the granules to dark brown (4). The internal structure of the pigment granules is normal (2, 5). In common with some other mutations that reduce pigmentation, *ru* causes reduced number of projections of retinal ganglion cells to the ipsilateral lateral geniculate nucleus (3). Kidney concentration of lysosomal enzymes is elevated, probably because of a low rate of secretion into the urine. Lysosomal morphology is normal (6). Ruby-eye mice have platelet storage pool deficiency characterized by prolonged bleeding time, normal platelet number, and low platelet dense granule number and dense granule serotonin content (7).

References

1. Dunn, L.C. 1945. A new eye color mutant in the mouse with asymmetrical expression. Proc. Natl. Acad. Sci. USA 31:343–346.

2. Hearing, V.J., P. Phillips, and M.A. Lutzner. 1973. The fine structure of melanogenesis in coat color mutants of the mouse. J. Ultrastruct. Res. 43:88–106.
3. LaVail, J.H., R.A. Nixon, and R.L. Sidman. 1978. Genetic control of retinal ganglion cell projections. J. Comp. Neurol. 182:399–422.
4. Markert, C.L., and W.K. Silvers. 1956. The effects of genotype and cell environment on melanoblast differentiation in the house mouse. Genetics 41:429–450.
5. Moyer, F.H. 1966. Genetic variation in the fine structure and ontogeny of mouse melanin granules. Am. Zool. 6:43–66.
6. Novak, E.K., F. Wieland, G.P. Jahrus, and R.T. Swank. 1980. Altered secretion of kidney lysosomal enzymes in the mouse pigment mutants ruby-eye, ruby-eye-2-J, and maroon. Biochem. Genet. 18:549–561.
7. Novak, E.K., S.-W. Hui, and R.T. Swank. 1984. Platelet storage pool deficiency in mouse pigment mutations associated with seven distinct genetic loci. Blood 63:536–544.

ru-2 locus, Chr 7

ru-2, ruby-eye-2, recessive. Arose in a substrain of C57BL. Homozygotes are identical in appearance to *ru/ru* mice. By use of Cattanach's insertion *Is(In7;X)1Ct*, Eicher (3) has produced *+/ru-2/ru-2* mice and shown that they are intermediate between wild type and *ru-2/ru-2* in color.

ru-2^hz, haze, recessive. Shown to be an allele of *ru-2* by Eicher and Fox (4). Arose in the DBA/2J strain. In homozygotes the hair tips are brownish and the underfur is white. At birth the eyes are light-colored (2).

ru-2^mr, maroon, recessive. Shown to be an allele of *ru-2* by Eicher and Fox (4). Arose spontaneously in a 'lactation' stock. Eyes are colorless at birth, but darken to a rich maroon. The coat color varies from near normal to extreme pallid, even among littermates, and darkens with age (1). Kidney concentrations of the lysosomal enzymes β-glucuronidase and β-galactosidase are almost twice as high in maroon mice as in controls (5). The enzymes are synthesized at a normal rate but secreted into the urine at a reduced rate. Lysosomal morphology is normal (6). Maroon mice have platelet storage pool deficiency characterized by prolonged bleeding time, normal platelet number, and low platelet dense granule number and dense granule serotonin content (7).

References

1. Bateman, N. 1957. Mouse News Lett. 16:7.
2. Dickie, M.M. 1965. Mouse News Lett. 32:44.
3. Eicher, E.M. 1970. The position of *ru-2* and *qv* with

respect to the *flecked* translocation in the mouse. Genetics 64:495–510.

4. Eicher, E.M., and S. Fox. 1977. Mouse News Lett. 56:42.
5. Meisler, M.H. 1978. Synthesis and secretion of kidney β-glucuronidase in mutant *le/le* mice. J. Biol. Chem. 253:3129–3134.
6. Novak, E.K., F. Wieland, G.P. Jahrus, R.T. Swank. 1980. Altered secretion of kidney lysosomal enzymes in the mouse pigment mutants ruby-eye, ruby-eye-2-J, and maroon. Biochem. Genet. 18:549–561.
7. Novak, E.K., S.-W. Hui, and R.T. Swank. 1984. Platelet storage pool deficiency in mouse pigment mutations associated with seven distinct genetic loci. Blood 63:536–544.

Rua locus, raffinose undecaacetate tasting, Chr 8?

This locus controls ability to taste raffinose undecaacetate (RUA). The allele Rua^a determines ability to taste RUA and occurs in strains BALB/c, DBA/2, and C3H; the allele Rua^b determines inability to taste RUA and occurs in C57BL/6 and probably many other strains. Heterozygotes are intermediate in sensitivity. Sensitivity to RUA probably involves sensitivity to all the component parts of the RUA molecule. *Rua* showed no recombination with *Qui* among 23 recombinant inbred strains (1). The location of *Qui* on Chr 8 is in some doubt. See *Qui*.

References

1. Lush, I.E. 1986. The genetics of tasting in mice. IV. The acetates of raffinose, galactose and β-lactose. Genet. Res. 47:117–123.

rv, rib-vertebrae, recessive

Arose spontaneously in the C57L/J strain. Extensive linkage tests have not revealed its chromosomal location. Homozygotes of both sexes are fertile, but females are often poor breeders. Affected mice have reduced body length and often distal tail kinks of varying degrees of severity. There are fusions of the ribs and irregularities and fusions of the neural arches and transverse processes, probably resulting from irregularities in somite formation. There is duplication of the caudal neural tube and sometimes unilateral absence of kidneys. Affected embryos can sometimes be recognized at 11 days of gestation by a characteristic budding of the tail tip, and often show irregularities of mesenchymal condensation in the anlage of the vertebral column and sprouting of the terminal gut (1).

References

1. Theiler, K., and D. Varnum. 1985. Development of rib-vertebrae: a new mutation in the house mouse with accessory caudal duplications. Anat. Embryol. 173:111–116.

Rv-1 locus

See *Fv-1* locus.

Rv-2, *Rv-3* loci

Rv-2 locus, Rauscher virus susceptibility-2, Chr 9
See also *Fv-2* locus.
Rv-3 locus, Rauscher virus susceptibility-3, Chr X.

These two loci influence susceptibility to the spleen focus-forming (SFFV) component of Rauscher leukemia virus (RLV). *Rv-2* has a major effect on susceptibility; *Rv-3* has a minor effect demonstrable only at low dose. The $Rv-2^r$ allele determining resistance to RLV is dominant; it occurs in the C57BL/6 and C57BL/10 strains. The $Rv-2^s$ allele determining susceptibility is recessive; it occurs in the BALB/c and CBA/Lac strains. The $Rv-3^r$ allele determining resistance to a low dose of RLV is recessive and occurs in the C57BL/6 strain; the $Rv-3^s$ allele determining susceptibility is dominant and occurs in the CBA/Lac strain. The effect of *Rv-2* on susceptibility to RLV is very similar to that of *Fv-2* on susceptibility to Friend leukemia virus. The two loci are located in the same region of Chr 9 and it is very likely that they are identical. No other locus similar to *Rv-3* has been described (1).

References

1. Heller, E., and D.H. Pluznik. 1984. Chromosomal assignment of two murine genes controlling susceptibility to spleen focus formation by Rauscher leukemia virus. Exp. Hematol. 12:645–649.

Rvil-1 locus, RadLV-induced leukemia-1, Chr 2

This locus controls susceptibility to leukemia induced by the Kaplan strain of radiation-induced leukemia virus (RadLV) obtained from a radiation-induced leukemia. The allele in C57BL/6By, $Rvil-1^s$, determines susceptibility; the allele in BALB/cBy, $Rvil-1^r$, determines resistance. Susceptibility was assessed, after intrathymic injection in 3- to 6-week old mice, by mortality curves ending at about 250 days of age. *Rvil-1* is probably located between *H-3* and the agouti locus (1).

References

1. Meruelo, D., M. Offer, and A. Rossomando. 1983. Induction of leukemia by both fractionated X-irradiation and radiation leukemia virus involves loci in the chromosome 2 segment *H-30–A*. Proc. Natl. Acad. Sci. USA 80:462–466.

Rw, rump-white, semidominant, Chr 5

Found in the progeny of a low-intensity neutron irradiated (C3H/He × 101/H)F1 male (1). Heterozygotes have white hindlegs and tails and white fur on the posterior part of the abdomen, with occasional pigmented spots in the white areas. The white area is more extensive ventrally than dorsally. Pigmented melanocytes occur in the epidermis, even in the area with white fur, and may be particularly noticeable in skin of the scrotum and tail. Homozygotes die *in utero* by mid-pregnancy. *Rw* is very closely linked to *Ph* and *W*, two other loci with semidominant alleles affecting pigmentation (2).

References

1. Batchelor, A.L., R.J.S. Phillips, and A.G. Searle. 1966. A comparison of the mutagenic effectiveness of chronic neutron- and γ-irradiation of mouse spermatogonia. Mutat. Res. 3:218–229.
2. Searle, A.G., and G.M. Truslove. 1970. A gene triplet in the mouse. Genet. Res. 15:227–235.

S

s locus, Chr 14

Radiation-induced mutation at this locus is very common (13).

s, piebald, recessive. This is a very old mutation of the mouse fancy; possibly some piebalds in existing stocks are of independent origin. Homozygotes show irregular white spotting, the amount of which is greatly influenced by minor modifying genes (5). Homozygotes have dark eyes. The white areas of the coat are completely lacking in melanocytes, and there is a reduction in the number of melanocytes in the choroid layer of the eye (2, 8). There may also be defects in the structure of the iris, suggesting that pigment cells make some structural or inductive contribution to normal development of the iris (6). Homozygotes may develop megacolon which is always associated with lack of ganglion cells in the distal portion of the colon. The incidence of megacolon is affected by minor modifying genes (1). Pigment cells and enteric ganglion cells of the colon are both derived from the neural crest, and Mayer (9) has shown by explantation of embryonic tissues that the defect leading to white spotting is in the neural crest rather than in the skin. The defect probably consists of failure of pigment cells to differentiate in certain tissue environments rather than in failure to migrate (10). The distribution of white areas in the skin and other organs is probably due to normal regional differences in these tissues in capacity to support pigmentation and not to regional heterogeneity among the pigment cells themselves (4, 10–12).

s^l, piebald-lethal, recessive. Found in the F2 generation of a cross between C3H/HeJ and C57BL/6J. Homozygotes are almost completely white with dark eyes and with only an occasional small pigmented spot on the head or rump. *s/s^l* mice resemble *s/s* in degree of spotting. All homozygotes develop megacolon with lack of enteric ganglion cells in the posterior end of the colon. They usually die at about 2 weeks of age, but some live a year or more and may breed (7). Functional studies of the colon and rectum have shown that inhibitory cholinergic innervation is absent in these mice (14, 15). Enteric ganglion cells in normal embryos enter the gut by way of the vagal outgrowth at 10 days of gestation and migrate down the gut. In *s^l/s^l* embryos migration is slower and does not keep up with elongation of the gut, so that the neuroblasts never reach the end of the gut, even though they migrate 6 to 7 days longer (16). In *s^l/s^l* mice the neural epithelium of the inner ear is abnormal, probably as a result of defects in the part of the acoustic ganglion derived from the neural crest (3).

References

1. Bielchowsky, M., and G.C. Schofield. 1962. Studies on megacolon in piebald mice. Austral. J. Exp. Biol. Med. Sci. 40:395–404.

2. Billingham, R.E., and W.K. Silvers. 1960. The melanocytes of mammals. Quart. Rev. Biol. 35:1–40.
3. Deol, M.S. 1967. The neural crest and the acoustic ganglion. J. Embryol. Exp. Morphol. 17:535–541.
4. Deol. M.S. 1971. Spotting genes and internal pigmentation patterns in the mouse. J. Embryol. Exp. Morphol. 26:123–133.
5. Dunn, L.C., and D.R. Charles. 1937. Studies on spotting patterns. I. Analysis of quantitative variations in the pied spotting of the mouse. Genetics 22:14–42.
6. Dunn, L.C., and J. Mohr. 1952. An association of hereditary eye defects with white spotting. Proc. Natl. Acad. Sci. USA 38:872–875.
7. Lane, P.W. 1966. Association of megacolon with two recessive spotting genes in the mouse. J. Hered. 57:29–31.
8. Markert, C.L., and W.K. Silvers. 1956. The effects of genotype and cell environment on melanoblast differentiation in the house mouse. Genetics 41:429–450.
9. Mayer, T.C. 1965. The development of piebald spotting in mice. Dev. Biol. 11:319–334.
10. Mayer, T.C. 1967. Pigment cell migration in piebald mice. Dev. Biol. 15:521–535.
11. Mayer, T.C. 1967. Temporal skin factors influencing the development of melanoblasts in piebald mice. J. Exp. Zool. 166:397–403.
12. Mayer, T.C. 1977. Enhancement of melanocyte development from piebald neural crest by a favorable tissue environment. Dev. Biol. 56:255–262.
13. Russell, W.L. 1965. Evidence from mice concerning the nature of the mutation process. Proc. XI Int. Congr. Genet. 2:257–264.
14. Ueki, S., E. Okamoto, K. Kuwata, A. Toyosaka, and K. Nagai. 1985. Qualitative and quantitative analysis of muscarinic acetylcholine receptors in the piebald lethal mouse model of Hirschsprung's disease. Gastroenterology 88:1834–1842.
15. Vaillant, C., A. Bu'lock, R. Demaline, and G.J. Dockray. 1982. Distribution and development of peptidergic nerves and gut endocrine cells in mice with congenital aganglionic colon, and their normal littermates. Gastroenterology 82:291–300.
16. Webster, W. 1973. Embryogenesis of the enteric ganglia in normal mice and in mice that develop aganglionic megacolon. J. Embryol. Exp. Morphol. 30:573–585.

sa, satin, recessive, Chr 13

Probably radiation induced. Homozygotes are viable and fertile. The hair texture is altered to produce a silky coat with a high sheen. All hair types are present, but they are thinner than normal and the medulla is a rarely interrupted cylindrical structure without air spaces and lacking normal keratinization. Development of the hair follicle appears normal except for the structurally abnormal medulla which becomes apparent as the hair is leaving the bulb (2). The *sa* gene in homozygotes suppresses natural killer cell activity slightly and suppresses alloimmune cytotoxic T-cell function more severely. It acts synergistically with *bg* to cause greater suppression of both these functions than in either homozygote alone (1).

References

1. McGarry, R.C., R. Walker, and J.C. Roder. 1984. The cooperative effect of the *satin* and *beige* mutations on the suppression of NK and CTL activities in mice. Immunogenetics 20:527–534.
2. Trigg, M.J. 1972. Hair growth in mouse mutants affecting coat texture. J. Zool. Lond. 168:165–198.

Saa complex, serum amyloid A, Chr 7

This complex codes for the acute phase protein, serum amyloid A (SAA), a precursor for a major component of amyloid fibrils. There are at least three *Saa* genes, *Saa-1* and *Saa-2*, which are about 95 per cent homologous (5) and *Saa-3*, which is about 70 per cent homologous with the other two (3, 6). *Saa-1* and *Saa-2* are adjacent and arranged in divergent transcriptional direction (6). Taylor and Rowe (4) detected genetic polymorphism by hybridization of DNA restriction fragments to a cDNA clone specific for *Saa-3* (3). By hybridization with four different restriction enzymes, 14 variants were detected among inbred strains. Five variants were shown to segregate as a single unit and were mapped to Chr 7 between *Gpi-1* and *p* by use of recombinant inbred and congenic strains. Probably all of the sequences detected by this probe are clustered on Chr 7. There are two major haplotypes: *Saa^a* in most inbred strains, and *Saa^b* in ABP, BDP, I/Ln, P, PRO/1Re, SJL, and 129. Seven other strains each have a rare haplotype (4). Four structural SAA variants demonstrated by two-dimensional electrophoresis of the cell-free translation product of liver mRNA were found by Baumann and colleagues (1, 2). The strain distribution of these variants correlates completely with that of four of the haplotypes found by Taylor and Rowe (4).

References

1. Baumann. H., and F.G. Berger. 1985. Genetics and evolution of the acute phase proteins in mice. 1985. Mol. Gen. Genet. 201:505–512.
2. Baumann, H., W.A. Held, and F.G. Berger. 1984. The acute phase response of mouse liver: genetic analysis of the major acute phase reactants. J. Biol. Chem. 259:566–573.
3. Stearman, R.S., C.A. Lowell, G. Peltzman, and J.M. Morrow. 1986. The sequence and structure of a new serum amyloid A gene. Nucl. Acids Res. 14:797–809.
4. Taylor, B.A., and L. Rowe. 1984. Genes for serum amy-

loid A proteins map to chromosome 7 in the mouse. Mol. Gen. Genet. 195:491–499.

5. Yamamoto, K., and S. Migita. 1985. Complete primary structure of two major serum amyloid A proteins deduced from cDNA sequences. Proc. Natl. Acad. Sci. USA 82:2915–2919.

6. Yamamoto, K., M. Shiroo, and S. Migita. 1986. Diverse gene expression for isotypes of murine serum amyloid A protein during acute phase reaction. Science 232:227–229.

Sac locus, saccharin preference

This locus influences preference for a 0.1 per cent solution of saccharin over water. The dominant allele Sac^b, present in the C57BL/6J strain, determines high saccharin preference; the recessive allele Sac^d, present in the DBA/2J strain, determines low saccharin preference. The mechanism of preference is probably related to the incentive value of the taste of saccharin rather than to the threshold for detection (1).

References

1. Fuller, J.L. 1974. Single-locus control of saccharin preference in mice. J. Hered. 65:33–36.

Sal-1 locus

See *Abp* locus.

Sap locus, serum amyloid P-component, Chr 1

This locus controls endogenous concentration of the acute phase reactant, serum amyloid P-component (SAP). The level of SAP in untreated mice varies widely among inbred strains. On the basis of SAP levels in the C57BL/6, DBA/2, C3H/He, and BALB/c inbred strains, BXD and BXH recombinant inbred strains, and the B6.C-H-25^c congenic strain, the following allelic symbols have been tentatively assigned: Sap^a determining high level of endogenous SAP in DBA/2 and BALB/c; Sap^b determining low level in C57BL/6; and Sap^c determining a weaker high level than Sap^a in C3H/He. *Sap* is closely linked to *Ly-9* on Chr 1 (1).

References

1. Mortensen, R.F., P.T. Le, and B.A. Taylor. 1985. Mouse serum amyloid P-component (SAP) levels controlled by a locus on chromosome 1. Immunogenetics 22:367–375.

Sas-1 locus, serum antigenic substance-1, Chr 1

This locus controls an antigenically detected substance in serum that migrates electrophoretically in the fast β region slightly anodal to hemolytic complement (3). It is probably identical with the *Mud* locus which controls a similar substance and has a similar strain distribution of alleles (1,2). The allele $Sas-1^a$ (*Mud1*) determines an antigenic determinant present in the C57 family of strains, AKR/J, BALB/c, and others. No allelic antigen determined by the $Sas-1^o$ allele has been found (3), but Naylor and Cinader (1) were able to produce a weak anti-Mud2 antiserum. $Sas-1^o$ (*Mud2*) occurs in strains RF, A/He, DBA/2, and others. *Sas-1* is in the same chromosomal region as complement factor H (*Cfh*) and may be identical with it. See *Cfh* locus.

References

1. Naylor, D.H., and B. Cinader. 1970. Inheritance, hormonal regulation and properties of polymorphic murine antigens Mud1 and Mud2. Int. Arch Allergy Appl. Immunol. 39:511–539.

2. Rosenstreich, D.L., M.G. Groves, H.A. Hoffman, and B.A. Taylor. 1978. Location of the *Sas-1* locus on mouse chromosome 1. Immunogenetics 7:313–320.

3. Wortis, H.H. 1965. A gene locus concerned with an antigenic serum substance in *Mus musculus*. Genetics 52:267–273.

Sas-2 locus, serum antigenic substance-2, Chr 1

This locus controls a serum protein that migrates immunoelectrophoretically in the β region of serum proteins. The protein was identified by reciprocal immunization between strains BALB/c and MOL-ANJ, a strain derived from wild *Mus musculus molossinus* in Japan. The allele $Sas-2^a$ determines an antigenic specificity that occurs in all standard inbred strains tested, including A/He, AKR, BALB/c, C3H/He, C57BL/6, DBA/2, and NZB; the allele $Sas-2^b$ determines a specificity that occurs in several inbred strains derived from *M. m. molossinus*. Heterozygotes have both antigens. *Sas-2* is on Chr 1 very close to *Pep-3*. This region also contains the *Sas-1* locus. Like Sas-2, the Sas-1 protein migrates in the β region but alleles at the two loci have different strain distributions and the two proteins give different precipitin lines in the Ouchterlony test. The two loci are therefore probably different (1).

References

1. Harada, Y.-N, J.-I. Hayakawa, E. Noda, and T. Tomita. 1987. A gene locus controlling a serum protein migrating electrophoretically in the β region of mice and detected by using a strain derived from the Japanese wild mouse (*Mus musculus molossinus*). J. Immunogenet. 14:33–41.

sb, stub, recessive

This mutant is extinct but may possibly have recurred in *pu* which it closely resembles. It arose in the inbred CF-1 stock at Carworth Farms. The embryology of *sb* was not studied, but the description of postnatal homozygotes differs in no way from that of *pu* (1).

References

1. Dunn, L.C., and S. Gluecksohn-Schoenheimer. 1942. Stub, a new mutation in the mouse. J. Hered. 33:235–239.

sc, screw-tail, recessive, extinct

Arose spontaneously in the inbred Bagg albino strain at Cold Spring Harbor. Homozygotes show widespread abnormalities of the skeleton including shortened kinked tails, defective vertebral centra which may result in sharp kinks in the thoracic or lumbar region, a short broad unsegmented sternum, abnormal teeth and jaws, and abnormalities of the calvarium. The number of presacral vertebrae is increased by about one (3). The abnormalities of the sternum appear to result from retarded growth of the ribs which fails to bring the bilateral sternal bands into close union (2). Abnormalities of dentition appear to result from retarded growth of the mandibular condyle, maxillary sutures, and connective tissue of the dental pulp (1).

References

1. Bhaskar, S.N., I. Schour, E.C. MacDowell, and J.P. Weinmann. 1951. The skull and dentition of screw tail mice. Anat. Rec. 110:199–229.
2. Bryson, V. 1945. Development of the sternum in screw tail mice. Anat. Rec. 91:119–141.
3. MacDowell, E.C., J.S. Potter, T. Laanes, and E.N. Ward. 1942. The manifold effects of the screw tail mouse mutation. J. Hered. 33:439–449.

scb, scabby, recessive, Chr 8

Arose in the C3H/HeH strain. Homozygotes are recognizable shortly after birth by transverse areas of scar tissue that appear on the skin, especially in the lumbar region, with similar scaly regions on legs and tail. The tail may be short and kinky and have constrictions. There is defective hair growth in scabby areas which leads to dark transverse areas in the agouti juvenile coat. The adult coat appears nearly normal. Penetrance is complete or nearly so (2). *scb* is on Chr 8 between *Os* and *e* (1).

References

1. Beechey, C.V., and A.G. Searle. 1982. Mouse News Lett. 66:64.

2. Searle, A.G., and C.V. Beechey. 1977. Mouse News Lett. 57:16.

sch, scant hair, recessive, Chr 9

Arose in strain C57BL/Ks-*db*. Homozygotes can be identified at 7 to 8 days by cessation of hair growth. Expression ranges from thin hair to nakedness and does not change with age. Vibrissae are normal. Both sexes are fertile (1).

References

1. Hummel, K.P., and D.B. Chapman. 1971. Mouse News Lett. 45:28.

scid, severe combined immune deficiency, recessive, Chr 16

Arose in the C.B-17 inbred strain (BALB/c.C57BL/Ka-*Igh-1^b*). Most homozygotes have no detectable IgM, IgG1, IgG2a, IgG2b, IgG3, or IgA, but a few have low levels of one to three of these immunoglobulin isotypes. The size of the lymphoid organs is only one-tenth or less that of normal. Thymus, lymph nodes, and splenic follicles are virtually devoid of lymphocytes. Homozygotes are deficient in both B- and T-cell function. Their spleen cells do not respond to either B- or T-cell mitogens and they are unable to reject skin grafts. They lack detectable B-cells and pre-B cells. In spite of the small thymus and lack of functional T-cells, the Thy-1 marker is present on a majority of cells recovered from the thymus, and T-cell lymphomas occur in 10 per cent or more of affected mice. *scid* specifically impairs differentiation of stem cells into mature lymphocytes. Myeloid cell differentiation is not affected. The basic defect in these mice appears to be in the lymphoid stem cells and not in the cellular environment, since functional T- and B-cells are found in mice reconstituted with normal bone marrow (2, 3). However, full reconstitution of the immune deficiency occurs only after irradiation of the recipients, indicating that *scid/scid* mice may have normal numbers of a radiation-sensitive stem cell that has defective proliferative capacity (4). The rearrangements of immunoglobulin and T-cell receptor genes that normally occur in B and T lymphocytes are not found in homozygous *scid* mice. However, in Abelson leukemia virus-transformed B-cells of these mice and in their occasional T-cell lymphomas, rearrangements, most of which are abnormal, are found. This suggests that *scid* may act through an effect on the recombinase system catalyzing the assembly of immunoglobulin and T-cell receptor genes, and that lymphocytes with these defects are not able to develop further (5). Homozygous *scid* mice are fertile and, under specific pathogen-free conditions, may survive a year or more (1). *scid* is on Chr 16 near *md* (H.O. Sweet, personal communication).

References

1. Bosma, G.C., R.P. Custer, and M.J. Bosma. 1983. A severe combined immune deficiency mutation in the mouse. Nature 301:527–530.
2. Custer, R.P., G.C. Bosma, and M.J. Bosma. 1985. Severe combined immune deficiency (SCID) in the mouse. Pathology, reconstitution, neoplasms. Am. J. Pathol. 120:464–477.
3. Dorshkind, K., G.M. Keller, R.A. Phillips, R.G. Miller, G.C. Bosma, M. O'Toole, and M.J. Bosma. 1984. Functional status of cells from lymphoid and myeloid tissues in mice with severe combined immunodeficiency disease. J. Immunol. 132:1804–1808.
4. Fulop, G.M., and R.A. Phillips. 1986. Full reconstitution of the immune deficiency in scid mice with normal stem cells requires low-dose irradiation of the recipients. J. Immunol. 136:4438–4443.
5. Schuler, W., I.J. Weiler, A. Schuler, R.A. Phillips, N. Rosenberg, T.W. Mak, J.F. Kearney, R.P. Perry, and M.J. Bosma. 1986. Rearrangement of antigen receptor genes is defective in mice with severe combined immune deficiency. Cell 46:963–972.

Scl-1, *Scl-2* loci

Scl-1 locus, susceptibility to cutaneous leishmaniasis-1, Chr 8. This locus controls susceptibility to intradermal inoculation of the protozoon parasite *Leishmania tropica major*. The allele *Scl-1^r* determines resistance and occurs in strains A/J, C3H/He, DBA/1, C57BL/6, and many others. At 16 weeks after infection, mice of these strains had completely resolved their skin lesions. The allele *Scl-1^s* determines susceptibility and occurs in strains SWR and BALB/c. At 16 weeks, mice of these strains had developed large non-healing lesions. All had died by 7.5 months after infection. (BALB/c × C57BL/6)F1 hybrids were intermediate in rate of expansion of cutaneous lesions, but all ultimately succumbed to the infection. The F2 generation could be clearly separated into resistant, intermediate, and susceptible groups (4). The susceptibility of BALB/c mice appears to be due to activity of a B-cell-dependent suppressor T-cell which suppresses the delayed type hypersensitivity response that would result in resolution of the infection (7). Lymph node cells of resistant C57BL/6 mice, when stimulated with *L. major* antigens, produce interferon-γ, but similarly stimulated lymph node cells of susceptible BALB/c mice do not (8). Susceptibility and resistance can be transmitted with hematopoie-

tic donor cells (5). The effects of cutaneous infection with *L. mexicana amazonensis* in BALB/c and A/J mice closely resemble those of *L. tropica major* (1). Strain P/J is also susceptible to cutaneous infection by *L. tropica major*, but in crosses with resistant C3H/He, resistance segregates as a single gene with resistance dominant. The locus responsible is thought to be different from *Scl-1* (6). Data from 13 recombinant inbred strains are consistent with location of *Scl-1* on Chr 8 (2).

Scl-2 locus. This locus is said to control a difference in susceptibility of susceptible BALB/c and resistant C3H/He to the *L. mexicana mexicana* strain of parasite (3).

References

1. Andrade, Z.A., S.G. Reed, S.B. Roters, and M. Sadigursky. 1984. Immunopathology of experimental cutaneous leishmaniasis. Am. J. Pathol. 114:137–148.
2. Blackwell, J.M., and B.A. Taylor. 1984. Mouse News Lett. 70:86.
3. Blackwell, J.M., B. Roberts, and J. Alexander. 1985. Responses of BALB/c mice to leishmanial infection. Curr. Top. Microbiol. Immunol. 122:97–106.
4. DeTolla, L.J. Jr., P.A. Scott, and J.P. Farrell. 1981. Single gene control of resistance to cutaneous leishmaniasis in mice. Immunogenetics 14:29–39.
5. Howard, J.G., C. Hale, and F.Y. Liew. 1980. Genetically determined susceptibility to *Leishmania tropica* infection is expressed by haematopoietic donor cells in mouse radiation chimaeras. Nature 288:161–162.
6. Nacy, C.A., M.S. Meltzer, and A.H. Fortier. 1984. Macrophage activation to kill *Leishmania tropica*: characterization of P/J mouse macrophage defects for lymphokine-induced antimicrobial activities against *Leishmania tropica* amastigotes. J. Immunol. 133:3344–3350.
7. Sachs, D.L., P.A. Scott, R. Asofsky, and F.A. Sher. 1984. Cutaneous leishmaniasis in anti-IgM-treated mice: enhanced resistance to functional depletion of a B cell-dependent T cell involved in the suppressor pathway. J. Immunol. 132:2072–2077.
8. Sodick, M.D., R.M. Locksley, C. Tubbs, and H.V. Raff. 1986. Murine cutaneous leishmaniasis: resistance correlates with the capacity to generate interferon-γ in response to leishmania antigens *in vitro*. J. Immunol. 136:655–661.

Sco locus, scopolamine modification of exploratory activity, Chr 17

Scopolamine reduces the activity level of mice with high basal short-term exploratory activity and increases the activity level of mice with low basal short-term exploratory activity. The *Sco* locus is said to modulate this effect of scopolamine, with the allele *Sco^a*, present in BALB/cBy, determining absence of dose-related activity change, and the allele *Sco^p*, present in C57BL/6By, determining presence of such activity change.

Analysis of CXB recombinant inbred strains shows that *Sco* is probably linked to *H-2* on Chr 17 (1).

References

1. Oliverio, A., B.E. Eleftheriou, and D.W. Bailey. 1973. Exploratory activity: genetic analysis of its modification by scopolamine and amphetamine. Physiol. Behav. 10:893–899.

Sd, Danforth's short tail, semidominant, Chr 2

Found as a spontaneous mutation by Danforth (1). Many heterozygotes and all homozygotes die shortly after birth from urogenital abnormalities. The surviving heterozygotes may have good viability and fertility. Heterozygotes have short tails with a reduced number of caudal vertebrae and some kinking. Tails may be absent and the third and fourth sacral vertebrae missing. The bodies of all the vertebrae are reduced (4). One or both kidneys may be reduced in size or absent. In the absence of a kidney the ureter may be short or absent. Homozygotes have similar but much more severe abnormalities. In addition, the anus is imperforate, and the rectum and sometimes the urethra and bladder are absent (2). The developmental effects can be traced to a structurally abnormal notochord, more severe toward the caudal end, leading to abnormal vertebrae and to reduction of the cloaca and tail gut (5). The notochord shows discontinuities at 9 days of gestation, and by 11 days mesenchymal organization around the notochord is abnormal (6). Organ culture experiments attempting to determine whether the lack of kidneys in *Sd/Sd* is due to defective kidney mesenchyme or to defective ureters indicated that both are quantitatively defective (3). A locus that modifies tail length in *Sd/+* mice was discovered by Wallace and found to be located close to *ln* on Chr 1 (7).

References

1. Dunn, L.C., S. Gluecksohn-Schoenheimer, and V. Bryson. 1940. A new mutation in the mouse affecting spinal column and urogenital system. J. Hered. 31:343–348.
2. Gluecksohn-Schoenheimer, S. 1943. The morphological manifestations of a dominant mutation in mice affecting tail and urogenital system. Genetics 28:341–348.
3. Gluecksohn-Waelsch, S., and T.R. Rota. 1963. Development in organ tissue culture of kidney rudiments from mutant mouse embryos. Dev. Biol. 7:432–444.
4. Grüneberg, H. 1953. Genetical studies on the skeleton of the mouse. VI. Danforth's short tail. J. Genet. 51:317–326.
5. Grüneberg, H. 1958. Genetical studies on the skeleton of

the mouse. XXII. The development of Danforth's short tail. J. Embryol. Exp. Morphol. 6:124–148.
6. Paavola, L.G., D.B. Wilson, and E.M. Center. 1980. Histochemistry of the developing notochord, perichordal sheath and vertebrae in Danforth's short tail (*Sd*) and normal C57BL/6 mice. J. Embryol. Exp. Morphol. 55:227–245.
7. Wallace, M.E. 1976. A modifier mapped in the mouse. Genetica 46:529.

Sdh-1 locus, sorbitol dehydrogenase-1, Chr 2

This locus controls variation in electrophoretic mobility of sorbitol dehydrogenase (SDH-1; E.C. 1.1.1.14), a tetrameric enzyme present in kidney, testis, and in lesser amounts in other tissues. The locus was discovered independently by use of isoelectric focusing and at first called *Sodh-1* (4). The allele *Sdh-1^a* determines a fast cathodally migrating band and is found in all inbred strains examined; the allele *Sdh-1^b* determines a slow variant found in the Peru stock (2, 3). The activity of SDH-1B is about 25 per cent of that of SDH-1A. Heterozygotes have faint parental bands and an intermediate band and intermediate activity (1). A third allele, *Sdh-1^c*, determining absence of enzyme activity was found in the C57BL/LiA strain. *Sdh-1^a/Sdh-1^c* mice have intermediate enzyme activity (3).

References

1. Andrews, S.J., and J. Peters. 1983. Linkage analysis and biochemical genetics of sorbitol dehydrogenase-1 (*Sdh-1*) in the mouse. Biochem. Genet. 21:809–817.
2. Holmes, R.S., J.T. Jones, and J. Peters. 1978. Genetic variation, cellular distribution and ontogeny of sorbitol dehydrogenase isozymes in male reproductive tissues of the mouse. J. Exp. Zool. 206:279–288.
3. Holmes, R.S., J.A. Duly, and J. Hilgers. 1982. Sorbitol dehydrogenase genetics in the mouse: a 'null' mutant in a 'European' C57BL strain. Anim. Bld. Grps. Biochem. Genet. 13:263–272.
4. Skow, L.C., and J.B. Whitney. 1978. Mouse News Lett. 59:26.

Sdr-1, *Sdr-2* loci

Sdr-1 locus, serine dehydratase regulator-1. This locus controls levels of the gluconeogenic enzymes serine dehydratase (SD; E.C. 4.1.1.13) and phosphoenolpyruvate carboxykinase (PEPCK; E.C. 4.1.1.32) induced in liver by feeding a high-protein low-carbohydrate diet. The allele *Sdr-1^a* determines high level of induction of both enzymes and occurs in the BALB/cJ strain; the allele *Sdr-1^b* determines low level of induction of both enzymes and occurs in the AKR, C57BL/6,

BALB/cBy, and most other inbred strains. F1 hybrids between BALB/cJ and AKR have intermediate levels of induction of both enzymes. *Sdr-1* is not linked to *Sdr-2* (1).

Sdr-2 locus, serine dehydratase regulator-2. This locus controls level of serine dehydratase, but not of phosphoenolpyruvate carboxykinase, induced in liver by a 48-hour fast. The allele *Sdr-2ᵃ* determines high level of induction and occurs in the BALB/cJ strain; the allele *Sdr-2ᵇ* determines low level of induction and occurs in the AKR and most other inbred strains. F1 hybrids between BALB/cJ and AKR have an intermediate level of induction. *Sdr-2* is not linked to *Sdr-1* (1).

References

1. Coleman, D.L. 1980. Genetic control of serine dehydratase and phosphoenolpyruvate carboxykinase in mice. Biochem. Genet. 18:969–979.

se locus, Chr 9

Many spontaneous and radiation- and chemical-induced mutations have been found at this locus, often with accompanying mutations at the closely linked dilute (*d*) locus. The spatial and functional relationships of these mutations have been studied by Russell (8) and the organization of the DNA in this region by Rinchik *et al.* (7). Four of the known mutations are listed below.

se, short-ear, recessive. Arose spontaneously in mice obtained from a commercial breeder (6). Homozygotes have short, slightly ruffled external ears recognizable at about 14 days of age. The defective ears are due to defective cartilage framework. The whole skeleton is abnormal, being slightly smaller than normal with numerous local defects including a reduced or bifurcated xiphisternum, reduced number of ribs and sternebrae, reduction or absence of the ulnar sesamoid bone of the wrist, the medial sesamoid bone of the knee, and the anterior tubercles of the sixth cervical vertebra, and other similar defects (3). The skeletal abnormalities are traceable to defective condensation of mesenchyme in the embryo (2). In adult *se/se* mice, rate of proliferation of periosteal cells in the healing of bone fractures is reduced (4). Homozygotes show variable frequency of abnormalities of the soft tissues, some of them probably secondary to crowding of the viscera caused by the defective skeleton, and including hydroureter and hydronephrosis, multiple lung cysts, medially displaced left ovary, displaced right renal artery, and multiple small giant-cell granulomas on the ventral surface of the liver

(5). The hydronephrosis and hydroureter have been further studied by Wallace (9) who found a correlation between incidence of hydronephrosis and more posterior position of the ureteral veins.

seˡ, short-ear lethal, recessive. Russell (8) described eight lethal short-ear mutations independently induced by radiation in (101 × C3H)F1 hybrids and collectively designated *seˡ*. Compounds with se (*se/seˡ*) resemble *se/se* and are viable and fertile. One of these mutations studied by Dunn (1) is probably a small deletion involving *se* and extending part way to *sv* (Snell's waltzer). Homozygotes die between 7 and 8 days postcoitum.

seˢᵛ, short-eared waltzer, recessive. Probably radiation-induced in a (101 × C3H)F1 hybrid. Homozygotes are viable and resemble *se sv/se sv* mice. The mutation involves changes at both the *se* and the closely linked *sv* loci and is probably a small rearrangement rather than a deletion (8).

seˣ, intermediate short-ear, recessive. Probably radiation-induced in a (101 × C3H)F1 hybrid. Homozygotes are viable and resemble *se/se* except that the ears are longer. *seˣ/se* mice have ears like *se/se* or possibly a little longer (8).

References

1. Dunn, G.R. 1972. Embryological effects of a minute deficiency in linkage group II of the mouse. J. Embryol. Exp. Morphol. 27:147–154.
2. Green, E.L, and M.C. Green. 1942. The development of three manifestations of the short ear gene in the mouse. J. Morphol. 70:1–19.
3. Green, M.C. 1951. Further morphological effects of the short ear gene in the mouse. J. Morphol. 88:1–22.
4. Green, M.C. 1958. Effects of the short ear gene in the mouse on cartilage formation in healing bone fractures. J. Exp. Zool. 137:75–88.
5. Green, M.C. 1968. Mechanism of the pleiotropic effects of the short ear gene in the mouse. J. Exp. Zool. 167:129–150.
6. Lynch, C.J. 1921. Short ears, an autosomal mutation in the house mouse. Am. Nat. 55:421–426.
7. Rinchik, E.M., L.B. Russell, N.G. Copeland, and N.A. Jenkins. 1986. Molecular genetic analysis of the *dilute-short ear* (*d-se*) region of the mouse. Genetics 112:321–342.
8. Russell, L.B. 1971. Definition of functional units in a small chromosomal segment of the mouse and its use in interpreting the nature of radiation-induced mutations. Mutat. Res. 11:107–123.
9. Wallace, M.E. 1976. Hydronephrosis in the mouse: the effects of the short-ear gene, sex and ureteral vascular system. Am. J. Anat. 147:19–32.

sea, sepia, recessive, Chr 1

Arose spontaneously in the C57BL/6J strain. Homozygotes closely resemble *bg/bg* mice but are a little darker, and the eyes are not light enough to be distinguished from those of wild type at birth. Homozygotes are viable and fertile (1).

References

1. Sweet, H.O., and P.W. Lane. 1977. Mouse News Lett. 57:19.

Segr loci, segregation reversal, Chr X

In somatic cell hybrids between mouse and Chinese hamster cells, the chromosomes of one of the parents are gradually lost while those of the other parent are not. In a particular set of hybrids between Chinese hamster E36 cells and BALB/c sarcoma lines CMS4 or MethA(s), only Chinese hamster chromosomes are lost. It was shown that if an X chromosome from a different BALB/c cell line or from normal embryo cells is introduced into the Chinese hamster E36 cells, the direction of segregation is reversed and only mouse chromosomes are lost. The mechanism by which the X chromosome produces this effect is not known (1).

References

1. Pravtcheva, D.D., and F.H. Ruddle. 1983. Normal X chromosome induced reversion in the direction of chromosome segregation in mouse–Chinese hamster somatic cell hybrids. Exp. Cell Res. 148:265–272.

Sep-1 locus

See *Apoa-1* locus.

Sev-1 locus

See *Emv-3* locus.

Sey locus, Chr 2

Sey, small eye, semidominant. Arose in a line previously selected for small body weight at the Institute of Animal Genetics at Edinburgh. Expression in heterozygotes is extremely variable, but penetrance is very nearly complete (6). The diameter of the eye ranges from 2 mm to almost normal. The orbit may also be small and result in asymmetrical skull and brain. The optic chiasma is sometimes absent. There is gross thickening of extracellular membranes in the eye, especially in the lens capsule, cellular degeneration, overgrowth and invasiveness, and mutual adhesiveness of tissues.

In embryos older than 13 days, lens cells may have grossly thickened membranes, and the inner and outer limiting retinal membranes and Descemet's membrane may be abnormally thick (1). Sialic acid concentration is 2.6-fold higher in the lens capsule of *Sey*/+ mice than in controls, and the heteropolysaccharide units are shorter than normal (5). Homozygotes can be identified at 10.5 days of gestation by absence of the lens and nasal placodes. By the time of birth, they have neither eyes nor nasal tissue but otherwise appear normal externally. They do not survive birth (3).

Sey^H, small eye-Harwell, semidominant. Found among offspring of an irradiated (C3H/HeH × 101/H)F1 female. Heterozygotes are more severely affected than *Sey*/+ mice. They are usually smaller at birth and may have a white belly spot. The eyes often have coloboma, visible in 14-day embryos. Sometimes one or both ears are small and low set. Both sexes are fertile, but females are poor breeders. There is probably some prenatal death. Homozygotes die prenatally, probably before or shortly after implantation. *Sey*^{H}/*Sey* embryos may survive at least to day 15, but are more severely affected than *Sey/Sey* embryos. *Sey*^{H} is on Chr 2, probably about 3 cM proximal to *pa* (3).

Sey^{dey} (formerly *Dey*), Dickie's small eye, semidominant. Arose spontaneously in the C3H/HeJ strain. It has the same chromosome location as *Sey* (2) and was shown to be an allele of *Sey* by Hogan *et al.* (4). *Sey*^{dey} is very similar to *Sey*^{H} in its effects in both homozygotes and heterozygotes. Its effects on the anatomy and embryological development of heterozygotes have been described by Theiler *et al.* (7).

References

1. Clayton, R.M., and J.C. Campbell. 1968. Small eye, a mutant in the house mouse apparently affecting the synthesis of extracellular membranes. J. Physiol. Lond. 198:74P–75P.
2. Davisson, M.T. 1986. Position of Dey on Chr 2. Mouse News Lett. 75:30–31.
3. Hogan, B.L.M., G. Horsburgh, J. Cohen, C.M. Hetherington, G. Fisher, and M.F. Lyon. 1986. *Small eyes (Sey): a homozygous lethal mutation on chromosome 2 which affects the differentiation of both lens and nasal placodes in the mouse. J. Embryol. Exp. Morphol. 97:95–110.
4. Hogan, B., C. Hetherington, and M.F. Lyon. 1987. Allelism of small eyes (Sey) with Dickie's small eye (Dey) on Chr 2. Mouse News Lett. 77:135–138.
5. Pritchard, D.J., R.M. Clayton, and W.L. Cunningham. 1974. Abnormal lens capsule carbohydrate associated with the dominant gene "small-eyes" in the mouse. Exp. Eye Res. 19:335–340.

6. Roberts, R.C. 1967. *Small-eyes*, a new dominant mutant in the mouse. Genet. Res. 9:121–122.
7. Theiler, K., D.S. Varnum, and L.C. Stevens. 1978. Development of Dickie's small eye, a mutation in the house mouse. Anat. Embryol. 155:81–86.

sf, scurfy, recessive, Chr X

Arose spontaneously. Heterozygous males can first be recognized at about 11 days of age by a reddening of the genital papilla. They develop scaliness, first of the tail and later of other parts of the body. The skin appears tight, and the eyelids open late. Scurfy males usually die before or shortly after weaning. The survivors are small and sterile (3). They have very small abdominal testes and lack a scrotum. This hypogonadism suggests that *sf* is probably homologous to X-linked ichthyosis with hypogonadism in man (1). Occasional scurfy females have occurred and proved to be X/O in sex chromosome type. These females resemble scurfy males in appearance and viability (3). Heterozygous females are indistinguishable from homozygous wild-type females (2).

References

1. Lyon, M.F. 1986. Hypogonadism in scurfy (sf) males. Mouse News Lett. 74:93.
2. Russell, L.B. 1964. Another look at the single-active-X hypothesis. Trans. NY Acad. Sci. Ser. II 26:726–736.
3. Russell, W.L., L.B. Russell, and J.S. Gower. 1959. Exceptional inheritance of a sex-linked gene in the mouse explained on the basis that the X/O sex-chromosome constitution is female. Proc. Natl. Acad. Sci. USA 45:554–560.

sg, staggerer, recessive, Chr 9

Occurred spontaneously in a stock of obese mice in 1955. Homozygotes usually die during the fourth week. Some survive to adulthood, and one male has bred. Homozygotes show a staggering gait, mild tremor, hypotonia, and small size (12). The cerebellar cortex is grossly underdeveloped with a deficiency of granule cells and Purkinje cells. The deficiency of granule cells in the external granular layer is already evident at birth. The remaining granule cells migrate inward from the external layer prematurely and then degenerate (15). Purkinje cells are much delayed in postnatal differentiation and lack the dendritic spines on which synapses with the parallel fibers from the granule cells normally occur (8). Golgi cells are not clearly distinguishable from Purkinje cells and it is possible that their number is also reduced (6). Examination of the cerebellum of *sg/sg* ⟷ +/+ chimeras using cellular markers for Purkinje cells and granule cells has shown that the *sg* effect is intrinsic to the Purkinje cells and that granule cells are affected secondarily (5, 7). Purkinje cells are probably defective as early as postnatal day 4 (10). The granule cell deficiency may result from failure of Purkinje cells to adequately stimulate granule cell genesis (13), as well as from later cell death due to failure of synapsis with Purkinje cells. Other effects of *sg* include persistence of multiple innervation of Purkinje cells by climbing fibers (2), reduction in size of deep cerebellar nuclei (9) and inferior olivary complex (1), and abnormal patterns of ganglioside composition and enzymatic activity (11). Cerebellar cells of *sg/sg* mice at 7 days postnatal have immature cell surface components of a type which are present in +/+ cells at late prenatal and neonatal stages (4, 14). In particular, the conversion of neural cell adhesion molecules (N-CAM) from embryonic to adult form which is normally complete by 21 days does not occur in *sg/sg* mice (3). Staggerer mice have been used as a source of an agranulate cerebellum in a number of investigations of the composition and function of granule cells.

References

1. Blatt, G.J., and L.M. Eisenman. 1985. A qualitative and quantitative light microscopic study of the inferior olivary complex in the adult staggerer mutant mouse. J. Neurogenet. 2:51–66.
2. Crepel, F., N. Delhaya-Bouchard, J.M. Guastavino, and I. Sampaio. 1980. Multiple innervation of cerebellar Purkinje cells by climbing fibers in *staggerer* mutant mice. Nature 283:483–484.
3. Edelman, G.M., and C.-M. Chuong. 1982. Embryonic to adult conversion of neural cell adhesion molecules in normal and staggerer mice. Proc. Natl. Acad. Sci. USA 79:7036–7040.
4. Hatten, M.E., and A. Messer. 1978. Postnatal cerebellar cells from staggerer mutant mice express embryonic cell surface characteristics. Nature 276:504–506.
5. Herrup, K. 1983. Role of staggerer gene in determining cell number in cerebellar cortex. I. Granule cell death is an indirect consequence of staggerer gene action. Dev. Brain Res. 11:267–274 (Brain Res. 313).
6. Herrup, K., and R.J. Mullen. 1979. Regional variation and absence of large neurons in the cerebellum of the staggerer mouse. Brain Res. 172:1–12.
7. Herrup, K., and R.J. Mullen. 1979. Staggerer chimeras: intrinsic nature of Purkinje cell defects and implications for normal cerebellar development. Brain Res. 178:443–457.
8. Landis, D.M.D., and R.L. Sidman. 1978. Electron microscopic analysis of postnatal histogenesis in the cerebellar cortex of staggerer mutant mice. J. Comp. Neurol. 179:831–863.
9. Roffler-Tarlov, S. and K. Herrup. 1981. Quantitative

examination of the deep cerebellar nuclei in the staggerer mutant mouse. Brain Res. 215:49–59.

10. Roffler-Tarlov, S., and M. Turey. 1982. The content of amino acids in the developing cerebellar cortex and deep cerebellar nuclei of granule cell deficient mutant mice. Brain Res. 247:65–73.

11. Schaal, H., C. Wille, and W. Wille. 1985. Changes of ganglioside pattern during cerebellar development of normal and staggerer mice. J. Neurochem. 45:544–551.

12. Sidman, R.L., P.W. Lane, and M.M. Dickie. 1962. Staggerer, a new mutation in the mouse affecting the cerebellum. Science 137:610–612.

13. Somnez, E., and K. Herrup. 1984. Role of staggerer gene in determining cell number in cerebellar cortex. II. Granule cell death and persistence of the external granule cell layer in young mouse chimeras. Dev. Brain Res. 12:271–283 (Brain Res. 314).

14. Trenkner, E. 1979. Postnatal cerebellar cells of *staggerer* mutant mice express immature components on their surface. Nature 277:566–567.

15. Yoon, C.H. 1972. Developmental mechanism for changes in cerebellum of "staggerer" mouse, a neurological mutant of genetic origin. Neurology 22:743–754.

Sgp-1 locus, serum glycoprotein-1, Chr 17

This locus controls the level of serum gp70, a 70-kDa glycoprotein similar to the envelope protein of retroviruses. The locus is closely linked to but not in the *H-2* complex. The allele *Sgp-1ᵃ* is associated with the *H-2ᵈ* haplotype and determines a high level of serum gp70 in C57BL/10 congenic strains. The allele *Sgp-1ᵇ* is associated with most other *H-2* haplotypes and determines a low serum gp70 in B10 congenic strains and in their F1 hybrids with NZB, a high gp70 producer. It also determines an increased gp70 response to lipopolysaccharide (LPS) stimulation in these mice. The allele *Sgp-1ᶜ* is associated with the *H-2ˢ* haplotype and determines a low serum gp70 and a lack of gp70 response to LPS stimulation. Production of gp70 is apparently affected by other genes not linked to *H-2* (1).

References

1. Maruyama, N., and C.O. Lindstrom. 1983. *H-2*-linked regulation of serum gp70 production in mice. Immunogenetics 17:507–521.

Sgp-2 locus, serum glycoprotein-2, Chr 7 (possibly identical to *Gv-2*)

This locus controls enhancement of the level of gp70 in serum after lipopolysaccharide (LPS) injection. The allele *Sgp-2ᵃ* determines high rate of enhancement and occurs in the 129 strain; the allele *Sgp-2ᵇ* determines low rate of enhancement and occurs in the C57BL/6 strain. Heterozygotes are intermediate. *Sgp-2* is linked to *Hbb* on Chr 7 with about 20 per cent recombination, which puts it in the vicinity of *Gv-2* (Gross virus-2). *Sgp-2* and *Gv-2* have similar effects on gp70 and may be identical (1).

References

1. Maruyama, N., C.O. Lindstrom, H. Sato, and F.J. Dixon. 1983. Serum gp70 production regulated by a gene on chromosome 7. Immunogenetics 18:365–371.

sh-1, shaker-1, recessive, Chr 7

Found by Lord and Gates (4) in a Bagg albino strain. Viability is normal, and breeding ability is high for a circling mutant. Homozygotes show the circling, head-tossing, deafness, and hyperactivity characteristic of mutants of this type. They can sometimes hear and swim well on the surface of water up to 4 weeks or more but lose the ability later. The degenerative changes of the labyrinth may occur a little later than in some of the other waltzing mutants. By light microscopy, the changes are seen to consist of degeneration of the organ of Corti, the spiral ganglion, and the stria vascularis in the cochlea, and of the saccular macula and the vestibular ganglion in the vestibular labyrinth (1). Kikuchi and Hilding (3) found virtual absence of efferent innervation of the hair cells of the organ of Corti at 12 days postnatal and postulated that the primary effect of *sh-1* was in the efferent innervation. However, Shnerson *et al.* (6) found abnormal spiral ganglion and outer hair cells at 3 days, followed later by arrival of a reduced number of efferent nerves which subsequently degenerated. Efferent innervation of the inner hair cells was normal. The eighth nerve action potential and cochlear potentials begin to develop but are never normal and have disappeared by 22 days of age (5). In females heterozygous for *sh-1* and Cattanach's insertion, *Is(In7;X)1Ct*, which are mosaic for shaker-1 and normal cells, the cochlea shows patchy distribution of mutant and normal hair cells, while the saccular macula appears uniformly affected throughout but with great variation between maculae in degree of effect. These results suggest that, in the cochlea, the action of *sh-1* is localized, but in the macula, there must be some averaging of the effects on individual cells (2). Tan Creti (7) found marked gliosis in the medial vestibular nucleus of *sh-1/sh-1* mice.

References

1. Deol, M.S. 1956. The anatomy and development of the mutants pirouette, shaker-1 and waltzer in the mouse. Proc. R. Soc. Lond. (Biol.) 145:206–213.

2. Deol, M.S., and M.C. Green. 1969. Cattanach's translocation as a tool for studying the action of the shaker-1 gene in the mouse. J. Exp. Zool. 170:301–309.
3. Kikuchi, K., and D.A. Hilding. 1965. The defective organ of Corti in shaker-1 mice. Acta Oto-Laryngol. 60:287–303.
4. Lord, E.M., and W.H. Gates. 1929. Shaker, a new mutation of the house mouse. Am. Nat. 63:435–442.
5. Mikaelian, D.O., and R.J. Ruben. 1964. Hearing degeneration in shaker-1 mouse. Arch. Otolaryngol. 80:418–430.
6. Shnerson, A., M. Lenoir, T.R. Van de Water, and R. Pujol. 1983. The pattern of sensorineural degeneration in the cochlea of the deaf shaker-1 mouse: ultrastructural observations. Dev. Brain Res. 9:305–315 (Brain Res. 285).
7. Tan Creti, D.M. 1969. Neuropathology of mutant mice with auditory and/or vestibular deficiencies. J. Neuropathol. Exp. Neurol. 28:159 (Abstr.).

sh-2, shaker-2, recessive, Chr 11

Discovered by Dobrovolskaïa-Zavadskaïa (3) in the descendants of an irradiated male. Viability and fertility are nearly normal. This mutant is very similar in behavior and pathology to *sh-1*, with the exception that the abnormalities are observed a little earlier in *sh-2*. Homozygotes appear to be deaf from the beginning, and the saccular macula is abnormal at birth (2). The stria vascularis appears normal by light microscopy at 2 weeks, but begins to show degenerative changes shortly thereafter. Many of the cells contain electron-dense inclusions filled with a fine granular material (1). At 2.5 months, the type I hair cells of the cristae ampullares and maculae utriculi contain rod-shaped inclusion bodies composed of actin filaments (4).

References

1. Anniko. M. 1982. Specific pathology of the stria vascularis in postnatal progressive genetic inner ear disorder. Hear. Res. 6:247–258.
2. Deol. M.S. 1954. The anomalies of the labyrinth of the mutants varitint-waddler, shaker-2 and jerker in the mouse. J. Genet. 52:562–588.
3. Dobrovolskaïa-Zavadskaïa, N. 1928. L'irradiation des testicules et l'hérédité chez la souris. Arch. Biol. 38:457–501.
4. Sobin, A., M. Anniko, and A. Flock. 1982. Rods of actin filaments in type I hair cells of the shaker-2 mouse. Arch. Otorhinolaryngol. 236:1–6.

Sha, shaven, semidominant, Chr 15

Arose in a stock of the Endocrinology Department, Edinburgh. Heterozygotes have a greasy coat and some waviness of the vibrissae. Homozygotes do not grow a first coat during the preweaning period and thereafter show cyclic growth and loss of very short sparse furry hairs. The vibrissae are short and wavy at birth and throughout life. Hairs are of normal size in heterozygotes, but the air spaces in the medulla are smaller than normal and filled with sudanophilic fluid. In homozygotes the hairs are thin and uneven in diameter, and the air spaces are very small and irregular and filled with sudanophilic fluid. *Sha* is closely linked to naked (*N*). The skin histology of *Sha/Sha* mice is similar to that of *N/N* mice. Keratinization of the hairs is defective, and they fail to penetrate the epidermis (1).

References

1. Flanagan, S.P., and J.H. Isaacson. 1967. Close linkage between genes which cause hairlessness in the mouse. Genet. Res. 9:99–100.

shi locus, Chr 18

The locus maps near the distal end of Chr 18 (21).

shi, shiverer, recessive. Arose in the SWV strain at the University of British Columbia. Homozygotes show a violent shiver when disturbed beginning at 12 days. Shivering increases in severity with age, and there is incoordination of the hindlimbs. Post-weaning mice have attacks in which they lie rigid and motionless for many seconds. Homozygotes are noticeably smaller than their littermates by 4 weeks and have a shortened life span. They normally die between 50 and 100 days, usually while in an attack. They are fertile but do not breed well (8). There is a severe myelin deficiency throughout the central nervous system (CNS) (4), and a moderate hypomyelination in the peripheral nervous system (PNS) (16). In the CNS there are occasional regions of normal appearing myelin (19). Most of the sparse myelin that forms in the CNS lacks the major dense line; in the PNS the major dense line is present (17). Immunochemical studies have shown that myelin basic protein (MBP) is lacking in both the CNS (11) and PNS (12,15). Heterozygous (*shi/+*) mice have normal behavior and structurally normal myelin but have only half the normal amount of MBP in both the CNS and PNS (2, 5, 6). Translation products of polyribosomes isolated from brains of *shi/shi* and *shi/+* mice produced 0 per cent and about 50 per cent of control (+/+) amounts of MBP, respectively (7). Co-culture experiments have shown that the effect of *shi* is intrinsic to the oligodendrocytes and Schwann cells (3, 20). The MBP gene was found to be on Chr 18 by use of mouse–hamster somatic cell hybrids and is probably identical

shi

to the *shi* gene. In the *shi* mutation there is a deletion of five of the six exons of the MBP gene (18).

shi^{mld}, myelin deficient, recessive. Arose spontaneously in the MDB/Dt strain (9, 10). Homozygotes closely resemble *shi/shi* mice in behavior, neuropathology, and MBP deficiency (13, 14). However, in the *shi^{mld}* mutation, a portion of the MBP gene is duplicated. The duplication is not immediately adjacent to the intact gene but may be responsible for repressing its expression (1).

References

1. Akowitz, A., K. Scheld, E. Barbarese, and J.H. Carson. 1986. A partial duplication of the myelin basic protein gene in *MLD* mice. Trans. Am. Soc. Neurochem. 17:108 (Abstr.).
2. Barbarese, E., M.L. Nielson, and J.H. Carson. 1983. The effect of the shiverer mutation on myelin basic protein expression in homozygous and heterozygous mouse brain. J. Neurochem. 40:1680–1686.
3. Billings-Gagliardi, S., A.L. Hall, G.B. Stanhope, R.J. Altschuler, R.L. Sidman, and M.K. Wolf. 1984. Cultures of shiverer mutant cerebellum injected with normal oligodendrocytes make both normal and shiverer myelin. Proc. Natl. Acad. Sci. USA 81:2558–2561.
4. Bird, T.D., D.F. Farrell, and S.M. Sumi. 1978. Brain lipid composition of the shiverer mouse: (genetic defect in myelin development). J. Neurochem. 31:387–391.
5. Cammer, W. 1982. Partial deficiencies in the myelin proteins of developing mice heterozygous for the shiverer mutation. Dev. Genet. 3:155–163.
6. Cammer, W., S. Kalin, and T. Zimmerman. 1984. Biochemical abnormalities in spinal cord myelin and CNS homogenates in heterozygotes affected by the shiverer mutation. J. Neurochem. 42:1372–1378.
7. Campagnoni, A.T., C.W. Campagnoni, J.-M. Bourre, C. Jacque, and N. Baumann. 1984. Cell-free synthesis of myelin basic proteins in normal and dysmyelinating mutant mice. J. Neurochem. 42:733–739.
8. Chernoff, G.F. 1981. Shiverer, an autosomal recessive mutant mouse with myelin deficiency. J. Hered. 72:128.
9. Doolittle, D.P., and K.M. Schweikart. 1977. Myelin deficient, a new neurological mutant in the mouse. J. Hered. 68:331–332.
10. Doolittle, D.P., N. Baumann, and G. Chernoff. 1981. Allelism of two myelin deficiency mutations in the mouse. J. Hered. 72:285.
11. Dupouey, P., C. Jacque, J.M. Bourre, F. Casselin, A. Privat, and N. Baumann. 1979. Immunochemical studies of myelin basic protein in shiverer mouse devoid of major dense line of myelin. Neurosci. Lett. 12:113–118.
12. Kirschner, D.A., and A.L. Ganser. 1980. Compact myelin exists in the absence of basic protein in the shiverer mutant mouse. Nature 283:207–210.
13. Matthieu, J.-M., H. Ginalski, R.L. Friede, S.R. Cohen, and D.P. Doolittle. 1980. Absence of myelin basic protein and major dense line in CNS myelin of the mld mutant mouse. Brain Res. 191:278–283.
14. Matthieu, J.-M., F.X. Omlin, H. Ginalski-Winkelman, and B.J. Cooper. 1984. Myelination in the CNS of *mld* mutant mice: comparison between composition and structure. Dev. Brain Res. 13:149–158 (Brain Res. 315).
15. Mikoshiba, K., K. Takamatsu, and Y. Tsukada. 1983. Peripheral nervous system of shiverer mutant mice: developmental change of myelin components and immunohistochemical demonstration of the absence of MBP and presence of P2 protein. Dev. Brain Res. 7:71–79 (Brain Res. 283).
16. Peterson, A.C., and G.M. Bray. 1984. Hypomyelination in the peripheral nervous system of shiverer mice and in *shiverer* ⟷ *normal* chimaera. J. Comp. Neurol. 227:348–356.
17. Privat, A., C. Jacque, J.M. Bourre, P. Dupouey, and N. Baumann. 1979. Absence of the major dense line in myelin of the mutant mouse "shiverer". Neurosci. Lett. 12:107–112.
18. Roach, A., N. Takahashi, D. Pravtcheva, F. Ruddle, and L. Hood. 1985. Chromosome mapping of mouse myelin basic protein gene and structure and transcription of the partially deleted gene in shiverer mutant mice. Cell 42:149–155.
19. Rosenbluth, J. 1980. Central myelin in the mouse mutant shiverer. J. Comp. Neurol. 194:639–648.
20. Shine, H.D., and R.L. Sidman. 1984. Immunoreactive myelin basic proteins are not detected when shiverer mutant Schwann cells and fibroblasts are co-cultured with normal neurons. J. Cell Biol. 98:1291–1295.
21. Sidman, R.L., C.S. Conover, and J.H. Carson. 1985. Shiverer gene maps near the distal end of chromosome 18 in the house mouse. Cytogenet. Cell Genet. 39:241–245.

shm, shambling, recessive, Chr 11

Found in a randombred population which had been exposed to X-irradiation in the four preceding generations. Homozygotes are smaller than normal and seldom fertile. They are first recognizable at 16 to 18 days by their small size and wobbly gait. Shambling mice walk with the hindlimbs held relatively stiff and take exaggerated steps so that a hindfoot often lands in front of the forefeet. They do not swim well, and when taken out of the water may undergo a spell of violent trembling lasting half an hour or more (1). Viability and longevity are somewhat reduced (2). In the central nervous system and in a number of visceral organs there are reported to be focal deposits of phospholipid-like material which has not been identified (3). There are no degenerative myelin changes, and brain gangliosides are normal in composition and amount (4). In brain microsomes, there is a 60 per cent decrease in rate of synthesis of phosphatidylserine (PS) in homozygotes and a 80 per cent decrease in heterozygotes, the loss probably being due to loss of the calmodulin-sensitive PS synthesis present in normal brain microsomes (5).

References

1. Green, E.L. 1967. Shambling, a neurological mutant of the mouse. J. Hered. 58:65–68.
2. Green, E.L. 1971. Fitness of heterozygotes of deleterious recessive mutations in the mouse. Mutat. Res. 12:281–289.
3. Meier, H. 1967. Pathological findings in shambling, a hereditary neuropathy of mice. J. Neuropathol. Exp. Neurol. 26:620–633.
4. Seyfried, T.N., E.J. Weber, and W.L. Daniel. 1976. Absence of brain ganglioside abnormalities in shambling mutant mice. J. Neurochem. 27:295–296.
5. Yamaguchi, T., Y. Matsumura, and M. Yamaguchi. 1984. Decreases in phosphatidylserine synthesis in brain microsomes of shambling mutant mice. Biochem-Int. 8:347–352.

sho, shorthead, recessive

Arose spontaneously in an inbred line at the Nevis Biological Station of Columbia University. Newborn homozygotes are characterized by small size, disproportionate shortening of the head and limbs, and a large median cleft palate. They die soon after birth. The forelimbs are relatively more shortened than the hindlimbs. The small intestine is about one-half normal length, and in females one ovary or oviduct may occur medial to the kidney on that side. At 14 days of gestation the chondrocranium is severely foreshortened, and the part of the basal plate that will become the presphenoid is widened. Mutants can be recognized at 12 days of gestation by their small size and short heads, showing that the gene acts before cartilage formation (1). The cleft palate can be traced to 14-day embryos, when the tongue and lower jaw are seen to be abnormally small and the palatal shelves are prematurely closed at the anterior end. They enclose the tip of the tongue, preventing it from descending and thereby preventing the shelves from closing (2).

References

1. Fitch, N. 1961. A mutation in mice producing dwarfism, brachycephaly, cleft palate and micromelia. J. Morphol. 109:141–149.
2. Fitch, N. 1961. Development of cleft palate in mice homozygous for the shorthead mutation. J. Morphol. 109:151–157.

si, silver, recessive, Chr 10

This mutant is widespread in the English fancy. Homozygotes are extremely variable in appearance. Individual hairs of the coat of nonagouti silvers may be all white, all black, black with white tips, or white with gray or black bands. Silvering results from a reduction in number of pigment granules. Nonagouti silvers heterozygous for brown (*si/si a/a b/+*) have very light underfur (1). The effect of *si* is so variable that it is often difficult to classify in crosses, and its usefulness as a genetic marker is therefore limited.

References

1. Dunn, L.C., and L.W. Thigpen. 1930. The silver mouse, a recessive color variation. J. Hered. 21:495–498.

Sig, sightless, semidominant, Chr 6

Found in a low-intensity neutron irradiation experiment. Heterozygotes have both eyes open at birth and slight hydrocephaly. Penetrance is complete. The presumed homozygotes die around birth with more pronounced hydrocephaly and sometimes with hindfoot anomalies, especially syndactyly (1).

References

1. Searle, A.G. 1965. Mouse News Lett. 33:29.

Sinc locus, scrapie incubation period, Chr 2

This locus controls length of incubation period of the scrapie slow-virus-like agent of sheep after intracerebral or intraperitoneal administration to mice (5). The allele *Sinc*^{s7} in homozygotes determines short incubation period of the ME7 group of scrapie agents and long incubation period of the 22A group of agents; it occurs in the C57BL and CW strains. The allele *Sinc*^{p7} in homozygotes determines prolonged incubation period of the ME7 group of agents and short incubation period of the 22A group; it occurs in the VM strain. For most of the ME7 group of agents, heterozygotes are intermediate between parental types; for the 22A group and for some of the ME7 group, heterozygotes have longer incubation periods than either of the parents, i.e. the allele determining the longer period is overdominant. This overdominance and the inversion of gene effect according to the strain of the agent suggest that the *Sinc* locus may determine a multimeric replication site for scrapie, with the two alleles contributing different subunits (3, 4). The type of neuropathology, consisting primarily of vacuolation and amyloidosis with a wide range of morphology, is characteristic for any particular

Sinc

scrapie agent and *Sinc* allele combination (1). The scrapie agent contains a protein, termed a prion, but no detectable nucleic acid. The prion is now known to be encoded in the host genome (2, 6). In a different strain combination, a gene determining incubation time (*Prn-i*) was shown to be closely linked to the prion structural gene *Prn-p* on Chr 2. It is likely that *Sinc* and *Prn-i* are identical. See *Prn* complex.

References

1. Bruce, M.E., A.G. Dickinson, and H. Fraser. 1976. Cerebral amyloidosis in scrapie in the mouse; effect of agent strain and mouse genotype. Neuropathol. Appl. Neurobiol. 2:471–478.
2. Chesbro, B., R. Race, K. Wehrly, J. Nishio, M. Bloom, D. Lechner, S. Bergstrom, K. Robbins, L. Mayer, J.M. Keith, C. Garon, and A. Haase. 1985. Identification of scrapie prion protein-specific mRNA in scrapie-infected and uninfected brain. Nature 315:331–333.
3. Dickinson, A.G. 1975. Host–pathogen interactions in scrapie. Genetics 79:s387–s395.
4. Dickinson, A.G., and G.W. Outram. 1979. The scrapie replication-site hypothesis and its implications for pathogenesis. *In* S.B. Prusiner and W.J. Hadlow, eds., Slow Transmissible Diseases of the Nervous System, Vol. 2, 13–31. Academic Press, New York.
5. Dickinson, A.G., V.M.H. Meikle, and H. Fraser. 1968. Identification of a gene which controls the incubation period of some strains of scrapie agent in mice. J. Comp. Pathol. 78:293–299.
6. Oesch, B., D. Westaway, M. Wälchli, M.P. McKinley, S.B.H. Kent, R. Aebersold, R.R. Barry, P. Tempst, D.B. Taplow, L.E. Hood, S.B. Prusiner, and C. Weissmann. 1985. A cellular gene encodes scrapie PrP 27-30 protein. Cell 40:735–746.

Sip locus, schedule-induced polydipsia

This locus controls susceptibility to polydipsia (elevated water consumption) induced in food-deprived mice by periodic food reinforcement at 150-s intervals. The allele Sip^d determines susceptibility to schedule-induced polydipsia and occurs in the DBA/2J strain; the allele Sip^b determines resistance and occurs in the C57BL/6J strain. Heterozygotes are fully susceptible. In susceptible mice, water intake in a 3-hour testing session may be threefold greater than the normal 24-hour intake (1).

References

1. Symons, J.P., and R.L. Sprott. 1976. Genetic analysis of schedule induced polydipsia. Physiol. Behav. 17:837–839.

Sis locus (c-*sis*), simian sarcoma oncogene, Chr 15

This is the cellular homolog of the transforming gene (v-*sis*) of the simian sarcoma virus originally found in a fibrosarcoma of a woolley monkey. A 20 kDa protein product of v-*sis*, p28*sis*, has been shown to be highly homologous with the human platelet-derived growth factor (PDGF) (2, 5), and the human c-*sis* oncogene is the structural gene for PDGF (1). It is therefore probable that the viral oncogene was derived from the cellular PDGF gene of the woolley monkey. The function of *Sis* in the mouse has not been reported. The gene is recognized in the Southern blot hybridization procedure by a probe made from the human c-*sis* gene. By screening mouse–Chinese hamster somatic cell hybrids with this probe, *Sis* was found to be located on Chr 15. A DNA restriction fragment length difference between C57BL/6 and four other strains (A/J, BALB/c, C3H/He, and DBA/2) was later found by Meruelo *et al.* (4). By use of recombinant inbred strains between C57BL/6 and the other strains, the authors showed that *Sis* was located near *Ly-6* on Chr 15. Chr 15 is involved in translocations or trisomies occurring in various mouse neoplastic diseases. While some of these translocation break points have been shown to occur at the locus of the *Myc* oncogene on Chr 15, it is possible that others involve the *Sis* locus (3).

References

1. Chiu, I.-M., E.P. Reddy, D. Givol, K.C. Robbins, S.R. Tronick, and S.A. Aaronson. 1984. Nucleotide sequence analysis identifies the human c-sis proto-oncogene as a structural gene for platelet-derived growth factor. Cell 37:123–129.
2. Doolittle, R.F., M.W. Hunkapiller, L.E. Hood, S.G. Devare, K.C. Robbins, S.A. Aaronson, and H.J. Antoniades. 1983. Simian sarcoma virus *onc* gene, v-*sis*, is derived from the gene (or genes) encoding a platelet-derived growth factor. Science 221:275–277.
3. Kozak, C.A., J.F. Sears, and M.D. Hoggan. 1983. Genetic mapping of the mouse proto-oncogene c-sis to chromosome 15. Science 221:867–868.
4. Meruelo, D., A. Rossomando, S. Scandalis, P. D'Eustachio, R.E.K. Fournier, D.R. Roop, D. Saxe, C. Blatt, and M.N. Nesbitt. 1987. Assignment of the *Ly-6–Ril-1–Sis–H-30–Pol-5/Xmmv-72–Ins-3–Krt-1–Int-1–Gdc-1* region to mouse chromosome 15. Immunogenetics 25:361–372.
5. Waterfield, M.D., G.T. Scrace, N. Whittle, P. Stroobant, A. Johnson, A. Wasteson, B. Westermark, C.-H. Heldin, J.S. Huang, and T.F. Deuel. 1983. Platelet-derived growth factor is structurally related to the putative transforming protein p28[sis] of simian sarcoma virus. Nature 304:35–39.

Lyon MF, Searle, AG, eds. Genetic Variants and Strains
of the Laboratory Mouse, Second Edition. Oxford:Oxford
University Press, 1989:333.

Sl

Sk, scaly, semidominant

Arose spontaneously. Penetrance is probably complete in heterozygotes. In mild cases they show scaliness on shoulders and back of head only; in extreme expression the scaliness may cover the entire dorsal surface and ears, producing almost immobile animals which occasionally die. Scaliness is first noticeable at about 4 days. The coat grows in sparsely at first but may in milder cases become indistinguishable from normal. The fate of the homozygotes in unknown (1).

References

1. Kelly, E.M., and J.W. Bangham. 1955. Mouse News Lett. 12:47.

Skn-1, *Skn-2* loci (formerly *Sk-1*, *Sk-2*), skin antigen-1, -2

These two loci were postulated to account for the rejection of strain A skin grafts by lethally irradiated C57BL/6 mice which had been restored with (C57BL/6 × A)F1 bone marrow or spleen. The antigens can be detected by skin grafting or serologically. By serological methods they were shown to be present on epidermal cells and brain, but not on thymocytes, lymph node cells, or spleen cells (3). Different allelic forms occur in strain A and C57BL/6. Original results of a backcross of (B6 × A)F1 to B6 gave evidence of segregation for a single controlling locus (2), but later evidence from the same cross showed two-locus segregation (5). An Skn antigen present on skin but not on lymphocytes was found in the B6-*H-29*^{c} congenic strain bearing the *H-29* allele from the BALB/cBy strain (1). The *Skn* gene determining this antigen is either identical with *H-24* or closely linked to it on Chr 7. The genetics of Skn antigens has been critically reviewed by Steinmuller and Wachtel (4).

References

1. Harrison, D.E., and L.E. Mobraaten. 1984. Skin graft rejection in mice repopulated with marrow of the skin donor type: a *Skn* gene in a congenic line. Immunogenetics 19:503–509.
2. Lance, E.M., E.A. Boyse, S. Cooper, and E.A. Carswell. 1971. Rejection of skin allografts by irradiation chimeras: evidence for a skin-specific transplantation barrier. Transplant. Proc. 3:864–868.
3. Scheid, M., E.A. Boyse, E.A. Carswell, and L.J. Old. 1972. Serologically demonstrable alloantigen of mouse epidermal cells. J. Exp. Med. 135:938–955.
4. Steinmuller, D., and S.S. Wachtel. 1980. Transplantation biology and immunogenetics of murine skin-specific (Sk) alloantigens. Transplant. Proc. 12 (3) Suppl. 1:100–106.
5. Wachtel, S.S., H.T. Thaler, and E.A. Boyse. 1977. A second system of alloantigens expressed selectively on epidermal cells of the mouse. Immunogenetics 5:17–23.

Sl locus, Chr 10

Mutations at the *Sl* locus, both spontaneous and induced, have been discovered frequently. These mutations, like mutations at the *W* locus, cause deficiencies in pigment cells, germ cells, and blood cells, the alleles varying in severity of their effects. In general, homozygotes or compounds of two alleles are black-eyed white, are usually sterile in one or both sexes, and have severe macrocytic anemia: heterozygotes have diluted coat color with a small amount of white spotting, are fertile, and may be slightly anemic. Numerous experiments have demonstrated that precursors of pigment cells and blood cells of homozygotes behave normally when transplanted to an environment of wild-type cells, and that the locus exerts its effect through action of cells in the microenvironment of the apparently affected cells. This relation has not been demonstrated for germ cells, but there has been no satisfactory experimental system for testing it. Most of the experiments have used *Sl* and *Sl*^{d} homozygotes or *Sl/Sl*^{d} compounds as the severely affected genotype.

By means of recombining mutant and normal embryonic skin and neural crest and growing the recombined tissue in chick embryos, it has been shown that the *Sl* locus acts in embryonic skin and not in melanoblasts from the neural crest. The inhibitory action on pigment cell development occurs in both dermis and epidermis and probably also in other tissue through which melanoblasts normally migrate (17). The *Sl* locus is active in skin from 13 to at least 18 days of gestation (16), but is inactive in adult skin (23).

Primordial germ cells are absent from the gonads of homozygotes (5). They migrate normally toward the gonadal ridges but fail to proliferate and to be maintained there (18).

The blood of *Sl*-locus homozygotes and compounds has reduced numbers of macrocytic red cells, reduced packed cell volume, and elevated reticulocyte percentage (25). Neutrophils in both blood and bone marrow are reduced (24) and blood monocyte count is less than 10 per cent of normal (29). The number of splenic natural killer cells is about one-third of normal (8). Platelet count is normal, but there is a reduced number of enlarged megakaryocytes (9). Mast cells are virtually absent in skin and other tissues (11, 15). The marrow cells are capable of forming normal spleen colonies when injected into irradiated hosts, they can restore the blood-forming capacity of lethally irradiated normal

333

hosts, and they can cure the anemia of W/W^v mice. Steel mice cannot themselves be cured by injection of normal marrow, but they are partially cured by implantation of whole spleens (1). It is not known which cells in the blood-forming tissue of steel mutants are responsible for preventing normal hematopoiesis, but there is evidence that the defect is caused by local action of a nonmigratory stromal component, possibly by cell–cell contact or short-range diffusion (22). Several allelic combinations (Sl/Sl^d, Sl^m/Sl^m) are known to cause gastrointestinal bleeding, and the mice may lose 2.5 to 3.5 per cent of their total blood volume per day. The cause of the blood loss is not known. The loss could account for the elevated reticulocyte count found in steel mice (13, 30). Mutant steel alleles cause resistance to induction of spleen foci by Friend virus, probably by producing an unfavorable environment for the target lymphopoietic cells (6). They also cause a marked increase in incidence of lymphocytic leukemia on an appropriate genetic background (20). Some of the *Sl*-locus mutations are listed below.

Sl, steel, semidominant. Arose spontaneously in the C3H inbred strain (26). Heterozygotes have a slight dilution of coat color, more extreme on the belly than on the back, light ears, feet, and tail, and occasionally a belly spot or blaze. They also have a slight macrocytic anemia but are viable and fertile. Homozygotes are severely anemic *in utero* and die usually at 15 to 16 days of gestation (26). Primordial germ cells are absent in Sl/Sl and deficient in $Sl/+$. There is no pigment-forming ability in the skin of Sl/Sl embryos (5).

Sl^{cg}, cloud gray, semidominant. Found in offspring of a neutron-irradiated male. Homozygotes are very light gray with white blotches. $Sl^{cg}/+$ mice are barely lighter than wild type. They are fertile, but fertility of homozygotes is very poor (21). About one-third of the homozygous females have imperforate vaginas. Sl^{cg}/Sl mice are off-white with dark ears and black eyes (14).

Sl^{con}, contrasted, semidominant. Found in a neutron irradiation experiment. Heterozygotes are recognizable soon after birth by dark pigmentation of the genital papilla. The adult coat is slightly lighter than normal. Homozygotes also have dark external genitalia but a markedly diluted coat. Males are fertile but females are usually sterile. Homozygotes have a mild macrocytic anemia (2).

Sl^d, steel-Dickie, semidominant. Arose spontaneously in the DBA/2J inbred strain (7). This allele is similar to *Sl* in its effect on coat color in heterozygotes, but homozygotes are viable. They are white with black eyes and severely anemic. Sl/Sl^d are also black-eyed white and

anemic and many live as long as a year. Both Sl/Sl^d and Sl^d/Sl^d are sterile with very few germ cells in the gonads. The germ-cell-deficient ovaries develop tumors in old age (20).

Sl^{gb}, grizzle-belly, semidominant. Arose spontaneously in a $bt/bt\ Mi^{wh}/+$ female. Heterozygotes have light bellies. Homozygotes are anemic and die during the first week of life (27, 28).

Sl^m, steel-Miller, semidominant. Arose in a C57BL/6 substrain exposed to low-level irradiation (12, 19). Homozygotes are black-eyed white, anemic, and may survive up to one year. Males are sterile, but an occasional female may have one litter. Heterozygotes have diluted coat color and are otherwise normal. Homozygotes have gastrointestinal bleeding and lose about 3.5 per cent of their blood volume daily (13).

Sl^{pan} (formerly Sl^{11H}), panda, semidominant. Heterozygotes have slightly diluted coat color with occasionally a white head dot and white tail tip. Homozygotes have a mild macrocytic anemia and are fertile in males but sterile with very small ovaries in females. They are white with black eyes and black ears. Sl^{pan}/Sl^{con} mice have light coats and are fertile in males and sterile in females (3, 4).

Sl^{so}, sooty, semidominant. Arose in the C57BL/6 strain. $Sl^{so}/+$ have a dilute coat and light tail. Homozygotes are black-eyed white but live 10 to 40 days (12, 19).

sl^{du}, dusty, recessive. Arose spontaneously in the C3H/HeJ strain. Homozygotes have slightly diluted coat color and small white patches that vary in number from none to extensive white speckling, and they are not noticeably anemic. Sl/sl^{du} mice have a moderate anemia at birth, producing a recognizable paleness, and they usually have more white spotting than $Sl/+$ mice (10).

References

1. Bannerman, R.M., J.A. Edwards, and P.H. Pinkerton. 1973. Hereditary disorders of the red cell in mammals. Prog. Hematol. 8:131–179.
2. Beechey, C.V., and A.G. Searle. 1983. Contrasted, a steel allele in the mouse with intermediate effects. Genet. Res. 42:183–191.
3. Beechey, C.V., and A.G. Searle. 1985. Mouse News Lett. 73:17.
4. Beechey, C.V., J.F. Loutit, and A.G. Searle. 1986. Panda, a new steel allele. Mouse News Lett. 74:92.
5. Bennett, D. 1856. Developmental analysis of a mutation with pleiotropic effects in the mouse. J. Morphol. 98:199–234.
6. Bennett, M., R.A. Steeves, G. Cudkowicz, E.A. Mirand,

and L.B. Russell. 1968. Mutant Sl alleles of mice affect susceptibility to Friend spleen focus-forming virus. Science 162:564–565.

7. Bernstein, S.E. 1960. Mouse News Lett. 23:33.

8. Clark, E.A., L.D. Shultz, and S.B. Pollack, 1981. Mutations in mice that influence natural killer (NK) cell activity. Immunogenetics 12:601–613.

9. Ebbe, S., E. Phalen, P. D'Amore, and H. Howard. 1978. Megakaryocyte responses to thrombocytopenia and thrombocytosis in Sl/Sl^d mice. Exp. Hematol. 6:201–212.

10. Green, M.C., and H.O. Sweet. 1973. Mouse News Lett. 49:32.

11. Hayashi, C., T. Sonoda, T. Nakano, N. Nakayama, and Y. Kitamura. 1985. Mast cell precursors in the skin of mouse embryos and their deficiency in embryos of Sl/Sl^d genotype. Dev. Biol. 109:234–241.

12. Hollander, W.F. 1964. Mouse News Lett. 30:29.

13. Kales, A.N., W. Fried, and C.W. Gurney. 1966. Mechanism of the hereditary anemia of Sl^m mutant mice. Blood 28:387–397.

14. Kelly, E.M. 1974. Mouse News Lett. 50:52.

15. Kitamura, Y., and S. Go. 1979. Decreased production of mast cells in Sl/Sl^d anemic mice. Blood 53:492–497.

16. Mayer, T.C. 1970. A comparison of pigment cell development in albino, steel, and dominant spotting mutant mouse embryos. Dev. Biol. 23:297–309.

17. Mayer, T.C. 1973. Site of gene action in steel mice: analysis of the pigment defect by mesoderm–ectoderm recombinations. J. Exp. Zool. 184:345–352.

18. McCoshen, J.A., and D.J. McCallion. 1975. A study of the primordial germ cells during their migratory phase in steel mutant mice. Experientia 31:589–590.

19. Miller, D.S. 1963. Coat color and behavior in mutations in inbred mice under chronic low-level γ-irradiation. Radiat. Res. 19:184–185.

20. Murphy, E.D. 1977. Effects of mutant steel alleles on leukemogenesis and life span in the mouse. J. Natl. Cancer Inst. 58:107–110.

21. Owens, J. 1972. Mouse News Lett. 47:61.

22. Ploemacher, R.E., W.J. Molendijk, N.H.C. Brons, and H. de Ruiter. 1986. Defective support of Sl/Sl^d splenic stroma for humoral regulation of stem cell proliferation. Exp. Hematol. 14:9–15.

23. Poole, T.W., and W.K. Silvers. 1979. Capacity of adult steel (Sl/Sl^d) and dominant spotting (W/W^v) mouse skin to support melanogenesis. Dev. Biol. 72:398–400.

24. Ruscetti, F.W., D.R. Boggs, B.J. Torok, and S.S. Boggs. 1976. Reduced blood and marrow neutrophils and granulocytic colony-forming cells in Sl/Sl^d mice. Proc. Soc. Exp. Biol. Med. 152:398–402.

25. Russell, E.S. 1970. Abnormalities of erythropoiesis associated with mutant genes in mice. In A.L. Gordon, ed., Regulation of Hematopoiesis, Vol. I, 649–675. Appleton-Century-Crofts, New York.

26. Sarvella, P.A., and L.B. Russell. 1956. Steel, a new dominant gene in the house mouse. J. Hered. 47:123–128.

27. Schaible, R.H. 1961. Mouse News Lett. 24:38.

28. Schaible, R.H. 1963. Mouse News Lett. 29:48–49.

29. Shibata, Y., and A. Volkman. 1985. The effect of hemopoietic microenvironment on splenic suppressor macrophages in congenitally anemic mice of genotype Sl/Sl^d. J. Immunol. 135:3905–3910.

30. Wolf, N.S. 1978. Dissecting the hematopoietic microenvironment. II. The kinetics of the erythron in the Sl/Sl^d mouse and the dual nature of its anemia. Cell Tissue Kinet. 11:325–334.

sla, sex-linked anemia, Chr X

Probably radiation induced (3). Hemizygous males and homozygous females can be recognized at birth by their pale color and small size. They have a severe hypochromic anemia which tends to disappear with age and which can be cured by parenteral administration of iron. The anemia is due to a defect in intestinal iron transport. Iron uptake by the absorbing cells of the mucosa is normal, but iron transfer out of the mucosal cells is defective. The nature of the defect is unknown, but it may be associated with deficiency of a transferrin-like protein. The intestinal transport defect is not present at birth but develops by 12 to 15 days. Paradoxically, the defect in absorption is greater on high-than on low-iron diets. It appears to affect absorption of inorganic iron more severely than that of organic iron compounds (1). The anemia of newborn mice is due to a defect in placental iron transport similar to that in the intestine (4). Control of rate of iron uptake may be influenced by stores of iron in the absorbing cells, which acquired the iron from the circulation while they were still undifferentiated residents of the deep crypts. In the crypt cells of *sla*/Y mice the amount of iron deposited after intravenous iron administration is greater than normal, either because of increased avidity or decreased return to plasma. These abnormal stores of iron might carry an abnormal message into the absorbing cells (2).

References

1. Bannerman, R.M. 1976. Genetic defects of iron transport. Fed. Proc. 35:2281–2285.

2. Bedard, Y.C., P.H. Pinkerton, and G.T. Simon. 1976. Uptake of circulating iron by the duodenum of normal mice and mice with altered iron stores, including sex-linked anemia. High resolution radioautographic study. Lab. Invest. 34:611–615.

3. Falconer, D.S., and J.H. Isaacson. 1962. The genetics of sex-linked anaemia in the mouse. Genet. Res. 3:248–250.

4. Kingston, P.J., C.E.M. Bannerman, and R.M. Bannerman. 1978. Iron deficiency anaemia in newborn *sla* mice: a genetic defect of placental iron transport. Br. J. Haematol. 40:265–276.

Slf

Slf, sex-linked fidget, semidominant, Chr X

Found in an experiment searching for sex-linked recessive lethals. Hemizygous males show a slight up and down head-shaking, most easily identifiable around weaning age. In a few heterozygous females, behavior resembles that of hemizygous males, but in most it is normal (1).

References
1. Phillips, R.J.S., and G. Fisher. 1978. Mouse News Lett. 58:44.

Slk

See *dl* locus.

Slp locus

See *H-2S*.

slt, slaty (slt was formerly used for *bg*), recessive, Chr 14

Arose in a non-inbred stock carrying *ld^J* and *mg*. In *a/a slt/slt* mice the coat color is slightly diluted, and the ears are slightly yellowish (1).

References
1. Green, M.C. 1972. Mouse News Lett. 47:36.

sm, syndactylism, recessive, Chr 12

Arose spontaneously in the A/Fa inbred strain (1). There is considerable preweaning mortality of homozygotes. Fertility is about normal in surviving males but low in females. Homozygotes have all four feet affected. The third and fourth digits are always syndactylous and the first and second sometimes so. Digits of the forefeet are usually joined by soft tissues only; digits of the hindfeet by fusions of cartilage and bone. Many homozygotes have tail kinks. The defect can be traced to hyperplasia of the apical ectodermal ridge of the limbs, of part of the limb epidermis, and of the epidermis of the distal part of the tail (2). Grüneberg concluded that the hyperplastic apical ectodermal ridge stimulated mesodermal overgrowth which led to enlargement and deformation of the footplate. This caused the middle digits to crowd together and become fused. Milaire (4) noted that hyperplasia of the apical ectodermal ridge was confined to the preaxial portions and found no evidence for enlargement of the footplate. He postulated that fusion of the digits was due to abnormal mechanical rigidity of the hyperplastic apical ectodermal ridge that failed to allow normal expansion of the growing mesoderm. *sm* is on Chr 12, 9 cM from *Pre-1* (3).

References
1. Grüneberg, H. 1956. Genetical studies on the skeleton of the mouse. XVIII. Three genes for syndactylism. J. Genet. 54:113–145.
2. Grüneberg, H. 1960. Genetical studies on the skeleton of the mouse. XXV. The development of syndactylism. Genet. Res. 1:196–213.
3. Hawes, N.L., S.H. Langley, and T.H. Roderick. 1980. Location of syndactylism on Chr 12. Mouse News Lett. 63:20.
4. Milaire, J. 1967. Histochemical observations on the developing foot of normal, oligosyndactylous (*Os/+*) and syndactylous (*sm/sm*) mouse embryos. Arch. Biol. 78:223–288.

smc, spondylo-metaphyseal chondrodysplasia, recessive

Arose at the MRC Radiobiology Unit at Harwell. Homozygotes are viable and the males are fertile, the females sterile. Expression in homozygotes is variable. They are generally slightly dwarfed with short thick tails and short limbs, and the toes of the hind feet may be flexed laterally. The long bones are bowed with enlarged epiphyses. Histologically, there is complete lack of alignment and ordering of calcified cells prior to ossification (2). The formation of growth-plate cartilage is disrupted, and secondary centers of ossification appear very late (3). The cartilage matrix stains darkly but unevenly with Alcian blue. The matrix is abnormally rich in collagen fibers which often form a sheath around the chondrocytes. Collagen balls—spherical inclusions of collagen fibers—are a unique and characteristic feature of the matrix. Chondrocytes contain large amounts of lipid. *smc* somewhat resembles cartilage matrix deficiency (*cmd*) and the two genes may be alleles (1).

References
1. Johnson, D.R. 1984. The ultrastructure of mandibular condylar and rib cartilage from mice carrying the spondylo-metaphyseal chondrodysplasia (*smc*) gene. J. Anat. 138:463–470.
2. Phillips, R.J.S., G. Fisher, and J.F. Loutit. 1980. Mouse News Lett. 62:47.
3. Thurston, M.N., D.R. Johnson, and N.F. Kember. 1985. Cell kinetics of growth cartilage in spondylo-metaphyseal chondrodysplasia (*smc*) mice. J. Anat. 140:435–445.

Smg-1 locus

See *Prt-4* locus.

Smg-2 locus

See *Prt-5* locus.

Smst locus, somatostatin, Chr 16

This locus codes for somatostatin (SST), the growth hormone release-inhibiting hormone. The locus was first found to be on Chr 16 using mouse–Chinese hamster somatic cell hybrids and a human genomic clone coding for SST (1). Later, a *Bst*EII DNA fragment length variant was found in the Czech II strain derived from wild *M. m. musculus* from Czechoslovakia. The variant was used in a cross of Czech II with BALB/c to locate *Smst* about 5 cM distal to *Igl-1* on Chr 16. The fragments in Czech II are 6.8 and 4.0 kb; in BALB/c they are 6.8 and 1.9 kb (2).

References

1. Naylor, S.L., A.Y. Sakaguchi, W.W. Chin, J. Jacobs, L.P. Shen, W.J. Rutter, and P.A. Lalley. 1984. Chromosomal mapping of mouse hormone genes: somatostatin, calcitonin, thyrotropin alpha subunit, and luteinizing hormone beta subunit. Cytogenet. Cell Genet. 37:551 (Abstr.).
2. Reeves, R.H., D. Gallahan, B.F. O'Hara, R. Callahan, and J.D. Gearhart. 1987. Genetic mapping of *Prm-1*, *Igl-1*, *Smst*, *Mtv-6*, *Sod-1* and *Ets-2*, and localization of the Down syndrome-region of mouse chromosome 16. Cytogenet. Cell Genet. 44:76–81.

sno, snubnose, recessive, Chr 4

Arose in the K/Gw strain inbred from Swiss albino mice. Homozygotes are recognizable at 2 weeks of age by their short noses. Spina bifida occulta occurs in the lumbar and often in the posterior thoracic and sacral vertebrae. There is great variation in severity, and occasionally kyphosis occurs (1).

References

1. Hollander, W.F. 1976. Genetic spina bifida occulta in the mouse. Am. J. Anat. 146:173–180.

Soa locus, sucrose octaacetate aversion

This locus controls presence or absence of aversion to sucrose octaacetate (SOA) (2, 7), strychnine (3), glucose pentaacetate (4), and other acetylated monosaccharides (5). The absence of aversion is due to inability to taste these substances, not to a preference for them (1). The threshold for response to SOA recorded in the glossopharyngeal and chorda tympani nerves is several orders of magnitude lower in tasters than in non-tasters (6). The allele Soa^a determines ability to taste these substances and aversion to consuming them in drinking water; it occurs in the SWR strain. The allele Soa^b determines inability to taste these substances and no aversion to them; it occurs in all other strains tested and in wild mice from two locations. Non-tasters may even show a slight preference for SOA at appropriate low concentrations. Heterozygotes are tasters, but may show slightly less aversion to these bitter substances than SWR mice (2).

References

1. Harder, D.B., G. Whitney, P. Frye, J.C. Smith, and M.E. Rashotte. 1984. Strain differences among mice in taste psychophysics of sucrose octaacetate. Chem. Senses 9:311–323.
2. Lush, I.E. 1981. The genetics of tasting in mice. I. Sucrose octaacetate. Genet. Res. 38:93–95.
3. Lush, I.E. 1982. The genetics of tasting in mice. II. Strychnine. Chem. Senses 7:93–98.
4. Lush, I.E. 1983. More about tasting in mice. Genet. Res. 41:306 (Abstr.).
5. Lush, I.E. 1986. The genetics of tasting in mice. IV. The acetates of raffinose, galactose and β-lactose. Genet. Res. 47:117–123.
6. Shingai, T., and L.M. Beidler. 1985. Interstrain differences in bitter taste responses in mice. Chem. Senses 10:51–55.
7. Warren, R.P., and R.C. Lewis. 1970. Taste polymorphism in mice involving a bitter sugar derivative. Nature 227:77–78.

soc, soft coat, recessive, Chr 3

Arose spontaneously in the BALB/cJ strain. Homozygotes are identifiable at birth by their sparse slightly curled whiskers. Eruption of the first coat may be slightly delayed. The hairs tend to stick together in small tufts and stand out untidily from the body surface. Hairs of all four types are thin with occasional caliber irregularities. The epidermis is morphologically abnormal at birth, the cells of the stratum germinativum being more flattened than normal and having more distinct cell boundaries. The thinning out of the epidermis which normally occurs at 6 or 7 days is delayed by 4 or 5 days, and there is delay in formation of the interruptions in the stratum germinativum giving rise to the hair canals. The hair bulbs are shorter than normal, and the dermal papillae are small. When the abnormally thin hairs, formed in the small hair bulbs, approach the surface, they are delayed in erupting by the lack of hair canals, and kinks and irregularities of the hair shafts result (1).

References

1. Trigg, M.J. 1972. Hair growth in mouse mutants affecting coat texture. J. Zool. Lond. 168:165–198.

Sod-1 locus, superoxide dismutase-1, Chr 16

This locus codes for the cytoplasmic form of superoxide dismutase (SOD-1; E.C. 1.15.1.1). It was first shown to be on the distal part of Chr 16 by use of mouse–Chinese hamster somatic cell hybrids in which the mouse and Chinese hamster enzymes could be distinguished by isoelectric focusing and the proximal and distal parts of the chromosome distinguished by means of a Chr 16 translocation (3). Womack and Ashley (6) used an allelic variant to confirm this location but did not describe the variant. Later, a *Bst*EII DNA fragment length variant was found in the Czech II strain derived from wild *M. m. musculus* from Czechoslovakia. Czech II mice have fragments of 4.4 and 2.0 kb; BALB/c mice have fragments of 2.3, 2.1, and 2.0 kb. This difference was used in a cross of Czech II with BALB/c to locate *Sod-1* about 24 cM distal to *Mtv-6* on Chr 16 (4). In wild populations in Denmark, Selander *et al.* (5) reported electrophoretic variation of an enzyme thought to be indophenol oxidase. However, it was later realized (2) that the detection method used had revealed variation in SOD-1, then a newly discovered enzyme. Heterozygotes for the electrophoretic variants did not have an intermediate dimeric band, in contrast to the dimeric structure reported for SOD-1 in humans. SOD-1 activity differences between inbred strains occur but no genetic studies of these differences have been reported (1).

References

1. Bloor, J.H., D. Holtz, J. Kaars, and D.J. Kosman. 1983. Characterization of superoxide dismutase (SOD-1 and SOD-2) activities in inbred mice: evidence for quantitative variability and possible nonallelic SOD-1 polymorphism. Biochem. Genet. 21:349–364.
2. Creagan, R., J. Tischfield, F. Ricciuti, and F.H. Ruddle. 1973. Chromosome assignments of genes in man using mouse–human somatic cell hybrids: mitochondrial superoxide dismutase (indophenol oxidase-B, tetrameric) to chromosome 6. Humangenetik 20:203–209.
3. Francke, U., and R.T. Taggart. 1979. Assignment of the gene for superoxide dismutase (*Sod-1*) to a region of chromosome 16 and of *Hgprt* to a region of the X chromosome in the mouse. 1979. Proc. Natl. Acad. Sci. USA 76:5230–5233.
4. Reeves, R.H., D. Gallahan, B.F. O'Hara, R. Callahan, and J.D. Gearhart. 1987. Genetic mapping of *Prm-1*, *Igl-1*, *Smst*, *Mtv-6*, *Sod-1* and *Ets-2*, and localization of the Down syndrome-region of mouse chromosome 16. Cytogenet. Cell Genet. 44:76–81.
5. Selander, R.K., W.G. Hunt, and S.Y. Yang. 1969. Protein polymorphism and genic heterozygosity in two European subspecies of the house mouse. Evolution 23:379–390.
6. Womack, J.E., and P.S. Ashley. 1984. Mouse News Lett. 70:100.

Sod-2 locus, superoxide dismutase-2, Chr 17

This locus codes for the mitochondrial form of superoxide dismutase (E.C. 1.15.1.1). No electrophoretically detectable genetic variants have been found among 40 inbred strains and 136 wild mice examined. The locus was found to be on Chr 17 by use of clones derived from mouse–Chinese hamster hybridoma cells, of which some carried Chr 17 and others did not. There was complete concordance between loss of Chr 17 and loss of SOD-2 activity (1).

References

1. Szymura, J.M., M.R. Wabl, and J. Klein. 1981. Mouse mitochondrial superoxide dismutase locus is on chromosome 17. Immunogenetics 14:231–240.

Sol locus

See *Hba* complex.

Sp locus, Chr 1

Mutations at this locus occur frequently (2).

Sp, splotch, semidominant. Arose spontaneously in the C57BL inbred strain (5). Heterozygotes show white spotting on the belly and occasionally on the back, feet, and tail. Homozygotes die at 13 days of gestation with malformations which include rachischisis in the lumbosacral region and frequently in the region of the hindbrain, overgrowth of neural tissue, reduction or absence of spinal ganglia and their derivatives, and abnormal tail morphology. Neural crest and adjacent regions of *Sp/Sp* embryos fail to develop any pigment when transplanted to chick coelom or to the anterior chamber of the mouse eye (1). Gap junctional vesicles occur much more frequently in cells of the neural tube of 9-day *Sp/Sp* embryos than in those of normal sibs (7). At 10 and 11 days of gestation, the mesencephalon of *Sp/Sp* embryos shows increased numbers of mitoses and a prolonged cell generation time (6). In mice bearing splotch combined with the gene fidget which causes decreased incidence of neural tube mitoses (*Sp/Sp fi/fi*), the two genes counteract each other and there is a decreased incidence of neural tube defects (4). The teratogen, retinoic acid, can cause an increase or decrease of neural tube defects in offspring of *Sp/+* × *Sp/+* matings depending on the day of administration, presumably through effects on the formation and rate of closure of the anterior and posterior neuropores (3).

Sp^d, delayed splotch, semidominant. Arose spontaneously in the C57BL/6J strain. Heterozygotes have a white belly spot. Homozygotes are similar to *Sp/Sp* but have caudal rachischisis only, and survive to birth. *Sp*^{d}/*Sp* mice resemble *Sp*^{d}/*Sp*^{d} (2).

References

1. Auerbach, R. 1954. Analysis of the developmental effects of a lethal mutation in the house mouse. J. Exp. Zool. 127:305–329.
2. Dickie, M.M. 1964. New splotch alleles in the mouse. J. Hered. 55:97–101.
3. Kapron-Bras, C.M., and D.G. Trasler. 1985. Reduction in the frequency of neural tube defects in splotch mice by retinoic acid. Teratology 32:87–92.
4. Konyukhov, B.V., and O.V. Mironova. 1979. Interaction of the mutant genes splotch and fidget in mice. Soviet Genet. 15:407–411.
5. Russell, W.L. 1947. Splotch, a new mutation in the house mouse, Mus musculus. Genetics 32:102 (Abstr.).
6. Wilson, D.B. 1974. Proliferation in the neural tube of the splotch (*Sp*) mutant mouse. J. Comp. Neurol. 154:249–256.
7. Wilson, D.B., and L.A. Finta. 1979. Gap junctional vesicles in the neural tube of the splotch (*Sp*) mutant mouse. Teratology 19:337–340.

spa, spastic, recessive, Chr 3

Found in a hybrid stock by Chai (1). Preweaning mortality is high and breeding performance is low. Homozygotes can usually be recognized at 14 days of age but sometimes not until 5 or 6 weeks. They show spastic symptoms which sometimes occur spontaneously and can always be induced by handling. The spasms consist of rapid tremor, stiffness of posture, and difficulty in righting when placed on the back (1). No anatomical pathology occurs in either muscle or central nervous system except that herniated intervertebral discs and cysts of the leptomeninges occur, most severely in the lumbar region, possibly as a result of traumatic injuries consequent on spasticity and tremor (3). The spastic symptoms can be markedly alleviated by intraperitoneal injection of aminooxyacetic acid, an inhibitor of γ-aminobutyric acid transaminase (GABA-T), but not by Dilantin or trimethadione (2, 4). There is a decrease

to less than 20 per cent of normal in the receptor for glycine, the major inhibitory neurotransmitter in the spinal cord and brainstem. Evidence from strychnine binding and the effects of strychnine administered to normal mice indicates that the glycine receptor deficiency is probably the cause of the behavioral symptoms (5, 6).

References

1. Chai, C.K. 1961. Hereditary spasticity in mice. J. Hered. 52:241–243.
2. Chai, C.K., E. Roberts, and R.L. Sidman. 1962. Influence of aminooxyacetic acid, a γ-aminobutyrate transaminase inhibitor, on hereditary spastic defect in the mouse. Proc. Soc. Exp. Biol. Med. 109:491–495.
3. Meier, H., and C.K. Chai. 1970. Spastic, an hereditary neurological mutation in the mouse characterized by vertebral arthropathy and leptomeningeal cyst formation. Exp. Med. Surg. 28:24–38.
4. Roberts, E., and K. Kuriyama. 1968. Biochemical–physiological correlations in studies of the γ-aminobutyric acid system. Brain Res. 8:1–35.
5. White, W.F. 1985. The glycine receptor in the mutant mouse spastic (*spa*): strychnine binding characteristics and pharmacology. Brain Res. 329:1–6.
6. White, W.F., and A.H. Heller. 1982. Glycine receptor alteration in the mutant mouse *spastic*. Nature 298:655–657.

Sparc locus, secreted acidic cysteine-rich glycoprotein, Chr 11

This locus codes for a 43 kDa glycoprotein (SPARC) homologous to a 43 kDa bovine secreted endothelial cell protein. The gene is expressed at high levels in all extraembryonic tissues and at lower levels in the fetus, at high levels in parietal endoderm differentiating from F9 teratocarcinoma cells, and at lower levels in adult lung, adrenal, testis, and ovary. SPARC is secreted by some cell lines but its production is markedly reduced after transformation. A DNA polymorphism was detected with three different restriction enzymes. The allele *Sparc*^{b} determines a DNA pattern found in C57BL/6, AKR/J, and C57L; the allele *Sparc*^{d} determines a pattern found in strains DBA/2, C3H/He, and SWR. By use of mouse–Chinese hamster and mouse–rat somatic cell hybrids and by *in situ* hybridization, *Sparc* was found to be near the center of Chr 11 (1).

References

1. Mason, I.J., D. Murphy, M. Münke, U. Francke, R.W. Elliott, and B.L.M. Hogan. 1986. Development and transformation-sensitive expression of the *Sparc* gene on mouse chromosome 11. EMBO J. 5:1831–1837.

spc, sparse coat, recessive, Chr 14

Arose spontaneously in a stock carrying *jg* and *bf*. Homozygotes have fur that is sparse and patchy. They are usually recognizable at about 2 weeks when the coat beings to appear slightly rougher than normal. The whiskers are sometimes short. By 3 weeks, affected animals are usually recognizable with certainty. Viability and fertility of homozygotes are often considerably reduced (1).

References

1. Green, M.C. 1974. Mouse News Lett. 51:23.

spf locus, Chr X

spf, sparse-fur, incompletely recessive. Found in a descendant several generations removed from an irradiated male. Hemizygous males and homozygous females are detectable at 5 to 7 days by their small size, absence or relative paucity of fur, and wrinkled skin. They are extremely variable in appearance thereafter, some being almost fully furred and of normal size, and some remaining dwarfed and hairless for several weeks after weaning. Most heterozygous females are normal in appearance, but a few have the mutant phenotype. Hemizygous males and homozygous females frequently have urinary bladder stones composed mostly of orotic acid. Affected males have severely reduced activity of liver ornithine transcarbamylase (ornithine carbamoyltransferase, OTC; E.C. 2.1.3.3). Heterozygous females have intermediate activity (3). At pH 8.0, the optimum pH for normal OTC, affected males have only 25 per cent of normal activity of OTC. At pH 9, OTC activity is 150 per cent of normal with an increase in amount of OTC protein and a higher specific activity, suggesting that *spf* codes for an abnormal OTC molecule (1). This was shown to be the case by Veres *et al*. (7) who found that the *spf* mutation is the result of a single C to A base substitution that changes a histidine to an asparagine residue at amino acid 117. Defective activity of OTC in mutant mice occurs in the colon and small intestine as well as in liver (5). The *spf* defect resembles one form of OTC deficiency in man (2).

spf^*ash*^, abnormal skin and hair. Occurred in a stock at Purdue University. It closely resembles *spf*, except that heterozygous females appear to be more variable than *spf*/+ females, with many showing some hair loss, and some resembling homozygotes (4). OTC activity in liver is reduced to 8 per cent of normal in affected males and to 55 per cent of normal in heterozygous females but the enzyme protein is identical to that of controls. OTC messenger RNA from mutant livers apparently synthesizes two precursor polypeptides, one normal and the other larger than normal. Both are taken up by the liver mitochondria but only the normal one is assembled into the active trimeric enzyme (6). The *spf*^*ash*^ defect resembles a form of OTC deficiency in man different from the form that *spf* resembles (2).

References

1. Briand, P., L. Cathelineau, P. Kamoun, D. Gigot, and M. Penninckx. 1981. Increase of ornithine transcarbamylase protein in sparse-fur mice with ornithine transcarbamylase deficiency. FEBS Lett. 130:65–68.
2. Briand, P., B. Francois, D. Rabier, and L. Cathelineau. 1982. Ornithine transcarbamylase deficiencies in human males: kinetic and immunochemical classification. Biochim. Biophys. Acta 704:100–106.
3. DeMars, R., S.L. LeVan, B.L. Trend, and L.B. Russell. 1976. Abnormal ornithine carbamoyltransferase in mice having the sparse-fur mutation. Proc. Natl. Acad. Sci. USA 73:1693–1697.
4. Doolittle, D.P., L.L. Hulbert, and C. Cordy. 1974. A new allele of the sparse fur gene in the mouse. J. Hered. 65:194–195.
5. Qureshi, I.A., J. Letarte, and R. Ouellet. 1985. Expression of ornithine transcarbamylase deficiency in the small intestine and colon of sparse-fur mutant mice. J. Pediatr. Gastroenterol. Nutr. 4:118–124.
6. Rosenberg, L.E., F. Kaloucek, and M.D. Orsulak. 1983. Biogenesis of ornithine transcarbamylase in *spf*^*ash*^ mutant mice: two cytoplasmic precursors, one mitochondrial enzyme. Science 222:426–428.
7. Veres, G., R.A. Gibbs, S.E. Scherer, and C.T. Caskey. 1987. The molecular basis of the sparse fur mouse mutation. Science 237:415–417.

sph locus, Chr 1

This is probably the structural locus for the α-subunit of spectrin, the major skeletal protein of red cell membranes. By use of a cDNA clone to screen somatic cell hybrids, the α-spectrin gene (*Spna-1*) was shown to be on Chr 1 (6), the known location of *sph* (4). See *Spna-1* locus. Several mutations causing spherocytic anemia have been described at this locus.

sph, spherocytosis, recessive. Arose in an unpedigreed C3H stock. Homozygotes are extremely pale at birth and acquire a yellow color during the first few hours thereafter. They die within the first 24 hours. Examination of the blood shows numerous spherocytes accompanied by all the evidence of hemolytic anemia (7). Erythropoiesis is accelerated, but erythrocyte differentiation is without histological abnormality (5). α-spectrin is virtually absent from the red cell membrane of *sph/sph* mice (4). Protoporphyrin concentration is higher in red cells of *sph*/+ than of +/+ mice, and about 80 per cent of heterozygotes can be distinguished from their +/+ sibs by this method (9).

sph^{ha} (formerly *ha*), hemolytic anemia, recessive. Arose spontaneously in the DBA/1J inbred strain. Homozygotes show neonatal hypochromic anemia, microcytosis, and jaundice. Many die within a few days after birth, but occasional animals survive 6 months or more. All are sterile (2). Affected embryos can be identified by their pale color as early as 14 days of gestation, but the jaundice appears after birth. Transplantation studies have shown that the defect resides in the erythropoietic cells (3). The red cell membrane contains very little α-spectrin; however, about six times the normal amount of α-spectrin is synthesized by *sph^{ha}/sph^{ha}* reticulocytes but fails to accumulate in the membrane. The *sph^{ha}* mutation may produce a defective molecule that is not capable of stable binding to the membrane skeleton (4). Heterozygotes show a mild and well compensated anemia with a moderate reduction in red cell survival (1, 8). They have a higher concentration of protoporphyrin in red cells than +/+ mice, and about 50 per cent of *sph^{ha}/+* mice are identifiable by this method (9).

sph^{2Bc}, spherocytosis-2Bc, recessive. Arose spontaneously is a stock of mixed origin at the University of British Columbia. Homozygotes are pale at birth and become yellow-orange within a few hours. About one-fourth survive to weaning and one-third of these to 6 months. A few males and females have bred (10). Like *sph*, *sph^{2Bc}* causes absence of α-spectrin in the red cell membrane skeleton (4).

sph^{2J}, spherocytosis-2J, recessive. Arose in the C3HeB/FeJ strain. It is like *sph* in the absence of α-spectrin in the red cell membrane skeleton (4).

References

1. Bannerman, R.M., J.A. Edwards, and P.H. Pinkerton. 1973. Hereditary disorders of the red cells in animals. Prog. Hematol. 8:131–179.
2. Bernstein, S.E. 1963. Analysis of gene action and characterization of a new hematological abnormality, hemolytic anemia. Proc. XI Int. Cong. Genet. 1:186 (Abstr.).
3. Bernstein, S.E. 1969. Hereditary disorders of the rodent erythron. *In* J.R. Lindsay, ed., Genetics in Laboratory Animal Medicine, Natl. Acad. Sci. Publ. 1697:9–33. Washington, DC.
4. Bodine, D.M. IV, C.S. Birkenmeier, and J.E. Barker. 1984. Spectrin deficient inherited hemolytic anemias in the mouse: characterization by spectrin synthesis and mRNA activity in reticulocytes. Cell 37:721–729.
5. Brookhoff, D., L. Maggio-Price, S. Bernstein, and L. Weiss. 1982. Erythropoiesis in *ha/ha* and *sph/sph* mice, mutants which produce spectrin-deficient erythrocytes. Blood 59:646–651.
6. Huebner, K., A.P. Palumbo, M. Isobe, C.A. Kozak, S. Monaco, G. Rovera, C.M. Croce, and P.J. Curtis. 1985. The α-spectrin gene is on chromosome 1 in mouse and man. Proc. Natl. Acad. Sci. USA 82:3790–3793.
7. Joe, M., J.M. Teasdale, and J.R. Miller. 1962. A new mutation (*sph*) causing neonatal jaundice in the house mouse. Can. J. Genet. Cytol. 4:219–225.
8. Russell, E.S., and S.E. Bernstein. 1966. Blood and blood formation. *In* E.L. Green, ed., Biology of the Laboratory Mouse, 2nd ed., 351–372. McGraw-Hill, New York.
9. Sassa, S., and S.E. Bernstein. 1978. Studies of erythrocyte protoporphyrin in anemic mutant mice: use of modified hematofluorometer for the detection of heterozygotes for hemolytic disease. Exp. Hematol. 6:479–487.
10. Unger, A.E., M.J. Harris, S.E. Bernstein, J.C. Falcone, and S.E. Lux. 1983. Hemolytic anemia in the mouse: report of a new mutation and clarification in its genetics. J. Hered. 74:88–92.

Spi-1, *Spi-2* gene families, Chr 12

These closely related multigene families contain the genes controlling the major plasma protease inhibitors, α₁-protease inhibitor (also called α₁-antitrypsin) and contrapsin, respectively. The families are very closely linked and are located on Chr 12 in the same position as the closely linked loci *Pre-1* and *Pre-2* (prealbumin-1 and -2) (1, 2). It is possible that the prealbumin proteins are identical with the protease inhibitors but the exact relationship of *Spi-1* and *Spi-2* to *Pre-1* and *Pre-2* is not known. See *Pre-1*, *Pre-2*.

Spi-1 family, serum protease inhibitor complex-1. This family contains the gene for α₁-protease inhibitor. Two variants, recognized by Southern blot patterns with various restriction enzymes, are known, one in strains BALB/c and DBA/2 and the other in strain C57BL/6. The family contains at least four genes in BALB/c and five in C57BL/6 (1).

Spi-2 family. This family contains a structural and a regulatory gene for contrapsin, an inhibitor of trypsin-like proteins.

Spi-2s locus, serum protease inhibitor complex-2, structural. There are two variants at this locus, recognized by Southern blot patterns with various restriction enzymes, by Northern blot patterns of liver RNA, and by polyacrylamide gel electrophoresis of *in vitro*-synthesized contrapsin polypeptide. One variant occurs in BALB/c, DBA/2, C57L, AKR, and a number of other strains, the other variant in C3H/He, C57BL/6, A/J, and several other strains. The family contains at least three genes in both BALB/c and C57BL/6 (1).

Spi-1, Spi-2

Spi-2r locus, serum protease inhibitor complex-2 regulator. The multiple RNA transcripts of *Spi-2s* in liver are synthesized constitutively in some strains and only after inflammation in others. The strain distribution pattern of this difference in 21 C57Bl/6 × DBA/2 recombinant inbred strains corresponds exactly to that of *Spi-2s*, indicating that this regulatory gene, *Spi-2r*, is closely linked to *Spi-2s* and is *cis*-acting (1).

References

1. Hill, R.E., P.H. Shaw, R.K. Barth, and N.D. Hastie. 1985. A genetic locus closely linked to a protease inhibitor gene complex controls level of multiple RNA transcripts. Mol. Cell. Biol. 5:2114–2122.
2. Wilcox, F.H. 1975. Genetic variation in plasma prealbumin of the house mouse. J. Hered. 66:19–22.

Spl locus, plasma serotonin level, Chr 1

This locus controls level of serotonin in plasma. The allele *Spl^h* determines high level (1.7-2.2 µg/ml) and occurs in strain C57BL/6By; the allele *Spl^l* determines low level (0.6-0.8 µg/ml) and occurs in strain BALB/cBy. *Spl* is closely linked to *H-35* (provisionally H[w56]) and to *H-25* (1), which are known to be on Chr 1 (D.W. Bailey, personal communication).

References

1. Eleftheriou, B.E., and D.W. Bailey. 1972. A gene controlling plasma serotonin levels in mice. J. Endocrinol. 55:225–226.

Spl-1 locus, spleen antigen-1

This locus controls presence of an antigen on adult spleen cells. The allele *Spl-1^a* determining presence of the antigen occurs in the CBA/J strain; the allele *Spl-1^b* determining absence of the antigen occurs in the C3H/HeJ strain. The antigen is recognized by its ability to stimulate proliferation of neonatal thymus cells of CBA and C3H mice in mixed lymphocyte cultures. The reactivity disappears from the thymus cells of CBA mice a few days after birth but not from the thymus cells of C3H mice. It is thought that the loss of reactivity in the CBA thymus to an antigen present in the CBA liver may involve the mechanism that determines self-tolerance (1).

References

1. Carter, J., and T.G. Wegmann. 1973. Mendelian segregation of a tolerance-inducing self-antigen in the spleen. Cell. Immunol. 7:402–409.

spm, sphingomyelinosis, recessive

Arose in the C57BL/Ks strain. On this strain background, homozygotes can be recognized at about 7 weeks of age when they begin to lose weight and show abnormal behaviour consisting of tremor and ataxic gait. They continue to lose weight and the tremor and ataxia become more severe until death, which occurs at about 12 to 14 weeks (1). There is enlargement of the liver and spleen which is present as early as 4 weeks on the C57BL/Ks background but does not occur on some other backgrounds (2, 3). Large areas of liver and spleen are replaced by foam cells. Purkinje cells in the cerebellum are severely depleted. Sphingomyelin and free cholesterol are markedly elevated in liver and spleen but not in brain. Sphingomyelinase activity is reduced to about 30 per cent of normal in liver, 50 per cent in spleen, and 70 to 80 per cent in brain. In liver, the altered levels of sphingomyelin, cholesterol, and sphingomyelinase are present from 4 weeks until death (2). Heterozygotes have normal lipid concentrations and sphingomyelinase activity (1). The *spm* condition closely resembles human type A or C Niemann–Pick disease (3).

References

1. Miyawaki, S., S. Matsuoka, T. Sakiyama, and T. Kitagawa. 1982. Sphingomyelinosis, a new mutation in the mouse. A model of Niemann–Pick disease in humans. J. Hered. 73:257–263.
2. Miyawaki, S., S. Matsuoka, T. Sakiyama, and T. Kitagawa. 1983. Time course of hepatic lipids accumulation in a strain of mice with an inherited deficiency of sphingomyelinase. J. Hered. 74:465–468.
3. Miyawaki, S., H. Yoshida, and S. Mitsuoka. 1986. A model for Niemann–Pick disease. Influence of genetic background on disease expression in *spm/spm* mice. J. Hered. 77:379–384.

Spna-1 locus, α spectrin-1, Chr 1

This locus codes for the α subunit of the erythrocyte membrane skeletal protein, spectrin. It is the locus of the spherocytosis (*sph*) mutant alleles. See *sph* locus. A restriction fragment length polymorphism was found among inbred strains by use of a cDNA clone of the mouse α spectrin gene hybridized to *Eco*RI restricted DNA. The *Spna-1^a* allele determines a 5.0-kb fragment and occurs in most inbred strains, including A/J, AKR, BALB/c, C3H/He, and NZB; the allele *Spna-1^b* determines a 7.0-kb fragment and occurs in strains C57BL/6, C57BR/cd, C57L, HRS, and MA/My; a third allele determining a 4.3-kb fragment was found in *Mus spretus*. By use of five sets of recombinant inbred strains and an interspecific backcross between C57BL/6 and *M. spretus*, *Spna-1* was found to be about 3 cM distal to the *Ly-17*, *Mls*, *Apoa-2* group of loci (1).

References

1. Seldin, M.F., H.C. Morse III, L. D'Hoostelaere, J.L. Britten, and A.D. Steinberg. 1987. Mapping of alpha spectrin on distal mouse chromosome 1. Cytogenet. Cell Genet. 45:52–54.

Spnb-1 locus, β spectrin-1, Chr 12

This locus codes for the β subunit of the erythrocyte membrane skeletal protein, spectrin. It was found to be on Chr 12 by use of mouse–Chinese hamster somatic cell hybrids screened with a probe of mouse erythrocyte β spectrin cDNA (1). The mutation jaundiced (*ja*) is known to cause absence of β spectrin in the cell membrane. Its chromosomal location has not been reported, but it may be a mutation at the *Spnb-1* locus. See *ja*.

References

1. Laurila, P., L Cioe, C.A. Kozak, and P.J. Curtis. 1987. Assignment of mouse beta-spectrin gene to chromosome 12. Somat. Cell Mol. Genet. 13:93–98.

sq

See *wa-1*.

sr, spinner, recessive, Chr 9

Arose spontaneously in the C57BL/How inbred strain. Homozygotes show the typical head tossing, circling, deafness, and hyperactivity of the shaker–waltzer mutants. The abnormal behavior can be recognized as early as 7 days. The anomalies of the inner ear consist of degeneration of the organ of Corti and spiral ganglion, of reduction in size of the stria vascularis of the cochlea, and of lesser degeneration of the vestibular part of the labyrinth limited to the saccular macula. Both sexes are fertile, but females are not good mothers (1). *sr* is on Chr 9, probably distal to *Mod-1* (2).

References

1. Deol, M.S., and M.W. Robins. 1962. The spinner mouse. J. Hered. 53:133–136.
2. Fox, S., E.M. Eicher, and S. Reynolds. 1978. Mouse News Lett. 59:50.

Sr-1 locus, spectrotype regulation-1

The antibody response to the inulin determinant of bacterial levan requires presence of the *Igh^a* haplotype (found in BALB/c) of the immunoglobulin heavy chain complex. The isoelectric focusing (IEF) pattern of the IgG anti-inulin antibodies produced in *Igh^a*-bearing mice is regulated by *Sr-1*, a locus not linked to *Igh*. The allele in BALB/c, *Sr-1^a*, determines a single spectrotype by IEF analysis. The allele of C57BL/6, *Sr-1^b*, when introduced into *Igh^a*-bearing mice, determines a more complex pattern. Heterozygotes have a pattern very similar to that of *Sr-1^b*/*Sr-1^b* (1).

References

1. Stein, K.E., C. Bona, R. Lieberman, C.C. Chien, and W.E. Paul. 1980. Regulation of the anti-inulin antibody response by a nonallotype-linked gene. J. Exp. Med. 151:1088–1102.

Src locus (*c-src*), Rous sarcoma oncogene, Chr 2

This is the cellular homolog of the transforming gene of the Rous avian sarcoma virus. The viral oncogene, v-*src*, has been shown to produce a 60 000 MW protein, p60*src*, which phosphorylates proteins, specifically the tyrosine residues (2, 3). Since mouse cells contain easily detectable phosphorylated tyrosine, it is probable that the cellular gene *Src* also has tyrosine phosphorylase activity (3). The *Src* locus was shown to be on Chr 2 by use of mouse–Chinese hamster somatic cell hybrids screened with a viral oncogene probe (4). A DNA restriction site difference at this locus was found between the C57BL/6 and DBA/2 strains. The allele in C57BL/6, *Src^b*, generates two *Hind*III fragments of 16.5 and 14 kb: the allele in DBA/2, *Src^d*, generates a single *Hind*III fragment of 13 kb. By use of BXD and other recombinant inbred strains, *Src* was found to be 2 cM from *Psp* on the distal half of Chr 2 (1) and within 2 cM of a (5).

References

1. Blatt, C., M.E. Harper, G. Franchini, M.N. Nesbitt, and M.I. Simon. 1984. Chromosome mapping of murine c-*fes* and c-*src* genes. Mol. Cell. Biol. 4:978–981.
2. Collet, M.S., and R.L. Erickson. 1978. Protein kinase activity associated with avian sarcoma virus *src* gene product. Proc. Natl. Acad. Sci. USA 75:2021–2024.
3. Hunter, T., and B.M. Sifton. 1980. Transforming gene product of Rous sarcoma virus phosphorylates tyrosine. Proc. Natl. Acad. Sci. USA 77:1311–1315.
4. Sakaguchi, A.Y., P.A. Lalley, B.U. Zabel, R.W. Ellis, E.M. Scolnick, and S.L. Naylor. 1984. Chromosome assignments of four mouse cellular homologs of sarcoma and leukemia virus oncogenes. Proc. Natl. Acad. Sci. USA 81:525–529.
5. Taylor, B.A. and L. Rowe. 1985. Mapping the murine

Src

homolog of the *Src* oncogene and a related DNA sequence. Jackson Lab. Ann. Rep. 56:45–46.

Srch locus, *Src*-homologous sequence, Chr 2

This DNA sequence is homologous with the seven most 3' exons of the v-*src* gene but not with the 5' end of the gene. It is linked to *Src* but at a distance of about 3 cM. *Srch* may have arisen by duplication of the *Src* sequence, but the distance between them is unusually large for such an occurrence (1).

References

1. Taylor, B.A. and L. Rowe. 1985. Mapping the murine homolog of the *Src* oncogene and a related DNA sequence. Jackson Lab. Ann. Rep. 56:45–46.

Srlv-1 locus, susceptibility to RadLV-1

This locus controls susceptibility to leukemogenesis induced by the radiation leukemia virus RadLV. The allele *Srlv-1^s* determines sensitivity to the virus and occurs in the B10.AQR(n4) congenic strain into which it was probably introduced from a stock carrying the translocation *T(2;9)138Ca*. The allele *Srlv-1^r* determines resistance to the virus and occurs in strain C57BL/10 and a number of its *H-2* congenic lines. Susceptibility is dominant, in contrast to an *H-2*-linked locus in which resistance to RadLV is dominant. The *Srlv-1^s* determining susceptibility overrides the resistance conferred by the *H-2*-linked gene. Preliminary data suggest that the *Srlv-1* locus controls leukemogenesis by affecting virus proliferation (1,2).

References

1. Meruelo, D., M. Lieberman, N. Ginzton, B. Deak, and H.O. McDevitt. 1977. Genetic control of radiation leukemia virus-induced tumorigenesis. I. Role of the major histocompatibility complex, *H-2*. J. Exp. Med. 146:1079–1087.
2. Meruelo, D., M. Lieberman, B. Deak, and H.O. McDevitt. 1977. Genetic control of radiation leukemia virus-induced tumorigenesis. II. Influence of *Srlv-1*, a locus not linked to *H-2*. J. Exp. Med. 146:1088–1095.

srn, siren, recessive

Found among descendants of a cross between the SM/J and BUA strains. Most homozygotes die shortly after birth, but some probably die *in utero* (2). Their abnormalities are confined to the posterior end of the body and consist of absence of the hindlimbs, a single median hindlimb, or two partly fused hindlimbs accompanied by abnormalities of the pelvic girdle, kinked or otherwise abnormal tails, atresia ani and underdeveloped intestines, and fusion and other abnormalities of the urogenital system. A small percentage of heterozygotes have twisted tails (1). Craniofacial anomalies, including micrognathia, microstomia, and cleft palate, also occur in many *srn/srn* mice, the frequency being greater in those sirens more severely affected caudally (3). Homozygous embryos were first identified at 11 days of gestation by more ventral than normal position of the hindlimb buds. At this stage there is a single median umbilical artery rather than the normal paired arteries (4).

References

1. Hoornbeek, F.K. 1970. A gene producing symmelia in the mouse. Teratology 1:7–10.
2. Hoornbeek, F.K., and C. Schreiner. 1973. Abnormal segregation and sex ratio and reduced litter size associated with the siren phenotype in mice. Teratology 7:195–198.
3. Orr, B.Y., S.Y. Long, and A.J. Steffek. 1982. Craniofacial, caudal, and visceral anomalies associated with mutant sirenomelic mice. Teratology 26:311–317.
4. Schreiner, C., and F.K. Hoornbeek. 1973. Developmental aspects of sirenomelia in the mouse. J. Morphol. 141:345–358.

Srv locus, sensitivity to RadLV

This locus modifies the slope (1- or 2-hit) and level of sensitivity of the titration curve when B-tropic RadLV leukemia virus is propagated in embryo cells of nonpermissive *Fv-1* type. The dominant allele, *Srv^s*, determines 1-hit kinetics and high sensitivity and occurs in the C57BL/Ka strain; the recessive allele, *Srv^r*, determines 2-hit kinetics and low sensitivity and occurs in the BALB/c and NIH Swiss strains. Control of hitness by the *Srv* locus is expressed only in *Fv-1^n/Fv-1^b* mice, but control of sensitivity is independent of *Fv-1* genotype (1).

References

1. Declève, A., O. Niwa, J. Kojola, and H.S. Kaplan. 1976. New gene locus modifying susceptibility to certain B-tropic murine leukemia viruses. Proc. Natl. Acad. Sci. USA 73:585–590.

Ss locus

See *H-2* complex.

Ssp locus (formerly also *Msp-1*), sex-limited saliva pattern, Chr 17?

This locus influences electrophoretic mobility of a protein in the saliva of adult male mice. The allele *Ssp^s*

determines a slowly migrating band and occurs in the A/J, AKR, CBA/Br, C3H/St, and C57BL/Ks strains; the allele Ssp^f determines a fast band and occurs in the BALB/c, DBA/2, C57BL/6, NZB, and other strains. All females show only the fast phenotype. Males of the slow strains show a fast phenotype before puberty, and both fast and slow bands during puberty. Ssp^s is dominant. In the slow strains, testosterone induces the slow phenotype in females and accelerates acquisition of it in males (2, 3). The protein affected by *Ssp* is a dimer that binds androgen and is produced in the submandibular gland. The change from F to S induced by androgen probably results from an increase in molecular weight of the larger subunit. *Ssp* is probably a regulatory locus for the androgen-binding protein (Abp) whose structure is determined by the *Abp* locus (1, 2). A locus called *Msp-1*, mouse salivary protein-1, which appears to be identical to *Ssp* was described by Ikemoto and Matsushima (4). They found that, in the *H-2* congenic strains B10.A and B10.BR, the *Ssp* allele is that of the donor strain, suggesting that *Ssp* is linked to *H-2* on Chr 17. See *Abp* locus.

References

1. Dlouhy, S.R., and R.C. Karn. 1983. The tissue source and cellular control of the apparent size of androgen binding protein (Abp), a mouse salivary protein whose electrophoretic mobility is under the control of *sex-limited saliva pattern* (*Ssp*). Biochem. Genet. 21:1057–1070.
2. Dlouhy, S.R., and R.C. Karn. 1984. Multiple gene action determining a mouse salivary protein phenotype: identification of the structural gene for androgen binding protein (Abp). Biochem. Genet. 22:657–667.
3. Karn, R., S.R. Dlouhy, K.R. Springer, J.P. Hjorth, and J.T. Nielsen. 1982. Sex-limited genetic variation in a mouse salivary protein. Biochem. Genet. 20:493–504.
4. Ikemoto, S., and Y. Matsushima. 1984. Sex and strain differences of mouse variant salivary proteins. Differentiation 26:55–58.

sst

See *dr*.

st, shaker-short, recessive, probably extinct

Arose in a stock of hairless mice (1). Viability is much reduced. Females were sterile, but a few males bred. Most newborn homozygotes have one or two cerebral hernias covered with skin near the median point of the parieto-occipital suture of the skull. The blebs later dry to a scab. Tail length varies from absence to three-quarters normal length. From 5 days on there is disturbed equilibrium with erratic circus movements and,

in adults, marked lack of co-ordination. In the development of the brain no foramen of Magendie or choroid plexus is formed, leading to disturbed circulation of the cerebrospinal fluid (2). By the 16th day of gestation the lumen of the brain is narrow, and the walls are abnormally thick. About the 17th day there is a rupture of the roof of the mesencephalon and cerebellum, which gives rise to the cerebral hernias. The ear vesicles, as in kreisler (*kr*), are very abnormal in morphology, possibly because of a separation of the vesicles from the inductive influence of the myelencephalon.

References

1. Dunn, L.C. 1934. A new gene affecting behavior and skeleton in the house mouse. Proc. Natl. Acad. Sci. USA 20:230–232.
2. Bonnevie, K. 1936. Abortive differentiation of the ear vesicles following a hereditary brain-anomaly in the "short-tailed waltzing mice." Genetica 18:105–125.

Sta, autosomal striping, semidominant

Found in several offspring of an irradiated female and therefore probably occurred spontaneously. Heterozygotes of both sexes resemble *Ta*/+ females. There are occasional normal overlaps. Heterozygotes are of normal size and fertility. Homozygotes probably die before birth (1).

References

1. Bangham, J.W. 1968. Mouse News Lett. 38:31.

stb, stubby, recessive, Chr 2

Found in a linkage cross. It resembles achondroplasia (*cn*) but is less severe. Homozygotes are recognizable at 4 or 5 days by a domed head and thick short tail. Adults have shorter heads, bodies, and legs than normal, but there is some variation in expression. Most females are fertile and bring up their litters, but males are probably sterile. They produce ample motile sperm, and the reason for their sterility is not clear (1). Cartilage histology is within normal limits, but chondrogenesis stops earlier than in normal animals (2).

References

1. Lane, P.W., and M.M. Dickie. 1968. Three recessive mutations producing disproportionate dwarfing in mice: achondroplasia, brachymorphic, and stubby. J. Hered. 59:300–308.
2. Miller, W.A., and K.L. Flynn-Miller. 1976. Achondroplastic, brachymorphic and stubby chondrodystrophies in mice. J. Comp. Pathol. 86:349–363.

stm, stumpy, recessive

Arose spontaneously in the CBA/FaCam strain. Viability and fertility are both reasonably good. Affected animals are recognizable at birth or a few days afterwards (1). All cartilage bones, as well as the secondary cartilage of the mandibular condyle (5), are affected to a greater or lesser degree. They are usually shortened, and may have widened areas of muscle attachment. The proximal long bones are more severely affected than the distal long bones (2). In cartilage of the epiphyseal plates there is an abnormally wide zone of hypertrophy, a slightly increased mitotic rate, and more chondrocytes per lacunae than normal. Electron microscopy shows many chondrocytes to be in close approximation, some with tight junctions, and there is much interdigitation and folding of the cell membranes of adjacent chondrocytes (3). Growth of the femur and tibia is fairly rapid at first but stops in the 16- to 21-day period (4). The main factor responsible for reduced growth of the bones is low height of the cells of the hypertrophic layer (6).

References

1. Ferguson, J.M., and M.E. Wallace. 1973. Mouse News Lett. 49:23.
2. Ferguson, J.M., M.E. Wallace, and D.R. Johnson. 1978. A new type of chondrodystrophic mutant in the mouse. J. Med. Genet. 15:128–131.
3. Johnson, D.R. 1977. Ultrastructural observations on *stumpy* (*stm*), a new chondrodystrophic mutant in the mouse. J. Embryol. Exp. Morphol. 39:279–284.
4. Johnson, D.R. 1978. The growth of femur and tibia in three genetically distinct chondrodystrophic mutants of the house mouse. J. Anat. 125:267–275.
5. Johnson, D.R. 1983. Abnormal cartilage from the mandibular condyle of stumpy (*stm*) mutant mice. J. Anat. 137:715–728.
6. Thurston, M.N., D.R. Johnson, N.F. Kember, and W.J. Moore. 1983. Cell kinetics of growth cartilage in stumpy: a new chondrodystrophic mutant in the mouse. J. Anat. 136:407–415.

Sto-1 locus (originally *Sto^I*), steroid hydroxylase-1

This locus controls activity in the liver of females of a 16α-hydroxylase with high affinity for testosterone and progesterone and low affinity for pregnenolone. Two alleles are known, *Sto-1^a* (*Sto^{Ia}*) determining low activity in the 129/J strain, and *Sto-1^b* (*Sto^I*) determining high activity in the C57BL/6, DBA/2, AKR, C3H/I, and BALB/c strains. Heterozygotes have intermediate activity. *Sto-1* is not expressed in adult males. Adult males have another 16α-hydroxylase, form II, with a high affinity for pregnenolone. No strain differences in this enzyme were detected. In 129/J females (*Sto-1^a*/*Sto-1^a*) with low STO-1 activity, only form II enzyme is produced, resulting in a phenotype of low activity of a 16α-hydroxylase with high affinity for pregnenolone (1).

References

1. Pasleau, F., C. Kolodcizi-Mehaignoul, and J.E. Gielen. 1984. Genetic segregation of hepatic steroid 16α-hydroxylase activities in inbred strains of mice. Endocrinology 115:1371–1379.

Str, striated, semidominant, Chr X

Found among the progeny of an X-rayed male. Heterozygous females show a transverse striping similar to that of *Ta*/+ females. The dark stripes are due to shortening of the hairs and not to lack of zigzags as in *Ta*/+. Viability appears to be normal, but there is a shortage of *Str*/+ in segregating generations probably due to misclassification. Hemizygous males die at about 11 to 13 days of gestation (3). There is a shortage of auchenes and guard hairs among the short hairs of *Str*/+ females, but zigzags and awls occur in approximately the normal ratio to each other. In double heterozygotes with *Ta*, the two loci were shown to act in the same cell type (1). Since *Ta* is known to act in the epidermis (2), this is presumptive evidence that *Str* also acts in the epidermis.

References

1. Lyon, M.F. 1963. Attempts to test the inactive-X theory of dosage compensation in mammals. Genet. Res. 4:93–103.
2. Mayer, T.C., and M.C. Green. 1978. Epidermis is the site of action of tabby (*Ta*) in the mouse. Genetics 90:125–131.
3. Phillips, R.J.S. 1963. *Striated*, a new sex-linked gene in the house mouse. Genet. Res. 4:151–153.

Sts locus, steroid sulfatase, Chrs X, Y

This locus controls activity level of steroid sulfatase (STS) (arylsulfatase C) in germ cells and somatic tissues. High activity determined by the allele here called *Sts^b* occurs in the C57BL/6, SM, and SWR strains; low or intermediate activity determined by the allele here called *Sts^a* occurs in the A/J, C3H/He, and DBA/2 strains. In crosses of A/J with C57BL/6 or SWR, the difference behaves as if controlled by an autosomal locus with codominant inheritance (2, 4). However, a gene controlling activity level of STS was found to be on the X chromosome by determining that the activity level in oocytes of X/X mice was twice that in oocytes of X/O mice (3). Results of a cross between a C3H male (*Sts^a*/Y) and an X/O female (*Sts^b*/0) showed that there must be a functional *Sts* allele on the Y chromosome and that the *Sts* alleles undergo a high frequency of crossing over with the sex-determining part of the Y chromosome and thus simulate autosomal inheritance (5). Cattanach and Crocker (1) found that, in female heterozygotes, *Sts* shows about 10 per cent recombination with *Crm* (cream), a locus near the distal end of the X chromosome, but not in the pairing region. Among male offspring of males heterozygous for both *Sts* and the *Sxr* transposition that marks the distal end of the Y chromosome, 6 per cent were recombinants between *Sts* and *Sxr* (6), indicating that the obligate chiasma between the X and Y chromosomes does not occur at a fixed location and can occur beyond the *Sts* locus. Males carrying *Sts^a* on the X chromosome and either *Sts^a* or *Sts^b* on the Y chromosome, when compared, gave no evidence that the Y-linked gene is expressed (1).

References

1. Cattanach, B.M., and A.J.M. Crocker. 1986. X chromosomal location of Sts. Mouse News Lett. 74:94–95.
2. Erickson, R.P., K. Harper, and J.M. Kramer. 1983. Identification of an autosomal locus affecting steroid sulfatase activity among inbred strains of mice. Genetics 105:181–189.
3. Gartler, S.M., and M. Rivest. 1983. Evidence for X-linkage of steroid sulfatase in the mouse: steroid sulfatase levels in oocytes of XX and XO mice. Genetics 103:137–141.
4. Keinanen, B.M., K. Nelson, W.L. Daniel, and J.M. Roque. 1983. Genetic analysis of murine arylsulfatase C and steroid sulfatase. Genetics 105:191–206.
5. Keitges, E., M. Rivest, M. Siniscalco, and S.M. Gartler. 1985. X-linkage of steroid sulfatase in the mouse is evidence for a functional Y-linked allele. Nature 315:226–227.
6. Keitges, E.A., D.F. Schorderet, and S.M. Gartler. 1987. Linkage of the steroid sulfatase gene to the *sex-reversed* mutation in the mouse. Genetics 116:465–468.

stu, stumbler, recessive

Arose spontaneously in the C3H/HeJ strain. Homozygotes have clinical features suggesting a cerebellar defect. They can be recognized at 12 days of age by a stumbling locomotion characterized by a high-stepping broad-based gait. They become less active with age and progressively smaller than their normal sibs and usually die before weaning. There are fewer than normal Purkinje cells and granule cells from about 10 days of age onward and a consequent smaller size of the cerebellum. The Purkinje cells have small dendritic arborizations and immature spines on their somata. They also have an increased number of mitochondrial profiles both in cell bodies and in swellings on dendrites. The morphology of the granule cells appears normal (1, 2).

References

1. Caddy, K.W.T., and R.L. Sidman. 1981. Purkinje cells and granule cells in the cerebellum of the stumbler mutant mouse. Dev. Brain Res. 1:221–236 (Brain Res. 227).
2. Caddy, K.W.T., R.L. Sidman, and E.M. Eicher. 1981. Stumbler, a new mutant mouse with cerebellar disease. Brain Res. 208:251–255.

su, surdescens, recessive

Found in mice of the C57BL/Gr inbred strain and in a strain carrying *Mi^{wh}*. Homozygotes hear normally when young but become deaf or hard of hearing between 2 and 5 months. Behavior is normal. In the inner ear there is reduction in size of the stria vascularis and degeneration of the organ of Corti and the spiral ganglion (2). More severe changes that included a decrease in size of the endolymphatic space in the scala media and the sacculus were thought by Kocher to be caused by *su*, but were later shown by Deol (1) to be effects of the *Mi^{wh}* gene segregating in the stock.

References

1. Deol, M.S. 1967. The neural crest and acoustic ganglion. J. Embryol. Exp. Morphol. 17:533–541.
2. Kocher, W. 1960. Untersuchungen zur Genetik und Pathologie der Entwicklung spät einsetzender hereditärer Taubheit bei der Maus (*Mus musculus*). Arch. Ohr. Nas. Kehlk. Heilk. 177:108–145.

Suc-1 locus, sucrase-1

This locus controls level of activity in the intestine of sucrase and isomaltase, but not of maltase or trehalase. Low activity determined by the *Suc-1^a* allele occurs in the CBA/Ca strain; high activity (threefold higher) determined by the *Suc-1^b* allele occurs in the C57BL/6 strain. All other strains tested have two- to threefold higher levels than CBA/Ca. *Suc-1^a*/*Suc-1^b* mice have intermediate activity. CBA/Ca mice are able to regulate sucrase activity in response to changes in diet, but to a lesser extent than C57BL/6 mice. No structural difference between the enzymes of the two strains was detected (1).

Suc-1

References

1. James, P.S., M.W. Smith, G.W. Butcher, D. Brown, and E.K. Lund. 1986. Evidence for a possible regulatory gene (Suc-1) controlling sucrase expression in mouse intestine. Biochem. Genet. 24:169–181.

sv, Snell's waltzer, recessive, Chr 9

Found in the B10.HA(33NX) stock of Snell. Viability is nearly normal. Breeding ability is reduced, and males are more reliable breeders than females. Homozygotes show to a marked degree the typical circling, head-tossing, deafness, and hyperactivity of other mutants of this type. They are recognizable by the age of 1 week. The abnormalities of the inner ear consist of degeneration of the entire neuroepithelium comprising the organ of Corti, the saccular and utricular maculae, and the cristae of all three semicircular canals (1).

References

1. Deol, M.S., and M.C. Green. 1966. Snell's waltzer, a new mutation affecting behavior and the inner ear of the mouse. Genet. Res. 8:339–345.

Svl

See Cat.

Svp-1 locus, seminal vesicle protein-1, Chr 2

This locus controls electrophoretic variation in a protein secreted by the seminal vesicle. The allele $Svp-1^a$ determines presence of two slow anodally migrating bands (V and VI) and occurs in strains 129/Re, AKR, C3H/He. and others: the allele $Svp-1^b$ determines presence of band V and a faster band, IV, and occurs in strains C57BL/6, BALB/c, CBA/J, and others. Heterozygotes have bands IV, V, and VI. The function of the protein is not known (1).

References

1. Platz, R.D., and H.G. Wolfe. 1969. Mouse seminal vesicle proteins. The inheritance of electrophoretic variants. J. Hered. 60:187–192.

Svp-2 locus, seminal vesicle protein-2, Chr 7

This locus controls electrophoretic variation in a protein secreted by the seminal vesicle that migrates just cathodal to the protein controlled by Svp-1 (bands 7 and 8). Three alleles are known: $Svp-2^a$ determines presence of band 8 and absence or reduced intensity of band 7 and occurs in strain C57BL/6; $Svp-2^b$ deter-mines presence of band 8 and intense staining of band 7 and occurs in strains AKR, CBA, and others; $Svp-2^c$ determines intermediate intensity of staining of bands 7 and 8 and occurs in strains C3H/He, DBA/2, BALB/c, and others. Heterozygotes have patterns resembling mixtures of the two parental types (1). Svp-2 is on Chr 7 very close to Gpi-1 (2).

References

1. Moutier, R., K. Toyama, and M.F. Charrier. 1971. Contrôle génétique des protéines de la sécrétion des glandes séminales chez la souris et le rat. Experimentation Animale 4:7–18.
2. Taylor, B.A. 1982. Mouse News Lett. 67:22.

Svp-3 locus, seminal vesicle protein-3, Chr 2

This locus controls electrophoretic variation of a protein in the secretion of the seminal vesicles that migrates just cathodal to the SVP-2 bands. The allele $Svp-3^a$ determines a slow anodally migrating enzyme band and occurs in the B10.AKM/Sn strain; the allele $Svp-3^b$ determines a faster band and occurs in the AKR.M/Sn, C3H/He, and C57BL/10Orl strains. Heterozygotes have the two parental bands. Svp-3 shows no recombination with Svp-1 on Chr 2 (1).

References

1. Moutier, R., and M.F. Bertrand. 1983. Svp-3, a third polymorphic locus for mouse seminal vesicle proteins. Biochem. Genet. 21:797–800.

sw, swaying, recessive, Chr 15

Arose spontaneously in a cross testing for linkage of mg and dt^J. Homozygotes sway to one side or the other when attempting to move and may then pivot clockwise or counterclockwise around their rear legs. They swim poorly and often roll over (1). Their marked ataxia and hypertonia are probably attributable to malformations of the anterior vermis of the cerebellum and of the colliculi. The anterior parts of the cerebellar cortex are badly scrambled, and the white matter is fused directly with the superior colliculus. Despite the severe cortical malformation, the individual cell types in the cortex bear normal anatomic relations to one another (2).

References

1. Lane, P.W. 1967. Mouse News Lett. 36:40.
2. Sidman, R.L. 1968. Development of interneuronal connections in brains of mutant mice. In F.D. Carlson, ed., Physiological and Biochemical Aspects of Nervous Integration, 163–193. Prentice-Hall, Englewood, NJ.

Swl, sprawling, dominant or semidominant (homozygotes not obtainable)

Found in descendants of mice used several generations earlier in an irradiation experiment. Heterozygous females are fertile, but males have not bred. Affected mice are detectable at 7 to 10 days by abnormal posture of the hindlegs when the mouse is held up by the tail. They are smaller than normal for several weeks but later grow to about normal size. They show abnormal posture and locomotion resulting from defective position sense mainly affecting the hindlimbs. The sensory roots, dorsal columns of the spinal cord, and peripheral nerves are deficient in myelinated axons, with the disorder more severe caudally than rostrally. No evidence of axonal degeneration was found. The number of sensory ganglion cells is greatly reduced, and the cell bodies show marked chromatolysis which persists throughout life. There is a severe deficiency of sensory receptors, and spindles are virtually absent from the hindlimb muscles (1–3).

References

1. Duchen, L.W. 1975. "Sprawling": a new mutant mouse with failure of myelination of sensory axons and a deficiency of muscle spindles. Neuropath. Appl. Neurobiol. 1:89–101.
2. Duchen, L.W., and F. Scaravilli. 1977. The structure and composition of peripheral nerves and nerve roots in the *sprawling* mouse. J. Anat. 123:763–775.
3. Duchen, L.W., and F. Scaravilli. 1977. Quantitative and electron microscopic studies of sensory ganglion cells of the sprawling mouse. J. Neurocytol. 6:465–481.

Sxa locus, sex chromosome association, Chrs X and Y

The frequency of X–Y chromosome dissociation is low (< 30 per cent) in strain BALB/c and in Japanese wild mice *M. m. molossinus* (MOL). In their F1 hybrids it is high (> 70 per cent). Male offspring of female BALB/c by male MOL were backcrossed to BALB/c. In male offspring, the frequency of X–Y dissociation was either high or low. In successive backcrosses, the proportion of low dissociation males decreased in each generation to 0 at N5. To explain these findings, the authors postulate the existence of a locus *Sxa* on the common (pairing) part of the X and Y chromosomes. The allele in BALB/c is designated Sxa^a, that in MOL is designated Sxa^b. Homozygotes of either allele have low dissociation, heterozygotes have high dissociation. The low dissociation males in offspring of the backcrosses result from crossing over between *Sxa* and the differential part of the X and Y chromosomes. The decline in proportion of low dissociation males in successive backcrosses to BALB/c is attributed to autosomal factors that, when homozygous, decrease the frequency of crossing over in Sxa^a/Sxa^b heterozygotes (1).

References

1. Matsuda, Y., H.T. Imai, K. Moriwaki, and K. Kondo. 1983. Modes of inheritance of X-Y dissociation in interspecies hybrids between BALB/c mice and *Mus musculus molossinus*. Cytogenet. Cell Genet. 35:209–215.

Sxr, sex-reversed, dominant, Chrs X and Y

Found among the offspring of a male carrying Cattanach's translocation, *Is(In7;X)1Ct* (formerly *T(X;7)1Ct*). This genetic variant causes chromosomal females (X/X or X/O) to develop as males with small testes but otherwise normal morphology. *Sxr*/+ males are normal and fertile. Sxr was originally thought to be an autosomal dominant, but is now known to be the effect of an abnormal Y chromosome in which a piece of the proximal end of the Y which bears the genes for the H-Y antigen and for testis determination has been transposed to the distal end of the Y (4, 8). Its chromosomal designation is *Tp(Y)1Ct*. The distal ends of the X and Y chromosomes consist of short pairing segments in which an obligate chiasma occurs at every meiosis. This results in transfer of the *Sxr* segment to one-half of the X-bearing gametes and loss of it from one-half of the Y-bearing gametes. Segregation of *Sxr* thus simulates autosomal dominant inheritance with expression limited to females. X/X^{Sxr} mice are sterile males and have small testes which contain no spermatogonia after 10 days of age. However, X/X^{Sxr} germ cells in X/X^{Sxr} ⟷ +/+ female chimeras can produce normally functional X^{Sxr}-bearing oocytes (1). X^{Sxr}/O males are also sterile but have testes of normal size in which spermatogenesis becomes abnormal at the spermatid stage and produces sperm that are all grossly abnormal. Males of both types have normal mating behavior (2). If the normal X of X/X^{Sxr} mice is replaced by the *T(X;16)16H (T16H)* translocation, preferential inactivation of the X^{Sxr} chromosome occurs and some of these mice are fertile females (3). All of them, however, produce H-Y antigen, demonstrating that inactivation of the X^{Sxr} chromosome is not complete (7). By mating *T16H*/X^{Sxr} females to X/Y^{Sxr} males, X^{Sxr}/Y^{Sxr} mice were produced and proved to be viable and fertile males (5). McLaren *et al.* (6) found an *Sxr* variant, tentatively called *Sxr'*, which retained the testis-determining gene, *Tdy*, but lacked the H-Y antigen-determining gene, *Hya*. See *Hya*, *Tdy*.

Sxr

References

1. Bradbury, M.W. 1983. Functional capacity of sex-reversed (XX, *Sxr*/+) mouse germ cells as shown by progeny derived from XX, *Sxr*/+ oocytes of a female chimera. J. Exp. Zool. 226:315–320.
2. Cattanach, B.M., C.E. Pollard, and S.G. Hawkes. 1971. Sex-reversed mice: XX and XO males. Cytogenetics 10:318–337.
3. Cattanach, B.M., E.P. Evans, M.D. Burtenshaw, and J. Barlow. 1982. Male, female and intersex development in mice of identical chromosome constitution. Nature 300:445–446.
4. Evans, E.P., M.D. Burtenshaw, and B.M. Cattanach. 1982. Meiotic crossing over between the X and Y chromosomes of male mice carrying the sex-reversing (*Sxr*) factor. Nature 300:443–445.
5. McLaren, A., and P.S. Burgoyne. 1983. Daughterless X*Sxr*/Y*Sxr* mice. Genet. Res. 42:345–349.
6. McLaren, A., E. Simpson, K. Tomonari, P. Chandler, and H. Hogg. 1984. Male sexual differentiation in mice lacking H-Y antigen. Nature 312:552–555.
7. Simpson, E., A. McLaren, P. Chandler, and K. Tomonari. 1984. Expression of H-Y antigen by female mice carrying Sxr. Transplantation 37:17–21.
8. Singh, L., and K.W. Jones. 1982. Sex reversal in the mouse (*Mus musculus*) is caused by a recurrent nonreciprocal crossover involving the X and an aberrant Y chromosome. Cell 28:205–216.

Sxv locus, susceptibility to xenotropic virus, Chr 1

This locus controls susceptibility of mouse embryo cells to infection with xenotropic viruses. All laboratory strains carry the allele *Sxv^r* and their cells cannot be productively infected with xenotropic viruses. The embryo cells of most wild-derived mice of various subspecies of *Mus musculus* and of various related species will support growth of these viruses. *M. spretus* and *M. m. praetextus* were shown, in crosses to laboratory mice, to carry the fully dominant allele *Sxv^s* determining susceptibility. *Sxv* was found to be on Chr 1 close to but distinct from *Bxv-1*, an integrated xenotropic proviral genome. *Sxv* is also in the same region as *Rmc-1*, a locus that probably codes for a receptor for MCF viruses, and it is possible that the two loci control allelic receptors for the two kinds of viruses (1).

References

1. Kozak, C.A. 1985. Susceptibility of wild mouse cells to exogenous infection with xenotropic leukemia viruses: control by a single dominant locus on chromosome 1. J. Virol. 55:690–695.

sy locus, Chr 18

sy, shaker-with-syndactylism, recessive. Found by Hertwig (4) among descendants of an irradiated male. Most homozygotes die within the first month, and none have lived to breed. Homozygotes may be syndactylous on all four feet; the forefeet may often be normal and possibly one or both hindfeet very occasionally so. The remainder of the skeleton shows many slight anomalies in addition to small size. The shafts of the long bones are considerably thinner than normal, and there are differences in shape of the sacral vertebrae and the scapula (2). The osseous skeleton is less densely constructed than normal (3). Abnormal behavior appears during the first week. It consists of head-tossing and some circling, and the affected mice are always deaf. Abnormalities of the labyrinth can be seen at 13 days of gestation. There is excessive amount of mesenchymal tissue from which develops an excessive amount of perilymphatic space. Partial collapse of the endolymphatic space follows, and eventually extensive degeneration of all parts of the membranous labyrinth occurs. A quantitative abnormality of the mesodermal tissue has been suggested as the basis of the whole syndrome (1).

sy^fp, fused phalanges, recessive. Arose spontaneously in strain C3HeB/FeHu. Homozygotes are viable and fertile. All homozygotes show fusion of the three central digits of the hind-feet and about half show the same abnormality of the forefeet. Fusions occur between the second and third or third and fourth digits, but not all three (5). Behavior is normal. *sy/sy^fp* mice resemble *sy^fp/sy^fp* and show no abnormal behavior (6). Because homozygous *sy^fp* mice are fertile, this allele is a good marker for Chr 18 (7).

References

1. Deol, M.S. 1963. The development of the inner ear in mice homozygous for *shaker-with-syndactylism*. J. Embryol. Exp. Morphol. 11:493–512.
2. Grüneberg, H. 1956. Genetical studies on the skeleton of the mouse. XVIII. Three genes for syndactylism. J. Genet. 54:113–145.
3. Grüneberg, H. 1962. Genetical studies on the skeleton of the mouse. XXXII. The development of shaker with syndactylism. Genet. Res. 3:157–166.
4. Hertwig, P. 1942. Neue Mutationen und Koppelungsgruppen bei der Hausmaus. Z. Indukt. Abstammungs-Vererbungsl. 80:220–246.
5. Hummel, K.P., and D.B. Chapman. 1971. Mouse News Lett. 45:28.
6. Lane, P.W., and K.P. Hummel. 1973. Mouse News Lett. 49:32.
7. Lane, P.W., A.G. Searle, C.V. Beechey, and E.M.

Eicher. 1981. Chromosome 18 of the mouse. J. Hered. 72:409–412.

Syn-1 locus, synapsin-1, Chr X

This locus codes for synapsin-1 (SYN-I), a neuron-specific phosphoprotein associated with the membrane of synaptic vesicles. It may play a role in regulation of neurotransmitter release. *Syn-1* was identified by a rat SYN-I cDNA clone and mapped to Chr X by use of mouse–Chinese hamster somatic cell hybrids. *In situ* hybridization showed it to be close to the centromere in the A1 to A4 region (1).

References

1. Yang-Feng, T.L., L.J. DeGennaro, and U. Francke. 1986. Genes for synapsin I, a neuronal phosphoprotein, map to conserved regions of human and murine X chromosomes. Proc. Natl. Acad. Sci. USA 83:8679–8683.

T

t-complex, Chr 17

See Bennett (4) and Sherman and Wudl (31) for comprehensive reviews of the earlier literature, Silver (35) and Frischauf (13) for reviews of more recent molecular genetic studies, and Frischauf (Chapter 5, this volume) for an up-to-date map of the complex. The *t* complex encompasses a region of about 12 cM on the proximal portion of Chr 17. The region has been identified by several semidominant (*T*) alleles and numerous recessive (*t^n*) haplotypes. In wild populations and probably also in some laboratory populations, *t* haplotypes may be very common. Most *t* haplotypes in wild populations are so-called 'complete' haplotypes. In general, *T/+* mice are short-tailed, and for the complete *t* haplotypes, *t^n/+* mice are normal, *T/t^n* mice are tailless, and *T/T* and *t^n/t^n* homozygotes are lethal. Compounds of two different *t* haplotypes are viable and normal tailed but males are sterile. Segregation ratios in males but not in females of the genotypes *+/t^n* and *T/t^n* are very abnormal, usually with a large excess of offspring from the *t*-bearing gametes. The sperm and spermatogenesis of *+/t^n* and *T/t^n* males appear normal, but in most cases the *t*-bearing sperm fertilize a disproportionately high proportion of eggs when such males are mated. The high transmission ratio is thought to be responsible for maintaining *t* haplotypes in the wild in spite of the lethality of homozygotes (10, 11). In *+/t^n* and *T/t^n* mice, recombination is greatly reduced in the region of Chr 17 between *T* and *H-2*.

In the laboratory, existing *t* haplotypes occasionally give rise to new so-called 'partial' *t* haplotypes by recombination with wild type in the region of crossover suppression (24). For a pedigree of all such new haplotypes known at that time, see Klein and Hammerberg (19). These recombinant haplotypes contain only a part of the *t* complex. They have varying degrees of lethality, recombination suppression, transmission ratio distortion, and male sterility. Polymorphism of the numerous genes that have been identified in the *t* haplotypes is quite limited. This and other evidence leads to the conclusion that the *t* complex arose only once and that all existing haplotypes have arisen by mutation or recombination from existing *t* haplotypes (35).

The lethality of homozygous *t* haplotypes is due to presence of one or more lethal mutations, which may be at different loci in different haplotypes. Compounds of two haplotypes having lethals at different loci are viable and the two haplotypes are said to fall into different complementation groups. There are now at least 16 complementation groups known (4, 20). Six different single-locus lethals and one determined by two loci were analyzed and shown to be distributed, with some clustering, over the whole length of the *t* complex (1, 32). Although recombination is suppressed in *t/+* heterozygotes, normal recombination can occur between different *t* haplotypes in their overlapping regions (36). Crosses involving various combinations of this kind were used to map the *t* lethals and have also showed that the distal part of the *t* complex from a point proximal to *tf* to beyond *H-2* is inverted with respect to that part of the wild-type chromosome (2, 33, 35). Other evidence shows that the proximal part of the *t* complex contains another non-overlapping inversion that includes the region from *T* to between *qk* and *tf* (16). The presence of these inversions explains the recombination suppression in *t/+* heterozygotes. Since

t-complex

suppression occurs over the whole t region, it is likely that another smaller inversion spans the region between the two larger ones (35).

The transmission ratio distortion that occurs in complete t haplotypes and may or may not occur in partial t haplotypes has been shown to be due to the cumulative action of three or more distorter (*Tcd*) genes on a responder gene *Tcr*. Distortion occurs when *Tcr* is present and heterozygous. The distorters act additively, either in *cis* or in *trans*, to raise the transmission of whatever chromosome carries the responder. In the absence of the responder, the distorters are transmitted at a normal ratio (22). How the distorter genes produce this effect is not known. Evidence from segregation ratios in $+/+ \longleftrightarrow +/t^{w73}$ chimeras supports the conclusion that inactivation of the +-bearing sperm is determined during meiosis (28). Shur (34) described a fourfold increase in galactosyltransferase activity on t-bearing sperm from mice of a number of complete t haplotype in either homozygous or heterozygous condition or in compounds of two complete haplotypes, but how this is related to transmission ratio distortion is unknown. The relationship of transmission ratio distortion to male sterility of viable compounds of complementing t haplotypes has been clarified by Lyon (23), who showed that male sterility is due to the interaction of three sterility factors that are carried on the same haplotypes as the three distorter genes and are probably identical with them. When heterozygous, the distorter/sterility genes act on the wild-type form of the responder gene to render sperm carrying it non-functional. In homozygotes, the harmful effect of these genes is stronger, affecting both forms of the responder gene and leading to sterility. This explains the paradoxical situation of high transmissibility of t haplotypes in heterozygotes with wild type and zero transmissibility in compounds with each other.

Two sets of molecular markers are becoming increasingly useful in mapping studies of the t complex. A set of eight testicular cell proteins specific to the t complex were detected by two-dimensional gel electrophoresis and the loci determining them (*Tcp-n*) mapped to positions over the whole length of the complex (38). Soon afterwards, a set of 15 independent DNA clones that hybridize to one or more sequences specific to the t complex were isolated by Röhme *et al.* (29) and six of these were mapped to diverse regions of the complex (12).

A very large number of t haplotypes have been described. They may be either complete or partial, and they can be grouped into 16 or more complementation groups depending on the particular lethal mutation

they carry. It is beyond the scope of this summary to list them here. Descriptions of the main complementation groups and their prototype haplotypes can be found in Bennett (4), Guénet *et al.* (15), Nizetić *et al.* (27), and Winking and Silver (40). Klein and Hammerberg (19) list all haplotypes known at that time.

The genetic variants of the t complex listed below are the individual genes contained in the complex. They include the dominant short-tailed (*T*) mutations; genes responsible for transmission ratio distortion, male sterility, and a maternal effect; lethal mutations; protein variants; and DNA restriction fragment variants.

T, brachyury, semidominant. Discovered by Dobrovolskaïa-Zavadskaïa (9) in a laboratory stock. Tail length of heterozygotes varies from nearly normal to absent. There tends to be one fewer presacral vertebrae than in +/+ sibs. Studies of development of heterozygotes have shown that the notochord in the tail region is abnormal. It is present in the whole tail but becomes incorporated in the neural tube or the tail gut (14). T/t^n mice are tailless. T/T embryos never develop a notochord or only a rudimentary one, and the portion of the body posterior to the forelimb buds is not formed. They die at 10 days of gestation (8). In organ culture, T/T neural tube can induce cartilage in normal somites, but cartilage cannot be induced in T/T somites (3). Several other developmental studies have been reported but none have revealed the basic cause of the defects.

T^c, curtailed, semidominant. Found among offspring of an irradiated (C3H/He × 101/H)F1 male mated to females of a multiple recessive stock. Heterozygotes have complete or near-complete absence of the tail, absence of the odontoid process of the axis, absence of the nucleus pulposi of the intervertebral discs, a tendency for rib and vertebral fusions, and a slight decrease in average number of presacral vertebrae. They occasionally show paralysis of the hindlimbs and atresia ani. Homozygotes die at the same time as T/T embryos and are similar to them but more severely affected. T^c/T embryos are variable and intermediate between T/T and T^c/T^c (30).

T^h, T-Harwell, semidominant. Found in the control series of an irradiation experiment. $T^h/+$ mice are indistinguishable from $T/+$, but T^h/T^h mice die at an earlier stage than T/T, being already highly abnormal at 8 days (21).

T^{hp}, hairpin tail, semidominant. Arose spontaneously in the AKR/J strain. It is probably a deletion that includes T and *qk*. Heterozygotes resemble $T/+$ mice phenotypically, and T^{hp}/t^n mice are tailless. Homozygotes die at 7 days of gestation (4). T^{hp} has a maternal effect: $T^{hp}/+$ mice whose T^{hp} gene was transmitted by their mother die at a late stage *in utero*, while $T^{hp}/+$ mice whose T^{hp} gene was transmitted by their father are normally viable even when the mother is also $T^{hp}/+$ (17, 18). $T^{hp}/+$ embryos with a maternally derived T^{hp} gene can be rescued if aggregated with normal embryos, but such chimeras, if female, fail to transmit T^{hp} to viable offspring (5). McGrath and Solter (25) transplanted pronuclei between $T^{hp}/+$ and $+/+$ ova and showed that the maternal effect was carried in the nucleus, not in the cytoplasm. The maternal effect of T^{hp} is probably due to deletion of the *Tme* locus (40). See *Tme* locus below.

T^{Or}, T-Oak Ridge, semidominant. This designation is used for a group of six mutations (T^{1Or}-T^{6Or}), probably radiation-induced in (101 × C3Hf)F1 hybrids. They behave genetically like *T* and cannot be distinguished from T^h on the basis of the phenotype of heterozygotes and homozygotes. Serological studies of T^{1Or} suggest that it may be a deletion which, however, does not extend as far as *qk* (6).

T^{Orl}, T-Orleans, semidominant. Arose spontaneously in a noninbred Swiss/Orl stock (26). $T^{Orl}/+$ *qk* mice are short-tailed and quaking, suggesting that T^{Orl} is a deletion involving both the *T* and *qk* loci. T^{Orl} is indistinguishable from T^{hp} except that it does not have a maternal effect (4). However, it is inseparably associated with a chromosomal region derived from a naturally occurring *t* haplotype. Its *t*-like properties are a consequence of closely linked *t*-haplotype genes, not of the deletion itself (37). Silver *et al.* (37) suggest symbol Tt^{Orl} for this mutant complex.

Tcd-1 to *-6* loci, T-complex ratio distorter-1 to -6. These loci act cumulatively to increase the transmission ratio of whatever chromosome carries the responder *Tcr* locus, presumably by a harmful effect on the Tcr^+-bearing chromosome or the sperm carrying it. The *Tcd* loci may be identical with the *Tcs* male sterility loci. They are distributed along the whole length of the *t* complex (22, 23, 35).

tcl-0, etc., t-complex lethals, recessive. These loci are designated by the number of the lowest numbered haplotype in the complementation group in which the lethal occurs. The *tcl* loci and some of the haplotypes in which they occur are shown in Table 2.8 (1, 2, 32). Both *tcl-12a* and *tcl-12b* are necessary for the lethal effect.

Tcp-1, -3 to *-9* loci, t-complex protein-1, -3 to -9 (Table 2.9). These loci code for testicular proteins. They were identified in two-dimensional electrophoretic gels as spots present in gels from testis of *t* haplotype-bearing

Table 2.8

Locus	Haplotype
tcl-0	t^0, t^6
tcl-4	t^4, t^9, t^{w18}
tcl-12a, -12b	t^{12}, t^{w32}
tcl-w1	t^{w1}, t^{w12}
tcl-w5	t^{w5}
tcl-w73	t^{w73}
tcl-wLub1	t^{wLub1}

mice and not present in gels from testis of wild-type mice. *Tcp-2* was found to be identical with *Pgk-2* and is not listed here. Allelic forms of five of the eight loci have been identified on wild-type Chr 17. The allele determining the more acidic form is designated with the superscript *a* and that determining the more basic form with the superscript *b* (38).

The *Tcp-1* gene was cloned by Willison *et al.* (39). The nucleotide differences between $Tcp-1^a$ and $Tcp-1^b$ produce six amino acid differences between the two proteins. A *Taq*I polymorphism of *Tcp-1* occurs among inbred strains. Strains C57BL/6, C57BL/10, NZB, and RIIIS have a fragment not present in all other strains examined.

Tcr, t-complex ratio responder. This mutant gene occurs in all *t* haplotypes that show transmission ratio distortion in heterozygotes with wild type. In combination with one or more *Tcd* genes, there is increased transmission of the chromosome carrying *Tcr*. In the absence of *Tcd* genes, as in the t^{low} haplotype, there is lowered transmission of *Tcr* (22).

Tcs-1 to *-6* loci, t-complex male sterility-1 to -6. These loci are probably identical to the *Tcd* loci. When homozygous, they cause male sterility, presumably by a harmful effect on both Chrs 17 mediated through the *Tcr* alleles, either *Tcr* or Tcr^+ (23, 35).

tct, t-complex tail length modifier, recessive. This gene interacts with *T* to produce taillessness. It is responsible for the tailless phenotype of T/t^n mice. No recombination with *T* has ever been observed and *tct* has therefore been considered to be an allele of *T*. However, Bode (7) found an ENU-induced mutation apparently identical to *tct* on a wild-type Chr 17 and showed that it could recombine with *T* and is located just distal to *T*.

Tme locus, T-associated maternal effect. This locus is deleted in the T^{hp} deletion and in the partial *t* haplotype, t^{wLub2}. This haplotype was recovered as a recombinant between the wild-derived t^{wLub1} haplotype and a wild-type chromosome 17. Embryos heterozygous for either of these deletions die at a late embryonic stage if

t-complex

Table 2.9

Locus	t-complex allele	+ allele	Molecular weight kDa
Tcp-1	a	b	63
Tcp-3	b	a	58
Tcp-4	b	a	52
Tcp-5	–	–	50
Tcp-6	–	–	45
Tcp-7	b	a	30
Tcp-8	b	a	15
Tcp-9	–	–	85

the deletion is inherited from the mother but not if it is inherited from the father (40). See T^{hp} locus for further description of the maternal effect. *Tme* must be inherited from the mother in order for normal development to proceed through late stages of gestation. How this effect is produced is not understood. *Tme* is located between *qk* and *tf* (40). The alleles present in wild-type and *t* chromosomes are presumably identical.

DNA restriction fragment variants. Röhme *et al.* (29) identified 15 sequences of Chr 17 DNA which showed restriction fragment length differences between *t* haplotypes and wild type. The sequences were designated T18, T48, T54, ... T180. Their proper full designation should be *D17Leh18*, *D17Leh48*, etc. Their position in the *t* complex according to the best current interpretation of the evidence is given by Silver (35) and Frischauf (Chapter 5, this volume) All complete *t* haplotypes examined are identical at all of these loci, although some of the loci are polymorphic among wild

Table 2.10

D17Leh18	D17Leh66C	D17Leh111
D17Leh48	D17Leh80	D17Leh119
D17Leh54	D17Leh89	D17Leh122
D17Leh66A	D17Leh94	D17Leh171
D17Leh66B	D17Leh108	D17Leh180

chromosomes (35). The 15 variants found by Röhme *et al.* are listed in Table 2.10. Others are likely to be found in the future.

References

1. Artzt, K. 1984. Gene mapping within the *T/t* complex of the mouse. III. *t*-lethal genes are arranged in three clusters on chromosome 17. Cell 39:565–572.

2. Artzt, K., P. McCormick, and D. Bennett. 1982. Gene mapping within the T/t complex of the mouse. I. *t*-lethal genes are nonallelic. Cell 28:463–470.

3. Bennett, D. 1958. *in vitro* study of cartilage induction in *T/T* mice. Nature 181:1286.

4. Bennett, D. 1975. The T-locus of the mouse. Cell 6:441–454.

5. Bennett, D. 1978. Rescue of a lethal *T/t* locus genotype by chimaerism with normal embryos. Nature 272:539.

6. Bennett, D., L.C. Dunn, M. Spiegelman, K. Artzt, J. Cookingham, and E. Schermerhorn. 1975. Observations on a set of radiation-induced dominant *T*-like mutations in the mouse. Genet. Res. 26:95–108.

7. Bode, V.C. 1984. Ethylnitrosourea mutagenesis and the isolation of mutant alleles for specific genes located in the *T* region of mouse chromosome 17. Genetics 108:457–470.

8. Chesley, P. 1935. Development of the short-tailed mutant in the house mouse. J. Exp. Zool. 70:429–459.

9. Dobrovolskaïa-Zavadskaïa, N. 1927. Sur la mortification spontanée de la queue chez la souris nouveau-née et sur l'existence d'un caractère héréditaire "non-viable". C. R. Soc. Biol. 97:114–116.

10. Dunn, L.C., A.B. Beasley, and H. Tinker. 1960. Polymorphisms in populations of wild house mice. J. Mammal. 41:220–229.

11. Dunn, L.C., D. Bennett, and A.B. Beasley. 1962. Mutation and recombination in the vicinity of a complex gene. Genetics 47:285–303.

12. Fox, H.S., G.R. Martin, M.F. Lyon, B. Herrmann, A.-M. Frischauf, H. Lehrach, and L.M. Silver. 1985. Molecular probes define different regions of the mouse *t* complex. Cell 40:63–69.

13. Frischauf, A.-M. 1985. The *T/t* complex of the mouse. Trends Genet. 1:100–103.

14. Grüneberg, H. 1958. Genetical studies on the skeleton of the mouse. XXIII. The development of brachyury and anury. J. Embryol. Exp. Morphol. 33:685–695.

15. Guénet, J.-L., H. Condamine, and J. Gaillard. 1981. t^{wPa-1}, t^{wPa-2}, t^{wPa-3}: three new *t* haplotypes in the mouse. Genet. Res. 36:211–217.

16. Herrmann, B., M. Bućan, P.E. Mains, A.-M. Frischauf, L.M. Silver, and H. Lehrach. 1986. Genetic analysis of the proximal portion of the mouse *t* complex: evidence for a second inversion within *t* haplotypes. Cell 44:469–476.

17. Johnson, D.R. 1974. Hairpin-tail: a case of post-reductional gene action in the mouse egg? Genetics 76:795–805.

18. Johnson, D.R. 1975. Further observations on the hairpin-tail (T^{hp}) mutation in the mouse. Genet. Res. 24:207–213.

19. Klein, J., and C. Hammerberg. 1977. The control of differentiation of the *T* complex. Immunol. Rev. 33:70–104.

20. Klein, J., P. Sipos, and F. Figueroa. 1984. Polymorphism of *t*-complex genes in European wild mice. Genet. Res. 44:39–46.

21. Lyon, M.F. 1959. A new dominant *T*-allele in the house mouse. J. Hered. 50:140–142.

22. Lyon, M.F. 1984. Transmission ratio distortion in mouse

t-haplotypes is due to multiple distorter genes acting on a responder locus. Cell 37:621–628.

23. Lyon, M.F. 1986. Male sterility of the mouse *t*-complex is due to homozygosity of the distorter genes. Cell 44:357–363.

24. Lyon, M.F., and K.B. Bechtol. 1977. Derivation of mutant *t*-haplotypes of the mouse by presumed duplication or deletion. Genet. Res.30:63–76.

25. McGrath, J. and D. Solter. 1984. Maternal *T^hp* lethality in the mouse is a nuclear, not cytoplasmic defect. Nature 308:550–551.

26. Moutier, R. 1973. Mouse News Lett. 49:42.

27. Nizetić, D., F. Figueroa, and J. Klein. 1984. Evolutionary relationships between the *t* and *H-2* haplotypes in the house mouse. Immunogenetics 19:311–320.

28. Olds-Clarke, P., and B. Peitz. 1986. Fertility of sperm from *t/+* mice: evidence that +-bearing sperm are dysfunctional. Genet. Res. 47:49–52.

29. Röhme, D., H. Fox, B. Herrmann, A.-M. Frischauf, J.-E. Edström, P. Mains, L.M. Silver, and H. Lehrach. 1984. Molecular clones of the mouse *t* complex derived from microdissected metaphase chromosomes. Cell 36:783–788.

30. Searle, A.G. 1966. Curtailed, a new dominant T-allele in the house mouse. Genet. Res. 7:86–95.

31. Sherman, M.I., and L.R. Wudl. 1977. T-complex mutations and their effects. *In* M.I. Sherman, ed., Concepts in Mammalian Embryogenesis, 136–234. MIT Press, Cambridge, Mass.

32. Shin, H.-S., D. Bennett, and K. Artzt. 1984. Gene mapping within the *T/t* complex of the mouse. IV. The inverted *MHC* is intermingled with several *t*-lethal genes. Cell 39:573–578.

33. Shin, H.-S., L. Flaherty, K. Artzt, D. Bennett, and J. Ravetch. 1983. Inversion in the *H-2* complex of *t*-haplotypes in mice. Nature 306:380–383.

34. Shur, B.D. 1981. Galactosyltransferase activities on mouse sperm bearing multiple *t^lethal* and *t^viable* haplotypes of the T/t complex. Genet. Res. 38:225–236.

35. Silver, L.M. 1985. Mouse *t* haplotypes. Ann. Rev. Genet. 19:179–208.

36. Silver, L.M., and K. Artzt. 1981. Recombination suppression of mouse *t*-haplotypes due to chromatin mismatching. Nature 290:68–70.

37. Silver, L.M., D. Lukralle, and J.I. Garrels. 1983. *T^Orl* is a novel, variant form of mouse chromosome 17 with a deletion in a partial *t* haplotype. Nature 301:422–424.

38. Silver, L.M., J. Uman, J. Danska, and J.I. Garrels. 1983. A diversified set of testicular proteins specified by genes within the mouse *t* complex. Cell 35:35–45.

39. Willison, K.R., K. Dudley, and J. Potter. 1986. Molecular cloning and sequence analysis of a haploid expressed gene encoding *t* complex polypeptide 1. Cell 44:727–738.

40. Winking, H., and L.M. Silver. 1984. Characterization of a recombinant mouse *t* haplotype that expresses a dominant lethal maternal effect. Genetics 108:1013–1020.

T3d, *T3e*, *T3g* loci, T3-δ, T3-ε, T3-γ, Chr 9

These loci code for three of the subunits of the invariable T3 portion of the T-cell receptor complex. Four such subunits are known in the mouse, γ, δ, ε, and μ. Their function may be to transmit a proliferation signal from the T-cell receptor α/β complex through the plasma membrane (3). No genetic variants of these subunits are known. The location of *T3d*, *T3e*, and *T3g* on Chr 9 was found by use of mouse–Chinese hamster somatic cell hybrids screened with cDNA clones coding for the T3-δ, T3-ε, and T3-γ chains (1, 2, 4).

References

1. Gold, V.P., J.J.M. van Dongen, C.C. Morton, G.A.P. Bruns, P. van den Elsen, A.H.M.G. van Kessel, and C. Terhorst. 1987. The gene encoding the ε subunit of the T3/T-cell receptor complex maps to chromosome 11 in humans and to chromosome 9 in mice. Proc. Natl. Acad. Sci. USA 84:1664–1668.

2. Krissansen, G.W., P.A. Gorman, C.A. Kozak, N.K. Spurr, D. Sheer, P.N. Goodfellow, and M.J. Crumpton. 1987. Chromosomal locations of the genes for the CD3(T3)γ subunit of the human and mouse CD3/T-cell antigen receptor complex. Immunogenetics 26:258–266.

3. Samuelson, L.E., J.B. Harford, and R.D. Klausner. 1985. Identification of the components of the murine T cell antigen receptor complex. Cell 43:223–231.

4. van den Elsen, P., G. Bruns, D.S. Gerhard, D. Pravtcheva, C. Jones, D. Housman, F.H. Ruddle, S. Orkin, and C. Terhorst. 1985. Assignment of the gene coding for the T3-δ subunit of the T3-T-cell receptor complex to the long arm of human chromosome 11 and to mouse chromosome 9. Proc. Natl. Acad. Sci. USA 82:2920–2924.

Ta locus, Chr X

Ta, tabby, semidominant. Arose spontaneously in a strain selected for large size. Tabby males breed satisfactorily, but homozygous females are often sterile. Heterozygous females are fully fertile. Hemizygous males and homozygous females are identical in phenotype with homozygous crinkled (*cr*) and downless (*dl*) mice and with homozygous or heterozygous sleek (*Dl^slk*) mice. They are characterized by absence of guard hairs and zigzags in the coat, a bald patch behind the ear, bald tail with a few kinks near the tip, reduced aperture of the eyelids, a respiratory disorder, and a modified agouti pattern (2). The number of vibrissae is reduced (3). The incisors may be reduced or absent, and the molars are usually smaller than normal with the third molar often absent (4, 12). There are defects of many endocrine glands. The structures affected by *Ta* all arise embryologically as downgrowths of solid epithelial cords, not by invagination with a lumen or by outgrowths from deep grooves (7). Heterozygous females are most easily recognized if they are agouti, in which case they show transverse stripes of light-colored normal and dark tabby hair. They have normal incisors but may have mutant- or intermediate-type molars (12). A small proportion of heterozygous females may show some slight defects of some of the exocrine glands (6). In the development of the coat of homozygotes and hemizygotes, hair follicle initiation begins at 17 days of gestation, 3 days later than normal, and ends 1 or 2 days after birth, several days earlier than normal. The hairs are of only one type and resemble abnormal awls (1, 5). By use of dermal–epidermal recombination grafts of embryonic flank skin, it was shown that *Ta* acts in the epidermis in its effects on structure of the hairs (9). By use of a similar technique with tail skin, it could not be shown that the effect of *Ta* in preventing growth of hair on the tail was confined to either the dermis or the epidermis. It is unclear whether *Ta* acts directly on the cells of the hair or through a diffusible product (10). The phenotype of *Ta*/+ females has been extensively studied because of its relevance to the X-inactivation theory of dosage compensation (4, 8).

Ta^c, tabby-c, semidominant. Arose in a line selected for body weight at the Institute of Animal Genetics at Edinburgh (11). It is indistinguishable from *Ta^J*. *Ta^c*/+ females are more nearly normal with respect to number of secondary vibrissae than *Ta*/+ females, but it is not certain whether this is an intrinsic effect of the *Ta* mutants or is due to a difference at the closely linked *Xce* locus controlling inactivation properties of the X chromosome (13).

Ta^J, tabby-J, semidominant. Arose spontaneously in the 129/Sv inbred strain. *Ta^J* resembles *Ta* except that hemizygous males and homozygous females have some hair on the tails, and the hairs are curved. *Ta/Ta^J* females have both hairless areas and areas with curved hairs on their tails (14).

References

1. Claxton, J.H. 1967. The initiation and development of the hair follicle population in tabby mice. Genet. Res. 10:161–171.
2. Falconer, D.S. 1953. Total sex-linkage in the house mouse. Z. Indukt. Abstammungs-Vererbungsl. 85:210–219.
3. Fraser, A.S., and B.M. Kindred. 1960. Selection for an invariant character, vibrissa number, in the house mouse. II. Limits to variability. Austral. J. Biol. Sci. 13:48–58.
4. Grüneberg, H. 1966. The molars of the tabby mouse, and a test of the "single-active X-chromosome" hypothesis. J. Embryol. Exp. Morphol. 15:223–244.
5. Grüneberg, H. 1969. Threshold phenomena versus cell heredity in the manifestation of sex-linked genes in mammals. J. Embryol. Exp. Morphol. 22:145–179.
6. Grüneberg, H. 1971. The glandular aspects of the tabby syndrome in the mouse. J. Embryol. Exp. Morphol. 25:1–19.
7. Grüneberg, H. 1971. The tabby syndrome in the mouse. Proc. R. Soc. Lond. (Biol.) 179:139–156.
8. Lyon, M.F. 1970. Genetic activity of sex chromosomes in somatic cells of mammals. Phil. Trans. R. Soc. Lond. (Biol.) 2509:41–52.
9. Mayer, T.C., and M.C. Green. 1978. Epidermis is the site of action of tabby (*Ta*) in the mouse. Genetics 90:125–131.
10. Pennycuik, P.R., and K.A. Raphael. 1984. The tabby locus (*Ta*) in the mouse: its site of action in tail and body skin. Genet. Res. 43:51–63.
11. Roberts, R.C. 1966. Mouse News Lett. 35:24.
12. Sofaer, J.A. 1969. Aspects of the tabby-crinkled-downless syndrome. I. The development of tabby teeth. J. Embryol. Exp. Morphol. 22:181–205.
13. Sofaer, J.A., and C.J. MacLean. 1970. Dominance in threshold characters. A comparison of two tabby alleles in the mouse. Genetics 64:273–290.
14. Stevens, L.C. 1963. Mouse News Lett. 29:40.

Tag locus, temporal α-galactosidase

Mice of the C57 and C58 family of strains have elevated levels of activity of α-galactosidase (E.C. 3.2.1.22) in liver beginning at about 25 days of age and reaching a twofold increase over that of the CBA/J, DBA/2, and C3H/He strains by 60 days. The enzyme levels in other organs are not affected (1). F1 hybrids between high and low strains have intermediate levels of liver enzyme activity. Segregation into distinct phenotypic classes does not occur in backcross and F2 progeny, but inbred strains and recombinant inbred strains in most cases separate into distinct high and low classes. The provisional symbol *Tag* has been assigned to the major locus responsible for this difference (2).

References

1. Lusis, A.J., and K. Paigen. 1975. Genetic determination of the α-galactosidase developmental program in mice. Cell 6:371–378.
2. Lusis, A.J., and J.D. West. 1978. X-linked and autosomal genes controlling mouse α-galactosidase expression. Genetics 88:327–342.

Tal, talipes, semidominant

Found among offspring of a male treated with combined hydroxyurea and X-rays. Heterozygotes cannot easily be identified at birth but can be classified with confidence at weaning. The feet are held in an abnormal position, there is soft tissue syndactylism involving digits 2, 3, and to a lesser extent 4, and there is often overgrowth and twisting of the toenails. The forefeet are less obviously affected than the hindfeet. Viability of heterozygotes at weaning is about 80 per cent. Homozygotes are more severely affected than heterozygotes with extreme shortening of the phalanges and significant reduction of the metatarsals. The tail shows dorsal and upward kinking to form a hook near the tip. Long bones are not affected. About 40 per cent of homozygotes survive to weaning and none have bred (1).

References

1. Cattanach, B.M., and C. Rasberry. 1987. A dominant mutation affecting the feet. Mouse News Lett. 77:123.

Tam-1 locus, tosyl arginine methylesterase-1, Chr 7

This locus controls electrophoretic mobility of a testosterone-induced esterolytic protease, tosyl arginine methylesterase, in submaxillary gland. The allele *Tam-1ᵃ* determines a fast cathodal double-banded pattern and occurs in strains A/J, BALB/c, NZB, and others; the allele *Tam-1ᵇ* determines a slower double band and occurs in DBA/2J, SJL/J, and a few others; the allele *Tam-1ᶜ* determines absence of enzyme activity and occurs in strains C57BL/6J, AKR/J, and others. *Tam-1ᵃ* and *Tam-1ᵇ* are codominant; *Tam-1ᶜ* is recessive to both the other alleles (1).

References

1. Skow, L.C. 1978. Genetic variation at a locus (*Tam-1*) for submaxillary gland protease in the mouse and its location on chromosome 7. Genetics 90:713–724.

TAP

See *Ly-6* locus.

Tas, T-associated sex reversal, dominant or semidominant, Chr 17

This mutation was found in a stock carrying hairpin tail (*Tʰᵖ*) produced by repeatedly crossing *Tʰᵖ* into the C57BL/6J strain. *Tʰᵖ* is a deletion extending as far as the *qk* locus. *Tʰᵖ/+* males from this stock are all partially or completely sex-reversed. When *Tʰᵖ* is crossed for several generations into the C3H/HeJ strain, no sex reversal of males occurs. One explanation of these results is that a locus, *Tas*, in the part of Chr 17 deleted in the *Tʰᵖ* deletion has an allele in the C57BL/6 strain

that must be present in double dose for normal male sexual differentiation; in the C3H/HeJ strain only a single dose of *Tas* is necessary, either because it is a different allele from that in C57BL/6 or because of differences in the effect of the background genome. The possibility that the sex-reversal effect is a pleiotropic effect of the *Tʰᵖ* gene or that it is an effect of a closely linked *de novo* mutation is not ruled out (1).

References

1. Washburn, L.L., and E.M. Eicher. 1983. Sex reversal in XY mice caused by dominant mutation on chromosome 17. Nature 303:338–340.

Tat locus, tyrosine aminotransferase, Chr 8

This locus codes for tyrosine aminotransferase (TAT; E.C. 2.6.1.5), a liver-specific enzyme that is developmentally and hormonally regulated. The gene contains 12 exons and extends over 9.2 kb. A DNA restriction fragment variant occurs in *Mus spretus*. The allele in DBA/2 and other laboratory strains (*Tatᵈ*) determines a *Taq*I fragment of 5.2 kb; the allele in *M. spretus* (*Tatˢ*) determines *Taq*I fragments of 5.2 and 4.0 kb. This polymorphism was used to map *Tat* to Chr 8 about 26 cM distal to *Es-1* (1). *Tat* was also shown to be on Chr 8 by use of mouse–rat and mouse–Chinese hamster somatic cell hybrids. Expression of TAT in liver cells can be prevented by introduction of a Chr 11 from kidney cells, an effect due to presence on Chr 11 of the *Tse-1* locus (tissue-specific extinguisher-1) (2).

References

1. Müller, G., G. Scherer, H. Zentgraf, S. Ruppert, B. Herrmann, H. Lehrach, and G. Schütz. 1985. Isolation, characterization and chromosomal mapping of the mouse tyrosine aminotransferase gene. J. Mol. Biol. 184:367–373.
2. Peterson, T.C., H.M. Killary, and R.E.K. Fournier. 1985. Chromosomal assignment and *trans* regulation of the tyrosine aminotransferase structural gene in hepatoma hybrid cells. Mol. Cell. Biol. 5:2491–2494.

tb, tumbler, recessive, Chr 1

Found in a special stock of C.P. Dagg. By 15 days homozygotes walk in a crab-like fashion. They may somersault, fall over, or jump when trying to go forward. They can swim but cannot hold on to a rope. Most homozygotes survive and can breed (1). No lesions have been found in the central nervous system, but the endocrine and lymphoid tissues show some abnormalities. The cells of the anterior pituitary have a diminished amount of cytoplasm (2). Growth hormone cells show signs of degeneration by 90 days of age (3). The thymus undergoes involution after the onset of neurological symptoms (2).

References

1. Dickie, M.M. 1965. Mouse News Lett. 32:45.
2. Dung, H.C. 1975. Growth retardation, high mortality, and low reproductivity of neurological mutant mice. Anat. Rec. 181:347–348 (Abstr.).
3. Lair, A.C. 1971. Observations on the fine structure of the alpha adenohypophyseal cell of the male tumbler mouse. Anat. Rec. 169:361 (Abstr.).

tc, truncate, recessive, Chr 6

Arose spontaneously in a Swiss albino stock at the Jackson Laboratory (1). This mutation was originally called boneless and was later shown to be allelic with two other mutations of independent origin, truncate and jointed, now extinct. Penetrance is incomplete. Homozygotes have short or absent tails, and many have missing sacral or lumbar vertebrae. In some cases several vertebrae in an intermediate position seem to be 'picked out'. The primary effect is an interruption of the notochord at $9\frac{1}{2}$ to 10 days of gestation. The sclerotomic cells, which normally migrate from the somites to the midline under the neural tube to form the vertebrae, degenerate in the region of the interrupted notochord. These defects lead eventually to interruption of the spine, paralysis of the hindlegs, difficult parturition in females, and absence of the median ventral fissure of the spinal cord (2).

References

1. Theiler, K. 1957. Boneless tail, ein recessives autosomales Gen der Hausmaus. Arch. Julius Klaus-Stift. Vererb. 32:474–481.
2. Theiler, K. 1959. Anatomy and development of the "truncate" (boneless) mutation in the mouse. Am. J. Anat. 104:319–343.

Tcd loci

See *t* complex.

tcl loci

See *t* complex.

Tcn-2 locus, transcobalamin-2, Chr 11

This locus controls electrophoretic variation in transcobalamin-2, the only vitamin B12-binding protein found in mouse serum. The protein has a molecular weight of 28 000 and was recognized by its ability to bind cyanocobalamin. The allele $Tcn-2^s$ determines a slow migrating form and occurs in serum from 25 strains including A/J, C57BL/6, BALB/c, and C3H/He; the allele $Tcn-2^f$ determines a more rapidly migrating form and was found only in the NZB, ST/b, and CPB-WV strains. Heterozygotes have both parental forms (1). *Tcn-2* is on Chr 11 about 7 cM proximal to *wa-2*, probably very close to the centromere (2).

References

1. Fràter-Schröder, M., O. Haller, R. Gmür, L. Kierat, and S. Anastasi. 1982. Allelic forms of mouse transcobalamin-2. Biochem. Genet. 20:1011–1014.
2. Fràter-Schröder, M., M. Prochazka, O. Haller, F. Arwert, H.J. Porch, L.C. Skow, L.-G. Lundin, J. Hilkens, and J. Hilgers. 1985. Localization of the gene for the vitamin B12 binding protein, transcobalamin II, near the centromere on mouse chromosome 11, linked to the hemoglobin alpha-chain locus. Biochem. Genet. 23:139–153.

Tcp loci

See *t* complex.

Tcr

See *t* complex.

Tcra complex, T-cell receptor α-chain, Chr 14

This complex contains the multiple genes that code for the components of the T-cell receptor α-chain, and that are rearranged and assembled into a complete gene during maturation of T lymphocytes. The α-chain combines with the T-cell receptor β-chain to form the antigen receptor that occurs on most T-cells. The *Tcra* complex contains, in order from 5' to 3', a large number of variable region (*V*) genes, a large number of joining region (*J*) genes, and a single constant region (*C*) gene (4, 8). An unusual feature of the *Tcra* complex is that it contains, in the region between the *V* and *J* regions, the *J* and *C* regions of the gene coding for *Tcrd*, the T-cell receptor δ-gene. See *Tcrd*. *Tcra* is transcribed at very low levels in the immature thymus of 16-day embryos and rises rapidly through birth to adulthood (6, 7). *Tcra* was first shown to be on Chr 14 by use of mouse–Chinese hamster somatic cell hybrids screened with a cDNA probe of the combined *V* and *C* sequences of a rearranged gene (5). A DNA restriction fragment length difference in the *Tcra-C* region was found between the C57BL/6 and DBA/2 strains: C57BL/6 has a *Bgl*I fragment of 9.4 kb; DBA/2 has a *Bgl*I fragment of 12 kb. BALB/c and C3H/He also differ from C57BL/6 in a *Tcra* DNA fragment. In 26 BXD, 12 BXH, and 7 CXB recombinant inbred strains, *Tcra* showed complete concordance with *Np-2* located in the proximal portion of Chr 14 (3). The C57BL/6 strain has also been shown to differ from other strains in the restriction fragment length pattern of several genes in the variable region of *Tcra* (*Tcra-V*) (1, 2).

References

1. Arden, B., J.L. Klotz, G. Siu, and L.E. Hood. 1985. Diversity and structure of genes of the α family of mouse T-cell antigen receptor. Nature 316:783–787.

2. Berman, J.W., A.J.D. Rocha, and R. Basch. 1986. Restriction length polymorphism in the variable region of the *Tcr* locus linked to histocompatibility antigen H-8 on murine chromosome 14. Immunogenetics 24:328–330.

3. Dembić, Z., W. Bannwarth, B.A. Taylor, and M. Steinmetz. 1985. The gene coding for the T-cell receptor α-chain maps close to the *Np-2* locus on mouse chromosome 14. Nature 314:271–273.

4. Hayday, A.C., D.J. Diamond, G. Tanigawa, J.S. Heilig, V. Folsom, H. Saito, and S. Tonegawa. 1985. Unusual organization and diversity of T-cell receptor α-chain genes. Nature 316:828–832.

5. Kranz, D.M., H. Saito, C.M. Disteche, K. Swinholm, D. Pravtcheva, F.H. Ruddle, H.N. Eisen, and S. Tonegawa. 1985. Chromosomal locations of the murine T-cell receptor alpha-chain gene and the T-cell gamma gene. Science 227:941–945.

6. Raulet, D.H., R.D. Garman, H. Saito, and S. Tonegawa. 1985. Developmental regulation of T-cell receptor gene expression. Nature 314:103–107.

7. Snodgrass, H.R., Z. Dembić, M. Steinmetz, and H. von Boehmer. 1985. Expression of T-cell antigen receptor genes during fetal development in the thymus. Nature 315:232–233.

8. Winoto, A., S. Mjolaness, and L. Hood. 1985. Genomic organization of the genes encoding mouse T-cell receptor α-chain. Nature 316:832–836.

Tcrb complex, T-cell receptor β-chain, Chr 6

This complex contains the multiple genes that code for the components of the T-cell receptor β-chain. The β-chain combines with the T-cell receptor α-chain to form the antigen receptor that occurs on most T-cells. The complex contains, in order from 5' to 3', a relatively small number of variable (*V*) genes (approximately 14 to 18) and two groups of genes, each comprising a diversity (*D*) gene, a cluster of about

seven joining (*J*) genes, and a single constant (*C*) gene (1, 5, 9). The genes are rearranged during maturation of T-cells to produce a functional gene composed of a *VDJC* combination. Most of the diversity among mature β-chains is thought to arise from variation in the *V–D–J* junctions formed during the rearranging process (1, 6). *Tcrb* gene expression begins before embryonic day-15. *D–J* joining occurs first and the more complex rearrangements are present by embryonic day-16, one day before expression of the receptor protein (2, 10). *Tcrb* polymorphisms have been detected by DNA restriction fragment length differences using several different endonucleases and by reaction of T-cells with a monoclonal antibody (allotype). A DNA fragment length variant occurs in strains SJL and SWR, all other strains including BALB/c and C57L having the major pattern (1, 4). The monoclonal antibody recognizes an antigen of T-cells in BALB/c and all other strains tested except SJL, SWR, C57BR, and C57L. Since C57L has the major DNA pattern and the variant allotype, the two variants must be due to different mutational events (4, 11). An additional variant occurs in the NZB strain. It consists of a deletion in the *Tcrb* gene extending from the first exon of the *C1* region through the *D2* and *J2* regions to the first exon of *C2*. Whether any relationship exists between the deletion and the autoimmune condition of the NZW strain is unknown (8). *Tcrb* was first shown to be on Chr 16 by use of mouse–Chinese hamster somatic cell hybrids (7), and later mapped to a position about 8 cM proximal to the *Igk* complex on that chromosome (3, 11).

References

1. Behlke, M.A., D.G. Spinella, H.S. Chou, W. Sha, D.L. Hartl, and D.Y. Loh. 1985. T-cell receptor β-chain expression: dependence on relatively few variable region genes. Science 229:566–570.

2. Born, W., J. Yagüe, E. Palmer, J. Kappler, and P. Marrack. 1985. Rearrangement of T-cell receptor β-chain genes during T-cell development. Proc. Natl. Acad. Sci. USA 82:2925–2929.

3. D'Hoostelaere, L.A., E. Jouvin-Marche, and K. Huppi. 1985. Localization of C_{Tβ} and C_x on mouse chromosome 6. Immunogenetics 22:277–283.

4. Epstein, R., N. Roehm, P. Marrack, J. Kappler, M. Davis, S. Hedrick, and M. Cohn. 1985. Genetic markers of the antigen-specific T cell receptor locus. J. Exp. Med. 161:1219–1224.

5. Gascoigne, N.R.J., Y.-H. Chien, D.M. Becker, J. Kavaler, and M.M. Davis. 1984. Genomic organization and sequence of T-cell receptor β-chain constant- and joining-region genes. Nature 310:387–391.

6. Hood, K., M. Kronenberg, and T. Hunkapiller. 1985. T

cell antigen receptors and the immunoglobulin supergene family. Cell 40:225–229.

7. Lee, N.E., P. D'Eustachio, D. Pravtcheva, F.H. Ruddle, S.M. Hedrick, and M.M. Davis. 1984. Murine T cell receptor beta chain is encoded on mouse chromosome 6. J. Exp. Med. 160:905–913.

8. Noonan, D.J., R. Kofler, P.A. Singer, G. Cardenas, F.J. Dixon, and A.N. Theofilopoulos. 1986. Delineation of a defect in T cell receptor β genes of NZW mice predisposed to autoimmunity. J. Exp. Med. 163:644–653.

9. Patten, P., T. Yokota, J. Rothbard, Y. Chien, K. Arai, and M.M. Davis. 1984. Structure, expression and divergence of T-cell β-chain variable regions. Nature 312:40–46.

10. Raulet, D.H., R.D. Garman, H. Saito, and S. Tonegawa. 1985. Developmental regulation of T-cell receptor gene expression. Nature 314:103–107.

11. Roehm, N.W., A. Carbone, E. Kushnir, B.A. Taylor, R.J. Riblet, P. Marrack, and J.W. Kappler. 1985. The major histocompatibility complex-restricted antigen receptor on T cells: the genetics of expression of an allotype. J. Immunol. 135:2176–2182.

Tcrd complex, T-cell receptor δ-chain, Chr 14

This complex contains a *C* region and at least one *J* region gene for the δ-chain of the γ/δ T-cell receptor. This receptor occurs on a small subpopulation of thymic and peripheral T-cells and on Thy-1$^+$ dendritic epidermal cells. The DNA of the *Tcrd* complex is located within the *Tcra* complex, between the V_α and J_α regions (2, 3). V_δ DNA sequences have not been identified and it is possible that the V_α sequences are used in constructing the mature *Tcrd* gene (1, 3). *Tcrd* rearrangements occur at least by 14 days of gestation, well before any *Tcra* message or protein is detected, and they may preclude *Tcra* rearrangements in the cells in which they occur (2, 3). No genetic variants of *Tcrd* have been described.

References

1. Born, W., C. Miles, J. White, R. O'Brien, J.H. Freed, P. Marrack, J. Kappler, and R.T. Kubo. 1987. Peptide sequences of T-cell receptor δ and γ chains are identical to predicted X and γ proteins. Nature 330:572–574.

2. Bonyhadi, M., A. Weiss, P.W. Tucker, R.E. Tigelaar, and J.P. Allison. 1987. Delta is the C_x-gene product in the γ/δ antigen receptor of dendritic epidermal cells. Nature 330:574–576.

3. Chien, Y.-H, M.Iwashima, K.B. Kaplan, J.F. Elliott, and M.M. Davis. 1987. A new T-cell receptor gene located within the alpha locus and expressed early in T-cell differentiation. Nature 327:677–682.

Tcrg complex, T-cell receptor γ-chain, Chr 13

This complex contains the multiple genes that code for components of the T-cell γ-chain. The γ-chain combines with the δ-chain to form a dimeric receptor that is expressed on a small subpopulation of thymic and peripheral T-cells. The *Tcrg* complex contains, in order from 5' to 3', three variable (*V*) genes and three pairs of joining (*J*) and constant (*C*) genes. It is likely that only one of the *V* and one of the *J* genes are productively recombined in mature T-cells. Most of the diversity is generated by variation at the *V–J* junction (1, 2, 4). *Tcrg* is expressed primarily in immature T-cells. The level declines as the cells mature at a time when expression of *Tcra* (α-chain) is increasing (7, 8). Five different DNA restriction fragment patterns among inbred strains are revealed by use of *Eco*RV endonuclease and a 700-bp probe containing the *C* and part of the *J* region of *Tcrg*. Haplotype *Tcrg*^{a} occurs in strains C3H/He, DBA/2, CE, CBA/J, and others; *Tcrg*^{b} occurs in C57BL/6, BALB/c, and others; *Tcrg*^{c} occurs in AKR, ST/b, NZB, and others; *Tcrg*^{d} occurs in I/Ln; and *Tcrg*^{e} occurs in 129/J and A/J. In addition, *Bam*II and *Eco*RI distinguish between the two *Tcrg*^{a} strains C3H/He and DBA/2 (6), and C57BL/10 lacks a *V–J–C* sequence present in BALB/c (3). *Tcrg* was found to be on Chr 13 in the proximal region by use of mouse–Chinese hamster somatic cell hybrids and *in situ* hybridization (5). It was later found to be closely linked to *bg* and *Hist1* near the centromere of that chromosome by use of recombinant inbred strains (6).

References

1. Hayday, A.C., H. Saito, S.D. Gillies, D.M. Kranz, G. Tanigawa, H.N. Eisen, and S. Tonegawa. 1985. Structure, organization, and somatic rearrangement of T cell gamma genes. Cell 40:259–269.

2. Hood, L., M. Kronenberg, and T. Hunkapiller. 1985. T cell antigen receptors and the immunoglobulin supergene family. Cell 40:225–229.

3. Iwamomoto, A., F. Rupp, P.S. Ohashi, C.L. Walker, H. Pircher, R. Joho, H. Hengartner, and T.W. Mak. 1986. T cell-specific γ genes in C57BL/10 mice: sequence and expression of new constant and variable region genes. J. Exp. Med. 163:1203–1212.

4. Kranz, D.M., H. Saito, M. Heller, Y. Takagaki, W. Haas, H.N. Eisen, and S. Tonegawa. 1985. Limited diversity of the rearranged T-cell γ gene. Nature 313:752–755.

5. Kranz, D.M., H. Saito, C.M. Disteche, K. Swinhelm, D. Pravtcheva, F.H. Ruddle, H.N. Eisen, and S. Tonegawa. 1985. Chromosomal location of the murine T-cell receptor alpha-chain gene and the T-cell gamma chain gene. Science 227:941–945.

6. Owen, F.L., B.A. Taylor, A. Zweidler, and J.G. Seidman. 1986. The murine γ-chain of the T cell receptor is closely linked to a spermatocyte specific histone gene and the beige coat color locus on chromosome 13. J. Immunol. 137:1044–1046.
7. Raulet, D.H., R.D. Garman, H. Saito, and S. Tonegawa. 1985. Developmental regulation of T-cell receptor gene expression. Nature 314:103–107.
8. Snodgrass, H.R., Z. Dembić, M. Steinmetz, and H. von Boehmer. 1985. Expression of T-cell antigen receptor genes during fetal development in the thymus. Nature 315:232–233.

Tcs loci

See *t* complex.

tct

See *t* complex.

Td, tattered, semidominant, Chr X

Found among offspring of an X-irradiated male. *Td*/Y males die *in utero*. *Td*/+ females show reduced viability (60 to 80 per cent of normal) with losses occurring both before and after birth. They can be recognized at 5 to 6 days of age by scarring of the skin in patches or streaks that may extend over the whole body. Hair growth is restricted to the unscarred regions. The more severely affected females have many small irregular bald areas, giving the coat a tattered appearance. At weaning, affected females tend to be smaller than normal (1). *Td* is located very close to *spf* at the centromeric end of Chr X (2).

References

1. Cattanach, B.M. 1982. Mouse News Lett. 66:61.
2. Cattanach, B.M. 1982. Mouse News Lett. 67:19.

td

See *du^td*.

Tda-1 locus, testis-determining autosomal-1, Chr 2?

This locus was detected after introduction of the X chromosome from *Mus poschiavinus* (a geographic variety of *M. domesticus*) into the C57BL/6J strain. The allele *Tda-1^b* occurs in the C57BL/6 strain. Homozygous *Tda-1^b*/*Tda-1^b* mice carrying the *M. poschiavinus* Y chromosome develop as true hermaphrodites or, more rarely, as females. The allele *Tda-1^d* occurs in *M. poschiavinus*, and also in the other laboratory strains DBA/2, C58, and BALB/c. Homozygous *Tda-1^d*/*Tda-1^d* or heterozygous *Tda-1^d*/*Tda-1^b* mice develop as normal males whether they carry the C57BL/6 or *M. poschiavinus* Y chromosome (1). The authors hypothesize that the *Tda-1* gene, when activated by the *Tdy* gene on the Y chromosome, commits gonad development to the testis development pathway. However, the combination of the *Tda-1^b* allele from C57BL/6 and the *Tdy^d* allele from *M. poschiavinus* either fails to achieve proper activation or fails to properly inactivate a female determining pathway (1). When females of the BXD recombinant inbred strains were crossed to C57BL/6 males consomic for the *M. poschiavinus* Y chromosome, half proved to be carrying *Tda-1^b* and half *Tda-1^d*. The strain distribution pattern suggested location of *Tda-1* on Chr 2, and also gave evidence for an additional autosomal testis-determining locus with different alleles in C57BL/6 and DBA/2 (2). See *Tdy* locus.

References

1. Eicher, E.M., and L.L. Washburn. 1983. Inherited sex reversal in mice: identification of a new primary sex-determining gene. J. Exp. Zool. 228:297–304.
2. Eicher, E.M., and L.L. Washburn. 1986. Identification and chromosomal position of genes involved in primary sex determination. Jackson Lab. Ann. Rep. 56:40–41.

Tdt locus, terminal deoxynucleotidyltransferase, Chr 19

This locus codes for TdT, an enzyme that catalyzes additions of deoxynucleotides to the 3' hydroxyl end of DNA primers without a template requirement. The enzyme is a nuclear protein of 58 to 60 kDa and is present in cortical thymocytes, primitive marrow lymphocytes, and in some leukemias. It may play a role in the insertion of nucleotides at joining junctions of immunoglobulin heavy-chain genes and possibly also of T-cell receptor genes. *Tdt* was found to be on Chr 19 by use of mouse–Chinese hamster somatic cell hybrids screened with a full-length mouse TdT cDNA probe (1).

References

1. Yang-Feng, T.L., N.R. Landau, D. Baltimore, and U. Francke. 1986. The terminal deoxynucleotidyltransferase gene is located on human chromosome 10 (10q23→q24) and on mouse chromosome 19. Cytogenet. Cell Genet. 43:121–126.

Tdy locus, testis determining-Y, Chr Y

This locus was detected after introduction of the Y chromosome from *Mus poschiavinus* into the C57BL/6J strain. The allele on the Y chromosome of *M. poschia-vinus*, Tdy^d, in combination with at least part of the remaining genome of the C57BL/6 strain, produces XY mice that develop as females or as true hermaphrodites (2). It was later found that presence in homozygous condition of the C57BL/6 allele at the autosomal locus ($Tda-1^b/Tda-1^b$), in addition to Tdy^b, was responsible for the observed sex reversal (1). The *Tdy* allele in the C57BL/6 strain, Tdy^b, produces normal males in combination with $Tda-1^b$. It is not known whether *Tdy* is a single gene, since the Y chromosome is transmitted intact between generations. See *Tda-1* locus.

References

1. Eicher, E.M., and L.L. Washburn. 1983. Inherited sex reversal in mice: identification of a new primary sex-determining gene. J. Exp. Zool. 228:297–304.
2. Eicher, E.M., L.L. Washburn, J.B. Whitney, III, and K.E. Morrow. 1982. *Mus poschiavinus* Y chromosome in the C57BL/6J murine genome causes sex reversal. Science 217:535–537.

ter, teratoma, incompletely recessive

Occurred in strain 129/Sv-$c^+ p^+$. In heterozygous males on this strain background, the incidence of spontaneous testicular teratomas is 33 per cent in comparison to 5 per cent in normal mice of the same strain. Heterozygous males and females are normally fertile. In homozygous males and females, the gonads at 15 days of gestation have many fewer primordial germ cells than normal. Adult females have small ovaries but produce a few litters. Adult males are sterile. On the strain 129 genetic background, 95 per cent of homozygous males have spontaneous teratomas, mostly bilateral. On the strain C57BL/6 genetic background, homozygous males and females are germ cell deficient but males do not develop teratomas (1).

References

1. Noguchi, T., and M. Noguchi. 1985. A recessive mutation (ter) causing germ cell deficiency and a high incidence of congenital testicular teratomas in 129/Sv-*ter* mice. J. Natl. Cancer Inst. 75:385–392.

tf, tufted, recessive, Chr 17

Probably arose spontaneously in a multiple recessive stock. Homozygotes show repeated waves of hair loss and regrowth which begin in the nose and pass posteriorly along the body. Viability and penetrance are good (1). *tf* is linked to the *T/t* complex and is used as a marker gene in balanced lethal *t*-haplotype stocks to monitor the occurrence of recombinant haplotypes. Two independent mutations to *tf* alleles have occurred within *t* haplotypes (2).

References

1. Lyon, M.F. 1956. Hereditary hair loss in the tufted mutant of the house mouse. J. Hered. 47:101–103.
2. Silver, L.M., L. Cisek, C. Jackson, and D. Lukralle. 1985. A new spontaneous mutation at the tufted locus within a mouse *t* haplotype. Genet. Res. 45:107–112.

Tfm, testicular feminization, Chr X

Found among offspring of a female homozygous for the X-linked genes *Ta* and Mo^{blo}. The literature up to 1981 has been extensively reviewed by Lyon *et al.* (9). Briefly, hemizygous males are outwardly female in appearance except that the vagina does not open until 3 months of age or not at all. Internally there are very small testes but no vas deferens or epididymis, and no male accessory glands. Spermatogonia and Sertoli cells are present in the testes, but spermatogenesis does not proceed past meiotic prophase (8). Fusion chimeras between *Tfm*/Y and +/Y embryos developed into fertile males and were used to produce *Tfm/Tfm* females, which proved to be fertile but with somewhat reduced fertility (7). *Tfm* acts by causing insensitivity to androgens. Almost all responses to androgens are absent or defective in *Tfm*/Y males and are not inducible by administration of androgens (3, 4). In *Tfm*/+ females they are inducible to an intermediate degree (10). The insensitivity of *Tfm*/Y mice to androgens is due to a defect in androgen-binding receptor present in the cytosol and the nucleus. The receptor is very much reduced in *Tfm*/Y mice and present in intermediate amounts in *Tfm*/+ females. The receptor has been demonstrated in kidney, submandibular gland (11), and brain (12). *Tfm*/Y males have about 20 to 25 per cent of normal testosterone-binding activity in brain (1) and 10 to 20 per cent in kidney. The residual receptors in kidney are indistinguishable by isoelectric focusing and DNA-cellulose eluting conditions from a minor component of normal androgen receptor (5). A second allele, Tfm^L, was found at the Laboratory Animal Centre, Carshalton. Preliminary experiments indicate that androgen receptors in mice bearing this allele are more severely reduced than in *Tfm*, but the residual receptors appear to be indistinguishable from wild-type receptors (6). *Tfm* has been used in recombinations of epithelium and mesenchyme from mammary gland and urogenital tissues of *Tfm*/Y and +/Y embryos to show that mesenchyme, not epithelium, is the target tissue for testosterone (2).

References

1. Attardi, B. and S. Ohno. 1978. Physical properties of androgen receptors in brain cytosol from normal and testicular feminized (Tfm/Y mice). Endocrinology 103:760–770.

2. Cunha, G.R., and B. Lung. 1978. Androgen and estrogen receptors in brain cytosol from male, female, and testicular feminized (Tfm/Y) mice. Endocrinology 98:864–874.

3. Dofuku, R., U. Tettenborn, and S. Ohno. 1971. Testosterone "regulon" in the mouse kidney. Nature New Biol. 232:5–7.

4. Dunn, J.F., and J.D. Wilson. 1975. Developmental study of androgen responsiveness in the submandibular gland of the mouse. Endocrinology 96:1571–1578.

5. Fox, T.O., and S.J. Wieland. 1981. Isoelectric focusing of androgen receptors from wild-type and *Tfm* mouse kidneys. Endocrinology 109:790–797.

6. Fox, T.O., D. Black, and J.A. Politch. 1983. Residual androgen binding in testicular feminization (TFM). J. Steroid Biochem. 19:577–581.

7. Lyon, M.F., and P.H. Glenister. 1980. Reduced reproductive performance in androgen-resistant *Tfm/Tfm* female mice. Proc. R. Soc. Lond. (Biol.) 208:1–12.

8. Lyon, M.F., and S.G. Hawkes. 1970. X-linked gene for testicular feminization in the mouse. Nature 227:1217–1219.

9. Lyon, M.F., B.M. Cattanach, and H.M. Charlton. 1981. Genes affecting sex differentiation in mammals. *In* C.R. Austin and R.G. Edwards, eds., Mechanisms of Sex Differentiation in Animals and Man, 329–386. Academic Press, New York.

10. Tettenborn, U., R. Dofuku, and S. Ohno. 1971. Noninducible phenotype exhibited by a proportion of female mice heterozygous for the X-linked testicular feminization mutation. Nature New Biol. 234:37–40.

11. Verhoeven, G., and J.D. Wilson. 1976. Cytosol androgen binding in submandibular gland and kidney of the normal mouse and the mouse with testicular feminization. Endocrinology 99:79–92.

12. Wieland, S.J., T.O. Fox, and C. Savakis. 1978. DNA-binding of androgen and estrogen receptors from mouse brain: behavior of residual androgen receptor from *Tfm* mutant. Brain Res. 140:159–164.

tg locus, Chr 8

tg, tottering, recessive. Arose spontaneously in the DBA/2 inbred strain. Viability is nearly normal. Homozygotes of both sexes are fertile, but breeding performance is low. Affected mice are characterized by a wobbly gait affecting particularly the hindquarters, which begins at about 3 to 4 weeks, and by intermittent spontaneous seizures (1). The seizures are of two kinds. (i) Sudden arrests of movement (behavioral absence seizures) begin at about 3 weeks and are accompanied by abnormal bursts of bilaterally synchronous spike waves in electrocorticograms (ECG); they last about 0.3 to 10 seconds, occur hundreds of times per day, and continue throughout life. (ii) Stereotyped partial motor seizures begin at about 4 weeks and are accompanied by abnormal ECG activity; they last 20 to 30 minutes, occur once or twice a day, and persist throughout life (8). The central nervous system appears normal by light microscopy (7). There is no discernible cerebellar hypoplasia (6). However, studies using fluorescent histochemistry have shown that *tg/tg* mice have a marked increase in number of noradrenergic fibers in the terminal fields innervated by locus ceruleus axons, the hippocampus, cerebellum, and dorsal lateral geniculate (3). Treatment of neonatal *tg/tg* mice with 6-hydroxydopamine, which selectively causes degeneration of distal noradrenergic axons from the locus ceruleus, almost completely abolishes the ataxic and seizure symptoms and the abnormal ECG patterns (7). These results suggest that the primary effect of *tg* may be in the neurons of the locus ceruleus.

tg^{la}, leaner, recessive. Arose spontaneously in the AKR/J strain. Many homozygotes die at weaning time, but some survive and females may breed. Homozygotes are recognized at 8 to 10 days of age. They show ataxia, stiffness, and retarded motor activity. Adults are characterized by instability of the trunk and hypertonia of the trunk and limb muscles. Seizures have not been observed. The cerebellum is reduced in size, particularly in the anterior region (11). There is loss of granule cells beginning at 10 days of age and loss of Purkinje and Golgi cells beginning after 1 month. Cell loss later slows but continues throughout life. Granule and Purkinje cells are more severely affected than Golgi cells and the anterior folia more severely affected than other parts of the cerebellum (2). *tg^{la}/tg* mice show ataxia, stiffness, and retarded motor activity at 15 to 17 days of age. Within a few days they develop a wobbly gait and intermittent focal seizures which continue throughout life. The cerebellum shows shrinkage and degenerative changes of the Purkinje cells (4).

tg^{rol}, rolling Nagoya, recessive. Found among descendants of a cross between the SIII and C57BL/6 strains. It apparently occurred in the SIII strain. Homozygotes show poor motor co-ordination of hindlimbs that may lead to falling and rolling, and sometimes stiffness of the hindlimbs and tail. No seizures have been observed. Symptoms are recognizable at 10 to 14 days. They appear a little earlier than those of *tg/tg* mice and are somewhat more severe. Viability is somewhat reduced. Females may breed, but males rarely do (6, 9). The cerebellum is grossly normal until 10 days of age but, after that, grows more slowly than normal (6). The size of the anterior part of the central lobe of the cerebellum is reduced with reduction in number of granule, basket, and stellate cells but normal numbers of Purkinje cells. There is reduced concentration of glutamate and increased concentration of glycine and taurine in the cerebellum and decreased activity of tyrosine hydroxylase in the cerebellum and other areas (5). tg^{rol}/tg mice resemble tg^{rol}/tg^{rol} in severity and time of appearance of wobbly gait and in absence of seizures (10).

References

1. Green, M.C., and R.L. Sidman. 1962. Tottering, a neuromuscular mutation in the mouse and its linkage with oligosyndactylism. J. Hered. 53:233–237.
2. Herrup, K., and S.L. Wilczynski. 1982. Cerebellar cell degeneration in the leaner mutant mouse. Neurosci. 7:2185–2196.
3. Levitt, P., and J.L. Noebels. 1981. Mutant mouse tottering: selective increase of locus ceruleus axons in a defined single-locus mutation. Proc. Natl. Acad. Sci. USA 78:4630–4634.
4. Meier, H., and A.D. MacPike. 1971. Three syndromes produced by two mutant genes in the mouse. J. Hered. 62:297–302.
5. Muramoto, O., I. Kanazawa, and K. Ando. 1981. Neurotransmitter abnormality in rolling mouse Nagoya, an ataxic mutant mouse. Brain Res. 215:295–304.
6. Nakane, K. 1976. Postnatal development of brain with congenital ataxia, rolling (rol) and tottering (tg). Teratology 14:248 (Abstr.).
7. Noebels, J.L. 1984. A single gene error of noradrenergic axon growth synchronizes central neurones. Nature 310:409–411.
8. Noebels, J.L., and R.L. Sidman. 1979. Inherited epilepsy: spike-wave and focal motor seizures in the mutant mouse tottering. Science 204:1334–1336.
9. Oda, S.I. 1973. The observation of rolling mouse Nagoya (rol), a new neurological mutant, and its maintenance. Exp. Anim. 22:281–288.
10. Oda, S. 1981. A new allele of the tottering locus, rolling mouse Nagoya, on chromosome no. 8 in the mouse. Jpn. J. Genet. 56:295–299.
11. Sidman, R.L., M.C. Green, and S.H. Appel. 1965. Catalog of the neurological mutants of the mouse. Harvard Univ. Press, Cambridge. 82 p.

Tgfb locus, transforming growth factor β, Chr 7

This locus codes for transforming growth factor β (TGF-β), one of two transforming factors designated TGF. TGF-β is a homodimer of about 25 kDa. It is structurally unrelated to TGF-α, but acts synergistically with it in inducing phenotypic transformation. TGF-β has been isolated from normal as well as transformed cells, including platelets, placenta, and kidney. *Tgfb* was shown to be on Chr 7 by use of mouse–Chinese hamster somatic cell hybrids screened with a mouse cDNA probe (1).

References

1. Fujii, D., J.E. Brissenden, R. Derynck, and U. Francke. 1986. Transforming growth factor β gene maps to human chromosome 19 long arm and to mouse chromosome 7. Somat. Cell Mol. Genet. 12:281–288.

Tgn locus, thyroglobulin, Chr 15

This locus codes for thyroglobulin (TG), the protein from which thyroxine is derived. Southern blots of DNA digested with various restriction endonucleases and probed with several rat TG cDNA clones showed multiple restriction fragments, some of which were polymorphic among inbred strains. Most inbred strains had one of three haplotypes. Haplotype Tgn^a occurs in strains A/J, AKR, RIIIS, and SWR; haplotype Tgn^b occurs in the C57–C58 family and in strains DBA/2, LG, NZB, SF/CAM, SJL, 129/J, and others; haplotype Tgn^c occurs in strains BALB/c, CBA/J, CE, and C3H/He. Five other haplotypes occur in various wild derived strains and in two inbred strains. *Tgn* is located near *Gpt-1* in the middle of Chr 15 and shows no recombination with *cog* (congenital goiter), suggesting that *cog* may be a mutation in the *Tgn* gene (1).

References

1. Taylor, B.A., and L. Rowe. 1987. The congenital goiter mutation is linked to the thyroglobulin gene in the mouse. Proc. Natl. Acad. Sci. USA 84:1986–1990.

Th locus, tyrosine hydroxylase, Chr 7

This locus codes for tyrosine hydroxylase (TH; E.C. 1.14.16.2), which catalyzes the rate-limiting step in the biosynthesis of catecholamines. Two restriction fragment patterns were found in *Hind*III- and *Taq*I-digested DNA probed with a rat genomic clone. The allele Th^n determines a 10.9-kb *Hind*III fragment and occurs in strains NZB/BlN, AKR, SWR, PL, C57BL/6, and DBA/2; the allele Th^s determines a 10.3-kb fragment and occurs in strains SM and C57L. Variants found with *Taq*I have a similar strain distribution and may define the same alleles as *Hind*III. In a backcross of (NZB × SM)F1 × NZB, *Th* was found to be located near the distal end of Chr 7 (1).

References

1. Brilliant, M.H., M.M. Neimann, and E.M. Eicher. 1987. Murine tyrosine hydroxylase maps to the distal end of

chromosome 7 within a region conserved in mouse and man. J. Neurogenet. 4:259–266.

Thb locus, thymus B-cell antigen, Chr 15

This locus controls the level of expression of the ThB antigen on B lymphocytes but not on thymocytes. The antigen is present on all B-cells and on immature thymocytes comprising about 50 per cent of lymphocytes of the thymus. It is present on these cells in all strains, but the degree of expression on B-cells differs between strains. The allele *Thb^h* determines high level of expression and occurs in the AKR, AL, C57BL/6, C57BL/10, DBA/2, SWR, and SJL strains; the allele *Thb^l* determines a low level of expression and occurs in the A/J, BALB/c, C3H.SW, CBA/Ca, CBA/J, and NZB strains. *Thb* has shown no recombination with loci determining antigens of the *Ly-6* complex on Chr 15 and shows the same strain distribution as these loci. It is thought to be otherwise unrelated to them because of the different cell distribution of ThB and because the effect of *Thb* is quantitative, unlike that of the *Ly-6* loci (1). The ThB antigen is an acidic protein of about 16 000 MW in contrast to the Ly-6.2 antigen which has a molecular weight of 33 500 (2).

References

1. Eckhardt, L.A., and L.A. Herzenberg. 1980. Monoclonal antibodies to ThB detect close linkage of *Ly-6* and a gene regulating ThB expression. Immunogenetics 11:275–291.
2. Matossian-Rogers, A., P. Rogers, J.A. Ledbetter, and L.A. Herzenberg. 1982. Molecular weight determination of two genetically linked cell surface murine antigens: ThB and Ly-6. Immunogenetics 15:591–599.

thd (formerly *th*), tilted head, recessive, Chr 1

Possibly radiation-induced. Homozygotes are viable and fertile. The head tilts to one side, the side being constant for an individual. The animals shake when held up by the tail (2). They swim with difficulty. Normal otoliths are lacking in both ears, but a limited number of crystals many times larger than normal are present (1). A very similar allele, now extinct, called unbalanced, *thd^ub*, occurred in the CBA/H strain (3). Treatment of pregnant mothers with manganese-supplemented water caused an increase in the number of abnormal otolith crystals in the *thd^ub*/*thd^ub* offspring and an improvement in swimming ability, but did not produce normal otoliths (1).

References

1. Erway, L.C., A.S. Fraser, and L.S. Hurley. 1971. Prevention of congenital otolith defect in pallid mutant mice by manganese supplementation. Genetics 67:97–108.
2. Kelly, E.M. 1958. Mouse News Lett. 19:37.
3. Lyon, M.F., and R. Meredith. 1965. Mouse News Lett. 32:38.

thf, thin fur, recessive, Chr 17

Arose spontaneously in a moderately inbred strain carrying the *T(2;4)1Sn* translocation. Homozygotes are first recognizable at about 5 weeks of age. At the first molt, the underfur is replaced only sparsely, and the adult coat consists chiefly of guard hairs. Viability and fertility are normal (1).

References

1. Key, M., and W.F. Hollander. 1972. Thin fur, a recessive mutant on chromosome 17 of the mouse. J. Hered. 63:97–98.

Tht, thick tail, semidominant, Chr 5

Arose in a female irradiated *in utero*. Heterozygotes can be recognized at birth by their shortened and thickened tails. With growth, the tail tends to become undulated (1). *Tht* is located near *Hm* in the proximal region of Chr 5 (2).

References

1. Beechey, C.V., and A.G. Searle. 1980. Mouse News Lett. 62:48.
2. Beechey, C.V., and A.G. Searle. 1982. Mouse News Lett. 67:19.

Thy-1 locus, thymus cell antigen-1 (theta), Chr 9

This locus determines an antigen present on cells of the thymus, a number of mouse leukemias, brain, and in lesser amounts on lymph node and spleen cells. The allele *Thy-1^a* determines an antigenic specificity, Thy-1.1, found in the AKR and RF strains; the allele *Thy-1^b* determines an antigenic specificity, Thy-1.2, found in the C3HeB/Fe and many other strains (4–6). The Thy-1 antigen is probably present on all T lymphocytes and absent from all B lymphocytes, and it thus serves as a valuable T-cell marker (4). It is very widely used in experiments designed to determine the distribution and function of T-cells. *Thy-1* specifies a T-cell surface glycoprotein, T25, with a molecular weight of 25 kDa (7). The protein appears to be anchored in the cell membrane by a lipid that is either phosphotidylinositol or closely related to it (2). Thy-1 may function in the cell membrane as a signal transduction molecule (1). The *Thy-1* locus, or possibly a gene closely linked to it, controls quantitative expression of a protein that is the same size as Thy-1 and is expressed on thymus and brain but not on lymph node and spleen cells (3).

References

1. Kroczek, R.A., K.C. Gunter, R.N. Germain, and E.M. Shevach. 1986. Thy-1 functions as a signal transduction molecule in T lymphocytes and transfected B lymphocytes. Nature 322:181–184.
2. Low. M.G., and P.W. Kincade. 1985. Phosphotidylinositol is the membrane-binding domain of the Thy-1 glycoprotein. Nature 318:62–64.
3. Oliveira, O.L.P., and D.B. Thomas. 1985. A new thymocyte membrane differentiation antigen for mouse and rat. Immunogenetics 21:559–568.
4. Raff, M.C. 1971. Surface antigenic markers for distinguishing T and B lymphocytes in mice. Transplant. Rev. 6:52–80.
5. Reif, A.E., and J.M. Allen. 1966. Mouse thymic iso-antigens. Nature 209:521–523.
6. Snell, G.D. 1971. The histocompatibility systems. Transplant. Proc. 3:1133–1138.
7. Trowbridge, I.S., and C. Mazauskas. 1976. Immunological properties of murine thymus-dependent lymphocyte surface glycoproteins. Eur. J. Immunol. 6:557–562.

Thy-2 locus, thymus cell antigen-2, Chr 17

This locus controls a minor cell-surface antigen present on the majority of thymocytes, on brain tissue, and at lower levels on bone marrow and spleen cells. It appears not to be present on liver, lymph-node, or red blood cells. The allele determining presence of the antigen, *Thy-2*^{b}, occurs in the A/J, C57BL/6, DBA/2, and other strains; the allele determining absence of the antigen, *Thy-2*^{a}, occurs in the 129/J, C57L, C57BR/cd, NZB, SWR, and SJL strains. Heterozygotes express half as much antigen as *Thy-2*^{b} homozygotes. The antigen is a 150-kDa polypeptide. *Thy-2* is located about 3 cM distal to *H-2* beyond *Pgk-2* on Chr 17 (1).

References

1. Siadak, A.W., and R.C. Nowinski. 1981. Thy-2: a murine thymocyte-brain alloantigen controlled by a gene linked to the major histocompatibility complex. Immunogenetics 12:45–58.

ti, tipsy, recessive, Chr 11

Found among the descendants of a cross between the C3H/H and 101/H inbred strains. Penetrance is complete, and homozygotes are viable and fertile. They have a rabbit-like gait from the age of 1 week, followed soon by a tendency to fall over. There is marked variation in expression and some amelioration in older animals (1).

References

1. Searle, A.G. 1961. Tipsy, a new mutant in linkage group VII of the mouse. Genet. Res. 2:122–126.

Timp locus, tissue inhibitor of metalloproteinases, Chr X

This locus codes for a glycoprotein (TIMP) that is a physiological inhibitor of collagenase, stromelysin, and gelatinase, and to a lesser extent of the polymorphonuclear leukocyte metalloproteinases. The mouse *Timp* gene has been isolated from an embryonic cDNA library by cross-hybridization to the human X-linked TIMP gene. A DNA restriction fragment length difference detected by hybridization to the mouse *Timp* probe occurs between *Mus musculus* and *M. spretus*. A cross between the two species showed *Timp* to be X-linked (1).

References

1. Jackson, I.J., T.D. LeCras, and A.J.P. Docherty. 1987. Assignment of the TIMP gene to the mouse X-chromosome using an inter-species cross. Nucl. Acids Res. 15:4357.

Tind, *Tpre*, *Tsu*, and *Tthy* loci, Chr 12

These four loci control antigens on T lymphocytes (4). They are closely linked to each other and to the immunoglobulin heavy chain constant region (*Igh-C*) on Chr 12, probably in the order *Tpre Tsu*, (*Tthy Tind*) *Igh-C* (5). The antigens controlled by the four loci have all been identified and distinguished by their reaction to various monoclonal antibodies prepared from BALB/c anti-C.AL-20 immune spleen cells. The antigens each show a characteristic distribution on T-cells of various maturational stage and function (2, 3). The *d* alleles of all four loci, which determine presence of the antigens, occur in the C.AL-20 strain and are mostly associated with the *Igh*^{d} and *Igh*^{e} haplotypes. The alleles determining absence of the antigens, which I have designated by an a superscript, occur in the BALB/c strain and are associated with the *Igh*^{a} and other haplotypes (5).

Tind locus, T peripheral cell antigen. The antigen controlled by this locus is present on mature peripheral lymphoid cells that are Ly-1$^+$2$^-$3$^-$ (8). The *Tind*^{d} allele occurs in strains C.AL-20, AKR, NZB, A/J, and C3H/He; the *Tind^a* allele occurs in strains BALB/c, C57BL/6, C57L, CBA/J, SWR, HRS, and MRL (5).

Tpre locus, pre-T-cell antigen. The antigen controlled by this locus is present on the thymus-repopulating T-cells of bone marrow and on normal fetal thymocytes. In the adult thymus, *Tpre* and *Tthy* are largely on overlapping populations of T-cells (3). The allele *Tpre^d* occurs in strains C.AL-20, AKR, NZB, and A/J; the allele *Tpre^a* occurs in strains BALB/c, C57BL/6, C57L, C3H/He, CBA/J, SWR, HRS, and MRL (5).

Tsu locus, T suppressor cell antigen. The antigen controlled by this locus is present on adult peripheral T suppressor cells that are Ly-1$^-$2$^+$3$^+$ (1, 6). The *Tsu^d* allele occurs in strains C.AL-20, AKR, A/J, and NZB; the *Tsu^a* allele occurs in strains BALB/c, DBA/2, C57BL/6, C3H/He, and others (5, 6).

Tthy locus, T thymocyte antigen. The antigen controlled by this locus is present on thymocytes that are Ly-1$^+$2$^+$3$^+$. The antigen is not recognized by anti-Tsu or anti-Tind antibodies. The cells bearing this antigen are descended from Tpre-bearing cells and are precursors of Tsu- and Tind-bearing cells (2, 7). The *Tthy^d* allele occurs in strains C.AL-20, A/J, AKR, NZB, and C3H/He; the *Tthy^a* allele occurs in strains BALB/c, C57BL/6, CBA/J, SWR, HRS, and MRL (5).

References

1. Owen, F.L. 1980. A mature T lymphocyte marker closely linked to Igh-1 that is expressed on the precursor for the suppressor T cell regulating a primary response to SRBC. J. Immunol. 124:1411–1417.
2. Owen, F.L. 1982. Products of the IgT-C region of chromosome 12 are maturational markers for T cells: sequence of apppearance in immunocompetent T cells parallels ontogenetic appearance of Tthy^d, Tind^d, and Tsu^d. J. Exp. Med. 156:703–718.
3. Owen, F.L. 1983. Tpre, a new alloantigen encoded in the IgT-C region of chromosome 12, is expressed on bone marrow of nude mice, fetal T cell hybrids, and fetal thymus. J. Exp. Med. 157:419–432.
4. Owen, F.L., and S.K. Keesee. 1983. Western blot analysis of products of a cluster of genes distal to immunoglobulin on chromosome 12 that encode T-cell antigens. Prog. Immunol. 5:881–890.
5. Owen, F.L., and R. Riblet. 1984. Genes for the mouse T cell alloantigens Tpre, Tthy, Tind, and Tsu are closely linked near *Igh* on chromosome 12. J. Exp. Med. 159:313–317.
6. Owen, F.L., R. Riblet, and B.A. Taylor. 1981. The T suppressor cell alloantigen Tsu^d maps near immunoglobulin allotype genes and may be a heavy chain constant region marker on a T cell receptor. J. Exp. Med. 153:801–810.
7. Owen, F.L., G.M. Spurll, and E. Panageas. 1982. Tthy^d, a new thymocyte alloantigen linked to Igh-1: implications for a switch mechanism for T cell antigen receptors. J. Exp. Med. 155:52–60.
8. Spurll, G.M., and F.L. Owen. 1981. A family of T-cell alloantigens linked to Igh-1. Nature 293:742–745.

tint (originally *t-int*, a non-standard symbol), t-interaction

Found in a stock homozygous for a partial t-haplotype (*t^{w111}*) and tufted (*tf*). It has no visible effect by itself in either heterozygotes or homozygotes. It is unlinked to the *t* complex on Chr 17, but interacts with the *t*-complex mutations *T*, *T^c*, *tct*, and *Fu^{ki}* to reduce tail length. The tail phenotype of the various combinations of *tint* with these mutations is shown in Table 2.11. There was no interaction with three dominant and two recessive non-chromosome 17 mutations affecting tail length.

Table 2.11

Tail length	Genotype
Normal	*tct/tct +/+*
Short	*tct/tct tint/+*, *tct/+ tint/tint*
Tailless	*tct/tct tint/tint*
Short	*T/+ +/+*
Tailless	*T/+ tint/+*
Tailless	*T/+ tint/tint*
Tailless	*T^c/+ +/+*
Tailless with spina bifida	*T^c/+ tint/+*
Kinky, long	*Fu^{ki}/+ +/+*
Kinky, long	*Fu^{ki}/+ tint/+*
Very short	*Fu^{ki}/+ tint/tint*

References

1. Artzt, K., J. Cookingham, and D. Bennett. 1987. A new mutation (*t-int*) interacts with the mutations of the mouse T/t complex that affect the tail. Genetics 116:601–605.

tip, tippy, recessive, Chr 9

Arose in a linkage cross in 1977. Homozygotes are small, hyperactive, cannot stand or walk without falling over, and die by 20 to 25 days of age. The gene is on Chr 9 near ducky, *du* (1).

References

1. Lane, P.W. 1984. Mouse News Lett. 71:31.

tk, tail kinks, recessive, Chr 9

Arose spontaneously in a BALB/c subline at the Chester Beatty Research Institute. Homozygotes are recognized by their shortened kinky tails. The tail vertebrae are very abnormal, and the cervical and upper thoracic vertebrae are also severely affected. The mice tend to have one more presacral vertebra than normal. The defects can be traced to the 10-day embryonic stage when the cervical sclerotomes in normal mice have differentiated into anterior and posterior halves which differ in density. No such differentiation occurs in *tk/tk* embryos (1). Penetrance is complete, and fertility is good. On some genetic backgrounds viability may be considerably reduced (2).

References

1. Grüneberg, H. 1955. Genetical studies on the skeleton of the mouse. XVI. Tail-kinks. J. Genet. 53:536–550.
2. Lilly, F. 1972. Mouse News Lett. 46:18.

Tk-1 locus, thymidine kinase-1, Chr 11

This is the structural locus for cytosol thymidine kinase (TK-1; E.C. 2.7.1.21). No genetic variants are known. The locus has been assigned to Chr 11 by use of somatic cell hybrids between human and mouse cells and Chinese hamster and mouse cells (1, 2), and to the B5 to D1.3 region of that chromosome by chromosome breakage analysis (3).

References

1. Kozak, C.A., and F.H. Ruddle. 1977. Assignment of the genes for thymidine kinase and galactokinase to *Mus musculus* chromosome 11, and the preferential segregation of this chromosome in Chinese hamster/mouse somatic cell hybrids. Somat. Cell Genet. 3:121–133.
2. McBreen, P., K.G. Orkwiszewski, C.J. Chern, W.J. Mellman, and C.M. Croce. 1977. Synteny of the genes for thymidine kinase and galactokinase in the mouse and their assignment to mouse chromosome 11. Cytogenet. Cell Genet. 19:7–13.
3. Sawyer, J.R., M.M. Moore, D. Clive, and J.C. Hozier. 1985. Regional sub-localization of the mouse thymidine kinase gene by breakage analysis. Cytogenet. Cell Genet. 40:738 (Abstr.).

tl, nonerupted teeth, recessive

Possibly radiation-induced. Homozygotes usually die at weaning but have been raised by special feeding to 3 months. Their teeth do not erupt, and there are bony spurs on the long bones and ribs. The tail is slightly shortened (1).

References

1. Kelly, E.M. 1955. Mouse News Lett. 12:48.

Tla locus, thymus leukemia antigen, Chr 17

This locus controls presence or absence of a cell surface antigen on normal thymus. The antigen is found in no other normal tissue but is found in a proportion of leukemias of probably all mouse strains (1, 3). The allele *Tla^a* determines presence of the antigenic specificities TL.1, 2, 3, 5, 6, 7 and occurs in strains A, C57BR, C58, SJL, and others; the allele *Tla^b* determines absence of the antigen (TL⁻) and occurs in strains C57BL/6, AKR, C3H/An, and many others; the allele *Tla^c* determines presence of specificities TL.2, 7 and occurs in strains BALB/c, DBA/2, 129/J, and others; the allele *Tla^d* determines presence of specificities TL.1, 2, 3 and occurs in strains A.CA and B10.M; the allele *Tla^e* determines presence of specificities TL.1, 2, 3, 5 and occurs in strains P/J and BDP/J; the allele *Tla^f* determines presence of specificity TL.2 (2, 8, 12). Heterozygotes with *Tla^b* have intermediate amounts of the antigens determined by the other alleles (3, 4). Leukemic cells of all strains have TL.1, 2; the leukemias of *Tla^a* and *Tla^d* also have TL.3; the leukemias of *Tla^b* and *Tla^c* have a different specificity, TL.4. This and other evidence is interpreted to mean that the *Tla* locus encompasses both structural and regulatory loci (2). The TL antigens are acquired by cells in the thymus. As the cells mature and leave the thymus, they completely lose TL (4). TL antigens are located in close proximity to H-2D antigens on the cell surface and show a reciprocal quantitative relationship to them in that the amount of H-2D antigen is greater on TL⁻ cells than on TL⁺ cells (4). Antibody to TL antigens can induce the loss of TL antigens from the surface of thymocytes or tumor cells, a process known as antigenic modulation (1). The cell surface component determined by the *Tla* locus is a class I antigen very similar to that determined by *H-2*, with subunits of molecular weight about 44 kDa (heavy chain) and 12 kDa (β2-microglobulin) (13). In mice of all the allelic types, thymocytes make the usual 45-kDa heavy chain, and all except *Tla^a* make an additional 53-kDa chain (6). Multiple mRNA transcripts coding for TL proteins are produced in thymocytes of all TL⁺ strains (*Tla^{a,c,d,e,f}*). Each allele produces a distinctive pattern of mRNA species and a distinctive pattern also occurs in mRNA from TL⁺ leukemias of TL⁻ (*Tla^b*) strains (5). The number of class I genes in the *Tla* region may differ in different alleles (10). Of the *T3* and *T13* genes, for example, *Tla^b* strains have only the *T3^b* allele, whereas *Tla^c* strains have both the $T3^c$ and $T13^c$ alleles (11). The number of genes has been estimated to be about 18 in BALB/c (Tla^a) and 13 in C57BL/10 (Tla^b) (7). Although only six haplotypes have been identified by serological methods, Obata *et al.* (10) have found, by analysis of DNA restriction fragments, three different patterns in Tla^b strains and two in Tla^c strains, making a total of nine haplotypes. In addition, Yokoyama *et al.* (14) found, by peptide mapping of TL antigens, two structurally distinct TL molecules among Tla^a strains. Two AKR proviral sequences were found in DNA clones of the *Tla* region isolated from the C57BL/10 strain (9).

References

1. Boyse, E.A., and L.J. Old. 1969. Some aspects of normal and abnormal cell surface genetics. Ann. Rev. Genet. 3:269–290.
2. Boyse, E.A., and L.J. Old. 1971. A comment on the genetic data relating to expression of TL antigens. Transplantation 11:561–562.
3. Boyse, E.A., L.J. Old, and S. Luell. 1964. Genetic determination of the TL (thymus leukemia) antigen in the mouse. Nature 201:779.
4. Boyse, E.A., L.J. Old, and E. Stockert. 1972. The relation of linkage group IX to leukemogenesis in the mouse. *In* P. Emmelot and P. Bentvelzen, eds., RNA Viruses and Host Genome in Oncogenesis, 171–175. North-Holland, Amsterdam.
5. Chen, Y.-T, Y. Obata, E. Stockert, and L.J. Old. 1985. Thymus-leukemia (TL) antigens of the mouse. Analysis of TL mRNA and TL cDNA from TL⁺ and TL⁻strains. J. Exp. Med. 162:1134–1148.
6. Chorney, M.J., F.-W. Shen, J.-S. Tung, and E.A. Boyse. 1985. Additional products of the *Tla* locus of the mouse. Proc. Natl. Acad. Sci. USA 82:7044–7047.
7. Fisher, D.A., S.W. Hunt III, and L. Hood. 1985. Structure of a gene encoding a murine thymus leukemia antigen, and organization of *Tla* genes in the BALB/c mouse. J. Exp. Med. 162:528–545.
8. Flaherty, L., and E. Rinchik. 1980. A new allele and antigen at the *Tla* locus. Immunogenetics 11:205–208.
9. Meruelo, D., R. Kornreich, A. Rossomando, C. Pampeno, A.L. Mellor, E.H. Weiss, R.A. Flavell, and A. Pellicer. 1984. Murine leukemia virus sequences are encoded in the murine major histocompatibility complex. Proc. Natl. Acad. Sci. USA 81:1804–1808.
10. Obata, Y., Y.-T. Chen, E. Stockert, and L.J. Old. 1985. Structural analysis of *TL* genes of the mouse. Proc. Natl. Acad. Sci. USA 82:5475–5479.
11. Pontarotti, P.A., H. Mashimo, R.A. Zeff, D.A. Fisher, L. Hood, A. Mellor, R.A. Flavell, and S.G. Nathenson. 1986. Conservation and diversity in the class I genes of the major histocompatibility complex: sequence analysis of a Tla^b gene and comparison with a Tla^c gene. Proc. Natl. Acad. Sci. USA 83:1782–1786.
12. Shen, F.-W., M. Chorney, and E.A. Boyse. 1982. Further

polymorphism of the *Tla* locus defined by monoclonal antibodies. Immunogenetics 15:573–578.
13. Vitetta, E.S., M.D. Poulik, J. Klein, and J.W. Uhr. 1976. Beta 2-microglobulin is selectively associated with H-2 and TL alloantigens on murine lymphoid cells. J. Exp. Med. 144:179–192.
14. Yokoyama, K., E. Stockert, L.R. Pease, Y. Obata, L.J. Old, and S.G. Nathenson. 1983. Polymorphism and diversity in the *Tla* gene system. Immunogenetics 18:445–451.

tlt, tilted, recessive, Chr 5

Arose in a stock carrying p^{6H}/p^d. Homozygotes often hold the head in a tilted position. They cannot swim and are probably deaf. When held up by the tail, they tuck the head toward the belly. Both sexes breed. *tlt* is on Chr 5, 15 cM from W^v (1).

References

1. Lane, P.W. 1986. Tilted (tlt). Mouse News Lett. 75:28.

tm, tremulous, recessive, extinct

Tremor is the most conspicuous symptom of *tm/tm* mice. They do not have convulsions. Both sexes are sterile. Affected mice have a higher concentration of creatine phosphate and ribose in the supernatant fraction of calcium-precipitated trichloracetic acid extracts of the brain than their normal sibs (2). They also have lower serum cholinesterase and serum β-globulins but higher serum albumins (1,3).

References

1. Yoon, C.H. 1961. Electrophoretic analysis of the serum proteins of neurological mutants in mice. Science 134:1009–1010.
2. Yoon, C.H., and D.J. Denuccio. 1963. Acid-soluble phosphorus in brains of neurological mice. J. Hered. 54:202–205.
3. Yoon, C.H., and S.R. Harris. 1962. Cholinesterase studies of neurological mutants in mice. I. Alterations in serum cholinesterase levels. Neurology 12:423–426.

Tme locus

See *t* complex.

tn, teetering, recessive, Chr 11

Arose spontaneously in the C3H/HeJ inbred strain (1). Homozygotes die between 5 and 6 weeks of age. They are first recognizable at 25 to 30 days by their stiff, slow, unstable movements and their growth retardation. They may assume and maintain unusual postures.

Occasionally they have running 'fits'. Just prior to death they look emaciated and lie on their sides with all limbs extended. In the brainstem and spinal cord there is dysgenesis of selective regions including the pons, trapezoid body, olivary and fastigial nuclei, and the pyramidal tract. There is also progressive Purkinje cell loss (2).

References

1. Lane, P.W. 1962. Mouse News Lett. 27:38.
2. Meier, H. 1967. The neuropathy of teetering, a neurological mutation in the mouse. Arch. Neurol. 16:59–66.

Tnfa, Tnfb loci, Chr 17

These loci code for products of mononuclear cells that are cytotoxic for tumors and cultured transformed cells. They were first found to be located on Chr 17 by use of mouse–human somatic cell hybrids screened with DNA clones for TNF-α from a mouse genomic library. The two genes are tandemly arranged and separated by about 1100 bp (2). Gardner *et al.* (1) found among 19 inbred strains at least five restriction fragment length variants detectable with a combination of four restriction endonucleases. By use of these variants in the BALB/c × STS and O20 × AKR recombinant inbred strains and in *H-2* congenic strains, they showed that the two genes are closely linked to the *H-2D* end of the *H-2* complex.

Tnfa locus, tumor necrosis factor-α. This factor (TNF-α) is identical to cachectin. In contrast to its cytotoxic effect on tumor cell lines, it acts as a cytokine for many types of normal cells. It is produced mostly, if not exclusively, by adherent cells of the monocyte/macrophage lineage (2).

Tnfb locus, tumor necrosis factor-β (also known as lymphotoxin). This factor (TNF-β) is secreted by T lymphocytes. There is evidence that it is delivered to its target cells by cytotoxic T-cells. The clones for this gene were distinguished from the *Tnfa* clones by hybridization with a human TNF-β probe. The intron–exon structure of *Tnfb* is very similarfto that of *Tnfa* and of the homologous genes in man (2).

References

1. Gardner, S.M., B.A. Mock, J. Hilgers, K.E. Huppi, and W.D. Roeder. 1987. Mouse lymphotoxin and tumor necrosis factor: structural analysis of the cloned genes, physical linkage, and chromosomal position. J. Immunol. 139:476–483.
2. Nedospasov, S.A., B. Hirt, A.N. Shakhov, V.N. Dobrynin, E. Kawashima, and C.V. Jongeneel. 1986. The genes for tumor necrosis factor (TNF-alpha) and lymphotoxin (TNF-beta) are tandemly arranged on chromosome 17 of the mouse. Nucl. Acids Res. 14:7713–7725.

To

See *Mo^{to}*.

Tol-1 locus, tolerance to BGG

This locus controls susceptibility to induction of tolerance to bovine γ-globulin. The allele *Tol-1^a* determines susceptibility to induction of tolerance and occurs in the DBA/2J strain; the allele *Tol-1^b* determines resistance to induction of tolerance and occurs in the BALB/cJ strain. *Tol-1^a* is dominant (2). Resistance can be transferred with macrophages and is probably due to ability of BALB/c macrophages to filter out a component of the immunogen that is responsible for inhibition of tolerance. The mechanism by which the uptake of the inhibiting component by macrophages results in resistance to tolerance induction is not known (1).

References

1. Cowing, C., D. Harris, G. Garabedian, M. Lakić, and S. Leskowitz. 1977. Immunologic tolerance to heterologous immunoglobulin: its relation to *in vitro* filtration by macrophages. J. Immunol. 119:256–262.
2. Lukić, M.L., H.H. Wortis, and S. Leskowitz. 1975. A gene locus affecting tolerance to BGG in mice. Cell. Immunol. 15:457–463.

tor, tortured, recessive

Arose in a non-inbred stock. Homozygotes may be recognized beginning at about 17 days of age by grossly abnormal postures and movements of all four limbs. After about 4 weeks, the abnormal behavior remains essentially unchanged. Affected animals may live for a year of more but do not breed. Severe pathological changes are widespread in the central and peripheral nervous systems. Degeneration in the trigeminal and dorsal root sensory ganglia is found at postnatal day 17. By 27 days there is degeneration in the white matter tracts at all levels of the spinal cord, the dorsal and ventral roots and peripheral nerves, neuron cell bodies of the dorsal and ventral horns of spinal gray matter, myelinated axons of several brainstem tracts and cerebellar cortex, and cerebellar granule cells (1).

References

1. Cowen, J., and R.L. Sidman. 1981. Mouse News Lett. 65:16.

tp, taupe, recessive, Chr 7

Arose spontaneously in the C57BL/10 inbred strain. Viability of homozygotes is normal, and fertility of males is normal, but females have difficulty gestating their young and do not nurse them. Their nipples are abnormal. Homozygotes have a diluted coat color, *a/a tp/tp* mice being slate gray (1).

References

1. Fielder, J.H. 1952. The taupe mouse. J. Hered. 43:75–76.

Tpi-1 locus, triosephosphate isomerase-1, Chr 6

This locus is probably the structural locus for triosephosphate isomerase (TPI-1; E.C. 5.3.1.1). The allele *Tpi-1^a* determines normal activity of TPI-1 in red cells and occurs in all inbred strains examined; the allele *Tpi-1^b* determines low activity in red cells (30 per cent of that in strain C57BL/10) and was found in a wild-caught mouse (3). The locus was first shown to be located on Chr 6 by use of mouse–Chinese hamster somatic cell hybrids (1, 2). By use of the *Tpi-1^b* variant, *Tpi-1* was mapped to the distal portion of Chr 6 (3).

References

1. Leinwand, L.A., C.A. Kozak, and F.H. Ruddle. 1978. Assignment of genes for triose phosphate isomerase to chromosome 6 and tripeptidase-1 to chromosome 10 in *Mus musculus* by somatic cell hybridization. Somat. Cell Genet. 4:233–240.
2. Minna, J.D., A.P. Bruns, A.H. Krinsky, P.A. Lalley, U. Francke, and P.S. Gerald. 1978. Assignment of a *Mus musculus* gene for triosephosphate isomerase to chromosome 6 and for glyoxylase-1 to chromosome 17 using somatic cell hybrids. Somat. Cell Genet. 4:241–252.
3. Peters, J., and G. Bulfield. 1984. Mouse News Lett. 70:82.

Tpre locus

See *Tind*, etc. loci.

Tr locus, Chr 11

Tr, trembler, dominant. Arose as a spontaneous mutation in a stock at the Institute of Animal Genetics, Edinburgh. Mortality of heterozygotes is high at 3 to 4 weeks but can be improved if moist food is provided within easy reach during this critical period. Females breed normally, but males are frequently sterile. Homozygotes are viable and cannot be distinguished from heterozygotes (3, 8). Affected animals are recognizable at 9 to 10 days by a rapid tremor which remains throughout life, a tendency to convulsions during the first 3 weeks, and spasticity in the muscles of the lower back and limbs. No pathology is found in the central nervous system or in the dorsal root ganglia. In the peripheral nervous system (PNS), myelin sheaths are thin or absent, and there is a marked deficiency of large myelinated axons. Myelination is delayed, many myelin sheaths are uncompacted, and much of the myelin degenerates. Schwann cells proliferate, digest the myelin debris, and engage in remyelination. These processes lead to the formation of 'onion bulbs', axons surrounded by Schwann cell cytoplasm and by layers of Schwann cell basement membrane (2, 10, 11). At the spinal cord–ventral root junction, some of the myelinated fibers of the spinal cord show the same type of abnormality found in the ventral roots and peripheral nerves (5). There is marked reduction in motor conduction velocities of the peripheral nerves (12). Transplantation experiments with sciatic nerves of normal and trembler mice have shown that the defect caused by *Tr* resides in the Schwann cells and not in the nerve cells (1). Cultures of dorsal root ganglia from 15- to 16-day *Tr/+* embryos develop abnormal myelination resembling that in *Tr/+* adults (13). Alterations in skeletal muscle of *Tr/+* mice have been described by Gale *et al.* (4) and in lipids and proteins of peripheral myelin by Yao and Bourre (14), Inuzuka *et al.* (9), and Heape *et al.* (6).

Tr^J, trembler-J, semidominant. Arose in the C57BL/6J strain. Heterozygotes are similar to *Tr/+* mice in behavior and neuropathology but less severely affected. Tremor cannot be reliably recognized before 20 to 25 days. There are no obvious seizures and only a mild gait abnormality. In the PNS, the myelin deficiency is considerably less severe than that of *Tr/+* mice. Homozygotes are recognizable by 8 days of age after which they become progressively disabled. They are unable to walk normally and can right themselves only with great difficulty. Most are dead by 18 days. The PNS is nearly devoid of myelin with myelinogenesis blocked in the promyelin stage. Schwann cells ensheathe individual axons but without formation of compact myelin. Severely affected offspring of a mating between *Tr^J/+* and *Tr/+* mice (probably *Tr^J/Tr*) were very severely myelin-deficient when examined at 16 days (7, 8).

References

1. Aguayo, A.J., M. Attiwell, J. Trecarten, S. Perkins, and G.M. Bray. 1977. Abnormal myelination in transplanted trembler mouse Schwann cells. Nature 265:73–75.
2. Ayers, M.M., and R.M. Anderson. 1976. Development

of onion bulb neuropathy in the trembler mouse. Morphometric study. Acta Neuropathol. 36:137–152.

3. Falconer, D.S. 1951. Two new mutants, "trembler" and "reeler", with neurological action in the house mouse (*Mus musculus* L.). J. Genet. 50:192–201.
4. Gale, A.N., S. Gomez, and L.W. Duchen. 1982. Changes produced by a hypomyelinating neuropathy in muscle and its innervation: morphological and physiological studies in the trembler mouse. Brain 105:373–393.
5. Harrison, B.M. 1977. The spinal cord–ventral root junction in the trembler mouse. Acta Neuropathol. 38:33–38.
6. Heape, A., H. Juguelin, M. Fabre, F. Boiron, and C. Cassagne. 1986. A quantitative developmental study of the peripheral nerve lipid composition during myelogenesis in normal and trembler mice. Dev. Brain Res. 25:181–189 (Brain Res. 390).
7. Henry, E.W., and R.L. Sidman. 1983. The murine mutation trembler-J: proof of semidominant expression by use of the linked vestigial tail marker. J. Neurogenet. 1:39–52.
8. Henry, E.W., J.S. Cowen, and R.L. Sidman. 1983. Comparison of trembler and trembler-J mouse phenotypes: varying severity of peripheral hypomyelination. J. Neuropathol. Exp. Neurol. 42:688–706.
9. Inuzuka, T., R.H. Quarles, J. Heath, and B.D. Trapp. 1985. Myelin-associated glycoprotein and other proteins in trembler mice. J. Neurochem. 44:793–797.
10. Low, P.A. 1976. Hereditary hypertrophic neuropathy in the trembler mouse. Part I. Histological studies: light microscopy. J. Neurol. Sci. 30:327–341.
11. Low, P.A. 1977. The evolution of "onion bulbs" in the hereditary hypertrophic neuropathy of the trembler mouse. Neuropathol. Appl. Neurol. 3:81–92.
12. Low, P.A., and J.G. McLeod. 1975. Hereditary demyelinating neuropathy in the trembler mouse. J. Neurol. Sci. 26:565–574.
13. Mithen, F.A., M. Cochran, C.J. Cornbrooks, and R.P. Bunge. 1982. Expression of the trembler mouse mutation in organotypic cultures of dorsal root ganglia. Dev. Brain Res. 4:407–415 (Brain Res. 256).
14. Yao, J.K., and J.-M. Bourre. 1985. Metabolic alterations of endoneurial lipids in developing trembler nerve. Brain Res. 325:21–27.

tra-1 locus

See *Dtc-1* locus.

Trf locus, transferrin, Chr 9

Alleles at this locus control variation in the electrophoretic properties of the iron-binding β-globulin component, transferrin, of the serum (2–4). Ashton and Braden (1) and Thompson *et al.* (5) have independently described variation in the electrophoretic properties of β-globulin which is probably controlled by the *Trf* locus, although this has not been tested genetically. Two codominant alleles are known, *Trf^a*, found only in the CBA and some related strains, and *Trf^b*, found in all other strains tested. The serum component of *Trf^a* is more negatively charged at pH 8.5 and migrates more rapidly than that of *Trf^b*.

References

1. Ashton, G.C., and A.W.H. Braden. 1961. Serum β-globulin polymorphism in mice. Austral. J. Biol. Sci. 14:248–253.
2. Cohen, B.L. 1960. Genetics of plasma transferrin in the mouse. Genet. Res. 1:431–438.
3. Cohen, B.L., and D.C. Shreffler. 1961. A revised nomenclature for the mouse transferrin locus. Genet. Res. 2:306–308.
4. Shreffler, D.C. 1960. Genetic control of serum transferrin type in mice. Proc. Natl. Acad. Sci. USA 46:1378–1384.
5. Thompson, S., J.F. Foster, J.W. Gowen, and O.E. Tauber. 1954. Hereditary differences in serum proteins of normal mice. Proc. Soc. Exp. Biol. Med. 87:315–317.

Trip locus

See *Pep-2* locus.

trm, tremor, recessive

Found by C.W. Emmens of Sydney, Australia. Homozygotes are viable and fertile. They have a lurching gait and appear to lift their hindlegs too high off the ground. After exertion they crouch in a corner of the cage and start trembling. They are relatively easy to classify when a month old. The behavior remains unchanged throughout life. Homozygotes are lighter in weight than their normal littermates throughout life. Histological sections of the brain showed no differences between mutant and normal animals. Orphenadrine and hyoscine in high doses reduced tremor, but there was no effect of α-methyldopa, atropine, barbiturates, or nethalide. Levels of several brain amines and acetylcholine were normal (1).

References

1. Schnieden, H., and G.M. Truslove. 1970. The effect of some drugs on genetic tremor in the mouse. Eur. J. Pharmacol. 11:33–37.

Trp53 loci

Trp53 locus, transformation-related protein 53, Chr 11. This locus codes for tumor antigen p53, a phosphoprotein found at elevated levels in a variety of transformed cells. In normal cells, p53 turns over very rapidly and the level is very low (2, 3). No genetic variants of *Trp53* are known. By use of mouse–Chinese hamster somatic cell hybrids screened with a monoclonal antibody to mouse p53, it was shown that *Trp53* is on Chr 11 (2).

Trp53-ps locus, transformation-related protein 53 pseudogene, Chr 14. Zakut-Houri *et al.* (3) showed that the mouse genome also contains this processed pseudogene for p53. Czosnek *et al.* (1) used mouse–Chinese hamster somatic cell hybrids screened with a mouse cDNA clone to show that the pseudogene is on Chr 14 and to confirm the location of *Trp53* on Chr 11.

References

1. Czosnek, H.H, B. Bienz, D. Givol, R. Zakut-Houri, D.D. Pravtcheva, F.H. Ruddle, and M. Oren. 1984. The gene and pseudogene for mouse p53 cellular tumor antigen are located on different chromosomes. Mol. Cell. Biol. 4:1638–1640.
2. Rotter, V., D. Wolf, D. Pravtcheva, and F.H. Ruddle. 1984. Chromosomal assignment of the murine gene encoding the transformation-related protein p53. Mol. Cell. Biol. 4:383–385.
3. Zakut-Houri, R., M. Oren, B. Bienz, V. Lavie, S Hazum, and D. Givol. 1983. A single gene and a pseudogene for the cellular tumor antigen p53. Nature 306:594–597.

Try-1 locus, trypsin-1, Chr 6

This locus codes for the endopeptidase trypsin-1 (E.C. 3.4.21.4). No genetic variants are known in laboratory inbred strains. The locus was first found to be on Chr 6 by use of mouse–Chinese hamster somatic cell hybrids and a cDNA probe isolated from a rat pancreatic cDNA library (2). A restriction fragment length difference between laboratory strains and *Mus spretus* in *Taq*I-digested DNA was later found and used to map *Try-1* close to *Tcrb* on Chr 6 (1). The *Prt-1* and *Prt-3* loci, which are closely linked, control electrophoretic variation in trypsinogen and trypsin. Their chromosomal location is not known but one of them may be identical with *Try-1*. See *Prt-1*, *Prt-3*.

References

1. Bućan, M., T. Yang-Feng, A.M. Colberg-Poley, D.J. Wolgemuth, J.-L. Guénet, U. Francke, and H. Lehrach. 1986. Genetic and cytogenetic localisation of the homeo box containing genes on mouse chromosome 6 and human chromosome 7. EMBO J. 5:2899–2905.
2. Honey, N.K., A.Y. Sakaguchi, P.A. Lalley, C. Quinto, R.J. MacDonald, C. Creik, G.I. Bell, W.J. Rutter, and S.L. Naylor. 1984. Chromosomal assignments of genes for trypsin, chymotrypsinogen B, and elastase in mouse. Somat. Cell Mol. Genet. 10:377–383.

Ts, tail-short, semidominant, Chr 11

Arose spontaneously in the BALB/c inbred strain (3). Homozygotes are detectable at 3.5 days post-coitum as retarded pale-staining morulae, and are all dead by 5.5 days. They also fail to develop in culture (4). Viability and expression of heterozygotes is strongly dependent on the genetic background (3). Heterozygotes are recognizable by their shortened kinked tails. They are smaller than normal and have numerous skeletal abnormalities including vertebral fusions and dyssymphyses, bilateral asymmetry of the length of the humerus and tibia, triphalangy of digit 1 of the forefoot, an additional pair of ribs, and often a shortened and highly abnormal skull. There is a prenatal anemia at 13 to 17 days which disappears before birth (2). It is probably due to a general retardation of development rather than to a specific defect in erythropoiesis (1).

References

1. Brotherton, T.W., D.H.K. Chui, E.C. McFarland, and E.S. Russell. 1979. Fetal erythropoiesis and hemoglobin ontogeny in tail-short (*Ts*/+) mutant mice. Blood 54:673–683.
2. Deol, M.S. 1961. Genetical studies on the skeleton of the mouse. XXVIII. Tail-short. Proc. R. Soc. Lond. (Biol.) 155:78–95.
3. Morgan, W.C. 1950. A new tail-short mutation in the mouse. J. Hered. 41:208–215.
4. Paterson, H.F. 1980. *in vivo* and *in vitro* studies on the early embryonic lethal tail-short (*Ts*) in the mouse. J. Exp. Zool. 24:247–256.

Tse-1, *Tse-2* loci

Tse-1 locus, tissue-specific extinguisher-1, Chr 11. This locus acts to suppress transcription of the gene for tyrosine aminotransferase (TAT; E.C. 2.6.1.5) in rat hepatoma cells. It was detected in hepatoma microcell hybrids containing single specific chromosomes from cultured mouse fibroblasts. In hepatoma hybrids containing fibroblast Chr 11, TAT expression was extinguished while expression of other hepatic enzymes was unaffected. *Tse-1* may normally act to suppress TAT production in fibroblasts and presumably also in other cells not producing TAT. It appears to be stably activated so that it survives and is active when transferred into other cells even those of other species (1).

Tse-2 locus, tissue-specific extinguisher-2. This locus acts to suppress expression of liver alcohol dehydrogenase (ADH-1; E.C. 1.1.1.1) in rat hepatoma cells. It was detected on a single but unidentified chromosome from mouse cultured kidney cells (1).

Tse-1, Tse-2

References

1. Killary, A.M., and R.E.K. Fournier. 1984. A genetic analysis of extinction: *trans*-dominant loci regulate expression of liver-specific traits in hepatoma hybrid cells. Cell 38:523–534.

Tse-3 locus, tissue-specific extinguisher-3, Chr 3?

This locus acts to suppress transcription of the gene for albumin. It was detected in microcell hybrids formed by introducing mouse chromosomes derived from a clone of mouse fibroblast L-cells into rat hepatoma cells. The hepatoma cells normally make albumin; introduction of a specific mouse chromosome, probably Chr 3, caused albumin mRNA production to cease. The *Tse-3* gene presumably is stably activated in all cells that do not produce albumin, and continues to be active when transferred into other cells, even those of different species (1).

References

1. Petit, C., J. Levilliers, M.-O. Ott, and M.C. Weiss. 1986. Tissue-specific expression of the rat albumin gene: genetic control of its extinction in microcell hybrids. Proc. Natl. Acad. Sci. USA 83:2561–2565.

Tsha, Tshb loci

Tsha locus, thyrotropin α, Chr 4. This locus codes for the α subunit of thyrotropin (thyroid-stimulating hormone, TSH-α). No genetic variants are known. The locus was found to be on Chr 4 by use of mouse–Chinese somatic cell hybrids and a mouse cDNA clone coding for TSH-α (1, 2).

Tshb locus, thyrotropin β, Chr 3. This locus codes for the β subunit of thyrotropin (thyroid-stimulating hormone, TSH-β). No genetic variants are known. By use of mouse–Chinese hamster somatic cell hybrids screened with a cloned cDNA probe, *Tshb* was shown to be on Chr 3 (1, 3).

References

1. Kourides, I.A., P.E. Barker, J.A. Gurr, D.D. Pravtcheva, and F.H. Ruddle. 1984. Assignment of the genes for the α and β subunits of thyrotropin to different mouse chromosomes. Proc. Natl. Acad. Sci. USA 81:517–519.
2. Naylor, S.L., W.W. Chin, H.M. Goodman, P.A. Lalley, K.-H. Grzeschik, and A.Y. Sakaguchi. 1983. Chromosome assignments of genes encoding α and β subunits of glycoprotein hormones in man and mouse. Somat. Cell Genet. 9:757–770.
3. Naylor, S.L., A.Y. Sakaguchi, L. McDonald, S. Todd,

P.A. Lalley, T.B. Shows, and W.W. Chin. 1986. Mapping thyrotropin β subunit gene in man and mouse. Somat. Cell Mol. Genet. 12:307–311.

Tsk, tight-skin, semidominant, Chr 2

Arose spontaneously in the B10.D2(58N) strain. Heterozygotes have tight skins detectable at 1 week. They have marked hyperplasia of the subcutaneous connective tissue, increased growth of cartilage and bone, and small tendons with hyperplasia of the tendon sheaths. In the loose connective tissue there are large accumulations of microfibrils in the intercellular spaces. Pronounced increase in size of the thoracic skeleton leads to pathological distention of the hollow thoracic viscera. Growth hormone concentration in the pituitary and plasma is normal. Homozygotes probably die at 7 to 8 days of gestation (2). The dermis is thicker than normal and lacks the regular weave characteristic of the collagen fibers of normal skin. Fibroblasts are abundant (4). Collagen content of the skin is increased, and cultured skin fibroblasts from *Tsk*/+ mice have been found to synthesize almost five times more Type I and Type III procollagen mRNA than fibroblasts of control mice (3). Glycoseaminoglycan content in the skin is higher than normal (5) and degranulated mast cells are increased in number (9). These characteristics resemble those of progressive systemic sclerosis in man. The distention of the lungs resulting from increased size of the thorax causes an emphysema-like condition which has been used as a model for investigating the pathology of human emphysema (6–8). DeLustro *et al.* (1) have reported that elastase-soluble lung peptide elicits delayed-type hypersensitivity in *Tsk*/+ but not in +/+ mice.

References

1. DeLustro, F.A., A.M. Mackel, and E.C. LeRoy. 1983. Delayed-type hypersensitivity to elastase-soluble lung peptides in the tight-skin (Tsk) mouse. Cell. Immunol. 81:175–179.
2. Green, M.C., H.O. Sweet, and L.E. Bunker. 1976. Tight-skin, a new mutation of the mouse causing excessive growth of connective tissue and skeleton. Am. J. Pathol. 82:493–512.
3. Jiminez, S.A., C.J. Williams, J.C. Myers, and R.I. Bashey. 1986. Increased collagen biosynthesis and increased expression of type I and type III procollagen genes in tight-skin (TSK) mouse fibroblasts. J. Biol. Chem. 261:657–662.
4. Menton, D.N., and R.A. Hess. 1980. The ultrastructure of collagen in the dermis of tight-skin (Tsk) mutant mice. J. Invest. Dermatol. 74:139–147.
5. Ross, S.C., T.G. Osborn, R.W. Dorner, and J. Zuchner.

374

1983. Glycoseaminoglycan content in the skin of the tight-skin mouse. Arth. Rheum. 26:653–657.

6. Rossi, G.A., G.W. Hunninghake, S.V. Szapiel, J.E. Gadek, J.D. Fulmer, O. Kawanami, V.J. Ferrans, and R.G. Crystal. 1980. The tight-skin mouse: an animal model of inherited emphysema. Bull. Eur. Physiopathol. Resp. 16:Suppl.157–165.
7. Rossi, G.A., G.W. Hunninghake, J.E. Gadek, S.V. Szapiel, O. Kawanami, J.V. Ferrans, and R.G. Crystal. 1984. Hereditary emphysema in the tight-skin mouse. Evaluation of pathogenesis. Am. J. Resp. Dis. 129:850–855.
8. Szapiel, S.V., J.D. Fulmer, G.W. Hunninghake, N.E. Nelson, O. Kawanami, V.J. Ferrans, and R.G. Crystal. 1981. Hereditary emphysema in the tight-skin (Tsk/+) mouse. Am. Rev. Resp. Dis. 123:680–685.
9. Walker, M., R. Harley, J. Maize, F. DeLustro, and E.C. LeRoy. 1984. Mast cells and their degranulation in the Tsk mouse model of scleroderma. Proc. Soc. Exp. Biol. Med. 180:323–328.

Tsk-2, tight skin-2, Chr 1

Found among offspring of a 101/H male treated with ENU. Heterozygotes can be recognized at 1 to 2 weeks of age by tightness of the skin over the shoulders when picked up. No other abnormality was detected and the mice are fully viable and fertile. Homozygotes probably die prenatally. *Tsk-2* is on Chr 1, probably between *fz* and *ln* (1).

References

1. Peters, J., and S.T. Ball. 1986. Tight skin-2 (Tsk-2). Mouse News Lett. 74:91–92.

Tsu locus

See *Tind*, etc. loci.

Tsz-1 locus, thymus size-1

This locus controls a twofold difference in thymus size relative to body size in mice 50 days of age or older. The allele *Tsz-1^l* determines large thymus size and occurs in the C57BL/6 and AKR strains; the allele *Tsz-1^s* determines small thymus size and occurs in the A/J, DBA/2, BALB/c, CBA/J, C3H/He, 129/J, C57BL/10, and MA/My strains. *Tsz-1^l* is dominant. Thymus size is also influenced by other genes. In the immediate postnatal period (0–23 days), the F1 mice of a cross between C57BL/6 and A/J resemble the small parent, and the difference at this stage appears to be controlled by two genes. *Tsz-1* shows loose linkage with *Ric* on Chr 5 in 20 BxD recombinant inbred strains but no linkage with other Chr 5 markers (1).

References

1. Peleg, L., and M.N. Nesbitt. 1984. Genetic control of thymus size in inbred mice. J. Hered. 75:126–130.

Tthy locus

See *Tind*, etc. loci.

tu, toe-ulnar

Mice with an extra digit on the ulnar side of one or both forefeet were found among descendants of a female treated with nitrogen mustard at the beginning of pregnancy. The trait may be due to a recessive mutation with reduced penetrance, but there are some irregularities in the segregation pattern that cast some doubt on this hypothesis (1).

References

1. Center, E.M. 1955. Postaxial polydactyly in the mouse. J. Hered. 46:144–148.

tub, tubby, recessive, Chr 7

Arose spontaneously in the C57BL/6J strain in 1977. Homozygotes are recognizable by increased body weight at 3 to 4 months in males and at 4 to 6 months in females. Both sexes are fertile. The increased weight is composed of excess adipose tissue. Blood glucose is normal, but plasma insulin is increased prior to obvious signs of obesity and may rise to 20 times normal by 6 months. The islets of Langerhans are moderately enlarged with signs of hyperactivity (1).

References

1. Coleman, D.L., E.M. Eicher, and J.L. Southard. 1978. Mouse News Lett. 59:25.

Tw, twirler, semidominant, Chr 18

Arose spontaneously in the PCS multiple recessive stock. Heterozygotes show head-shaking and circling behavior but are not deaf. Viability and fertility are normal except that adults tend to become obese and may then become sterile. Penetrance is incomplete. There are morphological abnormalities of the inner ear which consist of irregularities in the outline of the semicircular canals, sometimes amounting to branching, and reduction or absence of otoliths (1). Moderate astrocytosis in the vestibular nuclei and cerebellar white matter has been reported (2). Homozygotes have cleft lip and palate or cleft palate only. They die within 24 hours after birth (1).

References

1. Lyon, M.F. 1958. Twirler: a mutant affecting the inner ear of the house mouse. J. Embryol. Exp. Morphol. 6:105–116.
2. TanCreti, D.M. 1969. Neuropathology of mutant mice with auditory and/or vestibular deficiencies. J. Neuropathol. Exp. Neurol. 28:159 (Abstr.).

twi, twitcher, recessive, Chr 12

Arose spontaneously in the CE/J strain. Homozygotes can be recognized at about 3 weeks of age by a generalized tremor. Progressive weakness and wasting follow and death occurs by about 3 months. The principal pathological changes are degeneration of myelin sheaths in both the central and peripheral nervous systems, presence of multinucleated macrophages (globoid cells) containing a variety of inclusions in which there are crystalline and multi-angular structures and twisted tubules (1), and endoneurial edema (5). These abnormalities closely resemble those of human globoid cell leukodystrophy (Krabbe's disease) (1). In the brain and liver of homozygotes, there is a profound deficiency of galactosylceramidase and lactosylceramidase activity. Heterozygotes have intermediate levels of both enzymes (4). In the peripheral nervous system, the defect was shown to be intrinsic to the Schwann cells by means of nerve-grafting experiments (6). When sciatic nerves from affected mice were grafted into trembler (*Tr/+*) mice, which have defective Schwann cells, 1 to 4 months after grafting they showed normal myelin, no globoid cells, and very little endoneurial edema (7). Grafting to normal hosts also results in increased galactosylceramidase activity. These outcomes are attributed to enzyme replacement in the mutant Schwann cells (8). Transplantation of normal hematopoietic cells into twitcher mice gradually repairs the demyelination of the peripheral nerves, probably by enzyme transfer from normal macrophages (11). In spite of the deficiency of enzyme in affected mice and in patients with Krabbe's disease, there is no accumulation of the normal substrate, galactosylceramide, in the nervous system. However, in twitcher mice it was shown that, beginning at 7 days, there is a rapid and progressive accumulation of a toxic metabolite, galactosylsphingosine (psychosine), which is also a substrate for galactosylceramidase. No psychosine was detected in either homozygous normal or heterozygous mice. The accumulation of psychosine is thought to be responsible for the neurological disease in affected mice and probably also in patients with Krabbe's disease (2). In the kidneys of twitcher mice, in contrast to the nervous system, there is a large increase of galactosylceramide (50 × normal) and a smaller increase in liver and lung (3). This is accompanied by presence in the loop of Henle in the kidneys of large numbers of inclusions of the type found in the globoid cells of the nervous system (10). *twi* is on Chr 12 about 12 cM from *sm* (9).

References

1. Duchen, L.W., E.M. Eicher, J.M. Jacobs, F. Scaravilli, and F. Teixeira. 1980. Hereditary leucodystrophy in the mouse: the new mutant twitcher. Brain 103:695–710.
2. Igisu, H., and K. Suzuki. 1984. Progressive accumulation of toxic metabolite in a genetic leukodystrophy. Science 224:753–755.
3. Igisu, H., and K. Suzuki. 1984. Glycolipids in the spinal cord, sciatic nerve, and systemic organs of the twitcher mouse. J. Neuropathol. Exp. Neurol. 43:22–36.
4. Kobayashi, T., T. Yamanaka, J.M. Jacobs, F. Teixeira, and K. Suzuki. 1980. The twitcher mouse: an enzymatically authentic model of human globoid cell leukodystrophy (Krabbe disease). Brain Res. 202:479–483.
5. Powell, H.C., R.L. Knobler, and R.R. Myers. 1983. Peripheral neuropathy in the twitcher mutant: a new experimental model of endoneurial edema. Lab. Invest. 49:19–25.
6. Scaravilli, F., and J.M. Jacobs. 1981. Peripheral nerve grafts in hereditary leukodystrophic mutant mice (twitcher). Nature 290:56–58.
7. Scaravilli, F., and J.M. Jacobs. 1982. Improved myelination in nerve grafts from the leucodystrophic twitcher into trembler mice: evidence for enzyme replacement. Brain Res. 237:163–172.
8. Scaravilli, F., and K. Suzuki. 1983. Enzyme replacement in grafted nerve of *twitcher* mouse. Nature 305:713–715.
9. Sweet, H.O. 1986. Twitcher (twi) is on Chr 12. Mouse News Lett. 75:30.
10. Takahashi, H., H. Igisu, K. Suzuki, and K. Suzuki. 1984. Murine globoid cell leukocystrophy: the twitcher mouse. An ultrastructural study of the kidney. Lab. Invest. 50:42–50.
11. Yeager, A.M., S. Brennan, C. Tiffany, H.W. Moser, and G.W. Santos. 1984. Prolonged survival and remyelination after hematopoietic cell transplantation in the twitcher mouse. Science 225:1052–1054.

twt, twister, recessive, Chr 7

Arose in a C57BL/6 × C3Heb/Fe hybrid stock carrying *myd*. When picked up by the tail, all affected mice tuck their heads under. They may show circling behavior or tilted heads and none can swim. They are not deaf. *twt* is located very close to pink-eyed dilution, *p*, on Chr 7 (1).

References

1. Lane, P.W. 1981. Mouse News Lett. 64:59.

tx, toxic milk, recessive

Arose in the DL strain. Homozygotes show no overt distinguishing characteristics. Females are recognized by their litters, which show poor growth, hypopigmentation, tremors, and death at 2 weeks. The defects are due to a deficiency of copper in the offspring resulting from defective copper metabolism in the mother, such that there is greatly reduced gestational hepatic accumulation of copper as well as reduced copper concentration in the milk. Offspring of homozygous mothers can be rescued by fostering on normal females. Genotype with respect to *tx* of the offspring has no effect on presence or absence of the syndrome. Homozygous animals of both sexes at 7 weeks have normal copper levels of all tissues except liver and plasma. In liver, copper concentration is more than 10 times normal and increases progressively with age. Plasma copper is about half that of normal and remains low. Ceruloplasmin activity in plasma is also reduced (1).

References

1. Rauch, H. 1983. Toxic milk, a new mutation affecting copper metabolism in the mouse. J. Hered. 74:141–144.

ty, trembly, incompletely recessive, Chr X

Found among descendants of a cross between 129/J and NZB/BlNJ. Hemizygous males (*ty*/Y) exhibit tremors at about 2 weeks and seizures a little later. They do not survive beyond weaning. Histologically, they do not show the gross absence of myelin found in jimpy (*jp*/Y), another X-linked mutant with similar behavior. Some presumptive *ty*/+ females have mild tremors, poor coordination, and infertility. The heterozygous expression appears to be genotype-dependent (1).

References

1. Taylor, B.A., H. Meier, A. MacPike, and M. Williams. 1978. Mouse News Lett. 59:25.

U

U, umbrous, semidominant

Found among descendants of a stock carrying *wa-1*. *U* in homozygous condition causes a marked darkening in *A/A* mice. In heterozygotes the darkening is somewhat less. In *U/U* mice *A/a* is distinguishably darker than *A/A*. *U* is thus a modifier of dominance of *A* (1). No effect of *U* is recognizable in *a/a* mice. A similar mutant has been described by Robinson (2), but no tests for allelism have been made.

References

1. Mather, K., and S.B. North. 1940. Umbrous: a case of dominance modification in mice. J. Genet. 40:229–241.
2. Robinson, R. 1959. Sable and umbrous mice. Genetica 29:319–326.

Ucp locus, mitochondrial uncoupling protein, Chr 8

This locus codes for a cold-inducible mitochondrial uncoupling protein (UCP) found only in brown fat tissue. UCP is thought to be essential for heat production by uncoupling respiration from ATP production in the mitochondria. A 10-fold induction of UCP occurs after exposure of mice to 5°C. Two DNA restriction fragment patterns are found after digestion with *Bam*HI and hybridization to a UCP cDNA probe. The allele *Ucp^b* determines a 13-kb fragment and occurs in strains C57BL/6, C57L, C57BR/cd, and NZB; the allele *Ucp^d* determines a 23-kb fragment and occurs in strains DBA/2, BALB/c, AKR, C3H/He, SWR, SJL, and 129/J. By use of BXD, CXB, and AKXL recombinant inbred strains, *Ucp* was found to be on Chr 8 very close to *Es-1* (1).

References

1. Jacobsson, A., U. Stadler, M.A. Glotzer, and L.P. Kozak. 1985. Mitochondrial uncoupling protein from mouse brown fat: molecular cloning, genetic mapping, and mRNA expression. J. Biol. Chem. 260:16250–16254.

Udpgt-3 locus, UDP glucuronosyltransferase-3, Chr 5

This locus contains a DNA sequence or sequences homologous with a rat cDNA coding for the constitutive drug-metabolizing enzyme UDP glucuronosyltransferase-3 (UDPGT-3; E.C. 2.4.1.17). The locus was found to be on Chr 5 by use of mouse–Chinese hamster somatic cell hybrids screened with a rat UDPGT-3 cDNA clone. The same mouse sequences also hybridize with a rat UDPGT-4 cDNA clone, but the mouse gene or genes have not yet been characterized.

References

1. Krasnewich, D., C.A. Kozak, D.W. Nebert, and P.I. Mackenzie. 1987. Localization of UDP glucuronosyltransferase gene(s) on mouse chromosome 5. Somat. Cell Mol. Genet. 13:179–182.

Ul, ulnaless, dominant or semidominant, Chr 2

Arose in an irradiation experiment. Heterozygotes show extreme reduction of the radius and ulna and deformities of the tibia and fibula. Penetrance appears to be complete. Since males have not bred, homozygotes have not been reported (1).

References

1. Morris, T. 1967. Mouse News Lett. 36:34.

Umph-2 locus, uridine monophosphatase-2, Chr 11

This locus codes for UMPH-2, one of two isozymes of uridine monophosphatase found in most tissues of the mouse. UMPH-2 is found at low levels in heart and skeletal muscle an in larger amounts in most other tissues and in cultured cells (1). In mouse–Syrian hamster somatic cell hybrids, *Umph-2* cosegregated with galactokinase (*Glk*) on Chr 11 (2). No variants of the gene for the other isozyme, UMPH-1, have been described and its chromosome location is not known (1).

References

1. Swallow, D.M., V.S. Turner, and D.A. Hopkinson. 1983. Isozymes of rodent 5′-nucleotidase: evidence for two independent structural loci *Umph-1* and *Umph-2*. Ann. Hum. Genet. 47:9–17.
2. Wilson, D.E., D. Woodward, A. Sandler, J. Erickson, and A. Gurney. 1986. A gene for uridine monophosphatase-2 is on mouse chromosome 11. Am. J. Hum. Genet. 39:Suppl. A173 (Abstr.).

un locus, Chr 2

un, undulated, recessive. Found in mice obtained from a Cambridge fancier (7). Homozygotes have a shor-

tened and usually kinked tail. The caudal vertebrae are reduced in size but not in number. The kinks can easily be flattened out with the fingers but immediately return when released. Some homozygotes have marked kyphosis of the lower thoracic and upper lumbar region. The vertebrae are abnormally formed over the whole spine. The acromion process of the scapula is reduced or absent (3). The anomalies can be traced to the 11th day of gestation. The condensations of mesenchyme cranial to the sclerotomic fissure are smaller than normal. Instead of joining with the primitive centra in front of them to form the body of the vertebra, they remain with the material posterior to the sclerotomic fissure and enter into the intervertebral disk. The vertebrae are thus smaller and the disks larger than normal (4).

un^{ex}, undulated-extensive, recessive. Found among descendants of wild mice trapped in Peru. Homozygotes have an abnormal spine and tail and sometimes face and jaw defects, webbed feet, and polydactyly. Viability is good (5, 6).

un^m, undulated-minimal, recessive. Found among descendants of wild mice trapped in Peru. Homozygotes have good viability. They resemble *un/un* mice but are less severely affected (5, 6).

Un^s, undulated-short-tail, semidominant. Arose in the CC57BR/Y strain. Heterozygotes have wavy tails and are smaller than normal. Homozygotes die *in utero*. *Un^s/un* mice have short undulated tails and deformed spines and are very small. The two such mice observed died by 2 weeks of age (1, 2).

References

1. Blandova, Z.K., and I.K. Egorov. 1967. Mouse News. Lett. 36:57.
2. Blandova, Z.K., and I.K. Egorov. 1975. Mouse News. Lett. 52:43.
3. Grüneberg, H. 1950. Genetical studies on the skeleton of the mouse. II. Undulated and its "modifiers". J. Genet. 50:142–173.
4. Grüneberg, H. 1954. Genetical studies on the skeleton of the mouse. XII. The development of undulated. J. Genet. 52:441–455.
5. Wallace, M.E. 1979. Mouse News Lett. 61:29.
6. Wallace, M.E. 1980. Mouse News Lett. 63:10.
7. Wright, M.E. 1947. Undulated: a new genetic factor in *Mus musculus* affecting the spine and tail. Heredity 1:137–141.

Upg-1 locus, urinary pepsinogen-1, Chr 17

This locus controls electrophoretic mobility of the most anodal band of the group A pepsinogens found in mouse urine. The allele $Upg-1^s$ determines a slowly migrating band and occurs in wild mice and in about half the inbred strains tested, including C57BL/6, DBA/1, SJL, and 129/J; the allele $Upg-1^f$ determines a fast band and occurs in wild mice and in the A/J, AKR, BALB/c, C3H/He, DBA/2, and other strains. There are few other loci in the mouse with alleles so evenly distributed among inbred strains and with such widespread variation in natural populations (1). *Upg-1* is located on Chr 17 about 6 cM distal to *H-2* (2).

References

1. Szymura, J.M., and J. Klein. 1981. Linkage of a gene controlling urinary pepsinogen with the major histocompatibility complex of the mouse. Immunogenetics 13:267–271.
2. Szymura, J.M., and J. Klein. 1982. Mapping of the mouse *Upg-1* locus. Immunogenetics 16:89–90.

Upg-2 locus, urinary pepsinogen-2, Chr 1

This locus controls electrophoretic variation in the group B pepsinogens which migrate more slowly than the group A pepsinogens. The allele $Upg-2^s$ determines a single enzyme band and occurs in many strains, including A/J, BALB/c, CBA/Ca, C57BL/6, and NZB; the allele $Upg-2^d$ determines a double band and occurs in many other strains, including AKR, CBA/J, DBA/2, SJL, and SWR. Heterozygotes have double bands indistinguishable from those of homozygous $Upg-2^d$. *Upg-2* is located on Chr 1 very close to *Ren-1* and *Pep-3* (1).

References

1. Szymura, J.M., B.A. Taylor, and J. Klein. 1982. *Upg-2*: a urinary pepsinogen variant located on chromosome 1 of the mouse. Biochem. Genet. 20:1211–1219.

Ups locus, uroporphyrinogen I synthase, Chr 9

This locus controls structural variation revealed by isoelectric focusing of the enzyme uroporphyrinogen I synthase (UPS; E.C. 4.3.1.8). UPS is the third of eight enzymes in the biosynthetic pathway leading to heme production and is present in erythrocytes and all other tissues. In erythrocytes, the allele Ups^a determines a form with an isoelectric point (pI) of 5.6; it occurs in all inbred strains tested, in wild mice from California, and in *M. m. castaneus*. The allele Ups^b determines a variant with a pI of 5.4 found in wild *M. m. musculus* from Denmark. Heterozygotes have both parental bands. UPS isozymes in other tissues differ in pI from that in erythrocytes because of posttranslational modifications

(2). *Ups* is located on Chr 9 near the centromeric end (1).

References

1. Antonucci, T.K., V.C. Chapman, and M.H. Meisler. 1982. Linkage of the structural gene for uroporphyrinogen I synthase to markers on mouse chromosome 9 in a cross between feral and inbred mice. Biochem. Genet. 20:703–710.
2. Meisler, M.H., and M.L.C. Carter. 1980. Rare structural variants of human and murine uroporphyrinogen I synthase. Proc. Natl. Acad. Sci. USA 77:2848–2852.

ur, urogenital, recessive, extinct

Discovered in a balanced tailless stock (1). Homozygotes have short tails and are small at birth. They usually have cleft palates and die within a day or two. Some have normal palates and may survive for a month or more, but they remain small and sterile. Older animals are usually found to have hydronephrosis or polycystic kidneys. At birth the kidneys have only slight histological abnormalities but are deficient in alkaline phosphatase and probably do not function normally (3). The skeletal abnormalities consist of a reduction in length but not in width of midline structures, particularly at the anterior and posterior ends. There may be a small extra pair of ribs and fusions of ribs and sternebrae. The shortening of the skeleton can be seen in the tail at $10\frac{1}{2}$ days of gestation and in the head at 12 days (2). The origin of the kidney abnormalities is unknown.

References

1. Dunn, L.C., and S. Gluecksohn-Schoenheimer. 1947. A new complex of hereditary abnormalities in the house mouse. J. Exp. Zool. 104:25–51.
2. Fitch, N. 1957. An embryological analysis of two mutants in the house mouse, both producing cleft palate. J. Exp. Zool. 136:329–357.
3. Gluecksohn-Waelsch, S., and S.A. Kamell. 1955. Physiological investigations of a mutation in mice with pleiotropic effects. Physiol. Zool. 28:68–73.

us, urogenital syndrome, recessive

Occurred in a linkage cross in 1966. It resembles the extinct mutation *ur*. Homozygotes are small at birth and have a kinky tail. Skeletal abnormalities include split arches and fusions of cervical vertebrae, fusions of thoracic and lumbar vertebrae, and unilateral and bilateral fusions of one or more ribs. Delayed growth, sterility, and urogenital abnormalities also occur. Affected mice have a thinner coat with fewer zigzags and narrower awls than normal, and slightly abnormal behavior (1). *us* was thought to be located on Chr 13, but this was later shown to be in error (2).

References

1. Lane, P.W. 1973. Mouse News Lett. 49:32.
2. Lane, P.W. 1975. Mouse News Lett. 52:38.

uw, underwhite, recessive, Chr 15

Arose spontaneously in the C57BL/6J strain. The fur of homozygotes is a light buff color on top with very white underfur. Eyes are unpigmented at birth but darken to a dark reddish color at maturity (1).

References

1. Dickie, M.M. 1964. Mouse News Lett. 30:30.

V

v locus, Chr 10

v, waltzer, recessive. Probably originated in China many centuries ago (3). Viability and breeding ability are somewhat less than normal. Homozygotes show the typical circling, head-tossing, deafness, and hyperactivity of the circling mutants. Most of them are deaf from the beginning. Abnormalities of the inner ear include degeneration of the organ of Corti, spiral ganglion, stria vascularis, and saccular macula. Double heterozygotes with shaker-1 (*v*/+ *sh-1*/+) are deaf beginning at 3 to 6 months. They have changes similar to those of the homozygotes in the organ of Corti, stria vascularis, and spiral ganglion, but less severe and with much later onset (1,4).

v^{df}, deaf, recessive. First noticed in a laboratory stock through a somewhat anomalous position of the ears (2). Homozygotes may be deaf from the beginning or may hear for a few days before weaning but otherwise behave normally. Deafness is caused by degeneration of the organ of Corti, the spiral ganglion, and the stria vascularis. *v^{df}*/*v* mice are like *v^{df}*/*v^{df}* mice in hearing ability and behavior. Double heterozygotes with shaker-1 (*v^{df}*/+ *sh-1*/+) become deaf at a late stage (4).

References

1. Deol, M.S. 1956. The anatomy and development of the mutants pirouette, shaker-1 and waltzer in the mouse. Proc. R. Soc. Lond. (Biol.) 145:206–213.
2. Deol, M.S. 1956. A gene for uncomplicated deafness in the mouse. J. Embryol. Exp. Morphol. 4:190–195.
3. Keeler, C.E. 1931. The Laboratory Mouse. Its Origin, Heredity, and Culture. Harvard Univ. Press, Cambridge, Mass. 81p.
4. Kocher, W. 1960. Untersuchungen zur Genetik und Path-ologie der Enkwicklung von 8 Labyrinth-mutanten (*deaf-waltzer-shaker*-Mutanten) der Maus (*Mus musculus*). Z. Vererb. 91:114–140.

Va locus, Chr 3

Va, varitint-waddler, semidominant. Found by Cloudman and Bunker (1) in descendants of a cross of C57BL × C57BR strains. Viability of heterozygotes is nearly normal, but fertility is reduced. Mortality is very high in homozygotes, and very few of the survivors are fertile. Heterozygotes are deaf and show circling behavior, head-tossing, and hyperactivity. They circle somewhat less than some of the other circling mutants. Their coats are variegated with patches of normal-colored, diluted, and white fur. Homozygotes show more intense behavioral abnormalities than heterozygotes, and their coats are white, except for small patches of unaltered color near the ears and base of the tail (1). The pathological changes in heterozygotes include degeneration of the organ of Corti, stria vascularis, spiral ganglion, saccular macula, cristae ampullares, and vestibular ganglion. In homozygotes the degenerative changes are more severe and also include the utricular macula (3). In heterozygotes, in addition to the characteristic shaker–waltzer behavior, a number of behavioral defects involving reflexes and locomotor patterns and said to be characteristic of defects of the corpus striatum have been described (2). In the same study a close correlation was found between direction of laterality of circling and asymmetry of distribution of the wild-type patches of fur.

Va^J, varitint-waddler-Jackson, semidominant. Found in a linkage cross involving *Va* and probably arose by mutation from *Va*. *Va^l/*+ mice are more pigmented than *Va/*+ mice and behave normally although they are deaf. They have slightly diluted coat color, a large irregular belly spot, and white feet and tail tip. Homozygotes have much white spotting interspersed with patches of diluted color. They are deaf but behave normally and are fertile. *Va^l/Va* mice are somewhat similar to *Va^l/Va^l* but are smaller with more white spotting and abnormal behavior. They are deaf and circle vigorously. Their viability and fertility are considerably reduced (4).

References

1. Cloudman, A.M., and L.E. Bunker. 1945. The varitint-waddler mouse. J. Hered. 36:258–263.
2. Cools, A.R. 1972. Asymmetrical spotting and direction of circling in the varitint-waddler mouse. J. Hered. 63:167–171.
3. Deol, M.S. 1954. The anomalies of the labyrinth of the mutants varitint-waddler, shaker-2 and jerker in the mouse. J. Genet. 52:562–588.
4. Lane, P.W. 1972. Two new mutations in linkage group XVI of the house mouse. Flaky tail and varitint-waddler-J. J. Hered. 63:135–140.

vb, vibrator, recessive, Chr 11

Arose spontaneously in the DBA/2J strain. Homozygotes are recognizable at 10 to 12 days of age by a fine rapid tremor. This is followed by a degenerative phase in which there is progressive development of ascending motor paralysis and coarse cerebellar tremor, and finally a terminal phase in which there is loss of consciousness and death. Death occurs by 30 days of age in *vb/vb* mice with an inbred genetic background but many derived from outcrosses may live to 6 months. Selected neurons in spinal cord and later in brainstem and cerebellum show progressive degenerative changes featuring dilated cisternae of endoplasmic reticulum in cell bodies, dendrites, and axons, with eventual severe intracellular vacuolization and some cell death (1).

References

1. Weimer, W.B., P.W. Lane, and R.L. Sidman. 1982. Vibrator (*vb*): a spinocerebellar system degeneration with autosomal recessive inheritance in mice. Brain Res. 251:357–364.

vc, vacillans, recessive, Chr 4, probably extinct

Arose spontaneously in the DBA/1 strain. Homozygotes can be identified with certainty at 14 days of age by a violent tremor when walking and by swaying of the hindquarters. With age the instability lessens and a ducklike gait sets in. The tremor is less marked with age. Vacillans mice are less aggressive than normal. They are smaller, and females do not take good care of their litters. Muscular strength is about half that of normal mice. There is a peak mortality at weaning age, after which survival appears normal. Sexual maturity in males did not occur before $5\frac{1}{2}$ months. The central nervous system shows no gross anatomical abnormality on pathological examination (1).

References

1. Sirlin, J.L. 1956. Vacillans, a neurological mutant in the house mouse linked to brown. J. Genet. 54:42–48.

Ve, velvet coat, semidominant, Chr 15

Appeared in offspring of an irradiated male. Heterozygotes closely resemble *sa/sa* mice. The coat has a velvety sheen and on the ventral surface appears to be more luxuriant than normal. Heterozygotes are viable and fertile (2). Homozygotes die *in utero*. The gross morphology of the hairs of heterozygotes is largely unaltered. Internally the medulla is abnormal with widely dispersed pigment and extensive areas of fatty fluid in the air spaces. The epidermis is somewhat disorganized and less compact than normal, and the abnormalities of the hair follicles appear to be secondary to a general abnormality of the epidermal cells (4). This conclusion is reinforced by the observation that *Ve/Ve* embryos at 7 days of gestation have very little or no ectoderm. Trophoblast giant cells and parietal endoderm appear unaffected, but disorganization ensues and death occurs by 8 or 9 days (1, 3).

References

1. Diwan, S., and L.C. Stevens, 1974. Mouse News Lett. 51:25.
2. Maddux, S.C. 1964. Mouse News Lett. 31:41.
3. Rossant, J., and K.M. Vijh. 1981. *in vivo* and *in vitro* development of mouse embryos homozygous for the embryonic lethal velvet coat (*Ve*) mutation. J. Embryol. Exp. Morphol. 66:43–55.
4. Trigg, M.J. 1972. Hair growth in mouse mutants affecting coat texture. J. Zool. Lond. 168:165–198.

vi, visceral inversion, recessive

Found in descendants of a four-way cross. Probably extinct. Penetrance is probably not complete, and there may be some loss of homozygotes in segregating generations because of inviability. The mutant resembles *iv* but was not tested for allelism with it. Homozygotes show complete or partial reversal of the viscera as well as a general sickliness and often hydrocephalus (1).

References

1. Tihen, J.A., D.R. Charles, and T.O. Sippel. 1948. Inherited visceral inversion in mice. J. Hered. 39:29–31.

vl, vacuolated lens, recessive, Chr 1

Arose in the C3HeB/FeJ strain. Homozygotes have opaque white lenses, may have a white belly spot, and may have caudal spina bifida. Although the lens defect is the most consistent landmark, occasionally homozygotes have only a belly spot. Histologically the lens shows small vacuoles at birth. Spina bifida is present in all homozygous embryos at 11 to 12 days of gestation, but is visible in only one-third to one-half of them at birth. If this defect is not too severe, the lesion heals and the mice survive (1).

References

1. Dickie, M.M. 1967. Mouse News Lett. 36:39.

Vlm, vacuolated lens with microphthalmia, dominant or semidominant

Arose in a (101 × C3H)F1 hybrid after irradiation with ^{137}Cs gamma rays. In heterozygotes there are small and large vacuoles that fill up the whole lens. There is occasional rupture of the capsule and the lens substance extrudes into the anterior chamber. Pupillary dilation by atropine is poor. The cataract is associated with microphthalmia. Manifestation is bilateral and penetrance is complete. Homozygotes are viable and fertile but the condition of their cataracts was not described (1).

References

1. Kratochvilova, J. 1981. Dominant cataract mutations detected in offspring of gamma-irradiated male mice. J. Hered. 72:234–237.

vt, vestigial-tail, recessive, Chr 11

Arose spontaneously in the C57BR strain (3). Homozygotes have very short tails, varying from complete absence to about half normal length. They tend to have fewer presacral vertebrae than normal, and the bodies of lumbar vertebrae often ossify from bilateral twin centers rather than from a single center. At 10 days of gestation when there is a ventral ectodermal ridge at the zone of growth of the tail tip, the ridge is much reduced in *vt/vt* embryos. There is also a reduction in the tail gut. The neural tube in the tail bud is abnormal in the ventral region, tending to split off and round up to form accessory tubes (1). The primitive streak is abnormally small and fails to fill the space within the ectoderm. All the other abnormalities may be consequent on the abnormal primitive streak (2).

References

1. Grüneberg, H. 1957. Genetical studies on the skeleton of the mouse. XIX. Vestigial tail. J. Genet. 55:181–194.
2. Grüneberg, H., and G.A. DeS. Wickramaratne. 1974. A re-examination of two skeletal mutants of the mouse, vestigial tail (*vt*) and congenital hydrocephalus (*ch*). J. Embryol. Exp. Morphol. 31:207–222.
3. Heston, W.E. 1951. The "vestigial tail" mouse. J. Hered. 42:71–74.

W

W locus, Chr 5

Mutations at the *W* locus, like those at the *Sl* locus, cause reduction in pigmentation, sterility, and macrocytic anemia. They are all semidominant. Homozygotes or compounds of two different alleles are usually black-eyed white, sterile, and have severe macrocytic anemia, often causing death *in utero* or neonatally. Heterozygotes with wild-type have some white spotting with or without slightly diluted pigmentation, they are fertile, and they may be slightly anemic. Numerous experiments have shown that the anemia and pigment defects are due to intrinsic defects in progenitor cells of erythrocytes and melanocytes. This has not been shown for germ cells, probably because of lack of a suitable experimental system. Most of the experimental studies on the *W* locus have made use of the *W* and *W^v* alleles and their compound *W/W^v*.

By use of recombination grafts of embryonic neural crest and skin, Mayer and Green (26) showed that the pigment defect in *W^v/W^v* mice resides in the melanoblasts derived from the neural crest. The defect operates at some stage before proliferation of melanoblasts in the skin (24). Melanoblasts of heterozygotes (*W/+*) are also defective; unlike +/+ melanoblasts, they are unable to compete with albino (*c/c*) melanoblasts in fusion chimeras or in dermal–epidermal recombination grafts (14, 25).

The sterility of mice bearing two *W* alleles is due to a severe deficiency of primordial germ cells. Germ cells form normally in the yolk sac, but the rate of proliferation thereafter and of migration toward the germinal ridges is severely retarded (27). These mice tend to develop ovarian tumors, the lack of germ cells resulting in overproduction of pituitary gonadotropic hormone with excess stimulation of the gonad (28).

The anemia of *W*-locus homozygotes and compounds is detectable at about 12 days of gestation and persists throughout life. Red cell count and hemoglobin are considerably reduced and the red cells are macrocytic. Bone marrow megakaryocytes and neutrophils are also significantly reduced. The anemia can be cured by implantation of normal bone marrow or spleen cells without irradiation of the recipients, and bone marrow and spleen of the anemic mice have no life-sparing effect on irradiated normal mice. The genetic defect probably resides in erythropoietic cells at an erythropoietin-sensitive stage (2, 30). It seems likely that some lymphopoietic cells may also be abnormal, since normal lymphocytes implanted into unirradiated *W*-anemic mice will replace most host lymphocytes in the thymus and spleen (17). Severe mast cell deficiency in stomach and skin has been found in a number of *W*-locus homozygotes and compounds (11, 13, 37). *W/W^v* mice appear to have increased susceptibility to stomach ulcers under some conditions (18, 37). There is evidence that, in addition to the stem cell defect, a regulatory cell bearing Thy-1 antigen on its surface and necessary for differentiation of stem cells into erythrocytes is absent or defective in *W/W^v* mice (33). *W*-anemic mice are not susceptible to spleen focus formation induced by Friend leukemia virus, probably because they are deficient in the target cell for the virus (35). *W/W^v* mice have a mild deficiency of natural killer cells (6). Mutations at the *W* locus occur frequently.

W, dominant spotting, semidominant. This is an old mutant of the mouse fancy. Heterozygotes have variable amounts of white spotting depending on the genetic background. Colored areas may be interspersed with white hairs to produce a roan type of pattern (8). Heterozygotes have normal blood and are fully viable and fertile (15). Homozygotes are white with black eyes in which pigment is restricted to the retina. They have a severe macrocytic anemia beginning at 12 days of gestation and die within the first week after birth (30). Fetal liver cells of homozygotes differ from those of +/+ mice in the relative amounts of antigen detected by anti-erythroblast and by anti-RBC antisera (9).

W^{19H}, dominant spotting-19H, semidominant. Probably radiation induced in a (C3H/HeH × 101/H)F1 female. It is a deletion covering the *Ph* locus and a recessive lethal located 2 cM distal to *W*. Heterozygotes are not anemic, show minimal white spotting, and are not diluted, but many are small and runted or die pre- or postnatally. Homozygotes die before implantation (22).

W^{34J}, *W^{35J}*, *W^{37J}* to *W^{44J}*, dominant spotting-34J etc., semidominant. This series of 10 new alleles was studied by Geissler *et al.* (12) and Geissler and Russell (10). The alleles varied considerably in their effects on pigmentation in heterozygotes. In homozygotes, most were prenatal or neonatal lethals, but W^{39J} was viable, anemic, and of low fertility, W^{41J} was viable, anemic, and fertile, and W^{44J} was viable, non-anemic, and sterile in males and with greatly reduced fertility in females. $W^{42J}/+$ mice, which have diluted coats, may have spots of wild-type color in both the white and pigmented parts of the coat, suggesting that W^{42J} may often revert to wild-type (39).

W^a, Ames dominant spotting, semidominant. Found among offspring of an X-rayed male of the Z strain. Resembles W except that heterozygotes have a prominent blaze and more often have belly spotting. Homozygotes are anemic and die within a few days after birth (32).

W^b. Ballantyne's dominant spotting, semidominant. Arose in the C57/St strain. Heterozygotes are more extensively spotted then $W^v/+$ and the coat is more diluted. Homozygotes may survive to maturity. They are anemic, black-eyed white, and sterile (1).

W^{bd}, banded, semidominant. Found in a stock carrying Rw, go, and bf. Homozygotes are black-eyed white with slight pigmentation on the ears and snout and occasionally on the rump region and tail. They are viable and fertile and may be marginally anemic. Heterozygotes have a broad band lacking pigmentation in the trunk region. W^{bd}/W^v mice are viable and fertile and black-eyed white with occasional ear pigmentation. W^{bd} closely resembles W^{sh} but compounds of the two alleles have more pigmentation on the head than either homozygote. In addition, $W^{bd}/+$ and W^{bd}/W^{bd} mice are more sensitive to induction of anemia by irradiation than the corresponding genotypes of W^{sh} (3, 4).

W^{ct}, Cattanach's dominant spotting, semidominant. Found in a closed colony descended from C3H/HeH × 101/H hybrids. Heterozygotes resemble W^v and are mildly anemic. Homozygotes are anemic and black-eyed white and most die before 2 weeks (20).

W^e, extreme dominant spotting, semidominant. Found in a Robertsonian translocation stock maintained on a C3H/HeH × 101/H background. Heterozygotes have diluted pigment and extensive white markings on head, back, and belly. Homozygotes are anemic, black-eyed white, and die by 10 days. W^e/W^v mice are viable but infertile (5).

W^f, fertile dominant spotting, semidominant. Arose spontaneously in the C3H/He strain at the Institute Pasteur. Heterozygotes may have a small blaze on the forehead and a large belly spot, or on some backgrounds may be indistinguishable from wild type. They are slightly anemic. Homozygotes have very extensive

white spotting and are moderately anemic, but are fully fertile. Compounds with W^v are black-eyed white except for grayish patches at the roots of the ears and are also fertile (16). W^f homozygotes have defective delayed-type hypersensitivity (DTH) due to a deficiency of circulating DTH-mediating cells (23).

W^j, Jay's dominant spotting, semidominant, extinct. Arose spontaneously in the C3H strain. W^j resembles W in its effects in heterozygotes and homozygotes with the exception that it causes more white spotting in heterozygotes (31).

W^{pw}, panda-white, semidominant, probably extinct. Arose spontaneously in a Df(d se) stock at Oak Ridge National Laboratory. Heterozygotes have much white spotting with pigment restricted mainly to the head and base of tail, and occasionally on hips and shoulders. Homozygotes are anemic and die soon after birth. W^{pw}/W^v mice are black-eyed white and sterile (34).

W^s, Strong's dominant spotting, semidominant, probably extinct. Arose in a methylcholanthrene-treated subline of the Br-S strain. In the strain of origin, heterozygotes showed considerable white spotting but no dilution of the pigmented part of the coat, Homozygotes were anemic at 15 days of gestation and rarely survived to birth. W^s/W mice also die *in utero* (38). W^s/W^b are anemic and black-eyed white and may survive to maturity (1).

W^{sh}, sash, semidominant. Found among offspring of a C3H/HeH × 101/H cross. Heterozygotes have a broad white sash around the body in the lumbar region. Homozygotes are black-eyed white with full viability and fertility and no evidence of anemia. W^{sh}/W^v mice are black-eyed white with some pigment around the ears and eyes and are fertile and not anemic (21). $W^{sh}/W^{sh} \longleftrightarrow +/+$ embryo fusion chimeras resemble $W^{sh}/+$ mice in the distribution of pigment in the skin, a result unexpected if the distribution of melanocytes in the skin is random (36).

W^v, viable dominant spotting, semidominant. Discovered by Little and Cloudman (19) in the C57BL inbred strain. Heterozygotes have a variable amount of white spotting and also a slight dilution of coat color. They have a slight macrocytic anemia (29) but are fully viable and fertile. Homozygotes resemble W/W in color, anemia, and germ cells, but many of them survive to maturity. Marked abnormalities occur in the cochlea of the inner ear, an effect thought to be due to dependence of the neuroepithelium of the cochlea on the acoustic ganglion, which is probably a neural crest derivative and may therefore be defective in W^v/W^v mice (7).

W^x, a mutation indistinguishable from W, occurred in the C3H/HeJ strain in 1952 (31).

References

1. Ballantyne, J., F.G. Bock, L.C. Strong, and W.C. Quevedo. 1961. Another allele at the W locus of the mouse. J. Hered. 52:200–202.

2. Bannerman, R.M., J.A. Edwards, and P.H. Pinkerton. 1973. Hereditary disorders of the red cell in animals. Prog. Hematol. 8:131–179.

3. Beechey, C.V., and A.G. Searle. 1983. Two new W mutations. Mouse News Lett. 68:70.

4. Beechey, C.V., J.F. Loutit, and A.G. Searle. 1986. Banded—a new W allele. Mouse News Lett. 74:92.

5. Cattanach, B.M. 1978. Mouse News Lett. 59:18.

6. Clark, E.A., L.D. Shultz, and S.B. Pollack. 1981. Mutations in mice that influence natural killer (NK) cell activity. Immunogenetics 12:601–613.

7. Deol, M.S. 1970. The origin of the acoustic ganglion and effects of the gene dominant spotting (W^v) in the mouse. J. Embryol. Exp. Morphol. 23:773–784.

8. Dunn, L.C. 1937. Studies on spotting patterns. II. Genetic analysis of variegated spotting in the house mouse. Genetics 22:43–46.

9. Flaherty, L., L. Cantor, D. Zimmerman, and D. Bennett. 1977. Cell surface antigens on erythroid cells: a comparison of normal and anemic (W/W) mice. Dev. Biol. 59:237–240.

10. Geissler, E.N. and E.S. Russell. 1983. Analysis of the hematopoietic effects of new dominant spotting (W) mutations of the mouse. I. Influence upon hematopoietic stem cells. Exp. Hematol. 11:452–460.

11. Geissler, E.N., and E.S. Russell. 1983. Analysis of the hematopoietic effects of new dominant spotting (W) mutations of the mouse. II. Effects on mast cell development. Exp. Hematol. 11:461–466.

12. Geissler, E.N., E.C. McFarland, and E.S. Russell. 1981. Analysis of pleiotropism at the dominant white spotting (W) locus of the mouse: a description of ten new W alleles. Genetics 97:337–361.

13. Go, S., Y. Kitamura, and M. Nishimura. 1980. Effect of W and W^v alleles on production of tissue mast cells in mice. J. Hered. 71:41–44.

14. Gordon, J. 1977. Modification of pigmentation patterns in allophenic mice by the W gene. Differentiation 9:19–27.

15. Grüneberg, H. 1942. Inherited macrocytic anaemias in the house mouse. II. Dominance relationships. J. Genet. 43:285–293.

16. Guénet, J.-L., G. Marchal, G. Milon, P. Tambourin, and F. Wendling. 1979. Fertile dominant spotting in the house mouse: a new allele at the W locus. J. Hered. 70:9–12.

17. Harrison, D.E., and C.M. Astle. 1976. Population of lymphoid tissue in cured W-anemic mice by donor cells. Transplantation 22:42–46.

18. Kitamura, Y., M. Yokoyama, H. Matsuda, and M. Shimada. 1980. Coincidental development of forestomach

19. Little, C.C., and A.M. Cloudman. 1937. The occurrence of a dominant spotting mutation in the house mouse. Proc. Natl. Acad. Sci. USA 23:535–537.

20. Loutit, J.F., and B.M. Cattanach. 1983. Hematopoietic role for patch (Ph) revealed by a new W mutant (W^{cr}) in mice. Genet. Res. 42:29–39.

21. Lyon, M.F., and P.H. Glenister. 1982. A new allele sash (W^{sh}) at the W-locus and a spontaneous recessive lethal in mice. Genet. Res. 39:315–322.

22. Lyon, M.F., P.H. Glenister, J.F. Loutit, E.P. Evans, and J. Peters. 1984. A presumed deletion covering the W and Ph loci of the mouse. Genet. Res. 44:161–168.

23. Marchal, G., G. Milon, and J.-L. Guénet. 1980. Altered circulation of lymphocytes mediating delayed-type hypersensitivity when primed in W^f/W^f mice. Immunology 39:269–274.

24. Mayer, T.C. 1970. A comparison of pigment cell development in albino, steel, and dominant spotting mutant mouse embryos. Dev. Biol. 23:297–309.

25. Mayer, T.C. 1979. Interactions between normal and pigment cell populations mutant at the dominant spotting (W) and steel (Sl) loci in the mouse. J. Exp. Zool. 210:81–88.

26. Mayer, T.C., and M.C. Green. 1968. An experimental analysis of the pigment defect caused by the W and Sl loci in mice. Dev. Biol. 18:62–75.

27. Mintz, B., and E.S. Russell. 1957. Gene-induced embryological modifications of primordial germ cells in the mouse. J. Exp. Zool. 134:207–237.

28. Murphy, E.D. 1972. Hyperplastic and early neoplastic changes in the ovaries of mice after genic deletion of germ cells. J. Natl. Cancer Inst. 48:1283–1295.

29. Russell, E.S. 1949. Analysis of pleiotropism at the W-locus in the mouse: relationship between the effects of W and W^v substitution on hair pigmentation and on erythrocytes. Genetics 34:708–723.

30. Russell, E.S. 1970. Abnormalities of erythropoiesis associated with the W-locus in mice. In A.L. Gordon, ed., Regulation of Hematopoiesis, Vol. 4, 649–675. Appleton-Century-Crofts, New York.

31. Russell, E.S., F. Lawson, and G. Schabtach. 1957. Evidence for a new allele at the W-locus of the mouse. J. Hered. 48:119–123.

32. Schaible, R.H. 1963. Developmental genetics of spotting patterns in the mouse. PhD Dissertation. Iowa State Univ., Ames. 225 p.

33. Sharkis, S.J., W. Wiktor-Jedrzejazak, A. Ahmed, G.W. Santos, A. McKee, and K.W. Sell. 1978. Antitheta-sensitive cell (TSRC) and hematopoiesis: regulation of differentiation of transplanted cells in W/W^v anemic and normal mice. Blood 52:802–817.

34. Steele, M.S. 1974. Mouse News Lett. 50:52.

35. Steeves, R.A., M. Bennett, E.A. Mirand, and G. Cudkowicz. 1968. Genetic control by the W locus of suscepti-

bility to (Friend) spleen focus-forming virus. Nature 218:372–374.

36. Stephenson, R.A., P.H. Glenister, and J.E. Hornby. 1985. Site of beige (*bg*) and leaden (*ln*) pigment gene expression determined by recombinant embryonic skin grafts and aggregation mouse chimaeras employing sash (*W^{sh}*) homozygotes. Genet. Res. 46:193–205.

37. Stevens, J., and J.F. Loutit. 1982. Mast cells in spotted mutant mice (*W*, *Ph*, *mi*). Proc. R. Soc. Lond. B 215:405–409.

38. Strong, L.C., and W.F. Hollander. 1953. Two non-allelic mutants resembling "W" in the house mouse. J. Hered. 44:41– 44.

39. Whitney, J.B. III, and M.L. Lamoreux. 1982. Transposable elements controlling genetic instabilities in mammals. J. Hered. 73:12–18.

wa-1, waved-1, recessive, Chr 6

Found in a mixed mouse colony (2). Homozygotes are recognizable at 2 or 3 days of age by their curly whiskers. The first coat is strongly waved, but the hair becomes straight in later coats. Most of the whiskers also become straight, but the guard hairs are curved and shorter than normal. All hair types of the first coat show loose curves, possibly because of a defective internal root sheath which may prevent the hair shaft from moving upward at a normal rate (3). Bennett and Gresham (1) found that many *wa-1/wa-1* mice had eyelids open at birth and ascribed the condition to a closely linked gene, *eo*, but the possibility that open eyelids is a pleiotropic effect of *wa-1* was not excluded.

References

1. Bennett, J.H., and G.A. Gresham. 1956. A gene for eyelids open at birth in the house mouse. Nature 178:272–273.
2. Crew, F.A.E. 1933. Waved: an autosomal recessive coat form character in the mouse. J. Genet. 27:95–96.
3. Trigg, M.J. 1972. Hair growth in mouse mutants affecting coat texture. J. Zool. Lond. 168:165–198.

wa-2, waved-2, recessive, Chr 11

Found in an 'abnormal corpus callosum' stock (2). This mutant is very similar to *wa-1*. Homozygotes can be recognized at 2 to 3 days by their curly whiskers. The first coat is waved but later coats are not. The vibrissae usually remain curled and the guard hairs curved. Some homozygotes have eyelids open at birth (1). Butler and Robertson have ascribed the condition to an independent gene, squint (*sq*), but it is probably a pleiotropic effect of *wa-2*. The proportions of guard hairs and zigzags are about normal in *wa-2/wa-2* mice, but the hairs show irregularities and narrowing of the medulla, usually accompanied by thickening of the cortex. Bending occurs between these areas (3). In fusion chimeras between *wa-2/wa-2* and +/+ mice, waved and normal hairs occur in a patchy distribution, suggesting that *wa-2* acts locally in the hair follicle cells (4).

References

1. Butler, L., and D.A. Robertson. 1953. A new eye abnormality in the house mouse. J. Hered. 44:13–16.
2. Keeler, C.E. 1935. A second rexoid coat character in the house mouse. J. Hered. 26:189–191.
3. McLaren, A. 1971. The microscopic appearance of waved-2 hairs. Genet. Res. 17:257–260.
4. McLaren, A., and P. Bowman. 1969. Mouse chimaeras derived from fusion of embryos differing by nine genetic factors. Nature 224:238–240.

wal, waved alopecia, Chr 14

Arose spontaneously in the BALB/cLac strain. Homozygotes are recognized by their long waved fasciculated hair. After shedding of the juvenile coat, thin short hair grows in, but, by 2 months, the body appears hairless except for long waved hair on the head and tail. Vibrissae are normal. Some dermal cysts are found at 2 to 4 months. *wal* is on Chr 14 about 6 cM proximal to *s* (1).

References

1. Sorokina, J.D., and Z.K. Blandova. 1985. Mouse News Lett. 73:23.

Wap locus, whey acidic protein, Chr 11?

This locus controls variation revealed by electrophoresis and by isoelectric focusing of an acidic protein found in the whey fraction of mouse milk. The allele *Wap^a* determines a rapidly migrating form on polyacrylamide gel at pH 8.5 and occurs in all strains examined except YBR, which bears the allele *Wap^b* determining a slower migrating form. Heterozygotes have both parental forms (2). WAP-A and WAP-B have isoelectric points of 4.7 and 4.8, respectively, and differ in one amino acid. The protein has a molecular weight of 14 000 and appears to be unique to mouse milk (3). Gupta *et al.* (1) have isolated a mouse mammary gland cDNA clone homologous to a gene for a rat novel whey protein. Complete sequence analysis of the clone suggests that it may code for WAP. Evidence from somatic cell hybrids is consistent with location of the gene on Chr 11.

References

1. Gupta, P., J.M. Rosen, P. D'Eustachio, and F.H. Ruddle. 1982. Localization of the casein gene family to a single mouse chromosome. J. Cell Biol. 93:199–204.
2. Piletz, J.E., and R.E. Ganschow. 1981. Genetic variation in milk proteins in mice. Biochem. Genet. 19:1023–1030.
3. Piletz, J.E., M. Heinlen, and R.E. Ganschow. 1981. Biochemical characterization of a novel whey protein from murine milk. J. Biol. Chem. 256:11509–11516.

War, warfarin resistance, dominant in females, semidominant in males, Chr 7

Found in a wild population in the Plant Breeding Institute, Cambridge, where its frequency had been increasing because of exposure of the population to warfarin. It was discovered because it occurred in this population on the same chromosome as a closely linked color mutation, extreme dilution (c^e), whose frequency was noted to be increasing. Wild-type mice of both sexes were fully susceptible when tested at 4 months of age. Both *War/War* and *War/+* females were fully resistant when tested under the same conditions. Less than 100 per cent of similarly tested *War/War* males were resistant, and the proportion of resistant *War/+* males was intermediate between that and 0 per cent. Modifying genetic factors affected the degree of resistance in males but not in females. A similar mutation for warfarin resistance occurs in the rat and occupies an analogous position linked to the *p* and *c* loci in that species (1).

References

1. Wallace, M.E., and F.J. MacSwiney. 1976. A major gene controlling warfarin-resistance in the house mouse. J. Hyg. Camb. 76:173–181.

Wc, waved coat, semidominant, Chr 14

Found among descendants of a linkage cross carrying depilated (*dep*). Heterozygotes have a wavy coat and whiskers. They are recognizable at 4 or 5 days of age by a lighter color, wavy irregular whiskers, and short fuzzy hair on the head and upper back. On some genetic backgrounds, heterozygotes may be more severely affected; the skin becomes dry and scaly after birth, no hair develops, and the mice die before weaning. On an outbred background heterozygotea are normally viable and fertile (2). Homozygotes die *in utero*. After differentiation of embryonic and extraembryonic ectoderm, the embryonic ectoderm becomes pycnotic and degenerates, leaving an 'extraembryonic organism' consisting of relatively normal yolk sac, chorion, and allantois and devoid of an embryo proper (1).

References

1. Diwan, S., and L.C. Stevens. 1974. Mouse News Lett. 51:24–25.
2. Green, M.C. 1972. Mouse News Lett. 47:36.

wd, waddler, recessive, Chr 4, probably extinct

Arose spontaneously in a stock carrying furless (*fs*). Homozygotes can be identified at about 14 days of age. The hindquarters sway from side to side in a smooth arc, and the mice often fall over on their hips. No trembling or paralysis occurs. Homozygotes are smaller than their normal littermates from 3 weeks on, but viability is good and many waddlers are fertile. The abnormalities of behavior do not become more severe with age (1). Studies have been made of the serum proteins (2), serum cholinesterase (4), and acid-soluble phosphorus of the brain (3) in *wd/wd* mice, but the results are not easily interpretable.

References

1. Yoon, C.H. 1959. Waddler, a new mutation and its interaction with quivering. J. Hered. 50:238–244.
2. Yoon, C.H. 1961. Electrophoretic analysis of the serum proteins of neurological mutants in mice. Science 134:1009–1010.
3. Yoon, C.H., and D.J. Denuccio. 1963. Acid-soluble phosphorus in brains of neurological mice. J. Hered. 54:202–205.
4. Yoon, C.H., and S.R. Harris. 1962. Cholinesterase studies of neurological mutants in mice. I. Alterations in serum cholinesterase levels. Neurology 12:423–426.

we, wellhaarig, recessive, Chr 2

Arose as a spontaneous mutation in the stocks of Agnes Bluhm. Homozygotes are fully fertile. They have curly whiskers at 2 to 3 days of age and a wavy first coat, most strongly evident between 10 and 21 days. In later coats the waviness is lost. The hairs have a lower average diameter then those of normal mice (3). *we* has been shown to be on Chr 2 closely linked to pallid (*pa*) and has been bred in coupling with *pa* (1) with no interaction between the two genes. However, Graff *et al.* (2) reported a cross between *we/we* and *pa/pa* mice that produced F2 pallid wellhaarig offspring in much higher than expected frequency, all showing a severe skeletal abnormality and early death. The authors hypothesized that the effect was due to complementary action of genes in the *pa-we* segment and that the F2 pallid wellhaarig mice were *pa we/pa +* in genotype, accounting for the excess of this class in the F2. Why the results of this cross differed from those of similar previous crosses is not clear.

References

1. Falconer, D.S. 1954. Linkage in the mouse: the sex-linked genes and "rough". Z. Indukt. Abstammungs-Vererbungsl. 86:263–268.
2. Graff, R.J., D. Simmons, J. Meyer, D. Martin-Morgan, and M. Kurtz. 1986. Abnormal bone production associated with mutant mouse genes *pa* and *we*. J. Hered. 77:109–113.
3. Hertwig, P. 1942. Neue Mutationen und Koppelungsgruppen bei der Hausmaus. Z. Indukt. Abstammungs-Vererbungsl. 80:220–246.

wh, writher, recessive

This mutation is similar to *dt*, but has not been tested for allelism with it. Possibly radiation-induced. Affected mice die before weaning. They are unable to support themselves on their legs. Convulsions of the torso occur in later stages. The condition is first recognizable at 12 days (1).

References

1. Kelly, E.M. 1953. Mouse News Lett. 8 (Suppl.):15.

wi, whirler, recessive, Chr 4

Arose as a spontaneous mutation in a multiple recessive stock. Viability of homozygotes may be slightly reduced. Both sexes are fertile, but females do not always make good mothers. Homozygotes show the characteristic symptoms of the shaker syndrome, deafness, head-tossing, circling, and hyperactivity (1). They have defects of the membranous labyrinth similar to those of *sh-2* (M.S. Deol, personal communication). Body weight is decreased and metabolic rate and adrenocortical activity are increased, probably as a consequence of the hyperactivity (2, 3).

References

1. Lane, P.W. 1963. Whirler mice, a recessive behavior mutation in linkage group VIII. J. Hered. 54:263–266.
2. Sackler, A.M., and A.S. Weltman. 1967. Metabolic and endocrine differences between the mutation whirler and normal female mice. J. Exp. Zool. 164:133–140.
3. Weltman, A.S., A.M. Sackler, A.S. Lewis, and L. Johnson. 1970. Metabolism rate, biochemical and endocrine alterations in male whirler mice. Physiol. Behav. 5:17–22.

wl, wabbler-lethal, recessive, Chr 14

Arose in a stock carrying pirouette (*pi*). Homozygotes are first recognizable at 12 days of age and usually die at about 4 weeks. They have an abnormal wobbly gait and a pronounced tremor when walking. Difficulty in walking increases in severity until the animal dies (1). In an extensive study of behavioral development of this mutant, homozygotes were shown to be deficient in nearly all behaviors tested. From 20 days until death they were hypoglycemic (5). Histological examination showed myelin degeneration widely distributed throughout the central nervous system, particularly in the vestibulocerebellar and spinocerebellar systems (2). Electron microscopy showed widespread axonal degeneration in the medulla with secondary myelin dissolution. Similar abnormalities were present to a lesser extent in the basal ganglia, spinal cord, and cerebellum. Primary demyelination was found in the sciatic nerve (3). Aromatic amino acid metabolism is abnormal in affected mice with depressed activity of phenylalanine hydroxylase and elevated activity of several other enzymes. This condition characterizes several other neurological mutants and is probably secondary to their neurological deficit (4).

References

1. Dickie, M.M., J. Schneider, and P.J. Harman. 1952. A juvenile wabbler-lethal in the house mouse. J. Hered. 43:283–286.
2. Harman, P.J. 1954. Genetically controlled demyelination in the mammalian central nervous system. Ann. NY Acad. Sci. 58:546–550.
3. Luse, S.A., C. Chenard, and E.H. Finke. 1967. The wabbler-lethal mouse. An electron microscopic study of the nervous system. Arch. Neurol. 17:153–161.
4. Siegel, S.J., and H. Rauch. 1969. Aromatic amino acid metabolism n the wabbler-lethal mouse. Biochem. Genet. 2:311–318.
5. Thiessen, D.D. 1965. The wabbler-lethal mouse: a study in development. Anim. Behav. 13:87–100.

wr, wobbler, recessive

Arose in the C57BL/Fa strain. Homozygotes are recognizable at about 3 weeks of age by their smaller size, unsteady gait, and fine tremor of the head. They become progressively weaker, more severely in the forelimbs and anterior part of the body. Most affected mice die by 2 or 3 months of age, but some live for more than a year. None have bred. Histological examination shows degeneration in the cell bodies of motor nerve cells of the brainstem and spinal cord and denervation atrophy of skeletal muscle. The denervation atrophy is distributed similarly in all affected mice but does not involve all muscles, those of the anterior region being more affected than those of the posterior region, and proximal limb muscles more than distal (5).

Electron microscopy of the cervical anterior horns, cervical ventral roots, and brachial nerve shows vacuolar dilatation of the Golgi complex and endoplasmic reticulum, as well as other cytoplasmic changes, in the cell bodies, and wallerian degeneration with collapse and degeneration of myelin sheaths in the axons (1). There is loss of the larger myelinated fibers in the median nerve (2) and a reduction in both size and number of axons in the forelimb nerves. In the forelimb axons, there is a reduction in the rate of slow axonal transport affecting the neurofilament proteins more than it affects tubulin and actin (7). Biochemical studies have shown reduced cGMP in the cervical spinal cord (3), reduced protein synthesis in anterior horn neurons (8), and increased acid protease activity in spinal cord and muscle (4). Some minor systemic defects have been reported (6) but these may be secondary effects of the debilitated condition of the mice. The wobbler disease somewhat resembles amyotrophic lateral sclerosis and Werdnig–Hoffmann disease in man (6).

References

1. Andrews, J.M. 1975. The fine structure of the cervical spinal cord, ventral root and brachial nerves in the wobbler (wr) mouse. J. Neuropathol. Exp. Neurol. 34:12–27.
2. Bradley, W.G., and M. Jenkison. 1973. Abnormalities of peripheral nerves in murine muscular dystrophy. J. Neurol. Sci. 18:227–247.
3. Brooks, B.R., W.D. Lust, J.M. Andrews, and W.K. Engel. 1978. Decreased spinal cord cGMP in murine (wobbler) spontaneous lower motor neuron degeneration. Arch. Neurol. 35:590–591.
4. Chelmicka-Schorr, E., M. Sportiello, J.P. Antel, and B.G.W. Arnason. 1982. Acid protease activity in spinal cord and muscle in wobbler mice. J. Neurol. Sci. 56:141–145.
5. Duchen, L.W., and S.J. Strich. 1968. An hereditary motor neurone disease with progressive denervation of muscle in the mouse: the mutant "wobbler." J. Neurol. Neurosurg. Psychiat. 31:535–542.
6. Leestma, J.E. 1980. Animal model: motor neuron disease in the wobbler (wr/wr) mouse. Am. J. Pathol. 100:821–824.
7. Matsumoto, H., and P. Gambetti. 1986. Impaired slow axonal transport in wobbler mouse motor neuron disease. Ann Neurol. 19:36–43.
8. Murakami, T., F.L. Mastaglia, and W.G. Bradley. 1980. Reduced protein synthesis in spinal anterior horn neurons in wobbler mouse mutant. Exp. Neurol. 67:423–432.

wst, wasted, recessive, Chr 2

Arose spontaneously in the HRS/J strain. Homozygotes can be recognized at 20 days of age by tremor and uncoordinated body movements. They develop progressive paralysis and do not survive beyond 30 days. They have marked lymphoid hypoplasia in the spleen, thymus, lymph nodes, and peripheral blood. The relative proportion of T- and B-cells is not affected. In the central nervous system there is degeneration of Purkinje cells and focal demyelination in the cerebellar cortex and ventral columns of the spinal cord. Wasted mice show some similarities to human ataxia telangiectasia (AT) (7), but also numerous differences. There is a marked increase in spontaneous and γ-ray-induced chromosomal abnormalities in bone marrow as in AT (7, 9), but no increased sensitivity of spleen cells to killing by ultraviolet radiation (3), no increased sensitivity of cultured fibroblasts to killing by X- or γ-radiation (3, 10), and no increased post-irradiation inhibition of DNA synthesis (6). In addition, the immunological defects of wasted mice also differ considerably from those of the human disease (5). Wasted mice have a severely reduced number of IgA plasma cells in the entire large and small intestines, but the number of such cells in the spleen and the level of serum IgA are normal (4). *wst* is on Chr 2 very close to *Ra* near the distal end (8). Abbott *et al.* (1) found decreased activity of adenosine deaminase (ADA) in erythrocytes of *wst/wst* mice, with intermediate level of activity in heterozygotes. The locus coding for ADA, *Ada*, is also on the distal part of Chr 2 and Abbott *et al.* have postulated that *wst* is a mutation at the *Ada* locus. However, this conclusion is not supported by the work of Geiger and Nagy (2), who found no decrease in ADA activity in erythrocytes and an increase in ADA activity in spleen of wasted mice.

References

1. Abbott, C.M., C.J. Skidmore, A.G. Searle, and J. Peters. 1986. Deficiency of adenosine deaminase in the wasted mouse. Proc. Natl. Acad. Sci. USA 83:693–695.
2. Geiger, J.D., and J.I. Nagy. 1986. Lack of adenosine deaminase deficiency in the mutant mouse *wasted*. FEBS Lett. 208:431–434.
3. Inoue, T., K. Aikawa, H. Tezuka, T. Kada, and L.D. Shultz. 1986. Effect of DNA-damaging agents on isolated spleen cells and lung fibroblasts from mouse mutant "wasted", a putative animal model for ataxia telangiectasia. Cancer Res. 46:3979–3982.
4. Kaiserlian, D., D. Delacroix, and J.F. Bach. 1985. The wasted mutant mouse. I. An animal model of excretory IgA deficiency with normal serum IgA. J. Immunol. 135:1126–1131.
5. Kaiserlian, D., W. Savino, J. Uriel, J. Hassid, M. Dardenne, and J.-F. Bach. 1986. The wasted mutant mouse. II. Immunological abnormalities in a mouse described as a model of ataxia–telangiectasia. Clin. Exp. Immunol. 63:562–569.

6. Nordeen, S.K., V.G. Schaefer, M.H. Edgell, C.A. Hutchison III, L.D. Shultz, and M. Swift. 1984. Evaluations of wasted mouse fibroblasts and SV-40 transformed human fibroblasts as models of ataxia telangiectasia *in vitro*. Mutat. Res. 140:219–222.
7. Shultz, L.D., H.O. Sweet, M.T. Davisson, and D.R. Coman. 1982. 'Wasted,' a new mutant of the mouse with abnormalities characteristic of ataxia telangiectasia. Nature 297:402–404.
8. Sweet, H.O. 1984. Mouse News Lett. 71:31.
9. Tezuka, H., T. Inoue, T. Noguti, T. Kada, and L.D. Shultz. 1986. Evaluation of the mouse mutant "wasted" as an animal model for ataxia telangiectasia. I. Age-dependent and tissue-specific effects. Mutat. Res. 161:83–90.
10. Thacker, J. and W. Masson. 1984. Radiation sensitivity of cells cultured from 'wasted' mice. Mouse News Lett. 70:80.

Wt, waltzer-type, semidominant

Arose spontaneously at the Oak Ridge National Laboratory. Heterozygotes may show behavior varying from slight nervousness to rapid circling. They are not deaf. The posterior and lateral semicircular canals are abnormal, with defects ranging from slight shortening to complete absence distal to the ampulla. The rest of the labyrinth is normal. All $Wt/+$ mice appear to have morphological abnormalities of the labyrinth, but many of them do not show any behavioral abnormality. Circling behavior is associated with defective lateral rather than posterior canals. Both canals seem important in normal swimming, and a larger proportion of mice with defective canals can be detected on this basis. Wt/Wt mice die at about the 11th day of gestation (1).

References

1. Stein, K.F., and S.A. Huber. 1960. Morphology and behavior of waltzer-type mice. J. Morphol. 106:197–203.

wv, weaver, recessive, Chr 16

Arose spontaneously in the C57BL/6J strain (4). Homozygotes are recognizable in the second postnatal week by small size, instability of gait, weakness, and hypotonia. Many die at weaning age, but some survive to adulthood and females may breed. The cerebellum is very small and simple and almost devoid of granule cells, which degenerate during the second week. Purkinje cells are somewhat disarranged and have small dendritic trees (10). By electron microscopy, the Purkinje cells are seen to have normal dendritic spines with normal postsynaptic terminals, but their presynaptic mates from the parallel fibers of the granule cells are missing (3). Heterozygotes ($wv/+$) have normal behavior, but they have a smaller than normal cerebellum with a deficiency of granule cells, some of which have not migrated into the internal granule layer and remain scattered in the molecular layer (7). By autoradiographic labeling, Rezai and Yoon (7) showed that the defect in wv/wv and $wv/+$ mice first becomes apparent at about 5 days of age and consists of a reduced rate of migration of granule cells into the internal granule layer, the defect being more severe in wv/wv than in $wv/+$ mice. Granule cells migrate inward along radially distributed Bergmann glial fibers. Rakić and Sidman (5) found cytological abnormalities in some Bergmann fibers in wv/wv and concluded that the basic defect caused by wv might lie in the Bergmann fibers. Later evidence from cultures of mutant and normal cerebellum showed that granule cells of wv/wv and $wv/+$ mice have gene-dosage dependent abnormalities of morphology and cell behavior (12, 13). Furthermore, Goldowitz and Mullen (2), using $wv/+ \longleftrightarrow +/+$ chimeras and cellular markers for granule cells and Purkinje cells, showed that the migration defect of granule cells is intrinsic to the granule cells themselves. The disarrangement of Purkinje cells is not caused by intrinsic action of wv in these cells. Several secondary effects of the weaver defect on other cell types in the cerebellum and their interconnections have been described (1, 6, 11). Although the morphology of the cerebrum appears unaffected by wv, parts of the cerebrum of wv/wv mice show a marked deficiency of dopamine, particularly in the frontal cortex and the dorsal striatum, with a smaller deficiency in the olfactory tubercle, and normal level in the ventral striatum (8, 9). Weaver mice have been used as a source of an agranulate cerebellum in a number of studies of the composition and function of granule cells.

References

1. Crepel, F., and J. Mariani. 1976. Multiple innervation of Purkinje cells by climbing fibers in the cerebellum of the weaver mutant mouse. J. Neurobiol. 7:579–582.
2. Goldowitz, D., and R.J. Mullen. 1982. Granule cell as a site of gene action in the weaver mouse cerebellum: evidence from heterozygous mutant chimeras. J. Neurosci. 2:1474–1485.
3. Hirano, A., and H.M. Dembitzer. 1973. Cerebellar alterations in the weaver mouse. J. Cell Biol. 56:478–486.
4. Lane, P.W. 1964. Mouse News Lett. 30:32.
5. Rakić, P., and R.L. Sidman. 1973. Sequence of developmental abnormalities leading to granule cell deficit in cerebellar cortex of weaver mutant mice. J. Comp. Neurol. 152:103–132.
6. Rakić, P., and R.L. Sidman. 1973. Organization of cerebellar cortex secondary to deficit of granule cells in weaver mutant mice. J. Comp. Neurol. 153:133–162.

7. Rezai, Z., and C.H. Yoon. 1972. Abnormal rate of granule cell migration in the cerebellum of weaver mutant mice. Dev. Biol. 29:17–26.

8. Roffler-Tarlov, S., and A.M. Greybiel. 1984. Weaver mutation has differential effects on the dopamine-containing innervation of the limbic and nonlimbic striatum. Nature 307:62–66.

9. Schmidt, M.J., B.D. Sawyer, K.W. Perry, R.W. Fuller, M.M. Foreman, and B. Ghetti. 1982. Dopamine deficiency in the weaver mutant mouse. J. Neurosci. 2:376–380.

10. Sidman, R.L. 1968. Development of interneuronal connections in brains of mutant mice. *In* F.D. Carlson, ed.,

11. Sotelo, C. 1975. Anatomical, physiological and biochemical studies of the cerebellum from mutant mice. II. Morphological study of cerebellar cortical neurons and circuits in the weaver mouse. Brain Res. 94:19–44.

12. Willinger, M., and C. Haaksma, 1985. Cytoplasmic morphology of weaver (*wv*) mouse cerebellar neurons at the culture substratum. J. Neurosci. Res. 13:163–182.

13. Willinger, M., D.M. Margolis, and R.L. Sidman. 1981. Neuronal differentiation in culture of weaver mutant mouse cerebellum. J. Supramol. Struct. Cell Biochem. 17:79–86.

Physiological and Biochemical Aspects of Nervous Integration, 163–193. Prentice-Hall, Englewood, NJ.

X

Xcat, X-linked cataract, semidominant, Chr X

Found among offspring of an irradiated male. Hemizygous males and homozygous females have total lens opacity. Heterozygous females have variable expression ranging from totally opaque to totally clear lenses. No effects on other tissues were observed. *Xcat* is located about 22 cM distal to *Ta* (tabby) (1).

References

1. Favor, J., and W. Pretsch. 1987. Position of Xcat, a new X-linked cataract mutation. Mouse News Lett. 77:139.

Xce locus (formerly also called *Cg*, *O^hv*), X-chromosome controlling element, Chr X

Several investigators have found evidence for a locus on the X chromosome that influences the X-chromosome inactivation process by controlling interaction between the two X chromosomes of females in such a way as to modify the frequency with which one or the other X is inactivated (2, 3, 5, 7). The relative activity of the two chromosomes has been followed by examining the phenotype of females heterozygous for X-linked genes with visible effects on coat color or hair characteristics, or heterozygous for X-autosomal translocations carrying genes affecting coat color. A difference in relative activity of the X chromosomes has also been detected by use of electrophoretic patterns of the X-linked gene phosphoglycerate kinase (*Pgk-1*) (11).

Three alleles are known: *Xce^a* causes inactivation of a high proportion of chromosomes bearing it (low expression of genes on the same chromosome) and occurs in the C3H/HeH, 101/H, and CBA/Ca strains; *Xce^b* causes an intermediate level of inactivation (intermediate expression of genes on the same chromosome) and occurs in the JU/Fa, C57BL/6, and C57BL/Go strains; *Xce^c* causes a low level of inactivation (high expression of genes on the same chromosome) and was found in *M. m. musculus* captured in Denmark (1, 4, 6, 10). The inactive X chromosome can be detected at 6.5 days of gestation by differential staining. By use of a marked X chromosome, Rastan (8) showed that nonrandom inactivation associated with *Xce* alleles was already present at that time. *Xce* is located close to tabby (*Ta*) near the middle of the X chromosome (1). There is considerable evidence that X-inactivation proceeds from a single center. The inactivation center is also close to *Ta* and is very probably the same as the *Xce* locus (9).

References

1. Cattanach, B.M., and D. Papworth. 1981. Controlling elements in the mouse. V. Linkage tests with X-linked genes. Genet. Res. 38:57–70.

2. Cattanach, B.M., and C.E. Williams. 1972. Evidence of non-random X chromosome activity in the mouse. Genet. Res. 19:229–240.

3. Falconer, D.S., and J.H. Isaacson. 1972. Sex-linked variegation modified by selection in brindled mice. Genet. Res. 20:291–316.

4. Forrester, L.M., and J.D. Ansell. 1985. Parental influences on X chromosome expression. Genet. Res. 45:95–100.

5. Grahn, D., R.A. Lea, and J. Hulesch. 1970. Location of an X-inactivation controller gene on the normal X-chromosome of the mouse. Genetics 64:s25 (Abstr.).

6. Johnston, P.G., and B.M. Cattanach. 1981. Controlling elements in the mouse. IV. Evidence of non-random X-inactivation. Genet. Res. 37:151–160.

7. Ohno, S., L. Christian, B.J. Attardi, and J. Kan. 1973. Modification of expression of the *testicular feminization* (*Tfm*) gene of the mouse by a "controlling element" gene. Nature New Biol. 245:92–93.

8. Rastan, S. 1982. Primary non-random X-inactivation caused by controlling elements in the mouse demonstrated at the cellular level. Genet. Res. 40:139–147.

9. Rastan, S. 1983. Non-random X-chromosome inactivation in mouse X-autosomal translocation embryos—location of the inactivation centre. J. Embryol. Exp. Morphol. 78:1–22.

10. Rastan, S., and B.M. Cattanach. 1983. Interaction between the *Xce* locus and imprinting of the paternal X chromosome in mouse yolk-sac endoderm. Nature 303:635–637.

11. West, J.D., and V.M. Chapman. 1978. Variation for X chromosome expression in mice detected by electrophoresis of phosphoglycerate kinase. Genet. Res. 32:91–102.

xid, X-linked immune deficiency, recessive, Chr X

This mutation was found in the CBA/N strain derived from CBA/H mice obtained by the National Institutes of Health from Harwell in 1966. The locus lies about 7 cM distal to *Ta* in the central region of the X chromosome (2). The extensive literature up to 1982 has been reviewed by Scher (10). Briefly, hemizygous males and homozygous females have a defective immune response to type 2 thymus-independent (TI-2) antigens, a normal response to type 1 thymus-independent (TI-1) antigens, an impaired immune response to thymus-dependent (TD) antigens, and lymphocytes with impaired response to B-cell mitogens. Affected mice have about half as many spleen cells as normal controls. Heterozygous females have normal immune responses. The defect is expressed in B-cells which are characterized by low complement receptor, low Mls determinants, high surface IgM, low surface IgD, low IgG3 production, and are Lyb-3⁻, Lyb-5⁻, Lyb-7⁻, and Ia.W39⁻. Since cells of this type are present in immature normal mice, it was originally thought that *xid*/Y and *xid*/*xid* mice possessed only immature B-cells and were lacking normal adult B-cells (10). There is now considerable evidence that the B-cells of affected mice are different from any normal cells, either mature or immature (5, 8, 9, 14). Nor-

mal B-cells can be implanted in unirradiated *xid*/Y or *xid*/*xid* mice and will restore normal immune responsiveness. Such cells will not implant in normal mice unless the mice have been previously irradiated (16). B-cells of *xid*/Y or *xid*/*xid* mice are unable to compete with normal cells in double bone marrow chimeras made with a mixture of the two kinds of cells, and are gradually replaced by normal cells (13). Some B-cells in the Peyer's patches of adult *xid*/Y males resemble normal adult B-cells and are capable of producing a normal IgA response to orally administered TD and TI-2 antigens (6). Old affected mice (12 months) develop a normal response to these antigens and it is thought that these late maturing B-cells may be the mature B-cells of the Peyer's patches which emerge in advanced age (4). Normal B-cell response to TD and TI-2 antigens can be restored in young adult *xid*/Y and *xid*/*xid* mice by treatment with 8-mercaptoguanosine or with a B-cell stimulating factor produced by a cloned line of dendritic cells (1, 3). Differentiation of B-cells in *xid*/Y mice is dependent on presence of the thymus, in contrast to normal mice in which B-cell differentiation is unaffected by absence of the thymus (12). Most studies on *xid* have been performed with mice of the CBA/N strain or with offspring of a female CBA/N by male DBA/2 cross, which produces *xid*/Y males (affected) and *xid*/+ females (unaffected). When *xid* is transferred to the C3H/IIe background, *in vitro* responses to TI-1 antigens and B-cell mitogens are markedly lower than those on the other two backgrounds (7). The combination of *xid* with several inherited autoimmune diseases causes profoundly reduced manifestation of the autoimmunity (11, 15).

References

1. Ahmad, A., and J.J. Mond. 1986. Restoration of *in vitro* responsiveness of xid B cells to TNP-ficoll by 8-mercaptoguanosine. J. Immunol. 136:1223–1226.

2. Berning, A.K., E.M. Eicher, W.E. Paul, and I. Scher. 1980. Mapping the X-linked immune deficiency mutation (xid) of CBA/N mice. J. Immunol. 124:1875–1882.

3. Clayberger, C., R.H. DeKruyff, R. Fay, B. Haber, and H. Cantor. 1985. Evidence for defects in accessory and T cell subsets in mice expressing the *xid* defect. J. Mol. Cell. Immunol. 2:61–69.

4. Eldridge, J.H., H. Kiyono, S.M. Michalek, and J.R. McGhee. 1983. Evidence for a mature B cell subpopulation in Peyer's patches of young adult *xid* mice. J. Exp. Med. 157:789–794.

5. Hardy, R.R., K. Hayakawa, D.R. Parks, and L.A. Herzenberg. 1983. Demonstration of B-cell maturation in X-linked immunodeficient mice by simultaneous three-color immunofluorescence. Nature 306:270–272.

6. Kiyono, H., L.M. Mosteller, J.H. Eldridge, S.M. Micha-

lek, and J.R. McGhee. 1983. IgA responses in *xid* mice: oral antigen primes Peyer's patch cells for *in vitro* immune response and secretory antibody production. J. Immunol. 131:2616–2622.

7. Mond, J.J., G. Norton, W.E. Paul, I. Scher, F.D. Finkelman, S. House, M. Schaefer, P.K.A. Mongini, C. Hansen, and C. Bona. 1983. Establishment of an inbred line of mice that express a synergistic immune defect precluding *in vitro* responses to type 1 and type 2 antigens, B cell mitogens, and a number of T cell-derived helper factors. J. Exp. Med. 158:1401–1414.

8. Mosier, D.E. 1985. Are *xid* B lymphocytes representative of any normal B cell population?—a commentary. J. Mol. Cell. Immunol. 2:70.

9. Reid, G.K., and D.G. Osmond. 1985. B lymphocyte production in the bone marrow of mice with X-linked immunodeficiency (*xid*). J. Immunol. 135:2299–2302.

10. Scher, I. 1982. The CBA/N mouse strain: experimental model illustrating the influence of the X-chromosome on immunity. Adv. Immunol. 33:1–71.

11. Smith, H.R., T.M. Chused, and A.D. Steinberg. 1983. The effect of the X-linked immune deficiency gene (*xid*) upon the Y chromosome-related disease of BXSB mice. J. Immunol. 131:1257–1262.

12. Sprent, J. and J. Bruce. 1984. Physiology of B cells in mice with X-linked immunodeficiency. II. Influence of the thymus and mature T cells on B cell differentiation. J. Exp. Med. 160:335–340.

13. Sprent, J. and J. Bruce. 1984. Physiology of B cells in mice with X-linked immunodeficiency. III. Disappearance of *xid* B cells in double bone marrow chimeras. J. Exp. Med. 160:711–723.

14. Sprent, J. J. Bruce, Y. Ron, and S.R. Webb. 1985. Physiology of B cells in mice with X-linked immunodeficiency. I. Size, migratory properties, and turnover of the B cell pool. J. Immunol. 134:1442–1448.

15. Steinberg, B.J., P.A. Smathers, K. Frederiksen, and A.D. Steinberg. 1982. Ability of the *xid* gene to prevent autoimmunity in (NZB × NZW)F$_1$ mice during the course of their natural history, after polyclonal stimulation, or following immunization with DNA. J. Clin. Invest. 70:587–597.

16. Volf, D., L.L. Sensenbrenner, S.J. Sharkis, G.J. Elfenbein, and I Scher. 1978. Induction of partial chimerism in nonirradiated B-lymphocyte-deficient CBA/N mice. J. Exp. Med. 147:940–945.

Xld-1 locus, xylose dehydrogenase-1, Chr 7

This locus controls electrophoretic variation in xylose dehydrogenase, an NADP-dependent non-specific aldose oxidoreductase. The allele *Xld-1^a* determines a fast anodally migrating enzyme band and occurs in the IS/Cam strain and some wild populations; the allele *Xld-1^b* determines a slower band and occurs in most inbred strains as well as in wild populations; the allele *Xld-1^c* determines absence of a detectable band and occurs in the 101/H strain and a wild population in Scotland. *Xld-1^a/Xld-1^b* mice show a three-banded pattern, suggesting that the enzyme is a dimer. Heterozygotes of *Xld-1^c* with the other two alleles produce a band with decreased staining intensity. The preferred substrate of the enzyme is D-xylose, but it is also active on a variety of other sugars. The enzyme is most active in liver but is also active in brain, kidney, heart, and muscle, and very weakly active in spleen, lung, stomach, brown fat, testis, and colon. It is located primarily in the cytoplasm. *Xld-1* is on Chr 7, 4 cM distal to *Tam-1* (1).

References

1. Newton, M.F., H.R. Nash, J. Peters, and S.J. Andrews. 1982. Xylose dehydrogenase-1, a new gene on mouse chromosome 7. Biochem. Genet. 20:733–745.

Xlr complex, X-linked lymphocyte-regulated gene family, Chr X

This complex comprises a large gene family, all of whose members are on the X chromosome. In Southern blots, the genes are recognized as 10 to 14 bands that hybridize with a cDNA probe isolated from a T-cell-specific cDNA library. The complex was shown to be on Chr X by use of mouse–Chinese hamster somatic cell hybrids. A DNA restriction fragment polymorphism exists between the progenitor strains of two strains congenic for the X-linked *xid* mutation. Strain CBA/N (*xid/xid*) carries a *Pvu*II 5.2-kb fragment not found in strains MRL and BXSB. The congenic strains MRL-*xid* and BXSB-*xid* both carry the 5.2-kb fragment. A similar association was found between a *Bgl*I 9.6-kb fragment and *xid* in the MRL-*xid* strain. These results show that the *Xlr* complex is closely linked to the *xid* locus, which may be a member of the complex. It is possible that recombination can occur between members of the complex, since with *Bam*HI digestion, the BXSB-*xid* strain showed one fragment from BXSB and two from CBA/N (1). *Xlr* mRNA is expressed in T-cell hybridomas and other T-cell tumors, in some B-cell tumors, and in normal lymphoid tissue, but not in macrophage tumors, liver, or kidney. It is also expressed in late-stage but not in immature-stage plasmacytomas. The *xid* mutation causes a deficiency of mature B-cells and it is therefore of interest that *Xlr* was not expressed in three *xid*-bearing plasmacytomas examined (2).

Four X-linked immune response variants, three of which are associated with T- and/or B-cell antigens, have been described. See *LyX-1–LyX-3* loci. The relation of these antigens and immune responses with either the *Xlr* gene products or the *xid* locus is not known.

Table 2.12

Locus	Restriction enzyme	Chromosome	Linked marker	Strain	
				Present	Absent
Xmmv-2	*Eco*RI	9	*Thy-1*	C57BL/6	DBA/2, A/J, AKR
Xmmv-3	*Eco*RI	5(?)	*Afp*	C57BL/6	DBA/2, A/J, AKR
Xmmv-5	*Hind*III	5	*Pgm-1*	C57BL/6, A/J, AKR	DBA/2
Xmmv-6	*Hind*III	1	*Bxv-1*	DBA/2	C57BL/6, A/J, AKR
Xmmv-8	*Bgl*II	4	*Lyb-2*	C57BL/6	DBA/2
Xmmv-9	*Bgl*II	1	*Xmmv-6*	C57BL/6	DBA/2
Xmmv-15	*Pst*I	17(?)	*D17Leh66*	C57BL/6	DBA/2, A/J, AKR
Xmmv-21	*Pvu*II	12	*Pre-1*	DBA/2	C57BL/6
Xmmv-23	*Pvu*II	4	*Akp-2*	C57BL/6	DBA/2
Xmmv-25	*Pvu*II	12	*Pre-1*	–	–
Xmmv-27	*Xba*I	6	*Ly-2*	C57BL/6	DBA/2
Xmmv-29	*Xba*I	8	*Emv-2*	DBA/2	C57BL/6, AKR
Xmmv-31	*Hind*III	7	*Tam-1*	C57BL/6, A/J	DBA/2, AKR
Xmmv-34	*Hind*III	12	*Igh*	DBA/2	C57BL/6
Xmmv-35	*Bam*HI	7	*Coh*	C57BL/6	DBA/2, A/J, AKR
Xmmv-36	*Bam*HI	1	*Xmmv-6*	DBA/2	C57BL/6, A/J, AKR

References

1. Cohen, D.I., S.M. Hedrick, E.A. Nielsen, P. D'Eustachio, F. Ruddle, A.D. Steinberg, W.E. Paul, and M.M. Davis. 1985. Isolation of a cDNA clone corresponding to an X-linked gene family (XLR) closely linked to the murine immunodeficiency disorder *xid*. Nature 314:369–372.
2. Cohen, D.I., A.D. Steinberg, W.E. Paul, and M.M. Davis. 1985. Expression of an X-linked gene family (XLR) in late-stage B cells and its alteration by the *xid* mutation. Nature 314:372–374.

Xmmv-1 to *Xmmv-40* loci (formerly *env-1* to *env-40*), xenotropic-MCF leukemia virus-1 to -40

These loci are the sites of integration of the proviral genomes of xenotropic and mink cell focus-forming (MCF) murine leukemia viruses. They were found as differences between C57BL/6 and DBA/2 in DNA restriction fragment lengths identified using a cloned *env* gene of Moloney-MCF virus as a probe. Their chromosomal location was investigated by determining their strain distribution patterns in BXD recombinant inbred (RI) strains. Forty differences between the strains were found but only those whose probable chromosome location was determined are listed in Table 2.12 (1). *Xmmv-15* was originally thought to be on Chr 7, but this location has not been confirmed. The locus is probably on Chr 17 (B.A. Taylor, personal communication). *Xmmv-6* may be allelic with *Bxv-1*. However, *Bxv-1* determines inducibility in C57BL/6 and AKR, an effect that seems unlikely if these strains do not carry the proviral genome. See *Bxv-1* locus (1).

References

1. Blatt, C., K. Milcham, M. Haas, M.N. Nesbitt, M.E. Harper, and M.I. Simon. 1983. Chromosomal mapping of the mink cell focus-inducing and xenotropic *env* gene family in the mouse. Proc. Natl. Acad. Sci. USA 80:6298–6302.

Xmmv-42 to *Xmmv-70*, *Xmmv-76* loci (formerly *Xp-1* to *Xp-29*), xenotropic-MCF leukemia virus-42 to -70, -76

These loci are sites of integration of proviral genomes of xenotropic and mink cell focus-forming (MCF) murine leukemia viruses, identified with a xenotropic virus *env* gene probe. Digestion of DNA of many inbred strains with *Pvu*II revealed 29 different fragments whose strain distribution could be determined (Table 2.13). Fourteen of these fragments segregated in one or more of the eight sets of recombinant inbred strains examined, allowing identification of the chromosomal location of six loci (4). *Xmmv-55* (*Xp-14*) is now known to be on Chr 15 (1) and *Xmmv-65* (*Xp-24*) is on Chr 3 (2). *Xp-9* segregated as two distinct loci, the fragment in AKR (*Xmmv-50*) mapping to Chr 12 (4), and the fragment in DBA/2 (now designated *Xmmv-76*) mapping to Chr 7 (3). *Xmmv-61* may be the same as *Xmmv-9* (N.A. Jenkins, personal communication).

References

1. Hogarth, P.M., I.F.C. McKenzie, V.R. Sutton, K.M. Curnow, B.K. Lee, and E.M. Eicher. 1987. Mapping of the murine *Ly-6*, *Xp-14*, and *Gdc-1* loci to chromosome 15. Immunogenetics 25:21–27.

Table 2.13

Locus	Formerly	Fragment size (kb)	Chromosome	Strains
Xmmv-42	*Xp-1*	16.9	19	C57BR, C57L, C58
Xmmv-43	*Xp-2*	15.4	–	RIIIS
Xmmv-44	*Xp-3*	14.6	–	NZW/LacJ
Xmmv-45	*Xp-4*	13.5	–	I/Ln
Xmmv-46	*Xp-5*	10.4	–	NZW/LacJ, SWR
Xmmv-47	*Xp-6*	10-2	–	CBA/Ca, CE, C57BR, others
Xmmv-48	*Xp-7*	7.3	–	CBA/N, LG/J, NZB/Icr, SM
Xmmv-49	*Xp-8*	7.3	–	SJL
Xmmv-50	*Xp-9*	6.6	12	AKR
Xmmv-51	*Xp-10*	6.5	–	SJL, RFM/Un
Xmmv-52	*Xp-11*	6.3	5	Common, not in C57BR, C57L, SJL, 129/J, others
Xmmv-53	*Xp-12*	6.2	–	SJL, RFM/Un
Xmmv-54	*Xp-13*	6.0	–	Common, not in DA/Hu, RIIIS, SJL, ST/b, ST/a
Xmmv-55	*Xp-14*	5.6	15	AKR, C57 family, DBA/2, NZW/Lac, others
Xmmv-56	*Xp-15*	5.5	–	LG
Xmmv-57	*Xp-16*	5.3	–	DA/Hu
Xmmv-58	*Xp-17*	4.6	–	NZB
Xmmv-59	*Xp-18*	4.4	–	SM/J
Xmmv-60	*Xp-19*	3.7	–	A/J, BALB/c, NZB, SM, others
Xmmv-61	*Xp-20*	3.6	1	Common, not in DBA/2, NZB, others
Xmmv-62	*Xp-21*	3.3	4	C57BR, C57L, C58, 129/J, others
Xmmv-63	*Xp-22*	2.8	–	Common
Xmmv-64	*Xp-23*	2.4	–	AKR, BALB/cWt, CBA/J, C3H/He, C57 family, others
Xmmv-65	*Xp-24*	2.3	3	AKR, BALB/c, C57BL/Ks, DBA/2, others
Xmmv-66	*Xp-25*	2.1	–	C57BR, C57L, C58, LT
Xmmv-67	*Xp-26*	2.0	–	C57BL/Ks, C58, NZB, RF, others
Xmmv-68	*Xp-27*	1.9	–	CE, LP
Xmmv-69	*Xp-28*	1.3	–	Common, not in CBA/Ca, ST/b, SWR
Xmmv-70	*Xp-29*	1.0	–	Common
Xmmv-76	*Xp-9*	6.6	7	DBA/2

2. Paul, P.R., and R.W. Elliott. 1987. Analysis of the mouse *Amy* locus in recombinant inbred mouse strains. Biochem. Genet. 25:569–579.

3. Silver, J., and C.E. Buckler. 1986. A preferred region for integration of Friend murine leukemia virus in hematopoietic neoplasms is closely linked to the *Int-2* oncogene. J. Virol. 61:1156–1158.

4. Wejman, J.C., B.A. Taylor, N.A. Jenkins, and N.G. Copeland. 1984. Endogenous xenotropic murine leukemia virus-related sequences map to chromosomal regions encoding mouse lymphocyte antigens. J. Virol. 50:237–247.

Xmmv-71 to *Xmmv-74* loci, xenotropic-MCF leukemia virus-71 to -74

A large number of endogenous viral loci identified with xenotropic or MCF viral *env* or *pol* gene probes have been described by Meruelo *et al.* (1) and Rossomando and Meruelo, (3) Many of these are probably identical to some of the *Xmmv-1* to *Xmmv-70* loci. (See Kozak, Chapter 3, this volume.) Four of these loci are known

to be different from those previously described (M.T. Davisson, personal communication) and are listed in Table 2.14. The chromosomal locations of the loci were first determined by use of BALB/c × C57BL/6 recombinant inbred strains. The position of *Xmmv-72* near *Sis* and *Ly-6* on Chr 15 was confirmed by Meruelo *et al.* (2) using a different probe and three other sets of recombinant inbred strains.

References

1. Meruelo, D., A. Rossomando, M. Offer, J. Buxbaum, and A. Pellicer. 1983. Association of endogenous viral loci with genes encoding murine histocompatibility and lymphocyte differentiation antigens. Proc. Natl. Acad. Sci. USA 80:5032–5036.

2. Meruelo, D., A. Rossomando, S. Scandalis, P. D'Eustachio, R.E.K. Fournier, D.R. Roop, D. Saxe, C. Blatt, and M.N. Nesbitt. 1987. Assignment of the *Ly-6-Ril-1-Sis-H-30-Pol-5/Xmmv-72-Ins-3-Krt-1-Int-1-Gdc-1* region to mouse chromosome 15. Immunogenetics 25:361–372.

3. Rossomando, A., and D. Meruelo. 1986. Viral sequences

Table 2.14

Locus	Formerly	Restriction enzyme	Fragment size (kb)	Chromosome	Strains
Xmmv-71	–	*Hind*III	–	2	BALB/c
Xmmv-72	*Pol-5*	*Pvu*I *Kpn*I *Bam*HI	– 9.3	15	C57BL/6
Xmmv-73	–	*Hind*III	–	7	C57BL/6
Xmmv-74	–	*Pvu*II	–	1	C57BL/6

are associated with many histocompatibility genes. Immunogenetics 23:233–245.

Xmv-1 to *Xmv-5* loci, xenotropic murine leukemia virus-1 to -5

These loci are the sites of integration of xenotropic murine leukemia viruses in the BALB/c strain. They were recognized in *Hind*III digested genomic DNA hybridized to a probe specific for the *env* gene of xenotropic viruses. The chromosomal location of the loci was found by use of mouse–Chinese hamster somatic cell hybrids made with BALB/c mouse cells (1) (Table 2.15).

Table 2.15

Locus	Fragment size (kb)	Chromosome
Xmv-1	18.8	4
Xmv-2	16.7	4
Xmv-3	11.3	16
Xmv-4	9.9	11
Xmv-5	9.4	11

References

1. Hoggan, M.D., R.R. O'Neill, and C.A. Kozak. 1986. Nonecotropic murine leukemia viruses in BALB/c and NFS/N mice: characterization of the BALB/c *Bxv-1* provirus and the single NFS endogenous xenotrope. J. Virol. 60:980–986.

xn, exencephaly, recessive

Found in the random-bred CF/1 albino stock in the 11th generation of treatment of the stock with DDT. Homozygotes appear to be fully viable until birth but affected homozygotes die within a few hours postnatally. Penetrance is incomplete and dependent on the genetic background. It varies from 33 per cent to 84 per cent and is twice as high in females as in males. In the most severely affected mice there is extensive failure of the cephalic portion of the neural tube to close at 10 days of gestation, the bony vault of the cranium fails to develop, and the tissues of the brain are exposed. The failure to close can be mild or severe and can extend from the thalamus to the medulla. Unaffected *xn/xn* mice probably survive and reproduce, but they have not been examined to discover whether they have minor defects not detectable externally (1).

References

1. Wallace, M.E., P.J. Knight, and J.R. Anderson. 1979. Inheritance and morphology of exencephaly, a neonatal lethal recessive with partial penetrance, in the house mouse. Genet. Res. 32:135–149.

Xp-1 to *-29* loci

See *Xmmv-42* to *-70*.

Xpa locus, xeroderma pigmentosum A, Chr 4

In man, xeroderma pigmentosum is an inherited disease in which there is defective repair of DNA damage after exposure to ultraviolet (UV) light. Genetic studies have shown that excision-defective cells fall into seven complementation groups (A–G). In somatic cell hybrids between mouse embryo fibroblasts and a line of excision-defective human cells of the XP A group, there was a high correlation between presence of mouse Chr 4 and ability to repair DNA as measured by unscheduled DNA synthesis after UV irradiation. Chr 4, therefore, appears to contain a gene homologous to the human XP A gene (1).

References

1. Lin, P.-F., and F.H. Ruddle. 1981. Murine DNA repair gene located on chromosome 4. Nature 289:191–194.

Xpl, X-linked polydactyly, dominant, Chr X

Arose spontaneously in a balanced stock carrying *gl* and *dl^l*. Both homozygous and heterozygous females are viable and fertile. Hemizygous males are viable but most are sterile, with undescended testes and sometimes with hydronephrosis and missing kidneys. All these genotypes have preaxial polydactyly of the hindfeet, varying from a barely discernible thickening of the hallux on one foot to four extra toes on one or both feet, and may also have tibial hemimelia. Homozygous and heterozygous females are not phenotypically distinguishable. Penetrance is known to be incomplete in heterozygous females (about 80 to 90 per cent) and in hemizygous males (about 90 to 95 per cent) (1).

References

1. Sweet, H.O., and P.W. Lane. 1980. X-linked polydactyly (*Xpl*), a new mutation in the mouse. J. Hered. 71:207–209.

Xt locus, Chr 13

At least seven independent mutations at this locus have been found (6).

Xt, extra toes, semidominant. Arose spontaneously in the control series of an irradiation experiment, probably in either a (101 X C3H/He)F1 or a CBA/H mouse (6). Heterozygotes have varying numbers of extra digits on the preaxial side of forefeet and hindfeet. Occasionally there is a postaxial rudimentary digit. Heterozygotes may also have a belly spot and an interfrontal bone. They are fully viable and fertile. Homozygotes die *in utero* or at birth with multiple abnormalities including paddle-shaped feet with eight or nine digits, hemimelia, abnormalities of the brain, spinal cord, and sense organs, and edema. They can first be recognized at 9 days of gestation by excessively large pharyngeal arches and open neural tube (2).

*Xt^bph^*, brachyphalangy, semidominant. Found among the offspring of a (101 X C3H/He)F1 male who had been treated with low-intensity neutron irradiation and mated to a female of a multiple recessive stock (1). Heterozygotes have a thickening of the first digit on forefeet and hindfeet, and occasionally soft-tissue syndactylism of the other digits. The forefeet have a postaxial nubbin at the base of digit 5. The sternum is shortened and thickened, and the sternebrae are often fused. There may be a belly spot and an interfrontal bone. Homozygotes resemble *Xt/Xt* but are more severely affected, with a high incidence of exencephaly, unilateral and bilateral harelip, and more pronounced edema (3). There is a greater increase of fluid in the conceptus of the exencephalic than of the non-exencephalic Xt^{bph}/Xt^{bph} embryos, most of it in the amniotic sac (5). Trypan blue and Xt^{bph} have many effects on the embryo in common. Treatment of litters segregating for Xt^{bph} with trypan blue shows that the gene and teratogen act additively (4).

References

1. Batchelor, A.L., R.J.S. Phillips, and A.G. Searle. 1966. A comparison of the mutagenic effectiveness of chronic neutron- and γ-irradiation of mouse spermatogonia. Mutat. Res. 3:218–229.
2. Johnson, D.R. 1967. *Extra-toes*: a new mutant gene causing multiple abnormalities in the mouse. J. Embryol. Exp. Morphol. 17:543–581.
3. Johnson, D.R. 1969. Brachyphalangy, an allele of extra-toes in the mouse. Genet. Res. 13:275–280.
4. Johnson, D.R. 1970. Trypan blue and the extra-toes locus in the mouse. Teratology 3:105–110.
5. Johnson, D.R. 1971. Genes and genotypes affecting embryonic fluid relations in the mouse. Genet. Res. 18:71–79.
6. Lyon, M.F., T. Morris, A.G. Searle, and J. Butler. 1967. Occurrences and linkage relations of the mutant "extra-toes" in the mouse. Genet. Res. 9:383–385.

Y

Yaa, Y-linked autoimmune acceleration, Chr Y

This gene on the Y chromosome derived from the SB/Le inbred strain is postulated to account for the accelerated autoimmunity that occurs in males carrying this Y chromosome. It was discovered in the BXSB strain derived from a cross of an SB male to a C57BL/6 female. Mice of the BXSB strain develop spontaneous autoimmune disease characterized by moderate lymph node and spleen enlargement, hemolytic anemia, hypergammaglobulinemia, and immune complex glomerulonephritis (5). The disease process in this strain is strikingly accelerated in males, which live little more than a third as long as females. Similar acceleration occurs in males but not in females from an outcross of BXSB males to females of strain NZB, but does not occur in offspring of the reciprocal cross. The same effect of *Yaa* is seen in offspring of outcrosses to strains SJL, C57BL/6, and AKR (5). Although the development of autoimmunity is greatly accelerated by *Yaa*, the basic autoimmune disease occurring in these various strains and crosses appears to depend on autosomal factors (2, 3, 6). The accelerated autoimmunity of BXSB males and its absence in females can be reciprocally transferred by grafts of bone marrow and spleen cells (1). A dramatic increase in number of peripheral monocytes in BXSB males beginning at 2 months of age has been described (9). A study of B-cell development in BXSB mice showed that males become deficient in pre-B-cells (cytoplasmic μ-chain positive, surface immunoglobulin negative) after about 4 weeks and remain so throughout life. They presumably maintain their mature B-cell population by a self-renewal process, or else by a very rapid transition from the very early precursor stage to maturity (4). The accelerated autoimmunity is abolished in BXSB-*xid*/*Yaa* and (NZW × BXSB)-*xid*/*Yaa* male mice, suggesting that the effect of Yaa is dependent on the B-cell subset that is not present in mice bearing the X-linked immune deficiency gene, *xid* (6, 8). The *Yaa*-dependent progression of autoimmunity is also greatly retarded in beige homozygotes (*bg*/*bg*) as seen in SB/Le-*bg*/*bg* males (7). The effect of *Yaa* is not hormonally mediated (1, 5).

References

1. Eisenberg, R.A., S. Izui, P.J. McConahey, L. Hang, C.J. Peter, A.N. Theofilopoulos, and F.J. Dixon. 1980. Male determined accelerated autoimmune disease in BXSB mice: transfer by bone marrow and spleen cells. J. Immunol. 125:1032–1036.
2. Hudgins, C.C., R.T. Steinberg, D.M. Klinman, M.J. Patton Reeves, and A.D. Steinberg. 1985. Studies of consomic mice bearing the Y chromosome of the BXSB mouse. J. Immunol. 134:3849–3854.
3. Izui, S., K. Masuda, and H. Yoshida. 1984. Acute SLE in F₁ hybrids between SB/LE and NZW mice; prominently enhanced formation of gp70 immune complexes by a Y chromosome-associated factor from SB/Le mice. J. Immunol. 132:701–704.
4. Jyonouchi, H., P.W. Kincade, and R.A. Good. 1985. Age-dependent changes in B lymphocyte lineage cell populations of autoimmune-prone BXSB mice. J. Immunol. 134:858–864.
5. Murphy, E.D., and J.B. Roths. 1979. A Y chromosome associated factor in strain BXSB producing accelerated autoimmunity and lymphoproliferation. Arthr. Rheum. 22:1188–1194.
6. Rosenberg, Y.J., and A.D. Steinberg. 1984. Influence of X and Y chromosomes on B cell responses in autoimmune prone mice. J. Immunol. 132:1261–1264.
7. Roths, J.B., and E.D. Murphy. 1982. The beige mutation retards the progression of autoimmune disease. Jackson Lab. Ann. Rep. 53:65.
8. Smith, H.A., T.M. Chused, and A.D. Steinberg. 1983. The effect of the X-linked immune deficiency gene (*xid*) upon the Y chromosome-related disease of BXSB mice. J. Immunol. 131:1257–1262.
9. Wofsey, D., C.E. Kerger, and W.E. Seaman. 1984. Monocytosis in the BXSB model for systemic lupus erythematosis. J. Exp. Med. 159:629–634.

Ym, yellow mottled, semidominant, Chr X

Heterozygous females resemble *Mo*/+ females except that the mottling is yellowish and the vibrissae are not curly. Hemizygous males die *in utero*. *Ym* is near the *Mo* locus but recombines with it (1,2).

References

1. Hunsicker, P.R. 1968. Mouse News Lett. 38:31.
2. Hunsicker, P.R. 1969. Mouse News Lett. 40:42.

Unnamed loci

A number of variant traits in the mouse have been shown to be due to a difference at a single locus, but the locus has not been named. Most have not been mapped. Variants of this type known to me are listed below, grouped mainly by similarity of phenotype.

Y chromosome DNA sequences

Molecular genetic studies have revealed several classes of DNA sequences on the Y chromosome (1). The proximal region, which is responsible for sex determination, contains the *Tdy* locus, the gene coding for the H-Y antigen, and sequences isolated from the heterogametic female sex of snakes. This sequence is highly conserved in the heterogametic sex of vertebrates (8). A long region in the central part of the chromosome bears up to 100 copies of a sequence that hybridizes to a clone (M720) of an endogenous type C retrovirus isolated from *Mus cookii* (7). M720 also hybridizes to an endogenous murine retrovirus present in over 300 copies in this region. The region probably also contains a gene important in sperm motility (1). The distal region of the Y chromosome is homologous with the distal region of the X chromosome and forms an obligate chiasma with it at each meiosis. It contains the *Sts* locus, but no DNA sequences have been assigned to this region (1, 4).

A number of Y-chromosome DNA polymorphisms have been found among inbred strains. Workers in three laboratories have isolated different mouse DNA clones that recognize DNA sequences on the Y chromosome and that detect polymorphisms with a similar strain distribution pattern (2, 5, 6). All three groups found that strains BALB/c, C57BL/6, DBA/2, C3H/He, and A/J fall into one class and strain AKR into the other. Other strains in the first class are NZB (5, 6), YBR (5), CBA/J (2), 129 (2, 6), SEC, CE, and HRS (6), and others in the second class are SJL (2, 5), SWR, SWV, and RF (6). In addition, Bishop *et al.* (2) and Nishioka and Lamothe (6) found that their probes detected the same difference between the Y chromosomes of the subspecies *Mus musculus musculus* (like the first class) and *Mus musculus domesticus* (like the second class). The sequences detected by Lamar and Palmer (5) and Nishioka and Lamothe (6) were shown not to be located in the *H-Y*, *Tdy*, sex-determining

region, but their location is not known. A polymorphic sequence with a different strain distribution was found by Phillips *et al.* (7), using a probe composed of a long terminal repeat from an amphotropic retrovirus isolated from *M. m. musculus*. They found that a 5.5-kb *Eco*RI fragment present in the DNA of male mice of strains BALB/c, C57BL/10, NZB, and SM was not present in strains AKR and DBA/2. Finally, Blatt *et al.* (3), using a mink cell focus-forming *env* gene probe, found three sequences present on the Y chromosome of strains A/J and C57BL/6 and absent from DBA/2.

References

1. Eicher, E.M., and L.L. Washburn. 1986. Genetic control of primary sex determination in mice. Ann. Rev. Genet. 20:327–360.
2. Bishop, C.E., P. Boursot, B. Baron, F. Bonhomme, and D. Hatat. 1985. Most classical *Mus musculus domesticus* laboratory strains carry a *Mus musculus musculus* Y chromosome. Nature 315:70–72.
3. Blatt, C., K. Milcham, M. Haas, M.N. Nesbitt, M.E. Harper, and M.I. Simon. 1983. Chromosomal mapping of the mink cell focus-inducing and xenotropic *env* gene family in the mouse. Proc. Natl. Acad. Sci. USA 80:6298–6302.
4. Keitges, E., M. Rivest, M. Siniscalco, and S.M. Gartler. 1985. X-linkage of steroid sulfatase in the mouse is evidence for a Y-linked allele. Nature 315:226–227.
5. Lamar, E.E., and E. Palmer. 1984. Y-encoded, species-specific DNA in mice: evidence that the Y chromosomes exists in polymorphic form in inbred strains. Cell 37:171–177.
6. Nishioka, Y., and E. Lamothe. 1986. Isolation and characterization of a mouse *Y* chromosomal repetitive sequence. Genetics 113:417–432.
7. Phillips, S.J., E.H. Birkenmeier, R. Callahan, and E.M. Eicher. 1982. Male and female mouse DNAs can be discriminated using retroviral probes. Nature 297:241–243.
8. Singh, L., and K.W. Jones. 1982. Sex reversal in the mouse (*Mus musculus*) is caused by a recurrent nonreciprocal crossover involving the X and an aberrant Y chromosome. Cell 28:205–216.

Y chromosome loci

Jutley and Stewart (1) obtained marginally significant evidence for polymorphism of two loci on the Y chromosome. Comparisons were always between effects of different Y chromosomes on identical genetic backgrounds. One locus affects serum testosterone level, with the Y chromosome of the PHH strain causing low level and the Y chromosome of the PHL strain causing high level. Strain background also has a large effect. The other locus affects response of seminal vesicles to testosterone, with the Y chromosome of C57BL/6 causing low response and the Y of CBA/Fa causing high response. This effect occurs on the C57BL/6 but not on the CBA genetic background.

References

1. Jutley, J.K., and A.D. Stewart. 1986. Genetic analysis of the Y-chromosome of the mouse: evidence for two loci affecting androgen metabolism. Genet. Res. 47:29–34.

Activity of catecholamine biosynthetic enzymes

The activity level of three enzymes in this pathway, tyrosine hydroxylase (E.C. 1.14.16.2), dopamine-β-hydroxylase (E.C. 1.14.17.1), and phenylethanolamine N-methyltransferase (E.C. 2.1.1), is high in the BALB/cJ strain and about one-half as high in BALB/cN, two substrains that have been separated for more than 25 years. F1 mice have intermediate activity. Activities of all three enzymes segregate together in F2 and backcross generations, indicating that a single regulatory locus probably controls the phenotypic expression of the three enzymes (1). These two substrains also show a difference in fighting behavior, the BALB/cJ males being aggressive and the BALB/cN males not aggressive. The difference appears to be under single-locus control, but it has not been possible to test the mice of segregating generations for both enzyme level and behavior. It is therefore not known whether the same locus controls both traits (2).

References

1. Ciaranello, R.D., H.J. Hoffman, J.G.M. Shire, and J. Axelrod. 1974. Genetic regulation of catecholamine biosynthetic enzymes. II. Inheritance of tyrosine hydroxylase, dopamine-β-hydroxylase, and phenylethanolamine N-methyltransferase. J. Biol. Chem. 249:4528–4536.
2. Ciaranello, R.D., A. Lipsky, and J. Axelrod. 1974. Association between fighting behavior and catecholamine biosynthetic enzyme activity in two inbred mouse sublines. Proc. Natl. Acad. Sci. USA 71:3006–3008.

Manganese-independent carnosinase activity

Activity of Mn^{2+}-independent carnosinase (E.C. 3.4.13.3) in kidney and some other tissues is very low in most inbred strains, including BALB/c, SWR, DBA/2, C57BL/6, and A/J, and in some outbred CD-1 mice. Activity is high in inbred strain NZB/BlN and in other CD-1 mice. (BALB/c × NZB)F1 mice have intermediate levels and the F2 and backcross segregations indicate major control by a single locus, with possibly other factors playing a minor role. The strain distribution is different from that of all other peptidase loci reported (1).

References

1. Margolis, F.L., and M. Grillo. 1984. Inherited differences in mouse kidney carnosinase activity. Biochem. Genet. 22:441–451.

Colon carcinogenesis by DMH

Mice of the ICR/Ha strain are susceptible to induction of colon adenocarcinomas by 1,2-dimethylhydrazine (DMH); C57BL/Ha mice are resistant. F1 mice are all susceptible. Segregation in the F2 and backcross to C57BL/Ha indicates control by a single locus not linked to genetic markers on Chrs 1, 2, 4, 7, 8, 9, and 11 (2). The resistance of C57BL/Ha mice is associated with a smaller extent of initial methylation of colon DNA by the carcinogen than in susceptible ICR mice (1).

References

1. Cooper, H.K., J. Buecheler, and P. Kleihues. 1978. DNA alkylation in mice with genetically different susceptibility to 1,2-methylhydrazine-induced colon carcinogenesis. Cancer Res. 38:3063–3065.
2. Evans, J.T., T.B. Shows, E.E. Sproul, N.S. Paolini, A. Mittelman, and T.S. Hauschka. 1977. Genetics of colon carcinogenesis in mice treated with 1,2-dimethylhydrazine. Cancer Res. 37:134–136.

IgE immune response

Low doses of many antigens induce the production of IgE immunoglobulin (reagin), which triggers the anaphylactic response. The effectiveness of specific antigens in inducing IgE production is controlled by the *Ir-1* locus in the *H-2* complex (4, 9). Some strains produce only a very low and transient IgE response to any antigen, and this strain difference is controlled by one or two non-*H-2*-linked loci. Under at least two different immunization procedures, strain SJL produces a low or transient response and strain BALB/c a high response, and this difference is controlled by a single locus not linked to *H-2* or to *Igh-1* (6). The high response is dominant. The low response in SJL is due to the presence in this strain of a population of suppressor T-cells in the spleen which suppress IgE production. Elimination of these cells by X-irradiation or other treatment results in a normal IgE response (2, 7, 10). Production or suppression of IgE appears to depend on the induction in splenic T-cells of IgE-binding proteins which are of two kinds, IgE-potentiating and IgE-suppressing. In B6D2F1 mice, it was shown that an appropriate antigen-induced potentiating IgE-binding factor whereas in SJL mice the antigen-induced suppressive IgE-binding factor (5, 8). In the SJL.BALB/c-Igh^a (SJA) congenic strain, IgE production is even more severely suppressed than in SJL, but the reason for this is not clear (1, 3).

References

1. Adachi, M., K. Okumura, N. Watanabe, N. Noro, T. Masuda, and J. Yodoi. 1985. Lack of Fc receptor for IgE in STA9 mice. Immunogenetics 22:77–83.
2. Chiorazzi, N., D.A. Fox, and D.H. Katz. 1977. Hapten-specific IgE antibody responses in mice. VII. Conversion of IgE "non-responder" strains to IgE "responders" by elimination of suppressor T cell activity. J. Immunol. 118:48–54.
3. Hirano, T., Y. Kumaga, K. Okumura, and Z. Ovary. 1983. Regulation of murine IgE production: importance of a not-yet-described T cell for IgE secretion demonstrated in SJA9 mice. Proc. Natl. Acad. Sci. USA 80:3435–3438.
4. Ishizaka, K. 1976. Cellular events in the IgE antibody response. Adv. Immunol. 23:1–75.
5. Ishizaka, K. 1984. Regulation of IgE synthesis. Ann. Rev. Immunol. 2:158–182.
6. Itaya, T. K. Okumura, N. Watanabe, and Z. Ovary. 1978. Absence of correlation of Ig-1 allotype and IgE antibody suppression in SJL mice. J. Immunol. 120:1758–1759.
7. Ovary, Z., T. Itaya, N. Watanabe, and S. Kojima. 1978. Regulation of IgE in mice. Immunol Rev. 41:26–51.
8. Uede, T., and K. Ishizaka. 1984. IgE-binding factors from mouse T lymphocytes. II. Strain differences in the nature of IgE-binding factor. J. Immunol. 133:359–367.
9. Vaz, N.M., and B.B. Levine. 1970. Immune responses of inbred mice to repeated low doses of antigen: relationship to histocompatibility (*H-2*) type. Science 168:852–854.
10. Watanabe, N., S. Kojima, and S. Ovary. 1976. Suppression of IgE antibody production in SJL mice. I. Nonspecific suppressor T cells. J. Exp. Med. 143:833–845.

Immune response to H-2.'28' antigenic specificity

Strain C3H produces a strong cytotoxic response to this antigen; strain C57BL/10.BR does not. F1 mice are responsive. The response is specific for the H-2.'28' specificity. The response difference appears to be controlled by a single locus not linked to *a*, *H-2*, *Igh-1*, *Ly-1*, or *Ly-2* (1).

References

1. Hansen, T.H., S.E. Cullen, N. Shinohara, E. Schurko, and D.H. Sachs. 1977. Genetic control of the humeral response to an H-2 public specificity. J. Immunol. 118:1403–1408.

Immune response to H-2.32 antigenic specificity

Mice of the congenic strain B10.AKM-*H-2^m* produce a very low cytotoxic antibody response to the H-2D^k private specificity H-2.32; mice of strain AKR.M-*H-2^m* produce a strong response. The difference is predominantly controlled by a single non-*H-2* locus. The allele for strong response in AKR.M is dominant. The alleles segregate independently of *Igh*, *Ly-2*, and *Mls*. The defect in response of B10.AKM is detected at the level of B-cell function. This locus is quite similar to the locus controlling the immune response to H-2.'28', but it is not possible to test them for allelism (1).

References

1. Shinohara, N., T.H. Hansen, E. Schurko, I.F.C. McKenzie, and D.H. Sachs. 1978. Genetic control of the immune response to the H-2D^k private specificity, H-2.32. II. Genetic linkage analysis of a backcross generation. J. Immunol. 121:2472–2476.

Immune response to *Klebsiella pneumoniae* polysaccharide

BALB/c mice have a low primary immune response to this antigen (K47-PS), in contrast to C57BL/10.D2/nSn mice which have a normal response. The difference appears to be due to a single locus, with responsiveness dominant. The alleles segregate independently of sex, *a*, *b*, *c*, *Igh-1*, and probably *H-2* (1).

References

1. White-Scharf, M.E., and L.T. Rosenberg. 1978. Genetically controlled IgM hyporesponsiveness to a *K. pneumoniae* polysaccharide. Immunogenetics 6:81–89.

Immune response to Moloney virus-induced lymphoma

The cytotoxic antibody response against Moloney virus-induced cell surface antigen (MSCA) is low in strains A/Sn and C3H, and high in DBA/2, CBA, C57L, and C57BL. Responsiveness is dominant. The response difference appears to be controlled by a single locus which is not linked to *H-2* or *Igh-1*, or to isozyme loci on Chrs 1, 4, 7, 8, and 9 (1).

References

1. Asjö, B., R. Kiessling, G. Klein, and S. Povey. 1977. Genetic variation in antibody response and natural killer cell activity against a Moloney virus-induced lymphoma (YAC). Eur. J. Immunol. 7:554–558.

Resistance to polyoma virus, Chr 7

Chang and Hildemann (1) have described a runting syndrome produced in AKR mice by neonatal injection of polyoma virus. Mice of the C57BL/6 strain and of (AKR × C57BL/6)F1 hybrids are resistant. The backcross to AKR showed that the difference in susceptibility was controlled by a gene linked to or identical with the albino (*c*) locus and by at least one other independently segregating gene.

References

1. Chang, S.S., and W.H. Hildemann. 1964. Inheritance of susceptibility to polyoma virus in mice. J. Natl. Cancer Inst. 33:303–313.

Natural resistance to Sendai virus, Chr 1

Resistance to intranasal inoculation of a naturally occurring Sendai virus is high in C57BL/6 and low in DBA/2 mice. F1 hybrids are resistant. Resistance in F2 mice, in offspring of a backcross to DBA/2, and in 25 BXD recombinant inbred strains, segregated as if controlled by a single locus. Significant linkage with *Sas-1* on Chr 1 occurred in the recombinant inbred strains (1).

References

1. Brownstein, D.G. 1983. Genetics of natural resistance to Sendai virus infection in mice. Infect. Immun. 41:308–312.

Resistance to virus-induced diabetes mellitus

Some strains of mice, including SJL, SWR, and DBA/1, are susceptible to induction of diabetes mellitus by infection with encephalomyocarditis (EMC) virus. Other strains, including C57BL/6, AKR, A/J, and several others, are resistant (1). In crosses between susceptible and resistant strains, resistance is dominant, but the number of genes involved is uncertain (2, 3). Resistance appears to be related to the degree of viral replication in the beta cells of the pancreatic islets.

Virus replicates much faster in the beta cells of susceptible than of resistant mice (2). Infection with Coxsackie virus B4 that has been passaged in beta cell cultures also induces transient diabetes in strains SJL and SWR, but not in C57BL/6 and AKR (4).

References

1. Boucher, D.W., K. Hayashi, J. Rosenthal, and A.L. Notkins. 1975. Virus-induced diabetes mellitus. III. Influence of the sex and strain of the host. J. Infect. Dis. 131:462–466.
2. Notkins, A.L. 1977. Virus induced diabetes mellitus. Arch. Virol. 54:1–17.
3. Onodera, T., J.W. Yoon, K.S. Brown, and A.L. Notkins. 1978. Evidence for a single locus controlling susceptibility to virus-induced diabetes mellitus. Nature 274:693–696.
4. Yoon, J.W., T. Onodera, and A.L. Notkins. 1978. Virus-induced diabetes mellitus. XV. Beta cell damage and insulin-dependent hyperglycemia in mice infected with Coxsackie virus B4. J. Exp. Med. 148:1068–1080.

Resistance to pertussis HSF

A histamine-sensitizing factor in *B. pertussis* vaccine sensitizes some strains of mice, including SWR/J, and SJL/J, to the lethal effects of histamine. Other strains, including CBA/J, A/J, and AKR/J, are resistant to HSF-sensitization to histamine. Conflicting results for sensitivity of C57BL, BALB/c, and DBA/2 were found by Linthicum and Frelinger (1), Teuscher (2), and Wardlaw (3). Sensitization appeared to be controlled by a dominant allele at a single locus in segregating generations from crosses between SWR and CBA (3) and between C57BL/10 (sensitive) and DBA/2 (resistant) (2).

References

1. Linthicum, D.S., and J.A. Frelinger. 1982. Acute autoimmune encephalomyelitis in mice. II. Susceptibility is controlled by the combination of H-2 and histamine sensitization genes. J. Exp. Med. 135:31–40.
2. Teuscher, C. 1983. Experimental allergic orchitis in mice. II. Association of disease susceptibility with the locus controlling *Bordetella pertussis*-induced sensitivity to histamines. Immunogenetics 22:417–425.
3. Wardlaw, A.C. 1970. Inheritance of responsiveness to pertussis HSF in mice. Int. Arch. Allergy Appl. Immunol. 38:573–589.

Hybrid resistance to transplantable plasmacytoma MPC-11

This locus controls resistance to the BALB/c myeloma MPC-11 in hybrids between BALB/c mice and mice of other genotypes. The allele for resistance occurs in strains C57BL/10, C57BL/6, C57L, C57BL/Ks, and DBA/1. The allele for susceptibility occurs presumably in BALB/c and also in strains A/J, SJL, and DBA/2. Heterozygotes are mostly resistant, but about 25 per cent were susceptible in the F1 of a cross between BALB/c and C57BL/10. Resistance is slightly affected by alleles at the *H-2* locus, but the gene is not linked to *H-2*. It is also not linked to albino (*c*) or brown (*b*) (1). Resistance is abolished or reduced by total body irradiation but unaffected by silica injection, indicating that it probably depends on an active immune response (2).

References

1. Walker, M.C., and J.M. Phillips-Quagliata. 1979. Hybrid resistance to BALB/c plasmacytomas. I. Resistance to MPC-11 controlled by a locus not linked to H-2. J. Immunol. 122:1535–1543.
2. Walker, M.C., and J.M. Phillips-Quagliata. 1979. Hybrid resistance to BALB/c plasmacytomas. II. Radiation sensitivity and silica insensitivity of resistance to MPC-11. J. Immunol. 122:1544–1547.

SWR B-tropic virus

A B-tropic leukemia virus having a low degree of sequence homology with other endogenous type-C viruses is present in strain SWR mice and absent in other strains tested. It is inducible by IdU in cultured embryo cells. In crosses of SWR and NIH Swiss, inducibility and viral structural information segregate together as if controlled by a single locus, with inducibility dominant (1).

References

1. Robbins, K.C., J.R. Stephenson, C.D. Cabradilla, and S.A. Aaronson. 1977. Endogenous mouse type-C RNA virus of SWR cells: inducibility locus containing structural information for a new endogenous virus class. Virology 82:392–400.

Naturally occurring pancreatic virus

An endogenous type-C virus revealed by electron microscopy occurs in the pancreas of C57BL/He and C57BL/6J mice, and its incidence is increased by treatment with dexamethasone. The virus is not present in other organs or in BALB/c or C3Hf mice. Crosses of C57BL/He with BALB/c give evidence that viral production is controlled by a single locus with the BALB/c allele for restriction of virus production dominant (1).

References

1. Boiocchi, M., G. Della Torre, and G. Della Porta. 1975. Genetic control of endogenous C-type virus production in pancreatic acinar cells of C57BL/He and C57BL/6 mice. Proc. Natl. Acad. Sci. USA 72:1892–1894.

Ratio of Ly-2⁻/Ly-2⁺ T-cells

$Ly-2^+$ T-cells recognize antigen in association with class I major histocompatibility complex molecules and have cytotoxic and suppressor activity; $Ly-2^-$ T-cells recognize antigen in association with class II molecules and have helper or amplifier activity. The $Ly-2^-/Ly-2^+$ ratio is low (0.8–1.1) in C57BL/6 and high (1.5–2.0) in BALB/c mice. Heterozygotes resemble C57BL/6. In F2 mice and in CXB recombinant inbred strains, the difference behaves as if controlled by a single locus. Most of the difference is determined by selection or differential turnover in the peripheral lymphoid organs (1).

References

1. Kraal, G., I.L. Weissman, and E.C. Butcher. 1983. Genetic control of T-cell subset representation in inbred mice. Immunogenetics 18:585–592.

Cerebellar abnormality

A recessive mutation tentatively designated nm 544 arose at the Jackson Laboratory in strain BALB/cJ. Homozygotes are recognizable at about 10 days by their tendency to lean to one side or the other. Later they may assume a sprawling position with limbs extended to the side. They walk with a jerky forward movement but have no tremor. They may survive to adulthood but do not breed. In studies by Webster and Briggs (1), the sciatic nerve and brain were examined histologically and the only abnormality detected was displacement of Purkinje cells in the nodulus of the cerebellum. The Purkinje cells were scattered through the granule cell layer instead of being aligned at the junction of the granule and molecular layers as in normal controls (1).

References

1. Webster, W.S., and D. Briggs. 1981. A new mouse mutant showing cerebellar abnormalities. Brain Res. 218:412–416.

3 RETROVIRAL AND CANCER-RELATED GENES

CHRISTINE A. KOZAK

This mouse linkage map (Fig. 3.1) and the accompanying tables represent an updated version of a gene map published previously (42). It includes several classes of genes implicated in neoplasia and the transmission of mouse retroviruses. The first three tables list the virus integration sites which contain DNA copies of the genomes of the mouse mammary tumor viruses (MMTVs) (Table 3.1) or murine leukemia viruses (MuLVs) (Tables 3.2 and 3.3). Table 3.2 lists loci containing proviruses of MuLVs of the ecotropic host range class; Table 3.3 lists loci containing proviruses of the xenotropic or MCF (mink cell focus-forming) classes. These proviruses may be complete copies capable of expression as infectious virus, or defective copies detected by hybridization with retroviral probes. Since novel proviral integrations can be acquired following infection of germline cells, specific proviruses are often restricted to particular breeding lines or related inbred strains. Therefore, the common inbred strains which carry these proviruses are indicated in the first three tables. A more complete strain survey for endogenous MuLVs is available elsewhere (39,112), but for the MMTVs, not all inbred strains have been systematically typed (44).

The second group of genes included in this map are those which control virus replication and expression as well as resistance to virus-induced disease (Table 3.4). Few of these genes have been characterized at the molecular level. One notable exception, *Fv-4*, is included here as well as in Table 3.2 because, although it has been characterized as an ecotropic proviral gene, its expression is associated with resistance to infection by ecotropic MuLVs.

Third, mice contain proto–oncogenes identified most commonly by their homology to the transforming genes of the acute retroviruses or by their ability to transform NIH-3T3 cells (Table 3.5). Finally, tumor production is often associated with retroviral integration into specific chromosomal regions in somatic cells (Table 3.6). In some cases, such tumor-specific integrations disrupt known oncogenes such as *Myc* or *Myb* resulting in their altered expression. In other cases, such sites do not represent any previously identified proto-oncogenes and are therefore assumed to represent novel genes involved in tumor induction or maintenance.

The general conventions used in Fig. 3.1 are described elsewhere in this volume. References are given for the chromosomal localization only, not for the initial identification and characterization of these genes (see Chapter 2).

Table 3.1 Endogenous mouse mammary tumor viruses

Locus[1]	Mouse strains	Chromosome	References
Mtv-1[2]	DBA, C3H	7	105, 107
Mtv-2[2]	GRS	18	67
Mtv-3	GRS	11	79
Mtv-6	BALB/c, DBA, C3H	16	9, 88
Mtv-7	C3H/Bi, DBA, NFS, GRS	1	58, 67, 105
Mtv-8	various	6	89
Mtv-9	various	12	9
Mtv-11	DBA, C3H/He	14	9, 37, 58, 85
Mtv-13	DBA, NFS, A	4	71
Mtv-17	various	4	67
Mtv-21	BR6	8	84

[1] Nomenclature according to Reference 44.
[2] Expressed as infectious virus.

Table 3.2 Endogenous ecotropic murine leukemia viruses

Locus	Mouse strains	Chromosome	References
Akv-1 (*Emv-11*)[1]	AKR	7	92
Akv-2 (*Emv-12*)[1]	AKR/N	16	47
Akv-3 (*Emv-13*)	AKR	2	104
Akv-4 (*Emv-14*)[1]	AKR/J	11	104
Bbv[1]	B10.BR/SgLi	11	48
Bv (*Emv-2*)[1]	C57BL	8	48
C58v-1[1]	C58	8	48
C58v-4	C58	7	97
Cv (*Emv-1*)[1]	BALB/c, A, C3H	5	36, 45, 48
Emv-3 (*Dbv*, *Sev-1*)[1]	DBA, BDP, P	9	38, 48
Emv-15[1]	C57BL/6J-A^y, LT/Sv-A^y, 129/Sv-A^y	2	11
Emv-16[1]	RF/J	1	8
Emv-17[1]	RF/J	1	8
Emv-18[1]	B6.S (*Fv-2*$^{s/s}$)	9	73
Fgv-1[1]	C3H/FgLw	7	48
Fv-4	G (FRG)	12	19, 81
Mov-1[1]	BALB/Mo	6	6
Mov-7	129/Mo	1	75
Mov-9[1]	129/Mo	11	75
Mov-10	129/Mo	3	75
Mov-13[1]	C57BL/Mo	11	74, 75, 96
Mov-14[1]	C57BL/Mo	X	101
Mov-15[1]	BALB/Mo	X, Y	27
Mov-24[1]	(CFWxSWR)F$_1$	Y	100

[1] Expressed as infectious virus.

Table 3.3 Endogenous nonecotropic murine leukemia viral genes

Locus[1]	Mouse strains	Chromosome	References
Bxv-1[2]	C57BL, C57L, C58, BALB/c, AKR	1	46
Mmv-1	BALB/c	5	33
Mmv-2	BALB/c	3	33
Mmv-3	BALB/c	7	33
Mmv-4	BALB/c	7	33
Mmv-5	BALB/c	1	33
Mmv-6	BALB/c	12	33
Mmv-7	BALB/c	12	33
Mmv-9	BALB/c	1	33
Mmv-10	BALB/c	1	33
Mmv-12	BALB/c	3	33
Pol-7	C57BL/6	12	90
Pol-20	C57BL/6	5	90
Pol-23	BALB/c	2	90
Xmmv-2	C57BL	9	4
Xmmv-6	DBA	1	4
Xmmv-8	C57BL	4	4
Xmmv-9	C57BL	1	4
Xmmv-21	DBA	12	4
Xmmv-23	C57BL	4	4
Xmmv-27	C57BL	6	4
Xmmv-29	DBA	8	4
Xmmv-34	DBA	12	4
Xmmv-35	C57BL	7	4
Xmmv-36	DBA	1	4
Xmmv-42	C57L	19	112
Xmmv-50	AKR	12	112
Xmmv-52	DBA	5	112
Xmmv-55[3]	C57BL, 129	15	32, 65, 112
Xmmv-61[4]	C57L, AKR, C57BL, 129	1	112
Xmmv-62	C57L, 129, C58	4	112
Xmmv-71	BALB/c	2	65
Xmmv-72[3]	C57BL/6	15	32, 65, 90
Xmmv-74	C57BL/6	1	65
Xmv-1	BALB/c	4	33
Xmv-2	BALB/c	4	33
Xmv-3	BALB/c	16	33
XmmvY	A, C57BL6	Y	4

[1] The *env* hybridization probes used to detect *Xmmv* loci cannot distinguish the xenotropic and MCF host range classes of MuLVs. *Mmv* and *Xmv* loci were defined using MCF- or xenotropic-specific *env* probes. *Bxv-1* reacts with the xenotropic *env* probe. *Pol* loci react with a 3′ *pol* sequence and were not tested with *env* probes.

[2] Expressed as infectious xenotropic virus.

[3] *Xmmv-55* and *Xmmv-72* may be identical proviral loci.

[4] May be identical to *Xmmv-9*.

Table 3.4 Genes controlling virus replication, viral and tumor specific antigens, and resistance to virus-induced disease

Locus	Phenotype	Chromosome	References
Cxv-1	High levels of xentropic MuLV	17	113
Cxv-2	High xenotropic MuLV antigen (XenCSA)	4	72
Fv-1	Resistance to N- or B-tropic MuLVs	4	93
Fv-2	Resistance to Friend SFFV MuLV	9	53
Fv-4	Resistance to ecotropic MuLV	1	81
Gv-2	G_{IX} antigen expression	7	102
Hsp84-1	Tumor specific transplantation antigen related to heat shock protein	17	70
Hsp84-2	Tumor specific transplantation antigen related to heat shock protein	2	70
Hsp84-3	Tumor specific transplantation antigen related to heat shock protein	12	70
If-1	NDV induced circulating interferon	3	68
Ifa	α-interferon	4	16, 56
Ifb	β-interferon	4	15, 16
Ifg	γ-interferon	10	77
Ifgr	γ-interferon receptor	10	60
Ifrc	Interferon receptor	16	12
Ifx	NDV-induced interferon	X	18, 82
Inb-1	Enhanced MuLV induction	8	59
Inc-1	Enhanced MuLV induction	5	59
Mtvr-1	MMTV receptor	16	31
Ram-1	Amphotropic MuLV receptor	8	24
Rec-1	Ecotropic MuLV receptor	5	24
–	M813 ecotropic MuLV receptor	2	87
Ril-1	Resistance to radiation-induced leukemia	15	32, 52, 63
Rmc-1	MCF MuLV receptor	1	41
Rmcf	Resistance to MCF MuLVs	5	29
Rfv-1	Recovery from Friend virus-induced	7 (*H-2D*)	10
Rfv-2	splenomegaly	17 (*H-2K*)	10
Rgv-1	Resistance to Gross virus leukemogenesis	17 (*H-2K*)	54
Rmv-1	Resistance to Moloney virus	17 (*H-2I*)	17
Rmv-2	Resistance to Moloney virus	17 (*H-2I*)	17
Rmv-3	Resistance to Moloney virus	17 (*H-2D*)	17
Rrv-1	Resistance to radiation MuLV (RadLV)	17 (*H-2I*)	55
–	Resistance to RadLV	17 (*H-2I*)	62
Rvil	Resistance to RadLV induced leukemia	12	64
Sxv	Susceptibility to xenotropic MuLVs	11	43
Tla	Thymus-leukemia antigen	17	5
Trp53	Transformation-related protein	11	14, 91
Trp53-ps	Transformation-related protein (pseudogene)	11	14

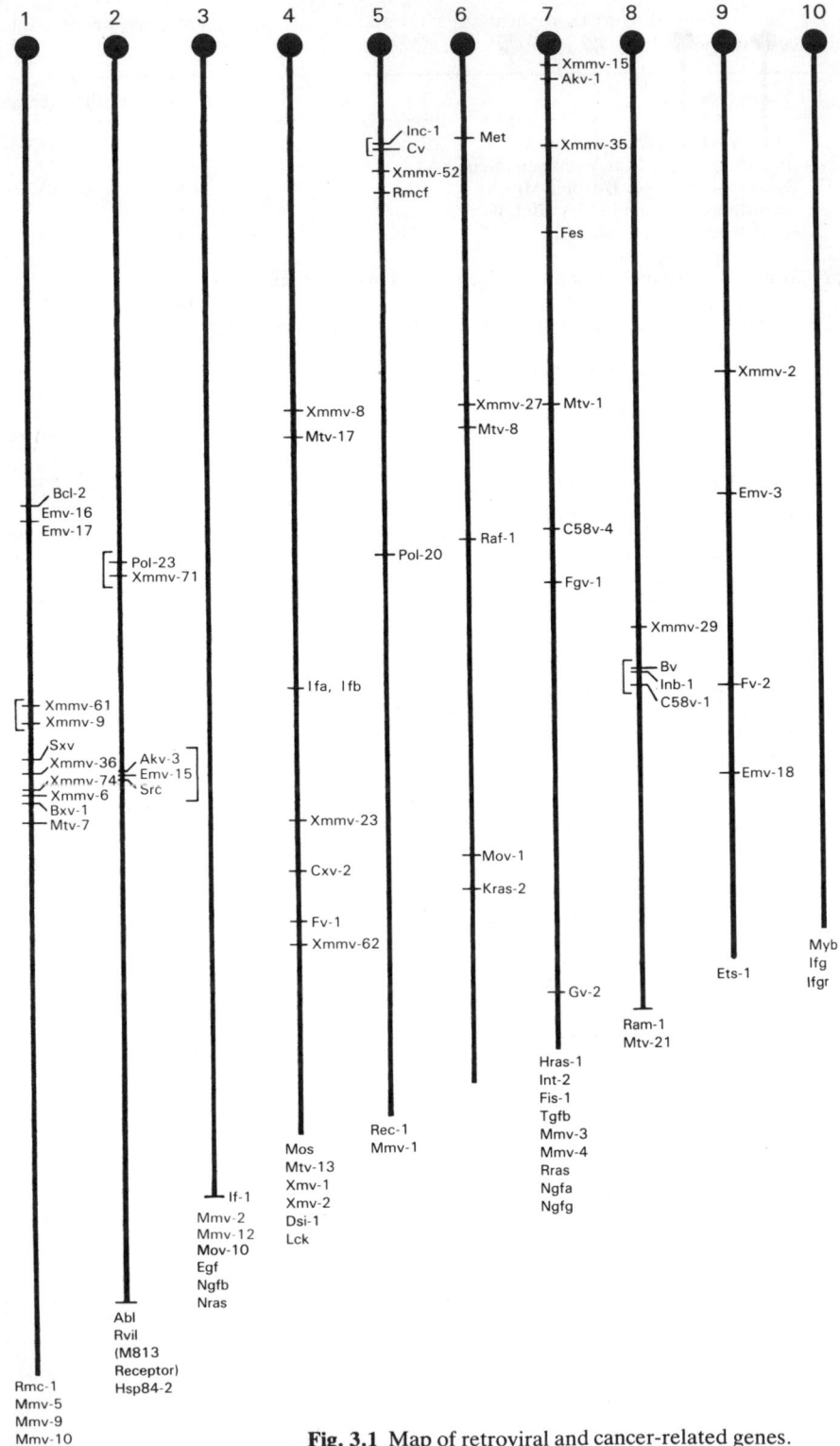

Fig. 3.1 Map of retroviral and cancer-related genes.

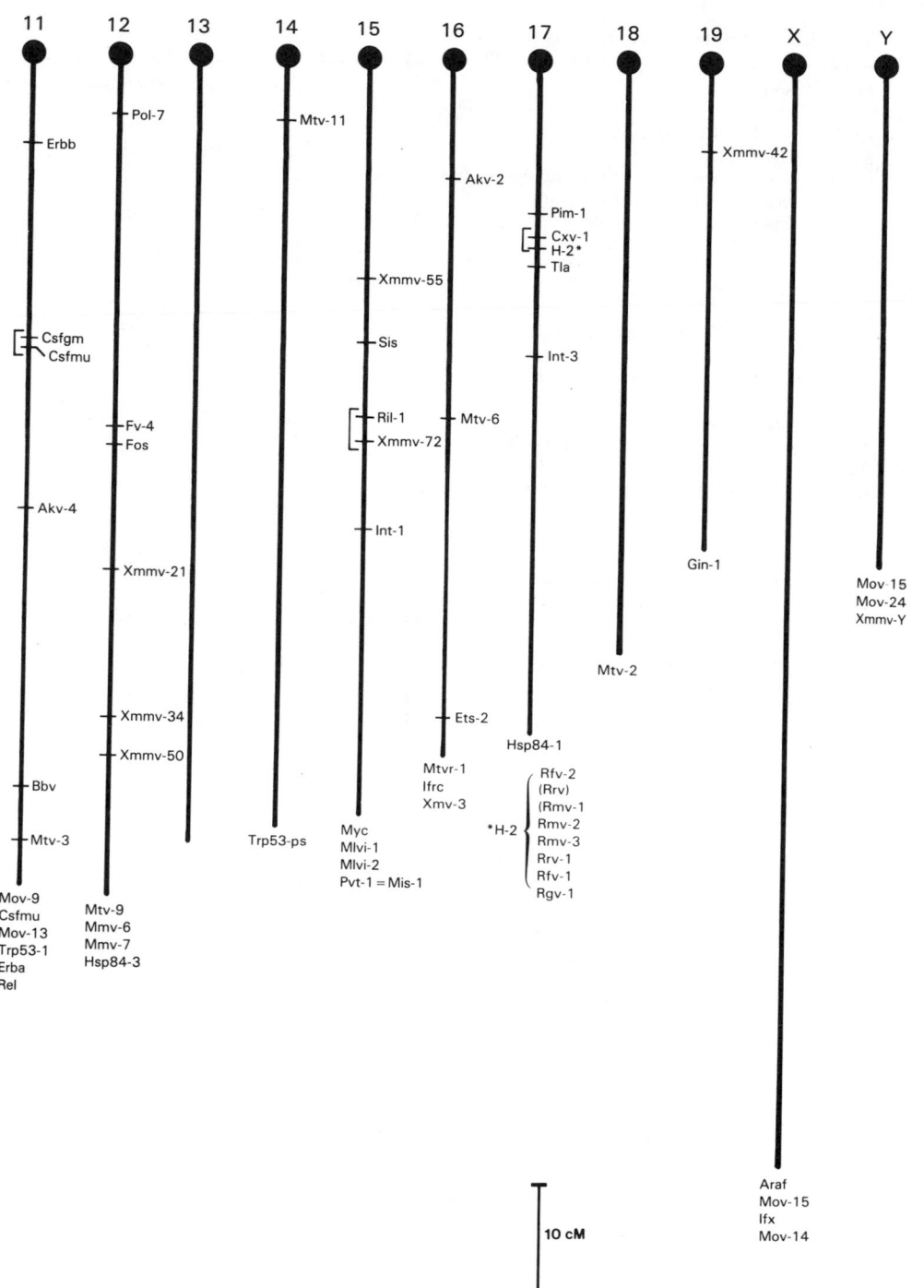

Table 3.5 Proto-oncogenes and growth factors

Locus	Source	Chromosome	References
Abl	Abelson murine leukemia virus	2	25
Bcl-2	B-cell leukemia	1	69, 78
Csfgm	B-cell macrophage colony stimulating factor	11	3
Csfm	Multi-colony stimulating factor	11	3
Egf	Epidermal growth factor	3	114
Erba	Avian erythroblastosis virus	11	115
Erbb	Avian erythroblastosis virus	11	99, 115
Ets-1	Avian retrovirus E26	9	110
Ets-2	Avian retrovirus E26	16	88, 111
Fes	Snyder–Theilen feline sarcoma virus	7	28,49
Fos	FBJ murine osteosarcoma virus	12	20
Lck	Lymphocyte-specific protein kinase	4	61
Mos	Moloney murine sarcoma virus	4	103
Myb	Avian myeloblastosis virus	10	95
Myc	Avian myelocytomalosis virus MC29	15	1, 13, 19, 52, 66
Ngfa	γ subunit of nerve growth factor	7	21,34
Ngfb	β subunit of nerve growth factor	3	114
Ngfg	γ subunit of nerve growth factor	7	21,34
Raf–1	3611 murine sarcoma virus, avian carcinoma virus MH2	6	89
Araf	*Raf*–related gene	X	2,35
Hras–1	Harvey murine sarcoma virus	7	49,86,95
Kras–2	Kirsten murine sarcoma virus	6	89,95
Nras	Transfection from human neuroblastoma SK–N–SH	3	26,94
Rras	Ras–related gene	7	57
Rel	Reticuloendotheliosis virus strain T	11	7
Sis	Simian sarcoma virus	15	1,50,52,66
Src	Rous sarcoma virus	2	28
Tgfb	Transforming growth factor β	7	22

Table 3.6 Tumor-specific viral integration sites

Locus	Inducing virus	Chromosome	References
Dsi-1	Moloney MuLV	4	108
Fis-1	Friend MuLV	7	98
Gin-1	Gross Passage A MuLV	19	109
Int-1	MMTV	15	52, 66, 80, 83
Int-2	MMTV	7	83
Int-3	MMTV	17	23
Mlvi-1	Moloney MuLV	15	51
Mlvi-2	Moloney MuLV	15	106
Pim-1	Moloney MuLV, MCF-MuLV	17	30, 76
Pvt-1 (*Mis-1*)	Moloney MuLV, MCF-MuLV	15	40, 110

References

1. Adolph, S., C.R. Bartram, and H. Hameister. 1987. Mapping of the oncogenes *Myc*, *Sis*, and *int-1* to the distal part of mouse chromosome 15. Cytogenet. Cell Genet. 44:65–68.

2. Avner, P., M. Bucan, D. Arnaud, H. Lehrach, and U. Rapp. 1987. *Araf* oncogene localizes on mouse X chromosome to region some 10–17 centimorgans proximal to hypoxanthine phosphoribosyltransferase gene. Somat. Cell Mol. Genet. 13:267–272.

3. Barlow, D.P., M. Bucan, H. Lehrach, B.L.M. Hogan, and N.M. Gough. 1987. Close genetic and physical linkage between the murine haemopoietic growth factor genes GM-CSF and Multi-CSF (IL3). EMBO J. 6:617–623.

4. Blatt, C., K. Mileham, M. Haas, M.N. Nesbitt, M.E. Harper, and M.I. Simon. 1983. Chromosomal mapping of the mink cell focus-inducing and xenotropic *env* gene family in the mouse. Proc. Natl. Acad. Sci. USA. 80:6298–6302.

5. Boyse, E.A., L.J. Old, and S. Luell. 1964. Genetic determination of the TL (thymus-leukemia) antigen in the mouse. Nature 201:779.

6. Breindl, M., J. Doehmer, K. Willecke, J. Dausman, and R. Jaenisch. 1979. Germ line integration of Moloney leukemia virus: Identification of the chromosomal integration site. Proc. Natl. Acad. Sci. USA. 76:1938–1942.

7. Brownell, E., C.A. Kozak, J.R. Fowle III, W.S. Modi, D. Pravtcheva, F.H. Ruddle, N.E. Rice, and S.J. O'Brien. 1986. Comparative genetic mapping of the *rel* proto-oncogene homologs in man, mouse and the domestic cat. Am. J. Hum. Genet. 39:194–202.

8. Buchberg, A.M., B.A. Taylor, N.A. Jenkins, and N.G. Copeland. 1986. Chromosomal localization of *Emv-16* and *Emv-17*, the two closely linked ecotropic proviruses of RF/J mice. J. Virol. 60:1175–1178.

9. Callahan, R., D. Gallahan, and C. Kozak. 1984. Two genetically transmitted BALB/c mouse mammary tumor virus genomes are located on chromosomes 12 and 16. J. Virol. 49:1005–1008.

10. Chesebro, B., and K. Wehrly. 1978. *Rfv-1* and *Rfv-2*, two *H-2*-associated genes that influence recovery from Friend leukemia virus-induced splenomegaly. J. Immunol. 120:1081–1085.

11. Copeland, N.G., N.A. Jenkins, and B.A. Lee. 1983. Association of the lethal (*A^y*) yellow coat color mutation with an ecotropic murine leukemia virus genome. Proc. Natl. Acad. Sci. USA. 80:247–249.

12. Cox, D.R., L.B. Epstein, and C.J. Epstein. 1980. Genes coding for sensitivity to interferon (*IfRec*) and soluble superoxide dismutase (*SOD-1*) are linked in mouse and man and map to mouse chromosome 16. Proc. Natl. Acad. Sci. USA. 77:2168–2172.

13. Crews, S., R. Barth, L. Hood, J. Prehn, and K. Calame. 1982. Mouse c-myc oncogene is located on chromosome 15 and translocated to chromosome 12 in plasmacytomas. Science 218:1319–1321.

14. Czosnek, H.H., B. Bienz, D. Givol, R. Zakut-Houri, D. Pravtcheva, F. H. Ruddle, and M. Oren. 1984. The gene and the pseudogene for mouse p53 cellular tumor antigen located in different chromosomes. Mol. Cell. Biol. 4:1638–1640.

15. Dandoy, F., E. DeMaeyer, F. Bonhomme, J.-L. Guenet, and J. DeMaeyer-Guignard. 1985. Segregation of restriction fragment length polymorphism in an interspecies cross of laboratory and wild mice indicates tight linkage of the murine IFN-β gene to the murine IFN-α genes. J. Virol. 56:216–220.

16. Dandoy, F., K.A. Kelley, J. DeMaeyer-Guignard, E. DeMaeyer, and P.M. Pitha. 1984. Linkage analysis of the murine interferon-α locus on chromosome 4. J. Exp. Med. 160:294–302.

17. Debre, P., S. Gisselbrecht, F. Pozo, and J.P. Levy. 1979. Genetic control of sensitivity to Moloney leukemia virus in mice. J. Immunol. 123:1806–1812.

18. DeMaeyer-Guignard, J., R. Zawatzky, F. Dandoy, and E. DeMaeyer. 1983. An X-linked locus influences early serum interferon levels in the mouse. J. Interferon Res. 3:241–252.

19. Erikson, J., J.F. Mushinski, and C. Croce. 1986. The locus for the serum prealbumin is proximal to the heavy chain locus on mouse chromosome 12q$^+$. J. Immunol. 136:3137–3139.

20. D'Eustachio, P. 1984. A genetic map of mouse chromosome 12 composed of polymorphic DNA fragments. J. Exp. Med. 160:827–838.

21. Evans, B.A., and R.I. Richards. 1985. Genes for the α and γ subunits of mouse nerve growth factor are contiguous. EMBO J. 4:133–138.

22. Fujii, D., J.E. Brissenden, R. Derynck, and U. Francke. 1986. Transforming growth factor β gene maps to human chromosome 19 long arm and to mouse chromosome 7. Somat. Cell Mol. Genet. 12:281–288.

23. Gallahan, D., C. Kozak, and R. Callahan. 1986. A new common integration region (*int-3*) for the mouse mammary tumor virus on mouse chromosome 17. J. Virol. 61:218–220.

24. Gazdar, A.F., H. Oie, P. Lalley, W. Moss, J.D. Minna, and U. Francke. 1977. Identification of mouse chromosomes required for murine leukemia virus replication. Cell 11:949–956.

25. Goff, S.P., P. D'Eustachio, F.H. Ruddle, and D. Baltimore. 1982. Chromosomal assignment of the endogenous proto-oncogene *C-abl*. Science 218:1317–1319.

26. Guerrero, I., A. Villasante, P. D'Eustachio, and A. Pellicer. 1984. Isolation, characterization, and chromosome assignment of mouse *N-ras* gene from carcinogen-induced thymic lymphoma. Science 225:1041–1043.

27. Harbers, K., P. Soriano, U. Muller, and R. Jaenisch. 1986. High frequency of unequal recombination in pseudoautosomal region shown by proviral insertion in transgenic mouse. Nature 324:682–685.

28. Harper, M.E., C. Blatt, S. Marks, M. Nesbitt, and M.I. Simon. 1984. Gene mapping in the mouse by analysis of

RFLP segregation in recombinant inbred strains. Cytogenet. Cell Genet. 37:488.

29. Hartley, J.W., R.A. Yetter, and H.C. Morse III. 1983. A mouse gene on chromosome 5 that restricts infectivity of mink cell focus-forming recombinant murine leukemia viruses. J. Exp. Med. 158:16–24.

30. Hilkens, J., H.T. Cuypers, G. Selten, V. Kroezen, J. Hilgers, and A. Berns. 1986. Genetic mapping of *Pim-1* putative oncogene to mouse chromosome 17. Somat. Cell Mol. Genet. 12:81–88.

31. Hilkens, J., B. van der Zeijst, F. Buijs, V. Kroezen, N. Bleumink, and J. Hilgers. 1983. Identification of a cellular receptor for mouse mammary tumor virus and mapping of its gene to chromosome 16. J. Virol. 45:140–147.

32. Hogarth, P.M., I.F.C. McKenzie, V.R. Sutton, K.M. Curnow, B.K. Lee, and E.M. Eicher. 1987. Mapping of the murine *Ly-6*, *Xp-14* and *Gdc-1* loci to chromosome 15. Immunogenetics 25:21–27.

33. Hoggan, M.D., R.R. O'Neill, and C.A. Kozak. 1986. Non-ecotropic murine leukemia viruses in BALB/c and NFS/N mice: characterization of the BALB/c *Bxv-1* provirus and the single NFS endogenous xenotrope. J. Virol. 60:980–986.

34. Howles, P.N., D.P. Dickinson, L.L. DiCaprio, M. Woodworth-Gutai, and K.W. Gross. 1984. Use of a cDNA recombinant for the γ-subunit of mouse nerve growth factor to localize members of this multigene family near the TAM-l locus on chromosome 7. Nucl. Acids Res. 12:2791–2805.

35. Huebner, K., A. ar-Rushdi, C.A. Griffin, M. Isobe, C. Kozak, B.S. Emanuel, L. Nagarajan, J.L. Cleveland, T.I. Bonner, M.D. Goldsborough, C.M. Croce, and U. Rapp. 1986. Actively transcribed genes in the *raf* oncogene group, located on the X chromosome in mouse and human. Proc. Natl. Acad. Sci. USA. 83:3934–3938.

36. Ihle, J.N., D.R. Joseph, and J.J. Domotor Jr. 1979. Genetic linkage of C3H/HeJ and BALB/c endogenous ecotropic C-type viruses to phosphoglucomutase-1 on chromosome 5. Science 204:71–73.

37. Jeffreys, A.J., K.V. Wilson, B.A. Taylor, and G. Bulfield. 1987. Mouse DNA "fingerprint": analysis of chromosome localization and germ-line stability of hypervariable loci in recombinant inbred strains. Nucl. Acids Res. 15:2823–2836.

38. Jenkins, N.A., N.G. Copeland, B.A. Taylor, and B.K. Lee. 1981. Dilute (*d*) coat colour mutation of DBA/2J mice is associated with the site of integration of an ecotropic MuLV genome. Nature 293:370–374.

39. Jenkins, N.A., N.G. Copeland, B.A. Taylor, and B.K. Lee. 1982. Organization, distribution, and stability of endogenous ecotropic murine leukemia virus DNA sequences in chromosomes of *Mus musculus*. J. Virol. 43:26–36.

40. Jolicoeur, P., E. Rassart, L. Villeneuve, and C. Kozak. 1985. Mouse chromosomal mapping of a murine leukemia virus integration region (*Mis-1*) first identified in rat thymic leukemia. J. Virol. 56:1045–1048.

41. Kozak, C.A. 1983. Genetic mapping of a mouse chromosomal locus required for mink cell focus-forming virus replication. J. Virol. 48:300–303.

42. Kozak, C.A. 1984. Retroviral and cancer-associated loci of the mouse (*Mus musculus*). Genet. Maps 2:294–297.

43. Kozak, C.A. 1985. Susceptibility of wild mouse cells to exogenous infection with xenotropic leukemia viruses: control by a single dominant locus on chromosome 1. J. Virol. 55:690–695.

44. Kozak, C., G. Peters, R. Pauley, V. Morris, R. Michalides, J. Dudley, M. Green, M. Davisson, O. Prakash, A. Vaidya, J. Hilgers, A. Verstraeten, N. Hynes, H. Diggelmann, D. Peterson, J.C. Cohen, C. Dickson, N. Sarkar, R. Nusse, H. Varmus, and R. Callahan. 1987. A standardized nomenclature for endogenous mouse mammary tumor viruses. J. Virol. 16:1651–1654.

45. Kozak, C.A., and W.P. Rowe. 1979. Genetic mapping of ecotropic murine leukemia virus-inducing locus of BALB/c mouse to chromosome 5. Science 204:69–71.

46. Kozak, C.A., and W.P. Rowe. 1980. Genetic mapping of xenotropic murine leukemia virus-inducing loci in five mouse strains. J. Exp. Med. 152:219–228.

47. Kozak, C.A., and W.P. Rowe. 1980. Genetic mapping of the ecotropic virus-inducing locus *Akv-2* of the AKR mouse. J. Exp. Med. 152:1419–1425.

48. Kozak, C.A., and W.P. Rowe. 1982. Genetic mapping of ecotropic murine leukemia virus-inducing loci in six inbred strains. J. Exp. Med. 155:524–534.

49. Kozak, C.A., J.F. Sears, and M.D. Hoggan. 1983. Genetic mapping of the mouse oncogenes *c-Ha-ras-1* and *c-fes* to chromosome 7. J. Virol. 47:217–220.

50. Kozak, C.A., J.F. Sears, and M.D. Hoggan. 1983. Genetic mapping of the mouse proto-oncogene *c-sis* to chromosome 15. Science 221:867–869.

51. Kozak, C.A., P.G. Strauss, and P.N. Tsichlis. 1985. Genetic mapping of a cellular DNA region involved in induction of thymic lymphomas (*Mlvi-1*) to mouse chromosome 15. Mol. Cell. Biol. 5:894–897.

52. LeClair, K.P., M. Rabin, M.N. Nesbitt, D. Pravtcheva, F.H. Ruddle, R.G.E. Palfree, and A. Bothwell. 1987. Murine *Ly-6* multigene family is located on chromosome 15. Proc. Natl. Acad. Sci. USA. 84:1638–1642.

53. Lilly, F. 1970. *Fv-2*: identification and location of a second gene governing the spleen focus response to Friend leukemia virus in mice. J. Natl. Cancer Inst. 45:163–169.

54. Lilly, F., E.A. Boyse, and L.J. Old. 1964. Genetic basis of susceptibility to viral leukemogenesis. Lancet ii:1207–1209.

55. Lonai, P., and N. Haran-Ghera. 1977. Resistance genes to murine leukemia in the *I* immune response gene region of the *H-2* complex. J. Exp. Med. 146:1164–1168.

56. Lovett, M., D.R. Cox, D. Yee, W. Boll, C. Weissmann, C.J. Epstein,, and L.B. Epstein. 1984. The chromosomal location of mouse interferon-α genes. EMBO J. 3:1643–1646.

57. Lowe, D.G., D.J. Capon, E. Delwart, A.Y. Sakaguchi, S.L. Naylor, and Goeddel, D.V. 1987. Structure of the human and murine R-*ras* genes, novel genes closely related to *ras* proto-oncogenes. Cell 48:137–146.

58. MacInnes, J.I., V.L. Morris, W.F. Flintoff, and C.A. Kozak. 1984. Characterization and chromosomal location of endogenous mouse mammary tumor virus loci in GR, NFS, and DBA mice. Virology 132:12–25.

59. McCubrey, J., and R. Risser. 1982. Allelism and linkage studies of murine leukemia virus activation genes in low leukemic strains of mice. J. Exp. Med. 155:1233–1238.

60. Mariano, T.M., C.A. Kozak, J.A. Langer, and S. Pestka. 1987. The mouse immune interferon receptor gene is located on chromosome 10. J. Biol. Chem. 262:5812–5814.

61. Marth, J.D., C. Disteche, D. Pravtcheva, F. Ruddle, E.G. Krebs, and R.M. Perlmutter. 1986. Localization of a lymphocyte-specific protein tyrosine kinase gene (*lck*) at a site of frequent chromosomal abnormalities in human lymphomas. Proc. Natl. Acad. Sci. USA. 83:7400–7404.

62. Meruelo, D., M. Lieberman, N. Ginzton, B. Deak, and H. McDevitt. 1977. Genetic control of radiation leukemia virus-induced tumorigenesis. J. Exp. Med. 146:1079–1087.

63. Meruelo, D., M. Offer, and N. Flieger. 1981. Genetics of susceptibility to radiation-induced leukemia. J. Exp. Med. 154:1201–1211.

64. Meruelo, D.M., M. Offer, and A. Rossomando. 1983. Induction of leukemia by both fractionated x-irradiation and radiation leukemia virus involves loci in the chromosome 2 segment H-30-A. Proc. Natl. Acad. Sci. USA. 80:462–466.

65. Meruelo, D., A. Rossomando, M. Offer, J. Buxbaum, and A. Pellicer. 1983. Association of endogenous viral loci with genes encoding murine histocompatibility and lymphocyte differentiation antigens. Proc. Natl. Acad. Sci. USA. 80:5032–5036.

66. Meruelo, D., A. Rossomando, S. Scandalis, P. D'Eustachio, R.E.K. Fournier, D.R. Roop, D. Saxe, C. Blatt, and M.N. Nesbitt. 1987. Assignment of the *Ly-6-Ril-1–Sis-H-30-Pol-5/Xmmv-72-Ins-3-Krt-1-Int-1-Gdc-1* region to mouse chromosome 15. Immunogenetics 25:361–372.

67. Michalides, R., R. Verstraeten, F.W. Shen, and J. Hilgers. 1985. Characterization and chromosomal distribution of endogenous mouse mammary tumor viruses of European mouse strains STS/A and GR/A. Virology 142:278–290.

68. Mobraaten, L.E., H.P. Bunker, J. DeMaeyer-Guignard, E. DeMaeyer, and D.W. Bailey. 1984. Location of histocompatibility and interferon loci on chromosome 3 of the mouse. J. Hered. 75:233–234.

69. Mock, B. 1987. In F. Melchers and M. Potter, eds, Mechanisms of B cell neoplasia. Editiones Roche, Basel, Switzerland. pp. 262–265.

70. Moore, S.K., C. Kozak, E.A. Robinson, S.J. Ullrich, and E. Appella. 1987. Cloning and nucleotide sequence of the murine *hsp84* cDNA and chromosome assignment of related sequences. Gene 56:29–40.

71. Morris, V.L., C. Kozak, J.C. Cohen, P.R. Shank, P. Jolicoeur, F. Ruddle, and H.E. Varmus. 1979. Endogenous mouse mammary tumor virus DNA is distributed among multiple mouse chromosomes. Virology 92:46–55.

72. Morse, H.C., III, T.M. Chused, J.W. Hartley, B.J. Mathieson, S.O. Sharrow, and B.A. Taylor. 1979. Expression of xenotropic murine leukemia viruses as cell-surface gp70 in genetic crosses between strains DBA/2 and C57BL/6. J. Exp. Med. 149:1183–1196.

73. Mowat, M., and A. Bernstein. 1983. Linkage of the *Fv-2* gene to a newly reinserted ecotropic retrovirus in *Fv-2* congenic mice. J. Virol. 47:471–477.

74. Münke, M., K. Harbers, R. Jaenisch, D.R. Cox, and U. Francke. 1985. Assignment of alpha 1 collagen gene to mouse chromosome 11. Cytogenet. Cell Genet. 40:706.

75. Münke, M., K. Harbers, R. Jaenisch and U. Francke. 1986. Chromosomal mapping of four different integration sites of Moloney murine leukemia virus including the locus for $\alpha 1$ (I) collagen in mouse. Cytogenet. Cell Genet. 43:140–144.

76. Nadeau, H.J., and S.J. Phillips. 1987. The putative oncogene *Pim-1* in the mouse: its linkage and variation among *t* hapolytpes. Genetics 117:533–541.

77. Naylor, S.L., P.A. Lalley, A.Y. Sakaguchi, and P.W. Gray. 1984. Mapping of mouse immune interferon (IFN-gamma) gene to chromosome 10. Cytogenet. Cell Genet. 37:550.

78. Negrini, M., E. Silini, C. Kozak, Y. Tsujimoto, and C. Croce. 1987. Molecular analysis of *mbcl-2*: Structure and expression of the murine gene homologous to the human gene involved in follicular lymphoma. Cell 49:455–463.

79. Nusse, R., J. De Moes, J. Hilkens, and R. Van Nie. 1980. Localization of a gene for expression of mouse mammary tumor virus antigens in the GR/Mtv-2⁻ mouse strain. J. Exp. Med. 152:712–719.

80. Nusse, R., A. Van Ooyen, D. Cox, Y.K.T. Fung, and H. Varmus. 1984. Mode of proviral activation of a putative mammary oncogene (*int-1*) on mouse chromosome 15. Nature 307:131–136.

81. Odaka, T., H. Ikeda, H. Yoshikura, K. Moriwaki, and S. Suzuki. 1981. *Fv-4*: gene controlling resistance to NB-tropic Friend murine leukemia virus. Distribution in wild mice, introduction into genetic background of BALB/c mice, and mapping of chromosomes. J. Natl. Cancer Inst. 67:1123–1127.

82. Pedersen, E.B., S. Haahr, S.C. Mogensen. 1983. X-linked resistance of mice to high doses of Herpes simplex virus type 2 correlates with early interferon production. Infect. Immun. 42:740–746.

83. Peters, G., C. Kozak, and C. Dickson. 1984. Mouse mammary tumor virus integration regions *int-1* and *int-2* map on different mouse chromosomes. Mol. Cell Biol. 4:375–378.

84. Peters, G., M. Placzak, S. Brookes, C. Kozak, R. Smith,

and C. Dickson. 1986. Characterization, chromosome assignment and segregation analysis of endogenous proviral units of mouse mammary tumor virus. J. Virol. 59:535–544.

85. Prakash, O., C. Kozak, and N.H. Sarkar. 1985. Molecular cloning, characterization, and genetic mapping of an endogenous murine mammary tumor virus proviral unit I of C3H/He mice. J. Virol. 54:285–294.

86. Pravtcheva, D.D., F.H. Ruddle, R.W. Ellis, and E.M. Scolnick. 1983. Assignment of murine cellular Harvey *Ras* gene to chromosome 7. Somat. Cell Genet. 9:681–686.

87. Rapp, U.R. and T.H. Marshall. 1980. Cell surface receptors for endogenous mouse type C viral glycoproteins and epidermal growth factor: tissue distribution in vivo and possible participation in specific cell-cell interaction. J. Supramol. Structure 14:343–352.

88. Reeves, R.H., D. Gallahan, B.F. O'Hara, R. Callahan, and J.D. Gearhart. 1987. Genetic mapping of *Prm-1*, *Igl-1*, *Smst*, *Mtv-6*, *Sod-1*, and *Ets-2* and localization of the Down syndrome region on mouse chromosome 16. Cytogenet. Cell Genet. 44:76–81.

89. Robbins, J.M., D. Gallahan, E. Hogg, C. Kozak, and R. Callahan. 1986. An endogenous mouse mammary tumor virus genome common in inbred mouse strains is located on chromosome 6. J. Virol. 57:709–713.

90. Rossomando, A., and D. Meruelo. 1986. Viral sequences are associated with many histocompatibility genes. Immunogenetics 23:233–245.

91. Rotter, V., D. Wolf, D. Pravtcheva, and F.H. Ruddle. 1984. Chromosomal assignment of the murine gene encoding the transformation-related protein p53. Mol. Cell. Biol. 4:383–385.

92. Rowe, W.P., J.W. Hartley, and T. Bremner. 1972. Genetic mapping of a murine leukemia virus-inducing locus of AKR mice. Science 178:860–862.

93. Rowe, W.P., and H. Sato. 1973. Genetic mapping of the *Fv-1* locus of the mouse. Science 180:640–641.

94. Ryan, J., C.P. Hart, and F.H. Ruddle. 1984. Molecular cloning and chromosome assignment of murine *N-ras*. Nucl. Acids Res. 12:6063–6072.

95. Sakaguchi, A.Y., P.A. Lalley, B.U. Zabel, R.W. Ellis, E.M. Scolnick, and S.L. Naylor. 1984. Chromosome assignments of four mouse cellular homologs of sarcoma and leukemia virus oncogenes. Proc. Natl. Acad. Sci. USA. 81:525–529.

96. Schnieke, A., K. Harbers, and R. Jaenisch. 1983. Embryonic lethal mutation in mice induced by retrovirus insertion into the $\alpha 1(I)$ collagen gene. Nature 304:315–320.

97. Silver, J. 1985. A defective ecotropic provirus closely linked to the albino locus. J. Virol. 55:494–496.

98. Silver, J., and Kozak, C. 1986. A common proviral integration region on mouse chromosome 7 in lymphomas and myelogenous leukemias induced by Friend murine leukemia virus (F-MuLV). J. Virol. 57:526–533.

99. Silver, J.S., J.B. Whitney III, C.A. Kozak, G. Hollis,

and I. Kirsch. 1985. *Erbb* is linked to the alpha-globin locus on mouse chromosome 11. Mol. Cell. Biol. 5:1784–1786.

100. Soriano, P., T. Gridley, and R. Jaenisch. 1987. Retroviruses and insertional mutagenesis in mice: proviral integration at the *Mov-34* locus leads to early embryonic death. Genes Dev. 1:366–375.

101. Stewart, C., K. Harbers, D. Jahner, and R. Jaenisch. 1983. X chromosome-linked transmission and expression of retroviral genomes microinjected into mouse zygotes. Science 221:760–762.

102. Stockert, E., H. Sato, K. Itakura, E.A. Boyse, and L.J. Old. 1972. Location of the second gene required for expression of the leukemia-associated mouse antigen G_{IX}. Science 178:862–863.

103. Swan, D., M. Oskarsson, D. Keithley, F.H. Ruddle, P. D'Eustachio, and G.F. Vande Woude. 1982. Chromosomal localization of the Moloney sarcoma virus mouse cellular (*c-mos*) sequence. J. Virol. 44:752–754.

104. Taylor, B.A., L. Rowe, N.A. Jenkins, and N.G. Copeland. 1985. Chromosomal assignment of two endogenous ecotropic murine leukemia virus proviruses of the AKR/J mouse strain. J. Virol. 56:172–175.

105. Traina, V.L., B.A. Taylor, J.C. Cohen. 1981. Genetic mapping of endogenous mouse mammary tumor viruses: locus characterization, segregation, and chromosomal distribution. J. Virol. 40:735–744.

106. Tsichlis, P.N., P.G. Strauss, and C.A. Kozak. 1984. A cellular region involved in the induction of thymic lymphomas (*Mlvi-2*) maps to mouse chromosome 15. Mol. Cell. Biol. 4:997–1000.

107. Van Nie, R., and A.A. Verstraeten. 1975. Studies of genetic transmission of mammary tumour virus by C3Hf mice. Int. J. Cancer 16:922–931.

108. Vijaya, S., D.L. Steffen, C. Kozak, and H.L. Robinson. 1987. *Dis-1*, a region with frequent proviral insertions in Moloney murine leukemia virus-induced rat thymomas. J. Virol. 1164–1170.

109. Villermur, R., Y. Monczak, E. Rassart, C. Kozak, and P. Jolicoeur. 1987. Identification of a new common provirus integration site in Gross passage A murine leukemia virus-induced mouse thymoma DNA. Mol. Cell. Biol. 7:512–522.

110. Villeneuve, L., E. Rassart, P. Jolicoeur, M. Graham, and J.M. Adams. 1986. Proviral integration site *Mis-1* in rat thymuses corresponds to the *pvt-1* translocation breakpoint in murine plasmacytomas. Mol. Cell. Biol. 6:1834–1837.

111. Watson, D.K., M.J. McWilliams-Smith, C. Kozak, R. Reeves, J. Gearhart, M.F. Nunn, W. Nash, J.R. Fowle III, P. Duesberg, T.S. Papas, and S.J. O'Brien. 1986. Conserved chromosomal positions of dual domains of the *ets* proto-oncogene in cats, mice and man. Proc. Natl. Acad. Sci. USA. 83:1792–1796.

112. Wejman, J.C., B.A. Taylor, N.A. Jenkins, and N.G. Copeland. 1984. Endogenous xenotropic murine leukemia virus-related sequences map to chromosomal regions

encoding mouse lymphocyte antigens. J. Virol. 50:237–247.

113. Yetter, R.A., J.W. Hartley, and H.C. Morse III. 1983. *H-2*-linked regulation of xenotropic murine leukemia virus expression. Proc. Natl. Acad. Sci. USA. 80:505–509.

114. Zabel, B.U., R.L. Eddy, P.A. Lalley, J. Scott, G.I. Bell, and T.B. Shows. 1985. Chromosomal locations of the human and mouse genes for precursors of epidermal growth factor and the β subunit of nerve growth factor. Proc. Natl. Acad. Sci. USA. 82:469–473.

115. Zabel, B.U., R.E.K. Fournier, P.A. Lalley, S.L. Naylor, and A.Y. Sakaguchi. 1984. Cellular homologs of the avian erythroblastosis virus *erb*-A and *erb*-B genes are syntenic in mouse but asyntenic in man. Proc. Natl. Acad. Sci. USA. 81:4874–4878.

4 LINKAGE MAP

MURIEL T. DAVISSON and THOMAS H. RODERICK

Assigning genes to chromosomes has been a continuing effort of geneticists and other scientists in a variety of fields since the first established linkage in any vertebrate was discovered in the mouse in 1915. With the advent of special linkage testing stocks, biochemical polymorphisms, DNA polymorphisms, and certain chromosomal rearrangements, and with a greater number of mapped genes and an increasing number of participating workers, the speed of mapping has accelerated rapidly. This linkage map (Fig. 4.1), showing the location of 965 loci, is the result of the cumulative efforts of many investigators in many institutions. The map is maintained in a computer at The Jackson Laboratory and is updated regularly. The programming that draws this map was done by Alan L. Hillyard.

The solid vertical bars depict the relative lengths of the chromosomes based on their proportional lengths (see Table 4.1) and an estimated total haploid length of 1600 centimorgans (cM) including the X chromosome. The centromeres are represented by knobs.

The names of the loci represented by the symbols to the right of the chromosomes can be found in Chapter 2. Nucleolus organizers are symbolized by NO, except on Chr 12 where a ribosomal RNA gene has been mapped using a DNA polymorphism and is symbolized *Rnr12*. The positions of the loci were determined from the amount of recombination in appropriate crosses. The numbers to the left of the chromosomes are recombination percentages and can be taken as a measure of

linear distance along the chromosomes. The map is based on data from classical genetic crosses, including male and female data, and from recombinant inbred strains. The recombination distances are given as distance from the centromere so the approximate distance between any two loci can be obtained by subtracting the value for the more proximal one from that for the more distal one. Distances between centromeres and proximal markers for most chromosomes have been determined using Robertsonian chromosomes and may be underestimated. Genes listed at the bottom of chromosomes have been assigned to specific chromosomes by parasexual methods or are just known to be linked to that chromosome. Please quote the original sources given in Chapter 6 (on recombination frequency) when referring to linkage data.

The data used to determine the position of the loci are given in Table 6.1 on recombination percentages. In many cases the recombinations between adjacent loci were determined directly; in others they were determined by subtraction. Some recombination values are based on very large samples with small sampling errors; others are based on very small samples and thus are much less reliable. In many cases the relative order of loci in the same vicinity is not known. The map therefore should be thought of as a summary of current information with some uncertainties.

We have used several conventions to indicate the relative certainty of the position of a locus and thus its

Table 4.1 Chromosome lengths and assignments of gene associations formerly designated as linkage groups. Chr = chromosome; Length = percentage of total female haploid length; LG = linkage group. The lengths given are an average of measurements made in standard inbred and random bred strains in different laboratories (1). See Table 11.1 for revised estimates

Chr	Length	LG	Chr	Length	LG	Chr	Length	LG	Chr	Length	LG
1	7.20	XIII	6	5.62	XI	11	4.63	VII	16	3.86	–
2	6.88	V	7	5.43	I	12	4.63	–	17	3.74	IX
3	6.10	XVI	8	5.00	XVIII	13	4.38	XIV	18	3.69	XV
4	5.89	VIII	9	4.96	II	14	4.31	III	19	2.65	XIII
5	5.80	XVII	10	4.83	IV,X	15	4.17	VI	X	6.18	XX
									Y	2.72	

value as a potential genetic marker for further mapping studies. The positions of many genes are well known from three-point crosses and extensive data. These 'anchor' loci are indicated by lines extending to the left of the vertical chromosomal bars beyond those of the other loci and the locus symbols are printed in bold-face. Sometimes two loci close together are both potential anchor markers but have not been tested together. In these cases we depicted as an anchor marker the locus which can be more readily tested by many laboratories. When the position of a locus is known with respect to one other locus, the line is drawn to, but not through, the chromosome and the locus with which it has been tested is appended to the symbol in parentheses. This means that the locus could just as easily be on the other side of the marker locus but at that same distance. When a locus is known to be close to another but recombination values are not known, the new locus is placed next to the linked locus but no line is drawn to the chromosome. When more than one locus maps to the same position, the loci are listed on the same line and, if one is an anchor locus, its symbol is listed first. An upward caret means the locus or loci following it belong in the line above. The symbols = > or <= indicate two loci linked closer than 1cM and the arrow points to the more distal locus. An expanded list of loci in the *H-2* to *Tla* region is given beside Chr 17, but maps of fine structure of this region are available elsewhere (2–4). Lines drawn across the expanded segment of chromosome denote loci whose order is well established.

In publishing linkage information, investigators should include the type of cross, the phase (coupling or repulsion) of the genes involved when it is appropriate, the sex of the F1 parent in backcrosses, the number of recombinants, and the total number examined. It is important for authors to indicate whether new information includes data formerly published or given in Mouse News Letter so that the compilers of the map will not utilize the same data twice.

We thank Priscilla W. Lane, Hope O. Sweet, Margaret C. Green, Christine A. Kozak, Peter A. Lalley, Frank H. Ruddle, and Benjamin A. Taylor for their significant help with the map. Maintenance of the map is supported by a contract from the Howard Hughes Medical Institute, grant BSR 84-18828 from the National Science Foundation, and grant CA 34196 from the National Cancer Institute.

References

1. Committee on Standardized Genetic Nomenclature for Mice. 1972. Standard karyotype of the mouse, *Mus musculus*. J. Hered. 63:69–72.
2. Hood, L.E., M. Steinmetz, and B. Malissen. 1983. Genes of the major histocompatibility complex of the mouse. Ann. Rev. Immunol. 1:529–568.
3. Klein, J. 1986. Natural History of the Major Histocompatibility Complex. John Wiley and Sons, New York.
4. Klein, J., F. Figueroa, and Z.A. Nagy. 1983. Genetics of the major histocompatibility complex: the final act. Ann. Rev. Immunol. 1:119–142.

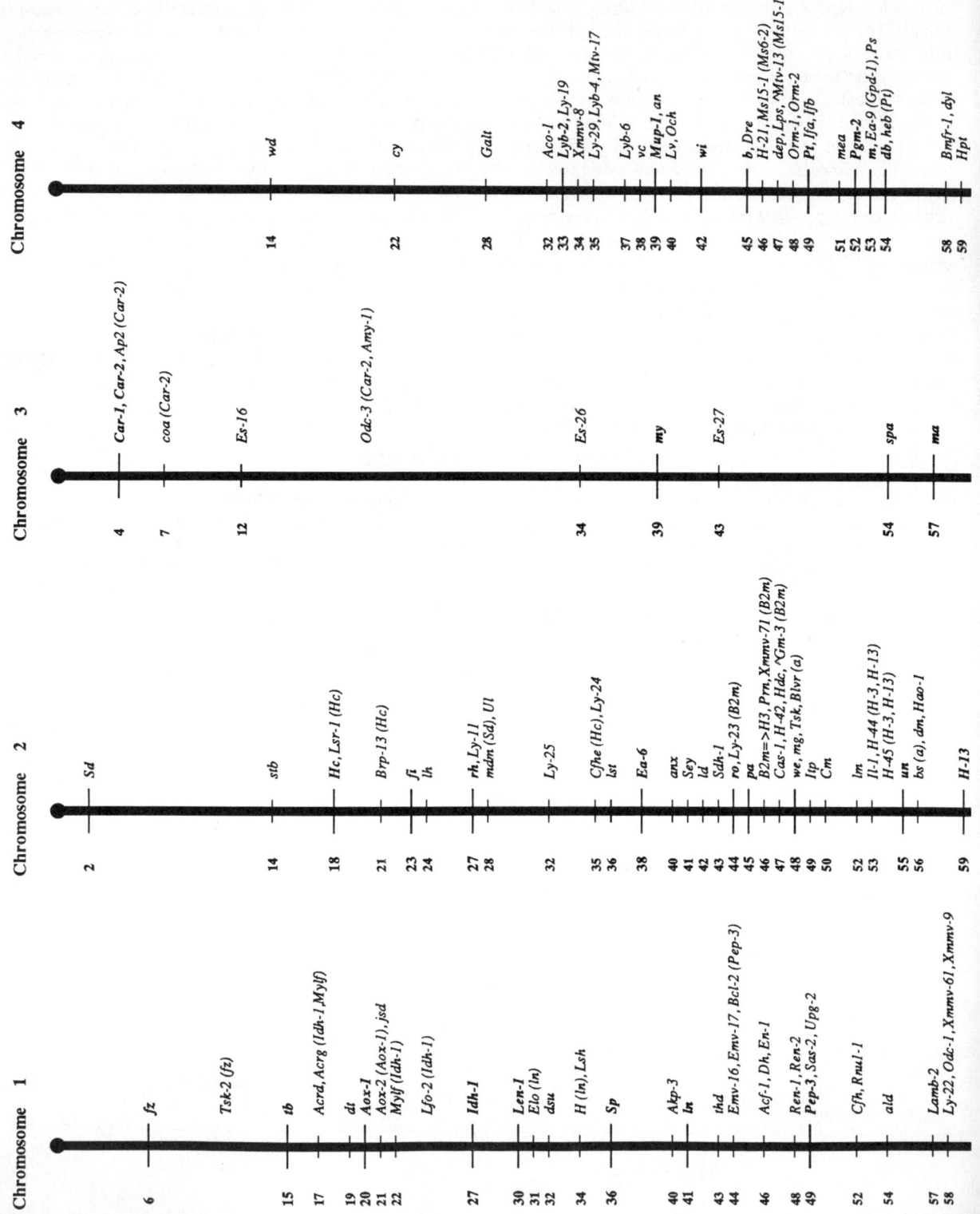

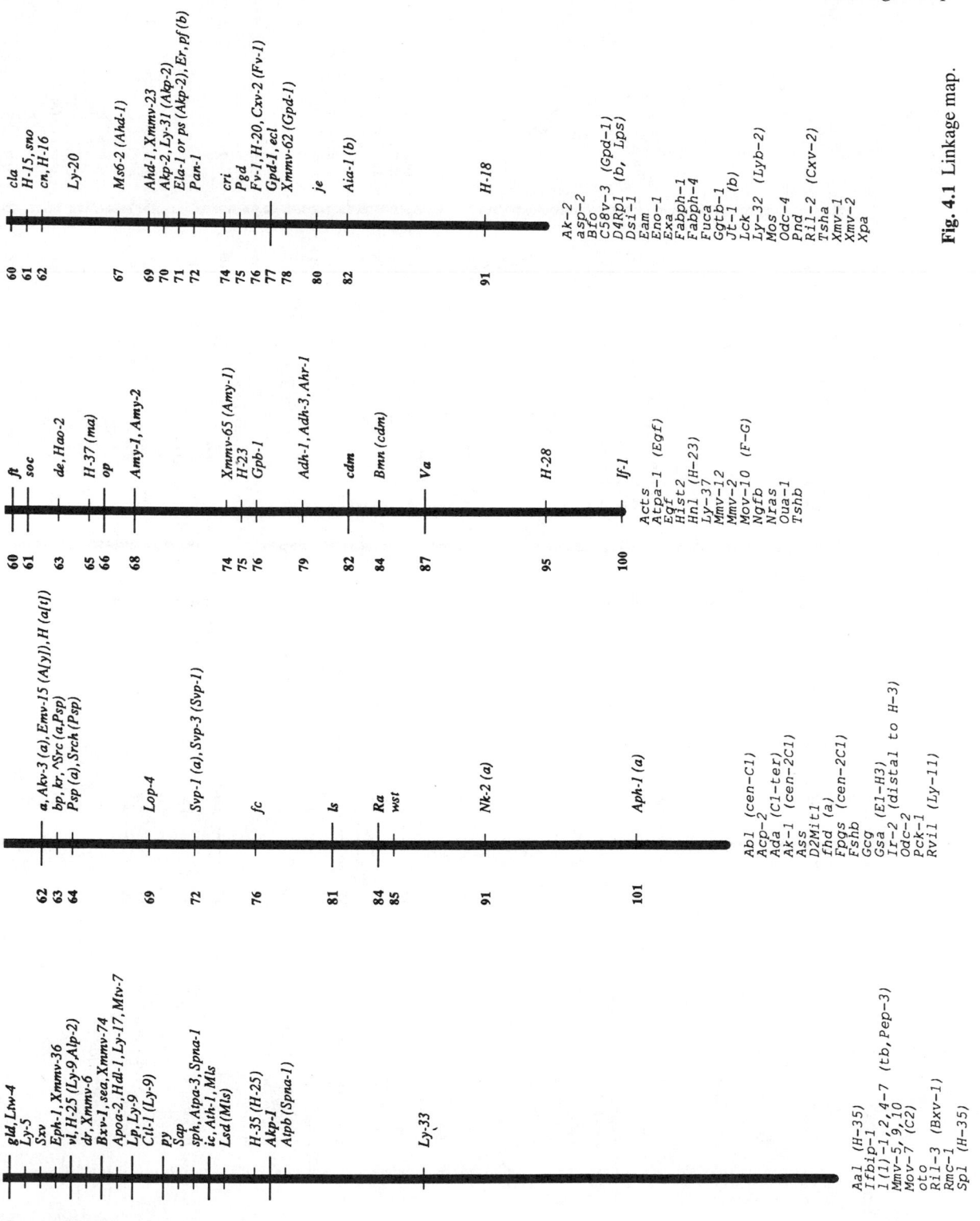

Fig. 4.1 Linkage map.

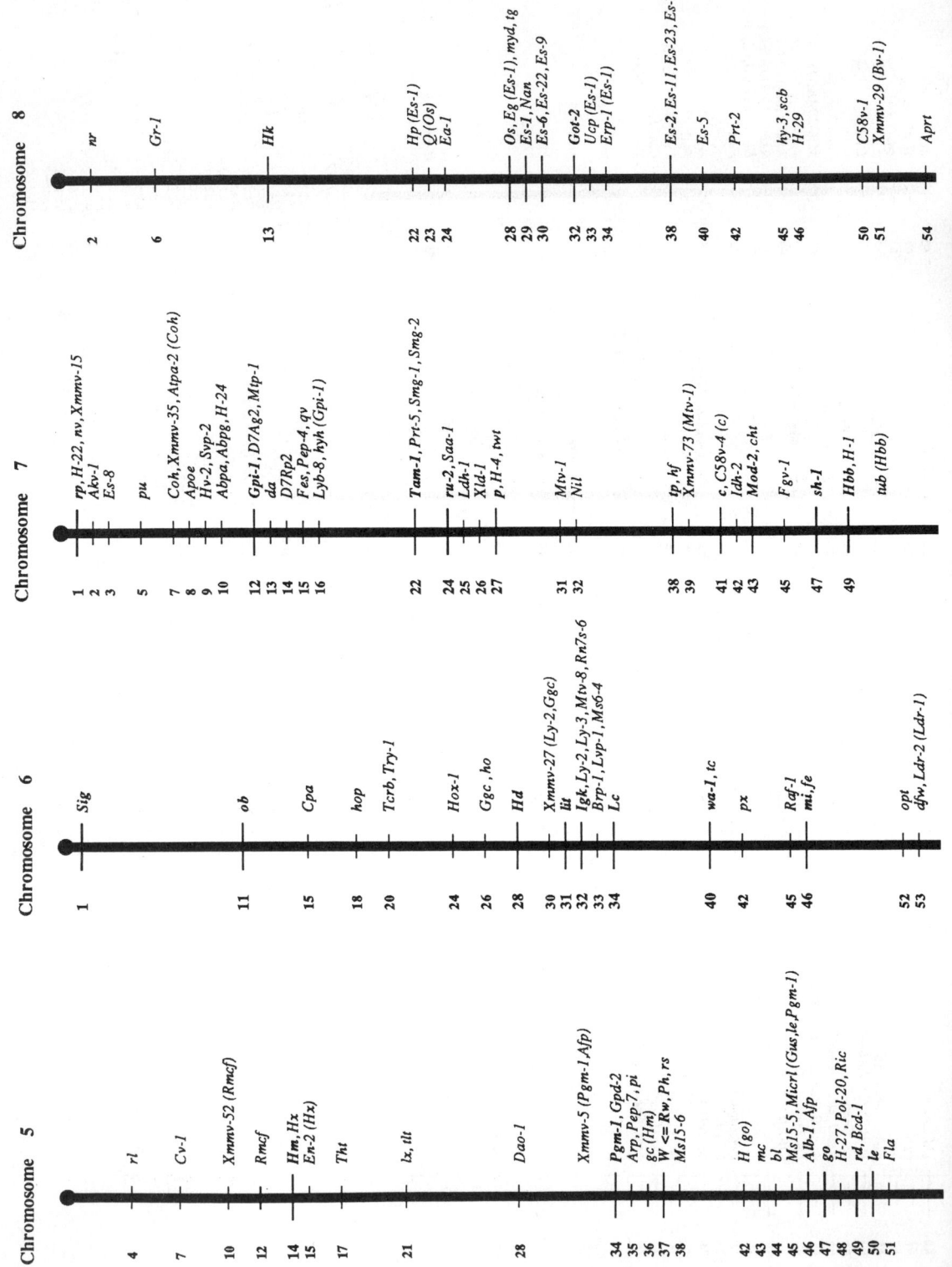

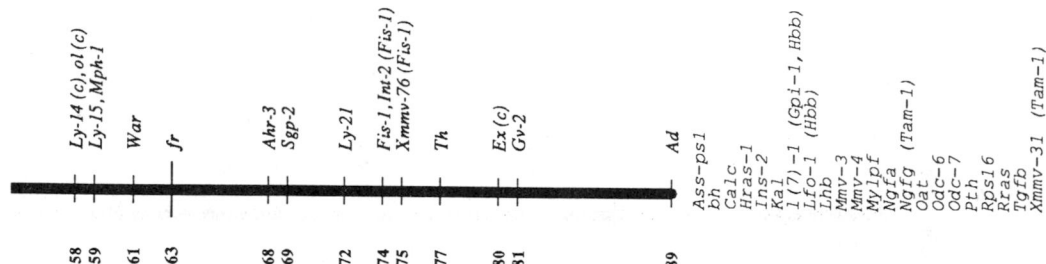

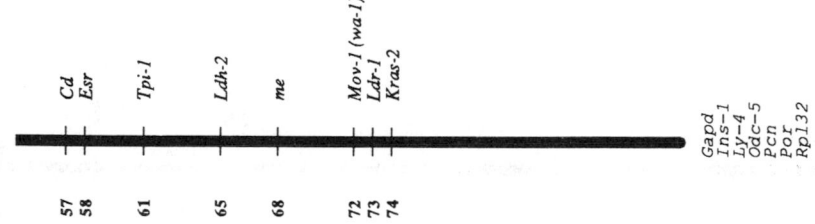

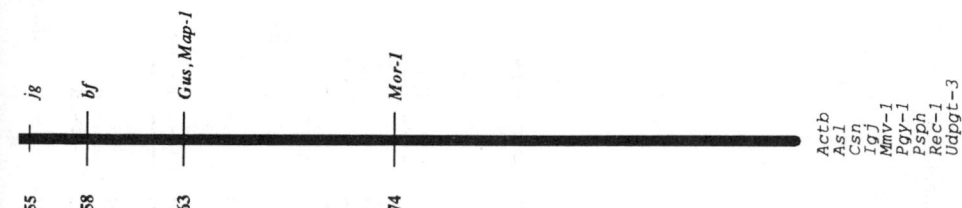

Fig. 4.1 *(Continued)*

421

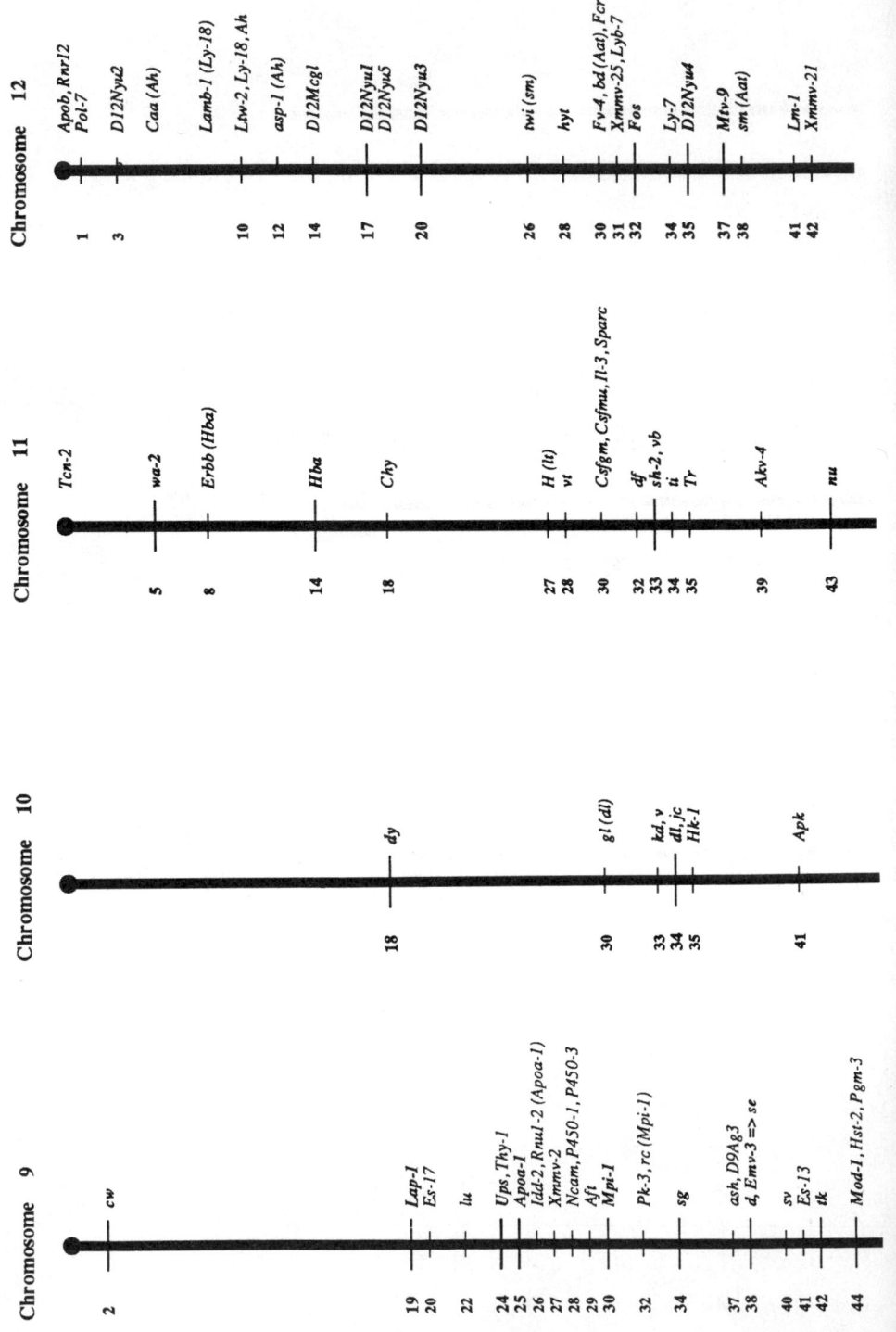

Chromosome 9

2 — *cw*

19 — *Lap-1*
20 — *Es-17*

22 — *lu*

24 — *Ups, Thy-1*
25 — *Apoa-1*
26 — *Idd-2, Rnu1-2 (Apoa-1)*
27 — *Xmmv-2*
28 — *Ncam, P450-1, P450-3*
29 — *Aft*
30 — *Mpi-1*

32 — *Pk-3, rc (Mpi-1)*

34 — *sg*

37 — *ash, D9Ag3*
38 — *d, Emv-3 => se*

40 — *sv*
41 — *Es-13*
42 — *tk*

44 — *Mod-1, Hst-2, Pgm-3*

Chromosome 10

18 — *dy*

30 — *gl (dl)*

33 — *kd, v*
34 — *dl, jc*
35 — *Hk-1*

41 — *Apk*

Chromosome 11

Tcn-2

5 — *wa-2*

8 — *Erbb (Hba)*

14 — *Hba*

18 — *Chy*

27 — *H (lt)*
28 — *vt*

30 — *Csfgm, Csfmu, Il-3, Sparc*

32 — *df*
33 — *sh-2, vb*
34 — *ii*
35 — *Tr*

39 — *Akv-4*

43 — *nu*

Chromosome 12

Apob, Rnr12
Pol-7

1 —
3 — *D12Nyu2*
    Caa (Ah)

    Lamb-1 (Ly-18)

10 — *Lrw-2, Ly-18, Ah*

12 — *asp-1 (Ah)*

14 — *D12Mcg1*

17 — *D12Nyu1*
     D12Nyu5

20 — *D12Nyu3*

26 — *twi (sm)*

28 — *hyt*

30 — *Fv-4, bd (Aat), Fcr*
31 — *Xmmv-25, Lyb-7*
32 — *Fos*

34 — *Ly-7*
35 — *D12Nyu4*

37 — *Mtv-9*
38 — *sm (Aat)*

41 — *Lm-1*
42 — *Xmmv-21*

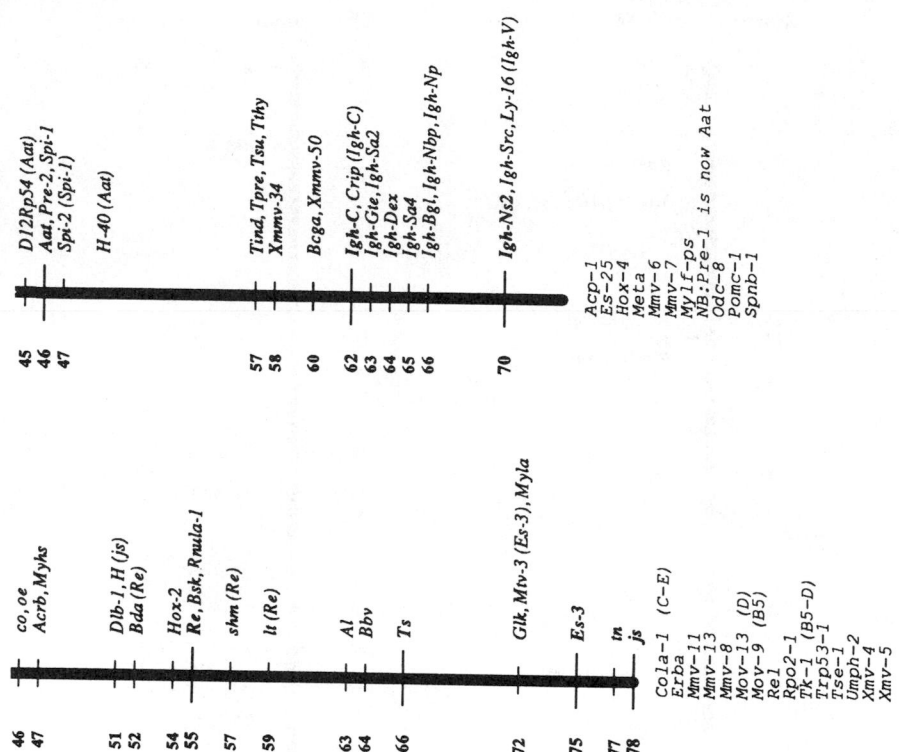

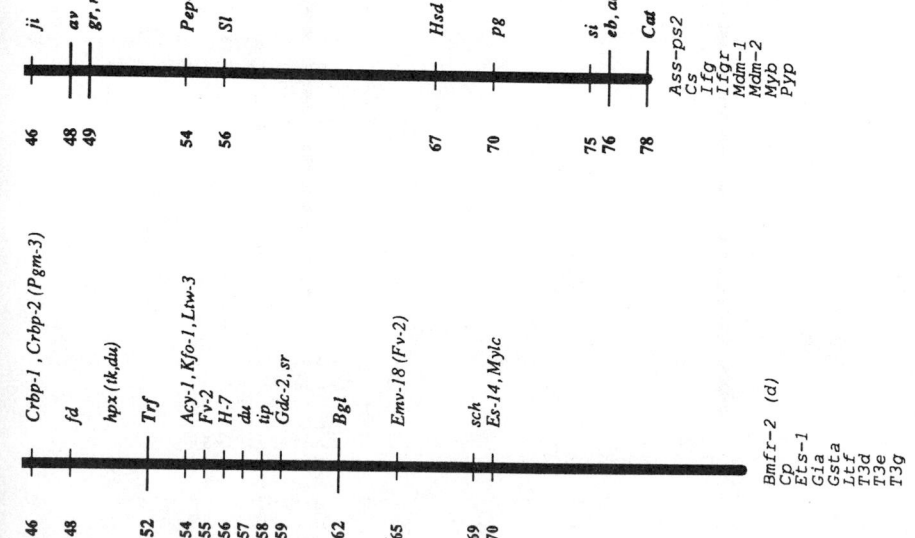

Fig. 4.1 (*Continued*)

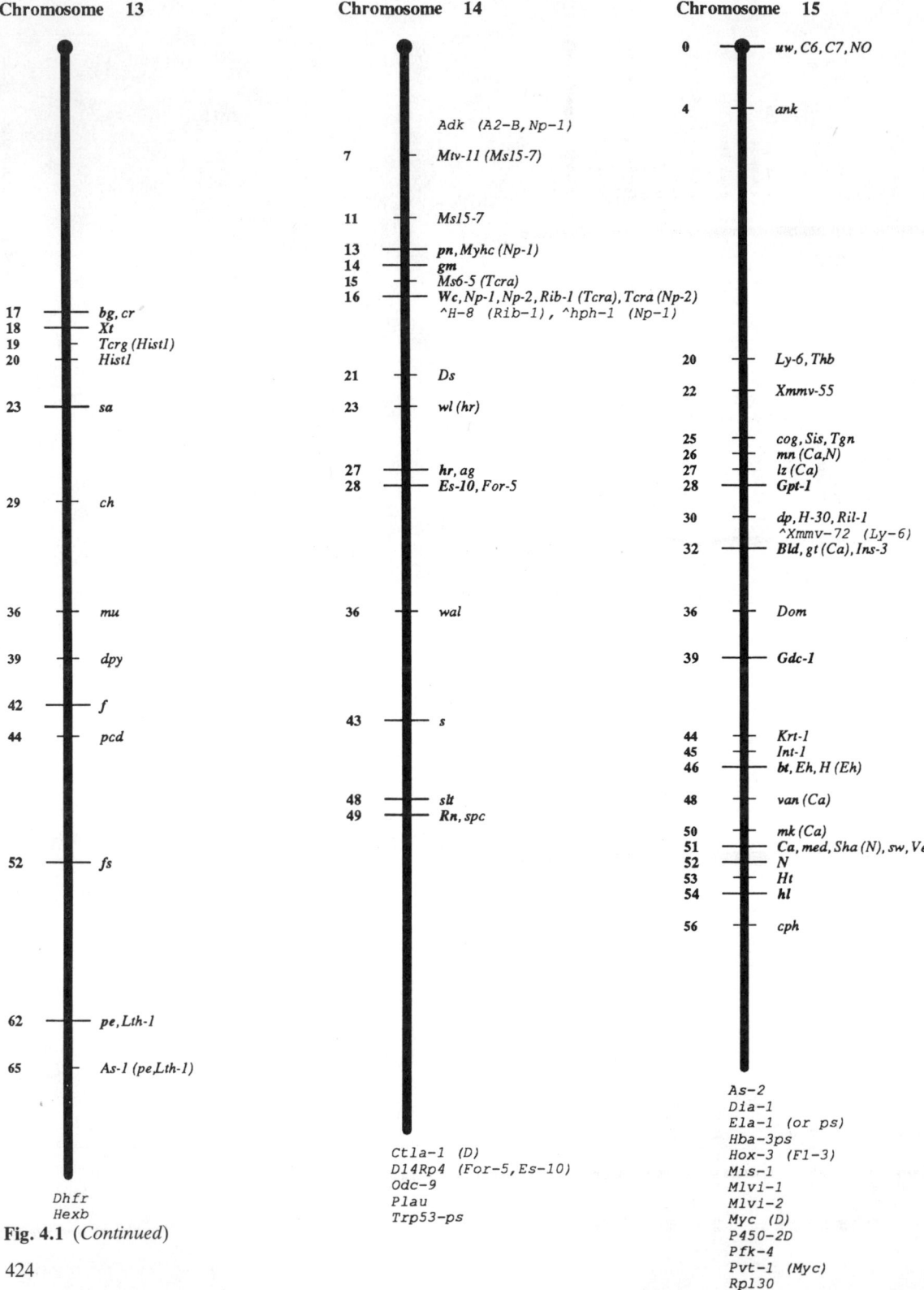

Chromosome 13

17	*bg, cr*
18	*Xt*
19	*Tcrg (Hist1)*
20	*Hist1*
23	*sa*
29	*ch*
36	*mu*
39	*dpy*
42	*f*
44	*pcd*
52	*fs*
62	*pe, Lth-1*
65	*As-1 (pe,Lth-1)*

Dhfr
Hexb

Fig. 4.1 (*Continued*)

Chromosome 14

	Adk (A2–B, Np–1)
7	*Mtv-11 (Ms15-7)*
11	*Ms15-7*
13	*pn, Myhc (Np-1)*
14	*gm*
15	*Ms6-5 (Tcra)*
16	*Wc, Np-1, Np-2, Rib-1 (Tcra), Tcra (Np-2)*
	^H-8 (Rib-1), ^hph-1 (Np-1)
21	*Ds*
23	*wl (hr)*
27	*hr, ag*
28	*Es-10, For-5*
36	*wal*
43	*s*
48	*slt*
49	*Rn, spc*

Ctla-1 (D)
D14Rp4 (For-5, Es-10)
Odc-9
Plau
Trp53-ps

Chromosome 15

0	*uw, C6, C7, NO*
4	*ank*
20	*Ly-6, Thb*
22	*Xmmv-55*
25	*cog, Sis, Tgn*
26	*mn (Ca,N)*
27	*lz (Ca)*
28	*Gpt-1*
30	*dp, H-30, Ril-1*
	^Xmmv-72 (Ly-6)
32	*Bld, gt (Ca), Ins-3*
36	*Dom*
39	*Gdc-1*
44	*Krt-1*
45	*Int-1*
46	*bt, Eh, H (Eh)*
48	*van (Ca)*
50	*mk (Ca)*
51	*Ca, med, Sha (N), sw, Ve*
52	*N*
53	*Ht*
54	*hl*
56	*cph*

As-2
Dia-1
Ela-1 (or ps)
Hba-3ps
Hox-3 (F1-3)
Mis-1
Mlvi-1
Mlvi-2
Myc (D)
P450-2D
Pfk-4
Pvt-1 (Myc)
Rpl30

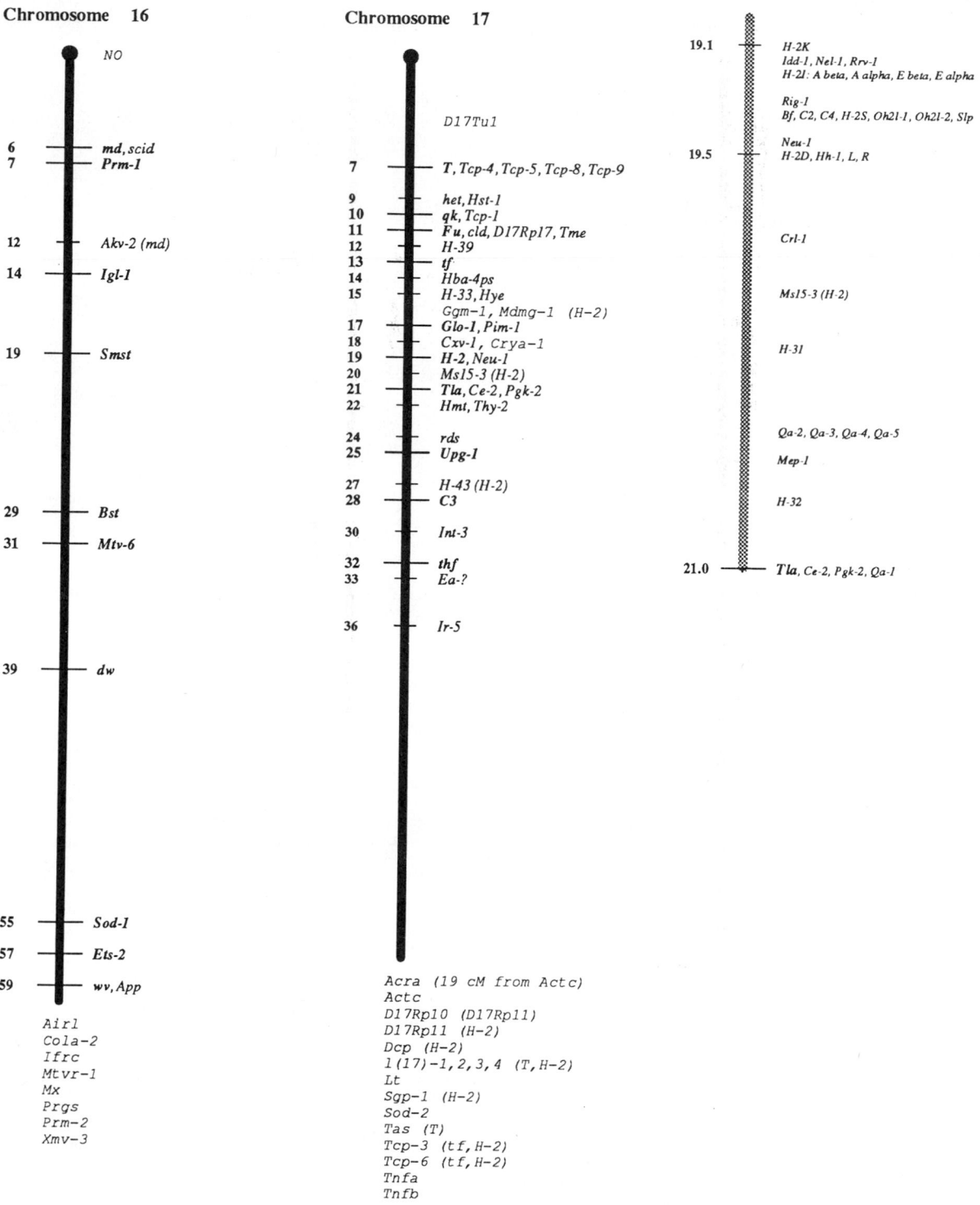

Chromosome 16

NO

6 — md, scid
7 — Prm-1

12 — Akv-2 (md)

14 — Igl-1

19 — Smst

29 — Bst

31 — Mtv-6

39 — dw

55 — Sod-1
57 — Ets-2
59 — wv, App

Airl
Cola-2
Ifrc
Mtvr-1
Mx
Prgs
Prm-2
Xmv-3

Chromosome 17

D17Tu1

7 — T, Tcp-4, Tcp-5, Tcp-8, Tcp-9

9 — het, Hst-1
10 — qk, Tcp-1
11 — Fu, cld, D17Rp17, Tme
12 — H-39
13 — tf
14 — Hba-4ps
15 — H-33, Hye
        Ggm-1, Mdmg-1 (H-2)
17 — Glo-1, Pim-1
18 — Cxv-1, Crya-1
19 — H-2, Neu-1
20 — Ms15-3 (H-2)
21 — Tla, Ce-2, Pgk-2
22 — Hmt, Thy-2

24 — rds
25 — Upg-1

27 — H-43 (H-2)
28 — C3

30 — Int-3

32 — thf
33 — Ea-?

36 — Ir-5

Acra (19 cM from Actc)
Actc
D17Rp10 (D17Rp11)
D17Rp11 (H-2)
Dcp (H-2)
l(17)-1,2,3,4 (T,H-2)
Lt
Sgp-1 (H-2)
Sod-2
Tas (T)
Tcp-3 (tf,H-2)
Tcp-6 (tf,H-2)
Tnfa
Tnfb

19.1 — H-2K
          Idd-1, Nel-1, Rrv-1
          H-2I: A beta, A alpha, E beta, E alpha
          Rig-1
          Bf, C2, C4, H-2S, Oh21-1, Oh21-2, Slp
          Neu-1
19.5 — H-2D, Hh-1, L, R

          Crl-1

          Ms15-3 (H-2)

          H-31

          Qa-2, Qa-3, Qa-4, Qa-5

          Mep-1

          H-32

21.0 — Tla, Ce-2, Pgk-2, Qa-1

Fig. 4.1 (*Continued*)

425

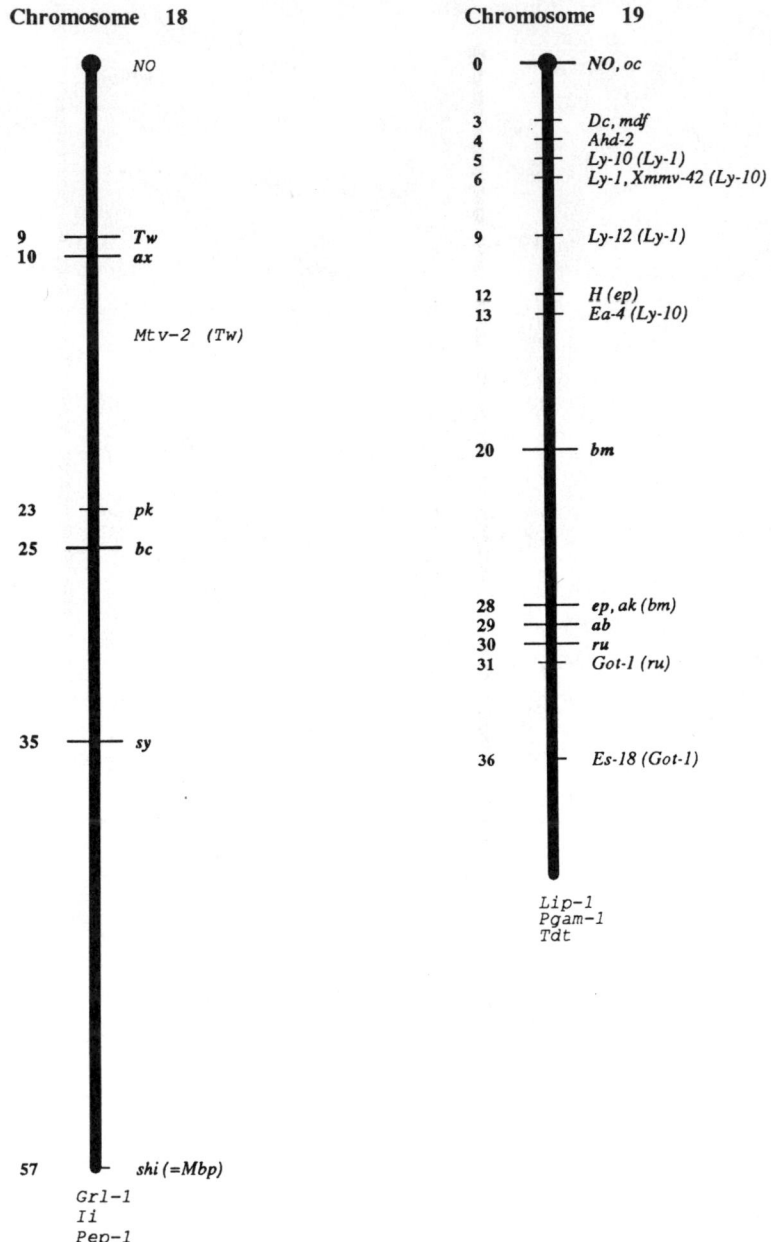

Chromosome 18

NO

9   *Tw*
10   *ax*

Mtv-2 (Tw)

23   *pk*

25   *bc*

35   *sy*

57   *shi (=Mbp)*
Grl-1
Ii
Pep-1

Chromosome 19

0   *NO, oc*

3   *Dc, mdf*
4   *Ahd-2*
5   *Ly-10 (Ly-1)*
6   *Ly-1, Xmmv-42 (Ly-10)*

9   *Ly-12 (Ly-1)*

12   *H (ep)*
13   *Ea-4 (Ly-10)*

20   *bm*

28   *ep, ak (bm)*
29   *ab*
30   *ru*
31   *Got-1 (ru)*

36   *Es-18 (Got-1)*

Lip-1
Pgam-1
Tdt

Fig. 4.1 (*Continued*)

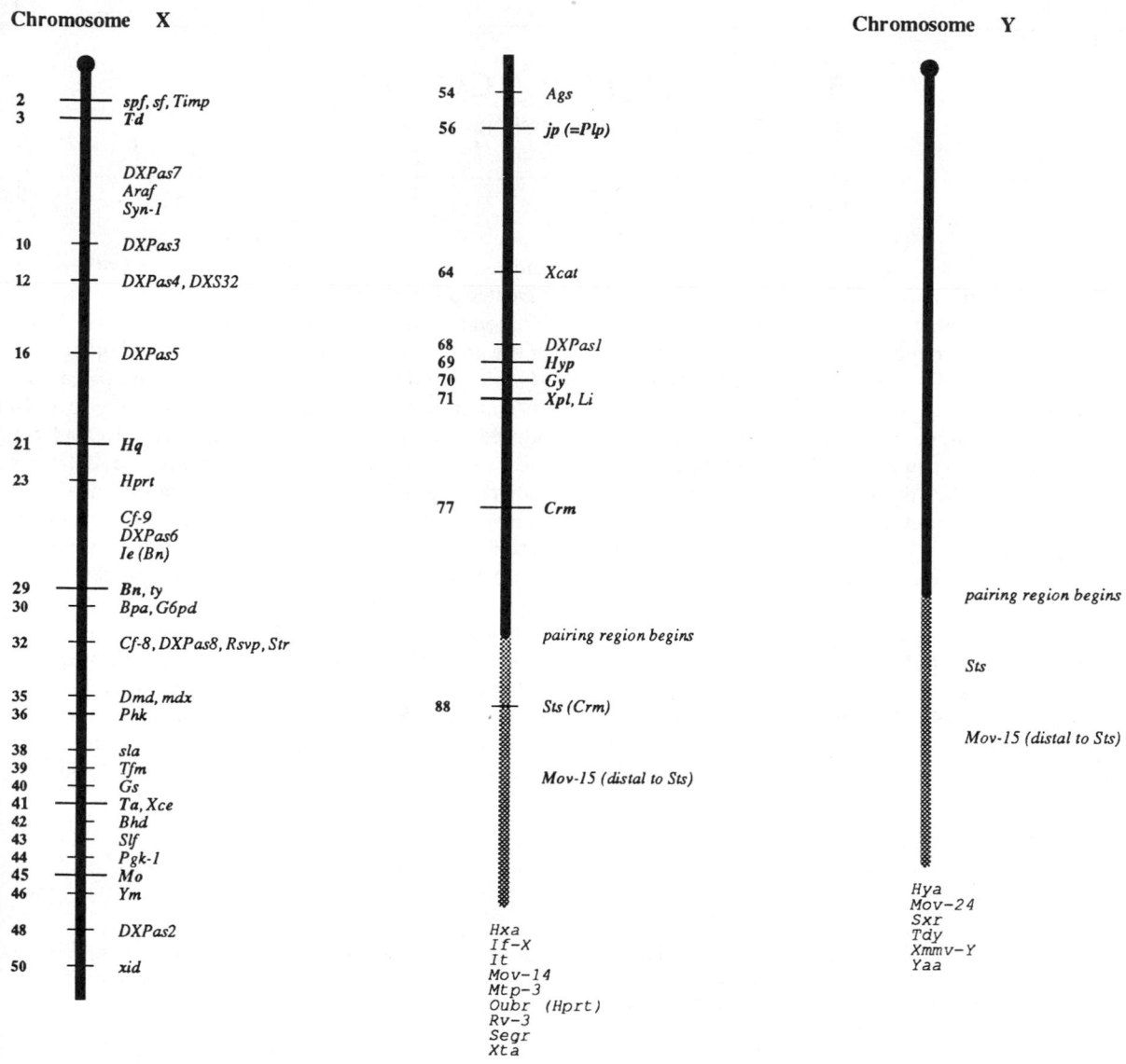

Fig. 4.1 (*Continued*)

5 MAP OF THE *t*-COMPLEX

ANNA-MARIA FRISCHAUF

This chapter shows maps of the relative locations of known genes and DNA markers in the segment of chromosome 17 occupied by the *t*-complex. Knowledge of the locations is derived from various sources. Conventional breeding tests and data from RI strains have been used to map morphological and biochemical markers in wild-type chromatin and information concerning these data can be found in Chapter 6 and Table 5.1. Genomic DNA markers for this segment of chromosome 17 have been obtained by microdissection of the chromosome (17). Many of these clones show RFLPs between *t* and wild-type chromatin. It has been possible to find the order of many of these polymorphic probes on the chromosome (Fig. 5.1) by studying partial *t*-haplotypes, derived by recombination between *t* and wild-type chromatin (7). Some of the partial haplotypes that have yielded this information concerning the proximal half of the region are shown schematically in Fig. 5.2. In turn, a combination of breeding techniques and use of the microdissected probes and others has revealed a variety of chromosomal changes. In particular, complete *t*-haplotypes carry two inversions, shown as the proximal (9) and distal (21) inversions in Fig. 5.1. Hence the order of loci differs between *t* and wild-type chromatin and both orders are shown in Fig. 5.1. In addition, use of the DNA markers has revealed the location of several small duplications and deletions either in partial *t*-haplotypes or in mutant alleles of brachyury, *T*, (Figs. 5.1 and 5.2). Most of these duplications and deletions appear to have arisen by crossing-over within the inversions (3, 10). Other markers in this segment of chromosome 17 are the testis proteins, designated TCP-1 to TCP-9, which were recognized as spots on 2D-gels, and which show polymorphism between *t* and wild-type (Chapter 2). These also are shown in Fig. 5.1. The effects of the *t*-complex on transmission ratio distortion and male sterility are mediated by three (or more) distorter genes, *Tcd-1* to *Tcd-3*, act-

Table 5.1 Recombinant inbred strain (BXD) segregation pattern of chromosome 17 loci

Loci	BXD strains																									
	0	0	0	0	0	0	1	1	1	1	1	1	1	1	2	2	2	2	2	2	2	2	2	3	3	3
	1	2	5	6	8	9	1	2	3	4	5	6	8	9	0	1	2	3	4	5	7	8	9	0	1	2
I	D	D	D	D	B	B	B	D	B	B	B	B	B	B	B	B	D	B	D	B	B	B	D	B	D	D
II	D	B	D	D	B	B	B	D	B	D	B	B	D	B	B	B	D	B	D	B	D	B	D	B	D	D
III	D	B	D	D	B	B	B	D	B	B	B	B	D	B	B	B	D	B	D	B	D	B	D	B	D	D
IV	D	B	D	D	B	B	B	D	B	B	B	B	D	B	B	B	D	B	D	B	D	B	D	B	D	D
V	D	B	D	D	B	B	B	D	B	–	B	B	D	B	B	B	D	B	D	B	D	B	D	B	D	D
VIa	D	B	D	D	B	D	D	D	B	B	B	D	D	B	B	D	D	B	D	D	B	D	B	D	D	D
VIb	D	B	D	D	B	D	D	B	B	B	B	D	D	B	B	D	D	B	D	D	B	D	B	D	D	D
VII	D	B	D	D	B	D	D	D	B	B	B	D	D	B	B	D	D	B	D	D	D	D	B	D	D	D

Loci in group		
I:	D17Leh66E, D17Leh119 (9)	
II:	D17Rp17 (15)	
III:	D17Leh66D (9) Tcp 1 (K. Willison, J. Potter, unpublished)	
IV:	D17Leh180, D17Leh94	
	D17Leh54M, D17Leh443 (3)	
V:	Hba-ps4 (13, 15)	
VIa:	Crya (12, 3, 22) Results in (3) and (22) disagree by one point. Given here as in (3).	
VIb:	Pim-1 (5, 16; F. Michiels, unpublished)	
VII:	H-2S (18), Tla (23, 3)	
	D17Leh525, D17Leh89 (3)	

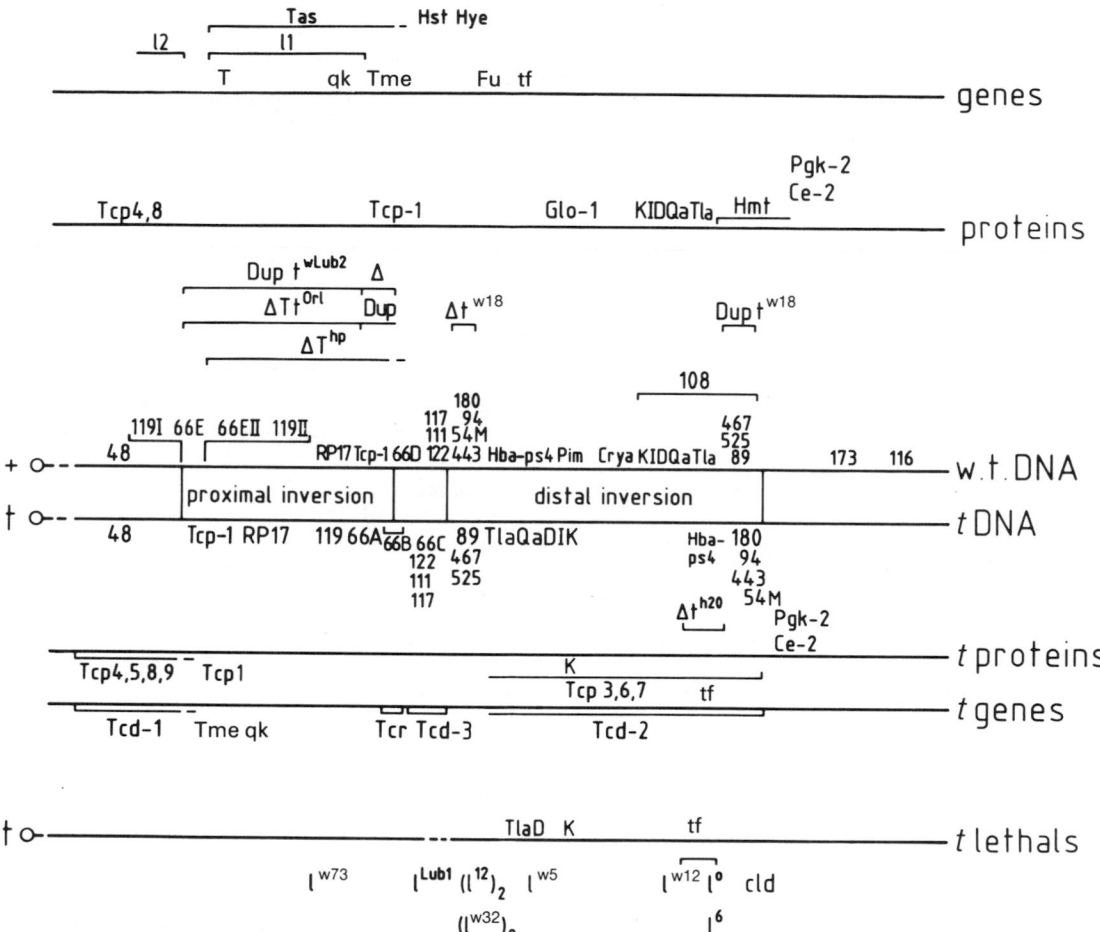

Fig. 5.1 Map of the *t*-complex. For references to mutant genes and proteins see Chapter 2, this book. Probes and mapping of DNA markers: Crya (12, 3, 22, 23), Hba-ps4 (6, 8, 15), Pim (5, 16, and F. Michiels, unpublished), D17Rp17 (RP17 on figure) (15), l(17)-1$_{Wis}$ (11 on figure), l(17)-2$_{Pas}$ (l2 on figure) (20).

DNA markers derived by microcloning (17) are given by numbers only. The full name is D17LehN (N (number) on figure). D17Leh48, D17Leh66A, D17Leh66B, D17Leh66C, D17Leh122, and D17Leh108 on *t* chromosomes (7); D17Leh66E, D17Leh66EII, D17Leh119l, D17Leh119ll, and D17Leh66D (9, 10); D17Leh180, D17Leh94, D17Leh54M, D17Leh443, D17Leh525, D17Leh89, and D17Leh467 (3); D17Leh117, D17Leh111, D17Leh173, and D17Leh116 (Bucan, Herrmann, Mains, Rappold, Silver, and Lehrach, unpublished).

The distances on the map do not reflect genetic distances. Where mapping has been done using partial *t* haplotypes or deletions, the position of a marker has been indicated by brackets. In general an attempt has been made to relate the other markers to the DNA probes, except for the last line (*t* lethals) (1, 21) where, except for the localization of the *H-2* genes with respect to the lethals, correlation to DNA markers is very difficult.

Δ (deletion) Dup (duplication) *Tcd 1, 2, 3* corresponding to *tcs 1, 2, 3* (14) have been indicated by the Tcd names only. Further mutations have been induced by ENU on wild-type (20) and *t* chromosomes (2, 11). The former have not been mapped in detail and are not included on the map, the latter are presumptive alleles of *T*, *qk*, and *tf*. Also, a lethal of the *t°* complementation group has been found on a wild-type chromosome (19).

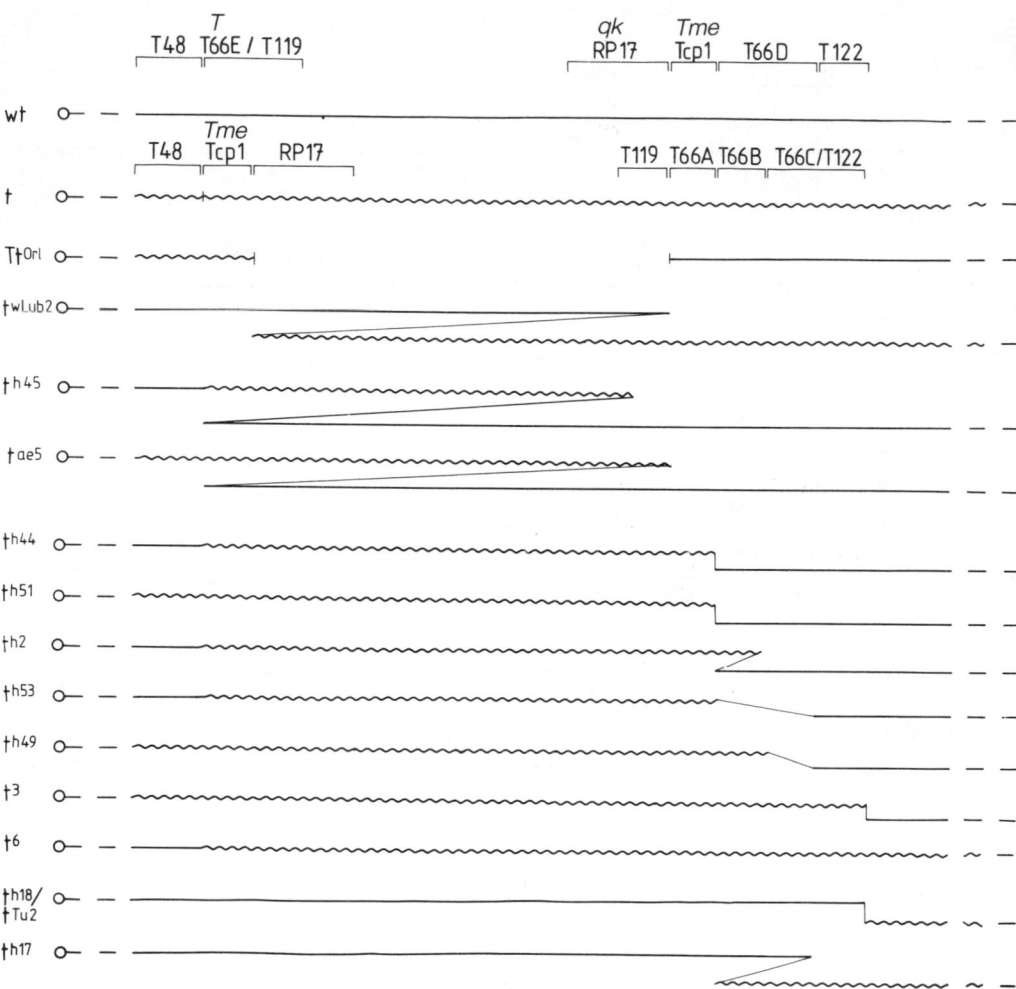

Fig. 5.2 Schematic representation of the structure of wild-type, complete *t*, and partial *t* chromosomes. Wild-type (wt) DNA is represented by a straight line, and *t* haplotype (*t*) DNA, by a wavy line. A broken line indicates an undefined distance. The distance between markers is not drawn to scale. Partial *t* haplotypes are drawn as two connected lines corresponding to their content of wt and *t* DNA as defined by the markers. Redrawn from (9).

ing on a responder, *Tcr* (14). Study of partial *t*-haplotypes, together with the information from DNA markers, has enabled the location of these genes with respect to the inversions (7, 14). Finally, work on recombination within the distal inversion, in female mice carrying two different *t*-haplotypes, has shown that various recessive lethals carried by *t*-haplotypes are non-allelic (1,4). These lethals have been mapped relative to the H-2 complex (21) but not in relation to the other markers, and their approximate positions are indicated on the bottom line of Fig. 5.1.

430

References

1. Artzt, K. 1984. Gene mapping within the T/t complex of the mouse. III: t-lethal genes are arranged in three clusters on chromosome 17. Cell 39:565–572.

2. Bode, V.C. 1984. Ethylnitrosourea mutagenesis and the isolation of mutant alleles for specific genes located in the t region of mouse chromosome 17. Genetics 108:457–470.

3. Bucan, M., B.G. Herrmann, A.M. Frischauf, V. Bautch, V. Boder, L.M. Silver, G.R. Martin, and H. Lehrach. 1987. Deletion and duplication of DNA sequences is associated with the embryonic lethal phenotype of the t^9

complementation group of the mouse *t*-complex. Genes and Development 1:376–385.

4. Condamine, H., J.-L. Guénet, and F. Jacob. 1983. Recombination between two mouse t-haplotypes (t^{w12}tf and t^{Lub-1}): segregation of lethal factors relative to centromere and tufted (tf) locus. Genet. Res., Camb. 42:335–344.

5. Cuypers, H.T., G. Selten, W. Quint, M. Zijlstra, E. Robanus-Maandag, W. Boelens, P. van Wezenbeek, C. Melief, and A. Berns. 1984. Murine leukemia virus-induced T cell lymphomagenesis: integration of proviruses in a distinct chromosomal region. Cell 37:141–150.

6. D'Eustachio, P., B. Fein, J. Michaelson, and B.A. Taylor. 1984. The α-globin pseudogene on mouse chromosome 17 is closely linked to H-2. J. Exp. Med. 159:958–963.

7. Fox, H.S., G.R. Martin, M.F. Lyon, B. Herrmann, A.-M. Frischauf, H. Lehrach, and L.M. Silver. 1985. Molecular probes define different regions of the mouse t-complex. Cell 40:63–69.

8. Fox, H.S., L.M. Silver, and G.R. Martin. 1984. An α-globin pseudogene is located within the mouse t-complex. Immunogenetics 19:125–130.

9. Herrmann, B., M. Bucan, P.E. Mains, A.-M. Frischauf, L.M. Silver, and H. Lehrach. 1986. Genetic analysis of the proximal portion of the mouse t-complex: evidence for a second inversion within t-haplotypes. Cell 44:469–476.

10. Herrmann, B.G., D.P. Barlow, and H. Lehrach. 1987. A large inverted duplication allows homologous recombination between chromosomes heterozygous for the proximal t-complex inversion. Cell 48:813–825.

11. Justice, M.J., and V.C. Bode. 1986. Induction of new mutations in a mouse t-haplotype using ethylnitrosourea mutagenesis. Genet. Res., Camb. 47:187–192.

12. King, C.R., T. Shinohara, and J. Piagtigorsky. 1982. αA-crystallin messenger RNA of the mouse lens: more noncoding than coding sequences. Science 215:985–987.

13. Leder, A., D. Swan, F. Ruddle, P. D'Eustachio, and P. Leder. 1981. Dispersion of α-like globin genes of the mouse to three different chromosomes. Nature 293:196–200.

14. Lyon, M.F. 1986. Male sterility of the mouse t-complex is due to homozygosity of the distorter genes. Cell 44:357–363.

15. Mann, E.A., L.M. Silver, and R.W. Elliott. 1986. Genetic analysis of a mouse t-complex locus that is homologous to a kidney cDNA clone. Genetics 114:993–1006.

16. Nadeau, J.N., and S.J. Phillips. 1987. The putative oncogene Pim-1 in the mouse: Its linkage and variation among *t* haplotypes. Genetics 117:533–541.

17. Röhme, D., H. Fox, B. Herrmann, A.-M. Frischauf, J.-E. Edström, P. Mains, L.M. Silver, and H. Lehrach. 1984. Molecular clones of the mouse t-complex derived from microdissected metaphase chromosomes. Cell 36:783–788.

18. Russell, J.H., C.B. Dobos, R.J. Graff, and B.A. Taylor. 1981. Genetic control of cross-reactive cytotoxic T-lymphocyte responses to a BALB/c tumor. Immunogenetics 14, 263–272.

19. Sanchez, E.R., and C. Hammerberg. 1984. A new t-complex embryonic lethal (tcl^0) has arisen in the T^{hp} chromosome of the mouse and is allelic to t^0. Genet. Res., Camb. 43:27–33.

20. Shedlovsky, A., J.-L. Guénet, L.L. Johnson, and W.F. Dove. 1986. Induction of recessive lethal mutations in the T/t-H-2 region of the mouse genome by a point mutagen. Genet. Res., Camb. 47:135–142.

21. Shin, H.-S., D. Bennett, and K. Artzt. 1984. Gene mapping within the T/t complex of the mouse. IV: The inverted MHC is intermingled with several t-lethal genes. Cell 39:573–578.

22. Skow, L.C., and M.E. Donner. 1985. The locus encoding αA-crystallin is closely linked to H-2K on mouse chromosome 17. Genetics 110:723–732.

23. Winoto, A., M. Steinmetz, and L. Hood. 1983. Genetic mapping in the major histocompatibility complex by restriction enzyme site polymorphisms: Most mouse class l genes map to the Tla complex. Proc. Natl. Acad. Sci. USA 80:3425–3429.

6 RECOMBINATION PERCENTAGES AND CHROMOSOMAL ASSIGNMENTS

MURIEL T. DAVISSON, THOMAS H. RODERICK, and DONALD P. DOOLITTLE

Genetic recombination percentages are the basic data with which the location and ordering of loci are determined in experimental organisms. Accumulation of such information for the mouse was started by Haldane *et al.* (395) with recombination data for pink eye (*p*) and albino (*c*) and has now reached a total of over 4700 independently derived recombination percentages based on over 2000 references, thus probably comprising the most extensive set of data of this kind for any organism. The data are now all contained and updated at The Jackson Laboratory in a computerized data base called 'Locusbase'.

Table 6.1 gives a summary of the recombination percentages used to construct the linkage map in Chapter 4. It includes most of the intervals for which linkage data have been published. Only a few very long intervals, where data existed for several well-defined intervening shorter intervals, have been omitted. The intervals between loci (columns 2 and 3) are arranged primarily in the order in which the more proximal (closer to the centromere) of the two loci appears on the linkage map. When there is more than one interval for a locus, the intervals are arranged in order of increasing recombination distances between the more distal loci and the same proximal locus. In most cases where two symbols have been published for the same locus, only one is used in the table and on the map. These include:

The recombination values shown are the combined estimates from all published data for that interval. Data were combined by weighting each estimate by the reciprocal of its variance (sum of pI/sum of I, where p is the recombination frequency; see Green, p. 64 (352a) or Mather, p. 56 (681a)). Some loci were assigned to chromosomes by use of somatic cell hybrid clones segregating mouse chromosomes or by other parasexual methods. These are given at the end of the list for each chromosome and identified by the word 'Syntenic' in the next-to-last column. The names and descriptions of the loci listed in column 2 are given in Chapter 2. The chromosomes contained in the Robertsonian and reciprocal translocations listed in Table 6.1 are given in Section 12.1.

This work was partially supported by grant BSR8418828 from the National Science Foundation, CA34196 from the National Cancer Institute, and a contract from the Howard Hughes Medical Institute. We thank Carolyn Blake, Linda Fournier, Alan Hillyard, and Moyha Lennon-Pierce for their help in putting the table and bibliography together.

Chr	Symbol used	Other symbols
8	*Bv-1*	*Emv-2, Inb-1*
5	*Cv-1*	*Emv-1, Inc-1*
1	*Lsh*	*Ity, Bcg*
12	*Aat*	*Pre-1*
1	*Cfh*	*Sas-1*
7	*Abpa*	*Sal-1, Abp*
7	*Emv-11*	*Akv-1*
16	*Emv-12*	*Akv-2*
2	*Emv-13*	*Akv-3*
11	*Emv-14*	*Akv-4*

Table 6.1 Summary of recombination data, given as per cent recombination ± the standard error, and chromosomal assignments

Chr	Loci Proximal	Distal	Female	Male	Combined or not stated	References
1	Hc1*	fz	6.00 ± 3.36			177
1	Hc1	ln	44.90 ± 7.10			177
1	Rb1Bnr	fz			1.00 ± 0.58	175
1	fz	tb			8.62 ± 3.81	977
1	fz	dt			19.74 ± 3.54	516, 597, 275
1	fz	Idh-1	21.40 ± 5.50	21.33 ± 2.42		1035, 864
1	fz	Lsh	26.80 ± 5.92			1035
1	fz	H(ln)			30.43 ± 9.59	20
1	fz	Sp	40.02 ± 0.85	32.66 ± 0.82		960, 208, 791
1	fz	ln	42.81 ± 0.81	36.47 ± 0.60	35.08 ± 1.96	516, 597, 275, 206, 960, 177, 200, 914, 791, 864
1	fz	Pep-3		30.77 ± 2.04		864, 861
1	fz	vl			50.95 ± 3.56	206
1	fz	Lp	50.00 ± 7.72	53.61 ± 3.58		960
1	fz	py	49.31 ± 0.94	49.40 ± 0.99		791
1	fz	Tsk-2			Linked	803
1	tb	Sp		27.18 ± 4.38		199
1	tb	ln			26.93 ± 3.57	977
1	tb	Lp			52.39 ± 3.64	977
1	Acrd	Acrg	≤7†			407
1	Acrd	Mylf	4.88 ± 3.36			407
1	Acrd	Idh-1	11.11 ± 5.24			407
1	Acrg	Mylf	4.76 ± 4.53			407
1	Acrg	Idh-1	10.81 ± 5.10			407
1	dt	ln			34.11 ± 3.06	516, 597, 275
1	Aox-1	Aox-2	0.70 ± 0.30			463, 458
1	Aox-1	Idh-1	8.06 ± 3.46		6.43 ± 1.55	463, 1088
1	Aox-1	Pep-3	30.12 ± 5.04		28.14 ± 3.48	463, 1088
1	jsd	Idh-1			4.00 ± 2.00	36
1	jsd	Pep-3			13.00 ± 2.00	36
1	dsu	Idh-1	6.25 ± 3.49		7.30 ± 3.50	1008, 714
1	dsu	Emv-17			14.20 ± 4.68	714
1	Mylf	Idh-1	5.41 ± 3.72		2.38 ± 1.84	857
1	Lfo-2	Idh-1			Near	260
1	Idh-1	Len-1	1.40 ± 1.40	5.40 ± 2.10		951
1	Idh-1	Lsh	8.90 ± 3.00		11.46 ± 3.59	1035, 77
1	Idh-1	Sp	9.09 ± 5.00			483
1	Idh-1	Akp-3		13.80 ± 3.10		1105
1	Idh-1	ln		12.94 ± 1.98	15.51 ± 4.99	864, 77, 483
1	Idh-1	Lsd			16.70 ± 9.64	769
1	Idh-1	Emv-17			17.44 ± 3.74	87, 714
1	Idh-1	En-1			19.15 ± 5.70	87
1	Idh-1	Acf-1		19.30 ± 5.20		1099
1	Idh-1	Pep-3	27.49 ± 1.79	20.89 ± 1.54	23.31 ± 1.81	866, 951, 130, 868, 68, 1099, 1105, 864, 898, 544
1	Idh-1	Cfh	25.00 ± 7.65		21.43 ± 12.91	866
1	Idh-1	Sas-2			24.04 ± 4.19	399
1	Idh-1	gld	39.00 ± 3.45			868
1	Idh-1	Bxv-1			38.01 ± 3.96	544
1	Idh-1	Mtv-7	5.80 ± 1.23			663
1	Idh-1	sph	46.20 ± 3.55			68
1	Idh-1	Mls			31.92 ± 6.01	898, 290, 769
1	Elo	ln	9.83 ± 0.93	9.70 ± 1.03		764
1	Len-1	Pep-3	25.30 ± 5.20	18.00 ± 5.20		951
1	Lsh	ln	10.70 ± 4.10		12.00 ± 10.00	1035, 77
1	Lsh	Pep-3	26.80 ± 5.92		20.00 ± 8.77	1035, 77, 947

Table 6.1—*cont.*

Chr	Loci		Female	Male	Combined or not stated	References
	Proximal	Distal				
1	*Lsh*	*Rnu1-1*			10.00	400
1	*Bcg*	*Lsh*			0	947
1	*Bcg*	*Ity*			4.00 ± 3.92	947
1	*H(ln)*	*ln*			8.69 ± 5.87	20
1	*Sp*	*ln*	7.07 ± 0.41	4.46 ± 0.35		960, 208, 607, 791
1	*Sp*	*py*	40.39 ± 0.92	27.59 ± 0.88		791
1	*Akp-3*	*Pep-3*		8.90 ± 2.60		1105
1	*ln*	*Emv-17*			3.62 ± 1.59	87
1	*ln*	*thd*	2.50 ± 0.68	1.32 ± 0.65		607
1	*ln*	*Dh*	4.71 ± 0.75	4.84 ± 0.41	1.94 ± 1.36	655, 914
1	*ln*	*Ren-1*			10.87 ± 2.65	87, 1021
1	*ln*	*Ren-2*			25.01 ± 15.31	1106
1	*ln*	*Upg-2*			33.31 ± 26.32	1021
1	*ln*	*Pep-3*	19.64 ± 5.29	6.97 ± 1.42	22.32 ± 6.81	1035, 864, 1021, 1034, 1106
1	*ln*	*ald*			26.55 ± 8.14	1034
1	*ln*	*Sxv*		25.00 ± 7.60	24.00 ± 10.90	549
1	*ln*	*vl*			15.52 ± 4.37	206
1	*ln*	*dr*			24.27 ± 4.22	655
1	*ln*	*Lp*	38.09 ± 7.49	35.05 ± 3.43		960
1	*ln*	*py*	37.89 ± 0.91	23.55 ± 0.84		791
1	*ln*	*En-1*			5.07 ± 1.87	87
1	*ln*	*Emv-16,17*			Near	498
1	*Emv-16*	*Emv-17*		≤0.7†		498
1	*Emv-16*	*En-1*			0.27 ± 0.27	508
1	*Emv-16*	*Ren-1*			5.87 ± 1.27	508
1	*Emv-17*	*Ren-1*			7.25 ± 2.21	87
1	*Emv-17*	*Pep-3*			7.13 ± 2.55	87
1	*Emv-17*	*Rnu1-1*			9.18 ± 4.15	87
1	*Emv-17*	*En-1*			1.60 ± 0.92	87
1	*En-1*	*Dh*			0.27 ± 0.27	508
1	*En-1*	*Ren-1,2*			2.38 ± 1.84	423
1	*En-1*	*Ren-1*			3.96 ± 1.35	87, 423
1	*En-1*	*Pep-3*			10.64 ± 4.50	87
1	*En-1*	*Lamb-2*			12.96 ± 7.33	423
1	*Acf-1*	*Pep-3*		3.15 ± 1.45		1099
1	*Acf-1*	*Lp*		20.69 ± 4.34		1099
1	*Dh*	*Pep-3*			3.80	447
1	*Dh*	*dr*			22.33 ± 4.10	655
1	*Ren-1*	*ald*			8.33 ± 5.51	1021, 1106
1	*Ren-1*	*Ren-2*			0	1106
1	*Ren-1*	*Upg-2*			≤2†	1021
1	*Ren-1*	*Pep-3*			0.51 ± 0.52	1106
1	*Ren-1*	*Ltw-4*			12.16 ± 4.20	264
1	*Ren-2*	*ald*			5.00 ± 2.99	1106
1	*Ren-1,2*	*Pep-3*			0.76 ± 0.55	1108
1	*Upg-2*	*Pep-3*			≤2†	1021
1	*Upg-2*	*ald*			8.33 ± 5.51	1021
1	*Pep-3*	*Bcl-2*			4.55 ± 3.09	712
1	*Pep-3*	*Rnu1-1*			3.49 ± 2.29	400, 87
1	*Pep-3*	*Cfh*	3.13 ± 3.08		0.59 ± 0.60	866, 1021
1	*Pep-3*	*ald*			9.52 ± 6.41	1048, 1034
1	*Pep-3*	*Lamb-2*			6.16 ± 2.55	261
1	*Pep-3*	*Ly-22*			7.70 ± 3.79	1027
1	*Pep-3*	*Odc-1*			8.20 ± 5.43	261
1	*Pep-3*	*gld*	9.42 ± 1.65		12.83 ± 1.92	868
1	*Pep-3*	*Ltw-4*			12.20 ± 4.22	264
1	*Pep-3*	*Ly-5*			5.00	903

Table 6.1—*cont.*

Chr	Loci Proximal	Distal	Female	Male	Combined or not stated	References
1	*Pep-3*	*Sxv*		15.79 ± 3.92	11.00 ± 1.70	549
1	*Pep-3*	*dr*			19.10 ± 4.80	1080
1	*Pep-3*	*Bxv-1*			18.09 ± 1.53	544, 538, 539
1	*Pep-3*	*Ly-17*			26.47 ± 4.81	898, 539
1	*Pep-3*	*Mtv-7*	23.02 ± 3.97		7.90 ± 5.96	663, 705
1	*Pep-3*	*Lp*	20.40 ± 4.07	18.40 ± 4.20	22.00 ± 2.87	868, 1099, 1100
1	*Pep-3*	*Ly-9*			17.20 ± 2.80	539
1	*Pep-3*	*sph*	22.80 ± 3.00			68
1	*Pep-3*	*Mls*			21.29 ± 4.48	898, 290
1	*Pep-3*	*Akp-1*			25.13 ± 4.18	1100
1	*Pep-3*	*Rmc-1*			Syntenic	548
1	*Pep-3*	*Sas-2*			0	399
1	*Sas-2*	*Akp-1*			26.83 ± 4.89	399
1	*Cfh*	*Sas-1*			≤1.6†	164
1	*Cfh*	*Lamb-2*			8.80 ± 4.82	261
1	*Cfh*	*Ly-22*			8.80 ± 4.82	1027
1	*Cfh*	*Ltw-4*			5.40 ± 2.66	264
1	*Cfh*	*Eph-1*			12.96 ± 7.33	631
1	*Cfh*	*Ly-17*	12.50 ± 3.90		25.28 ± 4.66	899, 209
1	*Cfh*	*Xmmv-74*			12.00 ± 6.63	702
1	*Cfh*	*Mls*	19.40 ± 4.70		17.13 ± 4.44	899, 209, 631, 866
1	*ald*	*Ltw-4*			7.10 ± 4.54	264
1	*ald*	*Mls*			7.00 ± 4.48	290
1	*Lamb-2*	*ald*			8.20 ± 5.43	261
1	*Lamb-2*	*Ly-22*			5.00 ± 2.99	261
1	*Lamb-2*	*Odc-1*			0.70 ± 0.72	261
1	*Lamb-2*	*Xmmv-61*			7.70 ± 5.00	261
1	*Ly-22*	*Xmmv-9*			≤1†	1091
1	*Ly-22*	*Ltw-4*			2.70 ± 1.76	1027
1	*Ly-22*	*Eph-1*			5.00 ± 2.99	1027
1	*Ly-22*	*Ly-17*			5.93 ± 2.03	526
1	*Ly-22*	*Mtv-7*			7.90 ± 5.96	705
1	*Ly-22*	*Ly-33*			24.44 ± 3.70	526
1	*Odc-1*	*ald*			8.20 ± 5.43	261
1	*Odc-1*	*Xmmv-61*			5.20 ± 3.61	261
1	*Ath-1*	*Ltw-4*			10.17 ± 4.81	785
1	*Ath-1*	*Apoa-2*			4.59 ± 1.34	782, 785
1	*Ath-1*	*Hdl-1*			6.00 ± 3.02	785
1	*Ath-1*	*Ly-9*			6.87 ± 3.37	785
1	*Ath-1*	*Akp-1*			8.30 ± 5.49	785
1	*Ath-1*	*H-25*			12.45 ± 5.83	785
1	*Ath-1*	*Akp-1*			6.00 ± 4.28	785
1	*gld*	*Lp*	5.08 ± 22.00		14.00 ± 3.50	868
1	*Ltw-4*	*Eph-1*			3.90 ± 2.58	631
1	*Ltw-4*	*Xmmv-6*			5.00	62
1	*Ltw-4*	*Xmmv-74*			6.52 ± 3.34	702
1	*Ltw-4*	*Apoa-2*			10.80 ± 6.85	629
1	*Ltw-4*	*Hdl-1*			10.80 ± 6.85	629
1	*Ltw-4*	*Mls*			4.30 ± 2.05	264
1	*Ly-5*	*Spna-1*			10.71 ± 5.85	918
1	*Sxv*	*Bxv-1*		4.00 ± 2.30		549
1	*Xmmv-36*	*Eph-1*			≤3.4†	62
1	*Xmmv-36*	*Xmmv-6*			1.00	62
1	*Xmmv-36*	*Mtv-7*			2.00	62
1	*Eph-1*	*Xmmv-74*			6.00 ± 4.15	702
1	*Eph-1*	*Mtv-7*			1.00 ± 1.05	1057
1	*Eph-1*	*Mls*			3.90 ± 2.58	631

Table 6.1—*cont.*

Chr	Loci Proximal	Loci Distal	Female	Male	Combined or not stated	References
1	*vl*	*Lp*			4.24 ± 0.96	368, 372
1	*vl*	*sea*			2.42 ± 1.04	1003
1	*Xmmv-6*	*Eph-1*			1.00	62
1	*Xmmv-6*	*Mtv-7*			2.00	62
1	*Xmmv-6*	*Mls*			6.00	62
1	*dr*	*Akp-1*			13.20 ± 4.10	1080
1	*Bxv-1*	*Ly-17*			3.40 ± 1.30	539
1	*Bxv-1*	*Lp*			5.00 ± 3.00	544
1	*Bxv-1*	*Ly-9*			3.30 ± 1.10	539
1	*Bxv-1*	*Ril-3*			Near	699
1	*Ly-17*	*Apoa-2*			≤8†	629
1	*Ly-17*	*Hdl-1*			≤4.7†	629
1	*Ly-17*	*Mtv-7*			4.60 ± 3.82	705
1	*Ly-17*	*Spna-1*			4.24 ± 1.56	918
1	*Ly-17*	*Ly-9*			0.89 ± 0.46	171, 539
1	*Ly-17*	*Mls*	6.90 ± 3.90		5.42 ± 2.19	899, 209, 898
1	*Ly-17*	*Mls*			Close	529
1	*Ly-17*	*Ly-33*			20.00 ± 3.44	526
1	*Apoa-2*	*Ath-1*			4.67 ± 1.72	784
1	*Apoa-2*	*Hdl-1*			0	629, 785
1	*Apoa-2*	*Ly-9*			3.10 ± 2.47	629
1	*Apoa-2*	*H-25*			4.50 ± 5.32	629
1	*Hdl-1*	*Ly-9*			3.30 ± 1.87	785, 629
1	*Hdl-1*	*Akp-1*			8.30 ± 5.49	785
1	*Hdl-1*	*H-25*			6.34 ± 3.82	785, 629
1	*Mtv-7*	*Ly-9*			2.40 ± 1.85	1057
1	*Mtv-7*	*Sap*			3.80 ± 2.53	717
1	*Mtv-7*	*Mls*			2.40 ± 1.85	1057
1	*Mtv-7*	*Akp-1*			4.60 ± 3.82	705
1	*Lp*	*sea*			≤3.2†	1003
1	*Lp*	*ic*	2.97 ± 1.69	4.60 ± 2.04	5.79 ± 1.29	368, 1001
1	*Lp*	*Akp-1*			11.40 ± 3.10	1100
1	*Ly-9*	*Sap*			≤3.2†	717
1	*Ly-9*	*Spna-1*			0.91 ± 0.94	918
1	*Ly-9*	*Mls*			3.20 ± 2.08	611
1	*Ly-9*	*Ctl-1*			9.90 ± 5.27	170
1	*Ly-9*	*H-25*			4.54 ± 5.35	433, 229
1	*Sap*	*Mls*			2.40 ± 1.85	717
1	*Spna-1*	*Akp-1*			6.25 ± 2.24	918
1	*Spna-1*	*Atpa-3*			0	518
1	*Spna-1*	*Atpb*			6.45 ± 4.41	518
1	*Mls*	*Akp-1*			3.60 ± 2.90	1100
1	*Lsd*	*Mls*			1.00 ± 1.05	769
1	*H-35*	*Aal*			Close	768
1	*H-35*	*Spl*			Close	255
1	*H-25*	*H-35*			12.00	25
1	*oto*	*tb,Pep-3*			Near In(1)1Rk	510
1	*l(1)-1*	*tb,Pep-3*			Near In(1)1Rk	865
1	*l(1)-2*	*tb,Pep-3*			In In(1)1Rk	865
1	*l(1)-4*	*tb,Pep-3*			In In(1)1Rk	865
1	*l(1)-5*	*tb,Pep-3*			In In(1)1Rk	865
1	*l(1)-6*	*tb,Pep-3*			In In(1)1Rk	865
1	*l(1)-7*	*tb,Pep-3*			In In(1)1Rk	865
1	Chr 1	*Mal-1*			Linked	624
1	Chr 1	*Atpb*			Syntenic	518
1	Chr 1	*Spna-1*			Syntenic	472
1	Chr 1	*Atpa-3*			Syntenic	518

Table 6.1—*cont.*

Chr	Loci		Female	Male	Combined or not stated	References
	Proximal	Distal				
1	Chr 1	Rmc-1			Syntenic	548
1	Chr 1	Ifbip-1			Syntenic	893
1	Chr 1	En-1			Syntenic	506
1	Chr 1	Mmv-5			Syntenic	438
1	Chr 1	Mmv-9			Syntenic	438
1	Chr 1	Mmv-10			Syntenic	438
1	Chr 1C3	Mov-7			Synt., *in situ*	725
2	Rb6Rma	Sd			1.90 ± 1.90	41
2	Sd	Hc			14.55 ± 1.69	164, 1044
2	Sd	stb	10.18 ± 5.13	11.30 ± 1.83		567
2	Sd	fi	20.55 ± 1.72	24.02 ± 1.82	25.00	1075, 1076
2	Sd	lh	24.11 ± 3.60			980
2	Sd	mdm			26.00 ± 3.90	605
2	Sd	ro			37.50	279
2	Sd	pa			40.43 ± 2.95	74
2	Sd	Cas-1			35.32 ± 3.24	197
2	Sd	un			44.80 ± 11.37	1129
2	Sd	a	45.32 ± 1.90	43.09 ± 1.91		106, 1075, 567
2	Sd	fc			40.60 ± 5.80	1068a
2	Sd	Ra	72.73 ± 13.43	56.10 ± 4.47	35.37 ± 5.28	106
2	Hc	Lsr-1			≤7.4†	333
2	Hc	Brp-13			3.33 ± 2.66	342
2	Hc	Cfhe			16.70 ± 4.60	742
2	Hc	Ly-24			28.60 ± 4.30	141
2	Hc	a			43.91 ± 3.13	141, 164
2	stb	fi			≤5†	567
2	stb	a	34.84 ± 5.40	40.91 ± 3.03		567
2	fi	rh	4.30 ± 1.27		4.63 ± 1.98	1065
2	fi	ro	21.24 ± 0.97	17.08 ± 1.13		1079
2	fi	pa	27.45 ± 1.57	24.79 ± 3.99	18.95 ± 3.58	104, 69
2	fi	a	36.22 ± 1.69	27.14 ± 1.90	33.39 ± 1.11	104, 1065, 1075, 99
2	lh	a			39.90	980
2	rh	a	34.89 ± 1.71			1065
2	Ly-11	Rvil-1			Near	700
2	Ul	pa	18.37 ± 3.91			180
2	Ul	a	32.65 ± 4.74			180
2	Gm-3	H-3			≤18†	416
2	Gm-3	B2m			≤18†	416
2	Ea-6	Ly-24			4.55 ± 5.36	141
2	Ea-6	pa			6.90 ± 3.20	618
2	Ea-6	a			25.36 ± 2.11	618
2	Ly-25	Ly-24			3.41 ± 1.72	436, 437
2	Ly-25	B2m			15.22 ± 5.30	436
2	Ly-24	B2m			10.69 ± 2.20	141, 436, 615
2	Ly-24	Cas-1			≤18†	141
2	Ly-24	a			14.57 ± 3.10	141
2	lst	anx			4.10 ± 1.90	665
2	lst	a	27.99	19.06		308
2	lst	Ra			47.30 ± 5.90	308
2	anx	pa			5.40	665
2	anx	a			24.70 ± 6.00	665
2	anx	Ra			39.20 ± 4.30	665
2	Sey	pa	4.60 ± 2.20	6.59 ± 2.60	2.67 ± 1.32	179, 431
2	Sey	we			3.33 ± 1.47	431
2	Sey	un			4.67 ± 1.72	431
2	Sey	mg			12.40 ± 2.40	431

Table 6.1—*cont.*

Chr	Loci		Female	Male	Combined or not stated	References
	Proximal	Distal				
2	*Sey*	*a*	34.48 ± 5.10	17.58 ± 3.99	13.85 ± 1.27	179, 431, 182, 1053
2	*ld*	*mg*			6.13 ± 1.40	382
2	*ld*	*a*			19.00 ± 3.10	382, 376
2	*ro*	*pa*			0.60	279
2	*ro*	*we*	2.87 ± 1.07	4.28 ± 1.16		279
2	*ro*	*a*			16.40 ± 2.60	279
2	*Sdh-1*	*pa*			1.90 ± 1.40	6
2	*Sdh-1*	*Hao-1*			12.60 ± 3.70	454
2	*Sdh-1*	*a*			18.09 ± 2.44	6, 454
2	*Sdh-1*	*bp*			20.20 ± 3.94	6
2	*Xmmv-71*	*H-3*			≤18†	702
2	*pa*	*Hdc*		3.28 ± 2.28	2.00 ± 1.40	674, 708
2	*pa*	*we*	3.88 ± 0.33	2.24 ± 0.30	0.66 ± 0.66	597, 1093, 279, 298, 431, 775
2	*pa*	*Tsk*	1.74 ± 1.00	2.34 ± 1.34		364
2	*pa*	*un*			3.33 ± 1.47	431
2	*pa*	*a*	19.16 ± 1.09	18.21 ± 1.26	15.57 ± 0.81	104, 106, 1042, 179, 180, 1093, 858, 431, 6, 74
2	*pa*	*bp*		21.31 ± 5.24	17.02 ± 2.62	676, 6, 708
2	*pa*	*Lop-4*	15.90 ± 5.50			1093
2	*pa*	*Ra*	39.40 ± 5.80	21.82 ± 4.19		106, 353
2	*pa*	*bp*		21.31 ± 5.24		708
2	*H-3*	*B2m*			≤3.3†	340, 835, 1025
2	*H-3*	*H-42*			*H-42* distal	347
2	*H-3*	*H-42*			Close	347
2	*H-3*	*we*	1.70 ± 0.69	5.00 ± 4.87		959, 956
2	*H-3*	*a*	13.84 ± 2.58	6.38 ± 2.52	20.19 ± 7.60	958, 965, 956
2	*H-3*	*Ir-2*			*Ir-2* distal	328
2	*B2m*	*Itp*			≤18†	1042
2	*B2m*	*Prn-p*			≤6.9†	96
2	*B2m*	*Ly-23*			0.32 ± 0.32	326
2	*B2m*	*pa*			3.10 ± 1.70	674
2	*B2m*	*Il-1*			4.41 ± 2.11	166a
2	*B2m*	*a*			6.76 ± 3.82	1025
2	*B2m*	*Psp*			11.29 ± 6.12	1025
2	*B2m*	*Ak-1*			Syntenic	151
2	*Prn-p*	*Prn-i(Sinc?)*			1.52 ± 1.51	97
2	*Prn-p*	*a*			11.00 ± 2.00	96
2	*H-42*	*we*			*H-42* proximal	347
2	*Cas-1*	*Hao-1*			7.40 ± 3.17	456
2	*Cas-1*	*a*			23.19 ± 1.89	429, 197
2	*Cas-1*	*Ra*			44.17 ± 4.53	197
2	*Hdc*	*Il-1*			1.52 ± 1.14	166a
2	*Hdc*	*bp*		21.31 ± 5.24	15.80 ± 3.60	674, 676, 708
2	*Hdc-c*	*B2m*			3.13 ± 1.76	676
2	*Hdc-c*	*pa*			1.98 ± 1.39	676
2	*Hdc-c*	*Hdc-e*			0	675
2	*Hdc-c*	*Hdc-s*			0	673
2	*we*	*mg*			0.12 ± 0.02	604
2	*we*	*Cm*			1.80	910103
2	*we*	*un*	7.50 ± 0.41	4.68 ± 0.39	2.67 ± 1.32	1093, 956, 959, 279, 298, 431, 775
2	*we*	*a*	14.65 ± 0.72	10.81 ± 0.79	11.17 ± 0.59	597, 412, 792, 431, 604
2	*we*	*Lop-4*	15.90 ± 5.50			1093
2	*we*	*Ra*	32.73 ± 1.64	33.25 ± 1.65		792
2	*Tsk*	*a*	12.27 ± 2.57	12.07 ± 1.56		364
2	*mg*	*dm*			5.99 ± 1.00	382
2	*mg*	*a*	13.15 ± 1.00	9.80 ± 1.16	11.58 ± 0.60	382, 572, 604
2	*Itp*	*pa*	5.90 ± 2.60			1042

Table 6.1—*cont.*

Chr	Loci		Female	Male	Combined or not stated	References
	Proximal	Distal				
2	*Itp*	*a*	18.00 ± 4.00		13.15 ± 8.84	1042
2	*Itp*	*Ada*			Syntenic	928
2	*Itp*	*Ak-1*			Syntenic	928
2	*lm*	*a*	9.62 ± 4.09			367
2	*lm*	*Ra*	36.84 ± 11.07			367
2	*Il-1a*	*Il-1b*			0	166a
2	*Il-1*	*Psp*			11.02 ± 4.35	166a
2	*Il-1*	*a*	12.12 ± 4.02	6.06 ± 4.15		166a
2	*H-44*				Between *H-3*	347a
2	*H-45*				and *H-13*	347a
2	*Cm*	*a*			12.90 ± 2.46	91, 957
2	*un*	*H-13*	5.61 ± 1.64			959
2	*un*	*a*	4.73 ± 0.35	4.56 ± 0.31	4.48 ± 0.48	300, 100, 1093, 956, 959, 879, 298, 431, 775
2	*un*	*bp*	5.04 ± 0.84			879
2	*un*	*Lop-4*	9.10 ± 4.30			1093
2	*dm*	*a*			4.55 ± 0.94	382, 664, 989
2	*dm*	*Ra*			29.00	664
2	*Hao-1*	*a*			5.95 ± 1.56	454, 456, 220
2	*H-13*	*a*	2.04 ± 1.01			959
2	*A^y*	*a*			0.12 ± 0.05	943
2	*a*	*Akv-3*		0		1037
2	*a*	*Emv-15*			0.55 ± 0.56	943
2	*a*	*bp*	0.33 ± 0.20	0	2.90 ± 1.60	879, 6
2	*a*	*kr*	0.97 ± 0.32		0.40	597, 412
2	*a*	*Psp*			3.00 ± 3.00	780
2	*a*	*bs*	5.60 ± 1.80			1066
2	*a*	*Lop-4*	6.80 ± 3.80			1093
2	*a*	*Svp-1*	9.28 ± 2.95		9.23 ± 3.59	828, 718
2	*a*	*fc*			13.70 ± 4.00	1068a
2	*a*	*Blvr*	13.50 ± 4.70			1149
2	*a*	*ls*			25.70	819
2	*a*	*wst*			22.90 ± 2.50	1015
2	*a*	*Ra*	20.62 ± 1.11	23.66 ± 1.00	24.34 ± 2.59	106, 367, 572, 792, 353, 376
2	*a*	*Nk-2*			28.60	829
2	*a*	*Aph-1*			38.69 ± 8.32	398a
2	*a*	*Emv-15*			Close	143
2	*a*	*Src*			Near	1039
2	*a*	*H(a^t)*			Near	691
2	*a*	*fhd*			Linked	1095b
2	*Akv-3*	*Psp*			1.00 ± 1.03	1037
2	*Akv-3*	*Svp-1*		4.10 ± 2.80		1037
2	*Src*	*Psp*			2.20 ± 1.68	400, 61
2	*Psp*	*Srch*			Near	1039
2	*Svp-1*	*Svp-3*	0			718
2	*fc*	*Ra*			6.40 ± 2.70	1068a
2	*ls*	*Ra*			3.14	819
2	*wst*	*Ra*			0.93 ± 1.30	1015
2	*Ada*	*Ak-1*			Syntenic	928
2	*Ir-2*	*a*			20.00 ± 5.96	327
2	*Ir-2*	*a*			*Ir-2* distal	328
2	Chr 2	*Abl*			Synt., *in situ*	890
2	Chr 2	*Acp-2*			Syntenic	556
2	Chr 2	*Ass*			Syntenic	733
2	Chr 2	*D2Mit1*			Syntenic	338
2	Chr 2	*Fshb*			Syntenic	338
2	Chr 2	*Gcg*			Syntenic	562

Table 6.1—*cont.*

Chr	Loci Proximal	Distal	Female	Male	Combined or not stated	References
2	Chr 2	*Gsa*			Syntenic	13
2	Chr 2	*Odc-2*			Linked	853
2	Chr 2	*Pck-1*			Syntenic	614
2	Chr 2cen-C1	*Abl*			Syntenic	703
2	Chr 2cen-C1	*Ak-1*			Syntenic	703
2	Chr 2cen-C1	*Fpgs*			Syntenic	703, 310
2	Chr 2C1-ter	*Itp*			Syntenic	703, 928
2	Chr 2C1-ter	*Ada*			Syntenic	703, 928
3	*Rb1Bnr*	*Car-1*	2.75 ± 1.25			243, 173
3	*Rb1Bnr*	*my*	30.88 ± 5.60			173
3	*Car-1*	*Car-2*	0			243
3	*Car-1*	*Es-16*	8.51 ± 2.90			1156
3	*Car-1*	*my*	33.33 ± 9.07			173
3	*Car-2*	*Ap-2*			Syntenic	415
3	*Car-2*	*coa*			2.70 ± 1.90	1006
3	*Car-2*	*Es-16*	6.20 ± 2.70		7.40 ± 2.00	1157, 1153
3	*Car-2*	*Odc-3*			5.56 ± 4.78	795
3	*Car-2*	*Amy-1*	40.74 ± 6.69		12.49 ± 13.07	71, 795
3	*Car-2*	*Adh-1*	43.99 ± 4.05			71
3	*Car-2*	*H-28*			Near	267
3	*Car-2*	*If-1*			Near	267
3	*Odc-3*	*Amy-1*			10.00 ± 8.00	795
3	*Es-16*	*Es-26*	21.00 ± 4.50		21.90 ± 3.30	1157, 1153
3	*Es-26*	*Es-27*			9.30 ± 2.33	1150, 1153
3	*Es-26*	*Amy-1*	13.60 ± 3.80			1157
3	*my*	*ma*	21.32 ± 2.38	15.62 ± 2.98	13.50 ± 5.20	236
3	*my*	*Va*	40.20 ± 2.89	39.70 ± 3.68	56.81 ± 16.00	236, 101
3	*Es-27*	*Amy-1*			31.10 ± 5.40	1153
3	*spa*	*ma*			5.70 ± 1.65	569, 599
3	*spa*	*ft*			3.81 ± 0.77	599, 569
3	*spa*	*Va*	31.00	25.00 ± 2.72		599
3	*ma*	*ft*	2.69 ± 0.66	4.84 ± 1.02	2.30 ± 0.30	599, 569
3	*ma*	*soc*	3.10 ± 1.00		3.06 ± 0.48	569
3	*ma*	*de*	7.10 ± 4.90		6.47 ± 1.03	569, 601
3	*ma*	*op*	11.80 ± 5.60		8.35 ± 1.00	569
3	*ma*	*Amy-1*		9.60 ± 2.60		569
3	*ma*	*H-23*			19.10 ± 4.90	710
3	*ma*	*cdm*	28.57 ± 8.54			1043
3	*ma*	*Va*	29.13 ± 1.42	26.74 ± 1.52	28.78 ± 1.92	569, 236, 599, 46, 710, 601
3	*ma*	*H-28*			37.50 ± 4.03	710
3	*ma*	*H-37*			7.80 ± 7.20	710
3	*ft*	*ma*			2.82 ± 0.94	596
3	*ft*	*soc*	0.90 ± 0.60		0.66 ± 0.24	569
3	*ft*	*op*			14.50 ± 4.30	569
3	*ft*	*Va*	30.57 ± 2.65	22.12 ± 2.03		599
3	*soc*	*ma*			3.77 ± 1.09	596
3	*soc*	*ft*			0.94 ± 0.54	596
3	*soc*	*Amy-1*		8.02 ± 1.44		236, 569
3	*soc*	*Va*		26.40 ± 2.80	29.22 ± 2.59	569, 978, 981
3	*de*	*op*			9.10 ± 2.90	569
3	*de*	*Adh-3*	17.40 ± 4.60	17.20 ± 5.00	18.00 ± 4.00	452, 453, 450
3	*de*	*Va*	25.63 ± 3.00		25.23 ± 1.75	452, 569, 601, 156
3	*Hao-2*	*Amy-1*	4.88 ± 2.38			462
3	*Hao-2*	*Adh-3*	17.57 ± 4.43			464
3	*op*	*Amy-1*		2.34 ± 1.25		569
3	*op*	*Va*	19.42 ± 2.54	22.82 ± 3.43	14.46 ± 1.81	670, 569, 594

Table 6.1—*cont.*

Chr	Loci		Female	Male	Combined or not stated	References
	Proximal	Distal				
3	*Amy-1*	*Amy-2*			⩽7†	66, 427
3	*Amy-1*	*Xmmv-65*			6.45 ± 2.85	795
3	*Amy-1*	*H-23*			4.55 ± 5.36	795
3	*Amy-1*	*Gbp-1*			2.38 ± 2.61	795
3	*Amy-1*	*Adh-1*	9.26 ± 3.94			71
3	*Amy-1*	*cdm*		17.00 ± 5.20	6.76 ± 3.82	1043, 795
3	*Amy-1*	*Va*		9.49 ± 0.90	18.11 ± 2.13	1043, 569, 236, 252
3	*H-23*	*Gbp-1*			4.55 ± 5.36	795
3	*H-23*	*Va*			13.23 ± 3.33	710, 24
3	*H-23*	*Hnl*			Close	256
3	*Gbp-1*	*Adh-3*			3.41 ± 1.12	843
3	*Gbp-1*	*Ahr-1*			⩽18†	843
3	*Gbp-1*	*Va*			11.70 ± 2.80	843
3	*Gbp-1*	*H-28*			4.55 ± 5.36	795
3	*Adh-3*	*Adh-1*		0	⩽2†	451, 457
3	*Adh-3*	*Ahr-1*	0		⩽1†	222, 221
3	*Adh-3*	*Adh-3t*	0	0.39 ± 0.39		460, 453
3	*Adh-3*	*Va*	9.85 ± 2.55	6.76 ± 1.27	7.13 ± 1.63	843, 460, 452, 453, 451, 221
3	*Adh-3*	*Bmn*	0			625
3	*Bmn*	*cdm*			2.50 ± 1.94	625
3	*Adh-1*	*Adh-1t*	0			455
3	*Adh-1*	*Va*		4.00 ± 1.70		451
3	*Ahr-1*	*Va*			4.80 ± 2.00	221
3	*cdm*	*Va*	8.05 ± 2.92	10.53 ± 3.21	7.32 ± 4.07	1043
3	*Va*	*H-28*			7.70 ± 2.20	710
3	*Va*	*If-1*			12.40 ± 3.00	710
3	*H-28*	*If-1*			5.90 ± 2.20	710, 267, 186
3	*Atpa-1*	*Egf*			28.12 ± 7.95	518
3	Chr 3	*Acts*			Syntenic	158, 157
3	Chr 3	*Adh-1*			Syntenic	125
3	Chr 3	*Atpa-1*			Syntenic	518
3	Chr 3	*Egf*			Syntenic	1138
3	Chr 3	*Hist2*			Syntenic	352, 146
3	Chr 3	*Ly-37*			Syntenic	919
3	Chr 3	*Mmv-2*			Syntenic	438
3	Chr 3	*Mmv-12*			Syntenic	438
3	Chr 3	*Mov-10*			Syntenic	725
3	Chr 3	*Ngfb*			Syntenic	1140
3	Chr 3	*Nras*			Syntenic	889, 393
3	Chr 3	*Oua-1*			Syntenic	540
3	Chr 3	*Tshb*			Syntenic	536, 749
4	*Rb2Bnr*	*Galt*			3.45 ± 1.96	729
4	*Rb2Bnr*	*Aco-1*			4.68 ± 2.64	729
4	*Rb2Bnr*	*Mup-1*			30.00 ± 10.25	729
4	*Rb2Bnr*	*b*		16.09 ± 3.94		114
4	*Rb2Bnr*	*m*		29.89 ± 4.91		114
4	*wd*	*b*	32.52 ± 3.67		29.40 ± 3.50	1137
4	*cy*	*b*	21.47 ± 2.97	24.11 ± 2.24	20.41 ± 5.76	500, 289
4	*cy*	*m*	26.18 ± 3.18	26.85 ± 2.32		500
4	*cy*	*Ps*	26.18 ± 3.18	27.12 ± 2.33	20.41 ± 5.76	500, 289
4	*Galt*	*Aco-1*			3.96 ± 1.89	729
4	*Galt*	*Mup-1*			9.20 ± 3.10	729
4	*Galt*	*b*			14.94 ± 3.82	729
4	*Aco-1*	*Mup-1*			7.01 ± 1.79	729
4	*Aco-1*	*b*			16.77 ± 4.43	729
4	*Lyb-2*	*Ly-19*			⩽5†	1024

Table 6.1—*cont.*

Chr	Loci		Female	Male	Combined or not stated	References
	Proximal	Distal				
4	Lyb-2	Xmmv-8			1	62
4	Lyb-2	Ly-32			Close	1027a
4	Lyb-2	Mtv-17			2.00 ± 1.53	705, 418
4	Lyb-2	Lyb-4			3.22 ± 1.72	468
4	Lyb-2	Ly-29			2.00 ± 2.16	343
4	Lyb-2	Lyb-6			4.00	519
4	Lyb-2	Mup-1			7.34 ± 2.08	519, 468, 1041, 897
4	Lyb-2	b			14.12 ± 3.54	1041, 897
4	Lyb-2	Pgm-2			20.59 ± 4.90	897
4	Ly-19	Orm-1			40.84 ± 34.16	33
4	Xmmv-8	Lyb-4			4.00 ± 2.58	62
4	Xmmv-8	Lyb-6			1.00 ± 1.07	62
4	Xmmv-8	Mup-1			5.00 ± 3.05	62
4	Xmmv-8	Xmmv-23			27.00 ± 17.24	62
4	Mtv-17	Ly-29			≤6.8†	343
4	Mtv-17	Mup-1			5.91 ± 2.77	705, 343, 418
4	Mtv-17	b			4.40 ± 3.70	418
4	Lyb-4	Mup-1			3.18 ± 1.29	468
4	Ly-29	Mup-1			7.80 ± 5.89	343
4	Ly-29	b			19.20 ± 14.82	343
4	vc	b	7.73 ± 1.44		5.40 ± 3.10	944, 945
4	vc	m	17.95 ± 2.75			944
4	Lyb-6	Mup-1			2.00	519
4	Mup-1	Lv			1.79 ± 1.35	728
4	Mup-1	b	1.79 ± 1.77		6.71 ± 1.04	929, 183, 1097, 1041, 1032, 728, 343, 897, 295, 470
4	Mup-1	Lps		11.11 ± 6.05	7.51 ± 1.93	761, 1087, 145, 1086
4	Mup-1	Orm-1	8.00 ± 5.40		10.71 ± 4.13	728, 32
4	Mup-1	m	21.59 ± 4.39			183
4	Mup-1	Ifa	9.37 ± 2.21		11.54 ± 5.55	728, 183
4	Mup-1	Ifb	16.00 ± 7.30			728
4	Mup-1	dyl			19.67 ± 2.94	896
4	Mup-1	Bmfr-1			21.20 ± 4.00	929
4	Mup-1	Gpd-1	50.00 ± 7.37	40.00 ± 5.22	43.30 ± 4.10	953, 781
4	an	b	8.46 ± 2.02	3.30 ± 1.29	9.50 ± 2.40	412, 689
4	Lv	b	4.81 ± 2.10		8.54 ± 4.29	482, 728
4	Lv	Ifa			12.86 ± 6.40	728
4	Och	b			4.62 ± 1.66	815, 820
4	Och	Ps			11.83 ± 3.35	820
4	wi	b	5.47 ± 0.64	1.82 ± 0.58	1.00 ± 0.61	600, 587
4	wi	db			12.03 ± 0.97	587
4	b	Dre			0	514
4	b	H-21			1.00	25
4	b	dep	2.27 ± 0.71	1.41 ± 0.70		687
4	b	Orm-1			3.00 ± 2.41	33
4	b	Pgm-2	7.84 ± 0.79			130, 459
4	b	Pt	4.60 ± 0.42	3.59 ± 0.55	5.17 ± 0.63	689, 439, 600, 1137, 687, 602, 295, 913, 140
4	b	mea	11.68 ± 1.94	8.28 ± 2.12		440
4	b	m	6.66 ± 0.69	10.83 ± 2.55	7.30 ± 1.76	483, 500, 944, 1126, 114, 913
4	b	Ps	8.77 ± 1.29	6.97 ± 1.01	6.38 ± 1.29	440, 500, 502, 592, 820, 289, 913
4	b	db			8.72 ± 0.97	587
4	b	H-15			16.00 ± 3.00	25
4	b	Ifa	4.54 ± 2.39		3.82 ± 1.97	168, 183, 169, 728
4	b	Ifb	16.70 ± 8.78			168
4	b	dyl			12.25 ± 1.33	896
4	b	Bmfr-1			14.00 ± 1.79	929

Table 6.1—*cont.*

Chr	Loci		Female	Male	Combined or not stated	References
	Proximal	Distal				
4	*b*	*Hpt*			14.11 ± 1.72	597
4	*b*	*cla*			14.10 ± 1.50	1012
4	*b*	*sno*	17.02 ± 1.06	14.97 ± 1.22		441
4	*b*	*cn*			16.59 ± 2.98	592
4	*b*	*Ly-20*			26.47 ± 17.55	698
4	*b*	*Ahd-1*	31.10 ± 3.30			459
4	*b*	*Er*		24.68 ± 3.47		390
4	*b*	*pf*			26.41 ± 6.87	696
4	*b*	*cri*			31.40 ± 4.00	360
4	*b*	*H-20*			25.00 ± 6.00	24
4	*b*	*Fv-1*			37.91 ± 2.16	872, 1032, 871
4	*b*	*Gpd-1*	31.73 ± 4.56		34.27 ± 3.24	872, 482, 1032, 140
4	*b*	*Aia-1*			36.67 ± 3.59	533
4	*b*	*Jt-1*			Linked	405
4	*b*	*D4Rp1*			Near	257
4	*H-21*	*H-15*			10.00	25
4	*dep*	*Pt*	2.72 ± 0.77	4.59 ± 1.24		687
4	*Lps*	*Ps*			12.90 ± 6.02	1086
4	*Lps*	*D4Rp1*			Near	257
4	*Orm-1*	*Orm-2*			≤5†	32, 31
4	*Orm-1*	*Ifa*	8.00 ± 5.40			728
4	*Orm-1*	*Ifb*	8.00 ± 5.40			728
4	*Pgm-2*	*Ahd-1*	23.42 ± 3.13		36.00 ± 11.00	459, 448
4	*Pgm-2*	*Pgd*			32.66 ± 5.99	132
4	*Pgm-2*	*Gpd-1*	29.00 ± 5.90	26.96 ± 4.14		459, 130
4	*Pgm-2*	*Tsha*			Syntenic	743
4	*Pt*	*m*	3.16 ± 0.60	4.06 ± 0.95	2.06 ± 1.02	600, 913
4	*Pt*	*Ps*			2.58 ± 1.14	913
4	*Pt*	*heb*			5.14 ± 2.01	1063, 313
4	*Pt*	*Hpt*			13.00 ± 1.60	602
4	*Pt*	*cla*			10.40 ± 1.00	1012
4	*Pt*	*cri*			24.70	385
4	*Pt*	*Gpd-1*			25.49 ± 4.32	140
4	*mea*	*Ps*	8.38 ± 2.00	5.41 ± 1.28		440
4	*m*	*Ps*	0.42 ± 0.42	0.32 ± 0.26	0.52 ± 0.52	500, 502, 913
4	*m*	*db*			1.08 ± 0.77	475
4	*m*	*Gpd-1*			24.00 ± 8.54	483
4	*Ps*	*cn*			15.56 ± 7.49	592
4	*H-15*	*H-16*			1.00	25
4	*H-15*	*Ifa*			4.55 ± 5.36	169
4	*Ms15-1*	*Ms6-2*			21.43 ± 12.91	496
4	*Ms15-1*	*Mtv-13*			0	496
4	*H-16*	*Ifa*			4.55 ± 5.36	169
4	*H-16*	*H-20*			1.90	25
4	*Ifa*	*m*	7.95 ± 2.88			183
4	*Ifa*	*Ifb*	≤7†			168, 728
4	*Ifa*	*H-20*			4.55 ± 5.36	169
4	*Ly-20*	*Fv-1*			11.11 ± 7.93	698
4	*Ly-20*	*Cxv-2*			7.14 ± 5.27	698
4	*Ms6-2*	*Ahd-1*			2.38 ± 1.84	496
4	*Ahd-1*	*Xmmv-23*			2.00 ± 1.62	62
4	*Ahd-1*	*Akp-2*			0.60 ± 0.40	1102
4	*Ahd-1*	*Pan-1*	4.71 ± 2.30	5.68 ± 2.47		953
4	*Ahd-1*	*Gpd-1*	8.65 ± 1.54			461, 459
4	*Ahd-1*	*je*	11.80 ± 3.90			461
4	*Xmmv-23*	*Akp-2*			2.00 ± 1.62	62
4	*Xmmv-23*	*Fv-1*			11.00 ± 6.08	62

Table 6.1—*cont.*

Chr	Loci		Female	Male	Combined or not stated	References
	Proximal	Distal				
4	Xmmv-23	Gpd-1			7.00 ± 4.01	62
4	Akp-2	Gpd-1			5.70 ± 1.90	1103, 1102
4	Akp-2	Ela-1			1.00	400
4	Akp-2	Ly-31			0	1026
4	Er	Gpd-1			5.00 ± 3.45	314
4	Pan-1	Gpd-1	7.32 ± 2.88	6.82 ± 2.69		953
4	Pgd	Tsha			Syntenic	743
4	H-20	Gpd-1			1.00	25, 22
4	Fv-1	Gpd-1			0.67 ± 0.50	872, 1032
4	Cxv-2	Ril-2			Near	699
4	Gpd-1	ecl			Linked	1045
4	Gpd-1	Xmmv-62			1.30 ± 0.95	1091
4	Gpd-1	je	1.50 ± 1.50		4.31 ± 1.89	461, 238
4	Gpd-1	H-18			14.00 ± 7.00	25, 22
4	Gpd-1	Ea-9		24.04 ± 4.19		312
4	Gpd-1	C58v-3			Near	499
4	Eam	Exa			Linked	767a
4	Chr 4	Exa			Linked	767b
4	Chr 4	Ak-2			Syntenic	559
4	Chr 4	asp-2			Linked	140
4	Chr 4	Dsi-1			Syntenic	1068
4	Chr 4	Eno-1			Syntenic	559, 161
4	Chr 4	Fabph-1			Syntenic	415
4	Chr 4	Fabph-4			Syntenic	415
4	Chr 4	Fuca			Syntenic	556
4	Chr 4	Ggtb-1			Syntenic	921
4	Chr 4	Lck			Synt., distal	672
4	Chr 4	Mos			Syntenic	996, 890
4	Chr 4	Mtv-13			Syntenic	716
4	Chr 4	Mtv-17			Syntenic	798
4	Chr 4	Odc-4			Linked	853
4	Chr 4	Pnd			Syntenic	1132
4	Chr 4	Tsha			Syntenic	748, 743, 536
4	Chr 4	Udpgt-3			Syntenic	552
4	Chr 4	Xmv-1			Syntenic	438
4	Chr 4	Xmv-2			Syntenic	438
4	Chr 4	Xpa			0	619
5	Rb1Wh	Tht	9.10 ± 3.50			43
5	Rb3Bnr	Rw		8.43 ± 3.05		114
5	rl	lx			16.00 ± 3.40	281
5	rl	W			29.00 ± 3.40	281
5	Cv-1	Hx			2.15 ± 1.50	508
5	Cv-1	Pgm-1		23.94 ± 5.06	25.20 ± 2.41	542, 485, 688
5	Cv-1	Gus		39.44 ± 5.80	43.00 ± 5.90	542, 688
5	Xmmv-52	Rmcf			2.10 ± 1.60	1091
5	Rmcf	Hm			2.20 ± 0.90	401
5	Rmcf	Pgm-1			20.25 ± 4.49	401
5	Hm	Hx			0.04 ± 0.04	369, 1010
5	Hm	lx			4.63 ± 2.02	357
5	Hm	tlt			6.94 ± 2.12	598
5	Hm	Pgm-1	20.41 ± 5.76			1119
5	Hm	gc			20.00 ± 5.16	1071
5	Hm	W	21.20 ± 1.27	23.11 ± 2.79	21.59 ± 0.62	203, 357, 204, 356, 382, 598, 373, 1010
5	Hm	mc			25.47 ± 3.43	373
5	Hm	go			25.36 ± 2.62	204
5	Hm	jg		44.68 ± 7.25	37.14 ± 6.18	382

Table 6.1—*cont.*

Chr	Loci		Female	Male	Combined or not stated	References
	Proximal	Distal				
5	*Hm*	*bf*	41.00 ± 6.00		38.59 ± 3.59	203, 1119
5	*Hm*	*Gus*	25.00 ± 10.00			1119
5	*Hm*	*Mor-1*	48.00 ± 6.00			1119
5	*Hx*	*W*			21.92 ± 0.66	978, 1010
5	*Hx*	*En-2*			1.36 ± 0.68	508
5	*Tht*	*Ph*			17.90 ± 2.90	40
5	*Tht*	*go*			27.90 ± 4.10	45
5	*Tht*	*bf*			39.30 ± 4.42	45
5	*lx*	*pi*	10.27 ± 2.03	12.03 ± 2.59		582
5	*lx*	*W*			17.07 ± 0.90	357, 103, 102
5	*tlt*	*W*			16.17 ± 2.08	598
5	*Ms15-6*	*Ms15-5*			20.58 ± 13.82	496
5	*Pgm-1*	*Gpd-2*	≤8.4†			737
5	*Pgm-1*	*W*			3.00 ± 1.71	483
5	*Pgm-1*	*Ms15-6*			4.05 ± 2.71	496
5	*Pgm-1*	*Xmmv-5*			7.00	62
5	*Pgm-1*	*Alb-1*	10.02 ± 3.42	14.41 ± 4.35		755
5	*Pgm-1*	*Afp*			9.00	400
5	*Pgm-1*	*Fla*	20.00 ± 5.20	13.64 ± 5.10		404a
5	*Pgm-1*	*Ric*			22.86 ± 4.76	386
5	*Pgm-1*	*rd*			17.65 ± 6.54	386
5	*Pgm-1*	*bf*	34.69 ± 6.88			1119
5	*Pgm-1*	*Gus*	25.00 ± 10.00	26.76 ± 5.25	36.68 ± 2.85	1119, 542, 688, 563
5	*Pgm-1*	*Gus-s*			36.70 ± 4.30	563
5	*Pgm-1*	*Mor-1*	43.00 ± 7.00			1119
5	*Pgm-1*	*Asl*			Syntenic	561
5	*Pgm-1*	*Rec-1*			Syntenic	767, 878
5	*Pep-7*	*Arp*			≤6.7†	683
5	*Pep-7*	*Rw*	2.25 ± 1.57	4.00 ± 3.92		807
5	*Pep-7*	*Asl*			Syntenic	561
5	*Pep-7*	*Rec-1*			Syntenic	767
5	*pi*	*W*	3.10 ± 0.60	1.09 ± 0.62		201, 582
5	*W*	*Ph*			0.08 ± 0.06	387
5	*W*	*Rw*		2.50 ± 1.80	≤0.5†	42, 912
5	*W*	*rs*			≤1.7†	972
5	*W*	*l(5)-1*			1.82 ± 0.42	635, 640
5	*W*	*mc*			6.21 ± 1.90	373
5	*W*	*go*			10.80 ± 1.14	204, 58
5	*W*	*bl*			11.51 ± 2.49	823
5	*W*	*rd*	11.63 ± 1.43	11.46 ± 1.11	15.57 ± 2.90	932, 934, 555
5	*W*	*le*	16.36 ± 0.94	9.25 ± 0.97		582, 571, 934, 356
5	*W*	*jg*		16.09 ± 3.94	20.56 ± 2.27	382
5	*W*	*bf*			21.74 ± 3.04	203
5	*Xmmv-5*	*Afp*			11.00	62
5	*Ph*	*Rw*			≤1†	912
5	*Rw*	*go*	10.11 ± 3.20	8.00 ± 5.43		807
5	*H(go)*	*go*			5.00 ± 3.00	20
5	*H(go)*	*H-27*			6.00	25
5	*Ms15-5*	*Ric*			4.07 ± 2.70	496, 1109a
5	*Ms15-5*	*Bcd-1*			4.07 ± 2.70	1109a
5	*Ms15-5*	*Fla*			8.04 ±	1109a
5	*go*	*bf*	32.58 ± 4.97	16.00 ± 7.33	8.20 ± 2.50	807, 45
5	*Pol-20*	*Ric*			≤9.3†	867
5	*Alb-1*	*Afp*			Tandem	488
5	*Alb-1*	*rd*			7.00 ± 2.55	129
5	*Alb-1*	*le,rd*			7.00 ± 2.55	129
5	*Afp*	*Ric*			5.00	400

Table 6.1—*cont.*

Chr	Loci		Female	Male	Combined or not stated	References
	Proximal	Distal				
5	*Afp*	*Igj*			Syntenic	1130
5	*Ric*	*rd*			3.22 ± 1.83	386
5	*Ric*	*Gus*			15.91 ± 5.51	386
5	*H-27*	*Pol-20*			≤18†	867
5	*H-27*	*H(go)*			6.00	22
5	*rd*	*Bcd-1*			≤6.8†	449
5	*rd*	*le*	0.95 ± 0.48	1.71 ± 0.61	≤3†	555, 934, 129
5	*rd*	*Gus*	7.60 ± 2.45	15.35 ± 2.26	29.00 ± 4.54	1119, 788, 932, 129
5	*rd*	*Mor-1*	23.63 ± 5.73			1119
5	*Bcd-1*	*Fla*			6.64 ± 2.90	1109a
5	*Bcd-1*	*Dao-1*	20.83 ± 4.14			917
5	*le*	*jg*		4.92 ± 2.77	4.53 ± 2.01	382
5	*le*	*Micrl*			*le* proximal	904
5	*le,rd*	*Gus*			29.00 ± 4.54	129
5	*jg*	*bf*			≤0.4†	1046
5	*bf*	*Gus*	7.15 ± 3.53			1119, 997
5	*bf*	*Gus-s*			12.05	997
5	*bf*	*Mor-1*	13.00 ± 4.00			1119
5	*Fla*	*Gus*			2.73 ± 2.20	1109a
5	*Gus*	*Micrl*			0	904
5	*Gus*	*Map-1*			3.80	212, 210
5	*Gus*	*Mor-1*	10.67 ± 3.56			1119
5	*Gus-e*	*Gus-t*			Close	627
5	*Gus-r*	*Gus-s*			0.67	997
5	Chr 5	*Actb*			Syntenic	157
5	Chr 5	*Asl*			Syntenic	556
5	Chr 5	*Csn*			Syntenic	394
5	Chr 5	*Igj*			Syntenic	1130
5	Chr 5	*Mmv-1*			Syntenic	438
5	Chr 5	*Pgy-1*			Synt., *in situ*	676a
5	Chr 5	*Psph*			Syntenic	558
5	Chr 5	*Rec-1*			Syntenic	331, 419, 767, 878
5	Chr 5	*Udpgt-3*			Syntenic	552
6	*Rb1Ald*	*Sig*		0		114
6	*Rb2Bnr*	*Sig*		0.76 ± 0.75		114
6	*Sig*	*hop*	21.95 ± 3.73	13.66 ± 1.85		442
6	*Sig*	*ho*			22.10 ± 3.10	973
6	*Sig*	*Hd*	32.37 ± 1.37	20.40 ± 1.47		442
6	*Sig*	*Lc*	36.34 ± 2.63	24.11 ± 4.04	34.10 ± 3.60	909, 800, 804, 908
6	*Sig*	*mi*	42.16 ± 3.32	33.93 ± 4.47	42.20 ± 3.75	909, 804, 908
6	*Sig*	*Tpi-1*	46.70 ± 4.26			804
6	*ob*	*Hd*			16.60 ± 4.40	478
6	*ob*	*wa-1*			23.50 ± 3.78	198
6	*ob*	*mi*	25.54 ± 6.88	25.15 ± 5.63		198
6	*Cpa*	*Hox-1*	10.81 ± 5.10			86
6	*Cpa*	*Tcrb*	5.41 ± 3.72			86
6	*Cpa*	*Igk*	21.62 ± 6.77			86
6	*Cpa*	*Raf-1*	29.73 ± 7.51			86
6	*hop*	*Hd*	8.74 ± 2.78	9.87 ± 1.71		442
6	*hop*	*mi*	20.08 ± 2.54	23.42 ± 1.28		442
6	*Mtv-14*	*Igk*			44.00 ± 5.10	856
6	*Mtv-14*	*Mtv-8*			44.00 ± 5.10	856
6	*Mtv-14*	*Raf-1*			39.00 ± 7.20	856
6	*Mtv-14*	*Kras-2*			54.00 ± 7.20	856
6	*Tcrb*	*Hd*		4.50 ± 2.21		167
6	*Tcrb*	*Igk*	16.22 ± 6.06			86
6	*Tcrb*	*Igk-C*		7.95 ± 2.88		167

Table 6.1—*cont.*

Chr	Loci		Female	Male	Combined or not stated	References
	Proximal	Distal				
6	*Tcrb*	*Raf-1*	29.73 ± 7.51			86
6	*Tcrb*	*Hox-1*	5.41 ± 3.72		4.55 ± 3.79	86, 711
6	*Tcrb*	*Try-1*	≤7.8†			86
6	*Hox-1*	*Hd*			0.85 ± 0.84	711
6	*Hox-1*	*Ggc*			0	86
6	*Hox-1*	*Igk*	10.81 ± 5.10		4.55 ± 3.79	86, 711
6	*Hox-1*	*Ly-2*			11.30 ± 6.12	86
6	*Ggc*	*Xmmv-27*			8.00 ± 4.70	62
6	*Ggc*	*Ly-2*			5.45 ± 1.62	86, 1057, 1059
6	*Ggc*	*Mtv-8*			13.40 ± 6.11	1057
6	*Ggc*	*fe*			28.57 ± 6.45	766
6	*ho*	*Hd*			2.20 ± 1.30	973
6	*Hd*	*lit*		3.33 ± 0.84		232
6	*Hd*	*Igk-C*		3.41 ± 1.97		167
6	*Hd*	*Ly-2*			3.33 ± 2.32	492
6	*Hd*	*Lvp-1*		7.44 ± 2.39		1101
6	*Hd*	*wa-1*			14.70 ± 4.00	478
6	*Hd*	*mi*	18.53 ± 1.26	15.57 ± 1.02	15.31 ± 3.42	381, 442, 1101, 232, 591, 492
6	*Hd*	*Cd*			23.53 ± 5.94	591
6	*Hd*	*me*	37.90 ± 6.30	43.30 ± 12.90	27.20 ± 6.20	359, 378
6	*lit*	*mi*		13.97 ± 1.63		232
6	*Xmmv-27*	*Ly-2*			2.00 ± 1.68	62
6	*Igk*	*Ly-3*			3.33 ± 3.28	1036
6	*Igk*	*Mtv-8*			≤3.5†	856
6	*Igk*	*Rn7s-6*			3.33 ± 3.28	1036
6	*Igk*	*Raf-1*	13.51 ± 5.62		10.00 ± 4.30	86, 856
6	*Igk*	*Kras-2*			33.00 ± 6.60	856
6	*Igk-V*	*Igk-C*		≤3.3†		167, 995
6	*Igk-Ef1*	*Igk-Ef2*			0.49 ± 0.49	1036
6	*Igk-Ef1*	*Igk-Pc*			≤8.3†	851
6	*Igk-Ef1*	*Ly-2*			≤8.3†	851
6	*Igk-Ef1*	*Ly-2,3*			Very close	336
6	*Igk-Ef1*	*Ly-3*			0.30 ± 0.30	335
6	*Igk-Ef1*	*Rn7s-6*			≤1.3†	1036
6	*Igk-Ef2*	*Rn7s-6*			0.49 ± 0.49	1036
6	*Igk-Pc*	*Ly-2*			≤8.3†	851
6	*Igk-Pc8*	*Ly-2,3*			≤5.3†	137
6	*Brp-1*	*Ly-2*			≤3.9†	342
6	*Brp-1*	*Lvp-1*			≤3.9†	342
6	*Ly-2*	*Ly-3*			≤0.5†	493
6	*Ly-2*	*Mtv-8*			9.57 ± 4.37	1057
6	*Ly-2*	*mi*			8.26 ± 1.94	493, 492
6	*Ly-2*	*Ldr-1*			37.50 ± 5.40	493
6	*Ly-3*	*Rn7s-6*			3.33 ± 3.28	1036
6	*Ly-3*	*Esr*	25.80 ± 5.70			1155
6	*Ly-3*	*Ldr-1*	43.10 ± 7.00			1155
6	*Mtv-8*	*Raf-1*			10.00 ± 4.30	856
6	*Mtv-8*	*Kras-2*			33.00 ± 6.60	856
6	*Lvp-1*	*mi*	13.22 ± 3.08			1101
6	*Ms6-4*	*Lvp-1*			0	496
6	*Lc*	*wa-1*	8.51 ± 2.04	7.77 ± 1.93	5.99 ± 1.21	818, 914
6	*Lc*	*px*			6.28	914
6	*Lc*	*mi*	11.76 ± 1.19	10.92 ± 1.09	12.72 ± 2.53	818, 800, 909, 804, 908
6	*Lc*	*Tpi-1*	22.60 ± 3.57			804
6	*wa-1*	*mi*	23.97 ± 1.08	4.85 ± 1.33	4.17 ± 1.02	90, 818, 914
6	*wa-1*	*Mov-1*		31.20 ± 3.10		80
6	*tc*	*wa-1*			0	597

447

Table 6.1—*cont.*

Chr	Loci		Female	Male	Combined or not stated	References
	Proximal	Distal				
6	*tc*	*mi*	6.87 ± 1.20	4.98 ± 1.08		597
6	*px*	*mi*			2.61 ± 0.26	914
6	*Raf-1*	*Kras-2*			29.00 ± 6.60	856
6	*Raf-1*	*Hox-1*	24.32 ± 7.05			86
6	*mi*	*opt*			5.59 ± 1.14	597, 591
6	*mi*	*dfw*			6.83 ± 1.06	581
6	*mi*	*Cd*			11.65 ± 3.16	375, 591
6	*mi*	*Tpi-1*	15.30 ± 3.00			804
6	*mi*	*me*	21.30 ± 4.60	25.40 ± 11.60		359
6	*mi*	*Ldh-2*	19.00 ± 3.90			800
6	*mi*	*Ldr-1*	28.70 ± 4.35			483
6	*fe*	*Ldr-1*			32.50 ± 7.41	766
6	*Ldr-2*	*Ldr-1*			20.46 ± 2.79	521
6	*Esr*	*Ldr-1*	15.10 ± 3.54			1155
6	*Tpi-1*	*Gapd*			Syntenic	85
6	*Pcn*	*Por*			Syntenic	942
6	Chr 6	*Gapd*			Syntenic	85
6	Chr 6	*Ins-1*			Syntenic	557
6	Chr 6	*Ly-4*			Syntenic	291
6	Chr 6	*Odc-5*			Linked	853
6	Chr 6	*Pcn*			Syntenic	942
6	Chr 6	*Por*			Syntenic	942
6	Chr 6	*Rpl32*			Syntenic	1098
7	*cen*	*Gpi-1*	11.30 ± 1.20			269
7	*H-22*	*H-24*			11.00	25, 22
7	*H-22*	*Gpi-1*			3.00 ± 3.00	21
7	*nv*	*c*			39.46 ± 2.31	1062
7	*rp*	*Gpi-1*			10.80 ± 2.20	334
7	*rp*	*Hbb*			46.60 ± 3.50	334
7	*Akv-1*	*Abpa*			4.76 ± 2.83	213
7	*Akv-1*	*Gpi-1*	12.80 ± 4.90	13.03 ± 2.50		1037, 870
7	*Akv-1*	*Gpi-1*			*Akv-1* prox.	874
7	*Akv-1*	*c*		33.30 ± 4.10		870
7	*Es-8*	*Gpi-1*			9.55 ± 1.94	128
7	*Es-8*	*Smg-1*			42.80 ± 2.79	773, 774
7	*pu*	*p*			21.61 ± 2.90	985
7	*pu*	*c*			33.54 ± 3.74	985
7	*Coh*	*Xmmv-35*			≤3.7†	62
7	*Coh*	*Atpa-2*			0	518
7	*Coh*	*Abpa*			3.85 ± 2.56	213
7	*Coh*	*Gpi-1*	5.80 ± 2.81		5.57 ± 2.87	1124
7	*Coh*	*D7Ag2*			3.49 ± 2.29	63
7	*Coh*	*Lyb-8*			5.70 ± 2.86	1016
7	*Coh*	*p*	27.54 ± 5.38		16.67 ± 8.78	1124
7	*Coh*	*c*			21.15 ± 5.66	1124
7	*Coh*	*Xmmv-31*			8.00 ± 4.70	62
7	*Xmmv-35*	*Gpi-1*			4.00 ± 2.69	62
7	*Xmmv-35*	*D7Ag2*			3.49 ± 2.29	63
7	*Xmmv-35*	*Xmmv-31*			7.60 ± 4.40	62
7	*Apoe*	*Gpi-1*			5.00 ± 4.23	628
7	*Apoe*	*Svp-2*			1.06 ± 1.11	628
7	*Apoe*	*Tam-1*			14.29 ± 11.04	628
7	*Apoe*	*Mtv-1*			16.67 ± 13.61	628
7	*Hv-2*	*Gpi-1*			3.90 ± 3.16	534
7	*Hv-2*	*Svp-2*			≤6.8†	534
7	*Hv-2*	*Tam-1*			12.50 ± 9.24	534
7	*Hv-2*	*p*	15.00 ± 8.00		27.60	955

Table 6.1—*cont.*

Chr	Loci Proximal	Distal	Female	Male	Combined or not stated	References
7	*Hv-2*	*c*	32.00 ± 6.00		37.60	955
7	*H-24*	*Gpi-1*			2.00	25
7	*Abpa*	*Gpi-1*			1.63 ± 1.14	213, 1051
7	*Abpa*	*Tam-1*			5.70 ± 2.32	213, 1051
7	*Abpa*	*Saa-1*			5.56 ± 6.76	213, 1038
7	*Abpa*	*Hbb*			31.71 ± 4.20	213
7	*Abpa*	*Abpg*			≤1†	213
7	*Gpi-1*	*D7Ag2*			≤3.3†	63
7	*Gpi-1*	*Svp-2*			≤3.4†	1049
7	*Gpi-1*	*Mtp-1*			≤4.9†	685
7	*Gpi-1*	*Fes*			2.60 ± 2.04	400, 61
7	*Gpi-1*	*Pep-4*	3.00 ± 1.70	4.00 ± 1.90		950
7	*Gpi-1*	*hyh*			3.90 ± 2.20	603
7	*Gpi-1*	*Tam-1*	13.87 ± 2.15		10.00 ± 5.66	753, 949
7	*Gpi-1*	*D7Rp2*			2.60 ± 2.04	1016
7	*Gpi-1*	*Lyb-8*			1.10 ± 1.15	1016
7	*Gpi-1*	*Prt-4*			12.43 ± 2.55	771
7	*Gpi-1*	*Prt-5*			2.68 ± 1.18	771
7	*Gpi-1*	*Saa-1*	12.50 ± 6.75		6.52 ± 4.72	33, 1038
7	*Gpi-1*	*ru-2*	12.13 ± 1.62	11.60 ± 2.38		174
7	*Gpi-1*	*Ldh-1*			15.50 ± 4.80	970
7	*Gpi-1*	*Xld-1*	16.87 ± 4.11		11.70 ± 3.70	753, 735
7	*Gpi-1*	*p*	16.19 ± 3.09		11.11 ± 7.41	1124, 130
7	*Gpi-1*	*Mtv-1*	19.00 ± 6.10		8.83 ± 4.84	663, 1057
7	*Gpi-1*	*cht*			33.80 ± 4.00	172
7	*Gpi-1*	*Idh-2*			21.10	489
7	*Gpi-1*	*c*	28.25 ± 2.41	19.87 ± 3.22	23.61 ± 2.45	517, 33, 685, 483, 130, 870, 21, 1124, 184, 128
7	*Gpi-1*	*Mod-2*	32.06 ± 4.08	34.78 ± 4.44	26.55 ± 3.07	130, 131, 924, 128
7	*Gpi-1*	*Hbb*	34.92 ± 2.79	30.08 ± 4.14	32.90 ± 1.83	517, 213, 172, 130, 483, 753, 484, 334, 489, 1124
7	*Gpi-1*	*Ly-21*			40.00	517
7	*Gpi-1*	*Ahr-3*	41.00 ± 5.60			682
7	*Gpi-1*	*l(7)-1*			In In(7)13Rk	863
7	*Gpi-1*	*Lhb*			Syntenic	743
7	*Gpi-1*	*Xmmv-31*			7.00 ± 4.18	62
7	*Gpi-1*	*D7Rp2*			D7Rp2 distal	257
7	*Gpi-1s*	*Gpi-1t*			≤2.8†	693
7	*Gpi-1s*	*c*			21.95 ± 6.46	693
7	*Gpi-1t*	*c*			21.95 ± 6.46	693
7	*D7Ag2*	*qv*			Near	63
7	*D7Ag2*	*Lyb-8*			1.02 ± 1.06	63
7	*da*	*p*			17.00 ± 8.00	283
7	*da*	*c*			23.00 ± 4.00	283
7	*qv*	*p*			12.40 ± 1.90	1135
7	*qv*	*c*			28.32 ± 3.17	1135
7	*Fes*	*Tam-1*			6.20 ± 3.86	61
7	*Fes*	*c*			8.06 ± 4.73	61
7	*Fes*	*c*			Close	339
7	*Pep-4*	*p*	23.00 ± 4.60	14.00 ± 3.50		950
7	*Tam-1*	*D7Rp2*			10.80 ± 6.37	1016
7	*Tam-1*	*Lyb-8*			7.60 ± 4.40	1016
7	*Tam-1*	*Prt-4*			≤3†	773, 771
7	*Tam-1*	*Prt-5*			≤1†	773, 954, 771
7	*Tam-1*	*Smg-2*			≤1†	954
7	*Tam-1*	*Saa-1*			≤5.3†	1038
7	*Tam-1*	*ru-2*	3.03 ± 1.33	≤4.1†		174

Table 6.1—*cont.*

Chr	Loci		Female	Male	Combined or not stated	References
	Proximal	Distal				
7	*Tam-1*	*Xld-1*	3.61 ± 2.05			753
7	*Tam-1*	*Mtv-1*			6.76 ± 3.82	1057
7	*Tam-1*	*Hbb*	27.66 ± 2.78		12.98 ± 7.34	949, 753
7	*Tam-1*	*Ngfg*			≤8.3†	469
7	*Tam-1*	*Xmmv-31*			14.00 ± 8.12	62
7	*Tam-1*	*Smg-1*			Near	952
7	*Tam-1*	*Smg-2*			Near	952
7	*Tam-1*	*D7Rp2*			*Tam-1* prox.	257
7	*D7Rp2*	*Lyb-8*			0.78 ± 0.79	1016
7	*D7Rp2-s*	*D7Rp2-r*			Very close	49
7	*Prt-4*	*Prt-5*			≤1.2†	771
7	*Prt-4*	*c*			22.90 ± 4.60	771
7	*Prt-5*	*Smg-2*			≤1†	954
7	*Prt-5*	*c*			22.90 ± 4.60	771
7	*Prt-5*	*Smg-1*			≤1†	773, 774
7	*Smg-1*	*c*			21.40 ± 2.31	773, 774
7	*Saa-1*	*c*	29.17 ± 9.28		10.73 ± 3.88	32, 33, 1038
7	*Saa-1*	*p*			26.31 ± 17.41	1038
7	*ru-2*	*p*	3.24 ± 1.20	1.72 ± 1.21		253
7	*ru-2*	*Mod-2*	20.30 ± 2.00	16.02 ± 2.73		174
7	*Ldh-1*	*Hbb*			25.30 ± 6.00	970
7	*Ldh-1*	*Lhb*			Syntenic	743
7	*Xld-1*	*Hbb*	27.71 ± 4.91			753
7	*p*	*H-4*			0	962
7	*p*	*twt*			≤4.1†	583
7	*p*	*Nil*			5.37 ± 1.45	1078
7	*p*	*hf*			13.24 ± 4.11	92
7	*p*	*tp*			7.69 ± 5.42	292, 1125
7	*p*	*c*	15.03 ± 0.38	9.14 ± 0.38	14.67 ± 0.34	108, 109, 196, 227, 130, 195, 277, 288, 831, 389, 388, 9
7	*p*	*sh-1*	16.92 ± 0.97	13.26 ± 1.82		834, 388
7	*p*	*Hbb*	15.00 ± 3.60	15.00 ± 3.60		950
7	*p*	*Mph-1*			26.16 ± 2.35	9
7	*p*	*Bsp*			Linked?	152
7	*Mtv-1*	*Xmmv-73*			≤14.7†	702
7	*Mtv-1*	*c*	25.87 ± 3.31	26.75 ± 2.81		1067, 1146
7	*Mtv-1*	*Mod-2r*			9.57 ± 4.37	1057
7	*Mtv-1*	*Hbb*			21.88 ± 10.66	1057
7	*Mtv-1*	*Fis-1*	50.00			938
7	*Nil*	*c*			10.33 ± 1.96	1078
7	*cht*	*Hbb*			7.70 ± 2.20	172
7	*hf*	*c*			2.94 ± 2.05	92
7	*tp*	*c*	2.26 ± 0.48			887
7	*Idh-2*	*Hbb*			5.30	489
7	*c*	*C58v-4*			≤22†	937
7	*c*	*Mod-2*	4.08 ± 2.83		0.49 ± 0.49	130, 241, 128
7	*c*	*Mod-2r*		1.59 ± 1.58		53
7	*c*	*Fgv-1*			6.01 ± 1.41	541
7	*c*	*sh-1*	4.09 ± 0.28	2.58 ± 0.45	5.67 ± 1.00	277, 389, 388, 329, 241, 234
7	*c*	*Hbb*	6.19 ± 0.65	3.46 ± 0.64	3.74 ± 0.70	517, 1112, 830, 130, 483, 685, 831, 53, 751, 937
7	*c*	*H-1*	8.09 ± 2.04	4.68 ± 3.46	9.01 ± 2.50	962, 965, 25
7	*c*	*Hma*			7.93 ± 2.11	751
7	*c*	*ol*	10.34 ± 3.20	14.10 ± 3.20	29.50 ± 3.54	412
7	*c*	*Ly-14*			17.14 ± 3.68	836
7	*c*	*Mph-1*			21.74 ± 4.97	9
7	*c*	*Ly-21*			27.00 ± 6.00	517

Table 6.1—*cont.*

Chr	Loci Proximal	Distal	Female	Male	Combined or not stated	References
7	c	Ahr-3	26.90 ± 5.00			682
7	c	Ex			39.00	923
7	C58v-4	Hbb			Near	499
7	Mod-2	sh-1			4.00	241
7	Mod-2	Hbb	6.44 ± 1.22	2.21 ± 1.09	4.66 ± 1.24	434, 174, 924, 128
7	Mod-2	Ly-15			15.70 ± 3.50	434
7	Mod-2c	Hbb			5.37 ± 1.57	128
7	Mod-2r	Hbb		19.00 ± 4.94	4.26 ± 2.94	53
7	Fgv-1	Hbb			8.04 ± 2.06	541, 873
7	Fgv-1	Hbb			Close	874
7	sh-1	Hbb	1.39 ± 0.62	2.02 ± 0.95		834
7	sh-1	War			17.00	1072
7	sh-1	fr	15.44 ± 2.96	16.39 ± 4.74		277
7	Hbb	H-1			≤1.2†	883, 267
7	Hbb	Hby		≤5.2†	≤1.8†	337, 988
7	Hbb	Hma			≤1.8†	751
7	Hbb	Ly-15			9.93 ± 2.51	434
7	Hbb	Ly-21			24.00 ± 5.00	517
7	Hbb	Sgp-2			19.48 ± 4.51	678
7	Hbb	Fis-1	24.60 ± 3.82			938
7	Hbb	Th			35.19 ± 4.50	81
7	Hbb	Gv-2			36.68 ± 3.83	992
7	Hbb	l(7)-1			In In(7)13Rk	863
7	Hbb	Lfo-1			Near	267
7	Hbb	tub			Close	242
7	Hbb	Hma			Linked?	181
7	H-1	Lfo-1			Near	267
7	War	fr			1.75 ± 0.01	1072, 1073
7	War	Ad			30.79 ± 4.39	1073
7	Ly-15	Th			23.15 ± 4.06	81
7	fr	Ad			26.46 ± 0.11	1072, 1073
7	Fis-1	Int-2			≤1.6†	935
7	Fis-1	Xmmv-76			1.01 ± 1.06	935
7	Fis-1	Th			3.33 ± 2.66	81
7	Ngfa	Ngfg			Synt., tandem	272
7	Chr 7	Ass-ps1			Syntenic	733
7	Chr 7	bh			Linked?	1077
7	Chr 7	Calc			Syntenic	748, 562
7	Chr 7	Hras-1			Syntenic	891, 545, 840, 890
7	Chr 7	Ins-2			Syntenic	557
7	Chr 7	Kal			Syntenic	680
7	Chr 7	Lhb			Syntenic	748, 743
7	Chr 7	Mmv-3			Syntenic	438
7	Chr 7	Mmv-4			Syntenic	438
7	Chr 7	Mylpf			Syntenic	158
7	Chr 7	Oat			Syntenic	762
7	Chr 7	Odc-6			0	853
7	Chr 7	Odc-7			0	853
7	Chr 7	Pth			Syntenic	562
7	Chr 7	Rps16			Syntenic	1098
7	Chr 7	Rras			Syntenic	892
7	Chr 7	Tgfb			Syntenic	321
8	Rb5Bnr	Os		26.76 ± 5.25		114
8	Rb11em	Aprt			44.44 ± 6.76	751a
8	nr	Os	26.61 ± 2.29		24.02 ± 1.21	382, 931
8	nr	e			50.80 ± 2.59	931
8	Gr-1	Es-1	15.48 ± 4.06	38.24 ± 8.33	30.23 ± 4.95	756, 720, 320

Table 6.1—*cont.*

Chr	Loci		Female	Male	Combined or not stated	References
	Proximal	Distal				
8	*Gr-1*	*Got-2*	25.20 ± 5.33	12.62 ± 4.18	32.56 ± 5.05	756, 320
8	*Gr-1*	*Tat*	35.29 ± 8.20			720
8	*Gr-1*	*Bv-1*			45.97 ± 3.07	541
8	*Gr-1*	*C58v-1*			42.59 ± 4.22	541
8	*Gr-1*	*e*			56.98 ± 5.34	320
8	*Gr-1*	*Aprt*			Syntenic	316, 537
8	*Gr-1*	*Ctrb*			Syntenic	465
8	*Gr-1*	*Mt-1*			Syntenic	150
8	*Hk*	*Os*	17.90 ± 2.38	7.61 ± 2.76	15.94 ± 4.41	362, 275
8	*Hp*	*Es-1*			6.94 ± 2.79	31, 32
8	*Q*	*Os*			4.73 ± 0.83	445, 446
8	*Ea-1*	*Es-1*			5.30 ± 3.60	309
8	*Ea-1*	*Es-5*			9.90 ± 6.30	309
8	*Os*	*tg*			≤1.7†	976, 361
8	*Os*	*Es-1*	2.00 ± 0.99	1.42 ± 1.00	0.46 ± 0.46	833, 1122, 877
8	*Os*	*Nan*	1.64 ± 1.63	0.51 ± 0.29	1.69 ± 1.69	634, 639
8	*Os*	*Es-6*		2.94 ± 2.05		1122
8	*Os*	*Es-2*		8.51 ± 2.35		1122
8	*Os*	*scb*			19.54 ± 4.25	47
8	*Os*	*hy-3*			16.84 ± 2.72	382
8	*Os*	*e*			30.46 ± 1.03	382, 275, 47, 907, 931, 284
8	*Eg*	*Es-1*			1.08 ± 1.07	511
8	*myd*	*Os*			≤3.5†	595
8	*myd*	*e*			30.00	595
8	*tg*	*Es-1*	5.83 ± 2.14		3.84 ± 1.98	1136, 1058
8	*Es-1*	*Es-6*		1.47 ± 1.46		1122
8	*Es-1*	*Es-9*	≤3.8†		0.18 ± 0.18	1158, 736
8	*Es-1*	*Es-22*			*Es-22* distal	254
8	*Es-1*	*Got-2*	1.59 ± 0.56		6.98 ± 3.88	806, 133, 185
8	*Es-1*	*Ucp*			4.30 ± 1.90	495
8	*Es-1*	*Erp-1*	8.86 ± 3.20	3.16 ± 1.75	6.40 ± 2.00	531
8	*Es-1*	*Es-2*	8.46 ± 0.78	9.27 ± 1.55	9.10 ± 1.09	832, 1084, 1158, 877, 1082, 1122, 736, 812
8	*Es-1*	*Es-7*	8.30 ± 2.80			1158
8	*Es-1*	*Es-23*	8.30 ± 2.80			1158
8	*Es-1*	*Es-5*	8.10 ± 0.91		9.50 ± 2.70	1082, 812
8	*Es-1*	*Prt-2*	12.75 ± 3.30	12.50 ± 11.69		1084
8	*Es-1*	*H-29*			17.00	25, 22
8	*Es-1*	*Bv-1*	24.93 ± 5.17			1037
8	*Es-1*	*Tat*	26.47 ± 7.57			720
8	*Es-1*	*Bv-1*	24.93 ± 5.17		30.23 ± 2.34	541, 688, 1037
8	*Es-1*	*C58v-1*			20.60 ± 3.25	541
8	*Es-1*	*Aprt*	29.27 ± 4.10	22.53 ± 2.63		751a
8	*Es-6*	*Es-9*			0.10 ± 0.09	1083
8	*Es-6*	*Es-2*		2.94 ± 2.05		1122
8	*Es-6, Es-9*	*Es-24*			Linked	1154a
8	*Es-9*	*Es-22*			≤2.3†	254
8	*Es-9*	*Es-2*	7.72 ± 2.14		9.27 ± 1.24	905, 1158, 736
8	*Es-9*	*Es-7*	8.30 ± 2.80			1158
8	*Es-9*	*Es-23*	8.30 ± 2.80			1158
8	*Got-2*	*Es-2*	6.57 ± 1.11		3.70	806, 133
8	*Got-2*	*Es-11*	6.57 ± 1.11			806
8	*Got-2*	*Bv-1*			27.81 ± 2.27	541
8	*Es-2*	*Es-11*	≤1†			806
8	*Es-2*	*Es-7*	≤3.8†		≤1.5†	1158, 133
8	*Es-2*	*Es-23*	≤3.8†			1158
8	*Es-2*	*Es-5*	≤1†		1.70 ± 1.30	1082, 806, 812

Table 6.1—*cont.*

Chr	Loci		Female	Male	Combined or not stated	References
	Proximal	Distal				
8	*Es-2*	*Prt-2*	3.92 ± 1.92	12.50 ± 11.69		1084
8	*Es-2*	*Es-23*			Near	1154
8	*Es-11*	*Es-5*	≤1†			806
8	*Es-7*	*Es-23*	≤3.8†			1158
8	*scb*	*e*			13.79 ± 3.70	47
8	*hy-3*	*e*			11.40 ± 3.20	382
8	*H-29*	*H-19*			18.00	25, 22
8	*Xmmv-29*	*Bv-1*			4.00 ± 2.69	62, 64
8	*C58v-1*	*e*			7.88 ± 3.02	541
8	*e*	*am*			30.00	697
8	*Mt-1*	*Mt-2*			Syntenic	916
8	Chr 8	*Aprt-ps*			Syntenic	286, 230
8	Chr 8	*Ctrb*			Syntenic	465
8	Chr 8	*Fabph-2*			Syntenic	415
8	Chr 8	*Mt-1*			Syntenic	150
8	Chr 8	*Mtv-21*			Syntenic	798
8	Chr 8	*Plat*			Syntenic	746, 845a
8	Chr 8	*Ram-1*			Syntenic	331
8	Chr 8	*Scl-1*			Linked?	57
9	*Rb163H*	*cw*			1.75 ± 1.23	632
9	*Rb6Bnr*	*cw*		2.50 ± 2.47		115
9	*Rb163H*	*se*	37.74 ± 6.67	37.50 ± 6.47	38.60 ± 4.56	632
9	*Rb163H*	*tk*			45.61 ± 4.66	632
9	*cw*	*Thy-1*	28.70 ± 3.03			217
9	*cw*	*Mpi-1*			36.00 ± 9.60	1123
9	*cw*	*d*	33.63 ± 3.16			217
9	*cw*	*se*	40.09 ± 2.72	36.55 ± 3.92	33.18 ± 1.97	273, 632
9	*cw*	*Mod-1*	40.67 ± 6.40			217
9	*lu*	*sg*			19.05 ± 3.24	358
9	*lu*	*d, se*	15.02 ± 3.52	22.91 ± 6.91	16.16 ± 2.85	384
9	*lu*	*tk*			23.77 ± 3.09	358
9	*lu*	*Mod-1*	12.50 ± 6.75			483
9	*lu*	*Trf*	23.81 ± 4.32	20.31 ± 3.56		92
9	*lu*	*sch*			46.82 ± 4.55	477
9	*Lap-1*	*Es-17*			1.20 ± 0.80	8
9	*Lap-1*	*Thy-1*	5.38 ± 1.98	3.64 ± 2.52		214
9	*Lap-1*	*Ups*			4.70 ± 1.60	8
9	*Lap-1*	*Apoa-1*		10.10 ± 3.23	6.73 ± 1.25	8, 244, 579, 629
9	*Lap-1*	*Xmmv-2*			8.00 ± 4.70	62
9	*Lap-1*	*Ncam*			8.20 ± 3.60	165
9	*Lap-1*	*P450-3*			10.00 ± 6.93	417
9	*Lap-1*	*Mpi-1*	12.31 ± 2.88	9.09 ± 3.88		214
9	*Lap-1*	*d*	21.25 ± 4.57		8.10 ± 4.75	1120
9	*Lap-1*	*Es-13*			17.54 ± 4.89	1121
9	*Lap-1*	*Mod-1*	28.05 ± 3.10	18.18 ± 5.20	18.30 ± 4.95	1120, 214, 1121
9	*Es-17*	*Ups*			2.95 ± 0.99	8
9	*Es-17*	*Apoa-1*			4.32 ± 1.17	8
9	*Es-17*	*Mpi-1*	8.70 ± 2.50			772
9	*Es-17*	*Mod-1*	18.90 ± 3.50			772
9	*Thy-1*	*Ncam*			5.30 ± 3.19	165
9	*Thy-1*	*P450-3*			6.60 ± 4.77	417
9	*Thy-1*	*Mpi-1*	8.15 ± 2.35	4.84 ± 2.73		214
9	*Thy-1*	*d*	12.29 ± 1.86	33.33 ± 10.29	19.57 ± 2.56	494, 217, 493, 365, 617
9	*Thy-1*	*Mod-1*	21.04 ± 2.38	20.29 ± 3.26	18.80 ± 4.40	217, 60, 214, 493
9	*Thy-1*	*fd*			25.00 ± 7.65	365
9	*Thy-1*	*Trf*	39.61 ± 3.90	36.67 ± 5.00	47.50 ± 10.40	467, 60
9	*Thy-1*	*Fv-2*			35.20 ± 4.27	617

Table 6.1—*cont.*

Chr	Loci		Female	Male	Combined or not stated	References
	Proximal	Distal				
9	*Ups*	*Apoa-1*			1.34 ± 0.65	8
9	*Ups*	*Mpi-1*	14.80 ± 5.00			7
9	*Ups*	*Mod-1*	37.00 ± 7.00			7
9	*Apoa-1*	*Xmmv-2*			2.00 ± 1.68	62
9	*Apoa-1*	*Ncam*			2.78 ± 1.26	165
9	*Apoa-1*	*Mpi-1*		15.65 ± 3.39		244
9	*Apoa-1*	*Pk-3*	10.60 ± 2.40			805
9	*Apoa-1*	*Idd-2*			*Idd-2* prox.	842
9	*Apoa-1*	*ash*			21.52 ± 4.62	579
9	*Apoa-1*	*D9Ag3*			19.57 ± 11.34	63
9	*Apoa-1*	*d*			10.00 ± 5.66	244
9	*Apoa-1*	*Es-13*		26.30 ± 10.10	10.00 ± 9.80	244
9	*Apoa-1*	*Mod-1*	23.03 ± 3.16	23.16 ± 4.33	15.00 ± 9.62	8, 244, 629
9	*Apoa-1*	*Acy-1*	35.96 ± 3.60	30.53 ± 4.72		8
9	*Rnu1-2*	*Apoa-1*			1.00	400
9	*Xmmv-2*	*Ncam*			2.40 ± 1.85	165
9	*P450-3*	*Pgm-3*			32.03 ± 26.44	417
9	*P450-3*	*Bgl*			32.03 ± 26.44	417
9	*Aft*	*se*			10.10 ± 4.10	580
9	*Aft*	*sch*			37.00 ± 6.60	580
9	*Mpi-1*	*rc*			1.87 ± 1.31	235
9	*Mpi-1*	*d*			13.30 ± 3.70	550
9	*Mpi-1*	*d,se*			10.89 ± 3.10	754
9	*Mpi-1*	*Emv-3*			26.08 ± 5.30	541
9	*Mpi-1*	*Mod-1*	15.78 ± 1.47	11.59 ± 1.52	18.81 ± 3.89	72, 754, 7, 269, 214, 772, 244, 215
9	*Mpi-1*	*Hst-2*	7.69 ± 7.39			72
9	*Mpi-1*	*Gdc-2*			32.53 ± 5.14	550
9	*sg*	*d,se*			4.49 ± 0.83	358
9	*Pk-3*	*d*			Close	503
9	*Pk-3*	*d,se*	7.90 ± 1.70			801
9	*Pk-3*	*Pgm-3*	15.90 ± 3.10			801
9	*Pk-3*	*Mod-1*	10.79 ± 1.27			805, 801
9	*Pk-3*	*Mod-1*			Near	503
9	*Idd-2*	*Mod-1*			*Idd-2* prox.	842
9	*ash*	*d*			0.91 ± 0.90	578
9	*ash*	*Mod-1*			6.93 ± 2.42	579
9	*D9Ag3*	*d*			1.10 ± 1.13	63
9	*D9Ag3*	*Pgm-3*			7.10 ± 4.06	63
9	*D9Ag3*	*Mod-1*			9.20 ± 5.12	63
9	*d*	*Emv-3*			≤1.5†	497
9	*d*	*se*	0.12 ± 0.90	0.74 ± 0.73	0.13 ± 0.05	330, 966, 344, 107, 801, 963
9	*d*	*Pgm-3*			6.85 ± 2.96	504
9	*d*	*tk*	4.00 ± 0.79	2.99 ± 0.74	4.29 ± 0.64	358, 575
9	*d*	*Mod-1*	8.48 ± 1.76		6.88 ± 1.04	1120, 483, 217, 504, 493, 244, 365, 550, 134, 787
9	*d*	*fd*	9.76 ± 1.72	8.08 ± 2.15	14.46 ± 3.73	575, 365
9	*d*	*Trf*	16.92 ± 2.62	8.51 ± 2.61		925
9	*d*	*Fv-2*	16.00 ± 7.33		17.91 ± 2.84	616, 617
9	*d*	*Gdc-2*			19.39 ± 2.96	550
9	*d*	*du*			21.26 ± 1.70	964, 358
9	*d*	*Bgl-s*	25.00 ± 4.51		20.78 ± 4.61	920, 134
9	*d*	*Mylc*			29.73 ± 7.51	30
9	*d*	*Bmfr-2*			Near	929
9	*d,se*	*Pgm-3*	6.80 ± 1.80			801
9	*d,se*	*Mod-1*	9.00 ± 1.80		7.92 ± 2.69	801, 754
9	*Emv-3*	*Mod-1*			18.17 ± 3.22	541
9	*se*	*sv*	2.65 ± 0.65	1.75 ± 1.74	≤3†	193

Table 6.1—*cont.*

Chr	Loci		Female	Male	Combined or not stated	References
	Proximal	Distal				
9	*se*	*tk*	5.19 ± 1.23	5.87 ± 1.92	7.89 ± 2.27	273, 632
9	*se*	*du*	21.56 ± 4.14	15.04 ± 7.00		384
9	*se*	*sch*			25.00 ± 3.13	474
9	*Es-13*	*Mod-1*			2.60 ± 2.60	1121, 244
9	*Pgm-3*	*Mod-1*	≤1.6†		≤1.5†	801, 504
9	*Pgm-3*	*Crbp-1*			1.95 ± 1.21	192
9	*Pgm-3*	*Crbp-2*			1.95 ± 1.21	192
9	*Crbp-1*	*Crbp-2*			≤2.8†	192
9	*tk*	*fd*	5.05 ± 1.27	6.83 ± 1.99		575
9	*tk*	*Fv-2*			30.80 ± 5.00	617
9	*tk*	*du*			11.99 ± 1.35	358
9	*tk*	*hpx*			Near	52
9	*hpx*	*du*			Near	52
9	*Mod-1*	*Hst-2*	≤20.6†			72
9	*Mod-1*	*Sep-2*			Linked	239
9	*Mod-1*	*fd*			5.26 ± 2.96	365
9	*Mod-1*	*Kfo-1*			5.90	211
9	*Mod-1*	*Acy-1*	12.92 ± 2.51	8.80 ± 1.47	5.17 ± 3.59	8, 732, 744
9	*Mod-1*	*Trf*	8.27 ± 1.63	9.36 ± 1.25	4.60	60, 805, 732, 126
9	*Mod-1*	*Ltw-3*			5.90 ± 3.63	263
9	*Mod-1*	*Fv-2*			12.96 ± 7.33	263
9	*Mod-1*	*H-7*			12.00 ± 5.00	25
9	*Mod-1*	*Gdc-2*			17.85 ± 2.95	550
9	*Mod-1*	*Bgl-e*		13.86 ± 3.40		215
9	*Mod-1*	*Bgl-s*			14.54 ± 1.40	134, 787
9	*Mod-1*	*sr*	20.44 ± 3.45			252a
9	*Mod-1*	*Es-14*	23.61 ± 5.00			71
9	*Kfo-1*	*Acy-1*			≤2.1†	744
9	*Kfo-1*	*Ltw-3*			≤1.5†	258
9	*Kfo-1*	*Fv-2*			5.70	211
9	*Kfo-1*	*Bgl-s*			11.11 ± 5.24	211
9	*Acy-1*	*Trf*		2.44 ± 0.83		732
9	*Acy-1*	*Ltw-3*			≤2.1†	744
9	*Acy-1*	*Fv-2*			6.36 ± 3.00	744
9	*Acy-1*	*sr*	6.92 ± 2.23			252a
9	*Acy-1*	*Bgl*			7.69 ± 3.53	744
9	*Trf*	*sr*	0.94 ± 0.66			252a
9	*Trf*	*Bgl-s*			7.70	126
9	*Ltw-3*	*Fv-2*			5.70 ± 2.64	263
9	*Ltw-3*	*H-7*			Close	267
9	*Fv-2*	*H-7*			Near	25
9	*Fv-2*	*Bgl*			*Fv-2* proximal	18
9	*Fv-2*	*Bgl-s*			5.18 ± 2.23	211, 263
9	*Fv-2*	*Emv-18*			8.23 ± 4.69	719
9	*Fv-2*	*Rv-2*			Near	409
9	*H-7*	*Bgl*			*H-7* proximal	18
9	*du*	*sch*			18.33 ± 6.40	477
9	*sr*	*sch*	10.97 ± 1.69			252a
9	*Bgl-e*	*Bgl-s*			0	78
9	*Bgl-s*	*Bgl-t*			≤1.8†	787
9	*sch*	*tip*			20.00	606
9	*Es-14*	*Mylc*			≤7.8†	30
9	Chr 9	*Cp*			Syntenic	27
9	Chr 9	*Ets-1*			Syntenic	1085
9	Chr 9	*Gia*			Syntenic	13
9	Chr 9	*Gsta*			Syntenic	159
9	Chr 9	*Ltf*			Syntenic	747

Table 6.1—*cont.*

Chr	Loci		Female	Male	Combined or not stated	References
	Proximal	Distal				
9	Chr 9	T3d			Syntenic	1148
9	Chr 9	T3e			Syntenic	341
9	Chr 9	T3g			Syntenic	553
10	Rb8Bnr	dy			17.60	93
10	dy	dl			16.17 ± 2.33	998
10	dy	av			26.89 ± 1.75	882
10	dy	gr			23.75 ± 2.51	998
10	dy	Sl			31.93 ± 2.17	1113
10	gl	dl			5.39 ± 1.53	597, 589, 577, 1011
10	gl	av			35.37 ± 5.47	597
10	gl	Sl	31.05 ± 4.77	17.65 ± 15.16		597
10	v	kd			≤32.8†	642
10	v	dl	1.87 ± 0.93		0.96 ± 0.48	370, 998, 153, 154
10	v	Hk-1			3.03 ± 2.11	799
10	v	Apk			6.98 ± 3.88	1115
10	v	gr	13.13 ± 2.61	12.80 ± 5.40	16.17 ± 2.62	65, 645, 998, 37, 799
10	v	ji	40.00 ± 15.49	10.20 ± 4.32	26.20 ± 8.20	967
10	v	Sl			28.48 ± 4.89	37, 282
10	v	Cat	40.71 ± 3.55	46.56 ± 3.74		645
10	kd	gr			16.40 ± 3.53	642
10	dl	Hk-1			1.04 ± 1.04	799
10	dl	Apk			≤6†	1115
10	dl	av			12.68 ± 1.01	597, 585
10	dl	gr			13.82 ± 0.79	366, 998
10	dl	Sl			22.42 ± 1.07	597, 910, 799
10	jc	dl			≤16.3†	251
10	jc	gr			16.60 ± 5.70	366
10	jc	Sl			28.71 ± 3.13	975
10	jc	eb			37.64 ± 4.07	251
10	Hk-1	gr			15.15 ± 4.41	799
10	Hk-1	Sl			18.75 ± 3.98	799
10	Apk	gr			6.25 ± 3.49	1115
10	Apk	Pep-2	21.82 ± 3.94		21.61 ± 3.92	1114, 1114a
10	Apk	Sl			26.83 ± 4.89	1114a
10	av	Sl	11.71 ± 2.43		15.33 ± 1.24	597, 382, 901, 734
10	av	si	32.14 ± 3.12	34.98 ± 3.35		902
10	av	eb			28.81 ± 3.10	123
10	mh	gr			≤1.2†	565
10	mh	Sl			20.10 ± 4.10	366
10	gr	ji			3.34	374
10	gr	Sl	8.14 ± 2.38	2.60 ± 2.50	16.36 ± 5.60	645, 884, 37
10	Sl	Pep-2			3.66 ± 2.07	1114, 1114a
10	Sl	his			4.30	1127
10	Sl	Hsd			11.10 ± 4.70	10
10	Sl	pg			14.20 ± 4.04	274
10	Sl	si	19.38 ± 2.99			382, 901
10	Sl	eb			20.19 ± 1.97	123
10	Sl	Cat	19.21 ± 3.98	22.63 ± 2.06		645, 644
10	Pep-2	Cs			Syntenic	149
10	Pep-2	Ifg			Syntenic	745
10	pg	si			20.10 ± 5.92	274
10	eb	at			7.14 ± 1.66	476, 1011
10	Ifg	Pyp			Syntenic	745
10	Chr 10	Ass-ps2			Syntenic	733
10	Chr 10	Cs			Syntenic	149
10	Chr 10	Ifg			Syntenic	745
10	Chr 10	Ifgr			Syntenic	669

Table 6.1—*cont.*

Chr	Loci		Female	Male	Combined or not stated	References
	Proximal	Distal				
10	Chr 10	Mdm-1			Syntenic	94
10	Chr 10	Mdm-2			Syntenic	94
10	Chr 10	Pyb			Syntenic	891, 890
10	Chr 10	Pyp			Syntenic	560
11	Rb4Bnr	wa-2			4.00 ± 2.77	113
11	Rb4Bnr	vt			28.00 ± 6.35	113
11	Rb4Bnr	Re			56.25 ± 6.20	113
11	Tcn-2	wa-2	6.92 ± 2.23			319
11	Tcn-2	Hba	18.63 ± 2.79	10.50 ± 3.70	7.10 ± 4.00	319
11	Tcn-2	Mtv-3			12.40 ± 9.17	705
11	Tcn-2	Es-3	48.60 ± 8.40			319
11	wa-2	Hba	12.31 ± 2.88		9.04 ± 1.57	319, 880
11	wa-2	Chy		12.50 ± 5.80		633
11	wa-2	vt	22.15 ± 1.69		26.99 ± 3.03	707, 915, 113
11	wa-2	sh-2	23.82 ± 0.84	29.49 ± 1.40	25.94 ± 1.17	299, 961, 98, 202, 1128, 413, 515
11	wa-2	ti	24.92 ± 1.85			915
11	wa-2	vb			31.70 ± 4.00	1090
11	wa-2	oe			36.08 ± 2.32	515
11	wa-2	Re	42.17 ± 2.54	43.60 ± 2.50	41.00 ± 4.45	105, 98, 280
11	wa-2	Al	39.46 ± 2.33			202
11	Erbb	Hba			5.37 ± 2.11	936
11	Erbb	Hox-2			33.30 ± 26.32	480a
11	Erbb	Il-3			20.00 ± 7.30	486
11	Hba	Hbx			Close	1096
11	Hba	sh-2			23.48 ± 3.69	880
11	Hba	Hox-2			26.90 ± 20.39	400a
11	Hba	Myhs	31.76 ± 8.30			857, 1094
11	Hba	Dlb-1	36.48 ± 5.60	40.37 ± 4.54		1061
11	Hba	Re	33.33 ± 11.11	16.67 ± 10.76		704
11	Hba	Rnu1a-1	30.77 ± 12.80	9.00 ± 11.11	2.08 ± 2.16	704
11	Hba	Myla	61.00 ± 6.19			857, 1094
11	Hba	Es-3		63.01 ± 4.48		1061
11	Chy	vt	5.70 ± 3.20	7.40 ± 2.50	0	633, 638
11	Chy	Re	35.47 ± 4.41	27.34 ± 2.40	26.65 ± 2.35	633, 638
11	vt	sh-2			1.72 ± 9.50	707
11	vt	ti	8.93 ± 1.16			915
11	vt	vb			0.90 ± 0.90	1090
11	vt	Re	28.71 ± 2.80	18.24 ± 1.89	24.00 ± 3.80	633, 706, 245
11	vt	Al	44.83 ± 5.33			202
11	Csfgm	Csfmu			≤7.8†	28
11	Sparc	Csfmu			≤7.8†	28
11	Sparc	Csfgm			≤7.8†	28
11	Sparc	Hox-2			21.43 ± 15.81	400a
11	H(lt)	lt			32.00 ± 7.00	20
11	df	Re	22.53 ± 2.24			29
11	sh-2	Tr	1.59 ± 1.11			278
11	sh-2	vb			35.04 ± 6.23	1090, 593, 597
11	sh-2	oe			17.15 ± 1.21	515
11	sh-2	Re	14.76 ± 2.18	19.23 ± 1.82	27.31 ± 2.89	280, 738, 98, 515
11	sh-2	Al	26.10 ± 1.55			597, 202
11	sh-2	tn			44.41 ± 4.92	597, 586
11	ti	Re	19.17 ± 2.12	20.45 ± 2.14		915
11	vb	Re	18.04 ± 1.92	17.59 ± 5.38		1090, 597
11	Tr	Akv-4	4.10 ± 2.80			1037
11	Tr	nu	7.41 ± 3.56			306
11	Tr	Hox-2	20.55 ± 4.73			400a
11	Tr	Re	21.97 ± 2.34			306, 278, 400a

Table 6.1—*cont.*

Chr	Loci		Female	Male	Combined or not stated	References
	Proximal	Distal				
11	*Akv-4*	*Hox-2*			8.94 ± 3.80	400a
11	*Akv-4*	*Re*	37.00 ± 7.00	12.00 ± 5.00		1037
11	*Akv-4*	*Es-3*	50.00 ± 7.14			1037
11	*nu*	*Re*	11.11 ± 4.28		19.40 ± 4.60	306
11	*oe*	*co*			Close	809
11	*oe*	*Re*			11.97 ± 1.08	55, 515
11	*oe*	*Es-3*			29.12 ± 4.41	55
11	*oe*	*tn*			42.83 ± 3.87	515
11	*Myhs*	*Acrb*	≤6.9†			407
11	*Myhs*	*Myla*	26.19 ± 6.78			857, 1094
11	*Myhs*	*Es-3*	27.03 ± 7.30			857, 1094
11	*H(js)*	*js*			27.00 ± 8.00	20
11	*Dlb-1*	*Re*	2.32 ± 1.31	3.45 ± 3.39		1061
11	*Dlb-1*	*Es-3*	23.08 ± 4.13	24.08 ± 3.02	7.70 ± 5.82	1061
11	*Bda*	*Re*			≤3.2†	1069
11	*Hox-2*	*Re*	1.09 ± 1.08			400a
11	*Hox-2*	*Glk*			31.82 ± 26.24	400a
11	*Hox-2*	*Es-3*			16.65 ± 11.10	400a
11	*Re*	*Rnu1a-1*			≤12.7†	704
11	*Re*	*Bsk*			≤1.8†	650, 651
11	*Re*	*shm*			1.50 ± 0.70	354
11	*Re*	*lt*			3.97 ± 1.74	205
11	*Re*	*Al*	22.15 ± 3.94	13.43 ± 2.08		202, 597
11	*Re*	*Ts*	10.58 ± 2.45			1007, 1002
11	*Re*	*Bbv*			9.15 ± 2.40	541
11	*Re*	*Es-3*	15.70 ± 2.48	18.00 ± 2.72	19.77 ± 2.40	1007, 1037, 1061, 860, 862, 55
11	*Re*	*tn*	22.83 ± 2.67	15.71 ± 6.17	21.00	597, 586
11	*Re*	*js*			16.67 ± 6.80	862
11	*Rnu1a-1*	*Mtv-3*			7.90 ± 5.96	704
11	*Rnu1a-1*	*Es-3*			12.50 ± 9.24	704
11	*Ts*	*Es-3*	5.17 ± 2.19			1007
11	*Bbv*	*Es-3*			12.50 ± 3.70	541
11	*Mtv-3*	*Es-3*			3.12 ± 0.91	758, 705
11	*Glk*	*Myla*			≤10.6†	857
11	*Glk*	*Es-3*			2.50 ± 1.60	709
11	*Glk*	*Umph-2*			Syntenic	1109
11	*Myla*	*Es-3*	8.11 ± 3.17		≤10.6†	857, 1094
11	*Es-3*	*js*			3.33 ± 3.28	862
11	Chr 11	*Sparc*			Syntenic	681
11	Chr 11	*Il-3*			Syntenic	486
11	Chr 11	*Csfgm*			Syntenic	346
11	Chr 11	*Erba*			Syntenic	723, 890, 1139
11	Chr 11	*Hox-2*			Synt., distal	723, 845, 507
11	Chr 11B3-C	*Mov-9*			Synt., *in situ*	725
11	Chr 11C-E	*Mov-13*			Synt., *in situ*	725
11	Chr 11	*Mmv-8*			Syntenic	438
11	Chr 11	*Mmv-11*			Syntenic	438
11	Chr 11	*Mmv-13*			Syntenic	438
11	Chr 11	*Rel*			Syntenic	83
11	Chr 11	*Rpo2-1*			Synt., *in situ*	838
11	Chr 11	*Trp53*			Syntenic	160, 869
11	Chr 11	*Tse-1*			Syntenic	810, 524
11	Chr 11C-E	*Cola-1*			Synt., *in situ*	724
11	Chr 11B5-D1	*Tk-1*			Syntenic	900
11	Chr 11	*Xmv-4*			Syntenic	438
11	Chr 11	*Xmv-5*			Syntenic	438
12	*Rb5Bnr*	*hyt*	27.66 ± 3.77			34

Table 6.1—*cont.*

Chr	Loci		Female	Male	Combined or not stated	References
	Proximal	Distal				
12	*Rnr12*	*Apob*			≤8.3†	628
12	*Rnr12*	*D12Nyu2*			2.38 ± 2.61	628
12	*Rnr12*	*Ly-18*			10.00 ± 8.00	628
12	*Rnr12*	*D12Nyu1*			16.67 ± 13.61	628
12	*Apob*	*D12Nyu2*			3.12 ± 2.48	628
12	*Apob*	*Ly-18*			10.24 ± 3.46	628
12	*Apob*	*Ah*			12.07 ± 6.67	628
12	*Apob*	*D12Nyu1*			16.67 ± 7.86	628
12	*Pol-7*	*D12Nyu2*			1.50 ± 1.60	867
12	*Pol-7*	*D12Nyu5*			3.30 ± 2.65	867
12	*Pol-7*	*D12Nyu1*			16.70 ± 11.14	867
12	*D12Nyu2*	*Ah*			6.82 ± 3.52	628
12	*D12Nyu2*	*Ly-18*			5.17 ± 3.59	628
12	*D12Nyu2*	*D12Nyu1*			13.70 ± 2.00	59, 166
12	*D12Nyu2*	*D12Nyu3*			12.80 ± 2.30	59
12	*D12Nyu2*	*Fos*			25.77 ± 2.56	59
12	*D12Nyu2*	*D12Nyu4*			26.47 ± 3.38	59
12	*D12Nyu2*	*Mtv-9*			30.21 ± 2.71	59
12	*D12Nyu2*	*Aat*			36.62 ± 2.86	59
12	*D12Nyu2*	*Igh-C*			49.31 ± 2.94	59
12	*Ah*	*D12Mcg1*			8.60 ± 3.28	139
12	*Ah*	*D12Nyu1*			8.11 ± 3.08	628
12	*Ah*	*Caa*			Linked	789
12	*Ly-18*	*Ah*			≤3.6†	628
12	*Ly-18*	*D12Nyu1*			4.39 ± 2.30	628
12	*Lamb-1*	*Ly-18*			Close	266
12	*Ltw-2*	*Ly-18*			0	1047
12	*D12Nyu1*	*D12Nyu3*			5.83 ± 1.63	59, 166
12	*D12Nyu1*	*D12Nyu5*			7.10 ± 4.10	166
12	*D12Nyu1*	*Fos*			17.31 ± 2.15	59, 166
12	*D12Nyu1*	*D12Nyu4*			19.76 ± 3.08	59
12	*D12Nyu1*	*Mtv-9*			21.71 ± 2.36	59
12	*D12Nyu1*	*Aat*			29.35 ± 2.66	59
12	*D12Nyu1*	*Igh-C*			43.14 ± 2.86	59
12	*D12Nyu5*	*D12Nyu3*			3.80 ± 4.39	166
12	*D12Nyu5*	*Fos*			21.43 ± 22.32	166
12	*D12Nyu3*	*D12Nyu4*			16.13 ± 2.95	59
12	*D12Nyu3*	*Fos*			14.38 ± 2.37	59, 166
12	*D12Nyu3*	*Mtv-9*			18.10 ± 2.66	59
12	*D12Nyu3*	*Aat*			28.57 ± 3.17	59
12	*D12Nyu3*	*Igh-C*			40.09 ± 3.37	59
12	*Fv-4*	*Aat*			17.65 ± 2.48	765, 487
12	*Fv-4*	*Igh-1*			23.66 ± 2.71	765, 487
12	*bd*	*Aat*			16.00 ± 7.33	1005
12	*Fcr*	*Igh-C*			32.05 ± 5.28	30a
12	*Fcr*				Near *Lm-1*	30a
12	*Xmmv-25*	*Aat*			12.00 ± 6.77	62
12	*Xmmv-25*	*Xmmv-21*			10.00 ± 5.66	62
12	*Lyb-7*	*Xmmv-25*			0	62
12	*Lyb-7*	*Igh-Np*			16.65 ± 11.10	993
12	*Lyb-7*	*Igh-1*			23.59 ± 4.56	993
12	*Fos*	*D12Nyu4*			2.35 ± 1.16	59, 166
12	*Fos*	*Mtv-9*			4.61 ± 1.20	59
12	*Fos*	*Aat*			12.58 ± 1.87	59, 166
12	*Fos*	*Igh-C*			27.06 ± 2.46	59, 166
12	*Ly-7*	*Aat*			11.30 ± 4.33	435
12	*Ly-7*	*Igh-C*			19.00 ± 6.96	435

Table 6.1—*cont.*

Chr	Loci		Female	Male	Combined or not stated	References
	Proximal	Distal				
12	D12Nyu4	Mtv-9			2.94 ± 1.30	59
12	D12Nyu4	Aat			11.04 ± 2.45	59
12	D12Nyu4	Igh-C			24.51 ± 3.23	59, 166
12	Mtv-9	Aat			10.77 ± 1.80	59
12	Mtv-9	Igh			12.00	95, 162
12	Mtv-9	Igh-C			25.74 ± 2.51	59
12	sm	Aat			8.57 ± 4.73	403
12	sm	twi			12.10 ± 3.07	1014
12	Lm-1	Aat			5.00 ± 2.35	763
12	Lm-1	Tsu			4.55 ± 2.19	763
12	Xmmv-21	Aat			4.00 ± 2.63	62
12	Xmmv-21	Xmmv-34			16.50 ± 9.51	62
12	D12Rp54	Aat			1.16 ± 1.21	422
12	Aat	Pre-2			≤2.3†	1104
12	Aat	Spi-1			≤3.4†	425
12	Aat	Tpre			10.00 ± 8.00	777
12	Aat	Xmmv-34			7.00 ± 4.09	62
12	Aat	Igh-C			9.00	400
12	Aat	Igh-C			16.13 ± 1.83	59, 166
12	Aat	Igh-1	1.90 ± 1.80	13.60 ± 4.50	8.04 ± 1.16	695, 765, 554, 487, 1031, 1089, 435
12	Aat	H-40			Close, near	307, 411
12	Spi-1	Spi-2			1.06 ± 1.11	425
12	Spi-2	Spi-2r			0	425
12	Spi-1	Igh-1			12.07 ± 6.67	425
12	Spi-2	Igh-1			15.38 ± 8.62	425
12	H-40	Igh-C			*Igh* distal	411
12	Tind	Tsu			Close	983
12	Tind	Tpre			Close	779
12	Tthy	Tpre			Close	779
12	Tsu	Tpre			2.37 ± 2.60	777
12	Tsu	Tpre			Close	779
12	Tind	Igh-C			50.00 ± 57.74	777
12	Tind	Igh-C			Near	983
12	Tthy	Igh-C			50.00 ± 57.74	777
12	Tsu	Igh-C			1.94 ± 1.43	777, 776
12	Tpre	Igh-C			Near	779
12	Xmmv-34	Igh-C			4.00 ± 2.63	62
12	Xmmv-34	Igh-Sa2			4.00 ± 2.63	62
12	Xmmv-34	Igh-Sa4			3.00 ± 2.15	62
12	Xmmv-50	Igh-C			1.60 ± 1.71	1091
12	Bcga	Aat			11.00 ± 6.34	906
12	Bcga	Igh-Src			8.00 ± 4.70	906
12	Bcga	Igh-C			2.50 ± 1.94	906
12	Igh-C	Crip			≤1.3†	56
12	Igh-1	Igh-7			≤9.2†	75
12	Igh-1	Igh-Src			7.60 ± 4.40	1089
12	Igh-1	Igh-Sa2			1.11 ± 1.16	1089
12	Igh-1	Igh-Gte			1.00 ± 0.74	509
12	Igh-1	Igh-Dex			2.02 ± 1.90	851
12	Igh-1	Igh-Np			3.80 ± 2.53	1089
12	Igh-1	Igh-Nbp			Linked	1089
12	Igh-1	Igh-Bgl			Linked	1089
12	Igh-C	Igh-Ns2			7.45 ± 2.03	825
12	Igh-Sa2	Igh-Sa4			2.40 ± 1.85	1089
12	Igh-Sa4	Igh-Np			1.11 ± 1.16	1089
12	Igh-Sa4	Igh-Nbp			1.11 ± 1.16	1089
12	Igh-Np	Igh-Src			1.11 ± 1.16	1089

Table 6.1—*cont.*

Chr	Loci		Female	Male	Combined or not stated	References
	Proximal	Distal				
12	*Igh-Nbp*	*Igh-Src*			1.11 ± 1.16	1089
12	*Igh-Np*	*Igh-Nbp*			0	1089
12	*Igh-V*	*Ly-16*			Near	296
12	Chr 12	*Acp-1*			Syntenic	316
12	Chr 12	*Es-25*			Linked	1048
12	Chr 12	*Lamb-1*			Syntenic	266
12	Chr 12	*Meta*			Syntenic	609, 839
12	Chr 12	*Mmv-6*			Syntenic	438
12	Chr 12	*Mmv-7*			Syntenic	438
12	Chr 12	*Odc-8*			0	853
12	Chr 12	*Pomc-1*			Syntenic	106 8
12	Chr 12	*Rnr12(NO)*			Syntenic	268
12	Chr 12	*Spnb-1*			Syntenic	608
13	*Rb4Bnr*	*Xt*			0	117
13	*T199H*	*bg*	17.19 ± 4.72			636
13	*cr*	*bg*			0.24 ± 0.17	590
13	*cr*	*Xt*	1.85 ± 0.65	0.62 ± 0.44	0.82 ± 0.82	648, 501
13	*cr*	*ch*			9.09 ± 1.87	383, 822
13	*cr*	*f*	33.79 ± 2.48		28.25 ± 2.44	530, 822, 355
13	*cr*	*fs*			32.86 ± 4.15	355
13	*bg*	*Xt*	0.44 ± 0.44	0.68 ± 0.68	0.83 ± 0.31	647, 590
13	*bg*	*Hist1*	5.00 ± 4.87	3.23 ± 3.18		176, 778
13	*bg*	*sa*	6.28 ± 1.40	3.72 ± 1.42	9.00 ± 2.75	986, 176, 636, 647
13	*bg*	*f*			27.15 ± 2.27	355
13	*bg*	*fs*			31.87 ± 2.00	355
13	*Xt*	*Hist1*	2.27 ± 2.24			176
13	*Xt*	*sa*	8.25 ± 1.15	2.70 ± 1.30		647, 443
13	*Xt*	*mu*	18.55 ± 2.61	10.55 ± 2.18		647
13	*Xt*	*dpy*	25.00 ± 2.33	15.48 ± 1.38		443
13	*Xt*	*f*	22.61 ± 2.40	18.76 ± 1.77		648
13	*Xt*	*pcd*			32.00	974
13	*Xt*	*pe*	40.94 ± 1.83	30.65 ± 1.95		648, 647
13	*Tcrg*	*Hist1*			1.50 ± 1.11	778
13	*Hist1*	*sa*	5.00 ± 4.87	3.23 ± 3.18		176
13	*Hist1-1s*	*Hist1-2bs*			≤5.8†	1142, 176
13	*sa*	*ch*			5.80 ± 1.80	383
13	*sa*	*dpy*	17.73 ± 2.06	10.27 ± 1.16		443
13	*sa*	*pe*	45.31 ± 6.22			636
13	*ch*	*f*			8.81 ± 2.51	822
13	*mu*	*pe*	29.41 ± 3.07	23.62 ± 3.01		647
13	*f*	*fs*			9.57 ± 1.30	355
13	*f*	*pe*			19.57 ± 4.10	355
13	*pcd*	*pe*			18.79 ± 4.80	974
13	*fs*	*pe*			11.48 ± 1.07	355
13	*pe*	*Lth-1*			0	262
13	*pe*	*As-1*			11.54 ± 6.30	262
13	*Lth-1*	*As-1*			2.17 ± 0.96	262
13	Chr 13	*Dhfr*			Syntenic	525
13	Chr 13	*Hexb*			Syntenic	525
14	*cen*	*Np-1*	15.80 ± 2.40			269
14	*Rb6Bnr*	*Np-1*	26.50 ± 3.60		8.89 ± 4.24	1117, 269
14	*Adk*	*Np-1*			Syntenic	613
14	*Adk*	*Es-10*			Syntenic	613
14	*Ms15-7*	*Mtv-11*			4.05 ± 2.71	496
14	*Ms15-7*	*Ms6-5*			4.84 ± 3.32	496
14	*pn*	*hr*	13.54 ± 1.26			523

Table 6.1—*cont.*

Chr	Loci		Female	Male	Combined or not stated	References
	Proximal	Distal				
14	pn	s	27.63 ± 1.02	24.39 ± 6.65	27.90 ± 3.00	522, 523
14	Myhc-a	Np-1	2.85 ± 2.81			1094
14	gm	Wc			2.02 ± 1.41	379, 380
14	gm	s			29.72 ± 6.92	250
14	gm	Rn			32.32 ± 4.70	380
14	Wc	Np-1			≤6†	1117
14	Wc	hr			10.53 ± 3.15	379
14	Wc	Es-10			6.25 ± 3.49	1117
14	Wc	Rn			30.30 ± 4.62	380
14	Np-1	Np-2			0	623
14	Np-1	hph-1	≤9.8†			67
14	Np-1	Es-10			9.67 ± 1.88	1117
14	Np-1	s			25.77 ± 4.44	1117
14	Np-2	Tcra			0	191
14	Np-2	Rib-1			0	265
14	Np-2	H-8			0	191
14	Np-2	Es-10	13.95 ± 3.73		8.17 ± 1.94	623, 1050
14	Np-2r	Es-10			34.38 ± 8.40	969
14	Ms6-5	Tcra			1.28 ± 1.35	496
14	Tcra	Rib-1			≤3.3†	265
14	Tcra	H-8			≤18†	191, 50
14	Rib-1	H-8			≤18†	265
14	Rib-1	Es-10			6.90 ± 3.30	265
14	Rib-1	For-5			6.90 ± 3.30	265
14	Ds	hr			5.67 ± 0.67	479
14	Ds	s			25.24 ± 1.23	479
14	wl	hr			3.79 ± 0.73	566
14	hr	ag			0	428
14	hr	wal			7.84 ± 0.93	971
14	hr	s	12.84 ± 0.86	2.39 ± 2.41	19.60 ± 0.77	597, 966, 377, 637, 523, 379, 479
14	hr	Rn	23.28 ± 1.73			377, 637
14	hr	For-5			Close	155
14	Es-10	For-5			≤3.7†	155
14	Es-10	s			15.46 ± 3.67	1117
14	Es-10	D14Rp4			Near	257
14	For-5	D14Rp4			Near	257
14	wal	s			5.62 ± 0.93	971
14	s	slt	3.03 ± 2.98	4.88 ± 1.50	7.52 ± 1.71	371, 379
14	s	Rn	5.35 ± 0.77	6.95 ± 1.29	8.79 ± 2.97	371, 377, 637, 379
14	s	spc	4.61 ± 1.42	7.10 ± 1.90	8.54 ± 2.18	371
14	slt	Rn	3.03 ± 2.98	1.95 ± 0.97		371
14	Rn	spc	0	0		371
14	Chr 14	Adk			Syntenic	613
14	Chr 14	Mtv-11			Syntenic	837
14	Chr 14	Ctla-1			Syntenic	84
14	Chr 14	Odc-9			Linked	853
14	Chr 14	Plau			Syntenic	746, 845a
14	Chr 14	Trp53-ps			Syntenic	160
14	Chr 14A2-B	Adk			Syntenic	895
14	Chr 14B-C	Np-1			Syntenic	895
14	Chr 14D2-E2	Es-10			Syntenic	895
15	Rb3Bnr	uw			0	114
15	Rb1Ald	uw			0	114
15	uw	ank	8.20 ± 2.80		2.92 ± 1.07	999, 1011
15	uw	Gpt-1			33.86 ± 3.14	436, 246
15	uw	Bld	41.67 ± 8.22	16.67 ± 5.75		1052
15	uw	bt	43.60 ± 2.34	38.26 ± 2.82	39.81 ± 2.82	1011, 207, 246

Table 6.1—*cont.*

Chr	Loci		Female	Male	Combined or not stated	References
	Proximal	Distal				
15	uw	Ca	40.54 ± 4.04	48.27 ± 4.64		207
15	C6	Gpt-1	25.97 ± 4.97			404
15	C6	Gdc-1	36.40 ± 5.50			404
15	C6a	C7		≤9.8†		770
15	C6b	C7		≤9.8†		770
15	ank	bt	28.90 ± 4.60		38.16 ± 4.44	999, 1011
15	ank	Ca	52.94 ± 12.11			999
15	Ly-6	Thb			0	231
15	Ly-6	Xmmv-55			1.83 ± 0.91	436, 1091
15	Ly-6	H-30			≤18†	701
15	Ly-6	Ins-3			3.66 ± 2.42	703
15	Ly-6	Xmmv-72			4.10 ± 3.39	702
15	Ly-6	Sis			4.84 ± 1.92	703
15	Ly-6	Ril-1			11.00 ± 2.00	699
15	Ly-6	Gpt-1			13.20 ± 8.88	436
15	Ly-6	Gdc-1			20.37 ± 3.88	436
15	Ly-6	Krt-1			39.45 ± 25.02	703
15	Xmmv-55	Gpt-1		≤2.4†	1.60 ± 1.71	436
15	Xmmv-55	Gdc-1			17.59 ± 3.66	436
15	cog	Tgn	≤13.3†		≤23†	1040
15	cog	Eh	0.45 ± 0.45	≤5.5†		35
15	cog	Ve	34.03 ± 4.04	36.09 ± 4.65		1040
15	Sis	Ins-3			1.02 ± 1.06	703
15	Sis	Xmmv-72			2.29 ± 1.12	703
15	Sis	Krt-1			17.27 ± 6.56	703
15	Sis	Int-1			30.01 ± 18.33	703
15	Sis	Gdc-1			32.94 ± 14.08	703
15	mn	Ca	24.90 ± 2.45	26.99 ± 2.95		1074
15	mn	N	22.87 ± 5.50	26.06 ± 5.30		1074
15	lz	Ca			23.88 ± 6.14	715
15	Gpt-1	Tgn	3.00 ± 3.00			1040
15	Gpt-1	Gdc-1	15.60 ± 4.10		9.01 ± 2.25	436, 404, 249
15	Gpt-1	bt			19.33 ± 3.62	246
15	dp	bt			31.00	1070
15	dp	Ca	25.86 ± 4.07			1070
15	dp	N	28.45 ± 4.19			1070
15	H-30	Xmmv-72			≤18†	702, 867
15	Xmmv-72	Ins-3			2.00 ± 1.53	703
15	Xmmv-72	Krt-1			22.41 ± 11.46	703
15	Xmmv-72	Int-1			30.96 ± 18.46	703
15	Xmmv-72	Gdc-1			41.66 ± 19.27	703
15	Bld	bt	17.65 ± 3.77	15.38 ± 2.30		1052
15	Bld	Ca	29.09 ± 3.11	20.28 ± 3.40		1052
15	Ins-3	Krt-1			15.39 ± 8.63	703
15	Ins-3	Int-1			16.67 ± 10.29	703
15	Ins-3	Gdc-1			19.23 ± 10.50	703
15	Dom	bt	10.32 ± 1.92			573
15	Dom	Eh	1.08 ± 0.49	0.36 ± 0.36		573
15	Dom	Ca	17.50 ± 2.01	23.77 ± 2.59		573
15	Gdc-1	Int-1			0.88 ± 0.91	703
15	Gdc-1	bt			7.63 ± 2.10	436
15	Gdc-1	Krt-1			2.54 ± 1.62	703
15	Krt-1	Int-1			0.98 ± 1.02	703
15	gt	Ca	17.01 ± 3.20	23.97 ± 4.90		933
15	bt	Eh			0.16 ± 0.11	391, 881
15	bt	hl	7.36 ± 0.50	8.62 ± 0.68		991, 444
15	bt	Ca	4.19 ± 0.41	5.20 ± 0.43		726, 1052, 305, 1074

Table 6.1—*cont.*

Chr	Loci		Female	Male	Combined or not stated	References
	Proximal	Distal				
15	*bt*	*Ve*	7.42 ± 1.21	12.47 ± 1.48		991
15	*bt*	*N*	4.38 ± 0.44	13.28 ± 0.72		726, 1074
15	*bt*	*cph*			13.37 ± 5.43	466a
15	*Eh*	*H(Eh)*			0	20
15	*Eh*	*Ca*	0	0		573
15	*Eh*	*Ve*			0	392
15	*Eh*	*med*			0	391
15	*Ht*	*Ca*	1.60 ± 1.60	6.60 ± 3.30		984
15	*hl*	*Ca*	1.36 ± 0.22	4.86 ± 0.50		444
15	*hl*	*Ve*	2.02 ± 1.41	4.35 ± 2.46		991
15	*Ca*	*sw*			≤18†	588
15	*Ca*	*Ve*			≤1.2†	991
15	*Ca*	*med*			0.27 ± 0.10	911, 44
15	*Ca*	*mk*	0.70 ± 0.40	2.63 ± 2.60		690
15	*Ca*	*N*	0.59 ± 0.15	1.75 ± 0.24	1.79 ± 0.72	142, 726, 1070, 305, 1074
15	*Ca*	*cph*	3.39 ± 2.36	8.82 ± 4.86		466a
15	*Sha*	*N*	1.43 ± 0.54	0.84 ± 0.38	0.77 ± 0.26	305
15	*Ve*	*N*	0.83 ± 0.58	1.92 ± 0.95		991
15	*Mlvi-2*	*Myc*			Syntenic	547
15	*Myc*	*Niard*			Synt.,close	668
15	*Myc*	*Pvt-1*			Syntenic *Myc* proximal	1111, 26
15	Chr 15	*NO*			Syntenic	410
15	Chr 15	*Ela-1* (or ps)			Syntenic	465
15	Chr 15	*As-2*			Syntenic	317
15	Chr 15	*Dia-1*			Syntenic	1028
15	Chr 15	*Hba-3ps*			Syntenic	612
15	Chr 15F1-3	*Hox-3*			Syntenic	17, 79, 844
15	Chr 15	*Int-1*			Syntenic	760, 797
15	Chr 15	*Mis-1*			Syntenic	505
15	Chr 15	*Mlvi-1*			Syntenic	547
15	Chr 15	*Mlvi-2*			Syntenic	547
15	Chr 15D2-3	*Myc*			Syntenic	890
15	Chr 15	*P450-2D*			Syntenic	343a
15	Chr 15	*Pfk-4*			Syntenic	14
15	Chr 15	*Pvt-1*			Syntenic	348, 144
15	Chr 15	*Rpl30*			Syntenic	1098
15	Chr 15	*Sis*			Syntenic	546
15	Chr 15D	*Gdc-1*			Syntenic	703
16	*Rb7Bnr*	*md*	1.00 ± 0.90		0	178, 814
16	*Hc16*	*md*	8.57 ± 3.34	4.08 ± 2.83		178
16	*md*	*Akv-2*			6.44 ± 2.47	543
16	*md*	*scid*			≤15†	180
16	*md*	*Igl-1*	8.58 ± 1.61	6.67 ± 2.20	10.90 ± 4.60	271, 270
16	*md*	*Bst*	28.80 ± 4.00	16.40 ± 4.40		270
16	*md*	*dw*			32.51 ± 1.88	576
16	*md*	*Sod-1*			42.86 ± 10.80	1114
16	*md*	*wv*			44.60 ± 5.65	576
16	*md*	*Mx*			55.20 ± 4.80	987
16	*scid*	*Igl-1*			7.28 ± 1.36	180
16	*Prm-1*	*Igl-1*	7.23 ± 2.84			848
16	*Akv-2*	*Igl-1*		≤5.8†		271
16	*Igl-1*	*Igl-1r*		≤3.1†		271
16	*Igl-1*	*Mtv-6*	17.20 ± 3.91			848
16	*Igl-1*	*Bst*	17.20 ± 3.90	13.60 ± 4.10		270
16	*Igl-1*	*Smst*	5.38 ± 2.34			848
16	*Prm-1*	*Smst*	13.08 ± 3.26			848

Table 6.1—*cont.*

Chr	Loci		Female	Male	Combined or not stated	References
	Proximal	Distal				
16	*Smst*	*Sod-1*	35.16 ± 5.01			848
16	*Smst*	*Est-2*	36.28 ± 4.52			848
16	*Mtv-6*	*Sod-1*	23.91 ± 4.45			848
16	*Mtv-6*	*Ets-2*	26.55 ± 4.15			848
16	*Mtv-6*	*Smst*	11.02 ± 2.88			848
16	*dw*	*wv*			20.23 ± 1.47	597
16	*Sod-1*	*App*	3.19 ± 1.81			850
16	*Sod-1*	*Ets-2*	1.16 ± 1.16		5.00 ± 3.45	848, 849
16	*Sod-1*	*Cola-2*			Syntenic	926
16	*Sod-1*	*Mtvr-1*			Syntenic	420
16	*Ets-2*	*App*	2.13 ± 1.49			850
16	*Cola-1*	*Cola-2*			Syntenic	926
16	Chr 16	*NO*			Syntenic	268
16	Chr 16C3-ter	*App*			Syntenic	622, 850
16	Chr 16	*Air1*			Syntenic	1
16	Chr 16	*Cola-2*			Syntenic	926
16	Chr 16	*Ifrc*			Syntenic	147, 620
16	Chr 16	*Mtvr-1*			Syntenic	420
16	Chr 16	*Mx*			Syntenic	987
16	Chr 16	*Prgs*			Syntenic	148
16	Chr 16	*Prm-2*			Syntenic	406
16	Chr 16	*Xmv-3*			Syntenic	438
17	*Rb7Bnr*	*T*	5.81 ± 2.52	1.96 ± 1.37		396, 114
17	*Rb7Bnr*	*tf*	9.83 ± 2.26	4.85 ± 1.67		396, 114
17	*Rb7Bnr*	*H-2*	13.01 ± 2.09	7.94 ± 3.41		396
17	*cen*	*T*			5.56 ± 2.34‡	11
17	*D17Tu1*				Between *cen* and *t* complex	512
17	*T*	*Hst-1*			4.50 ± 2.20	189
17	*T*	*het*			1.88 ± 0.71	1009
17	*T*	*qk*		2.19 ± 1.08	2.60 ± 0.60	982, 48, 979
17	*T*	*cld*	6.00	2.00		794
17	*T*	*H-39*			0	12
17	*T*	*Fu*	7.17 ± 2.23	2.97 ± 0.53	2.35 ± 0.42	226, 225, 3
17	*T*	*tf*	8.96 ± 0.71	4.81 ± 0.40	6.40 ± 1.55	826, 228, 224, 223, 646, 654, 363, 979, 827
17	*T*	*Hye*			*Hye* distal	551
17	*T*	*H-2*	14.72 ± 0.91	8.10 ± 0.71	8.53 ± 1.14	826, 968, 1018, 396, 363, 3, 324, 190, 1141
17	*T*	*H-2K*			12.10 ± 4.30	854
17	*T*	*Neu-1*			19.59 ± 2.78	1118
17	*T*	*Pgk-2*			19.28 ± 2.78	233, 1118
17	*T*	*rds*			20.00 ± 4.00	190
17	*T*	*Int-3*			18.52 ± 5.29	324
17	*T*	*thf*	27.12 ± 1.83	27.23 ± 1.65		396, 520
17	*T*	*Ea-?*			26.00 ± 2.77	827
17	*T*	*Ir-5*			29.73 ± 3.64	1141
17	*T*	*Tcd-1*			In *t* complex	653, 656
17	*T*	*Tcd-2*			In *t* complex	653, 656
17	*T*	*Tcd-3*			In *t* complex	656
17	*T*	*Tcr-1*			In *t* complex	653
17	*T*	*Tcp-4*			In *t* complex	940
17	*T*	*Tcp-5*			In *t* complex	940
17	*T*	*Tcp-8*			In *t* complex	940
17	*T*	*Tcp-1*			In *t* complex	940
17	*T*	*Tas*			In *t* complex	1081
17	*T*	*Tcp-2*			In *t* complex	940

Table 6.1—*cont.*

Chr	Loci		Female	Male	Combined or not stated	References
	Proximal	Distal				
17	*T*	*Tcp-3*			In *t* complex	940
17	*T*	*Tcp-6*			In *t* complex	940
17	*T*	*Tcp-7*			In *t* complex	940
17	*Tcd-1*				In *t* complex	656
17	*Tcd-2*				In *t* complex	656
17	*Tcd-3*				In *t* complex	656
17	*Hst-1*	*H-2*			13.50 ± 4.07	189
17	*het*	*Fu*	16.30 ± 2.34			1009
17	*qk*	*tf*		10.38 ± 2.30	5.26 ± 2.63	982, 979
17	*qk*	*Tcp-1*			*qk* distal	941
17	*qk*	*Tcp-1*			Near, *qk* dist	939
17	*H-39*				Between *qk* and *tf*	397
17	*Tcp-1*	*Hba-4ps*			*Tcp-1* prox.	311
17	*Tme*				Between *qk* and *tf*	1110
17	*D17Rp17*	*Hba-4ps*			3.30 ± 1.87	667
17	*D17Rp17*	*D17Rp10*			29.01 ± 18.19	667
17	*Fu*	*tf*	1.16 ± 0.47	0.36 ± 0.26		223
17	*Fu*	*H-2*	6.85 ± 2.96	1.83 ± 0.62	2.35 ± 0.63	2, 3, 968
17	*Fu*	*thf*	28.57 ± 4.41	18.69 ± 3.77		520
17	*Hba-4ps*	*tf*			Near	311
17	*Pim-1*	*Hba-4ps*			4.41 ± 2.11	731
17	*Hba-4ps*	*Crya-1*			7.80 ± 4.09	948
17	*Hba-4ps*	*H-2*			6.60 ± 3.14	163
17	*Hba-4ps*	*Upg-1*			9.60 ± 4.38	163
17	*Hba-4ps*	*D17Rp10*			19.60 ± 11.37	667
17	*tf*	*H-2*	3.47 ± 0.79	2.67 ± 0.70	5.60 ± 1.45	826, 363, 396, 827
17	*tf*	*C3*			24.00 ± 4.40	739
17	*tf*	*Ea-?*			20.40 ± 2.55	827
17	*Hye*				Proximal to *H-2*	551
17	*H-33*				Proximal to *H-2K*	304
17	*Mdmg-1*				Proximal to *H-2K*	23
17	*Pim-1*	*Crya-1*			1.06 ± 0.78	731
17	*Pim-1*	*H-2*			2.07 ± 1.00	731
17	*Crya-1*	*H-2*			1.05 ± 0.78	948
17	*Crya-1*				Proximal to *H-2K*	948
17	*Crya-1*	*Upg-1*			6.70 ± 3.47	948
17	*Crya-1*				Distal to *Glo-1*	948
17	*Ggm-1*	*H-2K*			Linked	402
17	*Glo-1*	*H-2*			1.87 ± 0.39	876, 875
17	*Glo-1*	*H-2(Ss)*		3.16 ± 1.79		694
17	*Glo-1*	*C4*		0.99 ± 0.70		894
17	*Glo-1*	*Ss*	2.53 ± 1.25	1.61 ± 1.60	1.81 ± 0.66	218, 216
17	*Glo-1*	*Neu-1*	2.04 ± 1.17	0.99 ± 0.70		1116, 894
17	*Glo-1*	*Ce-2*	0.86 ± 2.50	0.99 ± 0.70		730, 894
17	*Glo-1*	*Pgk-2*	2.82 ± 1.96	0.99 ± 0.70		216, 894
17	*Glo-1*	*Upg-1*		3.47 ± 1.29		894
17	*Glo-1*	*C3*	15.29 ± 2.89	6.88 ± 1.28	6.88 ± 1.37	730, 727, 875
17	*Glo-1*	*Pim-1*			Syntenic	421
17	*Glo-1r*	*Glo-1s*			≤11.7†	876
17	*Cxv-1*	*H-2K*			1.89 ± 1.89	1134
17	*H-2*	*Bf*			Within *H-2*	846, 790
17	*H-2*	*Hh-1*			Within *H-2*	138
17	*H-2*	*Tla*			1.51 ± 0.61	76
17	*H-2*	*Ce-2*			2.00	430
17	*H-2*	*Pgk-2*			≤22†	233

Table 6.1—*cont.*

Chr	Loci		Female	Male	Combined or not stated	References
	Proximal	Distal				
17	*H-2*	*Thy-2*			3.03 ± 2.11	927
17	*H-2*	*Hmt*	≤7.2†			297
17	*H-2*	*rds*			4.97 ± 1.55	1145, 190
17	*H-2*	*Upg-1*	6.06 ± 4.15	5.36 ± 1.38	6.00	1020, 1018, 1017, 1019
17	*H-2*	*H-43*			8.48 ± 2.17	491
17	*H-2*	*C3*			11.24 ± 1.95	875, 1143
17	*H-2*	*Int-3*			10.61 ± 3.79	324
17	*H-2*	*thf*	11.81 ± 2.10	20.66 ± 2.06		396
17	*H-2*	*Ea-?*			14.80 ± 2.25	827
17	*H-2*	*Ir-5*			17.09 ± 2.99	1141
17	*H-2*	*Agp-1*	27.96 ± 3.14			677
17	*H-2*	*Ads-3*			Within *H-2*	426
17	*H-2*	*Agp-3*			Within *H-2*	426
17	*H-2*	*Lpn-2*			Within *H-2*	426
17	*H-2*	*Dcp*			Linked	73
17	*H-2*	*C2*			Linked	1029
17	*H-2*	*Crl-1*			Linked	332
17	*H-2*	*C4bp*			Linked	1030
17	*H-2*	*Rrs*			Linked	1095a
17	*H-2(Ss)*	*C3*	14.00 ± 4.91	8.16 ± 3.91		740
17	*Sgp-1*	*H-2*			Linked	679
17	*Rfv-1*	*H-2*			Linked	136
17	*Rfv-2*	*H-2*			Linked	136
17	*D17Tu2*				In *H-2I*	293
17	*Rrv-1*				In *H-2I*	621
17	*Ads-3*	*Agp-3*			Close	426
17	*H-2K*	*Hmt*			4.60 ± 1.80	297
17	*H-2K*	*Hdd-1*			Close	842
17	*H-2K*	*Nel-1*			Linked	414
17	*Rrv-1*	*H-2K*			*Rrv-1* distal	621
17	*H-2L*	*H-2D* region			Linked	752
17	*Idd-1*	*H-2I*			*Idd-1* proximal	842
17	*H-2I*	*Neu-1*			0	294
17	*Rmv-1*	*H-2I*			Linked	188
17	*Rmv-2*	*H-2I*			Linked	188
17	*Lmp-2*	*Lmp-7*			0	713
17	*Lmp-2*				Between *H-2I*	713
17	*Lmp-7*				and *H-2D*	713
17	*Rig-1*	*H-2S*			*Rig-1* prox.	219
17	*C2*	*H-2S*			*C2* in *H-2S*	345
17	*Slp*	*Oh21-1*			Syntenic	1095
17	*Oh21-1*	*C4*			Syntenic	1095, 5
17	*C4*	*Oh21-2*			Syntenic	1095, 5
17	*C4*	*Neu-1*		≤1.5†		894
17	*C4*	*Ce-2*		≤1.5†		894
17	*C4*	*Pgk-2*		≤1.5†		894
17	*C4*	*Upg-1*		2.48 ± 1.10		894
17	*C4(Ss)*	*Bf*			Close	741
17	*Ss*	*Pgk-2*	1.41 ± 1.40	≤8.9†	0.97 ± 0.97	218, 216
17	*(Slp)*	*Bf*			Close	846
17	*Neu-1*	*H-2D*			In *H-2*	294
17	*Neu-1*	*Ce-2*		≤1.5†		894
17	*Neu-1*	*Pgk-2*	1.25 ± 0.77	≤1.5†	7.35 ± 2.24	1116, 808, 894, 1118
17	*Neu-1*	*Upg-1*		2.48 ± 1.10		894
17	*H-2*	*Ms15-3*			1.06 ± 1.11	496
17	*H-2D*	*Mep-1*			2.11 ± 0.85	847
17	*H-2D*	*Ce-2*	1.54 ± 0.68			793

Table 6.1—*cont.*

Chr	Proximal	Distal	Female	Male	Combined or not stated	References
17	*H-2D*	*Pgk-2*			2.19 ± 1.25	847
17	*H-2D*	*Hmt*			2.00 ± 1.20	297
17	*H-31*				Between	302
17	*Qa-2*				*H-2 & Tla*	303
17	*Qa-3*				" "	303
17	*H-32*				" "	302
17	*Qa-1*				" "	303
17	*H-2D*	*Mep-1*			*Mep-1* distal	70
17	*Rmv-3*	*H-2D*			Linked	188
17	*H-2T*				Between *Qa-2* & *Tla*	532
17	*Qa-4*				In *Qa* region	398
17	*Qa-5*				In *Qa* region	398
17	*Qa-6*				In *Qa* region	692
17	*Qa-6*	*Tla*			Linked	301
17	*Qa-7*				In *Qa* region	692
17	*Qa-8*				In *Qa* region	692
17	*Qa-11*				In *Qa* region	1147
17	*Q10*				In *Qa* region	946
17	*Mep-1*	*Pgk-2*			2.19 ± 1.25	847, 70
17	*Qb-1*				Distal to *Tla*	859
17	*Tla*	*Qa-9*			Linked	994
17	*Ce-2*	*Pgk-2*		0		894
17	*Ce-2*	*Upg-1*		2.48 ± 1.10		894
17	*Ce-2*	*C3*	12.75 ± 3.30			730
17	*Pgk-2*	*Upg-1*		2.48 ± 1.10		894
17	*Upg-1*	*thf*	6.06 ± 4.15	7.27 ± 2.48	6.90	1020, 1018
17	*C3*	*Pim-1*			Syntenic	421
17	*Acra*	*Actc*	19.05 ± 6.06			407
17	*D17Rp10*	*D17Rp11*			11.90 ± 7.75	667
17	*Pim-1*	*Sod-2*			Syntenic	421
17	*Tnfa*	*H-2D*			*Tnfa* prox.	721
17	*Tnfa*	*Tnfb*			Syntenic	750
17	Chr 17	*Kth-1*			Linked	258
17	Chr 17	*Lt*			Syntenic	325
17	Chr 17	*Sod-2*			Syntenic	1023, 1022
17	Chr 17	*Tnfa*			Syntenic	750, 325
17	Chr 17	*l(17)-1*			*T*-linked	922
17	Chr 17	*l(17)-2*			*T*-linked	922
17	Chr 17	*l(17)-3*			*T*-linked	922
17	Chr 17	*l(17)-4*			*T*-linked	922
18	*Dp(Hc18)*	*Tw*			8.92 ± 2.80	574, 38
18	*Rb6Rma*	*Tw*	19.23 ± 7.73	0		574
18	*Rb9Lub*	*pk*		1.50 ± 1.10		570
18	*Rb9Lub*	*sy*		10.50 ± 2.70		570
18	*ax*	*Tw*			≤1.5†	660, 657
18	*ax*	*sy*			19.87 ± 2.16	574
18	*Tw*	*pk*	20.00 ± 5.30	14.70 ± 1.81		570
18	*Tw*	*bc*		14.11 ± 3.51	17.20 ± 3.90	574
18	*Tw*	*sy*	33.22 ± 4.85	24.19 ± 1.75	29.00 ± 4.70	574, 570
18	*Tw*	*shi*	41.20 ± 6.00	47.70 ± 4.90		930
18	*Tw*	*Mtv-2*			Linked	705
18	*pk*	*sy*	22.29 ± 3.51	12.88 ± 1.48	16.05 ± 1.74	570
18	*bc*	*sy*			10.90 ± 2.20	574
18	*sy*	*shi*		21.35 ± 3.59	0	930
18	Chr 18	*NO*			Syntenic	268, 410
18	Chr 18	*Grl-1*			Syntenic	315

Table 6.1—*cont.*

Chr	Loci		Female	Male	Combined or not stated	References
	Proximal	Distal				
18	Chr 18	*Ii*			Syntenic	1131, 852, 855
18	Chr 18	*Pep-1*			Syntenic	316
19	*Rb163H*	*oc*			≤2.3†	671
19	*Rb163H*	*bm*	19.63 ± 3.87	22.50 ± 6.32		568
19	*Rb163H*	*ep*	30.84 ± 4.43	30.00 ± 7.23		568
19	*Rb163H*	*ru*	31.15 ± 3.39	34.31 ± 4.53		568
19	*oc*	*bm*			33.20 ± 6.90	671
19	*oc*	*ab*			34.50 ± 3.60	671
19	*oc*	*ru*			33.28 ± 2.03	671
19	*mdf*	*Dc*			0	1013
19	*mdf*	*ru*			37.98 ± 3.55	1013
19	*Dc*	*Ly-1*			0	135
19	*Dc*	*bm*	18.75 ± 3.78	14.12 ± 4.34		568
19	*Dc*	*ep*	25.00 ± 4.33	20.00 ± 4.97	34.74 ± 4.89	568, 194
19	*Dc*	*ru*	26.79 ± 4.33	23.53 ± 4.97		568
19	*Ahd-2*	*bm*	16.70 ± 4.80		16.44 ± 2.51	1054, 1056
19	*Ahd-2*	*ep*	20.80 ± 5.30		21.01 ± 2.75	1054, 1056
19	*Ahd-2*	*ru*	20.80 ± 5.30		21.04 ± 2.13	1054, 1056
19	*Ly-10*	*Ly-1*			2.51 ± 1.01	527, 528, 432
19	*Ly-10*	*Xmmv-42*			0	1091
19	*Ly-10*	*Ly-12*			3.60	432
19	*Ly-10*	*Ea-4*			8.10	432
19	*Ly-12*	*Ea-4*			4.20	432
19	*H(ep)*	*ep*			16.00 ± 5.00	20
19	*bm*	*ep*	6.17 ± 1.03	8.50 ± 1.27	7.72 ± 0.72	567, 568, 1054, 1056
19	*bm*	*ak*			8.09 ± 2.34	1064
19	*bm*	*ab*			9.43 ± 1.42	1002
19	*bm*	*ru*	7.23 ± 1.14	9.75 ± 1.55	8.99 ± 0.75	568, 1054, 1056, 1002
19	*ep*	*ru*	1.27 ± 0.33	2.18 ± 0.45	1.21 ± 0.41	571, 568, 1054, 1056
19	*ab*	*ru*			1.37 ± 0.39	1000, 1002
19	*ru*	*Got-1*			1.29 ± 0.74	238
19	*Got-1*	*Es-18*			4.97 ± 1.39	1151, 1152
19	Chr 19	*NO*			Syntenic	410
19	Chr 19	*Lip-1*			Syntenic	535, 556
19	Chr 19	*Pgam-1*			Syntenic	322
19	Chr 19	*Tdt*			Syntenic	1133
X	*Xcen-XD*	*Syn-1*	Syntenic			318
X	*HcX*	*spf*	2.41 ± 1.68			237
X	*spf*	*sf*	Close			112
X	*spf*	*DXS32*	10.00			666
X	*spf*	*Td*	0.60 ± 0.60			122
X	*spf*	*mdx*	19.00 ± 8.60			89
X	*Otc(spf)*	*Timp*	≤4.0†			722
X	*sf*	*Ta*	43.94 ± 4.32			1092
X	*Td*	*Ta*	38.40 ± 2.70			119
X	*Td*	*Mo*	45.50 ± 3.60			119
X	*DXPas3*	*DXPas4*	1.61 ± 1.60			4
X	*DXPas3*	*DXPas5*	5.66 ± 3.17			4
X	*DXPas3*	*Hprt*	10.00 ± 3.87			4
X	*DXPas3*	*Ta*	26.00 ± 6.20			4
X	*DXPas3*	*DXPas2*	32.79 ± 6.01			4
X	*DXPas3*	*DXPas1*	52.00 ± 7.00			4
X	*DXPas4*	*DXPas5*	5.36 ± 3.01			4
X	*DXPas4*	*Hprt*	8.82 ± 3.44			4
X	*DXPas4*	*Ta*	22.81 ± 5.56			4
X	*DXPas4*	*DXPas2*	30.77 ± 5.72			4

Table 6.1—*cont.*

Chr	Proximal	Distal	Female	Male	Combined or not stated	References
X	DXPas4	DXPas1	52.00 ± 6.00			4
X	DXPas5	Hprt	7.02 ± 3.38			4
X	DXPas5	Ta	15.56 ± 5.40			4
X	DXPas5	DXPas2	29.09 ± 6.12			4
X	DXPas5	DXPas1	44.00 ± 7.00			4
X	Hq	Dmd	19.51 ± 6.19			82
X	Hq	Bn	5.97 ± 2.89			630
X	Hq	G6pd	6.95 ± 2.15			82, 802
X	Hq	mdx	16.18 ± 2.58			88
X	Hq	Ta	20.65 ± 2.05			490
X	Hq	Phk	21.74 ± 8.60			82
X	Hq	Pgk-1	23.87 ± 3.85			757
X	Hq	Mo	25.21 ± 2.19			490
X	Hq	Ags	25.37 ± 5.32			630
X	Hq	Li	57.14 ± 5.10			110
X	Hprt	Cf-9	3.06 ± 1.74			722, 16
X	Hprt	DXPas6	6.54 ± 2.39			16
X	Hprt	G6pd	11.29 ± 2.32			16
X	Hprt	DXPas8	18.10 ± 3.76			16
X	Hprt	Ta	25.87 ± 3.66			16
X	Hprt	Pgk-1	21.49 ± 2.54			127
X	Hprt	Ags	28.60 ± 2.90			127
X	Hprt	mdx	5.60 ± 2.70			124
X	Hprt	Dmd	12.50 ± 3.70			408
X	Hprt	Ta	19.00 ± 5.00			4
X	Hprt	DXPas2	25.37 ± 5.32			4
X	Hprt	DXPas1	46.97 ± 6.14			4
X	Hprt	Oubr	Syntenic			610
X	Cf-9	DXPas6	3.16 ± 1.79			16
X	DXPas6	G6pd	8.57 ± 2.73			16
X	DXPas6	DXPas8	13.23 ± 4.11			16
X	Bn	Str	34.28 ± 8.02			821
X	Bn	Phk	16.05 ± 2.88			473
X	Bn	Ta	13.20 ± 1.01			247, 241, 584, 817, 240, 473, 15
X	Bn	Xce	3.33 ± 1.64			351
X	Bn	Mo	13.66 ± 1.57			15
X	Bn	Ags	19.40 ± 4.83			630
X	Bn	jp	27.75 ± 2.91			247, 817
X	Bn	Hyp	34.92 ± 3.54			241, 240
X	Bn	Gy	36.67 ± 3.11			659
X	Bn	Ie	Near			888, 480
X	Bn	ty	Near			1033
X	ty	Ta	12.00 ± 3.00			1033
X	Bpa	Ta	8.75 ± 3.20			816
X	Bpa	Mo	22.50 ± 6.60			816
X	G6pd	Cf-8	≤4.0†			722
X	G6pd	Dmd	4.76 ± 1.89			82, 408
X	G6pd	Phk	12.82 ± 5.35			82
X	G6pd	Ta	4.75 ± 1.24			16, 82, 802
X	Str	Ta	8.91 ± 1.06			821, 658, 652
X	Cf-8	Rsvp	≤4.0†			722
X	mdx	Ta	9.50 ± 6.40			89
X	mdx	Pgk-1	12.70 ± 3.90			124
X	mdx	Mo	10.07 ± 1.85			89, 88
X	mdx	Hyp	37.74 ± 6.70			89
X	G6pd	Dmd	4.62 ± 2.60			408
X	Dmd	Phk	5.75 ± 2.48			82, 408
X	Dmd	Ta	5.82 ± 2.31			82, 408

Table 6.1—*cont.*

Chr	Loci Proximal	Loci Distal	Female	Male	Combined or not stated	References
X	*Phk*	*Ta*	3.45 ± 1.14			473
X	*Phk*	*jp*	26.37 ± 4.62			473
X	*sla*	*Ta*	2.53 ± 0.63			276
X	*Tfm*	*Ta*	1.28 ± 1.27			641
X	*Tfm*	*Mo*	6.41 ± 2.77			641
X	*Gs*	*Ta*	Close			885
X	*Gs*	*Xce*	7.95 ± 0.80			349
X	*Gs*	*Mo*	4.76 ± 0.33			349, 350
X	*Ta*	*jp*	14.17 ± 2.19			597
X	*Ta*	*Xce*	≤3.7†			121, 116
X	*Ta*	*Bhd*	1.52 ± 1.50			813
X	*Ta*	*Slf*	2.17 ± 2.15			813
X	*Ta*	*Pgk-1*	2.70 ± 1.19			757
X	*Ta*	*Mo*	3.66 ± 0.39			641, 886, 490, 285, 1092, 276, 630, 15
X	*Ta*	*DXPas2*	7.00 ± 3.00			4
X	*Ta*	*xid*	6.60 ± 1.77			51
X	*Ta*	*Ags*	12.92 ± 1.94			630
X	*Ta*	*jp*	15.29 ± 1.26			817, 473, 247, 1004, 248
X	*Ta*	*Xcat*	22.94 ± 1.95			287
X	*Ta*	*DXPas1*	27.00 ± 6.00			4
X	*Ta*	*Hyp*	20.49 ± 2.53			241, 51
X	*Ta*	*Gy*	23.94 ± 1.89			39, 248, 659
X	*Ta*	*Xpl*	27.29 ± 1.74			1004
X	*Ta*	*Crm*	31.11 ± 6.90			39
X	*Xce*	*Mo*	11.82 ± 0.56			349, 118, 116
X	*Pgk-1*	*Ags*	9.12 ± 2.40			127
X	*Pgk-1*	*Mo*	≤2.7†			757
X	*Mo*	*Ym*	2.63 ± 1.06			481
X	*Mo*	*Ags*	8.89 ± 1.90			630
X	*Mo*	*Gy*	24.56 ± 4.03			659
X	*Mo*	*Li*	29.63 ± 5.07			110
X	*DXPas2*	*DXPas1*	26.15 ± 5.45			4
X	*xid*	*Hyp*	12.18 ± 2.33			51
X	*Ags*	*Hprt*	Syntenic			537
X	*jp*	*Gy*	13.77 ± 2.67			248
X	*jp*	*Xpl*	14.62 ± 2.11			1004
X	*Xcat*		Distal to *Ta*			287
X	*Hyp*	*Gy*	0.71 ± 0.35			649, 643
X	*Hyp*	*Li*	0.88 ± 0.35			120
X	*Gy*	*Bn*	36.67 ± 3.11			659
X	*Gy*	*Ta*	20.95 ± 2.37			659
X	*Gy*	*Mo*	24.56 ± 4.03			659
X	*Gy*	*Crm*	6.67 ± 3.72			39
X	*Crm*	*Sts*	10.87 ± 4.59			111
X	*Sts*	*Ta*	53.13 ± 8.82			513
X	Chr X	*Araf*	Syntenic			471
X	Chr X	*Hxa*	Linked			54
X	Chr X	*If-X*	Linked			187, 796
X	Chr X	*It*	Linked			824
X	Chr X	*Mov-14*	Linked			990
X	Chr X	*Mtp-3*	Linked			686
X	Chr X	*Rv-3*	Linked			409
X	Chr X	*Segr*	Linked			841
X	Chr X	*Tse-3*	Syntenic			811
X	Chr X	*Xta*	Linked			662
XY	*Sts*	*Sxr*			6.25 ± 4.28	513
XY	*Sxr*	*Ta*			59.38 ± 8.68	513

Table 6.1—*cont.*

Chr	Loci		Female	Male	Combined or not stated	References
	Proximal	Distal				
XY	Chr X,Y	*Sxa*			0	684
XY	Chr X,Y	*Mov-15*			0	725
Y	Chr Y	*Hya*		Linked		92a, 271a
Y	Chr Y	*Mov-24*		Syntenic		725
Y	Chr Y	*Tdy*		Linked		245a
Y	Chr Y	*Xmmv-Y*		Linked		62
Y	Chr Y	*Yaa*		Linked		725a
UN	*Prp*	*Cyx*		Linked		19, 626
UN	*Prp*	*Qui*		Linked		19, 626
UN	*Prp*	*Rua*		Linked		19, 626

* Hc followed by a number refers to a centromeric heterochromatin marker.

† Maximum value of recombination per cent (95% U.C.L.) (Mather, 681a, p.47). Formulas used to calculate values were $0.05 = (1-p)^N$ for backcrosses and $0.05 = (1-p^2/4)^N$ for repulsion intercrosses, when no recombinants were observed. Recombinant inbred strain values were taken from a table prepared by Earl L. Green. The statistical method for RI strains is given in Silver, 922a.

‡ The *t*-complex was genotyped using a cDNA probe for the MHC and assuming that an inversion associated with the *t*-complex inverts the MHC to become the most proximal marker for the *t*-complex.

References

1. Accolla, R.S., M. Jotterand-Bellomo, L. Scarpellino, A. Maffei, G. Carra, and J. Guardiola. 1986. *aIr-1*, A newly found locus on mouse chromosome 16 encoding a trans-acting activator factor for MHC class II gene expression. J. Exp. Med. 164:369–74.

2. Allen, S.L. 1955. *H-2ᶠ*, a tenth allele at the histocompatibility-2 locus in the mouse as determined by tumor transplantation. Cancer Res. 15:315–9.

3. Allen, S.L. 1955. Linkage relations of the genes histocompatibility-2 and fused tail, brachyury and kinky tail in the mouse, as determined by tumor transplantation. Genetics 40:627–50.

4. Amar, L.C., D. Arnaud, J. Cambrou, J-L. Guénet, and P.R. Avner. 1985. Mapping of the mouse X chromosome using random genomic probes and an interspecific mouse cross. EMBO J. 4:3695–700.

5. Amor, M., M. Tosi, C. Duponchel, C. Steinmetz, and T. Meo. 1985. Liver mRNA probes disclose two cytochrome P-450 genes duplicated in tandem with the complement *C4* loci of the mouse *H-2S* region. Proc. Natl. Acad. Sci. USA 82:4453–7.

6. Andrews, S.J., and J. Peters. 1983. Linkage analyses and biochemical genetics of sorbitol dehydrogenase-1 (*Sdh-1*) in the mouse. Biochem. Genet. 21:809–17.

7. Antonucci, T., V.C. Chapman, and M.H. Meisler. 1982. Linkage of the structural gene for uroporphyrinogen I synthase to markers on mouse chromosome 9 in a cross between feral and inbred mice. Biochem. Genet. 20:703–10.

8. Antonucci, T.K., O.H. von Deimling, B.B. Rosenblum, and M.H. Meisler. 1984. Conserved linkage within a 4-cM region of mouse chromosome 9 and human chromosome 11. Genetics 107:463–75.

9. Archer, J.R. 1975. Inheritance of the macrophage alloantigenic marker (Mph-1) in inbred mice. Genet. Res. 26:213–9.

10. Arfin, S.M., W.C. Hanford, and B.A. Taylor. 1979. Assignment of histidase-regulating locus to chromosome 10 of the mouse. Biochem. Genet. 17:529–35.

11. Artzt, K., C. Calo, E.N. Pinheiro, A. DiMeo-Talento, and F.L. Tyson. 1987. Ovarian teratocarcinomas in LT/Sv mice carrying t-mutations. Dev. Genet. 8:1–9.

12. Artzt, K., L. Hamburger, and L. Flaherty. 1977. H-39, a histocompatibility locus closely linked to the *T/t* complex. Immunogenetics 5:477–80.

13. Ashley, P.L., J. Ellison, K.A. Sullivan, H.R. Bourne, and D.R. Cox. 1987. Chromosomal assignment of the murine G_i and G_s genes. Am. J. Hum. Genet. 41:A155.

14. Ashley, P.L., and R.R. Flandermeyer. 1986. Identification of novel phosphofructokinase loci in mouse and man. Am. J. Hum. Genet. 39(Suppl.):A186 (Abstr.).

15. Auerbach, C., D.S. Falconer, and J.H. Isaacson. 1962. Test for sex-linked lethals in irradiated mice. Genet. Res. 3:444–7.

16. Avner, P., L. Amar, D. Arnaud, A. Hanauer, and J. Cambrou. 1987. Detailed ordering of markers localizing to the Xq26-Xqter region of the human X chromosome by the use of an interspecific *Mus spretus* mouse cross. Proc. Natl. Acad. Sci. USA 84:1629–33.

17. Awgulewitsch, A., M.F. Utset, C.P. Hart, W. McGinnis, and F.H. Ruddle. 1986. Spatial restriction in expression of a mouse homeo box locus within the central nervous system. Nature 320:328–35.

18. Axelrad, A. 1981. Personal communication.
19. Azen, A.A., I.E. Lush, and B.A. Taylor. 1986. Close linkage of mouse genes for salivary proline-rich proteins (PRPs) and taste. Trends Genet. 2:199–200.
20. Bailey, D.W., and H.P. Bunker. 1972. Located histocompatibility genes. Mouse News Lett. 47:18.
21. Bailey, D.W., T.H. Roderick, and H.P. Bunker. 1972. Personal communication.
22. Bailey, D.W. 1975. Genetics of histocompatibility in mice. I. New loci and congenic lines. Immunogenetics 2:249–56.
23. Bailey, D.W. 1986. Mandibular-morphogenesis gene linked to the *H-2* complex in mice. J. Craniofac. Genet. Dev. Biol. 2:33–9.
24. Bailey, D.W. 1976. Mapping new histocompatibility loci. Jackson Laboratory Annual Report 47:73.
25. Bailey, D.W. 1977. Personal communication.
26. Banerjee, M., F. Wiener, J. Spira, M. Babonits, M.G. Nilsson, J. Sumegi, and G. Klein. 1985. Mapping of the *c-myc*, *pvt-1* and immunoglobulin kappa genes in relation to the mouse plasmacytoma-associated variant (6;15) translocation breakpoint. EMBO J. 4:3183–8.
27. Baranov, V.S., A.L. Shwartzman, V.N. Gorbunova, V.S. Gaitskhoki, and S.A. Neifakh. 1985. Mapping of the ceruloplasmin gene on human and laboratory mouse chromosomes by direct in situ hybridization. Soviet Genet. 21:315–24.
28. Barlow, D.P., M. Bucan, H. Lehrach, B.L.M. Hogan, and N.M. Gough. 1987. Close genetic and physical linkage between the murine haemopoietic growth factor genes GM-CSF and MULTI-CSF (IL3). EMBO J. 6:617–33.
29. Bartke, A. 1965. Mouse News Lett. 32:52–3.
30. Barton, P.R.J., A. Cohen, B. Robert, M.Y. Fiszman, F. Bonhomme, J-L. Guénet, D.P. Leader, and M.E. Buckingham. 1985. The myosin alkali light chains of mouse ventricular and slow skeletal muscle are indistinguishable and are encoded by the same gene. J. Biol. Chem. 260:8578–84.
30a. Baum, C.M., J.P. McKearn, R. Riblet, and J.M. Davie. 1985. Polymorphism of FC receptor on murine B cells is Igh-linked. J. Exp. Med. 162:282–96.
31. Baumann, H., and F.G. Berger. 1984. Genetic analysis of the major acute phase reactants in the mouse: chromosomal location of the structural genes for haptoglobin and α-1-acid glycoprotein-2. Mouse News Lett. 71:49.
32. Baumann, H., and F.G. Berger. 1985. Genetics and evolution of the acute phase proteins in mice. Mol. Gen. Genet. 201:505–12.
33. Baumann, H., W.A. Held, and F.G. Berger. 1984. The acute phase response of mouse liver. Genetic analysis of the major acute phase reactants. J. Biol. Chem. 259:566–73.
34. Beamer, W.G., E.M. Eicher, L.J. Maltais, and J.L. Southard. 1981. Inherited primary hypothyroidism in mice. Science 212:61–3.

35. Beamer, W.G., L.J. Maltais, M.H. DeBaets, and E.M. Eicher. 1986. Inherited congenital goiter in mice. Endocrinology 120:838–40.
36. Beamer, W.G., K.L. Shultz, S.H. Langley, and T.H. Roderick. 1984-85. Juvenile spermatogonial depletion: genetic linkage. Jackson Laboratory Annual Report, p. 34.
37. Beechey, C.V., and A.G. Searle. 1976. Linkage of Sl^{con} and gr. Mouse News Lett. 55:15.
38. Beechey, C.V., M.D. Burtenshaw, B.B. Brown, E.P. Evans, and A.G. Searle. 1980. Chr 18: distance of Tw from centromere. Mouse News Lett. 63:17.
39. Beechey, C.V., and A.G. Searle. 1979. Crm at distal end of X. Mouse News Lett. 60:47.
40. Beechey, C.V., and A.G. Searle. 1980. Thick-tail, Tht. Mouse News Lett. 62:48.
41. Beechey, C.V., and A.G. Searle. 1980. Chr 2: distance of Sd from centromere. Mouse News Lett. 63:17.
42. Beechey, C.V., and A.G. Searle. 1986. Crossing over between the Rw and W loci. Mouse News Lett. 75:27.
43. Beechey, C.V., and A.G. Searle. 1985. Linkage between Tht and Rb1Wh. Mouse News Lett. 72:106.
44. Beechey, C.V., and A.G. Searle. 1984. Mouse News Lett. 71:28.
45. Beechey, C.V., and A.G. Searle. 1982. Position of Tht on chr 5. Mouse News Lett. 67:19.
46. Beechey, C.V., and A.G. Searle. 1980. Position of Va and ma on chr 3. Mouse News Lett. 62:52.
47. Beechey, C.V., and A.G. Searle. 1982. Position of scb on chr 8. Mouse News Lett. 66:64.
48. Bennett, D. 1974. Mouse News Lett. 50:31.
49. Berger, F., and R. Elliott. 1982. A gene complex that governs the structure of an androgen-regulated gene and the nature of the corresponding RNAs. Mouse News Lett. 67:34.
50. Berman, J.W., A.J.D. Rocha, and R. Basch. 1986. Restriction length polymorphism in the variable region of the *Tcr* locus linked to histocompatibility antigen *H-8* on murine chromosome 14. Immunogenetics 24:328–30.
51. Berning, A.K., E.M. Eicher, W.E. Paul, and I. Scher. 1980. Mapping of the X-linked immune deficiency mutation (*xid*) of CBA/N mice. J. Immunol. 124:1875–7.
52. Bernstein, S.E. 1986. Hypotransferrinemia with hemochromatosis (hpx). Mouse News Lett. 75:29.
53. Bernstine, E.G., C. Koh, and C.C. Lovelace. 1979. Regulation of mitochondrial malic enzyme synthesis in mouse brain. Proc. Natl. Acad. Sci. USA 76:6539–41.
54. Berryman, P.L., and W.K. Silvers. 1979. Studies on the *H-X* locus of mice. I. Analysis of polymorphism. Immunogenetics 9:363–7.
55. Biddle, F. 1972. Mouse News Lett. 46:20.
56. Birkenmeier, E.H., and J.I. Gordon. 1986. Developmental regulation of a gene that encodes a cysteine-rich intestinal protein and maps near the murine immuno-

globulin heavy chain locus. Proc. Natl. Acad. Sci. USA 83:2516–20.

57. Blackwell, J.M., J. Howard, and B.A. Taylor. 1984. Mapping of the gene controlling susceptibility to cutaneous leishmaniasis. Mouse News Lett. 70:86.

58. Blandova, Z.K., and A.M. Malashenko. 1985. Mouse News Lett. 73:23.

59. Blank, R.D., G.R. Campbell, M. Pollak, and P. D'Eustachio. 1987. Bayesian multilocus linkage mapping. *In* M. Potter, ed., Symposium on Immunoregulatory Genes.

60. Blankenhorn, E.P., and T.C. Douglas. 1972. Location of the gene for theta antigen in the mouse. J. Hered. 63:259–63.

61. Blatt, C., M.E. Harper, G. Franchini, M.N. Nesbitt, and M.I. Simon. 1984. Chromosomal mapping of murine c-fes and c-src genes. Mol. Cell. Biol. 4:978–81.

62. Blatt, C., K. Mileham, M. Haas, M.N. Nesbitt, M.E. Harper, and M.I. Simon. 1983. Chromosomal mapping of the mink cell focus-inducing xenotropic env gene family in the mouse. Proc. Natl. Acad. Sci. USA 80:6298–302.

63. Blatt, C., L. Weiner, J.G. Sutcliffe, M.N. Nesbitt, and M.I. Simon. 1985. Chromosomal mapping of murine brain specific genes. Cytogenet. Cell Genet. 40:583 (Abstr.).

64. Blatt, C. 1985. Personal communication.

65. Bloom, J.L., and D.S. Falconer. 1966. 'Grizzled,' a mutant in linkage group X of the mouse. Genet. Res. 7:159–67.

66. Bloor, J.H., and M.H. Meisler. 1980. Additional evidence for the close linkage of *Amy-1* and *Amy-2* in the mouse. J. Hered. 71:449–51.

67. Bode, V., F. Bonhomme, and J-L. Guénet. 1986. Mouse News Lett. 74:97.

68. Bodine, D.M., IV, C.S. Birkenmeier, and J.E. Barker. 1984. Spectrin deficient inherited hemolytic anemias in the mouse: characterization of four phenotypes by spectrin synthesis and mRNA activity in reticulocytes. Cell 37:721–9.

69. Bodmer, W.F. 1961. Viability effects and recombination differences in a linkage test with pallid and fidget in the house mouse. Heredity 16:485–95.

70. Bond, J.S., R.J. Beynon, J.F. Reckelhoff, and C.S. David. 1984. *Mep-1* gene controlling a kidney metalloendopeptidase is linked to the major histocompatibility complex in mice. Proc. Natl. Acad. Sci. USA 81:5542–5.

71. Bonhomme, F., F. Bemehdi, J. Britton-Davidian, and S. Martin. 1979. Analyse génétique de croisements interspécifiques *Mus musculus* L. *Mus spretus* Lataste: liaison de *Adh-1* avec *Amy-1* sur le chromosome 3 et de *Es-14* avec *Mod-1* sur le chromosome 9. C.R. Acad. Sci. Paris 289:545–8.

72. Bonhomme, F., J-L. Guénet, and J. Catalan. 1982. Présence d'un facteur de stérilité male, *Hst-2*, ségré-

geant dans les croisements interspécifiques. *M. musculus* L. X *M. spretus* Lastaste et lie a *Mod-1* et *Mpi-1* sur le chromosome 9. C.R. Acad. Sci. Paris 294:691–3.

73. Bonner, J.J., and M.L. Tyan. 1984. Glucocorticoid-induced cleft palate genes in chromosome 17: genetic linkage and mapping analyses. Immunogenetics 20:169–83.

74. Borger, R. 1950. Order of genes in the fifth linkage group of the house mouse. Nature 166:697.

75. Borges, M.S., Y. Kumagai, K. Okumura, N. Hirayama, Z. Ovary, and T. Tada. 1981. Allelic polymorphism of murine IgE controlled by the seventh immunoglobulin heavy chain allotype locus. Immunogenetics 13:499–507.

76. Boyse, E.A., L.J. Old, and E. Stockert. 1965. The TL (thymus leukemia) antigen: a review. *In* P. Grabar and P.A.M. Miescher, eds., Immunopathology, Fourth International Symposium, 23–40. Grune & Stratton, NY.

77. Bradley, D.J., B.A. Taylor, J. Blackwell, E.P. Evans, and J. Freeman. 1979. Regulation of Leishmania populations within the host. III. Mapping of the locus controlling susceptibility to visceral leishmaniasis in the mouse. Clin. Exp. Immunol. 37:7–14.

78. Breen, G.A.M., A.J. Lusis, and K. Paigen. 1977. Linkage of genetic determinants for mouse B-galactosidase electrophoresis and activity. Genetics 85:73–84.

79. Breier, G., M. Bucan, U. Francke, A.M. Colberg-Poley, and P. Gruss. 1986. Sequential expression of murine homeo box genes during F9 EC cell differentiation. EMBO J. 5:2209–15.

80. Breindl, M., J. Doehmer, K. Willecke, J. Dansman, and R. Jaenisch. 1979. Germline integration of Moloney leukemia virus: identification of the chromosomal integration site. Proc. Natl. Acad. Sci. USA 76:1938–42.

81. Brilliant, M.H., M.M. Neimann, and E.M. Eicher. 1987. Murine tyrosine hydroxylase maps to the distal end of chromosome 7 within a region conserved in mouse and man. J. Neurogenet. 4:259–66.

82. Brockdorff, N., G.S. Cross, J.S. Cavanna, E.M.C. Fisher, M.F. Lyon, K.E. Davies, and S.D.M. Brown. 1987. The mapping of a cDNA from the human X-linked Duchenne muscular dystrophy gene to the mouse X chromosome. Nature 328:166–8.

83. Brownell, E., C.A. Kozak, J.R. Fowle III, W.S. Modi, N.R. Rice, and S.J. O'Brien. 1986. Comparative genetic mapping of cellular rel sequences in man, mouse, and the domestic cat. Am. J. Hum. Genet. 39:194–202.

84. Brunet, J.F., M. Dosseto, F. Denizot, M.G. Mattei, W.R. Clark, T.M. Haqqi, P. Ferrier, M. Nabholz, A.M. Schmitt-Verhulst, M.F. Luciani, and P. Golstein. 1986. The inducible cytotoxic T-lymphocyte-associated gene transcript CTLA-1 sequence and gene localization to mouse chromosome 14. Nature 322:268–71.

85. Bruns, G., P.S. Gerald, P. Lalley, U. Francke, and J. Minna. 1979. Gene mapping of the mouse by somatic

cell hybridization. Cytogenet. Cell Genet. 25:139 (Abstr.).

86. Bucan, M., T. Yang-Feng, A.M. Colberg-Poley, D.J. Wolgemuth, J-L. Guénet, U. Francke, H. Lehrach. 1986. Genetic and cytogenetic localisation of the homeo box containing genes on mouse chromosome 6 and human chromosome 7. EMBO J. 5:2899–905.

87. Buchberg, A.M., B.A. Taylor, N.A. Jenkins, and N.G. Copeland. 1986. Chromosomal localization of *Emv-16* and *Emv-17*, two closely linked ecotropic proviruses of RF/J mice. J. Virol. 60:1175–8.

88. Bulfield, G., and J.H. Isaacson. 1985. X-linked muscular dystrophy. Mouse News Lett. 72:97.

89. Bulfield, G., W.G. Siller, P.A.L. Wight, and K.J. Moore. 1984. X-chromosome-linked muscular dystrophy (mdx) in the mouse. Proc. Natl. Acad. Sci. USA 81:1189–92.

90. Bunker, H., and G.D. Snell. 1948. Linkage of white and waved-1. J. Hered. 39:28.

91. Bunker, H. 1970. Personal communication.

92. Bunker, L.E., and Jr. 1959. Hepatic fusion, a new gene in linkage group I of the mouse. J. Hered. 50:40–4.

92a. Bunker, M.C. 1966. Y-chromosome loss in transplanted testicular teratomas of mice. Can J. Genet. Cytol. 8:312–27.

93. Butler-Browne, G., and J-L. Guénet. 1980. Evaluation of the genetic distance between the dy locus and the centromere of chromosome 10. Mouse News Lett. 62:72.

94. Cahilly-Snyder, L., T. Yang-Feng, U. Francke, and D.L. George. 1987. Molecular analysis and chromosomal mapping of amplified genes isolated from a transformed mouse 3T3 cell line. Somat. Cell Mol. Genet. 13:235–44.

95. Callahan, R., D. Gallahan, and C. Kozak. 1984. Two genetically transmitted BALB/c mouse mammary tumor virus genomes located on chromosomes 12 and 16. J. Virol. 49:1005–8.

96. Carlson, G., M. Lovett, C.J. Epstein, D. Westaway, P.A. Goodman, S.T. Marshall, and S.B. Prusiner. 1986. The prion gene complex: polymorphism of *Prn–p* and its linkage with agouti. 14th Mol. Biochem. Genet. Workshop (Abstr.).

97. Carlson, G.A., D.T. Kingsbury, P.A. Goodman, S. Coleman, S.T. Marshall, S. De Armond, D. Westaway, and S.B. Prusiner. 1986. Linkage of prion protein and scrapie incubation time genes. Cell 46:503–11.

98. Carter, T.C., and R.J.S. Phillips. 1953. The sex distribution of waved-2, shaker-2 and rex in the house mouse. Z. Indukt. Abstammungs.-Vererbungsl. 85:564–78.

99. Carter, T.C., and H. Grüneberg. 1950. Linkage between fidget and agouti in the house mouse. Heredity 4:373–6.

100. Carter, T.C. 1947. A new linkage in the house mouse: undulated and agouti. Heredity 1:367–72.

101. Carter, T.C. 1956. Genetics of the Little and Bagg X-rayed mouse stock. J. Genet. 54:311–26.

102. Carter, T.C. 1949. Position of 'luxate' in the third linkage group of the house mouse. Nature 164:1138.

103. Carter, T.C. 1951. The genetics of luxate mice. III. Linkage and independence. J. Genet. 50:300–6.

104. Carter, T.C. 1951. The position of fidget in linkage group V of the house mouse. J. Genet. 50:264–7.

105. Carter, T.C. 1951. Wavy-coated mice: phenotypic interactions and linkage tests between rex and (a) waved-1, (b) waved-2. J. Genet. 50:268–76.

106. Carter, T.C., and R.J.S. Phillips. 1954. Ragged, a semi-dominant coat texture mutant in the house mouse. J. Hered. 45:151–4.

107. Castle, W.E., W.H. Gates, S.C. Reed, and G.D. Snell. 1936. Identical twins in a mouse cross. Science 84:581.

108. Castle, W.E., and W.L. Wachter. 1924. Variations of linkage in rats and mice. Genetics 9:1–12.

109. Castle, W.E. 1919. Studies of heredity in rabbits, rats, and mice. Publ. Carnegie Inst. no. 288, p. 56.

110. Cattanach, B.M., A.J.M. Crocker, and J. Peters. 1984. A further new X-linked gene. Mouse News Lett. 70:80–1.

111. Cattanach, B.M., and A.J.M. Crocker. 1986. X chromosomal location of Sts. Mouse News Lett. 74:94–5.

112. Cattanach, B.M., and J.H. Isaacson. 1965. Genetic control over the inactivation of autosomal genes attached to the X-chromosome. Z. Vererbungsl. 96:313–23.

113. Cattanach, B.M., and H.J. Moseley. 1973. Assignment of LG VII to chromosome 11. Mouse News Lett. 48:31.

114. Cattanach, B.M., and H.J. Moseley. 1974. Mouse News Lett. 50:41–2.

115. Cattanach, B.M., and H.J. Moseley. 1974. Mouse News Lett. 50:42.

116. Cattanach, B.M., and D. Papworth. 1981. Controlling elements in the mouse. V. Linkage tests with X-linked genes. Genet. Res. 38:57–70.

117. Cattanach, B.M., C.E. Williams, and H.J. Bailey. 1972. Mouse News Lett. 47:34.

118. Cattanach, B.M. 1973. Location of Hq. Mouse News Lett. 49:29.

119. Cattanach, B.M. 1982. A new X-linked mutation. Mouse News Lett. 66:61–2.

120. Cattanach, B.M. 1985. Close linkage between Li and Hyp. Mouse News Lett. 73:17.

121. Cattanach, B.M. 1983. Location of Xce using Xce[a]/Xce[c] heterozygotes. Mouse News Lett. 69:24.

122. Cattanach, B.M. 1982. Location of tattered (Td). Mouse News Lett. 67:19.

123. Chapman, D.B. 1965. Personal communication.

124. Chapman, V., M. Murawski, D. Miller, and D. Swiatek. 1985. Linkage analyses of X-chromosome-linked muscular dystrophy, mdx. Mouse News Lett. 72:120.

125. Chapman, V.M., R.J. Blaird, M. Murawski, and B.A.

Quarantillo. 1979. Electrophoretic variation for alcohol dehydrogenase ADH-1. Mouse News Lett. 61:60.

126. Chapman, V.M., and W. Frels. 1975. Mouse News Lett. 53:61.

127. Chapman, V.M., P.G. Krater, and B.A. Quarantillo. 1983. Electrophoretic variation for X chromosome-linked hypoxanthine phosphoribosyl transferase (*Hprt*) in wild-derived mouse. Genetics 103:785–95.

128. Chapman, V.M., E.A. Nichols, and F.H. Ruddle. 1974. Esterase-8 (*Es-8*): characterization, polymorphism, and linkage of an erythrocyte esterase locus on chromosome 7 of *Mus musculus*. Biochem. Genet. 11:347–58.

129. Chapman, V.M., W.K. Noell, and D. Adler. 1975. Mouse News Lett. 53:61.

130. Chapman, V.M., F.H. Ruddle, and T.H. Roderick. 1971. Linkage of isozyme loci in the mouse. I. Phosphoglucomutase-2 (*Pgm-2*), mitochondrial NADP malate dehydrogenase (*Mod-2*), and dipeptidase-1 (*Dip-1*). Biochem. Genet. 5:l0l–10.

131. Chapman, V.M., and F.H. Ruddle. 1972. Glutamate oxaloacetate transaminase (GOT) genetics in the mouse: polymorphism of GOT-1. Genetics 70:299–305.

132. Chapman, V.M. 1975. 6-Phosphogluconate dehydrogenase (PGD) genetics in the mouse: linkage with metabolically related enzyme loci. Biochem. Genet. 13:849–56.

133. Chapman, V.M. 1973. Mouse News Lett. 48:45.

134. Chapman, V.M. 1973. Mouse News Lett. 49:46.

135. Cherry, M., and P.W. Lane. 1973. Mouse News Lett. 49:33.

136. Chesebro, B., and K. Wehrly. 1978. *Rfv-1* and *Rfv-2*, two H-2-associated genes that influence recovery from Friend leukemia virus induced splenomegaly. J. Immunol. 120:1081–5.

137. Claflin, J.L., B.A. Taylor, M. Cherry, and M. Cubberly. 1978. Linkage in mice of genes controlling an immunoglobulin kappa-chain marker and the surface alloantigen Ly-3 on T lymphocytes. Immunogenetics 6:379–87.

138. Clark, E.A., R.C. Harmon, and L.S. Wicker. 1977. Resistance of *H-2* heterozygous mice to parental tumors. II. Characterization of *Hh-1*-controlled hybrid resistance to syngeneic fibrosarcomas and the EL-4 lymphoma. J. Immunol. 119:648–56.

139. Cobb, R.R., T.A. Stoming, and J.B. Whitney III. 1987. The aryl hydrocarbon hydroxylase (*Ah*) locus and a novel restriction-fragment length polymorphism (RFLP) are located on mouse chromosome 12. Biochem Genet. 25:401–13.

140. Collins, R.L. 1970. A new genetic locus mapped from behavioral variation in mice: audiogenic seizure prone (*asp*). Behav. Genet. 1:99–109.

141. Colombatti, A., E.N. Hughes, B.A. Taylor, and J.T. August. 1982. Gene for a major cell surface glycoprotein of mouse macrophages and other phagocytic cells is on chromosome 2. Proc. Natl. Acad. Sci. USA 79:1926–9.

142. Cooper, C.B. 1939. A linkage between naked and caracul in the house mouse. J. Hered. 30:212.

143. Copeland, N.G., N.A. Jenkins, and B.K. Lee. 1983. Association of the lethal yellow (*A^y*) coat color mutation with an ecotropic murine leukemia virus genome. Proc. Natl. Acad. Sci. USA 80:247–9.

144. Cory, S., M. Graham, E. Webb, L. Corcoran, and J.M. Adams. 1985. Variant (6;15) translocations in murine plasmacytomas involve a chromosome 15 locus at least 72 kb from the c-myc oncogene. EMBO J. 4:675–81.

145. Coutinho, A., and T. Meo. 1978. Genetic basis for unresponsiveness to lipopolysaccharide in C57BL/10Cr mice. Immunogenetics 7:17–24.

146. Cox, D. Quoted in W.F. Marzluff and R.A. Graves. 1984. Organization and expression of mouse histone genes. *In* G.S. Stein, J.L. Stein, and W. Marzluff, eds., Histone Genes, 281–315. John Wiley & Sons, NY.

147. Cox, D.R., L.B. Epstein, and C.J. Epstein. 1980. Genes coding for sensitivity to interferon (*IfRec*) and soluble superoxide dismutase (SOD-1) are linked in mouse and man and map to mouse chromosome 16. Proc. Natl. Acad. Sci. USA 77:2168–72.

148. Cox, D.R., D. Goldblatt, and C.J. Epstein. 1982. Chromosomal assignment of mouse PRGS: further evidence for homology between mouse chromosome 16 and human chromosome 21. Am. J. Hum. Genet. 33:145A (Abstr.).

149. Cox, D.R., D. Goldblatt, and C.J. Epstein. 1982. Comparative gene mapping in man and mouse: assignment of the gene for citrate synthase (*Cis*) to mouse chromosome 10. Cytogenet. Cell Genet. 32:259.

150. Cox, D.R., and R.D. Palmiter. 1983. The metallothionein-I gene maps to mouse chromosome 8: implications for human Menkes disease. Hum. Genet. 64:61–4.

151. Cox, D.R., J.A. Sawicki, D. Yee, E. Appella, and C.J. Epstein. 1982. Assignment of the gene for beta-2-microglobulin (*B2m*) to mouse chromosome 2. Proc. Natl. Acad. Sci. USA 79:1930–4.

152. Crichton, D.N., and J.G.M. Shire. 1982. Genetic basis of susceptibility to splenic lipofuscinosis in mice. Genet. Res. 39:275–85.

153. Crocker, A.J.M. 1978. Location of sleek. Mouse News Lett. 59:20–1.

154. Crocker, M., and B.M. Cattanach. 1979. The genetics of sleek: a possible regulatory mutation of the tabby-crinkled-downless syndrome. Genet. Res. 34:231–8.

155. Cumming, R.B., M.F. Walton, J.C. Fuscoe, B.A. Taylor, J.E. Womack, and F.H. Gaertner. 1979. Genetics of formamidase-5 (brain formamidase) in the mouse: localization of the structural gene on chromosome 14. Biochem. Genet. 17:415–31.

156. Curry, G.A. 1959. Genetical and developmental studies on droopy-eared mice. J. Embryol. Exp. Morphol. 7:39–65.

157. Czosnek, H., U. Nudel, Y. Mayer, P.E. Barker, D.D.

Pravtcheva, F.H. Ruddle, and D. Yaffe. 1983. The genes coding for the cardiac muscle actin, the skeletal muscle actin and the cytoplasmic b-actin are located on three different mouse chromosomes. EMBO J. 2:1977–9.

158. Czosnek, H., U. Nudel, M. Shani, P.E. Barker, D.D. Pravtcheva, F.H. Ruddle, and D. Yaffe. 1982. The genes coding for the muscle contractile proteins, myosin heavy chain, myosin light chain 2, and skeletal muscle actin are located on three different mouse chromosomes. EMBO J. 1:1299–305.

159. Czosnek, H., S. Sarid, P.E. Barker, F.H. Ruddle, and V. Daniel. 1984. Glutathione S-transferase Ya subunit is coded by a multigene family located on a single mouse chromosome. Nucl. Acids Res. 12:4825–33.

160. Czosnek, H.H., B. Bienz, D. Givol, R. Zakut-Houri, D.D. Pravtcheva, F.H. Ruddle, and M. Oren. 1984. The gene and the pseudogene for mouse p53 cellular tumor antigen are located on different chromosomes. Mol. Cell. Biol. 4:1638–40.

161. D'Ancona, G.G., and C.M. Croce. 1977. Assignment of the gene for enolase to mouse chromosome 4 using somatic-cell hybrids. Cytogenet. Cell Genet. 19:1–6.

162. D'Eustachio, P., and C.A. Calabro. Genetic analysis of mouse chromosome 12. 14th Mol. Biochem. Genet. Workshop, and 1986 (Abstr.).

163. D'Eustachio, P., B. Fein, J. Michaelson, and B.A. Taylor. 1984. The α-globin pseudogene on mouse chromosome 17 is closely linked to *H-2*. J. Exp. Med. 159:958–63.

164. D'Eustachio, P., T. Kristensen, R.A. Wetsel, R. Riblet, B.A. Taylor, and B.F. Tack. 1986. Chromosomal location of the genes encoding complement components C5 and factor H in the mouse. J. Immunol. 137:3990–5.

165. D'Eustachio, P., G.C. Owens, G.M. Edelman, and B.A. Cunningham. 1985. Chromosomal location of the gene encoding the neural cell adhesion molecule (N-CAM) in the mouse. Proc. Natl. Acad. Sci. USA 82:7631–5.

166. D'Eustachio, P. 1984. A genetic map of mouse chromosome 12 composed of polymorphic DNA fragments. J. Exp. Med. 160:827–38.

166a. D'Eustachio, P., S. Jadidi, R.C. Fuhlbrigge, P.W. Gray, and D.D. Chaplin. 1987. Interleukin-1 α and β genes: linkage on chromosome 2 in the mouse. Immunogenetics 26:339–43.

167. D'Hoostelaere, L.A., E. Jouvin-Marche, and K. Huppi. 1985. Localization of CTβ and C$\varkappa$ on mouse chromosome 6. Immunogenetics 22:277–83.

168. Dandoy, F., E. De Maeyer, F. Bonhomme, J-L. Guénet, and J. De Maeyer-Guignard. 1985. Segregation of restriction fragment length polymorphism in an interspecies cross of laboratory and wild mice indicates tight linkage of the murine IFN-β gene to the murine IFN-α genes. J. Virol. 56:216–20.

169. Dandoy, F., K.A. Kelley, J. DeMaeyer-Guignard, E. DeMaeyer, and P.M. Pitha. 1984. Linkage analysis of the murine interferon-α locus on chromosome 4. J. Exp. Med. 160:294–302.

170. Davidson, W.F., B.J. Mathieson, C.A. Kozak, T.M. Chused, and H.C. Morse III. 1982. Chromosome 1 locus required for induction of CTL to H-2-compatible cells in NZB mice. Immunogenetics 15:321–5.

171. Davidson, W.F., H.C. Morse III, B.J. Mathieson, C.A. Kozak, and F-W. Shen. 1983. The B cell alloantigen Ly-17.1 is controlled by a gene closely linked to *Ly-20* and *Ly-9* on chromosome 1. Immunogenetics 17:325–9.

172. Davisson, M.T., and H.P. Bunker. 1986. Linkage of chocolate (cht) to chr 7. Mouse News Lett. 75:31.

173. Davisson, M.T., E.M. Eicher, and M.C. Green. 1976. Genes on chromosome 3 of the mouse. J. Hered. 67:155–6.

174. Davisson, M.T., S.H. Langley, T.H. Roderick, and L.C. Skow. 1979. Mouse News Lett. 60:50–1.

175. Davisson, M.T., and T.H. Roderick. 1975. Mouse News Lett. 52:38.

176. Davisson, M.T., and A. Zweidler. 1986. Histone gene mapping in the mouse. 14th Mol. Biochem. Genet. Workshop (Abstr.).

177. Davisson, M.T. 1980. Personal communication.

178. Davisson, M.T. 1984–85. Rb(16.17)7Bnr inhibits genetic recombination. The Jackson Laboratory 56th Annual Report, p. 40.

179. Davisson, M.T. 1986. Position of Dey on chr 2. Mouse News Lett. 75:30–1.

180. Davisson, M.T. 1987. Personal communication.

181. Day, C., and M. Nesbitt. 1977. Mouse News Lett. 57:10.

182. Day, T.H. 1986. Variation and evolution of the structural proteins with special reference to the lens proteins. PhD Thesis, Univ. of Edinburgh. Quoted B.L.M. Hogan *et al.*, J. Embryol. Exp. Morphol. 97:95–110.

183. De Maeyer, E., and F. Dandoy. 1987. Linkage analysis of the murine interferon alpha locus (*Ifa*) on chromosome 4. J. Hered. 78:143–6.

184. DeLorenzo, R.J., and F.H. Ruddle. 1969. Genetic control of two electrophoretic variants of glucosephosphate isomerase in the mouse (*Mus musculus*). Biochem. Genet. 3:151–62.

185. DeLorenzo, R.J., and F.H. Ruddle. 1970. Glutamate oxalate transaminase (GOT) genetics in *Mus musculus*: linkage, polymorphism, and phenotypes of the Got-2 and Got-1 loci. Biochem. Genet. 4:259–73.

186. DeMaeyer, E., J. De Maeyer-Guignard, and D.W. Bailey. 1975. Effect of mouse genotype on interferon production. I. Lines congenic at the *If-1* locus. Immunogenetics 1:438–43.

187. DeMaeyer-Guignard, J., R. Zawatsky, F. Dandoy, and E. DeMaeyer. 1983. An X-linked locus influences early serum interferon levels in mouse. J. Interferon Res. 3:241–52.

188. Debre, P., S. Gisselbrecht, F. Pozo, and J.P. Levy. 1979. Genetic control of sensitivity to Moloney leuke-

mia virus in mice. II. Mapping of three resistant genes within the *H-2* complex. J. Immunol. 123:1806–12.

189. Demant, P., J. Frejt, and J. Ivanyi. 1974. Hybrid sterility genetic system—mapping of *Hst-1* between T and H-2. Mouse News Lett. 51:29.

190. Demant, P.D., D. Ivanyi, and R. van Nie. 1979. The map position of the *rds* gene on the 17th chromosome of the mouse. Tissue Antigens 13:53–5.

191. Dembic, Z., W. Bannwarth, B.A. Taylor, and M. Steinmetz. 1985. The gene encoding the T-cell receptor α-chain maps close to the *Np-2* locus on mouse chromosome 14. Nature 314:271–3.

192. Demmer, L.A., E.H. Birkenmeier, D.A. Sweetser, M.S. Levin, S. Zollman, R.S. Sparkes, T. Mohandas, A.J. Lusis, and J.I. Gordon. 1987. The cellular retinol binding protein II gene. Sequence analysis of the rat gene, chromosomal localization in mice and humans, and documentation of its close linkage to the cellular retinol binding protein gene. J. Biol. Chem. 262:2458–67.

193. Deol, M.S., and M.C. Green. 1966. Snell's waltzer, a new mutation affecting behavior and the inner ear in the mouse. Genet. Res. 8:339–45.

194. Deol, M.S., and P.W. Lane. 1966. A new gene affecting the morphogenesis of the vestibular part of the inner ear in the mouse. J. Embryol. Exp. Morphol. 16:543–58.

195. Detlefsen, J.A., and L.S. Clemente. 1924. Linkage of a dilute color factor and dark eye in mice. Genetics 9:247–60.

196. Detlefsen, J.A. 1925. The linkage of dark-eye and color in mice. Genetics 10:17–32.

197. Dickerman, R.C., R.N. Feinstein, and D. Grahn. 1968. Position of the acatalasemia gene in linkage group V of the mouse. J. Hered. 59:177–8.

198. Dickie, M.M., and P.W. Lane. 1957. Mouse News Lett. 17:52. Plus letter to Roy Robinson 7/7/70.

199. Dickie, M.M., and J.L. Southard. 1970. Personal communication.

200. Dickie, M.M., and G.W. Woolley. 1950. Fuzzy mice. J. Hered. 41:193–6.

201. Dickie, M.M., and G.W. Woolley. 1946. Linkage studies with the pirouette gene in the mouse. J. Hered. 37:335–7.

202. Dickie, M.M. 1955. Alopecia, a dominant mutation in the house mouse. J. Hered. 46:31–4.

203. Dickie, M.M. 1965. Mouse News Lett. 32:46.

204. Dickie, M.M. 1966. Mouse News Lett. 34:30.

205. Dickie, M.M. 1969. Mouse News Lett. 40:29.

206. Dickie, M.M. 1969. Mouse News Lett. 40:29. Plus letter of J. Southard to Roy Robinson 9/15/70.

207. Dickie, M.M. 1968. Mouse News Lett. 39:27.

208. Dickie, M.M. 1964. New splotch alleles in the mouse. J. Hered. 55:97–101.

209. Dickler, H.B., D.L. Rosenstreich, A. Ahmed, and D.H. Sachs. 1980. Genetic linkage between *Lym-1*, *Sas-1* and *Mls* loci. Immunogenetics 10:93–6.

210. Dizik, M., and R. Elliott. 1977. Genes controlling the processing of α-mannosidase. Mouse News Lett. 56:58.

211. Dizik, M., R.W. Elliott, and F. Lilly. 1981. Construction of a D2.B6- *Fv-2ʳ* congenic strain. Non-lethality of the homozygous *Fv-2ʳ* genotype. J. Natl. Cancer Inst. 66:755–60.

212. Dizik, M., and R.W. Elliott. 1977. A gene apparently determining the extent of sialylation of lysosomal α-mannosidase in mouse liver. Biochem. Genet. 15:31–46.

213. Dlouhy, S.R., B.A. Taylor, and R.C. Karn. 1987. The genes for mouse salivary androgen-binding protein (ABP) subunits alpha and gamma are located on chromosome 7. Genetics 115:535–43.

214. Douglas, T.C., and P.E. Dawson. 1979. Location of the gene for theta antigen in the mouse. III. The position of *Thy-1* relative to *Lap-1* and *Mpi-1*. J. Hered. 70:250–4.

215. Douglas, T.C., and P.E. Dawson. 1981. Mouse News Lett. 65:23.

216. Douglas, T.C., and P.E. Dawson. 1979. The position of the gene for glyoxylase I on chromosome 17 of the mouse. Immunogenetics 8:367–71.

217. Douglas, T.C., T. Meo, and H. Skarvall. 1978. Location of the gene for theta antigen in the mouse. II. Three-point crosses place *Thy-1* in proximal region of chromosome 9. J. Hered. 69:224–8.

218. Douglas, T.C. 1979. Mouse News Lett. 60:58.

219. Dowsett, A.P., L.A. Herzenberg, and L.A. Herzenberg. 1981. Regulation of the production of murine IgGa by an *H-2*-linked gene and other unlinked genes. Immunogenetics 13:237–45.

220. Duley, J., and R.S. Holmes. 1974. α-Hydroxyacid oxidase genetics in the mouse: evidence for two genetic loci and a tetrameric subunit structure for the liver isozyme. Genetics 76:83–97.

221. Duley, J.A., and R.S. Holmes. 1982. Biochemical genetics of aldehyde reductase in the mouse: *Ahr-1*—a new locus linked to the alcohol dehydrogenase gene complex on chromosome 3. Biochem. Genet. 20:1067–83.

222. Duley, J.A., T.L. Seeley, and R.S. Holmes. 1983. Aldehyde reductase (Ahr-1). Mouse News Lett. 68:66.

223. Dunn, L.C., D. Bennett, and A.B. Beasley. 1962. Mutation and recombination in the vicinity of a comple gene. Genetics 47:285–303.

224. Dunn, L.C., and D. Bennett. 1971. Further studies of a mutation (*Low*) which distorts transmission ratios in the house mouse. Genetics 67:543–58.

225. Dunn, L.C., and E. Caspari. 1945. A case of neighboring loci with similar effects. Genetics 30:543–68.

226. Dunn, L.C., and E. Caspari. 1942. Close linkage between mutations with similar effects. Proc. Natl. Acad. Sci. USA 28:205–10.

227. Dunn, L.C. 1920. Linkage in mice and rats. Genetics 5:325–43.

228. Dunn, L.C. 1966. Mouse News Lett. 34:21.

229. Durda, P.J., S.C. Boos, and P.D. Gottlieb. 1979. T100:

a new murine cell surface glycoprotein detected by anti-Lyt-2.1 serum. J. Immunol. 122:1407–12.

230. Dush, M.K., J.A. Tischfield, S.A. Khan, E.Feliciano, J.M. Sikela, C.A. Kozak, and P.J. Stambrook. 1986. An unusual adenine phosphoribosyltransferase pseudogene is syntenic with its functional gene and is flanked by highly polymorphic DNAs. Mol. Cell Biol. 6:4161–7.

231. Eckhardt, L.A., and L.A. Herzenberg. 1980. Monoclonal antibodies to ThB detect close linkage of *Ly-6* and a gene regulating ThB expression. Immunogenetics 11:275–91.

232. Eicher, E.M., and W.G. Beamer. 1976. Inherited ateliotic dwarfism in mice: characteristics of the mutation, little, on chromosome 6. J. Hered. 67:87–91.

233. Eicher, E.M., M. Cherry, and L. Flaherty. 1978. Autosomal phosphoglycerate kinase linked to mouse major histocompatibility complex. Mol. Gen. Genet. 158:225–8.

234. Eicher, E.M., and D.L. Coleman. 1977. Influence of gene duplication and X-inactivation on mouse mitochondrial malic enzyme activity and electrophoretic patterns. Genetics 85:647–58.

235. Eicher, E.M., S. Fox, and S. Reynolds. 1977. Mouse News Lett. 56:42.

236. Eicher, E.M., and P.W. Lane. 1980. Assignment of LGXVI to chromosome 3 in the mouse. J. Hered. 71:315–8.

237. Eicher, E.M., J. May, J.L. Southard, and L. Washburn. 1977. Mouse News Lett. 56:42.

238. Eicher, E.M., S. Reynolds, and J.L. Southard. 1977. Mouse News Lett. 56:42.

239. Eicher, E.M., and S. Reynolds. 1978. Mouse News Lett. 58:48.

240. Eicher, E.M., J.L. Southard, C.R. Scriver, and F.H. Glorieux. 1976. Hypophosphatemia: a mouse model for human familial hypophosphatemic (vitamin D-resistant) rickets. Proc. Natl. Acad. Sci. USA 73:4667–71.

241. Eicher, E.M., and J.L. Southard. 1972. Mouse News Lett. 47:36–7.

242. Eicher, E.M., and J.L. Southard. 1979. Mouse News Lett. 60:51.

243. Eicher, E.M., R.H. Stern, J.E. Womack, M.T. Davisson, and T.H. Roderick. 1976. Evolution of mammalian carbonic anhydrase loci by tandem duplication: close linkage of *Car-1* and *Car-2* to the centromere region of chromosome 3 of the mouse. Biochem. Genet. 14:651–60.

244. Eicher, E.M., B.A. Taylor, S.C. Leighton, and J.E. Womack. 1980. A serum protein polymorphism determinant on chromosome 9 of *Mus musculus*. Mol. Gen. Genet. 177:571–6.

245. Eicher, E.M., and D. Varnum. 1986. Allelism of Den and Re. Mouse News Lett. 75:29–30.

245a. Eicher, E.M., L.L. Washburn, J.B. Whitney III, and K.E. Morrow. 1982. *Mus poschiavinus* Y chromosome in the C57BL/6J murine genome causes sex reversal. Science 217:535–7.

246. Eicher, E.M., and J.E. Womack. 1977. Chromosomal location of soluble glutamic-pyruvic transaminase-1 (*Gpt-1*) in the mouse. Biochem. Genet. 15:1–8.

247. Eicher, E.M. 1972. Mouse News Lett. 46:31.

248. Eicher, E.M. 1974. Mouse News Lett. 51:24.

249. Eicher, E.M. 1976. Mouse News Lett. 54:41.

250. Eicher, E.M. 1972. Personal communication.

251. Eicher, E.M. 1974. Personal communication.

252. Eicher, E.M. 1975. Personal communication.

253. Eicher, E.M. 1970. The position of *ru-2* and *qv* with respect to the flecked translocation in the mouse. Genetics 64:495–510.

254. Eisenhardt, E., and O. von Deimling. 1982. Interstrain variation of esterase-22, a new isozyme of the house mouse. Comp. Biochem. Physiol. 73B:719–24.

255. Eleftheriou, B.E., and D.W. Bailey. 1972. A gene controlling plasma serotonin levels in mice. J. Endocrinol. 55:225–6.

256. Eleftheriou, B.E. 1974. A gene influencing hypothalamic norepinephrine levels in mice. Brain Res. 70:538–40.

257. Elliott, R. 1981. Identification and mapping of new loci using cloned cDNA. Mouse News Lett. 64:87.

258. Elliott, R. 1979. Mouse variants studied by two-dimensional electrophoresis. Mouse News Lett. 61:59.

260. Elliott, R. 1978. Mouse News Lett. 59:59.

261. Elliott, R.W., D. Barlow, and B.L.M. Hogan. 1985. Linkage of genes for laminin B1 and B2 subunits on chromosome 1 in mouse. In Vitro Cell. Dev. Biol. 21:477–84.

262. Elliott, R.W., W.L. Daniel, B.A. Taylor, and E. Novak. 1985. Linkage of loci affecting a murine liver protein and arylsulfatase B to chromosome 13. J. Hered. 76:243–5.

263. Elliott, R.W., C. Hohman, C. Romijko, P. Louis, and and F. Lilly. 1978. Use of high resolution two dimensional electrophoresis of liver cytosol proteins in the discovery and mapping of a new mouse variant. *In* N. Catsimpoolas, ed., Electrophoresis, 261–74. Elsevier North Holland, Amsterdam.

264. Elliott, R.W., C. Romejko, and C. Hohman. 1980. Mapping the gene for LTW-4, a 26, molecular weight major protein of mouse liver and kidney. Mol. Gen. Genet. 180:17–22.

265. Elliott, R.W., L.C. Samuelson, M.S. Lambert, and M.H. Meisler. 1986. Assignment of pancreatic ribonuclease gene to mouse chromosome 14. Cytogenet. Cell Genet. 42:110–12.

266. Elliott, R.W. 1987. Map locations for laminin subunits B1 and B2. Mouse News Lett. 78:74.

267. Elliott, R.W. 1979. Use of two-dimensional electrophoresis to identify and map new mouse genes. Genetics 91:295–308.

268. Elsevier, S.M., and F.H. Ruddle. 1975. Location of genes coding for 18S and 28S ribosomal RNA within the genome of *Mus musculus*. Chromosoma 52:219–28.

269. Eppig, J.T., and E.M. Eicher. 1983. Application of the

ovarian teratoma mapping method in the mouse. Genetics 103:797–812.

270. Epstein, R., M. Davisson, K. Lehman, E.C. Akeson, and M. Cohn. 1986. Position of *Igl-1*, *md*, and *Bst* loci on chromosome 16. Immunogenetics 23:78–83.

271. Epstein, R., K. Lehmann, M. Cohn, C. Buckler, W. Rowe, and M.T. Davisson. 1984. Linkage of the *Igl-1* structural and regulatory genes to *Akv-2* on chromosome 16. Immunogenetics 19:527–37.

271a. Erickson, R.P. 1977. Androgen modified expression compared with Y linkage of male specific antigen. Nature 265:59–61.

272. Evans, B.A., and R.I. Richards. 1985. Genes for the α and γ subunits of mouse nerve growth factor are contiguous. EMBO J. 4:133–8.

273. Falconer, D.S., and J.H. Isaacson. 1966. Curly-whiskers and its linkage with tail-kinks in linkage group II of the mouse. Genet. Res. 8:111–3.

274. Falconer, D.S., and J.H. Isaacson. 1965. Mouse News Lett. 32:30–1.

275. Falconer, D.S., and J.H. Isaacson. 1965. Mouse News Lett. 32:31.

276. Falconer, D.S., and J.H. Isaacson. 1962. The genetics of sex-linked anemia in the mouse. Genet. Res. 3:248–50.

277. Falconer, D.S., and G.D. Snell. 1952. Two new hair mutants, rough and frizzy, in the house mouse. J. Hered. 43:53–7.

278. Falconer, D.S., and W.R. Sobey. 1953. The location of 'trembler' in linkage group VII. J. Hered. 44:159–60.

279. Falconer, D.S. 1954. Linkage in the mouse: the sex-linked genes and 'rough.' Z. Indukt. Abstammungs.-Vererbungsl. 86:263–8.

280. Falconer, D.S. 1947. Linkage of rex with shaker-2 in the house mouse. Heredity 1:133–5.

281. Falconer, D.S. 1952. Location of 'reeler' in linkage group III in the mouse. Heredity 6:255–7.

282. Falconer, D.S. 1956. Mouse News Lett. 14:13.

283. Falconer, D.S. 1957. Mouse News Lett. 17:40.

284. Falconer, D.S. 1962. Mouse News Lett. 27:30.

285. Falconer, D.S. 1953. Total sex-linkage in the house mouse. Z. Indukt. Abstammungs.-Vererbungsl. 85:210–19.

286. Farber, R.A., and D. Zielinski. 1986. Assignment of a processed mouse *Aprt* pseudogene to the same chromosome as the functional gene. Cytogenet. Cell Genet. 42:198–201.

287. Favor, J., and W. Pretsch. 1987. Position of Xcat, a new X-linked cataract mutation. Mouse News Lett. 77:139.

288. Feldman, H.W. 1924. Linkage of albino allelomorphs in rats and mice. Genetics 9:487–92.

289. Ferguson, J.M., and M.E. Wallace. 1975. Mouse News Lett. 52:31.

290. Festenstein, H., C. Bishop, and B.A. Taylor. 1977. Location of *Mls* locus on mouse chromosome 1. Immunogenetics 5:357–61.

291. Field, E.H., B. Tourvieille, P. D'Eustachio, and J.R. Parnes. 1987. The gene encoding the mouse T cell differentiation antigen L3T4 is located on chromosome 6. J. Immunol. 138:1968–70.

292. Fielder, J.H. 1952. The taupe mouse. J. Hered. 43:75–6.

293. Figueroa, F., M. Kasahara, H. Tichy, E. Neufeld, U. Ritte, and J. Klein. 1987. Polymorphism of unique non-coding DNA sequences in wild and laboratory mice. Genetics 117:101–8.

294. Figuerova, F., D. Klein, S. Tewarson, and J. Klein. 1982. Evidence for placing the *Neu-1* locus within the mouse H-2 complex. J. Immunol. 129:2089–92.

295. Finlayson, J.S., D.M. Hudson, and B.L. Armstrong. 1969. Location of the *Mup-a* locus on mouse linkage group VIII. Genet. Res. 14:329–31.

296. Finnegan, A., and F.L. Owen. 1981. Ly-18, a new alloantigen present on cytotoxic T cells and controlled by a gene(s) linked to the *Igh-V* locus. J. Immunol. 127:1947–53.

297. Fischer-Lindahl, K., B. Hausmann, and V.M. Chapman. 1983. A new *H-2*-linked class I gene whose expression depends on a maternally inherited factor. Nature 306:383–5.

298. Fisher, R.A., and W. Landauer. 1953. Sex differences of crossing-over in close linkage. Am. Nat. 87:116.

299. Fisher, R.A., M.F. Lyon, and A.R.G. Owen. 1947. The sex chromosomes in the house mouse. Heredity 1:355–66.

300. Fisher, R.A. 1949. A preliminary linkage test with agouti and undulated mice. I. The fifth linkage group. Heredity 3:229–41.

301. Flaherty, L., M. Karl, and C.L. Reinisch. 1982. Syngeneic tumor immunizations produce Qa antibodies. Discovery of a new Qa antigen, Qa-6. Immunogenetics 16:329–37.

302. Flaherty, L., and S.S. Wachtel. 1975. Identification of two new loci, *H-31* and *H-32*, and alleles. Immunogenetics 2:81–5.

303. Flaherty, L. 1978. Genes of the Tla system. The new Qa system of antigens. *In* H.C. Morse III, ed., Origins of Inbred Strains of Mice, 409–22. Academic Press, NY.

304. Flaherty, L. 1975. H-33—a histocompatibility locus to the left of the *H-2* complex. Immunogenetics 2:325–9.

305. Flanagan, S.P., and J.H. Isaacson. 1967. Close linkage between genes which cause hairlessness in the mouse. Genet. Res. 9:99–110.

306. Flanagan, S.P. 1966. 'Nude' a new hairless gene with pleiotropic effects in the mouse. Genet. Res. 8:294–309.

307. Forman, J., R. Riblet, K. Brooks, E.S. Vitetta, and L.A. Henderson. 1984. H-40, an antigen controlled by an *Igh* linked gene and recognized by cytotoxic T lymphocytes. I. Genetic analysis of H-40 and distribution of its product on B cell tumors. J. Exp. Med. 159:1724–40.

308. Forsthoefel, P.F., and T.E. Shenk. 1970. Linkage rela-

tionships of Strong's luxoid gene in the mouse. J. Hered. 61:64–6.

309. Foster, M., M.L. Petras, and P. Tomlin. 1972. Personal communication. *In* R. Robinson, ed., Gene Mapping in Laboratory Mammals, Part B, p. 363. Plenum, NY.

310. Fournier, R.E.K., and R.G. Moran. 1983. Complementation mapping in microcell hybrids: localization of *Fpgs* and *Ak-1* on *Mus musculus* chromosome 2. Somat. Cell Genet. 9:69–84.

311. Fox, H.S., L.M. Silver, and G.R. Martin. 1984. An alpha-globin pseudogene is located within the mouse *t* complex. Immunogenetics 19:125–30.

312. Fox, R.R., and P.M. Black. 1987. Erythrocyte antigen-9 (Ea-9). Mouse News Lett. 78:57.

313. Fox, S., E.M. Eicher, and S. Reynolds. 1977. Mouse News Lett. 57:21.

314. Fox, S., S. Reynolds, and E.M. Eicher. 1977. Mouse News Lett. 57:21.

315. Francke, U., and U. Gehring. 1980. Chromosome assignment of a murine glucocorticoid receptor gene (*Grl-1*) using intraspecies somatic cell hybrids. Cell 22:657–64.

316. Francke, U., P.A. Lalley, W. Moss, J. Ivy, and J.D. Minna. 1977. Gene mapping in *Mus musculus* by interspecific cell hybridization: assignment of the genes for tripeptidase-1 to chromosome 10, dipeptidase-2 to chromosome 18, acid phosphatase-1 to chromosome 12 and adenylate kinase-1 to chromosome 2. Cytogenet. Cell Genet. 19:57–84.

317. Francke, U., P. Tetri, R.T. Taggart, and N. Oliver. 1981. Conserved autosomal syntenic group on mouse (MMU) chromosome 15 and human (HSA) chromosome 22: assignment of a gene for arylsulfatase A to MMU15 and regional mapping of DIA1, ARSA, and ACO2 on HSA 22. Cytogenet. Cell Genet. 31:58–69.

318. Francke, U., T.L. Yang-Feng, and L.J. DeGennaro. 1986. Genes for synapsin I, aneural phosphoprotein, map to conserved regions of human and murine X chromosomes. Am. J. Hum. Genet. 39(Suppl.):A155 (Abstr.).

319. Frater-Schroder, M., M. Prochazka, O. Haller, F. Arwert, H.J. Porck, L.C. Skow, L-G. Lundin, J. Hilkens, and J. Hilgers. 1985. Localization of the gene for the vitamin B12 binding protein transcobalamin II, near the centromere on mouse chromosome 11, linked with the hemoglobin alpha-chain locus. Biochem. Genet. 23:139–53.

320. Frels, W., and V.M. Chapman. 1977. Linkage of glutathione reductase (Gr). Mouse News Lett. 56:58.

321. Fujii, D., J.E. Brissenden, R. Derynck, and U. Francke. 1986. Transforming growth factor β gene maps to human chromosome 19 long arm and to mouse chromosome 7. Somat. Cell Mol. Genet. 12:281–8.

322. Fundele, R., T. Bucher, A. Gropp, and H. Winking. 1981. Enzyme patterns in trisomy 19 of the mouse. Dev. Genet. 2:291–303.

323. Furukawa, F., N. Maruyama, H. Yoshida, Y. Hamashima, S. Hirose, and T. Shirai. 1985. Genetic studies on the skin lupus band test in New Zealand mice. Clin. Exp. Immunol. 59:146–52.

324. Gallahan, D., C. Kozak, and R. Callahan. 1987. A new common integration region (int-3) for mouse mammary tumor virus on mouse chromosome 17. J. Virol. 61:218–20.

325. Gardner, S.M., B.A. Mock, J. Hilgers, K.E. Huppi, and W.D. Roeder. 1987. Mouse lymphotoxin and tumor necrosis factor: structural analysis of the cloned genes, physical linkage, and chromosomal position. J. Immunol. 139:476–83.

326. Gasser, D.L., J. Ziebar, and K. Matsumoto. 1983. A lymphocyte antigen encoded by a locus closely linked to *H-3* on the second chromosome of the mouse. Proc. Natl. Acad. Sci. USA 80:6620–3.

327. Gasser, D.L. 1969. Genetic control of the immune response in mice. I. Segregation data and localization to the fifth linkage group of a gene affecting antibody production. J. Immunol. 103:66–70.

328. Gasser, D.L. 1976. Genetic studies of the H-3 region of mice and implications for polymorphism of histocompatibility loci. Immunogenetics 3:271–6.

329. Gates, W.H. 1931. Linkage of the factor shaker with albinism and pink-eye in the house mouse. Z. Indukt. Abst. Verebungsl. 59:220–6.

330. Gates, W.H. 1928. Linkage of the factors for short-ear and density in the house mouse (*Mus musculus*). Genetics 13:170–9.

331. Gazdar, A.E., H. Oie, P. Lalley, W.W. Moss, J.D. Minna, and U. Francke. 1977. Identification of mouse chromosomes required for murine leukemia virus reproduction. Cell 11:949–56.

332. Gelfand, M.C., D.H. Sachs, R. Lieberman, and W.E. Paul. 1974. Ontogeny of B lymphocytes. III. H-2 linkage of a gene controlling the rate of appearance of complement receptor lymphocytes. J. Exp. Med. 139:1142–53.

333. Gervais, F., M. Stevenson, and E. Skamene. 1984. Genetic control of resistance to Listeria monocytogenes: regulation of leukocyte inflammatory responses. J. Immunol. 132:2078–83.

334. Gibb, S., E.M. Hakansson, L-G. Lundin, and J.G.M. Shire. 1981. Reduced pigmentation (*rp*), a new coat colour gene with effects on kidney lysosomal glycosidases in the mouse. Genet. Res. 37:95–103.

335. Gibson, D.M., S.J. Maclean, and M. Cherry. 1983. Recombination between kappa chain genetic markers and the Lyt-3 locus. Immunogenetics 18:111–6.

336. Gibson, D.M., B.A. Taylor, and M. Cherry. 1978. Evidence for close linkage of a mouse light chain marker with the *Ly-2, 3* locus. J. Immunol. 121:1585–90.

337. Gilman, J.G., and O. Smithies. 1968. Fetal hemoglobin variants in mice. Science 160:885–6.

338. Glaser, T., D. Gerhard, C. Jones, L. Albritton, P. Lal-

ley, and D. Houseman. 1985. A fine structure deletion map of chromosome 11p. Cytogenet. Cell Genet. 40:643 (Abstr.).

339. Glaser, T. 1986. Personal communication.

340. Goding, J.W. 1981. Evidence for linkage of murine B2-microglobulin to *H-3* and *Ly-4*. J. Immunol. 126:1644–6.

341. Gold, D.P., J.J.M. Van Dongen, C.C. Morton, G.A.P. Bruns, P. Van Den Elsen, A.H.M. Geurts Van Kessell, and C. Terhorst. 1987. The gene encoding the epsilon subunit of the T3/T-cell receptor complex maps to chromosome 11 in humans and to chromosome 9 in mice. Proc. Natl. Acad. Sci. USA 84:1664–8.

342. Goldman, D., and H.J. Pikus. 1986. Fourteen genetically variant proteins of mouse brain: discovery of two new variants and chromosomal mapping of four loci. Biochem. Genet. 24:183–94.

343. Gonez, L.J., K.S. Sandrin, M.M. Henning, J. Hilgers, and I.F.C. McKenzie. 1985. *Ly-29*: a locus closely linked to *Mtv-20* on chromosome 4 codes for a new mouse lymphocyte surface alloantigen. Immunogenetics 22:305–8.

343a. Gonzalez, F.J., T. Matsunaga, K. Nagata, U.A. Meyer, D.W. Nebert, J. Pastewka, C.A. Kozak, J. Gillette, H.V. Gelboin, and J.P. Hardwick. 1987. Debrisoquine 4-hydroxylase: characterization of a new P450 gene subfamily, regulation, chromosomal mapping, and molecular analysis of the DA rat polymorphism. DNA 6:149–61.

344. Goodwins, I.R., and M.A.C. Vincent. 1955. Further data on linkage between short-ear and Maltese dilution in the house mouse. Heredity 9:413–4.

345. Gorman, J.C., R. Jackson, J.R. Desantola, D. Shreffler, and J.P. Atkinson. 1980. Development of a hemolytic assay for mouse C2 and determination of its genetic control. J. Immunol. 125:344–51.

346. Gough, N.M., J. Gough, D. Metcalf, A. Kelso, D. Grail, N.A. Nicola, A.W. Burgess, and A.R. Dunn. 1984. Molecular cloning of cDNA encoding a murine haematopoietic growth regulator, granulocyte-macrophage colony stimulating factor. Nature 309:763–7.

347. Graff, R.J., D. Martin-Morgan, and M.E. Kurtz. 1985. Allograft rejection defined antigens of the *B2m*, *H-3* region. J. Immunol. 135:2842–6.

347a. Graff, R.J., D. Martin-Morgan, and M.E. Kurtz. 1987. Multiplicity of chromosome 2 histocompatibility genes: new loci, *H-44* and *H-45*. Immunogenetics 26:111–4.

348. Graham, M., J.M. Adams, and S. Cory. 1985. Murine T lymphomas with retroviral inserts in the chromosomal 15 locus for plasmacytoma variant translocations. Nature 314:740–3.

349. Grahn, D., R.A. Lea, and J.H. Hulesch. 1970. Location of an X-inactivation controller gene on the normal X-chromosome of the mouse. Genetics 64:s25.

350. Grahn, D. 1972. Mouse News Lett. 47:20.

351. Grahn, D. 1973. Mouse News Lett. 48:21.

352. Graves, R.A., S.E. Wellman, I-M. Chiu, and W.F. Marzluff. 1985. Differential expression of two clusters of mouse histone genes. J. Mol. Biol. 183:179–94.

352a. Green, M.C. 1963. Methods for testing linkage. *In* W.J. Burdette, ed., Methodology in Mammalian Genetics, 56–82. Holden-Day, Inc, San Francisco, CA.

353. Green, E.L., and S.J. Mann. 1961. Opossum, a semidominant lethal mutation affecting hair and other characteristics of mice. J. Hered. 52:223–7.

354. Green, E.L. 1968. Linkage of shambling with rex in linkage group VII of the mouse. J. Hered. 59:59.

355. Green, E.L. 1969. Personal communication.

356. Green, M.C., and S.C. Fox. 1965. Mouse News Lett. 32:46.

357. Green, M.C., and S.C. Fox. 1968. Mouse News Lett. 39:28.

358. Green, M.C., and P.W. Lane. 1967. Linkage group II of the house mouse. J. Hered. 58:225–8.

359. Green, M.C., and L.D. Shultz. 1975. Motheaten, an immunodeficient mutant of the mouse. I. Genetics and pathology. J. Hered. 66:250–8.

360. Green, M.C., R.L. Sidman, and O.H. Pivetta. 1972. Cribriform degeneration (*cri*): a new recessive neurological mutation in the mouse. Science 176:800–3.

361. Green, M.C., and R.L. Sidman. 1962. Tottering—a neuromuscular mutation in the mouse. J. Hered. 53:233–7.

362. Green, M.C., G.D. Snell, and P.W. Lane. 1963. Linkage Group XVIII of the mouse. J. Hered. 54:245–7.

363. Green, M.C., and J.H. Stimpfling. 1966. Mouse News Lett. 35:32. Plus Green, M.C. 1970. Personal communication.

364. Green, M.C., H.O. Sweet, and L.E. Bunker. 1976. Tight skin, a new mutation of the mouse causing excessive growth of connective tissue and skeleton. Am. J. Pathol. 82:493–512.

365. Green, M.C., H.O. Sweet, M. Cherry, and E.M. Eicher. 1975. Mouse News Lett. 52:38.

366. Green, M.C., H.O. Sweet, P.W. Lane, and E.M. Eicher. 1975. Mouse News Lett. 52:37.

367. Green, M.C., and H.O. Sweet. 1973. Mouse News Lett. 49:32.

368. Green, M.C., and H.O. Sweet. 1974. Mouse News Lett. 51:24.

369. Green, M.C., and H.O. Sweet. 1975. Mouse News Lett. 52:38.

370. Green, M.C., and H.O. Sweet. 1975. Mouse News Lett. 53:34.

371. Green, M.C., and H.O. Sweet. 1976. Mouse News Lett. 55:17.

372. Green, M.C., and H.O. Sweet. 1979. Mouse News Lett. 60:50.

373. Green, M.C., and E.F. Woodworth. 1967. Mouse News Lett. 37:34.

374. Green, M.C., and E.F. Woodworth. 1968. Mouse News Lett. 39:28. Plus Green, M.C. 1970. Personal communication.

375. Green, M.C., and E.F. Woodworth. 1970. Mouse News Lett. 43:32. Plus letter to Roy Robinson 7/27/70.

376. Green, M.C. 1962. Mouse News Lett. 26:34–5.

377. Green, M.C. 1971. Mouse News Lett. 44:30.

378. Green, M.C. 1971. Mouse News Lett. 45:29.

379. Green, M.C. 1973. Mouse News Lett. 48:35.

380. Green, M.C. 1973. Mouse News Lett. 49:32. Plus Green, M.C. 1973. Personal communication.

381. Green, M.C. 1973. Mouse News Lett. 49:32.

382. Green, M.C. 1970. Personal communication.

383. Green, M.C. 1970. The developmental effects of congenital hydrocephalus (*ch*) in the mouse. Dev. Biol. 23:585–608.

384. Green, M.C. 1961. The position of luxoid in linkage group II of the mouse. J. Hered. 52:297–300.

385. Green, M.C. 1972. Mouse News Lett. 47:37.

386. Groves, M.G., D.L. Rosenstreich, B.A. Taylor, and J.V. Osterman. 1980. Host defenses in experimental scrub typhus: mapping the gene that controls natural resistance in mice. J. Immunol. 125:1395–9.

387. Grüneberg, H., and G.M. Truslove. 1960. Two closely linked genes in the mouse. Genet. Res. 1:69–90.

388. Grüneberg, H. 1935. A three-factor linkage experiment in the mouse. J. Genet. 31:157–62.

389. Grüneberg, H. 1936. Further linkage data on the albino chromosome of the house mouse. J. Genet. 33:255–65.

390. Guénet, J-L., B. Salzgeber, and M.T. Tassin. 1979. Repeated epilation: a genetic epidermal syndrome in mice. J. Hered. 70:90–4.

391. Guénet, J-L. 1976. Mouse News Lett. 55:21.

392. Guénet, J-L. 1980. Mouse News Lett. 62:72.

393. Guerrero, I., A. Villasante, P. D'Eustachio, and A. Pellicer. 1984. Isolation, characterization, and chromosome assignment of mouse N-ras gene from carcinogen-induced thymic lymphoma. Science 225:1041–3.

394. Gupta, P., J.M. Rosen, P. D'Eustachio, and F.H. Ruddle. 1982. Localization of the casein gene family to a single mouse chromosome. J. Cell Biol. 93:199–204.

395. Haldane, J.B.S., A.D. Sprunt, and N.M. Haldane. 1915. Reduplication in mice. J. Genet. 5:133–5.

396. Hammerberg, C., and J. Klein. 1975. Linkage relationships of markers on chromosome 17 of the house mouse. Genet. Res. 26:203–11.

397. Hammerberg, C. 1981. Mouse News Lett. 64:72.

398. Hammerling, G.J., U. Hammerling, and L. Flaherty. 1979. Qat-4 and Qat-5, new murine T-cell antigens governed by the Tla region and identified by monoclonal antibodies. J. Exp. Med. 150:108–16.

398a. Harada Y., J. Hayakawa, and T. Tomita. 1986. Antigenic polymorphism of a serum protein migrating electrophoretically in the alpha region detected by using a strain derived from the Japanese wild mouse (*Mus musculus molossinus*). Immunogenetics 24:47–50.

399. Harada, Y-N., J-I. Hayakawa, E. Noda, and T. Tomita. 1987. A gene locus controlling a serum protein migrating electrophoretically in the beta region of mice and detected by using a strain derived from the Japan-ese wild mouse (*Mus musculus molossinus*). J. Immunogenet. 14:33–41.

400. Harper, M.E., C. Blatt, S. Marks, M.N. Nesbitt, and M.I. Simon. 1984. Gene mapping the mouse by analysis of RFLP segregation in recombinant inbred strains. Cytogenet. Cell Genet. 37:488 (Abstr.).

400a. Hart, C.P., D.K. Dalton, L. Nichols, L. Hunihan, T.H. Roderick, S.H. Langley, B.A. Taylor, and F.H. Ruddle. 1988. The *Hox-2* homeobox gene complex on mouse chromosome 11 is closely linked to *Re*. Genetics 118:319–27.

401. Hartley, J.W., R.A. Yetter, and H.C. Morse III. 1983. A mouse gene on chromosome 5 that restricts infectivity of MCF-type recombinant murine leukemia viruses. J. Exp. Med. 158:16–24.

402. Hashimoto, Y., A. Suzuki, T. Yamakawa, N. Miyashita, and K. Moriwaki. 1983. Expression of GM1 and GD1a in mouse liver is linked to the H-2 complex on chromosome 17. J. Biochem. (Tokyo) 94:2043–8.

403. Hawes, N.L., S.H. Langley, and T.H. Roderick. 1980. Location of syndactylism (sm) on chr 12. Mouse News Lett. 63:20.

404. Hayakawa, J-I., H. Nikaido, and T. Koizumi. 1985. Assignment of the gene locus for the sixth component (C6) of complement in mice to chromosome 15. Immunogenetics 22:637–42.

404a. Hayakawa J., H. Nikaido. 1987. Two types of liver-specific F antigen are encoded by a locus located on chromosome 5 in mice. Immunogenetics 26:366–9.

405. Hayes, C.E., K.K. Klyczek, D.P. Krum, R.M. Whitcomb, D.A. Hullett, and H. Cantor. 1984. Chromosome 4 *Jt* gene controls murine T cell surface I-J expression. Science 223:559–63.

406. Hecht, N.B., K.C. Kleene, P.C. Yelick, P.A. Johnson, D.D. Pravtcheva, and F.H. Ruddle. 1986. Mapping of haploid expressed genes: genes for both mouse protamines are located on chromosome 16. Somat. Cell Mol. Genet. 12:203–8.

407. Heidmann, O., A. Buonanno, B. Geoffroy, B. Robert, J-L. Guénet, J.P. Merlie, and J-P. Changeus. 1986. Chromosomal location of muscle nicotinic acetylcholine receptor genes in the mouse. Science 234:866–8.

408. Heilig, R., C. Lemaire, J.L. Mandel, L. Dandolo, L. Amar, and P. Avner. 1987. Localization of the region homologous to the Duchenne muscular dystrophy locus on the mouse X chromosome. Nature 328:168–70.

409. Heller, E., and D.H. Pluznik. 1984. Chromosomal assignment of two murine genes controlling susceptibility to spleen focus formation by Rauscher leukemia virus. Exp. Hematol. 12:645–9.

410. Henderson, A.S., E.M. Eicher, M.T. Yu, and K.C. Atwood. 1974. The chromosomal location of ribosomal DNA in the mouse. Chromosoma 49:155–60.

411. Henderson, L.A., R. Ciavarro, R. Riblet, and J. Forman. 1986. H-40, an antigen controlled by an *Igh*-linked gene and recognized by cytotoxic T lymphocytes. II.

Recognition of H-40 as a tumor antigen in leukemic animals. J. Immunol. 133:2778–85.

412. Hertwig, P. 1942. Neue Mutationen and Koppelungsgruppen bei der Hausmaus. Zeit. ind. Abst. Vererb. 80:220–46.

413. Heston, W.E. 1941. A relationship between susceptibility to induced pulmonary tumors and certain known genes in mice. J. Natl. Cancer Inst. 2:127–32.

414. Hetherington, C.M., E.P. Blankenhorn, and E. Simpson. 1979. An H-2-restricted CML target antigen controlled by a gene linked to the *H-2* complex. Immunogenetics 9:255–60.

415. Heuckeroth, R.O., E.H. Birkenmeier, M.S. Levin, and J.I. Gordon. 1987. Analysis of the tissue-specific expression, developmental regulation, and linkage relationships of a rodent gene encoding heart fatty acid binding protein. J. Biol. Chem. 262:9709–17.

416. Hibbs, M.L., P.M. Hogarth, R.A. Harris, and I.F.C. McKenzie. 1985. Gm-3.2, a new granulocyte/macrophage alloantigen. Immunogenetics 21:67–70.

417. Hildebrand, C.E., F.J. Gonzalez, C.A. Kozak, and D.W. Nebert. 1985. Regional linkage analysis of the dioxin-inducible P-450 gene family on mouse chromosome 9. Biochem. Biophys. Res. Comm. 130:396–406.

418. Hilgers, J. and colleagues (not specified). 1984. Mouse News Lett. 71:14.

419. Hilkens, J., A. Colombatti, M. Strand, E. Nichols, F.H. Ruddle, and J. Hilgers. 1979. Identification of a mouse gene required for binding of Rauscher MuLV envelope gp70. Somat. Cell Genet. 5:39–49.

420. Hilkens, J., B. van der Zeijst, F. Buijs, V. Kroezen, N. Bleumink, and J. Hilgers. 1983. Identification of a cellular receptor for mouse mammary tumor virus and mapping of its gene to chromosome 16. J. Virol. 45:140-7. Also Mouse News Lett. 1983. 68:84.

421. Hilkens, J., H.T. Cuypers, G. Selten, V. Kroezen, J. Hilgers, and A. Berns. 1986. Genetic mapping of *Pim-1* putative oncogene to mouse chromosome 17. Somat. Cell Mol. Genet. 12:81–8.

422. Hill, R.E., R.K. Barth, and N.D. Hastie. 1982. Polymorphic expression of a mouse liver specific gene. Mouse News Lett. 67:36.

423. Hill, R.E., A.E. Hall, C.M. Sime, and N.D. Hastie. 1987. A mouse homeo box-containing gene maps near a developmental mutation. Cytogenet. Cell Genet. 44:171–4.

425. Hill, R.E., P.H. Shaw, R.K. Barth, and N.D. Hastie. 1985. A genetic locus closely linked to a protease inhibitor gene complex controls the level of multiple RNA transcripts. Mol. Cell. Biol. 5:2114–22.

426. Hirose, S., R. Nagasawa, I. Sekikawa, M. Hamaoki, Y. Ishida, H. Sato, and T. Shirai. 1983. Enhancing effect of *H-2*-linked NZW gene(s) on the autoimmune traits of (NZB x NZW)F1 mice. J. Exp. Med. 158:228–33.

427. Hjorth, J.P. 1979. Mouse News Lett. 61:25.

428. Hoecker, G., A. Martinez, S. Markovic, and O.

Pizarro. 1954. Agitans, a new mutation in the house mouse with neurological effects. J. Hered. 45:104.

429. Hoffman, H.A., and C.K. Grieshaber. 1974. Genetic studies of murine catalase. J. Hered. 65:277–9.

430. Hoffman, H.A., and C.K. Grieshaber. 1977. Mouse News Lett. 56:51.

431. Hogan, B.L.M., G. Horsburgh, J. Cohen, C.M. Hetherington, G. Fisher, and M.F. Lyon. 1986. Small eyes (*Sey*): a homozygous lethal mutation on chromosome 2 which affects the differentiation of both lens and basal placodes in the mouse. J. Embryol. Exp. Morphol. 97:95–110.

432. Hogarth, M., I.F.C. McKenzie, B.A. Taylor, and M. Cherry. 1983. Personal communication.

433. Hogarth, P.M., J. Craig, and I.F.C. McKenzie. 1980. A monoclonal antibody detecting the Ly-9.2 (Lgp100) cell-membrane alloantigen. Immunogenetics 11:65–74.

434. Hogarth, P.M., E.M. Eicher, and I.F.C. McKenzie. 1986. Mapping of the murine Ly-15 (LFA-1) locus to chromosome 7. Immunogenetics 23:348–9.

435. Hogarth, P.M., I.F.C. McKenzie, L. Lanier, D.W. Bailey, and B.A. Taylor. 1984. Location of *Ly-7* on mouse chromosome 12. Immunogenetics 19:539–43.

436. Hogarth, P.M., I.F.C. McKenzie, V.R. Sutton, K.M. Curnow, B.K. Lee, and E.M. Eicher. 1987. Mapping of the murine *Ly-6*, *Xp-14* and *Gdc-1* loci to chromosome 15. Immunogenetics 25:21–7.

437. Hogarth, P.M., A. Rigby, V.R. Sutton, I.F.C. McKenzie, and J. Hilgers. 1984. The Ly-25.1 specificity: definition with a monoclonal antibody. Immunogenetics 19:83–6.

438. Hoggan, M.D., R.R. O'Neill, and C.A. Kozak. 1986. Nonecotropic murine leukemia viruses in BALB/c and NFS/N mice: characterization of the BALB/c Bxv-1 provirus and the single NFS endogenous xenotrope. J. Virol. 60:980–6.

439. Hollander, W.F., and L.C. Strong. 1951. Pintail, a dominant mutation linked with brown in the house mouse. J. Hered. 42:179–82.

440. Hollander, W.F., and K.S. Waggie. 1977. Meander tail: a recessive mutant located in chromosome 4 of the mouse. J. Hered. 68:403–6.

441. Hollander, W.F. 1976. Genetic spina bifida occulta in the mouse. Am. J. Anat. 146:173–80.

442. Hollander, W.F. 1976. Hydrocephalic-polydactyl, a recessive pleiotropic mutant in the mouse and its location in chromosome 6. Iowa State J. Res. 51:13–23.

443. Hollander, W.F. 1971. Linkage relationships of dumpy, a recessive mutant on chromosome 13 of the mouse. J. Hered. 72:358–9.

444. Hollander, W.F. 1965. Personal communication.

445. Hollander, W.F. 1966. Mouse News Lett. 35:30.

446. Hollander, W.F. 1968. Mouse News Lett. 38:23.

447. Holmes, L.B. 1986. Identification of *Dh/+* and *Dh/Dh* embryos through close linkage of *Dh* and peptidase-3. Teratology 4:353–7.

448. Holmes, R., and G. Timms. 1979. Mitochondrial alde-

hyde dehydrogenase - positioning of *Ahd-1* on chromosome 4. Mouse News Lett. 61:35.

449. Holmes, R., and R. van Nie. 1984. Mouse News Lett. 70:65.

450. Holmes, R., S.J. Andrews, and C.V. Beechey. 1979. Gene order of Hao-2, Adh-3 and other loci on chromosome 3. Mouse News Lett. 6l:35.

451. Holmes, R.S., R. Albanese, F.D. Whitehead, and J.A. Duley. 1981. Mouse alcohol dehydrogenase isozymes: products of closely localized duplicated genes exhibiting divergent kinetic properties. J. Exp. Zool. 217:151–7.

452. Holmes, R.S., S.J. Andrews, and C.V. Beechey. 1981. Genetic regulation of alcohol dehydrogenase C2 in the mouse. Developmental consequences of the temporal locus (*Adh-3t*) and positioning of *Adh-3* on chromosome 3. Dev. Genet. 2:89–98.

453. Holmes, R.S., S.J. Andrews, and C.V. Beechey. 1981. Alcohol dehydrogenase C2 (stomach enzyme). Mouse News Lett. 64:54.

454. Holmes, R.S., J.A. Duley, and J. Hilgers. 1982. Sorbitol dehydrogenase genetics in the mouse: a 'null' mutant in a 'European' C57BL strain. Anim. Blood Grps. Biochem. Genet. 13:263–72.

455. Holmes, R.S., J.A. Duley, and S. Imai. 1983. Alcohol dehydrogenase temporal locus (Adh-lt). Mouse News Lett. 68:67.

456. Holmes, R.S. and J.A. Duley. 1975. Biochemical and genetic studies of peroxisomal multiple enzyme systems; α-hydroxyacid oxidase and catalase. *In* C.L. Markert, ed., Isozymes, Vol. I, Molecular Structure, 191–211. Academic Press, NY.

457. Holmes, R.S., and J.A. Duley. 1982. Alcohol dehydrogenase isozymes (Adh-1 and Adh-3 on chromosome 3). Mouse News Lett. 66:60.

458. Holmes, R.S., L.R. Leitjten, and J.A. Duley. 1981. Liver aldehyde oxidase and xanthine oxidase genetics in the mouse. Anim. Blood Grps. Biochem. Genet. 12:193–9.

459. Holmes, R.S., and G.P. Timms. 1981. Mouse aldehyde dehydrogenase genetics: positioning of *Ahd-1* on chromosome 4. Anim. Blood Grps. Biochem. Genet. 12:1–5.

460. Holmes, R.S. 1979. Genetics and ontogeny of alcohol dehydrogenase isozymes in the mouse: evidence for a cis-acting regulator gene (*Adt-1*) controlling C2 isozyme expression in reproductive tissues and close linkage of *Adh-3* and *Adt-1* on chromosome 3. Biochem. Genet. 17:461–72.

461. Holmes, R.S. 1978. Genetics and ontogeny of aldehyde dehydrogenase isozymes in the mouse: localization of *Ahd-1* encodings the mitochondrial isozyme on chromosome 4. Biochem. Genet. 16:1207–18.

462. Holmes, R.S. 1978. Genetics of hydroxyacid oxidase isozymes in the mouse. Localisation of *Hao-2* on linkage group XVI. Heredity 41:403–6.

463. Holmes, R.S. 1979. Genetics, ontogeny and testosterone inducibility of aldehyde oxidase isozymes in the mouse: evidence for two genetic loci (*Aox-1* and *Aox-2*) closely linked on chromosome 1. Biochem. Genet. 17:517–27.

464. Holmes, R.S. 1977. The genetics of α-hydroxy acid oxidase and alcohol dehydrogenase in the mouse: evidence for multiple gene loci and linkage between *Hao-2* and *Adh-3*. Genetics 87:709–16.

465. Honey, N.K., A.Y. Sakaguchi, P.A. Lalley, C. Quinto, R.J. MacDonald, C. Craik, G.I. Bell, W.J. Rutter, and S.L. Naylor. 1984. Chromosomal assignments of genes for trypsin, chymotrypsin B, and elastase in mouse. Somat. Cell Mol. Genet. 10:377–83.

466. Horowitz, J.M., and R. Risser. 1982. A locus that enhances the induction of endogenous ecotropic murine leukemia viruses is distinct from genome-length ecotropic proviruses. J. Virol. 44:950–7.

466a. Horton, C.E. Jr., M.T. Davisson, J.B. Jacobs, G.T. Bernstein, A.B. Retik, and J. Mandell. 1988. Congenital progressive hydronephrosis in mice: a new recessive mutation. J. Urol. 40: in press.

467. Horton, M.A., and C.M. Hetherington. 1980. Genetic linkage of *Ly-6* and *Thy-1* loci in the mouse. Immunogenetics 11:521–5.

468. Howe, R.C., A. Ahmed, T.J. Faldetta, J.E. Byrnes, K.M. Rogan, B.A. Taylor, and R.E. Humphreys. 1979. Mapping of the *Lyb-4* gene to different chromosomes in DBA/2J and C3H/HeJ mice. Immunogenetics 9:221–32.

469. Howles, P.N., D.P. Dickinson, L.L. DiCaprio, M. Woodworth-Gutai, and K.W. Gross. 1984. Use of a cDNA recombinant for the γ-subunit of mouse nerve growth factor to localize members of this multigene family near the TAM-1 locus on chromosome 7. Nucl. Acids Res. 12:2791–805.

470. Hudson, D.M., J.S. Finlayson, and M. Potter. 1967. Linkage of one component of the major urinary protein complex of mice to the brown coat color locus. Genet. Res. 10:195–8.

471. Huebner, K., A. Ar-Rushdi, C.A. Griffin, M. Isobe, C. Kozak, B.S. Emanuel, L. Ngarajan, J.L. Cleveland, T.I. Bonner, M.D. Goldsborough, C.M. Croce, and U. Rapp. 1986. Actively transcribed genes in the raf oncogene group, located on the X chromosome in mouse and human. Proc. Natl. Acad. Sci. USA 83:3934–8.

472. Huebner, K., A.P. Palumbo, M. Isobe, C.A. Kozak, S. Monaco, G. Rovera, C.M. Croce, and P.J. Curtis. 1985. The α-spectrin gene is on chromosome 1 in mouse and man. Proc. Natl. Acad. Sci. USA 82:3790–3.

473. Huijing, F., E.M. Eicher, and D.L. Coleman. 1973. Location of phosphorylase kinase (*Phk*) in the mouse X chromosome. Biochem. Genet. 9:193–6.

474. Hummel, K.P., and D. Chapman. 1973. Mouse News Lett. 48:35.

475. Hummel, K.P., and D.B. Chapman. 1970. Mouse News Lett. 42:29.

476. Hummel, K.P., and D.B. Chapman. 1971. Mouse News

Lett. 45:29. Plus Hummel, K.P. 1972. Personal communication.

477. Hummel, K.P., and D.B. Chapman. 1974. Mouse News Lett. 51:24.

478. Hummel, K.P. 1970. Hypodactyly, a semidominant lethal mutation in mice. J. Hered. 61:219–20.

479. Hummel, K.P. 1965. Personal communication.

480. Hunsicker, P. 1974. Mouse News Lett. 50:51–2.

481. Hunsicker, P.R. 1969. Mouse News Lett. 40:42.

482. Hutton, J.J., and D.L. Coleman. 1969. Linkage analysis using biochemical variants in mice. II. Levulinate dehydratase and autosomal glucose-6 phosphate dehydrogenase. Biochem. Genet. 3:517–23.

483. Hutton, J.J., and T.H. Roderick. 1970. Linkage analyses using biochemical variants in mice. III. Linkage relationships of eleven biochemical markers. Biochem. Genet. 4:339–50.

484. Hutton, J.J. 1969. Linkage analyses using biochemical variants in mice. I. Linkage of the hemoglobin beta-chain and glucosephosphate isomerase loci. Biochem. Genet. 3:507–15.

485. Ihle, J.N., D.R. Joseph, and J.J. Domotor Jr. 1979. Genetic linkage of C3H/HeJ and BALB/c endogenous ecotropic c-type viruses to phosphoglucomutase-1 on chromosome 5. Science 204:71–3.

486. Ihle, J.N., J. Silver, and C.A. Kozak. 1987. Genetic mapping of the mouse interleukin 3 gene to chromosome 11. J. Immunol. 138:3051–4.

487. Ikeda, H., H. Sato, and T. Odaka. 1981. Mapping of the Fv-4 mouse gene controlling resistance to murine leukemia viruses. Int. J. Cancer. 28:237–40.

488. Ingram, R.S., R.W. Scott, and S.M. Tilghman. 1981. α-Fetoprotein and albumin genes are in tandem in the mouse genome. Proc. Natl. Acad. Sci. USA 78:4694–8.

489. Institut des Sciences de L'Évolution. 1983. Mouse News Lett. 68:79.

490. Isaacson, J.H., J. Stewart, and D.S. Falconer. 1974. Location of Hq. Mouse News Lett. 50:33.

491. Ishikawa, H., E. Kubota, H. Suzuki, and K. Saito. 1985. The target minor H antigen for F1 cytotoxic T lymphocytes induced by Igh-congenic parental spleen cells is coded for by gene linked to H-2. J. Immunol. 134:2953–9.

492. Itakura, K., and E.A. Boyse. 1972. Personal communication.

493. Itakura, K., J.J. Hutton, E.A. Boyse, and L.J. Old. 1972. Genetic linkage relationships of loci specifying differentiation alloantigens in the mouse. Transplantation 13:239–43.

494. Itakura, K., J.J. Hutton, E.A. Boyse, and L.J. Old. 1971. Linkage groups of the theta and Ly-A loci. Nature New Biol. 230:126.

495. Jacobsson, A., U. Stadler, M.A. Glotzer, and L.P. Kozak. 1985. Mitochondrial uncoupling protein from mouse brown fat. Molecular cloning, genetic mapping and mRNA expression. J. Biol. Chem. 260:16250–4.

496. Jeffreys, A.J., V. Wilson, R. Kelly, B.A. Taylor, and G. Bulfield. 1987. Mouse DNA 'fingerprints': analysis of chromosome localization and germ-line stability of hypervariable loci in recombinant inbred strains. Nucl. Acids Res. 15:2823–36.

497. Jenkins, N.A., N.G. Copeland, B.A. Taylor, and B.K. Lee. 1982. Organization, distribution, and stability of endogenous ecotropic murine leukemia virus DNA sequences in chromosomes of *Mus musculus*. J. Virol. 43:26–36.

498. Jenkins, N.A., and N.G. Copeland. 1985. High frequency germline acquisition of ecotropic MuLV proviruses in SWR/J-RF/J hybrid mice. Cell 43:811–9.

499. Jenkins, N.A. and N.G. Copeland. 1987. Retroviral DNA content of the mouse genome. *In* V.A. McKusick, T.H. Roderick, J. Mori, and N.W. Paul, eds., Medical and Experimental Mammalian Genetics: A Perspective. March of Dimes Birth Defects Foundation, Birth Defects Original Article Series, Vol. 23, No. 3, 109–22. Alan R. Liss Inc, NY.

500. Johnson, D.R., and M.E. Wallace. 1979. Crinkly-tail, a mild skeletal mutant in the mouse. J. Embryol. Exp. Morphol. 53:327–33.

501. Johnson, D.R. 1967. Mouse News Lett. 36:44.

502. Johnson, D.R. 1969. Polysyndactyly, a new mutant gene in the mouse. J. Embryol. Exp. Morphol. 21:285–94.

503. Johnson, F.M., F. Chaslow, G. Anderson, P. MacDougal, R.W. Hendren, and S.E. Lewis. 1981. A variation in mouse kidney pyruvate kinase activity determined by a mutant gene on chromosome 9. Genet. Res. 37:123–31.

504. Johnson, F.M., R.W. Hendren, F. Chasalow, L.B. Barnett, and S.E. Lewis. 1981. A null mutation at the mouse phosphoglucomutase-1 locus and a new locus, *Pgm-3*. Biochem. Genet. 19:599–615.

505. Jolicoeur, P., L. Villeneuve, E. Rassart, and C. Kozak. 1985. Mouse chromosomal mapping of a murine leukemia virus integration region (Mis-1) first identified in rat thymic leukemia. J. Virol. 56:1045–8.

506. Joyner, A.L., T. Kornberg, K.G. Coleman, D.R. Cox, and G.R. Martin. 1985. Expression during embryogenesis of a mouse gene with a sequence homology to the Drosophila engrailed gene. Cell 43:29–37.

507. Joyner, A.L., R.V. Lebo, Y.W. Kan, R. Tjian, D.R. Cox, and G.R. Martin. 1985. Comparative chromosome mapping of a conserved homoeo box region in mouse and human. Nature 314:173–5.

508. Joyner, A.L., and G.R. Martin. Expression during embryogenesis of *En-1* and *En-2*, two unlinked mouse genes that have extensive sequence homology with the Drosophila engrailed and invected genes. Symposium talk presented by Martin at Genetics Society of America 56th Annual Meeting, and June 1987.

509. Ju, S-T., and M.E. Dorf. 1980. Idiotypic analysis of anti-gat antibodies. IX. Genetic mapping of the Gte idiotypic marker within the IghV locus. J. Immunol. 126:183–6.

510. Juriloff, D.M., K.K. Sulik, T.H. Roderick, and B.K. Hogan. 1985. Genetic and developmental studies of a new mouse mutation that produces otocephaly. J. Craniofacial Genet. Dev. Biol. 5:121–45.

511. Karl, T.R., and V.M. Chapman. 1974. Linkage and expression of the *Eg* locus controlling inclusion of β-glucuronidase into microsomes. Biochem. Genet. 11:367–72.

512. Kasahara, M., F. Figuerova, and J. Klein. 1987. Random cloning of genes from mouse chromosome 17. Proc. Natl. Acad. Sci. USA 84:3325–8.

513. Keitges, E.A., D.F. Schorderet, and S.M. Gartler. 1987. Linkage of the steroid sulfatase gene to the sex-reversed mutation in the mouse. Genetics 116:465–8.

514. Kelly, E.M. 1968. Mouse News Lett. 38:31.

515. Kelton, D.E., and H. Rauch. 1968. Linkage of open eyelids with linkage group VII of the mouse. J. Hered. 59:27–8.

516. Kelton, D.E. 1965. Mouse News Lett. 32:60.

517. Kennard, J., and D. Meruelo. 1982. Precise mapping of the gene for Ly-21.2 on the 7th chromosome. J. Immunogenet. 9:389–95.

518. Kent, R.B., D.A. Fallows, E. Geissler, T. Glaser, J.R. Emanuel, P.A. Lalley, R. Levenson, and D.E. Housman. 1987. Genes encoding α and β subunits of Na, K-ATPase are located on three different chromosomes in the mouse. Proc. Natl. Acad. Sci. USA 84:5369–73.

519. Kessler, S., A. Ahmed, and I. Scher. 1979. Murine B cell alloantigens Lyb 2, Lyb 4, and Lyb 6 are products of distinct genes clustered on chromosome 4. Fed. Proc. 38:1460.

520. Key, M., and W.F. Hollander. 1972. Thin fur, a recessive mutant on chromosome 17 of the mouse. J. Hered. 63:97–8.

521. Khlebodarova, T.M., and O.L. Serov. 1980. A new locus regulating the expression of the *Ldh-2* gene in mouse liver. Biochem. Genet. 18:1027–39.

522. Kidwell, J.F., J.W. Gowen, and J. Stadler. 1961. Pugnose—a recessive mutation in linkage group 3 of mice. J. Hered. 52:145–8.

523. Kidwell, J.F., J.W. Gowen, and J. Stadler. 1966. Pugnose linkage in the mouse. J. Hered. 57:229–30.

524. Killary, A.M., and R.E.K. Fournier. 1984. A genetic analysis of extinction: trans-dominant loci regulate expression of liver-specific traits in hepatoma hybrid cells. Cell 38:523–34.

525. Killary, A.M., R.J. Leach, R.G. Moran, and R.E.K. Fournier. 1986. Assignment of genes encoding dihydrofolate reductase and hexosaminidase B to *Mus musculus* chromosome 13. Somat. Cell Mol. Genet. 12:641–8.

526. Kimura, S., N. Tada, J. Furze, and H. Festenstein. 1987. Ly-33: a new lymphocyte alloantigen controlled by a gene linked to *Ly-17* locus on mouse chromosome 1. Immunogenetics 26:315–6.

527. Kimura, S., N. Tada, and U. Hammerling. 1980. A new lymphocyte alloantigen (Ly-10) controlled by a gene linked to the *Lyt-1* locus. Immunogenetics 10:363–72.

528. Kimura, S., N. Tada, Y. Liu-Lam, and U. Hammerling. 1983. Further genetic characterization of mouse Ly-m10. Immunogenetics 17:2112–14.

529. Kimura, S., N. Tada, E. Nakayama, Y. Liu, and U. Hammerling. 1981. A new mouse cell-surface antigen (Ly-m20) controlled by a gene linked to *Mls* locus and defined by monoclonal antibodies. Immunogenetics 14:3–14.

530. King, J.W.B. 1956. Linkage group XIV of the house mouse. Nature 178:1126–7.

531. Kirby, G.C. 1974. The genetics of an electophoretic variant of an erythrocytic protein in the house mouse (*Mus musculus*), Anim. Blood Grps. Biochem. Genet. 5:153–7.

532. Klein, J., and C. Chiang. 1978. A new locus (*H-2T*) at the *D* end of the *H-2* complex. Immunogenetics 6:235–43.

533. Knight, J.G., and D.D. Adams. 1981. Genes determining autoimmune disease in New Zealand mice. J. Clin. Lab. Immunol. 5:165–70.

534. Knobler, R.L., B.A. Taylor, M.K. Wooddell, W.G. Beamer, and M.B.A. Oldstone. 1984. Host genetic control of mouse hepatitis virus type-4 (JHM strain) replication. II. The gene locus for susceptibility is linked to *Svp-2* locus on mouse chromosome 7. Clin. Exp. Immunogenet. 1:217–22.

535. Koch, G., P.A. Lalley, M. McAvoy, and T.B. Shows. 1981. Assignment of LIPA, associated with human lipase acid deficiency, to human chromosome 10 and comparative assignment to mouse chromosome 19. Somat. Cell Genet. 7:345–58.

536. Kourides, I.A., P.E. Barker, J.A. Gurr, D.D. Pravtcheva, and F.H. Ruddle. 1984. Assignment of the genes for the α and β subunits of thyrotropin to different mouse chromosomes. Proc. Natl. Acad. Sci. USA 81:517–9.

537. Kozak, C., E. Nichols, and F.H. Ruddle. 1975. Gene linkage analysis in the mouse by somatic cell hybridization: assignment of adenine phosphoribosyl transferase to chromosome 8 and α-galactosidase to the X chromosome. Somat. Cell Genet. 1:371–82.

538. Kozak, C., and W.P. Rowe. 1978. Genetic mapping of xenotropic leukemia virus-inducing loci in two mouse strains. Science 199:1448–9.

539. Kozak, C.A., W.F. Davidson, and H.C. Morse III. 1984. Genetic and functional relationships of the retroviral and lymphocyte alloantigen loci on mouse chromosome 1. Immunogenetics 19:163–8.

540. Kozak, C.A., R.E.K. Fournier, L.A. Leinwand, and F.H. Ruddle. 1979. Assignment of the gene controlling cellular ouabain resistance to *Mus musculus* chromosome 3 using human/mouse microcell hybrids. Biochem. Genet. 17:23–34.

541. Kozak, C.A., and W.P. Rowe. 1982. Genetic mapping of ecotropic murine leukemia virus-inducing loci in six inbred strains. J. Exp. Med. 155:524–34.

542. Kozak, C.A., and W.P. Rowe. 1979. Genetic mapping

of the ecotropic murine leukemia virus-inducing locus of the BALB/c mouse to chromosome 5. Science 204:69–71.

543. Kozak, C.A., and W.P. Rowe. 1980. Genetic mapping of the ecotropic virus inducing locus *Akv-2* of the AKR mouse. J. Exp. Med. 152:1419–23.

544. Kozak, C.A., and W.P. Rowe. 1980. Genetic mapping of xenotropic murine leukemia virus-inducing loci in five mouse strains. J. Exp. Med. 152:219–28.

545. Kozak, C.A., J.F. Sears, and M.D. Hoggan. 1983. Genetic mapping of the mouse oncogenes c-Ha-ras-1 and c-fes to chromosome 7. J. Virol. 47:217–20.

546. Kozak, C.A., J.F. Sears, and M.D. Hoggan. 1983. Genetic mapping of the mouse proto-oncogene c-sis to chromosome 15. Science 221:867–9.

547. Kozak, C.A., P.G. Strauss, and P.N. Tsichlis. 1985. Genetic mapping of a cellular DNA region involved in induction of thymic lymphomas (Mlvi-1) to mouse chromosome 15. Mol. Cell. Biol. 5:894–7.

548. Kozak, C.A. 1983. Genetic mapping of a mouse chromosomal locus required for mink cell focus-forming virus replication. J. Virol. 48:300–3.

549. Kozak, C.A. 1985. Susceptibility of wild mouse cells to exogenous infection with xenotropic leukemia viruses: control by a single dominant locus on chromosome 1. J. Virol. 55:690–5.

550. Kozak, L.P., D.L. Burkart, and J.P. Hjorth. 1982. Unlinked structural genes for developmentally regulated isozymes of sn-glycerol-3-phosphate dehydrogenase in mice. Dev. Genet. 3:1–6.

551. Kralova, J., and A. Lengerova. 1979. H-Y antigen: genetic control of the expression as detected by host-versus-graft popliteal lymph node enlargement assay maps between the *T* and *H-2* complexes. J. Immunogenet. 6:429–38.

552. Krasnewich, D., C.A. Kozak, D.W. Nebert, and P.I. Mackenzie. 1987. Localization of UDP glucuronosyltransferase gene(s) on mouse chromosome 5. Somat. Cell Mol. Genet. 13:179–82.

553. Krissansen, G.W., P.A. Gorman, C.A. Kozak, N.K. Spurr, D. Shear, P.N. Goodfellow, and M.J. Crumpton. 1987. Chromosomal locations of the gene coding for the CD3 (T3) gamma subunit of the human and mouse CD3/T-cell antigen receptor complexes. Immunogenetics 26:258–66.

554. Kueppers, F., C.C. Lee, J. Mills, and R. Riblet. 1982. Linkage of loci for alpha-1-antitrypsin (*Pi*) and immunoglobulin heavy chain (*Igh-1*) in the mouse. Am. J. Hum. Genet. 34:56A (Abstr.).

555. LaVail, M.M., and R.L. Sidman. 1974. C57BL/6J mice with inherited retinal degeneration. Arch. Ophthalmol. 91:394–400.

556. Lalley, P.A. 1981. Mouse gene mapping in somatic cell hybrids. Mouse News Lett. 64:79–81.

557. Lalley, P.A., and J.M. Chirgwin. 1984. Mapping of mouse insulin genes. Cytogenet. Cell Genet. 37:515 (Abstr.).

558. Lalley, P.A., and J.A. Diaz. 1984. Comparative gene mapping in the mouse involving genes assigned to human chromosomes 7 and 20. Cytogenet. Cell Genet. 37:514 (Abstr.).

559. Lalley, P.A., U. Francke, and J.D. Minna. 1978. Homologous genes for enolase, phosphogluconate dehydrogenase, phosphoglucomutase, and adenylate kinase are syntenic on mouse chromosome 4 and human chromosome 1p. Proc. Natl. Acad. Sci. USA 75:2382–6.

560. Lalley, P.A., J.D. Minna, and U. Francke. 1978. Conservation of autosomal gene synteny groups in mouse and man. Nature 274:160–2.

561. Lalley, P.A., S.L. Naylor, and T.B. Shows. 1979. Gene assignment of argininosuccinate lyase to mouse chromosome 5. Cytogenet. Cell Genet. 25:178 (Abstr.).

562. Lalley, P.A., A.Y. Sakaguchi, R.L. Eddy, N.H. Honey, G.I. Bell, L-P. Shen, W.J. Rutter, J.W. Jacobs, G. Heinrich, W.W. Chin, and S.L. Naylor. 1987. Mapping polypeptide hormone genes in the mouse: somatostatin, glucagon, calcitonin, and parathyroid hormone. Cytogenet. Cell Genet. 44:92–7.

563. Lalley, P.A., and T.B. Shows. 1974. Lysosomal and microsomal glucuronidase genetic variant alters electrophoretic mobility of both hydrolases. Science 185:442–4.

565. Lane, P.W., and M.S. Deol. 1974. Mocha, a new coat color and behavior mutation on chromosome 10 of the mouse. J. Hered. 65:362–4.

566. Lane, P.W., and M.M. Dickie. 1961. Linkage of wabbler-lethal and hairless in the mouse. J. Hered. 52:159–60.

567. Lane, P.W., and M.M. Dickie. 1968. Three recessive mutations producing disproportionate dwarfing in mice. Achondroplasia, brachymorphic, and stubby. J. Hered. 59:300–8.

568. Lane, P.W., E.M. Eicher, and J.L. Southard. 1974. Chromosome 19 of the house mouse. J. Hered. 65:297–300.

569. Lane, P.W., and E.M. Eicher. 1979. Gene order in linkage group XVI of the house mouse. J. Hered. 70:239–44.

570. Lane, P.W., and E.M. Eicher. 1985. The location of plucked (*pk*) on chromosome 18 of the mouse. J. Hered. 76:476–7.

571. Lane, P.W., and E.L. Green. 1967. Pale ear and light ear in the house mouse, mimic mutations in linkage groups XII and XVII. J. Hered. 58:17–20.

572. Lane, P.W., and M.C. Green. 1960. Mahogany, a recessive color mutation in linkage group V of the mouse. J. Hered. 51:228–30.

573. Lane, P.W., and H.M. Liu. 1984. Association of megacolon with a new dominant spotting gene (*Dom*) in the mouse. J. Hered. 75:435–9.

574. Lane, P.W., A.G. Searle, C.V. Beechey, and E.M. Eicher. 1981. Chromosome 18 of the house mouse. J. Hered. 72:409–12.

575. Lane, P.W., and H.O. Sweet. 1973. Mouse News Lett.

48:34. Plus Green, M.C. 1973. Personal communication.

576. Lane, P.W., and H.O. Sweet. 1979. Mouse News Lett. 60:52.

577. Lane, P.W., and H.O. Sweet. 1987. Personal communication.

578. Lane, P.W., and H.O. Sweet. 1984. Ashen mapped. Mouse News Lett. 70:84.

579. Lane, P.W., and J.E. Womack. 1979. Ashen, a new coat color mutation in chromosome 9 of the mouse. J. Hered. 70:133–5.

580. Lane, P.W. 1987. Abnormal feet and tail (Aft). Mouse News Lett. 78:56–7.

581. Lane, P.W. 1987. Deaf waddler (dfw). Mouse News Lett. 77:129.

582. Lane, P.W. 1967. Linkage groups III and XVII in the mouse and the position of the light-ear locus. J. Hered. 58:21–4.

583. Lane, P.W. 1981. Mouse News Lett. 64:59.

584. Lane, P.W. 1960. Mouse News Lett. 23:36. Plus Lane, P.W. 1960. Personal communication.

585. Lane, P.W. 1966. Mouse News Lett. 34:32.

586. Lane, P.W. 1967. Mouse News Lett. 37:34.

587. Lane, P.W. 1968. Mouse News Lett. 38:24. Plus letter to Roy Robinson 7/7/70.

588. Lane, P.W. 1970. Mouse News Lett. 42:30.

589. Lane, P.W. 1971. Mouse News Lett. 44:30.

590. Lane, P.W. 1971. Mouse News Lett. 45:29.

591. Lane, P.W. 1972. Mouse News Lett. 47:36–7.

592. Lane, P.W. 1973. Mouse News Lett. 49:33.

593. Lane, P.W. 1974. Mouse News Lett. 50:44.

594. Lane, P.W. 1975. Mouse News Lett. 52:37.

595. Lane, P.W. 1975. Mouse News Lett. 53:34–5.

596. Lane, P.W. 1975. Mouse News Lett. 53:35.

597. Lane, P.W. 1970. Personal communication.

598. Lane, P.W. 1987. Tilted (tlt). Mouse News Lett. 77:129.

599. Lane, P.W. 1972. Two new mutations in Linkage Group XVI of the house mouse. Flaky tail and varitint-waddler-J. J. Hered. 63:135–40.

600. Lane, P.W. 1963. Whirler mice, a recessive behavior mutation in linkage group VIII. J. Hered. 54:263–6.

601. Lane, P.W. 1980. Chr 3 position of de. Mouse News Lett. 62:56.

602. Lane, P.W. 1983. Mouse News Lett. 68:72.

603. Lane, P.W. 1985. Hydrocephaly with hop gait (hyh). Mouse News Lett. 73:18.

604. Lane, P.W. 1971. Mouse News Lett. 45:29.

605. Lane, P.W. 1985. Muscular dystrophy with myositis (mdm). Mouse News Lett. 73:18.

606. Lane, P.W. 1984. Tippy (tip). Mouse News Lett. 71:31.

607. Larsen, M.M. 1965. Mouse News Lett. 33:68.

608. Laurila, P., L. Cioe, C.A. Kozak, and P.J. Curtis. 1987. Assignment of mouse beta-spectrin gene to chromosome 12. Somat. Cell Mol. Genet. 13:93–8.

609. Law, L.W. 1984. Assignment of the gene encoding for Meth A tumour rejection antigen (TATA) to chromosome 12 of the mouse. Br. J. Cancer 50:109–11.

610. Law, M.L., X. Mo, X. Zhang, and F.T. Kao. 1983. Genes coding for ouabain resistance (OUBR) and HPRT are syntenic in the mouse genome. Cytogenet. Cell Genet. 37:518 (Abstr.).

611. Ledbetter, J.A., J.W. Goding, T.T. Tsu, and L.A. Herzenberg. 1979. A new mouse lymphoid alloantigen (Lgp 100) recognized by a monoclonal rat antibody. Immunogenetics 8:347–60.

612. Leder, A., D. Swan, F. Ruddle, P. D'Eustachio, and P. Leder. 1981. Dispersion of α-like globin genes of the mouse to three different chromosomes. Nature 293:196–200.

613. Leinwand, L., R.E.K. Fournier, E.A. Nichols, and F.H. Ruddle. 1978. Assignment of the gene for adenosine kinase to chromosome 14 in *Mus musculus* by somatic cell hybridization. Cytogenet. Cell Genet. 21:77–85.

614. Lem, J., and R.E.K. Fournier. 1985. Assignment of the gene encoding cytosolic phosphoenol pyruvate carboxykinase (GTP) to *Mus musculus* chromosome 2. Somat. Cell Mol. Genet. 11:633–8.

615. Lesley, J., and I.S. Trowbridge. 1982. Genetic characterization of a polymorphic murine cell-surface glycoprotein. Immunogenetics 15:313–20.

616. Lilly, F. 1970. Identification and location of a second gene governing spleen focus response to Friend leukemia virus in mice. J. Natl. Cancer Inst. 45:163–9.

617. Lilly, F. 1972. Mouse News Lett. 46:18.

618. Lilly, F. 1967. The location of histocompatibility-6 in the mouse genome. Transplantation 5:83–5.

619. Lin, P-F., and F.H. Ruddle. 1981. Murine DNA repair gene located on chromosome 4. Nature 289:191–4.

620. Lin, P-F., D.L. Slate, F.C. Lawyer, and F.H. Ruddle. 1980. Assignment of the murine interferon sensitivity and cytoplasmic superoxide dismutase genes to chromosome 16. Science 209:285–7.

621. Lonai, P., and N. Haran-Ghera. 1977. Resistance genes to murine leukemia in the I immune response gene region of the H-2 complex. J. Exp. Med. 146:1164–8.

622. Lovett, M., D. Goldgaber, P. Ashley, D.R. Cox, D.C. Gajdusek, and C.J. Epstein. 1987. The mouse homolog of the human amyloid β protein (ADAP) gene is located on the distal end of mouse chromosome 16: further extension of the homology between human chromosome 21 and mouse chromosome 16. Biochem. Biophys. Res. Comm. 144:1069–75.

623. Lukey, T., K. Neote, J.F. Loman, A.E. Unger, F.G. Biddle, and F.F. Snyder. 1985. Purine nucleoside phosphorylase (Np) in the mouse: linkage relationship of Np-2 to esterase-10 (*Es-10*) and Np-1 on chromosome 14. Biochem. Genet. 23:347–56.

624. Lundin, L-G., K. Bergquist, and E.M. Hakansson. 1983. Malaria resistance. Mouse News Lett. 68:89.

625. Lundin. L-G. 1987. A gene (*Bmn*) controlling β-man-

nosidase activity in the mouse is located in the distal part of chromosome 3. Biochem. Genet. 25:603–10.

626. Lush, I., E. Azen, and B.A. Taylor. 1986. Mouse News Lett. 74:121.

627. Lusis, A.J., V.M. Chapman, R.W. Wangenstein, and K. Paigen. 1983. Trans-acting temporal locus within the β-glucuronidase gene complex. Proc. Natl. Acad. Sci. USA 80:4398–402.

628. Lusis, A.J., B.A. Taylor, D. Quon, S. Zollman, and R.C. LeBoeuf. 1987. Genetic factors controlling structure and expression of apolipoproteins B and E in mice. J. Biol. Chem. 262:7594–604.

629. Lusis, A.J., B.A. Taylor, R.W. Wangenstein, and R.C. LeBoeuf. 1983. Genetic control of lipid transport in mice. II. Genetic variations of high density lipoproteins. J. Biol. Chem. 258:5071–8.

630. Lusis, A.J., and J.D. West. 1978. X-linked and autosomal genes controlling mouse a-galactosidase expression. Genetics 88:324–42.

631. Lyman, S.D., A. Poland, and B.A. Taylor. 1980. Genetic polymorphism of microsomal epoxide hydrolase activity in the mouse. J. Biol. Chem. 255:8650–4.

632. Lyon, M.F., J.M. Butler, and R. Kemp. 1968. The positions of the centromeres in linkage groups II and IX of the mouse. Genet. Res. 11:193–9.

633. Lyon, M.F., and P.H. Glenister. 1986. Gene order of Chy-vt-Re on chromosome 11. Mouse News Lett. 74:96.

634. Lyon, M.F., and P.H. Glenister. 1986. Position of neonatal anemia (Nan) on chromosome 8. Mouse News Lett. 74:95.

635. Lyon, M.F., and P.H. Glenister. 1982. A new allele sash (W^{sh}) at the W locus and a spontaneous recessive lethal in mice. Genet. Res. 39:315–22.

636. Lyon, M.F., and P.H. Glenister. 1974. Position of translocation break in T(10;13)199H. Mouse News Lett. 51:21–2.

637. Lyon, M.F., and P.H. Glenister. 1978. Roan and freckled. Mouse News Lett. 59:18.

638. Lyon, M.F., and P.H. Glenister. 1985. Chylous ascites (Chy) on chromosome 11. Mouse News Lett. 72:106–7.

639. Lyon, M.F., and P.H. Glenister. 1985. Neonatal anemia (Nan) on chromosome 8. Mouse News Lett. 72:107.

640. Lyon, M.F., and P.H. Glenister. 1980. Position of a recessive lethal linked to sash, W^{sh}. Mouse News Lett. 62:53.

641. Lyon, M.F., and S.G. Hawkes. 1970. X-linked gene for testicular feminization in the mouse. Nature 227:1217–9.

642. Lyon, M.F., and E.V. Hulse. 1971. An inherited kidney disease of mice resembling human nephronophthisis. J. Med. Genet. 8:41–8.

643. Lyon, M.F., S. Jarvis, and Y. Goyder. 1981. Relationship of Gy and Hyp. Mouse News Lett. 64:56.

644. Lyon, M.F., and S. Jarvis. 1979. Lens opacity (Lop). Mouse News Lett. 61:37.

645. Lyon, M.F., S.E. Jarvis, I. Sayers, and R.S. Holmes. 1981. Lens opacity: a new gene for congenital cataract on chromosome 10 of the mouse. Genet. Res. 38:337–41.

646. Lyon, M.F., and R. Meredith. 1964. Investigations of the nature of t-alleles in the mouse. I. Genetic analysis of a series of mutants derived from a lethal allele. Heredity 19:301–12.

647. Lyon, M.F., and R. Meredith. 1969. Muted, a new mutant affecting coat color and otoliths of the mouse, and its position in linkage group XIV. Genet. Res. 14:163–6.

648. Lyon, M.F., T. Morris, A.G. Searle, and J. Butler. 1967. Occurrences and linkage relations of the mutant 'extra-toes' in the mouse. Genet. Res. 9:383–5.

649. Lyon, M.F., C.R. Scriver, L.R.I. Baker, H.S. Tenenhouse, J. Kronick, and S. Mandla. 1986. The Gy mutation: another cause of X-linked hypophosphatemia in mouse. Proc. Natl. Acad. Sci. USA 83:4899–903.

650. Lyon, M.F., and J.F. Zenthon. 1987. Relationship between bareskin, denuded and rex. Mouse News Lett. 77:125.

651. Lyon, M.F., and J.F. Zenthon. 1986. Close linkage of bareskin (Bsk) and rex (Re). Mouse News Lett. 74:96.

652. Lyon, M.F. 1963. Attempts to test the inactive-X theory of dosage compensation in mammals. Genet. Res. 4:93–103.

653. Lyon, M.F. 1982. Genetic basis of male sterility and segregation distortion due to t-haplotype in the mouse. Genet. Res. 40:102–3 (Abstr.).

654. Lyon, M.F. 1956. Hereditary hair loss in the tufted mutant of the house mouse. J. Hered. 47:101–3.

655. Lyon, M.F. 1961. Linkage relations and some pleiotropic effects of the dreher mutant in the house mouse. Genet. Res. 2:92–5.

656. Lyon, M.F. 1986. Male sterility of the mouse t-complex is due to homozygosity of the distorter genes. Cell 44:357–63.

657. Lyon, M.F. 1975. Mouse News Lett. 52:36.

658. Lyon, M.F. 1966. Order of loci on the X-chromosome of the mouse. Genet. Res. 7:130–3.

659. Lyon, M.F. 1966. Mouse News Lett. 35:28.

660. Lyon, M.F. 1958. Twirler: a mutant affecting the inner ear of the house mouse. J. Embryol. Exp. Morphol. 6:105–16.

662. Lyons, J.B. 1984. Personal communication to editor. Mouse News Lett. 71:5.

663. MacInnes, J.I., V.L. Morris, W.F. Flintoff, and C.A. Kozak. 1984. Characterization and chromosomal location of endogenous mouse mammary tumor virus loci in GR, NSF, and DBA mice. Virology 132:12–25.

664. Mackensen, J.A. 1962. Mouse News Lett. 27:38.

665. Maltais, L.J., P.W. Lane, and W.G. Beamer. 1984. Anorexia, a recessive mutation causing starvation in preweanling mice. J. Hered. 75:468–72.

666. Mandel, J.L., B. Arveiler, G. Camerino, A. Hanauer,

R. Heilig, M. Koenig, and I. Oberle. 1986. Genetic mapping of the human X chromosome: linkage analysis of the q26-q28 region that includes the fragile X locus and isolation of expressed sequences. Cold Spring Harbor Symp. Quant. Biol. 51:195–203.

667. Mann, E., R.W. Elliott, and C. Hohman. 1984. Of four loci spanning mouse chromosome 17, two map to the t complex. Mouse News Lett. 71:48.

668. Marcu, K.B., L.J. Harris, L.W. Stanton, J. Erikson, R. Watt, and C.M. Croce. 1983. Transcriptionally active c-myc oncogene is contained within NIARD, a DNA sequence associated with chromosome translocations in B-cell neoplasia. Proc. Natl. Acad. Sci. USA 80:519–23.

669. Mariano, T.M., C.A. Kozak, J.A. Langer, and S. Pestka. 1987. The mouse immune interferon receptor gene is located on chromosome 10. J. Biol. Chem. 262:5812–14.

670. Marks, S.C. Jr., and P.W. Lane. 1976. Osteopetrosis, a new recessive skeletal mutation on chromosome 12 of the mouse. J. Hered. 67:11–18.

671. Marks, S.C. Jr., M.F. Seifert, and P.W. Lane. 1985. Osteosclerosis, a recessive skeletal mutation on chromosome 19 in the mouse. J. Hered. 76:171–6.

672. Marth, J.D., C. Disteche, D. Pravtcheva, F. Ruddle, E.G. Krebs, and R.M. Perlmutter. 1986. Localization of a lymphocyte-specific protein tyrosine kinase gene (lck) at a site of frequent chromosomal abnormalities in human lymphomas. Proc. Natl. Acad. Sci. USA 83:7400–4.

673. Martin, S.A., and G. Bulfield. 1984. A structural gene (*Hdc-s*) for mouse kidney histidine decarboxylase. Biochem. Genet. 22:645–56.

674. Martin, S.A., B.A. Taylor, T. Watanabe, and G. Bulfield. 1984. Histidine decarboxylase phenotypes of inbred mouse strains: a regulatory locus (*Hdc*) determines kidney enzyme concentration. Biochem. Genet. 22:305–22.

675. Martin, S.A.M., and G. Bulfield. 1984. A regulatory locus, *Hdc-e*, determines the response of mouse kidney histidine decarboxylase to estrogen. Biochem. Genet. 22:1037–46.

676. Martin, S.A.M., B.A. Taylor, T. Watanabe, and G. Bulfield. 1984. Histidine decarboxylase phenotypes of inbred mouse strains a regulatory locus (*Hdc*) determines kidney enzyme concentration. Biochem. Genet. 22:305–22.

676a. Martinsson, T., and G. Levan. 1987. Localization of the multidrug resistance-associated 170 kDa P-glycoprotein gene to mouse chromosome 5 and to homogeneously staining regions in multidrug-resistant mouse cells by in situ hybridization. Cytogenet. Cell Genet. 45:99–101.

677. Maruyama, N., F. Furukawa, Y. Nakai, Y. Sasaki, K. Ohta, S. Ozaki, S. Hirose, and T. Shirai. 1983. Genetic studies of autoimmunity in New Zealand mice. IV. Contribution of NZB and NZW genes to the spontaneous occurrence of retroviral gp70 immune complexes in (NZB x NZW)F1 hybrid and the correlation to renal disease. J. Immunol. 130:740–6.

678. Maruyama, N., C.O. Lindstrom, H. Sato, and F.J. Dixon. 1983. Serum gp70 production regulated by a gene on murine chromosome 7. Immunogenetics 18:365–71.

679. Maruyama, N., and C.O. Lindstrom. 1983. H-2-linked regulation of serum gp70 production in mice. Immunogenetics 17:507–21.

680. Mason, A.J., B.A. Evans, D.R. Cox, J. Shine, and R.I. Richards. 1983. Structure of mouse kallikrein gene family suggests a role in specific processing of biologically active peptides. Nature 303:300–7.

681. Mason, I.J., D. Murphy, M. Münke, U. Francke, R.W. Elliott, and B.L.M. Hogan. 1986. Developmental and transformation-sensitive expression of the *Sparc* gene on mouse chromosome 11. EMBO J. 5:1831–7.

681a. Mather, K. 1938. The Measurement of Linkage in Heredity. John Wiley & Sons, NY.

682. Mather, P.B., and R.S. Holmes. 1985. Aldehyde reductase isozymes in the mouse: evidence for two new loci and localization of *Ahr-3* on chromosome 7. Biochem. Genet. 23:483–96.

683. Matossian-Rogers, A., L. DeGiorgi, and S. Povey. 1983. Alloimmune interactions of a lymphoproliferative disease-inducer gene *Arp* and linkage to *Pep-7*. Immunogenetics 18:639–48.

684. Matsuda, Y., H.T. Imai, K. Moriwaki, and K. Kondo. 1983. Modes of inheritance of X-Y dissociation in intersubspecies hybrids between BALB/c mice and *Mus musculus molossinus*. Cytogenet. Cell Genet. 35:209–15.

685. Matsushima, Y., and S. Ikemoto. 1984. Genetic analysis of mouse tear protein: linkage of the *Mtp-1* locus on chromosome 7. Jpn. J. Zootec. Sci. 55:197–8.

686. Matsushima, Y., T. Imai, and S. Ikemoto. 1984. Evidence for X linkage of the mouse tear protein system-3 (Mtp-3) with mosaic expression in heterozygous females. Biochem. Genet. 22:577–85.

687. Mayer, T.C., N.J. Kleiman, and M.C. Green. 1976. Depilated (*dep*), a mutant gene that effects the coat of the mouse and acts in the epidermis. Genetics 84:59–65.

688. McCubrey, J., and R. Risser. 1982. Allelism and linkage studies of murine leukemia virus activation in low leukemic strains of mice. J. Exp. Med. 155:1233–8.

689. McFarland, E. 1965. Personal communication.

690. McFarland, E.C., and E.S. Russell. 1975. Mouse News Lett. 53:35.

691. McKenzie, I.F.C., M.M. Henning, and K.M. Curnow. 1985. A new H gene associated with a coat color mutation. Transplant. Proc. 17:754–5.

692. McKenzie, I.F.C., P.M. Hogarth, M.S. Sandrin, V.R. Sutton, R. Harris, and D.G. Penington. 1983. Complexities of the Qa antigenic system in the mouse. Transplant. Proc. 15:149–51.

693. McLaren, A., and M. Buehr. 1981. *Gpi* expression in female germ cells of the mouse. Genet. Res. 37:303.

694. Meo, T., T. Douglas, and A.M. Rijnbeek. 1977. Glyoxalase I polymorphism in the mouse: a new genetic marker linked to H-2. Science 198:311–3.

695. Meo, T., J. Johnson, C.V. Beechey, S.J. Andrews, J. Peters, and A.G. Searle. 1980. Linkage analyses of murine immunoglobulin heavy chain and serum prealbumin genes establish their location on chromosome 12 proximal to the T(5;12)31H breakpoint in band 12F1. Proc. Natl. Acad. Sci. USA 77:550–3.

696. Meredith, R. 1965. Mouse News Lett. 32:39.

697. Meredith, R. 1971. Mouse News Lett. 45:31.

698. Meruelo, D., M. Offer, and N. Flieger. 1983. A new lymphocyte cell surface antigen, Ly-22.2, controlled by a locus on chromosome 4 and a second unlinked locus. J. Immunol. 130:946–50.

699. Meruelo, D., M. Offer, and N. Flieger. 1981. Genetics of susceptibility to radiation-induced leukemia. J. Exp. Med. 154:1201–11.

700. Meruelo, D., M. Offer, and A. Rossomando. 1983. Induction of leukemia by both fractionated X-irradiation and radiation leukemia virus involves loci in the chromosome 2 segment H-30-A. Proc. Natl. Acad. Sci. USA 80:462–6.

701. Meruelo, D., A. Paolino, N. Flieger, and M. Offer. 1980. Definition of a new T lymphocyte cell surface antigen, Ly 11.2. J. Immunol. 125:2713–8.

702. Meruelo, D., A. Rossomando, M. Offer, J. Buxbaum, and A. Pellicer. 1983. Association of endogenous viral loci with genes encoding murine histocompatibility and lymphocyte differentiation antigens. Proc. Natl. Acad. Sci. USA 80:5032–6.

703. Meruelo, D., A. Rossomando, S. Scandalis, P. D'Eustachio, R.E.K. Fournier, D.R. Roop, D. Saxe, C. Blatt, and M.N. Nesbitt. 1987. Assignment of the *Ly-6– Ril-1– Sis– H-30– Pol5/ Xmmv-72– Ins-3– Krt-1– Int-1– Gdc-1* region to murine chromosome 15. Immunogenetics 25:361–72.

704. Michael, S.K., J. Hilgers, C. Kozak, J.B. Whitney III, and E.F. Howard. 1986. Characterization and mapping of DNA sequence homologous to mouse U1a1 snRNA: localization on chromosome 11 near the *Dlb-1* and *Re* loci. Somat. Cell Mol. Genet. 12:215–23.

705. Michalides, R., R. Verstraeten, F.W. Shen, and J. Hilgers. 1985. Characterization and chromosomal distribution of endogenous mouse mammary tumor viruses of European mouse strains STS/A and GR/A. Virology 142:278–90.

706. Michie, D. 1955. Genetical studies with 'vestigial tail' mice. I. The sex difference in crossing-over between vestigial and rex. J. Genet. 53:270–9.

707. Michie, D. 1955. Genetical studies with 'vestigial tail' mice. II. The position of vestigial in the seventh linkage group. J. Genet. 53:280–4.

708. Middleton, R.J., S.A.M. Martin, and G. Bulfield. 1987. A new regulatory gene in the histidine decarboxy-lase gene complex determines the responsiveness of the mouse kidney enzyme to testosterone. Genet. Res. 49:61–7.

709. Mishkin, J.D., B.A. Taylor, and W.J. Mellman. 1976. *Glk*: a locus controlling galactokinase activity in the mouse. Biochem. Genet. 14:635–40.

710. Mobraaten, L., H.P. Bunker, J. DeMaeyer-Guignard, E. DeMaeyer, and D.W. Bailey. 1984. Location of histocompatibility and interferon loci on chromosome 3 of the mouse. J. Hered. 75:233–4.

711. Mock, B.A., L.A. D'Hoostelaere, R. Matthai, and K. Huppi. 1987. A mouse homeo box gene, *Hox-1.5*, and the morphological locus, *Hd*, map to within 1 cM on chromosome 6. Genetics 116:607–12.

712. Mock, B.A. 1987. Mapping of the mouse bcl-2 gene. Mouse News Lett. 78:63.

713. Monaco, J.J., and H.O. McDevitt. 1986. The LMP antigens: a stable MHC controlled multi subunit protein complex. Hum. Immunol. 15:416–26.

714. Moore, K.J., A.M. Buchberg, M.C. Strobel, P.K. Seperack, N.G. Copeland, and N.A. Jenkins. 1986. The interaction of the dilute suppressor gene (*dsu*) with dilute and other coat color mutations of the mouse. 14th Mol. Biochem. Workshop (Abstr.).

715. Morris, T. 1966. Mouse News Lett. 35:27.

716. Morris, V.L., C. Kozak, J.C. Cohen, P.R. Shank, P. Jolicoeur, F. Ruddle, and H.E. Varmus. 1979. Endogenous mouse mammary tumor virus DNA is distributed among multiple mouse chromosomes. Virology 92:46–55.

717. Mortensen, R.F., P.T. Le, and B.A. Taylor. 1985. Mouse serum amyloid P-component (SAP) levels controlled by a locus on chromosome 1. Immunogenetics 22:367–75.

718. Moutier, R., and M.F. Bertrand. 1983. *Svp-3*, a third polymorphic locus for mouse seminal vesicle proteins. Biochem. Genet. 21:797–800.

719. Mowat, M., and A. Bernstein. 1983. Linkage of the *Fv-2* gene to a newly reinserted ecotropic retrovirus in *Fv-2* congenic mice. J. Virol. 47:471–7.

720. Müller, G., G. Scherer, H. Zentgraf, S. Ruppert, B. Herrmann, H. Lehrach, and G. Schütz. 1985. Isolation, characterization and chromosomal mapping of the mouse tyrosine aminotransferase gene. J. Mol. Biol. 184:367–73.

721. Müller, U., C.V. Jongeneel, S.A. Nedospasov, K.F. Lindahl, and M. Steinmetz. 1987. Tumour necrosis factor and lymphotoxin genes map close to *H-2D* in the mouse major histocompatibility complex. Nature 325:265–7.

722. Mullins, L.J., S.G. Grant, J. Pazik, D.A. Stephenson, and V.M. Chapman. 1987. Assignment and linkage of four new genes to the mouse X chromosome. Mouse News Lett. 77:150–1.

723. Münke, M., D.R. Cox, I.J. Jackson, B.L.M. Hogan, and U. Francke. 1986. The murine *Hox-2* cluster of

homeo box containing genes maps distal on chromosome 11 near the tail-short (*Ts*) locus. Cytogenet. Cell Genet. 42:236–40.

724. Münke, M., K. Harbers, R. Jaenisch, D.R. Cox, and U. Francke. 1985. Assignment of the alpha 1(I) collagen gene to mouse chromosome 11. Cytogenet. Cell Genet. 40:706 (Abstr.).

725. Münke, M., K. Harbers, R. Jaenisch, and U. Francke. 1986. Chromosomal mapping of four different integration sites of Moloney murine leukemia virus including the locus for α(I) collagen in mouse. Cytogenet. Cell Genet. 43:140–9.

725a. Murphy, E.D., and J.B. Roths. 1979. A Y chromosome associated factor in strain BXSB producing accelerated autoimmunity and lymphoproliferation. Arthritis Rheumatism 22:1188–94.

726. Murray, J.M., and G.D. Snell. 1945. Belted, a new sixth chromosome mutation in the mouse. J. Hered. 36:266–8.

727. Nadeau, J. 1982. Personal communication.

728. Nadeau, J.H., F.G. Berger, K.A. Kelley, P.M. Pitha, C.L. Sidman, and N. Worrall. 1986. Rearrangement of genes located on homologous chromosomal segments in mouse and man. The location of genes for alpha and beta-interferon, alpha-1 acid glycoprotein-1 and -2, and aminolevulinate dehydratase on mouse chromosome 4. Genetics 104:1239–55.

729. Nadeau, J.H., and E.M. Eicher. 1982. Conserved linkage of soluble aconitase and galactose-1-phosphate uridyl transferase in mouse and man: assignment of these genes to mouse chromosome 4. Cytogenet. Cell Genet. 34:271–81.

730. Nadeau, J.H., S.J. Phillips, and I.K. Egorov. 1985. Recombination between the *t6* complex and linked loci in the house mouse. Genet. Res. 45:251–64.

731. Nadeau, J.H., and S.J. Phillips. 1987. The putative oncogene Pim-1 in the mouse: its linkage and variation among *t* haplotypes. Genetics 117:533–41.

732. Nadeau, J.H. 1986. A chromosomal segment conserved since divergence of lineages leading to man and mouse. The gene order of aminoacylase-1, transferrin, and β-galactosidase on mouse chromosome 9. Genet. Res. 48:175–8.

733. Nakamura, D., R. Popp, E.M. Eicher, and P.A. Lalley. 1985. Comparison of the arginosuccinate synthetase gene family in mouse and man. Cytogenet. Cell Genet. 40:710 (Abstr.).

734. Nash, D.J. 1964. Mouse News Lett. 30:53–4.

735. Nash, H.R., M. Newton, J. Peters, and S.A. Andrews. 1979. Mouse News Lett. 60:54.

736. Nash, H.R., and O. von Deimling. 1982. Kidney esterase of *Mus musculus*: further polymorphism of esterase-6, esterase-9, and a new esterase, esterase-20. Biochem. Genet. 20:537–54.

737. Nash, H.R. 1984. Mouse News Lett. 71:33.

738. Nasrat, G.E. 1956. Estimation of the recombination fraction between the two linked genes *Re* and *sh-2* in the house mouse when the female is the heterozygous parent. Proc. Zool. Soc. (Bengal) 9:85–7.

739. Natsuume-Sakai, S., J-I. Hayakawa, S. Amano, and M. Takahaski. 1979. Genetic mapping of the locus controlling structural variations of murine C3 in the chromosome 17. J. Immunol. 123:947–8.

740. Natsuume-Sakai, S., J-I. Hayakawa, and M. Takahashi. 1978. Genetic polymorphism of murine C3 controlled by a single co-dominant locus on chromosome 17. J. Immunol. 121:491–8.

741. Natsuume-Sakai, S., K. Moriwaki, S. Migita, K. Sudo, K. Suzuki, D-Y. Lu, C. Wang, and M. Takahashi. 1983. Structural polymorphism of murine factor B controlled by a locus closely linked to the H-2 complex and demonstration of multiple alleles. Immunogenetics 18:117–24.

742. Natsuume-Sakai, S., K. Sudoh, T. Kaidoh, H-I. Hayakawa, and M. Takahashi. 1985. Structural polymorphism of murine complement factor H controlled by a locus located between the *Hc* and the β2M locus on the second chromosome of the mouse. J. Immunol. 134:2600–6.

743. Naylor, S.L., W.W. Chin, H.M. Goodman, P.A. Lalley, K-H. Grzeschik, and A.Y. Sakaguchi. 1983. Chromosome assignment of genes encoding the α and β subunits of glycoprotein hormones in man and mouse. Somat. Cell Genet. 9:757–70.

744. Naylor, S.L., R.W. Elliott, J.A. Brown, and T.B. Shows. 1982. Mapping of aminoacylase-1 and B-galactosidase-A to homologous regions of human chromosome 3 and mouse chromosome 9 suggests location of additional genes. Am. J. Hum. Genet. 34:235–4.

745. Naylor, S.L., P.W. Gray, and P.A. Lalley. 1984. Mouse immune interferon (IFN-γ) gene is on chromosome 10. Somat. Cell Mol. Genet. 10:531–4.

746. Naylor, S.L., A. Marshall, A.M. Killary, P.A. Lalley, D. Belin, R. Rickles, S. Strickland, and B. Rajput. 1987. The genes for mouse plasminogen activator and urokinase map to chromosomes 8 and 14, respectively. (HGM 9) Cytogenet. Cell Genet., 47:669 (Abstr.).

747. Naylor, S.L., A. Marshall, A. Solomon, J.R. McGill, J. McCombs, V.L. Magnuson, C.M. Moore, P.A. Lalley, B.T. Pentecost, and C. Teng. 1987. Lactoferrin maps to human chromosome 3(q21-q23) and mouse chromosome 9. (HGM) Cytogenet. Cell Genet., 47:669 (Abstr.).

748. Naylor, S.L., A.Y. Sakaguchi, W.W. Chin, J. Jacobs, L.P. Shen, W.J. Rutter, and P.A. Lalley. 1984. Chromosomal mapping of mouse hormone genes: somatostatin, calcitonin, thyrotropin alpha subunit, and luteinizing hormone beta subunit. Cytogenet. Cell Genet. 37:551 (Abstr.).

749. Naylor, S.L., A.Y. Sakaguchi, L. McDonald, S. Todd, P.A. Lalley, T.B. Shows, and W.W. Chin. 1986. Mapping thyrotropin β subunit gene in man and mouse. Somat. Cell Mol. Genet. 12:307–11.

750. Nedospasov, S.A., B. Hirt, A.N. Shakhov, V.N. Dobrynin, E. Kawashima, and C.V. Jongeneel. 1986.

The genes for tumor necrosis factor (TNF-alpha) and lymphotoxin (TNF-beta) are tandemly arranged on chromosome 17 of the mouse. Nucl. Acids Res. 14:7713–25.

751. Nesbitt, M.M., B. Bakay, M.B. Gardner, and C. Day. 1979. Isoenzyme pattern of HPRT in murine erythrocytes: control by an autosomal locus. Biochem. Genet. 17:957–64.

751a. Nesterova, T.B., P.M. Borodin, S.M. Zakian, and O.L. Serov. 1987. Assignment of the gene for adenine phosphoribosyltransferase on the genetic map of mouse chromosome 8. Biochem. Genet. 25:563–8.

752. Neuport-Sautes, C., and P. Demant. 1978. The *H-2L* locus. Mouse News Lett. 57:8.

753. Newton, M.F., H.R. Nash, J. Peters, and S.J. Andrews. 1982. Xylose dehydrogenase-1, a new gene on mouse chromosome 7. Biochem. Genet. 20:733–45.

754. Nichols, E.A., V.M. Chapman, and F.H. Ruddle. 1973. Polymorphism and linkage for mannosephosphate isomerase in *Mus musculus*. Biochem. Genet. 8:47–53.

755. Nichols, E.A., F.H. Ruddle, and M.L. Petras. 1975. Linkage of the locus for serum albumin in the house mouse, *Mus musculus*. Biochem. Genet. 13:551–5.

756. Nichols, E.A., and F.H. Ruddle. 1975. Polymorphism and linkage of glutathione reductase in *Mus musculus*. Biochem. Genet. 13:323–9.

757. Nielson, J.T., and V.M. Chapman. 1977. Electrophoretic variation for X-chromosome-linked phosphoglycerate kinase (PGK-1) in the mouse. Genetics 87:319–25.

758. Nusse, R., J. de Moes, J. Hilkens, and R. van Nie. 1980. Localization of a gene for expression of mouse mammary tumor virus antigens in the GR/Mtv-2⁻ mouse strain. J. Exp. Med. 152:712–9.

760. Nusse, R., A. van Ooyen, D. Cox, Y.K. Fung, and H. Varmus. 1984. Mode of proviral activation of a putative mammary oncogene (*int-1*) on mouse chromosome 15. Nature 307:131–6.

761. O'Brien, A.D., D.L. Rosenstreich, I. Scher, G.H. Campbell, R.P. MacDermott, and S.B. Formal. 1980. Genetic control of susceptibility to *Salmonella typhimurium* in mice: role of the *Lps* gene. J. Immunol. 124:20–4.

762. O'Donnell, J.J., K.M. Vannas-Sulonen, T.B. Shows, and D.R. Cox. 1985. Ornithine aminotransferase (OAT) maps to human chromosome 10 and mouse chromosome 7. Cytogenet. Cell Genet. 40:716 (Abstr.).

763. O'Toole, M.M., R. Riblet, and M.J. Bosma. 1984. Identification and mapping of *Lm-1*, an *Igh*-linked locus for a murine alloantigen. Immunogenetics 20:265–75.

764. Oda, S-I., T. Watanabe, and K. Kondo. 1980. A new mutation, eye lens obsolescence, *Elo*, on chromosome 1 in the mouse. Jpn J. Genet. 55:71–5.

765. Odaka, T., H. Ikeda, K. Yoshikura, H. Moriwaki, and S. Suzuki. 1981. *Fv-4*: gene controlling resistance to NB-tropic Friend murine leukemia virus. Distribution

in wild mice, introduction into genetic background of BALB/c mice, and mapping of chromosomes. J. Natl. Cancer Inst. 67:1123–7.

766. Oh, Y-S., and T. Tomita. 1987. Linkage of faded gene (*fe*) to chromosome 6 of the mouse. Jikken Dobutsu, Exp. Anim. (Tokyo) 36:73–7.

767. Oie, H.K., A.F. Gazdar, P.A. Lalley, E.K. Russell, J.D. Minna, J. DeLarco, G.J. Todaro, and U. Francke. 1978. Mouse chromosome 5 codes for ecotropic murine leukemia virus cell-surface receptor. Nature 274:60–2.

767a. Oliverio, A., and B.E. Eleftheriou. 1976. Motor activity and alcohol; genetic analysis in the mouse. Physiol. Behav. 16:577–81.

767b. Oliverio, A., B.E. Eleftheriou, and D.W. Bailey. 1973. Exploratory activity: genetic analysis of its modification by scopolamine and amphetamine. Physiol. Behav. 10:893–9.

768. Oliverio, A., B.E. Eleftheriou, and D.W. Bailey. 1973. A gene influencing active avoidance performance in mice. Physiol. Behav. 11:497–501.

769. Opalka, B., and E. Kolsch. 1983. Evidence for a new lymphocyte-stimulating determinant (*Lsd*) detected by alloreactive T cell lines. Eur. J. Immunol. 13:24–30.

770. Orren, A., M.J. Hobart, H.R. Nash, and P.J. Lachmann. 1985. Close linkage between mouse genes determining the two forms of complement component C6 and component C7, and cis action of a C6 regulatory gene. Immunogenetics 21:591–9.

771. Otto, J., and O. von Deimling. 1981. *Prt-4* and *Prt-5*: new constituents of a gene cluster on chromosome 7 coding for esterproteases in the submandibular gland of the house mouse (*Mus musculus*). Biochem. Genet. 19:431–44.

772. Otto, J., and O.H. von Deimling. 1983. Esterase-17 (ES-l7): characterization and genetic location on chromosome 9 of a bis-p-nitrophenyl phosphate-resistant esterase of the house mouse (*Mus musculus*). Biochem. Genet. 21:37–48.

773. Otto, J. 1980. Submandibular gland esterproteases. Mouse News Lett. 63:12.

774. Otto, J. 1980. Submandibular gland esterproteases: new loci *Prt-4* and *Prt-5* linked to *c* on chromosome 7. Mouse News Lett. 62:46–7.

775. Owen, A.R.G. 1953. The analysis of multiple linkage data. Heredity 7:247–64.

776. Owen, F.L., R. Riblet, and B.A. Taylor. 1981. The T suppressor cell alloantigen *Tsu^d* maps near immunoglobulin allotype genes and may be a heavy chain constant region marker on a T cell receptor. J. Exp. Med. 153:801–10.

777. Owen, F.L., and R. Riblet. 1984. Genes for the mouse T cell alloantigens Tpre, Tthy, Tind, and Tsu are closely linked near *Igh* on chromosome 12. J. Exp. Med. 159:313–7.

778. Owen, F.L., B.A. Taylor, A. Zweidler, and J.G. Seidman. 1986. The murine gamma-chain of the T cell receptor is closely linked to a spermatocyte specific his-

tone gene and the beige coat color locus on chromosome 13. J. Immunol. 137:1044–6.

779. Owen, F.L. 1983. *Tpre*, a new alloantigen encoded in the IgT-C region of chromosome 12, is expressed in bone marrow of nude mice, fetal T cell hybrids, and fetal thymus. J. Exp. Med. 157:419–32.

780. Owerbach, D., and J.P. Hjorth. 1980. Inheritance of a parotid secretory protein in mice and its use in determining salivary amylase quantitative variants. Genetics 95:129–41.

781. Ozaki, S., H. Honda, N. Maruyama, S. Hirose, M. Hamaoki, H. Sato, and T. Shirai. 1983. Genetic regulation of erythrocyte autoantibody production in New Zealand black mice. Immunogenetics 18:241–54.

782. Paigen, B., D. Albee, P.A. Holmes, and D. Mitchell. 1987. Genetic analysis of murine strains C57BL/6J and C3H/HeJ to confirm the map position of *Ath-1*, a gene determining atherosclerosis susceptibility. Biochem. Genet. 25:501–11.

784. Paigen, B., D. Mitchell, and B. Ishida. 1987. *Ath-1*, a gene determining atherosclerosis susceptibility in mice, maps on chromosome 1 near *Alp-2*. Genetics 116:s22 (Abstr.).

785. Paigen, B., D. Mitchell, K. Reue, A. Morrow, A.J. Lusis, B.A. Taylor, and R.C. LeBoeuf. 1987. *Ath-1*, a gene determining atherosclerosis susceptibility and high density lipoprotein levels in mice. Proc. Natl. Acad. Sci. USA 84:3763–7.

787. Paigen, K., M. Meisler, J. Felton, and V. Chapman. 1976. Genetic determination of the β-galactosidase developmental program in mouse liver. Cell 9:533–9.

788. Paigen, K., and W.K. Noell. 1961. Two linked genes showing a similar timing of expression in mice. Nature 190:148–50.

789. Palayoor, S.T., and T.N. Seyfried. 1984. Genetic association between Ca2+ ATPase activity and audiogenic seizures in mice. J. Neurochem. 42:1171–4.

790. Paolucci, E.S., and D.S. Shreffler. 1983. *H-2*-linked murine factor B phenotypes. Immunogenetics 17:67–78.

791. Parsons, P.A. 1958. A balanced four-point linkage experiment for linkage group XIII in the house mouse. Heredity 12:77–95.

792. Parsons, P.A. 1958. Additional three-point data for linkage group V of the mouse. Heredity 12:357–62.

793. Passmore, H., and E. Ogin. 1982. Map position of Ce-2. Mouse News Lett. 66:75.

794. Paterniti J.R., Jr., W.V. Brown, H.N. Ginsberg, and K. Artzt. 1983. Combined lipase deficiency (*cld*): a lethal mutation on chromosome 17 of the mouse. Science 221:167–9.

795. Paul, P.R., and R.W. Elliott. 1987. Analysis of the mouse *Amy* locus in recombinant inbred mouse strains. Biochem. Genet. 25:569–79.

796. Pedersen, E.B., S. Haahr, and S.C. Mogensen. 1983. X-linked resistance of mice to high doses of Herpes simplex virus type 2 correlates with early interferon production. Infect. Immun. 42:740–6.

797. Peters, G., C. Kozak, and C. Dickson. 1984. Mouse mammary tumor virus integration regions *int-1* and *int-2* map on different mouse chromosomes. Mol. Cell. Biol. 4:375–8.

798. Peters, G., M. Placzak, S. Brookes, C. Kozak, R. Smith, and C. Dickson. 1986. Characterization, chromosomal assignment and segregation analysis of endogenous proviral units of mouse mammary tumor viruses. J. Virol. 59:535–44.

799. Peters, J., S.J. Andrews, and C.V. Beechey. 1980. Position of Hk-1 on chromosome 10. Mouse News Lett. 63:17.

800. Peters, J., and S.J. Andrews. 1985. Linkage of lactate dehydrogenase-2, *Ldh-2*. Biochem. Genet. 23:217–25.

801. Peters, J., and S.J. Andrews. 1984. The *Pk-3* gene determines both the heart, M1, and the kidney, M2, pyruvate kinase isozymes in the mouse; and a simple electrophoretic method for separating phosphoglucomutase-3. Biochem. Genet. 22:1047–63.

802. Peters, J., and S.T. Ball. 1985. Location of G6pd. Mouse News Lett. 73:17–8.

803. Peters, J., and S.T. Ball. 1986. Tightskin-2 (Tsk-2). Mouse News Lett. 74:91–2.

804. Peters, J., and G. Bulfield. 1984. Location of triose phosphate isomerase-1, Tpi-1. Mouse News Lett. 70:82.

805. Peters, J., H.R. Nash, E.M. Eicher, and G. Bulfield. 1981. Polymorphism of kidney pyruvate kinase in the mouse is determined by a gene, *Pk-3*, on chromosome 9. Biochem. Genet. 19:757–69.

806. Peters, J., and H.R. Nash. 1977. Polymorphism of esterase 11 in *Mus musculus*, a further esterase locus on chromosome 8. Biochem. Genet. 15:217–26.

807. Peters, J., S. Povey, S. Jeremiah, and L. De Giorgi. 1983. Linkage relationships of peptidase-7, *Pep-7*, in the mouse. Biochem. Genet. 21:801–7.

808. Peters, J., D.M. Swallow, S.J. Andrews, and L. Evans. 1981. A gene (*Neu-1*) on chromosome 17 of the mouse affects acid α-glucosidase and codes for neuraminidase. Genet. Res. 38:47–55.

809. Peterson, A., and F. Biddle. 1970. Mouse News Lett. 43:19.

810. Peterson, T.C., A.M. Killary, and R.E.K. Fournier. 1985. Chromosomal assignment and trans regulation of the tyrosine aminotransferase structural gene in hepatoma hybrid cells. Mol. Cell. Biol. 5:2491–4.

811. Petit, C., J. Levilliers, M-O. Ott, and M.C. Weiss. 1986. Tissue-specific expression of the rat albumin gene: genetic control of its extinction in microcell hybrids. Proc. Natl. Acad. Sci. USA 83:2561–5.

812. Petras, M.L., and F.G. Biddle. 1967. Serum esterases in the house mouse, *Mus musculus*. Can. J. Genet. Cytol. 9:704–10.

813. Phillips, R.J.S., and G. Fisher. 1978. Mouse News Lett. 58:43–4.

814. Phillips, R.J.S., and G. Fisher. 1979. Non-agouti curly: change of symbol. Mouse News Lett. 60:46–7.

815. Phillips, R.J.S., and S.G. Hawker. 1973. Mouse News Lett. 49:29.

816. Phillips, R.J.S., and M.H. Kaufman. 1974. Bare-patches, a new sex-linked gene in the mouse, associated with a high production of XO females. II. Investigations into the nature and mechanism of the XO production. Genet. Res. 24:27–41.

817. Phillips, R.J.S. 1954. Jimpy, a new totally sex-linked gene in the house mouse. Z. Indukt. Abstammungs.-Vererbungsl. 86:322–6.

818. Phillips, R.J.S. 1960. Lurcher, a new gene in linkage group XI of the house mouse. J. Genet. 57:35–42.

819. Phillips, R.J.S. 1966. Mouse News Lett. 34:27.

820. Phillips, R.J.S. 1974. Location of Och. Mouse News Lett. 50:42.

821. Phillips, R.J.S. 1963. Striated, a new sex-linked gene in the house mouse. Genet. Res. 4:151–3.

822. Phillips, R.J.S. 1956. The linkage of congenital hydro-cephalus in the house mouse. J. Hered. 47:302–4.

823. Phillips, R.J.S. 1970. Mouse News Lett. 42:26.

824. Phipps, E.L. 1969. Irregular teeth, It. Mouse News Lett. 40:41.

825. Pietsky, D.S., S.E. Riordan, and D.H. Sachs. 1979. Genetic control of the immune response to staphylococ-cal nuclease. IX. Recombination between genes deter-mining BALB/c antinuclease idiotypes and the heavy chain allotype locus. J. Immunol. 122:842–6.

826. Pizairo, O., and L.C. Dunn. 1970. A study of recombi-nation between the *H-2* (histocompatibility) locus and loci closely linked with it in the house mouse. Trans-plantation 9:207–18.

827. Pizairo, O., and U. Vergara. 1973. Relationship between locus R (*Ea-2*) and other loci of the ninth lin-kage group of the house mouse. Folia Biol. (Prague) 19:89–94.

828. Platz, R.D., and H.G. Wolfe. 1969. Mouse seminal vesicle proteins. The inheritance of electrophoretic variants. J. Hered. 60:187–92.

829. Pollack, S.B., and S.L. Emmons. 1982. NK-2.1: an NK-associated antigen detected with NZB anti-BALB/c serum. J. Immunol. 129:2277–81.

830. Popp, R.A., and W. St. Amand. 1964. A sex difference in recombination frequency in the albinism-hemoglobin interval of linkage group I in the mouse. J. Hered. 55:101–3.

831. Popp, R.A., and W. St. Amand. 1960. Studies on the mouse hemoglobin locus. I. Identification of hemoglo-bin types and linkage of hemoglobin with albinism. J. Hered. 51:141–4.

832. Popp, R.A. 1967. Linkage of *Es-1* and *Es-2* in the mouse. J. Hered. 58:186–8.

833. Popp, R.A. 1965. Loci linkage of serum esterase pat-terns and oligosyndactyly. J. Hered. 56:107–8.

834. Popp, R.A. 1962. Studies on the mouse hemoglobin loci. J. Hered. 53:73–80.

835. Potter, T.A., I.F.C. McKenzie, and M.C. Cherry. 1981. Murine lymphocyte alloantigens: the *Ly-4* locus. J. Immunogenet. 8:15–25.

836. Potter, T.A., and I.F.C. McKenzie. 1981. Identifi-cation of new murine lymphocyte alloantigens: antisera prepared between C57L 129 and related strains define new loci. Immunogenetics 12:351–69.

837. Prakash, O., C. Kozak, and N.H. Sarkar. 1985. Mole-cular cloning, characterization, and genetic mapping of an endogenous mammary tumor virus proviral unit I of C3H/He mice. J. Virol. 54:285–94.

838. Pravtcheva, D., M. Rabin, M. Bartolomei, J. Corden, and R.H. Ruddle. 1986. Chromosomal assignment of gene encoding the largest subunit of RNA polynmerase II in the mouse. Somat. Cell Mol. Genet. 12:523–8.

839. Pravtcheva, D.D., A.B. DeLeo, F.H. Ruddle, and L.J. Old. 1981. Chromosome assignment of the tumor-speci-fic antigen of a 3-methylcholanthrene induced mouse sarcoma. J. Exp. Med. 154:964–77.

840. Pravtcheva, D.D., F.H. Ruddle, R.W. Ellis, and E.M. Scolnick. 1983. Assignment of murine cellular Harvey ras gene to chromosome 7. Somat. Cell Genet. 9:681–6.

841. Pravtcheva, D.D., and F.H. Ruddle. 1983. Normal X chromosome induced reversion in the direction of chro-mosome segregation in mouse–Chinese hamster soma-tic cell hybrids. Exp. Cell Res. 148:265–72.

842. Prochazka, M., P.H. Le, D. Serreze, D. Coleman, and E.H. Leiter. 1986. Genetic control of insulin-dependent diabetes in non-obese diabetic (NOD) mice. Mouse News Lett. 75:32.

843. Prochazka, M., P. Staeheli, R.S. Holmes, and O. Haller. 1985. Interferon-induced guanylate-binding proteins: mapping of the murine *Gbp-1* locus to chro-mosome 3. Virology 145:273–9.

844. Rabin, M., A. Ferguson-Smith, C.P. Hart, and F.H. Ruddle. 1986. Cognate homeo-box loci mapped on homologous human and mouse chromosomes. Proc. Natl. Acad. Sci. USA 83:9104–8.

845. Rabin, M., C.P. Hart, A. Ferguson-Smith, W. McGin-nis, M. Levine, and F.H. Ruddle. 1985. Two homoeo box loci mapped in evolutionarily related mouse and human chromosomes. Nature 314:175–8.

845a. Rajput, B., A. Marshall, A.M. Killary, P.A. Lalley, S.L. Naylor, D. Belin, R.J. Rickles, and S. Strickland. 1987. Chromosomal assignments of genes for tissue plasminogen activator and urokinase in mouse. Somat. Cell Mol. Genet. 13:581–6.

846. Raos, M.H., and P. Demant. 1982. Murine comple-ment factor B (BF): sexual dimorphism and *H-2*-linked polymorphism. Immunogenetics 15:23–30.

847. Reckelhoff, J.F., J.S. Bond, R.J. Beynon, S. Savarir-ayan, and C.S. David. 1985. Proximity of the *Mep-1* gene to *H-2D* on chromosome 17 in mice. Immunoge-netics 22:617–23.

848. Reeves, R.H., D. Gallahan, B.F. O'Hara, R. Callahan, and J.D. Gearhart. 1987. Genetic mapping of *Prm-1*, *Igl-1*, *Smst*, *Mtv-6*, *Sod-1* and *Ets-2* and localization of

the Down syndrome region on mouse chromosome 16. Cytogenet. Cell Genet. 44:76–81.

849. Reeves, R.H., J.D. Gearhart, and J.W. Littlefield. 1986. Genetic basis for a mouse model of Down syndrome. Brain Res. Bull. 16:803–14.

850. Reeves, R.H., N.K. Robakis, M.L. Oster-Granite, H.M. Wisniewski, J.T. Coyle, and J.D. Gearhart. 1987. Genetic linkage in the mouse of genes involved in Down syndrome and Alzheimer's disease in man. Mol. Brain Res. 2:215–21.

851. Riblet, R., L. Claflin, D.M. Gibson, B.J. Mathieson, and M. Weigert. 1980. Antibody gene linkage studies in (NZB x C58) recombinant inbred lines. J. Immunol. 124:787–9.

852. Richards, J.E., D.D. Pravtcheva, C. Day, F.H. Ruddle, and P.P. Jones. 1985. Murine invariant chain gene: chromosomal assignment and segregation in recombinant inbred strains. Immunogenetics 22:193–9.

853. Richards-Smith, B., and R. Elliott. 1984. Mapping of a family of repeated sequences in the mouse genome. Mouse News Lett. 71:46–7.

854. Rinchik, E.M., and D.B. Amos. 1983. Genetic analysis of the *T- H-2* region in non- *t* chromosomes. I. Two new congenic strains isolate *Qglo-1* from *H-2*. Immunogenetics 17:445–55.

855. Roach, A., N. Takahashi, D. Pravtcheva, F. Ruddle, and L. Hood. 1985. Chromosomal mapping of mouse myelin basic protein gene and structure and transcription of the partially deleted gene in shiverer mutant mice. Cell 42:149–55.

856. Robbins, J.M., D. Gallahan, E. Hogg, C. Kozak, and R. Callahan. 1986. An endogenous mouse mammary tumor virus genome common in inbred mouse strains is located on chromosome 6. J. Virol. 57:709–13.

857. Robert, B., P. Barton, A. Minty, P. Daubas, A. Weydert, F. Bonhomme, J. Catalan, D. Chazottes, J-L. Guénet, and M. Buckingham. 1985. Investigation of genetic linkage between myosin and actin genes using an interspecific mouse back-cross. Nature 314:181–3.

858. Roberts, E., and J.H. Quisenberry. 1935. Linkage of the genes for non-yellow (*y*) (*a*) and pink-eye (*p2*) (*pa*) in the house mouse (*Mus musculus*). Am. Nat. 69:181–3.

859. Robinson, P.J. 1985. *Qb-1*, a new class I polypeptide encoded by the *Qa* region of the mouse *H-2* complex. Immunogenetics 22:285–9.

860. Roderick, T.H., J.J. Hutton, and F.H. Ruddle. 1970. Linkage of esterase-3 (*Es-3*) and rex (*Re*) on linkage group VII of the mouse. J. Hered. 61:278–9.

861. Roderick, T.H. 1971. Mouse News Lett. 44:30.

862. Roderick, T.H. 1972. Mouse News Lett. 47:37.

863. Roderick, T.H. 1986. Personal communication.

864. Roderick, T.H. 1979. Personal communication.

865. Roderick, T.H. 1983. Using inversions to detect and study recessive lethals and detrimentals in mice. *In* F.J. de Serres and W. Sheridan, eds., Utilization of Mammalian Specific Locus Studies in Hazard Evaluation and Estimation of Genetic Risk, 135–67. Plenum Publishing Corporation, New York.

866. Rosenstreich, D.L., M.G. Groves, H.A. Hoffman, and B.A. Taylor. 1978. Location of the Sas-1 locus on mouse chromosome 1. Immunogenetics 7:313–20.

867. Rossomando, A., and D. Meruelo. 1986. Viral sequences are associated with many histocompatibility genes. Immunogenetics 23:233–45.

868. Roths, J., E.D. Murphy, and E.M. Eicher. 1984. A new mutation, *gld*, that produces lymphoproliferation and autoimmunity in C3H/HeJ mice. J. Exp. Med. 159:1–20.

869. Rotter, V., D. Wolf, D. Pravtcheva, and F.H. Ruddle. 1984. Chromosomal assignment of the murine gene encoding the transformation-related protein p53. Mol. Cell. Biol. 4:383–5.

870. Rowe, W.P., J.W. Hartley, and T. Bremmer. 1972. Genetic mapping of a murine leukemia virus-inducing locus of AKR mice. Science 178:860–2.

871. Rowe, W.P., J.B. Humphrey, and F. Lilly. 1973. A major genetic locus affecting resistance to infection with murine leukemia viruses. III. Assignment of the *Fv-1* locus to linkage group VIII of the mouse. J. Exp. Med. 137:850–3.

872. Rowe, W.P., and H. Sato. 1973. Genetic mapping of the *Fv-1* locus of the mouse. Science 180:640–1.

873. Rowe, W.P. 1973. Genetic factors in the natural history of murine leukemia virus infection: G.H.A. Clowes memorial lecture. Cancer Res. 33:3061–8. Plus Rowe, W.P. 1974. Personal communication.

874. Rowe, W.P. 1978. Leukemia virus genomes in the chromosomal DNA of the mouse. Harvey Lectures 71:173–92.

875. Rubinstein, P., K. Vienne, and G.F. Hoecker. 1979. The location of the *C3* and *Glo* (glyoxylase I) loci of the IXth linkage group in mice. J. Immunol. 122:2584–9.

876. Rubenstein, P., and K. Vienne. 1982. Genetic control of the quantitative variation of erythrocyte glyoxylase-1 (GLO-1) in mice. Biochem. Genet. 20:153–63.

877. Ruddle, F.H., T.B. Shows, and T.H. Roderick. 1969. Esterase genetics in *Mus musculus*: expression, linkage and polymorhism of locus *Es-2*. Genetics 62:393–9.

878. Ruddle, N.H., B.S. Conta, L. Leinwand, C. Cozak, F. Ruddle, P. Besmer, and D. Baltimore. 1978. Assignment of the receptor for ecotropic murine leukemia virus to mouse chromosome 5. J. Exp. Med. 148:451–65.

879. Runner, M.N. 1959. Linkage of brachypodism. A new member of linkage group V of the house mouse. J. Hered. 50:81–4.

880. Russell, E.S., and E.C. McFarland. 1974. Genetics of mouse hemoglobins. Ann. NY Acad. Sci. 241:25–38.

881. Russell, E.S., and E.C. McFarland. 1977. Mouse News Lett. 56:42.

882. Russell, E.S., and J.L. Southard. 1966. Mouse News Lett. 35:32.

883. Russell, E.S., B. Whitney, and D. Bailey. 1979. Personal communication.

884. Russell, L.B., and N.L. Cacheiro. 1975. Superimposition of linkage groups X and IV. Mouse News Lett. 52:46–7.

885. Russell, L.B., and M.M. Larsen. 1965. Mouse News Lett. 33:69.

886. Russell, L.B. 1960. Mouse News Lett. 23:58.

887. Russell, L.B. 1963. Mouse News Lett. 29:73.

888. Russell, L.B. 1983. X-Autosome translocations in the mouse: their characterization and use as tools to investigate gene inactivation and gene action. *In* A.A. Sandberg, ed., Cytogenetics of the Mammalian X Chromosome, Part A Basic Mechanisms of X Chromosome Behavior, 205–50. Alan R. Liss, NY.

889. Ryan, J., C.P. Hart, and F.H. Ruddle. 1984. Molecular cloning and chromosome assignment of murine N-ras. Nucl. Acids Res. 12:6063–72.

890. Sakaguchi, A.Y., P.A. Lalley, B.U. Zabel, R. Ellis, E. Scolnick, and S.L. Naylor. 1984. Mouse proto-oncogene assignments. Cytogenet. Cell Genet. 37:573–4 (Abstr.).

891. Sakaguchi, A.Y., P.A. Lalley, B.U. Zabel, R.W. Ellis, E.M. Scolnick, and S.L. Naylor. 1984. Chromosome assignments of four mouse cellular homologs of sarcoma and leukemia virus oncogenes. Proc. Natl. Acad. Sci. USA 81:525–9.

892. Sakaguchi, A.Y., D.V. Lowe, D. Goeddel, P.A. Lalley, and S.L. Baylor. 1986. R-ras: a member of the ras family on human chromosome 19. Am. J. Hum. Genet. 39(Suppl.):A167.

893. Samanta, H., D.D. Pravtcheva, F.H. Ruddle, and P. Lengyel. 1984. Chromosomal location of mouse gene 202 which is induced by interferons and specifies a 56.5 kD protein. J. Interferon Res. 4:295–300.

894. Samollow, P.B., J.L. VandeBerg, A.L. Ford, T.C. Douglas, and C.S. David. 1986. Electrophoretic analysis of liver neuraminidase-1 variation in mice and additional evidence concerning the location of Neu-1. J. Immunogenet. 13:29–39.

895. Samuelson, L.C., and R.A. Farber. 1985. Cytological localization of adenosine kinase, nucleoside phosphorylase-1, and esterase-10 genes on mouse chromosome 14. Somat. Cell Mol. Genet. 11:157–65.

896. Sanyal, S., R. Van Nie, J. De Moes, and R.K. Hawkins. 1986. Map position of dysgenetic lens (*dyl*) locus on chromosome 4 in the mouse. Genet. Res. 48:199–200.

897. Sato, H., K. Itakura, and E.A. Boyse. 1977. Location of *Lyb-2* on mouse chromosome 4. Immunogenetics 4:591–5.

898. Sato, H., S. Kimura, and K. Itakura. 1981. Genetic and serologic reevaluation of LyM-1 antigen specified by a gene closely linked to *Mls*: establishment of LyM-1 antigen of the mouse. J. Immunogenet. 8:27–40.

899. Sato, H., S. Natsuume-Sakai, M. Katagiri, and K. Itak-

ura. 1981. Location of *LyM-1* locus on *Mus musculus* chromosome 1. J. Immunogenet. 8:185–9.

900. Sawyer, J.R., M.M. Moore, D. Clive, and J.C. Hozier. 1985. Regional sublocalization of the mouse thymidine kinase gene by breakage analysis. Cytogenet. Cell Genet. 40:738 (Abstr.).

901. Schaible, R. 1964. Personal communication.

902. Schaible, R.H. 1961. Mouse News Lett. 24:38.

903. Scheid, M.P., K.S. Landreth, J-S. Tang, and P.W. Kincade. 1982. Preferential but nonexclusive expression of macromolecular antigens on B-lineage cells. Immunol. Rev. 69:141–59.

904. Schlagel, C.J., and A. Ahmed. 1982. Evidence for genetic control of microwave-induced augmentation of complement receptor-bearing B lymphocytes. J. Immunol. 129:1530–3.

905. Schollen, J., K. Bender, and O. von Deimling. 1975. Esterase. XXI. *Es-9*, a possibly new polymorphic esterase in *Mus musculus* genetically linked to *Es-2*. Biochem. Genet. 13:369–77.

906. Schrier, D.J., J.L. Sternick, E.M. Allen, and V.L. Moore. 1983. Immunogenetics of BCG-induced anergy in mice: control by genes linked to the Igh complex. J. Immunol. 128:1466–9.

907. Searle, A.G., and C.V. Beechey. 1970. Mouse News Lett. 42:27.

908. Searle, A.G., and C.V. Beechey. 1970. Mouse News Lett. 43:29.

909. Searle, A.G., and C.V. Beechey. 1974. Sex difference in chr 6 recombination frequencies. Mouse News Lett. 50:40.

910. Searle, A.G., and C.V. Beechey. 1975. Mouse News Lett. 52:35–6.

911. Searle, A.G., and C.V. Beechey. 1975. Mouse News Lett. 52:35.

912. Searle, A.G., and G.M. Truslove. 1970. A gene triplet in the mouse. Genet. Res. 15:227–35.

913. Searle, A.G. 1968. Mouse News Lett. 38:22.

914. Searle, A.G. 1964. The genetics and morphology of two luxoid mutants in the house mouse. Genet. Res. 5:171–97.

915. Searle, A.G. 1961. Tipsy, a new mutant in linkage group VII of the mouse. Genet. Res. 2:122–6.

916. Searle, P.F., B.L. Davison, G.W. Stuart, T.M. Wilkie, G. Norstedt, and R.D. Palmiter. 1984. Regulation, linkage, and sequence of mouse metallothionein I and II genes. Mol. Cell. Biol. 4:1221–30.

917. Seeley, T-L., and R.S. Holmes. 1981. Genetics and ontogeny of butyryl CoA dehydrogenase in the mouse and linkage of *Bcd-1* with *Dao-1*. Biochem. Genet. 19:333–45.

918. Seldin, M.F., H.C. Morse III, L. D'Hoostelaere, J.L. Britten, and A.D. Steinberg. 1987. Mapping of alpha-spectrin on distal mouse chromosome 1. Cytogenet. Cell Genet. 45:52–4.

919. Sewell, W.A., M.H. Brown, M.J. Owen, P.J. Fink, C.A. Koack, and M.J. Crumpton. 1987. The murine

homologue of the T lymphocyte CD2-antigen: molecular cloning, chromosome assignment and cell surface expression. Eur. J. Immunol. 17:1015–20. Plus Morse H.C. III. 1987. Personal communication.

920. Seyedyazdani, I., and L.G. Lundin. 1973. Association between β-galactosidase activities and coat color in mice. J. Hered. 64:295–6.

921. Shaper, N.L., J.H. Shaper, G.F. Hollis, H. Chang, I.R. Kirsch, and C.A. Kozak. 1987. The gene for galactosyltransferase maps to mouse chromosome 4. Cytogenet. Cell Genet. 44:18–21.

922. Shedlovsky, A., J-L. Guénet, L.L. Johnson, and W.F. Dove. 1986. Induction of recessive lethal mutations in the T/ t- H-2 region of the mouse genome by a point mutagen. Genet. Res. 47:135–42.

922a. Silver, J. 1985. Confidence limits for estimates of genetic linkage based on analysis of recombinant inbred strains. J. Hered. 76:436–40.

923. Shire, J.M.G. 1970. Mouse News Lett. 42:40.

924. Shows, T.B., V.M. Chapman, and F.H. Ruddle. 1970. Mitochondrial malate dehydrogenase and malic enzyme: Mendelian inherited electrophoretic variants in the mouse. Biochem. Genet. 4:707–18.

925. Shreffler, D.C. 1963. Linkage of the mouse transferrin locus. J. Hered. 54:127–9.

926. Shupp Byrne, D.E., and R.L. Church. 1983. Assignment of the genes for mouse type I procollagen to chromosome 16 using mouse fibroblast Chinese hamster somatic cell hybrids. Somat. Cell Genet. 9:313–31.

927. Siadak, A.W., and R.C. Nowinski. 1981. Thy-2: a murine thymocyte-brain alloantigen controlled by a gene linked to the major histocompatibility complex. Immunogenetics 12:45–58.

928. Siciliano, M.J., R.E.K. Fournier, and R.L. Stallings. 1984. Regional assignment of ADA and ITPA to mouse chromosome 2 (C1—ter). J. Hered. 75:175–80.

929. Sidman, C.L., J.D. Marshall, W.G. Beamer, J.H. Nadeau, and E.R. Unanue. 1986. Two loci affecting B cell responses to B cell maturation factors. J. Exp. Med. 163:116–28.

930. Sidman, R.L., C.S. Conover, and J.H. Carson. 1985. Shiverer gene maps near the distal end of chromosome 18 in the house mouse. Cytogenet. Cell Genet. 39:241–5.

931. Sidman, R.L., and M.C. Green. 1970. 'Nervous disease' a new mutant mouse with cerebellar disease. Colloques Internat. du Centre Nationale de la Recherche Scientifique 924:69–79.

932. Sidman, R.L., and M.C. Green. 1965. Retinal degeneration in the mouse: location of the rd locus in linkage group XVII. J. Hered. 56:23–9.

933. Sidman, R.L., H.C. Kinney, and H.O. Sweet. 1985. Transmissible spongiform encephalopathy in the gray tremor mutant mouse. Proc. Natl. Acad. Sci. USA 82:253–7.

934. Sidman, R.L. 1968. Personal communication.

935. Silver, J., and C.E. Buckler. 1986. A preferred region for integration of Friend murine leukemia virus in hematopoietic neoplasms is closely linked to the Int-2 oncogene. J. Virol. 60:1156–8.

936. Silver, J., J.B. Whitney III, C. Kozak, G. Hollis, and I. Kirsch. 1985. Erbb is linked to the alpha-globin locus on mouse chromosome 11. Mol. Cell. Biol. 5:1784–6.

937. Silver, J. 1985. A defective ecotropic provirus closely linked to the albino locus. J. Virol. 55:494–6.

938. Silver, J. 1986. Personal communication.

939. Silver, L.M., K. Artzt, and D. Bennett. 1979. A major testicular cell protein specified by a mouse T/ t complex gene. Cell 17:275–84.

940. Silver, L.M., J. Ulman, J. Danska, and J.I. Garrels. 1983. A diversified set of testicular cell proteins specified by genes within the mouse t complex. Cell 35:35–45.

941. Silver, L.M., M. White, and K. Artzt. 1980. Evidence for unequal crossing over within the mouse T/ t complex. Proc. Natl. Acad. Sci. USA 77:6077–80.

942. Simmons, D.L., P.A. Lalley, and C.B. Kasper. 1985. Chromosomal assignments of genes coding for components of the mixed-function oxidase system in mice. Genetic localization of the cytochrome P-450PCN and P-450PB gene families and the NADPH-cytochrome P-45 9 oxidoreductase and epoxide hydratase genes. J. Biol. Chem. 260:515–21.

943. Siracusa, L.D., L.B. Russell, E.M. Eicher, D.J. Corrow, N.G. Copeland, and N.A. Jenkins. 1987. Genetic organization of the agouti region of the mouse. Genetics 117:93–100.

944. Sirlin, J.L. 1957. Location of vacillans in linkage group VIII of the house mouse. Heredity 11:259–60.

945. Sirlin, J.L. 1956. Vacillans, a neurological mutant in the house mouse linked with brown. J. Genet. 54:42–8.

946. Siwarski, D.F., Y. Barra, G. Jay, and M.J. Rogers. 1985. Occurrence of a unique MHC class I gene in distantly related members of the genus Mus. Immunogenetics 21:267–76.

947. Skamene, E., P. Gros, A. Forget, P.A.L. Kongshavn, C. St. Charles, and B.A. Taylor. 1982. Genetic regulation of resistance to intracellular pathogens. Nature 297:506–9.

948. Skow, L.C., and M.E. Donner. 1985. The locus encoding αA-crystallin is closely linked to H-2K on mouse chromosome 17. Genetics 110:723–32.

949. Skow, L.C. 1978. Genetic variation at a locus (TAM-1) for submaxillary gland protease in the mouse and its location on chromosome 7. Genetics 90:713–24.

950. Skow, L.C. 1981. Genetic variation for prolidase (PEP-4) in the mouse maps near the gene for glucosephosphate isomerase (GPI-1) on chromosome 7. Biochem. Genet. 19:695–700.

951. Skow, L.C. 1982. Location of a gene controlling electrophoretic variation in mouse γ-crystallins. Exp. Eye Res. 34:509–16.

952. Skow, L.C. 1979. Mouse News Lett. 61:55.

953. Skow, L.C. 1983. Mouse News Lett. 69:46–7.

954. Skow, L.C. 1983. Mouse News Lett. 69:47.

955. Smith, M.S., R.E. Click, and P.G.W. Plagemann. 1984. Control of mouse hepatitis virus replication in macrophages by a recessive gene on chromosome 7. J. Immunol. 133:428–32.

956. Snell, G.D., and H.P. Bunker. 1964. Histocompatibility genes of mice. IV. The position of *H-3* in the fifth linkage group. Transplantation 2:743–51.

957. Snell, G.D., and H.P. Bunker. 1967. Mouse News Lett. 37:34.

958. Snell, G.D., S. Counce, B. Smith, L. Dube, and D. Kelton. 1955. A 5th chromosome histocompatibility locus identified in the mouse by tumor transplantation. Proc. Am. Assoc. Cancer Res. 2:46.

959. Snell, G.D., G. Cudkowicz, and H.P. Bunker. 1967. Histocompatibility genes of mice. VII. H13, a new histocompatibility locus in the fifth linkage group. Transplantation 5:492–503.

960. Snell, G.D., M.M. Dickie, P. Smith, and D.E. Kelton. 1954. Linkage of loop-tail, leaden, splotch and fuzzy in the mouse. Heredity 8:271–4.

961. Snell, G.D., and L.W. Law. 1939. A linkage between shaker-2 and wavy-2 in the house mouse. J. Hered. 30:447.

962. Snell, G.D., and L.C. Stevens. 1961. Histocompatibility genes of mice. III. *H-1* and *H-4*, two histocompatibility loci in the first linkage group. Immunology 4:366–79.

963. Snell, G.D. 1928. A cross-over between the genes for short-ear and density in the house mouse. Proc. Natl. Acad. Sci. USA 14:926–8.

964. Snell, G.D. 1955. Ducky, a new second chromosome mutation in the mouse. J. Hered. 46:27–9.

965. Snell, G.D. 1958. Histocompatibility genes of the mouse. II. Production and analysis of isogenic resistant lines. J. Natl. Cancer Inst. 21:843–77.

966. Snell, G.D. 1931. Inheritance in the house mouse, the linkage relations of short-ear, hairless, and naked. Genetics 16:42–74.

967. Snell, G.D. 1945. Linkage of jittery and waltzing in the mouse. J. Hered. 36:279–80.

968. Snell, G.D. 1952. Preliminary data on crossing-over between *H-2*, *Fu*, *Ki* and *T* in the mouse. Heredity 6:247–54.

969. Snyder, F. 1981. Regulation of purine metabolism. Mouse News Lett. 65:10.

970. Soares, E.R. 1978. Mouse News Lett. 59:12.

971. Sorokina, J.D., and Z.K. Blandova. 1985. Mouse News Lett. 73:23.

972. Southard, J., and M.C. Green. 1971. Mouse News Lett. 45:29.

973. Southard, J. 1981. Mouse News Lett. 64:60.

974. Southard, J.L., and E.M. Eicher. 1977. Mouse News Lett. 57:20.

975. Southard, J.L. 1970. Mouse News Lett. 42:30.

976. Southard, J.L. 1970. Mouse News Lett. 43:33.

977. Southard, J.L. 1971. Mouse News Lett. 44:30.

978. Southard, J.L. 1971. Mouse News Lett. 45:29.

979. Southard, J.L. 1972. Mouse News Lett. 46:31. Plus Southard, J.L. 1973. Personal communication.

980. Southard, J.L. 1973. Mouse News Lett. 49:32.

981. Southard, J.L. 1971. Mouse News Lett. 45:29.

982. Southard, J.L. 1971. Personal communication.

983. Spurll, G.M., and F.L. Owen. 1981. A family of T-cell alloantigens linked to *Igh-1*. Nature 293:742–4.

984. St Amand, W., and M.B. Cupp. 1957. Mouse News Lett. 16:37.

985. St Amand, W., and M.B. Cupp. 1957. Mouse News Lett. 17:88.

986. St Amand, W., and M.B. Cupp. 1958. Mouse News Lett. 19:38.

987. Staeheli, P., D. Pravtcheva, L-G. Lundin, M. Acklin, F. Ruddle, J. Lindenmann, and O. Haller. 1986. Interferon-regulated influenza virus resistance gene *Mx* is localized on mouse chromosome 16. J. Virol. 58:967–9.

988. Stern, R.H., E.S. Russell, and B.A. Taylor. 1976. Strain distribution and linkage relationship of a mouse embryonic hemoglobin variant. Biochem. Genet. 14:373–81.

989. Stevens, L.C., and J.A. Mackensen. 1961. Mouse News Lett. 24:41.

990. Stewart, C., K. Harbers, D. Jähner, and R. Jaenisch. 1983. X chromosome linked transmission and expression of retroviral genomes microinjected into mouse zygotes. Science 221:760–2.

991. Stieler, C., and W.F. Hollander. 1972. Location of the velvet coat mutant in linkage group VI of the mouse. J. Hered. 63:212–4.

992. Stockert, E., H. Sato, K. Itakura, E.A. Boyse, L.J. Old, and J.J. Hutton. 1972. Location of the second gene required for expression of the leukemia-associated mouse antigen G(IX). Science 178:862–3.

993. Subbarao, B., A. Ahmed, W.E. Paul, I. Scher, R. Liebermann, and D.E. Mosier. 1979. *Lyb-7*, a new B cell alloantigen controlled by genes linked to the IgC(H) locus. J. Immunol. 122:2279–85.

994. Sutton, V.R., P.M. Hogarth, and I.F.C. McKenzie. 1983. Description of a new Qa antigenic specificity, 'Qa-m9,' whose expression is under complex genetic control. J. Immunol. 131:1363–7.

995. Swan, D., P. D'Eustachio, J. Leinwand, J. Seidman, D. Keithley, and F.H. Ruddle. 1979. Chromosomal assignment of the mouse kappa light chain genes. Proc. Natl. Acad. Sci. USA 76:2735–9.

996. Swan, D., M. Oskarsson, D. Keithley, F.H. Ruddle, P. D'Eustachio, and G.F. van de Wonde. 1982. Chromosomal localization of the Moloney sarcoma virus mouse cellular (c-*mos*) sequence. J. Virol. 44:752–4.

997. Swank, R.T., K. Paigen, and R.E. Ganschow. 1973. Genetic control of glucuronidase induction in mice. J. Mol. Biol. 81:225–43.

998. Sweet, H.O., and M.C. Green. 1977. Mouse News Lett. 57:21.

999. Sweet, H.O., and M.C. Green. 1981. Progressive anky-

losis, a new skeletal mutation in the mouse. J. Hered. 72:87–93.

1000. Sweet, H.O., and P.L. Lane. 1977. Mouse News Lett. 57:20 refs. RF. chapter.

1001. Sweet, H.O., and P.W. Lane. 1971. Mouse News Lett. 45:29.

1002. Sweet, H.O., and P.W. Lane. 1979. Mouse News Lett. 59:27.

1003. Sweet, H.O., and P.W. Lane. 1979. Mouse News Lett. 60:50.

1004. Sweet, H.O., and P.W. Lane. 1980. X-linked polydactyly (*Xpl*), a new mutation in the mouse. J. Hered. 71:207-9. Also Sweet, H.O. 1979. X-linked polydactyly (Xpl). Mouse News Lett. 61:42.

1005. Sweet, H.O., S.L. Langley, and M.C. Green. 1982. Mouse News Lett. 66:66.

1006. Sweet, H.O., and M. Prochazka. 1985. Mouse News Lett. 73:18.

1007. Sweet, H.O., T.H. Roderick, and M.T. Davisson. 1979. Mouse News Lett. 60:51.

1008. Sweet, H.O. 1983. Dilute suppressor, a new suppressor gene in the house mouse. J. Hered. 74:305–6.

1009. Sweet, H.O. 1980. Head tilt (het). Mouse News Lett. 63:19. Plus Sweet, H.O. 1981. Personal communication.

1010. Sweet, H.O. 1982. Mouse News Lett. 66:66.

1011. Sweet, H.O. 1987. Personal communication.

1012. Sweet, H.O. 1985. Mouse News Lett. 73:18.

1013. Sweet, H.O. 1983. Mouse News Lett. 68:72.

1014. Sweet, H.O. 1986. Twitcher (twi) is on chr 12. Mouse News Lett. 75:30.

1015. Sweet, H.O. 1984. Wasted linkage. Mouse News Lett. 71:31.

1016. Symington, F.W., B. Subbarao, D.E. Mosier, and J. Sprent. 1982. Lyb-8.2: a new B cell antigen defined and characterized with a monoclonal antibody. Immunogenetics 16:381–91.

1017. Szymura, J.M., and J. Klein. 1981. Linkage of a gene controlling urinary pepsinogen with the major histocompatibility complex of the mouse. Immunogenetics 13:267–71.

1018. Szymura, J.M., and J. Klein. 1982. Mapping of the mouse *Upg-1* locus. Immunogenetics 16:89–90. Plus Szymura, J.M. 1982. Personal communication.

1019. Szymura, J.M., and J. Klein. 1981. Mouse News Lett. 64:93–4.

1020. Szymura, J.M., and J. Klein. 1981. Mouse News Lett. 65:48.

1021. Szymura, J.M., B.A. Taylor, and J. Klein. 1982. *Upg-2*: a urinary pepsinogen variant located on chromosome 1 of the mouse. Biochem. Genet. 20:1211–9.

1022. Szymura, J.M., M.R. Wabl, and J. Klein. 1981. Mouse News Lett. 65:48.

1023. Szymura, J.M., M.R. Wabl, and J. Klein. 1981. Mouse mitochondrial superoxide dismutase locus is on chromosome 17. Immunogenetics 14:231–40.

1024. Tada, N., S. Kimura, and U. Hammerling. 1982. Immunogenetics of mouse B-cell alloantigen systems defined by monoclonal antibodies and gene cluster formation of these loci. Immunol. Rev. 69:99–126.

1025. Tada, N., S. Kimura, A. Hatzfeld, and U. Hammerling. 1980. Ly-m11: the H-3 region of mouse chromosome 2 controls a new surface alloantigen. Immunogenetics 11:441–9.

1026. Tada, N., S. Kimura, Y. Liu-Lam, and U. Hammerling. 1984. A new lymphocyte surface antigen (Ly-m31) controlled by a gene closely linked to the *Akp-2* locus on mouse chromosome 4. Immunogenetics 20:589–92.

1027. Tada, N., S. Kimura, Y. Liu-Lam, and U. Hammerling. 1983. Mouse alloantigen system Ly-m22 predominantly expressed on T lymphocytes and controlled by a gene linked to *Mls* region on chromosome 1. Hybridoma 2:29–38.

1027a. Tada, N., N. Tamaoki, S. Ikegami, S. Nakamura, K. Dairiki, and T. Fujimori. 1987. The *Lyb-2– Ly-19* region of mouse chromosome 4 controls a new cell-surface alloantigen: Ly-32. Immunogenetics 25:269–70.

1028. Taggart, R.T., P. Tetri, and U. Francke. 1980. Assignment of the gene for NADH diaphorase Dia-1 to mouse chromosome 15. Somat. Cell Genet. 6:769–76.

1029. Takahashi, S., Y. Fukuoka, K. Moriwaki, T. Okuda, T. Tachibana, S. Natsuume-Sakai, and M. Takahashi. 1984. Structural polymorphism of mouse complement C2 detected by microscale peptide mapping: linkage to *H-2*. Immunogenetics 19:493–501.

1030. Takahashi, S., M. Takahashi, T. Kaidoh, S. Natsuume-Sakai, and T. Takahashi. 1984. Genetic mapping of mouse C4-bp locus to the H-2D-Qa interval. J. Immunol. 132:6–8.

1031. Taylor, B.A., D.W. Barley, M. Cherry, R. Riblet, and M. Weigert. 1975. Genes for immunoglobulin heavy chain and serum prealbumin protein are linked in the mouse. Nature 256:644–6.

1032. Taylor, B.A., H.G. Bedigian, and H. Meier. 1977. Genetic studies of the *Fv-1* locus of mice: linkage with *Gpd-1* in recombinant inbred lines. J. Virol. 23:106–9.

1033. Taylor, B.A., H. Meier, A. Macpike, and M. Williams. 1978. Mouse News Lett. 59:25.

1034. Taylor, B.A., and H. Meier. 1975. Mapping the adrenal lipid gene of the AKR/J mouse strain. Genet. Res. 26:307–12.

1035. Taylor, B.A., and A.D. O'Brien. 1982. Position on mouse chromosome 1 of a gene that controls resistance to Salmonella typhimurium. Infect. Immun. 36:1257–60.

1036. Taylor, B.A., L. Rowe, D.M. Gibson, R. Riblet, R. Yetter, and P.D. Gottlieb. 1985. Linkage of a 7S RNA sequence and kappa light chain genes in the mouse. Immunogenetics 22:471–81.

1037. Taylor, B.A., L. Rowe, N.A. Jenkins, and N.G. Copeland. 1985. Chromosomal assignment of two endogenous ecotropic murine leukemia virus proviruses of the AKR/J mouse strain. J. Virol. 56:172–5.

1038. Taylor, B.A., and L. Rowe. 1984. Genes for serum

amyloid A proteins map to chromosome 7 in the mouse. Mol. Gen. Genet. 195:491–9.

1039. Taylor, B.A., and L. Rowe. 1984-85. Mapping the murine homology of the Src oncogene and a related DNA sequence. Jackson Laboratory Annual Report, p. 45.

1040. Taylor, B.A., and L. Rowe. 1987. The congenital goiter mutation is linked to the thyroglobulin gene in the mouse. Proc. Natl. Acad. Sci. USA 84:1986–90.

1041. Taylor, B.A., and F-W. Shen. 1977. Location of *Lyb-2* on chromosome 4: evidence from recombinant inbred strains. Immunogenetics 4:597–9.

1042. Taylor, B.A., D.M. Walls, and M.J. Wimsatt. 1987. Localization of the inosine triphosphatase locus (*Itp*) on chromosome 2 of the mouse. Biochem. Genet. 25:267–74.

1043. Taylor, B.A. 1976. Linkage of the cadmium resistance locus to loci on mouse chromosome 12. J. Hered. 67:389–90.

1044. Taylor, B.A. 1975. Mouse News Lett. 53:35.

1045. Taylor, B.A. 1976. Epistatic circling gene of C57L/J. Mouse News Lett. 55:17.

1046. Taylor, B.A. 1979. Personal communication.

1047. Taylor, B.A. 1985. Personal communication.

1048. Taylor, B.A. 1987. Personal communication.

1049. Taylor, B.A. 1982. Mouse News Lett. 67:22.

1050. Taylor, B.A. 1981. Mouse News Lett. 65:28.

1051. Taylor, B.A. 1982. Mouse News Lett. 66:66.

1052. Teicher, L.S., and E.W. Caspari. 1978. The genetics of blind — a lethal factor in mice. J. Hered. 69:86–90.

1053. Theiler, K., D.S. Varnum, and L.C. Stevens. 1978. Development of Dickie's small eye, a mutation in the house mouse. Anat. Embryol. 155:81–6.

1054. Timms, G.P., and R.S. Holmes. 1981. Genetics of aldehyde dehydrogenase isozymes in the mouse: evidence for multiple loci and localization of *Ahd-2* on chromosome 19. Genetics 97:327–36.

1056. Timms, G.P., and R.S. Holmes. 1982. Cytoplasmic aldehyde dehydrogenase (*Ahd-2* on chromosome 19). Mouse News Lett. 66:60.

1057. Traina, V.L., B.A. Taylor, and J.C. Cohen. 1981. Genetic mapping of endogenous mouse mammary tumor viruses: locus characterization, segregation, and chromosomal distribution. J. Virol. 40:735–44.

1058. Tsuji, S., and H. Meier. 1969. Linkage of serum esterase and tottering in the mouse. J. Hered. 60:221–2.

1059. Tulchin, N., and B.A. Taylor. 1981. γ-Glutamyl cyclotransferase: a new genetic polymorphism in the mouse (*Mus musculus*) linked to *Lyt-2*. Genetics 99:109–16.

1060. Uhler, M., E. Herbert, P. D'Eustachio, and F.D. Ruddle. 1983. The mouse genome contains two nonallelic pro-opiomelanocortin genes. J. Biol. Chem. 258:9444–53.

1061. Uiterdijk, H.G., B.A.J. Ponder, M.F.W. Festing, J. Hilgers, L. Skow, and R. Van Nie. 1986. The gene controlling the binding sites of Dolichos biflorus agglutinin, *Dlb-1*, is on chromosome 11 of the mouse. Genet. Res. 47:125–9.

1062. van Abeelen, J.H.F., and P.H.W. van der Kroon. 1967. Nijmegen waltzer — a new neurological mutant in the mouse. Genet. Res. 10:117–8.

1063. Varnum, D.S., and S.C. Fox. 1981. Head blebs: a new mutation on chromosome 4 of the mouse. J. Hered. 72:293.

1064. Varnum, D.S., and L.C. Stevens. 1975. Mouse News Lett. 53:35.

1065. Varnum, D.S., and L.C. Stevens. 1974. Rachiterata: a new skeletal mutation on chromosome 2 of the mouse. J. Hered. 65:91–3.

1066. Varnum, D.S. 1983. Blind-sterile: a new mutation on chromosome 2 of the house mouse. J. Hered. 74:206–7.

1067. Verstraeten, A.A., and R. van Nie. 1978. Genetic transmission of mammary tumor virus in the DBAf mouse strain. Int. J. Cancer 21:473–5.

1068. Vijaya, S., D.L. Steffen, C. Kozak, and H.L. Robinson. 1987. Dsi-l, a region with frequent proviral insertions in Moloney murine leukemia virus-induced rat thymomas. J. Virol. 61:1164–70.

1068a. Wallace, M.E. 1979. Mapping with mimic genes in the mouse. J. Hered. 70:216.

1069. Wallace, M.E., and J. Ferguson. 1984. Linkage for bald-arthritic, Bda. Mouse News Lett. 71:19.

1070. Wallace, M.E., and P.P. Green. 1974. Mouse News Lett. 50:29.

1071. Wallace, M.E., and C. Lowe. 1984. Tibia-less, a new Peru mutant. Mouse News Lett 71:18.

1072. Wallace, M.E., and F.J. MacSwiney. 1975. Warfarin resistance and a new gene for obesity. Mouse News Lett. 53:20.

1073. Wallace, M.E., and F.M. MacSwiney. 1979. An inherited mild middle-aged adiposity in wild mice. J. Hyg. 82:309–17.

1074. Wallace, M.E., and S.A. Mallyon. 1972. Unusual recombination values and the mapping of the lethal minature in the house mouse. Genet. Res. 20:257–62.

1075. Wallace, M.E. 1957. A balanced three-point experiment for linkage group V of the house mouse. Heredity 11:223–58.

1076. Wallace, M.E. 1950. Locus of the gene "fidget" in the house mouse. Nature 166:407.

1077. Wallace, M.E. 1963. Mouse News Lett. 29:22.

1978. Wallace, M.E. 1967. Personal communication.

1079. Wallace, M.E. 1972. Reduction in recombination due to a deletion in a colony of wild mice. J. Hered. 63:297–300.

1080. Washburn, L., and E.M. Eicher. 1986. A mutation at the dreher locus, (dr^{2J}). Mouse News Lett. 75:28–9.

1081. Washburn, L.L., and E.M. Eicher. 1983. Sex reversal in XY mice caused by dominant mutation on chromosome 17. Nature 303:338–40.

1082. Wassmer, B., S.M. de Looze, and O. von Deimling. 1985. Biochemistry and genetics of esterase-20 (ES-20), a second trimeric carboxylesterase of the house mouse (*Mus musculus*). II. A unique recombination reveals

ES-20 as a hybrid enzyme. Biochem. Genet. 23:759-70. Also de Looze S. 1985. Mouse News Lett. 72:101.

1083. Wassmer, B., and O. von Deimling. 1983. Mouse News Lett. 69:20.

1084. Watanabe, T., N. Ogasawara, and H. Goto. 1976. Genetic study of pancreatic proteinase in mice (*Mus musculus*): linkage of the *Prt-2* locus on chromosome 8. Biochem. Genet. 14:999–1002.

1085. Watson, D.K., M.J. McWilliams-Smith, C.F. Kozak, R. Reeves, J. Gearhart, M.F. Nunn, W. Nash, J.R. Fowle III, P. Duesberg, T.S. Papas, and S.J. O'Brien. 1986. Conserved chromosomal positions of dual domains of the ets protoncogene in cats, mice and humans. Proc. Natl. Acad. Sci. USA 83:1792–6.

1086. Watson, J., K. Kelley, M. Largen, and B.A. Taylor. 1978. The genetic mapping of a defective LPS response gene in C3H/HeJ mice. J. Immunol. 120:422–4.

1087. Watson, J., M. Largen, and K.P.W. McAdam. 1978. Genetic control of endotoxic response in mice. J. Exp. Med. 147:39–49.

1088. Watson, J.G., T.J. Higgins, P.B. Collins, and S. Chaykin. 1972. The mouse liver aldehyde oxidase locus (*Aox*). Biochem. Genet. 6:195–204.

1089. Weigert, M., and R. Riblet. 1978. The genetic control of antibody variable regions of the mouse. Springer Semin. Immunopathol. 1:133–69.

1090. Weimar, W.R., P.W. Lane, and R.L. Sidman. 1982. "Vibrator" (*vb*): autosomal recessive inheritance of spongiform spinocerebellar degeneration in mice. Brain Res. 251:357–64.

1091. Wejman, J.C., B.A. Taylor, N.A. Jenkins, and N.G. Copeland. 1985. Endogenous xenotropic murine leukemia virus-related sequences map to chromosomal regions encoding mouse lymphocyte antigens. J. Virol. 50:237–47.

1092. Welshons, W.J., and L.B. Russell. 1959. The Y-chromosome as the bearer of male determining factors in the mouse. Proc. Natl. Acad. Sci. USA 45:560–6.

1093. West, J.D., and G. Fisher. 1986. Further experience of the mouse dominant cataract mutation test from an experiment with ethylnitrosourea. Mutat. Res. 164:127–36.

1094. Weydert, A., P. Daubas, I. Lazaridis, P. Barton, I. Garner, D.P. Leader, F. Bonhomme, J. Catalan, D. Simon, J-L. Guénet, F. Gros, and M.E. Buckingham. 1985. Genes for skeletal muscle myosin heavy chains are clustered and are not located on the same mouse chromosome as a cardiac myosin heavy chain gene. Proc. Natl. Acad. Sci. USA 82:7183–7.

1095. White, P.C., D.D. Chaplin, J.H. Weis, B. Dupont, M.I. New, and J.G. Seidman. 1984. Two steroid 21-hydroxylase genes are located in the murine S region. Nature 312:465–7.

1095a. Whitmore, A.C., G.F. Babcock, and G. Haughton. 1978 Genetic control of susceptibility of mice to rous sarcoma virus tumorigenesis. II. Segregation analysis

of strain A.SW-associated resistance to primary tumor induction. J. Immunol. 121:213–20.

1095b. Whitmore, S. 1986. Mouse News Lett. 74:107.

1096. Whitney, J.B., III., and E.S. Russell. 1980. Linkage of genes for adult α-globin and embryonic α-like globin chains. Proc. Natl. Acad. Sci. USA 77:1087–90.

1097. Whitney, J.B., and T.H. Roderick. 1978. Mouse News Lett. 58:49.

1098. Wiedemann, L.M., P. D'Eustachio, D.E. Kelley, and R.P. Perry. 1987. Three functional ribosomal protein genes are unlinked in mouse genome. Somat. Cell Mol. Genet. 13:77–80.

1099. Wilcox, F., J.E. Womack, and T.H. Roderick. 1979. Linkage of the locus for conversion of albumin (*Acf-1*) in the house mouse, *Mus musculus*. Biochem. Genet. 17:159–62.

1100. Wilcox, F.H., L. Hirschhorn, B.A. Taylor, J.E. Womack, and T.H. Roderick. 1979. Genetic variation in alkaline phosphatase of the house mouse (*Mus musculus*) with emphasis on a manganese requiring isozyme. Biochem. Genet. 17:1093–107.

1101. Wilcox, F.H., and T.H. Roderick. 1982. Location on chromosome 6 of the locus for a major liver protein (*Lvp-1*) of the house mouse. Genet. Res. 40:213–5.

1102. Wilcox, F.H., and B.A. Taylor. 1981. Genetics of the Akp-2 locus for alkaline phosphatase of liver, kidney, bone, and placenta in the mouse. J. Hered. 72:387–90.

1103. Wilcox, F.H., J.E. Womack, and T.H. Roderick. 1976. Genetic variation in alkaline phosphatase isozymes in liver of the house mouse. Genetics 83:s82.

1104. Wilcox, F.H. 1975. Genetic variation in plasma prealbumin of the house mouse. J. Hered. 66:19–22.

1105. Wilcox, F.H. 1983. Genetics of alkaline phosphatase of the small intestine of the house mouse. Biochem. Genet. 21:641–52.

1106. Wilson, C.M., E.G. Erdos, J.D. Wilson, and B.A. Taylor. 1978. Location on chromosome 1 of *Rnr*, a gene that regulates renin in the submaxillary gland of the mouse. Proc. Natl. Acad. Sci. USA 75:5623–6.

1108. Wilson, C.M. 1984. Mouse News Lett. 71:51.

1109. Wilson, D.E., D. Woodard, A. Sandler, J. Erickson, and A. Gurney. 1987. Provisional assignment of the gene for uridine monophosphatase-2 (*Umph-2*) to mouse chromosome 11. Biochem. Genet. 25:1–6.

1109a. Winchester, G., N.A. Mitchison, and B.A. Taylor. 1987. The structural gene for F liver protein (*Flp*) maps to chromosome 5 of the mouse. Immunogenetics 26:356–8.

1110. Winking, H., and L.M. Silver. 1984. Characterization of a recombinant mouse *T* haplotype that expresses a dominant lethal maternal effect. Genetics 108:1013–20.

1111. Wirschubsky, Z., S. Ingvarsson, A. Carstenssen, F. Wiener, G. Klein, and J. Sumego. 1985. Gene localization on sorted chromosomes: definitive evidence on the relative positioning of genes participating in the mouse plasmacytoma-associated typical translocation. Proc. Natl. Acad. Sci. USA 82:6975–9.

1112. Wolfe, H.G., E.S. Russell, and S.O. Packer. 1963. Hemoglobins in mice. Segregation of genetic factors affecting electrophoretic mobility and solubility. J. Hered. 54:107–12.

1113. Wolfe, H.G. 1963. Mouse News Lett. 29:40.

1114. Womack, J.E., and P.S. Ashley. 1984. Mouse News Lett. 70:100.

1114a. Womack, J.E., S. Ashley, L.B. Barnett, and S.E. Lewis. 1986. Linkage of *Pep-2* and *Apk* on mouse chromosome 10. Biochem. Genet. 24:721–7.

1115. Womack, J.E., and S.B. Aucrbach. 1978. An acid phosphatase locus expressed in mouse kidney (*Apk*) and its genetic location. Biochem. Genet. 16:239–45.

1116. Womack, J.E., and C.S. David. 1982. Mouse gene for neuraminidase activity (*Neu-1*) maps to the *D* end of *H-2*. Immunogenetics 16:177–80.

1117. Womack, J.E., M.T. Davisson, E.M. Eicher, and D.A. Kendall. 1977. Mapping of nucleoside phosphorylase (*Np-1*) and esterase 10 (*Es-10*) on mouse chromosome 14. Biochem. Genet. 15:347–55.

1118. Womack, J.E., and E.M. Eicher. 1977. Liver specific lysosomal acid phosphatase deficiency (*Apl*) on mouse chromosome 17. Mol. Gen. Genet. 155:315–7.

1119. Womack, J.E., N.L. Hawes, E.R. Soares, and T.H. Roderick. 1975. Mitochondrial malate dehydrogenase (*Mor-1*) in the mouse: linkage to chromosome 5 markers. Biochem. Genet. 13:519–25.

1120. Womack, J.E., M.A. Lynes, and B.A. Taylor. 1975. Genetic variation of an intestinal leucine arylaminopeptidase (*Lap-1*) in the mouse and its location on chromosome 9. Biochem. Genet. 13:511–8.

1121. Womack, J.E., B.A. Taylor, and J.E. Barton. 1978. Esterase 13, a new mouse esterase locus with recessive expression and its genetic location on chromosome 9. Biochem. Genet. 16:1107–12.

1122. Womack, J.E. 1975. Esterase-6 (*Es-6*) in laboratory mice: hormone influenced expression and linkage relationship to oligosyntactylism (*Os*), esterase-1 (*Es-1*), and esterase-2 (*Es-2*) in chromosome 8. Biochem. Genet. 13:311–22.

1123. Womack, J.E. 1976. Mouse News Lett. 55:18.

1124. Wood, A.W., and B.A. Taylor. 1979. Genetic regulation of coumarin hydroxylase activity in mice. Evidence for single locus control on chromosome 7. J. Biol. Chem. 254:5647–51.

1125. Woodworth, E.F. 1960. Personal communication.

1126. Woolley, G.W. 1945. Misty dilution in the mouse. J. Hered. 36:269–70.

1127. Wright, A.F., G. Bulfield, S.M. Arfin, and H. Kacser. 1982. Comparison of the properties of histidine ammonia-lyase in normal and histidinemic mutant mice. Biochem. Genet. 20:245–64.

1128. Wright, M.E. 1947. Two sex-linkages in the house mouse with unusual recombination values. Heredity 1:349–54.

1129. Wright, M.E. 1947. Undulated: a new genetic factor in *Mus musculus* affecting the spine and tail. Heredity 1:137–41.

1130. Yagi, M., P. D'Eustachio, F.H. Ruddle, and M.E. Koshland. 1982. J chain is encoded by a single gene unlinked to other immunoglobulin structural genes. J. Exp. Med. 155:647–54.

1131. Yamamoto, K., G. Floyd-Smith, U. Francke, N. Koch, W. Lauer, B. Dobberstein, R. Schafer, and G.J. Hammerling. 1985. The gene encoding the Ia associated invariant chain is located on chromosome 18 in the mouse. Immunogenetics 21:83–90.

1132. Yang-Feng, T.L., G. Floyd-Smith, M. Nemer, J. Drouin, and U. Francke. 1985. The pronatriodilatin gene is located on the distal short arm of human chromosome 1 and on mouse chromosome 4. Am. J. Hum. Genet. 37:1117–28.

1133. Yang-Feng, T.L., N.R. Landau, D. Baltimore, and U. Francke. 1986. The terminal deoxynucleotidyltransferase gene is located on human chromosome 10 (10q23–q24) and on mouse chromosome 19. Cytogenet. Cell Genet. 43:121–6.

1134. Yetter, R.A., J.A. Hartley, and H.C. Morse III. 1983. *H-2*-linked regulation of xenotropic murine leukemia virus expression. Proc. Natl. Acad. Sci. USA 80:505–9.

1135. Yoon, C.H., and E.P. Les. 1957. Quivering, a new first chromosome mutation in mice. J. Hered. 48:176–80.

1136. Yoon, C.H. 1969. Disturbances in developmental pathways leading to a neurological disorder of genetic origin, "leaner", in mice. Dev. Biol. 20:158–81.

1137. Yoon, C.H. 1961. Linkage relationship of the waddler gene in mice with evidence for temperature effect on crossing over. J. Hered. 52:279–81.

1138. Zabel, B.U., R.L. Eddy, P.A. Lalley, J. Scott, G.I. Bell, and T.B. Shows. 1985. Chromosomal locations of the human and mouse genes for precursors of epidermal growth factor and the β subunit of nerve growth factor. Proc. Natl. Acad. Sci. USA 82:469–73.

1139. Zabel, B.U., R.E. Fournier, P.A. Lalley, S.L. Naylor, and A.Y. Sakaguchi. 1984. Cellular homologs of the avian erythroblastosis virus erb-A and erb-B genes are syntenic in mouse but asyntenic in man. Proc. Natl. Acad. Sci. USA 81:4874–8.

1140. Zabel, B.U., A.Y. Sakaguchi, P.A. Lalley, J. Scott, and S.L. Naylor. 1984. The nerve growth factor (*Ngf*) gene is on mouse chromosome 3. Cytogenet. Cell Genet. 37:614 (Abstr.).

1141. Zaleski, M., and J. Klein. 1974. Immune response of mice to Thy-1.1 antigen: genetic control by alleles at the *Ir-5* locus loosely linked to the H-2 complex. J. Immunol. 113:1170–7.

1142. Zweidler, A. 1984. Core histone variants of the mouse: primary structure and differential expression. *In* G.S. Stein, J.L. Stein, and W. Marzluff, eds., Histone Genes, 339–71. John Wiley & Sons, New York.

1143. daSilva, F.P., G.F. Hoecker, N.K. Day, K. Vienne,

504

and P. Rubinstein. 1978. Murine complement component 3: genetic variation and linkage to *H-2*. Proc. Natl. Acad. Sci. USA 75:963–5.

1145. van Nie, R., D. Ivanyi, and P. Demant. 1978. A new *H-2*-linked mutation, *rds*, causing retinal degeneration in the mouse. Tissue Antigens 12:106–8.

1146. van Nie, R., and A.A. Verstraeten. 1975. Studies of genetic transmission of mammary tumor virus by C3Hf mice. Int. J. Cancer 16:922–31.

1147. van de Meugheuvel, W., G. van Seventer, and P. Demant. 1985. A new Tla region antigen Qa-11, similar to Qa-2 and associated with B type beta-2-microglobulin. J. Immunol. 134:2507–12.

1148. van den Elsen, P., G. Bruns, D.S. Gerhard, D. Pravtcheva, C. Jones, D. Housman, F.H. Ruddle, S. Orkin, and C. Terhorst. 1985. Assignment of the gene coding for the T3-delta subunit of the T3-T-cell receptor complex to the long arm of human chromosome 11 and to mouse chromosome 9. Proc. Natl. Acad. Sci. USA 82:2920–4.

1149. von Deimling, O., J. Peters, and S. Povey. 1984. Linkage of biliverdin reductase, Blvr. Mouse News Lett. 70:81.

1150. von Deimling, O. 1985. Mouse News Lett. 73:15.

1151. von Deimling, O. 1985. Mouse News Lett. 73:15.

1152. von Deimling, O. 1986. Mouse News Lett. 75:25.

1153. von Deimling, O. 1986. Mouse News Lett. 74:89.

1154. von Deimling, O. 1982. Mouse News Lett. 67:16.

1154a. von Deimling, O.H., and J. Hilgers. 1982. ES-24. Mouse News Lett. 67:16.

1155. von Deimling, O.H., J. Otto, and A.B. Reske-Kunz. 1982. *Esr*, a second locus in the house mouse controlling esterase-5. Biochem. Genet. 20:351–8.

1156. von Deimling, O.H., P. Schupp, and J. Otto. 1981. Esterase-16 (*Es-16*): characterization, polymorphism and linkage to chromosome 3 of kidney esterase locus of the house mouse. Biochem. Genet. 19:1091–9.

1157. von Deimling, O.H., B. Wassmer, and M. Müller. 1984. Esterase-26 (ES-26): characterization and genetic location on chromosome 3 of an eserine-sensitive esterase of the house mouse (*Mus musculus*). Biochem. Genet. 22:1119–26.

1158. von Deimling, O.H. 1984. Esterase-23 (ES-23): characterization of a new carboxylesterase isozyme (EC 3.1.1.1) of the house mouse, genetically linked to ES-2 on chromosome 8. Biochem. Genet. 22:769–82.

505

7 LINKAGE AND SYNTENY HOMOLOGIES IN MOUSE AND MAN

JOSEPH H. NADEAU and ANDREW H. REINER

Criteria proposed by the Committee on Comparative Mapping (4) were used to identify homologous genes. Definitions proposed by the Committee on Comparative Mapping (4) and by Nadeau and Taylor (5) are followed here. The term *marker* designates homologous genes that have been assigned to specific chromosomes. Each gene is regarded as a marker for the chromosome or chromosomal segment on which it is located. Duplicated genes that are closely linked such as *Car-1* and *Car-2* were counted as a single marker. *Synteny conservation* is defined as the occurrence of two or more pairs of homologous markers on the same chromosome in two or more species; the markers need not occur in the same order. *Linkage conservation* is defined as conservation not only of synteny, but also of gene order. Symbols for mouse genes are defined in Mouse News Letter and in Chapter 2 of this book and for human genes in Human Gene Mapping (3). Searle *et al.* (6) provide a recent summary of comparative gene mapping data for mouse and man.

A total of 201 markers have been mapped in both mouse and man, 188 on autosomes and 13 on the X chromosome. A listing of the chromosomal location of these genes in mouse and man is provided in Table 7.1 and in Fig. 7.1. Autosomal markers consist of two groups, one group of 117 whose chromosomal locations have been determined in the mouse through identification of genetic variants and conventional gene mapping and another group of 71 that have been assigned to a specific chromosome but have not been precisely located within the associated linkage map.

Complete conservation of synteny was observed for markers on the X chromosome (Fig. 7.1). Despite complete synteny conservation, several chromosomal rearrangements that disrupt gene order have occurred since divergence of lineages leading to mouse and man. The most obvious is that the mouse X chromosome is acrocentric whereas the human X chromosome is metacentric. Inspection of the location of markers on the X chromosome of mouse and man (Fig. 7.1) reveals other examples of disrupted gene order (2, 7). As a result of

these rearrangements, there are only three examples of linkage conservation: *Hprt*, *Cf-9*, and *G6pd*; *Syn-1* and *spf*; *Pgk-1*, *Plp*, and *Ags* (Fig. 7.1). Other examples of linkage conservation for genes on the X chromosome are difficult to identify because chromosome arm assignments have not been determined for many of these genes in man (1).

Table 7.2 summarizes data for autosomal linkage conservation in mouse and man. There are 26 conserved autosomal segments involving 77 markers on 14 of the 19 autosomes. Segment lengths range from 1 cM for markers on mouse Chr 6 to 33 cM for markers on mouse Chr 8. Because recombination between *Np-1* (*Np-2*) and *Tcra* has not been detected, the length of the segment marked by *Np-1* (*Np-2*) and *Tcra* on Chr 14 is not known. (The 95 per cent upper confidence limit is 1.92 cM.) The mean empirical length of the 24 autosomal segments for which estimates of length are available is 11.0 cM (SD, 8.9 cM). There are five examples of linkage conservation across the centromere of an autosome in man.

Table 7.3 summarizes data for synteny conservation in mouse and man. The 117 autosomal markers define 52 homologous segments and the 71 markers that have been assigned to a specific chromosome but have not been precisely located within the associated linkage map define another 12 homologous segments. The minimum estimate of the total number of homologous autosomal segments is 65. There are nine examples of synteny conservation across the centromere, including the five examples of linkage conservation.

Table 7.4 summarizes data for genes whose order has not been conserved since divergence of lineages leading to mouse and man. These probably represent examples of linkage disruption. There are six examples involving three chromosomes. Three examples involve genes located on the same chromosome arm in man and two involve genes located on opposite arms. Although some chromosomes, e.g. Chr 9, have several segments representing linkage conservation without apparent disruption, other chromosomes, such as Chrs 4, 7, and

17, not only have several conserved segments but also have several apparently disrupted segments.

Nadeau and Taylor (5) proposed a number of measures of the extent of autosomal linkage conservation. We applied their methods to the present data and obtained the following results for autosomes:

1. The expected length (m) of each conserved segment was calculated by using eqn. 2 of Nadeau and Taylor (5) and is given in Table 7.2. Expected lengths are greater than the empirical distance (r) between the outermost markers because it is unlikely the markers for each conserved segment occur at the boundary of the segment. The expected length takes this problem into account. The mean of these expected lengths was 22.2 cM (SD, 15.2 cM).

2. We also estimated the mean and standard error of the lengths of all conserved autosomal segments in the genome. Estimates of the length of conserved segments given above are based on segments marked by two or more pairs of homologous markers; segments lacking identified markers and segments with a single identified marker were necessarily excluded from the analysis. Exclusion of these segments probably leads to an overestimate of the length of conserved segments. Equations 10 and 11 (5) resolve this problem. We estimate the mean length of all conserved autosomal segments in the genome to be 10.4 ± 2.4 cM.

3. The mean length of all conserved segments can be used to estimate the number of linkage disruptions that have occurred since divergence of lineages leading to mouse and man (Eqs. 15 and 17; Ref. 5). This number was 134 ± 35 disruptions. This corresponds to a rate of about 0.96 disruptions per million years (myr) and is equivalent to 0.48 reciprocal translocations per myr.

We maintain a Homology Database and Programs (HMDP) consisting of a database for mammalian gene homology data (cf. Appendix) and of programs for producing comparative maps (cf. Fig. 7.1). This database stores, retrieves, and sorts mammalian gene mapping information. Linkage and synteny data together with references are included for all mammals for which data are available. Each record in the database contains gene symbol, gene name, global symbol (the symbol that connects data files for each species), chromosome, regional location if known, and reference. Currently, the data files hold information for 20 species. The number of genes ranges from 24 for rat to over 200 for mouse. More than 700 references are included.

We welcome inquiries concerning comparative gene mapping data and information concerning errors of commission or omission.

Table 7.1 Chromosomal location of homologous markers in mouse and man. References for the chromosomal and linkage assignments for autosomal and X-linked genes in mouse and man are provided in the appendix to this chapter

Mouse			Human		
Gene	Chr	Reference	Gene	Chr	Reference
Aat	12	239, 550	PI	14q	122, 326, 476
Abl	2	223	ABL	9q	23, 139, 241, 242, 277
Aco-1	4	415	ACO1	9p	57, 103, 469, 585
Acp-1	12	196	ACP1	2p	181, 235, 236
Acp-2	2	221, 327	ACP2	11p	221, 450
Actb	5	119	ACTB	7	434
Actc	17	119	ACTC	15q	233
Acts	3	118, 119	ACTA	1	233
Acy-1	9	8, 414, 431	ACY1	3p	302, 431
Ada	2	330, 523	ADA	20q	37, 259, 395, 465, 558
Adh-1	3	41, 255	ADH1	4q	530
Adh-3	3	252, 253, 404	ADH3	4q	530
Adk	14	345	ADK	10q	304
Afp	5	124, 239	AFP	4q	237, 503, 567
Ags	X	200, 315, 455	GLA	Xq	189, 200
Ahd-1	4	254, 592	ALDH2	12	263
Ahd-2	19	555, 556	ALDH1	9q	263
Ak-1	2	187, 196	AK1	9q	4, 103, 165, 166, 469
Ak-2	4	334	AK2	1p	520
Akp-2	4	592	ALPL	1	541
Alb-1	5	124, 440	ALB	4q	237, 291, 567, 582
Amy-1	3	41, 249, 337, 442	AMY1	1p	408, 561, 614
Amy-2	3	249, 404, 442	AMY2	1p	408, 561, 614
Apoa-1	9	8, 168	APOA1	11q	292, 339, 508
Apoa-2	1	351, 353	APOA2	1	307, 325, 399
App	16	88	APP	21q	88
Aprt	8	315	APRT	16q	204, 338, 363, 557
As-1	13	132	ARSB	5	189, 244
As-2	15	197	ARSA	22q	197, 198
Asl	5	333	ASL	7	430
Ass	2	422	ASS	9q	70, 126
B2m	2	362, 385, 386, 488	B2M	15q	177, 225, 517
Bgl	9	34, 48, 78, 84, 431, 455	GLB1	3p	58, 431
Blvr	2	151	BLVR	7p	379, 456
C3	17	125, 205, 419, 424, 491	C3	19p	205, 269, 290, 515
Calc	7	221, 328, 384, 427	CALC	11p	215, 221, 260, 262, 274, 359, 384

Table 7.1—*cont.*

Mouse			Human			Mouse			Human		
Gene	Chr	Reference	Gene	Chr	Reference	Gene	Chr	Reference	Gene	Chr	Reference
Car-1	3	36, 172	*CA1*	8	63, 136	*Glo-1*	17	160, 382, 390	*GLO1*	6p	31, 234, 581
Car-2	3	172, 404	*CA2*	8	421, 573	*Got-1*	19	79, 146, 506	*GOT1*	10q	93, 94, 112, 243, 519
Cas-1	2	180, 250, 384	*CAT*	11p	214, 215, 262, 274, 287, 384, 423	*Got-2*	8	146	*GOT2*	16q	276, 559
						Gpi-1	7	147, 272	*GPI*	19q	54, 269, 290, 352
Cbs	17	412	*CBS*	21q	412						
Cf-9	X	411	*F9*	Xq	46, 76, 365	*Gpt-1*	15	86, 169	*GPT1*	8q	275, 298, 299, 378, 590
Coh	7	350, 603	*CYP1*	19q	135, 466	*Gr-1*	8	439	*GSR*	8p	140, 142
Cola-1	11	64, 406	*COL1A1*	17q	45, 98, 228, 266, 531	*Grl-1*	18	201	*GRL*	5q	212
						Gsta	9	117	*GST3*	11q	540
Cola-2	16	64	*COL1A2*	7q	43, 47, 228, 229, 266	*Gus*	5	455	*GUSB*	7q	32, 75, 91, 191, 230, 329
Cpa	6	258	*CPA*	7q	257						
Crbp	9	153	*RBP1*	3	100, 153	*H-2*	17	303	*HLA*	6p	3, 234, 335, 554
Crya-1	17	529	*CRYA1*	21q	472						
Cryg-1	1	349, 528	*CRYG*	2q	164, 349, 562	*Hba*	11	493, 524	*HBA*	16p	22, 49, 156, 157, 311
Cs	10	106	*CS*	12	111, 245, 568	*Hbb*	7	221, 272, 384, 493	*HBB*	11p	74, 155, 179, 214, 215, 221, 262, 274, 384, 403
Csfgm	11	12, 226	*GMCSF*	5q	265						
Ctrb	8	258, 576, 577	*CTRB*	16q	256, 499						
D21S16	16	88	*D21S16*	21q	88						
D21S52	16	88	*D21S52*	21q	88						
D21S56	17	88	*D21S56*	21q	88	*Hexb*	13	300	*HEXB*	5q	189, 219
D21S58	16	88	*D21S58*	21q	88	*Hk-1*	10	331	*HK1*	10	519
Dhfr	13	300	*DHFR*	5q	207	*Hox-1*	6	62, 373, 473	*HOX1*	7p	62, 67, 473
Dia-1	15	197, 546	*DIA1*	22q	182, 198, 261, 288	*Hox-2*	11	285, 405, 474	*HOX2*	17q	67, 285
						Hox-3	15	473	*HOX3*	12q	473
Egf	3	404, 613	*EGF*	4q	51, 402, 613	*Hp*	8	25, 26	*HP*	16q	204, 357, 489, 525
Ela-1	15	258	*ELA1*	12	256, 257						
Eno-1	4	120, 334	*ENO1*	1p	217, 377, 520	*Hprt*	X	12, 13, 82, 83, 199, 200, 355	*HPRT*	Xq	200, 454
Erba	11	497, 612	*ERBA1*	17q	496, 516, 536, 612						
Erbb	11	407, 524, 612	*ERBB*	7p	241, 496, 536, 538, 612	*Hras-1*	7	221, 384, 471, 497	*HRAS1*	11p	74, 179, 214, 215, 221, 241, 262, 274, 359, 384, 403, 447
Es-10	14	163, 458, 504, 601	*ESD*	13q	44, 85, 163, 569, 575						
Es-17	9	8, 41, 449	*ESA4*	11q	324, 375						
Ets-1	9	579	*ETS1*	11q	145, 154, 241, 579	*Hyp*	X	170	*HYP*	X	374
						Idh-1	1	272	*IDH1*	2q	190, 235, 236
Ets-2	16	88, 479, 480, 579	*ETS2*	21q	88, 241, 495, 579	*Idh-2*	7	53, 331	*IDH2*	15q	59, 231
Fes	7	39, 239	*FES*	15q	127, 240, 241, 242, 277	*Ifa*	4	66, 89, 130, 131, 148, 296, 348, 416, 571	*IFNA*	9p	154, 452, 453, 560
Fos	12	121	*FOS*	14q	18						
Fpgs	2	187	*FPGS*	9q	280	*Ifb*	4	130, 297, 416, 571	*IFNB*	9p	376, 452, 453, 560
Fshb	2	221	*FSHB*	11p	221, 222, 262, 578	*Ifg*	10	439	*IFNG*	12q	426, 560
Fuca	4	327	*FUCA1*	1p	68, 69, 188, 565	*Ifrc*	16	109	*IFNAR*	21q	77, 393, 547, 615
G6pd	X	12, 83, 361	*G6PD*	Xq	454	*Igh*	12	286, 550	*IGH*	14q	184
Galt	4	415	*GALT*	9p	2, 57, 394, 396, 409, 585	*Igk*	6	62, 218, 486, 544	*IGK*	2p	358, 372
Gapd	6	61	*GAPD*	12p	60, 245	*Igl*	16	122, 175, 193, 479	*IGL*	22q	6, 141, 176, 227
Gdc-1	15	251, 270, 319, 321, 383, 473	*GPD1*	12	375	*Ii*	18	138, 484, 605	*DHLAG*	5	99
						Ins-2	7	221, 332, 384	*INS*	11p	74, 179, 221, 238, 262, 274, 359, 384, 450
Ggc	6	62, 564	*GCTG*	7p	38						
Glk	11	316, 370, 391	*GALK*	17q	174						

Table 7.1—*cont.*

Mouse			Human			Mouse			Human		
Gene	Chr	Reference	Gene	Chr	Reference	Gene	Chr	Reference	Gene	Chr	Reference
Int-1	15	1, 383, 443, 457	*INT1*	12	444	*Pep-7*	5	331, 462	*PEPS*	4	40, 55, 509, 521
Int-2	7	457	*INT2*	11q	71	*Pfk-4*	15	10	*PFK4*	12	10
Itp	2	330, 523, 549	*ITPA*	20p	259, 380, 395	*Pgam-1*	19	208	*PGAMA*	10q	289
Kras-2	6	213, 486, 497	*KRAS2*	12p	241, 371, 447, 496, 500, 502	*Pgd*	4	334, 583	*PGD*	1p	284, 377, 586
						Pgk-1	X	83, 320	*PGK1*	Xq	580
Ldh-1	7	52, 221, 331, 445	*LDHA*	11p	194, 221, 262, 324, 607	*Pgk-2*	17	171, 572	*PGK2*	19	209, 210
						Pgm-1	5	272	*PGM2*	4	40, 534, 591
Ldh-2	6	52, 461	*LDHB*	12p	87, 245, 281, 283, 323	*Pgm-2*	4	80, 334	*PGM1*	1p	104, 284, 377, 435, 492, 520
Lhb	7	427, 432	*LHB*	19	432	*Pgm-3*	9	279, 417	*PGM3*	6q	284, 335
Lip-1	19	310	*LIPA*	10	308, 310, 436	*Phk*	X	268, 354	*PYK*	X	387
Ltf	9	553	*LTF*	3	553	*Pim-1*	17	248, 418	*PIM1*	6p	116, 420
Lv	4	271, 272, 416	*ALAD*	9q	5, 29, 166, 468	*Pk-3*	9	278, 459, 460	*PKM2*	15q	90, 92, 323, 569
Ly-2	6	62, 273	*CD8*	2p	539	*Plp*	X	134	*PLP*	Xq	593
Mag	7	20	*GMA*	19	20	*Pnd*	4	609	*PND*	1p	608, 609
Mbp	18	301, 485	*MBP*	18q	505, 533	*Pomc*	12	566	*POMC*	2p	451, 614
Met	6	150	*MET*	7q	28, 43, 178, 241, 428, 535, 574, 587	*Prgs*	16	105	*PRGS*	21q	19, 73, 107, 108, 398
						Prn	2	532, 584	*PRIP*	20p	532
Mod-1	9	8, 41, 272, 522	*ME1*	6q	87	*Prp*	8	14	*PRP*	12p	15, 232
						Psph	5	330	*PSP*	7	309
Mod-2	7	80, 522	*ME2*	6	312, 313	*Pyp*	10	331	*PP*	10q	95, 183, 367, 369, 438
Mor-1	5	522	*MDH2*	7p	32, 111, 518, 569						
Mos	4	543	*MOS*	8q	72, 241, 433, 470	*Raf-1*	6	13, 318, 486	*RAF1*	3p	42, 241
Mpi-1	9	441	*MPI*	15q	90 92, 323, 569	*Rb-1*	14	163	*RB1*	13q	163
						Rel	11	56	*REL*	2p	56
Mt-1	8	110, 220, 513	*MT1*	16q	293, 340, 507	*Ren-1*	1	96, 597	*REN*	1	97
Mt-2	8	513	*MT2*	16q	293, 340, 507	*Saa-1*	7	551	*SAA1*	11p	305
Myb	10	497	*MYB*	6q	127, 240, 241, 371, 496	*Sap*	1	400	*APCS*	1q	185, 186, 305
Myc	15	1, 114, 270, 497, 498	*MYC*	8q	128, 241, 360, 496, 498, 510, 548	*Sdh-1*	2	7	*SORD*	15	158
						Sis	15	1, 270, 317, 383	*SIS*	22q	24, 129, 241, 277
Myhs	11	118, 344, 407	*MYH1*	17p	344, 478, 511	*Smst*	16	328, 427, 479	*SST*	3q	614
Ncam	9	123	*NCAM*	11q	437	*Sod-1*	16	61, 88, 109, 199, 479	*SOD1*	21q	11, 88, 108, 398, 526, 547, 604
Ngfb	3	613	*NGFB*	1p	192, 408, 613						
Np-1	14	504, 601	*NP*	14q	195, 481, 483	*Sod-2*	17	545	*SOD2*	6q	113
Nras	3	404, 494	*NRAS*	1p	137, 143, 144, 241, 371, 408, 464, 467, 475	*Spna-1*	1	264, 514	*SPTA1*	1q	264
						Src	2	39, 239, 497	*SRC1*	20q	241, 496, 501
						Sts	X	211, 295	*STS*	Xp	115, 397
Oat	7	448	*OAT*	10q	392, 448, 477	*Syn-1*	X	610	*SYN1*	Xp	610
Orm-1	4	26, 27, 416	*ORM*	9q	4, 165, 166	*T3d*	9	173, 224	*T3D*	11q	173, 224, 537, 570
P450-1	9	246, 563	*CYP2*	15q	247						
Pep-1	18	196, 343, 512	*PEPA*	18q	112, 366	*Tat*	8	410, 463	*TAT*	16q	21
Pep-2	10	196, 602	*PEPB*	12q	87, 236, 245, 281, 283	*Tcn-2*	11	202, 203	*TC2*	22q	9, 167
						Tcp-1	17	596	*TCP1*	6q	594, 595
Pep-3	1	80	*PEPC*	1q	284, 435, 492, 520	*Tcra*	14	152	*TCRA*	14q	17, 102, 537
						Tcrb	6	62, 65, 322	*TCRB*	7q	16, 43, 65, 101, 341, 401, 537
Pep-4	7	331, 527	*PEPD*	19q	55, 135, 290, 352, 364, 368, 369	*Tcrg*	13	322	*TCRG*	7p	401, 537
						Tdt	19	611	*TDT*	10q	611
						Tfm	X	356	*DHTR*	X	388, 389, 589
						Tgfb	7	206	*TGFB*	19q	206

509

Table 7.1—*cont.*

Mouse			Human			Mouse			Human		
Gene	Chr	Reference	Gene	Chr	Reference	Gene	Chr	Reference	Gene	Chr	Reference
Tgn	15	552	*TG*	8q	33, 35, 336	*Try-1*	6	258	*TRY1*	7q	256, 257
Th	7	50	*TH*	11p	274, 403	*Tsha*	4	314, 427, 432	*CGA*	6q	432
Thy-1	9	161	*THY1*	11q	482	*Tshb*	3	314, 425	*TSHB*	1p	162, 425
Tk-1	11	316, 370	*TK1*	17q	324	*Umph-2*	11	542, 599, 600	*UMPH2*	17q	598
Tpi-1	6	342, 390	*TPI1*	12p	245, 281, 283, 323	*Ups*	9	8	*UPS*	11q	216, 381
						Xpa	4	346	*XPAC*	1q	294
Trf	9	78, 84	*TF*	3q	267, 606	*mdx*	X	81	*DMD*	Xp	413
Trp53	11	407, 490	*TRP53*	17p	30	*spf*	X	149, 159	*OTC*	Xp	347

Table 7.2 Linkage conservation of autosomal genes: a listing of segments whose linkage has been conserved since divergence of lineages leading to mouse and man. Markers are listed in order from the most proximal to the most distal on mouse chromosomes. Expected lengths were calculated by using the methods of Nadeau and Taylor (5)

Segment	Mouse Chr	Human Chr	Empirical length (cM)	Expected length (cM)
Idh-1–Cryg-1	1	2q	3	9
Ren-1–Pep-3–Apoa-2–Sap–Spna-1	1	1	24	36
Sdh-1–B2m	2	15	3	9
Itp–Prn–Src	2	20p and q	15	30
Galt–Aco-1–Lv	4	9p and q	12	24
Akp-2–Pgd	4	1	5	15
Pgm-1–Pep-7–Afp (Alb-1)	5	4	12	24
Gus–Mor-1	5	7	11	33
Met–Cpa–Tcrb–Try-1–Hox-1–Ggc	6	7p and q	15	21
Igk–Ly-2	6	2p	1	3
Tpi-1–Ldh-2–Kras-2	6	12p	13	26
Coh–Gpi-1	7	19q	5	15
Saa-1–Ldh-1	7	11p	3	9
Hbb–Th–Int-2	7	11p	29	58
Hp–Got-2–Ctrb–Tat	8	16q	33	55
Es-17–Thy-1–Ups–Apoa-1–Ncam	9	11q	8	12
P450-1–Mpi-1–Pk-3	9	15q	4	8
Pgm-3–Mod-1	9	6q	3	9
Crbp–Trf–Acy-1–Bgl	9	3p and q	16	27
Myhs–Hox-2–Glk	11	17p and q	25	50
Fos–Igh–Aat	12	14q	14	28
Np-1–Tcra	14	14q	Not known*	–
Rb-1–Es-10	14	13q	1	3
Gdc-1–Int-1	15	12	6	18
App–D21S16–D21S52–D21S58–Sod-1–Ets-2	16	21q	8	11
Pgk-2–C3	17	19	7	21

* The 95 per cent confidence limit is 1.92 cM.

Table 7.3 A listing of conserved autosomal syntenic segments. Markers are listed in order from the most proximal to the most distal on mouse chromosomes

Marker	Mouse Chr	Human Chr
Idh-1, Cryg-1	1	2q
Ren-1, Pep-3, Apoa-2, Sap,		
Spna-1	1	1
Abl, Ak-1, Ass, Fpgs	2	9q
Cas-1, Acp-2, Fshb	2	11p
Sdh-1, B2m	2	15
Prn, Itp, Src, Ada	2	20p and q
Amy-1 (Amy-2), Acts, Ngfb, Tshb	3	1
Adh-1 (Adh-3), Egf	3	4q
Pgm-2, Akp-2, Pgd, Ak-2, Eno-1,		
Fuca, Pnd, Xpa	4	1p and 1q
Galt, Aco-1, Lv, Orm-1, Ifa, Ifb	4	9p and 9q
Pgm-1, Pep-7, Afp (Alb-1)	5	4
Gus, Mor-1, Actb, Asl, Psph	5	7
Met, Cpa, Tcrb, Try-1, Hox-1,		
Ggc	6	7p and q
Igk, Ly-2	6	2p
Tpi-1, Ldh-2, Kras-2, Gapd	6	12p
Int-2, Saa-1, Ldh-1, Hbb, Th,		
Calc, Hras-1, Ins-2	7	11p and q
Fes, Idh-2	7	15q
Coh, Gpi-1, Pep-4, Lhb, Mag,		
Tgfb	7	19p and q
Hp, Got-2, Ctrb, Tat, Aprt, Mt-1		
(Mt-2)	8	16q
Crbp, Trf, Acy-1, Bgl, Ltf	9	3p and 3q
Pgm-3, Mod-1	9	6q
Es-17, Thy-1, Ups, Apoa-1,		
Ncam, Ets-1, Gsta, T3d	9	11q
P450-1, Mpi-1, Pk-3	9	15q
Hk-1, Pyp	10	10
Pep-2, Cs, Ifg	10	12
Myhs, Hox-2, Glk, Trp53, Cola-1,		
Erba, Tk-1, Umph-2	11	17p and 17q
Acp-1, Pomc	12	2p
Fos, Igh, Aat	12	14q
As-1, Dhfr, Hexb	13	5
Np-1, Tcra	14	14q
Rb-1, Es-10	14	13q
Tgn, Gpt-1, Myc	15	8q
Gdc-1, Int-1, Ela-1, Hox-3, Pfk-4	15	12
Sis, As-2, Dia-1	15	22q
Sod-1, Ets-2, Ifrc, Prgs, App,		
D21S16, D21S52, D21S58	16	21q
Tcp-1, Pim-1, Glo-1, H-2, Sod-2	17	6p and 6q
D21S56, Crya-1, Cba	17	21q
Pgk-2, C3	17	19
Grl-1, Ii	18	5
Mbp, Pep-1	18	18q
Got-1, Lip-1, Pgam-1, Tdt	19	10

Table 7.4 Examples of changes in gene order between mouse and man. The location of the human homologue is given in parentheses after each gene

Mouse Chr	
4	*Lv* (9q), *Orm-1* (9q), and *Pgm-2* (1p) are located between *Aco-1* (9p) and *Ifa, Ifb* (9b)
4	*Ifa* (9b), *Ifb* (9p), and *Ahd-1* (12) are located between *Pgm-2* (1p) and *Akp-2* (1)
7	*Fes* (15q) is located between *Gpi-1* (19q) and *Pep-4* (19p)
7	*Pep-4* (19p), *Saa-1* (11p), and *Ldh-1* (11p) are located between *Fes* (15q) and *Idh-2* (15q)
7	*Idh-2* (15q) and *Mod-2* (6) are located between *Ldh-1* (11p) and *Hbb* (11p)
17	*Crya-1* (21q) and *D21S56* (21q) are located between *Tcp-1* (6q) and *H-2* (6p)

Acknowledgements

We thank William Jenkins for his help in entering gene mapping data into HMDP and Drs Muriel Davisson and Benjamin A. Taylor for their many helpful comments on a draft of this chapter. This work was supported by NIH grants GM32461 and GM39414.

References

1. Goodfellow, P.N., K.E. Davies, and H.-H. Ropers. 1985. Report of the committee on the genetic constitution of the X and Y chromosomes. Cytogenet. Cell Genet. 40:296–352.
2. Francke, U., and R.T. Taggart. 1980. Comparative gene mapping: order of loci on the X chromosome is different in mice and humans. Proc. Natl. Acad. Sci. USA 77:3595–3599.
3. Human Gene Mapping 9 (1987). Ninth International Workshop on Human Gene Mapping. Cytogenet. Cell Genet. 46, Nos. 1–4, 1988.
4. Lalley, P.A., and V.A. McKusick. 1985. Report of the committee on comparative mapping. Human Gene Mapping 8 (1985): Eighth International Workshop on Human Gene Mapping. Cytogenet. Cell Genet. 40:536–566.
5. Nadeau, J.H., and B.A. Taylor. 1984. Lengths of chromosomal segments conserved since divergence of man and mouse. Proc. Natl. Acad. Sci. USA 81:814–818.
6. Searle, A.G., J. Peters, M.F. Lyon, E.P. Evans, J.H. Edwards, and V.J. Buckle. 1987. Chromosome maps of man and mouse, III. Genomics 1:3–18.
7. Yang-Feng, T.L., L.J. DeGennaro, and U. Francke. 1986. Genes for synapsin I, a neuronal phosphoprotein, map to conserved regions of human and murine X chromosomes. Proc. Natl. Acad. Sci. USA 83:8679–8683.

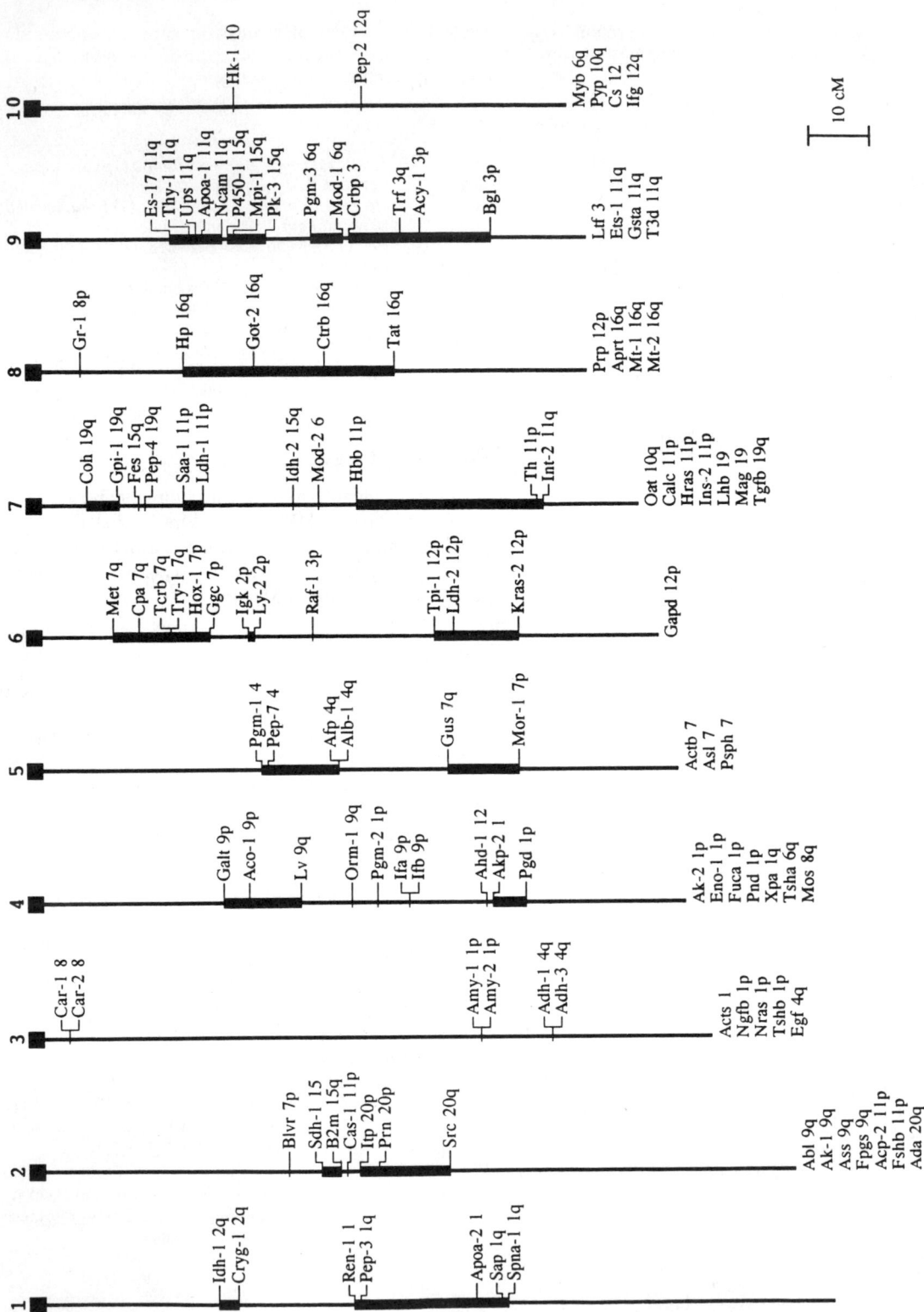

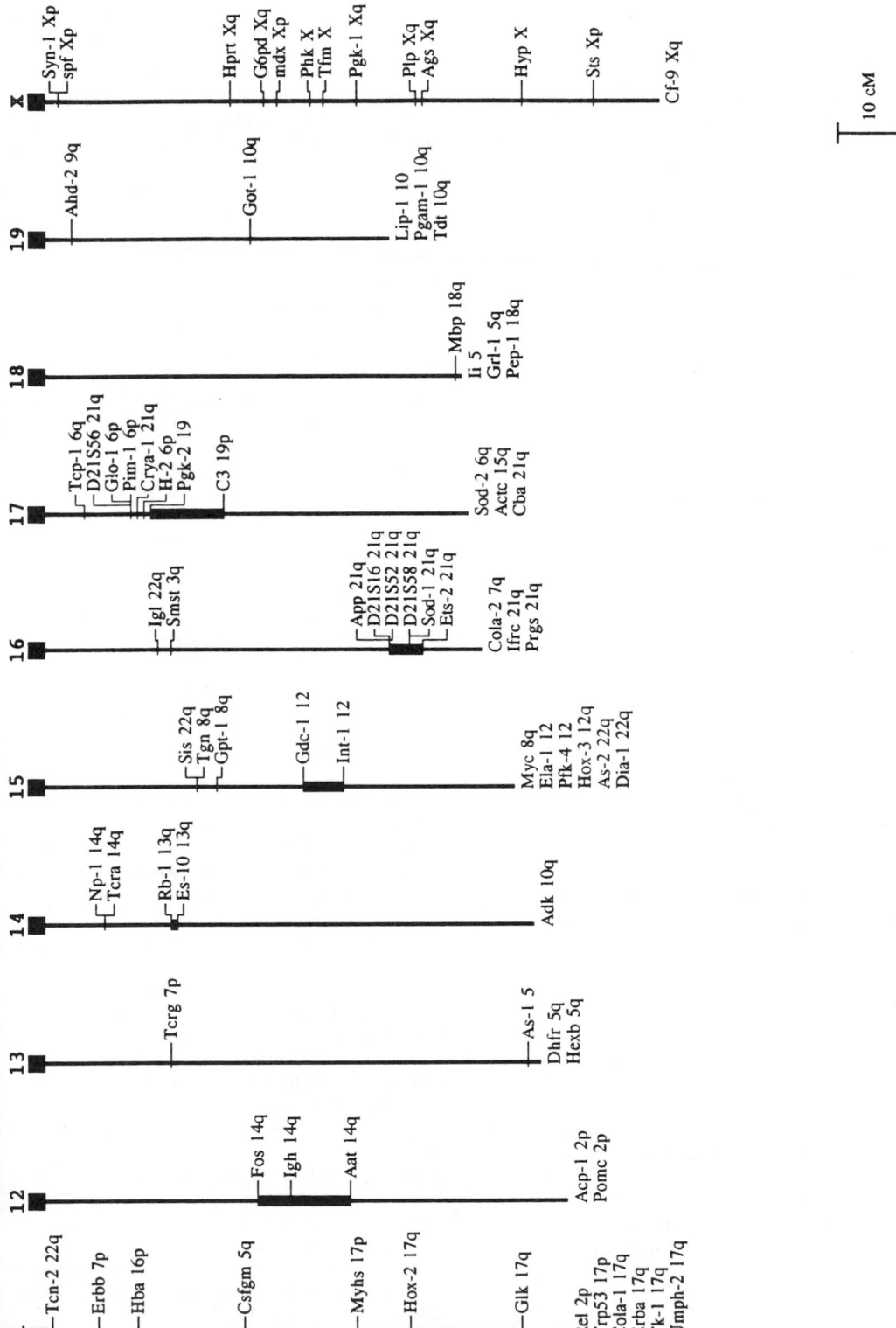

Fig. 7.1 Chromosome map of linkage and synteny homologies in mouse and man. All genes mapped in both mouse and man are given. The chromosomal location of the homologous gene in man is given after each mouse gene; chromosome arm assignment is also given, when known. Highlighted are chromosome segments that are marked by two or more genes whose homologues in man are located on the same chromosome and that are not interrupted by unrelated genes. Synteny assignments are given below each chromosome.

Appendix

References for linkage and synteny assignments listed in Table 1.

1. Adolph, S., C.R. Bartram, and H. Hameister. 1987. Mapping of the oncogenes *Myc*, *Sis*, and *int-1* to the distal part of mouse chromosome 15. Cytogenet. Cell Genet. 44:65–68.

2. Aitken, D.A., and M.A. Ferguson-Smith. 1979. Intrachromosomal assignment of the structural gene for *GALT* to the short arm of chromosome 9 by gene dosage studies. Cytogenet. Cell Genet. 25:131.

3. Albert, E.D., M.P. Baur, and W.R. Mayr. 1984. Histocompatibility Testing 1984. Springer Verlag, Berlin 764 p.

4. Allerdice, P.W., H. Kaita, M. Lewis, P.J. McAlpine, P. Wong, J. Anderson, and E.R. Giblett. 1986. Segregation of marker loci in families with an inherited paracentric insertion of chromosome 9. Am. J. Hum. Genet. 39:612–617.

5. Amorim, A., F. Schunter, H. Ritter, and J. Kompf. 1984. Linkage studies on the *ALAD* polymorphism. Cytogenet. Cell Genet. 37:400.

6. Anderson, M.L.M., M.F. Szanjnert, J.-C. Kaplan, L. McColl, and B.D. Young. 1984. The isolation of a human IgV lambda gene from a recombinant library of chromosome 22 and estimation of its copy number. Nucl. Acids Res. 12:6647–6661.

7. Andrews, S.J., and J. Peters. 1983. Linkage analyses and biochemical genetics of sorbitol dehydrogenase-1 (*Sdh-1*) in the mouse. Biochem. Genet. 21:809–817.

8. Antonucci, T.K., O.H. von Deimling, B.B. Rosenblum, L.C. Skow, and M.H. Meisler. 1984. Conserved linkage within a 4-cM region of mouse chromosome 9 and human chromosome 11. Genetics 107:463–475.

9. Arwert, F., H.J. Porck, M. Frater-Schroder, C. Brahe, A. Guerts van Kessel, A. Westerveld, P. Meera Khan, K. Zang, R.R. Frants, H.T. Kortbeek, and A.W. Erikson. 1986. Assignment of human transcobalamin II (*TC2*) to chromosome 22 using somatic cell hybrids and monosomic meningioma cells. Hum. Genet. 74:378–381.

10. Ashley, P.L., R.R. Flandermeyer, and D.R. Cox, 1986. Identification of noval phosphofructokinase loci in mouse and man. Am. J. Hum. Genet. 39:A186.

11. Aula, P., J. Leisti, and H. von Koskull. 1973. Partial trisomy 21. Clin. Genet. 4:241–245.

12. Avner, P., L. Amar, D. Arnaud, A. Hanauer, and J. Cambrou. 1987. Detailed ordering of markers localizing to the Xq26-Xqter region of the human X chromosome by the use of an interspecific *Mus spretus* mouse cross. Proc. Natl. Acad. Sci. USA 84:1629-1633.

13. Avner, P., M. Bucan, D. Arnaud, H. Lehrach, and U. Rapp. 1987. *A-raf* oncogene localizes on mouse X chromosome to region some 10–17 centimorgans proximal to hypoxanthine phosphoribosyltransferase gene. Somat. Cell Mol. Genet. 13:267–272.

14. Azen, E., D.M. Carlson, P.A. Lalley, and E. Vanin. 1984. Salivary proline-rich genes on chromosome 8 of mouse. Science 226:967–969.

15. Azen, E., P.A. Goodman, and P.A. Lalley. 1985. Human salivary proline-rich protein genes on chromosome 12. Am. J. Hum. Genet. 37:418–424.

16. Barker, P.E., F.H. Ruddle, H.-D. Royer, O. Acuto, and E.L. Reinherz. 1984. Chromosomal location of human T-cell receptor gene TiBeta. Science 23:348–349.

17. Barker, P.E., H.-D. Royer, F.H. Ruddle, and E.L. Reinherz. 1985. Human T cell receptor gene *TCRA* lies in region 14pter-q21. Cytogenet. Cell Genet. 40:576–577.

18. Barker, P.E., M. Rabin, M. Watson, W.R. Breg, F.H. Ruddle, and I.M. Verma. 1984. Human c-fos oncogene mapped within chromosomal region 14q21-q31. Proc. Natl. Acad. Sci. USA 81:5826–5830.

19. Bartley, J.A., and C.J. Epstein. 1980. Gene dosage effect for glycinamide ribonucleotide synthetase in human fibroblasts trisomic for chromosome 21. Biochem. Biophys. Res. Comm. 93:1286–1289.

20. Barton, D.E., M. Arquint, J. Roder, R. Dunn, and U. Francke. 1987. The myelin–associated glycoprotein gene: mapping to human chromosome 19 and mouse chromosome 7 and expression in quivering mice. Genomics 1:107–112.

21. Barton, D.E., T.L. Yang-Feng, and U. Francke. 1986. The human tyrosine aminotransferase gene mapped to the long arm of chromosome 16 (region 16q22-q24) mapped by somatic cell hybrid analysis and in situ hybridization. Hum. Genet. 72:221–224.

22. Barton, P., S. Malcolm, C. Murphy, and M.A. Ferguson-Smith. 1982. Localization of the human alpha-globin gene cluster to the short arm of chromosome 16 (16p12-16pter) by hybridization in situ. J. Mol. Biol. 156:269–278.

23. Bartram, C.R., A. De Klein, A. Hagemeijer, G. Grosveld, and D. Bootsma. 1984. Oncogene translocation in chronic myelocytic leukemia. Cytogenet. Cell Genet. 37:415.

24. Bartram, C.R., A. Klein, A. Hagemeijer, G. Grosveld, N. Heisterkamp, and J. Groffen. 1984. Localization of the human c-sis oncogene in Ph 1-positive and Ph 1-negative chronic myelocytic leukemia by in situ hybridization. Blood 63:223–225.

25. Baumann, H., and F.G. Berger. 1984. Genetic analysis of the major acute phase reactants in the mouse: chromosomal location of the structural genes for haptoglobin and alpha-1-acid glycoprotein-2. Mouse News Lett. 71:49.

26. Baumann, H., and F.G. Berger. 1985. Genetics and evolution of the acute phase reactant proteins in mice. Mol. Gen. Genet. 201:505–512.

27. Baumann, H., W.A. Held, and F.G. Berger. 1984. The acute phase response of mouse liver: genetic analysis of the major acute phase reactants. J. Biol. Chem. 259:566–573.

28. Beaudet, A., A. Bowcock, M. Buchwald, L. Cavalli-

Sforza, M. Farrall, M.-C. King, K. Klinger, J.-M. Lalouel, G. Lathrop, S. Naylor, and J. Ott. 1986. Linkage of cystic fibrosis to two tightly linked DNA markers: joint report from a collaborative effort. Am. J. Hum. Genet. 39:681–693.

29. Beaumont, C., C. Foubert, B. Grandchamp, D. Weil, V.C. N'Guyen, M.S. Gross, and Y. Nordman. 1984. Assignment of the human gene for delta aminolevulinate dehydrase to chromosome 9 by somatic cell hybridization and specific enzyme immunoassay. Ann. Hum. Genet. 48:153–159.

30. Benchimol, S., P. Lamb, L. Crawford, D. Sheer, E. Solomon, T. Shows, G. Bruns, and J. Peacock. 1985. Human p53 is on the short arm of chromosome 17. Cytogenet. Cell Genet. 40:580–581.

31. Bender, K., and K.-H. Grzeschik. 1982. Possible assignment of glyoxalase I (*GLO*) to chromosome 6 using man–mouse somatic cell hybrids. Hum. Genet. 61:127–134.

32. Benn, P., C.J. Chern, G. Bruns, I.W. Craig, and C.M. Croce. 1977. Assignment of the genes for human beta-glucuronidase and mitochondrial malate dehydrogenase to the region pter-q22 of chromosome 7. Cytogenet. Cell Genet. 19:273–280.

33. Berge-Lefranc, J.-L., G. Cartouzou, M.-G. Mattei, E. Passage, C. Malezet-Desmoulins, and S. Lissitzky. 1985. Localization of the thyroglobulin gene by in situ hybridization to human chromosomes. Hum. Genet. 69:28–31.

34. Berger, F.G., and K. Paigen. 1979. Cis-active control of mouse beta-galactosidase biosynthesis by a systemic regulatory locus. Nature 282:314–316.

35. Bernardi, F., P. Patracchini, A. Monticelli, S. Varrone, V. Aiello, E. Calzolari, G. Marchetti, and V.E. Avvedimento. 1985. Human thyroglobulin gene is located on the terminal part of the long arm of chromosome 8. Cytogenet. Cell Genet. 40:582–583.

36. Biddle, F.G., and M.L. Petras. 1967. The inheritance of a non-hemoglobin erythrocytic protein in *Mus musculus*. Genetics 57:943–949.

37. Bissbort, S., H.W. Hitzeroth, C.T. van den Berg, and T.F. Wienker. 1987. Linkage relationship between the genes for adenosine deaminase and S-adenosy-homocysteine hydrolase on human chromosome 20. Hum. Genet. 77:277–279.

38. Bissbort, S., K. Bender, and K.-H. Grzeschik. 1984. Assignment of the human gene for gamma-glutamyl-cyclotransferase (*GCTG*) to chromosome 7p. Cytogenet. Cell Genet. 37:421.

39. Blatt, C., M.E. Harper, G. Franchini, M.N. Nesbitt, and M.I. Simon. 1984. Chromosomal mapping of murine c-fes and s-src genes. Mol. Cell. Biol. 4:978–981.

40. Bocian, M., T. Mohandas, R.S. Sparkes, and S. Funderbunk. 1979. Peptidase S (*PEPS*) and phosphoglucomutase 2 (*PGM*) loci on the short arm of human chromosome 4. Cytogenet. Cell Genet. 25:138.

41. Bonhomme, F., F. Bemehdi, J. Britton-Davidian, and S. Martin. 1979. Analyse génétique de croisements inter-

spécifiques *Mus musculus* L. X *Mus spretus* Lataste: liaison de *Adh-1* avec *Amy-1* sur le chromosome 3 et de *Es-14* avec *Mod-1* sur le chromosome 9. C. R. Acad. Sci. Paris 289:545–548.

42. Bonner, T., S.J. O'Brien, W.G. Nash, U.R. Rapp, C.C. Morton, and P. Leder. 1984. Two human homologs of the *raf* (*mil*) oncogene are located on human chromosomes 3 and 4. Science 223:71–74.

43. Bowcock, A.M., J. Crandall, L. Daneshvar, G.M. Lee, B. Young, V. Zunzunegui, C. Craik, L.L. Cavalli-Sforza, and M.-C. King. 1986. Genetic analysis of cystic fibrosis: linkage of DNA and classical markers in multiplex families. Am. J. Hum. Genet. 39:699–706.

44. Bowcock, A.M., L.A. Farrer, L.L. Cavalli-Sforza, J.M. Hebert, K.K. Kidd, M. Frydman, and B. Bonne-Tamir. 1987. Mapping the Wilson disease locus to a cluster of linked polymorphic markers on chromosome 13. Am. J. Hum. Genet. 41:27–35.

45. Boyd, C.D., K. Weliky, S. Toth-Fejel, S.B. Deak, A.M. Chrisiano, J.W. McKenzie, L.J. Sandell, K. Tryggvason, and E. Magenis. 1986. The single copy gene coding for human alpha 1(IV) procollagen is located at the terminal end of the long arm of chromosome 13. Hum. Genet. 74:121–125.

46. Boyd, Y., I.W. Craig, E. Munro, and B.R. Migeon. 1984. Localization of the factor IX locus to Xq26-qter. Cytogenet. Cell Genet. 37:425–426.

47. Brebner, D., A. Grobler-Rabie, A. Bester, C. Mathew, and C. Boyd. 1985. Two new polymorphic markers in the human proalpha 2-1 collagen gene. Hum. Genet. 10:25–27.

48. Breen, G.A., A.J. Lusis, and K. Paigen. 1977. Linkage of genetic determinants for mouse beta-galactosidase electrophoresis and activity. Genetics 85:73–84.

49. Breuning M.H., K. Madan, M. Verjaal, J.T. Wijen, P. Meera Khan, and P.L. Pearson. 1987. Human alpha-globin maps to pter-p13.3 in chromosome 16 distal to *PGP*. Hum. Genet. 76:287–289.

50. Brilliant, M.H., M.N. Niemann, and E.M. Eicher. 1987. Murine tyrosine hydroxylase maps to the distal end of chromosome 7 within a linkage group conserved in mouse and man. J. Neurogenet. 4:259–266.

51. Brissenden, J.E., A. Ullrich, and U. Francke. 1984. Human chromosomal mapping of genes for insulin-like growth factors I and II and epidermal growth factor. Nature 310:781–784.

52. Britton-Davidian, J., A. Ruiz Bustos, and M. Topal. 1978. Lactate dehydrogenase polymorphism in *Mus musculus* L. and *Mus spretus* Latase. Experientia 34:1144–1145.

53. Britton-Davidian, J., and J. Pasteur. 1983. Personal communication. Mouse News Lett. 68:79.

54. Brook, J.D., D.J. Shaw, L. Meredith, G.A.P. Bruns, and P.D. Harper. 1984. Localization of genetic markers and orientation of the linkage group on chromosome 19. Hum. Genet. 68:282–285.

55. Brown, S., P.A. Lalley, and J.D. Minna. 1978. Assign-

ment of the gene for peptidase S (*PEPS*) to chromosome 4 in man and confirmation of peptidase D (*PEPD*) assignment to chromosome 19. Cytogenet. Cell Genet. 22:167–171.

56. Brownell, E., C.A. Kozak, J.R. Fowle, W.S. Modi, N.R. Rice, and S.J. O'Brien. 1986. Comparative genetic mapping of cellular rel sequences in man, mouse, and the domestic cat. Am. J. Hum. Genet. 39:194–202.

57. Bruns, G.A.P., A.C. Leary, R.E. Eisenman, C.W. Bazinet, V.M. Regina, and P.S. Gerald. 1978. Expression of *ACONs* and *GALT* in man–rodent somatic cell hybrids. Cytogenet. Cell Genet. 22:172–176.

58. Bruns, G.A.P., A.C. Leary, V.M. Regina, and P.S. Gerald. 1977. Lysosomal beta-D-galactosidase in man–hamster somatic cell hybrids. Human Gene Mapping 4:177–181.

59. Bruns, G.A.P., R.E. Eiseman, and P.S. Gerald. 1976. Human mitochondrial NADP-dependent isocitrate dehydrogenase in man–mouse somatic cell hybrids. Cytogenet. Cell Genet. 17:200–211.

60. Bruns, G., and P.S. Gerald. 1976. Human glyceraldehyde-3 phosphate dehydrogenase in man–rodent somatic cell hybrids. Science 192:54–56.

61. Bruns, G., P.S. Gerald, P. Lalley, U. Francke, and J. Minna. 1979. Gene mapping of the mouse by somatic cell hybridization. Cytogenet. Cell. Genet. 25:139.

62. Bucan, M., T. Yang-Feng, A.M. Colberg-Poley, D.J. Wolgemuth, J.-L. Guénet, U. Francke, and H. Lehrach. 1986. Genetic and cytogenetic localization of the homeo box containing genes on mouse chromosome 6 and human chromosome 7. EMBO J. 5:2899–2905.

63. Butterworth, P., J. Barlow, C. Konialis, S. Povey, and Y.H. Edwards. 1985. The assignment of human erythrocyte carbonic anhydrase CA1 to chromosome 8. Cytogenet. Cell Genet. 40:597.

64. Byrne, D.E.S., and R.L. Church. 1983. Assignment of the genes for mouse type 1 procollagen to chromosome 16 using mouse fibroblast–Chinese hamster somatic cell hybrids. Somat. Cell Genet. 9:313–331.

65. Caccia, N., M. Kronenberg, D. Saxe, R. Haars, G.A.P. Bruns, J. Goverman, M. Malissen, H. Willard, Y. Yoshikai, M. Simon, and L. Hood. 1984. The T cell receptor beta-chain genes are located on Chromosome 6 in mice and Chromosome 7 in humans. Cell 37:1091–1099.

66. Cahilly, L.A., D. George, B.L. Daugherty, and S. Petska. 1985. Subchromosomal localization of mouse IFN-alpha genes by in situ hybridization. J. Interferon Res. 5:391–395.

67. Cannizzaro, L.A., C.M. Croce, C.A. Griffen, A. Simeone, E. Boncinelli, and K. Huebner. 1987. Human homeo box-containing genes located at chromosome regions 2q31-2q37 and 12q12-12q13. Am. J. Hum. Genet. 41:1–15.

68. Carritt, B. 1985. Genomic cloning of the short arm of human chromosome 1. Cytogenet. Cell Genet. 40:599.

69. Carritt, B., J. King, and H.M. Welch. 1982. Gene order and localization of enzyme loci on the short arm of chromosome 1. Ann. Hum. Genet. 46:329–335.

70. Carritt, B., P.S.G. Goldfarb, M.L. Hooper, and C. Slack. 1977. Chromosome assignment of a human gene for argininosuccinate synthetase expression in Chinese hamster × human somatic cell hybrids. Exp. Cell Res. 106:71–78.

71. Casey, G., R. Smith, D. McGillivray, G. Peters, and C. Dickson. 1986. Characterization and chromosome assignment of the human homolog of *int-2*, a potential proto-oncogene. Mol. Cell. Biol. 6:502–510.

72. Caubet, J.-F., D. Mathieu-Mahul, A. Bernheim, C.-J. Larsen, and R. Berger. 1985. Human proto-oncogene c-mos (*MOS*) maps to 8q11. Cytogenet. Cell Genet. 40:602.

73. Chadefaux, B., D. Allard, M.O. Rethore, O. Raoul, M. Poissonier, S. Gilgenkrantz, C. Cheruy, and H. Jerome. 1984. Assignment of human phosphoribosylglycinamide synthetase locus to region 21q22.1. Hum. Genet. 66:190–192.

74. Chaganti, R.S.K., S.C. Jhanwar, S.E. Antonarakis, and W.S. Hayward. 1985. Germ-line chromosomal localization of genes in chromosome 11p linkage: parathyroid hormone, beta-globin, c-Ha-ras-1, and insulin. Somat. Cell. Mol. Genet. 11:197–202.

75. Chan, T.-S., M.P. Reardon, and R.M. Greenstein. 1976. Somatic cell hybrid assignment of a structural gene for human beta-glucuronidase to chromosome 7 by use of an X/7 translocation. Cytogenet. Cell Genet. 17:291–295.

76. Chance, P.F., K.A. Dyer, K. Kurachi, S. Yoshitake, H.H. Ropers, P. Wieacker, and S.M. Gartler. 1984. Regional assignment of the human factor IX gene by molecular hybridization. Cytogenet. Cell Genet. 37:435.

77. Chany, C., M. Vignal, P. Couillin, V.C. Nguyen, J. Boué, and A. Boué. 1975. Chromsomal localization of human gene governing the interferon-induced antiviral state. Proc. Natl. Acad. Sci. USA 72:3129–3133.

78. Chapman, V.M. 1973. Personal communication. Mouse News Lett. 49:46.

79. Chapman, V.M., and F.H. Ruddle. 1972. Glutamate oxaloacetate transaminase (GOT) genetics in the mouse: polymorphism of *Got-1*. Genetics 70:299–305.

80. Chapman, V.M., F.H. Ruddle, and T.H. Roderick. 1971. Linkage of isozyme loci in the mouse: phosphoglucomutase-2 (*Pgm-2*), mitochondrial NADP malate dehydrogenase (*Mod-2*) and dipeptidase-1 (*Dip-1*). Biochem. Genet. 5:101–110.

81. Chapman, V.M., M. Murawski, D. Miller, and D. Suriatek. 1985. Personal communication. Mouse News Lett. 72:120.

82. Chapman, V.M., P.G. Kratzer, and B.A. Quarantillo. 1983. Electrophoretic variation for X chromosome-linked hypoxanthine phosphoribosyl transferase (HPRT) in wild-derived mice. Genetics 103:785–795.

83. Chapman, V.M., and T.B. Shows. 1976. Somatic cell genetic evidence for X-chromosome linkage of three enzymes in the mouse. Nature 259:665–667.

84. Chapman, V.M., and W. Frels. 1975. Personal communication. Mouse News Lett. 53:61.

85. Chen, S.-H., R.P. Creagan, E.A. Nichols, and F.H. Ruddle. 1974. Assignment of human esterase-D gene to chromosome 13. Cytogenet. Cell Genet. 14:269–272.

86. Chen, S.-H., R.P. Donahue, and C.R. Scott. 1973. The genetics of glutamic-pyruvic transaminase in mice: Inheritance, electrophoretic phenotypes, and postnatal changes. Biochem. Genet. 10:23–28.

87. Chen, T.R., F.A. McMorris, R. Ceragan, F. Ricciuti, J. Tischfield, and F.H. Ruddle. 1973. Assignment of the genes for malate oxidoreductase decarboxylating to chromosome 6 and peptidase B and lactate dehydrogenase B to chromosome 12 in man. Am. J. Hum. Genet. 25:200–207.

88. Cheng, S.V., J.H. Nadeau, R.E. Tanzi, P.C. Watkins, J. Jagadesh, B.A. Taylor, N. Sacchi, and J.F. Gusella. 1988. Comparative mapping of DNA markers from the familial Alzheimer disease and Down syndrome regions of human chromosome 21 to mouse chromosomes 16 and 17. Proc. Natl. Acad. Sci. USA. 85:6032–6036.

89. Cheng Z.Y., M. Lovett, L.B. Epstein, and C.J. Epstein. 1986. The mouse IFN-alpha (*Ifa*) locus: correlation of physical and linkage maps by in situ hybridization. Cytogenet. Cell Genet. 41:101–106.

90. Chern, C.J., and C.M. Croce. 1975. Confirmation of the synteny of the human genes for mannose phosphate isomerase and pyruvate kinase and their assignment to chromosome 15. Cytogenet. Cell Genet. 15:299–305.

91. Chern, C.J., and C.M. Croce. 1976. Assignment of the structural gene for human beta-glucuronidase to chromosome 7 and tetrameric association of subunits in the enzyme molecule. Am. J. Hum. Genet. 28:350–356.

92. Chern, C.J., R. Kennet, E. Engel, W.J. Mellman, and C.M. Croce. 1977. Assignment of the structural genes for the alpha subunit of hexosaminidase A, mannose phosphate isomerase and pyruvate kinase to the region q22-qter of human chromosome 15. Somat. Cell Genet. 3:553–560.

93. Chern, C.J., W.J. Mellman, and C.M. Croce. 1975. Assignment of the structural gene for cytoplasmic glutamic-oxaloacetic transaminase (GOT) to region q24-qter of human chromosome 10. Am. J. Hum. Genet. 27:25A.

94. Chern, C.J., W.J. Mellman, and C.M. Croce. 1976. Assignment of the structural locus for cytoplasmic glutamic-oxaloacetic transaminase to region q24-qter on human chromosome 10. Somat. Cell Genet. 2:177–182.

95. Chern, C.J. 1976. Localization of the structural genes for hexokinase-1 and inorganic pyrophosphatase on region (pter-q24) of human chromosome 10. Cytogenet. Cell Genet. 17:338–342.

96. Chirgwin, J.M., I.M. Schaefer, J.A. Diaz, and P.A. Lalley. 1984. Mouse kidney renin gene is on chromosome 1. Somat. Cell Mol. Genet. 10:633–637.

97. Chirgwin, J.M., I.M. Schaefer, P.S. Rotwein, N. Piccini, K.W. Gross, and S.L. Naylor. 1984. Human renin gene is on chromosome 1. Somat. Cell Mol. Genet. 10:415–421.

98. Church, R.L., N. Sundar Raj, and J.K. McDougall. 1980. Regional chromosomal mapping of the human skin type 1 procollagen gene using adenovirus 12-fragmentation of human–mouse somatic cell hybridization. Cytogenet. Cell Genet. 27:24–30.

99. Claesson-Welsh, L., P.E. Barker, D. Larhammar, L. Rask, F.H. Ruddle, and P.A. Peterson. 1984. The gene encoding the human class II antigen-associated chain is located on chromosome 5. Immunogenetics 20:898–893.

100. Colantuoni, V., M. Rocchi, L. Roncuzzi, and G. Romeo. 1986. Mapping of human cellular retinol-binding protein to chromosome 3. Cytogenet. Cell Genet. 43:221–222.

101. Collins, M.K.L., P.N. Goodfellow, M.J. Dunne, N.K. Spurr, E. Solomon, and M.J. Owen. 1984. A human T-cell antigen receptor beta chain gene maps to chromosome 7. EMBO J. 3:2347–2349.

102. Collins, M.K.L., P.N. Goodfellow, N.K. Spurr, E. Solomon, G. Tanigawa, S. Tonegawa, and M.J. Owen. 1985. The human T-cell receptor alpha-chain gene maps to chromosome 14. Nature 314:273–274.

103. Cook, P.J.L., E.B. Robson, K.E. Buckton, C.A. Slaughter, J.E. Gray, C.E. Blank, F.E. James, M.A.C. Ridler, J. Insley, and M. Hulten. 1978. Segregation of *ABO*, *AK1* and *ACONS* in families with abnormalities of chromosome 9. Ann. Hum. Genet. 41:365–376.

104. Cousineau, A.J., J.V. Higgens, E. Hackel, D.F. Waterman, H. Toriello, P.A. Calile, and P.J.L. Cook. 1981. Cytogenetic recognition of chromosomal duplication dup(1)(p31.4-p22.1) and the detection of three different alleles at the *PGM1* locus. Ann. Hum. Genet. 45:337–340.

105. Cox, D.R., D. Goldblatt, and C.J. Epstein. 1982. Chromosomal assignment of mouse *PRGS*: further evidence for homology between mouse chromosome 16 and human chromosome 21. Am. J. Hum. Genet. 33:145A.

106. Cox, D.R., D. Goldblatt, and C.J. Epstein. 1982. Comparative gene mapping in man and mouse: assignment of the gene for citrate synthetase (*Cis*) to mouse chromosome 10. Cytogenet. Cell Genet. 32:259.

107. Cox, D.R., D. Goldblatt, and C.J. Epstein. 1982. Confirmation of the assignment of *PRGS* to human chromosome 21. Cytogenet. Cell Genet. 32:259.

108. Cox, D.R., H. Kawashina, S. Vora, and C.J. Epstein. 1984. Regional mapping of *SOD-1*, *PRGS*, and *PFK-L* on human chromosome 21. Cytogenet. Cell Genet. 37:441–442.

109. Cox, D.R., L.B. Epstein, and C.J. Epstein. 1980. Genes coding for sensitivity to interferon (*Ifrc*) and soluble superoxide dismutase (*SOD-1*) are linked in mouse and man and map to mouse chromosome 16. Proc. Natl. Acad. Sci. USA 77:2168–2172.

110. Cox, D.R., and R.D. Palmiter. 1983. The metallothionein-1 gene maps to mouse chromosome 8: implications for human Menkes' disease. Hum. Genet. 64:61–64.

111. Craig, I.W., V. van Heyningen, D. Finnegan, and W.F. Bodmer. 1974. Analysis of the mitochondrial enzymes citrate synthetase (EC 4.1.3.7) and malate dehydrogenase (EC 1.1.3.7) in human–mouse somatic cell hybrids. Cytogenet. Cell Genet. 13:76–78.

112. Creagan, R., J. Tischfield, F.A. McMorris, S. Chen, M. Hirschi, T.R. Chen, F. Ricciutti, and F.H. Ruddle. 1973. Assignment of the genes for human peptidase A to chromosome 18 and cytoplasmic glutamic oxaloacetate transaminase to chromosome 10 using somatic cell hybrids. Cytogenet. Cell Genet. 12:187–198.

113. Creagan, R., J. Tischfield, F. Ricciuti, and F.H. Ruddle. 1973. Chromosome assignments of genes in man using mouse–human somatic cell hybrids: mitochondrial superoxide dismutase (indophenol oxidase-B, tetrameric) to chromosome 6. Humangenetik 20:203–209.

114. Crews, S., R. Barth, L. Hood, J. Prehn, and K. Calame. 1982. Mouse c-myc oncogene is located on chromosome 15 and translocated to chromosome 12 in plasmacytomas. Science 218:1319–1321.

115. Curry, C.J.R., R.E. Magenis, M. Brown, J.T. Lawn, T. Tsai, P. O'Lague, P. Goodfellow, T. Mohandas, E.A. Bergner, and L.J. Shapiro. 1984. Inherited chondrodysplasia punctata due to a deletion of the terminal short arm of an X chromosome. N. Eng. J. Med. 298:1010–1015.

116. Cuypers, H.T., G. Selten, A. Berns, and A.H.M. Geurts van Kessel. 1986. Assignment of the human homologue of *Pim-1*, a mouse gene implicated in leukemogenesis to the pter-q12 region of chromosome 6. Hum. Genet. 72:262–265.

117. Czosnek, H., S. Sarid, P.E. Barker, F.H. Ruddle, and V. Daniel. 1984. Glutathione-S-transferase Ya subunit is coded by a multigene family located on a single mouse chromosome. Nucl. Acids Res. 12:4825–2833.

118. Czosnek, H., U. Nudel, M. Shani, P.E. Barker, D.D. Pravtcheva, F.H. Ruddle, and D. Yaffe. 1982. The gene coding for the muscle contractile proteins, myosin heavy chain, myosin light chain 2, and skeletal muscle actin are located on three different mouse chromosomes. EMBO J. 1:1299–1305.

119. Czosnek, H., U. Nudel, Y. Mayer, P.E. Barker, D.D. Pravtcheva, F.H. Ruddle, and D. Yaffe. 1983. The gene coding for the cardiac muscle actin, the skeletal muscle actin and the cytoplasmic beta-actin are located on three different mouse chromosomes. EMBO J. 2:1977–1979.

120. D'Ancona, G.G., and C.M. Croce. 1977. Assignment of the gene for enolase to mouse chromosome 4 using somatic cell hybrids. Cytogenet. Cell Genet. 19:1–6.

121. D'Eustachio, P. 1984. A genetic map of mouse chromosome 12 composed of polymorphic DNA fragments. J. Exp. Med. 160:827–838.

122. D'Eustachio, P., A.L. Bothwell, T.K. Takaro, D. Baltimore, and F.H. Ruddle. 1981. Chromosomal location of structural genes encoding murine immunoglobulin lambda light chains. J. Exp. Med. 153:793–800.

123. D'Eustachio, P., G.C. Owens, G.M. Edelman, and B.A. Cunningham. 1985. Chromosomal location of the gene encoding the neural cell adhesion molecule (N-CAM) in the mouse. Proc. Natl. Acad. Sci. USA 82:7631–7635.

124. D'Eustachio, P., R.S. Ingram, S.M. Tilghman, and F.H. Ruddle. 1981. Murine alpha-fetoprotein and albumin: two evolutionarily linked proteins encoded on the same mouse chromosome. Somat. Cell Genet. 7:289–294.

125. Da Silva, F.P., G.F. Hoecker, N.K. Day, K. Vienne, and P. Rubenstein. 1978. Murine complement component 3: genetic variation and linkage to H-2. Proc. Natl. Acad. Sci. USA 75:963–965.

126. Daiger, S.P., N.S. Hoffman, R.S. Wildin, and T.-S. Su. 1984. Multiple, independent restriction site polymorphisms in human DNA detected with a cDNA probe to argininosuccinate synthetase (AS). Am. J. Hum. Genet. 36:736–749.

127. Dalla-Favera, R., G. Franchini, S. Martinotti, F. Wong-Staal, R.C. Gallo, and C.M. Croce. 1982. Chromosomal assignment of the human homologues of feline sarcoma virus and avian myeloblastosis oncogenes. Proc. Natl. Acad. Sci. USA 79:4714–4717.

128. Dalla-Favera, R., M. Bregni, J. Erikson, D. Patterson, R.C. Gallo, and C.M. Croce. 1982. Human c-myc oncogene is located on the region of chromosome 8 that is translocated in Burkitt lymphoma cells. Proc. Natl. Acad. Sci. USA 79:7824–7827.

129. Dalla-Favera, R., R.C. Gallo, A. Giallongo, and C.M. Croce. 1982. Chromosomal localization of the human homolog (c-sis) of the Simian sarcoma virus onc gene. Science 218:686–688.

130. Dandoy, F., E. DeMaeyer, F. Bonhomme, J.-L. Guénet, and J. DeMaeyer-Guignard. 1985. Segregation of restriction fragment variants in an interspecies cross of laboratory and wild mice indicates tight linkage of the murine IFN-beta gene to the murine IFN-alpha gene. J. Virol. 56:216–220.

131. Dandoy, F., K.A. Kelley, J. DeMaeyer-Guignard, E. DeMaeyer, and P.M. Pitha. 1984. Linkage analysis of the murine interferon-alpha locus on mouse chromosome 4. J. Exp. Med. 160:294–302.

132. Daniel, W.L. 1976. Genetic control of heat sensitivity and activity level of murine arylsulfatase B. Biochem. Genet. 14:1003–1018.

133. Darlington, G.J., K.H. Astrin, S.P. Muirhead, R.J. Desnick, and M. Smith. 1982. Assignment of human alpha-1 antitrypsin to chromosome 14 by somatic cell hybrid analysis. Proc. Natl. Acad. Sci. USA 79:870–873.

134. Dautigny, A., M.G. Mattei, D. Morello, P.M. Alliel, D. Pham-Dinh, L. Amar, D. Arnaud, D. Simon, J.F. Mattei, J.-L. Guénet, and P. Jolles. 1986. The structural gene coding for myelin-associated proteolipid protein is mutated in jimpy mice. Nature 321:867–869.

135. Davis, M.B. 1987. Linkage between the loci for peptidase D and cytochrome P-450 (*CYP1*) on chromosome 19. Ann. Hum. Genet. 51:9–12.

136. Davis, M.B., L.F. West, J.H. Barlow, P.H.W. Butterworth, J.C. Lloyd, and Y.H. Edwards. 1987. Regional localization of carbonic anhydrase genes *CA1* and *CA3* on human chromosome 8. Somat. Cell Mol. Genet. 3:173–178.

137. Davis, M., S. Malcolm, A. Hall, and C.J. Marshall. 1983. Localization of the *N-RAS* oncogene to chromosome 1cen-p21 by in situ hybridization. EMBO J. 2:2281–2283.

138. Day, C.E., and P.P. Jones. 1983. The gene encoding the Ia antigen-associated invariant chain (*I$_i$*) is not linked to the *H-2* complex. Nature 302:157–159.

139. De Klein, A., A. Geurts Van Kessel, G. Grosveld, C.R. Bartram, A. Hagemeyer, D. Bootsma, N.K. Spurr, N. Heisterkamp, J. Groffen, and J.R. Stephenson. 1982. A cellular oncogene is translocated to the Philadelphia chromosome in chronic myelocytic leukemia. Nature 300:765–767.

140. De La Chapelle, A., A.W. Erikson, M. Kirjarinta, and F. Knutar. 1971. Glutathione reductase activity in haematological disorders associated with C trisomy. Eur. J. Clin. Invest. 1:366.

141. De La Chapelle, A., G. Lenoir, J. Boué, A. Boué, P. Gallano, C. Huerre, M.-F. Szajnert, M. Jeanpierre, J.-M. Lalouel, and J.-C. Kaplan. 1983. Lambda Ig constant region genes are translocated to chromosome 8 in Burkitt's lymphoma with t(8;22). Nucl. Acids Res. 11:1133–1142.

142. De La Chapelle, A., P. Vuopio, and A. Icen. 1976. Trisomy 8 in the bone marrow associated with high red cell glutathione reductase activity. Blood 47:815–826.

143. De Martinville, B., J.M. Cunningham, M.J. Murray, and U. Francke. 1983. The *N-ras* oncogene assigned to the short arm of human chromosome 1. Nucl. Acids Res. 11:5267–5275.

144. De Martinville, B., J.M. Cunningham, M.J. Murray, and U. Francke. 1984. The *N-RAS* oncogene assigned to chromosome 1 (p31-cen) by somatic cell hybrid analysis. Cytogenet. Cell Genet. 37:531.

145. De Taisne, C., A. Gegonne, D. Stehelin, A. Bernheim, and R. Berger. 1984. Chromosomal localization of the human proto-oncogene c-ets. Nature 310:581–583.

146. DeLorenzo, R.J., and F.H. Ruddle. 1970. Glutamate oxalate transaminase (GOT) genetics in *Mus musculus*: linkage, polymorphism, and phenotypes of the *Got-2* and *Got-1* loci. Biochem. Genet. 4:259–273.

147. DeLorenzo, R.J., and F.H. Ruddle. 1969. Genetic control of two electrophoretic variants of glucosephosphate isomerase in the mouse (*Mus musculus*). Biochem. Genet. 3:151–162.

148. DeMaeyer, E., and F. Dandoy. 1987. Linkage analysis of the murine interferon alpha locus (*Ifa*) on chromosome 4. J. Hered. 78:143–146.

149. DeMars, R., S.L. Le Van, B.L. Trend, and L.B. Russell. 1976. Abnormal ornithine carbamoyltransferase in mice having the sparse-fur mutation. Proc. Natl. Acad. Sci. USA 73:1693–1697.

150. Dean, M., C. Kozak, J. Robbins, R. Callahan, S. O'Brien, and G.F. Vande Woude. 1987. Chromosomal localization of the *met* proto-oncogene in the mouse and cat genome. Genomics 1:167–173.

151. Deimling, O. von, J. Peters, and S. Povey. 1984. Personal communication. Mouse News Lett. 70:81.

152. Dembic, Z., W. Bannwarth, B.A. Taylor, and M. Steinmetz. 1985. The gene encoding the T-cell receptor alpha-chain maps close to the *Np-2* locus on mouse Chromosome 14. Nature 314:271–273.

153. Demmer, L.A., E.H. Birkenmeier, D.A. Sweetser, M.S. Levin, S. Zollman, R.S. Sparkes, T. Mohandas, A.J. Lusis, and J.I. Gordon. 1987. The cellular retinol binding protein II gene: sequence analysis of the rat gene, chromosomal localization in mice and humans, and documentation of its close linkage to the cellular retinol binding protein gene. J. Biol. Chem. 262:2458–2467.

154. Diaz, M.O., M.M. LeBeau, P.M. Pitha, and J.D. Rowley. 1986. Interferon and c-ets-1 genes in the translocation (9:11)(p22;q23) in human acute monocytic leukemia. Science 231:265–267.

155. Diesseroth, A., A. Nienhuis, H. Lawrence, R. Giles, P. Turner, and F.H. Ruddle. 1978. Chromosomal localization of the human beta globin gene on human chromosome 11 by somatic cell hybrids. Proc. Natl. Acad. Sci. USA 75:1456–1460.

156. Diesseroth, A., A. Nienhuis, P. Turner, R. Velez, W.F. Anderson, F.H. Ruddle, J. Lawrence, R.P. Creagan, and R.S. Kucherlapati. 1977. Localization of the human alpha globin structural gene to chromosome 16 in somatic cell hybrids by molecular hybridization assay. Cell 12:205–218.

157. Diesseroth, A., and D. Hendrick. 1978. Human alphaglobin gene expression following chromosomal dependent gene transfer into mouse erythroleukemia cells. Cell 15:55–63.

158. Donald, L.J., H.S. Wang, and J.L. Hamerton. 1980. Assignment of the sorbitol dehydrogenase locus to human chromosome 16. Cytogenetics 32:268.

159. Doolittle, D.P., L.L. Hulbert, and C. Cordy. 1974. A new allele of the sparse fur gene in the mouse. J. Hered. 65:194–195.

160. Douglas, T.C., and P.E. Dawson. 1979. The position of the gene for glyoxalase 1 on chromosome 17 of the mouse. Immunogenetics 8:367–371.

161. Douglas, T.C., T. Meo, and H. Skarvall. 1978. Location of the gene for the theta antigen in the mouse. II. Three-point crosses place *Thy-1* in proximal region of chromosome 9. J. Hered. 69:224–228.

162. Dracapoli, N.C., W.J. Rettig, G.K. Whitfield, B.A. Spengler, J.L. Biedler, L.J. Old, and I.A. Kourides. 1985. Assignment of the structural gene for the beta subunit of thyroid stimulating hormone to human chromosome 1p22. Cytogenet. Cell Genet. 40:619.

163. Duncan, A.M.V., C. Morgan, B.L. Gallie, R.A. Phil-

lips, and J. Squire. 1987. Reevaluation of the sublocalization of esterase D and its relation to the retinoblastoma locus by in situ hybridization. Cytogenet. Cell Genet. 44:153–157.

164. Dunnen, J.T. den, R.J.E. Jongbloed, N.H. Lubsen, A.H.M. Geurts van Kessel, and A. Westerveld. 1985. Human lens gamma-crystallin genes are located in the p12-qter region of chromosome 2. Cytogenet. Cell Genet. 40:616.

165. Eiberg, H., J. Mohr, and L. Staubnielsen. 1982. Linkage of orosomucoid (ORM) and ABO and AK1. Cytogenet. Cell Genet. 32:372.

166. Eiberg, H., J. Mohr, and L. Staubnielsen. 1983. Delta-amino-levulinate dehydratase: synteny with *ABO-AK1-ORM* (and assignment to chromosome 9) Clin. Genet. 23:150–154.

167. Eiberg, H., N. Moller, J. Mohr, and L.S. Nielsen. 1986. Linkage of transcobalamin II (*TC2*) to the P blood group system and assignment to chromosome 22. Clin. Genet. 29:354–359.

168. Eicher, E.M., B.A. Taylor, S.C. Leighton, and J.E. Womack. 1980. A serum protein polymorphism determinant on chromosome 9 of *Mus musculus*. Mol. Gen. Genet. 177:571–576.

169. Eicher, E.M., and J.E. Womack. 1977. Chromosomal location of soluble glutamic-pyruvic transaminase-1 (*Gpt-1*) in the mouse. Biochem. Genet. 15:1–8.

170. Eicher, E.M., J.L. Southard, C.R. Schriver, and F.H. Glorieux. 1976. Hypophosphatemia: mouse model for human familial hypophosphatemic (vitamin D-resistant) rickets. Proc. Natl. Acad. Sci. USA 73:4667–4671.

171. Eicher, E.M., M. Cherry, and L. Flaherty. 1978. Autosomal phosphoglycerate kinase linked to mouse major histocompatibility complex. Mol. Gen. Genet. 158:225–228.

172. Eicher, E.M., R.H. Stern, J.E. Womack, M.T. Davisson, T.H. Roderick, and S.C. Reynolds. 1976. Evolution of mammalian carbonic anhydrase loci by tandem duplication: close linkage of *Car-1* and *Car-2* to the centromere region of chromosome 3 of the mouse. Biochem. Genet. 14:651–660.

173. Elsen, P. van den, G. Bruns, D.S. Gerhard, D. Pravetcheva, C. Jones, D. Housman, F.H. Ruddle, S. Orkin, and C. Terhorst. 1985. Assignment of the gene coding for the T3-delta subunit of the T3-T-cell receptor complex to the long arm of human chromosome 11 and to mouse chromosome 9. Proc. Natl. Acad. Sci. USA 82:2920–2924.

174. Elsevier, S.M., R.S. Kucherlapati, E.A. Nichols, R.P. Creagan, R.E. Giles, F.H. Ruddle, K. Willecke, and J.K. McDougall. 1974. Assignment of the gene for galactokinase to human chromosome 17 and its regional localization to band q21-q22. Nature 251:633–635.

175. Epstein, R., K. Lehmann, M. Cohn, C. Buckler, W. Rowe, and M. Davisson. 1984. Linkage of the *Igl-1* structural and regulatory gene to *Akv-2* on chromosome 16. Immunogenetics 19:527–537.

176. Erikson, J., J. Martinis, and C.M. Croce. 1981. Assignment of the genes for human lambda immunoglobulin chains to chromosome 22. Nature 294:173–175.

177. Faber, H.B., R.S. Kucherlapati, M.D. Poulik, F.H. Ruddle, and O. Smithies. 1976. Beta-2-microglobulin locus on human chromosome 15. Somat. Cell Genet. 2:141–153.

178. Farrall, M., E. Watson, G. Bates, G. Bell, J. Bell, K.A. Davies, X. Estivill, H. Kruyer, H.-Y. Law, N. Lench, and W. Lissens. 1986. Further data supporting linkage between cystic fibrosis and the *met* oncogene and haplotype analysis with *met* and pJ3.11. Am. J. Hum. Genet. 39:713–719.

179. Fearon, E.R., S.E. Antonarakis, D.A. Meyers, and M.A. Levine, 1984. c-Ha-ras-1 oncogene lies between beta-globin and insulin loci on human chromosome 11p. Am. J. Hum. Genet. 36:329–337.

180. Feinstein, R.N., R.D. Dickerman, and D. Grahn. 1967. Catalasemic mutant in mice. Genetics 56:559.

181. Ferguson-Smith, M.A., B.F. Newman, P.M. Ellis, D.M.G. Thompson, and I.D. Riley. 1973. Assignment by deletion of human red cell acid phosphatase gene locus to the short arm of chromosome 2. Nature New Biol. 242:271–274.

182. Fisher, R.A., S. Povey, M. Bobrow, E. Solomon, Y. Boyd, and B. Carritt. 1977. Assignment of the *DIA1* locus to chromosome 22. Ann. Hum. Genet. 41:150–155.

183. Fisher, R.A., W. Putt, S. Povey, K.E. Buckton, I.P. Gormley, and P. Perry. 1976. Assignment of the locus determining inorganic pyrophosphatase to chromosome 10 in man. Cytogenet. Cell Genet. 16:129–130.

184. Flanagan, J.G., and T.H. Rabbitts. 1982. Arrangement of human immunoglobulin heavy chain constant region genes implies evolutionary duplication of a segment containing gamma, epsilon, and alpha genes. Nature 300:709–713.

185. Floyd-Smith, G.A., A.S. Whitehead, R.H. Colten, and U. Franke. 1985. Human serum amyloid P component (SAP) is located on the proximal long arm of chromosome 1. Cytogenet. Cell Genet. 40:631.

186. Floyd-Smith, G., A.S. Whitehead, H.R. Colten, and U. Francke. 1986. The human C-reactive protein gene (*CRP*) and serum amyloid P component gene (*APCS*) are located on the proximal long arm of chromosome 1. Immunogenetics 24:171–176.

187. Fournier, R.E.K., and R.G. Moran. 1983. Complementation mapping in microcell hybrids: localization of *Fpgs* and *Ak-1* on *Mus musculus* chromosome 2. Somat. Cell Genet. 9:69–84.

188. Fowler, M.L., H. Nakai, M.G. Byers, H. Fukushima, R.L. Eddy, W.M. Henry, L.L. Haley, and J.S. O'Brien. 1986. Chromosome 1 localization of the human alpha-L-fucosidase structural gene with a homologous site on chromosome 2. Cytogenet. Cell Genet. 43:103–108.

189. Fox, M.F., D.C. Du Toit, L. Warnick, and A.E. Retief.

1984. Regional localization of alpha-galactosidase (*GLA*) to Xpter-Xq22, hexosaminidase B (*HEXB*) to 5q13-qter, and arylsulfatase B (*ARSB*) to 5pter-q13. Cytogenet. Cell Genet. 38:45–49.

190. Francke, U. 1974. Regional localization of the human genes for malate dehydrogenase-1 and isocitrate dehydrogenase-1 on chromosome 2 by interspecific hybridization using human cells with the balanced reciprocal translocation t(1;2)(q32;q12). Rotterdam Conference 1978:138–142.

191. Francke, U. 1976. The human gene for beta-glucuronidase is on chromosome 7. Am. J. Hum. Genet. 28:357–362.

192. Francke, U., B. de Martinville, L. Coussens, and A. Ullrich. 1983. The human gene for the beta subunit of nerve growth factor is located on the proximal short arm of chromosome 1. Science 222:1248–1250.

193. Francke, U., B. de Martinville, P. D'Eustachio, and F.II. Ruddle. 1982. Comparative gene mapping: murine lambda light chain genes are located in region cen–B5 of mouse chromosome 16 not homologous to human chromosome 21. Cytogenet. Cell Genet. 33:267–271.

194. Francke, U., D.L. George, and V.M. Riccardi. 1977. Gene dose effect: intraband mapping of *LDHA* locus using cells from four individuals with different interstitial deletions of 11p. Cytogenetics 19:197.

195. Francke, U., N. Busby, D. Shaw, S. Hansen, and M.G. Brown. 1976. Intrachromosomal gene mapping in man: assignment of nucleoside phosphorylase to region 14 cen-to-14q21 by interspecific hybridization of cells with a t(X;14)(p22;q21) translocation. Somat. Cell Genet. 2:27–40.

196. Francke, U., P.A. Lalley, W. Moss, J. Ivy, and J.D. Minna. 1977. Gene mapping in *Mus musculus* by interspecific cell hybridization: assignment of the gene for tripeptidase-1 to chromosome 10, dipeptidase-2 to chromosome 18, acid phosphatase-1 to chromosome 12, and adenylate kinase-1 to chromosome 2. Cytogenet. Cell Genet. 19:57–84.

197. Francke, U., P. Tetri, R.T. Taggart, and N. Oliver. 1981. Conserved autosomal syntenic group on mouse (MMU) chromosome 15 and human (HSA) chromosome 22: assignment of a gene for arylsulfatase A to MMU15 and regional mapping of *DIA1*, *ARSA*, *ACO2* on HSA22. Cytogenet. Cell Genet. 31:58–69.

198. Francke, U., P. Tetri, R.T. Taggart, and N. Oliver. 1982. Regional mapping of *DIA1*, *ARSA*, and *ACO2* on chromosome 22 using hybrids with a t(15;22)(q14;q13:31) translocation. Cytogenet. Cell Genet. 32:276–277.

199. Francke, U., and R.T. Taggart. 1979. Assignment of the gene for cytoplasmic superoxide dismutase (*Sod-1*) to a region of chromosome 16 and of *Hprt* to a region of the X chromosome in the mouse. Proc. Natl. Acad. Sci. USA 76:5230–5233.

200. Francke, U., and R.T. Taggart. 1980. Comparative gene mapping: order of loci on the X chromosome is different in mice and humans. Proc. Natl. Acad. Sci. USA 77:3595–3599.

201. Francke, U., and U. Gehring. 1980. Chromosome assignment of a murine glucocorticoid receptor gene (*Grl-1*) using intraspecies somatic cell hybrids. Cell 22:657–664.

202. Frater-Schroder, M., M. Prochzka, O. Haller, F. Arwert, H.J. Porck, L.C. Skow, L.-G. Lundin, J. Hilkens, and J. Hilgers. 1985. Localization of the gene for the vitamin B12 binding protein, transcobalamin II, near the centromere on mouse Chromosome 11, linked to the hemoglobin alpha-chain locus. Biochem. Genet. 23:139–153.

203. Frater-Schroder, M., O. Haller, R. Gmur, L. Kierat, and S. Anastasi. 1982. Allelic forms of transcobalamin 2. Biochem. Genet. 20:1001–1014.

204. Fratini, A., R.N. Simmers, D.F. Callen, V.J. Hyland, J.A. Tischfield, P.J. Stambrook, and G.R. Sutherland. 1986. A new location for the human adeninephosphoribosyltransferase gene (*APRT*) distal to the haptoglobin (*HP*) and fra (16)(q23) (RA16D) loci. Cytogenet. Cell Genet. 43:10–13.

205. Friedrich, A., H. Brunner, D. Smeets, E. Lambermon, and H.-H. Ropers. 1987. Three-point linkage analysis employing *C3* and 19cen markers assigns myotonic dystrophy to 19q. Hum. Genet. 75:291–293.

206. Fujii, D., J.E. Brissenden, R. Derynck, and U. Francke. 1986. Transforming growth factor beta gene maps to human chromosome 19 long arm and to mouse chromosome 7. Somat. Cell Mol. Genetics. 12:281–288.

207. Funanage, V.L., T.T. Myoda, P.A. Moses, and H.R. Cowell. 1984. Assignment of the human dihydrofolate reductase gene to the q11–q22 region of chromosome 5. Mol. Cell. Biol. 4:2010–2016.

208. Fundele, R., T. Bucher, A. Gropp, and H. Winking. 1981. Enzyme patterns in trisomy 19 of the mouse. Dev. Genet. 2:291–303.

209. Gartler, S.M., D.E. Riley, R.V. Lebo, M.-C. Cheung, R.L. Eddy, and T.B. Shows. 1986. Mapping of human autosomal phosphoglycerate kinase sequence to chromosome 19. Somat. Cell Mol. Genet. 12:395–401.

210. Gartler, S., D. Riley, R. Eddy, and T. Shows. 1985. A human phosphoglycerate kinase gene has been assigned to chromosome 19. Cytogenet. Cell Genet. 40:635–636.

211. Gartler, S., and M. Rivest. 1983. Evidence for X-linkage of steroid sulfatase in the mouse: steroid sulfatase levels in oocytes of XX and XO mice. Genetics 103:137–141.

212. Gehring, U., B. Segnitz, B. Foellmer, and U. Francke. 1985. Assignment of the human gene for the glucocorticoid receptor to chromosome 5. Proc. Natl. Acad. Sci. USA 82:3751–3755.

213. George, D.L., B. de Martinville, and U. Francke. 1984. Amplified DNA sequences, including a c-Ki-ras oncogene, present in Y1 mouse adrenal tumor cells originate from mouse chromosome 6. Cytogenet. Cell Genet. 37:477–488.

214. Gerhard, D.S., and D.E. Housman. 1985. Ordering of

15 markers on human chromosome 11p by meiotic recombination: analysis by three point linkage tests. Cytogenet. Cell Genet. 40:640.

215. Gerhard, D.S., T. Vulliamy, G. Bruns, P. Kramer, J. Gusella, J. Kidd, K. Kidd, and D.E. Housman. 1985. Linkage relationships among nine markers on the short arm of chromosome 11. Cytogenet. Cell Genet. 40:640–641.

216. Giampietro, P., A.-L. Wang, F. Arredondo-Vega, M. Smith, and R.J. Desnick. 1982. Assignment of genes for human porphobilinogen deaminase and esterase A4 to chromosomal region 11q23-qter. Cytogenet. Cell Genet. 32:280–281.

217. Giblett, E.R., S.-H. Chen, J.E. Anderson, and M. Lewis. 1974. A family study suggesting genetic linkage of phosphopyruvate hydratase (enolase) to the Rh blood group system. Cytogenet. Cell Genet. 13:91–92.

218. Gibson, D.M., S.J. Maclean, and M. Cherry. 1983. Recombination between kappa chain genetic markers and the *Lyt-3* locus. Immunogenetics 18:111–116.

219. Gilbert, F., R. Kucherlapati, R. Creagan, M.J. Murname, G.J. Darlington, and F.H. Ruddle. 1975. Tay-Sachs' and Sandhoffs' diseases: the assignment of genes for hexosaminidase A and B to individual human chromosomes. Proc. Natl. Acad. Sci. USA 72:263–267.

220. Glanville, N., D.M. Durnam, and R.D. Palmiter. 1981. Structure of mouse metallothionein-1 gene and its mRNA. Nature 292:267–269.

221. Glaser, T., D. Gerhard, C. Jones, L. Albritton, P. Lalley, and D. Housman. 1985. A fine structure map of chromosome 11p. Cytogenet. Cell Genet. 40:643.

222. Glaser, T., G. Bruns, P. Watkins, T. Shows, and D. Housman. 1985. The beta subunit of follicle stimulating hormone (*FSHB*) is deleted in WAGR patients: a further definition of the *WAGR* locus. Cytogenet. Cell Genet. 40:642.

223. Goff, S.P., P. D'Eustachio, F.H. Ruddle, and D. Baltimore. 1982. Chromosomal assignment of the endogenous proto-oncogene C-abl. Science 218:1317–1319.

224. Gold, D.P., J.M. van Dongen, C.C. Morton, G.A.P. Bruns, P. van den Elsen, H.M. Guerts van Kessel, and C. Terhorst. 1987. The gene encoding the epsilon subunit of the T3/T-cell receptor complex maps to chromosome 11 in humans and to chromosome 9 in mice. Proc. Natl. Acad. Sci. USA 84:1664–1668.

225. Goodfellow, P.N., E.A. Jones, V. van Heyningen, E. Solomon, M. Bobrow, V. Miggiano, and W.F. Bodmer. 1975. The beta-2-microglobulin gene is on chromosome 15 and not in the *HL-A* region. Nature 254:267–269.

226. Gough, N.M., J. Gough, D. Metcalf, A. Kelso, D. Grail, N.A. Nicola, A.W. Burgess, and A.R. Dunn. 1984. Molecular cloning of cDNA encoding a murine haemopoietic growth regulator, granulocyte-macrophage colony stimulating factor. Nature 309:763–767.

227. Goyns, M.H., B.D. Young, A. Geurts van Kessel, A. de Klein, G. Grosveld, C.R. Bartram, and D. Bootsma. 1984. Regional mapping of the human immunoglobulin lambda light chain to the Philadelphia chromosome in chronic myeloid leukemia. Leuk. Res. 8:547–553.

228. Griffen, C.A., B.S. Emanuel, J.R. Hansen, W.K. Cavenee, and J.C. Myers. 1987. Human collagen genes encoding basement membrane alpha 1(IV) and alpha 2(IV) chains map to the distal long arm of chromosome 13. Proc. Natl. Acad. Sci. USA 84:512–516.

229. Grobler-Rabie, A., D. Brebner, S. Vandenplas, G. Wallis, R. Dagliesh, R.E. Kaufman, J. Bester, C.G.P. Mathew, and C.D. Boyd. 1985. Polymorphism of a DNA sequence in the human proalpha 21 collagen gene. J. Med. Genet. 22:182–186.

230. Grzechik, K.-H. 1976. Assignment of a structural gene for beta-glucuronidase to human chromosome 7. Somat. Cell Genet. 2:401–410.

231. Grzeschik, K.-H. 1976. Assignment of a gene for human mitochondrial isocitrate dehydrogenase (ICD-M, EC 1.1.1.41) to chromosome 15. Hum. Genet. 34:23–38.

232. Grzeschik, K.-H., and H. Kazazian. 1985. Report of the committee on the genetic constitution of chromosomes 10, 11, and 12. Cytogenet. Cell Genet. 40:179–205.

233. Gunning, P., P. Ponte, H. Kedes, R. Eddy, and T. Shows. 1984. Chromosomal location of the co-expressed human skeletal and cardiac actin genes. Proc. Natl. Acad. Sci. USA 81:1813–1817.

234. Haines, J.L., and J.A. Trofatter. 1986. Multipoint linkage analysis of spinocerebellar ataxia and markers on chromosome 6. Genet. Epidemiol. 3:399–405.

235. Hamerton, J.L., T. Mohandas, P.J. McAlpine, and G.R. Douglas. 1975. Assignment of three human gene loci to regions of chromosome 2. Cytogenet. Cell Genet. 14:176–178.

236. Hamerton, J.L., T. Mohandas, P.J. McAlpine, and G.R. Douglas. 1975. Localization of human gene loci using spontaneous chromosome rearrangements in human–Chinese hamster somatic cell hybrids. Am. J. Hum. Genet. 27:595–608.

237. Harper, M.E., and A. Dugaiczyk. 1983. Linkage of the evolutionary related serum albumin and alpha fetoprotein genes within q11-22 of human chromosome 4. Am. J. Hum. Genet. 35:565–572.

238. Harper, M.E., A. Ullrich, and G.F. Saunders. 1981. Localization of the human insulin gene to the distal end of the short arm of chromosome 11. Proc. Natl. Acad. Sci. USA 78:4458–4460.

239. Harper, M.E., C. Blatt, S. Marks, M.N. Nesbitt, and M.I. Simon. 1984. Gene mapping in the mouse by analysis of RFLP segregation in recombinant inbred strains. Cytogenet. Cell Genet. 37:488.

240. Harper, M.E., G. Franchini, J. Love, M.I. Simon, R.C. Gallo, and F. Wong-Staal. 1983. Chromosomal sublocalization of human c-myb and c-fes cellular oncogenes. Nature 304:169–171.

241. Heim, S., and F. Mitelman. 1987. Nineteen of 26 cellular oncogenes precisely localized in the human genome map to one of 83 bands involved in primary cancer-specific rearrangements. Hum. Genet. 75:70–72.

242. Heisterkamp, N., J. Fraffen, J.R. Stephenson, N.K. Spurr, P.N. Goodfellow, E. Solomon, B. Carritt, and W.F. Bodmer. 1982. Chromosomal localization of human cellular homologues of two viral oncogenes. Nature 299:747–749.

243. Hellkuhl, B., and K.-H. Grzeschik. 1976. Localization of GOTs to the q24-qter region of human chromosome 10. Cytogenet. Cell Genet. 16:157–158.

244. Hellkuhl, B., and K.-H. Grzeschik. 1978. Assignment of a gene for arylsulfatase B to human chromosome 5 using human–mouse somatic cell hybrids. Cytogenet. Cell Genet. 22:203–206.

245. Herbschleb-Voogt, E., M. Monteba van Heuval, L.M.M. Wijnen, A. Westerveld, P.L. Pearson, and P. Meera Khan. 1978. Chromosomal assignment and regional localization of *CS, ENO2, GAPDH, LDH2, PEPB,* and *TPI* in man–rodent cell hybrids. Cytogenet. Cell Genet. 22:482–486.

246. Hildebrand, C.E., F.J. Gonzalez, C.A. Kozak, and D.W. Nebert. 1985. Regional linkage analysis of the dioxin-inducible *P-450* gene family on mouse chromosome 9. Biochem. Biophys. Res. Comm. 130:396–406.

247. Hildebrand, C.E., F.J. Gonzalez, O.W. McBride, and D.W. Nebert. 1985. Assignment of the human 2, 3, 7, 8-tetrachlorodibenzs-p-dioxin-inducible cytochrome P-450 gene to chromosome 15. Nucl. Acids Res. 13:2009–2015.

248. Hilkens, J., H.T. Cuypers, G. Selten, V. Kroezen, J. Hilgers, and A. Berns. 1986. Genetic mapping of *Pim-1* putative oncogene to mouse chromosome 17. Somat. Cell Mol. Genet. 12:81–88.

249. Hjorth, J.P. 1979. Genetic variation in mouse salivary amylase rate of synthesis. Biochem. Genet. 17:665–682.

250. Hoffman, H.A., and C.K. Grieshaber. 1974. Genetic studies of murine catalase. J. Hered. 65:277–279.

251. Hogarth, P.M., I.F.C. McKenzie, V.R. Sutton, K.M. Curnow, B.K. Lee, and E.M. Eicher. 1987. Mapping of the mouse *Ly-6, Xp-14,* and *Gdc-1* loci to Chromosome 1. Immunogenetics 25:21–27.

252. Holmes, R.S. 1977. The genetics of alpha-hydroxyacid oxidase and alcohol dehydrogenase in the mouse: evidence for multiple gene loci and linkage between *Hao-3* and *Adh-3* Genetics 87:709–716.

253. Holmes, R.S. 1979. Genetics and ontogeny of alcohol dehydrogenase isozymes in the mouse: evidence for a cis-acting regulator gene (*Adt-1*) controlling C2 isozyme expression in reproductive tissues and close linkage of *Adh-3* and *Adt-1* on chromosome 3. Biochem. Genet. 17:461–472.

254. Holmes, R.S. and J.A. Duley. 1982. Aldehyde dehydrogenase isozymes *Adh-1* and *Adh-3* on chromosome 3. Mouse News Lett. 66:60.

255. Holmes, R.S., R. Albanese, F.D. Whitehead, and J.A. Duley. 1981. Mouse alcohol dehydrogenase isozymes: products of closely localized duplicated genes exhibiting divergent kinetic properties. J. Exp. Zool. 217:151–157.

256. Honey, N.K., A.Y. Sakaguchi, C. Quinto, R.J. Mac-Donald, G.I. Bell, C. Craik, W.J. Rutter, and S.L. Naylor.

257. Honey, N.K., A.Y. Sakaguchi, C. Quinto, R.J. Mac-Donald, W.J. Rutter, and S.L. Naylor. 1984. Assignment of the human genes for elastase to chromosome 12, and for trypsin and carboxypeptidase A to chromosome 7. Cytogenet. Cell Genet. 37:492.

258. Honey, N.K., A.Y. Sakaguchi, P.A. Lalley, C. Quinto, R.J. MacDonald, W.J. Rutter, and S.L. Naylor. 1984. Assignment in mouse of the genes for chymotrypsinogen B, elastase, trypsin, and carboxypeptidase A. Cytogenet. Cell Genet. 37:492–493.

259. Hopkinson, D.A., C. Junien, S. Heuertz, J.F. Mattei, and J. Frezal. 1981. Confirmation of the assignment of the locus determining *ADA* to chromosome 20 in man: data on possible syntney of *ADA* and *ITP* in human–Chinese hamster somatic cell hybrids. Cytogenet. Cell Genet. 16:159–160.

260. Hoppener, J.W.M., P.H. Steenbergh, J. Zandberg, E. Bakker, P.L. Pearson, A.H.M. Guerts van Kessel, H.S. Jansz, and C.J.M. Lips. 1984. Localization of the polymorphic human calcitonin gene on chromosome 11. Hum. Genet. 66:309–312.

261. Hors-Cayla, M.C., C. Junien, S. Heuertz, J.F. Mattei, and J. Frezal. 1981. Regional assignment of arylsulfatase A, mitochondrial aconitase and NADH-cytochrome b5 reductase by somatic cell hybridization. Hum. Genet. 58:140–143.

262. Housman, D.E., D.S. Gerhard, T. Glaser, and C. Jones. 1985. Mapping of chromosome 11p by linkage and somatic cell genetic techniques: a comparison of relative map order derived by each method. Cytogenet. Cell Genet. 40:655–656.

263. Hsu, L.C., A. Yoshida, and T. Mohandas. 1986. Chromosomal assignment of the genes for human aldehyde dehydrogenase-1 and aldehyde dehydrogenase-2. Am. J. Hum. Genet. 38:641–648.

264. Huebner, K., A.P. Palumbo, M. Isobe, C.A. Kozak, S. Monaco, G. Rovera, C. Croce, and P.J. Curtis. 1985. The alpha-spectrin gene is on chromosome 1 in mouse and man. Proc. Natl. Acad. Sci. USA 82:3790–3793.

265. Huebner, K., M. Isobe, C.M. Croce, D.W. Golde, S.E. Kaufman, and J.C. Gasson. 1985. The human gene encoding GM-CSF is at 5q21-q32, the chromosome region deleted in the 5q- anomaly. Science 230:1282–1285.

266. Huerre, C., C. Junien, D. Weil, M.-L. Chu, M. Morabito, V.C. Nguyen, J.C. Myers, C. Foubert, M.-S. Gross, D.J. Prokop, and A. Boué. 1983. Human type 1 procollagen genes are located on different chromosomes. Proc. Natl. Acad. Sci. USA 79:6627–6630.

267. Huerre, C., G. Uzan, K.-H. Grzeschik, D. Weil, M. Levin, M.C. Hors-Cayla, J. Boué, A. Kahn, and C. Junien. 1984. The structural gene for transferrin *TF* maps to 3q21-3qter. Ann. Genet. 27:5–10.

268. Huijing F., E.M. Eicher, and D.L. Coleman. 1973.

Location of phosphorylase kinase (*Phk*) in the mouse X-chromosome. Biochem Genet. 9:193–196.

269. Hulsebos, T., H. Brunner, B. Weiringa, U. Friedrich Humphries, and O. Myklebost. 1985. Regional assignment of *C3*, *GPI*, *APOC2*, and beta-*HCG* and their linkage relationships with *DM* and 19cen. Cytogenet. Cell Genet. 40:658.

270. Huppi, K., R. Duncan, and M. Potter. 1988. *Myc-1* is centromeric to the linkage group *Ly-6–Sis–Gdc-1* on mouse chromosome 15. Immunogenetics 27:215–219.

271. Hutton, J.J., and D.L. Coleman. 1969. Linkage analyses using biochemical variants in mice. II. Levulinate dehydratase and autosomal glucose-6-phosphate dehydrogenase. Biochem. Genet. 3:517–523.

272. Hutton, J.J., and T.H. Roderick. 1970. Linkage analyses using biochemical variants in mice. III. Linkage relationships of eleven biochemical markers. Biochem. Genet. 4:339–350.

274. Jeanpiere, M., I. Henry, C. Turleau, A. Ullrich, J. Mallet, J. de Grouchy, and C. Junien. 1985. Beckwith–Wiedeman syndrome (BWS) and 11p15 markers. Cytogenet. Cell Genet. 40:661.

275. Jeremiah, S.J., C. Kielty, S. Povey, and T. McLellan. 1984. Immunological characterization of human GPT in hybrids supports assignment to chromosome 8. Cytogenet. 37:498–499.

276. Jeremiah, S.J., S. Povey, M.W. Burley, C. Kielty, M. Lee, G. Spowart, G. Corney, and P.J.L. Cook. 1982. Mapping studies on human mitochondrial glutamate oxaloacetate transaminase. Ann. Hum. Genet. 46:145–152.

277. Jhanwar, S.C., B.G. Neel, W.S. Hayward, and R.S.K. Changanti. 1984. Localization of the cellular oncogenes *ABL*, *SIS*, and *FES* on human germ-line chromosomes. Cytogenet. Cell Genet. 38:73–75.

278. Johnson, F.M., F. Chaslow, G. Anderson, P. MacDougal, R.W. Hendren, and S.E. Lewis. 1981. A variation in mouse kidney pyruvate kinase activity determined by a mutant gene on chromosome 9. Genet. Res. 37:123–131.

279. Johnson, F.M., R.W. Hendren, F. Chasalow, L.B. Barnett, and S.E. Lewis. 1981. A null mutation at the mouse phosphoglucomutase-1 locus and a new locus, *Pgm-3*. Biochem. Genet. 19:599–615.

280. Jones, C., and F.T. Kao. 1984. Regional mapping of the folylpolyglutamate synthetase gene (*FPGS*) to 9cen-q34. Cytogenet. Cell Genet. 37:499.

281. Jongsma, A.P.M., A. Hagemeijer, and P. Meera Khan. 1975. Regional mapping of *TPI*, *LDH-B*, and *PEP-B* on chromosome 12 of man. Cytogenet. Cell Genet. 14:189–191.

282. Jongsma, A., H. Someren, A. Westerveld, A. Hagemeijer, and P. Pearson. 1973. Localization of genes on human chromosomes by studies of human–Chinese hamster somatic cell hybrids. Humangenetik 20:195–202.

283. Jongsma, A.P.M., W.R.T. Los, and J. Hagemeijer. 1974. Evidence for synteny between the human loci for triose phosphate isomerase, lactate dehydrogenase-B, and peptidase-B and the regional mapping of these loci on chromosome 12. Cytogenet. Cell Genet. 13:106–107.

284. Jongsma, A., H. Someren, A. Westerveld, A. Hagemeijer, and P. Pearson. 1973. Localization of genes on human chromosomes by studies of human–Chinese hamster somatic cell hybrids. Humangenetik 20:195–202.

285. Joyner, A.L., R.V. Lebo, Y.W. Kan, R. Tjian, D.R. Cox, and G.R. Martin. 1985. Comparative mapping of a conserved homeo box region in human and mouse. Cytogenet. Cell Genet. 40:663.

286. Juneja, R.K., B. Gahne, J. Rapacz, and J. Hasler-Rapacz 1986. Linkage between the porcine genes encoding immunoglobulin heavy-chain allotypes and some serum alpha-protease inhibitors: a conserved linkage in pig, mouse, and human. Ann. Genet. 17:225–233.

287. Junien, C., C. Turleau, J. deGrouchy, R. Said, M.O. Rethore, R. Tenconi, and J.L. Dufier. 1980. Regional assignment of catalase (*CAT*) gene to band 11p13; Association with the *WAGR* complex. Ann. Genet. 23:165–168.

288. Junien, C., M. Vibert, D. Weil, V.C. Nguyen, and J.-C. Kaplan. 1978. Assignment of NADH-cytochrome b5 reductase (*DIA1* Locus) to human chromosome 22. Hum. Genet. 42:233–239.

289. Junien, C., S. Despoisse, C. Turleau, J. de Grouchy, T. Bucher, and R. Fundele. 1982. Assignment of phosphoglycerate mutase (*PGAMA*) to human chromosome 10. Ann. Genet. 25:25–27.

290. Kaneda, Y., H. Hayes, T. Uchida, M.C. Yoshida, and Y. Okada. 1987. Regional assignment of five genes on human chromosome 19. Chromosoma 95:8–12.

291. Kao, F.-T., J.W. Hawkins, M.L. Law, and A. Dugaiczyk. 1982. Assignment of the structural gene for albumin to human chromosome 4. Hum. Genet. 62:377–341.

292. Karathansis, S.K., J. McPherson, V. Zannis, and J.L. Breslow. 1983. Linkage of human apolipoprotein A-1 and C-111 genes. Nature 304:371–373.

293. Karin, M., and T.B. Shows. 1984. The human metallothionein gene family is clustered on chromosome 16. Proc. Natl. Acad. Sci. USA 81:5494–5498.

294. Keijzer, W., M. Stefanini, A. Westerveld, and D. Bootsma. 1984. Mapping of the XPAC gene involved in complementation of the defect in xeroderma pigmentosum group A cells. Cytogenet. Cell Genet. 37:508.

295. Keitges, E., M. Rivest, M. Siniscalco, and S.M. Gartler. 1985. X-linkage of steroid sulphatase in the mouse is evidence for a functional Y-linked allele. Nature 315:226–227.

296. Kelley, K.A., C.A. Kozak, F. Dandoy, F. Sor, D. Skup, J.D. Windass, J. DeMaeyer, P.M. Pitha, and E. DeMaeyer. 1983. Mapping of murine interferon-alpha gene to chromosome 4. Gene 26:181–188.

297. Kelley, K.A., C.A. Kozak, and P.M. Pitha. 1985. Localization of the mouse interferon-beta gene to chromosome 4. J. Interferon Res. 5:409–413.

298. Kielty, C.M., S. Povey, and D.A. Hopkinson. 1982. Regulation of expression of liver-specific enzymes II. Activation and chromosomal localization of soluble glutamate-pyruvate transaminase. Ann. Hum. Genet. 46:135–143.

299. Kielty, C.M., S. Povey, and D.A. Hopkinson. 1982. Regulation of expression of liver-specific enzymes III. Further analysis of a series of rat hepatomas × human somatic cell hybrids. Ann. Hum. Genet. 47:307–327.

300. Killary, A.M., R.J. Leach, R.G. Moran, and R.E. Fournier. 1986. Assignment of the genes encoding dihydrofolate reductase and hexosaminidase B to mouse chromosome 13. Am. J. Hum. Genet. A159:39.

301. Kimura, M., H. Inoko, M. Katsuki, A. Ando, T. Sato, T. Hirose, H. Takashima, S. Inayama, H. Ckano, and K. Takamatsu. 1987. Myelin-deficient mice: shiverer mutant mice show a deletion in gene(s) coding for myelin basic protein. J. Neurochem. 44:692–696.

302. Kit, S., M. Hazeb, H. Qavi, D.R. Dubbs, and S. Pathak 1980. Integration site(s) of herpes simplex virus type 1 thymidine kinase gene and regional assignment of the gene for aminoacylase-1 in human chromosomes. Cytogenet. Cell Genet. 26:93–103.

303. Klein, J. 1975. Biology of the Mouse Histocompatibility 2 Complex. Springer-Verlag, New York: 620p.

304. Klobutcher, L.A., E.A. Nichols, R.S. Kucherlapati, and F.H. Ruddle. 1976. Assignment of the gene for human adenosine kinase to chromosome 10 using somatic cell hybrid clone panel. Cytogenet. Cell Genet. 16:171–174.

305. Kluve-Beckerman, B., S.L. Naylor, A. Marshall, J.C. Gardner, T.B. Shows, and M.D. Benson. 1986. Localization of human *SAA* gene(s) to chromosome 11 and detection of DNA polymorphisms. Biochem. Biophys. Res. Comm. 137:1196–1204.

307. Knott, T.J., R.L. Eddy, M.E. Robertson, L.M. Priestly, J. Scott, and T.B. Shows. 1984. Chromosomal localization of the human apoprotein C1 gene and of a polymorphic apoprotein A11 gene. Biochem. Biophys. Res. Comm. 125:299–306.

308. Koch, G.A., M. McAvoy, S.L. Naylor, M.G. Byers, L.L. Haley, R.L. Eddy, J.A. Brown, and T.B. Shows. 1979. Assignment of lipase A (*LIPA*) to human chromosome 10. Cytogenet. Cell Genet. 25:174.

309. Koch, G.A., R.L. Eddy, L.L. Haley, M.G. Byers, M. McAvoy, and T.B. Shows. 1983. Assignment of the human phosphoserinephosphatase gene (*PSP*) to the pter-q22 region of chromosome 7. Cytogenet. Cell Genet. 35:67–69.

310. Koch, G., P.A. Lalley, M. McAvoy, and T.B. Shows. 1981. Assignment of *LIPA*, associated with human acid lipase deficiency, to human chromosome 10 and comparative assignment to mouse chromosome 19. Somat. Cell Genet. 7:345–358.

311. Koeffler, H.P., R.S. Sparkes, H. Stang, and T. Mohandas. 1981. Regional assignment of genes for human alpha-globin and phosphoglycolate phosphatase to the short arm of chromosome 16. Proc. Natl. Acad. Sci. USA 78:7015–7018.

312. Kompf, J., F. Schunter, P. Wernet, and H. Ritter. 1985. Linkage between loci for mitochondrial malic enzyme (*ME2*) and coagulation factor X111A subunit (*F13A*). Hum. Genet. 70:43–44.

313. Kompf, J., and H. Ritter. 1985. Linkage between the loci for mitochondrial malic enzyme (*ME2*) and coagulation factor XXIIIA subunit (*F13A*). Cytogenet. Cell Genet. 40:672.

314. Kourides, I.A., P.E. Barker, J.A. Gurr, D.D. Pravtcheva, and F.H. Ruddle. 1984. Assignment of the genes for the alpha and beta subunits of thyrotropin to different mouse chromosomes. Proc. Natl. Acad. Sci. USA 81:517–519.

315. Kozak, C.A., E. Nichols, and F.H. Ruddle. 1975. Gene linkage analysis in the mouse by somatic cell hybridization: assignment of adenine phosphoribosyltransferase to chromosome 8 and alpha-galactosidase to the X-chromosome. Somat. Cell Genet. 1:371–382.

316. Kozak, C.A., and F.H. Ruddle. 1977. Assignment of the genes for thymidine kinase and galactokinase to *Mus musculus* chromsome 11, and the preferential segregation of this chromosome in Chinese hamster/mouse somatic cell hybrids. Somat. Cell Genet. 3:121–133.

317. Kozak, C.A., J.F. Sears, and M.D. Hogan. 1983. Genetic mapping of the mouse proto-oncogene c-sis to chromosome 15. Science 221:867–868.

318. Kozak, C., M.A. Gunnell, and U.R. Rapp. 1984. A new oncogene, c-raf, is located on mouse chromosome 6. J. Virol. 49:297–299.

319. Kozak, L.P., D.L. Burkart, and J.P. Hjorth. 1982. Unlinked structural genes for the developmentally regulated isozymes of 'sn'-glycerol-3-phosphate dehydrogenase in mice. Dev. Genet. 3:1–6.

320. Kozak, L.P., G.K. McLean, and E.M. Eicher. 1974. X linkage of phosphoglycerate kinase in the mouse. Biochem. Genet. 11:41–47.

321. Kozak, L.P., and K.J. Erdelsky. 1975. The genetics and developmental regulation of L-glycerol 3-phosphate dehydrogenase. J. Cell. Physiol. 85:437–448.

322. Kranz, D.M., H. Saito, C.M. Disteche, K. Swisshelm, D. Pravtcheva, F.H. Ruddle, H.E. Eisen, and S. Tonegawa. 1985. Chromosomal locations of the murine T-cell receptor alpha-chain gene and the T-cell gamma gene. Science 227:941–944.

323. Kucherlapati, R.S., R.P. Creagan, E.A. Nichols, D.S. Borgaonkar, and F.H. Ruddle. 1975. Synteny relationships of four human genes: mannose phosphate isomerase to pyruvate kinase-3 and triose phosphate isomerase to lactate dehydrogenase B. Cytogenet. Cell Genet. 14:194–197.

324. Kucherlapati, R., J.K. McDougall, and F.H. Ruddle. 1974. Regional localization of the human genes for thymidine kinase, lactate dehydrogenase-A and esterase-A4. Cytogenet. Cell Genet. 13:108–110.

325. Lackner, N.J., S.W. Law, J. Brewer, A.Y. Sakaguchi, and S.L. Naylor. 1984. The human apolipoprotein A-II gene is located on chromosome 1. Biochem. Biophys. Res. Comm. 122:877–883.

326. Lai, E.C., F.-T. Kao, M.L. Law, and S.L.C. Woo. 1983. Assignment of the alpha1-antitrypsin gene and a sequence related gene to human chromosome 14 by

molecular hybridization. Am. J. Hum. Genet. 35:385–392.

327. Lalley, P. 1981. Personal communication. Mouse News Lett. 64:79–81.

328. Lalley, P.A., A.Y. Sakaguchi, R.L. Eddy, N.H. Honey, G.I. Bell, L.-P. Shen, W.J. Rutter, J.W. Jacobs, G. Heinrich, W.W. Chin, and S.L. Naylor. 1987. Mapping polypeptide hormone genes in the mouse: somatostatin, glucagon, calcitonin, and parathyroid hormone. Cytogenet. Cell Genet. 44:92–97.

329. Lalley, P.A., J.A. Brown, R.L. Eddy, L.L. Haley, M.G. Byers, A.P. Goggin, and T.B. Shows. 1977. Human beta-glucuronidase: assignment of the structural gene to chromosome 7 by using somatic cell hybrids. Biochem. Genet. 15:367–382.

330. Lalley, P.A., and J.A. Diaz. 1984. Comparative gene mapping in the mouse involving genes assigned to human chromosomes 7 and 20. Cytogenet. Cell Genet. 37:514–515.

331. Lalley, P.A., J.D. Minna, and U. Francke. 1978. Conservation of autosomal gene synteny groups in mouse and man. Nature 274:160–163.

332. Lalley, P.A., and J.M. Chirgwin. 1984. Mapping of mouse insulin genes. Cytogenet. Cell Genet. 37:515.

333. Lalley, P.A., S.L. Naylor, and T.B. Shows. 1979. Gene assignment of argininosuccinate lyase to mouse chromosome 5. Cytogenet. Cell Genet. 25:178.

334. Lalley, P.A., U. Francke, and J.D. Minna. 1978. Homologous genes for enolase, phosphogluconate dehydrogenase, phosphoglucomutase, and adenylate kinase are syntenic on mouse chromosome 4 and human chromosome 1p. Proc. Natl. Acad. Sci. USA 75:2382–2386.

335. Lamm, L.U., U. Friedrich, G. Brunn, S. Petersen, J. Jorgensen, J. Nielsen, A.J. Therkelsen, and F. Kissmeyer-Nielsen. 1974. Assignment of the major histocompatibility complex to chromosome no. 6 in a family with a pericentric inversion. Hum. Hered. 24:273–284.

336. Landegent, J.E., G.J.B. van Ommen, F. Baas, A.K. Raap, P. van Duijn, and M. van der Ploeg. 1985. High-sensitivity in situ hybridization of human single-copy genes using 2-acetyl-aminofluorene (AAF)-modified DNA probes and reflection contrast microscopy. Cytogenet. Cell Genet. 40:677.

337. Lane, P.W., and E.M. Eicher. 1979. Gene order in linkage group XVI of the house mouse. J. Hered. 70:239–244.

338. Lavinha, J., N. Morrison, L. Glasgow, and M.A. Ferguson-Smith. 1984. Further evidence for the regional localization of human APRT and DIA4 on chromosome 16. Cytogenet. Cell Genet. 37:517.

339. Law, S.W., G. Gray, H.B. Brewer, A.Y. Sakaguchi, and S.L. Naylor. 1984. Chromosomal localization of the apolipropotein A1 gene by in situ hybridization. Biochem. Biophys. Res. Comm. 118:934–942.

340. LeBeau, M.M., M.O. Diaz, M. Karin, and J.D. Rowley. 1985. Metallothionein gene cluster is split by chromo-some 16 rearrangements in myelomonocytic leukemia. Nature 313:709–711.

341. LeBeau, M.M., M.O. Diaz, J.D. Rowley, and T.W. Mak. 1984. Chromosomal localization of the human T cell receptor beta chain genes. Cell 41:335.

342. Leinwand, L.A., C.A. Kozak, and F.H. Ruddle. 1978. Assignment of genes for triose phosphate isomerase to chromosome 6 and tripeptidase-1 to chromosome 10 in *Mus musculus* by somatic cell hybridization. Somat. Cell Genet. 4:233–240.

343. Leinwand, L.H., and F.H. Ruddle. 1978. Assignment of the gene for dipeptidase 2 to *Mus musculus* chromosome 18 by somatic cell hybridization. Biochem. Genet. 16:477–484.

344. Leinwand, L., R.E.K. Fournier, B. Nadal-Girard, and T. Shows. 1983. Multigene family for sarcomeric myosin heavy chain in mouse and human DNA localization on a single chromosome. Science 221:766–769.

345. Leinwand, L., R.E.K. Fournier, E.A. Nichols, and F.H. Ruddle. 1978. Assignment of the gene for adenosine kinase to chromosome 14 in *Mus musculus* by somatic cell hybridization. Cytogenet. Cell Genet. 21:77–85.

346. Lin, P.-F., and F.H. Ruddle. 1981. Murine DNA repair gene located on chromosome 4. Nature 289:191–194.

347. Lindgren, V., B. de Martinville, A.L. Horwich, L.E. Rosenberg, and U. Francke. 1984. Human ornithine transcarbamylase locus mapped to band Xp21.1 near the Duchenne muscular dystrophy locus. Science 226:698–700.

348. Lovett, M., D.R. Cox, D. Yee, W. Boll, C. Weissman, C.J. Epstein, and L.B. Epstein. 1984. The chromosomal location of the mouse interferon-alpha genes. EMBO J. 3:1643–1646.

349. Lubsen, N.H., J.H. Renwick, L.-C. Tsui, M.L. Breitman, and J.G.G. Schoenmakers. 1987. A locus for a human hereditary cataract is closely linked to the gamma-crystallin gene family. Proc. Natl. Acad. Sci. USA 84:489–492.

350. Lush, I.E., and K.M. Andrews. 1978. Genetic variation between mice in their metabolism of coumarin and its derivatives. Genet. Res. 31:177–186.

351. Lusis, A.J., B.A. Taylor, R.W. Wangenstein, and R.C. LeBoeuf. 1983. Genetic control of lipid transport in mice. II. Genes controlling structure of high density lipoproteins. J. Biol. Chem. 258:5071–5078.

352. Lusis, A.J., C. Heinzman, R.S. Sparkes, R. Geller, M.C. Sparkes, and T. Mohandas. 1985. Regional mapping on human chromosome 19: apolipoprotein E, apolipoprotein C11, low density lipoprotein (LDL) receptor, peptidase D, glucose phosphate isomerase. Cytogenet. Cell Genet. 40:683.

353. Lusis, A.J., K.L. Reue, R.C. Le Boeuf, O. Ben-Zeev, M.C. Schotz, and B. Paigen. 1984. Genetic factors in plasma lipid transport and atherosclerosis. Cytogenet. Cell. Genet. 37:526–527.

354. Lyon, J.B. 1970. The X-chromosome and the enzymes

controlling muscle glycogen: phosphorylase kinase. Biochem. Genet. 4:169–185.

355. Lyon, M.F., J. Zenthon, M.D. Burtenshaw, and E.P. Evans. 1987. Localization of the *Hprt* locus by in situ hybridization and distribution of loci on the mouse X chromosome. Cytogenet. Cell Genet. 44:163–166.

356. Lyon, M.F., and S.G. Hawkes. 1970. X-linked gene for testicular feminization in the mouse. Nature 227:1217–1219.

357. Magenis, R.E., F. Hecht, and E.W. Lovrien. 1970. Heritable fragile site on chromosome 16: probable localization of haptoglobin locus in man. Science 170:85–87.

358. Malcolm, S., P. Barton, D.L. Bentley, M.A. Ferguson-Smith, C.S. Murphy, and T.H. Rabbitts. 1982. Assignment of a IGKV locus for immunoglobulin light chains to the short arm of chromosome 2 (2p13-cen) by in situ hybridization using a cRNA probe of Hx101lCh4A. Cytogenet. Cell Genet. 32:296.

359. Mannens, M., R.M. Slater, C. Heyting, A. Guerts van Kessel, E. Goedde-Salz, R.R. Frants, G.J.B. Van Ommen, and P.L. Pearson. 1987. Regional localization of DNA probes on the short arm of chromosome 11 using aniridia-Wilms' tumor-associated deletions. Hum. Genet. 75:180–187.

360. Marcu, K.B., L.J. Harris, L.W. Stanton, J. Erikson, R. Watt, and C.M. Croce. 1983. Transcriptionally active c-myc oncogene is contained within NIARD, a DNA sequence associated with chromosome translocations in B-cell neoplasia. Proc. Natl. Acad. Sci. USA 80:519–523.

361. Mareschal, J.C., L.R. Valcovic, and H.V. Malling. 1976. Personal communication. Mouse News Lett. 54:49.

362. Margulies, D.H., J.R. Parnes, N.A. Johnson, and J.G. Seidman. 1983. Linkage of beta-2 microglobulin and ly-m11 by molecular cloning and DNA-mediated gene transfer. Proc. Natl. Acad. Sci. USA 80:2328–2331.

363. Marimo, B., and F. Giannelli. 1975. Gene dosage effect in human trisomy 16. Nature 256:204–206.

364. Martiniuk, F., A. Ellenbogan, K. Hirschhorn, and R. Hirschhorn. 1985. Further regional localization of the genes for human acid alpha glucosidase (*GAA*), peptidase D (*PEPD*), and alpha mannosidase B (*MANB*) by somatic cell hybridization. Hum. Genet. 69:109–111.

365. Mattei, M.G., M.A. Bateman, R. Heilig, I. Oberle, K. Davies, J.L. Mandel, and J.F. Mattei. 1985. Localization by in situ hybridization of the coagulation factor IX gene and of two polymorphic DNA probes with respect to the fragile X site. Hum. Genet. 69:327–331.

366. McAlpine, P.J., A.E. Chudley, M. Ray, and J.L. Hamerton. 1973. The peptidase A gene locus and human chromosome 18. Am. J. Hum. Genet. 25:49A.

367. McAlpine, P.J., T. Mohandas, M. Ray, H. Wang, and J.L. Hamerton. 1975. Assignment of the inorganic pyrophosphatase gene locus (*PP*) to chromosome 10 in man. Cytogenet. Cell Genet. 14:201–203.

368. McAlpine, P.J., T. Mohandas, M. Ray, H. Wang, and J.L. Hamerton. 1976. Evidence for the assignment of the peptidase D (*PEPD*) gene locus to chromosome 19 in man. Cytogenet. Cell Genet. 16:204–205.

369. McAlpine, P.J., T. Mohandas, M. Ray, and J.L. Hamerton. 1975. Assignment of pyrophosphatase gene locus (*PP*) to chromosome 10 and peptidase D gene locus (*PEPD*) to chromosome 19 in man. Am. J. Hum. Genet. 27:61A.

370. McBreen, P., K.G. Orkwiszewski, C.J. Chern, W.J. Mellman, and C.M. Croce. 1977. Synteny of the genes for thymidine kinase and galactokinase in the mouse and assignment to mouse chromosome 11. Cytogenet. Cell Genet. 19:7–13.

371. McBride, O.W., D.C. Swan, S.R. Tronick, R. Gol, D. Klimanis, D.E. Moore, and S.A. Aaronson. 1983. Regional chromosomal localization of *N-ras*, *K-ras-1*, *K-ras-2* and *myb* oncogenes in human cells. Nucl. Acids Res. 11:8221–8226.

372. McBride, O.W., D. Swan, P. Leder, P. Hieter, and G. Hollis. 1982. Chromosomal location of human immunoglobulin light chain constant region genes. Cytogenet. Cell Genet. 32:297–298.

373. McGinnis, W., C.P. Hart, W.J. Gehring, and F.H. Ruddle. 1984. Molecular cloning and chromosome mapping of a mouse DNA sequence homologous to homeotic genes of Drosophila. Cell 38:675–680.

374. McKusick, V.A. 1983. Mendelian Inheritance in Man: Catalogs of Autosomal Dominant, Autosomal Recessive, and X-Linked Phenotypes. Johns Hopkins University Press, Baltimore, 1077 p.

375. McKusick, V.A. 1984. Human gene map. *In* S.J. O'Brien, ed., Genetic Maps 1984, 417–441. Cold Spring Harbor Press, Cold Spring Harbor, New York.

376. Meager, A., H. Graves, D.C. Burke, and D.M. Swallow. 1979. Involvement of a gene on chromosome 9 in human fibroblast interferon production. Nature 280:493–495.

377. Meera Khan, P., B.A. Doppert, A. Hagemeijer, and A. Westerveld. 1974. The human loci for phosphopyruvate hydratase and guanylate kinase are syntenic with the *PGD-PGM1* linkage group in man–Chinese hamster somatic cell hybrids. Cytogenet. Cell Genet. 13:130–131.

378. Meera Khan, P., and L.M.M. Wijnen. 1984. Further evidence assigning the red cell *GPT* gene to chromosome 16. Cytogenet. Cell Genet. 37:539.

379. Meera Khan, P., L.M.M. Wijnen, and K.-H. Grzeschik. 1984. Electrophoretic characterization and genetics of human biliverdin reductase (BLVR; EC.1.3.1.24); assignment of BLVR to the p14-1/2cen region of human chromosome 7 in mouse–human somatic cell hybrids. Biochem. Genet. 21:123–133.

380. Meera Khan, P., P.L. Pearson, L. Wijnen, B.A. Doppert, A. Westerveld, and D. Bootsma. 1976. Assignment of inosine triphosphatase gene to gorilla chromosome 13 and to human chromosome 20 in primate–rodent somatic cell hybrids. Cytogenet. Cell Genet. 16:420–421.

381. Meisler, M.H., L. Wanner, F.T. Kao, and C. Jones. 1981. Localization of the uroporphyrinogen I synthase locus to human chromosome region 11q13-qter and interconversion of enzyme isomers. Cytogenet. Cell Genet. 31:124–128.

382. Meo, T., T. Douglas, and A.-M. Rijnbeeck. 1977. Glyoxalase I polymorphism in the mouse: a new genetic marker linked to H-2. Science 198:311–313.

383. Meruelo, D., A. Rossomando, A. Scandalis, P.D'Eustachio, R.E.K. Fournier, D.R. Roop, D. Saxe, C. Blatt, and M.N. Nesbit. 1987. Assignment of the *Ly-6-Ril-1–Sis–H-30–Pol-5 Xmmv-72–Ins-3–Krt-1–Int-1-Gdc-1* region to mouse chromosome 15. Immunogenetics 25:361–372.

384. Meyers, D.A., T.H. Beaty, N.E. Maestri, S.D. Kittur, S.E. Antonarakis, and H.H. Kazazian. 1987. Multipoint mapping studies of six loci on chromosome 11. Hum. Hered. 37:94–101.

385. Michaelson, J. 1981. Genetic polymorphism of beta-2 microglobulin (*B2m*) maps to the H-3 region of chromosome 2. Immunogenetics 13:167–171.

386. Michaelson, J. 1983. Genetics of beta-2 microglobulin in the mouse. Immunogenetics 17:219–259.

387. Migeon, B.R., and F. Huijing. 1974. Glycogen storage disease associated with phosphorylase kinase deficiency: evidence for X-inactivation. Am. J. Hum. Genet. 26:360–368.

388. Migeon, B.R., H.W. Moser, A.B. Moser, J.R. Axelman, D. Sillence, and R.A. Norum. 1981. Adrenoleukodystrophy: evidence for X-linkage, inactivation and selection favoring the mutant allele in heterozygous cells. Proc. Natl. Acad. Sci. USA 78:5066–5070.

389. Migeon, B.R., T.R. Brown, J. Axelman, and C.J. Migeon. 1981. Studies of the locus for androgen receptor: localization on the human X chromosome and evidence for homology with the *Tfm* locus in the mouse. Proc. Natl. Acad. Sci. USA 78:6339–6343.

390. Minna, J.D., G.A.P. Bruns, A.H. Krinsky, P.A. Lalley, U. Francke, and P.S. Gerald. 1978. Assignment of a *Mus musculus* gene for triosephosphate isomerase to chromosome 6 and for glyoxalase-1 to chromosome 17 using somatic cell hybrids. Somat. Cell Genet. 4:241–252.

391. Mishkin, J.D., B.A. Taylor, and W.J. Mellman. 1976. *Glk*: a locus controlling galactokinase activity in the mouse. Biochem. Genet. 14:635–640.

392. Mitchell, G., D. Valle, H. Willard, G. Steel, M. Suchanek, and L. Brody. 1986. Human ornithine-delta-amino transferase (*OAT*): cross-hybridizing fragments mapped to chromosome 10 and Xp11.1-21.1. Am. J. Hum. Genet. 39:163.

393. Mogensen, K.E., F. Vignaux, and I. Gresser. 1982. Enhanced expression of cellular receptors for human interferon alpha on peripheral lymphocytes from patients with Down's syndrome. FEBS Lett. 140:285–287.

394. Mohandas, T., R.S. Sparkes, J.D. Shulkin, K.E. Toomey, and S.J. Funderburk. 1979. Regional localization of human gene loci on chromosome 9: studies of somatic cell hybrids containing human translocations. Am. J. Hum. Genet. 31:586–600.

395. Mohandas, T., R.S. Sparkes, M.B. Passage, M.C. Sparkes, J.H. Miles, and M.M. Kaback. 1980. Regional mapping of *ADA* and *ITP* on human chromosome 20: cytogenetic and cell studies in an X/20 translocation. Cytogenet. Cell Genet. 26:28–35.

396. Mohandas, T., R.S. Sparkes, M.C. Sparkes, and J.D. Shulkin. 1977. Assignment of the human gene for galactose-1-phosphate uridyl transferase to chromosome 9: studies with Chinese hamster–human somatic cell hybrids. Proc. Natl. Acad. Sci. USA 74:5628–5631.

397. Mondello, C., H.-H. Ropers, I.W. Craig, E. Tolley, and P.N. Goodfellow. 1987. Physical mapping of genes and sequences at the end of the human X chromosome short arm. Ann. Hum. Genet. 51:137–143.

398. Moore, E.E., C. Jones, F.T. Kao, and D.C. Oates. 1977. Synteny between glycinamide ribonucleotide synthetase and superoxide dismutase (soluble). Am. J. Hum. Genet. 29:389–396.

399. Moore, M.N., F.-T. Kao, Y.-K. Tsao, and L. Chan. 1984. Nucleotide sequence of a cloned cDNA and localization of its structural gene on human chromosome 1. Biochem. Biophys. Res. Comm. 123:1–7.

400. Mortensen, R.F., P.T. Le, and B.A. Taylor. 1985. Mouse serum amyloid P-component (SAP) levels controlled by a locus on chromosome 1. Immunogenetics 22:367–375.

401. Morton, C.C., A.D. Duby, R.L. Eddy, C. Murre, R.A. Waldman, T.B. Shows, and J.G. Seidman. 1985. The human T-cell receptor beta chain and the T-cell gamma chain gene are assigned to chromosome 7. Cytogenet. Cell Genet. 40:703.

402. Morton, C., G. Bell, and T. Shows. 1985. Epidermal growth factor (*EGF*) is located at q25-q27 on human chromosome 4. Cytogenet. Cell Genet. 40:702–703.

403. Moss, P.A.H., K.E. Davis, C. Boni, J. Mallet, and S.T. Reeders. 1986. Linkage of tyrosine hydroxylase to four other markers on the short arm of chromosome 11. Nucl. Acids Res. 14:9927–9932.

404. Mucenski, M.L., B.A. Taylor, N.G. Copeland, and N.A. Jenkins. 1989. Chromosomal location of *Evi-1*, a common site of ecotropic viral integration in AKXD murine myeloid tumors. Oncogene Res. (in press).

405. Muenke, M., D.R. Cox, I.J. Jackson, B.L.M. Hogan, and U. Francke. 1986. The murine *Hox-2* cluster of homeo box containing genes maps distal on chromosome 11 near the tail-short (*Ts*) locus. Cytogenet. Cell Genet. 42:236–240.

406. Muenke, M., K. Harbers, R. Jaenisch, and U. Francke. 1986. Chromosomal mapping of four different integration sites of Moloney murine leukemia virus including the locus for alpha 1 (1) collagen in mouse. Cytogenet. Cell Genet. 43:140–149.

407. Muenke, M., and U. Francke. 1987. The physical map of *Mus musculus* chromosome 11 reveals evolutionary rela-

tionships with different syntenic groups of genes in *Homo sapiens*. J. Mol. Evol. 25:134–140.

408. Muenke, M., V. Lindgren, B. De Martinville, and U. Francke. 1984. Comparative analysis of mouse–human hybrids with rearranged chromosomes 1 by in situ hybridization and Southern blotting: high resolution mapping of *NRAS*, *NGFB*, and *AMY* on chromosome 1. Somat. Cell Mol. Genet. 10:589–599.

409. Mulcahy, M.T., and R.G. Wilson. 1980. Where is the gene for *GALT*? Hum. Genet. 54:129–130.

410. Muller, G., G. Scherer, H. Zentgraf, S. Ruppert, B. Herrmann, H. Lehrach, and G. Schütz. 1985. Isolation, characterization and chromosomal mapping of the mouse tyrosine aminotransferase gene. J. Mol. Biol. 184:367–373.

411. Mullins, L.J., S.G. Grant, J. Pazik, D.A. Stephenson, and V.M. Chapman. 1987. Personal communication. Mouse News Lett. 77:150–151.

412. Munke, M., J.P. Kraus, T. Ohura, and U. Francke. 1988. The gene for cystathione beta-synthetase (*CBS*) maps to the subtelomeric region on human chromosome 21q and to proximal mouse chromosome 1. Am. J. Hum. Genet. 42:550–559.

413. Murray, J.M., K.E. Davies, P.S. Harper, L. Meredith, C.R. Mueller, and R. Williamson. 1982. Linkage relationship of a cloned DNA sequence on the short arm of the X chromosome to Duchenne muscular dystrophy. Nature 300:69–71.

414. Nadeau, J.H. 1986. A chromosomal segment conserved since divergence of lineages leading to man and mouse: the gene order of aminoacylase-1, transferrin and beta-galactosidase on mouse Chromosome 9. Genet. Res. 48:175–178.

415. Nadeau, J.H., and E.M. Eicher. 1982. Conserved linkage of soluble aconitase and galactose-1-phosphate uridyl transferase in mouse and man: assignment of these genes to mouse chromosome 4. Cytogenet. Cell Genet. 34:271–281.

416. Nadeau, J.H., F.G. Berger, K.A. Kelley, P.M. Pitha, C.L. Sidman, and N. Worrall. 1986. Rearrangement of genes located on homologous chromosomal segments in mouse and man: the location of alpha- and beta-interferon, alpha-1 acid glycoprotein-1 and -2, and aminolevulinate dehydratase on mouse chromosome 4. Genetics 114:1239–1255.

417. Nadeau, J.H., J. Kompf, G. Siebert, and B.A. Taylor. 1981. Linkage of *Pgm-3* in the house mouse and homologies of three phosphoglucomutase loci in mouse and man. Biochem. Genet. 19:465–474.

418. Nadeau, J.H., and S.J. Phillips. 1987. The putative oncogene *Pim-1* in the mouse: its linkage and variation among *t* haplotypes. Genetics 117:533–541.

419. Nadeau, J.H., S.J. Phillips, and I.K. Egorov. 1985. Recombination between the *t*[6] complex and linked loci in the house mouse. Genet. Res. 45:251–264.

420. Nagarajan, L., E. Louie, Y. Tsujimoto, A. Ar-Rushdi, K. Huebner, and C.M. Croce. 1986. Localization of the human *PIM* oncogene (*PIM*) to a region of chromosome 6 involved in translocations in acute leukemias. Proc. Natl. Acad. Sci. USA 83:2556–2560.

421. Nakai, H., M.G. Byers, P.J. Venta, R.E. Tashian, and T.B. Shows. 1987. The gene for human carbonic anhydrase II (*CA2*) is located at chromosome 8q22. Cytogenet. Cell Genet. 44:234–235.

422. Nakamura, D., R. Popp, E. Eicher, and P.A. Lalley. 1985. Comparison of the argininosuccinate synthetase gene family in mouse and man. Cytogenet. Cell Genet. 40:710.

423. Narahara, K., K. Kikkawa, and S. Kimira. 1984. Regional mapping of catalase and Wilms' tumor genitourinary abnormalities, and mental retardation triad loci to the chromosomal segment 11p1305–p1306. Hum. Genet. 66:181–185.

424. Natsuume-Sakai, S., J.-I. Hayakawa, and M. Takahashi. 1978. Genetic polymorphism of murine C3 controlled by a single co-dominant locus on chromosome 17. J. Immunol. 121:491–498.

425. Naylor, S.L., A.Y. Sakaguchi, L. McDonald, S. Todd, P.A. Lalley, T.B. Shows, and W.W. Chin. 1986. Mapping of thyrotropin beta subunit gene in man and mouse. Somat. Cell Mol. Genet. 12:307–311.

426. Naylor, S.L., A.Y. Sakaguchi, T.B. Shows, M.L. Law, D.V. Goeddel, and P.W. Gray. 1983. Human immune interferon gene is located on chromosome 12. J. Exp. Med. 57:1020–1027.

427. Naylor, S.L., A.Y. Sakaguchi, W.W. Chin, J. Jacobs, L.P. Shen, W.J. Rutter, and P.A. Lalley. 1984. Chromosomal mapping of mouse hormone genes: somatostatin, calcitonin, thyrotropin alpha subunit, and luteinizing hormone beta subunit. Cytogenet. Cell Genet. 37:55.

428. Naylor, S.L., D.R. Barnett, J.M. Buchanan, J. Latimer, K. Wieder, S. Marshall, J. Gardner, C.R. Denning, M. Gluckson, R. Pinero, and H. Rendon. 1986. Linkage of cystic fibrosis locus and polymorphic DNA markers in 14 families. Am. J. Hum. Genet. 39:707–712.

429. Naylor, S.L., P.A. Lalley, A.Y. Sakaguchi, and P.W. Gray. 1984. Mapping of mouse immune interferon (IFN-gamma) gene to mouse chromosome 10. Cytogenet. Cell Genet. 37:550.

430. Naylor, S.L., R.J. Klebe, and T.B. Shows. 1978. Argininosuccinic aciduria: assignment of the argininosuccinate lyase gene to the pter-q22 region of human chromosome 7 by bioautography. Proc. Natl. Acad. Sci. USA 75:6159–6162.

431. Naylor, S.L., R.W. Elliot, J.A. Brown, and T.B. Shows. 1982. Mapping of aminoacylase-1 and beta-galactosidase-A to homologous regions of human chromosome 3 and mouse chromosome 9 suggests location of additional genes. Am. J. Hum. Genet. 34:235–244.

432. Naylor, S.L., W.W. Chin, H.M. Goodman, P.A. Lalley, K.-H. Grzeschik, and A.Y. Sakaguchi. 1983. Chromosome assignments of genes encoding the alpha and beta subunits of glycoprotein hormones in man and mouse. Somat. Cell Genet. 9:757–770.

433. Neel, B.G., S.C. Jhanwar, R.S.K. Chaganti, and W.S. Hayward. 1982. Two human c-onc genes are located on the long arm of chromosome 8. Proc. Natl. Acad. Sci. USA 79:7842–7846.

434. Ng, S.-Y., P. Gunning, R. Eddy, P. Ponte, J. Leavitt, L. Kedes, and T. Shows. 1985. Chromosome 7 assignment of the human beta-actin functional gene (*ACTB*) and the chromosomal dispersion of pseudogenes. Cytogenet. Cell Genet. 40:712.

435. Nguyen, V.C., C. Billardon, J.Y. Picard, J. Feingold, and J. Frezal. 1971. Liaison probable (linkage) entre les locus *PGM* et peptidase C chez l'homme. C. R. Acad. Sci. 272:485–487.

436. Nguyen, V.C., D. Weil, M.C. Hors-Cayla, M.S. Gross, S. Heuertz, C. Foubert, and J. Frezal. 1980. Assignment of the genes for human lyosomal acid lipases A and B to chromosomes 10 and 16. Hum. Genet. 55:375–381.

437. Nguyen, V.C., M.G. Mattei, C. Goridis, J.F. Mattei, and B.R. Jordan. 1985. Localization of the human *N-CAM* gene to chromosome 11 by in situ hybridization with a murine N-CAM cDNA probe. Cytogenet. Cell Genet. 40:713.

438. Nguyen, V.C., R. Rebourcet, D. Weil, C. Pangalos, and J. Frezal. 1975. Localization d'un locus de structure de la pyrophosphatase inorganique 'erythrocytaire' sur le chromosome 10 chez l'homme par la methode d'hybridation cellulaire homme-hamster. C. R. Acad. Sci. D 281:435–438.

439. Nichols, E.A., and F.H. Ruddle. 1975. Polymorphism and linkage of glutathione reductase in *Mus musculus*. Biochem. Genet. 13:323–329.

440. Nichols, E.A., F.H. Ruddle, and M.L. Petras. 1975. Linkage of the locus for serum albumin in the house mouse, *Mus musculus*. Biochem. Genet. 13:551–559.

441. Nichols, E.A., V.M. Chapman, and F.H. Ruddle. 1973. Polymorphism and linkage for mannosephosphate isomerase in *Mus musculus*. Biochem. Genet. 8:47–53.

442. Nielsen, J.T., and K. Sick. 1975. Genetic polymorphism of amylase isoenzymes in feral populations of the house mouse. Hereditas 79:279–286.

443. Nusse, R., A. von Ooyen, D. Cox, Y.K.T. Fung, and H. Varmus. 1984. Mode of proviral activation of a putative mammary oncogene (*int-1*) on mouse chromosome 15. Nature 307:131–136.

444. Nusse, R., V.T.L. Veer, A.G. van Kessel, A. van Agthoven, D. Bootsma, and H. Varmus. 1984. Chromosomal localization of a human homologue of a putative mammary tumor oncogene. Cytogenet. Cell Genet. 37:556–557.

445. O'Brien, D., A. Linnenbach, and C.M. Croce. 1978. Assignment of the gene for lactic dehydrogenase A to mouse chromosome 7 using mouse–human hybrids. Cytogenet. Cell Genet. 21:72–76.

447. O'Brien, S.J., W.G. Nash, J.L. Goodwin, D.R. Lowy, and E.H. Chang. 1983. Dispersion of the *ras* family of transforming genes to four different chromosomes in man. Nature 302:839–842.

448. O'Donnell, J.J., K.M. Vannas-Sulonen, T.B. Shows, and D.R. Cox. 1985. Ornithine aminotransferase (*OAT*) maps to human chromosome 10 and mouse chromosome 7. Cytogenet. Cell Genet. 40:716.

449. Otto, J., and O.H. von Deimling. 1983. Esterase-17 (ES-17): characterization and genetic location on chromosome 9 of a bis-p-nitrophenyl phosphate-resistant esterase of the house mouse (*Mus musculus*). Biochem. Genet. 21:37–48.

450. Owerbach, D., G.I. Bell, W.J. Rutter, J.A. Brown, and T.B. Shows. 1981. Localization of the insulin gene (*INS*) on the short arm of chromosome 11. Diabetes 39:267.

451. Owerbach, D., W.J. Rutter, J.L. Roberts, P. Whitfield, J. Shine, P.H. Seeburg, and T.B. Shows. 1981. The pro-opiocortin (adrenocorticotropin/beta-lipotropin) gene is located on chromosome 2 in humans. Somat. Cell Genet. 7:359–369.

452. Owerbach, D., W.J. Rutter, T.B. Shows, P. Gray, D.V. Goeddel, and R.M. Lawn. 1981. Leukocyte and fibroblast interferon genes are located on human chromosome 9. Proc. Natl. Acad. Sci. USA 78:3123–3127.

453. Owerbach, D., W.J. Rutter, T.B. Shows, P. Gray, D.V. Goeddel, and R.M. Lawn. 1982. Both leukocyte (*IFL*) and fibroblast (*IFF*) interferon genes are on chromosome 9. Cytogenet. Cell Genet. 32:307.

454. Pai, G.S., J.A. Sprenkle, T.T. Do, C.E. Mareni, and B.R. Migeon. 1980. Localization of loci for hypoxanthine phosphoribosyl transferase and glucose-6-phosphate dehydrogenase and biochemical evidence of nonrandom X chromosome expression from studies of a human X-autosome translocation. Proc. Natl. Acad. Sci. USA 77:2810–2813.

455. Paigen, K. 1979. Acid hydrolases as models of genetic control. Ann. Rev. Genet. 13:417–466.

456. Parkar, M., S.J. Jeremiah, S. Povey, A.F. Lee, F.O. Finley, P.N. Goodfellow, and E. Solomon. 1984. Confirmation of the assignment of human biliverdin reductase to chromosome 7. Ann. Hum. Genet. 48:57–60.

457. Peters, G., C. Kozak, and C. Dickson. 1984. Mouse mammary tumor virus integration regions int-1 and int-2 map on different mouse chromosomes. Mol. Cell Biol. 4:375–378.

458. Peters, J., and H.R. Nash. 1976. Polymorphism of esterase-10 in *Mus musculus*. Biochem. Genet. 14:119–125.

459. Peters, J., H.R. Nash, E.M. Eicher, and G. Bulfield. 1981. Polymorphism of kidney pyruvate kinase in the mouse is determined by a gene, *Pk-3*, on Chromosome 9. Biochem. Genet. 19:757–769.

460. Peters, J., and S.J. Andrews. 1984. The *Pk-3* gene determines both the heart, M_1 and the kidney, M_2, pyruvate kinase isozymes in the mouse; and a simple electrophoretic method for separating phosphoglucomutase-3. Biochem. Genet. 22:1047–1063.

461. Peters, J., and S.J. Andrews. 1985. Linkage of lactate dehydrogenase-2, *Ldh-2*, in the mouse. Biochem. Genet. 23:217–225.

462. Peters, J., S. Povey, S. Jeremiah, and L. De Giorgi.

1983. Linkage relationships of peptidase-7, *Pep-7*, in the mouse. Biochem. Genet. 21:801–807.

463. Peterson, T.C., A.M. Killary, and R.E.K. Fournier. 1985. Chromosomal assignment and trans regulation of the tyrosine aminotransferase structural gene in hepatoma hybrid cells. Mol. Cell Biol. 5:2491–2494.

464. Pettenati, M.J., A. Milatovich, N.A. Heerema, and C.G. Palmer. 1985. Evidence for separate regions of localization for NRAS probes to chromosome 1. Cytogenet. Cell Genet. 40:722.

465. Philip, T., G. Lenoir, M.O. Rolland, I. Phillip, M. Hamet, B. Lauras, and J. Fraisse. 1985. Regional assignment of the *ADA* locus on 10q13.2-qter by gene dosage studies. Cytogenet. Cell Genet. 27:187–189.

466. Phillips, I.R., E.A. Shepard, S. Povey, M.B. Davis, G. Kelsey, M. Monteiro, L.F. West, and J. Cowell. 1985. A cytochrome *P-450* gene family mapped to human chromosome 19. Ann. Hum. Genet. 49:267–274.

467. Popescu, N.C., S.C. Amsbaugh, J.A. Di Paolo, S.R. Tronick, S.A. Aaronson, and D.C. Swan. 1985. Chromosomal localization of three human *ras* genes by in situ molecular hybridization. Somat. Cell Mol. Genet. 11:149–155.

468. Potluri, V.R., K.A. Astrin, J.G. Wetmur, D.F. Bishop, and R.J. Desnick. 1987. Human delta-aminolevulinate dehydratase chromosomal localization to 9q34 by in situ hybridization. Hum. Genet. 76:236–239.

469. Povey, S., C.A. Slaughter, D.E. Wilson, I.P. Gormley, K.E. Buckton, R. Perry, and M. Bobrow. 1976. Evidence for the assignment of the loci *AK1*, *AK3*, and *ACONS* to chromosome 9 in man. Ann. Hum. Genet. 39:413–422.

470. Prakash, K., O.W. McBride, D.C. Swan, S.G. Devare, S.R. Tronik, and S.A. Aaronson. 1982. Molecular cloning and chromosomal mapping of a human locus related to the transforming gene of Moloney murine sarcoma virus. Proc. Natl. Acad. Sci. USA 79:5210–5214.

471. Pravtcheva, D.D., F.H. Ruddle, R.W. Ellis, and E.M. Scolnick. 1983. Assignment of murine cellular Harvey Ras gene to chromosome 7. Somat. Cell Genet. 9:681–686.

472. Quax-Jeuken, Y., W. Quax, G. van Rens, P. Meera Khan, and H. Bloemendal. 1985. Assignment of the alphaA-crystallin gene (*CRYA1*) to chromosome 21. Cytogenet. Cell Genet. 40:727–728.

473. Rabin, M., A. Ferguson-Smith, C.P. Hart, and F.H. Ruddle. 1986. Cognate homeo-box loci mapped on homologous human and mouse chromosomes. Proc. Natl. Acad. Sci. USA 83:9104–9108.

474. Rabin, M., C.P. Hart, A. Ferguson-Smith, W. McGinnis, M. Levine, and F.H. Ruddle. 1985. Two homeo box loci mapped in evolutionarily related mouse and human chromosomes. Nature 314:175–178.

475. Rabin, M., M. Watson, P.E. Barker, J. Ryan, W.R. Breg, and F.H. Ruddle. 1984. *NRAS* transforming gene maps to region p11-p13 on chromosome 1 by in situ hybridization. Cytogenet. Cell Genet. 38:70–72.

476. Rabi, M., M. Watson, W.R. Breg, V. Kidd, S.L.C. Woo, and F.H. Ruddle. 1985. Human alpha1-antichymotrypsin (*AACT*) and alpha1-antitrypsin (*PI*) genes map to the same region on chromosome 14. Cytogenet. Cell Genet. 40:728.

477. Ramesh, V., R. Eddy, G.A. Bruns, V.E. Shih, T.B. Shows, and J.F. Gusella. 1987. Localization of the ornithine aminotransferase gene and related sequences on two human chromosomes. Hum. Genet. 76:121–126.

478. Rappold, C.A., and H.P. Vosberg. 1983. Chromosomal localization of a human myosin heavy chain gene by in situ hybridization. Hum. Genet. 65:195–197.

479. Reeves, R.H., D. Gallahan, B.F. O'Hara, R. Callahan, J.D. Gearhart, and 1987. Genetic mapping of *Prm-1*, *Igl-1*, *Smst*, *Mtv-6*, *Sod-1*, and *Ets-2*, and localization of the Down syndrome region on mouse chromosome 16. Cytogenet. Cell Genet. 44:76–81.

480. Reeves, R.H., J.D. Gearhart, and J.W. Littlefield. 1986. Genetic basis for a mouse model for Down syndrome. Brain Res. Bull. 16:803–814.

481. Remes, G.M., R.A. Fisher, E. Hackel, A.J. Cousineau, and J.V. Higgens. 1984. SRO refinement for nucleoside phosphorylase by deletion mapping of chromosome 14. Cytogenet. Cell Genet. 37:568.

482. Rettig, W.J., N.C. Dracopoli, B.A. Spengler, J.L. Biedler, and L.J. Old. 1985. Somatic cell genetic analysis of human cell surface antigens, including putative human *Thy-1*: eight distinct antigenic systems controlled by chromosome 11. Cytogenet. Cell Genet. 40:732.

483. Ricciuti, F., and F.H. Ruddle. 1973. Assignment of nucleoside phosphorylase to D-14 and localization of X-linked loci in man by somatic cell genetics. Nature New Biol. 241:180.

484. Richards J.E., D.D. Pravtcheva, C. Day, F.H. Ruddle, and P.P. Jones. 1985. Murine invariant chain gene: chromosomal assignment and segregation in recombinant inbred strains. Immunogenetics 22:193–199.

485. Roach, J., N. Takahashi, D. Pravtcheva, F.H. Ruddle, and L. Hood. 1985. Chromosomal mapping of mouse myelin basic protein gene and structure and transcription of the partially deleted gene in shiverer mutant mice. Cell 42:149–155.

486. Robbins, J.M., D. Gallahan, E. Hogg, C. Kozak, and R. Callahan. 1986. An endogenous mouse mammary tumor virus genome common in inbred mouse strains is located on chromosome 6. J. Virol. 57:709–713.

488. Robinson, P.J., L. Lundin, K. Sege, L. Graf, H. Wigzell, and P.A. Peterson. 1981. Location of the mouse beta-2 microglobulin gene *B2m* determined by linkage analysis. Immunogenetics 14:449–452.

489. Robson, E.B., P.E. Polani, S.J. Dart, P.A. Jacobs, and J.H. Renwick. 1969. Probable assignment of the alpha locus of haptoglobin to chromosome 16 in man. Nature 223:1163–1165.

490. Rotter, V., D. Wolf, D. Pravtcheva, and F.H. Ruddle.

1984. Chromosomal assignment of the murine gene encoding the transformation-related protein p53. Mol. Cell Biol. 4:383–385.

491. Rubenstein, P., K. Vienne, and G.F. Hoecker. 1979. The location of the *C3* and *Glo* (glyoxalase I) loci of the IXth linkage group in mice. J. Immunol. 122:2584–2589.

492. Ruddle, F.H., F. Ricciuti, F.A. McMorris, J. Tischfeld, R. Creagan, G. Darlington, and T. Chen. 1972. Somatic cell genetic assignment of peptidase C and the Rh linkage group to chromosome A-1 in man. Science 176:1429–1439.

493. Russell, E.S., and E.C. McFarland. 1974. Genetics of mouse hemoglobins. Ann. NY Acad. Sci. 241:25–38.

494. Ryan, J., C.P. Hart, and F.H. Ruddle. 1984. Molecular cloning and chromosome assignment of murine *N-ras*. Nucl. Acids Res. 12:6063–6072.

495. Sacchi, N., D.K. Watson, A.H.M. Guerts van Kessel, A. Hagemeijer, J. Kersey, H.D. Drabkin, D. Patterson, and T. Papas. 1986. Hu-ets-1 and Hu-ets-2 genes are transposed in acute leukemia with (4:11) and (8:21) translocations. Science 231:379–382.

496. Sakaguchi, A.Y., B.U. Zabel, K.-H. Grzeschik, M.L. Law, and S.L. Naylor. 1984. Human proto-oncogene assignments. Cytogenet. Cell Genet. 37:572–573.

497. Sakaguchi, A.Y., P.A. Lalley, B.U. Zabel, R. Ellis, E. Skolnick, and S.L. Naylor. 1984. Mouse proto-oncogene assignments. Cytogenet. Cell Genet. 37:573–574.

498. Sakaguchi, A.Y., P.A. Lalley, and S.L. Naylor. 1983. Human and mouse cellular myc protooncogenes reside on chromosomes involved in numerical and structural aberrations in cancer. Somat. Cell Genet. 9:391–405.

499. Sakaguchi, A.Y., S.L. Naylor, C. Quinto, W.J. Rutter, and T.B. Shows. 1982. The chymotrypsinogen B gene (*CTRB*) is on human chromosome 16. Cytogenet. Cell Genet. 32:313.

500. Sakaguchi, A.Y., S.L. Naylor, R.A. Weinberg, and T.B. Shows. 1982. Organization of human proto-oncogenes. Am. J. Hum. Genet. 34:175A.

501. Sakaguchi, A.Y., S.L. Naylor, and T.B. Shows. 1983. A sequence homologous to Rous sarcoma virus v-src on human chromosome 20. Prg. Nucl. Acids Res. Mol. Biol. 29:291–293.

502. Sakaguchi, A.Y., S.L. Naylor, T.B. Shows, J.J. Toole, M. McCoy, and R.A. Weinberg. 1983. Human c-Ki-ras2 proto-oncogene on chromosome 12. Science 219:1081–1082.

503. Sakai, M., T. Tamaoki, M. Watson, W.R. Breg, F.H. Ruddle, and M. Rabin. 1983. Human alpha-fetoprotein gene is located on chromosome 4. Am. J. Hum. Genet. 35:151A.

504. Samuelson, L.C., and R.A. Farber. 1985. Cytological localization of adenosine kinase, nucleoside phosphorylase-1, and esterase-10 genes on mouse chromosome 14. Somat. Cell Mol. Genet. 11:157–165.

505. Saxe, D.F., N. Takahashi, L. Hood, and M.I. Simon. 1985. Localization of the human myelin basic protein gene (*MBP*) to region 18q22-qter by in situ hybridization. Cytogenet. Cell Genet. 39:246–249.

506. Schafer, R., J. Doehmer, I. Rademacher, and K. Willecke. 1980. Assignment of the gene for cytoplasmic glutamate-oxaloacetate transaminase to mouse chromosome 19 using Chinese hamster × mouse somatic cell hybrids. Somat. Cell Genet. 6:709–717.

507. Schmidt, C.J., D.H. Hamer, and O.W. McBride. 1984. Chromosomal location of human metallothionein genes; implicaiton for Menkes' disease. Science 224:1104–1106.

508. Schroeder, W.T., and G.F. Saunders. 1987. Localization of human catalase and apolipoprotein A-I genes to chromosome 11. Cytogenet. Cell Genet. 44:231–233.

509. Schumutz S.M., and N.E. Simpson. 1984. Suggested assignment of peptidase S (*PEPS*) to 4p11-4q12 by exclusion using gene dosage, accounting for variability in fibroblasts. Hum. Genet. 64:134–138.

510. Schwab, M., H.E. Varmus, J.M. Bishop, K.-H. Grzeschik, S.L. Naylor, A.Y. Sakaguchi, G. Brodeur, and J. Trent. 1984. Chromosome localization in normal human cells and neuroblastomas of a gene related to c-myc. Nature 308:288–291.

511. Schwartz, C.E., E. McNally, L. Leinwand, and M.H. Skolnick. 1986. A polymorphic human myosin heavy chain locus is linked to an anonymous single copy locus (*D17S1*) at 7p13. Cytogenet. Cell Genet. 43:117–120.

512. Searle, A.G. 1978. Personal communication. Mouse News Lett. 59:1.

513. Searle, P.F., B.L. Davison, G.W. Stuart, T.M. Wilkie, G. Norstedt, and R.D. Palmiter. 1984. Regulation, linkage, and sequence of mouse metallothionein I and II genes. Mol. Cell Biol. 4:1221–1230.

514. Seldin, M.F., H.C. Morse, L. D'Hoostelaere, J.L. Britten, and A.D. Steinberg. 1987. Mapping of alpha-spectrin on distal mouse chromosome 1. Cytogenet. Cell Genet. 45:52–54.

515. Shaw, D.J., A.L. Meredith, J.D. Brook, M. Sarfarazi, H.G. Harley, S.M. Huson, G.I. Bell, and P.S. Harper. 1986. Linkage relationships of the insulin receptor gene with the complement component 3, LDL receptor, apolipoprotein C2, and myotonic dystrophy loci on chromosome 19. Hum. Genet. 74:267–269.

516. Sheer, D., D.M. Sheppard, M. LeBeau, J.D. Rowley, C. San Roman, and E. Solomon. 1985. Localization of the oncogene c-*erb*A1 immediately proximal to the acute promyelocytic leukemia breakpoint on chromosome 17. Ann. Hum. Genet. 49:167–171.

517. Sheer, D., L.R. Hiorns, K.F. Stanley, P.N. Goodfellow, D.M. Swallow, S. Povey, H. Heisterkamp, J. Gorffen, J.R. Stephenson, and E. Solomon. 1983. Genetic analysis of the 15;17 chromosome translocation associated with acute promyelocytic leukemia. Proc. Natl. Acad. Sci. USA 80:5007–5011.

518. Shimizu, N., Y. Shimizu, and F.H. Ruddle. 1978. Assignment of the human mitochondrial NAD-linked malate dehydrogenase gene to the p22-qter region of chromosome 7. Cytogenet. Cell Genet. 22:441–445.

519. Shows, T.B. 1974. Synteny of human genes for glutamic oxaloacetic transaminase and hexokinase in somatic cell hybrids. Cytogenet. Cell Genet. 13:143–145.

520. Shows, T.B., and J.A. Brown. 1975. Mapping chromosomes 1 and 2 employing a 1/2 translocation in somatic cell hybrids. Cytogenet. Cell Genet. 14:421–425.

521. Shows, T.B., J.A. Brown, R.L. Eddy, M.G. Byers, L.L. Haley, E.S. Cooper, and A.P. Goggin. 1978. Assignment of peptidase S (*PEPS*) to chromosome 4 in man using somatic cell hybrids. Hum. Genet. 43:119–125.

522. Shows, T.B., V.M. Chapman, and F.H. Ruddle. 1970. Mitochondrial malate dehydrogenase and malic enzyme: Mendelian inherited variants in the mouse. Biochem. Genet. 4:707–718.

523. Siciliano, M.J., R.E.K. Fournier, and R.L. Stallings. 1984. Regional assignment of *ADA* and *ITPA* to mouse chromosome 2 (C1-ter). J. Hered. 75:175–180.

524. Silver, J., J.B. Whitney, C. Kozak, G. Hollis, and I. Kirsch. 1985. *Erbb* is linked to the alpha-globin locus on mouse chromosome 11. Mol. Cell Biol. 5:1784–1786.

525. Simmers, R.N., I. Stupans, and G.R. Sutherland. 1985. The haptoglobin locus is distal to the fragile site at 16q22. Cytogenet. Cell Genet. 40:745–746.

526. Sinet, P.M., J. Couturier, B. Dutrillaux, M. Poissonier, O. Raoul, M.-O. Rethore, D. Allard, J. Lejeune, and H. Jerome. 1976. Trisomie 21 et superoxyde dismutase-1 (*IPO-A*): Tentative de localisation sur la sous-bande 21q22.1. Exp. Cell Res. 97:47–55.

527. Skow, L.C. 1981. Genetic variation for prolidase (*PEP-4*) in the mouse maps near the gene for glucosephosphate isomerase (*GPI-1*) on chromosome 78. Biochem. Genet. 19:695–700.

528. Skow, L.C. 1982. Location of a gene controlling electrophoretic variation in mouse gamma-crystallins. Exp. Eye Res. 34:509–516.

529. Skow, L.C., and M.E. Donner. 1985. The locus encoding alphaA-crystallin is closely linked to *H-2K* on mouse chromosome 17. Genetics 110:723–732.

530. Smith, M., G. Duester, L. Carlock, and J. Wasmuth. 1985. Assignment of *ADH1*, *ADH2*, and *ADH3* genes (class I ADH) to human chromosome 4q21-4q25. Cytogenet. Cell Genet. 40:748.

531. Solomon, E., L. Hiorns, D. Sheer, and D. Rowe. 1984. Confirmation that the type 1 collagen gene on chromosome 17 is *COL1A1* using a human genomic probe. Ann. Hum. Genet. 48:39–42.

532. Sparkes, R.S., M. Simon, V.H. Kohn, R.E.K. Fournier, J. Lem, I. Klisak, C. Heinzman, C. Blatt, M. Lucero, and T. Mohandas. 1986. Assignment of the human and mouse prion protein genes to homologous chromosomes. Proc. Natl. Acad. Sci. USA 83:7358–7362.

533. Sparkes, R.S., T. Mohandas, C. Heinzmann, H.J. Roth, I. Klisak, and A.T. Campagnoni. 1987. Assignment of the myelin basic protein gene to human chromosome 18q22-qter. Hum. Genet. 75:147–150.

534. Sparkes, R.S., T. Mohandas, M.C. Sparkes, and J.D. Shulkin. 1978. Regional localization of human phosphoglucomutase-2 locus on chromosome 4. Exp. Cell Res. 111:492–495.

535. Spence, J.E., C.L. Rosenbloom, W.E. O'Brien, D.K. Seilheimer, S. Cole, R.E. Ferrell, R.C. Stern, and A.L. Beaudet. 1986. Linkage of DNA markers to cystic fibrosis in 26 families. Am. J. Hum. Genet. 39:729–734.

536. Spurr, N.K., E. Solomon, M. Jansson, D. Sheer, P.N. Goodfellow, W.F. Bodmer, and B. Vennstrom. 1984. Chromosomal localization of the human homologues to the oncogenes erbA and B. EMBO J. 3:159–163.

537. Spurr, N.K., E. Solomon, P. Goodfellow, M.J. Owen, M.K.L. Collins, S. Tonegawa, A. Stinson, and T. Rabbitts 1985. Genetics of the T cell receptor. Cytogenet. Cell Genet. 40:754.

538. Spurr, N.K., P.N. Goodfellow, E. Solomon, M. Parkar, B. Vennstrom, and W.F. Bodmer. 1984. Mapping of cellular oncogenes; erb B on chromosome 7. Cytogenet. Cell Genet. 37:591.

539. Sukhatme, V.P., A.C. Vollmer, M. Erikson, M. Isobe, C. Croce, and J.R. Parnes. 1985. Gene for human T-cell differentiation antigen Leu-2/78 is closely linked to the K light chain locus on chromosome 2. J. Exp. Med. 161:409–434.

540. Suzuki, T., and P. Board. 1984. Glutathione-S-transferase gene mapped to chromosome 11 is GST3 not GST1. Somat. Cell Mol. Genet. 10:319–320.

541. Swallow, D.M., S. Povey, P.N. Goodfellow, P. Andrews, and H. Harris. 1985. The liver/bone/kidney isozyme of alkaline phosphatase (ALPL) is coded by a gene on chromsome 1. Cytogenet. Cell Genet. 40:756.

542. Swallow, D.M., V.S. Turner, and D.A. Hopkinson. 1983. Isozymes of rodent 5' nucleotidase: evidence for two independent structural loci *Umph-1* and *Umph-2*. Ann. Hum. Genet. 47:9–17.

543. Swan, D., M. Oskarsson, D. Keithley, F.H. Ruddle, P. D'Eustachio, and G.F. Vande Woude. 1982. Chromosomal localization of the Moloney sarcoma virus mouse cellular (c-mos) sequence. J. Virol. 44:752–754.

544. Swan, D., P. D'Eustachio, L. Leinwand, J. Seidman, D. Keithley, and F.H. Ruddle. 1979. Chromosomal assignment of the mouse kappa light chain genes. Proc. Natl. Acad. Sci. USA 76:2735–2739.

545. Szymura, J.M., M.R. Wabl, and J. Klein. 1981. Mouse mitochondrial dismutase locus is on chromsome 17. Immunogenetics 14:231–240.

546. Taggart, R.T., P. Tetri, and U. Francke. 1980. Assignment of the gene for NADH diaphorase *Dia-1* to mouse chromosome 15. Somat. Cell Genet. 6:769–776.

547. Tan, Y.-H., I. Tischfield, and F.H. Ruddle. 1973. The linkage of genes for the human interferon-induced antiviral protein and indophenol oxidase-B traits to chromosome G-21. J. Exp. Med. 137:317–330.

548. Taub, R., I. Kirsch, C. Morton, G. Lenoir, D. Swan, S. Tronick, S. Aaronson, and P. Leder. 1982. Translocation of the c-myc gene into the immunoglobulin heavy chain locus in human Burkitt lymphoma and murine

plasmacytoma cells. Proc. Natl. Acad. Sci. USA 79:7837–7841.

549. Taylor, B.A., D.M. Walls, and M.J. Wimsatt. 1987. Localization of the inosine triphosphatase locus (*Itp*) on chromosome 2 of the mouse. Biochem. Genet. 25:267–273.

550. Taylor, B.A., D.W. Bailey, M. Cherry, R. Riblet, and M. Weigert. 1975. Genes for immunoglobulin heavy chain and serum prealbumin protein are linked in mouse. Nature 256:644–646.

551. Taylor, B.A., and L. Rowe. 1984. Genes for serum amyloid A proteins map to chromosome 7 in the mouse. Mol. Gen. Genet. 195:491–499.

552. Taylor, B.A., and L. Rowe. 1987. The congenital goiter mutation is linked to the thyroglobulin gene in the mouse. Proc. Natl. Acad. Sci. USA 84:1986–1989.

553. Teng, C.T., B.T. Pentecost, A. Marshall, A. Solomon, B.H. Bowman, P.A. Lalley, and S.L. Naylor. 1987. Assignment of the lactotransferrin gene to human chromosome 3 and to mouse chromosome 9. Somat. Cell Mol. Genet. 13:689–693.

554. Terasaki, P. 1980. Histocompatibility Testing 1980. UCLA Tissue Typing Laboratory. UCLA Press, Los Angeles.

555. Timms, G.P., and R.S. Holmes. 1981. Genetics of aldehyde dehydrogenase isozymes in the mouse: evidence for multiple loci and localization of *Ahd-2* on chromosome 19. Genetics 97:327–336.

556. Timms, G.P., and R.S. Holmes. 1982. Personal communication. Mouse News Lett. 66:60.

557. Tischfield, J.A., and F.H. Ruddle. 1974. Assignment of the gene for adenosine phosphoribosyltransferase to human chromosome 16 by mouse human somatic cell hybridization. Proc. Natl. Acad. Sci. USA 71:45–49.

558. Tischfield, J.A., R.P. Creagan, E.A. Nichols, and F.H. Ruddle. 1974. Assignment of a gene for adenosine deaminase to human chromosome 20. Hum. Hered. 24:1–11.

559. Tolley, E., V. van Heyningen, R. Brown, M. Bobrow, and I.W. Craig. 1980. Assignment to chromosome 16 of a gene necessary for the expression of human mitochondrial glutamate oxaloacetate transaminase (aspartate aminotransferase) (EC 2.6.1.1). Biochem. Genet. 18:947–954.

560. Trent, J.M., S. Olsen, and R.M. Lawn. 1982. Chromosomal localization of human leukocyte, fibroblast, and immune interferon genes by means of *in situ* hybridization. Proc. Natl. Acad. Sci. USA 79:7809–7813.

561. Tricoli, J.V., and T.B. Shows. 1984. Assignment of alpha amylase genes to the p22.1-p21 region of chromosome 1. Cytogenet. Cell Genet. 37:597.

562. Tsui, L.-C., M.L. Breitman, S.O. Meakin, H.F. Willard, Y. Shiloh, T. Donlon, and G. Bruns. 1985. Localization of the human gamma-crystallin gene cluster (*CRYG*) to the long arm of chromosome 2, region q33-q35. Cytogenet. Cell Genet. 40:763–764.

563. Tukey, R.H., P.A. Lalley, and D.W. Nebert. 1984. Localization of cytochrome *P1-450* and *P3-450* genes to mouse chromosome 9. Proc. Natl. Acad. Sci. USA 81:3163–3166.

564. Tulchin, N., and B.A. Taylor. 1981. Gamma-glutamyl cyclotransferase: a new genetic polymorphism in the mouse (*Mus musculus*) linked to *Lyt-2*. Genetics 99:109–116.

565. Turner, V.S., B.M. Turner, R. Kucherlapati, F.H. Ruddle, and K. Hirshhorn. 1976. Assignment of the human alpha-L-fucosidase locus to chromosome 1 by use of a 'clone panel'. Cytogenet. Cell Genet. 16:238–240.

566. Uhler, M., E. Hebert, P. D'Eustachio, and F.H. Ruddle. 1983. The mouse genome contains two nonallelic pro-opiomelanocortin genes. J. Biol. Chem. 258:9444–9453.

567. Urano, Y., M. Sakai, K. Watanabe, and T. Tamaoki. 1984. Tandem arrangement of the albumin and alpha-fetoprotein genes in the human genome. Gene 32:255–261.

568. Van Heyningen, V., I. Craig, and W.F. Bodmer. 1973. Genetic control of mitichondrial enzymes in human–mouse somatic cell hybrids. Nature 242:183–187.

569. Van Heyningen, V., M. Bobrow, W.F. Bodmer, S.E. Gardiner, S. Povey, and D.A. Hopkinson. 1975. Chromosome assignment of some human enzyme loci: mitochondrial malate dehydrogenase to 7, mannose phosphate isomerase and pyruvate kinase to 15 and probably, esterase D to 13. Ann. Hum. Genet. 38:295–303.

570. Van den Elsen, P., G. Bruns, D.S. Gerhard, D. Pravtcheva, C. Jones, D. Housman, F.H. Ruddle, S. Orkin, and C. Terhorst. 1985. Assignment of the gene coding for the T3-delta subunit of the T3-T cell receptor complex to the long arm of human chromosome 11. Proc. Natl. Acad. Sci. USA 82:2920–2924.

571. Van der Korput, J.A.G.M., J. Hilkens, V. Kroezen, E.C. Zwarthoff, and J.Trapman. 1985. Mouse interferon alpha and beta genes are linked at the centromere proximal region of chromosome 4. J. Gen. Virol. 66:43–48.

572. VandeBerg, J.L., and J. Klein. 1978. Localization of mouse *Pgk-2* gene at the D end of the *H-2* complex. J. Exp. Zool. 203:319–324.

573. Venta, P.J., T.B. Shows, P.J. Curtis, and R.E. Tashian. 1983. Polymorphic gene for carbonic anhydrase II: Molecular disease marker on chromosome 8. Proc. Natl. Acad. Sci. USA 80:4437–4440.

574. Wainwright, B.J., P.J. Scambler, and R. Williamson. 1987. Regional localization of three probes closely linked to the cystic fibrosis locus by deletion analysis. Cytogenet. Cell Genet. 44:101–102.

575. Ward, P., S. Packman, W. Loughman, M. Sparkes, A. McMahon, T. Gregory, and A. Ablin. 1984. Location of the retinoblastoma susceptibility gene(s) and the human esterase D locus. J. Med. Genet. 21:92–95.

576. Watanabe, T., N. Ogasawara, and H. Goto. 1976. Genetic study of pancreatic proteinase in mice (*Mus muscu-*

lus): genetic variants of trypsin and chymotrypsin. Biochem. Genet. 14:697–707.

577. Watanabe, T., and T. Tomita. 1974. Genetic study of pancreatic and alpha-amylase in mice (*Mus musculus*). Biochem. Genet. 12:419–428.

578. Watkins, P., R. Eddy, A. Beck, V. Vellucci, J. Gusella, and T. Shows. 1985. Assignment of the human gene for the beta subunit of follicle stimulating hormone (*FSHB*) to chromosome 11. Cytogenet. Cell Genet. 40:773.

579. Watson, D.K., M.J. McWilliams-Smith, C. Kozak, R. Reeves, J. Gearhart, M.F. Nunn, W. Nash, J.R. Fowle, P. Duesberg, T.S. Papas, and S.J. O'Brien. 1986. Conserved chromosomal positions of dual domains of the ets protooncogene in cats, mice, and humans. Proc. Natl. Acad. Sci. USA 83:1792–1796.

580. Weil, D., Nguyen Van Cong, R. Rebourcet, M.S. Gross, and J. Frezal. 1979. Regional mapping of enzyme loci on human chromosomes 2, 17, 5, and X by use of somatic cell hybridization. Cytogenet. Cell Genet. 25:215–216.

581. Weitkamp, L.R. 1976. Linkage of GLO with HLA and Bf. Effect of population and sex on recombination frequency. Tissue Antigens 7:273–279.

582. Weitkamp, L.R., J.H. Renwick, J. Berger, D.C. Shreffler, O. Drachman, R. Wuhrmann, M. Braend, and G. Franglen. 1970. Additional data and summary for albumin-*Gc* linkage in man. Hum. Hered. 20:1–7.

583. Weitkamp, L.R., S.A. Guttormsen, and R.M. Gutendyke. 1971. Genetic linkage between a locus for *6-PGD* and *Rh* locus: evaluation of possible heterogeneity in the recombination fraction between sexes and among families. Am. J. Hum. Genet. 23:462–470.

584. Westaway, D., P.A. Goodman, C.A. Mirenda, M.P. McKinley, G.A. Carlson, and S.B. Prusiner. 1987. Distinct prion proteins in short and long scrapie incubation period mice. Cell 51:651–662.

585. Westerveld, A., H.M.A. Beyersbergen Van Henegouwen, and H. Van Someren. 1975. Evidence for synteny between the human loci for galactose-1-phosphate uridyl transferase and aconitase in man-Chinese hamster somatic cell hybrids, Cytogenet. Cell Genet. 14:283–284.

586. Westerveld, A., and P. Meera Khan. 1972. Evidence for linkage between human loci for 6-phosphogluconate dehydrogenase and phosphoglucomutase 1 in man-Chinese hamster somatic cell hybrids. Nature 236:30–32.

587. White, R., M. Leppert, P. O'Connell, Y. Nakamura, S. Woodward, M. Hoff, J. Herbst, M. Dean, G. Vande Woude, G.M. Lathrop, and J.-M. Lalouel. 1986. Further linkage data on cystic fibrosis: the Utah study. Am. J. Hum. 39:694–698.

588. Whitehead, A.S., E. Solomon, S. Chambers, W.F. Bodmer, S. Povey, and G. Fey. 1982. Assignment of a structural gene for the third component of human complement to chromosome 19. Proc. Natl. Acad. Sci. USA 79:5021–5025.

589. Wieacker, P., J.E. Griffen, T. Weinker, J.M. Lopez, J.D. Wilson, and M. Breckwoldt. 1987. Linkage analysis with RFLPs in families with androgen resistance syndromes: evidence for close linkage between the androgen receptor locus and the DXS1 segment. Hum. Genet. 76:248–252.

590. Wijnen, L.M.M., and P. Meera Khan. 1982. Assignment of *GPT1* to human chromosome 16. Cytogenet. Cell Genet. 32:327.

591. Wijnen, L.M.N., K.-H. Grzeschik, P.L. Pearson, and P. Meera Khan. 1977. The human *PGM-2* and its chromosomal location in man-mouse hybrids. Hum. Genet. 37:217–218.

592. Wilcox, F.H., and B.A. Taylor. 1981. Genetics of the *Akp-2* locus for alkaline phosphatase of liver, kidney, bone, and placenta in the mouse. J. Hered. 72:387–390.

593. Willard, H.F., and J.R. Riordan. 1985. Assignment of the gene for myelin proteolipid protein to the X chromosome: implications for X-linked myelin disorders. Science 230:940–942.

594. Willison, K., A. Kelly, K. Dudley, P. Goodfellow, N. Spurr, V. Groves, P. Gorman, D. Sheer, and J. Trowsdale. 1987. The human homologue of the mouse *t*-complex gene, *TCP1*, is located on chromosome 6 but is not near the *HLA* region. EMBO J. 6:1967–1974.

595. Willison, K., K. Dudley, N. Spurr, and P. Goodfellow. 1985. Chromosomal assignment of *TCP-1* the human homologue of a mouse *t*-complex locus. Cytogenet. Cell Genet. 40:779–780.

596. Willison, K.R., K. Dudley, and J. Potter. 1986. Molecular cloning and sequence analysis of a haploid expressed gene encoding *t* complex polypeptide 1. Cell 44:727–738.

597. Wilson, C.M., E.G. Erdos, J.D. Wilson, and B.A. Taylor. 1978. Location on chromosome 1 of *Rnr*, a gene that regulates renin in the submaxillary gland of the mouse. Proc. Natl. Acad. Sci. USA 75:623–626.

598. Wilson, D.E., D.M. Swallow, and S. Povey. 1986. Assignment of the human gene for uridine 5′-monophosphohydrolase (*UMPH2*) to the long arm of chromosome 17. Ann. Hum. Genet. 50:223–227.

599. Wilson, D.E., D. Woodard, A. Sandler, J. Erikson, and A. Gurney. 1986. The gene for uridine monophosphatase-2 is on mouse chromosome 11. Am. J. Hum. Genet. 39:A173.

600. Wilson, D.E., D. Woodard, A. Sandler, J. Erikson, and A. Gurney. 1987. Provisional assignment of the gene for uridine monophosphatase-2 (*UMPH-2*) to mouse chromosome 11. Biochem. Genet. 55:1–9.

601. Womack, J.E., M.T. Davisson, E.M. Eicher, and D.A. Kendall. 1977. Mapping of nucleoside phosphorylase (*Np-1*) and esterase 10 (*Es-10*) on mouse chromosome 14. Biochem. Genet. 15:347–355.

602. Womack, J.E., S. Ashley, L.B. Barnett, and S.E. Lewis. 1986. Linkage of *Pep-2* and *Apk* on chromosome 10. Biochem. Genet. 24:721–727.

603. Wood, A.W., and B.A. Taylor. 1979. Genetic regulation of coumarin hydroxylase activity in mice: evidence for single locus control on chromosome 7. J. Biol. Chem. 254:5647–5651.

604. Wulfsberg, E.A., R.E. Carrel, I.J. Klisak, T.J. O'Brien, J.A. Sykes, and R.S. Sparkes. 1983. Normal superoxide dismutase-1 (*SOD-1*) activity with deletion of chromosome band 21q21 support localization of *SOD-1* locus to 21q22. Hum. Genet. 64:271–272.

605. Yamamoto, K., G. Floyd-Smith, U. Francke, N. Koch, W. Lauer, B. Dobberstein, R. Shafer, and G.J. Hammerling. 1985. The gene encoding the Ia-associated invariant chain is located on chromosome 18 in the mouse. Immunogenetics 21:83–90.

606. Yang, F., J.B. Lumm, J.R. McGill, C.M. Moore, S.L. Naylor, P.H. van Bragt, W.D. Baldwin, and B.H. Bowman. 1984. Human transferrin: cDNA characterization and chromosomal localization. Proc. Natl. Acad. Sci. USA 81:2752–2756.

607. Yang-Feng, T.L., G.A.P. Bruns, A.J. Carroll, K.O.J. Simola, and U. Francke. 1986. Localization of the *LDHA* gene to 11p14-11p15 by in situ hybridization of an *LDHA* cDNA probe to two translocations with breakpoints in 11p13. Hum. Genet. 74:331–334.

608. Yang-Feng, T.L., G. Floyd-Smith, M. Nemer, J. Drouin, and U. Francke. 1985. Human pronatriodilatin is located on the distal short arm of chromosome 1. Cytogenet. Cell Genet. 40:783.

609. Yang-Feng, T.L., G. Floyd-Smith, M. Nemer, J. Drouin, and U. Francke. 1985. The pronatriodilatin gene is located on the distal short arm of human chromosome 1 and on mouse chromosome 4. Am. J. Hum. Genet. 37:1117–1128.

610. Yang-Feng, T.L., L.J. DeGennaro, and U. Francke. 1986. Genes for synapsin I, a neuronal phosphoprotein, map to conserved regions of human and murine X chromosomes. Proc. Natl. Acad. Sci. USA 83:8679–8683.

611. Yang-Feng, T.L., N.R. Landau, D. Baltimore, and U. Francke. 1986. The terminal deoxynucleotidyltransferase gene is located on human chromosome 10 (10q23-q24) and on mouse chromosome 19. Cytogenet. Cell Genet. 43:121–126.

612. Zabel, B.U., R.E.K. Fournier, P.A. Lalley, S.L. Naylor, and A.Y. Sakaguchi. 1984. Cellular homologs of the avian erythroblastosis virus *erb-A* and *erb-B* genes are syntenic in mouse but asyntenic in man. Proc. Natl. Acad. Sci. USA 81:4874–4878.

613. Zabel, B.U., R.L. Eddy, P.A. Lalley, J. Scott, G.I. Bell, and T.B. Shows. 1985. Chromosomal locations of the human and mouse genes for precursors of epidermal growth factor and the beta subunit of nerve growth factor. Proc. Natl. Acad. Sci. USA 82:469–473.

614. Zabel, B.U., S.L. Naylor, A.Y. Sakaguchi, G.I. Bell, and T.B. Shows. 1983. High resolution chromosomal localization of human genes for amylase, proopiomelanocortin, somatostatin and a DNA fragment (D391) by in situ hybridization. Proc. Natl. Acad. Sci. USA 80:6932–6936.

615. Zhang, Z.-X., E. de Clercq, H. Heremans, M. Verhaegen-Lewalle, and J. Content. 1982. Antiviral and anticellular activities of human and murine Type I and Type II interferons in human cells monosomic, disomic, and trisomic for chromosome 21. Proc. Soc. Exp. Biol. Med. 170:103–111.

8 DNA POLYMORPHISMS

ROSEMARY W. ELLIOTT

Table 8.1 contains information on probes that reveal variant fragments in mouse DNA. These variant fragments are popularly known as restriction fragment length polymorphisms (RFLPs). Polymorphism refers to a variant which is present at a frequency of greater than 1 per cent in an inbreeding population and this kind of information is not usually available for mouse populations. However, as it is likely that most of these fragments can also be found in DNA from trapped mice or from mice that have been laboratory bred for only a few generations, the use of the commonly accepted term 'polymorphism' is reasonable. In support of this, we have hybridized most of our probes to DNA from wild-derived *M. musculus* and *M. domesticus* as well as other related species and find the same sized DNA fragments as we found in DNA from inbred mice.

A list of probes for mouse DNA is published annually in the February Mouse News Letter, beginning in Feb. 1987 (Eppig *et al.* 1987). A subset of these probes are associated with fragment length variation and these are the subject of this table. For each probe, data are presented on the vector, insert length, and insertion site. This is followed by information on the hybridizing fragments in mouse DNA, the enzyme used, fragment sizes, both constant and variant, allele designations, and strains carrying each allele.

The information is organized by chromosome. Within each chromosome the loci are listed by probable order. An estimate of the distance from the centromere has been made for each locus. Four major sources of information have been used for this estimate. The first is the mouse linkage map (Davisson and Roderick 1986) that provides estimates of intervals between loci. A second source is the Mouse Chromosome Atlas (Lyon 1987) that presents map intervals as distances from the centromere. A third source is the publication referenced in the table. A fourth is the response to a request for information. Many contributors generously summarized their data on their probes, making the compiling task much easier, and sent data giving estimates of distances from nearby reference loci. The final result is a reasonable estimate of order and distance that should be, however, used with caution. In situ-

ations such as the X-chromosome and chromosome 17, where several laboratories have contributed information, the relative order of loci obtained from one laboratory is clearly much more reliable than that obtained by intermingling data from several laboratories. Until all probes are tested on a single set of animals, the order cannot be determined. However, it seemed to be of use to indicate when loci were likely to be clustered. At the end of each chromosome are those loci which could not be ordered.

In several instances, the probe used is not completely described. This is indicated by 'Not given'. At times a DNA fragment was used. This is indicated by 'fr'. When two enzymes were used, either at the time of insertion, or the time of excision of the fragment, a dash has been placed between the enzyme names.

In the five columns describing the restriction fragments in mouse genomic DNA, the following conventions were observed. Under ENZYME the indicated enzyme may be the only one used or it may be selected from several which gave similar information. If a second enzyme gave new alleles, this is indicated either by giving the second enzyme a new line, or by separating the enzyme names by a slash.

In the column headed CON. FRAG., a number, or set of numbers separated by commas indicates the size(s) of any constant fragments. 'No' indicates the absence of a constant fragment. 'Not given' indicates that no information is available. 'Several' or 'Many' indicates the presence of a number of fragments. Where two enzymes are reported, a slash has been used to separate the results for the two enzymes.

The column headed ALLELE gives allele designations for the polymorphic fragments. Allele designations were often made in the reference or were subsequently communicated by the authors. In several cases, allele designations were not given. They have been added, using the following guidelines. Where mapping used RI strain progenitors, the designations customary for inbred strains was used, i.e. if C57BL/6 was included in the strain list the allele designation was b. For an allele in DBA/2, d was chosen, h for C3H. For interspecific crosses, the designation customary for the inbred strain was used. Alleles in *M. spretus* were designated s while those in *M. musculus* were designated m. However, in crosses using CzechII mice, the authors had usually designated the alleles d and c.

Under the column LENGTH are listed the fragment lengths for each allele either as found in the reference or as supplied separately by the authors. Sometimes the approximate length can be calculated from a published figure. These estimates are indicated by 'approximate' symbols. Otherwise 'not given' is used. 'Absent' indicates that the phenotype of the designated allele is the absence of the fragment found in another allele. 'Present' is used if the size of this fragment cannot be determined from the data at hand. 'Several' and 'Many' are used as in the column for constant fragments while 'Various' indicates that when a number of enzymes are used, most enzymes produce one or more polymorphic fragments.

In the column headed STRAINS/SPEC., up to three inbred strains are listed for each allele. If the paper shows results for only one, a full designation of the strain is given. Otherwise, an abbreviated version of the name is used.

Initially, the intent was to include only DNA variation between inbred strains and preference has been given to these data. The table has, however, been extended to include published data using interspecific crosses because this method of mapping has been

shown to be very powerful and because backcrosses are available from a number of laboratories. Investigators should be aware, however, that DNA fragment length differences have been found between *M. spretus* and inbred strains for every locus tested. The same may well be true for crosses involving *M. musculus* and inbred strains.

The column headed CONTACT/AVAILABILITY lists the holder(s) of the probes used and some information on the availability. Numbers indicate that the probe is available (1) freely, (2) to collaborators, (3) in future, (4) not available, (5) not known.

Information on which Table 8.1 is based was obtained in response to a request for information during 1986. This was followed by a literature search. A computerized literature search using DNA polymorphism and mouse as the search criteria yielded papers which relate mainly to human genetics. Thus the search concentrated on journals likely to have relevant papers and authors likely to publish in this area. The author extends apologies to those whose papers were omitted.

Davisson, M., and T. Roderick. 1986. Mouse linkage map. Mouse News Lett. 75:11–15.

Eppig, J., S. Brown, R.W. Elliott, and K. Willison. 1987. List of mouse DNA clones and probes. Mouse News Lett. 77:81–89.

Lyon, M.F. 1987. Mouse chromosome atlas. Mouse News Lett. 78:12–33.

Table 8.1 DNA polymorphisms

LOCUS NAME	CHR.	DIST. CEN.	LOCUS	PROBE NAME	VECTOR	SITE	INS.LEN. (BP)	POLYMORPHISM ENZYME	CON.FRAG	ALLELE	LENGTH	STRAINS/SPEC	CONTACT/ AVAILABILITY	REFERENCE
d-Nicotinic acetyl-choline receptor	1	17	Acrd	dAchr	M13mp18	EcoRI-PvuII	300-600	PvuII	No	d s	5.0 4.2	DBA2 M.spretus	Robert/1	Heidmann et al. 1986
g-Nicotinic acetyl-choline receptor	1	17	Acrg	gAchr	M13mp18	EcoRI-BglI	300-600	EcoRI	No	d s	9.0 5.0	DBA2 M.spretus	Robert/1	Heidmann et al. 1986
Myosin alkali light chains 1,3	1	22	Mylf	pGLC450	pBR322	BamHI	470	PstI	Not given	b d	4.1 4.3	C57BL/6J DBA2J	Robert/1	Robert et al. 1985
B cell leukemia junction-2	1	44	Bcl-2	bcl-2 (Ncm4)	ab fr.	EcoRI-HindIII	1400	XbaI	7.4	a b	5.0 5.8	A,BALB,C57 DBA,STS,129	Potter/5	Mock et al. 1988
Endog.ecot.MULV-17	1	44	Emv-17	pPS1.25	pUC18	PstI-SstI	1250	EcoRI	No	a b c d	18.4 9.6 7.4 6.8	RF/J AKR,DBA,C3H SWRJ A,C57BL,BALB	Copeland/1	Buchberg et al. 1986
Engrailed-like-1	1	46	En-1	pL4.U6	pUC9	EcoRI	750	TaqI	1.5	b d	1.4 4.2	C57BL/J DBA2,AKR/J	Hill/3	Hill et al. 1987
Renin	1	48	Ren-1,-2	pRn1-3 pSM479	pBR322 pBR322	PstI PstI	1400 380	Many PvuII	Not given No	b s b s	Various Various 9.0 10.0,14.0	BALB,C3H,C57 AKR,SWR C57BL/6J DBA2J	Rougeon/1 Gross/1	Panthier et al. 1982 Piccini et al. 1982
Complement factor H	1	52	Cfh	Mu25IV	pBR322	PstI	4300	EcoRI	15 fr.	a b	4.6 Absent	AKR,BALB,C57 C3H,DBA,SJL	Tack/2	D'Eustachio et al. 1986
Laminin B2	1	57	Lamb-2	pPE9	pUC8	EcoRI-HindIII	1500	HinfI EcoRI	1.0,0.8 7.0	a,b d a b,d	1.6 1.3 5.0 3.7	Most strains DBA,O20,ST AKR,129,MOLD Most strains	Hogan/1	Elliott et al. 1985
Ornithine decarbox-ylase-1	1	58	Odc-1	pMK934	pBR322	PstI	1000	HindIII	Several	a b	2.6 2.4	AKR,SM,NZB Most strains	Elliott/1 Berger/1	Elliott et al. 1985
Lymphocyte antigen-5	1	61	Ly-5	pLy5-68	pCD	PstI-BamHI	4700	BamHI	5 fr.	a b c d,e,f,g,?	≈5.2 ≈10 ≈5.2,≈5 Various	RIII,SJL,STS Most strains ST/bJ,NH wild-derived	Seldin/5 Shen/5	Shen et al. 1985 Seldin et al. 1987b
Serum amyloid P	1	71	Sap	pmSAP3	M13mp18	EcoRI	1000	EcoRI	No	a b d	25.0 16.0,4.6 5.4	AKR,NZB,SJL C57,NC DBA,BALB,A	Maruyama/5	Ishikawa et al. 1987
Spectrin, alpha	1	72	Spna-1	pMaSp1	pBR322	PstI	750	EcoRI	No	a b s	5.0 7.0 4.3	Most strains C57,HRS,MA M.spretus	Curtis/5	Cioe & Curtis, 1985 Seldin et al. 1987c
Sodium, potassium ATPase a-3	1	72	Atpa-3	rb13c	Not given	Not given	2700	TaqI	Several	a s	Absent Present	BALB/c M.spretus	Herrera/5	Kent et al. 1987

Table 8.1 (*Continued*)

LOCUS NAME	CHR.	DIST. CEN.	LOCUS	PROBE NAME	VECTOR	SITE	INS.LEN. (BP)	POLYMORPHISM ENZYME	CON.FRAG	ALLELE	LENGTH	STRAINS/SPEC	CONTACT/ AVAILABILITY	REFERENCE
Complement-5	2	18	Hc	pC5	pBR322	PstI	5200	HindIII	10 fr.	a b	2.7 4.6,6.0	C57,C3H.SJL AKR,SWR,DBA	Tack/2	D'Eustachio et al. 1986
Interleukin-1a	2	53	Il-1a	pMull-1a	Lgt10	EcoRI- EcoRV	650	MspI/TaqI	No	a b c d e f	7.1,6.6/≈23 6.2/8.4 6.0/≈18 6.2/≈18 6.0/≈23 6.3,5.6/≈18	BALB/cJ C57BL,C57L AKR,SWR DBA2J C3H/HeJ SJL/J	Gray/5	Gray et al. 1986 D'Eustachio et al. 1987
Interleukin-1b	2	53	Il-1b	pMull-1b	Lgt10	HindIII- EcoRI	1350	TaqI	No	a b	4.2,3.8,1.9 5.6,1.7	BALB/cJ AKR,C57,DBA	Gray/5	Gray et al. 1986 D'Eustachio et al. 1987
Endog.ecotr.MULV-15	2	62	Emv-15	p15.4	pBR325	EcoRI	1100	KpnI	7.5	a b v	12.0 8.3 12.0,3.0	BALB,C3H,SJL A,C57,DBA YS	Copeland/1	Siracusa et al. 1987 Copeland et al. 1983
AKR virus-3 Endog.ecotr.MULV-13	2	62	Akv-3 Emv-13	pEnv	pBR322	EcoRI	400	PvuII	No	a b	3.9 Absent	AKR/J Most strains	Jenkins/1	Jenkins et al. 1981
Proto-oncogene src	2	62	Src	v-sarc	Not given	Not given	Not given	HindIII	No	b d	14.0,16.5 13.0	C57BL/J DBA2J	Simon/1	Blatt et al. 1984
Carbonic anhydraseII	3	4	Car-2	pMCAII	pBR322	PstI	1480	PstI/BglII	Yes	b d	5.2,3.2/3.1,2 8.4/5.1	C57BL/YBR A,BALB,DBA	Venta/1 Curtis	Venta et al. 1985 Curtis et al. 1983
Ecotr.vir.site-1	3	25	Evi-1	pUC1.OM3	pUC18	HindIII	1000	EcoRV	Not given	a b	23.8 22.5	BALB/cJ,C57 DBA,A,AK,C3H	Copeland/1	Mucenski et al. 1988
Nerve growth factorβ	3	64	Ngfb	pmngf6	pBR322- deriv.	PstI	Not given	TaqI	Not given	d s	5.4 2.5	C57BL/6J M.spretus	Rutter/5	Mucenski et al. 1988
Proto-oncogene Nras	3	64	Nras	Nras	SP6	Not given	Not given	TaqI	Not given	d s	1.9 3.3,2.3,1.1	C57BL/6J M.spretus	Pellicer/5	Mucenski et al. 1988
Salivary amylase Pancreatic amylase	3 3	68 68	Amy-1 Amy-2	pMSa104 pMPa21	pBR322 pBR322	PstI PstI	1630 1500	HindIII BglII	Several Several	b d d d a y	6.3 8.1 4.2 4.4 4.7 5.4	129,C57,O20 A,BALB,DBA 129,C57,O20 A,BALB,DBA A/J YBR/Ki	Hagenbüchle /1	Hagenbüchle et al. 1980 Paul & Elliott, 1987 Meisler et al. 1986
			Amy-1 promotor	Amy-1 promotor	fr.	HaeIII-PstI	95	EcoRI	No					
Epidermal growth factor	3	81	Egf	pmefg-10b	pBR322- deriv.	PstI	Not given	EcoRV TaqI	Not given Not given	a b c d	21.2 18.4 5.8 5.3	C57BL/6J A,AK,C3H,DBA A,AK,C57,BALB C3H,DBA	Rutter/5	Mucenski et al. 1988
Alcohol dehydrogenase	3	85	Adh-1	pHJEmYA2	pUC	BglI fr.	Not given	EcoRI	2.5	b d	6.4 5.3	C57,YBR BALB,DBA	Edenberg/5	Zhang et al. 1987
Sodium potassium ATPase a-1	3		Atpa-1	rb5	Not given	Not given	1200	Not given	Not given	a s	Not given Not given	Inbred strains M.spretus	Levenson/5 Housman/5	Kent et al. 1987

540

LOCUS NAME	CHR.	DIST. CEN.	LOCUS	PROBE NAME	VECTOR	SITE	INS.LEN. (BP)	ENZYME	POLYMORPHISM CON.FRAG	ALLELE	LENGTH	STRAINS/SPEC	CONTACT/ AVAILABILITY REFERENCE
alpha-Actin, skeletal muscle	3		Acts	pAM91	pBR322	PstI	1360	BglI	No	a / b	7.0 / 9.0	BALB/c AKR,C57,DBA	Buckingham/1 Minty et al. 1983
Moloney-MuLV int-10	3		Mov-10	p10	pBR322	PstI	1400	EcoRI	Yes	b / c	7.0 / 9.0	C57BL/6 BALB/c,AKR	Jaenisch/5 Münke et al. 1986
Mouse urinary protein	4	39	Mup-1	p1057	pBR322	PstI	800	Many	Many	a / b	Several / Several	BALB,C3H,DBA C57BL,C57L	Hastie/1 Held/1 Bennett et al. 1982
alpha-1 acid glyco-protein	4	48	Orm-1	pIRL-10	pBR322	PstI	850	EcoRI	6.8,6.2	a / b	8.2,2.1 / 13.0,5.4	AKR,C57L,SWR C57BL,DBA	Baumann/1 Berger/1 Baumann & Berger, 1985
Interferon alpha	4	49	Ifa	Mu1204IFN fr.-a2	IFN fr.	PstI-HincII	392	HindIII	Several	b / d	11.0,2.2 9.4,2.1	C57BL/6J DBA,BALB	Demaeyer-Guignard/5 Nadeau et al. 1986
				IFNa9	SV40-pBR322	HindIII	1800	EcoRI	Several	b / d	2.3 2.4	C57BL/6 BALB/c,HW13	Demaeyer-Guignard/5 Seif & DeMaeyer-Guignard. 1986
Interferon-ß	4	49	Ifb	pM3	fr.	PstI-BglII	300	HindIII	6.0	a / s	14.0 / 26.0	AKR/J M.spretus	Higashi/5 Nadeau et al. 1986
Random cDNA clone	4	56	D4Rp1	pMK1440 pMKA11	pBR322	PstI	1200	EcoRI	7.8,0.57	a b c m	6.3,3.9 5.7,3.9 6.3,4.5 4.6,4.5.3,9,2.5	SWR,CD C57,DBA,129 BALB,C3H,SJL MOL,CAS,YBR	Elliott/1 Berger/1 Elliott, 1981
Pronatriodilantin Atrial natriuretic factor	4	80	Pnd Anf	pCAR60	pUC9	PstI-AccI	786	PvuII	No	b / d	0.5 / 0.7	A,BALB,C57 AKR,C3H,DBA	Davies/5 Gross/5 Mullins et al. 1987
Proto-oncogene mos	4		Mos	MSH	pBR322	HindIII	8000	TaqI	31.4	b / d	4.7 / 5.6	AKR,C57,SWR BALB,C3H,DBA	VandeWoude/2
Random cDNAclone	4		D4Rp18	pMK174	pBR322	PstI	986	BamHI	No	c m h g	1.7 14.6 8.5 1.9	Most strains MOLD,129,LP M.hortulanus M.musculus	Elliott/1 Mann et al. 1986
Albumen	5	46	Alb-1	pMSA433	fr.	PstI	900	HindIII/ PvuII	3.3/No	a / b	5.2/8.0 4.2/7.0	BALB/c Most strains	Sala-Trepat /5 Jubier-Maurin et al. 1985 Bellis et al.1987
Alpha fetoprotein	5	46	Afp	HcII440	pUC9	EcoRI-PvuII	650	EcoRI	No	a / b	10.0 / 2.2	BALB,C3H C57BL/6J	Tilghman/1 Belayew & Tilghman, 1982
ß-Glucuronidase	5	63	Gus-s	pGus-1	pBR322	PstI	1850	HindIII	7.1,2.4	a / b,h	4.0 3.8	A,BALB,SM C57,C3H	Ganschow/1 Gallagher et al. 1987
				pGA-1	pBR322	PstI	1400	PstI	6.9	a / b,h	5.3 0.8,5.0	AJ C57,C3H	Paigen/1 Watson et al. 1985
Non-muscle ß-actin	5		Actb	pAL41	pBR322	PstI	1150	EcoRI	≈10,≈7	c / d	20.0 / 13.0	BALB/cJ DBA/2J	Buckingham/1 Minty et al. 1983

Table 8.1 (*Continued*)

LOCUS NAME	DIST. CEN. CHR.	LOCUS	PROBE NAME	VECTOR	SITE	INS. LEN. (BP)	POLYMORPHISM ENZYME	CON.FRAG	ALLELE	LENGTH	STRAINS/SPEC	CONTACT/ AVAILABILITY	REFERENCE
Proto-oncogene met	6	6	pmet-1	Met	pBR322	EcoRI	5000	EcoRV	30.0	b d	15.0 12.0	C57BL/6J DBA,C3H,BALB	VandeWoude/5	Dean et al. 1987
								MspI	2.0	b d	6.5 2.3	C57BL/6J BALBcJ		
Carboxypeptidase A	6	15	Il-3	Cpa	pBR322	PstI	500	EcoRI	Not given	b s	Not given Not given	M.domesticus M.spretus	Rutter/5	Búcan et al. 1986
Trypsin-1	6	20	pcXP4-78	Try-1	pBR322	PstI	850	TaqI	Several	b s	Not given Not given	M.domesticus M.spretus	Craik/5	Búcan et al. 1986
T-cell receptor-β	6	20	86T1(Vb11) fr.	Tcrb		BamHI-EcoRI	Not given	BamHI	6.0	a b	6.4 12.0	SJL/J Most strains	Davis/5	D'Hoostelaere et al. 1985 Behlke et al. 1985
			p4.1		fr.	Not given	TaqI	Not given	b s	Not given Not given	M.domesticus M.spretus	Steinmetz/5	Búcan et al. 1986
Homeobox cluster-1	6	24	p577 Hox-1.1 m6		fr.	EcoRI	577	BamHI	No	b s	45.0 >45	C57BL/6J M.spretus	Colberg-Poley/5	Búcan et al. 1986
Testis-specific homeo box	6	24	pHBT Hox-1.4 Hbt-1		Lgt10	PstI-HindIII	750	Xba-1	No	b d	Not given Not given	C57BL/6J DBA/2J	Wolgemuth/5	Búcan et al. 1986
	6	24	Mo10 Hox-1.5		fr.	EcoRI	1400	EcoRV	6,3,5,4	a b	12.5 16.4	A,BALB,C3H AKR,C57,STS	McGinnis/5	Mock et al. 1987a
	6	24	MH1-G5 Hox1.7 Hox-1.7		Lambda Lgt10	EcoRI fr. HindIII fr.	14900 1700	EcoRI	Several	b d	Not given Not given	C57B,BALB AKR,C3H,DBA	D'Eustachio /5	Rubin et al. 1987
Immunoglobulin kappa	6	32	pECk	Igk	pBR322	EcoRI-BamHI fr.	6600	BamHI	No	d b	10.0 12.7	SJL/J C3H	Huppi/5	D'Hoostelaere et al. 1985
			Vk24 M167VL Jk		Not given	Not given	Not given	BamHI	No	a b	7.2 20, 9.5	AKR/J NZB,C57	Gibson/5 Storb/5	D'Hoostelaere & Gibson, 1986 Yang et al. 1987
					fr.	Not given	XbaI	5.0	a b	3.9 3.4	C58/J C57BL/6J	Boyd/5	Yang et al. 1987
			Vk-Ser		fr	PstI-KpnI	200	EcoRI	No	a b	2.1 2.8	C58/J C57BL/6J	Boyd/5	Yang et al. 1987
			pT1		fr.	EcoRI-HindIII	2500	TaqI	Not given	b s	Not given Not given	M.domesticus M.spretus	Steinmetz/5	Búcan et al. 1986
Ornithine decarboxylase-5	6	32	pK217	Odc-5	pBR322	PstI	500	SstI	Several	a b	6 fr. 5 fr.	AKR Most strains	Elliott/1 Berger/1	Richards-Smith & Elliott, 1984
7s RNA locus 6	6	32	pA6	Rn7s-6	pAT153	Not given	190	BamHI	Several	b a	7.5 Absent	C57,DBA,NZB AKR,C58,PL	Balmain/5	Taylor et al. 1985
Mammary tumor virus	6	32	pUI3P	Mtv-8	fr.	HindIII	1100	EcoRI	No	b n	6.6 (4.9 flanking)	Most strains NZB/B1NJ	Hynes/5	Robbins et al. 1986
			MMTV env		Not given	Not given	Not given	EcoRI	Several	c a d	(4.9 plus 2 fr) 5.9 Absent	CzechII BALB/c C58/J	Dudley/5	Yang et al. 1987
Proto-oncogene raf	6	45	pl71	Raf	fr.	EcoRI-XmaI	1000	BamHI	No	a c	17.0 8.0	C3H CzechII	Rapp/5 ATTC/P	Robbins et al. 1986
			3611-MSV		fr.	XhoI-BstEII	1200	TaqI	Not given	b s	Not given Not given	M.domesticus M.spretus	Rapp/5	Búcan et al. 1986
Proto-oncogene Kras	6	74	pKMSV	Kras	fr.	HincII	1000	PstI	No	b c	23.0 12.0	C3H CzechII	Chang/5	Robbins et al. 1986
			pHiHi-3		pBR322	EcoRI	1000	EcoRI	Several	a b	0.55 0.7	A,BALB,SWR C57,DBA,C3H	Scolnick/5	Ryan et al. 1987

542

LOCUS NAME	CHR.	DIST. CEN.	LOCUS	PROBE NAME	VECTOR	SITE	INS. LEN. (BP)	ENZYME	CON. FRAG	ALLELE	LENGTH	STRAINS/SPEC	CONTACT/ AVAILABILITY	REFERENCE
AKR virus-1 Endog. ecotr. MULV-11	7	3	Akv-1 Emv-11	pEnv	pBR322	EcoRI	400	PvuII	No	a d	4.8 Absent	AKR/J Most strains	Jenkins/1	Jenkins et al. 1981
Phenobarb.-ind. cyt-P450	7	7	Coh	pP450b-5	pBR322	PstI	1870	EcoRI	Many	b d	5.7 6.0	A,AKR,C57BL 129,C57L,DBA	Simmons/5	Simmons & Kasper, 1983
Proto-oncogene fes	7	15	Fes	pN26	Lambda	EcoRI	13000	EcoRI	No	a b	13.0 12.0	A,AKR C57BL,DBA	Francini/5	Blatt et al. 1984
Random cDNA	7	20	D7Rp2	pMK908	pBR322	PstI	1150	HindIII	0.6	b d	1.9 2.2	C57B,AKR,C3H DBA,C57L,SWR	Elliott/1	Elliott & Berger, 1983
				pMAK-1E	pBR322	BglI-RsaI	457	HindIII	4.2	b d	1.9 2.1	C57BL/6J DBA/2J	Berger/1 Lingrel/5	King et al. 1986
Serum amyloid A	7	24	Saa-1	pRS48	pBR322	PstI	480	BamHI	Several	a b	10.5 9.4	Most strains SJL,129,I,P	Taylor/5	Taylor & Rowe, 1984
Random cDNA	7	24	D7Rp14	pMK1443	pBR322	PstI	720	BamHI	26.0,12.0	b d w n	5.3 9.2 14.6 5.3,4.9	C57,BALB,C3H DBA,MOL,M.spr DW/J NZB/B1NJ	Elliott/1	Elliott, Pers. Comm.
Ornithine decarboxylase-6	7	24	Odc-6	pMK3048	pBR322	PstI	900	HindIII	Several	b d	18.0 12.0	Most strains DBA,SM,I,MOL	Elliott/1 Berger/1	Richards-Smith & Elliott, 1984
Nerve growth factor-gamma	7	24	Ngfg	pSM676	pBR322	PstI	520	EcoRI	Several	b d	Several Several	C57BL,BALB DBA/2J	Gross/1	Howles et al. 1984
β-Hemoglobin	7	49	Hbb	pCR1bMG9 BA1-4 pBR&G2	pCR1	EcoRI	540	Several	Not given	d s	Not given Not given	BALB/c C57BL/10	Rougeon/1 Edgell/1 Jones/2	Rougeon & Mach, 1977 Weaver et al. 1981 Jones, Pers. Comm.
				bh3	pUC12	PstI-HindIII	387	RsaI	No	d4 d	1.0 2.0	101/H,C3H/He HBC		
			Hbbh3	bh3	fr.	HindIII-EcoRI	1200	PstI	No	d s	9.6 7.5	BALB/c C57	Martin/5	Bellis et al. 1987
Friend virus int site (tumor)	7	74	Fis-1	p1.8	pUC8	BamHI	1800	HincII	No	a b c	2.1 2.5 2.3	NFS/N C57BL,C57L AKR,BALB,DBA	Silver/1	Silver & Buckler, 1986
MMTV int. site-2	7	74	Int-2	Int-2a	fr.	EcoRI-SacI	1900	PstI	No	b a	2.1 4.4	AK,BALB,DBA C57BL,C57L	Peters/5	Moore et al. 1986
Sodium potassium ATPase a-2	7		Atpa-2	rb-2	fr.	EcoRI-StuI	2000	Not given	Not given	d s	Not given Not given	C57BL/6 M.spretus	Levenson/5	Kent et al. 1987
DNA fragment	7		H19	H19	pUC9	BamHI	1100	MspI	No	a b	1.1 1.5	BALB/c,C3H C57BL/6J	Tilghman/1	Pachnis et al. 1984
APRT pseudogene	8	10	Aprt-2	p15	pBR328 fr.	EcoRI SmaI-AvaII	3000 700	BglII	No	a c	3.3 6.2	A/J BALB/c	Stambrook/2	Dush et al. 1986
Mitochondrial uncoupling protein	8	33	Ucp	pCIN-1	pBR322	PstI	900	BamHI	No	b d	13.0 23.0	C57BL/6J DBA/2J	Kozak/1	Jacobsson et al. 1985

Table 8.1 (*Continued*)

LOCUS NAME	CHR.	DIST. CEN.	LOCUS	PROBE NAME	VECTOR	SITE	INS.LEN. (BP)	ENZYME	CON.FRAG	ALLELE	LENGTH	STRAINS/SPEC	CONTACT/ AVAILABILITY	REFERENCE
Uvomorulin	8	50	Um	F5H3	fr.	HindIII-EcoRI	900	EcoRI	3.5	d	3.4	C57BL/6J	Kemler/5	Eisstetter et al. 1988
										s	2.1,0.8,0.6	M.spretus		
Tyrosine aminotransferase	8	55	Tat	pmTAT-SH3pUC8	Sall-HindIII		3700	TaqI	5.2	b	Absent	DBA2	Schutz/5	Müller et al. 1985
										s	4.0	M.spretus		
Endog. ecotr. MULV-2	8	55	Emv-2 Blv	pEnv	pBR322	EcoRI	400	XbaI	No	b	11.5	C57BL	Jenkins/1	Jenkins et al. 1981
										d	Absent	Most strains		
Finger protein-2	8		Fp-2	mFP2	Lgt10	Not given	Not given	Not given	Not given	b	Not given	C57BL/6J	Willison/2	Ashworth & Willison, 1987
										d	Not given	DBA2J		
Neural cell adhesion molecule	9	28	N:	pEC501	pBR328	PstI	4500	MspI	3.2,2.6,2.2	a	2.7,1.7	BALB,C3H,DBA	Cunningham/2	D'Eustachio et al. 1985
										b	1.0,0.8,0.7	AKR,C57,SWR		
TCDD-ind. P-450	9	28	P450-3	pP3450FL	pBR322	BamHI	1894	PvuII	5.1,57,0.84	a	4.6	AKR/J	Nebert/1	Hildebrand et al. 1985
										b	4.8	C57L,DBA,C57		
Endog. ecotr. MULV-3 dilute	9	35	Emv-3 d,Dbv	pEnv	pBR322	EcoRI	400	PvuII	Various	b	Absent	Most strains	Jenkins/1	Jenkins et al. 1981
										d	5.4	DBA,I,P,HRS		
Myosin alkali light.chain, cardiac ventricle	9	68	Myl-3	pGLC450	pBR322	BamHI	470	BamHI	Not given	b	Not given	DBA2	Robert/1 Buckingham/1	Robert et al. 1985
										s	Not given	M.spretus		
Proto-oncogene myb	10		Myb	pG4M2b	Not given	EcoRI	2200	EcoRI	16,2.1,1,9,1	a	6.5	C57,DBA,STS	Reddy/1 Huppi/1	Mock et al. 1987c
										b	4.2	BALB,SJL,MA		
Proto-oncogene erbb	11	8	Erbb	Erbb	fr.	HindIII-EcoRI	2500	PstI	No	a	5.3	AKR,BALB,SEC	Vennstrom/5	Silver et al. 1985
										b	5.5	C57,129,SM,I		
					fr.	Sstl-PvuII	1700	BamHI	Not given	d	9.3	M.domesticus	Oncor/P	Buchberg et al. 1988
										s	11.0	M.spretus		
Proto-oncogene rel	11	11	Rel	Rel	fr.	EcoRI	900	TaqI	Not given	d	10.0	M.domesticus	Temin/5	Buchberg et al. 1988
										s	4.7	M.spretus		
Hemoglobin-alpha	11	14	Hba	pCR1aMG	pCR1	EcoRI	500	HindIII	Not given	d	Not given	DBA2J	Rougeon/1	Robert et al. 1985
										s	Not given	M.spretus		
Finger protein 1	11	20	Fp-1	mFP1	Lgt10	Not given	Not given	Not given	Not given	b	Not given	C57BL/6J	Willison/2	Ashworth & Willison, 1987
										d	Not given	DBA2J		
Granulocyte macroph. colony stim. factor	11	26	Csfgm	pEl-11	Not given	Not given	790	PstI/ BamHI	Not given	d	2.8/9.3	M.domesticus	Miyatake/5	Buchberg et al. 1988
										s	6.4/7.0	M.spretus		
Interleukin-3	11	26	Il-3	fpGVIL12	retrovirus	HincII-EcoRI	2500	MspI	Not given	d	2.1	M.domesticus	Fung/5	Buchberg et al. 1988
										s	2.4,1.4	M.spretus		

LOCUS NAME	CHR.	DIST. CEN.	LOCUS	PROBE NAME	VECTOR	SITE	INS.LEN. (BP)	POLYMORPHISM ENZYME	CON.FRAG	ALLELE	LENGTH	STRAINS/SPEC	CONTACT/ AVAILABILITY	REFERENCE
Secr.acidic,cysteine-rich glycoprotein	11	30	Sparc	pPE220	pUC8	EcoRI-HindIII	600	BglII	No	b d	4.7 3.8,1.2	C57,AKR DBA,C3H,SWR	Hogan/1	Mason et al. 1986
DNA fragment (HD-linked)	11	30	D11G8 HsaD4S10	pSC33	pUC9	EcoRI-HindIII	3300	MspI/TaqI	Several	b d	2.0/2.9 4.2/4.3	AKR,C57BL DBA,C3H	Gusella/5	Cheng et al. 1987
Transforming protein p53	11	43	Trp53-1	Not given	Not given	Not given	538	TaqI	Not given	d s	2.5,1.7 1.75	M.domesticus M.spretus	Benchimol/5	Buchberg et al. 1988
Ecotr.vir.int.site-2	11	47	Evi-2	p597.1	pUC18	PstI	500	HindIII/ TaqI	Not given	a b	1.9/4.4 2.4/2.0	AKR/J C57,C3H,DBA	Copeland/1	Buchberg et al. 1988
b-Nicotinic acetyl-choline receptor	11	47	Acrb	bAchr	M13mp18	EcoRI-PstI	300-600	HincII	No	d s	2.5 1.5	DBA/2J M.spretus	Merlie/5	Heidmann et al. 1986
Myosin heavy ch,, fast skeletal	11	47	Myh-1	pMLC1a	Not given	Not given	Not given	BglI	No	d s	Not given Not given	DBA/2 M.spretus	Robert/5	Robert et al. 1985
U1 snRNA	11	55	Rnu1a-1	pUE236	pAT153	Hha-Hinf fr.	550	HindIII	No	c s	5.1 4.5	BALB/cHeA STS/A	Whitney/5	Michael et al. 1986
Granulocyte colony stimulating factor	11	55	Csfg	pMG2	pBR327	EcoRI	1400	TaqI	Not given	d s	3.9 2.5	M.domesticus M.spretus	Tsuchiya/5	Buchberg et al. 1988
Proto-oncogene erba	11	55	Erba	Erba	fr.	PstI	478	TaqI	Not given	d s	4.4,4.1,1.5 2.3,1.5	M.domesticus M.spretus	Oncor/P	Buchberg et al. 1988
Proto-oncogene erbb-2 Proto-oncogene neu	11	55	Erbb-2	neuc(t)	pSP65	BamHI	420	PstI	Not given	d s	4.0 4.4	M.domesticus M.spretus	Barbacid/5	Buchberg et al. 1988
Homeobox cluster-2	11	55	Hox-2.1	pRRF5	Not given	EcoRI	3000	TaqI	4,6,3,9,2.8	a b	Absent 0.7	A,BALB,C57 RIII,STS,SWR	Joyner/5	Mock et al. 1987b
			Hox-2.2	pMo-4.6	pAT153	EcoRI	5300	HindIII	5.5	j k l	9.0,1.9,0.5 9.0,2.4 8.0,1.9,0.5	Most strains A,AKR,CBA M.spretus	McGinnis/5 Ruddle/5 Hart/5	Hart et al. 1988
			Hox-2.3	pMo-1.1	pAT153	EcoRI	2900	EcoRI	No	a b	3.0 15.0	Most strains A,AKR,CBA		
			Hox2.2 (mh22a)	p6	pUC19	BamHI-EcoRI	2000	Not given	Not given	a b	Not given	A/J C57BL/6J	Blatt/5	Lonai et al. 1987
			Hox2.3 (mh22b)	p4	pUC19	BamHI-EcoRI	700	Not given	Not given	a b	Not given	A/J C57BL/6J	Blatt/5	Lonai et al. 1987
			Hox2.1	RRF12	Not given	Not given	Not given	Not given	Not given	a b	Not given	A/J C57BL/6J	Tjian/5	Lonai et al. 1987
Myosin alkali lt.ch. cardiac atrial	11	72	Myla	pGLC260	pBR322	PstI	260	HinfI BamHI	No Not given	b h d s	Not given Not given Not given Not given	C57BL/6J C3H/HeJ DBA/J M.spretus	Barton/2	Robert et al. 1985

545

Table 8.1 (Continued)

LOCUS NAME	CHR.	DIST. CEN.	LOCUS	PROBE NAME	VECTOR	SITE	INS.LEN. (BP)	POLYMORPHISM ENZYME	CON.FRAG	ALLELE	LENGTH	STRAINS/SPEC	CONTACT/ AVAILABILITY	REFERENCE
DNA fragment	12	3	D12Nyu2	M13p19-25	M13	HincII	<500	TaqI	No	a b	4.1,2.7 3.9,3.6	Most strains C57BL/6J	D'Eustachio /1	D'Eustachio, 1984
D12RC2.3	12	3	D12RC2.3	pRC2.3	fr.	EcoRI	2300	HindIII	No	b d	7.6 11.2	C57BL Most strains	Whitney/1	Cobb et al. 1987
DNA fragment	12	17	D12Nyu5	M13p30-3	M13	HincII	<500	MspI	No	a b	4.4,3.7 3.8	AKR,BALB C57,SWR	D'Eustachio /1	D'Eustachio, 1984
DNA fragment	12	17	D12Nyu1	M13p7-97	M13	HincII	<500	EcoRI/MspI	No	a b c	7.9/3.7 12/33.7 12/3.9,3.6	BALB,C3H,DBA C57BL,C57L SJL,SWR	D'Eustachio /1	D'Eustachio, 1984
DNA fragment	12	20	D12Nyu3	M13p20-1	M13	HincII	<500	EcoRI/TaqI	No	a b	6.9/3.8,2.0 7.9/4.2,3.6	SWR,SJL,SM Most strains	D'Eustachio /1	D'Eustachio, 1984
Proto-oncogene fos	12	32	Fos	pfos	pBR322	PstI	1400	EcoRI/MspI	Yes	a b c	4.0/8.5 5.7/3.2 5.7/8.5	SWR/J Most strains SJL/J	Verma/5 ATCC/P	D'Eustachio, 1984 Curran et al. 1982
DNA fragment	12	35	D12Nyu4	M13p2615	M13	HincII	<500	TaqI	No	a b	7.7,1.8 9.2	SWR/J Most strains	D'Eustachio /1	D'Eustachio, 1984
alpha-1 Antitrypsin	12	46	Aat	pC1.2	pBR322	PstI	Not given	EcoRI	7.9,3.8	a b	8.9,2.5 7.5,5.9,5.2	AKR,BALB,DBA C57L,C3H,A	D'Eustachio /1	D'Eustachio et al. 1984
Serine protease inhibitor-1	12	46	Spi-1	pLv1796	pBR322	PstI	750	BglII (+others)	Several	a b	Several Several	BALB,DBA C57BL/6J	Hill/1	Hill et al. 1985
Serine protease inhibitor-2	12	47	Spi-2	pLv54	pBR322	PstI	800	HindIII (+others)	Several	a b c	Several Several Several	BALB,DBA C57BL,C3H 129	Hill/1	Hill et al. 1985
Cysteine-rich intest. protein	12	62	Crip	pMI3	pBR322	PstI	303	KpnI/TaqI	Not given	b d	Not given Not given	C57BL/6J Most strains	Birkenmeier/5	Birkenmeier & Gordon, 1986
Immunoglobulin heavy chain	12	62-70	Igh	V186 VAR34 VS107 V81X VAR104	fr. fr. fr. fr. fr.	PstI-HinfI PstI DdeI PvuII-PstI Sau3A	264 354 175 331 800	EcoRI EcoRI/HindIII EcoRI EcoRI PstI	Several Several Several Several Several	Several Several Several Several a b	Many Many Many Many 1.05 Absent	Various Various Various Various BALB/c C57BL/6	Blankenstein /5	Blankenstein et al. 1987
				pMu(3741)	pMB9	EcoRI	1110	EcoRI	No	a b h	12.5 14.5 11.2	BALB,C57L C57BL/6,SM C3H/HeJ	Arnheim/5	Marcu et al. 1980
T-cell receptor-gamma	13	19	Tcrg	pHDS4	pBR322	PstI	700	EcoRV	8.9,7.4	a b c d e	≈10 ≈30,≈15,6.6 Not given Not given Not given	C3H,DBA,SM C57,BALB,RF AKR,SWR,NZB I/LnJ A/J,129/J	Seidman/5	Owen et al. 1986
				gamma7.1	Lgt10	EcoRI	2400	EcoRI	10.5	b a c	13.4,7.5 16.0 16.0,6,6.6	BALB,C57BL/6 C3H,B10,B/Ks AKR/J	Tucker/5	Rathbun et al. 1988

LOCUS NAME	CHR.	DIST. CEN.	LOCUS	PROBE NAME	VECTOR	SITE	INS.LEN. (BP)	POLYMORPHISM ENZYME	CON.FRAG	ALLELE	LENGTH	STRAINS/SPEC	CONTACT/ AVAILABILITY	REFERENCE
T-cell receptor alpha	14	16	Tcra	T1.2	pUC9	EcoRV-AvaII fr.	200	BglI	No	d	12.0	BALB,C3H,DBA	Steimetz/5	Dembic et al. 1985
										b	9.4	C57BL/6J		
				pHSD58	pBR322	PstI	1370	EcoRI	6.2,3,9,2.2	c	20.0,15.0	Most strains	Tonegawa/5	Berman et al. 1986
										b	17.0,12.0,0.9	C57BL/6J		
Pancreatic ribo-nuclease	14	16	Rib-1	pMPR-1	pBR322	PstI	590	BamHI	No	a	3.7	Most strains	Meisler/1	Elliott et al. 1986
										b	5.8	C57,FS,MOLD		
Lymphocyte antigen-6	15	20	Ly-6	pKIy6.1-2R	pUC19	EcoRI	750	EcoRI	Several	6.2	Several	C57BL/6J	Bothwell/5	Meruelo et al. 1987
										6.1	Several	BALB/c		LeClair et al. 1986
Endog. xenotr. vir-55	15	22	Xmmv-55	pXenv	Not given	Not given	Not given	PvuII	Not given	b	Present	SM,C57BL,LT	Hogarth et al. 1987	
										c	Absent	NZB,CAST		
Thyroglobulin congenital goiter	15	25	Tgn cog	pRT57	Not given	Not given	Not given	PvuII/TaqI	Several	a	3.0/5.5	A,AKR,SWR	Di Lauro/5	Taylor & Rowe, 1987
										b	2.5/4.4	C57,DBA,SJL	Taylor/5	
										c	2.5/3.5	BALB,CBA,C3H		
										p	2.5/5.5	PL/J,PERU		
Proto-oncogene sis	15	25	Sis	pSSV11	fr.	PstI-XbaI	1000	EcoRI	No	a	20.0	C57BL/6	Robbins/5	Meruelo et al. 1987
										b	14.0	BALB/c		
Endog. xenotr. vir-72	15	30	Xmmv-72 Pol-5	pB65.5PV2	fr.	SmaI	358	BamHI	Several	b	9.3	C57BL/6	Meruelo et al. 1987	
										c	Absent	BALB/c		
Insulin-3	15	32	Ins-3	pI19	fr.	PstI	323	PstI	Not given	a	Several	C57BL/6	Lomedico/5	Meruelo et al. 1987
										b	Several	A/J		
Glycerol-3-phosphate dehydrogenase	15	39	Gdc-1	pH8,pC8	pBR322	PstI	990	HindIII	No	b	7.9	C57BL/6J	Kozak/1	Kozak & Birkenmeier, 1983
										d	5.9	M.castaneous		
								PstI	5.6	b	3.2	C57BL/6J		
				pC8	pBR322	PstI	350	BglI	Not given	c	2.9	BALB/c		Meruelo et al. 1987
										a	5.8	C57BL/6		
								PvuII	No	b	7.0	BALB/c	Birkenmeier/1	Hogarth et al. 1987
										n	3.2	NZB		
										s	3.0	SM		
Keratin b1	15	44	Krt-1	pK4.2(b1)	fr.	PstI	1580	PstI	Not given	a	Several	C57BL/6	Steinert/5	Meruelo et al. 1987
										b	Several	A/J		
				pK435(a1)	fr.	PstI	1850	PstI	Not given	a	Several	C57BL/6	Steinert/5	Meruelo et al. 1987
										b	Several	A/J		
Keratin b2	15	44	Krt-2	pK3.68(b2)	fr.	PstI	1650	HindIII	Not given	a	Several	C57BL/6	Steinert/5	Meruelo et al. 1987
										b	Several	A/J		
Protamine-1	16	7	Prm-1	pMP1-1	pUC8	SalI	404	BstEII	No	a	16.8	BALB/c	Hecht/2	Reeves et al. 1987
										c	6.8	CzechII,CBA		
AKR int. site-2 (infectious virus)	16	12	Akv-2	D242	pBR322	HindIII fr.	600	XbaI	No	a	16.0	AKR/J	Martin/5	Epstein et al. 1984
										b	8.0	Most strains		
Immunoglobulin lambda	16	14	Igl-1	MOPC 104E (PABL1)	pBR322	PstI	≈5000	EcoRI	Several	a	8.6	AKR/J	Potter/5	Reeves et al. 1987
										c	Absent	CzechII	Huppi/5	
				Igl-c1 (pCL14E)	pCRI	HaeIII	600	KpnI	Not given	a	4.3	AKR/J	Blomberg/5	Epstein et al. 1984
										b	7.9	SJL/J		Epstein et al 1986
			Igl-5	pCL5s	pUC18	BamHI-EcoRI fr.	500	HindIII	No	d	17.0	M.domesticus	Kindt/5	Mami et al.1988
										s	8.5	M.spretus		
										a	6.0	M.abotti		

Table 8.1 (*Continued*)

LOCUS NAME	CHR.	DIST. CEN.	LOCUS	PROBE NAME	VECTOR	SITE	INS.LEN. (BP)	POLYMORPHISM ENZYME	CON.FRAG	ALLELE	LENGTH	STRAINS/SPEC	CONTACT/ AVAILABILITY	REFERENCE
Somatostatin	16	19	Smst	p1-6-13	pBR322	PstI	550	BstEII	6.8	a / c	1.9 / 4.0	BALB/c / CzechII	Goodman/5	Reeves et al. 1987
MMTV int. site-6	16	31	Mtv-6	MMTVLTR	fr.	Not given	Not given	EcoRI	Several	a / c	16.0 / Absent	BALB/c / CzechII	Callahan/5	Reeves et al. 1987
Superoxide dismutase (cyt c	16	55	Sod-1	SOD4.1	pBR322	EcoRI-SalI	4100	BstEII	2.0	a / c	2.3.2.1 / 4.4	AKR,BALB,C57 / CzechII,CBA	Reeves/3	Reeves et al. 1987
Proto-oncogene ets-2	16	57	Ets-2	P1.27	pBR322	PstI	1270	BamHI	No	a / c	9.4 / 17.0	AKR,BALB,C57 / CzechII,CBA	Reeves/2	Reeves et al. 1987
Influenza virus sensitivity	16		Mx	pMx41	pHG327	BamHI	1650	EcoRI	10.0	a / b / c	7.5,5.0,2.6,0.75 / 4.0,2.6 / 107.5	BALB.A2G-Mx / BALB,C57B,A / CBA/J	Klein/5	Staeheli et al. 1986
DNA fragment	17	5	D17Tu1	D17Tu1 from 3.4.1	pUC8	EcoRI	9400	TaqI	No	a / b / c / d / e / f / g / h / i / j	5.1,3.9 / 5.7,3.9 / 4.4,2.8 / 4.5,3.9 / 5.1,2.8 / 6.5,5.1 / 5.1,4.5 / 5.1,4.9 / 6.0 / 4.5	PL/J,M.hort. / M.domesticus / CEJ / M.domesticus / M.domesticus / M.musculus / MA,129,P,I,RF / Most inbreds / M.caroli / M.castaneous	Klein/5	Figueroa et al. 1987
DNA fragment	17	5	D17Leh48	Tu48	pUC9	EcoRI	400	BamHI	No	a / t	6.2 / 5.9	Most strains / t haplotypes	Lehrach/1 Silver/1	Fox et al. 1985
DNA fragment	17	5	D17Leh508	Tu508	pUC9	EcoRI	2500	BamHI	Not given	b / s	7.0 / 2.0	Most inbreds / M.spretus	Lehrach/2	Lehrach, Pers. Comm.
DNA fragment	17	5	D17Leh119	p119A-R	pUC9	EcoRI	3200	TaqI	No	d / b	10.0 / Absent	DBA2J / C57BL/6J	Lehrach/1	Herrmann et al. 1986
DNA fragment	17	5	D17Leh66E	p66M-RT	pUC9	EcoRI-TaqI	2300	BamHI	Several	d / b / t	18.0 / 3.7 / 3.2	DBA,C3H,BALB / C57BL,AKR / t haplotypes	Lehrach/1	Herrmann et al. 1986
DNA fragment	17	5	D17Leh66B	p66M-RT	pUC9	EcoRI-TaqI	2300	XhaI	Several	d / b	Not given / Not given	DBA2J / C57BL/6J	Lehrach/1	Herrmann et al. 1986
DNA fragment	17	5	D17Leh122	Tu122	pUC9	EcoRI	400	TaqI	No	a / t	larger / 1.9	Most strains / t haplotypes	Lehrach/1 Silver/1	Fox et al. 1985
DNA fragment	17	5	D17Leh550	Tu550	pUC9	EcoRI	2000	MspI	5.0	a / t	Absent / 4.0	Most strains / t haplotypes	Lehrach/2	Lehrach, Pers. Comm.
DNA fragment	17	5	D17Leh443	Tu443	pUC8	EcoRI	1700	EcoRI / MspI	No / No	a / b / c / t	Smaller / Larger / 15.0 / 8.0	C57BL/6J / DBA2J / 129/J / tw5/tw5	Lehrach/2	Búcan et al. 1987
Protein kinase	17	10	D17Rp17	pMK174	pBR322	PstI	986	BamHI	No	b / d / t	6.3,4.1,2.8 / 6.1,4.1,2.8 / 6.3,5.4,1.6	C57,AKR,SWR / DBA,C3H,A,SM / t haplotypes	Elliott/1	Mann et al. 1986

LOCUS NAME	CHR.	DIST. CEN.	LOCUS	PROBE NAME	VECTOR	SITE	INS.LEN. (BP)	ENZYME	POLYMORPHISM CON.FRAG	ALLELE	LENGTH	STRAINS/SPEC	CONTACT/ AVAILABILITY	REFERENCE
t complex polypep-1	17	12	Tcp-1	pB1.4	Charon 4A	EcoRI	Not given	TaqI	No	a / b	4.1.2.8 / 4.3,3.3	AKR,BALB,C3H C57,NZB,RIII	Willison/5	Willison et al. 1986
DNA fragment	17	13	D17Leh66D	p66M-RT	pUC9	TaqI-EcoRI	2300	XbaI	One	d / b	Larger / Smaller	DBA/2J C57BL/6J	Lehrach/1	Herrmann et al. 1986
								TaqI	5.3,4.6,4.4	b / a / h	9.4.6.3(*2) / 9.4,6.3 / Absent	C57BL,SWR AKR,129 C3H/HeJ	Silver/5	Schimenti et al. 1987
DNA fragment	17	13	D17Leh54M	p5411-R2.9	pUC9	EcoRI	2900	TaqI/ EcoRI	Two/Two	b / d	Larger/Smaller Smaller/Larger	C57BL/6J DBA/2J	Lehrach/2	Bucan et al. 1987
alpha-Hemoglobin pseudogene	17	14	Hba4ps	apsi4	pBR322	EcoRI	2400	TaqI	Several	a / b / c / t	3.5,1.3 / 3.1 / 4.7 / 5.2	AKR,BALB,C3H C57,NZB,DBA* SM* t haplotypes	Leder/1	D'Eustachio et al. 1984
								MspI	Several	a / b / d	6.7,6.2 / 11.8,11.1 / 10.9	AKR,BALB,C3H C57,NZB,SM* DBA*,A.WB*P		Fox et al. 1984 Mann et al. 1986
Proto-oncogene pim-1	17	17	Pim-1	A	fr.	BamHI	923	BamHI/ HincII	No	a / b / c / d / e / f / t	2.2/13.8 / 0.9/10.2 / 5.1,12/10.2 / 2.2/13.8 / 2.2,10.2 / 2.2/18.0	A,AKR,C3H BALB,DBA,t NZB,RIII C57BL,SM,SWR CAST,MOLF,RF M.spretus t haplotypes	Nadeau/5 Cuypers/5	Nadeau & Philips, 1987 Warren et al. 1987
alpha A-crystallin	17	18	Crya-1	pMaCR2	pUC9	PstI	1000	HincII/BglII (12/14 enz. tested)	No	a / b / c / d / e / f / t	12.0/13.0 / 10.5/13.0 / 10.5/10.3,2.7 / 10.5/11.2.7 / 10.5/11.2.7 / 9.0/8.0 / 8.5,1.9/12.5,1.2t haplotypes	BALB,DBA,SWR AKR,C3H,C57 PLJ MOLD/Rk CASO/Rk SJL/J	Skow/5	Skow & Donner, 1985 Skow et al. 1987
Histocompatibility complex-2K	17	19	H-2K	0.6HindIII	fr.	HindIII	600	BamHI	No	H-2b / H-2d / H-2k / Several t12	6.3 / 6.7 / 7.6 / 2.5 / 3.0	C57BL/6J BALB/c AKRJ Most strains t12	Bennett/5 Flavell/5	Uehara et al. 1987
				LS1/1	fr.	BamHI	3000	BamHI	No					
Histocompatibility complex-2I	17	19	H-2I	pH-2d-37C	fr.	PstI	250	BamHI	No	H-2b,k,q / H-2d / H-2s	9.3,8.0 / 8.0,6.3 / 8.5	C57,AKR,DBA BALB/c SJL/J	Kourilsky	Lalanne et al. 1985
Histocompatibility complex-2I-A	17	19	H-2I-A	I-Aa I-Ab	fr. fr.	HindIII EcoRI	1200 5800	Several Several BglII	Varies Varies	Several Several a c	Various Various 5.0 9.0	Various Various C3H/OuJ CzechII	Seidman/5 Singer/5	McConnell et al. 1986 Gallahan et al. 1987
Histocompatibility complex-2Eβ2	17	19	H-2Eβ2	Eβ2-2 EβB1	fr. fr.	PstI-AccI EcoRI	300 2000	PstI Asp718/ BamHI	No No	H-2b / Several / H-2b / H-2k / H-2f / H-2p / H-2q,s	1.8 / 2.1 / 7.1/8.2 / 17.0/6.2 / 24.0/6.2 / 24.0/3.8 / 24.0/8.2	C57BL/6J Various C57BL AKR,CBA,C3H B10.M P,BRSUNT DBA/1,SWR	Braunstein/5 Steinmetz/5	Braunstein & Germain, 1986 Lafuse & David. 1986

Table 8.1 (*Continued*)

LOCUS NAME	CHR.	DIST. CEN.	LOCUS	PROBE NAME	VECTOR	SITE	INS.LEN. (BP?)	POLYMORPHISM ENZYME	CON.FRAG	ALLELE	LENGTH	STRAINS/SPEC	CONTACT/ AVAILABILITY	REFERENCE
Histocompatibility complex-2Ea	17	19	H-2Ea	Ea	fr.	SalI	3400	BamHI/SacI	6.6/20.0	H-2b,s H-2f,q	6.6/4.4 Absent/1.7	C57BL/10 B10.M,RBF	Hood/5	Lafuse et al. 1986
DNA fragment	17	19	D17Tu2	D17Tu2 from 9.4.3	pUC8	HindIII	3300	TaqI	2.2	a b c d	4.5 5.5 6.2 7.2	I,P,WB,WC,BDP Most inbreds CE,MA,RF,SM RIII,NZW	Klein/5	Figueroa et al. 1987
Complement protein-4	17	19	C4	pMC4/7	pBR322	PstI	1900	Most enz.	Yes	Several	Several	Various	Tosi/5	Levi-Strauss et al. 1985
Sex limited protein	17	19	Slp							Several	Several	Various		Stavenhagen et al. 1987
Steroid 21-hydroxyl-ase	17	19	Oh21	pM210H	pBR322	EcoRI	2700	TaqI	No	s c	3.8,0.9 3.8,4.3	BALB,AKR,C57 BALB.K	Meo/1	Amor et al. 1985
					fr.	BamHI-EcoRI	821	ApaI	No	a h	1.9,1.7 1.8	BALB,C57BL C3H/HeJ	Seidman/5	Chaplin et al. 1986
Complement Factor B	17	19	Bf	pBmB2	fr.	PstI	1300	MspI	No	Most f,r,v	3.0,2.6 3.9,3.5	Most A.CA,RIII,SM	Sackstein/5	Sackstein et al. 1984
Histocompatibility complex-Qa-2	17	20	Qa-2	Probe 1	fr.	BamHI	≈100	BamHI	Yes	b d	Several Several	C57BL/6 BALB/c	Hood/5	Mellor et al. 1985
				pH2IIa	pBR322	Not given	200	BamHII	Several	H-2k H-2d	5.4 6.2,2.7	CBA BALB/c		Steinmetz et al. 1981
DNA fragment	17	20	D17Leh89	Tu89	pUC8	EcoRI	1700	KpnI	One	d b	larger smaller	DBA/2J C57BL/6J	Lehrach/2	Búcan et al. 1987
DNA fragment	17	20	D17Leh525	T525	pUC8	EcoRI	2000	BgIII	One	d b	larger smller	DBA/2J C57BL/6J	Lehrach/2	Búcan et al. 1987
DNA fragment	17	23	D17Leh173	Tu173	pUC9	EcoRI	500	TaqI	Several	d b	larger smaller	DBA/2J C57BL/6J	Lehrach/2,3	Lehrach, Pers. Comm.
DNA fragment	17	24	D17Leh116	Tu116	pUC8	EcoRI	1000	PvuII	One	d b	smaller larger	DBA/2J C57BL/6J	Lehrach/2,3	Lehrach, Pers. Comm.
Random cDNA	17	26	D17Rp11	pMK1109	pBR322	PstI	900	BamHI	No	a s l u	18.0 10.0 12.5 25.0	C57B,DBA,NZB SWR,WB,129 C57L,DW,1P AU,MOL,SM	Elliott/1	Mann et al. 1984
MMTV int.site-3	17	30	Int-3	int-3	fr.	Xba-EcoRI	2900	PvuII	No	a c	4.0 2.3	C3H CzechII	Callahan/5	Gallahan et al. 1987
alpha-Actin, cardiac	17		Actc	pAF81	pBR322	PstI	1100	BclI	No	a b	23.0 20.0	BALB/c C57BL/6	Buckingham/1	Garner et al. 1986 Minty et al. 1983
a-Nicotinic acetyl-choline receptor	17		Acra	aAChR	M13mp18	EcoRI-XmnI	600	HincII	No	d s	6.0 3.0	DBA/2J M.spretus	Merlie/5	Heidmann et al. 1986
Invariant chain	18	57	Ii	pli5	Not given	PstI	Not given	XbaI	No	a d	10.0 9.0/1.1	AKR/J DBA,C57L	Jones/5	Richards et al. 1985
Random cDNA	19	5	D19Rp19	pMK1711	pBR322	PstI	800	EcoRI	Seven fr.	b d	8.5 4.4,4.0	Most strains DBA,C57L,MOLD	Elliott/1	Elliott &Ehrenreich, 1988

550

LOCUS NAME	CHR.	DIST. CEN.	LOCUS	PROBE NAME	VECTOR	SITE	INS.LEN. (BP)	POLYMORPHISM ENZYME	CON.FRAG	ALLELE	LENGTH	STRAINS/SPEC	CONTACT/ AVAILABILITY	REFERENCE
DNA fragment	X	1	DX44	MSDX44	pGEM4	EcoRI	1000	TaqI / MspI	2.0 / No	a / b / a,b / s	8.0 / 7.4 / 8.0 / 8.6	C57BL/10 SWR/J M.domesticus M.spretus	Brown/1	Fisher et al. 1985
DNA fragment	X	1	DX70	p70-38	pBR325	EcoRI	9000	Several	Yes	a / b	constant fr. / additional fr.	A,BALB,B6,DBA SWR/J NZB,SJL,B10	Disteche/2	Disteche et al. 1985
Ornithine transcarbamoylase.sparse fur	X	2	Otc spf	pOTC	pUC9	EcoRI	1400	BglII	Yes	b / m / s	4.2,3.6 / 3.9 / 5.7,3.4	C57BL M.musculus M.spretus	Caskey/5	Mullins et al. 1988
Chr.granulomatous dis. Cyochrome b245	X	2	Cgd Cybb	4CA	pUC19	EcoRI	1100	BglII	Yes	b / m / s	6.0 / Not given / 8.0,5.2	C57BL M.musculus M.spretus	Orkin/5	Chapman, Pers. Comm.
Tissue inhibitor of metalloproteases	X	2	Timp	p3/10	pSP64	HindIII	4300	BamHI	No	b / m / s	1.9,1.7 / 3.9 / 3.8	C57BL M.musculus M.spretus	Williams/5	Mullins et al. 1988
Synapsin	X	4	Syn-1	p5E2	M13mp8	EcoRI	1500	BglII	Many weak	b / m / s	5.7 / Not given / 6.7	C57BL M.musculus M.spretus	DeJerravo/5	Chapman, Pers. Comm.
DNA fragment	X	6	DX68	p68-36	pBR325	EcoRI	4000	Several	Yes	a / w	Several / Several	Inbred strains other species	Disteche/2	Disteche et al. 1985
DNA fragment	X	11	DXPas7 HsaDXS32	M2C	Not given	Not given	Not given	Not given	Not given	b / s	Not given / Not given	C57BL/6 M.spretus	Mandel/5	Avner et al. 1987b
DNA fragment	X	16	DX141	MDRX141	pSP64	EcoRI	1200	TaqI	No	m / s	Four fr. / Two fr.	M.domesticus M.spretus	Brown/1	Fisher et al. 1985
Proto-oncogene Araf	X	17	Araf	p19-1	pUC12	EcoRI	1200	XbaI	6.4,1.6	b / s	8.4 / 2.8	C57BL/6 M.spretus	Rapp/5	Avner et al. 1987b
DNA fragment	X	18	DXPas3	p66	LNM1149	EcoRI	3100	TaqI	Not given	d / s	Not given / Not given	M.domesticus M.spretus	Avner/2	Amar et al. 1985 / Avner et al. 1987a
DNA fragment	X	19	DXPas4	p87	LNM1149	EcoRI	2500	TaqI	Not given	d / s	Not given / Not given	M.domesticus M.spretus	Avner/2	Amar et al. 1985 / Avner et al. 1987a
DNA fragment	X	21	DX222	MDSX222	pGEM4	EcoRI	140	TaqI	No	m / s	4.7,3.5,2.8 / 4.9,2.0	M.domesticus M.spretus	Brown/1	Fisher et al. 1985
DNA fragment	X	21	DX10	MDSX10	pGEM4	EcoRI	500	MspI	No	m / s	14.0,5.0 / 5.5,2.5	M.domesticus M.spretus	Brown/1	Fisher et al. 1985
DNA fragment	X	21	DXPas5	p100	LNM1149	EcoRI	2700	TaqI	Not given	d / s	Not given / Not given	M.domesticus M.spretus	Avner/2	Amar et al. 1985 / Avner et al. 1987a
DNA fragment	X	22	DX219	MDSX219	pSP64	EcoRI	600	TaqI	5.1	m / s	2.4 / 2.1	M.domesticus M.spretus	Brown/1	Fisher et al. 1985

Table 8.1 (*Continued*)

552

LOCUS NAME	CHR.	DIST. CEN.	LOCUS	PROBE NAME	VECTOR	SITE	INS.LEN. (EP)	ENZYME	POLYMORPHISM CON.FRAG	ALLELE	LENGTH	STRAINS/SPEC	CONTACT/ AVAILABILITY	REFERENCE
X-linked lymphocyte regulated family	X	22	Xlr	pM1	pUC9	EcoRI	920	BamHI	Yes	b m s	3.3 Not given 1.7	C57BL M.musculus M.spretus	Davis/5	Chapman, Pers. Comm.
Hypoxanthine phosphoribosyl transferase	X	23	Hprt	pMEV	pcD-X	PstI-PvuII	1400	HindIII	Yes	b m s	12.8.7.5 11.7.4.8,2.8 12.8,11.7,4.8	C57BL M.musculus M.spretus	Johnson/5	Mullins et al. 1988
DNA fragment	X	25	DX36	MDRS36	pGEM4	EcoRI	400	TaqI	5.2.2.8	m s	3.4 Absent	M.domesticus M.spretus	Brown/1	Fisher et al. 1985
Clotting factor IX	X	25	Cf-9	pKT218	pBR322-de:	PstI	3100	EcoRI	Yes	b m s	6.2 20.0 5.8	C57 BL M.musculus M.spretus	McGraw/5	Mullins et al. 1988
Flank to G6PD	X	30	G6pd	Gdx	LEMBL3	BamHI	14400	TaqI	2.7	b s	2.3 6.2	M.domesticus M.spretus	Avner/2	Amar et al. 1985 Avner et al. 1987a
Glucose-6-phosphate dehydrogenase	X	30	G6pd	pGDP3	pBR322	PstI	2800	PstI HindIII	Yes, weak Yes, weak	b m b s	7.4 7.9 4.5 1.6	C57 BL M.musculus C57 BL M.spretus	Luzatto/5	Mullins et al. 1988
Clotting factor VIII	X	32	Cf-8	p61-51	pSP65	EcoRI	4700	EcoRI BamHI	Yes Yes	b m b s	13.0,3.6 13.5,3.1 13.0,11.0,2.3 13.5,10.0,2.0	C57 BL M.musculus C57 BL M.spretus	Knopf/5	Mullins et al. 1988
Red sensitive visual pigment	X	32	Rsvp	hs-7	pUC18	EcoRI	1100	BglII PvuII	No Yes	b m b s	14.8 12.5 6.6,4.8,2.8 6.5,5.5,5.1	C57BL/6J M.musculus C57 BL M.spretus	Nathans/5	Mullins et al. 1988 Nathans et al. 1986
Duchenne muscular dystrophy	X	33	Dmd	Mc2-6	Bluescript	EcoRI	2700	EcoRI	Several	b m s	Several Several Several	M.domesticus M.musculus M.spretus	Kunkel/5	Grant et al. 1988
DNA fragment	X	35	DX120	MDSX120	pGEM4	EcoRI	190	TaqI	No	m s	3.4 4.0	M.domesticus M.spretus	Brown/1	Fisher et al. 1985
Ornithine decarboxylase-13	X	43	Odc-13	pMK934	pBR322	PstI	1000	BglII	Several	b m	7.0 2.7	C57BL/6J M.musculus	Elliot/1 Berger/1	Stephenson et al. 1988
DNA fragment	X	48	DXPas2	p52	LNM1149	EcoRI	2800	TaqI	Not given	d s	Not given Not given	M.domesticus M.spretus	Avner/2	Amar et al. 1985 Avner et al. 1987a
Myelin proteolipid jimpy	X	56	Plp jp	P-23	pBR322	PstI	1720	EcoRI	No	b s	6.3 5.5	C57BL.CBA M.spretus	Dautigny/2	Dautigny et al. 1986
DNA fragment	X	63	DX225	MDRX225	pGEM4	EcoRI	180	TaqI	4.6	m s	5.6,4.8,4.2 4.4	M.domesticus M.spretus	Brown/1	Fisher et al. 1985
DNA fragment	X	68	DXPas1	p45	LNM1149	EcoRI	2900	TaqI	Not given	d s	Not given Not given	M.domesticus M.spretus	Avner/2	Amar et al. 1985 Avner et al. 1987a

LOCUS NAME	DIST. CEN. CHR. LOCUS	PROBE NAME	VECTOR	SITE	INS.LEN. (BP)	ENZYME	CON.FRAG	ALLELE	LENGTH	STRAINS/SPEC	CONTACT/ AVAILABILITY	REFERENCE
DNA fragment	X DX17A	p17A	pUC18	EcoRI	5800	EcoRI (+others)	No	a b	5.8 3.7,3.3	A,,BALB,C57B AKR,DBA,NZB	Disteche/2	Disteche et al. 1982
DNA fragment	X DX31A	p31A	pUC13	EcoRI	1000	HindIII		a b,c m	 7.2,2.7	C57BL/10 *Mspretus* BALB,C57,DBA	Disteche/2	Disteche et al. 1982
DNA fragment	Y DY2	pY2	pBR322	BamHI	500	PvuII	3.6,1.5	d m	3.8 4.8	AKR,SJL BALB,C57,DBA	Lamar/1	Lamar & Palmer, 1984
DNA fragment	Y DY3	pY3	pBR322	BamHI	500	BamHI	3.7,1.5,1.0	d m	4.9 7.5,5.0	AKR,SJL BALB,C57,DBA	Lamar/1	Lamar & Palmer, 1984
Genomic repeat	Y DYAc11	AC11	pBR322	BamHI	3.8	EcoRI	Several	d m		AKR,SJL,SWR Most strains	Nishioka/5	Nishioka, 1987
alpha-Actin,skeletal	U	Acta	fr.	BamHI	2.6	BgII	6.6,2.5	a b	4.6,1.7 4.4,2.0	BALB/c C57B,DBA,AKR	Davidson/5	Seldin et al. 1987a
Histone -3	U H3F2	pFO422	fr.	EcoRI	2.1	EcoRI	Nine fr.	a b	8.6,8.2,4.1 8.0,7.6	A,BALB,C3H AKR,C57B,DBA	Stein/5	Seldin & Steinberg, 1987

References

Amar, L.C., D. Arnaud, J. Cambrou, J.-L. Guénet, and P.R. Avner. 1985. Mapping of the mouse X chromosome using random genomic probes and an inter-specific mouse cross. EMBO J. 4:3695–3700.

Amor, M., M. Tosi, C. Duponchel, M. Steinmetz, and T. Meo. 1985. Liver mRNA probes disclose two cytochrome P-450 genes duplicated in tandem with the complement C4 loci of the mouse H-2S region. Proc. Natl. Acad. Sci. USA 82: 4453–4457.

Ashworth, A., and K. Willison. 1987. Chromosomal localization of mouse cDNAs homologous to the Drosophila gene *Kruppel*. First Intl. Workshop on Mouse Gene Mapping. Mouse News Lett. 79:84.

Avner, P., L. Amar, D. Arnaud, A. Hanauer, and J. Cambrou. 1987a. Detailed ordering of markers localizing to the Xq26-Xqter region of the human X chromosome by the use of an interspecific *Mus spretus* mouse cross. Proc. Natl. Acad. Sci. USA 84:1629–1633.

Avner, P., M. Búcan, D. Arnaud, H. Lehrach, and U. Rapp. 1987b. A-*raf* oncogene localizes on mouse X chromosome to region some 10–17 centimorgans proximal to hypoxanthine phosphoribosyltransferase gene. Somat. Cell Mol. Genet. 13: 267–272.

Baumann, H., and F.G. Berger. 1985. Genetics and evolution of the acute phase proteins in mice. Mol. Gen. Genet. 201:505–512.

Behlke, M.A., D.G. Spinella, H.S. Chou, W. Sha, D.L. Hartl, and D.Y. Loh. 1985. T-cell receptor β-chain expression: dependence on relatively few variable region genes. Science 229:566–570.

Belayew, A., and S.M. Tilghman. 1982. Genetic analysis of α-fetoprotein synthesis in mice. Mol. Cell. Biol. 2:1427–1435.

Bellis, M., V. Jubier-Maurin, B. Dod, F. Vanlergerghe, A.-M. Laurent, C. Senglat, F. Bonhomme, and G. Roizés. 1987. Distributions of two recently inserted long interspersed elements of the Ll repetitive family at the Alb and βh3 loci in wild mice populations. Mol. Biol. Evol. 4:351–363.

Bennett, K.L., P.A. Lalley, R.K. Barth, and N.D. Hastie. 1982. Mapping the structural genes coding for the major urinary proteins in the mouse: Combined use of recombinant inbred strains and somatic cell hybrids. Proc. Natl. Acad. Sci. USA 79:1220–1224.

Berman, J.W., A.J.D. Rocha, and R. Basch. 1986. Restriction length polymorphism in the variable region of the Tcr locus linked to histocompatibility antigen H-8 on murine chromosome 14. Immunogenetics 24:328–330.

Birkenmeier, E.H., and J.I. Gordon. 1986. Developmental regulation of a gene that encodes a cysteine-rich intestinal protein and maps near the murine immunoglobulin heavy chain locus. Proc. Natl. Acad. Sci. USA 83:2516–2520.

Blankenstein, T., F. Bonhomme, and U. Krawinkel. 1987. Evolution of pseudogenes in the immunoglobulin V_H-gene family of the mouse. Immunogenetics 26:237–248.

Blatt, C., M.E. Harper, G. Franchini, M.N. Nesbitt, and M.I. Simon. 1984. Chromosomal mapping of murine c-fes and c-src genes. Mol. Cell. Biol. 4:978–981.

Braunstein, N.S., and R.N. Germain. 1986. The mouse Eβ2 gene: a class II MHC β gene with limited intraspecies polymorphism and an unusual pattern of transcription. EMBO J. 5:2469–2476.

Búcan, M., T. Yang-Feng, A.M. Colberg-Poley, D.J. Wolgemuth, J.-L. Guénet, U. Francke, and H. Lehrach. 1986. Genetic and cytogenetic localization of the homeo box containing genes on mouse chromosome 6 and human chromosome 7. EMBO J. 5:2899–2905.

Búcan, M., B.G. Herrmann, A.-M. Frischauf, V.L. Bautch, V. Bode, L.M. Silver, G.R. Martin, and H. Lehrach. 1987. Deletion and duplication of DNA sequences is associated with the embryonic lethal phenotype of the t^9 complementation group of the mouse t complex. Genes Dev. 1:376–385.

Buchberg, A.M., B.A. Taylor, N.A. Jenkins, and N.G. Copeland. 1986. Chromosomal localization of Emv-16 and Emv-17, two closely linked ecotropic proviruses of RF/J mice. J. Virol. 60: 1175–1178.

Buchberg, A.M., H.G. Bedigian, B.A. Taylor, E. Brownell, J.N. Ihle, S. Nagata, N.A. Jenkins, and N.G. Copeland. 1988. Localization of *Evi-2* to chromosome 11: linkage to other proto-oncogene and growth factor loci using interspecific backcross mice. Oncogene Res. 2:149–165.

Chaplin, D.D., L.J. Galbraith, J.G. Seidman, P.C. White, and K.L. Parker. 1986. Nucleotide sequence analysis of murine 21-hydroxylase genes: mutations affecting gene expression. Proc. Natl. Acad. Sci. USA 83, 9601–9605.

Cheng, S.V., T.G. Lugo, R.E. Tanzi, J.B. Whitney, III, R.E.K. Fournier, and J.F. Gusella. 1987. Chromosomal localization of the mouse homolog of the Huntington's disease linked G8 (D4S10) marker. DNA 6:401–407.

Cioe, L., and P. Curtis. 1985. Detection and characterization of a mouse α-spectrin cDNA clone by its expression in *Escherichia coli*. Proc. Natl. Acad. Sci. USA 82:1367–1371.

Cobb, R.R., T.A. Stoming, and J.B. Whitney, III. 1987. The aryl hydrocarbon hydroxylase (Ah) locus and a novel restriction-fragment length polymorphism (RFLP) are located on mouse chromosome 12. Biochem. Genet. 25:401–413.

Copeland, N.G., N.A. Jenkins, and B.K. Lee. 1983. Association of the lethal yellow (A^Y) coat color mutation with an ecotropic murine leukemia virus genome. Proc. Natl. Acad. Sci. USA 80:247–249.

Curran, T., G. Peters, C. Van Beveren, N.M. Teich, and I.M. Verma. 1982. FBJ murine osteosarcoma virus: identification and molecular cloning of biologically active proviral DNA. J. Virol. 44:674–682.

Curtis, P.J., E. Withers, D. Demuth, R. Watt, P.J. Venta, and R.E. Tashian. 1983. The nucleotide sequence and derived amino acid sequence of cDNA coding for mouse carbonic anhydrase II. Gene 25:325–332.

Dautigny, A., M.-G. Mattei, D. Morello, P.M. Alliel, D. Pham-Dinh, L. Amar, D. Arnaud, D. Simon, J.-F. Mattei, J.-L. Guénet, P. Jollés, and P. Avner. 1986. The structural gene coding for myelin-associated proteolipid protein is mutated in *jimpy* mice. Nature 321:867–869.

Dean, M., C. Kozak, J. Robbins, R. Callahan, S. O'Brien, and G.F. Vande Woude. 1987. Chromosomal localization of the *met* proto-oncogene in the mouse and cat genome. Genomics 1:167–173.

Dembic, Z., W. Bannwarth, B. Taylor, and M. Steinmetz. 1985. The gene encoding the T-cell receptor α-chain maps close to the Np-2 locus on mouse chromosome 14. Nature 314:271–273.

D'Eustachio, P. 1984. A genetic map of mouse chromosome 12 composed of polymorphic DNA fragments. J. Exp. Med. 160:827–838.

D'Eustachio, P., B. Fein, J. Michaelson, and B.A. Taylor. 1984. The α-globin pseudogene on mouse chromosome 17 is closely linked to H-2. J. Exp. Med. 159:958–963.

D'Eustachio, P., G.C. Owens, G.M. Edelman, and B.A. Cunningham. 1985. Chromosomal location of the gene encoding the neural cell adhesion molecule (N-CAM) in the mouse. Proc. Natl. Acad. Sci. USA 82:7631–7635.

D'Eustachio, P., T. Kristensen, R.A. Wetzel, R. Riblet, B.A. Taylor, and B.F. Tack. 1986. Chromosomal location of the genes encoding complement components C5 and factor H in the mouse. J. Immunol. 137:3990–3995.

D'Eustachio, P., S. Jadidi, R.C. Fuhlbrigge, P.W. Gray, and D.D. Chaplin. 1987. Interleukin-1 α and β genes: linkage on chromosome 2 in the mouse. Immunogenetics 26:339–343.

D'Hoostelaere, L.A., and D.M. Gibson. 1986. The organization of immunoglobulin variable kappa chain genes on mouse chromosome 6. Immunogenetics 23:260–265.

D'Hoostelaere, L.A., E. Jouvin-Marche, and K. Huppi. 1985. Localization of $C_{T\beta}$ and C_K on mouse chromosome 6. Immunogenetics 22:277–283.

Disteche, C.M., L.M. Kunkel, A. Lojewski, S.H. Orkin, M. Eisenhard, E. Sahar, B. Travis, and S.A. Latt. 1982. Isolation of mouse X-chromosome specific DNA from an X-enriched lambda phage library derived from flow sorted chromosomes. Cytometry 2:282–296.

Disteche, C.M., U. Tantravahi, S. Gandy, M. Eisenhard, D. Adler, and L.M. Kunkel. 1985. Isolation and characterization of two repetitive DNA fragments located near the centromere of the mouse X chromosome. Cytogenet. Cell Genet. 39:262–268.

Dush, M.K., J.A. Tischfield, S. Khan, E. Feliciano, J.M. Sikela, C.A. Kozak, and P.J. Stambrook. 1986. An unusual adenine phosphoribosyltransferase pseudogene is syntenic with its functional gene and is flanked by highly polymorphic DNAs. Mol. Cell. Biol. 6:4161–4167.

Eistetter, H.R., S. Adolph, M. Ringwald, D. Simon-Chazottes, R. Schuh, J.-L. Guénet, and R. Kemler. 1988. Chromosomal mapping of the structural gene coding for the mouse cell adhesion molecule uvomorulin. Proc. Natl. Acad. Sci. USA 85:3489–3493.

Elliott, R.W. 1981. Identification and mapping of new loci using cloned cDNA. Mouse News Lett. 64:87.

Elliott, R.W., and F.G. Berger. 1983. DNA sequence polymorphism in an androgen-regulated gene is associated with alteration in the encoded RNAs. Proc. Natl. Acad. Sci. USA 80:501–504.

Elliott, R.W., and E. Ehrenreich. 1988. A cDNA probe for the proximal end of mouse chromosome 19. Mouse News Lett. 80:180.

Elliott, R.W., D. Barlow, and B.L.M. Hogan. 1985. Linkage of genes for laminin B1 and B2 subunits on chromosome 1 in mouse. In Vitro Cell. Dev. Biol. 21:477–484.

Elliott, R.W., L.C. Samuelson, M.S. Lambert, and M.H. Meisler. 1986. Assignment of pancreatic ribonuclease gene to mouse chromosome 14. Cytogenet. Cell Genet. 42:110–112.

Epstein, R., K. Lehmann, M. Cohn, C. Buckler, W. Rowe, and M. Davisson. 1984. Linkage of the Igl-1 structural and regulatory genes to Akv-2 on chromosome 16. Immunogenetics 19:527–537.

Epstein, R., M. Davisson, K. Lehmann, E.C. Akeson, and M. Cohn. 1986. Position of Igl-1, md, and Bst loci on chromosome 16 of the mouse. Immunogenetics 23:78–83.

Figueroa, F., M. Kasahara, H. Tichy, E. Neufeld, U. Ritte, and J. Klein. 1987. Polymorphism of unique noncoding DNA sequences in wild and laboratory mice. Genetics 117:101–108.

Fisher, E.M.C., J.S. Cavanna, and S.D.M. Brown. 1985. Microdissection and microcloning of the mouse X chromosome. Proc. Natl. Acad. Sci. USA 82:5846–5849.

Fox, H.S., L.M. Silver, and G.R. Martin. 1984. An alpha globin pseudogene is located within the mouse *t* complex. Immunogenetics 19:125–130.

Fox, H.S., G.R. Martin, M.F. Lyon, B. Herrmann, A.-M. Frischauf, H. Lehrach, and L.M. Silver. 1985. Molecular probes define different regions of the mouse *t* complex. Cell 40:63–69.

Gallagher, P.M., M.A. D'Amore, S.D. Lund, R.W. Elliott, J. Pazik, C. Hohman, T.R. Korfhagen, and R.E. Ganschow. 1987. DNA sequence variation within the β-glucuronidase gene complex among inbred strains of mice. Genomics 1:145–152.

Gallahan, D., C. Kozak, and R. Callahan. 1987. A new common integration region (int-3) for mouse mammary tumor virus on mouse chromosome 17. J. Virol. 61:218–220.

Garner, I., A.J. Minty, S. Alonso, P.J. Barton, and M.E. Buckingham. 1986. A 5' duplication of the α-cardiac actin gene in BALB/c mice is associated with abnormal levels of α-cardiac and α-skeletal actin mRNAs in adult cardiac tissue. EMBO J. 5:2559–2567.

Grant, S.G., L.J. Mullins, D.A. Stephenson, and V.M. Chapman. 1988. Manuscript in preparation.

Gray, P.W., D. Glaister, E. Chen, D.V. Goeddel, and D. Pennica. 1986. Two interleukin 1 genes in the mouse: cloning and expression of the cDNA for murine interleukin 1β. J. Immunol. 137:3644–3648.

Hagenbüchle, O., R. Bovey, and R.A. Young. 1980. Tissue-specific expression of mouse α-amylase genes: nucleotide sequence of isoenzyme mRNAs from pancreas and salivary gland. Cell 21:179–187.

Hart, C.P., D.K. Dalton, L. Nichols, L. Hunihan, T.H. Roderick, S.H. Langley, B.A. Taylor, and F.H. Ruddle. 1988. The

Hox-2 homeo box gene complex on mouse chromosome 11 is closely linked to *Re*. Genetics 118:319–327.

Heidmann, O., A. Buonanno, B. Geoffroy, B. Robert. J.-L. Guénet, J.P. Merlie, and J.-P. Changeux. 1986. Chromosomal localization of muscle nicotinic acetylcholine receptor genes in the mouse. Science 234:866–868.

Herrmann, B., M. Búcan, P.E. Mains, A.-M. Frischauf, L.M. Silver, and H. Lehrach. 1986. Genetic analysis of the proximal portion of the mouse *t* complex: evidence for a second inversion within *t* haplotypes. Cell 44:469–476.

Hildebrand, C.E., F.J. Gonzalez, C.A. Kozak, and D.W. Nebert. 1985. Regional linkage analysis of the dioxin-inducible P-450 gene family on mouse chromsome 9. Biochem. Biophys. Res. Comm. 130:396–406.

Hill, R.E., P.H. Shaw, R.K. Barth, and N.D. Hastie. 1985. A genetic locus closely linked to a protease inhibitor gene complex controls the level of multiple RNA transcripts. Mol. Cell. Biol. 5:2114–2122.

Hill, R.E., A.E. Hall, C.M. Sime, and N.D. Hastie. 1987. A mouse homeo box-containing gene maps near a developmental mutation. Cytogenet. Cell Genet. 44:171–174.

Hogarth, P.M., I.F.C. McKenzie, V.R. Sutton, K.M. Curnow, B.K. Lee, and E.M. Eicher. 1987. Mapping of the mouse Ly-6, Xp-14 and Gdc-1 loci to chromosome 15. Immunogenetics 25:21–27.

Howles, P.N., D.P. Dickinson, L.L. DiCaprio, M. Woodworth-Gutai, and K.W. Gross. 1984. Use of a cDNA recombinant for the γ-subunit of mouse nerve growth factor to localize members of this multigene family near the TAM-1 locus on chromosome 7. Nucl. Acids Res. 12:2791–2805.

Ishikawa, N., K. Shigemoto, and N. Maruyama. 1987. The complete nucleotide and deduced amino acid sequence of mouse serum amyloid P component. Nucl. Acids Res. 15:7186.

Jacobsson, A., U. Stadler, M.A. Glotzer, and L.P. Kozak. 1985. Mitochondrial uncoupling protein from mouse brown fat. J. Biol. Chem. 260:16250–16254.

Jenkins, N.A., N.G. Copeland, B.A. Taylor, and B.K. Lee. 1981. Dilute (d) coat colour mutation of DBA/2J mice is associated with the site of integration of an ecotropic MuLV genome. Nature 293:370–374.

Jubier-Maurin, V., B.J. Dod, M. Bellis, M. Piechadzyk, and G. Roizes. 1985. Comparative study of the L1 family in the genus *Mus*. Possible role of retroposition and conversion events in its concerted evolution. J. Mol. Biol. 184:547–564.

Kent, R.B., D.A. Fallows, E. Geissler, T. Glaser, J.R. Emanuel, P.A. Lalley, R. Levenson, and D.E. Housman. 1987. Genes encoding α and β subunits of Na,K-ATPase are located on three different chromosomes in the mouse. Proc. Natl. Acad. Sci. USA 84:5369–5373.

King, D., L.D. Snider, and J.B. Lingrel. 1986. Polymorphism in an androgen-regulated mouse gene is the result of the insertion of a B1 repetitive element into the transcription unit. Mol. Cell. Biol. 6:209–217.

Kozak, L.P., and E.H. Birkenmeier. 1983. Mouse sn-glycerol-3-phosphate dehydrogenase: molecular cloning and genetic

mapping of a cDNA sequence. Proc. Natl. Acad. Sci. USA 80:3020–3024.

Lafuse, W.P., and C.S. David. 1986. Recombination hot spots within the I region of the mouse H-2 complex map to the E_β and E_α genes. Immunogenetics 24:352–360.

Lafuse, W.P., S. Singh, S. Savarirayan, and C.S. David. 1986. DNA restriction fragment analysis of E_α genes in E-negative H-2 recombinant mouse strains. Immunogenetics 24:68–70.

Lalanne, J.-L., C. Transy, S. Guerin, S. Darche, P. Meulien, and P. Kourilsky. 1985. Expression of class I genes in the major histocompatibility complex: identification of eight distinct mRNAs in DBA/2 mouse liver. Cell 41:469–478.

Lamar, E.E., and E. Palmer. 1984. Y-encoded, species-specific DNA in mice: evidence that the Y chromosome exists in two polymorphic forms in inbred strains. Cell 37:171–177.

LeClair, K.P., R.G.E. Palfree, P.M. Flood, U. Hammerling, and A. Bothwell. 1986. Isolation of a murine Ly-6 cDNA reveals a new multigene family. EMBO J. 5:3227–3234.

Levi-Strauss, M., M. Tosi, M. Steinmetz, J. Klein, and T. Meo. 1985. Multiple duplications of complement C4 gene correlate with H-2-controlled testosterone-independent expression of its sex-limited isoform, C4-Slp. Proc. Natl. Acad. Sci. USA 82:1746–1750.

Lonai, P., E. Arman, H. Czosnek, F.H. Ruddle, and C. Blatt. 1987. New murine homeoboxes: structure, chromosomal assignment, and differential expression in adult erythropoiesis. DNA 6:409–418.

Mami, F., P.-A. Cazenave, and T.J. Kindt. 1988. Conservation of the immuno-globulin Cλ 5 gene in the *Mus* genus. EMBO J. 7:117–122.

Mann, E.A., L.M. Silver, and R.W. Elliott. 1986. Genetic analysis of a mouse *t* complex locus that is homologous to a kidney cDNA clone. Genetics 114:993–1006.

Marcu, K.B., J. Banerji, N.A. Penncavage, R. Lang, and N. Arnheim. 1980. 5' flanking region of immunoglobulin heavy chain constant region genes displays length heterogeneity in germlines of inbred mouse strains. Cell 22:187–196.

Mason, I.J., D. Murphy, M. Münke, U. Francke, R.W. Elliott, and B.L.M. Hogan. 1986. Developmental and transformation-sensitive expression of the Sparc gene on mouse chromosome 11. EMBO J. 5:1831–1837.

McConnell, T.J., B. Darby, and E.K. Wakeland. 1986. Restriction fragment length polymorphisms of class II gene sequences in mice expressing minor structural variants of I-A^k and I-A^P. J. Immunol. 136:3076–3084.

Meisler, M.H., T.K. Antonucci, L.O. Treisman, D.L. Gumucio, and L.C. Samuelson. 1986. Interstrain variation in amylase gene copy number and mRNA abundance in three mouse tissues. Genetics 113:713–722.

Mellor, A.L., J. Antoniou, and P.J. Robinson. 1985. Structure and expression of genes encoding murine Qa-2 class I antigens. Proc. Natl. Acad. Sci. USA 82:5920–5924.

Meruelo, D., A. Rossomando, S. Scandalis, P. D'Eustachio, R.E.K. Fournier, D.R. Roop, D. Saxe, C. Blatt, and M.N. Nesbitt. 1987. Assignment of the Ly-6–Ril-1–Sis–H-30–Pol-

5/Xmmv-72–Ins-3–Krt-1–Int-1–Gdc-1 region to mouse chromosome 15. Immunogenetics 25:361–372.

Michael, S.K., J. Hilgers, C. Kozak. J.B. Whitney III, and E.F. Howard. 1986. Characterization and mapping of DNA sequence homologous to mouse Ula1 snRNA; localization on chromosome 11 near the Dlb-1 and Re loci. Somat. Cell Mol. Genet. 12:215–223.

Minty, A.J., S. Alonso, J.-L. Guénet, and M.E. Buckingham. 1983. Number and organization of actin-related sequences in the mouse genome. J. Mol. Biol. 167:77–101.

Mock, B.A., L.A. D'Hoostelaere, R. Matthai, and K. Huppi. 1987a. A mouse homeo box gene, Hox-1.5, and the morphological locus, Hd, map to within 1 cM on chromosome 6. Genetics 116:607–612.

Mock, B., J. Hilgers, K. Huppi, and L. D'Hoostelaere. 1987b. A TaqI restriction fragment length polymorphism at the Hox-2.1 locus cosegregates with the Dlb-1 locus on mouse chromosome 11. Nucl. Acids Res. 15:2397.

Mock, B., R. Skurla, K. Huppi, L. D'Hoostelaere, D. Klinman, and J.F. Mushinski. 1987. A restriction fragment length polymorphism at the murine c-myb locus. Nucl. Acids Res. 15:4700.

Mock, B.A., D. Givol, L.A. D'Hoostelaere, K. Huppi, M.F. Seldin, N. Gurfinkel, T. Unger, M. Potter, and J.F. Mushinski. 1988. Mapping of the *bcl-2* oncogene on mouse chromosome 1. Cytogenet. Cell Genet. 47:11–15.

Moore, R., G. Casey, S. Brookes, M. Dixon, G. Peters, and C. Dickson. 1986. Sequence, topography and protein coding potential of mouse *int-2*: a putative oncogene activated by mouse mammary tumour virus. EMBO J. 5:919–924.

Mucenski, M.L., B.A. Taylor, N.G. Copeland, and N.A. Jenkins. 1988. Chromosomal location of *Evi-1*, a common site of ecotropic viral integration in AKXD murine myeloid tumors. Oncogene Res. 2:219–233.

Müller, G., G. Scherer, H. Zentgraf, S. Ruppert, B. Herrmann, H. Lehrach, and G. Schütz. 1985. Isolation, characterization and chromosomal mapping of the mouse tyrosine aminotransferase gene. J. Mol. Biol. 184:367–373.

Mullins, J.J., Q. Zeng, and K.W. Gross. 1987. Mapping of the mouse atrial natriuretic factor gene: evidence for tight linkage to the Fv-1 locus. Hypertension 9:518–521.

Mullins, L.J., S.G. Grant, D.A. Stephenson, and V.M. Chapman. 1988. Multilocus molecular mapping of the mouse X chromosome. Genomics 3:187–194.

Münke, M., K. Harbers, R. Jaenisch, and U. Francke. 1986. Chromosomal mapping of four different integration sites of Moloney murine leukemia virus including the locus for $\alpha1(I)$ collagen in mouse. Cytogenet. Cell Genet. 43:140–149.

Nadeau, J.H., and S.J. Phillips. 1987. The putative oncogene Pim-1 in the mouse: its linkage and variation among *t* haplotypes. Genetics 117:533–541.

Nadeau, J.H., F.G. Berger, K.A. Kelley, P.M. Pitha, C.L. Sidman, and N. Worrall. 1986. Rearrangement of genes located on homologous chromosomal segments in mouse and man: the location of genes for alpha- and beta-interferon, alpha-1 acid glycoprotein-1 and -2 and amino-levulinate dehydratase on mouse chromosome 4. Genetics 104:1239–1255.

Nathans, J., D. Thomas, and D.S. Hogness. 1986. Molecular genetics of human color vision: the genes encoding blue, green and red pigments. Science 232:193–202.

Nishioka, Y., 1987. Y-chromosomal DNA polymorphism in mouse inbred strains. Genet. Res. 50:69–72.

Owen, F.L., B.A. Taylor, A. Zweidler, and J.G. Seidman. 1986. The murine γ-chain of the T cell receptor is closely linked to a spermatocyte specific histone gene and the beige coat color locus on chromosome 13. J. Immunol. 137:1044–1046.

Pachnis, V., A. Belayew, and S.M. Tilghman. 1984. Locus unlinked to α-fetoprotein under the control of the murine raf and Rif genes. Proc. Natl. Acad. Sci. USA 81:5523–5527.

Panthier, J.-J., I. Holm, and F. Rougeon. 1982. The mouse Rn locus: S allele of the renin regulator gene results from a single structural gene duplication. EMBO J. 1:1417–1421.

Paul, P.R., and R.W. Elliott. 1987. Analysis of the mouse Amy locus in recombinant inbred mouse strains. Biochem. Genet. 25:569–579.

Piccini, N., J.L. Knopf, and K.W. Gross. 1982. A DNA polymorphism, consistent with gene duplication, correlates with high renin levels in the mouse submaxillary gland. Cell 30:205–213.

Rathbun, G.A., W. Born, W.A. Kuziel, and P.W. Tucker. 1988. Diversity of the mouse T cell receptor C γ 1 gene: structural analysis in C57BL/Ka. Immunogenetics 27:121–126.

Reeves, R.H., D. Gallahan, B.F. O'Hara, R. Callahan, and J.D. Gearhart. 1987. Genetic mapping of Prm-l, Igl-l, Smst, Mtv-6, Sod-l and Ets-2, and localization of the Down syndrome region on mouse chromosome 16. Cytogenet. Cell Genet. 44:76–81.

Richards, J.E., D.D. Pravtcheva, C. Day, F.H. Ruddle, and P.P. Jones. 1985. Murine invariant chain gene: chromosomal assignment and segregation in recombinant inbred strains. Immunogenetics 22:193–199.

Richards-Smith, B., and R.W. Elliott. 1984. Mapping of a family of repeated sequences in the mouse genome. Mouse News Lett. 7:46–47.

Robbins, J.M., D. Gallahan, E. Hogg, C. Kozak, and R. Callahan. 1986. An endogenous mouse mammary tumor virus genome common in inbred mouse strains is located on chromosome 6. J. Virol. 57:709–713.

Robert, B., P. Barton, A. Minty, P. Daubas, A. Weydert, F. Bonhomme, J. Catalan, D. Chazottes, J.-L. Guénet, and M. Buckingham. 1985. Investigation of genetic linkage between myosin and actin genes using an interspecific mouse back-cross. Nature 314:181–183.

Rougeon, F., and B. Mach. 1977. Cloning and amplification of α and β mouse globin gene sequences synthesised *in vitro*. Gene 1:229–239.

Rubin, M.R., W. King, L.E. Toth, I.S. Sawczuk, M.S. Levine, P. D'Eustachio, and M.C. Nguyen-Huu. 1987. Murine Hox-1.7 homeo-box gene: cloning, chromosomal location and expression. Mol. Cell. Biol. 7:3836–3841.

Ryan, J., P.E. Barker, M.N. Nesbitt, and F.H. Ruddle. 1987. KRAS2 as a genetic marker for lung tumor susceptibility in inbred mice. JNCI 79:1351–1357.

Sackstein, R., M.H. Roos, P. Démant, and H.R. Colten. 1984. Subdivision of the S region of the mouse major histocompatibility complex by identification of genomic polymorphisms of the Class III genes. Immunogenetics 20:321–330.

Schimenti, J., L. Vold, D. Socolow, and L.M. Silver. 1987. An unstable family of large DNA elements in the center of the mouse *t* complex. J. Mol. Biol. 194:583–594.

Seif, I., and J. DeMaeyer-Guignard. 1986. Structure and expression of a new murine interferon-alpha gene: MuIFN-$\alpha_1$9. Gene 43:111–121.

Seldin, M.F., and A.D. Steinberg. 1987. Human histone 3 F2 detects RFLPs in inbred mice, cosegregates with H4F2 and does not map to possible syntenic groups on mouse distal chromosome 1 and distal chromosome 3. Nucl. Acids. Res. 15:5900.

Seldin, M.F., L.A. D'Hoostelaere, and A.D. Steinberg. 1987a. Mouse skeletal alpha actin has limited restriction fragment length polymorphism and is not a member of the human l_q–mouse distal chromosome 1 syntatic group. Nucl. Acids Res. 15:1881.

Seldin, M.F., L.A. D'Hoostelaere, A.D. Steinberg, Y. Saga, and H.C. Morse III. 1987b. Allelic variants of Ly-5 in inbred and natural populations of mice. Immunogenetics 26:74–78.

Seldin, M.F., H.C. Morse III, L. D'Hoostelaere, J.L. Britten, and A.D. Steinberg. 1987c. Mapping of alpha–spectrin on distal mouse chromosome 1. Cytogenet. Cell Genet. 45:52–54.

Shen, F.-W., Y. Saga, G. Litman, G. Freeman, J.-S. Tung, H. Cantor, and E.A. Boyse. 1985. Cloning of Ly-5 cDNA. Proc. Natl. Acad. Sci. USA 82:7360–7363.

Silver, J., and C.E. Buckler. 1986. A preferred region for integration of friend murine leukemia virus in hematopoietic neoplasms is closely linked to the *Int-2* oncogene. J. Virol. 60:1156–1158.

Silver, J., J.B. Whitney III, C. Kozak, G. Hollis, and I. Kirsch. 1985. *Erbb* is linked to the alpha-globin locus on mouse chromosome 11. Mol. Cell. Biol. 5:1784–1786.

Simmons, D.L., and C.B. Kasper. 1983. Genetic polymorphisms for a phenobarbital-inducible cytochrome P-450 map to the Coh locus in mice. J. Biol. Chem. 258:9585–9588.

Siracusa, L.D., L.B. Russell, N.A. Jenkins, and N.G. Copeland. 1987. Allelic variation within the *Emv-15* locus defines genomic sequences closely linked to the *agouti* locus on mouse chromosome 2. Genetics 117:85–92.

Skow, L.C., and M.E. Donner. 1985. The locus encoding αA-crystallin is closely linked to H-2K on mouse chromosome 17. Genetics 110:723–732.

Skow, L.C., J.N. Nadeau, J.C. Ahn, H.-S. Shin, K. Artzt, and D. Bennett. 1987. Polymorphism and linkage of the αA-crystallin gene in *t*-haplotypes of the mouse. Genetics 116:107–111.

Staeheli, P., O. Haller, W. Boll, J. Lindermann, and C. Weissmann. 1986. Mx protein: constitutive expression in 3T3 cells transformed with cloned Mx cDNA confers selective resistance to influenza virus. Cell 44:147–158.

Stavenhagen, J., F. Loreni, C. Hemenway, M. Kalff, and D.M. Robins. 1987. Molecular genetics of androgen-dependent and -independent expression of mouse sex-limited protein. Mol. Cell. Biol. 7:1716–1724.

Steinmetz, M., K.W. Moore, J.G. Frelinger, B.T. Sher, F.-W. Shen, E.A. Boyse, and L. Hood. 1981. A pseudogene homologous to mouse transplantation antigens: transplantation antigens are encoded by eight exons that correlate with protein domains. Cell 25:683–692.

Stephenson, D.A., R.W. Elliott, V.M. Chapman, and S.G. Grant. 1988. Identification of an X-linked member of the Odc gene family in the mouse. Nucl. Acid Res. 16:1642.

Taylor, B.A., and L. Rowe. 1984. Genes for serum amyloid A proteins map to chromosome 7 in the mouse. Mol. Gen. Genet. 195:491–499.

Taylor, B.A., and L. Rowe. 1987. The congenital goiter mutation is linked to the thyroglobulin gene in the mouse. Proc. Natl. Acad. Sci. USA 84:1986–1990.

Taylor, B.A., L. Rowe, D.M. Gibson, R. Riblet, R. Yetter, and P.D. Gottlieb. 1985. Linkage of a 7S RNA sequence and kappa light chain genes in the mouse. Immunogenetics 22:471–481.

Uehara, H., K. Abe, C.-H.T. Park, H.-S. Shin, D. Bennett, and K. Artzt. 1987. The molecular organization of the H-2K region of two *t*-haplotypes: implications for the evolution of genetic diversity. EMBO J. 6:83–90.

Venta, P.J., J.C. Montgomery, D. Hewett-Emmett, K. Wiebauer, and R.E. Tashian. 1985. Structure and exon to protein domain relationships of the mouse carbonic anhydrase II gene. J. Biol. Chem. 260:12130–12135.

Warren, W., S.C. Jacobs, and C.S. Cooper. 1987. A BamHI RFLP at the Pim-1 locus on mouse chromosome 17. Nucl. Acids Res. 15:3195.

Watson, G., M. Felder, L. Rabinow, K. Moore, C. Labarca, C. Tietze, G.V. Molen, L. Bracey, M. Brabant, J. Cai, and K. Paigen. 1985. Properties of rat and mouse β-glucuronidase mRNA and cDNA, including evidence for sequence polymorphism and genetic regulation of mRNA levels. Gene 36:15–25.

Weaver, S., M.B. Comer, C. Jahn, C.A. Hutchison III, and M.H. Edgell. 1981. The adult β-globin genes of the 'single' type mouse C57BL. Cell 24:403–411.

Willison, K.R., K. Dudley, and J. Potter. 1986. Molecular cloning and sequence analysis of a haploid expressed gene encoding *t* complex polypeptide 1. Cell 44:727–738.

Yang, J.-N., R.T. Boyd, P.D. Gottlieb, and J.P. Dudley. 1987. The endogenous retrovirus Mtv-8 on mouse chromosome 6 maps near several kappa light chain markers. Immunogenetics 25:222–227.

Zhang, K., W.F. Bosron, and H.J. Edenberg. 1987. Structure of the mouse Adh-1 gene and identification of a deletion in a long alternating purine–pyrimidine sequence in the first intron of strains expressing low alcohol dehydrogenase activity. Gene 57:27–36.

9 HIGHLY REPEATED DNA FAMILIES IN THE GENOME OF *MUS MUSCULUS*

NICHOLAS D. HASTIE

Classical DNA reassociation experiments revealed that not all DNA sequences are represented equally in the eukaryotic genome (7). It was concluded that in the mouse genome, for example, DNA sequences fall into any one of three frequency classes: a highly repeated class thought to be composed entirely of centromeric satellite sequences which constitute 10 per cent of the genome; a moderately repetitive class comprising numerous families each with hundreds or many thousands of members at dispersed locations and constituting *in toto* 20 per cent of the genome; and, the single-copy component which made up the final 70 per cent of the genome and included most of the genes. Additional experiments showed that the bulk of moderately repetitive sequences were interspersed with single-copy sequences (12, 16, 17).

Since the advent of recombinant DNA technology it has been possible to study the structure and genomic distribution of individual cloned copies of DNA sequences belonging to these three classes. We now know that there is a considerable degree of overlap between the three components. Some dispersed repeated sequences may be more abundant in the genome than satellite DNA sequences. Also many genes are members of small families, the remaining members of which may be functional genes or, as is often the case, functionless pseudogenes which can be located at different chromosomal sites from the functional progenitor gene (2, 70).

For the purpose of this brief review I will concentrate, rather arbitrarily, on sequences which are present at a concentration of 10 000 copies or greater in the genome of *Mus musculus*. Essentially five such families, two satellites and three dispersed families, dominate the genome of *Mus musculus* (3). These are the most well characterized repeated sequences and may well act as paradigms for the structure and behaviour of less repeated elements.

One generalization which stems from studies of these highly repeated sequences is that the genomes of mammals are highly plastic. There are, for example, considerable differences in the representation of particular dispersed DNA families in the rat compared with the mouse (69). Also the major satellite of *Mus musculus* appears to be markedly different from the major satellites of *Mus caroli* and *Mus spretus* (8, 59, 62, 77). Differences of this nature, together with other considerations, have led to the notion that the majority of these repeated sequences are functionless or so-called selfish DNA (20, 56). However, I will discuss how these differences can be exploited to develop marker systems of use to the developmental biologist and geneticist.

For several reasons, which should become apparent, I have placed the discussion of satellites and dispersed DNA families into different sections.

9.1 Satellite DNA

The term 'satellite' DNA was coined originally to define families of highly repeated sequences which could be separated from main-band DNA by virtue of their density (38). This definition is no longer adequate and Singer has redefined satellites as those sequences which are centromerically located, tandemly repeated, and usually present in more than 10 000 copies in the genome (73). Many families of this nature would be overlooked by density criteria alone. For example, the minor satellite of *Mus musculus* (see Section 9.1.2) is not sufficiently abundant to be revealed as a separate density peak and would anyway be obscured by the major satellite which is of a similar density (59). Likewise there is evidence from molecular hybridization studies that *Mus spretus* has a major satellite component but this is not evident in density analysis (8, 59, 77).

Satellite DNAs comprise a sizeable portion (up to 50 per cent) of the genomes of higher eukaryotes but their function, if any, remains obscure. Satellite sequences are consistently associated with constitutive heterochromatin and a large body of evidence shows that they are not transcribed. This and their centromeric location

559

led to the proposal that they may play a structural role in chromosome segregation. Proposed functions for satellite DNA include: (i) to promote pairing of homologous chromosomes during meiosis; (ii) to provide specific attachment sites for binding of the chromosome to the spindle apparatus (14, 89); (iii) to reduce the level of meiotic recombination in the vicinity of the centromere (94). A more pessimistic viewpoint is that satellite DNAs provide no specific function but are merely selfish passengers which are maintained without exerting a deleterious effect on the genome. Certainly any function of satellite DNA cannot depend on absolute sequence content as this can vary considerably between two closely related rodent species. It might be that only a small proportion of the satellite cluster is functioning and that the remainder protects the functional core. We may only gain insights into this question of satellite DNA function when we can manipulate mammalian chromosomes in the way that has been achieved for yeast (6).

Two different satellite DNAs have been identified in the genome of *Mus musculus*. To reflect their relative abundance we shall call these the major and minor satellites.

9.1.1 The major satellite

M. musculus has been shown to contain one major satellite band with a density of 1.690 g/cm^3 (38, 84). From analysis of reassociation kinetics it was concluded that the satellite was a unit of 300–400 bp long repeated approximately 10^6 times and thus representing on the order of 10 per cent of the genome (92). *In situ* hybridization studies demonstrated that the major satellite is located at the centromeres of all the chromosomes with the exception of the Y chromosome (36, 57). Analysis of the major satellite by restriction endonuclease digestion first provided insights into its structure and evolution. Southern showed that the enzyme *Eco*RII digested mouse satellite into a ladder of small bands which were multimers of a basic repeating unit of 220–260 bp (81). This monomer unit was the most abundant fragment followed by the dimer and the trimer in an ever-decreasing series. This type of pattern, the A pattern, is also produced when satellite DNA is digested by *Ava*II and *Sau*96I. This is in contrast to the majority of restriction enzymes which produce a B-type pattern in which, characteristically, most of the DNA is undigested and just a few very large multimers are observed. Partial restriction with *Eco*RII revealed a ladder of high molecular weight multimers which decreased in intensity as digestion proceeded towards completion, thus confirming the tandem nature

of the repeat. Also produced in the *Eco*RII digestion products were fragments of length corresponding to $\frac{1}{8}$, $\frac{1}{4}$, and $\frac{1}{2}$ multiples of the moner unit. This and other considerations led Southern to propose that the present day satellite has evolved from smaller ancestral units 9–18 bp in length which have undergone alternating cycles of amplification and divergence. These conjectures have received strong support from DNA sequencing studies which showed that the exact length of the satellite is 234 bp (Table 9.1), consisting of two units of 116 and 118 bp (32). The ancestral sequence is thought to consist of three nonanucleotides GAAAAATGA, GAAAAAACT, and GAAAAACGT (Fig. 9.1).

It has been possible to study the evolution of the major satellite on individual mouse chromosomes which have been separated from the rest of the genome in somatic cell hybrids. This analysis showed that the basic organization of the repeat on specific chromosomes is very similar to the consensus but that there are chromosome-specific variants (9). Thus, mechanisms must exist to maintain inter- and intrachromosomal homogeneity of the repeat.

One provocative finding indicative, perhaps, of a functional role was that the major satellite can bind efficiently *in vitro* to the microtubule-associated protein (14, 89).

9.1.2 The minor satellite

The minor satellite was first detected by screening a small library of repetitive sequences inserted into a plasmid vector (59). The screen was designed to identify sequences which were highly repeated and which had evolved rapidly between different mouse species, properties characteristic of satellite DNA. Replicate filters plated with the library were hybridized to labelled DNA from *Mus musculus*, *Mus spretus*, and *Mus caroli*. Several clones gave a strong signal with *M. musculus* DNA but a weak or undetectable signal with *M. spretus* and *M. caroli* DNA. These clones turned out to be representatives of the major satellite. One clone on the other hand gave a strong signal with *M. spretus* DNA but a very weak signal with DNA from *M. caroli* or *M. musculus* even though the library was constructed from the latter. This clone corresponded to a new satellite, the minor *M. musculus* satellite.

This minor satellite is present in approximately 50 000 copies in *M. musculus* (Table 9.1) where it is centromerically located on most if not all chromosomes apart from the Y chromosome (59, 77). The restriction enzyme *Msp*I digests this satellite to produce a ladder based on a monomer unit of 120 bp. The sequence of the complete unit has not been determined but the

```
(a)
 α₁   G G A C C T G G A A T A T G G C G A G A A        A A C T G A A
 β₁   A A T C A C G G A A A A T G A    G A A A T A C A C A C T T T A
 α₂   G G A C G T G A A A T A T G G C G A GGA A        A A C T G A A
 β₂   A A A G G T G G A A A A T TTA    G A A A T G T C C A C T G T A
 α₃   G G A C G T G G A A T A T G G C A A G A A        C A C T A A A
 β₃   A A T C A T G G A A A A T G A    G A A A C A T C C A C T T G A
 α₄   C G A C T T G A A A A A T G A C G A A A T        C A C T G A A
 β₄   A A A C G T G A A A A A T G A    G A A A T G C A C A C T G A A

(b)
      A A   A C G T G G A A A A T G A   G A   A A T      C A C T G A A
      G G             A       T       G         G   A        A

      C     T G A C              T       A       C            T T
              T                                                A G
              C

(c)
      A A A C G T , G A A A A A T G A , G A A A A      A A C T , G A A
```

Fig. 9.1 *Internal repetitivity of the mouse major satellite DNA repeat unit, adapted from Hörz and Altenburger.* (a) The 234 bp of the repeat unit are arranged in eight subunits such as to obtain maximum homology. The subunits are numbered α_1 to β_4 as indicated on the left. Nucleotides identical in all eight repeats are underlined. The consensus sequence derived from the eight repeats is shown in (b) with nucleotides occurring only once or twice in a given position set in smaller print. A possible ancestral sequence of mouse satellite is shown in (c). It consists of three tandemly repeated nonanucleotides related to each other: GA_5TGA, GA_6CT, GA_5CGT.

sequence of a 90 bp stretch showed that the minor satellite is composed of two types of smaller internal repeat which are similar to those found in the major satellite, suggesting that both satellites evolved from a common ancestor (59, 77).

Variation of the minor satellite between inbred strains of Mus musculus

When genomic DNA is digested with the restirction enzyme *Hae*III a large number of bands is revealed following gel electrophoresis, Southern blotting, and hybridization to a minor satellite probe. Although the majority of these fragments is shared between all inbred strains, a minority of fragments is variant such that each strain tested had a unique combination of variant fragments. Several other restriction enzymes reveal such variant fragments (77). It has been possible to follow the segregation behaviour of some of these variants in mouse crosses and in somatic cell hybrids. This analysis has shown that each variant segregates as

an independent genetic locus. Such variants should be very useful for typing inbred strains, for following satellite evolution, and for studying recombination between centromeres and outside markers.

It is much more difficult to study variants of the major satellite as these are likely to be obscured due to the much higher frequency of this repeat relative to the minor satellite.

9.1.3 Evolution of satellite DNA within the genus *Mus*

Southern and dot blot analysis has been carried out to determine the abundance and repeating nature of the major and minor satellites in a range of *Mus* species (77). This analysis has confirmed that these satellites are evolving rapidly and in a manner consistent with the known phylogenetic relationships. In the commensal species, *M. m. domesticus*, *M. m. musculus*, *M. m. molossinus*, and *M. m. castaneus*, the absolute amounts of the major and minor satellites are similar, i.e. on the

561

order of 700 000 and 100 000 copies per haploid genome, respectively. Restriction endonuclease analysis showed that the basic repeating unit is the same for each satellite in all these species but there is a considerable degree of variation in the restriction patterns obtained for the minor satellite. An aboriginal species *M. hortulanus* also has a very similar representation of these satellites to the commensal species. However, in another aboriginal species, *M. spretus*, the pattern is reversed in that there are approximately 25 000 copies of the major satellite and 400 000 copies of the minor satellite. In the distantly related Asian species *M. cookii*, *M. cervicolor*, and *M. caroli* the levels of hybridization to both the major and minor satellites are reduced by at least 100-fold relative to those found within the commensal species. There appears to be no hybridization to the minor satellite at all. In spite of the very low copy number, digestion with *Sau*96I reveals the same repeating length and ladder pattern for the major satellite in the Asian species as in the commensal species. The low level of major satellite in *M. caroli* concurs with the density analysis which revealed no satellites with the same density as in *M. musculus*. However, there is convincing evidence that *M. caroli* has at least one major centromeric satellite (77, 84). In the most distantly related Asian species *M. pahari* it was not possible to demonstrate any sequences homologous to the major or minor satellite. It should be stressed that the reduction in signal is likely to be due to both a combination of lower copy number and sequence divergence. The combined data showed that the two satellites are not evolving in concert.

9.1.4 Satellite sequences can be used as markers for cell lineage analysis in interspecific chimaeras

The difference in level of hybridization of the major satellite probe to *M. musculus* and *M. caroli* DNA has formed the basis for an *in situ* hybridization procedure which makes it possible to distinguish the cells of the two species in interspecific chimaeras (66, 78). This should prove to be invaluable in the analysis of cell lineage during development. In many ways this constitutes the perfect marker system since it is ubiquitous, cell autonomous, independent of expression, and can be carried out on tissue sections. For technical reasons interspecific chimaeras are usually constructed by injecting *M. caroli* cells into blastocysts of *M. musculus* which are then implanted into *M. musculus* uteri. Hence the ideal marker probe would be one which

recognized the *M. caroli* component of the chimaeras, not the *M. musculus* component. There is ample evidence that a major centromeric satellite exists in *M. caroli* (77, 84). When cloned this should provide the appropriate probe.

9.1.5 Undermethylation of the minor and major satellites in germ cells

It is possible to study the overall state of methylation of a particular DNA sequence by using the restriction enzymes *Msp*I and *Hpa*II. These are so called isoschizomers, both recognizing the same site, CC↓GG, the difference being that *Msp*I will cut at this site if the second C residue is methylated, *Hpa*II will not. Greater than 90 per cent of the methylation in mammalian DNA takes place at the C residue of a CpG couplet so a proportion of these sites can be investigated for their state of methylation using both these enzymes. The significance of methylation is unclear (5). Genes which are actively transcribed in a specific tissue are likely to be undermethylated in this tissue relative to tissues in which they are transcriptionally silent. It might be expected that satellite DNAs, which are transcriptionally inert and located within constitutive heterochromatin, would be heavily methylated. Neither the major nor minor mouse satellite produces an A-type ladder when digested with *Msp*I which would reflect sites in the majority of monomers (68). However, the majority of the minor satellite does enter an agarose gel following cleavage with *Msp*I which results in a series of bands ranging from the monomer, which is most abundant, to units of 50 members or greater. In somatic tissues the majority of these sites are methylated. However, in both male and female germ cells there is a considerable reduction in the level of methylation of the minor satellite (68). By analysis of purified male germ cells it was possible to conclude that this undermethylated state exists in primary spermatogonia, several stages prior to meiosis. The major satellite is digested by *Msp*I to produce a B-type pattern so that only a small proportion will enter an agarose gel. The analysis is therefore less satisfactory than for the minor satellite but the conclusions remain the same (68). What is the significance of this germline undermethylation? Perhaps it reflects the need for relaxation of the centromeric heterochromatin to allow pairing of chromosome homologues during meiosis. Whatever the significance, these findings demonstrate that even though the satellite is transcriptionally silent throughout development its structure can vary.

9.2 Dispersed repetitive DNA families

The first mammalian dispersed repetitive DNA sequences to be isolated and sequenced were members of the human 'Alu' family, so called because the majority of these repeats contain the recognition sequence for the restriction endonuclease *Alu*I (33). It came as a surprise at the time to learn that there are more than 500 000 copies of this 300 bp element which in total comprise 5–6 per cent of the human genome. Sequencing studies of genomic copies of the Alu sequences revealed that they were flanked on either side by a direct repeat of a 10–20 bp stretch (18, 35). This feature was reminiscent of the situation found for a variety of transposable elements in bacteria, yeast, and Drosophila. Insertion of such elements is invariably associated with a duplication of a short stretch of the target site. This observation has led to a proposed mechanism for insertion. Since the discovery of the 'Alu' sequence several other highly repetitive DNA sequences have been identified in primates, rodents, and other mammals. These squences all appeared to fit into two categories in terms of size, those of the order of several hundred bp in length and those several thousand bp long. Singer coined the term SINE (short interspersed element) for the former and LINE (long interspersed element) for the latter (74). All the major SINEs appear to be homologous to well-characterized small cellular RNA species from which they were thought to derive. This and other considerations have led to the hypothesis that the SINEs arise from DNA intermediates which are formed by reverse transcription of the homologous RNA species (2, 34, 70, 87). The general term retroposon has been introduced to classify such sequences (63). Another characteristic of these SINEs is an A-rich region of variable length in the part corresponding to the 3' terminus of the cognate RNA. This is thought to reflect the ability of the RNA to fold over on itself and act as a primer for reverse transcription. The LINEs also have this feature and other characteristics of retroposons except that a template RNA of known function has yet to be described.

Perhaps the most compelling evidence for the functionless nature of these elements is that the major SINEs in the human, mouse, and rabbit are markedly different and are located at different positions relative to orthologous genes in the three species (13). In spite of this there is every reason to suspect that these repeated sequences have played a major role in the restructuring of the genome during evolution and in the generation of polymorphisms and rearrangements within genes. Both these areas are worthy of further study as is the mechanism of transposition of these sequences. The possibility also exists that a small minority of these elements have taken on functional roles. In addition some of these elements may provide mouse-specific markers to help in the isolation of genes following transfection of mouse DNA into cells of other species.

The following discussion will concern mainly the two major SINEs and the predominant LINE in the mouse genome. Other less well characterized sequences will receive only a superficial coverage.

9.2.1 The B1 SINE

The most highly repeated dispersed element in the mouse genome is the B1 sequence, first isolated and characterized by Krayev *et al.* (43). This sequence is of the order of 130–150 bp in length and is repeated 130 000–180 000 times, constituting altogether 0.7–1 per cent of the mouse genome (Table 9.1) (3). The B1 sequence bears strong homology to the human 'Alu' sequence but the two elements are structurally dissimilar (37, 43, 86). Both of these sequences in turn are homologous to regions of the 7SL RNA from which they are likely to have derived (1, 85, 86). The 7SL RNA is a highly conserved RNA species found associated with proteins to form the signal recognition particle (SRP) located in the cytoplasm (91). This particle plays a key role in processing proteins which are destined to be exported from the cell (91). The relationships between B1, Alu, and 7SL RNA are complex but worth considering because they shed light on the origins of these repetitive elements. Figure 9.2, derived from a scheme published by Ullu and Tschudi (86), summarizes these relationships clearly. The Alu sequence is a dimeric structure, the two halves being unequal and both dissimilar to the B1 which is a monomeric unit. The 7SL RNA molecule is approximately 300 nucleotides long but only the 5' 100 nucleotides and 3' 45 nucleotides are represented in the SINES, the central portion being 7SL-specific. The right-hand Alu monomer contains an extra 30 bp portion of the 7SL RNA relative to the left-hand monomer. The B1 sequence on the other hand is more like the left-hand monomer but with an internal duplication of a 30 bp stretch. How did these SINES derived from the 7SL RNA? When the SRP is subjected to limited nuclease digestion the 7SL RNA is cleaved into three fragments corresponding to the left- and right-hand portions which contribute to the SINES and the central 7SL-specific portion (86). It is reasonable to suppose that following such an event *in vivo* the right- and left-hand

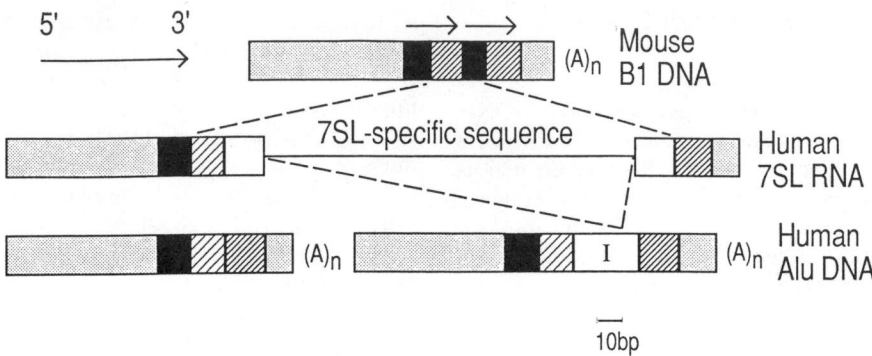

Fig. 9.2 The structural relationship of human 7SL RNA to the consensus sequences of human *Alu* and mouse B1 DNA. Adapted from Ullu and Tschudi (86). Homologous sequences are indicated by identical shading. Human Alu DNA is a head to tail dimer of two similar sequences ~130 bp long. The right monomer contains an insert (I) which is not present in the left half. The mouse B1 consensus sequence contains an internal tandem duplication of 30 bp between positions 62 and 121 as indicated by arrows. $(A)_n$ denotes an A-rich sequence which follows the Alu and B1 sequences at the 3′ end.

portions could ligate and then be reverse transcribed to produce the ancestral SINE progenitor. The B1 sequence could be generated by a deletion and duplication event within this progenitor. The Alu sequence could be formed by fusion of a deleted derivative of the progenitor to a non-deleted derivative.

The 7SL RNA gene is transcribed by RNA polymerase III and has the canonical internal bipartite promoters. Both B1 (A and B boxes Fig. 9.3) and the left Alu monomer have *pol*III promoters which will function *in vitro* (23, 95); hence it has been proposed that transcription of these SINES can lead to further rounds of transposition. However, it has not been possible to demonstrate *pol*III transcription of Alu sequences *in vivo* (58). Both sequences can however be transcribed as part of *pol*III transcripts of genes (3) and may be found in the 3′ untranslated region of mRNAs (47).

9.2.2 The mouse has a dimeric repeat remarkably similar to the human Alu

It has been proposed that Alu and B1 sequences amplified in human and mouse genomes respectively since these species diverged from a common ancestor (19). In support of this notion no B1-like sequence has been reported in human and until recently no dimeric Alu-like sequence had been detected in the mouse. However, in this laboratory we have isolated a mouse repeated sequence which is dimeric, nearly 300 nucleotides long, and shows greater than 90 per cent homology with Alu (Meehan, Hill, and Hastie, unpublished observations). There are approximately 10 000 copies of this sequence in the mouse genome. This degree of

conservation is surprisingly high for a functionless pseudogene formed prior to human–mouse divergence. It seems likely that if this element formed prior to mammalian radiation the sequence has been subject to constraint because it has evolved as the ideal retroposon. The second and less likely possibility is that the sequence is conserved because it does provide a selective advantage to the organism. Alternatively, this complex structure has formed and amplified independently within the two lineages again, perhaps, because it is selected as a tailor-made transposon.

9.2.3 The B2 SINE may have derived from a tRNA

The B2 sequence is approximately 190 nucleotides in length and repeated 80 000–100 000 times in the mouse genome (3, 44). B2 homologues in rat and hamster are given the general classification, 'type 2 Alu' sequences (28). It has not been possible to isolate a human counterpart to this family, again suggesting that these SINES amplified subsequent to mammalian radiation. Like B1 and Alu, B2 sequences have the A and B boxes of the RNA polymerase III promoter (44). Rogers first drew attention to the fact that parts of SINES bear homology to tRNAs (63). This has been confirmed by three independent groups (15, 46, 67). Figure 9.4(a) compares the sequence of a rat lysine tRNA with a stretch of 60 nucleotides close to the 5′ end of the B2 sequence (67). As shown in Fig. 9.4(b), it is possible, in principle, for this part of the B2 sequence to adopt a clover-leaf structure. Major SINES in rabbit, rat, and goat bear even more convincing homologies than B2 to

```
10                                               50
GAGTCTATTA GTTACCTTTT AAAAAATACA GAAATGGCCG GGCGAGGTGG CACACGCCTT TAATCCCAGC ACTCGGGAAG CAAAGGCAGG
   100                                                150
TGGATTTCTG AATTCGAGGC CAGCCTGGTC TACAAAGTGA GCTCCAGGAC AGCCAGGGCT ACACAGAGAA ACCCTGTCTC GAAAAACCAA
                 200
AAAAAAAATA AAACAAAATA AAATAAAAAA TACAGAAATG AAAACAACTT
```

Fig. 9.3 Sequence of a complete copy of a B1 element and flanking sequences adapted from Kalb *et al.* (37). The shaded regions is the B1 sequence. The flanking direct repeats are boxed. The A and B boxes of the *pol*III promoter are underlined.

particular tRNA species, making it very likely that these sequences are retroposon derivatives of tRNA (64). Obviously, in order for the B2 sequence to have arisen a tRNA copy must have first inserted into an unrelated stretch of DNA prior to transcription and reverse transcription.

B2 transcripts are expressed at high levels early in development and in some tumour cell lines

One prerequisite for inheritance of a retrotransposition event is that the RNA precursor must be transcribed early in development or in the germline. Small heterogeneous B2 transcripts have been detected in the cytoplasm of undifferentiated embryonal carcinoma and embryonal stem (EK) cells (3, 53). Following differentiation of these stem cells in culture the concentration of these transcripts reduced dramatically. Subsequently it was shown by *in situ* hybridization that cytoplasmic B2 transcripts are expressed at high levels in oocytes and early embryos (88). It remains to be seen whether these RNAs perform any function in development and whether they are transcribed from a small subset of the 100 000 B2 elements in the genome. The high level of expression early in development may be a reflection of the rapid cell turnover rate as B2 transcripts have been shown to be cell-cycle regulated and serum stimulated (21). In addition these small RNAs have been detected at high levels in various transformed cell lines where they have indeed been shown to be *pol*III transcripts (40, 41, 53, 76). Sequencing of DNA copies of these small RNAs from tumour cells has revealed that they derive only from the B2 sequence itself and that the heterogeneity is due to a variable length A-rich stretch at the 3′ end (42). Each of the transcripts sequenced was shown to derive from a different B2 element in the genome.

B2 insertion can be associated with DNA polymorphisms and altered gene expression

If the SINES have expanded in the genome since human–rodent divergence we may expect to detect recent insertions which might be associated with restriction fragment length polymorphisms or alteration to the structure or expression of genes. Several such cases have been documented. A B2 element has been detected in the spacer region of ribosomal RNA genes of mice of the BALB/c strain but not other strains examined (39). The B2 SINE is unusual in that it contains a sequence identical to the poly A addition site found at the 3′ end of protein-coding genes which are

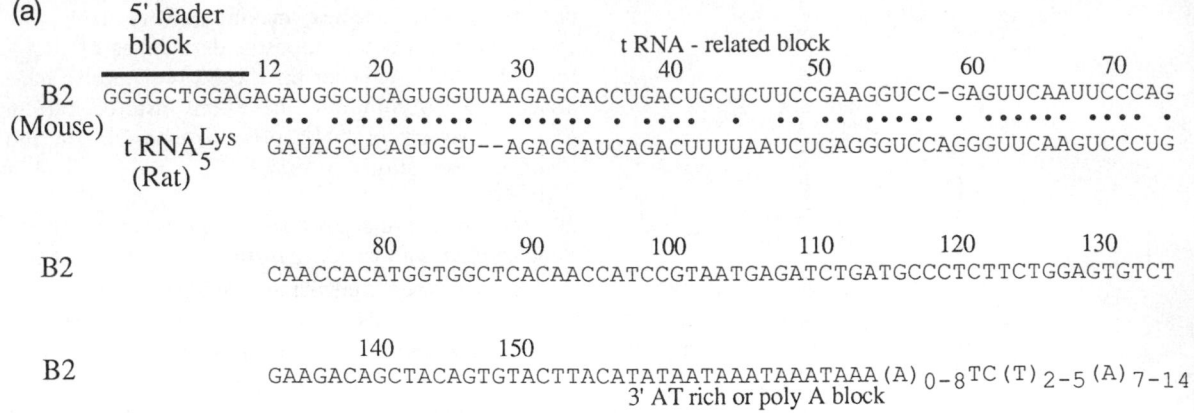

(a)

5' leader block ————— 12

t RNA - related block

B2 (Mouse)

tRNA$_5^{Lys}$ (Rat)

B2 80 90 100 110 120 130

B2 140 150

3' AT rich or poly A block

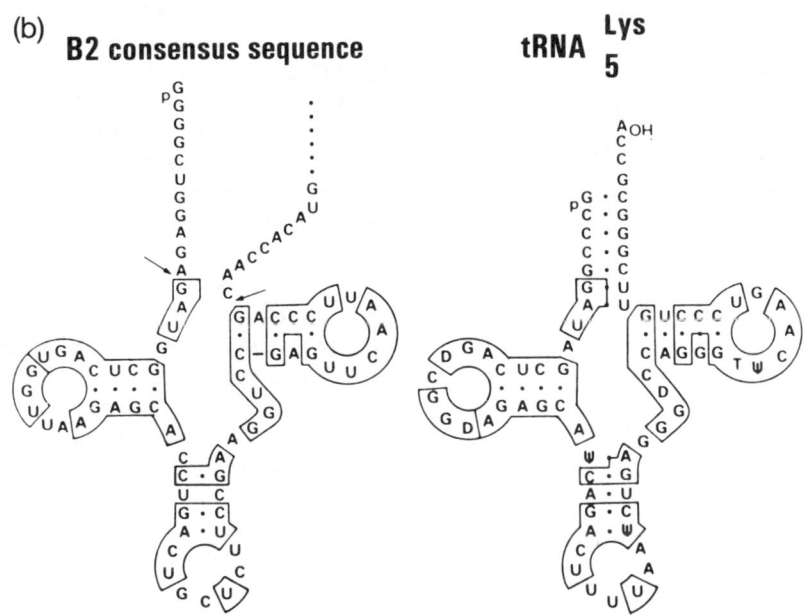

(b) **B2 consensus sequence**   **tRNA $_5^{Lys}$**

Fig. 9.4 (a) The structure of the mouse B2 repeat and its relationship with a rat lysine tRNA. Adapted from Sakamoto and Okada (67). (b) Sequence and structural homologies between the 5' part of the B2 consensus sequence and lysine tRNA$_5$. Adapted from Sakamoto and Okada (67). Identical sequences are boxed. Arrows indicate putative recombination sites between the tRNA-unrelated block.

transcribed by RNA polymerase II. If a B2 element is inserted into the region of a gene corresponding to the 3' untranslated region of mRNA the new B2-specific poly A addition site might be used instead of the gene's own poly A addition site. This is precisely what has happened in several genes within the MHC cluster (45). A B2 element is found in the 3' untranslated region of

H2L^d and *H2D^d* mRNAs but not the *H2K^d* transcript. Further studies have shown that the B2 poly A addition site is used at the expense of the natural site leading to altered *L^d* and *D^d* mRNAs which as a result could be less stable than the *K^d* transcript. These examples demonstrate the potential of retroposon insertion to affect the structure and expression of genes.

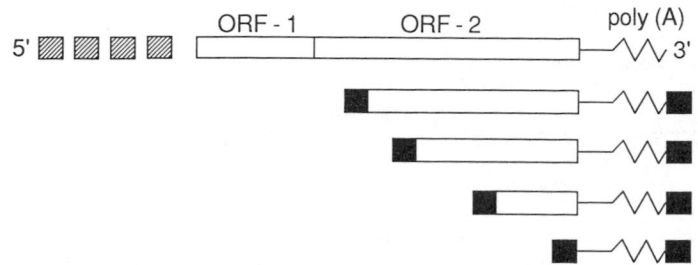

Fig. 9.5 Diagram of the arrangement of a full-length LINE-1 sequence (top line) and of 5'-truncated dispersed repeats (lower lines) flanked by short direct repeats (black squares). ORF, open reading frame. Hatched boxes, promoter-like motifs. Adapted from Rogers (65).

9.2.4 The major mouse LINE or L1md—a complex family of variable length elements

When mouse DNA is digested with certain restriction endonucleases such as *Eco*RI or *Msp*I, then electrophoresed on an agarose gel and visualized by ethidium bromide staining, prominent bands are revealed over a background smear. These bands are internal fragments of the major mouse LINE. The terminology used to describe this family of sequences has been rather confusing. Originally it was called the 1.3RI family to reflect the size in kbp of the internal fragment produced by the enzyme *Eco*RI in a large number of members (30). Subsequently this element was termed MIF-1 depicting the Major Insterspersed Fragment 1 (10, 52). It now seems sensible to adopt the terminology of Voliva *et al.* who advocated that this sequence be called L1md for LINE 1 of *Mus domesticus* (90). Full-length members of this family are on the order of 6–7 kbp (Table 9.1). However, family members are highly heterogeneous in size and only a fraction are full length (Fig. 9.5). The majority of members are variably truncated and missing the 5' end (4, 22, 25, 90). Fragments of the L1 repeat of the order of 500 nucleotides in length and representing the extreme 3' end are present in 100 000 copies in the genome (Table 9.1). Such sequences were originally thought to represent a distinct family of SINES called the R repeat before their relationship with the LINE was demonstrated (24, 49). The majority of family members start at the 3' end; there is almost an inverse relationship between the length of the members and their copy number and full-length copies are likely to be represented less than 10 000 times in the genome.

These sequences possess several features of retroposons. They are flanked by direct repeats and at the R end there is a long A or A-rich stretch indicating that this corresponds to the 3' end of a transcript. It seems reasonable to suppose that the variably truncated sequences are formed from incomplete reverse transcripts of a full-length RNA transcript.

The sequence of full-length L1 repeats has provided key insights into their likely origins (Fig. 9.5) (48, 65). There appear to be two open reading frames (ORFS) which overlap by 14 bp. Of particular interest is the finding that the long ORF bears regions of strong homology with reverse transcriptases of retroviruses and other retroposons (48). Hence it seems that the success of this retroposon lies in its ability to encode a self-copying enzyme.

Another interesting feature of the mouse L1 is a repeating motif of a 208 bp GC-rich stretch at the 5' end (48). This repeat element includes stretches similar to those found in the promoters of housekeeping genes. This feature suggests a mechanism to allow further rounds of transposition from a retroposon as downstream motifs could be included in the transcript and inserted as DNA copies into the genome where they could act as promoters.

All mammals investigated have LINES with homology to L1md (75). The well-characterized human L1 repeat is also variably truncated at the 5' end (75). It is gratifying that the most conserved regions are the ORFs (11, 50). The 5' repeating motif has not yet been detected in close to full length human LINES which may lack this feature altogether.

There are two reasons to suppose that the vast majority of these elements, constituting up to 5 per cent of the genome, are functionless. Firstly, as stated above, most members are truncated; secondly, stop codons may be found in the middle of the ORFs of a large number of members sequenced.

For the retroposon model to be correct it should be possible to detect full length RNA transcripts of these LINES. These have not been identified in mouse but have in humans and, more interestingly, in undifferen-

Table 9.1 The most highly repeated sequences in the genome of *Mus musculus*

Repeated sequence	Size of repeating unit in base pairs	Approximate copy number	Per cent of genome	Comments
Satellites				
Major	234	700 000	5.5	Tandemly arrayed at all centromeres except Y. Useful cell marker in interspecific chimaeras.
Minor	120	50 000–100 000	0.2–0.4	Tandemly arrayed at most centromeres except Y. Pattern variant between inbred strains. Undermethylated in germ cells.
Dispersed				
B1	130	130 000–180 000	0.6–0.8	Probably derived from 7SL RNA. Flanked by direct repeats; A-rich 3′ end; *pol*III promoter.
Alu	260–280	10 000	0.1	Dimeric derivative of 7SL RNA; high degree of homology with human Alu repeat.
B2	190	80 000–120 000	0.5–0.8	Probably derived from tRNA; flanked by direct repeats; *pol*III promoter; A-rich 3′ end; small cytoplasmic transcripts in embryos and tumour cells.
ID	80	10 000	0.025	Rat has 200 000 copies probably derived from tRNA; small brain-specific cytoplasmic *pol*III transcripts.
MT	>370	?	?	–
L1md	varies, up to 6800	Varies inversely with length 5000–100 000	4	Two ORFs in full length members; one ORF homology with reverse transcriptase; variably truncated at 5′ end. 3′ end equivalent to R sequence.
R	400–500	70 000–100 000	1.5	3′ end of L1md; flanked by direct repeats; presumed 3′ end A-rich.
EC	Variable up to several hundred	100 000	1	Simple stretches of alternating CT, GT, etc.

tiated embryonal carcinoma cell lines (79). These might be full-length transcripts which may encode functional proteins and act as templates for reverse transcription. Perhaps this putative reverse transcriptase acts in trans to copy the RNAs which are precursors of SINES.

The L1 sequence does not appear to be distributed in the mouse genome randomly with respect to the SINES

When a mouse genomic lambda phage library is screened for the presence of B1 and B2 sequences, 80 and 50 per cent of the plaques are positive, respectively (4). Likewise approximately 15–20 per cent of plaques are positive for the central portion of the L1 repeat. If these sequences are randomly located relative to each other we would expect approximately 80 and 50 per cent of L1-containing clones to include also B1 and B2 sequences respectively. However, in actuality, only 30 per cent of L1-containing clones contain B1 elements and 20 per cent contain B2 sequences (4). This discrepancy cannot just be explained by the L1 sequences displacing the SINES due to, for example, more recent insertion into the genome as the LINES are on average only 3–4 kbp within a genomic clone of 15 kbp on average. There are other data which also suggest that these

repeated sequences may be partially compartmentalized. When very high molecular weight DNA is subjected to density gradient fractionation and the various fractions then electro-phoresed and Southern blotted with the different repeated sequences as probes, the L1 sequence is revealed to be in a lower density component than the B1 sequence (80). The significance of these observations is unclear and may only become clarified when we have information about the long-term organization of the genome over several thousand kbp stretches of DNA.

9.2.5 Species-specific variation in dispersed repeated DNA sequences can be exploited for the isolation of mouse genes following transfection

The first human oncogene to be cloned, the H-*ras* gene, was isolated with the aid of Alu repeated sequences which acted as tags for any human DNA sequences introduced into mouse cells by transfection (55, 71). In these experiments DNA from a human bladder carcinoma cell line was shown to transform mouse cells neoplastically in culture at a frequency of approximately 1 in 10^5 cells following transfection using the calcium phosphate precipitation method (72). Greater than 95 per cent of human DNA fragments of the size used in the transfection experiments were likely to include at least one of the Alu sequences which are distributed on average at 5 kb intervals in the human genome. It is possible therefore to use the Alu sequence as a probe to identify human DNA fragments amongst a mouse DNA background when conditions of high stringency are used for the hybridization. Primary transformants took up hundreds of different human DNA fragments which produced a smear when the DNA was subjected to Southern hybridization using the Alu sequence as a probe (55, 71, 72). However, secondary transformants obtained by transfection with DNA from primary transformants contained only one or a few human fragments. Independent secondary transformants had Alu hybridizing fragments in common as well as different fragments. The common Alu-containing fragments are those which include the H-*ras* gene. A genomic DNA library from these secondary transformants was constructed in λ phage and the H-*ras* oncogene was isolated from this library using Alu as a probe in colony hybridization (72).

It would be desirable to be able to clone mouse genes which can confer a phenotype of interest following transfection using a similar approach. The situation in the mouse is not as convenient as no repeated sequence approaches the repetition frequency of the Alu sequence in the human genome (Table 9.1). The choice of repeated sequence to be used for gene tagging obviously depends on the complement of repetitive sequences in the species from which the recipient cells are derived. The most highly repeated mouse sequence, the B1 SINE, hybridizes strongly to human DNA and is highly conserved and abundant in rat and hamster genomes (4). Hence the B1 sequence is not suitable for the purpose.

There is no apparent human counterpart to the B2 SINE and the human LINE-1 does not hybridize to the R sequence. So a combination of these two probes could be used to isolate mouse fragments following transfection into human cells. Approximately 55 per cent of clones in a mouse genomic λ library contain a B2 sequence and 50 per cent an R sequence (3, 4). Thus about 75 per cent of clones will hybridize to either one of these probes. On the other hand, if the library from the secondary transfectant were to be constructed in a cosmid where the inserts are more than twice the size of those in lambdaphage, 90 per cent of mouse fragments should be identifiable. Of course, if a gene were very large, i.e. greater than 50 kb, the change of identifying at least part of this gene with one of these repeated sequences should be high. The B2 sequence is highly represented in both rat and hamster genomes but the mouse R sequence hybridizes very poorly to DNA of these species at high stringency (Meehan and Hastie, unpublished observations). Thus, following transfection of mouse DNA into hamster cells, it would be desirable to construct a cosmid library and probe this with the R sequence.

In addition repeated sequences can be used to isolate anonymous DNA fragments corresponding to specific mouse chromosomes or regions of specific chromosomes (26). For example, a genomic library could be constructed from a hamster–mouse cell hybrid containing only the X chromosome as its mouse complement. The R sequence could then be used as a probe to isolate mouse-specific clones which must have originated from the X chromosome. The same approach could be used to isolate clones from chromosome fragments introduced into cells of another species by chromosome-mediated gene transfer (54, 60, 61, 93).

9.3 Some other repeated sequences in the mouse genome

There are several reports in the literature of new families of highly repeated sequences which do not belong to the B1, B2, or L1 families. However, in some cases these may be subregions of the very large L1 repeat. For example, two families, LL rep 1 and LL rep 2, were described (31). Unfortunately, no sequence information was included but both these elements were found to be situated close to the R element more often than expected by chance. In hindsight these are likely to be subregions of the L1 repeat and not additional families as claimed. However, two additional mouse repeated sequence families have been reported, the MT repeat and the mouse equivalent of the rat ID repeated sequence.

9.3.1 The MT repeat

This sequence was first detected in several cDNA clones in a library constructed from poly A+ RNA of mouse cerebellum (29). This is the most highly represented repeated sequence in cerebellar mRNA. The MT repeat has also been located upstream of a cytochrome P-450 oxidoreductase gene and close to the site of integration of a polyomavirus genome. It has not been possible to define the ends of the sequence which is at least 370 bp long. Thus it is not yet known whether the typical features of a retroposon, namely flanking repeats and an A-rich stretch, are associated with the amplification mechanism. The repetition frequency of this sequence in the genome is unclear because the filter assay used to test this property gave a 50-fold difference in abundance between the B1 and R sequences which in actuality differ in copy number by only a factor of two (3). The MT repeat is likely to be in the range of 10 000–100 000 copies in the genome. The rat genome contains a repeated sequence with 90 per cent homology to the MT element (29).

9.3.2 The ID sequence

The ID or 'identifier' sequence was first recovered as an 80 bp repeat found in several rat brain-specific mRNAs (82). Of additional interest was the fact that this sequence hybridized to a small brain-specific *pol*III-transcribed cytoplasmic polyadenylated RNA species 160 nucleotides long. It was postulated that this sequence might play a role in the control of transcription of brain-specific genes (83). In this model the ID sequences within genes were first activated transcriptionally by RNA polymerase III. This rendered the surrounding chromatin accessible to RNA polymerase II so that the genes themselves would then be transcribed. This model has now fallen into disrepute for several reasons. Firstly there are of the order of 200 000 copies of the ID sequence which appear to be randomly distributed in the rat genome so that most genomic fragments 15 kb in length would contain one of these elements. Thus there are many more than would be expected to reside in brain-specific genes alone. This is borne out by the finding that several non-brain-specific mRNAs contain ID sequences. Secondly, it has been shown that the level of newly synthesized *pol*II transcripts containing ID sequence is no higher in the brain than other tissues (69). Finally, it was not possible to demonstrate ID sequences in the mouse at first but it is now known that there are approximately only 10 000 copies of this sequence in the mouse genome (69). Hence we must accept that greater than 95 per cent of the rat ID sequences perform no function and are likely to have appeared in the rat genome subsequent to rodent radiation. No ID sequence counterpart has been detected in the human genome. Recently it has been shown that the ID sequence shows strong homology to a phenylalanyl tRNA from which it is thought to have derived as a retroposon (46, 67). In accord with this the genomic ID sequences are flanked by direct repeats and have an A-rich stretch at the end corresponding to the 3′ terminus of the RNA.

Of interest is the fact that the ID sequence of the mouse is also copied by RNA polymerase III specifically in the brain to produce a 160 nucleotide cytoplas-mic transcript as in the rat (69); this transcript may yet prove to play some functional role.

9.3.3 Evolutionarily conserved sequences

A sequence was described which is highly repeated in the mouse genome and a variety of vertebrates and invertebrates (51). This was termed the evolutionarily conserved or EC sequence. Subsequently several such EC sequences were isolated from a small library of mouse repeated sequences (3). In all cases sequencing showed that these were comprised of stretches of poly-pyrimidines several hundred nucleotides in length. It is not surprising that such sequences are evolutionarily conserved due to their simplicity. It is likely that the conserved region of the original EC sequence also comprised CpT or GpT stretches (51). Similarly alternating GpC or ApT stretches have been detected at high levels in the genomes of many eukaryotes (27). These are likely to have arisen by a local replication 'slippage' mechanism or by unequal crossing-over rather than by a transposition event. It is perhaps incorrect to think of these as dispersed repeated DNA sequences. The functional significance of these simple repeated sequences remains obscure.

Acknowledgements

I am indebted to my colleagues, Richard Meehan, Robert Hill, and Linda Siracusa for valuable discussions, to Katie Rae for excellent secretarial help, and to Norman Davidson and Sandy Bruce for photographic work of the highest quality.

References

1. Balmain, A., R. Krumlauf, J.K. Vass, and G.D. Birnie. 1982. Cloning and characterization of the abundant cytoplasmic 7S RNA from mouse cells. Nucl. Acids Res. 10:4259–4227.

2. Baltimore, D. 1985. Retroviruses and retroposons: The role of reverse transcription in shaping the eukaryotic genome. Cell 40:481–482.

3. Bennett, K.L., R.E. Hill, D.F. Pietras, M. Woodworth-Gutai, C. Kane-Haas, J.M. Houston, J.K. Heath, and N.D. Hastie. 1984. Most highly repeated dispersed DNA families in the mouse genome. Mol. Cell. Biol. 4:1561–1571.

4. Bennett, K. L., and N.D. Hastie. 1984. Looking for relationships between the most repeated dispersed DNA sequences in the mouse; small R elements are found associated consistently with long MIF repeats. EMBO J. 3:467–472.

5. Bird, A.P. 1986. CpG-rich islands and the function of DNA methylation. Nature 321:209–213.

6. Blackburn, E.H. 1985. Artificial chromosomes in yeast. Trends Genet. 1:8–12.

7. Britten, R.J., and D.E. Kohne. 1968. Repeated sequences in DNA. Science 161:529–540.

8. Brown, S.D.M., and G.A. Dover. 1980. Conservation of segmental variants of satellite DNA of *Mus musculus* in a related species: *Mus spretus*. Nature 285:47–49.

9. Brown, S.D.M., and G.A. Dover. 1980. The specific organization of satellite DNA sequences on the X-chromosome of *Mus musculus*: partial independence of chromosome evolution. Nucl. Acids Res. 8:781–792.

10. Brown, S.D.M., and G.A. Dover. 1981. Organization and evolutionary progress of a dispersed repetitive family of sequences in widely separated rodent genomes. J. Mol. Biol. 150:441–466.

11. Burton, F.H., D.D. Loeb, C.F. Voliva, S.L. Martin, M.H. Edgell, and C.A. Hutchison, III. 1986. Conservation throughout mammalia and extensive protein-encoding capacity of the highly repeated DNA long interspersed sequence one. J. Mol. Biol. 187:291–304.

12. Cech, T.R., A. Rosenfeld, and J.E. Hearst. 1973. Characterization of the rapidly renaturing sequences in mouse main-band DNA. J. Mol. Biol. 81:299–325.

13. Cheng, J.F., R. Printz, T. Callaghan, D. Shuey, and R.C. Hardison. 1984. The rabbit C family of short, interspersed repeats. J. Mol. Biol. 176:1–20.

14. Corces, V.G., J. Salas, M.L. Salas, and J. Avila. 1978. Binding of microtubule proteins to DNA: specificity of the interaction. Eur. J. Biochem. 86:473–479.

15. Daniels, G.R., and P.L. Deininger. 1985. Repeat sequence families derived from mammalian tRNA genes. Nature 317:819–822.

16. Davidson, E., B. Hough, C. Amenson, and R. Britten. 1973. General interspersion of repetitive with non-repetitive sequence elements in the DNA of Xenopus. J. Mol. Biol. 77:1–23.

17. Davidson, E., B.R. Hough, C.S. Amenson, R.J. Britten, D.E. Graham, B.R. Neufeld, M.J. Smith, R.B. Goldberg, and G.A. Galau. 1973. Sequence organization in animal DNAs. *In* B.A. Hamkalo, and J. Papaconstantinou, eds., Molecular Cytogenetics, 9–30. Plenum Press, New York.

18. Deininger, P.L., D.J. Jolly, C.M. Rubin, T. Friedmann, and C.W. Schmid. 1981. Base sequence studies of 300 nucleotide renatured repeated human DNA clones. J. Mol. Biol. 151:17–33.

19. Deininger, P.L., and G.R. Daniels. 1986. The recent evolution of mammalian repetitive DNA elements. Trends Genet. 2:76–81.

20. Doolittle, W.F., and C. Sapienza. 1980. Selfish genes, the phenotype paradigm and genome evolution. Nature 284:601–603.

21. Edwards, D.R., C.L.J. Parfett, and D.T. Denhardt. 1985. Transcriptional regulation of two serum-induced RNAs in mouse fibroblasts: Equivalence of one species to B2 repetitive elements. Mol. Cell. Biol. 5:3280–3288.

22. Fanning, T.G. 1983. Size and structure of the highly repetitive BAM HI element in mice. Nucl. Acids Res. 11:5073–5091.

23. Fuhrman, S.A., P.L. Deininger, P. La Porte, T. Friedmann, and E.P. Geiduschek. 1981. Analysis of transcription of the human Alu family ubiquitous repeating element by eukaryotic RNA polymerase III. Nucl. Acids Res. 9:6439–6456.

24. Gebhard W., T. Meitinger, T. Hochtl, and H.G. Zachau. 1982. A new family of interspersed repetitive DNA sequences in the mouse genome. J. Mol. Biol. 157:453–471.

25. Gebhard, W., and H.G. Zachau. 1983. Organization of the R family and other interspersed repetitive DNA sequences in the mouse genome. J. Mol. Biol. 170:255–270.

26. Gusella, J.F., C. Keys, A. Varsanyi-Breiner, F.-T. Kao, C. Jones, T.T. Puck, and D. Housman. 1980. Isolation and localisation of DNA segments from specific human chromosomes. Proc. Natl. Acad. Sci. USA 77:2829–2833.

27. Hamada, H., M.G. Petrino, and T. Kakunaga. 1982. A novel repeated element with Z-DNA-forming potential is widely found in evolutionarily diverse eukaryotic genomes. Proc. Natl. Acad. Sci. USA 79:6465–6469.

28. Haynes, S.R., and W.R. Jelinek. 1981. Low molecular weight RNAs transcribed *in vitro* by RNA polymerase III from Alu-type dispersed repeats in Chinese hamster DNA are also found *in vivo*. Proc. Natl. Acad. Sci. USA 78:6130–6134.

29. Heinlein, U.A.O., R. Lange-Sablitzky, H. Schaal, and W. Wille. 1986. Molecular characterization of the MT-family of dispersed middle-repetitive DNA in rodent genomes. Nucl. Acids Res. 14:6403–6416.

30. Heller, R., and N. Arnheim. 1980. Structure and organization of the highly repeated and interspersed 1.3kb EcoR1-Bgl II sequence family in mice. Nucl. Acids Res. 8:5031–5042.

31. Heller, D., M. Jackson and L. Leinwand. 1984. Organization and expression of non-Alu family interspersed repetitive DNA sequences in the mouse genome. J. Mol. Biol. 173:419–436.

32. Hörz, W., and W. Altenburger. 1981. Nucleotide sequence of mouse satellite DNA. Nucl. Acids Res. 9:683–696.

33. Houck, C.M., F.P. Rinehard, and C.W. Schmid. 1979. A ubiquitous family of repeated DNA sequences in the human genome. J. Mol. Biol. 132:289–306.

34. Jagadeeswaran, P., B. Forget, and S.M. Weissman. 1981. Short interspersed repetitive DNA elements in eucaryotes: transposable DNA elements generated by reverse transcription of RNA pol III transcripts. Cell 26:141–142.

35. Jelinek, W.R., T.P. Toomey, L. Leinwand, C.H. Duncan, P.A. Biro, P.V. Choudary, S.M. Weissman, C.M. Rubin, C.M. Houck, P.L. Deininger, and C.W. Schmid. 1980. Ubiquitous interspersed repeated sequences in mammalian genomes. Proc. Natl. Acad. Sci. USA 77:1398–1402.

36. Jones, K.W. 1970. Chromosomal and nuclear location of

mouse satellite DNA in individual cells. Nature 225:912–915.

37. Kalb, V.F., S. Glasser, D. King, and J.B. Lingrel. 1983. A cluster of repetitive elements within a 700 base pair region in the mouse genome. Nucl. Acids Res. 11:2177–2184.

38. Kit, S. 1961. Equilibrium sedimentation in density gradients of DNA preparations from animal tissues. J. Mol. Biol. 3:711–716.

39. Kominami, R., M. Muramatsu, and K. Moriwaki. 1983. A mouse type 2 Alu sequence (B2) is mobile in the genome. Nature 301:87–89.

40. Kramerov, D.A., I.V. Lekakh, O.P. Samarina, and A.P. Ryskov. 1982. The sequences homologous to major interspersed repeats B1 and B2 of mouse gneome are present in mRNA and small cytoplasmic poly(A) tRNA. Nucl. Acids Res. 10:7477–7491.

41. Kramerov, D.A., S.V. Tillib, I.V. Lekakh, A.P. Ryskov, and G.P. Georgiev. 1985. Biosynthesis and cytoplasmic distribution of small poly(A)-containing B2 RNA. Bioehcm. Biophys. Acta 824:85–98.

42. Kramerov, D.A., S.V. Tillib, A.P. Ryskov, and G.P. Georgiev. 1985. Nucleotide sequence of small polyadenylated B2 RNA. Nucl. Acids Res. 13:6423–6437.

43. Krayev, A.S., D.A. Kramerov, K.G. Skryabin, A.P. Ryskov, A.A. Bayev, and G.P. Georgiev. 1980. The nucleotide sequence of the ubiquitous repetitive DNA sequence B1 complementary to the most abundant class of mouse fold-back RNA. Nucl. Acids Res. 8:1201–1215.

44. Krayev, A.S., T.V. Markusheva, D.A. Kramerov, A.P. Rysko, K.G. Skryabin, A.A. Bayev, and G.P. Georgiev. 1982. Ubiquitous transposon-like repeats B1 and B2 of the mouse genome: B2 sequencing. Nucl. Acids Res. 10:7461–7475.

45. Kress, M., Y. Barra, J.G. Seidman, G. Khoury, and G. Jay. 1984. Functional insertion of an Alu type 2 (B2 SINE) repetitive sequence in murine class I genes. Science 226:974–977.

46. Lawrence, C.B., D.P. McDonnell, and W.J. Ramsey. 1985. Analysis of repetitive sequence elements containing tRNA-like sequences. Nucl. Acids Res. 13:4239–4252.

47. Lehrman, M.A., W.J. Schnieder, T.C. Südhof, M.S. Brown, J.L. Goldstein, and D.W. Russell. 1985. Mutation in LDL receptor: Alu–Alu recombination deletes exons encoding transmembrane and cytoplasmic domains. Science 227:140–146.

48. Loeb, D.D., R.W. Padgett, S.C. Hardies, W.R. Shehee, M.B. Comer, M.H. Edgell, and C.A. Hutchison, III. 1986. The sequence of a large L1Md element reveals a tandemly repeated 5′ end and several features found in retrotransposons. Mol. Cell. Biol. 6:168–182.

49. Lueders, K.K., and B.M. Paterson. 1982. A short interspersed repetitive element found near some mouse structural genes. Nucl. Acids Res. 10:7715–7729.

50. Martin, S.L., C.F. Voliva, F.H. Burton, M.H. Edgell, and C.A. Hutchison. 1984. A large interspersed repeat found in mouse DNA contains a long open reading frame

that evolves as if it encodes a protein. Proc. Natl. Acad. Sci. USA 81:2308–2312.

51. Meisfeld, R., M. Krystal, and N. Arnheim. 1981. A member of a new repeated sequence family which is conserved throughout eucaryotic evolution is found between the human δ and β globin genes. Nucl. Acids Res. 9:5931–5946.

52. Meunier-Rotival, M., P. Soriano, G. Cuny, F. Strauss, and G. Bernardi. 1982. Sequence organization and genomic distribution of the major family of interspersed repeats of mouse DNA. Proc. Natl. Acad. Sci. USA 79:355–359.

53. Murphy, D., P.M. Brickell, D.S. Latchman, K. Willison, and P.W.J. Rigby. 1983. Transcripts regulated during normal embryonic development and oncogenic transformation share a repetitive element. Cell 35:865–871.

54. Murphy, P.D., and F.H. Ruddle. 1985. Isolation and regional mapping of random X sequences for distal human X chromosome. Somat. Cell Mol. Genet. 11:433–444.

55. Murray, M.J., B.-Z. Shilo, C. Shih, D. Cowing, H.W. Hsu, and R.A. Weinberg. 1981. Three different human tumor cell lines contain different oncogenes. Cell 25:355–361.

56. Orgel, L.E., and F.H.C. Crick. 1980. Selfish DNA: the ultimate parasite. Nature 284:604–607.

57. Pardue, M.L., and J.G. Gall. 1970. Chromosomal localization of mouse satellite DNA. Science 168:1356–1358.

58. Paulson, K., and C.W. Schmid. 1986. Transcriptional inactivity of Alu repeats in HeLa cells. Nucl. Acids Res. 14:6145–6158.

59. Pietras, D.F., K.L. Bennett, L.D. Siracusa, M. Woodworth-Gutai, V.M. Chapman, K.W. Gross, C. Kane-Hass, and N.D. Hastie. 1983. Construction of a small *Mus musculus* repetitive DNA library: identification of a new satellite sequence in *Mus musculus*. Nucl. Acids Res. 11:6965–6983.

60. Porteous, D.J., J.E.N. Morten, G. Cranston, J.M. Fletcher, A.R. Mitchell, V. van Heyningen, J.A. Fantes, P.A. Boyd, and N.D. Hastie. 1986. Molecular and physical arrangements of human DNA in HRAS1-selected, chromosome-mediated transfectants. Mol. Cell. Biol. 6:2223–2232.

61. Porteous, D.J. 1986. Rapid isolation and characterization of hybridization selected recombinants from lambda genomic libraries. Analyt. Biochem. 159:17–23.

62. Rice, N.R., and N.A. Straus. 1973. Relatedness of mouse satellite DNA to DNA of various *Mus* species. Proc. Natl. Acad. Sci. USA 70:3546–3550.

63. Rogers, J.H. 1985. Origin and evolution of retroposons. Int. Rev. Cytol. 93:187–279.

64. Rogers, J.H. 1985. Origins of repeated DNA. Nature 317:765–766.

65. Rogers, J.H. 1986. The origin of retroposons. Nature 319:725.

66. Rossant, J., M. Vijh, L.D. Siracusa, and V.M. Chapman. 1983. Identification of embryonic cell lineages in histologi-

cal sections of *M. musculus* ⟷ *M. caroli* chimaeras. J. Embryol. Exp. Morph. 73:179–191.

67. Sakamoto, K., and N. Okada. 1985. Rodent type 2 Alu family, rat identifier sequence, rabbit C-family, and bovine or goat 73-bp repeat may have evolved from tRNA genes. J. Mol. Evol. 22:134–140.

68. Sanford, J., L. Forrester, V. Chapman, A. Chandley, and N.D. Hastie. 1984. Methylation patterns of repetitive DNA sequences in germ cells of *Mus musculus*. Nucl. Acids Res. 12:2823–2834.

69. Sapienza, C., and B. St-Jacques. 1986. 'Brain-specific' transcription and evolution of the identifer sequence. Nature 319:418–420.

70. Sharp, P.A. 1983. Conversion of RNA to DNA in mammals: Alu-like elements and pseudogenes. Nature 301:471–472.

71. Shih, C., L.C. Padhy, M. Murray, and R.A. Weinberg. 1981. Transforming genes of carcinomas and neuroblastomas introduced into mouse fibroblasts. Nature 290:261–264.

72. Shih, C., and R.A. Weinberg. 1982. Isolation of a transforming sequence from a human bladder carcinoma cell line. Cell 29:161–169.

73. Singer, M.F. 1982. Highly repeated sequences in mammalian genomes. Int. Rev. Cytol. 76:67–112.

74. Singer, M.F. 1982. SINEs and LINEs: highly repeated short and long interspersed sequences in mammalian genomes. Cell 28:433–434.

75. Singer, M.F., and J. Skowronski. 1985. Making sense out of LINES: long interspersed repeat sequences in mammalian genomes. Trends Biochem. Sci. 10: 119–112.

76. Singh, K., M. Carey, S. Saragosti, and M. Botchan. 1985. Expression of enhanced levels of small RNA polymerase III transcripts encoded by the B2 repeats in simian virus 40-transformed mouse cells. Nature (London) 314:553–556.

77. Siracusa, L.D.A. 1985. A molecular and genetic analysis of satellite DNA in the mouse. PhD thesis, State University of New York at Buffalo.

78. Siracusa, L.D.A., V.M. Chapman, K.L. Bennett, N.D. Hastie, D.F. Pietras, and J. Rossant. 1983. Use of repetitive DNA sequences to distinguish *Mus musculus* and *Mus caroli* cells by *in situ* hybridisation. J. Embryol. Exp. Morph. 73:163–178.

79. Skowronski, J., and M.F. Singer. 1985. Expression of a cytoplasmic LINE-1 transcript is regulated in a human teratocarcinoma cell line. Proc. Natl. Acad. Sci. USA 82:6050–6054.

80. Soriano, P., M. Meunier-Rotival, and G. Bernardi. 1983. The distribution of interspersed repeats is nonuniform and conserved in the mouse and human genomes. Proc. Natl. Acad. Sci. USA 80:1816–1920.

81. Southern, E.M. 1975. Long range periodicities in mouse satellite DNA. J. Mol. Biol. 94:52–69.

82. Sutcliffe, J.G., R.J. Milner, R.E. Bloom, and R.A. Lerner. 1982. Common 82-nucleotide sequences unique to brain RNA. Proc. Natl. Acad. Sci. USA 79:4942–4946.

83. Sutcliffe, J.C., R.J. Milner, J.M. Gottesfeld, and R.A. Lerner. 1984. Identifier sequences are transcribed specifically in brain. Nature 308:237–241.

84. Sutton, W.D., and M. McCallum. 1972. Related satellite DNA's in the genus *Mus*. J. Mol. Biol. 71:633–656.

85. Ullu, E., S. Murphy, and M. Melli. 1982. Human 7SL RNA consists of a 140 nucleotide middle-repetitive sequence inserted in an Alu sequence. Cell 29:195–202.

86. Ullu, E., and C. Tschudi. 1984. Alu sequences are processed 7SL RNA genes. Nature 312:171–172.

87. Van Arsdell, S.W., R.A. Denison, L.B. Bernstein, A.M. Weiner, T. Manser, and R.F. Gesteland. 1981. Direct repeats flank three small nuclear RNA pseudogenes in the human genome. Cell 26:11–17.

88. Vasseur, M., H. Condamine, and P. Duprey. 1985. RNAs containing B2 repeated sequences are transcribed in the early stages of mouse embryogenesis. EMBO J. 4:1749–1753.

89. Villasante, A., V.G. Corces, R. Manso-Martinez, and J. Avila. 1981. Binding of microtubule protein to DNA and chromatin: possibility of simultaneous linkage of microtubule to nucleic acid and assembly of the microtubule structure. Nucl. Acids Res. 9:895–908.

90. Voliva, C.F., C.L. Jahn, M.B. Comer, C.A. Hutchison, and M.H. Edgell. 1983. The L1Md long interspersed repeat family in the mouse: almost all examples are truncated at one end. Nucl. Acids Res. 11:8847–8860.

91. Walter, P., and G. Blobel. 1982. Signal recognition particle contains 7S RNA essential for protein translocation across the endoplasmic reticulum. Nature 299:691–698.

92. Waring, M., and R.J. Britten. 1966. Nucleotide sequence repetition: a rapidly reassociating fraction of mouse DNA. Science 154:791–794.

93. Weis, J.H., J.G. Seidman, D.E. Housman, and D.L. Nelson. 1986. Eucaryotic chromosome transfer: production of a murine-specific cosmid library from a neo-linked fragment of murine chromosome 17. Mol. Cell. Biol. 6:441–451.

94. Yamamoto, M., and G.L. Miklos. 1978. Genetic sources of heterochromatin in *Drosophila melanogaster* and their implication for the function of satellite DNA. Chromosoma 66:71–98.

95. Young, P.R., R.W. Scott, D.H. Hamer, and S.M. Tilghman, S.M. 1982. Construction and expression *in vivo* of an internally deleted mouse α fetoprotein gene: presence of a transcribed Alu-like repeat within the first intervening sequence. Nucl. Acids Res. 10:3099–3116.

10 RULES FOR NOMENCLATURE OF CHROMOSOME ANOMALIES

COMMITTEE ON STANDARDIZED GENETIC NOMENCLATURE FOR MICE
Chairman: MARY F. LYON

10.1 Symbols for chromosome anomalies

(a) Initial letter symbols shall be adopted as follows:

T	Translocation
Rb	Robertsonian translocation
In	Inversion
Is	Insertion
Ts	Trisomy
Ms	Monosomy
Dp	Duplication
Df	Deficiency
Del	Deletion
Tp	Transposition

(b) Successive anomalies in a series shall be distinguished by a series symbol, consisting of a number followed by an abbreviation of the name of the person or laboratory who discovered the anomaly. This abbreviation should either be that already assigned for the designation of inbred substrains or sublines, or, where there is no pre-assigned abbreviation, it should follow the rules for such abbreviations and not duplicate an existing symbol in the standard list of abbreviations, e.g. T37H (the 37th translocation found at Harwell), In5Rk (the 5th inversion found by Roderick).

(c) The chromosome(s) involved in the anomaly shall be indicated by adding the appropriate Arabic numerals in parentheses, between the initial letter and the series symbol, e.g. In(2)5Rk (indicating that the inversion in this case involves chromosome 2).

Chromosomes shall be numbered and identified according to the system given by Nesbitt and Francke (1), and extended by Evans (Chapter 11) and by Sawyer *et al.* (2).

The X and Y chromosomes shall be indicated by capital letters rather than numbers.

(d) The two chromosomes involved in translocations and insertions should be separated by a semicolon, whereas in Robertsonian translocations the centromere

should be indicated by a point, e.g. T(4;X)37H; Rb(9.19)163H.

In the case of insertions, the number of the chromosome donating the inserted portion should be given first, e.g. Is(7;1)40H, an insertion of part of Chr 7 into Chr 1.

(e) When an animal carries two or more aberrations which are potentially separable by recombination the symbols for both (or all) anomalies should be given, e.g.

Rb(16.17)Bnr T(1;17)190Ca/++, an animal carrying a Robertsonian and a reciprocal translocation involving Chr 17, in 'coupling';

Rb(5.15)3Bnr +/+ In(5)9Rk, an animal carrying a Robertsonian and an inversion involving Chr 5 in repulsion.

(f) The symbols L and S should be used to denote the long and short arms of mouse chromosomes. In translocations, breaks in the short arm should be so designated, but the L for long arm may be omitted if the meaning is clear, e.g. T(4S;5)99H, a translocation involving a break in the short arm of chromosome 4 and the long arm of 5.

(g) When the position(s) of the chromosomal breakpoint(s) relative to G-bands are known these shall be indicated by adding the band numbers, as given in the standard karyotype of the mouse (1), after the appropriate numbers, e.g.

T(2H1;8A4)26H, a reciprocal translocation having breaks in band H1 of Chr 2 and band A4 of Chr 8. In(XA1;XE)H, an inversion involving breaks in bands A1 and E of the X chromosome;

Df(7E1)c8Rl, a deficiency of band 7E1 manifesting as a mutation to albino, *c*;

Is(7F1-7C;XF1)1Ct an inverted insertion of a segment 7F1-7C of Chr 7 into the X chromosome at band F1.

(h) For pericentric inversions the symbols LS and/or appropriate band numbers should be used, e.g.

In(YLS), pericentric inversion involving the Y chromosome;

In(8A2LS), pericentric inversion with break in band A2 of the long arm and in the short arm;

In(5C2.15E1)Rb3Bnr 1Ct, the first inversion found by Cattanach in Rb3Bnr, and involving bands 5C2 and 15E1.

(i) Trisomies and monosomies should be denoted by the appropriate initial symbol, followed by the chromosome(s) concerned in parentheses, and by the translocation number if the trisomy etc. is derived from a translocation, e.g. $Ts(1^{13})89H$, trisomy for the proximal end of 1 and the distal end of 13, derived from T(1;13)89H.

10.2 Variations in heterochromatin and chromosome banding

(a) The symbol NO should be reserved for nucleolus organizers. Different organizers should be distinguished by chromosome numbers.

(b) Centric heterochromatin. The symbol H should be used for heterochromatin, followed by a symbol indicating the chromosome region involved, in this case c for centromeric, and a number indicating the chromosome on which it lies, e.g. Hc14—centric heterochromatin on Chr 14. Variations in size, etc. of any block should be indicated by superscripts, using n for normal or standard, l for large, and s for small bands, e.g. $Hc14^n$—normal or standard centric heterochromatin on Chr 14. In describing a new variant a single inbred strain should be named as the prototype or standard strain.

The centromere itself (as opposed to centric heterochromatin) should be denoted by the symbol C or Cen.

(c) G-band polymorphisms. When a recognizable and heritable variant in size, staining density, etc. of a particular chromosomal G-band is discovered this should be indicated by giving the designation of the band affected, in accordance with the standard karyotype of the mouse, with a superscript to indicate the variant concerned, e.g. $17A2^s$—small A2 band in Chr 17.

When a supernumerary band becomes visible this may be due to a small duplication, and if so should be designated as such (see Section 10.1). If the supernumerary band is due not to a duplication but to a further resolution within a band then a new band should be designated, as a subdivision of the appropriate known band.

Thus, if the known band has a letter but no numbered subdivisions, then such numbered subdivisions should be created. If numbered subdivisions require to be further subdivided then a decimal system should be used, i.e. band 17B should be divided into 17B1, 17B2, etc.; further subdivision would lead to 17B1.1, 17B1.2, etc.

10.3 Use of human chromosome nomenclature

The system proposed here does not preclude the use of the type of nomenclature used for human chromosomes in cytogenetic papers dealing with whole arm changes or mosaics, e.g. 41,XY+13, or 39,X//41,XYY, using a double slash to separate the components of the mosaic.

References

1. Nesbitt, M.N., and U. Francke. 1973. A system of nomenclature for band patterns of mouse chromosomes. Chromosoma 41: 145–158.
2. Sawyer, J.R., M.M. Moore, and J.C. Hozier. 1987. High resolution G-banded chromosomes of the mouse. Chromosoma 95: 350–358.

11 STANDARD NORMAL CHROMOSOMES

11.1 Standard idiogram

EDWARD P. EVANS

Since the publication of the first edition, a high-resolution G-banded idiogram has been produced by Sawyer and Hozier (5). Although they claim that it reveals over 200 additional bands, its use may be limited because the resolution was achieved by the action of actinomycin D on cultured cells at the expense of producing a severe reduction in the mitotic index. The G-banded idiogram presented here (Fig. 11.1) only reveals a further 100 bands but it is based on the measurements made on conventionally banded chromosomes from 10 mitotic cells obtained directly from the livers of day 12 embryos of the C57BL/6J/Ola inbred strain. It follows the standard idiogram previously presented except for these differences:

1. Chromosome numbers and X and Y are placed at the centric ends to facilitate finding the chromosomes.

2. The Y chromosome is shown with a short arm (3).

3. The existing length anomalies between the drawings of the longer and shorter chromosomes, as compared to their proportional lengths estimated by Nesbitt and Francke (4), are corrected, e.g. Chromosome 1 in the idiogram measures 47 mm and has a proportional length of 7.20 per cent whereas Chromosome 16 measures 20 mm [$\leq \frac{1}{2}$] but has a proportional length of 3.86 per cent [$\geq \frac{1}{2}$]. The proportional lengths were re-estimated from measurements made in the 10 cells and did not differ widely (Table 11.1) from the previous estimates.

4. The A1 band sizes in certain chromosomes may largely differ as a result of centromeric heterochromatin variation (2) between the strains used in the past and here.

5. Some band positions are changed, e.g. band 10C2 in most, if not in all strains, is close to 10D1 rather than lying in the middle of 10C.

Table 11.1 Chromosome measurements expressed as percentage of total female haploid lengths; upper figures present estimates, lower figures previous estimates

Chr		Chr		Chr		Chr	
1	7.20	6	5.53	11	4.72	16	3.81
	7.20		5.62		4.63		3.86
2	6.95	7	5.19	12	4.88	17	3.86
	6.88		5.43		4.63		3.74
3	5.99	8	4.97	13	4.38	18	3.88
	6.10		5.00		4.38		3.69
4	5.89	9	4.79	14	4.46	19	2.73
	5.89		4.96		4.31		2.65
5	5.68	10	4.74	15	4.05	X	6.23
	5.80		4.83		4.17		6.18
						Y	2.86
							2.72

6. Within the limitations of illustrating band optical density by shading and hatching, some of the band densities are changed, e.g. 4A5.

7. In some of the chromosomes greater band resolution is achieved and is denoted according to the suggested extended nomenclature (1) but using decimal numerals rather than letters for better consistency with human nomenclature. Major regions already subdivided are further subdivided, for example 1C1 into 1C1.1, 1C1.2, and 1C1.3. Major regions not previously subdivided are more simply subdivided, for example XC into XC1, XC2, and XC3. Additional, distal terminal bands are denoted by additions to existing subdivisions, for example the band beyond XF4, which was previously the terminal band, becomes XF5.

References

1. Committee on Standardized Genetic Nomenclature for mice. 1985. Mouse News Lett. 72:13–15.
2. Davisson, M.T. 1981. Centromeric heterochromatin variants. *In* M.C. Green, ed., Genetic variants and strains of

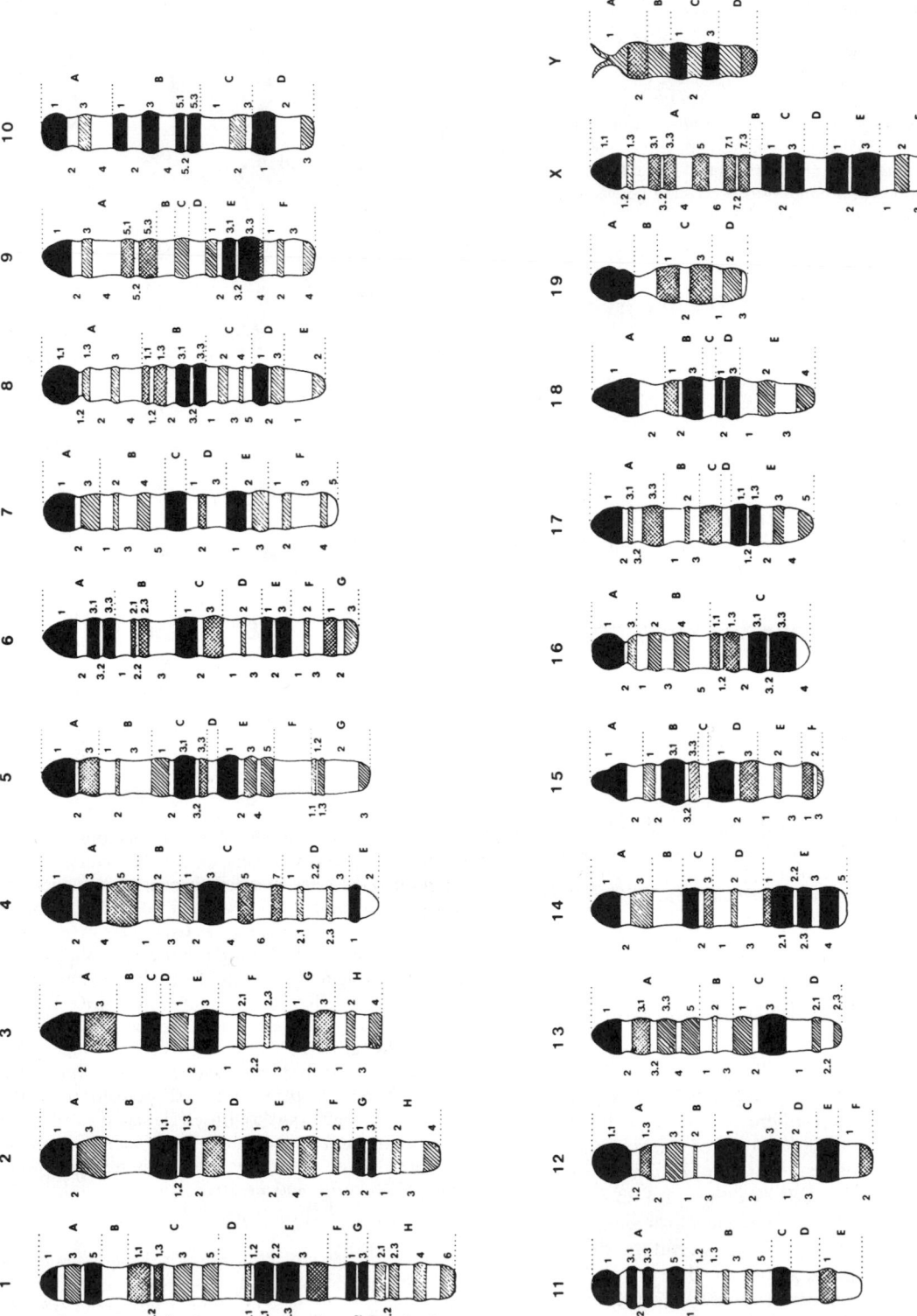

Fig. 11.1 Idiogram of the banding patterns in mouse chromosomes revealed by Giemsa staining after treatment with 2xSSC and trypsin.

577

the laboratory mouse, 357–358. Gustav Fischer Verlag, Stuttgart.

3. Ford, C.E. 1966. The murine Y chromosome as a marker. Transplantation 4:333–335.

4. Nesbitt, M.N., and U. Francke. 1973. A system of nomenclature for band patterns of mouse chromosomes. Chromosoma 41:145–158.

5. Sawyer, J.R., and J.C. Hozier. 1986. High resolution of mouse chromosomes: banding conservation between man and mouse. Science 232:1632–1635.

11.2 Standard G-banded karyotype

EDWARD P. EVANS

See Fig. 11.2.

References

1. Gallimore, P.H., and C.R. Richardson. 1973. An improved banding technique exemplified in the karyotypic analysis of two strains of rat. Chromosoma 41: 259–263.

11.3 Standard Q-banded karyotype

MURIEL T. DAVISSON

See Fig. 11.3 on p. 579.

References

1. Miller, O.J., D.A. Miller, R.E. Kouri, P.W. Allderdice, V.G. Dev, M.S. Grewal, and J.J. Hutton. 1971. Identification of the mouse karyotype by quinacrine fluorescence, and tentative assignment of seven linkage groups. Proc. Natl. Acad. Sci. USA 68:1530–1533.

11.4 Standard pachytene karyotype

EDWARD P. EVANS

The method used in the preparation is described in the first edition and the karyotype presented is the same apart from the substitition of some of the bivalents for clearer examples and the placing of chromosome numbers at the centric ends (see Fig. 11.4 on p. 580). As in other mammals, the pachytene chromomeres are seen to correspond to the G-bands of mitotic chromosomes. Since the publication of the first edition, another pachytene map of the mouse spermatocyte has been published by Fang and Jagiello (1).

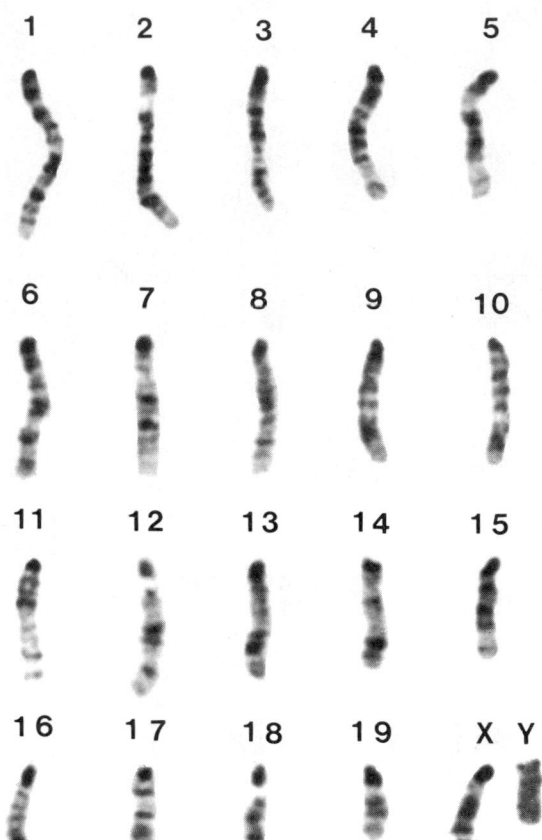

Fig. 11.2 G-banded example of a male haploid set from C57BL/6J/Ola embryos. Livers of day 12 embryos were macerated in MEM and the cell suspension obtained resuspended in hypotonic 0.56 per cent potassium chloride for 8 minutes at room temperature. Cells were centrifuged and fixed in 3 methanol : 1 glacial acetic acid. Slides were made by air-drying and stored for 7–10 days before banding. Banding was effected by a combined ASG/Trypsin method (1) which involved incubating the slides for 1–2 hours in 2xSSC at 60–65°C, cooling to room temperature and rinsing in 0.85 per cent saline. Slides were then flooded in a horizontal plane with 0.025 per cent trypsin in saline for 30 seconds, quickly rinsed in saline to stop the tryptic activity, and stained in 1–2 per cent Giemsa in pH6.8 buffer. On achieving adequate staining, slides were rinsed in pH6.8 buffer and air-dried.

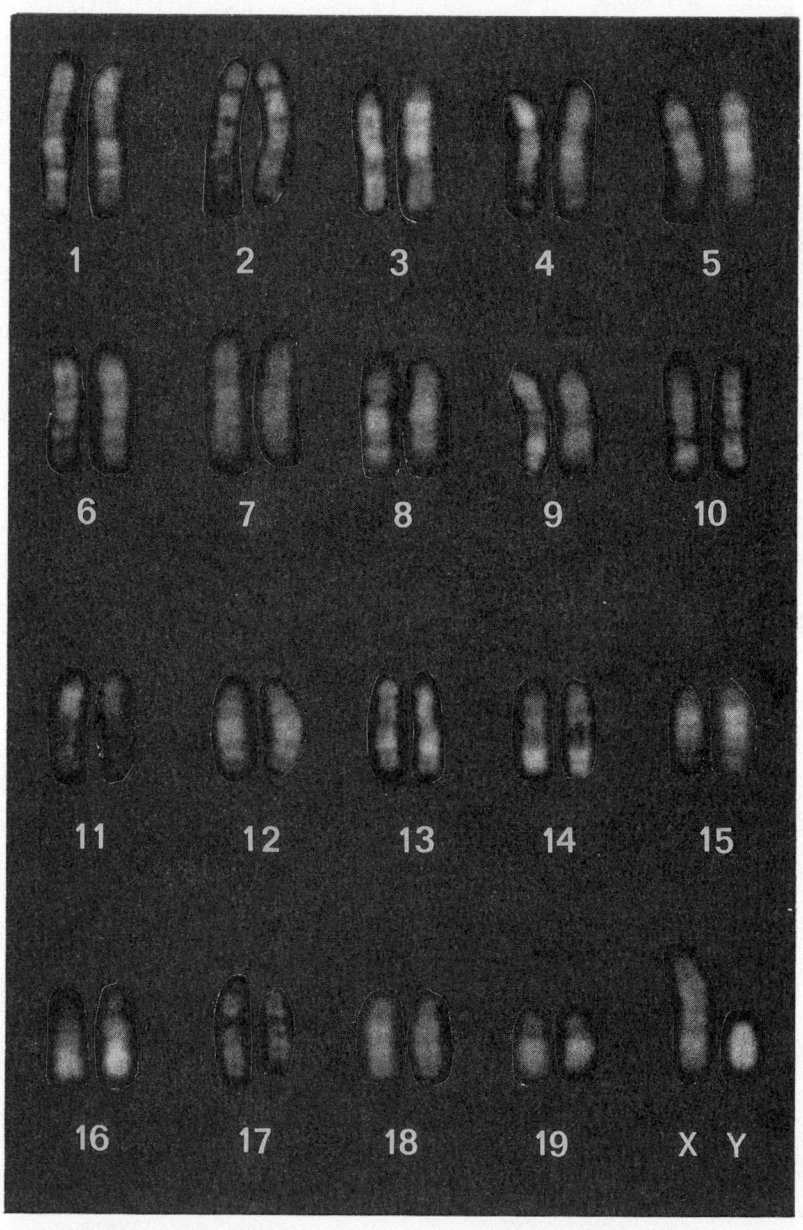

Fig. 11.3 Q-banded karyotype of chromosomes from cultured leucocytes of a C57BL/6J male. The chromosomes were stained with quinacrine mustard dihydrochloride by the method of Miller *et al.* (1) and photographed using an FITC excitation filter and barrier filter 50.

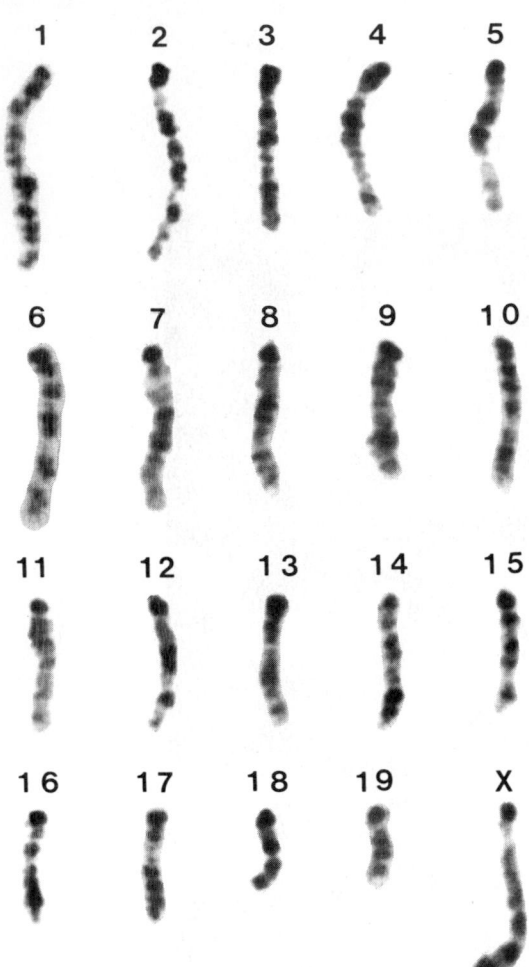

Fig. 11.4 Pachytene karyotype showing chromomere patterns.

References

1. Fang, J.S., and G. Jagiello. 1981. A pachytene map of the mouse spermatocyte. Chromosoma 82:437–445.

11.5 Standard karyotype of early replicating bands (RBG-banding)

IMRE E. SOMSSICH and

HORST HAMEISTER

Embryonic fibroblasts or spleen cells stimulated by 10 μg/ml concanavalin A were used for short-term culture. Seven to nine hours before harvest of the cells bromodeoxyuridine, 10 μg/ml, and fluorodeoxyuridine, 0.5 μg/ml, were added. The cells were treated for 1 h with colcemid (0.05 μg/ml), exposed to hypotonicity (0.075 M KCl) for 25 min, fixed, and spread on cold wet slides. Staining was done using the Hoechst–Giemsa technique (1). In brief: at least 24 h old preparations were stained with Hoechst 33258, 2.5 μg/ml, for 10 min, dipped in a phosphate-buffered salt solution, pH 5.5, and exposed to UV light for 30 min. Subsequently the slides were incubated in 2 × SSC (0.3 M NaCl, 0.03 M tri-sodium citrate) at 59°C for 1 h and thereafter stained in 5 per cent Giemsa for 10–20 min. RBG-banding is due to labelling of *in vivo* growing cells during part of the preceding DNA synthetic phase with bromodeoxyuridine. The intensity of staining depends upon the progression within the S-phase of the individual cell. Therefore the staining pattern is more variable than GTG- or QFQ-banding. This is best illustrated by the late replicating X and Y chromosome, (eX = early replicating and lX = late replicating X chromosome) (see Fig. 11.5 on p. 581). For more details see ref. 2.

References

1. Perry, P., and S. Wolff. 1974. New Giemsa method for the differential staining of sister chromatids. Nature 251:156–158.
2. Somssich, I.E., H. Hameister, and H. Winking. 1981. The pattern of early replicating bands in the chromosomes of the mouse. Cytogenet. Cell Genet. 30:222–231.

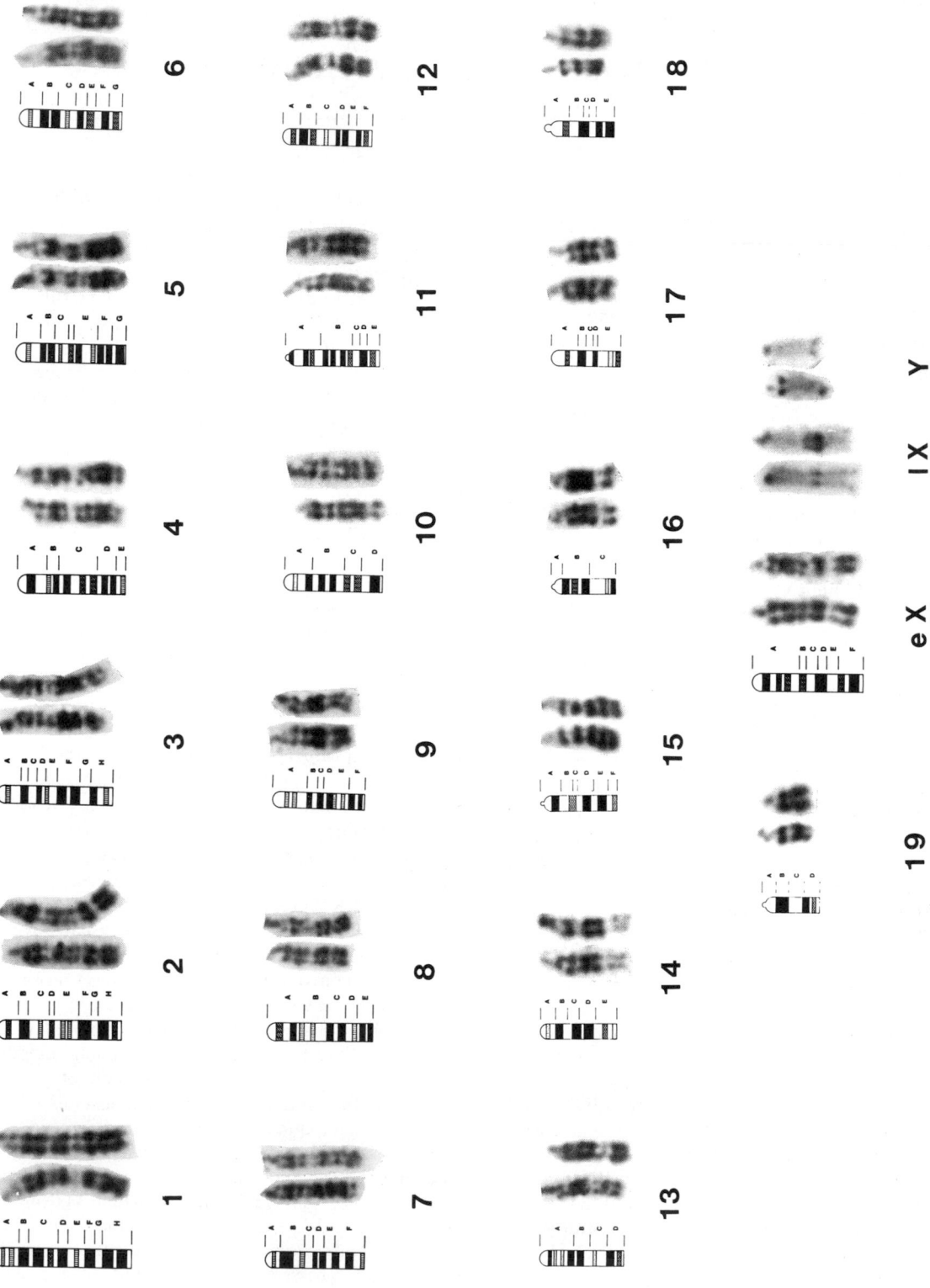

Fig. 11.5 Karyotype treated to show early replicating bands (RBG-banding) with idiograms on left. For the X chromosome, both early replicating (eX) and late replicating (lX) patterns are shown.

12 CHROMOSOMAL VARIANTS

12.1 Numerical variants and structural rearrangements

ANTONY G. SEARLE

This section lists those heritable chromosome re-arrangements in the mouse which are thought to exist at present, with brief descriptions of their origin and chief characteristics. Some account is also given of known variations in number of whole chromosomes and their effects. It seemed useful to include these since one, the XO female, is viable and fertile, while others may live for various lengths of time. Known holders of chromosomal variants are listed annually in Mouse News Letter (Oxford University Press).

Translocations are divided into reciprocal (with and without involvement of the X), Robertsonian (whole arm), and insertional. The Robertsonians are further subdivided into those which are laboratory- or wild-derived, the latter with reference to their place of origin. Within each group, the translocations are listed according to their serial number, as are the inversions. To assist the reader in locating translocations and inversions that involve particular chromosomes, Table 12.1 shows all the listed variants, arranged according to the chromosomes involved and identified by serial number. A map of chromosomal breakpoints in translocations and inversions is given in Section 13.1 and a genetic map of inversions in Section 13.2

Abbreviations used in the text are: Chr = chromosome; RF = recombination frequency.

12.1.1 Polyploidy

Beatty (23) discussed the generation of polyploids and reported spontaneously occurring triploid, tetraploid, and hexaploid embryos at $3\frac{1}{2}$ days p.c.; induced 8-ploid and 16-ploid embryos have been reported (289) at early cleavage stages. Some triploids are known to survive until at least the 10th day of embryonic life, usually with gross abnormalities, the nature of which seems to depend in part on the genetic background (309). Most tetraploids die at the egg cylinder stage, with similar abnormalities to those described in some triploids (for-mation of fetal membranes but complete cessation of

embryo development), but a few survive to late fetal life or even to term (289).

The incidence of triploidy can be increased by the use of heat shock or colchicine at the time of fertilization, or by delaying fertilization (255, 309), or by fertilization *in vitro* (154). It is also higher in certain stocks of mice carrying silver (*si*), although *si* itself does not seem responsible (23, 255). Tetraploidy can be induced by the treatment of embryos approaching the second clea-vage division with cytochalasin B (289).

12.1.2 Aneuploidy

Bond and Chandley (63) have reviewed the various types of defect which lead to aneuploidy. For trisomy, the most important of these is meiotic non-disjunction but chromosome loss, and therefore monosomy, can also arise by entirely separate mechanisms involving chromosome damage. It can be induced by irradiation of meiotic and post-meiotic germ-cell stages but the extent to which numerical non-disjunction can be induced by radiation is still under investigation. How-ever, it is known that radiation can increase the fre-quency of reciprocal (but probably not Robertsonian) translocations and thus lead to extra tertiary trisomics and monosomics.

Non-disjunction occurs spontaneously and its fre-quency increases with maternal age. Heterozygotes for wild-derived Robertsonians show high frequencies of non-disjunction, which are even higher (usually over 30 per cent) in compounds of two Robertsonians with an arm in common (monobrachial homology) and reach 100 per cent in a recently constructed compound of three Robertsonians with tribrachial homology (42, 162, 278). By use of suitable visible genetic markers, mice which arise from complementation of gametes with an extra, and a missing, chromosome can be detected. Compounds of two Robertsonians with monobrachial homology are particularly useful for the study of specific monosomics and trisomics for the com-mon arm, e.g. Chr 15 if the compound is Rb(4.15) 4Rma/Rb(6.15)1Ald, since the non-disjunctional gametes mainly carry both the metacentrics or neither (166). If intercrossed, such compounds also generate

582

nullisomics and tetrasomics for the chromosome concerned. However, such intercrosses are not always possible because many compounds of two Robertsonians with monobrachial homology are male-sterile.

12.1.2.1 Autosomal aneuploidy

Nullisomy

The only nullisomic investigated so far has been that for Chr 15. This seems to be lethal at a very early stage in development, since none have been found at the blastocyst stage (25, 45), when Chr 15 monosomics can still be detected.

Monosomy

The effects of all autosomal monosomics except those for Chrs 7, 8, and 13 have been investigated (14, 15, 112, 215, 216; see also ref. 183) and found to be very severe, with death at or before implantation. One group in which death is particularly early, i.e. almost entirely at the morula or early blastocyst stage, comprises Chrs 2, 3, 5, 6, 11, and 15. Monosomics for Chrs 1, 10, 14, and 16 may survive to be implanted but very few live beyond the neurula stage. The other monosomics studied seem to have intermediate effects, with survival usually to the late blastocyst stage without implantation. However, there are some discrepancies between the results of different laboratories, which suggest that differences in genetic background may be important in determining the length of survival.

Trisomy

All autosomal trisomics have now been studied to a greater or lesser extent (112, 123, 160, 183, 301, 302). All survive longer than the monosomics but resemble them in that they include a group which die early, i.e. soon after implantation. This comprises Chrs 2, 3, 4, 5, 6, 7, 8, 9, 11, 15, and 17 and is thus rather similar to the early lethal monosomic group. All in the later lethal group tend to show specific malformations, except trisomy–6, and there is a general tendency for retardation as compared to normal littermates. It is now realized that at least four of the trisomics (for Chrs 13, 14, 16, and 19; possibly also Chr 18 although no liveborn 18-trisomics have been reported yet) may survive to term, while a trisomy–19 female mouse has survived for up to 20 weeks, with production of one offspring (294). The most frequent malformations associated with trisomy are CNS and heart defects. Those of Chr 16 have been particularly thoroughly studied because they resemble those of human trisomy 21, which shows a homologous region with this mouse chromosome (123). Thus mouse trisomy–16 can be regarded, to some extent anyway, as

a mouse model of Down syndrome. Other lines of research on mouse trisomics include (i) attempts to 'rescue' particular trisomic cellular systems, by chimaera formation, to study their normality or otherwise, (ii) quantitative study of enzyme patterns in particular trisomics, so as to detect abnormal gene dosage effects which may help to reveal the chromosomal location of the gene concerned.

Tetrasomy

The only complete tetrasomies studied so far have been those for Chrs 15 (45) and 16 (109a). Both were more severely affected than the corresponding trisomies, with death on about the ninth day of gestation in tetrasomies for Chr 15 and the 12th day for Chr 16.

12.1.2.2 Sex-chromosomal aneuploidy

Monosomy

The XO condition is compatible with viability and fertility in mice although about 35 per cent die *in utero*. This contrasts with over 99 per cent intrauterine lethality of human XOs and sterility of survivors. Litter-size of XO mice is almost halved and reproductive lifespan is shortened (200), with numbers of postnatal oocytes halved as the result of excess atresia at the late pachytene stage (69a). The litter-size reduction is partly the result of pre-implantation death of OY zygotes, as well as extra lethality of XO (227), which also helps to explain the less than expected frequency of XO daughters. Various authors have reported a preferential segregation of the X chromosome into the nucleus of XO oocytes at meiosis rather than with the polar body but this was not confirmed in the latest investigation (66) neither was there evidence for a maternal age effect on this segregation.

Trisomy

1. XXY mice are sterile males in which there is complete lack of spermatogenesis as a rule, although occasional patches of almost normal spermatogenic activity may occur (84a). These are usually ascribed to the presence of an XY clone, but some spermatogonial metaphases with 41 chromosomes were found in an XXY mouse with XY/XXY germinal mosaicism (64).

2. XYY mice are also phenotypically normal males which suffer from spermatogenic impairment but to a lesser extent than XXY. Some spermatozoa are usually formed, enough for transient fertility in one male (135). As with XXY mice small populations of XY spermatogonia and spermatocytes may be seen but fail to become dominant over the XYY line, unlike the human situation.

Table 12.1 Translocations and paracentric inversions arranged to show the chromosomes involved. Recipocal translocations and insertaions are shown in the upper right, Robertsonian translocations in the lower left, and inversions on the main diagonal. The translocations and inversions are identified by a serial number and standard subline symbol. Some of the subline symbols are abbreviated as follows: A = Ald; B = Bnr; I = Iem; L = Lub; R = Rma; S = Sic

Second chromosome of reciprocal translocation

1st Chr \ 2nd Chr	1	2	3	4	5	6	7	8	9	10	11	12	13	14	15	16	17	18	19	X*
1	1Rk 12Rk 24Rk 38Rk 44H‡ 46Rk 49Rk	5Ca			44H‡		40H†		27H	11Ad			70H 1Wa				190Ca			
2	5H 18L	2H 5Rk 19Rk 40Rk 45Rk	24H	1Ca 1Sn 13H		7Ca		26H 2Wa	11H		30H			1Gso		28H				14RI
3	1B 1L		11Rk 55Rk	5Rk 11Rk					45H			30Rk								
4				1By 1Dn 28Rk 32Rk	46H			36H‡												1RI 7RI 8RI 37H
5			1ICg 2Tu		2Rk 9Rk 30Rk 33Rk			3Rk	9RI			31H	5Ad 264Ca		4Ad					
6		4I	3Tu	2B		47Rk 48Rk		2Ad		8Ad		32H	6Ad							
7	1R					13R	13Rk						7Ad	2Iem	9H			50H	145H	1Ct† 1Neu 2Neu 3Neu 2RI 3RI 5RI 6RI 18RI
8		2L	2R				14Rk 34Rk 36H‡									17H				

Reciprocal translocations and related variants. Rows = first chromosome (1st Chr); columns = second chromosome (2nd Chr).

1st \ 2nd	1	2	3	4	5	6	7	8	9	10	11	12	13	14	15	16	17	18	19	X
9			12L						26Rk 39Rk		9Ad						10Ad 138Ca			
10	10B			8Tu	3Nam	13Tu				6Rk 17Rk 42Rk		10Rl	3Ad 199H	8Rk			11Rl	12Rl 18H		
11				12R					14Tu	8B 5R	20Rk	25Rk						38H	42H	
12					9B 1Tu 1Nam	3S		5B 22L	10Tu 1Cam	17R 5L									13Rl	
13			7Tu		70L	3R 1H 9Tu	1Oxe	16R	6B	1Ad	4B 6L 6Tu	3Ct 15Tu	31Rk 43Rk 50Rk	22Rk 36Rk 54Rk	18Rk 21Rk 35Rk					
14				11Tu	21L						1Dn	8R			6Ca					
15		2Ct	2Rk	4R 1Rk	3B 15R 4L 12Tu	1A				23L										
16				13L	7R	24L		38L 6S 1I	9R		2H		1Mpl			51Rk	43H		16H	
17		4H 11R													64L	7B 8L 32L 54L 74L 75L			15Rl	
18	10R			3H			9L													
19		6R			1Wh	2Icg		1Ct	163H											37Rk
X*			2Ad																1H 3H	

Chr First chromosome of Robertsonian translocation

* In the symbols of X-autosomal translocations in the text, the X chromosome is given first.
† These are non-reciprocal insertions.
‡ An inversion and translocation combined (T36H and T44H).

Sex-chromosomal mosaics and chimaeras

Among the sex-chromosomal mosaics which have been described are: (i) XO/XYY, sterile male with only XYY in the germ-line (137); (ii) XO/XY which was a true hermaphrodite (193); (iii) XO/XX, one fertile and one sterile (82, 157); (iv) XO/XY/XYY male with normal testes, which probably arose at an early cleavage division but in which the Y clone was outgrowing the others (129). Levy and Burgoyne (189b) have shown that XO germ-cells are selectively eliminated from XO/XY and XO/XY/XYY mice and have postulated the existence of a spermatogenesis gene on the mouse Y chromosome. Their mosaics were derived from crosses with the BALB/cWt inbred strain which has a high frequency of spontaneous true hermaphrodites. These are XO/XY or XO/XY/XYY mosaics, in which the proportion of ovarian tissue seems to be related to the percentage of XO cells (122).

Many XX/XY chimaeras have been reported as the result of morula aggregation (221). Most of these are males, some are females, and a few are intersexes. This suggests that >30 per cent of XY cells in a gonad primordium leads to testis development (221a). In general, both male and female chimaeras produce functional gametes of one class only. However, Ford *et al.* (147) have reported the occurrence of a functional oocyte which was apparently of the XY constitution. Nevertheless, the authors show that there is very good evidence against the functioning of XX germ cells in the male environment.

12.1.3 X–Y separation

Although not an aneuploid condition, this is included here because it is a well-defined anomaly which involves the sex chromosomes and may be inherited. In its extreme form, i.e. 100 per cent dissociation of the X and Y chromosomes at meiosis so that no X–Y bivalent is found at metaphase I, it causes complete sterility, with absence of secondary spermatocytes and post-meiotic germ cells (24). This is one of a number of phenomena which suggest an association between failure of sex chromosomes to pair during meiosis and spermatocyte death (69). A 70 per cent frequency of X–Y dissociation is found in F_1 hybrids between Japanese wild mice (*Mus musculus molossinus*) and inbred laboratory mice (BALB/c and B.10BR) and high frequencies are also found in the DBA/2J, DDK, and WB/Re inbred strains as well as in some hybrids involving these (59, 177, 219, 220, 246). X–Y dissociation may reach frequencies of 70–90 per cent in XY, Sxr males, where it seems to be connected with a tendency for the aberrant

Y chromosome to pair on itself at meiotic prophase (93b). Evidence for genetic control of X–Y dissociation have been obtained by several authors (24, 51, 219, 292). Matsuda *et al.* (219) postulated the existence of at least one heritable factor (provisionally called *Sxa*) responsible for the end-to-end association of the sex chromosomes, while Biddle *et al.* (51) reported that, in the C57BL/6J and DBA/2J strains and crosses between them, sex-chromosomal univalency was controlled by (i) an unknown number of autosomal factors, (ii) the DBA Y chromosome, (iii) two X-chromosome factors which were not closely linked.

12.1.4 Reciprocal translocations

These are symmetrical exchanges of chromosomal material between non-homologous chromosomes. Their general properties in mammals have been reviewed by Ford and Clegg (142), and by Searle *et al.* (281) with special reference to meiotic behaviour in mice. Carter *et al.* (77) summarized early work on translocation induction in the mouse by Snell, Hertwig, and others and also described genetic methods for the detection of translocation heterozygotes and homozygotes. Heterozygotes typically show 'semi-sterility' or 'partial sterility' (really semifecundity); when mated to unaffected mice, they produce litters of reduced size, usually somewhat less than half normal. This is because of events at meiosis that lead to the production of 50 per cent unbalanced gametes from normal (alternate or adjacent–1) segregation as well as 100 per cent unbalanced gametes from the rarer non-disjunctional type of segregation (adjacent–2). The resultant zygotes usually die *in utero* at the time of implantation (as 'small moles' or deciduomata) but may survive longer and produce later deaths, fetuses with malformations (e.g. exencephaly), or even viable but often abnormal tertiary trisomics or tertiary monosomics (183). At metaphase I of meiosis, translocation heterozygotes form characteristic configurations by which they can be identified cytologically, namely rings of four elements (RIV), chains of four (CIV), trivalent + univalent (CIII + I), the last in particular leading to tertiary trisomy. The form of the configuration depends on whether chiasmata are formed in all four arms (RIV), three arms (CIV), two adjacent arms (CIII + I), or two opposite arms (bivalents only). The last type cannot be distinguished from normal unless the bivalent has unequal arms. Similar multivalent associations can be detected in the synaptonemal complex of translocation heterozygotes at pachytene (228). If the translocation is unequal (in terms of the lengths of parts exchanged), it may lead to the formation of somatic marker chromosomes appreciably longer or shorter than any in the usual set. A translocation involving the short arm of a chromosome can lead to a somatic marker in the form of a metacentric.

Homozygotes for some translocations are lethal, sometimes because of a closely linked lethal gene. Most are viable and 'fully fertile', i.e. with normal-sized litters, but all their offspring by normal males are semi-sterile. Males heterozygous for certain translocations may suffer from defective spermatogenesis, leading to infertility or complete sterility. This sterility of males is characteristic of reciprocal translocations involving the X chromosome (114, 259), but it also occurs in all known Y-autosome translocations, in tertiary trisomics and in many autosomal translocations (72, 202, 272, 280) especially those in which one breakpoint is proximal and the other distal on the chromosomes concerned. Such an arrangement often generates long and short marker chromosomes and is associated with failures of pairing at meiosis, which may be an important factor in generating the male sterility (60, 149). This chromosomally derived male sterility is also associated with a disruption of pachytene DNA metabolism (172). Moreover, some X-linked enzymes show higher activity at late meiotic prophase in these sterile males than in normal fertile ones, which suggests that some reactivation of the inactive X chromosome has occurred (171). The severity of the male spermatogenic defect may depend partly on the genetic background (150). Certain combinations of translocations may also lead to male sterility (78). Female heterozygotes for X-autosome translocations may be adversely affected in growth rate, etc. (265). They may also show variegation of the coat when heterozygous for certain coat-colour loci on the affected autosome if the normal autosome has the recessive allele for the locus concerned. These 'V-type position effects' (261) are now thought to be the result of a spread of inactivation from the X-chromosomal inactivation centre into contiguous autosomal material (84, 85). Females heterozygous for male-sterile translocations are fertile but there is recent evidence that oogenesis may be affected in these and in female tertiary trisomics, with a marked reduction in numbers of small oocytes in young mice, as compared with normal littermates (70, 226, 286).

Reciprocal translocations have been used extensively for assigning mouse linkage groups to chromosomes (223, 224) as well as for assigning genes to particular regions of chromosomes by: (i) duplication-deficiency mapping (121); (ii) comparing genetic and cytological breakpoint positions (234, 283); (iii) *in situ* hybridization. They have also been used for finding the position of centromeres relative to the linkage group concerned (281), for reducing litter-size (190), and for the study of genomic imprinting. This last use has stemmed from the discovery that, when genetically marked translocation heterozygotes are intercrossed, zygotes formed from gametes with complementary duplications and deficiencies can usually form perfectly normal and viable mice (281). However, when these duplications and deficiencies involve certain regions of Chrs 2, 6, 7, 8, 11, 17 (and possibly some others) and come in from one particular parent, the resultant zygotes may die before or after birth or suffer from certain abnormalities (87, 279). It is thought that particular genes in the regions concerned need to be imprinted in some way during germ-cell maturation. This cannot happen normally if the genes or regions concerned are transmitted by one parent rather than by both. This helps to explain why parthenogenetic embryos cannot survive to term (182, 218).

Translocations listed in the following sections are those for which locations of breakpoints are known and stocks are believed to exist at present. There are others described in the literature (usually only briefly) which are thought to be extinct.

12.1.4.1 Translocations involving the X-chromosome

The origins, nature, and uses of these have been reviewed by L.B. Russell (259). Rastan (247) has used some to find the chromosomal location of the X-inactivation centre (see also *Xce*) and to formulate a model for the initiation of inactivation.

T(X;7)1Neu. Arose as a variegated female after spermatozoal sampling of (102/El x C3H/El)F$_1$ males treated with 250 mg/kg ethylmethane sulphonate (EMS) in a specific locus experiment. Semi-sterile females show variegation when heterozygous for c^{ch} and p; males are sterile. Translocation breakpoints are at XF3 and 7B3, which give rise to long X^7 and short 7^X marker chromosomes (235).

T(X;4)1Rl. (R1). Arose in a post-meiotic germ cell of a (101 ♀ x C3H ♂)F$_1$ male after acute X-irradiation in a specific locus experiment, being detected because of variegation for brown in females heterozygous for the translocation with b on the intact Chr 4. In these animals, about 40 to 90 per cent of the fur is brown (261). However, no variegation is shown with m (268). Male carriers are sterile because of degeneration of spermatocytes in pachytene; female carriers are usually semi-sterile. This translocation is about 7 units from *Ta* and

9–14 units from *b* (259). Cytological positions of breakpoints are at XF2 and 4A5 (259), which makes the translocated product with the X centromere a somatic marker, 1.3 times as long as Chr 1 (114). This translocation and T6Rl show extensive non-homologous synapsis, which is thought to occur when a translocation breakpoint is in or very near to a dark-staining G-band (9).

T(X;7)2Neu. Arose after spermatid sampling and EMS treatment (175 mg/kg), like T1Neu. Females also show variegation with c^{ch} *p*. These are also semi-sterile, though with somewhat larger litter-sizes than T1Neu, while males are sterile. Translocation breakpoints are in XF3 and 7B3 as T1Neu but not in identical parts of the bands (235).

T(X;7)2Rl. (R2). Arose after X-irradiation of spermatozoa of a (101 ♀ x C3H ♂)F$_1$ hybrid male in a specific locus experiment (265). Semi-sterile heterozygous females have reduced viability and weigh less than normal sisters. They can show variegation with *p* and *tp* but not with *c* (266). This translocation has not recombined with *c* on Chr 7, but its breakpoint is presumed to lie between *c* and *tp* from other evidence; on the X chromosome the breakpoint lies about 17 units distal to *spf*. There may be some inhibition of crossing-over near the Chr–7 breakpoint. Cytological breakpoints are at XA4 and 7D3 (259). This makes the translocated chromosome with the centromere of 7 (7^X) a somatic marker, 1.2 times the length of Chr 1.

T(X;7)3Neu. Arose, like T2Neu, after spermatid sampling and EMS treatment (175 mg/kg) in a specific locus test (2). Breakpoints are at XF4 and 7D1 (130). Further details are not available yet.

T(X;7)3Rl. (R3). Arose after X-irradiation of spermatocytes of a (101 ♀ x C3H ♂)F$_1$ hybrid male in a specific locus experiment (265). Semi-sterile heterozygous females tend to have lower weaning weights than normal ones, lower survival to 7 days, and other deleterious effects. They can show variegation for *c* locus alleles but only slightly for *p* (266). Males are sterile because of spermatogenic breakdown at pachytene. Cytological breakpoints are at XA2 and 7F1 (259), which makes the 7^X product a somatic marker, about 1.3 times as long as Chr 1. T3Rl shows 0.2 per cent recombination with *c*, with breakpoint on its distal side and with no evidence for crossover suppression (266). The *Mod–2* locus is inactivated by T3Rl (50), so that the breakpoint probably lies between *Mod–2* and *sh–1*. T3Rl shows about 3.5 per cent recombination with *spf* (259) and is probably distal to that locus.

T(X;7)5Rl. (R5). Arose after X-irradiation of spermatozoa of a (101 ♀ x C3H ♂)F$_1$ hybrid male in a specific locus experiment (265). Semi-sterile heterozygous females have slightly lower weaning weights and survival than normal. They can show variegation of the *c*, *tp*, and *p* loci (266). Males are sterile because of spermatogenic breakdown at pachytene. Cytological breakpoints are at XF1 and 7A3, which leads to long and short somatic markers, the long X^7 marker being 1.4 times the length of Chr 1 (259). Only 20 per cent of pachytene spermatocytes from carrier males form quadrivalents, apparently because of premature desynapsis of the short 7^X from its Chr 7 partner (10). The Chr 7 breakpoint is about 7 units proximal to the *p* locus; there is no evidence for crossover suppression. The X-chromosomal breakpoint is 11 units distal to *Ta* (259).

T(X;7)6Rl. (R6). Of spontaneous origin, first appearing as a cluster of two from a (101 ♀ x C3H ♂)F$_1$ female in a specific locus experiment and probably originating in a strain 101 female. Heterozygous females tend to be lighter at weaning, to have a higher incidence of sterility than normal, and to show other deleterious effects (265). They can show high-grade variegation at both the *c* and *p* loci also to a lesser extent at *tp* and *ru–2* (266). Male carriers are completely sterile, with spermatogenesis ceasing after pachytene. Cytological breakpoints are at XF1 and 7B3 (259), which leads to the X^7 product being a somatic marker about 1.3 times the length of Chr 1. The T6Rl breakpoints lie about 6.5 units on the distal side of *Ta* and about 0.6 units on the proximal side of *ru–2*, with no signs of crossover suppression (259). Only 42 per cent of pachytene spermatocytes form quadrivalents, probably for the same reason as with T5Rl (8, 11).

T(X;4)7Rl. (R7). From neutron-irradiated spermatozoa of a (101 ♀ x C3H ♂)F$_1$ male in a specific locus experiment. It can show variegation for *b* and (to a lesser extent) for *m* in heterozygous females, which are semi-sterile (257, 259). T7Rl lies between *spf* and *Ta*, with about 11 per cent recombination between it and *Ta*; the breakpoint on Chr 4 is about 12 units distal to *m*, with no inhibition of crossing-over (259). Cytological breakpoints are at XA2 and 4D1 (259), making the translocated product with the Chr 4 centromere a somatic marker, 1.3 times as long as Chr 1. Viability is reduced in heterozygous females; male carriers are sterile, with cessation of spermatogenesis at pachytene (257).

T(X;4)8Rl. (R8). Induced by post-spermatogonial treatment with cyclophosphamide in a specific locus experiment and recognized by its variegation with *b* in females. Breakpoint positions are not yet known but give a 4^X product which is longer than the X. T8Rl shows about 1 per cent recombination with *m* and is probably distal to it (259). It shows about 13 per cent recombination with *Gs*, which is distal to it.

T(X;12)13Rl. (R13). This was found in a stock descended from an X-irradiated male in which postgonial germ-cell stages were sampled. Female heterozygotes are semi-sterile and small; males are sterile (73). No variegated phenotype has been found. Cytological breakpoints are in or near XA3 and in 12A giving translocation products of medium size (259). The translocated product 12^X is selectively unlabelled (i.e. inactive) in 82 per cent of adult cells, in contrast to the situation in T16H (see below). However, no phenotypic change has been detected in heterozygotes for sex-linked markers.

T(X;2)14Rl. (R14). Found after post-spermatogonial treatment of males with repeated doses of triethylenemelamine. Semi-sterile heterozygous females can show variegation for *pa*. Breakpoints are in bands XF and 2C, giving an X2 chromosome about the same length as Chr 1. T14Rl is proximal to the *pa* locus and about 14 units from it (259).

T(X;17)15Rl. (R15). Arose after triethylenemelamine treatment of post-spermatogonial stages (259). Breakpoints are in XA and 17A and do not give rise to somatic markers. Carrier males are sterile.

T(X;16)16H. (Searle's translocation). Arose in a (C3H ♀ x 101 ♂)F₁ female that had been mated to a *Ta* male and X-irradiated at the pronuclear stage. A phenotypically wild-type semi-sterile daughter proved to have the *Ta*/+ genotype, but to carry a translocation between the X chromosome and a small autosome in which the normal X behaved as if completely inactivated, while the translocated one remained active in all cells (212). Thus inactivation was non-random. Other sex-linked mutant genes also failed to show heterozygous expression if on the normal X, but behaved as if dominant (without variegation in the heterozygote) if introduced on to the translocated X-chromosome by crossing-over (192, 212). Males are sterile, but some spermatocytes reach metaphase I, 80–90 per cent forming chain quadrivalents (143). There is much degeneration of spermatocytes and spermatids (248). Ohno and Lyon (238) reported that the normal X chromosome showed positive heteropycnosis in nearly 90 per cent of somatic prophases of female carriers. Takagi (293) observed that although the 16^X product could become inactivated the X^{16} product did not, contrary to the results of Disteche *et al.* (111), but in line with present

views that the X-chromosomal material which is not in physical contact with the inactivation centre remains active (214, 247). Thus the non-random X inactivation in surviving T16H/+ females must be the result of the selective elimination of cells with both the X and X^{16} active. The T16H sex vesicle is enlarged and sometimes the 16^X and X^{16} products form an illegitimate synaptonemal complex in it (290, 291). Chromosomal breakpoints are at 16B5 and XD-distal (120) and do not give somatic markers. The breakpoint on the X is about three-quarters of the map distance from *Bn* to *Ta*; there is marked crossover suppression (191, 212). On Chr 16 the breakpoint is about 40 units from *md* (32). T16H has been given a somatic marker chromosome by combination with Rb(16.17)7Bnr (37). T16H has been used extensively for investigations in which complete inactivation of the normal X chromosome is an advantage.

T(X;7)18Rl. (R18). Arose after spermatogonial stem-cell exposure to ethylnitrosourea (ENU) in a specific locus experiment (262), but was probably the result of spontaneous mutation in a post-meiotic stage. Like T2Rl the Chr 7 breakpoint lies between the *c* and *p* loci, leading to variegation in semi-sterile or infertile female carriers with respect to *p* but not *c*. Females are small, of reduced prenatal viability and ability to rear offspring. Males are sterile.

T(X;4)37H. Induced by X-irradiation of spermatozoa of a (C3H ♀ x 101 ♂)F₁ male (285). Female heterozygotes are semi-sterile and can show variegation for *b*; males are sterile (282). Spermatogenesis mainly stops at pachytene, but occasional germ cells continue to diakinesis. The X-chromosomal breakpoint is proximal to *Ta* and about 2 cM distal to *spf* (214, 282); the Chr 4 breakpoint is about 17 units distal to *m* with no signs of crossover suppression. Cytological breakpoints are at XA2 and 4D3 to give very long and short somatic marker chromosomes (282). Size is reduced in female heterozygotes, by 10 per cent at birth and 30 per cent at weaning. T37H and T38H tend to generate XO mice.

T(X;11)38H. Induced by X-irradiation of spermatozoa of a (C3H ♀ x 101 ♂)F₁ male (285). Males are sterile with spermatogenic breakdown in meiotic prophase (282). Cytological breakpoints are at XA1 and 11E1 (136, 282). Long and short somatic markers are formed, with 60 per cent CIV and 40 per cent CIII + I configurations at MI in heterozygous females. Resultant tertiary trisomics, which are XO, XX, or XY plus the small X^{11} marker chromosome, may survive birth although with sterility in the male and some other abnormalities. However, some die *in utero* with exencephaly. T38H showed 4.8 ± 2.3 per cent recombination with *spf* but the breakpoint must be on the proximal side of the *spf* locus, since *in situ* hybridization studies have shown that this locus (homologous with human OTC) lies in the long marker of T38H but the short marker of T37H (214). The Chr 11 breakpoint is distal to the *Re* locus, with an RF of 3.7 ± 1.8 per cent (39, 282). No variegation has been detected in semi-sterile *Re^{wc}*/+ females with respect to wavy coat.

12.1.4.2 Translocations involving the Y chromosome

All Y-autosome translocations reported so far (72, 189, 272) have led to complete sterility in the males carrying them, with cessation of spermatogenesis usually at the spermatocyte or spermatid stage. Thus it has not proved possible so far to maintain any of these as a stock.

12.1.4.3 Translocations of the autosomes

T(2;4)1Ca. Induced in spermatozoa of a male carrying *Ca*, *Mi^{wh}*, and *T* (77). Cytological breakpoints are at 2B and 4C2, with no clear somatic markers, while configurations in metaphase I of the male are 81 per cent RIV, 9 per cent CIV, and 10 per cent with bivalents only (283). The Chr 2 breakpoint is between *Sd* and *fi* and about 9 units from *Sd*. The Ch 4 breakpoint is about 3 units on the proximal side of *b*. It is not known whether the homozygote is lethal.

T(2;14)1Gso. Derived from a triethylenemelamine-treated hybrid male, F_1(SEC x C57BL), mated to an F_1(C3H x C57BL) female (270). The original semi-sterile male had neurological symptoms which were also found in his translocation-carrying descendants. Affected mice could be distinguished from postnatal day 1 by a reduced ability to right themselves, followed by an inability to swim ('diver' phenotype). No histological defects were found in the ear or the CNS. Chromosomally unbalanced derivatives and translocation homozygotes were lethal *in utero*. Breakpoints of T1Gso are at 2E4 and 14E3; heterozygotes formed 77 per cent RIV and 16 per cent CIV configurations at meiotic metaphase I, the rest being 20 bivalents.

T(2;4)1Sn. (Snell's translocation). This is the earliest known translocation still in existence. It was derived from an X-irradiated male, and was originally described as T-F₁146, then as T(5;8)a (287). Cytological breakpoints are at 2H1 and 4D4 (130). Chromosomal configurations in spermatocytes consist of 12 per cent RIV, 1 per cent CIV, and 87 per cent without multivalents (283). Snell *et al.* (288) showed that heterozygotes for this translocation produced quite a high proportion

of lethal anencephalic embryos; some of these also have cardiac defects (183). Eicher and Washburn (120) found that these anencephalics, and some runted live-born without malformations, carried the 4^2 translocation product, but were monosomic for the normal Chr 4. Snell (287) showed that this translocation was very closely linked to the *a* locus with breakpoint about 1 unit from it. By duplication-deficiency mapping, Eicher and Washburn (121) concluded that the breakpoint must be distal to *bp*, which is itself about 0.3 units distal to *a*. Snell (287) showed that *a* and *b* were linked in mice carrying this translocation, the Chr 4 break being about 21 units distal to *m*, but proximal to *Gpd–1* (120). Homozygotes for T1Sn are viable (287). From intercrosses of T1Sn heterozygotes which were homozygous for *pa* to others which were homozygous for *b*, Snell obtained homozygotes for both *pa* and *b* and was thus able to show that: (i) *pa* and *b* lay on opposite arms of the translocation; (ii) unbalanced gametes could survive and complement each other to give balanced zygotes (see also ref. 281). This translocation has also been used to demonstrate parental origin effects associated with the distal region of Chr 2 and associated with abnormal phenotypes (87).

T(1;13)1Wa. Induced in a triple metacentric male (Rb1Ald/Rb4Bm/Rb163H) by 4-Gy acute X-irradiation. It was kept because its breakpoints resembled those of T70H (236) being 1C2 or 1C3 and 13D1 (58). When intercrossed with T70H the resultant T1Wa/T70H mice have two heteromorphic marker bivalents but no multivalents at meiosis (298). Such males are frequently sterile. In females, numbers of chiasmata in the translocation bivalents were severely reduced compared with T70H homozygotes because of the interstitial loops, but there was no change in males. However the frequency of non-disjunction was increased in both sexes; in the case of females this even included chromosomes not involved in the translocation.

T(6;8)2Ad. Recovered from a female mouse derived from a 3–Gy gamma-irradiated C3H/El male with sampling of spermatocytes (1). Male heterozygotes show about 54 per cent RIV and 46 per cent CIV in the 71 per cent of spermatocytes which are multivalent. Breakpoints are in 6F3 and 8A4, giving rise to long 6^8 and short 8^6 somatic markers.

T(7;14)2Iem. Arose spontaneously in an outbred stock of mice (12). Breakpoints are in 7F3 or F4 and in 14B. This translocation was used to show that deficiency for the part of Chr 7 distal to the breakpoint led to embryonic death by the blastocyst stage; thus this region must be of major importance in the genetic control of early development (15).

T(2;8)2Wa. Like T1Wa, this was induced in post-meiotic male germ-cells by 4-Gy acute X-irradiation of a triple metacentric male (Rb1Ald/Rb4Bnr/Rb163H) and was kept because it had similar breakpoints to those of a pre-existing translocation, namely T(2;8)26H (236). Breakpoints are at 2H3 and 8A4. The compound T2Wa/T26H leads to sterility in about 20 per cent of males (62). The degree of spermatogenic impairment was correlated with the frequency of proximity between the sex chromosomes and translocation multivalent and thus to lack of meiotic pairing within the univalent.

T(10;13)3Ad. Detected as a spontaneous event in a heritable translocation test (1). Breakpoints are at 10A4 and 13B. At metaphase I there are 44 per cent multivalents, 37 per cent of these being RIV and 63 per cent CIV.

T(5;8)3Rk. Occurred simultaneously with T(12;17)4Rk and In23Rk in a single male being tested for inversions (109). This male produced offspring carrying the three chromosomal aberrations in various combinations. T4Rk and In23Rk are now extinct.

T(5;15)4Ad. Arose spontaneously in a heritable translocation test (1). Breakpoints are at 5B and 15D2. There were only 34 per cent multivalents in metaphase spermatocytes, 82 per cent of these being CIV and 18 per cent RIV. Male carriers had a reduced testes weight.

T(5;13)5Ad. Recovered in a female after a heritable translocation test with spermatid sampling after 250 mg/kg ethylnitrosourea. Breakpoints are at 5B and 13Dl (2). Eighty-five per cent of MI spermatocytes were multivalent, 66 per cent being chains and 34 per cent rings.

T(1;2)5Ca. This was induced in a male of the *Ca Mi^{wh} T* phenotype (YX stock) by X-irradiation of spermatozoa (77). Cytological breakpoints are at 1H3 and 2H1 (120), leading at metaphase I of the male to 82 per cent RIV, 7 per cent CIV, and 11 per cent bivalents only (283). T5Ca is about 23 units distal to *ln*, the RF being significantly higher in female heterozygotes, namely 29 per cent against 17 per cent in males (78). In Chr 2, the breakpoint is very close to *a*, with no confirmed case of separation. There is no evidence for crossover suppression. Homozygotes for T5Ca seem to be lethal. Heterozygotes with T(1;2)83Ca were sterile in the male, with azoospermia (78).

T(3;4)5Rk. Detected in progeny of a DBA/2J male treated with triethylenemelamine (98, 105). Cytological breakpoints are at 3D and 4C2 (96, 98). T5Rk is about 8 units distal to *my* on Chr 3 and very close to *b* on Chr 4 (0.9 per cent recombination). This suggests that the *b* locus is in or very near 4C2 (98).

T(6;13)6Ad. Arose in a male after paternal treatment with 2.5 mg/kg mitomycin C and spermatocyte sampling in a heritable translocation test. Breakpoints are at 6B3 and 13Dl (2). Sixty-five per cent of metaphase spermatocytes have multivalent configurations, half being rings and half chains.

T(14;15)6Ca. This is the well-known marker translocation which has been used extensively in immunological and other studies. It was induced by X-irradiation of spermatozoa of males carrying *Ca*, *Mi^{wh}*, and *T* (77). Cytological breakpoints are at 14E3 and 15A2 (130a), giving rise to the distinctive small marker. E.P. Evans found that 37 per cent of metaphase I configurations were CIII + I, the remainder being RIV or CIV (118). Baranov and Dyban (16) showed that spermatogenesis was defective in heterozygous (but not homozygous) T6Ca males, leading to oligospermia or even absence of spermatozoa in some males derived from crosses to the C57BL strain. Forejt (149) associated this defective spermatogenesis in T6Ca/+ males with non-random association between the centromeric heterochromatin regions of Chrs X and 15 at diakinesis/metaphase I, with an H–2 linked gene also involved (150). Eicher and Green (118) found that the Chr 14 breakpoint of T6Ca was distal to *s*, but there were 0/335 recombinants between the breakpoint and the *s* locus. Carter *et al.* (78) also reported no confirmed cases of crossing-over between these loci. However, Eicher and Green considered it probable that T6Ca reduces recombination in the region of the breakpoint. On Chr 15 the breakpoint is very close to *uw* (118). Eicher and Green calculated that the frequency of adjacent–2 disjunction was 0.29, based on the rate of recovery of complementation offspring. T6Ca homozygotes are viable. Tertiary trisomics for this translocation are discussed in Section 12.1.4.5.

T(2;6)7Ca. Induced in spermatozoa of a male carrying *Ca*, *Mi^{wh}*, and *T* (77). Cytological breakpoints are at 2D and 6C2, not generating somatic marker chromosomes (232). Configurations in male heterozygotes at metaphase I are 89 per cent RIV, 9 per cent CIV, 1 per cent CIII + I, and 1 per cent with 20 bivalents only (283). The viability of the homozygote is not known. Carter *et al.* (78) showed that the Chr 2 breakpoint was 5 units from *pa* on the side away from *a* (i.e. proximal); on Chr 6 the breakpoint is about two-thirds of the map distance from *mi* to *Lc*. There may be some crossover suppression on the chromosome. Recombination frequencies are higher in female than in male heterozygotes.

T(7;13)7Ad. Arose in a male after paternal treatment with 250 mg/kg ethylnitrosourea and spermatocyte sampling in a heritable translocation test. Breakpoints are at 7F3 and 13C2 (2).

T(8;10)8Ad. Arose in a male after paternal treatment with 240 mg/kg ethylnitrosourea and spermatogonial sampling in a specific locus test. Breakpoints are at 8B2 and 10C (2).

T(10;14)8Rk. Arose in a female, with involvement of the Robertsonian translocation Rb(9.14)6Bnr, after paternal treatment with TEM (triethylenemelamine). It gives a tiny 10^{14} marker chromosome, also present in a tertiary trisomic female product (95). It now only exists separately from the Robertsonian (100).

T(9;11)9Ad. Induced in male spermatozoa by 3-Gy gamma-irradiation in a heritable translocation test. Breakpoints are at 9B and 11D (2).

T(7;15)9H. Induced by X-irradiation of spermatogonia of a (C3H ♀ x 101 ♂)F₁ male. T9H was found in a semi-sterile daughter (271). Chromosome breakpoints are at 7B3 and 15C (128). The spectrum of configurations at metaphase I in male heterozygotes is 62 per cent RIV, 23 per cent CIV, and 14 per cent bivalents only (273). The Chr 7 breakpoint is about 1.5 units proximal to the *p* locus, with no sign of crossover suppression (276). In Chr 15, the breakpoint is about two-fifths of the map distance from *uw* to *Ca^d* (28). Crossing-over seems to be enhanced in this region. T9H homozygotes are viable.

T(5;10)9Rl. Found in offspring of a (101 ♀ x C3H ♂)F₁ male given fission neutron irradiation, exposed stem-cell spermatogonia being sampled (71). Cytological positions of breakpoints are at 5E and 10D. Heterozygotes for T9Rl resemble *Sl*/+ in having slightly diluted coat colour and a small head-spot; it thus comprises one of nearly 30 independent 'steeloid' mutations which are kept at Oak Ridge (258). No recombination between the steeloid phenotype and the translocation (as judged by fertility criteria) has been observed in any of the four translocations involving Chr 10: T(5;10)9Rl, T(10;12)10Rl, T(10;17)11Rl, and T(10;18)12Rl, studied by Cacheiro and Russell (71). The steeloid phenotype is presumed to arise because one translocation breakpoint is at or very near the *Sl* locus. The T9Rl breakpoint in Chr 5 seems to be close to the *W* locus with recombination of 2.1 to 2.5 per cent between *W^v* and the steeloid phenotype.

T(9;17)10Ad. Arose in a female after paternal treatment with 20 mg/kg methylmethane sulphonate and spermatid sampling in a heritable translocation test. Breakpoints are at 9D and 17E2 (2).

T(10;12)10Rl. This translocation was discovered in offspring derived from X-irradiated spermatozoa (71) and is one of the 'steeloid' group (see T(5;10)9Rl) in having a phenotypic effect on the coat like that of steel (*Sl*). Cytological breakpoints were in bands 10D and 12A. No recombination has been observed between the T10Rl semi-sterility and the steeloid phenotype, and it seems clear that the Chr 10 breakpoint is at or very near the *Sl* locus. The Chr 12 breakpoint has not been genetically mapped yet.

T(1;10)11Ald. Recovered in a male after a heritable translocation test with spermatid sampling after 3-Gy gamma-irradiation. Breakpoints are at 1F and 10D2 (2).

T(2;9)11H. Induced by X-irradiation of spermatogonia of a (C3H ♀ x 101 ♂)F₁ male. T11H was found in a daughter (271). Cytological breakpoints are at 2E4 and 9A4 (231) and do not lead to somatic marker chromosomes. Meiotic configurations in males at metaphase I are 88 per cent RIV, and 12 per cent CIV (277). The estimated frequency of adjacent–2 disjunction is about 14 per cent. The Chr 2 breakpoint is about 2 units proximal to *pa*; the Chr 9 breakpoint is about 9 units from *se*, between it and *cw*. There are no signs of crossover suppression. Homozygotes are viable.

T(10;17)11Rl. This translocation arose spontaneously in non-irradiated controls (71) and is another member of the 'steeloid' group, with a phenotypic effect like that of steel *Sl*. Cytological breakpoints are at 10D and 17C. The steeloid phenotype is inseparable from the translocation, so the Chr 10 breakpoint is thought to be at the *Sl* locus. That for Chr 17 has not yet been determined.

T(10;18)12Rl. This translocation was discovered in the offspring of neutron-irradiated males in which exposed spermatogonia were sampled. Like the last, it is a member of the 'steeloid' group described by Cacheiro and Russell (71). Cytological breakpoints are at 10D and 18D. The Chr 10 breakpoint is presumably at the *Sl* locus; that for Chr 18 has not yet been determined.

T(2;4)13H. Induced by X-irradiation of spermatogonia of a (C3H ♀ x 101 ♂)F₁ male (271). Cytological breakpoints are at 2C1 and 4C7 and do not lead to clear somatic marker chromosomes. Meiotic configurations in spermatocytes at metaphase I are 87 per cent RIV and 13 per cent CIV (283). On Chr 2, T13H lies about 8 units from *fi* on its proximal side; the breakpoint on Chr

4 lies midway between the *b* and *m* loci. There is probably some crossover suppression in this region. Homozygotes for T13H are viable. Male double heterozygotes for T1Ca and T13H (both involving Chrs 2 and 4) appear to be early postnatal lethals (as judged by death of *Sd* males tagging T1Ca), although females are fully viable (27). However, male double heterozygotes for T13H and T1Sn are viable and form quadrivalent configurations. T13H has been used for studies on genetic imprinting phenomena which involve Chr 2 (87) in which anomalous phenotypes occur on intercrossing T13H/+.

T(8;16)17H. Induced by X-irradiation of spermatogonia of a (C3H ♀ x 101 ♂)F₁ male (271). Evans (128) reported breakpoints to be in 8E1 and 16C3; Eicher and Beamer (117) reported the latter to be 16C2. Configurations at metaphase I in the male are 48 per cent RIV, 33 per cent CIV, 3 per cent CIII + I, and 17 per cent bivalents only (277), with failures of association being mainly in the distal segments. T17H lies between the *Os* and *E^{so}* loci on Chr 8 and shows 7 per cent so recombination with *E^{so}*, but there are signs of some crossover suppression so the actual map distance may be greater (26a). It segregates independently of the *md* locus on Chr 16, confirming its distal location (35). Chromosome imbalance frequently leads to postimplantation death, often with exencephaly and other malformations (183).

T(10;18)18H. Induced by X-irradiation of spermatogonia of a (C3H ♀ x 101 ♂)F₁ male. T18H was found in a semi-sterile son (207). The breakpoints are in bands 10B4 and 18C (233); no somatic marker is produced. Configurations at metaphase I in males were 76 per cent RIV, 14 per cent CIV, 1 per cent CIII + I, and 9 per cent bivalents only (186, 277). This translocation is tagged by *v* on Chr 10, with which it shows 1.8 per cent recombination, and is only 4 units from *dl* between that locus and *Sl* (277). On Chr 18, T18H shows 34 per cent recombination with *Tw* and is only 3 units from *sy* (186), thus having a breakpoint very close to that for T50H. There is no evidence for crossover suppression on either chromosome. The homozygote is viable.

T(2;3)24H. Induced in spermatogonia of a (C3H ♀ x 101 ♂)F₁ male after X-irradiation (207). Cytological breakpoints are at 2E2 and 3H1, with the 3² product slightly longer than Chr 1 (283). Configurations at metaphase I of the male are 82 per cent RIV, 15 per cent CIV, and 4 per cent bivalents only. 12–21 per cent of the embryonic deaths in progeny of T24H heterozygotes are late (i.e. after the deciduomatal or small mole stage), and much of this is connected with exencephaly

(183), as in Snell's translocation, T(2;4)1Sn. T24H is between the *fi* and *we* loci on Chr 2 and about 8 units from *we*; there may be a slight crossover suppression (283). On Chr 3, the breakpoint is between *ma* and *Va*, about 4 units from *Va*, with very little crossover suppression (34). T24H has been used in genetic imprinting studies on Chr 2 (87) in which anomalous phenotypes occur on intercrossing T24H/+.

T(2;8)26H. Induced in spermatogonia of a (C3H ♀ x 101 ♂)F₁ male by fission neutron irradiation at low dose-rate, in a specific locus experiment (22). It was detected because it gave an agouti-umbrous phenotype when *a* was on the other chromosome. Cytological breakpoints are at 2H1 and 8A4 (54), giving rise to a long marker chromosome. Although Searle *et al.* (281) found 79 per cent RIV, 20.5 per cent CIV, and 0.5 per cent bivalent configurations in T26H/+ males at metaphase I, de Boer (53) found a lower proportion of rings (53 per cent) and a higher proportion of chains, presumably connected with a changed genetic background. The frequency of adjacent–2 disjunction at meiosis was estimated at about 17 per cent by Searle *et al.* (281) and 10 per cent by de Boer (53). The T26H breakpoint on Chr 2 is at the *a* locus; that on Chr 8 is 11 units from *Os* on its proximal side (277, 281). There is no evidence for crossover suppression. Homozygotes for T26H are viable and have a distinctive dark agouti phenotype. Embryonic death in T26H/+ progeny mainly occurs between the start of implantation and day 6; the numbers and types of abnormality correlate well with expectation in terms of chromosomal imbalance (57). Intercrosses of T26H/+ with distal or proximal markers led to the discovery of defective complementation phenomena which involve the distal regions of Chrs 2 and 8 (277, 278). These operate when both copies of the region concerned are maternal in origin and seem to be connected with genetic imprinting.

T(1;9)27H. Induced in spermatogonia of a (C3H ♀ x 101 ♂)F₁ male after fission neutron irradiation (22) in a specific locus experiment. Detected because it has the *se* phenotype with *se* on the intact Chr 9, it was therefore called *se^{5H}* at first. Cytological breakpoints are at 1C1 and 9E3 or 9E4, with the 9¹ product a long somatic marker, while 1⁹ is between Chrs 18 and 19 in size (230). Spectrum of configurations at metaphase I in a male was 56 per cent RIV, 43 per cent CIV, 1 per cent bivalents only (273). The T27H breakpoint is between *fz* and *ln* on Chr 1, about 13 units from *ln*, if crossover enhancement is allowed for (27). The breakpoint in Chr 9 is at *se*, but no recombination has been found with *tk* in a large number of tests. T27H homozygotes appear to be lethal *in utero* (273). T27H has been used to estimate that Chr 9 extends about 49 units beyond *se* (7).

T(2;16)28H. Induced in spermatogonia of a (C3H ♀ x 101 ♂)F₁ male by X-irradiation (146). It resembles Snell's translocation T(2;4)1Sn in several respects. Cytological breakpoints are at 2H4 (130) and at 16C3. Meiotic configurations at metaphase I in males are 3 per cent RIV, 5 per cent CIV, and 92 per cent bivalents only (283), showing that only short segments were exchanged because of the very distal positions of breakpoints. As in T1Sn, embryonic survival is high in progeny of females mated to carrier males, and embryonic death is frequently post-implantation and associated with exencephaly and cardiac disorders (183). T28H is about 7 units distal to *ls* on Chr 2 (274), probably with some crossover suppression, but the genetic breakpoint on Chr 16 has not been determined yet. Like other Chr 2 translocations T28H has been used in studies on genetic imprinting but does not give hyperkinetic and hypokinetic offspring on intercrossing, presumably because the Chr 2 anomaly factor is located proximal to the T28H breakpoint but distal to the T30H one (87).

T(2;11)30H. Arose after treatment of a (101 ♀ x C3H ♂)F₁ male with triethylenemelamine in a specific locus experiment (81). Like T26H it was detected because of a phenotypic effect at the *a* locus, since it gives a non-agouti phenotype when *a* is on the intact Chr 2. Cytological breakpoints are at 2H1 and 11B1 and do not lead to somatic marker chromosomes. Meiotic configurations in the male at metaphase I are 80 per cent RIV, 16 per cent CIV, 1 per cent CIII + I, and 3 per cent with 20 bivalents (283). The breakpoint in Chr 2 is at the *a* locus; that in Chr 11 is about 3 units from *vt* on its distal side (31), with some crossover suppression. The homozygote is a post-natal lethal, which usually dies before 2 weeks old with severe hydronephrosis but which can survive longer if urination is induced (89a). The compound with T26H (also with a breakpoint at *a*) is sometimes male-fertile (40). Cattanach and Kirk (89) have used T30H to show that the proximal region of Chr 11 is involved in a parental effect on size: a maternal duplication/paternal deficiency for this region in T30H/+ intercrosses gives offspring which are smaller than normal at birth. The absence of abnormally large young from the reciprocal cross (as expected from Rb(11.13)4Bnr results) suggested the presence of a lethal factor in the proximal region of Chr 2 (87). The authors also reported that maternal duplication/paternal deficiency for the distal part of Chr 2 in these intercrosses led to a lethal hypokinetic phenotype,

while paternal duplication/maternal deficiency gave a lethal hyperkinetic one.

T(3;12)30Rk. Found in the son of a DBA/2J male treated with 0.3 mg triethylenemelamine (97, 101). Cytological breakpoints are at 3A2 and 12C1. No multivalent configurations are found in heterozygous males.

T(5;12)31H. Induced in an oocyte of a (C3H ♀ x 101 ♂)F₁ female given acute X-irradiation (275). Male carriers are usually sterile with reduced sperm count, but some fertile ones have been recovered (47). About 17 per cent of spermatids are diploid (189a). Female carriers show a 73 per cent reduction in oocyte numbers (286a). Cytological breakpoints are at 5B and 12F1, giving rise to long and short marker chromosomes and to mainly CIV and CIII + I configurations at metaphase I in the male but there is more pairing in spermatocytes than oocytes (286a). Viable tertiary trisomics and monosomics are produced; they are sterile in males but fertile in females (47). However, ovarian volume was reduced in such females (226) and there was a 71 per cent reduction in numbers of oocytes (286). Synaptonemal complex analyses of these tertiary trisomics did not reveal any clear-cut relationship between pairing errors and gametogenic impairment (217). A double tertiary trisomic (N + 5¹² + 5¹²) male has been found, which was sterile and small, with a shortened head (46). T31H shows 31 per cent recombination with *Rw* in Chr 5, and the breakpoint is on the side opposite *go*, i.e. on the proximal side (47). On Chr 12 the breakpoint is very close to *Pre–1* in the order centromere, *Igh–1*, *Pre–1*, T31H, and there is marked crossover suppression (222). T31H seems lethal in the homozygote (33).

T(6;12)32H. Induced by X-irradiation of spermatozoa of a (C3H ♀ x 101 ♂)F₁ male (285). A semi-sterile daughter carried this translocation. It leads to sterility in males due to defective spermatogenesis, although a few spermatozoa may be produced (280). Chromosomal breakpoints are at 6G1 and 12B (134), giving long and short somatic marker chromosomes. Configurations at metaphase I of males are 12 per cent RIV, 68 per cent CIV, 18 per cent CIII + I, and 2 per cent bivalents only. Viable tertiary trisomics are produced, which have 20 bivalents and the small 12⁶ marker (43), and are fertile in the female (see Section 12.1.4.5). T32H is about 26 units distal to the *mi* locus on Chr 6 (36) but its position on the Chr 12 map is still unknown.

T(4;8S)In(8LS)36H. Induced by X-irradiation of spermatozoa of a (C3H ♀ x 101 ♂)F₁ male (285). Male carriers are sterile, with cessation of spermatogenesis mainly around metaphase I, although some spermatozoa are formed (280). The translocation, with Chr 4 breakpoint at 4D2, is associated with a pericentric inversion in Chr 8, with breakpoints at 8C1 and in the short arm of Chr 8, giving rise to a submetacentric marker chromosome with a reduced amount of C-band material (134). On Chr 4, T36H lies distal to *m* and about 14 units from it (30); on Chr 8 it shows 32 per cent recombination with the *Os* locus, with the breakpoint on the side opposite E^{so} (134). Presumably, this is the breakpoint in the short arm. T36H/+ females tend to produce fetuses with cardiac anomalies (183).

T(1;7)40H. Now called Is(7;1)40H. See Section 12.1.6.

T(11;19)42H. Induced by X-irradiation of spermatozoa of a (C3H ♀ x 101 ♂)F_1 male (285). Male carriers are sterile, with very small testes in which spermatogenesis stops at the pachytene stage of meiosis (280). Oogenesis is also impaired, with a retardation in ovarian growth (225a) and a 65 per cent reduction in oocyte numbers in 3–5 day female carriers (70). Cytological breakpoints are at 11D and 19B (134), which generate a small marker chromosome, while the 11^{19} product has a non-staining gap. This suggests that the breakpoint in Chr 19 is in a secondary constriction area near the centromere. On Chr 11, T42H is between the *vt* and *Re* loci and about 9 units from *Re* (38); on Chr 19 the breakpoint is about 20 units from *bm* on its proximal side, therefore very close to the centromere as expected (29). There is no evidence for crossover suppression in the distal part of Chr 19. Females heterozygous for T42H tend to produce anencephalic and exencephalic fetuses (183) as a result of chromosome imbalance.

T(16;17)43H. Induced by spermatozoal X-irradiation of a (C3H ♀ x 101 ♂)F_1 male (285). The mating produced a semi-sterile daughter who had sterile sons. Male sterility is caused by cessation of spermatogenesis around metaphase I, with very few post-meiotic germcells (280). However, male fertility can be restored by crossing to Rb(16.17)Bnr, so that T43H homozygotes can be produced (158). Cytological breakpoints are at 16A and 17B (134). Nearly all of Chr 16, including most of its centromeric heterochromatin, is transferred on to Chr 17, a parallel situation to that found in T(10;13)199H (q.v.). The small 16^{17} product is about the same size as the Y chromosome. Forejt and Gregorová (151) reported that there was non-random association between the C band of the X chromosome and the translocation configurations in T43H as well as in T199H and T(7;19)145H. In addition, the configurations tend to have positively heteropycnotic regions,

often associated with the allocyclic X chromosome, in some early diakineses. Configurations at male diakinesis/metaphase I are about 70 per cent CIV; by incorporating the t^{12} haplotype Forejt and Gregorová reduced the CIV frequency to 25 per cent in early diakinesis, which led to a higher incidence of centromeric heterochromatin contact between the X and the configuration. Gregorová *et al.* (158) have found viable and partially fertile products of T43H that carry the 17^{16} translocation product, but have only one normal Chr 16, i.e. they are Ts(17^{16})43H Ms16, thus having three doses of the region of Chr 17 up to the T43H break. T(10;13)199H (q.v.) produces similar unbalanced products. Forejt *et al.* (152) found that the T43H breakpoint on Chr 17 was about 23 units distal from *T* and close to the K (proximal) end of the *H–2* complex. This was confirmed by use of the trisomic described above. Agulnik *et al.* (6) placed T43H about 21 units from *T* and about 2 units from *tf*. Thus it leads to enhanced crossing-over in the *T* to *H–2* region. Beechey and Searle (35) found an RF of 6.7 ± 4.6 per cent between T43H and the *md* locus which is in agreement with the proximity of this locus to the centromere. Capková and Forejt (76) have produced a new congenic strain C57BL/10Sn–T43H//Ph which is homozygous for T43H and in which the *H–2* complex on 16^{17} is separated from the *T-t* complex on 17^{16}.

T(In1;5)44H. Found in the son of an F_1(C3H ♀ x 101 ♂) male given two doses of 5-Gy acute X-irradiation 24 h apart in a specific locus test, with spermatogonial sampling (47a). Kept because of its light coat colour, not associated with the specific loci and inseparable from the chromosome anomaly. Consists of a paracentric inversion of nearly all Chr 1, with breakpoints at A1 in centromeric heterochromatin and H6, followed by a reciprocal translocation between Chrs 1 and 5, with breakpoints at 1C2 and 5F. Both sexes breed but are semi-sterile. This compound shows 12.4 per cent recombination with the *W* locus (6.9 ± 2.9 per cent in coupling and 21.9 ± 6.5 per cent in repulsion) and the breakpoint is probably distal to it. At diakinesis/MI the heterozygote shows 88 per cent ring and 10 per cent chain quadrivalents, with one bivalent (130). Because of the paracentric inversion acentric and dicentric chromosomes are formed; in fact all 50 MI configurations examined included one or two dicentric chromosomes.

T(4;9)45H. Induced by spermatozoal X-irradiation of homozygous Rb(4.15)4Rma males (43a). Linkage tests have shown that its Chr 4 breakpoint is proximal to *b* with an RF of 1.8 ± 1.8 per cent between them, while it is close to *d, se* on Chr 9 (RF = 5.0 ± 3.4 per cent).

T(4;5)46H. Discovered in an F_1 male from a specific locus mutation experiment which involved combined treatment with hydroxyurea and X-rays (92b). Breakpoints are at 4A4 or 4A5 and 5B and multivalent frequencies are relatively low.

T(7;18)50H. Induced by X-irradiation of spermatogonia of a (C3H ♀ x 101 ♂)F_1 male (207). Breakpoints are within bands 7E2 to 7F2 and 18B3 to 18D (233). Configurations at metaphase I in the male are 47 per cent RIV, 47 per cent CIV, and 6 per cent bivalents only (277). The T50H breakpoint in Chr 7 is between *sh–1* and *fr* and about 4 units from the former, with no evidence for crossover suppression (26a). In Chr 18, there is about 32 per cent recombination between *Tw* and the breakpoint which is between *Tw* and *sy*, about 3 units from *sy* (186). Homozygotes for T50H are viable. The estimated frequency of adjacent–2 disjunction for this translocation is 29 per cent (277). Together with T9H it has been used to show that Chr 7 is involved in non-complementation phenomena, both with maternal duplication/paternal deficiency and the reverse in translocation intercrosses (279).

T(1;13)70H. Induced in spermatogonia of a (C3H ♀ x 101 ♂)F_1 male after X-irradiation (207). Cytological breakpoints are at 1A4 and 13D1, which generate both long and short markers (54). Male meiotic configurations are mainly CIV, but CIII + I is also found, leading to tertiary trisomic zygotes (see Section 12.1.4.5) and a few RIV (53, 281). The sperm count is slightly reduced in heterozygous males (53). This seemed to be associated with an overall reduction in chiasma frequency, a character showing high heritability (56). The frequency of adjacent–2 disjunction in heterozygotes was estimated to be about 13 per cent by Searle *et al.* (281) and 22 per cent by de Boer (53). T70H lies between the *fz* and *ln* loci on Chr 1, about one-third of the distance from *fz*. The translocation shows an RF of 8 per cent with *fz*, but there is some crossover suppression (281). On Chr 13 T70H lies between the *sa* and *pe* loci and about 14 units from *pe*, with no suppression of crossing over in this chromosome (274a). Homozygotes are viable (281).

T(9;17)138Ca. Induced by X-irradiation of spermatozoa of a male of the CWX stock, carrying *lx*, *s*, and W^v, mated to a CBA/H female (77). Cytological positions of breakpoints are at 9B and 17D (120), which do not lead to somatic marker chromosomes (48). The spectrum of configurations at male metaphase I is 41 per cent RIV, 41 per cent CIV, 18 per cent CIII + I or bivalents only, and 0.4 per cent with two univalents (148). Forejt (148) reported that presence of the t^{12}

allele significantly decreased chiasma frequency in the T138Ca configurations from 3.53 to 3.26, while Lyon and Glenister (199) found similar decreases with T/t^{h2} and T/t^6. In Chr 9, T138Ca lies between the *cw* and *d* loci, about 18 units from *d* (78, 199); Klein and Klein (184) reported that the breakpoint on Chr 17 was about 3 units from H–2 on its distal side. However, recombination over the *T-tf* and *T-H–2* regions is much enhanced, so that *T*-T138Ca recombination is about 35 per cent (184, 203). The t^6 haplotype reduces crossing-over between *T* and the translocation break in T138Ca, but does not abolish it completely (203). Lyon *et al.* (205) found large differences between reciprocal crosses, involving T138Ca heterozygotes, T/t^{h2}, and *cw/cw*, in frequencies of adjacent–2 disjunction. The anomalous results were probably connected with inviability of zygotes receiving only a paternal homologue of a certain region of Chr 17 (199). Homozygotes for T138Ca are viable.

T(7;19)145H. Induced by X-irradiation of a (C3H ♀ x 101 ♂) F_1 male mouse, with sampling of spermatozoa and spermatids (202). T145H males are sterile, with spermatogenic arrest during meiosis so that few if any spermatozoa are produced. Cytological breakpoints are at 7B3 and 19D1 (120) leading to production of a small marker chromosome (7^{19}). Tertiary trisomics with a normal chromosome set plus Chr 7^{19} are runted at birth (121). Configurations at metaphase I in males are 3 per cent RIV, 40 per cent CIV, 29 per cent CIII + I, 14 per cent with bivalents only, and 16 per cent with two univalents (202). T145H is about 0.8 units from *p* on Chr 7 (196), the order of loci being *Gpi-1*–Is1Ct–*ru-2*–T145H–*p* (121). On Chr 19, T145H is about 3 units from *ep* (196) and distal to *ru* (121). Forejt and Gregorová (151) have reported non-random contact between the C bands of the X-chromosome and the T145H configuration in diakinesis/metaphase I plates. They consider that the impairment of spermatogenesis found in this and other male-sterile translocations is related to this phenomenon. For T31H, T43H, and T145H, Forejt and co-workers (153) have also reported an association between the synaptonemal complexes of the translocations at pachytene and the XY pair.

T(1;17)190Ca. Induced by X-irradiation of spermatozoa of males carrying *lx*, *s*, and W^v (77). Cytological breakpoints are at 1H5 and 17B (120), which generate long and short somatic marker chromosomes (48). The T190Ca break in Chr 17 is very close to the *tf* locus (with complete absence of recombination) but seems to be proximal to it (209). When originally induced, this translocation carried the *t*-haplotype, t^6, which causes

596

crossover suppression in this area. Present stocks of T190Ca carry a mutant haplotype of t^6, t^{h17}, which permits recombination. This has revealed that T190Ca enhances recombination in the proximal segment of Chr 17, since *T–T190Ca* recombination is about 20 per cent, whereas normal *T–tf* recombination is 7 per cent (195). Carter *et al.* (78) found 28 per cent recombination between the t^6 tag and *ln* in T190Ca heterozygotes, with the break lying distal to *ln* on Chr 1. Crossing-over is not suppressed in this chromosome. Homozygotes cannot be obtained. Configurations at male metaphase I are 51 per cent RIV, 42 per cent CIV, and 7 per cent CIII + I or bivalents; frequencies of RIV are reduced in the presence of t^{12} (148). Tertiary trisomics, Ts(17^1)190Ca, survive for a few days after birth (204).

T(10;13)199H. This male-sterile translocation was induced by X-irradiation of a (C3H ♀ x 101 ♂)F$_1$ male (202). Males are sterile because of failure of spermatogenesis at the primary spermatocyte stage. Metaphase I may be reached but few if any later stages are seen. No ring configurations were found at metaphase I, but there were 85 per cent CIV, 3 per cent CIII + I, and 12 per cent bivalents only. Forejt and Gregorová (151) have reported a non-random association between X-chromosomal C bands and the translocation configuration in diakinesis/metaphase I plates of these sterile males, as well as positively heteropycnotic regions of the configuration, apparently in phase with the allocyclic X. Cytological breakpoints are at 10C1 and 13A1 (110), with virtually all of Chr 13, including C-banding material of the centromeric region, translocated on to the distal part of Chr 10. The translocated chromosome with interstitial C-band material is a distinctive mitotic marker. T199H lies about 4 units from *gr*, probably on the distal side (197). On Chr 13, the centromeric breakpoint is about 17 units from *bg* (198). T199H/+ females produce runted offspring, which have 40 chromosomes, but with the 10^{13} translocation product in place of Chr 13. Both sexes with this condition are fertile, and the condition behaves like a dominant (120).

T(5;13)264Ca. Induced by spermatozoal X-irradiation of a male of the CWX stock carrying *lx*, *s*, and W^v (77). Breakpoints are at 5E2 and 13B (128) and do not give somatic marker chromosomes. At metaphase I in the male, mainly RIV configurations are formed, with occasional CIV, CIII + I and bivalents only (281). This translocation is tagged by W^v on Chr 5, being about 4 units distal to it (78). On Chr 13 (LG XIV), the order is T264–*f*–*cr* with the breakpoint about 8 units from *f* (241), while recombination between the W^v tag and *pe* is about 22 per cent (281). There is no evidence for

crossover suppression. The frequency of adjacent–2 disjunction in T264Ca heterozygotes is about 9 per cent (281). Viable male homozygotes have been obtained recently (44).

12.1.4.4 Transposition

Tp(Y)1Ct. This transposition of a proximal part of the Y chromosome, including *Tdy* on the short arm, to a terminal position in the X–Y pairing region, is also known as sex-reversed, *Sxr*, and is described under that name in Chapter 2.

12.1.4.5 Tertiary monosomies and trisomies

Tertiary monosomy or trisomy, associated with counts of 39 or 41 chromosomes respectively, originates when there is numerical meiotic non-disjunction in a translocation heterozygote. This can lead to the production of gametes deficient in one of the translocated chromosomes, or having an extra translocated chromosome (202, 272). Viable zygotes may result if the amount of chromosomal material lost or gained is not too large. This tends to happen with certain male-sterile translocations in particular, when one breakpoint is near the centromeric end and the other more distal.

Lyon and Meredith (202) described tertiary trisomic mice in the offspring of female heterozygotes for the male-sterile translocations T158H and T194H. Male trisomics were sterile, with small testes; females showed somewhat reduced fertility. T158H trisomics tended to have abnormal postural reflexes. Monosomics for T194H were also found, again with sterility in the male and low fertility or sterility in the female. These monosomics usually showed an abnormal phenotype, with widening of the skull, though association of phenotype with abnormal karyotype was not complete. Tertiary monosomy and trisomy (with the small marker chromosome deficient or in excess) have been studied in the partially male-sterile translocation T(5;12)31H (q.v.). Male monosomics are sterile, with sperm-head abnormalities; litter-size is reduced in offspring of females, and it seems likely that some monosomics die before birth (47). However, a male-fertile tertiary monosomic for the pericentric inversion In1Wa (q.v.) has been found (26).

Tertiary trisomics for T(14;15)6Ca, with 20 bivalents and the 15^{14} marker chromosome, are viable but retarded. Cattanach reported a frequency of 5.8 per cent from XO T6Ca/+ females; they tended to show nervous movements and head-shaking (82). Eicher (116) reported that these trisomics were produced by female and not male translocation carriers. This was in line with Baranov and Dyban's (17) findings of more abnormal embryos at 11 to 12 days p.c. in offspring of female than of male T6Ca heterozygotes. Oshimura and Takagi (239) have explained Eicher's finding by their discovery that the frequency of non-disjunction at metaphase II involving the T6Ca small marker is 22 per cent in female heterozygotes but only 4 per cent in male heterozygotes. Kaufman (181) has confirmed the high frequency of aneuploid oocytes in these heterozygotes. Baranov and Dyban (20) showed that the frequency of trisomic progeny from T6Ca/+ females depended greatly on maternal genotype, being about three times as high in (C3H x CBA-T6)F_1 females as in (C57BL x CBA-T6)F_1 females. Tertiary trisomy, Ts(X^{11})38H, has also been found for the small translocation product of T(X;11)38H; males are sterile, but a female has bred, giving small litters (282).

Two trisomic stocks have been developed as follows:

Ts(12^6)32H. This arose from T(6;12)32H (q.v.) and has the normal chromosome complement plus the short 12^6 somatic marker. Only females are fertile but show growth retardation. In Ts32H males very few sperm are formed (43).

Ts(1^{13})70H. This arose from the reciprocal translocation T(1;13)70H (q.v.), which is normally fertile in the male but has long and short marker chromosomes. The trisomic mouse carries the short marker in addition to a normal set of chromosomes (52). Both sexes are fertile, though litter-size from intercrosses is low. Some of the trisomics show characteristic shortening and malformations of the anterior skull bones, often with bent nasals and associated tooth abnormalities. Further studies (55, 61) have shown that trisomic males are oligospermic (about 30 per cent normal sperm count in epididymis) which leads to delayed fertilization. Many spermatozoa are morphologically abnormal (77 per cent in uterus and 25 per cent in oviduct), but these abnormalities seem unrelated to the haploid karyotype. Meiotic examination showed that the telomeric part of Chr 13 had a greater potential for forming chiasmata than the proximal segment of Chr 1, which included centromeric heterochromatin. Fifty per cent of secondary spermatocytes from heterozygous males carried the small marker, showing that there was no loss of the marker at first meiotic division.

12.1.5 Robertsonian translocations

These translocations (otherwise known as centric fusions or whole arm translocations) were first reported

in the mouse in 1967. Since then over 60 have been discovered, mainly in wild populations but also increasingly in laboratory strains. Every autosome is involved in at least one separately maintained Robertsonian translocation and an X-autosome one has now been reported (2).

Gropp and Winking (162) have reviewed the properties and behaviour of Robertsonian translocations, especially those from feral populations. Chromosome combinations in these have been tabulated (295a). Those of the tobacco mouse (*Mus poschiavinus*) are the best known and show a number of interesting features that seem to reflect evolutionary divergence between this species and *Mus musculus*. Thus heterozygotes containing tobacco mouse metacentrics and house mouse chromosomes usually show high frequencies of non-disjunction, and some may exhibit crossover suppression for proximal chromosomal regions, as described later. The former property, not found in laboratory-derived Robertsonians apart from Rb1Ct (q.v.), has been used for the generation of zygotes monosomic and (especially) trisomic for particular chromosomes. Combinations of two Robertsonians sharing a common chromosome arm (i.e. with monobrachial homology) have proved particularly useful (166), since these have very high frequencies of non-disjunction, allow trisomies for specific chromosomes to be predetermined, and permit the study of nullisomics. However, male compounds of two Robertsonians with a common arm are frequently sterile (127). Gropp and Winking (163) have provided a list of those compounds which are fertile, sub-fertile, or sterile in the male, but pointed out that this depends partly on genetic background.

The metacentrics formed by Robertsonian translocations are useful somatic markers. They also allow the construction of stocks with karyotypes more similar to the human one, in which the induction of pericentric inversions can be studied, for instance.

12.1.5.1 Laboratory-derived

Rb(10.14)1Ad. Recovered after 200 mg/kg procarbazine treatment with spermatid sampling, in heritable translocation test. Has been made homozygous (1).

Rb(6.15)1Ald. This was first observed by Léonard and Deknudt (188) in their subline of the AKR inbred strain, all members of which were homozygous for it. Litter-size in the strain was low, but this was probably for other reasons. There is no evidence for crossover suppression in this translocation (86) and the frequency of non-disjunction in heterozygotes is known to be low

(165). Baranov and Dyban (21) have studied chromosomally abnormal embryos generated by Rb1Ald.

Rb(8.19)1Ct. Detected in a test for non-disjunction involving Rb(8.12)5Bnr (91). The translocation is fertile in both hetero- and homozygotes, but is male-sterile in combination with Rb5Bnr, with absence of sperm. It is the first Robertsonian translocation of house mouse origin to cause high levels of non-disjunction (88a).

Rb(11.14)1Dn. Arose in a backcross of T(3;12)20Rk to the AEJ/GnRk strain. It is maintained homozygous on the same AEJ background (101).

Rb(6.13)1H. Found in a mutant stock at Harwell. Heterozygotes and homozygotes are fertile in both sexes (245).

Rb(3.5)1Icg. Found in an outcross of a heterozygote for Rb(8.17)1Iem. Has normal fertility in the heterozygote and is being maintained in homozygous state (5).

Rb(8.17)1Iem. This translocation appeared in a colony of white mice kept at the Institute of Experimental Medicine, Leningrad (18). There was no noticeable effect on fertility in heterozygotes or homozygotes of either sex, but trisomic embryos are found in the progeny of heterozygous females, though not males (19). However, the frequency of non-disjunction in heterozygotes is fairly low (165).

Rb(7.13)1Oxe. First found in a cell from a *nu/nu* mouse (131) and is being established on this background, as it could prove a useful *nu* marker.

Rb(4.15)1Rk. Found in a male from an inversion stock homozygous for In(1)12Rk (106).

Rb(5.19)1Wh. White and Tjio (299) reported the presence of this Robertsonian in a randomly bred albino stock of unknown origin at the US National Institutes of Health. Fertility of homozygotes was normal; that of heterozygotes was somewhat reduced. There was a low incidence of aneuploid cells at meiotic metaphase II. White *et al.* (300) showed that Chrs 5 and 19 were involved and combined Rb1Wh with Rb(9.19)163H in order to generate trisomy–19. These mice sometimes survive to birth but those with one Rb1Wh and two Rb163H translocations frequently have cleft palate (301). Higher frequencies of trisomy 19 are found earlier in gestation, and those surviving weigh less than normal mice (302).

Rb(X.2)2Ad. Arose spontaneously in a heritable translocation test, as the son of a female carrying T(5;13)5Ad (2). Has now been made homozygous and is the first X-autosome Robertsonian to survive.

Rb(1.15)2Ct. Found in a stock segregating for Rb(16.17)7Bnr. Heterozygotes and homozygotes are fully fertile in both sexes (92). There is some evidence for raised levels of non-disjunction in heterozygous females (86b) but crossing-over with *fz* and *ln* is not suppressed.

Rb(11.16)2H. Arose spontaneously in a stock carrying Rb(4.6)2Bnr (206). Heterozygotes of both sexes are fully fertile.

Rb(6.19)2Icg. Found in a cross involving Rb(8.17)1Iem; is being maintained in homozygous state (4).

Rb(3.15)2Rk. Found in an F_1 male from a cross between a female carrying Rb(16.17)7Bnr and a T/t^6 male (106). Heterozygotes are fertile.

Rb(12.13)3Ct. This was first detected in a male heterozygous for Rb(9.14)6Bnr. Both hetero- and homozygotes are fully fertile, as are double heterozygotes with Rb6Bnr (88). Levels of non-disjunction are not raised in heterozygotes (88a).

Rb(4.18)3H. Arose spontaneously in a stock carrying Rb(16.17)7Bnr and T(1;17)190Ca. Hetero- and homozygotes of both sexes are fertile (201).

Rb(2.17)4H. Found in two F_1(C3H/HeH x 101/H) sibs (94).

Rb(2.6)4Iem. Arose spontaneously in laboratory mice (13).

Rb(1.2)5H. Arose in a stock of mice carrying the genes Sl^{con} and *dl*, with formation of a particularly large metacentric (156). A homozygous stock has been constructed (211).

Rb(9.19)163H. Evans *et al.* (138) found this translocation, giving a submetacentric chromosome, in a stock carrying Cattanach's translocation Is(In7;X)1Ct (see Section 12.1.6) and showed that homozygotes and heterozygotes of both sexes were phenotypically normal and fertile, although fertility of the male heterozygotes is somewhat reduced. The frequency of non-disjunction in translocation heterozygotes is very low (210). There is no evidence for crossover suppression (86).

12.1.5.2 Wild-derived

The *Rb1Bnr* to *Rb7Bnr* series ('Pos')

These comprise the most-studied group of Robertsonians. They were discovered in the dark-coloured tobacco mouse *Mus poschiavinus*, which has been caught in the wild only in the high-altitude Valle di Poschiavo in S.E. Switzerland (167, 168). This species has seven autosomal metacentrics and thus differs from the house mouse *Mus musculus domesticus* with respect to seven Robertsonian translocations, given below.

Fertility is markedly reduced in the F_1 between the tobacco and the laboratory mouse because of gametic aneuploidy (295) resulting from non-disjunction. When the individual Robertsonian translocations are isolated (90), some show crossover suppression in proximal chromosome regions (86). Cattanach concluded that this probably stemmed from minor structural or genic differences between the tobacco mouse and house mouse chromosome regions concerned and was not related to non-disjunction. Chromosomal banding patterns of the metacentrics have been determined by Zech *et al.* (310) and linkage groups by Cattanach and co-workers (86, 90, 93). Non-disjunction frequencies in heterozygotes for these Robertsonians have been estimated from metaphase II analysis (86), from frequencies of fetal aneuploidy at various stages (144, 144a, 165), and by genetic method (210, 293a). Gropp *et al.* (165) found evidence for a higher rate of non-disjunction in heterozygous females, with possibly some selection against unbalanced male germ cells.

Rb(1.3)1Bnr. Estimates of non-disjunction frequencies in Rb1Bnr heterozygotes, based on metaphase II chromosome counts, range from 14 to 20 per cent (86); Ford and Evans (144) obtained a figure of 22 per cent from trisomic implants. These high frequencies of non-disjunction, which occur even on a laboratory mouse background, are associated with increased zygotic loss (90), which has been studied both cytologically and morphologically (165). There is a higher frequency of non-disjunction in female Rb1Bnr heterozygotes than in males (304). The compound of Rb1Bnr and Rb2Ct, which have monobrachial homology for Chr 1, gives a very high frequency of non-disjunction for that chromosome, which has been used in the MBH method for detecting non-disjunction and chromosome loss (92a).

Testis weights are reduced in both homo- and heterozygotes, more so in the former, in which the sperm count is also reduced to about 70 per cent of controls (90). Rb1Bnr markedly suppresses crossing-over in proximal regions of Chr 1, with an extension of the suppression for some distance distal to *fz* (86).

Rb(4.6)2Bnr. Non-disjunction frequency estimates for heterozygous males are 14 and 34 per cent from metaphase II counts (86) and 25 per cent from trisomic implants (144). Lyon *et al.* (210) obtained a frequency of 15 per cent (increasing steadily with maternal age) for non-disjunction in Chr 4 with both sexes combined. There is marked zygotic loss in progeny even on a laboratory mouse background (86b), but only a slight reduction in testis weights of homo- and heterozygotes and in sperm counts of the former (90). There is crossover suppression in the Chr 4 region between the *b* locus and the centromere, but little evidence for it on Chr 6 (86). Evans (127) reported that the compounds of Rb2Bnr and Rb(6.15)1Ald were male-sterile; Gropp and Winking (164) found them male-fertile in 1978 but male-sterile in 1982.

Rb(5.15)3Bnr. Estimates of non-disjunction frequencies from metaphase II counts in heterozygous males range from 13 per cent to 28 per cent (86) with a frequency of 9 per cent from trisomic implants (144). Cattanach and Moseley (90) found a high level of zygotic loss with this translocation, which is also present on a laboratory mouse background (86b). They also reported testis weight reductions in Rb3Bnr homo- and heterozygotes, with sperm count in the former being about 60 per cent of normal. Rb3Bnr suppresses crossing-over proximal to the *W* locus in Chr 5, but there is no evidence for a similar phenomenon in Chr 15. Evans (127) reported that the compound of Rb3Bnr and Rb(6.15)1Ald showed reduced fertility in the male, although Gropp and Winking (163) found it fully fertile.

Rb(11.13)4Bnr. Non-disjunction frequencies seem to be particularly high in heterozygotes for this translocation, being estimated as 28 per cent by Gropp *et al.* (166) and 29 per cent by Cattanach and Moseley (90) from metaphase II counts. Ford and Evans (144) estimated non-disjunction frequency to be 36 to 39 per cent. Zygotic loss is also high (90), even on a laboratory mouse background (88a); testis weight is reduced in both homo- and heterozygotes. Gropp *et al.* (165) found 18 per cent aneuploid embryos in 8-day fetuses from Rb4Bnr heterozygous females, 97 per cent of these being trisomic. There is no good evidence for crossover suppression in either chromosome arm (86).

Rb(8.12)5Bnr. Non-disjunction frequencies are much lower than in Rb4Bnr, with estimates of 4 and 8 per cent from metaphase II counts (86), with zygotic loss also much lower, as expected (90), and with frequencies of aneuploid embryos of only about 1 per cent in progeny of heterozygotes of both sexes (165). However, testis weights are reduced to about 75 per cent of controls in both homo- and heterozygotes, while the sperm count in homozygotes is also markedly lower. There are no signs of crossover suppression (86). When Rb5Bnr is combined with Rb(8.17)1Iem, Rb(8.19)1Ct, or Rb(3.8)2Rma resultant males are sterile (91, 163).

Rb(9.14)6Bnr. Non-disjunction frequency estimates from metaphase II counts vary between 8 and 23 per cent for this translocation (90, 141, 166). Ford (141) also estimated non-disjunction frequencies of 39 and 38 per cent from pre- and post-implantation embryos. This variability has been discussed by Lyon *et al.* (210) who reported a frequency of 15 per cent in both sexes combined for the Chr 14 arm using genetic markers. Cattanach and Moseley (90) found a fairly low level of zygotic loss, and testis weights and sperm counts were somewhat reduced, as in other members of this series. There is no evidence for crossover suppression (51). Compounds of Rb(9.19)163H and this translocation are sterile (127) or subfertile (163) in the male.

Rb(16.17)7Bnr. Non-disjunction frequencies estimated from metaphase II counts vary from 4 to 12 per cent (86), with a figure of 10 per cent from post-implantation embryos (144a). The frequency of aneuploid embryos is much higher in progeny of female than of male heterozygotes (165). In heterozygotes, testis weights and sperm counts are similar to those of controls, but testis weight is reduced in the homozygote (90). There are, as yet, no data on crossover suppression in Chr 16 and no evidence for it on Chr 17 (86). The compound of Rb7Bnr with Rb(8.17)1Iem is male sterile (163), but that with Rb(2.17)1Rma is not (303).

The Rb8Bnr to Rb13Bnr series ('Alp')

Gropp *et al.* (169) described members of this series (mainly with different preliminary designations), and further information is given by Capanna *et al.* (75) and Gropp (159). The mice concerned were trapped in various predominantly Alpine areas of Switzerland and showed an array of different combinations of metacentric chromosomes. They are considered to be *Mus musculus*.

Rb(10.11)8Bnr. From mice caught in Vicosoprano, Val Bregaglia (Grisons). The same combination of chromosomes was found in mice from Poveredo (Val Mezolana) and from Chiavenna (169). Heterozygous females give low frequencies of aneuploid fetuses (165).

Rb(4.12)9Bnr. From mice caught at Mutten (Albula Valley, Grisons). The same metacentric was found in mice from Bottmingen (Basel area), Chiavenna (Italy), and Roveredo (Grisons) (169). Heterozygotes give about 3 per cent aneuploid fetuses (165), but there is no clear evidence for crossover suppression between the *fz* and *ln* loci in heterozygotes of either sex (86a).

Rb(1.10)10Bnr. This translocation, originally called 'Bondo', came from the place of that name in the Val Bregaglia (169). Female heterozygotes give about 6 per cent aneuploid fetuses (165), but there is no clear evi-

dence for crossover suppression between the *fz* and *ln* loci in heterozygotes of either sex (86a).

The Rb1Rma to Rb9Rma series ('CD')

Capanna *et al.* (74) reported a population of *Mus musculus*, living in the Central Apennines and stretching as far as the Adriatic coast, that had a diploid number of only 22, with nine pairs of metacentric chromosomes. Those introduced into the laboratory were trapped at Cittaducale (Northern Latium), hence 'CD'. Further information is given in Capanna *et al.* (75) and Gropp (159) with revision of some of the karyotypic identification and nomenclature. Eight of the nine metacentrics involve different combinations of chromosomes from those in the tobacco mouse series. Crosses between members of these two multimetacentric populations lead to the production of supermultivalent configurations of very regular appearance (chains or rings) in oocytes and spermatocytes, providing proof of the preservation of homology in the chromosome arms (75, 162). However, male offspring from such crosses were sterile, and there was a very high frequency of non-disjunction in female hybrids. Heterozygotes with laboratory mice (NMRI-Han) showed high frequencies of non-disjunction in both sexes (75).

The Robertsonians are as follows:

Rb(1.7)1Rma
Rb(3.8)2Rma
Rb(6.13)3Rma
Rb(4.15)4Rma
Rb(10.11)5Rma, having the same metacentric as Rb8Bnr
Rb(2.18)6Rma
Rb(5.17)7Rma
Rb((12.14)8Rma
Rb(9.16)9Rma.

The Rb10Rma to Rb18Rma series ('CB')

Like the previous series these are derived from the Central Apennines (Campobasso, Molise) from a population having nine pairs of metacentric chromosomes (75). However, only one metacentric, namely Rb(9.16)18Rma, has the same combination of chromosomes found in a 'CD' metacentric. Crosses of 'CB' mice with 'CD' or *poschiavinus* lead to complete male sterility in the offspring. Definitive designations of the Robertsonians maintained in this series are as follows:

Rb(1.18)10Rma
Rb(2.17)11Rma
Rb(4.11)12Rma
Rb(6.7)13Rma

Rb(5.15)15Rma, like *Rb3Bnr*
Rb(8.14)16Rma
Rb(10.12)17Rma.

The Rb1Lub to Rb75Lub series (Lübeck)

Four additional feral mouse populations from Italy have been collected by Dr E. Capanna; one has been collected by Dr B. Dulic in Yugoslavia. All have been studied at Lübeck (159, 305, 307).

The first population was collected from Alpie Orobie, near Bergamo in N. Italy. The following are maintained:

Rb(1.3)1Lub, like *Rb1Bnr*
Rb(2.8)2Lub
Rb(5.15)4Lub, like *Rb3Bnr* and *Rb15Rma*
Rb(10.12)5Lub, like *Rb17Rma* and *Rb16Lub*
Rb(11.13)6Lub, like *Rb4Bnr*
Rb(16.17)8Lub, like *Rb7Bnr*
Rb(7.18)9Lub.

The second population was collected near Ancona in Central Italy (159). Maintained members are:

Rb(3.9)12Lub, like *Rb20Lub*
Rb(4.17)13Lub.

The third population was collected in Lipari, Isole Eolie, Southern Italy (159). Maintained members are:

Rb(1.2)18Lub, like *Rb10Lub*
Rb(5.14)21Lub
Rb(8.12)22Lub, like *Rb5Bnr*
Rb(10.15)23Lub
Rb(6.16)24Lub, like *Rb14Lub.*

The fourth group stemmed from one male offspring of wild mice trapped near Milan (307). There were eight pairs of metacentrics, but the only survivor in the laboratory is:

Rb(16.17)32Lub, like *Rb7Bnr* etc.

The fifth group (305) originated from wild mice trapped in Zadar, Yugoslavia. Seven out of eight specimens showed a karyotype with 12 metacentrics; one animal had 11. The only maintained Robertsonian is:

Rb(8.17)38Lub, like *Rb1Iem.*

Other maintained Robertsonians in the Lübeck series are:

Rb(16.17)54Lub. From Cremona in Northern Italy (161).

Rb(15.17)64Lub. Recovered from a sample of 19 mice caught in a coastal region of NW Greece (306) with two

other Robertsonians which are not maintained. Rb64Lub is now bred on a *tf/tf* background.

Rb(5.13)70Lub. Isolated from wild mice caught in Luino, NE Lombardy in Italy (170, 305).

Rb(16;17)74Lub. Also isolated from the Luino population (170, 305). One male with Rb74Lub also carried a lethal t-allele, t^{Lub6}, associated with an unusually large block of heterochromatin on the Chr 17 arm.

Rb(16.17)75Lub. Isolated from a male mouse trapped near Pavia which was also homozygous for 9.14 and 16.17 metacentrics (305).

The Rb1Sic to Rb7Sic series (Sicily)

Wild mice were obtained from several locations near Palermo, and some were found to have high numbers of metacentrics though others had none (187). In all, several different metacentrics were present in the population, but some mice were heterozygous for one or more. The available Robertsonian translocations are as follows:

Rb(6.12)3Sic
Rb(8.17)6Sic, like *Rb1Iem.*

The Tubingen series

Adolph and Klein (3) have described wild populations of mice with Robertsonian translocations found in Southern Germany, Spain and Scotland. For the last see also refs. 67 and 68. The Robertsonians are as follows:

Rb(4.12)1Tu. Found homozygous in the Ravensburg area of Southern Germany, also in the Tubingen area.

Rb(2.5)2Tu. Found in the Ravensburg area only, usually in heterozygous state.

Rb(3.6)3Tu. Found in Ravensburg in homozygous or heterozygous state, with the same arrangement found also near Milan.

Rb(11.13)6Tu. Found commonly around Ravensbruck, with the same arrangement common in the Alpine system.

Rb(3.14)7Tu. Found on Eday, one of the Orkney islands in Scotland, but not on the mainland (Caithness).

Rb(4.10)8Tu. Present in wild populations both on Eday and on the Scotland mainland.

Rb(6.13)9Tu. Only found on the Scotland mainland.

Rb(9.12)10Tu. One of three pairs of metacentrics (Rb7Tu, Rb8Tu, Rb10Tu) recovered from Eday but also found on the Scottish mainland.

Rb(4.14)11Tu. Found in Spain south of a demarcation line which separates all-acrocentric and metacentric populations (3).

Rb(5.15)12Tu. From the same region of Spain, like those below:

> *Rb(6.10)13Tu*
> *Rb(9.11)14Tu*
> *Rb(12.13)15Tu.*

Other wild-derived Robertsonians

Rb(9.12)1Cam. Found in wild mice from Peru kept in laboratory conditions (297).

Rb(13.16)1Mpl. Found in four mice from Ibiza (Balearic Isles), but not present in Majorca or Minorca, which had all-acrocentric chromosomes (65).

Rb(4.12)1Nam. Found in wild mouse populations in four different localities in Belgium in heterozygous or homozygous state (173, 174). Has now been introduced into the BALB/c and C57BL strains (175).

Rb(5.10)3Nam. As above (173).

Chromosomal polymorphism has also been discovered in Tunisia (270a) with populations in and near Monastir being mainly homozygous for the following nine Robertsonians: Rb(1.11)2Mpl, Rb(2.16)3Mpl, Rb(3.12)4Mpl, Rb(4.6)5Mpl, Rb(5.14)6Mpl, Rb(7.18)7Mpl, Rb(8.9)8Mpl, Rb(10.17)9Mpl, and Rb(13.15)10Mpl. These mice also showed a high frequency of a distal duplication in Chr 1, found in other populations too.

12.1.6 Insertions

Is(In7;X)1Ct (Cattanach's translocation). This first appeared in a mutation experiment in which CBA males were treated with triethylenemalamine and then mated to females of a specific locus tester stock (PCT) homozygous for seven recessive genes including c^{ch} and p (79). An F_1 female, derived from a treated spermatozoon, showed a white variegated pattern on an otherwise wild-type background. This 'flecked' phenotype was inherited in crosses to $p\ c^{ch}$ homozygotes. The female carried a translocation whereby part of Chr 7, including the c, p, and ru–2 loci, was inserted into the X chromosome; variegation for each gene can be seen in appropriate genotypes. The variegated phenotype results primarily from a combination of two phenomena: (i) random X-inactivation (194); and (ii) a V-type position effect operating when the X chromosome carrying the translocated segment is inactivated, with spread of inactivation into this segment from both sides (83, 84). By this means, wild-type alleles on the translocated segment are inactive in some clones of pigment cells, so that recessive alleles on the non-translocated homologue (or homologues) can manifest their phenotypic effects on coat colour. Two types of flecked females occur, both fully viable: Type I with the balanced translocation and Type II with the transposed piece of autosome present as a duplication: *Dp(In7;X)1Ct*. Of the equivalent types of male, Type I is viable but tends to become sterile early in life, whereas only about one-quarter of Type II males are viable, but are fully fertile as a rule. The corresponding deficiency is lethal in both sexes.

The translocation is non-reciprocal (237), being an inverted insertion of about one-third of Chr 7 into the X chromosome (114) so as to produce a long somatic marker. No crossing-over has been detected within the inverted segment. The genetic positions of breaks on Chr 7 are between ru–2 and qv (115) and between sh–1 and Hbb (113, 308). In the X chromosome the breakpoint is between Mo and Gy, with 4 per cent recombination between it and Ta and 2 per cent between it and Mo^{br} (80). No crossovers have been found in several hundred offspring from crosses with jp (Cattanach, personal communication), so there is marked crossover suppression. The breakpoints are in the proximal one-third of 7C and at the distal edge of 7E3 or proximal edge of 7F1, while the X breakpoint is between XE1 and XF1 or in the proximal region of band XF1 (121).

Cattanach's translocation has been used extensively in many investigations. These include: (i) studies on the spreading effect and its two-directional nature, which has a bearing on the control of X-inactivation (85); (ii) position effect variegation and the reactivation of previously inactive loci (84) as well as the action of the X-chromosomal controlling element, *Xce* (83); (iii) studies on imprinting and X-inactivation (139, 237, 247a); (iv) comparisons of variegation in chimeras and heterozygotes for this insertion (93a); (v) biochemical studies on gene dosage (119); (vi) effects of X-inactivation during spermatogenesis (124).

Is(7;InX)1Neu. Arose after treatment of a female with methyl methanesulphonate in a specific locus experiment. An interstitial deletion of part of Chr 7, including the p locus, has been inserted into an inverted segment of the X-chromosome. Breakpoints are at XB, XD (the point of insertion), and XF, also at 7C and 7D (2). Both balanced and unbalanced progeny of this stock are viable and fertile.

Is(7;1)40H. Induced by X-irradiation of spermatozoa of a (C3H ♀ x 101 ♂)F₁ male (285). Male carriers are sterile, with cessation of spermatogenesis at or before pachytene and with testis weight reduced to 30 per cent of normal (282). This is an insertion of part of Chr 7 into Chr 1, with breakpoints in bands 1B, 7B1, and 7F1. One long marker is formed and one the same length as the Y. The breakpoint in Chr 1 is between *fz* and *ln*, the recombination with *fz* being 24 per cent. In Chr 7 there is linkage between the insertion and *fr*, with a breakpoint about 12 units proximal to this locus. Linkage tests with *ru–2* and *p* suggested that the other breakpoint on Chr 7 was 0.2 units distal to *ru–2*. However, later studies showed no recombination between the breakpoint and *Gpi–1s* (41) so this locus is now thought to be within the insertion; thus the Is40H breakpoint is proximal to that of Is1Ct, as expected from G-banding data.

Is(HSR;1C5)1Lub (provisional symbol). This apparent insertion of a homogeneously staining region (HSR) into Chr 1 has been found in a number of wild populations (296) from Spain to Russia. The size of the extra segment varies from 6–30 per cent of a standard Chr 1. It has now been introduced into laboratory mouse strains and has been identified in a subline of RBA/Dn (99), which will be called RBA1/Dn in future.

12.1.7 Inversions

While reciprocal translocations have been known in the mouse for over 50 years, chromosomal inversions have been detected only in the last 20, their production being a matter of deliberate policy for the development of tester stocks for studies in mutagenesis (249, 251). This followed considerable success of tester stocks carrying inversions in *Drosophila melanogaster* for facilitating the detection of recessive lethal mutations. Their great advantage from this point of view is that there is suppression of crossing-over between the inverted region and the structurally normal homologue (due to non-transmission of crossover gametes or non-survival of zygotes derived from them), which helps greatly in isolating any lethal mutation induced in these regions, especially if a linked genetic marker is present. Because of the dicentric chromatids, which result from crossing-over in a paracentric inversion, characteristic anaphase bridges (sometimes continuing to telophase) are produced at first meiotic division, the frequency of which gives an initial estimate of the map length of the inverted segments (253). Associated acentric fragments may also be seen. Inversion loops can be seen in synaptonemal complex preparations of early pachytene in

heterozygotes but these are eliminated in later pachytene by synaptic adjustment (229).

The additional embryonic mortality associated with paracentric inversions is generally small or negligible in male heterozygotes but may be quite high in female heterozygotes (145). Apparently, there is gametic selection in the male against the unbalanced products of anaphase bridge formation (including diploid sperm) which may involve the uterotubal junction functioning as a barrier (185). Some homozygotes for these inversions are lethal, but most have full fertility.

Heterozygotes for pericentric inversions never produce dicentric anaphase bridges or acentric fragments, because the centromere is within the inversion loop. Instead, they produce duplication-deficiency gametes, like heterozygotes for reciprocal translocations, if there is a chiasma within the loop. Thus an appreciable additional amount of embryonic lethality would be expected in heterozygotes for long pericentric inversions, formed in metacentric chromosomes arising from Robertsonian translocations.

A large number of paracentric inversions have now been studied cytologically by G-banding (104, 107), which has shown that breakpoints tend to be located in non-staining bands (see also Section 13.1). Male heterozygotes for two inversions on the same chromosome may be completely sterile (250). Two paracentric inversions are associated with the *t*-complex on Chr 17 (see Chapters 2 and 5).

12.1.7.1 Paracentric inversions

In(X)1H. This X-chromosome inversion was found among the descendants of a (C3H x 101)F₁ male that had received a fractionated X-ray dose to spermatogonia (132). It was first associated with the X-linked male-lethal *Bpa* mutation, since *Bpa*/+ females were found to produce many XO daughters, with a paternal X-chromosome. This property was found to be separable from *Bpa* (244) and proved to be a result of heterozygosity for the inversion (132). This was confirmed by the presence of characteristic bivalents in oocytes at diakinesis and by G-banding, which showed that the breakpoints were at A1 and F, so that the inversion stretches at least from *spf* to beyond *jp*, since there is pronounced crossover suppression within this region. However, *Crm* is outside the inversion (140). As expected, males carrying the inversion do not give any extra frequency of XO offspring. The inversion has been used as a crossover suppressor in the search for induced X-linked lethals (208). The homozygote is viable.

In(1)1Rk. This was induced in postmeiotic germ cells of a DBA/2J male by exposure to 900 R acute X-irradiation, and detected by the presence of meiotic anaphase bridges in testes of male offspring (252, 253). Breakpoints are in or just distal to band 1B and in band 1F, giving an estimated physical length of 42 per cent of the chromosome. The distal genetical breakpoint is just beyond *Pep–3* (formerly *Dip–1*) (103); the proximal is near to *tb* (254). The inversion is estimated to cover 37 per cent of the total genetic length. Heterozygotes gave 34 per cent of anaphase bridges and also a number of telophase bridges, which were 'consistently lumpy'. A homozygous line has been established. Heterozygotes may have a slight reduction of litter size (253).

In(2)2H. The mutant agouti-suppressor, A^s, is associated with this large inversion, with breakpoints in the region of 2D (or possibly 2E1) and 2H1, so that about 40 per cent of the chromosome is involved (133). The frequency of inversion bivalents in heterozygotes at metaphase I is remarkably low (0.2 per cent), and no bridges were found at anaphase I. No pachytene loops were seen, but the pairing showed signs of strain. Linkage studies suggested that the inversion stretched from the T7Ca breakpoint, distal to *fi*, to the *a* locus.

In(5)2Rk. This was induced in a postmeiotic germ cell of a DBA/2J male by an acute X-ray exposure of 850 R (252, 253). It has a frequency of about 19 per cent of anaphase bridges in carriers, with acentric fragments in about 2 per cent of these. Since the map length of an inversion should be at least half the frequency of anaphase bridges (253), the length of In2Rk would seem to be 10 cM or more. Its breakpoints are 5D and 5G2 (107), one being near *Pgm–1* on Chr 5 (254). A homozygous line has been established.

In(X)3H. Found in association with the presumably induced sex-linked lethal *sll–1* (242) in an experiment in which F1(C3H x 101) males were given a fractionated X-ray dose to spermatogonia. Breakpoints are in XA2 and XF1 (243).

In(2)5Rk. This inversion was also induced in DBA/2J males by 900 R of acute X-irradiation (253). It covers approximately 50 per cent of Chr 2 according to Ford *et al.* (145) and 37 per cent according to Davisson and Roderick (102), with breakpoints at 2D and 2H1 (107). Genetic breakpoints are distal to *fi* and near *un* (254). Heterozygotes give about 34 per cent of anaphase bridges, also with associated telophase bridges, which are lumpy like those of In1Rk. C-banding revealed four characteristic types of bivalent at diakinesis (145); diploid secondary spermatocytes, usually containing an intact dicentric chromatid, are common. Heterozygo-

sity for this inversion has little effect on male fertility but significantly reduces female fertility (145). The homozygote is lethal (251).

In(10)6Rk. This was induced by acute exposure of DBA/2J male postmeiotic stages to 900 R of X-rays. It is located in the proximal part of Chr 10, with its distal breakpoint near *gr* (254). Cytological breakpoints are at 10A4 and 10C1 (107). It gives about 26 per cent of anaphase bridges in heterozygous carriers. A homozygous line has been established (253).

In(5)9Rk. This long inversion was induced by intraperitoneal injection of C3D2F1 males with TEM, with sampling of postmeiotic germ cells (253). About 90 per cent of Chr 5 is covered by this medially located inversion (145), which generated about 75 per cent of anaphase bridges in heterozygotes (253). These bridges are uniformly solid in appearance and are frequently followed by telophase bridges. Breakpoints are at 5A2 and 5G2 (107). A homozygous line has been established. Fertility of In9Rk heterozygotes is reduced in both sexes (145), with a striking increase in numbers of diploid sperm (176).

A number of inversions have been induced with triethylenemelamine (TEM) and are listed in Table 12.2 with information about their origin, anaphase bridge frequency, and cytological breakpoints (107, 108, 251, 273). The known genetic breakpoints of these, as well as of the previously listed inversions, are shown in Fig. 13.2.

Of these, *In25Rk*, *In28Rk*, and probably *In21Rk* and *In34Rk* are homozygous lethal. *In(10)17Rk* homozygotes are also homozygous for pygmy (*pg*) on Chr 10, so the inversion breakpoint may be at the *pg* locus. *In20Rk* heterozygotes have reduced litter-size but no translocations seem to be involved; however, *In(1)24Rk* was induced simultaneously with a 1;13 translocation, in which the Chr 1 breakpoint is within the inverted segment (251).

The following three inversions were recovered after 1000 R (10 Gy) γ-irradiation was given to strain DBA/2J males (251):

> *In(14)36Rk*, with anaphase bridge frequency of 0.43
>
> *In(19)37Rk*, with anaphase bridge frequency of 0.59
>
> *In(1)38Rk*, with anaphase bridge frequency of 0.53.

In addition, one spontaneous inversion, *In(5)33Rk*, has been reported (251) which arose in the Chr 5 arm of Rb(5.15)3Bnr. There are fertility problems in heterozygotes. The anaphase bridge frequency is 59 per cent

Table 12.2 Inversions induced by triethylenemelamine

	Strain of origin	Dose of TEM (mg/kg)	Anaphase bridge frequency	Breakpoints
In(3)11Rk	C57BL/6J	0.2	0.29	3D, 3H1
In(1)12Rk	C57BL/6J	0.2	0.23	1D, 1H5
In(7)13Rk	C3D2F$_1$	0.2	0.45	7B, 7F1
In(8)14Rk	DBA/2J	0.2	0.35	8A2 (or 4) 8C3
In(10)17Rk	DBA/2J	0.3	0.65	10A4, 10D2
In(15)18Rk	DBA/2J	0.3	0.37	15A2, 15E
In(2)19Rk	DBA/2J	0.3	0.61	2B, 2H1
In(11)20Rk	DBA/2J	0.3	0.47	11A2, 11D
In(15)21Rk	DBA/2J	0.4	0.47	15A1, 15E
In(14)22Rk	DBA/2J	0.4	0.66	14B, 14E5
In(1)24Rk	DBA/2J	0.4	0.73	1A4, 1H5
In(12)25Rk	DBA/2J	0.4	0.46	12B, 12F1
In(9)26Rk	DBA/2J	0.4	0.24	9B, 9F1
In(4)28Rk	POS*	0.3	0.56	4C2, 4D3
In(5)30Rk	POS (F)*	0.4	0.27	5A2, 5E2
In(13)31Rk	POS (F)*	0.3	0.70	13A2, 13D1
In(4)32Rk	POS (P)*	0.5	0.21	4C2, 4E2
In(8)34Rk	POS (P)*	0.5	0.37	– –
In(15)35Rk	DBA/2J	0.5	0.11	15C, 15F4

* Stocks carrying Robertsonian translocations.

and breakpoints are provisionally located at 5A2 and 5F. Genetic information on the following inversions is also given in Fig. 13.2.:

In(4)1By, In(4)1Dn, In(9)39Rk, In(2)40Rk, In(10)42Rk, In(13)43Rk, In(2)45Rk, In(1)46Rk, In(6)47Rk, In(6)48Rk, In(1)49Rk, In(13)50Rk, In(16)51Rk, In(14)54Rk, and In(3)55Rk.

T(In1;5)44H is described in Section 12.1.4.3 on reciprocal translocations.

12.1.7.2 Pericentric inversions

In(6.15)Rb1Ald/1Wa. Nijhoff (236) has reported the induction of a pericentric inversion in Rb(6.15)1Ald after 400 rad of acute X-irradiation of a triple metacentric stock (homozygous for Rb4Bnr, Rb163H, and Rb1Ald). Probable breakpoints are 6A2 and 15F2. Progeny of heterozygous males showed a high level of postimplantation loss. The original stock has now been lost as a result of recombination with Chr 15 (26) being replaced with one having a 15^6 chromosome deficient for the proximal part of Chr 6 and the distal part of Chr 15. Nevertheless, testes weights and sperm counts are near normal and there is no sign of prenatal selection against this karyotype.

T(4;8S)In(8LS)36H. This combination of reciprocal translocation and pericentric inversion is described in Section 12.1.4.3.

In(Y)Lub. Winking (303) reported the presence of a small meta- to submetacentric chromosome among males from the NMRI stock. It proved to be a derivative Y chromosome, presumably the result of a pericentric inversion. Carriers are fully fertile.

12.1.8 Duplications and deficiencies
12.1.8.1 Duplications

These may arise *de novo* or as a secondary result of the formation of insertions or translocations. The fertile tertiary trisomy *Ts(1^{13})70H* (q.v.) comes in the latter category since it arose as the result of non-disjunction in T(1;13)70H and is a duplication of the proximal end of Chr 1 and the distal end of Chr 13. Other viable tertiary trisomies are male-sterile and are dealt with under the translocation concerned. Two viable duplications of parts of Chr 7, generated by insertions, are considered below. However, a third duplication of part of Chr 7 (bands 7B1–7F1), generated by Is(7;1)40H, leads to the production of exencephalic fetuses (284).

Dp(In7;X)1Ct. This is the type II form of Cattanach's translocation Is(In7;X)1Ct (q.v.) with a duplication of Chr 7 bands 7E3–7F1. About one-quarter of these are viable and fertile in males.

Dp(7;InX)1Neu. This duplication of Chr 7 bands 7C–7D, derived from Is(7;InX)1Neu, has not been described in detail but appears to be viable and fertile (2).

Dp(7)1Rl. This tandem duplication (268a, 269) was found in a female mouse from a cross of an irradiated SEC/Rl ♀ (*c^{ch}/c^{ch}*) and unirradiated 101/Rl, which was small and proved to have an unusual type of β-haemoglobin instead of the expected *Hbb^d/Hbb^s* phenotype.

The abnormal haemoglobin was consistent with over-production of β^{SEC}, and the coat colour, after crossing to albino, c/c, was consistent with the presence of two doses of c^{ch} and one of c. Thus the duplication includes both the c and Hbb loci, as well as Mod–2 and sh–1 (264). The length of Chr 7 is increased by approximately 20 per cent, and band 7E with adjacent sub-bands is duplicated in tandem. Recombination has been observed between the duplicated c loci and also between the duplicated Hbb loci, but is rare, indicating crossover inhibition. Heterozygotes weigh about 60 per cent of normal at weaning; survival to 10 days of age is about 40 per cent of normal. Litter-size in outcrosses of female heterozygotes was only 3.5, compared with 5.5 from male heterozygotes, indicating a possible effect of the duplication on maternal functions. Homozygotes die prenatally, probably around the time of implantation (264).

12.1.8.2 Deficiencies

The first deficiency to be reported in the mouse was by Painter (240) who cytologically examined the 'non-disjunction' (v-0) mice of Gates (155) and found that one chromosome was much reduced in size, presumably due to a large deficiency on Chr 10, including the v locus. Since then, a number of small deletions have been reported involving the closely linked d and se loci (256, 267). These were induced in male and female germ cells of (101 x C3H)F_1 mice by treatment with X-rays, γ-rays, or fission neutrons in specific locus experiments. By means of complementation tests L.B. Russell (256, 260) revealed the existence of over 20 complementation groups spanning 11–12 functional units. A similar type of analysis has now been carried out on c locus mutants (260, 268a) induced by radiation (including fission neutrons) and has revealed 13 complementation groups which involve eight functional units. The longest deficiencies also include the Mod–2 and sh–1 loci and cover 4–8 cM. These have been detected cytologically and shown to involve the loss of much of band 7E (263). By A- and G-banding methods, Miller *et al.* (225) showed that the radiation-induced recessive lethal c^{25H} had a deletion of about 8 per cent of the length of Chr 7. This deletion is known to cover Mod–2 as well as the c locus (125).

By cytological examination of the pachytene stage of oogenesis and spermatogenesis in a series of six X-ray induced lethal c locus 'alleles', Jagiello *et al.* (178) demonstrated the presence of diagnostic deficiency loops in all of them (c^{3H}, c^{6H}, and c^{25H} from Harwell; c^{14Cos}, c^{65K}, and c^{112K} from Oak Ridge). These mutants are described under the c locus in Chapter 2.

Some mutations in the t-complex are known to be deficiencies. Thus T^{Orl} is a deletion extending from T to some point distal to qk (126) and t^{h20} may include a small deletion covering the loci of tf, Fu^{kb}, and the t^6 lethal factor, though no visible chromosomal change has yet been detected (195). T^{hp} (179, 180) involves a deletion (49), as does W^{19H} (213). For further information, see Chapters 2 and 5. Tertiary monosomies, which are deficient for the distal end of one chromosome and the proximal end of another, are discussed in Section 12.1.4.5.

Acknowledgements

I am indebted to Dr Bruce Cattanach for his critical and constructive comments. I should also like to thank Mr Colin Beechey for helping me to assemble the information given here and Mrs Debbie Badger for preparing innumerable drafts with patience, care and skill. Drs Muriel Davisson and Tom Roderick kindly provided unpublished information on a number of inversions. Part of this work was carried out during tenure of a Leverhulme award.

References

1. Adler, I.D. 1984. Characterization of translocations. Mouse News Lett. 71:38–40.
2. Adler, I.D., and A. Neuhäuser-Klaus. 1987. Seventeen stocks of mice with reciprocal or Robertsonian translocations. Mouse News Lett. 77:139–142.
3. Adolph, S., and J. Klein. 1981. Robertsonian variation in *Mus musculus* from Central Europe, Spain and Scotland. J. Hered. 72:219–221.
4. Agulnik, S.I., and A.I. Agulnik. 1985. New Robertsonian translocation—Rb(6.19)2Icg. Mouse News Lett. 75:41.
5. Agulnik, S.I., A.I. Agulnik, and A.O. Ruvinsky. 1983. Mouse News Lett. 69:41.
6. Agulnik, S.I., A.I. Agulnik, and A.O. Ruvinsky. 1985. Recombination distances between the loci in the proximal part of chromosome 17 in females heterozygous for translocation Rb(8.17)1Iem, Rb(16.17)7Bnr and T(16;17)43H. Mouse News Lett. 75:40.
7. Alavantic, D., and A.G. Searle. 1984. Chr 9: distance to distal end. Mouse News Lett. 70:79.
8. Ashley, T. 1983. Nonhomologous synapsis of the sex chromosomes in the heteromorphic bivalents of the X–Y translocations in male mice: R5 and R6. Chromosoma 88:178–183.
9. Ashley, T., and L.B. Russell. 1986. A new type of non-homologous synapsis in T(X;4)1Rl translocation male mice. Cytogenet. Cell Genet. 43:194–200.
10. Ashley, T., L.B. Russell, and N.L.A. Cacheiro. 1982. Synaptonemal complex analysis of X–Y translocations in

male mice. I. R3 and R5 quadrivalents. Chromosoma 87:149–164.

11. Ashley, T., L.B. Russell, and N.L.A. Cacheiro. 1983. Synaptonemal complex analysis of two X–Y translocations in male mice. II. R2 and R6 quadrivalents. Chromosoma 88:171–177.

12. Baranov, V.S. 1979. New marker reciprocal translocations T(7;14)2Iem and T(7;14;15)3Iem in laboratory mice. Genetika 15:1651–1660.

13. Baranov, V.S. 1981. Rb(2.6)4Iem—a new marker translocation of the Robertsonian type in the laboratory mouse *Mus musculus*. Tsitologia 23:1362–1367.

14. Baranov, V.S. 1983. Chromosomal control of early embryonic development in mice. I. Experiments on embryos with autosomal monosomy. Genetica 61:165–177.

15. Baranov, V.S. 1983. Chromosomal control of early embryonic development in mice. II. Experiments on embryos with structural aberrations of autosomes 7, 9, 14 and 17. Genet. Res., Camb. 41:227–239.

16. Baranov, V.S., and A.P. Dyban. 1968. Analysis of spermatogenic and embryogenic abnormalities in mice heterozygous for the chromosome translocation T6. Genetika 4 (12):70–83.

17. Baranov, V.S., and A.P. Dyban. 1970. Correlation between normal embryonic development and the type of chromosome abnormalities arising during meiosis in mice that are heterozygous with regard to T6 translocations. Ontogenez 1:248–261.

18. Baranov, V.S., and A.P. Dyban. 1971. Embryogenesis and peculiarities of karyotype in mouse embryos with central fusion of chromosomes (Robertsonian translocation). Ontogenez 2: 164–176.

19. Baranov, V.S., and A.P. Dyban. 1972. Disturbances of embryogenesis in trisomy of autosomes arising in offspring of mice with Robertsonian translocations (centric fusion) of T1Iem. (In Russian) Arch. Anat. Histol. Embryol. 63:67–77.

20. Baranov, V.S., and A.P. Dyban. 1972. Effect of the maternal genotype on the frequency of trisomy among autosomes in embryogenesis. Bull. Exp. Biol. Med. 74:1566–1569.

21. Baranov, V.S., and A.P. Dyban. 1973. Cytogenetic analysis of early embryogenesis in mice heterozygous for Robertsonian translocation T1Ald. Ontogenez 4:587–594.

22. Batchelor, A.L., R.J.S. Phillips, and A.G. Searle. 1966. A comparison of the mutagenic effectiveness of chronic neutron- and γ-irradiation of mouse spermatogonia. Mutat. Res. 3:218–229.

23. Beatty, R.A. 1957. Parthenogenesis and polyploidy in mammalian development. Cambridge University Press, Cambridge, 132p.

24. Beechey, C.V. 1973. X–Y chromosome dissociation and sterility in the mouse. Cytogenet. Cell. Genet. 12:60–67.

25. Beechey, C.V. 1987. Time of death of nullisomy and monosomy 15 embryos. Mouse News Lett. 78:51.

26. Beechey, C.V., and P. de Boer. 1983. Mouse News Lett. 69:55.

26a. Beechey, C.V., and A.G. Searle. 1972. Mouse News Lett. 46:28.

27. Beechey, C.V., and A.G. Searle. 1975. Mouse News Lett. 53:30–31.

28. Beechey, C.V., and A.G. Searle. 1976. Mouse News Lett. 54:38–39.

29. Beechey, C.V., and A.G. Searle. 1977. Mouse News Lett. 56:39.

30. Beechey, C.V., and A.G. Searle. 1978. Mouse News Lett. 58:46.

31. Beechey, C.V., and A.G. Searle. 1978. Mouse News Lett. 59:19–20.

32. Beechey, C.V., and A.G. Searle. 1979. Mouse News Lett. 61:40.

33. Beechey, C.V., and A.G. Searle. 1980. Mouse News Lett. 62:50.

34. Beechey, C.V., and A.G. Searle. 1980. Mouse News Lett. 62:52.

35. Beechey, C.V., and A.G. Searle. 1980. Mouse News Lett. 63:17.

36. Beechey, C.V., and A.G. Searle. 1981. Mouse News Lett. 65:22.

37. Beechey, C.V., and A.G. Searle. 1983. Mouse News Lett. 68:69.

38. Beechey, C.V., and A.G. Searle. 1984. Mouse News Lett. 70:81.

39. Beechey, C.V., and A.G. Searle. 1985. Mouse News Lett. 72:103–104.

40. Beechey, C.V., and A.G. Searle. 1986. Compound of T(2;8)26H and T(2;11)30H. Mouse News Lett. 74:93.

41. Beechey, C.V., and A.G. Searle. 1986. Position of Is40H on Chr 7. Mouse News Lett. 74:95.

42. Beechey, C.V., and A.G. Searle. 1987. New non-disjunction tester stock established. Mouse News Lett. 77:127–128.

43. Beechey, C.V., and A.G. Searle. 1987. Tertiary trisomies for T(6;12)32H. Mouse News Lett. 77:128.

43a. Beechey, C.V., and A.G. Searle. 1987. T(4;9)45H. Mouse News Lett. 78:50–51.

44. Beechey, C.V., and A.G. Searle. 1987. Unpublished information.

45. Beechey, C.V., and A.G. Searle. 1988. Effects of 0–4 copies of chromosome 15 on mouse embryonic development. Cytogenet. Cell Genet. 47:66–71.

46. Beechey, C.V., and R. Speed. 1981. Mouse News Lett. 64:56.

47. Beechey, C.V., M. Kirk, and A.G. Searle. 1980. A reciprocal translocation induced in an oocyte and affecting fertility in male mice. Cytogenet. Cell. Genet. 276:129–146.

47a. Beechey, C.V., A.G. Searle, M.D. Burtenshaw, and E.P. Evans. 1987. Mouse News Lett. 77:126.

48. Bennett, D. 1965. The karyotype of the mouse, with identification of a translocation. Proc. Natl. Acad. Sci. USA 53:730–737.

49. Bennett, D. 1975. The T-locus of the mouse. Cell 6:441–454.
50. Bernstine, E.G., and L.B. Russell. 1977. Gene dosage and inactivation effects on the expression of mitochondrial malic enzyme activity in *Mus musculus*. Genetics 86:s6.
51. Biddle, F.G., B.G. MacDonald, and B.A. Eales. 1985. Genetic control of sex-chromosomal univalency in the spermatocytes of C57BL/6J and DBA/2J mice. Can. J. Genet. Cytol. 27:741–750.
52. Boer, P. de. 1973. Fertile tertiary trisomy in the mouse. Cytogenet. Cell Genet. 12:435–442.
53. Boer, P. de. 1976. Male meiotic behaviour and male and female litter size in mice with the T(2;8)26H and T(1;13)70H reciprocal translocations. Genet. Res., Camb. 27:369–387.
54. Boer, P. de, and M. van Gijsen. 1974. The location of the positions of the breakpoint involved in the T26H and T70H mouse translocations with the aid of Giemsa-banding. Can. J. Genet. Cytol. 16:783–788.
55. Boer, P. de, and A. Groen. 1974. Fertility and meiotic behaviour of male T70H tertiary trisomics of the mouse. Cytogenet. Cell Genet. 13:489–510.
56. Boer, P. de, and F.A. van der Hoeven. 1977. Son-sire regression based heritability estimates of chiasma frequency, using T70H mouse translocation heterozygotes, and the relation between univalence, chiasma frequency and sperm production. Heredity 39:335–343.
57. Boer, P. de, and P.H.M.D. de Maar. 1976. A histological study of embryonic death caused by heterozygosity for the T26H reciprocal mouse translocation. J. Embryol. Exp. Morphol. 35:595–606.
58. Boer, P. de, and J.H. Nijhoff. 1980. Mouse News Lett. 62:78–79.
59. Boer, P. de, and J.H. Nijhoff. 1981. Incomplete sex chromosome pairing in oligospermic male hybrids of *Mus musculus* and *M. musculus molossinus* in relation to the source of the Y chromosome and the presence or absence of a reciprocal translocation. J. Reprod. Fert. 62:235–243.
60. Boer, P. de, and A.G. Searle. 1980. Workshop on chromosomal aspects of male sterility in mammals. Summary and synthesis. J. Reprod. Fert. 60:257–265.
61. Boer, P. de, F. van der Hoeven, and J.A.P. Chardon. 1976. The production, morphology, karyotypes and transport of spermatozoa from tertiary trisomic mice and the consequences for egg fertilisation. J. Reprod. Fertil. 48:249–256.
62. Boer, P. de, A.G. Searle, F.A. van der Hoeven, D.G. de Rooij, and C.V. Beechey. 1986. Male pachytene pairing in single and double translocation heterozygotes and spermatogenic impairment in the mouse. Chromosoma 93:326–336.
63. Bond, D.J., and A.C. Chandley. 1983. Aneuploidy. Oxford University Press, Oxford.
64. Breckon, G. 1981. XXY mouse with germinal mosaicism. Mouse News Lett. 65:21.
65. Britton-Davidian, J. 1983. Mouse News Lett. 69:35.
66. Brook, J.D. 1983. X-chromosome segregation, maternal age and aneuploidy in the XO mouse. Genet. Res. 41:85–95.
67. Brooker, P.C. 1982. Robertsonian translocations in *Mus musculus* from N.E. Scotland and Orkney. Heredity 48:305–309.
68. Brooker, P.C., and R.J. Berry. 1981. Mouse News Lett. 64:65–66.
69. Burgoyne, P.S. 1979. Evidence for an association between univalent Y chromosomes and spermatocyte loss in XYY mice and man. Cytogenet. Cell Genet. 23:84–89.
69a. Burgoyne, P.S., and T.G. Baker. 1985. Perinatal oocyte loss in XO mice and its implications for the aetiology of gonadal dysgenesis in XO women. J. Reprod. Fert. 75:633–645.
70. Burgoyne, P.S., S. Mahadevaiah, and U. Mittwoch. 1985. A reciprocal autosomal translocation which causes male sterility in the mouse also impairs oogenesis. J. Reprod. Fertil. 75:647–652.
71. Cacheiro, N.L.A., and L.B. Russell. 1975. Evidence that linkage group IV as well as linkage group X of the mouse are in chromosome 10. Genet. Res., Camb. 25:193–195.
72. Cacheiro, N.L.A., L.B. Russell, and M.S. Swartout. 1974. Translocations, the predominant cause of total sterility in sons of mice treated with mutagens. Genetics 76:73–91.
73. Cacheiro, N.L.A., L.B. Russell, and J.W. Bangham. 1978. A new mouse X-autosome translocation with non-random inactivation. Genetics 88:s13.
74. Capanna, E., M. Cristaldi, P. Perticone, and M. Rizone. 1975. Identification of chromosomes involved in the 9 Robertsonian fusions of the Apennine mouse with a 22-chromosome karyotype. Experientia 31:294–296.
75. Capanna, E., A. Gropp, H. Winking, G. Noack, and M.V. Civitelli. 1976. Robertsonian metacentrics in the mouse. Chromosoma 58:341–353.
76. Capková, J., and J. Forejt. 1984. Research news. Mouse News Lett. 71:43.
77. Carter, T.C., M.F. Lyon, and R.J.S. Phillips. 1955. Gene-tagged chromosome translocations in eleven stocks of mice. J. Genet. 53:154–166.
78. Carter, T.C., M.F. Lyon, and R.J.S. Phillips. 1956. Further genetic studies of eleven translocations in the mouse. J. Genet. 54:462–473.
79. Cattanach, B.M. 1961. A chemically-induced variegated-type position effect in the mouse. Z. Vererbungsl. 92:165–182.
80. Cattanach, B.M. 1966. The location of Cattanach's translocation in the X-chromosome linkage map of the mouse. Genet. Res., Camb. 8:253–256.
81. Cattanach, B.M. 1966. Chemically induced mutations in mice. Mutat. Res. 3:346–353.
82. Cattanach, B.M. 1967. A test of distributive pairing between two specific non-homologous chromosomes of the mouse. Cytogenetics 6:67–77.
83. Cattanach, B.M. 1970. Controlling elements in the

mouse X-chromosome. III. Influence upon both parts of an X divided by rearrangement. Genet. Res., Camb. 16:293–301.

84. Cattanach, B.M. 1974. Position effect variegation in the mouse. Genet. Res., Camb. 23:291–306.

84a. Cattanach, B.M. 1974. Genetic disorders of sex determination in mice and other mammals. International Congress Series no. 310: Birth defects:129–141. Excerpta Medica, Amsterdam.

85. Cattanach, B.M. 1975. Control of chromosome inactivation. Ann. Rev. Genet. 9:1–18.

86. Cattanach, B.M. 1978. Crossover suppression in mice heterozygous for tobacco mouse metacentrics. Cytogenet. Cell Genet. 20:264–281.

86a. Cattanach, B.M. 1979. Linkage between fz and ln in heterozygotes for Robertsonian translocations. Mouse News Lett. 61:40–41.

86b. Cattanach, B.M. 1982. Non-disjunction tests with Robertsonian translocations. Mouse News Lett. 66:62–63.

87. Cattanach, B.M. 1986. Parental origin effects in mice. J. Embryol. Exp. Morphol. 97 Suppl:137–150.

88. Cattanach, B.M., and A.J.M. Crocker. 1979. Mouse News Lett. 60:44.

88a. Cattanach, B.M., and C. Jones. 1980. Non-disjunction tests with Robertsonian translocations. Mouse News Lett. 62:50–51.

89. Cattanach, B.M., and M. Kirk. 1985. Differential activity of maternally and paternally derived chromosome regions in mice. Nature 315:496–498.

89a. Cattanach, B.M., and M. Kirk. 1987. Lethality of T(2;11)30H homozygotes. Mouse News Lett. 78:52.

90. Cattanach, B.M., and H. Moseley. 1973. Non-disjunction and reduced fertility caused by the tobacco mouse metacentric chromosomes. Cytogenet. Cell Genet. 12:264–287.

91. Cattanach, B.M., and J.R.K. Savage. 1976. Mouse News Lett. 54:38.

92. Cattanach, B.M., I. Murray, and T.R.L. Bigger. 1977. Mouse News Lett. 56:37.

92a. Cattanach, B.M., D. Papworth, and M. Kirk. 1984. Genetic tests for autosomal non-disjunction and chromosome loss in mice. Mutation Res. 126:189–204.

92b. Cattanach, B.M., C. Rasberry, and E.P. Evans. 1987. T(4;5)46H. Mouse News Lett. 78:51.

93. Cattanach, B.M., C.E. Williams, and H. Bailey. 1972. Identification of the linkage groups carried by the metacentric chromosome of the tobacco-mouse (*Mus poschiavinus*). Cytogenetics 11:412–413.

93a. Cattanach, B.M., H.G. Wolfe, and M.F. Lyon. 1972. A comparative study of the coats of chimaeric mice and those of heterozygotes for X-linked genes. Genet. Res., Camb. 19:213–228.

93b. Chandley, A.C., and R.M. Speed. 1987. Cytological evidence that the Sxr fragment of XY, Sxr mice pairs homologously at meiotic prophase with the proximal testis-determining region. Chromosoma 95:345–349.

94. Crocker, A.J.M. 1980. A new Robertsonian translocation. Mouse News Lett. 63:14.

95. Davisson, M.T. 1979. Mouse News Lett. 60:49.

96. Davisson, M.T. 1980. Mouse News Lett. 63:19–20.

97. Davisson, M.T. 1984. Mouse News Lett. 70:85.

98. Davisson, M.T. 1985. Mouse News Lett. 73:19.

99. Davisson, M.T. 1986. A Chr 1 aberration in RBA/Dn. Mouse News Lett. 75:34.

100. Davisson, M.T. 1987. Personal communication.

101. Davisson, M.T., and E.C. Akeson. 1983. Mouse News Lett. 69:26.

102. Davisson, M.T., and T.H. Roderick. 1975. Chromosomal banding patterns of two paracentric inversions in mice. Cytogenet. Cell Genet. 12:398–403.

103. Davisson, M.T., and T.H. Roderick. 1975. Cytogenetics of mammalian paracentric inversions. Genetics 80:s25.

104. Davisson, M.T., and T.H. Roderick. 1976. Cytology of inversions in mice. Genetics 83:s19.

105. Davisson, M.T., and T.H. Roderick. 1977. Mouse News Lett. 57:20.

106. Davisson, M.T., and T.H. Roderick. 1978. Mouse News Lett. 58:48.

107. Davisson, M.T., and T.H. Roderick. 1988. Localization of cytological breakpoints in induced chromosome rearrangements in the mouse (*Mus musculus*). (Ms. in preparation).

108. Davisson, M.T., N. Hawes, and T.H. Roderick. 1977. Mouse News Lett. 57:19–20.

109. Davisson, M.T., T.H. Roderick, and M.N. Murphy. 1976. Mouse News Lett. 56:41.

109a. Debrot, S., and C.J. Epstein. 1986. Tetrasomy 16 in the mouse: a more severe condition than the corresponding trisomy. J. Embryol. Exp. Morph. 91:169–180.

110. Dev, V.G., D.A. Miller, J. Charen, and O.J. Miller. 1974. Translocation of centromeric heterochromatin in the T(10;13)199H stock of *Mus musculus* and localization of chromosome breakpoints. Cytogenet. Cell Genet. 13:256–267.

111. Disteche, C.M., E.M. Eicher, and S.A. Latt. 1981. Late replication patterns in adult and embryonic mice carrying Searle's X-autosome translocation. Exp. Cell Res. 133:357–362.

112. Dyban, A.P., and V.S. Baranov. 1987. Cytogenetics of mammalian embryonic development. Clarendon Press, Oxford.

113. Eicher, E.M. 1967. The genetic extent of the insertion involved in the flecked translocation in the mouse. Genetics 55:203–212.

114. Eicher, E.M. 1970. X-autosome translocations in the mouse: total inactivation versus partial inactivation of the X chromosome. Adv. Genet. 15:175–259.

115. Eicher, E.M. 1970. The position of *ru-2* and *qv* with respect to the flecked translocation in the mouse. Genetics 64:495–510.

116. Eicher, E.M. 1973. Translocation trisomic mice: production by female and not male translocation carriers. Science 180:81.

117. Eicher, E.M., and W.G. Beamer. 1979. Mouse News Lett. 60:51.
118. Eicher, E.M., and M.C. Green. 1972. The T6 translocation in the mouse: its use in trisomy mapping, centromere localization, and cytological identification of linkage group III. Genetics 71:621–632.
119. Eicher, E.M., and D.L. Coleman. 1977. Influence of gene duplication and X-inactivation on mouse mitochondrial malic enzyme activity and electrophoretic patterns. Genetics 85:647–658.
120. Eicher, E.M., and L.L. Washburn. 1977. Mouse News Lett. 56:43.
121. Eicher, E.M., and L.L. Washburn. 1978. Assignment of genes to regions of mouse chromosomes. Proc. Natl. Acad. Sci. USA 75:946–950.
122. Eicher, E.M., W.G. Beamer, L.L. Washburn, and W.K. Whitten. 1980. A cytogenetic investigation of inherited true hermaphroditism in BALB/cWt mice. Cytogenet. Cell. Genet. 28:104–115.
123. Epstein, C.J. 1985. Mouse monosomies and trisomies as experimental systems for studying mammalian aneuploidy. Trends Genet. 1:129–134.
124. Erickson, R.P. 1986. Cattanach's translocation [Is(7;X)Ct] corrects male sterility due to homozygosity for chromosome 7 deletions. Genet. Res., Camb. 43:35–41.
125. Erickson, R.P., E.M. Eicher, and S. Gluecksohn-Waelsch. 1974. Demonstration in mouse of X-ray induced deletions for a known enzyme structural locus. Nature 248:416–417.
126. Erickson, R.P., S.E. Lewis, and K.S. Slusser. 1978. Deletion mapping of the t complex of chromosome 17 of the mouse. Nature 274:163–164.
127. Evans, E.P. 1976. Male sterility and double heterozygosity for Robertsonian translocations in mouse. Chromosomes Today 5:75–81.
128. Evans, E.P. 1979. Mouse News Lett. 60:71.
129. Evans, E.P. 1982. A mosaic male mouse. Mouse News Lett. 67:29.
130. Evans, E.P. 1987. Unpublished information.
130a. Evans, E.P. 1989. Mouse News Lett. 83:164.
131. Evans, E.P., and M.D. Burtenshaw. 1984. Yet another Robertsonian translocation in the laboratory mouse. Mouse News Lett. 70:94.
132. Evans, E.P., and R.J.S. Phillips. 1975. Inversion heterozygosity and the origin of XO daughters of *Bpa*/+ female mice. Nature 256:40–41.
133. Evans, E.P., and R.J.S. Phillips. 1978. Mouse News Lett. 58:45.
134. Evans, E.P., C.V. Beechey, and A.G. Searle. 1977. Mouse News Lett. 57:17.
135. Evans, E.P., C.V. Beechey, and M.D. Burtenshaw. 1978. Meiosis and fertility in XXY mice. Cytogenet. Cell Genet. 20:249–263.
136. Evans, E.P., M.D. Burtenshaw, and M.F. Lyon. 1987. A repositioning of the X chromosome breakpoint of T(X;11)38H. Mouse News Lett. 77:147.
137. Evans, E.P., C.E. Ford, and A.G. Searle. 1969. A 39, X/41, XYY mosaic mouse. Cytogenetics 8:87–96.
138. Evans, E.P., M.F. Lyon, and M. Daglish. 1967. A mouse translocation giving a metacentric marker chromosome. Cytogenetics 6:105–119.
139. Evans, H.J., C.E. Ford, M.F. Lyon, and J. Gray. 1965. DNA replication and genetic expression in female mice with morphologically distinguishable X chromosomes. Nature 206:900–903.
140. Fisher, G., and M.F. Lyon. 1981. Mouse News Lett. 64:58.
141. Ford, C.E. 1975. The time in development at which gross genome unbalance is expressed. *In* M. Balls and A.E. Wild, eds., The early development of mammals, 285–304. Cambridge University Press, Cambridge.
142. Ford, C.E., and H.M. Clegg. 1969. Reciprocal translocations. Br. Med. Bull. 25:110–114.
143. Ford, C.E., and E.P. Evans. 1964. A reciprocal translocation in the mouse between the X chromosome and a short autosome. Cytogenetics 3:295–305.
144. Ford, C.E., and E.P. Evans. 1973. Robertsonian translocations in mice: segregational irregularities in male heterozygotes and zygotic unbalance. Chromosomes Today 4:387–397.
144a. Ford, C.E., and E.P. Evans. 1973. Non-expression of genome unbalance in haplophase and early diplophase of the mouse and incidence of karyotypic abnormality in post-implantation embryos. *In* A. Boué and C. Thibault, eds., Les accidents chromosomique de la reproduction. Institut National de la Santé et de la Recherche Medicale, Paris.
145. Ford, C.E., E.P. Evans, and M.D. Burtenshaw. 1976. Meiosis and fertility in mice heterozygous for paracentric inversions. *In* K. Jones and P.E. Brandham, eds., Current chromosome research, 123–132. Elsevier/North Holland Biomedical Press, Amsterdam.
146. Ford, C.E., A.G. Searle, E.P. Evans, and B.J. West. 1969. Differential transmission of translocations induced in spermatogonia of mice by irradiation. Cytogenetics 8:447–470.
147. Ford, C.E., E.P. Evans, M.D. Burtenshaw, H.M. Clegg, M. Tuffrey, and R.D. Barnes. 1975. A functional "sex-reversed" oocyte in the mouse. Proc. R. Soc. Lond. (Biol.) 190:187–197.
148. Forejt, J. 1972. Chiasmata and crossing-over in the male mouse (*Mus musculus*). Suppression of recombination and chiasma frequencies in the ninth linkage group. Folia Biol. 18:161–170.
149. Forejt, J. 1974. Non-random association between a specific autosome and the X chromosome in meiosis of the male mouse: possible consequence of the homologous centromeres' separation. Cytogenet. Cell Genet. 13:369–383.
150. Forejt, J. 1976. Spermatogenic failure of translocation heterozygotes affected by H–2 linked gene in mouse. Nature 260:143–145.
151. Forejt, J., and S. Gregorová. 1977. Meiotic studies of

translocations causing male sterility in the mouse. I. Autosomal reciprocal translocations. Cytogenet. Cell Genet. 19:55–56.

152. Forejt, J., Capková, and S. Gregorová. 1978. Mouse News Lett. 59:55–56.

153. Forejt, J., S. Gregorová, and P. Goetz. 1981. XY pair associates with the synaptonemal complex of autosomal male-sterile translocations in pachytene spermatocytes of the mouse. Chromosoma 82:41–54.

154. Fraser, L.R., H.M. Zanellotti, G.R. Paton, and L.M. Drury. 1976. Increased incidence of tripoidy in embryos derived from mouse eggs fertilized *in vitro*. Nature 260:39–40.

155. Gates, W.H. 1927. A case of non-disjunction in the mouse. Genetics 12:295–306.

156. Goyder, Y. 1983. New Robertsonian translocation. Mouse News Lett. 68:67.

157. Green, M.C. 1967. Mouse News Lett. 37:33.

158. Gregorová, S., V.S. Baranov, and J. Forejt. 1981. Partial trisomy (including T-t gene complex) of chromosome 17 of the mouse. The effect on male fertility and the transmission to progeny. Folia Biol. 27:171–177.

159. Gropp, A. 1977. Mouse News Lett. 57:27–28.

160. Gropp, A. 1982. Value of an animal model for trisomy. Virchows Archiv A 395:117–131.

161. Gropp, A., and H. Winking. 1980. Robertsonian translocations in mice from wild populations. Mouse News Lett. 62:59–60.

162. Gropp, A., and H. Winking. 1981. Robertsonian translocations: cytology, meiosis, segregation patterns and biological consequences of heterozygosity. *In* R.J. Berry, ed., Biology of the house mouse, 141–181. Academic Press, London.

163. Gropp, A., and H. Winking. 1981. Mouse News Lett. 65:33.

164. Gropp, A., and H. Winking. 1983. Mouse News Lett. 68:76–77.

165. Gropp, A., D. Giers, and U. Kolbus. 1974. Trisomy in the fetal backcross progeny of male and female metacentric heterozygotes of the mouse. I. Cytogenet. Cell Genet. 13:511–535.

166. Gropp, A., U. Kolbus, and D. Giers. 1975. Systematic approach to the study of trisomy in the mouse. II. Cytogenet. Cell Genet. 14:42–62.

167. Gropp, A., U. Tettenborn, and E. von Lehmann. 1969. Chromosomenuntersuchungen bei der Tabakmaus (*M. poschiavinus*) und bei Tabakmaus-Hybriden. Experientia 25:875–876.

168. Gropp, A., U. Tettenborn, and E. von Lehmann. 1970. Chromosomenvariationen vom Robertson'schen Typus bei der Tabakmaus, M. poschiavinus, und ihren Hybriden mit der Laboratoriumsmaus. Cytogenetics 9:9–23.

169. Gropp, A., H. Winking, L. Zech, and H. Müller. 1972. Robertsonian chromosomal variation and identification of metacentric chromosomes in feral mice. Chromosoma 39:265–288.

170. Gropp, A., H. Winking, C. Redi, E. Capanna, J. Brit-ton-Davidian, and G. Noack. 1982. Robertsonian karyotype variation in wild house mice from Rhaeto-Lombardia. Cytogenet. Cell Genet. 34:67–77.

171. Hotta, Y., and A.C. Chandley. 1982. Activities of X-linked enzymes in spermatocytes of mice rendered sterile by chromosomal alterations. Gamete Res. 6:65–72.

172. Hotta, Y., A.C. Chandley, H. Stern, A.G. Searle, and C.V. Beechey. 1979. A disruption of pachytene DNA metabolism in male mice with chromosomally-derived sterility. Chromosoma 73:287–300.

173. Hübner, R. 1988. Populations Robertsoniennes chez la souris 'sauvage' en Belgique. Ann. Soc. R. Zool. Belg. 118:69–75.

174. Hübner, R., L. Koulischer, and M. Lichterberger. 1985. Mouse News Lett. 72:116–117.

175. Hübner, R., L. Koulischer, and M. Lichterberger. 1985. Mouse News Lett. 73:27.

176. Hugenholtz, A.P. 1984. Mouse News Lett. 71:35.

177. Imai, H.T., Y. Matsuda, T. Shiroishi, and K. Moriwaki. 1981. High frequency of X-Y chromosome dissociation in primary spermatocytes of F_1 hybrids between Japanese wild mice and inbred laboratory mice. Cytogenet. Cell Genet. 29:166–175.

178. Jagiello, G., J.S. Fang, H.A. Turchin, S.E. Lewis, and S. Gluecksohn-Waelsch. 1976. Cytological observations of deletions in pachytene stages of oogenesis and spermatogenesis in the mouse. Chromosoma 58:377–386.

179. Johnson, D.R. 1974. Hairpin-tail: a case of post-reductional gene action in the mouse egg? Genetics 76:795–805.

180. Johnson, D.R. 1975. Further observations on the hairpin-tail (T^{hp}) mutation in the mouse. Genet. Res., Camb. 24:207–213.

181. Kaufman, M.H. 1976. Incidence of chromosomally unbalanced gametes in T(14;15)6Ca heterozygote mice. Genet. Res., Camb. 27:77–84.

182. Kaufman, M.H., S.C. Barton, and M.A.H. Surani. 1977. Normal post-implantation development of mouse parthenogenetic embryos to the forelimb bud stage. Nature 265:53–55.

183. Kirk, K.M., and A.G. Searle. 1988. Phenotypic consequences of chromosome imbalance in the mouse. *In* A. Daniel, ed., The cytogenetics of mammalian autosomal rearrangements, 739–768, Alan Liss, New York.

184. Klein, J., and D. Klein. 1972. Position of the translocation break T(2;9)138Ca in linkage group IX of the mouse. Genet. Res., Camb. 19:177–179.

185. Krzanowska, H. 1974. The passage of abnormal spermatozoa through the uterotubal junction of the mouse. J. Reprod. Fert. 38:81–90.

186. Lane, P.W., A.G. Searle, C.V. Beechey, and E.M. Eicher. 1981. Chromosome 18 of the house mouse. J. Hered. 72:409–412.

187. Lehmann, E., von, and A. Radbruch. 1977. Robertsonian translocations in *Mus musculus* from Sicily. Experientia 33:1025–1026.

188. Léonard, A., and G. Deknudt. 1967. A new marker for chromosome studies in the mouse. Nature 214:504–505.

189. Léonard, A., and G. Deknudt. 1969. Etude cytologique d'une translocation chromosome Y-autosome chez la souris. Experientia 25:876–877.

189a. Levy, E.R., and P.S. Burgoyne. 1986. Diploid spermatids: a manifestation of spermatogenic impairment in XO Sxr and T31H/+ male mice. Cytogenet. Cell Genet. 42:159–163.

189b. Levy, Y., and P.S. Burgoyne. 1986. The fate of XO germ cells in the testes of XO/XY and XO/XY/XYY mouse mosaics: evidence for a spermatogenesis gene on the mouse Y chromosome. Cytogenet. Cell Genet. 42:208–213.

190. Lyon, M.F. 1954. Stage of action of the litter-size effect on absence of otoliths in mice. Z. Indukt. Abstammungs-Vererbungsl. 86:289–292.

191. Lyon, M.F. 1966. Order of loci on the X-chromosome of the mouse. Genet. Res., Camb. 7:130–133.

192. Lyon, M.F. 1966. Lack of evidence that inactivation of the mouse X-chromosome is incomplete. Genet. Res., Camb. 8:197–203.

193. Lyon, M.F. 1969. A true hermaphrodite mouse presumed to be an XO/XY mosaic. Cytogenetics 8:326–331.

194. Lyon, M.F. 1972. X-chromosome inactivation and developmental patterns in mammals. Biol. Rev. 47:1–35.

195. Lyon, M.F., and K.B. Bechtol. 1977. Derivation of mutant t-haplotypes of the mouse by presumed duplication or deletion. Genet. Res., Camb. 30:63–76.

196. Lyon, M.F., and P.H. Glenister. 1971. Mouse News Lett. 45:24.

197. Lyon, M.F., and P.H. Glenister. 1972. Mouse News Lett. 46:29.

198. Lyon, M.F., and P.H. Glenister. 1974. Mouse News Lett. 51:22.

199. Lyon, M.F., and P.H. Glenister. 1977. Factors affecting the observed number of young resulting from adjacent–2 disjunction in mice carrying a translocation. Genet. Res., Camb. 29:83–92.

200. Lyon, M.F., and S.G. Hawker. 1973. Reproductive lifespan in irradiated and unirradiated chromosomally XO mice. Genet. Res., Camb. 21:185–194.

201. Lyon, M.F., and S. Jarvis. 1979. Mouse News Lett. 60:44.

202. Lyon, M.F., and R. Meredith. 1966. Autosomal translocation causing male sterility and viable aneuploidy in the mouse. Cytogenetics 5:335–354.

203. Lyon, M.F., and R.J.S. Phillips. 1959. Crossing-over in mice heterozygous for t-alleles. Heredity 13:23–32.

204. Lyon, M.F., and I.M. Sayers. 1977. Mouse News Lett. 57:17.

205. Lyon, M.F., P.H. Glenister, and S.G. Hawker. 1972. Do the H–2 and T-loci of the mouse have a function in the haploid phase of sperm? Nature 240:152–153.

206. Lyon, M.F., I.M. Mason, and T.R.L. Bigger. 1977. Mouse News Lett. 56:37.

207. Lyon, M.F., R.J.S. Phillips, and A.G. Searle. 1964. The overall rates of dominant and recessive lethal and visible mutation induced by spermatogonial X-irradiation of mice. Genet. Res. 5:448–467.

208. Lyon, M.F., R.J.S. Phillips, and G. Fisher. 1982. Use of an inversion to test for induced X-linked lethals in mice. Mutat. Res. 92:217–228.

209. Lyon, M.F., I.M. Sayers, and E.P. Evans. 1978. Mouse News Lett. 58:44.

210. Lyon, M.F., H.C. Ward, and G.M. Simpson. 1975. A genetic method for measuring non-disjunction in mice with Robertsonian translocations. Genet. Res., Camb. 26:283–295.

211. Lyon, M.F., S. Wall, and J.F. Zenthon. 1985. Mouse News Lett. 72:104.

212. Lyon, M.F., A.G. Searle, C.E. Ford, and S. Ohno. 1964. A mouse translocation suppressing sex-linked variegation. Cytogenetics 3:306–323.

213. Lyon, M.F., P.H. Glenister, J.F. Loutit, E.P. Evans, and J. Peters. 1984. A presumed deletion covering the W and Ph loci of the mouse. Genet. Res. 44:161–168.

214. Lyon, M.F., J. Zenthon, E.P. Evans, M.D. Burtenshaw, K.A. Wareham, and E.D. Williams. 1986. Lack of inactivation of a mouse X-linked gene physically separated from the inactivation centre. J. Embryol. Exp. Morphol. 97:75–85.

215. Magnuson, T., S. Smith, and C. Epstein. 1982. The development of monosomy 19 embryos. J. Embryol. Exp. Morphol. 69:223–236.

216. Magnuson, T., S. Debrot, J. Dimpfl, A. Zweig, T. Zamora, and C.J. Epstein. 1985. The early lethality of autosomal monosomy in the mouse. J. Exp. Zool. 236:353–360.

217. Mahadevaiah, S., and U. Mittwoch. 1986. Synaptonemal complex analysis in spermatocytes and oocytes of tertiary trisomic Ts(5^{12})31H mice with male sterility. Cytogenet. Cell Genet. 41:169–176.

218. Markert, C.L. 1982. Parthenogenesis, homozygosity and cloning in mammals. J. Hered. 73:390–397.

219. Matsuda, Y., H.T. Imai, K. Moriwaki, and K. Kondo. 1983. Modes of inheritance of X-Y dissociation in inter-subspecies hybrids between BALB/c mice and *Mus musculus molossinus*. Cytogenet. Cell Genet. 35:209–215.

220. Matsuda, Y., H.T. Imai, K. Moriwaki, and F. Bonhomme. 1982. X-Y chromosome dissociation in wild derived *Mus musculus* subspecies, laboratory mice, and their F$_1$ hybrids. Cytogenet. Cell Genet. 34:241–252.

221. McLaren, A. 1976. Mammalian chimaeras. Cambridge University Press, Cambridge, 154p.

221a. McLaren, A. 1983. Sex reversal in the mouse. Differentiation 23 (Suppl.):593–598.

222. Meo, T., J. Johnson, C.V. Beechey, S.J. Andrews, J. Peters, and A.G. Searle. 1980. Linkage analysis of murine immunoglobulin heavy chain and serum prealbumin genes establish their locations on chromosome 12 proximal to the T(5;12)31H breakpoint in band 12F1. Proc. Natl. Acad. Sci. USA 77:550–553.

223. Miller, D.A., and O.J. Miller. 1972. Chromosome mapping in the mouse. Science 178:949–955.

224. Miller, O.J., and D.A. Miller. 1975. Cytogenetics of the mouse. Ann. Rev. Genet. 9:285–303.

225. Miller, D.A., V.G. Dev, R. Tantravahi, O.J. Miller, M.B. Schiffman, R.A. Yates, and S. Glueksohn-Waelsch. 1974. Cytological detection of the c^{25H} deletion involving the albino (c) locus on chromosome 7 in the mouse. Genetics 78:905–910.

225a. Mittwoch, U., S. Mahadevaiah and M.B. Olive. 1981. Retardation of ovarian growth in male-sterile mice carrying an autosomal translocation. J. Med. Genet. 18:414–417.

226. Mittwoch, U., S. Mahadevaiah, and L.A. Setterfield. 1984. Chromosomal anomalies that cause male sterility in the mouse also reduce ovary size. Genet. Res., Camb. 44:219–224.

227. Morris, T. 1968. The XO and OY chromosome constitutions in the mouse. Genet. Res., Camb. 12:125–137.

228. Moses, M.J., L.B. Russell, and N.L.A. Cacheiro. 1977. Mouse chromosome translocations: visualization and analysis by electron microscopy of the synaptonemal complex. Science 196:892–894.

229. Moses, M.J., P.A. Poorman, T.H. Roderick, and M.T. Davisson. 1982. Synaptonemal complex analysis of mouse chromosomal rearrangements. IV. Synapsis and synaptic adjustment in two paracentric inversions. Chromosoma 84:457–474.

230. Nesbitt, M.N. 1973. Mouse News Lett. 49:23.

231. Nesbitt, M.N. 1975. Mouse News Lett. 52:31.

232. Nesbitt, M.N. 1975. Mouse News Lett. 53:68.

233. Nesbitt, M.N., and C.V. Beechey. 1979. Mouse News Lett. 60:45.

234. Nesbitt, M.N., and U. Francke. 1973. A system of nomenclature for band patterns of mouse chromosomes. Chromosoma (Berl.) 41:145–158.

235. Neuhäuser-Klaus, A., and I.-D. Adler. 1984. Mouse News Lett. 71:40.

236. Nijhoff, J.H. 1977. Mouse News Lett. 57:40–41.

237. Ohno, S., and B.M. Cattanach. 1962. Cytological study of an X-autosome translocation in *Mus musculus*. Cytogenetics 1:129–140.

238. Ohno, S., and M.F. Lyon. 1965. Cytological study of Searle's X-autosome translocation in *Mus musculus*. Chromosoma (Berl.) 16:90–100.

239. Oshimura, M., and N. Takagi. 1975. Meiotic disjunction in T(14;15)6Ca heterozygotes and fate of chromosomally unbalanced gametes in embryonic development. Cytogenet. Cell Genet. 15:1–16.

240. Painter, T. 1927. The chromosome constitution of Gates' 'non-disjunction' (v-O) mice. Genetics 12:379–392.

241. Phillips, R.J.S. 1961. Mouse News Lett. 24:34.

242. Phillips, R.J.S. 1979. Mouse News Lett 60:45–46.

243. Phillips, R.J.S., and E.P. Evans. 1979. A new X-linked inversion. In(X)3H. Mouse News Lett. 61:39.

244. Phillips, R.J.S., and M.H. Kaufman. 1974. Bare-patches, a new sex-linked gene in the mouse, associated with a high production of XO females. II. Investigations into the nature and mechanism of the XO production. Genet. Res., Camb. 24:27–41.

245. Phillips, R.J.S., and J.R.K. Savage. 1976. Mouse News Lett. 55:14.

246. Rapp, M.E., Therman, and C. Denniston. 1977. Non-pairing of the X and Y chromosomes in the spermatocytes of BDF$_1$ mice. Cytogenet. Cell Genet. 19:85–93.

247. Rastan, S. 1983. Non-random X-chromosome inactivation in mouse X-autosome translocation embryos—location of the inactivation centre. J. Embryol. Exp. Morphol. 78:1–22.

247a. Rastan, S., and B.M. Cattanach. 1983. Interaction between the *Xce* locus and imprinting of the paternal X chromosome in mouse yolk-sac endoderm. Nature, Lond. 303:635–637.

248. Reader, C.R., and A.J. Solari. 1969. The histology and cytology of the seminiferous epithelium of mice carrying Searle's X-autosome translocation. Acta Physiol. Lat. Am. 19:249–256.

249. Roderick, T.H. 1971. Producing and detecting paracentric chromosomal inversions in mice. Mutat. Res. 11:59–69.

250. Roderick, T.H. 1976. Inversions of mice in studies of mutagenesis. Genetics 83:s64.

251. Roderick, T.H. 1983. Using inversions to detect and study recessive lethals and detrimentals in mice. *In* F.J. de Serres and W. Sheridan, eds., Utilization of mammalian specific locus studies in hazard evaluation and estimation of genetic risk, 135–167. Plenum, New York.

252. Roderick, T.H., and N.L. Hawes. 1970. Two radiation-induced chromosomal inversions in mice (*Mus musculus*). Proc. Natl. Acad. Sci. USA 67:961–967.

253. Roderick, T.H., and N.L. Hawes. 1974. Nineteen paracentric chromosomal inversions in mice. Genetics 76:109–117.

254. Roderick, T.H., N.L. Hawes, and M.N. Murphy. 1976. Detecting and using paracentric chromosome inversions. Ann. Rep. Jackson Lab. 47:30–32.

255. Russell, L.B. 1962. Chromosome aberrations in experimental mammals. Prog. Med. Genet. 2:230–294.

256. Russell, L.B. 1971. Definition of functional units in a small chromosomal segment of the mouse and its use in interpreting the nature of radiation-induced mutations. Mutat. Res. 11:107–123.

257. Russell, L.B. 1972. A second T(X;8) in the mouse. Genetics 71:s53–s54.

258. Russell, L.B. 1978. Mouse News Lett. 59:45.

259. Russell, L.B. 1983. X-autosome translocations in the mouse: their characterization and use as tools to investigate gene inactivation and gene action. *In* A.A. Sandberg, ed., Cytogenetics of the mammalian X chromosome, Part A, 205–250. Alan Liss, New York.

260. Russell, L.B. 1983. Qualitative analysis of mouse specific-locus mutations: information on genetic organization, gene expression and the chromosomal nature of induced lesions. *In* F.J. de Serres, ed., Utilization of mammalian

specific locus studies in hazard evaluation and estimation of genetic risk, 241–258. Plenum, New York.

261. Russell, L.B., and J.W. Bangham. 1961. Variegated-type position effects in the mouse. Genetics 46:509–525.

262. Russell, L.B., and J.W. Bangham. 1986. A new T(X;7) with probable absence of inactivation in one autosomal segment. Mouse News Lett. 75:42–43.

263. Russell, L.B., and N.L.A. Cacheiro. 1977. The *c* locus region of the mouse: genetic and cytological studies of small and intermediate deficiencies. Genetics 86:s53–s54.

264. Russell, L.B., and S. Maddux. 1977. Mouse News Lett. 56:54.

265. Russell, L.B., and C.S. Montgomery. 1969. Comparative studies on X-autosome translocations in the mouse. I. Origin, viability, fertility and weight of five T(X;1)'s. Genetics 63:103–120.

266. Russell, L.B., and C.S. Montgomery. 1970. Comparative studies on X-autosome translocations in the mouse. II. Inactivation of autosomal loci, segregation, and mapping of autosomal breakpoints in five T(X;1)'s. Genetics 64:281–312.

267. Russell, L.B., and W.L. Russell. 1960. Genetic analysis of induced deletions and of spontaneous non-disjunction involving chromosome 2 of the mouse. J. Cell. Comp. Physiol. 56, Suppl. 1:169–188.

268. Russell, L.B., J.W. Bangham, and C.L. Saylors. 1962. Delimitation of chromosomal regions involved in V-type position effects from X-autosome translocations in the mouse. Genetics 47:981–982.

268a. Russell, L.B., W.L. Russell, R.A. Popp, C. Vaughan, and K.B. Jacobson. 1976. Radiation induced mutations at mouse haemoglobin loci. Proc. Natl. Acad. Sci. USA 73:2843–2846.

269. Russell, L.B., W.L. Russell, N.L.A. Cacheiro, C.M. Vaughan, R.A. Popp, and K.B. Jacobson. 1975. A tandem duplication in the mouse. Genetics 80:s71.

270. Rutledge, J.C., K.T. Cain, N.L.A. Cacheiro, C.V. Fornett, C.G. Wright, and W.M. Generoso. 1985. A balanced translocation in mice with a neurological defect. Science 231:395–397.

270a. Said, K., T. Jacquart, C. Montgelard, H. Sonjaya, A.N. Helal, and J. Britton-Davidian. 1986. Robertsonian house mouse populations in Tunisia: a caryological and biochemical study. Genetica 68:151–156.

271. Searle, A.G. 1964. Genetic effects of spermatogonial X-irradiation on productivity of F_1 female mice. Mutat. Res. 1:99–108.

272. Searle, A.G. 1974. Nature and consequences of induced chromosome damage in mammals. Genetics 78:173–186.

273. Searle, A.G. 1981. Numerical variants and structural rearrangements. *In* M.C. Green, ed., Genetic variants and strains of the laboratory mouse, 324–357. Fischer, Stuttgart.

274. Searle, A.G., and C.V. Beechey. 1970. Mouse News Lett. 43:29.

274a. Searle, A.G., and C.V. Beechey. 1973. Mouse News Lett. 48:32.

275. Searle, A.G., and C.V. Beechey. 1974. Cytogenetic effects of X-rays and fission neutrons in female mice. Mutat. Res. 24:171–186.

276. Searle, A.G., and C.V. Beechey. 1974. Mouse News Lett. 50:40.

277. Searle, A.G., and C.V. Beechey. 1978. Complementation studies with mouse translocations. Cytogenet. Cell Genet. 20:282–303.

278. Searle, A.G., and C.V. Beechey. 1982. The use of Robertsonian translocations in the mouse for studies on non-disjunction. Cytogenet. Cell Genet. 33:81–87.

279. Searle, A.G., and C.V. Beechey. 1985. Noncomplementation phenomena and their bearing on nondisjunctional effects. *In* V.L. Dellarco, P.E. Voytek, and A. Hollaender, eds., Aneuploidy, 363–376. Plenum, New York.

280. Searle, A.G., C.V. Beechey, and E.P. Evans. 1978. Meiotic effects in chromosomally derived male sterility of mice. Ann. Biol. Anim. Biochim. Biophys. 18:391–398.

281. Searle, A.G., C.E. Ford, and C.V. Beechey. 1971. Meiotic disjunction in mouse translocations and the determination of centromere position. Genet. Res., Camb. 18:215–235.

282. Searle, A.G., C.V. Beechey, E.P. Evans, and M. Kirk. 1983. Two new X-autosome translocations in the mouse. Cytogenet. Cell Genet. 35:279–292.

283. Searle, A.G., C.V. Beechey, E.M. Eicher, M.N. Nesbitt, and L.L. Washburn. 1979. Colinearity in the mouse genome: a study of chromosome 2. Cytogenet. Cell Genet. 23:255–263.

284. Searle, A.G., C.V. Beechey, P. de Boer, D.G. de Rooij, E.P. Evans, and M. Kirk. 1983. A male-sterile insertion in the mouse. Cytogenet. Cell Genet. 36:617–626.

285. Searle, A.G., C.E. Ford, E.P. Evans, C.V. Beechey, M.D. Burtenshaw, and H.M. Clegg. 1974. The induction of translocations in mouse spermatozoa. I. Kinetics of dose response with acute x-irradiation. Mutat. Res. 22:157–174.

286. Setterfield, L.A., and U. Mittwoch. 1986. Reduced oocyte numbers in tertiary trisomic mice with male sterility. Cytogenet. Cell Genet. 41:177–180.

286a. Setterfield, L.A., S. Mahadevaiah, and U. Mittwoch. 1988. Chromosome pairing and germ cell loss in male and female mice carrying a reciprocal translocation. J. Reprod. Fert. 82:369–379.

287. Snell, G.D. 1946. An analysis of translocations in the mouse. Genetics 31:157–180.

288. Snell, G.D., E. Bodemann, and W. Hollander. 1934. A translocation in the house mouse and its effect on development. J. Exp. Zool. 67:93–104.

289. Snow, M.H.L. 1973. Tetraploid mouse embryos produced by cytochalasin B during cleavage. Nature 244:513–414.

290. Solari, A.J. 1971. The behaviour of chromosomal axes in Searle's X-autosome translocation. Chromosoma 34:99–112.

291. Solari, A.J. 1974. The behaviour of the X-Y pair in mammals. Int. Rev. Cytol. 38:273–313.

292. Sotomayor, R.E., and R.B. Cumming. 1977. XY dissociation in mice: a model that may account for sex aneuploidy in humans? Genetics 86:560–561.

293. Takagi, N. 1980. Primary and secondary nonrandom X chromosome inactivation in early female mouse embryos carrying Searle's translocation T(X;16)16H. Chromosoma 81:439–459.

293a. Tease, C., and B.M. Cattanach. 1986. Mammalian cytogenetic and genetic tests for nondisjunction. *In* F.J. de Serres, ed., Chemical mutagens, Vol. 10, 215–283. Plenum Press, New York.

294. Tease, C., and G. Fisher. 1985. Mouse News Lett. 72:105.

295. Tettenborn, U., and A. Gropp. 1970. Meiotic disjunction in mice and mouse hybrids. Cytogenetics 9:272–283.

295a. Tichy, H., and I. Vircak. 1987. Chromosomal polymorphism in the house mouse (*Mus domesticus*) of Greece and Yugoslavia. Chromosoma 95:31–36.

296. Traut, W., H. Winking, and S. Adolph. 1984. An extra segment in chromosome 1 of wild *Mus musculus*: a C-band positive homogeneously staining region. Cytogenet. Cell Genet. 38:290–297.

297. Wallace, M.E. 1981. Wild mice heterozygous for a single Robertsonian translocation. Mouse News Lett. 64:49.

298. Wauben-Penris, P.J.J., F.A. van der Hoeven, and P. de Boer. 1983. Chiasma frequency and nondisjunction in heteromorphic bivalents: meiotic behaviour in T(1;13)70H/T(1;13)1Wa mice as compared to T(1;13)70H/T(1;13)70H mice. Cytogenet. Cell Genet. 36:547–553.

299. White, B.J., and J.H. Tjio. 1967. A mouse translocation with 38 and 39 chromosomes but normal NF. Hereditas 58:284–296.

300. White, B.J., J.-H. Tjio, L.C. Van de Water, and C. Crandall. 1972. Trisomy for the smallest autosome of the mouse and identification of the T1Wh translocation chromosome. Cytogenetics 11:363–378.

301. White, B.J., J.H. Tjio, L.C. Van de Water, and C. Crandall. 1974. Trisomy 19 in the laboratory mouse. I. Frequency in different crosses at specific developmental stages and relationship of trisomy to cleft palate. Cytogenet. Cell Genet. 13:217–231.

302. White, B.J., J.-H. Tjio, L.C. Van de Water, and C. Crandall. 1974. Trisomy 19 in the laboratory mouse. II. Intra-uterine growth and histological studies of trisomies and their normal litter-mates. Cytogenet. Cell Genet. 13:232–245.

303. Winking, H. 1978. Mouse News Lett. 58:53–54.

304. Winking, H., and A. Gropp. 1976. Meiotic disjunction of metacentric heterozygotes in oocytes versus spermatocytes. *In* P.G. Crosignani and D.R. Mishell, eds., Ovulation in the human, 47–56. Academic Press, London.

305. Winking, H., B. Dulic, and A. Gropp. 1979. Mouse News Lett. 60:55.

306. Winking, H., A. Gropp, and G. Bulfield. 1981. Robertsonian chromosomes in mice from North-Western Greece. Mouse News Lett. 64:69–70.

307. Winking, H., G. Noack, A. Gropp, and E. Capanna. 1979. Mouse News Lett. 60:55.

308. Wolfe, H.G. 1967. Mapping the hemoglobin locus in mice transmitting the flecked translocation. Genetics 55:213–218.

309. Wroblewska, J. 1971. Developmental anomaly in the mouse associated with triploidy. Cytogenetics 10:199–207.

310. Zech, L., E.P. Evans, C.E. Ford, and A. Gropp. 1972. Banding patterns in the mitotic chromosomes of tobacco mouse. Exp. Cell Res. 70:263–268.

12.2 Centromeric heterochromatin variants

MURIEL T. DAVISSON

The A-T rich satellite DNA of the mouse is localized in heterochromatin near the centromeric region of the chromosomes (9, 12). When fixed chromosomes are appropriately treated, this heterochromatin is visualized as C-bands, darkly stained regions just distal to the centromeres. These C-bands are equivalent to the A1 G-bands in the standard karyotype (see Section 11.2). Variation in the size of one or more of the C-bands can occur among inbred strains and species of *Mus musculus* and *domesticus*. The variants are transmitted intact in crosses and can serve as markers for the centromeric region. They are designated by the symbol *Hc*, for heterochromatin in the centromeric region, followed by the number of the chromosome involved. Variations in size are indicated by superscripts: *s* (small or undetectable), *n* (normal), *l* (large) with C57BL/6J usually the standard. Table 12.3 gives the published information on strain distribution of centromeric heterochromatin variants.

The chromosomes of *Mus cervicolor* fluoresce brightly in the pericentromeric region when stained for quinacrine fluorescence (Q-bands). This is in contrast to *Mus musculus* or *domesticus* chromosomes which show weak fluorescence in this region. *Mus cervicolor* chromosomes have C-bands comparable to *M. musculus* and *domesticus* and slightly larger than the brightly fluorescent centromeric regions (2).

Interstitial C-bands have been detected in at least two chromosomes in wild populations of *Mus domesticus* and *musculus* subspecies (7, 11, 14). An extra C-band positive segment between bands C5 and D of Chr 1 has been found in wild mice captured in central (14) and eastern (7) Europe. The C-band staining seems to be less intense than that of pericentromeric heterochromatin and the segment does not contain satellite DNA in detectable amounts (14). Some *M. m. molossinus* mice from Japanese wild populations have a positive C-band in the middle of Chr 18 (11). It is not clear whether this segment has detectable satellite DNA sequences.

This work was supported by grants HD17784 and RR01183 from the National Institutes of Health and grant BSR8418828 from the National Science Foundation.

Table 12.3 Centromeric heterochromatin variants in different strains. Only the small and large variants are shown, all others being normal, except in the case of *Mus musculus molossinus* which may have two types of a particular C-band. The *M. m. molossinus* stocks were fairly recently derived from the wild, and some stocks may have a particular marker while others may not. The Y chromosome shows no C-band in any of the mice typed. Parentheses indicate conflicting reports or only slight deviation from normal

Strain	Centromeric hetero-chromatin	Reference
A/HeHok	$Hc1^s$, $Hc4^s$, $Hc9^l$, $Hc13^l$, $Hc18^l$	16*
A/J	$Hc4^s$	1
AKR	$Hc3^s$, $Hc19^s$	1
AKR/Hok	$Hc1^s$, $Hc3^s$, $Hc6^s$, $Hc8^l$, $Hc12^l$, $Hc13^l$, $Hc16^l$, $Hc19^l$	16
BALB/cJ	$(Hc6^l)$, $Hc7^s$, $(Hc13^l)$, $Hc18^s$	6, 10
BALB/cHok, BALB/cAnHok	$Hc1^s$, $Hc7^s$, $Hc9^l$, $Hc13^l$, $Hc14^l$, $Hc16^l$	16
CBA/J, CBA/H	$(Hc1^s)$, $(Hc6^l)$, $Hc14^s$, $(Hc18^s)$	5, 6
CBA/JHok, CBA/H-T6Hok	$Hc1^s$, $Hc9^l$, $Hc12^l$, $Hc14^s$, $Hc16^l$, $Hc18^s$	16
C3H/Di	$(Hc1^s)$, $Hc14^s$, $(Hc18^s)$	5, 6
C3H/HeJ	$Hc1^s$, $Hc14^s$, $Hc18^s$	1, 6, 13
C3H/HeHok	$Hc1^s$, $Hc9^l$, $Hc12^l$, $Hc14^s$, $Hc16^l$, $Hc18^s$	16
C3H/HeHOxe†	$Hc18^s$, HcX^s	4
C57BL/6J		1
C57BL/6JHok	$Hc1^s$, $Hc8^l$, $Hc16^l$, $Hc18^l$	16
C57BL/10J	$Hc11^s$	6, 10
C57BL/10ScSnPh	$Hc11^s$	5, 6
B10.A	$Hc11^s$	5, 6
B10.D2	$Hc11^s$	5, 6
C57BR/cdJ	$(Hc10^s)$, $Hc12^l$, $Hc13^s$	10
C57L/J	$Hc3^s$, $Hc5^l$, $Hc10^s$, $Hc12^l$, $Hc13^s$	1
DBA/1J	$(Hc1^s)$, $(Hc10^s)$, $Hc14^s$	5, 10
DBA/2J	$Hc1^l$, $Hc14^s$	1
DDD/Hok, DRC/Hok	$Hc1^s$, $Hc4^l$, $Hc6^l$, $Hc8^l$, $Hc9^l$, $Hc10^s$, $Hc11^s$, $Hc12^l$, $Hc13^l$, $Hc15^s$, $Hc16^l$, $Hc18^l$	16
DHS/Hok	$Hc1^s$, $Hc4^s$, $Hc8^l$, $Hc12^l$, $Hc13^l$	16
HLS/Hok	$Hc1^s$, $Hc4^s$, $Hc5^l$, $Hc8^l$, $Hc9^l$, $Hc12^l$, $Hc13^l$	16
KFY/Hok	$Hc1^s$, $Hc8^l$, $Hc9^l$, $Hc12^l$, $Hc13^l$, $Hc14^s$	16
MT/Hok	$Hc1^s$, $Hc8^l$, $Hc9^l$, $Hc10^s$, $Hc12^l$, $Hc13^l$, $Hc14^s$	16
SJL/JHok	$Hc1^s$, $Hc3^l$, $Hc4^s$, $Hc9^l$, $Hc12^l$, $Hc13^l$, $Hc17^s$	16
SWJ/Hok	$Hc1^s$, $Hc7^s$, $Hc8^l$, $Hc10^s$, $Hc12^l$, $Hc16^l$	16
SM/J	$Hc16^s$	1
T(7;X)1Ct/Hok	$Hc1^s$, $Hc7^s$, $Hc13^l$	16
101/Rl	$Hc8^s$, $Hc13^l$	4
101/HOxe†	$Hc1^l$, $Hc2^l$, $Hc5^l$, $Hc7^s$, $Hc8^s$, $Hc13^l$, $Hc14^s$, $Hc15^l$, $Hc17^l$	4

Table 12.3—*cont.*

Strain	Centromeric hetero-chromatin	Reference
102/El‡	$Hc5^l$, $Hc14^s$, $Hc15^l$, $Hc17^s$	4
129	$Hc8^s$, $Hc14^s$	1
M. m. molossinus	$Hc1^{n,l}$, $Hc2^{n,l}$, $Hc3^s$, $Hc4^{n,l}$ $Hc5^{n,s}$, $Hc6^{n,l}$, $Hc7^s$, $Hc8^{n,s}$, $Hc9^{n,s}$, $Hc10^{n,s}$, $Hc11^{n,l}$, $Hc12^{s,n,l}$, $Hc13^s$, $Hc14^s$, $Hc15^s$, $Hc16^s$, $Hc17^{n,s}$, $Hc18^{n,l}$, $Hc19^{n,s}$, HcX^s	1, 3, 8, 11

* The classification system for The Hokkaido University substrains typed by Yoshida and Kodama (16) differs from that used by others. Some of their large (*l*) variants may be included as normal (*n*) by other investigators. Their small (*s*) and missing (*o*) are grouped in the *s* category here.

† The Harwell strains C3H/HeH and 101/H are probably the same as the respective Oxford substrains but the Harwell substrains were only characterized for *Hc8* and *Hc13* (4).

‡ Formerly 101/El and 101/Sl (15).

References

1. Dev, V.G., D.A. Miller, and O.J. Miller. 1973. Chromosome markers in *Mus musculus*: Strain differences in C-banding. Genetics 75:663–670.
2. Dev, V.G., D.A. Miller, O.J. Miller, J.T. Marshall, Jr, and T.C. Hsu. 1973. Quinacrine fluorescence of *Mus cervicolor* chromosomes. Bright centromeric heterochromatin. Exp. Cell. Res. 79:475–479.
3. Dev, V.G., D.A. Miller, R. Tantravahi, R.R. Schreck, T.H. Roderick, B.F. Erlanger, and O.J. Miller. 1975. Chromosome markers in *Mus musculus*: Differences in C-banding between the subspecies *M. m. musculus* and *M. m. molossinus*. Chromosoma 53:335–244.
4. Evans, E.P., M.D. Burtenshaw, and I.-D. Adler. 1985. Chromosome differences between substrains of the inbred 101 mouse strain. Genet. Res., Camb. 46:353–356.
5. Forejt, J. 1973. Centromeric heterochromatin polymorphism in the house mouse. Evidence from inbred strains and natural populations. Chromosoma 43:187–201.
6. Forejt, J. 1979. Centromeric heterochromatin variations: mouse. *In* P.L. Altman and D.D. Katz, eds., Inbred and genetically defined strains of laboratory animals, Part 1, p. 44. Federation of American Societies for Experimental Biology, Bethesda, Maryland.
7. Gorlov, I.P., S.I. Agulnik, and A.I. Agulnik. 1986. Polymorphism for duplicated heterochromatin block in chromosome 1 in natural mice population of Novosibirsk. Mouse News Lett. 75:41–42.
8. Ikeuchi, T. 1978. Notes on the centromeric heterochromatin of the Japanese wild mouse, *Mus musculus molossinus*. Chromosome Information Service No. 25:24–26.
9. Jones, K.W. 1970. Chromosomal and nuclear location of mouse satellite DNA in individual cells. Nature 225:912–915.
10. Miller, D.A., R. Tantravahi, V.G. Dev, and O.J. Miller. 1976. Q- and C-band chromosome markers in inbred strains of *Mus musculus*. Genetics 82:67–75.
11. Moriwaki, K., and M. Minezawa. 1976. Geographical distribution of No. 18 chromosome polymorphism in the Japanese feral mouse, *Mus musculus molossinus*. Ann. Rep. Natl. Inst. Genet. No. 27:46–47.
12. Pardue, M.L., and J.G. Gall. 1970. Chromosomal localization of mouse satellite DNA. Science 168:1356–1358.
13. Schnedl, W. 1971. The karyotype of the mouse. Chromosoma 35:111–116.
14. Traut, W., H. Winking, and S. Adolph. 1984. An extra segment in chromosome 1 of wild *Mus musculus*: a C-band positive homogeneously staining region. Cytogenet. Cell Genet. 38:290–297.
15. West, J.D., M.F. Lyon, J. Peters, and P.B. Selby. 1985. Genetic differences between substrains of the inbred mouse strain 101 and designation of a new strain 102. Genet. Res., Camb. 46:349–352.
16. Yoshida M.C., and Y. Kodama. 1983. C-band patterns of chromosomes in 17 strains of mice. Cytogenet. Cell Genet. 35:51–56.

12.3 Nucleolus organizer regions

MURIEL T. DAVISSON

Genes coding for ribosomal RNA (rRNA) have been detected on at least six different chromosomes in the mouse genome. Because detectable ribosomal DNA (rDNA) is located at the nucleolus organizer regions of chromosomes, the symbol NOR has been given to chromosomal sites of rDNA. These regions have been shown by hybridization *in situ* of rRNA to contain the genes for 5.8S, 18S, and 28S rRNA (3, 5–7). In all laboratory strains rDNA is located just distal to the centromeric heterochromatin, usually at the secondary constriction.

The sites of rDNA are cytologically revealed by either hybridization *in situ* of labelled rRNA or NOR-specific silver stains (3–5). The number of chromosome pairs with detectable rDNA and the distribution of rDNA sites varies among strains. The size of rDNA sites also varies, reflecting the number of rRNA genes at particular sites. The size of NOR regions can vary among NOR chromosomes within a strain, among different strains, and in at least one case between sublines of a single strain.

More recently, restriction fragment length polymorphism (RFLP) variants of rDNA have been detected among mouse strains and used to map rDNA to proximal Chr 12 (1, 8). The gene symbol *Rnr* (RNA, ribosomal) has been proposed for the loci detected by these RFLPs (8).

Table 12.4 Chromosomal locations of nucleolus organizers (NOR) in different strains, detected by labelled rRNA (3,5–7) or silver staining (2), or rDNA detected by RFLP variants (1)

Strain	\multicolumn{6}{c}{Chromosomes with NOR}	Reference

Strain	12	15	16	17	18	19	Reference
BALB/c[a]	+	+	+		+		2
CBA/CaH-T6J		+			+	+	5, 6
C3H/HeJ, C3H/St	+[b]	+	+		+		1, 2
C57BL/6J	+	+	+		+	+	6
R163H1ALD/Ei[c]		+			+[d]		5
R347BNR/Ei[e]	+	+	+		+	+	6
SEC/1ReJ	+		+		+		3
TF/Gn					+		6
M. m. molossinus[f]	+		+	+	+	+	2

[a] Chr 16 stained with Ag-staining in some but not all BALB/cJ cells. This probably indicates a small NOR.
[b] Mapped using RFLPs only.
[c] A stock of mice homozygous for Rb(9.19)163H and Rb(6.15)1Ald.
[d] A chromosome tentatively identified as Chr 18 had an NOR.
[e] An incipient inbred strain derived from a cross between *Mus musculus poschiavinus* and a European strain of mice.
[f] *Mus musculus molossinus* mice were not inbred when studied. The lines typed were segregating for the NOR in Chr 12.

Although rDNA is always found at secondary constrictions, it is sometimes possible to demonstrate rDNA in chromosomes without obvious secondary constrictions. Occasionally, as in Chr 16 of BALB/cJ, a chromosome binds NOR- specific label but not consistently in all cells. Also C3H/HeJ Chr 12 does not have a silver-detectable NOR (2) but has rDNA fragments (1). Probably these chromosomes carry a reduced number of rRNA gene copies, making the NOR sites more difficult to detect. These kinds of data suggest it is possible that many, perhaps even all, of the mouse chromosomes contain at least one rRNA gene near the centromere.

Additional chromosomal sites of rDNA have been identified in wild mice captured in Europe (9). These are not included in Table 12.4 because they are not readily available in laboratory strains. They are, however, potential new cytological markers in stocks recently derived from wild populations. In *Mus**

* Although the authors call these *Mus musculus* (9), it is likely mice from these localities are *Mus domesticus*.

species from Italy and Yugoslavia silver-detected NORs have been found in terminal sites on Chrs 4 and 13, in addition to the proximal locations on Chrs 12, 16, 18, and 19. Chromosomes from a single *Mus spretus* specimen from Portugal had terminal NORs on Chrs 4, 13, and 19, but no proximal NORs. These observations suggest that laboratory stocks of *Mus* recently derived from wild populations may show greater variability in NORs than is found among inbred laboratory strains.

This work was supported by grants HD17784 and RR01183 from the National Institutes of Health and grant BSR8418828 from the National Science Foundation.

References

1. Arnheim, N., D. Treco, B. Taylor, and E.M. Eicher. 1982. Distribution of ribosomal gene length variants among mouse chromosomes. Proc. Natl. Acad. Sci. USA 79:4677–4680.
2. Dev, V.G., R. Tantravahi, D.A. Miller, and O.J. Miller. 1977. Nucleolus organizers in *Mus musculus* subspecies and in the RAG mouse cell line. Genetics 86:389–398.
3. Elsevier, S.M., and F.H. Ruddle. 1975. Location of genes coding for 18S and 28S ribsomal RNA within the genome of *Mus musculus*. Chromosoma 52:219–228.
4. Goodpasture, C., and S.E. Bloom. 1975. Visualization of nucleolar organizer regions in mammalian chromosomes using silver staining. Chromosoma 53:37–50.
5. Henderson, A.S., E.M. Eicher, M.T. Yu, and K.C. Atwood. 1974. The chromosomal location of ribosomal DNA in the mouse. Chromosoma 49:155–160.
6. Henderson, A.S., E.M. Eicher, M.T. Yu, and K.C. Atwood. 1976. Variation in ribosomal RNA gene number in mouse chromosomes. Cytogenet. Cell Genet. 17:307–316.
7. Henderson, A.S., M.T. Yu, and C. Milcarek. 1979. On the chromosomal location of 5.8S DNA in people and mice. Cytogenet. Cell Genet. 23:201–207.
8. Taylor, B.A., E.M. Eicher, and N. Arnheim. 1987. Locus symbol for rRNA genes, Mouse News Lett. 77: 130.
9. Winking, H., K. Nielsen, and A. Gropp. 1980. Variable positions of NORs in *Mus musculus*. Cytogenet. Cell Genet. 26: 158–164.

13 MAPS OF CHROMOSOMAL VARIANTS

13.1 Map of reciprocal translocations, inversions, and insertions

ANTONY G. SEARLE and

COLIN V. BEECHEY

As in the last edition (12), breakpoint positions of existing chromosomal rearrangements are shown on both the G-banded map of Nesbitt and Francke (10) and the linkage map. This time, however, inversions are included as well as reciprocal translocations, insertions and the Sxr transposition (Tp) which is now known to involve the short arm of the Y (2). See Chapter 12.1. for other references to breakpoints and 13.2 for linkage maps of inversions. In addition, the positions of some mapped genes also located to a particular G-band by *in situ* hybridization or otherwise (refs. 1, 3, 5–9, 11) are shown by placing the same symbol beside the chromosome band concerned and the locus on the linkage map.

The abbreviated symbol of each translocation and pericentric inversion has been placed, as far as possible, opposite the G-band in which the breakpoint is thought to be, with lines to this band and to the estimated position on the linkage map. For other inversions and insertions vertical lines join the two breakpoints on a particular chromosome, with the symbol joined fairly centrally to this line. Dashed lines are used when there is uncertainty over the exact band positions of breakpoints or their location on the linkage map. Translocation symbols are bracketed when the translocations concerned share breakpoints with respect to bands or location on the linkage map. Unbanded inversions have been omitted.

Linkage maps have been constructed by (i) assuming a total map length of 1600 cM, as before, (ii) multiplying this by the percentages of the whole karyotype represented by each chromosome (see Table 4.1) to derive an estimate of the linkage map length of each chromosome. For the actual drawing, 1 cM equalled 1.3 mm, reduced to 0.9 mm for publication. The loci included in the maps are those actually used in linkage tests with the translocations, etc., with the addition of some end-markers; distances apart are drawn to the scale given above. Where there is evidence that a particular rearrangement is leading to crossover suppression, proportional map distances to marker loci have been used rather than those based on actual recombination frequencies. Where the distance from the proximal end-marker to the centromere is not accurately known the line between them is interrupted. Short arms have been placed on all the chromosomes except 1 and 7, as before (4).

Some general comments may be in order, in view of the increased amount of information in these maps as compared with the first edition. The marked tendency for the breakpoints of rearrangements to be located in light-staining G-bands is still found. Moreover, the concept of colinearity of genetic and cytological maps (13) still holds very well, since there are only two or three instances in which cytological and genetic positions appear to disagree in a total of 37 translocations for which relevant fairly precise data on positions of breakpoints are available.

There is still good evidence that in certain chromosomes (e.g. numbers 1, 4, 7, 8, 9, X) genetic breakpoints of translocations tend to be closer together than banding ones, with little signs of a tendency in the opposite direction, except possibly for Chrs 16 and 19. This suggests that on the chromosomes concerned, and possibly others, chiasmata are more likely to occur towards the ends of chromosomes than elsewhere. However, it should be remembered that the distribution of chiasmata in multivalents with separate proximal and distal regions may well be different from that in intact bivalents of the chromosomes concerned. The most striking example of differences between cytological and genetic map positions remains T(14;15)6Ca.

Note added in proof: E.P. Evans and M.D. Burtenshaw (Mouse News Lett. **83**, 164) have reported revised breakpoint positions for T6Ca, namely 14E3 and 15A2. These reduce the cytogenetic differences mentioned above.

620

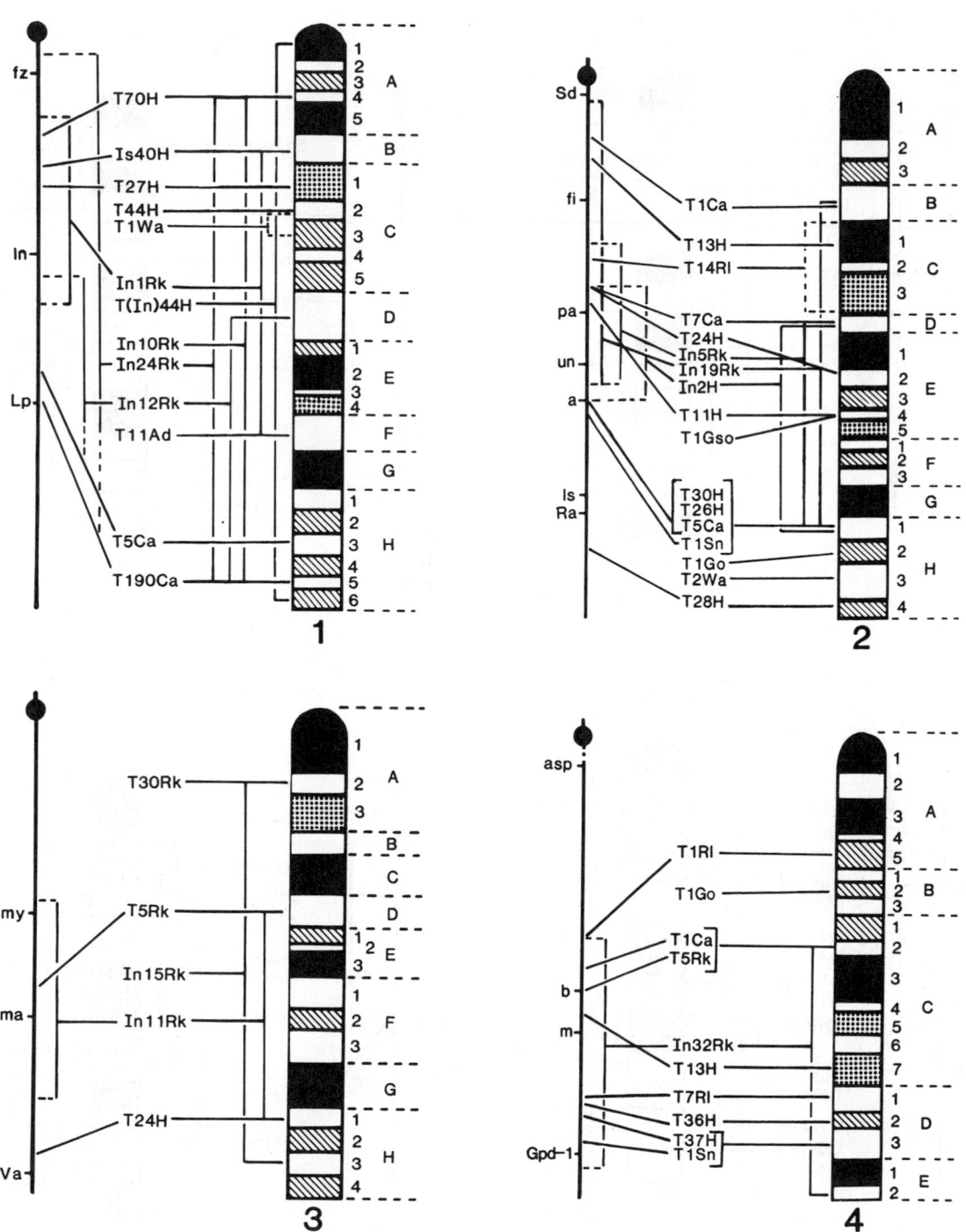

Fig. 13.1 Breakpoint positions of reciprocal translocations (T), inversions (In), and insertions (Is) on G-banded chromosomes and linkage maps.

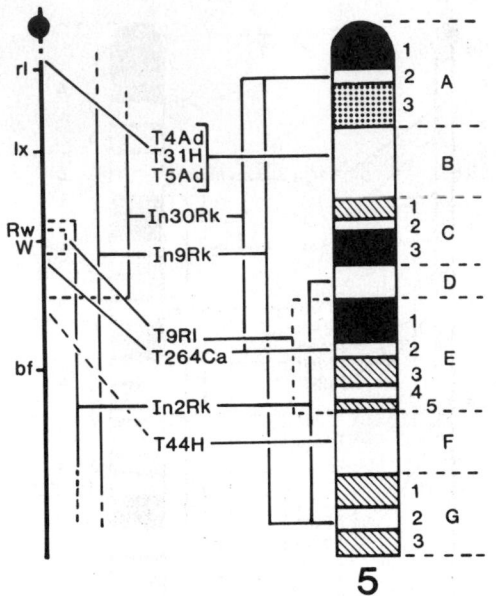

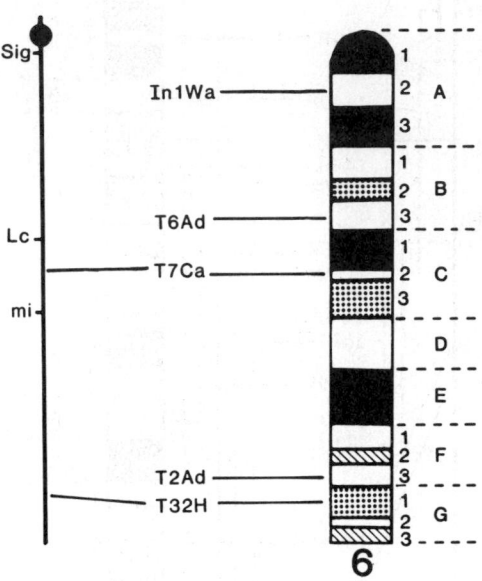

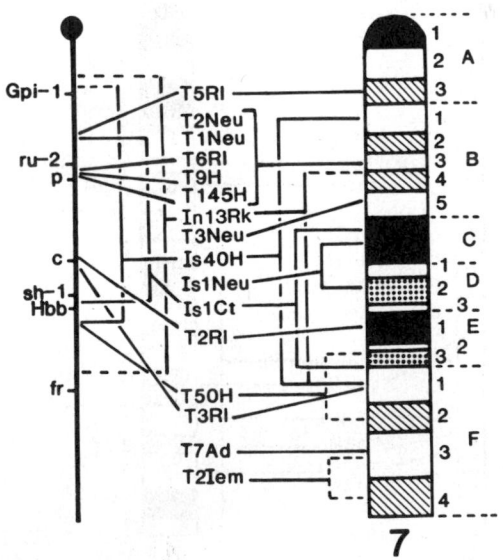

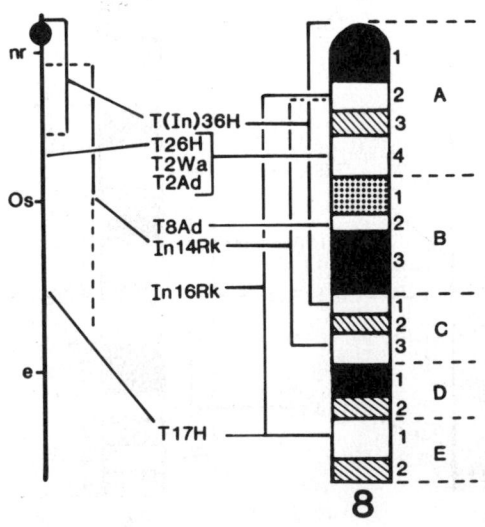

Fig. 13.1 (*Continued*)

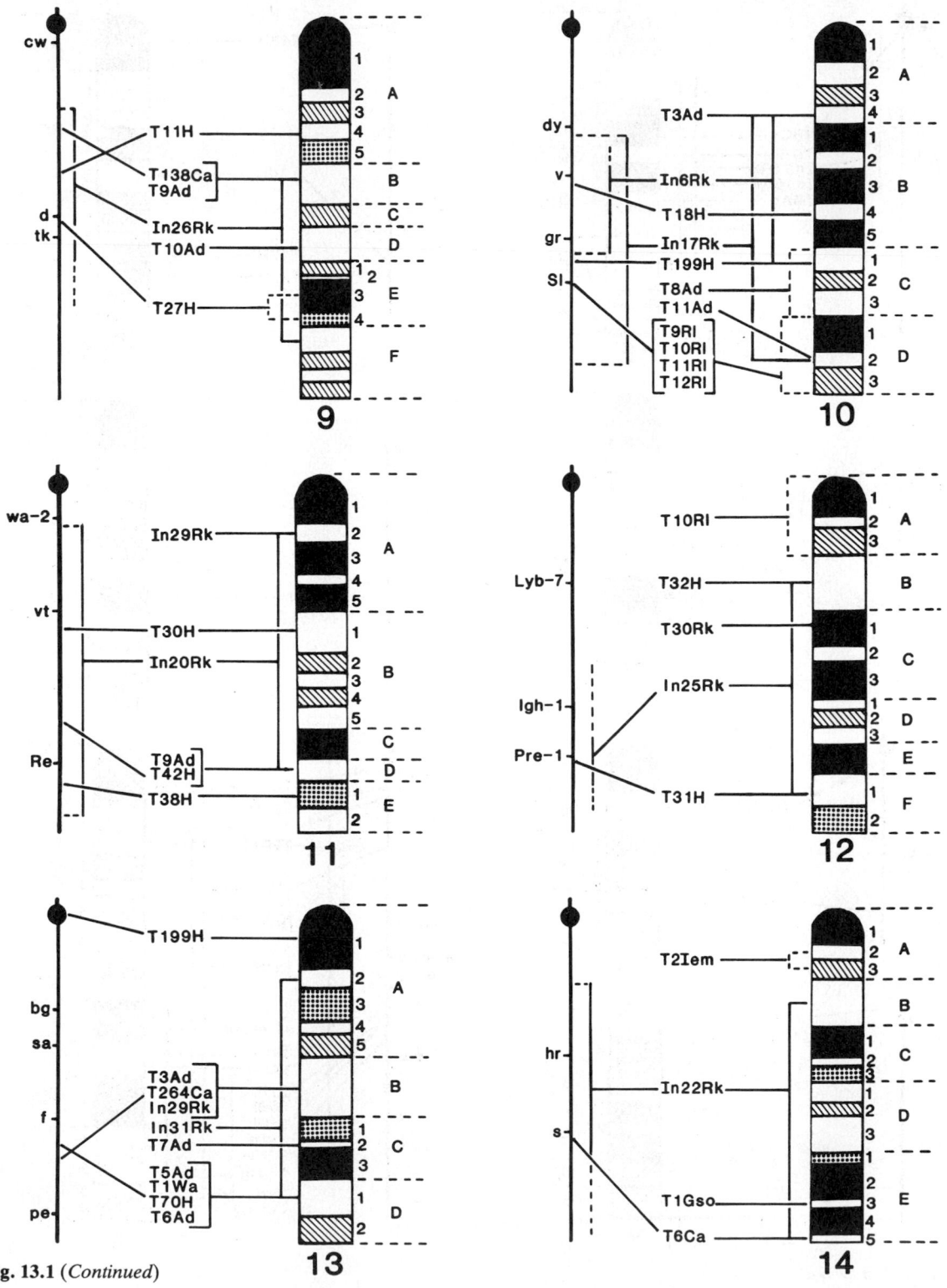

Fig. 13.1 (*Continued*)

623

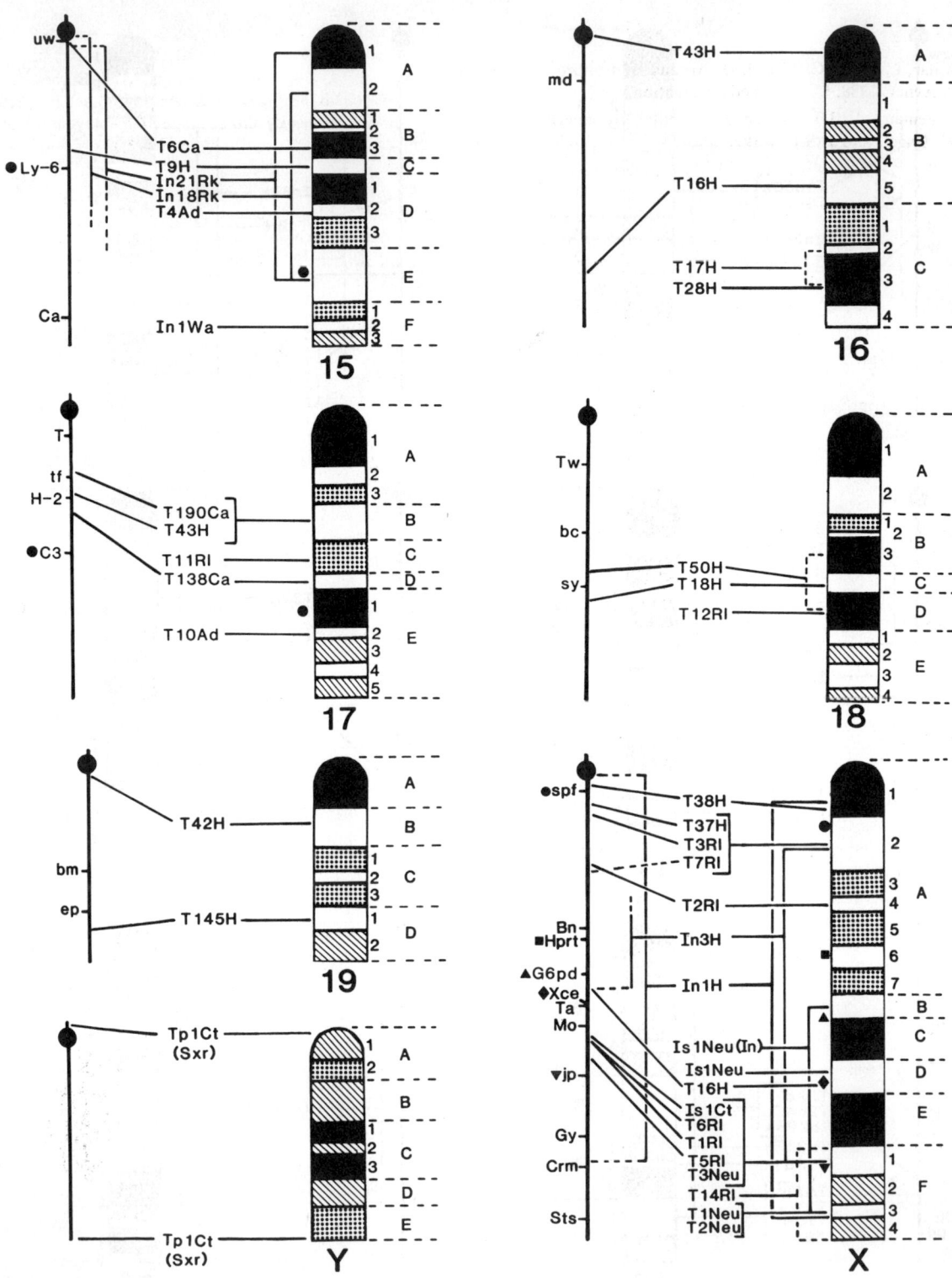

Fig. 13.1 (*Continued*)

References

1. Amar, L.C., M.-G. Mattei, D. Arnaud, J. Cambrou, and P. Avner. 1988. Unpublished information.

2. Arnemann, J., J.T. Epplen, H.J. Cooke, U. Sauermann, W. Engel, and J. Schmidtke. 1987. A human Y-chromosomal DNA sequence expressed in testicular tissue. Nucleic Acids Research 15:8713–8724.

3. Dautigny, A., M.-G. Mattei, D. Morello, P.M. Alliel, D. Pham-Dinh, L. Amar, D. Arnaud, D. Simon, J.-F. Mattei, J.-L. Guenet, P. Jollès, and P. Avner. 1986. The structural gene coding for myelin-associated proteolipid protein is mutated in jimpy mice. Nature 321:867–869.

4. Grzeschik, K.-H., and M.A. Kim. 1976. The structure of the paracentric region in mouse chromosomes. Chromosomes Today 5:361–365.

5. LeClair, K.P., M. Rabin, M.N. Nesbitt, D. Pravtcheva, F.H. Ruddle, R.G.E. Palfree, and A. Bothwell. 1986. Murine *Ly-6* multigene family is located on chromosome 15. Proc. Natl. Acad. Sci. USA 84:1638–1642.

6. Lyon, M.F., J. Zenthon, E.P. Evans, M.D. Burtenshaw, and K.R. Willison. 1988. Extent of the mouse *t* complex and its inversions shown by in situ hybridization. Immunogenet. 27:375–382.

7. Lyon, M.F., J. Zenthon, M.D. Burtenshaw, and E.P. Evans. 1987. Localization of the *Hprt* locus by in situ hybridization and distribution of loci on the mouse X-chromosome. Cytogenet. Cell Genet. 44:163–166.

8. Lyon, M.F., J. Zenthon, E.P. Evans, M.D. Burtenshaw, K.A. Wareham, and E.D. Williams. 1986. Lack of inactivation of a mouse X-linked gene physically separated from its inactivation centre. J. Embryol. Exp. Morphol. 97:75–85.

9. Münke, M., D.R. Cox, I.J. Jackson, B.L.M. Hogan, and U. Francke. 1986. The murine *Hox-2* cluster of homeobox containing genes maps distal on chromosome 11 near the tail-short (*Ts*) locus. Cytogenet. Cell Genet. 42:236–240.

10. Nesbitt, M.N., and U. Francke. 1973. A system of nomenclature for band patterns of mouse chromosomes. Chromosoma 41:145–158.

11. Rastan, S., and E. Robertson. 1985. X-chromosome deletions in embryo-derived (EK) cell lines associated with lack of X-chromosome inactivation. J. Embryol. Exp. Morphol. 90:379–388.

12. Searle, A.G., and C.V. Beechey. 1981. Map of reciprocal translocations and insertions. *In* M.C. Green, ed., Genetic variants and strains of the laboratory mouse, 360–365. Gustav Fischer, Stuttgart.

13. Searle, A.G., C.V. Beechey, E.M. Eicher, M.N. Nesbitt, and L.L. Washburn. 1979. Colinearity in the mouse genome: a study of chromosome 2. Cytogenet. Cell Genet. 23:255–263.

13.2 Map of inversions

THOMAS H. RODERICK, NORMAN L. HAWES,
MURIEL T. DAVISSON, and
ALAN L. HILLYARD

Forty seven paracentric inversions now are available in the mouse. The chromosomal locations of 45 of them are shown in Figure 13.2. Further information on chromosomal breakpoints are given in Chapter 12.1.7. Pericentric inversions found in Robertsonian chromosomes are not shown on this map but are also described in Chapter 12.1.7.

Paracentric inversions have been ascertained (1) cytologically by the inverted order of bands on properly stained chromosomes, (2) by their characteristic and generally high frequency of first meiotic anaphase bridges in inversion heterozygotes, and (3) by their effect of decreasing recombination among loci in the region of the inversion in inversion heterozygotes. The banded chromosomes can be prepared from blood cells; the bridges are most easily observed in histological sections of the testis.

Paracentric inversions have been located in three different ways (1) by linkage tests in standard backcrosses involving mapped loci such as are shown on the locus map in Chapter 4. (2) by cytological observation of the inverted order of bands on identified chromosomes, and (3) by studying F1 hybrids involving other chromosomally located inversions. In the last method a male heterozygous for two inversions in the same chromosome pair will generally show reduced meiotic activity, reduced testis size, and sterility when the inversions each overlap one breakpoint of the other. A male heterozygous for two or three inversions will show a high frequency of double anaphase bridges when the inversions are located on different chromosomes. This high frequency of double bridges is predictable from the product of the characteristic frequencies of single bridges for each inversion.

For an explanation of the general conventions of this map see Chapter 4. Here the inversions are shown against a map of the 'anchor' or well positioned loci. The inversion names are given to the right of the chromosome and are abbreviated to include the number and the laboratory. Laboratory symbols are: H = Harwell, By = Bailey, Dn = Davisson, and Rk = Roderick. The extent of each inversion is indicated by a vertical line to the left of the chromosome. A solid line spans the loci which have been tested for linkage with the inversion or which are presumed to be contained within the inversion from other linkage data. The inversions either do not recombine with those loci or their recombination with them is greatly reduced. The dashed extensions of the lines show the additional length estimated from the particular frequency of anaphase bridges. In general, the higher the frequency of bridges, the longer the inversion. Wavy extensions of the inversion lines show chromosomal regions known to be involved in the inversion by cytological examination of the bands. Inversions In(4)1By, In(6)48Rk and In(16)51Rk have been assigned to chromosomes by cytological or preliminary linkage data but their position on the chromosome is not yet known. These are listed below their respective chromosomes.

This work was supported by grant GM–19656 from the National Institutes of Health, contract DE-AC02-76EV03267 from the Department of Energy, and a contract with the Howard Hughes Medical Institute.

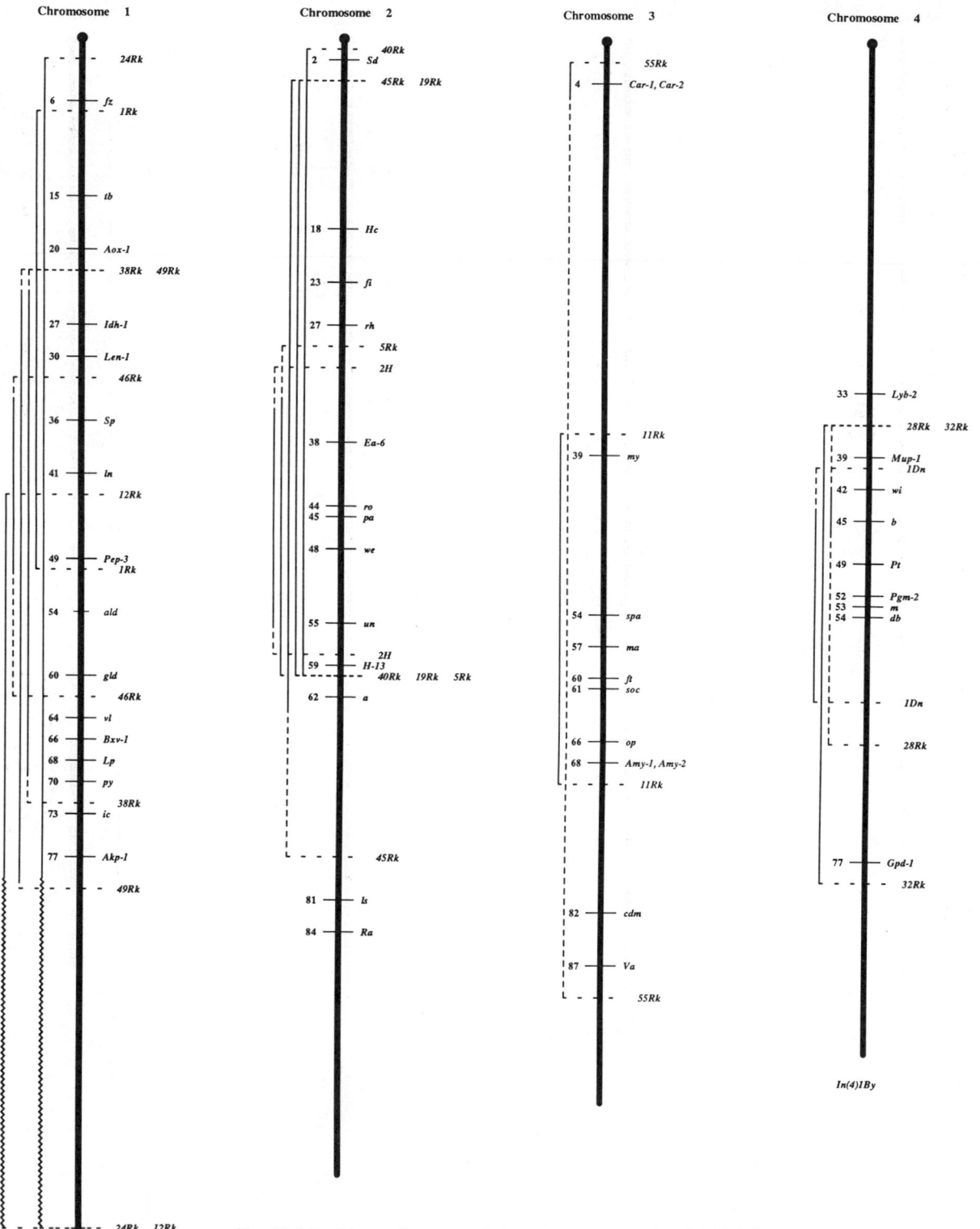

Fig. 13.2 Positions of paracentric inversions on an abbreviated linkage map.

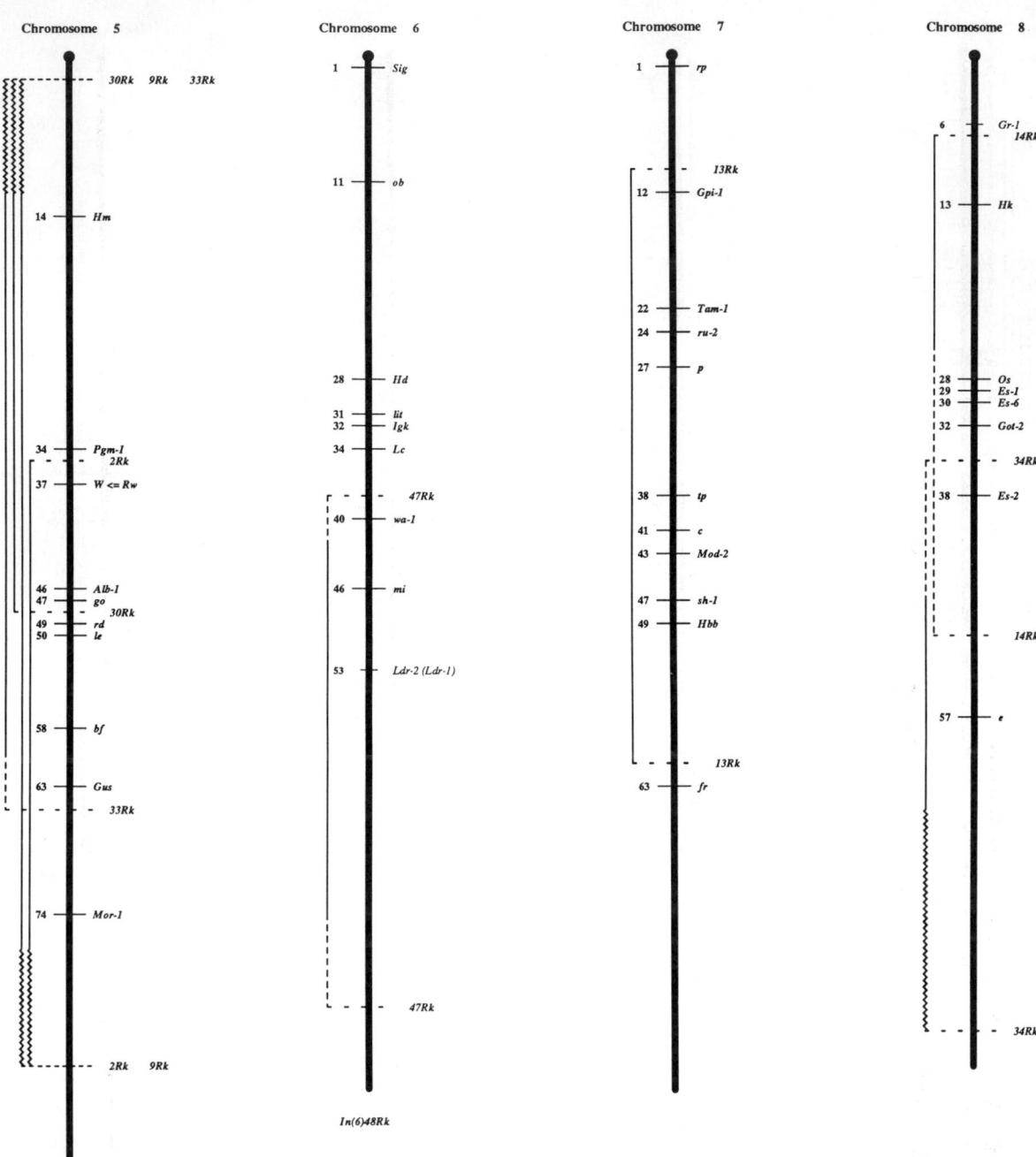

Fig. 13.2 (*Continued*)

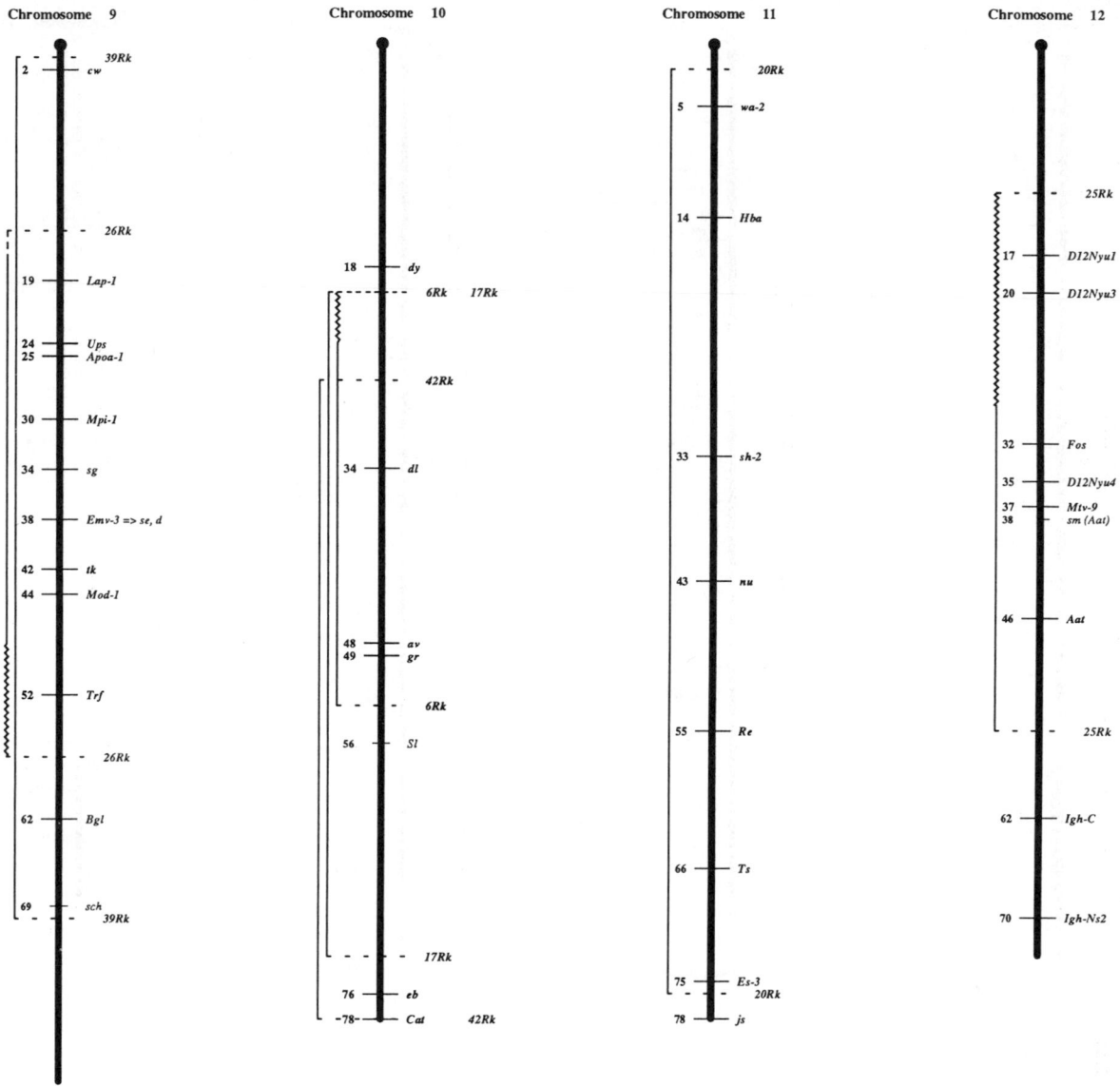

Fig. 13.2 *(Continued)*

629

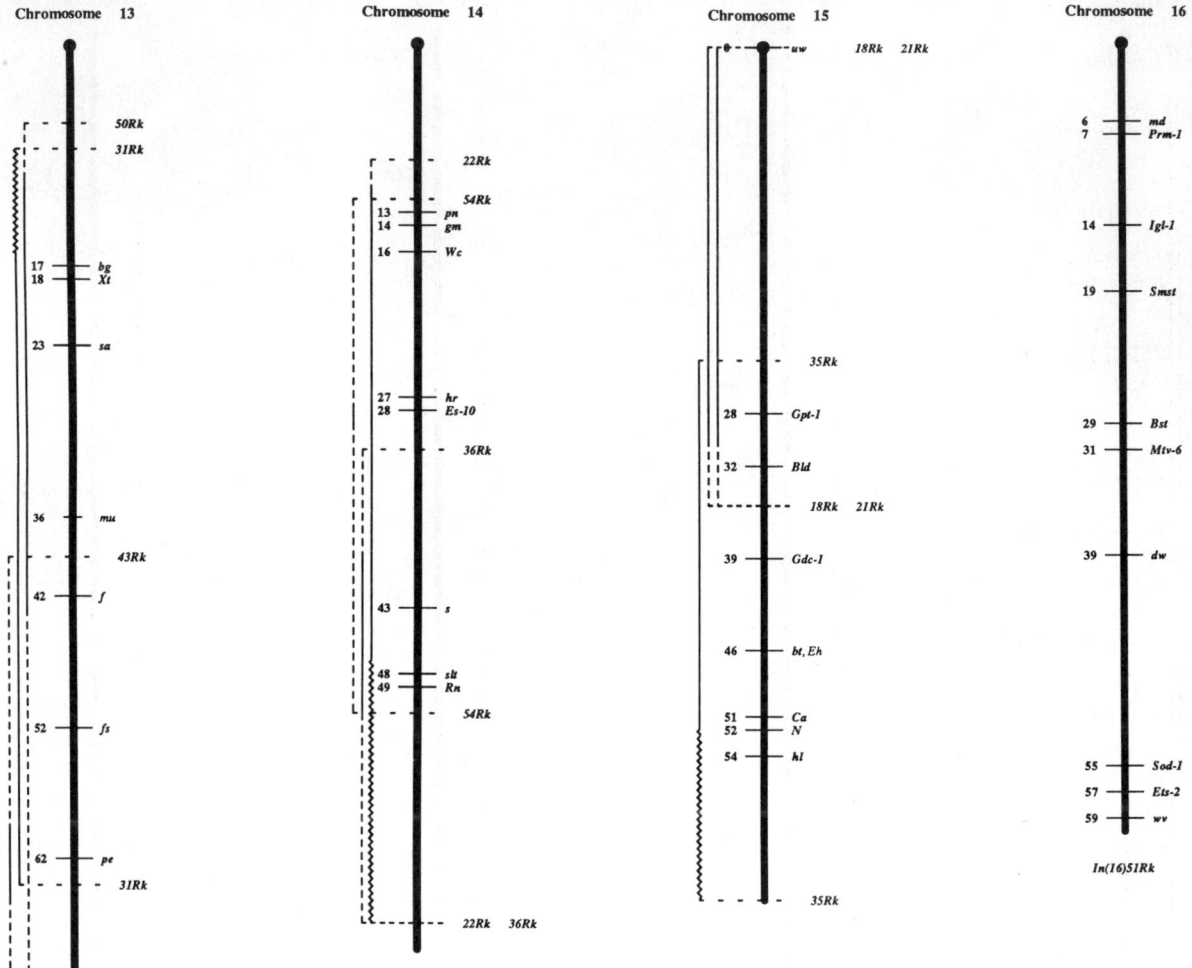

Fig. 13.2 (*Continued*)

630

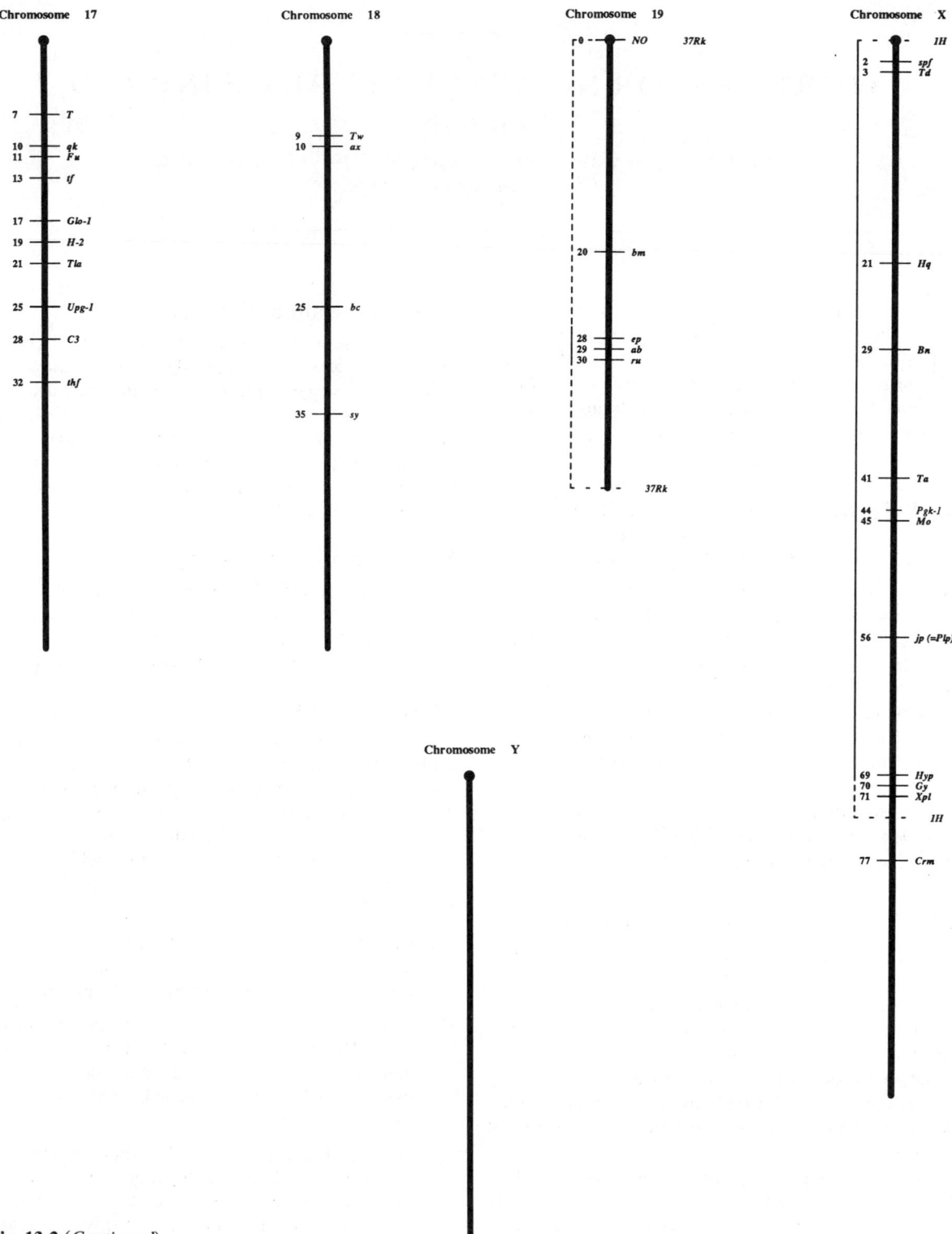

Fig. 13.2 (*Continued*)

14   RULES FOR NOMENCLATURE OF INBRED STRAINS

COMMITTEE ON STANDARDIZED GENETIC NOMENCLATURE FOR MICE
Chairman: MARY F. LYON

These new rules were adopted by the Committee in 1978. They are the most recent version of rules for nomenclature of inbred strains first published in 1952 (1) and revised in 1960 (2). They include a revision of the rules for nomenclature of inbred strains preserved by freezing, which appeared in 1976 (3).

14.1 Inbred strains

14.1.1 Definition of inbred strain

A strain shall be regarded as inbred when it has been mated brother × sister (hereafter called b × s) for 20 or more consecutive generations (F20), and can be traced to a single ancestral breeding pair in the 20th or subsequent generations. Parent × offspring matings may be substituted for b × s matings provided that, in the case of consecutive parent × offspring matings, the mating in each case is to the younger of the two parents.

Exceptionally, other breeding systems may be used, provided that the inbreeding coefficient achieved is at least equal to that at F20 (i.e. 0.99).

14.1.2 Symbols for inbred strains

Inbred strains shall be designated by a capital letter or letters in roman type or, less preferably, by a combination of capital letters and numbers, beginning with a letter. It is urged that anyone naming a new stock consult the most recent listing of inbred strains (4–6) to avoid duplication. Brief symbols are preferred.

An exception is allowed in the case of stocks already widely used and known by a designation that does not conform.

Any strains with a common origin separated before F20 shall be regarded as related inbred strains and be given symbols that indicate relationship, e.g. NZB, NZC, NZO.

14.1.3 Indication of inbreeding

When it is desired to indicate the number of generations of b × s inbreeding, this shall be done by appending, in parentheses, an F followed by the number of inbred generations. Example: (F87). If, because of incomplete information, the number given represents only part of the total inbreeding, this should be indicated by preceding it with a question mark and plus sign, e.g. YBL (F?+10).

14.1.4 Priority in strain symbols

If two inbred strains are assigned the same symbol, the symbol to be retained shall be determined by priority in publication. For this purpose, listing in *Mouse News Letter* shall be regarded as publication.

14.1.5 Substrains

Inbred strains may be subdivided into substrains and/or sublines (see below). An established inbred strain is considered to be divided into substrains when known or probable genetic differences become established in separate branches. Such differences could arise by: residual heterozygosity at the time of branching; mutation; or contamination. Hence, substrains should be considered to be formed:

(a) when branches are separated before F40 (i.e. after 20 to 40 generations of b x s matings). In such cases, residual heterozygosity may be present.

(b) when a branch is known to have been maintained separately from other branches for 100 or more generations from their common ancestor. The existence of differences arising by mutation is then highly probable.

(c) when genetic differences from other branches are discovered. Such differences could arise by any of the three means listed above. Contamination is likely to lead to numerous genetic differences, and may thus be distinguishable from mutation. If contamination is thought likely, the strain should be renamed, e.g. A2G, a strain arising by contamination of strain A, when maintained by Glaxo (G).

14.1.6 Designation of substrains

A substrain shall be known by the name of the parent followed by a slanted line and an appropriate substrain symbol. Substrain symbols may be:

(a) numbers, e.g. DBA/1, DBA/2;

(b) an abbreviation of the name of the person or laboratory originating the strain or substrain. The initial letter of this symbol should be a roman capital; all other letters should be lower case. Abbreviations should be checked with published lists (4–6) to avoid duplication, e.g. A/He, Heston substrain of strain A; CBA/J, Jackson Laboratory substrain of CBA;

(c) a combination of name abbreviation and number, beginning with a number, when one investigator or laboratory originates more than one substrain, e.g. C57BL/6J, C57BL/10J.

Exceptions, such as lower-case letters, are permitted for already well-known strains; e.g. BALB/c; C57BR/cd.

Where successive genetic differences arise in an inbred strain, leading to successive substrain formation, the substrain symbols should be accumulated, since the genetic differences too will be cumulative, e.g. CBA/HN, a substrain of CBA found by National Institutes of Health (N) to be genetically different from CBA/H, a substrain originated by Harwell, from which it arose.

14.1.7 Sublines

An inbred strain or substrain should be considered as divided into sublines when environmental, maternal, or cytoplasmic differences are certainly or probably present, or when genetic differences are possible rather than probable. Sublines are thus constituted either by:

(a) manipulative processes, such as fostering, egg transfer, hand-rearing, ovary transplant, or preservation by freezing; or

(b) transfer of a stock to another investigator.

14.1.8 Designation of sublines

Whereas correct designation of substrains is obligatory, since different substrains may have substantial genetic differences, designation of sublines is optional. This is because the manipulative processes of transfer of stock may not in all cases be relevant to the work being performed.

1. Sublines developed by manipulative processes, such as foster-nursing, may be indicated by appending an appropriate lower-case letter to the strain symbol, i.e.

Egg transfers	– e
Foster-nursing	– f
Hand-rearing	– h
Ovary transplant	– o
Freeze preservation	– p
Fostered on hand-reared	– fh

The strain used as foster parent or egg or ovary recipient may be indicated by the addition of its symbol or an abbreviation of it, e.g. C3Hf; C3HfC57BL or C3HfB, C3H fostered on C57BL; C3HeB, eggs of strain C3H transferred to C57BL.

Symbols for manipulative processes should be placed after any substrain symbols. If there is any risk of confusion, the symbol for fostering, etc. may be used in subscript position, e.g. A/He$_f$B, Heston substrain of A fostered on C57BL.

2. One significance of manipulative processes, such as fostering, is that they may add or eliminate vertically transmitted viruses. If the identity of an added or removed virus is known, or if a virus is added artificially, this should be indicated by appending a symbol for the virus, in capital letters, followed by a + or − sign according to whether it has been added or removed, e.g. C3H/HeN-MTV+, purified MTV inoculated into young caesarean-derived C3H/HeN.

3. Transfer of a strain to another investigator may introduce relevant environmental differences, and also genetic differences as the number of generations of separation increases. A subline separated by transfer should be indicated by an abbreviation of the name of the person or laboratory holding it, as for a substrain, but separated by a double slanted line. Symbols for sublines should follow those for substrains, e.g. C3H/He//H, a subline of C3H/He maintained at Harwell.

Subline symbols should *not* be accumulated, since environmental factors are unlikely to act cumulatively in this sense. Where a certain branch of a strain or substrain has been subjected to repeated manipulations or transfers, the symbol for the most recent event should be given, and the history of the subline should be recorded in *Mouse News Letter*.

Coisogenic and congenic strains (see below) may be divided into sublines, in which case the subline symbol should follow the complete symbol for the congenic strain, e.g. CBA/Ca-*se*//Gn, Carter's substrain of CBA, with mutation to *se*, maintained by Green.

14.2 Recombinant inbred strains

Strains formed by crossing two inbred strains, followed by 20 or more generations of b × s mating, shall be known as recombinant inbred (RI) strains. The names of such strains shall consist of an abbreviation of both parental strain names separated by a capital X (letter eks) with no intervening spaces, e.g. CXB, a set of recombinant inbred strains derived from a cross of BALB/c x C57BL.

Different RI strains in the same series should be distinguished by numbers (e.g. CXB2, CXB3, etc.) except for strains already well known by a different designation. For clarity, it is optional to use a hyphen to separate elements that distinguish strains in a series, e.g. CXB6-2.

14.3 Coisogenic, congenic, and segregating inbred strains

Two strains that are genetically identical (i.e. isogenic), except for a difference at a single locus, are called *coisogenic*. True coisogenicity can probably be achieved only by mutation within an existing inbred strain, while lines obtained by inbreeding with forced heterozygosis (segregating inbred), or by crossing on to an inbred strain (congenic), usually differ in a short chromosomal segment rather than a single gene. Congenic lines that differ at a histocompatibility locus and, therefore, resist each other's grafts are called *congenic resistant* (CR) lines.

Congenic or coisogenic strains are designated by a compound symbol consisting of two parts separated by a period or hyphen, the first part in all cases being the full or abbreviated symbol of the background or parent strain. The second part may be *either* the symbol of the differential locus and allele *or* an abbreviated symbol of the donor strain, *or* both of these.

1. *Coisogenic strains* formed by the occurrence of a mutation within an inbred strain shall be designated by the strain symbol and, where appropriate, the substrain symbol followed by a hyphen and the gene symbol of the differential allele (in italics in printed articles; e.g. DBA/Ha-d^+). When the mutant gene is maintained in the heterozygous condition, this should be indicated by including a + sign in the symbol; e.g. A/Fa-+/c; C3H/N-+/W^v. Such strains differ from the substrains described in Sections 14.1.5 and 14.1.6 in that in coisogenic strains the genetic difference is simple and precisely defined.

2. *Segregating inbred strains* developed by inbreeding with forced heterozygosis shall be designated, like coisogenic strains, by a strain symbol followed by a hyphen and the gene symbol of the segregating locus. However, indication of the segregating locus is optional, e.g. 129 or 129-c^{ch}/c (129 is customary); WB-W/+ or WB.

The minimal inbreeding requirement for such a strain shall be the same as that for an inbred strain, i.e. 20 generations of b × s matings.

3. *Congenic strains* produced by repeated crosses on to an inbred strain shall be designated by the full or abbreviated symbol of the background strain followed by: (i) a period and an abbreviated symbol of the donor strain, and (ii) a hyphen and the symbol of the differential locus or loci and allele(s) (in italics), e.g. B10.129-H-12^b, a strain with the genetic background of C57BL/10Sn (=B10), but which differs from that strain in a differential allele (H-12^b) derived from strain 129/J.

If several lines derived from the same donor strain are available, the individual lines are distinguished by a number and/or letter in parentheses, e.g. B10.129(12M)-H-12^b, B10.129(21M)-H-4^b p.

In cases where the use of such a full symbol is inappropriate, as when the donor strain is not inbred, or the genetic difference is complex, or where the strain is already widely known, a less complete symbol may be used, e.g. B10-*pa* H-3^e a^t; B10.129(5M); A.SW.

A strain developed by this method shall be regarded as congenic when a minimum of 10 backcross generations to the background strain have been made, counting the first hybrid or F_1 generation as generation 1. The number of backcross generations shall be indicated by N and the number in parentheses. In the case where it is necessary to employ more complex mating systems, the generations should be expressed as N equivalents (NE) and the strain regarded as congenic at a minimum of NE10. For segregating inbred strains developed by inbreeding with forced heterozygosis, the number of generations of such breeding should be indicated by FH and the number in parentheses. Where a coisogenic strain arises by mutation, the number of generations of inbreeding since the mutation should be indicated by intercalating M + in the indication of inbreeding, e.g. (N10 F6), 10 generations of backcrossing followed by six generations of b × s inbreeding; (NE12 F17), a complex system of backcrosses and intercrosses genetically equivalent to 12 backcrosses, followed by 17 generations of b × s matings; (F? + M + F23), 23 generations of b × s matings since the occurrence of a mutation; (FH27), a strain developed by 27 generations of b × s matings with forced heterozygosis.

14.4 Inbred strains preserved by freezing

The strain name, as normally used in accordance with standardized nomenclature, should be followed by the letter p to indicate preservation by freezing.

1. If the uterine foster mother is of the same strain as the thawed embryos transferred to her, no further designation is necessary.

2. If the uterine foster mother is of a different strain from the embryos the letter p should be followed by the usual nomenclature for egg transfer, i.e. the letter e followed by the strain designation of the foster mother, e.g. C57BL/6p, strain C57BL preserved by freezing and transferred to a C57BL/6 foster mother; C57BL/6peCBA/H, strain C57BL/6 preserved by freezing and transferred to a CBA/H foster mother.

In accordance with standardized nomenclature, abbreviations may be used for strain names in these complex symbols.

3. In denoting the generation of inbreeding, the years of freezing and of thawing should be indicated in parentheses, together with the letter P, e.g. F37 (P1975-1977) + F16, frozen at generation 37 in 1975, thawed in 1977, with 16 generations of inbreeding since thawing.

References

1. Committee on Standardized Nomenclature for Inbred Strains of Mice. 1952. Standardized nomenclature for inbred strains of mice. Cancer Res. 12:602–613.
2. Committee on Standardized Genetic Nomenclature for Mice. 1960. Standardized nomenclature for inbred strains of mice, second listing. Cancer Res. 20:145–169.
3. Committee on Standardized Genetic Nomenclature for Mice. 1976. Nomenclature for inbred strains of mice preserved by freezing. Mouse News Lett. 54:2–3.
4. Festing, M.F.W. 1979. Inbred strains in biomedical research, Macmillan Press, London; Oxford University Press, New York.
5. Festing, M.F.W. 1986. Inbred strains of mice, 9th listing. Inbred Strains of Mice 14:3–25.
6. Staats, J. 1986. Standardized nomenclature for inbred strains of mice: eighth listing. Cancer Res. 45: 945–977.

15 INBRED STRAINS OF MICE

MICHAEL F.W. FESTING

Lists of inbred strains of mice, their coat colour and mutant genes, and brief details of their characteristics have been published regularly under the aegis of the Committee on Standardized Genetic Nomenclature for Mice by Joan Staats. Her most recent eighth listing (6) was published in 1985. This listing is based largely on her eighth listing, but with substantial revision. Only very brief details of the name, coat colour, genes, origin, and strain characteristics are given due to limitations of space. Substrain symbols for strains where no major substrains are recognized have mostly been dropped, so SB/Le is listed as SB. Sets of recombinant inbred strains have been recorded by giving the set designation, but without listing individual strains. The list has 294 entries compared with 233 in the eighth listing.

Further details of strain characteristics may be obtained from a number of sources, including other chapters in this book. For example, many strains listed here carry mutant genes whose effects are given in detail by Margaret C. Green (Chapter 2). Other sources of information may be found in the general literature, including books specifically about strain characteristics (1, 3) and books which give details of strain characteristics indirectly (usually in a specialized field) (2, 4, 5).

References

1. Altman, P.L., and D.D. Katz (eds.). 1979. Inbred and genetically defined strains of laboratory animals, Part 1. Mouse and rat. Fed. Am. Soc. for Exp. Biol., Bethesda, Maryland.
2. Berry, R.J. (ed.). 1981. Biology of the house mouse. Symp. Zool. Soc. (Lond.) 47.
3. Festing M.F.W. 1979. Inbred strains in biomedical research. Macmillan Press, London.
4. Klein, J. 1975. Biology of the mouse histocompatibility-2 complex. Springer-Verlag, Berlin, Heidelberg.
5. Skamene, E. (ed.). 1985. Genetic control of host resistance to infection and malignancy. Alan R. Liss, New York.
6. Staats, J. 1985. Standardized nomenclature for inbred strains of mice: eighth listing. Cancer Res. 45:945–977.

15.1 List of inbred mouse strains

A Albino: *a, b, c*. Origin: Strong 1921 from a cross of Cold Spring Harbor albino and Bagg albino. Major substrains are:

A/Br Inbr ?. Strong to Little c 1927, to Br (Leeds, UK) c 1934.

A/Fa Inbr ? + 18 (Lac). Strong to Little c 1927, to Fa 1934.

A/He Inbr 195. Origin: Strong to Bi c 1929, to He c 1940.

A/J Inbr (J) 196. Strong to J c 1930.

AB Inbr (Jena) 70. Albino: *A, B, c, D*. From unspecified stock of Agnes Bluhm to P. Hertwig in 1934, to J. Fortner in 1945, to H. Rohrer in 1949, to Jena 1952.

ABJ Inbr (Le) 60. Cinnamon: *A, b*. Also carries *ab^J* (asebia-J). Origin: *ab^J* arose in BALB/cJ in 1968. One outcross to C3H/HeDi, then bxs.

ABP Inbr (Le) 64. Pink-eyed-fawn: *a, b, bt, p, se*. Origin: A.B. Griffen to Lane 1968. Sib-mated since then. Also carries *wa-1*.

ACR Inbr ?. Albino. Origin: AKR substrain to CIBA 1949. Known as AKR/FuA. Genetic contamination in 1953, renamed ACR.

AE Inbr (Nctr) 80. Pink-eyed fawn: *a^e, b, d, p, se*. Origin: stock 3H-220 from Hollander, Iowa to Inst. Cancer Research, PA in 1962 at about F3.

AEJ Inbr (Rk) 70. Black. *a^e*. Origin: Hollander, Iowa. Ames waltzer stock with miscellaneous markers. Crossed once to C57BL/10 followed by cross to C57BL/6-*A^{W-J}*, then bxs.

AG Inbr (Cam) c155. Colour depends on genotype: *A^y/a^e × A^y/a^e* with *A^w, A, A^s, a^t, a*, and *p* continuously backcrossed to the *A^y/a^e* line. Origin: R.A. Fisher, 'A' from Grüneberg in 1945 at F34 (a CBA substrain).

AKR Albino: *a, B, c*. Origin: Detwiler to Furth as a high-leukaemia strain from 1928 to 1936, then random bred at the Rockefeller Inst. for several generations. bxs by Mrs. Rhoades to F9 then C. Lynch to F21. Major substrains are:

AKR/J Inbr 155+?. To J 1940.

AKR/Lw Inbr ?. To Law 1940.

AKR/FuA see ACR/A.

AKR/Cum Inbr ?. Origin ?. Differs from other substrains in being *Mod-1ᵃ*.

AKXL- Set of 18 recombinant inbred strains. Inbr 15–40+. Origin: B.A. Taylor from a cross between AKR/J and C57L/J.

AL Inbr (N) 195. Albino: *a, b, c*. Origin: believed to have originated from a strain A outcross to unknown strain followed by bxs mating. Should not be considered a substrain of A.

AM Inbr (Nctr) 63. Mottled yellow, mottled agouti: A^{vy}/a^m, a^m/a^m. Origin: a^m/a^m males from L.B. Russell, Oak Ridge Natl. Lab. in 1965 at F5 to the Inst. of Cancer Research, PA. Mated by A^{vy}/a females (F12) of strain VY/Wf. To Nctr in 1972. Maintained by A^{vy}/a^m × a^m/a^m sib matings.

ATEB Inbr (Le) 42. Grey: *a, d*. Also carries *at/+* (atrichosis), *eb/+* (eye-blebs). Origin: *at* arose in DBA/1 at F15 in 1964, *eb* arose in a non-inbred hairless stock maintained by Hummel in 1960. bxs as balanced stock.

AU Inbr (J) ?+69. Black. *a, U*. Origin: R.A. Fisher, about 1937 from about 50 per cent CBA/Gr, 28.5 per cent Fisher's colour stocks, and 21.5 per cent Grüneberg's wavy stock, the latter two being unrelated stocks of American origin.

AX Inbr (Pa) 27. Black. a^x. Origin: L.B. Russell's balanced lethal stock A^y/a^x 6PB, to M.H.L. Snow in 1969. Backcrossed to C57BL/6 (N5). Sib-mated with forced heterozygosity for the embryonic lethal a^x.

AXB- Inbr c 20+. Set of 48 recombinant inbred strains developed by Nesbitt from A/J × C57BL/6 and C57BL/6 × A/J.

A2G Inbr (A) 107+. Albino *a, b, c*. Origin: illegitimate mating of strain A mice at Glaxo Labs. UK in 1942–50, followed by bxs mating.

BA Inbr (Nmg) 64. Sandy (?):*b*. Origin: from a non-inbred obese stock maintained by Falconer.

BALB/c Albino: *A, b, c*. Origin: stock acquired by H. Bagg in 1913, to MacDowell, to Snell in 1932 (who added the /c). Major substrains are:

    BALB/cHeAn. Inbr ?. To Snell c 1932, to He c 1935.
    BALB/cJ Inbr 145+. Snell to J after 1940.
    BALB/cRL Inbr 80+. Snell to Green c 1937, to W.L. and L.B. Russell 1948.

BBT Inbr (Cv) 65. Black belted: *a, bt*. Origin: cross of outbred *hr/hr, bt/bt* × C57BL/Ha.

BDP Inbr (J) 124. Pale fawn with short ears: *a, b, d, p, se*. Origin W.H. Gates, dilute brown female from Little × pink-eyed male from Strong. Inbred since 1926.

BLN Inbr (Nmg) 58. Black: *a*. Origin: Cpb non-inbred yellow stock. Inbred from 1965. Very good reproductive performance.

BN/a Inbr (W) 90. Albino: *A, B, c, D*. Origin: Anna Dux, Gliwice 1950 from unknown parents. Split into two substrains in 1957 when taken to Warsaw.

BN/b Inbr (W) 96. Albino, see above.

BNT Inbr (Le) 73. Black: *a*. Origin: mutation to *Bn* (bent-tail) discovered by E.D. Garber. From A.B. Griffen to Lane 1960. Inbred to F27, then one outcross to C57BL/6J × CBA/Ca F1, followed by bxs. (frozen embryos only).

BRSUNT Inbr (N) 158. Chocolate: *a, b*. Origin: Strong, a branch of BRS (from strain NH) continued without further methylcholanthrene treatment.

BRVR Inbr ?. Albino. c. Origin: Webster 1930 selected two lines resistant and susceptible to infection with *Bacillus enteritidis* (*Salmonella sp*), followed by sib mating. Later tests with louping ill and St. Louis encephalitis virus led to the development of three lines BSVS (bacteria susceptible, virus susceptible), BSVR (bacteria susceptible, virus resistant), and BRVS. Later BRVR was developed from a cross between BSVR and BRVS. There may have been some confusion between the lines. As they have been poorly characterized by genetic markers, authenticity should not always be assumed.

BRX58N- Inbr 20+. Set of 11 recombinant inbred strains developed by B.A. Taylor from C57BR/cdJ × B10.D2(58N)/Sn.

BSVR Inbr ?. Albino. See BRVR.

BSVS Inbr ?. Albino. See BRVR.

BUB Inbr (J) 122. Albino *a, c*. Origin: from albinos of unknown ancestry at Brown University, maintained by random mating until 1945.

BXA- see AXB.

BXD- Inbr 30+. Set of 24 recombinant inbred strains developed by B.A. Taylor from C57BL/6J × DBA/2J.

BXH- Inbr 30+. Set of 13 recombinant inbred strains developed by B.A. Taylor from C57BL/6J × C3H/HeJ.

BXSB Inbr (Mp) 49. Agouti: +. Origin: E.D. Murphy from a cross of C57BL/6J × SB, followed by selection of the satin, non-beige phenotype, followed by bxs mating. Develops lupus-like autoimmune syndrome.

CAT Inbr ?+26 (Lac). Albino: *c, Catᶠʳ*. Origin: Cat^{Fr} arose in strain of unknown origin. Inbred by A. Muggleton-Harris before 1986.

CBA Agouti: +. Origin: Strong 1920 from a cross of a Bagg albino female and a DBA male. Strain CBA was selected from low mammary tumour incidence, and C3H for a high incidence.

CBA/Br Inbr 150. J to Haldane and Grüneberg 1932, to Bonser 1933, to Amsterdam 1948. Carries the *H-2^g* allele in contrast with other substrains which are *H-2^k*.

CBA/Ca and CBA/CaH. Inbr (J) 150+. J to Haldane and Grüneberg 1932, to Carter via Royal Cancer Hospital and British Emp. Cancer Campaign. To Harwell c 1953.

CBA/CaH-T6. Inbr (J) N13 F88. Substrain of CBA/CaH carrying the T(14;15)6Ca translocation introduced by M.F. Lyon.

CBA/J Inbr (J) 194. Strong to Andervont 1947 to J 1948. This substrain may have been genetically contaminated prior to 1948. It differs from CBA/Ca at several loci.

CBA/N Inbr 36+?. To N 1966 from Harwell (CBA/H) heterozygous for *fm* (foam-cell reticulosis). *xid* (X-linked immune deficiency) mutation discovered in non-*fm* stock. Hemizygous males and homozygous females have a defective immune response to type III pneumococcal polysaccharide.

CBA/St Inbr (Man) 173. Strong to Mann 1962. Ki has two substrains one of which has a high incidence of mammary tumours and is *rd^+* and the other with a low incidence of tumours and with the *rd* genotype.

CC57BR Inbr (Y) 115. Chocolate *a*, *b*. Origin: Medvedev 1943 from a cross of BALB/c × C57BL/N.

CC57W Inbr (Y) 107. Albino: *a*, *b*, *c*. Origin: Medvedev 1943 from a cross of BALB/c × C57BL/N.

CE Inbr (J) 97. Black-eyed grey. *A^w*, *c^e*. Origin: from wild mice trapped by J.E. Knight in 1920. Inbred by Eaton to at least F15, to Woolley then to J in 1948. Probably crossed with laboratory mice.

CFCW Inbr (Rl) 100+. Albino: *c*, *Ca*. Origin: outbred mice from Carworth Farms to Rl 1948.

CFW Inbr (Jic) 58. Albino: *c*. Origin: outbred mice from Carworth Farms. Inbreeding started in 1964. Note that there are several other inbred CFW lines.

CHI Inbr (Man) 160. Agouti: +. Origin: Strong 1920 from a cross of the Bagg albino × DBA. strains C, CBA, C3H and C121 originated from the same cross.

CL Inbr(Fr) 70. Albino: *b*, *c*. Origin: stock from Morgan carrying "msl" (migratory spot lesion). Crossed with A/J then inbred with selection for cleft lip (incidence now 25 per cent in viable 17-day embryos).

CN Inbr(Nmg) 27. Cinnamon: *b*. Origin: Fa non-inbred obese stock. Inbred from 1976.

CPB-K Inbr(Nmg) 70+16. Steely blue: *a*, *B*, *d*. Origin: Cpb stock of unknown origin, inbred from 1979.

CS Inbr (Nga) 94. Albino: *a*, *b*, *c*, *D*, *s*. Origin: 1956 from a cross between NBC and SII, both now extinct. Unrelated to IVCS.

CT Inbr (Lac) ?+8. Agouti: +. Origin: *ct* (curly-tail) mutation arose spontaneously in GFF inbred mice. Backcrossed to CBA an unknown number of times, then bxs of presumed *ct/ct* homozygotes.

CWD Inbr (Le) 53. Non-agouti dilute: *a*, *d*, *cw* (curly-whiskers). Origin: *cw* arose in CBA/Cbi at Chester Beatty Res. Inst. To Harwell. Stock T11H from A.G. Searle to J in 1971. Crossed to B6C3Fe-*a*/*a*F1 then to DBA/2, then inbred.

CXB- Inbr (By) 91. Set of seven Recombinant Inbred Strains (CXBD, E, G, H, I, J, K) developed by Bailey from a cross of BALB/cBy × C57BL/6By.

CXS- Set of 14 recombinant inbred strains developed by A from a cross of BALB/cA × STS/A.

C3H Agouti. Origin. Strong 1920 from a cross between the Bagg albino stock and DBA (see also CBA), with selection for a high incidence of mammary tumours.

C3H/Bi Inbr(Ki) 160. Origin: Strong to Bittner 1931, to Kirschbaum 1952.

C3H/He Inbr (N) 160. Origin: Strong to Andervont 1930, to Heston 1941.

C3H/HeJ Inbr. (J) 180. Origin: as for He substrain. Heston to J 1947. Carries *Lps* (lipopolysaccharide) gene causing defective immune response to lipopolysaccharide and other antigens.

C3H/St. Inbr (Man) 190. Origin: Strong 1920 (see above).

C3H_eB/De Inbr (Dt) 125. Origin: Deringer, fertilized ova of C3H transferred to C57BL.

C3H_eB/FeJ Inbr(J) 123. Origin: Fekete 1948. C3H/HeJ embryos transferred to C57BL/6 foster mothers, to Hummel at F3, to J in 1950.

C3H_eB/FeJLe-*a*/*a*. Inbr (Le) N7F40. Black: *a*. Origin: non-agouti (*a*) from C57BL/6J.

C3HA Inbr (Y) 102. Agouti: + Origin: Pogosianz, C3H female × A male, followed by sib mating. Originally 30 per cent mammary tumours, but has declined.

C57BL Black: *a*. Origin: Little 1921 from the mating of female 57 with male 52 from Miss Abbie Lathrop's stock. The same cross gave rise to strains C57L and C57BR. Female 58 mated with the same male gave rise to strain C58. Major substrains are:

C57BL/A Inbr (A) ?+142. Origin: Little to A c 1932.

C57BL/An Little to Andervont 1932. Differs from /6 and /10 sublines at the *Ce-1* locus.

C57BL/GrFa. Origin: Little to Grüneberg 1932, to Falconer 1947.

C57BL/Ks. Origin: C57BL/6J to Biesele in 1947, then pen bred. To Kaliss in 1948. Ks resumed inbreeding. To J 1948. Carries *H-2^d* due to possible genetic contamination.

C57BL/6J Inbr (J) 150. Origin: substrains 6 and 10 were separated prior to 1937. Substrains 6 and 10 differ at the *H-9*, *Igh-2*, and *Lv* loci.

C57BL/10J Inbr (J) 158. Origin: see C57BL/6.

C57BL/10ScSn. Inbr (J) ? +136. Origin: Little to W.L. Russell to J.P. Scott at F26 as a separate substrain. To Snell at F35–36.

C57BR/cd Inbr (J) 178. Brown: *a*, *b*. Origin: Little from the same cross that gave rise to C57BL, C57BR/a, and C57L. Black and brown substrains were separated in the first generation. Substrain cd was established at F13 from a cross between two brown substrains, one of which had previously given rise to C57BR/a. To Heston 1938, to J 1947 at F66.

C57L Inbr (J) 162. Grey *a*, *b*, *ln*. Origin: J. Murray 1933, mutation to *ln* in F20 of a C57BR substrain which is now extinct. To Cloudman, to Heston 1938, to J 1947 at F45.

C57P Inbr (A): 115. Origin: R. Korteweg 1934. Cross of DBA × C57BL then N20 to the C57BL, followed by bxs mating. (Formerly listed as P/A.)

C58 Inbr (J) 198. Black: *a*. Origin: MacDowell 1921 from mating of litter mates 58 and 52 of Miss Abbie Lathrop's stock (see also C57BL).

DA Inbr (Mob): 125. Albino: *c*. Origin: Hummel 1948 from outbred 'Swiss' mouse with mammary tumour in the female. Low mammary tumour incidence.

DBA Grey: *a*, *b*, *d*. Origin: Little 1909 from stock segregating for coat colour. Oldest of all inbred strains of mice. In 1929–30 crosses were made between substrains, and several new substrains established, including the widely used substrains /1 and /2. Major substrains are:

DBA/LiA Inbr(A) ?+126. Origin: Little to Amsterdam c 1932.

DBA/1 Inbr (J) ?+117. Origin. Substrain maintained by Little at the Jackson Laboratory.

DBA/2 Inbr (J) 150. Origin: Substrain maintained by J.

DC Inbr (Le) 100. Agouti: +. Origin: mutation to dancer *Dc* arose in an obese stock outcrossed to a

BALB/c × C3H/He hybrid in 1956. One cross to C3H/HeJ, then inbred.

DD Inbr (A) 91. Albino: *A*, *B*, *c*, *S*. Origin: From non-inbred ddN stock of Cen. Lab. Exp. Animals, Osaka University 1956, to Heston 1957, to Nara 1959.

DDD Inbr.(Hok) 81. Albino: *A*, *B*, *c*, *S*. Origin: dd stock mice taken by Dr. S. Hata from Germany to Kitasato Inst. before 1920. (dd = Deutschland–Densenbyo). To Manchuria (K. Ando) and then back to Japan. Inbreeding begun by Suzuki in 1962 at the Univ. of Tokyo.

DDI Inbr. 83. Albino: *a*, *B*, *c*, *D*, *S*. Origin: animals from Lab. Infect. Dis. Tohoku Univ. in 1948 and 1953. Maintained by D. Takasuki (1948–63), by H. Nikaido (1960–65), and by Sadao Ishigaki since 1965. Inbreeding started in 1960, and reached F20 in 1968.

DDK Inbr 105. Albino: *A*, *B*, *c*, *D*, *S*. Origin: K. Kondo from dd stock mice from Inst. Infect. Disease. Tokyo in 1944. Fertility is reduced when females outcrossed to other strains, due to pre-implantation embryonic death.

DHS Inbr(Hok) 98. Albino: *c*. Origin: Germany to Kitasato Inst. 1910 to 1920, to Inst. Infect. Dis. (Univ. Tokyo) in 1944, to Hok 1957. bxs inbreeding started in 1957.

DK Inbr (Lm) 20. Black, yellow or sable: *A^y/a*, *b*. Origin: undocumented matings among mice from the MRC Radiobiology Unit, Harwell. *A^y/a* mice become sable in phenotype at the first moult, but at about 6 months of age revert to yellow.

DKI Inbr 92. Albino: *a*, *B*, *c*, *S*. Origin: outbred ddN mice of Cen. Lab. Exp. Anim. Inbreeding started by B. Kitasato. 1953.

DKI-R Inbr (Jko) N11F36. Origin: cross of DKI/Jko × C3H/He, followed by backcrossing to DKI with selection for a gene for relative resistance to *Salmonella enteritidis*.

DL Inbr (Ra) 105. Black, a. Origin: balanced *se+/+d^e* stock from Truslove to E.S. Russell, to Kelton 1958. Inbred by Kelton.

DLS Inbr (Le) 85. Black, *a*. Origin: as for DL, but independently inbred. To Le in 1968 at F32. Carries balanced lethal *se+/+d^l*.

DRC Inbr(Hok) ?+48. Albino (?). Origin: DDD mated to C57BL/6. N8 to DDD then bxs. To Hok 1972 from Univ. Tokyo.

DW Inbr(J) 110. Grey *ln*, Dwarf *dw*: Origin: *dw* mutation arose in a stock of silver mice obtained from an English fancier before 1929. Le to J 1966.

EBT Inbr. (Cv) 63. Yellow, belted (?) *a*, *bt*, *e*. Origin: cross of non-inbred *bt/bt* × C57BL/Ha-*e/e*. To Chapman 1974.

EL Inbr (Yok) 84. Albino: *c*. Origin: K. Imaizumi, *El* (epilepsy) mutation in non-inbred albino mice in 1954.

F Inbr 140. Grey with white head spot *a*, *b*, *c^{ch}*, *d*, *s*. Origin: Strong 1926 from a group of unpedigreed Bussey Inst. mice.

FB Inbr (Ki) 58. Colour ?. Origin: Kirschbaum 1942–52 from multiple crosses between A, F, and NH. About 50 per cent reticular tissue neoplasms.

FL/1 Inbr (Brk) 93. Black: *a*. Origin: *f*, flexed-tail gene from Snell's WA linkage testing stock. Seven rounds of cross-intercross to C3H by Jay and E.S. Russell. Outcross to WB/Re by Re, with bxs mating of *f/f* progeny started in 1956.

FL/4 Inbr(Mob): N30F46. Origin: see FL/1. Separated from FL/1Re-*f/+ W/+* at N30 in 1968. In 1980 fostered on to C57BL/6J.

FRG Inbr. (Tbr) 35: Albino (?). Origin: non-inbred dd mice Chugai Lab. to Tbr 1981 at F26. Resistant to Friend leukaemia virus.

FS Inbr (Dn) 77. Fawn: *b*, *c^{ch}*, *fr*, *p*. Origin: Snell linkage testing stock to M.C. Green to Eicher 1971, to Davisson 1983. Also carries *sh-1/sh-1*.

FSB Inbr (Dn) 98. Black: *a*. Origin: furless mutation *fs* arose in unpedigreed stock at Ohio State University in 1951. Mutant females mated with C57BL/10 males.

FTC Inbr (U) ?+77. Light-grey with pink eyes: *a*, *b*, *p*. Origin: formerly called CPB-FT. CPB-TNO (1950) to Vet. Fac. Utrecht in 1973.

GL Inbr (Le) 50. Agouti: +. Origin: mutation to grey-lethal (*gl*) discovered in a stock segregating for *c^e* by Grüneberg in 1935. From G to Jay to M. Dickie to P. Lane who inbred to F25, then one outcross to *dt^l/dt^l* and bxs mating as a balanced stock.

GLF Inbr(Y) 65. Agouti +. Origin: from J as bearers of grey-lethal mutation. Selected for normal genotype by progeny testing.

GRS Inbr (A) 101. Albino *a*, *c*. Origin: Mühlbock 1965 from outbred mice obtained from Grumbach in Zurich. High incidence of mammary tumours which are hormone-responsive.

GT Inbr (Le) 28. Breeders are agouti: *A*, *gt/+*, *gt/gt* are light grey with white belly spot and have tremors like myelin-deficient mice. Origin: developed by Lane. *gt* mutation arose in strain HYIII/Le in 1977 at F55. One outcross to C3H_eB/FeJ × C57BL/6-*A^{w-J}*F1, then bxs. Has a spontaneous spongiform encephalopathy

whose expression is determined by the interaction of genetic factors and an unconventional unrecognized transmissible agent.

HC Inbr (U) ?+82. Light grey (?): *b*, *d*, *p*. Origin: CPB-TNO (about 1950) to Vet. Fac. Utrecht in 1973.

HLC Inbr 47. Black: *a*. Origin: from hybrid stock derived from crosses involving C57BL/6J, C57BR/cd, A/J, BALB/c, LG, and SM. Selected for high leucocyte count (strain LLC was selected for low leucocyte count).

HRS Inbr (J) 79. Albino, *b*, *c*, *d*. Origin: hairless (*hr*) stock from Crew to Carnochan to Heston to Chase, to E.L. Green 1952, to Les 1956, to M.C. Green 1959, to J 1964. Maintained by mating +/*hr* females with *hr/hr* males.

HTG Inbr (Mob) 51. Cinnamon *A*, *b*. Origin: Gorer about 1937 as an H-2 locus recombinant from heterogeneous stock of unstated origin.

HTH Inbr (Ao) ?+53. Brown. *a*, *b*. Origin: as for HTG.

HTI Inbr (Ao) ?+54. Black: *a*. Origin: as for HTG.

HYIII Inbr (Le) 83. Agouti: +. Origin: mutation to hydrocephalus-3 discovered in heterogeneous stock by H. Grüneberg. To M.C. Green 1963, then b × s. To Lane 1975. *hy-3/hy-3* mice die with frank hydrocephalus by 4–5 weeks.

I Inbr (N) 143. Pink-eyed-fawn with variable white patches: *a*, *b*, *d*, *p*, *s*. Origin: Strong 1926 from unpedigreed mice. Also carries sex-linked *Phk^b* controlling the activity of phosphorylase kinase.

IC Inbr (Le) 59. Agouti: +. Origin: ichthyosis (*ic*) gene reported by Carter and Phillips in 1950, to Snell 1950, to M.C. Green 1970 to Lane 1975. Heterogeneous stock crossed once to CBA followed by bxs. Homozygous *ic/ic* mice show abnormal clumping of chromatin in nuclei.

ICR Inbr 29. Albino. Origin: BLU:Ha(ICR) outbred stock from the Institute of Cancer Research in 1958, to Arbor Scientific Co. Ltd, to UBC, Vancouver in 1977, followed by sib mating.

IF Inbr (A) 70. Black with yellow belly, *a^t*. Origin: Bonser (Leeds, UK) 1932–36. Selection for early papilloma development after treatment with carcinogens. Susceptible to several carcinogens.

IITES Inbr (Ng) 76. Colour ? *a*, *b*, *d*, *s*. Origin: from a cross involving CS, DBA/2, NBC, and ITES.

IS Inbr (Dn) Agouti: +. Origin: *Mus musculus praetextus* male caught in Israeli port × *M. m. musculus* female from laboratory stock carrying *bt*, *m*, *b*, and *a*. To Roderick, to Eicher 1971 at F42, to Dn 1983.

ITES Inbr (Nga): 87. Colour ?:*a*, *b*, *d*/+. Origin: from cross involving CS, DBA/2, and SII in 1962. Relatively radiation resistant.

IVCE Inbr (Csk) 73. Albino. Origin: separated during inbreeding and selection of IVCS. Regular 4-day oestrus cycle.

IVCS Inbr (Csk) 73. Albino: *a*, *B*, *c*, *S*. Origin: from ddN stock of Central Lab. of Exp. Anim., selected for regular oestrus cycle.

J Inbr (Glw) 92. Cinnamon: *A*, *b*. Origin: Falconer. High frequency of cleft-palate.

JBT Inbr (Jd) 86. Chocolate belted: *a*, *b*, *bt*. Origin: from Falconer's outbred JC stock. Frequent tail kinks.

JE Inbr (Le) 47. Slate ruby-eyed: *a*, *f*, *ru je*/+. Origin: Fisher to Snell 1948, to Lane 1969, bxs since 1972. Homozygous jerker (*je*/*je*) females are poor breeders.

JGBF Inbr (Ty) 55. Khaki: *a*, *jg* +/+ *bf*. Origin: mutation to jagged-tail occurred in C3H/HeJ in 1960. N1 to C57BL/10 and bxs, then N1 to C57BL/6J-*bf* and inbred as a balanced stock. Homozygous *jg*/*jg* mice are usually born dead. Ty maintains it as an RI line, Le as a balanced mutant strain.

JIGR Inbr (Dn) 65. Agouti: +. Segregates for *gr* (grizzled) and *ji* (jittery). Origin: *ji* arose in waltzing (*v*) stock of Snell before 1957. bxs to F15. To M.C. Green (1963) and outcrossed to *gr*/*gr* stock (F30). Balanced stock bxs. To Eicher 1972, then Davisson 1980. *gr* from Falconer to Snell 1950. Two crosses to CBA then bxs to F30. Cross to *ji*/+ to make the balanced stock. *ji*/*ji* die before weaning. *gr*/*gr* often poor breeders.

JU Inbr (Ct) 99. Albino *a*, *c*. Origin: Falconer 1952 from crosses involving Goodale's and MacArthur's large strains, Bateman's high lactation strain, and various mutant stocks with about 50 per cent C57BL/Fa ancestry.

KE Inbr (Kw) 94. Albino *a*, *b*, *c*, *P*. Origin: Kraznowska 1952 from mice of unknown origin. Cross also produced KP. High (17.6 per cent) incidence of spermatozoa with abnormal heads.

KF Inbr (Tbr) 48. Albino *a*, *B*, *c*, *S*. Origin: noninbred ddN stock of Central Lab. of Exp. Anim. to Nara 1964. Used as a recipient for a Leydig cell tumour.

KI Inbr (Glw) 141. Albino: *c*. Origin: *Fu^{ki}* (kinky) stock inbred by Dunn and Glueck"sohn-Waelsch. Heterozygotes have tail abnormalities; homozygotes die before birth.

KK Inbr 90. Albino: *a*, *B*, *c*, *D*, *S*. Origin: K. Kondo 1944 from Japanese dealer stock (Kasukabe group). Small litters, frequent diabetes mellitus, occasional obesity in old animals.

KP Inbr. (Kw) 91. Fawn (?) *a*, *b*, *p*. Origin: see KE. High embryonal and post-embryonal mortality, frequent sterile matings, low sperm production.

KR Inbr (Nga) 102. Albino: *A*, *B*, *c*, *D*, *S*. Origin: same as KK 1952. Good reproductive performance. Imperforate vagina 2 per cent.

KSB Inbr (Nga) 80. Black, white patches *a*, *B*, *D*, *s*. Origin: as KK 1944–48. Characteristics resemble KK.

LDJ Inbr (Le) 72. Black or very black: *a*, *mg*/+. Also carries *ld^J*/+. Origin: *ld^J* arose spontaneously in CBA/Ca-*se* stock (M.C. Green) 1959. One cross to C57BL/10Gn, bxs to F3, outcrossed to an inbred *mg*/*mg* stock (F27), and bxs as balanced stock. To Lane 1975. The *mg* mutation arose in a cross of C3H female × Swiss stock in 1950, with bxs until 1950.

LG Inbr (J) 104. Albino, *a*, *c*. Origin: developed by Goodale with selection for large body size beginning in 1931. Inbred by Runner. Wt at 28 days 20.1 g, at 60 days 47.4 g. Docile.

LIS Inbr (A) 101. Albino *c*. Origin: Mühlbock, from Hyg. Inst. Zurich in 1955. Closely related to STS and LTS.

LLC Inbr (Ckc) 46. Colour ?. Origin: Chai 1976. Selected for low leucocyte count from a cross involving C57BL/6J, C57BR/cdJ, A/J, BALB/cJ, LG/J, and SM/Ckc (see also HLC). Mean leucocyte count now 4000–5000/cu mm.

LM Inbr (Bc) 49. Agouti. Origin: 'C3H' mice from Rockland Farms in 1966, inbred by J.R. Miller, then outcrossed to SWV at F27, followed by sib mating. Resulting colony 1/8th SWV, and the rest from the original stock. Homozygous for the *lg^{Ml}* mutation resulting in 94 per cent open-eyes at birth.

LP Inbr (J) 125. Colour: white-bellied agouti with white patches *A^w*, *s*. Origin: Dunn 1928 from a chinchilla stock from Castle and some coat colour stocks from English fanciers. To Scott, to Dickie 1947, to J 1949. Longest life-span among 22 strains.

LPT Inbr (Le) 89. Colour: agouti. Origin: mutation to loop-tail (*Lp*) arose in strain A in 1949. From W. Hollander to Snell 1950. Crossed (N7) to C57BL/6J, once to C3H/He, then bxs. To Lane 1969. *Lp*/+ mice have crooked or looped tails and some head wobbling. Homozygotes die at birth.

LS Inbr (Le) 80. Black-and-tan, or black and tan with white spots: *a^t* *ls*/*a^t* +. Origin: mutation to lethal spotting (*ls*) occurred in C57BL-*a^t* mice at Harwell. Phillips to Lane as balanced stock in 1961. Homozygous *ls*/*ls* develop megacolon, but some survive and breed.

LST Inbr (Bc) 63. Albino. A tabby/circling stock from Oak Ridge c 1957. To Kathryn F. Stein, Mount Holyoke College, Mass. to U.B.C., Vancouver in 1966 at F9. Homozygous for the *lg^Stn* gene resulting in 99 per cent open-eyes at birth.

LT Inbr (Sv) 121. Colour: Light brown *a*, *B^lt*. Origin: MacDowell 1950 from a mutation at the brown locus in strain C58. Outcrossed to BALB/c. To Chase, to Re in 1957 at F28. About 50 per cent spontaneous ovarian teratomas.

LTS Inbr (A) 98. Albino: *c*. Origin: Mühlbock, from P. Loustalot, Ciba, Basel 1954. Closely related to LIS and STS. High mammary tumour incidence.

MA Inbr (J): 132. Albino: *c*. Origin: Marsh's strain 3. Mice from the Lathrop–Loeb colony (1903–1915). 32 generations of cousin mating by Marsh, to W.S. Murray who started bxs. Polydipsia–polyuria in 79 per cent virgin females, with gross kidney disease in males.

MAS Inbr (A) 98. Albino. Origin: Mühlbock from same stock as GRS, and is similar to it at 48/50 biochemical loci. Susceptible both to the C3H and GRS mammary tumour viruses.

MK Inbr (Ty) 78. Colour ?. Origin: mutation to *mk* (microcytic anaemia) in descendents of a B6D2F1 cross. From M.M. Dickie to Re in 1963. Segregates for *mk*, causing microcytic anaemia in *mk*/*mk* due to a generalized impairment of cellular iron uptake.

MOC Inbr (U) ?+61. Albino: *c*. Origin: Inst. Trop. Hygiene, Amsterdam (?) to CPB-TNO in about 1957. To Vet. Fac. Utrecht in 1973. Has skeletal deformities with brachypodism.

MOR Inbr (Cv) 44. Black: *a*. Origin: wild mice trapped in Ohio by Bruell, to Shows, to Chapman 1972. Carries *Mor-1^b* gene, a mitochondrial malate dehydrogenase variant.

MRL Inbr (J) 65. Albino: *a*, *c*. Origin: Murphy from crosses started in about 1960. Now estimated to have a composite genome of LG (75 per cent), AKR/J (12.6 per cent), C3H (12.1 per cent), and C57BL/6 (0.3 per cent). A mutation *lpr* (lymphoproliferation) was found at F12. *lpr*/*lpr* mice develop massive generalized enlargement of the lymph nodes and autoimmunity, and usually die at 14–16 weeks of age.

MS Inbr (A) 30. Albino. Origin: A.U. Aeschbacher before 1973, from four males of a Swiss CD-1 stock treated with methylurea. Selection based on high inci-

dence of fetal death. More susceptible than other strains to several mutagens/carcinogens.

MWT Inbr (Le) 59. Colour varies: carries *a^t*, *Mi^wh*, *W^v*, *T*. Origin: M.C. Green from a heterogeneous background about 1959.

MY Inbr (Le) 98. Chocolate: *a*, *b*. Origin: C57BR/cd female × line 85 F71 male from MacDowell in 1948. F8 male outcrossed to C3H_eB/HuJ; selected for *aa*, *bb* in F2. Carries *my*/*my*, but the only visible effect is missing eyes, though kidneys may also be missing.

MYD Inbr (Le) MYD is N7F12, MYD-+/+ is F71. Agouti: +. Origin: mutation to *myd* (myodystrophy) occurred in 1963 in *ls* (lethal spotting) stock from R. Phillips in 1961. Crossed to C57BL/6J-*A^wJ*, then bxs. *Os* from ROP crossed to +/+ and to *myd*/+ (N7). Maintained as *Os*+/+*myd* balanced stock. A +/+ substrain was derived at F35. *myd*/*myd* have a progressive and diffuse myopathy and die between 5 weeks and 5 months.

N Inbr (Ao): 119. Colour: dilute with white patches: *a*, *b*, *d*, *s*. Origin: Strong, about 1926 from a group of unpedigreed mice. Ancestral to strain PBR.

NBR Inbr (Nga) 90. Chocolate: *a*, *b*. Origin: from Japanese fancy mice (Nishiki–Nezumi group) between 1944 and 1949.

NC/Nga Inbr (Nga) 112. Cinnamon: *A*, *b*. Origin: K. Kondo, from Japanese fancy mice (Nishiki–Nezumi group). (Note designation NC has been used for two different strains, and this clash will have to be resolved, ed.)

NC/CpbU Inbr (U) ?+80. Light grey (?): *a*, *b*, *p*. Origin: CPB-TNO (1950) to Vet. Fac. Utrecht in 1973. Poor breeders, obese.

NFR Inbr (N) 59. Albino. Origin: from NIH outbred stock as multiple-branch inbred from 1936, then random breeding in the 1960s. Inbred as HSFR with selection for resistance to the action of histamine after treatment with *Bordetella pertussis* in 1966, then separated as NFR in 1972.

NFS Inbr (N) 57. Albino: *a*, *c*. Origin: as for NFR, but selectively bred for sensitivity to the action of histamine after treatment with *Bordetella pertussis*.

NGP Inbr (N) 69. Albino: *a*, *c*. Origin: from NIH general-purpose outbred stock N:GP(S).

NH Inbr (Ao) 109. Dilute spotted: *a*, *d*, *p*, *s*. Origin: Strong, from a cross involving CBA, N, and JK.

NIH Inbr (Ola) ?+27. Albino: *a*, *b*, *c*, *rd*. Origin: Pitman at NIH from N:NIH(S) stock to Natl. Inst. Biol. Standards, Hampstead, England in 1968. To Burroughs

Welcome 1970 and Ola in 1975. Good reproductive performance.

NMRI Inbr (Lac) 50. Albino: *c*. Origin: outbred Swiss mice from Lynch to S. Poiley at NIH in 1937. Inbred by Pl, known as NIH/Pl. To US Naval Medical Res. Inst. at F51. Known as NMRI. Lac maintains only frozen embryos. Note: there are also several outbred colonies of NMRI.

NOD Inbr (Komeda) 22. Albino: *c*. Origin: a substrain developed by selection for diabetes from F6 of the CTS strain, which was derived from JCL:ICR. About 80 per cent of females and 20 per cent of males develop insulin-dependent diabetes by the age of 30 weeks.

NON Inbr (Komeda) 27. Albino: *c*. Origin: a non-diabetic substrain with the same origin as NOD.

NRH Inbr (Nrs) 71. Albino. *a*, *b*, *c*. Origin: CF-1 from Carworth Farms to Takeda Chem. Industries Ltd. to Jmo to Nrs. Inbreeding started Sept. 1960 without selection. Foster nursed on germ-free NDII in 1972.

NXSM- Inbr c 20. Set of 19 recombinant inbred strains developed by Eva M. Eicher from NZB/BlNJ × SM/J.

NX129- Inbr c 20. Set of eight recombinant inbred strains developed by B.A. Taylor from NZB/BlNJ × 129/J.

NYLR Inbr (Nya) 89. Albino: *c*. Origin: outbred colony from a single pair of unknown origin obtained from a breeder in Albany, NY in 1930. bxs matings started in 1942.

NZB Inbr (J) 121. Black: *a*. Origin: outbred mice from the Imp. Cancer Res. Fund, London to Univ. of Otago Med. School 1930. Inbred by Bieslchowsky 1948. A number of other strains, including NZO, NZC, NZW, NZX, and NZY were developed from the same stock. Strain NZW was inbred independently by Hall. NZB develops autoimmune haemolytic anaemia.

NZBR Inbr 75+. Chocolate brown: *a*, *b*. Origin: presumed mutation to brown (*b*) in NZB/Bl at F73. Develops autoimmune anaemia like NZB.

NZC Inbr (Wehi) 137. Chocolate-brown: *a*, *b*. Origin: see NZB. Congenital cystic kidneys. High incidence of ovarian granulose cell tumours.

NZO Inbr (Wehi) 128. Agouti: +. Origin: see NZB. Very obese, hyperglycaemic but not hyperinsulinaemic. Defective pancreatic islets.

NZW Inbr (Umc) 100. Albino: *b*, *c*, *p*. Origin: see NZB. NZB × NZWF1 develop systemic lupus erythematosus, antinuclear antibody, haemolytic anaemia, and proteinuria.

NZX Inbr (Wehi) 90. Colour ?. Origin: from a NZC × NZY cross at F33 and F27, respectively. Some imperforate vagina and megacolon.

NZY Inbr (Wehi) 120. Colour: Cinnamon piebald (?): *b*, *s*. Origin: see NZB. A piebald male in F2 was mated to a tan sister. Subsequently inbred with selection for piebald gene *s*. Megacolon.

O20 Inbr (A) 194. Albino: *a*, *c*. Origin: R. Kortweg, 1931, from Amsterdam petshop mice. Mammary tumours 0 per cent in virgins, 5 per cent in breeders, 13 per cent in force-bred females.

P Inbr (J) 153. Light fawn, pink eyes: *a*, *b*, *s*, *p*. Also carries *se* (short-ear). Origin: Snell, following outcross of BDP outbred stock developed by Gates 1926.

P/A see C57P.

PAA Inbr 45. Colour: Agouti (?). Origin: wild pair trapped in Philadelphia factory (see also PAB, PAC, PAD).

PAB Inbr (42). Colour: Agouti, but 91 per cent white head spot. Origin: see PAA.

PAC Inbr (42). Colour: Agouti (?). Origin: see PAA.

PAD Inbr (45). Colour: Agouti (?). Origin: see PAA.

PC Inbr (U) ?+60. Colour ?:*a*, *b*, *d*, *p*. Origin: formerly CPB-P. CPB-TNO (1950) to Vet. Fac. Utrecht 1973.

PE Inbr (Rl) 90+. Colour: Agouti and pearl: +, *pe*. Origin: pearl mutation causing dilution of black and yellow pigments arose in C3H/HeRl mice before 1954. Maintained by *pe/+* × *pe/pe* matings.

PET Inbr (Wmr) 91. Colour ?. Origin: from a cross between C3H and black pet-shop mice. Med. Coll. Va 1956 to Wmr 1958.

PF Infr (Fo) 35. Colour: agouti piebald: *A*, *s*, *lst*. Origin: ?. Selected for high expression of polydactyly and tibial hemimelia in heterozygotes caused by *lst*.

PHH Inbr (We) 95. Colour: Grey: *a*, *ln*. Origin: Weir from MacArthur. Outbred stock obtained from Butler in 1949. bxs mating with selection for high blood pH (see also PHL). Blood pH now 7.48 ± 0.004.

PHL Inbr (We) 97. Colour (?): *aᵗ*, *b*, *ln*. Origin: see PHH, but selected for low blood pH. Blood pH now 7.43 ± 0.004.

PL Inbr (J) 130. Albino: *c*. Origin: non-inbred Princeton stock started in 1922 from 200 mice purchased from a dealer. Inbred by Lynch. High incidence of leukaemia.

PM Inbr 80. Colour ?. Origin: C. Biancifiori, Perugia, from C3Hb × BALB/cfC3H. Descendants bxs mated with selection for high incidence of mammary tumours.

High mammary tumour incidence and polycystic kidneys.

PN Ibr 52. Albino: *c*. Origin: albino mice from pet shop in New Zealand in 1948. Maintained by Wigley in Med. Res Dept. and inbred starting in 1964, with selection for positive ANA (anti nuclear antibody) in first three generations. ANA appear at about 5 months, and 80 per cent are positive by 10 months.

PT Inbr (Y) 45. Near-white: *a*, *b*, *c^{ch}*, *d*, *p*, *s*. Origin: seven-locus stock from Dr Mary Lyon, Harwell in 1969. Inbred since then. Also carries *se* and *T^y/+*.

PRO Inbr (Brk) 61. Colour ?: *a*, *c^{ch}*, *p*. Origin: E.S. Russell from a cross of 129/ReJ × C57BL/6J. bxs with selection for *a*, *c^{ch}*, and *p*. Mutation to *Pro-1^b* (proline oxidase-1). Activity of proline oxidase in liver, kidney, and brain is about 20 per cent of normal.

PUH Inbr (Cam) 50. Agouti. Origin: wild mouse from Rimach valley in Peru (Peru-Harland) × CBA. Inbred by M.E. Wallace. Many mutations present in this strain.

QC Inbr (U) ?+94. Off-white (?): *a*, *c^e*. Origin: Hagedoor to Hirschfeld (1937) to CPB (1949), to Vet. Fac. Utrecht (1973).

RBA Inbr (Dn) ?+25. Agouti. Origin: from wild mice captured near Mutten, Albula Valley, Grisons, S.E. Switzerland by A. Gropp and H. Winking. To Davisson and Roderick 1977, to Davisson 1981. Homozygous for Rb(4.12)9Bnr. Has never been outcrossed to a laboratory strain.

RBB Inbr (Dn) ?+35. Agouti. Origin: wild mice captured near Bondo, Val Bregaglia, Grisons, S.E. Switzerland by A. Gropp and H. Winking. To Davisson and Roderick 1977, to Davisson 1981. Homozygous for Rb(1.10)10Bnr. Probably has never been crossed to a laboratory strain, but is not as wild as RBA.

RBC Inbr (Dn) ?+33. Albino. Origin: Rb(8.17)1Iem arose in a colony of white mice at the Inst. Exp. Med., Leningrad. From Rappolovo Nursery of Acad. Med. Sci of USSR to Gropp, to Davisson and Roderick 1977, to Davisson 1981.

RBD Inbr (Dn) 60. Albino: *A*, *E^{tob}*, *c*. Origin: derived from crosses of Swiss mice (probably Han:NMRI, A. Gropp personal communication to Dn) with wild mice captured in the Valle di Poschiavo in SE Switzerland. To Fls from Gropp (probably via F. Ruddle) to T. Roderick in 1970, bxs ever since. To M. Davisson in 1981. Homozygous for the Robertsonian chromosomes Rb(5.15)3Bnr, Rb(11.13)4Bnr, and Rb(16.17)7Bnr.

RBE Inbr (Dn) 36. Albino: *A*, *c*. Origin: Davisson from a cross of two stocks derived from F1 hybrids

between wild mice captured in the Valle di Poschiavo in S.E. Switzerland and Swiss laboratory mice (probably Han:NMRI, A. Gropp, personal communication to Dn) or a B6D2F1 female. F1s from A. Gropp to F.H. Ruddle to T.H. Roderick in 1970. Stocks to M.T. Davisson in 1974. Homozygous for the Robertsonian chromosomes Rb(1.3)1Bnr, Rb(4.6)2Bnr, Rb(8.12)5Bnr, and Rb(9.14)6Bnr.

RBF Inbr (Dn) 60. Albino: *c*, *E^{tob}*. Origin: Swiss mice (probably Han:NMRI) crossed with wild mice captured in Valle di Poschiavo in S.E. Switzerland. Gropp to Roderick 1970 at F1. to Davisson 1981. Homozygous for Rb(1.3)1Bnr, Rb(8.12)5Bnr, and Rb(9.14)6Bnr.

RC Inbr (U) ?+75. Albino: *c*. Origin: formerly CPB-R. Rhodesfarm, S. Africa (?) to Compton to Hagedoorn to CPB-TNO (1949), to Vet. Fac. Utrecht (1973).

RF Inbr (J) 113. Albino: *a*, *c*. Origin: Furth 1928, from Rockefeller Inst. General purpose stock of domestic origin, transferred to Oak Ridge.

RFM Inbr 120 +. Albino: *a*, *c*. Origin: Furth, to Oak Ridge 1949. Strain has been found to be segregating for *H-2^f* and *H-2^k* and substrains also differ in other respects, so genetic contamination of some substrains must be suspected.

RHJ Inbr (Le) 72. Albino: *a*, *b*, *c*. Origin: mutation to rhino-J (*hr^{rh-J}*) in a Carworth Farms stock received by E.P. Nagler, then back to Carworth, then to M.M. Dickie 1951. To Lane 1960. N6 to BALB/cHu, bxs to F16, then outcrossed to albino stock followed by bxs. Homozygotes begin to lose hair at about 10 days and females do not nurse young.

RNC Inbr 50+. Colour ?. Origin: R.C. Roberts, Edinburgh 1968. Three pairs carrying *Re* (Rex), *Tr* (trembler), and *nu* (nude). *Tr* was lost. Recombinant *Re nu/+ +* were mated bxs. *nu/nu* mice are athymic with defective cell-mediated immune response.

ROP Inbr (Le) 67. Agouti: +. Origin: heterogeneous stock maintained by M.C. Green in 1960, to Lane 1975. Carries *Ra/+* (Ragged), *Os/+* (oligosyndactyly), and *Pt/+* (pintail).

RSV Inbr (Le) 66. Agouti: +. Origin: Carter's Stock 1 (Edinburgh linkage testing stock E1) to Woodworth 1950. Outcrossed to C57BL/6, CBA, and C3H. To Lane 1967. bxs thereafter. Carries *Re/Re* (Rex), *Sd/+* (Danforth's short-tail) and *Va/+* (varitint-waddler).

RIII Inbr 80+. Albino: *A*, *c*. Origin: Dobrovolskaia-Zavadskaia, Inst. du Radium, Paris 1928, then see below.

RIII/An. Dobrovolskaia-Zavadskaia to Andervont.

RIIIS. In 1967 both RIII and RIII/An maintained by J failed to breed. RIII/2J was developed from a cross between RIII/An and SEC/1Re, backcrossed to RIII/An, then to RIII/J. Similar to RIII/An except 1/8th of genes expected to be derived from SEC/1Re. Name later changed to RIIIS.

RIII/SeA From Severi (Perugia) to Mühlbock (Amsterdam) 1964. Differs from other substrains at the *Hbb* and *Mup* loci.

RIII-*ro*. Dickinson, UK from mutation to *ro*.

RW Inbr (W) 82. Albino: *a, B, c*, D. Origin: albino mice from the University of Wroclaw, Poland. Inbred since 1960. No known crosses with other strains.

SB Inbr (Le) 101. Pale grey: *sa, bg, A^w*. Origin: mutations to *sa* and *bg* occurred independently in stocks used for mutation rate studies at Oak Ridge. Rl to Le 1961, then bxs. Beige mice develop a Chediak–Higashi-like syndrome with giant lysosomal granules in the leucocytes, pigment dilution, and defective immune responses due to absence of natural killer cells. Mice develop progressive pneumonitis.

SC Inbr (U) ?+93. Albino: *c*. Origin: a 'Bagg Swiss' from Rockefeller Inst. to Laidlow (Hampstead, England), to Hagedoorn, to CPB-TNO (1949), to Vet. Fac. Utrecht in 1973. Probably same origin as BRVR and BRVS. Selected for 'bacillary resistant, virus susceptible' (like BSVS ?).

SD Inbr (Glw) 122. Albino: *c*. Origin: Dunn and Gluecksohn-Waelsch. Carries *Sd* (Danforth's short-tail) gene. Homozygotes and some heterozygotes develop urogenital abnormalities, and die shortly after birth. Heterozygotes have short tails and other abnormalities.

SEA Inbr (J) 142. Grey: *b, d* +/*d se*. Origin: EL and MC Green from BALB/c × P. Carries *se* (short-ear) gene causing short ears, skeletal abnormalities.

SEC Pale brown: *a, b, c^{ch}*. Origin: Green prior to 1946 from a cross between NB and BALB/c. Carries *se* (short-ear) gene causing short ears and skeletal abnormalities. Substrains as follows:

SEC/Rl Inbr (Rl) 100 +. Colour ?: *a, b, c^{ch}, d se*/++. Origin: E.L. Green 1948 to Rl at F20–F21.

SEC/1Gn Inbr (Le) 149. Gets multiple lung cysts, especially in *se/se*.

SEC/1Re Inbr (J) 138. From SEC/1Gn, but short-ear eliminated by E.S. Russell.

SF Inbr (Dn) 42+32. Agouti: +. Origin: wild mice trapped in coal mine and sib-mated by M. Barnawell (Berkeley CA) as 'Corte Madera' to Cambridge 1959.

SHI Inbr (Le) 78. Agouti chinchilla: *c^{ch}*. Origin: Snell's FS stock to M.C. Green. Outcrossed to C57BL/

10Gn then bxs. To Lane 1975. Carries *sh-1* (shaker-1) gene. Homozygotes show circling, head tossing, deafness, and hyperactivity.

SHN Inbr (Mei) 51. Albino: *A, b, c*. Origin: SWM/Ms from Natl. Inst. Genet. Misima 1963. bxs from 1964, with selection for high early incidence of mammary tumours (see also SLN). Mammary tumours 90–100 per cent in breeders at 6.6 m.

SHR Inbr (Dn) 59. Colour ?. Origin: E.L. Green from *shm* (shambling) mutation which arose in outbred stock in 1960. Crossed with C57BL/6J, C3H_eB/FeJ, and finally to a *Re* (rex) stock. *shm/shm* have abnormal gait, small body size, are infertile, and have phospholipid-like material deposited in the central nervous system.

SJL Inbr (J) 104. Albino: *c, p*. Origin: Swiss Webster mice from three sources to J 1938–1943. Pen-bred until 1955 when bxs started. High incidence of reticulum-cell sarcoma resembling Hodgkin's disease.

SK Inbr (Dn) ?+38. Agouti: +. Origin: three mice trapped on Skokholm Island off Pembrokeshire in 1962 by R.J. Berry. bxs (to F19) by M.E. Wallace, then to Rk, Dn, and Ei.

SL Inbr(A) ?+90. Albino: *A, B, c*. Origin: derived from SMA as a high-leukaemia strain by K. Tutikawa in Misima. However, leukaemia incidence is low.

SLN Inbr (Mei) 45. Colour ?. Origin: see SHN. This is the high-late mammary tumour line. Mammary tumour incidence is about 50–60 per cent in breeders, 9–10 per cent in virgins at 10 months.

SM Inbr (J) 112. White-bellied agouti or black: *A^w/a* or *a/a*. Origin: MacArthur 1939 by crossing seven stocks including DBA and selecting for small body size. To Runner 1948, who began bxs mating. Small body size which tends to disappear as the animals mature.

SRH Inbr (Nmg) 53. Dilute brown: *a, b, d*. Origin: van Abeelen from a cross of C57BL/6J × DBA/2J, N4 to DBA/2 followed by bxs (from 1966) with selection for behaviour traits. High exploratory rearing and locomotor activity.

SRL Inbreeding, colour, and origin as SRH, but low exploratory rearing and locomotor activity. Differs from SRH in developmental-age dependent behavioural characters.

SSIN Inbr (Utsp) 22+. Albino. Origin: 1970s, from outbred SENCAR (sensitive to carcinogens) selected for high skin tumour number following challenge for 10 generations with a topical application of dimethylbenz(a)anthracene followed by repetitive application of the tumour promoter 12-0-tetradecanoylphorbol-13-

acetate (TPA). Inbred in 1986. Extreme sensitivity to TPA.

SSL Inbr (Le) 60. White spotted, heavy spotting, or black-eyed white depending on genotype: *a*, *s/s* or *s/s^l*, or *s^l/s^l*. Origin: developed by Le. *s^l* arose in the F2 from a cross of C3H/HeSn × C57BL/6J. bxs to F18. In 1968 outcrossed to *a/a*, *s/s* stock (F19) and maintained by sib mating of *s/s^l* to present. The *s* gene came from a multiple recessive stock of Holman to Runner before 1955. Outcrossed to C57BL/6J and bxs to F19. In 1968 crossed to *a/a hr +/+ s^l* to make *s/s^l*. *s^l/s^l* get megacolon and generally die though a few live to breed.

ST Inbr. (J) 143. Albino: *a*, *b*, *c*. Origin: Englebreth-Holm from outbred Danish white mice in about 1940. To Heston in 1947 at F23. Two major substrains are known which differ at the *H-2* locus. These were separated after more than eight generations of sib-mating.

ST/a See above. This is the *H-2^b* substrain which is not so widely used.

ST/b See above. *H-2^k* substrain.

STAR Inbr (N) 83. Albino: *c*. Origin: Gotchevsky (Austria) who obtained a stock of albino mice from Bittner in the early 1930's, to NIH in 1972. Homozygous for a dominant cataract.

STR Inbr (N) 145. Brown: *a*, *b*. Origin: Strong from a stock treated with methylcholanthrene between generations F4 and F27.

STR/1 Inbr (N) 138. Piebald brown: *a*, *b*, *s*. Origin: piebald mutation in STR at F29 in 1961. Develops osteoarthropathy of the knee joints, and obstructive uropathy in males before 16 months.

STS Inbr (A) 100. Albino: *c*. Origin: Mühlbock, from albino mice from the Hyg. Inst. Zurich in 1955. Closely related to LIS and LTS. Lung tumours in both sexes.

STU Inbr 90. Albino. Origin: Gonnert, Bayer-Leverkusen from Swiss mice. To Max-Planck-Inst. Für Virusforsch. (Tubingen) in 1950.

STX Inbr (Le) 59. Black: *E^{so}*. Origin: sombre (*E^{so}*) mutation in C3H held by N. Bateman before 1961, to M. Foster, to M.C. Green 1966. Crossed with *Tw* (twirler) from M. Lyon in 1967, then bxs. Also carries *Xt^J* (extra-toes-J). Has been well characterized at polymorphic loci.

SUMS Inbr (Weller) ?+20. Colour ? Origin: spontaneous hydrocephalic mutation in a stock at Guy's Hospital, London. To R.O. Weller in 1973 and bxs mating. Carries autosomal recessive unnamed mutation causing hydrocephalus in homozygotes which is lethal by 2–6 weeks of age.

Swiss: Various inbred strains have been developed from outbred Swiss stock. It should not be assumed that they have similar characteristics.

SWR Inbr (J) 148. Albino: *c*. Origin: C. Lynch from outbred 'Swiss' mice from Albert de Coulon, Lausanne obtained in 1926. Develops polydipsia and polyuria on ageing.

SWV Inbr (Bc) 73. Albino: *A*, *c*, plus unknown dilution gene. Origin: outbred animals from Defense Research Bd. Suffield, Alberta to Cent. Animal Depot, Univ. Brit. Columbia. Inbreeding started in 1959. Develops nephrogenic diabetes insipidus with a progressive and unique kidney defect.

SWXJ- Inbr (28–32). Set of 14 recombinant inbred strains developed by Beamer from a cross of SWR × SJL.

TA1 Inbr. 90. Albino: *a*, *b*, *c*. Origin: outbred mice from Tianjin (China) in June 1956. Inbred since then. Well characterized at polymorphic loci.

TA2 Inbr. 66. Albino: *a*, *B*, *c*, *d*. Origin: outbred 'Kun Ming' mice from Bioproducts Institute, Peking in 1962. Inbred since then.

TF Inbr. (Le) 77. Black: *a*. Origin: M.C. Green from *Ttf/t^b* stock received from M.F. Lyon in 1961, then outcrossed to C57BL/10Gn to obtain *+/tf* chromosome in 1962, followed by bxs. To Lane 1975. Good chromosome 17 markers.

TFH Inbr (H) 33. Agouti: +. Origin: outbred *T tf/+ tf* stock. Sib mating started in 1972. Homozygous *tf/tf*, but forced segregation of *T/+*.

TH Inbr (Wl) 93. Black-and-tan: *a^t*. Origin: H.G. Wolfe before 1961 with selection for high blood pH. Good reproductive performance. Arterial blood pH 7.34.

TKDU Inbr (Dn) 69. Grey: *a*, *d*. Also carries *du* (ducky) and *tk* (tail kinks). Origin: *du* from C. Keeler to G.D. Snell 1948, to M.C. Green 1956, to Eicher 1972, to Davisson 1980. Cross to DBA then bxs to F10. crossed to BALB/c then bxs to F4. Crossed to DBA/1 then bxs to F2. Crossed to C57BL/10 then bxs to F2. Crossed to BALB/c-*tk* (from Grüneberg 1961) and maintained as balanced stock. *tk* arose in BALB/c (probably An substrain) in 1953 at the Chester Beatty R. Inst.

TL Inbr (Wl) 96. Brown with white-belly: *A^w*, *b*. Origin: as for TH, but with selection for low blood pH. Arterial blood pH 7.30.

TM Inbr 90+. Albino. Origin: Casas 1948 from unpedigreed Swiss mice.

TP Inbr (Rl) 68+. Black or slate-grey: *a*, *tp/+*. Origin: Jackson Lab. Taupe mutation in C57BL/10J in 1948. N6 to C57BL/10, then bxs. Homozygous *tp* mice viable, but females cannot rear young due to abnormal nipples.

TPS Inbr (Y) 38. Black: *a*. Origin: Heterogeneous stock synthesized from A/Sn, BALB/c, CBA/Ca, C57BL/6J, DBA/2, GLF, I/St, and 101/H, with selection for high sensitivity to cytogenetic effect of thiotepa in bone marrow cells.

TR Inbr (Dn) 81. Irregular patches of full-coloured and very lightly coloured fur: *Mo^to*. Origin: Dickie 1952 from tortoise (*Mo^to*) mutation in non-inbred obese stock. Maintained by matings of *Mo^to* females with normal males. Mottled females with much white in coat do not usually mature. Darker ones survive well.

TSI Inbr. (A) ?+72. Albino: *a*, *c*. Origin: Mühlbock, from stock supplied by P. Schafer (Tubingen) in 1958. High incidence of leukosis in both sexes.

TSJ Inbr (Le) 64. Cinnamon: *b*. Origin: mutation to *Ts* (tail-short) arose at NCI in 1946. From Morgan to G.D. Snell 1950. Crossed to C57BL/6, C57BR/cd, and BALB/cSn, then bxs. *Ts/Ts* lethal before 6 days of gestation. *Ts/+* have short kinked tails and skeletal abnormalities.

TT6 Inbr (Le) 54. Agouti. Origin: 'TR' stock from M.F. Lyon to M.C. Green in 1961, to Le 1970. One cross to jittery-grizzled stock, bxs to F7, one cross to B6CBAF1-*A^{W-J}/A* in 1972, then bxs. Carries *T* (brachyury), *t^6*, and *tf* (tufted).

V Inbr (Le) 49. Grey, and grey-piebald (?): *a*, *ln*, *s*. Origin: G.D. Snell from a stock carrying *v* (waltzer), *s* (piebald), and *ln* (leaden) from Ludwin in 1947. Crossed to C57BL/10, then to *fz* (fuzzy) stock in 1960 and non-sib mated. To Lane 1969, then bxs. *v/v* females are poor mothers.

VC Inbr (U) ?+86. Chinchilla (?):*c^{ch}*. Origin: formerly CPB-V. Hagedoorn (?) to CPB/TNO in 1949, to Vet. Fac. Utrecht in 1973. Relatively low concentration of major urinary protein in males.

VY Inbr (Nctr) 81. Mottled yellow, pseudoagouti and black: *A^{vy}/a*, *a/a*. Origin: the *A^{vy}* mutation occurred in C3H/HeJ in 1960. Backcrossed to C57BL/6J. One N3 male and two N1 females from M.M. Dickie, to Inst. Cancer Research in 1962. These were mated and the offspring sib mated. Maintained by *A^{vy}/a × a/a* matings since then. To the Nctr in 1972 at F35.

WB Inbr (J) 119. Black or black-eyed white: *a*, *W*. Origin: *W* (dominant spotting) heterozygotes from S.J. Holman and S.G. Waelsch. Sib-mating started in 1948.

W/W mice have macrocytic anaemia, are sterile, lack pigmentation in coat, and die at about 11 days.

WC Inbr (J) N58F37. Colour grey, white, or black depending on genotype: *a*, *Sl*. Origin: *Sl* (steel) mutation backcrossed to strain WB-+/+. Phenotype of *Sl/Sl* superficially similar to *W/W*, but defect is due to abnormal environment for stem-cell development.

WHT Inbr (Gy) 125+. Albino: *c*. Origin: Hewitt, Westminster Hospital London 1958. Host for a range of transplantable tumours.

WLHR Inbr (Le) 92. Chocolate: *a*, *b*. Carries *wl* and *hr*. Origin: mutation to *wl* (wabbler-lethal) occurred in pirouette stock in 1948. Crossed to *hr* (hairless) stock from K.P. Hummel and inbred as balanced stock. Homozygous *wl/wl* young die before weaning.

WLL Inbr (A): ?+112. Albino: *c*. Origin: Kreyberg 1929 from a commercial dealer in Oslo. In 1934 'White label' mice were sent to Leeds, becoming WLL, and some stayed in Oslo, becoming WLO (now extinct ?).

WN Inbr (Nga) 60. Colour depends on genotype: *a*, *W^n*. Origin: new mutation at the W locus. *W^n/W^n* die at about 16–18 days of gestation with anaemic syndrome. Heterozygotes normal but with more extensive white spotting than in *W^v/+*.

WR Inbr (Y) 36. Black spotted: *a*, *W^y/+*. Origin: developed by selection for dilution of pigmentation of *a/a*, *W^y/+* mice and a high incidence of somatic reversion of *W^y* to + in a cross of 129 and 129 × C57BL/6-*W^y/+*. Inbred since 1975. Anaemic *W^y/W^y* die at birth. *W^y/+* mice have light coat colour, 25 per cent with black spots.

X Inbr (Gf) 89. Albino: *a*, *c*. Origin: unknown. Inbred by Goldfeder since 1953. Complete absence of spontaneous tumours, and relatively resistant to most carcinogens.

XLII Inbr (Orl) 50+. Cinnamon (?): *b*. Origin: Dobrovolskaia-Zavadskaia from a cross between C57L and C57BL. Very low tumour incidence.

XVII Inbr (Rd) 80+ Albino: *c*. Origin: Dobrovolskaia-Zavadskaia 1928. bxs inbreeding since 1943. Very susceptible to lung oncogenesis by chemical agents.

YBR Inbr (Ki) 132. Colour brown or yellow: *A^y/a*, *b*. Origin: Little to Andervont 1936, to Heston 1946 at An's F30. Amyloid at least 50 per cent in both sexes and genotypes. Obesity in crosses with other strains.

YS Inbr (Nctr) 84. Mottled yellow, pseudoagouti, black, piebald, depending on genotype: *A^y/a*, or *a/a*, *s*.

Origin: Chase to Inst. of Cancer Research 1959 at F38–39. In 1962 outcrossed to an N3 male from a (C3H/ HeJ × C57BL/6) × C57BL/6 cross. A^{vy}/a mice were backcrossed to the YS strain to N35.

YT Inbr (Y) 45. Near-white: a, b, c^{ch}, p, d, ln. Origin: from crosses involving C57L, DBA/2, C57BL/Go, 129/ J. Inbred since 1967. Carries multiple recessive genes including go (angora), producing long-hair.

YX Inbr 50. Colour: Yellow: A^y/a^x. Origin: from an A^y/a^x stock of L.B. Russell crossed with an A^{vy}/a male in 1968. Homozygous a^x is lethal.

101 Inbr ?+79 (H). White-bellied agouti: A^w. Origin: L.C. Dunn before 1936 (see also 129).

102 Inbr (El) 93. Agouti: +. Origin: from 101/Rl to El, then presumed genetic contamination. Differs from 101 at the a, Car-2, Gpi-1^s, Hba, and lop-2 loci.

129 Inbr and colour depends on substrain (see below). Origin: Dunn 1928 from crosses of coat colour stocks from English fanciers and a chinchilla stock from Castle. This strain has a common origin with strain 101. A number of major substrains, which trace back to the J in 1948 are:

129/Re Inbr (J) 89. Pale yellow: A^w, c^{ch}, p. Non-dystrophic substrain of 129/Re-dy.

129/RrJ Inbr (J) 97. Pale yellow, or albino: A^w, c^{ch} (or c), p. Origin: J 1948.

129/Sv-ter/+ Inbr (Sv) N8 F49. Agouti with light belly: A^w, c^{ch}, p^+. Also carries a gene ter causing a high incidence of testicular teratomas. Origin: the W gene was backcrossed repeatedly to 129, and at generation N8 a female produced 8/38 male offspring with testicular teratomas. All subsequent members derived from that mating. The W gene has been eliminated.

201 Inbr (Rl) 75+. Colour: various depending on genotype A^y/a, pe. Origin: Russell, from PE/Rl and a non-inbred A^y/a stock. Maintained by matings of A^y/a pe/pe × a/a pe/pe. Occasional somatic reverse mutations.

615 Inbr 64. Chocolate: a, b. Origin ?: Tianjin, China. Well characterized at polymorphic loci.

16 THE WILD HOUSE MOUSE AND ITS RELATIVES

FRANÇOIS BONHOMME and JEAN-LOUIS GUÉNET

Although the house mouse has been used for years by investigators, its systematics have remained poorly documented and sometimes even controversial. The original users of the Linnaean classification, for example, termed every rodent with a mouse-like appearance '*Mus*' so that even the black rat was called *Mus rattus*. As more samples were collected from around the world, a more precise systematics of the genus *Mus* emerged based mainly on skull and teeth row patterns. Within this overall framework local samples with somewhat conspicuous peculiarities in body size, shape, or coat colour were often considered as belonging to distinct taxa. A total of around 130 named forms were counted by Ellerman (21) in his 1941 publication, with as many as 60 subspecies for *Mus musculus* alone. In reaction to this, Schwarz and Schwarz (47) considered that there was only one polytypic species for the genus *Mus* in Europe; a point of view which can still be found in recent texts (20).

Using more sophisticated approaches to, and appropriate tools for biosystematical inferences, modern taxonomists have been able to characterize unambiguously the natural units of the house mouse and its relatives.

The purpose of this chapter is to describe these units, and to stress their contribution to presently existing laboratory stocks.

16.1 Systematics within the genus *Mus*

16.1.1 The complex species *Mus musculus*

Genetic polymorphisms have greatly contributed to our understanding of intraspecific systematic relationships since they allow measurement of the amount of genetic exchange. In the field of *Mus musculus* systematics the most significant results have been made by groups of investigators studying nuclear protein polymorphisms (mostly changes in electrophoretic mobility). Among the groups involved were Selander *et al.* (48, 29), Sage (45), Marshall (32), Marshall and Sage (33), Berry and Peters (2), Moriwaki and co-workers (35, 36), and Thaler, Bonhomme, and co-workers (5, 10, 15, 49).

The contribution of studies on other types of markers to the systematics of the house mouse is exemplified by the work of Gropp, Winking, and co-workers (26, 27, 54), Capanna and co-workers (17, 18, 19), Moriwaki and co-workers (37), Nash and co-workers (41), Adolph and Klein (1), and Britton-Davidian, Said, and co-workers (16, 46) for karyological variation, Ferris and co-workers (23, 24), Yonekawa, Moriwaki, and co-workers (38, 55, 56), and Boursot, Bonhomme, and co-workers (13, 14, 25) for mitochondrial DNA variation, and Klein and co-workers (30, 31) and Nadeau *et al.* (40) for studies on the M.H.C. and the *T/t* complex.

Taken together, these studies have demonstrated the existence of *four main biochemical groups* distributed throughout the Old World, as shown in Fig. 16.1. These groups are distributed as follows:

- *Mus musculus domesticus* (Rutti, 1772) Western Europe and the Mediterranean basin, Africa, Arabia, and the Middle East. Transported by man to the New World and almost everywhere else;

- *Mus musculus musculus* (Linnaeus, 1758)—from eastern Europe to Japan across the USSR and northern China;

- *Mus musculus castaneus* (Waterhouse, 1843)— from Ceylon to south east Asia including the Indo-Malayan archipelago;

- *Mus musculus bactrianus* (Blyth, 1846)—from Iran to Pakistan and India.

Electrophoretic studies at 42 loci have established Nei's genetic distances between these four groups as ranging from the highly significant values of 0.15 to 0.35 (10). Whilst they are well differentiated as determined by protein variation, it is important to underline that these four main units exchange genes wherever they come into contact. It is these exchanges which justify the use of subspecific designations (latin trinomens). The best understood cases of gene exchange are those between *M. m. musculus* and *M. m. domesticus* in Europe, and between *M. m. musculus* and *M. m. castaneus* in Japan. The latter has been well studied by Moriwaki, Yonekawa, and colleagues (see for instance ref. 56, and

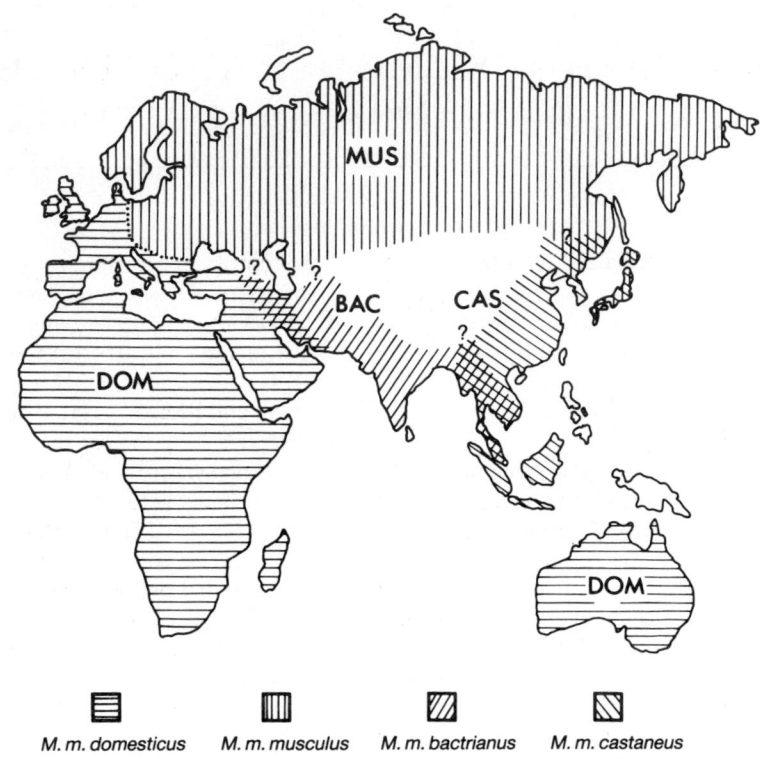

M. m. domesticus   M. m. musculus   M. m. bactrianus   M. m. castaneus

Fig. 16.1 Geographical distribution in the Old World of the four main taxonomic units making up the complex species *Mus musculus* (from (4)).

literature cited therein), who have shown that the colonization of the Japanese archipelago by mice was the result of at least two invasions paralleling those of man. The first one by *M. m. castaneus* came from the south; the second one by *M. m. musculus* from the west. The two subspecies have subsequently hybridized extensively, giving rise to a unique population often referred to as *M. m. molossinus*, which represents one of the clearest cases of reticulate evolution in mammals. The *musculus/castaneus* interaction is also known to occur in continental China.

At the other end of Eurasia, *M. m. domesticus* and *M. m. musculus* are also known to interact along a hybrid zone ranging from Denmark (48), to Bulgaria (9), where they exchange nuclear genes in a nonrandom fashion. Extensive mtDNA flow across this line occurred at several points of the zone (22, 13), and individuals with nearly 100 per cent *M. m. musculus* nuclear genome and a *M. m. domesticus* cytoplasm have been found in Sweden while the reverse can be found in Greece. However, gene flow is asymmetrical

and differential depending upon which part of the genome is considered. For instance, an entire chromosome such as the Y does not seem to cross the contact zone (50). The blending of the two genomes is thus not complete and there is some partial occurrence of male sterility in the northern part of the hybrid zone. Nevertheless, autosomal alleles characteristic of *M. m. domesticus* have been found in *M. m. musculus* populations more than 200 km north of the hybrid zone of east Bulgaria. It therefore appears unlikely that the process of gene exchange that started some 5000 years ago (29) following the deforestation of central Europe is about to stop.

The two situations described above provide evidence for east/west exchange opportunities occurring in the following fashion: *domesticus* ⟷ *musculus* ⟷ *castaneus* (see Fig. 16.1). Although it is certain that they do occur much less is known about the interactions of *M. m. bactrianus* with the remainder of the species which, given its distribution south of the Himalayas, probably present the sequence *domesticus* ⟷ *bactrianus* ⟷ *castaneus*. In addition to these natural interactions in the Old World, *M. m. domesticus* and, to a lesser extent *M. m. castaneus*, have been transported by man in recent times to almost all other regions of the world giving rise eventually to new hybrid populations testifying once again to the fact that the blending of different genomes is a reality in nature.

It has to be noted that our nomenclatorial choice does not preclude internal heterogeneity in the units thus defined. Because of a certain degree of intraspecific heterogeneity many morphotypes of the most extensively studied subspecies, *M. m. domesticus*, have often been designated with names of their own such as *brevirostris*, *praetextus*, *poschiavinus* etc. In fact, in spite of coat colour and karyotypic differences, these animals are almost indistinguishable from one another even using genetic techniques such as protein electrophoresis or mtDNA RFLP's. This reflects the fact that they all share a common and probably recent ancestry and that gene flow from say, Scotland to Egypt, is theoretically possible even if seriously impeded by geographical, ecological, or genetical factors. Even Robertsonian karyotypes which occur throughout almost the entire *M. m. domesticus* range*, do not seem to constitute absolute obstacles, since there is evidence of gene flow across the clinal zone linking the 40 acrocentrics to the fused karyotypes (J. Britton-Davidian, personal communication). It therefore seems to us legitimate to consider *Mus musculus domesticus* and all its local variations as a single unit. The same is probably true of the other subspecies, and many designations (e.g. *urbanus*, *homourus*, *tytleri*, *wagneri*, . . .) will probably fall into synonymy with either *musculus*, *castaneus*, or *bactrianus* as additional wild populations are studied.

Another point we feel should be discussed is the use of subspecific trinomens. Some authors (Sage (45), Marshall and Sage (33), Ferris *et al.* (24)) have preferred to consider all forms as distinct specific entities. Other authors (Selander *et al.* (48), Thaler *et al.* (49)) have spoken of 'semi-species' for the particular case of *domesticus* and *musculus*. Some others (Bonhomme *et al.* (5, 10)), have preferred provisionally not to make any nomenclatorial choice and have used numbered biochemical groups. It seems to us that we should choose now that there is a much better knowledge of the natural interactions displayed by these taxa *in natura*, as illustrated by the brief description already presented: *None of the four main units is completely genetically isolated from the other three, none is able to*

live sympatrically with any other. In those locations where they meet, there is evidence of exchanges ranging from differential introgression limited to selected genetic components to a complete blending. It is therefore necessary to keep all these taxonomical units, whose evolutionary fate is unpredictable, within a species framework. Our proposal is in accord with the recent trend in systematics tending to give to the subspecies category a wide meaning: that is an entity showing a differentiated gene pool whilst retaining the ability to intercross with other such entities.

16.1.2 The other species of the genus *Mus*

Over most of its range, the complex species *Mus musculus* coexists sympatrically with at least one other species belonging to the same genus.

In the western part of the Old World there are three such species (see Fig. 16.2) making five possible sympatric pairs with either *Mus musculus domesticus*, *Mus musculus musculus*, or even *Mus musculus bactrianus*.

These species are:

- the western Mediterranean short-tailed mouse *Mus spretus* (Lataste, 1883),

- the mound-building mouse *Mus spicilegus* (Petenyi, 1882), and

- the eastern Mediterranean short-tailed mouse that we shall provisionally refer to as biochemical group *Mus 4A*.

Biochemical criteria discriminate easily between these different species which, although they occur sympatrically with representatives of the complex species *Mus musculus*, never produce hybrids with them in nature.

Nomenclature

Here again an unavoidable nomenclatorial problem arises since on the one hand we have to use unambiguous designations having the same meaning for everyone and on the other we are obliged to respect the international nomenclature code and its rule of priority. This last rule is however hard to respect since all the European forms are both very close morphologically and variable from one place to the other and only recently have the specific identities of these forms been clearly and unambiguously established biochemically. In the southern part of its range, for instance, *Mus musculus musculus* has permanent outdoor populations with a white belly and a rather short tail which are virtually impossible to tell apart from mound-builder mice in a mixed sample (occurring frequently in the same

---

* To our knowledge, Robertsonian populations with different chromosomal combinations have been found in Belgium, Denmark, France, Germany, Greece, Italy, Scotland, Spain, Switzerland, Tunisia, and Yugoslavia.

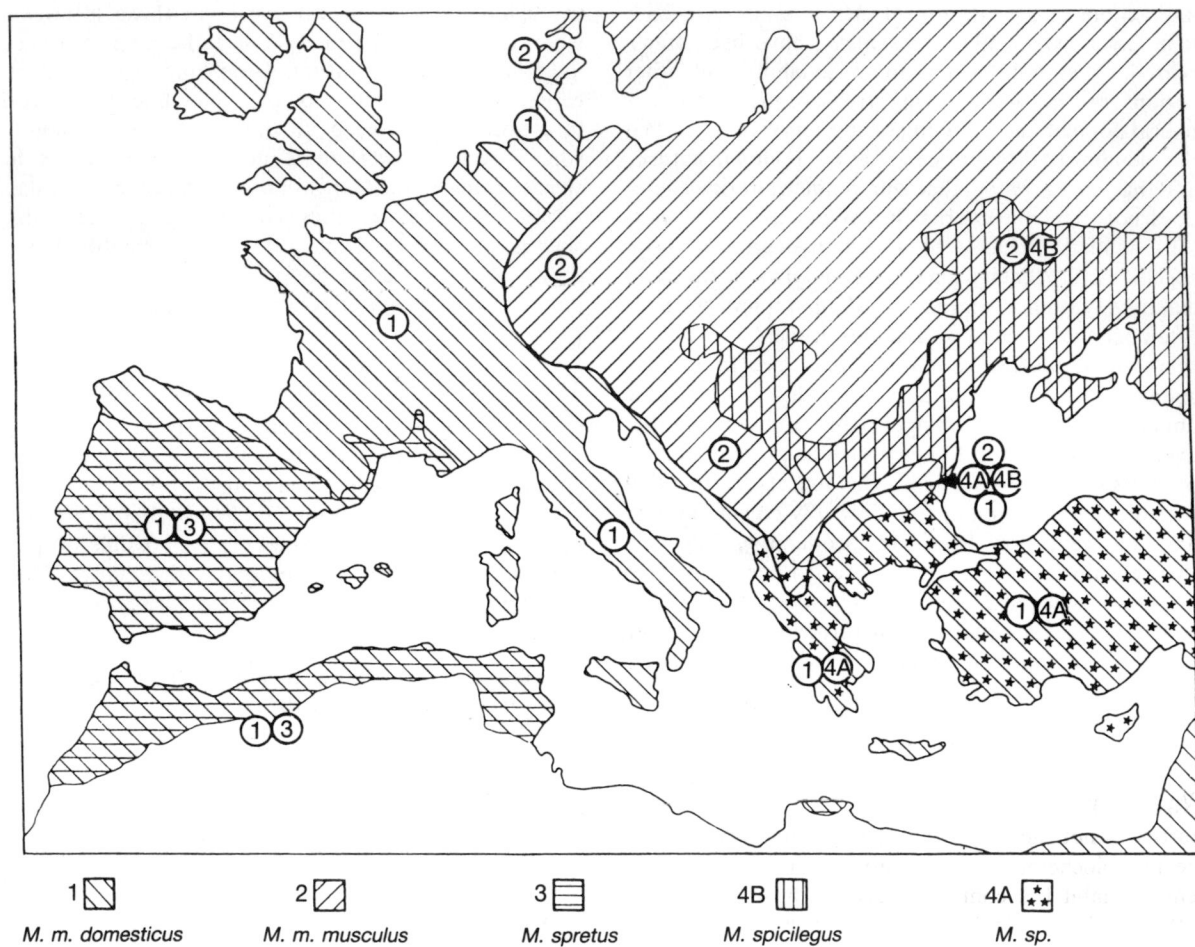

1 ⬚	2 ⬚	3 ⬚	4B ⬚	4A ⬚
M. m. domesticus	*M. m. musculus*	*M. spretus*	*M. spicilegus*	*M. sp.*

Fig. 16.2 Geographical distribution of the five taxa belonging to the genus *Mus* in Europe (from (10)).

trapping line). It is only now that biochemical analysis can differentiate them clearly that multicharacter discriminant functions are available to assist morphologists in telling them apart.

It is often hard to know exactly what animals have been described by authors in the past who were unaware of the biosystematical complexity of the taxon. Moreover, it is not always possible to recover the type specimens and *a posteriori* identifications are difficult. There are therefore a lot of potentially suitable names for the European species of *Mus*. However, it seems that the biochemical group *Mus* 3 should settle down as *Mus spretus* (Lataste, 1883); biochemical group *Mus* 4B as *Mus spicilegus* (Petenyi, 1882) also called *Mus hortulanus* (Nordmann, 1840) by Marshall and Sage (33) but accepted as *Mus*

spicilegus by Marshall (34): the type specimens of *hortulanus*, from a garden in Odessa, may well in fact belong to *M. m. musculus*. As to *Mus* 4A, it is unclear for the time being whether it should be called *Mus abbotti*, described by Waterhouse in 1837 from a juvenile trapped in Trebizond, Turkey, where both *Mus* 4A and forms related to *M. m. domesticus* or *M. m. bactrianus* occur, or even *Mus macedonicus* (if the animals described by Petrov and Ruzic (42) in Yugoslavia are in fact *Mus* 4A which is *a priori* hard to tell), or if it should receive a name of its own such as the *Mus spretoides* proposed by Thaler in Bonhomme *et al.* (10), but never described as a holotype. It is to be hoped that further multidimensional morphometric studies of type specimens will clarify the nomenclature of the European species.

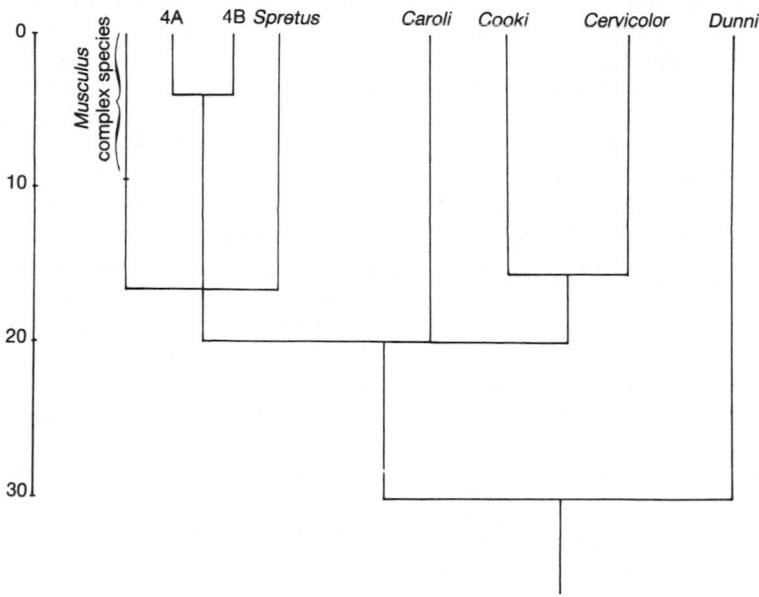

Fig. 16.3 Schematic dendrogram showing the biochemical relatedness of several species of the genus *Mus* based on protein electrophoretic polymorphism. The ordinate is a measure of the number of differentially fixed loci out of the 42 analysed. (Data from (10).)

Phylogeny

Europe is not the homeland of the genus *Mus*. All of the five biochemically differentiated groups that currently inhabit the continent seem to have entered it with man during Neolithic times, at least as far as the western part of Europe is concerned. Representatives of the genus have apparently inhabited India and Southeast Asia since their origins. These species are *Mus caroli*, *Mus cervicolor*, and *Mus cooki*, and in India representatives of the subgenus Leggada such as *Mus booduga* or *Mus dunni*. There are many other forms described as species belonging to the genus *Mus*, but only biochemical analysis will show whether or not they have been correctly assigned or if, as the authors suppose, they either fall out of the range of the genus or into synonymy with one of the above-mentioned species. Figure 16.3 shows a schematic phylogeny of all the taxa described so far in this chapter, based on protein electrophoresis. The multifurcations presented are points where alternative branching orders are possible due to the proximity of the splitting events and the limited resolving power of the technique. The species set described here corresponds to the group of species falling within the genus *Mus*, *sensu stricto*. Apart from Robertsonian variation, all these species have a basic

karyotype of 40 acrocentric chromosomes. Viable F1 offspring have been obtained between laboratory female mice and each of the European species by natural fertilization (summarized in Bonhomme *et al.* (10)). Using artificial insemination West *et al.* (53) recovered embryos sired by male *M. dunni* and *M. caroli* sperm, but failed to recover them when *M. cervicolor* sperm were used.

16.1.3 Relationships with other murid genera

Comparisons with *Rattus* have been made frequently in different laboratories, and attempts made to estimate the time of *Mus/Rattus* divergence. There are moreover at least three wild species not belonging to the genus *Mus sensu stricto* as defined above which are maintained in laboratory colonies and often used for comparison. These are *Nannomys minutoides*, *Coelomys pahari*, and *Pyromys platythrix*. These three genera were formerly considered as subgenera of the genus *Mus*, on the basis principally of their morphology. The reasons why they should be excluded from the genus *Mus* are explained elsewhere (Bonhomme *et al.* (11); Bonhomme (4)). Let us simply say that they seem no more related to *Mus* than they are to other well defined murid genera. Recent data obtained from single copy DNA hybridization (F. Catzeflis, personal communication) suggest the following order of dichotomies: starting with *Rattus* around 8–11 M.Y. then *Pyromys*, *Nannomys*, *Coelomys* at intermediate times, and finally the ancestors of the genus *Mus sensu stricto* with the splitting off of the Indian pigmy mice (*Mus dunni*, *Mus booduga*) occurring between 3.2 and 1.5 M.Y. All the above-mentioned genera have distinct karyotypic formulae of their own.

16.2 The laboratory mice and the wild species

16.2.1 Classical inbred strains

Started at the beginning of this century by a few pioneers, the establishment of duly characterized laboratory mice owes much to the existence of a fancy mouse tradition. The known pedigrees of these classical inbred strains (see for instance, Morse (39)) show that many intercrosses occurred in the early part of their development, together with a few introductions of unknown parentage, either by purchase from pet dealers or by wild animals. These laboratory lines were classically taken as being archetypes of *Mus musculus domesticus* and their closeness has been demonstrated for instance by the fact that almost all of them possess the same *domesticus* type mtDNA (Ferris *et al.* (22)). Further studies on the characterization of the Y chromosome (Bishop *et al.* (3)) and other genetic markers have however shown unambiguously that the bulk of classical inbred strains are in fact a mosaic of various genetic components derived from more than a simple taxon of the complex species *Mus musculus*. In particular, they almost all show a *M. m. musculus* component that could be of Asian origin and linked to the early introduction of Japanese fancy mice (waltzing mice for instance) into the European tradition. Irrespective of their origin it is however clear that the allelic diversity they show cannot be interpreted as resulting from high mutation rates during their hundred year history (Bonhomme *et al.* (12)). In some sense, today's classical inbred laboratory strains may be considered as recombinant inbred strains between *Mus musculus domesticus* and (at least) *Mus musculus musculus*.

16.2.2 Wild derived strains

Over the last 10 years, a variety of new strains from known geographical and taxonomical origins have been

created in various laboratories (Table 16.1). The list of those that are completely inbred or very nearly so is given in this chapter. Many useful randombred stocks of wild mice are moreover maintained in various laboratories, and a more complete description of these stocks has been given in a recent book (43).

Mus musculus domesticus is the most widely represented taxon amongst these stocks, followed by the *musculus* group (including *molossinus*) then by *castaneus*.

Beside these strains belonging to the *Mus musculus* species, representatives of *Mus* 4A, *Mus spicilegus*, and *Mus spretus* are now maintained in the laboratory, and some of them have already undergone 20 generations of sib matings. A special mention should be made of *Mus spretus*. It was crossed for the first time to *Mus musculus domesticus* and to BALB/c by Bonhomme *et al.* (6), and, because of the very large number of allelic differences involved, subsequently used to make chromosome assignments in first generation back-crosses (Bonhomme *et al.* (7, 8). The technique, used since then by others and summarized in Guénet (28), has proven to be exceedingly useful for otherwise unvariant proteins (44) when a DNA probe allows the detection of restriction fragment length polymorphisms between *Mus musculus* and *Mus spretus*. Because of its great power, and since the restriction maps are colinear, the *spretus* × *musculus* system is bound to become as important as the R.I. strains are for the detection of genetic linkage.

The same sort of crosses could be done with *Mus* 4A or *Mus spicilegus*. They however show a lesser degree of divergence from *Mus musculus*.

16.3 Inbred strains of wild origin

The following inbred strains have been derived from ancestral animals recently trapped in the feral state and have never been crossed to laboratory mice. All have been inbred for at least 20 generations at the time of publication.

16.3.1 Strains derived from *Mus m. domesticus* progenitors

BFM Established from a few wild breeders trapped by François Bonhomme on the University Campus of Montpellier in 1976.

At F12 the BFM line was split into two sublines: BFM/1 (F30) and BFM/2 (F31).

The BFM/1 line has the following constitution:

Table 16.1 List of laboratories breeding wild derived animals

François Bonhomme
Institut des Sciences de l'Evolution (L.A. 327)
Université de Montpellier II (U.S.T.L.)
Place Eugène Bataillon
F-34060 Montpellier Cedex, France

Verne M. Chapman
Roswell Park Memorial Institute
Department of Health—State of New York
666 Elm Street
Buffalo, New York 14263, USA

Muriel T. Davisson
The Jackson Laboratory
Bar Harbor
Maine 04609, USA

Otto von Deimling
Chemische Abteilung
Pathologisches Institut der
Universität Freiburg
Albertstrasse, 19
7800 Freiburg, FRG

Eva M. Eicher
The Jackson Laboratory
Bar Harbor
Maine 04609, USA

Jiri Forejt
Institute of Molecular Genetics
Czechoslovak Academy of Sciences
Videnska 1083/270
142 20 Praha-4 Krc, Czechoslovakia

Murray B. Gardner, MD
Medical Pathology, School of Medicine
University of California
Davis, California 95616, USA

Jean-Louis Guénet
Unité de Génétique des Mammifères
Institut Pasteur
25, rue du Docteur Roux
75724 Paris, Cedex 15, France

Yoshibumi Matsushima
Tohoku Dental University
Koriyama-shi Fukushima
963 Japan

Kazuo Moriwaki
National Institute of Genetics
Yata 1, 111 Misima
Sizuoka-ken
411 Japan

Joseph H. Nadeau
The Jackson Laboratory
Bar Harbor
Maine 04609, USA

Josephine Peters
MRC Radiobiology Unit
Chilton, Didcot
Oxon., OX11 0RD, UK

Michaël Potter
National Institutes of Health
National Cancer Institute
Bethesda
Maryland 20892, USA

Thomas H. Roderick
The Jackson Laboratory
Bar Harbor
Maine 04609, USA

Heinz Winking
Institut für Biologie
Medizinische Universität zu Lübeck
Ratzeburger Allee 160
D-2400 Lübeck 1, FRG

Acp/1ᵃ, Ada-1ᵃ, Adh-1ᵃ, Ak-1ᵃ, Alb-1ᵃ, Amy-1ᵃ, Car-2ᵇ, Es-1ᵇ, Es-2ᵇ, Es-3ᵇ, Es-5ᵇ, Es-6ᶜ, Es-9ᵈ, Es-10ᵃ, Es-11ᵃ, Es-14ʳ, Es-17ᵃ, Es-25ᵇ, Es-26ᵃ, Es-27ᵃ, Gapdᵃ, Gda-1ᵃ, Gdc-1ᵇ, Glo-1ᵃ, Got-1ᵃ, Got-2ᵃ, Gpd-1ᵃ, Gpi-1ᵃ, Gpt-1ᵃ, Idh-1ᵃ, Idh-2ᵃ, Ldh-1ᵃ, Ldh-2ʳ, Ldr-1ᵇ, Mod-1ᵃ, Mor-1ᵃ, Mor-2ᵃ, Mpi-1ᵇ, Np-1ᵃ, Pgd-1ᵇ, Pgk-1ᵇ, Pgm-1ᵃ, Pgm-2ᵇ, Pk-1ᵃ, Pk-3ᵃ, Sod-1ᵃ, Trfᵇ. BFM/1 also carries an electrophoretic variant allele at the *Sdh-1* locus which is neither *Sdh/1ᵃ* nor *Sdh-1ᵇ*

Holder: MPL.

The BFM/2 line displays the same distribution of alleles except for *Sdh-1* and *Ldr-1* where it carries *Sdh-1ᵃ* and *Ldr-1ᵃ*. The *Hbb* allele of BFM/2 is *Hbbˢ*.
Holder: PAS, MPL.

CLA F22 Original breeders of this line are the same as those of the WSA and WSB inbred lines of Michael Potter (see *infra*). They were trapped on a farm near Centreville, Maryland and propagated by brother ×

655

sister mating exclusively. The WSA and WSB lines were separated from the CLA line at generation 3.

This line has been typed for the following markers: *Idh-1^b*, *Pep-3^b*, *Es-3^c*, *Pgm-1^a*, *Ly1^b*, *Lys2^a*, and *Ly5^b*.

Holder: NIH (M. Potter).

MOR2/Cv F43 is a strain which was wild-trapped in Ohio in the early 1960s. It was maintained as a random bred stock by Jan Bruell at Case Western Reserve until it was shipped to Thomas Shows and Frank Ruddle at Yale University in 1968. In 1970 the strain was shipped to Roswell Park Memorial Institute where it started to be maintained as an inbred stock. MOR2/Cv has an electrophoretic variant for *Mor-1* and that trait has been selected and maintained since 1968.

Holder: ROS, JAX(Rk).

PERA group of strains

Seven breeders, trapped by Olga Atteck in 1961, in Peru, for Margaret Wallace at the University of Cambridge (51, 52), gave rise to a small closed colony. At about generation 15, in 1971, members of this colony were passed to Thomas H. Roderick and Eva M. Eicher who began numbering F generations. Rk and Ei stocks were separated in 1971 (51).

PERA/Cam Ei–F40
Akp-1^b, *Alp-1^a*, *Amy-1^b*, *Car-2^a*, *Es-1^b*, *Es-3^b*, *Es-10^b*, *Es-11^a*, *Glo-1^a*, *Got-2^b*, *Gpd-1^a*, *Gpi-1^b*, *Gpt-1^a*, *Gr-1^a*, *Hbb-^{d}*, *Idh-1^b*, *Mod-1^b*, *Mpi-1^b*, *Neu-1^b*, *Pep-3^b*, *Pgm-1^a*, *Pgm-2^a*, *Trf^a*.

Holder: JAX(Ei)

PERA/Cam Rk–F50
Alb-1^a, *Alp-1^a*, *Amy-1^b*, *Apk^a*, *C3^b*, *Car-1^a*, *Car-2^a*, *Es-1^b*, *Es-2^b*, *Es-3^b*, *Es-10^b*, *Gdc-1^b*, *Glo-1^a*, *Got-1^a*, *Got-2^b*, *Gpt-1^a*, *Hbb^d*, *Idh-1^b*, *Igh-4^a*, *Ldh-1^a*, *Mod-1^b*, *Mod-2^b*, *Mup-1^b*, *Np-1^a*, *Pep-3^b*, *Pgk-2^b*, *Pgm-1^a*, *Sdh-1^b*, *Tam-1^a*, *Trf^b*.

Holder: JAX(Rk).

RBA/Dn F26. Original breeders were captured near Mutten, Albula Valley, Grisons, S.E. Switzerland then sent to Muriel T. Davisson and Thomas H. Roderick in 1977. Animals are homozygous for Rb(4.12)9Bnr (26).

RBB/Dn F42. Derived from wild mice captured near Bondo, Val Bregaglia, Grisons, S.E. Switzerland then sent to Muriel T. Davisson and Thomas H. Roderick in 1977. Homozygous for Rb(1.10)10Bnr (26).

SF group of strains

SF/CamEi F10+32+36. Progenitors were trapped in California. From Barnawell to Margaret Wallace at Cambridge at F10, to Thomas H. Roderick and Eva

M. Eicher in 1971 at F42. Rk and Ei sublines separated in 1971 (51, 52).
Akp-1^b, *Alp-1^b*, *Amy-1^a*, *Car-2^a*, *Es-1^b*, *Es-3^c*, *Es-10^b*, *Es-11^b*, *Glo-1^a*, *Got-2^b*, *Gpd-1^a*, *Gpi-1^b*, *Gpt-1^b*, *Gr-1^b*, *Hbb^d*, *Idh-1^a*, *Mod-1^a*, *Mpi-1^b*, *Neu-1^b*, *Pep-3^b*, *Pgm-1^b*, *Pgm-2^a*, *Trf^b*.

Holder: JAX(Ei).

SF/CamRk F10+32+35.
Ahd-1^a, *Akp-1^b*, *Akp-2^b*, *Alb-1^a*, *Alp-1^b*, *Amy-1^a*, *Apk^a*, *As-1^a*, *C3^c*, *Car-1^a*, *Car-2^a*, *Es-1^b*, *Es-2^b*, *Es-3^c*, *Es-10^b*, *Ggc^a*, *Glo-1^a*, *Got-1^a*, *Got-2^b*, *Gpd-1^a*, *Gpi-1^b*, *Hba^d*, *Hbb^d*, *Hrt-1^b*, *Idh-1^a*, *Igh-4^b*, *Lap-1^b*, *Ldh-1^a*, *Ldr-1^a*, *Len-2^a*, *Mod-1^a*, *Mup-1^a*, *Np-1^a*, *Pan-1^b*, *Pan-2^b*, *Pep-3^b*, *Pgk-2^b*, *Pgm-1^b*, *Pgm-3^b*, *Pre-1^a*, *Rnr^b*, *Sas-1^o*, *Sdh-1^a*, *Tam-1^c*, *Trf^b*.

Holder: JAX(Rk).

SK group of strains

Trapped on Skokholm Island off Pembrokeshire by R.J. Berry, given to Margaret Wallace at Cambridge who sib mated to F19 then passed to Thomas H. Roderick and Eva M. Eicher in 1971. Rk and Ei sublines separated at F19 (51, 52).

SK/CamRk F19+31.
Ahd-1^b, *Akp-1^b*, *Akp-2^a*, *Alb-1^a*, *Amy-1^a*, *Apk^a*, *Apl^b*, *As-1^a*, *C-3^b*, *Car-1^b*, *Car-2^a*, *Es-1^b*, *Es-2^a*, *Es-3^c*, *Es-6^c*, *Es-10^b*, *Es-13^a*, *Fv-1^n*, *Ggc^b*, *Got-1^a*, *Got-2^a*, *Gpd-1^a*, *Gpi-1^a*, *Gr-1^a*, *Hba^d*, *Hbb^d*, *Idh-1^a*, *Igh-4^a*, *Ldh-1^a*, *Ldr-1^b*, *Mod-1^b*, *Mod-2^b*, *Neu-1^b*, *Np-1^a*, *Pep-3^b*, *Pgk-2^a*, *Pgm-1^a*, *Pgm-2^a*, *Pgm-3^c*, *Psp^b*, *Rnr^s*, *Sas-1^o*, *Sdh-1^a*, *Tam-1^c*, *Trf^b*.

Holder: JAX(Rk).

SK/CamEi F42.
Akp-1^b, *Alp-1^b*, *Amy-1^a*, *Car-2^a*, *Es-1^b*, *Es-3^c*, *Es-10^b*, *Es-11^b*, *Glo-1^a*, *Got-2^b*, *Gpd-1^a*, *Gpi-1^b*, *Gpt-1^b*, *Hbb^d*, *Idh-1^a*, *Mod-1^a*, *Mpi-1^b*, *Neu-1^b*, *Pep-3^b*, *Pgm-1^b*, *Pgm-2^a*, *Trf^b*.

Holder: JAX(Ei).

WLA F27. Original breeders of this strain were wild-trapped in 1976 near Toulouse (France) by Jean-Louis Guénet and brother × sister mated since this time. Original breeders were free of t haplotypes.
Acp-1^a, *Ada-1^a*, *Adh-1^a*, *Ak-1^a*, *Alb-1^a*, *Amy-1^a*, *Car-2^b*, *Es-1^b*, *Es-2^b*, *Es-3^b*, *Es-5^b*, *Es-10^a*, *Gda-1^a*, *Gdc-1^b*, *Glo-1^a*, *Got-1^a*, *Got-2^a*, *Gpd-1^a*, *Gpi-1^a*, *Hbb^s*, *Idh-1^a*, *Idh-2^a*, *Ldh-1^a*, *Ldh-2^p*, *Ldr-1^b*, *Mod-1^b*, *Mod-2^a*, *Mor-1^a*, *Mor-2^a*, *Mpi-1^b*, *Np-1^a*, *Pgd^b*, *Pgm-1^a*, *Pgm-2^a*, *Pk-1^a*, *Pk-2^a*, *Sdh-1^a*, *Sod-1^a*, *Trf^b*.

Holder: PAS.

WMP F20. Established from a few breeders trapped in Monastir (Tunisia) by G. Lefranc in 1980 and propagated since this time by brother × sister matings at the Institut Pasteur. Animals of this strain are homozygous for nine pairs of metacentric chromosomes:

Rb(1.11)2Mpl,   *Rb(2.16)3Mpl,*   *Rb(3.12)4Mpl,*
Rb(4.6)5Mpl,   *Rb(5.14)6Mpl,*   *Rb(7.18)7Mpl,*
Rb(8.9)8Mpl,   *Rb(10.17)9Mpl,*   *Rb(13.15)10Mpl.*

This strain is sometimes referred to as 22MO.

Holder: PAS.

WSA F20 (see CLA strain *supra*)
Holder: NIH (M. Potter).

WSB F20 (see CLA strain *supra*)
Holder: NIH (M. Potter).

16.3.2 Strains derived from *Mus m. castaneus* progenitors

CASA/Rk F ?+33. Original breeders were trapped in Thailand by Joe T. Marshall, then sent to Verne Chapman at Yale University who sent them to Thomas H. Roderick and Eva M. Eicher in 1971.

F generations numbered since 1971.

Akp-1^b, Alb-1^a, Alp-1^b, Amy-1^a, Apk^a, C3^b, Car-1^b, Car-2^a, Es-1^{ab}, Es-2^b, Es-3^b, Es-10^a, Gdc-1^d, Glo-1^a, Got-1^b, Got-2^b, Gpi-1$^{c,\ or\ a}$, Gpt-1^b, Gr-1^a, Hbb^d, Idh-1^c, Ldh-1^a, Ldr-1^a, Mod-1^a, Mod-2^a, Mup-1^b, Np-1^b, Pep-3^b, Pgk-1^b, Pgk-2^b, Pgm-1^a, Trf^b.

Holder: JAX(Rk).

CAST/Ei F ?+34. Same origin as CASA/Rk.

Akp-1^b, Alp-1^a, Amy-1^b, Car-2^a, Es-1^h, Es-2^i, Es-3^b, Es-5^a, Es-6^e, Es-7^c, Es-8^b, Es-9^b, Es-10^a, Es-11^b, Es-13^a, Es-14^r, Es-16^a, Es-17^b, Es-18^d, Es-19^a, Es-22^b, Es-23^b, Es-24^b, Es-25^b, Es-26^a, Es-27^a, Glo-1^a, Got-1^a, Got-2^b, Gpd-1^c, Gpi-1^a, Gpt-1^b, Gr-1^a, Hbb^d, Idh-1^c, Mod-1^a, Mpi-1^a, Neu-1^b, Pep-3^b, Pgm-1^b, Pgm-2^a, Trf^b.

Holder: JAX(Ei).

16.3.3 Strains derived from *Mus m. molossinus* progenitors

MOA F50. From wild animals trapped in Anjo-City, Aichi Prefecture, Japan in 1967.

Alb-1^a, Akp-1^b, Amy-1^b, Amy-2^b, Aph-1^b, Car-1^a, Gpd-1^b, Gpi-1^a, Es-1^a, Es-2^c, Es-5^a, Es-11^b, Hbb^p, Hc^l, Idh-1^b, Ldr-1^a, Ly-2^a, Mod-2^b, Mor-1^a, Mup-1^c, Pep-3^a, Pgd^b, Pgm-1^b, Pgm-2^b.

Holder: Moriwaki.

MOLC/Rk F46. Twenty-two mice of the *Mus musculus molossinus* species were received from Michael Potter on October 23, 1969, all mice being within one to three generations of capture from the wild. All stocks (Rk and Ei) descend from Potter's male 539 and a female daughter of Potter's cage 603. Two separate lines were

established immediately which led to MOLC/Rk and MOLD/Rk originally called MOL 3-I and MOL 3-III respectively.

MOLE/Rk, is a line selected under brother × sister mating for intensity of a white spotting phenotypic variant present from time to time. Split off from MOLD/Rk at F23.

Akp-1^b, Alb-1^a, Alp-1^b, Amy-1^b, Apk^b, C3^b, or C3^c, Car-2^a, Es-1^f, Es-2^h, Es-3^b, Es-5^a, Es-6^e, Es-7^e, Es-8^a, Es-9^c, Es-10^c, Es-11^c, Es-13^a, Es-14^r, Es-16^a, Es-17^b, Es-18^b, Es-19^a, Es-22^a, Es-23^a, Es-24^c, Es-25^b, Es-26^a, Es-27^a, Gdc-1^b, Glo-1^a, Got-1^a, Got-2^b, Gpi-1^a, Gpt-1^a, Hba^d, Hbb^p, Idh-1^c, Ldh-1^a, Ldr-1^a, Mod-1^a, Mod-2^a, Mup-1^c, Np-1^b, Pep-3^c, Pgm-1^b, Sod-1^a, Tam-1^b, Trf^b.

Holder: JAX(Rk).

MOLD/Rk F50. The genetic constitution of MOLD/Rk is identical to MOLC/Rk except for a few loci where the following alleles have been found: *Car-1^b, Es-1^f, Es-2^d, Es-3^a, Es-5^a, Es-6^e, Es-7^e, Es-8^a, Es-9^c, Es-10^c, Es-11^c, Es-13^a, Es-14^r, Es-16^a, Es-17^b, Es-18^b, Es-19^a, Es-22^a, Es-23^a, Es-24^c, Es-25^b, Es-26^a, Es-27^a, Gpt-1^c, Mpi-1^a, Pep-4^b, Pgk-1^b, Pgm-2^a, Sdh-1^a.*

Holder: JAX(Rk).

MOLE/Rk F49. The genetic constitution of MOLE/Rk is identical to MOLC/Rk except for a few loci where the following alleles have been found: *Apk^a, Car-1^b, Es-1^f, Es-2^d, Es-3^b, Es-5^a, Es-6^e, Es-7^e, Es-8^a, Es-9^c, Es-11^c, Es-13^a, Es-14^r, Es-16^a, Es-17^b, Es-18^b, Es-19^a, Es-22^a, Es-23^a, Es-24^c, Es-25^b, Es-26^a, Es-27^a, Gpt-1^c, Hbb^d, Pgk-1^b.*

The *Hba* alloform is not known. The *Es-10* and *Idh-1* loci have been found polymorphic for the two alloforms: *Es-10^b* and *Es-10^c*, *Idh-1^a* and *Idh-1^c*.

Holder: JAX(Rk).

MOLF/Ei F39.

Ahd-1^b, Akp-1^b, Akp-2^a, Alb-1^a, Amy-1^a, Apk^a, Apl^b, As-1^a, C3^b, Car-1^b, Car-2^a, Es-1^f, Es-2^d, Es-3^b, Es-5^a, Es-6^e, Es-7^e, Es-8^a, Es-9^c, Es-10^c, Es-11^c, Es-13^a, Es-14^r, Es-16^a, Es-17^b, Es-18^b, Es-19^a, Es-22^a, Es-23^a, Es-24^c, Es-25^b, Es-26^a, Es-27^a, Fv-1^n, Ggc^b, Got-1^a, Got-2^a, Gpd-1^a, Gpi-1^a, Gr-1^a, Hba^d, Hbb^d, Idh-1^a, Igh-4^a, Ldh-1^a, Ldr-1^b, Mod-1^b, Mod-2^b, Neu-1^b, Np-1^a, Pep-3^b, Pgk-2^a, Pgm-1^a, Pgm-2^a, Pgm-3^c, Psp^b, Rnr^s, Sas-1^o, Sdh-1^a, Tam-1^c, Trf^b.

Holder: JAX(Ei).

MOM F30. From wild animals trapped in Mizuho-District, Nagoya City, Aichi Prefecture, Japan in 1972.

Amy-1^b, Amy-2^b, Aph-1^b, Car-1^a, Gpd-1^b, Es-1^a, Es-2^c, Hbb^p, Hc^l, Idh-1^b, Ldr-1^a, Ly-2^a, Msp-3^c, Mtp-1^s, Mtp-5^o, Mtp-7^o.

Holder: Moriwaki.

MSM F21. From wild animals trapped in Mishima-City, Shizuoka Prefecture, Japan in 1978.
Akp-1^b, *Amy-2^b*, *Aph-1^b*, *Car-1^a*, *Gpd-1^b*, *Gpi-1^a*, *Es-1^a*, *Es-2^c*, *Hbb^p*, *Hc^l*, *Idh-1^b*, *Ldr-1^a*, *Ly-2^a*, *Msp-3^c*, *Mtp-1^s*, *Mtp-5^o*, *Mtp-7^o*, *Pep-3^a*, *Pgm-1^b*, *Pgm-2^a*.

Holder: Moriwaki.

16.3.4 Strains derived from *Mus m. musculus* progenitors

PWD F34. Inbred strain derived from a pair of wild animals trapped in cages of exotic birds in the Prague Zoo.
Akp-1^b, *Amy-1^a*, *Apk^a*, *Car-1^a*, *Car-2^a*, *Es-1^a*, *Es-2^h*, *Es-3^b*, *Es-5^a*, *Es-6^a*, *Es-7^b*, *Es-8^a*, *Es-9^g*, *Es-10^c*, *Es-11^c*, *Es-13^a*, *Es-14^r*, *Es-16^a*, *Es-17^b*, *Es-18^b*, *Es-22^i*, *Es-23^f*, *Es-25^b*, *Es-26^a*, *Es-27^a*, *Glo-1^a*, *Got-1^a*, *Got-2^b*, *Gpi-1^a*, *Gus^a*, *Idh-1^b*, *Mod-1^a*, *Pre-2^c*, *Prt-4^b*, *Prt-5^a*, *Tam-1^d*, *Trf^a*.

Holder: PRA, von Deimling.

PWB F34. Inbred strain derived from the same founder pair as PWD.

Holder: PRA.

PWK F38. Inbred strain derived from a pair of wild animals (*Mus musculus musculus*) trapped in Lhotka near Prague, Czechoslovakia in 1974. All biochemical characteristics are very similar to strain PWD except for the *Es-13* locus where the *Es-13^b* allele has been found.
Ada-1^a, *Acp-1^a*, *Adh-1^a*, *Ak-1^a*, *Akp-1^b*, *Alb-1^a*, *Amy-1^a*, *Apk^a*, *Car-1^a*, *Car-2^b*, *Es-1^a*, *Es-2^h*, *Es-3^b*, *Es-5^a*, *Es-6^a*, *Es-7^b*, *Es-8^a*, *Es-9^g*, *Es-10^c*, *Es-11^c*, *Es-13^b*, *Es-14^r*, *Es-16^a*, *Es-17^b*, *Es-18^b*, *Es-22^i*, *Es-23^f*, *Es-25^b*, *Es-26^a*, *Es-27^a*, *Gda-1^a*, *Gdc-1^b*, *Glo-1^a*, *Got-1^a*, *Got-2^a*, *Gpd-1^c*, *Gpi-1^a*, *Gus^a*, *Hbb^c*, *Idh-1^a*, *Idh-2^a*, *Ldh-1^a*, *Ldh-2^r*, *Ldr-1^b*, *Mod-2^b*, *Mor-1^a*, *Mor-2^a*, *Mpi-1^a*, *Pgd-1^b*, *Pgm-2^b*, *Pk-2^a*, *Pre-2^c*, *Prt-4^b*, *Prt-5^a*, *Sdh-1^a*, *Sod-1^b*, *Tam-1^d*, *Trf^b*. PWK is also homozygous for a 'slow' electrophoretic variant at both the *Pgm-1* and *Np-1* loci and for a 'fast' variant at the *Mod-1* locus.

Holders: PAS; PRA.

16.3.5 Strains derived from *Mus spretus* progenitors

SPE F23. Derived from wild animals trapped in Santa Fe (near Granada, Spain) by Louis Thaler in 1978 then propagated as closed colony in Montpellier by François Bonhomme for a few generations. Strictly inbred at the Institut Pasteur since 1981.
Adh-1^c, *Amy-1^e*, *Car-2^a*, *Es-2^a*, *Es-3^b*, *Es-5^a*, *Es-14^l*, *Gdc-1^b*, *Glo-1^a*, *Got-1^a*, *Got-2^b*, *Gpd-1^a*, *G6pd^a*, *Gpi-1^a*, *Hao-2^c*, *Hbb^s*, *Hk-1^a*, *Idh-1^a*, *Idh-2^a*, *Ldh-1^a*, *Mod-*

1^a, *Mod-2^b*, *Mor-1^a*, *Mor-2^a*, *Pgd-1^b*, *Pgm-2^a*, *Sdh-1^a*, *Sod-1^a*.

SPE is also homozygous for unique electrophoretic variants at the *Alb-1*, *Es-1*, *Es-10*, *Es-15*, *Np-1*, and *Pgm-1* loci.

Holder: PAS.

16.4 Mouse stocks not completely inbred (i.e. with less than 20 generations of unrelaxed brother × sister matings) which can be of interest with regards to their genetic make-up

16.4.1 Strains derived from *Mus m. domesticus* progenitors

PAC F18. Established from mice wild-trapped in a pasta factory in Philadelphia, PA. About 20 lines were inbred by James Conner for studies in animal behaviour who kept them with him after a move to Nebraska. Three lines have been shipped to Verne Chapman at Roswell Park Memorial Institute in 1974. PAC carries a novel allelic haplotype for β-glucuronidase.

Holder: ROS.

WLC F10. Established by Murray B. Gardner from wild breeders trapped near Lake Casitas (Southern California USA) and propagated by strict brother × sister matings since 1983. Animals of this inbred strain have been found negative when assayed with MTV DNA probes. The mammary tumor incidence is null. Mammary tissue can be transplanted.

Holder: Gardner.

16.4.2 Strains derived from *Mus m. molossinus* progenitors

KOR F7. From wild breeders trapped in Koriyama-shi, Fukushima Japan in 1983 by Yoshibumi Matsushima. Animals are:
Akp-1^b, *Car-2^a*, *Es-1^a*, *Es-2^c*, *Es-3^b*, *Es-10^c*, *Gpd-1^c*, *Gpi-1^a*, *Hbb^p*, *Idh-1^b*, *Ldr-1^a*, *Mod-1^a*, *Msp-3^c*, *Mtp-1^s*, *Mtp-2^o*, *Mtp-3^o*, *Mtp-4^o*, *Mtp-5^o*, *Mtp-6^o*, *Mtp-7^o*, *Mtp-8^a*, *Mtp-9^o*, *Mtp-10^{sf}*, *Mup-1^c*, *Pep-3^b*, *Pgm-1^b*, *Trf-1^b*.

Holder: Matsushima.

16.4.3 Strains derived from *M. m. musculus* progenitors

MAI F18. Derived from wild animals trapped at Illmitz (Austria) by François Bonhomme in 1981. First propagated as a closed colony, then brother × sister mated at the Institut Pasteur. Animals of this strain are homozygous for a recessive gene producing a yellow coat colour somewhat similar to *b/b*.

Acp-1^a, Alb-1^c, Amy-1^a, Car-2^a, Es-1^a, Es-14^r, Gdc-1^b, Glo-1^a, Got-1^a, Gpi-1^a, Idh-1^b, Idh-2^a, Ldh-1^a, Ldh-2^r, Mod-1^a, Mod-2^b, Mor-1^a, Mor-2^a, Pgd-1^b, Pk-1^a, Sdh-1^a, Sod-1^b.

Holder: PAS.

MBT F15. From wild animals trapped near General Toshevo in Bulgaria in 1980 by François Bonhomme then brother × sister mated at the Institut Pasteur.

Acp-1^a, Ak-1^a, Alb-1^c, Amy-1^a, Car-2^a, Es-1^a, Es-14^r, Gdc-1^b, Glo-1^a, Got-1^a, Got-2^b, Gpd-1^c, Gpi-1^a, Idh-1^b, Idh-2^a, Ldh-1^a, Ldh-2^r, Mod-2^b, Mor-1^a, Mor-2^a, Pgd-1^b, Sdh-1^a, Sod-1^b.

Holder: PAS.

16.4.4 Strains derived from *Mus caroli*

KAR F20. This strain was established from breeders trapped in Thailand by Joseph T. Marshall and sent to the Roswell Park Memorial Institute in 1975.

16.5 Strains (or stocks) carrying (or segregating for) alleles derived from the feral state

Since feral specimens represent a very important source of polymorphism, new alleles at specific loci have been found in the wild and several laboratories have derived *congenic strains or stocks* segregating for a given locus (or haplotype) of feral origin. For details of these polymorphic loci with alleles of feral origin readers must refer to specialized chapters (like the list of named mutant genes compiled by Margaret Green, Chapter 2, or to Jan Klein's contribution, Chapter 19).

In the same way many inbred strains have been established by continuous brother × sister matings from the initial cross of a single wild progenitor (usually male) to 'laboratory' counterparts with concomitant selection for a given locus (or haplotype). These stocks thus represent new inbred strains with variable and random contribution of two unrelated genomes. Although the genotypic constitution of the original breeders is generally not known these inbred strains represent a possible source of genetic polymorphism. We list here a few of these strains:

The 'Peru-Coppock' stocks
Original breeders of this group of strains were trapped in 1976 near Nana (22 km east of Lima) Peru by John Coppock. Three stocks are maintained in laboratories.

LH stock. F19. In F2 the wild ancestors of this strain were crossed to C57BL/Go, and in F12 to a hybrid (C3H/HeHx101/H). Animals are now homozygous *Ldh-2^b/Ldh-2^b*.

HK stock. F19. In F8 the wild ancestors of this strain were crossed to a Laboratory stock homozygous for *Df^{slk}/Df^{slk}* as a way of selecting for the *Hk-1^b* allele. The stock is now homozygous for *Hk-1^b* and segregates for *A*, and *a*.

SDH stock. F22. In F12 the wild ancestors of this strain were crossed to C3H/HeH. The stock is now homozygous for *Sdh-1^b* and segregates for *A^w*, and *A*.

Holder: HAR.

PUC/Fre F26. The original breeders of this inbred line were obtained from the Peru-Coppock stock of Dr Josephine Peters in 1977 prior to the derivation of *LH, HK,* and *SDH* stocks.

Apk^a, Akp-1^b, Amy-1^a, Car-1^b, Car-2^a, Es-1^b, Es-2^b, Es-3^b, Es-5^a, Es-6^b, Es-7^d, Es-8^b, Es-9^e, Es-10^b, Es-11^b, Es-13^a, Es-14^r, Es-16^c, Es-17^a, Es-18^a, Es-22^e, Es-23^c, Es-25^b, Es-26^a, Es-27^a, Got-1^a, Got-2^a, Gpd-1^a, Ldr-1^a, Mod-1^b, Mpi-1^b, Prt-4^a, Prt-5^b, Tam-1^a.

Holder: FRE (von Deimling).

WAZ F27. This strain is a recombinant between two unrelated genomes. A male wild-trapped in 1976 in the Amsterdam zoological garden was mated to females of the BTBR/Nev (*T tf/+ tf*) inbred strain. After selection for the 'wild' centromeric region of the chromosome 17 by Karen Artzt the animals have been given to Jean-Louis Guénet and brother × sister mated since this time. The original wild male breeder was free of t haplotype.

Holder: PAS.

IS/CamRk F25+50. Original breeders (identified as *M. m. praetextus*) were trapped in an Israeli port, then sent to Margaret Wallace at the University of Cambridge in 1961 where they were crossed with a strain of *M. m. domesticus*. They have been brother × sister mated since this time. Introduced in the Jackson Laboratory by Thomas H. Roderick and Eva M. Eicher at F25 (51).

Alb-1^a, Amy-1^a, Akp-1^b, Akp-2^b, Ahd-1^a, Apk^a, Apl^b, As-1^b, C3^c, Car-1$^{a \, or \, b}$, Car-2^a, Es-1^b, Es-2^a, Es-3^c, Es-10^a, Es-13^a, Fv-1^n, Gdc-1^b, Glo-1^a, Got-1^a, Got-2^a, Gpd-1^b, Gpi-1^a, Gpt-1^a, Hba^b, Hbb^d, Hrt-1^b, Idh-1^b, Ldh-1^a, Ldr-1^b, Len-2^a, Mod-1^a, Mod-2^b, Mup-1^a, Neu-1^b, Np-1^a, Pan-1^b, Pan-2^b, Pep-3^b, Pgk-1^b, Pgm-1^b, Pgm-2^a, Pgm-3^c, Pre-1^a, Psp^b, +rd, Rnr^s, Sas-1^o, Sdh-1^a, Tam-1^a, Trf^b.

Holder: JAX(Rk).

16.6 Chromosomal variants of feral origin

Many inbred stocks carrying Robertsonian chromosomal rearrangements have been established by brother × sister mating starting from an initial cross of a single wild male to several 'laboratory females'. These stocks sometimes represent new inbred strains with variable and random contribution of two unrelated genomes. Although the genotypic constitution of the original breeders is generally not known these inbred strains represent a possible source of genetic polymorphism.

In the same way other stocks have been repeatedly backcrossed to inbred laboratory strains so that only the regions close to the centromeres of the Robertsonian chromosomes can be expected to be of feral origin.

16.7 Miscellaneous stocks of wild rodents of interest in mouse genetics

Several investigators have bred under laboratory conditions various stocks derived from the feral state and belonging either to the *Mus* genus or to a closely related genus (*Coelomys*, *Nannomys*, *Pyromys*). A listing of these stocks has been published in 1986 by Dr Michael Potter (43).

Acknowledgements. The Authors wish to express gratitude to all of their colleagues who kindly sent information about their stocks. They are very grateful to Josette Catalan, Isabelle Fleurance, and Dominique Simon-Chazottes who helped in the preparation of the manuscript and files.

References

1. Adolph, S., and J. Klein. 1981. Robertsonian variation in *Mus musculus* from central Europe, Spain and Scotland. J. Hered. 72:219–221.
2. Berry, R.J., and J. Peters. 1981. Allozymic variation in house mouse populations. *In* M.H. Smith and J. Joule, eds., Mammalian population genetics, 242–253. University of Georgia Press, Athens, Georgia.
3. Bishop, C.E., P. Boursot, B. Baron, F. Bonhomme, and D. Hatat. 1985. Most classical *Mus musculus domesticus* laboratory mouse strains carry a *Mus musculus musculus* Y chromosome. Nature 325:70–72.
4. Bonhomme, F. 1986. Evolutionary relationships in the Genus *Mus*. *In* M. Potter, J. Nadeau, and M.P. Cancro, eds., The wild mouse in immunology, 19–34. Springer Verlag, New York.
5. Bonhomme, F., J. Britton-Davidian, L. Thaler, and C. Triantaphyllidis. 1978. Sur l'existence en Europe de quatre groupes de souris (genre *Mus* L.) du rang espèce et

semi-espèce, démontrée par la génétique biochimique. C.R. Acad. Sci. Paris 287:631–633.
6. Bonhomme, F., S. Martin, and L. Thaler. 1978. Hybridization between *M. musculus domesticus* L. and *Mus spretus* Lataste under laboratory conditions. Experientia 34 (9):1140–1141.
7. Bonhomme, F., F. Benmehdi, J. Britton-Davidian, and S. Martin. 1979. Analyse génétique de croisements interspécifiques *Mus musculus* L. × *Mus spretus* Lataste: liaison de *Adh-1* avec *Amy-1* sur le chromosome 3 et de *Es-14* avec *Mod-1* sur le chromosome 9. C.R. Acad. Sci. Paris 289:545–548.
8. Bonhomme, F., J.L. Guénet, and J. Catalan. 1982. Présence d'un facteur de stérilité mâle, *Hst-2*, ségrégeant dans les croisements interspécifiques *M. musculus* L. × *M. spretus* Lataste et lié à *Mod-1* et *Mpi* sur le chromosome 9. C.R. Acad. Sci. Paris 294:691–693.
9. Bonhomme, F., J. Catalan, P. Orsini, S. Guerassimov, and L. Thaler. 1983. Le complexe d'espèces du genre Mus en Europe Centrale. I. Génétique. Z. Säugetierkunde 48:78–86.
10. Bonhomme, F., J. Catalan, J. Britton-Davidian, V.M. Chapman, K. Moriwaki, E. Nevo, and L. Thaler. 1984. Biochemical diversity and evolution in the genus Mus. Biochem. Genet. 22:275–303.
11. Bonhomme, F., D. Iskandar, L. Thaler, and F. Petter. 1985. Electromorphs and phylogeny in muroïd rodents. *In* W.P. Luckett and J.L. Hartenberger, eds., Evolutionary relationships among rodents: A multidisciplinary analysis, 671–684. Plenum, New York.
12. Bonhomme, F., J.L. Guénet, B. Dod, K. Moriwaki, and G. Bulfield. 1987. The polyphyletic origin of laboratory inbred mice and their rate of evolution. Biol. J. Linn. Soc. Lond. 30:51–58.
13. Boursot, P., F. Bonhomme, J. Britton-Davidian, J. Catalan, H. Yonekawa, P. Orsini, S. Gerasimov, and L. Thaler. 1984. Introgression différentielle des génomes nucléaires et mitochondriaux chez deux semi-espèces de souris. C.R. Acad. Sci. Paris 299:365–370.
14. Boursot, P., T. Jacquart, F. Bonhomme, J. Britton-Davidian, and L. Thaler. 1985. Différenciation géographique du génome mitochondrial chez *Mus spretus* Lataste. C.R. Acad. Sci. Paris 301:161–166.
15. Britton-Davidian, J., and L. Thaler. 1978. Evidence for the presence of two sympatric species of mice (Genus *Mus* L.) in southern France based on biochemical genetics. Biochem. Genet. 16:213–225.
16. Britton-Davidian, J., F. Bonhomme, H. Croset, E. Capanna, and L. Thaler. 1980. Variabilité génétique chez les populations de souris (genre *Mus* L.) à nombre chromosomique réduit. C.R. Acad. Sci. Paris 290:195–198.
17. Capanna, E., M.V. Civitelli, M. Cristaldi, and G. Noack. 1977. Chromosomal rearrangement, reproductive isolation and speciation in mammals. The case of *Mus musculus*. Boll. Zool. 44:213–246.
18. Capanna, E., and E. Riscassi. 1978. Robertsonian karyotype variability in natural *Mus musculus* populations in

the Lombardy area of the Po Valley. Boll. Zool. 45:63–71.

19. Capanna, E. 1982. Robertsonian numerical variation in animal speciation: *Mus musculus*, an emblematic model. *In* Mechanisms of speciation, 155–177. A.P. Liss, New York.

20. Corbet, G.B. 1984. The mammals of the Palearctic Region: a taxonomic review. Suppl. Brit. Mus. (Nat. Hist.), London, 45 pp.

21. Ellerman, J.R. 1941. The families and genera of living rodents, Vol. II. British. Mus. (Nat. Hist.), London.

22. Ferris, S.D., R.D. Sage, and A.C. Wilson. 1982. Evidence from mtD.N.A. sequences that common laboratory strains of inbred mice are descended from a single female. Nature 295:163–165.

23. Ferris, S.D., R.D. Sage, C.M. Huang, J.T. Nielsen, U. Ritte, and A.C. Wilson. 1983. Flow of mitochondrial DNA across a species boundary. Proc. Natl. Acad. Sci. USA 80:2290–2294.

24. Ferris, S.D., R.D. Sage, E.M. Prager, U. Ritte, and A.C. Wilson 1983. Mitochondrial DNA evolution in mice. Genetics 105:681–721.

25. Fort, Ph., F. Bonhomme, P. Darlu, M. Piechaczyk, Ph. Jeanteur, and L. Thaler. 1984. Clonal divergence of mitochondrial DNA versus populational evolution of nuclear genome. Evol. Theor. 7 (2):81–90.

26. Gropp, A., H. Winking, L. Zech, and H.J. Müller. 1972. Robertsonian chromosomal variation and identification of metacentric chromosomes in feral mice. Chromosoma 39:265–288.

27. Gropp, A., H. Winking, C. Redi, E. Capanna, J. Britton-Davidian, and G. Noack. 1982. Robertsonian karyotype variation in wild house mice from Rhaeto-Lombardia. Cytogenet. Cell Genet. 34:67–77.

28. Guénet, J.L. 1986. The contribution of wild derived mouse inbred strains to gene mapping methodology. *In* M. Potter, J.H. Nadeau, and M.P. Cancro, eds., The wild mouse in immunology, 109–113. Springer Verlag, New York.

29. Hunt, W.G., and R.K. Selander. 1973. Biochemical genetics of hybridisation in European house mice. Heredity 31:11–33.

30. Klein, J., D. Gotze, J.H. Nadeau, and E.K. Wakeland. 1981. Population immunogenetics of murine H-2 and t systems. Symp. Zool. Soc. Lond. 47:439–453.

31. Klein, J., M. Golubic, O. Budimir, R. Schöpfer, M. Kasahara, and F. Figueroa. 1986. On the origin of t chromosomes. *In* M. Potter, J.H. Nadeau, and M.P. Cancro, eds., The wild mouse in immunology, 239–246. Springer Verlag, New York.

32. Marshall, J.T. 1981. Taxonomy. *In* L. Foster, D. Small, and G. Fox, eds., The mouse in biomedical research, 17–26. Academic Press, New York.

33. Marshall, J.T., and R.D. Sage. 1981. Taxonomy of the house mouse. Symp. Zool. Soc. Lond. 47:15–25.

34. Marshall, J.T. 1986. Systematics of the Genus *Mus*. *In* M. Potter, J.H. Nadeau, and M.P. Cancro, eds., The wild

mouse in immunology, 12–18. Springer Verlag, New York.

35. Minezawa, M., K. Moriwaki, and K. Kondo. 1981. Geographical survey of protein variation in wild populations of Japanese house mouse, *Mus musculus molossinus*. Jap. J. Genet. 56:27.

36. Miyashita, N., K. Moriwaki, M. Minezawa, Z. Yu, D. Lu, S. Migita, H. Yonekawa, and F. Bonhomme. 1985. Allelic constitution of hemoglobin beta chain in wild populations of the mouse, *Mus musculus*. Biochem. Genet. 23:975–986.

37. Moriwaki, K., H. Yonekawa, O. Gotoh, M. Minezawa, H. Winking, and A. Gropp. 1984. Implications of the genetic divergence between European wild mice with Robertsonian translocations from the viewpoint of mitochondrial DNA. Genet. Res. 43:277–287.

38. Moriwaki, K., N. Miyashita, H. Suzuki, Y. Kurihara, and H. Yonekawa. 1986. Genetic features of major geographical isolates of *Mus musculus*. *In* M. Potter, J.H. Nadeau, and M.P. Cancro, eds., The wild mouse in immunology, 55–61. Springer Verlag, New York.

39. Morse, H.C., ed., 1978. Origin of inbred mice. Academic Press, New York.

40. Nadeau, J.H., E.K. Wakeland, D. Götze, and J. Klein. 1981. The population genetics of the H-2 polymorphism in European and North African populations of the house mouse (*Mus musculus* L.). Genet. Res. 37:17–32.

41. Nash, H.R., P.C. Brooker, and S.M. Davis. 1983. The Robertsonian translocation house mouse populations of North East Scotland: a study of their origin and evolution. Heredity 50:303–310.

42. Petrov, B., and A. Ruzic. 1983. Preliminary report of the taxonomical status of the members of the genus *Mus* in Yugoslavia with description of a new subspecies *M. hortulanus macedonicus* (*sub. nova*). Proc. Drugi Simpozyum Osauni, 175–178, SP Serbie Belgrad.

43. Potter, M., J.H. Nadeau, and M.P. Cancro, eds. 1986. Current topics in microbiology and immunology, Vol. 127. The wild mouse in immunology. Springer Verlag, New York, 395 p.

44. Robert, B., P. Barton, A. Minty, P. Daubas, A. Weydert, F. Bonhomme, J. Catalan, D. Chazottes, J.L. Guénet, and M. Buckingham. 1985. Investigation of genetic linkage between myosin and actin genes using an interspecific mouse back-cross. Nature 314:181–183.

45. Sage, R.D. 1981. Wild mice. *In* L. Foster, D. Small, and G. Fox, eds., The mouse in biomedical research, Vol. 4 (1), 39–90. Academic Press, New York.

46. Saïd, K., T. Jacquart, C. Mongelard, H. Sonjaya, A.N. Helal, and J. Britton-Davidian. 1985. Robertsonian house mouse populations in Tunisia: a caryological and biochemical study. Genetica 68:151–156.

47. Schwarz, E., and H.K. Schwarz. 1943. The wild and commensal stocks of the house mouse, *Mus musculus*. J. Mammal. 24:59–72.

48. Selander, R.K., W.G. Hunt, and S.Y. Yang. 1969. Protein polymorphism and genetic heterozygosity in two

European subspecies of the house mouse. Evolution 23:379–390.

49. Thaler, L., F. Bonhomme, and J. Britton-Davidian. 1981. Processes of speciation and semi-speciation in the house mouse. Symp. Zool. Soc. Lond. 47:27–41.

50. Vanlerberghe, F., B. Dod, P. Boursot, M. Bellis, and F. Bonhomme. 1986. Absence of Y chromosome introgression across the hybrid zone between *Mus musculus domesticus* and *Mus musculus musculus*. Genet. Res., Camb. 48 191–197.

51. Wallace, M.E. 1970. Mouse News Lett. 42:20–22.

52. Wallace, M.E. 1971. An unprecedented number of mutants in a colony of wild mice. Environ. Pollution 1:175–184.

53. West, J.D., W.L. Frels, V.E. Papaioannou, J. Karr, and V.M. Chapman. 1977. Development of interspecific hybrids of *Mus*. J. Emb. Exp. Morphol. 41:233–243.

54. Winking, H. 1986. Some aspects of Robertsonian karyotype variation in European wild mice. *In* M. Potter, J.H. Nadeau, and M.P. Cancro, eds., The wild mouse in immunology, 68–74. Springer Verlag, New York.

55. Yonekawa, H., K. Moriwaki, O. Gotoh, I. Hayashi, Watanabe, N. Miyashita, M.L. Petras, and Y. Tagashira. 1981. Evolutionary relationships among five subspecies of *Mus musculus* based on restriction enzyme cleavage patterns of mitochondrial DNA. Genetics 98:801–816.

56. Yonekawa, H., O. Gotoh, Y. Tagashira, Y. Matsushima, L.I. Shi, W.S. Cho, N. Miyashita, and K. Moriwaki. 1986. A hybrid origin of Japanese mice. *In* M. Potter, J.H. Nadeau, and M.P. Cancro, eds., The wild mouse in immunology, 62–67. Springer Verlag, New York.

17 STRAIN DISTRIBUTION OF POLYMORPHIC VARIANTS

THOMAS H. RODERICK and JOHN N. GUIDI

The computerized genetic database called MATRIX at The Jackson Laboratory contains information on 426 polymorphic loci in 569 inbred strains and substrains. Table 17.1 gives the distribution of a selected subset of this data base, that is, alleles at 338 loci in 246 strains. Arbitrarily we have included only those strains typed for 20 or more loci and only those loci for which 15 or more strains have been typed. Strain symbols followed by a dash, as for example C57BL/6-, denote 'composite strains'. The rows or 'strain vectors' in the matrix for these composite strains contain data derived from all substrains of that original strain stem. When a composite strain is shown, we have included in this report only those substrains characterized for at least 50 loci. The exception is in the C3H/- collection of substrains where several allelic discrepancies exist probably due to residual heterozygosity before the lines were separated. Reference numbers for the loci in Table 17.1 are given in Table 17.2.

If in any case there is a discrepancy between substrains with respect to alleles at any locus, an asterisk appears in the composite strain vector. A discrepancy could be the result of: (i) incomplete homozygosity at the time of separation of the substrains; (ii) mutation in one or more lines after separation of the lines; or (iii) a mistake in allelic characterization by one laboratory or another. Each cell in the MATRIX is linked with its references so it is possible to determine the sources for each data point.

Because the information is accessible either by strain or locus, it can be used to compare loci with respect to the strain distributions of their alleles, or strains with respect to the loci at which they are alike or different. The first kind of comparison can be made by comparing 'locus vectors'. Identical strain distributions for two loci over several genetically independent strains raise the possibility that loci are identical, even though they may have been described by different methods. A comparison of the 'strain vectors' on the other hand can identify pairs of very different strains for maximizing the number of loci in a linkage test. Strain vectors are also useful in estimating genetic relationships among strains.

We thank M.T. Davisson, D.P. Doolittle, and M. Lennon-Pierce for considerable assistance in searching the literature, and A.L. Hillyard for help in computer programming. This work was supported by funds from the Jackson Laboratory and by a contract with the Howard Hughes Medical Institute.

Table 17.1. Distribution of alleles at 338 loci in 246 inbred strains. The loci and strains are arranged alphabetically according to the following plan on pages numbered as indicated.

	Loci													
Strains	*a–*	*Av-1–*	*D12Nyu3–*	*Es-5–*	*Fv-1–*	*H-4–*	*Igh-C–*	*Len-1–*	*Ly-11–*	*Lyb-3–*	*Pan-2–*	*rd–*	*Tpre–*	*Xmmv-57–*
101–BIR/A	664	665	666	667	668	669	670	671	672	673	674	675	676	677
BRSUNT/N–C58	678	679	680	681	682	683	684	685	686	687	688	689	690	691
C58/J–FS/Ei	692	693	694	695	696	697	698	699	700	701	702	703	704	705
FVB/NA–MRL/MpJ	706	707	708	709	710	711	712	713	714	715	716	717	718	719
MWT/Le–RIII/SeA	720	721	722	723	724	725	726	727	728	729	730	731	732	733
RIIIS/J–WC/ReJ	734	735	736	737	738	739	740	741	742	743	744	745	746	747
WK/Re–YS/Wf	748	749	750	751	752	753	754	755	756	757	758	759	760	761

The laboratory mouse

```
Locus     ->    a A A A A   A A A A A   A A A A A   A A A A A   A A A A A
(read down)     a b c c     d d d d g   h h h h k   k l m m o   o p p p s
                t p f o     h h h h s   d d r p     p b y y x   x h k o a
                a - -       - - - -     - - - -     - - - - -   - - a 1
                1 1         1 2 3 3     1 2 1 1     2 1 1 2 1   2 1 - s
                            e t                                 1

--------------------------------------------------------------------------
Chromosome      2 1 7 1 4   3 U 3 3 X   1 4 1 3 1   4 5 3 3 1   1 2 1 9 1
(read down)     2           N           2   9                   0   3
--------------------------------------------------------------------------

101/-           1 . . . .   a . a a h   . a . a .   . . . . a   a . . . .
129/-           1 a a b a   a a a a h   d a a a b   b a a a a   a . a b a
129/J           1 a a b a   . . a . h   d a . a b   b a a a 2   . . a b a
129/MA          1 . . . .   a a a a .   . a a . .   . . . . a   a . . . .
129/Rr-         1 . . . a   . . . . .   . . . . .   . . . . a   . . . . .

129/SvJ         1 a a . a   . . . . .   . . . . b   . . a . .   . . a b .
615/-           . . . . .   . . . . .   . . . . a   . . a a .   . . a a .
A2G/-           + . . . a   a a a a .   b b a . .   . . . . b   a . . . .
A/-             a o a a a   a a a a h   b a a a b   a a a a a   a a a b b
A/BrA           a . . . .   a a a a .   b a a . .   . . . . a   a . . . .

A/HeJ           a o a a a   . . . . h   b a . . b   a a a a .   . . a b b
A/J             a o a a a   . . a . h   b a a a b   a . a . 2   . . a b b
A/WySnA         a . . . .   a a a a .   b a a . .   . . . . a   a . . . .
A/WySnJ         . . . . a   . . . . .   . . . . b   . . . . .   . . a b b
ABJ/Le          . . . . .   . . . . .   . . . . b   . . a . .   . . . b .

ABP/LeJ         a . . . a   . . . . .   . . . . b   . . a . .   . . . a .
ACR/A           a o . . a   a a a a .   d b a . .   . . . . a   a . . . .
AEJ/GnRk        3 o . . a   . . . . .   . . . . .   . . a a .   . . a b .
AKR/-           a a a b a   a a a a h   d b a a b   a a a a a   a a a a a
AKR/FuRdA       a . . . .   a a a a .   d b a . .   . . . . a   a . . . .

AKR/J           a a a b a   . . . . h   d b . . b   a a a a 2   . a a a a
AKR/LwN         a . . . .   . . . . .   d . . . .   . . a . .   . . a . .
AL/N            a . . . .   . . . . .   b . . . .   . . . . .   . . . . .
ATEB/Le         . . . . .   . . . . .   . . . . a   . . a . .   . . . b .
AU/SsJ          a o a a a   . . . . h   d b . . a   a a a a .   . . a a a

BALB/c-         + a a b a   a a a a h   b b a a b   a a a a a   a a a b a
BALB/cAnN       + . . . .   . . . . .   b . . . .   . . . . .   . . . . .
BALB/cByA       + . a . .   a a a a .   b b a . .   . . . . a   a . . . .
BALB/cByJ       + . . a .   . . . . .   4 . . b .   . a a . .   x . a b .
BALB/cCdA       . . . . .   a a a a .   . b a . .   . . . . a   a . . . .

BALB/cCrglA     + . . . .   a a a a .   b b a . .   . . . . a   a . . . .
BALB/cGnWtBm    + . . . .   . . . . .   . . . . b   . a a . .   . . a . .
BALB/cGr-       + . . a .   . . . . .   . . . . b   . a a . a   . . a . .
BALB/cGrRk      + . . . .   . . . . .   . . . . b   . a a . .   . . a . .
BALB/cHeA       + . . . .   a a a a .   b b a . .   . . . . a   a . . . .

BALB/cJ         + a . b a   . . . . h   b b . . b   a a a a 2   . . a b a
BDP/J           a a . . a   . . . . h   b a . . b   b a a a 2   . . a a a
BFM2/A          . . . . .   . . . . .   . . . . .   . . . . .   . . . . .
BIMA/A          . . . . .   . . . . .   . . . . .   . . . . .   . . . . .
BIR/A           . . . . .   . . . . .   . . . . .   . . . . .   . . . . .

        1 = A*w    2 = a?    3 = a*e    4 = b?
```

```
Locus    ->   A A b B B    B B B B c    C C C C C    c C C C C    C d D D D
(read down)   v v 2 c      g g g v      3 4 6 a a    d e e f o    v 7 1 1
              - -  m d     l l l -          r r      m - - h h    - R 2 2
              1 2    -     - - - 1          - -      1 2 e        1 p N N
                     1     e s t            1 2                     2 y y
                                                                   - u u
                                                                   s 1 2
--------------------------------------------------------------------------------
Chromosome    U U 4 2 5    9 9 9 8 7    1 1 1 3 3    3 2 1 2 7    5 9 7 1 1
(read down)   N N                       7 7 5    7                        2 2
--------------------------------------------------------------------------------

101/-          . . . . a    b d d . +    . . . . .    . . . . .    . . . . .
129/-          . . + a b    b d d b 1    b . . a a    + . b . h    b + d b a
129/J          . . + a b    b d d b 1    b . . a a    + . a . h    b + d . .
129/MA         . . + . .    . . . . 1    . . . . .    + . a . .    . + . . .
129/Rr-        . . . . .    . . . . 2    . . . a .    . . . . .    . . . . .

129/SvJ        . . . . .    . . . b +    b . . . a    . . b . .    b + . b a
615/-          . . . . .    . . . . .    . . . a a    . . b . .    . . . . .
A2G/-          . . b . .    . . . . c    . h . . .    + . a . .    . + . . .
A/-            ? ? b a a    a h d b c    b h a a b    3 . a f l    a + b a a
A/BrA          . . b . .    . . . . c    . h . . .    3 . a . .    . + . . .

A/HeJ          . . b . .    a h d b c    b h . a b    3 . a . l    a + . . .
A/J            ? ? b . a    a h d b c    b h a a b    3 . a f .    a + b a a
A/WySnA        . . b . .    . . . . c    . h . . .    . . a . .    . + . . .
A/WySnJ        . . . a .    . . . b .    . . . . b    . . a . .    a + . . .
ABJ/Le         . . . . .    . . . . .    . . . . .    . . b . .    . . . . .

ABP/LeJ        ? ? b . .    . . . . +    b . . a a    . . . . .    . . . . .
ACR/A          . . + . .    . . . . c    . h . . .    + . b . .    . + . . .
AEJ/GnRk       . . + . .    . . . . +    c . a b .    . . . . .    . + . . .
AKR/-          . . + a .    b d d b c    b l b a a    + . b f l    b + b a a
AKR/FuRdA      . . + . .    . . . . c    . l . . .    + . b . .    . + . . .

AKR/J          . . + a .    b d d b c    b l . a a    + . b f l    b + b a a
AKR/LwN        . . + . .    . . . b c    . . . . a    . . b . .    b + . . .
AL/N           . . b . .    a h d . c    . . . . b    . . a . .    . . . . .
ATEB/Le        . . . . .    . . . . .    . . . . a    . . . . .    . . . . .
AU/SsJ         . . + . .    b d d b +    b . . a b    + . b . .    b + . . .

BALB/c-        s s b a b    a h d b c    b h a a b    3 b a f .    a + b a a
BALB/cAnN      s s b . .    . . . . c    b . . . b    . . a . .    . . . . .
BALB/cByA      . . b . .    . . . . c    . h . . .    . . a . .    . + . . .
BALB/cByJ      s s b a .    a h d . c    b . . a b    . . a . .    . + b . .
BALB/cCdA      . . . . .    . . . . c    . . . . .    . . a . .    . . . . .

BALB/cCrglA    . . b . .    . . . . c    . h . . .    3 . a . .    . + . . .
BALB/cGnWtBm   . . b . .    . . . . c    b . . . b    . . . . .    . + . . .
BALB/cGr-      . . b . .    . . . . c    b . . . b    . . . . .    . + . . .
BALB/cGrRk     . . b . .    . . . . c    b . . . b    . . . . .    . + . . .
BALB/cHeA      . . b . .    . . . . c    . h . . .    3 . a . .    . + . . .

BALB/cJ        . . b . .    a h d b c    b h . a b    3 . a . .    a + . a a
BDP/J          . . b a .    b d d b +    . . . a a    + . a . .    b d . . .
BFM2/A         . . . . .    . . . . .    . . . . .    . . . . .    . . . . .
BIMA/A         . . . . .    . . . . .    . . . . .    . . . . .    . . . . .
BIR/A          . . . . .    . . . . .    . . . . .    . . . . .    . . . . .

        1 = c*ch    2 = c*ch/c    3 = cdm
```

```
Locus    ->  D D D E E    E E E E    E E E E    E E E E    E E E E
(read down)  1 1 l a a    a a a g m   m m m m    m m m m    m p s s
             2 2 b - -    - - - a      v v v v    v v v v    v h - -
             N N - 2 4    5 6 7        - - - -    - - - -    - - 1 2 3
             y y 1                     4 5 6 7 8  9 1 1 1 1  1 1
             u u                                   0 1 2 3 4
             3 4
-----------------------------------------------------------------------
             Chromosome   1 1 1 U 1    U 2 U 8 U   U U U U U   U U 7 1 2   1 1 8 8 1
             (read down)  2 2 1 N 9    N   N   N   N N N N N   N N   6     1         1
-----------------------------------------------------------------------

101/-         . . b . .    . . a . .    . . . . .    . . . . .    . . . . .
129/-         b b b b a    a a a a h    b b b b b    b b b b b    b b b b c
129/J         . . . . .    . . . a h    b b b b b    b b b b b    b b b b c
129/MA        . . . . .    . . . . .    . . . . .    . . . . .    . . b b c
129/Rr-       . . b . .    . . a . .    . . . . .    . . . . .    . . . . .

129/SvJ       b b . . .    . . . . .    b b b b b    b b b b b    b . b b c
615/-         . . . . .    . . . . .    . . . . .    . . . . .    . . b b c
A2G/-         . . . . .    . . . . .    . . . . .    . . . . .    . . b b c
A/-           b b b b a    a a b a h    b b b b b    b b b b b    b d b b c
A/BrA         . . . . .    . . . . .    . . . . .    . . . . .    . . b b c

A/HeJ         . . . . .    . . . a .    b b b b b    b b b b b    b d b b c
A/J           b b . . .    . . . a h    b b b b b    b b b b b    b d b b c
A/WySnA       . . . . .    . . . . .    . . . . .    . . . . .    . . b b c
A/WySnJ       . . . b .    . . . . .    b b b b b    b b b b b    b . b b c
ABJ/Le        . . . . .    . . . . .    . . . . .    . . . . .    . . b . c

ABP/LeJ       . . . . .    . . . . .    . . . . .    . . . . .    . . a . a
ACR/A         . . . . .    . . . . .    . . . . .    . . . . .    . . b b b
AEJ/GnRk      . . . . .    . . . . .    . . . . .    . . . . .    . . b b c
AKR/-         b b b b a    . a b a h    b b b b b    b b a * a    * b b b c
AKR/FuRdA     . . . . .    . . . . .    . . . . .    . . . . .    . . b b c

AKR/J         b b . . .    . . . a h    b b b b b    b b a b a    a b b b c
AKR/LwN       . . . . .    . . . . .    b b b b b    b b a a a    b . b b c
AL/N          . . . . .    . . . . .    . . . . .    . . . . .    . . b . c
ATEB/Le       . . . . .    . . . . .    . . . . .    . . . . .    . . b . a
AU/SsJ        . . . . .    . . . . .    b b b b b    b b b b b    b b b b c

BALB/c-       b b b b a    . b b a l    b b b b b    b b b b b    b b b b a
BALB/cAnN     . . . . .    . . . . .    . . . . .    . . . . .    . . b . a
BALB/cByA     . . . . .    . . . . .    . . . . .    . . . . .    . . b b a
BALB/cByJ     . . . . .    . . . . .    . . . . .    . . . . .    . b b b a
BALB/cCdA     . . . . .    . . . . .    . . . . .    . . . . .    . . b b a

BALB/cCrglA   . . . . .    . . . . .    . . . . .    . . . . .    . . b b a
BALB/cGnWtBm  . . . . .    . . . . .    . . . . .    . . . . .    . . b b a
BALB/cGr-     . . . . .    . . . . .    . . . . .    . . . . .    . . b b a
BALB/cGrRk    . . . . .    . . . . .    . . . . .    . . . . .    . . b b a
BALB/cHeA     . . . . .    . . . . .    . . . . .    . . . . .    . . b b a

BALB/cJ       b b . . .    . . . a l    b b b b b    b b b b b    b . b b a
BDP/J         . . . b a    . . b a h    b b b b b    b b b b b    b . b b a
BFM2/A        . . . . .    . . . . .    . . . . .    . . . . .    . . b b b
BIMA/A        . . . . .    . . . . .    . . . . .    . . . . .    . . a b a
BIR/A         . . . . .    . . . . .    . . . . .    . . . . .    . . a b a
```

Locus → (read down)	Es-5	Es-6	Es-7	Es-8	Es-9	Es-10	Es-11	Es-12	Es-13	Es-14	Es-16	Es-17	Es-18	Es-22	Es-23	Es-24	Es-25	Es-26	Es-27	Esr	Fabpi	Fbp-1	Fbp-2	For	Fos5
Chromosome (read down)	8	8	8	7	8	14	8	UN	9	9	3	9	19	8	8	8	1	32	3	6	3	UN	UN	1	2
101/-	.	.	.	.	.	.	.	.	.	.	.	.	.	.	.	.	.	.	.	.	.	.	.	d	.
129/-	b	a	b	a	a	b	a	.	a	r	a	a	a	d	c	f	b	a	r	a	.	b	.	d	b
129/J	.	a	.	a	a	b	a	.	a	.	.	.	.	d	c	.	b	a	.	.	.	.	.	d	.
129/MA	b	a	b	a	a	b	a	.	a	r	a	a	a	d	c	f	b	a	r	a	.	b	.	.	.
129/Rr-	.	.	.	.	.	.	a	.	.	.	.	.	.	.	.	.	.	.	.	.	.	.	.	.	.
129/SvJ	.	.	.	.	.	b	a	.	a	.	.	a	.	.	.	.	.	.	.	.	.	.	.	.	b
615/-	.	.	.	.	.	a	a	.	.	.	.	.	.	.	.	.	.	.	.	.	.	.	.	.	.
A2G/-	b	a	b	a	a	a	a	.	a	r	a	a	a	d	c	a	a	a	r	a	.	.	.	.	.
A/-	b	a	b	a	a	a	a	.	a	r	a	a	a	d	c	h	a	b	r	a	.	b	.	b	b
A/BrA	b	.	b	a	.	.	.	.	.	.	.	.	.	.	c	.	.	b	.	.	.	b	.	.	.
A/HeJ	b	a	.	a	.	a	.	.	a	.	.	.	.	.	.	.	.	.	.	.	.	.	.	b	.
A/J	.	a	.	.	a	a	a	.	a	.	a	.	a	d	c	a	a	b	.	.	.	.	.	b	b
A/WySnA	b	d	b	a	a	a	a	.	a	r	a	a	a	d	c	h	a	b	r	a	.	.	.	.	.
A/WySnJ	.	.	.	.	.	a	a	.	.	.	.	.	.	.	.	.	.	.	.	.	.	.	.	.	.
ABJ/Le	.	.	.	.	.	b	a	.	.	.	.	.	.	.	.	.	.	.	.	.	.	.	.	.	.
ABP/LeJ	.	.	.	.	.	a	b	.	.	.	.	.	.	.	.	.	.	.	.	.	.	.	.	.	.
ACR/A	b	a	b	a	a	a	a	.	a	r	a	a	a	d	c	a	a	c	s	a	.	.	.	.	.
AEJ/GnRk	.	.	.	.	.	a	.	.	b	.	.	.	.	.	.	.	.	.	.	.	.	.	.	.	.
AKR/-	b	a	b	a	a	b	a	a	a	r	a	a	a	d	c	f	a	c	s	a	b	b	.	d	b
AKR/FuRdA	b	a	b	a	a	b	a	.	a	r	a	a	a	d	c	f	a	c	s	a	.	b	.	.	.
AKR/J	.	a	.	a	.	b	.	a	a	.	.	.	.	.	c	.	.	.	.	.	b	.	.	d	b
AKR/LwN	.	a	.	a	.	b	.	.	.	.	.	.	.	.	.	.	a	.	.	.	.	.	.	.	.
AL/N	.	.	.	.	.	a	.	.	.	.	.	.	.	.	.	.	.	.	.	.	.	.	.	.	.
ATEB/Le	.	.	.	.	.	b	a	.	.	.	.	.	.	.	.	.	.	.	.	.	.	.	.	.	.
AU/SsJ	.	a	.	a	.	b	a	.	.	.	.	.	.	.	.	.	.	.	.	.	.	b	a	d	.
BALB/c-	b	a	b	a	a	a	a	.	a	r	a	a	a	d	c	a	a	a	r	a	b	b	.	b	b
BALB/cAnN	b	.	.	.	.	a	.	.	.	.	.	.	.	.	.	.	.	.	.	.	.	.	.	.	.
BALB/cByA	b	a	b	a	a	a	a	.	a	r	a	a	a	d	c	a	a	a	r	a	.	.	.	.	.
BALB/cByJ	b	.	.	.	.	a	a	.	a	.	.	.	.	.	.	.	.	.	.	.	b	.	.	.	.
BALB/cCdA	b	a	b	.	a	a	a	.	a	r	a	a	a	d	c	a	a	a	r	a	.	b	.	.	.
BALB/cCrglA	b	a	b	a	a	a	a	.	a	r	a	a	a	d	c	a	a	a	r	a	.	.	.	.	.
BALB/cGnWtBm	.	.	.	.	.	a	.	.	.	.	.	.	.	.	.	.	.	.	.	.	.	.	.	.	.
BALB/cGr-	.	.	.	.	.	a	a	.	.	.	.	.	.	.	.	.	.	.	.	.	.	.	.	.	.
BALB/cGrRk	.	.	.	.	.	a	.	.	.	.	.	.	.	.	.	.	.	.	.	.	.	.	.	.	.
BALB/cHeA	b	a	b	a	a	a	a	.	a	r	a	a	a	d	c	a	a	a	r	a	.	b	.	.	.
BALB/cJ	b	a	.	a	.	a	a	.	a	.	.	.	.	.	c	.	a	.	.	.	.	.	.	b	b
BDP/J	.	a	.	a	.	a	b	.	a	.	.	.	.	.	.	.	.	.	.	.	.	.	.	b	.
BFM2/A	b	c	b	.	d	a	a	.	a	r	a	a	a	c	c	e	b	a	r	a	.	.	.	.	.
BIMA/A	b	a	b	.	a	a	a	.	a	r	a	a	a	d	c	a	b	a	r	a	.	.	.	.	.
BIR/A	b	a	b	.	a	a	a	.	a	r	a	a	a	d	c	a	b	a	r	a	.	.	.	.	.

The laboratory mouse

```
Locus    ->   F F G G G   G G G G G   G G G G G   G G G G G   G G H H H
(read down)   v v a b d   d d d g k   l l m o o   p p p r u   u v - - -
              - - l p c   c r r c     k o - t t   d i t - s   s - 1 2 3
              1 2 t - -   - - -       - 3 - -     - - - 1 -   - 1
                    1 1   2 1 2       1   1 2     1 1 1       r s

----------------------------------------------------------------------
Chromosome    4 9 4 3 1   9 U U 6 U   1 1 2 1 8   4 7 1 8 5   5 U 7 1 2
(read down)         5       N N N       1 7 9           5       N   7
----------------------------------------------------------------------

101/-         . . ; . .   . . . . .   . . . . .   . a . . .   . a . k .
129/-         n s b a b   b h l a l   a a a a b   a a a a b   * a b b b
129/J         n s b a b   . h l a l   a a . a b   a a a a b   b . . b b
129/MA        . . . a .   . . . . .   . a . a b   a b . a .   . . . b .
129/Rr-       . . . . .   . . . . .   . a . . .   . . a . .   . . . a .

129/SvJ       . . b . .   b . . . .   . a . a b   a a a a .   . a b b b
615/-         . . . . b   . . . . .   . a . a b   b a b . .   b . . . .
A2G/-         b s . b .   . . . . .   . a . a b   c a . b .   . . . a .
A/-           b s b a b   b h h a l   b a b a b   b a a a a   a a b a 2
A/BrA         b . . . .   . . . . .   . a . a b   b a . a .   . . . a .

A/HeJ         . . b . b   b h h a .   b a . a b   b a a a a   a . 3 a 2
A/J           b s b a b   b h h a l   b a . a b   b a a a a   a . . a .
A/WySnA       . s . . .   . . . . .   . a . a b   b a . a .   . . . a .
A/WySnJ       . . b . b   b . . . .   . a b a b   b a a a a   . . b a 2
ABJ/Le        . . . . .   . . . . .   . a . . b   b a a a .   . . . . .

ABP/LeJ       . . b . b   . . . a .   . a . . b   b b a b .   b . . b .
ACR/A         b s . . .   . . . . .   . a . a b   b a . a .   . . . k .
AEJ/GnRk      . . b . b   . . . . .   . a . a b   a b a . .   . . . . .
AKR/-         n s a b b   b h l a l   b a a a b   b a a a .   h a b k 2
AKR/FuRdA     n s . b .   . . . . .   . a . a b   b a . a .   h . . k .

AKR/J         n s a . b   b h l a l   b a a a b   b a a a .   h . . k 2
AKR/LwN       . . . . b   . . . . .   . a . a b   b a a a .   h . . k .
AL/N          . . . . .   . . . . .   . . . . .   b a . . .   a . . a .
ATEB/Le       . . . . .   . . . . .   . a . . b   b b a a .   . . . . .
AU/SsJ        . . b b .   . . . a .   . a . a b   b b a a a   b . . q .

BALB/c-       b s b a c   . h l a .   a a a a b   b a a a a   a b b d c
BALB/cAnN     b s . . .   . . . . .   . . . . .   b a . . .   a . . d .
BALB/cByA     . s . a .   . . . . .   . a . a b   b a . a .   . . . d .
BALB/cByJ     . . b . c   . . . a .   . o . a b   b a a a .   a . b d .
BALB/cCdA     . . . . .   . . . . .   . a . a b   b a . a .   . . . . .

BALB/cCrglA   b s . . .   . . . . .   . a . a b   b a . a .   . . . . .
BALB/cGnWtBm  . . . . .   . . . . .   . a . a b   b a a a .   . . . . .
BALB/cGr-     . . . . .   . . . . .   . a . a b   b a a a .   . . . . .
BALB/cGrRk    . . . . .   . . . . .   . a . a b   b a a a .   . . . . .
BALB/cHeA     b s . a .   . . . . .   . a . a b   b a . a .   . . . d .

BALB/cJ       . . b a c   . h l a .   a a . a b   b a a a a   a . b d c
BDP/J         b . a b b   b . . a .   a a . a b   a a a a b   b . a p 2
BFM2/A        . . . . .   . . . . .   . . . . .   . . . . .   . . . . .
BIMA/A        . . . . .   . . . . .   . . . . .   . . . . .   . . . . .
BIR/A         . . . . .   . . . . .   . . . . .   . . . . .   . . . . .
```

1 = c?   2 = -a,-b   3 = -a,-c

```
Locus    ->   H H H H H   H H H H H   H H H H H   H H I I I   I I I I I
(read down)   - - - - -   - a a b b   b c e m p   r s d f g   g g g g g
              4 7 8 9 1   1 o o a a   b x a       t d h - h   h h h h h
                      2   3 - -   -   b       -   - - - 1 -   - - - - -
                          1 2     4           1   1 1   1     2 3 4 5 A
                                  p                                   r
                                  s                                   s
-------------------------------------------------------------------------------
Chromosome    7 9 1 U U   2 2 3 1 1   7 2 U 7 8   U 1 1 3 1   1 1 1 1 1
(read down)       4 N N       1 7         N       N 0     2   2 2 2 2 2
-------------------------------------------------------------------------------

101/-         . . . . .   . . . . .   . 1 . . .   . . . . b   1 . . . .
129/-         b a b . b   b c a * c   d 1 . a .   b l a . a   a . a . o
129/J         b . . . .   . . . a .   d 1 . a .   b l a . a   a a . a .
129/MA        . . . . .   . c a . .   d . . . .   . a . . .   . . . . .
129/Rr-       . . . . .   b . . . .   . . a . .   . a . a .   . a . . .

129/SvJ       b a b . b   b . . a c   d . . . .   b . a . .   . . . . .
615/-         . . . . .   . . . . .   s . . . .   . a . . .   . . . . .
A2G/-         . . . . .   . a a . .   d o . . .   . a h . .   . . . . .
A/-           2 b 2 a *   a a a g .   d o * a a   . l a l e   d e a e a
A/BrA         . . . . .   . a a . .   d o . . .   . a . e .   . . . . .

A/HeJ         2 b 2 a 2   a . . g .   d o . . .   . . a l e   . . . . a
A/J           . . . . .   . . . g .   d o b a a   . l a l e   d e a e a
A/WySnA       . . . . .   . a a . .   d o . . .   . a . . .   . . . . .
A/WySnJ       2 b 2 a a   a . . . .   d o . . .   . a . . .   . . a . a
ABJ/Le        . . . . .   . . . . .   d . . . .   . a . . .   . . . . .

ABP/LeJ       . . . . .   . . . . .   d . . . .   . a . . .   . . . . .
ACR/A         . . . . .   . a a f .   d o . . .   . b l d .   . . . . .
AEJ/GnRk      . . . . .   . . . a .   s . . . .   . a . . .   . . b . .
AKR/-         a a 2 . 2   a a a f a   d o a a a   . l b h d   d d a a o
AKR/FuRdA     . . . . .   . a a . .   d o . . .   . . b h a   . . . . .

AKR/J         a . . . .   a . . f a   d o . a a   . l b h d   d d a a .
AKR/LwN       . . . . .   . . . . .   d o . . .   . . b . .   . . . . .
AL/N          . . . . .   . . . g .   d o a . .   . . a . d   1 . a . a
ATEB/Le       . . . . .   . . . . .   d . . . .   . . b . .   . . . . .
AU/SsJ        . . . . .   . . . d .   p o . a .   b l b . .   . . a . .

BALB/c-       2 b c b a   2 a a b b   d 1 a a a   b l a l a   a a a a o
BALB/cAnN     . . . . .   . . . b .   d 1 . . .   . . a . a   . . . . .
BALB/cByA     . . . . .   . a a . .   d 1 . . .   . . a . a   . . . . .
BALB/cByJ     . b c . .   . . b . .   d 1 . . a   b l a l a   . . . . .
BALB/cCdA     . . . . .   . a a . .   d . . . .   . . a . .   . . . . .

BALB/cCrglA   . . . . .   . a a . .   d 1 . . .   . . a l a   . . . . .
BALB/cGnWtBm  . . . . .   . . . . .   d . . . .   . . a . .   . . . . .
BALB/cGr-     . . . . .   . . . . .   d . . . .   . . a . .   . . . . .
BALB/cGrRk    . . . . .   . . . . .   d . . . .   . . a . .   . . . . .
BALB/cHeA     . . . . .   . a a . .   d 1 . . .   . . a l a   . . . . .

BALB/cJ       2 b . b a   2 . . b b   d 1 . a .   b l a . a   a a a a .
BDP/J         b a 2 2 a   2 . . h .   d 1 a . .   . b . h .   a . a . .
BFM2/A        . . . . .   . . . . .   . . . . .   . . . . .   . . . . .
BIMA/A        . . . . .   . . . a .   . . . . .   . . . . .   . . . . .
BIR/A         . . . . .   . . . a .   . . . . .   . . . . .   . . . . .
```

1 = -a,-c   2 = -a,-b

```
Locus    ->    I I I I I   I I I I I   I I I I I   I I K K K   L L L L L
(read down)    g g g g g   g g g g g   g g g g g   g g f t t   a a a d d
               h h h h h   h h h h h   h h k k k   l l o h h   m m p h r
               - - - - -   - - - - -   - - - - -   - - - - -   b b - - -
               C D I L N   P S S S S   S S E P T   1 1 1 1 2   - - 1 1 1
                 e n e p   c a a a a   a r f c r       r       1 2
                 x u v         1 2 3 4     5 c 1 p

Chromosome     1 1 1 1 1   1 1 1 1 1   1 1 6 6 6   1 U 9 1 U   1 1 9 7 6
(read down)    2 2 2 2 2   2 2 2 2 2   2 2         6 N   7 N   2
```

Strain	Igh-C	Igh-De	Igh-I	Igh-L	Igh-N	Igh-P	Igh-S1	Igh-S2	Igh-S3	Igh-S4	Igh-S5	Igh-Sr	Igk-Ef1	Igk-Pc	Igk-Trp	Igl-1	Igl-1	Kfo-1	Kth-1	Kth-2	Lamb-1	Lamb-2	Lap-1	Ldh-1	Ldr-1
101/-	.	.	.	.	a	.	.	.	.	.	.	.	.	.	.	.	.	.	.	.	.	.	.	.	.
129/-	.	a	.	a	.	a	o	o	a	.	.	.	b	b	b	.	a	.	.	.	a	a	b	.	a
129/J	.	.	.	a	.	.	.	.	a	.	.	.	b	b	b	.	a	.	.	.	a	a	b	.	a
129/MA	.	.	.	.	.	.	.	.	.	.	.	.	.	.	.	.	.	.	.	.	.	.	.	.	a
129/Rr-	.	.	.	.	.	.	.	.	.	.	.	.	.	.	.	.	.	.	.	.	.	.	.	.	a
129/SvJ	.	.	.	.	.	.	.	.	.	.	.	.	.	.	.	.	.	.	.	.	.	.	.	.	a
615/-	.	.	.	.	.	.	.	.	.	.	.	.	.	.	.	.	.	.	.	.	.	.	.	.	a
A2G/-	.	.	.	.	.	.	.	.	.	.	.	.	.	.	.	.	.	.	.	.	.	.	.	.	a
A/-	e	o	a	.	o	o	a	.	o	o	o	o	b	b	b	.	a	.	.	.	a	b	b	.	a
A/BrA	.	.	.	.	.	.	.	.	.	.	.	.	.	.	.	.	.	.	.	.	.	.	.	.	a
A/HeJ	e	o	a	.	.	o	a	.	.	.	.	o	.	.	b	.	.	.	.	.	.	.	b	.	a
A/J	e	o	.	.	o	.	a	.	o	o	o	o	b	b	b	a	.	.	.	.	a	b	b	.	a
A/WySnA	.	.	.	.	.	.	.	.	.	.	.	.	.	.	.	.	.	.	.	.	.	.	.	.	a
A/WySnJ	.	.	.	.	.	.	.	.	.	.	.	.	.	.	.	.	.	.	.	.	.	.	.	.	a
ABJ/Le	.	.	.	.	.	.	.	.	.	.	.	.	.	.	.	.	.	.	.	.	.	.	.	.	.
ABP/LeJ	.	.	.	.	.	.	.	.	.	.	.	.	.	.	.	.	.	.	.	.	a	b	.	.	.
ACR/A	.	.	.	.	.	.	.	.	.	.	.	.	.	.	.	.	.	.	.	.	.	.	.	.	a
AEJ/GnRk	.	.	.	.	.	.	.	.	.	.	.	.	.	.	.	.	.	.	.	.	.	.	b	a	b
AKR/-	d	o	o	o	o	o	o	o	o	o	a	a	a	a	a	.	.	b	o	a	a	a	b	a	a
AKR/FuRdA	.	.	.	.	.	.	.	.	.	.	.	.	.	.	.	.	.	.	.	.	.	.	.	.	a
AKR/J	d	.	.	.	.	.	.	.	o	o	a	.	a	a	a	.	.	b	o	a	a	a	b	.	a
AKR/LwN	.	.	.	.	.	.	.	.	.	.	.	.	.	.	.	.	.	.	.	.	.	.	.	a	.
AL/N	o	o	a	a	.	o	o	o	o	o	o	.	.	b	b	.	.	.	.	.	.	.	.	.	.
ATEB/Le	.	.	.	.	.	.	.	.	.	.	.	.	.	.	.	.	.	.	.	.	.	.	.	.	.
AU/SsJ	.	.	.	.	.	.	.	.	.	.	.	.	.	.	.	.	.	.	.	.	a	a	b	.	a
BALB/c-	a	a	a	a	o	a	o	o	a	.	o	o	b	b	b	a	a	b	a	a	a	b	b	a	a
BALB/cAnN	.	.	.	.	.	.	.	.	.	.	.	.	.	.	.	.	.	.	.	.	.	.	.	.	.
BALB/cByA	.	.	.	.	.	.	.	.	.	.	.	.	.	.	.	.	.	.	.	.	.	.	.	.	a
BALB/cByJ	.	.	.	.	.	.	.	.	.	.	.	.	.	.	.	.	.	b	a	a	a	b	b	a	a
BALB/cCdA	.	.	.	.	.	.	.	.	.	.	.	.	.	.	.	.	.	.	.	.	.	.	.	.	a
BALB/cCrglA	.	.	.	.	.	.	.	.	.	.	.	.	.	.	.	.	.	.	.	.	.	.	.	.	a
BALB/cGnWtBm	.	.	.	.	.	.	.	.	.	.	.	.	.	.	.	.	.	.	.	.	.	.	.	a	a
BALB/cGr-	.	.	.	.	.	.	.	.	.	.	.	.	.	.	.	.	.	.	.	.	.	.	.	.	a
BALB/cGrRk	.	.	.	.	.	.	.	.	.	.	.	.	.	.	.	.	.	.	.	.	.	.	.	.	a
BALB/cHeA	.	.	.	.	.	.	.	.	.	.	.	.	.	.	.	.	.	.	.	.	.	.	.	.	a
BALB/cJ	a	.	.	a	.	.	.	.	.	.	.	.	b	b	b	a	.	.	.	.	.	.	b	a	a
BDP/J	.	.	.	.	.	.	.	.	.	.	.	.	.	.	b	.	.	.	.	.	a	b	b	.	a
BFM2/A	.	.	.	.	.	.	.	.	.	.	.	.	.	.	.	.	.	.	.	.	.	.	.	.	.
BIMA/A	.	.	.	.	.	.	.	.	.	.	.	.	.	.	.	.	.	.	.	.	.	.	.	.	.
BIR/A	.	.	.	.	.	.	.	.	.	.	.	.	.	.	.	.	.	.	.	.	.	.	.	.	.

```
Locus   ->    L L L L   L L L L L   L L L L L   L L L L L   L L L L L
(read down)   e e f m   p s t t t   t t t t v   v y y y y   y y y y y
              n n o p   t h h h n   w w w w     p - - - -   - - - - -
              - - - -   - - - - -   - - - -     - 1 2 3 5   6 7 8 9 1
              1 2 1 1   1 1 1 2 2   2 3 4 6     1                     0
--------------------------------------------------------------------------
Chromosome    1 U 7 1   U 1 1 U U   1 9 1 U 4   6 1 6 6 1   1 1 U 1 1
(read down)   N   7 N   N 3 N N     2     N     9           5 2 N   9
--------------------------------------------------------------------------

101/-         a b . .   . . . . .   . . . . .   . . . . .   . . a . .
129/-         a b . o   a r a . b   . . a a c   a b b b b   b a a a b
129/J         a b . .   a . a . b   . . a a c   a b b b b   b a a a b
129/MA        a b . .   . . . . .   . . . . .   . . . . .   . . . . .
129/Rr-       . . . .   . . . . .   . . . . .   . . . . .   . . . . .

129/SvJ       . b . .   . . . . .   . . . . .   b b . . .   . . . . .
615/-         . . . .   . . . . .   . . . . .   . . . . .   . . . . .
A2G/-         a b . .   . r . . .   . . . . .   . b b b b   . . . . .
A/-           . . a a   a r o . a   . a b a a   a b b b b   a a b a b
A/BrA         . . . .   . . . . .   . . . . .   . b b b b   . . . . .

A/HeJ         . . . .   a . . . a   . . b a a   a . . . .   . . . . .
A/J           . . . a   a . o . a   . a b a .   a . . b .   . . . a .
A/WySnA       . . . .   . . . . .   . . . . .   . b . . .   . . . . .
A/WySnJ       . . . .   a . . . .   . . . . .   a b b b b   a a b a b
ABJ/Le        . . . .   . . . . .   . . . . .   . . . . .   . . . . .

ABP/LeJ       . . . .   . . . . .   . . . . .   . . . . .   . . . . .
ACR/A         a b . .   . . . . .   . . . . .   . b . . .   . . . . .
AEJ/GnRk      . . . .   . . . . .   . . . . .   . . . . .   . . . . .
AKR/-         a b b . a a r a o a   o a b a a   a b a a b   b a b a b
AKR/FuRdA     a b . .   . . . . .   . . . . .   . b a a b   . . . . .

AKR/J         a b b . . a . a o a   o a b a a   a b a a b   b a b a b
AKR/LwN       . . . .   . . . . .   . . . . a   . b a a .   . . . a .
AL/N          . . . .   . . . . .   . . . . .   . b . . .   . . . . .
ATEB/Le       . . . .   . . . . .   . . . . .   . . . . .   . . . . .
AU/SsJ        a b . .   . . a . .   . . . . .   a . . . .   . . . . .

BALB/c-       b b b a   a s a o b   o a a a a   a b b b b   a a b a b
BALB/cAnN     . . . .   . . . . .   . . . . a   . b b b .   . . . a .
BALB/cByA     b b . .   . . . . .   . . . . .   . b b b b   . . . . .
BALB/cByJ     . b b .   a . a o b   o a a a .   a b b b b   a a b a b
BALB/cCdA     b b . .   . . . . .   . . . . .   . . . . .   . . . . .

BALB/cCrglA   . . . .   . . . . .   . . . . .   . b b b b   . . . . .
BALB/cGnWtBm  . . . .   . . . . .   . . . . .   . . . . .   . . . . .
BALB/cGr-     . . . .   . . . . .   . . . . .   . . . . .   . . . . .
BALB/cGrRk    . . . .   . . . . .   . . . . .   . . . . .   . . . . .
BALB/cHeA     . . . .   . . . . .   . . . . .   . b b b b   . . . . .

BALB/cJ       b b . .   . . . . b   . . a a a   a . b b .   . . . . .
BDP/J         . . . a   . . . . .   . . a a .   a b a b b   b a . a .
BFM2/A        . . . .   . . . . .   . . . . .   . . . . .   . . . . .
BIMA/A        b b . .   . . . . .   . . . . .   . . . . .   . . . . .
BIR/A         . . . .   . . . . .   . . . . .   . . . . .   . . . . .
```

The laboratory mouse

Locus → (read down)	Ly-11	Ly-12	Ly-13	Ly-14	Ly-15	Ly-16	Ly-17	Ly-18	Ly-19	Ly-20	Ly-21	Ly-22	Ly-23	Ly-24	Ly-25	Ly-26	Ly-27	Ly-28	Ly-29	Ly-30	Ly-31	Ly-32	Ly-39	Ly-40	Lyb-2
Chromosome (read down)	2	19	UN	7	7	12	11	11	42	4	7	1	2	2	2	UN	UN	UN	4	UN	4	47	1	UN	4
101/-	.	.	.	.	.	.	.	.	.	.	.	.	.	.	.	.	.	.	.	.	.	.	.	.	.
129/-	b	b	b	a	*	.	a	a	b	b	b	b	b	b	b	b	b	b	b	b	b	b	.	1	b
129/J	.	b	b	a	a	.	a	a	b	.	b	.	.	b	.	b	.	.	.	.	b	.	.	1	b
129/MA	.	.	.	.	.	.	.	.	.	.	.	.	.	.	.	.	.	.	.	.	.	.	.	.	.
129/Rr-	.	.	.	.	.	.	.	.	.	.	.	.	.	.	.	.	.	.	.	.	.	.	.	.	.
129/SvJ	.	b	.	.	b	.	.	.	.	.	.	.	b	.	.	b	.	.	b	.	b	.	.	.	.
615/-	.	.	.	.	.	.	.	.	.	.	.	.	.	.	.	.	.	.	.	.	.	.	.	.	.
A2G/-	.	.	.	.	.	.	.	.	.	.	.	.	.	.	.	.	.	.	.	.	.	.	.	.	.
A/-	a	b	b	b	b	.	a	a	b	b	a	b	a	b	b	b	a	b	a	b	a	b	1	b	b
A/BrA	.	.	.	.	.	.	.	.	.	.	.	.	.	.	.	.	.	.	.	.	.	.	.	.	.
A/HeJ	.	.	.	.	.	.	.	.	.	.	.	.	a	.	.	.	.	.	.	.	a	.	.	.	.
A/J	.	.	.	.	.	.	a	.	.	.	.	.	a	b	.	.	.	.	.	.	.	.	.	.	b
A/WySnA	.	.	.	.	.	.	.	.	.	.	.	.	.	.	.	.	.	.	.	.	.	.	.	.	.
A/WySnJ	.	b	b	b	b	.	a	a	b	.	a	.	a	b	.	b	.	.	.	.	a	.	1	b	.
ABJ/Le	.	.	.	.	.	.	.	.	.	.	.	.	.	.	.	.	.	.	.	.	.	.	.	.	.
ABP/LeJ	.	.	.	.	.	.	.	.	.	.	.	.	.	.	.	.	.	.	.	.	.	.	.	.	.
ACR/A	.	.	.	.	.	.	.	.	.	.	.	.	.	.	.	.	.	.	.	.	.	.	.	.	.
AEJ/GnRk	.	.	.	.	.	.	.	.	.	.	.	.	.	.	.	.	.	.	.	.	.	.	.	.	.
AKR/-	b	b	b	a	b	.	b	a	a	b	b	b	a	b	b	b	b	a	a	b	a	a	1	b	c
AKR/FuRdA	.	.	.	.	.	.	.	.	.	.	.	.	.	.	.	.	.	.	.	.	.	.	.	.	.
AKR/J	.	b	b	a	b	.	2	a	a	.	b	.	.	b	.	b	.	.	.	.	a	.	1	b	3
AKR/LwN	.	.	.	.	.	.	.	.	.	.	.	.	.	.	.	.	.	.	.	.	.	.	.	.	c
AL/N	.	.	.	.	.	.	.	.	.	.	.	.	.	.	.	.	.	.	.	.	.	.	.	.	.
ATEB/Le	.	.	.	.	.	.	.	.	.	.	.	.	.	.	.	.	.	.	.	.	.	.	.	.	.
AU/SsJ	.	.	.	.	.	.	.	.	.	.	.	a	.	.	.	.	.	.	.	.	a	.	.	.	.
BALB/c-	a	b	b	a	a	a	b	a	b	b	b	b	a	a	b	b	a	b	b	b	a	b	1	b	b
BALB/cAnN	.	.	.	.	.	.	.	.	.	.	.	.	.	.	.	.	.	.	.	.	.	.	.	.	b
BALB/cByA	.	.	.	.	.	.	.	.	.	.	.	.	.	.	.	.	.	.	.	.	.	.	.	.	.
BALB/cByJ	.	b	b	a	a	.	b	a	b	.	b	.	.	a	.	b	.	.	.	.	a	.	1	b	.
BALB/cCdA	.	.	.	.	.	.	.	.	.	.	.	.	.	.	.	.	.	.	.	.	.	.	.	.	.
BALB/cCrglA	.	.	.	.	.	.	.	.	.	.	.	.	.	.	.	.	.	.	.	.	.	.	.	.	.
BALB/cGnWtBm	.	.	.	.	.	.	.	.	.	.	.	.	.	.	.	.	.	.	.	.	.	.	.	.	.
BALB/cGr-	.	.	.	.	.	.	.	.	.	.	.	.	.	.	.	.	.	.	.	.	.	.	.	.	.
BALB/cGrRk	.	.	.	.	.	.	.	.	.	.	.	.	.	.	.	.	.	.	.	.	.	.	.	.	.
BALB/cHeA	.	.	.	.	.	.	.	.	.	.	.	.	.	.	.	.	.	.	.	.	.	.	.	.	.
BALB/cJ	.	.	.	.	.	.	.	.	.	.	.	.	.	.	.	.	.	.	.	.	.	.	.	.	.
BDP/J	.	.	.	.	.	.	a	.	a	b	.	a	.	.	.	.	.	b	.	.	b	a	1	.	.
BFM2/A	.	.	.	.	.	.	.	.	.	.	.	.	.	.	.	.	.	.	.	.	.	.	.	.	.
BIMA/A	.	.	.	.	.	.	.	.	.	.	.	.	.	.	.	.	.	.	.	.	.	.	.	.	.
BIR/A	.	.	.	.	.	.	.	.	.	.	.	.	.	.	.	.	.	.	.	.	.	.	.	.	.

1 = -b   2 = b?   3 = (-)

```
Locus    ->    L L L L L   L M M M M   M M M M   N N N N   N O O p P
(read down)    y y y y y   y a e l o   o o p u   c e k k p   p r r a
               b b b b b   b p p s d   d r h i p   a u - - -   - m m n
               - - - - -   - - - -   - - - -   m - 1 2 1   2 - - -
               3 4 5 6 7   8 1 1 1   2 1 1 1   1           1 2   1

---------------------------------------------------------------------
Chromosome     U 4 U 4 1   7 5 1 1 9   7 5 7 9 4   9 1 1 2 1   1 4 4 7 4
(read down)    N   N   2       7                   7 7 4   4
---------------------------------------------------------------------

101/-          . . . . .   . b . . .   b . . . .   . b . . .   . . . . .
129/-          b b . b .   . b a . a   b a b b a   . b o o a   a . . p b
129/J          . b . . .   . . a . a   b a . b a   . b . . a   a . . p b
129/MA         . . . . .   . . . . a   b a . b .   . b . . a   . . . . .
129/Rr-        . . . . .   . b . . .   b . . . .   . b . . .   . . . p .

129/SvJ        . . . . .   . . . a .   . . . b .   . b . . .   . . . + b
615/-          . . . . .   . . . . b   b a . b b   . . . . a   . . . . .
A2G/-          . . . . .   . . . . a   b a . b a   . b . . a   . . . . .
A/-            b b b b b   b b a c a   b a b b a   . b o a a   . b a + .
A/BrA          . . . . .   . . . . a   b a . b a   . b . . a   . . . . .

A/HeJ          . . . . .   . b . . a   b a . b a   . b . o a   . . . + .
A/J            . b . . .   . b . c a   b . b b a   . b o a .   . b a + .
A/WySnA        . . . . .   . . . . a   b a . b a   . b . . a   . . . . .
A/WySnJ        . . . . .   . . . . a   . . . b .   . b . . .   . . . . .
ABJ/Le         . . . . .   . . . . a   . . . b .   . b . . .   . . . . .

ABP/LeJ        . . . . .   . . . . b   . . . b .   . b . . a   . . . p .
ACR/A          . . . . .   . . . . a   b a . b b   . b . . a   . . . . .
AEJ/GnRk       . . . . .   . . . . a   b a . b a   . . . . a   . . . + .
AKR/-          b b b b b   a * b a b   b a b b a   b b . . a   a a a + .
AKR/FuRdA      . . . . .   . . . . b   b a . b a   . b . . a   . . . . .

AKR/J          . b . . .   . h b . b   . a . b a   b b . . a   . a a . .
AKR/LwN        . . . . .   . . . . b   b a . b a   . . . . a   a . . . .
AL/N           . . . . .   . b . . a   . . . . a   . b . . a   . . . . .
ATEB/Le        . . . . .   . . . . a   . . . b .   . b . . .   . . . . b
AU/SsJ         . . . . .   . b a . b   b a . b a   . b . . .   . . . + b

BALB/c-        b b b b b   b b . b a   b a b b a   a b o a a   a b a + b
BALB/cAnN      . . . . .   . . . . a   . . . b a   . . . . .   a . . . .
BALB/cByA      . . . . .   . . . . a   b a . b a   . b . . a   . . . . .
BALB/cByJ      . . . . .   . . . . a   . a . b a   . b . . a   . b a + b
BALB/cCdA      . . . . .   . . . . a   b a . b .   . b . . a   . . . . .

BALB/cCrglA    . . . . .   . . . . a   b a . b a   . b . . a   . . . . .
BALB/cGnWtBm   . . . . .   . . . . a   b a . b a   . b . . a   . . . . .
BALB/cGr-      . . . . .   . . . . a   b a . b a   . b . . a   . . . + .
BALB/cGrRk     . . . . .   . . . . a   b a . b a   . b . . a   . . . + .
BALB/cHeA      . . . . .   . . . . a   b a . b a   . b . . a   . . . . .

BALB/cJ        . b . . .   . b . . a   b a . b a   a b . . a   . . . . b
BDP/J          . . . . .   . b . . b   . a . b b   . b . . .   . . . p .
BFM2/A         . . . . .   . . . . .   . . . . .   . . . . .   . . . . .
BIMA/A         . . . . .   . . . . .   . . . . .   . . . . .   . . . . .
BIR/A          . . . . .   . . . . .   . . . . .   . . . . .   . . . . .
```

```
Locus    ->    P P P P   P P P P   P P P P   P P P P Q   Q Q Q Q Q
(read down)    a c d e e   e e g g   g g g k r   r r r s a   a a a a e
               n a   p p   p p d k k m m m - e   n t t p -   - - - d
               - -   - -   - -       - - - 3 -   - - - 1     2 3 4 5 -
               2 1   2 3   4 7 1 2   1 2 3   2   p 1 2         1

-------------------------------------------------------------------------
Chromosome     U U U 1 1   7 5 4 X 1   5 4 9 9 1   2 U 8 2 1   1 1 1 1 1
(read down)    N N N 0             7             2   N             7   7 7 7 7 7
-------------------------------------------------------------------------

101/-          . . . . .   . . . . .   . . . . .   . . . . .   . . . . .
129/-          b b a . b   . . b b c   a a b a .   a . . b b   a a a a .
129/J          b . a . b   . . b . c   a a b . .   a . . b b   a a . . .
129/MA         . . . . b   . . b b .   a a . . .   . . . . .   . . . . .
129/Rr-        . . . . .   . . . . .   a . . a .   . . . . .   . . . . .

129/SvJ        b . . . b   . . . . c   a a . . .   . . . . .   . . . . .
615/-          . . . . a   . . . . .   1 a . . .   . . . . .   . . . . .
A2G/-          . . . . b   . . b b a   b a . a .   . . . . .   . . . . .
A/-            . a a . b   . . b b a   a a a . b   . a a a a   a a a b a
A/BrA          . . . . b   . . b b a   a a . . .   . . . . .   . . . . .

A/HeJ          . . . . b   . . b b a   a a a . b   . a a a .   . . a b .
A/J            . . a . b   . . . . a   a a a . b   . . . a a   a a a b a
A/WySnA        . . . . b   . . b b a   a a . . .   . . . . .   . . . . .
A/WySnJ        . . . . b   . . . . a   a a . . .   . . . . .   . . a b .
ABJ/Le         . . . . b   . . . . .   b a . . .   . . . . .   . . . . .

ABP/LeJ        . . . . a   . . . . a   b a . . .   . . . b .   . . . . .
ACR/A          . . . . b   . . b b b   a a . . .   . . . . .   . . . . .
AEJ/GnRk       . . . . a   . . . b a   a . c . .   . . . . .   . . . . .
AKR/-          . a . a b   b a b b b   a a a . a   . . . b b   b b b b b
AKR/FuRdA      . . . . b   . . b b b   a a . . .   . . . . .   . . . . .

AKR/J          . a . . b   b . b b b   a a a . a   . . . b .   . . . . b
AKR/LwN        . . . a b   b a b . .   a a . . .   . . . b .   . . . . .
AL/N           . . . . b   . . . . .   b . . . .   . . . . .   . . . . .
ATEB/Le        . . . . b   . . . . .   b a . . .   . . . . .   . . . . .
AU/SsJ         b . . . b   b . b b a   b a a . .   . . . b .   . . . . b

BALB/c-        b a a a a   b a b b a   a a a a a   a a a b b   * * b b b
BALB/cAnN      . . . . a   . . . . a   a . . . .   . . . . .   b b . . .
BALB/cByA      . . . . a   . . b b a   a a . . .   . . . . .   . . . . .
BALB/cByJ      b . . . a   . . . b a   a a m . .   . . . b b   b b b b b
BALB/cCdA      . . . . a   . . b b a   a a . . .   . . . . .   . . . . .

BALB/cCrglA    . . . . a   . . b b a   a a . . .   . . . . .   . . . . .
BALB/cGnWtBm   . . . . a   . . . . a   a a . . .   . . . . .   a a . . .
BALB/cGr-      . . . . a   . . . . a   a a . a .   . . . . .   . . . . .
BALB/cGrRk     . . . . a   . . . . a   a a . . .   . . . . .   . . . . .
BALB/cHeA      . . . . a   . . . b a   a a . . .   . . . . .   . . . . .

BALB/cJ        b a a a a   b a b b a   a a a . a   a a a b b   a a . b b
BDP/J          . b a . c   . . b b a   b a a . .   . . . a a   b b b b a
BFM2/A         . . . . .   . . . . .   . . . . .   . . . . .   . . . . .
BIMA/A         . . . . .   . . . . .   . . . . .   . . . . .   . . . . .
BIR/A          . . . . .   . . . . .   . . . . .   . . . . .   . . . . .
```

1 = ab

```
Locus    ->   r r R R R    S S S S S    S S S S S    S S S T T    T T T T T
(read down)   d d e i n    a a a a d    l o o o p    s v v a c    c g h i l
              s n b 7      a a s h r    p a d d i    p p p m n    r n y n a
              - - s        - - - -      - - -        - - - -      g   - d
              1 1 -        1 1 1 1      1 2 2        1 2 1 2      1     1
                  6                       r

--------------------------------------------------------------------------
Chromosome    5 1 1 1 6    7 7 1 2 U    1 U 1 1 1    U 2 7 7 1    1 1 9 1 1
(read down)     7 4                N    7 N 6 7 2    N       1    3 5   2 7
--------------------------------------------------------------------------

101/-         . . . . .    . . a . .    . . . . .    . . . . s    . . . . .
129/-         + . s a b    b . a a 1    . b a a .    . a . a s    e b b . c
129/J         + . s a b    b . a a 1    . . . . .    . a . a .    e b b . c
129/MA        . . . . .    . . a . .    . . . . .    . a . . s    . . . . .
129/Rr-       + . . . .    . . a . .    . b . . .    . . . . .    . . . . .

129/SvJ       + . . . .    b . . a .    . b a a .    . . . a .    . . . . c
615/-         . . . . .    . . a . .    . . . . .    . . . c .    . . . . .
A2G/-         + + . . .    . . . a .    . b . . .    . a . . s    . . b . a
A/-           + + b a b    a b o a 2    . . a a b    s b c a s    e a b d a
A/BrA         + + . . .    . . . a .    . . . . .    . b . . .    . . b . a

A/HeJ         + . . . .    a . o a .    . . . . .    . . . a .    . . . . a
A/J           + . b a b    a b o a 2    . . . . b    s . . a s    e a . d a
A/WySnA       . . . . .    . . a . .    . . . . .    . b . . .    . . b . a
A/WySnJ       + . . . .    . . a . .    . . a a .    . . . . .    . . b . .
ABJ/Le        . . . . .    . . . . .    . . . . .    . . . . .    . . . . .

ABP/LeJ       + . . a .    b . a . .    . . . . .    . a . . s    . . b . .
ACR/A         + + . . .    . . a . .    . . . . .    . a . . s    . . b . .
AEJ/GnRk      + . . . .    . . a . .    . . . . .    . . . c .    . . . . .
AKR/-         + + s a a    a b a a b    o b a a d    s a b c s    c a a d b
AKR/FuRdA     + + . . .    . . . a .    . . . . .    . a . . .    . . a . b

AKR/J         + . s a a    a b a a b    o . a a d    s a . c s    c a a d b
AKR/LwN       . . . . .    . . . . .    . . a a .    . . . . .    . . a . b
AL/N          . . . . .    . . a . .    . . . . .    . b . . .    . . . . .
ATEB/Le       . . . . .    . . . . .    . . . . .    . . . . .    . . . . .
AU/SsJ        + . s a .    . . o a .    . b . . .    . . . c .    . . . . .

BALB/c-       + + b a b    a b a a *    a b a a d    f b c a s    b c b o c
BALB/cAnN     . . . . .    . . . . .    . . . . .    . . . . .    . . b . c
BALB/cByA     . . . . .    . . . a .    . . . . .    f b . . .    . . b . .
BALB/cByJ     + . . a .    a b . a b    . b . . d    . . . a .    . c b . c
BALB/cCdA     . . . . .    . . . a .    . . . . .    . . . . .    . . . . .

BALB/cCrglA   + + . . .    . . . a .    . . . . .    . b . . .    . . b . c
BALB/cGnWtBm  + . . . .    . . . a .    . . . . .    . . . a .    . . . . .
BALB/cGr-     . . . . .    . . . a .    . b . . .    . . . a .    . . . . .
BALB/cGrRk    + . . . .    . . . a .    . . . . .    . . . a .    . . . . .
BALB/cHeA     + + . . .    . . . a .    . . . . .    . b . . .    . . b . c

BALB/cJ       + . b . b    a . a a a    a . a a .    . b . a .    b . . . c
BDP/J         3 . s a .    b . a a .    a b . . .    . . . . s    . . a . a
BFM2/A        . . . . .    . . . . .    . . . . .    . . . . s    . . . . .
BIMA/A        . . . . .    . . . . .    . . . . .    . . . . s    . . . . .
BIR/A         . . . . .    . . . . .    . . . . .    . . . . s    . . . . .

         1 = a?       2 = b?       3 = rd
```

```
Locus     ->   T T T T U   U U U W X   X X X X X   X X X X X   X X X X X
(read down)    p r s t c   p p p a m   m m m m m   m m m m m   m m m m m
               r f u h p   g g s p m   m m m m m   m m m m m   m m m m m
               e   y       - -   v     v v v v v   v v v v v   v v v v v
                           1 2         - - - - -   - - - - -   - - - - -
                               9       4 4 4 4 4   4 4 4 5 5   5 5 5 5 5
                                       2 3 4 5 6   7 8 9 0 1   2 3 4 5 6
---------------------------------------------------------------------------
Chromosome     1 9 1 1 8   1 1 9 U 1   1 U U U U   U U U 1 U   5 U U 1 U
(read down)    2   2 2     7     N     9 N N N N   N N N 2 N     N N 5 N
---------------------------------------------------------------------------

101/-          . . . . .   . . . . .
129/-          . b . . d   s d a a a   o o o o o   o o o o o   o o a a o
129/J          . b . . d   s d . a a   o o o o o   o o o o o   o o a a o
129/MA         . b . . .   . . b . .
129/Rr-        . . . . .   . . . a .

129/SvJ        . b . . .   s d a a a   o o o o o   o o o o o   o o a a o
615/-          . b . . .   . . . . .
A2G/-          . b . . .   . . a . .
A/-            d b d d .   f s a a a   o o o o o   o o o o o   a o a o o
A/BrA          . b . . .   . . . . .

A/HeJ          . b . . .   . . a a a   o o o o o   o o o o o   a o a o o
A/J            d b d d .   f s . a a   o o o o o   o o o o o   a o a o o
A/WySnA        . b . . .   . . . . .
A/WySnJ        . b . . .   f . . . a   o o o o o   o o o o o   a o a o o
ABJ/Le         . b . . .   . . . . .

ABP/LeJ        . b . . .
ACR/A          . b . . .   . . a . .
AEJ/GnRk       . b . . .   . d . . .
AKR/-          d b d d d   f d a a a   o o o o o   o o o a o   a o a a o
AKR/FuRdA      . b . . .

AKR/J          d b d d d   f d a a a   o o o o o   o o o a o   a o a a o
AKR/LwN        . b . . .
AL/N           . b . . .
ATEB/Le        . b . . .
AU/SsJ         . a . . .   s s a a a   o o o o o   o o o o o   a o a o o

BALB/c-        o b o o d   f s a a a   o o o o o   o o o o o   a o a o o
BALB/cAnN      . b . . .
BALB/cByA      . b . . .
BALB/cByJ      . b . . d   . . . a a   o o o o o   o o o o o   a o a o o
BALB/cCdA      . . . . .

BALB/cCrglA    . b . . .
BALB/cGnWtBm   . b . . .   . . . . a   o o o o o   o o o o o   a o a o o
BALB/cGr-      . b . . .
BALB/cGrRk     . b . . .
BALB/cHeA      . b . . .

BALB/cJ        . b . . .   f s a a a   o o o o o   o o o o o   a o a o o
BDP/J          . b . . .   f d a a o   o o o o o   o o o a o   o o a a o
BFM2/A         . . . . .
BIMA/A         . . . . .   . . a . .
BIR/A          . . . . .   . . a . .
```

```
Locus      ->    X X X X X   X X X X X   X X X
(read down)      m m m m m   m m m m m   m m m
                 m m m m m   m m m m m   m m m
                 v v v v v   v v v v v   v v v
                 - - - - -   - - - - -   - - -
                 5 5 5 6 6   6 6 6 6 6   6 6 7
                 7 8 9 0 2   3 4 5 6 7   8 9 0
---------------------------------------------------------------------
Chromosome       U U U U 4   U U 3 U U   U U U
(read down)      N N N N     N N   N N   N N N
---------------------------------------------------------------------

101/-            . . . . .   . . . .     . . .
129/-            o o o o a   a * o o o   o a a
129/J            o o o o a   a a o o o   o a a
129/MA           . . . . .   . . . . .   . . .
129/Rr-          . . . . .   . . . . .   . . .

129/SvJ          o o o o a   a o o o o   o a a
615/-            . . . . .   . . . . .   . . .
A2G/-            . . . . .   . . . . .   . . .
A/-              o o o a o   a o o o o   o a a
A/BrA            . . . . .   . . . . .   . . .

A/HeJ            o o o a o   a o o o o   o a a
A/J              o o o a o   a o o o o   o a a
A/WySnA          . . . . .   . . . . .   . . .
A/WySnJ          o o o a o   a o o o o   o a a
ABJ/Le           . . . . .   . . . . .   . . .

ABP/LeJ          . . . . .   . . . . .   . . .
ACR/A            . . . . .   . . . . .   . . .
AEJ/GnRk         . . . . .   . . . . .   . . .
AKR/-            o o o o o   a a a o o   o a a
AKR/FuRdA        . . . . .   . . . . .   . . .

AKR/J            o o o o o   a a a o o   o a a
AKR/LwN          . . . . .   . . . . .   . . .
AL/N             . . . . .   . . . . .   . . .
ATEB/Le          . . . . .   . . . . .   . . .
AU/SsJ           o o o a o   a o o o o   o a a

BALB/c-          o o o a o   a o a o o   o a a
BALB/cAnN        . . . . .   . . . . .   . . .
BALB/cByA        . . . . .   . . . . .   . . .
BALB/cByJ        o o o a o   a o a o o   o a a
BALB/cCdA        . . . . .   . . . . .   . . .

BALB/cCrglA      . . . . .   . . . . .   . . .
BALB/cGnWtBm     o o o a o   a o a o o   o a a
BALB/cGr-        . . . . .   . . . . .   . . .
BALB/cGrRk       . . . . .   . . . . .   . . .
BALB/cHeA        . . . . .   . . . . .   . . .

BALB/cJ          o o o a o   a o a o o   o a a
BDP/J            o o o o a   a o a o o   o a a
BFM2/A           . . . . .   . . . . .   . . .
BIMA/A           . . . . .   . . . . .   . . .
BIR/A            . . . . .   . . . . .   . . .
```

```
Locus   ->      a A A A A    A A A A A    A A A A A    A A A A A    A A A A A
(read down)     a b c c      d d d d g    h h h h k    k l m m o    o p p p s
                t p f o      h h h h s    d d r p      p b y y x    x h k o -
                a - -        - - - - -    - - - -      - - - - -    - -   a 1
                1 1          1 2 3 3      1 2 1 1      2 1 1 2 1    2 1   - s
                               e t                                        1
----------------------------------------------------------------------------
Chromosome      2 1 7 1 4    3 U 3 3 X    1 4 1 3 1    4 5 3 3 1    1 2 1 9 1
(read down)       2            N          2   9                        0   3
----------------------------------------------------------------------------

BRSUNT/N        a . . . .    . . . . .    d . . . .    . . . . .    . . . . .
BS              . . . . .    . . . . .    . . . . .    . a . a .    . . . . .
BUB/BnJ         a a . b a    . . . . h    . b . . b    a a a a 1    . . a b a
BXSB/MpJ        . . . . .    . . . . .    . . . b      . . a . .    . . . a .
C3H/-           + o a a a    a a a a h    b b a a b    a a a a a    a a a b a

C3H/An          . . . . .    . . . . .    . . . . .    . . . . .    . . . . .
C3H/Bi          + . . . .    . . . . .    . . . . .    . . . . .    . . . . .
C3H/BiU         . . . . .    . . . . .    . . . . .    . . . . .    . . . . .
C3H/Crgl        . . . . .    . . . . .    . . . . .    . . . . .    . . . b .
C3H/He-         + o a a a    a a a a h    b b a a b    a a a a a    a a a b a

C3H/HeA         + . . . .    a . a a .    b b . a .    . . . . a    a . . . .
C3H/HeAf        . . . . .    . . . . .    . . . . .    . . . . .    . . . . .
C3H/HeDiSnA     + . . . .    a a a a .    b b a . .    . . . . a    a . . . .
C3H/HeJ         + o a a a    . . . . h    b b a . b    a a a a 1    . . a b a
C3H/HeOuJ       . . . . .    . . . . .    . . . b      . . a . .    . . . b .

C3H/HeSnJ       + . a . .    . . . . .    . . . b      . . a . .    . a b .
C3H/St          + . a . .    . . . . .    . . . . .    . . . . .    . . . . .
C3H/StWi        + . . . .    . . . . .    . . . . .    . . . . .    . . . . .
C3HeB/FeJ       + o a a a    . . a . .    b b . . b    a a a a 1    . . a b .
C3HfB/HeN       + . . . .    . . . . .    . . . . .    . . . . .    . . . . .

C57BL/6-        a o a a a    a a b b h    b a a b a    b a a a a    a a a a b
C57BL/6ByA      a . . . .    a a b b .    b a a b .    . . . . a    a . . . .
C57BL/6ByJ      a . a . a    . . . . .    b . . a      . a . . .    . a a . .
C57BL/6J        a o a a a    a . b b h    b a a b a    b a a a a    . . a a b
C57BL/6JN       a . . . .    . . . . .    b . . . .    . . . . .    . . . . .

C57BL/10-       a o a . a    a a b b h    b a a b a    b a a . a    a a a a b
C57BL/10J       a o . . a    . . . . h    b a a . a    b a a . .    . . a a b
C57BL/10ScSnA   a . . . .    a a b b .    b a a b .    . . . . a    a . . . .
C57BL/10ScSnJ   a o a . .    . . . . .    b a . r .    . . a . a    a . a a .
C57BL/AnHf      a . . . .    . . . . .    . . . . .    . . . . .    . . . . .

C57BL/ImrHeA    a . . . .    a a b b .    b a a b .    . . . . a    a . . . .
C57BL/KaLwN     a . . . .    . . . . .    . . . . .    . . . . .    . . . . b
C57BL/KsJ       a o a . a    . . a . .    b b . . a    b a a a 2    . . a a b
C57BL/LiA       a . a . .    a . b b .    b . . b .    . . . . .    . . . . .
C57BR/cd-       a o a . a    . . . . h    b a . . a    b a a a a    . . a a b

C57BR/cdJ       a o a . a    . . . . h    b a . . a    b a a a 1    . . a a b
C57L/-          a o a . a    . . . . h    b a . . a    b a a a 1    . . a a b
C57L/J          a o a . a    . . . . h    b a . . a    b a a a 1    . . a a b
C57P/A          a . . . .    . . . . .    b . . . .    . . . . .    . . . . b
C58/-           a o a a a    . . a . h    b a a a a    b . a a 2    . . a a a
```

1 = a?   2 = b?

Strain distribution of polymorphic variants

Locus (read down) →	Av-1	Av-2	b2m	Bcd-1	B	Bgl-e	Bgl-s	Bgl-t	Bv-1	c	C3	C4	C6	Car-1	Car-2	cdm1	Ce-2	Cee	Cfh	Coh	Cv-1	d	D7Rp2-s	D12Nyu1	D12Nyu2
Chromosome (read down)	UN	UN	4	2	5	9	9	9	8	7	17	17	15	3	3	3	2	17	2	7	5	9	7	12	12
BRSUNT/N	.	.	b	.	.	b	d	d	.	+	.	.	.	.	a	.	.	a	.	.	.	.	.	.	.
BS	.	.	.	.	.	.	.	.	.	.	.	.	.	.	.	.	.	.	.	.	.	.	.	.	.
BUB/BnJ	.	.	.	.	.	b	d	d	b	c	b	.	.	a	b	+	.	a	.	l	b	+	b	.	.
BXSB/MpJ	.	.	.	.	.	.	.	.	.	.	.	.	.	.	a	.	.	.	.	.	.	+	.	.	.
C3H/-	r	r	+	a	.	*	*	d	b	+	b	l	a	a	b	1	b	b	f	l	a	+	b	a	a
C3H/An	.	.	.	.	.	.	.	.	.	.	b	.	.	.	.	.	.	.	.	.	.	.	.	.	.
C3H/Bi	.	.	+	.	.	.	.	.	.	+	b	.	.	.	.	.	.	.	.	.	.	.	.	.	.
C3H/BiU	.	.	.	.	.	.	.	.	.	.	.	.	.	.	.	.	.	.	.	.	.	.	.	.	.
C3H/Crgl	.	.	.	.	.	.	.	.	.	.	.	.	.	.	.	.	.	.	.	.	.	.	.	.	.
C3H/He-	r	r	+	a	.	a	h	d	b	+	b	l	a	a	b	1	b	b	f	l	a	+	b	a	a
C3H/HeA	.	.	+	.	.	.	.	.	.	+	.	l	.	.	.	1	.	b	.	.	.	+	.	.	.
C3H/HeAf	.	.	.	.	.	.	.	.	.	.	.	.	.	.	.	.	.	.	.	.	.	.	.	.	.
C3H/HeDiSnA	.	.	+	.	.	.	.	.	.	+	.	l	.	.	.	.	.	b	.	.	.	+	.	.	.
C3H/HeJ	.	.	+	a	.	a	h	d	b	+	b	l	a	a	b	1	.	b	f	l	a	+	b	a	a
C3H/HeOuJ	.	.	.	.	.	.	.	.	.	.	.	.	.	.	b	.	.	.	.	.	.	.	.	.	.
C3H/HeSnJ	.	.	+	a	.	.	.	.	b	+	b	.	.	.	b	.	.	b	.	.	a	+	.	.	.
C3H/St	.	.	+	.	.	b	d	d	.	+	.	.	.	.	.	.	.	.	.	.	.	.	b	.	.
C3H/StWi	.	.	+	.	.	.	.	.	.	+	.	.	.	.	b	.	.	b	.	.	.	.	.	.	.
C3HeB/FeJ	.	.	+	.	.	.	.	.	b	+	b	.	.	a	b	1	.	b	.	.	a	+	.	.	.
C3HfB/HeN	.	.	+	.	.	.	.	.	.	+	.	.	.	.	b	.	.	b	.	.	.	.	.	.	.
C57BL/6-	r	r	+	b	a	b	h	b	a	+	b	h	a	a	a	1	c	a	f	l	b	+	b	b	b
C57BL/6ByA	.	.	+	.	.	.	.	.	.	+	.	h	.	.	.	1	.	a	.	.	.	+	.	.	.
C57BL/6ByJ	r	r	+	b	.	b	h	b	.	+	.	.	.	a	a	1	.	a	.	.	.	+	b	.	.
C57BL/6J	r	r	+	b	a	b	h	b	a	+	b	h	.	a	a	1	c	a	.	l	b	+	b	b	b
C57BL/6JN	r	r	+	.	.	.	.	.	.	+	.	.	.	.	a	.	.	a	.	.	.	.	.	.	.
C57BL/10-	.	.	+	b	.	b	h	b	a	+	b	h	a	a	a	1	c	a	.	.	b	+	b	.	.
C57BL/10J	.	.	+	.	.	b	h	b	a	+	b	h	.	a	a	1	c	a	.	.	b	+	.	.	.
C57BL/10ScSnA	.	.	+	.	.	.	.	.	.	+	.	h	.	.	.	1	.	a	.	.	.	+	.	.	.
C57BL/10ScSnJ	.	.	+	b	.	.	.	.	a	+	b	.	.	a	a	.	.	a	.	.	b	+	b	.	.
C57BL/AnHf	.	.	+	.	.	.	.	.	.	+	.	.	.	.	a	.	b	a	.	.	.	.	.	.	.
C57BL/ImrHeA	.	.	+	.	.	.	.	.	.	+	.	h	.	.	.	1	.	a	.	.	.	+	.	.	.
C57BL/KaLwN	.	.	+	.	.	.	.	.	.	+	.	.	.	.	a	.	.	a	.	.	.	.	.	.	.
C57BL/KsJ	.	.	+	.	.	b	d	d	a	+	b	.	.	a	a	+	.	a	.	.	b	+	.	.	.
C57BL/LiA	.	.	+	.	.	.	.	.	.	+	.	h	.	.	.	1	.	a	.	.	.	+	.	.	.
C57BR/cd-	.	.	b	b	.	b	h	b	a	+	b	l	.	a	b	+	c	a	f	h	b	+	d	.	.
C57BR/cdJ	.	.	b	b	.	b	h	b	a	+	b	l	.	a	b	+	.	a	.	h	b	+	d	.	.
C57L/-	r	r	b	b	.	b	h	b	b	+	b	h	.	a	b	+	c	a	f	h	b	+	d	b	a
C57L/J	r	r	b	b	.	b	h	b	b	+	b	.	.	a	b	+	.	a	f	h	b	+	d	b	a
C57P/A	.	.	+	.	.	.	.	.	.	+	.	h	.	.	.	1	.	a	.	.	.	+	.	.	.
C58/-	r	r	+	a	.	b	h	b	b	+	b	.	.	a	a	+	c	a	.	.	b	+	d	.	.

1 = cdm

The laboratory mouse

```
Locus    ->    D D D E E   E E E E    E E E E   E E E E   E E E E
(read down)    1 1 l a a   a a a g m  m m m m   m m m m   m p s s
               2 2 b - -   - - - a    v v v v   v v v v   v h - -
               N N - 2 4   5 6 7      - - - -   - - - - -  - - 1 2 3
               y y 1                4 5 6 7 8   9 1 1 1 1   1 1
               u u                              0 1 2 3   4
               3 4

Chromosome     1 1 1 U 1   U 2 U 8 U   U U U U U   U U 7 1 2   1 1 8 8 1
(read down)    2 2 1 N 9   N   N   N   N N N N N   N N   6     1       1
-----------------------------------------------------------------------------

BRSUNT/N       . . . . .   . . . a .   . . . . .   . . . . .   . . b . a
BS             . . . . .   . . . . .   . . . . .   . . . . .   . . b b c
BUB/BnJ        . . . b a   . . a . .   b b b b b   b b b b b   b d b b c
BXSB/MpJ       . . . . .   . . . . .   . . . . .   . . . . .   . . a b c
C3H/-          b b . b .   * * a a h   b b b b b   b b b b b   b b b b c

C3H/An         . . . . .   . . . . .   . . . . .   . . . . .   . . . . .
C3H/Bi         . . b . .   . . . a .   . . . . .   . . . . .   . . b b c
C3H/BiU        . . . . .   . . . . .   . . . . .   . . . . .   . . b b c
C3H/Crgl       . . . . .   . . . . .   . . . . .   . . . . .   . . . . .
C3H/He-        b b b b .   b a a a h   b b b b b   b b b b b   b b b b c

C3H/HeA        . . . . .   . . . . .   . . . . .   . . . . .   . . b b c
C3H/HeAf       . . . . .   . . . . .   . . . . .   . . . . .   . . b b c
C3H/HeDiSnA    . . . . .   . . . . .   . . . . .   . . . . .   . . b b c
C3H/HeJ        b b . . .   . . . a h   b b b b b   b b b b b   b b b b c
C3H/HeOuJ      . . . . .   . . . . .   . . . . .   . . . . .   . . b . c

C3H/HeSnJ      . . . . .   . . . . .   b b b b b   b b b b b   b . b . c
C3H/St         . . . b .   a b . a .   . . . . .   . . . . .   . . b b c
C3H/StWi       . . . . .   . . . . .   . . . . .   . . . . .   . . b . c
C3HeB/FeJ      . . . . .   . . . . h   b b b b b   b b b b b   b . b b c
C3HfB/HeN      . . . . .   . . . . .   . . . . .   . . . . .   . . b . c

C57BL/6-       b b b b .   . a b a l   b b b b b   b b b b b   b b a b a
C57BL/6ByA     . . . . .   . . . . .   . . . . .   . . . . .   . . a b a
C57BL/6ByJ     . . . . .   . . . . .   . . . . .   . . . . .   . . a b a
C57BL/6J       b b b . .   . . . a l   b b b b b   b b b b b   b b a b a
C57BL/6JN      . . . . .   . . . . .   . . . . .   . . . . .   . . a . a

C57BL/10-      . . b . .   . . b . l   b b b b b   b b b b b   b . a b a
C57BL/10J      . . . . .   . . . . l   b b b b b   b b b b b   b . a b a
C57BL/10ScSnA  . . . . .   . . . . .   . . . . .   . . . . .   . . a b a
C57BL/10ScSnJ  . . b . .   . . . . .   b b b b b   b b b b b   b . a b a
C57BL/AnHf     . . . . .   . . . . .   . . . . .   . . . . .   . . a . a

C57BL/ImrHeA   . . . . .   . . . . .   . . . . .   . . . . .   . . a b .
C57BL/KaLwN    . . . . .   . . . . .   . . . . .   . . . . .   . . a . a
C57BL/KsJ      . . b b b   . . . . l   b b b b b   b b b b b   b b a b a
C57BL/LiA      . . . . .   . . . . .   . . . . .   . . . . .   . . a b a
C57BR/cd-      . . . b a   . . b . l   b b b b b   b b b b b   b b a b a

C57BR/cdJ      . . . . a   . . . . l   b b b b b   b b b b b   b b a b a
C57L/-         b b b b a   . . b . l   b b b b b   b b b b b   b . a b a
C57L/J         b b . . .   . . . . l   b b b b b   b b b b b   b . a b a
C57P/A         . . . . .   . . . . .   . . . . .   . . . . .   . . a b a
C58/-          . . . b a   b a a a l   b b b b b   b b b b b   b . b b c
```

Locus → (read down)	Es-5	Es-6	Es-7	Es-8	Es-9	Es-10	Es-11	Es-12	Es-13	Es-14	Es-16	Es-17	Es-18	Es-22	Es-23	Es-24	Es-25	Es-26	Es-27	Esr	Fabp-i	Fbp-1	Fbp-2	For	Fos-5
Chromosome (read down)	8	8	8	7	8	14	8	UN	9	9	3	9	19	8	8	8	12	3	3	6	3	UN	UN	14	12
BRSUNT/N	.	.	.	.	.	b	.	.	.	.	.	.	.	.	.	.	.	.	.	.	.	.	.	.	.
BS	b	.	.	.	.	.	a	.	.	.	.	.	.	.	.	.	.	.	.	.	.	.	.	.	.
BUB/BnJ	.	a	.	a	.	c	.	.	a	.	.	.	.	.	.	.	.	.	.	.	.	.	b	.	.
BXSB/MpJ	.	.	.	.	.	a	a	.	.	.	.	.	.	.	.	.	.	.	.	.	.	.	.	.	.
C3H/-	b	a	b	a	a	b	a	a	a	r	a	a	a	d	c	a	b	a	r	a	b	b	a	d	b
C3H/An	.	.	.	.	.	.	.	.	.	.	.	.	.	.	.	.	.	.	.	.	.	.	.	.	.
C3H/Bi	b	.	.	.	.	.	.	.	.	.	.	.	.	.	.	.	.	.	.	.	.	.	.	.	.
C3H/BiU	b	a	b	.	a	b	a	.	a	r	a	a	a	d	c	a	b	a	r	a	.	.	.	.	.
C3H/Crgl	.	.	.	.	.	.	.	.	.	.	.	.	.	.	.	.	.	.	.	.	.	.	.	.	.
C3H/He-	b	a	b	a	a	b	a	a	a	r	a	a	a	d	c	a	b	a	r	a	b	b	a	d	b
C3H/HeA	b	.	b	a	.	.	.	.	.	.	.	.	.	.	.	.	.	.	.	.	.	.	.	.	.
C3H/HeAf	b	a	b	.	a	b	a	.	a	r	a	a	a	d	c	a	b	a	r	a	.	.	.	.	.
C3H/HeDiSnA	b	a	b	a	a	b	a	.	a	r	a	a	a	d	c	a	b	a	r	a	.	.	.	.	.
C3H/HeJ	b	.	.	a	.	b	a	a	a	.	.	.	.	.	.	.	b	.	.	.	b	.	.	d	b
C3H/HeOuJ	.	.	.	.	.	b	a	.	.	.	.	.	.	.	.	.	.	.	.	.	.	.	.	.	.
C3H/HeSnJ	b	.	.	.	.	b	.	.	.	.	.	.	.	.	.	.	.	.	.	.	.	.	.	.	.
C3H/St	b	.	.	.	.	.	.	.	.	.	.	.	.	.	.	.	.	.	.	.	.	.	.	.	.
C3H/StWi	.	.	.	.	.	b	.	.	.	.	.	.	.	.	.	.	.	.	.	.	.	.	.	.	.
C3HeB/FeJ	.	a	.	.	.	b	a	.	a	.	.	.	.	.	.	.	.	.	.	.	.	.	.	.	.
C3HfB/HeN	.	.	.	.	.	b	.	.	.	.	.	.	.	.	.	.	.	.	.	.	.	.	.	.	.
C57BL/6-	b	a	b	a	a	a	a	a	a	r	a	a	a	d	c	a	b	a	r	a	b	b	.	b	b
C57BL/6ByA	b	a	b	a	a	a	a	.	a	r	a	a	a	d	c	a	b	a	r	a	.	.	.	.	.
C57BL/6ByJ	b	.	.	.	.	a	.	.	a	.	.	.	.	.	.	a	.	.	.	.	.	.	.	.	.
C57BL/6J	b	a	.	a	.	a	a	a	a	.	a	.	.	.	.	.	b	.	.	.	b	.	.	b	b
C57BL/6JN	.	.	.	.	.	a	.	.	.	.	.	.	.	.	.	.	.	.	.	.	.	.	.	.	.
C57BL/10-	b	a	b	a	a	a	a	.	a	r	a	a	a	d	c	a	b	a	r	a	.	.	.	b	.
C57BL/10J	.	.	.	.	.	a	a	.	a	.	.	.	.	.	.	.	.	.	.	.	.	.	.	b	.
C57BL/10ScSnA	b	a	b	a	a	a	a	.	a	r	a	a	a	d	c	a	b	a	r	a	.	.	.	o	o
C57BL/10ScSnJ	b	a	b	.	a	a	a	.	a	.	a	b	a	d	c	a	.	.	.	.	.	.	.	.	.
C57BL/AnHf	.	.	.	.	.	a	.	.	.	.	.	.	.	.	.	.	.	.	.	.	.	.	.	.	.
C57BL/ImrHeA	.	.	b	.	.	.	.	.	.	.	.	.	.	.	c	.	.	.	.	.	.	.	.	.	.
C57BL/KaLwN	.	.	.	.	.	a	.	.	.	.	.	.	.	.	.	.	.	.	.	.	.	.	.	.	.
C57BL/KsJ	.	.	.	.	.	a	a	.	a	.	a	.	.	.	.	.	.	.	.	.	.	.	.	b	.
C57BL/LiA	b	a	b	a	a	a	a	.	a	r	a	a	a	d	c	a	b	a	r	a	.	b	.	.	.
C57BR/cd-	.	a	.	a	.	a	a	b	a	.	.	.	.	.	.	.	.	.	.	.	.	.	.	b	.
C57BR/cdJ	.	a	.	a	.	a	.	b	a	.	.	.	.	.	.	.	.	.	.	.	.	.	.	b	.
C57L/-	.	a	.	a	.	a	a	b	a	.	.	.	.	.	.	.	b	.	.	.	b	.	.	b	b
C57L/J	.	a	.	a	.	a	a	b	a	.	.	.	.	.	.	.	b	.	.	.	b	.	.	b	b
C57P/A	b	.	b	.	.	.	.	.	.	.	.	.	.	.	.	.	.	.	.	.	.	.	.	.	.
C58/-	b	a	b	a	a	a	a	.	a	r	a	a	a	d	c	a	b	a	r	a	.	b	.	b	.

The laboratory mouse

```
Locus      ->    F F G G G   G G G G G   G G G G G   G G G G G   G G H H H
(read down)      v v a b d   d d d g k   l l m o o   p p p r u   u v - - -
                 - - l p c   c r r c     k o - t t   d i t - s   s - 1 2 3
                 1 2 t - -   - - - -     - 3 - -     - - - 1 -   - 1
                     1 1     2 1 2       1   1 2     1 1 1   r   s
```

```
Chromosome       4 9 4 3 1   9 U U 6 U   1 1 2 1 8   4 7 1 8 5   5 U 7 1 2
(read down)            5     N N   N     1 7   9           5     N   7
```

```
BRSUNT/N         . . . . .   . . . . .   . . . . .   b b . . .   a . . . .
BS               . . . . .   . . . . .   . . . a b   a b a . .   . . . . .
BUB/BnJ          n . c . b   b . . a .   . b . a b   b b a b b   b . . q .
BXSB/MpJ         . . . . .   . . . . .   . a . a b   b a a a .   . . . b .
C3H/-            n s b a b   b h . a a   b a a a b   b b a a .   * a a k b

C3H/An           . . . . .   . . . a .   . . . . .   . . . . .   . a . k .
C3H/Bi           n s . . .   . . . a .   . . . . .   . b . . .   . . a k b
C3H/BiU          . . . . .   . . . . .   . . . . .   . . . . .   . . . . .
C3H/Crgl         . . . . .   . . . . .   . . . . .   . . . . .   . . . . .
C3H/He-          n s b a b   b h . a a   b a a a b   b b a a .   h a a k b

C3H/HeA          n s . . .   . . . . .   . a . a b   b b . a .   . . . k .
C3H/HeAf         . . . . .   . . . . .   . . . . .   . . . . .   . . . . .
C3H/HeDiSnA      . s . . .   . . . . .   . a . a b   b b . a .   . . . k .
C3H/HeJ          . . b a b   b h . a a   b a a a b   b b a a .   h . . k .
C3H/HeOuJ        . . . . .   . . . . .   . a . . .   b b a a .   . . . k .

C3H/HeSnJ        . . . . b   . . . . .   . a . a b   b b a a .   . . a k b
C3H/St           . . . . .   . . . . .   . . . . .   . b . . .   b . . k .
C3H/StWi         . . . . .   . . . . .   . . . . .   b b . . .   b . . . .
C3HeB/FeJ        n . b a b   b . . a .   . a . a b   b b a a .   h . . k .
C3HfB/HeN        . . . . .   . . . . .   . . . . .   b b . . .   . . . . .

C57BL/6-         b r b b b   b h l b 1   a a b a b   a b a a b   b b c b a
C57BL/6ByA       . r . b .   . . . . .   . a . a b   a b . a .   . . . b .
C57BL/6ByJ       . . b . .   . . . . b   . a . a b   a b a a .   . . c b .
C57BL/6J         b r b b b   b h l b 1   a a . a b   a b a a b   b . c b a
C57BL/6JN        . . . . .   . . . . .   . . . . .   a b . . .   b . . b .

C57BL/10-        b r b b b   b . . . .   a a b a b   a b a a b   b b c b a
C57BL/10J        . . b . b   b . . . .   a a . a b   a b a a b   b b . b .
C57BL/10ScSnA    . r . b .   . . . . .   . a . a b   a b . a .   . . . b .
C57BL/10ScSnJ    b r . . .   b . . . .   . a . a b   a b a a .   b . c b a
C57BL/AnHf       . . . . .   . . . . .   . . . . .   a b . . .   . . . b .

C57BL/ImrHeA     . r . b .   . . . . .   . a . a b   a b . a .   . . . . .
C57BL/KaLwN      . . . . .   . . . . .   . . . . .   a b . . .   . . . . .
C57BL/KsJ        b . b b b   b . . b .   . a . a b   a b a a .   b . 2 d a
C57BL/LiA        b r . b .   . . . . .   . a . a b   a b . a .   . . . 3 .
C57BR/cd-        n r b . b   b l l a .   a a b a b   a a a a b   b b c k a

C57BR/cdJ        n r b . b   b l l a .   a a . a b   a a a a b   b b c k a
C57L/-           n r b . b   b l l a .   a a b a b   a a a a b   b . c b a
C57L/J           n r b . b   b l l a .   a a b a b   a a a a b   b . c b a
C57P/A           . r . . .   . . . . .   . a . a b   a b . a .   . . . 3 .
C58/-            n r b b b   b h l b 4   a a a a b   a a a a b   * a c k b
```

682   1 = c?   2 = -a,-c   3 = be   4 = b?

Strain distribution of polymorphic variants

Locus (read down) →	H-4	H-7	H-8	H-9	H-12	H-13	Hao-1	Hao-2	Hba-4ps	Hba	Hbb-1	Hcx	Hea	Hm	Hp	Hrt-1	Hsd-1	Idh-1	If-1	Igh-1	Igh-2	Igh-3	Igh-4	Igh-5	Igh-Ars
Chromosome (read down)	7	9	14	UN	UN	2	2	3	11	17	7	2	UN	7	8	UN	1	10	3	12	12	12	12	12	12
BRSUNT/N	.	.	.	.	.	.	.	.	f	.	d	o	a	.	.	.	l	b	.	a	.	.	.	.	.
BS	.	.	.	.	.	.	.	.	.	.	s	.	.	.	.	.	.	a	.	.	.	.	.	.	.
BUB/BnJ	.	.	.	.	.	.	.	.	f	.	d	1	b	.	.	.	b	.	a	.	a	.	a	.	.
BXSB/MpJ	.	.	.	.	.	.	.	.	.	.	s	.	.	.	.	.	.	a	.	.	.	.	.	.	.
C3H/-	a	*	1	*	*	*	a	a	c	a	d	1	a	a	a	.	l	*	l	j	a	.	a	.	o
C3H/An	.	.	.	.	.	.	.	.	.	.	.	.	.	.	.	.	.	.	.	.	.	.	.	.	.
C3H/Bi	.	1	1	a	a	a	.	.	.	.	d	.	.	.	.	.	b	.	.	.	.	.	.	.	.
C3H/BiU	.	.	.	.	.	.	.	.	.	.	.	.	.	.	.	.	.	.	.	.	.	.	.	.	.
C3H/Crgl	.	.	.	.	.	.	.	.	.	.	.	.	.	.	.	.	.	.	.	.	.	.	.	.	.
C3H/He-	a	b	1	a	a	c	a	a	c	c	d	1	.	a	a	.	.	a	l	j	a	.	a	.	.
C3H/HeA	.	.	.	.	.	.	.	.	.	.	d	1	.	.	.	.	.	a	.	a	.	.	.	.	.
C3H/HeAf	.	.	.	.	.	.	.	.	.	.	.	.	.	.	.	.	.	.	.	.	.	.	.	.	.
C3H/HeDiSnA	.	.	.	.	.	.	a	a	.	.	d	1	.	.	.	.	.	a	.	.	.	.	.	.	.
C3H/HeJ	.	.	.	.	.	.	.	.	c	c	d	1	.	a	a	.	.	a	l	j	.	.	a	.	.
C3H/HeOuJ	.	.	.	.	.	.	.	.	.	.	d	.	.	.	.	.	.	a	.	.	.	.	.	.	.
C3H/HeSnJ	a	b	1	a	a	c	.	.	.	.	d	1	.	.	.	.	.	a	.	j	.	.	a	.	.
C3H/St	.	.	.	.	.	.	.	.	.	.	d	.	.	.	.	.	.	a	.	.	.	.	.	.	.
C3H/StWi	.	.	.	.	.	.	.	.	.	.	d	.	.	.	.	.	.	b	.	.	.	.	.	.	.
C3HeB/FeJ	.	.	.	.	.	.	.	.	c	.	d	.	.	.	.	b	.	a	.	.	.	.	.	.	.
C3HfB/HeN	.	.	.	.	.	.	.	.	.	.	d	.	.	.	.	.	.	a	.	.	.	.	.	.	.
C57BL/6-	a	a	a	l	a	a	a	a	a	b	s	l	a	b	a	b	h	a	h	b	b	b	b	b	o
C57BL/6ByA	.	.	.	.	.	.	a	a	.	.	s	l	.	.	.	.	.	a	h	b	.	.	.	.	.
C57BL/6ByJ	.	.	.	.	.	.	.	.	.	.	s	l	.	a	.	.	.	a	.	b	.	.	.	.	.
C57BL/6J	a	a	a	l	a	a	a	a	a	b	s	l	a	b	a	b	h	a	h	b	b	b	b	b	.
C57BL/6JN	.	.	.	.	.	.	.	.	a	.	s	l	.	.	.	.	.	a	.	b	.	.	.	.	.
C57BL/10-	a	a	a	a	a	a	a	a	a	.	s	l	.	.	.	b	h	a	h	b	2	.	b	.	.
C57BL/10J	a	.	.	.	.	.	.	.	a	.	s	l	.	.	.	.	h	a	h	b	.	.	.	.	.
C57BL/10ScSnA	.	.	.	.	.	.	a	a	b	.	s	l	.	.	.	.	.	a	h	b	.	.	.	.	.
C57BL/10ScSnJ	a	a	a	a	a	a	.	.	a	.	s	l	.	.	.	b	.	a	h	b	.	h	.	b	.
C57BL/AnHf	.	.	.	.	.	.	.	.	.	.	s	l	.	.	.	.	.	a	.	.	.	.	.	.	.
C57BL/ImrHeA	.	.	.	.	.	.	a	a	.	.	s	l	.	.	.	.	.	a	h	b	.	.	.	.	.
C57BL/KaLwN	.	.	.	.	.	.	.	.	a	.	s	.	.	.	.	.	.	a	.	b	2	.	.	.	.
C57BL/KsJ	a	a	l	1	1	a	.	.	a	.	s	.	a	b	.	.	h	b	h	.	.	.	b	.	.
C57BL/LiA	.	.	.	.	.	.	.	.	.	.	s	l	.	.	.	.	.	a	h	b	.	.	b	.	.
C57BR/cd-	a	a	a	a	a	a	.	.	a	.	s	l	.	.	.	.	h	b	h	j	a	.	a	.	o
C57BR/cdJ	a	a	a	a	a	a	.	.	a	.	s	l	.	.	.	.	h	b	h	a	a	.	a	.	.
C57L/-	a	a	a	a	a	a	.	.	a	b	s	l	a	.	a	.	h	b	.	a	a	.	a	.	.
C57L/J	a	a	a	a	a	a	.	.	a	b	s	l	a	.	a	.	h	b	.	a	a	.	a	.	.
C57P/A	.	.	.	.	.	.	.	.	.	.	s	l	.	.	.	.	.	a	h	b	.	.	.	.	.
C58/-	a	a	a	.	a	b	.	.	b	b	s	l	a	b	.	.	h	a	h	a	a	.	.	.	.

1 = -a,-b   2 = -a,-c

```
Locus      ->    I I I I I   I I I I I   I I I I I   I I K K K   L L L L L
(read down)      g g g g g   g g g g g   g g g g g   g g f t t   a a a d d
                 h h h h h   h h h h h   h h k k k   l l o h h   m m p h r
                 - - - - -   - - - - -   - - - - -   - - - - -   b b - - -
                 C D I L N   P S S S S   S S E P T   1 1 1 1 2   - - 1 1 1
                   e n e p     c a a a a   a r f c r       r       1 2
                   x u v         1 2 3 4   5 c 1 p
----------------------------------------------------------------------------------
    Chromosome   1 1 1 1 1   1 1 1 1 1   1 1 6 6 6   1 U 9 1 U   1 1 9 7 6
    (read down)  2 2 2 2 2   2 2 2 2 2   2 2         6 N   7 N   2
----------------------------------------------------------------------------------

BRSUNT/N         . . . . .   . . . . .   . . . . .   . . . . .   . . . . .
BS               . . . . .   . . . . .   . . . . .   . . . . .   . . . . a
BUB/BnJ          . . . a .   . . . . .   . . b . .   . . . . .   a b b . a
BXSB/MpJ         . . . . .   . . . . .   . . . . .   . . . . .   . . . . a
C3H/-            . o a o o   o o o o o   . o b b b   . a b o o   a b b . a

C3H/An           . . . . .   . . . . .   . . . . .   . . . . .   . . . . .
C3H/Bi           . . . . .   . . . . .   . . . . .   . . . . .   . . . . .
C3H/BiU          . . . . .   . . . . .   . . . . .   . . . . .   . . . . .
C3H/Crgl         . . . . .   . . . . .   . . . . .   . . . . .   . . . . .
C3H/He-          . . . . .   . . . o o   . o b b b   . a b o o   a b b . a

C3H/HeA          . . . . .   . . . . .   . . . . .   . . . . .   . . . . a
C3H/HeAf         . . . . .   . . . . .   . . . . .   . . . . .   . . . . a
C3H/HeDiSnA      . . . . .   . . . . .   . . . . .   . . . . .   . . . . a
C3H/HeJ          . . . . .   . . . o o   . . b b b   . a b o o   a b b . a
C3H/HeOuJ        . . . . .   . . . . .   . . . . .   . . . . .   . . . . .

C3H/HeSnJ        . . . . .   . . . . .   . . . . .   . . . . .   . . . . .
C3H/St           . . . . .   . . . . .   . . . . .   . . . . .   . . . . a
C3H/StWi         . . . . .   . . . . .   . . . . .   . . . . .   . . . . .
C3HeB/FeJ        . . . . .   . . . . .   . . . . .   . . . o .   a b b . a
C3HfB/HeN        . . . . .   . . . . .   . . . . .   . . . . .   . . . . .

C57BL/6-         b o o a a   o o o o o   o a b b b   a a a a a   a b a a a
C57BL/6ByA       . . . . .   . . . . .   . . . . .   . . . . .   . . . . a
C57BL/6ByJ       . . . . .   . . . . .   . . . . .   . . . . .   . . a . a
C57BL/6J         b . . . .   . . . o o   o . b b b   a a a a a   a b a a a
C57BL/6JN        . . . . .   . . . . .   . . . . .   . . . . .   . . . . .

C57BL/10-        b o o . a   o o o . .   . a . . .   . a . . .   a b a . a
C57BL/10J        b . . . .   . . . . .   . a . . .   . a . . .   a b a . a
C57BL/10ScSnA    . . . . .   . . . . .   . . . . .   . . . . .   . . . . a
C57BL/10ScSnJ    h h . . .   . . . . .   . . . . .   . . . . .   . . . h a
C57BL/AnHf       . . . . .   . . . . .   . . . . .   . . . . .   . . . . .

C57BL/ImrHeA     . . . . .   . . . . .   . . . . .   . . . . .   . . . . a
C57BL/KaLwN      . o . . a   . . . . .   . . . . .   . . . . .   . . . . .
C57BL/KsJ        . . . . a   . . . . .   . a . . b   . . . . .   a b a . a
C57BL/LiA        . . . . .   . . . . .   . . . . .   . . . . .   . . . . a
C57BR/cd-        . a . a .   . . . . .   . a b . b   . . . . .   1 b a . a

C57BR/cdJ        . . . a .   . . . . .   . . b . b   . . . . .   1 b a . a
C57L/-           . a a a o   a . a a .   o . b b b   . a a a a   1 b a a a
C57L/J           . . . a .   . . . a .   o . b b b   . a a a a   1 b a a a
C57P/A           . . . . .   . . . . .   . . . . .   . . . . .   . . . . a
C58/-            a a a a o   a o o a .   o a a a a   . . . . .   1 b a a a
```

```
Locus    ->   L L L L L   L L L L L   L L L L L   L L L L L   L L L L L
(read down)   e e f m n   p s t t t   t t t t v   v y y y y   y y y y y
              n n o p a   t h h h n   w w w w     p - - - -   - - - - -
              - - - -     -   - - -   - - - -     - 1 2 3 5   6 7 8 9 1
              1 2 1 1     1   1 2 2   2 3 4 6     1                     0

--------------------------------------------------------------------------
Chromosome    1 U 7 1 U   U 1 1 U U   1 9 1 U 4   6 1 6 6 1   1 1 U 1 1
(read down)     N 7 N N   N 3 N N     2   N       9           5 2 N   9
--------------------------------------------------------------------------

BRSUNT/N      . . . . .   . . . . .   . . . . .   . a . . .   . . . . .
BS            . . . . .   . . . . .   . . . . .   . . . . .   . . . . .
BUB/BnJ       . . . . .   . . . b .   . a a b c   a b b b .   . . . . .
BXSB/MpJ      . . . . .   . . . . .   . . . . .   . b b b .   . . . . .
C3H/-         a b b a .   . r a o a   a * * a a   b a a b b   a a b a b

C3H/An        . . . . .   . . . . .   . . . . .   . a a b b   . . . a .
C3H/Bi        . . . . .   . . . . a   . b a a .   . a a b b   . . . . .
C3H/BiU       . . . . .   . . . . .   . . . . .   . . . . .   . . . . .
C3H/Crgl      . . . . .   . . . . .   . . a . .   . . . . .   . . . . .
C3H/He-       a b b . .   . r a o a   a a a a a   b a a b b   a a b a b

C3H/HeA       . . . . .   . . . . .   . . . . .   . a a b b   . . . . .
C3H/HeAf      a b . . .   . . . . .   . . . . .   . . . . .   . . . . .
C3H/HeDiSnA   . . . . .   . . . . .   . . . . .   . a a b b   . . . . .
C3H/HeJ       . . b . .   . . a o a   a a a a .   b a a b b   a a b a b
C3H/HeOuJ     . . . . .   . . . . .   . . . . .   . . . . .   . . . . .

C3H/HeSnJ     . . . . .   . . . . .   . . . . .   . a a b b   . . . . .
C3H/St        . . . . .   . . . . .   . . . . .   . . . . .   . . . . .
C3H/StWi      . . . . .   . . . . .   . . . . .   . . . . .   . . . . .
C3HeB/FeJ     a b . . .   b . . . a   . a a a a   b . . . .   . . . . .
C3HfB/HeN     . . . . .   . . . . .   . . . . .   . . . . .   . . . . .

C57BL/6-      b b a . b   . . o o b   a b b a b   a b b b b   b b b b b
C57BL/6ByA    . . . . .   . . . . .   . . . . .   . b b b b   . . . . .
C57BL/6ByJ    . . . . .   . . . . b   . . . . .   . b b b b   . . . . .
C57BL/6J      b b a . .   . . o o b   a b b a b   a b b b b   b b b b b
C57BL/6JN     . . . . .   . . . . .   . . . . b   . b b b .   . . . b .

C57BL/10-     b b . o .   a s o . b   . b b a c   a b b b b   . . b b .
C57BL/10J     . . . . .   a . o . b   . . b a c   a . . . b   . . . b .
C57BL/10ScSnA . . . . .   . . . . .   . . . . .   . b b b b   . . . . .
C57BL/10ScSnJ b b . . .   . s . . b   . b b a c   . b b b b   . . . b .
C57BL/AnHf    . . . . .   . . . . .   . . . . .   . . . . .   . . . . .

C57BL/ImrHeA  . . . . .   . . . . .   . . . . .   . b b b b   . . . . .
C57BL/KaLwN   . . . . .   . . . . .   . . . . .   . b . . .   . . . . .
C57BL/KsJ     . . . . .   . . o . b   . b b a .   . b b . .   . . . . .
C57BL/LiA     b b . . .   . . . . .   . . . . .   . b b b b   . . . . .
C57BR/cd-     . . . . b   . r o . .   . . a a a   a b b b b   b a b b a

C57BR/cdJ     . . . . .   . . o . .   . . a a a   a b b b b   b a b b a
C57L/-        a b a . b   a r o a b   a b a a a   a b b b b   b a b b a
C57L/J        a b a . .   a . o a b   a b a a a   a b b b b   b a b b a
C57P/A        . . . . .   . . . . .   . . . . .   . . . . .   . . . . .
C58/-         . . . . b   . . . . .   . . a a c   a b a a b   b b b b b
```

```
Locus    ->    L L L L L   L L L L L   L L L L L   L L L L L   L L L L L
(read down)    y y y y y   y y y y y   y y y y y   y y y y y   y y y y y
               - - - - -   - - - - -   - - - - -   - - - - -   - - - - b
               1 1 1 1 1   1 1 1 1 2   2 2 2 2 2   2 2 2 2 3   3 3 3 4 -
               1 2 3 4 5   6 7 8 9 0   1 2 3 4 5   6 7 8 9 0   1 2 9 0 2

Chromosome     2 1 U 7 7   1 1 1 4 4   7 1 2 2 2   U U U 4 U   4 4 1 U 4
(read down)      9 N         2   2                 N N N   N       7 N
```

Strain	1	2	3	4	5	6	7	8	9	10	11	12	13	14	15	16	17	18	19	20	21	22	23	24	25
BRSUNT/N	.	.	.	.	.	.	.	.	.	.	.	.	.	.	.	.	.	.	.	.	.	.	.	.	.
BS	.	.	.	.	.	.	.	.	.	.	.	.	.	.	.	.	.	.	.	.	.	.	.	.	.
BUB/BnJ	.	.	.	.	.	.	.	.	.	.	.	a	.	.	.	.	.	.	.	.	a	.	.	.	.
BXSB/MpJ	.	.	.	.	.	.	.	.	.	.	.	.	.	.	.	.	.	.	.	.	.	.	.	.	.
C3H/-	a	b	b	b	b	.	a	a	b	.	b	b	a	b	b	b	a	b	b	b	a	b	1	b	*
C3H/An	.	.	.	.	.	.	.	.	.	.	.	b	.	.	.	.	.	.	.	.	a	.	.	.	b
C3H/Bi	.	.	.	.	.	.	.	.	.	.	.	b	.	.	.	.	.	.	.	.	.	.	.	.	.
C3H/BiU	.	.	.	.	.	.	.	.	.	.	.	.	.	.	.	.	.	.	.	.	.	.	.	.	.
C3H/Crgl	.	.	.	.	.	.	.	.	.	.	.	.	.	.	.	.	.	.	.	.	.	.	.	.	.
C3H/He-	a	b	b	b	b	.	a	a	b	.	b	b	a	b	b	b	a	b	b	b	a	b	1	b	b
C3H/HeA	.	.	.	.	.	.	.	.	.	.	.	.	.	.	.	.	.	.	.	.	.	.	.	.	.
C3H/HeAf	.	.	.	.	.	.	.	.	.	.	.	.	.	.	.	.	.	.	.	.	.	.	.	.	.
C3H/HeDiSnA	.	.	.	.	.	.	.	.	.	.	.	.	.	.	.	.	.	.	.	.	.	.	.	.	.
C3H/HeJ	a	b	b	b	b	.	a	a	b	.	b	b	.	b	b	b	a	b	b	b	a	b	1	b	b
C3H/HeOuJ	.	.	.	.	.	.	.	.	.	.	.	.	a	.	.	.	.	.	.	.	a	.	.	.	.
C3H/HeSnJ	.	.	.	.	.	.	.	.	.	.	.	.	.	.	.	.	.	.	.	.	a	.	.	.	.
C3H/St	.	.	.	.	.	.	.	.	.	.	.	.	.	.	.	.	.	.	.	.	.	.	.	.	.
C3H/StWi	.	.	.	.	.	.	.	.	.	.	.	.	.	.	.	.	.	.	.	.	.	.	.	.	.
C3HeB/FeJ	.	.	.	.	.	.	.	.	.	.	.	.	.	.	.	.	.	.	.	.	a	.	.	.	.
C3HfB/HeN	.	.	.	.	.	.	.	.	.	.	.	.	.	.	.	.	.	.	.	.	.	.	.	.	.
C57BL/6-	b	b	b	b	b	b	b	b	b	b	b	b	b	b	b	b	b	b	b	b	b	b	b	b	b
C57BL/6ByA	.	.	.	.	.	.	.	.	.	.	.	.	.	.	.	.	.	.	.	.	.	.	.	.	.
C57BL/6ByJ	.	.	.	.	b	.	.	.	.	.	.	.	.	.	.	.	.	.	.	.	b	.	.	.	.
C57BL/6J	.	b	b	b	b	.	b	b	b	.	b	.	.	b	.	b	.	.	.	.	.	.	b	b	b
C57BL/6JN	.	.	.	.	.	.	.	.	.	.	.	.	.	.	.	.	.	.	.	.	.	.	.	.	b
C57BL/10-	.	b	b	b	.	.	a	.	.	a	.	.	b	.	b	b	.	.	b	.	b	.	.	.	b
C57BL/10J	.	.	.	.	.	.	a	.	.	.	.	.	.	.	.	.	.	.	.	.	b	.	.	.	b
C57BL/10ScSnA	.	.	.	.	.	.	.	.	.	.	.	.	.	.	.	.	.	.	.	.	.	.	.	.	.
C57BL/10ScSnJ	.	.	.	.	.	.	.	.	.	.	.	.	.	.	.	b	.	.	.	.	b	.	.	.	.
C57BL/AnHf	.	.	.	.	.	.	.	.	.	.	.	.	.	.	.	.	.	.	.	.	.	.	.	.	.
C57BL/ImrHeA	.	.	.	.	.	.	.	.	.	.	.	.	.	.	.	.	.	.	.	.	.	.	.	.	.
C57BL/KaLwN	.	.	.	.	.	.	.	.	.	.	.	.	.	.	.	.	.	.	.	.	.	.	.	.	.
C57BL/KsJ	.	b	.	.	b	.	.	.	.	.	.	b	.	.	b	b	.	.	b	.	b	.	.	.	.
C57BL/LiA	.	.	.	.	.	.	.	.	.	.	.	.	.	.	.	.	.	.	.	.	.	.	.	.	.
C57BR/cd-	b	a	a	b	b	.	b	b	b	.	b	b	b	b	a	b	.	.	a	b	b	a	2	1	a
C57BR/cdJ	.	a	a	b	b	.	b	b	b	.	b	b	.	b	.	b	.	.	a	.	b	.	2	1	a
C57L/-	b	a	a	b	b	a	b	b	b	b	b	b	b	b	b	b	b	.	a	b	b	.	b	1	a
C57L/J	.	a	a	b	b	a	b	b	b	.	b	b	b	b	.	b	.	.	.	.	.	.	b	1	a
C57P/A	.	.	.	.	.	.	.	.	.	.	.	.	.	.	.	.	.	.	.	.	.	.	.	.	.
C58/-	b	b	b	b	a	a	b	b	b	b	b	b	b	b	b	b	b	b	a	a	b	a	b	b	a

686   1 = -b   2 = b?

```
Locus   ->    L L L L L    L M M M M    M M M M M    N N N N N    N O O p P
(read down)   y y y y y    y a e l o    o o p p u    c e k k p    p r r a n
              b b b b b    b p p s d    d r h i p    a u - - -    - m m   n
              - - - - -    - - - - -    - - - - -    m - 1 2 1    2 - - -
              3 4 5 6 7    8 1 1 1 1    2 1 1 1 1    1           1 2 1

--------------------------------------------------------------------------------
Chromosome    U 4 U 4 1    7 5 1 1 9    7 5 7 9 4    9 1 1 2 1    1 4 4 7 4
(read down)   N   N   2      7              7 7   4      4
--------------------------------------------------------------------------------

BRSUNT/N      . . . . .    . b . . b    . . . . a    . b . . .    . . . . .
BS            . . . . .    . . . . a    b a . . .    . . . . a    . . . . .
BUB/BnJ       . . . . .    . b . . a    b a . b a    . b . . .    . . . . .
BXSB/MpJ      . . . . .    . . . . b    b . . b .    . b . . a    . . . . .
C3H/-         b a a b a    b * b c a    b a b b *    a b o a a    a b a + .

C3H/An        . . . . .    . . . . .    . . . . .    . . . . .    . . . . .
C3H/Bi        . . . . .    . . . . .    . . . b a    . . . . .    . . . . +
C3H/BiU       . . . . .    . . . . .    . . . . .    . . . . .    . . . . .
C3H/Crgl      . . . . .    . . . . .    . . . . .    . . . . .    . . . . .
C3H/He-       b a a b a    b h b c a    b a b b a    a b o a a    . b a + .

C3H/HeA       . . . . .    . . . . a    b a . b a    . b . . a    . . . . .
C3H/HeAf      . . . . .    . . . . .    . . . . .    . . . . .    . . . . .
C3H/HeDiSnA   . . . . .    . . . . a    . a . b a    . b . . a    . . . . .
C3H/HeJ       b a a b a    b h b c a    b a . b a    a b o a .    . b a + .
C3H/HeOuJ     . . . . .    . . . . a    . . . . .    . . . . .    . . . . .

C3H/HeSnJ     . . . . .    . . . . a    . . . b .    . b . . .    . . . . +
C3H/St        . . . . .    . b . . .    . . b b .    . b . . .    . . . . +
C3H/StWi      . . . . .    . . . . b    . . . . a    . b . . .    . . . . .
C3HeB/FeJ     . . . . .    . . . . a    . a . . a    . b . . a    . . . + b
C3HfB/HeN     . . . . .    . . . . a    . . . . a    . b . . .    . . . . .

C57BL/6-      b b b b b    b b a b b    b a . b b    b b . . a    b b a + b
C57BL/6ByA    . . . . .    . . . . b    b a . b b    . b . . a    . b a + .
C57BL/6ByJ    . . . . .    . . . . b    . . . b b    . b . . a    . b a + .
C57BL/6J      . b . . .    . b a . b    b a . b b    b b . . a    . b a + b
C57BL/6JN     . . . . .    . . . . b    . . . . b    . . . . .    . . . . .

C57BL/10-     . . . . .    . b a b b    b a b b b    . b . . a    b . . + b
C57BL/10J     . . . . .    . b . . b    . . . b b    . b . . a    . . . . +
C57BL/10ScSnA . . . . .    . . . . b    b a . b b    . b . . a    . . . . .
C57BL/10ScSnJ . . . . .    . b . . b    . d b b b    . b . . a    . . . + b
C57BL/AnHf    . . . . .    . . . . b    . . . . b    . . . . .    . . . . .

C57BL/ImrHeA  . . . . .    . . . . b    b a . b b    . b . . a    . . . . .
C57BL/KaLwN   . . . . .    . . . . b    . . . . b    . . . . .    . . . . .
C57BL/KsJ     . b . . .    . . a . b    . a . b b    . b . . .    . . . . +
C57BL/LiA     . . . . .    . . . . b    b a . a b    . b . . a    . . . . .
C57BR/cd-     b . . . .    . . a . b    b a . b a    . b a o a    . . . + .

C57BR/cdJ     . . . . .    . . a . b    b a . b a    . b a o a    . . . + .
C57L/-        b b . . .    . b a b b    c a . b b    b b . o a    . a a + .
C57L/J        . b . . .    . b a . b    c a . b b    b b . o a    . a a + .
C57P/A        . . . . .    . . . . b    b a . a b    . b . . a    . . . . .
C58/-         b . . a .    b b a . b    b a b b b    . b a o a    . . . + .
```

```
Locus    ->   P P P P   P P P P   P P P P   P P P P Q   Q Q Q Q
(read down)   a c d e e  e e g g   g g g k r  r r r s a  a a a a e
              n a   p p  p p d k k m m m - e  n t t p -  - - - - d
              - -   - -  - - - - - - - - 3 -  - - -   1  2 3 4 5 -
              2 1   2 3  4 7 1 2   1 2 3   2  p 1 2            1

--------------------------------------------------------------------
Chromosome    U U U 1 1  7 5 4 X 1  5 4 9 9 1  2 U 8 2 1  1 1 1 1 1
(read down)   N N N 0    7          2          N       7  7 7 7 7
--------------------------------------------------------------------

BRSUNT/N      . . . . b  . . . . .  b . . . .  . . . . .  . . . . .
BS            . . . . .  . . b b .  a a . . .  . b a . .  . . . . .
BUB/BnJ       . . b . b  . . b b a  a a a . .  a . . a .  a a b b a
BXSB/MpJ      . . . . b  . . . . .  a a . . .  . . . . .  . . . . .
C3H/-         . a a . b  . . b b b  * a a a b  a a a b b  b b b b b

C3H/An        . . . . .  . . . . .  . . . . .  . . . . b  b b b b .
C3H/Bi        . . . . .  . . . . .  a . . . .  . . . . .  b b . . .
C3H/BiU       . . . . .  . . . . .  . . . . .  . . . . .  . . . . .
C3H/Crgl      . . . . .  . . . . .  . . . . .  . . . . .  . . . . .
C3H/He-       . a a . b  . . b b b  b a a a b  a a a b .  . . . . b

C3H/HeA       . . . . b  . . b b b  b a . . .  . . . . .  . . . . .
C3H/HeAf      . . . . .  . . . . .  . . . . .  . . . . .  . . . . .
C3H/HeDiSnA   . . . . b  . . b b b  b a . . .  . . . . .  . . . . .
C3H/HeJ       . a . . b  . . b b b  b a a . b  . . . b .  . . . . b
C3H/HeOuJ     . . . . b  . . . . .  b . . . .  . . . . .  . . . . .

C3H/HeSnJ     . . . . b  . . . . b  b a . . .  a . . . .  . . . . .
C3H/St        . . . . .  . . . . .  b . . . .  . . . . .  . . . . .
C3H/StWi      . . . . b  . . . . .  a . . . .  . . . . .  . . . . .
C3HeB/FeJ     b . a . b  . . . . b  b a . . .  . . b . .  . . . . .
C3HfB/HeN     . . . . b  . . . . .  b . . . .  . . . . .  . . . . .

C57BL/6-      b b a a a  b a b b a  a a a a b  a a a b b  a a a a b
C57BL/6ByA    . . . . a  . . b b a  a a . . .  . . . . .  . . . . .
C57BL/6ByJ    . . . . a  . . . . a  a a . . .  . . . . .  . . . . .
C57BL/6J      b . a a a  b a b b a  a a a a b  a a a b .  . . . . b
C57BL/6JN     . . . . a  . . . . .  a . . . .  . . . . .  . . . . .

C57BL/10-     b . a . a  b . b b a  a a a . .  a . . . b  a a a a b
C57BL/10J     . . a . a  . . . . a  a a . . .  . . . . .  . . . . .
C57BL/10ScSnA . . . . a  . . b b a  a a . . .  . . . . .  . . . . .
C57BL/10ScSnJ b . . . a  b . . . a  a a . . .  a . . . .  . . . . b
C57BL/AnHf    . . . . c  . . . . .  a . . . .  . . . . .  . . . . .

C57BL/ImrHeA  . . . . a  . . b a .  a a . . .  . . . . .  . . . . .
C57BL/KaLwN   . . . . a  . . . . a  a . . . .  . . . . .  . . . . .
C57BL/KsJ     . . a . a  . . b a .  a a a a .  . . b . .  . . . . .
C57BL/LiA     . . . . a  . . b b .  a a . . .  . . . . .  . . . . .
C57BR/cd-     . b a . a  . . b . a  a a c . a  . . . . a  b b b b a

C57BR/cdJ     . . a . a  . . b . a  a a c . a  . . . . .  . . b b a
C57L/-        . b . a a  b a b b c  a a c a .  . . . b b  a a a a b
C57L/J        . . . a a  b a b b c  a a c . .  . . . b .  . . . . b
C57P/A        . . . . a  . . b b a  a a . . .  . . . . .  . . . . .
C58/-         . b a a a  b a b b a  a a c a .  . . . a a  b b a a a
```

```
Locus    ->      r r R R R   S S S S S   S S S S S   S S S T T   T T T T T
(read down)      d d e i n   a a a d d   l o o o p   s v v a c   c g h i l
                   s n b 7   a a s h r   p a d d i   p p p m n   r n y n a
                   - - s     - - - -     - - -       - - - -       g   - d
                   1 1 -     1 1 1 1     1 2 2       1 2 1 2       1
                       6                   r


                -------------------------------------------------------------
    Chromosome   5 1 1 1 6   7 7 1 2 U   1 U 1 1 1   U 2 7 7 1   1 1 9 1 1
    (read down)    7   4             N   7 N 6 7 2   N     1       3 5   2 7
                -------------------------------------------------------------


BRSUNT/N         . . . . .   . . . . .   . . . . .   . . . . .   . . . . .
BS               . . . . .   . . . . .   . . a . .   . . . . .   . . . . .
BUB/BnJ          1 . s a b   a . o a 2   . . . . .   . . . c .   . b a . a
BXSB/MpJ         . . . . .   . . . . .   . . . . .   . . . . .   . . b . .
C3H/-            1 . b a b   a b o a 2   o b a a b   s a c c s   a c b d b

C3H/An           1 . . . .   . . . . .   . . . . .   . a . . .   . . . . b
C3H/Bi           . . . . .   . . . . .   . . . . .   . . . . .   . . . . b
C3H/BiU          . . . . .   . . . . .   . . . . .   . . . . .   . . . . .
C3H/Crgl         . . . . .   . . . . .   . . . . .   . . . . .   . . . . .
C3H/He-          1 . b a b   a b o a 2   o b a a b   . a c c s   a c b d b

C3H/HeA          1 . . . .   . . . . .   . . . . .   . a . . .   . . b . b
C3H/HeAf         . . . . .   . . . . .   . . . . .   . . . . .   . . . . .
C3H/HeDiSnA      . . . . .   . . . a .   . . . . .   . a . . .   . . b . b
C3H/HeJ          1 . b a b   a b o a 2   . . a a b   . a . . s   a c b d b
C3H/HeOuJ        . . . . .   . . . . .   . . . . .   . . . . .   . . . . .

C3H/HeSnJ        1 . . . .   . . . a .   . . . . .   . . . . .   . . . . b
C3H/St           1 . . . .   . . . . .   . . . . .   s . . . .   . . b . .
C3H/StWi         . . . . .   . . . . .   . . . . .   . . . . .   . . . . .
C3HeB/FeJ        1 . . a .   a . o a .   . . . . .   . . c . .   . . . . .
C3HfB/HeN        . . . . .   . . . . .   . . . . .   . . . . .   . . . . .

C57BL/6-         + . b b b   a b a a 2   o b a a b   f b a c s   b b b o b
C57BL/6ByA       . . . . .   . . . a .   . . . . .   . b . . .   . . b . b
C57BL/6ByJ       + . . . .   . b . a .   . b a a .   . . . . .   . . b . b
C57BL/6J         + . b b b   a b a a 2   o . a a b   f b . c s   b b b o .
C57BL/6JN        . . . . .   . . . . .   . . . . .   . . . . .   . . b . b

C57BL/10-        + + b b .   a . a a 2   o b . . .   . b . c .   . . b . b
C57BL/10J        + . b b .   a . a a 2   o . . . .   . b . c .   . . . . b
C57BL/10ScSnA    + + . . .   a . . a .   . . . . .   . b . . .   . . b . b
C57BL/10ScSnJ    + . . . .   a . a s .   . b . . .   . . . c .   . . b . b
C57BL/AnHf       . . . . .   . . . . .   . . . . .   . . . . .   . . . . .

C57BL/ImrHeA     + + . . .   . . . a .   . . . . .   . b . . .   . . b . b
C57BL/KaLwN      . . . . .   . . a . .   . . . . .   . . . . .   . . . . .
C57BL/KsJ        + . . b .   a . a a .   . . . . .   s . . c .   . . b . .
C57BL/LiA        + + . . .   . . . . .   . . . . .   s b . . .   . . b . .
C57BR/cd-        + . b b b   a . a a 2   o . . . d   . b b c .   b b b . a

C57BR/cdJ        + . b b b   a . a a 2   o . . . d   . . . c .   b b b . a
C57L/-           + . b b b   a b a a .   o b a a d   . . . c s   b b b o c
C57L/J           + . b b b   a b a a .   . . a a d   . . . c s   b b b o c
C57P/A           + + . . .   . . . . .   . . . . .   . b . . .   . . b . .
C58/-            + . b b a   a . a a 2   o b a a b   . a . c s   a b b . a
           1 = rd     2 = b?
```

Locus (read down) →	Tpre	Trf	Tsuy	Tth	Ucp	Upg-1	Upg-2	Ups	Wap	Xmmv-9	Xmmv-42	Xmmv-43	Xmmv-44	Xmmv-45	Xmmv-46	Xmmv-47	Xmmv-48	Xmmv-49	Xmmv-50	Xmmv-51	Xmmv-52	Xmmv-53	Xmmv-54	Xmmv-55	Xmmv-56
Chromosome (read down)	12	9	12	12	8	17	1	9	UN	1	19	UN	UN	UN	UN	UN	UN	UN	12	UN	5	UN	UN	15	UN
BRSUNT/N	.	b	.	.	.	.	.	.	.	.	.	.	.	.	.	.	.	.	.	.	.	.	.	.	.
BS	.	b	.	.	.	.	.	.	.	.	.	.	.	.	.	.	.	.	.	.	.	.	.	.	.
BUB/BnJ	.	b	.	.	.	f	d	a	a	o	o	o	o	o	o	o	o	o	o	a	o	a	o	a	a	o
BXSB/MpJ	.	b	.	.	.	.	.	.	.	.	.	.	.	.	.	.	.	.	.	.	.	.	.	.	.
C3H/-	o	*	o	d	d	f	s	a	a	a	o	o	o	o	o	o	o	o	o	o	a	o	a	o	o
C3H/An	.	.	.	.	.	.	.	.	.	.	.	.	.	.	.	.	.	.	.	.	.	.	.	.	.
C3H/Bi	.	a	.	.	.	.	.	.	.	.	.	.	.	.	.	.	.	.	.	.	.	.	.	.	.
C3H/BiU	.	.	.	.	.	.	.	.	.	.	.	.	.	.	.	.	.	.	.	.	.	.	.	.	.
C3H/Crgl	.	.	.	.	.	.	.	.	.	.	.	.	.	.	.	.	.	.	.	.	.	.	.	.	.
C3H/He-	o	b	o	d	d	f	s	a	a	a	o	o	o	o	o	o	o	o	o	o	a	o	a	o	o
C3H/HeA	.	b	.	.	.	.	.	.	.	.	.	.	.	.	.	.	.	.	.	.	.	.	.	.	.
C3H/HeAf	.	.	.	.	.	.	.	.	.	.	.	.	.	.	.	.	.	.	.	.	.	.	.	.	.
C3H/HeDiSnA	.	.	.	.	.	.	.	.	.	.	.	.	.	.	.	.	.	.	.	.	.	.	.	.	.
C3H/HeJ	o	b	o	d	d	f	s	a	a	a	o	o	o	o	o	o	o	o	o	o	a	o	a	o	o
C3H/HeOuJ	.	b	.	.	.	.	.	.	.	.	.	.	.	.	.	.	.	.	.	.	.	.	.	.	.
C3H/HeSnJ	.	b	.	.	.	.	.	.	a	a	o	o	o	o	o	o	o	o	o	o	a	o	a	o	o
C3H/St	.	b	.	.	.	.	.	.	.	.	.	.	.	.	.	.	.	.	.	.	.	.	.	.	.
C3H/StWi	.	.	.	.	.	.	.	.	.	.	.	.	.	.	.	.	.	.	.	.	.	.	.	.	.
C3HeB/FeJ	.	b	.	.	.	f	.	.	a	a	o	o	o	o	o	o	o	o	o	o	a	o	a	o	o
C3HfB/HeN	.	.	.	.	.	.	.	.	.	.	.	.	.	.	.	.	.	.	.	.	.	.	.	.	.
C57BL/6-	o	b	o	o	b	s	s	a	a	a	o	o	o	o	o	o	o	o	o	o	a	o	a	a	o
C57BL/6ByA	.	b	.	.	.	.	.	a	.	.	.	.	.	.	.	.	.	.	.	.	.	.	.	.	.
C57BL/6ByJ	.	b	.	.	.	.	.	.	a	a	o	o	o	o	o	o	o	o	o	o	a	o	a	a	o
C57BL/6J	o	b	o	o	b	s	s	a	a	a	o	o	o	o	o	o	o	o	o	o	a	o	a	a	o
C57BL/6JN	.	b	.	.	.	.	.	.	.	.	.	.	.	.	.	.	.	.	.	.	.	.	.	.	.
C57BL/10-	.	b	.	.	.	s	s	a	a	a	o	o	o	o	o	o	o	o	o	o	a	o	a	a	o
C57BL/10J	.	b	.	.	.	.	.	a	a	a	o	o	o	o	o	o	o	o	o	o	a	o	a	a	o
C57BL/10ScSnA	.	b	.	.	.	.	.	a	.	.	.	.	.	.	.	.	.	.	.	.	.	.	.	.	.
C57BL/10ScSnJ	.	b	.	.	.	s	s	.	a	a	o	o	o	o	o	o	o	o	o	o	a	o	a	a	o
C57BL/AnHf	.	b	.	.	.	.	.	.	.	.	.	.	.	.	.	.	.	.	.	.	.	.	.	.	.
C57BL/ImrHeA	.	b	.	.	.	.	.	.	.	.	.	.	.	.	.	.	.	.	.	.	.	.	.	.	.
C57BL/KaLwN	.	b	.	.	.	.	.	.	.	.	.	.	.	.	.	.	.	.	.	.	.	.	.	.	.
C57BL/KsJ	.	b	.	.	.	f	.	.	a	a	o	o	o	o	o	o	o	o	o	a	o	a	o	a	a	o
C57BL/LiA	.	b	.	.	.	.	.	.	.	.	.	.	.	.	.	.	.	.	.	.	.	.	.	.	.
C57BR/cd-	.	b	.	.	b	s	s	.	.	a	a	o	o	o	o	a	o	o	o	o	o	o	a	a	o
C57BR/cdJ	.	b	.	.	b	s	s	.	.	a	a	o	o	o	o	a	o	o	o	o	o	o	a	a	o
C57L/-	o	b	o	o	b	s	s	a	a	a	a	o	o	o	o	a	o	o	o	o	o	o	a	a	o
C57L/J	o	b	o	o	b	s	s	a	a	a	a	o	o	o	o	a	o	o	o	o	o	o	a	a	o
C57P/A	.	b	.	.	.	.	.	.	.	.	.	.	.	.	.	.	.	.	.	.	.	.	.	.	.
C58/-	.	b	.	.	.	s	.	a	a	a	a	o	o	o	o	a	o	o	o	o	a	o	a	a	o

Locus → (read down)	Xmmv-57	Xmmv-58	Xmmv-59	Xmmv-60	Xmmv-62	Xmmv-63	Xmmv-64	Xmmv-65	Xmmv-66	Xmmv-67	Xmmv-68	Xmmv-69	Xmmv-70
Chromosome (read down)	UN	UN	UN	UN	4	UN	UN	3	UN	UN	UN	UN	UN
BRSUNT/N	.	.	.	.	.	.	.	.	.	.	.	.	.
BS	.	.	.	.	.	.	.	.	.	.	.	.	.
BUB/BnJ	o	o	o	o	o	a	o	a	o	o	o	a	a
BXSB/MpJ	.	.	.	.	.	.	.	.	.	.	.	.	.
C3H/-	o	o	o	o	o	a	*	o	o	o	o	a	a
C3H/An	.	.	.	.	.	.	.	.	.	.	.	.	.
C3H/Bi	.	.	.	.	.	.	.	.	.	.	.	.	.
C3H/BiU	.	.	.	.	.	.	.	.	.	.	.	.	.
C3H/Crgl	.	.	.	.	.	.	.	.	.	.	.	.	.
C3H/He-	o	o	o	o	o	a	*	o	o	o	o	a	a
C3H/HeA	.	.	.	.	.	.	.	.	.	.	.	.	.
C3H/HeAf	.	.	.	.	.	.	.	.	.	.	.	.	.
C3H/HeDiSnA	.	.	.	.	.	.	.	.	.	.	.	.	.
C3H/HeJ	o	o	o	o	o	a	a	o	o	o	o	a	a
C3H/HeOuJ	.	.	.	.	.	.	.	.	.	.	.	.	.
C3H/HeSnJ	o	o	o	o	o	a	o	o	o	o	o	a	a
C3H/St	.	.	.	.	.	.	.	.	.	.	.	.	.
C3H/StWi	.	.	.	.	.	.	.	.	.	.	.	.	.
C3HeB/FeJ	o	o	o	o	o	a	o	o	o	o	o	a	a
C3HfB/HeN	.	.	.	.	.	.	.	.	.	.	.	.	.
C57BL/6-	o	o	o	o	o	a	*	o	o	o	o	a	a
C57BL/6ByA	.	.	.	.	.	.	.	.	.	.	.	.	.
C57BL/6ByJ	o	o	o	o	o	a	o	o	o	o	o	a	a
C57BL/6J	o	o	o	o	o	a	a	o	o	o	o	a	a
C57BL/6JN	.	.	.	.	.	.	.	.	.	.	.	.	.
C57BL/10-	o	o	o	o	o	a	a	o	o	o	o	a	a
C57BL/10J	o	o	o	o	o	a	a	o	o	o	o	a	a
C57BL/10ScSnA	.	.	.	.	.	.	.	.	.	.	.	.	.
C57BL/10ScSnJ	o	o	o	o	o	a	a	o	o	o	o	a	a
C57BL/AnHf	.	.	.	.	.	.	.	.	.	.	.	.	.
C57BL/ImrHeA	.	.	.	.	.	.	.	.	.	.	.	.	.
C57BL/KaLwN	.	.	.	.	.	.	.	.	.	.	.	.	.
C57BL/KsJ	o	o	o	o	o	a	a	a	o	a	o	a	a
C57BL/LiA	.	.	.	.	.	.	.	.	.	.	.	.	.
C57BR/cd-	o	o	o	o	a	a	a	o	a	o	o	a	a
C57BR/cdJ	o	o	o	o	a	a	a	o	a	o	o	a	a
C57L/-	o	o	o	o	a	a	a	o	a	o	o	a	a
C57L/J	o	o	o	o	a	a	a	o	a	o	o	a	a
C57P/A	.	.	.	.	.	.	.	.	.	.	.	.	.
C58/-	o	o	o	o	a	a	a	o	a	a	o	a	a

```
Locus    ->      a A A A A     A A A A A     A A A A A     A A A A A     A A A A A
(read down)      a b c c       d d d d g     h h h h k     k l m m o     o p p p s
                 t p f o       h h h h s     d d r p       p b y y x     x h k o -
                 a - -         - - - - -     - - - -       - - - - -     - - a 1
                 1 1           1 2 3 3       1 2 1 1       2 1 1 2 1     2 1 - s
                               e t                                       1

-----------------------------------------------------------------------------------
Chromosome       2 1 7 1 4     3 U 3 3 X     1 4 1 3 1     4 5 3 3 1     1 2 1 9 1
(read down)      2             N             2 9                         0   3
-----------------------------------------------------------------------------------

C58/J       a o a a a   . . a . h   b a a a a   b . a a 1   . . a a a
C58/JA      l . . . .   . . . . .   . . . . .   . . . . .   . . . . .
CALB/Rk     . . . . .   . . . . .   . . . b     . a b . .   . . a b .
CASA/Rk     . . . . .   . . . . .   . . . b     . a a . .   . . a b .
CAST/Ei     . . . . b   . . . . .   . . . . .   . . . . .   . . . . .

CBA/-       + * a b a   a a a a h   b b a a *   a a b b b   a a a * a
CBA/BrA     + . a . .   a a a a .   b b a . .   . . . . b   a . . . .
CBA/Ca-     + o a . a   a . a a h   b b a a b   a a a a b   a . a a a
CBA/CaH-T6J + . a . a   . . . . .   . . . b     . a a a .   . . a a .
CBA/CaHN    + . . . .   . . . . .   b . a . .   . . . . .   . . . . .

CBA/CaJ     + o . . a   . . . . h   . b . . b   a . a . .   . . a a a
CBA/J       + a a b a   . . a . h   b b . . a   a a b b 1   . . a b a
CE/-        + a . a a   . . a . h   b b a a b   a a b b *   c . . a a
CE/J        + a . a a   . . a . h   b b a a b   a a b b a   c . . a a
CFO         . . . . .   . . . . .   . . . . .   . a . a .   . . . . .

CHMU/Le     . . . . .   . . . . .   . . . a     . . a . .   . . b .
CSB/-       . . . . .   . . . . .   . . . b     . a a . .   . . a .
CWD/Le      . . . . .   . . . . .   . . . a     . . a . .   . . b .
DA/HuSn     . . . . .   . . . . .   . . . .     . . a . .   . . . . .
DBA/1-      a a b . a   . . a . h   d b . a a   b a a a 1   . . . b a

DBA/1J      a a b . a   . . . . h   d b . . a   b a a a 1   . . . b a
DBA/2-      a a b b a   . . a a h   d b . a a   a a a a b   a a a b a
DBA/2DeJ    a . . . .   . . . . .   . . . a     . . a . .   . . a b .
DBA/2J      a a b b a   . . . . h   d b . . a   a a a a b   . . a b a
DBA/LiA     a . . . .   a a a a .   d b a . .   . . . b     a . . . .

DBA/LiAf    . . . . .   . . . . .   . . . .     . . . . .   . . . . .
DC/Le       . . . . .   . . . . .   . . . b     . . a . .   . . b .
DD/He       + . . . .   . . . . .   . . . .     . . . . .   . . . . .
DD/HeA      + . . . .   a a a a .   d b a . .   . . . . a   a . . . .
DD/HeAf     . . . . .   . . . . .   . . . .     . . . . .   . . . . .

DDD         + . . . .   a . b b .   . . . b b   . . . . .   . . . . .
DDK         + . . . .   . . . . .   . . . b     . . . . .   . . . . .
DE/J        + . . a .   . . . . h   b . . .     . . b b 2   . . a . .
DLS/Le      + . . . .   . . . . .   . . . b     . . a . .   . . a .
DW/J        + . . . a   . . . . h   . . . a     . a a . .   . . a .

F/St        a . . . .   . . . . h
FL/1Re      a . a . .   . . . . h                                      b .
FL/2Re      . . . . .   . . . . .                                      b .
FM/JmsA     . . . . .   a . a a .   . . . a     . . . . .
FS/Ei       . o . . .   . . . . .   . . . .     . a a . .   . . a .
```

1 = b?   2 = a?

```
Locus    ->    A A b B B    B B B B c    C C C C C    c C C C C    C d D D D
(read down)    v v 2 c d    g g g v      3 4 6 a a    d e e f o    v   7 1 1
               - - m   d    l l l -            r r    m - - h h    -   R 2 2
               1 2     1    - - - 1            1 2    1 2 e        1     N N
                             e s t                                  p   y y
                                                                   2   - u u
                                                                       s 1 2
---------------------------------------------------------------------------------
Chromosome     U U 4 2 5    9 9 9 8 7    1 1 1 3 3    3 2 1 2 7    5 9 7 1 1
(read down)    N N                       7 7 5              7            2 2
---------------------------------------------------------------------------------

C58/J          r r + a .    b h b b +    b . . a a    + . a . .    b + d . .
C58/JA         . . + . .    . . . . +    . . . . .    . . . . .    . + . . .
CALB/Rk        . . + . .    . . . . +    b . . b a    . . . . .    . + . . .
CASA/Rk        . . + . .    . . . . +    b . . b a    . . . . .    . + . . .
CAST/Ei        . . . . .    . . . . .    d . . . .    . . . . .    . . . . .

CBA/-          r r + a .    b d d b +    b * . a *    + . b f l    * + b . .
CBA/BrA        . . + . .    . . . . +    . h . . .    + . b . .    . + . . .
CBA/Ca-        r r + . b    . d . b +    b . . a a    + . b . l    b + . . .
CBA/CaH-T6J    . . + . .    . . . b +    b . . a a    + . b . .    b + . . .
CBA/CaHN       . . + . .    . . . b +    b . . . a    . . b . .    b + . . .

CBA/CaJ        r r + . .    . d . b +    b . . a a    + . . . l    b + . . .
CBA/J          . . + a .    b d d b +    b l . a b    + . b f .    a + b . .
CE/-           . . + a .    a h d b 1    b . . a b    + . b . .    b + d . .
CE/J           . . + a .    a h d b 1    b . . a b    + . b . .    b + d . .
CFO            . . . . .    . . . . .    . . . . .    . . . . .    . . . . .

CHMU/Le        . . . . .    . . . . .    . . . . a    . . . . .    . . . . .
CSB/-          . . . . .    . . . . .    . . a a .    . . b . .    . . . . .
CWD/Le         . . . . .    . . . . .    . . . . a    . . . . .    . . . . .
DA/HuSn        . . . a .    . . . b c    c . . . .    . . . . .    b + . . .
DBA/1-         . . b a .    b d d b +    b h . a a    + . b f h    b d . a a

DBA/1J         . . b a .    b d d b +    b h . a a    + . b . h    b d . a a
DBA/2-         r r b a .    b d d b +    b h a a b    + . a f h    b d d a a
DBA/2DeJ       . . b . .    . . . b +    . . . a b    . . . . .    b d . . .
DBA/2J         . . b a .    b d d b +    b h . a b    + . a . h    b d d a a
DBA/LiA        . . b . .    . . . . +    . h . . .    + . . . .    . d . . .

DBA/LiAf       . . . . .    . . . . .    . . . a .    . . . . .    . . . . .
DC/Le          . . . . .    . . . . .    . . . a .    . . . . .    . . . . .
DD/He          . . + . .    a h b . c    . . . a .    . . a . .    . . . . .
DD/HeA         . . + . .    . . . . c    . h . . .    + . a . .    . + . . .
DD/HeAf        . . . . .    . . . . .    . . . . .    . . . . .    . . . . .

DDD            . . + . .    . . . . c    b . . . b    . . . . .    . + . . .
DDK            . . + . .    . . . . c    . . . . b    . . . . .    . + . . .
DE/J           . . + . .    a h d . 1    . . . . .    . . . . .    . . . . .
DLS/Le         . . . . .    . . . . .    . . . . a    . . . . .    . . . . .
DW/J           . . + . .    a h d . +    b . . a b    . . . . .    . + . . .

F/St           . . b . .    b d d . 2    . . . . a    . . a . .    . d . . .
FL/1Re         . . + . .    . h . . c    b . . a b    . . . . .    . + . . .
FL/2Re         . . . . .    . . . . .    . . . a b    . . . . .    . . . . .
FM/JmsA        . . . . .    . . . . +    . . . . .    . . a . .    . . . . .
FS/Ei          . . b . .    . . . b 2    b . . a a    . . . . .    a + . . .
```

          1 = c*e   2 = c*ch

The laboratory mouse

```
Locus    ->    D D D E E   E E E E E   E E E E   E E E E E   E E E E
(read down)    1 1 l a a   a a a g m   m m m m   m m m m   m p s s
               2 2 b - -   - - - a     v v v v   v v v v   v h - -
               N N - 2 4   5 6 7       - - - -   - - - -   - - 1 2 3
               y y 1                   4 5 6 7 8 9 1 1 1 1   1 1
               u u                               0 1 2 3    4
               3 4
               ------------------------------------------------------
Chromosome     1 1 1 U 1   U 2 U 8 U   U U U U U   U U 7 1 2   1 1 8 8 1
(read down)    2 2 1 N 9   N   N   N   N N N N N   N N   6     1       1
               ------------------------------------------------------

C58/J          . . . . .   . . . . l   b b b b b   b b b b b   b . b b c
C58/JA         . . . . .   . . . . .   . . . . .   . . . . .   . . b b c
CALB/Rk        . . . . .   . . . . .   . . . . .   . . . . .   . . a a b
CASA/Rk        . . . . .   . . . . .   . . . . .   . . . . .   . 1 b b
CAST/Ei        . . . . .   . . . . .   . . . . .   . . . . .   . . . . .

CBA/-          . . b b a   . a a a h   b b b b b   b b b b b   b d b b c
CBA/BrA        . . . . .   . . . . .   . . . . .   . . . . .   . . b b c
CBA/Ca-        . . b . .   . . . . .   b b b b b   b b b b b   b d b b c
CBA/CaH-T6J    . . . . .   . . . . .   b b b b b   b b b b b   b . b b c
CBA/CaHN       . . b . .   . . . . .   b b b b b   b b b b b   b . b . c

CBA/CaJ        . . . . .   . . . . .   b b b b b   b b b b b   b d b b c
CBA/J          . . . b a   . a a a h   b b b b b   b b b b b   b d b b c
CE/-           . . . b a   . . b a .   b b b b b   b b b b b   b . b b c
CE/J           . . . b a   . . b a .   b b b b b   b b b b b   b . b b c
CFO            . . . . .   . . . . .   . . . . .   . . . . .   . . b b c

CHMU/Le        . . . . .   . . . . .   . . . . .   . . . . .   . . a . a
CSB/-          . . . . .   . . . . .   . . . . .   . . . . .   . . b b b
CWD/Le         . . . . .   . . . . .   . . . . .   . . . . .   . . b . c
DA/HuSn        . . . b a   . . b . .   b b b a b   b b b b b   b . . . .
DBA/1-         b b b b a   b b b . h   b b b b b   b b b b b   b d b b c

DBA/1J         b b . . .   . . . . h   b b b b b   b b b b b   b d b b c
DBA/2-         b b b b a   b b b a h   b b b b b   b b b b b   b d b b c
DBA/2DeJ       . . . . .   . . . . h   b b b b b   b b b b b   b . b b c
DBA/2J         b b . . .   . . . . h   b b b b b   b b b b b   b d b b c
DBA/LiA        . . . . .   . . . . .   . . . . .   . . . . .   . . b b c

DBA/LiAf       . . . . .   . . . . .   . . . . .   . . . . .   . . b b c
DC/Le          . . . . .   . . . . .   . . . . .   . . . . .   . . b . c
DD/He          . . . . .   . . . . .   . . . . .   . . . . .   . . b . c
DD/HeA         . . . . .   . . . . .   . . . . .   . . . . .   . . b b c
DD/HeAf        . . . . .   . . . . .   . . . . .   . . . . .   . . b b c

DDD            . . . . .   . . . . .   . . . . .   . . . . .   . . b b c
DDK            . . a . .   . . . . .   . . . . .   . . . . .   . . b b c
DE/J           . . . . .   . . . a .   . . . . .   . . . . .   . . b b c
DLS/Le         . . . . .   . . . . .   . . . . .   . . . . .   . . b . c
DW/J           . . . . .   . . . . .   . . . . .   . . . . .   . . b b a

F/St           . . . a a   a a . . .   . . . . .   . . . . .   . . b . .
FL/1Re         . . . . .   . . . . .   . . . . .   . . . . .   . . . . .
FL/2Re         . . . b a   . . b . .   . . . . .   . . . . .   . . . . .
FM/JmsA        . . . . .   . . . . .   . . . . .   . . . . .   . . b b c
FS/Ei          . . . . .   . . . . .   b b b b b   b b b b b   b . . . .
```

1 = ab

```
Locus    ->    E E E E   E E E E E   E E E E E   E E E E   F F F F F
(read down)    s s s s   s s s s s   s s s s s   s s s s   a b b o o
               - - - -   - - - - -   - - - - -   - - - r   b p p r s
               5 6 7 8 9 1 1 1 1 1   1 1 1 2 2   2 2 2 2   p - - -
                         0 1 2 3 4   6 7 8 2 3   4 5 6 7   i 1 2 5
-------------------------------------------------------------------------
    Chromosome  8 8 8 7 8   1 8 U 9 9   3 9 1 8 8   8 1 3 3 6   3 U U 1 1
    (read down)             4   N             9         2           N N 4 2
-------------------------------------------------------------------------

C58/J          . a . a .   a a . a .   . . . . .   . . . . .   . . . b .
C58/JA         b a b . a   a a . a r   a a a d c   a b a r a   . b . . .
CALB/Rk        . . . . .   b . . . .   . . . . .   . . . . .   . . . . .
CASA/Rk        . . . . .   a . . . .   . . . . .   . . . . .   . . . . .
CAST/Ei        . . . . .   . b . . .   . . . . .   . . . . .   . . . . .

CBA/-          b a b a a   b a . a r   a a a d c   a b a r a   . b . d .
CBA/BrA        b a b a a   b a . a r   a a a d c   a b a r a   . b . . .
CBA/Ca-        b a . . a   b a . a .   . . . d .   . . . . .   . . . . .
CBA/CaH-T6J    b . . . .   b a . . .   . . . . .   . . . . .   . . . . .
CBA/CaHN       . . . . .   b . . . .   . . . . .   . . . . .   . . . . .

CBA/CaJ        . . . . .   b a . a .   . . . . .   . . . . .   . . . . .
CBA/J          b a . a .   b a . a .   . . a . c   . b a . .   . . . d .
CE/-           b a b a a   b a . a r   a a a d c   a . a r .   . b a d .
CE/J           . a . a .   b . . . .   . . . . .   . . . . .   . b a d .
CFO            b . . . .   . a . . .   . . . . .   . . . . .   . . . . .

CHMU/Le        . . . . .   a a . . .   . . . . .   . . . . .   . . . . .
CSB/-          . . . . .   b a . . .   . . . . .   . . . . .   . . . . .
CWD/Le         . . . . .   b a . . .   . . . . .   . . . . .   . . . . .
DA/HuSn        . . . . .   . . . . .   . . . . .   . . . . .   . . . . .
DBA/1-         . . . . a   b a . a .   . . . . .   . . . . .   . . . d b

DBA/1J         . . . . .   b a . a .   . . . . .   . . . . .   . . . d b
DBA/2-         b a b a a   b a a a .   a a a d c   a b a . .   d b a d b
DBA/2DeJ       . . . . .   b a . . .   . . . . .   . . . . .   . . . . .
DBA/2J         b a . a a   b a a a .   a a a d c   a b a . .   d . . d b
DBA/LiA        b a b a a   b a . a r   a a a d c   a b a r a   . . . . .

DBA/LiAf       b a b . a   b a . a r   a a a d c   a b a r a   . . . . .
DC/Le          . . . . .   a a . . .   . . . . .   . . . . .   . . . . .
DD/He          . . . . .   b . . . .   . . . . .   . . . . .   . . . . .
DD/HeA         a . b a .   . . . . .   . . . . .   . . . . r   . b . . .
DD/HeAf        b a b . a   b a . b r   a a a d c   a b b r o   p . . . .

DDD            . . . . .   . . . . .   . . . . .   . . . . .   . . . . .
DDK            . . . . .   . . . . .   . . . . .   . . . . .   . . . . .
DE/J           . a . a .   . . . . .   . . . . .   . . . . .   . . . . .
DLS/Le         . . . . .   a a . . .   . . . . .   . . . . .   . . . . .
DW/J           b a . a .   a a . . .   . . . . .   . . . . .   . . . . .

F/St           . . . . .   . a . . .   . . . . .   . . . . .   . . . . .
FL/1Re         . . . . .   a . . . .   . . . . .   . . . . .   . . . . .
FL/2Re         . . . . .   a . . . .   . . . . .   . . . . .   . . . . .
FM/JmsA        . . . . .   . . . . .   . . . . .   . . . . .   . . . . .
FS/Ei          . . . . .   b . . . .   . . . . .   . . . . .   . . . . .
```

```
Locus     ->   F F G G G   G G G G G   G G G G G   G G G G G   G G H H H
(read down)    v v a b d   d d d g k   l l m o o   p p r u     u v - - -
               - - l p c   c r r c     k o - t t   d i t - s   s - 1 2 3
               1 2 t - -   - - -       - 3 - -     - - - 1 -   - 1
                     1 1   2 1 2       1   1 2     1 1 1  r    s

---------------------------------------------------------------------------
   Chromosome  4 9 4 3 1   9 U U 6 U   1 1 2 1 8   4 7 1 8 5   5 U 7 1 2
   (read down)       5       N N N     1 1   9           5       N     7
                                       7   9
---------------------------------------------------------------------------

C58/J          n r b . b   b h l b 1   a a a a b   a a a a b   b . c k b
C58/JA         . . . . .   . . . . .   . a . a b   a a . a .   . . . . .
CALB/Rk        . . . d .   . . . . .   . a . a b   . a b . .   . . . . .
CASA/Rk        . . . d .   . . . . .   . a . b b   . 2 b a .   . . . . .
CAST/Ei        . . b . .   . . . . .   . a . . .   . . . . .   . . . . .

CBA/-          n s b b c   b h l a 3   b a b a b   b b a a b   * b a * 4
CBA/BrA        n s . . .   . . . . .   . a . a b   b b . a .   . . . . q
CBA/Ca-        n s b . b   b . . a .   . a . a b   b b a a b   b b a k 4
CBA/CaH-T6J    n s b . c   b . . a .   . a . a b   b b a a .   . b . k .
CBA/CaHN       . . . . .   . . . . .   . . . . .   b b . . .   b . . k .

CBA/CaJ        n . b . b   . . . a .   . a . a b   b b a a b   h . a k 4
CBA/J          . . b b c   b h l a 3   b a . a b   b b a a .   h b a k 4
CE/-           n s . b b   b . . a .   a a a a b   a a a a b   b a f k b
CE/J           n s . b b   b . . a .   a a a a b   a a a a b   b . f k b
CFO            . . . . .   . . . . .   . . a b   b a b . .   . . . . .

CHMU/Le        . . . . .   . . . . .   . a . . b   a b a a .   . . . b .
CSB/-          . . . b .   . . . . .   . a . a b   a a a . .   . . . . .
CWD/Le         . . . b .   . . . . .   . a . a b   a b a a .   . . . . .
DA/HuSn        . . . . b   b . . . .   . . . . .   b . . . .   . . 5 6 4
DBA/1-         n s b . b   b . . . .   a a a a b   a a a a b   b b a q b

DBA/1J         n s b . b   b . . . .   a a a a b   a a a a b   b b a q b
DBA/2-         n s b b b   b h l a .   a a a a b   b a a a b   b a a d b
DBA/2DeJ       . . . . .   b . . . .   . a . a b   b a a a .   b . . . .
DBA/2J         n s b b b   b h l a .   a a a a b   b a a a b   b . a d b
DBA/LiA        n . . . .   . . . . .   . a . a b   a a . a .   . . . q .

DBA/LiAf       . . . b .   . . . . .   . . . . .   . . . . .   . . . . .
DC/Le          . . . . .   . . . . .   . a . . b   b b a a .   . . . d .
DD/He          . . . . .   . . . . .   . . . . .   b a . . .   b . . . .
DD/HeA         n s . . .   . . . . .   . a . a b   b a . a .   . . . 7 .
DD/HeAf        . . . b .   . . . . .   . . . . .   . . . . .   . . . . .

DDD            n s . . .   . . . . .   . . . . .   b a . . .   . . . . .
DDK            . . . . .   . . . . .   . . . . .   b a . . .   . . . . .
DE/J           . . . . .   . . . . .   . . . a b   b a a . .   a . . 8 .
DLS/Le         . . . . .   . . . . .   . a . . b   b b a a .   . . . b .
DW/J           . . b . .   . . . a .   . a . a b   a a a a .   b . . b .

F/St           . . . . .   . . . . .   . . . . .   b a . . .   . . . p .
FL/1Re         . . . . .   . . . a .   . . . . .   . . a . .   . . . k .
FL/2Re         . . . . .   . . . a .   . . . . .   . . a . .   . . 5 k 4
FM/JmsA        . . . . .   . . . . .   . a . a b   b a . b .   . . . . .
FS/Ei          . . . . .   . . . b .   . . . . .   . a a . .   b . . . 4
```

1 = b?	2 = c?	3 = a?	4 = -a,-b	5 = -a,-c
6 = qp	7 = bp	8 = k?		

```
Locus    ->   H H H H H   H H H H H   H H H H H   H H I I I   I I I I I
(read down)   - - - - -   - a a b b   b c e m p   r s d f g   g g g g g
              4 7 8 9 1   1 o o a a   b x a       t d h - h   h h h h h
                      2   3 - - -       -         - - 1 -     - - - - -
                          1 2 4         1         1 1 1       2 3 4 5 A
                              p                                       r
                              s                                       s
--------------------------------------------------------------------------
Chromosome    7 9 1 U U   2 2 3 1 1   7 2 U 7 8   U 1 1 3 1   1 1 1 1 1
(read down)     4 N N         1 7         N       N 0     2   2 2 2 2 2
--------------------------------------------------------------------------

C58/J         a a a . a   b . . b b   s l a b .   . h a h a   a . . . .
C58/JA        . . . . .   . . . . .   s . . . .   . . a . .   . . . . .
CALB/Rk       . . . . .   . . . . .   d . . . .   . . a . .   . . . . .
CASA/Rk       . . . . .   . . . . .   d . . . .   . . c . .   . . . . .
CAST/Ei       . . . . .   . . . . .   . . . . .   . . . . .   . . . . .

CBA/-         a 1 1 b 1   a a a d .   d 1 a a .   . l a h j   a a a a o
CBA/BrA       . . . . .   . a a . .   d 1 . . .   . b h a .   . . . . .
CBA/Ca-       a 1 1 b 1   a a a . .   d 1 . . .   . a . a .   a . a . .
CBA/CaH-T6J   . . . . .   . . . . .   d 1 . . .   . b . a .   a . a . .
CBA/CaHN      . . . . .   . . . . .   d 1 . . .   . b . . .   . . . . .

CBA/CaJ       a 1 1 b 1   a . . . .   d . . . .   . . b . .   . . . . .
CBA/J         a 1 1 b 1   a . . d .   d 1 a a .   . l b l j   a a a a .
CE/-          1 1 1 a a   b . . f .   s o . . .   b . a l f   f f a a o
CE/J          1 1 1 a a   b . . f .   s o a . .   b . a l f   f f a a .
CFO           . . . . .   . . . . .   d . . . .   . . b . .   . . . . .

CHMU/Le       . . . . .   . . . . .   s . . . .   . . a . .   . . . . .
CSB/-         . . . . .   . . . . .   s . . . .   . . a . .   . . . . .
CWD/Le        . . . . .   . . . . .   d . . . .   . . a . .   . . . . .
DA/HuSn       1 b 1 b 1   a . . . .   . . . . .   . . . . g   c . a . .
DBA/1-        1 a b . a   b . . . b   d 1 a . .   b . b l c   c . . . .

DBA/1J        1 a b . a   b . . . b   d 1 a . .   b . b l c   c . . . .
DBA/2-        1 a b . a   b . . g b   d o a a a   b . b h c   c a a a o
DBA/2DeJ      . . . . .   . . . . .   d . . . .   . . b . .   . . . . .
DBA/2J        1 a b . a   b . . g b   d o a a a   b . b h c   c a a a .
DBA/LiA       . . . . .   . c a . .   d 1 . . .   . . b l a   . . . . .

DBA/LiAf      . . . . .   . . . . .   . . . . .   . . . . .   . . . . .
DC/Le         . . . . .   . . . . .   d . . . .   . . a . .   . . . . .
DD/He         . . . . .   . . . . .   d o a . .   . . a . .   . . . . .
DD/HeA        . . . . .   . a a . .   d o . . .   . . a l a   . . . . .
DD/HeAf       . . . . .   . . . c .   . . . . .   . . . . .   . . . . .

DDD           . . . . .   . . . . .   s o . . .   . . a . .   . . . . .
DDK           . . . . .   . . . . .   s o . . .   . . a . .   . . . . .
DE/J          . . . . .   . . . c .   s o . . .   . . a . f   2 . a . .
DLS/Le        . . . . .   . . . . .   d . . . .   . . a . .   . . . . .
DW/J          . . . . .   . . . b .   s . a . .   . . a . .   . . . . .

F/St          . . . . .   . . . . .   s 1 a . .   . . a . .   . . . . .
FL/1Re        . . . . .   . . . c .   d . a . .   . . . . .   . . . . .
FL/2Re        1 b 1 a a   a . . c .   d . . . .   . . . . .   . . . . .
FM/JmsA       . . . . .   . . . . .   d . . . .   . . b . .   . . . . .
FS/Ei         . . . . .   . . . . .   d . . . .   . . . . .   . . b . .
```

        1 = -a,-b   2 = -a,-c

```
Locus     ->    I I I I I   I I I I I   I I I I I   I I K K K   L L L L L
(read down)     g g g g g   g g g g g   g g g g g   g g f t t   a a a d d
                h h h h h   h h h h h   h h k k k   l l o h h   m m p h r
                - - - - -   - - - - -   - - - - -   - - - - -   b b - - -
                C D I L N   P S S S S   S S E P T   1 1 1 1 2   - - 1 1 1
                  e n e p     c a a a a   a r f c r       r       1 2
                  x u v         1 2 3 4   5 c 1   p

    Chromosome  1 1 1 1 1   1 1 1 1 1   1 1 6 6 6   1 U 9 1 U   1 1 9 7 6
    (read down) 2 2 2 2 2   2 2 2 2 2   2 2             6 N   7 N   2
--------------------------------------------------------------------------------

C58/J           a . . a .   . . . a .   o . a a a   . . . . .   1 b a a a
C58/JA          . . . . .   . . . . .   . . . . .   . . . . .   . . . . a
CALB/Rk         . . . . .   . . . . .   . . . . .   . . . . .   . . . a a
CASA/Rk         . . . . .   . . . . .   . . . . .   . . . . .   . . . a a
CAST/Ei         . . . . .   . . . . .   . a . . .   . . . . .   . . . . .

CBA/-           j o a o o   o o o o o   o a b b b   . . . o .   a b b . *
CBA/BrA         . . . . .   . . . . .   . . . . .   . . . . .   . . . . b
CBA/Ca-         . . . . .   . . . . .   . . . . .   . . . o .   a b b . b
CBA/CaH-T6J     . . . . .   . . . . .   . . . . .   . . . . .   . . . . a
CBA/CaHN        . . . . .   . . . . .   . . . . .   . . . . .   . . . . .

CBA/CaJ         . . . . .   . . . . .   . . . . .   . . . o .   a b b . b
CBA/J           j . . . .   . . . o o   o . b b b   . . . o .   a b b . a
CE/-            f . . a o   o o o . .   . o b b b   a . . . .   a b b . a
CE/J            f . . a .   . . . . .   . . b b b   a . . . .   a b b . a
CFO             . . . . .   . . . . .   . . . . .   . . . . .   . . . . a

CHMU/Le         . . . . .   . . . . .   . . . . .   . . . . .   . . . . .
CSB/-           . . . . .   . . . . .   . . . . .   . . . . .   . . . . .
CWD/Le          . . . . .   . . . . .   . . . . .   . . . . .   . . . . .
DA/HuSn         . . . . .   . . . . .   . . . . .   . . . . .   . . . . .
DBA/1-          . . . . .   . o o . a   . o b . b   . . . . .   a d b . a

DBA/1J          . . . . .   . . . . a   . . b . b   . . . . .   a d b . a
DBA/2-          c o o . o   o . a . a   o o b b b   a . b a a   a d b a a
DBA/2DeJ        . . . . .   . . . . .   . . . . .   . . . . .   a d . . a
DBA/2J          c . . . .   . . . . a   o . b b b   a . b a a   a d b a a
DBA/LiA         . . . . .   . . . . .   . . . . .   . . . . .   . . . . a

DBA/LiAf        . . . . .   . . . . .   . . . . .   . . . . .   . . . . .
DC/Le           . . . . .   . . . . .   . . . . .   . . . . .   . . . . .
DD/He           . . . . .   . . . . .   . . . . .   . . . . .   . . . . .
DD/HeA          . . . . .   . . . . .   . . . . .   . . . . .   . . . . a
DD/HeAf         . . . . .   . . . . .   . . . . .   . . . . .   . . . . .

DDD             . . . . .   . . . . .   . . . . .   . . . . .   . . . . a
DDK             . . . . .   . . . . .   . . . . .   . . . . .   . . . . a
DE/J            . . . . .   . . . . .   . . . . .   b . . . .   . . . . b
DLS/Le          . . . . .   . . . . .   . . . . .   . . . . .   . . . . .
DW/J            . . . . .   . . . . .   . . . . .   . . . . .   a b . . b

F/St            . . . . .   . . . . .   . . . . .   . . . . .   . . . . .
FL/1Re          . . . . .   . . . . .   . . . . .   . . . . .   . . . . .
FL/2Re          . . . . .   . . . . .   . . . . .   . . . . .   . . . . a
FM/JmsA         . . . . .   . . . . .   . . . . .   . . . . .   . . . . a
FS/Ei           . . . . .   . . . . .   . . . . .   . . . . .   a b . . .
```

```
Locus      ->   L L L L L   L L L L L   L L L L L   L L L L L   L L L L L
(read down)     e e f m n   p s t t t   t t t t v   v y y y y   y y y y y
                n n o p a   t h h h n   w w w w     p - - - -   - - - - -
                - - -   -   -   - - -   - - - -       - 1 2 3 5   6 7 8 9 1
                1 2 1   1   1   1 2 2   2 3 4 6     1                     0
------------------------------------------------------------------------------
Chromosome      1 U 7 1 U   U 1 1 U U   1 9 1 U 4   6 1 6 6 1   1 1 U 1 1
(read down)       N 7 N       N 3 N N   2     N         9         5 2 N   9
------------------------------------------------------------------------------

C58/J           . . . . .   . . . . .   . . a a c   a b a a b   b b b b b
C58/JA          . . . . .   . . . . .   . . . . .   . . . . .   . . . . .
CALB/Rk         . . . . .   . . . . .   . . . . .   . . . . .   . . . . .
CASA/Rk         c b . . .   . . . . .   . . . . .   . . . . .   . . . . .
CAST/Ei         c b . . .   . . . . .   . . . . .   . . . a .   . . . . .

CBA/-           a b . . a   b r a . a   . . a a a   b a a b b   a a b a a
CBA/BrA         a b . . .   . . . . .   . . . . .   . a a b b   . . . . .
CBA/Ca-         . . . . a   . r a . a   . . a a a   . a a b b   a a . a a
CBA/CaH-T6J     . . . . .   . r . . .   . . . . .   . a a b b   . . . a .
CBA/CaHN        . . . . a   . . . . .   . . . . .   . a a b b   a a . a a

CBA/CaJ         . . . . .   . . a . a   . . a a .   . . . . .   . . . . .
CBA/J           a . . . a   b . a . a   . . a a a   b a a b b   a a b a a
CE/-            a b . . a   . s a . a   . a . a a   a b a b b   a a b a b
CE/J            a b . . .   . . a . a   . a a a a   a b a b b   a a b a b
CFO             . . . . .   . . . . .   . . . . .   . . . . .   . . . . .

CHMU/Le         . . . . .   . . . . .   . . . . .   . . . . .   . . . . .
CSB/-           . . . . .   . . . . .   . . . . .   . . . . .   . . . . .
CWD/Le          . . . . .   . . . . .   . . . . .   . . . . .   . . . . .
DA/HuSn         . . . . .   . . . . .   . . . . .   . b b b .   b . . . .
DBA/1-          a b . o .   b s a . b   . . a a a   b a a b b   b a b a a

DBA/1J          a b . o .   b . a . b   . . a a .   b a a b b   a a b a a
DBA/2-          a b b a a   b r a o b   o a a a a   b a a b b   b a b a a
DBA/2DeJ        . . . . .   . . . . b   . . a a .   . . . . .   . . . . .
DBA/2J          a b b a .   b . a o b   o a a a a   b a a b b   b a b a a
DBA/LiA         . . . . .   . . . . .   . . . . .   . a a b b   . . . . .

DBA/LiAf        . . . . .   . . . . .   . . . . .   . . . . .   . . . . .
DC/Le           . . . . .   . . . . .   . . . . .   . . . . .   . . . . .
DD/He           . . . . .   . . . . .   . . . . .   . b b b a   . . . . .
DD/HeA          . . . . .   . . . . .   . . . . .   . b b b a   . . . . .
DD/HeAf         a b . . .   . . . . .   . . . . .   . . . . .   . . . . .

DDD             . . . . .   . . . . .   . . . . .   . . . . .   . . . . .
DDK             . . . . .   . . . . .   . . . . .   . . . . .   . . . . .
DE/J            . . . . .   . . . . .   . a . a .   a . . . .   . . . . .
DLS/Le          . . . . .   . . . . .   . . . . .   . . . . .   . . . . .
DW/J            . . . . .   . . . . b   . . a a .   . . . . .   . . . . .

F/St            . . . . .   . r . . .   . . . . .   . . . . .   . . . b .
FL/1Re          . . . . .   . . . . .   . . . . .   . . . . .   . . . . .
FL/2Re          . . . . .   . . . . .   . . . . .   . a b b .   . . . . .
FM/JmsA         . . . . .   . . . . .   . . . . .   . . . . .   . . . . .
FS/Ei           . . . . .   . . . . b   . . b a .   . . . . .   . . . . .
```

Locus -> (read down)	Ly-11	Ly-12	Ly-13	Ly-14	Ly-15	Ly-16	Ly-17	Ly-18	Ly-19	Ly-20	Ly-21	Ly-22	Ly-23	Ly-24	Ly-25	Ly-26	Ly-27	Ly-28	Ly-29	Ly-30	Ly-31	Ly-32	Ly-39	Ly-40	Lyb-2
Chromosome (read down)	2	19	UN	7	7	1	1	12	4	42	7	1	2	2	2	UN	UN	UN	4	UN	4	4	17	UN	4
C58/J	.	b	b	b	a	a	b	b	b	.	b	.	b	b	.	b	.	.	.	.	b	.	b	b	.
C58/JA	.	.	.	.	.	.	.	.	.	.	.	.	.	.	.	.	.	.	.	.	.	.	.	.	.
CALB/Rk	.	.	.	.	.	.	.	.	.	.	.	.	.	.	.	.	.	.	.	.	.	.	.	.	.
CASA/Rk	.	.	.	.	.	.	.	.	.	.	.	.	.	.	.	.	.	.	.	.	.	.	.	.	.
CAST/Ei	.	.	.	.	.	.	.	.	.	.	.	.	.	.	.	.	.	.	.	.	.	.	.	.	.
CBA/-	a	b	b	b	b	b	a	a	b	b	b	b	a	b	b	b	.	b	b	b	a	a	1	.	*
CBA/BrA	.	.	.	.	.	.	.	.	.	.	.	.	.	.	.	.	.	.	.	.	.	.	.	.	.
CBA/Ca-	.	.	.	.	.	.	a	a	b	.	.	b	.	.	.	b	.	b	.	.	a	b	.	.	b
CBA/CaH-T6J	.	.	.	.	.	.	a	.	.	.	.	.	.	.	.	.	.	.	.	.	.	.	.	.	b
CBA/CaHN	.	.	.	.	.	.	a	a	b	.	.	b	.	.	.	.	.	b	.	.	a	b	.	.	b
CBA/CaJ	.	.	.	.	.	.	a	.	.	.	.	.	.	.	.	.	.	.	.	.	a	.	.	.	b
CBA/J	a	b	b	b	b	b	a	a	b	b	b	b	a	b	b	b	.	b	b	b	a	a	1	.	a
CE/-	a	b	b	b	b	.	a	b	a	b	.	b	a	b	a	a	a	.	a	a	a	.	1	.	c
CE/J	.	b	b	b	b	.	a	b	a	.	.	b	a	b	.	a	.	.	.	.	a	.	1	.	c
CFO	.	.	.	.	.	.	.	.	.	.	.	.	.	.	.	.	.	.	.	.	.	.	.	.	.
CHMU/Le	.	.	.	.	.	.	.	.	.	.	.	.	.	.	.	.	.	.	.	.	.	.	.	.	.
CSB/-	.	.	.	.	.	.	.	.	.	.	.	.	.	.	.	.	.	.	.	.	.	.	.	.	.
CWD/Le	.	.	.	.	.	.	.	.	.	.	.	.	.	.	.	.	.	.	.	.	.	.	.	.	.
DA/HuSn	.	.	.	.	.	.	.	.	.	.	.	.	.	.	.	.	.	.	.	.	.	.	.	.	.
DBA/1-	b	b	b	b	b	.	b	b	b	.	b	a	b	a	b	b	b	b	a	b	b	a	.	b	a
DBA/1J	.	b	b	b	b	.	b	b	b	.	b	a	.	a	.	b	.	.	.	.	b	.	.	b	a
DBA/2-	b	b	b	b	b	.	b	b	b	b	b	a	b	a	b	b	b	b	a	b	a	a	1	b	a
DBA/2DeJ	.	.	.	.	.	.	.	.	.	.	.	.	.	.	.	.	.	.	.	.	a	.	.	.	.
DBA/2J	.	b	b	b	b	.	b	b	b	.	b	a	.	a	.	b	.	.	.	.	a	.	1	b	a
DBA/LiA	.	.	.	.	.	.	.	.	.	.	.	.	.	.	.	.	.	.	.	.	.	.	.	.	.
DBA/LiAf	.	.	.	.	.	.	.	.	.	.	.	.	.	.	.	.	.	.	.	.	.	.	.	.	.
DC/Le	.	.	.	.	.	.	.	.	.	.	.	.	.	.	.	.	.	.	.	.	.	.	.	.	.
DD/He	.	.	.	.	.	.	.	.	.	.	.	.	.	.	.	.	.	.	.	.	.	.	.	.	.
DD/HeA	.	.	.	.	.	.	.	.	.	.	.	.	.	.	.	.	.	.	.	.	.	.	.	.	.
DD/HeAf	.	.	.	.	.	.	.	.	.	.	.	.	.	.	.	.	.	.	.	.	.	.	.	.	.
DDD	.	.	.	.	.	.	.	.	.	.	.	.	.	.	.	.	.	.	.	.	.	.	.	.	.
DDK	.	.	.	.	.	.	.	.	.	.	.	.	.	.	.	.	.	.	.	.	.	.	.	.	.
DE/J	.	.	.	.	.	.	.	.	.	.	.	.	.	.	.	.	.	.	.	.	.	.	.	.	.
DLS/Le	.	.	.	.	.	.	.	.	.	.	.	.	.	.	.	.	.	.	.	.	.	.	.	.	.
DW/J	.	.	.	.	.	.	.	.	.	.	.	.	.	.	.	.	.	.	.	.	.	.	.	.	.
F/St	.	.	.	.	.	.	.	.	.	.	.	.	.	.	.	.	.	.	.	.	.	.	.	.	.
FL/1Re	.	.	.	.	.	.	.	.	.	.	.	.	.	.	.	.	.	.	.	.	.	.	.	.	.
FL/2Re	.	.	.	.	.	.	.	.	.	.	.	.	.	.	.	.	.	.	.	.	.	.	.	.	.
FM/JmsA	.	.	.	.	.	.	.	.	.	.	.	.	.	.	.	.	.	.	.	.	.	.	.	.	.
FS/Ei	.	.	.	.	.	.	.	.	.	.	.	.	.	.	.	.	.	.	.	.	.	.	.	.	.

1 = -b

```
Locus    ->    L L L L L    L M M M M    M M M M M    N N N N N    N O O P P
(read down)    y y y y y    y a e l o    o o p p u    c e k k p    p r r a n
               b b b b b    b p p s d    d r h i p    a u - - -    - m m
               - - - - -    - - - - -    - - - - -    m - 1 2 1    2 - -
               3 4 5 6 7    8 1 1 1 1    2 1 1 1 1    1              1 2   1

---------------------------------------------------------------------------------
Chromosome     U 4 U 4 1    7 5 1 1 9    7 5 7 9 4    9 1 1 2 1    1 4 4 7 4
(read down)    N   N   2      7                          7 7   4    4
---------------------------------------------------------------------------------

C58/J          . . . . .    . b a . b    b a b b b    . b a o a    . . . + .
C58/JA         . . . . .    . . . . b    b a . b .    . . . . a    . . . . .
CALB/Rk        . . . . .    . . . . .    . . . 1 .    . . . b .    . . . + .
CASA/Rk        . . . . .    . . . . a    a . . . 2    . . . b .    . . . + .
CAST/Ei        . . . . .    . . . . .    . . . . .    . . . . .    . . . . .

CBA/-          a a b a b    b * b * b    b a b b a    . b o a a    . . . + .
CBA/BrA        . . . . .    . . . . b    b a . b a    . b . . a    . . . + .
CBA/Ca-        a . . b .    b b . b b    b a b b a    . b . . a    . . . + .
CBA/CaH-T6J    . . . . .    . . . b b    b a . b a    . . . . a    . . . + .
CBA/CaHN       a . . b .    b . . . b    b . . . a    . . . . .    . . . . .

CBA/CaJ        . . . . .    . b . b b    b . b b .    . b . . a    . . . + .
CBA/J          . a b a b    b h b d b    b a . b a    . b o a a    . . . + .
CE/-           b b a b a    b b b . a    b a b b a    . b . o a    . . . + a
CE/J           . b . . .    . b b . a    b a b b a    . b . o a    . . . + a
CFO            . . . . .    . . . . b    b a . . .    . . . . a    . . . . .

CHMU/Le        . . . . .    . . . . b    . . . b .    . b . . .    . . . . .
CSB/-          . . . . .    . . . . b    b a . b b    . . . . .    . . . . .
CWD/Le         . . . . .    . . . . a    . . . b .    . b . . .    . . . . .
DA/HuSn        . . . . .    . . . . a    . . . . .    . . . . .    . . . . .
DBA/1-         b . a a a    a b a a a    b a b b a    . b o o a    . . . + a

DBA/1J         . . . . .    . b a . a    b a . b a    . b . . a    . . . + a
DBA/2-         b a a a a    a b a a a    b a b b a    a b o a a    a b a + a
DBA/2DeJ       . . . . .    . . . . a    . . . b .    . b . . .    . . . + .
DBA/2J         . . . . .    . b . . a    b a . b a    a b o a a    a b a + a
DBA/LiA        . . . . .    . . . . a    b a . b a    . . . . a    . . . . .

DBA/LiAf       . . . . .    . . . . .    . . . . .    . . . . .    . . . . .
DC/Le          . . . . .    . . . . a    . . . b .    . b . . .    . . . . .
DD/He          . . . . .    . b . . a    o . p . a    . b . . .    a . . . a
DD/HeA         b b b b b    . . . . a    b a . b a    . b . . a    . . . . n
DD/HeAf        . . . . .    . . . . .    . . . . .    . . . . .    . . . . .

DDD            . . . . .    . . . . a    . . . . a    . . . . .    . . . . .
DDK            . . . . .    . . . . a    . . . . a    . . . . .    . . . + .
DE/J           . . . . .    . b . . a    b a . b .    . b . . .    . . . + .
DLS/Le         . . . . .    . . . . b    . . . b .    . b . . .    . . . . .
DW/J           . . . . .    . b . . a    . a . b b    . b . . .    . . . + .

F/St           . . . . .    . b . . a    . . a . a    . b . . .    . . . . .
FL/1Re         . . . . .    . . . . a    . . . . .    . . . . .    . . . . .
FL/2Re         . . . . .    . . . . a    . . . . .    . . . . .    . . . . .
FM/JmsA        . . . . .    . . . . a    b a . b b    . b . . a    . . . . .
FS/Ei          . . . . .    . . . . b    . . . b .    . . . . .    . . . p .
```

           1 = ?   2 = b?

701

```
Locus      ->    P P P P P   P P P P   P P P P P   P P P P Q   Q Q Q Q Q
(read down)      a c d e e   e e g g   g g g k r   r r r s a   a a a a e
                 n a   p p   p p d k k m m m - e   n t t p -   - - - - d
                 - -   - -   - - - -   - - - 3 -   - - - 1     2 3 4 5 -
                 2 1   2 3   4 7 1 2   1 2 3   2   p 1 2       1       1

-----------------------------------------------------------------------------
Chromosome       U U U 1 1   7 5 4 X 1   5 4 9 9 1   2 U 8 2 1   1 1 1 1 1
(read down)      N N N 0             7           2   N       7   7 7 7 7 7
-----------------------------------------------------------------------------

C58/J        . . a a a   b a b . a   a a c . .   . . . a   . . . . a
C58/JA       . . . . a   . . b b a   a a . .   . . . .   . . . . .
CALB/Rk      . . . . a   . . . . .   . . . .   . . . .   . . . . .
CASA/Rk      . . . . b   . . . b b   a . . .   . . . .   . . . . .
CAST/Ei      . . . . .   . . . . .   . . . .   . . . .   . . . . .

CBA/-        . . . . b   b . b b b   * a a a a   a a a b b   b b b b b
CBA/BrA      . . . . b   . . b b a   b a . .   . . . .   . . . . .
CBA/Ca-      . . . . b   . . b b b   b a a a   . a a b b   b b b b .
CBA/CaH-T6J  . . . . b   . . b b b   b a a .   . a a b .   . . . . .
CBA/CaHN     . . . . b   . . . . .   b . . .   . . . .   . . b b .

CBA/CaJ      . . . . b   . . . b b   b a a a   . . . b .   . . . . .
CBA/J        . . . . b   b . b . b   a a a . a   a a a . .   b b b b b
CE/-         b a . . b   b . b b a   a a b . b   a . . .   . . . . b
CE/J         b a . . b   b . b b b   a a b . b   a . . b .   . . . . b
CFO          . . . . .   . . b b .   a a . .   . b a .   . . . . .

CHMU/Le      . . . . a   . . . . .   a a . .   . . . .   . . . . .
CSB/-        . . . . b   . . . . .   b a . .   . . . .   . . . . .
CWD/Le       . . . . b   . . . . .   b a . .   . . . .   . . . . .
DA/HuSn      . . . . .   . . . . b   . . . .   . . b .   . . . . .
DBA/1-       b . . . b   . . . . a   b a . a   . . a b   a b . b b

DBA/1J       b . . . b   . . . . a   b a . .   . . . a .   a b . . b
DBA/2-       b b . a b   b a b b a   b a c a a   a a a a b   a a a b b
DBA/2DeJ     . . . . b   . . . . a   b a . .   . . . .   . . . . .
DBA/2J       b . . a b   b a b b a   b a c . a   a a a a .   . . . . b
DBA/LiA      . . . . b   . . b b a   b a . .   . . . .   . . . . .

DBA/LiAf     . . . . .   . . . . .   . . . .   . . . .   . . . . .
DC/Le        . . . . b   . . . . .   b a . .   . . . .   . . . . .
DD/He        . . . . c   . . . . a   b . . .   . . . .   . . . . .
DD/HeA       . . . . c   . . b b a   b a . .   . . . .   . . . . .
DD/HeAf      . . . . .   . . . . .   . . . .   . . . .   . . . . .

DDD          . . . . b   . . . . .   a . . .   . . . .   . . . . .
DDK          . . . . c   . . . . .   b . . .   . b a .   . . . . .
DE/J         . . . . b   . . b b .   a a . .   . . . .   . . . . .
DLS/Le       . . . . a   . . . . .   b a . .   . . . .   . . . . .
DW/J         . . . . b   . . . . .   a a b .   . . . b   . . . . .

F/St         . . . . b   . . . . .   b . . .   . . . .   . . . . .
FL/1Re       . . . . .   . . . . b   . . . .   . . . .   . . . . .
FL/2Re       . . . . .   . . . . b   . . . .   . . . .   . . . . .
FM/JmsA      . . . . b   . . b b a   b a . .   . . . .   . . . . .
FS/Ei        . . . . .   . . . . a   . . b .   . . b .   . . . . .
```

```
Locus    ->    r r R R R     S S S S S     S S S S S     S S S T T     T T T T T
(read down)    d d e i n     a a a d d     l o o o p     s v v a c     c g h i l
                 s n b 7     a a s h r     p a d d i     p p p m n     r n y n a
                 - - s       - - - -         - - -         - - - -     g   - d
                 1 1 -       1 1 1 1         1 2 2         1 2 1 2         1
                     6                           r

---------------------------------------------------------------------------------
  Chromosome   5 1 1 1 6     7 7 1 2 U     1 U 1 1 1     U 2 7 7 1     1 1 9 1 1
  (read down)    7     4                 N   7 N 6 7 2   N       1       3 5   2 7
---------------------------------------------------------------------------------

C58/J        +  .  b  b  a     a  .  a  a  l     o  .  a  a  b     .  .  .  c  s     a  b  b  .  a
C58/JA       .  .  .  .  .     .  .  .  .  .     .  .  .  .  .     .  a  .  .  .     .  .  .  .  .
CALB/Rk      .  .  .  .  .     .  .  b  .  .     .  .  .  .  .     .  .  .  .  .     .  .  .  .  .
CASA/Rk      .  .  .  .  .     .  .  .  .  .     .  .  .  .  .     .  .  .  .  .     .  .  .  .  .
CAST/Ei      .  .  .  a  .     .  .  .  .  .     .  .  a  a  .     .  .  .  .  .     .  .  .  .  .

CBA/-        *  +  *  a  b     a  .  o  a  l     o  b  a  .  d     s  b  *  c  s     a  c  b  o  b
CBA/BrA      +  +  .  .  .     .  .  .  a  .     .  .  .  .  .     s  b  .  .  .     .  .  b  .  b
CBA/Ca-      +  .  b  a  .     a  .  o  a  .     .  b  a  a  .     .  .  c  .  .     .  .  b  .  b
CBA/CaH-T6J  +  .  .  .  .     .  .  a  .        .  .  a  .  .     .  .  c  .  .     .  .  b  .  b
CBA/CaHN     .  .  .  .  .     .  .  .  .  .     .  .  .  .  .     .  .  .  .  .     .  .  b  .  b

CBA/CaJ      +  .  b  a  .     .  .  o  a  .     .  .  a  a  .     .  .  .  .  .     .  .  .  .  .
CBA/J        2  .  s  a  b     a  .  o  a  l     o  .  .  .  d     .  b  .  c  s     a  c  b  o  b
CE/-         +  .  b  a  b     .  .  o  a  .     .  b  .  .  .     .  .  a  s        a  c  b  .  b
CE/J         +  .  b  a  b     .  .  o  a  .     .  b  .  .  .     .  .  a  s        a  c  b  .  b
CFO          .  .  .  .  .     .  .  .  .  .     .  .  a  .  .     .  .  .  .  .     .  .  .  .  .

CHMU/Le      .  .  .  .  .     .  .  .  .  .     .  .  .  .  .     .  .  .  .  .     .  .  .  .  .
CSB/-        .  .  .  .  .     .  .  .  .  .     .  .  .  .  .     .  .  .  .  .     .  .  .  .  .
CWD/Le       .  .  .  .  .     .  .  .  .  .     .  .  .  .  .     .  .  .  .  .     .  .  .  .  .
DA/HuSn      2  .  .  .  .     .  .  o  .  .     .  .  .  .  .     .  .  .  c  .     .  .  b  .  b
DBA/1-       +  .  s  a  b     a  .  o  a  l     o  b  .  .  .     .  .  .  b  .     .  .  b  .  b

DBA/1J       +  .  s  a  b     a  .  o  a  l     o  .  .  .  .     .  .  .  b  .     .  .  b  .  b
DBA/2-       +  +  s  a  b     a  b  o  a  l     a  b  a  a  d     f  a  .  b  s     a  b  b  .  c
DBA/2DeJ     +  .  .  a  .     .  .  .  a  .     .  .  .  .  .     .  .  .  .  .     .  .  .  .  .
DBA/2J       +  .  s  a  b     a  b  o  a  l     a  .  a  a  d     f  .  .  b  s     a  b  b  .  c
DBA/LiA      +  +  .  .  .     .  .  .  a  .     .  .  .  .  .     .  a  .  .  .     .  .  b  .  c

DBA/LiAf     .  .  .  .  .     .  .  .  .  .     .  .  .  .  .     .  .  .  .  .     .  .  .  .  .
DC/Le        .  .  .  .  .     .  .  .  .  .     .  .  .  .  .     .  .  .  .  .     .  .  .  .  .
DD/He        +  .  .  .  .     .  .  o  .  .     .  .  .  .  .     .  .  .  .  c     .  g  b  .  c
DD/HeA       +  +  .  .  .     .  .  a  .  .     .  .  .  .  .     .  a  .  .  .     .  .  b  .  c
DD/HeAf      .  .  .  .  .     .  .  .  .  .     .  .  .  .  .     .  .  .  .  f     .  .  .  .  .

DDD          .  .  .  .  .     .  .  .  .  .     .  .  .  .  .     .  .  .  .  .     .  .  b  .  .
DDK          .  .  .  .  .     .  .  .  .  .     .  .  .  .  .     .  .  .  .  .     .  .  .  .  .
DE/J         +  .  .  .  .     .  .  o  .  .     .  .  .  .  .     .  .  .  .  .     .  .  .  .  .
DLS/Le       .  .  .  .  .     .  .  .  .  .     .  .  .  .  .     .  .  .  .  .     .  .  .  .  .
DW/J         +  .  .  a  .     .  .  .  .  .     .  .  .  .  .     .  .  a  s  .     .  .  .  .  .

F/St         .  .  .  .  .     .  .  .  .  .     .  .  .  .  .     .  .  .  .  .     .  .  .  .  a
FL/1Re       2  .  .  .  .     .  .  .  .  .     .  .  .  .  .     .  .  .  .  .     .  .  .  .  .
FL/2Re       2  .  .  .  .     .  .  .  .  .     .  .  .  .  .     .  .  .  .  .     .  .  b  .  b
FM/JmsA      .  .  .  .  .     .  .  .  .  .     .  .  .  .  .     .  .  .  .  .     .  .  .  .  .
FS/Ei        +  .  .  b  .     a  .  .  .  .     .  .  .  .  .     .  .  .  .  s     .  .  .  .  .

        1 = b?        2 = rd
```

703

Locus → (read down)	Tpre	Trf	Tsuy	Tth	Ucp	Upg-1	Upg-2	Ups	Wap	Xmmv-9	Xmmv-42	Xmmv-43	Xmmv-44	Xmmv-45	Xmmv-46	Xmmv-47	Xmmv-48	Xmmv-49	Xmmv-50	Xmmv-51	Xmmv-52	Xmmv-53	Xmmv-54	Xmmv-55	Xmmv-56
Chromosome (read down)	12	9	12	12	8	17	1	9	UN	1	19	UN	UN	UN	UN	UN	UN	UN	12	UN	5	UN	UN	15	UN
C58/J	.	b	.	.	.	s	.	a	a	a	a	o	o	o	o	a	o	o	o	o	a	o	a	a	o
C58/JA	.	b	.	.	.	.	.	.	.	.	.	.	.	.	.	.	.	.	.	.	.	.	.	.	.
CALB/Rk	.	b	.	.	.	.	.	.	.	.	.	.	.	.	.	.	.	.	.	.	.	.	.	.	.
CASA/Rk	.	b	.	.	.	.	.	.	.	.	.	.	.	.	.	.	.	.	.	.	.	.	.	.	.
CAST/Ei	.	.	.	.	.	.	.	.	.	.	.	.	.	.	.	.	.	.	.	.	.	.	.	.	.
CBA/-	o	a	o	o	.	f	*	a	a	a	o	o	o	o	o	*	*	o	o	o	a	o	a	o	o
CBA/BrA	.	a	.	.	.	.	.	.	.	.	.	.	.	.	.	.	.	.	.	.	.	.	.	.	.
CBA/Ca-	.	a	.	.	.	f	s	.	a	a	o	o	o	o	o	a	*	o	o	o	a	o	a	o	o
CBA/CaH-T6J	.	a	.	.	.	f	s	.	a	a	o	o	o	o	o	a	o	o	o	o	a	o	a	o	o
CBA/CaHN	.	a	.	.	.	.	.	.	.	a	o	o	o	o	o	a	a	o	o	o	a	o	a	o	o
CBA/CaJ	.	a	.	.	.	f	s	.	a	a	o	o	o	o	o	a	o	o	o	o	a	o	a	o	o
CBA/J	o	a	o	o	.	f	d	a	a	a	o	o	o	o	o	o	o	o	o	o	a	o	a	o	o
CE/-	.	b	.	.	.	.	s	a	.	a	o	o	o	o	o	a	o	o	o	o	o	o	a	o	o
CE/J	.	b	.	.	.	.	s	a	.	a	o	o	o	o	o	a	o	o	o	o	o	o	a	o	o
CFO	.	b	.	.	.	.	.	.	.	.	.	.	.	.	.	.	.	.	.	.	.	.	.	.	.
CHMU/Le	.	b	.	.	.	.	.	.	.	.	.	.	.	.	.	.	.	.	.	.	.	.	.	.	.
CSB/-	.	b	.	.	.	.	.	.	.	.	.	.	.	.	.	.	.	.	.	.	.	.	.	.	.
CWD/Le	.	b	.	.	.	.	.	.	.	.	.	.	.	.	.	.	.	.	.	.	.	.	.	.	.
DA/HuSn	.	.	.	.	.	.	.	.	.	o	o	o	o	o	o	o	o	o	a	o	o	o	o	a	o
DBA/1-	.	b	.	.	.	s	d	a	a	o	o	o	o	o	o	o	o	o	a	o	a	o	a	a	o
DBA/1J	.	b	.	.	.	s	d	a	a	o	o	o	o	o	o	o	o	o	a	o	a	o	a	a	o
DBA/2-	.	b	.	.	d	f	d	a	a	o	o	o	o	o	o	o	o	o	a	o	a	o	a	a	o
DBA/2DeJ	.	b	.	.	.	.	.	.	.	o	o	o	o	o	o	o	o	o	a	o	a	o	a	a	o
DBA/2J	.	b	.	.	d	f	d	a	a	o	o	o	o	o	o	o	o	o	a	o	a	o	a	a	o
DBA/LiA	.	b	.	.	.	.	.	.	.	.	.	.	.	.	.	.	.	.	.	.	.	.	.	.	.
DBA/LiAf	.	.	.	.	.	.	.	.	.	.	.	.	.	.	.	.	.	.	.	.	.	.	.	.	.
DC/Le	.	b	.	.	.	.	.	.	.	.	.	.	.	.	.	.	.	.	.	.	.	.	.	.	.
DD/He	.	b	.	.	.	.	.	.	.	.	.	.	.	.	.	.	.	.	.	.	.	.	.	.	.
DD/HeA	.	b	.	.	.	.	.	.	.	.	.	.	.	.	.	.	.	.	.	.	.	.	.	.	.
DD/HeAf	.	.	.	.	.	.	.	.	.	.	.	.	.	.	.	.	.	.	.	.	.	.	.	.	.
DDD	.	b	.	.	.	.	.	.	.	.	.	.	.	.	.	.	.	.	.	.	.	.	.	.	.
DDK	.	b	.	.	.	.	.	.	.	.	.	.	.	.	.	.	.	.	.	.	.	.	.	.	.
DE/J	.	.	.	.	.	.	.	.	a	.	.	.	.	.	.	.	.	.	.	.	.	.	.	.	.
DLS/Le	.	a	.	.	.	.	.	.	.	.	.	.	.	.	.	.	.	.	.	.	.	.	.	.	.
DW/J	.	b	.	.	.	.	s	.	.	.	.	.	.	.	.	.	.	.	.	.	.	.	.	.	.
F/St	.	.	.	.	.	.	.	.	.	.	.	.	.	.	.	.	.	.	.	.	.	.	.	.	.
FL/1Re	.	b	.	.	.	.	.	.	.	.	.	.	.	.	.	.	.	.	.	.	.	.	.	.	.
FL/2Re	.	b	.	.	.	.	.	.	.	.	.	.	.	.	.	.	.	.	.	.	.	.	.	.	.
FM/JmsA	.	b	.	.	.	.	.	.	.	.	.	.	.	.	.	.	.	.	.	.	.	.	.	.	.
FS/Ei	.	b	.	.	.	s	d	.	.	.	.	.	.	.	.	.	.	.	.	.	.	.	.	.	.

Locus -> (read down)	Xmmv-57	Xmmv-58	Xmmv-59	Xmmv-60	Xmmv-62	Xmmv-63	Xmmv-64	Xmmv-65	Xmmv-66	Xmmv-67	Xmmv-68	Xmmv-69	Xmmv-70
Chromosome (read down)	UN	UN	UN	UN	4	UN	UN	3	UN	UN	UN	UN	UN
C58/J	o	o	o	o	a	a	a	o	a	a	o	a	a
C58/JA	.	.	.	.	.	.	.	.	.	.	.	.	.
CALB/Rk	.	.	.	.	.	.	.	.	.	.	.	.	.
CASA/Rk	.	.	.	.	.	.	.	.	.	.	.	.	.
CAST/Ei	.	.	.	.	.	.	.	.	.	.	.	.	.
CBA/-	o	o	o	o	o	a	*	o	o	o	o	*	a
CBA/BrA	.	.	.	.	.	.	.	.	.	.	.	.	.
CBA/Ca-	o	o	o	o	o	a	*	o	o	o	o	o	a
CBA/CaH-T6J	o	o	o	o	o	a	o	o	o	o	o	o	a
CBA/CaHN	o	o	o	o	o	a	a	o	o	o	o	o	a
CBA/CaJ	o	o	o	o	o	a	o	o	o	o	o	o	a
CBA/J	o	o	o	o	o	a	a	o	o	o	o	a	a
CE/-	o	o	o	o	o	a	o	a	o	o	a	a	a
CE/J	o	o	o	o	o	a	o	a	o	o	a	a	a
CFO	.	.	.	.	.	.	.	.	.	.	.	.	.
CHMU/Le	.	.	.	.	.	.	.	.	.	.	.	.	.
CSB/-	.	.	.	.	.	.	.	.	.	.	.	.	.
CWD/Le	.	.	.	.	.	.	.	.	.	.	.	.	.
DA/HuSn	a	o	o	a	o	a	a	o	o	o	o	a	a
DBA/1-	o	o	o	o	o	a	o	a	o	o	o	a	a
DBA/1J	o	o	o	o	o	a	o	a	o	o	o	a	a
DBA/2-	o	o	o	o	o	a	o	a	o	o	o	a	a
DBA/2DeJ	o	o	o	o	o	a	o	a	o	o	o	a	a
DBA/2J	o	o	o	o	o	a	o	a	o	o	o	a	a
DBA/LiA	.	.	.	.	.	.	.	.	.	.	.	.	.
DBA/LiAf	.	.	.	.	.	.	.	.	.	.	.	.	.
DC/Le	.	.	.	.	.	.	.	.	.	.	.	.	.
DD/He	.	.	.	.	.	.	.	.	.	.	.	.	.
DD/HeA	.	.	.	.	.	.	.	.	.	.	.	.	.
DD/HeAf	.	.	.	.	.	.	.	.	.	.	.	.	.
DDD	.	.	.	.	.	.	.	.	.	.	.	.	.
DDK	.	.	.	.	.	.	.	.	.	.	.	.	.
DE/J	.	.	.	.	.	.	.	.	.	.	.	.	.
DLS/Le	.	.	.	.	.	.	.	.	.	.	.	.	.
DW/J	.	.	.	.	.	.	.	.	.	.	.	.	.
F/St	.	.	.	.	.	.	.	.	.	.	.	.	.
FL/1Re	.	.	.	.	.	.	.	.	.	.	.	.	.
FL/2Re	.	.	.	.	.	.	.	.	.	.	.	.	.
FM/JmsA	.	.	.	.	.	.	.	.	.	.	.	.	.
FS/Ei	.	.	.	.	.	.	.	.	.	.	.	.	.

```
Locus    ->   a A A A A   A A A A A   A A A A A   A A A A A   A A A A A
(read down)     a b c c   d d d d g   h h h h k   k l m m o   o p p p s
                t p f o   h h h h s   d d r p     p b y y x   x h k o -
              a - -       - - - -     - - - -     - - - - -   - - a 1
              1 1         1 2 3 3     1 2 1 1     2 1 1 2 1   2 1 - s
                          e t                                       1

--------------------------------------------------------------------------
Chromosome    2 1 7 1 4   3 U 3 3 X   1 4 1 3 1   4 5 3 3 1   1 2 1 9 1
(read down)   2             N             2   9               0       3
--------------------------------------------------------------------------

FVB/NA        . . . . .   a a a a .   . b a . .   . . . . a   a . . . .
GRS/A         + . a . .   a a a a h   b b a . .   . . . . a   a . . . .
GRS/N         + . . . .   . . . . h   b . . . .   . . . . .   . . . . .
HC/CpbU       . . . . .   . . . . .   . . . . .   . . . . .   . . . . .
HRS/J         + o a . a   . . . . h   b b . . a   a a a . .   . . . b a

HSFR/N        + . . . .   . . . . .   b . . . .   . . . . .   . . . . .
HSFS/N        + . . . .   . . . . .   b . . . .   . . . . .   . . . . .
HTG/-         + a . . .   . . . . .   . . . . .   . . . . .   . . . . .
HYIII/Le      . . . . .   . . . . .   . . a . .   . . a . .   . . . a .
I/-           a a . b .   a . a a h   d b . . b   a . a . a   a . a a .

I/LnReJ       . a . b .   . . . . .   1 b . . b   a . a . .   . . a a .
IC/Le         . . . . .   . . . . .   . . . b .   . . a . .   . . . a .
IF/BrA        2 . . . .   a . a a .   . . . a .   . . . . .   . . . . .
IS/Cam-       + a . . a   . . . . .   . a . . b   b a a . a   . . a a b
IS/CamRk      + a . . a   . . . . .   . a . . b   b a a . .   . . a a b

JE/Le         a . . . a   . . . . .   . . . . a   . a a . .   . . . a .
JGBF/Le       . . . . .   . . . . .   . . . . a   . a . . .   . . . a .
KC/CpbU       . . . . .   . . . . .   . . . . .   . . . . .   . . . . .
KK/-          a . . . .   . . . . .   . . . . b   . . . . .   . . . . .
KSB/-         a . . . .   . . . . .   . . . . .   . . . . .   . . . . .

LG/J          + o . b .   . . . . h   d b . . b   a a a a 3   . . a b b
LIS/A         + . . . .   a . a a h   b . . a .   . . . . .   . . . . .
LP/J          4 o . a a   . . . . h   d a . . b   b a a a a   . . a b a
LPT/Le        . . . . .   . . . . .   . . . . a   . . a . .   . . . a .
LS/Le         5 . . . .   . . . . .   . . . . b   . a a . .   . . . a .

LT/ChReSv     a . . . .   . . . . .   . . . . .   . . . . .   . . . a .
LTS/A         + . . . .   a . a a h   b . . a .   . . . . .   . . . . .
MA/J          + o . b .   . . . . h   b . . . .   . a a a 3   . . a . .
MA/MyJ        . o a . a   . . . . .   . a . . b   a a a . .   . . a a a
MAS/A         + . . . .   a a a a h   b b a . .   . . . . a   a . . . .

MOC/CpbU      . . . . .   . . . . .   . . . . .   . . . . .   . . . . .
MOLA          4 . . . .   . . . . .   . . . . .   . a b b .   . . . . .
MOLC/Rk       . . . . .   . . . . .   . . . . b   . a b . .   . . b b .
MOLD/Rk       . . . . .   . . . . .   . . . . b   . a b . .   . . b b .
MOLD/RkA      . . . . .   a . a a .   . b a . .   . . . . a   b . . . .

MOLE/Rk       . . . . .   . . . . .   . . . . b   . a b . .   . . a b .
MOLF/Ei       . . . . .   . . . . .   . . . . b   . . b . .   . . . b .
MOLN          6 . . . .   . . . . .   b . . . .   . a . a .   . . . . .
MOLT          6 . . . .   . . . . .   b . . . .   . a . b .   . . . . .
MRL/MpJ       a . . . .   . . . . .   . . . b .   . . a . .   . . . b .
```

1 = d?   2 = a*t   3 = a?   4 = A*w   5 = a*t/+
6 = A or A*w

Locus (read down)	Av‑1	Av‑2	b2m	Bcd‑1	B5	Bgl‑e	Bgl‑s	Bgl‑t	Bv‑1	c	C3	C4	C6	Car‑1	Car‑2	cdm	Cde‑1	Ce‑2	Cfh	Co	Cv‑1	D7Rp2‑s	D7Rp2	D12Nyu1	D12Nyu2
Chromosome (read down)	UN	UN	4	2	5	9	9	9	8	7	17	17	15	3	3	3	2	1	2	7	5	9	7	12	12
FVB/NA	.	.	.	.	.	.	.	.	.	c	.	.	.	.	.	.	.	.	.	.	.	.	.	.	.
GRS/A	.	.	+	.	.	b	d	d	.	c	c	h	.	.	.	+	.	a	.	.	.	+	.	.	.
GRS/N	.	.	+	.	.	b	d	d	.	c	c	.	.	.	a	.	.	a	.	.	.	+	.	.	.
HC/CpbU	.	.	.	.	.	.	.	.	.	.	.	.	.	.	.	.	.	.	.	.	.	.	.	.	.
HRS/J	.	.	b	.	.	.	h	.	b	c	b	.	.	a	b	1	.	b	.	.	a	d	.	.	.
HSFR/N	.	.	+	.	.	.	.	.	.	c	.	.	.	.	b	.	.	a	.	.	.	.	.	.	.
HSFS/N	.	.	+	.	.	.	.	.	.	c	c	.	.	.	b	.	.	a	.	.	.	.	.	.	.
HTG/-	.	.	b	a	.	.	.	.	.	+	b	.	.	.	.	.	.	.	.	.	.	.	.	.	.
HYIII/Le	.	.	.	.	.	.	.	.	.	.	.	.	.	.	a	.	.	.	.	.	.	.	.	.	.
I/-	.	.	b	.	.	b	d	d	b	+	b	.	.	a	a	1	.	a	.	.	a	d	d	.	.
I/LnReJ	.	.	.	.	.	.	.	.	b	.	b	.	.	a	a	1	.	a	.	.	a	d	d	.	.
IC/Le	.	.	.	.	.	.	.	.	.	.	.	.	.	.	a	.	.	.	.	.	.	.	.	.	.
IF/BrA	.	.	+	.	.	.	.	.	.	+	.	h	.	.	.	.	.	a	.	.	.	+	.	.	.
IS/Cam-	.	.	+	.	.	.	.	.	.	+	c	.	b	a	.	.	.	a	.	.	.	+	.	.	.
IS/CamRk	.	.	+	.	.	.	.	.	.	+	c	.	.	a	a	.	.	a	.	.	.	+	.	.	.
JE/Le	.	.	+	.	.	.	.	.	.	+	b	.	.	.	a	.	.	.	.	.	.	.	.	.	.
JGBF/Le	.	.	.	.	.	.	.	.	.	.	.	.	.	.	a	.	.	.	.	.	.	.	.	.	.
KC/CpbU	.	.	.	.	.	.	.	.	.	.	.	.	.	.	.	.	.	.	.	.	.	.	.	.	.
KK/-	.	.	+	.	.	.	.	.	.	c	.	.	.	.	a	.	.	a	f	.	.	+	.	.	.
KSB/-	.	.	+	.	.	a	h	d	.	+	.	.	.	.	.	.	.	.	.	.	.	.	.	.	.
LG/J	.	.	+	b	.	b	d	d	b	c	b	.	.	a	a	1	.	.	.	.	b	+	.	.	.
LIS/A	.	.	+	.	.	b	h	d	.	c	.	h	.	.	.	+	.	a	.	.	.	+	.	.	.
LP/J	?	?	+	a	.	b	d	d	b	+	b	.	.	a	a	+	.	b	f	.	b	+	.	b	a
LPT/Le	.	.	.	.	.	.	.	.	.	.	.	.	.	.	a	.	.	.	.	.	.	.	.	.	.
LS/Le	.	.	+	.	.	.	.	.	.	+	b	.	.	.	a	.	.	.	.	.	.	.	.	.	.
LT/ChReSv	.	.	2	a	.	.	.	.	b	.	.	.	.	a	b	.	.	.	.	.	b	+	b	.	.
LTS/A	.	.	+	.	.	b	d	d	.	c	.	h	.	.	.	+	.	a	.	.	.	+	.	.	.
MA/J	.	.	+	b	.	a	h	d	.	c	.	.	.	a	a	+	.	a	.	.	.	+	.	.	.
MA/MyJ	.	.	+	.	.	.	.	.	b	c	b	.	.	a	a	.	.	b	.	.	b	+	d	.	.
MAS/A	.	.	+	.	.	b	h	d	.	c	.	h	.	.	.	+	.	a	.	.	.	+	.	.	.
MOC/CpbU	.	.	.	.	.	.	.	.	.	.	.	.	.	.	.	.	.	.	.	.	.	.	.	.	.
MOLA	.	.	+	.	.	.	.	.	.	+	.	.	.	.	.	.	.	.	.	.	.	+	.	.	.
MOLC/Rk	.	.	+	.	.	.	.	.	.	+	c	.	.	.	a	.	.	.	.	.	.	+	.	.	.
MOLD/Rk	.	.	+	.	.	.	.	.	.	+	c	.	b	a	.	.	.	.	.	.	.	+	.	.	.
MOLD/RkA	.	.	.	.	.	.	.	.	.	+	.	h	.	b	.	.	.	a	.	.	.	+	.	.	.
MOLE/Rk	.	.	+	.	.	.	.	.	.	+	b	.	.	b	a	.	.	.	.	.	.	+	.	.	.
MOLF/Ei	.	.	.	.	.	.	.	.	.	.	.	.	.	.	a	.	.	.	.	.	.	+	.	.	.
MOLN	.	.	+	.	.	.	.	.	.	+	.	.	.	.	.	.	.	.	.	.	.	+	.	.	.
MOLT	.	.	+	.	.	.	.	.	.	+	.	.	.	.	.	.	.	.	.	.	.	+	.	.	.
MRL/MpJ	.	.	.	.	.	.	.	.	.	c	.	.	.	.	a	.	.	.	.	.	.	+	.	.	.

1 = cdm   2 = B*lt

The laboratory mouse

```
Locus   ->   D D D E E   E E E E   E E E E   E E E E   E E E E
(read down)  1 1 l a a   a a a g m   m m m m   m m m m   m p s s
             2 2 b - -   - - - a     v v v v   v v v v   v h - -
             N N - 2 4   5 6 7       - - - -   - - - -   - - 1 2 3
             y y 1               4 5 6 7 8   9 1 1 1 1   1 1
             u u                           0 1 2 3     4
             3 4
------------------------------------------------------------------
Chromosome   1 1 1 U 1   U 2 U 8 U   U U U U U   U U 7 1 2   1 1 8 8 1
(read down)  2 2 1 N 9   N   N   N   N N N N N   N N   6     1       1
------------------------------------------------------------------

FVB/NA       . . . . .   . . . . .   . . . . .   . . . . .   . . b b c
GRS/A        . . . . .   . . . . .   . . . . .   . . . . .   . . b b b
GRS/N        . . . . .   . . . . .   . . . . .   . . . . .   . . b . b
HC/CpbU      . . . . .   . . . . .   . . . . .   . . . . .   . . b c b
HRS/J        . . . b a   . . b . .   b b b b b   b b b b b   b b b b a

HSFR/N       . . . . .   . . . . .   . . . . .   . . . . .   . . b . c
HSFS/N       . . . . .   . . . . .   . . . . .   . . . . .   . . b . c
HTG/-        . . b b a   . . b . .   . . . . .   . . . . .   . . . . .
HYIII/Le     . . . . .   . . . . .   . . . . .   . . . . .   . . a . a
I/-          . . . b a   a a b a .   b b b b b   b b b b b   b . b b c

I/LnReJ      . . . . .   . . . . .   b b b b b   b b b b b   b . b b c
IC/Le        . . . . .   . . . . .   . . . . .   . . . . .   . . b . c
IF/BrA       . . . . .   . . . . .   . . . . .   . . . . .   . . a b a
IS/Cam-      . . . . .   . . . . .   . . . . .   . . . . .   . . b a c
IS/CamRk     . . . . .   . . . . .   . . . . .   . . . . .   . . b a c

JE/Le        . . . . .   . . . . .   . . . . .   . . . . .   . . a . a
JGBF/Le      . . . . .   . . . . .   . . . . .   . . . . .   . . a . a
KC/CpbU      . . . . .   . . . . .   . . . . .   . . . . .   . . b b b
KK/-         . . . . .   . . . . .   . . . . .   . . . . .   . . b a c
KSB/-        . . . . .   . . . . .   . . . . .   . . . . .   . . b a .

LG/J         . . . . .   . . . . .   a b b b b   b b b b b   b b b b c
LIS/A        . . . . .   . . . o .   . . . . .   . . . . .   . . b b c
LP/J         b b . b a   . . a a .   b a b b b   b b b b b   b . b b c
LPT/Le       . . . . .   . . . . .   . . . . .   . . . . .   . . b . a
LS/Le        . . . . .   . . . . .   . . . . .   . . . . .   . . b . c

LT/ChReSv    . . . . .   . . . . .   b b b b b   b b b b b   b . . . .
LTS/A        . . . . .   . . . o .   . . . . .   . . . . .   . . b b c
MA/J         . . . b a   . . b a .   . . . . .   . . . . .   . . b b c
MA/MyJ       . . . . .   . . . . .   b b b b a   a b b b b   b b b b c
MAS/A        . . . . .   . . . . .   . . . . .   . . . . .   . . b b c

MOC/CpbU     . . . . .   . . . . .   . . . . .   . . . . .   . . b b c
MOLA         . . . . .   . . . . .   . . . . .   . . . . .   . . a c a
MOLC/Rk      . . . . .   . . . . .   . . . . .   . . . . .   . . b a b
MOLD/Rk      . . . . .   . . . . .   . . . . .   . . . . .   . . b a c
MOLD/RkA     . . . . .   . . . . .   . . . . .   . . . . .   . . b d b

MOLE/Rk      . . . . .   . . . . .   . . . . .   . . . . .   . . b a b
MOLF/Ei      . . . . .   . . . . .   . . . . .   . . . . .   . . b . b
MOLN         . . . . .   . . . . .   . . . . .   . . . . .   . . a c a
MOLT         . . . . .   . . . . .   . . . . .   . . . . .   . . a c a
MRL/MpJ      . . . . .   . . . . .   . . . . .   . . . . .   . . b b c
```

```
Locus    ->     E E E E E   E E E E E   E E E E E   E E E E E   F F F F F
(read down)     s s s s s   s s s s s   s s s s s   s s s s s   a b b o o
                - - - - -   - - - - -   - - - - -   - - - - r   b p p r s
                5 6 7 8 9   1 1 1 1 1   1 1 1 2 2   2 2 2 2     p - -
                            0 1 2 3 4   6 7 8 2 3   4 5 6 7     i 1 2 5
----------------------------------------------------------------------------
Chromosome      8 8 8 7 8   1 8 U 9 9   3 9 1 8 8   8 1 3 3 6   3 U U 1 1
(read down)                 4   N       9           2           N N 4 2
----------------------------------------------------------------------------

FVB/NA          b a b . a   a a . b r   a a a d c   a b c s a   . a . . .
GRS/A           b a b a a   a a . a r   a a a d c   a b a r a   . . . . .
GRS/N           . . . . .   a . . . .   . . . . .   . . . . .   . . . . .
HC/CpbU         b a b . a   b a . a r   a a a d .   a a a r a   . . . . .
HRS/J           . a . a .   b a . a .   . . . . .   . . . . .   . . . d .

HSFR/N          . . . . .   a . . . .   . . . . .   . . . . .   . . . . .
HSFS/N          . . . . .   a . . . .   . . . . .   . . . . .   . . . . .
HTG/-           . . . . .   . . . . .   . . . . .   . . . . .   . . . . .
HYIII/Le        . . . . .   b a . . .   . . . . .   . . . . .   . . . . .
I/-             . . . . a   b . . . .   . . . . .   . . c . .   . a . . .

I/LnReJ         . . . . .   b . . . .   . . . . .   . . . . .   . . . . .
IC/Le           . . . . .   b a . . .   . . . . .   . . . . .   . . . . .
IF/BrA          b . b a .   . . . . .   . . . . .   . . . . .   . . . . .
IS/Cam-         . a b . a   a b . a .   . . a c c   . b a . .   . b a . .
IS/CamRk        . . . . .   a . a . .   . . . . .   . . . . .   . . . . .

JE/Le           . . . . .   a a . . .   . . . . .   . . . . .   . . . . .
JGBF/Le         . . . . .   a a . . .   . . . . .   . . . . .   . . . . .
KC/CpbU         b a b . a   b a . a r   a a a d c   a b a s a   . . . . .
KK/-            . . . . .   . . . . .   . . . . .   . . . . .   . . . . .
KSB/-           . . . . .   . . . . .   . . . . .   . . . . .   . . . . .

LG/J            . a . a .   b a . b .   . . . . .   . . . . .   . . . . .
LIS/A           b a b a b   a a . b r   a a a h c   a b a r a   . . . . .
LP/J            . a . a a   b a . a .   a b a d c   . b a . .   . . . d .
LPT/Le          . . . . .   a a . . .   . . . . .   . . . . .   . . . . .
LS/Le           . . . . .   b a . . .   . . . . .   . . . . .   . . . . .

LT/ChReSv       . . . . .   . . . . .   . . . . .   . . . . .   . . . . .
LTS/A           b a b a b   a a . a r   a a a h c   a b a r a   . . . . .
MA/J            . . . a .   a . . . .   . . . . .   . . . . .   . . . . .
MA/MyJ          . a . . .   a a . a .   . . . . .   . . . . .   . . . b .
MAS/A           b a b a a   a a . a r   a a a d c   a b a r a   . . . . .

MOC/CpbU        b a b . a   b a . a r   a a a d c   a b a s a   . . . . .
MOLA            a . . . .   . b . . .   . . . . .   . . . . .   . . . . .
MOLC/Rk         . . . . .   c a . . .   . . . . .   . . . . .   . . . . .
MOLD/Rk         . . c . .   c c . . .   . . . . .   . . . a .   . . . . .
MOLD/RkA        a e c . c   b c . a r   a b b a a   c b a r o   . . . . .

MOLE/Rk         . . . . .   b a . . .   . . . . .   . . . . .   . . . . .
MOLF/Ei         . . . . .   c 1 . . .   . . . . .   . . . . .   . . . . .
MOLN            b . . . .   . b . . .   . . . . .   . . . . .   . . . . .
MOLT            a . . . .   . b . . .   . . . . .   . . . . .   . . . . .
MRL/MpJ         . . . . .   b a . . .   . . . . .   . . . . .   . . . . .

              1 = b?
```

```
Locus     ->   H H H H H   H H H H H   H H H H H   H H I I I   I I I I I
(read down)    - - - - -   - a a b b   b c e m p   r s d f g   g g g g g
               4 7 8 9 1   1 o o a a   b   x a     t d h - h   h h h h h
                       2   3 - - -         -       - - 1 -     - - - - -
                           1 2   4         1       1 1   1     2 3 4 5 A
                                 p                                     r
                                 s                                     s
             -----------------------------------------------------------------
Chromosome     7 9 1 U U   2 2 3 1 1   7 2 U 7 8   U 1 1 3 1   1 1 1 1 1
(read down)        4 N N       1 7         N       N 0     2   2 2 2 2 2
             -----------------------------------------------------------------

FVB/NA         . . . . .   . c a c .   d . . . .   . . a . .   . . . . .
GRS/A          . a . . .   . a a 1 .   s o . . .   . . a h c   . . . . .
GRS/N          . . . . .   . . . c .   s o . . .   . l a . .   . . . . .
HC/CpbU        . . . . .   . . . . .   . . . . .   . . . . .   . . . . .
HRS/J          . . . . .   . . . b .   s . . . .   . . a . .   . . a . .

HSFR/N         . . . . .   . . . . .   s 1 . . .   . . a . .   . . . . .
HSFS/N         . . . . .   . . . . .   d . . . .   . . a . .   . . . . .
HTG/-          . . . . .   . . . . .   . . . . .   . . . a .   . . a . .
HYIII/Le       . . . . .   . . . . .   d . . . .   . . a . .   . . . . .
I/-            . . . . .   . . . f .   d o a . .   . . a . c   c . a . .

I/LnReJ        . . . . .   . . . f .   d o . . .   . . a . .   . . a . .
IC/Le          . . . . .   . . . . .   d . . . .   . . b . .   . . . . .
IF/BrA         . . . . .   . . . . .   d 1 . . .   . . a . a   . . . . .
IS/Cam-        . . . . .   . . . b .   d . . . .   b h b . .   . . . . .
IS/CamRk       . . . . .   . . . b .   d . . . .   b . b . .   . . . . .

JE/Le          . . . . .   . . . . .   s . . . .   . . b . .   . . . . .
JGBF/Le        . . . . .   . . . . .   d . . . .   . . a . .   . . . . .
KC/CpbU        . . . . .   . . . . .   . . . . .   . . . . .   . . . . .
KK/-           . . . . .   . . . . .   s o . . .   . . a . .   . . . . .
KSB/-          . . . . .   . . . . .   s . c . .   . . b . .   . . . . .

LG/J           . . . . .   . . b b .   d . a . .   . . a . .   . . . . .
LIS/A          . . . . .   . . . c .   d o a . .   . . a . a   . . . . .
LP/J           2 2 b b 2   b . . b .   d 1 a a .   b l a . b   3 . b . o
LPT/Le         . . . . .   . . . . .   s . . . .   . . a . .   . . . . .
LS/Le          . . . . .   . . . c .   s . . . .   . . a . .   . . . . .

LT/ChReSv      . . . . .   . . . . .   s . . b .   . . . . .   . . . . .
LTS/A          . . . . .   . . . c .   d o a . .   . . a h a   . . . . .
MA/J           . . . . .   . . . a .   s 1 . . .   . . a . j   a . . . .
MA/MyJ         . . . . .   . . . f .   s . . b .   . l a . a   a . a . .
MAS/A          . . . . .   . a a g .   s o a . .   . . a h a   . . . . .

MOC/CpbU       . . . . .   . . . . .   . . . . .   . . . . .   . . . . .
MOLA           . . . . .   . . . . .   p . . . .   . . b . .   . . . . .
MOLC/Rk        . . . . .   . . . d .   p . . . .   . . c . .   . . . . .
MOLD/Rk        . . . . .   . . . . .   p . . . .   . . c . .   . . . . .
MOLD/RkA       . . . . .   . b a a .   c . . . .   . . b . .   . . . . .

MOLE/Rk        . . . . .   . . . . .   d . . . .   . . 4 . .   . . . . .
MOLF/Ei        . . . . .   . . . . .   p . . . .   . . c . .   . . . . .
MOLN           . . . . .   . . . . .   d . . . .   . . b . .   . . . . .
MOLT           . . . . .   . . . . .   d . . . .   . . b . .   . . . . .
MRL/MpJ        . . . . .   . . . . .   d . . . .   . . a . .   . . . . .
```

1 = -c   2 = -a,-b   3 = -a,-c   4 = a,c

```
Locus    ->    F F G G G   G G G G G   G G G G G   G G G G G   G G H H H
(read down)    v v a b d   d d d g k   l l m o o   p p p r u   u v - - -
               - - l p c   c r r c     k o - t t   d i t - s   s - 1 2 3
               1 2 t - -   - - -       - 3 - -     - - - 1 -   - 1
                   1 1     2 1 2       1 1 2       1 1 1       r     s
-----------------------------------------------------------------------------
   Chromosome   4 9 4 3 1   9 U U 6 U   1 1 2 1 8   4 7 1 8 5   5 U 7 1 2
   (read down)      5       N N N       1 7 9           5       N   7
-----------------------------------------------------------------------------

FVB/NA         . . . b .   . . . . .   . . . a b   b b . . .   . . . . .
GRS/A          n s . b .   . . . . .   . a . a b   b b . a .   . . . 1 .
GRS/N          . . . . .   . . . . .   . . . . .   b b . . .   . . . . .
HC/CpbU        . . . . .   . . . . .   . . . . .   . . . . .   . . . . .
HRS/J          . . b . .   b . . a .   . a . . b   a a a a a   . . . k .

HSFR/N         . . . . .   . . . . .   . . . . .   b b . . .   . . . . .
HSFS/N         . . . . .   . . . . .   . . . . .   b b . . .   . . . . .
HTG/-          . . . . b   . . . . .   . . . . .   b . . . .   . . . g .
HYIII/Le       . . . . .   . . . . .   . a . . b   b b a a .   . . . . .
I/-            b s . . b   b . . a .   a a . . b   b a a a .   a a . j .

I/LnReJ        . . . . b   b . . a .   a a . . b   b a a a .   . . . j .
IC/Le          . . . . .   . . . . .   . a . . b   a b a a .   . . . k .
IF/BrA         . r . . .   . . . . .   . a . a b   a b . a .   . . . b .
IS/Cam-        n . b . b   . . . a .   . a . a a   b a a . .   . . . . .
IS/CamRk       n . . . b   . . . . .   . a . a a   b a a . .   . . . . .

JE/Le          . . b . b   . . . . .   . a . . b   a b a b .   . . . b .
JGBF/Le        . . . . .   . . . . .   . a . . b   a b a a .   . . . b .
KC/CpbU        . . . . .   . . . . .   . . . . .   . . . . .   . . . . .
KK/-           . . . . .   . . . . .   . . . . .   a b . . .   . . . b .
KSB/-          . . . . .   . . . . .   . . . . .   b . . . b   b . . . .

LG/J           b . . b b   b . . a .   a a . a b   b a a a a   a . . d .
LIS/A          n s . b .   . . . . .   . a . a b   a b . b b   b . . 1 .
LP/J           n . b a b   . . . a .   b a . a b   b a a . b   b . 2 3 b
LPT/Le         . . . . .   . . . . .   . a . . b   b b a a .   . . . b .
LS/Le          . . . . .   . . . . .   . a . . b   b b a a .   . . . . .

LT/ChReSv      . . . . .   . . . a .   . . . . .   . a . . .   . . . d .
LTS/A          n s . b .   . . . . .   . a . a b   b b . b b   b . . s .
MA/J           n . . . .   . . . . .   a b . a b   b a b a .   b b . k .
MA/MyJ         . . b b b   b . . a .   . b . a b   b a b a b   b . . k .
MAS/A          n s . b .   . . . . .   . a . a b   b b . b b   b . . s .

MOC/CpbU       . . . . .   . . . . .   . . . . .   . . . . .   . . . . .
MOLA           . . . . .   . . . . .   . . . a b   c a b . .   . . . . .
MOLC/Rk        . . . . b   . . . . .   . a . a b   c a a b .   . . . . .
MOLD/Rk        . . . . b   . . . . .   . a . a b   c a c b .   . . . . .
MOLD/RkA       . . . . .   . . . . .   . a . a b   d a . b .   . . . . .

MOLE/Rk        . . . . b   . . . . .   . a . a b   c a c a .   . . . . .
MOLF/Ei        . . . . .   . . . . .   . a . 4 b   c a a a .   . . . . .
MOLN           . . . . .   . . . . .   . a . a b   c a b . .   . . . . .
MOLT           . . . . .   . . . . .   . a . a b   c a b . .   . . . . .
MRL/MpJ        . . . b .   . . . . .   . a . a b   b a a a .   . . . k .
```

        1 = dx   2 = -a,-c   3 = bc   4 = a?

```
Locus      ->   I I I I I   I I I I I   I I I I I   I I K K K   L L L L L
(read down)     g g g g g   g g g g g   g g g g g   g g f t t   a a a d d
                h h h h h   h h h h h   h h k k k   l l o h h   m m p h r
                - - - - -   - - - - -   - - - - -   - - - - -   b b - - -
                C D I L N   P S S S S   S S E P T   1 1 1 1 2   - - 1 1 1
                  e n e p   c a a a a   a r f c r           r   1 2
                  x u v       1 2 3 4   5 c 1 p

Chromosome      1 1 1 1 1   1 1 1 1 1   1 1 6 6 6   1 U 9 1 U   1 1 9 7 6
(read down)     2 2 2 2 2   2 2 2 2 2   2 2           6 N 7 N   2
---------------------------------------------------------------------------

FVB/NA          . . . . .   . . . . .   . . . . .   . . . . .   . . . . b
GRS/A           . . . . .   . . . . .   . . . . .   . . . . .   a b . . a
GRS/N           . . . . .   . . . . .   . . . . .   . . . . .   . . . . .
HC/CpbU         . . . . .   . . . . .   . . . . .   . . . . .   . . . . .
HRS/J           . . . . .   . . . . .   . . . . .   . . . . .   . . b . a

HSFR/N          . . . . .   . . . . .   . . . . .   . . . . .   . . . . .
HSFS/N          . . . . .   . . . . .   . . . . .   . . . . .   . . . . .
HTG/-           . . . . .   . . . . .   . . . . .   . . . . .   . . . . .
HYIII/Le        . . . . .   . . . . .   . . . . .   . . . . .   . . . . .
I/-             . . . . .   . . . . .   . . b . .   . . . . .   a b . . a

I/LnReJ         . . . . .   . . . . .   . . b . .   . . . . .   a b . . a
IC/Le           . . . . .   . . . . .   . . . . .   . . . . .   . . . . .
IF/BrA          . . . . .   . . . . .   . . . . .   . . . . .   . . . . a
IS/Cam-         . . . . .   . . . . .   . . b . .   . . . . .   . . . a b
IS/CamRk        . . . . .   . . . . .   . . . . .   . . . . .   . . . a b

JE/Le           . . . . .   . . . . .   . . . . .   . . . . .   . . . . .
JGBF/Le         . . . . .   . . . . .   . . . . .   . . . . .   . . . . .
KC/CpbU         . . . . .   . . . . .   . . . . .   . . . . .   . . . . .
KK/-            . . . . .   . . . . .   . . . . .   . . . . .   . . . . a
KSB/-           . . . . .   . . . . .   . . . . .   . . . . .   . . . . .

LG/J            . . . . .   . . . . .   . . b . .   . . . . .   a b b . a
LIS/A           . . . . .   . . . . .   . . . . .   . . . . .   . . . . a
LP/J            . . a a .   . . . . .   . a . . .   . . a . .   a b b . b
LPT/Le          . . . . .   . . . . .   . . . . .   . . . . .   . . . . .
LS/Le           . . . . .   . . . . .   . . . . .   . . . . .   . . . . .

LT/ChReSv       . . . . .   . . . . .   . . . . .   . . . . .   . . a . .
LTS/A           . . . . .   . . . . .   . . . . .   . . . . .   . . . . b
MA/J            . a . . .   g g g g g   . . . b .   . . . . .   . . . . a
MA/MyJ          . . . . .   . . . . .   . . b b .   . . . . .   a b b . a
MAS/A           . . . . .   . . . . .   . . . . .   . . . . .   . . . . a

MOC/CpbU        . . . . .   . . . . .   . . . . .   . . . . .   . . . . .
MOLA            . . . . .   . . . . .   . . . . .   . . . . .   . . . . a
MOLC/Rk         . . . . .   . . . . .   . . . . .   . . . . .   . . . a a
MOLD/Rk         . . . . .   . . . . .   . . . . .   . . . . .   . . . a a
MOLD/RkA        . . . . .   . . . . .   . . . . .   . . . . .   a a . . a

MOLE/Rk         . . . . .   . . . . .   . . . . .   . . . . .   . . . a a
MOLF/Ei         . . . . .   . . . . .   . . a . .   . . . . .   . . . . .
MOLN            . . . . .   . . . . .   . . . . .   . . . . .   . . . . a
MOLT            . . . . .   . . . . .   . . . . .   . . . . .   . . . . a
MRL/MpJ         . . . . .   . . . . .   . . . . .   . . . . .   . . . . a
```

Strain distribution of polymorphic variants

```
Locus    ->    L L L L L    L L L L L    L L L L L    L L L L L    L L L L L
(read down)    e e f m n    p s t t t    t t t t v    v y y y y    y y y y y
               n n o p a    t h h h n    w w w w      p - - - -    - - - - -
               - - - -      -   - - -    - - - -      - 1 2 3 5    6 7 8 9 1
               1 2 1   1    1   1 2 2    2 3 4 6      1                     0

---------------------------------------------------------------------------

Chromosome     1 U 7 1 U    U 1 1 U U    1 9 1 U 4    6 1 6 6 1    1 1 U 1 1
(read down)    N   7 N      N 3 N N      2     N            9      5 2 N   9

---------------------------------------------------------------------------

FVB/NA         a b . . .    . . . . .    . . . . .    . . . . .    . . . . .
GRS/A          . . . . .    . . . . .    . . . . .    . b a b b    . . . . .
GRS/N          . . . . .    . . . . .    . . . . .    . b a b .    . . . . .
HC/CpbU        . . . . .    . . . . .    . . . . .    . . . . .    . . . . .
HRS/J          . . . . .    . . a . b    . . b a a    . . . . .    . . . . .

HSFR/N         . . . . .    . . . . .    . . . . .    . b . . .    . . . . .
HSFS/N         . . . . .    . . . . .    . . . . .    . b b a .    . . . . .
HTG/-          . . . . .    . . . . .    . . . . .    . b b . .    . . . a .
HYIII/Le       . . . . .    . . . . .    . . . . .    . . . . .    . . . . .
I/-            . . . . .    . . . . b    . a * . .    a b a b .    . . . . .

I/LnReJ        . . . . .    . . . . b    . a a a .    a . . b .    . . . . .
IC/Le          . . . . .    . . . . .    . . . . .    . . . . .    . . . . .
IF/BrA         . . . . .    . . . . .    . . . . .    . b a b b    . . . . .
IS/Cam-        a b . . .    . . . . b    . . a a .    . . . b .    . . . . .
IS/CamRk       . b . . .    . . . . .    . . . . .    . . . . .    . . . . .

JE/Le          . . . . .    . . . . .    . . . . .    . . . . .    . . . . .
JGBF/Le        . . . . .    . . . . .    . . . . .    . . . . .    . . . . .
KC/CpbU        . . . . .    . . . . .    . . . . .    . . . . .    . . . . .
KK/-           . . . . .    . . . . .    . . . . .    . . . . .    . . . . .
KSB/-          . . . . .    . . . . .    . . . . .    . . . . .    . . . . .

LG/J           a b . . .    . . . . a    . a b a .    b . . b .    . . . a .
LIS/A          a b . . .    . . . . .    . . . . .    . b b b a    . . . . .
LP/J           a b . . .    a . . . b    . a a a c    a b b b b    a a a a .
LPT/Le         . . . . .    . . . . .    . . . . .    . . . . .    . . . . .
LS/Le          . . . . .    . . . . .    . . . . .    . . . . .    . . . . .

LT/ChReSv      . . . . .    . . . . .    . . . . .    . . . . .    . . . . .
LTS/A          a b . . .    . . . . .    . . . . .    . b b b a    . . . . .
MA/J           . . . . .    . . . . .    . . . . a    b a a . .    . . . . .
MA/MyJ         . . . . .    . . . . b    . a b b .    . . . b .    . . . b .
MAS/A          a b . . .    . . . . .    . . . . .    . b a b b    . . . . .

MOC/CpbU       . . . . .    . . . . .    . . . . .    . . . . .    . . . . .
MOLA           . a . . .    . . . . .    . . . . .    . . . . .    . . . . .
MOLC/Rk        . a . . .    . . . . .    . . . . .    . . . . .    . . . . .
MOLD/Rk        c a . . .    . . . . .    . . . . .    . . . . .    . . . . .
MOLD/RkA       . . . . .    . . . . .    . . . . .    . . . . .    . . . . .

MOLE/Rk        . . . . .    . . . . .    . . . . .    . . . . .    . . . . .
MOLF/Ei        . . . . .    . . . . .    . . . . .    . . . a .    . . . . .
MOLN           . . . . .    . . . . .    . . . . .    . . . . .    . . . . .
MOLT           . . . . .    . . . . .    . . . . .    . . . . .    . . . . .
MRL/MpJ        . . . . .    . . . . .    . . . . .    . b a a .    . . . . .
```

Locus → (read down)	Ly-11	Ly-12	Ly-13	Ly-14	Ly-15	Ly-16	Ly-17	Ly-18	Ly-19	Ly-20	Ly-21	Ly-22	Ly-23	Ly-24	Ly-25	Ly-26	Ly-27	Ly-28	Ly-29	Ly-30	Ly-31	Ly-32	Ly-39	Ly-40	Lyb-42
Chromosome (read down)	2	19	UN	7	7	12	1	12	4	4	7	1	2	2	2	UN	UN	UN	4	UN	4	4	17	UN	4
FVB/NA	.	.	.	.	.	.	.	.	.	.	.	.	.	.	.	.	.	.	.	.	.	.	.	.	.
GRS/A	.	.	.	.	.	.	.	.	.	.	.	.	.	.	.	.	.	.	.	.	.	.	.	.	c
GRS/N	.	.	.	.	.	.	.	.	.	.	.	.	.	.	.	.	.	.	.	.	.	.	.	.	c
HC/CpbU	.	.	.	.	.	.	.	.	.	.	.	.	.	.	.	.	.	.	.	.	.	.	.	.	.
HRS/J	.	.	.	.	.	.	.	.	.	.	.	b	.	.	.	.	.	.	.	.	a	.	.	.	b
HSFR/N	.	.	.	.	.	.	.	.	.	.	.	.	.	.	.	.	.	.	.	.	.	.	.	.	.
HSFS/N	.	.	.	.	.	.	.	.	.	.	.	.	.	.	.	.	.	.	.	.	.	.	.	.	a
HTG/-	.	.	.	.	.	.	.	.	.	.	.	.	.	.	.	.	.	.	.	.	.	.	.	.	b
HYIII/Le	.	.	.	.	.	.	.	.	.	.	.	.	.	.	.	.	.	.	.	.	.	.	.	.	.
I/-	.	.	.	.	.	.	b	.	.	.	a	.	.	.	.	.	.	.	.	.	a	.	.	.	a
I/LnReJ	.	.	.	.	.	.	.	.	.	.	a	.	.	.	.	.	.	.	.	.	a	.	.	.	.
IC/Le	.	.	.	.	.	.	.	.	.	.	.	.	.	.	.	.	.	.	.	.	.	.	.	.	.
IF/BrA	.	.	.	.	.	.	.	.	.	.	.	.	.	.	.	.	.	.	.	.	.	.	.	.	.
IS/Cam-	.	.	.	.	.	.	.	.	.	.	.	.	.	.	.	.	.	.	.	.	.	.	.	.	.
IS/CamRk	.	.	.	.	.	.	.	.	.	.	.	.	.	.	.	.	.	.	.	.	.	.	.	.	.
JE/Le	.	.	.	.	.	.	.	.	.	.	.	.	.	.	.	.	.	.	.	.	.	.	.	.	.
JGBF/Le	.	.	.	.	.	.	.	.	.	.	.	.	.	.	.	.	.	.	.	.	.	.	.	.	.
KC/CpbU	.	.	.	.	.	.	.	.	.	.	.	.	.	.	.	.	.	.	.	.	.	.	.	.	.
KK/-	.	.	.	.	.	.	.	.	.	.	.	.	.	.	.	.	.	.	.	.	.	.	.	.	.
KSB/-	.	.	.	.	.	.	.	.	.	.	.	.	.	.	.	.	.	.	.	.	.	.	.	.	.
LG/J	.	.	.	.	.	.	.	.	.	.	.	b	.	.	.	.	.	.	.	.	a	.	.	.	.
LIS/A	.	.	.	.	.	.	.	.	.	.	.	.	.	.	.	.	.	.	.	.	.	.	.	.	.
LP/J	.	b	b	a	b	b	a	.	b	.	b	b	b	.	b	b	b	.	b	.	.	.	.	.	.
LPT/Le	.	.	.	.	.	.	.	.	.	.	.	.	.	.	.	.	.	.	.	.	.	.	.	.	.
LS/Le	.	.	.	.	.	.	.	.	.	.	.	.	.	.	.	.	.	.	.	.	.	.	.	.	.
LT/ChReSv	.	.	.	a	.	.	.	.	.	.	.	.	.	b	.	b	.	.	.	.	a	.	.	.	.
LTS/A	.	.	.	.	a	.	.	.	.	.	.	.	.	.	.	.	.	.	.	.	.	.	.	.	.
MA/J	.	b	.	.	b	.	.	.	.	.	.	.	.	.	.	b	.	.	b	.	.	.	.	.	.
MA/MyJ	.	.	.	.	b	.	b	.	.	.	.	b	.	.	b	b	.	.	.	.	a	.	.	.	.
MAS/A	.	.	.	.	.	.	.	.	.	.	.	.	.	.	.	.	.	.	.	.	.	.	.	.	.
MOC/CpbU	.	.	.	.	.	.	.	.	.	.	.	.	.	.	.	.	.	.	.	.	.	.	.	.	.
MOLA	.	.	.	.	.	.	.	.	.	.	.	.	.	.	.	.	.	.	.	.	.	.	.	.	.
MOLC/Rk	.	.	.	.	.	.	.	.	.	.	.	.	.	.	.	.	.	.	.	.	.	.	.	.	.
MOLD/Rk	.	.	.	.	.	.	.	.	.	.	.	.	.	.	.	.	.	.	.	.	.	.	.	.	.
MOLD/RkA	.	.	.	.	.	.	.	.	.	.	.	.	.	.	.	.	.	.	.	.	.	.	.	.	.
MOLE/Rk	.	.	.	.	.	.	.	.	.	.	.	.	.	.	.	.	.	.	.	.	.	.	.	.	.
MOLF/Ei	.	.	.	.	.	.	.	.	.	.	.	.	.	.	.	.	.	.	.	.	.	.	.	.	.
MOLN	.	.	.	.	.	.	.	.	.	.	.	.	.	.	.	.	.	.	.	.	.	.	.	.	.
MOLT	.	.	.	.	.	.	.	.	.	.	.	.	.	.	.	.	.	.	.	.	.	.	.	.	.
MRL/MpJ	.	.	.	.	.	.	.	.	.	.	.	.	.	.	.	.	.	.	.	.	.	.	.	.	.

```
Locus    ->    L L L L L   L M M M M   M M M M M   N N N N N   N O O p P
(read down)    y y y y y   y a e l o   o o p p u   c e k k p   p r r a n
               b b b b b   b p p s d   d r h i p   a u - - -   - m m   n
               - - - - -   - - - - -   - - - - -   m - 1 2 1   2 - -   -
               3 4 5 6 7   8 1 1 1 1   2 1 1 1 1   1           1 2   1

-------------------------------------------------------------------------
Chromosome     U 4 U 4 1   7 5 1 1 9   7 5 7 9 4   9 1 1 2 1   1 4 4 7 4
(read down)    N   N   2     7                       7 7   4   4
-------------------------------------------------------------------------

FVB/NA         . . . . .   . . . . a   b a . b a   . . . . a   . . . . .
GRS/A          . . . . .   . b . . a   b a . b b   . b . . a   . . . . .
GRS/N          . . . . .   . b . . a   . . . . b   . b . . .   . . . . .
HC/CpbU        . . . . .   . . . . .   . . . . .   . . . . .   . . . . .
HRS/J          . . . . .   . . . . a   . a . b a   . . . . a   . . . + .

HSFR/N         . . . . .   . . . . b   . . . . a   . . . . .   . . . . .
HSFS/N         . . . . .   . . . . a   . . . . a   . . . . .   . . . . .
HTG/-          . . . . .   . . . . a   . . . . .   . . . . .   . . . + .
HYIII/Le       . . . . .   . . . . b   . . . b .   . b . . .   . . . . .
I/-            . . . . .   . b . . b   . . a b a   . b . . a   . . . p .

I/LnReJ        . . . . .   . b . . b   . . . b .   . b . . a   . . . p .
IC/Le          . . . . .   . . . . b   . . . b .   . b . . .   . . . . .
IF/BrA         . . . . .   . . . . a   b a . b .   . b . . a   . . . + .
IS/Cam-        . . . . .   . . . . a   b . . . a   . b . . a   . . . + b
IS/CamRk       . . . . .   . . . . a   b . . . a   . b . . a   . . . + b

JE/Le          . . . . .   . . . . a   . . . b a   . b . . a   . . . + .
JGBF/Le        . . . . .   . . . . b   . . . b .   . b . . .   . . . . .
KC/CpbU        . . . . .   . . . . .   . . . . .   . . . . .   . . . . .
KK/-           . . . . .   . . . . a   . . . . b   . b . . .   . . . + .
KSB/-          . . . . .   . b . . a   . . . . .   . b . . .   . . . + .

LG/J           . . . . .   . b . . a   b a . b .   . b . . .   . . . . .
LIS/A          . . . . .   . b . . b   b a . b b   . b . . a   . . . . .
LP/J           . . . . .   . b a . a   b a . b a   . b o a a   . . . + b
LPT/Le         . . . . .   . . . . a   . . . b .   . b . . .   . . . . .
LS/Le          . . . . .   . . . . b   . . . b b   . b . . a   . . . + .

LT/ChReSv      . . . . .   . . . . a   . . . . .   . . . . .   . . . . .
LTS/A          . . . . .   . b . . b   b a . b a   . b . . a   . . . . .
MA/J           . . . . .   . . . . a   b a . a a   . b . . .   . . . . .
MA/MyJ         . . . . .   . . b . a   b . . a a   . b a o .   . . . . .
MAS/A          . . . . .   . b . . a   b a . b b   . b . . a   . . . . .

MOC/CpbU       . . . . .   . . . . .   . . . . .   . . . . .   . . . . .
MOLA           . . . . .   . . . . b   b a . . .   . . . . a   . . . + .
MOLC/Rk        . . . . .   . . . . a   a . . a c   . b . . b   . . . + .
MOLD/Rk        . . . . .   . . . . a   a . . a c   . b . . b   . . . + .
MOLD/RkA       . . . . .   . . . . a   a a . a .   . . . . b   . . . . .

MOLE/Rk        . . . . .   . . . . a   a . . a c   . b . . b   . . . + .
MOLF/Ei        . . . . .   . . . . a   a . . a .   . b . . .   . . . . .
MOLN           . . . . .   . . . . b   b a . . .   . . . . b   . . . + .
MOLT           . . . . .   . . . . b   a a . . .   . . . . b   . . . + .
MRL/MpJ        . . . . .   . . . . a   b . . b .   . b . . a   . . . . .
```

715

```
Locus     ->   P P P P P   P P P P P   P P P P P   P P P P Q   Q Q Q Q Q
(read down)    a c d e e   e e g g g   g g g k r   r r r s a   a a a a e
               n a   p p   p p d k k   m m m - e   n t t p -   - - - - d
               - -   - -   - -   - -   - - - 3 -   - - -   1   2 3 4 5 -
               2 1   2 3   4 7   1 2   1 2 3   2   p 1 2     1
               -------------------------------------------------------------
Chromosome     U U U 1 1   7 5 4 X 1   5 4 9 9 1   2 U 8 2 1   1 1 1 1 1
(read down)    N N N 0     7         7         2   N       7   7 7 7 7 7
               -------------------------------------------------------------

FVB/NA         . . . . b   . . b b b   a a . . .   . . . . .   . . . . .
GRS/A          . . . . b   . . b b a   b a c . .   . . . . .   b b . . .
GRS/N          . . . . b   . . . . .   b . . . .   . . . . .   . . . . .
HC/CpbU        . . . . .   . . . . .   . . . . .   . . . . .   . . . . .
HRS/J          . . . . a   . . b . b   a a a . .   . . . b .   . . . . .

HSFR/N         . . . . b   . . . . .   a . . . .   . . . . .   . . . . .
HSFS/N         . . . . b   . . . . .   a . . . .   . . . . .   . . . . .
HTG/-          . b . . .   . . . . a   . . . . .   . . . b .   . . . . .
HYIII/Le       . . . . b   . . . . .   b a . . .   . . . . .   . . . . .
I/-            . b . . b   . . . b a   b a c . .   b . . . b   . . . . ?

I/LnReJ        . . . . b   . . . b a   b a c . .   b . . . .   . . . . ?
IC/Le          . . . . c   . . . . .   b a . . .   . . . . .   . . . . .
IF/BrA         . . . . b   . . b b .   b a . . .   . . . . .   . . . . .
IS/Cam-        b . . . b   a . . b .   b a c b .   . . . b .   . . . . .
IS/CamRk       b . . . b   . . . b .   b a c . .   . . . b .   . . . . .

JE/Le          . . . . a   . . . b .   a a . . .   . . . . .   . . . . .
JGBF/Le        . . . . a   . . . . .   a a . . .   . . . . .   . . . . .
KC/CpbU        . . . . .   . . . . .   . . . . .   . . . . .   . . . . .
KK/-           . . . . b   . . . . .   a . . . .   . . . . .   . . . . .
KSB/-          . . . . .   . . . . .   . . . . .   . b a . .   . . . . .

LG/J           . . . . b   . . b . b   a a b . .   . . . b .   . . . . b
LIS/A          . . . . b   . . b b a   b a . . .   . . . . .   . . . . .
LP/J           b . a . a   . . b b c   a a . . a   a . . b .   . . . . b
LPT/Le         . . . . b   . . . . .   b a . . .   . . . . .   . . . . .
LS/Le          . . . . b   . . . . .   a a . . .   . . . . .   . . . . .

LT/ChReSv      . . . . .   . . . . a   . . a . .   a . . . .   . . . . .
LTS/A          . . . . b   . . b b b   b a . . .   . . . . .   . . . . .
MA/J           . a . . b   . . b b .   b a . . .   . . . . b   b b . . .
MA/MyJ         . . . . c   . . . . b   b a c . .   b . . . .   . . . . b
MAS/A          . . . . b   . . b b b   b a . . .   . . . . .   . . . . .

MOC/CpbU       . . . . .   . . . . .   . . . . .   . . . . .   . . . . .
MOLA           . . . . .   . . b b .   e b . . .   . b b . .   . . . . .
MOLC/Rk        . . . . b   . . . . .   b a . . .   . . . . .   . . . . .
MOLD/Rk        . . . . b   b . . b .   b a . . .   . . . . .   . . . . .
MOLD/RkA       . . . . b   . . b b b   b a . . .   . . . . .   . . . . .

MOLE/Rk        . . . . b   . . . b .   b a . . .   . . . . .   . . . . .
MOLF/Ei        . . . . a   . . . . .   b a . . .   . . . . .   . . . . .
MOLN           . . . . .   . . b b .   e a . . .   . b b . .   . . . . .
MOLT           . . . . .   . . b b .   e b . . .   . b a . .   . . . . .
MRL/MpJ        . . . . b   . . . . .   a a . . .   . . . . .   . . . . .
```

```
Locus    ->    r r R R R   S S S S S   S S S S S   S S S T T   T T T T T
(read down)    d d e i n   a a a d d   l o o o p   s v v a c   c g h i l
               s n b 7     a a s h r   p a d d i   p p p m n   r n y n a
               - - s       - - - -     - - -       - - - -     g   - d
               1 1 -       1 1 1 1     1 2 2       1 2 1 2     1
                   6                       r

-------------------------------------------------------------------------
Chromosome     5 1 1 1 6   7 7 1 2 U   1 U 1 1 1   U 2 7 7 1   1 1 9 1 1
(read down)      7   4               N   7 N 6 7 2   N       1   3 5   2 7
-------------------------------------------------------------------------

FVB/NA         . . . . .   . . . a .   . . . . .   . . . . s   . . . . .
GRS/A          + + . a .   . . . a .   . . . . .   f a . . f   . . b . c
GRS/N          . . . . .   . . . . .   . . . . .   . . . . .   . . b . c
HC/CpbU        . . . . .   . . . . .   . . . . .   . . . . .   . . . . .
HRS/J          + . . . .   a . a . .   . . . . .   . . . a .   . . . o b

HSFR/N         . . . . .   . . . . .   . . . . .   . . . . .   . . . . .
HSFS/N         . . . . .   . . . . .   . . . . .   . . . . .   . . b . a
HTG/-          + . . . .   . . . . .   . . . . .   . . . a .   . . b . b
HYIII/Le       . . . . .   . . . . .   . . . . .   . . . . .   . . . . .
I/-            + . s . b   b . o . .   . . . . d   . . . . .   d b b . b

I/LnReJ        + . s a b   b . o . .   . . . . d   . . . . .   d b . . .
IC/Le          . . . . .   . . . . .   . . . . .   . . . . .   . . . . .
IF/BrA         . . . . .   . . . . .   . . . . .   . a . . .   . . b . c
IS/Cam-        + . s . b   . . o a .   . b . . .   . . . a .   . . . . .
IS/CamRk       + . s . .   . . o a .   . . . . .   . . . a .   . . . . .

JE/Le          + . . . .   . . . . .   . . . . .   . . . . .   . . . . .
JGBF/Le        . . . . .   . . . . .   . . . . .   . . . . .   . . . . .
KC/CpbU        . . . . .   . . . . .   . . . . .   . . . . .   . . . . .
KK/-           . . . . .   . . . . .   . . . . .   . . . . .   . . b . .
KSB/-          . . . . .   . . . . .   . . . . .   . . . . .   . . . . .

LG/J           + . s a b   a . . a .   . . . . b   . . . c .   a b . . .
LIS/A          + + . . .   . . . . .   . . . . .   . a . . f   . . b . b
LP/J           + . b a .   a . a a 1   . . . . .   . . . . .   c . b . c
LPT/Le         . . . . .   . . . . .   . . . . .   . . . . .   . . . . .
LS/Le          + . . . .   . . . . .   . . . . .   . . . . .   . . . . .

LT/ChReSv      + . . . .   . . . . .   . . . . d   . . . a .   . . . . .
LTS/A          + + . . .   . . . . .   . . . . .   . a . . f   . . b . b
MA/J           + . . . .   . . a . .   o . . . .   . . . . .   c . a . b
MA/MyJ         + . s a b   a . a a .   . . a a b   . . p b s   a b . . b
MAS/A          + + . . .   . . . a .   . . . . .   . a . . f   . . b . b

MOC/CpbU       . . . . .   . . . . .   . . . . .   . . . . .   . . . . .
MOLA           . . . . .   . . . . .   . . b . .   . . . . .   . . . . .
MOLC/Rk        . . . . .   . . . . .   . . a . .   . . . b .   . . . . .
MOLD/Rk        . . . . .   . . . a .   . . a . .   . . . b .   . . . . .
MOLD/RkA       . . . b .   . . . a .   . . . . .   . . . . s   . . . . .

MOLE/Rk        . . . . .   . . . . .   . . a . .   . . . b .   . . . . .
MOLF/Ei        . . . . a   . . . . .   . . a . .   . . . . .   . . . . .
MOLN           . . . . .   . . . . .   . . b . .   . . . . .   . . . . .
MOLT           . . . . .   . . . . .   . . a . .   . . . . .   . . . . .
MRL/MpJ        . . . . .   . . . . .   . . . . .   . . . . .   . . b o .
```

        1 = b?

```
Locus    ->     T T T T U   U U U W X   X X X X X   X X X X X   X X X X X
(read down)     p r s t c   p p p a m   m m m m m   m m m m m   m m m m m
                r f u h p   g g s p m   m m m m m   m m m m m   m m m m m
                e   y       - -   v     v v v v v   v v v v v   v v v v v
                            1 2         - - - - -   - - - - -   - - - - -
                        9   4 4 4 4 4   4 4 4 5 5   5 5 5 5 5
                            2 3 4 5 6   7 8 9 0 1   2 3 4 5 6
---------------------------------------------------------------------------
Chromosome      1 9 1 1 8   1 1 9 U 1   1 U U U U   U U U 1 U   5 U U 1 U
(read down)     2   2 2     7     N     9 N N N N   N N N 2 N     N N 5 N
---------------------------------------------------------------------------

FVB/NA          . b . . .   . . a . .   . . . . .   . . . . .   . . . . .
GRS/A           . b . . .   . . a . .   . . . . .   . . . . .   . . . . .
GRS/N           . . . . .   . . . . .   . . . . .   . . . . .   . . . . .
HC/CpbU         . . . . .   . . . . .   . . . . .   . . . . .   . . . . .
HRS/J           o b o o .   s s . . a   o o o o o   o o o o o   a o a o o

HSFR/N          . b . . .   . . . . .   . . . . .   . . . . .   . . . . .
HSFS/N          . b . . .   . . . . .   . . . . .   . . . . .   . . . . .
HTG/-           . . . . .   . . . . a   o o o o o   o o o o o   a o a o o
HYIII/Le        . a . . .   . . . . .   . . . . .   . . . . .   . . . . .
I/-             . b . . .   f d a a o   o o o a o   o o o a o   o o a a o

I/LnReJ         . b . . .   f . a . o   o o o a o   o o o a o   o o a a o
IC/Le           . a . . .   . . . . .   . . . . .   . . . . .   . . . . .
IF/BrA          . b . . .   . . . . .   . . . . .   . . . . .   . . . . .
IS/Cam-         . b . . .   f d . . .   . . . . .   . . . . .   . . . . .
IS/CamRk        . b . . .   f . . . .   . . . . .   . . . . .   . . . . .

JE/Le           . b . . .   . . . . .   . . . . .   . . . . .   . . . . .
JGBF/Le         . b . . .   . . . . .   . . . . .   . . . . .   . . . . .
KC/CpbU         . . . . .   . . . . .   . . . . .   . . . . .   . . . . .
KK/-            . b . . .   . . . . .   . . . . .   . . . . .   . . . . .
KSB/-           . b . . .   . . . . .   . . . . .   . . . . .   . . . . .

LG/J            . b . . .   . d a . a   o o o o o   o a o o o   a o a o a
LIS/A           . b . . .   . . a . .   . . . . .   . . . . .   . . . . .
LP/J            . b . . .   . s b a a   o o o o o   o o o o o   o o a o o
LPT/Le          . b . . .   . . . . .   . . . . .   . . . . .   . . . . .
LS/Le           . b . . .   . . . . .   . . . . .   . . . . .   . . . . .

LT/ChReSv       . b . . .   f s . . a   o o o o o   a o o o o   a o a a o
LTS/A           . b . . .   . . . . .   . . . . .   . . . . .   . . . . .
MA/J            . . . . .   . . . . .   . . . . .   . . . . .   . . . . .
MA/MyJ          . b . . .   s d a a a   o o o o o   o o o o o   a o a o o
MAS/A           . b . . .   . . a . .   . . . . .   . . . . .   . . . . .

MOC/CpbU        . . . . .   . . . . .   . . . . .   . . . . .   . . . . .
MOLA            . b . . .   . . . . .   . . . . .   . . . . .   . . . . .
MOLC/Rk         . b . . .   . . . . .   . . . . .   . . . . .   . . . . .
MOLD/Rk         . b . . .   . . . . .   . . . . .   . . . . .   . . . . .
MOLD/RkA        . b . . .   . . b . .   . . . . .   . . . . .   . . . . .

MOLE/Rk         . b . . .   . . . . .   . . . . .   . . . . .   . . . . .
MOLF/Ei         . b . . .   . . . . .   . . . . .   . . . . .   . . . . .
MOLN            . b . . .   . . . . .   . . . . .   . . . . .   . . . . .
MOLT            . b . . .   . . . . .   . . . . .   . . . . .   . . . . .
MRL/MpJ         o b o o .   . . . . .   . . . . .   . . . . .   . . . . .
```

Locus -> (read down)	X m m v - 5 7	X m m v - 5 8	X m m v - 5 9	X m m v - 6 0	X m m v - 6 2		X m m v - 6 3	X m m v - 6 4	X m m v - 6 5	X m m v - 6 6	X m m v - 6 7		X m m v - 6 8	X m m v - 6 9	X m m v - 7 0
Chromosome (read down)	U N	U N	U N	U N	4		U N	U N	3 N	U N	U N		U N	U N	U N
FVB/NA	.	.	.	.	.		.	.	.	.	.		.	.	.
GRS/A	.	.	.	.	.		.	.	.	.	.		.	.	.
GRS/N	.	.	.	.	.		.	.	.	.	.		.	.	.
HC/CpbU	.	.	.	.	.		.	.	.	.	.		.	.	.
HRS/J	o	o	o	a	o		a	a	o	o	o		o	a	a
HSFR/N	.	.	.	.	.		.	.	.	.	.		.	.	.
HSFS/N	.	.	.	.	.		.	.	.	.	.		.	.	.
HTG/-	o	o	o	o	o		a	o	a	o	o		o	a	a
HYIII/Le	.	.	.	.	.		.	.	.	.	.		.	.	.
I/-	o	o	o	o	o		a	o	a	o	o		o	a	a
I/LnReJ	o	o	o	o	o		a	o	a	o	o		o	a	a
IC/Le	.	.	.	.	.		.	.	.	.	.		.	.	.
IF/BrA	.	.	.	.	.		.	.	.	.	.		.	.	.
IS/Cam-	.	.	.	.	.		.	.	.	.	.		.	.	.
IS/CamRk	.	.	.	.	.		.	.	.	.	.		.	.	.
JE/Le	.	.	.	.	.		.	.	.	.	.		.	.	.
JGBF/Le	.	.	.	.	.		.	.	.	.	.		.	.	.
KC/CpbU	.	.	.	.	.		.	.	.	.	.		.	.	.
KK/-	.	.	.	.	.		.	.	.	.	.		.	.	.
KSB/-	.	.	.	.	.		.	.	.	.	.		.	.	.
LG/J	o	o	o	o	o		a	o	a	o	o		o	a	a
LIS/A	.	.	.	.	.		.	.	.	.	.		.	.	.
LP/J	o	o	o	o	a		a	o	o	o	o		a	a	a
LPT/Le	.	.	.	.	.		.	.	.	.	.		.	.	.
LS/Le	.	.	.	.	.		.	.	.	.	.		.	.	.
LT/ChReSv	o	o	o	a	o		a	a	o	a	a		o	a	a
LTS/A	.	.	.	.	.		.	.	.	.	.		.	.	.
MA/J	.	.	.	.	.		.	.	.	.	.		.	.	.
MA/MyJ	o	o	o	o	o		a	a	o	o	o		o	a	a
MAS/A	.	.	.	.	.		.	.	.	.	.		.	.	.
MOC/CpbU	.	.	.	.	.		.	.	.	.	.		.	.	.
MOLA	.	.	.	.	.		.	.	.	.	.		.	.	.
MOLC/Rk	.	.	.	.	.		.	.	.	.	.		.	.	.
MOLD/Rk	.	.	.	.	.		.	.	.	.	.		.	.	.
MOLD/RkA	.	.	.	.	.		.	.	.	.	.		.	.	.
MOLE/Rk	.	.	.	.	.		.	.	.	.	.		.	.	.
MOLF/Ei	.	.	.	.	.		.	.	.	.	.		.	.	.
MOLN	.	.	.	.	.		.	.	.	.	.		.	.	.
MOLT	.	.	.	.	.		.	.	.	.	.		.	.	.
MRL/MpJ	.	.	.	.	.		.	.	.	.	.		.	.	.

```
Locus    ->   a A A A A    A A A A A    A A A A A    A A A A A    A A A A A
(read down)     a b c c      d d d d g    h h h h k    k l m m o    o p p p s
                t p f o      h h h h s    d d r p      p b y y x    x h k o -
                  a - -      - - - -      - - - -      - - - - -    - - a 1
                  1 1        1 2 3 3      1 2 1 1      2 1 1 2 1    2 1 - s
                     e t                                           1

---------------------------------------------------------------------------------
Chromosome    2 1 7 1 4    3 U 3 3 X    1 4 1 3 1    4 5 3 3 1    1 2 1 9 1
(read down)   2            N            2       9               0       3
---------------------------------------------------------------------------------

MWT/Le        1 o . . a    . . . . .    . . . . a    . a a . .    1 . a . a
MY/HuLe       . . . . .    . . . . .    . . . . b    . . a . .    . . a . a
MYD/Le        . . . . .    . . . . .    . . . . a    . . a . .    . . a . a
NBL/N         a . . . .    . . . . .    d . . . .    . . . . .    . . . . .
NC            + . . . .    . . . . .    . . . a .    . . . . .    . . . . .

NC/CpbU       . . . . .    . . . . .    . . . . .    . . . . .    . . . . .
NFR/N         + . . . .    . . . . .    b . . . .    . . . . .    . . . . .
NFS/N         + . . . .    . . . . .    b . . . .    . . . . .    . . . . .
NFS/NA        . . . . .    a a a a .    . b b a .    . . . . a    a . . . .
NGP/N         a . . . .    . . . . h    b . . . .    . . . . .    . . . . .

NH/LwN        a . . . .    . . . . h    b . . . .    . . . . .    . . . . .
NOD/Lt        + . . . .    . . a . .    . . . . b    . . b b .    . . . b .
NON/J         + . . . .    . . a . .    . . . . b    . . a a .    . . . a .
NU/J          . . . . .    . . . . .    . . . . b    . . a . .    . . . b .
NZB/-         a a b a a    . . a . h    d b b a b    a a a . a    a a a a a

NZB/B1NJ      a a b a a    . . . . h    d b . . b    a a a . .    . . a a a
NZW/-         + . . . a    . . a . h    d b b a b    . . a . a    a . . a .
NZW/LacJ      . . . . .    . . . . .    . . . . b    . . a . .    a . . a .
O20/-         a . . . .    a . a a h    b . . b .    . . . . .    . . . . .
O20/A         a . . . .    a . a a h    b . . b .    . . . . .    . . . . .

P/-           a a . b .    . . . . h    b a . . b    b a a a 2    . . a a a
P/J           a a . b .    . . . . h    b a . . b    b a a a 2    . . a a a
PC/CpbU       . . . . .    . . . . .    . . . . .    . . . . .    . . . . .
PERA/Rk       + . . . .    . . . . .    . . . . .    . a b . .    . . a a .
PH/Re         . . . . .    . . . . h    . . . . .    . . . . .    . . . a .

PL/J          + a . b .    . . . . h    b b . . b    a a a a 2    . a a a a
PRO/Re        a . . . .    . . . . .    . . . . .    . . . a .    . . a a .
QC/CpbU       . . . . .    . . . . .    . . . . .    . . . . .    . . . . .
RBA1/Dn       . . . . .    . . . . .    . . . . b    . . b . .    . . a . .
RBA/Dn        + 2 . . .    . . . . .    . . . a .    . a b . .    . a b . .

RBB/Dn        + a . . .    . . . . .    . . . . b    . a a . .    . a a . .
RBC/Dn        . o . . .    . . . . .    . . . . .    . a a . .    . a a . .
RBF/DnJ       . . . . .    . . . . .    . . . . b    . . a . .    . . b . .
RBG/Dn        . . . . .    . . . . .    . . . . b    . . a . .    . . b . .
RC/CpbU       . . . . .    . . . . .    . . . . .    . . . . .    . . . . .

RF/J          a a a b a    . . . . h    d b . . b    a a a a 2    . . a a a
RFM/-         a . . . .    . . . . h    . . . . .    . . . . .    . . . . .
RHJ/Le        . . . . .    . . . . .    . . . . b    . . a . .    . . . b .
RIII/-        + . . . .    a a a a .    b b a . .    . . . . a    a . . . .
RIII/SeA      + . . . .    a a a a .    b b a . .    . . . . a    a . . . .
```

720   1 = a*t   2 = a?

Strain distribution of polymorphic variants

```
Locus    ->   A A b B B    B B B B c    C C C C C    c C C C C    C d D D D
(read down)   v v 2 c      g g g v      3 4 6 a a    d e e f o    v 7 1 1
              - - m d      l l l -          r r      m - - h h    - R 2 2
              1 2 - 1      - - - 1          - -      1 2 e        1 p N N
                   1       e s t            1 2                     2 y y
                                                                   - u u
                                                                   s 1 2
              ----------------------------------------------------------------
Chromosome    U U 4 2 5    9 9 9 8 7    1 1 1 3 3    3 2 1 2 7    5 9 7 1 1
(read down)   N N                       7 7 5          7              2 2
              ----------------------------------------------------------------

MWT/Le        . . + . .    . . . . +    b . . a a    . . . . .    . . . . .
MY/HuLe       . . . . .    . . . . .    . . . . a    . . . . .    . . . . .
MYD/Le        . . . . .    . . . . .    . . . . a    . . . . .    . . . . .
NBL/N         . . + . .    . . . . +    . . . . a    . . a . .    . . . . .
NC            . . b . .    . . . . +    a . a . a    . . a f .    . + . . .

NC/CpbU       . . . . .    . . . . .    . . . . .    . . . . .    . . . . .
NFR/N         . . + . .    . . . . c    . . . . b    . . a . .    . . . . .
NFS/N         . . + . .    . . . . c    . . . . b    . . a . .    . . . . .
NFS/NA        . . . . .    . . . . c    . . . . .    . . a . .    . . . . .
NGP/N         . . . . .    a h b . c    . . . . b    . . a . .    . . . . .

NH/LwN        . . + . .    b d d . +    . . . . a    . . a . .    . d . . .
NOD/Lt        . . + . .    . . . . c    b . . a a    + . a . .    . + . . .
NON/J         . . + . .    . . . . c    b . . a b    + . a . .    . + . . .
NU/J          . . . . .    . . . . .    . . . . b    . . . . .    . . . . .
NZB/-         r r + a b    b d d b +    b . a a a    + . b f .    b + . b a

NZB/BlNJ      r r + a .    b d d b +    b . . a a    + . b . .    b + . b a
NZW/-         . . b . .    b d d . c    b . . . a    . . a . .    . + . . .
NZW/LacJ      . . . . .    . . . . .    b . . . a    . . . . .    . . . . .
O20/-         . . + . .    b d d . c    . h . . .    + . a . .    . + d . .
O20/A         . . + . .    b d d . c    . h . . .    + . a . .    . + . . .

P/-           . . b a .    b d d b +    b . . a a    + . a . .    b d . . .
P/J           . . b a .    b d d b +    b . . a a    + . a . .    b d . . .
PC/CpbU       . . . . .    . . . . .    . . . . .    . . . . .    . . . . .
PERA/Rk       . . + . .    . . . . +    b . . a a    . . . . .    . + . . .
PH/Re         . . . . .    b h d . .    b . . a b    . . . . .    . . . . .

PL/J          . . + a .    b d d b c    b . . a b    + . a f l    b + . b a
PRO/Re        . . . b .    . . . . 1    b . . a a    2 . . . .    . . . . .
QC/CpbU       . . . . .    . . . . .    . . . . .    . . . . .    . . . . .
RBA1/Dn       . . . . .    . . . . .    . . . . a    . . . . .    . . . . .
RBA/Dn        . . + . .    . . . . +    b . . 3 a    . . . . .    . + . . .

RBB/Dn        . . + . .    . . . . +    b . . a a    . . . . .    . + . . .
RBC/Dn        . . . . .    . . . . c    b . . a a    . . . . .    . . . . .
RBF/DnJ       . . . . .    . . . . .    . . . . a    . . . . .    . . . . .
RBG/Dn        . . . . .    . . . . .    . . . . a    . . . . .    . . . . .
RC/CpbU       . . . . .    . . . . .    . . . . .    . . . . .    . . . . .

RF/J          . . + b .    b d d b c    b l . a a    + . b . l    a + d . .
RFM/-         . . . . .    b d d . c    . . . . .    m . . . .    . . . . .
RHJ/Le        . . . . .    . . . . .    . . . . a    + . a . .    . . . . .
RIII/-        . . + . .    . . . . c    . h . . b    + . a . .    . + . . .
RIII/SeA      . . b . .    . . . . c    . h . . .    + . a . .    . + . . .
```

1 = c*ch   2 = cdm   3 = a,b

```
Locus      ->    D D D E E   E E E E E   E E E E E   E E E E E   E E E E E
(read down)      1 1 l a a   a a a g m   m m m m m   m m m m m   m p s s
                 2 2 b - -   - - - a     v v v v v   v v v v v   v h - -
                 N N - 2 4   5 6 7       - - - - -   - - - - -   - - 1 2 3
                 y y 1                   4 5 6 7 8   9 1 1 1 1   1 1
                 u u                               0 1 2 3   4
                 3 4
--------------------------------------------------------------------------------
Chromosome       1 1 1 U 1   U 2 U 8 U   U U U U U   U U 7 1 2   1 1 8 8 1
(read down)      2 2 1 N 9   N   N   N   N N N N N   N N   6       2 1
--------------------------------------------------------------------------------

MWT/Le           . . . . .   . . . . .   . . . . .   . . . . .   . . a b a
MY/HuLe          . . . . .   . . . . .   . . . . .   . . . . .   . . b . a
MYD/Le           . . . . .   . . . . .   . . . . .   . . . . .   . . b . a
NBL/N            . . . . .   . . . . .   . . . . .   . . . . .   . . b . a
NC               . . . . .   . . . . .   . . . . .   . . . . .   . . b b c

NC/CpbU          . . . . .   . . . . .   . . . . .   . . . . .   . . b b b
NFR/N            . . . . .   . . . . .   . . . . .   . . . . .   . . b . c
NFS/N            . . . . .   . . . . .   . . . . .   . . . . .   . . b . a
NFS/NA           . . . . .   . . . . .   . . . . .   . . . . .   . . b b a
NGP/N            . . . . .   . . . . .   . . . . .   . . . . .   . . b . c

NH/LwN           . . . . .   . . . . .   . . . . .   . . . . .   . . b . a
NOD/Lt           . . . . .   . . . . .   . . . . .   . . . . .   . . b b c
NON/J            . . . . .   . . . . .   . . . . .   . . . . .   . . b b c
NU/J             . . . . .   . . . . .   . . . . .   . . . . .   . . b . c
NZB/-            b b b b a   . . b . .   b b b b b   b b b b b   b . b b c

NZB/BlNJ         b b . . .   . . . . .   b b b b b   b b b b b   b . b . c
NZW/-            . . b . .   . . . . .   . . . . .   . . . . .   . . b b c
NZW/LacJ         . . . . .   . . . . .   . . . . .   . . . . .   . . b . c
O20/-            . . . . .   . . . . .   . . . . .   . . . . .   . . b b c
O20/A            . . . . .   . . . . .   . . . . .   . . . . .   . . b b c

P/-              . . . . .   . . . a .   b b b b b   b b b b b   b . b b a
P/J              . . . . .   . . . a .   b b b b b   b b b b b   b . b b a
PC/CpbU          . . . . .   . . . . .   . . . . .   . . . . .   . . b b b
PERA/Rk          . . . . .   . . . . .   . . . . .   . . . . .   . . b b b
PH/Re            . . . . .   . . . . .   . . . . .   . . . . .   . . b b c

PL/J             b b . . .   . . . a h   b b b b b   b b b b b   b b b c c
PRO/Re           . . . . .   . . . . .   . . . . .   . . . . .   . . . . .
QC/CpbU          . . . . .   . . . . .   . . . . .   . . . . .   . . b b c
RBA1/Dn          . . . . .   . . . . .   . . . . .   . . . . .   . . a . b
RBA/Dn           . . . . .   . . . . .   . . . . .   . . . . .   . . a c a

RBB/Dn           . . . . .   . . . . .   . . . . .   . . . . .   . . b b a
RBC/Dn           . . . . .   . . . . .   . . . . .   . . . . .   . . b b a
RBF/DnJ          . . . . .   . . . . .   . . . . .   . . . . .   . . b . b
RBG/Dn           . . . . .   . . . . .   . . . . .   . . . . .   . . a . c
RC/CpbU          . . . . .   . . . . .   . . . . .   . . . . .   . . b b c

RF/J             . . . a .   b b b a l   b b b b b   b b b b b   b b b b b
RFM/-            . . . a .   . . . . .   . . . . .   . . . . .   . . a . .
RHJ/Le           . . . . .   . . . . .   . . . . .   . . . . .   . . b . c
RIII/-           . . a a a   b . b . h   . . . . .   . . . . .   . . b b *
RIII/SeA         . . . . .   . . . . .   . . . . .   . . . . .   . . b b a
```

```
Locus   ->   E E E E E   E E E E E   E E E E E   E E E E E   F F F F F
(read down)  s s s s s   s s s s s   s s s s s   s s s s s   a b b o o
             - - - - -   - - - - -   - - - - -   - - - - r   b p p r s
             5 6 7 8 9   1 1 1 1 1   1 1 1 2 2   2 2 2 2     p - - -
                         0 1 2 3 4   6 7 8 2 3   4 5 6 7     i 1 2 5

---------------------------------------------------------------------------
Chromosome   8 8 8 7 8   1 8 U 9 9   3 9 1 8 8   8 1 3 3 6   3 U U 1 1
(read down)          4   N          9          2          N N 4 2
---------------------------------------------------------------------------

MWT/Le       . . . . .   b a . . .   . . . . .   . . . . .   . . . . .
MY/HuLe      . . . . .   a a . . .   . . . . .   . . . . .   . . . . .
MYD/Le       . . . . .   a a . . .   . . . . .   . . . . .   . . . . .
NBL/N        . . . . .   a . . . .   . . . . .   . . . . .   . . . . .
NC           . . . . .   . . . . .   . . . . .   . . . . .   . . . . .

NC/CpbU      b a b . a   b a . a r   a a a d c   a b a r a   . . . . .
NFR/N        . . . . .   a . . . .   . . . . .   . . . . .   . . . . .
NFS/N        . . . . .   a . . . .   . . . . .   . . . . .   . . . . .
NFS/NA       b a b . a   a a . b r   a a a d c   f b a r a   . b . . .
NGP/N        . . . . .   a . . . .   . . . . .   . . . . .   . . . . .

NH/LwN       . . . . .   b . . . .   . . . . .   . . . . .   . . . . .
NOD/Lt       . . . . .   a a . . .   a . . . .   . . . . .   . . . . .
NON/J        . . . . .   a a . . .   a . . . .   . . . . .   . . . . .
NU/J         . . . . .   a a . . .   . . . . .   . . . . .   . . . . .
NZB/-        b a b . .   b a . a .   . . . . .   . b . . .   . . . d b

NZB/BlNJ     . a . . .   b a . a .   . . . . .   . . . . .   . . . d b
NZW/-        b . . . a   a a . . .   . . . c .   . b . . .   . . . . .
NZW/LacJ     . . . . .   a a . . .   . . . . .   . . . . .   . . . . .
O20/-        b a b a a   b a . a r   a a a d c   a b c s a   . . . . .
O20/A        b a b a a   b a . a r   a a a d c   a b c s a   . . . . .

P/-          . a . a .   a b . a .   . . . . .   . . . . .   . a a b .
P/J          . a . a .   a b . a .   . . . . .   . . . . .   . a a b .
PC/CpbU      b a b . a   b a . a r   a a a d c   a b a r a   . . . . .
PERA/Rk      . . . . .   b . . . .   . . . . .   . . . . .   . . . . .
PH/Re        . a . a .   b . . . .   . . . . .   . . . . .   . . . . .

PL/J         . a . a .   b a . a .   . . . . .   . . . . .   . . . d b
PRO/Re       . . . . .   a . . . .   . . . . .   . . . . .   . . . . .
QC/CpbU      b a b . a   b a . a r   a a a d c   a a a s a   . . . . .
RBA1/Dn      . . . . .   b a . . .   . . . . .   . . . . .   . . . . .
RBA/Dn       . . . . .   1 b . . .   . . . . .   . . . . .   . . . . .

RBB/Dn       . . . . .   b a . . .   . . . . .   . . . . .   . . . . .
RBC/Dn       . . . . .   a . . . .   . . . . .   . . . . .   . . . . .
RBF/DnJ      . . . . .   b a . . .   . . . . .   . . . . .   . . . . .
RBG/Dn       . . . . .   a a . . .   . . . . .   . . . . .   . . . . .
RC/CpbU      b a b . a   b a . a r   a a a d c   a b a r a   . . . . .

RF/J         . a . a .   b a . a .   . . . . .   . . . . .   . . . d .
RFM/-        . . . . .   a . . . .   . . . . .   . . . . .   . . . . .
RHJ/Le       . . . . .   a a . . .   . . . . .   . . . . .   . . . . .
RIII/-       b a b a a   b a . a r   a a a d c   a b a r a   . . . . .
RIII/SeA     b a b a a   . a . a r   a a a d c   a b a r a   . . . . .
```

        1 = ab

Locus → (read down)	H-4	H-7	H-8	H-9	H-12	H-13	Hao-1	Hao-2	Hba-4ps	Hba	Hbb	Hc	He	Hm	Hp	Hrt-1	Hsd-1	Idh-1	If-1	Igh-1	Igh-2	Igh-3	Igh-4	Igh-5	Igh-Ars
Chromosome (read down)	7	9	14	UN	UN	2	2	3	11	17	7	2	UN	7	8	UN	10	1	3	12	12	12	12	12	12
MWT/Le	.	.	.	.	.	.	.	.	a	.	s	.	.	.	.	.	.	a	.	.	.	.	b	.	.
MY/HuLe	.	.	.	.	.	.	.	.	.	.	s	.	.	.	.	.	.	a	.	.	.	.	.	.	.
MYD/Le	.	.	.	.	.	.	.	.	.	.	d	.	.	.	.	.	.	a	.	.	.	.	.	.	.
NBL/N	.	.	.	.	.	.	.	.	.	.	d	o	.	.	.	.	.	b	.	b	.	.	.	.	.
NC	.	.	.	.	.	.	.	.	.	.	s	o	.	.	.	.	.	b	.	.	.	.	.	.	.
NC/CpbU	.	.	.	.	.	.	.	.	.	.	.	.	.	.	.	.	.	.	.	.	.	.	.	.	.
NFR/N	.	.	.	.	.	.	.	.	.	.	s	1	.	.	.	.	.	a	.	.	.	.	.	.	.
NFS/N	.	.	.	.	.	.	.	.	.	.	d	1	.	.	.	.	.	a	.	.	.	.	.	.	.
NFS/NA	.	.	.	.	.	.	c	a	c	.	d	.	.	.	.	.	.	a	.	.	.	.	.	.	.
NGP/N	.	.	.	.	.	.	.	.	.	.	d	o	a	.	.	.	l	b	.	.	.	.	.	.	.
NH/LwN	.	.	.	.	.	.	.	.	d	.	d	o	a	.	.	.	h	b	.	f	.	.	.	.	.
NOD/Lt	.	.	.	.	.	.	.	.	.	.	s	.	.	.	.	.	.	a	.	.	.	.	.	.	.
NON/J	.	.	.	.	.	.	.	.	.	.	s	.	.	.	.	.	.	b	.	.	.	.	.	.	.
NU/J	.	.	.	.	.	.	.	.	.	.	s	.	.	.	.	.	.	a	.	.	.	.	.	.	.
NZB/-	.	.	.	.	.	.	.	.	d	b	d	o	.	.	.	b	h	a	.	e	1	.	a	.	o
NZB/BlNJ	.	.	.	.	.	.	.	.	d	b	d	o	.	.	.	b	h	a	.	e	1	.	a	.	.
NZW/-	.	.	.	.	.	.	.	.	.	.	d	1	a	.	.	.	.	b	.	e	.	.	.	.	.
NZW/LacJ	.	.	.	.	.	.	.	.	.	.	d	.	.	.	.	.	.	b	.	.	.	.	.	.	.
O20/-	.	.	.	.	.	.	.	.	a	.	d	o	a	.	.	.	.	a	l	a	.	.	.	.	.
O20/A	.	.	.	.	.	.	.	.	a	.	d	o	.	.	.	.	.	a	l	a	.	.	.	.	.
P/-	.	.	.	.	.	.	.	.	h	.	d	1	.	.	.	.	l	b	.	h	a	.	a	.	.
P/J	.	.	.	.	.	.	.	.	h	.	d	1	a	.	.	.	l	b	.	h	a	.	a	.	.
PC/CpbU	.	.	.	.	.	.	.	.	.	.	.	.	.	.	.	.	.	.	.	.	.	.	.	.	.
PERA/Rk	.	.	.	.	.	.	.	.	.	.	d	.	.	.	.	.	.	b	.	.	.	.	a	.	.
PH/Re	.	.	.	.	.	.	.	.	d	.	s	.	a	.	.	.	.	b	.	.	.	.	.	.	.
PL/J	2	2	2	a	a	2	.	.	f	.	d	1	a	.	.	.	l	b	.	j	1	.	a	.	.
PRO/Re	.	.	.	.	.	.	.	.	a	.	d	.	.	.	.	.	l	.	.	.	.	.	.	.	.
QC/CpbU	.	.	.	.	.	.	.	.	.	.	.	.	.	.	.	.	.	.	.	.	.	.	.	.	.
RBA1/Dn	.	.	.	.	.	.	.	.	.	.	d	.	.	.	.	.	.	a	.	.	.	.	.	.	.
RBA/Dn	.	.	.	.	.	.	.	.	.	.	s	.	.	.	.	.	.	a	.	.	.	.	.	.	.
RBB/Dn	.	.	.	.	.	.	.	.	.	.	s	.	.	.	.	.	.	a	.	.	.	.	.	.	.
RBC/Dn	.	.	.	.	.	.	.	.	.	.	d	.	.	.	.	.	.	a	.	.	.	.	.	.	.
RBF/DnJ	.	.	.	.	.	.	.	.	.	.	d	.	.	.	.	.	.	b	.	.	.	.	.	.	.
RBG/Dn	.	.	.	.	.	.	.	.	.	.	s	.	.	.	.	.	.	b	.	.	.	.	.	.	.
RC/CpbU	.	.	.	.	.	.	.	.	.	.	.	.	.	.	.	.	.	.	.	.	.	.	.	.	.
RF/J	2	2	2	2	2	a	.	.	f	.	d	o	.	a	.	b	l	a	h	c	c	.	a	.	o
RFM/-	.	.	.	.	.	.	.	.	.	.	.	.	.	.	.	.	.	.	.	.	.	.	.	.	.
RHJ/Le	.	.	.	.	.	.	.	.	.	.	d	.	.	.	.	.	.	a	.	.	.	.	.	.	.
RIII/-	.	.	.	.	.	.	a	a	c	.	*	1	c	.	.	.	.	*	.	*	c	g	a	a	.
RIII/SeA	.	.	.	.	.	.	a	a	.	.	d	1	.	.	.	.	.	b	.	g	.	.	.	.	.

1 = -a,-c   2 = -a,-b

```
Locus     ->    F F G G G   G G G G G   G G G G G   G G G G G   G G H H H
(read down)     v v a b d   d d d g k   l l m o o   p p p r u   u v - - -
                - - l p c   c r r c     k o - t t   d i t - s   s - 1 2 3
                1 2 t - -   - - -       - 3 - -     - - - 1 -   - 1
                      1 1   2 1 2         1   1 2   1 1 1   r   s


------------------------------------------------------------------------------
Chromosome      4 9 4 3 1   9 U U 6 U   1 1 2 1 8   4 7 1 8 5   5 U 7 1 2
(read down)           5       N N   N   1 7   9         5           N     7
------------------------------------------------------------------------------

MWT/Le          . . b . b   b . . b .   . a . . b   a a a a .   . . . b .
MY/HuLe         . . . . .   . . . . .   . a . . b   a a a a .   . . . . .
MYD/Le          . . . . .   . . . . .   . a . . b   b b c a .   . . . . .
NBL/N           . . . . .   . . . . .   . . . . .   c b . . .   . . . d .
NC              . . . . .   . . . . .   . . . . .   b a . . .   . . . . .

NC/CpbU         . . . . .   . . . . .   . . . . .   . . . . .   . . . . .
NFR/N           . . . . .   . . . . .   . . . . .   b b . . .   . . . . .
NFS/N           . . . . .   . . . . .   . . . . .   b b . . .   . . . . .
NFS/NA          . . . . .   . . . . .   . a . a b   b b . . .   . . . . .
NGP/N           . . . . .   . . . . .   . . . . .   b b . . .   a . . . .

NH/LwN          . . . . .   . . . . .   . . . . .   b a . . .   . . . . .
NOD/Lt          n . . b b   . . . . .   . a . a b   b a b a .   . . . g .
NON/J           n . . b b   . . . . .   . a . a b   b a a a .   . . . b .
NU/J            . . . . .   . . . . .   . b . . a   b a a a .   . . . . .
NZB/-           n s a b b   b . . a .   a a a a b   b a b a b   b . . d .

NZB/B1NJ        n . a . b   b . . a .   a a . a b   b a b a b   b . . d .
NZW/-           . . . b .   . . . . .   . a . . b   b a a a .   . . . z .
NZW/LacJ        . . . b .   . . . . .   . a . . b   b a a a .   . . . z .
O20/-           n s . b .   . . . . .   . a . a b   a b . a b   b . . 1 .
O20/A           n s . b .   . . . . .   . a . a b   a b . a b   b . . 1 .

P/-             n . . b b   b . . a 2   a a . a b   a a a a b   b . d p .
P/J             n . . b b   b . . a 2   a a . a b   a a a a b   b . d p .
PC/CpbU         . . . . .   . . . . .   . . . . .   . . . . .   . . . . .
PERA/Rk         . . . . b   . . . . .   . a . a b   . . a . .   . . . . .
PH/Re           . . . . .   . . . a .   . . . . .   b a a a .   . . . k .

PL/J            n . . b b   b . . a .   a b . a b   a a a a b   b . 3 u 4
PRO/Re          b . . . .   b . . b .   . . . . .   . a . a .   . . . b .
QC/CpbU         . . . . .   . . . . .   . . . . .   . . . . .   . . . . .
RBA1/Dn         . . . . .   . . . . .   . a . a b   b a b a .   . . . . .
RBA/Dn          . . . b .   . . . . .   . a . a b   b 5 b a .   . . . 6 .

RBB/Dn          . . . b .   . . . . .   . a . a a   b a a a .   . . . b .
RBC/Dn          . . . . .   . . . . .   . a . a b   . b a . .   . . . . .
RBF/DnJ         . . . . .   . . . . .   . a . . b   b b a a .   . . . 6 .
RBG/Dn          . . . . .   . . . . .   . a . . b   a b a a .   . . . b .
RC/CpbU         . . . . .   . . . . .   . . . . .   . . . . .   . . . . .

RF/J            n s a b b   b h l a b   . a b a b   a a a a b   b b 3 k 4
RFM/-           . . . . .   . . . . .   . a . . .   . a . a .   b . . f .
RHJ/Le          . . . . .   . . . . .   . a . . b   b b a a .   . . . d .
RIII/-          b s . . .   . . . . .   . a . a b   b a . a .   . . . r .
RIII/SeA        b s . a .   . . . . .   . a . a b   b a . a .   . . . r .
```

1 = pz	2 = c?	3 = -a,-c	4 = -a,-b	5 = ab		
6 = wild						

The laboratory mouse

```
Locus     ->   I I I I I   I I I I I   I I I I I   I I K K K   L L L L L
(read down)    g g g g g   g g g g g   g g g g g   g g f t t   a a a d d
               h h h h h   h h h h h   h h k k k   l l o h h   m m p h r
               - - - - -   - - - - -   - - - - -   - - - - -   b b - - -
               C D I L N   P S S S S   S E P T     1 1 1 1 2   - - 1 1 1
                 e n e p   c a a a a   a r f c r   r           1 2
                 x u v         1 2 3 4 5 c 1   p

Chromosome     1 1 1 1 1   1 1 1 1 1   1 1 6 6 6   1 U 9 1 U   1 1 9 7 6
(read down)    2 2 2 2 2   2 2 2 2 2   2 2         6 N   7 N   2
-----------------------------------------------------------------------------
MWT/Le         . . . . .   . . . . .   . . . . .   . . . . .   . . . . a
MY/HuLe        . . . . .   . . . . .   . . . . .   . . . . .   . . . . .
MYD/Le         . . . . .   . . . . .   . . . . .   . . . . .   . . . . .
NBL/N          . . . . .   . . . . .   . . . . .   . . . . .   . . . . .
NC             . . . . .   . . . . .   . . . . .   . . . . .   . . . . a

NC/CpbU        . . . . .   . . . . .   . . . . .   . . . . .   . . . . .
NFR/N          . . . . .   . . . . .   . . . . .   . . . . .   . . . . .
NFS/N          . . . . .   . . . . .   . . . . .   . . . . .   . . . . .
NFS/NA         . . . . .   . . . . .   . . . . .   . . . . .   . . . . b
NGP/N          . . . . .   . . . . .   . . . . .   . . . . .   . . . . .

NH/LwN         . . o . .   o . . . .   . . . . .   . . . . .   . . . . .
NOD/Lt         . . . . .   . . . . .   . . . . .   . . . . .   . . . . a
NON/J          . . . . .   . . . . .   . . . . .   . . . . .   . . . . a
NU/J           . o . . .   . . . . .   . . . . .   . . . . .   . . . . .
NZB/-          n o . a o   o . . . .   . a b . b   . . . . .   a a b . a

NZB/BlNJ       n . . . .   . . . . .   . . b . .   . . . . .   a a b . .
NZW/-          . . . . .   . . . . .   . . . . .   . . . . .   . . . . a
NZW/LacJ       . . . . .   . . . . .   . . . . .   . . . . .   . . . . .
O20/-          . . . . .   . . . . .   . . . . .   . . . . .   a d . . a
O20/A          . . . . .   . . . . .   . . . . .   . . . . .   a d . . a

P/-            h . a . .   . . . . .   . . b . .   . . . . .   a b b . a
P/J            h . a . .   . . . . .   . . b . .   . . . . .   a b b . a
PC/CpbU        . . . . .   . . . . .   . . . . .   . . . . .   . . . . .
PERA/Rk        . . . . .   . . . . .   . . . . .   . . . . .   . . . a .
PH/Re          . . . . .   . . . . .   . . . . .   . . . . .   . . . . b

PL/J           j . a o .   . . . . .   . . a a a   . . . . .   a b b . a
PRO/Re         . . . . .   . . . . .   . . . . .   . . . . .   . . a . .
QC/CpbU        . . . . .   . . . . .   . . . . .   . . . . .   . . . . .
RBA1/Dn        . . . . .   . . . . .   . . . . .   . . . . .   . . . . .
RBA/Dn         . . . . .   . . . . .   . . . . .   . . . . .   . . . a b

RBB/Dn         . . . . .   . . . . .   . . . . .   . . . . .   . . . . a
RBC/Dn         . . . . .   . . . . .   . . . . .   . . . . .   . . . a a
RBF/DnJ        . . . . .   . . . . .   . . . . .   . . . . .   . . . . .
RBG/Dn         . . . . .   . . . . .   . . . . .   . . . . .   . . . . .
RC/CpbU        . . . . .   . . . . .   . . . . .   . . . . .   . . . . .

RF/J           . . a . .   . . a . a   o . a a a   . . . . .   a b b . a
RFM/-          . . . . .   . . . . .   . . . . .   . . . . .   . . . a .
RHJ/Le         . . . . .   . . . . .   . . . . .   . . . . .   . . . . .
RIII/-         . . a . .   o . . . .   . . . . b   . a . . .   . . . . a
RIII/SeA       . . . . .   . . . . .   . . . . .   . . . . .   . . . . a
```

Strain distribution of polymorphic variants

```
    Locus    ->   L L L L L   L L L L L   L L L L L   L L L L L   L L L L L
    (read down)   e e f m n   p s t t t   t t t t v   v y y y y   y y y y y
                  n n o p a   t h h h n   w w w w     p - - - -   - - - - -
                  - - -   -   -   - - -   - - - -     - 1 2 3 5   6 7 8 9 1
                  1 2 1   1   1   1 2 2   2 3 4 6     1                     0
    ---------------------------------------------------------------------------
    Chromosome    1 U 7 1 U   U 1 1 U U   1 9 1 U 4   6 1 6 6 1   1 1 U 1 1
    (read down)     N   7 N   N 3 N N       2   N         9           5 2 N   9
    ---------------------------------------------------------------------------

    MWT/Le        . . . . .   . . . . .   . . . . .   . . . . .   . . . . .
    MY/HuLe       . . . . .   . . . . .   . . . . .   . . . . .   . . . . .
    MYD/Le        . . . . .   . . . . .   . . . . .   . . . . .   . . . . .
    NBL/N         . . . . .   . . . . .   . . . . .   . b a b .   . . . . .
    NC            . . . . .   . . . . .   . . . . .   . . . . .   . . . . .

    NC/CpbU       . . . . .   . . . . .   . . . . .   . . . . .   . . . . .
    NFR/N         . . . . .   . . . . .   . . . . .   . b . . .   . . . . .
    NFS/N         . . . . .   . . . . .   . . . . .   . b . . .   . . . a .
    NFS/NA        a b . . .   . . . . .   . . . . .   . . . . .   . . . . .
    NGP/N         . . . . .   . . . . .   . . . . .   . b . . .   . . . . .

    NH/LwN        . . . . .   . . . . .   . . . . .   a . . . .   . . . . .
    NOD/Lt        . . . . .   . . . . .   . . . b .   . . . . .   . . . . .
    NON/J         . . . . .   . . . . .   . . . b .   . . . . .   . . . . .
    NU/J          . . . . .   . . . . .   . . . . .   . . . . .   . . . . .
    NZB/-         a b . a .   . r a . b   . . a a .   a b b b b   a a b a b

    NZB/BlNJ      a b . . .   . . a . .   . . . . .   a b b b b   a a b a b
    NZW/-         . . . a .   . s . . .   . . . . .   . b . . .   . . . . .
    NZW/LacJ      . . . . .   . . . . .   . . . . .   . . . . .   . . . . .
    O20/-         a a . . .   . . . . b   . a a a .   . b a b b   . . . . .
    O20/A         a a . . .   . . . . .   . a . . .   . b a b b

    P/-           . . . o .   . . a . b   . a . . .   a b . b .   . . . . .
    P/J           . . . o .   . . a . b   . a a a .   a . . b .   . . . . .
    PC/CpbU       . . . . .   . . . . .   . . . . .   . . . . .   . . . . .
    PERA/Rk       . . . . .   . . . . .   . . . . .   . . . . .   . . . . .
    PH/Re         . . . . .   . . . . .   . . . . .   . . . . .   . . . . .

    PL/J          . . . a a   a . a . .   . . b b c   a b a a b   b a . a b
    PRO/Re        . . . . .   . . . . .   . . . . .   . . . . .   . . . . .
    QC/CpbU       . . . . .   . . . . .   . . . . .   . . . . .   . . . . .
    RBA1/Dn       . . . . .   . . . . .   . . . . .   . . . . .   . . . . .
    RBA/Dn        . . . . .   . . . . .   . . . . .   . . . . .   . . . . .

    RBB/Dn        . . . . .   . . . . .   . . . . .   . . . . .   . . . . .
    RBC/Dn        . . . . .   . . . . .   . . . . .   . . . . .   . . . . .
    RBF/DnJ       . . . . .   . . . . .   . . . . .   . . . . .   . . . . .
    RBG/Dn        . . . . .   . . . . .   . . . . .   . . . . .   . . . . .
    RC/CpbU       . . . . .   . . . . .   . . . . .   . . . . .   . . . . .

    RF/J          a b . . .   a . a . a   . a b a b   a b a a b   b a . a .
    RFM/-         . . . . .   . . . . .   . . . . .   . . . . .   . . . . .
    RHJ/Le        . . . . .   . . . . .   . . . . .   . . . . .   . . . . .
    RIII/-        a b . . a   . . a . .   . . . . .   . b b b a   a a . a .
    RIII/SeA      a b . . .   . . . . .   . . . . .   . b b b a   . . . . .
```

Locus -> (read down)	Ly-11	Ly-12	Ly-13	Ly-14	Ly-15	Ly-16	Ly-17	Ly-18	Ly-19	Ly-20	Ly-21	Ly-22	Ly-23	Ly-24	Ly-25	Ly-26	Ly-27	Ly-28	Ly-29	Ly-30	Ly-31	Ly-32	Ly-39	Ly-40	Ly-42b
Chromosome (read down)	2	19	UN	7	7	12	1	12	4	4	7	1	2	2	2	UN	UN	UN	4	UN	4	4	17	UN	4
MWT/Le	.	.	.	.	.	.	.	.	.	.	.	.	.	.	.	.	.	.	.	.	.	.	.	.	.
MY/HuLe	.	.	.	.	.	.	.	.	.	.	.	.	.	.	.	.	.	.	.	.	.	.	.	.	.
MYD/Le	.	.	.	.	.	.	.	.	.	.	.	.	.	.	.	.	.	.	.	.	.	.	.	.	.
NBL/N	.	.	.	.	.	.	.	.	.	.	.	.	.	.	.	.	.	.	.	.	.	.	.	.	.
NC	.	.	.	.	.	.	.	.	.	.	.	.	.	.	.	.	.	.	.	.	.	.	.	.	.
NC/CpbU	.	.	.	.	.	.	.	.	.	.	.	.	.	.	.	.	.	.	.	.	.	.	.	.	.
NFR/N	.	.	.	.	.	.	.	.	.	.	.	.	.	.	.	.	.	.	.	.	.	.	.	.	.
NFS/N	.	.	.	.	.	.	a	.	.	.	.	.	.	.	.	.	.	.	.	.	.	.	.	.	.
NFS/NA	.	.	.	.	.	.	.	.	.	.	.	.	.	.	.	.	.	.	.	.	.	.	.	.	.
NGP/N	.	.	.	.	.	.	.	.	.	.	.	.	.	.	.	.	.	.	.	.	.	.	.	.	.
NH/LwN	.	.	.	.	.	.	.	.	.	.	.	.	.	.	.	.	.	.	.	.	.	.	.	.	.
NOD/Lt	.	.	.	.	.	.	.	.	.	.	.	.	.	.	.	.	.	.	.	.	.	.	.	.	.
NON/J	.	.	.	.	.	.	.	.	.	.	.	.	.	.	.	.	.	.	.	.	.	.	.	.	.
NU/J	.	.	.	.	.	.	.	.	.	.	.	.	.	.	.	.	.	.	.	.	.	.	.	.	.
NZB/-	a	b	b	.	a	.	a	.	b	b	b	a	a	b	b	a	b	b	b	b	a	b	.	1	.
NZB/BlNJ	.	b	b	.	a	.	.	.	b	.	b	a	.	b	.	a	.	.	.	.	a	.	.	1	.
NZW/-	.	.	.	.	.	.	.	.	.	.	.	.	a	.	.	.	.	.	.	.	a	.	.	.	.
NZW/LacJ	.	.	.	.	.	.	.	.	.	.	.	.	.	.	.	.	.	.	.	.	a	.	.	.	.
O20/-	.	.	.	.	.	.	.	.	.	.	.	.	.	.	.	.	.	.	.	.	.	.	.	.	.
O20/A	.	.	.	.	.	.	.	.	.	.	.	.	.	.	.	.	.	.	.	.	.	.	.	.	.
P/-	.	.	.	.	.	.	.	.	.	.	.	a	.	.	.	.	.	.	.	.	b	.	.	.	.
P/J	.	.	.	.	.	.	.	.	.	.	.	a	.	.	.	.	.	.	.	.	b	.	.	.	.
PC/CpbU	.	.	.	.	.	.	.	.	.	.	.	.	.	.	.	.	.	.	.	.	.	.	.	.	.
PERA/Rk	.	.	.	.	.	.	.	.	.	.	.	.	.	.	.	.	.	.	.	.	.	.	.	.	.
PH/Re	.	.	.	.	.	.	.	.	.	.	.	.	.	.	.	.	.	.	.	.	.	.	.	.	.
PL/J	b	b	.	.	a	.	a	.	b	.	b	b	b	.	b	b	b	.	b	.	a	.	1	.	b
PRO/Re	.	.	.	.	.	.	.	.	.	.	.	.	.	.	.	.	.	.	.	.	.	.	.	.	.
QC/CpbU	.	.	.	.	.	.	.	.	.	.	.	.	.	.	.	.	.	.	.	.	.	.	.	.	.
RBA1/Dn	.	.	.	.	.	.	.	.	.	.	.	.	.	.	.	.	.	.	.	.	.	.	.	.	.
RBA/Dn	.	.	.	.	.	.	.	.	.	.	.	.	.	.	.	.	.	.	.	.	.	.	.	.	.
RBB/Dn	.	.	.	.	.	.	.	.	.	.	.	.	.	.	.	.	.	.	.	.	.	.	.	.	.
RBC/Dn	.	.	.	.	.	.	.	.	.	.	.	.	.	.	.	.	.	.	.	.	.	.	.	.	.
RBF/DnJ	.	.	.	.	.	.	.	.	.	.	.	.	.	.	.	.	.	.	.	.	.	.	.	.	.
RBG/Dn	.	.	.	.	.	.	.	.	.	.	.	.	.	.	.	.	.	.	.	.	.	.	.	.	.
RC/CpbU	.	.	.	.	.	.	.	.	.	.	.	.	.	.	.	.	.	.	.	.	.	.	.	.	.
RF/J	.	.	.	.	b	.	a	a	a	b	b	a	b	b	.	b	b	a	a	b	a	a	.	.	c
RFM/-	.	.	.	.	.	.	.	.	.	.	.	.	.	.	.	.	.	.	.	.	.	.	.	.	.
RHJ/Le	.	.	.	.	.	.	.	.	.	.	.	.	.	.	.	.	.	.	.	.	.	.	.	.	.
RIII/-	.	.	.	.	.	.	a	.	b	.	b	a	.	.	.	b	.	.	.	.	b	b	1	.	.
RIII/SeA	.	.	.	.	.	.	.	.	.	.	.	.	.	.	.	.	.	.	.	.	.	.	.	.	.

1 = -b

```
Locus    ->    L L L L L   L M M M M   M M M M M   N N N N N   N O O p P
(read down)    y y y y y   y a e l o   o o p p u   c e k k p   p r r a
               b b b b b   b p p s d   d r h i p   a u - - -   - m m n
               - - - - -   - - - - -   - - - - -   m - 1 2 1   2 - - -
               3 4 5 6 7   8 1 1 1 1   2 1 1 1 1     1         2   1 2   1


Chromosome     U 4 U 4 1   7 5 1 1 9   7 5 7 9 4   9 1 1 2 1   1 4 4 7 4
(read down)    N   N   2       7             7 7   4   4
---------------------------------------------------------------------------
MWT/Le         . . . . .   . . . . a   . a . b b   . b . . a   . . . + .
MY/HuLe        . . . . .   . . . . a   . . . b .   . b . . .   . . . . .
MYD/Le         . . . . .   . . . . b   . . . b .   . b . . .   . . . . .
NBL/N          . . . . .   . . . . a   . . . . a   . . . . .   . . . + .
NC             . . . . .   . . . . a   . . . . b   . . . . .   . . . + .

NC/CpbU        . . . . .   . . . . .   . . . . .   . . . . .   . . . . .
NFR/N          . . . . .   . . . . b   . . . . a   . . . . .   . . . . .
NFS/N          . . . . .   . . . . a   . . . . a   . . . . .   . . . . .
NFS/NA         . . . . .   . . . . a   b a . b a   . b . . a   . . . . .
NGP/N          . . . . .   . b . b .   . . . . c   . b . . .   . . . . .

NH/LwN         . . . . .   . b . b .   . . . . a   . b . . .   b . . p .
NOD/Lt         . . . . .   . . a . b   . . . b a   . b . . .   . . . . .
NON/J          . . . . .   . . a . a   . . . b a   . b . . .   . . . . .
NU/J           . . . . .   . . . . a   . . . b .   . b . . .   . . . . .
NZB/-          . . . . .   a . . a b   b a b b a   . b . o a   a . . + b

NZB/BlNJ       . . . . .   . . . . b   . . . . a   . b . o a   a . . + b
NZW/-          . . . . .   . b . . a   . . b b a   . b . . .   . . . + .
NZW/LacJ       . . . . .   . . . . a   . . . b .   . b . . .   . . . . .
O20/-          . . . . .   . . . . a   b a . b b   . b . . a   . . . . .
O20/A          . . . . .   . . . . a   b a . b b   . b . . a   . . . . .

P/-            . . . . .   . . . . b   b a . b b   . b . . a   . . . p .
P/J            . . . . .   . . . . b   b a . b b   . b . . a   . . . p .
PC/CpbU        . . . . .   . . . . b   . . . . .   . . . . .   . . . . .
PERA/Rk        . . . . .   . . . . b   b . . . b   . . . . a   . . . + .
PH/Re          . . . . .   . b . b .   . a . b .   . b . . .   . . . + .

PL/J           . . . . .   a b . . a   b a . b b   . b . . a   . . . . .
PRO/Re         . . . . .   . . . . b   . . . . .   . . . . .   . . . p .
QC/CpbU        . . . . .   . . . . .   . . . . .   . . . . .   . . . . .
RBA1/Dn        . . . . .   . . . . b   . . . b .   . b . . .   . . . . .
RBA/Dn         . . . . .   . . . . b   b a . b b   . b . . a   . . . + .

RBB/Dn         . . . . .   . . . . a   a a . b b   . b . . a   . . . + .
RBC/Dn         . . . . .   . . . . b   b a . . b   . b . . a   . . . . .
RBF/DnJ        . . . . .   . . . . a   . . . b .   . b . . .   . . . . .
RBG/Dn         . . . . .   . . . . b   . . . b .   . b . . .   . . . . .
RC/CpbU        . . . . .   . . . . .   . . . . .   . . . . .   . . . . .

RF/J           b b b . b   b b b . a   . a . b a   . b . . a   . . . . a
RFM/-          . . . . .   . b . . a   . . . b .   . b . . .   . . . . .
RHJ/Le         . . . . .   . . . . a   . . . b .   . b . . .   . . . . .
RIII/-         . . . . .   . . . . b   b a . b *   . b . . a   . . . + .
RIII/SeA       . . . . .   . . . . b   b a . b a   . b . . a   . . . . .
```

```
Locus    ->    P P P P    P P P P    P P P P    P P P P Q    Q Q Q Q Q
(read down)    a c d e e   e e g g    g g g k r   r r r s a    a a a a e
               n a p p    p p d k k   m m m - e   n t t p -    - - - - d
               - -   - -   - -   - -   - - - 3 -   - - - 1    2 3 4 5 -
               2 1   2 3   4 7   1 2   1 2 3   2   p 1 2    1

------------------------------------------------------------------------
Chromosome     U U U 1 1   7 5 4 X 1   5 4 9 9 1   2 U 8 2 1   1 1 1 1 1
(read down)    N N N 0     7           2           N   7       7 7 7 7 7
------------------------------------------------------------------------

MWT/Le         . . . . a   . . . . a   a a .       . . . b .   . . . . .
MY/HuLe        . . . . c   . . . . .   a a .       . . . . .   . . . . .
MYD/Le         . . . . a   . . . . .   a a .       . . . . .   . . . . .
NBL/N          . . . . b   . . . . .   a .         . . . . .   . . . . .
NC             . . . . b   . . . . .   a .         . b a . .   . . . . .

NC/CpbU        . . . . .   . . . . .   . .         . . . . .   . . . . .
NFR/N          . . . . b   . . . b .   b .         . . . . .   . . . . .
NFS/N          . . . . b   . . . . a   a .         . . . . .   . . . . .
NFS/NA         . . . . b   . . b b a   a a .       . . . . .   . . . . .
NGP/N          . . . . b   . . . . .   b .         . . . . .   . . . . .

NH/LwN         . . . . b   . . . . .   b .         . . . . .   . . . . .
NOD/Lt         . . . . b   . . . b a   a a . a     . . . . .   . . . . .
NON/J          . . . . b   . . . b a   a a . a     . . . . .   . . . . .
NU/J           . . . . a   . . . . .   a a .       . . . . .   . . . . .
NZB/-          b a . . c   . . . b a   b a a .     . . . a a   a a . . a

NZB/BlNJ       b . . . c   . . . b a   b a a .     . . . . a   . . . . a
NZW/-          . . . . b   . . . . a   b a . a     a . . . b   b b b b .
NZW/LacJ       . . . . b   . . . . .   b a . a     a . . . .   . . . . .
O20/-          . . . . b   . . b b a   b a .       . . . . .   . . . . .
O20/A          . . . . b   . . b b a   b a .       . . . . .   . . . . .

P/-            . . . . c   . . b . a   b a a .     b . . . .   . . b b a
P/J            . . . . c   . . b . a   b a a .     b . . . .   . . b b a
PC/CpbU        . . . . .   . . . . .   . .         . . . . .   . . . . .
PERA/Rk        . . . . b   . . . . b   a .         . . . . .   . . . . .
PH/Re          . . . . b   . . b . b   b a .       . . . . .   . . . . .

PL/J           . a a . b   . . b b a   b a c . .   a . . b a   . . . . a
PRO/Re         . . . . .   . . . . a   . . a .     . . . b .   . . . . .
QC/CpbU        . . . . .   . . . . .   . .         . . . . .   . . . . .
RBA1/Dn        . . . . b   . . . . .   a a .       . . . . .   . . . . .
RBA/Dn         . . . . a   . . . b .   a a .       . . . . .   . . . . .

RBB/Dn         . . . . a   . . . b .   a a .       . . . . .   . . . . .
RBC/Dn         . . . . a   . . . b .   b .         . . . . .   . . . . .
RBF/DnJ        . . . . c   . . . . .   b a .       . . . . .   . . . . .
RBG/Dn         . . . . b   . . . . .   a a .       . . . . .   . . . . .
RC/CpbU        . . . . .   . . . . .   . .         . . . . .   . . . . .

RF/J           b a b . b   b . . b b   a a c . a   . . . a b   b b . . b
RFM/-          . . . a a   b a b . .   . a .       . . . . .   . . . . .
RHJ/Le         . . . . b   . . . . .   a a .       . . . . .   . . . . .
RIII/-         . a . . b   . . b b b   a a .       . . . . b   b b . . .
RIII/SeA       . . . . b   . . b b b   a a .       . . . . .   . . . . .
```

Strain distribution of polymorphic variants

```
Locus    ->     r r R R R   S S S S S   S S S S S   S S S T T   T T T T T
(read down)     d d e i n   a a a d d   l o o o p   s v v a c   c g h i l
                  s n b 7   a a s h r   p a d d i   p p p m n   r n y n a
                  - - s     - - - -       - - -     - - - -     g   - d
                  1 1 -     1 1 1 1       1 2 2     1 2 1 2       1
                      6                       r
                -------------------------------------------------------------
Chromosome      5 1 1 1 6   7 7 1 2 U   1 U 1 1 1   U 2 7 7 1   1 1 9 1 1
(read down)       7   4             N   7 N 6 7 2   N       1   3 5     2 7
                -------------------------------------------------------------

MWT/Le          + . . . .   . . a . .   . . . . .   . . . . .   . . . . .
MY/HuLe         + . . . .   . . . . .   . . . . .   . . . . .   . . . . .
MYD/Le          . . . . .   . . . . .   . . . . .   . . . . .   . . . . .
NBL/N           . . . . .   . . a . .   . . . . .   . . . . .   . . . . c
NC              . . . . .   . . . . .   . . . . .   . . . . .   . . b . .

NC/CpbU         . . . . .   . . . . .   . . . . .   . . . . .   . . . . .
NFR/N           . . . . .   . . . . .   . . . . .   . . . . .   . . b . .
NFS/N           . . . . .   . . . . .   . . . . .   . . . . .   . . b . .
NFS/NA          . . . . .   . . a . .   . . . . .   . . . s .   . . . . .
NGP/N           . . . . .   . . . . .   . . . . .   . . . . .   . . b . .

NH/LwN          . . . . .   . . . . .   . . . . .   . . . . .   . . . . .
NOD/Lt          . . . . .   . . . . .   . . . . .   . . . . .   . . b . .
NON/J           . . . . .   . . . . .   . . . . .   . . . . .   . . a . .
NU/J            . . . . .   . . . . .   . . . . .   . . . . .   . . . . .
NZB/-           + . s a b   a . o a 1   . b . . .   f . . . f   c b b d a

NZB/B1NJ        + . s a b   a . o a 1   . . . . .   . . a . .   c b b . .
NZW/-           . . . . .   . . . . .   . b . . .   . . . s .   . . . . b
NZW/LacJ        . . . . .   . . . . .   . . . . .   . . . . .   . . . . .
O20/-           + r . . .   . . . . .   . . . . .   . . a . s   . . b . c
O20/A           + r . . .   . . . . .   . . . . .   . . a . s   . . b . c

P/-             2 . s a b   b . a a 3   a . . . .   . . b s .   c . . . e
P/J             2 . s a b   b . a a 3   a b . . .   . . b s .   c . . . e
PC/CpbU         . . . . .   . . . . .   . . . . .   . . . . .   . . . . .
PERA/Rk         . . . . .   . . . b .   . . . . .   . . a . .   . . . . .
PH/Re           + . . . .   . . . . .   . . . . .   . . . . .   . . . . .

PL/J            2 . s a a   a . o a .   o . a a .   . . . c s   a p a . a
PRO/Re          + . . . .   b . . . .   . . . . .   . . . . .   . . . . .
QC/CpbU         . . . . .   . . . . .   . . . . .   . . . . .   . . . . .
RBA1/Dn         . . . . .   . . . . .   . . . . .   . . . . .   . . . . .
RBA/Dn          . . . . .   . . . . .   . . . . .   . . . . .   . . . . .

RBB/Dn          . . . . .   . . . . .   . . . . .   . . . . .   . . . . .
RBC/Dn          . . . . .   . . . . .   . . . . .   . . . . .   . . . . .
RBF/DnJ         . . . . .   . . . . .   . . . . .   . . . . .   . . . . .
RBG/Dn          . . . . .   . . . . .   . . . . .   . . . . .   . . . . .
RC/CpbU         . . . . .   . . . . .   . . . . .   . . . . .   . . . . .

RF/J            + . s a a   a . o a 3   o . a a d   . . . . .   b . a . b
RFM/-           . . . . .   . . . . .   . . a . .   . . . . .   . . a . .
RHJ/Le          . . . . .   . . . . .   . . . . .   . . . . .   . . . . .
RIII/-          + + . . .   . . * a .   . . . . .   . b . . .   . . b . b
RIII/SeA        + + . . .   . . . a .   . . . . .   . b . . .   . . b . b
```

     1 = a?   2 = rd   3 = b?

Locus → (read down)	Tpre	Trf	Tsuy	Tth	Ucp	Upg-1	Upg-2	Ups	Wap	Xmmv-9	Xmmv-42	Xmmv-43	Xmmv-44	Xmmv-45	Xmmv-46	Xmmv-47	Xmmv-48	Xmmv-49	Xmmv-50	Xmmv-51	Xmmv-52	Xmmv-53	Xmmv-54	Xmmv-55	Xmmv-56
Chromosome (read down)	12	9	12	12	8	17	1	9	UN	1	19	UN	UN	UN	UN	UN	UN	UN	12	UN	5	UN	UN	15	UN
MWT/Le	.	b	.	.	.	.	.	.	.	.	.	.	.	.	.	.	.	.	.	.	.	.	.	.	.
MY/HuLe	.	b	.	.	.	.	.	.	.	.	.	.	.	.	.	.	.	.	.	.	.	.	.	.	.
MYD/Le	.	b	.	.	.	.	.	.	.	.	.	.	.	.	.	.	.	.	.	.	.	.	.	.	.
NBL/N	.	b	.	.	.	.	.	.	.	.	.	.	.	.	.	.	.	.	.	.	.	.	.	.	.
NC	.	b	.	.	.	.	.	.	.	.	.	.	.	.	.	.	.	.	.	.	.	.	.	.	.
NC/CpbU	.	.	.	.	.	.	.	.	.	.	.	.	.	.	.	.	.	.	.	.	.	.	.	.	.
NFR/N	.	b	.	.	.	.	.	.	.	.	.	.	.	.	.	.	.	.	.	.	.	.	.	.	.
NFS/N	.	b	.	.	.	.	.	.	.	.	.	.	.	.	.	.	.	.	.	.	.	.	.	.	.
NFS/NA	.	b	.	.	.	.	.	a	.	.	.	.	.	.	.	.	.	.	.	.	.	.	.	.	.
NGP/N	.	b	.	.	.	.	.	.	.	.	.	.	.	.	.	.	.	.	.	.	.	.	.	.	.
NH/LwN	.	b	.	.	.	.	.	.	.	.	.	.	.	.	.	.	.	.	.	.	.	.	.	.	.
NOD/Lt	.	b	.	.	.	f	.	.	.	.	.	.	.	.	.	.	.	.	.	.	.	.	.	.	.
NON/J	.	b	.	.	.	s	.	.	.	.	.	.	.	.	.	.	.	.	.	.	.	.	.	.	.
NU/J	.	b	.	.	.	.	.	.	.	.	.	.	.	.	.	.	.	.	.	.	.	.	.	.	.
NZB/-	d	b	d	d	b	.	s	a	a	o	o	o	o	o	o	o	*	o	o	o	a	o	a	o	o
NZB/BlNJ	.	b	.	.	b	.	s	a	a	o	o	o	o	o	o	o	o	o	o	o	a	o	a	o	o
NZW/-	.	b	.	.	.	.	.	.	.	a	o	o	a	o	a	o	o	o	a	o	a	o	a	a	o
NZW/LacJ	.	b	.	.	.	.	.	.	.	a	o	o	a	o	a	o	o	o	a	o	a	o	a	a	o
O20/-	.	b	.	.	.	.	.	.	.	.	.	.	.	.	.	.	.	.	.	.	.	.	.	.	.
O20/A	.	b	.	.	.	.	.	.	.	.	.	.	.	.	.	.	.	.	.	.	.	.	.	.	.
P/-	.	b	.	.	.	f	s	a	a	o	o	o	o	o	o	o	o	o	a	o	o	o	a	a	o
P/J	.	b	.	.	.	f	s	a	a	o	o	o	o	o	o	o	o	o	a	o	o	o	a	a	o
PC/CpbU	.	.	.	.	.	.	.	.	.	.	.	.	.	.	.	.	.	.	.	.	.	.	.	.	.
PERA/Rk	.	b	.	.	.	.	.	.	.	.	.	.	.	.	.	.	.	.	.	.	.	.	.	.	.
PH/Re	.	b	.	.	.	.	.	.	.	.	.	.	.	.	.	.	.	.	.	.	.	.	.	.	.
PL/J	.	b	.	.	.	f	d	a	a	a	o	o	o	o	o	o	o	o	a	o	a	o	a	o	o
PRO/Re	.	b	.	.	.	s	s	.	.	.	.	.	.	.	.	.	.	.	.	.	.	.	.	.	.
QC/CpbU	.	.	.	.	.	.	.	.	.	.	.	.	.	.	.	.	.	.	.	.	.	.	.	.	.
RBA1/Dn	.	b	.	.	.	.	.	.	.	.	.	.	.	.	.	.	.	.	.	.	.	.	.	.	.
RBA/Dn	.	b	.	.	.	.	.	.	.	.	.	.	.	.	.	.	.	.	.	.	.	.	.	.	.
RBB/Dn	.	b	.	.	.	.	.	.	.	.	.	.	.	.	.	.	.	.	.	.	.	.	.	.	.
RBC/Dn	.	b	.	.	.	.	.	.	.	.	.	.	.	.	.	.	.	.	.	.	.	.	.	.	.
RBF/DnJ	.	b	.	.	.	.	.	.	.	.	.	.	.	.	.	.	.	.	.	.	.	.	.	.	.
RBG/Dn	.	b	.	.	.	.	.	.	.	.	.	.	.	.	.	.	.	.	.	.	.	.	.	.	.
RC/CpbU	.	.	.	.	.	.	.	.	.	.	.	.	.	.	.	.	.	.	.	.	.	.	.	.	.
RF/J	.	b	.	.	.	s	d	a	a	o	o	o	o	o	o	o	o	o	a	o	a	o	a	o	o
RFM/-	.	.	.	.	.	.	.	.	.	o	o	o	o	o	o	o	o	o	o	a	o	a	a	o	o
RHJ/Le	.	b	.	.	.	.	.	.	.	.	.	.	.	.	.	.	.	.	.	.	.	.	.	.	.
RIII/-	.	b	.	.	.	.	.	a	.	.	.	.	.	.	.	.	.	.	.	.	.	.	.	.	.
RIII/SeA	.	b	.	.	.	.	.	.	.	.	.	.	.	.	.	.	.	.	.	.	.	.	.	.	.

Locus -> (read down)	X m m v - 5 7	X m m v - 5 8	X m m v - 5 9	X m m v - 6 0	X m m v - 6 2	X m m v - 6 3	X m m v - 6 4	X m m v - 6 5	X m m v - 6 6	X m m v - 6 7	X m m v - 6 8	X m m v - 6 9	X m m v - 7 0
Chromosome (read down)	UN	UN	UN	UN	4	UN	UN	3	UN	UN	UN	UN	UN
MWT/Le	.	.	.	.	.	.	.	.	.	.	.	.	.
MY/HuLe	.	.	.	.	.	.	.	.	.	.	.	.	.
MYD/Le	.	.	.	.	.	.	.	.	.	.	.	.	.
NBL/N	.	.	.	.	.	.	.	.	.	.	.	.	.
NC	.	.	.	.	.	.	.	.	.	.	.	.	.
NC/CpbU	.	.	.	.	.	.	.	.	.	.	.	.	.
NFR/N	.	.	.	.	.	.	.	.	.	.	.	.	.
NFS/N	.	.	.	.	.	.	.	.	.	.	.	.	.
NFS/NA	.	.	.	.	.	.	.	.	.	.	.	.	.
NGP/N	.	.	.	.	.	.	.	.	.	.	.	.	.
NH/LwN	.	.	.	.	.	.	.	.	.	.	.	.	.
NOD/Lt	.	.	.	.	.	.	.	.	.	.	.	.	.
NON/J	.	.	.	.	.	.	.	.	.	.	.	.	.
NU/J	.	.	.	.	.	.	.	.	.	.	.	.	.
NZB/-	o	a	o	a	o	a	o	a	o	a	o	a	a
NZB/BlNJ	o	a	o	a	o	a	o	a	o	a	o	a	a
NZW/-	o	o	o	o	o	a	o	o	o	a	o	a	a
NZW/LacJ	o	o	o	o	o	a	o	o	o	a	o	a	a
O20/-	.	.	.	.	.	.	.	.	.	.	.	.	.
O20/A	.	.	.	.	.	.	.	.	.	.	.	.	.
P/-	o	o	o	o	a	a	o	a	o	o	o	a	a
P/J	o	o	o	o	a	a	o	a	o	o	o	a	a
PC/CpbU	.	.	.	.	.	.	.	.	.	.	.	.	.
PERA/Rk	.	.	.	.	.	.	.	.	.	.	.	.	.
PH/Re	.	.	.	.	.	.	.	.	.	.	.	.	.
PL/J	o	o	o	o	o	a	o	a	o	o	o	a	a
PRO/Re	.	.	.	.	.	.	.	.	.	.	.	.	.
QC/CpbU	.	.	.	.	.	.	.	.	.	.	.	.	.
RBA1/Dn	.	.	.	.	.	.	.	.	.	.	.	.	.
RBA/Dn	.	.	.	.	.	.	.	.	.	.	.	.	.
RBB/Dn	.	.	.	.	.	.	.	.	.	.	.	.	.
RBC/Dn	.	.	.	.	.	.	.	.	.	.	.	.	.
RBF/DnJ	.	.	.	.	.	.	.	.	.	.	.	.	.
RBG/Dn	.	.	.	.	.	.	.	.	.	.	.	.	.
RC/CpbU	.	.	.	.	.	.	.	.	.	.	.	.	.
RF/J	o	o	o	o	o	a	o	o	o	a	o	a	a
RFM/-	o	o	o	o	o	a	a	o	o	a	o	a	a
RHJ/Le	.	.	.	.	.	.	.	.	.	.	.	.	.
RIII/-	.	.	.	.	.	.	.	.	.	.	.	.	.
RIII/SeA	.	.	.	.	.	.	.	.	.	.	.	.	.

```
Locus    ->    a A A A A   A A A A A   A A A A A   A A A A A   A A A A A
(read down)      a b c c   d d d d g   h h h h k   k l m m o   o p p p s
                 t p f o   h h h h s   . d d r p   p b y y x   x h k o -
                 a - -     - - - - -   . . . . .   - - - - -   - - . a 1
                 1 1       1 2 3 3     1 2 1 1     2 1 1 2 1   2 1 . - s
                           e t                                       1
-----------------------------------------------------------------------------
Chromosome     2 1 7 1 4   3 U 3 3 X   1 4 1 3 1   4 5 3 3 1   1 2 1 9 1
(read down)    2           N           2 9                     0       3
-----------------------------------------------------------------------------

RIIIS/J      + a . . a   . . . . h   b a . . a   b a a a .   . . a b a
ROP/GnLe     + . . . a   . . . . .   . . . . a   . a a . .   . . . a .
RSV/Le       + o . . a   . . . . .   . . . . a   . a a . .   . . . b .
SB/Le        1 . . . .   . . . . .   . . . . .   . . . . .   . . . . .
SC/CpbU      . . . . .   . . . . .   . . . . .   . . . . .   . . . . .

SEA/GnJ      + a . . a   . . . . .   b . . . b   . a a a 2   . . a a a
SEC/1ReJ     a a a b a   . . . . h   b b . . b   a a a a .   . . a b a
SEC/-        a a a b a   . . . . h   b b . . b   a a a a .   . . a b a
SF/Cam-      + a . . a   a . b b .   . a . b b   a a a . a   . . a b a
SF/CamRk     + a . . .   . . . . .   . a . . b   a a a . .   . . a b a

SH1/Le       . . . . .   . . . . .   . . . . b   . . a . .   . . . b .
SJL/J        + a . b a   . . . . h   d b a . b   a a a a a   a a . a a
SJL/J-       + a a b a   a a a a h   d b a a b   a a a a a   a a . a a
SJL/JA       + . . . .   a a a a .   . b a a .   . . . . a   a . . . .
SK/Cam-      + 2 a . a   . . . . .   . b . . b   a a a . .   . a a b a

SK/CamRk     + 2 . . .   . . . . .   . b . . b   a a a . .   . . a . a
SL/NiA       . . . . .   a a a a .   . b a . .   . . . a .   a . . . .
SM/-         3 o . b a   . . . . h   b b a . a   a a a a a   . . a a a
SM/J         3 o . b a   . . . . h   b b a . a   a a a a a   . . a a a
SSL/Le       . . . . .   . . . . .   . . . a .   . a . . .   . . a . .

ST/a         . . . . .   . . . . .   . . . . .   . . . . .   . . . . .
ST/bJ        a a a b c   . . . . h   d b . . b   a a a . 2   . . a a a
ST/bJN       a . . . .   . . . . .   b . . . .   . . . . .   . . . . .
STR/1N       a . . . .   . . . . .   d . . . .   . . . . .   . . . . .
STR/N        a . . . .   . . . . h   b . . . .   . . . . .   . . . . .

STS/A        + . a . .   a a a a h   b b a . .   . . . . a   a . . . .
STS/N        . . . . .   . . . . .   . . . . .   . . . . .   . . . . .
STX/LeJ      + . . . a   . . . . .   h h . . .   . a a . .   . . . . .
SWR/-        + o a b a   a . a a h   d b . r b   a a a a a   b . a b a
SWR/J        + o a b a   a . a a h   d b . . b   a a a a a   b . a b a

TA1/-        . . . . .   . . . . .   . . . . b   . a a . .   . . a a .
TA2/-        . . . . .   . . . . .   . . . . b   . a . . .   . . a b .
TA3/-        . . . . .   . . . . .   . . . . b   . a a . .   . . b . .
TB1/-        . . . . .   . . . . .   . . . . a   . a a . .   . . . a .
TF/GnLe      a . . . a   . . . . .   . . . . a   . a a . .   . . . a .

TSI/A        a . a . .   a a a a h   . b a . .   . . . . a   a . . . .
TSJ/Le       . . . . .   . . . . .   . . . . b   . . a . .   . . . b .
V/Le         a . . . a   . . . . .   . . . . a   . a a . .   . . . a .
WB/ReJ       a a . . .   . . . . h   d . . . b   . . a . 2   . . . a .
WC/ReJ       a o . . .   . . . . h   . . . . b   . . a . .   . . . a .
```

1 = A[w]   2 = a?   3 = A*w

Strain distribution of polymorphic variants

```
Locus    ->   A A b B B    B B B B c    C C C C C    c C C C C    C d D D D
(read down)   v v 2 c      g g g v -    3 4 6 a a    d e e f o    v 7 1 1 1
              - -   m d    l l l - 1        r r      m - - h h    - R 2 2
              1 2     -    - - - 1          - -      1 2   e      1 p N N
                      1    e s t            1 2            1      2 y y
                                                                 - u u
                                                                 s 1 2
              -------------------------------------------------------------
Chromosome    U U 4 2 5    9 9 9 8 7    1 1 1 3 3    3 2 1 2 7    5 9 7 1 1
(read down)   N N                       7 7 5            7            2 2
              -------------------------------------------------------------

RIIIS/J       . . + . .    b d d b c    b . . a b    + . . . .    b + . . .
ROP/GnLe      . . + . .    . . . . +    b . . a a    + . . . .    . . . . .
RSV/Le        . . + . .    . . . . +    b . . a b    . . . . .    . . . . .
SB/Le         . . + . .    . . . . +    b . . a b    . . . . .    . . . . .
SC/CpbU       . . . . .    . . . . .    . . . . .    . . . . .    . . . . .

SEA/GnJ       s s b a .    . . . b +    b . . a a    + . . . .    a d b . .
SEC/1ReJ      . . b . .    b d d b 1    b . . a b    2 . a . .    a + b . .
SEC/-         . . b . .    b d d b 1    b . . a b    2 . a . .    a * b . .
SF/Cam-       . . + . .    . . . . +    c . . a a    . . b . .    . + . . .
SF/CamRk      . . + . .    . . . . +    c . . a a    . . b . .    . + . . .

SH1/Le        . . . . .    . . . . .    . . . a .    . . . . .    . . . . .
SJL/J         r r + a a    . d . b c    c . . a b    . . a . l    b + d c a
SJL/J-        r r + a a    . d . b c    c . . a b    . . a . l    b + d c a
SJL/JA        . . + . .    . . . . c    . . . . .    . . a . .    . + . . .
SK/Cam-       . . + . .    . . . . +    b . . b a    . . b . .    . + . . .

SK/CamRk      . . + . .    . . . . +    b . . b a    . . . . .    . + . . .
SL/NiA        . . . . .    . . . . c    . h . . .    + . a . .    . . . . .
SM/-          . . + . a    b d d b +    b . . b b    + . a f .    a + d b a
SM/J          . . + . a    b d d b +    b . . b b    + . a f .    a + d b a
SSL/Le        . . . . .    . . . . .    . . . a .    . . . . .    . . . . .

ST/a          . . . . .    . . . . c    . . . . .    . . . . .    . . . . .
ST/bJ         . . + . .    b d d b c    b l . a b    + . b . .    b + . . .
ST/bJN        . . b . .    . . . . c    . . . b .    . . b . .    . . . . .
STR/1N        . . b . .    b d d . +    . . . b .    . . a . .    . . . . .
STR/N         . . b . .    b d d . +    . . . a .    . . a . .    . . . . .

STS/A         . . + . .    b d d . c    c h . . .    + . a . .    . + . . .
STS/N         . . . . .    . . . . c    . . . . b    . . a . .    c . . . .
STX/LeJ       . . + . .    . . . . +    b . . . b    . . . . .    . . . . .
SWR/-         . . + a .    b d d b c    c . . a b    + . a . l    b + . c a
SWR/J         . . + a .    b d d b c    c . . a b    + . a . l    b + . c a

TA1/-         . . . . .    . . . . .    . . . a b    . . b . .    . . . . .
TA2/-         . . . . .    . . . . .    . . . a .    . . b . .    . . . . .
TA3/-         . . . . .    . . . . .    . . . a b    . . b . .    . . . . .
TB1/-         . . . . .    . . . . .    . . . a b    . . b . .    . . . . .
TF/GnLe       . . + . .    . . . . +    3 . . a .    . . a . .    . . . . .

TSI/A         . . + . .    b d d . c    . h . . .    + . a . .    . + . . .
TSJ/Le        . . . . .    . . . . .    . . . . b    . . . . .    . . . . .
V/Le          . . + . .    . . . . +    . . . a b    . . . . .    . . . . .
WB/ReJ        . . + b .    b h b . +    c h . b a    2 . a f .    . + . . .
WC/ReJ        . . b . .    . d . . .    b h . b a    + . . . .    . + . . .

              1 = c*ch     2 = cdm      3 = b?
```

Locus (read down) →	D12Nyu3	D12Nyu4	Dlb-1	Ea-2	Ea-4	Ea-5	Ea-6	Ea-7	Ega	Em	Emv-4	Emv-5	Emv-6	Emv-7	Emv-8	Emv-9	Emv-10	Emv-11	Emv-12	Emv-13	Emv-14	Eph-1	Es-1	Es-2	Es-3
Chromosome (read down)	12	12	11	UN	19	UN	2	UN	8	UN	UN	UN	UN	UN	UN	UN	UN	7	16	2	11	1	8	8	1
RIIIS/J	.	.	.	.	.	.	.	.	.	.	b	b	b	b	b	b	b	b	b	b	b	b	b	b	c
ROP/GnLe	.	.	.	.	.	.	.	.	.	.	.	.	.	.	.	.	.	.	.	.	.	.	a	b	a
RSV/Le	.	.	.	.	.	.	.	.	.	.	.	.	.	.	.	.	.	.	.	.	.	.	b	.	c
SB/Le	.	.	.	.	.	.	.	.	.	.	.	.	.	.	.	.	.	.	.	.	.	.	b	.	.
SC/CpbU	.	.	.	.	.	.	.	.	.	.	.	.	.	.	.	.	.	.	.	.	.	.	b	b	c
SEA/GnJ	.	.	b	.	.	.	.	.	.	.	b	b	b	b	b	b	b	b	b	b	b	.	b	b	a
SEC/1ReJ	.	.	.	b	.	.	.	b	.	.	b	b	b	b	b	b	b	b	b	b	b	b	b	b	a
SEC/-	.	.	.	b	.	.	.	b	.	.	b	b	b	b	b	b	b	b	b	b	b	b	b	b	a
SF/Cam-	.	.	.	.	.	.	.	.	.	.	.	.	.	.	.	.	.	.	.	.	.	.	b	b	c
SF/CamRk	.	.	.	.	.	.	.	.	.	.	.	.	.	.	.	.	.	.	.	.	.	.	b	b	c
SH1/Le	.	.	.	.	.	.	.	.	.	.	.	.	.	.	.	.	.	.	.	.	.	.	b	.	c
SJL/J	a	b	b	b	a	.	.	b	a	h	b	b	b	b	b	a	a	b	b	b	b	b	b	b	c
SJL/J-	a	b	b	b	a	.	.	b	a	h	b	b	b	b	b	a	a	b	b	b	b	b	b	b	c
SJL/JA	.	.	.	.	.	.	.	.	.	.	.	.	.	.	.	.	.	.	.	.	.	.	b	b	c
SK/Cam-	.	.	.	.	.	.	.	.	.	.	.	.	.	.	.	.	.	.	.	.	.	.	b	a	c
SK/CamRk	.	.	.	.	.	.	.	.	.	.	.	.	.	.	.	.	.	.	.	.	.	.	b	a	c
SL/NiA	.	.	.	.	.	.	.	.	.	.	.	.	.	.	.	.	.	.	.	.	.	.	b	b	c
SM/-	a	b	.	b	a	.	.	b	a	.	b	b	b	b	b	b	b	b	b	b	b	.	b	b	c
SM/J	a	b	.	b	a	.	.	b	a	.	b	b	b	b	b	b	b	b	b	b	b	.	b	b	c
SSL/Le	.	.	.	.	.	.	.	.	.	.	.	.	.	.	.	.	.	.	.	.	.	.	b	.	a
ST/a	.	.	.	.	.	.	.	.	.	.	.	.	.	.	.	.	.	.	.	.	.	.	b	b	b
ST/bJ	.	.	.	b	a	.	.	b	a	l	b	b	a	b	b	b	b	b	b	b	b	.	b	b	b
ST/bJN	.	.	.	.	.	.	.	.	.	.	.	.	.	.	.	.	.	.	.	.	.	.	b	.	b
STR/1N	.	.	.	.	.	.	.	.	.	.	.	.	.	.	.	.	.	.	.	.	.	.	b	.	a
STR/N	.	.	.	.	.	.	.	.	.	.	.	.	.	.	.	.	.	.	.	.	.	.	b	.	b
STS/A	.	.	.	.	.	.	.	.	o	.	.	.	.	.	.	.	.	.	.	.	.	.	b	b	c
STS/N	.	.	.	.	.	.	.	.	.	.	.	.	.	.	.	.	.	.	.	.	.	.	b	.	c
STX/LeJ	.	.	.	.	.	.	.	.	.	.	.	.	.	.	.	.	.	.	.	.	.	.	.	.	.
SWR/-	a	a	a	b	a	.	.	b	a	l	b	b	b	b	b	b	b	b	b	b	b	b	b	b	c
SWR/J	a	a	.	.	.	.	.	.	a	l	b	b	b	b	b	b	b	b	b	b	b	b	b	b	c
TA1/-	.	.	.	.	.	.	.	.	.	.	.	.	.	.	.	.	.	.	.	.	.	.	a	b	a
TA2/-	.	.	.	.	.	.	.	.	.	.	.	.	.	.	.	.	.	.	.	.	.	.	b	b	c
TA3/-	.	.	.	.	.	.	.	.	.	.	.	.	.	.	.	.	.	.	.	.	.	.	b	b	a
TB1/-	.	.	.	.	.	.	.	.	.	.	.	.	.	.	.	.	.	.	.	.	.	.	b	b	a
TF/GnLe	.	.	.	.	.	.	.	.	.	.	.	.	.	.	.	.	.	.	.	.	.	.	a	.	a
TSI/A	.	.	.	.	.	.	.	.	.	.	.	.	.	.	.	.	.	.	.	.	.	.	b	b	c
TSJ/Le	.	.	.	.	.	.	.	.	.	.	.	.	.	.	.	.	.	.	.	.	.	.	b	.	a
V/Le	.	.	.	.	.	.	.	.	.	.	.	.	.	.	.	.	.	.	.	.	.	.	a	.	a
WB/ReJ	.	.	b	a	.	.	.	b	.	.	.	.	.	.	.	.	.	.	.	.	.	.	b	.	a
WC/ReJ	.	.	.	.	.	.	.	.	.	.	.	.	.	.	.	.	.	.	.	.	.	.	b	.	a

```
Locus    ->     E E E E E    E E E E E    E E E E E    E E E E E    F F F F F
(read down)     s s s s s    s s s s s    s s s s s    s s s s s    a b b o o
                - - - - -    - - - - -    - - - - -    - - - - r    b p p r s
                5 6 7 8 9    1 1 1 1 1    1 1 1 2 2    2 2 2 2      p - - -
                             0 1 2 3 4    6 7 8 2 3    4 5 6 7      i 1 2 5
--------------------------------------------------------------------------------
  Chromosome    8 8 8 7 8    1 8 U 9 9    3 9 1 8 8    8 1 3 3 6    3 U U 1 1
  (read down)                4   N            9            2            N N 4 2
--------------------------------------------------------------------------------

RIIIS/J         . a . . .    b a . a .    . . . . .    . . . . .    . . . d .
ROP/GnLe        . a . a .    a a . . .    . . . . .    . . . . .    . . . . .
RSV/Le          . . . . .    b a . . .    . . . . .    . . . . .    . . . . .
SB/Le           . . . . .    . . . . .    . . . . .    . . . . .    . . . . .
SC/CpbU         b a b . a    b a . b r    a a a d c    a a a r a    . . . . .

SEA/GnJ         . a . a .    a a . a .    . . . . .    . . . . .    . a a b .
SEC/1ReJ        . a . a .    a a . a .    . . . . .    . . . . .    . . . b .
SEC/-           . a . a .    a a . a .    . . . . .    . . . . .    . . . b .
SF/Cam-         . . . . .    b . . . .    . . . . .    . . . . .    . . . . .
SF/CamRk        . . . . .    b . . . .    . . . . .    . . . . .    . . . . .

SH1/Le          . . . . .    a a . . .    . . . . .    . . . . .    . . . . .
SJL/J           . a . a .    b a . b .    . . . . c    . b c . .    . . . d c
SJL/J-          b a b a a    b a . b r    a a a d c    f b c s a    . b . d c
SJL/JA          b a b . a    b a . b r    a a a d c    f b c s a    . b . . .
SK/Cam-         . a b . b    b b . a .    a b a d d    g b a . .    . . . . .

SK/CamRk        . c . . .    b . . a .    . . . . .    . . . . .    . . . . .
SL/NiA          b a b . a    b a . b r    a a a d c    a b a r a    . . . . .
SM/-            b a b . a    b a . a r    a a a d c    a b a r a    . b a d b
SM/J            . a . . a    b a . a .    . . . . .    . . . . .    . b a d b
SSL/Le          . . . . .    b a . . .    . . . . .    . . . . .    . . . . .

ST/a            . . . . .    . . . . .    . . . . .    . . . . .    . . . . .
ST/bJ           . a . . .    a a . a .    . . . . .    . . . . .    . . . b .
ST/bJN          . . . . .    a . . . .    . . . . .    . . . . .    . . . . .
STR/1N          . . . . .    b . . . .    . . . . .    . . . . .    . . . . .
STR/N           . . . . .    b . . . .    . . . . .    . . . . .    . . . . .

STS/A           b a b a b    a a . b r    a a a h c    a b a r a    . b . . .
STS/N           . . . . .    a . . . .    . . . . .    . . . . .    . . . . .
STX/LeJ         . . . . .    a . . . .    . . . . .    . . . . .    . . . . .
SWR/-           . a . a a    a a a b .    . . . . .    . b a . .    b a a b a
SWR/J           . a . a .    a a a b .    . . . . .    . . . . .    b a a b a

TA1/-           a a b . a    b b . a r    a a a d c    a b b r .    . . . . .
TA2/-           b a b . a    a a . a r    a a b d c    a b a r a    . . . . .
TA3/-           . . . . .    b a . . .    . . . . .    . . . . .    . . . . .
TB1/-           . . . . .    b a . . .    . . . . .    . . . . .    . . . . .
TF/GnLe         . . . . .    a a . . .    . . . . .    . . . . .    . . . . .

TSI/A           b a b a a    b a . a r    a a a d c    a a a r a    . a . . .
TSJ/Le          . . . . .    a a . . .    . . . . .    . . . . .    . . . . .
V/Le            . . . . .    b a . . .    . . . . .    . . . . .    . . . . .
WB/ReJ          . . . . .    a a . . .    . . . . .    . . . . .    . . . . .
WC/ReJ          . . . . .    a a . . .    . . . . .    . . . . .    . . . . .
```

```
Locus     ->   F F G G   G G G G G   G G G G G   G G G G G   G G H H H
(read down)    v v a b d   d d d g k   l l m o o   p p p r u   u v - - -
               - - l p c   c r r c °   k o - t t   d i t - s   s - 1 2 3
               1 2 t - -   - - -       - 3 - -     - - - 1 -   - 1
                     1 1   2 1 2       1   1 2     1 1 1   r     s

Chromosome     4 9 4 3 1   9 U U 6 U   1 1 2 1 8   4 7 1 8 5   5 U 7 1 2
(read down)          5       N N   N   1 2   9           5       N     7

------------------------------------------------------------------------

RIIIS/J        b . b . b   b . . a .   a a . a b   b a b a .   b . . r .
ROP/GnLe       . . b . .   b . . b .   . a . . b   a a a a .   . . . . .
RSV/Le         . . b . .   . . . a .   . a . . b   a b a a .   . . . k .
SB/Le          . . . . b   . . . . .   . a . . .   . . a . .   . . . b .
SC/CpbU        . . . . .   . . . . .   . . . . .   . . . . .   . . . . .

SEA/GnJ        . . a b b   b . . a .   . a . a b   a b a a a   a . . d .
SEC/1ReJ       b . b a b   b h l a .   . a . a b   b a a a a   a . . d .
SEC/-          b . b a b   b h l a .   . a . a b   b a a a a   a . . d .
SF/Cam-        . . c . .   . . . a .   . a . a b   a b . . .   . . . . .
SF/CamRk       . . . . .   . . . a .   . a . a b   a b . . .   . . . . .

SH1/Le         . . . . .   . . . . .   . a . . b   b b a a .   . . . . .
SJL/J          . . b . b   b h l a .   a a a a b   b a a b .   b a l s 2
SJL/J-         . . b b b   b h l a .   a a a a b   b a a b .   b a l s 2
SJL/JA         . . . . .   . . . . .   . a . a b   b a . b .   . . . . .
SK/Cam-        n . b . .   . . . b .   . a . a a   a a a a -   . . . . .

SK/CamRk       n . . . .   . . . b .   . . . a a   a a . a .   . . . . .
SL/NiA         . . . b .   . . . . .   . b . a b   b a . a .   . . . . .
SM/-           n . b a .   b . . a 3   a a . a b   b a a a a   a . . v 2
SM/J           n . b a b   b . . a 3   a a . a b   b a a a a   a . . v 2
SSL/Le         . . . . .   . . . . .   . a . . b   a b a a .   . . . . .

ST/a           . . . . .   . . . . .   . . . . .   b . . . .   . . . b .
ST/bJ          n s a b b   b . . a .   b a . a b   b a a a .   b . . k .
ST/bJN         . . . . .   . . . . .   . . . . .   b a . . .   a . . k .
STR/1N         . . . . .   . . . . .   . . . . .   b b . . .   . . . . .
STR/N          . . . . .   . . . . .   . . . . .   b b . . .   . . . . .

STS/A          n s . b .   . . . . .   . a a a b   b b . b b   b . . 4 .
STS/N          . . . . .   . . . . .   . . . . .   b b . . .   . . . . .
STX/LeJ        . . b . b   . . . . .   l . . . .   . b a . .   b . . . .
SWR/-          n . b b b   b h l a .   a b a a a   b b a b b   b b 1 q b
SWR/J          n . b . b   b h l a .   a b . a a   b b a b b   b . 1 q b

TA1/-          . . . . b   . . . . .   . a . a b   b 5 a . .   b . . . .
TA2/-          . . . . b   . . . . .   . . . a b   b b b . .   b . . . .
TA3/-          . . . . b   . . . . .   . a . a b   b a a . .   . . . . .
TB1/-          . . . . b   . . . . .   . a . a b   b a a . .   . . . . .
TF/GnLe        . . b . b   . . . . .   . a . . b   b b a a .   . . . q .

TSI/A          b s . b .   . . . . .   . a . a b   b a . a .   . . . 4 .
TSJ/Le         . . . . .   . . . . .   . a . . b   b a a a .   . . . d .
V/Le           . . b . b   . . . . .   . a . . b   a a c a .   . . . b .
WB/ReJ         . . . . .   . . . . .   . a . . b   a b a a .   b . e 6 a
WC/ReJ         . . . . .   . . . . .   . a . . b   a b a a .   . . a 6 a
```

738   1 = -a,-c   2 = -a,-b   3 = a?   4 = dx   5 = ab
       6 = ja

```
Locus    ->    H H H H H     H H H H H     H H H H H     H H I I I     I I I I I
(read down)    - - - - -     - a a b b     b c e m p     r s d f g     g g g g g
               4 7 8 9 1     1 o o a a     b x a         t d h - h     h h h h h
                       2     3 - - -       -             - - 1 -       - - - - -
                             1 2   4       1             1   1         2 3 4 5 A
                                 p                                             r
                                 s                                             s
---------------------------------------------------------------------------------
Chromosome     7 9 1 U U     2 2 3 1 1     7 2 U 7 8     U 1 1 3 1     1 1 1 1 1
(read down)        4 N N           1 7         N           N 0   2       2 2 2 2 2
---------------------------------------------------------------------------------

RIIIS/J        . . . . .     . . . f .     s . . . .     b . a . .     . . a . .
ROP/GnLe       . . . . .     . . . d .     s . . . .     . . b . .     . . b . .
RSV/Le         . . . . .     . . . a .     s . . . .     . . a . .     . . b . .
SB/Le          . . . . .     . . . . .     d . . . .     . . . . .     . . . . .
SC/CpbU        . . . . .     . . . . .     . . . . .     . . . . .     . . . . .

SEA/GnJ        . . . . .     . . . h .     d l . a .     . . a . j     a a a a .
SEC/1ReJ       . . . . .     . . . b .     s . a . .     b . a . h     . . a . .
SEC/-          . . . . .     . . . b .     s l a . .     b . a . *     . . a . .
SF/Cam-        . . . . .     . . . d .     d . . . .     b l a . .     . . b . .
SF/CamRk       . . . . .     . . . d .     d . . . .     b . a . .     . . b . .

SH1/Le         . . . . .     . . . . .     d . . . .     . . b . .     . . . . .
SJL/J          a 1 1 1 1     1 . . c .     s l a b .     b l b . b     2 . b . o
SJL/J-         a 1 1 1 1     1 a a c .     s l a b .     b l b . b     2 . b . o
SJL/JA         . . . . .     . a a . .     s . . . .     . . b . .     . . . . .
SK/Cam-        . . . . .     . . . d .     d . . . .     . l a . .     . . a . .

SK/CamRk       . . . . .     . . . d .     d . . . .     . . a . .     . . a . .
SL/NiA         . . . . .     . a a c .     d . . . .     . . a . .     . . . . .
SM/-           . . . . .     d a a d .     s l . . .     b l b . b     2 . b . o
SM/J           . . . . .     d a a d .     s l . . .     b l b . b     2 . b . o
SSL/Le         . . . . .     . . . . .     d . . . .     . . a . .     . . . . .

ST/a           . . . . .     . . . . .     . . . . .     . . a . .     . . . . .
ST/bJ          . . . . .     . . . c .     d o . a .     b l a l a     a . 3 . .
ST/bJN         . . . . .     . . . c .     d o . . .     . . a . .     . . . . .
STR/1N         . . . . .     . . . f .     s l . . .     . . b . b     . . . . .
STR/N          . . . . .     . . . . .     s l a . .     . . b . a     a . . . .

STS/A          . . . . .     . c a g .     d o a . .     . . a h a     . . . . .
STS/N          . . . . .     . . . c .     d . a . .     . . a . .     . . . . .
STX/LeJ        . . . . .     . . . c .     s . . . .     . . a . .     . . a . .
SWR/-          1 1 1 . b     a . . c b     s o a b a     a l a . c     c . a . o
SWR/J          1 1 1 . b     a . . c b     s o a b a     a l a . c     c . a . o

TA1/-          . . . . .     . . . . .     4 . . . .     . . a . .     . . . . .
TA2/-          . . . . .     . . . . .     d . . . .     . . a . .     . . . . .
TA3/-          . . . . .     . . . . .     s . . . .     . . a . .     . . . . .
TB1/-          . . . . .     . . . . .     s . . . .     . . a . .     . . . . .
TF/GnLe        . . . . .     . . . a .     s . . . .     . . a . .     . . . . .

TSI/A          . . . . .     . a a d .     d l a . .     . . a h a     . . . . .
TSJ/Le         . . . . .     . . . . .     d . . . .     . . a . .     . . . . .
V/Le           . . . . .     . . . . .     s . . . .     . . b . .     . . . . .
WB/ReJ         1 1 1 a 1     1 . . c .     d o . . .     . . a . b     2 . b . .
WC/ReJ         1 1 1 1 1     a . . c .     d . . . .     . . a . b     2 . . . .

        1 = -a,-b    2 = -a,-c    3 = b?        4 = sd
```

```
Locus    ->      I I I I I   I I I I I   I I I I   I I K K K   L L L L L
(read down)      g g g g g   g g g g g   g g g g g   g g f t t   a a a a d d
                 h h h h h   h h h h h   h h k k k   l l o h h   m m p h r
                 - - - - -   - - - - -   - - - - -   - - - - -   b b - - -
                 C D I L N   P S S S S   S S E P T   1 1 1 1 2   - - 1 1 1
                   e n e p   c a a a a   a r f c r       r         1 2
                   x u v       1 2 3 4   5 c 1 p
-------------------------------------------------------------------------------
Chromosome       1 1 1 1 1   1 1 1 1 1   1 1 6 6 6   1 U 9 1 U   1 1 9 7 6
(read down)      2 2 2 2 2   2 2 2 2 2   2 2           6 N   7 N   2
-------------------------------------------------------------------------------

RIIIS/J          . . . . .   . . . . .   . . b . .   . . . . .   . . a . a
ROP/GnLe         . . . . .   . . . . .   . . . . .   . . . . .   . . . . a
RSV/Le           . . . . .   . . . . .   . . . . .   . . . . .   . . . . .
SB/Le            . . . . .   . . . . .   . . . . .   . . . . .   a b . . .
SC/CpbU          . . . . .   . . . . .   . . . . .   . . . . .   . . . . .

SEA/GnJ          . . a . .   . . . . .   . . . b .   . . . . .   . . . . a
SEC/1ReJ         . . a . .   . . . . .   . . . b .   . . . . .   a b b . a
SEC/-            . . a . .   . . . . .   . . . b .   . . . . .   a b b . a
SF/Cam-          . . . . .   . . . . .   . . a . .   . . . . .   . . b a a
SF/CamRk         . . . . .   . . . . .   . . a . .   . . . . .   . . b a a

SH1/Le           . . . . .   . . . . .   . . . . .   . . . . .   . . . . .
SJL/J            b o . a o   o o o . .   . a b . b   b b . . .   a b b . a
SJL/J-           b o . a o   o o o . .   . a b . b   b b . . .   a b b . a
SJL/JA           . . . . .   . . . . .   . . . . .   . . . . .   . . . . a
SK/Cam-          . . . . .   . . . . .   . . b . .   . . . . .   . . . a b

SK/CamRk         . . . . .   . . . . .   . . . . .   . . . . .   . . . a b
SL/NiA           . . . . .   . . . . .   . . . . .   . . . . .   . . . . b
SM/-             . . . . o   . . . . .   . . b . .   . . . . .   a a b . a
SM/J             . . . . o   . . . . .   . . b . .   . . . . .   a a b . a
SSL/Le           . . . . .   . . . . .   . . . . .   . . . . .   . . . . .

ST/a             . . . . .   . . . . .   . . . . .   . . . . .   . . . . .
ST/bJ            . a . a o   a . . . .   . . b b b   . . . . .   a d b . a
ST/bJN           . . . . .   . . . . .   . . . . .   . . . . .   . . . . .
STR/1N           . . . . .   . . . . .   . . . . .   . . . . .   . . . . .
STR/N            . . . . .   . . . . .   . . . . .   . . . . .   . . . . .

STS/A            . . . . .   . . . . .   . . . . .   . . . . .   a b . . a
STS/N            . . . . .   . . . . .   . . . . .   . . . . .   . . . . .
STX/LeJ          . . . . .   . . . . .   . . . . .   . . . . .   a b . . .
SWR/-            . . a . .   . o o . .   . o b . b   . . b a a   a b b . b
SWR/J            . . a . .   . . . . .   . . b . b   . . b a a   a b b . b

TA1/-            . . . . .   . . . . .   . . . . .   . . . . .   . . . . a
TA2/-            . . . . .   . . . . .   . . . . .   . . . . .   . . . . a
TA3/-            . . . . .   . . . . .   . . . . .   . . . . .   . . . . .
TB1/-            . . . . .   . . . . .   . . . . .   . . . . .   . . . . .
TF/GnLe          . . . . .   . . . . .   . . . . .   . . . . .   . . . . .

TSI/A            . . . . .   . . . . .   . . . . .   . . . . .   . . . . b
TSJ/Le           . . . . .   . . . . .   . . . . .   . . . . .   . . . . .
V/Le             . . . . .   . . . . .   . . . . .   . . . . .   . . . . .
WB/ReJ           . . . . .   . . . . .   . . . . .   . . . . .   a b . . .
WC/ReJ           . . . . .   . . . . .   . . . . .   . . . . .   . . . . .
```

Strain distribution of polymorphic variants

```
Locus    ->     L L L L L   L L L L L   L L L L L   L L L L L   L L L L L
(read down)     e e f m n   p s t t t   t t t t v   v y y y y   y y y y y
                n n o p a   t h h h n   w w w w     p - - - -   - - - - -
                - - - - -   - - - - -   - - - - -   - - - - -   - - - - -
                1 2 1   1   1   1 2 2   2 3 4 6     1 1 2 3 5   6 7 8 9 1
                                                                        0

-------------------------------------------------------------------------
Chromosome      1 U 7 1 U   U 1 1 U U   1 9 1 U 4   6 1 6 6 1   1 1 U 1 1
(read down)       N 7   N   N 3 N N     2     N     9             5 2 N 9
-------------------------------------------------------------------------

RIIIS/J         a b . o .   a . . . b   . a b a .   a . . b .   . . . a .
ROP/GnLe        . . . . .   . . . . .   . . . . .   . . . . .   . . . . .
RSV/Le          . . . . .   . . . . .   . . . . .   . . . . .   . . . . .
SB/Le           . . . . .   . . . . .   . . . . .   . . . . .   . . . . .
SC/CpbU         . . . . .   . . . . .   . . . . .   . . . . .   . . . . .

SEA/GnJ         . . . . .   . . a . b   . . a a .   a . . . .   . . . a .
SEC/1ReJ        b b . . .   a . a . b   . . a a c   a . b . .   . . . . .
SEC/-           b b . . .   a . a . b   . . a a c   a . b . .   . . . . .
SF/Cam-         a a . . .   . . . . b   . . a a .   . . . b .   . . . . .
SF/CamRk        . a . . .   . . . . .   . . . . .   . . . . .   . . . . .

SH1/Le          . . . . .   . . . . .   . . . . .   . . . . .   . . . . .
SJL/J           a b . . .   . . o . b   . a . . a   b b b b a   b a b a b
SJL/J-          a b . . a   . . o . b   . a . . a   b b b b a   b a b a b
SJL/JA          . . . . .   . . . . .   . . . . .   . . . . .   . . . . .
SK/Cam-         . . . . .   . . . . b   . . a a .   . . . b .   . . . . .

SK/CamRk        . . . . .   . . . . .   . . . . .   . . . . .   . . . . .
SL/NiA          . . . . .   . . . . .   . . . . .   . . . . .   . . . . .
SM/-            a b . a .   . . a . b   . . b a b   b b a b .   b a . a .
SM/J            a b . a .   . . a . .   . . . a b   b b a b .   b a . a .
SSL/Le          . . . . .   . . . . .   . . . . .   . . . . .   . . . . .

ST/a            . . . . .   . . . . .   . . . . .   . . . . .   . . . . .
ST/bJ           a b . . .   . . a . b   . a b a a   a . b b b   . . . . .
ST/bJN          . . . . .   . . . . .   . . . . a   . b b b .   . . . . .
STR/1N          . . . . .   . . . . .   . . . . .   . b . . .   . . . . .
STR/N           . . . . .   . . . . .   . . . . .   . . . . .   . . . . .

STS/A           a b . . .   . . . . .   . . . . .   . b b b a   . . . . .
STS/N           . . . . .   . . . . .   . . . . .   . b b b .   . . . . .
STX/LeJ         . . . . .   . . . . .   . . . . .   . . . . .   . . . . .
SWR/-           a b b . a   a . a o b   a a b b a   a b b b b   b a b a .
SWR/J           a b b . .   a . a o b   a a b b a   a b b b b   b a b a .

TA1/-           . . . . .   . . . . .   . . . . .   . . . . .   . . . . .
TA2/-           . . . . .   . . . . .   . . . . .   . . . . .   . . . . .
TA3/-           . . . . .   . . . . .   . . . . .   . . . . .   . . . . .
TB1/-           . . . . .   . . . . .   . . . . .   . . . . .   . . . . .
TF/GnLe         . . . o .   . . . . .   . . . . .   . . . . .   . . . . .

TSI/A           . . . . .   . . . . .   . . . . .   . b b b b   . . . . .
TSJ/Le          . . . . .   . . . . .   . . . . .   . . . . .   . . . . .
V/Le            . . . . .   . . . . .   . . . . .   . . . . .   . . . . .
WB/ReJ          . . . . .   . . . . .   . . a a .   . b b b .   b . . . .
WC/ReJ          . . . . .   . . . . .   . . . . .   . . . . .   . . . . .
```

```
Locus    ->   L L L L L   L L L L L   L L L L L   L L L L L   L L L L L
(read down)   y y y y y   y y y y y   y y y y y   y y y y y   y y y y y
              - - - - -   - - - - -   - - - - -   - - - - -   - - - - b
              1 1 1 1 1   1 1 1 1 2   2 2 2 2 2   2 2 2 3     3 3 3 4 -
              1 2 3 4 5   6 7 8 9 0   1 2 3 4 5   6 7 8 9 0   1 2 9 0 2

--------------------------------------------------------------------------
Chromosome    2 1 U 7 7   1 1 1 4 4   7 1 2 2 2   U U U 4 U   4 4 1 U 4
(read down)     9 N          2   2                 N N N   N       7 N
--------------------------------------------------------------------------

RIIIS/J       . . . . .   . . . . .   . a . . .   . . . . .   b . . . .
ROP/GnLe      . . . . .   . . . . .   . . . . .   . . . . .   . . . . .
RSV/Le        . . . . .   . . . . .   . . . . .   . . . . .   . . . . .
SB/Le         . . . . .   . . . . .   . . . . .   . . . . .   . . . . .
SC/CpbU       . . . . .   . . . . .   . . . . .   . . . . .   . . . . .

SEA/GnJ       . . . . .   . a . . .   . a a . .   . . . . .   b . . . .
SEC/1ReJ      . . . . .   . . . . .   . b . . .   . . . . .   a . . . .
SEC/-         . . . . .   . . . . .   . b . . .   . . . . .   a . . . .
SF/Cam-       . . . . .   . . . . .   . . . . .   . . . . .   . . . . .
SF/CamRk      . . . . .   . . . . .   . . . . .   . . . . .   . . . . .

SH1/Le        . . . . .   . . . . .   . . . . .   . . . . .   . . . . .
SJL/J         . b b a a   . a b a .   b . b b .   b . . . .   a . 1 1 c
SJL/J-        b b b a a   b a b a b   b a b b a   b b b a a   a a 1 1 c
SJL/JA        . . . . .   . . . . .   . . . . .   . . . . .   . . . . .
SK/Cam-       . . . . .   . . . . .   . . . . .   . . . . .   . . . . .

SK/CamRk      . . . . .   . . . . .   . . . . .   . . . . .   . . . . .
SL/NiA        . . . . .   . . . . .   . . . . .   . . . . .   . . . . .
SM/-          . b . . b   . . . . .   . a b . .   b . . . .   b . . . .
SM/J          . b . . b   . . . . .   . a b . .   b . . . .   b . . . .
SSL/Le        . . . . .   . . . . .   . . . . .   . . . . .   . . . . .

ST/a          . . . . .   . . . . .   . . . . .   . . . . .   . . . . .
ST/bJ         . . . . .   . . . . .   . a . . .   . . . . .   a . . . .
ST/bJN        . . . . .   . . . . .   . . . . .   . . . . .   . . . . .
STR/1N        . . . . .   . . . . .   . . . . .   . . . . .   . . . . .
STR/N         . . . . .   . . . . .   . . . . .   . . . . .   . . . . .

STS/A         . . . . b   . . . . .   . a . . a   b . . a .   a . . . 2
STS/N         . . . . .   . . . . .   . . . . .   . . . . .   . . . . .
STX/LeJ       . . . . .   . . . . .   . . . . .   . . . . .   . . . . .
SWR/-         a a b a a   . a b b b   b a a b a   b b b a b   a a 1 1 a
SWR/J         . a b a a   . a b b .   b a a b .   b . . . .   a . 1 1 a

TA1/-         . . . . .   . . . . .   . . . . .   . . . . .   . . . . .
TA2/-         . . . . .   . . . . .   . . . . .   . . . . .   . . . . .
TA3/-         . . . . .   . . . . .   . . . . .   . . . . .   . . . . .
TB1/-         . . . . .   . . . . .   . . . . .   . . . . .   . . . . .
TF/GnLe       . . . . .   . . . . .   . . . . .   . . . . .   . . . . .

TSI/A         . . . . .   . . . . .   . . . . .   . . . . .   . . . . .
TSJ/Le        . . . . .   . . . . .   . . . . .   . . . . .   . . . . .
V/Le          . . . . .   . . . . .   . . . . .   . . . . .   . . . . .
WB/ReJ        . . . . .   . . . . .   . . . . .   . . . . .   b . . . .
WC/ReJ        . . . . .   . . . . .   . . . . .   . . . . .   b . . . .
```

          1 = -b   2 = -a,-b,-c

Strain distribution of polymorphic variants

```
Locus      ->   L L L L L   L M M M M   M M M M M   N N N N   N O O p P
(read down)     y y y y y   y a e l o   o o p p u   c e k k p   p r r a
                b b b b b   b p p s d   d r h i p   a u - - -   - m m n
                - - - - -   - - - - -   - - - - -   m - 1 2 1   2 - - -
                3 4 5 6 7   8 1 1 1 1   2 1 1 1 1   1           1 2 1
                              7                     7 7   4     4

Chromosome      U 4 U 4 1   7 5 1 1 9   7 5 7 9 4   9 1 1 2 1   1 4 4 7 4
(read down)     N   N   2   7                       7 7   4     4
                              7

RIIIS/J         . . . . .   . b . . b   b a . b b   . b . . .   . . . . b
ROP/GnLe        . . . . .   . . . . b   . a . b a   . b . . a   . . . + .
RSV/Le          . . . . .   . . . . b   . . . b a   . b . . .   . . . + .
SB/Le           . . . . .   . . . . a   . . . . a   . . . . .   . . . + .
SC/CpbU         . . . . .   . . . . .   . . . . .   . . . . .   . . . . .

SEA/GnJ         . . . . .   . . . . b   . a . b a   . b . . .   . . . + .
SEC/1ReJ        . . . . .   . b . . a   b a . b a   . b . . a   . . . + b
SEC/-           . . . . .   . b . . a   b a . b a   . b . . a   . . . + b
SF/Cam-         . . . . .   . . . . *   . . . . a   . . . . a   . . . + b
SF/CamRk        . . . . .   . . . . a   . . . . a   . . . . a   . . . + b

SH1/Le          . . . . .   . . . . a   . . . b .   . b . . .   . . . . .
SJL/J           . b . . .   . b . c a   b a b b a   b b a o a   a . . p b
SJL/J-          b b b b b   b b . c a   b a b b a   b b a o a   a . . p b
SJL/JA          . . . . .   . . . . a   b a . b .   . b . . a   . . . . .
SK/Cam-         . . . . .   . . . . *   b . . . .   . b . . a   . . . + .

SK/CamRk        . . . . .   . . . . b   b . . . .   . b . . a   . . . + .
SL/NiA          . . . . .   . . . . a   b a . b .   . b . . a   . . . . .
SM/-            . a . . .   . . . . a   a a . b a   . a a o a   . . . + b
SM/J            . a . . .   . . . . a   a a . b a   . a a o a   . . . + b
SSL/Le          . . . . .   . . . . b   . . . b .   . b . . .   . . . . .

ST/a            . . . . .   . . . . .   . . . . .   . . . . .   . . . . .
ST/bJ           . . . . .   . . . . b   . a . b a   . b . . a   . . . . a
ST/bJN          . . . . .   . . . . b   . . . . a   . . . . .   . . . . .
STR/1N          . . . . .   . . . . b   . . . . a   . . . . .   . . . . .
STR/N           . . . . .   . b . . b   . . . . a   . b . . .   . . . . .

STS/A           . . . . .   . b . . b   b a . b b   . b . . a   . . . . .
STS/N           . . . . .   . . . . b   . . . . b   . b . . .   . . . . .
STX/LeJ         . . . . .   . . . . .   . . . . a   . . . . .   . . . + .
SWR/-           b a a a .   b b . . a   b a . b a   b b . . a   a a a . a
SWR/J           . a . . .   . b . . a   b a . b a   b b . . a   a a a . a

TA1/-           . . . . .   . . . . a   b a . b a   . . . . a   . . . . .
TA2/-           . . . . .   . . . . b   . . . . b   . . . . a   . . . . .
TA3/-           . . . . .   . . . . a   b a . b b   . . . . .   . . . . .
TB1/-           . . . . .   . . . . b   a a . b b   . . . . .   . . . . .
TF/GnLe         . . . . .   . . . . b   . . . b b   . b . . a   . . . + .

TSI/A           . . . . .   . b . . a   b a . b b   . b . . a   . . . . .
TSJ/Le          . . . . .   . . . . a   . . . b .   . b . . .   . . . . .
V/Le            . . . . .   . . . . b   . . . b b   . b . . a   . . . + .
WB/ReJ          . . . . .   . b . . b   . . . b a   . b . . a   . . . + .
WC/ReJ          . . . . .   . b . . b   . . . b a   . b . . a   . . . + .
```

```
Locus     ->   P P P P P   P P P P P   P P P P P   P P ·P P Q   Q Q Q Q Q
(read down)    a c d e e   e e g g g   g g g k r   r r r s a   a a a a e
               n a   p p   p p d k k   m m m - e   n t t p -   - - - - d
               - -   - -   - -   - -   - - - 3 -   - - -   1   2 3 4 5 -
               2 1   2 3   4 7   1 2   1 2 3   2   p 1 2             1

---------------------------------------------------------------------------
Chromosome     U U U 1 1   7 5 4 X 1   5 4 9 9 1   2 U 8 2 1   1 1 1 1 1
(read down)    N N N 0         7           2     N       7   7 7 7 7 7
---------------------------------------------------------------------------

RIIIS/J        b . . . c   . . . b b   a a a a .   b . . b .   . . . . c
ROP/GnLe       . . . . a   . . b b .   a a . . .   . . . b .   . . . . .
RSV/Le         . . . . b   . . . . b   a a . . .   . . . b .   . . . . .
SB/Le          . . . . .   . . . . .   . . . . .   . . . . .   . . . . .
SC/CpbU        . . . . .   . . . . .   . . . . .   . . . . .   . . . . .

SEA/GnJ        . . . . a   . . b . a   a a a . .   a . . . .   . . . . b
SEC/1ReJ       b . . . a   b . b b a   a a a . a   a . . b .   . . . . b
SEC/-          b . . . a   b . b b a   a a a . a   a . . b .   . . b b b
SF/Cam-        b . . . b   b . . . b   b . b . .   . . . b .   . . . . .
SF/CamRk       b . . . b   . . . . b   b . b . .   . . . . .   . . . . .

SH1/Le         . . . . a   . . . . .   b a . . .   . . . . .   . . . . .
SJL/J          b a b . b   . . b b a   b a c . .   a . . b a   . . a b a
SJL/J-         b a b . b   . . a b a   b a c . .   a . . b a   . . a b a
SJL/JA         . . . . b   . . a b a   b a . . .   . . . . .   . . . . .
SK/Cam-        . . . . b   b . . . a   a a c . .   . . . b .   . . . . .

SK/CamRk       . . . . b   . . . . a   a a c . .   . . . b .   . . . . .
SL/NiA         . . . . b   . . b b a   a a . . .   . . . . .   . . . . .
SM/-           a . . . b   . . . b a   a b a a .   a . . . .   . . . . b
SM/J           a . . . b   . . . b a   a b a a .   a . . . .   . . . . b
SSL/Le         . . . . b   . . . . .   b a . . .   . . . . .   . . . . .

ST/a           . . . . .   . . . . .   a a . . .   . . . . .   . . . . .
ST/bJ          b . . . c   . . . b b   a a a . .   a . . b .   b b . . b
ST/bJN         . . . . b   . . . . .   a . . . .   . . . . .   . . . . .
STR/1N         . . . . b   . . . . .   b . . . .   . . . . .   . . . . .
STR/N          . . . . b   . . . . .   b . . . .   . . . . .   . . . . .

STS/A          . . . . b   . . b b a   b a . . .   . . . . .   . . . . .
STS/N          . . . . b   . . . . .   b . . . .   . . . . .   . . . . .
STX/LeJ        . . . . a   . . b a .   a . . . .   . . . . .   . . . . .
SWR/-          b a a . b   b . b b a   b a c a a   a . . b a   a b b b a
SWR/J          b . a . b   b . b b a   b a c a a   a . . b a   a b . . a

TA1/-          . . . . c   . . . . .   a a . . .   . . . . .   . . . . .
TA2/-          . . . . b   . . . . .   b a . . .   . . . . .   . . . . .
TA3/-          . . . . b   . . . . .   a a . . .   . . . . .   . . . . .
TB1/-          . . . . a   . . . . .   a a . . .   . . . . .   . . . . .
TF/GnLe        . . . . b   . . . . a   a a . . .   . . . . .   . . . . .

TSI/A          . . . . b   . . b b a   b a . . .   . . . . .   . . . . .
TSJ/Le         . . . . a   . . . . .   a a . . .   . . . . .   . . . . .
V/Le           . . . . b   . . . . a   a a . . .   . . . . .   . . . . .
WB/ReJ         . . . . b   . . . . a   a a a . .   . . . . .   . . . . .
WC/ReJ         . . . . b   . . . . a   a a . . .   a . . . .   . . . . .
```

744

```
Locus   ->   r r R R R   S S S S S   S S S S S   S S S T T   T T T T T
(read down)  d d e i n   a a a a d   l o o o p   s v v a c   c g h i l
             s n b 7     a a a s h   p a d d i   p p p m n   r n y n a
             - -   s     - - - -     - -         - - - -       -   d
             1 1   -     1 1 1 1     1 2 2       1 2 1 2       1
                   6                     r
------------------------------------------------------------------------
Chromosome   5 1 1 1 6   7 7 1 2 U   1 U 1 1 1   U 2 7 7 1   1 1 9 1 1
(read down)    7   4             N   7 N 6 7 2   N       1   3 5   2 7
------------------------------------------------------------------------

RIIIS/J      + . s . b   . . o a .   . . . . .   . . . a s   a a . . .
ROP/GnLe     + . . . .   . . a . .   . . . . .   . . . . .   . . . . .
RSV/Le       . . . . .   . . o . .   . . . . .   . . . . .   . . . . .
SB/Le        . . . . .   . . . . .   . a a . .   . . . . .   . . . . .
SC/CpbU      . . . . .   . . . . .   . . . . .   . . . . .   . . . . .

SEA/GnJ      + . b . .   . . a a .   . b . . .   . . . . s   . . . . .
SEC/1ReJ     + . b a .   a . a a .   . . a a d   . . . a s   . . b . .
SEC/-        + . b a .   a . a a .   . . a a d   . . . a s   . . b . .
SF/Cam-      . . b . a   a . o a .   . . . . .   . . . c .   . b . . .
SF/CamRk     . . b . .   . . o a .   . . . . .   . . . c .   . . . . .

SH1/Le       . . . . .   . . . . .   . . . . .   . . . . .   . . . . .
SJL/J        1 . s a b   b . o a 2   . . . . .   . . . b s   . b b . a
SJL/J-       1 . s a b   b . o a 2   . . . . .   . . . b s   . b b . a
SJL/JA       . . . . .   . . . a .   . . . . .   . . . . .   . . . . .
SK/Cam-      + . s . a   . . o a .   . . . . .   . . . c .   . . . . .

SK/CamRk     + . s . .   . . o a .   . . . . .   . . . c .   . . . . .
SL/NiA       . . . . .   . . . a .   . . . . .   . . . . s   . . . . .
SM/-         + . s a b   a . o a 3   . b . . .   . . . a s   a . b . b
SM/J         + . s a b   a . o a 3   . b . . .   . . . a s   a . b . b
SSL/Le       . . . . .   . . . . .   . . . . .   . . . . .   . . . . .

ST/a         . . . . .   . . . . .   . . . . .   . . . . .   . . . . .
ST/bJ        1 . s . b   a . a a 2   . b . . .   . . . a f   c b b . b
ST/bJN       . . . . .   . . . . .   . . . . .   . . . . .   . . b . b
STR/1N       . . . . .   . . . . .   . . . . .   . . . . .   . . . . .
STR/N        . . . . .   . . a . .   . . . . .   . . . . .   . . . . .

STS/A        1 + . a .   . . . a .   . . . . .   f a . . f   . . b . c
STS/N        . . . . .   . . . . .   . . . . .   . . . . .   . . b . c
STX/LeJ      . . b . .   . . . . .   . . . . .   . . . . .   . . . . .
SWR/-        1 . s a b   a b a a .   . a a a d   f . . c s   c a b o a
SWR/J        1 . s a b   a b a a .   . . a a d   f . . c s   c a b o a

TA1/-        . . . . .   . . a . .   . . . . .   . . . c .   . . . . .
TA2/-        . . . . .   . . a . .   . . . . .   . . . a .   . . . . .
TA3/-        . . . . .   . . . . .   . . . . .   . . . . .   . . . . .
TB1/-        . . . . .   . . . . .   . . . . .   . . . . .   . . . . .
TF/GnLe      + . . . .   . . . . .   . . . . .   . . . . .   . . . . .

TSI/A        1 + . . .   . . a . .   . . . . .   f a . . s   . . b . b
TSJ/Le       . . . . .   . . . . .   . . . . .   . . . . .   . . . . .
V/Le         + . . . .   . . . . .   . . . . .   . . . . .   . . . . .
WB/ReJ       1 . . a .   . . a . .   . . . . .   . . . c .   . . b . b
WC/ReJ       1 . . . .   . . a . .   . . . . .   . . . c .   . . . . .
```

          1 = rd   2 = b?   3 = a?

Locus (read down)	Tpre	Trf	Tsuy	Tth	Ucp	Upg-1	Upg-2	Ups	Wap	Xmmv-9	Xmmv-42	Xmmv-43	Xmmv-44	Xmmv-45	Xmmv-46	Xmmv-47	Xmmv-48	Xmmv-49	Xmmv-50	Xmmv-51	Xmmv-52	Xmmv-53	Xmmv-54	Xmmv-55	Xmmv-56
Chromosome (read down)	12	9	12	12	8	17	1	9	UN	19	19	UN	UN	UN	UN	UN	UN	UN	12	UN	5	UN	UN	15	UN
RIIIS/J	.	b	.	.	.	s	s	a	a	o	o	a	o	o	o	o	o	o	o	o	a	o	o	o	o
ROP/GnLe	.	a	.	.	.	.	.	.	.	.	.	.	.	.	.	.	.	.	.	.	.	.	.	.	.
RSV/Le	.	a	.	.	.	.	.	.	.	.	.	.	.	.	.	.	.	.	.	.	.	.	.	.	.
SB/Le	.	b	.	.	.	.	.	.	.	.	.	.	.	.	.	.	.	.	.	.	.	.	.	.	.
SC/CpbU	.	.	.	.	.	.	.	.	.	.	.	.	.	.	.	.	.	.	.	.	.	.	.	.	.
SEA/GnJ	.	b	.	.	.	f	s	a	a	a	o	o	o	o	o	o	o	o	a	o	o	o	a	o	o
SEC/1ReJ	.	b	.	.	.	f	d	a	.	a	o	o	o	o	o	o	o	o	o	o	a	o	a	o	o
SEC/-	.	b	.	.	.	f	d	a	.	a	o	o	o	o	o	o	o	o	o	o	a	o	a	o	o
SF/Cam-	.	b	.	.	.	s	s	.	.	.	.	.	.	.	.	.	.	.	.	.	.	.	.	.	.
SF/CamRk	.	b	.	.	.	.	.	.	.	.	.	.	.	.	.	.	.	.	.	.	.	.	.	.	.
SH1/Le	.	b	.	.	.	.	.	.	.	.	.	.	.	.	.	.	.	.	.	.	.	.	.	.	.
SJL/J	.	b	.	.	d	s	d	a	a	o	o	o	o	o	o	o	o	a	o	a	o	a	o	a	o
SJL/J-	.	*	.	.	d	s	d	a	a	o	o	o	o	o	o	o	o	a	o	a	o	a	o	a	o
SJL/JA	.	b	.	.	.	.	.	.	.	.	.	.	.	.	.	.	.	.	.	.	.	.	.	.	.
SK/Cam-	.	b	.	.	.	s	s	.	.	.	.	.	.	.	.	.	.	.	.	.	.	.	.	.	.
SK/CamRk	.	b	.	.	.	.	.	.	.	.	.	.	.	.	.	.	.	.	.	.	.	.	.	.	.
SL/NiA	.	b	.	.	.	.	.	a	.	.	.	.	.	.	.	.	.	.	.	.	.	.	.	.	.
SM/-	.	b	.	.	.	.	d	a	a	o	o	o	o	o	o	o	a	o	o	o	a	o	a	a	o
SM/J	.	b	.	.	.	.	d	a	.	o	o	o	o	o	o	o	a	o	o	o	a	o	a	a	o
SSL/Le	.	b	.	.	.	.	.	.	.	.	.	.	.	.	.	.	.	.	.	.	.	.	.	.	.
ST/a	.	b	.	.	.	.	.	.	.	o	o	o	o	o	o	o	o	o	o	o	a	o	o	a	o
ST/bJ	.	b	.	.	.	s	d	a	.	o	o	o	o	o	o	o	o	o	a	o	a	o	o	a	o
ST/bJN	.	b	.	.	.	.	.	.	.	.	.	.	.	.	.	.	.	.	.	.	.	.	.	.	.
STR/1N	.	.	.	.	.	.	.	.	.	.	.	.	.	.	.	.	.	.	.	.	.	.	.	.	.
STR/N	.	a	.	.	.	.	.	.	.	.	.	.	.	.	.	.	.	.	.	.	.	.	.	.	.
STS/A	.	b	.	.	.	.	.	a	.	.	.	.	.	.	.	.	.	.	.	.	.	.	.	.	.
STS/N	.	.	.	.	.	.	.	.	.	.	.	.	.	.	.	.	.	.	.	.	.	.	.	.	.
STX/LeJ	.	b	.	.	.	.	.	.	.	.	.	.	.	.	.	.	.	.	.	.	.	.	.	.	.
SWR/-	o	b	o	o	d	s	d	a	a	o	o	o	o	o	a	o	o	o	o	o	a	o	a	a	o
SWR/J	o	b	o	o	d	s	d	a	a	o	o	o	o	o	a	o	o	o	o	o	a	o	a	a	o
TA1/-	.	b	.	.	.	.	.	.	.	.	.	.	.	.	.	.	.	.	.	.	.	.	.	.	.
TA2/-	.	b	.	.	.	.	.	.	.	.	.	.	.	.	.	.	.	.	.	.	.	.	.	.	.
TA3/-	.	b	.	.	.	.	.	.	.	.	.	.	.	.	.	.	.	.	.	.	.	.	.	.	.
TB1/-	.	b	.	.	.	.	.	.	.	.	.	.	.	.	.	.	.	.	.	.	.	.	.	.	.
TF/GnLe	.	1	.	.	.	.	s	.	.	.	.	.	.	.	.	.	.	.	.	.	.	.	.	.	.
TSI/A	.	b	.	.	.	.	.	a	.	.	.	.	.	.	.	.	.	.	.	.	.	.	.	.	.
TSJ/Le	.	b	.	.	.	.	.	.	.	.	.	.	.	.	.	.	.	.	.	.	.	.	.	.	.
V/Le	.	b	.	.	.	.	.	.	.	.	.	.	.	.	.	.	.	.	.	.	.	.	.	.	.
WB/ReJ	.	b	.	.	.	.	.	.	.	.	.	.	.	.	.	.	.	.	.	.	.	.	.	.	.
WC/ReJ	.	b	.	.	.	.	.	.	.	.	.	.	.	.	.	.	.	.	.	.	.	.	.	.	.

1 = b?

Locus → (read down)	Xmmv-57	Xmmv-58	Xmmv-59	Xmmv-60	Xmmv-62	Xmmv-63	Xmmv-64	Xmmv-65	Xmmv-66	Xmmv-67	Xmmv-68	Xmmv-69	Xmmv-70
Chromosome (read down)	UN	UN	UN	UN	4N	UN	UN	3N	UN	UN	UN	UN	UN
RIIIS/J	o	o	o	a	o	a	a	o	o	a	o	a	a
ROP/GnLe	.	.	.	.	.	.	.	.	.	.	.	.	.
RSV/Le	.	.	.	.	.	.	.	.	.	.	.	.	.
SB/Le	.	.	.	.	.	.	.	.	.	.	.	.	.
SC/CpbU	.	.	.	.	.	.	.	.	.	.	.	.	.
SEA/GnJ	o	o	o	o	a	a	o	a	o	o	o	a	a
SEC/1ReJ	o	o	o	a	o	a	o	a	o	o	o	a	a
SEC/-	o	o	o	a	o	a	o	a	o	o	o	a	a
SF/Cam-	.	.	.	.	.	.	.	.	.	.	.	.	.
SF/CamRk	.	.	.	.	.	.	.	.	.	.	.	.	.
SH1/Le	.	.	.	.	.	.	.	.	.	.	.	.	.
SJL/J	o	o	o	o	o	a	o	o	o	o	o	a	a
SJL/J-	o	o	o	o	o	a	*	o	o	o	o	a	a
SJL/JA	.	.	.	.	.	.	.	.	.	.	.	.	.
SK/Cam-	.	.	.	.	.	.	.	.	.	.	.	.	.
SK/CamRk	.	.	.	.	.	.	.	.	.	.	.	.	.
SL/NiA	.	.	.	.	.	.	.	.	.	.	.	.	.
SM/-	o	o	a	a	o	a	o	a	o	o	o	a	a
SM/J	o	o	a	a	o	a	o	a	o	o	o	a	a
SSL/Le	.	.	.	.	.	.	.	.	.	.	.	.	.
ST/a	o	o	o	a	o	a	o	o	o	a	o	a	a
ST/bJ	o	o	o	o	o	a	a	a	o	o	o	o	a
ST/bJN	.	.	.	.	.	.	.	.	.	.	.	.	.
STR/1N	.	.	.	.	.	.	.	.	.	.	.	.	.
STR/N	.	.	.	.	.	.	.	.	.	.	.	.	.
STS/A	.	.	.	.	.	.	.	.	.	.	.	.	.
STS/N	.	.	.	.	.	.	.	.	.	.	.	.	.
STX/LeJ	.	.	.	.	.	.	.	.	.	.	.	.	.
SWR/-	o	o	o	o	o	a	a	o	o	a	o	o	a
SWR/J	o	o	o	o	o	a	a	o	o	a	o	o	a
TA1/-	.	.	.	.	.	.	.	.	.	.	.	.	.
TA2/-	.	.	.	.	.	.	.	.	.	.	.	.	.
TA3/-	.	.	.	.	.	.	.	.	.	.	.	.	.
TB1/-	.	.	.	.	.	.	.	.	.	.	.	.	.
TF/GnLe	.	.	.	.	.	.	.	.	.	.	.	.	.
TSI/A	.	.	.	.	.	.	.	.	.	.	.	.	.
TSJ/Le	.	.	.	.	.	.	.	.	.	.	.	.	.
V/Le	.	.	.	.	.	.	.	.	.	.	.	.	.
WB/ReJ	.	.	.	.	.	.	.	.	.	.	.	.	.
WC/ReJ	.	.	.	.	.	.	.	.	.	.	.	.	.

```
Locus    ->    a A A A A    A A A A A    A A A A A    A A A A A    A A A A A
(read down)      a b c c    d d d d g    h h h h k    k l m m o    o p p s
                 t p f o    h h h h s    d d r p    p b y y x    x h k o -
                   a - -    - - - -      - - - -    - - - - -    - - a 1
                   1 1    1 2 3 3      1 2 1 1    2 1 1 2 1    2 1 - s
                        e t                                          1
--------------------------------------------------------------------------------
Chromosome     2 1 7 1 4    3 U 3 3 X    1 4 1 3 1    4 5 3 3 1    1 2 1 9 1
(read down)    2            N            2   9                     0       3
--------------------------------------------------------------------------------

WK/Re          a o . . .    . . . h      . . . .      . . a . .    . . a .
WLHR/Le        a . . . .    . . . .      . . . a      . a a . .    . . b .
WLL/BrA        + . . . .    . . . h    b . . .        . . . .      . . . .
WLL/BrAf       . . . . .    . . . .      . . . .      . . . . .    . . . a
YBR/-          1 . . . a    . . . .      . . . .      . a a . .    . . . a

YS/Wf          1 . . . .    . . . h      . . . .      . . . . .    . . . .
          1 = A*y/a
```

Strain distribution of polymorphic variants

```
Locus      ->    A A b B B    B B B B c    C C C C C    c C C C C    C d D D D
(read down)      v v 2 c      g g g v      3 4 6 a a    d e e f o    v 7 1 1
                 - -   m d    l l l -            r r    m - - h h    - R 2 2
                 1 2     -    - - - 1            - -    1 2 e        1 p N N
                       1      e s t              1 2                 2 y y
                                                                    - u u
                                                                    s 1 2
--------------------------------------------------------------------------------
Chromosome       U U 4 2 5    9 9 9 8 7    1 1 1 3 3    3 2 1 2 7    5 9 7 1 1
(read down)      N N                       7 7 5                7            2 2
--------------------------------------------------------------------------------

WK/Re            . . . b .    . d . . .    b . . b a    1 . . . .    . . . . .
WLHR/Le          . . b . .    . . . . +    c . . a a    . . . . .    . . . . .
WLL/BrA          . . + . .    b d d . c    . h . . .    + . a . .    . + . . .
WLL/BrAf         . . . . .    . . . . .    . . . . .    . . . . .    . . . . .
YBR/-            . . b . .    b h d . .    . h . . a    . b . . .    . + d . .

YS/Wf            . . . . .    b h b . .    . . . . a    . . a . .    . . . . .
                 1 = cdm
```

749

```
Locus     ->    D  D  D  E  E     E  E  E  E  E     E  E  E  E  E     E  E  E  E  E     E  E  E  E  E
(read down)     1  1  l  a  a     a  a  a  g  m     m  m  m  m  m     m  m  m  m  m     m  p  s  s
                2  2  b  -  -     -  -  -     a     v  v  v  v  v     v  v  v  v  v     v  h  -  -  -
                N  N  -  2  4     5  6  7           4  5  6  7  8     9  1  1  1  1     1  1     1  2  3
                y  y  1                                               0  1  2  3       4
                u  u
                3  4
                -----------------------------------------------------------------------------------------
Chromosome      1  1  1  U  1     U  2  U  8  U     U  U  U  U  U     U  U  7  1  2     1  1  8  8  1
(read down)     2  2  1  N  9     N     N     N     N  N  N  N  N     N  N     6        1           1
                -----------------------------------------------------------------------------------------

WK/Re           .  .  .  .  .     .  .  .  .  .     .  .  .  .  .     .  .  .  .  .     .  .  .  b  .  .
WLHR/Le         .  .  .  .  .     .  .  .  .  .     .  .  .  .  .     .  .  .  .  .     .  .  b  .  c
WLL/BrA         .  .  .  .  .     .  .  .  .  .     .  .  .  .  .     .  .  .  .  .     .  .  b  b  c
WLL/BrAf        .  .  .  .  .     .  .  .  .  .     .  .  .  .  .     .  .  .  .  .     .  .  b  b  c
YBR/-           .  .  .  b  .     a  a  .  o  .     .  .  .  .  .     .  .  .  .  .     .  .  .  .  .

YS/Wf           .  .  .  .  .     .  .  .  .  .     .  .  .  .  .     .  .  .  .  .     .  .  a  .  c
```

```
Locus   ->   E E E E E   E E E E E   E E E E E   E E E E E   F F F F F
(read down)  s s s s s   s s s s s   s s s s s   s s s s s   a b b o o
             - - - -     - - - -     - - - -     - - - - r   b p p r s
             5 6 7 8 9   1 1 1 1 1   1 1 1 2 2   2 2 2 2     p - - -
                         0 1 2 3 4   6 7 8 2 3   4 5 6 7     i 1 2 5
-----------------------------------------------------------------------------
Chromosome   8 8 8 7 8   1 8 U 9 9   3 9 1 8 8   8 1 3 3 6   3 U U 1 1
(read down)              4   N           9           2        N N 4 2
-----------------------------------------------------------------------------

WK/Re        . . . . .   b . . . .   . . . . .   . . . . .   . . . . .
WLHR/Le      . . . . .   c . . . .   . . . . .   . . . . .   . . . . .
WLL/BrA      b a b a a   a a . a r   a a c d c   a b a r a   . b . . .
WLL/BrAf     b a b . a   a a . a r   a a c d c   a b a . a   . . . . .
YBR/-        . . . . .   b . . . .   . a . . .   . . . . .   . . . . .

YS/Wf        . . . . .   . . . . .   . . . . .   . . . . .   . . . . .
```

751

The laboratory mouse

```
Locus    ->    F F G G G   G G G G G   G G G G G   G G G G G   G G H H H
(read down)    v v a b d   d d d g k   l l m o o   p p p r u   u v - - -
               - - l p c   c r r c     k o - t t   d i t - s   s - 1 2 3
               1 2 t - -   - - -       - 3 - -     - - - 1 -   - 1
                   1 1     2 1 2       1   1 2     1 1 1   r   s

--------------------------------------------------------------------------
Chromosome     4 9 4 3 1   9 U U 6 U   1 1 2 1 8   4 7 1 8 5   5 U 7 1 2
(read down)            5     N N   N   1 7 9             5       N   7
--------------------------------------------------------------------------

WK/Re          . . . . .   . . . . .   . . . . .   . . a . .   . . 1 2 3
WLHR/Le        . . . . .   b . . . .   . a . . b   a b a b .   . . . . .
WLL/BrA        n s . b .   . . . . .   . a . a b   a a . a .   . . . 4 .
WLL/BrAf       . . . . .   . . . . .   . . . . .   . . . . .   . . . . .
YBR/-          . . b . .   . . . . .   . a . . .   . a a . .   b . . d f

YS/Wf          . . . . .   . . . . .   . . . . .   b b . . .   b . . . .
         1 = -a,-c   2 = ja     3 = -a,-b   4 = bs
```

752

Strain distribution of polymorphic variants

Locus (read down)	H-4	H-7	H-8	H-9	H-12	H-13	Hao-1	Hao-2	Hba-4ps	Hba	Hbb	Hcx	Hea-1	Hm	Hp	Hrt-1	Hsd-1	Idh	If-1	Igh-1	Igh-2	Igh-3	Igh-4	Igh-5	Igh-Ars
Chromosome (read down)	7	9	14	UN	UN	2	2	3	11	17	7	2	UN	7	8	UN	10	1	3	12	12	12	12	12	12
WK/Re	a	a	1	1	1	a	.	.	f	.	s	.	.	.	.	.	.	.	.	b	2	.	b	.	.
WLHR/Le	.	.	.	.	.	.	.	.	d	.	d	.	.	.	.	.	.	b	.	.	.	.	.	.	.
WLL/BrA	.	.	.	.	.	.	.	.	.	.	d	1	a	.	.	.	.	a	h	a	.	.	.	.	.
WLL/BrAf	.	.	.	.	.	.	.	.	.	.	.	.	.	.	.	.	.	.	.	.	.	.	.	.	.
YBR/-	.	.	.	.	.	.	.	.	.	.	d	o	a	.	.	.	.	.	.	c	.	.	a	.	.
YS/Wf	.	.	.	.	.	.	.	.	.	.	s	.	a	.	.	.	l	b	.	.	.	.	.	.	.

1 = -a,-b   2 = -a,-c

```
Locus    ->    I I I I I   I I I I I   I I I I I   I I K K K   L L L L L
(read down)    g g g g g   g g g g g   g g g g g   g g f t t   a a a d d
               h h h h h   h h h h h   h h k k k   l l o h h   m m p h r
               - - - - -   - - - - -   - - - - -   - - - - -       b b - - -
               C D I L N   P S S S S   S S E P T   1 1 1 1 2   - - 1 1 1
                 e n e p     c a a a a   a r f c r     r           1 2
                 x u v         1 2 3 4   5 c 1   p
-------------------------------------------------------------------------------------
Chromosome     1 1 1 1 1   1 1 1 1 1   1 1 6 6 6   1 U 9 1 U   1 1 9 7 6
(read down)    2 2 2 2 2   2 2 2 2 2   2 2         6 N   7 N   2
-------------------------------------------------------------------------------------

WK/Re          . . . . .   . . . . .   . . . . .   . . . . .   . . . . .
WLHR/Le        . . . . .   . . . . .   . . . . .   . . . . .   . . . . .
WLL/BrA        . . . . .   . . . . .   . . . . .   . . . . .   . . . . .  a
WLL/BrAf       . . . . .   . . . . .   . . . . .   . . . . .   . . . . .
YBR/-          . . . . .   . . . . .   . . . . .   b . . . .   a b . . .

YS/Wf          . . . . .   . . . . .   . . . . .   . . . . .   . . . . .
```

Strain	Len-1	Len-2	Lfo-1	Lmp	Lna-1	Lpt-1	Lsh	Lth-1	Lth-2	Ltn-2	Ltw-2	Ltw-3	Ltw-4	Ltw-6	Lv	Lvp-1	Ly-1	Ly-2	Ly-3	Ly-5	Ly-6	Ly-7	Ly-8	Ly-9	Ly-10
Chromosome	1	UN	7	17	UN	UN	1	13	UN	UN	12	9	1	UN	4	6	19	6	6	1	15	12	UN	1	19
WK/Re	.	.	.	.	.	.	.	.	.	.	.	.	.	.	.	.	.	.	.	.	.	.	.	.	.
WLHR/Le	.	.	.	.	.	.	.	.	.	.	.	.	.	.	.	.	.	.	.	.	.	.	.	.	.
WLL/BrA	a	b	.	.	.	.	.	.	.	.	.	.	.	.	.	.	a	b	b	b	.	.	.	.	.
WLL/BrAf	.	.	.	.	.	.	.	.	.	.	.	.	.	.	.	.	.	.	.	.	.	.	.	.	.
YBR/-	a	b	.	.	.	.	.	a	.	b	.	a	b	a	.	.	.	.	.	.	.	.	.	.	.
YS/Wf	.	.	.	.	.	.	.	.	.	.	.	.	.	.	.	.	.	.	.	.	.	.	.	.	.

```
Locus    ->    L L L L L   L M M M M   M M M M M   N N N N N   N O O p P
(read down)    y y y y y   y a e l o   o o p p u   c e k k p   p r r   a
               b b b b b   b p p s d   d r h i p   a u - - -   - m m   n
                                                   m - 1 2 1   2 - - -
               3 4 5 6 7   8 1 1 1 1   2 1 1 1 1   1           1 2   1

---------------------------------------------------------------------------
Chromosome     U 4 U 4 1   7 5 1 1 9   7 5 7 9 4   9 1 1 2 1   1 4 4 7 4
(read down)    N   N   2       7                   7 7 4       4
---------------------------------------------------------------------------

WK/Re          . . . . .   . . . . b   . . . . b   . . . . a   . . . + .
WLHR/Le        . . . . .   . . . . a   . . . b 1   . b . . a   . . . + .
WLL/BrA        . . . . .   . b . . a   b a . b a   . b . . a   . . . . .
WLL/BrAf       . . . . .   . . . . .   . . . . .   . b . . .   . . . . .
YBR/-          . . . . .   . b . . a   . . . . b   . b . . .   . . . . .

YS/Wf          . . . . .   . b . . b   . . . . a   . b . . .   . . . . .
        1 = b?
```

Strain distribution of polymorphic variants

Locus -> (read down)	L	L	L	L	L	L	L	L	L	L	L	L	L	L	L	L	L	L	L	L	L	L	L	L	L
	y	y	y	y	y	y	y	y	y	y	y	y	y	y	y	y	y	y	y	y	y	y	y	y	y
	-	-	-	-	-	-	-	-	-	-	-	-	-	-	-	-	-	-	-	-	-	-	-	-	b
	1	1	1	1	1	1	1	1	1	2	2	2	2	2	2	2	2	2	2	3	3	3	3	4	-
	1	2	3	4	5	6	7	8	9	0	1	2	3	4	5	6	7	8	9	0	1	2	9	0	2
Chromosome (read down)	2	1	U	7	7	1	1	1	4	4	7	1	2	2	2	U	U	U	4	U	4	4	1	U	4
	9		N			2		2								N	N	N		N			7	N	
WK/Re	.	.	.	.	.	.	.	.	.	.	.	.	.	.	.	.	.	.	.	.	.	.	.	.	.
WLHR/Le	.	.	.	.	.	.	.	.	.	.	.	.	.	.	.	.	.	.	.	.	.	.	.	.	.
WLL/BrA	.	.	.	.	.	.	.	.	.	.	.	.	.	.	.	.	.	.	.	.	.	.	.	.	.
WLL/BrAf	.	.	.	.	.	.	.	.	.	.	.	.	.	.	.	.	.	.	.	.	.	.	.	.	.
YBR/-	.	.	.	.	.	.	.	.	.	.	.	.	b	.	.	.	.	.	.	.	.	.	.	.	.
YS/Wf	.	.	.	.	.	.	.	.	.	.	.	.	.	.	.	.	.	.	.	.	.	.	.	.	.

Locus → (read down)	Pan-2	Pca-1	Pd	Pep-2	Pep-3	Pep-4	Pep-7	Pgd	Pgk-1	Pgk-2	Pgm-1	Pgm-2	Pgm-3	Pk-3	Pre-2	Prn-p	Prt-1	Prt-2	Psp	Qa-1	Qa-2	Qa-3	Qa-4	Qa-5	Qed-1
Chromosome (read down)	UN	UN	UN	10	1	7	5	4	X	17	5	4	9	9	12	2	UN	8	2	17	17	17	17	17	17
WK/Re	.	.	.	.	b	.	.	.	b	.	b	a	.	.	.	.	.	.	.	.	.	.	.	.	.
WLHR/Le	.	.	.	b	.	.	.	.	b	.	b	a	.	.	.	.	.	.	.	.	.	.	.	.	.
WLL/BrA	.	.	.	.	c	.	.	b	b	a	b	a	.	.	.	.	.	.	.	.	.	.	.	.	.
WLL/BrAf	.	.	.	.	.	.	.	.	.	.	.	.	.	.	.	.	.	.	.	.	.	.	.	.	.
YBR/-	.	.	.	.	.	.	.	.	b	.	a	.	.	.	.	a	.	.	.	.	.	.	.	.	.
YS/Wf	.	.	.	b	.	.	.	.	.	.	a	.	.	.	.	.	.	.	.	.	.	.	.	.	.

Strain distribution of polymorphic variants

```
Locus     ->    r r R R R     S S S S S     S S S S S     S S S T T     T T T T T
(read down)     d d e i n     a a a d d     l o o o p     s v v a c     c g h i l
                  s n b 7     a a s h r     p a d d i     p p p m n     r n y n a
                  - - s       - - - -       - - -         - - - -       g   - d
                  1 1 -       1 1 1 1       1 2 2         1 2 1 2       1     1
                      6                         r

----------------------------------------------------------------------------------
Chromosome      5 1 1 1 6     7 7 1 2 U     1 U 1 1 1     U 2 7 7 1     1 1 9 1 1
(read down)       7   4                 N     7 N 6 7 2   N       1     3 5   2 7
----------------------------------------------------------------------------------

WK/Re           + . . . .     . . . . .     . . . . .     . . . . .     . . . . .
WLHR/Le         . . . . .     . . . . .     . . . . .     . . . . .     . . . . .
WLL/BrA         + + . . .     . . . . .     . . . . .     . a . s .     . . b . .
WLL/BrAf        . . . . .     . . . . .     . . . . .     . . . . .     . . . . .
YBR/-           . . a . .     a . . . .     a . . . .     . . . . .     . . . . .

YS/Wf           . . . . .     . . . . .     . . . . .     . . . . .     . . . . .
```

759

Locus → (read down)	Tpre	Trf	Tsuy	Tth	Ucp	Upg-1	Upg-2	Ups	Wap	Xmmv-9	Xmmv-42	Xmmv-43	Xmmv-44	Xmmv-45	Xmmv-46	Xmmv-47	Xmmv-48	Xmmv-49	Xmmv-50	Xmmv-51	Xmmv-52	Xmmv-53	Xmmv-54	Xmmv-55	Xmmv-56
Chromosome (read down)	12	9	12	12	8	17	1	9	UN	1	19	UN	UN	UN	UN	UN	UN	UN	12	UN	5	UN	UN	15	UN
WK/Re	.	b	.	.	.	.	.	.	.	.	.	.	.	.	.	.	.	.	.	.	.	.	.	.	.
WLHR/Le	.	b	.	.	.	.	.	.	.	.	.	.	.	.	.	.	.	.	.	.	.	.	.	.	.
WLL/BrA	.	b	.	.	.	.	.	.	.	.	.	.	.	.	.	.	.	.	.	.	.	.	.	.	.
WLL/BrAf	.	.	.	.	.	.	.	.	a	.	.	.	.	.	.	.	.	.	.	.	.	.	.	.	.
YBR/-	.	b	.	.	.	.	.	a	b	.	.	.	.	.	.	.	.	.	.	.	.	.	.	.	.
YS/Wf	.	.	.	.	.	.	.	.	.	.	.	.	.	.	.	.	.	.	.	.	.	.	.	.	.

Locus -> (read down)	X	X	X	X	X		X	X	X	X	X		X	X	X
	m	m	m	m	m		m	m	m	m	m		m	m	m
	m	m	m	m	m		m	m	m	m	m		m	m	m
	v	v	v	v	v		v	v	v	v	v		v	v	v
	-	-	-	-	-		-	-	-	-	-		-	-	-
	5	5	5	6	6		6	6	6	6	6		6	6	7
	7	8	9	0	2		3	4	5	6	7		8	9	0
Chromosome (read down)	U	U	U	U	4		U	U	3	U	U		U	U	U
	N	N	N	N			N	N		N	N		N	N	N

WK/Re	.	.	.	.	.		.	.	.	.	.		.	.	.
WLHR/Le	.	.	.	.	.		.	.	.	.	.		.	.	.
WLL/BrA	.	.	.	.	.		.	.	.	.	.		.	.	.
WLL/BrAf	.	.	.	.	.		.	.	.	.	.		.	.	.
YBR/-	.	.	.	.	.		.	.	.	.	.		.	.	.
YS/Wf	.	.	.	.	.		.	.	.	.	.		.	.	.

Table 17.2 Loci and their reference numbers

Loci	Reference number
a	74, 83, 87, 89, 117, 136, 178, 199, 222, 230
Aat	31, 161, 178, 214, 241
Abpa	35, 213
Acf-1	240
Aco-1	144, 148
Adh-1	83, 97
Adh-2	83
Adh-3e	36, 83, 95, 97, 117
Adh-3t	83, 95, 97
Ags	124, 126
Ah	18, 82, 83, 153, 198, 210, 220
Ahd-1	83, 94, 99, 243
Ahd-2	83, 221
Ahr-1	36
Akp-1	59, 178, 189, 204, 222, 242, 255
Akp-2	243
Alb-1	157, 178, 231
Amy-1	59, 86, 101, 104, 117, 140, 141, 158, 178, 189, 204, 230, 255
Amy-2	59, 104, 117, 189, 231
Aox-1	83, 96, 98, 122, 233
Aox-2	83, 96, 98
Aph-1	77
Apk	101, 158, 178, 247, 255
Apoa-1	40, 45, 59, 68, 117, 125, 178, 189, 204, 255, 256
As-1s	26, 27, 49
Av-1	176
Av-2	176
b	74, 83, 87, 89, 117, 136, 178, 199, 222, 230
B2m	67, 107, 130, 131, 197
Bcd-1	187
Bgl-e	13
Bgl-s	13, 52
Bgl-t	13
Bv-1	103
c	74, 83, 87, 89, 117, 178, 199, 222, 230
C3	29, 108, 117, 143, 150, 178, 236
C4	83, 198
C6	78, 79
Car-1	39, 117, 178, 189
Car-2	39, 45, 59, 87, 88, 89, 117, 138, 140, 141, 158, 178, 189, 222, 255
cdm	83, 117, 216
Ce-1	174
Ce-2	83, 87, 88, 89, 108, 117, 138, 143, 189, 222, 224, 255
Cfhe	151
Coh	253
Cv-1	103
d	22, 83, 103, 111, 117, 178, 199, 222, 230, 246
D7Rp2-s	48
D12Nyu1	31
D12Nyu2	31
D12Nyu3	31
D12Nyu4	31
Dlb-1	170
Ea-2	107, 198
Ea-4	107, 198
Ea-5	107, 198
Ea-6	107, 198

Table 17.2—*cont*.

Loci	Reference number
Ea-7	107, 172, 198
Eg	18, 64
Ema	182
Emv-4	103
Emv-5	103
Emv-6	103
Emv-7	103
Emv-8	103
Emv-9	103
Emv-10	103
Emv-11	103
Emv-12	103
Emv-13	103
Emv-14	103
Eph-1	127
Es-1	10, 59, 76, 83, 84, 87, 88, 89, 117, 138, 140, 141, 158, 177, 178, 180, 183, 189, 198, 222, 231, 255, 256
Es-2	76, 83, 84, 117, 140, 141, 158, 177, 178, 180, 183, 189, 198, 222, 226, 231, 255, 256
Es-3	59, 76, 83, 84, 87, 88, 89, 117, 140, 141, 158, 177, 178, 180, 183, 189, 198, 222, 231, 255, 256
Es-5	76, 82, 83, 84, 198, 231
Es-6	10, 42, 84, 149, 158, 165, 198
Es-7	83, 84, 226
Es-8	17, 83, 158
Es-9	10, 42, 84, 146, 149, 165, 185, 198, 238
Es-10	59, 84, 87, 89, 101, 113, 140, 141, 158, 178, 189, 198, 204, 248, 255
Es-11	59, 84, 164, 189, 226, 231, 255
Es-12	211
Es-13	40, 84, 252
Es-14	84
Es-16	84, 117, 228
Es-17	3, 84, 159, 160
Es-18	84, 225
Es-22	10, 42, 84
Es-23	84, 226
Es-24	10, 84, 225
Es-25	84, 227
Es-26	84, 229
Es-27	84
Esr	84
Fabpi	203
Fbp-1	147
Fbp-2	147
For-5	24
Fos	31
Fv-1	83, 117, 119, 198, 201, 215
Fv-2	83, 119, 198
Galt	144
Gbp-1	117, 173, 200
Gdc-1	101, 110, 117, 158, 178, 179, 189, 255
Gdc-2	111
Gdr-1	100
Gdr-2	100
Ggc	223
Gk	19
Glk	135
Glo-1	59, 83, 101, 108, 113, 133, 158, 178, 189, 256

Table 17.2—*cont.*

Table 17.2—*cont.*

Loci	Reference number
Gm-3	81
Got-1	30, 59, 83, 101, 117, 136, 140, 141, 158, 178, 179, 183, 189, 231, 255
Got-2	30, 59, 83, 117, 140, 141, 158, 178, 179, 180, 183, 189, 231, 255
Gpd-1	59, 83, 87, 88, 89, 138, 140, 141, 146, 158, 177, 178, 180, 183, 189, 198, 215, 222, 231, 255
Gpi-1	59, 76, 83, 87, 88, 89, 113, 138, 140, 141, 158, 177, 178, 180, 183, 189, 198, 222, 231, 255, 256
Gpt-1	41, 59, 101, 117, 136, 158, 178, 189, 204, 231, 255
Gr-1	59, 83, 113, 140, 141, 156, 158, 178, 204
Gus-r	18
Gus-s	18, 63, 87, 89, 114, 255
Gv-1	202
H-1	70, 71, 72, 198
H-2	1, 38, 55, 59, 70, 83, 87, 106, 108, 117, 138, 140, 141, 152, 186, 198, 209, 222, 224, 245
H-3	65, 70, 72, 198
H-4	70, 72, 198
H-7	70, 198
H-8	70, 198
H-9	70, 198
H-12	70, 198
H-13	70, 73, 198
Hao-1	83, 93
Hao-2	83, 93
Hba	61, 87, 89, 178, 180, 198, 236
Hba-4ps	32, 58
Hbb	59, 76, 82, 83, 87, 88, 89, 138, 140, 141, 154, 158, 177, 178, 180, 183, 189, 198, 222, 231, 236, 255, 256
Hc	33, 82, 83, 87, 89, 152, 175, 198, 222
Hex-1	44
Hma	154
Hp	8
Hrt-1	193
Hsd	6
Idh-1	59, 76, 83, 87, 88, 89, 117, 122, 138, 140, 141, 158, 177, 178, 180, 183, 189, 198, 222, 231, 255
If-1	83, 198
Igh-1	13, 55, 80, 83, 118, 134, 152, 161, 186, 198
Igh-2	75, 80, 198
Igh-3	75
Igh-4	75, 134, 178
Igh-5	75
Igh-Ars	234
Igh-C	7
Igh-Dex	75
Igh-Inu	75
Igh-Lev	234
Igh-Np	75
Igh-Pc	75
Igh-Sa1	75
Igh-Sa2	75
Igh-Sa3	75
Igh-Sa4	75
Igh-Sa5	75
Igh-Src	75
Igk-Ef1	75, 218
Igk-Pc	75
Igk-Trp	75
Igl-1	75
Igl-1r	75

Loci	Reference number
Kfo-1	45
Kth-1	45
Kth-2	45
Lamb-1	47
Lamb-2	47
Lap-1	40, 54, 178, 251
Ldh-1	113, 158, 178
Ldr-1	76, 83, 117, 140, 177, 178, 180, 183, 198, 222, 231, 255, 256
Len-1	195, 196
Len-2	193, 196
Lfo-1	45
Lmp	137
Lna-1	139
Lpt-1	142
Lsh	12
Lth-1	45, 49
Lth-2	45
Ltn-2	45
Ltw-2	45
Ltw-3	45, 50
Ltw-4	51, 68
Ltw-6	43
Lv	87, 89, 117, 180, 198
Lvp-1	142, 239
Ly-1	1, 82, 83, 87, 89, 107, 112, 130, 131, 139, 140, 141, 198
Ly-2	1, 82, 83, 87, 89, 107, 130, 131, 139, 140, 141, 198
Ly-3	1, 82, 83, 87, 89, 107, 130, 131, 139, 140, 141, 198, 218
Ly-5	82, 83, 130, 131, 139, 198
Ly-6	129, 130, 131, 139
Ly-7	130, 131, 139
Ly-8	130, 131, 139, 172
Ly-9	37, 109, 116, 128, 130, 131, 139, 172
Ly-10	130, 131, 139
Ly-11	139
Ly-12	130, 131, 139, 172
Ly-13	130, 131, 139, 172
Ly-14	130, 131, 139, 172
Ly-15	92, 130, 131, 139
Ly-16	55, 139
Ly-17	109, 130, 131, 139, 184
Ly-18	131, 139
Ly-19	131, 139
Ly-20	139
Ly-21	131, 139
Ly-22	139, 207
Ly-23	66, 139
Ly-24	21, 131, 139
Ly-25	91, 139
Ly-26	90, 131, 139
Ly-27	139
Ly-28	139
Ly-29	69, 131, 139
Ly-30	139
Ly-31	139, 208
Ly-32	139
Ly-39	131
Ly-40	131
Lyb-2	87, 89, 139, 184, 188
Lyb-3	139

Table 17.2—*cont.*

Loci	Reference number
Lyb-4	62, 139
Lyb-5	139
Lyb-6	139
Lyb-7	139
Lyb-8	139
Map-1	18, 101
Mep-1	11, 117
Mls-1	53, 198
Mod-1	40, 59, 83, 87, 88, 89, 111, 113, 117, 138, 140, 141, 158, 177, 178, 180, 183, 189, 198, 222, 231, 255
Mod-2	59, 83, 101, 136, 140, 141, 158, 178, 180, 183, 189, 198, 231
Mor-1	83, 158, 178, 183, 189, 190, 231
Mph-1	4, 5
Mpi-1	59, 76, 83, 111, 113, 155, 158, 178, 183, 189, 204
Mup-1	9, 45, 76, 83, 87, 88, 89, 117, 138, 143, 171, 178, 189, 198, 222, 255
Ncam	34
Neu-1	28, 59, 83, 108, 115, 167, 178, 180, 249, 250
Nk-1	169
Nk-2	169
Np-1	83, 140, 141, 158, 178, 183, 231, 248, 255
Np-2	14
Orm-1	8
Orm-2	8
P	87, 178, 198, 199, 222, 230
Pan-1	193
Pan-2	193
Pca-1	139, 198
Pd	25
Pep-2	113, 158
Pep-3	59, 83, 87, 88, 89, 109, 113, 138, 140, 141, 146, 158, 177, 178, 180, 189, 198, 212, 222, 255, 256
Pep-4	158, 194
Pep-7	113, 158
Pgd	16, 83, 113, 158, 231
Pgk-1	83, 101, 117, 178, 231
Pgk-2	38, 83, 108, 117, 178, 224
Pgm-1	59, 76, 83, 87, 88, 89, 136, 138, 140, 141, 146, 158, 177, 178, 180, 183, 189, 198, 204, 222, 231, 255
Pgm-2	59, 83, 113, 140, 141, 158, 177, 178, 180, 183, 189, 198, 204, 231, 255
Pgm-3	145
Pk-3	117, 163, 166
Pre-2	241
Prn-p	15
Prt-1	231, 232
Prt-2	231, 232
Psp	86
Qa-1	106, 108
Qa-2	106, 108, 257
Qa-3	106, 108, 257
Qa-4	56, 106
Qa-5	56, 106
Qed-1	120, 121
rd	82, 83, 177, 178, 179, 191, 199
rds	83
Ren-1	244
Rib-1	46
Rn7s-6	218
Saa	217
Saa-1	8

Table 17.2—*cont.*

Loci	Reference number
Sas-1	33, 152, 181, 254
Sdh-1	83, 178, 179, 237, 255
Sdr-1	20
Slp	198
Soa	123
Sod-1	23, 113, 143, 158, 231
Sod-2	143, 158
Spi-2r	85
Ssp	35, 105
Svp-1	83, 101, 198
Svp-2	198
Tam-1	178, 192, 255
Tcn-2	60, 61
Tcrg	162
Tgn	219
Thy-1	1, 82, 83, 87, 89, 112, 117, 130, 139, 140, 141, 198, 222
Tind	161
Tla	56, 57, 83, 87, 89, 106, 108, 198, 224
Tpre	161
Trf	40, 59, 76, 82, 83, 87, 89, 117, 136, 178, 189, 198, 222, 231, 255, 256
Tsu	161
Tthy	161
Ucp	102
Upg-1	32, 117, 204, 205
Upg-2	206
Ups	2, 3, 132
Wap	168
Xmmv-9	235
Xmmv-42	235
Xmmv-43	235
Xmmv-44	235
Xmmv-45	235
Xmmv-46	235
Xmmv-47	235
Xmmv-48	235
Xmmv-49	235
Xmmv-50	235
Xmmv-51	235
Xmmv-52	235
Xmmv-53	235
Xmmv-54	235
Xmmv-55	235
Xmmv-56	235
Xmmv-57	235
Xmmv-58	235
Xmmv-59	235
Xmmv-60	235
Xmmv-62	235
Xmmv-63	235
Xmmv-64	235
Xmmv-65	235
Xmmv-66	235
Xmmv-67	235
Xmmv-68	235
Xmmv-69	235
Xmmv-70	235

References

1. Acton, R.T., E.P. Blankenhorn, T.C. Douglas, R.D. Owen, J. Hilgers, H.A. Hoffman, and E.B. Boyse. 1973. Variations among sublines of inbred AKR mice. Nature New Biol. 245:8–10.

2. Antonucci, T.K., V.C. Chapman, and M.H. Meisler. 1982. Linkage of the structural gene for uroporphyrinogen I synthase to mark on mouse chromosome 9 in a cross between feral and inbred mice. Biochem. Genet. 20:703–710.

3. Antonucci, T.K., O.H. von Deimling, B.B. Rosenblum, L.C. Skow, and M.H. Meisler. 1984. Conserved linkage within a 4-cM region of mouse chromosome 9 and human chromosome 11. Genetics 107:463–475.

4. Archer, J.R. 1975. Inheritance of the macrophage alloantigenic marker (Mph-1) in inbred mice. Genet. Res. 26:213–219.

5. Archer, J.R. 1979. (SJL/J, CE/J, and C58/J typed for H-2 and Mph-1). Mouse News Lett. 60:33. (Personal communication)

6. Arfin, S.M., W.C. Hanford, and B.A. Taylor. 1979. Assignment of histidase regulating locus to chromosome 10 of the mouse. Biochem. Genet. 17:529–535.

7. Baum, C.M., J.P. McKearn, R. Riblet, and J.M. Davie. 1985. Polymorphism of Fc receptor on murine B cells is Igh-linked. J. Exptl. Med. 162:282–296.

8. Baumann, H., W.A. Held, and F.G. Berger. 1984. The acute phase response of mouse liver. Genetic analysis of the major acute phase reactants. J. Biol. Chem. 259:566–573.

9. Berger, F.G., and P. Szoka. 1981. Biosynthesis of the major urinary proteins in mouse liver: A biochemical genetic study. Biochem. Genet. 19:1261–1273.

10. Berning, W., S.M. de Looze, and O. von Deimling. 1985. Identification and development of a genetically closely-linked carboxylesterase family of the mouse liver. Comp. Biochem. Physiol. 80B:859–865.

11. Bond, J.S., R.J. Beynon, J.F. Reckelhoff, and C.S. David. 1984. Mep-1 gene controlling a kidney metalloendopeptidase is linked to the major histocompatibility complex in mice. Proc. Natl. Acad. Sci. USA 81:5542–5545.

12. Bradley, D.J. 1977. Regulation of Leishmania populations within the host II. Genetic control of acute susceptibility of mice to Leishmania donovani infection. Clin. Exptl. Immunol. 30:130–140.

13. Breen, G.A.M., A.J. Lusis, and K. Paigen. 1977. Linkage of genetic determinants for mouse beta-galactosidase electrophoresis and activity. Genetics 85:73–84.

14. Brenmer, T.A., E. Premkumar-Reddy, K. Nayar, and R.E. Kouri. 1978. Nucleoside phosphorylase 2 (Np-2) of mice. Biochem. Genet. 16:1143–1151.

15. Carlson, G.A., D.T. Kingsbury, P.A. Goodman, S. Coleman, S.T. Marshall, S. DeArmond, D. Westaway, and S.B. Prusiner. 1986. Linkage of prion protein and scrapie incubation time genes. Cell 46:503–511.

16. Chapman, V.M. 1975. 6-phosphogluconate dehydrogenase (PGD) genetics in the mouse: Linkage with metabolically related enzyme loci. Biochem. Genet. 13:849–856.

17. Chapman, V.M., E.A. Nichols, and F.H. Ruddle. 1974. Esterase-8 (Es-8): Characterization, polymorphism, and linkage of an erythrocyte esterase locus on chromosome 7 of *Mus musculus*. Biochem. Genet. 11:347–358.

18. Chapman, V.M., K. Paigen, L. Siracusa, and J.E. Womack. 1979. Biochemical variation, Mouse: Part II. Phenotypes and strain distribution. In P.L. Altman and D.D. Katz (eds.), Biolog. Handbooks III, Inbred and Genetically Defined Strains of Lab. Anims., Fed. Amer. Soc. Exptl. Biol. pp. 77–100.

19. Coleman, D.L. 1977. Genetic control of glucokinase activity in mice. Biochem. Genet. 15:297–305.

20. Coleman, D.L. 1980. Genetic control of serine dehydratase and phosphoenolpyruvate carboxykinase in mice. Biochem. Genet. 18:969–979.

21. Colombatti, A., E.N. Hughes, B.A. Taylor, and J.T. August. 1982. Gene for a major cell surface glycoprotein of mouse macrophages and other phagocytic cells is on chromosome 2. Proc. Natl. Acad. Sci. USA 79:1926–1929.

22. Corrow, D. 1981. The Jackson Laboratory, Bar Harbor, ME. (Gene distribution pattern of d locus). (Personal communication).

23. Cox, D.R. 1984. University of California, San Francisco. (Personal communication).

24. Cumming, R.B., M.F. Walton, J.C. Fuscoe, B.A. Taylor, J.E. Womack, and F.H. Gaertner. 1979. Genetics of formamidase-5 (brain formamidase) in the mouse: Localization of the structural gene on chromosome 14. Biochem. Genet. 17:415–431.

25. Dagg, C.P., D.L. Coleman, and G.M. Fraser. 1964. A gene affecting the rate of pyrimidine degradation in mice. Genetics 49:979–989.

26. Daniel, W.L. 1976. Genetic control of heat sensitivity and activity level of murine arylsulfatase B. Biochem. Genet. 14:1003–1018.

27. Daniel, W.L., and M.S. Caplan. 1980. Comparative biochemistry of murine arylsulfatase B. Biochem. Genet. 18:625–642.

28. Daniel, W.L., J.E. Womack, and P.S. Henthorn. 1981. Murine liver arylsulfatase B processing influenced by region on chromosome 17. Biochem. Genet. 19:211–225.

29. daSilva, F.P., G.H. Hoecker, N.K. Day, K. Vienne, and P. Rubinstein. 1978. Murine complement 3: Genetic variation and linkage to H-2. Proc. Nat. Acad. Sci. USA 75:963–965.

30. DeLorenzo, R.J., and F.H. Ruddle. 1970. Glutamate oxalate transaminase (GOT) genetics in *Mus musculus*: linkage, polymorphism, and phenotypes of the Got-2 and Got-1 loci. Biochem. Genet. 4:259–273.

31. D'Eustachio, P. 1984. A genetic map of mouse chromosome 12 composed of polymorphic DNA fragments. J. Exptl. Med. 160:827–838.

32. D'Eustachio, P., B. Fein, J. Michaelson, and B.A. Taylor. 1984. The alpha-globin pseudogene on mouse chromosome 17 is closely linked to H-2. J. Exptl. Med. 159:958–963.

33. D'Eustachio, P., T. Kristensen, R.A. Wetsel, R. Riblet, B.A. Taylor, and B.F. Tack. 1986. Chromosomal location of the genes encoding complement components C5 and factor H in the mouse. J. Immunol. 137:3990–3995.

34. D'Eustachio, P., G.C. Owens, G.M. Edelman, and B.A. Cunningham. 1985. Chromosomal location of the gene encoding the neural cell adhesion molecule (N-CAM) in the mouse. Proc. Natl. Acad. Sci. USA 82:7631–7635.

35. Dlouhy, S.R., and R.C. Karn. 1984. Multiple gene action determining a mouse salivary protein phenotype: Identification of the structural gene for androgen binding protein (Abp). Biochem. Genet. 22:657–667.

36. Duley, J.A., and R.S. Holmes. 1982. Biochemical genetics of aldehyde reductase in the mouse: Ahr-1—a new locus linked to the alcohol dehydrogenase gene complex on chromosome 3. Biochem. Genet. 20:1067–1083.

37. Durda, P.J., S.C. Boos, and P.D. Gottlieb. 1979. T100: a new murine cell surface glycoprotein detected by anti-Lyt-2.1 serum. J. Immunol. 1222:1407–1412.

38. Eicher, E.M., M. Cherry, and L. Flaherty. 1978. Autosomal phosphoglycerate kinase linked to mouse major histocompatibility locus. Molec. Gen. Genet. 158:225–228.

39. Eicher, E.M., R.H. Stern, J.E. Womack, M.T. Davis son, and T.H. Roderick. 1976. Evolution of mammalian carbonic anhydrase loci by tandem duplication: Close linkage of Car-1 and Car-2 to the centromere region of chromosome 3 of the mouse. Biochem. Genet. 14:651–660.

40. Eicher, E.M., B.A. Taylor, S.C. Leighton, and J.E. Womack. 1980. A serum protein determined on chromosome 9 of *Mus musculus*. Molec. Gen. Genet. 177:571–576.

41. Eicher, E.M., and J.E. Womack. 1977. Chromosomal location of soluble glutamic-pyruvic transaminase-1 (Gpt-1) in the mouse. Biochem. Genet. 15:1–8.

42. Eisenhardt, E., and O. von Deimling. 1982. Interstrain variation of Esterase-22, a new isozyme of the house mouse. Comp. Biochem. Physiol. 73B:719–724.

43. Elliott, R.W. 1979. (Ltw-6). Mouse News Lett. 61:59. (And personal communication, 1981).

44. Elliott, R.W., (cited by Chapman, Paigen, Siracusa & Womack). 1979. Hex-1; hexoseaminidase-1. In P.L. Altman and D.D. Katz, eds., Inbred and Genetically Defined Strains of Laboratory Animals. Fed. Amer. Soc. Exptl. Biol. Biological Handbook, 88. Bethesda, MD.

45. Elliott, R.W. 1981. Strain distribution patterns of several loci described by 2-D electrophoresis. (Personal communication).

46. Elliott, R.W. 1986. Assignment of pancreatic ribo-nuclease gene to mouse chromosome 14. Cytogenet. Cell Genet. 42:110–112.

47. Elliott, R.W., D. Barlow, and G.L.M. Hogan. 1985. Linkage of genes from Laminin B1 and B2 subunits on chromosome 1 in mouse. In Vitro Cell. & Devel. Biol. 21:477–484.

48. Elliott, R.W., and F.G. Berger. 1983. DNA sequence polymorphism in an androgen-regulated gene is associated with alteration in the encoded RNAs. Proc. Natl. Acad. Sci. USA 80:501–504.

49. Elliott, R.W., W.L. Daniel, B.A. Taylor, and E.K. Novak. 1985. Linkage of loci affecting a murine liver protein and arylsulfatase B to chromosome 13. J. Hered. 76:243–246.

50. Elliott, R.W., C. Hohman, C. Romejko, P. Louis, and F. Lilly. 1978. Use of high resolution two dimensional electrophoresis of liver cytosol proteins in the discovery and mapping of a new mouse variant. In Catsimpoolas (ed.) Electrophoresis '78, p 261–274. Elsevier, Amsterdam.

51. Elliott, R.W., C. Romejko, and C. Hohman. 1980. Mapping the gene for LTW-4, a 26,000 molecular weight major protein of mouse liver and kidney. Molec. Gen. Genet. 180:17–22.

52. Felton, J., M. Meisler, and K. Paigen. 1974. A locus determining beta-galactosidase activity in the mouse. J. Biol. Chem. 249:3267–3272.

53. Festenstein, H., and P. Demant. 1973. Workshop summary on genetic determinants of cell-mediated immune reactions in the mouse. Transplant. Proc. 5:1321–1327.

54. Finlay, M.F., and L.L. Huang. 1985. A new variant of serum leucine aminopeptidase in the mouse: its development and possible regulation. Biochem. Genet. 23:169–180.

55. Finnegan, A.L., and F.L. Owen. 1981. Ly-18, a new alloantigen present on cytotoxic T cells and controlled by gene(s) linked to the Igh-V locus. J. Immunol. 127:1947–1953.

56. Flaherty, L., and E. Rinchik. 1980. A new allele and antigen at the Tla locus. Immunogenetics 11:205–208.

57. Flaherty, L., K. Sullivan, and D. Zimmerman. 1977. The Tla locus: a new allele and antigenic specificity. J. Immunol. 119:571–575.

58. Fox, H.S., L.M. Silver, and G.R. Martin. 1984. An alpha globin pseudogene is located within the mouse t complex. Immunogenetics 19:125–130.

59. Fox, R.R., and R.F. Norberg. 1983. (Allelic characterization of Jackson Laboratory strains. Personal communication through Jackson Laboratory reports to date).

60. Fräter-Schröder, M., O. Haller, R. Gmür, L. Kierat, and S. Anastasi. 1982. Allelic forms of mouse transcobalamin 2. Biochem. Genet. 20:1001–1014.

61. Fräter-Schröder, M., M. Prochazka, O. Haller, F. Arwert, H.J. Porck, L.C. Skow, L.-G. Lundin, J. Hilkens, and J. Hilgers. 1985. Localization of the gene for vitamin B12 binding protein, transcobalamin II, near the

centromere on mouse Ch. 11, linked with the hemoglobin alpha chain locus. Biochem. Genet. 23:139–153.

62. Freund, J.G., A. Ahmed, R.E. Budd, M.E. Dorf, K.W. Sell, W.E. Vannier, and R.E. Humphreys. 1976. The L1210 leukemia cell bears a B Lymphocyte specific, non-H-2 alloantigen. J. Immunol. 117:1903–1905.

63. Gallagher, P.M., M.A. D'Amore, S.D. Lund, R.W. Elliott, J. Pazik, C. Hohman, T.R. Korfhagen, and R.E. Ganschow. 1987. DNA sequence variation with the beta-glucuronidase gene complex among inbred strains of mice. Genomics 1:145–152.

64. Ganschow, R., and K. Paigen. 1968. Glucuronidase phenotypes of inbred mouse strains. Genetics 9:335–349.

65. Gasser, D.L. 1976. Genetic studies on the H-3 region of mice and implications for polymorphism of histocompatibility loci. Immunogenetics 3:271–276.

66. Gasser, L., J. Ziebur, and K. Matsumoto. 1983. A lymphocyte antigen encoded by a locus closely linked to H-3 on the second chromosome of the mouse. Proc. Natl. Acad. Sci. USA 80:6620–6623.

67. Goding, J.W. 1981. Evidence for linkage of murine beta-2-microglobulin to H-3 and Ly-4. J. Immunol. 126:1644–1646.

68. Goldman, D., and H.J. Pikus. 1986. Fourteen genetically variant proteins of mouse brain: discovery of two new variants and chromosomal mapping of four loci. Biochem. Genet. 24:183–194.

69. Gonez, L.J., M.S. Sandrin, M.M. Henning, J. Hilgers, and I.F.C. McKenzie. 1985. Ly-29: A locus closely linked to Mtv-20 on chromosome 4 codes for a new mouse lymphocyte surface alloantigen. Immunogenetics 22:305–308.

70. Graff, R.J. 1970. Polymorphism of histocompatibility genes in the mouse. Transplant. Proc. 2:15–23.

71. Graff, R.J. 1979. Concerning H-1 and H-4 of certain strains of mice. (Personal communication).

72. Graff, R.J., and D.W. Bailey. 1973. The non-H-2 histocompatibility loci and their antigens. Transplant. Rev. 15:26–49.

73. Graff, R.J., D.H. Brown, and G.D. Snell. 1978. The alleles at the H-13 locus. Immunogenetics 7:413–423.

74. Green, E.L., (ed). 1968. Handbook on Genetically Standardized Jax Mice, 2nd Edition. The Jackson Laboratory. Bar Harbor, ME.

75. Green, M.C. 1979. Genetic nomenclature for the immunoglobulin loci of the mouse. Immunogenetics 8:89–97.

76. Groen, A. 1977. Identification and genetic monitoring of mouse inbred strains using biochemical polymorphisms. Lab Anim. 11:209–214.

77. Harada, Y.-N., J.-I. Hayakawa, and T. Takeshi. 1986. Antigenic polymorphism of a serum protein migrating electrophoretically in the alpha region detected by using a strain derived from the Japanese wild mouse (*Mus musculus molossinus*). Immunogenetics 24:47–50.

78. Hayakawa, J.-I., H. Nikaido, and T. Koizumi. 1984. Genetic polymorphism of the sixth component of complement (C6) in mice. Immunogenetics 20:633–638.

79. Hayakawa, J.-I., H. Nikaido, and T. Koizumi. 1985. Assignment of the gene locus for the sixth component (C6) of complement in mice to chromosome 15. Immunogenetics 22:637–642.

80. Herzenberg, L.A. 1964. A chromosome region for gamma-2a and beta-2a globulin H chain isoantigens in the mouse. Cold Spring Harbor Symp. Quant. Biol. 29:455–462.

81. Hibbs, M.L., P.M. Hogarth, R.A. Harris, and I.F.C. McKenzie. 1985. Gm-3.2, a new granulocyte/macrophage alloantigen. Immunogenetics 21:61–70.

82. Hilgers, J. 1976. (Inbred strains and markers.) Mouse News Lett. 54:30–31. (Personal communication)

83. Hilgers, J. 1982. (Distribution of alleles at marker loci in the strains maintained in the Amsterdam Colony). Mouse News Lett. 66:52. (Personal communication)

84. Hilgers J. and R. Poort-Keesom. 1986. (Strain distribution patterns). Mouse News Lett 76:14–26. (Personal communication)

85. Hill, R.E., P.H. Shaw, R.K. Barth, and N.D. Hastie. 1985. A genetic locus closely linked to a protease inhibitor gene complex controls the level of multiple RNA transcripts. Mol. Cell. Biol. 5:2114–2122.

86. Hjorth, J.P. 1979. (Research news). Mouse News Lett. 62:41. (Personal communication)

87. Hoffman, H.A. 1978. Genetic quality control of the laboratory mouse. In HC Morse III (ed.), Origins of Inbred Mice, 217–234. Academic Press, NY.

88. Hoffman, H.A. 1981. (Personal communication to George Wolff concerning his typing of strains VY/Wf and YS/Wf).

89. Hoffman, H.A., K.T. Smith, J.S. Crowell, T. Nomura, and T. Tomita. 1980. Genetic quality control of laboratory animals with emphasis on genetic monitoring. 7th ICLAS Symp., Utrecht 1979, p 307–317. Gustav Fischer Verlag, Stuttgart, NY.

90. Hogarth, P.M., S.E. Latham, and I.F.C. McKenzie. 1984. A new murine alloantigen: Ly-26.1. Immunogenetics 19:355–358.

91. Hogarth, P.M., A. Rigby, V.R. Sutton, I.F.C. McKenzie, and J. Hilgers. 1984. The Ly-25.1 specificity: definition with a monoclonal antibody. Immunogenetics 19:83–86.

92. Hogarth, P.M., I.D. Walker, I.F.C. McKenzie, and T.A. Springer. 1985. The Ly-15 alloantigenic system: a genetically determined polymorphism of the murine lymphocyte function-associated antigen-1 molecule. Proc. Natl. Acad. Sci. USA 82:526–530.

93. Holmes, R.S. 1977. The genetics of alpha-hydroxyacid oxidase and alcohol dehydrogenase in the mouse: Evidence for multiple gene loci and linkage between Hao-2 and Adh-3. Genetics 87:709–716.

94. Holmes, R.S. 1978. Genetics and ontogeny of aldehyde dehydrogenase isozymes in the mouse: Localization of

Ahd-1 encoding the mitochondrial isozyme on chromosome 4. Biochem. Genet. 16:1207–1218.

95. Holmes, R.S. 1979. Genetics and ontogeny of alcohol dehydrogenase isozymes in the mouse: Evidence for a cis-acting regulator gene (Adt-1) controlling C2 isozyme expression in reproductive tissues and close linkage of Adh-3 and Adt-1 on chromosome 3. Biochem. Genet. 17:461–472.

96. Holmes, R.S. 1979. Genetics, ontogeny, and testosterone inducibility of aldehyde oxidase isozymes in the mouse: Evidence for two genetic loci (Aox-1 and Aox-2) closely linked on chromosome 1. Biochem. Genet. 17:517–527.

97. Holmes, R.S., J.A. Duley, and S. Imai. 1982. Alcohol dehydrogenase isozymes in the mouse: genetic regulation, allelic variation among inbred strains and sex differences of liver and kidney A2 isozyme activity. Blood Groups and Biochem. Genet. 13:97–108.

98. Holmes, R.S., L.R. Leijten, and J.A. Duley. 1981. Liver aldehyde oxidase and xanthine oxidase genetics in the mouse. Anim. Blood Groups and Biochem. Genet. 12:193–199.

99. Holmes, R.S., and G.P. Timms. 1981. Mouse aldehyde dehydrogenase genetics: Positioning of Ahd-1 on chromosome 4. Anim. Blood Groups and Biochem. Genet. 12:1–5.

100. Hutton, J.J. 1971. Genetic regulation of glucose 6-phosphate dehydrogenase activity in the inbred mouse. Biochem. Genet. 5:315–331.

101. Hutton, J.J. 1978. Biochemical polymorphisms—detection, distribution, chromosomal location, and applications. p. 235–254, *In* HC Morse III (ed.), Origins of Inbred Mice. Academic Press, NY.

102. Jacobsson, A., U. Stadler, M.A. Glotzer, and L.P. Kozak. 1985. Mitochondrial uncoupling protein from mouse brown fat. J. Biol. Chem. 260:16250–16254.

103. Jenkins, N.A., N.G. Copeland, B.K. Lee, and B.A. Taylor. 1982. Organization, distribution, and stability of endogenous ectropic murine leukemia virus DNA sequences in chromosomes of *Mus musculus*. J. Virol. 43:26–36.

104. Kaplan, R.D., V.M. Chapman, and F.H. Ruddle. 1973. Electrophoretic variation of alpha-amylase in two inbred strains of *Mus musculus*. J. Hered. 64:155–157.

105. Karn, R.C., S.R. Dlouhy, K.R. Springer, J.P. Hjorth, and J.T. Nielsen. 1982. Sex-limited genetic variation in mouse salivary protein. Biochem. Genet. 20:493–504.

106. Kincade, P.W., L. Flaherty, G. Lee, T. Watanabe, and J. Michaelson. 1980. Distribution and ontogeny of Qa antigen expression. J. Immunol. 124:2879–2885.

107. Klein, J. 1975. Biology of the Mouse Histocompatibility-2 Complex. Springer-Verlag, NY.

108. Klein, J., L. Flaherty, J.L. VandeBerg, and D.C. Shreffler. 1978. H-2 haplotypes, genes, regions and antigens: First listing. Immunogenetics 6:489–512.

109. Kozak, C.A., W.F. Davidson, and H.C. Morse III. 1984. Genetic and functional relationships of the retroviral and lymphocyte alloantigen loci on mouse chromosome 1. Immunogenetics 19:163–168.

110. Kozak, L.P. 1972. Genetic control of alpha-glycerolphosphate dehydrogenase in mouse brain. Proc. Nat. Acad. Sci. USA 69:3170–3174.

111. Kozak, L.P., D.L. Burkart, and J.P. Hjorth. 1982. Unlinked structural genes for the developmentally regulated isozymes of sn-glycerol-3-phosphate dehydrogenase in mice. Dev. Genet. 3:1–6.

112. Krog, H.-H., M. Wysocka, P. Kisielow, and C.Z. Radzikowski. 1977. Genetic characterization of important mouse strains used in cancer research. Hereditas 87:201–204.

113. Lalley, P.A. 1983. Allelic variants of alpha-L-fucosidase in the mouse. (Personal communication)

114. Lalley, P.A., and T.B. Shows. 1974. Lysosomal and microsomal glucuronidase: genetic variant alters electrophoretic mobility of both hydrolases. Science 185:442–444.

115. Lalley, P.A., and T.B. Shows. 1977. Lysosomal acid phosphatase deficiency: Liver specific variant in the mouse. Genetics 87:305–317.

116. Ledbetter, J.A., J.W. Goding, T.T. Tsu, and L.A. Herzenberg. 1979. A new mouse lymphoid alloantigen (Lgp 100) recognized by a monoclonal rat antibody. Immunogenetics. 8:347–360.

117. Leiter, E.H. 1986. Genetic typing of 49 loci for strains NOD/J and NON/J. (Personal communication).

118. Lieberman, R. 1978. Genetics of IgCH (allotype) locus in the mouse. Springer Seminars Immunopath. 1:7–30.

119. Lilly, F., and T. Pincus. 1973. Genetic control of murine viral leukemogenesis. Advan. Cancer Res. 17:231–277.

120. Lindahl, K.F. 1979. (A new locus in the Tla region). Mouse News Lett. 60:33–34. (Personal communication).

121. Lindahl, K.F. 1981. Strain distribution pattern for Qed-1. (Personal communication).

122. Lush, I.E. 1978. Genetic variation of some aldehyde-oxidizing enzymes in the mouse. Anim. Blood Groups Biochem. Genet. 9:85–96.

123. Lush, I.E. 1981. The genetics of tasting in mice. I. Sucrose octaacetate. Genet. Res. 38:93–95.

124. Lusis, A.J., and K. Paigen. 1975. Genetic determination of the alpha-galactosidase developmental program in mice. Cell 6:371–378.

125. Lusis, A.J., B.A. Taylor, R.W. Wangenstein, and R.C. LeBeuf. 1983. Genetic control of lipid transport in mice. II. Genes controlling structure of high density lipoproteins. J. Biol. Chem. 258:5071–5078.

126. Lusis, A.J., and J.D. West. 1976. X-linked inheritance of a structural gene for alpha-galactosidase in *Mus musculus*. Biochem. Genet. 14:849–855.

127. Lyman, S.D., A. Pland, and B.A. Taylor. 1980. Genetic polymorphism of microsomal epoxide hydrolase activity in the mouse. J. Biol. Chem. 18:8650–8654.

128. Mathieson, B.J., S.O. Sharrow, K. Bottomly, and B.J. Fowlkes. 1980. Ly9, an alloantigenic marker of lymphocyte differentiation. J. Immunol. 125:2127–2136.

129. McKenzie, I.F.C., J. Gardiner, M. Cherry, and G.D. Snell. 1977. Lymphocyte antigens: Ly-4, Ly-6, and Ly-7. Transplant. Proc. 9:667–669.

130. McKenzie, I.F.C., P.M. Hogarth, and T.A. Potter. 1981. Further description of Ly allo-antigens. In N.L. Warner and D. Metcalf (eds.)., Workshop in Leukaemias. UICC Technical Report Series, 61/G:14.

131. McKenzie, I.F.C., and P.M. Hogarth. 1983. Strain distribution patterns for loci controlling lymphocyte antigens. (Personal communication).

132. Meisler, M.H., and M.L.C. Carter. 1980. Rare structural variants of human and murine uroporphyrinogen I synthase. Proc. Natl. Acad. Sci. USA 77:2848–2852.

133. Meo T., T. Douglas, and A.-M. Rijnbeek. 1977. Glyoxalase 1 polymorphism in the mouse: a new genetic marker linked to H-2. Science 198:311–312.

134. Minna, J.D., G.M. Michaelson, and L.A. Herzenberg. 1967. Identification of a gene locus for gamma G1 immunoglobulin H chains and its linkage to the H chain chromosome region in the mouse. Proc. Nat. Acad. Sci. USA 58:188–194.

135. Mishkin, J.D., B.A. Taylor, and W.J. Mellman. 1976. Glk: a locus controlling galactokinase activity in the mouse. Biochem. Genet. 14:635–640.

136. Mobraaten, L.E. 1979–1980. Concerning strains RIIIS/J, formerly RIII/2J. (Personal communication).

137. Monaco, J.J., and H.O. McDevitt. 1986. The LMP antigens: a stable MHC-controlled multisubunit protein complex. Human Immunol. 15:416–426.

138. Morse, H.C. III. 1981. Genetic characterization of strain F/St. (Personal communication).

139. Morse, H.C. III, F.-W. Shen, and U. Hämmerling. 1987. Genetic nomenclature for loci controlling mouse lymphocyte antigens. Immunogenetics 25:71–78.

140. Murphy, E.D., and J.B. Roths. 1978. (A new congenic inbred strain, MRL/Mp). Mouse News Lett. 58:51. (Personal communication).

141. Murphy, E.D., and J.B. Roths. 1978. (A new recombinant inbred strain, BXSB/Mp). Mouse News Lett. 58:51–52. (Personal communication).

142. Nadeau, J.H. 1981. (Personal communication).

143. Nadeau, J.H. 1982–1983. Typing of various loci and strains. (Personal communication).

144. Nadeau, J.H., and E.M. Eicher. 1982. Conserved linkage of soluble aconitase and galactose-1-phosphate uridyl transferase in mouse and man: Assignment of these genes to mouse chromosome 4. Cytogenet. Cell Genet. 4:271–281.

145. Nadeau, J.H., J. Kompf, G. Siebert, and B.A. Taylor. 1981. Linkage of Pgm-3 in the house mouse and homologies of three phosphoglucomutase loci in mouse and man. Biochem. Genet. 19:465–474.

146. Nash, H.R. 1981. Information on various isozymes. (Personal communication through J. Staats).

147. Nash, H.R., and P.R. Challinor. 1985. Fructose bisphosphatase isozymes of the mouse. I. Inheritance. Biochem. Genet. 23:499–510.

148. Nash, H.R., and M.F. Newton. 1980. (Aconitase). Mouse News Lett. 62:57–58. (Personal communication).

149. Nash, H.R., and O. von Deimling. 1982. Kidney esterases of *Mus musculus*: Further polymorphism of Esterase-6, Esterase-9, and a new esterase, Esterase-20. Biochem. Genet. 20:537–554.

150. Natsuume-Sakai, J., J.-I. Hayakawa, and M. Takahaski. 1978. Genetic polymorphism of murine C3 controlled by a single co-dominant locus on chromosome 17. J. Immunol. 121:491–498.

151. Natsuume-Sakai, S., K. Sudoh, T. Kaidoh, J.-I. Hayakawa, and M. Takahashi. 1985. Structural polymorphism of murine complement factor H controlled by a locus located between the Hc and B2m locus on the second chromosome of the mouse. J. Immunol. 134:2600–2606.

152. Naylor, D.H., and B. Cinader. 1970. Inheritance, hormonal regulation and properties of polymorphic murine antigens Mud1 and Mud2. Intern. Arch. Allergy Appl. Immunol. 39:511–539.

153. Nebert, D.W., N.M. Jensen, H. Shinozuka, H.W. Kunz, and T.J. Gill III. 1982. The Ah phenotype survey of forty-eight rat strains and twenty inbred mouse strains. Genetics 100:79–87.

154. Nesbitt, M.N., B. Bakay, M.B. Gardner, and C. Day. 1979. Isoenzyme pattern of HPRT in murine erythrocytes: Control by an autosomal locus. Biochem. Genet. 17:957–964.

155. Nichols, E.A., V.M. Chapman, and F.H. Ruddle. 1973. Polymorphism and linkage of mannosephosphate isomerase in *Mus musculus*. Biochem. Genet. 8:47–53.

156. Nichols, E.A., and F.H. Ruddle. 1975. Polymorphism and linkage of glutathione reductase in *Mus musculus*. Biochem. Genet. 13:323–329.

157. Nichols, E.A., F.H. Ruddle, and M.L. Petras. 1975. Linkage of the locus for serum albumin in the house mouse, *Mus musculus*. Biochem. Genet. 13:551–555.

158. O'Brien, S.J., J.L. Moore, M.A. Martin, and J.E. Womack. 1982. Evidence for the horizontal acquisition of murine AKR virogenes by recent horizontal infection of the germ line. J. Exptl. Med. 155:1120–1132.

159. Otto, J., and O. von Deimling. 1981. (Esterase-17 on chromosome 9). Mouse News Lett. 64:53. (Personal communication).

160. Otto, J., and O.H. von Deimling. 1983. Esterase-17 (Es-17): Characterization and genetic location on chromosome 9 of a bis-p-nitrophenyl phosphate-resistant esterase of the house mouse (*Mus musculus*). Biochem. Genet. 21:37–48.

161. Owen, F.L., and R. Riblet. 1984. Genes for the mouse T cell allo-antigens Tpre, Tthy, Tind, and Tsu are closely linked near Igh on chromosome 12. J. Exptl. Med. 159:313–317.

162. Owen, F.L., B.A. Taylor, A. Zweidler, and J.G. Seidman. 1986. The murine gama-chain of the T cell receptor is closely linked to a spermatocyte specific histone gene and the beige coat color locus on chromosome 13. J. Immunol. 137:1044–1046.

163. Peters, J., and S.J. Andrews. 1984. The Pk-3 gene determines both the heart, M1, and the kidney, M2, pyruvate kinase isozymes in the mouse; and a simple electrophoretic method for separating phosphoglucomutase-3. Biochem. Genet. 22:1047–1063.

164. Peters, J., and H.R. Nash. 1977. Polymorphism of Esterase 11 in *Mus musculus*, a further esterase locus on chromosome 8. Biochem. Genet. 15:217–226.

165. Peters, J., and H.R. Nash. 1978. Esterases of *Mus musculus*: Substrate and inhibition characteristics, new isozymes, and homologies with man. Biochem. Genet. 16:553–569.

166. Peters, J., H.R. Nash, E.M. Eicher, and G. Bulfield. 1981. Polymorphism of kidney pyruvate kinase in the mouse is determined by a gene, Pk-3, on chromosome 9. Biochem. Genet. 19:757–769.

167. Peters, J., D.M. Swallow, S.J. Andrews, and L. Evans. 1981. A gene (Neu-1) on chromosome 17 of the mouse affects acid alpha-glucosidase and codes for neuraminidase. Genet. Res. 38:47–55.

168. Piletz, J.E., and R.E. Ganschow. 1981. Genetic variation of milk proteins in mice. Biochem. Genet. 19:1023–1029.

169. Pollack, S.B., and S.L. Emmons. 1982. NK-2.1: an NK-associated antigen detected with NZB anti-BALB/c serum. J. Immunol. 129:2277–2281.

170. Ponder, B.A.J., M.F.W. Festing, and M.M. Wilkinson. 1985. An allelic difference determines reciprocal patterns of expression of binding sites for Dolichos biflorus lectin in inbred strains of mice. J. Embryol. Exp. Morph. 87:229–239.

171. Potter, M., J.S. Finlayson, D.W. Bailey, E.B. Mushinski, and B.L. Reamer, J.L. Walters. 1973. Major urinary protein and immunoglobulin allotypes of recombinant inbred mouse strains. Genet. Res. 22:325–328.

172. Potter, T.A., and I.F.C. McKenzie. 1981. Identification of new murine lymphocyte alloantigens: antisera prepared between C57L, 129 and related strains define new loci. Immunogenetics 12:351–369.

173. Prochazka, M., P. Staeheli, R.S. Holmes, and O. Haller. 1985. Interferon-induced guanylate-binding proteins: mapping of the murine Gbp-1 locus to chromosome 3. Virology 145:273–279.

174. Rechcigl, M. Jr., and W.E. Heston. 1963. Tissue catalase activity in several C57BL substrains and in other strains of inbred mice. J. Nat. Cancer Inst. 30:855–864.

175. Rhodes, J.C., L.S. Wicker, and W.J. Urba. 1980. Genetic control of susceptibility to Cryptococcus neoformans in mice. Infect. and Immun. 29:494–499.

176. Risser, R., M. Potter, and W.P. Rowe. 1978. Abelson virus-induced lymphomagenesis in mice. J. Exptl. Med. 148:714–726.

177. Roderick, T.H. 1978. Further information on subline differences. *In* H.C. Morse III (ed.), Origins of Inbred Mice, p 485–494. Academic Press, NY.

178. Roderick, T.H., S.H. Langley and E. Holt. 1978–1987. (Personal communication).

179. Roderick, T.H., S.H. Langley, and E. Akeson. 1979. (Further genetic characterization of inbred strains of mice). Mouse News Lett. 61:43–45. (Personal communication).

180. Roderick, T.H., F.H. Ruddle, V.M. Chapman, and T.B. Shows. 1971. Biochemical polymorphisms in feral and inbred mice (*Mus musculus*). Biochem. Genet. 5:457–466.

181. Rosenstreich, D.L., M.G. Groves, H.A. Hoffman, and B.A. Taylor. 1978. Location of the Sas-1 locus on mouse chromosome 1. Immunogenetics 7:313–320.

182. Rubinstein, P., N. Liu, E.W. Streun, and F. Decary. 1974. Electrophoretic mobility and agglutinability of red blood cells: A 'new' polymorphism in mice. J. Exptl. Med. 139:313–322.

183. Ruddle, F.H. 1973. (Biochemical markers of some of the Amsterdam inbred strains). Mouse News Lett. 48:20. (Personal communication).

184. Sato, H., S. Kimura, and K. Itakura. 1981. Genetic and serologic re-evaluation of LyM-1 antigen specified by a gene closely linked to Mls: establishment of LyM-1 antigen of the mouse. J. Immunogenet. 8:27–40.

185. Schollen, J., K. Bender, and O. von Deimling. 1975. Esterase. XXI. Es-9, a possibly new polymorphic esterase in *Mus musculus* genetically linked to Es-2. Biochem. Genet. 13:369–377.

186. Schrier, D.J., J.L. Sternick, E.M. Allen, and V.L. Moore. 1982. Immunogenetics of BCG-induced anergy in mice: Control by genes linked to the Igh complex. J. Immunol. 128:1466–1469.

187. Seeley, T.-L., and R.S. Holmes. 1981. Genetics and ontogeny of butytyl CoA dehydrogenase in the mouse and linkage of Bcd-1 with Dao-1. Biochem. Genet. 19:333–345.

188. Shen, F.-W., M. Spanondis, and E.A. Boyse. 1977. Multiple alleles of the Lyb-2 locus. Immunogenetics 5:481–484.

189. Shi, S. 1987. Genetic profile of Chinese inbred mouse strains. Proc. CAMS and PUMC 2(3):168–171.

190. Shows, T.B., V.M. Chapman, and F.H. Ruddle. 1970. Mitochondrial malate dehydrogenase and malic enzyme: Mendelian inherited electrophoretic variants in the mouse. Biochem. Genet. 4:707–718.

191. Sidman, R.L., and M.C. Green. 1965. Retinal degeneration in the mouse. J. Hered. 56:23–29.

192. Skow, L.C. 1978. Genetic variation at a locus (Tam-1) for submaxillary gland protease in the mouse and its location on chromosome 7. Genetics 90:713–724.

193. Skow, L.C. 1979. (Six new genes identified by isoelectric focusing.) Mouse News Lett. 61:55. (And personal communication).

194. Skow, L.C. 1981. Genetic variation for prolidase (Pep-4) in the mouse maps near the gene for glucosephosphate isomerase (Gpi-1) on chromosome 7. Biochem. Genet. 19:695–700.

195. Skow, L.C. 1982. Experimental location of a gene con-

trolling electrophoretic variation in mouse gamma crystallins. Exptl. Eye Res. 4:509–516.

196. Skow, L.C., M.D. Donner, R.A. Popp, and E.G. Bailiff. 1985. A second polymorphic lens crystallin (Len-2) in the mouse: genetic and biochemical analysis of Len-1 and Len-2. Biochem. Genet. 23:181–189.

197. Snell, G.D., M. Cherry, I.F.C. McKenzie, and D.W. Bailey. 1973. Ly-4, a new locus determining a lymphocyte cell-surface alloantigen in mice. Proc. Natl. Acad. Sci. USA 70:1108–1111.

198. Staats, J. 1976. Standardized nomenclature for inbred strains of mice: Sixth listing. Cancer Res. 36:4333–4377.

199. Staats, J. 1981. List of inbred strains. *In* M.C. Green (ed.), Genetic Variants and Strains of the Laboratory Mouse. Gustav Fischer Verlag, Stuttgart, p 373–376.

200. Staeheli, P., M. Prochazka, P.S. Steigmeier, and O. Haller. 1984. Genetic control of interferon action: mouse strain distribution and inheritance of an induced protein with guanylate-binding property. Virology 137:135–142.

201. Stephenson, J.R., and S.A. Aaronson. 1979. Induction of an endogenous B-tropic type C RNA virus from SWR/J mouse embryo cells in tissue culture. Virology 70:352–359.

202. Stockert, E., L.J. Old, and E.A. Boyse. 1971. The G(IX) system, a cell surface allo-antigen associated with murine leukemia virus: Implications regarding chromosomal integration of the viral genome. J. Exptl. Med. 133:1334–1344.

203. Sweetser, D.A., E.H. Birkenmeier, I.J. Klisak, S. Zollman, R.S. Sparkes, T. Mohandas, A.J. Lusis, and J.I. Gordon. 1987. The human and rodent intestinal fatty acid binding protein genes. J. Biol. Chem. 262:16060–16071.

204. Szymura, J.M., and J. Klein. 1981. (Mouse Urinary pepsinogen: a new locus Upg-1 linked to the H-2 complex on chromosome 17). Mouse News Lett. 64:93–94. (Personal communication).

205. Szymura, J.M., and J. Klein. 1981. Linkage of a gene controlling urinary pepsinogen with the major histocompatibility complex of the mouse. Immunogenetics 13:267–271.

206. Szymura, J.M., B.A. Taylor, and J. Klein. 1982. Upg-2: A urinary pepsinogen variant located on chromosome 1 of the mouse. Biochem. Genet. 20:1211–1219.

207. Tada, N., S. Kimura, Y. Liu-Lam, and U. Hämmerling. 1983. Mouse alloantigen system Ly-m22 predominantly expressed on T lymphocytes and controlled by a gene linked to Mls region on chromosome 1. Hybridoma 2:29–38.

208. Tada, N., S. Kimura, Y. Liu-Lam, and U. Hämmerling. 1984. A new lymphocyte surface antigen (Ly-m31) controlled by a gene closely linked to the Akp-2 locus on mouse chromosome 4. Immunogenetics 20:589–592.

209. Takahashi, S., Y. Fukuoka, K. Moriwaki, T. Okuda, T. Tachibana, S. Natsuume-Sakai, and M. Takahashi. 1984. Structural polymorphism of mouse complement C2

detected by microscale peptide mapping: linkage to H-2. Immunogenetics 19:493–501.

210. Taylor, B.A. 1971. Strain distribution and linkage tests of 7,12-dimethylbenzanthracene (DMBA) inflammatory response in mice. Life Sci. 10:1127–1134.

211. Taylor, B.A. 1977. (New esterase variant, Es-12). Mouse News Lett. 56:41. (Personal communication).

212. Taylor, B.A. 1979. (Dip-1 data, now Pep-3). Mouse News Lett. 61:42. (Personal communication).

213. Taylor, B.A. 1982. (Salivary protein variant, Sal-1 on chromosome 7). Mouse News Lett. 66:66. (Personal communication).

214. Taylor, B.A., D.W. Bailey, M. Cherry, R. Riblet, and M. Weigert. 1975. Genes for immunoglobulin heavy chain and serum prealbumin protein are linked in mouse. Nature 256:644–646.

215. Taylor, B.A., H.G. Bedigian, and H. Meier. 1977. Genetic studies of the Fv-1 locus of the mouse: linkage with Gpd-1 in recombinant inbred lines. J. Virol. 23:106–109.

216. Taylor, B.A., H.J. Heiniger, and H. Meier. 1973. Genetic analysis of resistance to cadmium-induced testicular damage in mice (37380). Proc. Soc. Exptl. Biol. Med. 143:629–633.

217. Taylor, B.A., and L. Rowe. 1984. Genes for serum Amyloid A proteins map to chromosome 7. Mol. Gen. Genet. 195:491–499.

218. Taylor, B.A., L. Rowe, D.M. Gibson, R. Riblet, R. Yetter, and P.D. Gottlieb. 1985. Linkage of a 7S RNA sequence and kappa light chain genes in the mouse. Immunogenetics 22:471–481.

219. Taylor, B.A., and L. Rowe. 1987. The congenital goiter mutation is linked to the thyroglobulin gene in the mouse. Proc. Nat. Acad. Sci. USA 84:1986–1990.

220. Thomas, P.E., J.J. Hutton, and B.A. Taylor. 1973. Genetic relationship between aryl hydrocarbon hydroxylase inducibility and chemical carcinogen induced skin ulceration. Genetics 74:655–659.

221. Timms, G.P., and R.S. Holmes. 1981. Genetics of aldehyde dehydrogenase isozymes in the mouse: evidence for multiple loci and localization of Ahd-2 on chromosome 19. Genetics 97:327–336.

222. Tomita, T. 1983. Genetic monitoring. *In* K. Kondo, T. Tomita, K. Esaki, and J. Hayakawa (eds.) Genetic Control of Laboratory Animals, p 75–94. Soft Science, Inc., Tokyo, Japan. (In Japanese).

223. Tulchin, N., and B.A. Taylor. 1981. Gamma-glutalmyl cyclotransferase: A new genetic polymorphism in the mouse (*Mus musculus*) linked to Lyt-2. Genetics 99:109–116.

224. VandeBerg, J.L., and J. Klein. 1978. Localization of mouse Pgk-2 gene at the D end of the H-2 complex. J. Exptl. Zool. 203:319–324.

225. von Deimling, O. 1982. (Strain distribution patterns of alleles for Es-18 and Es-24). Mouse News Lett. 67:15–16. (Personal communication).

226. von Deimling, O.H. 1984. Esterase-23 (Es-23): Charac-

terization of a new carboxylesterase isozyme (EC 3.1.1.1) of the house mouse, genetically linked to Es-2 on chromosome 8. Biochem. Genet. 22:769–782.

227. von Deimling, O.H., and B.A. Taylor. 1987. Esterase-25 (Es-25): Identification and characterization of a new kidney arylesterase of the house mouse, genetically linked to Ly-18 on chromosome 12. Biochem. Genet. 25:639–646.

228. von Deimling, O.H., P. Schupp, and J. Otto. 1981. Esterase-16 (Es-16): Characterization, polymorphism, and linkage to 3 of a kidney esterase locus of the house mouse. Biochem. Genet. 19:1091–1099.

229. von Deimling, O.H., B. Wassmer, and M. Muller. 1984. Esterase-26 (Es-26): characterization and genetic location on chromosome 3 of an eserine-sensitive esterase of the house mouse (*Mus musculus*). Biochem. Genet. 22:1119–1126.

230. Watanabe, T. 1982. (Personal communication).

231. Watanabe, T., T. Ito, and N. Ogasawara. 1982. Biochemical markers of three strains derived from Japanese wild mouse (*Mus musculus molossinus*). Biochem. Genet. 20:385–393.

232. Watanabe, T., and T. Tomita. 1974. Genetic study of pancreatic proteinase and alpha-amylase in mice (*Mus musculus*). Biochem. Genet. 12:419–428.

233. Watson, J.G., T.J. Higgins, P.B. Collins, and S. Chaykin. 1972. The mouse liver aldehyde oxidase (Aox) locus. Biochem. Genet. 6:195–204.

234. Weigert, M., and M. Potter. 1977. Antibody variable-region genetics: Summary and abstracts of the Homogeneous Immunoglobulin Workshop VII. Immunogenetics 4:401–435.

235. Wejman, J.C., B.A. Taylor, N.A. Jenkins, and N.G. Copeland. 1984. Endogenous xenotropic murine leukemia virus-related sequences map to chromosomal regions encoding mouse lymphocyte antigens. J. Virol. 50:237–247.

236. Whitney, J.B. III. 1978. (Personal communication).

237. Whitney, J.B. III, L.C. Skow, S.H. Langley, T.H. Roderick, and P.A. Lalley. 1980. Genetic mapping of sorbitol dehydrogenase (Sdh-1) to chromosome 2 in the mouse. (Personal communication).

238. Wienker, Th. 1980. (Typing of Esterase-9 in different laboratory strains and wild stocks). Mouse News Lett. 63:12. (Personal communication).

239. Wilcox, F.H. 1972. Genetic variation of a major liver protein in the mouse. J. Hered. 63:60–62.

240. Wilcox, F.H. 1973. Genetic differences in alteration of albumin in the house mouse, *Mus musculus*. Biochem. Genet. 10:69–78.

241. Wilcox, F.H. 1975. Genetic variation in plasma pre-albumin of the house mouse. J. Hered. 66:19–22.

242. Wilcox, F.H., L. Hirschhorn, B.A. Taylor, J.E. Womack, and T.H. Roderick. 1979. Genetic variation in

alkaline phosphatase of the house mouse (*Mus musculus*) with emphasis on a manganese-requiring isozyme. Biochem. Genet. 17:1093–1107.

243. Wilcox, F.H., and B.A. Taylor. 1981. Genetics of the Akp-2 locus for alkaline phosphatase of liver, kidney, bone, and placenta in the mouse. J. Hered. 7:387–390.

244. Wilson, C.M., M. Cherry, B.A. Taylor, and J.D. Wilson. 1981. Genetic and endocrine control of renin activity in the submaxillary gland of the mouse. Biochem. Genet. 19:509–523.

245. Witham, B.A. (ed.) 1988. The H-2 complex of the mouse. Jax Notes (No. 433), (Jackson Laboratory, Bar Harbor ME 04609).

246. Wolff, G.L. 1986. Allelic characterizations of BALB/cStCrlNctr. (Letter from George L. Wolff to T.H. Roderick).

247. Womack, J.E., and S.B. Auerbach. 1978. An acid phosphatase locus expressed in mouse kidney (Apk) and its genetic location on chromosome 17. Biochem. Genet. 16:239–245.

248. Womack, J.E., M.T. Davisson, E.M. Eicher, and D.A. Kendall. 1977. Mapping of nucleoside phosphorylase (Np-1) and esterase 10 (Es-10) on mouse chromosome 14. Biochem. Genet. 15:347–355.

249. Womack, J.E., and E.M. Eicher. 1977. Liver-specific lysosomal acid phosphatase deficiency (Apl) on mouse chromosome 17. Molec. Gen. Genet. 155:315–317.

250. Womack, J.E., D. Lu Shun Yan, and M. Potier. 1981. Gene for neuraminidase activity on mouse chromosome 17 near H-2: pleiotropic effects on multiple hydrolases. Science 212:63–65.

251. Womack, J.E., M.A. Lynes, and B.A. Taylor. 1975. Genetic variation of an intestinal leucine arylaminopeptidase (Lap-1) in the mouse and its location on chromosome 9. Biochem. Genet. 13:511–518.

252. Womack, J.E., B.A. Taylor, and J.E. Barton. 1978. Esterase-13, a new mouse esterase locus with recessive expression and its genetic location on chromosome 9. Biochem. Genet. 16:1107–1112.

253. Wood, A.W., and B.A. Taylor. 1979. Genetic regulation of coumarin hydroxylase activity in mice: Evidence for single locus control on chromosome 7. J. Biol. Chem. 254:5647–5651.

254. Wortis, H.H. 1965. A gene locus concerned with an antigenic serum substance in *Mus musculus*. Genetics 52:267–273.

255. Xing, R.-C. 1986. A survey of 30 biochemical markers in three Chinese inbred strains of mice. Acta Genetica Sinica 13:383–390.

256. Xing, R.-C., and T.H. Roderick. 1984. Typing of inbred strains and of mice (unpublished material). The Jackson Laboratory, Bar Harbor ME.

257. Zimmerman, D., and T.H. Hansen. 1978. Further serological analysis of the Qa antigens: analysis of an anti-H-2.28 serum. Immunogenetics 6:245–251.

18 RECOMBINANT INBRED STRAINS

BENJAMIN A. TAYLOR

Recombinant inbred (RI) strains are derived by systematic inbreeding from the cross of two pre-existing progenitor strains (3). Such strains possess particular attributes that make them especially useful for segregation analysis, gene mapping, and analysis of pleiotropism (3,170). In Table 18.1 extant sets of RI strains are listed with information about inbreeding and holders. In Tables 18.2–18.16 the published strain distribution patterns (SDPs) are given for those sets that have been characterized appreciably. Loci are arranged according to their known or apparent gene order so far as possible and arbitrarily thereafter. Thus, the indicated gene order may not always be correct. Markers that exhibit identical SDPs, and are apparently closely linked, are listed together to save space in the larger RI sets. A few tentative linkage assignments have been indicated by enclosing the chromosome number within parentheses. Unmapped loci are arranged alphabetically. In some cases chromosomal assignments have been made that were not claimed in the cited publication(s). In other cases linkage assignments have been changed to reflect new information about linkage relationships, chiefly based upon the addition of new markers in the RI system. In a number of cases locus symbols have been changed from those used in the initial publication.

Readers in doubt about the identity of a particular locus should consult Chapter 2 of this volume or the periodical *Mouse News Letter*. In a few cases in which no locus symbol was given initially, provisional locus symbols have been adopted. Other, non-standard, identifying symbols have been used without italics to indicate their unofficial status. Certain inferred SDPs based on the analysis of complex phenotypes and fragmentary SDPs have been excluded from the tables. Readers should also recognize that not all SDPs have been established with the same degree of certainty and that typing errors may have arisen from a variety of sources. Certain loci or gene complexes have been typed independently by multiple investigators using different methods and reagents (e.g. the immunoglobulin heavy chain locus). In such cases not all typings have been cited. A few RI strains have become extinct since the previous listing. In instances in which extensive genetic typings and a supply of genomic DNA are available, extinct strains have been included. It should be emphasized that these tables include only published SDPs and the holders of the various RI sets may possess extensive unpublished typings.

I wish to acknowledge support for this work by NIH research grants GM18684 and CA33093.

Table 18.1 Established and incipient recombinant inbred strains of mice

Progenitors		Designation	Number of strains	Generations of inbreeding	Holder
C57BL/6J	DBA/2J	BXD	26	63–88	Taylor
AKR/J	C57L/J	AKXL	17	41–83	Taylor
C57BL/6J	C3H/HeJ	BXH	12	71–86	Taylor
BALB/cBy	C57BL/6By	CXB	7	76–97	Mobraaten
AKR/J	DBA/2J	AKXD	25	33–63	Taylor
SWR/J	C57L/J	SWXL	6	56–84	Taylor
NZB/BlNJ	129/J	NX129	7	20–62	Taylor
NZB/Icr	C58/J	NX8	6	50–52	Riblet
B10.D2(58N)/Sn	C57L/J	58NXL	4	56–65	Taylor
C57L/J	C57BL/6J	LXB	3	60–67	Taylor
C57BR/cdJ	B10.D2(58N)/Sn	BRX58N	9	52–69	Taylor
C57L/J	PL/J	LXPL	5	17–50	Taylor
A/J	C57BL/6J	AXB	24	28–50	Nesbitt
C57BL/6J	A/J	BXA	22	28–50	Nesbitt
BALB/cHeA	STS/A	CXS	14	~40	Hilgers
O20/A	AKR/FuRdA	OXA	14	~30	Hilgers
BALB/cKe	SJL/J	CXJ	7	30–35	Cohn
129/Sv-Sl^{J}, C, P	A/HeJ	9XA	8	33–41	Stevens
LT/Sv	C57BL/6J	LTXB	3	33–40	Stevens
BALB/cJPas	DBA/2JPas	CXD	9	47–58	Guénet
129/SvPas-C, P, Sl/+	C57BL/6JPas	129XB	13	35–55	Guénet
C57BL/6N	AKR/N	B6NXAKN	15	35–60	Nebert
C57BL/6N	C3H/HeN	B6NXC3N	20	35–60	Nebert

Table 18.2 Strain distribution patterns of loci typed in BXD RI strains

Locus	Chromosome	1	2	5	6	8	9	11	12	13	14	15	16	18	19	20	21	22	23	24	25	27	28	29	30	31	32	References
																												BXD
Aox-1	1	B*	B	B	D*	D	B	B	B	B	B	D	D	D	D		B	B	B	B	D	D	D	D	D	B	D	74
Idh-1	1	B	B	B	D	D	B	B	B	B	B	D	D	B	D		B	B	B	B	D	D	D	D	D	B	D	14
Mylf	1	B	B	B	D	D	B	B	B	B	B	D	D	D	B		B	B	D	B	B	D	D	D	D	B		138
Lsh, Bcg	1	D	B	B	D	D	B	B	B	B	B	D	B	B	B	B	B	B	D	B	B	B	D	B	D	B	D	14, 154
Bcl-2	1	B	B	D	D	B	B	B	B	B	B	D	D	D	B	B	B	B	D	B	B	B	D	D	D	B	D	115
Emv-17	1	B	B	D	D	B	B	D	B	D	B	D	D	D	B	D	B	B	D	D	B	B	B	D	B	B	D	19
En-1	1	B	B	D	D	D	B	D	B	D	D	B	D	B	B	B	B	B	D	D	B	D	D	D	D	B	D	81, 67
Upg-2, Pep-3	1	B	B	D	D	B	B	D	B	D	B	B	D	D	B	B	B	B	D	D	B	D	D	D	D	B	D	164, 196
Ren-2	1	B	B	D	D	B	B	D	B	D	B	B	B	D	B	B	B	B	B	D	B	D	D	D	D	B	D	196
Sas-1, Cfh	1	B	B	D	B	B	B	D	B	D	D	B	B	D	B	D	B	B	D	D	B	D	D	D	D	B	D	140, 31
Lamb-2	1	B	B	D	B	B	D	D	B	D	B	D	B	D	B	B	B	D	D	D	B	D	D	D	D	B	B	42
Ly-22, Xmmv-61, Xmmv-9	1	B	B	D	B	B	D		B	D		D	B	D	B	B	B	D	D		D	D	D	D	D	B	D	168, 189, 12
Ltw-4	1	B	B	D	B	B	D	D	B	B		B	D	B	B	B	B	D	D	D	D	D	D	D	D	B	D	47
E	1	B	B	D	B	B	D	D	B	B		B	B	D	B	B	D	D	D	D	D	D	D	D	D		D	132
Sap	1	B	B	D	D	B	D	D	B	D		B	B	D	B		D	B	D	D	D	D	D	B	D	B		118
Ly-9	1	B	D	D	D	B	D	D	B	B	B	B	B	B	B	B	D	B	D	B	B	D	D	B	B	B		89
Mls	1	D	B	D	B	D	D	D	B	D	D	D	B	B	B	D	D	B	B	D	B	D	D	D	D	B	D	53, 98
Lsd	1	B	D	B	D	D	D	D	B	D	B	B	B	B	B	B	D	B	D	B	D	D	D	D	D	B	D	123
Mtv-7	1	B	B	B	B	D	D		B	B	B	D	B	B	B	B	B	B	D	B	B	B	D	D	B	B	D	184
Eph-1, Xmmv-36,	1	B	B	D	B	D	D	D	B	D	B	B	B	D	B	D	D	B	D	B	D	D	B	D	D	B	D	97, 12
Xmmv-6	1	B	B	D	D	D	D	D	B	B	B	D	B	B	D	B	B	D	B	B	D	D	D	D	B	B	D	12
Hc (C5)	2	B	B	B	B	D	D	B	B	D	B	B	D	D	D	B	D	D	D	B	D	B	B	D	D	D	D	31
Brp-13	2	B	D	B	B	B		D	B	D	D	D	B	B	B	B	B	B	D	D	B	D	B	B	D	B	D	57
Ly-24	2	B	D	D	B	D	B	B	D	B	B	B	B	D	D	B	D	D	D	B	B	D	D	D	D	B	D	23
B2m	2	D	D	B	B	D	B	B	D	D	B	D	D	B	D	D	D	B	B	D	B	D	D	D	D	B	B	20, 166
Hdc	2	B	B	B	B	D	B	B	D	D	B	D	D	D	D	D	D	B	D	D	B	B	D	D	B	D	B	102
Il-1a	2	B	B	D	D	D	D	D	D	D	D	B	D	D	B			B	D	B	B	B	B	D	B	D	B	30
Src-1	2	B	B	B	D	D	B	D	D	B	B	D	B	D		B	D	B	D	B	D	D	B.	D	D	D	B	11
Psp	2	D	D	B	D	D	D	D	D	B	B	B	B	B	D	D	D	B	D	D	D	D	D	D	D	D	D	69
Odc-2	2	B	B	D	D	B	D	B	D	D	B	D	D	D	D	D	D	B	D	B	B	D	D	D	D	D	D	136
Svp-1	2	D	D	D	D	B	D	D	B	D	B	B	B	D	B	D	D	D	D	B	B	D	D	D	D	D	D	172
Car-2, Ap2	3	D	B	B	B	D	B	D	B	D	B	D	D	D	D	B	B	B	B	B	B	B	B	B	D	B	B	126, 64
Xmmv-65	3	B	B	B	B	B	B	B	D	B		B	B	B	D	B	B	B	D	B	D	D	B	B	B	B	B	189
Amy-1,2	3	B	D	B	B	B	B	B	B	D	B	D	B	D	D	D	B	D	D	D	D	B	D	B	B	B	B	126
Adh-3	3	B	D	B	B	D	D	D	B	B	B	D	D	B	D		B	B	D	B	B	B	D	D	B	B	D	74
Bmn	3	B	B	B	B	B	B	B	B	B	D	B	B	D	D	D	D	B	D	D	D	D	D	D	B	B	D	94
cdm	3	B	D	B	D	D	B	D	D	B	B	D	B	D	D	B	B	B	D	B	D	B	D	B	D	B	B	176, 173
Xmmv-8	4	D	B	B	B	B	D	B	B	D	B	B	D	D	B	B	D	D	D		D	D	B	B	D	B	D	12
Lyb-2	4	D	B	B	B	B	B	B	D	B	B	B	D	B	B	B	D	B	D	D	D	D	B	B	D	B		182
Lyb-4	4	D	B	B	B	B	B	B	B	D	D	B	B	B	B	D	D	B	D	D	D	B	B	B	B	B		75
Mup-1	4	D	B	B	B	B	B	B	D	D	D	B	D	D	B	B	B	B	B	D	D	D	B	B	B	B	B	182
Lv	4	D	B	B	B	D	D	B	D	D	B	B	B	D	B	B	D	B	D	D	D	D	D	B	B	B	B	120
b	4	D	B	B	B	B	D	D	B	D	D	D	B	D	B	B	B	D	D		D	D	D	B	D	B	B	182
Ifa	4	D	B	B	B	B	D	B	B	D	D	D	B	B	B	B	D	B	D		D	D	D	B	D	B	B	120

* B and D are used here as generic symbols of alleles inherited from C57BL/6J and DBA/2J, respectively.

Table 18.2—*cont.*

Locus	Chromosome	1	2	5	6	8	9	11	12	13	14	15	16	18	19	20	21	22	23	24	25	27	28	29	30	31	32	References
Ly-20	4	D	B	B	D	B	B	D	D	D	D	D	D	D	B	B	D	B	D	D	D	B	D	D	D			108
Xmmv-23	4	D	B	B		B	B	D	D		D	D	B	D	D	D	D	D	D	D	D	D	D	D	D	B	D	12
Mtv-13	4	D	B	B	B	D	D	D	B	D	D	D	D	D	D	B	B	D		B	B	D	D	D	D	B	B	184
Ms15-1	4	D	B	D	B	B	D	D	D	D	D	D	D	D	D	D	B	D	B	B	B	D	D	D	D	B	D	78
Ms6-2	4	D	B	D	B	D	D	D	D	D	D	D	D	D	D	B	D	B	D	B	B	D	D	D	D	B	D	78
Ahd-1, Akp-2	4	D	B	D	B	D	B	D	D	D	D	D	D	D	B	D	D	B	B	D	B	B	D	B	D	B	D	193
Ly-31	4	D	B	B	B	D	B	D	D	D	D	B	D	D	B	B	D	B	B	D	B	D	D	D	D	B	D	169
Fv-1	4	D	B	B	B	D	B	D	D	D	D	B	B	D	B	B	D	D	B	D	B	D	D	D	D	B	D	175
Gpd-1	4	D	B	D	B	D	B	D	D	D	D	B	D	D	B	B	D	D	D	D	B	B	D	B	D	D	D	175
Xmmv-52	5	D	D	D	D	B	B	D	D	D	D	B	D	D	D	B	B	D		B	B	D	D	D	B	D	D	189
Xmmv-5	5	B	D	D	B	B	B	D	D	D	D	D	D	D	D	D	D	D		D	D	D	D	B	D	D	D	12
Pgm-1	5	B	D	D	B	D	D	D	B	B	D	B	D	B	D	D	D	D	B	D	B	D	D	D	D	D	D	59
Ms15-6	5	B	D	D	B	B	B	D	B	B	D	B	D	B	D	D	D	D		D	D	D	D	B	D	D	D	78
Ric	5	B	D	D	B	B	B	D	B	B	D	B	D	B	D	B	D	D	D	D	B	B	B	D	D	D	D	59
Tsz-1	5	B	D	D	D	D	D	D	D	D	D	B	D	B	D	D	D	D		B	B	D	B	B	D	B	D	127
Ms15-5	5	D	D	D	D	D	B	D	D	D	D	B	D	D	D	B	D	D	B	D	D	D	B	D	D	D	D	78
Bcd-1	5	B	D	D	B	D	B	D	B	D	D	B	D	D	B	D	D	D		D	B	B	B	D	D	D	B	197
C	5	B	D	D	B	D	B	D	B	D	D	B	D	D	D	D	D	D	B	D	D	B	B	D	B	B	D	132
Fla	5	B	D	B	B	D	B	D	B	D	D	B	D	D	D	D	D	D		D	D	B	B	D	B	B	D	197
Met	6	B	B	B	D	D	B	D	D	D	D	B	D	D	B	D	B	D		B	B	D	B	B	D	B	B	27
Hbt-1	6	B	B	B	D	D	B	D	D	D	D	B	B	D	B	D	B	D		D	B	D	B	B	D	B	B	18
Hox-1	6	B	B	B	D	D	B	D	D	D	D	B	B	D	B	D	D	D	B	B	D	D	B	B	D	B	B	142
Ggc	6	B	B	B	D	D	B	D	D	D	D	B	B	B	D	B	D	D		B	D	D	B	B	D	B	B	185
Xmmv-27	6	D	B	D	D	D	D	D	D	D	D	B	D	D	D	D	D	D		B	D	D	D	B	D	B	B	12
Ly-2, Lvp-1	6	D	B	D	D	D	B	B	D	D	D	B	D	B	D	D	D	D	B	D	D	D	D	B	D	B	B	185, 192
Ms6-4	6	D	D	D	D	D	B	B	B	B	B	B	D	D	D	B	D	D		D	D	B	D	B	D	B	B	78
P450PB	7	B	D	B	D	D	B	B	D	D	D	B	D	D	B	D	B	B		B	B	D	B	B	B	B	D	152
Coh	7	B	D	B	D	B	B	B	D	D	D	D	B	D	B	B	B	B		B	B	D	B	B	B	B	D	200
Xmmv-35	7	B	D	B	D	B	B	D	D	D	D	D	B	D	B	D	B	D		D	B	D	D	B	B	B	D	12
Svp-2	7	B	D	B	D	B	B	B	D	D	D	D	B	D	B	B	B	B		B	D	D	D	B	D	B	D	172
Abpa, Gpi-1	7	D	B	B	D	D	D	B	D	D	D	D	B	D	B	B	B	D		B	B	D	B	B	B	B	D	38, 200
Lyb-8	7	B	B	B	D	B	D	D	D	D	D	D	B	D	B	B	B	B		B	B	D	D	B	B	B	B	163
D7Rp2	7	B	B	B	D	B	D	D	D	D	D	D	B	D	B	B	B	B		B	B	D	B	B	B	B	B	43
Odc-6	7	B	D	B	D	B	D	D	D	D	D	D	D	D	B	D	B	B		B	B	D	B	B	B	B	D	136
Tam-1, Ngfg	7	B	D	B	D	B	B	D	D	D	D	D	B	D	B	B	B	B		D	B	D	D	B	B	D	B	157, 76
Mtv-1	7	B	D	B	D	B	D	D	D	D	B	B	B	D	B	B	B	B		D	B	D	B	B	B	B	D	184
Xmmv-31	7	B	D	B	D	B	D	D	D	D	B	B	B	D	B	B	B	B		D	B	D	B	B	B	D	B	12
B	7	B	B	D	D	B	D	B	D	D	B	D	B	D	B	B	B	B		B	B	D	D	B	B	D	D	132
Mod-2	7	B	B	B	D	B	D	D	D	D	D	D	B	D	B	B	B	B		D	B	B	D	B	B	B	D	184
Hbb	7	B	B	B	D	D	D	D	D	D	B	D	D	D	B	B	B	D		B	B	D	B	B	B	D	D	200
Odc-7	7	B	B	B	D	D	D	D	B	B	D	D	B	D	B	B	B	B		B	B	D	D	B	B	B	D	136
Fis-1	7	D	D	B	D	D	D	D	D	B	B	D	D	D	B	D	B	D		B	D	D	D	D	B	D	D	150
Int-2, Xmmv-76	7	D	D	B	D	B	D	D	B	B	B	D	B	D	B	D	B	B		B	D	D	D	D	B	D	D	150, 189
Es-1	8	D	B	B	D	D	D	D	B	D	B	B	D	D	D	D	B	B		B	B	B	B	B	B	B	D	172
Xmmv-29	8	D	B	B	D	D	D	D	B	B	B	D	D	D	B	D	D	D		D	D	D	D	D	B	B	D	12
Emv-2 (Bv-1)	8	D	B	D	D	D	D	D	D	B	B	B	D	B	B	B	D	B		D	D	D	B	D	B	D	D	80
Lap-1	9	B	D	D	D	D	D	D	D	B	B	B	D	D	D	B	D	D		D	D	B	B	D	D			198

Table 18.2—*cont.*

Locus	Chromosome	1	2	5	6	8	9	11	12	13	14	15	16	18	19	20	21	22	23	24	25	27	28	29	30	31	32	References
Apoal	9	B	D	D	D	D	D	B	D	D	D	B	D	D	D	B	D	B	D	D	D	B	B	D	D	B	D	41
Ncam	9	B	D	D	D	D	D	B	D	D	D	B	D	D	D		D	B	D	D	D	B	B	D	D	B	D	32
Xmmv-2	9	B	B	D	B	D	D	B	D	D	D	B	D	D	D		D	B	D	D	D	B	D	D	D	B	D	12
Cdt	9		B	B	B	B	D	B	D	B	B	B	D	D	D	B	D	D	D	B	D	B	D	D	B	B	D	144
d (Emv-3)	9	D	B	D	D	D	B	B	B	B	B	B	D	D	B	D	D	D	D	B	D	B	B	B	D	B	B	198, 80
Pgm-3	9	D	B	B	D	D	B	B	B	B	B	B	D	D	B	D	D	D	D	B	D	B	B	B	D	B	B	121
Mod-1	9	D	B	B	D	D	B	B	B	B	B	B	D	B	B	D	D	D	D	B	D	B	D	B	D	B	D	198
Crbp-2	9	D	B	B	D	D	B	B	D	B	B	B	D	B	B	D	B	D	D	B	D	B	B	B	B	B		37
Ltw-3	9	D	B	B	B	D	B	B	B	B	B	B	D	B	B	D	B	D	D	B	D	B	B	B	D	B		45
Fv-2	9	D	B	B	B	D	B	B	D	B	B	B	D	D	B	D	B	D	D	B	D	B	B	B	B	B		45
Bgl-s	9	B	D	B	D	D	B	B	D	B	B	B	B	D	B	D	D	D	B	D	B	B	D	B	B	D	D	107
Sparc	11	D	D	D	D	B	B	D	D	D	D	B	D	D	D	D	D	D	D	B	B	D	B	D	D	D	D	103
Es-3	11	D	D	D	D	D	D	B	B	B	B	D	B	B	B	D	D	B	D	D	B	D	D	D	D	B	D	172
D12Mcg1	12	B	B	D	B	B	B	B	B	D	D	B	B	B	D	D	B	D	D	D	B	B	D	B	B	B	D	22
D12Nyu2	12	B	B	D	B	B	D	B	B	D	B	D	B	B	B	D	B	D	D	D	D	D	D	B	B	D	D	33
Ah	12	B	B	B	B	B	B	B	D	B	B	D	B	B	D	D	B	D	D	D	B	D	D	D	B	D	D	90
D12Nyu1	12	B	B	B	B	B	B	B	B	D	D	D	B	B	D	D	B	D	D	B	B	D	D	D	B	D	D	33
Mtv-9	12	D	B	B	B	D	D	B	D	D	D	D	D	D	D	B	B	D	D	D	D	D	D	D	D	D	D	184
Spi-2	12	B	B	B	B	D	D	B	D	B	B	D	D	D	D	D	B	D	D	D	D	D	D	D	D	D	D	68
Aat	12	B	B	B	B	B	D	B	D	B	B	D	D	D	B	D	B	B	D	D	B	D	D	D	D	D	D	174, 33
Xmmv-34	12	B	B	B	B	B	D	B	D	B	B	D	D	D	B	B	B	B	D	D	B	D	B	B	D	B	D	12
Crip	12	D	B	D	B	B	D	D	B	D	B	D	D	D	D	D	B	D	D	B	D	D	D	B	D	D	D	10
Igh-C	12	D	B	B	B	B	B	D	B	D	D	B	D	D	B	D	B	B	D	D	D	D	D	B	B	D	D	174, 33
Igh-Gie, Odc-8	12	D	B	B	B	D	D	D	B	D	D	B	D	D	B	B	B	B	D	D	B	D	D	B	B	D	D	82, 136
Igh-Np	12	B	B	B	B	B	D	D	B	B	B	D	D	D	B	B	B	B	D	D	B	B	D	B	B	D	D	99
Tcrg	13	D	B	B	B	B	D	D	B	B	D	B	D	D	D	B	B	D	B	D	B	B	B	D	B	B	B	125
Hist1	13	D	B	B	B	D	D	D	B	B	B	B	D	D	D	B	B	D	B	D	B	D	B	D	B	D		125
As-1	13	D	D	B	D	B	D	D	D	D	D	D	D	D	B	B	D	B	B	B	D	D	D	D	D	D	D	44
Lth-1	13	D	D	B	D	B	D	D	D	D	D	D	D	D	B	B	D	B	B	B	B	D	D	D	D	D	D	44
Odc-9	14	D	D	B	B	B	B	D	B	B	B	D	B	D	B	B	B	D	D	D	D	B	D	B	B	D	D	136
Ms15-7	14	B	D	B	D	B	B	B	D	B	B	D	D	D	D	B	D	D	D	D	D	B	B	D	B	D	D	78
Mtv-11	14	B	B	B	D	D	D	B	B	B	D	B	D	D	D	B	B	D	D	B	D	B	B	B	B	D	D	184
Ms6-5	14	B	B	B	D	B	D	B	D	B	B	D	D	D	D	B	B	B	D	B	B	B	B	B	B	D	D	78
Np-2, Tcra,	14	B	D	B	D	B	D	B	D	B	B	B	D	D	B	B	B	B	B	D	B	D	D	B	D	D	D	36, 48
Rib-1	14																											
Es-10, For-5	14	B	D	D	D	D	D	D	B	D	D	D		B	D	B	D	D	B	D	B	B	B	D	D	B	D	24
Sis	15	B	D	D	B	D	B	D	B	D	B	D	B	D	D	B	D	D	D	D	D	D	D	D	D	D		110
Xmmv-15	(17)	D	D	D	B	B	B	B	B	B	B	B	B	D	B	B	B	D	B	D	B	B	D	B	B	B	D	12
T66E, T119	17	D	D	D	B	B	D	B	D	B	B	B	D	D	D	D	B	D	B	D	D	D	D	B	B	D	D	63
D17Rp17	17	D	D	D	B	D	D	D	B	B	D	B	B	D	B	B	B	D	B	D	B	B	D	B	B	D	D	101
T66D	17	D	B	D	D	D	D	D	B	B	B	B	D	D	B	B	B	D	B	D	D	B	D	B	B	D	D	63
D17Leh443,	17	D	B	B	B	B	D	D	D	D	B	B	D	D	D	D	B	D		B	D	D	D	B	B	D		17
D17Leh94, D17Leh180,	17																											
D17Leh54M	17																											
Hba-4ps	17	D	B	B	B	B	B	D	D	B	B	B	B	D	B	B	D	D	B	D	D	B	D	B	D	D	D	101
Ms15-3	17	D	B	D	D	B	D	D	B	B	B	B	D	D	B	B	B	D	B	D	D	B	D	B	D			78
Pim-1	17	D	B	D	D	B	D	D	D	B	B	B	D	D	B	D	D	D	D	D	D	D	D	D	D	D	D	122

Table 18.2—*cont.*

Locus	Chromosome	BXD 1	2	5	6	8	9	11	12	13	14	15	16	18	19	20	21	22	23	24	25	27	28	29	30	31	32	References
Crya-1	17	D	B	D	D	B	D	D	D	B	B	B	D	D	B	B	D	D	B	D	D	B	D	B	D	D	D	156
H-2S	17	D	B	D	D	B	D	D	D	B	B	B	D	D	B	B	D	D	B	B	B	D	B	B	D	D	D	144
Tla, D17Leh89, D17Leh525	17	D	B	D	D	B	D	D	D	B	B	B	D	D	B	B	D	D	B	D	D	D	D	B	D	D	D	17
Upg-1	17	D	D	D	D	B	B	D	D	D	B	B	D	D	D	D	B	B	D	D	D	D	D	B	B	D	D	164
D17Rp10	17	D	B	B	B	B	B	D	B	D	D	B	D	D	D	D	B	B	D	D	B	D	D	B	D	D	D	100, 46
Ly-10	19	D	B	D	D	D	B	B	B	D	B	B	B	B	D	B	B	B	D	D	B	B	B	B	B	B	B	165
Ly-1	19	B	B	B	B	D	B	B	B	D	B	B	B	B	D	D	B	B	D	D	B	D	B	D	D	B	B	165
Brp-8	—	B	D	B	D	B	B	B	B	D	D	D	B	D				B		B			B		B		D	57
Brp-12	—	B	B	B	D	B	B	B	B	D	D	B	B	D				B		D			D		D		D	57
Brp-14	—	B	B	D	B	B	D	D	D	B	D	D	B	D	B		B	B	B	D	B	D	B	D	D	D	D	57
GL-Y	—	B	B	B	B	B	B	D	B	B	D	D	D	B	D		B	B		B	B	B	B	D	D	B	D	145
Irgt	—	B	B	B	B	B	D	D	B	B	D	B	B	B	B	D	B	D		D	D	B	B	D	D	B	B	187
Ly-7	—	B	B	B	B	D	B	B	B	D	D	B	D	B	B	D	B	B	B	D	D	D	D	D	D	D	D	72
Ms6-1	—	B	D	B	B	B	B	B	D	B	D	B	D	D	D	D	B	B		B	B	D	B	B	B	B	B	78
Ms6-3	—	B	D	D	D	B	D	D	D	B	D	D	D	D	D	D	B	D		D	B	B	D	D	D	D	D	78
Ms15-2	—	B	D	B	D	B	D	B	B	B	D	D	D	D	B	B	B	B		D	D	B	B	B	B	B	D	78
Ms15-4	—	B	B	D	D	B	D	B	B	B	D	D	B	B	D	D	D	B		D	D	B	D	B	B	B	D	78
Ms15-8	—	B	B	D	D	B	B	D	B	D	D	D	B	B	D	B	B	B		D	D	B	B	B	B	B	B	78
Odc-10	—	B	D	D	D	B	D	D	D	D	D	D	D	D	D	D	B	D	B	D	D	B	D	D	D	D	D	136
Odc-11	—	B	B	B	D	B	D	B	D	B	D	D	D	B	D	B	B	D	B	B	B	D	B	B	B	D	D	136
Tpmt	—	D	B	D	D	B	D	D	B	D	B	D	D	D	D	D	D	D	D	D	D	B	D	D	D	D	D	124
Xmmv-3	—	B	D	D	D	B	D	D	B	D	B	D	D	D	D	B	B	B	D	D	D	B	D	D	B	B	D	12
Xmmv-7, Xmmv-54	—	D	D	D	B	D	B	D	D	D	D	D	D	D	D	D	D	B	D	D	D	D	D	D	D	B	D	12, 189
Xmmv-13	—	B	B	B		B	D	D	B	D	B	D	D	D	D	D	B	B	D	D	D	D	D	D	B	D	D	12
Xmmv-21	—	B	B	B		B	D	D	D	B	B	D	D	B	D	D	D	D	D	B	D	D	D	D	D	D	D	12
Xmmv-25	—	B	B	B		D	B	D	D	B	B	D	D	B	D	D	B	D	D	D	D	B	D	D	B	D	D	12

Table 18.3 Strain distribution patterns of loci typed in AKXL RI strains

Locus	Chromo-some	AKXL																		References
		5	6	7	8	9	12	13	14	16	17	19	21	24	25†	28	29	37	38	
ln	1	L*	L	A*	A	A	L	A	A	L	A	L	L	A	A	L	A	L	L	177, 196
Upg-2, Ren-2, Pep-3	1	A	A	L	A	A	A	L	L	L	A	L	L	A	A	A	A	L	A	164, 196
Odc-1, Lamb-2	1	A	A	L	L	A	A	L	L	L	A	L	L	A	A	L	A	L	A	42
ald	1	A	A	A	A	A	L	L	L	L	A	L	L	L	A	L	A	L	A	177
Ltw-4	1	L	L	L	A	A	L	L	L	L	A	L	L	L	A	L	A	L	A	47
Spna-1	1	L	L	L	L	A	L	L	L	L	A	L	L	L	A	A	A	A	A	147
Mls	1	L	L	L	L	A	L	L	L	L	A	L	L	L	A	A	A	L	A	53
Akp-1	1	A	L	L	L	L	L	L	L	L	A	L	L	L	A	L	A	A	L	191
Hc (C5)	2	A	A	L	L	A	A	A	L	L	A	L	L	A	A	A	L	A	L	31
B2m	2	L	L	L	A	L	A	L	L	L	A	L	L	A	A	A	L	A	L	20, 166
Ly-23	2	L	L	L	A	L	A	L	L	L	A	L	L	A	L	A	L	A	L	56
Emv-13 (Akv-3)	2	L	L	L	A	A	A	L	L	A	L	A	L	A	A	A	L	A	L	159
Svp-1	2	A	A	A	L	A	A	A	L	A	L	L	L	A	A	A	L	A	L	179
Car-2, Ap2	3	L	A	L	L	L	A	L	A	L	A	A	A	L	A	L	L	A	A	126, 64
Xmmv-47	3	L	L	L	L	L	A	A	A	L	L	A	L	A	A	L	L	L	L	189
Xmmv-65	3	A	L	A	L	L	A	A	A	L	L	L	L	A	L	L	L	A	A	189
Amy-1,2	3	A	L	A	L	L	A	A	A	L	L	L	L	A	A	A	L	L	A	126
Ly-19	4	L	A	A	A	L	A	A	L	A	A	L	A	A	A	A	A	L	A	167
Mup-1	4	L	A	L	A	A	A	L	L	A	L	L	L	A	A	L	L	L	L	7
b	4	L	A	L	A	A	L	L	L	L	A	L	L	L	A	L	L	L	L	173
Ahd-1, Akp-2	4	L	A	A	L	A	A	L	A	L	A	A	L	L	L	L	A	L	A	193
Ly-31	4	L	A	A		A	A	L	L	L	A	A	L	L	L	L	A	L	A	169
Gpd-1, Xmmv-62	4	L	A	A	L	A	A	L	L	L	A	A	A	L	L	L	L	A	A	193, 189
Bcd-1	5	A	A	L	L	L	A	A	A	A	L	L	L	L	A	L	L	L	L	197
Fla	5	L	A	L	A	A	L	A	A	A	A	L	L	L	L	L	L	L	L	197
Gus	5	L	A	L	A	A	L	A	A	A	A	L	L	L	L	L	L	L	L	197
MyK-103	6	L	L	L	L	L	L	A	A	L	A	L	A	L	A	A	L	L	L	194
Tcrb	6	L	L	L	L	L	L	A	A	L	A	A	A	L	A	A	L	L	L	139
Igk-Efl	6	L	L		L		L	A	L	L	L	L	L	A	A	A	L	L	L	178
Rn7s-6, Igk-J, Ly-2, Ly-3, Odc-5, Fabpl	6	L	L	L	L	L	L	A	L	L	L	L	L	A	A	A	L	L	L	178, 13, 136, 162
Emv-11 (Akv-1)	7	L	A	A	A	A	L	L	L	A	A	L	A	A	A	L	L	A	L	159
Lyb-8	7	A	L	A	L	A	L	L	L	A	A	A	A	A	A	L	L	A	A	163
D7Rp2	7	A	L	A	L	A	L	L	L	A	A	A	L	A	A	L	L	A	A	43
c	7	A	A	A	A	A	L	L	L	A	A	A	L	L	L	A	A	A	A	173
Odc-7	7	A	A	A	A	A	L	L	L	A	A	A	L	L	L	A	A	A	A	136
Fis-1, Int-2	7	L	L	L	L	L	A	A	L	L	L	L	L	L	A	A	A	L	A	150
Th	7	L	A	L	L	L	A	A	L	L	A	L	L	L	A	A	A	L	A	16
P450-3	9		L		A	A	L	A	L	A	A	L	L	A	L	L	A	A	A	65
Pgm-3	9	A	L	A	A	A	A	A	L	L	L	L	A	A	A	A	L	A	A	121
Crbp-2	9	L	L	A	A	A	A	A	L	L	L	L	A	A	A	A	L	A	A	37
Ltw-3	9	A	A	A	A	A	A	A	L	L	L	L	A	A	A	A	A	A	L	45
Fv-2	9		A		A	A	A	A	L	L	L	L	L	A	A	A	A	L	L	45
Bgl-s	9		A		A	A	A	A	L	L	A	L	L	A	A	A	A	L	L	45
Erbb	11	A	L	L	L	A	L	L	A	A	A	L	A	L	A	A	A	L	A	151
Hba	11	A	L		L	A	L	A	A	A	A	L	L	L	A	L	L	L	A	151
Emv-14 (Akv-4)	11	A	A	A	A	L	L	A	A	L	A	A	L	A	L	L	L	A	A	159
Es-3	11	A	L	L	A	L	L	L	L	L	A	A	L	L	L	L	L	A	A	114
Glk	11	A	L	A	A	L	L	A	L	L	A	A	L	L	L	L	L	A	L	114
D12Nyu5	12	L	A	L	L	A	A	L	L		L	L	L	A	A	L	A	L	A	33
Apob	12	A	L	L	A	A	L	L	A	L	A	L	L	L	A	L	A	L	L	95
Ly-18	12	A	A	L	L	A		L	A	A	A	L	L	A	A	L	A	L	A	84
D12Nyu1	12	A	A	L	L	A	A	L	A		A	L	L	L	A	L	A	L	A	33

779

Table 18.3—*cont.*

Locus	Chromo-some	AKXL																		References
		5	6	7	8	9	12	13	14	16	17	19	21	24	25†	28	29	37	38	
Aat	12	A	L	L	A	L	A	A	A	L	A	L	L	A	A	A	A	A	A	174, 33
Igh-1, Crip, Odc-8	12	A	L	L	A	A	A	L	L	L	A	L	L	A	A	A	L	A	A	174, 10, 136
Igh-Dex	12		L		A		A	L	L	L	A	L	L	A	A	A	L	A	A	133
Xmmv-50	12	A	L	A	A	A	L	L	L	L	A	L	L	A	A	A	L	A	A	189
As-1	13	A	A	L	A	L	L	L	A	A	A	L	A	A	L	L	A	A	A	44
Lth-1	13	L	A	L	A	L	L	L	A	A	A	L	A	A	A	L	L	A	A	44
Rib-1	14	A	A	A	L	L		A	L	L	L	L	A	A	L	L	L	L	A	48
Tgn	15	A	L	L	L	L	L	L	L	L	L	L	L	A	A	A	A	A	A	181
Hba-4ps	17	L	L	L	A	L	L	A	L	L	L	L	A	A	L	A	A	L	A	29
H-2, Tla, Pgk-2	17	L	A	L	A	L	L	A	A	L	L	L	A	L	L	A	L	L	A	21, 149
Ce-2, Thy-2	17	L	A	L	A	L	L	A	A	L	L	L	A	L	L	A	L	L	L	149
D17Rp11	17	L	A	L	A	L	L	A	A	L	L	L	A	L	L	A	L	L	L	100
Upg-1	17	L	A	L	A	L	L	A	A	L	L	L	A	L	L	A	L	L	L	164
D17Rp10	17	L	L	L	A	A	L	A	A	L	L	A	A	L	L	L	L	L	L	100, 46
Ii	18	A	A	L	L	L	L	A	A	L	L	L	L	L	A	L	L	L	L	135
Ly-10	19	A	A	A	L	A		L	A	L	L	L	L	A	A	A	A	A	L	165
Xmmv-42	19	A	A	A	L	A	A	L	A	L	L	L	L	A	A	A	A	A	L	189
Xp-11	–	A	A	L	A	L	A	A	L	A	L	A	L	L	L	L	A	L	A	189
Xmmv-66	–	L	A	A	L	L	L	A	L	A	L	A	A	A	L	L	L	A	A	189

* L and A are used here as generic symbols of alleles from C57L/J and AKR/J, respectively.
† Extinct strain.

Table 18.4 Strain distribution patterns of loci typed in BXH RI strains

Locus	Chromo-some	BXH												References
		2	3	4	6	7	8	9	10	11	12	14	19	
Lsh, Bcg	1	H*	B*	H	H	H	H	H	B	B	B	H	H	14, 155
Emv-17	1	H	H	H	H	H	B	H	H	H	H	H	H	19
Pep-3	1	B	H	B	H	H	H	H	H	B	B	H	B	140
Sas-1, Cfh	1	B	H	B	H	B	B	H	H	B	B	H	B	140, 31
Ltw-4	1	B	H	B	H	B	B	H	B	B	B	H	B	47
Ly-17, Apoa2	1	B	B	B	H	B	B	H	B	B	H	H	B	85, 96
Ly-9, Spna-1	1	B	H	B	H	B	B	H	B	B	H	H	B	89, 147
Akp-1	1	B	H	H	B	B	B	H	B	B	H	H	B	191
Ly-11	2	B	H	B	B	H	H	H	B	B	H	H		109
B2m, Ly-23	2	B	B	B	B	B	H	B	H	B	H	H	B	20, 56
Hdc	2	B	B	H	B	B	H	B	H	B	H	H	B	102
Il-1a	2	B	B	H	H	B	H	B	H	B	H	H	B	30
a, Emv-15	2	B	H	B	H	B	H	H	B	B	H	H	B	93, 153
Car-2	3	H	B	H	B	H	B	B	B	B	B	B	B	126
Odc-3	3	H	B	H	B	H	B	H	B	H	B	B	B	136
Amy-1,2	3	H	B	B	B	H	B	H	H	H	B	B	B	126
Adh-3, Gbp-1	3	B	B	B	B	H	B	B	H	H	B	B	B	131
Mup-1, Lps	4	H	H	H	H	H	H	H	B	B	H	B	B	188
Odc-4	4	H	B	H		H	H	H	H	H	B	H	H	136
Ahd-1, Ly-31, Akp-2	4	B	B	B	H	H	B	H	H	H	B	B	H	193, 169
Fv-1, Gpd-1	4	B	B	B	H	B	B	B	H	B	B	B	H	175

Table 18.4—*cont.*

Locus	Chromo-some	BXH												References
		2	3	4	6	7	8	9	10	11	12	14	19	
Emv-1 (Cv-1), *En-2*	5	H	H	B	H	B	H	H	B	B	H	B	H	79, 81
Pgm-1	5	H	H	H	B	B	H	H	B	B	B	H	H	59
Pol-20	5	H	H	H	B	H	H	H	B	B	B	H		141
Ric, rd	5	H	H	H	B	H	H	H	B	B	B	H	H	59
Fla	5	B	H	H	B	H	H	H	H	B	H	H	H	61
Gus	5	B	H	H	B	B	H	H	H	B	B	B	H	59
Met	6	H	B	B	B	B	H	H	H	B	H	B	B	27
Hox-1	6	H	B	B	B	B	B	B	H	B	B	B	B	142
Ggc	6	H	B	B	B	B	B	B	B	B	B	B	B	185
Ly-2, Lvp-1, Fabpl	6	B	B	B	B	B	B	B	B	B	B	B	B	185, 192, 162
Apoe	7	B	H	H	B	B	B	B	B	H	H	B	B	95
Svp-2	7	H	H	H	B	B	B	B	B	H	H	B	B	95
Mtv-1	7	H	H	H	H	B	H		H	H	H	H	H	184
Mod-2	7	H	H	H	H	B	H	H	H	H	H	B	H	184
Hbb	7	B	B	H	H	B	H	H	H	H	B	B	B	160
Es-1	8	B	H	H	B	B	B	H	B	H	H	B	H	79, 173
Pol-6	8	B	H	H	B	B	B	H	H	B	H	B		141
Emv-2 (Bv-1)	8	B	B	H	H	H	H	B	B	B	B	B	B	79
Lap-1	9	B	B	H	B	B	B	H	B	H	H	B	H	198
Apoa1	9	B	B	H	H	B	B	H	H	H	H	B	H	41
Ncam	9	B	B	H	B	H	B	H	H	H	H	B	H	32
Mod-1	9	H	H	H	H	H	B	H	H	B	H	B	H	41, 171
Fv-2	9	H	H	B	H	B	H	H	B	B	H	B	H	79
Hsd	10	B	H	B	H	H	B	H	H	B	H	B	H	1
Sparc	11	B	H	B	H	H	H	B	B	B	B	B		103
Myla	11		H	B	H	H	H	B	B	H	B	H	B	138
Glk	11	B	B	H	B	H	H	B	B	H	B	H	B	114
Es-3	11	B	B	H	B	H	H	B	B	H	B	B	B	114
Rnr12	12	B	B	B	H	H	B	H	H	H	B	H	H	2
Apob	12	B	B	B	H	H	B	H	H	H	B	H	H	95
D12Nyu2	12	B	B	B	H	H	B	H	B	H	B	H	H	33
Pol-7	12	B	B	B	H	H	B	H	B	H	B	H		141
Ly-18, Ah	12	B	B	B	B	H	B	H	B	H	H	H	H	84, 128
D12Nyu1	12	H	B	B	B	H	H	H	H	H	H	H	H	33
Mtv-9	12	B	B	B	B	H	B		H	B	H	H	B	184
Igh-C, Crip, Odc-8	12	H	B	H	H	H	B	H	B	B	H	B	H	188, 10, 136
Igh-Gte	12	H	B	H	H	H	B	H	B	B	H		H	82
Tcrg	13	B	B	B	H	H	H	B	H	B	H	B	B	125
Hist1	13	B	B	B	H	H	H	B	H	B		B	B	125
As-1	13	H	H	B	B	B	H	H	H	H	B	B	H	44
Lth-1	13	H		B	B	B	H	H	H	H	B	B	H	44
Odc-9	14	H	H	H	B	B	B	H	B	B	H	H	H	136
Mtv-11	14	H	B	H	B	B	B		B	H	H	H	H	184
Rib-1	14	H	B	H	B	B	H	B	B	B	B	H	H	48
Es-10, For-5	14	B	B	H	B	B	H	B	B	H	B	H	H	24
Tgn	15	B	B	B	H	B	B	H	B	B	B	B	H	181
Ly-6	15	B	B	B	B	H	H	H	B	B	B	B	H	52
Pol-5	15	H	H	B	B	H	H	H	B	B	B	B	H	141
Xmmv-55	15	H	H	B	B	H	H	H	B	B	B	B	H	189
Sis	15	H	H	B	B	H	H	H	B	H		B		110
Gdc-1	15	B	H	B	B	B	B	B	B	H		B		110

Table 18.4—*cont.*

Locus	Chromosome	BXH												References
		2	3	4	6	7	8	9	10	11	12	14	19	
RB-G	16	H	B	H	H	H	B	B	B	H	B	B	H	2
Mtv-6	16	H	B	H	H	B	H		B	H	H	H	H	184
D17Rp17	17	H	H	H	B		B	H	B	H	H	H	B	101
Hba-4ps	17	H	H	B	H	H	B	H	B	H	H	H	B	29
H-2	17	H	H	B	H	H	B	B	B	B	H	H	B	188
Upg-1	17	B	H	B	H	H	B	B	B	B	H	H	B	29
Ahd-1t	–		B	B		H	B	B	H	B	H	B	H	5
E16OH	–	B	H	H	H	B	B	B	H	B	H	H	B	15
H3F2	–	B	B	B	B	H	B	B	H	H	H	B	B	148
Ly-7	–	B	B	H	H	H	H	B	B	B	H	B	H	72
Mcr	–	H	H	H	H	H	B	H	H	B	H	B	H	87
Mups	–		H	B	H	B	H	B	B	H				8
Pca-1	–	H	B	H	B	H	H	B	B	B	H	H	H	40
Pol-15	–	B	H	H	H	H	H	B	B	H	H	B		141
RB-B	–	B	H	B	B	B	B	H	B	B	H	B	B	2
RB-C	–	B	H	B	B	B	B	B	B	B	H	B	B	2

* H and B are used here as generic symbols of alleles inherited from C3H/HeJ and C57BL/6J, respectively.

Table 18.5 Strain distribution patterns of loci typed in CXB RI strains

Locus	Chromosome	CXB							References
		D	E	G	H	I	J	K	
Ltw-4	1	C*	B*	B	C	C	B	C	49
Apoa2, Ly-9	1	C	B	C	B	B	C	C	96, 70, 89
H-25, H-35	1	C	B	C	B	B	B	C	4
Xmmv-60	1	C	B	C	C	B	B	C	189
Ly-11	2	B	B	C	C	B	C	B	109
Ea-6	2	C	B	C	B	B	C	B	158
Ly-24, Pol-23	2	C	B	C	B	B	B	B	23, 141
H-3, B2m, Ly-23, Itp, Il-1b	2	C	B	B	C	B	B	B	4, 20, 56, 183, 30
a, Emv-15	2	C	C	C	B	B	B	B	3, 93, 153
Car-2	3	C	C	B	B	C	C	B	49
Xmmv-65	3	B	C	B	B	C	B	C	189
H-37	3	B	C	C	B	C	B	C	4
Amy-1,2	3	C	C	C	B	C	C	B	126
H-23	3	C	C	C	C	C	C	B	4
Adh-3, Ahr-1, Gbp-1	3	C	C	C	B	C	C	B	39, 131
H-28, If-1	3	C	C	B	B	C	C	B	4, 34
Mtv-17	4	C	C	B	C	C	C	B	112
Mup-1	4	B	C	C	C	B	C	B	129
b, H-21	4	B	B	C	C	C	C	B	4
Ifa	4	C	B	C	C	C	C	B	25
H-15	4	C	B	C	C	B	C	B	4
H-16	4	C	B	C	C	C	B	B	4
H-18, Ahd-1, Akp-2, Ly-31, H-20	4	B	B	C	B	C	B	B	4, 193, 169
Gpd-1	4	C	C	C	B	C	B	B	130
Emv-1 (Cv-1)	5	B	C	B	C	C	C	C	105
En-2	5	B	C	B	C	C	B	C	81

Table 18.5—*cont.*

Locus	Chromo-some	CXB							References
		D	E	G	H	I	J	K	
H-27, *Pol-20*	5	B	B	C	B	C	B	C	4, 141
Gus-s	5	C	B	C	C	B	B	B	161
Ggc, *Hox-1*	6	B	B	C	B	C	C	C	185, 117, 142
Svp-2	7	C	C	C	C	C	B	B	119
H-22, *Gpi-1*, *H-24*	7	B	C	B	C	C	C	B	4, 130
c, *Lfo-1*, *Hbb*, *H-1*	7	B	B	C	B	C	B	B	4, 49, 143
Ly-15	7	B	C	C	B	B	B	C	71
Fis-1, *Int-2*	7	B	B	B	B	B	B	B	150
H-19	8	C	B	B	B	B	B	B	4
Pol-6	8	B	B	C	B	B	C	C	141
Es-1	8	C	B	C	C	B	C	C	119
H-29	8	C	C	B	C	B	C	B	4
Emv-2 (Bv-1)	8	B	B	B	C	B	B	B	105
Lap-1, *Apoa1*	9	C	C	C	B	B	B	C	41
Ncam	9	C	B	C	B	B	B	C	32
Mod-1	9	C	C	C	B	B	B	C	130
H-7, *Fv-2*, *Ltw-3*	9	C	C	C	B	B	C	C	4, 49
Bgl-e	9	C	C	C	C	B	B	C	49
Hsd	10	B	C	B	C	B	C	C	1
Hba, *Erbb*	11	C	B	B	C	B	B	C	143, 151
Apob	12	B	C	C	C	C	B	C	95
D12Nyu2	12	B	C	C	C	C	B	B	33
Es-25	12	B	C	B	C	C	B	C	28
Ly-18, *H-17*, *H-34*, *H-38*, *Ltw-2*, *Pol-7*, *Pol-17*, *Pol-1*, *Ah*	12	B	C	C	C	C	B	C	165, 4, 49, 141, 128
D12Nyu1	12	C	C	C	C	C	B	C	33
Aat	12	B	B	C	C	B	C	C	174
Igh-C, *-Dex*, *-Pc*, *-Np*, *Crip*	12	B	B	C	B	B	C	B	129, 133, 91, 77, 10
As-1, *Lth-1*	13	B	C	C	C	B	B	C	44
Rib-1, *Np-2*, *Tcra*, *H-8*	14	C	C	B	B	C	C	C	48, 36, 9, 4
Tgn	15	B	B	C	B	B	C	B	181
Ly-6, *Pol-5*	15	B	B	C	C	B	C	B	106, 141
Sis	15	C	B	C	C	B	C	B	110
Gdc-1	15	C	B	C	C	B	B	C	88
Mtv-6	16	B	C	B	C	B	C	C	112
Hba-4ps, *Crya-1*	17	C	B	C	C	B	B	B	29, 156
H-2, *Upg-1*	17	C	B	B	C	B	B	B	4, 29
D17Rp10	17	B	C	B	C	B	B	B	100
Ii	18	C	B	B	C	B	B	C	135
Ea-4	19	C	B	B	C	B	B	B	158
Acta	–	B	C	B	C	B	B	B	146
Av-1	–	C	B	B	C	B	C	B	137
Av-2	–	C	C	B	C	C	B	C	137
H-26	–	C	B	B	C	C	C	B	4
H-30	–	C	B	C	C	B	C	B	4
H-36	–	C	B	B	C	B	C	B	4
If-2	–	C	B	B	C	B	B	C	35

Table 18.5—*cont.*

Locus	Chromosome	D	E	G	H	I	J	K	References
		CXB							
Ly-7	–	B	C	B	C	B	C	C	106
Ly-27	–	B	C	C	B	C	B	C	104
Ly-34	–	B	C	C	C	B	B	B	83
Pca-1	–	C	B	B	B	B	B	C	62
Pol-8	–	C	C	C	C	C	B	B	141
Pol-10	–	C	C	C	C	B	B	C	141
Pol-15	–	C	C	C	B	B	B	B	141
Pol-16	–	C	B	B	B	B	B	B	141
Pol-19	–	C	C	C	B	B	B	C	141
Pol-26	–	C	B	C	C	B	C	B	141

* C and B are used here as generic symbols of alleles inherited from BALB/cBy and C57BL/6By, respectively.

Table 18.6 Strain distribution patterns of loci typed in AKXD RI strains

Locus	Chromosome	1	2	3	6	7	8	9	10	11	12	13	14	15	16	17	18	20	21	22	23	24	25	26	27	28	References
Lamb-2	1	A*	D*	A	D	D	A	D			D	A	A	A	D		D	D	A	D	D		D		A		42
Odc-1	1	A	D	A	D	D	A	D			D	A	A	A	D		A	D	A	D	D		D		A		42
Xmmv-61	1	A	D	A	D	D	A	A	D	A	D	A	A	A	D	D	A	D	A	A	A	A	D	A	A	A	189
Il-1a	2	D	D	D	A	D	A	A	D	D	A	D	A	D	A	D	D	A	D	A	D	A	D	A	A	A	30
Emv-13 (Akv-3)	2	D	A	D	A	D	A	A	D	A	A	A	A	D	D	D	D	D	D	A	D	A	D	D	A	A	179
Ly-19	4	A	A	D	D	D	A	A	A	A	A	A	D	D	A	D	D	A	A	D	D	A	D	A	D	A	165
b	4	D	D	A	D	D	A	A	A	D	A	A	A	D	D	A	D	D	A	D	A	A	D	D	D	D	173
Orm-1	4	D	D	A	D	D	A	A			A	A	A	D	D		D	D	D	A	A	A	D		D		6
Rmcf	5	D	D	D	A	D	A	D	A	A	A	D	A	D	A	A	D	A	A	D	A	A	D	A	A	A	60
Xmmv-52	5	D	D	D	A	D	A	D	A	A	A	D	A	D	A	A	A	A	A	A	A	A	A	A	A	A	189
Pgm-1	5	D	A	D	A	D	D	D	A	D	D	A	A	A	D	A	A	A	A	A	A	D	D	A	D	A	60
Igk-Ef1, Rn7s-6, Ly-3	6	D	A	D	A	D	A	A	D	D	A	A	D	D	A	A	A	D	A	D	D	A	D	D	A	D	178
Igk-J	6	D	A	D	A	D	A	A	D	D	A	A	D	D	A	A	A	D	A	D	D	A	D	D	A		13
Emv-11 (Akv-1)	7	A	D	A	A	D	A	D	A	D	A	D	A	A	A	D	A	D	A	A	A	A	A	D	D	D	179
Abpa	7	A	D	A	A	D	A	D	A	D	A	A	A	A	A	D	A	A	A	A	A	A	D	A	D	D	38
Tam-1	7	A	D	D	A	D	D	D	A	D	D	A	A	A	A	D	A	A	A	A	D	A		A	D	D	38
c	7	D	A	D	A	A	D	A	A	D	D	A	D	A	A	D	A	D	A	A	A	A	A	A	D	D	173
Apoa1	9	D	D	A	A	D	A	D	A	A	A	D	D	D	D	D	A	D	D	D	A	A	D	A	D	A	121
Ncam	9	D	D	D	D	D	D	A	A	A	A	D	D	D	D	D	A	D	D	D	A	A	D	A	D	A	32
d (Emv-3)	9	D	D	D	A	D	D	A	A	D	D	D	D	D	A	D	D	A	D	D	A	A	D	D	D	A	80
Pgm-3	9	D	D	D	D	D	D	A	A	D	D	D	D	D	A	D	D	A	D	D	A	A	D	D	D	D	121
Mod-1	9	D	D	D	D	D	D	A	A	D	D	D	D	D	A	D	D	D	D	D	A	A	D	D	D	D	121
Sparc	11	D	D	D	A	A	A	A			A	D	A	D			A	D	D		A		D				103
Emv-14 (Akv-4)	11	A	A	D	A	A	D	A	D	D	A	A	A	D	D	D	A	D	D	D	A	A	D	A	A	D	179
Apob	12	D	D	D	A	D	D	A	D	D	A	A	A	A	D	A	D	A	A	A	A	D	D	D	D	D	95
Es-25	12	D	D	A	D	D	D	A	D	A	A	A	A	A	D	A	D		A	A	A	D	D	D	D	D	28
Ly-18	12	D	D	A	D	A	D	A	D	D	A	A	A	D	D	A	D	A	A	A	A	A	D	D	A	A	165
Tgn	15	D	D	D	A	D	A	A	D	A	D	A	D	A	D	D	A	D	D	D	A	D	A	D	A	D	181
Crya-1	17	A	D	A	A	A	D	D	A	A	A	D	D	A	D	A	D	D	A	D	D	D	A	A	A	A	156
li	18	D	A	A	A	A	A	D	A	A	D	D	A	D	A	D	A	A	A	D	A	D	A	D	A	A	135
Ly-10	19	D	D	D	A	A	D	A	D	A	A	A	D	D	D	A	A	A	A	D	A	A	D	A	A	D	165
Ly-1	19	D	D	D	A	A	D	A	D	A	D	A	D	D	A	A	A	A	A	D	A	A	D	A	A	D	165
Tpmt	–	D	A	A	A	D	A	A	A	A	A	A	D	D	A	D	A	A	A	A	A	A		D		A	124
Xmmv-54	–	D	D	D	D	A	D	D	A	D	D	A	D	A	A	A	A	D	D	A	D	A	A	A	D	A	189

* A and D are used here as generic symbols of alleles inherited from AKR/J and DBA/2J, respectively.

Table 18.7 Strain distribution patterns of loci typed in SWXL RI strains

Locus	Chromo-some	SWXL							References
		4	7†	12	14	15	16	17	
ln	1	S*	S	L*	S	S	L	S	196
Ren-2, Pep-3	1	L	S	L	S	L	S	S	196
Xmmv-61, Ly-22	1	L	S	L	S	S	S	L	189, 168
Ly-17, Akp-1	1	S	S	L	S	S	S	L	85, 191
Hc	2	L	S	S	S	L	S	S	31
Ly-11	2	S	S	L	L	L	S	L	109
B2m, Ly-23, a, Emv-15	2	S	L	L	L	L	S	L	20, 56, 93, 153
Amy-1,2	3	L	L	L	L	S	S	L	126
b	4	S	S	S	L	S	L	S	173
Ahd-1	4	L	L	S	L	L	L	S	193
Ly-31, Akp-2	4	L	L	S	L	S	L	S	169, 193
Gpd-1, Xmmv-62	4	L	S	S	L	S	L	S	193, 189
Coh, Gpi-1	7	L	L	S	L	S	S	L	200
Odc-6	7	L	S	S	L	S	S	L	136
c	7	L	S	S	S	S	S	L	173
Odc-7	7	S	S	S	L	S	S	L	136
Ly-15	7	L	S	S	S	L	S	S	71
Lap-1	9	L	S	S	S	S	S	S	41
Apoa1	9	L	L	L	S	S	S	S	41
Es-13, Mod-1	9	S	L	S	S	S	S	S	199, 41
Hsd	10	S	S	L	S	L	S	S	1
Sparc	11	S	L	L	L		S	S	103
Dlb-1	11	S	L	S	S	S	S	S	186
Ah	12	S	S	L	L	S	L	L	200
D12Nyu3, D12Nyu1	12	S	S	S	L	S	S	L	33
D12Nyu4, Fos	12	L	L	L	L	S	S	L	33
Igh-C	12	L	L	L	L	L	L	L	33
Odc-8	12	S	L	L	L	L	L	L	136
As-1	13	L	S	S	S	L	L	S	44
Lth-1	13	L	S	S	S	S			44
Tgn	15	S	S	S	S	S	L	L	181
D17Rp11	17	L	L	S	L	S	S	S	100
Xmmv-42	19	L	L	S	S	S	S	L	189
Lamb-1	–		S	L	S	L	S	S	42
Soa	–	S		L	S	S	S	L	190
Xmmv-66	–	S	S	L	L	L	L	S	189
Xmmv-67	–	L	S	S	S	S	L	S	189
Xp-11	–	L	S	L	S	L	L	S	189

* S and L are used here as generic symbols of alleles inherited from SWR/J and C57L/J, respectively.

† Extinct strain.

Table 18.8 Strain distribution patterns of loci typed in NX129 RI strains

Locus	Chromo-some	NX129						References
		1	2	5	7†	10	12	
Xmmv-61, Xmmv-60	1	N*	9*	N	9	N	N	189
a, Emv-15	2	N	9	N	N	9	9	153
Xmmv-65	3	N	9	N	9	9	9	189
Gbp-1	3	N	N	9	9	N	9	131
Ly-31	4	9	N	N		N	9	169
Ahd-1, Akp-2, Gpd-1, Xmmv-62	4	9	N	N	N	N	9	193, 189
Abpa	7	N	N	N	N	N	9	38
Saa, p	7	N	N	N	N	N	N	180
c	7	N	N	N	N	9	N	180
Ly-15	7	9		9	N	N	N	71
Pgm-3, Mod-1	9	9	9	9	N	9	9	121
Apob	12	9	N	9	9	9	9	95
Ly-18	12	9	N	N	9	9	9	165
Xmmv-55, Ly-6	15	9	9	N	N	9	N	189, 165
Xmmv-58	–	9	N	9	N	N	9	189
Xmmv-67	–	9	9	9	N	N	N	189

* N and 9 are used here as generic symbols of alleles inherited from NZB/BINJ and 129/J, respectively.　　　　† Extinct strain.

Table 18.9 Strain distribution patterns of loci typed in NX8 RI strains

Locus	Chromo-some	NX8					13 A	13† B	15†	16	17†	18	19†	20	References
		3†	4†	5†	6†	9									
Pep-3	1	N*	8*	N	8	8	N	N	8	N	N	8	8	8	26
Xmmv-61	1	N	8	8	8	N	8	8	N	8	N	N	N	8	189
Xmmv-60	1	N	8	N	8	N	N	N	N	8	N	N	N	8	189
Xmmv-65	3	N	8	8	N	N	8	8	N	8	N	8	8	N	189
Mup-1	4	8	8	8	8	N	N	8	8	N	N	N	8	N	26
Xmmv-62	4	8	8	8	8	N	8	8	8	N	N	N	8	N	189
Gpd-1	4	N	8	8	8	N	8	8	8	N	N	N	N	N	26
Pgm-1	5		N	N	8	N	N	N	8	N		8	8		26
Igk-V, Rn7s-6,	6	8	8	8	N	8	8	8	N	8	8	N	8	8	134, 178
Ly-2, Hbb	7	8	N	N	8	N	N	N	N	8	8	8	8	N	26
Lap-1	9	8	8	N	N	N	8	8	N	8	8	N	N	8	26
Tcn-2	11	N	N	N	8	8	8	N	8	N	N	N	N	8	54
Ah	12	N	N	N	8	8	N	N	N	N	N	8	N	N	26
Aat	12	N	8	8	N	8	8	8	8	N	8	8	N	N	134
Igh-C, Igh-V	12	N	8	8	N	8	8	8	N	8	8	N	N	8	134
Es-10	14	N	8	N	N	N	N	8	8	8	8	8	N	N	26
Gpt-1	15	8	8	8	8	N	N	8	8	8	8	8	N	8	26
H-2	17	8	8	N	N	8	8	8	N	8	8	N	N	N	134
Xmmv-42	19	N	N	N	N	N	N	N	N	8	8	N	8	8	189
Xmmv-47	–	8	8	N	N	N	8	8	N	8	N	8	8	N	189
Xmmv-48	–	N	N	8	8	N	N	N	8	8	8	8	8	N	189
Xmmv-58	–	N	8	8	8	N	N	N	N	N	8	N	N	N	189
Xmmv-66	–	8	8	N	8	8	N	N	N	8	N	8	8	8	189

* N and 8 are used here as generic symbols of alleles inherited from NZB/Icr and C58/J, respectively.
† Extinct strains.

Table 18.10 Strain distribution patterns of loci typed in CXS RI strains

Locus	Chromosome	CXS														References
		1	2	3	4	5	6	7	8	9	10	11	12	13	14	
Len-1	1	S*	C*	S	S	C	C	C	C	S	C	C	S	C	S	66
Ren-2	1	S	C	S	S	C	S	S	S	S	S	S	C	C	C	195
Pep-3	1	S	C	S	S	C	S	S	S	S	C	S	C	C	S	113
Ly-5	1	S	S	S	S	C	S	S	S	S	C	S	C	C	S	66
Mtv-7	1	S	C	S	S	C	C	S	C	S	C	S	C	S	S	113
Ly-22, Ly-17	1	S	C	C	S	C	C	S	S	S	C	S	C	S	S	113
Akp-1	1	S	C	C	S	C	C	S	C	S	S	S	C	S	S	113
Hc	2	C	S	C	C	C	S	C	C	S	C	C	S	C	C	66
Ly-25	2	S	S	S	C	C	C	C	S	C	S	C	C	S	C	73
Ly-24	2	S	S	S	C	C	C	S	S	C	S	C	C	S	C	73
Hao-1	2	S	S	S	C	C	S	C	S	C	S	C	C	S	C	66
Gbp-1	3	S	S	C	S	S	S	C	S	S	C	S	C	C	S	131
Lyb-2	4	C	S	C	C	S	C	C	C	C	S	C	S	C	S	113
Ly-29, Mtv-20	4	C	S	C	C	S	C	C	C	C	S	S	S	C	S	58, 113
Mup-1	4	C	S	C	C	S	C	C	S	C	S	S	C	C	S	58
b	4	C	S	C	C	S	C	C	S	C	S	S	C	S	S	58
Emv-1 (Cv-1)	5	S	S	S	C	S	C	C	C	S	C	S	C	S	C	66
Pgm-1	5	S	S	S	S	C	C	C	C	S	C	C	C	C	C	66
Dao-1, rd	5	S	S	C	S	C	C	C	S	C	C	S	S	C	C	66
Gus-r	5	C	C	C	S	C	S	C	C	C	S	C	S	S	C	66
Gpi-1, Mtp-1	7	S	S	S	S	S	S	C	C	C	S	S	C	S	S	66
Prt-4, Prt-5	7	S	S	S	S	S	C	C	C	C	S	S	S		S	66
Gr-1	8	C	S	C	S	S	S	S	S	C	C	C	S	S	C	66
Eg	8	C	C	C	S	S	S	S	S	C	C	S	S	S	C	66
Es-9	8	C	C	C	S	S	S	S	S	S	C	S	S	S	C	66
Mod-1	9	S	S	S	S	S	S	C	C	C	C	S	C	C	S	66
Tcn-2	11	C	S	S	C	S	C	S	C	C	S	C	C	C	S	55
Hba	11	C	S	C	C	C	S	S	C	C	S	C	C	C	S	55
Dlb, Hox-2	11	C	C	S	S	S	S	S	C	S	S	S	S	C	S	186, 116
Rnul-4	11	C	C	S	S	S	S	S	C	S	C	S	S	C	S	111
Mtv-3	11	C	C	S	C	S	C	S	C	S	S	S	S	C	S	113
Es-3	11	C	S	S	C	S	C	S	C	S	S	S	S	C	S	111
Mtv-9	12	C	C	C	S	C	C	C	S	C	S	C	S	S	S	113
RB-F, RB-G	16	S	C	S	S	C	S	C	S	C		S	S	C	S	66
Mtv-6	16	S	C	C	S	S	C	C	C	C	C	S	S	S	C	113
H-2	17	C	S	S	C	S	S	C	C	C	S	C	S	C	S	66
Acta	–	S	S	S	S	C	S	S	S	C	C	S	S	S		146

* S and C are used here as generic symbols of alleles inherited from STS/A and BALB/cHeA, respectively.

Table 18.11 Strain distribution patterns of loci typed in 58NXL and LXB RI strains

Locus	Chromosome	58NXL					LXB			References
		1	2†	3	4	8	2	3	4	
Idh-1	1	L*	B*	B	B	B	B	L	B	14
Lsh	1	L	B	B	B	L	B	L	B	14
ln	1	B	B	B	B	L	B	L	B	14
b	4	L	B	B	B	L	L	B	L	173
Ggc	6	B	B	B	L	B	B	L	L	185
Aat	12	B	B	B	B	L	B	L	L	72
Igh-1	12	B	B	B	B	L	B	L	L	72
Ly-7	–	L		L	B	L	B	B	L	72

* L is used here as a generic symbol of alleles inherited from the C57L/J strain. B is used as a symbol of alleles inherited from B10.D2(58N)/Sn in the 58NXL RI strains and from C57BL/6J in the LXB RI strains.
† Extinct strain.

Table 18.13 Strain distribution patterns of loci typed in LXPL RI strains

Locus	Chromosome	LXPL				References
		1†	2	4	6	
ln	1	P*		P	L*	173
Itp	2	L	L	P	L	183
a	2	L	P	P	L	183
b	4	P		L	L	173
Ahd-1	4	P	L	P	L	193
Akp-2	4	P	L	P	P	193
Ly-31	4	L		P	P	169
Igk-Ef1, Rn7s-6	6	L	L	L	L	178
Igk-J	6			L	L	13
c	7	L	P	L	L	173
Tgn	15	L	P	L	L	181

* P and L are used here as generic symbols of alleles inherited from PL/J and C57L/J, respectively. † Extinct strain.

Table 18.12 Strain distribution patterns of loci typed in BRX58N RI strains

Locus	Chromosome	BRX58N										References
		1	3†	4	7	8	9	10	11	12	13	
Idh-1	1	R*	R	B*	B	R	R	B	R	B	B	14
Lsh	1	R	R	R	B	R	R	R	R	B		14
b	4	R	B	B	B	B	R	B	R	B	R	173
Ggc	6	R	R	R	B	B	B		R	R	R	185
Aat	12	B	R	R	R	R	B	R	R	R	R	72
Igh-1	12	B	B	B	R	B	B	R	B	R	R	72
Ly-7	–	R	B	B	R	R	R	R	R	B	R	72

* R and B are used here as generic symbols of alleles inherited from C57BR/cdJ and B10.D2(58N)/Sn, respectively. † Extinct strain.

Table 18.14 Strain distribution patterns of loci typed in AXB RI strains

Locus	Chromosome	AXB																									References
		1	2	3	4	5	6	7	8	9	10	11	12	13	14	15	17	18	19	20	21	22	23	24	25	
Lsh	1	B*	A*		A		B	A									A									155
B2m	2	A	B			A	A	B		A			A	B		A	A									92
Gus	5	B	A			A	A	B		A			A	B			A									92
Met	6	B	B	A	A	B	A	A	B	A	A	B	A	B	B		B	B		B		B	B	B	A	27
Hox-1.5	6	B	B	A	A		A	A	B	B	A	B	A	B	B	B	A	B	B	B	B	B	A	B		117
Fes	7	B	A	A	B	A	A		A	B		A		A	A			B	A					B		11
Gpi-1	7	B	A	A	B	A	A		A	B		A		A	A			B	A					B		11
Tam-1	7	B	B	A	B	A	A		A	B		A		B	A			B	A							11
c	7	B	A	A	B	B	A		A	B		A		A	B			B	B					B		11
Hbb	7	B	A	A	B	A	A		A	B		A		A	B			B	B					B		11
Ly-6	15	A	A	B	B	B	A	B	B	A	B	B	A	B	B	B	B	B	B	B	B	B	A	A	B	110
Sis	15	A	A	B	B	A	A	B	B	A	B	B	A	B	B	A	B	A	A	A	B	B	A	A	B	110
Pol-5	15	A	A	B	B	B	A	B	B	A	B	B	A	B	B	B	B	A	A	A	B	B	A	A	B	110
Ins-3	15	A	A		B		A	B	B		B	B	A	B	A		A		A		A		A			110
Krt-1	15	A	B	B	B	A	B	B	A	A	B	B	A	A	A	A	A		A		A		B		B	110
Int-1	15	A	B	B	B	A	B	B		B	B	A		A		A	A	A	A	A		B	B			110
Gdc-1	15	A	A	B	B	B	B	B	A	A	B	B	A	A	A	A	A	A	A	A		B	B	A		110
H-2	17	A	A		A		B	A		A			A	B		B	A									92

* B and A are used here as generic symbols of alleles inherited from C57BL/6J and A/J, respectively.

Table 18.15 Strain distribution patterns of loci typed in BXA RI strains

Locus	Chromosome	1	2	3†	4	5†	6	7	8	9	10	11	12	13	14	16	17	18	19	20	22	23	24	25	References
Lsh	1						B*		A*					A											155
B2m	2	A	A		B		A		B		A	B			B										92
Gus	5	A	A		A		B				A	A			B										92
Met	6	B	A		B		A	A	B	A	A	B	B	A	A			A	A	B	A	B	B	A	27
Hox-1.5	6	A	B		A		B	A	A	A	A	A	A	A	B	A		A	B	B	A	A	A	B	117
Fes	7	B		B	A		A	A						A	A				A			A			11
Gpi-1	7	B		B	B		A	A						A					B			A			11
Tam-1	7	B		B	A		A	B						A	A				B			A			11
c	7	B		B	A		B	A						A	A				A			B			11
Hbb	7	B		B	A		B	A						A	A				A			A			11
Ly-6	15	B	B		B		B	B	A	B	B	B	B	B	A	B		B	B	B	B	B	B	B	110
Sis	15	B	B	B	B		B	A	A	B	B	B	B	B	A	B	A	B	B	B	B		B	B	110
Pol-5	15	B	B	B	B	A	B	B	A	B	B	B	B	B	A	B	A	B	B	B	B	B	B	B	110
Ins-3	15	B		B	B	A	B		A			B	B	B	A	B	A				B	B			110
Krt-1	15	B		B	A	A		B				B	A	B	A	B		B	A		A	B			110
Int-1	15			B	A	A						B	A		A	B	A	B	A		A	B			110
Gdc-1	15	A	A	B	A	A	B	A	B	A	A	B	A	B	A	B	A	B	A	A	A	B	B		110
H-2	17	B	B		A		A		A			B	B		B										92

* See footnote to Table 18.14.
† Extinct strains.

Table 18.16 Strain distribution patterns of loci typed in CXJ RI strains

Locus	Chromosome	1	3	4	6	8	9	10†	11†	15	References
Cfh	1	C*	J*		C	J	J	C	J	C	31
Il-1a	2	C	J		C	C	J	J	J	C	30
Itp	2	C	J	J	C	C	J	J	J	C	183
Tcrb	6	C	J	J	J	J	J	J	C	J	50
Igk	6	C	C	J	J	J	J	J	J	C	50
Hv-2	7	J	J	C	C	C	C	J	C	J	86
Hbb	7	C	J	C	C	C	C	J	C	C	51
Ncam	9	J	J		J	C	C	C	J	J	32
D12Nyu5	12	C	C		J	C	J	C	J	C	33
D12Nyu3, D12Nyu1	12	C	C		J	C	J	J	J	C	33
Fos	12	C	C		C	C	J	J			33
Igh	12	C	J	C	C	J	J	J	J	J	51
Tcra-V2, Tcra-C	14	C	C	C	C	J	C	C	C	J	51
Tcra-V1	14	C	C	J	C	J	C	C	C	J	51
Es-10	14	C	C	J	J	J	C	C	C	J	51
Igl	16	C	J	C	J	C	J	J	J	J	51
H-2	17	C	C	J	J	C	C	C	C	J	51

* C and J are used here as generic symbols of alleles inherited from BALB/cKe and SJL/J, respectively.
† Extinct strains.

References

1. Arfin, S.M., W.C. Hanford, and B.A. Taylor. 1979. Assignment of histidase-regulating locus to chromosome 10 of the mouse. Biochem. Genet. 17:529–35.

2. Arnheim, N., D. Treco, B. Taylor, and E.M. Eicher. 1982. Distribution of ribosomal length variants among mouse chromosomes. Proc. Natl. Acad. Sci. USA 79:4677–80.

3. Bailey, D.W. 1971. Recombinant-inbred strains, an aid to finding identity, linkage, and function of histocompatibility and other genes. Transplantation 11:325–7.

4. Bailey, D.W. 1975. Genetics of histocompatibility in mice. I. New loci and congenic lines. Immunogenetics 2:249–56.

5. Balak, K.J., R.H. Keith, and M.R. Felder. 1982. Genetic and developmental regulation of mouse liver alcohol dehydrogenase. J. Biol. Chem. 257:15000–7.

6. Baumann, H., W.A. Held, and F.G. Berger. 1984. The acute phase response of mouse liver. Genetic analysis of the major acute phase reactants. J. Biol. Chem. 259:566–73.

7. Bennett, K.L., P.A. Lalley, and N.D. Hastie. 1982. Mapping of a multigene family: The structural genes coding for the major urinary proteins in the mouse are on chromosome 4. Proc. Natl. Acad. Sci. USA 79:1220–4.

8. Berger, F. 1983. Studies on genetic variation in major urinary protein synthesis in mouse liver. Biochem. Genet. 2:15-23.

9. Berman, J.W., A.J.D. Rocha, and R. Basch. 1986. Restriction length polymorphism in the variable region of the Tcr locus linked to histocompatibility antigen H-8 on murine chromosome 14. Immunogenetics 24:328–30.

10. Birkenmeier, E.H., and J.I. Gordon. 1986. Developmental regulation of a gene that encodes a cysteine-rich intestinal protein and maps near the murine immunoglobulin heavy chain locus. Proc. Natl. Acad. Sci. USA 83:2516–20.

11. Blatt, C., M.E. Harper, G. Franchini, M.N. Nesbitt, and M.I. Simon. 1984. Chromosomal mapping of murine c-fes and c-src genes. Mol. Cell. Biol. 4:978–81.

12. Blatt, C., K. Mileham, M. Haas, M.N. Nesbitt, M.E. Harper, and M.I. Simon. 1983. Chromosomal mapping of the mink cell focus-inducing xenotropic env gene family in the mouse. Proc. Natl. Acad. Sci. USA 80:6298–302.

13. Boyd, R.T., M.M. Goldrick, and P.D. Gottlieb. 1986. Genetic polymorphism at the mouse immunoglobulin J_x locus (Igk-J) as demonstrated by Southern hybridization and nucleotide sequence analysis. Immunogenetics 24:150–7.

14. Bradley, D.J., B.A. Taylor, J. Blackwell, E.P. Evans, and J. Freeman. 1979. Regulation of Leishmania populations within the host. III. Mapping of the locus controlling susceptibility to visceral leishmaniasis in the mouse. Clin. Exp. Immunol. 37:7–14.

15. Bradlow, N.L., R.J. Hershcopf, C.P. Martucci, and J. Fishman. 1985. Estradiol 16a-hydroxylation in mouse correlates with mammary tumor incidence and presence of murine mammary tumor virus: a possible model for the hormonal etiology of breast cancer in humans. Proc. Natl. Acad. Sci. USA 82:6295–99.

16. Brilliant, M.H., M.M. Neimann, and E.M. Eicher. 1987. Murine tyrosine hydroxylase maps to the distal end of chromosome 7 within a linkage group conserved in mouse and man. J. Neurogenet. 4:259–66.

17. Bucan, M., B.G. Herrmann, A.-M. Frischauf, V.L. Bautch, V. Bode, L.M. Silver, G.R. Martin, and H. Lehrach. 1987. Deletion and duplication of DNA sequences is associated with the embryonic lethal phenotype of the t complementation group of the mouse t complex. Genes Dev. 1:376–85.

18. Bucan, M., T. Yang-Feng, A.M. Colberg-Poley, D.J. Wolgemuth, J-L. Guénet, U. Francke, and H. Lehrach. 1986. Genetic and cytogenetic localisation of the homeo box containing genes on mouse chromosome 6 and human chromosome 7. EMBO J. 5:2899–905.

19. Buchberg, A.M., B.A. Taylor, N.A. Jenkins, and N.G. Copeland. 1986. Chromosomal localization of Emv-16 and Emv-17, two closely linked ecotropic proviruses of RF/J mice. J. Virol. 60:1175–8.

20. Chorney, M., F.-W. Shen, J. Michaelson, and E.A. Boyse. 1982. Monoclonal antibody to an alloantigenic determinant on β2-microglobulin (β2m) of the mouse. Immunogenetics 16:91–3.

21. Claflin, J.L., B.A. Taylor, M. Cherry, and M. Cubberly. 1978. Linkage in mice of genes controlling an immunoglobulin kappa-chain marker and the surface alloantigen Ly-3 on T lymphocytes. Immunogenetics 6:379–87.

22. Cobb, R.R., T.A. Stoming, and J.B. Whitney III. 1987. The aryl hydrocarbon hydroxylase (Ah) locus and a novel restriction-fragment length polymorphism (RFLP) are located on mouse chromosome 12. Biochem. Genet. 25:401–13.

23. Colombatti, A., E.N. Hughes, B.A. Taylor, and J.T. August. 1982. Gene for a major cell surface glycoprotein of mouse macrophages and other phagocytic cells is on chromosome 2. Proc. Natl. Acad. Sci. USA 79:1926–9.

24. Cumming, R.B., M.F. Walton, J.C. Fuscoe, B.A. Taylor, J.E. Womack, and F.H. Gaertner. 1979. Genetics of formamidase-5 (brain formamidase) in the mouse: localization of the structural gene on chromosome 14. Biochem. Genet. 17:415–31.

25. Dandoy, F., K.A. Kelley, J. DeMaeyer-Guignard, E. DeMaeyer, and P.M. Pitha. 1984. Linkage analysis of the murine interferon-α locus on chromosome 4. J. Exp. Med. 160:294–302.

26. Datta, S.K., F.L. Owen, J.E. Womack, and R.J. Riblet. 1982. Analysis of recombinant inbred lines derived form 'autoimmune' (NZB) and 'high leukemia' (C58) strain: independent multigenic systems control B cell hyperactivity, retrovirus expression, and autoimmunity. J. Immunol. 129:1539–44.

27. Dean, M., C. Kozak, J. Robbins, R. Callahan, S.

O'Brien, and G.F. Vande Woude. 1987. Chromosomal localization of the *met* proto-oncogene in the mouse and cat genome. Genomics 1:167–73.

28. Deimling, O.H.v., and B.A. Taylor. 1987. Esterase-25 (*Es-25*): identification and characterization of a new kidney arylesterase of the house mouse, genetically linked to *Ly-18* on chromosome 12. Biochem. Genet. 25:639–46.

29. D'Eustachio, P., B. Fein, J. Michaelson, and B.A. Taylor. 1984. The α-globin pseudogene on mouse chromosome 17 is closely linked to H-2. J. Exp. Med. 159:958–63.

30. D'Eustachio, P., S. Jedidi, R.C. Fuhlbrigge, P.W. Gray, and D.D. Chaplin. 1987. Interleukin-1 α and β genes: linkage on chromosome 2 in the mouse. Immunogenetics 26:339–43.

31. D'Eustachio, P., T. Kristensen, R.A. Wetsel, R. Riblet, B.A. Taylor, and B.F. Tack. 1986. Chromosomal location of the genes encoding complement components C5 and factor H in the mouse. J. Immunol. 137:3990–5.

32. D'Eustachio, P., G.C. Owens, G.M. Edelman, and B.A. Cunningham. 1985. Chromosomal location of the gene encoding the neural cell adhesion molecule (N-CAM) in the mouse. Proc. Natl. Acad. Sci. USA 82:7631–5.

33. D'Eustachio, P. 1984. A genetic map of mouse chromosome 12 composed of polymorphic DNA fragments. J. Exp. Med. 160:827–38.

34. De Maeyer, E., J. De Maeyer-Guignard, and D.W. Bailey. 1975. Effect of mouse genotype on interferon production. I. Lines congenic at the *If-1* locus. Immunogenetics 1:438–43.

35. DeMaeyer, E., J. De Maeyer-Guignard, W.T. Hall, and D.W. Bailey. 1974. A locus affecting circulating interferon levels induced by mouse mammary tumour virus. J. Gen. Virol. 23:209–11.

36. Dembic, Z., W. Bannwarth, B.A. Taylor, and M. Steinmetz. 1985. The gene encoding the T-cell receptor α-chain maps close to the *Np-2* locus on mouse chromosome 14. Nature 314:271–3.

37. Demmer, L.A., E.H. Birkenmeier, D.A. Sweetser, M.S. Levin, S. Zollman, R.S. Sparkes, T. Mohandas, A.J. Lusis, and J.I. Gordon. 1987. The cellular retinol binding protein II gene. Sequence analysis of the rat gene, chromosomal localization in mice and humans, and documentation of its close linkage to the cellular retinol binding protein gene. J. Biol. Chem. 262:2458–67.

38. Dlouhy, S.R., B.A. Taylor, and R.C. Karn. 1987. The genes for mouse salivary androgen-binding protein (ABP) subunits alpha and gamma are located on chromosome 7. Genetics 115:535–43.

39. Duley, J.A., and R.S. Holmes. 1982. Biochemical genetics of aldehyde reductase in the mouse: *Ahr-1* – A new locus linked to the alcohol dehydrogenase gene complex on chromosome 3. Biochem. Genet. 20:1067–83.

40. Dumont, F.J., R.C. Habbersett, L.Z. Coker, E.A. Nichols, and J.A. Treffinger. 1985. High level expression of the plasma cell antigen PC.1 on the T-cell subset expanding in the MRL/MpJ- *lpr/ lpr* mice: detection with a xenogeneic monoclonal antibody and alloantisera. Cell. Immunol. 96:327–37.

41. Eicher, E.M., B.A. Taylor, S.C. Leighton, and J.E. Womack. 1980. A serum protein polymorphism determinant on chromosome 9 of *Mus musculus*. Mol. Gen. Genet. 177:571–6.

42. Elliott, R.W., D. Barlow, and B.L.M. Hogan. 1985. Linkage of genes for laminin B1 and B2 subunits on chromosome 1 in mouse. In Vitro Cell. Dev. Biol. 21:477–84.

43. Elliott, R.W., and F.G. Berger. 1983. DNA sequence polymorphism in an androgen-regulated gene is associated with alteration in the encoded RNAs. Proc. Natl. Acad. Sci. USA 80:501–4.

44. Elliott, R.W., W.L. Daniel, B.A. Taylor, and E. Novak. 1985. Linkage of loci affecting a murine liver protein and arylsulfatase B to chromosome 13. J. Hered. 76:243–5.

45. Elliott, R.W., C. Hohman, C. Romijko, P. Louis, and F. Lilly. 1978. Use of high resolution two dimensional electrophoresis of liver cytosol proteins in the discovery and mapping of a new mouse variant. *In* N. Catsimpoolas, ed., Electrophoresis, 261–74. Elsevier North Holland, Amsterdam.

46. Elliott, R., E. Mann, and S. Berger. 1985. Correction. Mouse News Lett. 72:119.

47. Elliott, R.W., C. Romejko, and C. Hohman. 1980. Mapping the gene for LTW-4, a 26,000 molecular weight major protein of mouse liver and kidney. Mol. Gen. Genet. 180:17–22.

48. Elliott, R.W., L.C. Samuelson, M.S. Lambert, and M.H. Meisler. 1986. Assignment of pancreatic ribonuclease gene to mouse chromosome 14. Cytogenet. Cell Genet. 42:110–12.

49. Elliott, R.W. 1979. Use of two-dimensional electrophoresis to identify and map new mouse genes. Genetics 91:295–308.

50. Epstein, R., N. Roehm, P. Marrack, J. Kappler, M. Davis, S. Hedrick, and M. Cohn. 1985. Genetic markes of the antigen-specific T cell receptor locus. J. Exp. Med. 161:1219–24.

51. Epstein, R., G. Sham, J. Womack, J. Yague, E. Palmer, and M. Cohn. 1986. The cytotoxic T cell response to the male-specific histocompatibility antigen (H-Y) is controlled by two dominant immune response genes, one in the MHC, the other in the Tara-locus. J. Exp. Med. 163:759–73.

52. Feeney, A.J. 1978. Expression of Ly-6 on activated T and B cells: possible identity with Ala-1. Immunogenetics 7:537–43.

53. Festenstein, H., C. Bishop, and B.A. Taylor. 1977. Location of *Mls* locus on mouse chromosome 1. Immunogenetics 5:357–61.

54. Frater-Schroder, M., O. Haller, R. Gmur, L. Kierat, and S. Anastasi. 1982. Allelic forms of mouse transcobalamin 2. Biochem. Genet. 20:1001–14.

55. Frater-Schroder, M., M. Prochazka, O. Haller, F. Arwert, H.J. Porck, L.C. Skow, L-G. Lundin, J. Hilkens, and J. Hilgers. 1985. Localization of the gene for the vitamin B12 binding protein transcobalamin II, near the centromere on mouse chromosome 11, linked with the hemoglobin alpha-chain locus. Biochem. Genet. 23:139–53.

56. Gasser, D.L., J. Ziebar, and K. Matsumoto. 1983. A lymphocyte antigen encoded by a locus closely linked to *H-3* on the second chromosome of the mouse. Proc. Natl. Acad. Sci. USA 80:6620–3.

57. Goldman, D., and H.J. Pikus. 1986. Fourteen genetically variant proteins of mouse brain: discovery of two new variants and chromosomal mapping of four loci. Biochem. Genet. 24:183–94.

58. Gonez, L.J., K.S. Sandrin, M.M. Henning, J. Hilgers, and I.F.C. McKenzie. 1985. Ly-29: a locus closely linked to Mtv-20 on chromosome 4 codes for a new mouse lymphocyte surface alloantigen. Immunogenetics 22:305–8.

59. Groves, M.G., D.L. Rosenstreich, B.A. Taylor, and J.V. Osterman. 1980. Host defenses in experimental scrub typhus: mapping the gene that controls natural resistance in mice. J. Immunol. 125:1395–9.

60. Hartley, J.W., R.A. Yetter, and H.C. Morse III. 1983. A mouse gene on chromosome 5 that restricts infectivity of MCF-type recombinant murine leukemia viruses. J. Exp. Med. 158:16–24.

61. Hayakawa, J., and H. Nickaido. 1987. Two types of liver-specific F antigen are encoded by a locus located on chromosome 5 in mice. Immunogenetics 26:366–9.

62. Herberman, R.B., and T. Aoki. 1972. Immune and natural antibodies to syngeneic murine plasma cell tumors. J. Exp. Med. 136:94–111.

63. Herrmann, B., M. Bucan, P.E. Mains, A.-M. Frischauf, L.M. Silver, and H. Lehrach. 1986. Genetic analysis of the proximal portion of the mouse *t* complex: Evidence for a second inversion within *t* haplotypes. Cell 44:469–76.

64. Heuckeroth, R.O., E.H. Birkenmeier, M.S. Levin, and J.I. Gordon. 1987. Analysis of the tissue-specific expression, developmental regulation, and linkage relationships of a rodent gene encoding heart fatty acid binding protein. J. Biol. Chem. 262:9709–17.

65. Hildebrand, C.E., F.J. Gonzalez, C.A. Kozak, and D.W. Nebert. 1985. Regional linkage analysis of the dioxin-inducible P-450 gene family on mouse chromosome 9. Biochem. Biophys. Res. Commun. 130:396–406.

66. Hilgers, J., and J.W.A. Arends. 1985. A series of recombinant inbred strains between the BALB/cHeA and STS/A strains. Curr. Top. Microbiol. Immunol. 122:31–7.

67. Hill, R.E., A.E. Hall, C.M. Sime, and N.D. Hastie. 1987. A mouse homeo box-containing gene maps near a developmental mutation. Cytogenet. Cell Genet. 44:171–4.

68. Hill, R.E., P.H. Shaw, R.K. Barth, and N.D. Hastie. 1985. A genetic locus closely linked to a protease inhibi-

tor gene complex controls the level of multiple RNA transcripts. Mol. Cell. Biol. 5:2114–22.

69. Hjorth, P. 1980. Mouse News Lett. 62:41.

70. Hogarth, P.M., J. Craig, and I.F.C. McKenzie. 1980. A monoclonal antibody detecting the Ly-9.2 (Lgp100) cell-membrane alloantigen. Immunogenetics 11:65–74.

71. Hogarth, P.M., E.M. Eicher, and I.F.C. McKenzie. 1986. Mapping of the murine *Ly-15* (*LFA-1*) locus to chromosome 7. Immunogenetics 23:348–9.

72. Hogarth, P.M., I.F.C. McKenzie, L. Lanier, D.W. Bailey, and B.A. Taylor. 1984. Location of *Ly-7* on mouse chromosome 12. Immunogenetics 19:539–43.

73. Hogarth, P.M., A. Rigby, V.R. Sutton, I.F.C. McKenzie, and J. Hilgers. 1984. The Ly-25.1 specificity: definition with a monoclonal antibody. Immunogenetics 19:83–6.

74. Holmes, R.S., P.B. Mather, and J.A. Duley. 1985. Gene markers for alcohol-metabolizing enzymes among recombinant inbred strains of mice with differential behavioural responses toward alcohol. Anim. Blood Groups Biochem. Genet. 16:51–9.

75. Howe, R.C., A. Ahmed, T.J. Faldetta, J.E. Byrnes, K.M. Rogan, B.A. Taylor, and R.E. Humphreys. 1979. Mapping of the *Lyb-4* gene to different chromosomes in DBA/2J and C3H/HeJ mice. Immunogenetics 9:221–32.

76. Howles, P.N., D.P. Dickinson, L.L. DiCaprio, M. Woodworth-Gutai, and K.W. Gross. 1984. Use of a cDNA recombinant for the γ-subunit of mouse nerve growth factor to localize members of this multigene family near the TAM-1 locus on chromosome 7. Nucl. Acids Res. 12:2791–805.

77. Imanishi, T., and O. Makela. 1974. Inheritance of antibody specificity. I. Anti-(4-hydroxy-3-nitrophenyl)acetyl of the mouse primary response. J. Exp. Med. 140:1498–1510.

78. Jeffreys, A.J., V. Wilson, R. Kelley, B.A. Taylor, and G. Bulfield. 1987. Mouse DNA 'fingerprints': analysis of chromosome localization and germ-line stability of hypervariable loci in recombinant inbred strains. Nucl. Acids Res. 15:2823–36.

79. Jenkins, N.A., N.G. Copeland, B.A. Taylor, H.G. Bedigian, and B.K. Lee. 1982. Ecotropic murine leukemia virus DNA content of normal and lymphomatous tissues of BXH-2 recombinant inbred mice. J. Virol. 42:379–88.

80. Jenkins, N.A., N.G. Copeland, B.A. Tylor, and B.K. Lee. 1981. Dilute (*d*) coat colour mutation in DBA/2J mice is associated with the site of integration of an ecotropic MuLV genome. Nature 293:370–4.

81. Joyner, A.L., and G.R. Martin. 1987. *En-1* and *En-2*, two mouse genes with sequence homology to the Drosophila engrailed gene: expression during embryogenesis. Genes Dev. 1:29–38.

82. Ju, S-T., and M.E. Dorf. 1980. Idiotypic analysis of anti-GAT antibodies. IX. Genetic mapping of the Gte idiotypic marker within the *Igh-V* locus. J. Immunol. 126:183–6.

83. Kimura, S., J. Furze, and H. Festenstein. 1987. Ly-34: A new mouse lymphocyte alloantigen. Immunogenetics 26:381–2.

84. Kimura, S., N. Tada, Y. Liu, and U. Hammerling. 1981. A new mouse cell-surface antigen (Ly-m18) defined by a monoclonal antibody. Immunogenetics 13:547–54.

85. Kimura, S., N. Tada, E. Nakayama, Y. Liu, and U. Hammerling. 1981. A new mouse cell-surface antigen (Ly-m20) controlled by a gene linked to Mls locus and defined by monoclonal antibodies. Immunogenetics 14:3–14.

86. Knobler, R.L., D.S. Linthicum, and M. Cohn. 1985. Host genetic regulation of acute MHV-4 viral encephalomyelitis and acute experimental autoimmune encephalomyelitis in (BALB/cKe × SJL/J) recombinant-inbred mice. J. Neuroimmunol. 8:15–28.

87. Koizumi, T., and J. Hawakawa. 1987. Single locus control of the mast cell population in mouse skin. Immunogenetics 26:36–9.

88. Kozak, L.P. 1985. Interacting genes control glycerol-3-phosphate dehydrogenase expression in developing cerebellum of the mouse. Genetics 110:123–43.

89. Ledbetter, J.A., J.W. Goding, T.T. Tsu, and L.A. Herzenberg. 1979. A new mouse lymphoid alloantigen (Lgp 100) recognized by a monoclonal rat antibody. Immunogenetics 8:347–60.

90. Legraverend, C., S.O. Karenlampi, S.W. Bigelow, P.A. Lalley, C.A. Kozak, J.E. Womack, and D.W. Nebert. 1984. Aryl hydrocarbon hydroxylase induction by benzo[a]anthracene: regulatory gene localized to the distal portion of mouse chromosome 17. Genetics 107:447–61.

91. Lieberman, R., M. Potter, E.B. Mushinski, W. Humphrey, and S. Rudikoff. 1974. Genetics of a new Ig V_H (T15 idiotype) marker in the mouse regulating natural antibody to phosphorylcholine. J. Exp. Med. 139:983–1001.

92. Lovell, D.P., and R.P. Erickson. 1986. Genetic variation in the shape of the mouse mandible and its relationship to glucocorticoid-induced cleft palate analyzed by using recombinant inbred lines. Genetics 113:755–64.

93. Lovett, M., Z. Cheng, E.M. Lamela, T. Yokoi, and C.J. Epstein. 1987. Molecular markers for the agouti coat color locus of the mouse. Genetics 115:747–54.

94. Lundin, L.-G. 1987. A gene Bmn controlling β-mannosidase activity in the mouse is located in the distal part of chromosome 3. Biochem. Genet. 25:603–10.

95. Lusis, A.J., B.A. Taylor, D. Quon, S. Zollman, and R.C. LeBoeuf. 1987. Genetic factors controlling structure and expression of apolipoproteins B and E in mice. J. Biol. Chem. 262:7594–604.

96. Lusis, A.J., B.A. Taylor, R.W. Wangenstein, and R.C. LeBoeuf. 1983. Genetic control of lipid transport in mice. II. Genetic variations of high density lipoproteins. J. Biol. Chem. 258:5071–8.

97. Lyman, S.D., A. Poland, and B.A. Taylor. 1980. Genetic polymorphism of microsomal epoxide hydrolase activity in the mouse. J. Biol. Chem. 255:8650–4.

98. Lynch, D.H., R.E. Gress, B.W. Needleman, S.A. Rosenberg, and R.J. Hodes. 1985. T cell responses to Mls determinants are restricted by cross-reactive MHC determinants. J. Immunol. 134:2071–8.

99. Makela, O., K. Karjalainen, T. Imanishi-Kari, and B.A. Taylor. 1981. Linkage and recombination of V_H gene markers, Np and Nb, in recombinant inbred strains of mice. Immunol. Lett. 3:169–72.

100. Mann, E., R.W. Elliott, and C. Hohman. 1984. Of four loci spanning mouse chromosome 17, two map to the t-complex. Mouse News Lett. 71:48.

101. Mann, E.A., L.M. Silver, and R.W. Elliott. 1986. Genetic analysis of a mouse t complex locus that is homologous to a kidney cDNA clone. Genetics 114:993–1006.

102. Martin, S.A.M., B.A. Taylor, T. Watanabe, and G. Bulfield. 1984. Histidine decarboxylase phenotypes of inbred mouse strains: a regulatory locus (Hdc) determines kidney enzyme concentration. Biochem. Genet. 22:305–22.

103. Mason, I.J., D. Murphy, M. Munke, U. Francke, R.W. Elliott, and B.L.M. Hogan. 1986. Developmental and transformation-sensitive expression of the Sparc gene on mouse chromosome 11. EMBO J. 5:1831–7.

104. Matsushima, G.K., R.C. Harmon, and J.A. Frelinger. 1986. A new murine lymphocyte alloantigen, Ly-27.2. Immunogenetics 23:406–8.

105. McCubrey, J., and R. Risser. 1982. Allelism and linkage studies of murine leukemia virus activation in low leukemic strains of mice. J. Exp. Med. 155:1233–8.

106. McKenzie, I.F.C., J. Gardiner, M. Cherry, and G.D. Snell. 1977. Lymphocyte antigens: Ly-4, Ly-6, and Ly-7. Transplant. Proc. 9:667–9.

107. Meisler, M.H. 1976. Effects of the Bgs locus on mouse β-galactosidase. Biochem. Genet. 14:921–32.

108. Meruelo, D., M. Offer, and N. Flieger. 1983. A new lymphocyte cell surface antigen, Ly-22.2, controlled by a locus on chromosome 4 and a second unlinked locus. J. Immunol. 130:946–50.

109. Meruelo, D., M. Offer, and A. Rossomando. 1982. Evidence for a major cluster of lymphocyte differentiation antigens on murine chromosome 2. Proc. Natl. Acad. Sci. USA 79:7460–4.

110. Meruelo, D., A. Rossomando, S. Scandalis, P. D'Eustachio, R.E.K. Fournier, D.R. Roop, D. Saxe, C. Blatt, and M.N. Nesbitt. 1987. Assignment of the Ly-6– Ril-1– Sis– H-30– Pol5/ Xmmv-72– Ins-3– Krt-1– Int-1– Gdc-1 region to murine chromosome 15. Immunogenetics 25:361–72.

111. Michael, S.K., J. Hilgers, C. Kozak, J.B. Whitney III, and E.F. Howard. 1986. Characterization and mapping of DNA sequence homologous to mouse U1a1 snRNA: localization on chromosome 11 near the Dlb-1 and Re loci. Somat. Cell Mol. Genet. 12:215–23.

112. Michalides, R., and J. Hilgers. 1984. Mouse News Lett. 70:63–4.

113. Michalides, R., R. Verstraeten, F.W. Shen, and J. Hilgers. 1985. Characterization and chromosomal distribution of endogenous mouse mammary tumor viruses of

European mouse strains STS/A and GR/A. Virology 142:278–90.

114. Mishkin, J.D., B.A. Taylor, and W.J. Mellman. 1976. *Glk*: a locus controlling galactokinase activity in the mouse. Biochem. Genet. 14:635–40.

115. Mock, B.A. 1987. Mapping of mouse bcl-2 gene. Mouse News Lett. 78:63.

116. Mock, B.A., J. Hilgers, K. Huppi, and L. D'Hoostelaere. 1987. A TaqI restriction fragment length polymorphism at the Hox-2.1 locus cosegregates with the Dlb-1 locus on mouse chromosome 11. Nucl. Acids Res. 15:2397.

117. Mock, B.A., L.D. D'Hoostelaere, R. Matthai, and K. Huppi. 1987. A mouse homeo box gene, *Hox-1.5*, and the morphological locus, *Hd*, map within 1 cM on chromosome 6. Genetics 116:607–12.

118. Mortensen, R.F., P.T. Le, and B.A. Taylor. 1985. Mouse serum amyloid P-component (SAP) levels controlled by a locus on chromosome 1. Immunogenetics 22:367–75.

119. Moutier, R. 1978. Mouse News Lett. 58:64–5.

120. Nadeau, J.H., F.G. Berger, K.A. Kelley, P.M. Pitha, C.L. Sidman, and N. Worrall. 1986. Rearrangement of genes located on homologous chromosomal segments in mouse and man. The location of genes for alpha and beta-interferon, alpha-1 acid glycoprotein-1 and -2, and aminolevulinate dehydratase on mouse chromosome 4. Genetics 104:1239–55.

121. Nadeau, J.H., J. Kompf, G. Siebert, and B.A. Taylor. 1981. Linkage of *Pgm-3* in the house mouse and homologies of three phosphoglucomutase loci in mouse and man. Biochem. Genet. 19:465–74.

122. Nadeau, J.H., and S.J. Phillips. 1987. The putative oncogene *Pim-1* in the mouse: its linkage and variation among *t* haplotypes. Genetics 117:533–41.

123. Opalka, B., and E. Kolsch. 1983. Evidence for a new lymphocyte-stimulating determinant (*Lsd*) detected by alloreactive T cell lines. Eur. J. Immunol. 13:24–30.

124. Osterness, D.M., and R.M. Weinshilboum. 1987. Mouse thiopurine methyltransferase pharmacogenetics: biochemical studies and recombinant inbred strains. J. Pharmacol. Exp. Ther. 243:180–6.

125. Owen, F.L., B.A. Taylor, A. Zweidler, and J.G. Seidman. 1986. The murine γ-chain of the T cell receptor is closely linked to a spermatocyte specific histone gene and the beige coat color locus on chromosome 13. J. Immunol. 137:1044–6.

126. Paul, P.R., and R.W. Elliott. 1987. Analysis of the mouse *Amy* locus in recombinant inbred mouse strains. Biochem. Genet. 25:569–79.

127. Peleg, L., and M.N. Nesbitt. 1984. Genetic control of thymus size in inbred mice. J. Hered. 75:126–30.

128. Poland, A., E. Glover, and B.A. Taylor. 1987. The murine *Ah* locus: a new allele and mapping to chromosome 12. Mol. Pharmacol. 32:471–8.

129. Potter, M., J.S. Finlayson, D.W. Bailey, E.B. Mushinski, B.L. Reamer, and J.L. Walters. 1973. Major urinary protein and immunoglobulin allotypes of recombinant inbred mouse strains. Genet. Res. 22:325–8.

130. Potter, M., J.G. Pumphrey, and D.W. Bailey. 1975. Genetics of susceptibility to plasmacytoma induction. I. BALB/cAn (C), C57BL/6N (B6), C57BL/Ka (BK), (C × B6)F1, (C × BK)F1, and CXB recombinant-inbred strains. J. Natl. Cancer Inst. 54:1413–7.

131. Prochazka, M., P. Staeheli, R.S. Holmes, and O. Haller. 1985. Interferon-induced guanylate-binding proteins: mapping of the murine *Gbp-1* locus to chromosome 3. Virology 145:273–9.

132. Racine, R.R., and C.H. Langley. 1980. Genetic analysis of protein variations in *Mus musculus* using two-dimensional electrophoresis. Biochem. Genet. 18:185–97.

133. Riblet, R., B. Blomberg, M. Weigert, R. Lieberman, B.A. Taylor, and M. Potter. 1975. Genetics of mouse antibodies. I. Linkage of the dextran response locus, V_H-DEX, to allotype. Eur. J. Immunol. 5:775–7.

134. Riblet, R., L. Claflin, D.M. Gibson, B.J. Mathieson, and M. Weigert. 1980. Antibody gene linkage studies in (NZB x C58) recombinant inbred lines. J. Immunol. 124:787–9.

135. Richards, J.E., D.D. Pravtcheva, C. Day, F.H. Ruddle, and P.P. Jones. 1985. Murine invariant chain gene: chromosomal assignment and segregation in recombinant inbred strains. Immunogenetics 22:193–9.

136. Richards-Smith, B., and R. Elliott. 1984. Mapping of a family of repeated sequences in the mouse genome. Mouse News Lett. 71:46–7.

137. Risser, R., M. Potter, and W.P. Rowe. 1978. Abelson virus-induced lymphomagenesis in mice. J. Exp. Med. 148:714–26.

138. Robert, B., P. Barton, A. Minty, P. Daubas, A. Weydert, F. Bonhomme, J. Catalan, D. Chazottes, J-L. Guénet, and M. Buckingham. 1985. Investigation of genetic linkage between myosin and actin genes using an interspecific mouse back-cross. Nature 314:181–3.

139. Roehm, N.W., A. Carbone, E. Kushner, B.A. Taylor, R.J. Riblet, P. Marrack, and J.W. Kappler. 1985. The major histocompatibility complex-restricted antigen receptor on T cells: the genetics of an allotype. J. Immunol. 135:2176–82.

140. Rosenstreich, D.L., M.G. Groves, H.A. Hoffman, and B.A. Taylor. 1978. Location of the *Sas-1* locus on mouse chromosome 1. Immunogenetics 7:313–20.

141. Rossomando, A., and D. Meruelo. 1986. Viral sequences are associated with many histocompatibility genes. Immunogenetics 23: 233–45.

142. Rubin, M.R., W. King, L.E. Toth, I.S. Sawczuk, M.S. Levine, P. D'Eustachio, and M. Nguyen-Huu. 1987. Murine *Hox-1.7* homeo-box gene: cloning, chromosomal location, and expression. Mol. Cell. Biol. 7:3836–41.

143. Russell, E.S., S.L. Blake, and E.C. McFarland. 1972. Characterizaton and strain distribution of four alleles at the hemoglobin α-chain structural locus in the mouse. Biochem. Genet. 7:313–30.

144. Russell, J.H., C.B. Dobos, R.J. Graff, and B.A. Taylor.

1981. Genetic control of cross-reactive cytotoxic-T-lymphocyte responses to a BALB/c tumor. Immunogenetics 14:263–72.

145. Sekine, M., T. Yamakawa, and A. Suzuki. 1987. Genetic control of the expression of two extended globoglycolipids carrying either the stage specific embryonic antigen-1 or -3 determinant in mouse kidney. J. Biochem. 101:563–8.

146. Seldin, M.F., L.A. D'Hoostelaere, and A.D. Steinberg. 1987. Mouse skeletal alpha actin has limited restriction fragment length polymorphism and is not a member of the human 1q - mouse distal chromosome 1 syntenic group. Nucl. Acids Res. 15:1881.

147. Seldin, M.F., H.C. Morse III, L. D'Hoostelaere, J.L. Britten, and A.D. Steinberg. 1987. Mapping of alpha-spectrin on distal mouse chromosome 1. Cytogenet. Cell Genet. 45:52–54.

148. Seldin, M.F., and A.D. Steinberg. 1987. Human histone 3 F2 detects RFLPs in inbred mice, cosegregates with H4F2 and does not map to possible syntenic groups on distal chromosome 1 and distal chromosome 3. Nucl. Acids Res. 15:5900.

149. Siadak, A.W., and R.C. Nowinski. 1981. Thy-2: a murine thymocyte-brain alloantigen controlled by a gene linked to the major histocompatibility complex. Immunogenetics 12:45–58.

150. Silver, J., and C.E. Buckler. 1986. A preferred region for integration of Friend murine leukemia virus in hematopoietic neoplasms is closely linked to the *Int-2* oncogene. J. Virol. 60:1156–8.

151. Silver, J., J.B. Whitney III, C. Kozak, G. Hollis, and I. Kirsch. 1985. *Erbb* is linked to the alpha-globin locus on mouse chromosome 11. Mol. Cell. Biol. 5:1784–6.

152. Simmons, D.L., and C.B. Kasper. 1983. Genetic polymorphisms for a phenobarbital-inducible cytochrome P-450 map to the *Coh* locus in mice. J. Biol. Chem. 258:9585–8.

153. Siracusa, L.D., L.B. Russell, E.M. Eicher, D.J. Corrow, N.G. Copeland, and N.A. Jenkins. 1987. Genetic organization of the *agouti* region of the mouse. Genetics 117:93–100.

154. Skamene, E., P. Gros, A. Forget, P.A.L. Kongshavn, C. St. Charles, and B.A. Taylor. 1982. Genetic regulation of resistance to intracellular pathogens. Nature 297:506–10.

155. Skamene, E., P. Gros, A. Forget, P.J. Patel, and M.N. Nesbitt. 1984. Regulation of resistance to leprosy by chromosome 1 in the mouse. Immunogenetics 19:117–24.

156. Skow, L.C., and M.E. Donner. 1985. The locus encoding αA-crystallin is closely linked to *H-2K* on mouse chromosome 17. Genetics 110:723–32.

157. Skow, L.C. 1978. Genetic variation at a locus (*TAM-1*) for submaxillary gland protease in the mouse and its location on chromosome 7. Genetics 90:713–24.

158. Snell, G.D., M. Cherry, I.F.C. McKenzie, and D.W. Bailey. 1973. *Ly-4*, a new locus determining a lymphocyte cell-surface alloantigen in mice. Proc. Natl. Acad. Sci. USA 70:1108–11.

159. Steffen, D.L., B.A. Taylor, and R.A. Weinberg. 1982. Continuing germ line integration of AKV proviruses during the breeding of AKR mice and derivative recombinant inbred strains. J. Virol. 42:165–75.

160. Stern, R.H., E.S. Russell, and B.A. Taylor. 1976. Strain distribution and linkage relationship of a mouse embryonic hemoglobin variant. Biochem. Genet. 14:373–81.

161. Swank, R.T., and D.W. Bailey. 1973. Recombinant inbred lines, value in the genetic analysis of biochemical variants. Science 181:1249–52.

162. Sweetser, D.A., E.H. Birkenmeier, I.J. Klisak, S. Zollman, R.S. Sparkes, T. Mohandas, A.J. Lusis, and J.I. Gordon. 1987. The human and rodent intestinal fatty acid binding protein genes. Comparative analysis of their structure, expression, and linkage relationships. J Biol. Chem. 262:16060–71.

163. Symington, F.W., B. Subbarao, D.E. Mosier, and J. Sprent. 1982. Lyb-8.2: a new B cell antigen defined and characterized with a monoclonal antibody. Immunogenetics 16:381–91.

164. Szymura, J.M., B.A. Taylor, and J. Klein. 1982. *Upg-2*: a urinary pepsinogen variant located on chromosome 1 of the mouse. Biochem. Genet. 20:1211–9.

165. Tada, N., S. Kimura, and U. Hammerling. 1982. Immunogenetics of mouse B-cell alloantigen systems defined by monoclonal antibodies and gene cluster formation of these loci. Immunol. Rev. 69:99–126.

166. Tada, N., S. Kimura, A. Hatzfeld, and U. Hammerling. 1980. Ly-m11: the *H-3* region of mouse chromosome 2 controls a new surface alloantigen. Immunogenetics 11:441–9.

167. Tada, N., S. Kimura, Y. Liu, B.A. Taylor, and U. Hammerling. 1981. Ly-m19: the *Lyb-2* region of mouse chromosome 4 controls a new surface alloantigen. Immunogenetics 13:539–46.

168. Tada, N., S. Kimura, Y. Liu-Lam, and U. Hammerling. 1983. Mouse alloantigen system Ly-m22 predominantly expressed on T lymphocytes and controlled by a gene linked to Mls region on chromosome 1. Hybridoma 2:29–38.

169. Tada, N., S. Kimura, Y. Liu-Lam, and U. Hammerling. 1984. A new lymphocyte surface antigen (Ly-m31) controlled by a gene closely linked to the *Akp-2* locus on mouse chromosome 4. Immunogenetics 20:589–92.

170. Taylor, B.A. 1978. Recombinant inbred stains: use in gene mapping. *In* H. Morse, ed., Origins of inbred mice, 423–438. Academic Press, New York.

171. Taylor, B.A. 1984. *Mod-1* strain distribution pattern. Mouse News Lett. 71:32.

172. Taylor, B.A. 1987. RI strain distribution patterns. Mouse News Lett. 77:130.

173. Taylor, B.A. (unpublished).

174. Taylor, B.A., D.W. Bailey, M. Cherry, R. Riblet, and M. Weigert. 1975. Genes for immunoglobulin heavy

chain and serum prealbumin protein are linked in the mouse. Nature 256:644–6.

175. Taylor, B.A., H.G. Bedigian, and H. Meier. 1977. Genetic studies of the *Fv-1* locus of mice: linkage with *Gpd-1* in recombinant inbred lines. J. Virol. 23:106–9.

176. Taylor, B.A., H.J. Heiniger, and H. Meier. 1973. Genetic analysis of resistance to cadmium-induced testicular damage in mice. Proc. Soc. Exp. Biol. Med. 143:629–33.

177. Taylor, B.A., and H. Meier. 1975. Mapping the adrenal lipid gene of the AKR/J mouse strain. Genet. Res. 26:307–12.

178. Taylor, B.A., L. Rowe, D.M. Gibson, R. Riblet, R. Yetter, and P.D. Gottlieb. 1985. Linkage of a 7S RNA sequence and kappa light chain genes in the mouse. Immunogenetics 22:471–81.

179. Taylor, B.A., L. Rowe, N.A. Jenkins, and N.G. Copeland. 1985. Chromosomal assignment of two endogenous ecotropic murine leukemia virus proviruses of the AKR/J mouse strain. J. Virol. 56:172–5.

180. Taylor, B.A., and L. Rowe. 1984. Genes for serum amyloid A proteins map to chromosome 7 in the mouse. Mol. Gen. Genet. 195:491–9.

181. Taylor, B.A., and L. Rowe. 1987. The congenital goiter mutation is linked to the thyroglobulin gene in the mouse. Proc. Natl. Acad. Sci. USA 84:1986–90.

182. Taylor, B.A., and F-W. Shen. 1977. Location of *Lyb-2* on chromosome 4: evidence from recombinant inbred strains. Immunogenetics 4:597–9.

183. Taylor, B.A., D.M. Walls, and M.J. Wimsatt. 1987. Localization of the inosine triphosphatase locus (*Itp*) on chromosome 2 of the mouse. Biochem. Genet. 25:267–74.

184. Traina, V.L., B.A. Taylor, and J.C. Cohen. 1981. Genetic mapping of endogenous mouse mammary tumor viruses: locus characterization, segregation, and chromosomal distribution. J. Virol. 40:735–44.

185. Tulchin, N., and B.A. Taylor. 1981. γ-Glutamyl cyclotransferase: a new genetic polymorphism in the mouse (*Mus musculus*) linked to *Lyt-2*. Genetics 99:109–16.

186. Uiterdijk, H.G., B.A.J. Ponder, M.F.W. Festing, J. Hilgers, L. Skow, and R. Van Nie. 1986. The gene controlling the binding sites of Dolichos biflorus agglutinin, *Dlb-1*, is on chromosome 11 of the mouse. Genet. Res. 47:125–9.

187. Vidovic, D., J. Klein, and Z.A. Nagy. 1985. Recessive T cell response to poly [glu50 tyr50] possibly caused by self tolerance. J. Immunol. 134:3563–8.

188. Watson, J., R. Riblet, and B.A. Taylor. 1977. The response of recombinant inbred strains of mice to bacterial lipopolysaccharides. J. Immunol. 118:2088–93.

189. Wejman, J.C., B.A. Taylor, N.A. Jenkins, and N.G. Copeland. 1985. Endogenous xenotropic murine leukemia virus-related sequences map to chromosomal regions encoding mouse lymphocyte antigens. J. Virol. 50:237–47.

190. Whitney, G., and D.B. Harder. 1986. Single-locus control of sucrose octacetate tasting among mice. Behav. Genet. 16:559–74.

191. Wilcox, F.H., L. Hirschhorn, B.A. Taylor, J.E. Womack, and T.H. Roderick. 1979. Genetic variation in alkaline phosphatase of the house mouse (*Mus musculus*) with emphasis on a manganese requiring isozyme. Biochem. Genet. 17:1093–107.

192. Wilcox, F.H., and T.H. Roderick. 1982. Location on chromosome 6 of the locus for a major liver protein (*Lvp-1*) of the house mouse. Genet. Res. 40:213–5.

193. Wilcox, F.H., and B.A. Taylor. 1981. Genetics of the *Akp-2* locus for alkaline phosphatase of liver, kidney, bone, and placenta in the mouse. J. Hered. 72:387–90.

194. Wilkie, T.M., and R.D. Palmiter. 1987. Analysis of the integrant in Myk-103 transgenic mice in which males fail to transmit the integrant. Mol. Cell. Biol. 7:1646–55.

195. Wilson, C.M. 1984. Ren-2 genotypes of inbred strains maintained at the Netherlands Cancer Institute. Mouse News Lett. 71:51.

196. Wilson, C.M., E.G. Erdos, J.D. Wilson, and B.A. Taylor. 1978. Location on chromosome 1 of *Rnr*, a gene that regulates renin in the submaxillary gland of the mouse. Proc. Natl. Acad. Sci. USA 75:5623–6.

197. Winchester, G., N.A. Mitchison, and B.A. Taylor. 1987. The structural gene for F liver protein (*Flp*) maps to chromosome 9 of the mouse. Immunogenetics 26:356–8.

198. Womack, J.E., M.A. Lynes, and B.A. Taylor. 1975. Genetic variation of an intestinal leucine arylaminopeptidase (*Lap-1*) in the mouse and its location on chromosome 9. Biochem. Genet. 13:511–8.

199. Womack, J.E., B.A. Taylor, and J.E. Barton. 1978. Esterase 13, a new mouse esterase locus with recessive expression and its genetic location on chromosome 9. Biochem. Genet. 16:1107–12.

200. Wood, A.W., and B.A. Taylor. 1979. Genetic regulation of coumarin hydroxylase activity in mice. Evidence for single locus control on chromosome 7. J. Biol. Chem. 254:5647–51.

19 CONGENIC AND SEGREGATING INBRED STRAINS

19.1 Immunologically important loci

JAN KLEIN

The tables that follow contain over 800 congenic strains, some 500 of which carry as their differential locus the *H-2* complex. Altogether over 500 *H-2* haplotypes are listed. This, undoubtedly, is the largest collection of congenic strains and haplotypes of any vertebrate, if not of any animal species altogether. The compilation of this list would not have been possible without the co-operation of the holders of the congenic lines, whom I would like to thank for the information they sent me. I have tried to standardize the nomenclature used in this list but unfortunately some investigators, for reasons I do not comprehend, have refused to change symbols that do not comply with the established rules.

The tables require a few explanations. For the 'inbred partner'—and the 'donor strain'—entries I used full substrain designations rather than giving the symbol for the last holder only after the slanted line. Although the former practice is now discouraged, I used it nevertheless because it is important to know exactly which substrain of the inbred strain was used in the production of a particular congenic strain. I have also included strains that are not fully congenic (N3 to N6) in the list with the expectation that these strains will be completed by the time this book reaches the reader. The '+' symbol in the column specifying the number of backcrossing generations indicates that the particular congenic strain is derived from two strains that are already congenic and have the same inbred partner. Such 'second generation' congenic strains are usually maintained by brother × sister mating once they have become homozygous for the differential locus. For the second generation congenic strains, the original partner is given first and the congenic strain used for the derivation of the new strain is given in parentheses. I have not indicated the number of brother × sister generations of mating for the simple reason that it obviously changes continually.

In Table 19.3, I made no distinction between recombinant and so-called 'independent' *H-2* haplotypes. With the isolation of so many *H-2* haplotypes from wild mice, this distinction has become meaningless. It is clear that most, if not all, *H-2* haplotypes are genetic mosaics derived from other haplotypes, in other words, they are *all* 'recombinant'.

The allelic symbols used for the *H-2* haplotypes indicate the origin of the genes, except for the *Qa-1*, *Qa-2*, and *Tla* loci. For example, in the *H-2^{al}* haplotype, the *K^k* designation indicates that this allele is derived from the *H-2^k* haplotype. However, the *Qa-1^a* designation does not indicate derivation from the *H-2^a* haplotype; rather it is an arbitrary symbol that reflects only the sequence of description of alleles at this locus. This inconsistency will have to be rectified eventually; however, no agreement has as yet been reached on the nomenclature of *Qa* and *Tla* loci. I have tried to compensate for the lack of information in the sequential allelic designations by indicating in the superscript the strain of origin of the particular allele. This usage must, however, be considered as temporary; it does not comply with the rules for mouse genetic nomenclature and it should be discontinued once an agreement on consistent nomenclature of *Qa* and *Tla* loci is reached.

For completeness, I have included in the *H-2* haplotypes the two complement loci *C4* (*Ss*) and *C4S* (*Slp*) although, strictly speaking, they are not part of the *H-2* complex proper. The allelic designations indicate the origin of these genes rather than true allelic differences. Some of the genes that now bear different allelic designations because of different origin may, in fact, turn out to be identical.

Up to now, there have not been any official rules formalized for the designations of haplotypes that have the same *K*, *A*, *E*, *D*, and *L* but different *Qa* and *Tla* alleles. I have used the next available designation for such haplotypes. For example, I designated the haplotype borne by the *A-Tla^b* strain *H-2^{a16}*, that borne by the *C57BL/6-Tla^a* strain *H-2^{b2}*, and so on. This practice has not been officially agreed on, although I hope that eventually it will be. In newer haplotype designations, I started every new letter combination with numeral one (e.g. *pz1* instead of *pz*). I left, however, symbols that have been in use for some time (e.g. b, d, f, etc.) without numerals.

19.1.1 Symbols for holders and producers

The symbols given for holders and producers are as in Chapter 20, except for the following:

Han — Dr Ted H. Hansen, Department of Genetics, Washington University in St. Louis, 4566 Scott Avenue, St. Louis, MO 63110, USA.

Hk — Dr Kim J. Hasenkrug, The McLaughlin Research Institute, 1625 3rd Avenue North, Great Falls, MO 59401, USA.

Jax — The Jackson Laboratory, Bar Harbor, ME 04609, USA.

Mrf — Dr Donal B. Murphy, Department of Pathology, Yale University School of Medicine, New Haven, CT 06510, USA.

Mrw — Dr Kazuo Moriwaki, National Institute of Genetics, Department of Cytogenetics, Yata 1111, Misima, Sizuoka-ken, 411 Japan.

Psm — Dr Howard C. Passmore, Department of Zoology, Rutgers University, New Brunswick, NJ 08903, USA.

Shi — Dr T. Shiroishi, Department of Cytogenetics, National Institute of Genetics, Misima, Yata 1111, Sizuoka-ken, 411 Japan.

Sta — Dr T.H. Stanton, Department of Microbiology and Immunology, University of Washington, Seattle, WA 98195, USA.

Wo — Dr Henry Wortis, Department of Pathology, Tufts University School of Medicine, Boston, MA 02111, USA.

Wr — Dr Carol M. Warner, Department of Biochemistry and Biophysics, Iowa State University, Ames, Iowa 50011, USA.

Za — Dr Marek Zaleski, Department of Microbiology, State University of New York at Buffalo, Buffalo, NY 14214, USA.

Table 19.1 *H-2* congenic lines

Differential *H-2* haplotype	Congenic strain (synonym)	Inbred partner	Donor strain	Producer	N
a	BALB.A	BALB/cAn	A/J	Lil	14
	B10.A	C57BL/10SnSg	A/WySnSg	Sg	11
	C3H.A	C3H/HeHa	A/HeHa	Ha	9
	020.A(353)	020/A	B10.A/SgSnA	Dem	6
a1	A.AL	A/SnSf	C3H/HeJ, DBA/2J	Sf	11
	B10.AL	C57BL/10SnSf	A.AL/Sf	Sf	10
a2	B10.KDR3	C57BL/10SnSf, B10.S	B10.D2/n, B10.K	Sf	+2
a3	B10.KDR4	C57BL/10SnSf	B10.D2/n, B10.K	Sf	+2
a4	B10.RKD1	C57BL/10SnDv	B10.A(2R), B10.HTT	Dv	+
a5	B10.AIR1	C57BL/10SnDv	A.AIR1	Dv	+
a6	B10.HIR2	C57BL/10SnSg (B10.S)	B10.A(4R), B10.A(5R)	Sf	+2
a7	B10.RKD2	C57BL/10SnDv	B10.A(2R), B10.T(6R)	Dv	+
a8	B10.RKD3	C57BL/10SnDv	B10.A(2R), B10.HTT	Dv	+
a9	B10.RKD5	C57BL/10SnDv	B10.A(4R), B10.S(9R)	Dv	+
a10	BALB.A10	BALB/cKh	BALB.K, BALB/cKh	Sx	+
a11	BALB.A11	BALB/cKh	BALB.K, BALB/cKh	Sx	+
a13	B10.RKD4	C57BL/10SnDv	B10.A(4R), B10.PL	Dv	+
a14	B10.RKD6	C57BL/10SnDv	B10.A(2R), B10.T(6R)	Dv	+
a15	B10.KDR19	C57BL/10SnSf (B10.TBR6)	B10.KRR1, B10.TL	Sf	10
a16	A-*Tla*^b	A/JBoy	C57BL/6JBoy	Boy	15
a17	B10.A(R149)-*Tla*^b	C57BL/10Sn	B10.A, C57BL/10Sn	Mrf	10
a18	B10.A(R410)-*Tla*^b	C57BL/10Sn	B10.A, C57BL/10Sn	Mrf	10
a19	B10.RKD7	C57BL/10SnDv	B10.A(2R), B10.T(6R)	Dv	+
ak1	B6.AK1	C57BL/6JBoyFla	C57BL/6-*H-2*^k	Fla	+
an1	A.TFR1	A/SnSf	A.CA, A.TL	Sf	13
	B10.TFR1	C57BL/10SnSf	A.TFR1	Sf	10
an3	B10.RSF2	C57BL/10SnDv	B10.TBR2, B10.M	Dv	+
an4	B10.RSF4	C57BL/10SnDv	B10.RSB6, B10.M	Dv	+
an5	B10.RSF3	C57BL/10SnDv	B10.TBR4, B10.M	Dv	+

Table 19.1—*cont.*

Differential *H-2* haplotype	Congenic strain (synonym)	Inbred partner	Donor strain	Producer	N
an6	B10.RSF5	C57BL/10SnDv	B10.RSB12, B10.M	Dv	+
an7	B10.TBFR4	C57BL/10Sn (B10.D2/n)	B10.TBR2, B10.M	Sf	10
an8	B10.TBFR6	C57BL/10SnSg (B10.A)	B10.TBR2, B10.M	Sf	10
an9	B10.TBFR8	C57BL/10Sn (B10.D2/n)	B10.TBR2, B10.M	Sf	10
an10	B10.TBFR9	C57BL/10SnSg (B10.A)	B10.TBR2, B10.M	Sf	10
ap1	B10.M(11R)	C57BL/10SnSg	B10.A, B10.M	Sg	12
ap2	A.TFR2	A/SnSf	A.CA, A.TH	Sf	11
	B10.TFR2	C57BL/10SnSf	A.TFR2	Sf	10
ap3	A.TFR3	A/SnSf	A.CA, A.TH	Sf	11
	B10.TFR3	C57BL/10SnSf	A.TFR3	Sf	10
ap4	A.TFR4	A/SnSf	A.CA, A.TH	Sf	11
	B10.TFR4	C57BL/10SnSf	A.TFR4	Sf	10
ap5	A.TFR5	A/SnSf	A.CA, A.TL	Sf	12
	B10.TFR5	C57BL/10SnSf	A.TFR5	Sf	11
ap6	B10.RFD1	C57BL/10SnDv	B10.M, B10.RKD3	Dv	+
ap7	B10.RFD2	C57BL/10SnDv	B10.M, B10.RKD3	Dv	+
ap8	B10.RFD4	C57BL/10SnDv	B10.M, B10.RPD1	Dv	+
ap9	B10.RFD5	C57BL/10SnDv	B10.M, B10.RKD5	Dv	+
ap10	B10.RFD6	C57BL/10SnDv	B10.M, B10.RPD1	Dv	+
ap11	B10.RFD7	C57BL/10SnDv	B10.M, B10.RPD1	Dv	+
ap12	B10.RFD3	C57BL/10SnDv	B10.M, B10.RPD1	Dv	+
aq1	B10.M(17R)	C57BL/10SnSg	B10.A, B10.M	Sg	+
aq2	B10.RKF1	C57BL/10SnDv	B10.A, B10.RSF3	Dv	+
aq3	B10.RKF2	C57BL/10SnDv	B10.A(4R), B10.M	Dv	+
aq4	B10.RKF4	C57BL/10SnDv	B10.A(4R), B10.M	Dv	+
aq5	B10.RKF3	C57BL/10SnDv	B10.RKD7, B10.M	Dv	+
ar1	B10.LG	C57BL/10SnSg	LG/Ckc	Sf	10
ar2	B10.RIII(34R)	C57BL/10SnSg	B10.RIII(72NS)	Hk	12
as1	B10.S(8R)	C57BL/10SnSg	A.SW, B10.A	Sg	10
as2	B10.ASR2	C57BL/10SnSf	B10.A, B10.S	Sf	+1
as3	B10.ASR7	C57BL/10SnSf	B10.A, B10.S	Sf	+1
as4	B10.BASR1	C57BL/10Sn (B10.D2/n)	B10.BAR4, B10.S	Sf	+1
as5	B10.BASR2	C57BL/10SnSg [B10.F(13R)]	B10.BAR6, B10.S	Sf	+1
as6	B10.S(41R)	C57BL/10SnSg	B10.KPQ(38R), B10.S	Hk	13
as7	B10.RKS1	C57BL/10SnDv	B10.A(4R), B10.S	Dv	+
at1	A.TBR1	A/SnSf	A.BY, A.TL	Psm	+
at2	A.TBR2	A/SnSf	A.TL, C57BL/6J	Psm	10
	B10.TBR2	C57BL/10SnSg (B10.M)	A.TL, C57BL/6J	Sf, Psm	10
at3	A.TBR3	A/SnSf	A.TL, C57BL/6J	Psm	10
	B10.TBR3	C57BL/10SnSg (B10.M)	A.TL, C57BL/6J	Sf, Psm	10
at4	A.TBR4	A/SnSf	A.TL, C57BL/6J	Psm	10
	B10.TBR4	C57BL/10SnDv	A.TBR4	Dv	3
at5	A.TBR5	A/SnSf	A.TL, C57BL/6J	Psm	10
at6	B10.TBR6	C57BL/10SnSg (B10.M)	A.TL, C57BL/10SnSf	Sf	11
at8	B10.TBR11	C57BL/10Sn (B10.D2/n)	B10.BAR6, B10.S	Sf	+1
at9	B10.RSB1	C57BL/10SnDv	B10.S(9R), B10.A(4R)	Dv	+
at10	B10.RSB2	C57BL/10SnDv	B10.S(9R), B10.A(4R)	Dv	+
at11	B10.RSB3	C57BL/10SnDv	B10.HTT, B10.A(2R)	Dv	+
at12	A.TBR12	A/SnSf	A.TL, C57BL/6J	Psm	10
at13	A.TBR13	A/SnSf	A.BY, A.TL	Psm	+
at14	A.TBR14	A/SnSf	A.BY, A.TL	Psm	+
at15	A.TBR15	A/SnSf	A.TL, C57BL/6J	Psm	10
at16	A.TBR16	A/SnSf	A.BY, A.TL	Psm	+
at17	A.TBR17	A/SnSf	A.TL, C57BL/6J	Psm	+
at18	B10.THR2	C57BL/10SnSg (B10.M)	B10.A(4R), B10.HTT	Sf	+2
at19	B10.RSB4	C57BL/10SnSg	B10.S(9R), B10.A(4R)	Dv	+
at20	B10.RSB5	C57BL/10SnSg	B10.S(9R), B10.F(13R)	Dv	+
at21	B10.RSB6	C57BL/10SnSg	B10.S(9R), B10.F(13R)	Dv	+

Table 19.1—*cont.*

Differential *H-2* haplotype	Congenic strain (synonym)	Inbred partner	Donor strain	Producer	N
at22	B10.RSB7	C57BL/10SnSg	B10.HTT, B10.A(2R)	Dv	+
at23	B10.RSB8	C57BL/10SnSg	B10.S(9R), B10.F(13R)	Dv	+
at24	B10.RSB10	C57BL/10SnSg	B10.S(9R), B10.A(4R)	Dv	+
at25	B10.RSB11	C57BL/10SnSg	B10.TL, B10.F(13R)	Dv	+
at26	B10.RSB12	C57BL/10SnSg	B10.S(9R), B10.F(13R)	Dv	+
at27	B10.RSB13	C57BL/10SnSg	B10.TL, B10.F(13R)	Dv	+
at28	B10.RSB14	C57BL/10SnSg	B10.TL, B10.F(13R)	Dv	+
at29	B10.RSB15	C57BL/10SnSg	B10.TL, B10.F(13R)	Dv	+
at30	B10.TRB18	C57BL/10Sn (B10.D2/n)	B10.BAR6, B10.S	Sf	+1
aw1	B10.RIII(20R)	C57BL/10SnSg	B10.S(7R), B10.RIII	Sg	12
aw2	B10.A(R202)	C57BL/10Sn	Wild *Mus molossinus*, B10.A	Mrw, Shi	4
aw3	B10.A(R203)	C57BL/10Sn	Wild *Mus molossinus*, B10.A	Mrw, Shi	4
aw4	B10.A(R204)	C57BL/10Sn	Wild *Mus molossinus*, B10.A	Mrw, Shi	4
aw6	B10.A(R206)	C57BL/10Sn	Wild *Mus molossinus*, B10.A	Mrw, Shi	4
aw7	B10.A(R207)	C57BL/10Sn	Wild *Mus molossinus*, B10.A	Mrw, Shi	4
aw8	B10.A(R208)	C57BL/10Sn	Wild *Mus molossinus*, B10.A	Mrw, Shi	4
aw9	B10.A(R209)	C57BL/10Sn	Wild *Mus molossinus*, B10.A	Mrw, Shi	4
aw10	B10.A(R201)	C57BL/10Sn	Wild *Mus molossinus*, B10.A	Mrw, Shi	4
aw11	B10.A(R211)	C57BL/10Sn	Wild *Mus molossinus*, B10.A	Mrw, Shi	4
aw12	B10.A(R212)	C57BL/10Sn	Wild *Mus molossinus*, B10.A	Mrw, Shi	3
aw13	B10.A(R213)	C57BL/10Sn	Wild *Mus molossinus*, B10.A	Mrw, Shi	4
aw14	B10.A(R214)	C57BL/10Sn	Wild *Mus molossinus*, B10.A	Mrw, Shi	3
aw17	B10.A(R217)	C57BL/10Sn	Wild *Mus molossinus*, B10.A	Mrw, Shi	4
az1	B10.SM(22R)	C57BL/10SnSg	B10.A, SM/J	Sg	14
b	ABY/Kl	A/WySn	A.BY/Sn	Kl	11
	A.BY/Sn	A/WySn	Brachyury, non-inbred	Sn	11
	C3H.B10	C3H/HeJSf	C57BL/10J	Sf	10
	C3H/Bi-*H-2^b*	C3H/Bi	C57BL/6J	Lil	18
	C3H.SW/Sn	C3H/HeDiSn	Swiss, non-inbred	Sn	11
	D1.LP	DBA/1J	LP/J	Sn	11
b1	B10.HIR1	C57BL/10Sn (B10.S)	B10.A(4R), B10.A(5R)	Sf	+3
b2	C57BL/6-*Tla^a*	C57BL/6JBoy	A/JBoy	Boy	27
b3	B6.K1	C57BL/6JBoyFla	C57BL/6-*H-2^k*	Fla	+
b4	B6.K2	C57BL/6JBoyFla	C57BL/6-*H-2^k*	Fla	+
b5	B6.KB2	C57BL/6JBoyFla	C57BL/6-*H-2^k*	Rn	+
b6	B6.AC2	C57BL/6JBoyFla	A.CA/Sn	Fla	10
b7	B6.AC3	C57BL/6JBoyFla	A.CA/Sn	Fla	10
b8	B10.A(R297)-*Tla^a*	C57BL/10Sn	B10.A, C57BL/10Sn	Mrf	10
b9	B10.A(R310)-*Tla^a*	C57BL/10Sn	B10.A, C57BL/10Sn	Mrf	10
bb1	B10.AF(35R)	C57BL/10SnSg	B10.A(5R), B10.F(13R)	Hk	13
bf1	B10.RBF	C57BL/10Sn	C57BL/10SnDv, B10.M	Dv	+
bf2	B10.TFBR12	C57BL/10SnSf (B10.K, B10.TL)	A.TFR1, C57BL/10SnSf	Sf	+
bm1	B6.C-*H-2^bm1* (Hz1)	C57BL/6By	BALB/cBy	By	10
bm2	B6.C-*H-2^bm2* (Hz49)	C57BL/6By	BALB/cBy	By	11
bm3	C57BL/6-*H-2^bm3* (M505)	C57BL/6YEg	–	Eg	3
bm4	B6.C-*H-2^bm4* (Hz179)	C57BL/6By	BALB/cBy	By	9
bm5	C57BL/6-*H-2^bm5*	C57BL/6Kh	–	Kh, Mel	–
bm6	C57BL/6-*H-2^bm6*	C57BL/6Kh	–	Kh, Mel	–
bm7	B6.C-*H-2^bm7*	C57BL/6Kh	BALB/cKh	Kh, Mel	10
bm8	C57BL/6-*H-2^bm8*	C57BL/6Kh	–	Kh, Mel	–
bm9	B6.C-*H-2^bm9*	C57BL/6Kh	BALB/cKh	Kh, Mel	11
bm10	B6.C-*H-2^bm10*	C57BL/6Kh	BALB/cKh	Kh, Mel	10
bm11	B6.C-*H-2^bm11*	C57BL/6Kh	BALB/cKh	Kh, Mel	10
bm12	B6.C-*H-2^bm12*	C57BL/6Kh	BALB/cKh	Kh, Mel	10
	C3H-*H-2^bm12*	C3H/HeJ	B6.C-*H-2^bm12*	Han, Sf	7
bm13	B6.C-*H-2^bm13*	C57BL/6Kh	BALB/cKh	Kh, Mel	10
bm14	B6.C-*H-2^bm14* (HzW42)	C57BL/6By	BALB/cBy	By	14
bm16	C57BL/6-*H-2^bm16*	C57BL/6By	–	Kh, Mel	–

Table 19.1—*cont.*

Differential *H-2* haplotype	Congenic strain (synonym)	Inbred partner	Donor strain	Producer	N
bm17	C57BL/6-*H-2*^{bm17}	C57BL/6By	–	Kh, Mel	–
bm18	B6.C-*H-2*^{bm18}	C57BL/6By	BALB/cBy	Kh, Mel	10
bm19	B6.C-*H-2*^{bm19}	C57BL/6Kh	BALB/cKh	Kh, Mel	10
bm20	B6.C-*H-2*^{bm20}	C57BL/6Kh	BALB/cKh	Kh, Mel	10
bm21	B6.C-*H-2*^{bm21}	C57BL/6Kh	BALB/cKh	Kh, Mel	10
bm22	C57BL/6-*H-2*^{bm22}	C57BL/6Kh	–	Mel	–
bm23	B10.D2-*H-2*^{bm23}	C57BL/10Sn (B10.D2)	C57BL/10Sn	Eg	5
bm24	C57BL/10-*H-2*^{bm24}	C57BL/10ScNCr	–	Mel	–
bp1	020.DD	020/A	DD/HeA	Dem	5
bp2	B10.P(27R)	C57BL/10SnSg	B10.P	Hk	11
bp3	B10.P(39R)	C57BL/10SnSg	B10.P(10R)	Hk	10
bp4	B10.P(40R)	C57BL/10SnSg	B10.P(10R)	Hk	10
bp5 (bw1)	B10.F(14R)	C57BL/10SnSg	F/St	Sg	10
bq1	A.MBR	A.SW/Sn	B10.AKM, C57BL/10J	Sx	7
	B10.MBR	C57BL/10Sn	B10.AKM, C57BL/10Sn	Sx	+2
bq2	B10.Tf	C57BL/10SnSg	Brachyury, non-inbred	Sg	11
bq3	B10.RBQ1	C57BL/10Sn	B10.RBD, B10.Q	Dv	+
bq4	B10.QBR	C57BL/10J	C57BL/10J, B10.Q	Sx	+1
bq5	B10.D1(R108)	C57BL/10Y	C57BL/10Y, DBA/1J	Y	.
bs1	020.WLL	020/A	WLL/BrA	Dem	5
bv1	020.129	020/A	129/SvS1A	Dem	5
bw1	B10.(R231)	C57BL/10Sn	Wild *Mus molossinus*, C57BL/10Sn	Mrw, Shi	3
bw3	B10.(R233)	C57BL/10Sn	Wild *Mus molossinus*, C57BL/10Sn	Mrw, Shi	4
bw6	B10.(R236)	C57BL/10Sn	Wild *Mus molossinus*, C57BL/10Sn	Mrw, Shi	3
bw7	B10.(R237)	C57BL/10Sn	Wild *Mus molossinus*, C57BL/10Sn	Mrw, Shi	3
bw9	B10.(R239)	C57BL/10Sn	Wild *Mus molossinus*, C57BL/10Sn	Mrw, Shi	4
by1	B10.BYR1	C57BL/10Sn	B10.AQR	Klj	12
by2	B10.BYR2	C57BL/10Sn	B10.AQR	Klj	12
bz1	B10.SBR	C57BL/10J	C57BL/10J, B10.S	Sx	5
bz2	A.SB(1R)	A.SW/Sn	B10.MBR, A.SW/Sn	Sx	4
bz3	A.SB(2R)	A.SW/Sn	B10.MBR, A.SW/Sn	Sx	5
d	A.D2	A.SW/Sn	B10.D2/n	Sx	4
	C.D2 (C.D2Qa2$^+$)	BALB/cAnPt	DBA/2N	Pt	16
	BALB/c-*Qa-2*^{a} (Pt.JQa2$^+$)	BALB/cAnPt	BALB/cJ	Pt	6
	C57BL/6-*H-2*^{d}/a (HW19)	C57BL/6By	BALB/cBy	By	15
	C57BL/6-*H-2*^{d}/b (HW41)	C57BL/6By	BALB/cBy	By	16
	C57BL/6-*H-2*^{d}/c (HW101)	C57BL/6By	BALB/cBy	By	15
	B10.D2/n	C57BL/10Sn	B10.D2/o	Sn	+4
	B10.D2/o	C57BL/10Sn	DBA/2J	Sn	6
	B10.Rb7D	C57BL/10Sn	B10.D2/n, Wild *Mus domesticus*	Klj	10
	D1.C	SBA/1J	BALB/cJ	Sn	12
df1	B10.DFR1	B10.K	B10.TBFR4, B10.D2	Sf	10
dm1	B10.D2-*H-2*^{dm1} (M504)	C57BL/10SnEg	B10.D2/n	Eg	7
	BALB/c-*H-2*^{dm1}	BALB/cKhMel	B10.D2-*H-2*^{dm1}	Mel	9
dm2	BALB/c-*H-2*^{dm2}	BALB/cKh	–	Kh, Mel	–
dm4	C.B6-*H-2*^{dm4}	BALB/cKh	C57BL/6Kh	Kh, Mel	12
dm5	C.B6-*H-2*^{dm5}	BALB/cKh	C57BL/6Kh	Kh, Mel	15
dm6	C.B6-*H-2*^{dm6}	BALB/cBy	C57BL/6Kh	Mel	10
dr1	B10.RDR1	C57BL/10SnSf (B10.TL, B10.M)	B10.RIII, FM	Sf	10
ds1	B10.DSR1	C57BL/10Sn	B10.A	Mrf	+
ds2	B10.DSR2	C57BL/10Sn	B10.A	Mrf	+
ds3	B10.DSR3	C57BL/10Sn	B10.A	Mrf	+
dx1	B10.GR	C57BL/10J	GRS-*Mtv-2*$^-$	Dem	7
	B10.GSR	C57BL/10SnSf	GRS/A	Sf	10
	020.GR(705)	020/A	GRS-*Mtv-2*$^-$	Dem	6
dx2	020.TSI	020/A	TSI/A	Dem	5

Table 19.1—*cont.*

Differential H-2 haplotype	Congenic strain (synonym)	Inbred partner	Donor strain	Producer	N
f	A.CA/Sn	A/WySn	Caracul, non-inbred	Sn	11
	A.CA/Kl	A/WySn	A.CA/Sn	Kl	11
	B10.M	C57BL/10Sn	Non-Inbred	Sn	8
	020.F	020/A	F (non-inbred)	Dem	5
	020.M	020/A	B10.M/Sn	Dem	5
f2	B6.AC1	C57BL/6JBoyFla	A.CA/Sn	Fla	10
fb2	B10.RFB2	C57BL/10SnDv	B10.M, B10.RSB10	Dv	+
fb3	B10.RFB3	C57BL/10SnDv	B10.M, B10.RSB12	Dv	+
fb4	B10.TBFR7	C57BL/10SnSg (B10.A)	B10.TBR3, B10.M	Sf	10
fd1	B10.RFDB	C57BL/10SnDv	B10.M, B10.RQDB	Dv	+
fm1	A.CA-*H-2^{fm1}* (M506)	A/SnEg	A.CA	Eg	11
fm3	B10.M-*H-2^{fm2}*	C57BL/10Sn	B10.M	Mob, By	8
g	BALB.HTG (BALB.G)	BALB/cAn	HTG(C57BL, BALB/c)	Lil	10
	B10.HTG	C57BL/10Sn	HTG (C57BL, BALB/c)	Lil	12
	C3H.HTG(82NS)	C3H/HeJ	HTG(C57BL, BALB/c)	Sn	+5
g1	B10.D2(R101)	C57BL/10SnEg	B10.D2/n	Eg	4
g2	B10.GD	C57BL/10SnSf	D2.GD	Sf	10
	D2.GD	DBA/2J	C57BL/6J, DBA/2	Lil	13
g3	B10.D2(R103)	C57BL/10SnEg	B10.D2(M504)	Eg	+4
g4	B10.BDR1	C57BL/10J (B10.K)	C57BL/6J, DBA/2	Psm	11
g5	B10.BDR2	C57BL/10 (B10.K)	C57BL/6J, DBA/2	Psm	11
g6	B6.C-*H-2^{g6}*	C57BL/6By	BALB/cBy	By	22
h1	B10.A(1R)	C57BL/10SnSg	A/WySn	Sg	11
h2	BALB.2R	BALB/cAn	B10.A(2R)	Lil	10
	B10.A(2R)	C57BL/10SnSg	A/WySn	Sg	11
h3	B10.AM	C57BL/10SnSg	C57BL/6J, C3H/HeJ	Sg	11
	C3H.KBR	C3H/JSf (C3H.OH)	C57BL/6J, C3H/JSf	Sf	10
h4	B10.A(4R)	C57BL/10SnSg	A/WySn	Sg	11
h6	B10.BAR4	C57BL/10SnSg (B10.S)	C57BL/10SnSf, B10.A	Sf	+1
h8	B10.BAR10	C57BL/10SnSg (B10.S)	C57BL/10SnSf, B10.A	Sf	+4
h9	B10.BAR11	C57BL/10SnSg (B10.S)	C57BL/10SnSf, B10.A	Sf	+2
h10	B10.RKB	C57BL/10SnSg (B10.M)	B10.K, C57BL/6J	Dv	+
h11	B10.BAR6	C57BL/10SnSg (B10.S)	C57BL/10SnSf, B10.A	Sf	+1
h12	B10.RKB2	C57BL/10Sn	B10.A, B10.RFB1	Dv	+
h13	B10.RKB3	C57BL/10Sn	B10.A, B10.RFB1	Dv	+
h15	B10.A(15R)	C57BL/10SnSg	B10.A	Sg	8
h16	B10.BAR28	C57BL/10Sn	B10.A	Mrf	+
h17	B10.BAR29	C57BL/10Sn	B10.A	Mrf	+
h18	B10.BAR30	C57BL/10Sn	B10.A	Mrf	+
h19	B10.BAR31	C57BL/10Sn	B10.A	Mrf	+
h20	B10.BAR32	C57BL/10Sn	B10.A	Mrf	+
h21	B10.BAR33	C57BL/10Sn	B10.A	Mrf	+
h22	B10.BAR34	C57BL/10Sn	B10.A	Mrf	+
h23	B10.BAR35	C57BL/10Sn	B10.A	Mrf	+
h24	B10.BAR15	C57BL/10SnSf (B10.S, B10.ASR11)	C57BL/10SnSf, B10.A	Sf	+6
h25	C3H.BK(1R)	C3H/HeJ	C3H/HeJ, B6.C-*H-2^{bm10}*	Sx	1
i1	B10.BAR5	C57BL/10SnSg (B10.S)	C57BL/10SnSf, B10.A	Sf	+1
i2	B10.RBD	C57BL/10SnSg (B10.G)	C57BL/6J, B10.S(24R)	Dv	+
i3	B10.A(3R)	C57BL/10SnSg	A/WySn, C57BL/10Sn	Sg	12
i4	B10.BAR7	C57BL/10SnSf (B10.S)	C57BL/10SnSf, B10.A	Sf	+3
i5	B10.A(5R)	C57BL/10SnSg	A/WySn, C57BL/10SnSg	Sg	12
i6	B10.BAR8	C57BL/10Sn (B10.S)	C57BL/10SnSf, B10.A	Sf	+3
i7	B10.D2(R107)	C57BL/10SnEg	B10.D2/n	Eg	+4
i8	A.BTR1	A/WySn	A.BY, A.TL	Psm	+
i10	A.BTF3	A/WySn	A.TL, C57BL/6J	Psm	12
i11	A.BTR4	A/WySn	A.TL, C57BL/6J	Psm	10
i12	A.BTR5	A/WySn	A.TL, C57BL/6J	Psm	10
i13.	A.BTR6	A/WySn	A.BY, A.TL	Psm	+

Table 19.1—*cont.*

Differential *H-2* haplotype	Congenic strain (synonym)	Inbred partner	Donor strain	Producer	N
i14	A.BTR7	A/WySn	A.BY, A.TL	Psm	+
i15	B10.BAR12	C57BL/10SnSg (B10.S)	C57BL/10SnSf, B10.A	Sf	+2
i17	B10.YBR	C57BL/10J	C57BL/10J, B10.AQR	Sx	+1
i19	B10.BAR18	C57BL/10Sn	B10.A, C57BL/10Sn	Mrf	+
i20	B10.BAR19	C57BL/10Sn	B10.A, C57BL/10Sn	Mrf	+
i21	B10.S(21R)	C57BL/10SnSg	B10.S(7R), C57BL/10Sn	Sg	11
i22	B10.BAR20	C57BL/10Sn	B10.A, C57BL/10Sn	Mrf	+
i23	B10.BAR21	C57BL/10Sn	B10.A, C57BL/10Sn	Mrf	+
i24	B10.BAR22	C57BL/10Sn	B10.A, C57BL/10Sn	Mrf	+
i25	B10.BAR23	C57BL/10Sn	B10.A, C57BL/10Sn	Mrf	+
i26	B10.BAR24	C57BL/10Sn	B10.A, C57BL/10Sn	Mrf	+
i27	B10.BAR25	C57BL/10Sn	B10.A, C57BL/10Sn	Mrf	+
i28	B10.BAR26	C57BL/10Sn	B10.A, C57BL/10Sn	Mrf	+
i29	B10.TBFR13	C57BL/10SnSg (B10.TBR6)	A.TFR2, C57BL/10SnSf	Sf	10
i30	B10.BAR14	C57BL/10SnSg (B10.S)	C57BL/10SnSf, B10.A	Sf	+2
i31	A.DB(1R)	A.SW/Sn	A.BY, A.D2	Sx	+4
i32	A.DB(2R)	A.SW/Sn	A.BY, A.D2	Sx	+5
i33	B10.BAR27	C57BL/10Sn	B10.A, C57BL/10Sn	Mrf	+
i34	B10.S(30R)	C57BL/10SnSg	B10.S(7R) C57BL/10SnSg	Sg	11
i35	B10.PL(42R)	C57BL/10SnSg	B10.PL, C57BL/10SnSg	Hk	12
i36	B10.A(45R)	C57BL/10SnSg	B10.P(27R), B10.A, C57BL/10SnSg	Hk	9
ial	B10.D2(R106)	C57BL/10SnEg	B10.D2-*H-2^{dm2}*	Eg	+8
j	BALB.I	BALB/cAn	I/St	Lil	11
	C3H.JK	C3H/HeDiSn	JK/St	Sn	8
	B10.WB(69NS)	C57BL/10Sn	WB/Re	Sn	10
k	A.BR	A.SW/Sn	B10.BR	Sx	C5
	BALB.K (BALB/c-*H-2^k*)	BALB/cAn	C3H/An	Lil	12
	B6.C3H (C57BL/6-*H-2^k*)	C57BL/6J	C3H/An	Lil	18
	B6-*H-2^k*	C57BL/6JBoy	AKR/JBoy	Boy	14
	B10.AKR (C57BL/10-*H-2^k*)	C57BL/10J	AKR/J	Lil	20
	B10.BR	C57BL/10Sg	C57BR/c	Sn	9
	B10.CBA	C57BL/10SnPh	CBA/Ph	Ph	12
	B10.K	C57BL/10J	CBA/J	Sf	10
k3	B6.KTL1	C57BL/6JBoy	C57BL/6-*Tla^a*, C57BL/6-*H-2^k*	Sta	+
k4	B6.KTL2	C57BL/6JBoy	C57BL/6-*Tla^a*, C57BL/6-*H-2^k*	Sta	+
k5	B6.KTL3	C57BL/6JBoy	C57BL/6-*Tla^a*, C57BL/6-*H-2^k*	Sta	+
k6	B6.K3	C57BL/6JBoyFla	C57BL/6-*H-2^k*	Fla	−
k7	B6.K4	C57BL/6JBoyFla	C57BL/6-*H-2^k*, C57BL/6-*Tla^a*	Fla	−
k8	B6.K5	C57BL/6JBoyFla	C57BL/6-*H-2^k*, C57BL/6-*Tla^a*	Fla	−
k9	B6.KB1	C57BL/6JBoyFla	C57BL/6-*H-2^k*, C57BL/6J	Rn	−
k10	C3H.RKK	C3H/HeJ	C3H.A, C3H.OH	Dv	+
km1	CBA-*H-2^{km1}* (M523)	CBA/CaLacSto	−	Eg	−
kp1	B10.P(10R)	C57BL/10SnSg	B10.A, P/J	Sg	9
kr1	B10.RKR1	C57BL/10Sn	B10.A(4R), B10.RIII	Dv	+
kr2	B10.KRR2	C57BL/10SnSf (B10.TL)	B10.RDR1, B10.AM	Sf	10
m	AKR.M	AKR/J	Non-inbred	Sn	6
	B10.AKM	C57BL/10SnSg	AKR.M/Sn	Sg	10
m1	B10.QAR	C57BL/10Sn	B10.A, B10.G	Klj	11
m2	B10.RKQ1	C57BL/10Sn	B10.A(4R), B10.Q	Dv	+
m3	B10.RKQ2	C57BL/10Sn	B10.A(4R), B10.Q	Dv	+
m4	B10.KPQ(38R)	C57BL/10SnSg	B10.P(10R), B10.Q	Hk	12
o1	B10.OL	C57BL/10SnSf	C3H.OL	Sf	9
	C3H.OL	C3H/HeJ	DBA/2J, C3H/HeJ	Sf	10
o2	B10.OH	C57BL/10SnSf	C3H.OH	Sf	12
	C3H.OH	C3H/HeJSf	DBA/2, C3H/HeJSf	Sf–Sn	11
	C3H-*H-2°*	C3H/DiSn	DBA/2, C3H/J	Sf	12

Table 19.1—*cont.*

Differential *H-2* haplotype	Congenic strain (synonym)	Inbred partner	Donor strain	Producer	N
oz2	AKR.L-*H-2^{oz2}*	AKR/J	C57L/J	Cy	14
oz3	C3H.KBR	C3H/HeJ	C3H.SW, C3H/HeJ	Sx	+2
p	B10.CNB	C57BL/10SnEg	C3H.NB	Eg	9
	B10.F/Ao	C57BL/HaAo	F/StAo	Ao	14
	B10.F/Eg	C57BL/10SnEg	F/St	Eg	+5
	B10.F/Sg	C57BL/10SnSg	F/St	Sg	10
	B10.F/Y	C57BL/10Y	F/St	Y	7
	B10.NB	C57BL/10SnPh	C3H.NB	Ph	10
	B10.P	C57BL/10Sn	P/J	Sg	10
	B10.Y	C57BL/10Sn	Non-inbred	Sn	9+
	C3H.NB	C3H/HeDiSn	NB	Sn	12
	020.Y	020/A	B10.Y	Dem	5
pb1	B10.F(13R)	C57BL/10SnSg	F/St	Sg	12
pb2	B10.P(33R)	C57BL/10SnSg	B10.P	Hk	11
pb3	B10.P(36R)	C57BL/10SnSg	B10.P	Hk	11
pf1	B10.RPF1	C57BL/10Sn	B10.RPD2, B10.M	Dv	+
pf2	B10.RPF2	C57BL/10Sn	B10.M, B10.RPD1	Dv	+
pz1	B10.020	C57BL/10J	020/A	Dem	7
q	B10.D1/Ph	C57BL/10SnPh	DBA/1J	Ph	10
	B10.D1/Y	C57BL/10Y	DBA/1J	Y	+5
	B10.G	C57BL/10SnSg	Grey lethal stock	Sg	11
	B10.Q	C57BL/10SnSg	DBA/1J	Sg	11
	C3H.Q	C3H/JSf	STOLI/Lw	Sf	10
qb1	B10.RQB1	C57BL/10Sn	B10.T(6R), B10.A(2R)	Dv	+
qb2	B10.RQB2	C57BL/10Sn	B10.T(6R), B10.A(2R)	Dv	+
qb3	B10.RQB3	C57BL/10Sn	B10.T(6R), B10.A(2R)	Dv	+
qb4	B10.RQB4	C57BL/10SnDv	B10.Q, B10.RUB1	Dv	+
qb5	B10.RQB5	C57BL/10SnDv	B10.T(6R), C57BL/10SnDv	Dv	+
qb6	B10.RQB6	C57BL/10SnDv	B10.T(6R), B10.A(4R)	Dv	+
qf1	B10.RQF1	C57BL/10Sn	B10.RQB1, B10.M	Dv	+
qp1	B10.DA(80NS)	C57BL/10Sn	DA/HuSn	Sn	10
r	B10.RIII	C57BL/10SnSg	RIII/WyJ	Sg	11
	B10.RIII(71NS)	C57BL/10Sn	RIII/WyJ	Sn	8
	LP.RIII	LP/J	RIII/WyJ	Sn	6
ra1	B10.RIII(31R)	C57BL/10SnSg	B10.RIII	Hk	12
ra2	B10.RIII(32R)	C57BL/10SnSg	B10.RIII	Hk	12
rb2	B10.RIII(37R)	C57BL/10SnSg	B10.RIII(32R)	Hk	13
rs1	B10.RIII(43R)	C57BL/10SnSg	B10.RIII(37R), B10.S	Hk	14
s	A.SW/Kl	A/WySn	A.SW/Sn	Kl	6
	A.SW/Sn	A/WySn	Swiss, non-inbred	Sn	14
	BALB.S	BALB/cBy (BALB/cJ)	SJL/J	Mrf	10
	B10.ASW	C57BL/10SnPh	A.SW/Sn	Ph	11
	B10.S	C57BL/10SnSg	A.SW/Sn	Sg	10
	020.MA	020/A	MA/A	Dem	5
s2	B10.RSS	C57BL/10Sn	B10.RSD2, B10.S	Dv	+
sb1	B10.S(26R)	C57BL/10SnSg	B10.S(24R)	Sg	13
sb2	B10.BSV(25R)	C57BL/10SnSg	B10.BSVS	Sg	13
sb3	B10.S(44R)	C57BL/10SnSg	B10.S, B10.RIII(37R)	Hk	14
sq1	A.QSR1	A/J	A.SW, Grey lethal stock	Mcd	11
sq2	B10.QSR2	C57BL/10SnSg (B10.M)	A.SW, non-inbred	Mcd, Sf	13
sq3	B10.SQR	C57BL/10SnSg	B10.S, B10.MBR	Sx	+2
sx1	B10.DDD	C57BL/10SnSg (B10.D2)	DDD	Sf	10
t1	A.TL	A/WySnSf	A.AL, A.SW	Sf	11
	B10.TL	C57BL/10SnSf	A.TL	Sf	10
t2	A.TH	A/WySnSf	B10.S(7R)	Sf	11
	B10.S(7R)	C57BL/10SnSg	A.SW, B10.A	Sg	10
t3	B10.HTT	C57BL/10SnPh	Non-inbred line HTT (A.AL, A.SW)	Ph	10
t4	B10.S(9R)	C57BL/10SnSg	A.SW, B10.A	Sg	10

Table 19.1—*cont.*

Differential *H-2* haplotype	Congenic strain (synonym)	Inbred partner	Donor strain	Producer	N
t6	B10.S(24R)	C57BL/10SnSg	B10.A, B10.S	Sg	12
t7	B10.ASR1	C57BL/10SnSg	B10.A, B10.S	Sf	+1
t8	B10.ASR8	C57BL/10SnSg	B10.A, B10.S	Sf	+1
t9	B10.ASR11	C57BL/10SnSg	B10.KDR3, B10.S	Sf	+1
t10	B10.ASR12	C57BL/10SnSf	B10.KDR3, B10.S	Sf	+3
t11	B10.RSD1	C57BL/10SnSg	B10.S, B10.RFD5	Dv	+
t12	B10.RSD2	C57BL/10SnSg	B10.S, B10.RFD5	Dv	+
t13	B10.DSR6	C57BL/10Sn	B10.D2/n, B10.S	Mrf	+
t14	B10.DSR7	C57BL/10Sn	B10.D2/n, B10.S	Mrf	+
t15	B10.DSR4	C57BL/10Sn	B10.D2/n, B10.S	Mrf	+
t16	B10.DSR5	C57BL/10Sn	B10.D2/n, B10.S	Mrf	+
t17	B10.S(23R)	C57BL/10SnSg	B10.A, B10.S	Sg	12
t18	BALB.S(3R)	BALB/cBy (BALB/cJ)	SJL/J, BALB/cBy	Mrf	9
td1	B10.RPD1	C57BL/10SnSg	B10.F(13R), B10.S(9R)	Dv	+
td2	B10.RPD2	C57BL/10SnSg	B10.F(13R), B10.S(9R)	Dv	+
td3	B10.RPD3	C57BL/10SnSg	B10.F(13R), B10.TL	Dv	+
td4	B10.RPD4	C57BL/10SnSg	B10.RPB, B10.D2	Dv	+
td5	B10.RPD5	C57BL/10SnDv	B10.F(13R), B10.TL	Dv	+
td6	B10.RPD6	C57BL/10SnDv	B10.F(13R), B10.TL	Dv	+
td7	B10.RPD7	C57BL/10SnDv	B10.F(13R), B10.TL	Dv	+
u	B10.PL	C57BL/10SnSg	PL/J	Sg	11
	B10.PL(73NS)	C57BL/10SnSg	PL/J	Sn	8
ub1	B10.RUB1	C57BL/10SnDv	B10.PL, B10.A(4R)	Dv	+
ub2	B10.RUB2	C57BL/10SnDv	B10.PL, B10.A(4R)	Dv	+
v	B10.SM/Sg	C57BL/10SnSg	SM/J	Sg	11
	B10.SM(70NS)	C57BL/10Sn	SM/J	Sn	10
vb1	B10.RVB	C57BL/10SnSg	B10.A(4R), B10.SM	Dv	+
w1	B10.KPA42	C57BL/10SnKlj	Wild *Mus domesticus*	Klj	9
	B10.KPA132	C57BL/10SnKlj	Wild *Mus domesticus*	Klj	8
w2	B10.KPB68	C57BL/10SnKlj	Wild *Mus domesticus*	Klj	
w3	B10.SAA48	C57BL/10SnKlj	Wild *Mus domesticus*	Klj	11
w4	B10.GAA20	C57BL/10SnKlj	Wild *Mus domesticus*	Klj	11
w5	B10.KEA5	C57BL/10SnKlj	Wild *Mus domesticus*	Klj	12
w6	B10.T7WF	C57BL/10SnKlj	Wild *Mus domesticus*	Klj	12
w7	C3H.WSlp (C3H.WOA1)	C3H/JSf	Wild *Mus domesticus*	Klj, Sf	13
w8	B10.SNA70	C57BL/10SnKlj	Wild *Mus domesticus*	Klj	9
w9	B10.BUA19	C57BL/10SnKlj	Wild *Mus domesticus*	Klj	13
	B10.KEA2	C57BL/10SnKlj	Wild *Mus domesticus*	Klj	14
w10	B10.DRB62	C57BL/10SnKlj	Wild *Mus domesticus*	Klj	13
	B10.SNA57	C57BL/10SnKlj	Wild *Mus domesticus*	Klj	8
	B10.WOA105	C57BL/10SnKlj	Wild *Mus domesticus*	Klj	9
w11	B10.CAA2	C57BL/10SnKlj	Wild *Mus domesticus*	Klj	10
w12 (C4ᶠ)	B10.MOL1F	C57BL/10SnKlj	Wild *Mus domesticus*	Klj	11
w12 (C4ˢ)	B10.MOL1S	C57BL/10SnKlj	Wild *Mus domesticus*	Klj	11
w13	B10.LIB55	C57BL/10SnKlj	Wild *Mus domesticus*	Klj	11
	B10.STA10	C57BL/10SnKlj	Wild *Mus domesticus*	Klj	8
	B10.STA12	C57BL/10SnKlj	Wild *Mus domesticus*	Klj	12
w14	B10.STC77	C57BL/10SnKlj	Wild *Mus domesticus*	Klj	9
w15	B10.STC90	C57BL/10SnKlj	Wild *Mus domesticus*	Klj	9
	B10.STC90	C57BL/10SnKlj	Wild *Mus domesticus*	Klj	8
w16	B10.BUA1	C57BL/10SnKlj	Wild *Mus domesticus*	Klj	12
w17	B10.CAS2	C57BL/10SnKlj	Wild *Mus domesticus*	Klj	10
w18	B10.CHR51	C57BL/10SnKlj	Wild *Mus domesticus*	Klj	12
w19	B10.KPB128	C57BL/10SnKlj	Wild *Mus domesticus*	Klj	9
w20	B10.LIB18	C57BL/10SnKlj	Wild *Mus domesticus*	Klj	8
w21	B10.GAA37	C57BL/10SnKlj	Wild *Mus domesticus*	Klj	13
w22 (C4ᶠ)	B10.BUA16F	C57BL/10SnKlj	Wild *Mus domesticus*	Klj	18
w22 (C4ˢ)	B10.BUA16S	C57BL/10SnKlj	Wild *Mus domesticus*	Klj	18

Table 19.1—*cont.*

Differential H-2 haplotype	Congenic strain (synonym)	Inbred partner	Donor strain	Producer	N
w22 (*C4^f*)	B10.BUA19F	C57BL/10SnKlj	Wild *Mus domesticus*	Klj	13
w22 (*C4^s*)	B10.BUA19S	C57BL/10SnKlj	Wild *Mus domesticus*	Klj	13
w23	B10.CAS1	C57BL/10SnKlj	Wild *Mus domesticus*	Klj	12
w24	B10.KPA44	C57BL/10SnKlj	Wild *Mus domesticus*	Klj	11
w25	B10.STA39	C57BL/10SnKlj	Wild *Mus domesticus*	Klj	11
w26 (*C4^f*)	B10.CHA2F	C57BL/10SnKlj	Wild *Mus domesticus*	Klj	12
w26 (*C4^s*)	B10.CHA2S	C57BL/10SnKlj	Wild *Mus domesticus*	Klj	12
w27	B10.STA62	C57BL/10SnKlj	Wild *Mus domesticus*	Klj	10
w28	B10-*t^{12}*	C57BL/10SnKlj	Wild *Mus domesticus*	Klj	19
w29	B10-*t^{w2}*	C57BL/10SnKlj	Wild *Mus domesticus*	Klj	16
w34	B10.STU	C57BL/10SnSf (B10.MBR, B10.TBR6)	STU	Sf	10
wm1	B10.MOL-TEN1	C57BL/10Sn	Wild *Mus molossinus*	Mrw	12
wm2	B10.MOL-TEN2	C57BL/10Sn	Wild *Mus molossinus*	Mrw	10
wm3	B10.MOL-NSB	C57BL/10Sn	Wild *Mus molossinus*	Mrw	12
wm4	B10.MOL-OHM	C57BL/10Sn	Wild *Mus molossinus*	Mrw	12
wm5	B10.MOL-MSM	C57BL/10Sn	Wild *Mus molossinus*	Mrw	12
wm7	B10.MOL-SGR	C57BL/10Sn	Wild *Mus molossinus*	Mrw	10
wm8	B10.MOL-OKB	C57BL/10Sn	Wild *Mus molossinus*	Mrw	13
wm9	B10.MOL-YNG	C57BL/10Sn	Wild *Mus molossinus*	Mrw	13
wr7	B10.WR7	C57BL/10SnSf	B10.BR, Wild *Mus domesticus*	Klj, Sf	8
y1	B10.AQR	C57BL/10SnKlj	B10.A, T138 (non-inbred)	Klj	12
y2	B10.T(6R)	C57BL/10SnSg	Grey lethal stock, A/WySn	Sg	11
yb1	B10.RQDB	C57BL/10Sn	B10.T(6R), B10.A(2R)	Dv	+
z	B10.NZW	C57BL/10SnKlj	NZW	Klj	12
zq1	020.Q(R15)	020/A	R15/Dem	Dem	5
zq2	020.Q(R26)	020/A	R26/Dem	Dem	5
zq3	020.Q(R314)	020/A	R314/Dem	Dem	5
zx1	020.GR(R15150)	020/A	GRS-*Mtv-2$^-$*	Dem	8

Table 19.2 Double congenic strains in which one of the differential loci is the *H-2* complex

Differential locus	Chromosome	Congenic strain	Synonym	Inbred partner	Donor strain	Producer	N or F generation
H-2^a, *H-4^d*	17,7	B10-*H-2^{a}H-4^b*	2^{a}4^b	C57BL/10SnSg (B10.A/Sg)	B10.129(21M)	Wts	F24
H-2^d, *H-4^b*	17,7	B10-*H-2^{d}H-4^b*	2^{d}4^b	C57BL/10Sn (B10.D2/oSn)	B10.129(21M)	Wts	F34
H-2^{h2}, *H-4^b*	17,7	B10-*H-2^{h2}H-4^b*	2^{h2}4^b	C57BL/10SnSg [B10.A(2R)/Sg]	B10.129(21M)	Wts	F17
H-2^{i5}, *H-4^b*	17,7	B10-*H-2^{i5}H-4^b*	2^{i5}4^b	C57BL/10SnSg [B10.A(5R)/Sg]	B10.129(21M)	Wts	F9
H-2^b, *Fv-1^b*	17,–	AKR-*H-2^{b}Fv-1^b*		AKR/JBoy (AKR-*H-2^b*)	AKR-*Fv-1*	Boy	+
H-2^{b2}, *Ce-2^b*	17,17	C57BL/6-*Tla^{a}Ce-2^b*		C57BL/6JBoy	C57BL/6-*Tla^a*, C57BL/6-*H-2^k*	Boy	–
H-2^{bm1}, *H-4^b*	17,7	B6.C-*H-2^{bm1}H-4^b*	2^{bm1}4^b	C57BL/6By (B6.C-*H-2^{bm1}*)	B10-*H-2^{a}H-4^b*	Wts	F10
H-2^{bm3}, *H-4^b*	17,7	B6.C-*H-2^{bm3}H-4^b*	2^{bm3}4^b	C57BL/6YEg (B6.C-*H-2^{bm3}*)	B10-*H-2^{a}H-4^b*	Wts	F7
H-2^{bm4}, *H-4^b*	17,7	B6.C-*H-2^{bm4}H-4^b*	2^{bm4}4^b	C57BL/6By (B6.C-*H-2^{bm4}*)	B10-*H-2^{a}H-4^b*	Wts	F8
H-2^{bm5}, *H-4^b*	17,7	B6.C-*H-2^{bm5}H-4^b*	2^{bm5}4^b	C57BL/6Kh (B6.C-*H-2^{bm5}*)	B10-*H-2^{a}H-4^b*	Wts	F10
H-2^{bm6}, *H-4^b*	17,7	B6.C-*H-2^{bm6}H-4^b*	2^{bm6}4^b	C57BL/6Kh (B6.C-*H-2^{bm6}*)	B10-*H-2^{a}H-4^b*	Wts	F9
H-2^{bm8}, *H-4^b*	17,7	B6.C-*H-2^{bm8}H-4^b*	2^{bm8}4^b	C57BL/6Kh (B6.C-*H-2^{bm8}*)	B10-*H-2^{a}H-4^b*	Wts	F9
H-2^{bm11}, *H-4^b*	17,7	B6.C-*H-2^{bm11}H-4^b*	2^{bm11}4^b	C57BL/6Kh (B6.C-*H-2^{bm11}*)	B10-*H-2^{a}H-4^b*	Wts	F6
H-2^a, *H-7^b*	17,9	B10-*H-2^{a}H-7^b*	2^{a}7^b	C57BL/10SnSg (B10.A/Sg)	B10.C(47N)	Wts	F27
H-2^d, *H-7^b*	17,9	B10-*H-2^{d}H-7^b*	2^{d}7^b	C57BL/10Sn (B10.D2/oSn)	B10.C(47N)	Wts	F28
H-2^{h4}, *H-7^b*	17,9	B10-*H-2^{h4}H-7^b*	2^{h4}4^b	C57BL/10SnSg [B10.A(4R)/Sg]	B10.C(47N)	Wts	F17
H-2^{i5}, *H-7^b*	17,9	B10-*H-2^{i5}H-7^b*	2^{i5}7^b	C57BL/10SnSg [B10.A(5R)/Sg]	B10.C(47N)	Wts	F25
H-2^a, *H-3^b*	17,2	B10-*H-2^{a}H-3^b*	2^{a}3^b	C57BL/10SnSg (B10.A/Sg)	B10.LP-*H-3^b*	Wts	F17
H-2^d, *H-3^b*	17,2	B10-*H-2^{d}H-3^b*	2^{d}3^b	C57BL/10Sn (B10.D2/oSn)	B10.LP-*H-3^b*	Wts	F28
H-2^d, *Hc^o*	17,2	B10.D2/o		C57BL/10Sn	DBA/2J	Sn	N6
H-2^a, *rds*	17,17	020.A(334)		020/A	B10.A/Sn	Dem	N6
H-2^{dx}, *rds*	17,17	020.GR(710)		020/A	Grs-*Mtv-2$^-$*	Dem	N6
H-2^{pz}, *rds*	17,17	020.A(336)		020/A	B10.A/Sn	Dem	N6
H-2^{pz}, *rds$^+$*	17,17	020/Q(462)		020/A	DBA/LiAf	Dem	N8
H-2^q, *rds$^+$*	17,17	020.Q(450)		020/A	DBA/LiAf	Dem	N8

Table 19.3 Alleles at *H-2* loci in the known haplotypes

H-2 Haplotype	Origin	Alleles at *H-2* loci												Strain	Reference
		K	*Aβ*	*Aα*	*Eβ*	*Eα*	*C4*	*C4S*	*D*	*L*	*Qa-2*	*Tla*	*Qa-1*		
a	k/d?	k	k	k	k	k	d	d	d	d	a	a	a	A/J, B10.A	Common
a1	k/d	k	k	k	k	k	k	k	d	d	a	c	b	A.AL, B10.AL	83
a2	k/d	k	k	k	k	k	k	k	d	d	.	.	.	B10.KDR3	80
a3	k/d	k	k	k	k	k	d	k	d	d	.	.	.	B10.KDR4	80
a4	h2/i3	k	k	k	k	k	b	d	d	d	.	.	.	B10.RKD1	17
a5	a1/i9	k	k	k	k	k	b	b	d	d	.	.	.	A.AIR1	6
a6	h4/i5	k	k	k	.	b	b	b	d	.	.	.	.	B10.HIR2	81
a7	h2/y2	k	k	k	k	b	q	q	d	d	.	.	.	B10.RKD2	18
a8	h2/i3	k	k	k	k	b	b	b	d	d	.	.	.	B10.RKD3	18
a9	h4/i4	k	k	k	k	b	q	q	d	d	.	.	.	B10.RKD5	C.S. David, u.d.
a10	k/d	k	k	k	k	b	k	k	d	d	b	c	b	BALB.A10	D.H. Sachs, u.d.
a11	k/d	k	k	k	k	b	b	b	d	d	b	c	b	BALB.A11	D.H. Sachs, u.d.
a13	h4/u	k	k	k	k	b	q	q	d	d	.	.	.	B10.RKD4	C.S. David, u.d.
a14	h2/y2	k	k	k	k/q	q	q	q	d	d	.	.	.	B10.RKD6	C.S. David, u.d.
a15	kr2/h1	k	k	k	k	k	d	k	d	d	.	.	.	B10.KDR19	82
a16	a/b	k	k	k	k	k	d	d	d	d	a^A	b^{B6}	b^{B6}	A-Tla^b	10
a17	a/b	k	k	k	k	k	d	d	d	d	a	b^{B6}	b^{B6}	B10.A(R149)-Tla^b	74
a18	a/b	k	k	k	k	k	d	d	d	d	a	b^{B6}	b^{B6}	B10.A(R410)-Tla^b	74
a19	h2/y2	k	k	k	k	q	q	q	d	d	.	.	.	B10.RKD7	C.S. David, u.d.
ak1	b/k	b	b	b	b	b	b	b	k	k	b	a	a	B6.AK1	14
an1	t1/f	s	k	b	k	b	b	b	f	f	b	d	b	A.TFR1	83
an3	a2/f	s	s	s	s	b	b	b	f	f	.	.	.	B10.RSF2	C.S. David, u.d.
an4	a21/f	s	s	s	s	k/p	k	k	f	f	.	.	.	B10.RSF4	C.S. David, u.d.
an5	a4/f	s	s	s	s	k	b	b	f	f	.	.	.	B10.RSF3	19
an6	a26/f	s	k	k	.	k/p	b	b	f	f	.	.	.	B10.RSF5	19
an7	a2/f	s	k	k	.	b	f	f	f	.	.	.	.	B10.TBFR4	82
an8	a2/f	s	k	k	.	b	b	b	f	f	.	.	.	B10.TBFR6	82
an9	a2/f	s	k	k	.	b	b	b	f	f	.	.	.	B10.TBR8	82
an10	a2/f	s	k	k	.	f	f	f	f	d	.	.	.	B10.TBR9	82
ap1	f/a	f	f	f	f	f	f	f	d	d	a	a	a	B10.M(11R)	97
ap2	f/t2	f	f	f	f	f	f	f	d	d	(a)	a	(a)	A.TFR2	D.C. Shreffler and C.S. David, u.d.
ap3	f/t2	f	f	f	f	.	s	s	d	d	(a)	a	(a)	A.TFR3	D.C. Shreffler and C.S. David, u.d.
ap4	f/t2	f	f	f	f	.	s	s	d	d	(a)	a	(a)	A.TFR4	C.S. David, u.d.
ap5	f/t1	f	f	f	f/k	k	k	k	d	d	a	c	b	A.TFR5	84
ap6	f/a8	f	f	f	f	f	f	f	d	d	.	.	.	B10.RFD1	18
ap7	f/a8	f	f	f	f	f	f	f	d	d	.	.	.	B10.RFD2	18
ap8	f/td1	f	f	f	f	f	d	d	d	d	.	.	.	B10.RFD4	C.S. David, u.d.
ap9	f/a9	f	f	f	f	f	f	f	d	d	.	.	.	B10.RFD5	C.S. David, u.d.
ap10	f/td1	f	f	f	f	f	d	d	d	d	.	.	.	B10.RFD6	C.S. David, u.d.
ap11	f/td1	f	f	f	f	k	d	d	d	d	.	.	.	B10.RFD7	C.S. David, u.d.

Table 19.3—*cont.*

H-2 Haplotype	Origin	K	Aβ	Aα	Eβ	Eα	C4	C4S	D	L	Qa-2	Tla	Qa-1	Strain	Reference
ap12	fltd1	f	f	f	f	f	d	d	d	d	.	.	.	B10.RFD3	19
aq1	a/f	k	k	k	k	k	d	d	f	f	b	d	a	B10.M(17R)	97
aq2	a/an5	k	k	k	k	k	d	d	f	f	.	.	a	B10.RKF1	19
aq3	h4/f	k	k	k	k	b	b	b	f	f	.	.	.	B10.RKF2	19
aq4	h4/f	k	k	k	k	b	b	b	f	f	.	.	.	B10.RKF4	C.S. David, u.d.
aq5	a19/f	k	k	k	k	f	f	f	f	f	.	.	.	B10.RKF3	C.S. David, u.d.
ar1	dl?	d	.	.	.	.	.	.	.	.	.	.	.	B10.LG	C.S. David, u.d.
ar2	a/r	k	k	k	k/s	.	r	r	r	r	b	.	b	B10.RIII(34R)	K.J. Hasenkrug, u.d.
as1	a/s	k	k	k	k/s	s	s	s	s	s	a	b	b	B10.S(8R)	94
as2	a/s	k	k	k	k/s	s	d	d	s	.	.	.	.	B10.ASR2	79
as3	a/s	k	k	k	k/s	s	s	s	s	.	.	.	.	B10.ASR7	79
as4	h6/s	k	k	k	k/s	s	s	s	s	.	.	.	.	B10.BASR1	79
as5	h11/s	k	k	k	.	k	s	s	s	.	.	.	.	B10.BASR2	80
as6	m4/a	k	.	.	k	.	.	s	.	.	.	.	.	B10.S(41R)	K.J. Hasenkrug, u.d.
as7	h4/s	k	k	k	k	b	b	s	s	s	.	.	.	B10.RKS1	C.S. David, u.d.
at1	t1/b	s	k	k	k/b	b	b	b	b	△	.	.	.	A.TBR1	69
at2	t1/b	s	k	k	k/b	b	b	b	b	△	(a)	b	(b)	A.TBR2	69
at3	t1/b	s	k	k	k/b	b	b	b	b	△	a	b	b	A.TBR3	69
at4	t1/b	s	k	k	k/b	b	b	b	b	△	a	b	b	A.TBR4	69
at5	t1/b	s	k	k	k	k	b	b	b	△	.	.	.	A.TBR5	68
at6	h11/s	s	s	s	.	b	b	b	b	.	.	.	.	B10.TBR6	79
at8	t4/h4	s	s	s	s/b	k	d	d	b	.	.	.	.	B10.TBR11	82
at9	t4/h4	s	s	s	s	k	b	b	b	.	.	.	.	B10.RSB1	17
at10	t3/h2	s	s	s	s	k	d	d	b	.	.	.	.	B10.RSB2	17
at11	t1/b	s	s	s	s	k	k	k	b	.	.	.	.	B10.RSB3	17
at12	t1/b	s	k	k	k/b	b	b	b	b	△	.	.	.	A.TBR12	68
at13	t1/b	s	k	k	k	k	k	k	b	△	.	.	.	A.TBR13	68
at14	t1/b	s	k	k	k	k	k	k	b	△	.	.	.	A.TBR14	68
at15	t1/b	s	k	k	k	k	k	k	b	△	.	.	.	A.TBR15	68
at16	t1/b	s	k	k	k/b	b	b	b	b	△	.	.	.	A.TBR16	68
at17	t1/b	s	k	k	k/b	b	b	b	b	.	.	.	.	A.TBR17	68
at18	t3/h4	s	s	s	s	b	b	b	b	.	.	.	.	B10.THR2	81
at19	t4/h4	s	s	s	.	k	d	d	b	.	.	.	.	B10.RSB4	18
at20	t4/pb1	s	s	s	s	k/p	d	d	b	.	.	.	.	B10.RSB5	18
at21	t4/pb1	s	s	s	s	k/p	b	b	b	.	.	.	.	B10.RSB6	18
at22	t3/h2	s	s	s	s	k	b	b	b	.	.	.	.	B10.RSB7	18
at23	t4/pb1	s	s	s	s/b	k/p	b	b	b	.	.	.	.	B10.RSB8	18
at24	t4/h4	s	s	s	s/k	k	d	d	b	.	.	.	.	B10.RSB10	18
at25	t1/pb1	s	k	k	s	k/p	b	b	b	.	.	.	.	B10.RSB11	C.S. David, u.d.
at26	t4/pb1	s	s	s	s	k/p	b	b	b	.	.	.	.	B10.RSB12	C.S. David, u.d.
at27	t1/pb1	s	k	k	k	k	k	k	b	.	.	.	.	B10.RSB13	19
at28	t1/pb1	s	k	k	k	k	b	b	b	.	.	.	.	B10.RSB14	19
at29	t1/pb1	s	k	k	k	k/p	b	b	b	.	.	.	.	B10.RSB15	19

Designation	Allele	1	2	3	4	5	6	7	8	9	Strain	Ref.
at30	*h11/s*	s	s	s	s	s	b	.	.	.	B10.TBR18	82
aw1	*t2/r*	s	s	r	r	b	.	.	b	.	B10.RIII(20R)	6
aw2	*a/w*	k	k	w	w	w	w	w	r	b	B10.A(R202)	78
aw3	*a/w*	k	k	d	d	w	w	w	w	.	B10.A(R203)	78
aw4	*a/w*	w	w	d	d	w	w	w	w	.	B10.A(R204)	78
aw6	*a/w*	w	k	d	d	w	w	w	w	.	B10.A(R206)	78
aw7	*a/w*	k	k	d	d	w	w	w	w	.	B10.A(R207)	78
aw8	*a/w*	w	w	w	w	w	w	w	w	.	B10.A(R208)	78
aw9	*a/w*	k	k	d	d	w	w	w	w	.	B10.A(R209)	78
aw10	*a/w*	w	w	w	w	w	w	w	w	.	B10.A(R201)	78
aw11	*a/w*	k	w	d	d	w	w	w	w	.	B10.A(R211)	78
aw12	*a/w*	w	w	d	d	w	w	w	w	.	B10.A(R212)	78
aw13	*a/w*	w	w	d	d	k	w	w	w	.	B10.A(R213)	78
aw14	*a/w*	w	w	w	w	w	w	w	w	.	B10.A(R214)	78
aw17	*a/w*	k	k	d	d	w	w	k	w	.	B10.A(R217)	78
az1	*a/v*	k	k	v	v	v	▽	v	b	.	B10.SM(22R)	6
b	*—*	b	b	a	a	a	△	b	b	b	C57BL/10	Common
b1	*h4i5*	b	b	a^{B6}	a^{B6}	a^{B6}	.	b	b	b	B10.HIR1	81
b2	*b/a*	b/k	b	a^{AKR}	a^{AKR}	b^{AKR}	△	b	b	a^{A}	C57BL/6-Tla^a	10
b3	*b/k*	b	b	a^{AKR}	a^{AKR}	a^{B6}	△	b	b	a^{AKR}	B6.K1	28
b4	*b/k*	b	b	a^{AKR}	a^{AKR}	a^{B6}	△	b	b	a^{AKR}	B6.K2	28
b5	*b/k*	b	b	$a^{A,CA}$	d^{CA}	a^{B6}	△	b	b	$a^{A,CA}$	B6.KB2	72
b6	*b/f*	b	b	$a^{A,CA}$	d^{CA}	a^{B6}	△	b	b	$a^{A,CA}$	B6.AC2	14
b7	*b/f*	b	b	a^{A}	a^{A}	a?	△	b	b	a^{A}	B6.AC3	14
b8	*b/a*	b	b	a^{A}	a^{A}	a?	△	b	b	a^{A}	B10.A(R297)-Tla^a	74
b9	*b/a*	b	b	a^{A}	a^{A}	.	△	b	b	a^{A}	B10.A(R310)-Tla^a	74
bb1	*i5/av1*	b	b	.	.	b	.	b	b	.	B10.A(35R)	74 — Hasenkrug and Stimpfling, u.d.
bf1	*b/f*	b	b	f	f	f	f	f	f	.	B10.RBF	82
bf2	*an1/b*	b	b	b	b	b	.	f	f	.	B10.TFBR12	5 — C.S. David, u.d.
bm1	*b*	bm1	b	b	b	b	△	b	b	b	B6.C-H-2^{bm1}	5
bm2	*b*	bm2	b	b	b	b	△	b	b	b	B6.C-H-2^{bm2}	22
bm3	*b*	bm3	b	b	b	b	△	b	b	b	C57BL/6-H-2^{bm3}	4
bm4	*b*	bm4	b	b	b	b	△	b	b	b	B6.C-H-2^{bm4}	47
bm5	*b*	bm5	b	b	b	b	△	b	b	b	C57BL/6-H-2^{bm5}	47
bm6	*b*	bm6	b	b	b	b	△	b	b	b	C57BL/6-H-2^{bm6}	47
bm7	*b*	bm7	b	b	b	b	△	b	b	b	B6.C-H-2^{bm7}	?
bm8	*b*	bm8	b	b	b	b	△	b	b	b	C57BL/6-H-2^{bm8}	47
bm9	*b*	bm9	b	b	b	c	△	b	b	b	B6.C-H-2^{bm9}	57
bm10	*b*	bm10	b	b	b	b	△	b	b	b	B6.C-H-2^{bm10}	57
bm11	*b*	bm11	b	b	b	b	△	b	b	b	B6.C-H-2^{bm11}	57
bm12	*b*	b	bm12	b	b	b	△	b	b	b	B6.C-H-2^{bm12}	51
bm13	*b*	b	bm13	bm13	b	b	△	b	b	b	B6.C-H-2^{bm13}	57
bm14	*b*	b	bm14	bm14	b	b	f	b	b	.	B6.C-H-2^{bm14}	57 — D.W. Bailey, u.d.
bm16	*b*	bm16	b	b	b	b	△	b	b	b	C57BL/6-H-2^{bm16}	54
bm17	*b*	bm17	b	b	b	b	△	b	b	b	C57BL/6-H-2^{bm17}	54
bm18	*b*	bm18	b	b	b	b	△	b	b	b	B6.C-H-2^{bm18}	54
bm19	*b*	bm19	b	b	b	b	△	b	b	b	B6.C-H-2^{bm19}	54
bm20	*b*	bm20	b	b	b	b	△	b	b	b	B6.C-H-2^{bm20}	54

809

Table 19.3—*cont.*

H-2 Haplotype	Origin	K	Aβ	Aα	Eβ	Eα	C4	C4S	D	L	Qa-2	Tla	Qa-1	Strain	Reference
bm21	b	bm21	b	b	b	b	b	b	b	Δ	a	b	b	B6.C-H-2^{bm21}	54
bm22	b	bm22	b	b	b	b	b	b	b	Δ	a	b	b	C57BL/6-H-2^{bm22}	54
bm23	b	bm23	b	b	b	b	b	b	b	Δ	a	b	b	B10.D2-H-2^{bm23}	24
bm24	b	b	b	b	b	b	b	b	bm24	Δ	a	b	b	C57BL/10-H-2^{bm24}	56
bp1	–	bp1	.	.	.	.	.	.	bp1	.	.	c	.	BN/a	15
bp2	b/p	b	b	b	b	b	b	b	p	p	b	.	.	B10.P(27R)	K.J. Hasenkrug and J.H. Stimpfling, u.d.
bp3	b/kp1	b	.	.	.	.	.	.	p	.	.	.	.	B10.P(39R)	K.J. Hasenkrug and J.H. Stimpfling, u.d.
bp4	b/kp1	b	.	.	.	.	.	P	P	.	.	.	.	B10.P(40R)	K.J. Hasenkrug and J.H. Stimpfling, u.d.
bp5 (bw1)	b/p	b	b	.	b	b	.	P	P	P	b	.	.	B10.F(14R)	K.J. Hasenkrug and J.H. Stimpfling, u.d.
bq1	b/m	b	k	b	k	k	k	k	q	q	a	a	a	B10.MBR	76
bq2	b/q?	b	b	b	b	.	q	q	q	q	a	a	a	BQ2, B10.Tf	25, K.J. Hasenkrug and J.H. Stimpfling, u.d.
bq3	i2/q	b	b	b	b	b	.	.	q	q	.	.	b	B10.RBQ1	C.S. David, u.d.
bq4	b/q	b	b	b	.	.	.	q	q	q	a	b	.	B10.QBR	D.H. Sachs, u.d.
bq5	b/q	b	b	b	.	.	q	q	q	q	.	.	.	B10.D1(R108)	8
bs1	–	.	.	.	.	.	.	.	bs1	Δ	b	.	.	WLL/BrA	P. Démant, u.d.
bv1	b	b	b	b	b	b	b	b	b	Δ	a	c	b	129/ReJ	Common
bw1	b/w	b	w	w	w	w	w	w	w	w	a	.	.	B10.(R231)	78
bw3	b/w	b	w	w	w	w	w	w	w	w	a	.	.	B10.(R233)	78
bw6	b/w	b	w	w	w	w	b	w	w	w	.	.	.	B10.(R236)	78
bw7	w/b	w	b	b	b	b	b	b	b	Δ	.	.	.	B10.(R237)	78
bw9	w/b	w	b	b	b	b	b	b	b	Δ	a	b	b	B10.(R239)	78
by1	y/l/b	q	k	k	k	k	d	b	b	Δ	a	b	b	B10.BYR1	J. Klein and F. Figueroa, u.d.
by2	y/l/b	q	k	k	k	k	d	d	b	Δ	a	b	b	B10.BYR2	J. Klein and F. Figueroa, u.d.
bz1	b/s	b	s	s	s	s	s	s	s	.	a	b	b	B10.SBR	D.H. Sachs, u.d.
bz2	bq1/s	b	k	k	k	k	k	k	k	.	a	b	b	A.SB(1R)	D.H. Sachs, u.d.
bz3	bq1/s	b	k	k	k	k	k	k	k	d	a	b	b	A.SB(2R)	D.H. Sachs, u.d.
d	–	d	d	d	d	d	d	d	d	d	b	c	b	DBA/2, BALB/cJ	Common
	d	d	d	d	d	d	d	d	d	d	b	c	b	BALB/cBy	30
dm1	d	d	d	d	d	d	d	d	dm1	dm1	.	.	.	B10.D2-H-2^{dm1}	21
dm2	d	d	d	d	d	d	d	d	d	Δ	b	c	b	BALB/c-H-2^{dm2}	57
dm3	d	d	d	d	d	d	d	d	dm3	d	.	.	.	B6.C-H-2^{dm3}	57
dm4	d	dm4	d	d	d	d	d	d	d	d	.	.	.	C.B6-H-2^{dm4}	59
dm5	d	dm5	d	d	d	d	d	d	d	d	b	c	b	C.B6-H-2^{dm5}	60
dm6	d	d	d	d	d	d	d	d	dm6	d	.	.	.	CB6-H-2^{dm6}	.
dr1	d/r	d	d	d	d	d	d	d	r	d	b	c	.	B10.RDR1	82
ds1	d/s	d	d	d	d	d	d	d	s	d	.	.	.	B10.DSR1	63

Symbol	Allele												Strain	Reference
ds2	*d/s*	*d*	*d*	*d*	*d*	*d*	*s*	·	·	·	·	·	B10.DSR2	63
ds3	*d/s*	*d*	*d*	*d*	*d*	*s*	*s*	*s*	·	·	·	·	B10.DSR3	63
dx1	*d/w3*	*d*	*f*	*f*	*f*	*f*	*w3*	*f*	·	·	·	·	GSR/Hs, LG/Ckc	101, 93
dx2	*d/w3*	*d*	*f*	*f*	*f*	*f*	*w3*	*f*	*b*	*b*	·	·	020.TSI	P. Démant, pers. comm.
f	–	*f*	*f*	*f*	*f*	*f*	*f*	*f*	*b*	*b*	*b*	*b*	A.CA, B10.M	Common
f1	*f/b*	*f*	*f*	*f*	*f*	*f*	*f*	*f*	*a^B6*	*b^B6*	*b^B6*	*b^B6*	B6.ACA	14
fb2	*flat24*	*k*	*f*	*f*	*f*	*f*	*b*	*f/k*	*d*	*d*	·	·	B10.RFB2	C.S. David, u.d.
fb3	*flat26*	*p*	*f*	*f*	*f*	*f*	*b*	*f*	*b*	*b*	·	·	B10.RFB3	19
fb4	*a3lf*	*f*	*f*	*f*	*f*	*f*	*b*	*f*	*f*	*f*	*f*	·	B10.TBFR7	82
fd1	*flyb1*	*f*	*f*	*f*	*f*	*f*	*d*	*f*	*f*	*f*	*b*	*d*	B10.RFDB	C.S. David, u.d.
fm1	*f*	*fm1*	*f*	*f*	*f*	*f*	*f*	*f*	*f*	*f*	·	*a*	A.CA-H-2^*fm1*	23
fm2	*fm2*	*f*	*f*	*f*	*f*	*fm2*	*fm2*	*f*	*f*	*f*	*f*	*f*	B10.M-H-2^*fm2*	62
g	*d/b*	*d*	*d*	*d*	*d*	*d*	*b*	*d*	*d*	*d*	◁	*b*	HTG, B10.HTG	1
g1	*d/b*	*d*	*d*	*d*	*d*	*d*	*b*	*a*	*d*	*d*	◁	*b*	B10.D2(R101)	100
g2	*d/b*	*d*	*d*	*d*	*d*	*d/b*	*b*	*a*	*b*	*b*	◁	*b*	D2.GD, B10.GD	49
g3	*d/b*	*d*	*d*	*d*	*d*	*d*	*b*	*a*	*d*	*d*	◁	*b*	B10.D2(R013)	100
g4	*d/b*	*d*	*d*	*d*	*d*	*d*	*b*	*a*	*d*	*d*	·	·	B10.BDR1	H.C. Passmore, u.d.
g5	*d/b*	*d*	*d*	*d*	*d*	*d*	*b*	*a*	*d*	*d*	·	·	B10.BDR2	H.C. Passmore, u.d.
g6	*d/b?*	*k/b*	*d*	*d*	*d*	*d*	*b*	*a*	*k*	*k*	·	·	B6.C-H-2^*g6*	41
g7	*a/b*	*d*	*d*	*d*	*d*	*b*	*b*	*a*	*d*	*d*	◁	*b*	DBR7	25
h	*a/b*	*k*	*k*	*k*	*k*	*b*	*b*	*(a)*	◁	*b*	*b*	*b*	HTH	32
h1	*a/b*	*k*	*k*	*k*	*k*	*b*	*b*	*a*	◁	*b*	*b*	*b*	B10.A(1R)	95
h2	*a/b*	*k*	*k*	*k*	*k*	*b*	*b*	*a*	◁	*b*	*b*	*b*	B10.A(2R)	95
h3	*k/b*	*k*	*k*	*k*	*k*	*b*	*b*	*a*	◁	*b*	*b*	*b*	B10.AM	D.B. Amos, u.d.
h4	*a/b*	*k*	*k*	*k*	*k*	*b*	*b*	*a*	◁	*b*	*b*	*b*	B10.A(4R)	95
h6	*k/b*	*k*	*b*	*b*	*b*	*b*	*b*	·	◁	*b*	*b*	·	B10.BAR4	79
h7	*a/b*	·	·	·	·	·	·	*a*	·	·	·	·	L.AKR(A234)	13
h8	*a/b*	*k*	*k*	*k*	*k*	*b*	*b*	·	◁	·	·	*b*	B10.BAR10	80
h9	*k/b*	*k*	*k*	*k*	*k*	*b*	*b*	·	◁	·	·	*b*	B10.BAR11	80
h10	*k/b*	*k*	*k*	*k*	*k*	*b*	*b*	·	◁	·	·	*b*	B10.RKB	17
h11	*a/b*	*k*	*k*	*k*	*k*	*b*	*b*	·	◁	·	·	*b*	B10.BAR6	81
h12	*a/fb1*	*k*	*k*	*k*	*k*	*b*	*b*	·	◁	·	·	·	B10.RKB2	C.S. David, u.d.
h13	*a/fb1*	*k*	*k*	*k*	*k*	*b*	*b*	·	◁	·	·	·	B10.RKB3	C.S. David, u.d.
h15	*a/b*	*k*	*k*	*k*	*k*	*b*	*b*	*a*	◁	·	·	·	B10.A(15R)	97
h16	*a/b*	*k*	*k*	*k*	*k*	*b*	*b*	·	◁	·	·	·	B10.BAR28	63
h17	*a/b*	*k*	*k*	*k*	*k*	*b*	*b*	·	◁	·	·	·	B10.BAR29	63
h18	*a/b*	*k*	*k*	*k*	*k*	*b*	*b*	·	◁	·	·	·	B10.BAR30	63
h19	*a/b*	*k*	*k*	*k*	*k*	*b*	*b*	·	◁	·	·	·	B10.BAR31	63
h20	*a/b*	*k*	*k*	*k*	*k*	*b*	*b*	·	◁	·	·	·	B10.BAR32	63
h21	*a/b*	*k*	*b*	*b*	*b*	*b*	*b*	·	◁	·	·	·	B10.BAR33	63
h22	*a/b*	*b*	*b*	*b*	*b*	*b*	*b*	·	*d*	·	·	·	B10.BAR34	63
h23	*a/b*	*k*	*k*	*b*	*b*	*b*	*b*	·	◁	·	*b*	·	B10.BAR35	63
h24	*a/b*	*k*	*k*	*b*	*b*	*b*	*b*	·	*d*	·	·	·	B10.BAR15	82
h25	*k/bm10*	*k*	*k*	*k*	*k*	*b*	*b*	*(a)*	·	*b*	*b*	·	C3H.BK(1R)	75
i	*b/a*	*b*	*b*	*b*	*b*	*d*	*d*	*a*	*d*	*b*	*b*	*a*	HTI	32
i1	*b/a*	*b*	*b*	*b*	*b*	*d*	*d*	*a*	*d*	*b*	*b*	*a*	B10.BAR5	79
i2	*b/t6*	*b*	*b*	*b*	*b*	*b*	*d*	*a*	*d*	*b*	*b*	*a*	B10.RBD	17
i3	*b/a*	*b*	*b*	*b*	*b*	*d*	*d*	*a*	*d*	*b*	*b*	·	B10.A(3R)	95
i4	*b/a*	*b*	*b*	*b*	*b*	*d*	*b*	·	*d*	*b*	*b*	·	B10.BAR7	80

Table 19.3—*cont.*

H-2 Haplotype	Origin	Alleles at H-2 loci												Strain	Reference
		K	*Aβ*	*Aα*	*Eβ*	*Eα*	*C4*	*C4S*	*D*	*L*	*Qa-2*	*Tla*	*Qa-1*		
i5	*b/a*	*b*	*b*	*b*	*b/k*	*k*	*d*	*d*	*d*	*d*	*a*	*a*	*a*	B10.A(5R)	95
i6	*b/a*	*b*	*b*	*b*	*b*	*b*	*b*	*b*	*d*	*d*	*a*	*a*	*a*	B10.BAR8	80
i7	*b/d*	*b*	*b*	*b*	*b*	*b*	*b*	*b*	*d*	.	*a*	*c*	*b*	B10.D2(R107)	100
i8	*b/t1*	*b*	*b*	*b*	*b*	*b*	*b*	*b*	*d*	*d*	*a*	*c*	.	A.BTR1	69
i9	*b/t1*	*b*	*b*	*b*	*b*	*b*	*b*	*b*	*d*	*d*	*a*	*c*	*b*	BTR2/c	69
i10	*b/t1*	*b*	*b*	*b*	*b*	*b*	*b*	*b*	*d*	*d*	.	.	.	A.BTR3	68
i11	*b/t1*	*b*	*b*	*b*	*b*	*b*	*b*	*b*	*d*	*d*	.	.	.	A.BTR4	68
i12	*b/t1*	*b*	*b*	*b*	*b*	*b*	*b*	*b*	*d*	*d*	.	.	.	A.BTR5	68
i13	*b/t1*	*b*	*b*	*b*	*b*	*b*	*b*	*b*	*d*	*d*	.	.	.	A.BTR6	68
i14	*b/t1*	*b*	*b*	*b*	*b*	*b*	*b*	*b*	*d*	*d*	.	.	.	A.BTR7	68
i15	*b/a*	*b*	*b*	*b*	*b*	*b*	*b*	*b*	*d*	*d*	*a*	.	.	B10.BAR12	81
i18	*b/a*	*b*	*b*	*b*	*b*	*b*	*b*	*b*	*d*	*d*	*a*	.	*a*	B10.A(18R)	K.J. Hasenkrug and J.H. Stimpfling, u.d.
i19	*b/a*	*b*	*b*	*b*	*b*	*b*	*b*	*b*	*d*	.	.	.	.	B10.BAR18	63
i20	*b/a*	*b*	*b*	*b*	*b*	*b*	*b*	*b*	*d*	.	.	.	.	B10.BAR19	63
i21	*b/t2*	*b*	*b*	*b*	*b*	*b*	*d*	*b*	*d*	*d*	*a*	.	*a*	B10.A(21R)	K.J. Hasenkrug and J.H. Stimpfling, u.d.
i22	*b/a*	*b*	*b*	*b*	*b*	*b*	.	.	*d*	.	.	.	.	B10.BAR20	63
i23	*b/a*	*b*	*b*	*b*	*b*	*b*	.	.	*d*	.	.	.	.	B10.BAR21	63
i24	*b/a*	*b*	*b*	*b*	*k*	*b*	*d*	.	*d*	.	.	.	.	B10.BAR22	63
i25	*b/a*	*b*	*b*	*b*	*b*	*b*	.	.	*d*	.	.	.	.	B10.BAR23	63
i26	*b/a*	*b*	*b*	*b*	*b*	*b*	.	.	*d*	.	.	.	.	B10.BAR24	63
i27	*b/a*	*b*	*b*	*b*	*k*	*b*	*d*	*d*	*d*	.	.	.	.	B10.BAR25	63
i28	*b/a*	*b*	*b*	*b*	*b*	*b*	.	*d*	*d*	.	.	.	.	B10.BAR26	63
i29	*b/ap2*	*b*	*b*	*b*	*b*	*b*	*b*	*b*	*d*	.	.	.	.	B10.TBFR13	82
i30	*b/a*	*b*	*b*	*b*	*b*	*b*	*b*	*b*	*d*	.	.	.	.	B10.BAR14	82
i31	*b/d*	*b*	*b*	*b*	*d*	*d*	*d*	*d*	*d*	*d*	.	.	.	A.DB(1R), (A.DBRa)	75
i32	*b/d*	*b*	*b*	*b*	*b*	*b*	.	.	*d*	*d*	.	.	.	A.DB(2R), (A.DBRb)	75
i33	*b/a*	*b*	*b*	*b*	*k*	*k*	*d*	*d*	*d*	*d*	.	.	.	B10.BAR27	63
i34	*b/t2*	*b*	*b*	*b*	*b*	*b*	*b*	*b*	*d*	*d*	*a*	.	.	B10.S(30R)	K.J. Hasenkrug and J.H. Stimpfling, u.d.
i36	*bp2/a*	*b*	.	.	.	.	.	.	*d*	.	.	.	.	B10.A(45R)	K.J. Hasenkrug and J.H. Stimpfling, u.d.
ia1	*b/dm1*	*b*	*b*	*b*	*b*	*b*	*b*	*b*	*dm1*	*dm1*	*a*	*c*	*b*	B10.D2(R106)	100
j	*j/b?*	*j*	*j*	*j*	*j*	*j*	*j*	*j*	*b*	*b*	*a*	*b*	*b*	I/St, WB/J. B10.WB	87
k	–	*k*	*k*	*k*	*k*	*k*	*k*	*k*	*k*	*k*	*b*	*b*	*b*	C3H, CBA, AKR	Common
k2	–	*k*	*k*	*k*	*k*	*k*	*k*	*k*	*k*	*k*	*b*	*a*	*a*	C58, B10.BR	Common
k3	*k/a*	*k*	*k*	*k*	*k*	*k*	*k*	*k*	*k*	.	b^{AKR}	d^A	d^A	B6.KTL1	92
k4	*k/a*	*k*	*k*	*k*	*k*	*k*	*k*	*k*	*k*	.	b^{AKR}	b^{AKR}	d^A	B6.KTL2	92
k5	*k/a*	*k*	*k*	*k*	*k*	*k*	*k*	*k*	*k*	.	b^{AKR}	b^{AKR}	d^A	B6.KTL3	92

Symbol	Combination												Strain	Reference
k6	*k/b2*	k	k	k	k	k	k	k	k	a^{B6}	a^{A}	a^{A}	B6.K3	77
k7	*k/b2*	k	k	k	k	k	k	k	k	b^{AKR}	a^{A}	a^{A}	B6.K4	77
k8	*k/b2*	k	k	k	k	k	k	k	k	b^{AKR}	a^{A}	a^{A}	B6.K5	L. Flaherty, u.d.
k9	*k/b*	k	k	k	k	k	k	k	k	b^{AKR}	b^{B6}	b^{B6}	B6.KB1	72
k10	*a/o2*	k	k	d	d	d	k	k	k	.	.	.	C3H.RKK	C.S. David, u.d.
km1	*k*	km1	k	k	k	k	k	k	k	b	b	b	CBA-H-2^{km1}	9
kp1	*a/p*	p	p	p	p	p	p	p	p	b	b	b	B10.P((10R)	K.J. Hasenkrug and J.H. Stimpfling, u.d.
kr1	*h4/r*	r	b	r	r	r	b	r	r	.	.	.	B10.RKR1	C.S. David, u.d.
kr2	*h3/dr1*	r	k	k	k	k	k	k	k	.	.	.	B10.RKR2	82
m	*k/q?*	q	g	k	k	k	k	k	k	a	a	a	AKR.M, B10.AKM	86
m1	*a/q*	q	d	d	q	q	q	q	q	a	a	a	B10.QAR	44
m2	*h4/q*	.	b	b	.	.	.	.	.	a	a	a	B10.RKQ1	18
m3	*h4/q*	q	.	.	b	b	b	.	.	.	.	.	B10.RKQ2	C.S. David, u.d.
m4	*kp1/q*	q	p	p	p	p	p	p	p	a	b	b	B10.KPQ(38R)	K.J. Hasenkrug and J.H. Stimpfling, u.d.
o1	*d/k*	k	d	d	d	d	d	k	k	b	b	b	C3H.OL, B10.OL	83
o2	*d/k*	k	d	d	d	d	d	k	k	b	b	b	C3H.OH, B10.OH	85
oz1	*b/k*	.	b	b	b	b	b	b	b	b	b	b	C57BL6-H-2^{oz1}	99
oz2	*b/k*	.	.	.	.	.	.	.	.	.	.	.	AKR.L-H-2^{oz2}	13
oz3	*b/k*	p	b	b	b	b	b	b	b	b	b	b	C3H.KBR	D.H. Sachs, u.d.
p	*—*	p	p	p	p	p	p	p	p	b	e	b	P/J, B10.P, C3H.NB	Common
pb1	*p/b*	b	b	b	b	b	b	b	b	a	b	b	B10.F(13R)	K.J. Hasenkrug and J.H. Stimpfling, u.d.
pb2	*p/b*	.	p	p	p	p	p	p	p	a	b	.	B10.P(33R)	K.J. Hasenkrug and J.H. Stimpfling, u.d.
pb3	*p/b*	p	b	b	b	b	p	p	p	a	b	b	B10.P(36R)	K.J. Hasenkrug and J.H. Stimpfling, u.d.
pf1	*td2/f*	f	f	f	f	f	f	f	f	.	.	.	B10.RPF1	C.S. David, u.d.
pf2	*td1/f*	f	f	f	f	f	f	f	f	.	.	.	B10.RPF2	C.S. David, u.d.
pz1	*—*	pz1	.	.	.	.	.	.	.	.	.	.	020/A	16
q	*—*	q	q	q	q	q	q	q	q	.	b	b	DBA/1, B10.Q	Common
qb1	*y2/h2*	b	b	q	q	q	q	q	q	.	.	.	B10.RQB1	18
qb2	*y2/h2*	b	b	q	q	q	q	q	q	.	.	.	B10.RQB2	18
qb3	*y2/h2*	d	d	q	q	q	q	q	q	.	.	.	B10.RQB3	19
qb4	*q/ub1*	k	.	q/k	q	q	q	q	q	b	.	.	B10.RQB4	C.S. David, u.d.
qb5	*y2/b*	q	q	q	q	q	q	q	q	b	.	.	B10.RQB5	C.S. David, u.d.
qb6	*y2/h4*	q	q	q	q	q	q	q	q	b	.	.	B10.RQB6	C.S. David, u.d.
qf1	*qb1/f*	f	f	q	q	q	q	q	q	f	f	f	B10.RQF1	18
qp1	*q/s*	s	s	q	q	q	q	q	q	s	s	s	B10.DA(80NS)	G.D. Snell, u.d.
r	*—*	r	r	r	r	r	r	r	r	r	r	b	RIII, B10.RIII	Common
ra1	*r/a*	d	d	d	r	r	r	r	r	a	d	d	B10.RIII(31R)	K.J. Hasenkrug and J.H. Stimpfling, u.d.
ra2	*r/a*	.	.	.	.	.	.	.	.	a	.	.	B10.RIII(32R)	K.J. Hasenkrug and J.H. Stimpfling, u.d.
rb2	*ra2/b*	d	d	r	r	r	r	r	r	a	b	b	B10.RIII(37R)	K.J. Hasenkrug and J.H. Stimpfling, u.d.

Table 19.3—cont.

H-2 Haplotype	Origin	K	Aβ	Aα	Eβ	Eα	C4	C4S	D	L	Qa-2	Tla	Qa-1	Strain	Reference
rs1	rb2/s	r	.	.	.	.	.	.	s	.	.	.	.	B10.RIII(43R)	K.J. Hasenkrug and J.H. Stimpfling, u.d.
s	—	s	.	s	.	s	.	.	s	s	a	b	b	A.SW, B10.S	Common
s2	t2/s	s	s	s	s	s	s	s	s	s	.	b	.	B10.RSS	C.S. David, u.d.
sb1	t6/b	s	s	s	s	s	s	s	b	Δ	a	.	.	B10.S(26R)	K.J. Hasenkrug and J.H. Stimpfling, u.d.
sb2	t5/b	s	s	.	s	.	.	d	b	Δ	a	.	.	B10.BSVS(25R)	K.J. Hasenkrug and J.H. Stimpfling, u.d.
sb3	s/rb2	s	.	.	.	.	.	.	b	.	.	.	.	B10.S(44R)	K.J. Hasenkrug and J.H. Stimpfling, u.d.
sq1	s/q	s	s	s	s	s	q	q	q	.	.	.	.	A.QSR1	50
sq2	s/q	s	s	s	.	.	q	q	q	q	.	.	.	B10.QSR2	50
sq3	s/bq1	s	s	s	s	s	s	s	q	q	a	a	a	B10.SQR	76
sq4	s/q?	s	s	s	s	s	s	s	q	q	.	.	.	NFS/N	25
sx1	—	s	s	s	s	s	s	s	sx	.	.	.	.	DDD, B10.DDD	82
t1	s/a1	s	k	k	k	k	k	k	d	d	a	c	b	A.TL, B10.TL	83
t2	s/a	s	s	s	s	s	s	s	d	d	a	a	a	B10.S(7R), A.TH	94
t3	s/t1	s	s	s	s/k	k	k	k	d	d	a	c	b	B10.HTT	61
t4	s/a	s	s	s	s/k	k	s	d	d	d	a	a	a	B10.S(9R)	94
t5	s/d	s	s	s	s	s	d	d	d	d	a	.	.	BSVS	73, 84
t6	s/a	s	s	s	s	s	d	d	d	d	a	.	.	B10.S(24R)	K.J. Hasenkrug and J.H. Stimpfling, u.d.
t7	s/a	s	s	s	s	k	d	d	d	.	.	.	.	B10.ASR1	79
t8	s/a	s	s	s	.	s	s	s	d	.	.	.	.	B10.ASR8	80
t9	s/a2	s	s	s	s/k	k	k	k	d	.	.	.	.	B10.ASR11	80
t10	s/a2	s	s	s	s	k	k	k	d	.	.	.	.	B10.ASR12	80
t11	s/ap9	s	s	s	s	s	s	s	d	d	.	.	.	B10.RSD1	19
t12	s/ap9	s	s	s	s	s	s	s	d	.	.	.	.	B10.RSD2	19
t13	s/d	s	s	s	s	s	.	.	d	d	.	.	.	B10.DSR6	63
t14	s/d	s	s	s	s	s	d	.	d	d	.	.	.	B10.DSR7	63
t15	s/d	s	s	s	d	d	.	.	d	d	.	.	.	B10.DSR4	63
t16	s/d	s	s	s	d	d	d	d	d	d	.	.	.	B10.DSR5	63
t17	s/a	s	s	s	s	s	s	s	d	d	.	.	.	B10.S(23R)	K.J. Hasenkrug and J.H. Stimpfling, u.d.
t18	s/d	s	s	s	s	s	.	.	d	d	b	.	.	BALB.S(3R)	D.B. Murphy, u.d.
td1	pb1/t4	p	p	p	p	p/k	d	d	d	d	.	.	.	B10.RPD1	18
td2	pb1/t4	p	p	p	p	p/k	d	d	d	d	.	.	.	B10.RPD2	18
td3	pb1/t1	p	p	p	p	p	b	b	d	d	.	.	.	B10.RPD3	C.S. David, u.d.
td4	bu2/d	p	p	p	p	p	b	b	d	d	.	.	.	B10.RPD4	C.S. David, u.d.
td5	pb1/t1	p	p	p	p	p	b	b	d	d	.	.	.	B10.RPD5	C.S. David, u.d.
td6	pb1/t1	p	p	p	p	p	b	b	d	d	.	.	.	B10.RPD6	C.S. David, u.d.
td7	pb1/t1	p	p	p	p	p	u	u	d	d	.	.	.	B10.RPD7	C.S. David, u.d.
u	—	u	u	u	u	u	u	u	d	d	a	a	a	B10.PL(73NS)	89 or 90

Marker	Strain(s)	Ref.											
ub1	B10.RUB1	C.S. David, u.d.	·	·	·	b	b	u	u				u/h4
ub2	B10.RUB2	C.S. David, u.d.	·	·	·	b	u	u	u				u/h4
v	SMJ, B10.SM	89 or 90	b	b	a	v	v	v	v	v	v	v	—
w1	B10.KPA42, B10.KPA132	102	b	b	b	w1	w1	v	v	v	v	v	—
w2													
w2	B10.KPB68, $t^{Tuw10}/+$, $t^{Tuw18}/+$, $t^{Tuw30}/+$	42, 66	·	·	·	w2	w2	w2	w2	w2	w2	w2	—
w3	B10.SAA48	102	c	c	b	w3	w3	w3	w3	w3	w3	w3	—
w4	B10.GAA20	42	b	b	b	w4	w4	w4	w4	w4	w4	w4	—
w5	B10.KEA5	42	a	h	a	w5	w5	w5	w5	p	p	dv1	—
w6	B10.T7WF	41	e	a	a	w6	w6	w6	p	w6	w6	w6	—
w7	C3H.WOA1	43	·	c	a	w7	w7	w7	w7	w7	w7	w7	k/w
w8	B10.SNA70	38	c	c	a	w8	w8	w8	·	k	k	d	—
w9	B10.KEA2	20	b	b	a	w9	w9	w9	w9	d	d	w9	—
w10	B10.SNA57	20	·	·	b	w10	w10	v	v	v	v	v	w10
w10.1	B10.DRB62	20	b	b	c	w10	w10	v	v	v	v	v	w10
w10.2	B10.WOA105	220	c	c	c	w10	w10	v	v	v	v	v	—
w11	B10.CAA2	102	c	c	b	·	w7	w7	w7	p	p	fv1	—
w12	B10.MOL1	102	c	a	b	w12	w7	w12	w12	w12	pv1	k	—
w13	B10.STA10, B10.STA12, B10.LIB55	102	a	g	a	w13	w13	w13	w13	w13	w13	w13	—
w14	B10.STC77	102	c	c	a	·	z	p	p	p	pv1	dv2	—
w15	B10.STC90	102	b	b	b	w15	w15	·	w15	w15	kv1	w15	—
w16	B10.BUA1	102	c	c	a	w16	w16	w16	w16	w16	w16	w16	—
w17	B10.CAS2	102	a?	g	a	w3	w3	w3	w17	w17	w17	w17	—
w18	B10.CHR51	102	c	c	a	w5	w5	w5	w19	p	p	d	—
w19	B10.KPB128	102	c	b	a	Δ	b	·	·	s	s	s	—
w20	B10.LIB18	102	b	b	a	w8	w8	w16	w16	w16	w16	w16	—
w21	B10.GAA37	20	b	g	a	w16	w16	w16	w16	s	s	s	—
w22	B10.BUA16, B10.BUA19	20	b	b	a	k	k	w16	w16	w16	w16	w16	—
w23	B10.CAS1	45	a	a	a	w23	w23	w23	w23	w23	w23	w23	—
w24	B10.KPA44	20	c	c	b	z	z	d	w16	w16	d	d	—
w25	B10.STA39	20	c	c	b	w25	w25	·	·	k	w13	w3	—
w26	B10.CHA2	20	c	g	a	w26	w26	w26	k	k	kv2	q	—
w27	B10.STA62	20	c	c	b	w27	w27	w27	w27	w27	b	w27	—
w28	$t^{l2}/+$	98				w28	w28	·	w28	w31	w28	w29.7	—
w29	$t^l/+$, $t^{w2}/+$	98				w29	w29	w29	w29	w29	w29	w29	—
w30	$t^{w1}/+$, $t^l/+$, $t^{Tuw11}/+$	98				w30	w30	·	w2	w2	w30	w30	—
w31	$t^{w5}/+$, $t^{Tuw23}/+$, $t^{w94}/+$	98, 66				w31	w31	w31	w31	w31	w31	w31	w31
w32	$t^{w73}/+$	98, 66				·	w30	·	w2	w2	w30.2	w32	—
w33	$t^{ub1}/+$	98, 66				w30	w30	·	w2	w2	w30	w29.2	—
w34	STU	25	a			w3	w3	w3	w35	k	k	d	d(k)/w3
w35	TO2	25	b	b	b	w3	w3	w35	w35	w35	w35	w7	w7/w3
w36	$t^{Tuw2}/+$, $t^{Tuw25}/+$	66	b	b	b	w29.1	w29.5	w28.2	w36	·	w36	w29.5	—

Table 19.3—*cont.*

H-2 Haplotype	Origin	K	Aβ	Aα	Eβ	Eα	C4	C4S	D	L	Qa-2	Tla	Qa-1	Strain	Reference
w37	–	w30.4	w36.1		w31.1	w28			w37		.	.	.	$t^{Tuw7}/+$	98, 66
w38	–	w38	w38		w38	w28			w38		.	.	.	$t^{Tuw20}/+$	98, 66
w39	w7/b	w7	w35	w35	w35	w35	w35	w35	b	Δ	a	b	b	TO1	25
w40	–	w40	dv3	w40	w40	w40	w40	w40	w40	w40				W12A	103
w41	–	w7	w41	w41	w41				w16	w16					103
w42	w27/w	w27	w42	w42	w42	w42	w42	w42	w42	w42				CTC695	103
w43	–	w43	w43	w43	w43				w7v	w43				WRS936	103
w44	–	w44	w44	w44	w44	w44	w44	w44	w44	w44				BNK26	103
w45	–	w45	w45	w45	w45	w45	w45	w45	w45	w45				OBL984	103
w46	–	w46	w46	w46	w46				w3	w46				ISL1054	103
w47	w3/w	w3	w47	w47	w47	w47	w47	w47	w47	w47				GRL74	103
w48	–	w48	w48	w48	w48	w48	w48	w48	w48	w48				HOV162	103
w49	w/w4	w49	w49	w49	w49				w4	w4				EYB616	103
w50	w/w4	w50	w16	w50	w50	w50	w50	w50	w4	w4				BRY560	103
w51	–	w51	w51	w51	w51	w51	w51	w51	w51	w51				BNS570	103
w52	–	w52	w52	w52	w52	w52	w52	w52	w52	w52				GZA687	103
w53	–	w53	w53	w53	w53	w53	w53	w53	w53	w53				GPC882	103
w54	–	w54	w54	w54	w54	w54	w54	w54	w54	w54				GPC151	103
w55	–	w55	w55	w55	w55	w55	w55	w55	w55	w55				GRL1048	103
w56	–	w29.4	w30.3		w21	w28			dv					IS1025	103
w57	–	w57	w31.2		w31	w29.1			w57					$t^{Tuw32}/+$	66
w58	–	w29.1	w31.3		w31.2	w2			w28.1					$t^{Tuw8}/+$	66
w59	–	w59	w36		w36	w28.2			w59					$t^{Tuw6}/+$	66
w60	–	w29.6	w36		w36	w28.2			w60					$t^{Tuw26}/+$	66
w61	–	w29.3	w61		w61	w2			w61					$t^{Tuw28}/+$	66
w63	–	w30.3	w31.1		w31	w28.1			w29.1					$t^{Tuw29}/+$	66
w64	–	w30.3	w30		w2	w2			w64					$t^{Lub7}/+$	66
w65	–	w30	w30		w2	w2			w30					$t^{Lub9}/+, t^{Lub4}/+$	66
w66	–	w30.1	w30		w2	w2			w30.1					$t^{Tuw15}/+$	66
w67	w/s	w67	w67	w67	w67				s	s				$t^{Tuw12}/+$	66
w68	–	w68	w68	w68	w68	w68	w68	w68	w68	w68			.	BNT69	12
w69	–	w69	w69	w69	w69	w69	w69	w69	w69	w69			.	CRO433	12
w70	–	w70	w70	w70	w70	w70	w70	w70	w70	w70			.	GPC32	12
w71	–	w71	w71	w71	w71	w71	w71	w71	w71	w71			.	GPC881	12
w72	–	w72	w72	w72	w72	w72	w72	w72	w72	w72			.	GPC885	12
w73	–	w73	w73	w73	w73	w73	w73	w73	w73	w73			.	GRL1043	12
w74	–	w74	w74	w74	w74	w74	w74	w74	w74	w74			.	ISL1007	12
w75	–	w75	w75	w75	w75	w75	w75	w75	w75	w75			.	RTB962	12
w76	–	w76	w76	w76	w76	w76	w76	w76	w76	w76			.	RTB985	12
w77	–	w77	w77	w77	w77	w77	w77	w77	w77	w77			.	RTB987	12
w78	–	w78	w78	w78	w78	w78	w78	w78	w78	w78			.	THG629	12
w79	–	w79	w79	w79	w79	w79	w79	w79	w79	w79			.	WLA76	12
w80	–	w80	w80	w80	w80	w80	w80	w80	w80	w80			.	WPA76	12
														RTB991	12

Strain	Haplotype	Reference	Source
HUJ20	—	65	
HUJ22	—	65	
HUJ24	—	65	
B10.MOL-TEN1	—	78	
B10.MOL-TEN2	—		K. Moriwaki, u.d.
B10.MOL-NSB	—		K. Moriwaki, u.d.
B10.MOL-OHM	—		K. Moriwaki, u.d.
B10.MOL-MSM	—		K. Moriwaki, u.d.
B10.MOL-SGR	—	78	
B10.MOL-OKB	—		K. Moriwaki, u.d.
B10.MOL-YNG	—	78	
B10.WR7	w7/k	37	
B10.AQR	q/a	46	
B10.T(6R)	q/a	94	
B10.RQDB	y2/h2		C.S. David, u.d.
B10.NZW	z	25	
020.Q(R15)	pz/q		P. Démant, u.d.
020.Q(R26)	pz/q		P. Démant, u.d.
020.Q(R314)	pz/q		P. Démant, u.d.
020.GR(R15150)	pz/dx		P. Démant, u.d.
CAS3(RO)	c3/b?	26	
CAS3(R1)	c3/k	26	
CAS4(R2)	c4/k	26	
CAS4(R4)	c3/k	26	
CAS4(R5)	c4/k	26	
CAS3(R7)	c3/b	26	
CAS3(R8)	c3/RO	26	
CAS3(R9)	c3/b	26	
CAS3(R10)	R3/b	26	
CAS3(R11)	R2/b	26	
CAS4(R12)	c4/b	26	
CAS4(R13)	c4/k	26	
CAS4(R14)	c3/b	26	
CAS3(R16)	R7/b	26	
CAS3(R18)	R4/b	26	
CAS3(R23)	R8/k	26	
CAS3(R24)	RO/k	26	
CAS3(R25)	R7/b	26	
CAS4(R28)	c4/b	26	
CAS4(R29)	c4/b	26	
CAS3(R30)	R4/b	26	

Δ, Deletion; parentheses indicate presumed but not proved allele; .., undetermined or undesignated allele or haplotype; u.d., unpublished data.

Table 19.4 Congenic and coisogenic strains for mutant *H-2* haplotypes

Strain		Old *H-2* haplotype symbol	Mutated *H-2* locus	Amino acid interchange at position*	Reference
Standard designation	Synonym				
B6.C-*H-2^{bm1}*	B6.C(HZ1)	*ba*	*K*	152(Glu→Ala), 155(Arg→Tyr), 156(Leu→Tyr)	5
B6.C-*H-2^{bm2}*	B6.C(HZ49)	*bb*	*K*	155(Arg→Tyr), 156(Leu→Tyr)	5
C57BL/6-*H-2^{bm3}*	B6.M505	*bd*	*K*	77(Asp→Ser), 89(Lys→Ala)	22
B6.C-*H-2^{bm4}*	B6.C(HZ170)	*bf*	*K*	173(Lys→Gly), 174(Asn→Leu)	4
C57BL/6-*H-2^{bm5}*		*bg1*	*K*	116(Tyr→Phe)	53
C57BL/6-*H-2^{bm6}*		*bg2*	*K*	116(Tyr→Phe), 121(Cys→Arg)	53
B6.C-*H-2^{bm7}*		*bg3*	*K*	116(Tyr→Phe), 121(Cys→Arg)	57
C57BL/6-*H-2^{bm8}*		*bh*	*K*	22(Tyr→Phe), 23(Met→Ile), 24(Glu→Ser)	53
B6.C-*H-2^{bm9}*		*bi*	*K*	116(Tyr→Phe), 121(Cys→Arg)	57
B6.C-*H-2^{bm10}*		*bj*	*K*	163(Thr→Ala), 165(Val→Met), 173(Lys→Glu), (Asn→Leu)	57
B6.C-*H-2^{bm11}*		*bk*	*K*	77(Asp→Ser), 80(Thr→Asn)	57
B6.C-*H-2^{bm12}*		*bm*	*Aβ*		51
B6.C-*H-2^{bm13}*		*bn*	*D*		57
C57BL/6-*H-2^{bm14}*		*bo*	*D*		D.W. Bailey, u.d.
C57BL/6-*H-2^{bm16}*			*K*	116(Tyr→Phe)	58
C57BL/6-*H-2^{bm17}*			*K*		
B6.C-*H-2^{bm18}*			*K*		
B6.C-*H-2^{bm19}*			*K*		
B6.C-*H-2^{bm20}*			*K*		58
B6.C-*H-2^{bm21}*			*K*		
C57BL/6-*H-2^{bm22}*			*K*		55
B10.D2-*H-2^{bm23}*			*K*	75(Arg→His), 77(Asp→Ser)	24
C57BL/10-*H-2^{bm24}*			*D*		56
B10.D2-*H-2^{dm1}*	B10.D2(M504)	*da*	*D & L*	Deletion	21
BALB/c-*H-2^{dm2}*		*db*	*L*	Deletion	57
B6.C-*H-2^{dm2}*		*dc*	*D?*		57
C.B6-*H-2^{dm4}*		*dd*	*K*		57
C.B6-*H-2^{dm5}*			*K*		59
C.B6-*H-2^{dm6}*			*D*		60
A.CA-*H-2^{fm1}*	A.CA(M506)	*fa*	*K*		23
B10.M-*H-2^{fm2}*		*fb*	*D*		62
CBA-*H-2^{km1}*	CBA(M523)	*ka*	*K*		9

* See ref. 64.
u.d., unpublished data.

Table 19.5 Congenic strains for histocompatibility loci other than *H-2*

Differential locus	Chromosome	Congenic strain	(Synonym)	Inbred partner	Donor strain	N generations	Producer	Reference
H-1	7	B6.C-*H-1^b*	(HW80)	C57BL/6By	BALB/cBy	15	By	2
		B10.C3H-*H-1^a*	(40NX)	C57BL/10Sn	C3H/ScSn	9	Sn	35
		B10.D2-*H-1^{a}Hbb^b Mlv-1^a*	(58N)	C57BL/10Sn	DBA/2J	9	Sn	35
		B6.C-*H-1^{b}c Hbb^d*		C57BL/6By	BALB/cBy	15	By	2
		B10.A-*H-1^b*	(67NX)	C57BL/10Sn	A/Sn	9	Grf	35
		B10.AK-*H-1^b*	(68NX)	C57BL/10Sn	AKR/SnGrf	12	Grf	35
		B10.C-*H-1^b Hbb^d*	(41N)	C57BL/10Sn	BALB/cJ	8	Sn	35
		B10.K-*H-1^b*	(69NX)	C57BL/10Sn	C3H.K/Sn	12	Grf	35
		B10.129-*H-1^{b}c Hbb^d*	(5M)	C57BL/10Sn	129/J	11	Sn	35
		C3H.K-*H-1^{b}c*		C3H/ScSn	Non-inbred	11	Sn	35
		B10.P-*H-1^d*	(61NX)	C57BL/Sn	P/J	11	Grf	35
		B10.WB-*H-1^e*	(66NX)	C57BL/10Sn	WB/Re	11	Grf	35
		B10.CE-*H-1^f*	(62NX)	C57BL/10Sn	CE/J	11	Grf	35
		B10.SM-*H-1?*	(65NX)	C57BL/10Sn	SM/J	12	Sn	35
H-3	2	B10-*H-3^{b}we*		C57BL/10Sn	B10.UW	8	Sn	34
		B10.LP-*H-3^b H-13^b A^w*		C57BL/10Sn	LP/J	9	Sn	34
		B10.LP-*a H-3^b*		C57BL/10Sn	LP/J	13	Sn	34
		B10.LP-*H-3^b H-13^b*	(36NX)	C57BL/10Sn	LP/J	11	Sn	34
		B10.UW-*H-3^{b}we un a^t*	(69NS)	C57BL/10Sn	Non-inbred	8	Sn	34
		B10.C-*H-3^c*		C57BL/10Sn	BALB/cJ	11	Sn	34
		B10.C-*H-3^c H-13$^?$A*	(28NX)	C57BL/10Sn	BALB/10Sn	8	Sn	34
		B10.KR-*H-3^d*		C57BL/10Sn	Non-inbred	11	Sn	34
		B10.pa-*H-3^{e}a^t*		C57BL/10Sn	BALB/cJ	11	Sn	34
		B10.UW-*H-3^b*		C57BL/10Sn	Non-inbred	8	Sn	33
		C57BL/10-*pa a^t H-3^b*		C57BL/10Sn	C57BL/6-*pa*	11	Sn	?
		B10.SM-*H-3^c H-13^d*		C57BL/10Sn	SM/J	11	Grf	33
		B10.SM-*aH-3^c*	(91NX)	C57BL/10Sn	SM/J	12	Grf	33
		B10.KR-*aH-3^c*		C57BL/10Sn	Non-inbred	11	Grf	33
		B10.KR-*A^{w}H-3^{c}H-13^c*		C57BL/10Sn	Non-inbred	11	Grf	33
		B10.YBR-*H-3^c*		C57BL/10Sn	YBR-Ki	10	Grf	Graff *et al.*, pers. comm.
H-4	7	B10.129-*H-4^{b}p*	(21M)	C57BL/10Sn	129/J	15	Sn	36
H-7	9	B6.C-*H-7^b*	(HW23)	C57BL/6By	BALB/cBy	15	By	2
		B10.C-*H-7^b*	(47N)	C57BL/10Sn	BALB/cJ	7	Sn	36
H-8	–	B10.D2-*H-8^b*	(57N)	C57BL/10Sn	DBA/2J	8	Sn	36
		B6.C-*H-8^c*	(HW9)	C57BL/6By	BALB/cBy	15	By	2
H-9	–	B10.C-*H-9^b*	(45N)	C57BL/10Sn	BALB/cJ	9	Sn	36
H-10	–	B10.129-*H-10^b*	(9M)	C57BL/10Sn	129/J	7	Sn	36
H-11	–	B10.129-*H-11^b*	(10M)	C57BL/10Sn	129/J	8	Sn	36
		B10.D2-*H-11$^?$*	(55N)	C57BL/10Sn	DBA/2J	10	Sn	36
H-12	–	B10.129-*H-12^b*	(12M)	C57BL/10Sn	129/J	7	Sn	?
H-13	2	B10.CE-*H-13^b A^w*	(30NX)	C57BL/10Sn	CE/J	9	Sn	88
		B10.LP-*H-13^b H-13^b A^w*		C57BL/10Sn	LP/J	9	Sn	88
		B10.LP-*H-13^b A^w*		C57BL/10Sn	LP/J	11	Sn	88
		B10.FS-*H-13^b*	(92NX)	C57BL/10Sn	FS/Ei	11	Grf	33
		B10.C3H-*H-13^c*	(29NX)	C57BL/10Sn	C3H/ScSn	12	Grf	R.J. Graff, u.d.
		B10.KR-*H-13^c A*		C57BL/10Sn	Non-inbred	11	Sn	88
		B10.SM-*A^{w}H-13^d*		C57BL/10Sn	SM/J	11	Grf	33
H-15	4	B6.C-*H-15^c*	(HW13J)	C57BL/6By	BALB/cBy	9	By	2
		B6.C-*H-15^{a}H-16^{c}H-21^c*	(HW13)	C57BL/6By	BALB/6By	8	By	2
		B6.C-*H-15^{c}H-16^{c}H-21^c*	(HW13F)	C57BL/6By	BALB/6By	9	By	2
H-16	4	B6.C-*H-16^c*	(HW13K)	C57BL/6By	BALB/cBy	9	By	2
H-17	–	B6.C-*H-17^c*	(HW14)	C57BL/6By	BALB/cBy	17	By	2
H-18	4	B6.C-*H-18^c*	(HW17)	C57BL/6By	BALB/cBy	12	By	2
H-19	8	B6.C-*H-19^c*	(HW20)	C57BL/cBy	BALB/cBy	11	By	2
H-20	4	B6.C-*H-20^c*	(HW21)	C57BL/6By	BALB/cBy	17	By	2
H-21	4	B6.C-*H-21^c*	(HW35)	C57BL/6By	BALB/cBy	17	By	2
H-22	7	B6.C-*H-22^c*	(HW38)	C57BL/6By	BALB/cBy	12	By	2
H-23	3	B6.C-*H-23^c*	(HW53)	C57BL/6By	BALB/cBy	15	By	2
H-24	7	B6.C-*H-24^c*	(HW54)	C57BL/6By	BALB/cBy	16	By	2
H-25	1	B6.C-*H-25^c*	(HW65)	C57BL/6By	BALB/cBy	15	By	2
H-26	–	B6.C-*H-26^c*	(HW72)	C57BL/6By	BALB/cBy	18	By	2
H-27	5	B6.C-*H-27^c*	(HW77A)	C57BL/6By	BALB/cBy	15	By	2

Table 19.5—*cont.*

Differential locus	Chromosome	Congenic strain	(Synonym)	Inbred partner	Donor strain	N generations	Producer	Reference
H-28	3	B6.C-*H-28^c*/a	(HW81)	C57BL/6By	BALB/cBy	16	By	2
		B6.C-*H-28^c*/b	(HW94)	C57BL/6By	BALB/cBy	15	By	2
		B6.C-*H-28^c*/c	(HW97)	C57BL/6By	BALB/cBy	15	By	2
		B6.C-*H-28^c*/d	(HW110)	C57BL/6By	BALB/cBy	15	By	2
H-29	8	B6.C-*H-29^c*	(HW88)	C57BL/6By	BALB/cBy	15	By	2
H-30	–	B6.C-*H-30^c*	(HW105)	C57BL/6By	BALB/cBy	14	By	2
H-31	17	C57BL/6-*Tla^a*		C57BL/cBoy	A/JBoy	26	Boy	29
		A-*Tla^b*		A/JBoy	C57BL/6Boy	15	Boy	29
		B6.K-*H-2^k Tla^b*		C57BL/6J/Boy	AKR/JBoy	14	Boy	29
H-32	17	A-*Tla^b*		A/JBoy	C57BL/6JBoy	15	Boy	29
		C57BL/6-*Tla^a*		C57BL/6Boy	A/JBoy	26	Boy	29
H-33	17	BALB.TTF-*T H-33^a*/+*H-33^b*		BALB/c	Non-inbred	10	Fla	227
H-34	–	B6.C-*H-34^c*	(HW22)	C57BL/6By	BALB/cBy	16	By	2
H-35	1	B6.C-*H-36^c*	(HW56)	C57BL/6By	BALB/cBy	16	By	2
H-36	–	B6.C-*H-36^c*	(HW59)	C57BL/6By	BALB/cBy	16	By	2
H-37	3	B6.C-*H-37^c*	(HW106)	C57BL/6By	BALB/cBy	14	By	2
H-38	–	B6.C-*H-38^c*	(HW119)	C57BL/6By	BALB/cBy	18	By	3
H-40	12	BALB/c-*Igh^b*		BALB/cJ	C57BL/Ka	20	Pt	3
H-41	–	B10.STA12		C57BL/10Sn	Wild *Mus domesticus*	8	Klj	40
H-42	–	C3H-*Igh-1^{b}H-42^b*	(CWB)	C3H.SW/Sn	C57BL/6	20	Hz	7
H(Eh)	15	C57BL/6-*Eh*/+		C57BL/6By	Non-inbred	21	Jax-By	3
H(ep)	19	C57BL/6-*ep*/*ep*		C57BL/6J	C3Heb/FeJ	20	Jax-By	3
H(go)	5	C57BL/6-*go*/*go*		C57BL/6J	BALB/cJ	12	Jax-By	3
H(js)	11	C57BL/6-*js*/+		C57BL/6J	A/J	10	Jax/By	3
H(ln)	1	C57BL/6-*ln*/*ln*		C57BL/6J	C57BL/J	9	Jax/By	3
H(lt)	11	C57BL/6-*lt*/+		C57BL/6J	DBA/2J	12	Jax/By	3
H(pi)	5	C57BL/6-*pi*/+		C57BL/6J	C3H	14	Jax/By	3
H(sm)	–	C57BL/6-*sm*/+		C57BL/6J	A/Fa	5	Jax/By	3
H(tn)	11	C57BL/6-*tn*/+		C57BL/6J	C3H/HeJ	5	Jax-By	3

u.d., Unpublished data.

Table 19.6 Congenic strains for cellular alloantigens

Differential locus (synonym)	Chromosome	Congenic line	Inbred partner	Donor strain	N generation	Producer	Reference
Erythrocyte alloantigen loci							
Ea-2ᵃ (R, rho, H-14)	–	B6.RIII (76NS)	C57BL/6J	RIII/WyJ	11	Sn	88
		B10.F-*Ea-2ᵃ*	C57BL/10	F/St	15	Pp	70
		B10.RIII-*Ea-2ᵃ*	C57BL/10SnSg	RIII/WyJ	11	Sg	96
		B10.RFM-*Ea-2ᵃ*	C57BL/10SnSg	RFM/Pp	11	Sg	K.J. Hasenkrug and J.H. Stimpfling, u.d.
		B10.RIII-*Ea-2ᵃH-2ʳ*	C57BL/10SnSg	RIII/WyJ	11	Sg	K.J. Hasenkrug and J.H. Stimpfling, u.d.
		B10.RIII (72NS)	C57BL/10Sn	RIII/WyJ	8	Sn	91
		C57BL/6-*Ea-2ᵃ*	C57BL/6JBoy	B10.F-*Ea-2ᵃ*	8	Boy	E.A. Boyse, u.d.
		C57BL/10-*Ea-2ᵃEa-7ᵃ*	C57BL/10Sn	RIII/WyJ, 129J	8	Sn	88
Ea-3ᵃ (λ)	–	B10.L	C57BL/10SnEg	C57L/J	7	Eg	21
Ea-6ᵃ (H-6ᵃ)	2	C3H/Bi-*H-6ᵃ*	C3H/Bi	C3H/An	13	Lil	48
Ea-6ᵇ (H-6ᵇ)	2	C3H/Bi-*H-6ᵇ*	C3H/Bi	C57BL/6J	18	Lil	48
Ea-7ᵃ (T)	–	B10.129(5M)	C57BL/10Sn	129/J	11	Sn	96
		C57BL/10-*Ea-2ᵃEa-7ᵃ*	C57BL/10Sn	RIII/WyJ, 129/J	8	Sn	96
Lymphocyte alloantigen loci							
Ly-1ᵃ (Ly-A, Lyt-1, mu)	19	C57BL/6-*Ly-1ᵃ*	C57BL/6JBoy	C3H/Bi (?)	18	Boy	11
		C57BL/6-*Ly-1ᵃ*	C57BL/6J	Mixed	12	Sn	G.D. Snell, u.d.
Ly-2ᵃ (Ly-B, Lyt-2)	6	B6.PL(75NS)	C57BL/6J	PL/J	11	Sn	G.D. Snell, u.d.
		C57BL/6-*Ly-2ᵃLyt-3ᵃIgk*	C57BL/6JBoy	RF	7	Boy	11
Ly-5 (Lyt-4)	1	C57BL/6-*Ly-5ᵇ*	C57BL/6JBoy	SJL/J	22	Boy	E.A. Boyse, u.d.
		C57BL/6-*Ly-5ᶜ*	C57BL/6JBoy	ST/bJ	10	Boy	E.A. Boyse, u.d.
Ly-6	2(?)	020.AKR-*Ly-6ᵃ*	020/A	AKR/FuRdA	6	Dem	P. Démant, u.d.
Lyb-2	4	C57BL/6-*Lyb-2ᵃ*	C57BL/6JBoy	I/St	8	Boy	E.A. Boyse, u.d.
		C57BL/6-*Lyb-2ᶜ*	C57BL/6JBoy	CE	20	Boy	E.A. Boyse, u.d.
		C57BL/6-*Lyb-2ᶜMup-1ᵃ*	C57BL/6JBoy	CE	12	Boy	E.A. Boyse, u.d.
Thy-1	9	A-*Thy-1ᵃ*	A/JBoy	A.SW, AL/N	?	Boy	E.A. Boyse, u.d.
		A.AL-*Thy-1ᵃ*	A/JSf	A.SW	?	Sf	D.C. Shreffler, u.d.
		B6.PL(74NS)-*Thy-1ᵇ*	C57BL/6J	PL/J	?	Sn	G.D. Snell, u.d.
		AKR-*Thy-1ᵇ*/a (AKR-TH)	AKR/Sn	A.TH	12	Za	71
		C57BL/6-*Thy-1ᵃ* (B6.D-*Thy-1ᵃ*)	C57BL/6J (B6.C-*H-2ᵈ*)	B6.PL(74NS)	+	Za	71
		B10.S(7R)-*Thy-1ᵃ*	C57BL/10Sn [B10.S(7R)]	AKR/Sn	12	Za	71
		AKR-*Thy-1ᵇ*/b (AKR.D2)	AKR/Sn	DBA/2J	12	Za	71
		A.TL-*Thy-1ᵃ*	A/J(A.TL)	A/J (A.AL)	+	Za	71
Pca-1	–	C57BL/6-*Pca-1ᵃ*	C57BL/6JBoy	BALB/cJBoy	11	Boy	E.A. Boyse, u.d.
		BALB/c-*Pca-1ᵇ*	BALB/cJBoy	C57BL/6JBoy	20	Boy	E.A. Boyse, u.d.
Retrovirus-related and linked loci							
Gv-1	–	129-G⁻ₗₓ	129/JBoy	C57BL/6JBoy	11	Boy	E.A. Boyse, u.d.
Gv-1, Gv-2	–	C57BL/6-G⁺ₗₓ	C57BL/6JBoy	129/JBoy	11	Boy	E.A. Boyse, u.d.
Fv-1	4	BALB/c-*Fv-1ⁿ* (BALB.nD2)	BALB/cAn	DBA/2J	14	Lil	F. Lilly, u.d.
Fv-1, Gpd-1	4,4	C57BL/6-*Fv-1ⁿ*	C57BL/6JBoy	AKR/JBoy	11	Boy	E.A. Boyse, u.d.
		AKR-*Fv-1ᵇ*	AKR/JBoy	C57BL/6JBoy	11	Boy	E.A. Boyse, u.d.
Fv-1, Gpd-1, Gv-1	4,4,–	AKR-*Fv-1ᵇ*G⁻ₗₓ	AKR-*Fv-1ᵇ*	C57BL/6JBoy	?	Boy	E.A. Boyse, u.d.
Fv-1, H-2	4,17	AKR-*H-2ᵇFv-1ᵇ*	AKR-JBoy (AKR-*H-2ᵇ*)	AKR/JBoy (AKR-*Fv-1*)	+	Boy	E.A. Boyse, u.d.
Fv-2		DBA/2-*Fv-2ⁿ* (D2.RB)	DBA/2J	C57BL/6J	16	Lil	F. Lilly, u.d.
	–	C57BL/6-G⁺ₗₓM	C57BL/6JBoy	(Mutant)	+	Boy	E.A. Boyse, u.d.

u.d., Unpublished data.

Table 19.7 Congenic strains for immunoglobulin haplotypes

Haplotype	Congenic strain	Synonym	Inbred partner	Donor strain	N generation	Producer	Reference
Igh-1^a	C57BL/Ka-*Igh-1^a*	B.C-17	C57BL/Ka	BALB/cAn	17	Pt	M. Potter, u.d.
	C57BL/Ka-*Igh-1^a*	BC-8	C57BL/Ka	BALB/cAn	10	Pt	M. Potter, u.d.
	SJL-*Igh-1^a*	SJA/20	SJL/J	BALB/cJ	20	Hz	39
	B10.MBR-*Igh-1^a*	B10.MBR-a	C57BL/10Sn (B10.MBR)	BC-8	8	Sx	D.H. Sachs, u.d.
	B10.AKM-*Igh-1^a*	B10.AKM-a	C57BL/10Sn (B10.AKM)	BC-8	7	Sx	D.H. Sachs, u.d.
Igh-1^b	BALB/c-*Igh-1^b*	C.B-20	BALB/cAnPt	C57BL/Ka	20	Pt	M. Potter, u.d.
	BALB/c-*Igh-1^{b}CC BB*	CBB-22	BALB/cAnPt	C57BL/Ka	22	Pt	M. Potter, u.d.
	BALB/c-*Igh-1^{b}CC*	Cbb-22	BALB/cAnPt	C57BL/Ka	22	Pt	M. Potter, u.d.
	BALB/c-*Igh-C^{b}Igh-V^a*	BAB-14	BALB/cJ	C57BL/Ka	14	Hz	39
	BALB/c-*Igh-1^b*		BALB/cJ	C57BL/6	20	Wr	N.L. Warner, u.d.
	NZB-*Igh-1^b*		NZB	C57BL/6	8	Wr	N.L. Warner, u.d.
	C3H-*Igh-1^b*	CWB-13	C3H.SW	C57BL/6	9	Sx	D.H. Sachs, u.d.
	A.SW-*Igh-1^b*	A.SW-b	A.SW	B10.MBR	8	Sx	D.H. Sachs, u.d.
	A.TH-*Igh-1^b*	A.TH-b	A.TH	A.SW-b	3	Sx	D.H. Sachs, u.d.
	C3H-*Igh-1^b*	C3H-b	C3H-HeJ	CWB-13	5	Sx	D.H. Sachs, u.d.
	CBA-*Igh-1^b*	CBA.Igh^b	CBA/Tufts	C57BL/6J	14	Wo	67
Igh-1^c	BALB/c-*Igh-1^c*	C.D-11	BALB/cAnPt	DBA/2N	11	Pt	M. Potter, u.d.
	BALB/c-*Igh-1^c*		BALB/cJ	DBA/2J	10	Wr	N.L. Warner, u.d.
Igh-1^d	BALB/c-*Igh-1^d*	CAL/9	BALB/c	AL/N	9	Pt	M. Potter, u.d.
	BALB/c-*Igh-1^d*	CAL-20	BALB/c	AL/N	20	Pt	M. Potter, u.d.
	BALB/*Igh-1^d*		BALB/cJ	AKR/J	10	Wr	N.L. Warner, u.d.
Igh-1^f	BALB/c-*Igh-1^f*	C.CE-6	BALB/cAnPt	CE/J	6	Pt	M. Potter, u.d.
	BALB/c-*Igh-1^f*		BALB/cJ	CE/J	10	Wr	N.L. Warner, u.d.
Igh-1^g	BALB/c-*Igh-1^g*		BALB/cJ	RIII	10	Wr	N.L. Warner, u.d.
Igh-1^n	B10.D2-*Igh-1^n*		C57BL/10Sn	NZB	20	Wr	N.L. Warner, u.d.
	C57BL/6-*Igh-1^n*		C57BL/6J	NZB	20	Wr	N.L. Warner, u.d.
Igh-1°	BALB/c-*Igh-1°*	C.AL-20	BALB/cAnPt	AL/N	20	Pt	M. Potter, u.d.
Igh-1^{molo}	BALB/c-*Igh-1^{molo}*	C.MOLO-10	BALB/cAnPt	*Mus molossinus*	10	Pt	M. Potter, u.d.
Igh-1^{cla}	BALB/c-*Igh-1^{cla}*	C.CLA-7	BALB/cAnPt	CLA	7	Pt	M. Potter, u.d.

u.d., Unpublished data.

References

1. Amos, D., P.A. Gorer, and Z.B. Mikulska. 1955. The antigenic structure and genetic behaviour of a transplanted leukosis. Br. J. Cancer 9:209–215.

2. Bailey, D.W. 1975. Genetics of histocompatibility in mice. I. New loci and congenic lines. Immunogenetics 2:249–256.

3. Bailey, D.W., and H.P. Bunker. 1972. Mouse News Lett. 47:18.

4. Bailey, D.W., and M. Cherry. 1975. A new H-2 mutation. *In* Forty-sixth Annual Report, p. 73. Jackson Laboratory, Bar Harbor, Maine.

5. Bailey, D.W., and H.I. Kohn. 1965. Inherited histocompatibility changes in progeny of irradiated and unirradiated inbred mice. Genet. Res. (Camb.) 6:330–340.

6. Beisel, K.W., W.P. Lafuse, J.H. Stimpfling, and C.S. David. 1980. Description of the *H-2K-*, *H-2I-* and *H-2D-* gene products of the *H-2^v* haplotype. Immunogenetics 11:407–411.

7. Black, S., J. Goding, G. Gutman, L.A. Herzenberg, M. Loken, B. Osborne, W. van der Loo, and N. Warner. 1978. Immunoglobulin isoantigens (allotypes) in the mouse. V. Characteristics of IgM allotypes. Immunogenetics 7:213–230.

8. Blandova, Z.K. 1986. Mouse News Lett. 74:73.

9. Blandova, Z., Y.A. Mnatsakanyan, and I.K. Egorov. 1975. Study of *H-2* mutations in mice. VI. M523, a new *K*-end mutant. Immunogenetics 2:291–295.

10. Boyse, E.A., L. Flaherty, E. Stockert, and L.J. Old. 1972. Histoincompatibility attributable to genes near H-2 that are not revealed by hemagglutination or cytotoxicity tests. Transplantation 13:431–432.

11. Boyse, E.A., K. Itakura, E. Stockert, C.A. Iritani, and M. Miura. 1971. Ly-C: A third locus specifying alloantigens expressed only on thymocytes and lymphocytes. Transplantation 11:351–353.

12. Bukara, M., V. Vincek, F. Figueroa, and J. Klein. 1985. How polymorphic are class II loci of the mouse *H-2* complex? Immunogenetics 21:569–579.

13. Cherry, M. 1980. Mouse News Lett. 62:38.

14. Cook, R.G., R.N. Jenkins, L. Flaherty, and R.R. Rich. 1983. The Qa-1 alloantigens. II. Evidence for the expression of two Qa-1 molecules by the Qa-1^d genotype and for cross-reactivity between Qa-1 and H-2K. J. Immunol. 130:1293–1299.

15. Czarnomska, A., and P. Démant. 1975. A new *H-2* haplotype carrying possibly a mutation of *H-2^b*. Folia Biol. (Praha) 21:319–420.

16. Czarnomska, A., and P. Démant. 1980. H-2 antigenic specificities controlled by the translocation chromosome T190. Transplantation 30:69–72.

17. David, C.S. 1980. Mouse News Lett. 63:5.

18. David, C.S. 1982. Mouse News Lett. 66:46.

19. David, C.S. 1986. Mouse News Lett. 74:73.

20. Duncan, W.R., and J. Klein. 1980. Histocompatibility-2 system in wild mice. IX. Serological analysis of 13 new B10.W congenic lines. Immunogenetics 10:45–65.

21. Egorov, I.K. 1967. A mutation of the histocompatibility-2 locus in the mouse. Genetika (Moskva) 3:136–144 (in Russian).

22. Egorov, I.K., and Z.K. Blandova. 1968. The genetic homogeneity of the inbred mice bred at the Stolbovaya farm. II. Skin grafting tests. Genetika (Moskva) 12:63–69 (in Russian).

23. Egorov, I.K., and Z.K. Blandova. 1972. Histocompatibility mutations in mice: Chemical induction and linkage with the *H-2* locus. Genet. Res. (Camb.) 19:133–143.

24. Egorov, O.S., and I.K. Egorov. 1984. *H-2^{bm23}*—a new K^b mutant similar to, but not identical with *H-2^{bm3}*. Immunogenetics 20:83–87.

25. Figueroa. F., S. Tewarson, E. Neufeld, and J. Klein. 1982. *H-2* haplotypes of strains DBR7, B10.NZW, NFS, BQ2, STU, TO1 and TO2. Immunogenetics 15:431–436.

26. Fischer Lindahl, K. 1986. Genetic variants of histocompatibility antigens from wild mice. Curr. Top. Microbiol. Immunol. 127:272–278.

27. Flaherty, L. 1975. *H-33*—histocompatibility locus to the left of the *H-2* complex. Immunogenetics 2:325–329.

28. Flaherty, L. 1976. The *Tla* region of the mouse: Identification of a new serologically defined locus, Qa-2. Immunogenetics 3:533–539.

29. Flaherty, L., and S.S. Wachtel. 1975. *H(Tla)* system: Identification of two new loci, *H-31* and *H-32*, and alleles. Immunogenetics 2:81–85.

30. Forman, J., and L. Flaherty. 1978. Identification of a new CML target antigen controlled by a gene associated with the *Qa-2* locus. Immunogenetics 6:227–233.

31. Forman, J., R. Riblet, K. Brooks, E.S. Vitetta, and L.A. Henderson. 1984. H-40, an antigen controlled by an *Igh* linked gene and recognized by cytotoxic T lymphocytes. I. Genetic analysis of *H-40* and distribution of its products on B-cell tumors. J. Exp. Med. 159:1724–1740.

32. Gorer, P.A., and Z.B. Mikulska. 1959. Some further data on the H-2 system of antigens. Proc. R. Soc. Lond. (Biol.) 151:57–69.

33. Graff, R.J. 1978. Minor histocompatibility genes and their antigens. *In* H.C. Morse III ed., 371–389. Origins of inbred mice, Academic Press, New York.

34. Graff, R.J., and B.W. Bailey. 1973. The non-H-2 histocompatibility loci and their antigens. Transplant. Rev. 15:26–49.

35. Graff, R.J., D.H. Brown, and G.D. Snell. 1979. Thirteen new chromosome-7 congenic lines. Immunogenetics 8:245–256.

36. Graff, R.J., W.H. Hildemann, and G.D. Snell. 1966. Histocompatibility genes of mice. VI. Allografts of mice congenic at various non-*H-2* histocompatibility loci. Transplantation 4:425–437.

37. Hansen, T.H., and D.C. Shreffler. 1976. Characterization of a constitutive variant of the murine serum protein allotype, Slp. J. Immunol. 117:1507–1513.

38. Hauptfeld, M., and J. Klein. 1976. The *H-2* complexes of inbred and wild mice are organized in a similar fashion. Immunogenetics 3:603–607.

39. Herzenberg, L.A., and L.A. Herzenberg. 1977. Mouse immunoglobulin allotypes: Description and special methodology. *In* Handbook of experimental immunology. Blackwell, Oxford.

40. Juretić, A., I. Vucak, B. Malenica, Z.A. Nagy, and J. Klein. 1984. *H-41*, a new minor histocompatibility locus. I. Histogenetic analysis. J. Immunol. 133:2950–2954.

41. Klein, J. 1971. Cytological identification of the chromosome carrying the IXth linkage group (including H-2) in the house mouse. Proc. Natl. Acad. Sci. USA 68:1594–1597.

42. Klein, J. 1972. Histocompatibility-2 system in wild mice. I. Identification of five new *H-2* chromosomes. Transplantation 13:291–299.

43. Klein, J. 1975. A case of no sex-limitation of *Slp* in the murine *H-2* complex. Immunogenetics 2:297–299.

44. Klein, J. 1975. Biology of the mouse histocompatibility-2 complex. Springer-Verlag, New York.

45. Klein, J., and F. Figueroa. 1981. Polymorphism of the mouse *H-2* loci. Immunol. Rev. 60:23–57.

46. Klein, J., D. Klein, and D. Shreffler. 1970. H-2 types of translocation stocks T(2;9)138Ca, T(9;13)190Ca and *H-2* recombinant. Transplantation 10:309–320.

47. Kohn, J.I., and R.W. Melvold. 1974. Spontaneous histocompatibility mutations detected by dermal grafts: Significant changes in rate over a 10-year period in the mouse H-system. Mutation Res. 24:163–169.

48. Lilly, F. 1967. The location of histocompatibility-6 in the mouse genome. Transplantation 5:83–85.

49. Lilly, F., and J. Klein. 1973. An *H-2^g*-like recombinant in the mouse. Transplantation 16:530–532.

50. McDevitt, H.O., B.D. Deak, D.C. Shreffler, J. Klein, J.H. Stimpfling, and G.D. Snell. 1972. Genetic control of the immune response. Mapping of the *Ir-1* locus. J. Exp. Med. 135:1259–1287.

51. McKenzie, I.F.C., G.M. Morgan, M.S. Sandrin, M. Michaelideis, R.W. Melvold, and H.I. Kohn. 1979. A new *H-2* mutation in the *I* region in the mouse. J. Exp. Med. 150:1323–1338.

52. McKenzie, I.F.C., T. Pang, and R.V. Blanden. 1977. The use of H-2 mutants as models for the study of T cell activation. Immunol. Rev. 35:181–230.

53. Melief, C.J.M., R.S. Schwartz, H.I. Kohn, and R.W. Melvold. 1975. Dermal histocompatibility and *in vitro* lymphocyte reactions of three new *H-2* mutants. Immunogenetics 2:337–348.

54. Melvold, R.W. 1979. Mouse News Lett. 61:22.

55. Melvold, R.W. 1981. Mouse News Lett. 64:41.

56. Melvold, R.W. 1986. Mouse News Lett. 74:75.

57. Melvold, R.W., and H.I. Kohn. 1976. Eight new histocompatibility mutations associated with the *H-2* complex. Immunogenetics 3:185–191.

58. Melvold, R.W., G. Dunn, and H.I. Kohn. 1979. Mouse News Lett. 61:21.

59. Melvold, R.W., M.S. Sandrin, and I.F.C. McKenzie. 1982. The C.B6-*H-2^{dm5}* mutant. Immunogenetics 15:331–332.

60. Melvold, R.W., P.M. Stuart, A.E. Busker, and B. Beck-Maier. 1983. A new loss mutant in the H-2^d haplotype. Transplant. Proc. 15:2045–1047.

61. Meo, T., C.S. David, M. Nabholz, V. Miggiano, and D.C. Shreffler. 1973. Demonstration by MLR test of a previously unsuspected intra-*H-2* crossover on the B10.HTT strain: Implications concerning location of MLR determinants in the *Ir* region. Transplant. Proc. 5:1507–1510.

62. Mobraaten, E., and D.W. Bailey. 1973. Mouse News Lett. 48:17.

63. Murphy, D.B. 1986. Mouse News Lett. 74:73.

64. Nathenson, S.G., J. Geliebter, G.M. Pfaffenbach, and R.A. Zeff. 1986. Murine major histocompatibility class I-mutants. Molecular analysis and structure–function implications. Ann. Rev. Immunol. 4:471–502.

65. Neufeld, E., U. Ritte, F. Figueroa, and J. Klein. 1986. Low number of *H-2* haplotypes in some Israeli wild mouse populations. Immunogenetics 24:374–380.

66. Nižetíc, D., F. Figueroa, and J. Klein. 1984. Evolutionary relationships between the *t* and *H-2* haplotypes in the house mouse. Immunogenetics 19:311–320.

67. Nutt, N., J. Haber, and H.H. Wortis. 1984. Influence of Igh-linked gene products on the generation of T helper cells in the response to sheep erythrocytes. J. Exp. Med. 153:1225–1235.

68. Passmore, H.C. 1981. Mouse News Lett. 64:41.

69. Passmore, H.C., and K.W. Beisel. 1977. Intra-*H-2*-recombinants in *H-2^b/H-2^{tl}* heterozygotes in the mouse. I. Four cases of crossing over between the *S* and *D* regions. Tissue Antigens 9:121–130.

70. Popp, D.M. 1969. Histocompatibility-14: Correlation of the isoantigen rho and R-Z locus. Transplantation 7:233–241.

71. Reichner, J.S., T.J. Gorzynski, and M.B. Zaleski. 1983. New *Thy-1-* and *H-2* congenic strains of mice and their application in studies on the mechanism of anti-Thy-1.1 response. Immunol. Commun. 12:501–508.

72. Rinchik, E.M., and D.B. Amos. 1986. A mutation in a non-MHC murine cell surface antigen detectable by cytotoxic T lymphocytes. J. Immunol. 136:1992–1998.

73. Rose, N.R., A.O. Vladutiu, C.S. David, and D.C. Shreffler. 1973. Autoimmune murine thyroiditis. V. Genetic influence on the disease in BSVS and BRVR mice. Clin. Exp. Immunol. 15:281–287.

74. Rucker, J., M. Horowitz, E.A. Lerner, and D.B. Murphy. 1983. Monoclonal antibody reveals *H-2*-linked quantitative and qualitative variation in the expression of a Qa-2 region determinant. Immunogenetics 17:303–316.

75. Sachs, D.H. 1986. Mouse News Lett. 74:73.

76. Sachs, D.H., J.S. Arn, and T.H. Hansen. 1979. Two new recombinant *H-2* haplotypes, one of which juxtaposes K^b and I^k alleles. J. Immunol. 123:1965–1969.

77. Sharrow, S.O., L. Flaherty, and D.H. Sachs. 1984. Serologic cross-reactivity between class I MHC molecules and an H-2-linked differentiation antigen as detected by monoclonal antibodies. J. Exp. Med. 159:21–40.

78. Shiroishi, T., T. Sagai, and K. Moriwaki. 1982. A new wild-derived *H-2* haplotype enhancing *K-IA* recombination. Nature 300:370–372.

79. Shreffler, D.C. 1978. Mouse News Lett. 58:32.

80. Shreffler, D.C. 1980. Mouse News Lett. 62:38.

81. Shreffler, D.C. 1981. Mouse News Lett. 64:41.

82. Shreffler, D.C. 1986. Mouse News Lett. 74:73.

83. Shreffler, D.C., and C.S. David. 1972. Studies on recombination within the mouse *H-2* complex. I. Three recombinants which position the *Ss* locus within the complex. Tissue Antigens 2:232–240.

84. Shreffler, D.C., and C.S. David. 1975. The *H-2* major histocompatibility complex and the *I* immune response region: Genetic variation, function, and organization. Adv. Immunol. 20:125–195.

85. Shreffler, D.C., D.B. Amos, and R. Mark. 1966. Serological analysis of a recombination in the *H-2* region of the mouse. Transplantation 4:300–322.

86. Snell, G.D. 1958. Histocompatibility genes of the mouse. II. Production and analysis of isogenic resistant lines. J. Natl. Cancer Inst. 14:691–700.

87. Snell, G.D., and J.F. Stimpfling. 1966. Genetics of tissue transplantation. *In* E.L. Green, ed., Biology of the laboratory mouse, 457–491. McGraw Hill, New York.

88. Snell, G.D., G. Cudkowicz, and H.P. Bunker. 1967. Histocompatibility genes of mice. VII. *H-13*, a new histocompatibility locus in the fifth linkage group. Transplantation 5:492–503.

89. Snell, G.D., P. Démant, and M. Cherry. 1971. Hemagglutination and cytotoxic studies of H-2.I, H-2.1 and related specificities in the EK crossover regions. Transplantation 11:210–237.

90. Snell, G.D., R.J. Graff, and M. Cherry. 1971. Histocompatibility genes of mice. XI. Evidence establishing a

new histocompatibility locus, *H-12*, and a new *H-2* allele, *H-2^{bc}*. Transplantation 11:525–530.

91. Snell, G.D., G. Hoecker, D.B. Amos, and J.H. Stimpfling. 1967. Evidence that the 'R' and 'Z' blood group specificities of mice are allelic and distinct from H-2. Transplantation 5:481–491.

92. Stanton, T.H., S. Carbon, and M. Maynard. 1981. Recognition of alternate alleles and mapping of the *Qa-1* locus. J. Immunol. 127:1640–1643.

93. Stimpfling, J.H., and C.S. David. 1977. Mouse News Lett. 56:29.

94. Stimpfling, J.H., and A.E. Reichert. 1970. Strain C57BL/10ScSn and its congenic resistant sublines. Transplant. Proc. 2:39–47.

95. Stimpfling, J.H., and A. Richardson. 1965. Recombination within the histocompatibility-2 locus of the mouse. Genetics 51:831–846.

96. Stimpfling, J.H., and G.D. Snell. 1968. Detection of a non-H-2 blood group system with the aid of B10.129(5M) mice. Transplantation 6:468–475.

97. Stimpfling, J.H., A.E. Reichert, and P. Hudson. 1971. The serological properties of some *H-2* recombinant alleles. *In* A. Lengerová and M. Vojtišková, eds., Immunogenetics of the H-2 system, 10–17. Karger, Basel.

98. Sturm, S., F. Figueroa, and J. Klein. 1982. The relationship between *t* and *H-2* complexes in wild mice. I. The *H-2* haplotypes of 20 *t*-bearing strains. Genet. Res. (Camb.) 40:73–88.

99. Sullivan, K., and L. Flaherty. 1979. A new *H-2* recombinant strain, B6.AK1-*H-2^{ozl}*. Skin graft and MLC experiments. Immunogenetics 90:101–104.

100. Vedernikov, A.A., and I.K. Egorov. 1973. Study of *H-2* mutations in mice. II. Recombination analysis of the mutation 504. Genetika (Moskva) 9:60–66 (in Russian).

101. Zachařová, G., L. Rencková, A. Dux, and P. Démant. 1975. *I* region associated histoincompatibility with absence of accelerated rejection of second set skin grafts detected in tests with a new haplotype, *H-2^{dx}*. J. Immunogenet. 2:323–335.

102. Zaleska-Rutczynska, Z., and J. Klein. 1977. Histocompatibility-2 system in wild mice. V. Serologic analysis of sixteen B10.W congenic lines. J. Immunol. 119:193–1911.

103. Zaleska-Rutczynska, Z., F. Figueroa, and J. Klein. 1983. Sixteen new *H-2* haplotypes derived from wild mice. Immunogenetics 18:189–203.

19.2 Mutant genes and biochemical loci

PRISCILLA W. LANE and MARY F. LYON

This catalogue consists of three tables: Table 19.8, genes maintained at The Jackson Laboratory (PWL); Table 19.9, genes maintained elsewhere in the Western Hemisphere and Japan (PWL); and Table 19.10, genes maintained in Europe and the Near East (MFL). The information contained in Tables 19.8 and 19.9 was obtained by personal communication and is complete through 1988. The information in Table 19.10 was obtained from *Mouse News Letter* or from personal communication.

Data are for three kinds of stocks: (i) those with genes that occurred as mutations in an already inbred strain and have been maintained within that strain, either by repeated backcrossing or by brother × sister matings (coisogenic); (ii) those with mutations that occurred on other backgrounds and have been crossed into an inbred strain (congenic); (iii) those inbred by brother × sister matings with forced heterozygosity for a mutation (segregating). For the first two kinds, the strain name is separated from the gene symbol by a hyphen. For the third, if there is a strain designation, the genotype follows the strain name; when there is no strain name the genotype of the strain is listed. M indicates that the mutant gene arose by mutation in the

inbred strain. N designates the number of generations of crosses when the cross–intercross system has been used; NE is the equivalent number of backcross or cross generations when some irregular sequence of mating types has been used. F is the number of generations of brother × sister matings. In congenic strains, when the introduced mutation arose in another inbred strain, the strain of origin is indicated in some cases by a period and an abbreviated symbol for the donor strain following the symbol for the background strain.

Brother × sister matings (indicated by F) in congenic or coisogenic strains can be made either with forced heterozygosis or with homozygosis for the introduced gene. This information, when known, is indicated by the genotype of the strain. When the strain is being maintained by continued backcrossing, or its equivalent (indicated by N or NE), only the symbols for the introduced genes were given.

Some strains are included in which the level of inbreeding at the time of this report did not fully meet the criteria of the rules for nomenclature of inbred strains (i.e. 20 generations of full sib mating, or 10 backcrosses to an inbred strain). However, the inbreeding in these strains was sufficiently far advanced to make them useful, and further inbreeding could be expected to occur by the time of appearance of this catalogue.

For Tables 19.9 and 19.10 in some cases the source did not provide full information on the mode of inbreeding, the number of generations, or whether a congenic strain was maintained with forced heterozygosis or was made homozygous for the introduced gene. When it was not possible to supply the missing information it is omitted or its lack is indicated by a dash or a question mark. The symbol * added to the generation indicates that a strain is available only from a frozen embryo repository.

19.2.1 Strain holders

The scientists or laboratories identified by the symbols for Holders are given in Chapter 20, except for the following:

Alb   New York State Department of Health, Division of Laboratories and Research, Empire State Plaza, Albany, NY. 12201, USA. (Dr Anne Messer)

Gr   Dr Gillian Truslove, Galton Laboratory, University College London, Wolfson House, 4 Stephenson Way, London NW1 2HE, UK.

BUL   Dr G. Bulfield, A.F.R.C. Institute for Physiology and Genetics Research, Roslin, Midlothian EH25 9PS, UK.

MM   Dr Miriam Meisner, University of Michigan Medical School, Department of Human Genetics, Ann Arbor, MI 48109, USA.

NICHD   National Institute of Child Health and Human Development, Developmental Pharmacology Branch, Bethesda, MD 20014, USA. (Dr Daniel W. Nebert)

PVRC   Purdue Vision Research Center, Department of Biological Sciences, West Layfayette, IN 47907, USA. (Dr Lawrence H. Pinto)

Table 19.8 Genes maintained at The Jackson Laboratory

Gene symbol	Gene name	Strain and genotype	Generations inbred
A or $+$	Agouti	C57BL/6J- $A/a\ md$	NE13*
A^i	Intermediate agouti	C57BL/6J- A^i/a	M F24N2F1N1*
A^{iy}	Intermediate yellow	B6.C3H- A^{iy}/a	N14F21N1*
A^{sy}	Sienna yellow	C57BL/6J- A^{sy}/a	M N17F1N1*
A^{vy}	Viable yellow	C3HeB/FeJ- A^{vy}	N29*
		C57BL/6J- A^{vy}	N23*
A^w	White-bellied agouti	129/Sv- $A^y/A^w\ c^+\ p^+$	N9F57
A^{w-J}	White-bellied agouti-J	AEJ/GnLe a^e/A^{w-J}	F77
		C3H/HeJ- A^{w-J}	N11F28
		C57BL/6J- A^{w-J}	M F80
A^{w-15J}	White-bellied agouti-15J	DBA/2J- A^{w-15J}	M F20
A^y	Yellow	C57BL/6J- A^y	N125
a	Non-agouti	C3HeB/FeJLe- a	N7F47
a^e	Extreme non-agouti	AEJ/Gn- $a^e\ +/a\ bp^H$	N5F50
		AEJ/GnLe- a^e/A^{w-J}	F77
a^{t-33J}	Black and tan-33J	C57BL/6J- a^{t-33J}	M N27*
a^{td}	Tanoid	C57BL/6J- a^{td}	N91*
ab^J	Asebia-J	ABJ/Le $ab^J/+\ b/b$	F68
adr^{mto}	myotonia	SWR/J- $adr^{mto}/+$	M F42*
ak	Aphakia	$a/a\ ak/+$	F48
Aft	Abnormal feet and tail	129/Sv- $p^+\ c^{ch}\ Aft/+$	N14F7
an	Anemia	C57BL/6J- $an\ B^{lt}$	N73
		WB/Re- $an\ B^{lt}$	N47
ash	Ashen	C3H/HeSn- $ash/+$	M F33
at	Atrichosis	ATEB/Le $a/a\ d/d\ at\ +/+\ eb$	F40
av^J	Ames waltzer-J	B6.BKs- $av^J/+$	N1F38N3F1N1*
ax^J	Ataxia-J	C57BL/6J- ax^J	N20
b	Brown	C57BL/6J- $b\ m/+\ +$	N42F1N1*
		C57BL/6J- $b\ wi/+\ +$	N15F7N1*
b^{c-J}	Cordovan-J	C57BL/6J- b^{c-J}	N20F1N1
b^J	Brown-J	C57BL/6J- b^J	M N35F1N1*
B^{lt}	Light	C57BL/6J- $an\ B^{lt}$	N73
		WB/Re- $an\ B^{lt}$	N47
bc^{3J}	Bouncy-3J	C3H/HeSn- $bc^{3J}/+$	M F18

Table 19.8—*cont.*

Gene symbol	Gene name	Strain and genotype	Generations inbred
bf	Buff	C57BL/6J- *bf*/+	M N10F4N1*
		JGBF/Le *a*/ *a bf* +/+ *jg*	F46
bg	Beige	C57BL/6J- *bg*/+	N13F12*
bg^J	Beige-J	C57BL/6J- *bg^J*/+	M NE7F8
bg^{2J}	Beige-2J	C3H/HeJ- *bg^{2J}*/+	M F49
Bn	Bent-tail	BNT/Le *a*/ *a Bn*/+	F83*
bp^H	Brachypodism-H	AEJ/Gn- *a^e* +/ *a bp^H*	N5F50
bp^J	Brachypodism-J	A/J- *bp^J*/+	M N2F28
Bst	Belly spot and tail	C57BL/Ks- *Bst*/+	M N1F29
bt^{2J}	Belted-2J	C57BL/6J- *uw bt^{2J}*/+ +	M N10F1N1*
c	Albino	B10.C(41N)/Sn *c H-1^b Hbb^d*	NE8F65
		B10.129(SM)/nSn *c H-1^b Hbb^d*	NE11F52
		B6.C- *H-1^b*/By *c Hbb^d*	N15F47
		C3H.K/nSn *c H-1^b*	NE7F41
c^+	Albino (wild type)	C.B6- *c^+ Hbb^s*	N11F19
c^a	acromelanin	C3H/HeJ- *c^a*/+	M F12
c^{ch}	Chinchilla	129/J *A^w*/ *A^w c^{ch} p*/ *c p*	F105
c^h	Himalayan	C57BL/6J- *c^h*	N40*
c^J	Albino-J	C57BL/6J- *c^J*	M N40F1N1*
c^{2J}	Albino-2J	C57BL/6J- *c^{2J}*/+	M F43
c^{4J}	Albino-4J	B10.D2/nSn- *c^{4J}*/+	M F49N1*
c^{9J}	Albino-9J	C3H/HeJ- *c^{9J}*	M F6*
c^p	Platinum	C57BL/6J- *c^p*	N22F1N1*
Ca	Caracul	C57BL/6By- *Ca Sl*	N26F6N2*
Ca^J	Caracul-J	C3HeB/FeJLe- *a*/ *a- Ca^J Sl Hm*	N38
ch	Congenital hydrocephalus	*ch* +/+ *mu*	F26
cht	chocolate	C57BL/6J- *cht*/+	M F21
cla	Clasper	BALB/cBy- *cla*/+	M F15
cn	Achondroplasia	*a*/ *a cl*/ *c cn*/+	F59
coa	Cocoa	C57BL/10J- *coa*	M N1F15
cog	Congenital goiter	C.AKR- *cog*	N9
cph	Congenital progressive hydronephrosis	C57BL/6J- *cph*/+	N9
cpk	Congenital polycystic kidneys	C57BL/6J- *cpk*	M N33
		DBA/2J- *cpk*	N19
cw	Curly whiskers	CWD/Le *cw d*/+ *d a*/ *a*	F53
d	Dilute	C57BL/6J- *sg* + + +/+ *d se*	NE9F8
		a/ *a d fd*/+ +	F67
d^{+2J}	Dilute (wild type)	DBA/2J- *d^{+2J}*/ *d*	M N84*
d^{+17J}	Dilute (wild type)-17J	DBA/2J- *d^{+17J}*/ *d*	M F2N1F3N1*
d^{+18J}	Dilute (wild type)-18J	DBA/2J- *d^{+18J}*/ *d*	M F5N1*
d^l	Dilute-lethal	DLS/Le *a*/ *a d^l* +/+ *se*	F99
Dac	Dactylaplasia	SM/Ckc- *Dac*/+	N7F69
db	Diabetes	C3H.SW- *db*/+	N6F4
		C3HeB/FeJ- *db*/+	N6F47
		C57BL/6J- *db* +/+ *m*	N5F35
		C57BL/6J- *db m*/+ +	N6F29
		C57BL/KsJ- *db* +/+ *m*	M(*db*) N5F44
		C57BL/KsJ- *db m*/+ +	M(*db*) N6F41
		C57BL/KsJ.B6- *H-2^b db* +/+ *m*	N7F3
		C57BL/KsJ.B6- *H-2^b db*/+	N9
		CBA/J- *db m*/+ +	N7F19
db^{3J}	Diabetes-3J	129/J- *db^{3J}*/+	M N7F17*
Dc	Dancer	DC/Le *Dc*/+	F106
dfw	Deaf waddler	C3H/HeJ- *dfw*/+	M F12
dl^J	Downless-J	GL/Le *dl^J* +/+ *gl*	F56
dr^{2J}	Dreher-2J	AKR/J- *dr^{2J}*	M N5
Ds	Disorganization	*Ds*/+	F34
du	Ducky	TKDU/Dn *a*/ *a d* + *du*/ *d tk* +	F67

Table 19.8—*cont.*

Gene symbol	Gene name	Strain and genotype	Generations inbred
dw	Dwarf	B6.DW- *dw/+*	N11F9N1*
		DW/J *dw/+ ln/ ln*	F77
dw^J	Dwarf-J	C3H/HeJ- *dw^J/+*	M F16N6F8
dy	Dystrophia muscularis	129/ReJ- *dy/+*	M F66N3
dy^2J	Dystrophia muscularis-2J	C57BL/6J- *dy^2J*	N20
e	Recessive yellow	C57BL/6J- *e*	N20F1N1*
Ea-2	Erythrocyte antigen	B10.RIII(72NS)/Sn *Ea-2^a*	N8F58
		B6.RIII(76NS)/Sn *Ea-2^a*	NE11F42
Ea-7	Erythrocyte antigen-7	B10.129(5M)/oSn *H-1^b Ea-7^a*	NE9F76
eb	Eye blebs	ATEB/Le *a/ a d/ d at +/+ eb*	F40
Eh	Hairy ears	C57BL/6By- *Eh/+*	N30F2
ep	Pale ear	C57BL/6J- *ep/+*	N20F37N1*
f	Flexed-tail	C57BL/6J- *f*	N16
		WB/Re- *f*	N15
fat	Fat	C57BL/Ls- *fat/+*	N8F2
fd	Fur deficient	*a/ a d fd/+ +*	F67
fi	Fidget	*stb + a/+ fi a*	F67
fs	Furless	FSB/GnEi *a/ a fs/+*	F104
Fu	Fused	129/Rr- *Fu/+*	N31F64*
fz	Fuzzy	C57BL/6J- *fz ln/+ +*	N9F102
gl	Grey-lethal	GL/Le *dl^J +/+ gl*	F56
gld	Generalized lymphopro-liferative disease	C3H/HeJ- *gld*	M F38
		C57BL/6J- *gld*	N10
gm	Gunmetal	C57BL/6J- *gm/+*	M F33N1*
go	Angora	B6.C- *go*	N12F19N1*
Gpi-ls	Glucose phosphate isomerase-1 structural	B6.CAST- *Gpi-1^a*/Ei	N10F15
gr	Grizzled	JIGR/Dn *gr +/+ ji*	F64
		gr +/+ mh	F49
Gus-s	Beta-glucuronidase structural	C57BL/6J- *rd le Gus-s^h*	N30F48
Hba	Hemoglobin alpha-chain	C57BL/6J- *Hba^b*	N6*
		C57BL/6J- *Hba^d*	N7*
Hba^th-J	Alpha-thalessemia-J	C57BL/6- *Hba^th-J*	N29
		SEC/1Re- *Hba^th-J*	N25
		WB/Re- *Hba^th-J*	N23
Hbb	Hemoglobin beta-chain	C.B6- *c^+ Hbb^s*	N11F19
		C57BL/6J- *Hbb^d*	N9
		C57BL/6J- *Hbb^p*	N10F43
		B10.C(41N)Sn *c H-1^b Hbb^d*	NE8F65
		B10.129(5M)/nSn *c H-1^b Hbb^d*	NE11F52
		B6.C- *H-1^b*/By *c Hbb^d*	N15F47
		B10.D2(58N)/Sn *H-1^a Hbb^d Mlv-1^a*	NE9F78
Hbb^th	Beta-thalassemia	C57BL/6J- *Hbb^th*	N10
hbd	Hemoglobin deficient	C57BL/6J- *hbd*	N8
heb	Head blebs	AKR/J- *heb/+*	M N1F29
Hk	Hook	*Hk/+ c/ c*	F59*
Hm	Hammer-toe	C3HeB/FeJLe- *a/ a- Ca^J Sl Hm*	N38
ho^4J	Hotfoot-4J	DBA/2J- *ho^4J/+*	M F30
ho^5J	Hotfoot-5J	C57BL/6J- *ho^5J/+*	M F12N1*
Hpt	Hair patches	B6C3Fe- *a/ a- Hpt/+*	N6F23
hpx	Hypotransferrinemia	C57BL/6J- *hpx*	N12
		WB/Re- *hpx*	N12
hr	Hairless	HRS/J *b/ b c/ c d/ d hr/+*	F76
		WLHR/Le *a/ a b/ b hr +/+ wl*	F103
hr^rh-J	Rhino-J	RHJ/Le *a/ a b/ b c/ c hr^rh-J/+*	F76
hr^rh-7J	Rhino-7J	C57BL/6J- *Hbb^p hr^rh-7J*	M N10
Hx	Hemimelic extra toes	B10.D2/nSn- *Hx/+*	M F52
hy-3	Hydrocephalus-3	HYIII/Le *hy-3/+*	F86

Table 19.8—*cont.*

Gene symbol	Gene name	Strain and genotype	Generations inbred
Hyp	Hypophosphatemia	C57BL/6J- *Hyp*	N30
hyt	Hypothyroid	C.RF- *hyt*/+	N10F12
ic	Ichthyosis	AKR/J- *ic*	N12
		IC/Le *ic*/+	F63
ic⁺	Ichthyosis (wild type)	IC/Le *ic*⁺	F66
ic^J	Ichthyosis-J	C57BL/6J- *ic^J*/+	M F2N3F1N1*
jc	Jackson circler	C57BL/6J- *jc*/+	M N16F1N1*
je	Jerker	JE/Le *a*/ *a fl f je*/+ *ru*/ *ru*	F50
jg	Jagged-tail	JGBF/Le *a*/ *a bf* +/+ *jg*	F46
ji	Jittery	JIGR/Dn *gr* +/+ *ji*	F64
jpk	Juvenile polycystic kidneys	see *cph*	
js	Jackson shaker	C57BL/6J- *js*/+	N17F1N1*
jsd	Juvenile spermatogonial depletion	C57BL/6J- *jsd*/+	M F8
jv	Jackson waltzer	C57BL/6J- *jv*/+	M F25N1F1N1*
ld^J	Limb-deformity-J	LDJ/Le *ld^J* + *a*/+ *mg a*	F72
le	Light ear	C57BL/6J- *rd le*	N39F3N1*
		C57BL/6J- *rd le Gus-s^h*	N30F48
lit	Little	C57BL/6J- *lit*/+	M N4F13
lm	Lethal milk	C57BL/6J- *lm*	M F62
ln	Leaden	C57BL/6J- *fz ln*/+ +	N9F102
Lp	Loop tail	LPT/Le *Lp*/+	F94
lpr	Lymphoproliferation	AKRSmn.MRL- *lpr*	N14F6
		B10Sn.MRL- *lpr*	N11F?
		B6.MRL- *lpr*	N11F21
		MRL/Mp- *lpr*	N12F21
lpr⁺	Lymphoproliferation (wild type)	MRL/Mp *a c lpr*⁺	F55
ls	Lethal spotting	LS/Le *a^t ls*/ *a^t* +	F86
lt	Lustrous	B6.D2- *lt*/+	N12F20N1*
lu	Luxoid	C57BL/6J- *lu*	N37*
lx	Luxate	C57BL/6J- *lx W^v*	N108*
Ly-1	Lymphocyte antigen-1 (formerly Lyt-1)	C57BL/6J- *Ly-1^a*/Cy	N12F43
Ly-2	Lymphocyte antigen-2 (formerly Lyt-2)	B6.PL- *Ly-2^a Ly-3^a*/Cy	N11F39
Ly-3	Lymphocyte antigen-3 (formerly Lyt-3)	B6.PL- *Ly-1^a Ly-3^a*/Cy	N11F39
m	Misty	C57BL/6J- *b m*	N42F1N1*
		C57BL/6J- *db* +/+ *m*	N5F35
		C57BL/6J- *db m*/+ +	N6F29
		C57BL/KsJ- *db* +/+ *m*	M(*db*) N5F44
		C57BL/KsJ- *db m* +/+	M(*db*) N6F41
ma	Matted	C57BL/6J- *ma*	N22F1N1*
md	Mahoganoid	B6.C3- *md*/+	N7F14N1*
		C3H/HeJ- *md*/+	M F110
		C57BL/6J- *A*/ *a md*	NE13*
md^2J	Mahoganoid-2J	C3H/HeJ- *md^2J*	M F15N1F2*
mdm	Muscular dystrophy with myositis	C57BL/6J- *mdm*	M N8
Mdmg-1	Mandibular morpho-genesis-1	B6.C(HW19)/aBy- *H-2^d* *Mdmg-1*/ *Mdmg-1*	F75
mdx	X-linked muscular dystrophy	C57BL/10- *mdx*	M F?+F10
me	Motheaten	C3HeB/FeJLe- *a*/ *a- me*	N24
		C57BL/6J- *me*	M N14
me^v	Viable motheaten	C57BL/6J- *me^v*	M N6
mea^2J	Meander-tail-2J	C57BL/KsJ- *mea^2J m*/+ +	M (in BL/Ks- *db* +/+ *m*) F3
mg	Mahogany	LDJ/Le *ld^J* + *a*/+ *mg a*	F72

Table 19.8—*cont.*

Gene symbol	Gene name	Strain and genotype	Generations inbred
mg^{3J}	Mahogany-3J	C3H/HeB/FeJ- mg^{3J}	M N14*
mh	Mocha	gr +/+ mh	F49
mh^{2J}	Mocha-2J	C3H/HeJ- mh^{2J}/+	M N11
mi	Microphthalmia	C57BL/6J- Mi^{wh}/ mi	N93*
		C57BL/6J- mi	N97
mi^{rw}	Red eyed white	C57BL/6J- Mi^{wh}/ mi^{rw}	NE24*
mi^{sp}	Mi-spotted	C57BL/6J- Mi^{wh}/ mi^{sp}	M N68*
Mi^{wh}	White	C57BL/6By- Mi^{wh} W^v T	N29F2N1*
		C57BL/6J- Mi^{wh}	N148*
		MWT/Le a^t/ a^t Mi^{wh}/+ W^v/+ T/+	F70
mk	Microcytic anemia	MK/Re mk/+	F71
		WB/Re- mk	N26
Mo^{blo}	Blotchy	C57BL/6J- Mo^{blo}	N28
Mo^{br-J}	Brindled-J	C3H/HeJ- Mo^{br-J}/+	N15F20
Mo^{pew}	Pewter	CBA/J- Mo^{pew}/+	M N1F30
Mo^{to}	Tortoiseshell	C57BL/6J- Mo^{to}	N81*
		TR/DiEi a/ a Mo^{to}/+	F84
mto	Myotonia	see adr^{mto}	
mu	Muted	CHMU/Le ch +/+ mu	F26
myd	Myodystrophy	MYD/Le- Os +/+ myd	N7F21
myd^+	Myodystrophy (wild type)	MYD/Le myd^+	F71
N	Naked	C57BL/6J-N Sp	N18
nr	Nervous	C3HeB/FeJ- nr	N29
nu	Nude	BALB/cByJ- nu (N7 to BALB/cN)	N6
		C57BL/6J- nu	N14
nu^{str}	Streaker	AKR/J- nu^{str}/+	M N9
		C57BL/6J- nu^{str}	N13
ob	Obese	C57BL/6J- ob	N51
		C57BL/Ks- ob/+	N5F27
or^J	Ocular retardation-J	129/Sv- or^J/+	M F69
Os	Oligosyndactylism	C57BL/6By- Ra Os Pt	N26*
		C57BL/6J- Os +/+ tg^{la}	N14F4
		MYD/Le- Os +/+ myd	N7F21
		ROP/GnLe Ra/+ Os/+ Pt/+	F70
ot	Oscillator	C57BL/6J- ot/+	M N10
p^J	Pink-eyed dilution-J	C3H/HeJ- p^J/+	M F108
p^{un}	Pink-eye unstable	C57BL/6J- p^{un}/+	M N48F17*
pa	Pallid	C57BL/6- Tsk +/+ pa	N35F47
		C57BL/6J- pa	N45F1N1*
pcd	Purkinje cell degeneration	C57BL/6J- pcd	N15
pe	Pearl	C3HeB/FeJ- pe	N28
		C57BL/6J- pe	N24*
pg	Pygmy	C3H/HeNIcrWf- pg/+	N21F16
$Pgk-1$	Phosphoglycerate kinase-1	C57BL/6J- $Pgk-1^a$	N10
Ph	Patch	C57BL/6J- Ph	N88
pi	Pirouette	C57BL/6J- pi/+	N15F1N1*
pk	Plucked	B6.D2- pk	N19
Ps	Polysyndactyly	C57BL/6J- Ps (N28 B6/By)	N11 B6/JEi
Pt	Pintail	C57BL/6By- Ra Os Pt	N26*
		ROP/GnLe Ra/+ Os/+ Pt/+	F70
pu	Pudgy	pu + c^{ch}/+ p c^{ch}	F24
Ra	Ragged	C57BL/6By- Ra Os Pt	N26*
		C57BL/6J- Ra	N32*
		ROP/GnLe Ra/+ Os/+ Pt/+	F70
rd	Retinal degeneration	C57BL/6J- rd le	N39F3N1*
		C57BL/6J- rd le $Gus-s^h$	N30F48
Re	Rex	C57BL/6By- Re Sd Va^J	N12*
		C57BL/6J- Tr^J Re	N24
		SHR/GnEi Re +/+ shm	F62

Table 19.8—*cont.*

Gene symbol	Gene name	Strain and genotype	Generations inbred
Re^{den}	Denuded	C57BL/6By- Re^{den}/+	NE17*
Rf	Rib fusions	Rf/+	F80*
rh	Rachiterata	rh/+	F31
Rn	Roan	C57BL/10Gn- Rn/+	N10F66*
rs	Recessive spotting	C57BL/6J- rs	N29
ru	Ruby-eye	C57BL/6J- ru	N44*
ru-2^{hz}	Haze	B6.D2- ru-2^{hz}/+	N14F21N1*
ru-2^J	Ruby-eye-2-J	C57BL/6J- ru-2^J/+	M F51
rv	Rib vertebrae	B6.L- rv/+	N9F12
s	Piebald spotting	C3HeB/FeJLe- a/ a- s	N17
		SSL/Le a/ a s^l/ s	F58
s^l	Piebald lethal	SSL/Le a/ a s^l/ s	F58
sch	Scant hair	C57BL/Ks- sch/+	M F56
$scid$	Severe combined immune deficiency	C.B-17/IcrJ- $scid$	M F?
Sd	Danforth's short tail	C57BL/6By- Re Sd Va^J	N12*
		RSV/Le Re/ Re Sd/+ Va/+	F60
se	Short-ear	C57BL/6J- se	N30F3N1*
		C57BL/6J- sg + +/+ d se	NE9F8
		DLS/Le a/ a d^l +/+ se	F99
		SEA/GnJ b/ b d se/ d +	F146
		SEC/1GnLe a/ a b/ b c^{ch}/ c^{ch} se/+	F148
sea	Sepia	C57BL/6J- sea	M F67
sg	Staggerer	C57BL/6J- sg + +/+ d se	NE9F8
sh-1	Shaker-1	SHI/Le c^{ch} sh-1/ c^{ch} +	F82
sh-2	Shaker-2	sh-2/+	F49
shi	Shiverer	shi/+	F1
shm	Shambling	SHR/GnEi Re +/+ shm	F62
Sig	Sightless	C57BL/6By- Sig/+	N27F16
Sl	Steel	C3HeB/FeJLe- a/ a- Ca^J Sl Hm	N38
		C57BL/6By- Ca Sl	N26F6N2*
		WC/ReJ- Sl/+	N58F22
Sl^d	Steel Dickie	C57BL/6J- Sl^d/+	NE54
		WB/Re- Sl^d	N48
slt	Slaty	C57BL/6J- slt/+	N10F55
sm	Syndactylism	sm/+	F59
Sp	Splotch	C57BL/6J- N Sp	N18
Sp^d	Delayed-splotch	C57BL/6J- Sp^d/+	M N17F1N1*
spa	Spastic	C57BL/6J- spa	N16F1*
sr	Spinner	C57BL/6J- sr/+	N6F1N1*
stb	Stubby	stb + a/+ fi a	F67
stu	Stumbler	C3HeB/FeJLe- a/ a- stu	N5
Sxr	Sex reversed	C57BL/6J- A^{w-J}- Y^{Sxr}	N10
sy^{fp}	Fused phalanges	C3HeB/FeJ- sy^{fp}/+	N10F14
T	Brachyury	C3H/HeSn- T tf/+ tf	N7F?N2F1
		C57BL/10ScSn- T/+	F63
		C57BL/6By- Mi^{wh} W^v T	N29F2N1*
		MWT/Le a^t/ a^t Mi^{wh}/+ W^v/+ T/+	F70
		TF/GnLe a/ a T tf/+ tf	F81
		TT6/Le T tf/ t^6 +	F52
T^{hp}	Hairpin tail	C3H/HeSnJ- T^{hp}	N8
T^{2J}	Brachyury-2J	C57BL/6J- T^{2J}	M F3N5F1N1*
T^{4J}	Brachyury-4J	BALB/cBy- T^{4J}	M N5F1N1*
T^{Or}	T-Oak Ridge	C57BL/6J- T^{Or}	N10
t^0	t (complementation group)	LT.MA- Glo-1^b.B6- Glo-1^c- t^0	N7
t^{w32}		LT.MA- Glo-1^b.B6- Glo-1^a- t^{w32}	N8
t^{w71}		LT.MA- Glo-1^b.B6- Glo-1^a- t^{w71}	N10
Ta	Tabby	C57BL/6J- A^{w-J}- Ta +/+ Tfm	N3F28
Ta^{By}	Tabby-Bailey	B6.C- A^{w-J}- Ta^{By}/+ (N18 to B6/By)	B6/J N7F5*

Table 19.8—*cont.*

Gene symbol	Gene name	Strain and genotype	Generations inbred
Tdy	Testis determining-Y	C57BL/6J- *Tdy*	N16
tf	Tufted	C3H/HeSn- *T tf*/+ *tf*	N7F?N1F2
Tfm	Testicular feminization	C57BL/tJ- *A*^{w-J}- *Ta* +/+ *Tfm*	N3F28
tg	Tottering	C57BL/6J- *tg* (N13 to B10)	B6/J N28
tg^{la}	Leaner	C57BL/6J- *Os* +/+ *tg*^{la}	N14F4
tk	Tail-kinks	TKDU/Dn *a*/ *a d* + *du*/ *d tk* +	F67
tp	Taupe	*a*/ *a tp*/+	F62*
tp^{3J}	Taupe-3J	C57BL/6J- *tp*^{3J}/+	M N1F7
Tr^J	Trembler-J	C57BL/6J- *Tr*^J *Re*	N24
Ts	Tail-short	TSJ/Le *b*/ *b Ts*/+	F58
Tsk	Tight skin	C57BL/6J- *Tsk* +/+ *pa*	N35F47
tub	Tubby	C57BL/6J- *tub Hbb*^s/+ *Hbb*^p	M N1F25
twi	Twitcher	C57BL/6J- *twi*	N12
uw	Underwhite	C57BL/6J- *uw*	M NE17*
		C57BL/6J- *uw bt*^{2J}/+ +	M N10F1N1*
v	Waltzer	V/Le *a*/ *a fz ln*/ *fz ln s*/ *s v*/+	F54
Va	Varitint-waddler	C57BL/6J- *Va*	N67*
		RSV/Le *Re*/ *Re Sd*/+ *Va*/+	F60
Va^J	Varitint-waddler-J	C57BL/6By- *Re Sd Va*	N12*
Ve	Velvet coat	C57BL/6J- *Ve*	N10
vl	Vacuolated lens	C3H/HeSn- *vl*/+	M N2F71
W	Dominant spotting	C57BL/6J- *W*	N134
		FL/1Re- *W*	N98
		WB/ReJ *a*/ *a W*/+	F126
W^{17J}	Dominant spotting-17J	C57BL/6J- *W*^{17J}/+	M F53*
W^{20J}	Dominant spotting-20J	C57BL/6J- *W*^{20J}/+	M F54*
W^{24J}	Dominant spotting-24J	C57BL/6J- *W*^{24J}	N50*
W^{25J}	Dominant spotting-25J	C57BL/6J- *W*^{25J}	N42*
W^{35J}	Dominant spotting-35J	C57BL/6J- *W*^{35J}	M N57*
W^{37J}	Dominant spotting-37J	C57BL/6J- *W*^{37J}	M N54
W^{39J}	Dominant spotting-39J	C3H/HeJ- *W*^{39J}	N42
		C57BL/6J- *W*^{39J}	M N63*
W^{41J}	Dominant spotting-41J	C57BL/6J- *W*^{41J}	M N26
W^{42J}	Dominant spotting-42J	C57BL/6J- *W*^{42J}	M N56
W^{44J}	Dominant spotting-44J	C57BL/6J- *W*^{44J}	M N47
W^{45J}	Dominant spotting-45J	C57BL/6J- *W*^{45J}	M N52
W^{54J}	Dominant spotting-54J	129/Sv- *p*⁺ *c*^{ch} *W*^{54J}	M N18
W^{59J}	Dominant spotting-59J	NZB/BlNJ- *W*^{59J}	M N11
W^v	Viable dominant spotting	C57BL/6By- *Mi*^{wh} *W*^v *T*	N29F2N1*
		C57BL/6J- *W*^v	N139
		C57BL/6J- *lx W*^v	N108*
		MWT/Le *a*^t/ *a*^t *Mi*^{wh}/+ *W*^v/+ *T*/+	F70
W^x	Dominant spotting-x	C3H/HeJ- *W*^x/+	F108
		CBA/CaJ- *W*^x	N20*
wa-2	Waved-2	*a*/ *a wa-2*/+	F56
we	Wellhaarig	B10.UW/Sn *H-3*^b *we un a*^t	NE8F62
we^{Bkr}	Wellhaarig-Bunker	B10.129- *we*^{Bkr}/Cy	NE12F42*
wi	Whirler	C57BL/6J- *wi b*/+ +	N15F7N1*
wl	Wabbler-lethal	WLHR/Le *a*/ *a b*/ *b hr* +/+ *wl*	F103
xid	X-linked immune deficiency	CBA/CaHN- *xid*	F68
Xt^J	Extra toes-J	C57BL/6J- *Xt*^J	N27*
Y^{MET}	Y Chromosome, metacentric	BALB/cBy- *Y*^{MET}	N17
		C57BL/6J- *Y*^{MET}	N19
Yaa	Y-linked accelerated auto-immunity	BXSB/Mp- *Yaa*	F55
		C57BL/6J- *Yaa*	N9F12
Yaa⁺	Y-linked accelerated auto-immunity (wild type)	BXSB/Mp- *Yaa*⁺	N10F9

Table 19.9 Genes maintained in the Western Hemisphere (Jackson Laboratory excluded) and in Japan

Gene symbol	Gene name	Strain and genotype	Generations inbred	Holder
A	Agouti	C57BL/6JCj- A md	NE13pN2[†]	Cj
A^i	Intermediate agouti	C57BL/6JCj- A^i	M F24N1F1+N19	Cj
A^{iy}	Intermediate yellow	C57BL/6JCj- A^{iy}	N16F1+N18	Cj
A^{sy}	Sienna yellow	C57BL/6JCj- A^{sy}	M N17F1+N20	Cj
A^{vy}	Viable yellow	A/HeN- A^{vy}	N6	N
		AM/Wf$_r$C3H/Nctr A^{vy}/ a^m	F63	Nctr
		BALB/cAnN- A^{vy}	N13	N
		C3H/HeN- A^{vy}	N6	N
		C57BL/6N- A^{vy}	N15	N
		C57BL/6J- A^{vy}/ a	N65F40+15	Wak
		GRS/SN- A^{vy}	N8	N
		VY/Wf$_r$C3H/Nctr A^{vy}/ a	F81	Nctr
		VY/WfL A^{vy}/ a	F?+43	L
A^{w-23J}	White-bellied agouti-23J	C57BL/6JCj- A^{w-23J}/ A^{w-23J}	M F5	Cj
A^y	Yellow	201/Rl A^y/ a pe/ pe	F75+	Rl
		C57BL/6JCj- A^y	N100+N8	Cj
		DK/Lm A^y/ a b/ b	F22	Gb
a	Non-agouti	201/Rl A^y/ a pe/ pe	F75+	Rl
		DK/Lm A^y/ a b/ b	F22	Gb
		C3HeB/FeJLeCj- a/ a s/ s	N16F1+F6	Cj
		VY/Wf$_r$C3H/Nctr A^{vy}/ a	F81	Nctr
		VY/WfL A^{vy}/ a	F?+43	L
		SWR/JCj- a c^+	N10	Cj
a^e	Extreme non-agouti	AEJ/GnCj- a^e +/ a bp^H	N5F51+F3	Cj
a^m	Mottled agouti	AM/Wf$_r$C3H/Nctr A^{vy}/ a^m	F63	Nctr
a^{t-42J}	Black and tan	C57BL/6J- a^{t-42J}/ a^{t-42J}	M F7	Cj
Ags	Alpha galactosidase	C3H.MOL- Ags^m	N10	Cv
		C3H/HeHa- Pgk-1^a Ags^m	N10	Cv
Ah	Aromatic hydrocarbon responsiveness	AKR.B6- Ah^b	N?F26	NICHD
		B6.AKR[‡]- Ah^k	N?F31	NICHD
		B6.D2- Ah^d	NE12F20	NICHD
		C3H.D2- Ah^d	NE12F38	NICHD
		D2.B6- Ah^b	N?F32	NICHD
		D2.C3H- Ah^h	N?F31	NICHD
ank	Progressive ankylosis	C3HeB/FeJ- ank	N13	Ksb
ash	Ashen	C3H/HeSnCj- ash/ ash	M F30+11	Cj
b	Brown	C57BL/10J- b/ b	N10F6	Hir
b^J	Brown-J	C57BL/6JCj- b^J/ b^J	M N35+N2F3pF5[†]	Cj
bg^J	Beige-J	A/HeN- bg^J	N6	N
		BALB/cAnN- bg^J	N9	N
		CBA/N- bg^J	N6	N
		C3H/HeN- bg^J	N12	N
		C57BL/6J- bg^J	M F13	Ms
		C57BL/6N- bg^J	N10F13	N
		C57BL/6J- bg^J	M F?	Csk
		DBA/2N- bg^J	N6	N
		NFR/N- bg^J	N8	N
		NFS/N- bg^J	N6	N
		NZB/N- bg^J	N13	N
		P/N- bg^J	N9	N
		SJL/N- bg^J	N6	N
Bgl	Beta galactosidase	B6.CBA- Bgl^d Trf^a	N13	Cv
bp^H	Brachypodism-H	AEJ/GnCj- a^e +/ a bp^H	N5F51+3	Cj
c	Albino	C57BL/6J- c/ c	N10F4	Hir
c^{2J}	Albino-2J	C57BL/6J- c^{2J}/ c^{2J}	M F31	Glw
c^{3H}	Albino-3H	3H/Glw a/ a b/ b c^{ch}/ c^{3H}	F60	Glw
c^{6H}	Albino-6H	6H/Glw b/ b c^{ch}/ c^{6H}	F59	Glw
c^{25H}	Albino-25H	25H/Glw c^{ch}/ c^{25H}	F40	Glw
c^{112K}	Albino-112K	112K/Glw a/ a c^{ch}/ c^{112K}	F70	Glw

Table 19.9—*cont.*

Gene symbol	Gene name	Strain and genotype	Generations inbred	Holder
c^{ch}	Chinchilla	129/Rl A^w/ A^w c^{ch} p/ c p	F73	Rl
c^+	c-wild type	SWR/JCj- a c^+	*N10*	*Cj*
C-3	Complement component-3	C57BL/6J- C-3^a	N16	Boy
		B10.NC- C-3^a	N12F16+	Kuj
Cd	Crooked	Cd/+ a/ a	F27+14+NBS[§]	MTL
cla	clasper	BALB/cBy- cla/+	M F15	Sid
cmd	Cartilage matrix deficiency	a^t/ a^t cmd/+	F25	Ksb
cmd^{Bc}	Cartilage matrix deficiency-Bc	BALB/cGaBc- cmd^{Bc}/+	M F5	Bc
		BALB/cGaBc- cmd^{Bc}/+	M F3	Ksb
coa	Cocoa	C57BL/10JCj- coa/ coa	M N1F15+5	Cj
crn	Cranioschisis	crn/+	F53	Ksb
d	Dilute	C57BL/10J- d/ d	N10F1	Hir
		DL/Ra- d se^+/ d se^+	N10F?	Ra
		SEC/Rl a/ a b/ b c^{ch}/ c^{ch} d se/+ +	F85	Rl
d^l	Dilute-lethal	DL/Ra a/ a d^l +/+ se	F96+	Ra
d^{l-J}	Dilute-lethal-J	DL/Ra- d^{l-J} +/+ se	N10F?	Ra
d^{l-16J}	Dilute-lethal-16J	C57BL/6JCj- d^{l-16J} +/ d se	N15	Cj
d^{l-18J}	Dilute-lethal-18J	C57BL/6J- lpr?- d^{l-18J} +/ d se	M(d^l) N11F11+2	Cj
d^{15}	Dilute-15	DL/Ra- d^{15} +/+ se	N10F?	Ra
d^+	d-wild type	DL/Ra- d^+ se^+/ d^+ se^+	N10F?	Ra
db	Diabetes	C57BL/KsJ- db +/+ m	M(db) N7F?+31	Wak
		C57BL/KsJ- db +/+ m	M(db) N5F?	Ksb
		C57BL/KsJ- db +/+ m	M(db) N5F?	Csk
Dc	Dancer	DC/Le Dc/+	F64+10+NBS [§]	MTL
df	Ames dwarf	NFR/N- df	N12	N
Dh	Dominant hemimelia	BALB/c- Dh	N36	Boy
Dmm	Disproportionate micromelia	C57BL/6J- Dmm	N?F14	Ksb
dt	Dystonia musculorum	C57BL/6J- dt	N9	Cu
dt^{Alb}	Dystonia musculorum-Albany	BALB/cByJ- dt^{Alb}	M N22	Alb
dw	Dwarf	C3H/HeN- dw	N14	N
		NFR/N- dw	N12	N
dy	Dystrophia muscularis	C57BL/6J- dy	N?F22+22	Wak
Emv-16	Ecotropic murine leukemia virus-16	SWR/JCj- Emv-16 Emv-17	N13	Cj
Emv-17	Ecotropic murine leukemia virus-17	SWR/JCj- Emv-16 Emv-17	N11	Cj
Eo	Eye opacity	C3H$_f$/Rl- Eo	N20+	Rl
ep	Pale ear	C57BL/6J- ep/ ep	N?F15+2	Cj
		C57BL/6- ep/ ep le/ le	N?F24	MM
Es-1	Esterase-1	B6.YBR- Es-1^a Es-22°	N16	Cv
Es-22	Esterase-22	B6.YBR- Es-1^a Es-22°	N16	Cv
far	First arch	BALB/cGaBc- far/+	M F33	Bc
Frl^a	Fur loss-a	C57BL/10Rl- Frl^a	N20+	Rl
Frl^b	Fur loss-b	C57BL/10Rl- Frl^b	N33	Rl
Frl^c	Fur loss-c	C3H$_f$/Rl- Frl^c	N28	Rl
fro	Fragilitas ossium	C57BL/6JPas- fro/+	N?F16	Ksb
Fv-1	Friend virus susceptibility	AKR.B6- Fv-1^b	N11	Boy
		B6.AKR- Fv-1^n	N11	Boy
Glo-1	Glyoxylase-1	C57BL/6- Glo-1^b	N20	Boy
Gpd-1	Glucose-6 phosphate dehydrogenase	B6.AKR- Gpd-1^b	N20	Boy
Gpi-1	Glucose phosphate isomerase	C57BL/6- Gpi-1^a	N10?	Cv
gr	Grizzled	GR/Rl A/ A +/ gr	F55	Rl
gt	Gray tremor	GT/LeSid gt/+	F30	Sid
Gus	Beta glucuronidase	B6.A- Gus^a	N15	Cv
		B6.CAS- Gus^{cl}	N7	Cv

834

Table 19.9—*cont.*

Gene symbol	Gene name	Strain and genotype	Generations inbred	Holder
		B6.CAS- Gus^{cs}	N16	Cv
		B6.PAC- Gus^{n}	N12	Cv
		C57BL/6J- Gus^{s}	N9	Cv
		B6.MOR- Gus^{w5}	N14	Cv
		C57BL/6J- Gus^{w12}	N11	Cv
		C57BL/6J- Gus^{w16}	N13	Cv
		C57BL/6J- Gus^{w17}	N12	Cv
		C57BL/6J- Gus^{w18}	N11	Cv
		C57BL/6J- Gus^{w26}	N10	Cv
		C57BL/6J- Gus^{w28}	N10	Cv
Gv-1	Gross virus antigen-1	B6.129- *Gv-1 Gv-2*	N11	Boy
Gv-2	Gross virus antigen-2	B6.129- *Gv-1 Gv-2*	N11	Boy
H-2	Histocompatibility-2	B6.AKR- $H\text{-}2^{k}$	N15	Boy
		AKR.B6- $H\text{-}2^{b}$	N15	Boy
Hc	Hemolytic complement	B10.NC- Hc^{o}	N12F3+	Kuj
		NC.B10- Hc^{1}	N12F3+	Kuj
ho^{4J}	Hotfoot-4J	DBA/2J- ho^{4J}	M F30N2	Sid
Hprt	Hypoxanthine phospho-ribosyl transferase	C3H/HeHa- $Hprt^{a}$ $Pgk\text{-}1^{a}$ Ags^{m}	N9	Cv
hr	Hairless	C57BL/6JCu- *hr*	N28	Cu
		C57BL/10J- *hr/ hr*	N10F5	Hir
		C3H/HeN- *hr*	N10	N
Hx	Hemimelic extra toes	B10.D2/nSn- *Hx/+*	M F?+?	Ksb
hyt	Hypothyroid	C.RF- *hyt/+*	N10F12	Sid
Igh-C	Immunoglobulin heavy chain constant region	B10.NC- *Igh-C*	N12F13+	Kuj
Ie	Eye-ear reduction	C3H$_f$/Rl- *Ie*	N12+	Rl
		BALB/cRl- *Ie*	N12+	Rl
Lc	Lurcher	BALB/cByJ- *Lc/+*	N11F3	Alb
		C57BL/6J- *Lc*	N14	Sid
le	Light ear	C57BL/6J- *le/ le ep/ ep*	N?F24	MM
lg^{Ga}	Lid gap-Gates	AEJ/GnRk- lg^{Ga}	N10	Bc
Lp	Loop tail	LPT/Le *Lp/+*	F87+6	MTL
lpr	Lymphoproliferation	MRL/Mp- *lpr*	N10F?+5	Wak
		MRL/MpJ- *lpr*	N10?F?	Csk
		C57BL/6J- *lpr*	N11F10+3	Kuj
Ly-1	Lymphocyte antigen-1	B6.C3H- $Ly\text{-}1^{a}$	N18	Boy
Ly-2	Lymphocyte antigen-2	B6.C3H- $Ly\text{-}2^{a}$	N17	Boy
		B6.RF- $Ly\text{-}2^{a}$ $Ly\text{-}3^{a}$	N17	Boy
Ly-3	Lymphocyte antigen-3	B6.RF- $Ly\text{-}2^{a}$ $Ly\text{-}3^{a}$	N17	Boy
Ly-5	Lymphocyte antigen-5	B6.SJL- $Ly\text{-}5^{b}$	N22	Boy
Ly-9	Lymphocyte antigen-9	B6.C- $Ly\text{-}9^{a}$	N6	Boy
Ly-15	Lymphocyte antigen-15	B6.C- $Ly\text{-}15^{a}$	N10	Boy
Ly-17	Lymphocyte antigen-17	B6.C3H- $Ly\text{-}17^{a}$	N19	Boy
Lyb-2	B lymphocyte antigen-2	B6.I- $Lyb\text{-}2^{a}$	N8	Boy
		B6.CE- $Lyb\text{-}2^{c}$	N20	Boy
		B6.CE- $Lyb\text{-}2^{c}$ $Mup\text{-}1^{a}$	N12	Boy
m	Misty	C57BL/KsJ- *db +/+ m*	M(*db*) N5F?+31	Wak
		C57BL/KsJ- *db +/+ m*	M(*db*) N5F?	Csk
		C57BL/6JCj- db^{+} *m/+ m*	N5F30+1	Cj
mdx	Muscular dystrophy X linked	C57BL/10ScSn- *mdx/ mdx*	M F4+6	Cj
me	Motheaten	C3H/HeN- *me*	N16	N
		C57BL/6N- *me*	N6	N
		NFS/N- *me*	N15	N
		NZB/N- *me*	N6	N
		P/N- *me*	N6	N
mea	Meander tail	C57BL/6J- *mea*	N5	Sid
		DBA/2J- *mea*	N6	Sid

Table 19.9—*cont.*

Gene symbol	Gene name	Strain and genotype	Generations inbred	Holder
mh^{2J}	Mocha-2J	C57BL/6J- mh^{2J}	N11	Sid
mi^{ew}	Eyeless white	NAW/Wl- mi^{ew}	N20F18	Wl
Mi^{wh}	White	C57BL/6JCj- $Mi^{wh}/+$	N153F3	Cj
Mo^{blo}	Blotchy	C3H$_f$/Rl- Mo^{blo}	N49	Rl
Mo^{to}	Tortoise shell	NAW/Wl- Mo^{to}	N52F23	Wl
Mup-1	Major urinary protein	B6.CE- Lyb-2^c Mup-1^a	N11	Sid
myd	Myodystrophy	MYD/Le- Os $+/+$ myd	N7F22	Wak
nu	Nude	A/HeN- nu	N8	N
		AKR/N- nu	N14	N
		AL/N- nu	N15	N
		BALB/cAnN- nu	N11	N
		CBA/CaHN- nu	N15	N
		CBA/CaH-T6/J- nu	N9	N
		C3H/HeN- nu	N17	N
		C3H/HeJN- nu	N6	N
		C57BL/6N- nu	N16	N
		C57BL/10ScN- nu	N15	N
		C57L/N- nu	N10	N
		DBA/1N- nu	N8	N
		DBA/2N- nu	N14	N
		GRS/SN- nu	N8	N
		I/StN- nu	N14	N
		ICR/HaN- nu	N6	N
		NFR/N- nu	N10	N
		NFS/N- nu	N16	N
		NZB/N- nu	N17	N
		NZW/N- nu	N6	N
		P/N- nu	N6	N
		RIII/AnN- nu	N6	N
		SJL/N- nu	N8	N
		STR/N- nu	N7	N
		WB/ReN- W^v- nu	N14	N
nr	Nervous	BALB/cGrSid- $nr/+$	M F10	Sid
ob	Obese	C57BL/6J- $ob/+$	N40F8	Wak
Oel	Open eyelids with cleft palate	$Oel/+$	F55	Ksb
Os	Oligosyndactyly	MYD/Le- Os $+/+$ myd	N7F22	Wak
p	Pink-eyed dilution	C57BL/6ByCj- p/p $Ve^+/+$	N22F20+3	Cj
		C57BL/10J- p/p	N10F5	Hir
p^{un}	Pink-eye unstable	C57BL/6JCj- p^{un}/p^{un}	N48F20+5	Cj
		C57BL/6J- p^{un}/p^{un}	N10?F?	Cv
		C57BL/6JWl- p^{un}	N24F13N3	Wl
p^{25H}	P-25H	NAW/Wl- p^{25H}	N14	Wl
pa	Pallid	C57BL/6JCj- pa/pa	N45N1F3+3	Cj
Pca-1	Plasma cell antigen-1	C.B6- Pca-1^-	N20	Boy
		B6.C- Pca-1^+	N11	Boy
pcd	Purkinje cell degeneration	BALB/cBy- pcd	N9	Sid
		C57BL/6J- pcd	N14	Sid
pcd^{2J}	Purkinje cell degeneration-2J	SM/J- pcd^{2J}	M N3	Sid
		C57BL/6J- pcd^{2J}	N9	Sid
pe	Pearl	C57BL/6JCj- pe/pe	N23F1pN3F13[†]	Cj
		C57BL/6J- pe/pe	N20F12	PVRC
		C57BL/6J- pe/pe	N24	Sid
		PE/Rl A/A $pe/+$	F74	R1
		C57BL/6J- pe^H/pe^H	N9F4	PVRC
pe^{+P}	Pearl revertant-Purdue	C57BL/6J- pe^{+P}/pe^{+P}	M F5	PVRC
pe^{+2P}	Pearl revertant-2 Purdue	C57BL/6J- pe^{+2P}/pe^{+2P}	M F5	PVRC
		C57BL/6J- pe^{+2P}/pe^{+2P}	M F?	Sid

Table 19.9—*cont.*

Gene symbol	Gene name	Strain and genotype	Generations inbred	Holder
Pgk-1	Phosphoglycerate kinase-1	C3H/HeHa- *Pgk-1ᵃ Agsᵐ*	N10	Cv
qv³ᴶ	Quivering-3J	C57BL/6J- *qv³ᴶ*	M N1	Sid
ru-2ᴶ	Ruby-eye-2-J	C57BL/6JCj- *ru-2ᴶ/ ru-2ᴶ*	M F98+49+4	Cj
s	Piebald	C3HeB/FeJLeCj- *a/ a- s/ s*	N16F1+6	Cj
scid	Severe combined immunodeficiency	C.B-17/IcrCj- *scid/ scid*	M F8+4	Cj
Sd	Short Danforth	SD/Glw *c/ c Sd/+*	F129	Glw
se	Short-ear	C57BL/6JCj- *d se/ d se*	N8F1+14	Cj
		SEC/1Rl *a/ a b/ b cᶜʰ/ cᶜʰ d se/+ +*	F85	Rl
sf	Scurfy	129/Rl- *sf*	N74	Rl
shi	Shiverer	BALB/cBy- *shi*	N5	Sid
		C3H/HeJ- *shi*	N6	Sid
		C57BL/6J- *shi*	N9	Sid
		DBA/2J- *shi*	N8	Sid
		NFR/N- *shi*	NE6	N
shiᵐˡᵈ	Myelin deficient	BALB/cBy- *shiᵐˡᵈ*	N9	Sid
		C3H/HeJ- *shiᵐˡᵈ*	N9	Sid
		C57BL/6J- *shiᵐˡᵈ*	N9	Sid
		DBA/2J- *shiᵐˡᵈ*	N11	Sid
shm	Shambling-3J	C3H/HeSnJ- *shm³ᴶ/+*	M N3F9	Sid
Sl	Steel	SL/Rl *A/ A Sl/+*	F87	Rl
Sl³⁹ᴸᴸ	Steeloid-39LL	C3Hf/Rl- *Sl³⁹ᴸᴸ*	N12+	Rl
Sl¹ˣʸ	Steeloid-1XY	C3Hf/Rl- *Sl¹ˣʸ*	N12+	Rl
Sp	Splotch	SP/Rl *a/ a Sp/+*	F92	Rl
Spᵈ	Splotch-delayed	C57BL/6J- *Spᵈ/+*	M N14F11	MTL
stu	Stumbler	C57BL/6J- *stu*	N8	Sid
T	Brachyury	TF/GnLe *a/ a T tf/+ tf*	F55+18	Kuj
		T/Glw *b/ b T/+*	F55	Glw
		T12/Glw *a/ a b/ b T/ t¹²*	F70	Glw
t³	t-3	T3/Glw *b/ b t³/ t³*	F112	Glw
t¹²	t-12	T12/Glw *a/ a b/ b T/ t¹²*	F70	Glw
Ta	Tabby	C3H/Rl- *Ta*	N47	Rl
		TABR *Aʷ/ Aʷ b/ b Ta/+*	F46	Rl
thf	Thin fur	C57BL/6- *thf*	N20	Boy
Thy-1	Thymus cell antigen-1	B10.NRH- *Thy-1ᵃ*	N12F3+16	Kuj
Tla	Thymus leukemia antigen	B6.A- *Tlaᵃ*	N27	Boy
		A.B6- *Tlaᵇ*	N16	Boy
		B6.B10.A- *Tlaᵃ*	N15	Boy
tp³ᴶ	Taupe-3J	C57BL/6JCj- *tp³ᴶ/ tp³ᴶ*	M F1N1FY+3	Cj
Trf	Transferrin	B6.CBA- *bglᵈ Trfⁿ*	N13	Cv
tx	Toxic-milk	C57BL/6J- *tx*	N6	Ra
		DL/Ra- *d⁺ se⁺- tx*	N6	Ra
wl	Wabbler-lethal	C57BL/6JCu- *wl*	N30	Cu
W	Dominant spotting	WB/ReJ *W/+*	F?+11+6	Kuj
Wᵛ	Viable dominant spotting	C57BL/6J- *Wᵛ/+*	N?F47	Nrs
		C57BL/6J- *Wᵛ/+*	N?F36+9	Kuj
		C57BL/6JRl- *Wᵛ/+*	N15F72	Rl
wr	Wobbler	NFR/N- *wr*	N14	N
		C57BL/6JCu- *wr*	N10	Cu
wv	Weaver	BALB/cBy- *wv*	N8	Sid
		C57BL/6J- *wv*	N12	Sid
xid	X-linked-immune deficiency	A/HeN- *xid*	NE6	N
		AKR/N- *xid*	NE7	N
		AL/N- *xid*	NE7	N
		BALB/cAnN- *xid*	NE11	N
		CBA/CaHN- *xid*	NE10	N
		C3H/HeN- *xid*	NE10	N
		C3H/HeJN- *xid*	NE6	N

Table 19.9—*cont.*

Gene symbol	Gene name	Strain and genotype	Generations inbred	Holder
		C57BL/6N- *xid*	NE10	N
		C57BL/10ScN- *xid*	NE7	N
		C57L/N- *xid*	NE6	N
		DBA/1N- *xid*	NE6	N
		DBA/2N- *xid*	NE8	N
		GRS/SN- *xid*	NE6	N
		NFR/N- *xid*	NE9	N
		NFS/N- *xid*	NE10	N
		NZB/N- *xid*	NE8	N
		NZW/N- *xid*	NE6	N
		P/N- *xid*	NE6	N
		SJL/N- *xid*	NE6	N
		WB/ReN- W^v- *xid*	NE8	N
Xt^J	Extra toes-J	C57BL/6J- *Xt^J*	N21F15	MTL

†p = removed from freeze preservation and generations bred after removal follow p.

‡ = AKR/N not AKR/J.

§ NBS = not brother × sister mated, but not outcrossed.

Table 19.10 Genes maintained in Europe and the Near East

Gene symbol	Gene name	Strain and genotype	Generations inbred	Holder
A^{Pas1}	Agouti	C57BL/6/Pas-A^{Pas1}/a	F?+15	Pas
A^{Pas2}	Agouti	C57BL/6-H-2^k//Pas-A^{Pas2}/a	F?+25	Pas
A^{Pas3}	Agouti	STR/N//Pas-A^{Pas3}/a; b/b	F?+18	Pas
A^{vy}	Viable yellow	C3H/Icrf-A^{vy}	F?+99	Icrf
A^{wJ}	White-bellied agouti	C57BL/6J-A^{wJ}	MF?	Han
A^y	Yellow	129/Sv-A^y/+	N42	Pas, Ev
		C57BL/6J-A^y/+	F?+32	Pas, Hl
a	Non-agouti	HT/Pas-a bp pa; fz ln; pe	F39	Pas
		LT3/Pas-a; ep; fz ln; +/v	F?+33	Pas
		ABP/Pas-a; b; p^d/p; se; bt; wa-1	F34	Pas
		ADP/Pas-a; d; p	F30	Pas
		ZRDCT/Pas-a; d; p; b; ey	F62	Pas
		PT/Pas-a; d se; $c^{ch}p$; b; s	F35	Pas
a^t	Black and tan	BTBR/Pas-T tf/+ tf; a^t/a^t	F53	Pas
		C57BL/a^t	M+F?	H, Icrf, Han, Orl
Amy^{w1}	Amylase	C3H.Amy^{w1}	N10+	As
Amy^{w2}	Amylase	C3H.Amy^{w2}	N10+	As
Amy^{w3}	Amylase	C3H.Amy^{w3}	N10+	As
Amy^{w4}	Amylase	C3H.Amy^{w4}	N10+	As
Amy^{w5}	Amylase	C3H.Amy^{w5}	N10+	As
Amy^{w9}	Amylase	C3H.Amy^{w9}	N10+	As
Amy^{w11}	Amylase	C3H.Amy^{w11}	N10+	As
Amy^{w13}	Amylase	C3H.Amy^{w13}	N10+	As
Amy^{w15}	Amylase	C3H.Amy^{w15}	N10+	As
Amy^{w16}	Amylase	C3H.Amy^{w16}	N10+	As
Amy^{w17}	Amylase	C3H.Amy^{w17}	N10+	As
Amy^{w19}	Amylase	C3H.Amy^{w19}	N10+	As
Amy^{w21}	Amylase	C3H.Amy^{w21}	N10+	As
Amy^{w22}	Amylase	C3H.Amy^{w22}	N10+	As
Amy^{w33}	Amylase	C3H.Amy^{w33}	N10+	As
Amy^{w34}	Amylase	C3H.Amy^{w34}	N10+	As
Amy^{CE}	Amylase	C3H.CE	N10+	As
Amy^{YBR}	Amylase	C3H.YBR	N10+	As
ax^J	Ataxia	C57BL/6J//Pas-+/ax^J	F63	Pas
azh	Abnormal spermatozoan head shape	C57BL/6J-azh		Lac
B^{lt}	Light	129/Sv//Pas-T/+; B^{lt}/+	N27	Pas
b	Brown	ABP/Pas-a; b; p^d/p; se; bt; wa-1	F34	Pas
		ZRDCT/Pas a; d; p; b; ey	F62	Pas
		PT/Pas-a; d se; $c^{ch}p$; b; s	F35	Pas
		C57BL/6J//Pas-+/b	F?+15	Pas
bg	Beige	C57BL/6J-+/bg	F41	Orl, Pas, Dk, Max, Nimr, Ess, Ola
		NIH-nu; bg; xid		Ibm
bp	Brachypodism	CSB/Pas-c; se; bp	F19	Pas
		HT/Pas-a bp pa; fz ln; pe	F39	Pas
bp^Y		B10.RIII-bp^Y	M+F15	Y
bt	Belted	ABP/Pas-a; b; p^d/p; se; bt; wa-1	F34	Pas
c	Albino	CSB/Pas-c; se; bp	F19	Pas
		C3H/He-c		Lac
		C57BL/10ScSn-c		Lac
c^{2J}	Albino-2J	C57BL/6Orl-c^{2J}		Orl
c^{Orl}	Albino-Orleans	DBA/2Orl-c^{Orl}		Orl
c^{25H}	Lethal albino c^{25H}	C25H/Pas-c^{25H}/c^{ch}	F?+15	Pas
c^{ch}	Chinchilla	PT/Pas-a; d se; $c^{ch}p$; b; s	F35	Pas
c^e	Extreme dilution	RUEF/Y-c^e; ru; f	F30	Y
Ca	Caracul	129/Sv//Pas-Va/+; Ca/+	N29	Pas
Ca^d	Directional caracul	101C3H R.I. med +/+ Ca^d	F32	Pas
Cat	Dominant cataract	MORO/Ibm-Cat	F11	Ibm
Cm	Coloboma	C57BL/6J//Pas-Cm/+	N?+F22	Pas
Cs^a	Catalase	C57BL/AnlOrl-Cs^a		Orl
d	Dilute	ADP/Pas-a; d; p	F30	Pas
		ZRDCT/Pas-a; d; p; b; ey	F62	Pas
		PT/Pas-a; d se; $c^{ch}p$; b; s	F35	Pas
d^l	Dilute lethal	DBA1DLS R.I.-a; b; se+/+dl	F?+31	Pas
db	Diabetes	C57BL/6J-++/mdb	F31	Hl
		C57BL/6J-m+/+db	F39	Hl

839

Table 19.10—*cont.*

Gene symbol	Gene name	Strain and genotype	Generations inbred	Holder
		C57BL/KsJ-*db*	F32	Hl, Orl, Ola
		C57BL/KsJ-++/*mdb*	F42	Hl
		C57BL/KsJ-*m*+/+*db*	F?+F5	Hl, Y
db^{Pas}	Diabetes Pasteur	DW/J//Pas-+*db^{Pas}*; *ln*	F?+19	Pas
Dh	Dominant hemimelia	B6CBA R.I.-*Dh*/+	F34	Pas
dt^{Orl}	Dystonia musculorum	DBA/2Orl//Pas-+/*dt^{Orl}*	F?+25	Pas, Orl
dw	Dwarf	DW/J//Pas-+/*dw*; *ln*	F21	Pas
dy	Dystrophia muscularis	C57BL/6JOrl-*dy*		Orl
		129/Re-*c^{ch}p*; +/*dy*	F13	Pas, Orl, Y
E^{so}	Sombre	129/Sv//Pas-*E^{so}Os*/++	N4	Pas
		C3H/He//Pas-*E^{so}*/+; *Sl^{d-H}*/+	N32	Pas
Eh	Hairy ears	C57BL/6J//Pas-*Eh*/+	F55	Pas
		129/Sv//Pas-*Eh*/+	N42	Pas
		EHC/Pas-*Eh*/+	F14	Pas
		C57BL/6J//Pas-*Ve* +/+ *Eh*	F41	Pas
Er	Repeated epilation	129/Sv//Pas-*Er*/+	N35	Pas
ey	Eyeless	ZRDCT/Pas-*a*; *d*; *p*; *b*; *ey*	F62	Pas
f	Flexed tail	FRU/Pas-*f*; *ru*	N26-129; F19	Pas
fro	Fragilitas ossium	FRO-1/Pas-*a*; *b*; *fro*/*fro*	F6	Pas
fz	Fuzzy	LT3/Pas-*a*; *ep*; *fz ln*; +/*v*	F?+33	Pas
		HT/Pas-*a bp pa*; *fz ln*; *pe*	F39	Pas
fz^{Orl}	Fuzzy	C57BL/6JOrl-*fz^{Orl}*		Orl
fz^{Y}	Fuzzy	C57BL/10SnY-*fz^{Y}*	M+N18	Y
go^{Y}	Angora	C57BL-*go^{Y}*	M+F46	Y
Gpi-1s^{a}	Glucose phosphate isomerase	C57BL/Ola.129-*Gpi-1s^{a}*/Ws	N8F19	Ws
		C57BL/Ola.AKR-*Gpi-1s^{a}c*/Ws	N8F9	Ws
		C57BL/6-*Gpi-1^{a}*		Lac
Gpi-1s^{c}	Glucose phosphate isomerase	C57BL/Ola.*Gpi-1s^{c}*/Ws	N8F14	Ws
Gpi-1s^{a-m1H}		C57BL/Ola.*Gpi-1s^{a-m1H}*/Ws	NE8	Ws
Gpi-1^{c}		129/Sv//Ev-*Gpi-1^{c}*	N10+	Ev
Hba^{th}	Alpha-thalassemia	BALB/cAnCrlRij-*Hba^{th}*/+	N18	Rij
his	Histidinemia	C57BL/6-*his*	N8	BUL
hpc	Hyperspiny Purkinje cell	HPC/Pas-*ru*; +/*hpc*	F19	Pas
hph-1	Hyperphenylalaninemia-1	PKU/Pas-*hph-1*/*hph-1*	F?+12	Pas
hr	Hairless	A2G/Lac-+/*hr*		Tox
		C3H/TifBom-*hr*	N3	Bom
		WLHR/Le//Pas-*wl* +/+ *hr*	F90	Pas
		CBA-*hr*		Lee
		C3H/HeH-*hr*		Lee
		HRS/J	F?+22	Y
hr^{Pas2}	Hairless Pas 2	CB-20/Pas-+/*hr^{Pas2}*	F?+14	Pas
hr^{rhJ}	Rhino	RHJ-*hr^{rhJ}*		Ibm
hr^{rhY}		C57BL/10SnY-*hr^{rhY}*	M+N19	Y
hyt	Hypothyroid	BALB/cByEss-+/*hyt*	?+N4	Ess
je	Jerker	CJE/Pas-*c^{e}*; *je*	F19	Pas
jp	Jimpy	B6CBA R.I.-*Ta jp*/++	F?+30	Pas
kd	Kidney disease	CBA/CaH-*kd*		H
Lc	Lurcher	B6CBA R.I.-*Lc*/+	F4	Pas
lit	Little	C57BL/6J//Pas-+/*lit*	F?+9	Pas
ln	Leaden	LT3/Pas-*a*; *ep*; *fz ln*; +/*v*	F?+33	Pas
		HT/Pas-*a bp pa*; *fz ln*; *pe*	F39	Pas
lpr	Lymphoproliferation	C57BL/6J-*lpr*	N?F17	Pas, Ibm, Orl
		MRL/MPJ-*lpr*		Max, Ola
Lps^{d}	Lipopolysaccharide	C3H/HeJ-*Lps^{d}*		Ola
m	Misty	C57BL/6Pas-+/*m*	F29	Pas
mdx	X-linked muscular dystrophy	C57BL/10ScSn-*mdx*	MF?	BUL
me	Motheaten	C57BL/6Orl//Pas-+/*me*	F12	Pas
med	Motor end-plate disease	101C3H R.I.-+*Ca^{d}*/*med*+	F32	Pas
mi	Microphthalmia	G-*mi*	F40	Csr
		CBA/CaH-*T6mi*	F40	Csr
mi^{di}	Microphthalmia-defective iris	G-*mi^{di}*	F32	Csr
		CBA/CaH-*T6mi^{di}*	F22	Csr
min	Minuscule	C3H/HeH Pas-+/*min*	F19	Pas
Mo^{blo}	Blotchy	129/Sv//Pas-*Mo^{blo}*/+	N9F3	Pas
my	Blebs	MY/J//Pas-*my*/*my*	F103	Pas
N	Naked	NSP/Y-*N*/+; *Sp*/+	F58	Y

Table 19.10—*cont.*

Gene symbol	Gene name	Strain and genotype	Generations inbred	Holder
nr	Nervous	C3HeB/J//Pas-+/*nr*	F23	Pas
nu	Nude	BALB/c-+/*nu*		Many
		NIH-*nu*; *bg*; *xid*		Ibm
		AB/Jena-*nu*		Jena
		C57BL/10-*nu*		Crc, Nii
		CBA/Ca-*nu*		Crc, Nimr
		DBA/2-*nu*		Jena
		C3H/HeHan-*nu*		Han
		AKR/Bln-*nu*		Bln
		ST-*nu*		Fib
		N2B/Icrf-+/*nu*	F13	Icrf
		C57BL/6J-*nu*	N3	Bom, Ibm, Jena, Han, Orl, Nimr
nu^{str}	Streaker	AKR/J//Pas-+/*nu^{str}*	F?+8	Pas
nu^Y	Nude-Yurlovo	101/HY-+/*nu^Y*	F42	Y
ob	Obese	C57BL/6-+/*ob*	F?+11	Pas, Ibm, Hl, Ola
olt	Oligotriche	C3H/HeOrl//Pas-+/*olt*	F42	Pas, Orl
Os	Oligosyndactyly	C57BL/6-*Os*		Lac, Orl
p	Pink-eyed dilution	ABP/Pas-*a*; *b*; *p^d/p*; *se*; *bt*; *wa-1*	F34	Pas
		ADP/Pas-*a*; *d*; *p*	F30	Pas
		ZRDCT/Pas-*a*; *d*; *p*; *b*; *ey*	F62	Pas
		CBA/Ca-+/*p*		H
		PT/Pas-*a*; *d se*; *c^{ch}p*; *b*; *s*	F35	Pas
		C57BL/10-+/*p*		H
p^d	Dark pink eye	ABP/Pas-*a*; *b*; *p^d/p*; *se*; *bt*; *wa-1*	F34	Pas
pa	Pallid	HT/Pas-*a bp pa*; *fz ln*; *pe*	F39	Pas
par	Paralysé	PAR/Pas-+/*par*	F22	Pas
pcd	Purkinje cell degeneration	PCD/Pas-+*pe*/*pcd*+	F17	Pas
pe	Pearl	*pe/pe*; +/*wa-2*	F?+6	Pas
		HT/Pas-*a bp pa*; *fz ln*; *pe*	F39	Pas
Pgk-1^a	Phosphoglycerate kinase 1	PGK-1/Pas-*Pgk-1^a/Pgk-1^a*	F?+21	Pas
		C3H/He-*Pgk-1^a*/Ws	N8F29	Ws
qk	Quaking	C57BL/6J//Pas-*T^{2J}* +/+ *qk*		Pas
Ra	Ragged	129/Sv//Pas-*Re*/+; *Ra*/+	N30	Pas
Re	Rex	129/Sv//Pas-*Re*/+; *Ra*/+	N30	Pas
		C57BL/6Orl-*Re*		Orl
rl	Reeler	C57BL/6J//Pas-+/*rl*	F30	Pas
ro	Rough	RIII-*ro*		Lac, Dk
rp	Reduced pigmentation	C57BL/10ScSnEss-+/*rp*	N12	Ess
ru	Ruby eye	FRU/Pas-*f*; *ru*	N6(129); F19	Pas
		RUEF/Y-*c^e*; *ru*; *f*	F30	Y
Rw	Rump white	129/Sv//Pas-*Rw*/+; *Xt*/+	N35	Pas
s	Piebald	PT/Pas-*a*; *d se*; *c^{ch}p*; *b*; *s*	F35	Pas
		SLS/J//Pas-*a/a*; *s^l/s*	F?+44	Pas
s	Piebald lethal	SLS/J//Pas-*a/a*; *s^l/s*	F?+44	Pas
scid	Severe combined immunodeficiency	CB-17 Pas-+/*scid*	F?+5	Pas, Nimr
Sd	Danforth short tail	C3H/He Pas-*Sd*/+	N26	Pas
se	Short-ear	ABP/Pas-*a*; *b*; *p^d/p*; *se*; *bt*; *wa-1*	F34	Pas
		CSB/Pas-*c*; *se*; *bp*	F19	Pas
		PT/Pas-*a*; *d se*; *c^{ch}p*; *b*; *s*	F35	Pas
		DBA1DLS R.I.-*a/a*; *b/b*; +*d^l/se*+	F?+31	Pas
sg	Staggerer	C57BL/Pas-+/*sg*	F67	Pas
Sinc^{p7}	Scrapie incubation	VM/Dk-*Sinc^{p7}*	F83	Dk
		MB/Dk-*Sinc^{p7}*	F51	Dk
		IM/Dk-*Sinc^{p7}*	F57	Dk
Sinc^{s7}	Scrapie incubation	VM/Dk-*Sinc^{s7}*	F15	Dk
Sl^{d-H}	Steel Dickie Harwell	129/Sv//Pas-*E^{so}*/+; *Sl^{d-H}*/+	N31	Pas
		C3H/He Pas-*E^{so}*/+; *Sl^{d-H}*/+	N32	Pas
Sl^J	Steel-Jackson	129/Sv//Ev-*Sl^J*/+		Ev
Sl^{Pas}	Steel Pasteur	C3H/He Pas-*Sl^{Pas}*/+	F56	Pas
Sp	Splotch	NSP/Y-*N*/+; *Sp*/+	F58	Y
spa	Spastic	B6CBA R.I.-+/*a*; +/*spa*	F20	Pas
Sts^n	Steroid sulfatase	C3H/An//Pas-*Sts^n/Sts^n*	F?+45	Pas
Sxr^a	Sex reversal	129/Sv//Pas-*Sxr^a*/+; *Ta*/+	N4	Pas
		C57BL/6Mcl-*Sxr^a*/+	N11	Mcl
Sxr^b	(= Sxr-prime)	C57BL/6Mcl-*Sxr^b*	N12	Mcl

Table 19.10—*cont.*

Gene symbol	Gene name	Strain and genotype	Generations inbred	Holder
T	Brachyury	BTBR/Pas-*T tf*/+ *tf*; a^t	F53	Pas
		129/Sv//Pas-T/+; B^{lt}/+	N27	Pas
		129-*T*	F?+31	Ph
		B10.T/SnY-*T*/+	N62	Y
t^0	t haplotype	*T tf*/t^0+; *a*; *b*	F26	Pas
		129/Sv/Pas-t^0	N29	Pas, Ph
t^{12}	t haplotype	*T tf*/t^{12}+; *b*	F40	Pas
T^{2J}	Brachyury Jackson 2	C57BL/6J//Pas-T^{2J} +/+ *qk*	F27	Pas
T^{hp}	Hairpin-tail	C57BL/10-T^{hp}	N4	Ph
T^{Orl}	Brachyury Orleans	BRO/Pas-T^{Orl} *tf*/+ *tf*; a^t; *b*; *p*	F25	Pas
T^Y	Brachyury Yurlovo	PT/Y-*a*; *b*; $c^{ch}p$; *d se*; *s*; T^Y/+	F50	Y
t^{p4}	t haplotype	129-t^{p4}	N16	Ph
t^{p12}	t haplotype	129-t^{p12}	N12	Ph
t^{p14}	t haplotype	129-t^{p14}	N10	Ph
t^{w1}	t haplotype	*T tf*/t^{w1}+; *a*	F?+76	Pas
t^{w12}	t haplotype	*T tf*/t^{w12}*tf*; *a*; *p*	F?+37	Pas
		129/SvPas-t^{w12}	N29	Pas, Ph
t^{w18}	t haplotype	*T tf* or + +/t^{w18}+; a^t; *b*	F?+10	Pas
		129/SvPas-t^{w18}	N10	Pas
t^{w2}	t haplotype	*T tf*/t^{w2}+; *a*; *b*; *s*	F30	Pas
		129/SvPas-t^{w2}	N30	Pas
t^{w32}	t haplotype	*T tf*/t^{w32}+; *a*	F46	Pas
		129/SvPas-t^{w32}	N29	Pas, Ph
t^{w5}	t haplotype	*T tf*/t^{w5}+; a^t	F40	Pas
		129/SvPas-t^{w5}	N28	Pas, Ph
t^{w73}	t haplotype	*T tf*/t^{w73}+; *a*; *s*	F40	Pas
t^{wPas1}	t haplotype	*T tf*/t^{wPas1}+; a^t	F29	Pas
t^{wPas2}	t haplotype	*T tf*/t^{wPas2}+; a^t	–	Pas
t^{wPas3}	t haplotype	*T tf*/t^{wPas3}+; a^t	–	Pas
tf	Tufted	BTBR/Pas-*T tf*/+ *tf*; a^t	F53	Pas
		TFH/H-*T tf*/+ *tf*	F36	H
		TF/Y-*a*; *tf*	F48	Y
Tfm	Testicular feminization	TFM/Pas-*Tfm* +/+ *Ta*;	F?+14	Pas
tk	Tail-kinks	BALB/c-*tk*	MF74	Gr
		CXBP/GrEss-*tk*	F?+15	Ess
tp^{Orl}	Taupe	CXBJ/Orl-tp^{Orl}		Orl
Ts	Tail short	TS/J//Pas-*Ts*/+	F42+10	Pas
Un^s	Undulated short tail	C57BL/6JY-a^t; Un^s/+	N23F27	Y
v	Waltzer	LT3/Pas-*a*; *ep*; *fz ln*; +/*v*	F?+33	Pas
Va	Varitint-waddler	129/Sv//Pas-*Va*/+; *Ca*/+	N22	Pas
Ve	Velvet coat	C57BL/6J//Pas-*Ve*/+	F38	Pas
		129/Sv//Pas-*Ve*/+	N16	Pas
		C57BL/6J//Pas-*Ve* +/+ *Eh*	F41	Pas
W	Dominant spotting	C57BL/KaLwRij-*W*/+	N44	Rij
		DBA/2LwMblRij-*W*/+	N22	Rij
W^{Pas2}	Dominant spotting	C57BL/6 Pas-W^{Pas2}/+	F?+10	Pas
W^e	Extensive dominant spotting	129/Sv//Ev-W^e/+	N10+	Ev
W^{Pas3}	Dominant spotting	C3H/He Pas-W^{Pas3}/+	F?+10	Pas
W^f	Dominant spotting fertile	C3H/He Pas-W^f/+	F30	Pas
W^{Han}	Dominant spotting Han	C3H/He-W^{Han}/+	MF6	Han
W^{Tox}	Dominant spotting Tox	C57BL/10Sn-W^{Tox}/+		Tox
W^{sh}	Sash dominant spotting	129/Sv//Ev-W^{sh}		Ev
W^v	Viable dominant spotting	C3H/He Pas-W^v/+	N26	Pas
		129/Sv//Pas-W^v/+	N23F6	Pas, Ev
		C57BL/KaLwRij-W^v/+	N43	Rij
		C57BL/6Orl-W^v		Orl, Y
		DBA/2LwMblRij-W^v/+	N24	Rij
W^Y	Dominant spotting-Y	C57BL/6JY-W^Y/+	M+F57	Y
		WR/Y-*a*; W^Y/+	F40	Y
wa-1	Waved-1	ABP/Pas-*a*; *b*; p^d/*p*; *se*; *bt*; *wa-1*	F34	Pas
wa-2	Waved-2	*pe*; +/*wa-2*	F?+6	Pas
wal	Waved alopecia	BALB/c-*wal*	M+F45	Y
wl	Wabbler lethal	WLHR/Le//Pas-*wl*+/+*hr*	F90	Pas
xid	X-linked immunodeficiency	CBA/N-*xid*	F36	Rij, Nimr, Ola, Nii
		NIH-*nu*; *bg*; *xid*		Lac
Xt	Extra toes	129/Sv//Pas-*Rw*/+; *Xt*/+	N35	Pas
		C57BL/6-*Xt*/+		Ibm

The following is a list of holders and producers of inbred or other genetically defined mice who have been assigned subline codes by the International Committee on Standardized Genetic Nomenclature for Mice. The Institute of Laboratory Animal Resources, Washington, DC, which maintains these codes, distributed questionnaires between July and November 1986 to obtain current addresses. Where up-to-date information could not be elicited, a notation of 'address unknown' or 'last known address' is made. The list includes codes assigned through July 1988.

A   Dr Rob ten Berg, The Netherlands Cancer Institute, Plesmanlaan 121, 1066 CX Amsterdam, The Netherlands

Aa   Dr Peter Ebbesen, The Institute of Cancer Research, Radiumstationen, DK-8000 Aarhus C, Denmark

Aai   Ms Ruth Cinaglia, Ace Animals, Inc, P.O. Box 122, Boyertown, PA 19512, USA

Abg   Dr Bruce Dudek, SUNY Albany, Department of Psychology, 1400 Washington Avenue, Albany, NY 12222, USA

Adr   Mr John Cuthbertson, Moredun Research Institute, 408, Gilmerton Road, Edinburgh EH17 7JH, Scotland, UK

Ae   Dr Hans-Ulrich Aeschbacher, NESTEC Ltd, Research Department, Case postale 88, CH-1814 La Tour de Peilz, Switzerland

Agl   Dr Joe M. Angel, University of Texas System Cancer Center, Science Park, Research Division, P.O. Box 389, Smithville, TX 78957, USA

Ags   Last known address: Division of Animal Genetics, C.S.I.R.O., P.O. Box 90, Epping, New South Wales, Australia

Al   Dr R. C. Allen, Department of Pathology, Medical University of South Carolina, Charleston, SC 29425, USA

Alt   Dr David H. Neil, Health Sciences, Laboratory Animal Services, University of Alberta, B-109 Medical Sciences Building, Edmonton, Alberta T6G 2H7, Canada

Am   Dr J. L. Ambrus, Roswell Park Memorial Institute, 666 Elm Street, Buffalo, NY 14263, USA

An   H. B. Andervont. Deceased

Anl   Dr Thomas E. Fritz, Argonne National Laboratory, Division of Biological and Medical Research, 9700 South Cass Avenue, Argonne, IL 60439, USA

Ao   Dr D. B. Amos, Department of Microbiology and Immunology, Division of Immunology, Box 3010, Duke Medical Center, Durham, NC 27710, USA

Ar   Last known address: A.R.S.A.L., Casella Postale 39, Via di Valle Caia, 00040 Pomezia (Roma), Italy

Arc   Director, Animal Resources Centre, P.O. Box l80, Willetton, Western Australia 6155

Arl   Dr Martha A. Mann, University of Texas at Arlington, Department of Psychology, Room 313, Life Science Building, Box 19528, Arlington, TX 76119, USA

As   Dr J. Tønnes Nielsen, University of Aarhus, Institute of Molecular Biology, C. F. Møllers Alle 130, DK-8000 Aarhus C, Denmark

Asl   Animal Suppliers (London) Ltd. Code retired

Au   Dr Robert Auerbach, University of Wisconsin, Department of Zoology, 1117 West Johnson Street, Madison, WI 53706, USA

Ba   Dr Yolanda Piana de Bonaparte, Bioterio y Cáncer Experimental, Instituto de Oncología 'Angel H. Roffo', Avenida San Martín 5481 (1417), Buenos Aires, Argentina

Baz   Last known address: Centro per lo Studio e la Cura dei Tumori, Busto Arsizio, Italy

Bc   Drs Muriel J. Harris and Diana M. Juriloff, University of British Columbia, Department of Medical Genetics, Vancouver, British Columbia V6T 1W5, Canada

Bcr   Last known address: Department of Cancer Studies, The Medical School, University of Birmingham, Birmingham 15, England

Bd   Dr Gene A. Bingham, Biology Division, Oak Ridge National Laboratory, Y-12 Plant, P.O. Box Y, Building 9207, Oak Ridge, TN 37831, USA

Be   R. A. Beatty. Retired

Ben   Dr Dorothea Bennett, Department of Zoology, The University of Texas at Austin, Austin, TX 78712, USA

Bg   Drs Benson E. Ginsburg and Steven Maxson, U-154, Behavioral Genetics Laboratory, Biobehavioral Sciences Graduate Degree Program, The University of Connecticut, Storrs, CT 06268, USA

Bi   J. J. Bittner. Deceased

Bii   Mr C. Gentry and Mr E. Ackerley, Basel Institute for Immunology, Grenzacherstrasse 487, CH-4058 Basel, Switzerland

Bim   Dr R. Schmidt, Martin-Luther-Universität Halle-Wittenberg, Bereich Medizin, Institut für Biologie, Universitätsplatz 7, 4020 Halle/Saale, GDR

Bir   Dr Edward H. Birkenmeier, The Jackson Laboratory, Bar Harbor, ME 04609, USA

Bk   Dr Sidney L. Beck, Department of Biology, Ball State University, Muncie, IN 47306, USA

Bkl   Mr G. C. Bantin, Bantin and Kingman Ltd, The Field Station, Grimston, Aldbrough, Hull HU11 4QE, England *and* Mr Robert Schultz, Bantin and Kingman, Inc, 3403/3421 Yale Way, Fremont, CA 94538, USA

Bkr   Helen Bunker. Retired

Bl   M. Bielschowsky. Deceased

Bln   Dr Wolfgang Arnold, Academy of Sciences of GDR, Central Institute of Cancer Research, Lindenberger Weg 80, 1115 Berlin, GDR

Bm   Dr Wesley G. Beamer, The Jackson Laboratory, Bar Harbor, ME 04609, USA

Bn   Dr S. E. Bernstein, The Jackson Laboratory, Bar Harbor, ME 04609, USA

Bnr   A. Gropp. Deceased

Bog   Last known address: Jerzy Bogajewski, The Breeding Center for Experimental Animals, ul. Poznanska 10, 62081 Przezmierowo near Poznan, Poland

Bom   P. Schjødtz Hansen, Bomholtgård Breeding and Research Centre, 10 Bomholtvej, DK-8680 Ry, Denmark

Boy   Dr E. A. Boyse, Memorial Sloan-Kettering Cancer Center, 1275 York Avenue, New York, NY 10021, USA

Br   G. M. Bonser. Deceased

Brb   Barry R. Barber, Farmitalia Carlo Erba, Via Carlo Imbonati, 24, 20159 Milan, Italy

Bre   Dr Edward J. Breyere, The American University, 4400 Massachusetts Avenue, NW, Washington, DC 20016, USA

Brh   P. L. Broadhurst. Retired

Bri   Dr Ralph L. Brinster, School of Veterinary Medicine, University of Pennsylvania, 3800 Spruce Street, Philadelphia, PA 19104, USA

Brk   Dr Jane Barker, The Jackson Laboratory, Bar Harbor, ME 04609, USA

Bsc   Paul C. Bailey. Retired

Bsp   Last known address: Bioterio Central of the University of São Paulo Medical School, Av. Dr. Arnaldo, 455, 01246 São Paulo, Brazil

Bt   N. Bateman. Retired

Bts   Last known address: Radiobiology Department, Medical College of St. Bartholomew's Hospital, London EC1M 6BQ, England

Bua   J. C. Nightingale, University of Birmingham, The Medical School, Department of Anatomy, Animal Resources, Birmingham B15 2TJ, England

Buk   Ms Patricia France, Buckshire Corporation, P.O. Box 155, Perkasie, PA 18944, USA

Bul   Dr Grahame Bulfield, A.F.R.C. Institute for Physiology and Genetics Research, Roslin, Midlothian EH25 9PS, UK

Bw   Dr Russell V. Brown, Department of Biology, Virginia Commonwealth University, Richmond, VA 23220, USA

By   Dr Donald W. Bailey, The Jackson Laboratory, Bar Harbor, ME 04609, USA

Ca   T. C. Carter. Retired

Cag   University of Cambridge, School of Agriculture. Code retired

Cam   M. E. Wallace. Retired

Can   Mr Jocelyn Goyer, Charles River Canada Inc, 188 Lasalle Street, St-Constant, Quebec J0L 1X0, Canada

Cbi   James Wallace, Chester Beatty Research Labs, Institute of Cancer Research, 237 Fulham Road, London SW3 6JB, England

Cd   Albert Claude, Deceased

Cft   Dr P. Saibaba, Experimental Animal Production Facility (Animal House), Central Food Technological Research Institute, Mysore, 570013, India

Cg   Last known address: Mrs A. Cohen, Experimental Oncology Laboratory, Radiation Therapy Department, Johannesburg General Hospital, Johannesburg, South Africa

Ch   Herman B. Chase. Retired

Chp   Dorothy R. Chapman. Retired

Ci   Ernst Caspari. Retired

Cib   Dr G. B. Puttannaiah, Hindustan Ciba-Geigy Limited, Research Centre, Near Check Naka, Aarey Rd., Goregaon East, Bombay 400 063, India

Cit   Dr Alexander S. Apt, Central Institute for Tuberculosis, Moscow 107564, USSR

Ciu   Last known address: Unidad de Investigaciónes Cancerológicas, Avenida Cuauhtemoc 240, Ciudad de México, México

Cj   Drs Neal Copeland and Nancy Jenkins, Mammalian Genetics Laboratory, NCI Frederick Cancer Research Facility, P.O. Box B, Building 539, Frederick, MD 21701, USA

Ckc   C. K. Chai. Retired

Cl   Last known address: Mrs Ruth Clayton, Institute of Animal Genetics, West Mains Road, Edinburgh EH9 3JN, Scotland, UK

Clb   Dr P. C. van Mourik, Central Laboratory of the Bloodtransfusion Service, P.O. Box 9190, 1006 AD Amsterdam, The Netherlands

Cm   Dr R. B. Cumming, Biology Division, Oak Ridge National Laboratory, P.O. Box Y, Oak Ridge, TN 37831, USA

Cn   Dr A. L. Chapman, University of Kansas Medical Center, Kansas City, KS 66103, USA

Cnb   Dr Vet. J. Vankerkom, S.C.K./C.E.N., Boeretang 200, B-2400 Mol, Belgium

Cne   Dr V. Monaco, Laboratory of Experimental Animals, ENEA Casaccia, Casella Postale 2400, 00100 Roma A.D., Italy

Co   Dr George A. Carlson, The Jackson Laboratory, Bar Harbor, ME 04609, USA

Col   Dr Robert L. Collins, The Jackson Laboratory, Bar Harbor, ME 04609, USA

Cox   Laboratory Supply Co, Inc. Code retired

Cp   Dr A. B. Chapman, University of Wisconsin, College of Agricultural and Life Sciences, Department of Meat and Animal Science, 1675 Observatory Drive, Madison, WI 53706-1284, USA

Cpb   Ir. Ed. J. Gubbels, Harlan Olac CPB, P.O. Box 167, 3700 AD Zeist, The Netherlands

Cr   Dr Joseph Mayo and Mr Clarence Reeder, Biological Testing Branch, Frederick Cancer Research Facility, Building 428, Room 63, Frederick, MD 21701, USA

Crc   Dr S. Rastan, Division of Comparative Medicine, MRC Clinical Research Centre, Watford Road, Harrow, Middlesex HA1 3UJ, England

Crgl   Mrs Kay Yoshiura, Cancer Research Laboratory, 3510 Life Sciences Building, University of California, Berkeley, CA 94720, USA

Cri   Drs S. N. Naik and M. G. Deo, Cancer Research Institute, Tata Memorial Centre, Parel, Bombay-400 012, India

Crj   Dr Yukio Takei, Charles River Japan, Inc, 795, Shimofurusawa, Atsugishi, Kanagawa 243-02, Japan

Crl   Mr. Wayne Bolen, The Charles River Breeding Laboratories, Inc, 251 Ballardvale Street, Wilmington, MA 01887, USA, *and* Dr Yves Pasternak, Charles River France, B.P. 29, 76410 Cleon, St. Aubin-les-Elbeuf, France

Cs   R. T. Charles. Address unknown

Csk   Dr Jiro Adachi, Research Laboratories, Chugai Pharmaceutical Co, Ltd, 41-8 Takada 3-Chome, Toshima-ku, Tokyo 171, Japan

Csr   Dr M. J. Marshall, Charles Salt Research Centre, The Robert Jones and Agnes Hunt Orthopaedic Hospital, Oswestry, Shropshire SY10 7AG, England

Ct   Dr Bruce M. Cattanach, MRC Radiobiology Unit, Chilton, Didcot, Oxon OX11 0RD, England

Cu   Dr R. L. Curtis, Department of Anatomy and Cellular Biology, Medical College of Wisconsin, 8701 Watertown Plank Road, Milwaukee, WI 53226, USA

Cub   Last known address: Charles University, Faculty of General Medicine, Department of Biology, Albertov 4, Prague 2, Czechoslovakia

Cum   Dr James C. Kile, Jr and Mr Kenneth E. Kile, Sr, Cumberland View Farms, Route 3, Box 299, Clinton, TN 37716, USA

Cv   Dr Verne M. Chapman, Roswell Park Memorial Institute, Department of Molecular and Cellular Biology, 666 Elm Street, Buffalo, NY 14263, USA

Cvl   Dr Richard H. Civil, Department of Neurology, University Hospital, Cleveland, OH 44106, USA

Cwr   Mr Robert C. Voigt, Case Western Reserve University, Animal Resource Center, 2119 Abington Road, Cleveland, OH 44106, USA

Cy   Marianna Cherry. Retired

Da   Last known address: Dr D. A. Davies, Searle Research Laboratories, High Wycombe, Buckinghamshire HP12 4HL, England

Db   Dr David E. Briles, University of Alabama at Birmingham, 224 Tumor Institute, University Station, Birmingham, AL 35294, USA

De   Margaret K. Deringer. Retired

Dem   Dr Peter Démant, The Netherlands Cancer Institute, Plesmanlaan 121, 1066 CX Amsterdam, The Netherlands

Dg   Dr Charles P. Dagg, Department of Biology, University of Alabama at Birmingham, Birmingham, AL 35294, USA

Di   M. M. Dickie. Deceased

Dk   Dr H. Fraser, AFRC and MRC Neuropathogenesis Unit, Ogston Building, West Mains Road, Edinburgh EH9 3JF, Scotland, UK

Dm   L. Dmochowski. Deceased

Dn   Dr Muriel T. Davisson, The Jackson Laboratory, Bar Harbor, ME 04609, USA

Do   Drs Charles W. Mays and Glenn M. Taylor, University of Utah, Radiobiology Division, Building 351, Salt Lake City, UT 84112, USA

Dp   Dr Giuseppe Della Porta, Division of Experimental Oncology A, Istituto Nazionale Tumori, Via G. Venezian 1, 20133 Milano, Italy

Dt   Dr Donald P. Doolittle, Department of Animal Sciences, Purdue University, West Lafayette, IN 47907, USA

Du   Wilhelmina F. Dunning. Deceased

Dub   Mr Wayne Farmer, Dominion Laboratories, Inc, P.O. Box 459, Dublin, VA 24084, USA

Dv   Dr Chella S. David, Department of Immunology, Mayo Clinic, Rochester, MN 55905, USA

Ed   Last known address: Centre for Laboratory Animals, Edinburgh University, The Bush, Milton Bridge, Penicuik, Midlothian EH26 0PL, Scotland, UK

Eg   Dr Igor K. Egorov, The Jackson Laboratory, Bar Harbor, ME 04609, USA

Eh   J. Engelbreth-Holm. Deceased

Ei   Dr Eva M. Eicher, The Jackson Laboratory, Bar Harbor, ME 04609, USA

El   Dr U. H. Ehling, Gesellschaft für Strahlen- und Umweltforschung, 8042 Neuherberg bei München, FRG

Ep   Last known address: Dr. Carlos Epper, Atomic Energy National Commission, Radiobiology Laboratory, Avenida del Libertador Gen. San Martín 8250, Buenos Aires, Argentina

Epg   Dr Janan T. Eppig, The Jackson Laboratory, Bar Harbor, ME 04609

Er   Dr Robert P. Erickson, Department of Human Genetics, 4708 Medical Sciences II, Box 015, University of Michigan, Ann Arbor, MI 48109, USA

Eru   Experimental Radiopathology Unit. Code retired

Ess   Professor J. G. M. Shire, Biology Department, University of Essex, Colchester CO4 3SQ, England

Eup   Mrs Jean Hunter, University of Edinburgh Medical School, Department of Pharmacology, 1 George Square, Edinburgh EH8 9JZ, Scotland, UK

Ev   Dr Martin Evans, Department of Genetics, University of Cambridge, Downing Street, Cambridge CB2 3EH, England

Fa   D. S. Falconer. Retired

Fce   Barry R. Barber, Farmitalia Carlo Erba, Via Carlo Imbonati, 24, 20159 Milan, Italy

Fe   Elizabeth Fekete. Deceased

Ffm   Dr J. Svejcar, Universität Frankfurt, Institut für Humangenetik, Paul-Ehrlich-Strasse 41, D-6000 Frankfurt/Main, FRG

Fg   Frank H. J. Figge. Deceased

Fhc   Dr Barbara Johnston, Fred Hutchinson Cancer Research Center, 1124 Columbia Street, Seattle, WA 98104, USA

Fib   Dr Jørgen Kieler, The Fibiger Institute, Ndr. Frihavnsgade 70, DK-2100 Copenhagen Ø, Denmark

Fis   Last known address: Fisons Agrochemical Division, Chesterford Park Research Station, Nr. Saffron Walden, Essex, England

Fla   Dr Lorraine Flaherty, Wadsworth Center for Laboratories and Research, New York State Department of Health, Empire State Plaza, Albany, NY 12237, USA

Fn   Paul F. Fenton. Deceased

Fnn   Dr Richard H. Finnell, Department of Veterinary and Comparative Anatomy, Pharmacology, and Physiology (VCAPP), College of Veterinary Medicine, Washington State University, Pullman, WA 99164-6520, USA

Fo   P. Forsthoefel. Retired

Fp   Last known address: Istituto di Patologia Generale della Università di Ferrara, Via Fossato di Mortara, 64/a, 44100 Ferrara, Italy

Fr   F. Clarke Fraser. Retired

Fre   Dr Jeffrey A. Frelinger, Department of Microbiology and Immunology, University of North Carolina, Chapel Hill, NC 27514, USA

Fs	Morris Foster. Deceased
Fu	Jacob Furth. Deceased
Fx	Dr Thomas O. Fox, E. K. Shriver Center for Mental Retardation, 200 Trapelo Road, Waltham, MA 02254, USA
G	Last known address: Glaxo Laboratories Ltd, Greenford, Middlesex, England
Ga	Dr Allen H. Gates, University of Rochester Medical Center, Environmental Health Sciences Center (Toxicology), P.O. Box 641, Rochester, NY 14642, USA
Gb	Dr Donald B. Galbraith, Trinity College, Biology Department, Hartford, CT 06106, USA
Gc	Dr Gerald F. Chernoff, Washington State University, Department of Veterinary and Comparative Anatomy, Pharmacology, and Physiology (VCAPP), College of Veterinary Medicine, Pullman, WA 99164-6520, USA
Gd	Dr C. M. Goodall, New Zealand Institute for Cancer Research, P. O. Box 5190, Dunedin, New Zealand
Gen	Last known address: Università di Genova, Istituto di Farmacologia, Viale Benedetto XV, N.2 Genova, Italy
Gf	Dr Anna Goldfeder, Department of Biology, New York University, 952 Brown Building, Room 754, Washington Square, New York, NY 10003, USA
Gh	Dr Douglas Grahn, Argonne National Laboratory, Division of Biological and Medical Research, 9700 South Cass Avenue, Argonne, IL 60439, USA
Gib	Guiseppe Bianchi, Farmitalia Carlo Erba, Via Carlo Imbonati, 24, 20159 Milan, Italy
Gif	Code retired. Superseded by Orl
Gl	A. Glücksmann. Retired
Gli	Dr Halina Cios, Institute of Oncology, Department of Tumor Biology, 44-100 Gliwice, Poland
Glw	Dr Salome G. Waelsch, Department of Genetics, Albert Einstein College of Medicine, Eastchester Road and Morris Park, Bronx, NY 10461, USA
Gm	J. P. W. Gilman. Retired
Gn	E. L. and M. C. Green. Retired
Go	Peter Gorer. Deceased
Gr	Dr G. M. Truslove, Galton Laboratory, University College London, Wolfson House, 4 Stephenson Way, London NW1 2HE, England
Grf	Dr Ralph J. Graff, Cochran VA Hospital, 915 North Grand, St. Louis, MO 63106, USA
Gro	Dr G. A. van Oortmerssen, Center for Biology, State University of Groningen, Kerklaan 30, 9751 AE, HAREN (Gr), The Netherlands
Gs	Dr Ludwik Gross and Yolande Dreyfuss, Cancer Research Unit, VA Medical Center, 130 West Kingsbridge Road, Bronx, NY 10468, USA
Gw	John W. Gowen. Deceased
Gy	Dr J. F. Fowler, Gray Laboratory, Mount Vernon Hospital, Northwood, Middlesex HA6 2RN, England
H	Dr Mary F. Lyon, MRC Radiobiology Unit, Harwell, Didcot, Oxon, OX11 0RD, England
Ha	T. S. Hauschka. Retired
Hab	Dr Raúl Castillo Menéndez, Department of Experimental Animals, National Institute of Oncology and Radiobiology, 29 y E, Vedado, La Habana-4, Cuba
Han	Dr H. J. Hedrich, Zentralinstitut für Versuchstierzucht, Hermann-Ehlers-Allee 57, D-3000 Hannover 91, FRG
Hb	Dr H. Everett Hrubant, California State University, Department of Biology, Long Beach, CA 90804, USA
Hbl	Dr Gordon J. Eaton, Honeybrook Laboratories, 1125 West Lincoln Highway, Coatesville, PA 19320, USA
He	Walter E. Heston. Retired
Hf	Dr Harold A. Hoffman, Animal Genetic Systems, Inc, 628-G Lofstrand Lane, Rockville, MD 20850, USA
Hg	Paula Hertwig. Retired
Hi	Dr Jo Hilgers, The Netherlands Cancer Institute, Plesmanlaan 121, 1066 CX Amsterdam, The Netherlands
Hir	Dr T. Hirobe, Department of Biology, Faculty of Education, Iwate University, Ueda 3 Chome, Morioka 020, Japan
Hl	Professor Dr Lieselotte Herberg, Diabetes Forschungsinstitut an der Universität Düsseldorf, Auf'm Hennekamp 65, D-4000 Düsseldorf 1, FRG
Hla	Mr Thomas Miedel, Hilltop Lab Animals, Inc, P.O. Box 183, Scottdale, PA 15683, USA
Hlr	Dr Felix G. Garcia, Hoffman-La Roche, Inc, 340 Kingsland Street, Nutley, NJ 07110, USA
Hm	Dr Herbert Morse III, Building 7, Room 304, National Institutes of Health, Bethesda, MD 20892, USA
Hn	Dr F. K. Hoornbeek, University of New Hampshire, Zoology Department, Spaulding Building, Durham, NH 03824, USA

Hok   Professor Dr Motomichi Sasaki, Center for Experimental Plants and Animals, Chromosome Research Unit, Faculty of Science, Hokkaido University, North 10, West 8, Sapporo 060, Japan

Hop   Dr Peter C. Hoppe, The Jackson Laboratory, Bar Harbor, ME 04609, USA

How   Alma Howard. Deceased

Hr   Dr G. Hoecker, Departamento de Biologia Celular y Genetica, Facultad de Medicina, Universidad de Chile, Casilla 70061, Correo 7, Santiago, Chile

Hsd   Mr Jack McGinley, Harlan Sprague Dawley, Inc, P.O. Box 29176, Indianapolis, IN 46229, USA

Ht   H. B. Hewitt. Retired

Hu   Katharine P. Hummel. Retired

Huc   Dr Eliezer Sekeles, The Animal House, Hebrew University, Hadassah Medical School, P.O. Box 1172, Jerusalem 91010, Israel

Hul   University of Hull. Code retired

Hz   Drs Leonard A. Herzenberg and F. Timothy Gadus, Department of Genetics, Stanford University, Stanford, CA 94305, USA

Ibg   Mr Eugene Thomas, Institute for Behavioral Genetics, Campus Box 447, University of Colorado, Boulder, CO 80309, USA

Ibm   Professor W. Rossbach, Institute for Bio-Medical Research Ltd, CH-4414 Füllinsdorf, Switzerland

Ibr   Mr Chaim Amani, Israel Institute for Biological Research, P.O. Box 19, Ness-Ziona, Israel

Ico   Dr Alain Perrot, Iffa-Credo, Saint-Germain-sur-l'Arbresle, 69210 l'Arbresle, France

Icr   Dr Timothy Poole, Institute for Cancer Research, 7701 Burholme Avenue, Fox Chase, Philadelphia, PA 19111, USA

Icrc Code retired. Superseded by Cri

Icrf Mr J. D. Menzies, Clare Hall Laboratories, Imperial Cancer Research Fund, Blanche Lane, South Mimms, Herts EN6 3LD, England

Ig   Last known address: Dr Rigoberto Iglesias, Instituto de Medicina Experimental, Avenida Irarrázaval 849, Casilla 3401, Santiago, Chile

Imr   Ms Jane A. Holben, Coriell Institute for Medical Research, Copewood Street, Camden, NJ 08103, USA

Ino   Professor Dr Takayoshi Ino, Department of Animal Breeding, Faculty of Agriculture, Okayama University, Naka 1-1-1, Tsushima, Okayama 700, Japan

Irms Dr G. Mahouy, Institut de Recherches sur les Maladies du Sang, Laboratoire d'Expérimentation Animale, Hôpital St. Louis, 2 place du Dr. Fournier, 75010 Paris, France

Irr   Dr T. C. Anand Kumar, Institute for Research in Reproduction, Jehangir Merwanji Street, Parel, Bombay-400 012, India

Ivv   Dr Werner Wilk, Institut für Versuchstierkunde und Versuchstierkrankheiten, WE11 der Freien Universität Berlin, Krahnerstrasse 6, D-1000 Berlin 45, FRG

J   John L. Dorey, The Jackson Laboratory, Bar Harbor, ME 04609, USA

Jac   Dr Barbara Jacobs. Address unknown

Jah   Dr Nobuo Goto, National Institute of Animal Health, 1-1, Kannondai 3-chome, Yatabe-machi, Tsukuba-gun, Ibaraki-ken, 305, Japan

Jb   Dr Jan Bruell, Department of Psychology, University of Texas, Austin, TX 78712, USA

Jcl   Mr Toshio Kamon, CLEA JAPAN, Inc, No. 2 Inari Building 20-14, Aobadai-2, Meguro-ku, Tokyo, Japan

Jd   Last known address: A. Jurand, University of Edinburgh, Department of Genetics, West Mains Road, Edinburgh 9, Scotland, UK

Jem   Code retired. Superseded by Huc

Jena Professor Dr H. Heinecke, Zentralinstitut für Mikrobiologie und experimentelle Therapie der Akademie der Wissenschaften der DDR, Beutenbergstrasse 11, DDR-6900 Jena, GDR

Jic   Central Institute for Experimental Animals, 1430 Nogawa, Miyamae–ku, Kawasaki 213, Japan

Jko   Dr Daizo Ushiba, Keio University School of Medicine, Department of Microbiology, Shinano-machi, Shinjuku-ku, Tokyo, Japan

Jms   Dr Akio Matsuzawa, Laboratory Animal Research Center, Institute of Medical Science, University of Tokyo, 4-6-1 Shirokanedai, Minato-ku, Tokyo 108, Japan

Js   Dr Judy M. Strum, University of Maryland School of Medicine, Department of Anatomy, John Eager Howard Hall, 655 West Baltimore Street, Baltimore, MD 21201, USA

Jv   Dr John L. VandeBerg, Southwest Foundation for Biomedical Research, Genetics Department, P.O. Box 28147, San Antonio, TX 78284, USA

Jwg   Dr Jon W. Gordon, Department of Obstetrics and Gynecology, Mount Sinai School of Medicine, 1 Gustave L. Levy Place, New York, NY 10029, USA

Ka   H. S. Kaplan. Deceased

Kam   Kathryn A. Mason, Department of Radiation Oncology, School of Medicine, University of California, Los Angeles, CA 90024, USA

Kb   Berenice Kindred. Deceased

Kch   Dr Walter Kocher, Institut für Toxikologie und Embryonalpharmakologie, Freie Universitat Berlin, Garystrasse 5, D-1000 Berlin 33, FRG

Kd   James F. Kidwell. Retired

Kfl   Dr Kirsten Fischer-Lindahl, Howard Hughes Medical Institute, University of Texas Southwest Medical Center, 5323 Harry Hines Boulevard, Y5-310, Dallas, TX 75235-9050, USA

Kfm   H. Madörin, Kleintierfarm Madörin AG, Wölferstrasse 18, CH-4414 Füllinsdorf, Switzerland

Kga   Last known address: Kagoshima University, Laboratory of Animal Breeding, Faculty of Agriculture, Kagoshima, Japan

Kh   Henry I. Kohn. Retired

Ki   Professor Dr Masaki Ohmori, Kirschbaum Memorial Mouse Colony, Department of Pathology, Kagawa Medical School, Ikenobe, Miki, Kagawa 761-07, Japan

Kl   Drs G. and E. Klein, Department of Tumor Biology, Karolinska Institutet, Box 60400, S-104 01 Stockholm, Sweden

Klj   Professor Dr Jan Klein, Max-Planck-Institut für Biologie, Department of Immunogenetics, Corrensstrasse 42, 7400 Tubingen, FRG

Km   Drs Robert F. Kallman and J. Martin Brown, Department of Therapeutic Radiology, Stanford University Medical Center, Stanford, CA 94305, USA

Kn   Fr. Kröning. Deceased

Ko   N. Kobozieff. Retired

Kon   Dr Franz Gruber, Universität Konstanz, Tierforschungsanlage, Postfach 5560, D-7750 Konstanz, FRG

Ks   N. Kaliss. Retired

Ksb   Dr Kenneth S. Brown, Laboratory of Developmental Biology and Anomalies (LDBA), National Institute of Dental Research, Building 30, Room 412, NIH, Bethesda, MD 20892, USA

Kt   Dr Harold Kalter, Children's Hospital Research Foundation, Elland Avenue and Bethesda, Cincinnati, OH 45229, USA

Kuj   Dr J. Hayakawa, Institute for Experimental Animals, Kanazawa University School of Medicine, 13-1 Takara-machi, Kanazawa 920, Japan

Kun   Mr F. G. J. Janssen, Central Animal Laboratory, Catholic University, Geert Grooteplein N 29, P.O. 9101, 6500 HB Nijmegen, The Netherlands

Kw   Dr Halina Krzanowska, Jagiellonian University, Department of Genetics and Evolution, Institute of Zoology, Karasia 6, 30-060 Kraków, Poland

Kx   W. Eugene Knox. Deceased

L   Dr Terence T. Yen, Lilly Research Laboratories, Lilly Corporate Center, Indianapolis, IN 46285, USA

Lac   Dr David Whittingham, Experimental Embryology and Teratology Unit, MRC Laboratories, Woodmansterne Road, Carshalton, Surrey SM5 4EF, England

Lati Dr Gyula Györffy, LATI Közös Vállalat (Joint Enterprise), Táncsics M. Road, 2100.Gödöllő, Hungary

Lav   Dr Matthew M. LaVail, Department of Anatomy, University of California, San Francisco, CA 94143, USA

Lay   Dr W. M. Layton, Department of Anatomy, Dartmouth Medical School, Hanover, NH 03755, USA

Lb   Dr Miriam Lieberman, Cancer Biology Research Laboratory, Department of Therapeutic Radiology, Stanford University Medical Center, Stanford, CA 94305, USA

Ld   Dr J. Russell Lindsey, Department of Comparative Medicine, Schools of Medicine and Dentistry, University of Alabama at Birmingham, University Station, Birmingham, AL 35294, USA

Le   Mrs Priscilla W. Lane, The Jackson Laboratory, Bar Harbor, ME 04609, USA

Lee   Manager, Laboratory Animal Services, Worsley Building, University of Leeds, Clarendon Way, Leeds LS2 9JT, England

Lei   Last known address: Research Laboratories, Department of Rheumatology, University Hospital, Leiden, The Netherlands

Lhm   Professor H. Festenstein, Department of Immunology, London Hospital Medical College, 56–76 Ashfield Street, Whitechapel, London E1 2AD, England

Lhr   London Hospital Research Laboratories. See Lhm

Li   Clarence C. Little. Deceased

Lid   Last known address: Liverpool University, School of Dental Surgery, Boundary Place, Liverpool 7, England

Lil   Dr Frank Lilly, Department of Genetics, Albert Einstein College of Medicine, Bronx, NY 10461, USA

Ll   Last known address: Laboratory Animals Centre, Polish Academy of Sciences, Lomna Las k. Warsaw, Poland

Lm   Dr M. Lynn Lamoreux, Texas A&M University, Department of Biology, College Station, TX 77843-3258, USA

Ln   Dr J. B. Lyon, Department of Biochemistry, Emory University, Atlanta, GA 30322, USA

Lo   Dr L. M. Holland, Los Alamos National Laboratory, P.O. Box 1663, Los Alamos, NM 87545, USA

Lp   Dr Wayne S. Lapp, Department of Physiology, McGill University, Montreal, Quebec H3A 1B1, Canada

Ls   Dr Edwin P. Les, The Jackson Laboratory, Bar Harbor, ME 04609, USA

Lsh   Last known address: London School of Hygiene and Tropical Medicine, 395 Hatfield Road, St. Albans, Hertfordshire, England

Lt   Dr Edward H. Leiter, The Jackson Laboratory, Bar Harbor, ME 04609, USA

Lub   Dr Heinz Winking, Institut für Biologie, Medizinische Hochschule Lübeck, Ratzeburger Allee 160, D-2400 Lübeck, FRG

Lw   Dr Lloyd W. Law, National Cancer Institute, Laboratory of Cell Biology, Building 37, Room 1B22, National Institutes of Health, Bethesda, MD 20892, USA

Lws   Dr Susan E. Lewis, Life Science and Toxicology Division, Research Triangle Institute, P.O. Box 12194, Research Triangle Park, NC 27709, USA

Ly   Clara Lynch. Retired

M   Dr David D. Myers, Memorial Sloan-Kettering Cancer Center, 1275 York Avenue, New York, NY 10021, USA

Ma   Dr Edwin A. Mirand, Roswell Park Memorial Institute, 666 Elm Street, Buffalo, NY 14263, USA

Mac   Dr M. F. Tarttelin, Small Animal Production Unit, Massey University, Private Bag, Palmerston North, New Zealand

Mai   Microbiological Associates and M. A. Bioproducts. Code retired

Man   Dr Stanley J. Mann, Temple University Health Sciences Center, Central Animal Facilities, 3400 North Broad Street, Philadelphia, PA 19140, USA

Mar   Last known address: Royal Marsden Hospital, Radiotherapy Research Unit, Downs Road, Sutton, Surrey, England

Max   Dr H. Mossmann, Max-Planck-Institut für Immunbiologie, Stübeweg 51, D-7800 Freiburg, FRG

Mb   Dr Martin Hornberger, Max-Planck-Institut für Biochemie, D-8033 Martinsried, FRG

Mcd   Dr Hugh O. McDevitt, Department of Medical Microbiology, Fairchild Building, Stanford University School of Medicine, Stanford, CA 94305, USA

Mcl   Dr Anne McLaren, MRC Mammalian Development Unit, Wolfson House, University College London, 4 Stephenson Way, London NW1 2HE, England

Me   Sir Peter B. Medawar. Deceased

Mei   Dr Hiroshi Nagasawa, Experimental Animal Research Laboratory, Meiji University, Tama-ku, Kawasaki, Kanagawa 214, Japan

Mel   Dr Roger Melvold, Department of Microbiology and, Immunology, Northwestern University School of Medicine, 303 East Chicago Avenue, Chicago, IL 60611, USA

Mg   Walter Morgan. Retired

Mib   Dr B. V. Konyukhov, Institute of General Genetics, USSR Academy of Sciences, Gubkin St., 3, Moscow B-333, USSR 117809 GSP-1

Mk   Code retired. Superseded by Hok

Ml   Dr James R. Miller, Takeda Chemical Industries, Ltd, Central Research Division, 2-17-85 Jusohonmachi, Yodogawa-ku, Osaka 532, Japan

Mo   R. H. Mole. Retired

Mob   Dr Larry E. Mobraaten, The Jackson Laboratory, Bar Harbor, ME 04609, USA

Mp   Edwin D. Murphy. Deceased

Mq   Professor Jacqueline Mouriquand, Faculté de Médécine, Université Scientifique et Médicale de Grenoble, Domaine de la Merci, 38700, La Tronche, France

Mr   E. W. Miller. Address unknown

Mrk   Dr Louis DeTolla, Merck Sharp & Dohme Research Laboratories, RY80, A-31, P.O. Box 2000, Rahway, NJ 07065, USA

Mrp   Dr Donal B. Murphy, Department of Pathology, Yale University School of Medicine, 112 Lauder Hall, P.O. Box 3333, New Haven, CT 06510-8023, USA

Ms   Dr Kazuo Moriwaki, Genetic Stock Research Center, National Institute of Genetics, Yata 1, 111, Mishima, Shizuoka-ken 411, Japan

Mtl   Dr Daphne G. Trasler, McGill University, Department of Biology, 1205 Avenue Docteur Penfield, Montreal, Quebec H3A 1B1, Canada

Mts   Marc H. Tayman and Henry H. Matsunaga, MTS Laboratories, 7396 Trade Street, San Diego, CA 92121, USA

Mu   Irving Mauer. Address unknown

Mv   N. N. Medvedev. Retired

My   W. S. Murray. Deceased

Mza   Last known address: Dr J. Echave Llanos, Instituto de Patología General y Experimental, Facultad de Ciencias Medicas, Universidad Nacional de Cuyo, Mendoza, Argentina

N   Dr Carl T. Hansen, Veterinary Resources Branch, Division of Research Services, National Institutes of Health, Building 14G, Room 101, Bethesda, MD 20892, USA

Na   Dr Joseph H. Nadeau, The Jackson Laboratory, Bar Harbor, ME 04609, USA

Nar   Dr Yoshihiko Tsubura, Second Department of Pathology, Nara Medical University, 840 Shijo-cho, Kashihara-city, Nara-ken 634, Japan

Nav   Director, Laboratory Animal Sciences Branch, Naval Medical Research Institute, Building 21, Bethesda, MD 20814-5055, USA

Ncr   Last Known Address: Cancer Research Campaign, University of Nottingham, Nottingham, England

Nctr   Dr William F. McCallum, Division of Animal Husbandry, National Center for Toxicological Research, Jefferson, AR 72079, USA

Nem   Dr Sen-ichi Oda, Nagoya University, Research Institute of Environmental Medicine, Furo-cho, Chikusa-ku, Nagoya 464, Japan

Nga   Professor Dr Takeshi Tomita, School of Agriculture, Nagoya University, Nagoya 464, Japan

Ni   Dr Robertha van Nie, The Netherlands Cancer Institute, Plesmanlaan 121, 1066 CX Amsterdam, The Netherlands

Nia   Dr DeWitt G. Hazzard, National Institute on Aging, Building 31, Room 5C19, National Institutes of Health, Bethesda, MD 20892, USA

Nii   Dr Rajesh K. Anand, National Institute of Immunology Campus, JNU Complex, Saheed Jeet Singh Marg, New Delhi 10067, India

Nimr   Dr C. M. Hetherington, National Institute for Medical Research, Division of Biological Services, Mill Hill, London NW7 1AA, England

Nin   Last known address: National Institute of Nutrition, Indian Council of Medical Research, Jamei-Osmania, Hyderabad-7, India

Nip   Last known address: Nippon Rat Co, Ltd, 608-3, Negishi, Urawa, Saitama, Japan

Nmg   Dr J. H. F. van Abeelen, Department of Zoology, University of Nijmegen, Toernooiveld, 6525 ED Nijmegen, The Netherlands

No   D. J. Nolte. Deceased

Nrs   Head, Section of Laboratory Animals and Plants, National Institute of Radiological Sciences, 4-9-1, Anagawa, Chiba, 260 Japan

Ns   Dr Muriel Nesbitt, Department of Biology B-022, University of California at San Diego, La Jolla, CA 92093, USA

Nvs   Nevis Biological Station. Code retired

Nya   Dr Ronald F. Gruhn, Wadsworth Center for Laboratories and Research, New York State Department of Health, Empire State Plaza, Albany, NY 12201, USA

O   Professor Kristen Arnesen, University of Oslo, Department of Pathology, Ulleval Hospital, 0407 Oslo 4, Norway

Oci   Mr. Sydney Raper, Ontario Cancer Institute, University of Toronto, 500 Sherbourne Street, Toronto, Ontario M4X 1K9, Canada

Odn   Per Svendsen, Biomedical Laboratory, Odense University, Campusvej 55, DK-5230 Odense M, Denmark

Ok   Dr Yuzuru Kurabayashi, Animal Center for Medical Research, Okayama University Medical School, 2-5-1, Shikata-cho, Okayama-shi, Okayama, Japan

Ola   Harlan Olac Ltd, Shaw's Farm, Blackthorn, Bicester, Oxon OX6 0TP, England

Onl   Oncological Institute Bucharest, Bd.1 Mai II, P3 5916 Bucharest 12, Romania

Orl   Roland Motta, Centre de Sélection et d'Élevage d'Animaux de Laboratoire, CNRS, 3 B Rue de la Ferollerie, 45071 Orleans Cedex 2, France

Orr   Dr Nathaniel Revis, Oak Ridge Research Institute, 113 Union Valley Road, Oak Ridge, TN 37830, USA

Ort   Last known address: Royal National Orthopaedic Hospital, Brockley Hill, Stanmore, Middlesex HA7 4LP, England

Os   Last known address: Osaka University, Institute for Cancer Research, Dojima-Hamadori, Fukushima-ku, Osaka, Japan

Osb   Last known address: Osaka University, Department of Experimental Chemotherapy, Research Institute for Microbial Diseases, Yamadakami, Suita, Osaka-ku, Japan

Ott	Dr J. Nagai, Animal Research Centre, Agriculture Canada, Ottawa, Ontario K1A 0C6, Canada
Ou	Dr Henry C. Outzen, Institute for Medical Research, 2260 Clove Drive, San Jose, CA 95128, USA
Ove	Dr Paul Overbeek, Department of Cell Biology, Howard Hughes Medical Institute, Baylor College of Medicine, Houston, TX 77030, USA
Pa	Dr Virginia Papaioannou, Tufts University School of Medicine, Department of Pathology, 136 Harrison Avenue, Boston, MA 02111, USA
Pap	Stavros Kottaridis, Papanicolaou Research Center of Oncology and Experimental Surgery, Hellenic Anticancer Institute, 171, Alexandras Av., 115 22 Athens, Greece
Pas	Dr Jean-Louis Guénet, Institut Pasteur, Unité de Génétique des Mammiféres, Départment d'Immunologie, 25 rue du Docteur Roux, 75015 Paris, France
Pat	Dr C. S. Potten, Paterson Institute for Cancer Research, Christie Hospital, Manchester M20 9BX, England
Pbi	Dr Rui-chang Xing, National Institute for the Control of Pharmaceutical and Biological Products, Ministry of Health, Temple of Heaven, Beijing, Peoples Republic of China
Pe	P. R. Peacock. Retired
Per	Professor Emilio Bucciarelli, Divisione di Ricerche sul Cancro, Università degli Studi Policlinico, 06100 Perugia, Italy
Pfd	Dr ir. R. Leyten, Proefdierencentrum Universiteit Leuven, de Croylaan 34, B-3030, Heverlee (Leuven), Belgium
Ph	Drs Michael Boubelik and J. Forejt, Institute of Molecular Genetics, Czechoslovak Academy of Sciences, 14220 Praha 4, Videňská 1083, Czechoslovakia
Pha	Last known address: Institute for Experimental Pharmacology, Slovak Academy of Sciences, Dobrá Voda pri Trnave, Czechoslovakia
Pi	H. I. Pilgrim. Retired
Pin	Dr Lawrence H. Pinto, Department of Neurobiology and Physiology, Hogan Hall, 2153 Sheridan Road, Northwestern University, Evanston, IL 60201, USA
Pm	Dr Paul C. Montgomery, Wayne State University, School of Medicine, 7374 Scott Hall, Department of Immunology and Microbiology, 540 East Canfield Avenue, Detroit, MI 48201, USA
Po	Professor Dr Helen E. Pogosianz, Institute of Carcinogenesis, All-Union Cancer Research Center, Academy of Medical Sciences of the USSR, Kashirskoje sh.24, Moscow 115478, USSR
Pp	Drs Raymond and Diana Popp, Oak Ridge National Laboratory, Biology Division, P.O. Box Y, Oak Ridge, TN 37831, USA
Pr	Dr Louis J. Pierro, The University of Connecticut, Storrs Agricultural Experiment Station, 1376 Storrs Road, U-10, Room 215, Storrs, CT 06268, USA
Ps	Dr Donald D. Porteous. Address unknown
Pt	Dr Michael Potter, Laboratory of Genetics, National Cancer Institute, Building 37, Room 2B03, National Institutes of Health, Bethesda, MD 20892, USA
Pu	B. D. Pullinger. Retired
Ra	Dr Harold Rauch, Zoology Department, University of Massachusetts, Amherst, MA 01003, USA
Rap	Last known address: Breeding Farm Rappolova, Laboratory of Inbred Animals, Academy of Medical Science, Solyanka 14, Moscow, USSR
Rb	R. G. Busnel. Retired
Rbt	Dr Roy Riblet, Medical Biology Institute, 11077 North Torrey Pines Road, LaJolla, CA 92037, USA
Rd	G. Rudali. Deceased
Re	Elizabeth S. Russell. Retired
Rg	Last known address: University of Reading, Department of Physiology and Biochemistry, Reading, Berkshire, England
Rho	Last known address: Rhone-Poulenc Santé, Department of Biology, Research Centre, B.P. 14, F94407 Vitry sur Seine Cedex, France
Rib	Mr Kris Dyszynski, RIBI Immunochem Research, Inc, P.O. Box 1409, Hamilton, MT 59840, USA
Rij	Dr D. W. van Bekkum, REP Institutes TNO, P.O. Box 5815, NL-2280 HV Rijswijk, The Netherlands
Rj	Mrs Baurier-Margraim, Centre D'Élevage R. Janvier, Animaux de Laboratoire, Route des Chênes-Secs, Le Genest-St-Isle 53940, France

Rk   Dr Thomas H. Roderick, The Jackson Laboratory, Bar Harbor, ME 04609, USA

Rku   Dr Dennis M. Stark, Laboratory Animal Research Center, The Rockefeller University, 1230 York Avenue, Box 2, New York, NY 10021, USA

Rl   Drs W. L. and L. B. Russell, Mammalian Genetics Section, Biology Division, Oak Ridge National Laboratory, P.O. Box Y, Oak Ridge, TN 37831, USA

Rma   Last known address: Istituto di Anatomia Comparata, Università di Roma, Via Alfonso Borelli, 50, Rome, Italy

Rml   Mr Bryan Hestekin, Rocky Mountain Laboratories, Hamilton, MT 59840, USA

Rn   Dr Eugene M. Rinchik, Oak Ridge National Laboratory, Biology Division, P.O. Box Y, Oak Ridge, TN 37831, USA

Ro   U. Roth. Deceased

Ros   Ms Barbara A. Holdridge, West Seneca Laboratory, Health Research, Inc, Roswell Park Memorial Institute, 2863 Clinton Street, West Seneca, NY 14224, USA

Rp   Dr Rosemary W. Elliott, Department of Molecular Biology, Roswell Park Memorial Institute, 666 Elm Street, Buffalo, NY 14263, USA

Rr   M. N. Runner. Retired

Rt   Mr John B. Roths, The Jackson Laboratory, Bar Harbor, ME 04609, USA

Rw   Dr Sara Rockwell, Department of Therapeutic Radiology, Yale University School of Medicine, Hunter Radiation Therapy, P.O. Box 3333, New Haven, CT 06510-8040, USA

S   Last known address: Dr Kurt Stern, Department of Life Sciences, Bar-Ilan University, Ramat-Gan, Israel

Saf   Southern Animal Farms. Code retired

Sas   Mr Stephen F. Dwyer, Sasco, Inc, P.O. Box 66-DTS, Omaha, NE 68101, USA

Sbh   Professor O. Makela, Department of Serology and Bacteriology, University of Helsinki, Haartmaninkatu 3, 00290 Helsinki 29, Finland

Sc   J. P. Scott. Retired

Sch   ARS/Sprague-Dawley. Code retired

Scm   Dr Sandy C. Marks, Jr, Department of Anatomy, University of Massachusetts Medical School, 55 Lake Avenue, Worcester, MA 01605, USA

Scr   Dr Michael Goodman and Ms Sandie Urban, Research Institute of Scripps Clinic, 10666 North Torrey Pines Road, IMM 20, La Jolla, CA 92037, USA

Sd   K. L. Sydnor. Address unknown

Sda   Dr Kuniji Yamaki, Animal Breeding, Faculty of Agriculture, Tohoku University, 1-1 Tsutsumidori-Amamiyamachi, Sendai 980, Japan

Se   Lucio Severi. Retired

Sed   Dr Robert S. Sedlacek, Department of Radiation Medicine, Massachusetts General Hospital, Boston, MA 02114, USA

Sel   Last known address: Laboratorio Bioterapico Milanese, Selvi and C., Milan, Italy

Sf   Dr Donald C. Shreffler, Department of Genetics, P.O. Box 8031, Washington University School of Medicine, St. Louis, MO 63110, USA

Sfd   Dr Thomas E. Hamm, Jr, Division of Laboratory Animal Medicine, Quad 7, Building 330, Stanford University, Stanford, CA 94305, USA

Sg   Dr Jack H. Stimpfling, The McLaughlin Research Institute, 1625 Third Avenue North, Great Falls, MT 59401, USA

Shi   Shionogi and Company, Ltd, Koka-cho, Shiga, Japan

Shoe   Mr E. Anders, VEB Versuchstierproduktion, Hauptstrasse 62, 1291 Schönwalde, GDR

Si   Last known address: Dr Morten Simonsen, Queen Victoria Hospital, East Grinstead, Sussex RH19 3DZ, England

Sid   Drs Richard L. Sidman and Paul Neumann, Department of Neuroscience, Children's Hospital Medical Center, 300 Longwood Avenue, Boston, MA 02115, USA

Sim   Dr James D. Russell, Simonsen Laboratories, Inc, 1180-C Day Road, Gilroy, CA 95020, USA

Skh   Dr Stanley J. Mann, Temple University, Health Services Center, Central Animal Facility, 3400 North Broad Street, Philadelphia, PA 19140, USA

Slc   Dr Toshio Tanaka, Department of Laboratory Animal Medicine, Shizuoka Laboratory Animal Center, 1616 Koike-cho, Hamamatsu, Shizuoka, 435 Japan

Sm   Dr David Steinmuller, Tissue Typing Laboratory, 1520MSRB I, University of Michgan Medical School, 1150 West Medical Center Drive, Ann Arbor, MI 48109, USA

Smh   Curator and Dr S. D. M. Brown, Animal Department, St. Mary's Hospital Medical School, Paddington, London W2 1PG, England

Smn   Dr Charles L. Sidman, The Jackson Laboratory, Bar Harbor, ME 04609, USA

Sn   George D. Snell. Retired

Sp   William L. Simpson. Retired

Sr   Howard A. Schneider. Retired

Ss   Dr Willys K. Silvers, Department of Human Genetics, University of Pennsylvania School of Medicine, Philadelphia, PA 19104, USA

Ssc   Dr K. L. Fennestad, Statens Seruminstitut, Amager Boulevard 80, 2300 Copenhagen S, Denmark

St   Leonell C. Strong. Deceased

Sto   Last known address: Laboratory of Inbred Animals, Breeding Farm Stolbovaya, Academy of Medical Science, Solyanka 14, Moscow, USSR

Sv   Dr Leroy C. Stevens, The Jackson Laboratory, Bar Harbor, ME 04609, USA

Sw   Dr Sarah Walker, VA Hospital, 111F 800 Hospital Drive, Columbia, MO 65201, USA

Sx   Dr David H. Sachs, Chief, Transplantation Biology Section, Immunology Branch, National Cancer Institute, Building 10, Room 4B17, NIH, Bethesda, MD 20892, USA

Sy   A. Symeonidis. Deceased

Sz   Dr Leonard D. Shultz, The Jackson Laboratory, Bar Harbor, ME 04609, USA

T   L. Butler. Retired

Ta   Dr Hisashi Iwatsuka, Takeda Chemical Industries, Ltd, Central Research Division, 6-10-1 Himuro-cho, Takatsuki, Osaka 569, Japan

Tac   Mr Joseph W. Phelan, Taconic Farms, Inc, Hover Avenue, Germantown, NY 12526, USA

Tach   Dr Takehiko Tachibana. Address unknown

Tar   Dr A. K. Tarkowski, Department of Embryology, Institute of Zoology, University of Warsaw, 00-927/1 Warsaw, Poland

Tb   Ms Vicki Smith, Faculties Animal Care Facility, Australian National University, G.P.O. Box 4, Canberra A.C.T. 2601, Australia

Tbi   Last known address: Tokyo Biochemical Research Institute, 41-8 Takada 3 Toshima, Tokyo, Japan

Tbr   Dr Yoshihiko Tsubura, 2nd Department of Pathology, Nara Medical University, 840 Shijocho, Kashihara-city, Nara-ken 634, Japan

Ten   Dr Joseph E. Fuhr, University of Tennessee Medical Center/Knoxville, 1924 Alcoa Highway, Knoxville, TN 37920, USA

Tif   Dr R. Beglinger, Ciba-Geigy AG, Versuchstierzucht, CH-4332 Stein, Switzerland

Tohi   Dr Toshima Nobunaga, Institute for Experimental Animals, Tohoku University School of Medicine, 2-1 Seiryo-machi, Sendai 980, Japan

Ton   Dr Susan Tonkonogy, School of Veterinary Medicine, North Carolina State University, 4700 Hillsborough Street, Raleigh, NC 27606, USA

Tor   Last known address: Dr S. Tores, Instituto Nacional do Cáncer Sezao de Pesquisas, Pca. da Druz Vermelha 23, Rio de Janeiro, Guanabara, Brazil

Tox   Dr J. B. Greig, MRC Toxicology Unit, MRC Laboratories, Woodmansterne Road, Carshalton, Surrey SM5 4EF, England

Tr   Dr John J. Trentin, Division of Experimental Biology, Baylor University College of Medicine, Houston, TX 77025, USA

Tru   Mr David P. Kirstein, Trudeau Institute, P.O. Box 59, Saranac Lake, NY 12983, USA

Tu   Dr Felipe Figueroa, Max-Planck-Institut für Biologie, Department of Immunogenetics, Corrensstrasse 42, 7400 Tübingen, FRG

Tw   Professor Noboru Takasugi, Department of Biology, Yokohama City University, Kanazawa-ku, Seto 22-2, Yokohama 236, Japan

Ty   Dr Benjamin A. Taylor, The Jackson Laboratory, Bar Harbor, ME 04609, USA

U   Professor Dr L. F. M. van Zutphen, Department of Laboratory Animal Science, Veterinary Faculty, University of Utrecht, Yalelaan 1, P.O. Box 80.166, 3508 TD Utrecht, The Netherlands

Ucsd   Code Retired. Superseded by Gc.

Uct   M. E. Howard-Tripp, University of Cape Town Medical School, Animal Unit, Observatory, 7925, Cape Town, South Africa

Uip   Dr Paule Echinard-Garin, Animalerie Specialisée, US 71 I.R.S.C., CNRS, 7, rue Guy Moquet, B.P. No. 8, 94802 Villejuif Cedex, France

Umc   Mr Robert Ehlenfeldt, Department of Laboratory Medicine and Pathology, Box 724, Mayo Memorial Building, University of Minnesota, Minneapolis, MN 55455, USA

Umm   Dr Ronald M. McLaughlin, Department of Laboratory Animal Medicine, M154 Medical Science Building, University of Missouri, Columbia, MO 65212, USA

Un   Dr A. C. Upton, Institute of Environmental Medicine, NYU Medical Center, New York, NY 10021, USA

Unl   Dr S. McCormick, Unilever Research Laboratories, Biology Division, Colworth House, Sharnbrook, Bedford, England

Up   Dr Delta E. Uphoff, National Cancer Institute, Building 10, Room 6B18, National Institutes of Health, Bethesda, MD 20892, USA

Upj   Dr A. D. Ojerio, The Upjohn Company, 7000 Portage Road, Kalamazoo, MI 49001, USA

Urd   Ms Susan H. M. Poulton, Box 3062, Duke University Medical Center, Durham, NC 27710, USA

Utsp Dr Thomas J. Slaga, University of Texas System Cancer Center, Science Park, Research Division, P.O. Box 389, Smithville, TX 78957, USA

Vir   Last known address: Institute of Virology, Church Street, Glasgow, Scotland, UK

W   Dr Alina Czarnomska, Institute of Oncology, Department of Tumour Biology, Wawelska 15, 00-973 Warszawa, Poland

Wa   Dr P. de Boer, Department of Genetics, Agricultural University, Gen. Foulkesweg 53, 6703 BM Wageningen, The Netherlands

Wah   Professor Douglas Wahlsten, Department of Psychology, University of Waterloo, Waterloo, Ontario N2L 3G1, Canada

Wak   Dr Shigekatsu Tsuji, Department of Physiology, Wakayama Medical College, 9-bancho, Wakayama 640, Japan

Wal   Dr D. C. Thurley, Wallaceville Animal Research Centre, Ministry of Agriculture and Fisheries, Private Bag, Upper Hutt, New Zealand

We   John A. Weir. Retired

Wehi Last known address: Walter and Eliza Hall Institute of Medical Research, Melbourne 3050, Australia

Wf   Dr George L. Wolff, Division of Toxicology, National Center for Toxicological Research, Jefferson, AR 72079, USA

Wg   H. E. Walburg, Jr. Retired

Wh   Abraham White. Deceased

Wi   J. W. Wilson. Deceased

Wil   Dr Jean Wilson, The University of Texas, Southwestern Medical School, 5323 Harry Hines Boulevard, Dallas, TX 75235, USA

Wim   Drs Richard and Cynthia Wimer, Division of Neurosciences, Beckman Research Institute of City of Hope, Duarte, CA 91010, USA

Wits Dr J. C. Austin, Central Animal Service, University of the Witwatersrand Medical School, York Road, Parktown 2193, Republic of South Africa

Wl   Dr H. Glenn Wolfe, Department of Physiology and Cell Biology, University of Kansas, Lawrence, KS 66045, USA

Wm   Dr William H. Murphy, University of Michigan, Department of Microbiology, Medical Sciences II, Ann Arbor, MI 48109, USA

Wmr   Dr Willie M. Reams, Jr, Department of Biology, University of Richmond, Richmond, VA 23173, USA

Wms   Dr R. Michael Williams, 1 American Plaza, Suite 210, Evanston, IL 60201, USA

Wn   Dr A. Meshorer, Experimental Animal Center, Weizmann Institute of Science, P.O. Box 26, Rehovot 76 100, Israel

Wo   Dr James E. Womack, Department of Veterinary Pathology, Texas A&M University, College Station, TX 77843, USA

Wq   Dr Walter C. Quevedo, Jr, Division of Biology and Medicine, Brown University, Providence, RI 02912, USA

Wr   Dr Kelly H. Clifton, University of Wisconsin, Clinical Cancer Center, K4/312 C.S.C., 600 North Highland, Madison, WI 53792, USA

Wrm   Director, Department of Animal Resources, Walter Reed Army Institute of Research, Washington, DC 20307, USA

Ws   Dr John D. West, Department of Obstetrics and Gynaecology, University of Edinburgh, Centre for Reproductive Biology, 37 Chalmers Street, Edinburgh EH3 9EW, Scotland, UK

Wsl   Professor H. Bazin, University of Louvain, Faculty of Medicine, Experimental Immunology Unit, Clos Chapelle aux Champs 30.56, Brussels 1200, Belgium

Wt   Dr Wesley K. Whitten, The University of Tasmania, Department of Zoology, G.P.O. Box 252C, Hobart, Tasmania, Australia, 7001

Wts   Dr Peter J. Wettstein. Address unknown

Ww   E. F. Woodworth. Retired

Wy   George W. Woolley. Retired

Y   Dr A. M. Malashenko, Research Laboratory of Experimental Biological Models, Academy of Medical Sciences of the USSR, 143412 p/o Otradnoe, Krasnogorsk town, Moscow region, USSR

Yok   Drs Masaro Nakagawa and Toshihiko Asano, Department of Veterinary Science, The National Institute of Health, 10-35, Kamiosaki, 2-Chome, Shinagawa-ku, Tokyo, 141, Japan

Zal   Dr M. B. Zaleski, Department of Microbiology, State University of New York, Buffalo, NY 14214, USA

Zbz Dr med. Wolf H. Weihe, Universitätsspital Zürich, Biologisches Zentrallaboratorium, Ramistrasse 100, 8091 Zürich, Switzerland

Zgm Professor Dr Filip Culo, Medical Faculty, Department of Physiology, Salata 3, 41000 Zagreb, Yugoslavia

Zgr Dr Lidija Suman, Institut 'Ruder Boskovic', 41001 Zagreb, Bijenicka 54, Yugoslavia

Zur Professor Dr P. E. Thomann, Abteilung Labortierkunde der Universität Zürich, Winterthurerstrasse 190, CH-8057 Zürich, Switzerland

Index

Index

Index

Index

Index